PREFACE

Since 1922 there have been appearing in the pages of "Oil, Paint and Drug Reporter" a series of surveys covering the uses, potential applications and sales possibilities of chemicals and related raw materials.

At first these surveys appeared under the title "Where You Can Sell." Subsequently the title was changed to "Industrial Uses of Chemicals and Related Materials."

The early research was conducted by the late Ismar Ginsburg; and since his death in 1933 has been continued and expanded by Thomas C. Gregory, one of the editors of "The Condensed Chemical Dictionary" and a nationally-known expert on chemical marketing.

Recognizing the great permanent value of this information, an arrangement was effected between the publishers of "Oil, Paint and Drug Reporter" and Reinhold Publishing Corporation whereby the material could be made available in book form after rearrangement, amplification and complete editing by Mr. Gregory.

In its present form we believe it makes available to sales executives, research directors, manufacturers, dealers, and all others interested in the uses of chemical products, a mass of information unavailable elsewhere and of tremendous practical value.

Supplementary surveys by Mr. Gregory covering additional products and new uses will continue to appear in the pages of "Oil, Paint and Drug Reporter" and it is hoped that the welcome afforded this volume will justify the publication of supplementary volumes at suitable intervals.

REINHOLD PUBLISHING CORPORATION

June 1939.

WHAT THIS BOOK CONTAINS

It contains surveys on the uses, potential applications and sales possibilities of 5167 chemicals and related materials.

In making these surveys over a period of more than fifteen years the prime criterion has at all times been the commercial utility of the information.

For that reason the items dealt with have not been selected on a basis of chemical grouping; but rather because the substance was of commercial importance; or, for some reason, general interest has been aroused in it.

That is to say, all the aluminum compounds were not covered first; then all the calcium compounds, and so on. However, as the work has proceeded so many products have been dealt with that the effect has been much the same: the reader who checks the contents by chemical grouping will find few items of importance missing.

If the book does not contain information on a product in which you are interested, please refer to a file of "Oil, Paint and Drug Reporter" in which supplementary material is appearing weekly. This paper is available at most large libraries or may be obtained from Oil, Paint and Drug Publishing Co., Inc., 59 John Street, New York.

For information about products not included in this volume, or not yet covered in "Oil, Paint and Drug Reporter" communicate with Information Bureau, Reinhold Publishing Corporation, 330 West 42nd St., New York.

There is present a splendid representation of the following:

- Inorganic chemicals
- Organic chemicals
- Mineral acids
- Organic acids
- Fine chemicals
- Drugs and pharmaceutical products
- Related materials such as metals, petroleum products, greases, minerals, ores, animal and vegetable oils, essential oils, waxes, etc.

In addition there are many new products recently made available to the manufacturer for improving his product or solving difficulties encountered in processing operations.

Your attention is also directed to the fact that the name of each item is given not only in all variants commonly encountered in English but also in the more important foreign languages; and that in connection with many uses there is a patent reference.

HOW TO USE THIS BOOK

All products are arranged in *strict alphabetical order* and not by chemical classification. They are also titled by common name and not by strict chemical name. Experience has shown that many non-technical people handle chemicals and they look for products in this manner. Technical men are more apt to search by chemical classification.

Therefore:

Look for—

Beta-naphthol	not for	Naphthol, beta
Solvent Naphtha	" "	Naphtha, solvent
Sulphuretted hydrogen	" "	Hydrogen, sulphuretted
Soybean lecithin	" "	Lecithin, soybean
Soluble Prussian Blue	" "	Prussian blue, soluble
Potato Starch	" "	Starch, potato
Red Hematite	" "	Hematite, red
Red lead	" "	Lead oxide, red
S-Benzylisothiourea hydrochloride	" "	Benzylisothiourea hydrochloride, s-

Consult page vi for instructions "How to use this book"

Uses and Applications of Chemicals and Related Materials

Abelmoschus
Synonyms: Amber seed, Musk mallow, Musk okra, Musk seed.
Latin: Hibiscus abelmoschus.
French: Ambrette.
German: Abelmosch, Bisamkoerner.

Food
Ingredient of—
Food preparations of various sorts, to lend aromatic flavor and odor.

Perfumery
Ingredient of—
Perfume, toilet and cosmetic preparations of various sorts.
Substitute for—
Musk in making perfumes and the like.

Textile
——, *Miscellaneous*
To protect woolens against moths.

Abietene
French: Abiétene.
German: Abieten.

Miscellaneous
Ingredient of—
Compositions used for the removal of various kinds of stains from different materials.

Paint and Varnish
Ingredient of—
Paint and varnish removers.

Textile
——, *Miscellaneous*
Stain remover for—
Cleansing textile fabrics of all sorts and freeing them from stains and grease spots.

Abietic Acid
Synonyms: Abietenic acid.
French: Acide abiétique.
German: Tannensäure.

Chemical
Starting point in making—
Abietene, aluminum resinate, barium resinate, cadmium resinate, calcium resinate, copper resinate, hydroabietic acid, iron resinate, malonic acid, manganese resinate, nickel resinate, potassium resinate, silver resinate, sodium resinate, strontium resinate.

Miscellaneous
Assist in—
Promoting the growth of lactic and butyric acid ferments.
Preventive in—
Prohibiting infection and decomposition of raw material in the fermentation industry.

Abietic Acid Ester of Grapeseed Alcohol

Bituminous
Solvent (Brit. 445223) for—
Asphalt and other bituminous bodies

Dye
Solvent (Brit. 445223) for—
Dyestuffs, particularly oil-soluble coaltar dyes.

Fats, Oils, and Waxes
Solvent (Brit. 445223) for—
Fats, oils, waxes.

Resins
Solvent (Brit. 445223) for—
Oil-soluble glycerol-phthalic acid resins, polymerized vinyl compounds, synthetic resins.

Rubber
Solvent (Brit. 445223) for—
Rubber.

Abietic Acid Ester of Ricinoleic Alcohol

Bituminous
Solvent (Brit. 445223) for—
Asphalt and other bituminous bodies.

Dye
Solvent (Brit. 445223) for—
Dyestuffs, particularly oil-soluble coaltar dyes.

Fats, Oils, and Waxes
Solvent (Brit. 445223) for—
Fats, oils, waxes.

Resins
Solvent (Brit. 445223) for—
Oil-soluble glycerol-phthalic acid resins, polymerized vinyl compounds, synthetic resins.

Rubber
Solvent (Brit. 445223) for—
Rubber.

Abrus
Synonyms: Indian licorice, Jequirity jumble beads, Prayer beads, Wild licorice.
Latin: Abrus precatorius.

Chemical
Starting point in extracting—
Abrin

Pharmaceutical
In compounding and dispensing practice.

Absinthium
Synonyms: Alsei, Wormwood.
Latin: Artemisia absinthium.
German: Madderwort, Wermuth.

Chemical
Raw material for the production of—
Absinthin.

Fats and Oils
Raw material for the production of—
Essential oils.

Food
As a flavor in—
Beverages and condiments.

Perfumery
As an ingredient of—
Aromatic waters and lotions.

Pharmaceutical
In compounding and dispensing practice.

Aburachan Seed Oil
French: Huile de semences d'aburachan.
German: Aburachansamenoel.

Fuel
As an illuminant.

Soap
As a soapstock.

Acajou Balsam
Synonyms: Anacardia, Balsam, Cardol.
French: Baume d'acajou.

Ink
Ingredient of—
Indelible inks, ink for die-sinking work, stamp-pad ink.

Insecticide
Ingredient of—
Preparations used to combat ants.

Petroleum
Pigment for—
Paraffin in making black candles.

Pharmaceutical
In compounding and dispensing practice.

Accroides Gum
Synonyms: Black boy gum, Xanthorrhea resin, Yacca gum.
French: Gomme accroide.
German: Akaroidharz.

Leather
Finishing agent for—
Leather and leather goods.

Miscellaneous
Ingredient of—
Sealing wax compositions.

Paint and Varnish
Ingredient of—
Paint and varnish specialties.

Paper and Pulp
Finishing agent for—
Paper products.

Soap
Ingredient of—
Toilet soaps.

Acenaphthaquinone
German: Acenaphtachinon.

Dye
Starting point in making—
Ciba orange, vat dyestuffs.

Acenaphthene
Synonyms: Ethylenenaphthene, Naphacetene.
French: Acénaphtène.
German: Acenaphten, Azenaphten.

Chemical
Starting point in making—
Acenaphthene carboxylic acids (Brit. 274894), acenaphthenone, intermediate chemicals used in the preparation of drugs and perfumes, naphthaldehydic acid (U. S. 1439500).

Dye
Intermediate in making—
Sulphur dyes.
Starting point in making—
Intermediate chemicals used in the preparation of synthetic dyestuffs, Red R.

Acenaphthene-5-carboxylic Acid

Dye
Intermediate (Brit. 432885) in making—
Dyestuffs.

Acenaphthene-5:6-dicarboxylic Acid

Chemical
Starting point (Brit. 400069) in making—
Betanaphthol-1:4:5:8-tetracarboxylic acid, chloracenaphthenedicarboxylic acid, dinitronaphthalene-1:4:5:8-tetracarboxylic acid.

Dye
Intermediate (Brit. 432885) in making—
Dyestuffs.

Acenaphthenequinone

Dye
Starting point in making—
Anthra 2G, ciba orange G paste.

Acenaphthenone
French: Acénaphthènone.
German: Acenaphtenon.

Chemical
Starting point in making various intermediates.

Dye
Starting point (German 237719) in making—
Vat colors with isatin derivatives or naphthisatin derivatives.

Acetal

Perfume
Ingredient of—
Cosmetics, perfumes.

Acetaldehyde
Synonyms: Acetic aldehyde, Aldehyde, Acetaldehyd, Ethyl aldehyde.
French: Acétaldéhyde.
German: Acetaldehyd, Azetaldehyd.

Chemical
Starting point in making—
Acetals, acetic acid (synthetic), acetic anhydride, aldol, alpha-amino-propionic acid, ammonium derivative, butadiene: 1-3, chloroform, croton aldehyde, diethylbenzaldehyde acetal, dimethyl acetate, ethyl acetate, ethyl diacetate, methyl normal propyl carbinol, paraldehyde, synthetic acetic acid.

Dye
Starting point in making—
Aldehyde blue, Anthrapurpurin (1:2:7-trioxyanthraquinone), various other dyestuffs.

Insecticide
Ingredient of—
Disinfecting compositions.

Miscellaneous
Ingredient of—
Silvering compositions for mirrors.
Reagent in making—
Yeast albumen.

Petroleum
Ingredient of—
Motor fuel compositions.

Pharmaceutical
In compounding and dispensing practice.

Photography
Developing agent in—
Treating photographs.
Hardening agent in—
Treating dry gelatin films.

Plastics
Reagent in making—
Phenol-formaldehyde condensation products.

Rubber
Reagent in making—
Accelerator with the aid of amines, synthetic rubber.

Acetaldehydecyanohydrin
German: Azetaldehydcyanhydrin.

Chemical
Starting point in making—
Ethyl lactate (Brit. 264143).

Acetaldol

Chemical
Starting point (Brit. 249113) in making vulcanization accelerators with—
Anilin, diethylamine, ethylamine, ethylanilin, ethylenediamine, guanidin, methylamine, methylanilin, normalbutylamine, orthotoluidin, orthotolyldiguanide.

Acetaldoxime

Fuel
Primer (Brit. 429763) for—
Diesel engine fuel oils produced by the hydrogenation of coal.

Petroleum
Primer (Brit. 429763) for—
Diesel oils containing a high proportion of aromatic bodies.

Acetamide
Synonyms: Acetamid, Acetic acid amine.
French: Acétamide.
German: Acetamid, Azetamid.

Chemical
Reagent in making—
Alphanaphthylamine, phthalimide.
Solvent in—
Miscellaneous processes.
Stabilizer in making—
Hydrogen peroxide solutions.
Starting point in making—
Acetonitrile, methylamine, methylcyanide.

Acetamide Sulphate
French: Sulphate d'acétamide.
German: Acetamidsulfat, Schwefelsaeuresacetamid.

Chemical
Starting point in making—
Benzyl acetate (Brit. 255887).

1-Acetamido-4-aminoanthraquinone
Miscellaneous
Dyestuff (U.S. 1989133) for—
Cellulose acetate products (imparts shades of red).
Textile
Dyestuff (U.S. 1989133) for—
Cellulose acetate products (imparts shades of red).

13-Acetamidodiethylisorosindulin-1:6:11-trisulphonic Acid
Dye
Starting point (Brit. 431708 and 431709) in making—
Blue dyestuffs with 2:5-diaminometaxylene.

4-Acetamido-3-hydroxyquinaldin
Dye
Starting point (Brit. 429176) in making—
Acid dyestuffs for wool.

1-Acetamido-4-methylaminoanthraquinone
Synonyms: Alpha-acetamido-4-methylaminoanthraquinone.
French: Alpha-acétamido-4-méthyleaminoanthraquinone.
German: Alpha-acetamido-4-methylaminoanthrachinon.
Chemical
Starting point in making various derivatives.
Dye
Starting point in making various derived dyestuffs.
Textile
Reagent (Brit. 263260) in dyeing—
Cellulose acetate rayon yarns and fabrics.

4-Acetamido-1-naphthylamine-6-sulphonic Acid
French: Acide de 4-acétamido-1-napthylamine-6-sulphonique.
German: 4-Acetamido-1-napthylamin-6-sulphonsaeure.
Dye
Starting point (US1655550-1) in making—
Tetrakosazo dyestuffs, Trisazo dyestuffs.

5-Acetamino-2-amino-1:4-dimethoxybenzene
Synonyms: 5-Acetamino-2-amino-1:4-dimethoxybenzol.
French: 5-Acétamino-2-amino-1:4-diméthoxyebenzène.
Chemical
Starting point in making—
Aromatics, intermediates, pharmaceuticals.
Dye
Starting point (Brit. 307303) in making monoazo dyestuffs with—
N-Acetyl-H acid, N-betachloropropionyl-H acid, N-carbethoxy-H acid, N-chloroacetyl-H acid, N-phenylacetyl-H acid.

6-Acetaminobetanaphthol-4-sulphonic Acid
Dye
In dye syntheses
Starting point (Brit. 445999) in making—
Chromable orthohydroxy azo dyes by coupling with orthohydroxydiazonium compounds, such as those derived from 6-nitro-2-aminoparacresol or 4-chlor-2-aminophenol-6-sulphonic acid.

Acetanilide
Synonyms: Phenylacetamide, Antifebrin.
French: Antifébrine.
German: Acetanilid.

Chemical
Stabilizer in making—
Hydrogen peroxide compositions.
Reagent in making—
Paranitranilin, paranitroacetanilide, paraphenylenediamine.
Starting point in making—
Antispesin, pharmaceutical chemicals.
Dye
Starting point in making—
Para red, sulphur dyestuffs, various dyestuffs of other groups.
Miscellaneous
Reagent in making—
Synthetic camphor for use in the manufacture of celluloid.
Paint and Varnish
Ingredient of—
Cellulose ester dopes and lacquers.
Pharmaceutical
In compounding and dispensing practice.

Acetanisidin (Para)
French: Acétanisidine.
Chemical
Starting point in making—
Metanitroanisidin.
Dye
Starting point in making—
Tussaline dyestuffs.
Pharmaceutical
In compounding and dispensing practice.

Acetic Acid
Synonyms: Vinegar acid.
French: Acide acétique.
German: Essigsaeure.

Chemical
Reagent in making—
Acenaphthenone, acetnaphthenequinone, acetnaphthylidide, acetoacetic acid, acetoaminophenol, acetylalphanaphthylamine-5-sulphonic acid, acetylanthranilic acid, acetyl-1-naphthylamine-6-sulphonic acid, acetylorthotoluidin, acetylparatoluidin, alpha-amino-2-naphthol, alpha-anthrol, alphabromobetagammadiacetylglycerol, alphanaphthylamine, alphanaphthylglycin, alphanaphthylamine-6-sulphonic acid, alphanaphthylamine-7-sulphonic acid, aminopara-acetanilide, anthraquinone, anthraquinone-2-carboxylic acid, anthraquinonedicarboxylic acid, antisepsin, antithermin, benzanthrone, benzidinsulphonemonosulphonic acid, benzidin-3:3-disulphonic acid, benzidin-3-sulphonic acid, benzylidene acetate, betachloroanthraquinone-alphacarboxylic acid, 4-chloro-1-methylanthraquinone, 1-chloro-4-methranol, 5-chloroisatin, dichlorobenzidin, 1:3-diamino-2-methylanthraquinone, 5:8-dichloro-2-hydroxy-1-methylanthraquinone, dichlorophenylanthranol, diethyl acetate, diethylene hydrazin (asymetrical), dichloroacetic acid, 4:6-dihydroxy-2-methylanthraquinone, 9:10-dihydroxy-4-chloro-1-methylanthraquinone, dimethyl acetate, 2:2-dimethyl-1:1′-dianthraquinonyl, 2:2-dimethyl cyanurate (normal), erucic acid, ethylhydrazine, ethyleneacetobromohydrin, ethyleneacetochlorohydrin, ethylene diacetate, ethyleneiodochlorohydrin, 4-hydroxy-1-methylanthraquinone, 1-hydroxy-2-methylanthraquinone, 3-hydroxy-1-methylanthraquinone, meta-amino-azo benzoic acid, 4-nitro-1:3-phenylenediamine, 5-nitroanthraquinone-1-carboxylic acid, nitromethylanthraquinone, 2-nitroquinazarin, phenanthrenequinone, phenanthranaphthazin sulphonic acid, phenylhydrazone, phenylhydroxylamine, quinazarin, tanstuffs synthesized by the condensation of phenol with formaldehyde in the presence of sodium sulphite (German 426424), tetrachlorodiphenylamine, tetramethyldiaminobenzyhydrol, 1:3:6-trimethylanthraquinone, 1:4:7-trimethylanthraquinone, triphenyldihydroxyanthraquinone.
Reagent and starting point in making—
Acetals, acetanilide, acetanisidin, acetic anhydride, acetamide, acetins, acetphenetidin, acetyl bromide, acetyl chloride, acetyl iodide, acetone, acetochloro-

Acetic Acid (Continued)
form, acetophenone, aluminum acetate, amyl acetate, ammonium acetate, antipyrin, barium acetate, benzaldehyde, benzoic acid, benzoic acid anhydride, benzosalin, bergamiol, bismuth acetate, bismuth-aluminum acetonitrate, bismuth basic dibromohydroxynaphthene, bismuth basic gallate, bismuth iodosalicylate, bismuth oxyiodide, bornyl acetate, bromobehenolic acid, butyl acetate, butyl acetate (secondary), butyl acetate (tertiary), cadmium acetate, caesium acetate, calcium acetate, cetylic alcohol, chromium acetate, chrysarobin, chrysophanic acid, cinnamyl acetate, citronellol, cobalt acetate, copper acetate, copper acetate (basic), coumarone, coumaryl acetate, diacetin, dicalcium phosphate, ethyl acetate, ethylene bromide, ethylene diacetate, ethylene glycol, ethylene monoacetate, eugallol, euguform, ferric acetate, ferrous acetate, geranyl acetate, indol, iodophen, ionone, iononyl acetate, iron acetate (basic), isobutyl acetate, isobornyl acetate, lead acetate, lead acetate (tribasic), magnesium acetate, manganese acetate, mercury acetate, methyl benzoate, methyl cinnamate, methyl cyanide, monoacetin, nickel acetate, potassium acetate, saccharic acid, sodium acetate, strontium acetate, tetraiodophenolphthalein, thymyl acetate, tin acetate, triacetin, trichloroacetic acid, zinc acetate.
Solvent in making—
Compounds from hippuric acid or betanaphthalene sulphonic acid (Brit. 251651), phenanthraphenazin.

Dye
Reagent in making—
Alizarin yellow C, aurantia aureoline, brilliant azurine 3G, dimethylaminonaphthenazin, lanacyl blue BB, methyl violet B.
Solvent in making—
Azo dyestuffs from paranitranilin and 3-nitroparanitranilin for use on acetate rayon fiber and fabric (French 600106).

Food
Ingredient of—
Curing compositions for the treatment of fish, such as herring, and meats, such as hams and bacon.
Food flavoring compositions.
Reagent in making—
Fruit esters, vinegar.

Glues and Adhesives
Ingredient of—
Adhesive compositions (U. S. 1594522).

Insecticide
Ingredient of—
Insecticidal and germicidal compositions for killing ova

Leather
Reagent in making—
Artificial leathers.

Miscellaneous
Reagent in making—
Egg albumen.
Solvent in treating—
Vegetable fibers in making hair (French 600476).

Paint and Varnish
Ingredient of—
Special lacquers for airplanes.
Reagent in making—
Copper acetoarsenite (Schweinfurt green) varnishes, verdigris, white lead and other lead whites.

Pharmaceutical
In compounding and dispensing practice.

Photography
Reagent in making—
Moving picture films, noninflammable cellulose-acetate films.

Plastics
Ingredient of—
Compositions for cleaning and polishing celluloid.
Reagent in making—
Cellulose acetates, phenol-formaldehyde condensation products.

Rubber
Reagent in making—
Raw rubber by coagulation of rubber latex.

Textile
——, *Dyeing*
Assist in—
Dyeing wool, dyeing silk with acid and alizarin colors.

——, *Finishing*
Ingredient of—
Compositions used in scrouping silks.

——, *Manufacturing*
Reagent in making—
Acetate rayon.

——, *Printing*
Ingredient of—
Printing paste for calicos.
Solvent in—
Printing imitation embroidery or plaited effects on woven fabrics.

Acetic Acid Ester of Grapeseed Alcohol

Bituminous
Solvent (Brit. 445223) for—
Asphalt and other bituminous bodies.

Dye
Solvent (Brit. 445223) for—
Dyestuffs, particularly oil-soluble coaltar dyes.

Fats, Oils, and Waxes
Solvent (Brit. 445223) for—
Fats, oils, waxes.

Resins
Solvent (Brit. 445223) for—
Oil-soluble glycerol-phthalic acid resins.
Polymerized vinyl compounds.
Synthetic resins.

Rubber
Solvent (Brit. 445223) for—
Rubber.

Acetic Acid Ester of Ricinoleic Alcohol

Bituminous
Solvent (Brit. 445223) for—
Asphalt and other bituminous bodies.

Dye
Solvent (Brit. 445223) for—
Dyestuffs, particularly oil-soluble coaltar dyes.

Fats, Oils, and Waxes
Solvent (Brit. 445223) for—
Fats, oils, waxes.

Resins
Solvent (Brit. 445223) for—
Oil-soluble glycerol-phthalic acid resins, polymerized vinyl compounds, synthetic resins.

Rubber
Solvent (Brit. 445223) for—
Rubber.

Acetic Anhydride
Synonyms: Acetic acid anhydride.
French: Anhydride acétique.
German: Essigsaeureanhydrid.

Chemical
Ingredient of—
Compositions used in the nitration of hydrocarbons.
Reagent in making—
Alpha-acetylamino-8-naphthol-3:6 disulphonic acid, alphabetadichloroacetin, acetoeugenol, acetphenetidin, acetylaminoanthraquinone, acetylaminoazobenzene, acetyl-1:8-aminonaphthol-3:6-disulphonic acid, acetylanthranilic acid, acetylbenzidin, acetylcresotinic acid, acetylized methylenediguaiacol (Euguform), acetylmetatoluylenediamine, acetylmorphine, acetylnaphthylamine-5-sulphonic acid, acetyl-1:4-naphthylamine-7 sulphonic acid, acetyl-1:4-naphthylamine-4-sulphonic acid, acetylorthonitrobenzaldehyde, acetylorthotoluidin, acetylpara-aminophenol salicylate, acetylparaoxyphenylurethane (Neurodin), acetylparatoluidin, acetylphenacetin, acetylquinine, acetylsalicylic acid, acetyl xylidide, alkyl acetates of all sorts, benzoic acid anhydride, benzoylacetyl peroxide (Acetozone, Benzozone), benzyl acetate, benzylidene diacetate, butyl acetate, butylene acetate, butylidene diacetate, butyric anhydride, coumarin, diacetylbromacetamid (Neuronal), diacetylmorphine (Heroin), 2-2:dimethyl-1:1-dianthraquinone, ethyl acetate, ethylacetylsalicylic acid, ethylene acetate, ethylidene diacetate, guaiacol acetate (Eucol), hexamethyl rufigallate (Exodin), 3-hydroxy-1-methylanthraquinone, ionone, isatin, ketene, lead subacetate, linalyl acetate, menthyl acetate, menthylacetylsalicylic acid, menthylchloromethyl ether, methyl acetate, methylacetylsalicylic acid, methylene acetate, monochloroacetic acid, monoiodoacetic acid, 4-nitro-1:3-phenyl-

Acetic Anhydride (Continued)
ene diamine, orthoaminoazotoluene (Azodermin), ortho-orthodibromobenzidine, para-aminoacetophenone, phenyl acetate, phenylacetylsalicylic acid, pyracoll, pyrogallol monoacetate (Eugallol), rhein, rhein diacetate.

Dye
Reagent in making—
Various dyestuffs of different groups.

Paint and Varnish
Reagent in making—
Airplane dopes, airplane lacquers, varnishes.

Photographic
Reagent in making—
Noninflammable motion-picture films.

Plastics
Reagent in making—
Cellulose acetate.

Textile
Reagent in making—
Acetate rayon.

Aceticarsenic Acid

Chemical
Starting point in making—
Esters and salts, particularly the disodium salt (aricyl).
Various derivatives, particularly pharmaceuticals.

Miscellaneous
In veterinary medical practice.

Pharmaceutical
In compounding and dispensing practice.

Acetin

Explosives
Gelatinizing agent in making—
Smokeless powders.
Ingredient of—
Low-freezing dynamites, gelatines and other permissibles.
Reagent in making—
Dinitroacetylglycerin.

Leather
Reagent in—
Tanning of various leathers.

Textile
——, *Dyeing*
Ingredient of—
Basic dyestuffs liquors, indulin dyeing liquors, Perkin's violet dyeing liquors.

Acetoacetanilide
French: Acétoacétanilide.
German: Acetoacetanilid.

Dye
Starting point in making—
Hansa yellow G.

Acetoacetotoluidin

Dye
Starting point in making—
Pigment fast yellow G R extra.

Acetomesitylene

Anaylsis
Laboratory reagent.

Chemical
Reagent in—
Chemical synthesis.

Acetone
Synonyms: Aceton, Dimethyl ketal, Dimethyl ketone, Ketopropane, Methylacetal, 2-Propanone, Pyroacetic ether, Pyroacetic spirit.
Latin: Acetonum, Spiritus pyroaceticus.
French: Acétone, Ésprit pyroacétique, Éther pyroacétique.
German: Aceton, Essiggeist, Mesitalkohol.
Italian: Acetone, Chetone, Metilacetone.

Analysis
Reagent in various processes.

Chemical
Absorbent for—
Gases.
Dehydrating and desiccating reagent in making various products.
Denaturant for industrial alcohol.
Extracting medium in the treatment of—
Gallnuts for the removal of their tannin content.
Ingredient of—
Filling mass for storing explosive gases (U. S. 1591397).
Reagent in making—
Benzylidene acetone, chloroacetone, ionone, isoprene, ketene, methylacetone, 3-methyl butanol, methylheptenone, quinaldine, sodium acetone sulphoxylate, 1:3:5-trimethylbenzene.
Selective solvent in making various chemical products.
Solvent for coupling agents in synthesis.
Solvent for extracting—
Drugs, digestive ferments.
Solvent in preparation of—
Platinum contact catalysts.
Solvent in making—
Dimethylanthragallol, pyrogallol monoacetate (Eugallol).
Starting point in making—
Acetone bisulphite, acetone berberin, acetone chloroform, acetone oil, benzaldehyde, bromoform, benzylideneacetone, carbon tetrabromide, chloroform, chlorobutanol.
Condensation products, such as acetonedioxalic ester.
Compounds with ammonia, such as diacetoneamine.
Compounds with alkaline sulphites.
Compounds with chlorine, bromine, and other halogens, such as trichloroacetone, pentachloroacetone, monobromoacetone.
Compounds with chloroform.
Compounds with hydrocyanic acid, such as acetonecyanhydrol.
Compounds with mercury sulphate and mercury oxide.
Dimethylbutadiene, diacetone alcohol, iodoform, isopropyl alcohol, ketones, lepidin, mesityl oxide, methylisobutylketone, mesitylene, pinacone, sodium lygrosin, sulphonethylmethane (Trional), sulphonmethane (Sulphonal).

Dye
Reagent in making—
Indigo, isoquinolin dyestuffs, quinolin dyestuffs, nigrosin (condensation product of acetone and 1:8 naphthylene-4-sulphonic acid).
Starting point in making—
Azo dyestuffs, diphenylamine dyestuffs, dyestuffs with the aid of diazobenzene chloride.

Explosives
Reagent in making—
Trinitrotoluene (TNT).
Solvent in making—
Cordite, smokeless powder.

Fats and Oils
Extracting medium in obtaining fats and oils from oilseeds and other materials.

Glues and Adhesives
Solvent in making—
Adhesive compositions from nitrocellulose.

Gums
Solvent in preparing—
Gum arabic, gum senegal, gum tragacanth, Indian gum.

Leather
Reagent in—
Drying leather previous to the application of impregnating reagents.
Treating hides during the tanning process.
Solvent in making—
Artificial leathers, oak tannins.

Mechanical
Ingredient of—
Lubricating compositions.
Solvent in making—
Compositions for removing the carbon deposit in the cylinders of internal combustion engines.

Miscellaneous
Degreasing agent in the treatment of furs.
Ingredient of—
Cleansing preparations for fabrics and garments.
Disinfectant composition in combination with resorcinol.
Film cement compositions (U. S. 1596965).
Sealing wax compositions.
Reagent in—
Freezing of microtomes.
Making pepsin.

Acetone (Continued)
Solvent for—
Extracting products from asphalts and bitumens.
Storing acetylene.
Solvent in making—
Spotting fluids.

Paint and Varnish.
Ingredient in making—
Cellulose acetate solvent preparations, lacquers, varnishes.
Ingredient of—
Paint and varnish removers.
Solvent for—
Varnish gums.
Solvent in making—
Airplane dopes, airplane varnishes, bituminous paints, nitrocellulose lacquers, shellac preparations, varnishes of the resin type.

Perfume
Solvent for—
Essential oils.

Petroleum
Ingredient of—
Lubricating compositions, motor fuel compositions.
Reagent in—
Purification of paraffin.
Revivification of spent decolorizing and clarifying anhydrous magnesium silicates.
Solvent in—
Deodorizing crude oil, dewaxing paraffin-base oils.
Solvent in making—
Petroleum solutions.
Petroleum distillate solutions.
Petroleum products of improved quality, by removing the heavy and high-flashpoint constituents from mineral oils of moderately low specific gravity.

Pharmaceutical
In compounding and dispensing practice.

Photographic
Ingredient of—
Developing compositions in place of alkali.
Solvent in making—
Films, plates.

Plastics
Ingredient of—
Plastic compositions containing cellulose acetate (Brit. 252999).
Reagent in making—
Arabitol, artificial amber from copal resin, camphor substitutes, condensation products from casein, condensation products from glycerin.
Solvent in making—
Celluloid and celluloid preparations, horn substitutes, mannitol, pyroxylin plastics.

Printing
Ingredient of—
Compositions used in photomechanical flat printing processes.

Refrigeration
As a refrigerant.

Rubber
Ingredient of—
Rubber cements.
Reagent in making—
Synthetic rubber.
Solvent in—
Purification of crude rubber.

Soap
Ingredient of—
Liquid soaps.

Starch
Reagent in making—
Soluble starches from starch or flour by precipitation.
Dextrins from starch or flour by precipitation.

Textile
——, *Dyeing*
Impregnating agent in—
Treating raw cotton for dyeing with anilin black by oxidation.
Ingredient of—
Vat liquors, to increase the dispersion of the dyestuffs and to increase the stability of the liquor.

——, *Finishing*
Ingredient of—
Cleansing compositions for removing stains.
Solvent for—
Dissolving out rayon threads in cotton and rayon fabrics to produce pattern effects on woven goods (Brit. 237909).

——, *Manufacturing*
Ingredient of—
Compositions used for various preliminary treatments of fibers.
Solvent in—
Degreasing wool.
Degumming raw silk (in admixture with textile soap).
Manufacturing rayons.

——, *Miscellaneous*
Ingredient of—
Compositions used for improving the quality of cuprammonium rayon by purifying it and increasing its stability.
Solvent in making—
Rayon from casein.

Waxes and Resins
Reagent in making—
Synthetic resins, such as those made from casein by condensation with the aid of formaldehyde.
Solvent in preparing—
Bayberry wax, beeswax, candellila wax, carnauba wax, copal resins, coumarone, Japan wax, montan wax, ozokerite, resins, resin mixtures, rosin, waxes, wax mixtures.

Woodworking
Solvent in—
Artificial seasoning of wood.
Dyeing wood with green PLX and naphthol yellow S and fuchsin powder (German 422124).

Acetone Bisulphite
French: Bisulphite d'acétone.
German: Acetonbisulfit.

Chemical
Starting point in making—
Pure acetone.

Photographic
Reagent in—
Various processes.
Starting point in making—
Para-aminophenol developer.

Textile
——, *Dyeing*
Assist in—
Fabric dyeing, yarn dyeing.

——, *Printing*
Assist in—
Color pastes.

Acetone Chloroform
Synonyms: Acetochloretone, Acetoform, Anesin, Aneson, Chlorbutanol, Chloretone.
Latin: Chloretonum.
French: Acétonechloroforme, Alcool trichlorobutylique tertiare, Chlorétone, Chloroforme d'acétone.
German: Chlorbutanol, Chloreton.
Spanish: Cloreton.
Italian: Cloretone.

Chemical
Starting point in making—
Pharmaceutical derivatives.

Pharmaceutical
In compounding and dispensing practice.

Acetonecyanhydrin
Synonyms: Oxyisobutyric nitrile.
French: Cyanhydrine d'acétone, Nitrile oxyeisobutyrique.
German: Acetoncyanhydrin, Oxyisobutyronitril.

Chemical
Starting point in making—
Pharmaceuticals and other compounds.

Insecticide
Insecticide in 1 per cent or less aqueous solutions (U. S. 1559961).

Acetone Oils
French: Huiles d'acétone, Huiles acétonique.
German: Acetonoel.
Spanish: Aceites de acetone.
Italian: Olios di acetone, olios di chetone.

These substances are oily mixtures of high-boiling ketones containing other substances.

Analysis
Solvent for—
Cellulose derivatives, gums, natural resins, oils, synthetic resins.

Cellulose Products
Solvent for—
Cellulose acetate, cellulose ethers, nitrocellulose.

Ceramic
Solvent in—
Compositions, containing natural or synthetic resins, cellulose acetate, nitrocellulose, or other cellulose esters or ethers, used as coatings for protecting and decorating ceramic products.

Chemical
Denaturant for—
Industrial alcohol.
Solvent for—
Cellulose acetate, cellulose ethers, nitrocellulose.
Various raw and intermediate materials.
Starting point in making—
Methylethyl ketone.

Cosmetic
Solvent for—
Gums, oils.
Solvent in—
Nail enamels and lacquers containing natural or synthetic resins, nitrocellulose, cellulose acetate, or other cellulose esters or ethers as base material.
Perfumes.

Dry-Cleaning
Spotting agent for—
Greasy stains, gums, oils, resins.

Dye
Solvent in—
Purifying crude anthracene for the manufacture of dyestuffs.

Electrical
Solvent in—
Insulating compositions, containing natural or synthetic resins, nitrocellulose, cellulose acetate, or other cellulose esters or ethers, used for covering wire and in making electrical machinery and equipment.

Fats and Oils
Solvent for—
Oils.

Glass
Solvent in—
Compositions, containing natural or synthetic resins, nitrocellulose, cellulose acetate, or other cellulose esters or ethers, used in the manufacture of nonscatterable glass and as coatings for decorating and protecting glassware.

Glue and Adhesives
Solvent in—
Adhesive compositions containing gums, natural or synthetic resins, nitrocellulose, cellulose acetate, or other cellulose esters or ethers.

Gums
As a solvent.

Leather
Solvent in—
Compositions, containing natural or synthetic resins, nitrocellulose, cellulose acetate, or other cellulose esters or ethers, used in the manufacture of artificial leathers and as coatings for decorating and protecting leathers and leather goods.

Metal Fabrication
Solvent in—
Compositions, containing natural or synthetic resins, nitrocellulose, cellulose acetate, or other cellulose esters or ethers, used as coatings for protecting and decorating metallic articles.

Miscellaneous
Solvent for various materials in many manufacturing operations.
Solvent in—
Coating compositions, containing natural or synthetic resins, nitrocellulose, cellulose acetate, or other cellulose esters or ethers, used for protecting and decorating various articles.

Paint and Varnish
Ingredient of—
Dopes, enamels, lacquers, paints, paint removers, varnishes.
Solvent for—
Cellulose acetate, cellulose ethers, copals, cumar, natural resins, nitrocellulose, synthetic resins.

Paper
Solvent in—
Compositions, containing natural or synthetic resins, nitrocellulose, cellulose acetate, or other cellulose esters or ethers, used in the manufacture of coated papers and as coatings for decorating and protecting articles made of paper or pulp.

Pharmaceutical
Solvent miscible with—
Oils.

Photographic
Solvent in making—
Films from cellulose acetate, nitrocellulose, or other esters or ethers of cellulose.

Plastics
Gelatinizing agent in various compositions.
Solvent in making—
Laminated fiber products, molded products, plastics from or containing natural or synthetic resins, cellulose acetate, nitrocellulose or other cellulose esters or ethers.

Resins
Solvent for—
Copals, cumar, natural resins, synthetic resins.
Solvent in making—
Artificial resins from or containing cellulose acetate, nitrocellulose, or other cellulose esters or ethers.

Rubber
Solvent in—
Compositions, containing natural or synthetic resins, cellulose acetate, nitrocellulose, or other cellulose esters or ethers, used as coatings for protecting and decorating rubber goods.

Soap
Solvent for—
Oils.

Stone
Solvent in—
Compositions, containing natural or synthetic resins, cellulose acetate, nitrocellulose, or other cellulose esters or ethers, used as coatings for decorating and protecting artificial and natural stone.

Textile
Solvent in—
Compositions, containing gums, natural or synthetic resins, cellulose acetate, nitrocellulose, or other cellulose esters or ethers, used in the manufacture of coated fabrics.

Wood
Solvent in—
Compositions, containing natural or synthetic resins, cellulose acetate, nitrocellulose, or other cellulose esters or ethers, used as protective and decorative coatings on woodwork.

Acetone Peroxide
Synonyms: Peroxide of dimethyl ketone.

Fuel
Ignition improver (Brit. 444544) for—
Diesel engine fuels.
Reducer (Brit. 444544) of—
Inflammability hazards in diesel engine fuels.

Acetonitrile

Analysis
As an inert medium in physical-chemical investigations.

Chemical
Starting point in making—
Ethylamine, intermediates, organic chemicals, synthetic aromatics.
Used as a denaturant for ethyl alcohol.

Acetophenoneparaphenetidin
Synonyms: Malarin, Malarine.
German: Acetophenonparaphenetidid, Acetophenonparaphenetidin.
Chemical
Starting point in making various derivatives.
Pharmaceutical
In dispensing practice.

Acetophenonephenylmethyl Ketone
Cellulose Products
Plasticizer for—
Cellulose acetate.
For uses, see under general heading: "Plasticizers."

Acetopurpurin
Synonyms: Acetopurpine.
Textile
——, *Dyeing*
Dyestuff for—
Cotton yarns and fabrics, cotton-silk fabrics, cotton-wool fabrics, silk fabrics, silk-wool fabrics, wool fabrics, wool-cotton-silk fabrics.

Acetoxime
Fuel
Primer (Brit. 429763) for—
Diesel engine fuel oils produced by the hydrogenation of coal.
Petroleum
Primer (Brit. 429763) for—
Diesel oils containing a high proportion of aromatic bodies.

Acetoxylidide
Cellulose Products
Plasticizer (U. S. 1567343) for—
Cellulose esters or ethers.
For uses, see under general heading: "Plasticizers."

Acetozone
Pharmaceutical
Suggested for use as antiseptic and local anesthetic.
Sanitation
General disinfectant.

Acetphenetidin
Synonyms: Acetoparaphenetidin, Para-acetphenetidin, Paraethoxyacetanilide, Phenacetine.
Latin: Acetphenetidinum, Phenacetinum.
French: Acét-phénétidine, Amide acétique de l'amidophénétol, Para-acétophénétidine.
German: Acetphenetidin, Azetphenetidin.
Spanish: Fenacetina.
Italian: Fenacetina.
Pharmaceutical
In compounding and dispensing practice.
Suggested for use as—
Analgesic, antipyretic, sedative.

Acetylacetone
Analysis
Reagent in testing for—
Carbon disulphide.
Chemical
Starting point in making—
Aromatics, intermediates, methylheptenone, organic chemicals, pharmaceuticals.
Textile
Ingredient (Brit. 182166 and French 501700) of—
Solvent mixtures used in making a spinning solution containing cellulose acetate (added in the proportion of 50 to 100 per cent on the weight of cellulose acetate).

Acetylaminoacetparaphenetidin
German: Azetylaminoacetparaphenetidin.
Chemical
Starting point in making—
Acetylsalicylic acid acetamide-paraphenetidin (Aspirophen).

4-Acetylaminocoumaronone
French: 4-Acétyleaminocumaronone.
German: 4-Acetylaminocumaronon, 4-Azetylaminocumaronon.
Spanish: 4-Acetilaminocumaronona.
Italian: 4-Acetilaminocumaronona.
Chemical
Starting point in making—
Pharmaceutical derivatives.
Pharmaceutical
Suggested for use as antirheumatic and antithermic.

3-Acetylamino-4-oxyphenylarsenic Acid
French: Acide de 3-acétyleamino-4-oxyphénylearsinique.
German: 3-Acetylamino-4-oxyphenylarsinigsaeure.
Chemical
Starting point in making—
Ammonium, sodium and potassium salts used as medicines (Brit. 264797).
Derivatives with quinine hydrochloride and other cinchona alkaloids and their derivatives.

Acetylanisidin
Cellulose Products
Plasticizer (Brit. 312606) for—
Cellulose acetate, cellulose esters or ethers, cellulose nitrate (nitrocellulose).
For uses, see under general heading: "Plasticizers."

Acetylbenzoyl
Glue and Adhesives
Hardening agent (Brit. 444289) for—
Gelatin (the hardening effect is greatest at a pH value of 8).
Photographic
Hardening agent (Brit. 444289) for—
Gelatin (the hardening effect is greatest at a pH value of 8).

Acetyl Bromide
French: Bromure d'acétyle.
German: Acetylbromid.
Chemical
Reagent and starting point in making various intermediate materials for dye, pharmaceutical and perfume products.
Dye
Reagent in making various dyestuffs.

Acetylcaproyl
Glue and Adhesives
Hardening agent (Brit. 444289) for—
Gelatin (the hardening effect is greatest at a pH value of 8).
Photographic
Hardening agent (Brit. 444289) for—
Gelatin (the hardening effect is greatest at a pH value of 8).

Acetyl Chloride
French: Chlorure d'acétyle.
German: Acetylchlorid.
Analysis
Reagent in various processes.
Chemical
Reagent in making—
Acetic anhydride, acetophenone, acetoresorcinol, acetoxyphenylmethane, diacetylmorphine (Heroin), ethylideneacetoacetic ether, ketone musk, maleic acid, para-aminoacetophenone, paramethoxyacetophenone, quinolinic acid, quinolinic anhydride.
Starting point in making—
Acetamide.
Dye
Reagent and starting point in making various synthetic dyestuffs.

Acetylcyclohexylamine, Normal
German: N-acetylcyklohexylamin.
Chemical
Starting point in making—
Intermediates, pharmaceuticals, various derivatives.
Dye
Starting point (Brit. 340495) in making dyestuffs for rayons with the aid of—
Alphachloro-2:4-dinitrobenzene.
Alphachloro-2:4-dinitrobenzene-6-sulphonic acid.
Alphachloro-2:6-dinitrobenzene-4-sulphonic acid.
Alphachloro-2:4-dinitronaphthalene.
Alphachloro-2:4-dinitronaphthalene-sulphonic acid.
Alphachloro-4-nitrobenzene-2-carboxylic acid.
Alphachloro-2-nitrobenzene-4-sulphonic acid.
Alphachloro-4-nitrobenzene-2-sulphonic acid.
1:4-Dichloro-2-nitrobenzene.

Acetyldicyclohexylamine

Cellulose Products.
Plasticizer (French 506574) for—
Cellulose acetate, cellulose nitrate.
For uses, see under general heading: "Plasticizers."

Chemical
Starting point in making various derivatives.

Acetyldiphenylamine

Chemical
In organic syntheses.

Electrical
Stabilizer (Brit. 423938) for—
Transformer oils.

Fats and Oils
Stabilizer (Brit. 423938) for—
Vegetable oils.

Fuel
Stabilizer (Brit. 423938) for—
Coal-carbonization spirits.

Lubricant
Stabilizer (Brit. 423938) for—
Lubricants, lubricating oils.

Petroleum
Stabilizer (Brit. 423938) for—
Petroleum oils, shale oils.

Acetylene

French: Acétylène, Éthine.
German: Acetylen, Azetylen, Aethin, Steingas.

Analysis
Reagent in various analytical processes.

Chemical
Starting point in making—
Acetals, acetaldehyde, acetaldehydedisulphonic acid.
Acetaldehyde derivatives, such as sulphonic and carboxylic acids.
Acetic acid, acetic anhydride, acetone, acetylene dichloride, acetylene tetrabromide, acetylene tetrachloride, barium acetylide, caesium acetylide, calcium acetylide, copper acetylide, ethane, ethylene, ethylene acetate, ethylidene diacetate, formic acid, hydrogen, linalool, lithium acetylide, magnesium acetylide, manganese acetylide, mercury acetylide, metallic acetylides, 3-methylbutanol, nickel acetylide, potassium acetylide, propylene, pyrrol, rubidium acetylide, silver acetylide, sodium acetylide, strontium acetylide, synthetic tannins, tellurium acetylide, tetranitromethane, tetrachloromethane, thiophene, tin acetylide, titanium acetylide, tungsten acetylide, trichloroethylene, vinyl alcohol, zinc acetylide, zirconium acetylide.

Dye
Reagent in making various synthetic dyestuffs.
Starting point in making—
Indigoes.

Explosives
As a commercial explosive.

Fuel
For illuminating isolated buildings, in miner's lamps and outdoor lights of various sorts.
For various heating purposes.
For various lighting and illuminating purposes, such as in marine lights and signals.
Fuel in internal combustion engines.

Gas
Ingredient of—
Coal gas, water gas, mixed gas, and coke-oven gas (added to increase the calorific power).

Glass
Reagent in making—
Glass and glass products.

Ink
Reagent in making—
Inks.

Metallurgical
Reagent in—
Autogeneous welding, blowpipe work, cutting metals (used along with oxygen).
Reagent in making—
Black metal.

Miscellaneous
As an anesthetic.

Paint and Varnish
Reagent in making—
Graphite.
Starting point in making—
Acetylene black.

Photographic
Reagent in making—
Photographic papers.

Rubber
Reagent in making—
Synthetic rubber.

Textile
Reagent in printing—
Calicoes.

Waxes and Resins
Reagent in making—
Synthetic resins.

Acetylene Black

Synonyms: Acetylene lampblack.
French: Noir d'acétylène.
German: Azetylenruss.

Construction
Black pigment for coloring—
Mortars, stucco, and cements.

Electrical
Ingredient of—
Compositions used in making electrical insulating parts and for insulation of electrical machinery.

Ink
Ingredient of—
Black printing inks.

Leather
Black pigment for coloring—
Natural or artificial leather.

Linoleum and Oilcloth
As a black pigment.

Metallurgical
Black pigment for coating—
Fine mechanical appliances.

Miscellaneous
Black pigment in making—
Crayons, dressings for automobile tops, shoe polishes, stove polishes.
Ingredient of—
Compositions used in making phonograph records.
Compositions used in coating optical instruments.

Paint and Varnish
Black pigment in making—
Automobile lacquers, enamels, glue and casein paints, lacquers, oil paints, polishing compositions, paints for scenery, varnishes.

Plastics
As a black pigment

Printing
Black pigment in—
Lithography and in process engraving.

Rubber
As a black pigment and filler.

Stone
Black pigment for coloring—
Artificial building stone.

Textile
Black pigment in making—
Dark-colored waxed cloth.

Woodworking
Black pigment for—
Coating, impregnating.

Acetylene-Generator Waste

Synonyms: Calcium carbide residue, Carbide of lime.

Construction
Cheaper substitute for the lime element in—
Interior and exterior plasters (the plasticity, or spreading quality, of such plasters is said to be superior to that of other plasters; also, said (1) to work more smoothly under the trowel, (2) to impart "non-buckling" or "pitting" properties).
Starting point in making—
Fire-retardant whitewashes (one formula said to be recommended by insurance companies because of its fire-retarding qualities consists of carbide residue, rye flour, common salt, dissolved and mixed according

Acetylene-Generator Waste (Continued)
to a recommended manner; a whitewash of a type said to be used extensively by the United States Lighthouse Board consists of carbide residue, salt, ground rice, glue, Spanish whiting, dissolved and mixed according to a recommended manner).
Water-tightening medium in—
Concrete (plasticity, great workability, and a resulting reduction of placing and finishing costs are said to result from its use; also, said (1) to prevent segregation or unmixing of concrete, thus reducing stone pockets or honeycombing (2) to impart greater density).

Acetylene Tetrabromide
Synonyms: Muthmann's liquid, Tetrabromoethane.
French: Liqueur de muthmann, Tétrabromure d'acétylène.
German: Tetrabromacetylen, Tetrabromazetylen.
Spanish: Tetrabromuro de acetilena.
Italian: Tetrabromuro di acetilene.

Analysis
Solvent for—
Fats, oils, waxes.
Solvent in—
Separating minerals by order of specific gravity.

Chemical
Solvent for—
Fats, oils, waxes.
Solvent miscible with most common solvents.

Cosmetic
Solvent for—
Fats, oils, waxes.

Dry-Cleaning
Spotting agent for—
Fats, oils, waxes.

Fats, Oils, and Waxes
Solvent, characterized by pleasant odor, for—
Fats, oils, waxes.

Food
Solvent, characterized by pleasant odor, for—
Fats, oils, waxes.

Miscellaneous
Solvent for—
Fats, oils, waxes.
Solvent miscible with most common solvents.

Pharmaceutical
Solvent for—
Fats, oils, waxes.

Soap
Solvent for—
Fats, oils.

Acetylene Tetrachloride
Synonyms: Acetosol, Bonoform, Cellon.
French: Tétrachlorure d'acétylène, Tétrachlorure acétylènique.
German: Acetylentetrachlorid.
Spanish: Tetracloruro de acetilene.
Italian: Tetracloruro di acetilano.

Ceramics
Solvent in—
Compositions containing cellulose acetate or other esters or ethers of cellulose used for the decoration and protection of ceramic ware.

Chemical
Denaturant in—
Certain varieties of industrial alcohol.
Reagent in making—
Carbon tetrachloride, chlorinated derivatives of ethane and ethylene, dichloroethylene, trichloroethylene, various chlorinated organic compounds.
Solvent for—
Camphor and camphor derivatives, cellulose acetate, halogens, such as chlorine, bromine, and iodine.
Organic acids and organic compounds, phosphorus.
Sulphur and sulphur compounds, such as sulphur chloride.
Various purposes (in admixture with oil of turpentine and alcohol).
Starting point in making—
Intermediates and other organic chemicals.

Electrical
Solvent in making—
Insulating compositions containing cellulose acetate or other esters or ethers of cellulose and other ingredients, such as resins.

Fats and Oils
Reagent in—
Extracting various animal and vegetable fats and oils.
Solvent for various animal and vegetable fats and oils.

Food
Reagent in—
Extracting edible oils and fats from seeds and the like.
Solvent for—
Edible oils and fats.

Gas
Solvent for—
Coaltar.
Solvent in—
Extracting sulphur from spent oxide from the dry purifiers.

Glass
Solvent in—
Compositions containing cellulose acetate or other esters or ethers of cellulose and other ingredients, used in the manufacture of non-scatterable glass and for the decoration and protection of glassware.

Glues and Adhesives
Solvent in making—
Special adhesive preparations containing cellulose acetate or other esters or ethers of cellulose and other ingredients.

Gums
Solvent for—
Dammar, elemi, sandarac, shellac, soft copals.

Insecticide
As a mild insecticide
Ingredient of—
Insecticidal and germicidal compositions.

Jewelry
Solvent in making—
Artificial pearls.

Leather
Solvent in—
Compositions containing cellulose acetate or other esters or ethers of cellulose and other ingredients, used in the manufacture of artificial leather and for the decoration and protection of leather goods.

Metallurgical
Solvent for—
Degreasing and cleansing metals.
Preparing metallic surfaces for electroplating.
Solvent in—
Compositions containing cellulose acetate or other esters or ethers of cellulose and other ingredients, used for the decoration and protection of metallic articles.
Detinning processes.

Miscellaneous
As a general solvent
Solvent in making—
Compositions containing cellulose acetate or other esters or ethers of cellulose and other ingredients, used for the decoration and protection of miscellaneous compositions of matter.
Smoke-screen preparations, shoe polishes.

Paint and Varnish
Ingredient of—
Paint and varnish removers.
Solvent in making—
Enamels.
Paints, varnishes, enamels, lacquers, and dopes containing cellulose acetate or other esters or ethers of cellulose and other ingredients.
White varnishes.

Paper
Reagent in—
Removing printing ink from old newsprint.
Solvent in—
Compositions containing cellulose acetate or other esters or ethers of cellulose, used in the manufacture of coated paper and for the decoration and protection of pulp and paper products.

Petroleum
Solvent for—
Petroleum tar.

Photographic
Solvent in making—
Films containing cellulose acetate or other esters or ethers of cellulose.

Acetylene Tetrachloride (Continued)

Plastics
Solvent in making—
Compositions containing cellulose acetate or other esters or ethers of cellulose.

Printing
Solvent for—
Cleansing lithographic plates.

Resins and Waxes
Solvent for—
Artificial and natural resins.
Solvent in—
Extracting various resins and waxes.

Rubber
As a solvent.

Soap
Ingredient of—
Dry-cleansing soaps, stain-removing compositions, textile soaps.

Stone
Solvent for—
Cleansing stone surfaces.
Solvent in—
Compositions containing cellulose acetate or other esters or ethers of cellulose and other ingredients, used for the decoration and protection of natural and artificial stone.

Textile
—, *Dyeing*
Ingredient of—
Baths for dyeing wool, cotton, rayon, or mercerized cotton.
—, *Finishing*
Ingredient of—
Finishing baths for rayon, cotton, and wool fabrics.
—, *Manufacturing*
Ingredient of—
Silk-degumming baths.
Reagent in making—
Rayon
Solvent in—
Compositions containing cellulose acetate or other esters or ethers of cellulose, used in the manufacture of coated fabrics.

Woodworking
Solvent in—
Compositions containing cellulose acetate or other esters or ethers of cellulose, used for the decoration and protection of woodwork.

Acetyleugenol

Perfume
Fixative in making—
Various perfume preparations and toiletries.

Soap
Fixative in making—
Perfumed toilet soaps.

Acetyl-H Acid

Chemical
Starting point in making various derivatives.

Dye
Starting point in making—
Aurin, acetyl red, aminonaphthol red G, aminonaphthol red 6 B, azonaphthol red, naphthol reds.

Acetylhydroquinone

Petroleum
Stabilizing agent (Brit. 406195) for—
Cracked gasolines and other motor fuels.

Acetyl Iodide
French: Iodure acétylique, Iodure d'acétyle.
German: Acetyljodid, Azetyljodid, Jodacetyl, Jodazetyl.

Chemical
Starting point in making—
Intermediate chemicals, organic chemicals.

Acetylisoeugenol
French: Acétyleisoeugénole, Isoeugénole d'acétyle.

Chemical
Starting point in making—
Vanillin.

Acetylmethylcarbinol
Synonyms: Acetoin, 2-Butanolone-3, Dimethylketol, Methyl 1-hydroxyethyl ketone.

Beverages
Aroma carrier in—
Flavors, essences.

Food
Aroma carrier in—
Essences, flavors.

Perfumery
Aroma carrier.

Acetyl-1:4-naphthylenediamine-6-sulphonic Acid
Synonyms: 8-Acetamido-5-amino-2-naphtholsulphonic acid.
French: Acide 8-acétamido-5-amino-2-naphtholsulphonique, Acide d'acétyle-1:4-naphthylènediamine-6-sulphonique.
German: 8-Acetamido-5-amido-2-naphtolsulfonsäure, Acetyl-1:4-naphtylendiamin-6-sulfonsäure.

Chemical
Starting point in making—
Intermediates and other derivatives.

Dye
Starting point in making—
Diaminogen blue B, diaminogen blue BB.

Acetyl-1:4-naphthylenediamine-7-sulphonic Acid
French: Acide d'acétyle-1:4-naphthylènediamine-7-sulphonique.
German: Acetyl-1:4-naphthylendiamin-7-sulfonsäure.

Chemical
Starting point in making—
Intermediates and other derivatives.

Dye
Starting point in making—
Diaminogen blue B, diaminogen blue BB, various dyestuffs of the diaminogen series.

Acetylorthoaminophenol

Chemical
Starting point in making—
Intermediates, organic chemicals, pharmaceuticals.

Dye
Starting point (Brit. 347099) in making azo dyestuffs with the aid of—
Sulphanilic acid.

Acetylorthotoluidin

Chemical
Starting point in making—
Acetylanthranilic acid.

Acetylpara-aminophenol
French: Acétyle-paraaminophénol.
German: Acetparaaminophenol, Azetparaaminophenol, Acetylparaaminophenol, Azetylparaaminophenol.
Spanish: Acetilpara-aminofenol, Para-acetaminofenol.
Italian: Acetilpara-aminofenolo, Para-acetaminofenolo.

Chemical
Stabilizer (German 242324) for—
Hydrogen peroxide solutions.
Starting point in making—
Acetylpara-aminophenol salicylate (salophene), acetylparaphenetidin (phenacetin), various other pharmaceutical derivatives.

Acetylpara-aminophenyl Salicylate
Synonyms: Acetamidosalol, Acetoaminosalol, Acetylpara-aminosalol, Salophene, Salophen.
French: Salicylate d'acétylepara-aminophényle.
German: Salicylsaeuresacetylpara-aminophenolester.

Perfumery
Ingredient of cosmetic preparations.

Pharmaceutical
In compounding and dispensing practice.

Acetylparaphenylenediamine
Synonyms: Para-aminoacetanilide.
French: Acétyleparaphénylènediamine.
German: Azetylparaphenylendiamin.

Chemical
Starting point in making—
Intermediates, pharmaceuticals, and other derivatives.

Acetylparaphenylenediamine (Continued)

Dye
Starting point in making—
Azo dyestuffs, azo acid red, azo corellin, azo grenadine L, cotton yellow G, coomassie wool black R, coomassie wool black S, chromotrope 6B, lanafuchsin, leather brown, orange G, thiocatechin, thiophor yellow, victoria black, victor violet 4BS.
Starting point (Brit. 285096) in making dyestuffs in the presence of dimethylanilin, nitrobenzene, orthodichlorobenzene or naphthalene, with the aid of—
1-Allyleneoxy-4-aminoanthraquinone.
1-Allyloxy-4-aminoanthraquinone.
1-Amyleneoxy-4-aminoanthraquinone.
1-Amyloxy-4-aminoanthraquinone.
1-Butyleneoxy-4-aminoanthraquinone.
1-Butyloxy-4-aminoanthraquinone.
1-Ethyleneoxy-4-aminoanthraquinone.
1-Ethyloxy-4-aminoanthraquinone.
1-Heptyleneoxy-4-aminoanthraquinone.
1-Heptyloxy-4-aminoanthraquinone.
1-Hexyleneoxy-4-aminoanthraquinone.
1-Hexyloxy-4-aminoanthraquinone.
1-Isoallyloxy-4-aminoanthraquinone.
1-Isoamyloxy-4-aminoanthraquinone.
1-Isobutyloxy-4-aminoanthraquinone.
1-Isopropyloxy-4-aminoanthraquinone.
1-Methyleneoxy-4-aminoanthraquinone.
1-Methyloxy-4-aminoanthraquinone.
1-Pentyleneoxy-4-aminoanthraquinone.
1-Pentyloxy-4-aminoanthraquinone.
1-Propyleneoxy-4-aminoanthraquinone.
1-Propyloxy-4-aminoanthraquinone.

Acetylphenylhydrazin

Chemical
In organic syntheses.

Electrical
Stabilizer (Brit. 423938) for—
Transformer oils.

Fats and Oils
Stabilizer (Brit. 423938) for—
Vegetable oils.

Fuels
Stabilizer (Brit. 423938) for—
Coal-carbonization spirits.

Lubricant
Stabilizer (Brit. 423938) for—
Lubricants, lubricating oils.

Pharmaceutical
Suggested for use as—
Antipyretic.

Petroleum
Stabilizer (Brit. 423938) for—
Petroleum oils, shale oils.

Acetylphloroglucinol

Petroleum
Stabilizing agent (Brit. 406195) for—
Cracked gasolines and other motor fuels.

Acetylpyrocatechol

Petroleum
Stabilizing agent (Brit. 406195) for—
Cracked gasolines and other motor fuels.

Acetylpyrogallol

Petroleum
Stabilizing agent (Brit. 406195) for—
Cracked gasolines and other motor fuels.

Acetylquinine

German: Acetylchinin.

Chemical
Starting point in making—
Pharmaceutical derivatives.

Pharmaceutical
In compounding and dispensing practice.

Acetylresorcinol

Petroleum
Stabilizing agent (Brit. 406195) for—
Cracked gasolines and other motor fuels.

Acetylsalicylamide

Chemical
Starting point in making—
Pharmaceutical and other derivatives.

Pharmaceutical
In compounding and dispensing practice.

Acetylsalicylic Acid

Synonyms: Aspirin.
Latin: Acidum acetylosalicylicum, Acidum acetylsalicylicum.
French: Acide d'acétyle salicylique.
German: Acetyliertsalicylsäure, Azetyliertsalizylsäure, Acetylsäuressalizyl, Azetylsäuressalizyl, Salicylsäuresacetylester, Salizylsäuresazetylester, Salicyliertacetylsäure, Salizyliertazetylsäure.

Pharmaceutical
In compounding and dispensing practice.
Suggested for use (either alone or in combination) in treating—
Colds, headaches, nervous conditions, rheumatic condition.

Acetyltribromophenyl Salicylate

Synonyms: Acetyltribromsalol, Cordyl.

Chemical
Starting point in making various derivatives.

Pharmaceutical
In dispensing practice.

Acid Orange G

French: Orange d'acide.
German: Saeureorang.

Miscellaneous
Coloring matter (Brit. 279942) for—
Bone, furs, horn, ivory, and the like.

Photographic
Coloring matter (Brit. 279942) for—
Films made from celluloid.

Resins and Waxes
Coloring matter (Brit. 279942) for—
Resins.

Textile
——, Dyeing and Printing
Coloring matter for—
Woolens and other textiles.

Acriflavin Base

Synonyms: 3:6-Diamino-10-methylacridinium chloride, Euflavin, Euflavine, Neutral acriflavine, Neutral trypaflavine.
French: Acriflavine neutre, Chlorure de 3:6-diamino-10-méthyleacridinium.
German: 3:6-Diamino-10-methylacridiniumchlorid.
Spanish: Cloruro de 3:6-diamino-10-metilacridinio.
Italian: Cloruro di 3:6-diamino-10-metilacridinio.

Pharmaceutical
As an antiseptic, as a bacteriostatic.

Acrolein

Synonyms: Acraldehyde, Allyl aldehyde, Propenal.
French: Acroléine, Aldehyde d'allyle, Aldehyde allylique.
German: Akrolen, Allylaldehyd.

Chemical
Reagent (Brit. 325669) in making synthetic perfumes with the aid of—
1:3-Cyclohexadiene, 1:1-dimethylbutadiene, 1:3-dimethylbutadiene, 1:4-dimethylbutadiene, 2:3-dimethylbutadiene, 2:4-dimethylbutadiene, 1-methylbutadiene.
Reagent in making compounds with—
Dextrins, gums, proteins, starches.
Reagent in making—
Metallic colloids of various sorts.
Starting point in making—
Allyl alcohol, catalytic hydrogenation products, dichloroacrolein, monochloroacrolein, products with anilin, propionaldehyde, electrical insulating products.

Military
Ingredient of—
Poison gas preparations.

Metallurgical
Reagent in making colloidal forms of—
Osmium, rhodium, ruthenium.

Pharmaceutical
In compounding and dispensing practice.

Acrolein (Continued)

Plastics
Reagent in making—
Condensation products from phenol by reaction with formaldehyde.
Condensation products from urea by reaction with formaldehyde (Brit. 260288 and U. S. 1654215).

Refrigeration
Warning agent in methyl chloride for use in refrigeration.

Sanitation
Disinfectant in treatment of—
Water and sewage.
Ingredient of—
Disinfecting compositions (in admixture with formaldehyde).

Acroleinoxime

Fuel
Primer (Brit. 429763) for—
Diesel engine fuel oils produced by the hydrogenation of coal.

Petroleum
Primer (Brit. 429763) for—
Diesel oils containing a high proportion of aromatic bodies.

Acrylic Acid
Synonyms: Ethylenecarboxylic acid, Propene acid.
French: Acide d'acryle, Acide acrylique, Acide de éthylènecarboxylique, Acide de propène.
German: Acrylsäure, Akrylsäure, Propenesäure.

Adhesives
Starting point in making—
Polymerized esters used for making adhesives for gluing paper to metal (Brit. 311339).

Chemical
Starting point (in organic synthesis) in making—
Additive compounds with hydrogen, halogens, halogen hydrides, hydrogen cyanide.
Amides, esters, salts.
Polymerized products used for enhancing the properties of cellulose derivatives with respect to adhesive power, ductility, elasticity, stability to light; these products are colorless, odorless, noninflammable, and practically transparent.

Electrical
Starting point (Brit. 340677) in making—
Polymerized products, such as polymerized acrylic acid ester, halogenides, and nitriles, which, in admixture with paper, mica, or the like, or polymerized oils or oil preparations such as linoxyn, may be used as insulating materials for covering wires and cables.

Miscellaneous
Breaker (U. S. 1964444) of—
Emulsoid stability.

Paint and Varnish
Ingredient (Brit. 311339) of—
Paints (used for the purpose of improving the covering power, facilitating their use, and allowing the incorporation of a greater proportion of pigment).
Ingredient (Brit. 404819) of—
Compositions for cleansing metal prior to painting or otherwise coating; includes also rust-removing agents, removers of oil and grease, a soap-like glucoside of vegetable origin.
Starting point in making—
Polymerized products used for enhancing the properties of cellulose derivatives with respect to adhesive power, ductility, elasticity, stability to light; these products are colorless, odorless, noninflammable and practically transparent.
Polymerized products employed in finely pulverulent form for the production of lacquer coatings or films by directly applying the voluminous powder and then brushing or spraying with an organic solvent or heating in a furnace; may be employed as mixtures with (1) other polymerizable substances, (2) polymerization products of acrylic acid homologous, or salts thereof, or those esters thereof in which the alcohol group of the monomeric ester contains at least two carbon atoms less than the acid group (Brit. 404504).

Plastics
Starting point in making—
Polymerized products used for enhancing the properties of cellulose derivatives with respect to adhesive power, ductility, elasticity, stability to light; these products are colorless, odorless, noninflammable and practically transparent.

Rubber
Starting point (Brit. 340004) in making—
Synthetic rubber from polymerized hydrocarbons, in the form of coatings, films, discs, and threads, with the aid of iron carbonyl, nickel carbonyl, cobalt carbonyl, molybdenum carbonyl, tungsten carbonyl, or chromium carbonyl.

Soap
Ingredient (Brit. 407039) of—
Antiseptic washing and cleansing agents prepared by incorporating water-soluble salts of mercury, silver, or gold, which dissociate into metal ions, with aliphatic compounds (esters) having strong wetting and washing power, the metal salts thus formed being water-soluble.

Water and Sanitation
Breaker (U. S. 1964444) of—
Stability of emulsoids in sewage.

Acrylic Chloride
French: Chlorure d'acryle, Chlorure acrylique.
German: Akrylchlorid, Chlorakryl.
Spanish: Chloruro de acril.
Italian: Chloruro di acrile.

Miscellaneous
Starting point (French 697437) in making—
Polymerized artificial masses that give solutions which are viscous, elastic, vulcanizable, and stretchable.

Adenin
Synonyms: 6-Aminopurin.

Chemical
In organic synthesis.

Photographic
Defogging agent (Brit. 442731) for—
Gelatin having a strong tendency to cause fog.

Adeps Lanae (Anhydrous and Hydrous)
Synonyms: Anhydrous lanolin, Anhydrous wool fat, Hydrous wool fat, Lanolin, Lanoline, Wool fat, Wool grease.
Latin: Adeps lanae anhydricus, Adeps lanae cum aqua, Adeps lanae hydrosus, Lanolinum, Lanolinum anhydricum, Lanum.
French: Lanoline, Suinte de laine.
German: Wasserhaltiges wollfett, Wollfett.
Spanish: Lanolina.
Italian: Lanolina.

Construction
Rust-preventive coating and paint for—
Girders, steel structures.

Ink
Emulsifying agent (Brit. 388072) for—
Water-in-oil emulsion suitable for use as a printing ink vehicle.
Ingredient of—
Printing inks.

Leather
Alone or in combination for—
Dressing, finishing, softening.

Mechanical
As a special lubricant.
Ingredient of—
Belt-dressing compositions, cotton spinning oils.

Metallurgical
Rust-preventive for—
Iron, steel.

Miscellaneous
Protective coating for—
Ship bottoms.

Paint and Varnish
Emulsifying agent (Brit. 383238) for—
Paints of various kinds, quick-drying paints, varnishes.
Ingredient of—
Paints of various kinds, varnishes.

Perfume
As a skin-feeding medium for the hands.

Adeps Lanae (Continued)
Base for—
Cosmetics, creams, ointments, pomades.
Ingredient of—
Preparations for the hair.
Pharmaceutical
Base for—
Emulsions, ointments, salves.
Soap
Base for—
Shaving soaps, toilet soaps.

Adipic Acid
Synonyms: Adipinic acid, Hexane di-acid.
French: Acide adipinique, Acide adipique, Acide hexanedioique.
German: Adipinsäure, Hexandisäure.
Spanish: Acido adipico, Acido adipinico.
Italian: Acido adipico, Acido adipinico.
Analysis
Making standard solutions for various analytical purposes, particularly in volumetric analysis.
Chemical
Reagent in treating—
Yeast to remove the bitter matters contained therein.
Starting point in making—
Adipyl dichloride (German 318222), amyl adipate, calcium adipate, cetyl adipate, cyclohexyl adipate, esters with hydroaromatic alcohols, esters with hydrogenated crude cresol, isopropyl adipate, methylcyclohexanol adipate, octodecyl adipate, orthomethylcyclohexyl adipate, perfume fixatives (German 373219), pharmaceutical chemicals, products for protecting vegetables, sodium salts of derivatives of adipic acid and alcohols.
Dye
Ingredient of—
Various dyestuffs.
Food
Ingredient of—
Mineral yeast (used in the place of tartaric acid, cream of tartar, and biphosphates for the purpose of making a more stable product and one that is non-hygroscopic).
Reagent for—
Removing bitter matters from pressed yeast.
Leather
Resist in—
Dyeing leather goods.
Metallurgical
Reagent in treating—
Metals to color them and to produce bronze effects.
Miscellaneous
Ingredient of—
Fatty compositions made with glycollic acid (German 318922).
Polishing liquids.
Pharmaceutical
In compounding and dispensing practice.
Photographic
Reagent in making—
Photographic papers.
Plastics
Reagent in—
Galvanoplastic work.
Starting point (German 318922) in making—
Plastic resinous products with glycerin.
Resins and Waxes
Starting point (German 318922) in making—
Artificial wax sizes with glycerin and stearic, palmitic, and montanic acids.
Textile
Mordant and resist in—
Dyeing and printing cottons and rayons.

Adipic Acid Ester of Grapeseed Alcohol
Bituminous
Solvent (Brit. 445223) for—
Asphalt and other bituminous bodies.
Dye
Solvent (Brit. 445223) for—
Dyestuffs, particularly oil-soluble coaltar dyes.
Fats, Oils, and Waxes
Solvent (Brit. 445223) for—
Fats, oils, waxes.
Resins
Solvent (Brit. 445223) for—
Oil-soluble glycerol-phthalic acid resins, polymerized vinyl compounds, synthetic resins.
Rubber
Solvent (Brit. 445223) for—
Rubber.

Adipic Acid Ester of Ricinoleic Alcohol
Bituminous
Solvent (Brit. 445223) for—
Asphalt and other bituminous bodies.
Dye
Solvent (Brit. 445223) for—
Dyestuffs, particularly oil-soluble coaltar dyes.
Fats, Oils, and Waxes
Solvent (Brit. 445223) for—
Fats, oils, waxes.
Resins
Solvent (Brit. 445223) for—
Oil-soluble glycerol-phthalic acid resins, polymerized vinyl compounds, synthetic resins.
Rubber
Solvent (Brit. 445223) for—
Rubber.

Adolalphanaphthylamine
Petroleum
Preventive of—
Antiknock property deterioration (on storage) in gasoline containing di- and triolefins.
Discoloration (on storage) of gasoline containing di- and triolefins.
Gumming (on storage) of gasoline containing di- and triolefins.
Stabilizer for—
Gasoline containing di- and triolefins.

Adrenalin
Synonyms: Epinephrine, Suprarenalin, Suprarenin.
Latin: Adrenalinum.
French: Adrénaline.
German: Adrenalin.
Pharmaceutical
In compounding and dispensing practice.
Soap
Ingredient (U. S. 1744061) of—
Shaving preparations.

Agar-Agar
Synonyms: Ceylon Agar-Agar [chiefly gracilaria lichenoides, Ag.], Japanese Agar-Agar [Japanese isinglass, Sphaerococcus compressus, Ag., Gloiopeltis tenax, J. Ag., Gelidium corneum lam., Gelidium cartilogineum Gaill.], Macassar Agar-Agar [chiefly impure Eucheuma spinosum Ag., incrusted with salt], Vegetable isinglass.
French: Mousse de chine.
German: Wurmmoss, Wurmtang.
Spanish: Agar-Agar.
Italian: Agar-Agar.
Adhesives
Ingredient of—
Glue preparations, various adhesive preparations.
Substitute for—
Gelatin.
Chemical
Nutrient in—
Preparation of bacteriological culture media (plain agar-agar, glucose agar-agar, glycerin agar-agar, lactose-litmus agar-agar).
Food
Ingredient of various vegetarian foods.
Raw material in manufacture of—
Sausage casings.
Substitute for—
White of egg.
Thickening agent in—
Cake mixes, cream, ice cream, milk, pudding preparations, soups, various foods.
Miscellaneous
Suspending agent (U. S. 1724134) in—
Wire-drawing lubricant containing graphite and sulphonated oil.

Agar-Agar (Continued)

Perfume
Ingredient of—
Greaseless creams, lotions.

Pharmaceutical
Ingredient of various pharmaceutical preparations.
Suggested for use as a laxative.

Photographic
Reagent in—
Sensitized emulsions.

Textile
Sizing for—
Silk.
Thickener in—
Dyeing and printing of fabrics.

Agaric Acid
Latin: Acidum agaricinicum.
French: Acide agaricique.
German: Agaricinsaeure.

Chemical
Starting point in making—
Esters and salts, pharmaceuticals.

Pharmaceutical
In compounding and dispensing practice.

Akebi Seed Oil
French: Huile de semences d'akebi.
German: Akebisamenoel.

Chemical
Starting point in making—
Turkey red oil.

Food
As an edible fat.

Fuel
As an illuminant.

Soap
As a soapstock.

Alanin Hydrochloride
French: Hydrochlorure d'alanine.
German: Alaninchlorhydrat.

Chemical
Starting point in making—
Pharmaceutical product with hexamethylenetetramine (U. S. 1588753).

Pharmaceutical
In compounding and dispensing practice.

Albumen (Blood, Egg and Milk)
Synonyms: Albumine, Egg white.
Latin: Albumen, Ovi albumen.
French: Albumine de lait, Albumine d'oeuf, Albumine de sang.
German: Eiweisz.
Spanish: Albumina.

Analysis
Reagent in testing for—
Foreign oil in olive oil, sugar colors, wine colors.

Beverage
Reagent in clarifying wines and distilled liquors.

Chemical
As a clarifying agent.
Starting point in making—
Albuminates of various sorts, arsenic compounds, brominated derivatives for pharmaceutical use, chlorinated derivatives for pharmaceutical use, iron albuminates, iodated derivatives for pharmaceutical use, ichthyol albuminate, manganese albuminate, pharmaceutical products with albumoses and peptones, tannin albuminate.

Dye
Ingredient of—
Dye preparations.

Food
Clarifying agent in treating—
Cider, soft drinks.
Ingredient of—
Bakery products, confectionery, food preparations.

Glues and Adhesives
Ingredient of—
Cements, glues and veneers (French 563379).
Insolubilized glue preparations (French 493561).
Waterproof glue preparations with hexamethylenetetramine.

Leather
Ingredient of—
Compositions used for fixing gold on leather.

Metallurgical
Ingredient of—
Compositions used for fixing gold on metals.

Miscellaneous
Activating agent (Brit. 378888) for—
Enzymes in certain processes.
Clarifying agent in treating various liquids.
Ingredient of—
Compositions used for stamping products and materials with gold or bronze powder.
Compositions used in making porous insulating bodies (French 628701).
Varnishes used for treating skins.

Paint and Varnish
Ingredient of—
Colors for paint and varnish (used in admixture with sodium resinate) (French 545047).
Fine color preparations.
Lacquers (with the addition of sodium resinate) (French 545047).
Varnishes.
Varnishes (with the addition of sodium resinate) (French 545047).

Paper
Ingredient of—
Compositions used in sizing.

Pharmaceutical
In compounding and dispensing practice.

Photographic
Ingredient of—
Compositions used in making papers.
Compositions used for coating plates.

Plastics
Ingredient of—
Artificial ivory.
Infusible compositions made with the addition of glycerin (French 530145).

Printing
In printing and lithography.

Sugar
Reagent in clarifying—
Sugar and cane juices.

Textile
——, *Dyeing*
As a mordant.
Reagent in rendering dyes fast.

——, *Finishing*
Ingredient of—
Compositions used for fixing gold on fabrics.
Finishing lacquers.

Alcohol
Synonyms: Anhydrous ethyl alcohol, Chemically pure ethyl alcohol, Cologne spirits, Completely denatured alcohol, Dehydrated alcohol, Denatured alcohol, Ethanol, Ethyl alcohol, Ethyl hydroxide, Ethylic alcohol, Fermentation alcohol, Grain alcohol, Industrial alcohol, Proof spirit, Pure alcohol, Pure ethyl alcohol, Rectified spirit, Specially denatured alcohol, Spirits of wine, Synthetic ethyl alcohol.
Latin: Alcohol absolutum, Alcohol absolutus, Alcohol dehydratum, Alcohol dilutum, Alcohol ethylicum, Alcohol vini, Spiritus, Spiritus dilutus, Spiritus rectificatus, Spiritus vini rectificatissimus, Spiritus tenuior.
French: Alcool, Alcool absolu, Alcool dilué, Esprit de vin.
German: Absoluter alkohol, Aethyl alkohol, Alkohol, Rectificirter weingeist, Verdünnter spiritus, Verdünnter weingeist.
Spanish: Alcool anhidro, Alcool assoluto, Espiritu rectificado de vino.
Italian: Acquavite rectificata.

Note:—The uses enumerated below include all grades and varieties of pure and denatured alcohol.

Adhesives
Solvent in making—
Cements of various kinds.

Analysis
Preservative, reagent, and solvent in—
Analytical processes involving control and research in science and industry.

Alcohol (Continued)

Aviation

Indicator in—

Compasses, gauges, indicating devices of various kinds.

Cellulose Products

Solvent for—

Nitrocellulose in making various products, such as rayon, collodion and the like.

Chemical

Ethylating agent in—

Organic syntheses.

Extractant in—

Manufacturing processes.

Process material in making—

Acetaldehyde, acetanilide, acetic ether, acetone, acetparamidophenolsalol, acetphenetidin, acetylsalicylic acid, adalin, adeps lanae, aldehydes, aletrin, aloin, ammonium salts of lauric and oleic acids, anesthetic ether, antipyrine, apocynin, arabinose, arbutin, artificial musks, arsphenamine, asclepiadin, atophan, avenin, barbitol, benzaldehyde, benzene cyanide, benzidin, benzoic acid, benzoin, benzonaphthol, benzonaphthol benzoate, betanaphthol, betanaphthol benzoate, betanaphthol centralites, betanaphthol salicylate, bluing (laundry), butyric ether, calcium acetate, camphor (synthetic), caproic acid, caproic ether, certified food colors, chloral hydrate, chloroform, cinchona alkaloids, cinchophen, citronellol, cocaine salicylate, codeine, collodions, coumarin, creosote carbonate, dextrin, dextrose, diacetylmorphine, diamidophenol, diastase, diethyl phthalate, diethylbarbituric acid, digestive ferments, digitalis active principles, dimethylglyoxime, dimethyl sulphate, dinitrotoluene, emulsions, ethers, ethyl acetate, ethyl benzoate, ethyl bromide, ethyl butyrate, ethyl chloride, ethyl esters of cinnamic, lauric, benzoic, and pelargonic acids, ethyl lactate, ethyl laevulinate, ethyl nitrite, ethyl oxidehydrate, ethyl palmitate, ethyl propionate, ethyl sulphate, ethyl valerate, ethylene, ethylmorphine, formaldazone, formaldehyde, formic ether, galactose, glycerophosphates, guaiacol, guaiacol carbonate, heliotropin, hexachlorobenzene, homatropin, hydrastis alkaloids, hydrazonanisol, hydroquinone, insulin, inulin, intermediates, iodine, iriscin, isinglass, laevulose, maltose, mannose, melibiose, mercurochrome, Micheler's ketone, monobromated camphor, morphine, neoarsphenamine, neosalvarsan, nitrobenzene, nitrosobetanaphthol, nitrose ether, orthotoluolsulphamide, paramidophenol, paranitrophenatol, paraphenetidin, pectin, pepsin and similar products, phenacetin, phenolphthalein, phenylacetic acid, phenylcinchoninic acid, potassium cyanate, potassium hydroxide, potassium sodiumxanthate, propionic ether, protargentum, raffinose, resorcinol, rhamnose, saccharine, saccharose, salicylic acid, salicylic aldehyde, salol, salophen, salvarsan, santonin, saponin, sodium, sodium benzoate, sodium ethylsulphate, sodium hydrosulphate, sodium hydroxide, sodium hyposulphite, solidified alcohol, solvents, sorbit, strontium, strychnine, sulphonic acids, sulphuric ether, tannic acid, terpin hydrate, theobromine, toluidin, urea, uric acid, veronal, xanthates, xylose.

Reaction medium and solvent in—

Crystallization processes, extraction processes, purification processes.

Reducing agent in—

Organic syntheses.

Solvent for—

Acids, such as abietic, acetic, adipic, angelic, citric, tannic, tartaric.

Ammonia.

Ammonium compounds, such as the benzoate, bromide, chloride, iodide, nitrate, picrate, salicylate, sulphide, sulphocyanate, valerianate.

Balsams.

Castor oil (very soluble).

Delinquescent salts (except potassium carbonate).

Fine chemicals, such as acetanilide, acetylsalicylic acid, camphor, antipyrine, antipyrine acetylsalicylate, antipyrine salicylate, acetphenetidin, acetophenone, potassium iodide, iodine, menthol, thymol, and the like.

Fixed oils (sparingly).

Gums.

Inorganic chemicals, such as sodium hydroxide, potassium hydroxide, lithium hydroxide, aluminum bromide, aluminum chloride, antimony chloride, arsenic chloride, various other chlorides, calcium nitrate, silver nitrate, sulphur, phosphorus.

Intermediates.

Many other chemicals, raw materials, semifinished products, and materials in course of processing.

Nitrocellulose.

Organic chemicals, such as acetamide, acetic anhydride, acetin, acetone, acetylene, acetylene tetrabromide and tetrachloride, acrolein, aldol, allyl alcohol, aminoazotoluene hydrochloride, amyl acetate, amyl alcohols, amyl nitrate, amyl ether, amyl salicylate, amyl valeriate, anethole, anisaldehyde, anisole.

Pharmaceutical and medicinal chemicals, such as absinthin, acetal, acetylphenylhydrazin, acitrin, adonipin, alphol, aluminum phenolsulphonate, amylene, anemonin, anhalonidine, anhalonine, apiol.

Resins.

Soap.

Vegetable alkaloids, such as aconitine, apomorphine, apoatropine, apocodeine, codeine, morphine, hyoscine.

Volatile oils.

Starting point in making—

Alcoholates, such as those of soda and potash.

Esters with acids, such as acetic, benzoic, butyric, hydrochloric, formic, nitric, oxalic, succinic, propionic, sulphuric.

Miscellaneous chemicals, such as ethylmercaptan, ether, ethylene, dichloroethane, acetaldehyde, acetic acid, chloral hydrate, chloroform, iodoform.

Cosmetic

Diluent in—

Lotions, perfumes.

Extractant for—

Odorant principles.

Solvent for—

Cosmetic ingredients, hair tonic ingredients, odorants, perfume ingredients, shampoo ingredients, toilet preparation ingredients, toilet water ingredients, waxes.

Disinfectant

Antiseptic.

Solvent medium for—

Antiseptics, deodorants, disinfectants, germicides.

Dye

Process material in making—

Anilin and other dyestuffs, diethylanilin, dye intermediates, dye solutions, eosin, ethylanilin, gallocyanin, postal card colors.

Reaction medium in—

Reduction processes.

Solvent for—

Acenaphthene, acridin, alizarin, alphanaphthol, alphanaphthylamine, alphanitronaphthalene, aminoanthraquinone (sparingly), aminoazobenzene and hydrochloride, aminobenzoic acid, aminonaphtholdisulphonic acid (H acid), aminonaphtholsulphonic acid (Gamma acid), aminophenol (slightly) and hydrochloride, aminosalicylic acid, anilin and hydrochloride, anthracene, anthragallol, anthranilic acid, anthrapurpurin, anthraquinone, anthrarobin, benzanthrone, benzidin, betanaphthol, betanaphthylamine, diaminoazotoluene, diaminodiphenic acid, diaminodiphenylamine, diaminodiphenylmethane, diaminostilbene, dimethylanilin, intermediates, metanitroparatoluidin, naphthalene, naphthols, naphtholdisulphonic acid (R acid), naphtholsulphonic acid (Schaeffer's acid), naphtholsulphonic acid (F acid), naphthylamines, naphthylaminesulphonic acids, naphthylenediaminesulphonic acids, nitroanisole, nitrobenzene, nitronaphthalenesulphonic acid, nitroparacresol, nitrophenol, orthoanisidin, para-aminoacetanilide, para-aminodimethylanilin, paranitroorthotoluidin, paranitrophenetole, paranitrosophenol, paranitrotoluene.

Explosives and Matches

Reactant in making—

Fulminate of mercury.

Solvent for nitrocellulose in making—

Gelatin dynamites, smokeless powders.

Solvent in making—

T.N.T.

Fats, Oils, and Waxes

Extractant for—

Essential oils.

Alcohol (Continued)
Solvent for—
Castor oil (very soluble), fixed oils (sparingly), volatile oils, waxes (some).

Food
Preservative for—
Condiments, foodstuffs.
Process material in making—
Food preparations, sweetmeats.
Solvent for—
Condiments, food colors.
Starting point in making—
Vinegars.

Fuel and Light
Fuel for—
Chafing dishes, cigarlighters, small stoves, soldering torches, Welsbach mantles.
Process material in making—
Gas mantles.
Starting agent for use in—
Gasoline lamps (either gasoline only or lamps for burning either gasoline or kerosene).
Starting point in making—
Solid alcohol fuels.

Glass
Cleansing agent for—
Special glassware, such as thermometer and barometer tubes.
Solvent for—
Gums and resins used in coating mirrors.
Materials used in making nonscatterable glass.

Gums
Solvent for—
Certain gums.

Ink
Solvent for—
Dyes, ink ingredients.

Insecticide and Fungicide
Solvent in making—
Cattle dips, fumigants, insecticides, insect powders, moth-repellants, plantwashes, sheepdips.

Laundry
Solvent in making—
Laundry starches, washing preparations.

Leather
Solvent for—
Dressings, dyes, nitrocellulose, waxes (some).
Solvent in making—
Artificial leather.

Lubricant
Solvent for—
Mineral oils, lubricating ingredients.
Solvent in making—
Lubricants, soluble cutting oils.

Marine
Indicator in—
Compasses, gauges, indicating devices of various kinds.

Mechanical
Antifreeze for—
Automobile engines.
Carbon-remover for—
Automobile engines, marine engines.
Fuel for—
Automobile engines, marine engines.
Ingredient of—
Antifreeze preparations, engine fuels.

Metal Fabrication
Cleansing agent in the manufacture of—
Bronzeware, cutlery, metal articles, silverware.

Metallurgical
Cleansing agent.
Polishing medium.
Solvent in—
Coloring processes, etching solutions, lacquering processes, soldering fluxes.

Miscellaneous
Cleansing agent in—
Households, industrial factories.
Cleansing agent in the manufacture of—
Jewelry, watches.
Diluent in—
Furniture polishes, metal polishes, shoe polishes, various other special polishes.
Dye solvent in making—
Artificial flowers, hats, postcards.
General solvent for—
Household purposes, industrial use.
Gum solvent in making—
Brushes, felt hats, hats of various kinds, straw hats.
Indicator in—
Compasses, gauges, indicating devices of various kinds.
Lacquer solvent in making—
Hats.
Polishing medium.
Preservative for—
Anatomical specimens, botanical specimens, microscopical specimens, various products.
Solvent in making—
Carbon lamp filaments.

Oilcloth and Linoleum
Solvent.

Paint and Varnish
Blending agent, diluent, thinner, and solvent in—
Dopes, enamels, lacquers, paints, paint-removers, stains, varnishes, water colors.
Solvent for—
Colors, gums, nitrocellulose, resins.

Paper
Solvent for—
Nitrocellulose (used in coatings and decorative effects).

Petroleum
Solvent in—
Refining processes.

Pharmaceutical
Extractant for—
Alkaloids, other plant principles.
Extractant in making—
Fluid extracts, solid extracts, tinctures.
In compounding and dispensing practice.
Preservative for—
Anatomical specimens, microscopical specimens, various products.
Process material in making—
Anesthetics, antiseptics, drugs, liniments, lotions.
Rubbing agent.
Solvent for—
Drugs of many kinds.
Solvent and starting point in making—
Alcoholatures, alcoholates, spirits, tinctures.

Photographic
Dehydrating agent in making—
Negatives, prints, plates.
Solvent for—
Nitrocellulose.

Plastics
Solvent in making—
Celluloid, plastics of various types.
Products of various kinds, such as billiard balls, novelties, and other articles.

Printing
Solvent in making—
Photoengravings.

Rayon
Solvent.

Resins
Solvent for—
Natural resins, synthetic resins (some).

Rubber
Solvent.

Soap
Solvent in making—
Disinfecting soaps, liquid soaps, special soaps, textile soaps, transparent soaps.

Textile
Solvent for—
Dyes, nitrocellulose.
Solvent in—
Dyeing processes, printing processes.

Tobacco
Solvent for—
Flavorings, nicotine.

Aldehyde Ammonia

Synonyms: Aldamine, Aldehydate of ammonia, Ammoniated ethylic acetaldehyde, Ammonium aldehydate.
French: Aldéhydate d'ammoniaque, Aldéhydate d'ammonium.
German: Aldehydammoniak.
Spanish: Aldehidato de amonia.
Italian: Aldeidato di ammonio.

Chemical
Reagent (Brit. 312233) in extracting—
Humic acid from peat, humic colloids from peat.
Starting point in making—
Pure aldehyde.

Metallurgical
Ingredient (Brit. 309029) of—
Soldering fluids.

Plastics
Plasticizer (Brit. 281223) in making—
Casein plastics.

Alginic Acid

Synonyms: Algin.
French: Acide alginic, Acide alginique, Tangacide.
German: Alginsäure.
Italian: Alginico.

Ceramics
Ingredient of—
Compositions used for waterproofing various ceramic products.

Chemical
Emulsifying agent in making—
Emulsions of hydrocarbons of various groups of the aromatic and aliphatic series.
Emulsions of various chemicals, terpene emulsions.
Textile lubricants in emulsified form.
Wetting compositions in emulsified form.
Ingredient of—
Various chemical liquids (added for the purpose of increasing their viscosity).
Reagent for—
Purifying various pharmaceutical solutions.
Treating various chemical liquids for the purpose of clarifying and purifying them (French 570636).
Stabilizer in—
Emulsions of various chemicals and chemical products.
Starting point in making—
Ammonium alginate by reaction with ammonium hydroxide.
Calcium alginate by reaction with a calcium salt.
Copper-ammonium alginate, iodinated pharmaceutical products, potassium alginate, sodium alginate, zinc-ammonium alginate.

Construction
Ingredient of—
Compositions used for treating cement and concrete for the purpose of preventing deterioration when exposed to the action of alkalies and seawater.
Waterproofing compositions used for treating plaster of Paris, wallboard, cement, stucco, and concrete.

Disinfectant
Emulsifying agent in making—
Emulsified germicides and disinfectants.

Fats and Oils
Dispersing agent in making—
Emulsified fatty acids of animal and vegetable origin.
Emulsified fats and oils of animal and vegetable origin.
Emulsified fat-splitting compositions.
Emulsified greasing and lubricating compositions.
Emulsified sulphonated oils.
Reagent in—
Purifying various vegetable and animal fats and oils.
Stabilizer in—
Emulsions of various animal and vegetable fats and oils.

Fuel
Binder in—
Compositions for fuel briquettes, containing coal dust (used in the place of pitch).
Non-smoking fuel briquettes (used to avoid the large amount of smoke given off by briquettes made with the usual type of binder).

Glues and Adhesives
Ingredient (French 563726) of—
Adhesive preparations.
Reagent in—
Treating solutions of gelatin, glue, and other adhesives for the purpose of purifying and clarifying them.

Ink
Dispersing agent in making—
Emulsified printing and writing inks.
Ingredient of—
Printing ink (French 563726) (added for the purpose of thickening the product).
Various inks.

Insecticide
Dispersing agent in making—
Emulsified insecticidal and fungicidal preparations.

Leather
Dispersing agent in making—
Emulsified dressing compositions.
Emulsified fat-liquoring baths.
Emulsified soaking compositions.
Emulsified softening compositions.
Emulsified waterproofing compositions.
Ingredient (French 563726) of—
Compositions used for sizing leathers (used in place of starch and gum tragacanth).
Ingredient (French 533465) of—
Compositions containing various fatty substances, used in the preparation of emulsions for tanning and tawing leather.

Mechanical
Ingredient of—
Compositions used for covering steel tubes.
Compositions containing sodium carbonate, used as boiler compounds (added for the purpose of improving the water-softening properties of the sodium carbonate).

Metallurgical
Binder (French 518037) in—
Compositions containing graphite, lampblack, and antiseptics, used for repairing metallurgical furnaces and ovens.

Miscellaneous
Binder in making—
Compositions containing powdered mica, asbestos, coal, carbon, graphite, minerals, and the like.
Compositions used for sizing purposes (used in place of starches and gum tragacanth, and alleged to give a size of improved elasticity and greater transparency).
Preparations containing graphite, lampblack, and antiseptics, used for repairing stoves (French 518037).
Dispersing agent in making—
Automobile polishes in emulsified form.
Compositions for cleansing paint and metal surfaces.
Emulsions of various substances, such as coaltar, coaltar pitch, and asphalt.
Furniture polishes in emulsified form.
Metal polishes in emulsified form.
Scouring compositions in emulsified form, for various purposes.
Waterproofing compositions in emulsified form.
Ingredient of—
Antigrease compositions (French 563726).
Compositions used for treating rope and twine.
Compositions used for various waterproofing purposes.
Reagent in—
Treating various liquid preparations for the purpose of clarifying and purifying them.
Stabilizer in making—
Emulsions of various substances.

Paint and Varnish
Dispersing agent in making—
Emulsified asphaltic paints and varnishes.
Emulsified paints and varnishes.
Shellac emulsions.
Waterproofing compositions in emulsified form.
Ingredient (French 563726) of—
Compositions used for treating interior walls and ceilings.
Various paints, lacquers, and enamels.

Paper
Binder (French 563726) in making—
Sizing compositions (used in place of starches and gum tragacanth to give a more elastic and more transparent product).
Dispersing agent in making—
Sizing compositions in emulsified form, for use on pulp and paper.

Alginic Acid (Continued)

Various emulsions containing pitches, fats, oils, and the like for treating paper and pulp products.
Waterproofing compositions in emulsified form, for treating paper and pulp compositions and paperboard.
Waxing compositions in emulsified form, for treating paper and cardboard.

Ingredient of—
Compositions used for finishing paper.
Compositions used for waterproofing pulp and paper products.

Reagent in—
Treating waste liquors and the like for the purpose of clarifying and purifying them.

Petroleum

Dispersing agent in making—
Emulsified mineral cutting oils.
Emulsions of petroleum and petroleum distillates in stabilized form.
Kerosene emulsions.
Naphtha emulsions.
Soluble greases in emulsified form.
Solubilized emulsified oils.
Petroleum pitch emulsions.
Petroleum tar emulsions.

Ingredient of—
Emulsions of petroleum and petroleum distillates (added for the purpose of securing better dispersion)

Stabilizer in—
Emulsions of petroleum and petroleum distillates.

Plastics

Binder in making—
Various plastic compositions containing such substances as horn, ebonite, celluloid, ivory, bone, shell, galalith, formaldehydephenol condensation products, and other artificial resins.

Resins and Waxes

Dispersing agent in making—
Emulsions of various natural and artificial resins and waxes.

Stabilizing agent in making—
Emulsions of waxes and resins.

Rubber

Ingredient of—
Products of rubber latex.

Soap

Dispersing agent in making—
Hand-cleansing compositions in emulsified form.
Textile scouring soaps in emulsified form.
Various emulsified detersive preparations.

Ingredient of—
Bleaching compositions.

Sugar

Defacating agent in—
Refining sugar.

Reagent in—
Clarifying and purifying liquors in beet sugar refining.

Textile

——, *Bleaching*

Dispersing agent in making—
Emulsified bleaching baths.

——, *Dyeing*

Dispersing agent in making—
Dye baths in emulsified form.

Ingredient of various dye baths (added for the purpose of increasing the dispersion of the dyestuff).

Mordant in various dyeing processes.

——, *Finishing*

Dispersing agent in making—
Emulsified compositions for coating fabrics.
Emulsified sizing compositions.
Emulsified washing compositions.
Emulsified waterproofing compositions.

Ingredient of—
Compositions used for the waterproofing of fabrics (this treatment being followed by one in a solution of a metallic salt).
Compositions used for treating woolen fabrics to protect them against decomposition (French 518059).
Compositions used for sizing yarns and fabrics (French 563726) (used in place of starch and gum tragacanth for the purpose of obtaining a more elastic and more transparent size).

——, *Manufacturing*

Dispersing agent in making—
Emulsified baths for bowking operations.
Emulsified baths for fulling operations.
Emulsified baths for the carbonization of wool.
Emulsified baths for degreasing raw wool.
Emulsified baths for soaking silk.
Emulsified baths for degumming and boiling-off raw silk.
Emulsified compositions used in spinning operations.
Emulsified preparations for kier-boiling cotton.
Emulsified mercerizing baths.
Oiling emulsions for treating fabrics.

——, *Printing*

Mordant in printing various fabrics.

Thickener in making—
Printing pastes (used in place of gum tragacanth and British gum).

Water and Sanitation

Reagent in—
Treating waste waters and the like for the purpose of purifying and clarifying them.

Wine

Clarifying agent for—
Treating wines.

Alizarin

Synonyms: Alphabetadihydroxyanthraquinone, 1:2-Dihydroxyanthraquinone.
French: Alphabétadihydroxyeanthraquinone.
German: Alizarinsäure, Alphabetadihydroxyanthrachinon, 1:2-Dihydroxyanthrachinon, Krapport.

Analysis

As an indicator in volumetric analysis.

Reagent in testing for—
Aluminum and compounds, such as alum.
Aragonite.
Milk (to determine keeping qualities).

Chemical

Starting point in making—
Alizarinimide.
Various intermediates and other derivatives.

Dye

Starting point in making—
Alizarin astrol, alizarin Bordeaux B, alizarin cyanin R, alizarin cyanin G, alizarin garnet R, alizarin maroon W, alizarin orange marks, alizarin powder, alizarin red marks, alizarin red S, alizarin viridone G, anthrarubin, anthrapurpurin (1:2:7-trioxyanthraquinone), blue black B, color lakes with various metallic oxides, diacetyl alizarin, erweco alizarin red BS, hydroxyanthrarufin (1:2:5-trihydroxanthraquinone), leucoalizarin (1:2-hydroxanthranol), lead alizarate.

Photographic

Reagent in—
Developing and toning pictures.

Textile

——, *Dyeing*

Coloring matter in dyeing—
Cotton in red shades and also wool and silk.

——, *Printing*

Coloring matter in printing—
Wool and silk.

Allyl Alcohol

French: Alcool d'allyle, Alcool allylique.
German: Allylalkohol.

Chemical

Starting point in making—
Acrolein, allyl esters, allyl ether, allyl mercaptan, allyl mustard oil, allyl salts of acids and halogens, allylacetic acid, allylamine, allylmalonic acid, allylurea, betadichlorohydrin, diallylbarbituric acid (Dial).

Fats and Oils

Reagent (Brit. 277357) in making—
Emulsions, lubricants.

Fuel

Reagent in making—
Emulsified fuel compositions (Brit. 277357).

Insecticide

Ingredient of—
Compositions used in destroying soil nematodes.

Leather

Reagent in making—
Emulsified dressings (Brit. 277357).

Allyl Alcohol (Continued)

Military
Starting point in making—
Poison gas.

Petroleum
Reagent in making—
Motor fuel compositions.
Stable emulsions of petroleum and petroleum distillates (Brit. 277357).

Pharmaceutical
In compounding and dispensing practice.

Soap
Reagent in making—
Emulsified cleansing and detergent compositions (Brit. 277357).

Textile
——, *Finishing*
Reagent in making—
Emulsified cleansing and washing compositions (Brit. 277357).

Allyl Alphacrotonate
Synonyms: Alphacrotonic allyl ester.
French: Alphacrotonate d'allyle, Alphacrotonate allylique.
German: Allylalphacrotonat, Alphacrotonsäureallylester, Alphacrotonsäuresallyl.

Miscellaneous
Solvent and plasticizer (Brit. 321258) for—
Cellulose acetate, cellulose esters and ethers, cellulose nitrate, rubber.
For uses, see under general heading: "Solvents."

Allyl Carbamide
French: Carbamide allylique, Carbamide d'allyle.
German: Allylcarbamid.

Chemical
Starting point in making—
Intermediates, pharmaceuticals, salts and esters.

Dye
Starting point in making various synthetic dyestuffs.

Resins and Waxes
Starting point (Brit. 292912) in making synthetic resins with—
Acetylsalicylic acid, aliphatic dibasic acids, ammonium salicylate, anthranilic acid, benzoic acid, gallic acid, hydroxynaphthoic acid, magnesium salicylate, oxalic acid, phenolic acids, phthalic acid, salicylamide, strontium salicylate, succinic acid.

Allyl Dimethyldithiocarbamate

Disinfectant
As a bactericide (Australian 8103/32, Brit. 406979, U. S. 1972961).

Insecticide and Fungicide
As a fungicide (claimed effective against barley spores and pinewood fungi) (Australian 8103/32, Brit. 406979, U. S. 1972961).
As an insecticide (claimed effective against aphids) (Australian 8103/32, Brit. 406979, U. S. 1972961).

Allylenedinaphthol
French: Dinaphthol allylènique.
German: Allylendinaphtol.
Spanish: Allilendinaftolina.
Italian: Allilenedinaftolina.

Chemical
Starting point in making—
Intermediates, pharmaceuticals, various other derivatives.

Rubber
Reagent (U. S. 1841342) in preserving—
Rubber goods by vulcanizing them in the presence of this reagent.

1-Allyleneoxy-4-aminoanthraquinone

Chemical
Starting point in making various intermediates.

Dye
Starting point (Brit. 285096) in making dyestuffs in the presence of dimethylanilin, nitrobenzene, orthodichlorobenzene, naphthalene, and the like, with the aid of—
Acetylparaphenylenediamine, 5-amino-2-methylbenzimidazole, benzidin and derivatives and homologs, dimethylparaphenylenediamine, metanaphthylenediamine, metaphenylenediamine, metatoluylenediamine, metaxylidenediamine, orthonaphthylenediamine, orthophenylenediamine, orthotoluylenediamine, orthotoxylidenediamine, paranaphthylenediamine, paraphenylenediamine, paratoluylenediamine, paraxylidenediamine.

Allylenethiourea
Synonyms: Allylenesulphourea.
French: Sulphourée d'allylène, Sulphourée allylènique, Thiourée d'allylène, Thiourée allènique.
German: Allylensulfoharnstoff, Allylenthioharnstoff.

Chemical
Starting point in making various derivatives.
Starting point (Brit. 310534) in making rubber vulcanization accelerators with the aid of—
Alphanaphthylamine, anilin, betanaphthylamine, cyclohexylanilin, diphenylamine, ethylanilin, meta-anisidin, metacresidin, metanaphthylenediamine, metaphenylenediamine, metatoluidin, metatoluylenediamine, metaxylenediamine, metaxylidin, monomethylanilin, orthoanisidin, orthocreasidin, orthonaphthylenediamine, orthophenylenediamine, orthotoluidin, orthotoluylenediamine, orthoxylenediamine, orthoxylidin, para-anisidin, paracresidin, paranaphthylenediamine, paraphenylenediamine, paratoluidin, paratoluylenediamine, paraxylenediamine, paraxylidin.

Allyl Iodide
French: Iodure d'allylique, Iodure d'allyle.
German: Allyljodid, Jodallyl.

Chemical
Starting point in making—
Allylanilin, allylbenzene, allyl chloride, allyl ether, allyl mustard oil, allyl tribromide, croton nitrile, monoallylamine, tetra-allylammonium iodide.

Allylisopropylbarbituric Acid
French: Acide d'allylisopropylebarbiturique.
German: Allylisopropylbarbiturinsaeure.

Chemical
Starting point in making—
Allylisopropyl-lactic acid ureide (Brit. 264804).

Pharmaceutical
In compounding and dispensing practice.

Allyl Mandelate
French: Mandélate allylique, Mandélate d'allyle.
German: Mandelsaeuresallyl, Mandelsaeureallylester.

Paint and Varnish
Plasticizer (Brit. 270650) in making—
Cellulose ester lacquers, cellulose ester varnishes.

Plastics
Plasticizer in making—
Nitrocellulose plastics.

Allyl Mercaptan
French: Mercaptane d'allyle, Mercaptane allylique.
German: Allylmerkaptan.

Chemical
Starting point in making—
Intermediates, pharmaceuticals.
Starting point (Brit. 286749) in making rubber vulcanization accelerators with the aid of—
Dibenzylamine, diethylamine, diethylguanylthioureas, diphenylbiguanide, ditolylbiguanide, ethanolamines, guanylureas, isothioureas, isoureas, monophenylbiguanide, monophenylguanylthioureas, monotolylbiguanide, pentatolylbiguanide, piperidin, piperazin, tetramethylammonium hydroxide, tetraphenylbiguanide, tetratolylbiguanide, thioureas, trimethylsulphonium hydroxide.

Allyl Sulphide
French: Sulfure d'allyle, Sulfure allylique.
German: Allylsulfid, Schwefelallyl, Schwefelwasserstoffallyl, Schwefelwasserstoffsaeureallylester, Schwefelwasserstoffsaeuresallyl.

Chemical
Reagent and starting point in making—
Intermediates, pharmaceuticals, salts and esters.
Reagent (Brit. 298511) in treating—
Albumens and albumenoids.

Glues and Adhesives
Reagent (Brit. 298511) in treating—
Vegetable proteins, such as soya bean flour, linseed protein, and peanut protein, to make adhesives.

Miscellaneous
Reagent (Brit. 298511) in treating—
Vegetable proteins to make sizes and finishes.

Allylsulphuric Acid Chloride
French: Chlorure d'allylesulfurique.
German: Allylschwefelsaeureschlorid.

Dye
Reagent (Brit. 271533) in making soluble vat dyestuffs from—
Anthraquinone-1:2, indanthrone, flavanthrene, naphtacridine, thioindigo.

Allyl Thiosalicylate
French: Sulphosalicylate d'allyle, Sulphosalicylate allylique, Thiosalicylate d'allyle, Thiosalicylate allylique.
German: Allylsulfosalicylat, Allylthiosalicylat, Sulfosalicylsäureallylester, Sulfosalicylsäuresallyl, Thiosalicylsäureallylester, Thiosalicylsäuresallyl.
Spanish: Solfosalicylato de allil, Tiosalicylato de allil.
Italian: Sulfosalicilato di allile, Tiosalicylato di allile.

Chemical
Starting point (Brit. 262427) in making—
Synthetic drugs with the aid of oxides and other salts of antimony, arsenic, bismuth, gold, and silver.

Allyltriphenyl Chloride
French: Chlorure d'allyletriphénylé, Chlorure allylique et triphénylique.
German: Allyltriphenylchlorid, Chlorallyltriphenyl.

Chemical
Starting point in making various derivatives.

Miscellaneous
Mothproofing and moldproofing agent for the treatment of furs and hair.

Textile
Mothproofing and moldproofing agent in the treatment of wool and felt.

Almond Oil, Expressed
Synonyms: Almond oil, Oil of sweet almond.
Latin: Oleum amygdaloe dulcis, Oleum amygdaloe, Oleum amygdaloe, expressum.
French: Huile d'amande.
German: Mandeloel, Sussmandeloel.
Spanish: Aceite de almendra.
Italian: Olio di mandorla.

Miscellaneous
Ingredient of—
Special lubricating compositions.
Lubricant for—
Delicate machinery, watches, firearms.

Perfume
Ingredient of—
Creams, hair oils, lotions, pomades.

Pharmaceutical
Ingredient of—
Ointments and emulsions.
Suggested for use as demulcent and mild laxative.

Soap
Stock in making—
Fine toilet soaps, shaving creams.

Almond Shell

Explosives and Matches
Combustible ingredient (U. S. 1880116) in—
Dynamite composition (used in crushed form).

Aloe
Synonyms: Barbados aloe, Cape aloes, Curacoa aloes, Hepatic aloes.
Latin: Aloe barbadensis, Aloe socotrina.
French: Aloès socotrin, Ou sucotrin, Aloès hépatique des Barbades.
German: Socotra oder socotinische aloe, Barbados aloe.
Spanish: Acibar sucotrino.

Chemical
Starting point in making—
Aloin, aloetic acid, chrysamic acid.

Dye
Ingredient of—
Archil and certain anilin dyestuffs (added for the purpose of increasing their fastness to light).
Starting point in making—
Brown coloring matter.

Insecticide
Ingredient of—
Agricultural insecticidal and parasiticidal compositions.
Insecticidal and parasiticidal compositions used in the household.

Leather
Coloring matter in dyeing—
Leather and skins.

Miscellaneous
In veterinary medicine practice to heal wounds and sores.

Paint and Varnish
Coloring matter in making—
Paints and varnishes.

Pharmaceutical
Suggested for use as a laxative and cathartic.

Textile
Coloring matter in dyeing—
Cotton and woolen yarns and fabrics.

Woodworking
Coloring matter in dyeing—
Wood brown.

Alpha-acetamidoanthraquinone
German: Alpha-acetamidoanthrachinon.

Chemical
Starting point in making—
Intermediates.

Dye
Starting point in making—
Indanthrene copper R.
Other synthetic dyestuffs.

Alpha-acetamido-4-methoxyanthraquinone
German: Alpha-acetamido-4-methoxyanthrachinon.

Textile
——, *Dyeing*
Dyestuffs for rayon (Brit. 263260).

Alpha-acetaminoanthraquinone
German: Alpha-acetaminoanthrachinon.

Dye
Starting point in making—
Anthrene orange RT.
Indanthrene copper R.

Textile
——, *Dyeing*
Dyestuff for—
Acetate rayon (Brit. 263260).

Alpha-alphabetabetatetramethylbutylguaiacol

Pharmaceutical
Suggested (Brit. 431487) for use as—
Bactericide of high power.

Alpha-alphabetabetatetramethylbutylpyrogallol

Pharmaceutical
Suggested (Brit. 431487) for use as—
Bactericide of high power.

Alpha-alphabetabetatetramethylbutylquinol

Pharmaceutical
Suggested (Brit. 431487) for use as—
Bactericide of high power.

Alpha-alphabetabetatetramethylbutylresorcinol

Pharmaceutical
Suggested (Brit. 431487) for use as—
Bactericide of high power.

Alpha-alphadiaminopyridin

Chemical
Starting point (Brit 311349) in making pharmaceuticals with—
Anilin metatoluidin, metaxylidin, orthotoluidin, orthoxylidin, paratoluidin, paraxylidin.

Alpha-alphaparadihydroxydiphenylethane

Rubber
Age-resisting agent (U. S. 1958929).

Alpha-amino-4-anilidoanthraquinone
German: Alpha-amino-4-anilidoanthrachinon.

Dye
Starting point (Brit. 282854) in making dyestuffs with—
Acetaldehyde, benzaldehyde, butyraldehyde, cinnamyldehyde, crotonaldehyde, formaldehyde, heptaldehyde, hexaldehyde, paraformaldehyde, propionaldehyde, succinaldehyde.

Alpha-aminoanthraquinone
Synonyms: 1-Aminoanthraquinone.
German: Alpha-aminoanthrachinon, 1-Aminoanthrachinon.
Chemical
Starting point in making—
Alpha-aminoanthraquinone-1-carboxylic acid.
Intermediates, pharmaceuticals, synthetic aromatics.
Dye
Starting point in making—
Algol gray B, algol orange R, algol yellow WG, alizarin blue, alizarin cyanone, benzanthrone colors, corinth R, cyananthrol, indanthrene red R, indanthrene red G.

Alpha-aminoazonaphthalene
French: Alpha-aminoazonaphtalène.
German: Alpha-aminoazonaphtalin.
Chemical
Starting point in making various derivatives.
Dye
Starting point in making—
Fast pink for silk, magdala red, paraphenylene violet, rose magdala.

Alpha-amino-2-cyano-5-chlorobenzene
Synonyms: 1-Amino-2-cyano-5-chlorobenzene.
German: Alpha-amino-2-cyano-5-chlorbenzol, Alpha-amino-2-zyano-5-chlorbenzol, 1-Amino-2-zyano-5-chlorbenzol.
Chemical
Starting point in making—
Intermediates, pharmaceuticals.
Dye
Starting point (Brit. 301410) in making azo lakes and dyestuffs with the aid of—
2:3-Oxynaphthoic anilide.
2:3-Oxynaphthoic 4′-chloro-2′-toluidide.
2:3-Oxynaphthoic 5′-chloro-2′-anisidide.
2:3-Oxynaphthoic 4′-methoxyanilide.
2:3-Oxynaphthoic naphthylamide.
2:3-Oxynaphthoic 2′-toluidide.

Alpha-amino-2-ethoxynapthalene-6-sulphonic Acid
Synonyms: 1-Amino-2-ethoxynaphthalene-6-sulphonic acid.
French: Acide d'alpha-amino-2-éthoxyenaphthalène-6-sulphonique, Acide de 1-amino-2-éthoxyenaphthalène-6-sulphonique.
German: Alpha-amino-2-æthoxynaphtalin-6-sulfonsäure, 1-Amino-2-aethoxynaphtalin-6-sulfonsäure.
Chemical
Starting point in making—
Esters and salts, intermediates.
Dye
Starting point (Brit. 298518) in making azo dyestuffs with the aid of—
Alpha-aminonaphthalene, alpha-aminonaphthalene-6-sulphonic acid, alpha-aminonaphthalene-7-sulphonic acid, anilin, anilin-3-chloro-6-sulphonic acid, anilin-2:4-disulphonic acid, anilin-2:5-disulphonic acid, anilin-4-nitro-2:5-disulphonic acid, anilin-3-sulphonic acid, beta-amino-1-methoxybenzene-4-sulphonic acid, beta-amino-5-sulphobenzoic acid, 1:3-dioxyquinolin, methylketol, methylketolsulphonic acid, orthocresotinic acid, 1-phenyl-3-carboxy-5-pyrazolon, 1-phenyl-3-methyl-5-pyrazolon, salicylic acid, sulphazone.

Alpha-amino-2-mercapto-6-methyl-4-phenylamino-2′-carboxylic Acid
French: Acide d'alpha-amino-2-mercapto-6-méthyle-4-phényleamino-2′-carbonique.
German: Alpha-amino-2′-mercapto-6-methyl-4-phenylamino-2′-carbonsaeure.
Dye
Starting point (Brit. 265641) in making acid dyestuffs with—
Chloroanil, dichloroquinone, monochloroquinone, toluquinone, trichloroquinone.

Alpha-amino-2-methoxynaphthalene-6-sulphonic Acid
Synonyms: 1-Amino-2-methoxynaphthalenesulphonic acid.
French: Acide d'alpha-amino-2-méthoxyenaphthalène-6-sulphonique, Acide de 1-amino-2-méthoxyenaphthalène-6-sulphonique.
German: Alpha-amino-2-methoxynaphtalinsulfonsäure, 1-Amino-2-methoxynaphtalinsulfonsäure.
Chemical
Starting point in making—
Esters and salts, intermediates.
Dye
Starting point (Brit. 298518) in making azo dyestuffs with the aid of—
Alpha-aminonaphthalene, Alpha-aminonaphthalene-6-sulphonic acid, alpha-aminonaphthalene-7-sulphonic acid, anilin, anilin-3-chloro-6-sulphonic acid, anilin-2:4-disulphonic acid, anilin-2:5-disulphonic acid, anilin-4-nitro-2:5-disulphonic acid, anilin-3-sulphonic acid, beta-amino-1-methoxybenzene-4-sulphonic acid, beta-amino-5-sulphobenzoic acid, 1:3-dioxyquinolin, methylketol, methylketolsulphonic acid, orthocreosotinic acid, 1-phenyl-3-carboxy-5-pyrazolon, 1-phenyl-3-methyl-5-pyrazolon, salicylic acid, sulphazone.

Alpha-amino-2-methylanthraquinone
French: Alpha-amino-2-méthyleanthraquinone, 1-Amino-2-méthyleanthraquinone.
German: Alpha-amino-2-methylanthrachinon, 1-Amino-2-methylanthrachinon.
Chemical
Starting point in making—
Derivatives, such as carboxylic and sulphonic acids.
Intermediates, pharmaceuticals.
Dye
Starting point in making—
Algol orange red R, cyanthrol, leucol dark green B paste.
Textile
——, *Dyeing and Printing*
Component (Brit. 310827) of dyeing, printing, and stenciling compositions used on materials which contain cellulose esters and ethers, with the aid of—
Alpha-amino-2-methoxynaphthalene, alphanaphthylamine, betanaphthylamine, dimethylmeta-aminophenol, gammachlorobetaoxypropyl-1-naphthylamine, meta-anisidin, meta-aminophenol, metacresidin, metatoluidin, monoacetylmeta-aminophenol, metaphenylenediamine, nitrometaphenylenediamine, orthoanisidin, orthocresidin, omegaoxyethyl-1-naphthylamine, para-anisidin, paracresidin, paraxylidin, phenol.

Alpha-amino-2-naphthol-4-sulphonic Acid
Synonyms: 1-Amino-2-naphthol-4-sulphonic acid.
French: Acide d'alpha-amino-2-naphthole-4-sulphonique, Acide de 1-amino-2-naphthole-4-sulphonique.
German: Alpha-amino-2-naphtol-4-sulfonsäure, 1-Amino-2-naphtol-4-sulfonsäure.
Analysis
Reagent in the determination of—
Calcium, phosphates.
Chemical
Starting point in making—
Esters and salts, intermediates.
Dye
Starting point in making—
Chrome black BN, chrome blue black NB, chrome blue black NR, chrome palatin black 6 B, erichrome B, erichrome black T, erichrome blue black B, erichrome red B, nichrome black NT.
Soluble chromium compounds of azo dyestuffs (Brit. 260830).

Alpha-amino-2-naphthol-6-sulphonic Acid
Synonyms: 1-Amino-2-naphthol-6-sulphonic acid.
French: Acide d'alpha-amino-2-naphthole-6-sulphonique, Acide de 1-amino-2-naphthole-6-sulphonique.
German: Alpha-amino-2-naphtol-6-sulfonsäure, 1-amino-2-naphtol-6-sulfonsäure.
Chemical
Starting point in making—
Esters and salts, intermediates, sodium alpha-amino-2-naphthol-6-sulphonate.
Dye
Starting point in making various synthetic dyestuffs.

Alpha-amino-4-oxyanthraquinone
Synonyms: 1-Amino-4-oxyanthraquinone.
French: Alpha-amino-4-oxyeanthraquinone.
German: Alpha-amino-4-hydroxyanthrachinon, 1-Amino-4-hydroxyanthrachinon.
Chemical
Starting point in making—
Intermediates, pharmaceuticals.

Alpha-amino-4-oxyanthraquinone (Continued)

Dye
Starting point in making various synthetic dyestuffs.
Textile
Component (Brit. 310827) of dyeing, printing, and stenciling compositions used on materials containing cellulose esters or cellulose ethers (acetate rayon for example), with the aid of—
Alpha-amino-2-methoxynaphthalene, alphanaphthylamine, betanaphthylamine, dimethylmeta-aminophenol, gammachlorobetaoxypropyl-1-naphthylamine, meta-aminophenol, meta-anisidin, metacresidin, metatoluidin, monoacetylmeta-aminophenol, nitrometaphenylenediamine, omegaoxyethyl-1-naphthylamine, ortho-anisidin, orthocresidin, orthotoluidin, para-anisidin, paracresidin, paraxylidin, phenol.

Alpha-anthraquinonylmethane
German: Alpha-anthrachinonylmethan.
Dye
Starting point in making—
Dyestuffs of the alpha-alphadi and trianthraquinonylurea series for cellulose acetate rayon (Brit. 248858).

Alpha-b′-dodecyloxyethylglyceryl Ether
Soap
Emulsifying agent (Brit. 421490 and 411295) in—
Shaving creams, superfatted soaps, and the like.

Alphabenzoylaminoanthraquinone
Oils, Fats, Waxes
Coloring agent (Brit. 432867) for—
Paraffin and other mineral waxes, stearic acid, tallow and other solid triglycerides, beeswax, carnauba wax, and others.

Alphabenzylcinnamic Acid
French: Acide d'alphabenzylecinnamique.
German: Alphabenzylcinnzimtsäure.
Chemical
Starting point in making—
Esters, intermediates, pharmaceuticals, salts.
Miscellaneous
Ingredient (Brit. 319273) of—
Compositions used as wetting agents for special purposes.
Soap
Ingredient (Brit. 319273) of—
Detergent compositions.
Textile
——, *Dyeing*
Ingredient (Brit. 319273) of—
Dye liquors.
——, *Finishing*
Ingredient (Brit. 319273) of—
Finishing compositions.

Alphabetadiphenylethylbetaphenylethylpropylamine
Chemical
Claimed (U. S. 2006114) as—
Substitute for papaverine.

Alphabetadiphenylethylbetaphenylethylpropylamine Hydrochloride
Chemical
Claimed (U. S. 2006114) as—
Substitue for papaverine.

Alphabromo-2-amino-3-chloroanthraquinone
German: Alphabrom-2-aminochloranthrachinon.
Dye
Starting point (Brit. 278417) in making dyestuffs for wool, silk, and acetate rayon with the aid of—
Allylamine, amylamine, anilin, benzylamine, butylamine, metaphenylenediamine, metatoluidin, metaxylidin, monoethylamine, monoethylanilin, monomethylamine, naphthylamine, orthophenylenediamine, orthotoluidin, orthoxylidin, paraphenylenediamine, paratoluidin, paraxylidin.

Alphabromolauric Acid Cyclohexylester
Detergent
Starting point (Brit. 408754) in making—
Saponaceous products by reaction with tertiary amines, which may be used alone or with other soaps, fillers, or compounds giving off oxygen.

Alphabromolauric Acid Dodecylester
Detergent
Starting point (Brit. 408754) in making—
Saponaceous products by reaction with tertiary amines, which may be used alone or with other soaps, fillers, or compounds giving off oxygen.

Alphabromolauric Acid Hexadecylester
Soap
Starting point (Brit. 403883) in making—
Saponaceous products by reaction with amines, such as anilin, piperidin bases, hydroxyethylanilin, dihydroxyethylanilin, paratoluidin (these products may be used alone, or with soaps, fillers, or compounds giving off oxygen).

Alphabromolauric Acid Octadecylester
Detergent
Starting point (Brit. 408754) in making—
Saponaceous products by reaction with tertiary amines, which may be used alone or with other soaps, fillers, or compounds giving off oxygen.

Alphabromolauric Acid Tetradecylester
Soap
Starting point (Brit. 403883) in making—
Saponaceous products by reaction with amines, such as anilin, piperidin bases, hydroxyethylanilin, dihydroxyethylanilin, paratoluidin (these products may be used alone, or with soaps, fillers, or compounds giving off oxygen).

Alphabromoquinolingammacarboxylic Acid
French: Acide alphabromoquinolinegammacarbonique, Acide d'alphabromoquinolinegammacarboxyle, Acide alphabromoquinolinegammacarboxylique.
German: Alphabromchinolingammacarbonsaeure.
Chemical
Starting point (Brit. 270339) in making pharmaceutical chemicals with—
Benzylamine, diallylamine, diethylamine, diethylethylenediamine, di-isoamylamine, dimethylamine, dipropylamine, monomethylamine, normal ethylanilin, piperidin.

Alpha-b′-tetradecyloxyethylglyceryl Ether
Soap
Emulsifying agent (Brit. 421490 and 411295) in—
Shaving creams, superfatted soaps, and the like.

Alphabutylpyrrolidin
Chemical
Starting point in making various salts and other derivatives.
Insecticide
As an insecticide
Ingredient (U. S. 1748633) of—
Insecticidal compositions.
Sanitation
Ingredient (U. S. 1748633) of—
Germicidal preparations.

Alphabutylpyrrolidin Sulphate
Chemical
Starting point in making various derivatives.
Insecticide
As an insecticide.
Ingredient (U. S. 1748633) of—
Insecticidal compositions.
Sanitation
Ingredient (U. S. 1748633) of—
Germicidal compositions.

Alphachloro-2-amino-3-chloroanthraquinone
German: Alphachlor-2-amino-3-chloranthrachinon.
Dye
Starting point (Brit. 278417) in making dyestuffs for wool, silk, and acetate rayon with—
Allylamine, amylamine, anilin, benzylamine, butylamine, cresidin, diallylamine, diamylamine, ethylamine, ethylanilin, methylamine, methylanilin, metaphenylenediamine, metatoluidin, metaxylidin, naphthylamine, orthophenylenediamine, orthotoluidin, orthoxylidin, paraphenylenediamine, paratoluidin, paraxylidin, propylamine, propylanilin.

Alphachloroanthraquinone-2-carboxylic Acid
French: Acide d'alphachloroanthraquinone-2-carbonique.
German: Alphachloranthrachinon-2-carbonsaeure.

Dye
Starting point (Brit. 260588) in making anthraquinone-acridin dyestuffs with—
Alphanaphthylamine, anilin, benzylamine, benzidin, betanaphthylamine, dianisidin, dibenzylamine, dimethylanilin, diphenylamine, ethylanilin, meta-anisidin, metachloroanilin, metanitranilin, metanitroxylidin, metaphenylenediamine, metatoluidin, metatoluylenediamine, metaxylidin, methylanilin, orthoanisidin, orthochloroanilin, orthonitranilin, orthonitroxylidin, orthophenylenediamine, orthotoluidin, orthotoluylenediamine, orthoxylidin, para-anisidin, parachloroanilin, paranitranilin, paranitrotoluidin, paranitroxylidin, paraphenylenediamine, paratoluidin, paratoluylendiamine, paraxylidin, phenylamine, phenyldimethylamine, phenylmethylamine, xylylenediamines.

Alphachloro-2:4-dinitrobenzene-6-sulphonic Acid
French: Acide d'alphachloro-2:4-dinitrobenzène-6-sulfonique.
German: Alphachlor-2:4-dinitrobenzol-6-sulfonsaeure.

Dye
Starting point (Brit. 285504) in making dyestuffs with—
Benzidin, 2:4'-diaminodiphenyl, 4:4'-diamino-2-nitrodiphenyl, orthodianisidin.

Alphachloro-2:6-dinitrobenzene-4-sulphonic Acid
French: Acide d'alphachloro-2:6-dinitrobenzène-4-sulfonique.
German: Alphachlor-2:6-dinitrobenzol-4-sulfonsaeure.

Dye
Starting point (Brit. 285504) in making nitro dyestuffs with—
Benzidin, 2:4'-diaminodiphenyl, 4:4'-diamino-2-nitrodiphenyl, orthodianisidin.

Alphachloroquinolingammacarboxylic Acid Chloride
French: Chlorure d'alphachloroquinolinegammacarbonique.
German: Alphachlorchinolingammacarbonsaeurechlorid.

Chemical
Starting point (Brit. 270339) in making pharmaceuticals with—
Asymetrical diethylenediamine, benzylamine, diallylamine, diethylamine, di-isoamylamine, dimethylamine, dipropylamine, monoethylamine, normal ethylanilin, piperidin.

Alphacyclohexyloxyquinolinegammacarboxylic Acid
French: Acide d'alphacyclohexyleoxyquinolinegammacarbonique.
German: Alphacyclohexyloxychinolingammacarbonsaeure.

Chemical
Starting point (Brit. 270339) in making synthetic drugs with—
Benzylamine, diallylamine, diethylamine, diethylethylenediamine, di-isoamylamine, dimethylamine, dipropylamine, monobutylamine, monoethylamine, monomethylamine, monopropylamine, normal ethylanilin, piperidin.

Alphadiethylaminoanthraquinone
German: Alphadiaethylaminoanthrachinon.

Chemical
Starting point in making—
Alphadiethylanthramine (Brit. 260000).

Alphadiethylaminoethoxyquinolingammacarboxylic Acid
French: Acide d'alphadiéthyleaminoéthoxyquinoléinegammacarbonique, Acide d'alphadiéthyleaminoéthoxyquinoléinegammacarboxylique.
German: Alphadiaethylaminoaethoxychinolingammacarbonsaeure.

Chemical
Starting point (Brit. 270339) in making synthetic drugs with—
Benzylamine, diallylamine, diamylamine, dibutylamine, diethylamine, diethylethylenediamine, di-isoamylamine, di-isobutylamine, di-isoproplyamine, dimethylamine, dipropylamine, monoethylamine, monomethylamine, monopropylamine, normal ethylanilide, piperidin.

Alphadinaphthol

Dye
Reagent in making—
Dyes, intermediates.

Alphaethoxyquinolingammacarboxylic Acid
French: Acide d'alphaéthoxyquinolinecarbonique, Acide d'alphaéthoxyquinolinegammacarbonyle, Acide d'alphaéthoxyquinolinegammacarboxyle.
German: Alphaaethoxychinolingammacarbonsaeure.

Chemical
Starting point (Brit. 270339) in making therapeutic agents with—
Benzylamine, diallylamine, diamylamine, diethylamine, diethylethylenediamine, di-isoamylamine, dimethylamine, dipropylamine, monoethylamine, normal ethylanilide, piperidide.

Alphaethylbetapropylacrolein

Chemical
Starting point (Brit. 264673-4) in making rubber vulcanization accelerators with—
Anilin, benzanilin, benzidin, diphenylamine, metatoluidin, metaxylidin, monometachloroanilin, monometaethylanilin, mono-orthochloroanilin, mono-paramethylanilin, mono-orthoethylanilin, mono-orthomethylanilin, monoparaethylanilin, monometamethylanilin, monoparachloroanilin, naphthylamine, naphthylene-ethylamine, orthotoluidin, orthoxylidin, paratoluidin, paraxylidin, phenylamine.

Alphaethyloxy-4-aminoanthraquinone
German: Alphaaethyloxy-4-aminoanthrachinon.

Dye
Starting point (Brit. 285096) in making dyestuffs in the presence of dimethyianilin, nitrobenzene, orthodichlorobenzene or naphthalene, with the aid of—
Acetylparaphenylenediamine, 5-amino-2-methylbenzindazole, benzidin and derivatives and homologs, dimethylparaphenylenediamine, metaphenylenediamine, metatoluylenediamine, naphthylenediamine and derivatives and homologs, orthophenylenediamine, orthotoluylenediamine, paraphenylenediamine, paratoluylenediamine.

Alphaethylpropylorthocresol

Disinfectant
Germicide (U. S. 2073995).

Alphaiodoquinolingammacarboxylic Acid
French: Acide alphaiodoquinolinegammacarbonique. Acide d'alphaiodoquinolinegammacarbonyle.
German: Alphajodchinolingammacarbonsaeure.

Chemical
Starting point (Brit. 270339) in making synthetic drugs with—
Benzylamine, diallylamine, diethylamine, diethylethylenediamine, di-isoamylamine, dimethylamine, ethylanilin, monoethylamine, piperidin.

Alphaisatinanilide
Synonyms: 1-Isatinanalide, Isatin-1-phenylimide.
French: Anilide d'alphaisatine.
German: Alphaisatinanilid.

Chemical
Starting point in making various intermediates.

Dye
Starting point in making—
Alizarin indigo B, alizarin indigo 3R, ciba gray G, ciba violet B, ciba violet 3B, helindon blue 3GN, helindon brown 2R, helindon brown 5R, indigoid dyestuffs, thioindigo violet K.

Alphaisopropylbetaisopropylacrolein

Chemical
Starting point (Brit. 264673-4) in making rubber vulcanization accelerators with—
Anilin, benzanilin, benzidin, diphenylamine, monochloroanilin, monoethylanilin, monomethylanilin, naphthylamine, naphthylenethylamine, phenylamine, toluidin, xylidin.

Alphamethylglycerin Ether

Cellulose Products
Solvent for—
Cellulose nitrate (nitrocellulose).
For uses, see under general heading: "Solvents."

Alphamethylnaphthalene
German: Alphamethylnaphtalin.

Chemical
Starting point (Brit. 267940) in making cyclic ketones with—
Betabromobutyryl bromide, betabromobutyryl chloride, betabromobutyryl iodide, betabromopropionyl bromide, betabromopropionyl chloride, betabromopropionyl iodide, betachlorobutyryl bromide, betachlorobutyryl chloride, betachlorobutyryl iodide, betachloropropionyl bromide, betachloropropionyl chloride, betachloropropionyl iodide, betaiodobutyryl bromide, betaiodobutyryl chloride, betaiodobutyryl iodide, betaiodopropionyl bromide, betaiodoproprionyl chloride, betaiodopropionyl iodide.
Starting point in making—
3-Methylindole (Scatol).

Fats and Oils
Starting point in making—
Viscous oils with benzyl dichloride.

Alpha-naphthalene Sulphoamide

Cellulose Products
Plasticizer (Canada 340994, Brit. 417871) for—
Cellulose acetate.
For uses, see under general heading: "Plasticizers."

Alphanaphthalide
German: Alphanaphtalid.

Dye
Starting point (Brit. 274128) in making azo dyes with—
1:3-Dimethyl-4-amino-6-bromobenzene, 1:3-dimethyl-4-amino-6-chlorobenzene, 1:3-dimethyl-4-amino-2:6-dibromobenzene, 1:3-dimethyl-4-amino-2:6-dichlorobenzene.

Alphanaphthol-3:8-disulphonic Acid
Synonyms: Epsilon acid.
French: Acide d'alphanaphthol-3:8-disulphonique.
German: Alphanaphtol-3:8-disulfonsaeure.

Dye
Starting point (Brit. 265203) in making azo dyestuffs with—
Anilinmethyleneorthocresotinic acid, orthotoluidinmethyleneorthocresotinic acid.
Starting point in making.
Columbia blue G, Columbia blue R, congo fast blue, eosamin BG, erica 2GN, heliotropin 2B.

Alphanaphthylamine-4:6:8-trisulphonic Acid
Synonyms: 1-Naphthylamine-4:6:8-trisulphonic acid.
French: Acide d'alphanaphthylamine-4:6:8-trisulfonique.
German: Alphanaphtylamin-4:6:8-trisulfonsaeure.

Chemical
Starting point in making—
Alpha-amino-8-naphthol-4:6-disulphonic acid (K acid).
Urea derivatives.
Starting point (Brit. 278037) in making pharmaceutical chemicals with—
Alphanitronaphthalene-5-sulphochloride, bromonitrobenzoyl chlorides, chloronitrobenzoyl chlorides, iodonitrobenzoyl chlorides, nitroanisoyl chlorides, nitrobenzene sulphochlorides, nitrobenzoyl chlorides, 2-nitrocinnamyl chloride, 3-nitrocinnamyl chloride, 4-nitrocinnamyl chloride, 1:5-nitronaphthoyl chloride, 2-nitrophenylacetyl chloride, 4-nitrophenylacetyl chloride, nitrotoluyl chlorides.

Dye
Starting point in making—
Sulphur colors.

Alphanaphthylaminopropionic Acid

Rubber
Antioxidant (U. S. 1958469) for—
Rubber batches.

Alphanaphthylmethyl Ether

Chemical
Starting point in—
Organic synthesis.

Lubricant
Starting point (Brit. 440916) in making—
Products useful as lubricating oils or as pour-point depressors for paraffin-base lubricating oils by condensation with halogenated derivatives of aliphatic hydrocarbons, such as paraffin oils, paraffin, petrolatum, ceresin, ozokerite, or others contained in the middle to higher fractions of petroleum.

Alphanaphthylpiperidin

Petroleum
Stabilizer (Brit. 347916) for—
Transformer oils.

Rubber
Age resister (Brit. 347916).

Alphanitroanthraquinone-6-sulphonic Acid
Synonyms: 1-Nitroanthraquinone-6-sulphonic acid.
French: Acide sulphonique d'alphanitroanthraquinone (6).
German: Alphanitroanthrachinon-6-sulfonsäure.

Chemical
Starting point in making—
1:6-Dihydroxyanthraquinone.

Textile
——, *Dyeing*
Reserve in dyeing various fibers and fabrics.

——, *Printing*
Reserve in printing fabrics.

Alphanitro-2:4-dimethylbenzene
Synonyms: Alphanitro-2:4-dimethylbenzol.
French: Alphanitro-2:4-diméthylebenzène.

Chemical
Starting point (Brit. 278761) in making—
5-Chloroalpha-amino-2:4-dimethylbenzene.
1:3-Dichloroalpha-amino-2:4-dimethylbenzene.
3:5-Dichloroalpha-amino-2:4-dimethylbenzene.

Alphanitronaphthalene-5-sulphochloride
French: Sulfochlorure d'alphanitronaphthalène, Sulfochlorure alphanitronaphthalènique.
German: Alphanitronaphtalin-5-sulfochlorid, Alphanitronaphtalin-5-sulfonchlorid.

Chemical
Starting point (Brit. 278037) in making synthetic drugs with—
Alkoxynaphthaleneaminesulphonic acid.
Alphanaphthylamine-4:8-disulphonic acid.
Alphanaphthylamine-3:6:8-trisulphonic acid.
Alphanaphthylamine-4:6:8-trisulphonic acid.
4-Aminoacenaphthene-3:5-disulphonic acid.
4-Aminoacenaphthene-3-sulphonic acid.
4-Aminoacenaphthene-5-sulphonic acid.
4-Aminoacenaphthenetrisulphonic acid.
1:5-Aminonaphthol-3:6-disulphonic acid.
1:8-Aminonaphthol-3:6-disulphonic acid.
1:5-Aminonaphthol-7-sulphonic acid.
Bromonaphthylaminesulphonic acid.
Chloronaphthylaminesulphonic acid.
Iodonaphthylaminesulphonic acid.
Starting point in making—
Alphanitronaphthalene-5-sulphonic acid.

Alphaorthotolyl Diguanide

Resins
Increaser (U. S. 2010224 and 2010227) of—
Softening point of shellac.

Alphaoxyanthraquinone

Oils, Fats, Waxes
Coloring agent (Brit. 432867) for—
Paraffin and other mineral waxes, stearic acid, tallow and other solid triglycerides, beeswax, carnauba wax, and others.

Alphaoxy-4-chlorobenzoic Acid
French: Acide d'alphaoxy-4-chlorbenzoique.
German: Alphaoxy-4-chlorbenzoesaeure.

Leather
Mothproofing agent (Brit. 274425) in treating—
Skins.

Miscellaneous
Mothproofing agent in treating—
Furs, feathers, felt, hair.

Textile
——, *Finishing*
Mothproofing agent in treating—
Wool.

Alphaoxy-2-chloro-4-benzylaminoanthraquinone
German: Alphaoxy-2-chlor-4-benzylaminoanthrachinon.

Dye
Starting point (Brit. 268542) in making wool dyestuffs with—
Ammonium sulphite, potassium sulphite, sodium sulphite.

Alphaoxy-2-chloro-4-cresylaminoanthraquinone
German: Alphaoxy-2-chlor-4-kresylaminoanthrachinon.

Dye
Starting point (Brit. 268542) in making wool dyestuffs with—
Ammonium sulphite, potassium sulphite, sodium sulphite.

Alphaoxy-2-chloro-4-phenylaminoanthraquinone
German: Alphaoxy-2-chlor-4-phenylaminoanthrachinon.

Dye
Starting point (Brit. 268542) in making wool dyestuffs with—
Ammonium bisulphite, potassium bisulphite, sodium bisulphite.

Alphaoxy-2-chloro-4-tolylaminoanthraquinone
German: Alphaoxy-2-chlor-4-tolylaminoanthrachinon.

Dye
Starting point (Brit. 268542) in making anthraquinone dyestuffs with—
Ammonium sulphite, potassium sulphite, sodium sulphite.

Alphaoxy-2-chloro-4-xylylaminoanthraquinone
German: Alphaoxy-2-chlor-4-xylylaminoanthrachinon.

Dye
Starting point (Brit. 268542) in making wool dyestuffs with—
Ammonium sulphite, potassium sulphite, sodium sulphite.

Alphaoxy-4:6-dimethyl-2-benzoic Acid
French: Acide d'alphaoxy-4:6-diméthyle-2-benzoique.
German: Alphaoxy-4:6-dimethyl-2-benzoesaeure.

Leather
Mothproofing agent (Brit. 274425) for—
Skins.

Miscellaneous
Mothproofing agent (Brit. 274425) for—
Hair.

Textile
——, *Finishing*
Mothproofing agent (Brit. 274425) for—
Woolens.

Alphaoxy-6-methyl-4-bromo-2-benzoic Acid
French: Acide d'alphaoxy-6-méthyle-4-bromo-2-benzoique.
German: Alphaoxy-6-methyl-4-brom-2-benzoesaeure.

Leather
Mothproofing agent (Brit. 274425) for—
Hides and skins.

Miscellaneous
Mothproofing agent (Brit. 274425) for—
Hair, felt, and the like.

Textile
——, *Miscellaneous*
Mothproofing agent (Brit. 274425) for—
Wool and woolen fabrics.

Alphaphenoxyacetylamino-8-hydroxynaphthalenedisulphonic Acid

Chemical
Starting point in making—
Intermediates and other derivatives.

Dye
Starting point (Brit. 313710) in making dyestuffs with—
Anilin derivatives, meta-acetamino-5-aminoanisol, 4-aminoacetanilide, beta-aminobenzoic acid, paraxylidin.

Alphaphenoxyquinolingammacarboxylic Acid
French: Acide alphaphénoxyquinolinegammacarbonique, Acide d'alphaphénoxyquinolinegammacarboxyle, Acide alphaphénoxyquinolinegammacarboxylique.
German: Alphaphenoxychinolingammacarbonsaeure.

Chemical
Starting point (Brit. 270339) in making synthetic drugs with—
Benzylamine, diallylamine, diethylamine, diethylenediamine, di-isoamylamine, dimethylamine, dipropylamine, n-ethylanilide, monoethylamine, piperidin.

Alphaphenylcinchoninic Acid Chloride
French: Chlorure d'alphaphénylecinchoninique acide.
German: Alphaphenylcinchoninsaeureschlorid.

Chemical
Starting point (Brit. 304655) in making pharmaceuticals with—
Allylurethane, amylurethane, butylurethane, caprylurethane, ethylurethane, heptylurethane, hexylurethane, isoallylurethane, isoamylurethane, isobutylurethane, isopropylurethane, lactylurethane, methylurethane, propylurethane.

Alphapicolin

Chemical
Reagent in making—
Pharmaceutical chemicals.

Alphapinene

Chemical
Starting point in making—
Synthetic camphor, terpin hydrate, terpineol.

Alphapolymethylstyrene

Chemical
Starting point in making various derivatives.

Miscellaneous
Ingredient (Brit. 367126) of—
Compositions used for impregnating and stiffening felt.

Alphapropionamidoanthraquinone
German: Alphapropionamidoanthrachinon.

Textile
——, *Dyeing*
Dyestuff for—
Acetate rayon (Brit. 263260).

Alphathionaphthol

Petroleum
Antioxidant (Brit. 425569) for—
Lubricating, transformer, and switch oils, particularly solvent-extracted oils and others of a paraffinic nature, in which the natural inhibitor content may have been reduced during refining.

Alumina, Activated
French: Alumine activée, aluminium activée.
German: Aktivierte aluminum, Aktivierte tonerde.

Analysis
General dehydrating agent in analytical work, for drying gases and filling drying tubes for use in ultimate analysis of organic compounds, gas analyses, in desiccators, and the like.

Chemical
Atmospheric conditioning and dehumidifying agent in—
Process buildings, storage buildings.
Dehydrating agent for—
Chemicals, compressed gases.
Gases, such as acetylene, ammonia, argon, butane, butylene, carbon dioxide, chlorine, ethane, ethylchloride, ethylene, helium, hydrogen, hydrogen sulphide, isobutane, methyl chloride, neon, nitrogen, nitrous oxide, oxygen, phosgene, propane, propylene, resusiten, sulphur dioxide.
Dehydrating agent (by direct contact) for—
Organic liquids, such as benzene, butyl acetate, carbon bisulphide, carbon tetrachloride, ethyl acetate, methyl chloride, toluene, xylene.
Solids in powdered form that are difficult to dry because of the deleterious effect of elevated temperatures.
Vapors and liquids in refrigeration installations in chemical plants.

Explosives
Atmospheric conditioning agent in—
Process buildings, storage buildings.
Dehumidifying agent.
Dehydrating agent for—
Process materials.
Vapors and liquids in refrigerating installations.

Alumina, Activated (Continued)

Fats and Oils
Dehydrating agent for—
Vapors and liquids in refrigerating installations.

Fertilizer
Dehydrating agent for—
Ammonia.

Food
Atmospheric conditioning agent.
Dehydrating agent for—
Meat packed in boxes.
Various foods, such as fruits and vegetables.
Vapors and liquids in refrigerating installations.

Fuel
Dehydrating agent for—
Vapors and liquids in refrigerating installations in candle factories.

Gas
Dehydrating agent for—
Ammonia, benzene, coal gas, coke-oven gas, naphtha, toluene, xylene.

Glass
Air-conditioning and dehumidifying agent for—
Storerooms and warehouses where plate glass is stored, to prevent the etching action of condensed moisture.

Ice
Dehydrating agent for—
Vapors and liquids in refrigerating installations.

Metallurgical
Air-conditioning and dehumidifying agent for—
Storerooms and warehouses, to prevent corrosion and surface deterioration of steel products in moist atmospheres.
Dehydrating agent for—
Air (in connection with the Gayley process).
Ammonia, benzene, coke-oven gas, naphtha, toluene, xylene.

Miscellaneous
Air-conditioning and dehumidifying agent for—
Industrial buildings, storerooms, warehouses.
Drying agent in various industrial processes.

Petroleum
Dehydrating agent for—
Petroleum distillates.
Vapors and liquids in refrigerating installations.

Photographic
Air-conditioning and dehumidifying agent.

Refrigeration
Dehydrating agent for—
Vapors and liquids in industrial and domestic refrigerating installations.

Tobacco
Air-conditioning and dehumidifying agent.
Dehydrating agent for—
Vapors and liquids in refrigerating installations.

Wine
Air-conditioning and dehumidifying agent in—
Wine cellars and caves.

Aluminum

Synonyms: Aluminium.
French: Alumine, Aluminium.
German: Aluminium, Thonerde, Tonerde.
Spanish: Aluminio.
Italian: Alluminio.

In Common Commercial Forms

(Bars, Busbar, Cables, Ingots, Pipes, Plates, Rods, Sheets, Tubing, Wires, and Others)

Brewing
Metal for making—
Cans, kegs, piping, tanks, valves.
Various equipment formerly requiring linings or protective coatings to avoid adverse effects on color, limpidity, taste.

Building Construction
Metal for making—
Balustrades, columns, conduits, coping tiles, cornices, cresting doors, downspouts and gutters, electric light fixtures, entrances, fascias, floodlight brackets, frames and pendants, grille work, hardware, kick plates, marquises, newel posts, ornamental fences, panels, pilasters, radiators and covers, risers and treads, saddles, scuppers, sheet roofing, shingles, skylight frames, spandrels, stair railing, statuary, store front work and facing, thresholds, venetian blinds, ventilators, ventilating ducts, window casements, window frames, window sash, window sills.

Automotive
Metal for making—
Accessories, body members, fittings, power plant parts, underframes.

Aviation
Metal for making—
Fittings and accessories for wings, fuselage, power plants.
Structural members, coverings.

Chemical
Basic material in making various aluminum salts.
Metal for making—
Agitators, autoclaves, baffles, balances, belts, blowcases, bottles, cans, caps and seals, chutes, coils, collapsible tubes, condensers, containers, conveyors, coolers, crystallizers, culture pans, curing pans, dephlegmators, digesters, distilled water distributing systems, dryer belts, dryers, evaporators, fans, filling machines, filters, fittings, funnels, gas-scrubbers, heating coils, hoppers, jacketed kettles, kettles, laboratory apparatus, linings, mixers, pipelines, pots, pumps, rayon equipment, retorts, scale pans, scoops, screens, shipping drums, sieves, stills, tankcars, tanks, tanktrucks, trays, trucks, vacuum pans, valves, vats.

Dairy
Metal for making—
Ageing equipment, cans, coils, coolers, cream separators, dump tanks, filling machinery, filters, fittings, funnels, forewarming tanks, heaters, holding tanks.
Metal for making—
Hoppers, milker pails, milking machines, pasteurizing equipment, pipelines, pumps, ripeners, tanks, tank trucks, vacuum pans, valves, vats, weigh cans.

Electrical
Metal for making—
Battery eliminator parts, busbar, cable and connections, contacts, discs, electrode holders, electrolytic condensers, fixed condensers, fixtures, fuse wire, lamp receptacles, lead wires, lightning arrester trays, loud speaker diaphragms, loud speakers, magnetic coils, meter covers, radio chassis, radio panels, radio shields, rectifier parts, switch buttons, telephone equipment, variable condensers, wire.

Food
Metal for making—
Bins, canning equipment, caps and seals, chutes, collapsible tubes, containers, conveyors, cooking coils, cooking kettles, evaporators, filling machines, filters, fittings, foil, funnels, hoppers, jacketed kettles, jar caps, kettles, meat-cutting machines, mixers, pans, percolators, pipelines, pumps, range accessories, ranges, refrigeration equipment, screens, storage tanks, stovepipe, strainers, tabletops, trays, trucks, utensils, vacuum pans, valves, vats, waffle irons, weighing machine parts.

Metallurgical
Component of—
Alloys of various kinds, such as duraluminum, aluminum brasses and bronzes, lightweight alloys.
Deoxidizing agent in various processes.
Metal for making—
Burner parts, flask equipment, match plates, oven linings, recording instruments.
Precipitating agent in various processes.
Reducing agent in various processes.

Petroleum
Metal for making—
Bubble caps, busbar, cable, coatings, condensers, conduit, drums, gages, heat exchangers, nails, pipelines, pumps, screws, swing lines, tankcars, tanks, tanktrucks, tubing.

Paper
Metal for making—
Conduits, conveyors, dryers, rolls for fourdrinier tables, section boxes, tubing.

Pharmaceutical
Metal for making various apparatus, containers, and utensils (see under Chemical).

Aluminum (Continued)

Rubber
Filler.
Metal for making—
Bins, conduits, curing pans, electrodes, heat manifolds, mandrels, markers, molds, vulcanizing pans.

Railroading
Metal for making—
Accessories, bodies, body members, coaches and parts, power plant parts, structural members, underframing, trucks.

In Dust, Filings, Flakes, and Powder

Aviation
Coating agent for—
Airship fabrics.
Fireproofing material.
Ingredient of—
Airplane dopes.

Building Construction
Fireproofing agent.
Ingredient of—
Building material (U. S. 1357375).
Cement (U. S. 1269116).
Luting compositions (U. S. 1140760).

Ceramics
Ingredient of—
Decorations, glazes.

Chemical
Catalyst in making—
Ammonia from atmospheric nitrogen.
Organic acids from soda-cellulose waste liquors.
Starting point in making—
Aluminates.
Aluminum salts of acids and halogens.

Electrical
Construction material in—
Wire, conducters, and parts of electrical machinery.

Explosives
Ingredient of—
Blasting cartridges, detonators, explosives, pyrotechnic compositions.

Glass
Coating and decorating agent.
Ingredient of—
Special glass batches.

Leather
Process material (U. S. 1493062) in making—
Artificial leather.

Metal Fabricating
Decorating agent for—
Aluminum ware, enamelware.

Metallurgical
Coating agent for—
Copper, iron, steel.
Deoxidizer in—
Heat treatment of iron and steel.
Heating medium in—
Aluminic-thermic welding.
Ingredient of—
Aluminum alloys, aluminum bronze, chrome alloys, duralumin, ferrotungsten, iron alloys, manganese alloys, magnalium, plastic metal compositions.
Reagent in various metallurgical processes.
Reducing agent in making—
Chromium metal from chromium oxide.
Starting point in making—
Aluminum foil.

Miscellaneous
Coating agent for various materials.
Decorating agent for various materials.
Ingredient of—
Composition used for the generation of foam in fighting fires (German 430137).
Compositions used for stopping leaks in automobile radiators and other hot-water circulating systems (U. S. 1613055).
Metallic transfer product (U. S. 1444345).
Soldering compositions.
Paint base in—
Protective paints for structures and equipment exposed to corrosive fumes, temperature fluctuations, moisture, steam.
Light-reflecting paints used to increase the degree of illumination in interiors of buildings and in darkened areas.
Heat-reflecting paints.

Paper
Coating agent for—
Paper, pasteboard.

Paint and Varnish
Basic pigment in—
Aluminum paints used for priming, decorative, and light and heat-reflecting purposes.
Basic pigment in—
Lustrous paints (used plain or colored to simulate gilt or bronze or dyed with anilin colors).

Pharmaceutical
In compounding and dispensing practice.

Plastics
Ingredient (U. S. 1160362 and 1160365) of—
Billiard balls, buttons, knife handles, and various other plastic products.

Photographic
Ingredient of—
Flashlight powders.

Printing
Pigment in—
Decorative printing, printer's blankets (U. S. 1210375), printer's plates.

Rubber
Coating agent.
Decorative agent.
Ingredient of—
Rubber compositions.
Rubber-mending composition (U. S. 1389084).

Soap
Reagent in—
Floating soaps (used to retard rancidity).

Textile
Pigment in—
Printing calicoes.

In Foil and Ribbon

Aviation
Fireproofing material for—
Aircraft.

Electrical
Component of—
Condensers.

Food
Metal for making—
Containers.
Packaging material.

Miscellaneous
Component (U. S. 1506729) of—
Motion picture screen.
Decorative medium.
Fireproof insulation.
Packaging material.

Aluminum Acetate
Synonyms: Acetate of alumina.
French: Acétate d'alumine, Acétate d'aluminium, Mordant au roughe des indiennes.
German: Azetataluminium, Essigsaeurer alaun, Essigsäurealuminium, Essigsaeurestonerde.
Spanish: Acetato de aluminio.
Italian: Acetato di alluminio.

Chemical
Ingredient of catalytic preparations used in making—
Acenaphthylene, acenaphthaquinone, bisacenaphthylidenedione, naphthaldehydic acid, naphthalic anhydride and hemimellitic acid from acenaphthene (Brit. 295270).
Acetaldehyde from ethyl alcohol (Brit. 281307).
Acetic acid from ethyl alcohol (Brit. 281307).
Alcohols from aliphatic hydrocarbons (Brit. 281307).
Aldehydes or alcohols by reduction of esters (Brit. 306471).
Alphacampholide by reduction of camphoric acid (Brit. 306471).
Aldehydes and acids from toluene, orthochlorotoluene, orthobromotoluene, orthonitrotoluene, parachlorotoluene, parabromotoluene, paranitrotoluene, metachlorotoluene, metabromotoluene, metanitrotoluene, dinitrotoluenes, dibromotoluenes, dichlorotoluenes, chlorobromotoluenes, chloronitrotoluenes, bromonitrotoluenes (Brit. 295270).

Aluminum Acetate (Continued)
Aldehydes and acids from xylenes, pseudocumene, mesitylene, and paracymene (Brit. 295270).
Alphanaphthaquinone from naphthalene (Brit. 281307).
Anthraquinone from naphthalene (Brit. 295270).
Benzaldehyde and benzoic acid from toluene (Brit. 281307).
Benzoquinone from phenanthraquinone (Brit. 281307).
Benzyl alcohol by reduction of benzaldehyde (Brit. 306471).
Benzyl alcohol or benzaldehyde or phthalide from phthalic anhydride (Brit. 306471).
Butyl alcohol from crotonaldehyde (Brit. 306471).
Chloroacetic acid from ethylenechlorohydrin (Brit. 295270).
Diphenic acid from ethyl alcohol (Brit. 281307).
Ethyl alcohol from acetaldehyde (Brit. 306471).
Fluorenone from fluorene (Brit. 295270).
Formaldehyde from methanol or methane (Brit. 295270).
Formaldehyde from carbon monoxide or carbon dioxide (Brit. 306471).
Isopropyl alcohol by the reduction of acetone (Brit. 306471).
Maleic acid and fumaric acid by the oxidation of toluene, benzene, phenols, tar phenols, or furfural, or from benzoquinone or phthalic anhydride (Brit. 295270).
Methane by the reduction of carbon dioxide or carbon monoxide (Brit. 306471).
Methanol by reduction from carbon dioxide or carbon monoxide (Brit. 306471).
Naphthaldehydic acid, acenaphthaquinone, or bisacenaphthylidenedione from acenaphthylene (Brit. 281307).
Phenanthraquinone from phenanthrene or diphenic acid (Brit. 295270).
Phthalic acid and maleic acid from naphthalene (Brit. 295270).
Primary alcohols from aldehydes by reduction (Brit. 306471).
Propionic acid and butyric acid and higher alcohols, ketones, and acids from carbon dioxide or carbon monoxide (Brit. 306471).
Reduction of anthraquinone, benzoquinone, and the like to corresponding hydroxyl compounds (Brit. 306471).
Reduction of carbon dioxide and carbon monoxide (Brit. 306471).
Reduction of ketones, aldehydes, acids, esters, alcohols, ethers, and other organic compounds containing oxygen (Brit. 306471).
Salicylic acid and salicyl aldehyde from cresol (Brit. 295270).
Secondary butyl alcohol by the reduction of methylethyl ketone (Brit. 306471).
Valeryl alcohol by the reduction of valeraldehyde (Brit. 306471).
Vanillin and vanillic acid from eugenol and isoeugenol (Brit. 295270).
Starting point in making—
Acetonal, aluminum acetotartrate, aluminum tannate, potassium aluminum acetate.

Dye
Ingredient in making—
Dense color lakes.

Miscellaneous
Ingredient in making—
Embalming fluid compositions.

Paint and Varnish
Ingredient of—
Colored lacquers.

Pharmaceutical
In compounding and dispensing practice.

Textile
——, *Dyeing*
Assistant in dyeing with—
Alizarin colors, anilin black, madder red and pinks, red colors.
Mordant in—
Calico printing, textile dyeing, general practice.
Textile dyeing with a sulphoricinoleate or rank olive oil.
——, *Finishing*
Ingredient of—
Fireproofing compositions for yarns and fabrics.
Waterproofing compositions for yarns and fabrics.

Aluminum Albuminate
French: Albuminate d'aluminium.
German: Aluminiumalbuminat, Albuminsaeuresaluminium.

Rubber
Reagent (U. S. 1640817) in—
Reclaiming rubber.

Aluminum-Alphanaphthylamine-4-sulphonate
French: Alphanaphthylamine-4-sulphonate d'aluminium.
German: Alphanaphtylamin-4-sulphosaeuresaluminium, Aluminium-alphanaphtylamin-4-sulphonat.

Resins and Waxes
Ingredient of—
Phenol-aldehyde condensation products, added for the purpose of facilitating their separation by centrifuging (German 432727).

Aluminum-Ammonium Chloride
French: Chlorure d'aluminium et ammonium.
German: Aluminiumammoniumchlorid.

Miscellaneous
Carroting agent (Brit. 271026) in treating—
Felts, furs.

Aluminum Amyloxide
French: Amyloxyde d'aluminium.
German: Aluminiumamyloxyd.

Chemical
Reducing agent (German 434728) in making the following dihalogen and trihalogen alcohols—
Amyl, allyl, butyl, ethyl, isoallyl, isoamyl, isobutyl, isopropyl, methyl, propyl.

Aluminum Arsenate
French: Arsénate d'aluminium.
German: Aluminiumarsenat, Arsensaeuresaluminium.

Insecticide
Ingredient of insecticidal compositions in admixture with calcium arsenate (U. S. 1626872).

Aluminum Benzoate
French: Benzoate aluminique, Benzoate d'aluminium.
German: Aluminumbenzoat, Benzoesäuresaluminium, Benzoesäurestonerde, Tonerdebenzoat.
Spanish: Benzoato de aluminio.
Italian: Benzoato di alluminio.

Rubber
Retarding agent (U. S. 1929561) in—
Vulcanizing processes employing an ultra-accelerator.

Aluminum Betabenzoylpropionate

Plastics
Starting point (U. S. 2001380) in making—
Films.

Aluminum Betanaphtholdisulphonate

Perfume
Deodorant astringent in—
Cosmetic preparations.

Aluminum Borate
Synonyms: Aluminium borate.
French: Borate d'alumine.
German: Borsaeuresaluminium.

Ceramics
Ingredient of flux coatings for—
Chinaware, porcelains, potteries.

Chemical
Starting point in making—
Aluminum borotartrate.

Glass
Raw material in making special glasses.

Metallurgical
Ingredient of compositions used for coating enameled ware.

Aluminum Borotartrate
French: Borotartrate aluminique, Borotartrate d'aluminium.
German: Borsäuresaluminiumtartrat.
Spanish: Boratotartrato de aluminio.
Italian: Boratotartrato di alluminio.

Pharmaceutical
Suggested for use as—
Astringent in affections of nose and throat.

Aluminum Bromide
French: Bromure aluminique, Bromure d'aluminium.
German: Aluminiumbromid, Bromaluminium.
Spanish: Bromuro de aluminio.
Italian: Bromuro di alluminio.

Chemical
Reagent in—
Organic synthesis.

Miscellaneous
Ingredient (French 672746) of—
Soldering paste, containing also borax, zinc chloride, and sodium or ammonium bromide.

Rubber
As a thermo-softening agent (French 615195).

Aluminum Butyloxide
French: Butyloxyde d'aluminium.
German: Aluminiumbutyloxyd.

Chemical
Reducing agent (German 434728) in making the following dialkyl and trialkyl alcohols:
Dialkyl amyl, allyl, butyl, ethyl isoallyl, isoamyl, isobutyl, isopropyl, methyl, propyl.

Aluminum Carbide
French: Carbure aluminique, Carbure d'aluminium.
German: Aluminiumcarbid.
Spanish: Carburo de aluminio.
Italian: Carburo di alluminio.

Chemical
Catalyst (French 642391) in making—
Pyridin and its derivatives from aldehydes and ammonia.
Starting point in generating—
Methane.
Starting point in making—
Aluminum chloride, aluminum nitride.

Metallurgical
Addition agent (French 609829) to—
Molten bath of cryolite, fluorspar, and alumina in electrolytic production of aluminum.
Desulphurizing agent (French 573866) for—
Iron, steel.
Restrainer (Australian 106982) of—
Iron attack by sulphuric acid pickling baths.

Aluminum Carbonate
French: Carbonate aluminique, Carbonate d'aluminium.
German: Aluminiumcarbonat.
Spanish: Carbonato de aluminio.
Italian: Carbonato di alluminio.

Pharmaceutical
Suggested for use as—
Antiseptic styptic in croup, ocular affections, diarrhea, and other afflictions.

Aluminum Cetylacetate

Food
Starting point (U. S. 1955527) in making—
Chewing gum ingredients by heating with paraffin, vegetable or animal waxes, fats, resins, or polymerization products of a similar nature.

Miscellaneous
Starting point (U. S. 1955527) in making—
Substitutes for paraffin wax for waterproofing purposes (giving thin films which are more adhesive and possess a higher melting point than paraffin wax) by heating with paraffin, vegetable or animal waxes, fats, resins, or polymerization products of a similar nature.

Aluminum Chlorate
French: Chlorate aluminique, Chlorate d'aluminium.
German: Aluminiumchlorat, Chlorsäuresaluminium.
Spanish: Clorato de aluminio.
Italian: Clorato di alluminio.

Pharmaceutical
In compounding and dispensing practice.
Suggested for use as—
Nontoxic antiseptic.

Aluminum Chloride
French: Chlorure aluminique, Chlorure d'aluminium.
German: Aluminiumchlorid, Chlorwasserstoffsäures-aluminium, Chlorwasserstoffsäurestonerde.
Spanish: Cloruro de aluminio.
Italian: Cloruro di alluminio.

Analysis
Laboratory reagent in carrying out syntheses by the Friedel and Craft reaction.
Reagent in the testing for naphthalene.

Chemical
Augmenter (French 700511) of—
Steam-absorbing properties of activated carbons.
Catalyst in carrying out various reactions and syntheses.
Catalyst in—
Hydrogenating rubber to produce synthetic oils suitable for use in paints and for various impregnating purposes (Brit. 397136).
Production of acetophenone
Esters from lower aliphatic acids and olefins (Brit. 398527).
Ethylidene chloride from ethylene chloride (U. S. 1900276).
Catalyst in making—
Alkyl benzoates from benzotrichloride (U. S. 1866849).
Alkyl-substituted aromatic hydroxy compounds (U. S. 1892990).
Alkyl chlorides from olefin hydrocarbons.
Anilin from benzene (Brit. 250897).
Anilin by reduction of nitrobenzene with iron filings (Brit. 616274).
Aromatic alcohols, especially phenylethyl alcohols and their homologs, by the action of alkylene oxides on aromatic hydrocarbons, the alkylene oxides being diluted with an inert gas such as air, nitrogen, or carbon dioxide (Brit. 398136).
Aromatic aldehydes from benzenoid hydrocarbons, or halogenated derivatives, carbon monoxide, benzaldehyde, and water.
Benzaldehyde from benzene (Brit. 397124).
Benzaldehyde from benzene and carbon monoxide.
Benzoic acid from benzotrichloride (U. S. 1866849).
Benzoic acid from benzene, and cyanuric chloride (U. S. 1734029).
Bornyl esters of various sorts (Brit. 251147).
Chlorinated benzenes from benzene, or a partially chlorinated benzene, and chlorine (Brit. 388818).
Condensations in the benzanthrone series (French 612367 and 615831).
2:4-Dimethylbenzaldehyde from metaxylene (Brit. 397124).
High molecular polymerization products from iso-olefins (Brit. 401297).
Hydrocarbons, boiling between 120° and 200° C., by reacting ethylene and methyl chloride under pressure (French 695125).
Metal benzoates from benzotrichloride (U. S. 1866849).
Orthoamylbenzoylbenzoic acid (U. S. 1889347).
Organic acids from soda cellulose pulp waste liquors.
Orthobromotoluene.
Paratolualdehyde from toluene (Brit. 397124).
Perylene from 2:2'-dinaphthyl (Brit. 425363).
Phthalyl chloride, or its homologs, from phthalic anhydride, or its homologs, or its nuclear substitution products, and benzylidene chloride or benzyl chloride (Brit. 414570).
Reduction products from nitrobenzene, alphanitronaphthalene, orthonitrotoluene, chlorosulphonic acids, oxysulphonic acid, aminosulphonic acid, polynitrosulphonic acids, and nitrosulphonic acids (Brit. 263376).
Secondary aromatic amines useful as retardants of rubber deterioration (U. S. 1902115).
Vinyl compounds by condensation of vinyl chloride with a phenol (Brit. 409132).
Decolorizing agent (French 619857) for—
Acetone oils and methylene.
Ingredient of catalytic mixtures used in the manufacture of—
Acenaphthylene, acenaphthaquinone, bisacenaphthylidenedione, naphthaldehydic acid, naphthalic anhydride, and hemimellitic acid from acenaphthene (Brit. 295270).
Acetaldehyde from ethyl alcohol (Brit. 281307).
Acetic acid from ethyl alcohol (Brit. 281307).
Alcohols from aliphatic hydrocarbons (Brit. 281307).
Aldehydes or alcohols by the reduction of the corresponding esters (Brit. 306471).
Aldehydes and acids from toluene, orthochlorotoluene, orthonitrotoluene, orthobromotoluene, parachlorotoluene, parabromotoluene, paranitrotoluene, metachlorotoluene, metabromotoluene, metanitrotoluene, dichlorotoluenes, dibromotoluenes, dinitrotoluenes, bromonitrotoluenes, bromochlorotoluenes, chloronitrotoluenes (Brit. 295270).

Aluminum Chloride (Continued)

Aldehydes and acids from xylenes, pseudocumene, mesitylene, and paracymene (Brit. 281307).
Alphanaphthaquinone from naphthalene (Brit. 281307).
Anthraquinone from naphthalene (Brit. 295270).
Alphacampholide by the reduction of camphoric acid (Brit. 306471).
Benzaldehyde and benzoic acid from toluene (Brit. 281307).
Benzoquinone from phenanthraquinone (Brit. 281307).
Benzyl alcohol by the reduction of benzaldehyde (Brit. 306471).
Benzyl alcohol or benzaldehyde or phthalide by the reduction of phthalic anhydride (Brit. 306471).
Butyl alcohol by the reduction of crotonaldehyde (Brit. 306471).
Chloracetic acid from ethylenechlorohydrin (Brit. 295270).
Diphenic acid from ethyl alcohol (Brit. 281307).
Ethyl alcohol by the reduction of acetaldehyde (Brit. 306471).
Fluorenone from fluorene (Brit. 295270).
Formaldehyde by the reduction of methane or methanol (Brit. 306471).
Formaldehyde by the reduction of carbon dioxide or carbon monoxide (Brit. 306471).
Hydroxyl compounds by the reduction of anthraquinone, benzoquinone and similar compounds (Brit. 306471).
Isopropyl alcohol by the reduction of acetone (Brit. 306471).
Lubricating oils by polymerization of olefins, such as ethylene (Brit. 363846).
Maleic acid and fumaric acid by the oxidation of toluene, benzene, phenols, tar phenols, or furfural, or from benzoquinone or phthalic anhydride (Brit. 295270).
Methane by the reduction of carbon dioxide or carbon monoxide (Brit. 306471).
Naphthaldehydic acid, acenaphthaquinone, or bisacenaphthylidenedione from acenaphthylene (Brit. 281307).
Phenanthraquinone from phenanthrene or diphenic acid (Brit. 295270).
Phthalic acid and maleic acid from naphthalene (Brit. 295270).
Primary alcohols by the reduction of the corresponding aldehydes (Brit. 306471).
Propionic acid and butyric acid and higher alcohols, ketones, and acids by the reduction of carbon dioxide and carbon monoxide (Brit. 306471).
Reduction products of ketones, aldehydes, acids, esters, alcohols, ethers, and other organic compounds which contain oxygen (Brit. 306471).
Salicylic acid and salicylic aldehyde from cresol (Brit. 295270).
Secondary butyl alcohol by the reduction of methylethyl ketone (Brit. 306471).
Valeryl alcohol by the reduction of valeraldehyde (Brit. 306471).
Vanillin and vanillic acid from eugenol and isoeugenol (Brit. 295270).

Ingredient (Brit. 306471) of catalytic preparations which are used in the production of aromatic and aliphatic compounds, including—
Alphanaphthylamine from alphanitronaphthalene.
Amines from aliphatic nitro compounds, such as allyl nitriles, or nitromethane.
Amylamine from pyridin.
Anilin, azo-oxybenzene, azobenzene, and hydrazobenzene from nitrobenzene by reduction.
Aminophenols from nitrophenols.
3-Aminopyridin from 3-nitropyridin.
Amino compounds from the corresponding nitroanisoles.
Amines from oximes, Schiff's base, and nitriles.
Cyclohexamine, dicyclohexamine, and cyclohexylanilin from nitrobenzene.
Piperidin from pyridin.
Pyrrolidin from pyrrol.
Tetrahydroquinolin from quinolin.

Process material (Brit. 398136) in making—
Phenylethyl alcohol and other aromatic alcohols.

Reagent in—
Carrying out organic syntheses, in the manufacture of aromatic chemicals, pharmaceuticals and intermediate chemicals.
Carrying out the Friedel and Craft reaction in effecting organic syntheses.
Treating potassium ferrocyanide for making a catalyst which is to be used in the synthesis of ammonia (Brit. 253122).
Treatment of various organic compounds for the purpose of removing the hydrogen from aromatic compounds and for coupling aromatic radicles.

Reagent in making—
Acenaphthene, acetophenone, anthraquinone, carbazoles (Brit. 278771).
Aromatic aldehydes by reacting an aromatic hydrocarbon, or an ether of a mono or polyhydric phenol, or a halogenated aromatic hydrocarbon having one or several lateral chains, with hydrocyanic acid, the proportion of aluminum chloride being in excess of a molecule for each CN group of the acid or of a metallic cyanide (French 750842).
Benzoic acid (U. S. 1734029).
Benzoyl chloride.
Betachloroanthraquinone.
Betamethylanthraquinone.
5:8-Chloro-4-hydroxyl-1-methylanthraquinone.
Compounds with hydrocyanic acid.
Compounds of aluminum and fluorine by reacting upon fluorspar or sodium fluosilicate (French 695573).
Condensation products, such as 1:5-dibenzoylnaphthalene, diarylacenaphthenes, dichloroacetylacenaphthene (Brit. 291347).
Dimethyldiphenylanthrone.
Ethyl benzene.
Fluorene.
Ketone musk.
Isobutyl xylol.
Oily products, for use with chlorinated rubber, from sulphur and aliphatic homologs of benzene; for example, reactions with sulphur and toluene, sulphur and xylene, sulphur and ethylbenzene (Brit. 429764).
Pickling-bath inhibitors by reaction with quinoidin, sulphur, or sulphur chloride, in the presence or absence of a solvent; the end products are sulphurized quinoidin bodies (Brit. 411907).
Propylene dichloride.
Tetramethylanthrone.
Trichlorobutyl alcohol from butyl chloral (Brit. 251890).
Urea.
Urea salts.

Stabilizing agent (Brit. 250747) in treating—
Liquid hydrocyanic acid.

Starting point in making—
Pure aluminum sulphate (German 424129).
Various aluminum compounds.

Starting point (French 666122) in making—
Granulated aluminum oxide.

Bituminous

Catalyst in—
Thermic dissociation of tars of schist or lignite into saturated hydrocarbons of low boiling point (French 614229).
Transforming hydrocarbons of high boiling point, freed of constituents soluble in sulphuric acid, into hydrocarbons of low boiling point (French 630174).
Viscous oil production from tars and pitches (French 650799).
Viscous oil production from lignite tar or pitch or from their derivatives (French 669518).

Ingredient (French 669793) of—
Catalytic mixtures, with metals and metalloids, used in the conversion of tar oils into lubricating oils.

Starting point (French 691303) in making—
Hydrogenated products of charcoal, lignite, tar, pitch, cellulose, and wood.

Ceramics

Coloring agent (Brit. 410651) for—
Surfaces of unglazed, fired brick, tiles, earthenware, and the like. Coloration is effected by introducing into the kiln during the firing process, in the absence of any alkali chloride, at a stage when the contents of the kiln have reached a temperature at which the chloride sublimes substantially immediately on to the surface of the article being fired, and there decomposes with the deposition of the corresponding oxide; the colors may be modified by a prolonged firing in the absence of air by the action of the reducing gases of the kiln; the actual sublimation of the metallic chloride may be effected outside the kiln, the vapors being blown in through the firing holes, or the material may be preheated outside the kiln and introduced into the kiln before sublimation has taken place.

Aluminum Chloride (Continued)

Construction

Ingredient (Brit. 403230) of—
Plaster or cement made of (1) Portland cement or calcined gypsum and, (2) alcohol or triacetin.
Reagent (Brit. 371257) in making—
Cement plaster, drying with a glazed surface.
Reagent (Brit. 278788) in treating—
Cements and concretes, which contain wood, sawdust, shavings, cork, sisal, paper pulp and the like, in order to render them waterproof.
Setting accelerator (French 666763) for—
Cements.

Dye

Catalyst (French 709862) in—
Condensing halogen derivatives of squalene with phenol in making dye intermediates.
Condensing agent (Brit. 399241, addition to 381920) in making—
Vat dyes of the anthraquinone series.
Condensing agent (U. S. 1912378) in making—
Fast golden brown vat dyes.
Fractional crystallization agent (Brit. 403862, 341357, 237294, 265232, and 265964) in separating—
Isomeric vat dyes (from naphthalene tetracarboxylic acid and diamines).
Process material in making—
Indigoid dyes using 6-chloro-orthotoluidin as a starting material (French 617997).
Vat dyestuffs (French 596484).
Reagent in making—
Condensed nitroperylene colors.
Indanthrene yellow G.
Tetrabenzoylperylenes.
Vat colors.

Electrical

Catalyst (French (770848) in making—
Dielectics from polyphenyl derivatives of methane, polyphenyl derivatives of ethane, halogenated in an aromatic nucleus, and derivatives of benzophenone (such dielectics are said to be chemically stable under all conditions which may be met with in electrical apparatus and equipment and not to produce any gases capable of exploding in air when subjected to electric arcing).
Filler and adhesive (U. S. 1908792) in—
Crystallized alumina insulating agent for thermionic tube heater elements.
Ingredient (U. S. 1597165) of—
Electrolytic pastes for making dry cells.

Fats and Oils

Condensing agent (Brit. 394073) in making—
Lubricating oils (by converting animal or vegetable fatty substances into unsaturated products practically free from oxygen and polymerizing or condensing these products).

Fertilizer

Ingredient (U. S. 1880058) of—
Corrective agents (in admixture with zinc sulphate and copper sulphate) for the treatment of unproductive lands, such as the sawgrass soil of the Florida Everglades.

Fuel

Combustion catalyst (Brit. 408951) in—
Synthetic fuels consisting of suspensions of coal in oil (obtained by the destructive distillation or hydrogenation of coal or shale), stabilizers, water- or oil-soluble combustion accelerators, and oxidizing agents.

Fungicide and Insecticide

As an anticryptogamic (French 620941).
Ingredient (Brit. 396365) of—
Insecticidal mixture with sodium arsenate.

Leather

Tanning agent for—
Hides (French 614535).
Hides (used in conjunction with tin chlorides) (French 552161).

Lubricant

Catalyst in making—
Fluorescence promoters for lubricants (such promoters consist of hydrogenated or nonhydrogenated condensation products obtained by condensing olefins of purely aliphatic constitution which are liquid at ordinary temperature and substantially free from diolefins with between 20 and 70 percent of polynuclear organic substances, in which none of the nuclei is saturated with hydrogen at an elevated temperature) (Brit. 409696).
Lubricants by polymerization of olefins (Brit. 363846).
Lubricating oils by polymerization of cracked mineral oil distillates containing substantial amounts of unsaturated hydrocarbons (Brit. 414237).
Lubricating oils by polymerization of flash-distilled vapors, such vapors being produced from unvaporized oil separated after cracking heavy hydrocarbon oil (U. S. 1960625).
Polymerized products of nonsaturated hydrocarbons (obtained by cracking processes), such products being useful as lubricating oils (French 690966).
Pour-point improvers by condensing acid chlorides of molecular weight above 200 and derived from an aliphatic hydrocarbon; the chloride may be condensed alone or with other substances (Brit. 407956).
Setting-point reducers, viscosity index improvers, and stabilizers for lubricating oils and greases by polymerization or condensation of olefins, mineral or tar oils, waxes, fats, high molecular acids, or alcohols (Brit. 435597 and 435548).
Synthetic lubricants by condensing ethylene with naphthalene or tetralin.
Viscous oils from lignite tars or pitches or from their derivatives (French 669518).
Viscous oils from tars and pitches, or mineral oils, or lower olefins, such as ethylene, propylene, butylene (French 650799).
Ingredient (French 669793) of—
Catalytic mixtures, with metals and metalloids, used in the conversion of tar oils and mineral oils into lubricants.
Refining agent in—
Purifying and regenerating used lubricating oil by simultaneously separating low-flash fractions and precipitating asphaltic matter (Brit. 413537).
Removing refractory sulphonic acids and soaps from lubricants intended as airplane engine oils, ash-free and of specially low-temperature coefficient, prepared from mixed or paraffin-base stock (U. S. 1980377).

Metallurgical

Coating agent (U. S. 1920244) for—
Molds used in making castiron stair treads.
Ingredient of—
Electrolyte used in galvanoplastic soldering of aluminum to aluminum (French 523696)
Flux (in admixture with zinc chloride, ammonium chloride, sodium or potassium fluoride, and sodium chloride) used in hot-galvanizing iron articles by means of zinc or alloys of zinc, especially alloys of zinc and aluminum (French 701259).
Reagent (Brit. 412430) for—
Preparation of aluminum, or an alloy thereof, for electroplating (acts so that the film of oxide normally covering its surface is converted into aluminum hydride; reaction may be effected by immersion of the article and rapid withdrawal, or by spraying or by humidified vapor).
Starting point in making—
Aluminum metal by admixture with aluminum hydroxide in the electric furnace (French 634727).
Anhydrous alumina by low-temperature thermic decomposition (French 657444).

Miscellaneous

Ingredient (German 519062) of—
Heat-transfer medium "NS," in admixture with sodium chloride and iron chloride.
Polymerizing agent (U. S. 1932525) for—
Indene, styrene, coumarone, and other gum-forming constituents extracted from fuel gases by scrubbing with oil.
Reagent (in admixture with hydrogen peroxide) (French 697439) for—
Treating waste gases to prevent formation of carbon monoxide and assist in the formation of benzene.
Reagent (U. S. 1720487) in making—
Infusible asphaltic masses of high elasticity.

Paint and Varnish

Catalyst in making—
Antimony sulphide pigments.
Condensing agent (U. S. 1934033) in making—
Driers from phthalic anhydride and an alkylated aromatic hydrocarbon, the alkyl group of which contains at least three C atoms in a normal straight chain.

Paper

Reagent in making—
Parchmentized paper.

Aluminum Chloride (Continued)

Perfumery
Solvent (U. S. 1907424) for—
Cerium oxalate as a fixative of perfumes.
Cerium oxalate as a deodorant for perspiration.

Petroleum
Catalyst (U. S. 1909587) in—
Treating petroleum oils.
Catalyst in making—
Cracked oil products (French 616173).
Fluorescence promoters, for fuels and lubricants, consisting of hydrogenated or nonhydrogenated condensation products obtained by condensing olefins of purely aliphatic constitution which are liquid at ordinary temperature and substantially free from diolefins with between 20 and 70 percent of polynuclear organic substances, in which none of the nuclei is saturated with hydrogen, at an elevated temperature (these promoters may be added to hydrocarbon oils, that is, heavy oils, or middle oils, or lubricating oils, or liquid motor fuels, such as benzins, or other liquid hydrocarbon fractions, whether paraffine or naphthenic) (Brit. 409696).
Low-boiling products, such as gasoline, by conversion of high-boiling hydrocarbons (U. S. 1952898 and 1953612).
Lubricating oils by polymerization of cracked mineral oil distillates containing substantial amounts of unsaturated hydrocarbons (Brit. 414237).
Lubricating oils by polymerization of flash-distilled vapors, such vapors being produced from unvaporized oil separated after cracking heavy hydrocarbon oil (U. S. 1960625).
Lubricating oils, kerosene, and other oils from heavy petroleum stocks by treatment with hydrogen chloride or chlorine gas and addition of hydrocarbons of lower molecular weight either prior to or during the reaction (Brit. 398032).
Mineral oil products by transformation into hydrocarbons of lower boiling point, with simultaneous decoloration (French 671035).
Polymerized products of nonsaturated hydrocarbons (obtained by cracking processes) such products being useful as lubricating oils (French 690966).
Refined products from cracked oils by condensation of unsaturated constituents (French 630828).
Viscous oils from distillates of cracked hydrocarbons (French 608425).
Viscous oils from mineral oils or lower olefins, such as ethylene, propylene, butylene (French 650799).
Condensing agent (Brit. 397169) in making—
Condensation products of high molecular paraffin hydrocarbons, used to facilitate the separation of waxes from hydrocarbon oils.
Ingredient (French 669793) of—
Catalytic mixtures, with metals and metalloids, used in the conversion of mineral oils into lubricants.
Reagent in—
Refining.
Reagent (U. S. 1915206) in—
Stevens cracking process.

Pharmaceutical
In compounding and dispensing practice.

Plastics
Reagent (Brit. 250607) in treating—
Artificial horn products made from tendons and sinews.

Refractories and Abrasives
Starting point in making—
Anhydrous alumina (French 657444).
Granulated alumina (French 666122).

Resins
Catalyst in making—
Conversion products of ester gums, consisting of more saturated products of higher molecular weight and melting point, which do not give the Storch-Morawski reaction (a diluent may be present, and the products may be esterified with alcohols; products specified are dammar and mastic) (Brit. 399206).
Conversion products of resins, consisting of more saturated products of higher molecular weight and melting point which do not give the Storch-Morawski reaction (a diluent may be present, and the products may be esterified with alcohols; products covered are rosin and the benzene-soluble resin from rosin, glycerin, and phthalic anhydride) (Brit. 399206).
Neutral hydrocarbon resins suitable for varnishes by reacting (1) a diolefin and an olefin, (2) a diolefin and a substituted benzene from the unsaturated hydrocarbons of cracked distillate, (3) an olefin, a diolefin, and a substituted benzene (Brit. 391093).
Pale-colored varnish resins by polymerization of terpenes, with or without an alkylbenzene (U. S. 1939932).
Plastic materials which are resistant to abrasion and to the action of water, acid, alkali, and many organic liquids (these products, which may be chlorinated, may be milled and mixed with fillers, pigments, fibrous materials or rubber or rubber compounds, are used in the manufacture of heat-resisting covers for conveyor belts and flexible linings for tanks, barrels, ball mills, and pipelines) (Brit. 407948).
Shellac substitutes by halogenation of hydroxy, or polyhydroxy, carboxylic acids of an aliphatic or hydroaromatic character (U. S. 1903598).
Synthetic resins from polyvinyl alcohol, its esters, or its ethers (French 656151).
Synthetic resins by condensation of propylene with carbazol, its homologs, or its halogenated derivatives (French 666718).
Condensing agent (Brit. 397699) in making—
Resins and plastic resinous products from residual tars obtained in petroleum distillation, which according to their character may be employed as constituents of paints, varnishes, lacquers, rubber compounds, or as paving or coating materials, or for molding.
Reagent (German 420443) in making—
Resinous products by condensation of crude anthracene and phenanthrene.
Reagent (Brit. 397699) in making—
Artificial resins and resinous products from petroleum tar.

Rubber
Catalyst (French 696008) in making—
Synthetic materials, having the properties of either soft rubber or ebonite, by condensation of polyvinylic esters with nonsaturated aldehydes.
Catalyst (Brit. 397136) in making—
Paint and varnish oils and impregnating mediums for paper, cloth, leather, and other substances, by hydrogenation of natural or synthetic rubber in the presence of solvents, such as petroleum hydrocarbons.
Coagulant (Brit. 414205) for—
Coating transfer bases (rubber stamping devices and the like) prior to spray-coating with rubber.
Dispersing agent (Brit. 399870) in making—
Rubber-bonded asbestos products.
Reagent (Brit. 278395) in treating—
Rubber latex, used as a protective dispersive agent.
Thermoplasticizing agent (French 615195) for—
Rubber.

Sanitation
As a disinfectant for various purposes.

Textile
Delustring agent for—
Cellulose acetate fibers (U. S. 1927412).
Linen, rayon, or silk (applied simultaneously or subsequently with a solution of a salt of an aromatic orthodicarboxylic acid, such as the sodium or potassium salt of phthalic acid or its chloro derivatives; fastness to washing and dyeing is claimed) (Brit. 425418).
Flameproofing agent (French 600852) for—
Nitrocellulose threads
Impregnating agent in—
Wool carbonizing.
Preserver (French 601297) of—
Luster, transparency, and appearance of cellulose acetate products when subjected to the action of hot or boiling liquids.
Modifier (French 644636) of—
Viscosity of cellulose dipalmitate.

——, *Dyeing*
Ingredient of—
Dye liquors containing indigo blue.

——, *Manufacturing*
Reagent for—
Carbonizing wool.
Removing cotton and other vegetable fibers from wool and cotton mixtures.

Woodworking
Reagent for—
Preservation of wood.

Aluminum Chlorosulphonate
French: Chlorosulfonate d'aluminium.
German: Aluminiumchlorsulfonat, Chlorosulfonsaeurealuminium.

Petroleum
Reagent in treating—
Decolorized and deodorized cracked gasoline (U. S. 1649384).

Aluminum Dinaphthylnaphthenate

Lubricant
Addition agent (Brit. 433257) to—
Lubricating oils or greases, especially for use at high temperatures, such as cylinder oils, hydrogenated oils, or oils refined by treatment with sulphuric acid, clay, or extraction solvents.

Aluminum Ethyloxide
French: Éthyleoxyde d'aluminium.
German: Aluminiumaethyloxyd.

Chemical
Reducing agent (German 434728) in making dichlorinated, dibrominated, di-iodinated, trichlorinated, tribrominated and tri-iodinated derivatives of the following alcohols:—
Amyl, butyl, ethyl, isoamyl, isobutyl, isopropyl, methyl, propyl.

Aluminum Fluoride
French: Fluorure aluminique, Fluorure d'aluminium.
German: Fluoraluminium.

Ceramics
Ingredient of—
White enamels for porcelains, potteries, and other ceramic ware.

Chemical
Repressant in—
Controlling side fermentations in alcoholic fermentation.

Fermentation
Repressant in—
Controlling side fermentations in alcoholic fermentations.

Aluminum Formate
French: Formiate aluminique, Formiate d'aluminium.
German: Formiataluminium.

Paper
Impregnating agent for—
Paper.

Pharmaceutical
Suggested for use as—
Astringent, disinfectant.

Textile
Ingredient (Brit. 274611, 311885, 400244) of—
Delustering agent for artificial silk, composed of a solution with a sulphonated alyklated petroleum fraction.

Aluminum Gluconate

Leather
Pretanning agent (U. S. 1941485) in making—
White leathers which can be dyed any color and are similar to chrome-tanned leather in properties.

Aluminum-Hexamethylenetetramine Acetocitrate

Perfume
Ingredient of—
Powders and pomades for use on the skin.

Pharmaceutical
In compounding and dispensing practice.

Aluminum Hydroxide
Synonyms: Alumina hydrate, Alumina hydroxide, Aluminium hydrate, Aluminium hydroxide, Aluminium trihydrate, Aluminum hydrate, Aluminum trihydrate, Argilla, Hydrated alumina, Precipitated aluminum oxide, Precipitated oxide of alumina, Refined bauxite.
Latin: Alumini hydroxidum.
French: Alumine, Alumine hydratée, Hydrate aluminique, Hydrate d'aluminium, Hydroxyde aluminique, Hydroxyde d'aluminium.
German: Aluminiumhydrat, Aluminiumhydroxyd, Aluminiumtrihydrat, Aluminiumtrihydroxyd, Hydrasierte alumina, Hydratisierte alumina, Tonerdehydrat, Hydrasiertetonerde, Hydratisiertetonerde.
Spanish: Hidrato de aluminio.
Italian: Idrato di alluminio.

Ceramics
Ingredient of—
Glazing mixtures for potteries, porcelains, and chinaware.
Raw material in making—
Potteries.

Chemical
General filtering medium.
Ingredient of catalytic preparations used in making—
Acenaphthylene, acenaphthaquinone, bisacenaphthylidenedione, naphthaldehydic acid, naphthalic anhydride, and hemimellitic acid from acenaphthene (Brit. 295270).
Acetaldehyde from ethyl alcohol (Brit. 281307).
Acetic acid from ethyl alcohol (Brit. 281307).
Alcohols from aliphatic hydrocarbons (Brit. 281307).
Aldehydes or acids by the reduction of esters (Brit. 306471).
Alphacampholid from camphoric acid by reduction (Brit. 306471).
Aldehydes and acids from toluene, orthochlorotoluene, orthobromotoluene, orthonitrotoluene, parachlorotoluene, parabromotoluene, paranitrotoluene, metachlorotoluene, metabromotoluene, metanitrotoluene, dichlorotoluenes, dinitrotoluenes, dibromotoluene, chlorobromotoluene, chloronitrotoluene, bromonitrotoluene (Brit. 295270).
Aldehydes and acids from xylenes, pseudocumene, mesitylene, and paracymene (Brit. 281307).
Alphanaphthaquinone from naphthalene (Brit. 281307).
Anthraquinone from naphthalene (Brit. 295270).
Benzaldehyde and benzoic acid from toluene (Brit. 281307).
Benzoquinone from phenanthraquinone (Brit. 281307).
Benzyl alcohol by the reduction of benzaldehyde (Brit. 306471).
Benzyl alcohol or benzaldehyde or phthalide by the reduction of phthalic anhydride (Brit. 306471).
Butyl alcohol by the reduction of crotonaldehyde (Brit. 306471).
Chloroacetic acid from ethylenechlorohydrin (Brit. 295270).
Diphenic acid from ethyl alcohol (Brit. 281307).
Ethyl alcohol by the reduction of acetaldehyde (Brit. 306471).
Fluorenone from fluorene (Brit. 295270).
Formaldehyde by the reduction of methanol or methane (Brit. 306471).
Formaldehyde by the reduction of carbon dioxide or carbon monoxide (Brit. 306471).
Hydroxyl reduction compounds of anthraquinone, benzoquinone, and the like (Brit. 306471).
Isopropyl alcohol by the reduction of acetone (Brit. 306471).
Maleic acid and fumaric acid by the oxidation of toluene, benzene, phenols, tar phenols, or furfural, or from benzoquinone or phthalic anhydride (Brit. 295270).
Methane by the reduction of carbon dioxide or carbon monoxide (Brit. 306471).
Methanol by the reduction of carbon dioxide or carbon monoxide (Brit. 306471).
Naphthaldehydic acid, acenaphthaquinone, or bisacenaphthylidenedione from acenaphthylene (Brit. 281307).
Phenanthraquinone from phenanthrene or diphenic acid (Brit. 295270).
Phthalic acid and maleic acid from naphthalene (Brit. 295270).
Primary alcohols by the reduction of the corresponding aldehydes (Brit. 306471).
Propionic acid and butyric acid and higher alcohols, ketones, and acids by the reduction of carbon dioxide or carbon monoxide (Brit. 306471).
Reduction products of ketones, aldehydes, acids, esters, alcohols, ethers, and other organic compounds containing oxygen (Brit. 306471).
Salicylic acid and salicylic aldehyde from cresol (Brit. 295270).
Secondary butyl alcohol by the reduction of methylethyl ketone (Brit. 306471).
Valeryl alcohol by the reduction of valeraldehyde (Brit. 306471).
Vanillin and vanillic acid from eugenol or isoeugenol (Brit. 295270).
Ingredient (Brit. 304640) of catalytic preparations used in the production of various aromatic and aliphatic amines, including—

Aluminum Hydroxide (Continued)
Alphanaphthylamine from alphanitronaphthalene.
Amines from aliphatic nitro compounds, such as alkyl nitriles or nitromethane.
Amylamine from pyridin.
Anilin, azo-oxybenzene, azobenzene, and hydrazobenzene from nitrobenzene by reduction.
Aminophenols from nitrophenols.
3-Aminopyridin from 3-nitropyridin.
Amino compounds from the corresponding nitroanisoles.
Amines from oximes, Schiff's base, and nitriles.
Cyclohexamine, dicyclohexamine, and cyclohexylanilin from nitrobenzene.
Piperidin from pyridin.
Pyrrolidin from pyrrol.
Tetrahydroquinolin from quinolin.
Reagent in making—
Artificial mustard oil.
Various intermediates, pharmaceutical chemicals, and aromatics.
Starting point in making—
Aluminum sulphate, calcined alumina, aluminum acetate, aluminum borate, aluminum fluoride, aluminum oleate, aluminum palmitate, aluminum resinate, aluminum stearate.

Construction
Ingredient of—
Fire-resistant mortars.

Dye
Ingredient of—
Color lakes, dye preparations.

Fats and Oils
Filtering medium for purifying—
Animal and vegetable fats and oils.

Glass
Ingredient of—
Glass batches.

Ink
Ingredient of—
Lithographic inks, printing inks.

Mechanical
Ingredient of—
Lubricating compositions.

Miscellaneous
Filtering medium for—
Various products.
Mothproofing agent (Brit. 313771) for—
Furs, feathers, and the like.

Paint and Varnish
Reagent in making—
Lacquers, paints, varnishes.

Petroleum
Filtering medium for mineral oils.

Paper
Precipitating agent in—
Sizing with rosin.

Pharmaceutical
In compounding and dispensing practice.

Soap
Ingredient of various detergents.

Textile
——, *Dyeing*
Mordant in dyeing—
Yarns, warps, and fabrics.
——, *Finishing*
Reagent in waterproofing—
Delicate fabrics.
——, *Miscellaneous*
Mothproofing agent for wool.

Refractories
Ingredient of—
Fire-clay compositions.

Water and Sanitation
Filtering medium in—
Purifying water.

Aluminum Isoamyloxide
French: Isoamyleoxyde d'aluminium.
German: Aluminiumisoamyloxyd.

Chemical
Reagent (German 434728) in making the following alcohols:—
Dibromoamyl, dibromobutyl, dibromoethyl, dibromoisoamyl, dibromoisobutyl, dibromoisopropyl, dibromomethyl, dibromopropyl, dichloroamyl, dichlorobutyl, dichloroethyl, dichloroisoamyl, dichloroisobutyl, dichloroisopropyl, dichloromethyl, dichloropropyl, diiodoamyl, di-iodobutyl, di-iodoethyl, di-iodoisoamyl, di-iodoisobutyl, di-iodoisopropyl, di-iodomethyl, diiodopropyl, tribromoamyl, tribromobutyl, tribromoethyl, tribromoisoamyl, tribromoisobutyl, tribromoisopropyl, tribromomethyl, tribromopropyl, trichloroamyl, trichlorobutyl, trichloroethyl, trichloroisoamyl, trichloroisobutyl, trichloroisopropyl, trichloromethyl, trichloropropyl.

Aluminum Isopropylnaphthalenesulphonate
French: Isopropylenaphthalènesulphonate d'aluminium.
German: Aluminiumisopropylnaphtalinsulfonat, Isopropylnaphtalinsulphonsaeuresaluminium.

Textile
——, *Printing*
Ingredient (Brit. 269917) of—
Paste to enhance the saturation of the textile with the color and equalize the printed color design.

Aluminum Isopropyloxide
French: Isopropyleoxyde d'aluminium.
German: Aluminiumisopropyloxyd.

Chemical
Reducing agent (German 434728) in making dichlorinated, dibrominated, di-iodinated, trichlorinated, tribrominated and tri-iodinated derivatives of the following alcohols:—
Amyl, butyl, ethyl, isoamyl, isobutyl, isopropyl, methyl, propyl.

Aluminum Methylate
French: Méthylate d'aluminium.
German: Aluminiummethylat.

Chemical
Catalyst in making—
Acetic acid (Brit. 259641).

Aluminum 4-Methyl-6-isopropylenesulphonate
French: 4-Méthyle-6-isopropylènesulfonate d'aluminium.
German: Aluminium-4-methyl-6-isopropylensulfat, 4-Methyl-6-isopropylensulfonsaeuresaluminium.

Chemical
Catalyst (Brit. 276070) in making condensation products with—
Metacresol and acetone, orthocresol and acetone, paracresol and acetone.

Aluminum Methyloxide
French: Méthyleoxyde d'aluminium.
German: Aluminiummethyloxyd.

Chemical
Reducing agent (German 434728) in making the following alcohols:—
Dichloroalkyl, dibromoalkyl, di-iodoalkyl, tribromoalkyl, trichloroalkyl, tri-iodoalkyl.

Aluminum Naphthenate
French: Naphthène aluminique, Naphthène d'aluminium.
German: Naphtenaluminium.

Insecticide.
Oil-releasing agent (U. S. 1949799) in—
Insecticidal oil spray for application to sensitive plant foliage, comprising a mineral oil compounded with a small amount of partially esterified glycerol oleate.

Aluminum Nitrate
Synonyms: Aluminium nitrate, Nitrate of alumina.
French: Azotate d'alumine, Azotate aluminique, Azotate d'aluminium, Nitrate d'alumine, Nitrate aluminique, Nitrate d'aluminium.
German: Aluminiumnitrat, Salpetersäuresalaun, Salpetersäuresaluminium, Salpetersäurestonerde.
Spanish: Nitrato de aluminio.
Italian: Nitrato di alluminio.

Chemical
Reagent in making—
Acetal.
Starting point in making—
Aluminum acetate.
Aluminum borate.
Aluminum fluoride.
Aluminum oleate.
Aluminum palmitate.
Aluminum resinate.

Aluminum Nitrate (Continued)
Aluminum stearate.
Aluminum salts in general.
Ingredient of catalytic mixtures used in making—
Acenaphthylene, acenaphthaquinone, bisacenaphthylidenedione, naphthaldehydic acid, naphthalic anhydride, and hemimellitic acid from acenaphthene (Brit. 295270).
Acetaldehyde from ethyl alcohol (Brit. 281307).
Acetic acid from ethyl alcohol (Brit. 281307).
Alcohols from aliphatic hydrocarbons (Brit. 281307).
Aldehydes or acids by the reduction of the corresponding esters (Brit. 306471).
Aldehydes and acids from toluene orthochlorotoluene, orthobromotoluene, orthonitrotoluene, parachlorotoluene, paranitrotoluene, parabromotoluene, metachlorotoluene, metanitrotoluene, metabromotoluene, dichlorotoluenes, dibromotoluenes, dinitrotoluenes, bromochlorotoluenes, chloronitrotoluenes, bromonitrotoluenes (Brit. 295270).
Aldehydes and acids from xylenes, pseudocumenes, mesitylene, and paracymene (Brit. 281307).
Alphacampholide from camphoric acid by reduction (Brit. 306471).
Alphanaphthaquinone from naphthalene (Brit. 281307).
Anthraquinone from naphthalene (Brit. 281307).
Benzaldehyde and benzoic acid from toluene (Brit. 281307).
Benzoquinone from phenanthrene (Brit. 281307).
Benzyl alcohol by the reduction of benzaldehyde (Brit. 306471).
Benzyl alcohol or benzaldehyde or benzyl phthalide by the reduction of phthalic anhydride (Brit. 306471).
Butyl alcohol by the reduction of crotonaldehyde (Brit. 306471).
Diphenic acid from ethyl alcohol (Brit. 281307).
Ethyl alcohol by the reduction of acetaldehyde (Brit. 306471).
Fluorenone from fluorene (Brit. 295270).
Formaldehyde by the reduction of methane or methanol (Brit. 306471).
Formaldehyde by the reduction of carbon dioxide or carbon monoxide (Brit. 306471).
Hydroxyl compounds by the reduction of anthraquinone, benzoquinone, and the like (Brit. 306471).
Isopropyl alcohol by the reduction of acetone (Brit. 306471).
Maleic acid and fumaric acid by the oxidation of toluene, benzene, phenols, tar phenols, furfural, or from benzoquinone or phthalic anhydride (Brit. 295270).
Methane by the reduction of carbon dioxide or carbon monoxide (Brit. 306471).
Methanol by the reduction of carbon dioxide or carbon monoxide (Brit. 306471).
Naphthaldehydic acid, acenaphthaquinone, bisacenaphthylidenedione from acenaphthylene (Brit. 281307).
Phenanthraquinone from phenanthrene or diphenic acid (Brit. 295270).
Primary alcohols by the reduction of the corresponding aldehydes (Brit. 306471).
Propionic acid, butyric acid, and higher alcohols, ketones, and acids by the reduction of carbon dioxide or carbon monoxide (Brit. 306471).
Reduction products of ketones, aldehydes, acids, esters, alcohols, and ethers, and other organic compounds, which contain oxygen (Brit. 306471).
Salicylic acid and salicylic aldehyde from cresol (Brit. 295270).
Secondary butyl alcohol by the reduction of methylethylketone (Brit. 306471).
Valeryl alcohol by the reduction of valeric aldehyde (Brit. 306471).
Vanillin and vanillic acid from eugenol or isoeugenol (Brit. 295270).
Ingredient (Brit. 304640) of catalytic preparations used in the production of various aromatic and aliphatic amines, including—
Alphanaphthylamine from alphanitronaphthalene.
Amines from aliphatic nitro compounds, such as alkyl nitriles, or nitromethane.
Aminophenols from nitrophenols.
Amylamine from pyridin.
Anilin, azo-oxybenzene, azobenzene, and hydrazobenzene from nitrobenzene by reduction.
3-Aminopyridin from 3-nitropyridin.
Cyclohexamine, dicyclohexamine and cyclohexylanilin from nitrobenzene.
Amino compounds from the corresponding nitroanisoles.
Piperidin from pyridin.
Pyrrolidin from pyrrol.
Tetrahydroquinolin from quinolin.

Gas
Ingredient of—
Compositions used in the manufacture of gas-lamp mantles (used for the purpose of rendering the mantle more resistant in transportation and handling).
Reagent in—
Treating gas-lamp mantles for the purpose of hardening the tops of them.

Leather
Ingredient of—
Compositions used for the purpose of finishing leathers.
Reagent in—
Tawing.

Textile
Mordant in—
Dyeing and printing yarns and fabrics with alizarin red.

Aluminum Oleate
French: Oléate d'aluminium.
German: Aluminiumoleat, Oleinsaeuresaluminium, Oelsaeuresaluminium.

Miscellaneous
Thickening agent in—
Lubricating compositions (U. S. 1625969).

Paint and Varnish
Drier and flattener in—
Enamels, lacquers, paints, varnishes.

Textile
——, *Finishing*
Waterproofing agent in treating various fabrics.

Aluminum Palmitate
French: Palmitate aluminique, Palmitate d'aluminium.
German: Palmitinsäuresaluminium.

Leather
Glossing agent in—
Finishes.
Ingredient of—
Waterproofing compounds.

Mechanical
Thickener for—
Lubricating compositions, lubricating greases, lubricating oils.

Paint and Varnish
Flatting and suspensory agent in—
Paints, varnishes.

Paper
Glossing agent in—
Coatings.
Ingredient of—
Waterproofing compounds.

Textile
Ingredient of—
Waterproofing compounds.

Aluminum Palmitobenzenesulphonate
French: Palmitobenzènesulphonate d'aluminium.
German: Aluminiumpalmitobenzolsulfonat, Palmitobenzolsulfonsaeuresaluminium.

Textile
——, *Printing*
Ingredient of pastes for the purpose of enhancing the saturation of the textile with the color and equalizing the printing.

Aluminum-Phenyl Acetate

Petroleum
Addition agent (Brit. 433257) in—
Lubricating oils or greases, especially for use at high temperatures, such as cylinder oils, hydrogenated oils, or oils refined by treatment with sulphuric acid, clay, or extraction solvents.

Aluminum Phosphate
Synonyms: Phosphate of alumina.
French: Phosphate d'aluminium.
German: Aluminiumphosphat, Phosphorsaeuresaluminium.

Ceramics
Ingredient of flux in making—
Chinaware, porcelains, potteries, stoneware.

Paint and Varnish
Catalyst in making—
Light yellow lead chromate pigment of good covering power.

Aluminum Phosphate-Sulphates

Chemical
Claimed (Brit. 440400) as—
New commercial chemicals consisting of solutions of aluminum phosphate-sulphates prepared by bringing the three components, alumina, phosphoric acid, and sulphuric acid together under conditions of concentration and temperature to form clear and stable solutions containing more than one molecule of aluminum phosphate, $Al_2(PO_4)_2$, to two molecules of aluminum sulphate; that is, a molecular ratio of phosphoric acid to alumina higher than 0.33 molecule P_2O_5 to one molecule Al_2O_3; by evaporating the solutions, water-soluble solids (for example, in the form of small fragments) containing the higher ratio of aluminum phosphate to aluminum sulphate are obtained; by adding further phosphoric acid the molecular ratio of the phosphoric acid to alumina may be increased to 0.75 molecule P_2O_5 to one molecule Al_2O_3.
The new compounds have the property of decomposing in aqueous solution, without the addition of alkali, with precipitation of a very high proportion of their contained alumina and phosphoric acid when clear solutions thereof are heated at suitable strengths and temperatures, increased dilution and temperature tending toward increased precipitation; the decomposition may be assisted by adding a neutralizing reagent, for example, sodium carbonate; the precipitates formed by such decomposition normally consist of basic aluminum phosphates with varying ratios of Al_2O_3 to P_2O_5, according to the original composition and the conditions of decomposition, usually containing a small proportion of SO_3, and may be in bulky flocculent state, eminently suitable for use in arts.

Dye
Base (Brit. 440400) in—
Color lakes.

Leather
Water-treating agent (Brit. 440400) in—
Tanning processes.

Paper
Precipitator (Brit. 440400) of—
Aluminum phosphate in sizing processes.

Textile
Claimed (Brit. 440400) as—
Mordant for textile fibers.

Water and Sanitation
Claimed (Brit. 440400) as—
Water-purifying agent introducing novel advantageous effects.

Aluminum Resinate
Synonyms: Resinate of alumina, Aluminum soap.
French: Résinate d'alumine.
German: Aluminiumresinat.

Paint and Varnish
Drier in—
Enamels, lacquers, paints, varnishes.

Textile
——, *Finishing*
Ingredient of—
Finishing compositions, especially for treating fabrics for covering rails of billiard and pool tables.

Aluminum Stannate
French: Stannate d'aluminium.
German: Aluminiumstannat, Zinnsaeuresaluminium.

Ceramics
Ingredient of—
White, opaque enamels.

Metallurgical
Ingredient of—
White enamels.

Aluminum Stearate
Synonyms: Stearate of alumina.
French: Stéarate d'alumine.
German: Stearinsaeuresaluminium.

Mechanical
Ingredient of—
Cutting compositions, lubricating compositions.

Paint and Varnish
Flattening agent in—
Enamels, lacquers, paints, varnishes.

Petroleum
Ingredient of—
Mineral oil lubricating compositions.

Textile
——, *Finishing*
Ingredient of—
Waterproofing compositions for yarns and fabrics.

Aluminum Stearotoluenesulphonate
French: Stéarotoluènesulphonate d'aluminium.
German: Aluminiumstearotoluolsulfonat, Stearotoluolsulfonsaeuresaluminium.

Textile
——, *Printing*
Ingredient of pastes for the purpose of enhancing the saturation of the textile with the color and equalizing the printing.

Aluminum Sulphate
Synonyms: Sulphate of alumina, Papermaker's alum.
French: Sulfate d'alumine.
German: Aluminiumsulfat.

Cement
Ingredient in making—
Insulating cements.

Ceramics
Ingredient in making—
General ceramic ware, porcelains, potteries and the like.

Chemical
Catalyst in making—
Ethane gas.
Inert diluent in making—
Diazotizing preparations from paranitranilin (German 426033).
Reagent in making—
Ammonium palmitate, aluminum-betanaphthol disulphonate, synthetic tannins.
Starting point in making—
Alums and aluminum salts, aluminum hypochlorite, aluminum permanganate, aluminum sulphite.

Construction
Ingredient of—
Concrete waterproofing compositions.
Heat insulating compositions (Brit. 253919 and U. S. 1597093).
Insulating cements and mortars.

Dye
Reagent in making—
Azophor blue D, Azophorrosa A.
Greenish blue lakes and other color lakes.

Electrical
Ingredient of—
Accumulator and storage battery electrolytes.

Fats and Oils
Clarifying agent in purifying—
Fats and oils, especially in stearin candle manufacture.

Glass
Ingredient in making—
Glass resistant to sudden temperature variations and to chemical influences of all sorts.

Insecticide
Ingredient of—
Fungicides for combatting root rot of trees.
General insecticidal and germicidal compositions.

Leather
Ingredient of—
Tanning mixtures, especially for the tanning of white leathers.

Aluminum Sulphate (Continued)

Miscellaneous
Ingredient of—
Fire extinguishing compositions (Brit. 251334).
Polishing cloths (Brit. 256788).

Paint and Varnish
Ingredient of—
Paint and varnish removers.
Reagent in making—
Satin white.

Petroleum
Deodorizing and decolorizing agent in treating—
Petroleum, gasoline, kerosene and so on.
Ingredient of—
Lubricating compositions with mineral oil base.

Pulp and Paper
Size in making—
Newsprint, packing paper, writing paper.

Sugar
Reagent in treatment of—
Sugar juices to remove impurities, coloring matters and the like.

Textile
——, *Dyeing*
Assistant in dyeing with—
Alizarin and similar dyestuffs on woolen yarns and fabrics.
Ingredient of—
"Red liquor."
Mordant in—
General textile yarns and fabric dyeing.

Water and Sanitation
Reagent in—
Precipitation of sewage.
Purification of water to make it potable.

Aluminum Sulphate-Acetate
French: Sulfate et acétate aluminique, Sulfate et acétate d'aluminium.
German: Aluminiumsulfatacetat, Aluminiumsulfatazetat, Essigsäuresschwefelsäuresaluminium.
Italian: Acetatosolfato di alluminio.

Textile
——, *Miscellaneous*
Ingredient (Brit. 274611, 311885, 400244) of—
Delustering agent for artificial silk, composed of a solution with a sulphonated alkylated petroleum fraction.
——, *Dyeing*
Ingredient of—
Dye liquors containing alizarin red.
Mordant in—
Dyeing with anilin black.
——, *Finishing*
Ingredient of—
Compositions used in fireproofing yarns and fabrics.

Aluminum Tannate
French: Aluminium tannique, Tannate aluminique, Tannate d'aluminium.
German: Tanninaluminium.

Fats and Oils
Precipitant (U. S. 1745367) in—
Purifying vegetable oils.

Pharmaceutical
Suggested for use as—
Astringent in diseases of nasal and laryngeal mucous membranes.

Aluminum Tartrate
French: Tartrate d'alumine, Tartrate aluminique, Tartrate d'aluminium.
German: Aluminiumtartrat, Weinsaeuresaluminium, Weinsaeurestonerde.

Chemical
Catalyst (Brit. 291419) in purifying—
Anthracene, coaltar ammonia.
Stabilizer (Brit. 291419) in catalytic mixtures used in making—
Acenaphthylene, acenaphthaquinone, bisacenaphthylidenedione, naphthaldehydic acid, naphthalic anhydride, and hemimellitic acid from acenaphthene.
Acetic acid from ethyl alcohol.
Aldehydes and the corresponding acids from xylenes, pseudocumenes, mesitylene, paracymene.
Aldehydes and the corresponding acids from toluene, orthochlorotoluene, orthobromotoluene, orthonitrotoluene, parachlorotoluene, parabromotoluene, paranitrotoluene, metachlorotoluene, metanitrotoluene, metabromotoluene, dichlorotoluenes, dibromotoluenes, dinitrotoluenes, nitrochlorotoluenes, nitrobromotoluenes, chlorobromotoluenes.
Anthraquinone from anthracene.
Chloroacetic acid from ethylenechlorohydrin.
Fluorenone from fluorene.
Formaldehyde from methanol or methane.
Maleic acid from naphthalene.
Maleic acid and fumaric acid from benzol, toluol, phenols, furfural, or phthalic anhydride.
Naphthoquinone from naphthalene.
Phenanthraquinone from phenanthrene.
Phthalic anhydride from naphthalene.
Salicylic acid and salicylic aldehyde from cresols.
Vanillin and vanillic acid from eugenol and isoeugenol.

Textile
Assist in—
Dyeing with alizarin pink.

Aluminum Tungate
French: Tungate d'aluminium.
German: Aluminiumtungat, Tungsaeuresaluminium.

Paint and Varnish
Starting point (Brit. 270387) in making—
Paint driers, varnish driers.

Photographic
Starting point in making—
Light-sensitive varnishes.

Aluminum Uranate
French: Uranate d'alumine, Uranate aluminique, Uranate d'aluminium.
German: Aluminiumuranat, Uransaeuresaluminium, Uransaeurestonerde.

Chemical
Ingredient of catalytic preparations used in making—
Acenaphthylene, acenaphthaquinone, bisacenaphthylidenedione, naphthaldehydic acid, naphthalic anhydride, and hemimellitic acid from acenaphthene (Brit. 295270).
Acetaldehyde from ethyl alcohol (Brit. 281307).
Acetic acid from ethyl alcohol (Brit. 281307).
Alcohols from aliphatic hydrocarbons (Brit. 281307).
Aldehydes or alcohols by the reduction of esters (Brit. 306471).
Alpha-campholid by reduction of camphoric acid (Brit. 306471).
Aldehydes and acids from toluene, orthochlorotoluene, orthobromotoluene, orthonitrotoluene, metachlorotoluene, metabromotoluene, metanitrotoluene, parachlorotoluene, parabromotoluene, paranitrotoluene, dinitrotoluenes, dichlorotoluenes, dibromotoluenes, chlorobromotoluenes, chloronitrotoluenes, bromonitrotoluenes (Brit. 295270).
Aldehydes and acids from xylenes, pseudocumene, mesitylene, and paracymene (Brit. 295270).
Alphanaphthaquinone from naphthalene (Brit. 281307).
Anthraquinone from naphthalene (Brit. 281307).
Benzaldehyde and benzoic acid from toluene (Brit. 281307).
Benzoquinone from phenanthraquinone (Brit. 281307).
Benzyl alcohol by reduction of benzaldehyde (Brit. 306471).
Benzyl alcohol or benzyl aldehyde or phthalide by the reduction of phthalic anhydride (Brit. 306471).
Butyl alcohol by the reduction of croton aldehyde (Brit. 306471).
Chloroacetic acid from ethylenechlorohydrin (Brit. 295270).
Diphenic acid from ethyl alcohol (Brit. 281307).
Ethyl alcohol by the reduction of acetaldehyde (Brit. 306471).
Fluorenone from fluorene (Brit. 295270).
Formaldehyde from methane or methanol (Brit. 295270).
Formaldehyde by the reduction of carbon dioxide or carbon monoxide (Brit. 306471).
Isopropyl alcohol by the reduction of acetone (Brit. 306471).
Maleic acid and fumaric acid by the oxidation of benzene, toluene, phenol, tar phenols, or furfural, or from benzoquinone or phthalic anhydride (Brit. 295270).

Aluminum Uranate (Continued)
Methane by the reduction of carbon dioxide or carbon monoxide (Brit. 306471).
Methanol by the reduction of carbon dioxide or carbon monoxide (Brit. 306471).
Naphthaldehydic acid, acenaphthanone or bisacenaphthylidenedione from acenaphthylene (Brit. 281307).
Phenanthraquinone from phenanthrene or diphenic acid (Brit. 295270).
Phthalic acid and maleic acid from naphthalene (Brit. 295270).
Primary alcohols by the reduction of aldehydes (Brit. 306471).
Propionic acid and butyric acid and higher alcohols, ketones, and acids by the reduction of carbon dioxide or carbon monoxide (Brit. 306471).
Reduction of anthraquinone, benzoquinone, and the like to corresponding hydroxyl compounds (Brit. 306471).
Reduction of ketones, aldehydes, acids, esters, alcohols, ethers, and other organic compounds which contain oxygen (Brit. 306471).
Salicylic acid and salicylic aldehyde from cresol (Brit. 295270).
Secondary butyl alcohol by the reduction of methylethylketone (Brit. 306471).
Valeryl alcohol by the reduction of valeraldehyde (Brit. 306471).
Vanillin or vanillic acid from eugenol or isoeugenol (Brit. 295270).

Ambergris
Synonyms: Amber.
Latin: Ambra, Ambra grisea (cinera).
French: Ambre, Ambre gris, Ambre vrai.
German: Amber, Graue amber.

Food
Ingredient of—
Condiments and flavorings.

Perfume
Fixative in making—
Fine perfumes.
To fix odors in cosmetics.

Soap
To fix the odor in toilet soaps.

Pharmaceutical
In compounding and dispensing practice.

Amber Oil
French: Huile d'ambre.
German: Bernsteinoel.

Paint and Varnish
Ingredient of—
Paints, varnishes.

Perfumery
Ingredient of—
Perfume preparations in the place of artificial musk.
Ingredient (Brit. 279575) of—
Bath salts, cosmetics, dentrifices.

Pharmaceutical
In compounding and dispensing practice.

4-Aminoacenaphthene-3:4-disulphonic Acid

Chemical
Starting point in making—
Esters and salts, intermediates, pharmaceuticals.
Starting point (Brit. 314909) in making derivatives with—
3-Carboxyphenylthiocarbimide, diphenylurea-3:3′-dicarboxylic acid, 4-quinolylphenylurea-3:6′-dicarboxylic acid, symmetrical diphenylurea-3:3′-dicarboxylic acid, thiourea, thiourea-3:3′-dicarboxylic acid, urea.

4-Aminoacenaphthene-3:5-disulphonic Acid
French: Acide de 4-aminoacénaphthène-3:5-disulfonique.
German: 4-Aminoacenaphten-3:5-disulfonsaeure.

Chemical
Starting point (Brit. 278037) in making pharmaceuticals with—
Alphanitronaphthalene-5-sulphochloride, bromonitrobenzoyl chlorides, chloronitrobenzoyl chlorides, iodonitrobenzoyl chlorides, nitroanisoyl chlorides, nitrobenzene sulphochlorides, nitrobenzoyl chlorides, nitrocinnamyl chlorides, 1:5-nitronaphthoyl chloride, 2-nitrophenylacetyl chloride, 4-nitrophenylacetyl chloride, nitrotoluyl chlorides.

4-Aminoacenaphthene-3-sulphonic Acid
French: Acide de 4-aminoacénaphthène-3-sulfonique.
German: 4-Aminoacenaphten-3-sulfonsaeure.

Chemical
Starting point (Brit. 278037) in making synthetic drugs with—
Alphanitronaphthalene-5-sulphochloride, bromonitrobenzoyl chlorides, chloronitrobenzoyl chlorides, iodonitrobenzoyl chlorides, nitroanisoyl chlorides, nitrobenzene sulphochlorides, 2-nitrocinnamyl chloride, 3-nitrocinnamyl chloride, 4-nitrocinnamyl chloride, 1:5-nitronaphthoyl chloride, 2-nitrophenylacetyl chloride, 4-nitrophenylacetyl chloride, nitrotoluyl chlorides.

4-Aminoacenaphthene-5-sulphonic Acid
French: Acide de 4-aminoacénaphthène-5-sulfonique.
German: 4-Aminoacenaphten-5-sulfonsaeure.

Chemical
Starting point (Brit. 278037) in making pharmaceutical chemicals with—
Alphanitronaphthalene-5-sulphochloride, bromonitrobenzoyl chlorides, chloronitrobenzoyl chlorides, iodonitrobenzoyl chlorides, nitroanisoyl chlorides, nitrobenzene sulphochlorides, nitrobenzoyl chlorides, nitrocinnamyl chlorides, 1:5-nitronaphthoyl chloride, nitrophenylacetyl chlorides, nitrotoluyl chlorides.

1-Amino-8-acetamido-4:4′-acetamido-3′:6′-dimethoxyanilinoanthraquinone-2-sulphonic Acid

Textile
Blue dye (Brit. 432647) for—
Woolen fabrics.

1-Amino-4:4′-acetamidodiphenylamino-5-anthraquinone-2-sulphonic Acid

Textile
Blue dye (Brit. 432647) for—
Woolen fabrics.

6-Amino-4-acetylamino-1:3-dimethoxybenzene
Synonyms: 6-Amino-4-acetylamino-1:3-dimethoxybenzol.
French: 6-Amino-4-acétyleamino-1:3-diméthoxyebenzène.

Chemical
Starting point in making—
Aromatics, intermediates, pharmaceuticals.

Dye
Starting point (Brit. 307303) in making monoazo dyestuffs with—
N-Acetyl-H acid, N-betachloropropionyl-H acid, N-betachloroethanesulpho-H acid, N-benzoyl-H acid, N-carbethoxy-H acid, N-chloroacetyl-H acid, N-phenylacetyl-H acid, N-toluenesulpho-H acid.

2-Amino-4-acetyl-1:1-diphenyl Ether

Textile
Starting point (Brit. 453953) in making dyed or printed red colors of fine purity of shade and fastness to kier-boiling with—
2:3-Hydroxynaphthoic-2′-methylanilide.
2:3-Hydroxynaphthoic-2′:5-dimethoxyanilide.
2:3-Hydroxynaphthoic-4′-methoxy-2′-methylanilide.
2:3-Hydroxynaphthoic-3′:4′-dichloroanilide.
2:3-Hydroxynaphthoic-5′-chloro-2′-methylanilide.
2:3-Hydroxynaphthoic-4′-chloro-2′-methoxyanilide.
Starting point (Brit. 453953) in making dyed or printed scarlet-red colors of fine purity of shade and fastness to kier-boiling with—
2:3-Hydroxynaphthoic-3′-methylanilide.
2:3-Hydroxynaphthoic-4′-methylanilide.
2:3-Hydroxynaphthoic-3′-chloro-2′-ethoxyanilide.
2:3-Hydroxynaphthoic-2′-methoxy-5′-methylanilide.
2:3-Hydroxynaphthoic-4′-chloro-2′-methoxyanilide.

2-Amino-4-acetyl-3-methyl-1:1′-diphenyl Ether

Textile
Starting point (Brit. 453953) in making—
Dyed or printed red colors of fine purity of shade and fastness to kier-boiling with—
2:3-Hydroxynaphthoic-4′-chlor-2′-methoxyanilide.
2:3-Hydroxynaphthoic-5′-chlor-2′-methylanilide.
2:3-Hydroxynaphthoic-3′:4′-dichloroanilide.
2:3-Hydroxynaphthoic-2′:5′-dimethoxyanilide.
2:3-Hydroxynophthoic-4′-methoxy-2′-methylanilide.
2:3-Hydroxynaphthoic-2′-methylanilide.

2-Amino-4-acetyl-3-methyl-1:1'-diphenyl Ether (Continued)
Dyed or printed scarlet-red colors of fine purity of shade and fastness to kier-boiling with—
2:3-Hydroxynaphthoic-3'-chlor-2'-ethoxyanilide.
2:3-Hydroxynaphthoic-4'-chlor-2'-methoxyanilide.
2:3-Hydroxynaphthoic-2'-methoxy-5'-methylanilide.
2:3-Hydroxynaphthoic-3'-methylanilide.
2:3-Hydroxynaphthoic-4'-methylanilide.

2-Amino-4-acetyl-4'-methyl-1:1'-diphenyl Ether
Textile
Starting point (Brit. 453953) in making—
Dyed or printed red colors of fine purity of shade and fastness to kier-boiling with—
2:3-Hydroxynaphthoic-2'-methylanilide.
2:3-Hydroxynaphthoic-2':5'-dimethoxyanilide.
2:3-Hydroxynaphthoic-4'-methoxy-2'-methylanilide.
2:3-Hydroxynaphthoic-3':4'-dichloroanilide.
2:3-Hydroxynaphthoic-5'-chlor-2'-methylanilide.
2:3-Hydroxynaphthoic-4'-chlor-2'-methoxyanilide.
Dyed or printed scarlet-red colors of fine purity of shade and fastness to kier-boiling with—
2:3-Hydroxynaphthoic-3'-methylanilide.
2:3-Hydroxynaphthoic-4'-methylanilide.
2:3-Hydroxynaphthoic-3'-chlor-2'-ethoxyanilide.
2:3-Hydroxynaphthoic-2'-methoxy-5'-methylanilide.
2:3-Hydroxynaphthoic-4'-chlor-2'-methoxyanilide.

3-Aminoalizarin
Synonyms: Beta-aminoalizarin.
Dye
Starting point in making—
Alizarin blue WX, alizarin blue S, alizarin green X, alizarin indigo blue S.

4-Aminoalizarin
Synonyms: Alpha-aminoalizarin.
Dye
Starting point in making—
Alizarin garnet R, alizarin green S.

4-Aminoalphabutoxypyridin
Chemical
Condensing agent (Brit. 396778) in making—
Triazoles from 3:5-dimethylfurodiazole.

8-Amino-6-amyloxyquinolin
Chemical
Starting point (Brit. 399818) in making—
Compounds, said to be effective against malaria, by diazotizing and coupling with hydrocuprein or a substituted hydrocuprein.

4-Amino-1-anilino-8-hydroxyanthraquinone
Textile
Dyestuff (Brit. 402391, 402392, and 402393) for—
Producing blue colors on acetate rayon.

1-Amino-4-anilino-4-para-acetamidoanilino-5-anthraquinone-2-sulphonic Acid
Textile
Blue dye (Brit. 432647) for—
Woolen fabrics.

4-Amino-1:9-anthrapyrimidin
Chemical
In organic syntheses.
Photographic
Defogging agent (Brit. 442731) for—
Gelatin having a strong tendency to cause fog.

2-Aminoanthraquinone
Synonyms: Beta-aminoanthraquinone.
German: 2-Aminoanthrachinon, Beta-aminoanthrachinon.
Chemical
Starting point in making—
Beta-aminoanthracene (Brit. 260000).
Intermediates, beta-aminoanthraquinone-1-carboxylic acid, pharmaceuticals, synthetic aromatic chemicals.
Dye
Starting point in making—
Algol yellow 3G, algol red B, flavanthrene, helindone yellow 3GN, indanthrene blue RS, indanthrene dark blue BT, indanthrene brown B, indanthrene yellow G.

1-Aminoanthraquinone-2-carboxylic Acid
French: Acide d'alpha-aminoanthraquinone-2-carboxylique, Acide de 1-aminoanthraquinone-2-carboxylique.
German: Alpha-aminoanthrachinon-2-carbonsaeure, 1-Aminoanthrachinon-2-carbonsaeure.
Chemical
Starting point in making—
Esters and salts, intermediates, pharmaceuticals.
Dye
Starting point in making various synthetic dyestuffs.
Textile
——, *Dyeing and Printing*
Component (Brit. 310827) of dyeing, printing, and stenciling compositions used on materials containing cellulose esters and cellulose ethers, with the aid of—
Alpha-amino-2-methoxynaphthalene, alphanaphthylamine, betanaphthylamine, dimethylmeta-aminophenol, gammachlorobetaoxypropyl-1-naphthylamine, meta-anisidin, meta-aminophenol, metacresidin, metatoluidin, monoacetylmeta-aminophenol, metaphenylenediamine, nitrometaphenylenediamine, orthoanisidin, orthocresidin, omegaoxyethyl-1-naphthylamine, para-anisidin, paracresidin, paraxylidin, phenol.

1:8-Aminoanthraquinonesulphonic Acid
Chemical
Starting point in making—
Esters and salts, intermediates, pharmaceuticals, synthetic aromatics.
Dye
Starting point in making—
Alizarin direct blue B.

1-(1'Amino-2'-anthraquinonyl) benzothiazole-5-carboxyl Chloride
Dye
Starting point (Brit. 439570) in making—
Red vat dyestuffs by condensing with 1-aminoanthraquinone.
Reddish-brown vat dyestuffs by condensing with 1-amino-5-benzoamidoanthraquinone.
Reddish-violet vat dyestuffs by condensing with 1:4-diaminoanthraquinone.

Aminoazobenzenedisulphonic Acid
Synonyms: Aminoazobenzoldisulphonic acid.
French: Acide d'aminobenzènedisulphonique.
German: Aminoazobenzoldisulfonsäure.
Chemical
Starting point in making—
Intermediates.
Dye
Starting point in making—
Acid yellow, acid yellow J, acid yellow SS, biebrich scarlet, crocein scarlet O, crocein scarlet O extra, double scarlet, ponceau J extra, scarlet EC, wool black.
Textile
Coloring matter for—
Silk fabrics and yarns, wool fabrics and yarns.
Ingredient of—
Coloring mixtures used in the place of turmeric and fustic.
Dye baths containing fast red, fuchsin S, indigo carmine.

Aminoazobenzenesulphonic Acid
Synonyms: Para-aminoazobenzenesulphonic acid.
French: Acide d'aminoazobenzènesulphonique.
German: Aminoazobenzolsulfonsäure.
Chemical
Starting point in making various derivatives.
Dye
Starting piont in making—
Acid yellow, alizarin red 2B, clothdiazo brown N3JO, cloth scarlet G, crocein scarlet, crocein scarlet 3B, fast chloroazol red K, fast scarlet B, milling orange, salicin orange G.

1-Amino-4-benzamidoanthraquinone
Dye
In dye syntheses.
Starting point (Brit. 449477) in making—
Orange to red vat dyes by condensing with 4:6-dichlor-1:3:5-triazins carrying a halogenophenyl or alkoxyphenyl substituent in position 2.

1-Amino-5-benzamidoanthraquinone

Chemical
In organic syntheses.

Dye
In dye syntheses.
Starting point (Brit. 449263) in making—
Yellow vat dyes with—
4-(2':4'-dicarboxyphenyl)-7:8-phthaloyl-2-acridone acid chloride.
Paracarboxylphenyl-4-paratolyl-7:8-phthaloyl-2-quinolone acid chloride, normal.
4-Paracarboxylphenyl-7:8-phthaloyl-2-quinolin acid chloride.
7:8-Phthaloyl-2-quinolone-3-carboxylic acid.
7:8-Phthaloyl-2-quinolone-5-carboxylic acid.
Starting point (Brit. 449477) in making—
Orange to red vat dyes by condensing with 4:6-dichlor-1:3:5-triazins carrying a halogenophenyl or alkoxyphenyl substituent in position 2.

2-Aminobenzotrifluoride-5-sulphonic Acid

Dye
Intermediate (Brit. 446532) in making dyestuffs.

4-Aminobenzotrifluoride-3-sulphonic Acid

Dye
Intermediate (Brit. 446532) in making dyestuffs.

5-Aminobenzotrifluoride-4-sulphonic Acid

Dye
Intermediate (Brit. 446532) in making dyestuffs.

1-Amino-4-benzoylamino-2:5-dimethoxybenzene

Textile
Starting point (Brit. 396859) in—
Dyeing wool, either red or blue, by "Ingrain" process.

3-Amino-4-benzyloxybenzotrifluoride-6-sulphonic Acid

Dye
Intermediate (Brit. 446532) in making dyestuffs.

4-Aminobenzylsulphonic Acid
French: Acide d'amino-4-benzylsulphonique.
German: 4-Aminobenzylsulfosaeure.

Dye
Starting point (Brit. 265767) in making azo dyestuffs with—
Diphenylamine, ethylbenzylanilin, methylbenzylanilin, methylbetanaphthylamine, methyldiphenylamine.

2-Amino-1:4-bistrifluoromethylbenzene-6-sulphonic Acid

Dye
Intermediate (Brit. 446532) in making various dyestuffs.

5-Amino-1:3-bistrifluoromethylbenzene-2-sulphonic Acid

Dye
Intermediate (Brit. 446532) in making various dyestuffs.

1-Amino-4-bromoanthraquinone-2-methanesulphonic Acid

Dye
Starting point (Brit. 440208) in making—
Acid wool dyes by condensation with organic bases having at least one hydrogen atom attached to the nitrogen atom.

1-Amino-2-bromo-4-paratoluidinoanthraquinone

Oils, Fats, Waxes
Coloring agent (Brit. 432867) for—
Paraffin and other mineral waxes, stearic acid, tallow and other solid triglycerides, beeswax, carnauba wax, and others.

2-Amino-5-carbethoxyl-4'-oxy-3'-carboxydiphenyl-sulphone
French: 2-Amino-5-carbéthoxyle-4'-oxye-3'-carboxyediphénylesulphone.
German: 2-Amino-5-carbaethoxyl-4'-oxy-3'-carboxydiphenylsulfon.

Chemical
Starting point in making various intermediates.

Dye
Starting point (Brit. 288788) in making azo dyestuffs with—
Acetoaceticanilide-4-carboxylic acid.
2:8:6-Aminonaphtholsulphonic acid.
Diacetoacetylorthotoluidin.
2-Ethylaminonaphthalene-6-sulphonic acid.
Ethylbenzylanilin.
Ethylparasulphobenzylanilin.
2-Methylaminonaphthalene-7-sulphonic acid.
2-Methylaminonaphthalene-6-sulphonic acid.
2-Methylamino-8-naphthol-6-sulphonic acid.
2:6-Naphthylaminesulphonic acid.
2:7-Naphthylaminesulphonic acid.
2:3-Oxynaphthoic anilides.
2:3-Oxynaphthoic naphthalides.
2:3-Oxynaphthoic phenetidides.
2:3-Oxynaphthoic toluidides.
2:3-Oxynaphthoic xylidides.
2-Paratolylamino-8-naphthol-6-sulphonic acid.
Sodium methyldiketolsulphonate.

5-Amino-2-carbobetadiethylaminoethoxydiphenyl

Pharmaceutical
Claimed (U. S. 1976940) as—
Anesthetic.

5-Amino-2-carbogammadinormal-butylaminopropoxy-diphenyl

Pharmaceutical
Claimed (U. S. 1976940) as—
Anesthetic.

1-Amino-4-chloroanthraquinone-2-methanesulphonic Acid

Dye
Starting point (Brit. 440208) in making—
Acid wool dyes by condensation with organic bases having at least one hydrogen atom attached to the nitrogen atom.

3-Amino-4-chlorobenzoylbenzoic Acid
French: Acide de 3-amino-4-chlorobenzoylebenzoique.
German: 3-Amino-4-chlorbenzoylbenzosaeure.

Chemical
Starting point (Brit. 264916) in making—
1-Amino-2-chloro-4-bromoanthraquinone.
1-Amino-2-methylanthraquinone.
1-Bromo-2-amino-3-methylanthraquinone.
1-Dibromo-2-aminoanthraquinone.
1:3-Dichloro-2-aminoanthraquinone.

Dye
Starting point (Brit. 264916) in making—
Flavanthrene dyestuffs, indanthrene dyestuffs.

3-Amino-4-chlorodiphenylsulphone
German: 3-Amino-4-chlordiphenylsulfon.

Dye
Starting point (Brit. 279146) in making dyestuffs with—
2:3-Oxynaphthoicbetanaphthalide, 2:3-oxynaphthoic-3-nitranilide, 2:3-oxynaphthoic-2-phenetidide, 2:3-oxynaphthoic-3-toluidide.

2-Amino-3-chloro-1:4-naphthoquinone

Chemical
Starting point in making—
2-Amino-3-mercapto-1:4-naphthoquinone (Brit. 262141).

1-Amino-2-cyano-4-chlorobenzene
Synonyms: 1-Amino-2-cyano-4-chlorbenzol, Alpha-amino-2-cyano-4-chlorobenzene, Alpha-amino-2-cyano-4-chlorbenzol.
German: Alpha-amino-2-zyano-4-chlorbenzol, 1-Amino-2-zyano-4-chlorbenzol.

Chemical
Starting point in making—
Intermediates, pharmaceuticals.

Dye
Starting point (Brit. 301410) in making azo lakes and dyestuffs with—
2:3-Oxynaphthoic anilide.
2:3-Oxynaphthoic 4'-chloro-2'-toluidide.
2:3-Oxynaphthoic 5'-chloro-2'-anisidide.
2:3-Oxynaphthoic 4'-methoxyanilide.
2:3-Oxynaphthoic naphthylamide.
2:3-Oxynaphthoic 2'-toluidide.

1-Amino-3-cyano-6-chlorobenzene
Synonyms: Alpha-amino-3-cyano-6-chlorbenzol, Alpha-amino-3-cyano-6-chlorobenzene.
German: Alpha-amino-3-zyano-6-chlorbenzol, 1-Amino-3-cyano-6-chlorbenzol, 1-Amino-3-zyano-6-chlorbenzol.

Chemical
Starting point in making—
Intermediates, pharmaceuticals.

Dye
Starting point (Brit. 301410) in making azo dyestuffs with—
2:3-Oxynaphthoic alphanaphthylamide.
2:3-Oxynaphthoic anilide.
2:3-Oxynaphthoic betanaphthylamide.
2:3-Oxynaphthoic 5′-chloro-2′-anisidide.
2:3-Oxynaphthoic 4′-chloro-2′-anisidide.
2:3-Oxynaphthoic 4′-methoxyanilide.
2:3-Oxynaphthoic 2′-toluidide.

1-Amino-2-cyano-5-methylbenzene
Synonyms: Alpha-amino-2-cyano-5-methylbenzene, 1-Amino-2-cyano-5-methylbenzol.
German: Alpha-amino-2-zyano-5-methylbenzol.

Chemical
Starting point in making—
Aromatics, intermediates, pharmaceuticals.

Dye
Starting point (Brit. 301410) in making azo dyestuffs with the aid of—
2:3-Oxynaphthoic alpha-anilide.
2:3-Oxynaphthoic alphanaphthylamide.
2:3-Oxynaphthoic 4-anisidide.
2:3-Oxynaphthoic anthroxyanilide.
2:3-Oxynaphthoic benzoxyanilide.
2:3-Oxynaphthoic betanaphthalide.
2:3-Oxynaphthoic betanaphthylamide
2:3-Oxynaphthoic 2-chloroanilide.
2:3-Oxynaphthoic 4-chloro-2-anisidide.
2:3-Oxynaphthoic 4′-chloro-2-anisidide.
2:3-Oxynaphthoic 5′-chloro-2-anisidide.
2:3-Oxynaphthoic 4′-chloro-2′-toluidide.
2:3-Oxynaphthoic cresoxyanilide.
2:3-Oxynaphthoic diacetoaceticanilide.
2:3-Oxynaphthoic 2′:5′-dimethoxy-1′-anilide.
2:3-Oxynaphthoic 4-methoxyanilide.
2:3-Oxynaphthoic naphthoxyanilide.
2:3-Oxynaphthoic orthotoluidide.
2:3-Oxynaphthoic phenoxyanilide.
2:3-Oxynaphthoic toloxyanilide.
2:3-Oxynaphthoic 3-toluidide.
2:3-Oxynaphthoic 4-toluidide.

1-Amino-3-cyano-2-methylbenzene
Synonyms: Alpha-amino-3-cyano-2-methylbenzene, Alpha-amino-3-cyano-2-methylbenzol, 1-Amino-3-cyano-2-methylbenzol.
French: Alpha-amino-3-cyano-2-méthylebenzène, 1-Amino-3-cyano-2-méthylebenzène.
German: Alpha-amino-3-zyano-2-methylbenzol, 1-Amino-3-zyano-2-methylbenzol.

Chemical
Starting point in making—
Intermediates, pharmaceuticals.

Dye
Starting point (Brit. 301410) in making azo colors and lakes with—
2:3-Oxynaphthoic anilide.
2:3-Oxynaphthoic 4′-chloro-2′-toluidide.
2:3-Oxynaphthoic 5′-chloro-2′-anisidide.
2:3-Oxynaphthoic 4′-methoxyanilide.
2:3-Oxynaphthoic naphthylamide.
2:3-Oxynaphthoic 2′-toluidide.

1-Amino-4-cyclohexylamino-5-acetamidoanthraquinone-2-sulphonic Acid

Textile
Blue dye (Brit. 432647) for—
Woolen fabrics.

Aminocymene

Mechanical
Antiknock agent for—
Gasoline motor fuel (stated to be almost equal in efficiency to tetraethyl lead).

Aminodibenzanthrone
French: Aminodibenzanthrone.
German: Aminodibenzanthron.

Chemical
Starting point in making—
Intermediates, pharmaceuticals.

Dye
Starting point in making various synthetic dyestuffs.
Starting point (Brit. 307847) in making dyestuffs with the aid of—
Alpha-amino-4-methoxyanthraquinone, alpha-aminoanthraquinone, alphachloroanthraquinone, betachloroanthraquinone, chloranil, cyanuric chloride, dibromoanthraquinone, dibromoisodibenzanthrone, dichloroallo-ms-naphthodianthrone, 1:5-dinitroanthraquinone, nitrodibenzanthrone, tetrabromopyranthrone, tribromopyranthrone, trichloroanthraquinoneacridin.

2-Aminodibenzfuran

Chemical
In organic syntheses.

Dye
Starting point (Brit. 437283) in making—
Reddish-violet dyestuffs by condensing with chloranil or other parabenzoquinones.

Aminodibenzodioxan

Chemical
In organic syntheses.

Dye
Starting point (Brit. 437283) in making—
Bluish-violet dyestuffs by condensing with chloranil or other parabenzoquinones.

1-Amino-2:4-dibenzoylbenzene

Dye
Starting point (Brit. 441855) in making—
Water-insoluble red dyes by coupling in the fiber with alphanaphthalide, 5-chlor-2:4-dimethoxybenzene, 2:3-hydroxynaphthoic, orthoanisidide, 2:3-hydroxynaphthoicorthophenetidide, or orthotoluidide.

1-Amino-2:5-dibenzoylbenzene

Dye
Starting point (Brit. 441855) in making—
Water-insoluble red dyes by coupling in the fiber with 4-chlor-2:5-dimethoxybenzene, 5-chlor-2:4-dimethoxybenzene, 2:5-dimethoxyanilide, or 2:3-hydroxy-naphthoic orthoanisidide.

1-Amino-3:5-dibenzoylbenzene

Dye
Starting point (Brit. 441855) in making—
Water-insoluble red dyes by coupling in the fibre with 2:3-hydroxynaphthoicorthophenetidide or 2:3-hydroxynaphthoicparaphenetidide.

4-Amino-3′:2′-dichlorodiphenylamine
French: 4-Amino-3′:2′-dichlorodiphényleamine.
German: 4-Amino-3′:2′-dichlordiphenylamin.

Chemical
Starting point in making various intermediates.

Dye
Starting point in making various synthetic dyestuffs.

Textile
Reagent (Brit. 313865) in dyeing silk, cotton and other textiles with the aid of—
2:3-Oxynaphthoic alphanaphthylamide.
2:3-Oxynaphthoic anilide.
2:3-Oxynaphthoic 2-anisidide.
2:3-Oxynaphthoic 3-anisidide.
2:3-Oxynaphthoic 4-anisidide.
2:3-Oxynaphthoic 4-benzyloxy-1-anilide.
2:3-Oxynaphthoic betanaphthylamide.
2:3-Oxynaphthoic 2-chloroanilide.
2:3-Oxynaphthoic 3-chloroanilide.
2:3-Oxynaphthoic 4-chloroanilide.
2:3-Oxynaphthoic 5-chloro-2-anisidide.
2:3-Oxynaphthoic 4-chloro-2-anisidide.
2:3-Oxynaphthoic dianisidide.
2:3-Oxynaphthoic 2:5-dimethoxy-1-anilide.
2:3-Oxynaphthoic 2-ethyl-1-anilide.
2:3-Oxynaphthoic 4-ethyl-1-anilide.
2:3-Oxynaphthoic 2-ethyl-5-chloroanilide.
2:3-Oxynaphthoic 3-methoxy-2-naphthylamide.
2:3-Oxynaphthoic 3-nitranilide.
2:3-Oxynaphthoic 2-phenetidide.

4-Amino-3′:2′-dichloro diphenylamine (Cont'd)
2:3-Oxynaphthoic 3-phenetidide.
2:3-Oxynaphthoic 4-phenetidide.
2:3-Oxynaphthoic 2-phenoxy-1-anilide.
2:3-Oxynaphthoic 5-chloro-2-toluidide.
2:3-Oxynaphthoic 2-toluidide.
2:3-Oxynaphthoic 3-toluidide.
2:3-Oxynaphthoic 4-toluidide.

4-Amino-3′:4′-dichlorodiphenylamine
German: 4-Amino-3′:4′-dichlordiphenylamin.

Chemical
Starting point in making—
Intermediates and other derivatives.

Dye
Starting point in making various synthetic dyestuffs.

Textile
Reagent (Brit. 313865) in dyeing silk and cotton yarns and fabrics and other textiles with the aid of—
2:3-Oxynaphthoic alphanaphthylamide.
2:3-Oxynaphthoic anilide.
2:3-Oxynaphthoic 2-anisidide.
2:3-Oxynaphthoic 3-anisidide.
2:3-Oxynaphthoic 4-anisidide.
2:3-Oxynaphthoic 4-benzyloxy-1-anilide.
2:3-Oxynaphthoic betanaphthylamide.
2:3-Oxynaphthoic 2-chloroanilide.
2:3-Oxynaphthoic 3-chloroanilide.
2:3-Oxynaphthoic 4-chloroanilide.
2:3-Oxynaphthoic 4-chloro-2-anisidide.
2:3-Oxynaphthoic 5-chloro-2-anisidide.
2:3-Oxynaphthoic 5-chloro-2-toluidide.
2:3-Oxynaphthoic dianisidide.
2:3-Oxynaphthoic 2:5-dimethoxy-1-anilide.
2:3-Oxynaphthoic 2-ethyl-1-anilide.
2:3-Oxynaphthoic 4-ethyl-1-anilide.
2:3-Oxynaphthoic 2-ethyl-5-chloroanilide.
2:3-Oxynaphthoic 3-methoxy-2-naphthylamide.
2:3-Oxynaphthoic 3-nitranilide.
2:3-Oxynaphthoic 2-phenetidide.
2:3-Oxynaphthoic 3-phenetidide.
2:3-Oxynaphthoic 4-phenetidide.
2:3-Oxynaphthoic 2-phenoxy-1-anilide.
2:3-Oxynaphthoic 2-toluidide.
2:3-Oxynaphthoic 3-toluidide.
2:3-Oxynaphthoic 4-toluidide.

3-Amino-4:6-dimethoxybenzotrifluoride-2-sulphonic Acid

Dye
Intermediate (Brit. 446532) in making various dyestuffs.

4-Amino-3:2′-dimethoxydiphenylamine
French: 4-Amino-3:2′-diméthoxyediphényleamine.
German: 4-Amino-3:2′-dimethoxydiphenylamin.

Chemical
Starting point in making—
Intermediates, pharmaceuticals.

Dye
Starting point in making various synthetic dyestuffs.

Textile
Reagent (Brit. 313865) in dyeing silk, cotton, and other textiles with—
2:3-Oxynaphthoic alphanaphthylamide.
2:3-Oxynaphthoic anilide.
2:3-Oxynaphthoic 2-anisidide.
2:3-Oxynaphthoic 3-anisidide.
2:3-Oxynaphthoic 4-anisidide.
2:3-Oxynaphthoic 4-benzyloxy-1-anilide.
2:3-Oxynaphthoic betanaphthylamide.
2:3-Oxynaphthoic 2-chloroanilide.
2:3-Oxynaphthoic 3-chloroanilide.
2:3-Oxynaphthoic 4-chloroanilide.
2:3-Oxynaphthoic 4-chloro-2-anisidide.
2:3-Oxynaphthoic 5-chloro-2-anisidide.
2:3-Oxynaphthoic 5-chloro-2-toluidide.
2:3-Oxynaphthoic dianisidide.
2:3-Oxynaphthoic 2:5-dimethoxy-1-anilide.
2:3-Oxynaphthoic 2-ethyl-1-anilide.
2:3-Oxynaphthoic 4-ethyl-1-anilide.
2:3-Oxynaphthoic 2-ethyl-5-chloroanilide.
2:3-Oxynaphthoic 3-methoxy-2-naphthylamide.
2:3-Oxynaphthoic 3-nitranilide.
2:3-Oxynaphthoic 2-phenetidide.
2:3-Oxynaphthoic 3-phenetidide.
2:3-Oxynaphthoic 4-phenetidide.
2:3-Oxynaphthoic 2-phenoxy-1-anilide.
2:3-Oxynaphthoic 2-toluidide.
2:3-Oxynaphthoic 3-toluidide.
2:3-Oxynaphthoic 4-toluidide.

1-Amino-2:7-dimethoxynaphthalene
Synonyms: Alpha-amino-2:7-dimethoxynaphthalene.
French: Alpha-amino-2:7-diméthoxyenaphthalène, 1-Amino-2:7-diméthoxyenaphthalène.
German: Alpha-amino-2:7-dimethoxynaphtalin, 1-Amino-2:7-dimethoxynaphtalin.

Chemical
Starting point in making—
Intermediates, pharmaceuticals.

Dye
Starting point (Brit. 298518) in making azo dyestuffs with—
Alpha-aminonaphthalene, alpha-aminonaphthalene-6-sulphonic acid, alpha-aminonaphthalene-7-sulphonic acid, anilin, anilin-3-chloro-6-sulphonic acid, anilin-3:4-disulphonic acid, anilin-2:4-disulphonic acid, anilin-2:5-disulphonic acid, anilin-4-nitro-2:5-disulphonic acid, anilin-3-sulphonic acid, beta-amino-1-methoxybenzene-4-sulphonic acid, beta-amino-5-sulphobenzoic acid, 1:3-dioxyquinolin, methylketol, methylketolsulphonic acid, orthocresotinic acid, 1-phenyl-3-carboxy-5-pyrazolone, 1-phenyl-3-methyl-5-pyrazolone, salicylic acid, sulphazone.

2-Amino-5-dimethylaminoanisole-3-sulphonic Acid

Dye
Starting point (Brit. 447905, 447906, and 448016) in making—
Monoazo dyes for leather, particularly chrome leather.

1-Amino-2:4-dimethyl-5-cyanobenzene
Synonyms: Alpha-amino-2:4-dimethyl-5-cyanobenzene, Alpha-amino-2:4-dimethyl-5-cyanobenzol.
French: Alpha-amino-2:4-diméthyle-5-cyanobenzène, 1-Amino-2:4-diméthyle-5-cyanobenzène.
German: Alpha-amino-2:4-dimethyl-5-cyanbenzol, 1-Amino-2:4-dimethyl-5-cyanbenzol, 1-Amino-2:4-dimethyl-5-zyanbenzol.

Chemical
Starting point in making—
Intermediates, pharmaceuticals.

Dye
Starting point (Brit. 301410) in making azo dyestuffs with—
2:3-Oxynaphthoic alphanaphthylamide.
2:3-Oxynaphthoic anilide.
2:3-Oxynaphthoic betanaphthylamide.
2:3-Oxynaphthoic 5′-chloro-2′-anisidide.
2:3-Oxynaphthoic 4′-chloro-2′-toluidide.
2:3-Oxynaphthoic 4′-methoxyanilide.
2:3-Oxynaphthoic 2′-toluidide.

1-Amino-2:3-dioxypropane
Synonyms: Alpha-amino-2:3-dioxypropane.
French: 1-Amino-2:3-dioxyepropane.
German: 1-Amino-2:3-dioxypropan.

Chemical
Reagent (Brit. 295024) in making dispersing preparations with—
Castor oil, cottonseed oil, linseed oil, oleic acid, olive oil, palmitic acid, ricinoleic acid, sulphoricinoleic acid, stearic acid.
Starting point in making—
Intermediates, salts and esters.

1-Amino-2:5-diparachlorobenzoylbenzene

Dye
Starting point (Brit. 441855) in making—
Water-insoluble red dyes by coupling in the fiber with betanaphthalide or 2:3-hydroxynaphthoic-5-chloro-orthotoluidide.

1-Amino-2:5-diparatoluolbenzene

Dye
Starting point (Brit. 441855) in making—
Water-insoluble red dyes by coupling in the fiber with alphanaphthalide, 2:3-hydroxynaphthoicorthophenetidide, or orthotoluidide.

4-Aminodiphenylamine

Chemical

Starting point in making—
Intermediates and other derivatives.

Textile

Reagent (Brit. 313865) in dyeing silk, cotton and other fibers and yarns with the aid of—
2:3-Oxynaphthoic anilide.
2:3-Oxynaphthoic 2-anisidide.
2:3-Oxynaphthoic 3-anisidide.
2:3-Oxynaphthoic 4-anisidide.
2:3-Oxynaphthoic 4-benzyloxy-1-anilide.
2:3-Oxynaphthoic betanaphthylamide.
2:3-Oxynaphthoic 2-chloroanilide.
2:3-Oxynaphthoic 3-chloroanilide.
2:3-Oxynaphthoic 4-chloroanilide.
2:3-Oxynaphthoic 4-chloro-2-anisidide.
2:3-Oxynaphthoic 5-chloro-2-anisidide.
2:3-Oxynaphthoic 5-chloro-2-toluidide.
2:3-Oxynaphthoic dianisidide.
2:3-Oxynaphthoic 2:5-dimethoxy-1-anilide.
2:3-Oxynaphthoic 2-ethyl-1-anilide.
2:3-Oxynaphthoic 4-ethyl-5-chloroanilide.
2:3-Oxynaphthoic 3-methoxy-2-naphthylamide.
2:3-Oxynaphthoic 3-nitranilide.
2:3-Oxynaphthoic 2-phenetidide.
2:3-Oxynaphthoic 3-phenetidide.
2:3-Oxynaphthoic 4-phenetidide.
2:3-Oxynaphthoic 2-phenoxy-1-anilide.
2:3-Oxynaphthoic 2-toluidide.
2:3-Oxynaphthoic 3-toluidide.
2:3-Oxynaphthoic 4-toluidide.
2:3-Oxynaphthoic 2-xylidide.
2:3-Oxynaphthoic 3-xylidide.
2:3-Oxynaphthoic 4-xylidide.

4-Aminodiphenylaminesulphonic Acid

French: Acide de 4-aminodiphénylеaminesulphonique.
German: 4-Aminodiphenylaminsulfonsaeure.

Dye

Starting point (Brit. 274999) in making dinitrophenylamine dyestuffs with—
1-Chloro-2:6-dinitrobenzene.
1-Chloro-2:6-dinitro-4-benzenesulphonic acid.
1-Chloro-2:4-dinitronaphthalene.
1-Chloro-2:4:6-trinitrobenzene.

2-Aminodiphenylene Oxide

Rubber

Antiaging agent (Brit. 422191).

4-Aminodiphenyl-4′-Sulphonanilide

Dye

As an intermediate.
Starting point (Brit. 399583) in making—
Sulphur dyes.

3-Amino-4-ethoxybenzotrifluoride-6-sulphonic Acid

Dye

Intermediate (Brit. 446532) in making dyestuffs.

4-Amino-2′-ethoxydiphenylamine

French: 4-Amino-2′-éthoxyediphényleamine.
German: 4-Amino-2′-aethoxydiphenylamin.

Chemical

Starting point in making—
Intermediates and other derivatives.

Dye

Starting point in making various synthetic dyestuffs.

Textile

Reagent (Brit. 313865) in dyeing silk and cotton yarns and fabrics and other textiles, with the aid of—
2:3-Oxynaphthoic alphanaphthylamide.
2:3-Oxynaphthoic anilide.
2:3-Oxynaphthoic 2-anisidide.
2:3-Oxynaphthoic 3-anisidide.
2:3-Oxynaphthoic 4-anisidide.
2:3-Oxynaphthoic 4-benzyloxy-1-anilide.
2:3-Oxynaphthoic betanaphthylamide.
2:3-Oxynaphthoic 2-chloroanilide.
2:3-Oxynaphthoic 3-chloroanilide.
2:3-Oxynaphthoic 4-chloroanilide.
2:3-Oxynaphthoic 4-chloro-2-anisidide.
2:3-Oxynaphthoic 5-chloro-2-anisidide.
2:3-Oxynaphthoic 5-chloro-2-toluidide.
2:3-Oxynaphthoic dianisidide.
2:3-Oxynaphthoic 2:5-dimethoxy-1-anilide.
2:3-Oxynaphthoic 2-ethyl-1-anilide.
2:3-Oxynaphthoic 4-ethyl-1-anilide.
2:3-Oxynaphthoic 2-ethyl-5-chloroanilide.
2:3-Oxynaphthoic 3-methoxy-2-naphthylamide.
2:3-Oxynaphthoic 3-nitranilide.
2:3-Oxynaphthoic 2-phenetidide.
2:3-Oxynaphthoic 3-phenetidide.
2:3-Oxynaphthoic 4-phenetidide.
2:3-Oxynaphthoic 2-phenoxy-1-anilide.
2:3-Oxynaphthoic 2-toluidide.
2:3-Oxynaphthoic 3-toluidide.
2:3-Oxynaphthoic 4-toluidide.

3-Amino-4-ethoxydiphenylsulphone

German: 3-Amino-4-aethoxydiphenylsulfon.

Dye

Starting point (Brit. 279146) in making dyestuffs with—
2:3-Oxynaphthoic-2-anisidide.
2:3-Oxynaphthoic-5-chloro-2-anisidide.
2:3-Oxynaphthoic-2-phenetidide.
2:3-Oxynaphthoic-5-chloro-2-toluidide.

8-Amino-6-ethoxyquinolin

Chemical

Starting point (Brit. 399818) in making—
Compounds, said to be effective against malaria, by diazotizing and coupling with hydrocuprein or a substituted hydrocuprein.

5-Amino-2-ethylaminotoluene-4-sulphonic Acid

Dye

Starting point (Brit. 447905, 447906, and 448016) in making—
Monoazo dyes for leather, particularly chrome leather.

5-Amino-4-ethylsulphonylbenzotrifluoride-2-sulphonic Acid

Dye

Intermediate (Brit. 446532) in making various dyestuffs.

2-Aminofluorene

Chemical

In organic syntheses.

Dye

Starting point (Brit. 437283) in making—
Violet dyestuffs by condensing with chloranil or other parabenzoquinones.

Aminoformic Acid

French: Acide d'aminoformique.
German: Aminoameisinsaeure.

Paint and Varnish

Ingredient of—
Cellulose acetate lacquers and varnishes, added for stabilizing purposes (Brit. 243722).

5-Amino-2-hydroxybenzoic Acid

French: Acide de 5-amino-2-hydroxyebenzoique, Acide de 5-amino-2-oxyebenzoique.
German: 5-Amino-2-hydroxybenzoesaeure, 5-Amino-2-oxybenzoesaeure.

Chemical

Starting point in making—
Aromatics, intermediates, pharmaceuticals, salts and esters.
Starting point (Brit. 305487) in making—
2:6-Dichloro-4(3′-nitrophenyl)-pyrimidin.
6-Nitro-2:4-dichloroquinazolin.

Dye

Starting point in making various synthetic dyestuffs.

Aminohydroxybenzoylaminoanthraquinone

French: Aminohydroxyebenzoyleaminoanthraquinone.
German: Aminohydroxybenzoylaminoanthrachinon.

Chemical

Starting point in making—
Intermediates, pharmaceuticals.

Dye

Starting point (Brit. 298696) in making anthraquinone vat dyestuffs with—
Aminoallylbenzoylaminoanthraquinone.
Aminoamylbenzoylaminoanthraquinone.
Aminobutylbenzoylaminoanthraquinone.
Aminoethylbenzoylaminoanthraquinone.
Aminomethylbenzoylaminoanthraquinone.
Aminopentylbenzoylaminoanthraquinone.
Aminopropylbenzoylaminoanthraquinone.

Aminohydroxybenzoylaminoanthraquinone (Cont'd)
Bromobenzoylaminoanthraquinone.
Bromobutylbenzoylaminoanthraquinone.
Bromoethylbenzoylaminoanthraquinone.
Bromomethylbenzoylaminoanthraquinone.
Bromopropylbenzoylaminoanthraquinone.
Chlorobenzoylaminoanthraquinone.
Chlorobutylbenzoylaminoanthraquinone.
Chloroethylbenzoylaminoanthraquinone.
Chloromethylbenzoylaminoanthraquinone.
Chloropropylbenzoylaminoanthraquinone.

Aminohydroxynitrodiphenylamine

Dye
Starting point in making—
Pyrogene black G, pyrogene blue, pyrogene direct blue, thion blue B.

Aminohydroxynitrodiphenylmethane

Dye
Starting point in making—
Pyrogene black G, pyrogene blue, pyrogene direct blue, thion blue B.

Aminoisodibenzanthrone

Dye
Starting point (Brit. 252903) in making dibenzanthrone dyestuffs with—
Benzoyl chloride.
Orthochlorobenzoyl chloride.
Paratoluene sulphochloride.
Paratoluene sulphonicethyl ester.
Paratoluene sulphonicmethyl ester.

1-Amino-2-mercapto-6-methyl-4-phenylamino-3′-carboxylic Acid
French: Acide de 1-amino-2-mercapto-6-méthylephényleamino-3′-carboxyle.
German: 1-Amino-2-mercapto-6-methyl-4-phenylamino-3′-carbonsaeure.

Dye
Starting point (Brit. 265641) in making acid dyestuffs with—
Dichloroquinone, chloranil, monochloroquinone, toluquinone, trichloroquinone.

2-Amino-2-mercapto-1:4-napthoquinone.
German: 2-Amino-2-mercapto-1:4-naphthochinon.

Dye
Starting point (Brit. 262141) in making dyestuffs with—
Alphanaphthaldehyde, aryltetrahydronaphthalene-1-aldehyde, 4-aminobenzaldehyde, 4-dimethylaminobenzaldehyde.

1-Amino-4-metatoluidinoanthraquinone

Textile
Dyestuff (Brit. 402391, 402392, and 402393) for—
Producing blue colors on acetate rayon.

1-Amino-4-methoxyanthraquinone

Dye
In dye syntheses.
Starting point (Brit. 449477) in making—
Orange to red vat dyes by condensing with 4:6-dichlor-1:3:5-triazins carrying a halogenophenyl or alkoxyphenyl substituent in position 2.

1:4-Amino-4-methoxyanthraquinone
German: 1:4-Amino-4-methoxyanthrachinon.

Dye
Starting point (French 604347) in making anthraquinone dyestuffs with—
Metabenzamidobenzoic acid, metamethoxybenzoyl chloride, meta-m′-diphenyldicarboxylic acid, 5-methylisophthalicbenzoic acid, 3-methylthiolbenzoic acid.

2-Amino-1-methoxybenzene-4-sulphonic Acid
Synonyms: Beta-amino-1-methoxybenzene-4-sulphonic acid.
French: Acide de 2-amino-1-méthoxyebenzène-4-sulphonique, Acide de béta-amino-1-methoxyebenzène-4-sulphonique.
German: 2-Amino-1-methoxybenzol-4-sulfonsäure, Beta-amino-1-methoxybenzol-4-sulfonsäure,

Chemical
Starting point in making—
Esters and salts, intermediates.

Dye
Starting point (Brit. 298518) in making azo dyestuffs with the aid of—
Alpha-amino-2-ethoxynaphthalene-6-sulphonic acid.
Alpha-amino-2:7-dimethoxynaphthalene.
Alpha-amino-2:7-dioxynaphthaleneglycollate.
Alpha-amino-2-methoxynaphthalene.
Alpha-amino-2-naphthoxybetapropionic acid.
Alpha-amino-2-oxyethoxynaphthalene sulphonate.
Alpha-aminonaphthalene.
Alpha-aminonaphthalene-6-sulphonic acid.
Alpha-aminonaphthalene-7-sulphonic acid.
1:3-Dioxyquinolin, methyl ketol, methylketolsulphonic acid, orthocresotinic acid, 1-phenyl-3-carboxy-5-pyrazolon, 1-phenyl-3-methyl-5-pyrazolon, phenyl-3-methyl-5-pyrazolon, salicylic acid, sulphazone.

1-Amino-4-methoxybenzothiazole

Dye
Starting point (Brit. 440112 and 440113) in making—
Blue dyes for acetate rayon by coupling with 5-betahydroxyethylaminoalphanaphthol.
Blue-green dyes for acetate rayon by coupling with 1-betahydroxyethyl-tetrahydroalphanaphthaquinolin.
Blue-green dyes for acetate rayon by coupling with 3:7-dihydroxytetrahydroalphanaphthaquinolin.
Pink dyes for acetate rayon by coupling with metatoluidin.
Red dyes for acetate rayon by coupling with beta-b′-dihydroxydiethylanilin.
Red dyes for acetate rayon by coupling with 1-phenylpiperazin.
Red-blue dyes for acetate rayon by coupling with 3-hydroxytetrahydroalphanaphthaquinolin.
Red-violet dyes for acetate rayon by coupling with 3-hydroxy-7-methyl-1-butyltetrahydroquinolin.

3-Amino-2-methoxybenzotrifluoride-4-sulphonic Acid

Dye
Intermediate (Brit. 446532) in making dyestuffs.

3-Amino-4-methoxybenzotrifluoride-6-sulphonic Acid

Dye
Intermediate (Brit. 446532) in making dyestuffs.

4-Amino-2′-methoxy-4′-chlorodiphenylamine
French: 4-Amino-2′-méthoxye-4′-chlorodiphénylamine.
German: 4-Amino-2′-methoxy-4′-chlordiphenylamin.

Chemical
Starting point in making—
Intermediates, pharmaceuticals, and other derivatives.

Dye
Starting point in making various synthetic dyestuffs.

Textile
Reagent (Brit. 313865) in dyeing silk, cotton, and other products with the aid of—
2:3-Oxynaphthoic alphanaphthylamide.
2:3-Oxynaphthoic anilide.
2:3-Oxynaphthoic 2-anisidide.
2:3-Oxynaphthoic 3-anisidide.
2:3-Oxynaphthoic 4-anisidide.
2:3-Oxynaphthoic betanaphthylamide.
2:3-Oxynaphthoic 2-chloroanilide.
2:3-Oxynaphthoic 3-chloroanilide.
2:3-Oxynaphthoic 4-chloroanilide.
2:3-Oxynaphthoic 4-chloro-2-anisidide.
2:3-Oxynaphthoic 5-chloro-2-anisidide.
2:3-Oxynaphthoic 4-benzyloxy-1-anilide.
2:3-Oxynaphthoic 5-chloro-2-toluidide.
2:3-Oxynaphthoic dianisidide.
2:3-Oxynaphthoic 2:5-dimethoxy-1-anilide.
2:3-Oxynaphthoic 2-ethyl-1-anilide.
2:3-Oxynaphthoic 4-ethyl-1-anilide.
2:3-Oxynaphthoic 2-ethyl-5-chloroanilide.
2:3-Oxynaphthoic 3-methoxy-2-naphthylamide.
2:3-Oxynaphthoic 3-nitranilide.
2:3-Oxynaphthoic 2-phenetidide.
2:3-Oxynaphthoic 3-phenetidide.
2:3-Oxynaphthoic 4-phenetidide.
2:3-Oxynaphthoic 2-phenoxy-1-anilide.
2:3-Oxynaphthoic 2-toluidide.
2:3-Oxynaphthoic 3-toluidide.
2:3-Oxynaphthoic 4-toluidide.

4-Amino-3′-methoxy-6′-chlorodiphenylamine
French: 4-Amino-3′-méthoxye-6′-chlorodiphénylamine.
German: 4-Amino-3′-methoxy-6′-chlordiphenylamin.

Chemical
Starting point in making—
Intermediates, pharmaceuticals.

Dye
Starting point in making various synthetic dyestuffs.

Textile
Reagent (Brit. 313865) in dyeing silk, cotton, and other textiles with the aid of—
2:3-Oxynaphthoic anilide.
2:3-Oxynaphthoic 2-anisidide.
2:3-Oxynaphthoic 3-anisidide.
2:3-Oxynaphthoic 4-anisidide.
2:3-Oxynaphthoic 4-benzyloxy-1-anilide.
2:3-Oxynaphthoic betanaphthylamide.
2:3-Oxynaphthoic 2-chloroanilide.
2:3-Oxynaphthoic 3-chloroanilide.
2:3-Oxynaphthoic 4-chloroanilide.
2:3-Oxynaphthoic 4-chloro-2-anisidide.
2:3-Oxynaphthoic 5-chloro-2-anisidide.
2:3-Oxynaphthoic 5-chloro-2-toluidide.
2:3-Oxynaphthoic dianisidide.
2:3-Oxynaphthoic 2:5-dimethoxy-1-anilide.
2:3-Oxynaphthoic 2-ethyl-1-anilide.
2:3-Oxynaphthoic 4-ethyl-1-anilide.
2:3-Oxynaphthoic 2-ethyl-5-chloroanilide.
2:3-Oxynaphthoic 3-methoxy-2-naphthylamide.
2:3-Oxynaphthoic 3-nitranilide.
2:3-Oxynaphthoic 2-phenetidide.
2:3-Oxynaphthoic 3-phenetidide.
2:3-Oxynaphthoic 4-phenetidide.
2:3-Oxynaphthoic 2-phenoxy-1-anilide.
2:3-Oxynaphthoic 2-toluidide.
2:3-Oxynaphthoic 3-toluidide.
2:3-Oxynaphthoic 4-toluidide.

3-Amino-4-methoxydiphenylsulphone

Dye
Starting point (Brit. 435817) in making—
Bluish-red dyestuffs by diazotizing and coupling with 3-orthoxylidide.

2-(4′-Amino-3′-methoxyphenyl)-6-methylbenzothiazole-sulphonic Acid

Dye
Starting point (Brit. 439372) in making—
Bluish-red cotton dyes by diazotizing and coupling with benzoyl-J-acid.
Bluish-red cotton dyes by diazotizing and coupling with carbonyl-J-acid.
Bluish-violet cotton dyes (when coppered) by diazotizing and coupling with phenyl-J-acid.
Reddish-violet cotton dyes by diazotizing and coupling with phenyl-J-acid.

8-Amino-6-methoxyquinolin

Chemical
Starting point (Brit. 399818) in making—
Compounds, said to be effective against malaria, by diazotizing and coupling with hydrocuprein or a substituted hydrocuprein.

1-Amino-4-methylamino-5-acetamidoanthraquinone-2-sulphonic Acid

Textile
Blue dye (Brit. 432647) for—
Woolen fabrics.

3-Amino-4-methylbenzophenone
French: 3-Amino-4-méthylebenzophénone.
German: 3-Amino-4-methylbenzophenon.

Chemical
Starting point in making—
Aromatics, intermediates, pharmaceuticals.

Dye
Starting point (Brit. 279146) in making azo dyestuffs with—
2:3-Oxynaphthoic 4-chloroanilide.
2:3-Oxynaphthoic 4-chloro-2-anisidide.

1-Amino-6-methylbenzothiazole

Dye
Starting point (Brit. 440112 and 440113) in making—
Yellow-red dyes for acetate rayon by coupling with beta-b′-dihydroxydiethylanilin.

3′-Amino-4′-methylbenzoylalphanaphthylamine
French: 3′-Amino-4′-méthylebenzoylealphanaphthyle-amine.
German: 3′-Amino-4′-methylbenzoylalphanaphtylamin.

Dye
Starting point (Brit. 279146) in making azo dyestuffs with—
2:5-Dimethoxyanilide.
2:3-Oxynaphthoicalphanaphthalide.
2:3-Oxynaphthoicanilide.
2:3-Oxynaphthoicbeta-anisidide.
2:3-Oxynaphthoic-3-nitranilide.
2:3-Oxynaphthoic-2-phenetidide.
2:3-Oxynaphthoic-2-toluidide.

3′-Amino-4′-methylbenzoyl-2-anisidin
French: 3′-Amino-4′-méthylebenzoyle-2-anisidine.

Dye
Starting point (Brit. 279146) in making dyestuffs with—
2:3-Oxynaphthoicalphanaphthalide.
2:3-Oxynaphthoic-3-anisidide.
2:3-Oxynaphthoicbetanaphthalide.
2:3-Oxynaphthoic-4-chloro-2-anisidide.
2:3-Oxynaphthoic-5-chloro-2-anisidide.
2:3-Oxynaphthoicdianisidide.
2:3-Oxynaphthoic-2:5-dimethoxyanilide.

3′-Amino-4′-methylbenzoylbetanaphthylamine

Dye
Starting point (Brit. 279146) in making dyestuffs with—
3-Anisidide, 4-anisidide, 4-chloroanilide, 4-chloro-2-anisidide, 5-chloro-2-toluidide, 3-phenetidide, 4-phenetidide.

3′-Amino-4′-methylbenzoyl-2-chloroanilide
French: Chloroanilide de 3′-amino-4′-méthylebenzoyle.
German: 3′-Amino-4′-methylbenzoyl-2-chloranilid.

Chemical
Starting point in making—
Aromatics, intermediates, pharmaceuticals.

Dye
Starting point (Brit. 279146) in making azo dyestuffs with—
2:3-Oxynaphthoic alphanaphthalide.
2:3-Oxynaphthoic 2-anisidide.
2:3-Oxynaphthoic betanaphthalide.
2:3-Oxynaphthoic 2-chloroanilide.

3′-Amino-4′-methylbenzoyl-5-chloro-2-anisidin
French: 3′-Amino-4′-méthylebenzoyle-5-chloro-2-anisidine.

Dye
Starting point (Brit. 279146) in making azo dyestuffs with—
2:3-Oxynaphthoicalphanaphthalide.
2:3-Oxynaphthoic-4-anisidide.
2:3-Oxynaphthoicbetanaphthalide.
2:3-Oxynaphthoic-4-chloro-2-anisidide.
2:3-Oxynaphthoic-5-chloro-2-anisidide.
2:3-Oxynaphthoicdianisidide.
2:3-Oxynaphthoicdimethoxyanilide.

3′-Amino-4′-methylbenzoyl-4-chloro-2-toluidin

Dye
Starting point (Brit. 279146) in making azo dyestuffs with—
2:3-Oxynaphthoicalphanaphthalide.
2:3-Oxynaphthoicbetanaphthalide.
2:3-Oxynaphthoic-2-chloroanilide.
2:3-Oxynaphthoicdianisidide.
2:3-Oxynaphthoic-3-methoxy-2-naphthalide.
2:3-Oxynaphthoic-3-toluidide.
2:3-Oxynaphthoic-4-toluidide.

3′-Amino-4′-methylbenzoyl-3-toluidin
French: 3′-Amino-4′-méthylebenzoyle-3-toluidine.

Dye
Starting point (Brit. 279146) in making azo dyestuffs with—
2:3-Oxynaphthoicalphanaphthalide.
2:3-Oxynaphthoic-3-anisidide.
2:3-Oxynaphthoicbetanaphthalide.
2:3-Oxynaphthoic-2-phenetidide.
2:3-Oxynaphthoic-2-toluidide.

3-Amino-4-Methyldiphenylsulphone

Dye
Starting point (Brit. 435817) in making—
Reddish-orange dyestuffs by diazotizing and coupling with paraxylidide.

4-Amino-2′-methyl-4′-methoxydiphenylamine
French: 4-Amino-2′-méthyle-4′-méthoxyediphenyle-amine.

Chemical
Starting point in making—
Intermediates and other derivatives.

Dye
Starting point in making various synthetic dyestuffs.

Textile
Reagent (Brit. 313865) in dyeing silk, cotton, and other fibers, yarns, and fabrics with the aid of—
2:3-Oxynaphthoic anilide.
2:3-Oxynaphthoic 2-anisidide.
2:3-Oxynaphthoic 3-anisidide.
2:3-Oxynaphthoic 4-anisidide.
2:3-Oxynaphthoic 4-benzyloxy-1-anilide.
2:3-Oxynaphthoic betanaphthylamide.
2:3-Oxynaphthoic 2-chloroanilide.
2:3-Oxynaphthoic 3-chloroanilide.
2:3-Oxynaphthoic 4-chloroanilide.
2:3-Oxynaphthoic 4-chloro-2-anisidide.
2:3-Oxynaphthoic 5-chloro-2-anisidide.
2:3-Oxynaphthoic 5-chloro-2-toluidide.
2:3-Oxynaphthoic dianisidide.
2:3-Oxynaphthoic 2:5-dimethoxy-1-anilide.
2:3-Oxynaphthoic 2-ethyl-1-anilide.
2:3-Oxynaphthoic 4-ethyl-5-chloroanilide.
2:3-Oxynaphthoic 3-methoxy-2-naphthylamide.
2:3-Oxynaphthoic 3-nitranilide.
2:3-Oxynaphthoic 2-phenetidide.
2:3-Oxynaphthoic 3-phenetidide.
2:3-Oxynaphthoic 4-phenetidide.
2:3-Oxynaphthoic 2-phenoxy-1-anilide.
2:3-Oxynaphthoic 2-toluidide.
2:3-Oxynaphthoic 3-toluidide.
2:3-Oxynaphthoic 4-toluidide.

4-Amino-3′-methyl-6′-methoxydiphenylamine

Chemical
Starting point in making—
Intermediates and other derivatives.

Dye
Starting point in making various synthetic dyestuffs.

Textile
Reagent (Brit. 313865) in dyeing silk, cotton, and other fibers, yarns, and fabrics, with the aid of—
2:3-Oxynaphthoic anilide.
2:3-Oxynaphthoic 2-anisidide.
2:3-Oxynaphthoic 3-anisidide.
2:3-Oxynaphthoic 4-anisidide.
2:3-Oxynaphthoic 4-benzyloxy-1-anilide.
2:3-Oxynaphthoic betanaphthylamide.
2:3-Oxynaphthoic 2-chloroanilide.
2:3-Oxynaphthoic 3-chloroanilide.
2:3-Oxynaphthoic 4-chloroanilide.
2:3-Oxynaphthoic 4-chloro-2-anisidide.
2:3-Oxynaphthoic 5-chloro-2-anisidide.
2:3-Oxynaphthoic 5-chloro-2-toluidide.
2:3-Oxynaphthoic dianisidide.
2:3-Oxynaphthoic 2:5-dimethoxy-1-anilide.
2:3-Oxynaphthoic 2-ethyl-1-anilide.
2:3-Oxynaphthoic 4-ethyl-5-anilide.
2:3-Oxynaphthoic 2-ethyl-5-chloroanilide.
2:3-Oxynaphthoic 3-methoxy-2-naphthylamide.
2:3-Oxynaphthoic 3-nitranilide.
2:3-Oxynaphthoic 2-phenetidide.
2:3-Oxynaphthoic 3-phenetidide.
2:3-Oxynaphthoic 4-phenetidide.
2:3-Oxynaphthoic 2-phenoxy-1-anilide.
2:3-Oxynaphthoic 2-toluidide.
2:3-Oxynaphthoic 3-toluidide.
2:3-Oxynaphthoic 4-toluidide.

1-Amino-2-methyl-4-paratoluidinoanthraquinone

Oils, Fats, Waxes
Coloring agent (Brit. 432867) for—
Paraffin and other mineral waxes, stearic acid, tallow and other solid triglycerides, beeswax, carnauba wax, and others.

5-Amino-2-methylsulphonylbenzotrifluoride-4-sulphonic Acid

Dye
Intermediate (Brit. 446532) in making various dyestuffs.

1:8-Aminonaphthoic Acid
French: Acide 1:8-aminonaphthoique, Acide de 1:8-aminonaphthoique.
German: 1:8-Aminonaphtoesaeure.

Chemical
Starting point (Brit. 278100) in making—
4:4′-Dibromo-1:1′-dinaphthyl-8:8′-dicarboxylic acid.
4:4′-Dichloro-1:1′-dinaphthyl-8:8′-dicarboxylic acid.
1:1′-Dichloro-2:2′-dinaphthyl-3:3′-dicarboxylic acid.
1:1′-Dinaphthyl-8:8′-dicarboxylic acid.
2:2′-Dinaphthyl-3:3′-dicarboxylic acid.
1:1′-Dinaphthyl-2:2′-dicarboxylic acid.
4:4′-Disulpho-1:1′-dinaphthyl-8:8′-carboxylic acid.
Alkoxy derivatives of these acids.

Dye
Starting point in making various synthetic dyestuffs.

2:3-Aminonaphthoic Acid
French: Acide de 2:3-aminonaphthoique.
German: 2:3-Aminonaphtoesaeure.

Dye
Starting point (Brit. 275307) in making azo dyestuffs with—
Betanaphthol, betanaphthylamine, diazotized alphanaphthylamine, diazotized 2:4-dinitranilin, diazotized orthoanisidin, diazotized para-aminoacetanilide, diazotized paracresidin, diazotized paranitranilin, diazotized paranitro-orthoanisidin, ethylbetanaphthylamine, resorcinol.

2:6-Aminonaphthoic Acid
French: Acide de 2:6-aminonaphthoique.
German: Aminonaphtoesaeure.

Dye
Starting point (Brit. 275307) in making azo dyestuffs with—
Alphanaphthylamine, aminohydroquinonedimethyl ether, anilin, betanaphthol, betanaphthylamine, diazotized alphanaphthylamine, diazotized betanaphthylamine, diazotized orthochloroanilin, diazotized metachloroanilin, 2:5-dichloroanilin, dimethylanilin, dimethylmetatoluidin, ethylbetanaphthylamine, metaphenylenediamine, metatoluylenediamine, 1:5-naphthylenediamine, orthoanisidin, orthotoluidin, 2:3-oxynaphthoic acid, 2:3-oxynaphthoylbetanaphthylamine, paranitranilin, paranitro-orthoanisidin, phenylmethylpyrazolone.

1:2-Aminonaphthol Ether
French: Éther de 1:2-aminonaphthole, Éther 1:2-aminonaphtholique.
German: 1:2-Aminonaphtolaether.

Dye
Starting point (Brit. 270428) in making dyestuffs with—
Alphanaphthylamine, anilin, cresidin, 4:5-dinitroalphanaphthylamine, 2:4-dinitranilin, metaphenylenediamine, metanitranilin, metatoluidin, paranitranilin, picramic acid.
Starting point (Brit. 252957) in making—
Disazo dyestuffs.
Starting point in making—
Alphyl blue black O and OK.

1-Amino-2-naphtholethylether-6-sulphonic Acid
Synonyms: Alpha-amino-2-naphtholethylether-6-sulphonic acid.
French: Acide d'alpha-amino-2-naphthole-éthyle-éther-6-sulfonique.
German: Alphaamino-2-naphtolaethylaether-6-sulfonsaeure, 1-Amino-2-naphtolaethylaether-6-sulfonsaeure.

Chemical
Starting point in making—
Esters and salts, intermediates.

Dye
Starting point (Brit. 308958) in making trisazo dyestuffs with—
Acetyl-1-amino-8-hydroxynaphthalene-3:6-disulphonic acid.
Acetyl-1-amino-8-naphthalene-3:6-disulphonic acid.
Acetyl-1-amino-8-naphthalene-4:6-disulphonic acid.
Aminoazotoluenesulphonic acid.
Anilin-w-methanesulphonic acid.

1-Amino-2-naphtholethylether-6-sulphonic Acid (Continued)
Benzoyl-1-amino-8-hydroxynaphthalene-3:6-disulphonic acid.
Betanaphthylamine-8-sulphonic acid.
Orthochlorobenzoyl-1-amino-8-hydroxynaphthalene-3:6-disulphonic acid.
Orthochlorobenzoyl-1-amino-8-naphthalene-3:6-disulphonic acid.
Orthochlorobenzoyl-1-amino-8-naphthalene-4:6-disulphonic acid.
Orthotoluidin.
Parasulphanilic acid.
S-xylidin.

1-Amino-8-naphthol-3-sulphonic Acid
Synonyms: Alpha-amino-8-naphthol-3-sulphonic acid.
French: Acide d'alpha-amino-8-naphthole-3-sulphonique, Acide de 1-amino-8-naphthole-3-sulphonique.
German: Alpha-amino-8-naphtol-3-sulfonsäure, 1-Amino-8-naphtol-3-sulfonsäure.
Chemical
Starting point in making—
Alpha-amino-8-naphthol-3:5-disulphonic acid.
Esters and salts of the acid.
Dye
Starting point in making various synthetic dyestuffs.

1-Amino-8-naphthol-4-sulphonic Acid
Synonyms: Alpha-amino-8-naphthol-4-sulphonic acid, S acid.
French: Acide d'alpha-amino-8-naphthole-4-sulphonique, Acide de 1-amino-8-naphthole-4-sulphonique, acide de S.
German: Alpha-amino-8-naphtol-4-sulfonsäure, 1-Amino-8-naphtol-4-sulfonsäure, S-sæure.
Chemical
Starting point in making—
Esters and salts, intermediates.
Dye
Starting point in making—
Azidin blue for wool G, benzo blue R, benzo blue, reddish G, benzocyanin B, Chicago blue B, Chicago blue 2R, Chicago blue 4R, Columbia green black D, Columbia blue R, diazo olive G, direct brown J, solid brown ONT, zambesi black BR.

1-Amino-2-naphthoxyalphapropionic Acid
Synonyms: Alpha-amino-2-naphthoxyalphapropionic acid.
French: Acide de 1-amino-2-naphthoxyealphapropionique, Acide d'alphaamino-2-naphthoxyealphapropionique.
German: Alpha-amino-2-naphtoxyalphapropionsäure, 1-Amino-2-naphtoxyalphapropionsäure.
Chemical
Starting point in making—
Esters, intermediates, salts.
Dye
Starting point (Brit. 298518) in making azo dyestuffs with—
Alpha-aminonaphthalene, alpha-aminonaphthalene-6-sulphonic acid, alpha-aminonaphthalene-7-sulphonic acid, anilin, anilin-3-chloro-6-sulphonic acid, anilin-2:4-disulphonic acid, anilin-2:5-disulphonic acid, anilin-4-nitro-2:5-disulphonic acid, anilin-3-sulphonic acid, beta-amino-1-methoxybenzene-4-sulphonic acid, beta-amino-5-sulphobenzoic acid, 1:3-dioxyquinolin, methylketol, methylketolsulphonic acid, orthocresotinic acid, 1-phenyl-3-carboxy-5-pyrazolone, 1-phenyl-3-methyl-5-pyrazolone, salicylic acid, sulphazone.

1-Amino-2-naphthoxybetapropionic Acid
French: Acide d'alpha-amino-2-naphthoxyebétapropionique.
German: Alpha-amino-2-naphtoxybetapropionsaeure.
Chemical
Starting point in making various intermediates.
Dye
Starting point (Brit. 298518) in making azo dyestuffs with—
Alpha-aminonaphthalene, alpha-aminonaphthalene-6-sulphonic acid, alpha-aminonaphthalene-7-sulphonic acid, anilin, anilin-3-chloro-6-sulphonic acid, anilin-2:4-disulphonic acid, anilin-2:5-disulphonic acid, anilin-4-nitro-2:5-disulphonic acid, anilin-3-sulphonic acid, beta-amino-1-methoxybenzene-4-sulphonic acid, beta-amino-5-sulphobenzoic acid, 1:3-dioxyquinolin, methyl ketol, methyl ketol-sulphonic acid, orthocresotinic acid, 1-phenyl-3-carboxy-5-pyrazolone, 1-phenyl-3-methyl-5-pyrazolone, salicylic acid, sulphazone.

Amino-5-nitro-2-aminobenzyl Sulphonate
French: Sulphonate d'amino-5-nitro-2-aminobenzyle.
German: Amino-5-nitro-2-aminobenzylsulfonat.
Dye
Starting point (Brit. 265767) in making monoazo dyestuffs with—
Betamethylaminonaphthalene-7-sulphonic acid.
Ethylbenzylanilin.

4-Amino-normal-butyl-normal-betasulphopropylanilin
Dye
Starting point (Brit. 417388 and 435479) in making—
Greenish-blue dyes for wool by coupling with 1:3-dianilinonaphthalene-8-sulphonic acid.
Blue dyes for wool by coupling with diethylrosindulindisulphonic acid.

1-Amino-4-orthoanisidinoanthraquinone
Textile
Dyestuff (Brit. 402391, 402392, and 402393) for—
Producing blue colors on acetate rayon.

1:4-Amino-oxyanthraquinone
German: 1:4-Amino-oxyanthrachinon.
Dye
Starting point in making—
Algol pink R paste.
1:4-Dimethylaminoanthraquinone (Brit. 268891).
Oils, Fats, Waxes
Coloring agent (Brit. 432867) for—
Paraffin and other mineral waxes, stearic acid, tallow and other solid triglycerides, beeswax, carnauba wax, and others.

4:4'-Amino-oxydiphenylamine
German: 4:4'-Aminohydroxydiphenylamin.
Dye
Starting point (Brit. 282111) in making dyestuffs for pelts, animal fibers, and acetate rayon, with the aid of—
Alphanaphthol, alphanaphthylamine, betanaphthol, betanaphthylamine, 1:5-dioxynaphthalene, 2:7-dioxynaphthalene.

2-Amino-1-oxynaphthalene-4:8-disulphonic Acid
French: Acide 2-amino-1-oxynaphthalène-4:8-disulphonique.
German: 2-Amino-1-oxynaphtalin-4:8-disulphonsaeure.
Dye
Starting point (Brit. 249884) in making azo dyestuffs with betanaphthol—
1-oxynaphthalene-4:8-disulphonic acid, 1-oxynaphthalene-3:8-disulphonic acid, 1-phenyl-3-methyl-5-pyrazolone.

3-Amino-4-oxyphenyldichloroarsin Hydrochloride
German: 3-Amino-4-oxyphenylchloroarsinchlorhydrat.
Chemical
Starting point in making—
Aminoarylarsenious oxides (Brit. 260382), aminoaryldichloroarsines.

6-Amino-oxythionaphthene
Synonyms: 6-Amino-oxysulphonaphthene.
German: 6-Amino-oxythionaphten.
Dye
Starting point (Brit. 285389) in making 2-thionaphthene-3-indoleindigoid dyestuffs with—
5:7-Dichloroisatin, 5:7-dibromoisatin, 5:7-di-iodoisatin.

5-Amino-2-oxytoluene
Synonyms: 5-Amino-2-oxytoluol.
Dye
Starting point (Brit. 267366) in making dyestuffs with cresotinic sulphinic acid by treatment with—
Acetic anhydride, benzoyl chloride, chlorocarbonic esters, paramethoxybenzoyl chloride, phthalic anhydride, phosgene, salicyl sulphochloride, toluene sulphochloride.

2-Aminophenanthraquinone
German: 2-Aminophenanthrachinon.
Photographic
Starting point (German 436161) in making desensitizers with—
Orthoaminodiphenylamine and substitutes.

3-Aminophenanthrene

Chemical
In organic syntheses.

Dye
Starting point (Brit. 437283) in making—
Bluish-violet dyestuffs by condensing with chloranil or other parabenzoquinones.

2-Aminophenetol-4-sulphodimethylamide
Synonyms: 2-Aminophenetole-4-thiodimethylamide.
French: 2-Aminophénétole-4-sulphodiméthyleamide, 2-Aminophénétole-4-thiodiméthyleamide.
German: 2-Aminophenetol-4-sulfodimethylamid, 2-Aminophenetol-4-thiodimethylamid.

Chemical
Starting point in making—
Aromatics, intermediates, pharmaceuticals.

Dye
Starting point (Brit. 279146) in making azo dyestuffs with—
2:3-Oxynaphthoic alphanaphthalide, 2:3-oxynaphthoic betanaphthalide, 2:3-oxynaphthoic 4-chloro-2-anisidide.

2-Amino-1-phenol-4:6-disulphonic Acid

Dye
Intermediate in making various dyestuffs.
Starting point (Brit. 404198) in making—
Dyestuffs (for coloring bones and bone objects orange tints) by reaction with 1-phenyl-3-methyl-5-pyrazolone and a chromium salt.
Dyestuffs (for coloring bones and bone objects red tints) by reaction with 2:4-dioxyquinolin and a chromium salt.

2-Amino-1-phenol-4-sulphonic Acid

Dye
Intermediate in making—
Dyes of Schultz No. 154, 155, 156, and 157.

5-Amino-4-phenoxy-2-acetylamino-1-methoxybenzene
French: 5-Amino-4-phénoxye-2-acétyleamino-1-méthoxyebenzène.
German: 5-Amino-4-phenoxy-2-acetylamino-1-methylbenzol.

Chemical
Starting point in making—
Aromatics, intermediates, pharmaceuticals.

Dye
Starting point (Brit. 307303) in making monoazo dyestuffs with—
N-Acetyl-H acid, N-benzoyl-A acid, N-betachloropropionyl-H acid, N-betachloroethanesulpho-H acid, N-carbethoxy-H acid, N-chloroacetyl-H acid, N-phenylacetyl-H acid, N-toluenesulpho-H acid.

3-Amino-2-phenoxybenzotrifluoride-4-sulphonic Acid

Dye
Intermediate (Brit. 446532) in making various dyestuffs.

2(4-Aminophenylamino)-6-oxynaphthalene Hydrochloride
French: Chlorhydrate de 2(4′-aminophényleamino)-6-oxyenaphthalène, Hydrochlorure de 2(4′-aminophényleamino)-6-oxyenaphthalène.
German: 2(4′-Aminophenylamino)-6-oxynaphtalinchlorydrat, 2(4′-Aminophenylamino)-6-oxynaphtalinhydrochlorid, Chlorwasserstoffsaeures-2(4′-aminophenylamino)-6-oxynaphtalin.

Leather
Reagent (Brit. 290126) in—
Dyeing.

Miscellaneous
Reagent (Brit. 290126) in—
Dyeing furs, hair and feathers.

3-Amino-4-phenylbenzophenone
French: 3-Amino-4-phénylebenzophénone.
German: 3-Amino-4-phenylbenzophenon.

Chemical
Starting point in making—
Aromatics, intermediates, pharmaceuticals.

Dye
Starting point (Brit. 279146) in making azo dyestuffs with—
2:3-Oxynaphthoic betanaphthalide.

5-Amino-2-phenylbenzthiazole

Petroleum
Reagent for—
Imparting fluorescence to hydrocarbon oils oı liquids.

6-Amino-2-phenylbenzthiazole

Petroleum
Reagent for—
Imparting fluorescence to hydrocarbon oils or liquids.

4-Aminophenylbetanaphthylamine
French: 4-Aminophénylebétanaphthyleamine.
German: 4-Aminophenylbetanaphtylamin.

Chemical
Starting point in making—
Intermediates, pharmaceuticals.

Dye
Starting point in making various synthetic dyestuffs.

Textile
Reagent (Brit. 313865) in dyeing silk, cotton and other textiles with the aid of—
2:3-Oxynaphthoic alphanaphthylamide.
2:3-Oxynaphthoic anilide.
2:3-Oxynaphthoic 2-anisidide.
2:3-Oxynophthoic 3-anisidide.
2:3-Oxynaphthoic 4-anisidide.
2:3-Oxynaphthoic 4-benzyloxy-1-anilide.
2:3-Oxynaphthoic betanaphthylamide.
2:3-Oxynaphthoic 2-chloroanilide.
2:3-Oxynaphthoic 3-chloroanilide.
2:3-Oxynaphthoic 4-chloroanilide.
2:3-Oxynaphthoic 4-chloro-2-anisidide.
2:3-Oxynaphthoic 5-chloro-2-anisidide.
2:3-Oxynaphthoic 5-chloro-2-toluidide.
2:3-Oxynaphthoic diansidide.
2:3-Oxynaphthoic 2:5-dimethoxy-1-anilide.
2:3-Oxynaphthoic 2-ethyl-1-anilide.
2:3-Oxynaphthoic 4-ethyl-1-anilide.
2:3-Oxynaphthoic 2-ethyl-5-chloroanilide.
2:3-Oxynaphthoic 3-methoxy-2-naphthylamide.
2:3-Oxynaphthoic 3-nitranilide.
2:3-Oxynaphthoic 2-phenetidide.
2:3-Oxynaphthoic 3-phenetidide.
2:3-Oxynaphthoic 4-phenetidide.
2:3-Oxynaphthoic 2-phenoxy-1-anilide.
2:3-Oxynaphthoic 2-toluidide.
2:3-Oxynaphthoic 3-toluidide.
2:3-Oxynaphthoic 4-toluidide.

2-Amino-5-phenylsulphonylbenzotrifluoride-4-sulphonic Acid

Dye
Intermediate (Brit. 446532) in making various dyestuffs.

5-Amino-4-phenylsulphonylbenzotrifluoride-2-sulphonic Acid

Dye
Intermediate (Brit. 446532) in making various dyestuffs.

Aminophenyl Thiocyanate

Lubricant
Starting point (Brit. 440175) in making—
Addition agents for high-pressure lubricating oils or greases by mixing and reacting with organometallic compounds.

8-Aminoquinolin

Chemical
Starting point (Brit. 399818) in making—
Compounds, said to be effective against malaria, by diazotizing and coupling with hydrocuprein or a substituted hydrocuprein.

3-Aminosalicylic Acid
French: Acide de 3-aminosalicyle, Acide 3-aminosalicylique.
German: 3-Amino-salicylsaeure.

Dye
Starting point (Brit. 276540) in making dyestuffs with—
Alpha-amino-2-alkyloxynaphthalene-6-sulphonic acid.
Alpha-amino-2-alkyloxynaphthalene-7-sulphonic acid.
Alpha-amino-2-naphthol ethers.
Alphanaphthylamine.
Alphanaphthylamine-6-sulphonic acid.
Alphanaphthylamine-7-sulphonic acid.
Aminohydroquinonedimethyl ether.
Meta-aminoparacresolmethyl ether.

5-Aminosalicylic Acid
French: Acide 5-aminosalicylique, Acide de 5-amino-salicyle.
German: 5-Amino-salicylsaeure.

Chemical
Starting point in making—
Esters and salts, pharmaceuticals, intermediates.
Starting point in making compounds used as drugs.

Dye
Starting point in making—
Anthracene acid black DSF, anthracene acid brown B, azo dyestuffs, chrome bordeaux, diamond black F, diamond green B.
Starting point (Brit. 276757) in making dyestuffs for viscose rayon with—
Alpha-amino-2-alkyloxynaphthalene-6-sulphonic acid.
Alpha-amino-2-alkyloxynaphthalene-7-sulphonic acid.
Alpha-amino-2-naphthol ethers.
Alphanaphthol-3:6-disulphonic acid.
Alphanaphthol-4-sulphonic acid.
Alphanaphthol-5-sulphonic acid.
Alphanaphthylamine.
Alphanaphthylamine-2-sulphonic acid (Cleve's acid).
Alphanaphthylamine-4-sulphonic acid.
Alphanaphthylamine-5-sulphonic acid.
Alphaphenylaminonaphthalene-8-sulphonic acid.
Aminohydroquinonedimethyl ether.
Betamethylaminonaphthalene-7-sulphonic acid.
Betanaphthol-6-sulphonic acid.
Betanaphthol-7-sulphonic acid.
Betanaphthylamine-6-sulphonic acid.
Betanaphthylamine-7-sulphonic acid.
Meta-aminoparacresolmethyl ether.
Paraxylidene and betanaphthol-6-sulphonic acid.
Starting point in making—
Diamond black F, diamond green, oxazin dyestuffs, sulphonated dyestuffs.

Paper
Reagent in making—
Transfer paper.

Pharmaceutical
In compounding and dispensing practice.

2-Amino-4-sulphobenzoic Acid
French: Acide de 2-amino-4-sulphobenzoique, Acide de 2-amino-4-thiobenzoique.
German: 2-Amino-4-sulfobenzoesäure, 2-Amino-4-thiobenzoesäure.

Chemical
Starting point in making—
Esters, intermediates, salts.
Starting point (Brit. 324041) in making insecticides with the aid of—
6-Benzoylamino-4-chloro-3-amino-5-methoxybenzene.
4-Chloro-2-anisidin, 4-chloro-2-toluidin, 3-chloro-2-toluidin, 5-chloro-2-toluidin, 6-chloro-2-toluidin, 3-chloro-4-toluidin, 4:5-dichloro-2-toluidin, 4:6-dichloro-2-toluidin, 3:6-dichloro-4-toluidin, 2:5-dichloroanilin, metachloroanilin, 6-nitro-4-methoxy-3-toluidin, 4-nitro-2-toluidin, 5-nitro-4-toluidin, 3-nitro-4-toluidin, sulphanilic acid.

2-Amino-5-Sulphobenzoic Acid
Synonyms: Beta-amino-5-sulphobenzoic acid.
French: Acide 2-amino-5-sulphobenzoique, Acide beta-amino-5-sulphobenzoique.
German: 2-Amino-5-sulfobenzoesäure, Beta-amino-5-sulfobenzoesäure.

Chemical
Starting point in making—
Esters, intermediates, salts.

Dye
Starting point (Brit. 298518) in making azo dyestuffs with the aid of—
Alpha-amino-2:7-dimethoxynaphthalene.
Alpha-amino-2-ethoxynaphthalene-6-sulphonic acid.
Alpha-amino-2-methoxynaphthalene.
Alpha-aminonaphthalene.
Alpha-aminonaphthalene-6-sulphonic acid.
Alpha-aminonaphthalene-7-sulphonic acid.
Alpha-amino-2-naphthoxybetapropionic acid.
Alpha-amino-2-oxyethoxynaphthalenesulphonic acid.
1:3-Dioxyquinolin, methyl ketol, methylketolsulphonic acid, orthocresotinic acid, 1-phenyl-3-carboxyl-5-pyrazolon, 1-phenyl-3-methyl-5-pyrazolon, phenyl-3-methyl-5-pyrazolon, salicylic acid, sulphazone.

3-Amino-5-sulphobenzoic Acid
Synonyms: 3-Amino-5-thiobenzoic acid.
French: Acide 3-amino-5-sulphobenzoique, Acide 3-amino-5-thiobenzoique.
German: 3-Amino-5-sulfobenzoesäure, 3-Amino-5-thiobenzoesäure.

Chemical
Starting point in making—
Esters and salts, intermediates, pharmaceuticals.
Starting point (Brit. 324041) in making intermediates and disinfectants with the aid of—
6-Benzoylamino-4-chloro-3-amino-5-methoxybenzene, 4-chloro-2-anisidin, 4-chloro-2-toluidin, 3-chloro-2-toluidin, 6-chloro-2-toluidin, 3-chloro-4-toluidin, 3:6-dichloro-4-toluidin, 4:6-dichloro-2-toluidin, 4:5-dichloro-2-toluidin, 2:5-dichloroanilin, metachloroanilin, 6-nitro-4-methoxy-3-toluidin, 4-nitro-2-toluidin, 5-nitro-2-toluidin, 3-nitro-4-toluidin, sulphanilic acid.

2-Aminotoluene-4-sulphodimethylamide
Synonyms: 2-Aminotoluene-4-thiodimethylamide.
French: Amide de 2-aminotoluène-4-sulphodiméthyle, Amide de 2-aminotoluène-4-thiométhyle.
German: 2-Aminotoluol-4-sulfodimethylamid, 2-aminotoluol-4-thiodimethylamid.

Chemical
Starting point in making—
Intermediates, pharmaceuticals.

Dye
Starting point (Brit. 279046) in making azo dyestuffs with—
2:3-Oxynaphthoic alphanaphthalide.
2:3-Oxynaphthoic 4-chloro-2-anisidide.
2:3-Oxynaphthoic betanaphthalide.
2:3-Oxynaphthoic 2-phenetidide.

2-Aminotoluene-4-sulpho-N-methylanilide.
French: Anilide de 2-aminotoluène-4-sulpho-N-méthyle, Anilide-2-aminotoluène-4-sulpho-N-méthylique, Anilide de 2-aminotoluène-4-thio-N-méthyle.
German: 2-Aminotoluol-4-sulfo-N-methylanilid, 2-Aminotoluol-4-thio-N-methylanilid.

Chemical
Starting point in making various intermediates.

Dye
Starting point (Brit. 279146) in making dyestuffs with—
2:3-Oxynaphthoic 4-chloroanilide.
2:3-Oxynaphthoic 5-chloro-2-anisidide.
2:3-Oxynaphthoic 5-chloro-2-toluidide.
2:3-Oxynaphthoic 3-toluidide.

4-Amino-1-toluidino-8-hydroxyanthraquinone

Textile
Dyestuff (Brit. 402391, 402392, and 402393) for—
Producing blue colors on acetate rayon.

2-Aminotolyl-4-ethylsulphone

Chemical
Starting point in making—
Intermediates, pharmaceuticals.

Dye
Starting point (Brit. 279146) in making azo dyestuffs with—
2:3-Oxynaphthoic 5-chloro-2-anisidide.

Aminotriazinsulphonic Acid

Chemical
Starting point in making—
Derivatives with various bases.

Pharmaceutical
In compounding and dispensing practice.

Ammonia
Synonyms: Alkaline air, Ammoniacal gas, Anhydrous ammonia, Volatile alkali.
French: Ammoniaque, Ammoniaque liquide.
German: Ammoniak, Ammoniakflüssigkeit, Ammoniakgas, Salmiakgeist.
Spanish: Amonio.
Italian: Ammonio.

The term, "ammonia," refers to a colorless gas conforming to the chemical formula NH_3. When subjected to a certain pressure this gas is converted into a colorless, mobile liquid, generally known as anhydrous ammonia, and sometimes called liquid ammonia. Aqua ammonia (ammonium hydroxide) is a solution of ammonia in water; household ammonia is a weak aqua

Ammonia (Continued)
ammonia. In its chemical combining form ammonia has the formula NH_4 and is properly called ammonium.
(The uses given below are those of anhydrous ammonia or aqua ammonia; they are not segregated according to the physical state of ammonia.)

Analysis
Reagent in—
Analytical processes involving control and research in science and industry.

Beverage
Ammoniating agent for—
Glycyrrhizin used as a foam producer.
Extractant for—
Bitter principles, saponins.

Bituminous
Process material (U. S. 1899314) in—
Desulphurizing low-boiling point hydrocarbon oils.
Source of—
Hydrogen.
Temperature reaction controller (Brit. 444936) for—
Destructive hydrogenation treatments of carbonaceous materials.

Building Construction
Reagent for—
Removing stains from marble.

Cellulose Products
Process material in various operations in the rayon industry (see "Rayon" below).
Process material (Brit. 409008) in making—
Nitrocellulose of reduced viscosity.

Chemical
Alkali in—
Chemical processing and manufacturing.
Extractant for—
Alkaloids of coffee, bitter principles, saponins, sugar.
Hydrogen ion regulating agent (U. S. 1875368) in—
Securing higher yields of volatile acids from fermentation of corncob residues.
Hydrogenating agent for—
Unsaturated hydrocarbons.
Neutralizing agent for—
Acids.
Process material in—
Concentrating mixtures of butane and butylene (U. S. 1866800).
Purification of arylamides of 2:3-hydroxynaphthoic acid (U. S. 1890201).
Purifying hydrogen by removal of carbon dioxide, carbon monoxide, carbonyl sulphide.
Process material in making—
Basic magnesium carbonate (Brit. 413869).
Calcium chloride (Solvay process).
4:4'-Diaminodiphenyl ether (U. S. 1890256).
Dispersing agents, emulsifying agents, hydrocyanic acid, inorganic chemicals, light fluffy dolomite (U. S. 1975213), metallo-organic compounds, organic chemicals, sodamide, sodium bicarbonate, sodium carbonates, sodium hydroxide (Solvay process).
Suspending agents, thiourea (U. S. 1889959), uranium oxide, urea, wetting agents.
Reactant in making—
Aldehyde-ammonias, amides, amides of higher fatty acids (Brit. 406691), amidines, amines, amino acids, amino compounds, aminoaryl alkyl and aminodiaryl ethers (U. S. 1932653), cyanogen compounds, diamines, nitriles, solvay process alkalies.
Source of—
Hydrogen, nitrogen.
Starting point in making—
Ethanolamines from ethylene oxide (U. S. 1904013), nitric acid, nitrogen oxides, nitrous anhydride for use in sulphuric acid manufacture.
Starting point in making ammonium compounds such as the—
Acetate, arsenate, benzoate, bicarbonate, bichromate, bifluoride, binoxalate, bisulphate, bisulphite, bitartrate, borate, bromide, camphorate, carbamate, carbonate, chloride, chromate, citrate, cuprate, cyanate, fluoride, hypophosphite, iodide, metavanadate, molybdate, nitrate, oxalate, perchlorate, persulphate, phosphate, phosphomolybdate, phosphotungstate, picrate, salicylate, sulphate, sulphate-nitrate, sulphide, sulphocyanate, tartrate, thiocyanate, thiosulphate, tungstate, valeriate.

Temperature reaction controller (Brit. 444936) for—
Destructive hydrogenation treatments of carbonaceous materials.

Cosmetic
Ingredient of—
Permanent wave preparations.
Reagent in—
Treatment of hair with hydrogen peroxide.
Saponification agent in—
Greaseless creams and lotions.

Dye
Process material in making—
Anilin colors, azo dyes (Brit. 388332), lakes.
Triaminohydroxyanthraquinone from leuco-1:4:5:8-tetrahydroxyanthraquinones or 1:4-diamino-5:8-dihydroxyanthraquinone (Brit. 396976).
Purifying agent in making—
Paranitroanilin from nitrochlorobenzene (U. S. 1903030).
Reagent for—
Separating paranitroanilin from nitrodiphenylamines (U. S. 1903030).

Dry Cleaning
Stain and spot remover.

Explosives and Matches
Starting point in making—
Ammonium nitrate, ammonium picrate.

Fats, Oils, and Waxes
Extractant for—
Bitter principles, saponins.
Neutralizing agent for—
Acids.
Reactant in making—
Amides of higher fatty acids (Brit. 406691).
Remover of free fatty acids from—
Fats, oils, waxes.
Source of—
Hydrogen, nitrogen.

Fertilizer
Fertilizing agent.
Ingredient of—
Fertilizer mixtures.
Starting point in making—
Ammonium sulphate-nitrate, ammoniated ammonium nitrate, ammoniated mixed fertilizers, ammoniated sodium nitrate, ammoniated superphosphates, ammoniated urea, ammonium nitrate, ammonium phosphate, ammonium sulphate, fertilizer materials, urea.

Firefighting
Phosgene formation preventer (Brit. 319320) in—
Carbon tetrachloride fire-extinguishers.
Starting point in making—
Ammonium phosphate and other salts used as fireproofing agents.

Food
Extractant for—
Alkaloids of coffee, bitter principles, saponins, sugar.
Fumigant for—
Cheese storerooms (miticide).
Ingredient of antiseptic ice (Brit. 408696) for preserving—
Fish, fruit, vegetable.
Process material in—
Bakery and confectionery operations.

Glue and Gelatin
Hydrolyzing agent for—
Gelatin, glue.

Ink
Ingredient of—
Inks, laundry marking inks.
Starting point in making—
Ink preparations.
Tannate and gallotannate and vanadate salts.

Insecticide and Fungicide
Exterminant for—
Texas cotton root rot (sclerotia).

Leather
Ammoniating agent in making—
Mothproofing agents for pelts and furs.
Protectives (polychlorocresols, polychloronaphthols, polychlorophenols) against parasites, molds, and red spots on hides.
Curing agent in making—
Leather.
Mold preventive in—
Tan liquors.
Slime preventive in—
Tan liquors.

Ammonia (Continued)

Lubricant
Ammoniating agent in making—
Emulsifying or suspending agents for lubricants.

Mechanical
Rust inhibitor for—
Idle steam boilers (reacts with the remaining moisture to form compounds which act as rustproofing agents).
Softening agent (with disodium phosphate) for—
Boiler feed waters (claimed to give a higher softening effect at a lesser cost than by using sodium carbonate and trisodium phosphate and to eliminate the undesirably high alkalinity of the water in the boiler and the acidity of the condensate).

Metallurgical
Annealing gas in—
Bright-annealing coldrolled strip metal (used in cracked form).
Case-hardening agent for—
Iron, steel.
Corrosion-resistance augmenter for—
Aluminum and its alloys.
Light metals and their alloys.
Magnesium and its alloys.
Decoppering agent for—
Iron.
Denickeling agent for—
Iron.
Nitriding agent for—
Steel.
Solvent for metallic salts in making—
Electrolytes for electrolytic recovery of metals.

Miscellaneous
Cleansing agent for—
Factory purposes, household purposes.
Ingredient of—
Coating solution for capsules, tablets, and the like (U. S. 1907203).
Sealing compositions (U. S. 1904445).
Remover of—
Tackiness of coating in oilskin manufacture (used in conjunction with shellac and water).
Reagent for—
Removing ink from printed paper by alternate treatments with oleic acid.
Source of—
Hydrogen, nitrogen.
Thawing agent for—
Frozen water pipes (a cylinder of ammonia is hooked up to taps; the heat generated by the dissolution of the ammonia and the rapid diffusion of the gas into the ice greatly assists the thawing).

Mining
Flotation reagent for—
Lead sulphide ores.

Paint and Varnish
Ammoniating agent in making—
Mercury antifouling paints.

Petroleum
Process material (U. S. 1899314) in—
Desulphurizing low-boiling point hydrocarbon oils.
Purifying agent for—
Hydrocarbons.
Source of—
Hydrogen, nitrogen.
Temperature reaction controller (Brit. 444936) for—
Destructive hydrogenation treatments of carbonaceous materials.

Pharmaceutical
In compounding and dispensing practice.
Reagent in making—
Insulin.

Photographic
Developing agent (Brit. 390616) in making—
Light-sensitive films for multicolor pictures.
Reactant in—
Development of latent images.
Solubilizing agent (U. S. 1901441) for—
Silver sulphate in making "dull emulsions."

Rayon
Buffer in—
Bleaching acetate rayon with hydrogen peroxide.
Contamination restrainer (U. S. 1932789) in—
Desulphurizing of rayon in package form.
Starting point in making—
Cuprammonium solutions in rayon manufacture.

Refrigeration
Refrigerant.

Resins
Accelerator in—
Phenol condensation process.
Catalyst in making—
Urea-furfural resins.
Process material in making—
Resins.
Remover (U. S. 1900132) of—
Free fatty acids from natural resins.

Rubber
Coagulation preventer (Brit. 393844) of—
Concentrated latex in making dispersions or emulsions of rubber latex.
Extractant (U. S. 1931002) for—
Water-soluble material in rubber products.
Improver (U. S. 1931002) of rubber's resistance to—
Electricity, water.
Reactant (U. S. 1896054) in—
Coagulation of rubber latex.
Viscosity promoter (Brit. 399394) in making—
Highly viscous rubber dispersions capable of being shaped, sprayed, or brush-applied (used with sodium silicate).

Soap
Alkali base in—
Ammonia soaps.
Extractant for—
Saponins.
Neutralizing agent for—
Acids.
Reactant in making—
Amides of higher fatty acids (Brit. 406691).
Source of—
Hydrogen.

Tobacco
Extractant for—
Nicotine.

Textile
Coherence increaser (U. S. 1899513) for—
Silk filaments.
Elasticity increaser (U. S. 1899513) for—
Silk filaments.
Extractant for—
Saponins.
Process material in—
Bleaching operations, calico printing, degumming processes, desizing operations, dyeing processes, retting operations, scouring processes, waterproofing processes.
Reagent (U. S. 1899513) for—
Cleansing raw silk in the cocoon.

Water and Sanitation
Process material (U. S. 1892972) in—
Removing aldehydes and ketones.
Water-purification agent.

Ammoniacal Magnesium Cyanide

Agriculture
Fumigant for—
Plants.

Ammoniated Soap of Palm and Olive Oils

Miscellaneous
As a wetting agent (Brit. 411908).
For uses, see under general heading: "Wetting agents."

Ammonium Acetate

Synonyms: Acetate of ammonia.
Latin: Acetas ammoniae, Ammonium aceticum.
French: Acétate ammoniaque, Acétate d'ammonium.
German: Ammoniumacetat, Ammoniumazetat, Essigsäuresammoniak, Essigsäuresammonium.

Analysis
Reagent in determining—
Iron and lead.
Reagent in separating—
Lead sulphate and calcium sulphate from one another and from barium sulphate and strontium sulphate.

Chemical
Reagent in making—
Chloralamide, methyl cyanide, succinimide.
Various other organic chemicals.

Ammonium Acetate (Continued)
Starting point in making—
Ammonium carbonate, ammonium chloride.
Other ammonium salts.

Insecticide
Reagent (Brit. 258623) in making—
Chloramine insecticides and germicides.

Pharmaceutical
In compounding and dispensing practice.

Textile
Mordant in—
Dyeing and printing.
Liquors and pastes used in dyeing and printing fast shades on textiles (Brit. 247694).

Ammonium Acid Adipinate

Leather
Buffer (Brit. 444184) in—
Obtaining level dyeings with acid or substantive dyes (the dyed effects are said to have great resistance to soap and alkalies).

Ammonium Acid Diglycollate

Leather
Buffer (Brit. 444184) in—
Obtaining level dyeings with acid or substantive dyes (the dyed effects are said to have great resistance to soap and alkalies).

Ammonium Acid Phthalate

Leather
Buffer (Brit. 444184) in—
Obtaining level dyeings with acid or substantive dyes (the dyed effects are said to have great resistance to soap and alkalies).

Ammonium Acid Saccharate

Leather
Buffer (Brit. 444184) in—
Obtaining level dyeings with acid or substantive dyes (the dyed effects are said to have great resistance to soap and alkalies).

Ammonium Alginate
Synonyms: Alginate of ammonia.
French: Alginate ammoniaque, Alginate d'ammonium.
German: Alginsäuresammoniak, Alginsäuresammonium, Ammoniumalginat.
Spanish: Alginato de amoniaco.
Italian: Alginato d'ammonio.

Ceramics
Ingredient of—
Compositions used for waterproofing of various ceramic wares.

Chemical
Emulsifying agent in making—
Dispersions of various chemicals.
Ingredient of—
Various chemical liquids (added for the purpose of increasing their viscosity).
Reagent in treating—
Various chemical liquids, as well as solutions of pharmaceutical products for the purpose of purifying and clarifying them (French 570636).
Stabilizer in—
Emulsions of various chemicals and chemical products.
Starting point in making—
Iodinated pharmaceutical products.

Construction
Ingredient of—
Compositions used for treating cement and concrete for the purpose of preventing the deterioration when exposed to the action of alkalies and seawater.
Waterproofing compositions, used for treating plaster of Paris, wallboard, cement, stucco, concrete.

Fats and Oils
Stabilizer in—
Emulsions of various animal and vegetable fats and oils.

Fuel
Binder in—
Compositions used for fuel briquettes and containing coal dust (used in place of pitch).
Non-smoking fuel briquettes (burns without developing large amounts of smoke, as do the usual binders employed for this purpose).

Glues and Adhesives
Ingredient (French 563726) of—
Adhesive preparations.
Reagent in treating—
Solutions of gelatin, glue, and other adhesives for the purpose of purifying and clarifying them.

Ink
Ingredient (French 563726) of—
Printing inks (added for the purpose of thickening the product).
Various inks.

Leather
Ingredient of—
Compositions used for sizing leathers (added to replace starch and gum tragacanth) (French 563726).
Compositions, containing various fatty substances, used in the preparation of emulsions for tanning and tawing leather (French 533465).

Mechanical
Ingredient of—
Compositions used for covering steel tubes.
Compositions, containing sodium carbonate, used as boiler compounds (added for the purpose of improving the water-softening properties of the sodium carbonate).

Metallurgical
Binder (French 518037) in—
Compositions, containing graphite, lampblack, and antiseptics, used for the purpose of repairing metallurgical furnaces and ovens.

Miscellaneous
Binder (French 518037) in—
Preparations, containing graphite, lampblack, and antiseptics, used for repairing stoves.
Emulsifying agent in making—
Emulsions of various products.
Ingredient of—
Antigrease coatings (French 563726).
Compositions used for treating rope and twine.
Compositions used for waterproofing purposes.
Binder in—
Compositions containing powdered mica, asbestos, coal, carbon, graphite, minerals, and the like.
Compositions used for sizing purposes (used in place of starches and gum tragacanth, giving a size of improved elasticity and more transparent).
Stabilizer in—
Emulsions of various substances.

Paint and Varnish
Ingredient (French 563726) of—
Compositions, used for treating interior walls and ceilings.
Various paints, lacquers, and enamels.

Paper
Binder (French 563726) in—
Sizing compositions (used in place of starches and gum tragacanth to give a more elastic and more transparent product).
Ingredient of—
Compositions used for finishing paper.
Compositions used for waterproofing pulp and paper products.
Compositions containing wood flour.
Reagent in treating—
Waste liquors and the like for the purpose of purifying and clarifying them.

Petroleum
Ingredient of—
Emulsions of petroleum and petroleum distillates (added for the purpose of securing better dispersion).
Stabilizer in—
Emulsions of petroleum and petroleum distillates.

Plastics
Binder in—
Various plastic compositions containing such substances as horn, ebonite, celluloid, ivory, bone, shell, galalith, formaldehyde-phenol condensation products, ureaformaldehyde condensation products, and other artificial resins.

Rubber
Ingredient of—
Products obtained with rubber latex.

Soap
Ingredient of—
Bleaching preparations, detergent preparations.

Ammonium Alginate (Continued)
Sugar
Defecating agent in—
Refining sugar.
Reagent in—
Clarifying and purifying liquors in beet sugar refining.
Textile
——, *Dyeing*
Ingredient of—
Various dye baths (added for the purpose of increasing the dispersion of the dyestuff).
Mordant in—
Various dyeing processes.
——, *Finishing*
Ingredient of—
Compositions used for the waterproofing of fabrics, this treatment being followed by one in a solution of a metallic salt.
Compositions used for treating woolen fabrics to protect them against decomposition (French 518059).
Compositions used for sizing yarns and fabrics (added in place of starch and gum tragacanth for the purpose of obtaining a more elastic and more transparent size) (French 563726).
——, *Printing*
Mordant in—
Printing various fabrics.
Thickener in—
Printing pastes (used in place of gum tragacanth and British gum).
Waxes and Resins
Emulsifying agent in making—
Dipsersions of waxes and resins, both artificial and natural (added for the purpose of increasing the dispersion of these substances).
Stabilizer in—
Emulsions containing both natural and artificial resins and waxes.
Water and Sanitation
Reagent in—
Treating waste waters and the like for the purpose of purifying and clarifying them.
Wine
Clarifying agent for—
Treating wines.

Ammonium Alum
Synonyms: Alum, Aluminum-ammonium sulphate, Ammonia alum.
Latin: Alumen ammoniacale, Alumen.
French: Alun d'ammoniaque, Sulfate d'aluminium et d'ammonium, Sulfate double d'alumine et d'ammoniaque, Sulfate double d'aluminium et d'ammonium.
German: Ammoniakolaun.
Spanish: Allume ammoniacle, Alumbre de ammoniaco, Solfato de amoniaco y de aluminio.
Italian: Solfato di alluminio e d'ammonio.
Analysis
Reagent for—
Identifying coloring matters, making staining solutions.
Cement
Ingredient (French 666186) of—
Sorrel cement.
Reagent in—
Hardening plaster casts.
Ceramics
Reagent in making—
Ceramic products.
Chemicals
Clarifying agent in chemical processes and in purifying organic and inorganic chemical products.
Ingredient of catalytic mixtures used in the manufacture of—
Acenaphthylene, acenaphthaquinone, bisacenaphthylidenedione, naphthaldehydic acid, naphthalic anhydride, and hemimellitic acid from acenaphthene (Brit. 295270).
Acetaldehyde from ethyl alcohol (Brit. 281307).
Acetic acid from ethyl alcohol (Brit. 281307).
Alcohols from aliphatic hydrocarbons (Brit. 281307).
Aldehydes or acids by the reduction of the corresponding esters (Brit. 306471).
Aldehydes and acids from toluene, orthochlorotoluene, orthonitrotoluene, orthobromotoluene, parachlorotoluene, paranitrotoluene, parabromotoluene, metachlorotoluene, metabromotoluene, metanitrotoluene, dichlorotoluenes, dibromotoluenes, dinitrotoluenes, chloronitrotoluenes, chlorobromotoluenes, bromonitrotoluenes (Brit. 295270).
Aldehydes and acids from xylenes, pseudocumenes, mesitylenes, and paracymene (Brit. 281307).
Alphacampholide from camphoric acid by reduction (Brit. 306471).
Alphanaphthaquinone from naphthalene (Brit. 281307).
Anthraquinone from naphthalene (Brit. 281307).
Benzaldehyde and benzoic acid from toluene (Brit. 281307).
Benzoquinone from phenanthraquinone (Brit. 281307).
Benzyl alcohol from benzaldehyde by reduction (Brit. 306471).
Benzyl alcohol or benzaldehyde or benzyl phthalide by the reduction of phthalic anhydride (Brit. 306471).
Butyl alcohol by the reduction of crotonaldehyde (Brit. 306471).
Diphenic acid from ethyl alcohol (Brit. 281307).
Ethyl alcohol by the reduction of acetaldehyde (Brit. 306471).
Fluorenone from fluorene (Brit. 295270).
Formaldehyde by the reduction of methane or methanol (Brit. 306471).
Formaldehyde by the reduction of carbon dioxide or carbon monoxide (Brit. 306471).
Hydroxyl compounds by the reduction of anthraquinone, benzoquinone, and the like (Brit. 306471).
Isopropyl alcohol by the reduction of acetone (Brit. 306471).
Maleic acid and fumaric acid by the oxidation of toluene, benzene, phenols, tar phenols, or furfural, or from benzoquinone or phthalic anhydride (Brit. 295270).
Methane by the reduction of carbon dioxide or carbon monoxide (Brit. 306471).
Methanol by the reduction of carbon dioxide or carbon monoxide (Brit. 306471).
Naphthaldehydic acid, acenaphthaquinone, bisacenaphthylidenedione from acenaphthylene (Brit. 281307).
Phenanthraquinone from phenanthrene or diphenic acid (Brit. 295270).
Primary alcohols by the reduction of the corresponding aldehydes (Brit. 306471).
Propionic acid and butyric acid and higher alcohols, ketones, and acids by the reduction of carbon dioxide or carbon monoxide (Brit. 306471).
Reduction products of ketones, aldehydes, esters, ethers, alcohols, and other organic compounds which contain oxygen (Brit. 306471).
Salicylic acid and salicylic aldehyde from cresol (Brit. 306471).
Secondary butyl alcohol by the reduction of methylethyl ketone (Brit. 306471).
Valeryl alcohol by the reduction of valeraldehyde (Brit. 306471).
Vanillin and vanillic acid from eugenol or isoeugenol (Brit. 295270).
Ingredient (Brit. 304640) of catalytic preparations used in the manufacture of various aromatic and aliphatic amines, such as—
Alphanaphthylamine from alphanitronaphthalene.
Amines from aliphatic nitro compounds, such as alkyl nitriles, or nitromethane.
Amylamine from pyridin.
Anilin, azo-oxybenzene, azobenzene, and hydrazobenzene from nitrobenzene by reduction.
Aminophenols from nitrophenols.
3-Aminopyridin from 3-nitropyridin.
Amino compounds from the corresponding nitroanisoles.
Amines from oximes, Schiff's base, and nitriles.
Cyclohexamine, dicyclohexamine, and cyclohexylanilin from nitrobenzene.
Piperidin from pyridin.
Pyrrolidin from pyrrol.
Tetrahydroquinolin from quinolin.
Starting point in making—
Aluminum acetoformate.
Aluminum sulphoacetate.
Pure alumina for making rubber appliances and artificial corundum and artificial precious stones.
Construction
Hardening agent in—
Plastering.

Ammonium Alum (Continued)
Ingredient of—
Concrete mixtures and plastering compositions containing Portland cement (Brit. 258320).

Dye
Reagent in making—
Color lakes, various dyestuffs.

Electrical
Ingredient (Brit. 249016) of—
Electrolytes for lead plate storage batteries.

Explosives
Ingredient of—
Composition used in making matches.

Food
Ingredient of—
Baking powders.
Reagent in making—
Confectionery, oleomargarin.

Glues and Adhesives
Ingredient of—
Marble cements, porcelain cements.
Vegetable glues and adhesive preparations.

Ink
Reagent in making—
Writing inks.

Jewelry
Reagent in making—
Artificial precious stones.

Leather
Reagent in—
Tanning, tawing.

Miscellaneous
For hardening specimens for microscopy.
Ingredient of—
Coating compositions (U. S. 1787425), fire-extinguishing compositions, fireproofing compositions, linings for safes, polishing compositions (U. S. 1774221), waterproofing compositions.
Reagent in—
Galvanostegy.
Taxidermy.
Treating objects of art or articles used for decorating purposes, which are made from plaster mixed with a solution of dextrin and molded into shape and then treated.

Paint and Varnish
Reagent in making—
Pigments.

Paper
Ingredient of—
Sizing compositions.

Pharmaceutical
Suggested for use as astringent, emetic, styptic, irritant, purgative, and diuretic.

Photographic
Reagent for—
Hardening the gelatin coating of plates, films, and papers.
Recovering silver from dilute photographic emulsions (French 483100).

Printing
Reagent in—
Process engraving and lithography.

Sugar
Reagent in—
Clarifying sugar juices.

Textile
——, *Dyeing*
As a mordant.

——, *Finishing*
Ingredient of—
Fireproofing compositions.
Preservative compositions for treating knitted fabrics.
Waterproofing compositions.

——, *Manufacturing*
Reagent (French 601297) for—
Treating acetate rayon to preserve its luster.

——, *Printing*
As a mordant.

Woodworking
Ingredient of—
Fireproofing compositions, waterproofing compositions.

Ammonium Betatetrahydronaphthalenesulphonate
French: Bétatétrahydronaphthalènesulphonate ammoniaque, Bétatétrahydronaphthalènesulphonate d'ammonium.
German: Ammoniumbetatetrahydronaphtalinsulfonat, Betatetrahydronaphtalinsulfonsäuresammoniak, Betatetrahydronaphtalinsulfonsäuresammonium.

Miscellaneous
As an emulsifying agent (Brit. 371293).
For uses, see under general heading: "Emulsifying agents."

Ammonium Bicarbonate
Synonyms: Ammonium acid carbonate.
French: Bicarbonate ammoniaque, Bicarbonate d'ammonium.
German: Kohlensaeuressaeureammonium, Kohlenstoffsaeuressaeureammonium.

Chemical
Starting point in making—
Ammonium carbonate and other ammonium salts.

Dye
Reagent in making—
Various dyestuffs.

Food
Leavening agent in making—
Pastries and in baking generally.

Miscellaneous
Ingredient of—
Fire-extinguishing preparations.

Pharmaceutical
In compounding and dispensing practice.

Textile
——, *Miscellaneous*
Defatting agent in treating—
Yarns and fabrics.

Ammonium Borate
French: Borate ammoniaque, Borate d'ammonium.
German: Borsäuresammoniak, Borsäuresammonium.
Spanish: Borato de amoniaco.
Italian: Borato d'ammonio.

Chemical
Starting material (Brit. 253623) in—
Manufacture of chloramines.

Electrical
Ingredient of—
Electrolytic condensers.

Miscellaneous
As a general fire-retarding agent.

Pharmaceutical
In compounding and dispensing practice.
Suggested for use in the treatment of urinary calculi.

Textile
Ingredient of—
Compositions for impregnating and fireproofing fabrics.

Woodworking
Ingredient of
Compositions for impregnating and fireproofing wood.

Ammonium Bromide
Latin: Ammonium bromatum.
French: Bromhydrate d'ammonium, Bromure d'ammonium.
German: Ammoniumbromid, Bromammon.
Spanish: Bromuro amoniaco, Bromuro de amonio.
Italian: Bromuro di ammonio.

Analysis
As a reagent.

Metallurgical
Ingredient of—
Soldering flux, containing also borax, zinc chloride, sodium bromide, and aluminum bromide (French 672746).
Soldering flux for aluminum, containing also anhydrous zinc chloride and sodium fluoride (French 642778).

Miscellaneous
As a fireproofing medium (German 355107, 390840; Australian 100696; U. S. 1612104).
Ingredient of—
Fireproofing composition for theater curtains and scenery, containing also ammonium sulphate and other products (French 665464).

Ammonium Bromide (Continued)
Mixture with ammonium phosphate for fireproofing all kinds of products (French 594784).

Pharmaceutical
Suggested for use as—
Sedative.

Photographic
Precipitant for—
Silver salts.
Reagent in making—
Gelatin-bromide emulsions.

Printing
Reagent in—
Lithography, process engraving.

Textile
Reagent (French 602297) for—
Conserving luster, transparency, and general appearance of cellulose acetate fabrics when subjected to hot or boiling liquids.

Ammonium Butylnaphthalenesulphonate
French: Butylenaphthalène-sulfonate d'ammoniaque, Butylenaphthalène-sulfonate d'ammonium.
German: Ammonbutylnaphtalinsulfonat, Ammoniumbutylnaphtalinsulfonat, Butylnaphtalinschwefelsäuresammoniak, Butylnaphtalinschwefelsäuresammonium, Butylnaphtalinsulfonsäuresammoniak, Butylnaphtalinsulfonsäuresammonium.
Spanish: Butilnaftalene-sulfonato de amonio.
Italian: Butilnaftalenesolfonato di ammonio.

Fire Prevention
Ingredient (French 665464; Brit. 302172) of—
Fire-extinguishing compositions (aqueous emulsion with carbon tetrachloride).
Fire-extinguishing composition, containing also aluminum sulphate and sodium bicarbonate.

Ammonium Butyraldehydrate
Synonyms: Ammonium-butyraldehyde hydrate.
French: Butyraldéhydrate d'ammoniaque, Butyraldéhydrate d'ammonium.
German: Butyraldehyd-ammoniak.
Spanish: Butiraldehidato de amoniaco.
Italian: Butiraldeidato di ammoniaco.

Rubber
Accelerator (French 567987) in—
Vulcanizing processes (distinguishing itself by its liquid state).

Ammonium Butyrate
French: Butyrate ammoniaque, Butyrate d'ammonium.
German: Ammoniumbutyrat, Buttersaeuresammonium.

Fats and Oils
Emulsifying agent (Brit. 277357) for various fats and oils.

Leather
Emulsifying agent (Brit. 277357) in making—
Dressings containing mineral oils and aliphatic alcohols, lubricating compositions.

Petroleum
Emulsifying agent (Brit. 277357) in making—
Motor fuel compositions containing mineral oil distillates and aliphatic alcohols.
Stable emulsions with mineral oils and aliphatic alcohols.

Soap
Emulsifying agent (Brit. 277357) in making—
Detergent compositions, soap compositions, textile soaps.

Textile
——, *Finishing*
Emulsifying agent (Brit. 277357) in making—
Scouring compositions.

Ammonium Carbamate
Synonyms: Ammonium carbaminate.
French: Carbamate d'ammonium, Carbamate ammonique.
German: Ammoniumcarbaminat, Carbaminsäuresammonium.
Spanish: Carbaminato de amonio.
Italian: Carbamato di ammonio.

Chemical
Starting point in making—
Barium carbamate (French 687091).
Calcium carbamate (French 687091).
Carbamates from alkalies and alkaline earths (French 756653).
Cyanamides from alkaline earths (French 702754).
Disodium carbamide (French 753038).
Metallic oxides (French 696374).
Urea (French 687188).
Zinc carbamate (French 687091).

Fertilizer
Starting point (French 685276) in making—
Two-plantfood fertilizer from phosphoric anhydride.

Ammonium Carbonate
Synonyms: Ammonia crystal, Ammonium sesquicarbonate, Carbonate of ammonia, Hartshorn salts, Sal volatile, Sesquicarbonate of ammonia, Volatile alkali, Volatile salt.
Latin: Ammonium carbonatum.
French: Carbonate ammoniaque, Carbonate d'ammonium, Sel volatile d'angelterre, Sesquicarbonate ammoniaque.
German: Ammoniumkarbonat, Ammoniumsesquikarbonat, Kohlensäuresammoniak, Kohlenstoffsäuresammoniak, Salmiaksalz.
Spanish: Carbonato de amoniaco.
Italian: Carbonato d'ammonio.

Analysis
Reagent in separating—
Arsenic from antimony.
Chlorine from bromine and iodine.
Group of alkali earths from magnesium and the alkalies.
Magnesium from lithium.

Chemical
Reagent in making—
Acetoacetic acid, aminoacetic acid, betaresorcylic acid, 2:6-dihydroxybenzoic acid, dimethylpiperazin, protocatechuic acid, pyrocatecholorthocarboxylic acid, intermediates, synthetic aromatic chemicals, synthetic organic and pharmaceutical chemicals.
Reagent in treating—
Uranium salts to obtain radio-active preparations.
Starting point in making—
Ammonium acetate, ammonium iodide, ammonium sulphocyanide, ammonium thiosulphate.
Barium carbonate and carbonates of other metals and alkaline earth metals.
Ingredient of catalytic mixtures containing mixed catalysts and used in the manufacture of—
Acenaphthylene, acenaphthaquinone, bisacenaphthylidenedione, naphthaldehydic acid, naphthalic anhydride, and hemimellitic acid from acenaphthene (Brit. 292570).
Acetaldehyde from ethyl alcohol (Brit. 281307).
Acetic acid from ethyl alcohol (Brit. 281307).
Alcohols from aliphatic hydrocarbons (Brit. 281307).
Aldehydes or acids by the reduction of the corresponding esters (Brit. 306471).
Aldehydes and acids from toluene, orthochlorotoluene, orthonitrotoluene, orthobromotoluene, metachlorotoluene, metabromotoluene, metanitrotoluene, parachlorotoluene, paranitrotoluene, parabromotoluene, dichlorotoluenes, dibromotoluenes, dinitrotoluenes, chlorobromotoluenes, chloronitrotoluenes, bromonitrotoluenes (Brit. 295270).
Aldehydes and acids from xylenes, pseudocumenes, mesitylenes, and paracymene (Brit. 281307).
Alphacampholide from camphoric acid by reduction (Brit. 306471).
Alphanaphthaquinone from naphthalene (Brit. 281307).
Benzaldehyde and benzoic acid from toluene (Brit. 281307).
Anthraquinone from naphthalene (Brit. 281307).
Benzoquinone from phenanthraquinone (Brit. 281307).
Benzyl alcohol from benzaldehyde by reduction (Brit. 306471).
Benzyl alcohol or benzaldehyde or benzyl phthalide by the reduction of phthalic anhydride (Brit. 306471).
Butyl alcohol by the reduction of crotonaldehyde (Brit. 306471).
Diphenic acid from ethyl alcohol (Brit. 306471).
Ethyl alcohol by the reduction of acetaldehyde (Brit. 306471).
Fluorenone from fluorene (Brit. 295270).
Formaldehyde by the reduction of methane or methanol (Brit. 306471).
Formaldehyde by the reduction of carbon dioxide or carbon monoxide (Brit. 306471).

Ammonium Carbonate (Continued)

Hydroxyl compounds by the reduction of anthraquinone, benzoquinone, or the like (Brit. 306471).
Isopropyl alcohol by the reduction of acetone (Brit. 306471).
Maleic acid and fumaric acid by the oxidation of toluene, benzene, phenols, tar phenols, or furfural, or from benzoquinone or phthalic anhydride (Brit. 295270).
Methane by the reduction of carbon dioxide or carbon monoxide (Brit. 306471).
Methanol by the reduction of carbon dioxide or carbon monoxide (Brit. 306471).
Naphthaldehydic acid, acenaphthaquinone, biacenaphthylidenedione from acenaphthylene (Brit. 281307).
Phenanthraquinone from phenanthrene or diphenic acid (Brit. 295270).
Primary alcohols by the reduction of the corresponding aldehydes (Brit. 306471).
Propionic acid and butyric acid and higher alcohols, ketones, and acids by the reduction of carbon dioxide or carbon monoxide (Brit. 306471).
Reduction products of ketones, aldehydes, esters, ethers, alcohols, and other organic compounds, which contain oxygen (Brit. 306471).
Salicylic acid and salicylic aldehyde from cresol (Brit. 306471).
Secondary butyl alcohol by the reduction of methylethyl ketone (Brit. 306471).
Valeryl alcohol by the reduction of valeraldehyde (Brit. 306471).
Vanillin and vanillic acid from eugenol or isoeugenol (Brit. 295270).

Ingredient (Brit. 304640) of catalytic preparations used in the manufacture of various aromatic and aliphatic amines, including—
Alphanaphthylamine from alphanitronaphthalene.
Amines from aliphatic nitro compounds, such as alkyl nitriles or nitromethane.
Amylamine from pyridin.
Anilin, azo-oxybenzene, azobenzene, and hydrazobenzene from nitrobenzene by reduction.
Aminophenols from nitrophenols.
3-Aminopyridin from 3-nitropyridin.
Amino compounds from the corresponding nitroanisoles.
Amines from oximes, Schiff's base, and nitriles.
Cyclohexamine, dicyclohexamine, and cyclohexylanilin from nitrobenzene.
Piperidin from pyridin.
Pyrrolidin from pyrrol.
Tetrahydroquinolin from quinolin.

Ceramics
Reagent in making various ceramic products.

Construction
Ingredient (Brit. 258320) of—
Concrete mixtures and plastic compositions containing Portland cement.

Dye
Ingredient of—
Casein colors.
Reagent in making—
Carriers for dyestuffs (French 599868).
Various dyestuffs, including fuchsin S.

Explosives
Ingredient of—
Explosive mixtures containing picric acid.
Smokeless powders (U. S. 1589237).

Food
Ingredient of—
Baking powders.
Reagent in separating—
Constituents of the cocoa bean.
Source of—
Carbon dioxide in baking (used in place of sodium bicarbonate).

Glass
Ingredient of—
Compositions for cleaning glass and glass articles (U. S. 1643251).

Glues and Adhesives
Ingredient of—
Casein glues and other adhesive preparations.

Leather
Ingredient of—
Mordant compositions used in the dyeing of glove leather and other types of leather.
Tanning compositions.

Miscellaneous
Ingredient of—
Fire-extinguisher compositions.
Mordant compositions for dyeing various materials.

Paper
Ingredient of—
Mordant preparations used in dyeing paper, cardboard, and paper and pulp compositions.

Perfume
Ingredient of—
Cosmetics, smelling salts.

Pharmaceutical
In compounding and dispensing practice.

Rubber
Ingredient of—
Rubber batches (added for the purpose of producing a gas at the temperature of vulcanization and inflating the rubber mass, in the manufacture of rubber balloons, balls, inflated toys, and similar articles).

Soap
Ingredient of—
Cleansing powders, dry-cleaning soaps.

Textile
——, *Dyeing*
Mordant in—
Baths used in dyeing various fabrics.

——, *Finishing*
Ingredient of—
Compositions for degreasing woolens.
Washing compositions.

Wine
Accelerator of—
Fermentation of grapes in making wines.

Ammonium Cetylsulphate

Metallurgical
Ingredient (U. S. 1974436) of—
Flux for use in soft soldering, containing also zinc chloride and alcohol.

Ammonium Chloride

Synonyms: Ammonium hydrochloride, Chloride of ammonia, Hydrochlorate of ammonia, Muriate of ammonia, Sal ammoniac.
Latin: Ammoniae hydrochloras, Ammoniae murias, Ammonium chloratum, Ammonium hydrochloratum, Ammonium muriaticum, Chloruretum ammoniacum, Sal ammoniacum.
French: Chlorhydrate d'ammoniaque, Chlorure d'ammonium, Hydrochlorate d'ammonique, Muriate d'ammoniaque, Sel ammoniac.
German: Ammoniumchlorid, Chlorammonium, Reiner salmiak, Salmiak.
Spanish: Cloruro amoniaco, Sal amoniaco.
Italian: Chlorurio d'ammonio, Chloruro ammonico, Sale ammoniaco.

Chemical
Catalyst in making—
Allyl alcohol from glycerin.
Aromatic and aliphatic chemicals by reduction or oxidation.
Process material in making—
Ammonium chlorostannate, ammonium nitrate, ammonium thiosulphate, ammonium-platinum chloride, chloroamine insecticides and germicides (Brit. 253623), sterilizing agents (U. S. 1589237), urea.
Starting point (Brit. 396760) in making organic amines with the aid of—
Alcohol, metacresol, meta-5-xylenol.
Reagent in making—
Alpha-aminopropionic acid (alpha-alanin), betanaphthylamine-7-sulphonic acid, bismuth salicylate, chromic acid, diethylbenzaldehydeacetal, formamide, intermediate chemicals, isatin, ketone musk, pharmaceutical chemicals, phenylhydroxylamine, synthetic aromatics, triethylamine.

Dye
Reagent in making—
Auramin, auramin G, various dyestuffs.

Ammonium Chloride (Continued)

Electrical
Ingredient of—
Dry batteries, flashlight and annunciator batteries.
Reagent in making—
Electric appliances of various sorts.

Explosives
Ingredient of various explosives and blasting powders (but principally European powders).

Fertilizer
Ingredient of various fertilizer mixtures.

Fuel
As a fuel economizer (U. S. 1618465).
Reagent in—
Manufacture of candles.

Leather
Process material in—
Tanning.

Metallurgical
Flux in—
Soldering, galvanizing, tinning.
Ingredient of—
Baths for nickel, zinc, and platinum plating.

Miscellaneous
Ingredient of—
Compositions for generating heat in waterless hot-water bags.
Rust cement for pipe joints.

Paint and Varnish
Ingredient (U. S. 1580914) of—
Paint and varnish removers.
Reagent in making—
Mars yellow pigment.

Petroleum
Reagent in—
Analysis of oilfield water (determination of lime by the Macfayden method).

Pharmaceutical
Suggested for the treatment of acute hepatitis, bronchitis, catarrhal jaundice, dysmenorrhea, gastric catarrh, glandular enlargements, sciatica.

Soap
Ingredient of—
Scouring powders, washing powders.

Resins and Waxes
Ingredient (Brit. 255692) of—
Solutions used in the manufacture of special grades of phenol-aldehyde resins.

Textile
Protective ingredient of—
Hot liquors used in the treatment of acetate rayon.
Reagent in—
Dyeing and printing of fabrics.

Ammonium Chloroiodide
French: Chloroiodure d'ammonium.
German: Ammoniumchlorjodid, Chlorjodammonium.

Food
Reagent in improving—
Flour (U. S. 1630143).

Ammonium 1:5-Chloronaphthalenesulphonate
French: 1:5-Chloronaphthalènesulfonate ammoniaque, 1:5-Chloronaphthalènesulfonate d'ammonium.
German: Ammonium-1:5-chlornaphtalinsulfonate, 1:5-Chlornaphtalinsulfonsaeuresammonium.

Chemical
Reagent (Brit. 263873) in making—
Aromatic hydrocarbon emulsions, terpene emulsions.

Fats and Oils
Reagent (Brit. 263873) in making—
Emulsions.

Leather
Reagent (Brit. 263873) in making—
Emulsified tanning compositions.

Miscellaneous
Reagent (Brit. 263873) in making—
Washing and cleansing compositions.

Paper
Reagent (Brit. 263873) in increasing the wetting and absorbing properties of—
Blotting paper, cardboard, duplicating paper.

Petroleum
Reagent (Brit. 263873) in making—
Emulsions of oils and distillates.

Resins and Waxes
Reagent (Brit. 263873) in making—
Emulsions.

Textile
——, *Dyeing*
Reagent (Brit. 263873) in making—
Emulsified dye baths.
——, *Finishing*
Reagent (Brit. 263873) in making—
Washing and finishing compositions.
——, *Manufacturing*
Reagent (Brit. 263873) in making—
Wool-carbonizing compositions.
Reagent (Brit. 263873) in increasing—
Absorptive powers of cotton wadding.

Ammonium Chloroplatinnate
Synonyms: Platinum-ammonium chloride.
French: Chloroplatinate d-ammonium.
German: Ammoniumplatinchlorid, Platinsalmiak.
Spanish: Chloroplatinato de amonio.
Italian: Chloroplatinato di ammonio.

Chemical
Catalyst in—
Organic synthesis.

Metallurgical
Electrolyte (in conjunction with citric acid) in—
Platinum plating.
Intermediate product in—
Extraction of platinum from its salts.

Photographic
Toning agent in printing processes.

Ammonium Citrate
French: Citrate ammoniaque, Citrate d'ammonium.
German: Ammoniumcitrat, Ammoniumzitrat, Citronsaeuresammonium, Zitronsaeuresammonium.

Chemical
Reagent (Brit. 253623) in making—
Chloramines.

Fats and Oils
Reagent (Brit. 277357) in making—
Emulsions, lubricating compositions.

Leather
Reagent (Brit. 277357) in making—
Dressing and finishing compositions containing mineral oils and aliphatic alcohols.

Miscellaneous
Ingredient of—
Rustproofing and preventive compositions.

Petroleum
Reagent (Brit. 277357) in making—
Motor fuel compositions containing mineral oils and aliphatic alcohols.
Stable mineral oil emulsions with aliphatic alcohols.

Pharmaceutical
In compounding and dispensing practice.

Soap
Reagent (Brit. 277357) in making—
Detergent compositions, soap compositions, textile soaps.

Textile
——, *Finishing*
Reagent (Brit. 277357) in making—
Finishing and washing preparations.

Ammonium Dibutylnaphthalenesulphonate
French: Dibutylenaphthalènesulfonate ammoniaque, Dibutylenaphthalènesulfonate d'ammonium.
German: Ammoniumdibutylnaphtalinsulfonat, Dibutylnaphtalinsulfonsaeuresammoniak, Dibutylnaphtalinsulfonsaeuresammonium.

Fats and Oils
Starting point (Brit. 279877) in making—
Solvent.

Miscellaneous
Ingredient (Brit. 279877) of—
Bleaching and cleansing compositions for parquet floors.
Washing compositions for various purposes.

Soap
Ingredient (Brit. 279877) of—
Washing and detergent compositions.

Ammonium Dibutylnaphthalenesulphonate (Cont'd)

Textile

——, *Dyeing*

Assist (Brit. 279877) in dyeing—
Woolen yarns and fabrics.

——, *Finishing*

Ingredient (Brit. 279877) of—
Compositions used in finishing textiles.

Ammonium Fluoride

French: Fluorure d'ammoniaque, Fluorure d'ammonium.
German: Ammonfluorid, Fluorammoniak, Fluorammonium.

Agriculture

Improver for—
Sandy soils.

Analysis

Reagent in—
Analytical methods for control, research, and other purposes.

Beverage

As an antiseptic.

Chemical

Reactant in—
Decomposing minerals.
Starting point in making—
Other fluoride salts, various fluoride salts (U. S. 1911004), various fluosilicates (U. S. 1911004 and 1899607).

Glass

As an etching agent.

Miscellaneous

Reagent in—
Carrotting furs.

Textile

Mordant in—
Printing and dyeing.

Ammonium Fluotitanate

Synonyms: Ammonium titanofluoride.
French: Fluotitanate d'ammonium.
German: Ammoniumfluotitanat, Titanfluorwasserstoffsaeuresammonium.

Leather

Ingredient of dyeing liquor used with tinctorial woods.

Ammonium Hypochlorite

French: Hypochlorite d'ammonium.
German: Ammoniumhypochlorit, Unterchloridsaeuresammonium.

Chemical

Reagent in making—
Orthoaminocinnamic acid from orthocyanocinnamic acid (German 440052).

Sanitation

Disinfectant in the treatment of potable waters.

Ammonium Lactate

French: Lactate ammoniaque, Lactate d'ammonium.
German: Ammoniumlactat, Milchsaeuresammonium, Milchsäuresammoniak.
Spanish: Lactico de ammoniaco.
Italian: Lactico d'ammonio.

Chemical

Starting point in making—
Chloramine derivatives (Brit. 253623).

Fats and Oils

Reagent (Brit. 277357) in making—
Emulsified lubricants, emulsions (Brit. 277357).

Leather

Ingredient of—
Dressings containing mineral oils and aliphatic alcohols in emulsified condition (Brit. 277357).
Finishing compositions, tanning compositions.

Metallurgical

Ingredient of—
Bath used for plating lead, tin, and Britannia metal articles.
Reagent in—
Plating nickel on zinc.

Petroleum

Reagent (Brit. 277357) in making—
Stable emulsions containing mineral oils and aliphatic alcohols.
Stable emulsified motor fuel compositions containing mineral oils and aliphatic alcohols.

Soap

Reagent (Brit. 277357) in making—
Emulsified detergents, emulsified soap compositions, emulsified textile soaps.

Ammonium Laurylsulphonate

Miscellaneous

As an emulsifying agent (Brit. 353475).
For uses, see under general heading: "Emulsifying agents."

Ammonium Linoleate

Synonyms: Linoleate of ammonia.
French: Linoléate d'ammoniaque, Linoléate d'ammonium.
German: Leinoelammonium, Leinoelammoniak.

Miscellaneous

As an emulsifying agent.
For uses, see under general heading: "Emulsifying agents."

Ammonium Nitrate

Synonyms: Nitrate of ammonia.
Latin: Ammonii nitras, Ammonio nitras, Nitras ammonicus.
French: Azoate ammoniaque, Azoate d'ammonium, Azotate ammoniaque, Azotate d'ammonium, Nitrate ammoniaque, Nitrate d'ammonium.
German: Ammoniaknitrat, Ammoniumnitrat.
Italian: Nitrato di ammoniaco, Nitrato di emonie.

Chemical

Absorbent for—
Oxides of nitrogen.
Solvent in making—
Red-colored zinc oxide.
Starting point in making—
Nitrous oxide.

Explosives and Matches

Ingredient of—
Dynamites, explosive composition (Brit. 397600), fuel-igniting substances (Brit. 315232), low-density explosive composition (U. S. 1901126), military explosives, pyrotechnic compositions, safety explosives, U. S. permissible explosives of class 1.

Fertilizer

Alone as a fertilizer.
Ingredient of—
Fertilizer compositions.

Refrigeration

Ingredient of—
Freezing mixtures.

Ammonium Octodecylsulphonate

Miscellaneous

As an emulsifying agent (Brit. 360539).
For uses, see under general heading: "Emulsifying agents."

Ammonium Oleate

Synonyms: Oleate of ammonia.
French: Oléate d'ammonium.
German: Ammoniumoleat, Oelsaeuresammonium.

Chemical

Solidifying agent in solid alcohol compositions.

Insecticide

Emulsifying agent in—
Insecticides and weed-killers (Brit. 261241).
Reagent in making—
Insecticides and germicides containing chloramines.

Leather

Ingredient of—
Vermin-proofing applications for leather (Brit. 253993).

Lumbering

Ingredient of—
Vermin-proofing applications for wood (Brit. 253993).

Miscellaneous

Ingredient of—
Mothproofing applications for furs, fabrics and rugs (Brit. 253993).

Oils and Fats

Thickening agent in—
Compounded mineral oil lubricants.

Soap

Ingredient of—
Detergent preparations, dry-cleaner's soaps.

Ammonium Oxalate
Synonyms: Oxalate of ammonia.
French: Oxalate ammoniaque, Oxalate d'ammonium.
German: Oxalsäuresammonium, Oxalsäuresammoniak.

Analysis
Reagent in various processes.

Chemical
Starting point in making—
Antimony oxalate, barium oxalate, calcium oxalate, copper oxalate, iron oxalate, lead oxalate, magnesium oxalate, manganese oxalate, nickel oxalate, potassium oxalate, silver oxalate, sodium oxalate, strontium oxalate, tin oxalate, titanium oxalate, tungsten oxalate, zinc oxalate.

Explosives
Ingredient of—
Permissible dynamites, safety explosives.

Fats and Oils
Reagent (Brit. 277357) in making—
Emulsified lubricants, emulsions of various sorts.

Leather
Reagent in making—
Emulsified finishes and dressings (Brit. 277357).

Petroleum
Reagent in making—
Emulsified motor fuels containing mineral oils and aliphatic alcohols.
Mineral oil emulsions with aliphatic alcohols.

Soap
Reagent in making—
Detergent compositions in emulsified form, emulsified soaps, textile soaps.

Textile
——, *Dyeing and Printing*
Ingredient of—
Padding liquor used in the dyeing and printing of fabrics and yarns with the aid of indigosol O to produce indigo black.

——, *Finishing*
Ingredient of—
Washing and scouring compositions.

Metallurgical
Ingredient of—
Electroplating baths.
Oxidizing agent in the metallurgy of copper.

Miscellaneous
Ingredient of—
Disinfectant preparations used on the hands.
Oxidizing agent in various processes.
Reagent for—
Washing infected yeasts.

Perfumery
Ingredient of—
Hair dyes.

Photographic
Hypo eliminator and reducer.
Ingredient of—
Compositions used in softening prints.

Plastics
Stabilizing agent in making—
Solutions of cellulose in ammoniacal copper oxide.

Soap
Bleaching agent in making—
Fine grades of white soap.

Textile
——, *Bleaching*
Bleach for—
Cottons and other textiles.

——, *Manufacturing*
Stabilizing agent in—
Solutions containing cellulose dissolved in ammoniacal copper oxide, used in the manufacture of cuprammonium rayon.

Ammonium Persulphate
Synonyms: Persulphate of ammonia.
French: Persulfate ammoniaque, Persulfate d'ammonium.
German: Ammoniakpersulfat, Ammoniumpersulfat, Perschwefelsaeuresammoniak, Perschwefelsaeuresammonium.

Analysis
Reagent in the chemical laboratory.

Chemical
Oxidizing agent in making various chemicals.
Reagent in making—
Aldehyde from alcohol, alizarin from hydroxyanthraquinone, anthion, dihydroxybenzoic acid from salicylic acid, ferric salts from ferrous salts, nitrohydroquinone from orthonitrophenol, purpurin from alizarin.
Reagent in—
Introducing hydroxyl group into benzene nucleus.
Starting point in making—
Potassium persulphate, sodium persulphate.

Dye
Oxidizing agent in making—
Synthetic dyestuffs of various groups, such as resoflavin W.

Electrical
Ingredient of—
Battery compositions to produce depolarizing effect.

Fats and Oils
Bleaching agent.
Deodorizer for—
Fish oils, seal oil, stearin, train oil, whale oil.

Food
Bleaching agent in treating—
Flour and meal.
Reagent in—
Improving the quality of flour and meal.
Making starch conversion products for breadmaking (Brit. 282178).

Glues and Adhesives
Bleaching agent.

Ammonium Phosphate
Synonyms: Diammoniumorthophosphate.
French: Phosphate d'ammonium, Phosphate ammoniaque.
German: Phosphorsaeuresammonium, Zweicbasichesammoniumphosphat.

Building
Ingredient of—
Mixtures for fireproofing structures.

Chemical
Catalyst in making—
Dimethylpiperazin, guaiacol from diazo-orthoanisol (U. S. 1623949).
Ingredient of—
Compositions of food for promoting micro-organism fermentation, yeast fermentation, fermentation with acetic acid bacteria.
Contact masses for use in chemical processes.
Reagent in making—
Ammonium phosphotungstate, ammonium phosphomolybdate.
Reagent in making—
Acetic acid by addition to molasses, syrups, grape sugar, and the like.

Explosives
Ingredient of—
Impregnating bath for the treatment of wood for matches.

Fertilizer
Alone or in mixtures as a plant food.

Miscellaneous
Ingredient of—
Compositions for fireproofing straw, pickling baths for making candle wicks.

Metallurgical
Flux in—
Soldering metals.
Ingredient of—
Platinum electroplating liquors.
Baths for plating aluminum with silver and nickel.
Baths for plating platinum and nickel alloys.
Baths for coloring lead.
Baths for coating metals with lead.
Hard soldering compositions.

Paint and Varnish
Reagent in making—
Green pigment (Arnaudon's green) by calcination with dichromates.

Paper
Ingredient of—
Compositions for fireproofing cardboard, paper pulp, paper products.

Ammonium Phosphate (Continued)

Pharmaceutical
In compounding and dispensing practice.

Sugar
Reagent in refining.

Textile
——, *Finishing*
Ingredient of—
Compositions for fireproofing fabrics and yarns.

Woodworking
Ingredient of fireproofing agents.

Ammonium Phosphomolybdate
French: Ammonium molybdophospate, Molybdophospate ammoniaque, Molybdophospate d'ammonium, Phosphomolybdate ammoniaque, Phosphomolybdate d'ammonium.
German: Ammoniummolybdaenphosphat, Ammoniumphosphomolybdat, Molybdaenphosphorsaeureammonium, Phosphomolybdaensaeuresammonium.

Analysis
Reagent in various processes.

Dye
Reagent (Brit. 275943) in making dye lakes with—
Para-aminobenzaldehyde.
4:4′-Tetramethyldiaminobenzaldehyde.
4:4′-Tetramethyldiaminobenzophenone.
4:4′-Tetramethyldiaminodiphenylmethane.

Paint and Varnish
Ingredient (Brit. 275969) of—
Oil or spirit lacquers, colored with basic dyestuffs and containing cellulose ester or ether bases.

Ammonium Polysulphide
Synonyms: Polysulphide of ammonia.
French: Polysulfure ammoniaque, Polysulfure d'ammonium.
German: Ammoniakpolysulfid, Ammoniumpolysulfid.
Spanish: Polisulfurato de amoniaco.
Italian: Polisulfurato di ammonio.

Analysis
Precipitating reagent in chemical analysis of metals and other substances.

Chemical
Precipitating agent in various processes.
Reducing agent in making various inorganic and organic chemicals, intermediates, pharmaceuticals, and synthetic aromatic chemicals.

Dye
Reducing agent in making various dyestuffs.

Fats and Oils
Reagent (Brit. 271553) in making—
Vulcanized oils.

Insecticide
As an insecticide and fungicide.
Ingredient of—
Insecticidal and fungicidal compositions.
Preparations for combatting powdery mildew.

Leather
Reagent in—
Dehairing hides.

Paper
Ingredient (Brit. 271553) of—
Compositions, containing rubber latex, used for treating paper and pulp.

Rubber
Reagent (Brit. 271553) in treating—
Rubber latex.

Textile
Reagent in—
Denitrating nitro rayons, treating viscose rayon filament to remove sulphur.

Ammonium-Potassium Borofluoride

Fuel
Ingredient (Brit. 463218) of—
Automotive fuels consisting of gasoline and ethyl alcohol (added to inhibit corrosion of magnesium metal, magnesium alloys, or other metal parts).
Automotive fuels consisting of gasoline, benzol, and methanol (added to inhibit corrosion of magnesium metal, magnesium alloys, or other metal parts).

Ammonium Propionate
French: Propionate ammoniaque, Propionate d'ammonium.
German: Ammoniumpropionat, Propionsaeuresammoniak, Propionsaeuresammonium.

Oils and Fats
Ingredient (Brit. 277357) of—
Lubricating compositions.

Leather
Ingredient (Brit. 277357) of—
Dressing and finishing compositions.

Petroleum
Ingredient (Brit. 277357) of—
Motor fuels containing mineral oils and aliphatic alcohols.
Stable mineral oil emulsions containing aliphatic alcohols.

Soap
Ingredient (Brit. 277357) of—
Detergent preparations, textile soaps.

Ammonium Salicylate
French: Salicylate ammoniaque, Salicylate d'ammonium.
German: Ammoniumsalicylat, Salicylsäuresammoniak, Salicylsäuresammonium.

Pharmaceutical
In compounding and dispensing practice.

Sanitation
As a germicide.
Ingredient of—
Germicidal preparations.

Resins and Waxes
Reagent (Brit. 292912) in making synthetic resins with the aid of the following carbamides:—
Acetyl, allyl, benzoyl, benzyl, butyl, citryl, cresyl, ethyl, formyl, gallyl, heptyl, hexyl, isoallyl, isoamyl, isobutyl, isopropyl, lactyl, methyl, naphthyl, pentyl, phenyl, phthalyl, picryl, propionyl, propyl, salicyl, succinyl, tolyl, valeryl, xylyl.

Ammonium Selenate
French: Séléniate ammoniaque, Séléniate d'ammonium.
German: Ammoniumselenat, Selensäuresammoniak, Selensäuresammonium.

Analysis
Reagent in testing for—
Alkaloids, codeine.

Glass
Reagent in making—
Red glass.

Miscellaneous
Mothproofing agent (Brit. 340318) in treating—
Furs, feathers, and the like.

Textile
Mothproofing agent (Brit. 340318) in treating—
Wool and felt.

Ammonium Selenite
French: Sélénite ammoniaque, Sélénite d'ammonium.
German: Ammoniumselenit, Selenigsäuresammoniak, Selenigsäuresammonium.

Miscellaneous
Mothproofing agent (Brit. 340318) in treating—
Furs, feathers, skins and other articles.

Textile
Mothproofing agent (Brit. 340318) in treating—
Felt and wool.

Ammonium Silicofluoride
Synonyms: Ammoniumfluorsilicate.
French: Fluosilicate d'ammonium, Silicofluorure d'ammonium.
German: Fluorsiliciumstoffsaeuresammonium, Silicofluorstoffsaeuresammonium, Silicofluorwasserstoffsaeuresammonium.

Chemical
Starting point in making—
Ammonium chlorate.

Metallurgical
Reagent in the treatment of—
Difficultly decomposable minerals, especially rare earth minerals, titanium minerals, zirconium minerals, and monazite sand (German 440274).
Ingredient of—
Solutions used for electroplating copper on zinc and iron.

Pharmaceutical
In compounding and dispensing practice.

Ammonium Silicomolybdate
French: Ammonium molybdosilicate, Molybdosilicate ammoniaque, Molybdosilicate d'ammonium, Silicomolybdate ammoniaque, Silicomolybdate d'ammonium.
German: Ammonium-silicomolybdat, Molybdaenkieselsaeuresammonium, Siliciummolybdaensaeuresammonium.

Dye
Reagent (Brit. 275943) in making color lakes with—
Para-aminobenzaldehyde.
4:4'-Tetramethyldiaminobenzhydrol.
4:4'-Tetramethyldiaminobenzophenone.
4:4'-Tetramethyldiaminodiphenylmethane.

Paint and Varnish
Ingredient (Brit. 275969) of—
Cellulose ester or ether oil or spirit lacquers containing basic dyestuffs.

Ammonium Silicotungstate
French: Ammonium tungstosilicate, Silicotungstate ammoniaque, Silicotungstate d'ammonium, Tungstosilicate ammoniaque, Tungstosilicate d'ammonium.
German: Ammoniumsilicotungstat, Ammoniumwolframsilicate, Siliciumwolframsaeuresammoniak, Siliciumwolframsaeuresammonium, Wolframkieselsaeuresammoniak, Wolframkieselsaeuresammonium.

Dye
Reagent (Brit. 275943) in making color lakes with—
Para-aminobenzaldehyde.
4:4'-Tetramethyldiaminobenzhydrol.
4:4'-Tetramethyldiaminobenzophenone.
4:4'-Tetramethyldiaminodiphenylmethane.

Paint and Varnish
Ingredient of—
Cellulose ester or ether lacquers and varnishes containing basic dyestuffs (Brit. 275969).

Ammonium Succinate
French: Succinate d'ammonium, Succinate ammoniaque.
German: Ammoniumsuccinat, Bernsteinsaeuresammonium.

Chemical
Reagent in making—
Succinamide.

Ammonium Sulphate
Synonyms: Sulphate of ammonia.
French: Sulfate d'ammoniaque, Sulfate d'ammonium.
German: Ammoniumsulfat, Schwefelsaeuresammoniak, Schwefelsaeuresammonium.

Analysis
Reagent for generating pure nitrogen gas.

Chemical
Activating agent in enhancing—
Fermentation processes, used in admixture with superphosphate.
Nitrogenous food in making—
Dry yeast.
Precipitant in purifying—
Protein matters and ferments, such as pepsin.
Reagent in making—
Antimony lactate, antimony salt, betanaphthylamine, iron alum, methylamine, lactic acid by fermentation of glucose, sodium nitride, sulphur trioxide when treated with sodium sulphate (German 298491 and 301791).
Reagent in enhancing—
Electrolytic oxidation of sulphuric acid to persulphuric acid.
Starting point in making—
Ammonia alum, ammonia salts of acids and halogens, mohr salt, sal ammoniac.

Dye
Reagent in making—
Indigo.

Electrical
Ingredient of—
Battery charges.

Fertilizer
As a plant food.
Ingredient of—
Mixed fertilizers containing phosphate.

Food
Ingredient of—
Leavened bread (U. S. 1593977).
Substitute for—
Tartaric acid and cream of tartar in making baking powders.

Fuel
Ingredient of—
Saline mixtures used in impregnating candle wicks.

Leather
Reagent in neutralizing—
Hides during tanning by means of bichromate, used along with sodium carbonate.

Metallurgical
Ingredient of—
Compositions which are used for coloring metals by the deposition of metallic zinc.
Compositions for producing patina on copper or bronze objects.
Nickel-plating baths, soldering liquors.
Reagent in—
Electrolytic recovery of metallic cobalt.
Extraction of uranium from pitchblende.
Galvanizing iron.
Treating oxidized zinc ores to obtain the metallic content.

Miscellaneous
Ingredient of—
Fireproofing compositions used in the treatment of abrasive sheet materials, such as emery paper, sand cloth, sand paper, garnet paper, emery cloth (Brit. 252165).
Fireproofing sizes in admixture with starches.

Paper
Ingredient of—
Fireproofing compositions.

Textile
——, *Dyeing and Printing*
Mordant in dyeing and printing fabrics and yarns.
——, *Finishing*
Ingredient of—
Fireproofing compositions.
——, *Manufacturing*
Ingredient of—
Precipitating liquors in spinning viscose filaments from cellulose xanthate solutions (French 602711).

Woodworking
Ingredient of—
Fireproofing compositions.

Ammonium Sulphocyanate
Synonyms: Ammonium rhodanate, Ammonium rhodanide, Ammonium sulphocyanide, Ammonium thiocyanate, Ammonium thiocyanide.
French: Rhodanate ammoniaque, Rhodanate d'ammonium, Rhodanure ammoniaque, Rhodanure d'ammonium, Sulphocyanate ammoniaque, Sulphocyanate d'ammonium, Thiocyanate ammoniaque, Thiocyanate d'ammonium.
German: Ammoniumrhodanat, Ammoniumrhodanid, Ammoniumrhodanuer, Ammoniumsulfocyanat, Ammoniumsulfocyanid, Ammoniumsulfocyanuer, rhodansäuresammoniak, Rhodansäuresammonium, Sulfocyansäuresammoniak, Sulfocyansäuresammonium, Thiocyansäuresammoniak, Thiocyansäuresammonium.
Spanish: Sulfocianato de amonio.

Analysis
Reagent in forensic and other analytical work for the determination and detection of—
Arsenic, antimony, copper, halogens, mercury, mustard oils, silver.
Reagent for the determination of—
Alcohol, copper sulphate, iron.

Chemical
Reagent in making—
Acetamide, guanidin, guanidin sulphocyanide from thiourea, thiocarbamide, thiourea from guanidin.
Starting point in making—
Calcium sulphocyanide, carbon bisulphide, cyanides, lead sulphocyanide, potassium cyanide, potassium ferrocyanide, sodium cyanide, sodium ferricyanide, various ferricyanides, sulphocyanides and cyanides, various intermediates and other organic chemicals.

Explosives
Ingredient of—
Fulminating agents, match-head compositions.
Reagent in making—
Explosive compositions.

Ammonium Sulphocyanate (Continued)

Fertilizer
Ingredient of—
Fertilizing compositions.

Insecticide
Ingredient of—
Weed-killers.

Metallurgical
Reagent in producing—
Greyish-black coatings on zinc.

Miscellaneous
Reagent in—
Double staining specimens for examination under the microscope.

Pharmaceutical
In compounding and dispensing practice.

Photographic
Reagent in various processes.

Refrigeration
Ingredient of—
Freezing compositions containing potassium nitrate, sodium nitrate, and ammonium nitrate.

Textile
——, *Dyeing*
Assist in various processes.
Ingredient of—
Dye baths, which are contained in copper vessels (added for the purpose of preventing the contamination of the dyestuff with copper salts and spoiling the shade of the dyed materials).
Dye liquors containing alizarin.
Padding liquor in dyeing cotton with various indigosol dyestuffs by the slop-padding methods.

——, *Finishing*
Ingredient of—
Solutions used (Brit. 260289) for stripping dyestuffs from cellulose esters and ethers, fabrics, films, regenerated cellulose rayons, threads, yarns.
Weighting solutions containing tin salts (added for the purpose of increasing the strength of the tin-weighted silks and also to prevent the formation of red spots on these silks).

——, *Printing*
Assist in fabric printing with various dyestuffs.

Ammonium Telluride
French: Tellurure ammoniaque, Tellurure d'ammonium.
German: Ammoniaktellurid, Ammoniumtellurid, Tellurammoniak, Tellurammonium.

Chemical
Catalyst (Brit. 263877) in making—
Acetone from isopropyl alcohol.
Dehydrogenated products from cyclohexane.
Isobutyraldehyde from isobutyl alcohol.
Isoveraldehyde from isoamyl alcohol (Brit. 262120).
Naphthalene from tetrahydronaphthalene.
Paracymene from turpentine.
Reagent (Brit. 292222) in making organic tellurium compounds from—
Pentamethylene alphaepsilondibromide.
Pentamethylene alphaepsilondichloride.
Pentamethylene alphaepsilondifluoride.
Pentamethylene alphaepsilondi-iodide.

Ammonium-Tin Carbonate

Textile
Delustring agent (Brit. 454968) for—
Cellulose acetate yarn, rayon in the spun form, regenerated cellulose yarn.

Ammonium-Tin Citrate

Textile
Delustring agent (Brit. 454968) for—
Cellulose acetate yarn, rayon in the spun form, regenerated cellulose yarn.

Ammonium-Tin Oxalate

Textile
Delustring agent (Brit. 454968) for—
Cellulose acetate yarn, rayon in the spun form, regenerated cellulose yarn.

Ammonium-Tin Tartrate

Textile
Delustring agent (Brit. 454968) for—
Cellulose acetate yarn, rayon in the spun form, regenerated cellulose yarn.

Ammonium-Titanium Carbonate

Textile
Delustring agent (Brit. 454968) for—
Cellulose acetate yarn, rayon in the spun form, regenerated cellulose yarn.

Ammonium-Titanium Citrate

Textile
Delustring agent (Brit. 454968) for—
Cellulose acetate yarn, rayon in the spun form, regenerated cellulose yarn.

Ammonium-Titanium Oxalate

Textile
Delustring agent (Brit. 454968) for—
Cellulose acetate yarn, rayon in the spun form, regenerated cellulose yarn.

Ammonium-Titanium Tartrate

Textile
Delustring agent (Brit. 454968) for—
Cellulose acetate yarn, rayon in the spun form, regenerated cellulose yarn.

Ammonium Uranate
French: Uranate ammoniaque, Uranate d'ammonium.
German: Uranhaltigammoniak, Uranhaltigammonium, Uraniammoniak, Uraniammonium.

Chemical
Reagent (Brit. 397515 and 397516) in removing—
Albuminous substances from therapeutic sera and other liquids.

Ammonium Vanadate
French: Vanadate ammoniaque, Vanadate d'ammonium.
German: Ammoniumvanadat, Vanadinsaeuresammoniak, Vanadinsaeuresammonium.

Analysis
Reagent in pharmaceutical assays.

Chemical
Catalyst (Brit. 265959) in making the following acids:—
Adipic, allyladipic, amyladipic, butyladipic, diethyladipic, dimethyladipic, ethyladipic, hexyladipic, heptyladipic, isoallyladipic, isoamyladipic, isobutyladipic, isopropyladipic, methyladipic, propyladipic.
Starting point in making—
Iron vanadochromate, vanadates of metallic and alkaline bases.

Dye
Reagent in making—
Various synthetic colorings.

Glass
Ingredient of—
Crystal glass.

Ink
Ingredient of—
Printing inks, writing inks.

Paint and Varnish
Drier for linseed oil in paints and varnishes.
Reagent in making—
Dry colors.

Textile
——, *Dyeing*
Ingredient of—
Padding liquor in dyeing cotton fabrics by the slop-padding method with indigosol O.
Mordant in—
Dyeing fabrics and yarns with anilin black.

——, *Printing*
Mordant in—
Printing indigo black with the aid of indigosol O.

Amyl Acetate
Synonyms: Amyl acetic ether, Banana oil, Pear oil, Isoamyl acetate.
Latin: Amylum aceticum.
French: Acétate d'amyle, Acétate amylique, Acétate d'isoamyle, Éther amylacétique, Éther d'amyle-acétyle, Essence de poivres.

Amyl Acetate (Continued)
German: Amylacetat, Amylazetat, Bananeoel, Essigsäureamylester, Essigsäureisoamylester, Essigsäuresamyl, Essigsäureisoamyl, Isoamylacetat, Isoamylazetat.

Adhesives
Solvent in making—
Film cement (U. S. 1596965), special cements.

Ceramics
Solvent in—
Compositions containing nitrocellulose and the like for coating ceramic products.

Chemical
Solvent in making—
Collodion preparations.

Explosives
Solvent in making—
Nitrocellulose, gun cotton, smokeless powder.

Food
Flavoring in—
Beverages, candies, desserts, jellies.
Ingredient of—
Artificial flavorings, fruit essences, jargonelle pear essence.

Glass
Solvent in—
Compositions containing nitrocellulose and the like for coating glass and making non-scatterable glass.

Insecticide
Ingredient of—
Animal insecticidal compositions.
Insecticides and the like for treating plants (German 421833).

Leather
Solvent for—
Tannins.
Solvent in making—
Artificial leathers.

Metallurgical
Solvent in—
Compositions containing nitrocellulose and other esters and ethers of cellulose for coating metals.
Electroplating.

Miscellaneous
Combustible in—
Photometric lamps.
Ingredient of—
Fireproofing compositions, lubricating compositions (U. S. 1603086).
Solvent in making—
Artificial pearls.
Spectacles and other general optical supplies.

Oilcloth and Linoleum
Solvent in making—
Coatings.

Paint and Varnish
Solvent in—
Bronze varnishes.
Nitrocellulose lacquers, varnishes, dopes, enamels, and paints.
Waterproof varnishes.

Paper
Solvent in making—
Stencils (U. S. 1719926).
Solvent in—
Compositions containing nitrocellulose or other cellulose esters or ethers for making coated paper and paper products.

Perfume
General solvent.
Solvent for camphor.
Solvent in making—
Cosmetics (Brit. 255148).

Photographic
Solvent in making—
Coatings for plates, films, and papers.
Motion picture films.

Plastics
Ingredient of—
Compound solvents (French 601546).
Solvent in making—
Celluloid products, celluloid cements.
Compounds of nitrocellulose or other cellulose esters or ethers.

Printing
Solvent in—
Color printing, photoengraving.

Rubber
Solvent in—
Coating compositions containing nitrocellulose or other esters or ethers of cellulose.

Soap
Ingredient of—
Detergent preparations (Brit. 255148).
Solvent in making—
Special soaps.

Stone
Solvent in—
Coating compositions containing cellulose esters or ethers.

Textile
——, *Dyeing and Printing*
Solvent in—
Dye liquors and printing pastes.

——, *Finishing*
Solvent in—
Coating compositions containing cellulose esters or ethers.
Fireproofing compositions, waterproofing compositions.

——, *Manufacturing*
Solvent in making—
Rayon.

Woodworking
Solvent in—
Coating compositions containing nitrocellulose or other esters or ethers of cellulose.

Amyl alcohol, active. See: Butyl carbinol, secondary.

Amyl alcohol, Fermentation. See: Fusel oil.

Amyl alcohol, primary normal. See Butyl carbinol, normal.

Amyl alcohol, secondary normal. See: Diethyl carbinol.

Amyl alcohol, tertiary. See: Dimethylethylcarbinol.

Amyl Alphacrotonate
Synonyms: Alphacrotonic amyl ester.
French: Alphacrotonate de amyle, Alphacrotonate amylique, Éther d'alphacrotoniqueamylique.
German: Alphacrotonamylester, Alphacrotonsäureamylester, Alphacrotonsäuresamyl, Amylalphacrotonat.

Miscellaneous
Solvent (Brit. 321258) for—
Cellulose acetate, cellulose esters and ethers, cellulose nitrate, rubber.
For uses, see under general heading: "Solvents."

Amylalphanaphthol, Normal

Disinfectant
Claimed (U. S. 2073996 and 2073997) to be—
Germicide combining high efficiency toward *staphylococcus aureus* and low toxicity.

Amyl Carbamide
French: Carbamide d'amyle, Carbamide amylique.

Chemical
Starting point in making—
Intermediates, pharmaceuticals.

Resins and Waxes
Starting point (Brit. 292912) in making synthetic resins with—
Acetylsalicylic acid, aliphatic dibasic acids, ammonium salicylate, anthranilic acid, benzoic acid, gallic acid, hydroxynaphthoic acid, magnesium salicylate, oxalic acid, phenolic acids, phthalic acid, salicylamide, salicylic acid, succinic acid, strontium salicylate.

Amyl Chloride
French: Chlorure d'amyle, Chlorure amilique.
German: Amylchlorid, Chloramyl.

Chemical
Solvent for various purposes.

Miscellaneous
Solvent for various purposes.

Amyl Chloride (Continued)

Paint and Varnish
Solvent in making—
Nitrocellulose and cellulose acetate lacquers, dopes, and varnishes.
Plastics
Solvent in making—
Nitrocellulose and cellulose acetate compounds.
Textile
——, *Finishing*
Solvent in making—
Waterproofing compositions of cellulose acetate used in the treatment of collars, cuffs, shirt fronts, tablecloths and other linen fabrics.

Amyl Chloroacetate

French: Chloroacétate de amyle, Chloroacétate amylique.
German: Amylchloracetat, Chloressigsaeureamylester.
Dye
Reagent in making—
Stable, water-soluble vat dyestuffs derivatives (Brit. 263898).

Amyl Cinnamate

French: Cinnamate d'amyle, Cinnamate amylique, Cinnamate de pentyle, Cinnamate pentylique, Pentylcinnamate.
German: Amylcinnamat, Amylzinnamat, Pentylcinnamat, Pentylzinnamat, Zimtsäureamylester, Zimtsäurepentylester, Zimtsäuresamyl, Zimtsäurespentyl.
Perfume
Ingredient of—
Perfumes.
Perfume in—
Cosmetic preparations of various sorts.
Soap
Perfume in—
Toilet soaps.

Amylcinnamic Aldehyde

French: Aldéhyde de amylecinnamyle, Aldéhyde amylique et cinnamique.
German: Amylcinnamylaldehyd.
Chemical
Starting point in making various derivatives.
Perfume
Ingredient of—
Hyacinth perfumes, jasmine base preparations, lilac perfumes.
Perfume in—
Cosmetics.
Soap
Perfume in making—
Toilet soaps.

Amylcresol

Chemical
Starting point (Brit. 444351) in making—
Fat-splitting catalysts and emulsifying agents, useful in dyeing, laundering, bleaching, and various other purposes, by reacting with formaldehyde and nonaromatic secondary amines (the salts of the products with water-soluble acids, or water-insoluble acids, or the quaternary ammonium salts, are claimed to be valuable for the purposes named).

Amylene

Synonyms: Betaisoamylene, Pental, Pentene, Trimethylethylene, Trimethylene, Valerene.
French: Bétaisoamylène, Pental, Pentène, Triméthyleéthylène, Triméthyle d'éthylène.
German: Betaisoamylen, Pental, Penten, Trimethylaethylen, Valeren.
Chemical
Solvent for general purposes.
Starting point in making—
Dimethylethylcarbinol.
Miscellaneous
Solvent for various purposes.
Paint and Varnish
Solvent in making—
Dopes, lacquers, and varnishes from cellulose esters and ethers.
Pharmaceutical
In compounding and dispensing practice.
Plastics
Solvent in making—
Cellulose ester and ether compounds.
Resins and Waxes
Starting point in making—
Sulphur resins.

Amylene Dichloride

French: Dichlorure d'amylène, Dichlorure amylènique.
German: Amylendichlorid, Dichloramylen.
Chemical
Solvent for various purposes.
Miscellaneous
Solvent for various purposes.
Paint and Varnish
Solvent in making—
Cellulose ester and ether varnishes and lacquers.
Plastics
Solvent in making—
Cellulose ester and ether compounds.

Amylenedinaphthol

French: Dinaphthol amylènique.
German: Amylendinaphtol.
Spanish: Amilendinaftolina.
Italian: Amilenedinaftolina.
Chemical
Starting point in making—
Intermediates, pharmaceuticals.
Various other derivatives.
Rubber
Reagent (U. S. 1841342) in preserving—
Rubber goods by vulcanizing them in the presence of this reagent).

1-Amyleneoxy-4-aminoanthraquinone

French: 1-Amylèneoxye-4-aminoanthraquinone.
German: 1-Amylenoxy-4-aminoanthrachinon.
Chemical
Starting point in making various intermediates.
Dye
Starting point (Brit. 285096) in making dyestuffs in the presence of dimethylanilin, nitrobenzene, orthodichlorobenzene, naphthalene, and the like, with the aid of—
Acetylparaphenylenediamine, 5-amino-2-methylbenzimidazole, benzidin and derivatives and homologs, dimethylparaphenylenediamine, metanaphthylenediamine, metaphenylenediamine, metatoluylenediamine, metaxylidenediamine, orthonaphthylenediamine, orthophenylenediamine, orthotoluylenediamine, orthoxylidenediamine, paranaphthylenediamine, paraphenylenediamine, paratoluylenediamine, paraxylidenediamine.

Amylenethiourea

Synonyms: Amylenesulphourea.
French: Sulphourée d'amylene, Sulphourée amylenique, Thiourée d'amylene, Thiourée amylenique.
German: Amylensulfoharnstoff, Amylenthioharnstoff.
Chemical
Starting point in making—
Intermediates, pharmaceuticals.
Starting point (Brit. 310534) in making rubber vulcanization accelerators with the aid of—
Alphanaphthylamine, anilin, betanaphthylamine, cyclohexylanilin, diphenylamine, ethylanilin, meta-anisidin, metacresidin, metanaphthylenediamine, metaphenylenediamine, metatoluidin, metatoluylenediamine, metaxylenediamine, metaxylidin, monomethylanilin, orthoanisidin, orthocresidin, orthonaphthylenediamine, orthophenylenediamine, orthotoluidin, orthotoluylenediamine, orthoxylenediamine, orthoxylidin, para-anisidin, paracresidin, paranaphthylenediamine, paraphenylenediamine, paratoluidin, paratoluylenediamine, paraxylenediamine, paraxylidin, phenylamine.

Amyl Ether

Synonyms: Amyl oxide, Diamyl ether, Diamyl oxide, Di-isoamyl ether.
French: Éther d'amyle, Éther amylique, Éther de diamyle, Éther diamylique, Éther de di-isoamyle, Éther di-isoamylique.
German: Amylaether, Amyloxyd, Diamylaether, Diamyloxyd, Di-isoamylaether.

Amyl Ether (Continued)

Chemical
Reagent in making—
Aromatics, intermediates, pharmaceuticals.
Solvent in—
Alkaloid extraction, grignard reaction, various processes.

Fats and Oils
Solvent for—
Extracting fats and oils.
Various fats and oils.

Miscellaneous
General solvent.
General extracting medium.

Paint and Varnish
Solvent in making—
Lacquers, paints, varnishes.

Perfume
Solvent for—
Extracting odoriferous matters.
Perfume bases.

Pharmaceutical
In compounding and dispensing practice.

Amyl Furoate
French: Furoate d'amyle, Furoate amylique.
German: Amylfurat, Furoesaeureamylester, Furoesaeuresamyl.

Paint and Varnish
Solvent in making—
Lacquers and varnishes.

Plastics
Solvent in making various products.

Amyl Gallate

Petroleum
Antioxidant (U. S. 1970339) for—
Vapor-phase-cracked hydrocarbon distillates (inhibits usual deterioration, loss of antiknock properties, gum development on storage).

Amyl Lactate

Cellulose Products
Solvent and plasticizer for—
Cellulose esters or ethers.
For uses, see under general heading: "Solvents."

Amyl Mandelate
French: Mandélate d'amyle, Mandélate amylique.
German: Amylmandelat, Mandelsaeureamylester, Mandelsaeuresamyl.

Paint and Varnish
Plasticizer (Brit. 270650) in making—
Cellulose ester and cellulose ether lacquers.

Plastics
Plasticizer (Brit. 270650) in making—
Nitrocellulose plastics.

Amylmetacresol, Normal
Synonyms: Amyl metacresol.
German: Amylmetakresol, Kresylsäuresmetamylester, Metakresylsäuresamylester.

Disinfectant
As a germicide.
Ingredient of—
Liquid germicidal compositions for moulds, bacteria, and other organisms.

Amylnaphthylaminesulphonic Acid
French: Acide d'amylnaphthylaminesulfonique.
German: Amylnaphtylaminsulfonsaeure.

Miscellaneous
Ingredient (Brit. 271071) of—
Bleaching compositions.

Soap
Ingredient of—
Detersive compositions.

Textile
——, *Dyeing*
Ingredient of—
Dye bath in various processes.
——, *Finishing*
Ingredient of—
Finishing compositions, wetting compositions.

Amylolamine
German: Amylolamin.

Chemical
Starting point in making—
Pharmaceuticals and other derivatives.

Electrical
Dispersive agent (Brit. 340294) in making—
Special lubricating compositions for use in electrical switches.

Fats and Oils
Dispersive agent (Brit. 340294) in making—
Nonfreezing lubricating compositions containing animal and vegetable oils and fats, as well as ethyleneglycol or its esters, borax, benzyl alcohol.
Special lubricating compositions of the above type, for use on locomotive axles, railway switches, hydraulic presses and hydraulic brakes.
Ingredient (Brit. 340294) of—
Compositions containing vegetable, animal, and mineral oils and greases, used as rust preventives.

Petroleum
Ingredient (Brit. 340294) of—
Special lubricating compositions containing mineral oils and greases.

Amylphenol

Chemical
Starting point (Brit. 444351) in making—
Fat-splitting catalysts and emulsifying agents, useful in dyeing, laundering, bleaching, and various other purposes, by reacting with formaldehyde and nonaromatic secondary amines (the salts of the products with water-soluble acids, or water-insoluble acids, or the quaternary ammonium salts, are claimed to be valuable for the purposes named).

Amylphenol, Paratertiary

Chemical
In organic synthesis.

Disinfectant
As a germicide (phenol coefficient of approximately 60—Hygienic Laboratory Method).

Insecticide
Suggested as—
Fumigant, insecticide.

Resins and Waxes
Reagent in making—
Oil-soluble varnish resins of the phenol-formaldehyde type.

Amyl Phthalate

Cellulose Products
Plasticizer for—
Cellulose nitrate (nitrocellulose).
For uses, see under general heading: "Plasticizers."

Amyl Propionate
French: Propionate d'amyle, Propionate amylique.
German: Amylpropionat, Propionsäuresamyl, Propionsäuresamylester.

Cellulose Products
Solvent for—
Cellulose nitrate.
For uses, see under general heading: "Solvents."

Amylresorcinol

Chemical
Starting point (Brit. 444351) in making—
Fat-splitting catalysts and emulsifying agents for use in dyeing, laundering, bleaching, and various other purposes, by reacting with formaldehyde and nonaromatic secondary amines (the salts of the products with water-soluble acids or water-insoluble acids, and the quaternary ammonium salts, are claimed to be valuable for the purposes named).

Amyl Salicylate
Synonyms: Amylenol, Orchidie, Trefol, Ulmarene.
French: Salicylate d'amyle.
German: Amylsalicylat, Salicylsaeuresamyl, Salicylsaeureamylester.

Foodstuffs
Ingredient of—
Beverages, flavoring extracts, fruit essences, prepared foods.

Amyl Salicylate (Continued)

Perfumery
Fixative in making various preparations.
Odorous agent in—
Cosmetics, perfumes.
Pharmaceutical
In compounding and dispensing practice.

Amyl Stearate
French: Stéarate d'amyle, Stéarate amylique.
German: Amylstearat, Stearinsäureamylester, Stearinsäuresamyl.
Cellulose Products
Plasticizer for—
Cellulose acetate, cellulose esters and ethers, cellulose nitrate.
For uses, see under general heading: "Plasticizers."
Chemical
Solvent for—
Benzyl abietate.
Starting point in making various derivatives.
Gums
Solvent for various gums, such as ester gum.
Paint and Varnish
Solvent for—
Shellac.
Resins and Waxes
Solvent for—
Copal esters, cumarone resins, mastic.

Amyl Sulphide
French: Sulfure d'amyle, Sulfure amylique.
German: Amylsulfid.
Chemical
Reagent in various operations.
Reagent (Brit. 298511) in treating—
Albumens, albuminoids.
Glues and Adhesives
Reagent (Brit. 298511) in making adhesive preparations from—
Flaxseed proteins, peanut proteins, soybean proteins, various other vegetable proteins.
Miscellaneous
Reagent (Brit. 298511) in making sizes and finishing compositions from—
Flaxseed proteins, peanut proteins, soybean proteins, various other vegetable proteins.

Amyl-4-Sulphophthalate
Miscellaneous
As an emulsifying agent (Brit. 418334).
For uses, see under general heading: "Emulsifying agents."

Amylsulphuric Acid Chloride
French: Chlorure d'amylesulphurique.
German: Amylschwefelsaeureschlorid, Chloramylschwefelsaeure.
Dye
Starting point (Brit. 271533) in making soluble vat dyestuffs from—
Anthraquinone-1:2, flavanthrene, indanthrene, naphthacridone, thioindigo.

Amyltetrahydronaphthalenecarboxylic Acid
French: Acide d'amyletétrahydronaphthalènecarbonique.
German: Amyltetrahydronaphtalincarbonsaeure.
Chemical
Ingredient of—
Emulsifying and solvent media used for various purposes (German 432942).
Fats and Oils
Emulsifying agent (German 432942).
Miscellaneous
Ingredient of—
Emulsifying and solvent agents used for various purposes (German 432942).

Amylthiocyan Acetate
French: Acétate d'amylethiocyane, Acétate amylethiocyanique.
German: Amylthiozyanacetat, Amylthiozyanazetat, Essigsäureamylthiozyanester.
Chemical
Starting point in making various derivatives.
Insecticide
Ingredient (Brit. 361900) of—
Insecticidal preparations (used in solution in water or an organic solvent, such as kerosene).

Amyl Thiocyanpropionate
Chemical
Starting point in making various derivatives.
Disinfectant
Ingredient (Brit. 361900) of—
Disinfectants and germicides (used in solution in water or in an organic solvent, such as kerosene).
Insecticide
Ingredient (Brit. 361900) of—
Insecticidal compositions (used in solution in water or in an organic solvent, such as kerosene).

Amyl Xanthate
French: Amyle xanthate, Xanthate d'amyle, Xanthate amylique.
German: Amylxanthogenat, Xanthogensäuresamyl, Xanthogensäureamylester.
Mining
Flotation agent in—
Ore concentration processes.

Anacardic Acid
French: Acide anacardique.
German: Anacardsäure.
Chemical
Starting point in making—
Salts, particularly lead anacardate.
Insecticide
For vermifuge purposes.

Angelic Acid
Synonyms: Angelicic acid, Angelinic acid.
Food
Ingredient of—
Flavoring extracts.
Water and Sanitation
Breaker (U. S. 1964444) of—
Emulsoids in sewage.

Anhydromethylenecitric Acid
Chemical
Starting point in making—
Esters and salts, hexamethylenetetramine anhydromethylenecitrate, pharmaceuticals, sodium anhydromethylenecitrate.
Pharmaceutical
In compounding and dispensing practice.

Anilin
Synonyms: Aminobenzene, Aniline, Aniline oil, Phenylamine.
Aviation
Ingredient (U. S. 1389084) of—
Airplane fabric mending composition.
Building Construction
Ingredient of—
Coating compositions for concrete.
Cellulose Products
Gelatinizer for—
Nitrocellulose.
Ingredient (U. S. 1421974) of—
Celluloid solvent.
Process material in making—
Celluloid substitutes, cellulose acetate, cellulose benzoate, cellulose butyrate, cellulose palmitate, nitrocellulose.
Solvent for—
Cellulose esters and ethers.
Chemical
Precipitant (U. S. 1337192) for—
Metal hydroxides.
Process material in making—
Acetanilide, 3-acetyl-4:5-diketo-2-furyl-1-phenylpyrolidin, alkyl derivatives, aluminum anilide, amines, 1-amino-4-anilinoanthraquinone, 1-amino-4-anilino-2-methylanthraquinone, N-(4′-amino-1′-alphaethyl)paratoluenesulphonamide, 2-aminoanthraquinone, aminohexamethylene, 4-amino-1-(paratolylsulphonamido)-naphthalene-2-sulphonic acid, 4-amino-1-(paratolyl-sulphonamido)naphthalene-6 (or 7 or 8)-sulphonic acid, anilides, anilin hydrochloride, 1-anilinobenzothiazole,

Anilin (Continued)

2-anilino-7-chloroanthraquinone, anilinogallocyanin, anilinomethanesulphonic acid, 1-anilino-2-naphthol, 3-anilino-1-naphthol, 4-anilino-1-naphthol, 7-anilino-2-naphthol, anthraquinone derivatives, arsanilic acid, N-arylaminonaphthols, atoxyl, azobenzene, betanaphthoquinonedianilide, bis(aminonaphthyl)metabenzenedisulphonamides, bis(aminonaphthyl)naphthalenedisulphonamides, (bis-aminosulphonaphthyl)metabenzenedisulphonamides, bis(aminosulphonaphthyl)naphthalenedisulphonamides, bromoanilins, carbodiphenylimide, 4-chlor-3-(5′-keto-3′-methyl-4′-phenylazo-1′-pyrazolylbenzene)sulphonic acid, chloranilins, diaminotriphenylmethanesulphonic acid, 2:5-dianilino-3:6-dichloroquinones, diazoaminobenzene, diazobenzene, dichloronitrodiphenylamine, diethylanilin, dimethylanilin, dimethyldiphenylurea, dinitrodiphenylamine, diphenylamine, diphenylguanidin, ethylanilin, ethylstarch, formanilide, halogen derivatives, indanthrene, iodoanilins, magnesium anilide, methylanilin, N:N′-methylene-bis-anilin, 2-(1′-naphthyl) cinchoninic acid, nitroanilins, orthoaminophenol-4-sulphonanilide, orthosulphamidobenzoylanilide, N-(para-arsenophenyl)glycinanilide, para-N-dimethylaminothiocarbanilide, paradisulphatoethylaminoazobenzene, pharmaceutical compounds, phenylglycin, N-phenylphthalimide, phenylstibnic acid and its sodium salt, phenylstibnic oxychloride, phthalimides, quinone, salvarsan, sodium hyposulphite (anhydrous), N-sulphatoethylanilin, thiocarbanilide and derivatives, thiourea, toluidin, triphenylguanidin.

Solvent for—
Chromyl chloride, indigo, mercury compounds (U. S. 1457675), sulphur, various chemicals.

Starting point in making various chemicals, such as halogen derivatives and other (for specific products see under "Process material in making").

Disinfectant
Germicide.

Dye
Purifying agent for—
Anthracene.
Solvent for—
Anthraquinone.
Starting point in making—
Anilin black, anilin colors, azo dyestuffs, magenta, safranins, various other dyestuffs.

Electrical
Solvent for—
Tungsten hexachloride in making tungsten lamp filaments.

Explosives
Gelatinizer for—
Nitrocellulose.
Solvent for—
Cellulose esters and ethers.
Stabilizer for—
Nitrostarch.
Trinitrotoluene (in carbon bisulphide solution).

Fats and Oils
Solidification preventer for—
Chinawood oil.

Ink
Ingredient of—
Indelible inks, inks of various types, inks for checks, documents, and the like.

Mechanical
Ingredient of—
Carbon remover for internal combustion engines.

Mining and Smelting
Solvent for—
Flotation oils.

Miscellaneous
Ingredient of—
Compositions for mending small boats (U. S. 1389084).
Floor dressing.
Preservative for—
Cork.

Paint and Varnish
Ingredient of—
Antifouling paints, paint removers, paints, varnishes.

Paper
Preservative for—
Papers, safety papers.

Perfume
Process material in—
Synthesis of perfumes.

Photographic
Coloring agent in—
Color photography.

Plastics
Process material in making—
Molded articles, plastics.

Printing
Process material in making—
Printing pastes, rollers.

Resins
Process material in making—
Resins of various types, sulphur-phenol resins.
Softening agent for—
Phenol-aldehyde condensates.
Solvent for—
Phenol-aldehyde condensates.
Starting point in making—
Resins with formaldehyde and phthalic anhydride.

Rubber
Accelerator of—
Vulcanization.
Antiager for—
Balata, isoprene, rubber.

Textile
Coloring agent in—
Dyeing processes, printing processes.
Preservative for—
Textiles.

Wood
Ingredient of—
Wood-mending compositions.
Preservative for—
Wood.

Anilin Acetate

French: Acétate d'aniline, Acétate anilique.
German: Anilacetat, Anilanazetat, Essigsaeureanilinester, Essigsaeuresanilin.

Analysis
Reagent in testing and determining furfural.

Chemical
Starting point in making various intermediates.

Anilinazoalphanaphthylamine

Dye
Starting point in making—
Azo carmine G, phenylrosindulin, rosindon, rosindulin.

Food
Color for—
Butter, food oils, oleomargarin.

Anilin Black

French: Noir d'aniline.
German: Anilin schwarz, Schwarzanilin.

Ink
Ingredient of various ink compositions.

Leather
Pigment for dyeing.

Miscellaneous
Pigment for various purposes.

Paint and Varnish
Pigment in making—
Lacquers, enamels, varnishes.

Textile
——, *Dyeing*
Dyestuff for—
Cotton yarns and fabrics, half-silk fabrics, silk yarns and fabrics, wool yarns and fabrics.

Anilin Butyrate

French: Butyrate d'aniline.
German: Anilinbutyrat.
Spanish: Butirato de anilina.
Italian: Butirato di anilina.

Metallurgical
Ingredient (French 526640) of—
Rust preventive or removal agent, containing also fusel oil, mineral oil, or liquid petrolatum.

Military
Ingredient (French 526640) of—
Mixture with fusel oil, mineral oil, or liquid petrolatum for removing gun barrel stains caused by chemical corrosion or reaction during firing.
Rust preventive or removal agent, containing also fusel oil, mineral oil, or liquid petrolatum, used for treating metallic parts of guns.

Anilinmethyleneorthocresotinic Acid
French: Acide d'anilineméthylèneorthocrésotinique.
German: Anilinmethylenorthokresotinsaeure.

Dye
Starting point (Brit. 265203) in making azo dyestuffs and lakes with—
Acetyl H acid, alphanaphthol-3:6-disulphonic acid, alphanaphthol-3:8-disulphonic acid, betachloro-5-sulphophenylmethylpyrazolon.

Anilin Nitrate
French: Nitrate d'aniline.
German: Anilinnitrat, Saltpetersaeuresanilin.

Chemical
Starting point in making—
Metanitranilin.

Anilin Thiocyanate

Lubricant
Starting point (Brit. 440175) in making—
Addition agents for high-pressure lubricating oils or greases by mixing and reacting with organometallic compounds.

Anisic Acid
Synonyms: Badianic acid, Methyl-para-oxybenzoic acid, Paramethoxybenzoic acid, Umbellic acid.
Latin: Acidum anisicum.
French: Acide anisique, Acide badianique, Acide dragonique, Acide méthyleparaoxybenzoique, Acide paraméthoxybenzoique, Acide umbellique.
German: Anissäure, Badiansäure, Dragonsäure, Methylparaoxybenzoesäure, Paramethoxybenzoesäure, Umbellasäure.
Spanish: Acido anisico, Acido metilparaoxibenzoico.
Italian: Acido anisico, Acido metilparaossibenzoico.

Chemical
Starting point in making—
Amyl anisate by reaction with amyl alcohol.
Benzyl anisate by reaction with benzyl alcohol.
Ammonium anisate by reaction with ammonium hydroxide.
Bismuth anisate by reaction with a bismuth salt.
Ethyl anisate by reaction with ethyl alcohol.
Isoallyl anisate by reaction with isoallyl alcohol.
Isoamyl anisate by reaction with isoamyl alcohol.
Methyl anisate by reaction with methanol.
Naphthyl anisate by reaction with betanaphthol.
Nickel anisate by reaction with a nickel salt.
Phenyl anisate by reaction with phenol.
Potassium anisate by reaction with a potassium salt.
Propyl anisate by reaction with propyl alcohol.
Resorcinol anisate by reaction with resorcinol.
Sodium anisate by reaction with sodium bicarbonate.
Strontium anisate by reaction with a strontium salt.
Zinc anisate by reaction with a zinc salt.
Various other esters and salts, pharmaceutical compounds, intermediates, and aromatics.

Dye
Reagent in making various dyestuffs.

Miscellaneous
Antiseptic for various purposes.

Paint and Varnish
Ingredient of—
Compositions used for obtaining a mellowing coat in decorating and finishing wood.

Perfume
As a fixative.
Ingredient of various cosmetics.

Pharmaceutical
Suggested for use as antithermic, antipyretic, antiseptic, antirheumatic in the place of salicylic acid, and analgesic.

Soap
Fixative in making—
Fine perfumed toilet soaps.

4-Anisidin

Chemical
Starting point in making—
Intermediates and other derivatives.
Starting point (Brit. 353537) in making acridin derivatives with the aid of—
2-Chloro-4-bromobenzoic acid, 2-Chloro-4-iodobenzoic acid, 2:4-Dichlorobenzoic acid.

Dye
Starting point in making various synthetic dyestuffs.

Anisolesulphonamide

Cellulose Products
Plasticizer (Canada 340993) for—
Cellulose acetate.
For uses, see under general heading: "Plasticizers."

Anisyl Acetate
French: Acétate d'anisyle, Acétate anisylique.
German: Anisylacetat, Anisylazetat, Essigsäureanisylester, Essigsäuresanisyl.

Chemical
Starting point in making—
Aromatics.

Perfume
Ingredient of—
Perfume preparations.
Perfume in—
Cosmetics.

Anisyl Carboxethylate
French: Carboxéthylate d'anisyle.
German: Anisylcarboxaethylat.
Spanish: Carboxetilato de anisil.
Italian: Carbossietilato di anisile.

Perfume
Ingredient (French 650100) of—
Perfumes.

Annatto
Synonyms: Achiotte, Annotta, Arnotta, Orlean, Orleana, Orellana, Racou, Terre de la Nouvelle-Orléans.

Dye
Ingredient of—
Color lakes with various bases.

Food
Coloring agent for—
Butter, cheese, oils.

Miscellaneous
Pigment in making—
Wall paper.

Paint and Varnish
Ingredient of—
Oil colors, water colors.
Pigment in—
Wood stains, varnishes.

Paper
Coloring agent.

Textile
——, *Dyeing*
Coloring matter in dyeing cotton, silk or wool by direct processes.

Anthracene
Synonyms: Paranaphthene.
French: Anthracine.
German: Anthracen, Paranaphten.
Spanish: Antraceno, Paranafteno.
Italian: Antracene, Paranaftene.

Analysis
Reagent in—
Testing for lignin.

Automotive
Ingredient of—
Motor fuel compositions (added for the purpose of preventing knocking in the engine) (French 630326).

Ceramics
Ingredient (Brit. 371901) of—
Compositions, containing various simple esters of cellulose or mixed inorganic-aliphatic cellulose esters, used for the decoration and protection of ceramic wares (added to increase the life of the film by absorbing ultra-violet rays).

Chemical
Reagent (Brit. 263873) in making—
Emulsions of hydrocarbons of various groups of the aliphatic and aromatic series.
Emulsions of various chemicals.
Fat solvents in emulsified form.
Terpene emulsions.
Starting point in making—
Alphamethylanthracene, anthracenecarboxylic acid, anthracenesulphonic acids, anthracenetetrone, anthradiol, anthramine, anthraquinone and derivatives, betamethylanthracene, carbazole, dibromoanthracene, dicloroanthracene, dihydroanthracene, hexahydroanthracene.

Anthracene (Continued)

Hydrogenated products by treatment with hydrogen in the presence of alkali and alkaline earth hydrides used as catalysts (French 649976).
Intermediates and other organic chemicals.
Phenanthrene, pharmaceutical chemicals, nitroanthracene.
Tanning agents by sulphonation and then condensation with oxybenzylic alcohol (French 610661).
Tetrahydroxyanthracene, trihydroxyanthracene, wetting agents.

Disinfectant
Reagent (Brit. 263873) in making—
Emulsified germicides and disinfectants.

Dye
Starting point in making—
Alizarin, alizarin dyestuffs, anilin dyestuffs.

Fats and Oils
Reagent (Brit. 263873) in making—
Emulsified fatty acids, emulsified fats and oils, emulsified fat-splitting compositions, emulsified greasing and lubricating compositions, emulsified sulphonated oils.

Glass
Ingredient (Brit. 371901) of—
Compositions, containing various esters of cellulose, such as cellulose acetate, nitrocellulose, and also various mixed inorganic-aliphatic esters, such as cellulose acetopropionate, used for the production of decorative and protective coatings on glassware and in the manufacture of non-scatterable glass (added for the purpose of increasing the life of the film by absorbing ultra-violet rays).

Ink
Ingredient (French 551615) of—
Printing inks.
Inks used for printing on bank notes, checks and other fiscal paper, secret numbers and marks which are rendered visible by the action of ultra-violet light rays and rays from the x-ray lamp.

Insecticide
Reagent (Brit. 263873) in making—
Emulsified insecticidal and fungicidal preparations.

Leather
Ingredient (Brit. 371901) of—
Compositions, containing various simple esters of cellulose or mixed inorganic-aliphatic esters of cellulose, used in the manufacture of artificial leather and for the production of decorative and protective coatings on leather (added for the purpose of increasing the life of the film by absorbing ultra-violet rays).
Reagent (Brit. 263873) in making—
Emulsified dressing compositions, emulsified fat-liquoring baths, emulsified soaking compositions, emulsified impregnating compositions, emulsified softening compositions, emulsified tanning compositions, emulsified waterproofing compositions.

Metallurgical
Ingredient of—
Compositions, containing various simple esters of cellulose or mixed inorganic-aliphatic esters of cellulose, used for the production of decorative and protective coatings on metallic goods (added for the purpose of prolonging the life of the film by the absorption of ultra-violet rays) (Brit. 371901).
Pickling baths containing sulphuric acid to protect iron against too sharp action of the acid (anthracene is used in the crude form).

Miscellaneous
Ingredient (Brit. 371901) of—
Compositions, containing various simple or mixed inorganic-aliphatic esters of cellulose, used for the production of decorative and protective coatings on various fibrous compositions of matter (added for the purpose of increasing the life of the film by the absorption of ultra-violet rays).
Reagent (Brit. 263873) in making—
Emulsified compositions for impregnating fibrous compositions.
Emulsified cleansing compositions, emulsified automobile polishes, emulsified furniture polishes, emulsified metal polishes, emulsified scouring compositions.
Emulsified waterproofing compositions.
Emulsified compositions for treating rope and twine.
Emulsions of various products, such as asphalts and bitumens.

Paint and Varnish
Ingredient of—
Lacquers, paints, varnishes, dopes, and enamels containing various simple esters of cellulose, such as cellulose acetate and nitrocellulose, or various mixed inorganic-aliphatic esters of cellulose, such as cellulose acetopropionate (added for the purpose of increasing the life of the film by the absorption of ultra-violet rays) (Brit. 371901).
Emulsified asphaltic paints and varnishes (Brit. 263873).
Emulsified paints and varnishes (Brit. 263873).
Shellac emulsions (Brit. 263873).
Waterproofing compositions in emulsified form (Brit. 263873).

Paper
Ingredient (Brit. 371901) of—
Compositions, containing various simple or mixed inorganic-aliphatic esters of cellulose, used for the manufacture of coated paper and also for the production of decorative and protective coatings on paper and pulp products (added for the purpose of increasing the life of the film by the absorption of ultra-violet rays).
Reagent (Brit. 263873) in making—
Emulsified impregnating compositions for treating paper, cardboard, paperboard.
Emulsified waterproofing compositions, emulsified waxing compositions, emulsified sizing compositions, emulsified compositions for finishing paper and pulp products.

Petroleum
Ingredient (U. S. 1734990) of—
Compositions, containing sulphuric acid, nitrobenzene, sodium hydroxide, and naphthalene, used for treating oil wells.
Reagent (Brit. 263873) in making—
Emulsified cuttings oils, emulsions of petroleum and petroleum distillates, kerosene emulsions, naphtha emulsions, soluble greases in emulsified form, solubilized emulsified oils, petroleum pitch emulsions, petroleum tar emulsions.
White oil and paraffin and paraffin oil emulsions.

Pyrotechnic
Ingredient of—
Color lights, signal rockets.

Resins and Waxes
Reagent (Brit. 263873) in making—
Emulsified compositions containing various natural and artificial resins.
Emulsified compositions containing various natural and artificial waxes.
Starting point in making—
Resinous products.
Resins by treatment of crude anthracene with chlorine after suspension in carbon bisulphide or benzene and heating, the products being used for varnishes, and impregnating wood (German 420443).

Rubber
Ingredient (Brit. 371901) of—
Compositions, containing various simple or mixed inorganic-aliphatic esters of cellulose, used for the production of decorative and protective coatings on rubber goods (added for the purpose of increasing the life of the film by the absorption of ultra-violet rays).

Soap
Reagent (Brit. 263873) in making—
Emulsified cleansing compositions, emulsified textile-scouring soaps, emulsified detergent preparations of various kinds.

Stone
Ingredient (Brit. 371901) of—
Compositions, containing various simple or mixed inorganic-aliphatic esters of cellulose, used for the production of protective and decorative coatings on natural and artificial stone (added for the purpose of increasing the life of the film by the absorption of ultra-violet rays).

Textile
——, *Bleaching*
Ingredient (Brit. 263873) of—
Emulsified bleaching baths.

——, *Dyeing*
Assist (Brit. 263873) in—
Baths containing acid dyestuffs.
Ingredient (Brit. 263873) of—
Dye baths in emulsified form.

Anthracene (Continued)

——, *Finishing*
Ingredient (Brit. 263873) of—
Emulsified compositions for coating fabrics.
Emulsified sizing compositions.
Emulsified wetting compositions.
Emulsified washing compositions.
Emulsified waterproofing compositions.

——, *Manufacturing*
Ingredient (Brit. 263873) of—
Emulsified bowking baths.
Emulsified fulling baths.
Emulsified baths for the carbonization of wool.
Emulsified baths for degreasing raw wool.
Emulsified baths for soaking silk.
Emulsified baths for degumming and boiling-off raw silk.
Emulsified compositions used in spinning operations.
Emulsified preparations for kier boiling cotton.
Emulsified mercerizing baths.
Oiling emulsions for treating fabrics.

——, *Printing*
Ingredient of—
Paste for printing calico.

Woodworking
Ingredient (Brit. 371901) of—
Compositions, containing various simple or mixed inorganic-aliphatic esters of cellulose, used for the production of decorative and protective coatings on woodwork (added for the purpose of increasing the life of the film by the absorption of ultra-violet rays).

Anthranilic Acid
Synonyms: Anthranil, Orthoaminobenzoic acid.
Latin: Acidum anthranilicum.
French: Acide anthranilique, Acide d'orthoaminobenzoique.
German: Anthranilsäure, Orthoaminobenzoesäure.

Automotive
Used in automobile radiators to prevent corrosion by water or aqueous solutions of alcohols, glycols, and glycerin.

Chemical
Starting point in making—
Anisylaminoanthraquinone.
Anthranylaminoanthraquinone.
Benzylaminoanthraquinone.
Bornylaminoanthraquinone.
Camphylaminoanthraquinone.
Cinnamylaminoanthraquinone.
Cresylaminoanthraquinone.
Crotonylaminoanthraquinone.
Formylaminoanthraquinone.
Gallylaminoanthraquinone.
Metanylaminoanthraquinone.
Methyl anthranilate.
Methyl methylanthranilate.
Naphthylaminoanthraquinone.
Phenylaminoanthraquinone.
Phenylglycinorthocarboxylic acid with the aid of chloroacetic acid.
Phthalylaminoanthraquinone.
Picrylaminoanthraquinone.
Pyridylaminoanthraquinone.
Resorcinylaminoanthraquinone.
Salicylaminoanthraquinone.
Styrylaminoanthraquinone.
Sulphanylaminoanthraquinone.
Thiosalicylic acid and derivatives.
Tolylaminoanthraquinone.
Various esters and salts.
Xylaminoanthraquinone.
Reagent (Brit. 319794) in making pyracridin drugs with the aid of—
2-Bromopyridin, 2-bromopyridinsulphonic acid, 2-bromopyridinsulphonic acid esters, 2-chloropyridin, 2-chloropyridinsulphonic acid, 2-chloropyridinsulphonic acid esters, 2-iodopyridin, 2-iodopyridinsulphonic acid, 2-iodopyridinsulphonic acid esters.

Dye
Starting point in making—
Alizarin red B, azo dyestuffs, diamond yellow R, lake red D, indanthrene red violet 2RN, indanthrene violet RN, indigo, methyl red, orange dyestuffs, pigment scarlets, pigment scarlet G, scarlets, thioindigo brown, thioindigo pink, thioindigo red, vat dyestuffs, violet dyestuffs.

Miscellaneous
Ingredient of—
Cleaning fluids.

Perfume
As a synthetic perfume base.

Pharmaceutical
In compounding and dispensing practice.

Resins and Waxes
Reagent (Brit. 292912) in making synthetic resins with the aid of—
Acetylcarbamide, allylcarbamide, amylcarbamide, benzoylcarbamide, benzylcarbamide, butylcarbamide, citrylcarbamide, cresylcarbamide, cyanamide, ethylcarbamide, formylcarbamide, gallylcarbamide, heptylcarbamide, hexylcarbamide, isoallylcarbamide, isoamylcarbamide, isobutylcarbamide, isopropylcarbamide, lactylcarbamide, methylcarbamide, naphthylcarbamide pentylcarbamide, phenylcarbamide, phthalylcarbamide, picrylcarbamide, propionylcarbamide, propylcarbamide, salicylcarbamide, tolylcarbamide, xylylcarbamide.

Anthranilic Acid Ester of Ricinoleic Alcohol

Bituminous
Solvent (Brit. 445223) for—
Asphalt and other bituminous bodies.

Dye
Solvent (Brit. 445223) for—
Dyestuffs, particularly oil-soluble coaltar dyes.

Fats, Oils, and Waxes
Solvent (Brit. 445223) for—
Fats, oils, waxes.

Resins
Solvent (Brit. 445223) for—
Oil-soluble glycerol-phthalic acid resins, polymerized vinyl compounds, synthetic resins.

Rubber
Solvent (Brit. 445223) for—
Rubber.

Anthranol

Chemical
Starting point (Brit. 277342) in making benzanthrone derivatives with the aid of—
Diallyl fumarate, diallyl malate, diamyl fumarate, diamyl malate, dibenzyl fumarate, dibenzyl malate, dibutyl fumarate, dibutyl malate, dimethyl fumarate, dimethyl malate.
Starting point in making—
Benzanthrone, diphenyl fumarate, diphenyl malate, diphthalyl fumarate, diphthalyl malate, dipropyl fumarate, dipropyl malate, fumaric acid, maleic acid.

Anthraquinone
French: Anthraquinone.
German: Anthrachinon.

Chemical
Starting point in making—
Alpha-acetaminoanthraquinone.
Alpha-aminoanthraquinone.
Alpha-aminoanthraquinonesulphonic acid, 5 or 8 (usually used in admixture).
Alpha-amino-2-bromanthraquinone.
Alpha-amino-2-brom-4-hydroxyanthraquinone.
Alpha-amino-2-brom-6-hydroxyanthraquinone.
Alpha-amino-4-chloranthraquinone.
Alpha-amino-6-chloranthraquinone.
Alpha-amino-2:4-dibromanthraquinone.
Alpha-amino-4-hydroxyanthaquinone.
Alpha-amino-4:5:8-trihydroxyanthraquinone.
Alpha-anthramine.
Alpha-anthrol.
Alphachloranthraquinone.
Alphahydroxyanthranol.
Alphahydroxyanthraquinone (erythroxyanthraquinone).
Alphamethylanthraquinone.
Alphanitroanthraquinone.
Alphanitroanthraquinone-6-sulphonic acid.
Aminobenzanthrone.
Anthraquinone-2-aldehyde.
Anthraquinonealphasulphonic acid sodium salt (sometimes called "gold salt").
Anthraquinonebetaaldehyde.
Anthraquinonebetasulphonic acid sodium salt (commonly known as "silver salt").
Anthraquinonedisulphonic acids, 1:5 and 1:8 (used either separately or in admixture).

Anthraquinone (Continued)
Anthraquinonedisulphonic acids, 2:6 and 2:8 (used either separately or in admixture).
Anthraquinonylbetaurea chloride.
Anthrone (tautomeric form, anthranol).
Benzanthrone.
Benzanthronequinolin.
Beta-acetaminoanthraquinone.
Beta-aminoanthraquinone.
Betachloranthraquinone.
Betahydroxyanthraquinone.
Chlorbenzanthrone.
Diacetaminoanthraquinone, 1:6.
Diacetaminoanthraquinone, 1:7.
Diaminoanthraquinone, 1:4.
Diaminoanthraquinone, 1:5.
Diaminoanthraquinone, 1:8.
Diaminoanthraquinone, 2:6.
Diaminoanthraquinone, 2:7.
Dibenzanthrone.
Dichloranthraquinone, 1:5 or 1:8 (used either separately or in admixture).
Dichloranthraquinone, 2:6 or 2:7 (used either separately or in admixture).
Dihydroxyanthraquinone, 1:4 (quinizarin).
Dihydroxyanthraquinone, 1:5 (Anthrarufin).
Dihydroxyanthraquinone, 1:8 (chrysazin).
Dihydroxyanthraquinone, 2:6 (anthraflavic acid).
1:5-Dihydroxy-4:8-diaminoanthraquinone.
Dinitroanthraquinone, 1:5 and 1:8 (usually used as a mixture).
Nitrobenzanthrone.
Oxyanthrone.
Various other intermediates.

Dye
Starting point in making—
Alizarin bordeaux B (1:2:5:8-tetrahydroxyanthraquinone).
Alizarin blue-black B (1:2:4-trianilinoanthraquinone, sulphonated).
Alizarin brown R (1:2:3-trihydroxyanthraquinone).
Alizarin cyanin green G (1:4-Ditoluidinoanthraquinone, sulphonated).
Alizarin irisol R (toluidino-4-hydroxyanthraquinone, sulphonated).
Alizarin orange G (1:2-dihydroxy-3-nitroanthraquinone).
Alizarin red, blue shade (1:2-dihydroxyanthraquinone).
Alizarin R G, yellowish (1:2:6-trihydroxyanthraquinone).
Alizarin sapphire B and SE (1:5-dihydroxy-4:8-diaminoanthraquinonesulphonic acids).
Alizarin S X, red shade (1:2:7-trihydroxyanthraquinone).
Anthracene blue WG (polyhydroxylated anthraquinone).
Anthraquinone vat black B (nitroviolanthrone).
Anthraquinone vat blue GCD (chlorinated idanthrone).
Anthraquinone vat blue R (indanthrone).
Anthraquinone vat brown B (condensation product of beta-aminoanthrone).
Anthraquinone vat dark blue B (violanthrone).
Anthraquinone vat green B (nitroviolanthrone).
Anthraquinone vat orange GRN.
Anthraquinone vat orange R (1:2-dianthraquinonylamine).
Anthraquinone vat red G (di[alpha-anthraquinonlylamino]anthraquinone 2:6).
Anthraquinone vat red RN (anthraquinonediachridone).
Anthraquinone vat scarlet G (1:4-di[benzoylamino] anthraquinone).
Anthraquinone vat violet 2R (chlorinated isoviolanthrone).
Anthraquinone vat violet R (isoviolanthrone).
Anthraquinone vat yellow G (flavanthrone).
Anthraquinone vat yellow 3G (succinoylalpha-aminoanthraquinone).
Anthraquinone vat yellow 3GN (2:2-dianthraquinonylurea).
Purpurin (1:2:4-trihydroxyanthraquinone).

Anthraquinone-2-aldehyde
German: Anthrachinon-2-aldehyd.

Chemical
Starting point in making—
Intermediates, pharmaceuticals.
Starting point (Brit. 314020) in making—
Anthraquinonebetacarboxylic acid.
Ethylanthraquinonebetacarboxylate.
Methylanthraquinonebetacarboxylate.
Pharmaceutical derivatives.

Dye
Starting point (Brit. 298545) in making dyestuffs with—
2:3-Aminohydroxyanthraquinone.

2-Anthraquinonylbeta-9-carbazolylethyl Sulphide

Chemical
Intermediate (Brit. 444262 and 444501) in—
Organic syntheses.

Pharmaceutical
Claimed (Brit. 444262 and 444501) to have—
Value for pharmaceutical purposes.

Rubber
Accelerator (Brit. 444262 and 444501) in—
Vulcanizing.

1-Anthraquinonylbetaparatoluenesulphonylethyl Sulphide

Chemical
Intermediate (Brit. 444262 and 444501) in—
Organic syntheses.

Insecticide
Insecticide (Brit. 444262 and 444501) for—
Animal pests, vegetable pests.

Textile
As a dyestuff (when employing suitable initial materials) (Brit. 444262 and 444501).
Assistant (Brit. 444262 and 444501) in—
Textile processing.

2-Anthraquinonylbetaparatoluenesulphonylethyl Sulphide

Chemical
Intermediate (Brit. 444262 and 444501) in—
Organic syntheses.

Insecticide
Insecticide (Brit. 444262 and 444501) for—
Animal pests, vegetable pests.

Textile
As a dyestuff (when employing suitable initial materials) (Brit. 444262 and 444501).
Assistant (Brit. 444262 and 444501) in—
Textile processing.

Anthrone

Chemical
Starting point (Brit. 268830) in making substituted benzanthrones, such as—
Benzyl:benzyl-2-dimethylbenzanthrone.
Benzyl:benzyl-3-dimethylbenzanthrone.
Benzylethylbenzyl-2-methylbenzanthrone.
Benzyldiphenylbenzanthrone.
Benzylhydroxybenzyl-2-acetylbenzanthrone.
Benzylhydroxybenzyl-2-benzoylbenzanthrone.
Benzylhydroxybenzyl-2-benzanthronecarboxylic acid ester.
Benzylhydroxybenzyl-2-phenylbenzanthrone.
Benzylmethylbenzanthrone.
Benzylmethylphenylbenzanthrone.
Benzylparamethoxyphenylbenzanthrone.
Phenylbenzanthrone.

Antimony
Synonyms: Antimony regulus, Star antimony, Star metal.
Latin: Stibium.
French: Antimoine.
German: Antimon, Spiessglanzmetall.
Spanish: Antimonio.
Italian: Antimonia.

Ceramics
Ingredient of—
Bathtub enamels and the like, enamels, glazes.

Chemical
Starting point in making—
Antimony-anilin tartrate, antimony arsenate, antimony arsenite, antimony chloride, antimony crocus, antimony fluorides, antimony glass, antimony iodide, antimony lactate, antimony oxides, antimony oxychloride, antimony potassium tartrate, antimony pyrocatachin, antimony red, antimony saccharate, antimony sulphate, antimony sulphide, antimony sulphurated, antimony tribromide, antimony white, lead antimoniate, sodium antimoniate, zinc antimonide.

Antimony (Continued)

Explosives and Matches
Ingredient of—
Shrapnel metal.

Metallurgical
Hardening agent for—
Lead, tin.
Ingredient of—
Alloys for household utensils and other articles.
Bearing metals, brasses, copper-tin alloys, lead-tin alloys, pewter, stereotype metal, tin-antimony alloys, type metal.

Antimony Betabenzoylpropionate

Plastics
Starting point (U. S. 2001380) in making—
Films.

Antimony Crocus

Synonyms: Saffron of antimony.

Chemical
Starting point in making—
Tartar emetic.

Antimony, Crude

Synonyms: Antimony matte, Antimony trisulphide, Black antimony, Concentrated stibnite, Crude antimony sulphide, Liquated sulphide, Needle antimony, Refined stibnite.
Latin: Antimonii sulphidum, Antimonium crudum, Stibium sulfuratum crudum, Stibium sulfuratum nigrum, Sulfuretum stibicum.
French: Antimoine cru, Sulfure d'antimoine du commerce.
German: Schwefelantimon, Schwefelspiessglanz.
Spanish: Antimonio crudo.
Italian: Solfuro di antimonio.

Chemical
Starting point in making—
Antimony chemicals.

Explosives and Matches
Ingredient of—
Fireworks.
Military shell charges to produce, on explosion, a dense white smoke which aids in range finding.
Military shell primers.
Safety matches.

Fuel
Ingredient of—
Candles.

Metallurgical
Source of—
Antimony metal.

Miscellaneous
As a veterinary medicine.

Antimony Dinaphthylnaphthenate

Lubricant
Addition agent (Brit. 433257) to—
Lubricating oils or greases, especially for use at high temperatures, such as cylinder oils, hydrogenated oils, or oils refined by treatment with sulphuric acid, clay, or extraction solvents.

Antimony Fluoride

French: Fluorure d'antimoine.
German: Fluorantimon, Fluorspiessglanz.

Ceramics
Ingredient of—
Glazes.

Textile
Mordant in—
Cotton dyeing and printing.

Antimony Fluorochloride

Chemical
Active fluorinating agent (U. S. 1978840) for—
Hydrocarbons (requires no catalyst).

Antimony Lactate

French: Lactate d'antimoine.
German: Milchsäueresantimon, Milchsäueresspiessglanz.

Textile
As a mordant in dyeing and printing.

Antimony Oxychloride

Synonyms: Algaroth powder, Antimony chloride (basic).
Latin: Mercurius vitae.
French: Oxychlorure d'antimoine.
German: Chloroxospiessglanz.

Chemical
Starting point in making—
Antimony chemicals, tartar emetic.
As a smoke-producing substance.

Pharmaceutical
In compounding and dispensing practice.

Antimony Pentachloride

French: Pentachlorure d'antimoine.
German: Antimonpentachlorid, Pentachlorantimon.
Spanish: Pentacloruro de antimonio.
Italian: Pentacloruro di antimonio.

Chemical
Catalyst in making—
Tetrachloroethane from acetylene and chlorine.
Chlorinating agent in—
Organic synthesis.

Dye
Chlorinating agent in making—
Dyestuffs.

Antimony Pentoxide

Synonyms: Antimonic acid, Antimonic anhydride, Stibnic acid.
French: Acide d'antimoine, Acide antimonique, Acide stibique, Anhydride antimonique, Anhydride stibique, Pentoxyde d'antimoine, Pentoxyde antimonique.
German: Antimonanhydrid, antimonpentoxyd, antimonsäure.

Ceramics
Reagent in making—
Chinaware and porcelains.
Ingredient of—
Enamels for use on fine ceramic ware.

Chemical
General decolorizing and refining agent.
Reagent in making—
Antimony soaps used for the purpose of rendering wool, woolen materials, felt, feathers, furs, and the like mothproof.
Organic antimony pharmaceuticals.
Substitute for tartar emetic.
Reagent in separating—
Alcohols.
Benzyl alcohol from mixtures with benzyl acetate (Brit. 252570).
Geraniol from terpineol (Brit. 252570).
Phenols (Brit. 252570).
Terpineol from borneol (Brit. 252570).
Terpineol from borneol and geraniol (Brit. 252570).
Starting point in making—
Antimony nitrate, antimony lactate, antimony sulphate, antimony tartrate, antimony tribromide, butyl antimonate, ethyl antimonate, methyl antimonate, propyl antimonate, tartar emetic.

Glass
Decolorizing agent in making—
Glass and glass products.
Ingredient of—
Colorless barium optical glass.
Refining agent in making—
Glass and glass products.

Metallurgical
Ingredient of—
Enamels used on iron and steel ware, to render them opaque.
Reagent in producing—
Metal coatings of antimony on iron, steel, copper, and nickel.

Paint and Varnish
Pigment in making—
Enamels, paints, varnishes, lacquers.

Pharmaceutical
In compounding and dispensing practice.

Rubber
Reagent (German 326819) in—
Vulcanization.

Textile
——, *Dyeing*
Mordant in dyeing yarns and fabrics.

Antimony Pentoxide (Continued)

——, *Printing*
Ingredient of—
Paste for printing colored resists on colored grounds, both being obtained by means of the basic dyestuffs.
Mordant in printing—
Calico and other fabrics.

Antimony-Phenyl Acetate

Petroleum
Addition agent (Brit. 433257) in—
Lubricating oils or greases, especially for use at high temperatures, such as cylinder oils, hydrogenated oils, or oils refined by treatment with sulphuric acid, clay, or extraction solvents.

Antimony-Potassium Oxalate
French: Oxalate d'antimoine et de potasse.
German: Oxalsaeureskaliumantimon, Oxalsaeureskaliumspiessglanz.

Textile
Mordant on—
Silk, wool.

Antimony-Potassium Tartrate
Synonyms: Potassio-tartrate of antimony, Potassium antimonyl tartrate, Tartar emetic, Tartarized antimony, Tartrated antimony.
Latin: Antimonii-potassii tartras, Antimonii-potassio tartras, Antimonium tartaratum, Antimonium tartarizatum, Stibio-kali tartaricum, Tartarus emeticus, Tartaras stibiatus.
French: Émétique tatré stibié, Tartrate d'antimoine et de potasse.
German: Brechweinstein, Spiessglanzkaliumtartrat.
Spanish: Tartrato antimonico potasico.
Italian: Tartrato di antimonico e di potassio.

Leather
Mordant in—
Dyeing.

Perfume
Ingredient of various cosmetics.

Pharmaceutical
In compounding and dispensing practice.
Suggested for use as a diaphoretic and expectorant.

Textile
Mordant in—
Dyeing, printing.

Antimony Red
Synonyms: Antimony oxysulphide, Antimony vermilion.

Paint and Varnish
Pigment in—
Oil colors, water colors.

Rubber
Pigment in—
Rubber goods.

Antimony Saccharate
French: Saccharate d'antimoine.
German: Antimonsaccharat.

Chemical
Ingredient of—
Baking powders (Brit. 252695).

Sugar
Agent in the recovering of sugar from molasses.

Antimony Salts
Synonyms: Antimony salt, de Haen's salt, Sodium antimony trifluoride.
French: Sel d'antimoine, Sel de de Haen.
German: Antimonsalz, de Haensalz.

Miscellaneous
Mordant in dyeing—
Various materials and products.

Textile
Mordant in dyeing and printing—
Textile materials.
Substitute for tartar emetic in—
Dyeing cotton goods, printing calico.

Antimony Sulphate
Synonyms: Antimony trisulphate.
French: Sulfate d'antimoine, Sulfate antimoinique, Trisulfate d'antimoine, Trisulfate antimoinique.
German: Antimonsulfat, Antimontrisulfat, Schwefelsaeuresantimon, Schwefelsaeuresantimontrioxyd.

Chemical
Starting point in making—
Antimony lactate by reaction with calcium lactate.

Explosives and Matches
Ingredient of—
Explosive compositions, pyrotechnic compositions.

Paper
Weighing agent in—
Paper stock.

Textile
——, *Dyeing*
As an assist.

Sanitation
Reagent in treating—
Sewage.

Antimony Sulphides
Synonyms: Antimony pentasulphide, Antimony sulphuret, Antimony trisulphide, Artificial sulphuret of antimony, Crimson antimony, Golden antimony, Red antimony.
Latin: Antimonii sulphidum.
French: Sulfure d'antimoine.
German: Schwefelantimon.
Italian: Solfuro di antimonio.

Chemical
Starting material in making—
Antimony chemicals.

Explosives and Matches
Ingredient of—
Matches.
Military shell charges to produce, on explosion, a dense white smoke which aids in range finding.
Military shell primers.
Military shell priming composition (Brit. 393449).
Military shell priming compositions (U. S. 1779851).
Percussion pellets for cartridges.
Pyrotechnic compositions.

Glass
Ingredient of—
Ruby glass.

Paint and Varnish
Pigment in—
Paints, varnishes.

Pharmaceutical
In veterinary practice.

Rubber
Pigment in—
Rubber goods.

Antimony Tartrolactate
French: Lactate et tartrate d'antimoine.
German: Spiessglanzmilchsäurestartrat.

Textile
Mordant in—
Dyeing, printing.

Antimony Tetroxide
French: Tétroxyde d'antimoine, Tétroxyde antimonique.
German: Antimontetroxyd.

Ceramics
Ingredient of—
Enamels for use on fine ceramic ware.
Reagent in making—
Chinaware and porcelains.

Chemical
General decolorizing and refining agent.
Reagent in making—
Antimony soaps used for mothproofing wool, woolens, felt, feathers, furs, and the like.
Organic antimony pharmaceuticals.
Reagent in separating—
Alcohols.
Benzyl alcohol from mixtures with benzyl acetate (Brit. 252570).
Geraniol from terpineol (Brit. 252570).
Phenols (Brit. 252570).
Terpineol from borneol (Brit. 252570).
Terpineol from borneol and geraniol (Brit. 252570).
Starting point in making—
Antimony nitrate, antimony lactate, antimony sulphate, antimony tribromide, butyl antimonate, ethyl antimonate, methyl antimonate, propyl antimonate, tartar emetic.

Antimony Tetroxide (Continued)

Glass

Decoloring agent in making—
 Glass and glass products.
Ingredient of—
 Colorless barium optical glass.
Refining agent in making—
 Glass and glass products.

Metallurgical

Ingredient of—
 Enamels used on iron and steel ware, to render them opaque.
Reagent in producing—
 Metal coatings of antimony on iron, steel, copper, and nickel.

Paint and Varnish

Pigment in making—
 Enamels, paints, lacquers, varnishes.

Pharmaceutical

In compounding and dispensing practice.

Rubber

Reagent (German 326819) in—
 Vulcanization.

Textile

——, Dyeing

Mordant in dyeing yarns and fabrics.

——, Printing

Ingredient of—
 Paste for printing colored resists on colored grounds, both being obtained by means of the basic dyestuffs.
Mordant in printing—
 Calicoes and other fabrics.

Antimony Thiocellobiose

French: Thiocéllobiose d'antimoine.
German: Antimonthiocellobiose.

Chemical

Starting point (Brit. 398020) in making—
 Complex double compounds of organic heavy metal mercapto compounds.

Antimony Tribromide

Synonyms: Antimony bromide.
French: Bromure d'antimoine, Bromure antimoinique, Tribromure d'antimoine, Tribromure antimoinique.
German: Antimonbromid, Antimontribromid, Tribromantimon.

Analysis

Reagent in mineralogical analysis and in general analytical work.

Antimony Trichloride

Synonyms: Antimonous chloride, Antimony chloride, Butter of antimony, Caustic antimony, Mineral butter.
French: Beurre d'antimoine, Chlorure d'antimoine, Chlorure antimoneux.
German: Antimonbutter, Antimonchlorid, Antimontrichlorid, Kaustisches antimon.

Analysis

Reagent in analyzing—
 Chloral, aromatic hydrocarbons.
Reagent in testing for—
 Sodium.

Chemical

Catalyst in preparing various organic compounds.
Ingredient of catalytic preparations used in making—
 Acenaphthylene, acenaphthaquinone, bisacenaphthylidenedione, naphthalydehydic acid, naphthalic anhydride, and hemimellitic acid from acenaphthene (Brit. 295270).
 Acetaldehyde from ethyl alcohol (Brit. 281307).
 Acetic acid from ethyl alcohol (Brit. 281,307).
 Alcohols from aliphatic hydrocarbons (Brit. 281307).
 Aldehydes and acids by the reduction of the corresponding esters (Brit. 306471).
 Aldehydes and acids from toluene, orthochlorotoluene, orthobromotoluene, orthonitrotoluene, parachlorotoluene, paranitrotoluene, parabromotoluene, metachlorotoluene, metanitrotoluene, metabromotoluene, dichlorotoluenes, dibromotoluenes, dinitrotoluenes, chlorobromotoluenes, bromonitrotoluenes, chloronitrotoluenes (Brit. 295370).
 Aldehydes and acids from xylenes, pseudocumenes, mesitylene and paracymene (Brit. 281307).
 Alphacampholide by the reduction of camphoric acid (Brit. 306471).
 Alphanaphthaquinone from naphthalene (Brit. 281307).
 Anthraquinone from naphthalene (Brit. 295270).
 Benzaldehyde and benzoic acid from toluene (Brit. 281307).
 Benzoquinone from phenanthraquinone (Brit. 281307).
 Benzyl alcohol by the reduction of benzaldehyde (Brit. 306471).
 Benzyl alcohol or benzaldehyde or phthalide by the reduction of phthalic anhydride (Brit. 306471).
 Butyl alcohol by the reduction of crotonaldehyde (Brit. 306471).
 Chloroacetic acid from ethylenechlorohydrin (Brit. 306471).
 Diphenic acid from ethyl alcohol (Brit. 281307).
 Ethyl alcohol by the reduction of acetaldehyde (Brit. 306471).
 Fluorenone from fluorene (Brit. 295270).
 Formaldehyde by the reduction of methane or methanol (Brit. 306471).
 Formaldehyde by the reduction of carbon dioxide or carbon monoxide (Brit. 306471).
 Hydroxyl reduction compounds of anthraquinone, benzoquinone, and the like (Brit. 306471).
 Isopropyl alcohol by the reduction of acetone (Brit. 306471).
 Maleic acid and fumaric acid by the oxidation of toluene, benzene, phenols, tar phenols, or furfural, or from benzoquinone or phthalic anhydride (Brit. 295270).
 Methane by the reduction of carbon dioxide or carbon monoxide (Brit. 306471).
 Methanol by the reduction of carbon dioxide or carbon monoxide (Brit. 306471).
 Naphthaldehydic acid, acenaphthaquinone, bisacenaphthylidenedione from acenaphthylene (Brit. 281307).
 Phenanthraquinone from phenanthrene or diphenic acid (Brit. 295270).
 Phthalic acid and maleic acid from naphthalene (Brit. 295270).
 Primary alcohols by the reduction of the corresponding aldehydes (Brit. 306471).
 Propionic acid and butyric acid and higher alcohols, ketones, and acids by the reduction of carbon dioxide or carbon monoxide (Brit. 306471).
 Reduction products of ketones, aldehydes, acids, esters, alcohols, ethers and other organic compounds which contain oxygen (Brit. 306471).
 Salicylic acid and salicylic aldehyde from cresol (Brit. 295270).
 Secondary butyl alcohol by the reduction of methylethylketone (Brit. 306471).
 Valeryl alcohol by the reduction of valeraldehyde (Brit. 306471).
 Vanillin and vanillic acid by the oxidation of eugenol or isoeugenol (Brit. 295270).
Ingredient (Brit. 304640) of catalytic preparations used in the production of various aromatic and aliphatic amines, including—
 Alphanaphthylamine from alphanitronaphthalene.
 Amines from aliphatic nitro compounds, such as alkyl nitriles, or nitromethane.
 Amines from oximes, Schiff's base, and nitriles.
 Amino compounds from the corresponding nitroanisoles.
 Aminophenols from nitrophenols.
 3-Aminopyridin from 3-nitropyridin.
 Amylamine from pyridin.
 Anilin, azo-oxybenzene, azobenzene, and hydrazobenzene by the reduction of nitrobenzene.
 Cyclohexamine, dicyclohexamine and cyclohexylanilin from nitrobenzene.
 Piperidin from pyridin.
 Pyrrolidin from pyrrol.
 Tetrahydroquinolin from quinolin.
Reagent in making—
 Acetyl tetrachloride.
 Antimonic organic pharmaceuticals.
Starting point in making—
 Antimony oxalate, antimony oxychloride, antimony trioxide, antimony-potassium tartrate (tartar emetic).
 Other antimony salts.

Dye

Reagent in making—
 Lakes, particularly from dyewood extracts.

Antimony Trichloride (Continued)

Explosives
Ingredient of—
Match head compositions, percussion cap compositions, pyrotechnical preparations.

Leather
Mordant in coloring—
Patent leathers.
Reagent in producing—
Special color on finished leathers.

Metallurgical
Reagent in—
Bronzing gun barrels and other metallic articles.
Coloring zinc black.
Staining iron and copper articles.

Miscellaneous
Ingredient of—
Furniture polishes (U. S. 1739322).

Paint and Varnish
Reagent in making—
Antimony vermilion.

Petroleum
Ingredient of—
Motor fuel (U. S. 1753294).
Reagent in preventing and removing—
Discolorations from petroleum products, such as kerosene and gasoline.

Pharmaceutical
In compounding and dispensing practice.

Textile
Mordant in—
Dyeing and printing yarns and fabrics.

Woodworking
Ingredient of—
Preparations used for the preservative treatment of wood and wood products (U. S. 1602959).

Antimony Trioxide

Synonyms: Antimonius acid, Antimony oxide, Flowers of antimony, Oxide of antimony, White oxide of antimony.
Latin: Antimonum oxidum.
French: Acide antimonieux, Fleurs argentines d'antimoine, Oxide d'antimoine.
German: Antimonoxyd, Antimonsaeureanhydrid, Antimontrioxyd.

Ceramics
Ingredient of enamels.

Chemical
Reagent in separating—
Alcohols.
Benzyl alcohol from mixtures with benzyl acetate (Brit. 252570).
Geraniol from terpineol (Brit. 252570).
Phenols (Brit. 252570).
Terpineol from borneol (Brit. 252570).
Terpineol from borneol and geraniol.
Starting point in making—
Antimony nitrate, antimony lactate, antimony sulphate, antimony tartrate, antimony tribromide, butyl antimonate, ethyl antimonate, methyl antimonate, propyl antimonate, tartar emetic.

Glass
Decolorizing agent in making—
Glass and glass products.
Ingredient of—
Colorless barium optical glass.
Refining agent in making—
Glass and glass products.

Metallurgical
Ingredient of—
Enamels for iron and steel (to render them opaque).

Paint and Varnish
Pigment in—
Enamels, paints, varnishes.

Pharmaceutical
In compounding and dispensing practice.

Textile
——, *Dyeing*
Mordant in dyeing yarns and fabrics.
——, *Printing*
Ingredient of—
Paste for printing colored resists on colored grounds, both being obtained by means of the basic dyestuffs.
Mordant in printing fabrics.

Antipyrine

Synonyms: Analgesine, Anodynine, Dihydrodimethylphenylpyrazine, Methozine, Parodyne, Phenazone, Phenyldimethylisopyrazolone, Phenyldimethylpyrazolon.
Latin: Antipyrina, Phenazonum, Pyrazolonum phenyldimethylicum.
French: Analgésine, Antipyrine, Parodyne, Anodynine.
German: Antipyrin, Phenyldimethylpyrazolon.
Spanish: Antipirina.
Italian: Antipirina.

Chemical
Starting point in making—
Amidopyrine, nitrosoantipyrine, various derivatives.

Cosmetic
Ingredient of—
Liquid preparation for giving skin a white appearance, containing also witchhazel extract, rose water, alcohol, glycerin, tallow, magnesium carbonate, and magnesium stearate.

Pharmaceutical
In compounding and dispensing practice.
Suggested for use as—
Analgesic, antipyretic.
Suggested for use in treating—
Gout, neuralgic conditions, rheumatism and other painful afflictions.
Temperature reducing agent in—
Febrile diseases.

Arbutin

Analysis
Reagent for—
Detection and determination of nitric acid and nitrate ions.

Pharmaceutical
In compounding and dispensing practice.

Rubber
Preservative (U. S. 1823119) in treating—
Rubber latex.

Argon

Analysis
Inert gas for laboratory work.

Electrical
Gaseous filler for—
Filament lamp bulbs used for all kinds of illuminating purposes, for general indoor and outdoor lighting as well as for other medium and high-power needs (frequently used in admixture with nitrogen).
Ingredient of—
Gas mixtures used in the so-called "Neon Signs" (sometimes used in admixture with mercury).

Miscellaneous
Commonly said to be used in place of hydrogen in gas thermometers (not so used).

Radio
Often recorded in use as a gaseous filler in radio tubes (not so used now).

Arsenic Acid

Synonyms: Orthoarsenic acid, True arsenic acid.
Latin: Acidum arsenicum.
French: Acide arsénique, Acide orthoarsénique.
German: Arseniksäure, Arsensäure, Orthoarseniksäure, Orthoarsensäure.
Spanish: Acido arsenico, Acido ortoarsenico.
Italian: Acido arsenico, Acido ortoarsenico.

Cement
Ingredient of—
Lime cement (added for the purpose of increasing resistance to acid solutions).

Chemical
General oxidizing agent in carrying out chemical reactions and in making intermediates, organic chemicals, pharmaceuticals and the like.
Reagent in making—
Beta-aminoanthraquinone.
Beta-aminoanthraquinone, beta-aminoanthraquinone derivatives (German 107046), benzyl bromide, neosalvarsan, atoxyl, salvarsan, various organic arsenicals.
Starting point in making—
Ammonium arsenate by reaction with ammonia.
Barium arsenate by reaction with a barium salt, for example, barium chloride.

Arsenic Acid (Continued)
Cadmium arsenate by reaction with a cadmium salt.
Cobaltic arsenate by reaction with a cobalt salt.
Ferric arsenate.
Ferrous arsenate.
Lead arsenate by reaction with litharge.
Lithium arsenate by reaction with lithium carbonate.
Magnesium arsenate by reaction with magnesium hydroxide.
Manganese arsenate.
Mercury arsenate by reaction with a mercury salt.
Nickel arsenate by reaction with a nickel salt.
Potassium arsenate by calcination with potassium nitrate.
Silver arsenate by reaction with a silver salt.
Sodium arsenate by reaction with a sodium salt.
Strontium arsenate by reaction with strontium chloride.
Zinc arsenate by reaction with zinc chloride.

Dye
General oxidizing agent in making various synthetic dyestuffs.
Reagent in making—
Brilliant fern blue, fuchsin NB, magenta red, parafuchsin, quinolin derivatives (German 87334).

Glass
Ingredient of—
Glass batch.

Stone
Reagent in—
Hardening stone.

Textile
Mordant in—
Dyeing and printing.

Arsenic Disulphide
Synonyms: Arsenic bisulphide, Arsenic rouge, Realgar, Red arsenic glass, Red orpiment, Red sulphide of arsenic, Red sulphuret of arsenic, Ruby arsenic.
French: Arsénic rouge, Orpin, Sulfure d'arsénic rouge.
German: Arsenbisulfid, Arsenblende, Arsendisulfid, Arsenrot, Rauschrot, Roterarsenik, Rotesarsenglas, Rotglas.

Explosives
In pyrotechnics.
Ingredient of—
Blue fire, white bengal fire.

Leather
Reagent for—
Removing wool from sheepskins prior to tanning.

Metallurgical
Ingredient of—
Lead compositions used in making bullets.

Paint and Varnish
Pigment in making—
Fixed colors.

Textile
——, *Dyeing*
Ingredient of—
Liquors for dyeing.
——, *Printing*
Ingredient of—
Pastes for printing calicoes.

Arsenic Trioxide
Synonyms: Arsenic, Arsenious acid, Arsenious oxide, Arsenous anhydride, White arsenic.
Latin: Acidum arsenicosum, Acidum arseniosum, Arsenicum album, Arseni trioxidum.
French: Acide arsénieux, Arsénic blanc, Fleurs d'arsénic.
German: Arsenictesäure, Arsenigesäure, Arsenik, Weisserarsenik.
Spanish: Acido arsenioso, Anhidrido arsenioso, Arsenico blanco.
Italian: Acido arsenioso, Anidride arseniosa, Arsenico bianco.

Agriculture
Weed killer.

Analysis
Reagent in—
Analytical processes involving control and research work.

Building Construction
Adherance promoter in—
Cement coatings for wooden piles used in piers, landing places, foundations, reclaimed land, and the like.
Elasticity promoter in—
Cement coatings for wooden piles used in piers, landing places, foundations, reclaimed land, and the like.
Insecticide in—
Cement coatings for wooden piles used in piers, landing places, foundations, reclaimed land, and the like.
Peeling preventer in—
Cement coatings for wooden piles used in piers, landing places, foundations, reclaimed land, and the like.
Starting point (Brit. 435015) in making—
Aluminum arsenate used in new hydraulically binding cements.

Ceramics
Ingredient of—
Enamels.

Chemical
Catalyst (Brit. 402438) in making—
Ethylene oxide.
Source of—
Arsenic.
Starting point in making—
Aluminum arsenate, arsanilic acid, arsenic acid (orthoarsenic acid), arsenic bromide, arsenic chloride, arsenic iodide, arsenic pentasulphide, arsenic pentoxide, calcium arsenate, calcium arsenite, chemical pigments, lead arsenate, lead arsenite, organic arsenicals used as drugs, Paris green, soda arsenate, soda arsenite.

Dye
Process material in making—
Chemicals for the fixation of anilin colors.

Glass
Clarifying agent in making—
Crystal glass.
Decolorizing agent in making—
Low-grade glass.
Remover of—
Yellowish tints imparted to glass by iron oxide.

Insecticide and Fungicide
Ingredient of—
Ant poisons, cattle dips, exterminants for rodents and the like, fly poisons (for flypaper), fungicidal compositions, insecticidal compositions, pesticidal compositions.

Leather
Ingredient of—
Skin preservatives.
Process material in—
Tanning pelts.

Mechanical
Ingredient of—
Boiler compounds.

Metallurgical
Fluidity promoter in—
Lead used in making shot and the like.
Hardening agent for—
Lead used in making shot and the like.
Ingredient of—
Alloys used in making locomotive firebox cases.

Paint and Varnish
Starting point in making—
Emerald green, emperor green, imperial green, kaiser green, king's green, meadow green, mitis green, moss green, new green, Paris green, parrot green, patent green, Scheele's green, Schweinfurth green, Vienna green.

Pharmaceutical
In compounding and dispensing practice.
Source of arsenic in making—
Antisyphiletics, such as arsphenamine and the like.

Textile
Fixing agent for—
Anilin dyes.
Mordant in—
Dyeing processes, printing processes.

Wood
Ingredient of—
Perservative and protective compositions.

Ascorbic Acid
French: Acide ascorbique.

Photographic
As a developing agent (Brit. 430264).

Aseptic Acid

Chemical
Starting point in making—
Esters and salts, pharmaceuticals.

Pharmaceutical
In compounding and dispensing practice.

Asphalt

Synonyms: Asphaltum, Bitumen, Gilsonite, Jew's pitch, Judean pitch, Manjak, Mineral pitch, Petroleum pitch.
French: Asphalte.
German: Erdharz, Erdpech.
Italian: Asfalto.

Construction
Ingredient of various waterproofing cements for coating structures where there is apt to be an inseepage of water, such as cellar walls, for joints, for lining purposes and the like.
Waterproofing agent for—
Bridge piers, dock timbers, general purposes, industrial tanks, piling.

Electrical
Ingredient of—
Insulating compositions.

Fuel
As a briquetting binder.
As a fuel.

Gas
Coating material for—
Inside of gas mains and pipes, acting as a waterproofing agent, an anticorrosive, and an antiseepage agent.

Glass
Reagent in—
Glass etching.

Ink
Ingredient of—
Transfer inks.

Mechanical
Impregnating agent for—
Power-transmission belting, conveyor belting.

Miscellaneous
Ingredient of—
Chewing gums.

Paint and Varnish
Ingredient of—
Acidproof paint or varnish, black varnishes, enamels, japans, roof cements, waterproof paint or varnish.

Paper
Impregnating agent for—
Asphaltic papers, roofing papers, special insulating papers.

Petroleum
Ingredient of—
Asphalt-clay compositions used for protecting oil pipelines from corrosion (usually in 50-50 mixture).
Mix in Cozzborough process for lubricant production.

Plastics
Binder in various plastic manufactures.
Ingredient of various plastic compositions.

Printing
Reagent in—
Etching processes, lithography.

Road Building
As a binder in sand roads.
As a surfacing material.
Ingredient of—
Road-making and paving compositions.

Rubber
Filler for—
Rubber compositions.
Ingredient of—
Mineral rubber.

Stone
Ingredient of—
Asphaltic factitious stone.

Water Supply
Coating material for—
Inside of water mains and pipes, acting as a waterproofing agent, an anticorrosive and an antiseepage agent.

Atropinesulphuric Acid

Chemical
Starting point in making—
Esters and salts, pharmaceuticals.

Atroscine Hydrobromide

French: Bromhydrate d'atroscine.
German: Atroscinbromhydrat, Atroscinhydrobromid, Bromwasserstoffsäuresatroscin.
Spanish: Bromhidrato de atroscina.
Italian: Bromidrato di atroscina.

Pharmaceutical
In general compounding and dispensing practice.
Suggested for use as—
Mydriatic, narcotic, sedative.

Avocado Oil

Cosmetic
Basic ingredient of—
Cosmetic creams.

Ball Clay

Ceramics
Ingredient of—
Architectural terra cotta.
Art pottery.
Chlorine-resistant lute comprising mixture with 40° caustic soda.
Enameling for granite ware.
High-grade tile.
White-bodied ware, including china, porcelain, general ware, chemical porcelain, porcelain electrical supplies, sanitary ware.

Miscellaneous
Ingredient of—
Artificial abrasives, asbestos products.

Paper
As a filler (very small amounts only as compared with kaolin).

Refractories
Raw material in making—
Glass factory pots and tanks, pins, stilts, and spurs for potter's use, saggers for potters, wads, porous bodies comprising refractory granules with a ceramic bond (Brit. 404306).

Barium Acetate

Latin: Baryum aceticum.
French: Acétate de baryte, Acétate barytique, Acétate de baryum.
German: Essigsäuresbaryum, Baryumacetat, Baryumazetat.

Analysis
Reagent for—
Determining calcium and the alkalies.
Precipitating sulphates and chromates.
Ingredient of catalytic preparations used in making—
Acenaphthylene, acenaphthaquinone, bisacenaphthylidenedione, naphthaldehydic acid, naphthalic anhydride, and hemimellitic acid from acenaphthylene (Brit. 295270).
Acetaldehyde from ethyl alcohol (Brit. 281307).
Acetic acid from ethyl alcohol (Brit. 281307).
Alcohols from aliphatic hydrocarbons (Brit. 281307).
Aldehydes and acids by the reduction of esters (Brit. 306471).
Alphacampholid from camphoric acid by reduction (Brit. 306471).
Aldehydes and acids from toluene, orthochlorotoluene, orthonitrotoluene, orthobromotoluene, parachlorotoluene, parabromotoluene, paranitrotoluene, metachlorotoluene, metabromotoluene, metanitrotoluene, dichlorotoluenes, dibromotoluenes, dinitrotoluenes, chlorobromotoluenes, chloronitrotoluenes, bromonitrotoluenes (Brit. 295270).
Aldehydes and acids from xylenes, pseudocumene, mesitylene, and paracymene (Brit. 281307).
Alphanaphthaquinone from naphthalene (Brit. 281307).
Anthraquinone from naphthalene (Brit. 295270).
Benzaldehyde and benzoic acid from toluene (Brit. 281307).
Benzoquinone from phenanthraquinone (Brit. 281307).
Benzyl alcohol by the reduction of benzaldehyde (Brit. 306471).
Benzyl alcohol or benzaldehyde or phthalide by the reduction of phthalic anhydride (Brit. 306471).

Barium Acetate (Continued)

Butyl alcohol by the reduction of crotonaldehyde (Brit. 306471).
Chloroacetic acid from ethylenechlorohydrin (Brit. 295270).
Diphenic acid from ethyl alcohol (Brit. 281307).
Ethyl alcohol by the reduction of acetaldehyde (Brit. 306471).
Fluorenone from fluorene (Brit. 295270).
Formaldehyde by the reduction of methanol or methane (Brit. 306471).
Formaldehyde by the reduction of carbon dioxide or carbon monoxide (Brit. 306471).
Hydroxyl reduction compounds of anthraquinone, benzoquinone, and the like (Brit. 306471).
Isopropyl alcohol by the reduction of acetone (Brit. 306471).
Maleic acid and fumaric acid by the oxidation of toluene, benzene, phenols, tar phenols, or furfural, or from benzoquinone or phthalic anhydride (Brit. 306471).
Methane by the reduction of carbon dioxide or carbon monoxide (Brit. 295270).
Methanol by the reduction of carbon dioxide or carbon monoxide (Brit. 306471).
Naphthaldehydic acid, acenaphthaquinone, or bisacenaphthylidenedione from acenaphthene (Brit. 281307).
Phenanthraquinone from phenanthrene or diphenic acid (Brit. 295270).
Phthalic acid and maleic acid from naphthalene (Brit. 295270).
Primary alcohols by the reduction of the corresponding aldehydes (Brit. 306471).
Propionic acid and butyric acid and higher alcohols, ketones, and acids by the reduction of carbon dioxide or carbon monoxide (Brit. 306471).
Reduction products of ketones, aldehydes, acids, esters, alcohols and other organic compounds containing oxygen (Brit. 306471).
Salicylic acid and salicylic aldehyde from cresol (Brit. 295270).
Secondary butyl alcohol by the reduction of methylethyl ketone (Brit. 306471).
Valeryl alcohol by the reduction of valeraldehyde (Brit. 306471).
Vanillin and vanillic acid from eugenol and isoeugenol (Brit. 295270).

Ingredient (Brit. 306471) of catalytic preparations used in the production of various aromatic and aliphatic amines, including—
Alphanaphthylamine from alphanitronaphthalene.
Amines from aliphatic nitro compounds, such as alkyl nitriles, or nitromethane.
Amylamin from pyridin.
Anilin, azo-oxybenzene, azobenzene, and hydrazobenzene from nitrobenzene by reduction.
Aminophenols from nitrophenols.
3-Aminopyridin from 3-nitropyridin.
Amino compounds from the corresponding nitroanisoles.
Amines from oximes, Schiff's base, and nitriles.
Cyclohexamine, dicyclohexamine, and cyclohexylanilin from nitrobenzene.
Piperidin from pyridin.
Pyrrolidin from pyrrol.
Tetrahydroquinolin from quinolin.

Starting point in making—
Acetone, barium pyrophosphate, barium stearate, barium sulphate, various barium salts.

Paint and Varnish

Reagent in making—
Verdigris.

Textile

Mordant for turkey-red dyeing.
Mordant in dyeing and printing cottons.

Barium Albuminate

French: Albuminate de baryum.
German: Albuminsaeuresbarium, Bariumalbuminat.

Rubber

Reagent (U. S. 1640817) in—
Reclaiming rubber.

Barium Aluminate

Abrasive

As an abrasive agent.
Ingredient of—
Compositions used in the preparation of abrasive wheels, stones, and the like.

Chemical

Ingredient of catalytic mixtures used in making—
Acenaphthylene, acenaphthaquinone, bisacenaphthylidenedione, naphthaldehydic acid, naphthalic anhydride, and hemimellitic acid (Brit. 295270).
Acetaldehyde from ethyl alcohol (Brit. 281307).
Acetic acid from ethyl alcohol (Brit. 281307).
Alcohols from aliphatic hydrocarbons (Brit. 281307).
Aldehydes or acids by the reduction of the corresponding esters (Brit. 306471).
Aldehydes and acids from toluene, orthochlorotoluene, orthonitrotoluene, orthobromotoluene, metachlorotoluene, metabromotoluene, metanitrotoluene, parachlorotoluene, parabromotoluene, paranitrotoluene, dinitrotoluenes, dibromotoluenes, dichlorotoluenes, chlorobromotoluenes, chlorotoluenes, bromonitrotoluenes (Brit. 295270).
Aldehydes and acids from xylenes, pseudocumenes, mesitylene, and paracymene (Brit. 281307).
Alphacampholide from camphoric acid by reduction (Brit. 306471).
Alphanaphthaquinone from naphthalene (Brit. 281307).
Anthraquinone from naphthalene (Brit. 281037).
Benzaldehyde and benzoic acid from toluene (Brit. 281307).
Benzoquinone from phenanthrene (Brit. 281307).
Benzyl alcohol by the reduction of benzaldehyde (Brit. 306471).
Benzyl alcohol, benzaldehyde, or benzyl phthalide by the reduction of phthalic anhydride (Brit. 306471).
Butyl alcohol by the reduction of crotonaldehyde (Brit. 306471).
Diphenic acid from ethyl alcohol (Brit. 281307).
Ethyl alcohol by the reduction of acetaldehyde (Brit. 306471).
Fluorenone from fluorene (Brit. 295270).
Formaldehyde by the reduction of methane or methanol (Brit. 306471).
Formaldehyde by the reduction of carbon dioxide or carbon monoxide (Brit. 306471).
Hydroxyl compounds by the reduction of anthraquinone, benzoquinone, and the like (Brit. 306471).
Isopropyl alcohol by the reduction of acetone (Brit. 306471).
Maleic acid and fumaric acid by the oxidation of toluene, benzene, phenols, tar phenols, or furfural, or from benzoquinone or phthalic anhydride (Brit. 295270).
Methane by the reduction of carbon dioxide or carbon monoxide (Brit. 306471).
Methanol by the reduction of carbon dioxide or carbon monoxide (Brit. 306471).
Naphthaldehydic acid, acenaphthaquinone, bisacenaphthylidenedione from acenaphthylene (Brit. 281307).
Phenanthraquinone from phenanthrene or diphenic acid (Brit. 295270).
Primary alcohols by the reduction of the corresponding aldehydes (Brit. 306471).
Propionic acid, butyric acid, and higher alcohols, ketones, and acids by the reduction of carbon dioxide or carbon monoxide (Brit. 306471).
Reduction products of ketones, aldehydes, acids, esters, alcohols, ethers, and other organic compounds which contain oxygen (Brit. 306471).
Salicylic acid and salicylic aldehyde from cresol (Brit. 295270).
Secondary butyl alcohol by the reduction of methylethyl ketone (Brit. 306471).
Valeryl alcohol by the reduction of valeraldehyde (Brit. 306471).
Vanillin and vanillic acid from eugenol or isoeugenol (Brit. 295270).

Ingredient (Brit. 304640) of catalytic preparations used in the production of various aromatic and aliphatic amines, including—
Alphanaphthylamine from alphanitronaphthalene.
Amines from aliphatic nitro compounds, such as alkyl nitriles, or nitromethane.
Amylamine from pyridin.
Anilin, azo-oxybenzene, azobenzene, and hydrazobenzene from nitrobenzene by reduction.
Aminophenols from nitrophenols.
3-Aminopyridin from 3-nitropyridin.
Amino compounds from the corresponding nitroanisoles.
Cyclohexamine, dicyclohexamine, and cyclohexylanilin from nitrobenzene.
Piperidin from pyridin.

Barium Aluminate (Continued)
Pyrrolidin from pyrrol.
Tetrahydroquinolin from quinolin.

Leather
Ingredient (French 594524) of—
Compositions used for mordanting, loading, and impermeabilizing leather.

Miscellaneous
Ingredient (French 594524) of—
Compositions used for mordanting, loading, and impermeabilizing various animal and vegetable substances.

Paper
Ingredient (French 594524) of—
Compositions used for mordanting, loading, and impermeabilizing paper and pulp products.

Textile
Ingredient (French 594524) of—
Compositions used for mordanting, loading, and impermeabilizing animal and vegetable textile materials.

Water
Reagent for—
Softening water.
Ingredient (Brit. 316023) of—
Water-softening compositions containing lime and sodium carbonate.

Barium-Aluminum-Iron Cyanide

Chemical
Catalyst (Brit. 446411) in—
Halogenating unsaturated hydrocarbons.

Barium-Anilin

Dye
Reagent (German 436533) in making anthracene dyestuffs from—
3:9-Dichlorobenzanthrone, 11:3-dichlorobenzanthrone.

Barium Betanaphtholdisulphonate
French: Bétanaphtoldisulphonate de barium.
German: Bariumbetanaphtoldisulfonat, Betanaphtoldisulfonsaeuresbarium.

Chemical
Starting point in making—
Aluminum betanaphtholdisulphonate.

Barium Borate
French: Borate de barium.
German: Borsaeures barium.

Chemical
Reagent in making—
Amydricaine, pentaborate, amylocaine borate, benzamine pentaborate, benzocaine borate, butyl pentaborate, cocaine borate, ethocaine pentaborate, glycocaine borate, orthocaine borate, phenocaine borate.

Barium Bromide
French: Bromure de baryum.
German: Bariumbromid, brombarium.
Spanish: Bromuro de bario.
Italian: Bromuro di bario.

Chemical
Reagent in making—
Bromides.

Barium Camphorate
French: Camphorate de barium.
German: Baryumcamphorat, Kamphersaeuresbaryum, Kamphorsaeuresbaryum.

Chemical
Starting point in making—
Hyoscyamine camphorate (Brit. 269498).

Pharmaceutical
In compounding and dispensing practice.

Barium Caprinate
French: Caprinate de baryum.
German: Capricsaeuresbarium.

Chemical
Reagent in making—
Decylic aldehyde.

Barium Carbide
French: Carbure de baryum.
German: Bariumcarbid.
Spanish: Carburo de bario.
Italian: Carburo di bario.

Abrasive
Ingredient in making—
Bariundum.

Chemical
Starting point in making—
Barium hydroxide.
Synthetic alcohol (with recovery of barytes) as by-product of sugar-making (French 521048).

Metallurgical
Restrainer (Austria 106982) of—
Iron attack by sulphuric acid pickling baths.

Barium Chlorate
French: Chlorate de barium, chlorate de baryte.
German: Bariumchlorat, Chlorsäuresbarium, Chlorsäuresbaryt.
Spanish: Clorato de bario.
Italian: Clorato di bario.

Chemical
Starting point in making—
Other chlorates.

Explosives and matches
Ingredient of—
Explosive compositions, green-fire compositions, pyrotechnic compositions.

Textile
Mordant in dyeing.

Barium Cyanamide
French: Cyanamide de baryum.
German: Bariumcyanamid.
Spanish: Cianamida de bario.
Italian: Cianamide di bario.

Chemical
Catalyst in making—
Ammonia (U. S. 1352177, 1352179, and 1473543).
Hydrocyanic acid (U. S. 1352176).
Drying agent (U. S. 1454591) for—
Gases.
Starting point in making—
Ammonia catalysts (U. S. 1352178).
Sodium cyanamide (U. S. 1318258).
Sodium cyanide (U. S. 1318258).
Sodium ferrocyanide (U. S. 1318258).

Barium Ethylxanthate
Synonyms: Barium ethylxanthogenate.
French: Éthyle-xanthogénate de baryum.
German: Aethylxanthogensaeuresbaryum, Bariumaethylxanthogenat.

Chemical
Starting point in making—
Accelerator of rubber vulcanization in combination with sulphur monochloride (Brit. 265169).

Barium Fluoride
French: Fluorure de baryum.
German: Baryumfluorid, Fluorwasserstoffsaeuresbaryum.

Ceramics
Ingredient of enamel compositions for—
Chinaware, porcelains, potteries.

Jewelry
Ingredient of batch in making—
Artificial rubies.

Miscellaneous
Ingredient of—
Antiseptic preparations, embalming fluids.

Barium Hexanitride

Electrical
Source (U. S. 1931647) of—
Barium and nitrogen (by thermal dissociation) in vacuum tubes.

Barium Iodate
French: Iodate barytique, Iodate de baryum.
German: Baryumjodat, Jodsaeuresbaryum.

Food
Preservative (Brit. 274164) in treating—
Butter, cream, eggs, fish, fruit preserves, margarin, milk, meat.

Barium Pentasulphide
French: Pentasulfide de baryum.
German: Baryumpentasulfid.

Chemical
Starting point in making—
Carbon bisulphide.

Barium Percarbonate
French: Percarbonate de baryum.
German: Perkohlensaeuresbaryum.

Chemical
Starting point in making—
Hydrogen peroxide.

Barium Permanganate
French: Permanganate de baryum.
German: Übermangansaueresbaryum.

Chemical
Starting point in making—
Mercury permanganate, silver permanganate.

Barium Phenolsulphonate
Synonyms: Barium sulphocarbolate, Barium sulphophenate, Barium sulphophenolate, Barium sulphophenylate.
French: Phénolsulphate de baryum, sulfophénate de baryum.
German: Bariumphenolsulfonat, Bariumsulfocarbolat, Bariumsulfophenat, Bariumsulfophenolat, Phenolsulphosaeuresbaryum.

Chemical
Starting point in making phenolsulphonate of—
Aluminum, bismuth, calcium, cadmium, copper, lead, lithium, magnesium, manganese, mercury, nickel, potassium, silver, sodium, strontium, zinc.

Barium Polysulphide
French: Polysulfure barique, Polysulfure de baryum.
German: Baryumpolysulfid.
Spanish: Polisulfurato de bario.
Italian: Polisulfurato di bario.

Fats and Oils
Reagent (Brit. 271553) in making—
Vulcanized oils.

Insecticide
As an insecticide and fungicide.
Ingredient (U. S. 1578520) of—
Insecticidal and fungicidal preparations.

Metallurgical
Reagent in—
Sulphiding oxidized ores for separation by flotation.

Paper
Ingredient (Brit. 271553) of—
Compositions, containing rubber latex, used for treating paper and pulp.

Rubber
Reagent (Brit. 271553) in treating—
Rubber latex.

Barium Silicate
Synonyms: Silicate of barium.
German: Bariumsilicat, Kieselsaeuresbarium.

Glass
Ingredient of batch in making glass.

Miscellaneous
Softening agent in treating hard waters.

Sugar
Reagent in treating—
Molasses to recover sugar content.
Plant juices to recover sugar content (Brit. 249759).

Barium Silicofluoride
Synonyms: Barium fluosilicate.
French: Fluosilicate de baryum, Silicofluorure de baryum.
German: Bariumsiliciumfluorid, Bariumflorsilikat, Fluosiliciumbarium, Siliciumfluorstoffsaeuresbarium, Siliciumfluorwasserstoffsaeuresbarium.

Chemical
Starting point (German 426735) in making—
Barium peroxide through the intermediate formation of barium fluoride and barium nitrate.

Construction
Preservative (Brit. 271293) in treating—
Brickwork, stucco, and other structural work.

Ceramics
Ingredient of various ceramic products.

Stone
Preservative in treating—
Artificial stone, natural stone.

Woodworking
As a preservative.

Barium Telluride
French: Tellurure de baryum.
German: Bariumtellurid, Tellurbarium.

Chemical
Catalyst (Brit. 263877) in making—
Acetone from isopropyl alcohol.
Dehydrogenated products from cyclohexane.
Isobutyraldehyde from isobutyl alcohol.
Isoveraldehyde from isoamyl alcohol (Brit. 262120).
Naphthalene from tetrahydronaphthalene.
Paracymene from turpentine.
Reagent (Brit. 292222) in making organic tellurium compounds from—
Pentamethylene alphaepsilondibromide.
Pentamethylene alphaepsilondichloride.
Pentamethylene alphaepsilondifluoride.
Pentamethylene alphaepsilondi-iodide.

Barium Titanate
French: Titanate barique, Titanate de baryte, Titanate de baryum.
German: Barium titanat, Titansäuresbaryt, Titansäuresbaryterde, Titansäuresbaryum.
Spanish: Titanato de bario.
Italian: Titanato di bario.

Chemical
Ingredient of catalytic mixtures used in the manufacture of—
Acenaphthylene, acenaphthaquinone, bisacenaphthylidenedione, naphthaldehydic acid, naphthalic anhydride, and hemimellitic acid from acenaphthene (Brit. 295270).
Acetaldehyde from ethyl alcohol (Brit. 281307).
Acetic acid from ethyl alcohol (Brit. 281307).
Alcohols from aliphatic alcohols (Brit. 281307).
Aldehydes and acids by the reduction of the corresponding aldehydes (Brit. 306471).
Aldehydes and acids from toluene, orthochlorotoluene, orthonitrotoluene, orthobromotoluene, parachlorotoluene, paranitrotoluene, parabromotoluene, metachlorotoluene, metabromotoluene, metanitrotoluene, dichlorotoluenes, dibromotoluenes, dinitrotoluenes, chlorobromotoluenes, chloronitrotoluenes, bromonitrotoluenes (Brit. 295270).
Aldehydes and acids from xylenes, pseudocumenes, mesitylene, and paracymene (Brit. 281307).
Alphanaphthaquinone from naphthalene (Brit. 281307).
Anthraquinone from naphthalene (Brit. 295270).
Benzaldehyde and benzoic acid from toluene (Brit. 281307).
Benzoquinone from phenanthraquinone (Brit. 281307).
Benzyl alcohol from benzaldehyde by reduction (Brit. 306471).
Benzyl alcohol, benzaldehyde, benzyl phthalide by the reduction of phthalic anhydride (Brit. 306471).
Butyl alcohol by the reduction of crotonaldehyde (Brit. 306471).
Chloroacetic acid from ethylenechlorohydrin (Brit. 295270).
Diphenic acid from ethyl alcohol (Brit. 281307).
Ethyl alcohol by the reduction of acetaldehyde (Brit. 306471).
Fluorenone from fluorene (Brit. 295270).
Formaldehyde by the reduction of methane or methanol (Brit. 306471).
Formaldehyde by the reduction of carbon dioxide or carbon monoxide (Brit. 306471).
Hydroxyl compounds by the reduction of anthraquinone, benzoquinone, and similar compounds (Brit. 306471).
Isopropyl alcohol by the reduction of acetone (Brit. 306471).
Maleic acid and fumaric acid by the oxidation of toluene, benzene, phenols, tar phenols, or furfural, or from benzoquinone or phthalic anhydride (Brit. 295270).
Methane by the reduction of carbon dioxide or carbon monoxide (Brit. 306471).
Methanol by the reduction of carbon dioxide or carbon monoxide (Brit. 306471).

Barium Titanate (Continued)

Naphthaldehydic acid, acenaphthaquinone, or bisacenaphthylidenedione from acenaphthylene (Brit. 281307).
Phenanthraquinone from phenanthrene or diphenic acid (Brit. 295270).
Phthalic acid and maleic acid from naphthalene (Brit. 295270).
Primary alcohols by the reduction of the corresponding aldehydes (Brit. 306471).
Propionic acid and butyric acid and higher alcohols, ketones, and acids by the reduction of carbon dioxide or carbon monoxide (Brit. 306471).
Reduction products of ketones, aldehydes, acids, esters, alcohols, ethers, and other organic compounds which contain oxygen (Brit. 306471).
Salicylic acid and salicylic aldehyde from cresol (Brit. 295270).
Secondary butyl alcohol by the reduction of methylethylketone (Brit. 306471).
Valeryl alcohol by the reduction of valeraldehyde (Brit. 306471).
Vanillin and vanillic acid by the oxidation of eugenol or isoeugenol (Brit. 295270).

Ingredient (Brit. 306460) of catalytic preparations used in the production of various aliphatic and aromatic compounds, including—
Alphanaphthylamine from alphanitronaphthalene.
Amines from aliphatic nitro compounds, such as allyl nitriles or nitromethane.
Amino compounds from the corresponding nitroanisoles.
Amylamine from pyridin.
Anilin, azo-oxybenzene, azobenzene, and hydrazobenzene from benzene by reduction.
Aminophenols from nitrophenols.
3-Aminopyridin from 3-nitropyridin.
Cychlohexamine, dicyclohexamine, and cyclohexylanilin from nitrobenzene.
Piperidin from pyridin.
Pyrrolidin from pyrrol.
Tetrahydroquinolin from quinolin.

Paint and Varnish

Ingredient of—
Paints and varnishes.
White pigment (used in admixture with zinc white).

Plastics

White pigment in—
Plastic compositions containing phenol-formaldehyde resins.

Barium Vanadate

French: Vanadate de baryum.
German: Baryumvanadat, Vanadinsaeuresbaryta, Vanadinsaeuresbaryum.

Chemical

Ingredient of catalytic preparations used in making—
Acenaphthylene, acenaphthaquinone, bisacenaphthylidenedione, naphthaldehydic acid, naphthalic anhydride, and hemimellitic acid from acenaphthene (Brit. 295270).
Acetaldehyde from ethyl alcohol (Brit. 281307).
Acetic acid from ethyl alcohol (Brit. 281307).
Alcohols from aliphatic hydrocarbons (Brit. 281307).
Aldehydes or alcohols by reduction of esters (Brit. 306471).
Alphacampholide by the reduction of camphoric acid (Brit. 306471).
Aldehydes and acids from toluene, orthochlorotoluene, orthobromotoluene, orthonitrotoluene, parachlorotoluene, parabromotoluene, paranitrotoluene, metanitrotoluene, metabromotoluene, metachlorotoluene, dichlorotoluenes, dibromotoluenes, dinitrotoluenes, chlorobromotoluenes, chloronitrotoluenes, bromonitrotoluenes (Brit. 295270).
Aldehydes and acids from xylenes, pseudocumenes, mesitylene, and paracymene (Brit. 295270).
Alphanaphthaquinone from naphthalene (Brit. 281307).
Anthraquinone from naphthalene (Brit. 281307).
Benzaldehyde and benzoic acid from toluene (Brit. 281307).
Benzoquinone from phenanthraquinone (Brit. 281307).
Benzyl alcohol or benzaldehyde or phthalide by the reduction of phthalic anhydride (Brit. 306471).
Butyl alcohol by the reduction of crotonaldehyde (Brit. 306471).
Chloroacetic acid from ethylenechlorohydrin (Brit. 295270).
Diphenic acid from ethyl alcohol (Brit. 281307).
Ethyl alcohol by the reduction of acetaldehyde (Brit. 306471).
Fluorenone from fluorene (Brit. 295270).
Formaldehyde by the reduction of methanol or methane (Brit. 306471).
Formaldehyde by the reduction of carbon monoxide or carbon dioxide (Brit. 306471).
Hydroxyl reduction products of anthraquinone, benzoquinone, and the like (Brit. 306471).
Isopropyl alcohol by the reduction of acetone (Brit. 306471).
Maleic acid and fumaric acid by the oxidation of toluene, benzene, phenols, tar phenols, or furfural, or from benzoquinone or phthalic anhydride (Brit. 295270).
Methane by the reduction of carbon dioxide or carbon monoxide (Brit. 306471).
Methanol by the reduction of carbon dioxide or carbon monoxide (Brit. 306471).
Naphthaldehydic acid, acenaphthaquinone, or bisacenaphthylidenedione from acenaphthylene (Brit. 281307).
Phenanthraquinone from phenanthrene or diphenic acid (Brit. 295270).
Phthalic acid and maleic acid from naphthalene (Brit. 295270).
Primary alcohols by the reduction of aldehydes (Brit. 306471).
Propionic acid, butyric acid, and higher alcohols, ketones, and acids from carbon dioxide or carbon monoxide (Brit. 306471).
Reduction products of carbon dioxide and carbon monoxide (Brit. 306471).
Reduction products of ketones, aldehydes, acids, esters, alcohols, ethers, and organic compounds containing oxygen (Brit. 306471).
Salicylic acid and salicylic aldehyde from cresol (Brit. 295270).
Secondary butyl alcohol by the reduction of methylethylketone (Brit. 306471).
Valeryl alcohol from valeraldehyde by reduction (Brit. 306471).
Vanillin and vanillic acid from eugenol or isoeugenol (Brit. 295270).

Reagent for various purposes.

Barytes

Synonyms: Barium sulphate, Heavy spar, Tiff.
Latin: Barii sulphas.
French: Barite, Baryte, Barytine, Barytine broyée, Sulfate de barium, Sulfate de baryum.
German: Baryt, Schwefelsaueresbaryum.

Abrasives

Ingredient (U. S. 1276509) of—
Grinding composition.

Adhesives

Ingredient of—
Aromatic cementing composition (U. S. 1455598).
Glass-to-metal cementing composition (U. S. 1132721).

Aviation

Ingredient (U. S. 1389084) of—
Adhesive for mending airplane fabrics.

Beverage

Reagent (U. S. 1912473) in determining—
Caustic strength of alkaline beverages.

Building Construction

Acid-proofing agent (U. S. 1495138) for—
Concrete.
Bonding agent (U. S. 1495138) for—
Concrete.
Filler (U. S. 1336178) in—
Composition stone floorings.
Improver (U. S. 1758026) in making—
Scum-free bricks.
Oilproofing agent (U. S. 1495138) for—
Concrete.
Waterproofing agent (U. S. 1495138) for—
Concrete.

Cement

Process material (U. S. 1194344) in making—
Portland cement.

Ceramics

Ingredient of—
Lead glazes.
Improver (U. S. 1758026) in making—
Scum-free ceramic products.

Barytes (Continued)

Chemical
Carrier for—
Catalysts, catalytic mixtures.
Ingredient of—
Catalytic mixtures.
Process material in making—
Aluminum nitride (U. S. 1268240).
Aluminum-potassium sulphate (U. S. 1296457).
Calcium sulphate (U. S. 1146491).
Carbon dioxide (U. S. 1360312).
Starting point in making—
Barium chemicals, such as chloride, carbonate, sulphide.
Blanc fixe (reduction to sulphide).
Lithopone (reduction to sulphide).

Dye
Inert base for—
Colors in making lakes.

Electrical
Process material in making—
Arc light electrodes (various patents).
Insulations (various patents).
Resistances (U. S. 1507379).
Storage battery electrodes (U. S. 1182513 and 1164464).
Storage battery separators (U. S. 1228368, 1495568, and 1262228).

Explosives and Matches
Luminophore (Brit. 391914) in—
Luminous compositions.

Fats and Oils
Carrier for—
Fish oil hydrogenation catalysts (U. S. 1222660).
Vegetable oil hydrogenation catalysts (U. S. 1222660).
Neutralizing agent (Brit 393108) in—
Purifying processes for fats and oils.
Process material (Brit. 380065 and 380052) in making—
Stable emulsions of fats, oils, paraffin, and organic solvents.
Starting point (Brit. 404874) in making—
Coatings for interior surfaces of oil tanks (in admixture with waterglass and kieselguhr), said coatings being removable with hot water.

Firefighting
Ingredient (German 458400) of—
Chemical fire-extinguisher.

Glass
Ingredient of—
Glazier's cement.
Producer of—
Iridescent effects in glass.

Insecticide
Ingredient of—
Insecticidal compositions.

Leather
Ingredient of—
Coating compositions, dressing compositions.
Process material (U. S. 1241950) in making—
Quebracho tanning compound.

Linoleum and Oilcloth
Filler in—
Linoleum, oilcloth.

Lubricant
Filler (U. S. 1276509) in—
Lubricant composition.
Starting point (U. S. 1881542) in making—
Lubricating composition (with colloidal clay) for machining operations.

Metallurgical
Closing agent (U. S. 1214630) for—
Pores in bronze castings.
Flux in—
Brass smelting.
Ingredient (U. S. 1286061) of—
Case-hardening composition.

Miscellaneous
Decolorizing agent (with carbon) (U. S. 1447461).
Improver (Brit. 393488, 393505, and 386161) of—
Visibility of polyvinyl alcohol threads under Roentgen rays.
Ingredient of—
Boat-mending composition (U. S. 1389084).
Bottle cap composition (U. S. 1351438).
Buffing composition (U. S. 1276509).
Copying pad (U. S. 1348812).
Dental cement (U. S. 1507379).
Dry hair cleansers.
Joint-filling compound for aluminum articles (U. S. 1223458).
Joint-filling compound for brass articles (U. S. 1223458).
Mending composition for artificial teeth (U. S. 1389084).
Metal-cleansing composition (U. S. 1471466).
Mine-ventilating tubing (U. S. 1432585).
Motion picture screen (U. S. 1231727).
Phonograph records (various patents).
Polishing compound (U. S. 1276509).

Paint and Varnish
Filler.
Ingredient of—
Dark-colored paints, distemper colors, silica paints, water paints.
Pigment.
Process material in making—
Light-stable chrome yellows (Brit. 403762).
Pigments (many patents).
Starting point (Brit. 444110) in making—
New blue pigments with manganates.

Paper
Filler in—
Bristol board, cardboard, paperboard, stiff papers, wallpaper.

Pharmaceutical
Reagent in—
X-ray photographing of intestinal tract.

Photographic
Filler and pigment in—
Photographic papers.
Process material in making—
Films (U. S. 1507174), X-ray films.

Plastics
Filler in—
Artificial ivory.

Printing
Filler in making—
Printing plates (U. S. 1377519).
Printing plate matrices (U. S. 1398142).

Resins
Filler
Shortening agent (U. S. 1894731) in making—
Phenol-tung oil-formaldehyde resinous coatings for stencil paper.

Rubber
Filler.

Soap
Carrier for—
Fish oil hydrogenation catalysts (U. S. 1222660).
Neutralizing agent (Brit. 393108) in—
Purifying processes for fats and oils.
Starting point (Brit. 404874) in making—
Coatings for interior surfaces of oil tanks (in admixture with waterglass and kieselguhr), said coatings being removable with hot water.

Textile
Delustring agent (U. S. 1932734) for—
Bemberg silk (used with aluminum hydroxide).
Filler in—
Textile fabrics.
Ingredient of—
Collar-waterproofing composition (U. S. 1453764).
Dressing compositions.

Batyl Alcohol
French: Alcool de batyl, Alcool batylique.
German: Batylalkohol.

Chemical
Starting point (Brit. 398818) in making—
Sulphonated derivatives valuable as detergents.

Bauxite
Synonyms: Beauxite, Ferruginous hydrate of alumina, Natural alumina hydrate.

Abrasive
Raw material in making—
Abrasive paper and cloth.
Starting point in making—
Artificial corundum and emery.

Building
As a building stone.
Ingredient of—
Mortars.

Bauxite (Continued)

Cement
For lining cement kilns.
Starting point in making—
Aluminous or molten cements.

Chemical
Catalyst in various chemical reactions of an organic nature.
Catalyst, precipitated on silica gel, in making—
Ether from alcohol, ethylene from ethyl alcohol.
Ingredient of catalytic preparations used in making—
Acenaphthylene, acenaphthaquinone, bisacenaphthylidenedione, naphthaldehydic acid, naphthalic anhydride, and hemimellitic acid from acenaphthene (Brit. 295270).
Acetaldehyde from ethyl alcohol (Brit. 281307).
Acetic acid from ethyl alcohol (Brit. 281307).
Alcohols from aliphatic hydrocarbons (Brit. 281307).
Aldehydes and acids from toluene, orthochlorotoluene, orthobromotoluene, orthonitrotoluene, parachlorotoluene, parabromotoluene, paranitrotoluene, metanitrotoluene, metachlorotoluene, metabromotoluene, dibromotoluenes, dinitrotoluenes, dichlorotoluenes, chloronitrotoluenes, chlorobromotoluenes, bromonitrotoluenes (Brit. 295270).
Aldehydes and acids from xylenes, pseudocumene, mesitylene and paracymene (Brit. 295270).
Alphanaphthaquinone from anthracene (Brit. 281307).
Anthraquinone from anthracene (Brit. 295270).
Benzaldehyde and benzoic acid from toluene (Brit. 281307).
Benzoquinone from phenanthraquinone (Brit. 281307).
Chloroacetic acid from ethylenechlorohydrin (Brit. 295270).
Diphenic acid from ethyl alcohol (Brit. 295270).
Fluorenone from fluorene (Brit. 295270).
Formaldehyde from methanol or methane (Brit. 295270).
Maleic acid and fumaric acid by the oxidation of benzene, toluene, phenol, tar phenols, or furfural, or from benzoquinone or phthalic anhydride (Brit. 295270).
Naphthaldehydic acid, acenaphthaquinone, or bisacenaphthylidenedione from acenaphthylene (Brit. 281307).
Phenanthraquinone from phenanthrene or diphenic acid (Brit. 295270).
Phthalic acid and maleic acid from naphthalene (Brit. 295270).
Salicylic acid and salicylic aldehyde from cresol (Brit. 295270).
Vanillin or vanillic acid from eugenol or isoeugenol (Brit. 295270).
Starting point in making—
Alum, alumina, aluminum salts, metallic aluminates.

Jewelry
Starting point in making—
Artificial rubies, sapphires, oriental amethysts.

Metallurgical
Starting point in making—
Aluminum metal.

Miscellaneous
Road-making material.

Paint and Varnish
Filler in—
Paints.
Reagent in making—
Pigments.

Petroleum
Reagent in refining—
Crude petroleum, cracked motor fuels, paraffin.
Reagent in desulphurizing—
Fuel oils.

Refractory
Starting point in making—
Aluminous products.

Sugar
Reagent in clarifying—
Molasses.

Bayberry
Synonyms: Laurel berries, Noble laurel, Sweet bay.
Latin: Fructus lauri, Laurus.
French: Baies de laurier.
German: Lorbeeren.

Fats and Oils
Starting point in extracting—
Bayberry wax, laurel oil.

Pharmaceutical
In compounding and dispensing practice.

Beechwood Flour
French: Farine de fayard, Farine de fouteau, Farine de hêtre.
German: Rothbuchemehl.
Italian: Farina di faggio.

Miscellaneous
Filler (U. S. 1902914) in—
Duplicating stencil compositions containing a protein (used to improve distribution of the softeners).

Beeswax (Yellow and Bleached)
Synonyms: White wax, Yellow wax.
Latin: Cera alba, Cera citrina, Cera flava.
French: Cire d'abeilles, Cire blanche, Cire jaune.
German: Gelbes wachs, Weisses wachs.
Spanish: Cera blanca, Cera amarilla.
Italian: Cera bianca, Cera gialla, Cera vergine.

Adhesives
Ingredient of—
Adhesive compositions.

Chemical
Raw material in making—
Bottles used for holding and shipping hydrofluoric acid.

Construction
Waterproofing agent (Brit. 287514), used alone or in combination with other substances for treating—
Brickwork, concrete, masonry, porous structural materials.

Electrical
Ingredient of—
Insulating compositions for various purposes.
Insulating compositions containing rubber.
Insulating agent in making—
Apparatus, cables, wires.

Food
Ingredient of—
Chewing gums.
Compositions for decorating fancy food products.
Raw material in making—
Artificial honeycombs.

Ink
Ingredient of—
Lithographic inks, printing inks, stamping inks, writing inks.

Leather
Ingredient of—
Dressings (U. S. 1847629), finishing preparations, polishing compositions.

Metallurgical
Ingredient of—
Compositions used for covering metals to provide protection against moisture, acids, alkalies, and other corrosive substances.
In various electroplating processes.
Protective agent in—
Acid etching.

Miscellaneous
Ingredient of—
Cleaning and polishing fluid (U. S. 1730654).
Compositions for making dental impressions (U. S. 1897034).
Compositions for making anatomical specimens.
Compositions for lining barrels and kegs.
Compositions used in the manufacture of incandescent gas mantles.
Floor polishes.
Polishes of various sorts.
Preparations for making imitation alabaster statues.
Shoe polishes.
Raw material in making—
Candles, grease crayons, imitation fruit and flowers, toys, wax figures.
Substitute (U. S. 1895527) for—
Paraffin as coating for mutiple boxboard food containers.
Waterproofing agent (Brit. 287514), used either alone or in compositions, for the treatment of—
Asbestos, porous materials of various kinds, strawboard.

Beeswax (Yellow and Bleached) **(Continued)**

Oils and Fats
Base of various lubricating compositions.
Ingredient of—
Axle greases, gun oils.
Lubricating grease compounded of castor oil, mineral oil, and aluminum stearate (U. S. 1881591).
Special lubricants.

Paint and Varnish
Ingredient of—
Paints, preparations containing dry colors, special floor waxes, varnishes, wood-fillers.

Paper
Ingredient of—
Compositions used in the manufacture of carbon paper.
Preparations used in making waxed paper.
Sizings for high-gloss paper.
Emulsified sizing compositions (Brit. 287514).
Shortening agent (U. S. 1894731) for—
Phenol-tung oil-formaldehyde resinous coating for stencil paper.

Perfume
Raw material in making—
Creams, pencils, pomades.

Pharmaceutical
In compounding and dispensing practice.

Printing
Ingredient of—
Compositions used for the preparation of acidproof coatings for plates in the electrotyping process.
Compositions used for making matrices in galvanoplastic work.
Process material in—
Lithography, photoengraving, process engraving.

Resin and Wax
Ingredient of—
Electrotypers' wax, sealing wax, shoemaker's wax.

Rubber
Filler in making—
Rubber compositions.

Stone
Waterproofing agent (Brit. 287514), used either alone or in admixture with other substances, for treating—
Artificial stone, natural stone.

Textile
Assistant agent (Brit. 397881) in—
Stretching cellulose acetate filaments.
Ingredient of—
Compositions used for finishing.
Compositions used for sizing.
Compositions used in the manufacture of waxed cloth.
Emulsified dressing (Brit. 287514).
Waterproofing coating, composed of blown asphalt, rubber, and wax, for cellulose fibers (U. S. 1880036).
Water proofing compositions (Brit. 287514).
Waterproofing agent in treating—
Yarns and fabrics.

Woodworking
Ingredient of—
Compositions used in the finishing of furniture and of lumber used for parquet floorings.

Behenolic Acid Chloride

Chemical
Starting point (Brit. 407956) in making pour-point improvers for machine oils, gear oils, and other lubricants by condensing with—
Anilin, anthracene oil.
Aromatics obtained by destructive hydrogenation or by dehydrogenation.
Benzene.
Cracking gases containing gaseous olefins (ethylene, propylene, and butylene).
Cyclic terpenes, ethylnaphthalene, liquid olefins, middle oil, naphthalene, naphthols, naphthylamines, nitrated aromatics, phenols, tars, toluene, xylene.

Ben Oil

Synonyms: Behen oil, Behn oil, Ben nut oil, Sorinja oil.
Latin: Oleum belanium, Oleum been, Oleum behen.
French: Huile de ben, Huile de ben, aile; Huile de binj.
German: Behenoel, Bennussoel, Benoel, Moringaoel.
Spanish: Olio de ben.
Italian: Olio di ben.

Food
Ingredient of special food preparations.

Mechanical
Lubricant, alone or in mixtures, for—
Delicate machinery, clocks, watches, precision instruments.

Perfumery
Ingredient of—
Creams, lotions, pomades.
Extractive medium for obtaining floral odorous principles by maceration.
Fixative for—
Fugative perfumes.
Ingredient of—
Jasmine perfumes, oil antique, tuberose perfumes, violet perfumes.
Perfume in various cosmetics.
Reagent in—
Enfleurage processes.

Pharmaceutical
Suggested for use as rubefacient in rheumatism, as a laxative and bland.

Bentonite

Abrasives
Ingredient of—
Abrasive wheels, stones, and the like, added so as to cut down on the amount of binding clay required.

Agricultural
Ingredient of—
Cattle dips, added in order to increase the wetting power of the dipping fluid and used in place of soap.
Seed-disinfecting compositions (Brit. 267968).

Analysis
Desiccating agent in laboratory processes.

Ceramics
Ingredient of—
Compositions for the manufacture of electrical and chemical porcelain, added to cut down on the amount of binding clay required.
Plasticizing agent in—
Correcting short products, making various products.
Suspending agent in making—
Glazes.

Cement
Ingredient of—
Cements, added so as to increase their mechanical strength and quicken their setting.
Retarding agent in making—
Gypsum plasters.

Chemical
Absorbent in general use.
Accelerator in making—
Emulsions of various sorts.
Dehydrating agent in general use.
Ingredient of—
Asphalt residue emulsions.
Coaltar residue emulsions.
Compositions containing sticky and tacky materials at ordinary temperatures, that ball during grinding, added to facilitate the operation and prevent balling.
Pitch residue emulsions.
Stabilizer in making—
Emulsions.
Suspending agent for—
Solids in liquid mediums.

Dye
Base in making—
Lake colors.

Explosives and Matches
Filler in making—
Nitroglycerin dynamites and permissibles.
Nitrostarch explosives.

Fats and Oils
Clarifying and bleaching agent.
Dehydrating agent.
Ingredient of—
Compound lubricants.

Fertilizer
Filler in various compositions.

Glues and Adhesives
Ingredient of various preparations.

Ink
Ingredient of—
Printer's ink.

Bentonite (Continued)

Insecticide
Sticking or spreading agent in—
Sprays and dusts.

Metallurgical
Bonding agent for—
Molding sand compositions.
Ingredient of—
Core washers.
Ground coats in dry enameling processes.
Mixtures used in metal enameling, resulting in improved suspension of fine ingredients.
Wet enamel compositions, to cut down amount of clay required.

Mining
Ingredient of—
Compositions containing soap, used for laying the dust in coal mines.

Miscellaneous
Extracting agent in removing—
Asphalt content of tar sands.
Ingredient of—
Crayons, in place of clay; indelible leads, pastel colors, pencils, in place of clay; shoe polishes.
Sizing agent for—
Cordage.
Stabilizing agent in—
Roofing compositions.

Paint and Varnish
Filler in—
Enamels, lacquers, paints, varnishes.
Ingredient of—
Cold-water paints, distempers, kalsomines.
Substitute for whiting in making—
Putties.

Paper
Filler in—
Cardboard, paper.
Ingredient of—
Waterproofed paperboard.
Reagent in—
Overcoming gumming of wire used on the papermaking machine.
Promoting retention of china clay used as a loading agent.
Removing carbon black in reworking old newsprint.

Petroleum
Cleansing agent in treating—
Gasoline, lubricating oils, kerosene.
Dehydrating agent in treating—
Crude oil, gasoline.
Refining and clarifying agent.

Perfumery
Ingredient of—
Dusting powders, facial clay packs, toilet creams, toilet powders.

Pharmaceutical
In compounding and dispensing practice.

Refractories
Ingredient of—
Graphite compositions for crucibles, furnace linings, and the like.

Rubber
Filler in various compounds.
Ingredient of—
Bentonite-rubber emulsions.

Soap
Ingredient of—
Detergent compositions, scouring soaps, special soaps.

Textile
——, *Finishing*
Filler for various fabrics.
Size for—
Cotton fabrics and yarns.
Scouring agent in treating—
Fabrics and yarns.
——, *Dyeing*
Mordant for—
Fabrics and yarns.
Reagent in—
Bottom dyeing yarns.
——, *Printing*
Ingredient of—
Color pastes.

Water and Sanitation
Softening agent in treating water.

Benzaldehyde
Synonyms: Artificial oil of bitter almond, Benzoic aldehyde.
Latin: Benzaldehydum.
French: Aldéhyde benzoique.
German: Benzaldehyd, Künstliches bittermandeloel.

Agricultural
Antiseptic (Brit. 278818) in—
Animal foods.
Flavoring agent (Brit. 278818) in—
Animal foods.

Beverage
As a flavoring agent.
As a substitute for oil of bitter almond.
Ingredient of—
Various mixtures of essential oils, essences, ethers, tinctures, and other substances, used in making flavors for cordials, liqueurs, and other alcoholic and nonalcoholic beverages; typical of such flavors is ethereal cherry oil flavor, composed of benzaldehyde, amyl acetate, amyl butyrate, oil of lemon, oil of sweet orange, oil of clove, and oil of cassia.

Chemical
Reagent in making—
Acetylbenzyl peroxide, (acetozone benzoxate, benzozone).
Diethylbenzaldehydeacetal.
Oxyphenylbenzyl ketone.
Phenylbenzoylcarbinol.
2-Phenylquinolin-4-carboxylic acid (atophan, cinchophen).
Starting point in making—
Benzaldoxime, benzaldehydephenylhydrazone, benzoic acid, benzoyl chloride, benzyl alcohol, benzyl benzoate, benzyl dichloride, benzylidene acetone, benzylideneanthrone, benzylideneazin, cinnamic acid, cinnamic aldehyde.
Condensation products with (a) primary amines; for example, formation of benzylideneanilins; (b) tertiary amines; for example, formation of substituted diamino derivatives of triphenylmethane; (c) sodium salts of fatty acids; for example, formation of unsaturated acids; (d) fatty aldehydes, ketones, and the like; for example, formation of unsaturated aldehydes or ketones.
Halogenated derivatives, mandelic acid.
Nitro derivatives, such as meta-, ortho-, or paranitrobenzaldehyde.
Polymerization products, sulphonated derivatives.

Dry Cleaning
Spotting agent for—
Mercurochrome stains—(1) Boil 45 minutes in soapy water; (2) spot; (3) spot with 25 percent hydrochloric acid; (4) rinse.

Dye
Starting point in making—
Acid green, acidiol green B, acridin dyes, acridin orange R, benzoflavin, carmine blue, carmine green, ethyl green, guinea green B, malachite green, triphenylmethane dyes.

Food
As a flavoring agent.
As a substitute for oil of bitter almond.
Ingredient of—
Flavoring extracts.
Imitation almond flavor, containing also glycerin, water, and glycopon XS.
Nonalcoholic almond flavors of the paste type.

Glues and Adhesives
Improver (U. S. 1895433) of—
Water-resistant qualities of casein glues, by addition to the wet glue base in conjunction with a retarder which may be an aromatic sulphonic acid or a soluble copper, nickel or calcium salt.

Laundering
Spotting agent for—
Mercurochrome stains (for method see under "Dry Cleaning").

Perfume
Ingredient of—
Cosmetics.
Medicated perfume, containing also oil of lavender, camphor, menthol, thymol, oil of rosemary, methyl salicylate, and terpeneless oil of bay.
Perfumes.

Benzaldehyde (Continued)
Skin lotion (milky), containing also tincture of benzoin, rose oil, borax, glycerin in rose water, and a paste composed of crushed almonds and rose water.
Pharmaceutical
As a flavoring agent.
In compounding and dispensing practice.
Ingredient of—
Acne lotion, containing also rose water, alcohol, glycerin, menthol, phenol, methyl salicylate, zinc oxide, calamine, and boric acid.
Codliver oil emulsion, containing also codliver oil, water, glycerin, gum tragacanth, oil of sassafras, oil of coriander, oil of cardamom, and tincture of vanilla.
Flavoring for pharmaceutical purposes, containing also oil of cassia, guaiacol, oil of sassafras, and oil of wintergreen.
Plastics
Starting point (French 755316) in making—
Plastics by condensing with a polymerized vinyl alcohol.
Soap
As an odorant.
Ingredient (Swedish 70883) of—
Soap composition containing a calcium carbonate detergent.

Benzaldehydecyanohydrin
German: Benzaldehydcyanhydrin.
Chemical
Starting point in making—
Ethyl ester of mandelic acid (Brit. 243143).

Benzamide Sulphate
French: Sulphate de benzamide.
German: Benzamidsulfat, Schwefelsaeuresbenzamid.
Chemical
Starting point in making—
Propyl benzoate (Brit. 255887).

6-Benzamidocresidin
Dye
As an intermediate.
Starting point (Brit. 396859) in making—
Fast violet colors on wool.

4-Benzamido-2:5-diethoxyanilin
Dye
Coupling agent (Brit. 434243) in making—
Water-insoluble blue dyestuffs with 2:3-hydroxynaphthoicdodecylamide.

4-Benzamido-2:5-dimethyloxyanilin
Dye
Starting point in making—
Fast reddish-blue colors for wool (Brit. 396859).
Water-insoluble dyes of reddish-dark blue shades (Brit. 397016).

4-Benzamido-3-hydroxyquinaldin
Dye
Starting point (Brit. 429176) in making—
Acid dyestuffs for wool.

6-Benzaminobetanaphthol-4-sulphonic Acid
Dye
In dye syntheses.
Starting point (Brit. 445999) in making—
Chromable orthohydroxy azo dyes by coupling with orthohydroxydiazonium compounds, such as those derived from 6-nitro-2-aminoparacresol or 4-chlor-2-aminophenol-6-sulphonic acid.

Benzanthrone
German: Benzanthrone.
Chemical
Starting point in making—
Benzanthrone (Brit. 260000), benzanthronesulphonic acid, bromobenzanthrone, chlorobenzanthrone, dibromobenzanthrone, dichlorobenzanthrone, diethylbenzanthrone, dimethylbenzanthrone, ethylbenzanthrone, halogenated derivatives, methylbenzanthrone.
Dye
Starting point in making—
Anthraquinone dyestuffs, indranthrene dark blue BO, indranthrene green B, indranthrene violet B, indranthrene violet 2R, indranthrene violet R extra, indranthrene violet RT, irridanthrene B, violanthrene.

Benzanthroneperidicarboxylic Anhydride
French: Anhydride de benzanthroneperidicarbonique, Anhydride de benzanthroneperidicarboxylique.
German: Benzanthronperidicarbonsaeuresanhydrid.
Chemical
Starting point in making—
Intermediates, pharmaceuticals.
Dye
Starting point (Brit. 308651) in making anthraquinone vat dyestuffs with—
2-Chloro-4:5-diaminoanisole.
2-Chloro-4:5-diaminotoluene.
2:3-Diamino-5-chlorotoluene.
2:3-Diaminotoluene.
4:5-Diaminoveratrol.
1:2-Naphthylenediamine sulphate.
Orthophenylenediamine.
Parachloro-orthophenylenediamine.
Paraethoxyorthophenylenediamine.

Benzanthronyl Sulphide
French: Sulfure benzanthronylique, Sulfure de benzanthronyle.
German: Benzanthronylsulfid, Schwefelbenzanthronyl.
Chemical
Starting point (Brit. 275271) in making—
Chlorinated derivatives of thiobenzanthrone.
Diaminobenzanthronyl derivatives.
Monoaminobenzanthronyl derivatives.

Benzene
Synonyms: Benzol, Benzole, Coal naphtha, Motor benzol, Phenyl hydride.
Latin: Benzenum.
French: Benzole, Benzène, Benzine.
German: Benzol, Phenylwasserstoff, Steinkohlenbenzin.
Spanish: Benzolo.
Italian: Benzolo.
Adhesives
Solvent in making—
Cements.
Analysis
Solvent for—
Alkaloids, camphor, fats, iodine, oils, phosphorus, resins, rubber, sulphur, various organic substances, waxes.
Solvent in—
Analytical processes involving control and research in pure science, and in school, hospital, and industrial laboratory work.
Aviation
Fuel for—
Internal combustion motors.
Ingredient of—
Fuels for internal combustion motors.
Cellulose Products
Diluent in—
Nitrocellulose solutions used for various purposes in industry.
Ingredient of—
Softening agents for cellulose esters.
Solvent for—
Camphor used as plasticizer for pyroxylin.
Ceramics
Diluent in—
Solutions containing nitrocellulose used for the production of decorative and protective coatings on ceramic ware.
Chemical
Solvent for—
Alkaloids, camphor, essential oils, fats, fixed oils, iodine, phosphorus, resins, rubber, sulphur, various organic substances, waxes.
Solvent in—
General manufacturing processes.
Starting point in making—
Derivatives of various kinds.
Synthetic organic chemicals used as such or in the manufacture of drugs, perfume ingredients, pharmaceuticals, or other products.
Cosmetic
Solvent for—
Fats, odorants, oils, resins, waxes.
Starting point in making—
Derivatives and synthetic organic chemicals used as odorants.

Benzene (Continued)

Disinfectant
Starting point in making—
Chlorobenzenes, disinfectants, germicides, pharmaceutical chemicals, phenol.

Dry Cleaning
Ingredient of—
Dry-cleaning agents.
Solvent for—
Fats, waxes, oils, resins.
Solvent in—
Dry cleaning.

Dye
Solvent in—
General manufacturing processes.
Starting point (directly or indirectly) in making—
Anilin, azobenzene, chlorobenzene, derivatives used in manufacturing processes, diphenylamine, intermediates, nitrobenzene, phenol, various other chemicals used in dye synthesis.

Electrical
Diluent in—
Solutions, containing nitrocellulose, used in the manufacture of cables, electrical wire and for coating machinery.

Explosives and Matches
Solvent for—
Camphor, phosphorus, sulphur.
Solvent in—
Manufacturing processes.
Starting point in making—
Chlorobenzene, phenol.

Fats, Oils, and Waxes
Extractant for—
Fats, oils, waxes.
Solvent.

Fuel and Heat
Enricher for—
Gases.
Fuel.

Glue and Gelatin
Degreasing agent for—
Bone used in making glue and gelatin.

Glass
Diluent in—
Solutions, containing nitrocellulose, used in the manufacture of nonscatterable glass and for decorative and protective coatings on glass.

Ink
Solvent for—
Ink ingredients.

Leather
Degreasing agent for—
Hides.
Diluent in—
Solutions, containing nitrocellulose, used in the manufacture of artificial leather and for the production of decorative and protective coatings on leather goods.
Solvent for—
Fats, oils, waxes and other dressing and polishing ingredients.

Linoleum and Oilcloth
Solvent in—
Manufacturing processes.

Mechanical
Cleansing solvent.
Fuel for—
Internal combustion motors.
Ingredient of—
Fuels for internal combustion motors.

Metal Fabrication
Diluent in—
Solutions, containing nitrocellulose, used for the production of decorative and protective coatings on metallic articles.

Metallurgical
Substitute for—
Acetylene in welding and cutting.

Miscellaneous
Diluent in—
Solutions, containing nitrocellulose, used for the production of decorative and protective coatings on various compositions of matter.
General solvent.
Ingredient of—
Solvent admixtures with chlorinated solvents.
Solvent for—
Fats, resins, waxes, oils, sulphur.

Paint and Varnish
Diluent or solvent in making—
Airplane dopes, enamels, lacquers, paints, stains, varnishes.
Ingredient of—
Paint removers.
Paint remover.
Solvent for—
Camphor, oil, resins, rubber, waxes.

Paper
Diluent in—
Solutions, containing nitrocellulose, used in the manufacture of coated paper and for the production of decorative and protective coatings on paper and pulp products.
Solvent for—
Waxes, resins, and camphor used in the manufacture of coated paper and for the production of decorative and protective coatings on paper and pulp products.

Petroleum
Process material in—
Purifying paraffin.
Solvent for—
Sulphur, waxes, oils.

Pharmaceutical
Solvent for—
Alkaloids, fats, iodine, oils, waxes.

Plastics
Diluent in—
Solutions containing nitrocellulose.
Solvent for—
Camphor, resins.

Printing
Solvent in—
Lithography, process engraving.

Resins
Solvent.

Rubber
Diluent in—
Solutions, containing nitrocellulose, used for the production of decorative and protective coatings on rubber goods.
Solvent for—
Caoutchouc, gutta-percha.

Soap
Solvent for—
Fats, oils.
Solvent in—
Special soaps and cleansing compounds.
Starting point in making—
Dry-cleaning soaps, special textile soaps.

Stone
Diluent in—
Solutions, containing nitrocellulose, used for the production of decorative and protective coatings on artificial and natural stone.

Textile
Cleansing agent for—
Fabrics.
Degreasing agent for—
Fibers, fabrics.
Diluent in—
Solutions, containing nitrocellulose, used in the production of coated fabrics.
General solvent.
Preservative for—
Sizing agents, such as starch and albumen.

Wood
Diluent in—
Solutions, containing nitrocellulose, used for the production of decorative and protective coatings on woodwork.

Benzeneazoalphanaphthylamine

Dye
Starting point in making—
Neutral gray G.

Benzene Sulphochloride
French: Sulphochlorure de benzène.
German: Benzol sulfochlorid.

Analysis
Reagent in detecting—
Primary, secondary and tertiary amines.

Chemical
Protective agent for the amino group in making nitro compounds.
Reagent in making—
Acetic anhydride, picryl chloride.

Dye
Starting point in making—
Brilliant sulphon red B.

Benzenesulphonamide
French: Sulphonamide de benzène.
German: Benzolsulfonamid.

Cellulose Products
Solvent for—
Cellulose acetate, cellulose esters and ethers.
For uses, see under general heading: "Solvents."

Chemical
Starting point in making various derivatives.

Resins and Waxes
Solvent for—
Natural and artificial resins.
Starting point (Brit. 340101) in making synthetic resins with the aid of—
Benzaldehyde.

Benzidin
Synonyms: Benzidin base, Paradiaminodiphenyl.
German: Bianilin.

Analysis
Reagent for—
Detection of sulphates in water, identification of blood.

Chemical
Starting point in making—
Accelerators of rubber vulcanization in combination with heptaldehyde (Brit. 259933).
Aminodichlordiphenyl (Brit. 253763).
Benzidin sulphate, orthonitrophenol, synthetic aromatics, synthetic pharmaceuticals.

Dye
Starting point in making—
Alkali yellow R, azidin violet, benzidin brown 3GO, chloramine black N, chlorazol deep brown B, columbia green, congo red, congo series dyes, diamine beta black, diamine bronze G, diazo blue black RS, oramine maroon, oxamine maroon, pyramine orange 2R.

Miscellaneous
Reagent in—
Blood stain detection in microscopical work.

Paper
Reagent in—
Determination of degree of lignification of wood.

Benzidinsulphonedisulphonic Acid
French: Acide de benzidinsulphonedisulphonique.
German: Benzidinsulfonbisaeure.

Dye
Starting point in making—
Sulphone azurin D.

Benzimidododecyl Ether Hydrochloride

Textile
Dispersing agent (Brit. 446976) in making—
Waterproof and crease-resisting finishes on natural and synthetic fibers (used in conjunction with sulphonated fats, albuminous derivatives, and formaldehyde or a substance yielding it).
Delustring agent (Brit. 446976) for—
Natural and synthetic fibers.
Wetting agent (Brit. 446976) in making—
Waterproof and crease-resisting finishes on natural and synthetic fibers (used in conjunction with sulphonated fats, albuminous derivatives, and formaldehyde, or a substance yielding it).

Benzimidophenyl Monosulphide Hydrochloride
Synonyms: Benzimidothiophenyletherhydrochloride.

Insecticide and Fungicide
Larvicide for—
Culicine mosquito larvae.

Benzoazolon-5-arsinic Acid
French: Acide de benzoxazolon-5-arsenieux.
German: Benzoxazolon-5-arsinigsaeure.

Chemical
Starting point in making—
4-Amino-3-oxybenzene-1-arsinic acid (German 439607).

Benzoic Acid
Synonyms: Flowers of benzoin, Phenylformic acid, Phenylmethanic acid.
Latin: Acidum benzoicum, Flores benzoes.
French: Acide benzoique, Acide phénylеformique, Acide phénylеméthanoique, Fleurs de benjoin.
German: Benzoeblumen, Benzoesäure, Phenylameisensäure, Phenylmethansäure.
Spanish: Acido benzoico, Acido fenilformico, Acido fenilmetanico.
Italian: Acido benzoico, Acido fenileformico, Acido phenilemetanico.

Analysis
Standard for—
Calorimetry, preparing volumetric solutions of alkalies.

Chemical
Reagent (Brit. 310869) in making—
Iodized pharmaceutical derivatives.
Starting point in making—
Ammonium benzoate by reaction with ammonium hydroxide.
Amyl benzoate by reaction with amyl alcohol.
Anthragallol.
Benzyl benzoate by reaction with benzyl alcohol.
Benzoyl chloride.
Benzoic anhydride.
Benzylidene chloride.
Butyl benzoate by reaction with butyl alcohol.
Bornyl benzoate (Brit. 251147).
Bismuth benzoate by reaction with a bismuth salt.
1:3 Dihydroxy-2-methylanthraquinone.
Ethyl benzoate by reaction with ethyl alcohol.
Isoallyl benzoate by reaction with isoallyl alcohol.
Isoamyl benzoate by reaction with isoamyl alcohol.
Linalyl benzoate by reaction with linalyl alcohol.
Metanitrobenzoic acid.
Mercury benzoate.
Methyl benzoate by reaction with methyl alcohol.
Naphthyl benzoate by reaction with betanaphthol.
Nickel benzoate.
Orthonitrobenzioc acid.
Paranitrobenzoic acid.
Phenylacridin.
Phenyl benzoate by reaction with phenol.
Potassium benzoate by reaction with a potassium salt.
Propyl benzoate by reaction with propyl alcohol.
Resorcinol benzoate by reaction with resorcinol.
Sodium benzoate by reaction with sodium bicarbonate.
Strontium benzoate by reaction with a strontium salt.
Succinimide.
Various esters and salts in addition to the above.
Various pharmaceutical compounds.
Various intermediates and aromatic chemicals.
Zinc benzoate by reaction with a zinc salt.

Dye
Reagent in making—
Alizarin brown, alizarin yellow A, anilin blue, anilin dyestuffs of various sorts, anthraquinone brown, anthraquinone dyestuffs; bright blue, superfine, spirit-soluble; diphenylamine blue, extra opal blue 6B, spirit blue.

Food
Ingredient of—
Bleaching compositions containing ammonium persulphate (added for the purpose of increasing the bleaching and preservative powers of the latter in the treatment of flour).
Preservative for various foods.

Miscellaneous
Preservative for various purposes.

Paint and Varnish
Ingredient of—
Compositions used to obtain a mellowing coat in decorating and finishing wood.

Paper
Reagent in—
Treating cellulose for the manufacture of paper.

Benzoic Acid (Continued)

Perfume
Fixative in perfumes (used in place of benzoin).
Ingredient of—
Antiseptic mouth washes, various cosmetics and toilet articles.
Preservative in—
Fatty substances used for recovering odoriferous constituents.

Pharmaceutical
Suggested for use as an antithermic, antipyretic, antiseptic, expectorant.
Used in admixture with insulin for treating diabetes.

Resins and Waxes
Reagent (Brit. 292912) for making synthetic resins with the aid of—
Acetylcarbamide, allylcarbamide, amylcarbamide, benzoylcarbamide, butylcarbamide, citrylcarbamide, cyanamide, ethylcarbamide, formylcarbamide, heptylcarbamide, hexylcarbamide, isoallylcarbamide, isoamylcarbamide, isobutylcarbamide, isopropylcarbamide, methylcarbamide, propylcarbamide.

Soap
Fixative in making—
Perfumed toilet soaps.
Reagent for—
Neutralizing slightly alkaline toilet soaps.

Textile
—, *Bleaching*
As a bleaching reagent.
—, *Printing*
Mordant in printing—
Calicoes and other fabrics.

Tobacco
Ingredient of—
Compositions used for improving the aroma and taste of tobacco.

Benzoic Acid Ester of Grapeseed Alcohol

Bituminous
Solvent (Brit. 445223) for—
Asphalt and other bituminous bodies.
Dye
Solvent (Brit. 445223) for—
Dyestuffs, particularly oil-soluble coaltar dyes.
Fats, Oils, and Waxes
Solvent (Brit. 445223) for—
Fats, oils, waxes.
Resins
Solvent (Brit. 445223) for—
Oil-soluble glycerol-phthalic acid resins.
Polymerized vinyl compounds, synthetic resins.
Rubber
Solvent (Brit. 445223) for—
Rubber.

Benzoic Acid Ester of Ricinoleic Alcohol

Bituminous
Solvent (Brit. 445223) for—
Asphalt and other bituminous bodies.
Dye
Solvent (Brit. 445223) for—
Dyestuffs, particularly oil-soluble coaltar dyes.
Fats, Oils, and Waxes
Solvent (Brit. 445223) for—
Fats, oils, waxes.
Resins
Solvent (Brit. 445223) for—
Oil-soluble glycerol-phthalic acid resins.
Polymerized vinyl compounds.
Synthetic resins.
Rubber
Solvent (Brit. 445223) for—
Rubber.

Benzolethylmethyl Ketone
German: Benzolaethylmethylketon.

Perfumery
Ingredient of—
Hair restorers, pomades.

Benzothiazyl 1-Thioacetate

Rubber
Delayed-action accelerator in—
Vulcanization processes.

Benzo Trifluoride

Oils, Fats, and Waxes
Starting point (Brit. 440175) in making—
Addition agents for high-pressure lubricating oils or greases, by mixing and reacting with organo-metallic compounds.

Benzoxazolone-5-arsenic Acid

Chemical
Starting point in making—
Pharmaceuticals and other derivatives.
Pharmaceutical
In compounding and dispensing practice.

4-Benzoylamino-3-hydroxy-2-methylquinolin

Dye
Starting point (Brit. 429176) in making—
Greenish-yellow dyes for wool, by fusing with phthalic anhydride and sulphonating the product.

2-Benzoylamino-5-naphthol-7-sulphonic Acid
French: Acide de 2-benzoyleamino-5-naphthole-7-sulfonique.
German: 2-Benzoylamin-5-naphtol-7-sulfonsaeure.

Dye
Starting point (Brit. 280320) in making viscose dyestuffs with the diazo derivatives of—
Anilin, betanaphthylamine, meta-aminobenzoic acid, 2-naphthylamine-6:8-disulphonic acid, parachloroanilin, paranitranilin.

Benzoyl Benzoate
French: Benzoate de benzoyle.
German: Benzoesaeuresbenzoyl.

Resins and Waxes
Solvent in making—
Resinous compositions (Brit. 235595).

Benzoylhydroquinone

Petroleum
Stabilizing agent (Brit. 406195) for—
Cracked gasolines and other motor fuels.

Benzoyl Peroxide
French: Peroxyde de benzoyle, Peroxyde benzoylique.
German: Benzoylperoxyd.

Analysis
Reagent in detecting—
Cholesterin.
Formaldehyde.
Chemical
Accelerating agent in—
Treating yeast for the purpose of increasing its activity in the fermentation process.
Catalyst in—
Polymerization processes.
Cosmetic
Bleaching agent for—
Fats, oils, waxes.
Fats and Oils
Reagent in bleaching—
Fats, greases, oils.
Food
Reagent in bleaching—
Flours of various sorts (used in admixture with dicalcium phosphate).
Miscellaneous
Reagent in bleaching various products.
Reagent in fixing—
Microscopical samples.
Pharmaceutical
In compounding and dispensing practice.
Waxes and Resins
Catalyst in—
Polymerization processes in making synthetic resins.
Reagent in bleaching—
Waxes.
Rubber
Emulsifying agent (Brit. 312949) in—
Producing synthetic rubber from diolefines.
Reagent in—
Rubber compounding, rubber vulcanizing.

Benzoylphloroglucinol

Petroleum
Stabilizing agent (Brit. 406195) for—
Cracked gasolines and other motor fuels.

Benzoylpyrocatechol

Petroleum
Stabilizing agent (Brit. 406195) for—
Cracked gasolines and other motor fuels.

Benzoylpyrogallol

Petroleum
Stabilizing agent (Brit. 406195) for—
Cracked gasolines and other motor fuels.

Benzoylresorcinol

Petroleum
Stabilizing agent (Brit. 406195) for—
Cracked gasolines and other motor fuels.

Benzyl Acetate

French: Acétate benzylique, Acétate de benzyle.
German: Benzylazetat, Essigsaeurebenzylester, Essigsaeuresbenzyl.

Food
Ingredient of—
Various fruit essences.

Glues and Adhesives
Solvent (Brit. 273290) in making—
Cements for laminated mica and other purposes.

Miscellaneous
Solvent (Brit. 273290) in making—
Insulating preparations for wires and electrical apparatus.

Paint and Varnish
Solvent in making—
Cellulose acetate varnishes and lacquers.
Cellulose ester-resin composite varnishes and lacquers.
Cellulose nitrate varnishes and lacquers.
Solvent (Brit. 273290) in making—
Insulating enamels, varnish bases.

Perfumery
Ingredient of—
Artificial coreopsis, artificial jasmine, artificial jonquille, artificial tuberose, lavender water.

Plastics
Solvent (Brit. 273290) in making various compositions.

Resins and Waxes
Solvent (Brit. 273748) in treating artificial resins of—
Phenol-aldehyde type.
Polyhydric alcohol-polybasic acid type.
Urea-aldehyde type.

Soap
Perfume in making—
Toilet soaps.

Benzyl Acetylsalicylate

Chemical
Starting point in making various derivatives.

Pharmaceutical
In compounding and dispensing practice.

Benzyl Alcohol

French: Alcool de benzyle, Alcool benzylique.
German: Benzylalkohol.

Ceramics
Plasticizer in—
Coating compositions containing cellulose esters or ethers, such as cellulose acetate and nitrocellulose.

Chemical
Solvent for—
Benzyl abietate, cellulose acetate, nitrocellulose.
Starting point in making various derivatives.

Dye
Ingredient (Brit. 319249) of—
Dye mixtures.

Fats and Oils
Ingredient of—
Linseed and castor oil mixtures.

Glass
Plasticizer in—
Compositions containing cellulose esters or ethers, such as cellulose acetate and nitrocellulose, used in the manufacture of none-scatterable glass and for coating and decorating glassware.

Gums
Solvent for—
Ester gums.

Leather
Plasticizer in—
Compositions containing cellulose esters or ethers, such as cellulose acetate and nitrocellulose, used in the manufacture of artificial leathers and for coating and decorating leather goods.

Miscellaneous
Ingredient (Brit. 319249) of—
Washing compositions containing alcohols and amyl carboxylic acid or sulphonic acids.
Plasticizer in—
Compositions containing cellulose acetate, nitrocellulose, or other esters or ethers of cellulose, used for coating various products.

Paint and Varnish
Grinding medium for—
Pigments.
Plasticizer and solvent in making—
Paints, varnishes, lacquers, dopes, and enamels containing cellulose acetate or nitrocellulose, along with various gums and resins.
Solvent for—
Shellac.

Paper
Plasticizer in—
Compositions containing cellulose esters or ethers, such as cellulose acetate and nitrocellulose, used for coating paper and for decorating paper and pulp products.

Plastics
Plasticizer in—
Compositions containing cellulose esters or ethers, such as cellulose acetate and nitrocellulose, as well as various gums and resins, such as cumarone resin and ester gum.

Resins and Waxes
Solvent for—
Copal esters, cumarone resins, glyceryl phthalate resins, mastic.

Rubber
Plasticizer in—
Coating compositions containing cellulose acetate, nitrocellulose, or other esters or ethers of cellulose, as well as gums and resins.

Stone
Plasticizer in—
Compositions containing cellulose acetate, nitrocellulose, or other esters or ethers of cellulose, used for coating artificial and natural stone.

Textile
Ingredient (Brit. 319249) of—
Compositions containing alcohols and amylcarboxylic acid or sulphonic acids, used for dyeing rayons, cotton, and wool with vat dyestuffs.
Finishing compositions.
Plasticizer in—
Compositions containing cellulose acetate, nitrocellulose, or other esters or ethers of cellulose, used in the manufacture of coated textiles.

Woodworking
Plasticizer in—
Compositions containing cellulose acetate, nitrocellulose, or other esters or ethers of cellulose, and gums and resins, such as cumarone resin, mastic, and ester gum.

Benzylamine

Chemical
Starting point in making various organic compounds.

Resins and Waxes
Catalyst in making—
Resinous condensation products from formaldehyde and tar or crude tar oils containing phenols (French 607655).

Benzylanilin

Ceramics
Ingredient (Brit. 343288) of—
Compositions, containing cellulose ethers, such as benzylcellulose, butycellulose, and the like, used for decorating and protecting ceramic products (added to prolong the life of the coating.)

Chemical
Starting point in making—
Benzaldehyde, intermediates, pharmaceuticals, various other derivatives.

Benzylanilin (Continued)

Dye
Reagent in making—
Fine suspensions of indigo dyestuffs.
Starting point in making—
Guinea green, triphenylmethane dyes.

Glass
Ingredient (Brit. 343288) of—
Compositions, containing cellulose ethers, such as butylcellulose, benzylcellulose, and the like, used in the manufacture of non-scatterable and for the decoration and protection of glassware (added to prolong the life of the coating).

Leather
Ingredient (Brit. 343288) of—
Compositions, containing cellulose ethers, such as butylcellulose, benzylcellulose, and the like, used in the manufacture of artificial leather and for the decoration and protection of leather goods (added to prolong the life of the coating).

Metallurgical
Ingredient (Brit. 343288) of—
Compositions, containing cellulose ethers, such as butylcellulose, benzylcellulose, and the like, used for the decoration and protection of metal goods (added to prolong the life of the coating).

Miscellaneous
Ingredient (Brit. 343288) of—
Compositions, containing cellulose ethers, such as butylcellulose, benzylcellulose, and the like, used for the decoration and protection of various fibrous compositions of matter (added to prolong the life of the coating).

Paint and Varnish
Ingredient (Brit. 343288) of—
Compositions, containing cellulose ethers, such as benzylcellulose, butylcellulose, and the like, used as lacquers, paints, varnishes, dopes and enamels (added to prolong the life of the coating).

Paper
Ingredient (Brit. 343288) of—
Compositions, containing cellulose ethers, such as benzylcellulose, butylcellulose, and the like, used in the manufacture of coated paper and for the decoration and protection of paper and pulp products (added to prolong the life of the coating).

Plastics
Ingredient (Brit. 343288) of—
Compositions, containing cellulose ethers, such as benzylcellulose, butylcellulose, and the like (added to prolong the life of the product).

Rubber
Ingredient (Brit. 343288) of—
Compositions, containing cellulose ethers, such as benzylcellulose, butylcellulose, and the like, used for the protection and decoration of rubber goods (added to prolong the life of the coating).

Stone
Ingredient (Brit. 343288) of—
Compositions, containing cellulose ethers, such as benzylcellulose, butylcellulose, and the like, used for the decoration and protection of natural and artificial stone (added to prolong the life of the coating).

Textile
Ingredient (Brit. 343288) of—
Compositions, containing cellulose ethers, such as benzylcellulose, butylcellulose, and the like, used for the manufacture of coated fabrics (added to prolong the life of the coating).

Woodworking
Ingredient (Brit. 343288) of—
Compositions, containing cellulose ethers, such as benzylcellulose, butylcellulose, and the like, used for the decoration and protection of woodwork (added to prolong the life of the coating).

Benzylanisol
French: Anisole de benzyle, Anisole benzylique.

Chemical
Starting point in making—
Aromatics, intermediates, pharmaceuticals.

Miscellaneous
Ingredient (Brit. 319273) of—
Compositions used as wetting agents for special purposes.

Soap
Ingredient (Brit. 319273) of—
Detergent preparations.

Textile
——, *Dyeing*
Ingredient (Brit. 319273) of—
Dye liquors.

——, *Finishing*
Ingredient (Brit. 319273) of various finishing compositions.

Benzylbenzanthrone

Dye
Starting point (Brit. 261888) in making dyestuffs with the following alcohols:—
Butyl, ethyl, isobutyl, isopropyl, methyl, propyl.
Starting point (Brit. 266130) in making dyestuffs with the following anilides:—
Barium, butyl, calcium, ethyl, methyl, magnesium, potassium, propyl, sodium, strontium.
Starting point (Brit. 275283) in making isoviolanthrone dyestuffs with—
Alphanaphthylamine, anilin, benzylamine, betanaphthylamine, diphenylamine, meta-anisidin, metaphenylenediamine, metatoluidin, metaxylidin, orthoanisidin, orthophenylenediamine, orthotoluidin, orthoxylidin, para-anisidin, paraphenylenediamine, paratoluidin, paraxylidin, phenylamine.

Benzyl Benzoate
Synonyms: Benzylbenzoic ether, Ergol, Rhodazil, Spasmodine.
French: Benzoate de benzyle, Benzoate benzylique, Éther benzylbenzoique.
German: Benzoesäurebenzylester, Benzoesäuresbenzyl, Benzylbenzoat.
Spanish: Benzoato de benzil.
Italian: Benzoato di benzile.

Ceramics
Solvent and plasticizer (Brit. 371901) in—
Compositions, containing various esters or ethers of cellulose, added for the purpose of lengthening the life of the film and used for the decoration and protection of ceramic ware.

Chemical
Solvent for—
Cellulose acetate, nitrocellulose.
Starting point in making various derivatives.

Food
Solvent and fixative in making—
Chewing gum, confectionery flavors, other flavoring compositions.

Glass
Solvent and plasticizer (Brit. 371901) in—
Compositions, containing various esters or ethers of cellulose, added for the purpose of lengthening the life of the film and used for the decoration and protection of glassware and in the manufacture of non-scatterable glass.

Leather
Solvent and plasticizer (Brit. 371901) in—
Compositions, containing various esters or ethers of cellulose, added for the purpose of lengthening the life of the film and used for the decoration and protection of leather goods and in the manufacture of artificial leather.

Metallurgical
Solvent and plasticizer (Brit. 371901) in—
Compositions, containing various esters or ethers of cellulose, added for the purpose of lengthening the life of the film and used for the decoration and protection of metallic articles.

Miscellaneous
Solvent and plasticizer (Brit. 371901) in—
Compositions, containing various esters or ethers of cellulose, added for the purpose of lengthening the life of the film and used for the decoration and protection of various compositions of matter.

Paint and Varnish
Solvent and plasticizer (Brit. 371901) in making—
Paints, varnishes, dopes, enamels, and lacquers containing esters or ethers of cellulose, such as cellulose acetate and nitrocellulose (added for the purpose of lengthening the life of the film).

Benzyl Benzoate (Continued)

Paper

Plasticizer and solvent (Brit. 371901) in—
 Compositions, containing various esters or ethers of cellulose, added for the purpose of lengthening the life of the film and used for the decoration and protection of paper and pulp compositions as well as in the manufacture of coated paper.

Perfume

Fixative for—
 Attos in perfume compositions.
Solvent for—
 Xylene and ketone musk and essential oils.
Solvent in making various toilet preparations.

Plastics

Plasticizer and solvent (Brit. 371901) in making—
 Plastic compositions containing esters or ethers of cellulose, such as cellulose acetate and nitrocellulose.
Substitute for camphor in making—
 Celluloid, pyroxylin plastic compositions.

Pharmaceutical

Suggested for use as antispasmodic and in treating diarrhea, dysentery, intestinal catarrh, asthma, and angina pectoris.

Resins and Waxes

Solvent (Brit. 235595) in making—
 Resinous compositions.

Rubber

Plasticizer and solvent (Brit. 371901) in—
 Compositions, containing various esters or ethers of cellulose, added for the purpose of lengthening the life of the film and used for the decoration and protection of rubber goods.

Stone

Plasticizer and solvent (Brit. 371901) in—
 Compositions, containing various esters or ethers of cellulose, added for the purpose of lengthening the life of the film and used for the decoration and protection of natural and artificial stone.

Textile

Plasticizer and solvent (Brit. 371907) in—
 Compositions, containing esters or ethers of cellulose, such as cellulose acetate and nitrocellulose, added for the purpose of lengthening the life of the film and used in the manufacture of coated textiles.

Woodworking

Plasticizer and solvent (Brit. 371901) in—
 Compositions, containing various esters or ethers of cellulose, added for the purpose of lengthening the life of the film and used for the decoration and protection of woodwork.

Benzylbetadiethylaminoethyldiphenylacetamide

Pharmaceutical

Claimed (Brit. 438659) to possess—
 Physiological properties resembling those of atropine.

Benzyl Bromide

French: Bromure de benzyle.
German: Benzylbromid.
Spanish: Bromuro de benzyl.
Italian: Bromuro di benzile.

Chemical

Starting point (French 588933) in making—
 Foaming and frothing agents with alkalies and naphthalenesulphonic acids.

Food

Antiseptic (French 580481 and 580482) for—
 Yeast.

Military

As a tear gas.
Ingredient of—
 German military gas known as "T. Stoff" (in admixture with xylyl bromide).

Benzyl-4-bromo-1:2-benzanthraquinone

German: Benzyl-4-brom-1:2-benzanthrachinon.

Chemical

Starting point in making—
 Intermediates, pharmaceuticals.

Dye

Starting point (Brit. 340524) in making dyestuffs with—
 Alpha-aminoanthraquinone.
 Alpha-amino-4-benzoylaminoanthraquinone.
 1:5-Diaminoanthraquinone.

Benzylbromobenzanthrone

German: Benzylbromanthron.

Chemical

Starting point in making—
 Intermediates, pharmaceuticals, various other derivatives.

Dye

Starting point (Brit. 345728) in making dyestuffs with the aid of—
 8-Aminopyrazolanthrone, 3-bromopyrazolanthrone, 8-chloropyrazolanthrone, 4-chloropyrazolanthrone, 5-dimethylaminopyrazolanthrone, 4-dimethylaminopyrazolanthrone.
Starting point (Brit. 261888) in making dyestuffs with the following alcohols:—
 Butyl, ethyl, isobutyl, isopropyl, methyl, propyl.
Starting point (Brit. 266030) in making isodibenzanthrone dyestuffs with the following anilides:—
 Barium, butyl, calcium, ethyl, magnesium, methyl, potassium, propyl, sodium, strontium.

Benzylbutyl Tartrate

French: Tartrate benzylbutylique, Tartrate de benzyle et butyle.
German: Weinsaeuresbenzylester, Weinsaeuresbenzylbutyl.

Miscellaneous

Ingredient (U. S. 1639080) of—
 Stencil sheets of cellulose acetate.

Benzyl Carboxethylate

French: Carboxéthylate de benzyle.
German: Benzylcarboxaethylat.
Spanish: Carboxetilato de benzil.
Italian: Carbossietilato di benzile.

Perfume

Ingredient (French 650100) of—
 Perfumes.

Benzyl Cellulose

French: Cellulose de benzyle, Cellulose benzylique.
German: Benzylzellulose.

Ceramics

Ingredient (Brit. 330895) of—
 Coating compositions containing artificial resins and used for the decoration and protection of ceramic products.

Construction

Ingredient (Brit. 330895) of—
 Coating compositions, containing artificial resins and the like, used for the protection of brickwork, etc.

Electrical

Ingredient (Brit. 330895) of—
 Coating compositions, containing artificial resins and the like, used for coating electrical apparatus, wire, and other articles.

Glass

Ingredient (Brit. 330895) of—
 Compositions, containing artificial resins and the like, used for producing non-scatterable glass and for coating glassware.

Leather

Ingredient (Brit. 330895) of—
 Compositions, containing artificial resins and the like, used in the manufacture of artificial leathers and for coating leather goods.

Metallurgical

Ingredient (Brit. 330895) of—
 Coating compositions, containing artificial resins and the like, used for decorating and protecting metal ware.

Paper

Ingredient (Brit. 330895) of—
 Compositions, containing artificial resins and the like, used in the manufacture of coated papers and in the coating and decorating of paper and pulp products.

Paint and Varnish

Ingredient (Brit. 330895) (used together with artificial resins and the like) of—
 Dopes, lacquers, enamels, paints, priming compositions, varnishes.

Plastics

Ingredient (Brit. 330895) of—
 Compositions containing artificial resins and the like.

Benzyl Cellulose (Continued)

Rubber
Ingredient (Brit. 330895) of—
Compositions, containing artificial resins and the like, used for coating rubber merchandise.

Stone
Ingredient (Brit. 330895) of—
Compositions, containing artificial resins and the like, used for coating artificial and natural stones.

Textile
Ingredient (Brit. 330895) of—
Compositions, containing artificial resins and the like, used for making coated textiles.

Woodworking
Ingredient (Brit. 330895) of—
Compositions, containing artificial resins and the like, used for coating and protecting wood.

Benzylchlor-4-carboxylic Acid Benzylester

Detergent
Starting point (Brit. 408754) in making—
Saponaceous products by reaction with tertiary amines, which may be used alone or with other soaps, fillers, or compounds giving off oxygen.

Benzylchlor-4-carboxylic Acid Betaphenylethylester

Detergent
Starting point (Brit. 408754) in making—
Saponaceous products by reaction with tertiary amines, which may be used alone or with other soaps, fillers, or compounds giving off oxygen.

Benzylchlor-4-carboxylic Acid Dodecylester

Soap
Starting point (Brit. 403883) in making—
Saponaceous products by reaction with amines, such as anilin, piperidin bases, hydroxyethylanilin, dihydroxyethylanilin, para-toluidin (these products may be used alone or with other soaps, fillers, or compounds giving off oxygen).

Benzylchlor-4-carboxylic Acid Hexadecylester

Soap
Starting point (Brit. 403883) in making—
Saponaceous products by reaction with amines, such as anilin, piperidin bases, hydroxyethylanilin, dihydroxyethylanilin, para-toluidin (these products may be used alone or with other soaps, fillers, or compounds giving off oxygen).

Benzylchlor-4-carboxylic Acid Tetradecylester

Soap
Starting point (Brit. 403883) in making—
Saponaceous products by reaction with amines, such as anilin, piperidin bases, hydroxyethylanilin, dihydroxyethylanilin, para-toluidin (these products may be used alone or with other soaps, fillers, or compounds giving off oxygen).

Benzylchlorobenzanthrone
German: Benzychlorbenzanthron.

Dye
Starting point (Brit. 261888) in making isobenzanthrone dyestuffs with—
Butyl alcohol, ethyl alcohol, isobutyl alcohol, isopropyl alcohol, methyl alcohol, propyl alcohol.
Starting point (Brit. 266030) in making dyestuffs with—
Bariumanilide, calciumanilide, butylanilide, ethylanilide, methylanilide, magnesiumanilide, potassiumanilide, propylanilide, sodiumanilide, strontiumanilide.

Benzylcresol

Chemical
Starting point (Brit. 444351) in making—
Fat-splitting catalysts and emulsifying agents for use in dyeing, laundering, bleaching, and various other processes, by reacting with formaldehyde and non-aromatic secondary amines (the salts of the products with water-soluble acids, or water-insoluble acids, or the quaternary ammonium salts, are claimed to be valuable for the purposes named).

Benzylcyclohexyl Phthalate
Synonyms: Cyclohexylbenzyl phthalate.
French: Phthalate de benzyle et cyclohexyle, Phthalate benzylique et cyclohexylique, Phthalate de cyclohexyle et benzyle, Phthalate cyclohexylique et benzylique.
German: Benzylcyklohexylphtalat, Cyklohexylbenzylphtalat, Phtalsaeurebenzylcyklohexylester, Phtalsaeurecyklohexylbenzylester, Phtalsaeuresbenzylcyklohexyl.

Paint and Varnish
Solvent and plasticizer in making—
Cellulose acetate varnishes and lacquers.
Solvent and plasticizer (Brit. 302961) in making—
Nitrocellulose varnishes and lacquers containing dammar, manila gum, copal, elemi, sandarac, mastic, ester resins, resins obtained from cyclic ketones, coumarone resins, resins of the indene series, vinyl resins, urea-formaldehyde condensation products.

Plastics
Solvent and plasticizer (Brit. 302961) in making—
Nitrocellulose plastics containing dammar, manila gum, copal, elemi, sandarac, mastic, ester resins, resins obtained from cyclic ketones, coumarone resins, resins of the indene series, vinyl resins, urea-formaldehyde condensation products.

Benzyldeltaphenylbutylmethylamine

Chemical
Claimed (U. S. 2006114) as—
Substitute for papaverine.

Benzyldeltaphenylbutylmethylamine Hydrochloride

Chemical
Claimed (U. S. 2006114) as—
Substitute for papaverine.

Benzyl Dichloride
Synonyms: Benzal chloride, Benzyl bichloride, Benzylene chloride, Benzylidene chloride, Chlorobenzal, dichlorbenzyl.
French: Dichlorure de benzil, Dichlorure de benzyle, Dichlorure benzylique.
German: Benzyldichlorid.

Chemical
Starting point in making—
Benzyl compounds.

Dye
Source of benzyl group in making—
Dyes.

Military
Substitute for—
Mustard gas in chemical warfare shell experiments, on account of the similarity of physical properties.

6-Benzyldichlorobenzanthrone

Dye
Starting point (Brit. 261888) in making dyestuffs with—
Butyl alcohol, ethyl alcohol, isobutyl alcohol, isopropyl alcohol, methyl alcohol, propyl alcohol.
Starting point (Brit. 266030) in making dyestuffs with—
Bariumanilide, butylanilide, calciumanilide, ethylanilide, magnesiumanilide, methylanilide, potassiumanilide, propylanilide, sodiumanilide, strontiumanilide.

Benzyldiethylbetadodecylthioethyl-Ammonium Bromide

Disinfectant
Claimed (Brit. 436725 and 436726) to be—
Bactericide, disinfectant.

Insecticide and Fungicide
Claimed (Brit. 436725 and 436726) to be—
Fungicide.

Benzyldiethylbetahydroxygammadodecoxypropyl-Ammonium Chloride

Disinfectant
Claimed (Brit. 436725 and 436726) to be—
Bactericide, disinfectant.

Insecticide and Fungicide
Claimed (Brit. 436725 and 436726) to be—
Fungicide.

Benzyldiethyl-normaldodecyl-Ammonium Chloride

Disinfectant
Claimed (Brit. 436725 and 436726) to be—
Bactericide, disinfectant.

Insecticide and Fungicide
Claimed (Brit. 436725 and 436726) to be—
Fungicide.

Benzyldiethyl-normaloctyl-Ammonium Chloride

Disinfectant
Claimed (Brit. 436725 and 436726) to be—
Bactericide, disinfectant.
Insecticide and Fungicide
Claimed (Brit. 436725 and 436726) to be—
Fungicide.

Benzyldimethyldodecyl-Ammonium Bromide

Disinfectant
Claimed (Brit. 436725 and 436726) to be—
Bactericide, disinfectant.
Insecticide and Fungicide
Claimed (Brit. 436725 and 436726) to be—
Fungicide.

Benzyldimethyldodecyl-Ammonium Chloride

Disinfectant
Claimed (Brit. 436725 and 436726) to be—
Bactericide, disinfectant.
Firefighting
Basic ingredient (Brit. 460649) in—
Air-foaming compositions for fire-extinguishing purposes.
Insecticide and Fungicide
Claimed (Brit. 436725 and 436726) to be—
Fungicide.

Benzyldimethyldodecyl-Ammonium Iodide

Disinfectant
Claimed (Brit. 436725 and 436726) to be—
Bactericide, disinfectant.
Insecticide and Fungicide
Claimed (Brit. 436725 and 436726) to be—
Fungicide.

Benzylethylanilin

French: Aniline de benzyle et d'éthyle, Aniline benzylique-éthylique.
German: Benzylaethylanilin.
Chemical
Starting point in making—
Intermediates and other derivatives.
Dye
Starting point in making—
Erioglaucin, light green SF, yellowish; patent blue A, xylene blue A.

Benzylethylanilindisulphonic Acid

Synonyms: Ethylbenzylanilindisulphonic acid.
French: Acide de benzyle-éthyleanilinedisulphonique, Acide d'éthylebenzyleanilinedisulphonique.
German: Aethylbenzylanilindisulfonsäure, Benzylaethylanilindisulponsäure.
Chemical
Starting point in making—
Esters and salts, intermediates, pharmaceuticals.
Dye
Starting point in making—
Acid violet N 10B, fast acid violet 10B.

Benzylethylanilinmonosulphonic Acid

French: Acide de benzyle-éthyleanilinemonosulphonique.
German: Benzylaethylanilinmonosulphonsäure.
Chemical
Starting point in making—
Esters, salts and other derivatives.
Dye
Starting point in making—
Acid violet 5B extra, acid violet 6B, azo cardinal G, benzyl green B, benzyl violet 4B, erioglaucin A, fast acid violet 6B, formyl violet 5BN, guinea green B, night green B.

Benzyl Formate

French: Formiate de benzyle, Formiate benzylique.
German: Ameisensäuresbenzylester, Formylsäuresbenzylester.
Spanish: Formico de benzil.
Italian: Formico di benzile.
Analysis
Solvent for—
Cellulose derivatives, natural resins, synthetic resins.
Solvent miscible with—
Alcohols, aromatic hydrocarbons, aliphatic hydrocarbons, halogenated hydrocarbons, ketones, oils.
Cellulose Products
Solvent for—
Cellulose acetate (some types), cellulose ethers, nitrocellulose.
Chemical
Intermediate in—
Organic syntheses.
Solvent for—
Cellulose acetate, cellulose ethers, nitrocellulose.
Solvent miscible with—
Alcohols, aromatic hydrocarbons, aliphatic hydrocarbons, halogenated hydrocarbons, ketones, oils.
Cosmetic
Solvent in—
Perfumes.
Dry-Cleaning
Spotting agent for—
Resins.
Miscellaneous
See also "Solvents."
Paint and Varnish
Ingredient of—
Dopes, enamels, lacquers, paints, paint removers, varnishes.
Reagent for—
Imparting great strength to films.
Solvent for—
Benzyl abietate, cellulose acetate, cellulose ethers, copals, cumar, dammar, elemi, ester gum, ethylcellulose, glyptols (with alcohol), mastic, natural resins, nitrocellulose, shellac, synthetic resins.
Pharmaceutical
Solvent miscible with—
Alcohols, oils, ethers, hydrocarbons.
Resins
Solvent for—
Benzyl abietate, copals, cumar, dammar, elemi, ester gum, glyptols (with alcohol), mastic, natural resins, sandarac, shellac, synthetic resins.
Solvent in making—
Artificial resins from or containing cellulose acetate, nitrocellulose, or other cellulose esters or ethers.

Benzylgammaphenylpropylmethylamine

Chemical
Claimed (U. S. 2006114) as—
Substitute for papaverine.

Benzylgammaphenylpropylmethylamine Hydrochloride

Chemical
Claimed (U. S. 2006114) as—
Substitute for papaverine.

Benzylglucose

French: Benzyle de glucose, Benzyle de glycose, Glucose benzylique, Glycose benzylique, Glucose benzylé, Glycose benzylé.
German: Benzylglykose.
Cellulose Products
Plasticizer (Brit. 415764) for—
Benzylcellulose, cellulose derivatives, nitrocellulose.
For uses, see under general heading: "Plasticizers."

Benzylhydrocarvone

French: Hydrocarvone de benzyle, Hydrocarvone benzylique.
Cellulose Products
Plasticizer for—
Cellulose acetate, cellulose esters and ethers, cellulose nitrate.
For uses, see under general heading: "Plasticizers."
Chemical
Starting point in making—
Aromatics, intermediates, organic chemicals, pharmaceuticals.

Benzylideneglycerol

Cellulose Products
Solvent for—
Cellulose esters and ethers.
For uses, see under general heading: "Solvents."

Benzyl Mandelate
French: Mandélate de benzyle, Mandélate benzylique.
German: Benzylmandelat, Mandelsaeurebenzylester, Mandelsaeurebenzyl.

Paint and Varnish
Plasticizer (Brit. 270650) in making—
Lacquers, varnishes.

Plastics
Plasticizer in making—
Nitrocellulose plastics.

Benzylmethyldodecylsulphonium Methosulphate

Textile
Mordant (Brit. 436592) in—
Dyeing natural or regenerated cellulose textile materials with chrome dyestuffs.

Benzylnaphthalene
German: Benzylnaphtalin.

Chemical
Starting point (German 431899) in manufacturing reagent for protecting animal fibers in mordant dyeing by treatment with—
Allyl alcohol, allyl chloride, amyl alcohol, amyl chloride, butyl alcohol, butyl chloride, ethyl alcohol, ethyl chloride, isoallyl alcohol, isoallyl chloride, isoamyl alcohol, isoamyl chloride, isobutyl alcohol, isobutyl chloride, isopropyl alcohol, isopropyl chloride, methyl alcohol, methyl chloride, propyl alcohol, propyl chloride.

Benzylnaphthalenesulphonic Acid
French: Acide de benzylenaphthalènesulfonique.
German: Benzylnaphtalinsulfonsaeure.

Soap
Ingredient of—
Detergent preparations (Brit. 280110).

Textile
——, *Bleaching*
Wetting agent in making—
Bleach liquors (Brit. 280110).

——, *Dyeing*
Wetting agent in making—
Dye liquors (Brit. 280110).

——, *Finishing*
Wetting agent in making—
Fulling baths.
Liquors for wetting felt-like fabrics (Brit. 280110).

——, *Manufacturing*
Wetting agent in making—
Wool-carbonizing liquors.
Wool-degreasing liquors (Brit. 280110).

Benzylnaphthylmethane

Chemical
Starting point in making—
Chrysene.

Benzyl Paraoxybenzoate
French: Paraoxybenzoate de benzyle, Paraoxybenzoate benzylique.
German: Benzylparaoxybenzoat, Paraoxybenzoesäurebenzylester, Paraoxybenzoesäuresbenzyl.

Chemical
Starting point in making various derivatives.

Miscellaneous
As a preservative.

Pharmaceutical
In compounding and dispensing practice.

Sanitation
As a general disinfectant.

Benzylpentaerythritol

Cellulose Products
Solvent, softener, and plasticizer (Brit. 358393) for—
Cellulose acetate, cellulose esters or ethers, nitrocellulose.
For uses, see under general heading: "Plasticizers."

Benzyl Phosphate
French: Phosphate de benzyle, Phosphate benzylique.
German: Benzylphosphat, Phosphorsäurebenzylester, Phosphorsäuresbenzyl.
Spanish: Fosfato de benzil.
Italian: Fosfato di benzile.

Miscellaneous
Mothproofing agent (U. S. 1748675) in treating—
Feathers, furs, skins, plumes, and similar articles.

Textile
Mothproofing agent (U. S. 1748675) in treating—
Woolens and felts.

Benzyl Resinate
Synonyms: Benzyl abietate, Resin ether L.
French: Abiétate de benzyle, Abiétate benzylique, Éther résinique L, Résinate de benzyle, Résinate benzylique.
German: Abietinsaeurebenzylester, Abietinsaeuresbenzyl, Benzylabietat.

Paint and Varnish
Plasticizer in making—
Lacquers, paints, varnishes.

Plastics
Plasticizer in making various compositions.

Benzylresorcinol

Chemical
Starting point (Brit. 444351) in making—
Fat-splitting catalysts and emulsifying agents for use in dyeing, laundering, bleaching, and various other purposes, by reacting with formaldehyde and nonaromatic secondary amines (the salts of the products with water-soluble acids or water-insoluble acids, and the quaternary ammonium salts, are claimed to be valuable for the purposes named).

Benzylsorbitol
Synonyms: Benzylsorbite.

Miscellaneous
Plasticizer (U. S. 1936093) for—
Cellulose acetate, cellulose esters and ethers, cellulose nitrate, natural resins, synthetic resins.
For uses, see under general heading: "Plasticizers."

Benzylsucrose
French: Benzyle de sucre de canne, Benzyle sucré, Sucre de canne benzylique, Sucre de canne benzylé.
German: Benzylzucker.

Cellulose Products
Plasticizer (Brit. 415764) for—
Benzylcellulose, cellulose derivatives, nitrocellulose.
For uses, see under general heading: "Plasticizers."

Benzylsulphonic Acid
French: Acide de benzylesulfonique.
German: Benzylsulfonsaeure.

Chemical
Starting point in making various organic compounds.

Textile
——, *Dyeing and Printing*
Solubilizing agent (Brit. 276100) in making printing pastes and dye liquors containing—
Acridin dyestuffs, aminoanthraquinones, reduced or unreduced, anthraquinones, reduced or unreduced, azines, azo dyestuffs, basic diarylmethane dyestuffs, basic triarylmethane dyestuffs, benzoquinoneanilides, chrome mordant dyestuffs, indigoids, naphthoquinones, reduced and unreduced, naphthoquinoneanilides, nitroarylamines, nitroarylphenols, nitrodiarylamines, nitrodiarylphenols, oxazines, pyridines, quinolines, quinoneimides, reduced and unreduced, sulphur dyestuffs, thiazines, xanthenes.

Benzyl 1:2:3:6-Tetrahydrophthalate

Cosmetic
Fixative (U. S. 2015239) in—
Perfumes.

Benzyl Thiocyanate
Synonyms: Benzyl ester of thiocyanic acid, Benzyl sulphocyanate.
French: Benzyle thiocyanique, Thiocyanate de benzyle, Thiocyanate benzylique.
German: Benzilester aus thiocyansäure, Benzilrhodanid, Rhodanbenzil, Rhodanbenzilester, Rhodanwasserstoffsäuresbenzilester, Thiocyansäuresbenzilester.

Insecticide and Fungicide
Ingredient of—
Dusting agent for groundfleas (German 520330).
Dusting agent for killing the larvae and imagines of *Piesma quadrata.*
Spraying agent for exterminating insects which winter in cellars or the like.
Spraying agent for bugs and lice (French 654416).
Spraying agent for green plantlice (German 501135).

Benzyl Thiocyanate (Continued)
Spraying agent containing also pyridin and turkey red oil (German 520330).
Spraying agent for permanently driving ants away from trees (German 520330).
Spraying agents for *Aphis rumicis.*

Benzylthioglycollic Acid
Synonyms: Benzylsulphoglycollic acid.
French: Acide de benzylesulfoglycollique, Acide de benzylethioglycollique.
German: Benzylsulfoglykolsaeure, Benzylthioglykolsaeure.
Dye
Reagent (Brit. 284288) in making thioindigoid dyestuffs with--
Acenaphthenequinone, alphaisatinanilide, 5:7-dibromoisatin, isatin, isatin homologs and substitution products, orthodiketones.

Benzyl Thiosalicylate
Synonyms: Benzylsulphosalicylate.
French: Sulfosalicylate de benzyle, Sulfosalicylate benzylique, Thiosalicylate de benzyle, Thiosalicylate benzylique.
German: Benzylsulfosalicylat, Benzylthiosalicylat, Sulfosalicylsaeurebenzylester, Sulfosalicylsaeuresbenzyl, Thiosalicylsaeurebenzylester, Thiosalicylsaeuresbenzyl.
Chemical
Starting point (Brit. 282427) in making synthetic drugs with—
Oxides and salts of silver, gold, arsenic, antimony, and bismuth.

Benzyltolyl Sulphide
Synonyms: Thiocresylbenzyl ether.
Fungicide and Insecticide
As an insecticide (German 363656 and 496281, Brit. 326803, French 684447).
As a fungicide (German 363656 and 496281, Brit. 326803, French 684447).
Ingredient (Brit. 326803, French 684447, German 496281) of—
Bird-lice eradicator, containing also cyclohexanone, tetrahydronaphthalene, and talc.
Dog-flea eradicator, containing also cyclohexanone, tetrahydronaphthalene, and alcohol.

Benzyltriethyl Chloride
French: Chlorure de benzyletriéthyle, Chlorure benzylique et triéthylique.
German: Benzyltriaethylchlorid, Chlorbenzyltriaethyl.
Chemical
Starting point in making various derivatives.
Miscellaneous
Reagent (Brit. 312613) for treating—
Hair, feathers, furs, and other animal products to render them resistant to moths.
Textile
Reagent (Brit. 312613) for treating—
Wool and felt to render them mothproof.

Benzyltriphenyl Nitrate
French: Nitrate de benzyletriphényle, Nitrate bnezylique et triphénylique.
German: Benzyltriphenylnitrat, Salpetersäurebenzyltriphenylester, Salpetersäuresbenzyltriphenyl.
Chemical
Starting point in making various derivatives.
Miscellaneous
Mothproofing and moldproofing agent in the treatment of furs and hair.
Textile
Mothproofing and moldproofing agent in the treatment of wool and felt.

Benzyl Violet
French: Violette de benzyle.
German: Benzylveilchen.
Chemical
Ingredient (Brit. 295605) of bacteriological, therapeutic, and biological stain preparations, with—
Cresol, guaiacol, hydroquinone, phenol, phloroglucinol, pyrocatechol, pyrogallol, resorcinol.
Textile
Coloring matter in—
Dyeing and printing cotton, wool, and silk yarns and fabrics.

Beta-acetamidoanthraquinone
German: Beta-acetamidoanthrachinon.
Chemical
Starting point in making—
Intermediates.
Dye
Starting point in making—
Indianthrene orange RT, other synthetic dyestuffs.

Beta-aminoethyl Alcohol
French: Alcool de béta-aminoéthyle.
German: Beta-aminonaethylalkool.
Dye
Starting point (Brit. 285968) in making dyestuffs for cellulose ethers and esters with—
Alphachloroanthraquinone, alphaoxy-4-aminoanthraquinone, dibromobenzanthrone, dibromoindigo.

Beta-aminononadecane
Rubber
Activating agent (Brit. 412635) for—
Vulcanization accelerators, particularly such as the arylenethiazole mercaptans and disulphides and thiuramsulphides.

Beta-aminophenol-4-chloro-5-sulphonic Acid
French: Acide de béta-aminophénol-4-chloro-5-sulphonique.
German: Beta-aminophenol-4-chlor-5-sulfonsaeure.
Dye
Starting point (Brit. 271897) in making dyestuffs with—
Metatoluidide, orthotoluidide, 1-oxynaphthalene-4-phenylketone.

Beta-aminophenol-4-sulpho-6-carboxylic Acid
French: Acide de béta-aminophénol-4-sulpho-6-carbonique.
German: Beta-aminophenol-4-sulfo-6-carbonsaeure.
Dye
Starting point (Brit. 271897) in making cotton dyestuffs with—
Acetoaceticanilide, alphanaphthylamine, orthoanisidide, orthotoluidide, parachloroanilide, paradichloroanilide.

Beta-aminopyridin
Synoynms: 2-Aminopyridine.
Chemical
Starting point (Brit. 265167) in making—
2-Acetylaminopyridin, 2-allylaminopyridin, 2-amylaminopyridin, 2-butylaminopyridin, 2-cetylaminopyridin, 2-ethylaminopyridin, 2-formylaminopyridin, 2-heptylaminopyridin, 2-hexylaminopyridin, 2-isoallylaminopyridin, 2-isoamylaminopyridin, 2-isobutylaminopyridin, 2-isopropylaminopyridin, 2-lactylaminopyridin, 2-methylaminopyridin, 2-octylaminopyridin, 2-pentylaminopyridin, 2-propionylaminopyridin, 2-propylaminopyridin.

Beta-aminotridecane
Rubber
Activating agent (Brit. 412635) for—
Vulcanization accelerators, particularly such as the arylenethiazole mercaptans and disulphides and thiuramsulphides.

1-Beta-anthraquinonylbenzothiazole-5-carboxyl Chloride
Dye
Starting point (Brit. 439570) in making—
Yellow vat dyestuffs by condensing with 1-aminoanthraquinone.
Yellow vat dyestuffs by condensing with 1-amino-5-benzoamidoanthraquinone.

1-Beta-anthraquinonylbenzothiazole-3:5-dicarboxyl Chloride
Dye
Starting point (Brit. 439570) in making—
Yellow vat dyestuffs by condensing with 1-aminoanthraquinone.

2′-Beta-anthraquinonylbetanaphthoxazole-3-carboxyl Chloride
Dye
Starting point (Brit. 439570) in making—
Yellow vat dyestuffs by condensing with 1-aminoanthraquinone.

Beta-b'-bis-3-aminometatolyloxydiethyl Ether

Dye

Starting point (U. S. 1978783) in making—
Bis-2:3-hydroxynaphthoyl derivatives useful as coupling components in making dyestuffs of good fastness to light, chlorine, and washing.

Beta-b'-bis-5-amino-orthoanisoxydiethyl Ether

Dye

Starting point (U. S. 1978783) in making—
Bis-2:3-hydroxynaphthoyl derivatives useful as coupling components in making dyestuffs of good fastness to light, chlorine, and washing.

Beta-b'-bis-4-chlor-2-aminophenoxydiethyl Ether

Dye

Starting point (U. S. 1978783) in making—
Bis-2:3-hydroxynaphthoyl derivatives useful as coupling components in making dyestuffs of good fastness to light, chlorine, and washing.

Beta-b'-bisorthoaminophenoxydiethyl Ether

Dye

Starting point (U. S. 1978783) in making—
Bis-2:3-hydroxynaphthoyl derivatives useful as coupling components in making dyestuffs of good fastness to light, chlorine, and washing.

Beta-b'-bispara-aminophenoxydiethyl Ether

Dye

Starting point (U. S. 1978783) in making—
Bis-2:3-hydroxynaphthoyl derivatives useful as coupling components in making dyestuffs of good fastness to light, chlorine, and washing.

Beta-b'-dichloroethyl Sulphide

French: Sulfure de béta-b'-dichloroéthyle, Sulfure béta-b-dichloroéthylique.
German: Beta-b'-dichloraethylsulfid, Schwefelbeta-b'-dichloraethyl.

Military

As a poison gas.

Paint and Varnish

Ingredient (Russian 28267) of—
Paints and varnishes used for protecting the bodies of ships against deposits of shells.

Beta-b'-dichloroisopropyl Monochloroacetate

French: Monochloracétate de béta-b'-dichlorisopropyle.
German: Beta-b'-dichlorisopropylmonochloracetat.
Spanish: Monocloracetato de beta-b'-diclorisopropil.
Italian: Monocloracetato di beta-b'-diclorisopropile.

Disinfectant

Bactericide (French 667633) for—
Furs, hair, pelts, skins, cheese, meat, bread (applied as 10 percent solution in beef extract or other protein medium).

Insecticide

Insecticide and parasiticide (French 667633) for—
Furs, hair, pelts, skins, cheese, meat, bread (applied as 10 percent solution in beef extract or other protein medium).

Betabetabis-3-nitro-4-hydroxyphenylpropane

Rubber

Antiscorching agent (Brit. 418376 and 418445) in—
Rubber compounding.

Betabetadiethylhydroxylamine

German: Betabetadiaethylhydroxylamin.

Chemical

Starting point in making—
Triethyloxamine.

Betabetadihydroxytriphenylene

Dye

Starting point (Brit. 445862) in making—
Brown dyes by coupling, either in substance or on the fiber, with aromatic diazo compounds.

Betabutoxyethylfluorene 9-Carboxylate

Cellulose Products

Plasticizer (U. S. 1975697) for—
Cellulose acetate, cellulose esters and ethers, cellulose nitrate.
For uses, see under general heading: "Plasticizers."

Betabutoxyethyl Oleate

Miscellaneous

Plasticizer (U. S. 2010560) for—
Cellulose nitrate, phenol-formaldehyde resins.
For uses, see under general heading: "Plasticizers."

Betachlorethyl Monochloroacetate

French: Monochloracétate de bétachloréthyle.
German: Betachloraethylmonochloracetat.
Spanish: Monocloracetato de betacloretil.
Italian: Monocloracetato di betacloretile.

Disinfectant

Bactericide (French 667633) for—
Furs, hair, pelts, skins, cheese, meat, bread (applied as 10 percent solution in beef extract or other protein medium).

Insecticide

Insecticide and parasiticide (French 667633) for—
Furs, hair, pelts, skins, cheese, meat, bread (applied as 10 percent solution in beef extract or other protein medium).

Betachloroethanealphasulphondodecylamide

Miscellaneous

As a wetting agent (Brit. 436862).
For uses, see under general heading: "Wetting agents."

Betachloroethyl Acetate

Synonyms: Chloroethyl acetate.
French: Acétate de bétachloréthyle, Acétate bétachloroéthylique.
German: Chloraethylessigsaeureseste r, Essigsaeureschloraethylester, Essigsaeureschloraethyl.

Cellulose Products

Solvent (German 391667) for—
Cellulose acetate, cellulose nitrate, natural resins, synthetic resins.
For uses, see under general heading: "Solvents."

Chemical

Starting point in making various derivatives.

Betachloroethyl Naphthenate

Miscellaneous

As a softening agent (Brit. 435864).
For uses, see under general heading: "Softening agents."

Betachloromethylnaphthalenesulphonic Acid

French: Acide bétachloronaphthalènesulphonique.
German: Betachlormethylsulfonsäure.
Spanish: Acido betaclormetilnaftalinsulfonico.
Italian: Acido betaclormetilnaftalinsulfonico.

Miscellaneous

As an emulsifying agent (Brit. 362016).
For uses, see under general heading: "Emulsifying agents."

4-Betadiethylaminoethoxy-3-carbobetadiethylaminoethoxydiphenyl

Pharmaceutical

Claimed (U. S. 1976921, 1976922 and 1976924) as—
Anesthetic.

Betadiethylaminoethylacetylbenzylamide

Pharmaceutical

Claimed (Brit. 438659) to possess—
Physiological properties resembling those of atropine.

Betadiethylaminoethylacetyltropamide

Pharmaceutical

Claimed (Brit. 438659) to possess—
Physiological properties resembling those of atropine.

Betadiethylaminoethylalphaphenylalphamethoxyacetamide

Pharmaceutical

Claimed (Brit. 438659) to possess—
Physiological properties resembling those of atropine.

Betadiethylaminoethyldiphenoxyacetamide

Pharmaceutical

Claimed (Brit. 438659) to possess—
Physiological properties resembling those of atropine.

Betadiethylaminoethylmandelamide

Pharmaceutical

Claimed (Brit. 438659) to possess—
Physiological properties resembling those of atropine.

Betadiethylaminoethyl Para-aminobenzoate Hydrochloride
Pharmaceutical
Suggested for use (Brit. 439168) as—
Anesthetic in conjunction with 3:4-dihydroxyphenylalphapropanolamine hydrochloride (greater stability and less toxicity than adrenalin are claimed).

Betadiethylaminoethylphenyldiethylacetamide
Pharmaceutical
Claimed (Brit. 438659) to possess—
Physiological properties resembling those of atropine.

Betadiethylaminoethyltriphenylacetamide
Pharmaceutical
Claimed (Brit. 438659) to possess—
Physiological properties resembling those of atropine.

Betadimethylaminoethyldiphenylacetamide
Pharmaceutical
Claimed (Brit. 438659) to possess—
Physiological properties resembling those of atropine.

Betadimethylaminoethylphenylacetamide
Pharmaceutical
Claimed (Brit. 438659) to possess—
Physiological properties resembling those of atropine.

Betadinaphthol
German: Betadinaphtol.
Chemical
Starting point in making—
Perylene (U. S. 1639658).

Betaethoxyethyl Oleate
Miscellaneous
Plasticizer (U. S. 2010560) for—
Cellulose nitrate, phenol-formaldehyde resins.
For uses, see under general heading: "Plasticizers."

Betaethoxyethyl Stearate
Cellulose Products
Plasticizer (Brit. 393619) for—
Cellulose esters or ethers, cellulose nitrate (nitrocellulose).
For uses, see under general heading: "Plasticizers."

Betaethylhydroxylamine
German: Betaaethylhydroxylamin.
Chemical
Starting point in making various intermediates.

Betagammadibromopropyl Ester of Coconut Oil Acids
Miscellaneous
As a softening agent (Brit. 435864).
For uses, see under general heading: "Softening agents."

Betagammadichlorobutane
Chemical
As a chlorinating medium.
Miscellaneous
Solvent for—
Cellulose acetate, cellulose esters and ethers, cellulose nitrate, fats and resins.
For uses, see under general heading: "Solvents."

Betagammadihydroxypropylanilin, Normal
Dye
In dye syntheses.
Starting point (Brit. 449498) in making—
Violet-blue dyes for acetate rayon, with diazotized 2:5-dichlor-4:6-dinitroanilin.

Betagammadihydroxypropyldodecyl Sulphide
Chemical
Starting point (Brit. 422937) in making—
Textile assistants by oxidation and subsequent sulphonation.

Betagammadihydroxypropylnaphthylamine, Normal
Dye
In dye syntheses.
Starting point (Brit. 449498) in making—
Violet-blue dyes for acetate rayon, with diazotized 2:5-dichlor-4:6-dinitroanilin.

Betagammadistearoxypropyl Chloride
Miscellaneous
As a softening agent (Brit. 435864).
For uses, see under general heading: "Softening agents."

Betaglucoside
Mechanical
Inhibitor of—
Scale formation in boilers.
Sludge in high-pressure boilers.
Foaming in boilers.
Promoter of—
Even boiling in boiler operation.
Remover of—
Oxygen, carbon dioxide, and hydrogen sulphide in boiler waters.
Retarder of—
Precipitation in inorganic treatments of boiler waters.
Treating agent for—
Very hard waters and alkali waters in boiler plants.

Beta-4-hydroxy-3-carboxyphenyloctane
Fungicide
As a fungicide (U. S. 2001767).

Betahydroxyethylaminoanthraquinone
Miscellaneous
Dyestuffs (U. S. 1989133) for—
Cellulose acetate products (imparts shades of red).
Textile
Dyestuffs (U. S. 1989133) for—
Cellulose acetate products (imparts shades of red).

Betahydroxyethylnaphthenyl Sulphide
Chemical
Starting point (Brit. 422937) in making—
Textile assistants by oxidation and subsequent sulphonation.

Betahydroxyethyl-N:N-butylcresidin, Normal
Chemical
Reagent in—
Organic synthesis.
Dye
Coupling agent (Brit. 421975) in making—
Light-fast and readily discharged navy-blue dyestuffs for acetate rayon with diazotised 6-bromo-2:4-dinitroanilin or 6-chloro-2:4-dinitroanilin.

Betahydroxyethyl-N:N-butylmetatoluidin, Normal
Chemical
Reagent in—
Organic synthesis.
Dye
Coupling agent (Brit. 421975) in making—
Light-fast and readily discharged blue-violet dyestuffs for acetate rayon with diazotised 6-bromo-2:4-dinitroanilin or 6-chloro-2:4-dinitroanilin.

Betahydroxyisoallylamine
French: Bétahydroxyeisoallyleamine.
Chemical
Starting point in making—
Intermediates, pharmaceuticals.
Dye
Starting point (Brit. 289807) in making dyestuffs with—
Alphachloroanthraquinone.
Leuco 1:4:5:8-tetrahydroxyanthraquinone.
Quinazarin.
1:4:5-Trihydroxyanthraquinone.

Betahydroxyisoamylamine
Chemical
Starting point in making—
Intermediates, pharmaceuticals.
Dye
Starting point (Brit. 289807) in making dyestuffs with—
Alphachloroanthraquinone.
Leuco 1:4:5:8-tetrahydroxyanthraquinone.
Quinazarin.
1:4:5-Trihydroxyanthraquinone.

Betahydroxyisobutylamine
French: Bétahydroxyeisobutyleamine.
German: Betahydroxyisobutylamin.
Chemical
Starting point in making—
Aromatics, intermediates, pharmaceuticals.
Dye
Starting point (Brit. 289807) in making dyestuffs with—
Alphachloroanthraquinone.
Leuco 1:4:5:8-tetrahydroxyanthraquinone.
Quinazarin.
1:4:5-Trihydroxyanthraquinone.

Betahydroxyisopropylamine
French: Bétahydroxyeisopropyleamine.
German: Betahydroxyisopropylamin.
Chemical
Starting point in making—
Aromatics, intermediates, pharmaceuticals.
Dye
Starting point (Brit. 289807) in making dyestuffs with—
Alphachloroanthraquinone.
Leuco 1:4:5:8-tetrahydroxyanthraquinone.
Quinazarin.
1:4:5-Trihydroxyanthraquinone.

Betahydroxypentylamine
Chemical
Starting point in making—
Intermediates and other derivatives.
Dye
Starting point (Brit. 289807) in making dyestuffs with the aid of—
Alphachloroanthraquinone.
Leuco 1:4:5:8-tetrahydroxyanthraquinone.
Quinazarin.
1:4:5-Trihydroxyanthraquinone.

4-Betahydroxypropylaminoanthraquinone
Textile
Dyestuffs (Brit. 447090 and 447037) for imparting—
Deep-blue shades to acetate rayon, either by dyeing or printing.

Betahydroxytriphenylene
Dye
Starting point (Brit. 445862) in making—
Brown dyes by coupling, either in substance or on the fiber, with aromatic diazo compounds.

Betahydroxytriphenyleneorthocarboxylic Acid
Dye
Starting point (Brit. 445862) in making—
Brown dyes by coupling, either in substance or on the fiber, with aromatic diazo compounds.

Betahydroxytriphenylenesulphonic Acid
Dye
Starting point (Brit. 445862) in making—
Brown dyes by coupling, either in substance or on the fiber, with aromatic diazo compounds.

Betaine Hydrobromide
French: Hydrobromure de bétaine.
German: Betainhydrobromid, Bromwasserstoffsaeuresbetain.
Chemical
Starting point in making—
Pharmaceutical products with hexamethylenetetramine (U. S. 1588753).
Insecticide
Starting point in making—
Hydrocyanic acid for application as an insecticide.
Pharmaceutical
In compounding and dispensing practice.
Photographic
Ingredient of—
Bleaching solution used in the bromoil process (German 426661).

Betaine Hydrochloride
French: Hydrochlorure de bétaine.
German: Betainchlorhydrat, Betainchlorid.
Chemical
Starting point in making—
Pharmaceutical product with hexamethylenetetramine (U. S. 1588753).
Insecticide
Starting point in making—
Hydrocyanic acid for application as an insecticide.
Pharmaceutical
In compounding and dispensing practice.
Photographic
Ingredient of—
Bleaching solution used in the bromoil process (German 426661).

4-Betamethoxybutyramido-2:5-diethoxyanilin
Dye
Starting point (Brit. 435711) in making—
Navy-blue dyestuffs by coupling with alphanapththylamide.

Betamethoxyethyl Oleate
Miscellaneous
Plasticizer (U. S. 2010560) for—
Cellulose nitrate, phenol-formaldehyde resins.
For uses, see under general heading: "Plasticizers."

Betamethylaminoalphapara-aminophenylpropanol
Pharmaceutical
Claimed (Brit. 440968) as—
Less excitant in action than the bases from which it is derived.

Betamethylaminoalphaparahydroxyphenylethanol
Pharmaceutical
Claimed (Brit. 440968) as—
Less excitant in action than the bases from which it is derived.

Betamethylanthraquinone
Synonyms: 2-Methylanthraquinone.
German: 2-Methylanthrachinon.
Chemical
Starting point in making—
Betamethylanthramine (Brit. 260000).
Metamethylanthracene.
Alphanitrobetamethylanthraquinone.
Dye
Starting point in making—
Anthraflavone, cibanone yellow, indanthrene golden-orange.

Betamethylnaphthalene
Synonyms: 2-Methylnaphthalene.
German: Betamethylnaphtalin, 2-Methylnaphtalin.
Chemical
Starting point in making—
Acetyl derivative of 4-chloro-1-aminobetanaphthalene.
Aminobetanaphthalene.
Betamethylnaphthol.
Bromo-1-nitrobetanaphthalene.
Chloroaminobetanaphthalene.
4-Chloro-1-aminobetanaphthalene.
4-Chloro-1-aminobetanaphthalene hydrochloride.
4-Chloro-1-aminobetanaphthalene sulphate.
4-Chloro-2-methylnaphthylphthalimide.
Dinitrobetanaphthalene.
1-Nitrobetanaphthalene.
Dye
Starting point in making—
Azo dyestuffs, dyestuffs from methylnaphthol, oxyazo dyestuffs.

Betanaphthalenesulphonamide
Cellulose Products
Plasticizer (Brit. 417871; Canada 340994) for—
Cellulose acetate.
For uses, see under general heading: "Plasticizers."

Betanaphthalide
German: Betanaphtalid.
Dye
Reagent (Brit. 274128) in making azo dyestuffs with—
1:3-Dimethyl-4-amino-6-bromobenzene.
1:3-Dimethyl-4-amino-6-chlorobenzene.
1:3-Dimethyl-4-amino-2:6-dibromobenzene.
1:3-Dimethyl-4-amino-2:6-dichlorobenzene.

Betanaphtholamyl Ether
Chemical
Starting point (Brit. 264860) in making dispersive agents with the following chlorides—
Allyl, amyl, benzyl, butyl, cetyl, ethyl, isoallyl, isoamyl, isobutyl, isopropyl, hexyl, methyl, naphthyl, phenyl, propyl, tolyl, xylyl.

Betanaphtholbutyl Ether
German: Beta-naphtolbutylaether.
Chemical
Starting point (Brit. 264860) in making dispersive agents with the following chlorides:—
Allyl, amyl, benzyl, butyl, cetyl, ethyl, isoallyl, isoamyl, isobutyl, isopropyl, hexyl, methyl, phenyl, propyl, naphthyl, tolyl, xylyl.

Betanaphtholethyl Ether
French: Éther bétanaphthole-éthylique, Éther de bétanaphthole-éthyle.
German: Betanaphtolaethylaether.

Chemical
Starting point in—
Organic synthesis.
Starting point (Brit. 277098) in making derivatives with—
Benzyl chloride, naphthyl chloride, phenyl chloride, phthalyl chloride, tolyl chloride, xylyl chloride.

Lubricant
Starting point (Brit. 440916) in making—
Products useful as lubricating oils or as pour-point depressors for paraffin base lubricating oils by condensation with halogenated derivatives of aliphatic hydrocarbons, such as paraffin oils, paraffin, petrolatum, ceresin, ozokerite, or others contained in the middle to higher fractions of petroleum.

Perfumery
Ingredient of—
Cosmetics, perfumes.

Soap
Perfume in—
Toilet soaps.

Betanaphtholisoallyl Ether
French: Éther bétanaphtholisoallique, Éther de bétanaphtholisoallyle.
German: Betanaphtolisoallylaether.

Chemical
Starting point (Brit. 277098) in making derivatives with—
Benzyl chloride, naphthyl chloride, phenyl chloride, phthalyl chloride, tolyl chloride, xylyl chloride.

Betanaphtholisoamyl Ether
French: Éther de bétanaphtholisoamyle, Éther bétanaphtholisoamylique.
German: Betanaphtolisoamylaether.

Chemical
Starting point (Brit. 277098) in making derivatives with—
Benzyl chloride, naphthyl chloride, phenyl chloride, phthalyl chloride, tolyl chloride, xylyl chloride.

Betanaphtholpropyl Ether
French: Éther de bétanaphtholpropyle, Éther bétanaphtholpropylique.
German: Betanaphtolpropylaether.

Chemical
Starting point (Brit. 277098) in making derivatives with—
Benzyl chloride, naphthyl chloride, phenyl chloride, phthalyl chloride, tolyl chloride, xylyl chloride.

Betanaphthoquinone
Synonyms: 2-Naphthoquinone.

Dye
Starting point in making—
Alizarin green B, naphthaphenazin.

Betanaphthoyl Chloride
French: Chlorure de bétanaphthoyl, Chlorure bétanaphthoylique.

Dye
Reagent (German 432579) in making anthraquinone vat dyestuffs with—
Alpha-aminoanthraquinone, 1:4-diaminoanthraquinone, 1:5-diaminoanthraquinone, 1:6-diaminoanthraquinone, 1:7-diaminoanthraquinone.

Betanaphthoylpropyl Ether

Chemical
Starting point (Brit. 264860) in making dispersive agents with—
Acetyl chloride, allyl chloride, amyl chloride, anthranyl chloride, benzoyl chloride, benzyl chloride, butyl chloride, caproyl chloride, capryl chloride, cetyl chloride, cinnamyl chloride, dibenzoyl chloride, dibenzyl chloride, diethyl chloride, dimethyl chloride, diphenyl chloride, ethyl chloride, formyl chloride, glyceryl chloride, heptyl chloride, hexyl chloride, isoamyl chloride, isobutyl chloride, isopropyl chloride, lactyl chloride, methyl chloride, naphthyl chloride, octyl chloride, phenyl chloride, phthalyl chloride, propenyl chloride, propyl chloride, resorcyl chloride, succinyl chloride, sulphanyl chloride, thionyl chloride, tolyl chloride, valeryl chloride, xylyl chloride.

Betanaphthylamine Phthalate

Rubber
Prevulcanization inhibitor (Brit. 422308).

2-Betanaphthylaminodiphenylene Oxide

Rubber
Antiaging agent (Brit. 422191).

Betanaphthyl Salicylate
Synonyms: Betol, Naphthalol, Salinaphthol, Salicylic naphthyl ester.
French: Salicylate de bétanaphtyle, Salicylate bétanaphtylique.
German: Salicylsaeuresbetanaphtyl, Salicylsaeurebetanaphtylester.

Chemical
Starting point in making—
Water-soluble tanning agents with chlorosulphonic acid (Brit. 266697).

Pharmaceutical
In compounding and dispensing practice.

2-Betanaphthylthiolquinolin Ethiodide

Dye
Process material (Brit. 454687) in making—
Cyanin dyes.

2-Betanaphthylthiolquinolin Methiodide

Dye
Process material (Brit. 454687) in making—
Cyanin dyes.

Betaoctadecylsulphoxyethylthioglycollic Acid

Chemical
Intermediate (Brit. 444262 and 444501) in—
Organic syntheses.

Insecticide
Insecticide (Brit. 444262 and 444501) for—
Animal pests, vegetable pests.

Textile
As a dyestuff (when employing suitable initial materials) (Brit. 444262 and 444501).
Assistant (Brit. 444262 and 444501) in—
Textile processing.

Betaparaisoamylbenzoylpropionic Acid

Miscellaneous
As a wetting agent (Brit. 449865).
For uses, see under general heading: "Wetting agents."

Betaparaisoamylphenylbutyric Acid

Miscellaneous
As a wetting agent (Brit. 449865).
For uses, see under general heading: "Wetting agents."

Betapara-normal-butylbenzoylpropionic Acid

Miscellaneous
As a wetting agent (Brit. 449865).
For uses, see under general heading: "Wetting agents."

Betapara-normal-butylphenylbutyric Acid

Miscellaneous
As a wetting agent (Brit. 449865).
For uses, see under general heading: "Wetting agents."

Betapara-normal-cyclohexylbenzoylpropionic Acid

Miscellaneous
As a wetting agent (Brit. 449865).
For uses, see under general heading: "Wetting agents."

Betapara-normal-cyclohexylphenylbutyric Acid

Miscellaneous
As a wetting agent (Brit. 449865).

Betapara-normal-hexylbenzoylpropionic Acid

Miscellaneous
As a wetting agent (Brit. 449865).
For uses, see under general heading: "Wetting agents."

Betapara-normal-hexylphenylbutyric Acid

Miscellaneous
As a wetting agent (Brit. 449865).
For uses, see under general heading: "Wetting agents."

Betapara-normal-laurylbenzoylpropionic Acid

Miscellaneous
As a wetting agent (Brit. 449865).
For uses, see under general heading: "Wetting agents."

Betapara-normal-laurylphenylbutyric Acid
Miscellaneous
As a wetting agent (Brit. 446568).
For uses, see under general heading: "Wetting agents."

Betapara-normal-octylbenzoylpropionic Acid
Miscellaneous
As a wetting agent (Brit. 449865).
For uses, see under general heading: "Wetting agents."

Betapara-normal-octylphenylbutyric Acid
Miscellaneous
As a wetting agent (Brit. 446568).
For uses, see under general heading: "Wetting agents."

Betapara-secondary-butylbenzoylpropionic Acid
Miscellaneous
As a wetting agent (Brit. 449865).
For uses, see under general heading: "Wetting agents."

Betapara-secondary-butylphenylbutyric Acid
Miscellaneous
As a wetting agent (Brit. 449865).
For uses, see under general heading: "Wetting agents."

Betaparatoluenesulphonylethylthioglycollic Acid
Chemical
Intermediate (Brit. 444262 and 444501) in—
Organic syntheses.
Insecticide
Insecticide (Brit. 444262 and 444501) for—
Animal pests, vegetable pests.
Textile
As a dyestuff (when employing suitable initial materials) (Brit. 444262 and 444501).
Assistant (Brit. 444262 and 444501) in—
Textile processing.

Betaphenoxyethyllauramide
Chemical
Starting point (Brit. 443902) in making—
Sulphonated sodium salts, stable to calcium chloride and acids, which are used as scouring agents for raw wool.

Betaphenoxyethylstearamide
Chemical
Starting point (Brit. 443902) in making—
Sulphonated sodium salts, stable to calcium chloride and acids, which are used as scouring agents for raw wool.

Betaphenylethylamine
German: Betaphenylaethylamin.
Chemical
Starting point (German 423027) in making—
Normal benzenesulphophenylethylglycin.
Normal benzenesulphotetrahydroisoquinolin.
Phenylethylglycin.
Tetrahydroisoquinolin.

(Betaphenylethylbetamethoxybetaphenylethyl)methylamine
Chemical
Claimed (U. S. 2006114) as—
Substitute for papaverine.

(Betaphenylethylbetamethoxybetaphenylethyl)methylamine Hydrochloride
Chemical
Claimed (U. S. 2006114) as—
Substitute for papaverine.

(Betaphenylethylbetaorthomethoxyphenylethyl)-ethylamine
Chemical
Claimed (U. S. 2006114) as—
Substitute for papaverine.

(Betaphenylethylbetaorthomethoxyphenylethyl)ethylamine Hydrochloride
Chemical
Claimed (U. S. 2006114) as—
Substitute for papaverine.

Betaphenylethylbetaphenylisopropylmethylamine
Chemical
Claimed (U. S. 2006114) as—
Substitute for papaverine.

Betaphenylethylbetaphenylisopropylmethylamine Hydrochloride
Chemical
Claimed (U. S. 2006114) as—
Substitute for papaverine.

Beta-4-phenylethyl Piperidinoethylbenzoate
Pharmaceutical
Claimed (U. S. 1997828) as—
Local anesthetic.

Betaphenylethylsulphonic Acid
Dye
Intermediate (Brit. 447067) in making—
Dyes containing one or more aryl residues carrying one or more alkysulphonic groups directly combined to the nucleus.

Betaphenylmethylglycidic Aldehyde
French: Aldéhyde bétaphénylemétbyleglycidique.
German: Betaphenylmethylglycidinaldehyd.
Spanish: Aldehido betafenilmetilglicidico.
Italian: Aldeide betafenilmetilglyicidio.
Perfume
Ingredient of—
Perfume compositions to give them a lilac and hyacinth odor.
Perfume in various toiletries.
Soap
Perfume in—
Toilet soaps.

Betapiperidinodiphenylacetamide
Pharmaceutical
Claimed (Brit. 438659) to possess—
Physiological properties resembling those of atropine.

Betapiperidinoethanol
Chemical
Reagent in making—
Monochloride of normal betamethoxyethylanthranilic acid betapiperidine ethylester (Brit. 260605).

Betapiperidinoethylphenylacetamide
Pharmaceutical
Claimed (Brit. 438659) to possess—
Physiological properties resembling those of atropine.

Betapyridylalphapiperidin
Insecticide
Ingredient (U. S. 1925225) of—
Insecticidal compound with tannic acid.

Betaresorcylic Anilide
French: Anilide de bétarésorcyle, Anilide bétarésorcylique.
German: Betaresorcylanilid.
Agricultural
Reagent in treating—
Seeds to protect them against mildew and other fungi.
Chemical
Starting point in making various derivatives.
Food
Reagent in treating—
Grains to protect them against mildew and other fungi.
Leather
Reagent in treating—
Leather to protect it against mildew and other fungi.
Paper
Reagent in treating—
Paper and paper products to protect them against mildew and other fungi.
Rubber
Reagent in treating—
Rubber and rubber products to protect them against mildew and other fungi.
Textile
Reagent in treating—
Cotton yarns and fabrics to protect them against mildew and other fungi.
Woodworking
Reagent in treating—
Wood and wood products to protect them against mildew and other fungi.

Betaspodumene

(Spodumene—lithium-aluminum silicate—ore which has been heated in a lime kiln and freed of associated minerals by tumbling and sifting.)

Ceramic
Ingredient of—
Pottery batches.

Chemical
Starting point in making—
Lithium chloride, other lithium salts.

Glass
Ingredient of batches in making—
Extremely tough glass.

Betasulphoethyl Oleate

Rubber
Stabilizer (Brit. 411478) for—
Rubber latex.

Betathionaphthol

Petroleum
Antioxidant (Brit. 425569) for—
Lubricating, transformer, and switch oils, particularly solvent-extracted oils and others of a paraffinic nature, in which the natural inhibitor content may have been reduced during refining.

Betatolyloxyethyllauramide

Chemical
Starting point (Brit. 443902) in making—
Sulphonated sodium salts, stable to calcium chloride and acids, which are used as scouring agents for raw wool.

Betatolyloxyethylstearamide

Chemical
Starting point (Brit. 443902) in making—
Sulphonated sodium salts, stable to calcium chloride and acids, which are used as scouring agents for raw wool.

Betatrichloroethane

Synonyms: Chloroethylene chloride, Ethylene chlorochloride, Ethylene monochlorochloride, Monochlorinated dutch liquid, Monochloroethylene chloride, Vinyl trichloride.
French: Bétatrichlorure d'éthane, Monochlorochlorure d'éthylène, Trichlorure de vinyle.
German: Trichloraethan.
Spanish: Tricloretano.
Italian: Tricloretano.

Analysis
Solvent for—
Alkaloids, fats, oils, waxes.

Cellulose Products
Solvent for—
Cellulose acetate (used in admixture with alcohol).
Solvent miscible with—
Alcohols, esters, ethers, ketones.

Ceramic
Solvent (in admixture with alcohol) in—
Compositions, containing natural or synthetic resins or cellulose acetate, used as coatings for protecting and decorating ceramic products.

Chemical
Solvent for—
Alkaloids, fats, oils, tar, waxes.
Solvent miscible with—
Alcohols, esters, ethers, ketones.

Coal Processing
Solvent for—
Tar.

Cosmetic
Fats, oils, waxes.
Solvent (in admixture with alcohol) in—
Nail enamels and lacquers containing natural or synthetic resins or cellulose acetate as base material.

Dry-Cleaning
Spotting agent for—
Fats, oils, resins, tars, waxes.

Electrical
Solvent (in admixture with alcohol) in—
Insulating compositions, containing natural or synthetic resins or cellulose acetate, used for covering wire and in making electrical machinery and equipment.

Fats, Oils, and Waxes
Solvent for—
Fats, oils, waxes.

Food
Solvent for—
Fats, oils, waxes.

Glass
Solvent (in admixture with alcohol) in—
Compositions, containing natural or synthetic resins or cellulose acetate, used in the manufacture of nonscatterable glass and as coatings for decorating and protecting glassware.

Glue and Adhesives
Solvent (in admixture with alcohol) in—
Adhesive compositions containing natural or synthetic resins or cellulose acetate.

Leather
Solvent (in admixture with alcohol) in—
Compositions, containing natural or synthetic resins or cellulose acetate, used in the manufacture of artificial leathers and as coatings for decorating and protecting leathers and leather goods.

Metal Fabricating
Solvent (in admixture with alcohol) in—
Compositions, containing natural or synthetic resins or cellulose acetate, used as coatings for protecting and decorating metallic articles.

Miscellaneous
Degreasing agent for various purposes.
Solvent for—
Fats, oils, tar, waxes.
Solvent miscible with—
Alcohols, esters, ethers, ketones.
Solvent (in admixture with alcohol) in—
Coating compositions, containing natural or synthetic resins, or cellulose acetate, used for protecting and decorating various articles.

Paint and Varnish
Ingredient of—
Paint removers.
Solvent (in admixture with alcohol) in—
Paints, varnishes, lacquers, enamels, and dopes containing natural or synthetic resins or cellulose acetate.

Paper
Solvent (in admixture with alcohol) in—
Compositions, containing natural or synthetic resins or cellulose acetate, used in the manufacture of coated papers and as coatings for decorating and protecting products made of paper or pulp.

Pharmaceutical
Solvent for—
Fats, oils, waxes.

Plastics
Solvent (in admixture with alcohol) in making—
Plastics from or containing natural or synthetic resins or cellulose acetate.

Resins
Solvent for—
Natural resins, synthetic resins.
Solvent (in admixture with alcohol) in making—
Artificial resins from or containing cellulose acetate.

Rubber
Solvent (in admixture with alcohol) in—
Compositions, containing natural or synthetic resins or cellulose acetate, used as coatings for protecting and decorating rubber goods.

Soap
Ingredient of—
Special soaps.
Solvent for—
Fats, oils.

Stone
Solvent (in admixture with alcohol) in—
Compositions, containing natural or synthetic resins or cellulose acetate, used as coatings for decorating and protecting artificial and natural stone.

Textile
Solvent (in admixture with alcohol) in—
Compositions, containing natural or synthetic resins or cellulose acetate, used in the manufacture of coated fabrics.

Wood
Solvent (in admixture with alcohol) in—
Compositions, containing natural or synthetic resins or cellulose acetate, used as protective and decorative coatings on woodwork.

Birch Tar
Synonyms: Dagget, Doggert, Litauer balsam.
French: Goudron de bouleau.
German: Birkenoel, Birkenteer.
Leather
Ingredient of—
Compositions for finishing Russian leathers.
Pharmaceutical
In compounding and dispensing practice.

Bird Lime
French: Glu.
German: Fliegenleim, Raupenlein, Vogel-leim.
Agricultural
Ingredient of—
Compositions for protecting trees, vineyards, and growing plants against caterpillars, animal pests, insects.
Insecticide
Ingredient of—
Fly-destroying compositions, insect-destroying compositions.

1:4-Bisacetamideanthraquinone
German: 1:4-Bisacetamid-anthrachinon.
Chemical
Starting point in making various intermediates.
Dye
Starting point in making various synthetic dyestuffs.
Textile
Dyestuff (Brit. 263260) for—
Acetate rayon.

Bisacetoacetylorthotoluidin
Dye
Starting point (German 431773) in making azo dyestuffs with—
Diazotized sulphanilic acid.
Diazotized 6-chlorometatoluidin-4-sulphonic acid.

Bis(2-aminophenyl) Disulphide
Insecticide
Exterminant for—
Culicine mosquito larvae.

Bis(beta-9-carbazolylethylthiol)ethane
Chemical
Intermediate (Brit. 444262 and 444501) in—
Organic syntheses.
Pharmaceutical
Claimed (Brit. 444262 and 444501) to have—
Value for pharmaceutical purposes.
Rubber
Accelerator (Brit. 444262 and 444501) in—
Vulcanizing.

Bis(betamethoxyethyl)methylene Ether
Cellulose Products
Plasticizer (Brit. 424837) for—
Cellulose acetate.
For uses, see under general heading: "Plasticizers."

Bis(betaphenylethyl)allylamine
Chemical
Claimed (U. S. 2006114) as—
Substitute for papaverine.

Bis(betaphenylethyl)allylamine Hydrochloride
Chemical
Claimed (U. S. 2006114) as—
Substitute for papaverine.

Bis(betaphenylethyl)ethylamine
Chemical
Claimed (U. S. 2006114) as—
Substitute for papaverine.

Bis(betaphenylethyl)ethylamine Hydrochloride
Chemical
Claimed (U. S. 2006114) as—
Substitute for papaverine.

Bisbetapiperindinoethyldiphenylacetamide
Pharmaceutical
Claimed (Brit. 438659) to possess—
Physiological properties resembling those of atropine.

Bis-3-brom-2-hydroxy-5-methylphenyl Sulphide
Pharmaceutical
Bactericide (Brit. 349004) for—
Staphylococci.

Bis-5-brom-2-hydroxyphenyl Sulphide
Fungicide and Insecticide
As a fungicide (Brit. 349004).
Ingredient (Brit. 349004) of—
Fungicidal admixture with talcum for combatting mildew on roses.
Fungicidal solution in normal caustic soda for impregnating dead wood.
Fungicidal composition for treating seed grain, contains also talcum and isobutylnaphthalene sodium-sulphonate.

Bis-5-chlor-2-hydroxyphenyl Sulphide
Fungicide and Insecticide
As an animal pesticide (Brit. 349004).
As an insecticide (Brit. 349004).
As a mothproofing agent (Brit. 349004).
Inhibitor (Brit. 349004) of—
Bacillus pyocyaneus development.

Bis-1′:4′-diamino-2′-anthraquinonylalphabetaanthraquinonethiazole
Dye
Starting point (Brit. 436951) in making—
Bluish-green dyestuffs by acylation with benzyl chloride.

Bis-1′:4′-diamino-2′-anthraquinonylbetabetaanthraquinoneiminazole
Dye
Starting point (Brit. 436951) in making—
Gray dyestuffs by acylation with anthraquinone beta-carboxylic-chloride.

Bis-1′:4′-diamino-2′-anthraquinonylbetabetaanthraquinoneoxazole
Dye
Starting point (Brit. 436951) in making—
Blue dyestuffs by acylation with benzyl chloride.
Bluish-grey dyestuffs by acylation with 1:9-thiazolanthrone 2-carboxylicchloride.

Bis-1′:5′-diamino-2′-anthraquinonylbetabetaanthraquinoneoxazole
Dye
Starting point (Brit. 436951) in making—
Ruby-red dyestuffs by acylation with benzyl chloride.

Bis-3:5-dibrom-2-hydroxyphenyl Sulphide
Sanitation
Bactericide (Brit. 349004) for—
Staphylococci.

Bis-3:5-dibrom-4-hydroxyphenyl Sulphide
Fungicide and Insecticide
As a mothproofing agent (Brit. 349004).
Sanitation
Bactericide (Brit. 349004) for—
Staphylococci.

4:4′-Bisdiethylamino-8-methyl-1:1′-diethylbenzthiocarbocyanin
Photographic
Sensitizer (Brit. 400951 and 418157) for—
Silver halide emulsion layers.

5:5′-Bisdiethylamino-8-methyl-1:1′-diethylbenzthiocarbocyanin
Photographic
Sensitizer (Brit. 400951 and 418157) for—
Silver halide emulsion layers.

5:5′-Bisdiethylamino-8-methyl-1:8:1′-triethylbenzthiocarbocyanin
Photographic
Sensitizer (Brit. 400951 and 418157) for—
Silver halide emulsion layers.

Bis(3:4-dimethoxybetaphenylethyl)methylamine
Chemical
Claimed (U. S. 2006114) as—
Substitute for papaverine.

Bis(3:4-dimethoxybetaphenylethyl)methylamine Hydrochloride

Chemical
Claimed (U. S. 2006114) as—
Substitute for papaverine.

5:5′-Bisdimethylamino-2:2′-dimethyloxacarbocyanin Iodide

Photographic
Sensitizing agent (Brit. 430357) for—
Emulsions.

Bis(gammaphenylpropyl)ethylamine

Chemical
Claimed (U. S. 2006114) as—
Substitute for papaverine.

Bis(gammaphenylpropyl)ethylamine Hydrochloride

Chemical
Claimed (U. S. 2006114) as—
Substitute for papaverine.

Bis-2-hydroxy-5-bromophenyl Oxide

French: Oxyde de bis-2-hydroxye-5-bromophényle, Oxyde bis-hydroxye-5-bromophénylique.
German: Bis-2-hydroxy-5-bromphenyloxyd.

Agriculture
Ingredient (Brit. 358508) of—
Compositions, containing talc, soap, glycerin, wool-fat, petroleum jelly, paraffin, and waxes, used for treating pests on domestic animals and also for disinfecting and preserving seeds and plants.

Chemical
Starting point in making various derivatives.

Insecticide
Ingredient (Brit. 358508) of—
Insecticidal and fungicidal preparations containing talc, soap, wool-fat, glycerin, petroleum jelly, paraffin, and waxes.

Miscellaneous
Ingredient (Brit. 358508) of—
Compositions, containing talc, soap, wool-fat, glycerin, petroleum jelly, paraffin, and waxes, used as polishes and preservatives; for example, in the treatment of catgut and in the mothproofing of feathers, furs, skins.

Perfume
Ingredient (Brit. 358508) of—
Cosmetics, toothwashes.

Resins and Waxes
Ingredient (Brit. 358508) of—
Antiseptic wax compositions.

Textile
Ingredient (Brit. 358508) of—
Mothproofing compositions containing talc, soap, wool-fat, glycerin, petroleum jelly, paraffin, and waxes.

Woodworking
Ingredient (Brit. 358508) of—
Preservative compositions containing talc, soap, wool-fat, glycerin, petroleum jelly, paraffin, and waxes.

Bis-2-hydroxy-5-chlorophenyl Oxide

French: Oxyde de bis-2-hydroxye-5-chlorophényle, Oxyde bis-2-hydroxye-5-chlorophénylique.
German: Bis-2-hydroxy-5-chlorophenyloxyd.

Chemical
Starting point in making—
Intermediates and other derivatives.

Insecticide
As an insecticide, germicide, and fungicide.
Ingredient (Brit. 358508) of—
Compositions, containing talc, soaps, glycerin, wool-fat, petrolatum, paraffin, waxes, and other components, used for treating domestic animals to remove pests.
Compositions for treating plants and seeds to disinfect them.
Insecticidal, germicidal, and fungicidal preparations containing waxes, paraffin, soaps, talc, petrolatum, glycerin, wool-fat, and other components.

Miscellaneous
Ingredient (Brit. 358508) of—
Preparations, containing talc, paraffin, waxes, petrolatum, soaps, wool-fat, and other components, used for treating catgut and other articles to preserve them.
Preservative preparations.
Preservatives for treating skins.
Special polishing compositions.

Perfume
Ingredient (Brit. 358508) of—
Cosmetic ointments, dentifrices.

Resins and Waxes
Ingredient (Brit. 358508) of—
Antiseptic wax preparations.

Textile
Ingredient (Brit. 358508) of—
Compositions, containing waxes, paraffin, petrolatum, glycerin, talc, soaps, wool-fat, used for treating fabrics in order to preserve them.

Woodworking
Ingredient (Brit. 358508) of—
Compositions used for preserving wood.

Bis(4-hydroxy-5-isopropyl-2-methylphenyl) Sulphide

Disinfectant
Bactericide (Brit. 349004) for—
Staphylococci.

Bis(2-hydroxy-5-methylphenyl) Sulphide

Animal Husbandry
As an animal pesticide (Brit. 349004).

Bis(4-hydroxy-3-methylphenyl) Sulphide

Animal Husbandry
As an animal pesticide (Brit. 349004).

Bis(2-hydroxynaphthyl-1) Sulphide

Disinfectant
As a bactericide (Brit. 349004).

Fungicide
As a fungicide (Brit. 349004).

Insecticide
As an insecticide (Brit. 349004).

Bis-2-hydroxyphenylamine

Chemical
Starting point in making—
Intermediates and other derivatives.

Disinfectant
As a germicidal agent.
Ingredient (Brit. 358508) of—
Germicidal compositions, containing petrolatum, soap, glycerin, talc, wool-fat, paraffin or other waxes, and other components.

Insecticide
As an insecticide and fungicide.
Ingredient (Brit. 358508) of—
Compositions, containing talc, soap, glycerin, wool-fat, petrolatum, paraffin or other waxes, and other components, used for treating domestic animals to remove pests and for general insecticidal and fungicidal purposes.
Compositions for treating plants and seeds to disinfect them.

Bis-4-hydroxyphenylamine

French: Bis-4-hydroxyephényleamine.
German: Bis-4-hydroxyphenylamin.

Chemical
Starting point in making—
Intermediates and other derivatives.

Insecticide
As an insecticide, germicide, and fungicide.
Ingredient (Brit. 358508) of—
Compositions, containing talc, soap, glycerin, wool-fat, petrolatum, paraffin, waxes, and other components, used for treating domestic animals to remove pests.
Compositions for treating plants and seeds to disinfect them.
Insecticidal, germicidal, and fungicidal preparations containing petrolatum, glycerin, soap, talc, wool-fat, paraffin, waxes, and other components.

Miscellaneous
Ingredient (Brit. 358508) of—
Preparations, containing talc, paraffin, waxes, petrolatum, soaps, wool-fat, and other components, used for treating catgut and other articles to preserve them.
Preservative preparations.
Preservatives for treating skins.
Special polishing compositions.

Bis-4-hydroxyphenylamine (Continued)

Perfume
Ingredient (Brit. 358508) of—
Cosmetic ointments, dentifrices.

Resins and Waxes
Ingredient (Brit. 358508) of—
Antiseptic wax preparations.

Textile
Ingredient (Brit. 358508) of—
Compositions, containing waxes, paraffin, petrolatum, glycerin, soaps, talc, and wool-fat, used for the treatment of fabrics in order to preserve them.

Woodworking
Ingredient (Brit. 358508) of—
Compositions used for preserving wood.

Bis(4-hydroxyphenyl-2-arsenic Acid) Sulphide

Fungicide
Fungicide (Brit. 349004) for—
Mildew on roses.

Bis-2-hydroxyphenyl Oxide
French: Oxyde de bis-2-hydroxyephényle.
German: Bis-2-hydroxyphenyloxyd.

Chemical
Preservative for various chemical purposes (Brit. 368508).
Starting point in making—
Intermediates and other derivatives.

Insecticide
As an insecticide.
Ingredient (Brit. 358508) of—
Fungicidal preparations, insecticidal preparations.

Miscellaneous
Ingredient (Brit. 358508) of—
Disinfecting preparations, preservative preparations.

Perfume
Ingredient (Brit. 358508) of—
Cosmetics, dentifrices.

Pharmaceutical
In compounding and dispensing practice.

Sanitation
Ingredient (Brit. 358508) of—
Disinfecting preparations.

Bis-3-hydroxyphenyl Oxide
French: Oxyde de bis-3-hydroxyephényle, Oxyde 3-hydroxyephénylique.
German: Bis-3-hydroxyphenyloxyd.

Chemical
Starting point in making—
Intermediates and other derivatives.

Insecticide
As an insecticide, germicide, and fungicide.
Ingredient (Brit. 358508) of—
Compositions, containing talc, soap, glycerin, wool-fat, petrolatum, paraffin, waxes, and other components, used for treating domestic animals to remove pests.
Insecticidal, germicidal, and fungicidal preparations, containing waxes, paraffin, soap, talc, petrolatum, glycerin, wool-fat, and other components.

Miscellaneous
Ingredient (Brit. 358508) of—
Preparations, containing talc, paraffin, waxes, petrolatum, and other components, used for treating plants, seeds, catgut, and other articles to preserve them.
Preparations for the preservation of skins.
Preservative preparations.
Special polishing compositions.

Perfume
Ingredient (Brit. 358508) of—
Cosmetic ointments, dentifrices.

Resins and Waxes
Ingredient (Brit. 358508) of—
Antiseptic wax preparations.

Textile
Ingredient (Brit. 358508) of—
Preservative compositions, containing waxes, paraffin, soap, talc, glycerin, wool-fat, and petrolatum, used for treating textile fabrics.

Woodworking
Ingredient (Brit. 358508) of—
Compositions used for treating wood to preserve it.

Bis-4-hydroxyphenyl Oxide
French: Oxyde de bis-4-hydroxyephényle, Oxyde bis-4-hydroxyephénylique.
German: Bis-4-hydroxyphenyloxyd.

Chemical
Starting point in making—
Intermediates and other derivatives.

Insecticide
As an insecticide, germicide, and fungicide.
Ingredient (Brit. 358508) of—
Compositions, containing talc, soap, glycerin, wool-fat, petrolatum, paraffin, waxes, and other components, used for treating domestic animals to remove pests.
Compositions for treating plants and seeds to disinfect them.
Insecticidal, germicidal, and fungicidal preparations, containing waxes, paraffin, soap, talc, petrolatum, glycerin, wool-fat, and other components.

Miscellaneous
Ingredient (Brit. 358508) of—
Preparations, containing talc, paraffin, waxes, petrolatum, soap, wool-fat, and other components, used for treating catgut, and other articles to preserve them.
Preservative preparations.
Preservatives for treating skins.
Special polishing compositions.

Perfume
Ingredient (Brit. 358508) of—
Cosmetic ointments, dentifrices.

Resins and Waxes
Ingredient (Brit. 358508) of—
Antiseptic wax preparations.

Textile
Ingredient (Brit. 358508) of—
Compositions, containing waxes, paraffin, soap, talc, glycerin, wool-fat, and petrolatum, used for treating fabrics to preserve them.

Woodworking
Ingredient (Brit. 358508) of—
Compositions used for preserving wood.

Bis(4-hydroxyphenyl) Sulphide

Disinfectant
As a bactericide (Brit. 349004).

Insecticide and Fungicide
As a fungicide (Brit. 349004).
As an insecticide (Brit. 349004).

Bis(2-hydroxy-3:5:6-tribromophenyl) Sulphide

Disinfectant
Bactericide (Brit. 349004) for—
Staphylococci.

Bismarck Brown
Synonyms: Anilin brown, Cinnamon brown, English brown, Gold brown, Manchester brown, Phenylene brown.
French: Brun d'Anglais, Brun d'aniline, Brun Bismarck, Brun de cannelle, Brun de Manchester, Brun d'or, Brun phénylene.
German: Anilinbraun, Bismarck braun, Englisches braun, Goldbraun, Manchesterbraun, Phenylenbraun, Zimtbraun.

Chemical
Ingredient (Brit. 303932) (with arsenic acid, arsenious acid, or their salts) of—
Bactericides, disinfectants, fungicides, insecticides, vermin-destroying compounds.

Leather
Coloring matter in dyeing—
Leather reddish brown, when used with tannin mordant.

Textile
Coloring matter in dyeing—
Cotton reddish brown, when used with tannin mordant.
Silk, wool.

Bismark Brown R
French: Brune de Bismark R.
German: Bismarkbraun R.

Chemical
Ingredient (Brit. 295605) of—
Bacteriological preparations, biological stains and therapeutic compositions containing—
Cresol, guaiacol, hydroquinone, phenol, phloroglucinol, pyrocatechol, pyrogallol, resorcinol.

Bismarck Brown R (Continued)

Miscellaneous
Dyestuff for various products.
Textile
Ingredient of dyeing and printing compositions for use on various textiles.

Bismuth
Latin: Bismuthum.
French: Étain de glace, Bismuth.
German: Wismut, Wismuth.
Spanish: Bismuto.
Italian: Bismuto.
Ceramics
Ingredient of—
Colorings for porcelain and chinaware.
Chemical
Starting point in making—
Bismuth salts of acids and halogens, bismuth albuminate, bismuth-ammonium salts, bismuth-methylene digallate, bismuth nucleinate, bismuth-oxyiodomethylgallol, bismuth-quinine iodide, bismuth-triparatolyl, bismuth-triphenyl.
Ink
Ingredient (Brit. 387844) of—
Writing inks (added to prevent corrosive action on pens).
Metallurgical
Coating metal for—
Iron, steel.
Ingredient of—
Bearing alloys, britannia metal, dental alloys.
Low melting point alloys, such as Newton's alloy, Rose's alloy, Wood's alloy.

Bismuth-Ammonium Chloride
French: Chlorure de bismuth et d'ammonium.
German: Wismuthammoniumchlorid.
Chemical
Starting point and reagent in various processes.
Miscellaneous
Carrotting agent in treating—
Furs, felt, and the like (Brit. 271026).

Bismuth Bromide
French: Bromure de bismuth.
German: Bromwismut, Wismutbromid.
Spanish: Bromuro de bismuto.
Italian: Bromuro di bismuto.
Chemical
Catalyst (Brit. 398527) in making—
Esters from lower aliphatic acids and olefins.

Bismuth Camphenilanate
Pharmaceutical
As an oil-soluble bactericide (Brit. 428147).

Bismuth Carbonate
Synonyms: Bismuth subcarbonate, Bismuthyl carbonate, Oxycarbonate of bismuth.
Latin: Bismuthi subcarbonas, Bismuthum subcarbonicum, Subcarbonas bismuthicus.
French: Souscarbonate de bismuth.
German: Basisches kohlensäureswismutoxyd, Wismutsubcarbonat.
Chemical
Starting point in making—
Bismuth salts.
Cosmetics
Ingredient of—
Creams, face paints, face powders, lotions.
Pharmaceutical
In compounding and dispensing practice.
Suggested for use as—
Intestinal opacifying agent in x-ray work.

Bismuth Delta^3-tetrahydro-2:5-endomethylene-6-methylbenzoate
Pharmaceutical
As an oil-soluble bactericide (Brit. 428147).

Bismuth Dinaphthylnaphthenate
Petroleum
Addition agent (Brit. 433257) in—
Lubricating oils or greases, especially for use at high temperatures, such as cylinder oils, hydrogenated oils, or oils refined by treatment with sulphuric acid, clay, or extraction solvents.

Bismuth Isocamphenilanate
Pharmaceutical
As an oil-soluble bactericide (Brit. 428147).

Bismuth Nitrate
Synonyms: Bismuth ternitrate, Bismuth trinitrate.
French: Azotate de bismuth.
German: Salpetersäureswismut, Wismutnitrat, Wismuttrinitrat.
Spanish: Nitrato de bismuto.
Italian: Nitrato di bismuto.
Chemical
Starting point in making—
Various salts of bismuth, such as bismuth benzoate, bismuth oleate, bismuth oxychloride, bismuth resorcinate, bismuth sulphite, bismuth valerate, bismuth chromate, bismuth phenate, bismuth subgallate, bismuth subnitrate, bismuth trioxide, and pharmaceutical compounds.
Metallurgical
Source of—
Bismuth in producing bismuth luster on metals.
Paint and Varnish
Ingredient of—
Luminous paints and enamels.
Pharmaceutical
In compounding and dispensing practice.

Bismuth Oleate-Gallate
Pharmaceutical
Claimed (Brit. 443860) to be—
Assimilable organic bismuth salt.

Bismuth Oleate-Quinate
Pharmaceutical
Claimed (Brit. 443860) to be—
Assimilable organic bismuth salt.

Bismuth Oleate-Salicylate
Pharmaceutical
Claimed (Brit. 443860) to be—
Assimilable organic bismuth salt.

Bismuth Oxycamphenilanate
Pharmaceutical
As an oil-soluble bactericide (Brit. 428147).

Bismuth Oxychloride
Synonyms: Bismuth subchloride, Bismuthyl chloride, Cosmetic bismuth, Pearl white.
Latin: Bismuthi oxychloridum, Bismuthi subchloridum.
French: Blanc de fard, Oxychlorure de bismuth.
German: Wismutoxychlorid.
Cosmetic
Ingredient of—
Creams, face powders, lotions.
Miscellaneous
Pigment in—
Artificial pearls.
Paint and Varnish
As a pigment.
Pharmaceutical
In compounding an dispensing practice.
Substitute for—
Bismuth subnitrate.

Bismuth Oxyiodide
French: Oxyiodure de bismuth.
German: Wismuthoxyjodid.
Ceramics
Starting point (Brit. 444740) in developing—
Pearl or mother-of-pearl effects in compositions, containing organic derivatives of cellulose, used as coatings for decorating ceramic products.
Chemical
Starting point in making—
Various bismuth preparations used as pharmaceuticals.
Cosmetic
Starting point (Brit. 444740) in developing—
Pearl or mother-of-pearl effects in nail enamels and lacquers containing an organic derivative of cellulose as a base material.
Electrical
Starting point (Brit. 444740) in developing—
Pearl or mother-of-pearl effects in insulating compositions containing organic derivatives of cellulose.

Bismuth Oxyiodide (Continued)

Glass
Starting point (Brit. 444740) in developing—
Pearl or mother-of-pearl effects in compositions, containing organic derivatives of cellulose, used as coatings for decorating glassware.

Leather
Starting point (Brit. 444740) in developing—
Pearl or mother-of-pearl effects in compositions, containing organic derivatives of cellulose, used for decorating leathers and leather goods.

Metallurgical
Starting point (Brit. 444740) in developing—
Pearl or mother-of-pearl effects in compositions, containing organic derivatives of cellulose, used as coatings for decorating metallic articles.

Miscellaneous
Starting point (Brit. 444740) in developing—
Pearl or mother-of-pearl effects in compositions, containing organic derivatives of cellulose, used for decorating various products.

Paint and Varnish
Starting point (Brit. 444740) in developing—
Pearl or mother-of-pearl effects in paints, varnishes, lacquers, enamels, and dopes containing organic derivatives of cellulose.

Paper
Starting point (Brit. 444740) in developing—
Pearl or mother-of-pearl effects in compositions, containing organic derivatives of cellulose, used as coatings for decorating products made of paper and pulp.

Pharmaceutical
In compounding and dispensing practice.
Suggested for use as—
Iodoform substitute, local antiseptic.

Plastics
Starting point (Brit. 444740) in developing—
Pearl or mother-of-pearl effects in plastics made from organic derivatives of cellulose.

Resins
Starting point (Brit. 444740) in developing—
Pearl or mother-of-pearl effects in synthetic resin products made from or containing organic derivatives of cellulose.

Rubber
Starting point (Brit. 444740) in developing—
Pearl or mother-of-pearl effects in compositions, containing organic derivatives of cellulose, used as coatings for decorating rubber products.

Stone
Starting point (Brit. 444740) in developing—
Pearl or mother-of-pearl effects in compositions, containing organic derivatives of cellulose, used as coatings for decorating artificial and natural stone.

Textile
Starting point (Brit. 444740) in developing—
Pearl or mother-of-pearl effects in compositions, containing organic derivatives of cellulose, used for decorating fabrics.

Woodworking
Starting point (Brit. 444740) in developing—
Pearl or mother-of-pearl effects in compositions, containing organic derivatives of cellulose, used for decorative coatings on woodwork.

Bismuth-Phenyl Acetate

Petroleum
Addition agent (Brit. 433257) in—
Lubricating oils or greases, especially for use at high temperatures, such as cylinder oils, hydrogenated oils, or oils refined by treatment with sulphuric acid, clay, or extraction solvents.

Bismuth Phenyldibromide
French: Dibromure phénylique de bismuth.
German: Dibromphenylwismuth, Wismuthphenyldibromid.

Miscellaneous
Mothproofing agent (Brit. 303092) in treating—
Hair, feathers, furs.

Textile
Mothproofing agent (Brit. 303092) in treating—
Wool and felt.

Bismuth Phosphate
French: Phosphate de bismuth.
German: Phosphorsauereswismut, Wismutphosphat.
Spanish: Fosfato de bismuto.
Italian: Fosfato di bismuto.

Chemical
Catalyst (Brit. 398527) in making—
Esters from lower aliphatic acids and olefins.

Bismuth-Quinine Iodide
French: Iodure de bismuth-quinine.
German: Jodwismuthchinin, Wismuthchininjodid.

Chemical
Starting point in making—
Lecithin compound (Brit. 257912).

Pharmaceutical
In compounding and dispensing practice.

Bismuth Resinate
Synonyms: Resinate of bismuth.
French: Résinate de bismuth.
German: Wismuthresinat.

Ceramics
Pigment in producing lustrous coatings on—
Chinaware, porcelains, potteries.

Glass
Pigment in producing lustrous colored effect on various kinds of glassware.

Paint and Varnish
Drier in making—
Enamels, lacquers, paints, varnishes.

Bismuth Ricinoleate-Gallate

Pharmaceutical
Claimed (Brit. 443860) to be—
Assimilable organic bismuth salt.

Bismuth Ricinoleate-Quinate

Pharmaceutical
Claimed (Brit. 443860) to be—
Assimilable organic bismuth salt.

Bismuth Ricinoleate-Salicylate

Pharmaceutical
Claimed (Brit. 443860) to be—
Assimilable organic bismuth salt.

Bismuth Sesquioxide
Synonyms: Bismuth oxide, Bismuth trioxide, Bismuth yellow, Bismuthous oxide, Protoxide of bismuth.
Latin: Bismuthi trioxidum.
French: Oxyde de bismuth.
German: Wismuthoxyd, Wismuthsesquioxyd, Wismuthtrioxyd.

Chemical
Catalyst (German 439150) in making—
Potassium nitrate from calcium cyanamid.
Starting point in making—
Bismuth alphanaphtholate, bismuth-ammonium citrate, bismuth betanaphtholate, bismuth camphorate, bismuth carbolate, bismuth glycollate, bismuth phenate, bismuth phenolsulphonate, bismuth salts of acids and halogens.

Ceramics
Ingredient of—
Red enamels for tiles (Brit. 245748).
Lustrous coatings for chinaware, porcelains and potteries.

Glass
Ingredient of—
High refractive lustrous glass (used in the place of lead oxide).
Red glass.

Pharmaceutical
In compounding and dispensing practice.

Bismuth Subgallate
Synonyms: Basic bismuth gallate, Dermatol.
French: Gallate basique de bismuth, Gallate de bismuth.
German: Gallussäuresbasischwismuth, Gallussäureswismuthoxyd, Wismuthbasischgallat, Wismuthgallat.
Spanish: Gallate basico de bismuto, Gallato de bismuto.
Italian: Gallato basico di bismuto, Gallato di bismuto.

Bismuth Subgallate (Continued)

Chemical

Starting point in making—
Airol, various pharmaceutical products.

Perfume

Ingredient of—
Skin creams possessing medicinal properties.

Pharmaceutical

Suggested for use as astringent and in treating diarrhea, in making surgical gauze and bandages.

Bismuth Subnitrate

Synonyms: Bismuth oxynitrate, Bismuthyl nitrate, Magistery of bismuth, Subnitrate of bismuth, White bismuth.
Latin: Bismuthi subnitras, Bismuthum album, Bismuthum hydriconitricum, Bismuthum subnitricum, Magisterium bismuthi, Subazotas bismuthicus, Subnitras bismuthicus.
French: Azotate(sous) de bismuth, Sousnitrate de bismuth.
German: Basisches salpetersäureswismutoxyd, Basisches wismutnitrat, Wismut subnitrat.
Spanish: Magisterio de bismuto, Nitrico(sub) bismutico.
Italian: Nitrato basico di bismuto.

Ceramics

Ingredient of—
Enamel compositions used for coating porcelains and chinaware.
Gilding compositions.

Chemical

Starting point in making—
Various salts of bismuth, such as bismuth-ammonium citrate, bismuth citrate, bismuth lactate, bismuth oxide, and pharmaceutical compounds.

Cosmetic

Ingredient of—
Face paints, powders, greases, and other cosmetic preparations.

Metallurgical

Source of—
Bismuth in producing bismuth luster on metals.

Pharmaceutical

In compounding and dispensing practice.

Bismuth Subsalicylate

Pharmaceutical

Suggested for use as local antiseptic, as intestinal antiseptic, and for treating typhoid.

Bismuth Sulphate

French: Sulfate de bismuth.
German: Schwefelsäureswismut, Wismutsulfat.
Spanish: Solfato de bismuto.
Italian: Solfato di bismuto.

Chemical

Catalyst (Brit. 398527) in making—
Esters from lower aliphatic acids and olefins.

Bismuth Trichloride

French: Trichlorure de bismuth.
German: Trichlorwismut, Wismutchlorid, Wismuttrichlorid.
Spanish: Cloruro de bismuto.
Italian: Cloruro di bismuto.

Chemical

Catalyst (Brit. 398527) in making—
Esters from lower aliphatic acids and olefins.
Starting point in making—
Bismuth salts
Bismuth organic arsenicals by reaction with iso-oxypropyldiarsinic acid (French 648325).

Coaltar

Agent (French 633643) for—
Treating coaltar.

Rubber

Thermoplasticizing agent (French 615195).

Bismuth Trioxide

Synonyms: Bismuth oxide, Bismuth yellow.
French: Oxyde de bismuth, Trioxyde de bismuth.
German: Wismutoxyd, Wismuttrioxyd.
Spanish: Oxido de bismuto, Trioxido de bismuto.
Italian: Oxido di bismuto, Trioxido di bismuto.

Ceramics

Ingredient of—
Glazes.

Chemical

Catalyst (Brit. 402438) in making—
Ethylene oxide from ethylene as an intermediate step in making ethyleneglycol and its derivatives.
Ingredient (Brit. 405282) of—
Catalytic mixture used in making acetic acid from carbon monoxide, methanol, and steam.
Starting point in making—
Bismuth salts.
Hard, granular, porous gels having catalytic or adsorbent properties by peptizing with an organic acid, such as formic, acetic, or chloracetic acid, and nearly dehydrating the peptized mass (Brit. 398517).

Glass

Ingredient of—
Red glass batches.

Miscellaneous

Ingredient (Brit. 403233) of—
Mineral oxide mixtures used in the production of weatherproof luminous substances.

Pharmaceutical

In compounding and dispensing practice.

Rubber

Ingredient (U. S. 1844306) of—
Vulcanizable rubber compound.

Bismuth Triparatolyl

French: Triparatolyle de bismuthe.
German: Wismuthtriparatolyl.

Miscellaneous

Reagent (Brit. 303092) in treating—
Furs, feathers, hair, and other animal products in order to render them resistant to the clothes moth.

Textile

Reagent (Brit. 303092) in treating—
Wool and felt in order to render them resistant to the action of the clothes moth.

Bismuth Triphenyl

French: Triphényle de bismuth, Triphényle bismuthique.
German: Wismuthtriphenyl.

Chemical

Starting point in making various derivatives.

Miscellaneous

Reagent in treating—
Fur, hair, feathers, and the like to render them mothproof.

Textile

Reagent in treating—
Mothproofing felt and wool.

Bismuth-Triphenyl Dibromide

French: Dibromure de bismuth triphénylé, Dibromure de bismuth triphénylique.
German: Dibromtriphenylwismuth, Wismuttriphenyldibromid.
Spanish: Dibromuro de bismuto trifenil.
Italian: Dibromuro di bismuto trifenilato.

Miscellaneous

Mothproofing agent (Brit. 303092) in treating—
Hair, feathers, furs, felt.

Textile

Mothproofing agent (Brit. 303092) in treating—
Woolen fabrics and yarns.

1:2:5:6-Bis-5′-phenyl-1′:2′-indoloanthraquinone

Textile

As a brown vat dye (Brit. 443958 and 443959).

1:2:1′:2′-Bisphthaloyl-6:6′-dicarbazolyl

Textile

As a brown vat dye (Brit. 443958 and 443959).

Bitter Almond Oil

Food

Flavoring agent in making—
Confectionery, liqueurs.
Ingredient of—
Confectionery, food preparations.

Perfume

Odor for—
Various cosmetics and toilet preparations.

Pharmaceutical

Suggested for use as nerve sedative.

Soap

Perfume for—
Toilet soaps.

Bittersweet
Synonyms: Woody nightshade.
Latin: Dulcamara, Stipites dulcamarae, Solanum
French: Douce-amère, Tiges de douce-amère, Tiges de dulcamara.
morelle grimpante, Vigne de Judée.
German: Alpranken, Bittersuess, Hindischkraut.

Chemical
Starting point in extracting—
Dulcamarin, solanidine, solanine.

Pharmaceutical
In compounding and dispensing practice.

Black Dammar
Synonyms: Black gum dammar, Black dammar resin.
French: Résine de dammar noir.
German: Schwarzdammar.

Gums
Substitute for burgundy pitch.

Pharmaceutical
In compounding and dispensing practice.

Blanc Fixe
Synonyms: Artificial barytes, Artificial heavy spar, Permanent white, Precipitated barium sulphate, Terra ponderosa.
Latin: Barii sulfas, Terra ponderosa.
French: Sulfate de barium, Sulfate de baryum, Sulfate de baryum precipité.
German: Barytweiss, Schwefelsaeuresbaryum.

Dye
Inert base for—
Colors in making lakes.

Glass
Constituent of—
Special glasses.

Ink
Ingredient of—
Antismearing composition for inks (U. S. 1439623).
Lithographing inks, printing inks.

Leather
Filler in making—
White leather.

Linoleum and Oilcloth
Filler in—
Linoleum, oil cloth.

Miscellaneous
In X-ray work.

Paint and Varnish
Filler in—
Paints and lacquers.
Pigment in—
Paints and lacquers.
Starting point (Brit. 444110) in making—
New blue pigments with manganates.

Paper
Filler in—
Paper of various kinds and quality.
Paperboard and cardboard of the better grades.
Wallpaper.

Photographic
Ingredient of—
Coatings for paper used as carrier for gelatin-bromides.

Rubber
Filler.

Textile
Filler for—
Textile fabrics.
Ingredient of—
Dressing compositions.

Bone Ash

Fertilizer
As a manure.
Ingredient of—
Compounded fertilizers.
Source of—
Lime, magnesia, phosphates.
Starting point in making—
Superphosphates.

Miscellaneous
Ingredient of—
Cleaning compounds, polishing compounds.

Bone Black
Synonyms: Animal black, Animal charcoal, Bone char, Bone charcoal, Char, Drop black, Ivory black, Ivory drop black, Virgin drop black.
Latin: Carbo animalis.
French: Charbon animal, Charbon d'ivoire, Charbon d'os, Noir animal, Noir d'ivoire, Noir d'os.
German: Beinschwarz, Elfenbeinschwarz, Knockenkohle, Schwarzes spodium, Tierkohle.
Spanish: Carbon animal, Carbon de huesos.
Italian: Carbone animale.

Analysis
Filtering medium for treating various liquids in the chemical laboratory for the purpose of purifying, decolorizing, and deodorizing them.

Chemical
Absorbent for—
Gases, chemicals.
Mineral salts.
Catalyst for various chemical reactions.
Filtering medium for—
Decolorizing glycerin.
Decolorizing and purifying intermediates, organic chemicals, synthetic pharmaceuticals, and synthetic aromatic chemicals.
General decolorizing and deodorizing medium.
General filtering medium.

Fats and Oils
Filtering medium in treating—
Animal and vegetable fats and oils to purify them, remove the color and odor.

Fertilizer
Ingredient of—
Fertilizing compositions.

Food
Filtering medium for purifying various food products.

Ink
Pigment in—
Plate printing ink, various printing inks.

Leather
Pigment in making—
Black leathers.

Linoleum and Oilcloth
As a pigment in coatings.

Metallurgical
Reagent in—
Cementation of steel.

Military
For filling gas masks.

Miscellaneous
Filtering medium for—
Improving wines and distilled liquors.
General filtering medium.
General decolorizing and deodorizing agent.
General pigmenting agent in making various compositions of matter.
Ingredient of—
Compositions for making crayons, shoe polishes, stove polishes.

Paint and Varnish
Ingredient of—
Fine colors.
Pigment in—
Paints, varnishes, and enamels.

Petroleum
Filtering medium for—
Decolorizing paraffin and white oils.

Plastics
As a pigment.

Resins and Waxes
Reagent in—
Purifying waxes and resins.

Sanitation
Filtering medium for—
Purifying and deodorizing water.

Sugar
Decolorizing agent for—
Sugar and molasses syrups.
Refining agent in—
Processing sugar.

Bone Fat
Synonyms: Bone grease, Bone tallow.
French: Graisse d'os, Petit suif, Suif d'os.
German: Knockenfett.

Bone Fat (Continued)

Fats and Oils
Ingredient of—
Lubricating compositions.
Fuel
Constituent of—
Candles.
Mechanical
Ingredient of—
Lubricating compositions.
Soap
Raw material in making—
Cheap colored soaps.

Bone Meal, Raw

Animal Husbandry
Mineral supplement in—
Chicken feeds.
Fertilizer
Ingredient of—
Compounded fertilizers.
Source of—
Ammonia, bone phosphate, phosphoric acid.

Bone Meal, Steamed

Animal Husbandry
Mineral supplement in—
Cattle feeds, hog feeds.
Fertilizer
Ingredient of—
Compounded fertilizers.
Source of—
Ammonia, bone phosphate.

Borax

Synonyms: Biborate of soda, Biborate of sodium, Borate of soda, Borate of sodium, Purified borax, Pyroborate of soda, Sodium biborate, Sodium borate, Sodium pyroborate, Sodium tetraborate, Tetraborate of soda.
Latin: Boras sodicus, Borax purificatus, Natrium biboracicum, Natrium biboricum, Sodae biboras, Sodii boras.
French: Bauracon, Borate de soude, Borate sodique, Borax, Sel de Perse, Soude boré.
German: Borax, Borsäuresnatrium, Borsäuresnatron.
Spanish: Borato de sosa, Borato sodico, Borax.
Italian: Borace, Borato di sodio.

Abrasives
Etching agent for—
Abrasives, corundum, emery, garnet, quartz, sand, silica.
Ingredient of—
Abrasive compositions.
Adhesives
Preservative for—
Glues.
Process material in making—
Adhesives for waxed paper.
Fireproof adhesives.
Solvent for—
Casein.
Agriculture
Thinning agent for—
Bird-lime.
Animal Remedies
Suggested for use as—
Antiseptic, diuretic, germicide.
Aviation
Fireproofing agent in—
Airplane dopes, airplane fabrics, balloon fabrics.
Analysis
Reagent in—
Blowpipe analysis.
Reagent in—
Volumetric analysis.
Building Construction
Acidproofing agent for—
Cement, concrete.
Bonding agent in—
Cement, concrete.
Hardening agent for—
Plaster of Paris.
Ingredient of—
High-grade, highly-polishable cements.
Hydraulic and other cements.
Lutes, magnesium cements, marble substitutes, mortars, waterproofed cements.
Oilproofing agent for—
Cement, concrete.
Setting retarder for—
Plaster of Paris.
Waterproofing agent for—
Cement, concrete.
Ceramics
As a flux.
Ingredient of—
Enamels, fluxes, frits.
Process material in making glazed products, such as—
Brick, chinaware, porcelains, potteries, tile.
Chemical
Contact agent in—
Chlorination of anthraquinone derivatives.
Process material in making—
Acetaldehyde, arsenic, benzene, 2-chlorethanol, chlorhydrins, filter-aid, formates, formic acid, hypochlorous acid, methyl borates, nickel borate, potassium salts from silicates, salicylic aldehyde, sodium silicate.
Starting point in making—
Boric acid, boric anhydride.
Borates, such as those of aluminum, ammonia, copper, lead, magnesium, sodium, and the like.
Boron, boron carbide, boron nitride, boron trichloride.
Cosmetic
Ingredient of—
Hair remedies.
Preservative in—
Cosmetics, creams.
Dentistry
Ingredient of—
Cements, fillings.
Dry Cleaning
Reagent in—
Ink removing, spot removing.
Dye
Process material in making—
Dyestuffs, especially those of the anthracene series.
Electrical
Ingredient of—
Arc-light electrodes, condenser electrolytes, electrical insulations, electrolytes, lightning arrester electrolytes, rectifier electrolytes.
Explosives
Coolant and retardant in—
Safety explosives.
Preservative in—
Explosives.
Fire-fighting
Fire-proofing agent for—
Excelsior substitutes, fabrics, lining safes.
Ingredient of—
Chemical fire-extinguishers, fireproofings.
Food
As a general preservative.
Ingredient of—
Baking compound, shortening.
Preservative for—
Bacon, fish, fruit, hams.
Fuel
As a fuel in admixture with peat.
Impregnating agent in making—
Candle wicks.
Germicides
Bactericide, germicide.
Glass
Batch ingredient in making—
Borosilicate glass, chemical glassware, colored glass, electrical glassware, electro-chemical glassware, food-cooking glassware, industrial glassware, lamp glass, milk glass, opaque glass, optical glass, ruby glass, silk and rayon (manufacturing) glassware, strass glass, thermal glass, thermometer glass, translucent glass.
Glue and Gelatin
Preservative for—
Glue, gelatin.
Gums
Ingredient of—
Shellac solvents.
Process material in making—
Substitutes for British gum.

Borax (Continued)

Purifying agent for—
Gums.
Solvent for—
Gums, shellac.

Inks
Preservative in—
Printing inks.

Insecticides
Ingredient of—
Ant repellant, croton bug repellant, fly repellant, insecticides, roach repellant, rodent repellant.

Jewelry
Ingredient of—
Anchoring cements in diamond polishing.
Jewellers' solders.

Laundries
Detergent in—
Wash waters.
Washing operations for delicate products such as laces and other materials sensitive to alkalies.
Softening agent for—
Rinsing waters, wash waters.

Leather
Preservative for—
Hides, pelts, skins.
Process material in—
Tanning processes.

Lubrication
Ingredient of—
Lubricants.

Metal Fabrication
Flux in—
Brazing operations, enamelling operations, metal coating, soldering operations, welding operations.
Ingredient of—
Brazing compounds, fluxes, soldering compounds, welding compounds.
Process material in—
Enamelling ironware.

Metallurgical
Case-hardening agent for—
Alloys, chrome alloys, iron, manganese alloys, molybdenum alloys.
Ingredient of—
Electrolytes for etching of brass, copper, nickel, zinc, etc.
Tempering agents.
Neutralizing agent for—
Pickling solutions.
Process material in—
Gold refining.

Mining
Process material in making—
Flotation agent.

Miscellaneous
Deodorant for—
Various purposes.
Detergent in—
Dish washing, glass cleaning compositions, household laundering.
Ingredient of—
Bath salts, belt-dressing compositions, binders of various kinds, chewing gum compositions, evaporation-preventing foams, fillers of various kinds, gasket compositions, metal polishes, shoe dressings, shoe fillers, shoe polishes, soldering compounds, soldering fluxes, stencil compositions, stiffening compositions, waterproofing compositions, waterproofing compositions for canvas.
Preservative for—
Cork.
Remover for—
Grease.
Solvent for—
Casein.
Gums in polishing compositions of various kinds.
Shellac in making stiffening compositions for hats and millinery.

Oral Hygiene
Ingredient of—
Dentifrices.

Paint and Varnish
Process material in making—
Calcimines, calcimine binders.
Solvent for—
Casein, shellac.
Starting point in making—
Driers for paints, varnishes, enamels and the like (consisting of the borate salts of various metals, such as of manganese, lead, etc.)
Guignet's green.

Paper
Ingredient of—
Coatings and glazes for paper, cards, and the like.

Pharmaceutical
In compounding and dispensing practice.
Suggested for use as—
Antiepileptic, antiseptic, diuretic, germicide.

Power and Heat Generation
Ingredient of—
Heat insulations.

Printing
Binder and preservative in—
Printers' roller compositions.

Refractories
Ingredient of—
Crucibles.

Resins
Process material in making—
Aldehyde-phenol condensate, formaldehyde-urea condensate.

Rubber
Coagulant for—
Rubber latex.
Coating agent for—
Molds.
Ingredient of—
Mold coating compositions.
Process material in making—
Rubber substitutes.

Soap
Ingredient of—
Cleaning compositions, detergent compositions, dishwashing compositions, hand-cleansing compositions, scouring compositions, washing agents for delicate fabrics, window cleaning agents.

Textile
As a solvent bleach.
Bleaching agent for—
Fabrics.
Detergent in—
Degumming silk and other fibres, scouring operations, washing operations for wool.
Washing operations for delicate products such as laces and other materials sensitive to alkalies.
Fireproofing agent for—
Fabrics.
Fixative for—
Alumina, iron oxide.
Ingredient of—
Sizes, stiffening compositions.
Lustring agent for—
Starched goods (added to the starch).
Mordant in—
Calico printing, dyeing processes.

Wood Processing
Fireproofing agent for—
Wood.
Hardening agent in making—
Imitation hardwood from soft wood.
Oilproofing agent for—
Wood.
Preservative for—
Wood.

Water and Sanitation
Ingredient of—
Water softeners.

Boric Acid

Synonyms: Boracic acid, Hydrogen borate, Orthoboric acid.
Latin: Acidum boracicum, Acidum boricum, Sal sedativum hombergii.
French: Acide borique.
German: Borsäure.
Spanish: Acido borico.
Italian: Acido borico.

Analysis
Reagent in detecting—
Acetanilide, turmeric.
Reagent in determining—
Ammonia.

Boric Acid (Continued)

Cement
Flux in making—
Well-burnt Portland cement, manufactured at low temperature.
Ingredient of—
High-grade cements which are capable of taking a high polish.
Plaster of paris compositions.

Ceramics
Ingredient of—
Compositions used in the manufacture of potteries.
Compositions used in the manufacture of fire-resistant articles.
Glaze preparations used in the manufacture of porcelains, potteries, tile, glazed brick, glazed earthenware, and other ceramic products.

Chemical
Catalyst in making—
Quinol from benzoquinone.
Condensing agent in making—
Intermediates, organic chemicals, pharmaceutical chemicals, synthetic aromatic chemicals.
Reagent (Brit. 252570) in treating—
Camphor oil fraction for the recovery of terpineol and borneol.
Reagent (Brit. 260346) in making—
Amydricaine pentaborate, amylocaine pentaborate, benzamine pentaborate, benzocaine pentaborate, butyn pentaborate, cocaine borate, ethocaine pentaborate, glycocaine borate, orthocaine borate, phenocaine borate.
Reagent in making—
Acrolein, arylaminoanthraquinones, benzylaminoanthraquinone, cresylaminoanthraquinone, 5:8-dichloro-4-hydroxy-1-methylanthraquinone, 5:8-dichloro-2-hydroxy-1-methylanthraquinone, 1:2-dihydroxyanthraquinone-3:5-disulphonic acid, 4:6-dihydroxy-2-methylanthraquinone, diphenylaminoanthraquinone, formylaminoanthraquinone, gallylaminoanthraquinone, 4-hydroxy-1-methylanthraquinone, mesitylaminoanthraquinone, naphthylaminoanthraquinone.
Nitric acid from sodium nitrate, with the simultaneous production of borax.
Oxyanthraquinone, oxynaphthanenequinone, phenylaminoanthraquinone, phthalylaminoanthraquinone, purpurin-3:8-disulphonic acid, pyrylaminoanthraquinone, quinazarinbetacarboxylic acid, resorcylaminoanthraquinone, salicylaminoanthraquinone, succinylaminoanthraquinone, tolylaminoanthraquinone, xylylaminoanthraquinone.
Starting point in making—
Aluminum borate, antimony borate, ammonium metaborate, amyl borate, barium metaborate, benzyl borate, bismuth borate, borax, bornyl borate, butyl borate, cadmium borate, calcium borate, chromium borate, cobalt borate, copper borate, ethyl borate, heptyl borate, hexyl borate, lead borate, magnesium borate, magnesium pyroborate, manganese borate, menthol borate, methyl borate, nickel borate, phenyl borate, potassium borate, strontium borate, tin borate, zinc perborate.
Various perborates of alkali, alkaline earth, and metallic elements.
Water glass.

Dye
Catalyst in making—
Various dyestuffs, such as alizarin brilliant green B and SE.
Condensing agent in making—
Dyestuffs of different groups.
Reagent in making—
Dichloroquinazarin, anthracene blue WG, hydroxyanthrarufin (1:2:5-trihydroxyanthraquinone), indanthrene blue 3G, quinazarin.

Fats and Oils
Preservative for various fats and oils.

Food
As a general preservative.
Preservative in treating—
Butter and cheese.
Reagent in—
Curing meats.

Fuel
Ingredient of—
Solutions used for the impregnation of candle wicks in the manufacture of stearin candles.

Glass
Ingredient of—
Glass pastes.
Raw material in making—
Special optical glass.
Reagent in making—
Colored glass.

Glues and Adhesives
Preservatives in—
Various glues and adhesive preparations of animal and vegetable origin that tend to become putrid.

Ink
Ingredient of—
Printing inks.

Insecticide
Ingredient of—
Insecticidal preparations.

Jewelry
Ingredient of—
Compositions used in the manufacture of artificial gems.

Leather
Ingredient of—
Dressing compositions.
Reagent in—
Deliming calf, sheep, and goat skins prior to tanning them into light leathers.
Preparing hides prior to tanning.
Tanning hides by means of iron salts (French 533850).
Treating pelts for the purpose of removing lime prior to tanning.

Metallurgical
As a general flux for various welding and other operations carried out on iron and other metals and alloys.
Ingredient of—
Compositions used for brazing copper.
Compositions used for coloring gold.
Compositions used for decorating iron and steel.
Compositions used for the production of coatings of enamel on cast and wrought iron.
Electrolyte, containing nickel sulphate, nickel chloride, citric acid, and basic lead carbonate, used for production of bright nickel coatings (U. S. 183835).
Flux compositions, containing one part of boric acid and three parts of sodium bicarbonate, a non-hygroscopic product of greater absorptive and protective action.
Nickel-plating baths (added in the form of crystals and used for the purpose of preventing lagging of cathodic efficiency behind anodic efficiency).
Starting point (French 535303, additional patent 24836) in making—
Boron by electrolysis.

Mechanical
As a special lubricant (used in fine powder form).

Miscellaneous
Ingredient of—
Bleaching preparations.
Compositions, containing shellac, used for stiffening hats.
Compositions used for fireproofing various fibrous products.
Compositions for lining safes.
Porous compositions in 5 percent alcohol solution for preserving skins (French 512654).
Lubricant for—
Dance floors and the like (used in fine powder form).

Paint and Varnish
Ingredient of—
Airplane dopes and varnishes made with nitrocellulose base, and containing magnesium chloride (French 560341).
Enamels.
Lacquers containing gold bronze (added in the proportion of 0.25 to 1 percent for the purpose of preventing the gelling of the products).
Paints.
Starting point in making—
Borated ultramarine pigment, Guignet's green pigment.

Paper
Ingredient of—
Preparations used in the production of glazed paper.
Preparations used for fireproofing fibrous compositions containing paper or pulp.

Perfume
Ingredient of—
After-shaving lotions, cuticle pomades, nail polishes, soapless shavings creams, sunburn preparations.

Boric Acid (Continued)

Petroleum
Ingredient of—
Motor fuel compositions containing petroleum distillates (Brit. 252018).
Reagent in—
Refining crude oil.

Pharmaceutical
Recommended for use as disinfectant, astringent, and antiseptic.

Photographic
Reagent in various processes.

Refractories
Ingredient of—
Solutions used for moistening mixtures of graphite and clay or the like, used for the manufacture of refractory crucibles, muffle furnaces, and similar apparatus.

Rubber
Reagent in—
Compounding rubber.

Soap
Ingredient of—
Special toilet soaps.

Starch
Ingredient of—
Starch glazes for treating linens.

Textile
——, *Bleaching*
Ingredient of—
Bleaching preparations.

——, *Dyeing*
Mordant in—
Dyeing yarns and fabrics.

——, *Finishing*
Ingredient of—
Compositions used for finishing linens.
Compositions, containing 30 percent of boric acid and 70 percent of borax, used for fireproofing cottons.
Compositions for fireproofing rayons (Brit. 251227).

——, *Manufacturing*
Ingredient of—
Compositions used in the manufacture of carpets.

——, *Printing*
Mordant in—
Printing pastes.

Wine
As a preservative.

Woodworking
Ingredient of—
Compositions, containing soft wood, used to imitate hard wood.
Compositions used to render wood weatherproof and fireproof.

Boric Acid Ester of Grapeseed Alcohol

Bituminous
Solvent (Brit. 445223) for—
Asphalt and other bituminous bodies.

Dye
Solvent (Brit. 445223) for—
Dyestuffs, particularly oil-soluble coal-tar dyes.

Fats, Oils, and Waxes
Solvent (Brit. 445223) for—
Fats, oils, waxes.

Resins
Solvent (Brit. 445223) for—
Oil-soluble glycerol-phthalic acid resins.
Polymerized vinyl compounds, synthetic resins.

Rubber
Solvent (Brit. 445223) for—
Rubber

Boric Acid Ester of Ricinoleic Alcohol

Bituminous
Solvent (Brit. 445223) for—
Asphalt and other bituminous bodies.

Dye
Solvent (Brit. 445223) for—
Dyestuffs, particularly oil-soluble coal-tar dyes.

Fats, Oils, and Waxes
Solvent (Brit. 445223) for—
Fats, oils, waxes.

Resins
Solvent (Brit. 445223) for—
Oil-soluble glycerol-phthalic acid resins.
Polymerized vinyl compounds, synthetic resins.

Rubber
Solvent (Brit. 445223) for—
Rubber

Boric Anhydride

Synonyms: Anhydrous boric acid, Boric acid anhydride, Boron oxide, Boron sesquioxide, Boron trioxide, Fused boric acid.
French: Anhydride borique, Oxyde de bore, Oxyde borique, Sesquioxyde de bore, Sesquioxyde borique, Trioxyde de bore, Trioxyde borique.
German: Boroxyd, Borsäuresanhydrid, Borsesquixoyd, Bortrioxyd.
Spanish: Anhidrio borico, Oxido borico, Sesquioxido borico, Trioxido borico.
Italian: Anidrido borica, Ossido borica, Sesquiossido borica, Triossido de borica.

Analysis
Reagent for—
Disintegrating silicates to determine silica and alkalies.
Reagent in blowpipe analysis.

Chemical
Reagent (German 401870 and 406768) in making—
Bornyl acetate and isobornyl acetate from pinene.
Starting point in making—
Boron carbide, boron chloride, boron nitride, zinc borate.

Glass
Reagent in making—
Borosilicate glass (French 547090, 547091, 547092).
Ultramarine-boron glass with the aid of sodium sulphide.

Metallurgical
For making high-speed alloyed tool steel (French 514763).
Reagent for—
Decomposing silicates in metallurgical processes.
Starting point in making—
Metallic boron.

Miscellaneous
Reagent (U. S. 1399216 and 1399217) for—
Disintegrating potassium silicate rocks and clays.

Borneol Cinnamate

Synonyms: Bornyl cinnamate.
French: Cinnamate de bornéole, Cinnamate de bornyle, Cinnamate bornylique.
German: Borneolcinnamat, Bornylcinnamat, Zimtsäureborneolester, Zimtsaeurebornylester, Zimtsaeuresborneol, Zimtsaeuresbornyl,

Chemical
Starting point in making—
Pharmaceuticals and other derivatives.

Pharmaceutical
In compounding and dispensing practice.

Borneol Formate

Synonyms: Bornyl formate.
French: Bornylformiate, Formiate de bornéole, Formiate de bornyl, Formiate bornylique.
German: Ameisensaeureborneolester, Ameisensaeuresborneol, Ameisensaeuresbornyl, Borneolformiat.

Paint and Varnish
Solvent (Brit. 283619) in making—
Cellulose ester and ether varnishes, lacquers, and dopes.

Perfumery
Ingredient of—
Synthetic essential oil compounds.

Plastics
Solvent (Brit. 283619) in making—
Cellulose ester and ether compounds.

Borneol Salicylate

Synonyms: Bornyl salicylate.
French: Salicylate de bornéole, Salicylate bornylique.
German: Borneolsalicylat, Bornylsalicylat, Salicylsaeureborneolester, Salicylsaeuresborneol, Salicylsaeuresbornyl.

Paint and Varnish
Solvent (Brit. 283619) in making—
Cellulose ester and ether varnishes, lacquers, and dopes.

Pharmaceutical
In compounding and dispensing practice.

Plastics
Solvent (Brit. 283619) in making—
Cellulose ester and ether compounds.

Bornyl Chloride
French: Chlorure de bornyle, Chlorure bornylique.
German: Bornylchlorid.

Chemical
Starting point in making—
Bornylanilin, camphene (German 439695).

Bornylcresol

Chemical
Starting point (Brit. 444351) in making—
Fat-splitting catalysts and emulsifying agents, useful in dyeing, laundering, bleaching, and various other purposes, by reacting with formaldehyde and non-aromatic secondary amines (the salts of the products with water-soluble acids, or water-insoluble acids, or the quaternary ammonium salts, are claimed to be valuable for the purposes named).

Bornylphenol

Chemical
Starting point (Brit. 444351) in making—
Fat-splitting catalysts and emulsifying agents, useful in dyeing, laundering, bleaching, and various other purposes, by reacting with formaldehyde and non-aromatic secondary amines (the salts of the products with water-soluble acids, or water-insoluble acids, or the quaternary ammonium salts, are claimed to be valuable for the purposes named).

Bornylresorcinol

Chemical
Starting point (Brit. 444351) in making—
Fat-splitting catalysts and emulsifying agents, for use in dyeing, laundering, bleaching, and various other purposes, by reacting with formaldehyde and non-aromatic secondary amines (the salts of the products with water-soluble acids, or water-insoluble acids, or the quaternary ammonium salts, are claimed to be valuable for the purposes named).

Borodisalicylic Acid

Chemical
Starting point in making—
Esters and salts, pharmaceuticals.

Pharmaceutical
In compounding and dispensing practice.

Borofluoric Acid
Synonyms: Fluorboric acid.
French: Acide de borofluorique, Acide de fluorborique.
German: Borfluorsaeure, Fluorborsaeure.

Chemical
Starting point in making various salts and other derivatives.

Miscellaneous
Reagent for various purposes.

Resins and Waxes
Catalyst (Brit. 314810) in making synthetic resins from—
Arylalkyl ethers, crude naphtha, metacresylmethyl ether, metaxylenylethyl ether, metaxylenylmethyl ether.

Boron (Amorphous or Crystalline)
French: Bore.
German: Bor.

Metallurgical
Addition agent in making—
Boron steels and alloys.

Boron Carbide
French: Carbure de bore.
German: Borocarbid.
Spanish: Carburo de boro.
Italian: Carburo di boro.

Abrasives
Cutting material for—
Abrasives, carborundum.

Electrical
As an electrical resistance material.
Cutting material for—
Molded insulation material.
Electrode (German 206177) in—
Arc lamps.
Terminal material (German 206177) in—
Arc lamps.

Glass
Cutting material for—
Glass of all kinds.

Metallurgical
Draw plate metal in—
Wire-drawing.
Protective coating (French 631193) for—
Carbon crucibles.

Miscellaneous
Cutting material in—
Diamond working.

Plastics
Cutting material for—
Molded products.

Stone
Cutting material for—
Asbestos, stone of all kinds.
Cutting material in—
Rock drilling.

Boron Fluoride
French: Fluorure de boron.
German: Fluorboron.

Chemical
Catalyst in—
Polymerizing gaseous olefines.
Starting point in making—
Borofluoric acid.

Petroleum
Catalyst in—
Synthesizing petroleum oils.

Boronia Oil
Synonyms: Oil of boronia megastigma.
French: Huile de boronia megastigma.
German: Boroniaoel.
Spanish: Aciete de boronia megastigma.
Italian: Olio di boronia megastigma.

Perfume
Ingredient of—
Perfumes, toilet vinegars.
Perfume in—
Cosmetics.

Boron Trichloride
French: Trichlorure de bore.
German: Borchlorid, Trichlorbor.

Metallurgical
Refining agent (Brit. 420694) for—
Aluminum, zinc, copper, and their alloys (used by blowing or bubbling through the molten metal in a gaseous form, the metal being maintained in a non-oxidizing atmosphere and afterward cast in a mould in which a nonoxidizing atmosphere is maintained, the treatment of the metal being effected under a pressure greater than atmospheric).

Refrigeration
Refrigerant (German 574562) in—
Compression refrigerating machines (claimed to have advantages of noncombustibility, nontoxicity, noncorrosive, presence of leaks is immediately detected, as the boron trichloride forms a mist on coming into contact with moisture owing to hydrolysis).

Borophenolic Acid
Note: Mixture of phenyl metaborate and phenyl triborate.
French: Acide, borophénique, Acide borophénolique.
German: Phenolborsäure.

Chemical
Starting point in making—
Esters and salts, pharmaceuticals.

Miscellaneous
As a strong bactericide for various purposes.

Pharmaceutical
In compounding and dispensing practice.

2-Brom-1-alphanaphthylaminoanthraquinone

Chemical
Starting point in—
Organic syntheses.

Dye
Starting point (Brit. 443958 and 443959) in making—
Vat dyestuffs.

2-Brom-1-aminoanthraquinone

Chemical
Starting point in—
Organic syntheses.
Starting point (U. S. 1999996) in making—
Seleno ethers by reacting with 1:1'-dibenzanthronyl diselenide.

4-Brom-5-aminobenzotrifluoride-2-sulphonic Acid
Dye
Intermediate (Brit. 446532) in making dyestuffs.

3-Brom-1-benzamido-4:2′-methylalphanaphthylamino-anthraquinone
Chemical
Starting point in—
Organic syntheses.
Dye
Starting point (Brit. 443958 and 443959) in making—
Vat dyestuffs.

1-Brom-2-chlorobutanol-3
Petroleum
Solvent (Brit. 435096) in—
Refining mineral oils.

3-Brom-10-deltadiethylaminoalphamethylbutylamino-acridin Dihydrochloride
Pharmaceutical
Claimed (Brit. 441007, 441132, and addition to 363392) as—
New pharmaceutical.

Bromhydrin
Fuel
Primer (Brit. 461320) for—
Diesel fuels.

3-Brom-4-hydroxydiphenyl
Synonyms: 3-Brom-4-phenylphenol.
Disinfectant
As a germicide.

5-Brom-2-hydroxydiphenyl
Disinfectant
As a bactericide (U. S. 1989081).

Bromic Acid
Latin: Acidum bromicum.
French: Acide bromique.
German: Bromsäure.
Spanish: Acido bromico.
Analysis
As an oxidizing agent.
Chemical
Oxidizing agent in making—
Acetic acid from ethyl alcohol.
Reagent in making—
Bromates of various metals, inorganic compounds, intermediate chemicals, organic compounds, pharmaceutical chemicals.
Dye
Oxidizing agent in making various synthetic dyestuffs.

Bromine
Synonyms: Brominium.
Latin: Bromum.
French: Brome.
German: Brom.
Spanish: Bromo.
Italian: Bromo.
Analysis
Reagent in—
Phenols detection, various chemical analyses.
Chemical
As a general oxidizing agent.
As a halogenating agent.
Brominating agent in making—
Inorganic chemicals, such as aluminum bromide, barium bromate, barium bromide, ferric bromide, ferrous bromide, iodine bromide, sodium bromate, strontium bromide.
Organic chemicals, such as acetylene tetrabromide, alphabromobutyric acid, alphabromonaphthalene, bromoacetic acid, bromopropionic acid, bromosuccinic acid, dibromoanthracene, ethyl bromide, ethyl monobromoacetate, ethylene bromide, ethylene chlorobromide, ethylene dibromide, methyl bromide, monobromobenzene, parabromobenzoic acid, paradibromobenzene.
Pharmaceutical chemicals, such as adalin, alkaloid bromides, bromikin, bromantipyrin.
Photographic chemicals, chiefly bromides of metals and alkali-metals.
Starting point in making—
Hydrobromic acid.
Disinfectant
As a disinfectant
Ingredient of—
Solid disinfectant (by admixture with kieselguhr).
Dye
As a halogenating agent.
As an oxidizing agent.
Brominating agent in making—
Anilin dyes, vat dyes.
Electrical
Depolarizing agent in—
Galvanic batteries.
Ink
In the manufacturing process.
Leather
In the manufacturing process.
Metallurgical
Reagent in—
Gold extraction in mining processes.
Silver removal in the purification of platinum.
Military
As a poisonous gas.
Brominating agent in making—
Poisonous and tear gases, such as xylyl bromide, benzyl bromide, bromoacetone, orthobromobenzyl cyanide, brominated methylethyl ketone, dibromo ketone.
Pharmaceutical
In compounding and dispensing practice, chiefly for sedative effects.
Resins
Starting point in making—
Sodium hyperbromite solution, used for bleaching shellac for water solution.
Rubber
Ingredient of—
Cement for adhering rubber to metal, containing also crepe rubber and benzene.
Water and Sanitation
As a disinfectant.

5-Brom-7-methoxy-4-methylisatin Alphachloride
Dye
Starting point (Brit. 441548) in making—
Dyestuffs by condensing with 4-chlor-2-hydroxy-6-methoxy-3-methylthionaphthen.

2-Brom-1:2′-methylalphanaphthylaminoanthraquinone
Chemical
Starting point in—
Organic syntheses.
Dye
Starting point (Brit. 443958 and 443959) in making—
Vat dyestuffs.

Bromoacetanilide
French: Acétylbromanilide, Bromacétanilide.
German: Acetylbromanilid, Azetylbromanilid, Bromacetanilid, Bromazetanilid.
Cellulose Products
Plasticizer for—
Cellulose esters and ethers.
For uses, see under general heading: "Plasticizers."

Bromoacetic Acid Dodecylester
Soap
Starting point (Brit. 403883) in making—
Saponaceous products by reaction with amines, such as anilin, piperidin bases, hydroxyethylanilin, dihydroxyethylanilin, paratoluidin (these products may be used alone, or with soaps, fillers, or compounds giving off oxygen).

Bromoacetic Acid Hexadecylester
Soap
Starting point (Brit. 403883) in making—
Saponaceous products by reaction with amines, such as anilin, piperidin bases, hydroxyethylanilin, dihydroxyethylanilin, paratoluidin (these products may be used alone, or with soaps, fillers, or compounds giving off oxygen).

Bromoacetic Acid Tetradecylester
Soap
Starting point (Brit. 403883) in making—
Saponaceous products by reaction with amines, such as anilin, piperidin bases, hydroxyethylanilin, dihydroxyethylanilin, paratoluidin (these products may be used alone, or with soaps, fillers, or compounds giving off oxygen).

Bromoacetone
French: Bromacétone.
German: Bromaceton.
Spanish: Bromoacetona.
Italian: Bromoacetona.
Military
As a tear gas.
Ingredient of—
Tear gas, in admixture with chloroacetone.
Tear gas, in admixture with benzyl bromide.
Tear gas, in admixture with xylyl bromide.

2-Bromoalphagammabromodeltabetabutenylphenol
Disinfectant
Claimed (Brit. 443113 and 389514) to be—
Disinfectant free of odor.

Bromoalphanaphthol
Pharmaceutical
Suggested for use (Brit. 351605) as—
Antiseptic.

4-Bromo-1-aminoanthraquinone-2-sulphonic Acid
German: 4-Brom-1-aminoanthrachinon-2-sulfonsaeure.
Dye
Starting point (Brit. 401132) in making blue acid dyes by condensation with—
2-Amino-5:6:7:8-tetrahydro-4-naphthylmethylsulphone.
5-Amino-2-acetamidophenylmethylsulphone.
Meta-aminophenylaceticsulphone.
Meta-aminophenylmethylsulphone.
Para-aminophenylmethylsulphone.

4-Bromo-1-aminoanthraquinone-2-sulphonic-cyclohexylamide
Dye
Starting material (Brit. 399095) in making—
Anthraquinone derivatives by condensation with ammonia or an amine.

2-Bromoanthraquinone
Chemical
Starting point in—
Organic syntheses.
Starting point (U. S. 1999996) in making—
Seleno ethers by reacting with 1:1′-dibenzanthronyl diselenide.

Bromobenzanthrone
Chemical
Starting point (Brit. 256059) in making—
Benzylbenzanthronylmercaptan, benzylbenzanthronyl sulphide.

Bromobenzoylaminoanthraquinone
French: Bromobenzoyleaminoanthraquinone.
German: Brombenzoylaminoanthrachinon.
Chemical
Starting point in making—
Intermediates, pharmaceuticals.
Dye
Starting point (Brit. 298696) in making anthraquinone vat dyestuffs with—
Aminohydroxybenzoylaminoanthraquinone.
Benzoylaminoanthraquinone.

Bromobenzoyl Chloride
French: Chlorure de bromobenzoyle, Chlorure bromobenzoylique.
German: Brombenzoylchlorid, Chlorbrombenzoyl.
Fats and Oils
As a bleaching agent (used in conjunction with caproyl chloride, oleyl chloride, and lauryl chloride.
Reagent for bleaching—
Various oilseed meals (used in conjunction with oleyl chloride, lauryl chloride, and caproyl chloride).
Food
Reagent for bleaching—
Various foodstuffs, such as flour, milling products of various sorts, egg yolk, and other vegetable and animal food products (used in conjunction with oleyl chloride, lauryl chloride, and caproyl chloride).
Soap
As a bleaching agent (used in conjunction with oleyl chloride, caproyl chloride, and lauryl chloride).
Waxes and Resins
Reagent for bleaching—
Various resins (used in conjunction with oleyl chloride, caproyl chloride, and lauryl chloride).

Bromobenzyl Cyanide
French: Cyanobromure de benzyle, Cyanure de bromobenzyle.
German: Brombenzylcyanid, Cyansäuresbrombenzylester.
Spanish: Cianuro de bromobenzil.
Italian: Cianuro di bromobenzile.
Military
As a tear gas agent.

Bromocyclohexane
German: Bromcyclohexan.
Chemical
Starting point (Brit. 261764) in making cyclohexylamines with—
Alpha-aminoanthraquinone, anilin, beta-aminoanthraquinone, beta-aminochloroanthraquinone, carbazole, chloranilin, diaminoanthraquinone, 1:4-diaminochloroanthraquinone, diphenylamine, monoethylanilin, monomethylanilin, naphthylamine, toluidin, xylidin.

2-Bromocyclohexane-1:4-dicarboxylic Acid
Cellulose Products
Plasticizer (Brit. 390541) for—
Cellulose esters and ethers.
For uses, see under general heading: "Plasticizers."

4-Bromodiethylacetylamino-1-phenyl-2:3-dimethyl-5-pyrazolone
Pharmaceutical
Claimed (U. S. 1906200) as—
Febrifuge.

6-Bromo-2:4-dinitroanilin
Chemical
Starting point in—
Organic synthesis.
Dye
Starting point in making various dyes, including—
Light-fast and readily discharged violet dyestuffs for acetate rayon by diazotizing and coupling with di(betahydroxyethyl)metatoluidin (Brit. 421975).
Light-fast and readily discharged navy-blue dyestuffs for acetate rayon by diazotizing and coupling with normal-betahydroxyethyl-N-butylcresidin (Brit. 421975).
Light-fast and readily discharged red-violet dyestuffs for acetate rayon by diazotizing and coupling with normal-betahydroxyethyl-N-N-butylmetatoluidin (Brit. 421975).
Light-fast and readily discharged blue-violet dyestuffs for acetate rayon by diazotizing and coupling with normal-betahydroxyethyl-N-butylmetatoluidin (Brit. 421975).
Starting point (Brit. 429936 and 430079) in making—
Blue dyestuffs for acetate rayon and animal fibers by diazotizing and coupling with 2:5-dimethoxybutyl-betasulphatoethylanilin.
Navy-blue dyestuffs for acetate rayon and animal fibers by diazotizing and coupling with the alphabutyl, betabutyl, or betapropyl derivative of 3-betasulphatoethylaminoparatolylmethyl ether.
Red-violet dyestuffs for acetate rayon and animal fibers by diazotizing and coupling with ethylbetasulphatoethylanilin.
Violet dyestuffs for acetate rayon and animal fibers by diazotizing and coupling with methylbetasulphatoethylanilin.

Bromoform
Synonyms: Bromoformum, Bromoformium, Formyl tribromide, Tribromomethane.
Analysis
Solvent in—
Analytical processes involving control or research work.
Chemical
Intermediate in—
Organic syntheses.
Solvent miscible with—
Alcohols, esters, ethers.
Fire-Fighting
Ingredient of—
Chemical fire extinguishers.
Mining
Solvent in—
Geological assays.
Miscellaneous
Solvent miscible with—
Alcohols, esters, ethers.
Pharmaceutical
Suggested for use as a—
Local anesthetic, nervine, sedative.

5-Bromo-8-hydroxyquinolin
Pharmaceutical
Suggested for use (Brit. 351605) as—
Antiseptic.

5-Bromo-7-iodo-8-hydroxyquinolin
Pharmaceutical
Suggested for use (Brit. 351605) as—
Antiseptic.

Bromomesitylene
Analysis
Reagent.
Chemical
Reagent in—
Chemical syntheses.

Bromonitrobenzyl Chloride
French: Chlorure de bromonitrobenzoyle, Chlorure bromonitrobenzoylique.
German: Bromnitrobenzoylchlorid.
Chemical
Starting point (Brit. 314909) in making derivatives with—
3-Carboxyphenylthiocarbamide.
Diphenylurea-3:3′-dicarboxylic acid.
4-Quinolylphenylurea-3:6′-dicarboxylic acid.
Symmetrical diphenylurea-3:3′-dicarboxylic acid.
Thiourea.
Thiourea-3:3′-dicarboxylic acid.
Urea.

4-Bromo-2-normalamylphenol
Pharmaceutical
As a bactericide (U. S. 1969801).

4-Bromo-2-Normalhexylphenol
Pharmaceutical
As a bactericide (U. S. 1969801).

Bromopinene Nitrate
Petroleum
Primer (Brit. 436027) for—
Diesel engine fuels (lowers ignition point).

Bromopropylene
Refrigeration
Refrigerant (U. S. 2014496) in—
Centrifugal compression and expansion systems.

5-Bromosalicylaldehyde
Photographic
Purification agent (U. S. 1973472) for—
Methylpara-aminophenol (developing agent).

Bromostearic Acid Nitrate
Petroleum
Primer (Brit. 436027) for—
Diesel engine fuels (lowers ignition point).

4-Bromo-2-styrylquinolin
French: 4-Bromo-2-styrylquinoléine.
German: 4-Brom-2-styrylchinolin.
Chemical
Starting point (Brit. 282143) in making pharmaceuticals with—
Allylamine, allylenediamine, alphanaphthylamine, ammonia, amylamine, amylenediamine, benzylamine, benzylenediamine, betanaphthylamine, butylamine, butylenediamine, ethylamine, ethylenediamine, heptylamine, heptylenediamine, hexylamine, hexylenediamine, metaphenylenediamine, metatoluylenediamine, methylamine, methylenediamine, orthophenylenediamine, orthotoluylenediamine, paraphenylenediamine, paratoluylenediamine, propylamine, propylenediamine, toluylamine.

Bromosuccinic Acid Cyclohexylester
Detergent
Starting point (Brit. 408754) in making—
Saponaceous products by reaction with tertiary amines, which may be used alone or with other soaps, fillers, or compounds giving off oxygen.

Bromosuccinic Acid Dodecylester
Detergent
Starting point (Brit. 408754) in making—
Saponaceous products by reaction with tertiary amines, which may be used alone or with other soaps, fillers, or compounds giving off oxygen.

Bromosuccinic Acid Hexadecylester
Soap
Starting point (Brit. 403883) in making—
Saponaceous products by reaction with amines such as anilin, piperidin bases, hydroxyethylanilin, dihydroxyethylanilin, paratoluidin (these products may be used alone or with other soaps, fillers, or compounds giving off oxygen).

Bromosuccinic Acid Octadecylester
Detergent
Starting point (Brit. 408754) in making—
Saponaceous products by reaction with tertiary amines, which may be used alone or with other soaps, fillers, or compounds giving off oxygen.

Bromosuccinic Acid Tetradecylester
Soap
Starting point (Brit. 403883) in making—
Saponaceous products by reaction with amines such as anilin, piperidin bases, hydroxyethylanilin, dihydroxyethylanilin, paratoluidin (these products may be used alone or with other soaps, fillers, or compounds giving off oxygen).

Bromothymol Blue
Analysis
As an indicator.
Miscellaneous
Reagent for—
Rapid detection of suffocating and vesicatory gases in dangerous concentrations.

6-Brom-1:2-phthaloylcarbazole
Textile
As a vat dye (Brit. 443958 and 443959).

4-Brom-1:2-phthaloyl-6-methylcarbazole
Textile
As a vat dye (Brit. 443958 and 443959).

Brucine
Synonyms: Dimethoxystrychnine.
Chemical
Denaturant for special industrial alcohol.
Starting point in making—
Yohimbine-brucine sulphate (German 437923).
Pharmaceutical
In compounding and dispensing practice.

Brucine-Anilide Acetate
Insecticide and Fungicide
Ingredient (U. S. 2015533) of—
Mildewproofing composition, comprising admixture with sodium sulphate, tartar emetic, and saponin, for treating animal fibers.
Mothproofing composition, comprising admixture with sodium sulphate, tartar emetic, and saponin, for treating animal fibers.

Brucine-Anilide Formate
Insecticide and Fungicide
Ingredient (U. S. 2015533) of—
Mildewproofing composition, comprising admixture with anhydrous sodium sulphate, zinc sulphate, and saponin, for treating animal fibers.
Mothproofing composition, comprising admixture with anhydrous sodium sulphate, zinc sulphate, and saponin, for treating animal fibers.

Brucine-Anilide Hydrochloride
Insecticide and Fungicide
Ingredient (U. S. 2015533) of—
Mildewproofing composition, comprising admixture with anyhydrous sodium sulphate, aluminum sulphate, and saponin, for treating animal fibers.
Mothproofing composition, comprising admixture with anhydrous sodium sulphate, aluminum sulphate and saponin, for treating animal fibers.

1:3-Butadiene
Synonyms: Divinyl, Erythrene, Vinylethylene.
French: 1:3-Butadiène, Divinyl, Erythrène, Vinyléthylène.
German: 1:3-Butadien, Divinyl, Erythren, Vinyläthylen, Vinylaethylen.
Spanish: 1:3-Butadieno, Divinil, Eritreno, Viniletileno, Vinil de etileno.
Italian: 1:3-Butadieno, Divinile, Eritreno, Viniletileno, Vinile di etileno.

1:3-Butadiene (Continued)

Chemical
Starting point in making—
Organic bases, useful as accelerators of rubber vulcanization, by reaction with ammonia, anilin, cyclohexylamine, ethylenediamine, methylamine, methylanilin, or piperidin (French 662431).
Polymerization products in the presence of sodium and sodium chloride in an atmosphere of nitrogen (French 687808).
Quinolin bases by reaction with dimethylanilin (French 685569).
Soluble condensation products (French 687773).
Soluble polymerization products (French 686934).

Electrical
Starting point (French 667955) in making—
Electrical insulators from polymerization products.

Fats and Oils
Starting point in making—
Oils for adhesive and cement mixtures (French 679539).
Polymerization products useful as oils (French 683284).
Varnish oils (French 679539).

Miscellaneous
Starting point in making—
Compositions of polymerization products and coaltar derivatives (French 690543).
Polymerized products used as base materials for all sorts of products (French 690484).
Polymerization products useful as very stable diaphragms for electrolytic tanks (French 668045).
Polymerization products for impregnating tissues (French 667955-7).
Polymerization products useful as cementing agents (French 667299).
Ribbons by oxidation of its polymerization products (French 684990).
Ribbons from polymerization products (French 692027).
Threads by oxidation of its polymerization products (French 684990).
Threads from polymerization products (French 692027).
Twisted fibers from polymerization products (French 692027).

Paint and Varnish
Starting point in making—
Polymerization products used as dispersing agents for pigments (French 689711).

Plastics
Starting point in making—
Elastic foils and sheets from polymerization products (French 667955-7).
Films by oxidation of its polymerization products (French 684990).
Nozzles and such products from polymerization products (French 667955-7).
Plastics from polymerization products (French 667955-7).
Plastics by oxidation of its polymerization products (French 684990).
Plastics with great adhesive properties and easily soluble from polymerization products (French 688592-3).

Resins and Waxes
Starting point in making—
Varnish and other useful resins by condensation with tetrahydronaphthalene, benzene, xylene, or other hydrocarbons, in the presence of aluminum chloride (French 676508).
Varnish and other coating resins by oxidation of its polymerization products (French 684990).

Rubber
Starting point in making—
Condensation products with toluene, xylene, mesitylene, or tetrahydronaphthalene (French 677748).
Emulsions of mixed polymerisates with sorbic acid ethylester, sorbic acid methyl ketone, cinnamic acid ethylester, betamethylgammachlorobutadiene, betachlorobutadiene, or 3-chlorostyrene (Brit. 387381).
Hydrocarbon liquids by condensation with toluene (Brit. 315312).
Plastic products analogous to rubber by polymerization in presence of alkaline metals or alkaline earth metals and organic halides, such as benzyl chloride, dibromocyclohexane, ethylene chloride, vinyl chloride, alphachloronaphthalene, alphabromonaphthalene (French 702784).
Polymerization products (French 677415, 691901, 693920, 708807, 715204, 721361).
Polymerization products by means of an alkali metal in presence of small amounts of ammonia, or of primary, secondary, or tertiary bases, or of their mixtures (French 678305).
Polymerization products in presence of alkali metals water (French 691662).
Polymerization products in presence of alkali metals or alkaline earth metals and ethylvinyl ether, or isopropylvinyl ether, or allylethyl ether (French 695299).
Polymerization products in presence of alkali metals or alkaline earth metals and acetals of crotonic aldehyde, acetophenone, or other organic chemicals containing one atom of carbon of which two valances at least are saturated by oxygen (French 695441).
Polymerization products using metals such as sodium as catalysts (French 696149).
Polymerization products in presence of accelerators, such as hydrogen peroxide, benzoyl peroxide (French 708808).
Polymerization products in the presence of alkali metals, alkaline earth metals, hydrates of alkali metals, or organic derivatives of metals (French 710791).
Polmerization products in presence of salts of bile acids or of animal biles (French 663995).
Synthetic materials by polymerization (French 697693).
Synthetic rubber by polymerization (French 633575).
Synthetic rubber latex by conversion of water-in-oil-emulsion (French 646704).
Synthetic rubber from polymerization products (French 664419, 697679).
Synthetic rubber by polymerization of its aqueous soap emulsions (French 667256).
Synthetic rubber mixtures (French 669942).
Synthetic rubber using nitrobenzene, dinitrobenzene, trinitrobenzene, or dinitronaphthalene as a plastifier (French 689070).
Synthetic rubber mixtures with isoprene and dimethylbutadiene (French 695745).
Synthetic rubber (Brit. 301515, 308755, addition to 283841).
Synthetic rubber by polymerization (Brit. 307708, 307937-8).
Synthetic rubber by polymerization with sodium hydroxide (Brit. 315356).

Butane

Note: Butane, according to the purpose, may be used either alone or in admixture with propane or air.

Agriculture
Fuel for—
Farming machinery.
Gas refrigerators.
Heating and cooking equipment.
Orchard heating equipment used to prevent damage by frosting of citrous fruits and other crops.
Poultry equipment, such as incubators, brooders, disinfecting burners.
Stationary engines, running pumps, lighting units, power units.

Analysis
As an extractant.
As a solvent.
Fuel for—
Burners, hot-plates, water stills, flash-point testers, sterilizers, ovens, and other heating and heated equipment in laboratories.

Animal Products
Fuel for—
Cooking equipment in packing plants.
Low-boiling extractant for—
Food products, glandular constituents, hormones, vitamins.

Automotive
Internal combustion fuel for—
Automobile engines in block testing and running-in operations.

Aviation
Ingredient of—
Zeppelin engine fuels, in admixture with hydrogen (U. S. 1936155).
Zeppelin engine fuels, in admixture with hydrogen or natural gas.

Bituminous
Precipitating agent (Brit. 409278) for—
Asphalts in hydrogenation residues obtained from coal, tars, and other materials.

Brewing
Fuel for—
De-pitching burners, keg-branding irons.

Butane (Continued)

Ceramics
Fuel for—
China kilns, testing furnaces.

Chemical
As a low-boiling extracting solvent.
Starting point in making—
Organic chemicals.

Construction
Internal-combustion fuel for—
Ditch-diggers, excavating machinery, hauling equipment, hoisting equipment, power shovels, road-graders, trucks.

Fats and Oils
Extracting agent for—
Vegetable oils (claim is made that high yields of good quality, pale oils are obtained and the meal is easily freed from solvent).

Food
As a low-boiling extractant.
Fuel in—
Bakery plant equipment, canning plant equipment, coffee roasters, confectionery apparatus, cooking equipment in homes, hotels, restaurants, roadstands; dairy equipment, peanut roasters, ripening heaters for bananas and other fruits.

Fuel
Fuel for—
Battery and radiator torches, bench furnaces.
Burners of various types, such as ring burners, bar burners, jet burners, ribbon burners, cluster burners, furnace burners, furnace kindlers.
Industrial or domestic heating where artificial or natural gas is not available or where the supply is limited or of high cost, or not used for various reasons; can also be used as standby fuel or temporary substitute because the same burners or burning equipment is adaptable to all these fuels.
Industrial or domestic heating where dust and dirt incidental to the use of coal is not desirable.
Industrial or domestic heating where adequate coal-storage space is not available or must be avoided for various reasons.
Internal-combustion engines.
Internal-combustion power equipment operating mostly on full throttle.
Water-heaters of various kinds.

Glass
Fuel for—
Burners, glass furnaces, glassworking machinery.

Hospitals
Fuel for—
Bandage incinerators, coffee urns, constant burning devices, diet kitchen equipment, hot-plates, main kitchen equipment, steam-tables, sterilizers, stoves.

Laundering
Fuel for—
Dryers, irons, mangles, pressing machines, small steam boilers.

Mechanical
Fuel for—
Stationary engines connected direct to generators as sources of regular power, boosters, or standby units.
Stationary engines running compressors, lighting units, pumps.

Metallurgical
Fuel for—
Blow torches, brazing torches, crucible furnaces, cutting torches, enameling ovens, japanning ovens, lead-melting pots, preheating torches, welding torches.
Gaseous fuel in—
Annealing processes, carburizing processes, heat-treating processes.
Inhibitor of—
Oxidizing of nickel and monel metal in annealing processes carried out in electric furnaces.

Miscellaneous
Fuel for—
Barber shop equipment, beauty-shop equipment, cleaning, pressing, and tailoring equipment; dental equipment, doctor's office equipment, household incinerators, illuminating equipment, such as household lights, portable lanterns, gas floodlights.

Motor Transportation
Combination internal-combustion fuel and refrigerant for—
Refrigerated trucks.
Internal combustion fuel for—
Buses, industrial plant jitneys, trackless vehicles, trucks.

Paint and Varnish
Fuel for—
Varnish kettles.
Solvent in—
Lacquer formulation.
Starting point in making—
Black pigment by incomplete combustion.

Petroleum
Fuel for—
Internal-combustion engines running pumps on pipelines.
Stationary engines connected direct to generators as sources of regular power, boosters, or standby units.
Stationary engines running compressors or lighting units.
Precipitating agent for—
Asphalts in crude petroleum, or distillation, cracking, or destructive hydrogenation residues obtained from tars or mineral oils (Brit. 409278, U. S. 1948163 and 1948164).
Solvent for—
High molecular weight constituents in making high-quality lubricating oils (Brit. 422471).
Paraffinic fractions in refining mineral oils and making lubricating oils (Brit. 421123).
Paraffin in refining mineral oils (Brit. 390222, 408947, 408948, and 423303; U. S. 1977054, 1977055, 1948346, and 1943236).
Standby gas for—
Fuel purposes (in admixture with air).

Pharmaceutical
Low-boiling extractant for—
Glandular principles, hormones, vitamins.

Printing
Fuel for—
Glue pots, intertype burners, linotype burners, monotype burners, typemetal melting pots.

Railroading
Fuel for—
Brazing torches, buffet stoves, building heating, cooking and dining-car equipment, cutting equipment, engine-driven lighting and power generators, gas-fired switch heaters, gas refrigerators, gas service in restaurants and lunch rooms, heating passenger sections in cold weather, prime-movers, soldering torches.
Stationary engines operating electric generators, air compressors, water pumps, shop shafting.
Stationary power units on switching locomotives, construction locomotives, rail cars, trains, and locomotives (propane is especially suitable and is used as refrigerant in air-conditioning trains).
Steaming-up locomotives in terminals and stations where use of oil burners for this purpose is objectionable or impracticable and where high-pressure steam is not available around the yard or powerhouse.
Thawing torches, water heaters.

Refractories
Fuel for—
Furnaces.

Refrigeration
As a refrigerant.
Fuel in—
Gas refrigerators.

Rubber
Fuel for—
Burners for cleaning tire molds, vulcanizing equipment.

Textile
Fuel in—
Calendering processes, drying processes, singeing processes.

Utilities
——, *Gasmaking*
Enrichener for—
Manufactured gas in recarburation of domestic and industrial gases.
Heating agent for—
Underfired cokeovens (to reduce coke production).
Increaser of—
Gas production in coalgas, watergas, or oilgas plants.
Standby gas (in admixture with air) for—
Peak loads, utilities supplying natural gas.
Starting point in making—
Gas by reforming process.
Substitute for—
Gas oil for the carburation of water-gas.

Butane (Continued)

——, *Power*
Fuel for—
Stationary engines connected direct to generators.
Stationary engines running compressors, lighting units.
Stationary engines as sources of regular power, boosters, or standby units.

Butenylpiperidin, Normal
French: N-butenylepiperidine.
Chemical
Starting point in making various derivatives.
Insecticide
As an insecticide.
Ingredient (Brit. 313934) of—
Insecticidal compositions.
Soap
Ingredient (Brit. 313934) of—
Insecticidal soaps.

Butoxydiphenylamine, Secondary
Rubber
Aging retardant (Brit. 424461).
Promoter (Brit. 424461) of—
Resistance to crack formation on flexing.

Butyl Acetate, Normal
Synonyms: Butanol acetate, N-Butyl acetate.
French: Acétate butylique normale, Acétate de butanole, Acétate de butyle normale.
German: Butanolacetat, Essigsaeurenormalbutylester, Essigsaeurebutanolester, Essigsaeuresnormalbutyl, Essigsaeuresbutanol.
Dye
Solvent in making various products.
Food
Ingredient of—
Flavoring extracts, fruit essences.
Leather
Solvent in making—
Artificial leathers, patent leathers.
Miscellaneous
Solvent in making—
Celluloid cements, polishes.
See also: "Solvents."
Paint and Varnish
Solvent and plasticizer in making—
Lacquers, varnishes, paints, and dopes containing nitrocellulose and other esters and ethers of cellulose.
Plastics
Solvent in making—
Plastics containing paper pulp pitch.
Solvent and plasticizer in making—
Compounds of cellulose acetate, nitrocellulose, pyroxylin, and other cellulose direvatives.

Butylacetyl Ricinoleate
French: Ricinoléate butyleacétylique, Ricinoléate de butyle et d'acétyle.
German: Butylacetylricinoleat, Butylacetylrizinoleat, Butylazetylricinoleat, Butylazetylrizinoleat, Ricinoelsäurebutylacetylester, Ricinoelsäuresbutylacetyl, Ricinusoelsäurebutylacetylester, Ricinusoelsäuresbutylacetyl, Rizinoelsäurebutylacetylester, Rizinoelsäuersbutylacetyl, Rizinoelsäuerbutylacetylester, Rizinusoelsäuresbutylacetyl.
Spanish: Ricinoleato de butile y de acetile.
Italian: Ricinoleato di butil e d'acetil.
Ceramics
Plasticizer in—
Compositions containing various derivatives of cellulose, such as nitrocellulose, used for the production of protective and decorative coatings on ceramic ware.
Chemical
Dispersing agent in making—
Dispersions of hydrocarbons of various groups of the aromatic and aliphatic series.
Dispersions of halogenated hydrocarbons of various aromatic and aliphatic groups.
Dispersions and emulsions of various chemicals.
Terpene emulsions.
Wetting compositions in emulsified form.
Disinfectant
Dispersing agent in making—
Emulsified disinfecting and deodorizing compositions.
Dye
Dispersing agent in making—
Emulsified lakes, emulsions of dyestuffs.
Fats and Oils
Dispersing agent in making—
Emulsified boring oils, emulsified drilling oils.
Emulsified fats and oils of both animal and vegetable origin.
Emulsified fatty acids of both animal and vegetable origin.
Emulsified sulphonated oils.
Greasing compositions in emulsified form.
Lubricating compositions in emulsified form, containing various vegetable and animal fats and oils.
Solvents for fats in emulsified form.
Stabilized emulsions of vegetable and animal fats and oils.
Wetting compositions in emulsified form, containing animal and vegetable fats and oils.
Wire-drawing oils in emulsified form.
Glass
Plasticizer in—
Compositions containing various esters or ethers of cellulose, such as nitrocellulose, used in the manufacture of nonshatterable glass and for the production of decorative and protective coatings on glassware.
Glues and Adhesives
Dispersing agent in making—
Emulsified adhesive preparations.
Ink
Dispersing agent in making—
Ink emulsions for writing and printing purposes.
Insecticide
Dispersing agent in making—
Emulsified insecticidal and fungicidal preparations.
Orchard sprays in emulsified form.
Vermin exterminators in emulsified form.
Leather
Dispersing agent in making—
Emulsified dressing compositions.
Emulsified fat-liquoring baths.
Emulsified finishing compositions.
Emulsified soaking compositions containing various animal and vegetable oils.
Emulsified waterproofing compositions.
Plasticizer in—
Compositions containing various derivatives of cellulose, such as nitrocellulose, used in the manufacture of artificial leather and for the production of decorative and protective coatings on leather goods.
Metallurgical
Plasticizer in—
Compositions containing various esters or ethers of cellulose, such as nitrocellulose, used for the production of decorative and protective coatings on metal ware.
Miscellaneous
Dispersing agent in making—
Automobile polishes in emulsified form.
Cleansing compositions in emulsified form, for use on painted and metallic surfaces.
Degreasing compositions in emulsified form.
Emulsified compositions containing various substances, such as tars and pitches.
Greasing compositions in emulsified form.
Furniture polishes in emulsified form.
Metal polishes in emulsified form.
Scouring compositions in emulsified form.
Special emulsified cleansing compositions.
Various emulsified compositions containing fats, oils, and miscellaneous substances, used for wetting, washing, and dispersing processes.
Waterproofing compositions for treating various fibrous and other compositions of matter.
Plasticizer in—
Compositions containing various esters or ethers of cellulose, such as nitrocellulose, used as coatings on various compositions.
Paint and Varnish
Dispersing agent in making—
Emulsified paints, varnishes, and other coating compositions.
Pigment emulsions, shellac emulsions.
Waterproofing compositions in emulsified form.
Plasticizer in making—
Lacquers, enamels, varnishes, and paints containing various esters or ethers of cellulose, such as nitrocellulose.

Butylacetyl Ricinoleate (Continued)

Paper
Dispersing agent in making—
Emulsified preparations used for the treatment of paper and pulp and various products made therefrom.
Sizing compositions in emulsified form.
Waterproofing compositions in emulsified form.
Waxing compositions in emulsified form.
Plasticizer in—
Compositions containing various esters or ethers of cellulose, used for the production of decorative and protective coatings on paper and pulp products and compositions, and in the manufacture of coated paper.

Perfume
Dispersing agent in making—
Emulsified creams, emulsified lotions, emulsified lanolin preparations, emulsified ointments, emulsified perfume preparations, emulsified shaving creams, emulsified sunburn preparations.

Petroleum
Dispersing agent in making—
Emulsified cutting compositions containing various mineral oil distillates.
Emulsified preparations containing kerosene.
Naphtha emulsions.
Petroleum distillate and residue emulsions.
Rayon oils in emulsified form.
Soluble greases in emulsified form.
Soluble oils in emulsified form, for the lubrication of textile and other machinery.
Various textile oils in emulsified form.

Photographic
Plasticizer in making—
Films from various esters or ethers of cellulose, such as nitrocellulose.

Plastics
Plasticizer in making—
Plastic compositions containing various esters or ethers, of cellulose, such as nitrocellulose.

Resins and Waxes
Dispersing agent in making—
Emulsified compositions containing various waxes, both artificial and natural.
Emulsified compositions containing various resins, both artificial and natural.

Rubber
Dispersing agent in making—
Emulsified rubber compositions, emulsified rubber cements.

Soap
Dispersing agent in making—
Emulsified detergents for various purposes.
Hand-cleansing compositions in emulsified form.
Textile scouring soaps in emulsified form.

Stone
Plasticizer in—
Compositions containing various esters or ethers of cellulose, such as nitrocellulose, used for the production of decorative and protective coatings on natural and artificial stone.

Textile
——, *Dyeing*
Dispersing agent in making—
Dye baths in emulsified form.

——, *Finishing*
Dispersing agent in making—
Emulsified coating compositions, emulsified dressing compositions, emulsified finishing compositions, emulsified impregnating compositions, emulsified scouring compositions, emulsified sizing compositions, emulsified washing compositions containing soaps, emulsified waterproofing compositions.

——, *Manufacturing*
Dispersing agent in making—
Dispersions used for the carbonization of wool.
Emulsified compositions for greasing operations.
Emulsified compositions for degreasing operations.
Emulsified compositions used in fulling operations.
Emulsified compositions for lubrication purposes in spinning and weaving.
Emulsified compositions for the mercerization of cotton.
Emulsified compositions for degumming silk.
Emulsified preparations for soaking silk.
Emulsified preparations for kier-boiling cotton.
Emulsified preparations for milling purposes.
Emulsified preparations for washing wool.

——, *Printing*
Dispersing agent in making—
Emulsified printing pastes.

Woodworking
Plasticizer in—
Compositions containing various esters or ethers of cellulose, such as nitrocellulose, used for the production of decorative and protective coatings on woodwork.

Butyl Acrylate
Synonyms: Acrylic acid butyl ester.
French: Acrylate de butyle, Acrylate butylique.
German: Acrylsäurebutylester, Acrylsäuresbutyl, Butylacrylat.

Miscellaneous
Solvent (Brit. 321258) for—
Cellulose acetate, cellulose esters and ethers, cellulose nitrate, rubber.
For uses, see under general heading: "Solvents."

Butyl Alcohol
Synonyms: Butanol, Butylic alcohol, Normal butyl alcohol, Primary butyl alcohol.
French: Alcool butylique.
German: Butanol, Butylalkohol, Buttersäuresalkohol.

Chemical
Extracting agent and solvent in producing—
Alkaloids, resorcinol.
Reagent in making—
Acetyl butyrate, artificial musk, benzyl butyrate, benzoyl butyrate, butenes, butyl acetate, butyl aldehyde, butylanilin, butyl anthanilate, butyl benzoate, butyl borate, butyl bromate, butyl bromide, butyl butylacetate, butyl butyrate, butyl cacodylate, butyl camphorate, butyl caproate, butyl caprylate, butyl chloride, butyl chlorophthalate, butyl chromate, butyl cinnamate, butyl citrate, butyl cyanacetate, butyl cyanide, butyl dichloroacetate, butyl dichloroarsine, butyl dichlorophthalate, butyl dioxystearate, butyl ether, butyl ethylmalonate, butyl formate, butyl gallate, butyl glutartrate, butyl glycolate, butyl iodate, butyl iodide, butyl lactate, butyl malate, butyl maleate, butyl malonate, butyl mercaptan, butyl molybdate, butyl monobromoacetate, butyl monobromobenzoate, butyl monobromobutyrate, butyl monobromopropionate, butyl monobromosuccinate, butyl monochloroacetate, butyl mucate, butyl nitrate, butyl nitrobenzoate, butyl nitrosalicylate, butyl oleate, butyl oxalate, butyl oxybenzoate, butyl palmitate, butyl para-aminobenzoate, butyl para-aminosalicylate, butyl phenylacetate, butyl phenylpropionate, butyl phosphate, butyl phthalate, butyl picrate, butyl proprionate, butyl salicylate, butyl stearate, butyl succinate, butyl sulphanilate, butyl sulphate, butyl tartrate, butyl tetrachlorophthalate, butyl thioacetate, butyl trichloroacetate, butyl valerate, butylene, butylene benzoate, butylene bromide, butylene butyrate, butylene chloride, butylene cinnamate, butylene citrate, butylene formate, butylene iodide, butylene lactate, butylene mucate, butylene phthalate, butylene proprionate, butylene salicylate, butylene succinate, butylene tartrate, butylene valerate, butyric acid, butyric anhydride, cinnamyl butyrate, dibutyl acetate, dibutyl benzoate, dibutyl bromide, dibutyl chloride, dibutyl cinnamate, dibutyl citrate, dibutyl formate, dibutyl gallate, dibutyl iodide, dibutyl lactate, dibutyl malonate, dibutyl mucate, dibutyl oleate, dibutyl palmitate, dibutyl phthalate, dibutyl proprionate, dibutyl salicylate, dibutyl sulphanilate, dibutyl tartrate, dibutyl valerate, ethyl butyrate, ethylene butyrate, formyl butyrate, lactyl butyrate, methyl butyrate, methylene butyrate, phenyl butyrate, phthalyl butyrate, propyl butyrate, propylene butyrate, salicyl butyrate, succinyl butyrate, toluyl butyrate, tributyrin, valeryl butyrate, wetting agents from anthracene and naphthalene.

Dye
Solvent in making—
Vat dyestuffs from benzanthrone.

Explosives
Solvent in making—
Nitrocellulose explosives.
Reagent in treating—
Nitrocellulose explosives, to render them non-explosive for the purpose of storage or transportation (Brit. 252382).

Butyl Alcohol (Continued)
Food
Reagent in making—
Various fruit essences.
Glass
Solvent in making—
Nonshatterable glass.
Leather
Solvent in making—
Artificial leathers, patent leather dopes.
Miscellaneous
Extracting agent for various purposes.
Solvent in making—
Preparations for cleansing old paintings.
Oils and Fats
Extracting agent in producing various fats and oils.
Paint and Varnish
Ingredient of paint and varnish removers.
Solvent in making—
Dip and flow lacquers.
Nitrocellulose lacquers and varnishes.
Shellac compositions.
Spirit varnishes.
Spray and brush lacquers.
Petroleum
Blending agent in oil refining.
Ingredient of—
Motor fuels.
Pharmaceutical
In compounding and dispensing practice.
Photographic
Solvent in making—
Films from cellulose acetate, motion picture films.
Plastics
Solvent in making—
Celluloid, collodion, fiberloid, pyralin, viscoloid.
Solvent in admixture with amyl alcohol and fusel oil in making nitrocellulose plastics.
Perfume
Solvent in extracting various perfume materials.
Textile
Solvent in making—
Rayon.
Waxes and Resins
Extracting agent and solvent.

Butylallylbarbituric Acid, Normal
French: Acide de n-butyleallylebarbiturique.
German: N-butylallylbarbitursaeure.
Chemical
Starting point (Brit. 301727) in making pharmaceutical chemicals with—
1-Phenyl-2:3-dimethyl-4-diamylamino-5-pyrazolone.
1-Phenyl-2:3-dimethyl-4-diethylamino-5-pyrazolone.
1-Phenyl-2:3-dimethyl-4-diheptylamino-5-pyrazolone.
1-Phenyl-2:3-dimethyl-4-dihexylamino-5-pyrazolone.
1-Phenyl-2:3-dimethyl-4-di-isoallylamino-5-pyrazolone.
1-Phenyl-2:3-dimethyl-4-di-isoamylamino-5-pyrazolone.
1-Phenyl-2:3-dimethyl-4-di-isobutylamino-5-pyrazolone.
1-Phenyl-2:3-dimethyl-4-di-isopropylamino-5-pyrazolone.
1-Phenyl-2:3-dimethyl-4-dimethylamino-5-pyrazolone.
1-Phenyl-2:3-dimethyl-4-dipentylamino-5-pyrazolone.
1-Phenyl-2:3-dimethyl-4-dipropylamino-5-pyrazolone.

Butyl Alphacrotonate
Synonyms: Alphacrotonic butyl ester.
French: Alphacrotonate de butyle, Alphacrotonate butylique.
German: Alphacrotonsäurebutylester, Alphacrotonsäuresbutyl, Butylalphacrotnat, Butylalphacrotonester.
Miscellaneous
Solvent (Brit. 321258) for—
Cellulose acetate, cellulose esters and ethers, cellulose nitrate, rubber.
For uses, see under general heading: "Solvents."

Butylamine Selenate, Normal
French: Séléniate de N-butyleamine.
German: N-butylaminselenat, Selensäure-N-butylaminester.
Miscellaneous
Reagent (Brit. 340318) for—
Mothproofing furs, feathers, hair.
Textile
Reagent (Brit. 340318) for—
Mothproofing wool and felt.

Butyl Bromide
Chemical
Reagent in—
Organic syntheses.
Fuel
Primer (Brit. 404682) in—
Diesel engine fuels (used in conjunction with alkyl nitrates, having two to four atoms in the molecule, whose function is that of reducing the delay period).
Reducer (Brit. 404682) of—
Spontaneous ignition temperature of diesel engine fuels.

Butylbromocresol
Chemical
Starting point (Brit. 444351) in making—
Fat-splitting catalysts and emulsifying agents, useful in dyeing, laundering, bleaching, and various other purposes, by reacting with formaldehyde and nonaromatic secondary amines (the salts of the products with water-soluble acids, or water-insoluble acids, or the quaternary ammonium salts, are claimed to be valuable for the purposes named).

Butylbromophenol
Chemical
Starting point (Brit. 444351) in making—
Fat-splitting catalysts and emulsifying agents, useful in dyeing, laundering, bleaching, and various other purposes, by reacting with formaldehyde and nonaromatic secondary amines (the salts of the products with water-soluble acids, or water-insoluble acids, or the quaternary ammonium salts, are claimed to be valuable for the purposes named).

Butylbromoresorcinol
Chemical
Starting point (Brit. 444351) in making—
Fat-splitting catalysts and emulsifying agents for use in dyeing, laundering, bleaching, and various other purposes, by reacting with formaldehyde and nonaromatic secondary amines (the salts of the products with water-soluble acids, or water-insoluble acids, and the quaternary ammonium salts, are claimed to be valuable for the purposes named).

Butyl Carbinol, Normal
Synonyms: 1-Pentanol, Primary normal amyl alcohol.
French: Carbinole normale de butyle, Carbinole, normale butylique.
German: N-butylcarbinol.
Chemical
As a solvent for various purposes.
Miscellaneous
As a solvent for various purposes.
Paint and Varnish
Solvent in making—
Cellulose ester and ether varnishes, lacquers, and dopes.
Plastics
Solvent in making—
Cellulose ester and ether compounds.

Butyl Carbinol, Secondary
Synonyms: Amyl alcohol, active; Amyl hydrate.
French: Alcool d'amyle, actif; Alcool amylique, actif; Carbinole de butyle, sécondaire; Hydrate amylique.
German: Aktiv amylalkohol, Amylhydrat, Sekundaer butylcarbinol.
Chemical
Solvent for various purposes.
Starting point in making—
3-Nitrophthalate, phenyl carbamate, urethane.
Miscellaneous
Solvent for various purposes.
Paint and Varnish
Solvent in making—
Cellulose ester and ether lacquers.
Plastics
Solvent in making—
Cellulose ester and ether compounds.

Butylchloral Hydrate
Synonyms: Butyl chloral, Trichlorobutylideneglycol.
French: Chloralhydrate butylique.
German: Butylchloralhydrat.
Chemical
Starting point (German 438983) in making drugs with—
Pyramidon.

Butylchloral Hydrate (Continued)

Paint and Varnish
Solvent in making—
Cellulose ester and ether lacquers.

Pharmaceutical
In compounding and dispensing practice.

Plastics
Solvent in making—
Cellulose nitrate plastics.

Butyl Chloroacetate
French: Chloroacétate de butyle, Chloroacétate butylique.
German: Butylchloracetate, Chloressigsaeuresbutyl.

Dye
Reagent in making—
Stable, water-soluble vat dyestuffs derivatives (Brit. 263898).

Butylchlorocresol

Chemical
Starting point (Brit. 444351) in making—
Fat-splitting catalysts and emulsifying agents, useful in dyeing, laundering, bleaching, and various other purposes, by reacting with formaldehyde and non-aromatic secondary amines (the salts of the products with water-soluble acids, or water-insoluble acids, or the quaternary ammonium salts, are claimed to be valuable for the purposes named).

Butylchlorophenol

Chemical
Starting point (Brit. 444351) in making—
Fat-splitting catalysts and emulsifying agents, useful in dyeing, laundering, bleaching, and various other purposes, by reacting with formaldehyde and non-aromatic secondary amines (the salts of the products with water-soluble acids, or water-insoluble acids, or the quaternary ammonium salts, are claimed to be valuable for the purposes named).

Butylchlororesorcinol

Chemical
Starting point (Brit. 444351) in making—
Fat-splitting catalysts and emulsifying agents, for use in dyeing, laundering, bleaching, and various other purposes, by reacting with formaldehyde and non-aromatic secondary amines (the salts of the products with water-soluble acids, or water-insoluble acids, or the quaternary ammonium salts, are claimed to be valuable for the purposes named).

Butyl Cinnamate
French: Cinnamate de butyle, Cinnamate butilique.
German: Butylcinnamat, Zimtsäurebutylester, Zimtsäuresbutyl.

Chemical
Starting point in making various derivatives.

Miscellaneous
Plasticizer and solvent (Brit. 321258) for—
Cellulose acetate, cellulose esters and ethers, cellulose nitrate, rubber.
For uses, see under general heading: "Plasticizers."

Butylcresol

Chemical
Starting point (Brit. 444351) in making—
Fat-splitting catalysts and emulsifying agents, useful in dyeing, laundering, bleaching, and various other purposes, by reacting wtih formaldehyde and non-aromatic secondary amines (the salts of the products with water-soluble acids, or water-insoluble acids, or the quaternary ammonium salts, are claimed to be valuable for the purposes named).

3-Butyl-2′:4′-diaminobenzene

Disinfectant
Bactericide and bacteriostatic (U. S. 2030897).

Butyldi-isoamyl Phosphate, Normal

Cellulose Products
Plasticizer for—
Nitrocellulose.
For uses, see under general heading: "Plasticizers."

Butyl Disulphide
Synonyms: Butyl bisulphide.
French: Bisulfure de butyle, Bisulfure butylique, Disulfure de butyle, Disulfure butylique.
German: Butylbisulfid, Butyldisulfid, Schwefelbutyl, Schwefelwasserstoffsäurebutylester.

Chemical
Reagent (Brit. 298511) in treating—
Albumenoids and albumens to convert them into compounds suitable for adhesive, sizing, and similar purposes.
Starting point in making various derivatives.

Glues and Adhesives
Reagent (Brit. 298511) in treating—
Vegetable proteins, such as soybean flour, linseed protein, peanut protein, and the like, for making glues and adhesives.

Miscellaneous
Reagent (Brit. 298511) in treating—
Vegetable proteins, such as soybean flour, linseed protein, peanut protein, and the like, for making sizing and finishing compositions.

1:3-Butyleneglycol
German: 1:3-butylenglykol.

Fats and Oils
Starting point (Brit. 279877) in making—
Solvents for fats and oils.

Miscellaneous
Ingredient (Brit. 279877) of—
Detergent compositions.
Cleansers and bleaches for parquet floors.

Soap
Ingredient (Brit. 279877) of—
Washing compositions.

Textile
——, *Dyeing*
Assist in making—
Wool-dyeing liquors (Brit. 279877).
——, *Finishing*
Ingredient of—
Detergent and washing compositions.

2:3-Butylene Glycol
Synonyms: Beta butylene glycol, 2:3-Dihydroxybutane, 2:3-Butanediol, Pseudo butylene glycol, Symmetrical dimethylethylene glycol.

Chemical
Substitute for—
Glycerin, where its modified properties offer an advantage.
Glycols, where its modified properties offer an advantage.

Miscellaneous
Substitute for—
Glycerin, where its modified properties offer an advantage.
Glycols, where its modified properties offer an advantage.

1:3-Butyleneglycol Diformate

Miscellaneous
As an emulsifying agent (Brit. 311795).
For uses, see under general heading: "Emulsifying agents."

1-Butyleneoxy-4-aminoanthraquinone
French: 1-Butylèneoxye-4-aminoanthraquinone.
German: 1-Butylenoxy-4-aminoanthrachinon.

Chemical
Starting point in making various intermediates.

Dye
Starting point (Brit. 285096) in making dyestuffs in the presence of dimethylanilin, nitrobenzene, orthodichlorobenzene, naphthalene, and the like, with the aid of—
Acetylparaphenylenediamine.
5-Amino-2-methylbenzimidazole.
Benzidin and derivatives and homologs.
Dimethylparaphenylenediamine.
Metanaphthylenediamine.
Metaphenylenediamine.
Metatoluylenediamine.
Metaxylidenediamine.
Orthonaphthylenediamine.
Orthophenylenediamine.
Orthotoluylenediamine.
Orthoxylidenediamine.
Paranaphthylenediamine.
Paraphenylenediamine.
Paratoluylenediamine.
Paraxylidenediamine.

Butylenethiourea
Synonyms: Butylenesulphonurea.
French: Sulphourée de butylène, Sulphourée butylènique, Thiourée de butylène, Thiourée butylènique.
German: Butylensulfonharnstoff, Butylenthioharnstoff.

Chemical
Starting point in making—
Intermediates, pharmaceuticals.
Starting point (Brit. 314909) in making derivatives with—
Alkoxyalphanaphthalenesulphonic acid.
Alpha-amino-5-naphthol-7-sulphonic acid.
Alphanaphthylamine-4:8-disulphonic acid.
Alphanaphthylamine-4:6:8-trisulphonic acid.
4-Aminoacenaphthene-3:5-disulphonic acid.
4-Aminoacenaphthene-5-sulphonic acid.
4-Aminoacenaphthene-3-sulphonic acid.
Aminoarylcarboxylic acids.
Aminoheterocyclic chlorides.
1:8-Aminonaphthol-3:6-disulphonic acid.
Bromonitrobenzoyl chlorides.
Chloroalphanaphthalenesulphonic acids.
Chloronitrobenzoyl chlorides.
Iodonitrobenzoyl chlorides.
Nitroanisoyl chlorides.
2-Nitrocinnamyl chloride.
3-Nitrocinnamyl chloride.
4-Nitrocinnamyl chloride.
1-Nitronaphthalene 5-sulphochloride.
2-Nitrophenylacetyl chloride.
4-Nitrophenylacetyl chloride.
Nitrotoluyl chlorides.
Starting point (Brit. 310534) in making rubber vulcanization accelerators with—
Alphanaphthylamine, anilin, benzylamine, cyclohexylanilin, meta-anisidin, metacresidin, metanaphthylenediamine, metaphenylamine, metaphenylenediamine, matatoluidin, metatoluylenediamine, metaxylenediamine, metaxylidin, monoethylanilin, monomethylanilin, orthoanisidin, orthocresidin, orthonaphthylenediamine, orthophenylamine, orthophenylenediamine, orthotoluidin, orthotoluenediamine, orthoxylenediamine, orthoxylidin, para-anisidin, paracresidin, paranaphthylenediamine, paraphenylamine, paraphenylenediamine, paratoluidin, paratoluenediamine, paraxylidin, paraxylenediamine.

Butylether Ethyleneglycol
German: Ethylenglykolbutyl ether.

Paint and Varnish
Ingredient of—
Preparations for removing lacquers and lacquer-enamels. (U. S. 1618482).

Butyl Ether, Tertiary

Petroleum
Blending agent and improver (Brit. 445503) for—
Gasoline motor fuels (the blended fuel can also contain a small amount of tetraethyl lead or tetramethyl lead).

Butylethylbarbituric Acid, Normal
French: Acide normale-butyléthylebarbiturique.
German: Normal-butylaethylbarbiturinsaeure.

Chemical
Starting point in making—
Hypnotic drug with diethylamine (U. S. 1621094).

Butylethyl Carbonate
French: Carbonate de butyle et d'éthyle, Carbonate butylique et éthylique.
German: Butylæthylkarbonat, Kohlenstoffsäurebutylæthylester, Kohlenstoffsäuresbutylæthyl.

Cellulose Products
Solvent for—
Cellulose acetate, cellulose esters and ethers, cellulose nitrate.
For uses, see under general heading: "Solvents."

Chemical
Starting point in making—
Intermediates and other derivatives.

Resins
Solvent for—
Artificial resins, natural resins.

Butyl Formate
French: Formiate butilique, Formiate de butyle.
German: Ameisensäuresbutylester, Formylsäuresbutylester.
Spanish: Formico de butil.
Italian: Formico di butile.

Analysis
Solvent for—
Cellulose derivatives, natural resins, synthetic resins.
Solvent miscible with—
Alcohols, ethers, hydrocarbons, oils.

Cellulose Products
Solvent for—
Cellulose acetate (some types), cellulose ethers, nitrocellulose.

Chemical
Intermediate in—
Organic syntheses.

Solvent for—
Cellulose acetate, cellulose ethers, nitrocellulose.
Solvent miscible with—
Alcohols, ethers, hydrocarbons, oils.

Cosmetic
Solvent in—
Perfume compositions.

Dry-Cleaning
Spotting agent for—
Resins.

Miscellaneous
See also: "Solvents."

Paint and Varnish
Ingredient of—
Dopes, enamels, lacquers, paint removers, paints, varnishes.
Reagent for—
Imparting great strength to films.
Solvent for—
Benzyl abietate, cellulose acetate, cellulose ethers, copals, cumar, dammar, elemi, ester gum, glyptols (with alcohols), mastic, natural resins, nitrocellulose, shellac, synthetic resins.

Pharmaceutical
Solvent miscible with—
Alcohols, oils, ethers, hydrocarbons.

Resins
Solvent for—
Benzyl abietate, copals, cumar, dammar, elemi, ester gum, glyptols (with alcohol), mastic, natural resins, sandarac, shellac, synthetic resins.
Solvent in making—
Artificial resins from or containing cellulose acetate, nitrocellulose, or other cellulose esters or ethers.

Butylfurol
French: Furole de butyle, Furole butylique.

Chemical
General solvent.
Starting point in making—
Intermediates, pharmaceuticals.

Gums
Solvent for—
Pontianak and other varnish gums.

Cellulose Products
Solvent for—
Cellulose acetate, cellulose esters and ethers, cellulose nitrate.
For uses, see under general heading: "Solvents."

Butyl Glycolate
French: Glycolate de butyle, Glycolate butylique.
German: Butylglykolat, Glykolsaeurebutylester, Glykolsaeuresbutyl.

Cellulose Products
Plasticizer (Brit. 311664) for—
Cellulose esters and ethers.
For uses, see under general heading: "Plasticizers."

Chemical
Starting point in making—
Intermediates, pharmaceuticals.

Butylidene Iodide, Normal

Chemical
Starting point in making intermediates.
Starting point (Brit. 353477) in making contrast mediums for x-ray photography with the aid of—
Ammonium sulphite, magnesium sulphite, monomethylamine sulphite, piperidin sulphite, piperazin sulphite, sodium sulphite.

Butyliodocresol

Chemical
Starting point (Brit. 444351) in making—
Fat-splitting catalysts and emulsifying agents, useful in dyeing, laundering, bleaching, and various other purposes, by reacting with formaldehyde and non-aromatic secondary amines (the salts of the products with water-soluble acids, or water-insoluble acids, or the quaternary ammonium salts, are claimed to be valuable for the purposes named).

Butyliodophenol

Chemical
Starting point (Brit. 444351) in making—
Fat-splitting catalysts and emulsifying agents, useful in dyeing, laundering, bleaching, and various other purposes, by reacting with formaldehyde and non-aromatic secondary amines (the salts of the products with water-soluble acids, or water-insoluble acids, or the quaternary ammonium salts, are claimed to be valuable for the purposes named).

Butyliodoresorcinol

Chemical
Starting point (Brit. 444351) in making—
Fat-splitting catalysts and emulsifying agents, for use in dyeing, laundering, bleaching, and various other purposes, by reacting with formaldehyde and non-aromatic secondary amines (the salts of the products with water-soluble acids, or water-insoluble acids, and the quaternary ammonium salts, are claimed to be valuable for the purposes named).

Butyl Lactate
French: Lactate de butanole, Lactate butilique, Lactate de butyle.
German: Milchsäurebutanolester, Milchsäurebutylester, Milchsäuresbutanol, Milchsäurebutyl.

Chemical
As a solvent.
Solvent for—
Nitrocelullose.

Dye
Solvent for various dyes.

Explosives
Solvent for—
Nitrocellulose.

Fats and Oils
Solvent for various oils.

Glass
Additional agent in—
Cellulose acetate solutions used to coat glass to prevent fogging by condensed moisture (used to promote clearness of the film deposited on evaporation).

Gums
Solvent for various gums.

Inks
Ingredient of—
Lithographic inks, printing inks.

Jewelry
Additional agent in—
Cellulose acetate solutions used for heightening the luster of artificial pearls (added to promote clearness of the film deposited on evaporation).

Leather
Additional agent in—
Cellulose acetate solutions used for rendering leather noninflammable and impermeable (added to promote clearness of the film deposited on evaporation).

Miscellaneous
Ingredient of—
Stencil lacquers, stencil enamels.

Paint and Varnish
Additional agent in—
Aeroplane dopes, varnishes and lacquers formulated around cellulose acetate (added to promote clearness of the film deposited on evaporation).
Decreasing viscosity and improving flow and leveling qualities of lacquers, enamels, and varnishes.
Nitrocellulose lacquers to give them the brushing qualities of oleoresinous finishes.
Adhesion promoter in—
Spray lacquers.
Antiskinning agent in—
Oil enamels, oil varnishes.
Glossing agent in—
Spray lacquers.
Ingredient of—
Cellulose acetate lacquers.
Nitrocellulose lacquers.
Nitrocellulose lacquers formulated particularly for interior architectural uses.

Photographic
Additional agent in—
Cellulose acetate solutions used in making noninflammable film (added to promote clearness).

Resins and Waxes
Solvent for—
Synthetic resins.

Butyl Mandelate
French: Mandélate butylique, Mandélate de butyle.
German: Mandelsaeuresbutyl, Mandelsaeuresbutylester.

Paint and Varnish
Plasticizer (Brit. 270650) in making—
Cellulose ester lacquers, cellulose ester varnishes.
See also: "Plasticizers."

Plastics
Plasticizer in making—
Nitrocellulose plastics.

Butylmercaptan, Normal

Insecticide and Fungicide
Fumigant and insecticide for—
Rice weevils (*Sitophilus oryza* L.).
Flour weevils (*Tribolium confusum* Fab.)
Granary weevils (*Sitophilus granarius* L.).
Larvacide for—
Larvae of the Indian-meal moth (*Pledia interpunctella* Hbn.).
Repellent to—
House flies (*Musca domestica* L.).
Green bottle flies (*Lucilia spp.*).
Black blowflies (*Phormia regina* Meig.).
Screw-worm flies (*Cochliomyia macelloria* Fab.).

4-Butylmercaptoalphanaphthol

Chemical
Starting point in making—
Intermediates, pharmaceuticals.

Dye
Starting point (Brit. 291825) in making indigoid dyestuffs with—
Isatin anilide, isatin chloride, reactive alpha derivatives of isatin.

Butylnaphthalenesulphonic Acid
French: Acide de butylnaphthalènesulphonique.
German: Butylnaphtalinsulfonsaeure.

Fats and Oils
Emulsifying agent in making—
Compositions containing various fats and oils (Brit. 266746).

Leather
Emulsifying agent in preparing—
Finishing agents.

Miscellaneous
See also under: "Emulsifying Agents."

Resins and Waxes
Emulsifying agent in making—
Compositions containing various resins and waxes.

Textile
——, *Dyeing*
Ingredient of—
Dyeing liquors, as a wetting out agent.
——, *Finishing*
Ingredient of—
Compositions used in cleansing and washing textile materials.
——, *Preliminary Treatment*
Ingredient of—
Compositions used in degreasing wool (Brit. 253105).

Butylnaphthenate, Normal
French: N-naphthénate de butanole, N-naphthénate butilique, N-naphthénate de butyle, N-naphthénate butylé.
German: N-butylnaphtenat, N-naphtensäurebutylester, N-naphtensäuresbutyl.
Spanish: N-naftenato de butile.
Italian: N-naftenato di butile.

Miscellaneous
As an emulsifying agent (Brit. 359116).
For uses, see under general heading: "Emulsifying agents."

Butyl Nitrate

Chemical
Reagent in—
Organic syntheses.

Fuel
Primer (Brit. 404682) in—
Diesel engine fuels (used in conjunction with other primers consisting of organic bromides or organic copper compounds whose function is that of reducing the spontaneous ignition temperature).
Reducer (Brit. 404682) of—
Delay period in diesel engine fuels.

2-Butyloctyl Acetate

Gums and Resins
Plasticizer (Brit. 442643) for—
Natural and artificial gums and resins.

Butylolamine

Chemical
Starting point in making—
Pharmaceuticals and other derivatives.

Fats and Oils
Dispersive agent (Brit. 340294) in making—
Non-freezing lubricating compositions, containing animal or vegetable oils and fats, as well as ethyleneglycol or its esters, borax, and benzyl alcohol.
Special lubricating compositions of the above type, for use on locomotive axles, railway switches, hydraulic presses, and hydraulic brakes.
Special lubricating compositions for use in electric switches.

Miscellaneous
Ingredient (Brit. 340294) of—
Compositions, containing vegetable, animal, or mineral oils and greases, used as rust preventives.

Petroleum
Ingredient (Brit. 340294) of—
Special lubricating compositions containing mineral oils and greases.

Butyl Oleate

French: Oléate de butyle, Oléate butylique.
German: Butyloleat, Oleinsäurebutylester, Oleinsäuresbutyl.

Cellulose Products
Plasticizer for—
Cellulose acetate, cellulose esters and ethers, celullose nitrate.
For uses, see under general heading: "Plasticizers."

Gums, Resins and Waxes
Solvent for—
Copal esters, cumarone, ester gums, various resins, waxes.

Butyl Orthosilicate

Glue and Adhesives
Ingredient (Brit. 428548) of—
Cellulose acetate or nitrocellulose base adhesives for safety glass.

Butyloxyacetanilide

Cellulose Products
Plasticizer (Brit. 312606) for—
Cellulose acetate, cellulose esters or ethers, cellulose nitrate (nitrocellulose).
For uses, see under general heading: "Plasticizers."

Butyl-4-oxy-2-quinolin, Normal

Dye
Starting point (Brit. 431649) in making—
Dyestuffs with aniline or 2:5-dichloranilin, 4-nitroanilin, paratoluene sulphonic ester of 2-aminophenol, halogen anilins, toluidins, xylidins, and the like, for coloring organic solvents, lacquers, fats, oils, resins, and waxes, in clear yellow, greenish-yellow, or reddish shades, fast to sublimation and other influences.

Butyl Paraoxybenzoate

Synonyms: Butyl parahydroxybenzoate.
French: Parahydroxyebenzoate de butyle, Parahydroxyebenzoate butylique, Paraoxyebenzoate de butyle, Paraoxyebenzoate butylique.
German: Butylparaoxybenzoat, Butylparahydroxybenzoat, Parahydroxybenzoesäurebutylester, Parahydroxybenzoesäuresbutyl, Paraoxybenzoesäurebutylester, Paraoxybenzoesäuresbutyl.

Chemical
Starting point in making various derivatives.

Food
As a preservative.

Miscellaneous
As a general preservative and disinfectant.

Pharmaceutical
In compounding and dispensing practice.

Sanitation
As a disinfectant.

Butylpentaerythritol

Cellulose Products
Solvent, softener, and plasticizer (Brit. 358393) for—
Cellulose acetate, cellulose esters or ethers, nitrocellulose.
For uses, see under general heading: "Plasticizers."

Butylphenetidin, Normal

Chemical
Antioxidant and stabilizer (Brit. 430335) for—
Unstable organic substances.

Fats, Oils, and Waxes
Antioxidant and stabilizer (Brit. 430335) for—
Fats, oils, waxes.

Petroleum
Antioxidant and stabilizer (Brit. 430335) for—
Petroleum products.
Inhibitor (Brit. 430335) of—
Gumming in petroleum products.

Rubber
As an antioxidant (Brit. 430335).

Butyl Propionate

French: Propionate de butyle, Propionate butylique.
German: Butylpropionat, Propionsäurebutylester, Propionsäuresbutyl.

Cellulose Products
Solvent for—
Cellulose esters and ethers, cellulose nitrate.
For uses, see under general heading: "Solvents."

Paint and Varnish
Ingredient of—
Brushing lacquers.
Retarder in making—
Lacquers.
Solvent in making—
Lacquers, paints, varnishes, dopes, and enamels containing nitrocellulose or other esters or ethers of cellulose.

Butyl Resinate, Normal

French: Abiétate de N-butyle, Abiétate N-butylique, Résinate de N-butyle, Résinate de N-butylique.
German: Abietinsaeure-N-butylester, Abietinsaeures-N-butyl, N-Butylabietat, N-Butylresinat.

Paint and Varnish
Plasitcizer in making—
Paints and varnishes, lacquers and dopes, containing nitrocellulose, cellulose acetate, or other cellulose esters or ethers.

Plastics
Plasticizer in making—
Compositions containing nitrocellulose, cellulose acetate or other cellulose esters or ethers.

Butylresorcinol, Normal

French: N-butylerésorcinol.
German: N-butylresorcin.

Miscellaneous
Ingredient of—
Antiseptic solutions in alcohol, benzol, and vegetable oils (U. S. 1649672).

Pharmaceutical
In compounding and dispensing practice.

Textile
Inhibitor (Brit. 446404) of—
Acidity and mould development in textile lubricants during storage of the lubricant or fabric.

Butyl Stearate

French: Stéarate de butyle, Stéarate butilique.
German: Butylstearat, Stearninsäurebutylester, Stearinsäuresbutyl.

Cellulose Products
Plasticizer and solvent for—
Cellulose esters and ethers.
For uses, see under general heading: "Plasticizers."

Butyl Stearate (Continued)
Resins and Waxes
Solvent for—
Copal ester, cumarone, ester gum.
Rubber
As a solvent.
Soap
Solvent in—
The manufacture of various soaps.
Detergent preparations and dry-cleaning soaps.

Butylsulphuric Acid Chloride
French: Chlorure d'acide butylesulphurique.
German: Butylschwefelsaeureschlorid.
Dye
Starting point (Brit. 271533) in making soluble vat dyestuffs with—
Anthraquinone-1:2, flavanthrone, indanthrone, naphthacridone, thioindigo.

Butyltetrahydronaphthalenecarboxylic Acid
French: Acide de butyletétrahydronaphthalènecarboxyle, Acide butyletétrahydronaphthalènecarboxylique.
German: Butyltetrahydronaphtalincarbonsaeure.
Chemical
Ingredient in making—
Emulsifying and dissolving mediums (German 432942).

Butyl Thiocyanpropionate
Chemical
Starting point in making various derivatives.
Disinfectant
Ingredient (Brit. 361900) of—
Disinfectants and germicides (used in solution in water or in an organic solvent, such as kerosene).
Insecticide
Ingredient (Brit. 361900) of—
Insecticidal compositions (used in solution in water or in an organic solvent, such as kerosene).

Butyl Thiosalicylate, Normal
French: Thiosalicylate de butyle, normale; Thiosalicylate butylique, normale.
German: N-Butylthiosalicylat, Thiosalicylsaeurenormalbutylester, Thiosalicylsaeuresnormalbutyl.
Chemical
Starting point (Brit. 282427) in making synthetic drugs with oxides and other salts of—
Antimony, arsenic, bismuth, gold, silver.

Butyl Undecylenate
French: Undecylenate de butyle, Undecylenate butylique.
German: Butylundecylenat, Undecylensäurebutylester, Undecylensäuresbutyl.
Spanish: Undecilenato de butil.
Italian: Undecylenato di butile.
Chemical
Starting point (French 615959) in making—
Aluminum, zinc, manganese, and bismuth undecylenates.
Leather
Reagent (French 615959) for—
Weighting and polishing leather.

Butyl Xanthate
French: Butyle xanthate, Xanthate butilique.
German: Butylxanthogenat, Xanthogensäuresbutyl, Xanthogensäurebutylester.
Metallurgical
Flotation agent in—
Ore concentration processes.

Butylxylamine
Dye
Coupling agent (Brit. 429618) in making—
Dyestuffs with diazotized arylamines (color being developed on the fiber by acid treatment).

Butyn
Chemical
Starting point (Brit. 260346) in making—
Butyn hydrobromide, butyn hydrochloride, butyn pentaborate, butyn salicylate, butyn sulphate.
Pharmaceutical
In compounding and dispensing practice.

Butyraldehydecyanohydrin
German: Butyraldehydcyanhydrin.
Chemical
Starting point in making—
Ethyl ester of alphahydroxy-n-valeric acid (Brit. 264143).

Butyraldoxime
Fuel
Primer (Brit. 429763) for—
Diesel engine fuel oils produced by the hydrogenation of coal.
Petroleum
Primer (Brit. 429763) for—
Diesel oils containing a high proportion of aromatic bodies.

Butyric Acid Ester of Grapeseed Alcohol
(Uses same as those given for Butyric Acid Ester of Ricinoleic Alcohol.)

Butyric Acid Ester of Ricinoleic Alcohol
Bituminous
Solvent (Brit. 445223) for—
Asphalt and other bituminous bodies.
Dye
Solvent (Brit. 445223) for—
Dyestuffs, particularly oil-soluble coaltar dyes.
Fats, Oils, and Waxes
Solvent (Brit. 445223) for—
Fats, oils, waxes.
Resins
Solvent (Brit. 445223) for—
Oil-soluble glycerol-phthalic acid resins.
Polymerized vinyl compounds.
Synthetic resins.
Rubber
Solvent (Brit. 445223) for—
Rubber.

Butyric Acid, Normal
Synonyms: Propylformic acid.
French: Acide(normal)de butyrique.
German: Normalbuttersaeure.
Chemical
Catalyst in making—
Vulcanization accelerator of the aldehyde-amine condensation type (Brit. 265931).
Starting point in making—
Alkyl and aryl esters, ammonium butyrate, amyl butyrate, butyric anhydride, caproic acid.
Salts of alkalies, alkaline metals, and metals.
Synthetic perfume material.
Food
Ingredient of—
Butter substitutes, confectionery, flavoring compositions, fruit essences.
Glues and Adhesives
Ingredient of—
Starch glues and pastes.
Leather
Reagent in tanning.
Tanning, dehairing.
Paint and Varnish
Ingredient of—
Lacquers, varnishes.
Plastics
Reagent in making—
Cellulose butyrate plastics.
Pharmaceutical
In compounding and dispensing practice.
Sanitation
Disinfectant in treating—
Water.

Butyrylanisidin
Cellulose Products
Plasticizer (Brit. 312606) for—
Cellulose esters or ethers.
For uses, see under general heading: "Plasticizers."

Butyryl Carbamide
French: Carbamide de butyryle, Carbamide butyrylique.
German: Butyrylcarbamid.
Chemical
Starting point in making—
Intermediates, pharmaceuticals.

Butyryl Carbamide (Continued)

Dye
Starting point in making various synthetic dyestuffs.

Resins and Waxes
Starting point (Brit. 292912) in making synthetic resins with—
Acetylsalicylic acid, aliphatic dibasic acids, ammonium salicylate, anthranilic acid, benzoic acid, gallic acid, hydroxynaphthoic acid, magnesium salicylate, oxalic acid, phenolic acids, phthalic acid, salicylamide, salicylic acid, strontium salicylate, succinic acid.

Butyrylhydroquinone

Petroleum
Stabilizing agent (Brit. 406195) for—
Cracked gasolines and other motor fuels.

Butyrylphenetidin

Cellulose Products
Plasticizer for—
Cellulose esters or ethers.
For uses, see under general heading: "Plasticizers."

Butyrylphloroglucinol

Petroleum
Stabilizing agent (Brit. 406195) for—
Cracked gasolines and other motor fuels.

Butyrylpyrocatechol

Petroleum
Stabilizing agent (Brit. 406195) for—
Cracked gasolines and other motor fuels.

Butyrylpyrogallol

Petroleum
Stabilizing agent (Brit. 406195) for—
Cracked gasolines and other motor fuels.

Butyrylresorcinol

Petroleum
Stabilizing agent (Brit. 406195) for—
Cracked gasolines and other motor fuels.

Cacao Shell

Dairying
Ingredient of—
Cattle feeds (said to increase the vitamin D content of butter and milk from the Winter to the Summer level).

Food
Starting material in making—
Tea-like beverage.

Cade Oil

Synonyms: Kade oil, Oil of juniper tar.
French: Huile de cade.

Perfumery
Ingredient of creams and pastes.

Pharmaceutical
In compounding and dispensing practice.

Soap
Raw material in making special and medicated soaps.

Cadmium Acetate

French: Acétate de cadmium, Acétate cadmique.
German: Cadmiumacetat, Cadmiumazetat, Essigsaeurescadmium.

Analysis
Reagent for various purposes.

Ceramics
Ingredient of—
Coatings to produce iridescent, vitrifiable colored effects.

Chemical
Ingredient of catalytic preparations used in making—
Acenaphthylene, acenaphthaquinone, naphthaldehydic acid, naphthalic anhydride, and hemimellitic acid from acenaphthene (Brit. 295270).
Acetaldehyde from ethyl alcohol (Brit. 281307).
Acetic acid from ethyl alcohol (Brit. 281307).
Alcohols from aliphatic hydrocarbons (Brit. 281307).
Aldehydes or alcohols by the reduction of esters (Brit. 306471).
Alphacampholide by the reduction of camphoric acid (Brit. 306471).
Aldehydes and acids from toluene, orthochlorotoluene, orthonitrotoluene, orthobromotoluene, parachlorotoluene, parabromotoluene, paranitrotoluene, metachlorotoluene, metabromotoluene, metanitrotoluene, dinitrotoluenes, dibromotoluenes, dichlorotoluenes, chlorobromotoluenes, chloronitrotoluenes, bromonitrotoluenes (Brit. 295270).
Aldehydes and acids from xylenes, pseudocumenes, mesitylene, and paracymene (Brit. 295270).
Alphanaphthaquinone from naphthalene (Brit. 281307).
Anthraquinone from naphthalene (Brit. 295270).
Benzaldehyde and benzoic acid from toluene (Brit. 281307).
Benzoquinone from phenanthraquinone (Brit. 281307).
Benzyl alcohol or benzaldehyde or phthalide by the reduction of phthalic anhydride (Brit. 306471).
Benzyl alcohol by the reduction of benzaldehyde (Brit. 306471).
Butyl alcohol by the reduction of crotonaldehyde (Brit. 306471).
Chloroacetic acid from ethylenechlorohydrin (Brit. 295270).
Diphenic acid from ethyl alcohol (Brit. 281307).
Ethyl alcohol by the reduction of acetaldehyde (Brit. 306471).
Fluorenone from fluorene (Brit. 295270).
Formaldehyde from methanol or methane (Brit. 295270).
Formaldehyde by the reduction of carbon dioxide or carbon monoxide (Brit. 306471).
Hydroxyl reduction compounds of anthraquinone, benzoquinone, and the like (Brit. 306471).
Isopropyl alcohol by the reduction of acetone (Brit. 306471).
Maleic acid and fumaric acid by the oxidation of toluene, benzene, phenols, tar phenols, or furfural, or from benzoquinone or phthalic anhydride (Brit. 295270).
Methane by the reduction of carbon dioxide or carbon monoxide (Brit. 306471).
Methanol by the reduction of carbon dioxide or carbon monoxide (Brit. 306471).
Naphthaldehydic acid, acenaphthaquinone, or bisacenaphthylidenedione from acenaphthylene (Brit. 281307).
Phenanthraquinone from phenanthrene or diphenic acid (Brit. 295270).
Phthalic acid and maleic acid from naphthalene (Brit. 295270).
Primary alcohols by the reduction of aldehydes (Brit. 306471).
Propionic acid and butyric acid and higher alcohols, ketones, and acids by the reduction of carbon dioxide or carbon monoxide (Brit. 306471).
Reduction products of carbon dioxide and carbon monoxide (Brit. 306471).
Reduction products of ketones, aldehydes, acids, esters, alcohols, ethers, and other organic compounds containing oxygen (Brit. 306471).
Salicylic acid and salicylic aldehyde from cresol (Brit. 295270).
Secondary butyl alcohol by the reduction of methylethyl ketone (Brit. 306471).
Valeryl alcohol by the reduction of valeraldehyde (Brit. 306471).
Vanillin and vanillic acid from eugenol and isoeugenol (Brit. 295270).
Reagent for various purposes.

Cadmium Antimonide

French: Antimoinure cadmique, Antimoinure de cadmium.
German: Antimoncadmium, Cadmiumantimonid.

Chemical
Catalyst (Brit. 263877) in making—
Acetone from isopropyl alcohol.
Dehydrogenated products from cyclohexane.
Isobutyraldehyde from isobutyl alcohol.
Isobutyronitrile from isobutylamine.
Naphthalene from tetrahydronaphthalene.
Paracymene from turpentine.
Catalyst (Brit. 262120) in making—
Isovaleraldehyde from isoamyl alcohol.
General chemical reagent.

Cadmium Arsenide

Chemical
Catalyst (Brit. 263877) in making—
Acetone from isopropyl alcohol.
Dehydrogenated products from cyclohexane.
Isobutyraldehyde from isobutyl alcohol.
Isobutyronitrile from isobutylamine.
Naphthalene from tetrahydronaphthalene.
Paracymene from turpentine.
Catalyst (Brit. 262120) in making—
Isovaleraldehyde from isoamyl alcohol.
General chemical reagent.

Cadmium Bismuthide
French: Bismuthide de cadmium.
German: Cadmiumbismuthid.

Chemical
Catalyst in making—
Acetone from isopropyl alcohol.
Isobutyraldehyde from isobutyl alcohol.
Isobutyronitrile from isobutylamine.
Naphthalene from tetrahydronaphthalene.
Paracymene from turpentine oil.

Cadmium Borate
French: Borate de cadmium, Borate cadmique.
German: Borsaeurescadmium, Cadmiumborat.

Chemical
Reagent for various purposes.
Ingredient of catalytic preparations used in making—
Acenaphthylene, acenapthaquinone, bisacenaphthylidenedione, naphthaldehydic acid, naphthalic anhydride, and hemimellitic acid from acenaphthene (Brit. 295270).
Acetaldehyde from ethyl alcohol (Brit. 281307).
Acetic acid from ethyl alcohol (Brit. 281307).
Alcohols from aliphatic hydrocarbons (Brit. 281307).
Aldehydes or alcohols by the reduction of esetrs (Brit. 306471).
Alphacampholid by reduction of camphoric acid (Brit. 306471).
Aldehydes and acids from toluene, metachlorotoluene, metanitrotoluene, metabromotoluene, dichlorotoluenes, dibromotoluenes, dinitrotoluenes, chlorobromotoluenes, chloronitrotoluenes, bromonitrotoluenes, parachlorotoluene, paranitrotoluene, parabromotoluene, orthochlorotoluene, orthobromotoluene, orthonitrotoluene (Brit. 295270).
Aldehydes and acids from xylenes, pseudocumene, mesitylene and paracymene (Brit. 295270).
Alphanaphthaquinone from naphthalene (Brit. 281307).
Anthraquinone from naphthalene (Brit. 295270).
Benzaldehyde and benzoic acid from toluene (Brit. 281307).
Benzoquinone from phenanthraquinone (Brit. 281307).
Benzyl alcohol by reduction of benzaldehyde (Brit. 306471).
Benzyl alcohol or benzyl aldehyde or phthalide by the reduction of phthalic anhydride (Brit. 306471).
Butyl alcohol by the reduction of crotonaldehyde (Brit. 306471).
Chloroacetic acid from ethylenechlorohydrin (Brit. 295270).
Diphenic acid from ethyl alcohol (Brit. 281307).
Ethyl alcohol by the reduction of acetaldehyde (Brit. 306471).
Fluorenone from fluorene (Brit. 295270).
Formaldehyde from methane or methanol (Brit. 295270).
Formaldehyde by the reduction of carbon dioxide or carbon monoxide (Brit. 306471).
Isopropyl alcohol by the reduction of acetone (Brit. 306471).
Maleic acid and fumaric acid by the oxidation of toluene, benzene, phenol, tar phenols, or furfural, or from benzoquinone or phthalic anhydride (Brit. 295270).
Methane by the reduction of carbon dioxide or carbon monoxide (Brit. 306471).
Methanol by the reduction of carbon dioxide or carbon monoxide (Brit. 306471).
Naphthaldehydic acid, acenaphthaquinone, or bisacenaphthylidenedione from acenaphthylene (Brit. 281307).
Phenanthraquinone from phenanthrene or diphenic acid (Brit. 295270).
Phthalic acid and maleic acid from naphthalene (Brit. 295270).
Primary alcohols by the reduction of aldehydes (Brit. 306471).
Propionic acid and butyric acid and higher alcohols, ketones, and acids by the reduction of carbon dioxide or carbon monoxide (Brit. 306471).
Reduction of anthraquinone, benzoquinone, and the like to the corresponding hydroxyl compounds (Brit. 306471).
Reduction of ketones, aldehydes, acids, esters, alcohols, ethers, and other organic compounds containing oxygen (Brit. 306471).
Salicylic acid and salicylic aldehyde from cresol (Brit. 295270).
Secondary butyl alcohol by the reduction of methylethylketone (Brit. 306471).
Valeryl alcohol by the reduction of valeraldehyde (Brit. 306471).
Vanillin and vanillic acid from eugenol or isoeugenol (Brit. 295270).

Cadmium Bromide
French: Bromure de cadmium.
German: Cadmiumbromid, Kadmiumbromid.
Spanish: Bromuro de cadmio.
Italian: Bromuro di cadmio.
Analysis
As a reagent.
Chemical
Catalyst (Brit. 398527) in making—
Esters from lower aliphatic acids and olefins.
Starting point in making—
Cadmium sulphide.
Metallurgical
Ingredient of—
Galvanic plating baths.
Photographic
As a metallic bromide.
As a toning agent.
Printing
Reagent in—
Lithography, process engraving.
Textile
Reagent in—
Dyeing processes, printing processes.

Cadmium Butylxanthogenate
Synonyms: Cadmium butylxanthate.
French: Butylexanthogénate de cadmium, Xanthate butylique de cadmium.
German: Butylxanthogensäurescadmium, Butylxanthogensäureskadmium.
Spanish: Butilxantogenato de cadmio.
Italian: Butilxantogenato di cadmio.
Rubber
Accelerator (French 563397) in—
Vulcanizing processes.

Cadmium Cyanide
Agriculture
Disinfectant (U. S. 1998092) for—
Seeds.
Metallurgical
Ingredient of—
Cadmium-plating baths.

Cadmium Di-isopropyldithiophosphate
Agriculture
Disinfectant (U. S. 1998092) for—
Seeds.

Cadmium Dinaphthylnaphthenate
Lubricant
Addition agent (Brit. 433257) in—
Lubricating oils or greases, especially for use at high temperatures, such as cylinder oils, hydrogenated oils, or oils refined by treatment with sulphuric acid, clay, or extraction solvents.

Cadmium Dipentamethylenethiuramdisulphide
Rubber
Secondary activator in—
Vulcanizing processes (for use with mercaptabenzthiazole).

Cadmium Dipentamethylenethiurammonosulphide
Rubber
Secondary activator in—
Vulcanizing processes (for use with mercaptabenzthiazole).

Cadmium Dipentamethylenethiuramtetrasulphide
Rubber
Secondary activator in—
Vulcanizing processes (for use with mercaptabenzthiazole).

Cadmium Iodide
French: Iodure de cadmium.
German: Cadmiumjodid, Jodcadmium, Jodkadmium, Jodwasserstoffsaeurescadmium, Kadmiumjodid.
Pharmaceutical
In compounding and dispensing practice.
Photographic
Reagent in—
Collodion process photography.
Printing
Reagent in—
Process engraving and the litho trades.

Cadmium Methylethyldithiocarbamate

Disinfectant
As a bactericide (Australian 8103/32, Brit. 406979, U. S. 1972961).

Insecticide and Fungicide
As a fungicide (claimed effective against *Aspergillus niger* and *Fomes Annonsus*) (Australian 8103/32, Brit. 406979, U. S. 1972961).
As an insecticide (Australian 8103/32, Brit. 406979, U. S. 1972961).

Cadmium Molybdate
French: Molybdate cadmique, Molybdate de cadmium.
German: Cadmiummolybdat, Kadmiummolybdat, Molybdaensaeurescadmium, Molybdaensaeureskadmium.

Chemical
Reagent for various purposes.
Ingredient of catalytic preparations used in making—
Acenaphthylene, acenaphthaquinone, bisacenaphthylidenedione, naphthaldehydic acid, naphthalic anhydride and hemimellitic acid from acenaphthene (Brit. 295270).
Acetaldehyde from ethyl alcohol (Brit. 281307).
Acetic acid from ethyl alcohol (Brit. 281307).
Alcohols from aliphatic hydrocarbons (Brit. 281307).
Aldehydes or alcohols by the reduction of esters (Brit. 306471).
Alphacampholid by the reduction of camphoric acid (Brit. 306471).
Aldehydes and acids from toluene, metachlorotoluene, metanitrotoluene, metabromotoluene, parachlorotoluene, parabromotoluene, paranitrotoluene, orthochlorotoluene, orthobromotoluene, orthonitrotoluene, dichlorotoluenes, dibromotoluenes, dinitrotoluenes, chlorobromotoluene, chloronitrotoluene, bromonitrotoluene (Brit. 295270).
Aldehydes and acids from xylenes, pseudocumene, mesitylene, and paracymene (Brit. 295270).
Alphanaphthaquinone from naphthalene (Brit. 281307).
Anthraquinone from naphthalene (Brit. 295270).
Benzaldehyde and benzoic acid from toluene (Brit. 281307).
Benzoquinone from phenanthraquinone (Brit. 281307).
Benzyl alcohol by the reduction of benzaldehyde (Brit. 306471).
Benzyl alcohol or benzaldehyde or phthalide by the reduction of phthalic anhydride (Brit. 306471).
Butyl alcohol by the reduction of crotonaldehyde (Brit. 306471).
Chloroacetic acid from ethylenechlorohydrin (Brit. 295270).
Diphenic acid from ethyl alcohol (Brit. 281307).
Ethyl alcohol by the reduction of acetaldehyde (Brit. 306471).
Fluorenone from fluorene (Brit. 295270).
Formaldehyde by the reduction of methane or methanol (Brit. 306471).
Formaldehyde by the reduction of carbon dioxide or carbon monoxide (Brit. 306471).
Hydroxyl reduction compounds of anthraquinone, benzoquinone, and the like (Brit. 306471).
Isopropyl alcohol by the reduction of acetone (Brit. 306471).
Maleic acid and fumaric acid by the oxidation of toluene, benzene, phenols, tar phenols, or furfural, or from benzoquinone or phthalic anhydride (Brit. 295270).
Methane by the reduction of carbon dioxide or carbon monoxide (Brit. 306471).
Methanol by the reduction of carbon dioxide or carbon monoxide (Brit. 306471).
Naphthaldehydic acid, acenaphthaquinone, or bisacenaphthylidenedione from acenaphthylene (Brit. 281-307).
Phenanthraquinone from phenanthrene or diphenic acid (Brit. 295270).
Phthalic acid and maleic acid from naphthalene (Brit. 295270).
Primary alcohols by the reduction of aldehydes (Brit. 306471).
Propionic acid and butyric acid and higher alcohols, ketones, and acids by the reduction of carbon monoxide or carbon dioxide (Brit. 306471).
Reduction products of ketones, aldehydes, acids, esters, alcohols, ethers, and other organic compounds containing oxygen (Brit. 306471).
Salicylic acid and salicylic aldehyde from cresol (Brit. 295270).
Secondary butyl alcohol by the reduction of methylethyl ketone (Brit. 306471).
Valeryl alcohol by the reduction of valeraldehyde (Brit. 306471).
Vanillin and vanillic acid from eugenol or isoeugenol (Brit. 295270).

Cadmium Oleate
French: Oléate cadmique, Oléate de cadmium.
German: Kadmiumoleat, Oleinsäureskadmium.

Ceramics
Impregnating agent in treating—
Porous ceramic ware.

Chemical
Starting point in making various derivatives.

Construction
Reagent in waterproofing—
Brick and cement, plaster and stucco.

Leather
Reagent in waterproofing—
Leather and artificial leather.

Miscellaneous
Reagent in waterproofing—
Felt.

Paper
Impregnating agent in waterproofing—
Paper, pulp, and their products.

Plastics
Ingredient of—
Plastic compositions (added for the purpose of securing a waterproofing effect).

Stone
Reagent in waterproofing—
Natural and artificial stone.

Textile
Reagent in waterproofing—
Various fabrics.

Woodworking
Reagent in waterproofing—
Wood and wood products.

Cadmium Palmitate
French: Palmitate cadmique, Palmitate de cadmium.
German: Kadmiumpalmitat, Palmitinsäures kadmium.

Ceramics
Impregnating agent in treating—
Porous ceramic ware.

Chemical
Starting point in making various derivatives.

Construction
Reagent in waterproofing—
Brick, cement, plaster, stucco.

Leather
Reagent in waterproofing—
Leather and artificial leather.

Miscellaneous
Reagent in waterproofing—
Felt.

Paper
Impregnating agent in waterproofing—
Paper, pulp, and their products.

Plastics
Ingredient of—
Plastic compositions (added for the purpose of securing a waterproofing effect).

Stone
Reagent in waterproofing—
Natural and artificial stone.

Textile
Reagent in waterproofing—
Various fabrics.

Woodworking
Reagent in waterproofing—
Wood and wood products.

Cadmium Pentamethylenedithiocarbamate

Rubber
Secondary activator in—
Vulcanizing processes (for use with mercaptabenzthiazole).

Cadmium-Phenyl Acetate

Petroleum

Addition agent (Brit. 433257) in—

Lubricating oils or greases, especially for use at high temperatures, such as cylinder oils, hydrogenated oils, or oils refined by treatment with sulphuric acid, clay, or with extraction solvents.

Cadmium Phosphide

French: Phosphure cadmique, Phosphure de cadmium.
German: Cadmiumphosphid, Phosphorcadmium.

Chemical

Catalyst (Brit. 263877) in making—

Acetone from isopropyl alcohol.
Dehydrogenated products from cyclohexane.
Isobutyraldehyde from isobutyl alcohol.
Isobutyronitrile from isobutylamine.
Naphthalene from tetrahydronaphthalene.
Paracymene from turpentine.

Catalyst (Brit. 262120) in making—

Isovaleraldehyde from isoamyl alcohol.

General chemical reagent.

Cadmium Resinate

French: Résinate de cadmium.
German: Cadmiumresinat.

Ceramics

Pigment in admixture with lavender oil for obtaining reddish-yellow colors on—

Chinaware, porcelains, potteries.

Cadmium Selenide

French: Sélénide de cadmium.
German: Cadmiumselenid, Selencadmium.

Chemical

Catalyst in making—

Acetone from isopropyl alcohol.
Isobutyraldehyde from isobutyl alcohol.
Isobutyronitrile from isobutylamine.
Naphthalene from tetrahydronaphthalene.
Paracymene from turpentine oil.

Cadmium Stearate

French: Stéarate cadmique, Stéarate de cadmium.
German: Kadmiumstearat, Stearinsäureskadmium.

Ceramics

Impregnating agent in treating—

Porous ceramic ware.

Chemical

Starting point in making various derivatives.

Construction

Reagent in waterproofing—

Brick, cement, plaster, stucco.

Leather

Reagent in waterproofing—

Leather and artificial leather.

Miscellaneous

Reagent in waterproofing—

Felt.

Paper

Impregnating agent in waterproofing—

Paper, pulp, and their products.

Plastics

Ingredient of—

Plastic compositions (added for the purpose of securing a waterproofing effect).

Stone

Reagent in waterproofing—

Natural and artificial stone.

Textile

Reagent in waterproofing—

Various fabrics and yarns.

Wood

Reagent in waterproofing—

Wood and wood products.

Cadmium Sulphide

Synonyms: Cadmium yellow, Sulphide of cadmium.
French: Sulphure de cadmium.
German: Cadmiumsulfid.

Ceramics

Pigment in making—

Glazes for potteries, porcelains and other wares.

Explosives

Ingredient of—

Fireworks and pyrotechnic preparations, added for the purpose of producing blue flames.

Glass

Ingredient of—

Yellow glazes used in making fine glassware.

Pigment in producing—

Deep-yellow tones in glassware.

Ink

Ingredient of—

Lithographic and engraving inks.

Paints and Varnishes

Ingredient of—

Ultramarine green, white lead mixtures.

Pigment in making—

Luminous paints, oil paints, water paints.

Paper and Pulp

Ingredient of—

Pigmenting mixtures containing barium sulphate.

Perfumery

Ingredient of—

Depilatory preparations.

Photographic

Pigment in producing—

Yellow image in color photography.

Rubber

Pigment in making—

Yellow-colored rubber goods.

Soap

Color for—

Toilet soaps.

Textile

——, *Dyeing*

Pigment in compositions for coloring yarns and fabrics.

Cadmium Sulphoselenide

Glass

Addition agent (U. S. 1983151) for—

Cadmium yellow preparations for glass batches (said to eliminate use of additional reducing agents and to develop the color at once on melting; by the addition of mixtures of cadmium and cadmium selenide a color range is available from yellow, through orange and pink to red in various depths, up to approximately black).

Cadmium Tantalate

French: Tantalate cadmique, Tantalate de cadmium.
German: Cadmiumtantalat, Tantalsaeurescadmium.

Chemical

Ingredient of catalytic preparations used in making—

Acenaphthylene, acenaphthaquinone, bisacenaphthylidenedione, naphthaldehydic acid, naphthalic anhydride, and hemimellitic acid from acenaphthene (Brit. 295270).
Acetaldehyde from ethyl alcohol (Brit. 295270).
Acetic acid from ethyl alcohol (Brit. 295270).
Alcohols from aliphatic hydrocarbons (Brit. 281307).
Aldehydes and acids from toluene, orthochlorotoluene, metachlorotoluene, parachlorotoluene, orthobromotoluene, metabromotoluene, parabromotoluene, orthonitrotoluene, metanitrotoluene, paranitrotoluene, dichlorotoluenes, dibromotoluenes, dinitrotoluenes, chlorobromotoluenes, chloronitrotoluenes, bromonitrotoluenes (Brit. 295270).
Aldehydes and acids from xylenes, pseudocumene, mesitylene, and paracymene (Brit. 295270).
Alphanaphthaquinone from naphthalene (Brit. 281307).
Anthraquinone from anthracene (Brit. 295270).
Benzaldehyde and benzoic acid from toluene (Brit. 281307).
Benzoquinone from phenanthraquinone (Brit. 281307).
Chloroacetic acid from ethylenechlorohydrin (Brit. 295270).
Diphenic acid from ethyl alcohol (Brit. 281307).
Fluorenone from fluorene (Brit. 295270).
Formaldehyde from methanol or methane (Brit. 295270).
Maleic acid and fumaric acid by the oxidation of benzene, toluene, phenol, tar phenols, or furfural, or from benzoquinone or phthalic anhydride (Brit. 295270).
Naphthaldehydic acid, acenaphthaquinone, or bisacenaphthylidenedione from acenaphthylene (Brit. 281-307).
Phenanthraquinone from phenanthrene or diphenic acid (Brit. 295270).
Phthalic acid and maleic acid from naphthalene (Brit. 295270).
Salicylic acid and salicylic aldehyde from cresol (Brit. 295270).
Vanillin and vanillic acid from eugenol or isoeugenol (Brit. 295270).

Cadmium Tantalate (Continued)
Ingredient (Brit. 304640) of catalytic preparations used in reduction reactions in making—
Alphanaphthylamine from alphanitronaphthalene.
Amines from aliphatic nitro bodies, such as allyl nitrites, or nitromethane.
Amines from oximes, Schiff's bases, and nitrites.
Aminoanisole from nitroanisole.
Aminophenols from nitrophenols.
Amylamine from pyridin.
Anilin from nitrobenzene.
Azobenzene from nitrobenzene.
Azoxybenzene from nitrobenzene.
Cyclohexamine, dicyclohexamine, and cyclohexylanilin from nitrobenzene.
Hydrazobenzene from nitrobenzene.
Piperidin from pyridin.
Pyrrolidin from pyrrol.
Tetrahydroquinolin from quinolin.

Cadmium Telluride
French: Tellurure cadmique, Tellurure de cadmium.
German: Cadmiumtellurid, Tellurcadmium.
Chemical
Catalyst (Brit. 263877) in making—
Acetone from isopropyl alcohol.
Dehydrogenated products from cyclohexane.
Isobutyraldehyde from isobutyl alcohol.
Iosbutyronitrile from isobutylamine.
Naphthalene from tetrahydronaphthalene.
Paracymene from turpentine.
Catalyst (Brit. 262120) in making—
Isovaleraldehyde from isoamypl alcohol.
General chemical reagent.

Cadmium Tungstate
Electrical
Luminous agent in—
Cathode-ray tubes used in television.

Cadmium Xanthate
Agriculture
Disinfectant (U. S. 1998092) for—
Seeds.

Caesium Silicate
Synonyms: Cesium silicate.
French: Silicate de césium.
German: Caesiumsilikat, Kieselsaeurescaesium.
Ceramics
Raw material in making ceramic wares.

Caesium Sulphate
Synonyms: Cesium sulphate.
French: Sulfate de césium.
German: Caesiumsulfat, Schwefelsaeurescaesium.
Brewing
Reagent in—
Making beer.
Food
Ingredient of—
Manufactured mineral waters.

Caffeine Hydrobromide
French: Bromhydrate de cafeine.
German: Bromwasserstoffsäurescoffein, Coffeinhydrobromid.
Spanish: Bromhidrato de cafeina.
Italian: Bromidrato di cafeina.
Pharmaceutical
Suggested for use as—
Antineuralgic, cardiatic, diuretic, sedative, tonic.

Calabar Bean
Synonyms: Chop nut, Ordeal bean, Split nut.
Latin: Physostigma, Semen physostigmatic.
French: Fèves de calabar.
German: Eseresamen, Gottesurteilbohnen, Kalabarbohnen.
Spanish: Haber del calabar, Eseve.
Chemical
Starting point in extracting—
Eserine, physostigmine.
Pharmaceutical
In compounding and dispensing practice.

Calcium
French: Calcium.
German: Calcium, Kalk.
Spanish: Calcio.
Italian: Calcio.
Analysis
Reagent in—
Carrying out reduction reactions in organic synthesis.
Producing vacuums for experimental laboratory purposes.
Chemical
Absorbent for various gases.
Reagent in—
Carrying out reduction reactions and the like in the manufacture of intermediates, organic chemicals, synthetic aromatic chemicals, synthetic pharmaceuticals.
Dehydrating alkalies (used in place of sodium and potassium).
Making hydrogen for filling balloons.
Purifying various inert gases.
Starting point in making—
Pure grades of calcium carbide.
Fats and Oils
Reagent in dehydrating—
Fats and oils of animal or vegetable origin.
Fertilizer
Reagent in making—
Artificial fertilizers by the fixation of atmospheric nitrogen.
Metallurgical
Deoxidizer and gasifier in treating—
Molten steel (possesses the special property of leaving no residue behind in the steel after treatment).
Deoxidizer in treating—
Copper, without affecting the mechanical properties and electrical resistivity of the metal.
Deoxidizing agent in treating—
Cast iron and steel (added in the form of briquettes with sponge iron for greensand castings in the proportion of 0.5 percent to produce a finer distribution of the graphite and to reduce the content of graphitic carbon, sulphur, and total insoluble residue, to improve impact value, to produce uniform grain structure and increase the transverse and tensile strength).
Hardening agent in making—
Antifriction metallic compositions.
Ingredient of—
Frary metal, containing 2 percent of barium.
Lead alloys for sheathing cables.
Lead alloys of the types Pb_3Ca, $PbCa$, and $PbCa_2$.
Light aluminum alloys.
Reagent in—
Decarburization, desulphurization, and purification of iron and iron alloys and various other metals and alloys.
Reducing agent in making—
Metals, for example chromium, manganese, and alloys, from oxides and halides.

Calcium Acetate
Synonyms: Acetate of lime, Calcic acetate, Calcium pyrolignite.
French: Acétate calcique, Acétate de calcium, Acétate de chaux, Pyrolignite calcique, Pyrolignite de calcium, Pyrolignite de chaux, Terre folice calcaire.
German: Calciumacetat, Calciumazetat, Essigsäurescalcium, Essigsäureskalk.
Chemical
Reagent in making—
Ethyl acetate, methyl acetate, methylallyl ketone, methylamyl ketone, methylbenzyl ketone, methylbutyl ketone, methylcresyl ketone, methylethyl ketone, methylheptyl ketone, methylhexyl ketone, methyllactyl ketone, methylnonyl ketone, methylphenyl ketone, methylpropyl ketone, methyltolyl ketone, methylxylyl ketone.
Starting point in making—
Acetic acid, acetone, aluminum sulphoacetate, aluminum acetate, cobalt acetate, copper acetate, lithium acetate, magnesium acetate, nickel acetate, potassium acetate, sodium acetate, various acetates by double decomposition with sulphates, zinc acetate.
Dye
Precipitant in making—
Color lakes.
Food
Reagent in making—
Artificial vinegar.
Insecticide
As a fungicide and insecticide.
Ingredient of—
Insecticidal and fungicidal compositions.

Calcium Acetate (Continued)

Leather
Reagent in—
Tanning.

Miscellaneous
Ingredient of—
Fireproofing compositions used for various purposes.
Reagent in—
Preparing fur skins.

Paper
Ingredient of—
Compositions used for the impregnation of paper used for packing soft soaps (Brit. 319517).
Fireproofing compositions.

Pharmaceutical
In compounding and dispensing practice.

Photographic
Reagent in various processes.

Textile
——, *Dyeing and Printing*
Ingredient of—
Aluminum mordants, chrome mordants, nitrate mordants used in dyeing textiles with alizarin red.
Printing pastes containing various alizarin dyestuffs, such as dark bordeaux, light violets, dark violets, puces, reds with alizarol, roses, and roses with alizarol.

——, *Finishing*
Ingredient of—
Fireproofing compositions.

Woodworking
Ingredient of—
Fireproofing compositions.

Calcium Adipate
French: Adipate calcique, Adipate de calcium, adipate de chaux.
German: Adipinsaeurescalcium, Adipinsaeureskalk, Calciumadipat, Kalkadipat.

Food
Ingredient (Brit. 312088) of—
Condiments, flavorings.

Pharmaceutical
In compounding and dispensing practice.

Calcium Albuminate
Synonyms: Lime albuminate.
French: Albuminate de chaux.
German: Albuminsaeureskalk, Kalkalbuminat.

Rubber
Reagent for—
Reclaiming rubber (U. S. 1640807).

Calcium Aluminate
French: Aluminate calcique, Aluminate de calcium, Aluminate de chaux.
German: Calciumaluminat, Kalciumaluminat.
Spanish: Aluminato de calcio.
Italian: Aluminato di calcio.

Chemical
Ingredient of catalytic mixtures used in the manufacture of—
Acenaphthylene, acenaphthaquinone, bisacenaphthylidenedione, naphthaldehydic acid, naphthalic anhydride, and hemimellitic acid from acenaphthene (Brit. 295270).
Acetaldehyde from ethyl alcohol (Brit. 281307).
Acetic acid from ethyl alcohol (Brit. 281307).
Alcohols from aliphatic hydrocarbons (Brit. 281307).
Aldehydes and acids by the reduction of the corresponding esters (Brit. 306471).
Aldehydes and acids from toluene, orthochlorotoluene, orthonitrotoluene, orthobromotoluene, parachlorotoluene, parabromotoluene, paranitrotoluene, metachlorotoluene, metanitrotoluene, metabromotoluene, dichlorotoluenes, dibromotoluenes, dinitrotoluenes, chloronitrotoluenes, chlorobromotoluenes, nitrobromotoluenes (Brit. 295270).
Aldehydes and acids from xylenes, pseudocumenes, mesitylene, and paracymene (Brit. 281307).
Alphanaphthaquinone from naphthalene (Brit. 295270).
Anthraquinone from naphthalene (Brit. 281307).
Benzaldehyde and benzoic acid from toluene (Brit. 281307).
Benzoquinone from phenanthraquinone (Brit. 281307).
Benzyl alcohol from benzaldehyde by reduction (Brit. 306471).
Benzyl alcohol or benzaldehyde or benzyl phthalide by the reduction of phthalic anhydride (Brit. 306471).
Butyl alcohol by the reduction of crotonaldehyde (Brit. 306471).
Chloroacetic acid from ethylenechlorohydrin (Brit. 295270).
Diphenic acid from ethyl alcohol (Brit. 281307).
Ethyl alcohol by the reduction of acetaldehyde (Brit. 306471).
Fluorenone from fluorene (Brit. 295270).
Formaldehyde by the reduction of carbon dioxide or carbon monoxide (Brit. 306471).
Formaldehyde by the reduction of methane or methanol (Brit. 306471).
Hydroxyl compounds by the reduction of anthraquinone, benzoquinone and similar compounds (Brit. 306471).
Isopropyl alcohol by the reduction of acetone (Brit. 306471).
Maleic acid and fumaric acid by the oxidation of toluene, benzene, phenols, tar phenols, or furfural, or from benzoquinone or phthalic anhydride (Brit. 295270).
Methane from carbon dioxide or carbon monoxide by reduction (Brit. 306471).
Methanol from carbon dioxide or carbon monoxide by reduction (Brit. 306471).
Naphthaldehydic acid, acenaphthaquinone, or bisacenaphthylidenedione from acenaphthylene (Brit. 281307).
Phenanthraquinone from phenanthrene or diphenic acid (Brit. 295270).
Phthalic acid and maleic acid from naphthalene (Brit. 295270).
Primary alcohols by the reduction of the corresponding aldehydes (Brit. 306471).
Propionic acid and butyric acid and higher alcohols, ketones, and acids by the reduction of carbon dioxide or carbon monoxide (Brit. 306471).
Reduction products of ketones, aldehydes, acids, esters, alcohols, ethers, and other organic compounds, which contain oxygen (Brit. 306471).
Salicylic acid and salicylic aldehyde from cresol (Brit. 295270).
Secondary butyl alcohol by the reduction of methylethyl ketone (Brit. 306471).
Valeryl alcohol by the reduction of valeraldehyde (Brit. 306471).
Vanillin and vanillic acid by the oxidation of eugenol or isoeugenol (Brit. 295270).
Ingredient (Brit. 306460) of catalytic preparations which are used in the production of various aromatic and aliphatic compounds, including—
Alphanaphthylamine from alphanitronaphthalene.
Amines from aliphatic nitro compounds, such as alkyl nitriles or nitromethane.
Amino compounds from the corresponding nitroanisoles.
Amylamine from pyridin.
Anilin, azo-oxybenzene, azobenzene, and hydrazobenzene from benzene by reduction.
Aminophenols from nitrophenols.
3-Aminopyridin from 3-nitropyridin.
Cyclohexamine, dicyclohexamine, and cyclohexylanilin from nitrobenzene.
Piperidin from pyridin.
Pyrrolidin from pyrrol.
Tetrahydroquinolin from quinolin.

Calcium-Aluminum-Iron Cyanide

Chemical
Catalyst (Brit. 446411) in—
Halogenating unsaturated hydrocarbons.

Calcium-Anilin
French: Aniline calcique, Aniline et calcium.
German: Kalkanilin.

Dye
Starting point (German 436533) in making anthracene dyestuffs from 3:9-dichlorobenzanthrone.
11:3-Dichlorobenzanthrone.

Calcium Benzoate
Synonyms: Benzoate of lime.
Latin: Calcii benzoas.
French: Benzoate calcique, Benzoate de calcium.
German: Benzoesäurescalcium, Benzoesäureskalzium, Calciumbenzoat, Kalziumbenzoat.
Spanish: Benzoato de calcio.
Italian: Benzoato di calcio.

Calcium Benzoate (Continued)

Pharmaceutical

In compounding and dispensing practice.

Suggested for use as—
Alterative, antiseptic.

Rubber

Retarding agent (U. S. 1929561) in—
Vulcanizing processes employing an ultra-accelerator.

Calcium Betanaphthol Alphasulphonate

Synonyms: Abrastol, Asaprol.

Fermentation

Preservative in making wines.

Pharmaceutical

In compounding and dispensing practice.

Calcium Bicarbonate

Synonyms: Bicarbonate of lime, Calcium acid carbonate.
French: Bicarbonate de chaux.
German: Doppeltekohlensaeureskalk, Kalkbicarbonat, Kohlensaeuresaeureskalk.

Construction

Ingredient in making—
Mortars for various ornamental purposes.

Metallurgical

Ingredient of—
Compositions which are used for producing various color effects on metals by electrolysis.

Calcium Boride

French: Bore calcique, Bore de calcium, Bore de chaux, Chaux boré.
German: Borcalcium, Kalkhaltigbor.

Metallurgical

Degasifying and oxidizing agent for—
Metals (principally nonferrous metals).

Calcium Bromide

Synonyms: Bromide of lime.
Latin: Calcii bromidum, Calcium bromatum.
French: Bromure de calcium, Bromure de chaux.
German: Bromcalcium, Bromkalk, Calciumbromid, Kalkbromid.
Spanish: Bromuro de calcio.
Italian: Bromuro di calcio.

Beverage

Ingredient of—
Effervescent mineral waters.

Chemical

Starting point (Brit. 395296) in making—
Soluble organic calcium salts useful in medicine.

Substitute for—
Potassium bromide.

Pharmaceutical

In compounding and dispensing practice.

Suggested for use as—
Nerve sedative.

Suggested as substitute for other bromides in treating—
Asthma, epilepsy, hysteria, tetanus.

Photographic

Substitute for other bromides.

Calcium Carbide

French: Carbure calcique, Carbure de calcium.
German: Calciumcarbid.
Spanish: Carburo de calcio.
Italian: Carburo di calcio.

Agriculture

As an anticryptogamic agent.

Chemical

Catalyst (French 678742) in making—
Diphenylurea from anilin and carbonic anhydride.

Dehydrating agent for—
Alcohol, various purposes.

Source of acetylene in making—
Acetals, acetaldehyde, acetaldehydedisulphonic acid, acetaldehyde derivatives, such as sulphonic and carboxylic acids, acetic acid, acetic anhydride, acetone, acetylene black, acetylene dichloride, acetylene tetrabromide, acetylene tetrachloride, barium acetylide, caesium acetylide, calcium acetylide, copper acetylide, ethane, ethylene, ethylene acetate, ethylidene diacetate, formic acid, hydrogen, linalool, lithium acetylide, magnesium acetylide, manganese acetylide, mercury acetylide, metallic acetylides, 3-methylbutanol, nickel acetylide, potassium acetylide, propylene, pyrrol, rubidium acetylide, silver acetylide, sodium acetylide, strontium acetylide, synthetic tanning agents, tellurium acetylide, tetranitromethane, tetrachloromethane, thiophene, tin acetylide, titanium acetylide, tungsten acetylide, trichloroethylene, vinyl alcohol, zinc acetylide, zirconium acetylide.

Disinfectant

Ingredient (French 628931) of—
Disinfecting composition, containing also acetic acid, cresol solution, and nicotine.

Distilled Liquor

Dehydrating agent for—
Alcohol.

Dye

Source of acetylene in making—
Indigos, various synthetic dyestuffs.

Electrical

Dehydrating agent in—
Electrostatic work.

Fertilizer

Ingredient of—
Fertilizer compositions.

Starting point in making—
Calcium cyanamide.

Food

Dehydrating agent in making—
Desiccated foods.

Fuel

Source of acetylene for—
Illumination purposes, in signal fires, harbor and channel buoys, trucks, yachts, tug boats, and other water vessels.
Illuminating purposes in isolated buildings, outdoor lights of various kinds, nongaseous mines.
Various heating purposes.

Gas

Source of acetylene for—
Compressing (storage) in cylinders for use in oxy-acetylene processes of welding and cutting used in many industries — metal, construction, wrecking, scrapping, reclamation, shipbuilding, railroading, building, boiler, tank and general steel plate construction, repairing.
Increasing calorific power of coal gas, water gas, mixed gas, coke-oven gas, and other gases.

Metallurgical

Deoxidant in—
Copper refining (French 668312).
Iron and steel making (French 517815).

Desulphurizing agent in—
Iron and steel making (French 495073; 573866).

Hardening agent for—
Steel.

Reducing agent in processing—
Calamine in a current of nitrogen to distil zinc and recover calcium cyanamide (French 624916).
Magnesium compounds and ores in making metallic magnesium (French 488735).
Metallic oxides and salts.
Ores and compounds to recover sodium and alkaline-earth metals (French 524804; 743123).
Sulphide ores of copper.

Restrainer (Austrian 106982) in—
Sulphuric acid pickling baths (for reducing attack on iron).

Miscellaneous

Ingredient (French 555893) of—
Composition for cleansing old paintings, containing also caustic potash or soda, salt, and water.

Calcium Caseinate

French: Caseinate de chaux.
German: Kalkkaseinat.

Insecticide

Ingredient of—
Insecticidal emulsions (U. S. 1646149).

Calcium Chlorate

French: Chlorate de chaux, Chlorate calcique.
German: Calciumchlorat, Chlorsäurescalcium, Chlorsäureskalk.

Agricultural

As a weed-killer.

Chemical

Reagent (Brit. 335203) in making—
Weed-killers with the aid of acids, such as hydrochloric, sulphuric, nitric, boric, oxalic, and tartaric acids; acid salts, such as sodium bisulphate, potas-

Calcium Chlorate (Continued)
sium bitartrate, calcium dihydrogen phosphate; acid-reacting salts, chlorides of ammonium, aluminum, iron, copper, zinc; mercuric chloride, sodium bichromate, and sodium fluosilicate.

Explosives
Ingredient of—
Pyrotechnic compositions.

Food
Reagent in making—
Mineral waters, soda water.

Photographic
Reagent in making—
Papers and film and in developing work.

Calcium Chloride
French: Chlorure de calcium, Chlorure de chaux.
German: Calciumchlorid, Chlorcalcium, Chlorkalzium, Chlorwasserstoffsäurescalcium, Chlorwasserstoffsäureskalk, Kalziumchlorid.
Spanish: Cloruro de calcio.
Italian: Cloruro di calcio.

Agriculture
For killing weeds.
Ingredient of—
Compositions for treating soil to remove growths choking crops.
Compositions used for feeding stock.

Analysis
General drying agent in analytical work, for drying gases and filling drying tubes for use in ultimate analysis of organic compounds, gas analysis, in desiccators, and the like.
Ingredient of—
Solutions, added for the purpose of raising the boiling point.
Maintaining constant high-temperature baths.
Reagent for detecting and determining—
Alcohol, alcohol in esters and volatile oils, bile pigments in bile and organic products, carbon monoxide in blood, fusel oil, malic acid, organic acids of both aliphatic and aromatic series, oxalic acid, pyrocatechin, sulphuric acid, tartaric acid.
Reagent in—
Analysis of the soil.
Separating various organic acids from one another, of both aromatic and aliphatic series.
Testing dyed wood for fastness to seawater.
Yield of citric acid from calcium citrate and sulphuric acid.

Automotive
Ingredient of—
Antifreeze solutions for use in radiators of automobiles, trucks, and stationary internal combustion engines.

Brewing
Reagent for—
Treating water, used in brewing beers and ales, for the purpose of removing its acidity prior to use.

Cement
Ingredient of—
Alumina cements (Brit. 251618).
Bore hole cements (added for the purpose of accelerating the rate of setting).
Slag cements.

Ceramics
Ingredient of—
Glazes for potteries, porcelains, chinaware, and chemical stoneware.

Chemical
Catalyst in making—
Acetal.
Compounds of both aliphatic and aromatic series by condensing the organic molecule.
Cyanamide.
Esters of acetic acid (German 232818).
Paratolylalphanaphthylamine.
Various organic compounds, obtained by the reaction between naphthols and ammonia.
Ingredient of—
Contact mass used in the manufacture of contact sulphuric acid (added for the purpose of counteracting the poisonous action of arsenic on the platinum catalyst).
Ingredient of catalytic mixtures used in the manufacture of—
Acenaphthylene, acenaphthaquinone, bisacenaphthylidenedione, naphthaldehydic acid, naphthalic anhydride, and hemimellitic acid from acenaphthene (Brit. 295270).
Acetaldehyde from ethyl alcohol (Brit. 281307).
Acetic acid from ethyl alcohol (Brit. 281307).
Alcohols from aliphatic hydrocarbons (Brit. 281307).
Aldehydes and acids by the reduction of the corresponding esters (Brit. 306471).
Aldehydes and acids from toluene, orthochlorotoluene, orthobromotoluene, orthonitrotoluene, parachlorotoluene, parabromotoluene, paranitrotoluene, metachlorotoluene, metabromotoluene, metanitrotoluene, dichlorotoluenes, dibromotoluenes, dinitrotoluenes, chlorobromotoluenes, chloronitrotoluenes, bromonitrotoluenes (Brit. 295270).
Aldehydes and acids from xylenes, pseudocumenes, mesitylene, and paracymene (Brit. 281307).
Alphanaphthaquinone from naphthalene (Brit. 295270).
Anthraquinone from naphthalene (Brit. 295270).
Benzaldehyde and benzoic acid from toluene (Brit. 281307).
Benzoquinone from phenanthraquinone (Brit. 281307).
Benzyl alcohol from benzaldehyde by reduction (Brit. 306471).
Benzyl alcohol, benzaldehyde, or benzyl phthalide by the reduction of phthalic anhydride (Brit. 306471).
Butyl alcohol by the reduction of crotonaldehyde (Brit. 306471).
Chloroacetic acid from ethylenechlorohydrin (Brit. 295270).
Diphenic acid from ethyl alcohol (Brit. 295270).
Ethyl alcohol by the reduction of acetaldehyde (Brit. 306471).
Fluorenone from fluorene (Brit. 295270).
Formaldehyde by the reduction of carbon dioxide or carbon monoxide (Brit. 306471).
Formaldehyde by the reduction of methane or methanol (Brit. 306471).
Hydroxyl compounds by the reduction of anthraquinone, benzoquinone, and similar compounds (Brit. 306471).
Isopropyl alcohol by the reduction of acetone (Brit. 306471).
Maleic acid and fumaric acid by the oxidation of toluene, benzene, phenols, tar phenols, or furfural, or from benzoquinone or phthalic anhydride (Brit. 295270).
Methane by the reduction of carbon dioxide or carbon monoxide (Brit. 306471).
Methanol by the reduction of carbon dioxide or carbon monoxide (Brit. 306471).
Naphthaldehydic acid, acenaphthaquinone, or bisacenaphthylidenedione from acenaphthylene (Brit. 281307).
Phenanthraquinone from phenanthrene or diphenic acid (Brit. 295270).
Phthalic acid and maleic acid from naphthalene (Brit. 295270).
Primary alcohols by the reduction of the corresponding aldehydes (Brit. 306471).
Propionic acid and butyric acid and higher alcohols, ketones, and acids by the reduction of carbon dioxide or carbon monoxide (Brit. 306471).
Reduction products of ketones, aldehydes, acids, esters, alcohols, ethers, and other organic compounds which contain oxygen (Brit. 306471).
Salicylic acid and salicylic aldehyde from cresol (Brit. 295270).
Secondary butyl alcohol by the reduction of methylethyl ketone (Brit. 306471).
Valeryl alcohol by the reduction of valeraldehyde (Brit. 295270).
Vanillin and vanillic acid by the oxidation of eugenol or isoeugenol (Brit. 295270).
Ingredient (Brit. 206460) of catalytic preparations which are used in the production of various aromatic and aliphatic compounds, including—
Alphanaphthylamine from alphanitronaphthalene.
Amines from aliphatic nitro compounds, such as allyl nitriles or nitromethane.
Amino compounds from the corresponding nitrophenols.
3-Aminopyridin from 3-nitropyridin.
Amylamine from pyridin.
Anilin, azo-oxybenzene, azobenzene, and hydrazobenzene from benzene by reduction.
Cyclohexamine, dicyclohexamine, and cyclohexylanilin from nitrobenzene.
Piperidin from pyridin.
Pyrrolidin from pyrrol.
Tetrahydroquinolin from quinolin.

Calcium Chloride (Continued)
Ingredient of—
Calcaona (pharmaceutical containing cocoa).
Kalzine (pharmaceutical containing sterilized gelatin, administered by injection).
Mixtures containing cupric hydroxide (added to aid in retaining the color and other physical properties when heated to 100 deg. C.)
Mixtures containing magnesium oxide or magnesium oxychloride used for the production of metallic magnesium by electric furnace heating under a vacuum.
Mugotan (pharmaceutical containing gum arabic).
Noridal suppositories.
Normosal.
Various pharmaceutical preparations.
Reagent in making—
Acetal.
Acetamide compounds by addition.
Alkyl chlorides.
Allyl chloride.
Alphanaphthylamine.
Anthraquinonemetadicarboxylic chloride.
Ammonium chloride by reaction with ammonia.
Barium chloride by treating solution of barium sulphide.
Barium chloride from heavy spar and carbon (German 154498).
Cerium salts.
Diparatolylmetaphenylenediamine.
Dextromannose compound.
Ethyl bromide, ethyl butyrate, ethyl chloride, ethyl iodide, ethylene chloride.
Formochlor by reaction with formaldehyde.
Glycerin compound.
Hexamethylenetetramine compound.
Hydrochloric acid, commercial grades.
Hydrosilicofluoric acid (German 191820).
Magnesium oxide in highly purified form from dolomite (French 454162).
Malonic acid, methyl acetate, methyl chloride, methyl iodide, methylal, oenanthol, pharmaceuticals, phenylhydroxylamine, silver permanganate, urea.
Various intermediates and organic chemicals.
Reagent in—
Drying industrial gases.
Heating and purifying enzymes from extracts of malt and pancreas gland (Brit. 251455).
Preventing volatilization of ammonia.
Purifying glycerin.
Reducing nitrobenzene to phenylhydroxylamine.
Starting point in making—
Artificial gypsum.
Calcium acetate from pyroligneous acid.
Calcium arsenate by reaction with sodium arsenate.
Calcium arsenite by reaction with sodium arsenite.
Calcium carbonate in precipitated form (precipitated chalk).
Calcium chlorate by electrolysis.
Calcium fluoride by reaction with sodium fluoride.
Calcium glycerophosphate by reaction with glycerophosphoric acid.
Calcium iodobehenate by reaction with erucic acid and hydriodic acid.
Calcium linoleate by reaction with sodium linoleate.
Calcium metaborate.
Calcium molybdate.
Calcium oleate by reaction with sodium oleate.
Calcium peroxide by reaction with sodium peroxide.
Calcium phosphate, dibasic, by reaction with disodium phosphate.
Calcium phosphate, tribasic, by reaction with tri-sodium phosphate with excess of ammonia.
Calcium silicide.
Calcium stearate by reaction with sodium stearate.
Calcium tartrate by reaction with crude cream of tartar.
Calcium thiosulphate.
Calcium tungstate by reaction with sodium tungstate.
Chlorine.
Double salts of calcium by reaction with solution of calcium acetate.
Metallic calcium.
Normalin (tasteless calcium chloride preparation (German 283649).
Substitute for glycerin where it is desired to make use of its water-absorbent properties.

Construction
Ingredient of—
Cement mortars.
Concrete compositions.
Concrete mixtures used in the building of highways.
Concrete mixtures (added for the purpose of protecting cement against frost).
Concrete mixtures, added to aid in their curing.
Mortars and wall plasters (added to increase the cementing power of the lime).

Dye
Precipitating agent in making—
Color lakes from brilliant lake red R, permanent red, litholrubin, and other dyes.
Reagent in making—
Alizarin, calcium alizarinate, chrysazin dyestuffs.
Reagent in purifying various dyestuffs.

Explosives
Ingredient of—
Gunpowder compositions.

Fats and Oils
Catalyst in—
Decomposition of fatty acids.
Reagent in—
Clarifying fats and oils of animal and vegetable origin.

Fertilizer
Fertilizer for septic soils, which contain lime and soda.
Ingredient of—
Fertilizer compositions.

Food
Reagent in—
Drying various foods, such as fruits and vegetables.
Making cheese, mineral waters, preserving meats in boxes.

Fuel
Ingredient of—
Solutions used for washing coal (Austrian 103892).
Solutions for treating peat before removing part of the water content.

Gas
Drying agent in—
Treating manufactured coal gas and coke-oven gas.
Ingredient of—
Water used in wet gas meters (added to prevent freezing in cold weather).
Reagent for—
Treating coal in order to improve its coking properties (German 233892).

Glues and Adhesives
Ingredient of—
Casein glues (U. S. 1604307), dextrin adhesives, library pastes, starch glues.

Mechanical
Ingredient of—
Solutions for high-pressure work.

Metallurgical
Ingredient of—
Compositions used for tempering metals.
Mixtures, containing silver nitrate, citric acid, and collodion, for burning a silver coating on aluminum.
Preparations for annealing and pickling meats (Brit. 321638).
Solutions containing cupric nitrate used for coloring copper brown.
Reagent in—
Leaching copper ores.
Recovering metallic molybdenum from molybdenum chloride, barium chloride, and calcium fluoride solution used as electrolyte in electrolytic process.
Nickel from ores by electrolytic process.
Smelting copper ores by the chloriding roast process.

Military
Ingredient of—
Compositions used in gas masks.

Mining
Reagent in—
Preventing coal-mine explosions (used as a dust-laying solution).

Miscellaneous
Binding agent for various purposes.
Drying agent for various purposes.
Ingredient of—
Compositions used for disinfectant purposes.
Compositions used as dust preventive.
Compositions, containing sand, cinders, coke breeze, etc., used for treating ice-bound roads.
Compositions, containing graphite and dextrin, used for sealing purposes (U. S. 1744348).
Compositions used for removing snow.

Calcium Chloride (Continued)
Porous absorbent compositions, containing an ammonium salt, an iron salt, and silicates (U. S. 1852029).
Sealing wax preparations.
Sizing compositions containing starch paste.
Solutions for use as fire extinguishers.
Solutions used in automatic sprinkler installations (added to prevent their freezing in cold weather).
Solutions for filling fire buckets.
Solutions for sprinkling on railway rails in snowfalls.
Solutions for making roads.
Solutions for laying dust and reducing destructive action of freezing on highways.
Solutions for preventing wind-blowing of farm soil.
Reagent in making—
Packaging material impervious to soft soap (Brit. 329517).

Paint and Varnish
Ingredient of—
Fireproof paints.
Reagent in making—
Dry colors, tungsten yellow from metallic tungsten, ultramarine, yellow ultramarine.

Paper
Ingredient of—
Compositions used for softening horny parchment paper.
Sizing compositions.

Petroleum
Ingredient of—
Lubricating greases.
Reagent in—
Dehydrating petroleum distillates.
Crude oil (used in conjunction with sodium chloride).

Pharmaceutical
Ingredient of—
Hay-fever medicine, medicinal baths.
Suggested for use as hemostatic, diuretic, blood coagulant, and cathartic.

Photographic
Reagent in making—
Silver chloride collodion paper (celloidin paper).

Refrigeration
Ingredient of—
Cold mixtures.
In making ice.
In meat packing by cold storage.

Resins and Waxes
Reagent in making—
Artificial resins of the phenol-formaldehyde condensation type (used in conjunction with formaldehyde) (French 563777).

Rubber
Coagulant for—
Rubber latex.

Starch
Reagent for—
Treating starches.
Reagent in making—
Soluble starch.

Stone
Binder in making—
Quartz stone.
Preservative for—
Artificial and natural stone.
Reagent in—
Hardening gypsum.

Sugar
In the refining process.

Textile
——, *Finishing*
Ingredient of—
Fireproofing compositions.
Sizing compositions containing starches.
Sizing compositions for cotton fabrics.

——, *Manufacturing*
Reagent in making—
Rayon filament resistant to water.

——, *Printing*
Ingredient of
Printing pastes foı cotton fabrics.

Water
Reagent in—
Purifying water.

Woodworking
Ingredient of—
Compositions, containing zinc chloride, copper hydroxide, copper sulphate, calcium oxide, magnesium hydroxide, sodium chloride, and magnesium chloride, used for preserving wood (U. S. 1852090).
Compositions used for fireproofing wood.
Compositions used for preserving wood.

Calcium Citrate
Synonyms: Citrate of lime.
French: Citrate de chaux.
German: Citronensaeurescalcium, Kalkzitrat, Zitronsaeurescalcium, Zitronsaeureskalk.

Chemical
Starting point in making—
Citric acid.

Perfumery
Ingredient of—
Toothpowders, toothpastes.

Pharmaceutical
In compounding and dispensing practice.

Calcium Cresylate
French: Crésylate de chaux, Crésylate de calcium.
German: Cresylsaeureskalk, Kalkcresylat.

Petroleum
Reagent in treating—
Water-in-oil emulsions and in breaking up petroleum emulsions (U. S. 1606698).

Calcium Ethylxanthate
Synonyms: Calcium ethylxanthogenate.
French: Éthyle-xanthogénate de chaux.
German: Aethylxanthogensaeureskalk, Calciumaethylxanthogenat.

Chemical
Starting point in making—
Accelerator of rubber vulcanization, in combination with sulphur monochloride (Brit. 265169).

Calcium Glutonate
French: Glutonate calcique, Glutonate de calcium, Glutonate de chaux.
German: Calciumglutonat, Glutonsäurescalcium, Glutonsäureskalk, Kalkglutonat.

Pharmaceutical
Ingredient (Brit. 332840) of preparations containing—
Alkali compounds, alkali earth compounds, alkaloids, cadmium glutonate, caffeine hydrochloride, caffeine, caffeine-sodium salicylate, camphor, codeine hydrochloride, colamine hydrochloride, gelatin, glycerin, glucose, glucosides, 2-ethoxy-6:9-diaminoacridum lactate, hexamethylenetetramine, iron glutonate, methylsulphonic acid salts of para-aminobenzoic acid diethylamineleucinol ester, methylene blue, nickel glutonate, sodium chloride, sodium salicylate, sterols, strontium glutonate, tartar emetic, thyroxin, trypan blue.
In compounding and dispensing practice.

Calcium Glycerinophosphate
Synonyms: Calcium glycerophosphate, Calcium phosphoglycerate, Glycerophosphate of lime.
French: Glycérinophosphate calcique, Glycérinophosphate de chaux.
German: Glycerinphosphorsaeurescalcium, Glycerinphosphorsaeureskalk, Kalkglycerinphosphat.

Chemical
Starting point in making—
Magnesium glycerinophosphate.
Potassium glycerinophosphate.
Sodium glycerinophosphate.

Pharmaceutical
In compounding and dispensing practice.

Calcium Hypochlorite
Synonyms: Bleaching powder, Calcium oxymuriate, Chloride of lime, Chlorinated calcium oxide, Chlorinated lime, Hypochlorite of lime, Oxymuriate of lime.
Latin: Calcaria chlorata, Calcii hypochloris, Calcis chloridum, Calx chlorin, Calx chlorinata, Chloris calcicus, Chloruretum calcis.
French: Chlorure de chaux, Poudre de knox, Poudre de tennant.
German: Bleichkalk, Chlorkalk.
Italian: Cloruro di calcio.

Calcium Hypochlorite (Continued)

Chemical
Oxidizing agent in—
Organic synthesis.
Reagent in making—
Chloroform, various chemicals.

Dry Cleaning
Deodorizing and spotting agent for—
White goods.

Explosives and Matches
Cotton-bleaching agent in making—
Gelatin dynamites, gun cotton, smokeless powders.

Fats and Oils
Bactericide.
Bleaching agent.
Deodorant.
Deodorant for—
Tankcars.
Germicide.
Rancidity retardant.

Foods
Bactericide.
Bleaching agent.
Deodorant.
Disinfectant.
Germicide.
Sterilizing agent.

Gas
Reagent in—
Purification of acetylene.

Laundering
Bleaching agent in—
Washroom waters and soap solutions.
Germicide in—
Washroom waters and soap solutions.

Leather
In tanning processes.

Miscellaneous
Bactericide.
Bleaching agent.
Deodorant.
Disinfectant.
Drying agent.
Germicide.
Stabilizer in—
Ink eradicators.
Sterilizing agent.

Paper
Bleaching agent for—
Paper stock of all kinds.
Digesting agent (U. S. 1894501) in making—
Wood pulp from poplar.
Oxidizing agent (U. S. 1894620) in making—
White filler from sulphate pulp lime mud.

Pharmaceutical
In compounding and dispensing practice.
Starting material in making—
Carrel-Dakin solution.
Suggested for use in treatment of—
Adynamic dysentery, angina, burns, chilblains, hospital gangrene, itch, snake bite, typhoid fever, ulcerated gums, ulcers, wounds.

Plastics
Bleaching agent for—
Cotton and pulp in making cellulose base plastics.

Soap
Ingredient of—
Detergent having disinfectant properties (Brit. 391407).
Detergent, in combination with sodium silicate and trisodium phosphate (U. S. 1894207).
Rancidity retardant for—
Fats, oils, soap powders.

Sanitation
Bactericide, deodorant, and sterilizing washing agent for—
Hospital walls and floors, hospital lavatories, hospital utensils, industrial buildings, industrial equipment, public and domestic convenience stations, public buildings.
Germicide and deodorant in—
Earth closets, sewage systems.

Textile
As "chemick" in bleaching processes.

Water
Bactericide, deodorant, and sterilizing agent in—
Emergency water supply systems.
Isolated water storage systems.
Municipal water storage and supply systems.
Ship water storage systems.
Swimming pools.
Water mains under construction.
Destructive agent for—
Algae in condenser water for power plants and refrigerating plants.

Calcium Iodate
Synonyms: Lime iodate.
French: Iodate de chaux.
German: Calciumjodat, Jodsaeurescalcium, Jodsaeureskalk, Kalkjodat.

Food
Preservative (Brit. 274164) in treating—
Butter, cream, eggs, fish, fruit preserves, margarin, milk.

Calcium Lactate
French: Lactate de calcium, Lactate de chaux.
German: Milchsäurescalcium, Milchsäureskalk.
Italian: Lactico di calcio.

Metallurgical
Suggested as a reagent in—
Plating nickel on zinc.

Pharmaceutical
In compounding and dispensing practice.
Suggested as a blood coagulant in the treatment of hemorrhages; also for administration prior to dental operations to inhibit bleeding.
Suggested source of lime where lime salts are indicated in medical treatment.

Leather
Suggested ingredient of—
Tanning and finishing compositions.

Textile
Suggested as a mordant.

Calcium Lactobionate
Synonyms: Lactobinoate of lime.

Chemical
Starting point (Brit. 395296) in making—
Soluble organic calcium salts useful in medicine.

Calcium Maltobionate
Synonyms: Maltobionate of lime.

Chemical
Starting point (Brit. 395296) in making—
Soluble organic calcium salts useful in medicine.

Calcium Methylsulphate
French: Méthylesulphate de chaux.
German: Calciummethylsulfat, Methylschwefelsaureskalk.

Chemical
Starting point in making—
Methyl thiocyanate.

Calcium Mucate
French: Mucate de calcium.
German: Schleimsaeurescalcium.

Chemical
Ingredient of—
Baking powders (Brit. 252695).

Calcium Naphthenate
Synonyms: Naphthenate of lime.
French: Naphtènate calcique, Naphtènate de calcium, Naphtènate de chaux.
German: Kalziumnaphtenat, Naphtensäureskalk, Naphtensäureskalzium.

Linoleum and Oilcloth
Drier (Brit. 353783) in making—
Compositions for application in the manufacture of linoleum.

Miscellaneous
Drier (Brit. 353783) in making—
Compositions of various drying oils, such as linseed oil, chinawood oil.

Paint and Varnish
Drier (Brit. 353783) in making—
Paints, varnishes, and enamels.

Calcium Oleate
French: Oleate calcique, Oleate de calcium.
German: Calciumoleat, Kalkoleat, Oleinsäurescalcium, Oleinsäureskalk.
Spanish: Oleato de calcio.
Italian: Oleato di calcio.

Chemical
Reagent in making—
Emulsions of various chemicals.

Fats and Oils
Reagent in making—
Emulsions, emulsified lubricating compositions.

Miscellaneous
Ingredient of—
Modeling waxes (added for the purpose of varying the hardness of the preparations).
Reagent in making—
Emulsions of various products.

Petroleum
Ingredient of—
Emulsions containing petroleum and petroleum distillates.
Reagent in making—
Lubricating compositions containing petroleum and petroleum distillates.

Textile
Ingredient of—
Softening compositions used in finishing fibers and fabrics.

Woodworking
Ingredient (Brit. 340101) of—
Compositions, containing cellulose acetate and natural or artificial gums and resins, used for decorating and coating woodwork.

Calcium Oxide
Synonyms: Burned lime, Lime, Quicklime.
Latin: Calcaria usta, Calcii oxidum, Calcium oxidatum, Calx usta, Calx viva, Oxydum calcium.
French: Chaux, Chaux comune, Chaux vive.
German: Aetzkalk, Gebrannter kalk, Kalk.
Spanish: Cal viva.
Italian: Calce, Ossidio di calcio.

For most uses other than those in which a caustic effect is desired calcium oxide is slaked by exposure to the air or with water before using.

Agriculture
——, *Livestock*
As an animal and poultry medicine for—
Conditioning.
Increasing resistance to abortion and tuberculosis.
Neutralizing acidity.
Bone-building agent in—
Animal and poultry feeds.
Metabolizing agent in—
Animal and poultry feeds.
Shell-forming agent in—
Poultry feeds.
Tooth-building agent in—
Animal feeds.

——, *Soils and Crops*
Carrier for—
Plant foods.
Detoxicating agent for—
Field, truck, and orchard soils (by precipitation of aluminum and iron salts).
Disease regulant for—
Growing Crops.
Granulating agent in the flocculation of—
Humus, soils.
Neutralizing agent for acidity of—
Fertilizers, soils.
Physiological regulating agent for plants, through its effect on cell rigidity, food transfer, protoplasm activity.
Plant food for—
Field, truck, and orchard crops.

Analysis
Dehydating agent.
Gas absorbent.
Neutralizing agent.
Reagent.

Ceramics
As a flux.
Ingredient of—
Glazes.

Chemical
Absorbent for—
Carbon dioxide in phenol manufacture.
Catalyst in—
Esterification of glycerin with tung oil.
Nitrogen fixation.
Peroxidation of alkalies.
Reduction of chromium oxide with carbide.
Catalyst in making—
Calcium cyanamide, chlorine (by Weldon process).
Dehydrating agent for—
Alcohol.
Neutralizing agent for excess of inorganic acids in making—
Phenol.
Sulphonated naphthalene intermediates.
Oxidizing agent for carbon in making—
Calcium carbide, calcium silicide.
Precipitant for—
Atropine, berberine, brucine, cocaine, codeine, corydaline, cryptopine, ecgonine, emetine, eucaine, hydrastine, hyoscine (scopolamine), hyoscyamine, laudanine, laudanosine, morphine, narceine, narcotine, nicotine, protopine, quinine, strychnine, thebaine, tropinone.
Precipitant in processing—
Ceria, didymia, dysprosia, erbia, europia, gadolina, holmia, lanthana, lutecia, neodymia, praseodymia, samaria, scandia, terbia, thoria, thulia, ytterbia, yttria, zirconia.
Reagent in extracting—
Potash from greensand and feldspar.
Reagent in making—
Acetic acid in wood distillation.
Acetone in wood distillation.
Acetone from sulphite paper waste liquor.
Alcohol from molasses.
Alcohol from sulphite paper waste liquor.
Ammonia from aluminum nitride.
Ammonia from beet sugar residues.
Ammonia from gas-works liquors.
Ammonia from oil-shale distillation products.
Benzaldehyde from benzene chloride.
Calcium acetate in wood distillation.
Calcium citrate.
Calcium ferrocyanide from spent iron oxide used in purifying illuminating gas.
Calcium saccharate from molasses.
Citric acid.
Decolorizing carbon from sulphite paper waste.
Hydrocarbons from aromatic acids.
Hydrocyanic acid from waste ore liquors.
Methanol from sulphite paper waste liquor.
Methanol in wood distillation.
Potassium ferrocyanide from spent iron oxide used in purifying illuminating gas.
Pyridin from quinolic acid.
Sodium acetate as by-product in soda paper pulp making.
Sodium bichromate.
Sodium ferrocyanide from spent iron oxide used in purifying illuminating gas.
Trichlorethylene.
Water-clarifying agents.
Reagent in—
Purifying ammonium sulphate.
Anthraquinone, borax, caffeine, glauber's salts, intermediates, magnesium sulphate, sodium chloride.
Treating chemical plant waste waters.
Starting point (direct or indirect) in making—
Calcium metal.
Calcium salts of acids and halogens.
Sodium bichromate.

Construction
Ingredient of—
Asphaltic concrete (cooling, hardening, and filling agent).
Cement concrete (hydrating, plasticizing, stabilizing, strengthening, water-tightening, and whitening agent).
Magnesia insulating coatings.
Mortars for brick, stone, and tile (bonding, plasticizing, and toughening agent).
Sand-lime brick (bonding, hydrating, neutralizing, strengthening, and whitening agent).
Sorel cement (bonding, chemical, plasticizing, toughening, and water-tightening agent).
Stuccos (bonding, plasticizing, toughening, and water-tightening agent).
Wall plasters (bonding, plasticizing, sound-deadening, and whitening agent).

Calcium Oxide (Continued)

Dye
Neutralizing agent for excess of inorganic acids in making—
Sulphonated naphthalene intermediates.
Reagent for—
Purification of intermediates.
Saponifying organic salts in the synthesis of dyes.

Electrical
Coating agent for—
Electric arc-welding electrodes.

Explosives
Neutralizing agent for excess acid in making—
Nitroglycerin, smokeless powders.

Fertilizer
Ingredient of—
Fertilizers.
Reagent in making fertilizer compositions from—
Molasses, quarry wastes, tannery wastes.

Food
Fat conserving agent for—
Butter.
Neutralizing agent and corrective in—
Butter, milk.
Plumping and swelling agent in making—
Gelatin.
Preservative for—
Butter, eggs.
Reagent in—
Grain classification.

Fuel
Component of—
Briquets with coal, peat, tar, and waste products.
Fat-splitting agent in—
Candle making.

Gas
Absorbent in—
Purification of illuminating gas.
Admixed with coal in—
Water-gas enrichment.

Glass
Raw material in making—
Bottle glass, lime glass, lime-flint glass, window glass.

Glues and Adhesives
Ingredient of—
Casein glues.
Vegetable glue made from calcium oxide, powdered ivory nut, casein, soda ash, trisodium phosphate, and sodium fluoride (U. S. 1895979).
Plumping and swelling agent in making—
Glue.
Reagent in making—
Siccatives with naphthenates.

Insecticide
Adhesive agent in—
Sprays.
Carrier for—
Disinfectants, fungicides.
Compounding agent in—
Bordeaux mixture, calcium arsenate, fungicides, insecticides, lead arsenate, lime-sulphur mixtures.
Inhibiting agent in—
Dusts, sprays, washes.
Preventive of insect birth and growth.
Repellent for—
Weevils in garnered crops.

Leather
As a depilatory.
Neutralizing agent for acids in—
Hardening of patent leather.

Linoleum and Oilcloth
As a filler.

Metallurgical
As a fluxing agent in—
Smelting and refining.
Ingredient of—
Iron ore briquets.
Lubricant for dies in—
Steel wire drawing.
Neutralizing agent for—
Acid mine waters, excess leaching acids, excess pickling acids, ores in flotation processes.
Neutralizing agent in—
Disposing of waste pickling acid.
Reagent in purification of—
Ferrochrome.
Regulating agent in—
Electric welding.
Rust-retarding agent for—
Iron.
Scouring agent in—
Electroplating.
Settling agent for—
Ore slimes in refining gold and silver ores.

Miscellaneous
Cleansing agent.
Ingredient of—
Buffing compositions, magnesium flashlight powders, phosphorescent mixtures, polishing compositions.
Neutralizing agent for excess acid in making—
Cattle feed by saccharifying sawdust with acids.
Neutralizing agent for general purposes.
Precipitant for—
Aluminum salts in making cleaning compounds.
Raw material in making—
Crucibles, limelight pencils.
Rust resistant.
Scouring agent.

Oils and Fats
Reagent in—
Deodorizing vegetable oils.
Saponifying agent in making—
Lubricants.

Paint and Varnish
Cementing agent.
Chemical reagent.
Fire preventive.
Hardening agent.
Ingredient of—
Cold-water paints, whitewash.
Neutralizing agent for excess acidity in making—
Lithopone.
Neutralizing agent (for resinous acids) in making—
Enamels, varnishes.
Pigment.
Precipitant in making—
Colloidal pigments, such as satin white.
Reagent in making—
Limed rosin.
Rust preventive.
Saponifier.
Weather-resistant.

Paper
As a filler.
Causticizing reagent in—
Rag paper making, soda process papermaking, sulphite process papermaking.
Digestant in making—
Pulp from poplar wood (U. S. 1894501).
Strawboard.
Reagent in making paper from—
Bagasse, corncobs, cotton linters, oat hulls, old newsprint.
Reagent in—
Removing dextrin from cellulose.
Scouring agent for—
Rags.

Petroleum
Dehydrating agent for—
Greases, petroleum.
Desulphurizing agent for—
Petroleum.
Saponifying agent in making—
Lubricating greases, various petroleum products.

Pharmaceutical
Ingredient of—
Mineral oil-base salves.
Reagent in making—
Milk of magnesia.
Starting point in making—
Lime syrup, limewater.
Suggested for use in treatment of acid stomach, diarrhea, dyspepsia, nausea, pseudo-membranous croup, vomiting.

Photographic
Reagent in making—
Sensitizers.

Plastics
Reagent in making—
Phenol condensation products.
Reagent (U. S. 1897977) to—
Lower solidification temperature in production of glyptal resins.

Calcium Oxide (Continued)

Refractories
Bond in—
Silica refractories.

Refrigeration
Coagulating agent in—
Clarifying turbid water used in the manufacture of ice.

Rubber
Carrier for—
Sulphur in vulcanizing processes.
Reagent in making—
Flocculated clay rubber filler.
Hydrolized glue for use in the prevulcanization of rubber.

Sanitation
Disinfectant for—
Barns, cesspools, chicken houses, drains, outhouses, stables.
Reagent in—
Sewage treatment by direct chemical and electrolytic lime processes.
Waste water disposal.

Soap
Neutralizing agent in—
Making rosin soaps.
Twitchell process (for glycerin water).
Twitchell process (for soluble acids).
Reagent in making—
Lime soaps.
Saponifying agent for—
Fats, greases, and oils.

Sugar
Precipitant in—
Steffens lime process for extracting sugar from molasses.
Reagent for—
Coagulating and neutralizing beet juice.
Cane juice, sorghum juice.

Textile
Mercerizing agent.
Mordant in certain dyeing processes.
Neutralizing agent in—
Carbonizing.
Scouring agent.

Tobacco
Reagent in—
Extraction of nicotine.

Water Supply
Neutralizing agent for—
Acidity.
Reagent in—
Deferrization, deodorization, filtration, phenol removal, sedimentation, softening by lime-soda process.

Calcium Oxybromide

Chemical
Oxidizing agent (Brit. 395296) in making—
Soluble organic calcium salts used in medicine.

Calcium Paratoluolsulphamide

Agriculture
Ingredient of—
Weed-killing compositions.

Chemical
Starting point in making various derivatives.

Insecticide
As an insecticide.

Calcium Phenolsulphonate
Synonyms: Calcium sulphocarbolate, Calcium sulphophenate, Calcium sulphophenolate, Calcium suplhophenylate.
French: Phénolsulphonate de chaux, Sulphophénate de chaux.
German: Kalkphenolsulfonat, Phenolsulfonsaeureskalk, Sulfocarbolsaeureskalk.

Animal Remedies
Ingredient of—
Chicken remedies.

Chemical
Denaturant for—
Alcohol.
Starting point in making—
Aluminum phenolsulphonate, bismuth phenolsulphonate, cadmium phenolsulphonate, copper phenolsulphonate, lead phenolsulphonate, lithium phenolsulphonate, magnesium phenolsulphonate, manganese phenolsulphonate, mercury phenolsulphonate, nickel phenolsulphonate, potassium phenolsulphonate, sodium phenolsulphonate, strontium phenolsulphonate, zinc phenolsulphonate.

Insecticide and Fungicide
Process material in making—
"Bouillie Lyonnaise" for destroying *Oidium* on vines.

Sanitation
Disinfectant for various purposes.

Pharmaceutical
In compounding and dispensing practice.
Suggested for use as—
Antiseptic, astringent.

Calcium Plumbate
French: Plumbate de chaux.
German: Calcium plumbat.

Chemical
Ingredient of—
Colloidal compounds of arsenic (U. S. 1573375).
Reagent in making—
Potassium ferricyanide.
Starting point in making—
Oxygen (rare process).

Explosives
Ingredient of—
Compositions for making heads of matches (added to moderate rate of combustion).

Calcium Polysulphide
Synonyms: Polysulphide of lime.
French: Foie de soufre calcaire, Polysulfure calcique, Polysulfure de calcium, Polysulfure de chaux.
German: Kalkpolysulfid.
Spanish: Polisulfurato de calcio.
Italian: Polisulfurato di calcio.

Fats and Oils
Reagent (Brit. 271553) in making—
Vulcanized oils.

Fertilizer
Ingredient of—
Fertilizing compositions used as top dressing.

Insecticide
As an insecticide and fungicide.
Ingredient (U. S. 1388678) of—
Insecticidal and germicidal compositions.

Leather
Reagent in—
Dehairing hides.

Metallurgical
Flotation agent for—
Separating ores.

Paper
Ingredient (Brit. 271553) of—
Compositions, containing rubber latex, used for treating paper and pulp.

Rubber
Reagent (Brit. 271553) in treating—
Rubber latex.

Calcium Resinate
Synonyms: Lime soap, Resinate of lime.
French: Résinate de chaux.
German: Calciumresinat.

Paper and Pulp
Ingredient of—
Waterproofing compositions for treating paper and pulp.

Resins and Waxes
Hardening agent for—
Rosin to be used in admixture with chinawood oil.

Textile
——, *Finishing*
Ingredient of waterproofing compositions for fabrics and yarns.

Woodworking
Ingredient of—
Waterproofing compositions for treating woods.

Calcium Ricinoleate

Pharmaceutical
Claimed (U. S. 2019933) to be—
Intestinal detoxification agent suitable for oral administration.

Calcium Silicide
French: Siliciure de calcium.
German: Calciumsilicid, Siliciumcalcium.

Chemical
Reagent in making—
Sodium metal.

Metallurgical
Reagent in—
Aluminothermic work in the place of aluminum.
Welding iron and steel.

Miscellaneous
Ingredient of—
Ignition pellets.
Priming agent in making—
Marine smoke bombs.

Calcium Silicofluoride
Synonyms: Calcium fluosilicate.
French: Fluosilicate de calcium, Fluosilicate de chaux.
German: Calciumfluorsilikat, Fluorstoffkieselsaeureskalk, Fluorwasserstoffkieselsaeurescalcium, Kalksilicofluorid, Siliciumfluorsaeurescalcium, Siliciumfluorwasserstoffsaeureskalk.

Ceramics
Reagent in making—
Chinaware, porcelains, potteries, stoneware.

Construction
Preservative in treating—
Bricks, stone, stucco, and other construction material.

Woodworking
As a preservative.

Calcium Stearate
French: Stéarate de chaux.
German: Kalkstearat, Stearinsaeureskalk, Stearinsaeurescalcium.

Textile
——, *Finishing*
Reagent in—
Waterproofing textiles.

Woodworking
Reagent in—
Waterproofing.

Calcium Sulphate
Synonyms: Alabaster, Anhydrite, Gypsum, Land plaster, Plaster of Paris, Sulphate of lime, Terra alba.
Latin: Calcii sulphas, Calcis sulphas, Calcium sulfuricum.
French: Gypse, Sulphate de calcium, Sulphate calcique, Sulphate de chaux.
German: Gebrannter gyps.

Agriculture
As a land-dressing.

Building and Construction
As a lathing material.
As a roofing (form of slabs).
As a tile for various purposes.
As a wallboard.
As an acoustic plaster.
As an insulating medium.
Ingredient of—
Artificial flooring compositions.
Artificial stone flooring.
Artificial marble.
Fireproofing products.
Industrial floorings.
Insulations.
Keene's cement.
Paste for filling cracks in floors, containing also silica, yellow dextrin, and water.
Sound absorbents.
Special cements.
Special plasters.
Starting point in making—
Decorative effects, ornamental work.

Ceramics
Setting accelerator (U. S. 1897667) for—
Cellular clay body.

Chemical
Catalyst (Brit. 397187) in making—
Aliphatic alcohols and ethers by absorbing olefins at elevated temperatures and pressure in aqueous solutions of acids which are weaker than sulphuric acid and hydrolyzing the product after addition of the appropriate amount of water or water vapor.
Reagent (Brit. 376080) in—
Process for decolorizing barytes.

Cosmetic
Ingredient of—
Dentifrices, perfumed artificial seasalt.

Explosives and Matches
Ingredient of—
Matchhead compositions.
Reducer of—
Exploding temperatures.

Fats and Oils
Purifying agent for—
Oils.

Fertilizer
Ingredient (U. S. 1894587) of—
Fertilizer containing also peat, finely divided iron, and a source of nitrogen.

Fire Extinguisher
As a fire-extinguishing medium.

Food
Ingredient of—
Bakery products.
Molding agent in making—
Candies.

Glass
Ingredient of—
Glass cements for various purposes, typical of some of these are the following: (1) Glass to brass, (2) aquarium glass to glass or metal.
Reagent in making—
Translucent glass.

Inks
Ingredient of—
Metallic lustrous inks.

Insecticide and Fungicide
Diluent for—
Paris green and other arsenical compositions.
Ingredient of—
Cockroach exterminant, in admixture in equal parts with fine dry oatmeal.
Fungicide, in admixture with copper oxalate (U. S. 1785472).
Insecticide for cabbage maggots, in admixture with calomel.
Nonpoisonous rat exterminant, in admixture with rye flour and oil of anise.
Seed disinfectant, in admixture with copper oxalate (U. S. 1785472).
Seed disinfectant, in admixture with alpha-mercury-dithienyl and lime (U. S. 1934803).

Linoleum and Oilcloth
As a filler.

Mechanical
Ingredient of—
Heat insulation containing also aluminum sulphate, limestone, soap, talc, and water.
Insulating pipe-covering compositions containing asbestos.

Metallurgical
Dusting agent for—
Foundry molds in making special castings.
Flux for—
Garnierite in smelting New Caledonian ores to obtain a nickel-iron matte.
Ingredient of—
Cement for iron castings, containing also iron filings, whiting, gum arabic, carbon black, and portland cement.
Polishing agent for—
Tinplate.

Miscellaneous
As a cement and adhesive.
As a general dehydrating agent.
As a molding agent for various purposes.
Filler in—
Buttons, electro-plate ornaments, jewelry, phonograph records.
Starting point (U. S. 1746717) in making—
Artificial snow by reacting with a boiling dilute sulphuric acid solution, filtering, and crystallizing .

Paint and Varnish
As a filler.
Color improver (U. S. 1857274) for—
Titanium dioxide.
Hiding power improver (U. S. 1857274) for—
Titanium dioxide.
Ingredient of—
Titanium pigments (Brit. 405340).

Calcium Sulphate (Continued)
Titanium pigments (U. S. 1857274).
Titanium pigments (Brit. 407674).
Water paints.
Precipitation accelerator (U. S. 1857274) for—
Titanium dioxide.
Reagent (Brit. 403762) in making—
Chrome yellows which are stable to light.

Pharmaceutical
Component of—
Surgical bandages.
Material for—
Plaster casts.
Suggested for use as—
Absorbent dressing for wounds, foul ulcers, and similar conditions.

Plastics
As a filler.
Dehydrating agent (French 755316) in making—
Plastics from polymerized vinyl alcohol and aldehydes.

Rubber
As a filler.
Molding agent in making—
Rubber stamps.

Stone
As an artificial stone.
As a filler for cracks and pits.
As a polishing agent.

Textile
Dusting agent (Brit. 399599) in cleansing—
Processed wool, raw wool.
Filler in—
Cotton goods.
Fireproofing agent for—
Fabrics.

Veterinary Medicine
Ingredient of—
Lice and mite tablets for poultry, containing also calcium sulphide, silica sand, sugar, and starch.
Worm-expeller containing also epsom salt, calcium silicate, venetian red, sand, and nicotine.

Calcium Sulphide
Synonyms: Calcic liver of sulphur.
French: Foie de soufre calcaire, Sulphure de calcium, Sulphure calcique.
German: Calciumsulfid, Kalkschwefelleber, Kalksulfid, Schwefelcalcium, Schwefelkalk, Schwefelwasserstoffsaeurescalcium.

Chemical
Starting point in making—
Calcium sulphydrate, calcium thiosulphate, sulphur, sulphuretted hydrogen.

Gas
Reagent for removing—
Carbon bisulphide from gas.

Leather
Ingredient of—
Compositions for removing hair from hides before tanning.

Metallurgical
Ingredient of—
Flotation oils for separating the constituents of nonsulphide ores.
Reagent for—
Precipitating silver from solutions in sodium thiosulphate and calcium triosulphate in the wet metallurgy of silver.
Treating the sulphur dioxide fumes from furnaces in refineries to produce catalysis by means of reduction with the aid of mineral oils.

Miscellaneous
Ingredient of—
Germicidal preparations.

Paint and Varnish
Ingredient of—
Luminous paints, luminous varnishes.
Starting point in making—
Sulphophone, a substitute for lithopone.

Perfumery
Ingredient of—
Depilatories.

Pharmaceutical
In compounding and dispensing practice.

Calcium Tannate
French: Tannate de chaux.
German: Calciumtartrat, Gerbsaeurescalcium.

Adhesives
Ingredient of—
Casein glue compositions (U. S. 1604310).

Pharmaceutical
In compounding and dispensing practice.

Calcium Tartrate
Synonyms: Tartrate of lime.
French: Tartrate de calcium, Tartrate de chaux.
German: Weinsaeurecalcium.

Chemical
Starting point in making—
Tartaric acid.

Food
Ingredient of—
Baking powders (Brit. 252695).

Calcium Thiocyanate
Synonyms: Calcium rhodanide, Calcium sulphocyanate, Calcium sulphocyanide.
French: Rhodanure calcique, Rhodanure de calcium, Rhodanure de chaux, Sulfocyanate calcique, Sulfocyanate de calcium, Sulfocyanate de chaux, Sulfocyanure calcique, Sulfocyanure de calcium, Sulfocyanure de chaux, Thiocyanate calcique, Thiocyanate de calcium, Thiocyanate de chaux.
German: Calciumrhodanuer, Calciumsulfocyanat, Calciumsulfocyanur, Calciumthiocyanat, Kalkrhodanuer, Kalksulfocyanuer, Kalkthiocyanat, Rhodansäureskalk, Schwefelcyancalcium, Schwefelzyancalcium, Sulfocyansäurescalcium, Sulfocyansäureskalk, Thiocyansäurescalcium, Thiocyansäureskalk.
Spanish: Sulfocianato de calcio, Thiocianato de calcio.
Italian: Sofocianato di calcio, Thiocianato di calcio.

Analysis
Reagent in—
Analytical methods involving control and research operations.

Cellulose Products
Ingredient (U. S. 1301652 and 1482076) of—
Cellulose solvent.
Process material (U. S. 1465994) in making—
Cellulose acetate filaments.
Solvent for—
Cellulose (said to have advantages over other solvents at particular concentrations).

Chemical
Process material in—
Chemical manufacture.
Starting point in making—
Ferricyanides of various metals.
Sulphocyanides of various metals.

Fuel and Gas
Ingredient (German 624842 and 625418) of—
Drying agent for gas, in admixture with solution of calcium nitrate.

Miscellaneous
Ingredient of—
Compositions used for making vulcanized fibers.
Suggested for use as a highly soluble, nonpoisonous, noninflammable salt in miscellaneous processes.

Paper
Ingredient of—
Compositions used for the production of parchmentized paper.
Ingredient (German 590326) of—
Parchmentizing solution, with formaldehyde, used in making vulcanized fiber.
Parchmentizing agent for—
Paper.
Process material in making—
Parchmentized paper (U. S. 1333465).
Vulcanized fiber (U. S. 1333465).

Textile
——, *Dyeing*
Ingredient of—
Bath in dyeing madder colors on wool.

——, *Finishing*
Ingredient of—
Compositions used for weighting fabrics.

——, *Manufacturing*
Ingredient of—
Bath in the production of the rayon filament.

Calcium Thiocyanate (Continued)

——, *Miscellaneous*
Mercerizing agent (U. S. 1482076) for—
Cotton fabric.
Stiffening agent for—
Textiles.

Calcium Tungstate

Electrical
Luminous agent in—
Intensifying screens for x-ray work.

Calophyllum Oil

Synonyms: Alexandrian laurel oil, Calaba oil, Dilo oil, Domba oil, Laurel nut oil, Ndilo oil, Njamplung oil, Pinnay oil, Poonseed oil, Tacamahac fat, Udilo oil.
French: Huile de calophyllum.
German: Kalophyllumoel.
Spanish: Aceite de calofilluma.
Italian: Olio di calofilluma.

Fuel
As an illuminant (by natives in Africa).

Pharmaceutical
As a native medicine.
Proposed as an antirheumatic (the oil, particularly its resinous component, is poisonous).

Soap
As a soapstock.

Cameline Oil

Synonyms: German sesame oil.
French: Huile de caméline, Huile de sesame allemand,
German: Deutsches sesamoel, Dotteroel, Leindotteroel.

Fats and Oils
Ingredient of—
Colza oil mixtures.

Food
Ingredient of various preparations.

Fuel
As a burning oil.
Ingredient of fuel compositions.

Paint and Varnish
Vehicle in making—
Enamels, lacquers, paints, varnishes.

Soap
Raw material in making—
Soft soaps.

Camphene Cinnamate

French: Cinnamate de camphène, Cinnamate camphènique.
German: Camphencinnamat, Zimtsäurecamphenester, Zimtsäurescamphen.

Chemical
Starting point in making—
Pharmaceuticals and other derivatives.

Pharmaceutical
In compounding and dispensing practice.

Camphor

Latin: Camphora.
French: Camphre.
German: Campher, Kampher, Kampfer.
Spanish: Alcanfor.
Italian: Canfora.

Note: Covers uses of Chinese, Japanese and Formosa natural camphor, and of synthetic camphor; "Dutch Camphor" and "Tub Camphor" are archaic names for Japanese camphor.

Adhesives
Plasticizer and preservative in—
Adhesive compositions containing pastes, cellulose acetate, nitrocellulose, or other esters or ethers of cellulose.

Aviation
Plasticizer in—
Dopes and cementing agents, containing cellulose acetate, nitrocellulose, or other esters and ethers of cellulose, used for treating and processing aviation fabrics.

Cellulose Products
Plasticizer in—
Cellulose acetate, cellulose esters and ethers, nitrocellulose.

Ceramics
Plasticizer in—
Compositions, containing cellulose acetate, nitrocellulose, or other esters or ethers of cellulose, used as coatings for protecting and decorating ceramic ware.

Chemical
Ingredient (U. S. 1234381 and 1241738) of—
Acidproofing compositions, alkali-proofing compositions, chlorine-proofing compositions.
Preserver (U. S. 1486468) for—
Opium solutions.
Starting point in making—
Campholenic acid, camphoric acid, camphoronic acid, camphor oxime, camphylamine, carvacrol (with iodine).
Chloro, bromo, nitro, and amino derivatives.
Cymene (with phosphoric anhydride).

Cosmetic
Ingredient of—
Hair-restorer and loss preventive (French 694297).
Hair-washing and curling preparation (French 620213).
Plasticizer in—
Nail enamels containing nitrocellulose or other esters or ethers of cellulose.

Electrical
Plasticizer in—
Compositions, containing cellulose acetate, nitrocellulose, or other esters or ethers of cellulose, used in making electrical insulations and for insulating electrical machinery and equipment.

Explosives and Matches
Plasticizer in making—
Explosives containing nitrocellulose, night lights, pyrotechnic compositions, smokeless powders.

Fats and Oils
Preservative for—
Oils.

Firefighting
Ingredient of—
Chemical fire-extinguishing mixtures with carbon tetrachloride (various patents).
Fireproofing composition (U. S. 1241738).

Glass
Ingredient of—
Lubricant (with turpentine) used in cutting, boring and grinding glass.
Plasticizer in—
Compositions, containing cellulose acetate, nitrocellulose, or other esters or ethers of cellulose, used in the manufacture of nonscatterable glass and as coatings for decorating and protecting glassware.

Insecticide
As a moth repellant.
Ingredient of—
Insecticidal preparations for moths.
Termite repellant (French 606215).

Leather
Plasticizer in—
Compositions, containing cellulose acetate, nitrocellulose, or other esters or ethers of cellulose, used in making artificial leather and as coatings for decorating and protecting leather goods.

Mechanical
Ingredient (U. S. 1372639) of—
Composition for removing carbon from internal-combustion engines.
Fuels for internal-combustion engines.

Metal Fabricating
Plasticizer in—
Compositions, containing cellulose acetate, nitrocellulose, or other esters or ethers of cellulose, used as coatings for decorating and protecting metal ware.

Miscellaneous
Ingredient of—
Artificial hair (U. S. 1505043, 1350820, and 1217028).
Automobile polish (French 670760).
Boat-mending composition (U. S. 1389084).
Deodorants (U. S. 1346337 and 1515364).
Embalming preparations.
Fat-reducing compound (U. S. 1369997).
Fluorescent screen (U. S. 1480896).
Furniture polish (U. S. 1363419).
Liquid fuel (U. S. 1496260).
Waterproofings (various patents).

Camphor (Continued)
Plasticizer in—
Compositions, containing cellulose acetate, nitrocellulose, or other esters or ethers of cellulose, used as coatings for decorating and protecting various fibrous products.
Oral Hygiene
Ingredient of—
Dentifrices.
Paint and Varnish
Plasticizer in—
Compositions, containing cellulose acetate, nitrocellulose, or other esters or ethers of cellulose, used as paints, varnishes, enamels, dopes, and lacquers.
Paper
Plasticizer in—
Compositions, containing cellulose acetate, nitrocellulose, or other esters or ethers of cellulose, used in the manufacture of coated paper and as coatings for decorating and protecting paper and pulp products.
Petroleum
Deteriorating reducer (U. S. 1930248) for—
Antiknock gasoline during storage.
Improver (Brit. 404046) of—
Exhaust odors of gasoline fuels.
Pharmaceutical
In compounding and dispensing practice.
Photographic
Plasticizer in making—
Films from cellulose acetate, nitrocellulose, or other esters or ethers of cellulose.
Plastics
Plasticizer in making—
Plastic compositions from cellulose acetate, nitrocellulose, or other esters or ethers of cellulose.
Resins
Process material in making—
Resins.
Rubber
Plasticizer in—
Compositions, containing cellulose acetate, nitrocellulose, or other esters or ethers of cellulose, used as coatings for protecting and decorating rubber goods.
Soap
Ingredient of—
Degreasing composition (U. S. 1219967).
Detergent (U. S. 1219967).
Preservative for—
Oils.
Stone
Plasticizer in—
Compositions, containing cellulose acetate, nitrocellulose, or other esters or ethers of cellulose, used as coatings for decorating and protecting artificial and natural stone.
Textile
Plasticizer in—
Collar-waterproofing composition (U. S. 1453764).
Compositions, containing cellulose acetate, nitrocellulose, or other esters or ethers of cellulose, used in the manufacture of coated textiles.
Wood
Plasticizer in—
Compositions, containing cellulose acetate, nitrocellulose, or other esters or ethers of cellulose, used as coatings for protecting and decorating woodwork.
Plastic compositions of cellulose esters or ethers used to decorate, fill, and repair woodwork.

Camphor-Betasulphonic Acid
Chemical
Catalyst (Brit. 440888) in making—
Monoglycerides or mixtures of glycerides rich in monoglycerides, by esterifying—
Lauric acid, oleic acid, palmitic acid, stearic acid.

Camphoric Acid
French: Acide camphorique, Acide de camphoryle.
German: Kamphersaeure.
Chemical
Starting point (Brit. 269498) in making derivatives of—
Atropine, hyoscyamine, scopolamine.
Starting point in making the following camphorates—
Allyl, amyl, antipyrin, butyl, benzyl, cinnamyl, ethyl, formyl, glyceryl, guaiacyl, hexamethylenetetramine (Amphotropine), lactyl, methyl, phenyl, propyl, pyramidon, salicylyl, santalyl, valeryl.
Pharmaceutical
In compounding and dispensing practice.
Plastics
Plasticizing agent in making—
Celluloid, cellulose plastics.

Camphor, Monobromated
Synonyms: Bromated camphor, Brominated camphor, 3-Bromocamphor, Camphor monobromate, Camphor monobromide, Monobrominated camphor.
Latin: Camphora monobromata, Camphorae monobromidum.
French: Bromure de alcanfor, Camphre monobromé.
German: Bromcamphor, Kamphermonobromid, Monobrom-camphor, Orthomonobromcamphor.
Spanish: Alcanfor monobromado.
Italian: Canfora monobromata.
Pharmaceutical
In compounding and dispensing practice.
Suggested for use as—
Antineuralgic, antispasmodic, sedative.

Camphor Oil, Heavy
Latin: Oleum camphorae.
French: Huile de camphre, lourde.
German: Dickes camphoroel, Dickes kampferoel.
Chemical
Denaturant for—
Alcohol.
Starting point in making—
Safrol.
Fuel
As an illuminant.
Insecticide
Ingredient of—
Insecticidal preparations (Brit. 278816).
Paint and Varnish
Diluent in making—
Oil colors, paints, varnishes.
Soap
Ingredient of—
Special products.

Camphor Oil, Light
Latin: Oleum camphorae.
French: Huile de camphre, légère.
German: Duennes camphoroel, Duennes kampferoel.
Chemical
Denaturant for—
Alcohol.
Starting point in making—
Safrol.
Fuel
As an illuminant.
Insecticide
Ingredient of—
Insecticidal preparations (Brit. 278816).
Paint and Varnish
Diluent in making—
Oil colors, paints, varnishes.
Substitute for turpentine.
Pharmaceutical
In compounding and dispensing practice.
Resins and Waxes
Ingredient of—
Phenol-formaldehyde condensation products, added for the purpose of increasing their elasticity.

Caproylhydroquinone
Petroleum
Stabilizing agent (Brit. 406195) for—
Cracked gasolines and other motor fuels.

Caproyl Peroxide
French: Peroxyde de caproyle, Peroxyde caproylique.
German: Caproylperoxyd.
Chemical
Starting point in making—
Bactericidal preparations, intermediates, internal antiseptics, organic chemicals, pharmaceuticals.
Fats and Oils
Bleaching agent in treating—
Animal fats and oils, vegetable fats and oils.
Food
Bleaching agent in treating various foodstuffs.

Caproyl Peroxide (Continued)
Miscellaneous
Bleaching agent for various purposes.
Perfume
Ingredient of—
Skin-bleaching creams, toothpastes, tooth powders.
Pharmaceutical
In compounding and dispensing practice.
Resins and Waxes
As a bleaching agent.
Soap
As a bleaching agent.

Caproylphloroglucinol
Petroleum
Stabilizing agent (Brit. 406195) for—
Cracked gasolines and other motor fuels.

Caproylpyrocatechol
Petroleum
Stabilizing agent (Brit. 406195) for—
Cracked gasolines and other motor fuels.

Caproylpyrogallol
Petroleum
Stabilizing agent (Brit. 406195) for—
Cracked gasolines and other motor fuels.

Caproylresorcinol
Petroleum
Stabilizing agent (Brit. 406195) for—
Cracked gasolines and other motor fuels.

Caprylhydroquinone
Petroleum
Stabilizing agent (Brit. 406195) for—
Cracked gasolines and other motor fuels.

Caprylic Alcohol, Primary
Synonyms: Normal octyl alcohol, Normal octylic alcohol.
French: Alcool de capryle, primaire; Alcool caprylique, primaire; Alcool octylique, normale.
German: Caprylalkohol, primaer; n-Octylalkohol.
Chemical
Starting point in making—
Caprylic acid, caprylic acetate, caprylic formate, capronic acid, octaldehyde, various esters of caprylic acid, various synthetic compounds.
Perfume
Ingredient of—
Rose perfumes, special compound odors.
Perfume in—
Cosmetics.
Soap
Perfume in—
Toilet soaps.

Caprylphloroglucinol
Petroleum
Stabilizing agent (Brit. 406195) for—
Cracked gasolines and other motor fuels.

Caprylpyrocatechol
Petroleum
Stabilizing agent (Brit. 406195) for—
Cracked gasolines and other motor fuels.

Caprylpyrogallol
Petroleum
Stabilizing agent (Brit. 406195) for—
Cracked gasolines and other motor fuels.

Caprylresorcinol
Petroleum
Stabilizing agent (Brit. 406195) for—
Cracked gasolines and other motor fuels.

Caprylylhydroquinone
Petroleum
Stabilizing agent (Brit. 406195) for—
Cracked gasolines and other motor fuels.

Caprylylphloroglucinol
Petroleum
Stabilizing agent (Brit. 406195) for—
Cracked gasolines and other motor fuels.

Caprylylpyrocatechol
Petroleum
Stabilizing agent (Brit. 406195) for—
Cracked gasolines and other motor fuels.

Caprylylpyrogallol
Petroleum
Stabilizing agent (Brit. 406195) for—
Cracked gasolines and other motor fuels.

Caprylylresorcinol
Petroleum
Stabilizing agent (Brit. 406195) for—
Cracked gasolines and other motor fuels.

Caramel
Synonyms: Burnt sugar, Sugar coloring.
Latin: Saccharum ustum.
French: Caramel, Couleur.
German: Gebrannterzucker, Karamel, Zuckercouleur, Zuckerfarbe, Zuckertinktur.
Spanish: Azucar quemado, Colores de azucar.
Italian: Caramelle.
Beverage
Coloring agent for—
Alcoholic beverages, carbonated beverages, cider, malt beverages.
Food
Coloring agent for—
Culinary products, such as soups, jellies, and sauces.
Vinegars.
Ingredient of—
Cake fillings, cake icings, candies, pastries.
Tobacco
As a coloring and flavoring agent.

Carbon Bisulphide
Synonyms: Carbon disulphide, Carbon sulphide.
Latin: Alcohol sulfuris, Carboneum sulfuratum, Carbonii bisulphidum, Carbonis bisulphidum.
French: Sulfure de carbone.
German: Kohlensulfid, Schwefelalkohol, Schwefelkohlenstoff.
Spanish: Sulfuro de carbono.
Italian: Solfuro di carbonio.
Agriculture
As a fungicide, insecticide, and verminicide.
Analysis
As a solvent and reagent.
Chemical
Process material (U. S. 1886587) in—
Production of xanthates from terpene alcohols (fenchyl alcohol, borneol, and terpineol).
Reagent in making—
Acetophenone, alphanitronaphthalene-8-sulphonic acid, aluminum chloride, aluminum sulphide from aluminum oxide, ammonium sulphocarbonate, ammonium sulphocyanide, barium sulphide, carbon tetrachloride, chrysene, dimethyldiphenylanthrone, ethyl isothiocyanide, methyl isothiocyanate, monobromodibenzylanthracene, paraminoacetophenone, paraparadiaminodiphenylurea, perchloromethyl sulphide, potassium-antimony tartrate, potassium xanthate, sodium xanthate, sulphides from oxides, sulphocyanides and cyanides, tetrabenzoylperylene, thiobisorthoaminoparaoxydiphenylamine, thiourea and derivatives, trichloromethylsulphuric chloride, trimethylene sulphide.
Reagent in—
Wood distillation (to provide an atmosphere during distillation so as to produce a material of high conductive power for both heat and electricity).
Solvent for—
Iodine, phosphorus, quinine and other alkaloids, sulphur.
Solvent in—
Extracting uncombined sulphur and bitumen from minerals.
Making (Brit. 291347) diacetylacenaphthene.
1:5-Dibenzoylnaphthalene.
Dichloroacetylacenaphthene.
Other diacidyl derivatives of the naphthalene and acenaphthene series.
Starting point in making—
Synthetic hydrocarbons and arsenicals by distillation over arsenic.
Construction
Ingredient (U. S. 1602726) of—
Compositions used in waterproofing cement and concrete.

Carbon Bisulphide (Continued)

Dye
Reagent in making—
Azidin orange, indigo, sulphur black, thion blue B.
Explosives and Matches
Reagent in making—
Matches.
Fats and Oils
Solvent in—
Extracting fats, extracting oils from seeds, extracting second quality oils (foots).
Food
Preservative for—
Foods, especially meat.
Gas
Solvent for—
Extracting sulphur from spent oxide recovered in purifying gas.
Gums and Waxes
As a solvent.
Insecticide
Alone or in combinations as—
Ant destroyer, chicken lice destroyer, corn fumigant, flea exterminant, mole exterminant, moth destroyer, nit destroyer, phylloxera exterminant in viticulture, plant insecticide, rodent exterminant, soil insecticide, vermin exterminant, weevil expellant in grain storage.
Mechanical
Antiknock agent (U. S. 1741206) in—
Motor fuels.
Diluent for—
Motor fuels.
Metallurgical
Solvent for—
Phosphorus in the electroplating of delicate objects, such as feathers, flowers, and grasses.
Miscellaneous
As a general solvent and preservative.
As a general cleansing agent, particularly in degreasing and dry-cleaning.
In spectroscopy.
Reagent for—
Filling glass prisms to make them highly light-refractive.
Solvent for—
Fats and waxes in the manufacture of candles.
Paint and Varnish
Ingredient of—
Paint and varnish removers.
Solvent in—
Manufacture of varnishes.
Perfume
Extractant for—
Aromatic principles, essential oils, flower odors.
Petroleum
Reagent in—
Purifying paraffin oil.
Photographic
Ingredient (with carbon monoxide) of—
Illuminant in the Sell lamp used to produce a very luminous and highly actinic light.
Pharmaceutical
Reagent in—
Pharmacopeial tests.
Suggested for use as an external counter-irritant and local anesthetic and in veterinary medicine.
Refrigeration
As a refrigerant.
Rubber
Solvent for—
Guttapercha.
Sulphur chloride in the cold process of vulcanizing rubber.
Vulcanized rubber in the manufacture of waterproofed goods by the deposition of a thin layer of the dissolved rubber on the fabric.
Solvent in—
Rubber cements, vulcanizing rubber.
Resins and Waxes
Solvent for—
Extracting residual wax from residues obtained in the refining of beeswax.
Resins and waxes.
Resinous products obtained from crude anthracene.
Textile
Reagent (U. S. 1736713) for—
Improving cotton.
Solvent in—
Dry cleansing, manufacture of viscose rayon, wool degreasing.

Carbon Black

Synonyms: Gas black.
French: Noire de carbone.
German: Kohlenschwarz.

Cement
Black pigment in making—
Dark-colored cement mixtures.
Ceramics
Black pigment in making—
Tile and other ceramic products.
Construction
Black pigment in making—
Mortars, stuccos, concretes.
Electrical
Ingredient of—
Insulating compositions used in the manufacture of electrical machinery and equipment, as well as cables and wiring.
Explosives
Ingredient of—
Liquid air explosive compositions.
Preparations used for making matches.
Pyrotechnic preparations.
Fuel
Stabilizer (U. S. 1902866) for—
Emulsion of water and heavy tar used to render coal or coke nondusting.
Ink
Black pigment in making—
Chinese inks, India inks, lithographic inks, marking inks, offset inks, printing ink (Brit. 388072), printing inks, stenciling inks, typewriter-ribbon inks.
Jewelry
Black pigment in—
Coloring artificial stones.
Leather
Black pigment in making—
Artificial leather, black leather, patent leather.
Linoleum and Oilcloth
Black pigment in making—
Oilcloth and linoleum.
Metallurgical
Black pigment in making—
Compositions for coating mechanical apparatus.
Case-hardening agent.
Crucible material (in admixture with graphite).
Ingredient of—
Furnace lutes.
Reagent in treating—
Steel by the cementation process.
Miscellaneous
Ingredient of—
Auto-top dressings.
Compositions for making black buttons.
Compositions for making phonograph records.
Crayons, shoe polishes, stove polishes.
Paint and Varnish
Filler (Brit. 395478) in—
Lacquers, varnishes.
For shading oil colors.
Pigment in making—
Automobile lacquers, black paints, black varnishes, black enamels, black lacquers, casein paints, glue paints, Japan varnishes, oil paints, paints for scenery.
Paper
Pigment in making—
Black coated paper, bookbinders' board, carbon copying paper, glazed paper, gray coated paper, paperboard products.
Petroleum
Ingredient of—
Lubricating compositions containing mineral oil distillates or mixtures of the same with other oils (added in place of graphite to increase the viscosity).
Plastics
Black pigment in making—
Colored cellulose and other plastic compositions.

Carbon Black (Continued)

Printing
In process engraving and the litho trades.

Rubber
Ingredient of—
Automobile tires, rubber goods.

Stone
Black pigment in making—
Artificial stone.

Textile
Black pigment in making—
Carriage cloth, tarpaulins, waxed colored cloth.

Woodworking
Black pigment for impregnating—
Furniture, ornamental work, musical instruments, picture frames, tops of desks and the like.

Carbon Chlorofluoride
French: Chlorofluorure de carbone, Chlorofluorure carbonique.
German: Chlorfluorkohlenstoff, Kohlenstoffchlorfluorid.

Refrigeration
As a refrigerating agent in domestic and industrial mechanical refrigeration, possessing special property of non-toxicity.

Carbon Dioxide
Synonyms: Carbonic acid, Carbonic acid gas, Carbonic anhydride.
Latin: Acidum carbonicum.
French: Acide carbonique, Anhydride carbonique, Bioxyde de carbone, Bioxyde carbonique.
German: Kohlendioxyd, Kohlensäure, Kohlensäureanhydrid.
Spanish: Acido carbonico, Anhidrido carbonico, Bioxido carbonico.
Italian: Anidrido carbonio, Biossido carbonio.

Agriculture
For activating the growth of plants in cultivated fields and greenhouses.
Ingredient of—
Compositions, containing sodium carbonate, used for washing sheep.
Compositions for preventing smut in wheat.
Compositions used as weed-killers.
Compositions used for dipping cattle.
Plantfood for plants growing in greenhouses and truck gardens and cultivated fields.

Analysis
For making freezing mixtures in the laboratory.
Reagent for—
Freezing samples in the analysis of rubber.

Automotive
Used in liquefied form as motive power in compression motors.

Brewing
Antiseptic in treating—
Beer kegs and various equipment around the brewery.
Brewed beer and beer during the course of brewing.
Antioxidant in treating—
Beer during the progress of manufacture.
Various equipment around the brewery.
Medium for conveying beers around the brewery.
Partial disinfectant in the manufacture of beer.
Reagent for treating—
Beers and ales for the purpose of intensifying the taste.

Chemical
Antioxidizing agent in—
Making chemicals which are easily oxidized on contact with the oxygen in the atmosphere.
Making phosphorus compounds (used as a medium during the distillation process).
Various chemical processes.
Diluent in making—
Acetyl chloride from acetylene and chlorine.
Acetylene tetrachloride by the actinic ray process.
Various other organic chemicals in whose manufacture an inert medium must be used to control the progress of the reaction and prevent the formation of secondary compounds.
Drying agent in making various chemicals and in various chemical processes.
Gaseous medium in the distillation of various organic compounds (employed to prevent decomposition by the action of the oxygen in the atmosphere).
Ingredient of—
Mixtures containing nitrogen gas used for the purpose of filling the space above benzene and alcohol and other highly inflammable liquids in tanks and containers to prevent their catching fire and exploding (French 519132).
Mixtures, containing carbon bisulphide vapors, methyl formate, ethyl formate, propylene oxide, ethyl acetate, and similar organic chemicals (added for the purpose of producing non-inflammable mixtures).
Reagent for—
Digesting oxidizable substances.
Displacing oxygen in atmosphere in various chemical operations.
Neutralizing excess alkalinity in various chemical processes and in various finished chemicals.
Precipitating aluminum hydroxide.
Recovering inflammable solvents in various processes.
Starting liquid air machines.
Reagent in making—
Alphanaphthol-2-carboxylic acid.
Ammonium carbonate by reaction between carbon dioxide and dry ammonia gas.
Ammonium carbonate.
Ammonium sulphate by the lime process.
Aspirin (acetylsalicylic acid).
Barium carbonate by passing a current of the gas into a solution of barium sulphide.
Barium nitrate.
Benzaldehyde.
Betanaphthol-1-carboxylic acid.
Bismuth subcarbonate.
Calcium chloride.
Calcium carbamate.
Calcium permanganate by dissolving calcium manganate and passing a current of the gas through the melt.
Caesium carbonate by passing a current of the gas into a solution of caesium oxide.
Cobaltous carbonate by passing a current of the gas into a solution of cobaltous acetate.
Copper carbonate by passing a current of the gas through a solution of copper sulphate.
9:10-Dibenzylanthracene.
Ferrous carbonate by the precipitation of a solution of a ferrous salt by passing through a current of the gas.
Heptinecarboxylic acid.
Hydrocyanic acid.
2-Hydroxynaphthalene-6-carboxylic acids (U. S. 1594608).
Lanthanum carbonate by passing a current of the gas through a solution of lanthanum nitrate.
Lead carbonate by passing a current of the gas through a solution of lead nitrate.
Lithium carbonate by passing a current of the gas through a solution of lithium chloride.
Magnesium carbonate by passing a current of the gas through a solution of magnesium sulphate.
Magnesium carbonate, basic.
Manganese carbonate by passing a current of the gas through a solution of a manganese salt.
Monobromodibenzylanthracene.
Nickel carbonate by passing a current of the gas through a solution of nickel sulphate.
Nickel carbonate, basic.
Parahydroxybenzoic acid.
Phenylglycol (U. S. 1594608).
Pure carbonates of various metals.
Potassium permanganate by passing a current of the gas into a solution obtained by extracting with water a melt of potassium hydroxide, manganese dioxide, and potassium chlorate.
Sal ammoniac.
Salicylic acid by the treatment of a hot solution of sodium phenate.
Soda ash and sodium bicarbonate by the Le Blanc, Claus, Chance, and Solvay processes.
Sodium permanganate by dissolving sodium manganate in water and passing in a current of chlorine, carbon dioxide, or ozone.
Thiourea from calcium cyanamide (French 630883).
Tetramethyldiarsin (cacodyl).
Zinc carbonate by passing a current of the gas through a solution of a zinc salt.
Zinc oxide in very fine granules (U. S. 1442265).
Starting point in—
Synthesis of urea.

Carbon Dioxide (Continued)

Vehicle for—
Gaseous distillation of tar, phthalic anhydride, and other organic substances.

Construction
Used by plumbers for loosening stuck pipe joints and the like and also for cleansing plumbing.

Dye
Reagent in—
Extracting logwood (used as a medium for producing an inert atmosphere in the process so that oxidation of the extract is avoided).
Reagent in making—
Black V extra, various synthetic dyestuffs.

Electrical
Reagent for—
Extinguishing fires in electrical equipment, such as generators, transformers, high-tension fuses, electric ovens, telephone switchboards.
Maintaining inert atmosphere in oil transformers and thus preventing possible burning of the oil.
Testing tightness of lead coverings on electric cables and the like.

Explosives
Ingredient of—
White fire.
Reagent in making—
Pyrotechnics.

Food
Antiseptic in protecting—
Coconut, eggs, fruit, fruit drinks and extracts, meats, milk, various food preparations.
Anticorrosive agent in the canning industry.
Antioxidizing agent for—
Protecting and treating various foods and food preparations.
Protecting flavor, vitamin content, and other useful characteristics of fruit drinks, fruit extracts, fruit juices, and other food preparations.
For producing clean atmosphere in baking bread, cake, and the like.
Leavening agent in making—
Bread, cake, and other baked products.
Partial disinfectant in making—
Butter and cheese, fruit drinks, ginger ales, ice cream, soda water, various food preparations.
Reagent for—
Eliminating sulphur dioxide from bleached beverages, fruits, vegetables and other food products (used as an inert medium which does not introduce any undesirable product into the food during the course of its action).
Intensifying taste of canned goods, fruit drinks, ginger ales, carbonated milk and other food preparations, the original taste being preserved.
Reagent for preserving—
Butter, cheese.
Eggs by maintaining them in an atmosphere of the gas of sufficient concentration to maintain the pH of the white of the eggs at 7.6 (U. S. 1922143).
Fruit juices by removing the pulp first and then subjecting the clarified juice of normal acidity to the action of ultra-violet rays and finally charging the juice with the gas (French 483422).
Ice cream.
Meats in fresh state, fresh fruits, and vegetables by a mixture of gases containing carbon dioxide, carbon monoxide, and vapors of carbon bisulphide and chlorine (French 517191).
Reagent for—
Preventing explosions of dusts in the milling of cereal flours and other food products.
Ripening citrous fruits.
Reagent in—
Canning coffee, coconut, nuts, and other food products (used as a substitute for vacuum canning process).
Reagent in making—
Butter, cream, and cheese (used to prevent spoilage during and after manufacture).
Carbonated water by the reaction between barium dioxide and carbon dioxide gas under pressure (French 628630).
Carbonated soft drinks, carbonated milk, carbonated waters.
Ice cream (used as an inert atmosphere which cause the ice cream to become firmer and more tasty; also avoids oxidation processes and prevents contamination of the ice cream with bacteria).
Shatterproof grapes by treating them with the gas before they are placed in refrigerator cars.
Reagent in storing and shipping—
Apples, butter, cheese, fruits and fruit preparations, grapes, grain, eggs, whole and loose; meats, milk, various food preparations.

Fuel
Ingredient of—
Acetylene fuel compositions (added for the purpose of preventing the acetylene from burning with a smoky flame).
Reagent for—
Preventing spontaneous combustion of coal by storing the coal in an atmosphere rich in carbon dioxide.
Producing high intensity of flame by the atomization of petroleum and fine distribution of the particles fed to the burner nozzle.

Gas
Vehicle in—
Distillation of coaltar.

Glues and Adhesives
Reagent in making—
Glues, gelatins, and adhesive preparations (used for the purpose of neutralizing the excess of alkali used in the manufacturing process).

Insecticide
Ingredient of—
Mixtures containing various fumigating substances (used for the purpose of increasing the rate of penetration of the poisons into the tracheal system of insects).
Suggested for use as insecticide and fungicide.

Leather
Reagent for—
Recovering various solvents used in the manufacture of artificial leather.

Mechanical
Reagent for—
Removing certain types of boiler scale.

Metallurgical
Reagent for—
Preventing blowholes in the making of large steel castings.
Refining molten ferrochromium, ferromanganese, ferromolybdenum, ferrotungsten, and ferrovanadium (French 562351).
Repairing gasoline tanks and containers (used for mixing with the gasoline vapors still remaining in the tanks after the liquid gasoline has been removed, enough being added so that the mixture of gasoline and carbon dioxide vapors from the tanks no longer ignites; the tank then being welded in the usual manner with oxyacetylene or electric equipment).
Repairing and processing tanks containing casinghead gasoline, for example to burn holes through such tanks.
Welding containers of hydrocarbon gases and liquids.
Reagent in—
Cementation process.

Military
Reagent for—
Furnishing motive power for propelling torpedoes.

Miscellaneous
Ingredient of—
Fire-extinguishing compositions, the carbon dioxide being in solution in carbon tetrachloride (French 631980).
Reagent for—
Atomizing gasoline to produce a mixture burning with intense flame.
Congealing sandy soils to facilitate excavation.
Extinguishing fires in coal piles.
Humane killing of animals.
Inflating motor vehicle tires.
Operating bells and other signals on railways.
Preventing explosives of inflammable liquids.
Preventing and extinguishing fires in coal mines, ships, and electrical equipment, also fires caused by electricity.
Raising sunken ships.
Recovering volatile and inflammable solvents.
Safeguarding inflammable liquids from ignition by gases from internal combustion engines (French 519362).
Spray painting.
Testing for leaks in pipelines.
Tightners of bottles and other containers.

Carbon Dioxide (Continued)

Throwing water on fires (used in liquefied form).
Transporting inflammable and otherwise hazardous liquids, such as coaltar solvents and petroleum distillates.

Mining

Reagent for—
Preventing and fighting fires in mines.

Paint and Varnish

Diluent in making—
Carbon black pigment.
Zinc oxide and lithopone pigments in an extremely fine state of subdivision.
Reagent in making—
White lead by the wet process.

Perfume

Ingredient of—
Carbonated bath preparations.

Petroleum

Ingredient of—
Mixtures containing nitrogen used for filling empty spaces in gasoline tanks (French 519132).
Reagent for—
Moving gasoline and other inflammable distillates around the refinery.
Purifying and fractionating crude oil and petroleum distillates (Brit. 277946).
Vehicle in—
Distillation of petroleum products.

Pharmaceutical

In frigotherpary for treating certain types of skin diseases.
In artificial respiration (used in admixture with oxygen).
Suggested for use as a refrigerating agent in medicine and surgery.
Suggested for use as local anesthetic.

Plastics

Reagent for—
Recovering volatile and inflammable solvents.

Refrigeration

Active agent in—
Refrigerating installations on board ship, in milk plants, market places, abattoirs, chocolate plants, and in all locations where gas that may accidentally escape from the pipes and other parts of the refrigerating equipment must not be dangerous.
Refrigerant in making—
Ice.

Resins and Waxes

Reagent in making—
Light-colored rosin.

Rubber

Reagent for—
Inflating air bags used in the manufacture of rubber goods of various sorts.
Making cellular rubber products.
Various mechanical rubber goods.
Providing an atmosphere in the dry curing of rubber and also for maturing rubber.

Soap

Reagent in making—
Disinfectant soaps.

Sugar

Reagent for—
Eliminating lime from sugar juices in the carbonation process.

Textile

——, *Finishing*

Reagent for—
Boiling out cotton and woolen textiles.
Fireproofing textile fibers and fabrics.

——, *Manufacturing*

Reagent for—
Recovering volatile and inflammable solvents in the manufacture of Chardonnet or nitro rayon.

Tobacco

Reagent in—
Packing tobacco in tins.

Water

Reagent for—
Removing residual carbonate.
Treating water softened for use in boilers with soda and lime.

Wine

Reagent for—
Clarifying wines.
Making carbonated "sparkling" wines.
Moving wines about the plant.
Protecting wines against molds.

Carbon Dioxide (Solidified)

Synonyms: Carbon dioxide snow, Carbon dioxide ice, Dry ice.
French: Acide carbonique, solidfée; Dioxyde de carbone, solidfée.
German: Kohlensäureeis, Kohlenstoffsäureeis, Kohlensäureschnee, Kohlenstoffsäureschnee.

Abrasive

Ingredient of—
Compositions containing liquid condensation products of phenol and formaldehyde used for abrasive purposes (U. S. 1901324).
Various abrasive compositions in granular form.

Analysis

For making freezing mixtures in laboratory work.
For separating mixtures by freezing.
Reagent for—
Freezing samples in the analysis of rubber.
Source of carbon dioxide for laboratory purposes.

Brewing

Substitute for cylinder gas in carbonating beer and in treating beer equipment, such as beer kegs, vats, treating beer during brewing, partially disinfecting beer and improving the taste of beers and ales.
Used as motive power for moving beer in the bottling process.

Cement

For curing Portland cement.

Chemical

Reagent for making—
Carbonates, such as ammonium carbonate, barium carbonate, calcium carbonate, copper carbonate, lanthanum carbonate, lead carbonate, lithium carbonate, magnesium carbonate, manganese carbonate, nickel carbonate, pure carbonate of various metals and alkaline earth metals, zinc carbonate.
Acetylsalicylic acid, alphanaphthol-2-carboxylic acid, barium nitrate, benzaldehyde, betanaphthol-1-carboxylic acid, bismuth subcarbonate, calcium chloride.
Calcium permanganate by dissolving calcium manganate and passing the gas obtained from the solid carbon dioxide through the solution.
Calcium carbamate, hydrocyanic acid, heptincarboxylic acid, parahydroxybenzoic acid, potassium permanganate, synthetic urea, thiourea, tetramethyldiarsin, sodium perborate, salicylic acid.
Zinc oxide in fine granules (U. S. 1442265).
Reagent for neutralizing alkalies.
Source of gaseous carbon dioxide used for various chemical purposes, such as antioxidizing agent, diluent, drying agent, gaseous medium for distillations, precipitating agent, recovering inflammable solvents.

Construction

Source of carbon dioxide gas for use in loosening stuck pipe joints and connections and also for cleansing plumbing.

Dye

Source of carbon dioxide for extracting logwood so that oxidation is prevented in the process; also for the manufacture of various synthetic dyestuffs.

Electrical

For cooling the vacuum trap in valves and neon signs.
For maintaining inert atmosphere in transformers and preventing possible ignition of the oil.
Source of carbon dioxide for extinguishing fires in electrical equipment, such as generators, transformers, high tension fuses, electric ovens, and telephone switchboards.

Explosives

Source of carbon dioxide gas in making pyrotechnics and white fire.

Food

For modifying the atmosphere in cold storage rooms for eggs.
Refrigerating medium in—
Shipping frozen meats, fruits, vegetables, various foodstuffs in trucks, railroad refrigerator cars, ships.
Source of carbon dioxide in preserving eggs, fruit, milk.
Source of carbon dioxide for use as antioxidizing agent, leavening agent in baking bread and cake, improving taste of foods, making carbonated drinks, freezing-canning operations of various kinds, making ice cream, eliminating sulphur dioxide used in bleaching,

Carbon Dioxide (Solidified) (Continued)
ripening citrous fruits, fumigating grain and grain elevators.
Used for preserving fish on trawlers and for preserving ice cream in the frozen state.
Fuel
Source of carbon dioxide gas for making fuel compositions, such as acetylene mixtures, preventing spontaneous combustion of coal.
Insecticide
As an insecticide alone or in admixture with ethylene oxide.
Leather
Source of carbon dioxide gas for recovering solvents in the manufacture of artificial leather.
Mechanical
Reagent for—
Stopping flow in pipelines in an emergency.
Metallurgical
Reagent for—
Assembling light alloy aeroplane parts with air-hardened aluminum alloy rivets which have been held in refrigerated boxes at the assemblers' benches, so as to prevent premature hardening.
Chilling castiron cylinder linings and valve sleeves.
Hardening chromium steel, nickel steel, and nickel-silicon steel by chilling after machining, thus preventing changes in surface composition and the formation of scale due to heat.
Shrink-fitting machined parts.
Source of carbon dioxide gas in the cementation process, for preventing blowholes in castings and refining operations.
Mining
As an explosive in coal mining.
Refrigeration
General freezing agent.
Refrigerant in long-distance hauling of perishable products.
Rubber
Reagent for—
Processing golf balls, which are chilled by the solidified carbon dioxide to a consistency favoring neat trimming, this process being applicable to rubber and gutta-percha balls.
Sugar
Source of carbon dioxide in the carbonation of sugar juices.
Miscellaneous
Ingredient of—
Materials that are to be ground or mixed in the dry state (added for the purpose of preventing balling).
Reagent for—
Controlling fires in cellars, manholes, ships' holds, coal piles.
Freezing for repair purposes sections of piping carrying such liquids as sulphuric acid and the like.
Fumigating rooms, houses.
Making rain by distribution from airplanes above the clouds.
Refrigerant for—
Shipping flowers in trucks.
Refrigerating agent in refrigerator cars.
Source of carbon dioxide gas for various operations, such as fire extinguishing, inflating tires.
Water
For cleaning water wells so as to increase the flow.

Carbon Monoxide
French: Monoxyde de carbone.
German: Kohlenoxyd, Kohlenstoffoxyd.
Chemical
Reagent in making—
Ammonium cyanate, ammonium formate, benzaldehyde, carbon oxychloride with chlorine, carbon oxysulphide, formic acid, iron pentacarbonyl, phosgene, potassium formate, sodium formate, urea.
Starting point in making—
Methanol, ethylene.
Starting point (Brit. 269302) in making derivatives of formamide with—
Alphanaphthylamine, anilin, benzidin, benzylamine, betanaphthylamine, dianisidin, dibenzylamine, diethylanilin, dimethylanilin, diphenylamine, metaphenylenediamine, metatoluidin, methylethylanilin, monoethylanilin, monomethylanilin, naphthylenediamine, orthophenyldiamine, orthotoluidin, paraphenylenediamine, paratoluidin, phenylamine, phenyldimethylamine, phenylmethylamine, toluylenediamine, xylidin, xylylenediamine.
Fuel
Fuel gas used alone or in mixtures as water gas and producer gas.
Metallurgical
Reagent in making—
Special steels.
Reagent in reducing—
Refractory oxides.
Reagent in refining—
Nickel by the Bond process.
Paint and Varnish
Reagent in making—
High grade zinc white pigment.

Carbon Tetrachloride
Synonyms: Perchlormethane, Tetrachloromethane.
French: Chlorure de méthyle perchloré, Tétrachlorure de carbone.
German: Benzinform, Chlorkohlenstoff, Kohlenstofftetrachlorid, Perchlormethan, Tetrachorlkohlenstoff, Tetrachlormethan.
Spanish: Tetracloruro de carbono.
Italian: Tetracloruro di carbonio.
Analysis
Reagent in analyzing and testing—
Coffee, hops, ashes, mineral phosphates, palm oil, rosin, rosin oil.
Solvent in making toxicological examinations for the determination of strychnine and atropine.
Reagent in making color tests for—
Bromine, iodine.
Solvent for—
Alkaloids, bromine, iodine, fats, oils, resins, waxes.
Solvent in extracting—
Fats.
Solvent for the extraction and assay of drugs.
Solvent in isolating alkaloids.
Ceramic
Solvent in—
Compositions, containing nitrocellulose, cellulose acetate, or other esters of cellulose, as well as resins, waxes, and gums, used for coating and decorating ceramic ware.
Chemical
Ingredient of—
Mixed solvents, containing benzin, benzene and other inflammable substances (added for the purpose of decreasing their inflammability and making a non-inflammable mixture).
Reagent for—
Introducing chlorine in the manufacture of inorganic and organic compounds.
Reagent in making—
Ammonium carbonate, aromatics, carbineol, chloroform, chlorinated hydrocarbons, hexachloroethane, methane, novoiodine, paraoxybenzoic acid, pharmaceuticals, tetrachloroethylene, various intermediates and other organic compounds, viscin.
Solvent for extracting—
Atropine, strychnine.
Solvent for
Purifying organic pharmaceuticals and other compounds.
Solvent in making—
Acetic anhydride, alphapyrrolcarboxylic acid, 1:4-dichloronaphthalene, 1:5-dichloronaphthalene, various other organic compounds.
Solvent for—
Acetone, alkali cellulose, aluminum palmitate, aluminum stearate, benzene, benzin, bitumen, butyl alcohol, camphor, cellulose acetate, cellulose dinaphthenate, chloroform, chlorinated hydrocarbons and the like, coaltar naphthas, cumarone, ethyl acetate, ethyl alcohol, ether, methanol, nitrocellulose, propyl alcohol, rubber heptachloride, trichloroethylene.
Construction
Solvent for washing—
Tiles and tiled fronts of buildings.
Dye
Solvent in making—
Parafuchsin, various other dyestuffs.
Electrical
Solvent in—
Compositions, containing cellulose acetate, nitrocellulose, or other esters or ethers of cellulose, and at

Carbon Tetrachloride (Continued)
times resins, gums, and the like, used for insulating, cables, wiring and electrical machinery and equipment.
Substitute for oil in—
Electric transformers.
High-tension switches of high electrical resistivity.
Fats and Oils
Solvent for—
Fats, greases, oils.
Solvent in extracting fats and oils from—
Fatty materials, meals, oilseeds, press cakes, waste products.
Solvent in recovering—
Oils from fuller's earth and other substances used in bleaching the oils.,
Solvent in recovering—
Tallow.
Food
Preservative for—
Grain.
Solvent in extracting—
Caffeine from coffee.
Glass
Solvent in—
Compositions, containing cellulose acetate, nitrocellulose, or other esters or ethers of cellulose, and artificial or natural resins and waxes and gums, used for the manufacture of nonscatterable glass and for the decoration and protection of glassware.
Glues and Adhesives
Ingredient of—
Special adhesive compositions containing cellulose acetate, nitrocellulose, or other esters or ethers of cellulose.
Ingredient (U. S. 1594522) of—
Adhesive preparations.
Solvent for decreasing—
Bones and hides for the manufacture of bone and hide glue and gelatin.
Gums
General solvent.
Solvent for—
Dammar, mastic, sandarac.
Ink
Reagent in making—
Printing inks.
Insecticide
As a cereal insecticide.
For fumigating weevil-infected wheat.
Ingredient of—
Insecticidal preparations (in admixture with ethylene dichloride).
Insecticide for controlling—
Grain weevil, peach-tree borer, phylloxera, San Jose scale.
Leather
Ingredient of—
Shoe polishes.
Solvent for—
Cleansing spotted leathers.
Removing natural oils and greases from hides before tanning so as to prevent staining thereafter and insure evenness of the leather finish and tan.
Solvent in—
Compositions, containing cellulose acetate, nitrocellulose, or other esters or ethers of cellulose, as well as artificial or natural resins, gums, and waxes, used in the manufacture of artificial leather and for the protection and decoration of leather goods.
Mechanical
Solvent for—
Cleansing machinery.
Recovering oil from cotton and wool waste.
Removing oils and greases from leather belting and the like.
Metallurgical
Solvent for—
Cleansing metals preparatory to further treatment.
Degreasing metal parts and castings preparatory to plating, varnishing, galvanizing, shellacking.
Solvent in—
Compositions, containing cellulose acetate, nitrocellulose, or other esters or ethers of cellulose, as well as artificial or natural resins, waxes, and gums, used for the protection and decoration of metal ware.
Metal polishes.
Miscellaneous
As a general sovlent.
As a delousing agent.
For the standardization of thermometers.
Ingredient and solvent in making—
Compositions for rendering fibrous materials transparent or translucent.
Compositions for repelling moth (in admixture with ethylene dichloride).
Compositions, containing clay, for cleansing ivory, horn, and bone.
Compositions, containing waxes, etc., used for polishing furniture.
Compositions for the fumigation of furs.
Compositions used as fire extinguishers.
Preparations for cleansing internal combustion engines.
Preparations for cleansing electric motors.
Preparations used for waxing purposes.
Preparations used for the removal of stains from celluloid articles.
Preparations used for cleansing typewriters.
Solvent compositions, containing ethylene dichloride, used for a variety of purposes.
Reagent for—
Detecting watermarks in stamps and paper.
Solvent in—
Compositions, containing cellulose acetate, nitrocellulose, or other esters or ethers of cellulose, with gums, waxes, and artificial or natural resins, used for the decoration and protection of fibrous compositions of matter.
Oilcloth and Linoleum
Reagent in making—
Coating compositions.
Paint and Varnish
Ingredient of—
Paint, lacquer, and varnish removers.
Stains.
Thin staining lacquers.
Viscous dipping lacquers.
Reagent in making—
Dry colors.
Solvent in making—
Fat lacquers and varnishes.
Lacquers, varnishes, enamels, and dopes containing cellulose acetate, nitrocellulose, or other esters or ethers of cellulose, with waxes, gums, and artificial or natural resins.
Paper
Solvent for—
Removing oil from paperstock, reworking newsprint.
Solvent in—
Compositions, containing cellulose acetate, nitrocellulose, or other esters or ethers of cellulose, with gums, waxes, and natural or artificial resins, used in the manufacture of coated paper and for coating and decorating paper and pulp products.
Perfumery
Solvent in extracting—
Perfumes and essential oils from flowers.
Petroleum
Solvent for—
Gasoline, paraffin, petroleum.
Pharmaceutical
In compounding and dispensing practice.
Photographic
Reagent in treating—
Edges of sound films.
Solvent in removing—
Stains from films.
Plastics
Solvent in making—
Compositions, containing cellulose acetate, nitrocellulose, or other esters or ethers of cellulose, with gums, waxes, and artificial or natural resins.
Printing
Solvent for cleaning—
Engraved plates, type, printing machinery, lithographic stones.
Refrigeration
As a refrigerating medium.
Resins and Waxes
Solvent for—
Kauri, shellac (when used with alcohol), soft copal.
Solvent in making—
Liquid wax preparations.
Solvent in extracting—
Waxes from raw materials.

Carbon Tetrachloride (Continued)

Rubber
Ingredient of—
Rubber cements possessing non-inflammable properties.
Rubber compositions used in the manufacture of rubberized cloth.
Solvent for—
Rubber, splicing acid.
Solvent in—
Compositions, containing cellulose acetate, nitrocellulose, or other esters or ethers of cellulose, with gums, waxes, and artificial or natural resins, used for the decoration and coating of rubber goods.

Sanitation
Solvent in—
Degreasing garbage.

Soap
Ingredient of—
Cleansing compositions, dry-cleaning compositions, spotting fluids.
Solvent in making—
Gelatinous water-soluble soaps from sulphonated oils and resins.
Paste soaps for removing grease stains.
Soaps with sodium ricinoleate.
Textile soaps from linseed oil and castor oil.

Stone
Solvent in—
Compositions, containing cellulose acetate, nitrocellulose, or other esters or ethers of cellulose, with artificial or natural resins, gums, and waxes, used for the decoration and protection of artificial and natural stone.

Textile
——, *Bleaching*
Solvent in—
Linen bleaching process, carried out in kiers.
——, *Finishing*
Ingredient of—
Scouring compositions containing sulphonated oil soaps.
Solvent in—
Coating compositions containing cellulose acetate, nitrocellulose, or other esters or ethers of cellulose.

——, *Manufacturing*
Solvent in—
Scouring wool.
Reagent in making—
Cellulose compounds which are indifferent to substantive colors.

Woodworking
Solvent in—
Compositions, containing cellulose acetate, nitrocellulose, or other esters or ethers of cellulose, used for decorating and protecting woodwork.

Carbonyl Chloride
Synonyms: Phosgene.
French: Chlorure de carbonyle, Chlorure carbonylique, Phosgène.
German: Carbonylchlorid, Phosgen.

Chemical
General chlorinating reagent.
Reagent in making—
Acetic anhydride, acetyl chloride, anthraquinone-10-carboxylic acid, antipyrin derivatives, aristochin, benzoxazolonearsinic acid (Brit. 439605), chlorocarbonic acid esters, creosote carbonate (creosotal), diethylbarbituric acid (Barbital, Veronal), diphenyl carbonate, dipropaesin, guaiacol carbonate (Duotal), methyl chloride, methylorthoaminophenol, para-p′-tetramethyldiaminobenzophenone, phenyl isocyanate, phenyl salicylate (Salol), quinine carbonic acid ethyl ester, santalol carbonate (Blenal), symmetrical dimethyldiphenylurea, thionyl chloride, thyresol, urea.

Dye
Reagent in making—
Azo dyestuffs, benzo fast orange, benzo fast red, benzo fast rose red, benzo fast scarlet, benzo fast yellow, benzo scarlet, brilliant sulphon red B, cotton yellow, ethyl violet, helindon yellow 3GN, methyl violet.
Soluble vat dyestuffs with the aid of dimethoxydibenzanthrone (Brit. 277398).
Triphenylmethane dyestuffs.

Glass
Bleaching agent in treating—
Sand for making fine glass.

Military
Poison gas.

Miscellaneous
Poison gas for various industrial and agricultural purposes.
Ingredient (Brit. 255101) of—
Cleansing and polishing compositions for floors, linoleum, and the like.

3-Carboxyphenylthiocarbimide

Chemical
Starting point (Brit. 314909) in making derivatives with—
Alkoxylalphanaphthalenesulphonic acid.
Alpha-amino-5-naphthol-7-sulphonic acid.
Alphanaphthylamine-4:8-disulphonic acid.
Alphanaphthylamine-4:6:8-trisulphonic acid.
4-Aminoacenaphthene-3:5-disulphonic acid.
4-Aminoacenaphthene-3-sulphonic acid.
4-Aminoacenaphthene-5-sulphonic acid.
4-Aminoacenaphthenetrisulphonic acid.
Aminoarylcarboxylic acids.
Aminoheterocyclic carboxylic acids.
1:8-Aminonaphthol-3:6-disulphonic acid.
Bromonitrobenzoyl chlorides.
Chloroalphanaphthalenesulphonic acids.
Chloronitrobenzoyl chlorides.
Iodonitrobenzoyl chlorides.
Nitroanisoyl chlorides.
Nitrobenzene sulphochlorides.
Nitrobenzoyl chlorides.
2-Nitrocinnamyl chloride.
3-Nitrocinnamyl chloride.
4-Nitrocinnamyl chloride.
1-Nitronaphthalene-5-sulphochloride.
1:5-Nitronaphthoyl chloride.
2-Nitrophenylacetyl chloride.
4-Nitrophenylacetyl chloride.
Nitrotoluyl chlorides.

Carnauba Wax
Synonyms: Brazil wax.
French: Cire de brasil, Cire de carnauba.
German: Brasilienwachs, Carnaubawachs.

Disinfectant
Ingredient (Brit. 358508) of—
Antiseptic compositions, containing such active substances as bis-2-hydroxyphenyl oxide and 4-hydroxyphenylamine.

Electrical
Ingredient of—
Compositions used in the manufacture of electric cables and wires.
Insulating compositions used in motors and other electric machinery and apparatus.
Waterproofing compositions used on electric appliances.

Fuel
Ingredient of—
Compositions used for making candles (added for the purpose of making the candles harder and more durable).

Insecticide
Ingredient (Brit. 358508) of—
Compositions, containing active ingredients of the type of bis-2-hydroxyphenyl oxide, 4-hydroxyphenylamine, and the like, used for treating cattle and other domestic animals to rid them of pests and also for disinfecting and preserving seeds and plants.

Leather
Ingredient of—
Compositions for cleaning white leather.
Compositions for dressing various leathers.
Compositions for polishing leather and leather goods.
Compositions for applying waterproofed coatings to leather goods.

Linoleum and Oilcloth
Ingredient of—
Compositions used for finishing linoleum.

Miscellaneous
Ingredient of—
Automobile polishes.
Compositions, containing active ingredients of type of bis-2-hydroxyphenyl oxide, 4-hydroxyphenylamine, and the like, used as polishes and as preservatives, for example in the treatment of catgut and in the mothproofing of feathers, furs, skins (Brit. 358508).
Compositions used for making heel balls.
Compositions used for making phonograph cylinders and graphaphone records.

Carnauba Wax (Continued)
Compositions for marking cloth (U. S. 1622353).
Compositions for waterproofing purposes.
Furniture polishes, shoe polishes.
Reagent in making—
Physical apparatus.
Substitute for beeswax in various compositions.
Paint and Varnish
Ingredient of—
Wax varnishes, enamels, lacquers, and the like used for special purposes.
Wood-finishing waxes.
Paper
Ingredient of—
Compositions used for stiffening cardboard containers used in place of tin cans.
Compositions for making carbon paper.
Compositions for making waxed colored paper.
Perfume
Ingredient (Brit. 358508) of—
Cosmetic preparations, containing active ingredients of the type of bis-2-hydroxphenyl oxide and 4-hydroxyphenylamine.
Resins and Waxes
Ingredient of—
Compositions containing other natural and synthetic waxes (added for the purpose of hardening the product and rendering it highly lustrous).
Textile
Ingredient (Brit. 358508) of—
Mothproofing compositions containing the active ingredients bis-2-hydroxyphenyl oxide and 4-hydroxyphenylamine.
Woodworking
Ingredient (Brit. 358508) of—
Preservative compositions containing the active ingredients bis-2-hydroxyphenyl oxide and 4-hydroxyphenylamine.

Carob Bean
Synonyms: Locust bean, St. John's bread.
French: Fèves de caroube, Fèves de carouge, Fèves de locuste.
German: Johanisbrotbohnen.
Spanish: Carruba.
Italian: Carruba.
Chemical
Starting point in making—
Ethyl alcohol.
Food
As an article of food (the whole fruit).
Used in place of coffee in certain countries.
Ingredient (U. S. 1150607) of—
Food compositions containing chestnuts, potato flour, salep, fats, gum arabic, and vanilla flavoring.
Gums
Starting point in making—
Carob gum, tragasol.
Leather
Ingredient of—
Compositions used for the purpose of increasing the weight of leather and also to accelerate the tanning process.
Miscellaneous
For making stockfeed.
Pharmaceutical
Ingredient of—
Cough mixtures.
Suggested for use as nutrient.

Carob Bean Gum
Synonyms: Industrial gum, Locust bean gum, Locust kernel gum, Tragasol.
French: Gomme de caroube, Gomme de carouge, Gomme de locuste.
German: Johanisbrotgummi.
Italian: Gomma di carruba, Gomma di locusta, Gomma di tragasola.
Chemical
Ingredient of—
Colloidal preparations of chemicals, metals, and the like, such as selenium (used as a protective colloid in place of gum tragacanth).
Food
Ingredient (Brit. 24877-1894) of—
Confectionery.
Glues and Adhesives
Ingredient of—
Mucilages and other adhesive preparations.
Leather
Ingredient of—
Dressing compositions.
Preparations for accelerating tanning action (Collegium 1924, 137).
Weighting compositions.
Miscellaneous
Binder in making—
Various compositions of matter.
Emulsifying agent in making—
Various dispersed product (used in place of gum tragacanth).
Nutrient medium in bacteriological work.
Perfume
Used in place of gum tragacanth in cosmetic preparations.
Textile
——, *Dyeing and Printing*
Ingredient (Brit. 8793-1893) of—
Dye baths and printing pastes.
——, *Manufacturing*
Ingredient (Brit. 8793-1893) of—
Compositions used in weaving cloth.

Carvacrolphthalein
Synonyms: Carvacrol-phtaleine.
French: Phthaleine de carvacrole.
German: Carvacrolphtalein.
Spanish: Carvacrolftaleina.
Italian: Carvacrolftaleina.
Pharmaceutical
Suggested for use as—
Laxative.

Carvone
Synonyms: Carvol.
Beverage
Aromatic material for—
Liqueurs (to impart a caraway aroma).
Flavoring material for—
Liqueurs (to impart a caraway taste).
Soft drinks.
Food
Flavoring material for—
Baker's products, confectionery.
Miscellaneous
Flavoring material for—
Chewing gum, dental preparations.
Perfume
Aromatic material for—
Cosmetics, perfumes.
Pharmaceutical
In dispensing and compounding practice.
Soap
Aromatic material for—
Toilet soaps.

Cascarilla
Synonyms: Cascarilla bark, Sweet bark, Sweet wood bark, Eleuthera bark.
Latin: Cascarillae cortex, Cortex eluteriae, Cortex thuris, Quina aromatica.
French: Écorce de cascarilla, Écorce de bois, douce; Chacrille, Écorce éleuthérienne, cascarill.
German: Cascarillabast, Cascarillaborke, Cascarillohe, Cascarillarinde, Kaskarillrinde, Sussholzbast, Sussholzrinde.
Spanish: Chacarilla, Quina aromatica.
Italian: Cascarilla, Cascariglia.
Fats and Oils
Starting point in extracting—
Cascarilla oil.
Food
Ingredient of—
Flavoring compositions.
Insecticide
Ingredient of—
Insecticidal preparations.
Miscellaneous
Ingredient of—
Fumigating preparations.
Perfume
Ingredient of—
Dentifrices, pastilles.

Cascarilla (Continued)

Pharmaceutical
In compounding and dispensing practice.

Tobacco
Flavoring for—
Chewing and smoking tobaccos.

Casein
Synonyms: Lactarene.
Latin: Caseinum.
French: Caillebotte, Caséogomme, Caséine.
German: Casein, Käsestoff, Milchcasein.
Spanish: Caseina.
Italian: Caseina.

Abrasive
Ingredient of—
Compositions used as backing for cloth before the application of the glue in the manufacture of emery cloth.

Agriculture
As a spread in various insecticidal and fungicidal sprays.
Used in treating vine diseases.

Analysis
Reagent in—
Determining effectiveness of various digestive ferments, such as pepsin and trypsin.
Testing for formaldehyde.

Ceramics
Ingredient of—
Compositions used for the manufacture of potteries and porcelains (added for the purpose of increasing the hardness of the finished ware).

Chemical
Starting point or reagent in making—
Acrolein compound, albumen, alkaloid compounds, aluminum caseinate, ammonium caseinate (Eucasin), argonin (silver caseinate) (pharmaceutical), arkase (pharmaceutical), biosan (casein-iron preparation), bismuth caseinate, biformic iodide preparation, bromine compounds, calcium caseinate (Protolac), cargel (pharmaceutical), cargento (pharmaceutical), casein citrate, casein compositions containing arsenic, casein compositions containing cocaine, casein compositions containing tannic acid, casein-formaldehyde compositions used as antiseptics and the like, caseinhydrol, casein-hydrobromic acid, casein-hydriodic acid, casein-hydrofluoric acid, casein-iodide, casein oxalate, casein phosphate, caseinphosphorol, copper caseinate, eucasin (ammonium caseinate) (German 84682), ferric caseinate, iodocasein, iron-casein preparations, mercury caseinate, nutrose (sodium caseinate, special nutrient), odda (pharmaceutical nutrient), peptones, periodocasein, plasmon, proferrin (pharmaceutical), protan (pharmaceutical), protolac (calcium caseinate, special nutrient), saccharated casein, sanose (containing albumens), santogen (containing sodium glycerinophosphate), silver caseinate (Argonin), sodium caseinate (nutrose), triferrin (iron-casein composition), various casein compositions containing opium alkaloids.

Construction
Ingredient of—
Compositions for damproofing walls, insulating cements, mortars, plaster lath compositions.

Dye
Ingredient of—
Color lakes, nonpoisonous rhodamine and eosin lakes.

Electrical
Ingredient of—
Insulating compositions for coating wires and parts of electrical machinery and equipment.

Food
Component of various food compositions.
Filler in—
Ice cream.
Ingredient of—
Artificial butters, artificial compositions used in the place of eggs, albuminous milks, baker's wares, baking powders, cheeses, children's foods, cocoas, diabetic foods, diaprotein preparations made with casein flour, dyspeptic foods, easily digestible and nutrient foods, such as soups, coffee, tea, and the like, the casein being used in the soluble form (caseinogen), infants' foods, meat extracts, malted milks, milk chocolates, modified milks, oleomargarin, reconstituted and synthetic foods, reconstructed milks, sausages, soup tablets.
Reagent for—
Decolorizing and clarifying fruit juices.

Explosives
Ingredient of—
Compositions used for the manufacture of matches.

Fats and Oils
Ingredient of—
Emulsifying compositions.

Fuel
Binder in making—
Fuel briquettes from coal dust and the like.

Glues and Adhesives
As a gelatin substitute.
Ingredient of—
Borax glues.
Compositions for fastening paper bags over mature flowers of plants to prevent uncontrolled pollination in breeding studies.
Compositions for mending glass, china, porcelain, meerschaum.
Compositions for making cardboard boxes.
Compositions for cementing cork or paper discs to metal shells in bottle caps.
Compositions for sealing paper on cigarets.
Glues for attaching linoleum to wood and cement.
Glues for making plywood and veneered panels and furniture.
Glues for attaching heels to shoes.
Glues for cementing metals.
Glues for cementing stone.
Glues for attaching paper labels to tin cans and glassware.
Latex glues, liquid glues, washable cements for boards, water-resistant glues, wood glues.
Reagent in—
Clarifying glues and gelatins.

Ink
Ingredient of—
Common inks, intaglio inks, and printing inks (U. S. 1621541-3).
Printing inks, containing borax, glycerin, oil of citronella, carbolic acid, and borax in aqueous medium (U. S. 1724603).

Leather
General finishing reagent in treating leather goods.
Ingredient of—
Compositions used in the manufacture of artificial leathers.
Dressing compositions for leather and leather goods.
Finishing compositions for treating light leathers, such as sheepskins, heavy grades of stock, and heavy splits of cowhide.
Finishing compositions for coating heavy goods colored and embossed to imitate leather.
Pigment finishing compositions.
Seasoning compositions for treating leather and leather goods.
Tanning compositions (used in the place of blood albumen).
Reagent for—
Decolorizing tanning extracts.

Linoleum and Oilcloth
Ingredient of—
Compositions with linseed oil used for making linoleum and oilcloth.

Mechanical
Ingredient of—
Anticorrosion preparations.
Antiradiation coverings for steam pipes and other equipment.
Asbestos compositions used for the manufacture of high-pressure steam gaskets.
Brake-shoe fillings and linings containing cement, blood, asbestos (U. S. 1724718).
Facings for brake linings, with admixture of asbestos.

Miscellaneous
Ingredient of—
Anticorrosion compositions, antiradiation compositions.
Compositions for making cartridge boxes and cases, buckets, bags, and the like from paper (used to wearproof and strengthen the paper).
Compositions containing cork (used as binder).
Compositions with formaldehyde for glazing casks.
Compositions containing plaster of paris, used for cooperage luting.
Compositions for priming artists' canvas.
Compositions used for making picture moldings.

Casein (Continued)
Compositions containing colored micas.
Compositions containing various ingredients and used as a substitute for cork.
Compositions for treating straw to render it impermeable.
Compositions containing bituminous substances used for treating and surfacing roads (Brit. 251098).
Emulsions containing woodtar and bitumens.
Liquid court plasters.
Shoe polishes and creams.
Reagent in making—
Artificial horse hair.
Substitute for gelatin, gums, shellac, and albumen in various compositions.

Paint and Varnish
Ingredient of—
Anilin dye paint for marking bags, iron barrels, and cases.
Asbestos paints for fireproofing wood and canvas.
Black casein paints, blue casein paints, boiled oil substitutes, calcimine washes, casein distempers, casein enamels, casein facade paints, casein lime paints, cold water paints, casein cement paints, encaustic paints, external washable cold water paints, interior paints, fireproofing paints for use on stage curtains and scenery, gloss enamels.
Latex casein paints for use on paper, cloth, leather, concrete and brickwork.
Marble lime colors for outside work.
Milk paints, containing soap, slaked lime, and turpentine.
Moulders' paints for making steel castings with clean surfaces.
Oleo-casein paints for use on wood, metal, stone, and stucco.
Paint and varnish removers.
Paints containing satin white.
Paints for marking bags, barrels, cases.
Putties, quick-drying paints, roofing pulps, sanitary calcimines.
Sodium silicate paints for painting very damp rooms, stone, brick and fresh dry plaster (lime or cement).
Stenciling paints, street-marking paints, stucco water paints, water color paints, waterproof paints, water-white casein varnishes (German 200919), wax color binding compositions.
Zinc white casein paints for use on paper, cloth, leather, wood and stone.
Reagent in making—
Formolactin, a formaldehyde product used for making antiseptic paints and varnishes for use in hospitals, dairies, etc.
Reagent in—
Treating ultramarine and similar pigments to make them usuable with oil in the manufacture of oil paints (Brit. 224273).
Substitute for shellac and linseed oil.

Paper
Assistant in rosin sizing process.
Ingredient of—
Compositions containing magnesium oxide and lime, used for treating paper slates and drawing paper to make them erasable.
Compositions for mothproofing paper bags for storing clothing and the like.
Compositions for enamelling paper and pulp.
Compositions for coating photographic papers.
Compositions for making paper and pulp and compositions resistant to tearing and proof against water, oil, rust, and grease and suitable for making sacks, cartons, wrappers, blue prints, photographic and lithographic papers, posters, documents, sand and emery papers, parchment substitutes, wrappers for bread, tobacco, and sugar, bags for lime, cement, flour, paints, and various hygroscopic commodities.
Compositions for making washable and antiseptic wallpaper.
Compositions for sizing halftone printing paper.
Compositions for sizing art printing papers.
Compositions for making transfer papers.
Compositions for applying the finishing coat on heavy wallpaper to imitate leather.
Compositions (used as a size) in varnishing tile paper so as to provide a better support for the varnish.
Compositions, containing clays, alum, and lime, used for sizing high-grade half-tone paper.
Compositions for making strong durable, waterproof, and fireproof asbestos paper, board (casein used in the place of fish glue).
Compositions containing shellac used for forming the top varnish on playing cards or applying top coat on waterproofed paper.
Sizing compositions for coated or enamelled paper, onion-skin, writing paper, oilproof and waterproof art paper, metachrome papers.
Sizing compositions for surface sizing writing paper (casein used in the place of glue).
Reagent in—
Sizing paper pulp in the beater by the rosin sizing process.

Perfume
Ingredient of—
Casein creams, massage creams.
Perfume compositions (added for the purpose of retaining the perfume).

Petroleum
Ingredient of—
Solidified compositions containing petroleum oils or distillates.

Pharmaceutical
Added to pharmaceutical preparations for the purpose of promoting their toleration without diminishing the medical action of the drug itself.
As an emulsifying agent.
Ingredient of—
Dermatological applications.
Vehicle for pharmaceutical preparations containing heavy metals, tannins, alkaloids, salicylates, iodides, and the like.

Photographic
Ingredient of—
Compositions used in the manufacture of films and plates.
Reagent in making—
Photographic prints and plates (French 151014, German 202108, and Brit. 19297-1908).
Sensitizing solutions for making casein pigment prints.

Plastics
Ingredient of—
Artificial horn preparations.
Casein-phenol-formaldehyde plastics.
Casein plastics for making buttons, buckles, electrical insulators, fountain pens, pencils, combs, beads, brush backs, manicure sets, cuticle sticks, paper knives, teething rings, cigaret holders, millinery ornaments, chessmen, checkers, dominos, cane and umbrella handles, novelties.
Casein-cellulose plastics.
Compositions for making covers for floors and walls.
Fireproof cellulose substitutes.
Galalith.
Imitation ivory, mother of pearl, shell, bone.
Thermoplastics.
Reagent in—
Treating celluloid to reduce its inflammability (used in place of camphor).

Printing
In bookbinding, in lithography.

Rubber
Ingredient of—
Hard rubber.
Rubber, gutta-percha, or balata latex compositions (Brit. 253740).

Soap
Ingredient of—
Buttermilk soaps, detersive compositions, milk soaps.
Toilet soaps (added for the purpose of increasing the firmness and lathering properties).

Textile
——, *Dyeing*
Reagent—
Fixing insoluble dyestuffs.
Fixing zinc white on cotton with the aid of formaldehyde.
Mordanting cotton yarns and fabrics so that they can be dyed with acid dyestuffs.

——, *Finishing*
Ingredient of—
Compositions used for giving cloth a metallic finish.
Compositions for making coated airplane cloth (U. S. 1521055-6).
Compositions for giving high gloss to fabrics.
Compositions for fixing mineral pigments so that they are fast to washing.
Compositions for finishing and waterproofing cloth in general.

Casein (Continued)
Compositions for making mercerized crepes.
Dressing compositions for linens.
Loading compositions for silk and cotton.
Detergent compositions (added to increase the detergent action, the casein being used in alkaline solution).
Sizing compositions.
Softening compositions.

——, *Manufacturing*
Ingredient of—
Adhesive preparations used in the manufacture of double cloths.
Reagent in making—
Rayon.
Starting point in making—
Threads or fibers in coagulated form (French 356508).

——, *Printing*
Thickener in—
Pastes for printing calico.

Wine
As a clarifying agent.

Woodworking
Ingredient of—
Artificial wood compositions.
Compositions used as adhesives in the manufacture of built-in wooden propellers and other parts of airplanes.

Cashew Nut Shell Oil
French: Huile de coque de noir d'acajou.
German: Elephantenlausrindeoel, Kaschunussrindeoel.

Electrical
Starting point (Brit. 272510) in making—
Compositions for insulating purposes.

Fats and Oils
Starting point (Brit. 272509) in making—
Violet coloring matter for fats and oils.

Glues and Adhesives
Starting point (Brit. 272510) in making—
Cement and adhesive ingredients.

Gums
Starting point (Brit. 272509) in making—
Pigments for gums.

Ink
Starting point (Brit. 272509) in making—
Ink pigments.
Printing inks with linseed oil, oleic acid, tung oil.

Miscellaneous
Starting point (Brit. 272510) in making—
Waterproofing reagents.

Oilcloth and Linoleum
Starting point (Brit. 272509) in making—
Pigments.

Paper
Starting point (Brit. 272510) in making—
Cardboard finishing reagents, paper finishing reagents.

Plastics
Starting point (Brit. 272509) in making—
Coloring matter for dyeing cellulose acetate and cellulose nitrate plastic compositions.

Paint and Varnish
Starting point (Brit. 272509) in making pigments used in—
Enamels, cellulose lacquers, lacquers, paints, stains, varnishes.

Resins and Waxes
Starting point (Brit. 272509) in making—
Pigment for coloring coumarone resin.

Rubber
Starting point (Brit. 272509) in making—
Pigment.

Castor Oil
Synonyms: Palma christi seed oil, Ricinus oil.
Latin: Oleum palmae christi, Oleum ricini, Oleum e semine ricini.
French: Huile de castor, Huile de ricin.
German: Ricinusoel, Rizinusoel.
Spanish: Aceite de ricino.
Italian: Olio di ricino.

Chemical
Starting point in making—
Pimelic acid, normal.
Oenanthol, octin-1, sulphoricinoleates.
Starting point (Brit. 310941) in making emulsifying agents for—
Alcohols, chlorohydrin, hydrogenated phenols, ketones.

Dye
Starting point (Brit. 310941) in making emulsifying agents for—
Anilin dye pastes.

Electrical
Ingredient of—
Insulating compositions.

Fats and Oils
Ingredient (Brit. 310941) of—
Boring oils containing xylene.
Starting point (Brit. 310941) in making—
Splitting agents, emulsifying agents.
Starting point in making—
Sulphonated oils, turkey red oil.

Glues and Adhesives
Ingredient of—
Casein glue compositions.

Illumination
Illuminant in lamps for special purposes.

Insecticide
Ingredient of—
Fly oils for cattle, fly "dope" for outers, fly-paper coatings, fungicidal compositions (French 566406), insecticidal compositions (French 566406).

Leather
Ingredient (Brit. 310941) of—
Treating compositions containing xylene.
Ingredient of—
Leather varnishes, tanning compositions, various leather dressings.
Preservative in treatment of—
Boots and shoes, harness leather, leather belting.
Softening agent in treatment of—
Boots and shoes, harness leather, leather belting.
Reagent in making—
Artificial leather.

Mechanical
Ingredient of—
Lubricants for automobile and airplane motors.
Lubricants for fine machinery, especially those operated at high speeds or at low temperatures.
Lubricants for racing sulkies and light horse-drawn vehicles.

Miscellaneous
Ingredient (Brit. 310941) of—
Cleaning compositions, in combination with xylene.
Ingredient of—
Waterproofing compositions for fibrous substances (Brit. 251961).

Paint and Varnish
Ingredient (Brit. 310941) of—
Mineral pigment pastes, in combination with xylene.
Ingredient of—
Pyroxylin lacquers, to prevent cracking.
Starting point in making—
Nitrated product used in acetone solution as a lacquer.

Paper and Pulp
Ingredient of—
Waterproofing compositions for treating paper and paper products (Brit. 251961).

Perfumery
Ingredient of—
Hair dressings, shampoos, toilet creams.

Petroleum
Ingredient (Brit. 310941) of—
Mineral oil emulsions, in combination with xylene.

Pharmaceutical
In compounding and dispensing practice.

Rubber
Ingredient of—
Rubber substitutes.

Sanitation
Ingredient of—
Cleansing and disinfecting compositions (French 566406).

Soap
Reagent in making—
Liquid soaps, shaving soaps, special toilet soaps, textile soaps, transparent soaps.

Textile
——, *Bleaching*
Ingredient (Brit. 310941) of—
Bleaching baths, in combination with xylene.

Castor Oil (Continued)

——, *Dyeing*
Ingredient (Brit. 310941) of—
Dye baths, in combination with xylene.
Assist in dyeing—
Cotton yarns, cotton fabrics, with alizarin.

——, *Finishing*
Ingredient of—
Waterproofing compositions for textile fabrics (Brit. 251961).
Ingredient (Brit. 310941) of—
Finishing compositions, in combination with xylene.
Washing compositions, in combination with xylene.

——, *Manufacturing*
Ingredient (Brit. 310941) (in combination with xylene) of—
Carbonizing liquors, mercerizing liquors, spinning oils.

Waxes and Resins
Ingredient (Brit. 310941) of—
Emulsions with waxes, in combination with xylene.

Castor Oil Fatty Acids
French: Acides grasses d'huile de ricin.
German: Ricinoelfettsaeure, Ricinusoelfettsaeure, Rizineoelfettsaeure, Rizinusoelfettsaeure.

Chemical
Starting point in making—
Esters and salts of the acids.

Dye
Emulsifying agent in making—
Color lakes and oil colors.

Fats and Oils
Ingredient (Brit. 313453) of—
Fat and oil splitting compositions.
Lubricating and greasing compositions.

Ink
Ingredient of various products.

Insecticide
Ingredient of—
Insecticidal and germicidal compositions.

Leather
Ingredient (Brit. 313453) of—
Treating and finishing compositions.

Miscellaneous
Ingredient (Brit. 313453) of—
Bleaching composition, cleansing compositions, emulsifying compositions, purifying compositions, washing compositions, wetting compositions.
Starting point in making—
Polishing compositions.

Paper
Ingredient (Brit. 313453) of—
Compositions used in the treatment and coating of paper.

Perfume
Ingredient of—
Cosmetics.

Pharmaceutical
As a coating for pills.
In compounding and dispensing practice.

Plastics
Ingredient of various compositions.

Resins and Waxes
Ingredient (Brit. 313453) of—
Wax-splitting compositions.
Ingredient of—
Resin and wax compositions.

Soap
Starting point in making—
Special soaps.

Textile
——, *Dyeing and Printing*
Fixing agent (Brit. 313453) in—
Dyeing with basic dyestuffs.
Ingredient of—
Dye baths and printing pastes.
Stabilizing agent (Brit. 313453) in—
Dyeing with vat dyestuffs.

——, *Finishing*
Ingredient of—
Finishing compositions, wetting baths.

——, *Manufacturing*
Ingredient of—
Oiling compositions.

Catechu (Black)
Synonyms: Black catechu, Cutt, Pegu catechu, Cutch, Pegu catechu.
Latin: Catechu nigrum, Terra japonica.
French: Cachou de pégu, Cutch, Cachou.
German: Phgu, Katechu.
Spanish: Catecu.
Italian: Catto, Caticu.

Chemical
Starting point in making—
Cutch.

Leather
Finishing agent for special grades of leather.

Pharmaceutical
In compounding and dispensing practice.

Textile
——, *Dyeing*
Assist in dyeing fabrics and yarns.

Cedarwood Oil
Synonyms: Cedar oil, Oil of red cedar wood.
Latin: Oleum ligni cedri, Oleum juniperi Virginianae.
French: Essence de bois de cédre, Huile de cédre.
German: Cederholzoel, Zedernoel.
Italian: Olio di cedro.

Analysis
Ingredient (Brit. 306119) of—
Spectroscopic fluids.
Reagent in—
Microscopic work.

Chemical
Source of—
Cedrene, cedrol.

Insecticide
Ingredient of—
Dusting compounds, moth repellants, sprays.
Various insecticidal compositions.

Miscellaneous
Ingredient of—
Carbon remover (U. S. 1878245).
Carbon remover containing also acetone, benzene, camphorated oil, denatured alcohol, and turpentine (U. S. 1869310).
Cleansing and polishing liquid (U. S. 1758317).
Furniture polish (U. S. 1739332).
Sweeping compounds (U. S. 1758735).

Perfume
Ingredient of—
Perfumes.
Odorant in—
Toilet preparations.

Pharmaceutical
In compounding and dispensing practice.

Soap
Ingredient of—
Disinfecting soaps.
Odorant in—
Toilet soaps.

Woodworking
As a polishing and finishing agent for fine woods.

Celestite
Synonyms: Celestine, Coelestin, Coelestine.
French: Célestine.
German: Cölestin.
Spanish: Celestina.
Italian: Celestina.

Chemical
Starting point in making—
Strontium salts.

Paint and Varnish
As a pigment.

Cellulose Acetate
Synonyms: Acetylated cellulose, Acetylcellulose, Celanese.
French: Acétate de cellulose, Acétate cellulosique, Cellulose acetylée.
German: Azetylcellulose, Cellulosacetat, Zellulosazetat.

Ceramics
Ingredient of—
Coating compositions used for protecting and decorating ceramic products.

Electrical
Insulating material in making—
Condensers.

Cellulose Acetate (Continued)
Ingredient of—
Coating compositions used for insulating, protecting, and decorating electric wires and apparatus.

Glass
Ingredient of—
Compositions used for coating glass to prevent fogging by condensed moisture.
Raw material in making—
Intermediate layer between the plates of nonscatterable glass.

Glue and Adhesives
Ingredient of—
Adhesive preparations containing also gums, resins, and other substances.

Jewelry
Ingredient of—
Compositions used to increase the luster of artificial pearls.

Leather
Ingredient of—
Compositions used in the manufacture of artificial leather and as coatings for protecting and decorating leather goods.

Metal Fabricating
Ingredient of—
Compositions used as coatings for the decoration and protection of metalware.

Miscellaneous
General sizing and finishing agent.
Ingredient of—
Compositions used as coatings for the decoration and protection of various fibrous and other products.
Compositions used in coating skins.
Raw material in making—
Filaments, phonograph records.
Used for various purposes in dentistry.

Paint and Varnish
Raw material in making—
Bronzing varnishes, cements, compositions for treating dirigible fabrics, dopes, enamels, lacquers, varnishes.

Paper
Ingredient of—
Compositions used as coatings for the decoration and protection of products made from paper and pulp and in the manufacture of coated paper.

Photographic
Ingredient of—
Compositions used to make noninflammable photographic and cinematographic films.

Plastics
Ingredient of—
Noninflammable plastics used in place of celluloid.

Rayon
Base of—
Celanese.

Rubber
Ingredient of—
Compositions used in place of rubber and gutta-percha.
Compositions used as coatings for the decoration and protection of rubber goods.

Stone
Ingredient of—
Compositions used as coatings for the decoration and protection of artificial and natural stone.

Textile
As the rayon fabric commonly known as celanese.
Ingredient of—
Compositions used in gilding lace.
Compositions used for producing decorative effects on fabrics.
Fireproofing compositions used in treating textile fabrics, especially linen.
Self-ironing fabrics.

Wood
Ingredient of—
Compositions used as coatings for the decoration and protection of wood products.
Plastic compositions used for filling and decorating woodwork.

Cellulose Acetate-Sodium Phthalate

Adhesives
Ingredient of—
Adhesive compositions.

Dye
Vehicle for carrying—
Dyestuffs.

Miscellaneous
Ingredient of—
Sizing compositions.

Paint and Varnish
Vehicle for carrying—
Pigments.

Paper
Ingredient of—
Sizing compositions.

Photographic
Antihalation backing for—
Photographic film.

Textile
Ingredient of—
Sizing compositions.

Cellulose Butyrate

Miscellaneous
Ingredient (Brit. 406011) of—
Ester mixture with cellulose acetate, used in making coating compositions by solution in an alkylene chloride, together with a solubilizing agent, a plasticizer, and gums, fats, waxes, resins, or the like.
Ester mixture with cellulose acetate and nitrate, used in making coating compositions by solution in an alkylene chloride, together with a solubilizing agent, a plasticizer, and gums, fats, waxes, resins, or the like.
Also see "Cellulose acetate" for complete list of uses of the more widely used cellulose esters.

Cellulose Crotonate
French: Crotonate de cellulose.
German: Krotonzellulose ester, Zellulosekrotonat.

Ceramics
Ingredient of—
Compositions, containing resins or gums, used for decorating and protecting ceramic products.

Electrical
Ingredient of—
Insulating compositions.

Glass
Ingredient of
Compositions, containing resins or gums, used in the manufacture of nonscatterable glass and for decorating and protecting glassware.

Glues and Adhesives
Ingredient of—
Adhesive preparations containing gums, resins, and other substances.

Leather
Ingredient of—
Compositions, containing resins or gums, used in the manufacture of artificial leathers and for coating and decorating leathers and leather goods.

Metallurgical
Ingredient of—
Compositions, containing resins or gums, used for decorating and protecting metallic ware.

Miscellaneous
Ingredient of—
Compositions, containing resins or gums, used for decorating and protecting various articles.

Paint and Varnish
Ingredient of—
Paints, varnishes, enamels, dopes, and lacquers containing resins or gums.

Paper
Ingredient of
Compositions, containing resins or gums, used in the manufacture of coated papers and also for decorating and protecting paper and pulp products.

Plastics
Ingredient of—
Compositions, containing resins or gums.

Rubber
Ingredient of—
Compositions, containing resins or gums, used for decorating and protecting rubber merchandise.

Stone
Ingredient of—
Compositions, containing resins or gums, used for decorating and protecting artificial and natural stone.

Cellulose Crotonate (Continued)

Textile
Ingredient of—
Coating compositions containing resins or gums.

Woodworking
Ingredient of—
Compositions, containing resins or gums, used for decorating and protecting woodwork.

Cellulose Nitrate
See: Nitrocellulose.

Cellulose Palmitate

Petroleum
Thickener (Brit. 416513) for—
Mineral oils.
Also see "Cellulose acetate" for complete list of uses of the more widely used cellulose esters.

Cellulose Propionate

Miscellaneous
Ingredient (Brit. 406011) of—
Ester mixture with cellulose acetate, used in making coating compositions by solution in an alkylene chloride, together with a solubilizing agent, a plasticizer, and gums, fats, waxes, resins, or the like.
Ester mixture wtih cellulose acetate and butyrate, used in making coating compositions by solution in an alkylene chloride, together with a solubilizing agent, a plasticizer, and gums, fats, waxes, resins, or the like.
Ester mixture with cellulose acetate and nitrate, used in making coating compositions by solution in an alkylene chloride, together with a solubilizing agent, a plasticizer, and gums, fats, waxes, resins, or the like.
Also see "Cellulose acetate" for complete list of uses of the more widely used cellulose esters.

Cellulose Stearate

Petroleum
Thickener (Brit. 416513) for—
Mineral oils.
Also see "Cellulose acetate" for complete list of uses of the more widely used cellulose esters.

Ceratain

Textile
Ingredient of—
Settling bath, used in the manufacture of filaments and other products from viscose (U. S. 1774712).

Ceresin
Synonyms: Earth wax, Fossil wax, Mineral wax, Purified ozokerite, Refined ozokerite.
French: Céresine, Cérésite, Cire de cérésine, Cire de cérésite, Cire d'ozokérite, purifiée; Cire minérale, ozocérite réfinée, ozokérite réfinée.
German: Ceresin, Ceresinwachs, Cerin, Cerosin, Erdwachs, Mineralwachs, Refinierte ozocerite.

Building and Construction
Ingredient of—
Emulsifying composition for bitumen, especially suitable for addition to bitumen employed in road construction to produce an emulsion in wet weather, containing also saponified resin, a binding agent, a filling material, and concentrated soda-potash lye (Brit. 387825).
Emulsions used as waterproofing agents by dispersion in cement mixes (U. S. 1906276).
Waterproofing compositions for brickwork, concrete, masonry, piles, shingles, and other porous structural materials.

Chemical
Ingredient of—
Coating compositions for acid tanks and chemical apparatus.
Solution used in obtaining polymerized products of acrylic acid, its esters, salts, or homologs; polymerides are insoluble in the wax (Brit. 404504).
Purifying agent (Brit. 398136) in making—
Aromatic alcohols, such as phenylethyl alcohols and their homologs, by the action of alkylene oxides on aromatic hydrocarbons.

Cooperage
Material for—
Lining and impregnating packages of various kinds.

Cosmetic
Raw material in making—
Creams, lipsticks, pastes, pencils, pomades.

Electrical
As a general insulating agent.
Binding, coating, and insulating agent in—
Electrical condensers.
Boiling-out agent for—
Treating cables and other materials to remove moisture and improve their electrical properties.
Coating and insulating agent for—
Dry-cell batteries.
Household light wires, radio wires, telephone wires, wires in all kinds of domestic electrical appliances.
Industrial electrical cables and industrial electrical machinery.
Radio coils and other electrical coils.
Utility cables and machinery.
Filler for
Cable junctions, instrument transformers, terminal boxes.
Ingredient of—
Insulating compositions containing rubber.
Insulating compositions for wires of all kinds.
Insulating compositions for industrial electrical cables and industrial electrical machinery.
Insulating compositions for electric utility cables and machinery.
Insulating and sealing compositions for dry-cells.
Molded insulations.
Sealing agent (Brit. 402967) for—
Electrolytic cells, such as condensers, to prevent the escape of liquid from exhaust ports provided for the escape of gases under pressure.
Waterproofing agent for—
Electrical instruments, electrical machinery.

Explosives
Coating agent for—
Stems of paper of vesta matches, stems of wooden matches.
Ingredient of—
Matchhead compositions.
Waterproofing agent for—
Explosives, matches.

Fuel
Coating agent (U. S. 1912697) for—
Treating coal to reduce the tendency to heat or to disintegrate because of oxidation.
Ingredient of—
Coating composition for coal; consisting also of colored cellulose pulp and benzene (U. S. 1902642).
Wax in—
Candle-making.

Food
Coating agent for—
Display molds for products such as artificial jellies, chocolates, foods of all kinds.
Ingredient of—
Candies, chewing gums, decorative compositions.
Preservative and coating agent for—
Eggs.
Raw material in making—
Artificial honeycombs.
Sealing agent for—
Bottled and jarred goods.

Forestry
Ingredient of—
Compositions for curing brown bast in rubber trees.
Grafting dressings (mixed with rosin).

Ink
Ingredient of—
Lithographic inks, non-offset compounds, offset compounds, printing inks, stamping inks.

Laundering
Lubricant for—
Flatirons and ironing machines.
Polishing and stiffening agent for—
Collars, cuffs, shirt fronts.

Leather
Ingredient of—
Dressings, finishing preparations, military paste polishes, polishing compositions, waterproofing agents.

Lubricant
Basis of various lubricating compositions.
Ingredient of—
Axle greases.
Lubricating grease, containing also castor oil, mineral oils, and aluminum stearate (U. S. 1881591).
Special lubricants.

Ceresin (Continued)

Mechanical
As a coating against rust.
Ingredient of—
Drawing oils, belt dressings.

Metallurgical
Ingredient of—
Compositions used for covering metals to provide protection against moisture, acids, alkalies, and other corrosive substances.
Corrosion-resisting coating compositions, containing also petrolatum, oxidized petroleum bitumen, asbestos, and powdered shale.

Miscellaneous
Ingredient of—
Automobile polish, containing also carnauba wax, rosin, turpentine substitute, and potash solution.
Automobile polish, containing also turpentine, beeswax, paraffin, and carnauba wax.
Compositions for making dental impressions.
Compositions for making anatomical specimens.
Compositions for painting old timber to prevent attack of death watch beetle.
Compositions for waterproofing automobile tops and tarpaulins.
Floor polishes, furniture polishes.
Furniture polish, containing also beeswax, raw linseed oil, turpentine, paraffin oil, potassium carbonate, animal-fat soap chips, and water.
Furniture polish, containing also bleached carnauba wax, paraffin, turpentine, white curd soap, pale rosin, water, and an aromatic oil.
Linoleum polishes.
Phosphorescent compounds made from dehydrated quinine sulphate, zinc sulphide, thorium phenolsulphonate or oleate, glycerin, zinc, or antimony powder; such mixtures are electrically conducting (Brit. 402-777).
Polishes of various sorts.
Preparations for making imitation alabaster statues.
Shoe polishes, ski waxes, wood polishes.
Raw material in making—
Grease crayons, imitation fruit and flowers, oil crayons.
Spotting pencils, containing also stearic acid and oil dyes, for dry cleaners and textile manufacturers, used for restoring original shades to textiles which have been decolored by stain-removing chemicals.
Toys and dolls.
Wax figures for exhibition purposes and for window display.
Waterproofing agent for—
Cloth liners for automobile tires.
Pasteboard signs exposed to the weather.
Soda-water straws.

Oils, Fats, and Waxes
Ingredient of—
Beeswax substitute, containing beeswax and glyceryl stearate.
Belt dressings, compounded waxes, electrotypers' wax, sealing wax, shoemakers' wax, wire-drawing oils.
Substitute for—
More expensive waxes.

Paint and Varnish
Absorbent in—
Paint and varnish removers.
Ingredient of—
Antifouling paints.
Lacquers for flexible materials, based on a soluble polymerized vinyl compound, a plasticizer, and a solvent (Brit. 389914).
Special floor waxes.
Varnish, containing also rosin, barytes or other pigments, and alcohol, used for bottles and also for cork capping.
Wood fillers.

Paper
Coating, impregnating, or sizing material in making—
Glassine paper (U. S. 1914798 and 1914799).
Sized pulp, waxed paper products.
Ingredient of—
Coating compositions.
Coating compositions for regenerated cellulose products (Brit. 414911).
Coating composition, used in making a washable and greaseproof wallpaper, containing also a cellulose derivative, such as cellulose nitrate or acetate, or ethylcellulose or benzylcellulose, and solvents, plasticizers, and natural or synthetic resins (Brit. 394974).
Compositions used in the manufacture of carbon paper.
Moisture-proof, transparent lacquer for coating wrapping paper, containing also a plasticizer, nitrocellulose or other cellulose derivative, and suitable solvents, but no castor oil, gum, or resin (Brit. 412687).
Preparations used in making waxed paper.
Sizing emulsions for paper (Brit. 395155 and 404386).
Sizings for high-gloss paper.
Waterproofing composition, containing also chlorinated rubber and a plasticizer (French 740013).
Waterproofing agent for—
Boxboard, cardboard, cartons, paper, paper drinking cups.

Pharmaceutical
In compounding and dispensing practice.

Photographic
Coating for—
Photographic papers.
Finishing agent for—
Glossy prints.

Plastics
Ingredient of—
Moulded products made from a solution of a cellulose derivative and a polyvinyl resin together with cork, leather dust, or wood pulp, a plasticizer, and other synthetic resins (Brit. 416412).
Phonograph disks.

Printing
Ingredient of—
Compositions used for the preparation of acidproof coatings for plates in the electrotyping process.
Compositions used for making matrices in galvanoplastic work.
Rosin-sulphur mixes for making printing forms.
Process material in—
Lithography, photoengraving, process engraving.

Rubber
Coating agent for—
Molds (to prevent sticking of the article molded).
Filler in making—
Rubber compositions.
Ingredient of—
Rubber compositions (added to give the rubber a polished or finished appearance).

Shipbuilding
Ingredient of—
Mixtures with tallow for greasing ships' slipways to facilitate launching operations.

Soap
Ingredient (Brit. 407039) of—
Antiseptic washing and cleansing agents prepared by incorporating water-soluble mercury silver, or gold salts, which dissociate into metal ions, with aliphatic compounds having strong wetting and washing power, containing at least 8 carbon atoms, having an acid sulphuric or phosphoric ester group or sulphonic acid group in an end position and forming water-soluble salts with said metals.

Textile
Glazing agent in—
Hot calendering.
Ingredient of—
Compositions used in the manufacture of waxed cloth.
Finishing compositions.
Impregnating or coating agents for fabrics made from a solution of a cellulose derivative and a polyvinyl resin, a plasticizer, and other synthetic resins (Brit. 416412).
Sizing compositions.
Softening compositions.
Viscose solution for producing dull-lustered rayon (U. S. 1902529).
Waterproofing coating, containing also castor oil, rubber, and petrolatum.
Waterproofing composition consisting of emulsion with a dispersion agent, an organic amine, pine oil, oleic acid, a synthetic wax, aluminum acetate, and sodium silicate (Brit. 401282).
Polishing agent for—
Weaving machine rollers.
Stiffening ("starching") agent for—
Linen.
Waterproofing agent in—
Treating yarns and fabrics.
Wax for—
Hosiery stitching threads.

Tobacco
Waterproofing agent for—
Packagings.

Ceresin (Continued)

Woodworking
Coating and impregnating agent for—
Artificially dried wood (to prevent reabsorption of moisture).
Log ends (to prevent splitting and infection by borers).
Ingredient of—
Compositions used in the finishing of furniture and of lumber used for parquet flooring.

Ceric Sulphate
Synonyms: Cerium sulphate.
French: Sulphate cérique, Sulphate de cérium.
German: Cerisulfat, Schwefelsäurescerioxyd.

Analysis
Reagent in determination of nitrogen dioxide.
Reagent in testing for—
Phenols, santonin.

Chemical
Catalyst (Germany 149677) in making—
Sulphuric acid by the contact process.
Ingredient of catalytic preparations used in making—
Acenaphthylene, acenaphthaquinone, bisacenaphthylidenedione, naphthaldehydic acid, naphthalic anhydride, and hemimellitic acid from acenaphthene (Brit. 295270).
Acetaldehyde from ethyl alcohol (Brit. 281307).
Acetic acid from ethyl alcohol (Brit. 281307).
Alcohols from aliphatic hydrocarbons (Brit. 281307).
Aldehydes and acids by the reduction of the corresponding esters (Brit. 306471).
Alphacampholide by the reduction of camphoric acid (Brit. 306471).
Aldehydes and acids from toluene, orthochlorotoluene, orthonitrotoluene, orthobromotoluene, parabromotoluene, paranitrotoluene, parachlorotoluene, metachlorotoluene, metabromotoluene, metanitrotoluene, dichlorotoluenes, dibromotoluenes, dinitrotoluenes, chloronitrotoluenes, chlorobromotoluenes, bromonitrotoluenes (Brit. 295270).
Aldehydes and acids from xylenes, pseudocumenes, mesitylene, and paracymene (Brit. 281307).
Alphanaphthaquinone from naphthalene (Brit. 281307).
Anthraquinone from naphthalene (Brit. 295270).
Benzaldehyde and benzoic acid from toluene (Brit. 281307).
Benzoquinone from phenanthraquinone (Brit. 281307).
Benzyl alcohol by the reduction of benzaldehyde (Brit. 306471).
Benzyl alcohol or benzaldehyde or phthalide by the reduction of phthalic anhydride (Brit. 306471).
Butyl alcohol by the reduction of crotonaldehyde (Brit. 306471).
Chloroacetic acid from ethylenechlorohydrin (Brit. 306471).
Diphenic acid from ethyl alcohol (Brit. 281307).
Ethyl alcohol by the reduction of acetaldehyde (Brit. 306471).
Fluorenone from fluorene (Brit. 295270).
Formaldehyde by the reduction of carbon dioxide or carbon monoxide (Brit. 306471).
Formaldehyde by the reduction of methane or methanol (Brit. 306471).
Hydroxyl reduction compounds of anthraquinone, benzoquinone, and the like (Brit. 306471).
Isopropyl alcohol by the reduction of acetone (Brit. 306471).
Maleic acid and fumaric acid by the oxidation of toluene, benzene, phenols, tar phenols, or furfural, or from benzoquinone or phthalic anhydride (Brit. 295-270).
Methane by the reduction of carbon dioxide or carbon monoxide (Brit. 306471).
Methanol by the reduction of carbon dioxide or carbon monoxide (Brit. 306471).
Naphthaldehydic acid, acenaphthaquinone, or bisacenaphthylenedione from acenaphthylene (Brit. 281307).
Phenanthraquinone from phenanthrene or diphenic acid (Brit. 295270).
Phthalic acid and maleic acid from naphthalene (Brit. 295270).
Primary alcohols by the reduction of the corresponding aldehydes (Brit. 306471).
Propionic acid and butyric acid and higher alcohols, ketones, and acids by the reduction of carbon dioxide or carbon monoxide (Brit. 306471).
Reduction products of ketones, aldehydes, acids, esters, alcohols, ethers, and other organic compounds containing oxygen (Brit. 306471).
Salicylic acid and salicylic aldehyde from cresol (Brit. 295270).
Secondary butyl alcohol by the reduction of methylethyl ketone (Brit. 306471).
Valeryl alcohol by the reduction of valeraldehyde (Brit. 306471).
Vanillin and vanillic acid from eugenol or isoeugenol (Brit. 295270).
Ingredient (Brit. 304640) of catalytic preparations used in the production of various aromatic and aliphatic amines, including—
Alphanaphthylamine from alphanitronaphthalene.
Amines from aliphatic nitro compounds, such as alkyl nitriles, or nitromethane.
Amines from oximes, Schiff's base, and nitriles.
Amino compounds from the corresponding nitroanisoles.
Aminophenols from nitrophenols.
3-Aminopyridin from 3-nitropyridin.
Amylamine from pyridin.
Anilin, azo-oxybenzene, azobenzene, and hydrazobenzene, by the reduction of nitrobenzene.
Cyclohexamine, dicyclohexamine, and cyclohexylanilin from nitrobenzene.
Piperidin from pyridin.
Pyrrolidin from pyrrol.
Tetrahydroquinolin from quinolin.
Oxidizing agent in making—
Intermediates and organic chemicals, used in the place of potassium permanganate.
Starting point in making—
Cerium salts.

Dye
Oxidizing agent in making—
Synthetic dyestuffs, used in the place of potassium permanganate.

Electrical
In electric storage batteries (Brit. 21566-1900).

Miscellaneous
Oxidizing agent in making various products.

Photographic
Reagent (German 123017) for—
Reducing intensity of negatives.
Reagent in weakening or strengthening silver image.
Reagent in making—
Photographic paper.
Reducing agent in making—
Flashlight powders.

Textile
Developer in—
Dyeing and printing with anilin black.

Cerium-Ammonium Nitrate
Synonyms: Cerous-ammonium nitrate.
French: Azotate double de cérium et d'ammonium.
German: Ceroammoniumnitrat, Cerammoniumnitrat, Salpetersaeuresceroammonium.

Chemical
Reagent in making—
Acetal.

Leather
Reagent in—
Tanning.

Lighting
Ingredient of—
Compositions used in making gas mantles.

Textile
——, *Dyeing and Printing*
Mordant for—
Alizarin colors on fabrics and yarns.

Cerium Hydroxide
Synonyms: Ceric hydroxide.
French: Hydroxyde cérique, Hydroxyde de cérium.
German: Cerihydroxyd.

Ceramics
Reagent in—
Coloring porcelains and potteries.

Chemical
As a strong reducing agent.
Reducing agent in making—
Cuprous salts from cupric salts.
Mercurous salts from mercuric salts.
Starting point in making—
Cerium chloride, cerium nitrate, other cerium salts.

Textile
Mordant in dyeing—
Yarns and fabrics.

Cerium Methylcyclohexylphthalate
Miscellaneous
Preventer (U. S. 1965608) of—
Nitrocellulose coatings discoloration by ultraviolet light.

Cerium Nitrate
Synonyms: Cerous nitrate.
French: Azotate de cérium, Azotate cérreux.
German: Cernitrat, Ceronitrat, Salpetersaeuresceroxyd.
Chemical
Reagent in making—
Acetal.
Starting point in making—
Cerrous-ammonium nitrate, ceric oxide.
Leather
Reagent in—
Tanning.
Lighting
Ingredient of—
Compositions used in making gas mantles.
Textile
——, *Dyeing and Printing*
Mordant for fabrics and yarns.

Cerium Titanofluoride
French: Titanofluorure cérique, Titanofluorure de cérium.
German: Ceritítanofluorid, Titanofluorcer.
Metallurgical
Ingredient (Brit. 13988 year 1912) of—
Pyrophoric electrodes.

Cerium Tungstate
French: Tungstate cérique, Tungstate de cérium, Wolframate de cérique, Wolframate de cérium.
German: Cerwolframat, Wolframsäurescer, Wolframsäuresceroxyd.
Spanish: Tungstato de cerio, Wolframato de cerio.
Italian: Tungstato di cerio, Wolframato di cerio.
Chemical
Ingredient of catalytic mixtures used in making—
Acenaphthylene, acenaphthaquinone, bisacenaphthylidenedione, naphthaldehydic acid, naphthalic anhydride, and hemimellitic acid (Brit. 295270).
Acetaldehyde from ethyl alcohol (Brit. 281307).
Acetic acid from ethyl alcohol (Brit. 281307).
Alcohols from aliphatic hydrocarbons (Brit. 281307).
Aldehydes or acids by the reduction of the corresponding esters (Brit. 306471).
Aldehydes and acids from toluene, orthochlorotoluene, parachlorotoluene, metachlorotoluene, orthonitrotoluene, paranitrotoluene, metanitrotoluene, orthobromotoluene, parabromotoluene, metabromotoluene, dichlorotoluenes, dinitrotoluenes, dibromotoluenes, chloronitrotoluenes, chlorobromotoluenes, bromonitrotoluenes (Brit. 295270).
Aldehydes and acids from xylenes, pseudocumenes, mesitylene, and paracymene (Brit. 281307).
Alphacampholide from camphoric acid by reduction (Brit. 306471).
Alphanaphthaquinone from naphthalene (Brit. 281307).
Anthraquinone from naphthalene (Brit. 281307).
Benzaldehyde and benzoic acid from toluene (Brit. 281307).
Benzoquinone from phenanthrene (Brit. 281307).
Benzyl alcohol by the reduction of benzaldehyde (Brit. 306471).
Benzyl alcohol or benzaldehyde or benzyl phthalide by the reduction of phthalic anhydride (Brit. 306471).
Butyl alcohol by the reduction of crotonaldehyde (Brit. 306471).
Diphenic acid from ethyl alcohol (Brit. 281307).
Ethyl alcohol by the reduction of acetaldehyde (Brit. 306471).
Fluorenone from fluorene (Brit. 295270).
Formaldehyde by the reduction of methane or methanol (Brit. 306471).
Formaldehyde by the reduction of carbon dioxide or carbon monoxide (Brit. 306471).
Hydroxyl compounds by the reduction of anthraquinone, benzoquinone, and the like (Brit. 306471).
Isopropyl alcohol by the reduction of acetone (Brit. 306471).
Maleic acid and fumaric acid by the oxidation of toluene, benzene, phenols, tar phenols, furfural, or from benzoquinone or phthalic anhydride (Brit. 295270).
Methane by the reduction of carbon dioxide or carbon monoxide (Brit. 306471).
Methanol by the reduction of carbon dioxide or carbon monoxide (Brit. 306471).
Naphthaldehydic acid, acenaphthaquinone, bisacenaphthylidenedione from acenaphthylene (Brit. 281307).
Phenanthraquinone from phenanthrene or diphenic acid (Brit. 295270).
Primary alcohols by the reduction of the corresponding aldehydes (Brit. 306471).
Propionic acid and butyric acid and higher alcohols, ketones, and acids by the reduction of carbon dioxide or carbon monoxide (Brit. 306471).
Reduction products of ketones, aldehydes, acids, esters, and acids by the reduction of carbon dioxide or carbon monoxide (Brit. 306471).
Salicylic acid and salicylic aldehyde from cresol (Brit. 295270).
Secondary butyl alcohol by the reduction of methylethyl ketone (Brit. 306471).
Valeryl alcohol by the reduction of valeraldehyde (Brit. 306471).
Vanillin and vanillic acid from eugenol or isoeugenol (Brit. 295270).
Ingredient (Brit. 304640) of catalytic preparations used in the production of various aromatic and aliphatic amines, including—
Alphanaphthylamine from alphanitronaphthalene.
Amines from aliphatic nitro compounds, such as alkyl nitriles or nitromethane.
Amylamine from pyridin.
Anilin, azo-oxybenzene, azobenzene, and hydrazobenzene from nitrobenzene by reduction.
Amines from oxides, Schiff's base, and nitriles.
Amino compounds from the corresponding nitroanisoles.
Aminophenols and nitrophenols.
3-Aminopyridin from 3-nitropyridin.
Cyclohexamine, dicyclohexamine, and cyclohexylanilin from nitrobenzene.
Piperidin from pyridin.
Pyrrolidin from pyrrol.
Tetrahydroquinolin from quinolin.
Electrical
Reagent in treating—
Arc carbons to improve their capacity for giving a brilliant light.

Cetraric Acid
Synonyms: Cetrarin.
French: Acide cétrarique.
German: Cetrarsäure.
Chemical
Starting point in making—
Esters, pharmaceuticals, salts.
Pharmaceutical
In compounding and dispensing practice.

Cetyl Adipate
Paint and Varnish
Gelatinizing or softening agent (Brit. 387534) in making—
Varnishes and similar compositions having a base of cellulose esters or ethers, in particular nitrocellulose and cellulose acetate.

Cetyl Alcohol
Synonyms: Cetylic alcohol, Ethal, Hexadecanol, Palmityl alcohol, Primary hexadecyl alcohol.
Latin: Alcohol cetylicum.
French: Alcool de cétyle.
German: Cetylalkohol.
Chemical
Starting point in making—
Antiseptic washing and cleansing agents from water-soluble salts of mercury, silver or gold, and acid sulphuric esters of unsaturated or saturated alcohols (Brit. 407039).
Cetyl paramethoxycarbanilate, used as a plasticizer for cellulose esters.
Cetylamine, used as a dispersing medium and a fat-splitting agent.
Sulphonated cetylbenzyl ethers used as wetting and detergent agents (Brit. 393937).
Wetting and dispersing agents with coconut alcohol, octadecyl alcohol, sulphuric acid monohydrate, and alkali (Brit. 418846).
Wetting, cleansing, emulsifying, and bleaching agents with boric acid and a sulphonating or phosphatizing agent (Brit. 409598).
Cosmetic
Base for—
Cosmetic creams, hand creams, massage creams.

Cetyl Alcohol (Continued)
Fixative for—
Essential oils, synthetic aromatics.
Suggested as an agent for producing a velvety condition of the skin.
Disinfectant
Solvent for—
Oil-soluble germicides.
Starting point in making—
Cetyl bisulphide, used as a fumigant.
Food
Starting point (U. S. 1475574) in making—
Butter substitute.
Insecticide
Starting point in making—
Cetyl paranitrophenylisocyanate, used as an insecticide and fungicide.
Miscellaneous
Starting point (U. S. 1951593) in making—
Synthetic esters used for producing films.
Paint and Varnish
Solvent for—
Cellulose acetate, nitrocellulose.
Starting point in making—
Cetyl acetate, used in nitrocellulose lacquers and dopes.
Petroleum
Sludging inhibitor (U. S. 1841070) for—
Transformer oils.
Pharmaceutical
In compounding and dispensing practice.
Ingredient of—
Lanolin substitute used as a base in salves and the like suggested for the treatment of dermatologic eruptions.
Starting point in making—
Cetyl bromide, cetyl iodide.
Suggested for use in treating—
Prurige, weeping eczema.

Cetyl Alcohol Boric Ester
Fats, Oils, and Waxes
Starting point (Brit. 448668) in making—
Emulsifying agents for fats, oils, and waxes by condensing, in the presence of a sulphonating agent, with boric acid esters of the cholesterols of woolfat and neutralizing the products.

Cetylamine
German: Cetylamin.
Chemical
Starting point in making various derivatives.
Starting point (Brit. 343899) in making—
Dispersing and emulsifying agents for producing emulsions of various chemicals.
Dye
Ingredient (Brit. 343899) of—
Dispersing preparations used in the production of emulsions of dyestuffs.
Fats and Oils
Ingredient (Brit. 343899) of—
Dispersing preparations used for the production of emulsions of vegetable and animal oils and fats.
Miscellaneous
Ingredient (Brit. 343899) of—
Dispersing, emulsifying, cleansing, and washing compositions used for various purposes.
Paint and Varnish
Ingredient (Brit. 343899) of—
Dispersing preparations used for the production of emulsions of pigments.
Soap
Ingredient (Brit. 343899) of—
Dispersing agents used for the production of emulsions of alkaline earth soaps.
Detergent and cleansing compositions (added to produce the dispersion of the soap).
Textile
Ingredient (Brit. 343899) of—
Scouring, washing, wetting, and cleansing compositions used for various textile purposes (added for the purpose of effecting emulsification and dispersion).

Cetylbenzyl Ether
French: Benzyle éther de cétyle, Benzyle éther cétylique, Éther benzilique de cétyle.
German: Cetylbenzilaether.
Soap
Starting point (Brit. 378454) in making—
Sulphonated derivatives used as cleansing agents.

Cetylbetagammadihydroxypropyl Sulphide
Chemical
Starting point (Brit. 435039) in making—
Hydrogen sulphates (sodium salts) for use as wetting, cleansing, and emulsifying agents.

Cetylbetahydroxyethyl Sulphide
Chemical
Starting point (Brit. 435039) in making—
Hydrogen sulphates (sodium salts) for use as wetting, cleansing, and emulsifying agents.

Cetylene
Miscellaneous
As an emulsifying agent (Brit. 360602).
For uses, see under general heading "Emulsifying agents."

Cetyl Ester of Betaine Chloride
Metallurgical
Frothing agent in—
Flotation concentration of minerals (said to closely approach the ideal properties of a reagent for these purposes; namely:—(1) the formation of an abundant froth, but one not too persistent, at low concentrations; (2) as effective in acid mediums as in alkaline mediums; (3) insensitive to salts, even in high concentrations; (4) absolutely inert as a collector in regard to both sulphurized and nonsulphurized minerals; (5) its froth-forming properties should not be affected by the collecting agents, including the soap; (6) it should emulsify rapidly and have a dispersive action on all collecting reagents that are usually employed; by the use of this reagent the employment of new collectors, such as the insoluble paraffin oils and butyl sulpholeate, is practicable).

Cetyl Ether of N-Oxymethylpyridinium Chloride
Textile
Reagent (Brit. 390553) for—
Increasing fastness to water of cellulosic materials dyed with substantive colors.

Cetyl Hexahydrophenylenediacetate
Paint and Varnish
Gelatinizing or softening agent (Brit. 387534) in making—
Varnishes and similar compositions having a base of cellulose esters or ethers, in particular nitrocellulose and cellulose acetate.

Cetyl Hydrophthalate
Paint and Varnish
Gelatinizing or softening agent (Brit. 387534) in making—
Varnishes and similar compositions having a base of cellulose esters or ethers, in particular nitrocellulose and cellulose acetate.

Cetyl Iodide
Analysis
Reagent.
Chemical
Reagent in—
Chemical syntheses.

Cetyl Isoselenocyanate
Disinfectant
Claimed (U. S. 1993040) to be—
Parasiticide.

Cetyl Isotellurocyanate
Disinfectant
Parasiticide (U. S. 1993040).

Cetyl Isothiocyanate
Disinfectant
Parasiticide (U. S. 1993040).

Cetyl Phthalate
Paint and Varnish
Gelatinizing or softening agent (Brit. 387534) in making—
Varnishes and similar compositions having a base of cellulose esters or ethers, in particular nitrocellulose and cellulose acetate.

Cetyl-Potassium Sulphate

Metallurgical

Frothing agent in—
Flotation concentration of minerals (said to closely approach the ideal properties of a reagent for these purposes; namely:—(1) the formation of an abundant froth, but one not too persistent, at low concentrations; (2) as effective in acid mediums as in alkaline mediums; (3) insensitive to salts, even in high concentrations; (4) absolutely inert as a collector in regard to both sulphurized and nonsulphurized minerals; (5) its froth-forming properties should not be affected by the collecting agents, including the soap; (6) it should emulsify rapidly and have a dispersive action on all collecting reagents that are usually employed. By the use of this reagent the employment of new collectors, such as the insoluble paraffin oils and butyl sulpholeate, is practicable).

Cetylpyridinium Bromide

French: Bromure de cétylepyridinium.
German: Bromcetylpyridinium, Cetylpyridiniumbromid.
Spanish: Bromuro de cetilpyridinium.
Italian: Bromuro di cetilepyridinium.

Dry-Cleaning

Addition agent (Brit. 453523) to—
Solvents, such as trichloroethylene, carbon tetrachloride, and benzene.

Leather

Waterproofing agent (Brit. 424410) for—
Leather.

Metallurgical

Inhibitor (Brit. 397553) of—
Corrosion of metal by sulphuric acid in pickling baths for steel.

Textile

Addition agent (Brit. 453523) to—
Dry-cleaning solvents for textile fabrics.
Mordant (Brit. 436592) in—
Dyeing natural or regenerated cellulosic textile materials with chrome dyestuffs.

Cetyl Salicylate

Cellulose Products

Plasticizer for—
Cellulose acetate, cellulose esters or ethers, cellulose nitrate (nitrocellulose).
For uses, see under general heading: "Plasticizers."

Cetyl Sebacate

Paint and Varnish

Gelatinizing or softening agent (Brit. 387534) in making—
Varnishes and similar compositions having a base of cellulose esters or ethers, in particular nitrocellulose and cellulose acetate.

Cetyl Selenocyanate

Disinfectant

Parasiticide (U. S. 1993040).

Cetyl-Sodium Phosphate

Metallurgical

Frothing agent in—
Flotation concentration of minerals (said to closely approach the ideal properties of a reagent for these purposes; namely:—(1) the formation of an abundant froth, but one not too persistent, at low concentrations; (2) as effective in acid mediums as in alkaline mediums; (3) insensitive to salts, even in high concentrations; (4) absolutely inert as a collector in regard to both sulphurized and non-sulphurized minerals; (5) its froth-forming properties should not be affected by the collecting agents, including the soap; (6) it should emulsify rapidly and have a dispersive action on all collecting reagents that are usually employed. By the use of this reagent the employment of new collectors, such as the insoluble paraffin oils and butyl sulpholeate, is practicable).

Cetyl Succinate

Paint and Varnish

Gelatinizing or softening agent (Brit. 387534) in making—
Varnishes and similar compositions having a base of cellulose esters or ethers, in particular nitrocellulose and cellulose acetate.

Cetylsulphobenzyl Ether

Rubber

Stabilizer (Brit. 411478) for—
Rubber latex.

Cetylsulphobenzyl Ether Sodium Salt

Soap

Starting point (Brit. 378454) in making—
Detergent by admixture with sodium chloride.
Detergent, containing also sodium sulphate and sodium salt of dodecylsulphobenzyl ether.

Cetylsulphoethyl Ether

Rubber

Stabilizer (Brit. 411478) for—
Rubber latex.

Cetyl-1-sulphuric Acid (Normal) Ester

Chemical

As an emulsifying agent.
Reagent in—
Organic synthesis.
Starting point (Brit. 440575) in making—
Emulsifying agents with salts of lead, aluminum, iron, tin, or barium (such emulsifying agents are said to form water-in-oil emulsions and are, preferably, produced in situ by (1) dissolving the sulphuric acid ester in the oil and (2) agitating with an aqueous solution of the metallic salt, for example, lead acetate; they are said to be useful for treating medicinal paraffin oil, neatsfoot oil, olive oil, castor oil, cottonseed oil, and petroleum lubricating oil; a heavy paraffin oil, so treated on the basis of 50 parts by weight of oil to 48.75 parts of water, is said to yield a heavy grease that has good lubricating properties and may readily be extended with oil; water-linseed oil type emulsion is offered as suitable for use as a paint base).

Cetyl Tellurocyanate

Disinfectant

Parasiticide (U. S. 1993040).

Cetyl Thiocyanate

Disinfectant

Parasiticide (U. S. 1993040).

Cetyltrimethylammonium Bromide

French: Bromure de cétyletriméthyleammonium.
German: Bromcetyldreifachmethylammoniak, Bromcetyldreifachmethylammonium.
Spanish: Bromuro de cetiltrimetailammonio.
Italian: Bromuro di cetiletrimetileammonio.

Metallurgical

Inhibitor (Brit. 397553) of—
Corrosion of metal by sulphuric acid in pickling baths for steel.

Charcoal, Activated

French: Charbon, activé.
German: Aktivierte kohle.

Chemical

General decolorizing and purifying agent for the treatment of various chemicals and chemical products.
Absorbent in—
Producing high vacuums used for various chemical purposes.
Recovering various solvents, such as acetone, butyl alcohol, ethyl alcohol, methanol, petroleum distillates of various sorts, benzene, and ethyl acetate.
Recovering sulphur dioxide and nitrogen oxides in various chemical processes.
Storing compressed gases of various sorts, such as sulphur dioxide and acetylene.
Carrier for catalysts, such as metallic salts and oxides, used in making various chemicals and chemical products.
Catalyst in—
Converting sulphuretted hydrogen into water and sulphur.
Catalyst in making—
Chlorinated hydrocarbons.
Fatty acids and other chemical compounds by the decomposition and oxidation of mineral oils.
Nitric oxide and nitric acid from nitrogen of the air, as well as from ammonia by oxidation.
Reagent in purifying and decolorizing—
Acetanilide and derivatives.
Alcohols, both aromatic and aliphatic, such as ethyl alcohol, methanol, benzyl alcohol, and higher fatty alcohols.

Charcoal, Activated (Continued)
Alkaloids, such as morphine, caffeine, quinine, cocaine, strychnine, codeine, and their salts.
Benzene
Boric acid, borax, and other salts of boron.
Citric acid and citrates.
Gallic acid and gallates.
Glycerin for pharmaceutical purposes.
Magnesium sulphate and other alkaline earth sulphates.
Pharmaceutical products.
Photographic chemicals.
Salicylic acid and other aromatic acids.
Salicylates of organic and inorganic bases.
Sodium sulphate and other sodium salts.
Tartaric acid and tartrates.

Fats and Oils
Catalyst carrier in making—
Solid fats from oils by hydrogenation.
Reagent in—
Bleaching various edible oils, such as coconut oil, palm kernel oil, and cottonseed oil.
Decolorizing various fats and oils of both animal and vegetable origin.

Food
General decolorizing and bleaching agent in the treatment of food products.
Reagent in—
Bleaching liquid food products.
Decolorizing and purifying beverages, edible oils, fruit juices, vinegars.
Deodorizing and purifying carbon dioxide for use in carbonating beverages.

Explosives
Absorbent in—
Recovering volatile solvents used in the manufacture of cordite.

Gases
Absorbent in—
Extracting benzene and other light oils from city gas, coal gas, coke oven gas.
Reagent in—
Purifying and removing obnoxious odors from various industrial gases.

Glues and Adhesives
Reagent in—
Treating crude gelatin liquor to obtain the pure product.

Leather
Absorbent in—
Recovering volatile solvents used in the manufacture of artificial leathers.

Military
Filler for gas masks.

Miscellaneous
As a bleaching agent—
As a decolorizing and deodorizing agent.
As a filtering medium.
Reagent in—
Deodorizing refrigerators, storage tanks, submarine vessels, and other confined spaces.
Recovering volatile solvents used in dry cleaning and in various manufacturing processes.

Paint and Varnish
Reagent in purifying—
Paint oils.

Petroleum
Absorbent for recovering—
Gasoline from casinghead gas.
Gasoline from natural gas.
Gasoline vapors which escape during the process of the cracking of heavy oils.
Gasoline vapors from storage tanks.
Reagent in—
Decolorizing and purifying dry-cleaning solvents and other distillates.

Plastics
Reagent in—
Recovering volatile solvents used in manufacturing products, such as celluloid.

Resins and Waxes
Decolorizing agent for—
Resins and waxes.

Rubber
Reagent in—
Recovering volatile solvents used in the manufacture of rubberized cloth, rubber cement and other products.

Sugar
Decolorizing agent for—
Molasses, cane juices.

Textile
Reagent in—
Recovering volatile solvents used in the manufacture of nitro rayon and acetate rayon.

Water
Absorbent for—
Chlorine in the purification and sterilization of water and other liquids.

Wine
As a decolorizing agent.

Chebulic Acid
French: Acide chébulique.
German: Chebulinsäure.

Chemical
Starting point in making—
Esters, pharmaceuticals, salts.

Pharmaceutical
In compounding and dispensing practice.

Cherry Gum
French: Gomme de cérisier.
German: Kirschgummi.

Food
Ingredient of—
Confectionery, pastries.

Glues and Adhesives
Ingredient of special adhesive preparations.

Miscellaneous
Starting point in making—
Various emulsion preparations.

Paint and Varnish
Ingredient of—
Bronze color compositions, water color compositions.

Paper
Size in making various grades of paper.

Pharmaceutical
In compounding and dispensing practice.

Printing
In process engraving and the litho trades.

Textile
——, *Finishing*
Ingredient of sizes for—
Laces, silks, twills.
——, *Printing*
Ingredient of pastes for various processes.

Cherry Oil
Synonyms: Cherrypit oil.
French: Huile de cérise, Huile de noyau de cérise.
German: Kirschenoel, Kirschkernoel.

Food
As a frying oil.
As a salad oil.
As a shortening.

Perfume
Ingredient of—
Cosmetics.

Cherrypit Meal
French: Farine de noyau de cérise.
German: Kirschkernmehl.

Agriculture
As a cattle food.
Ingredient of—
Animal foods.

Fertilizer
As a fertilizer for different purposes.
Ingredient of—
Special fertilizing compositions used for lawns and gardens.

Food
As a flour in the baking industry.
Ingredient of—
Flour (to give it an almond flavor).

Cherrypit Shell
French: Coque de noyau de cérise.
German: Kirschkernhuelsen, Kirschkernschalen.

Fuel
As a fuel for use in mechanically fired furnaces.

Chimyl Alcohol

Miscellaneous
Starting point (Brit. 398818) in making—
Detergents by sulphonation with sodium pyrosulphate.

Chinawood Oil

Synonyms: Chinese wood oil, Japanese wood oil, Tung oil, Wood oil.
Latin: Oleum dryandrae, Oleum elaecoccae verniciae.
French: Huile de abrasin, Huile de bois, Huile de bois de chine, Huile de bois du Japon, Huile de Canton, Huile d'eloeococca, Huile de Hankow, Huile de Tung.
German: Chinesiches holzoel, Elaekokkoel, Holzoel, Japanisches holzoel, Oelfirnisbaumoel.
Spanish: Aceite de madera Chino.
Italian: Olio di legno di giappone.

Ceramics
Ingredient of—
Compositions used for the production of a protective film on chinaware, earthenware, stoneware, and other ceramic products.

Construction
Ingredient of—
Compositions used to produce a waterproof film on concrete, stucco, masonry, and other porous building materials.

Explosives
Ingredient (U. S. 1738628) of—
Compositions, containing also manganese resinate and lead oxide, digested in carbon tetrachloride, used for the waterproofing of paper shotgun shells.

Fuel
As an illuminant and burning oil.

Glues and Adhesives
Ingredient (Brit. 332257) of—
Special adhesive compositions.

Ink
Ingredient of—
Chinese inks.

Insecticide
Ingredient of—
Insecticidal preparations of great potency used for application to the roots of plants or by fumigation.

Leather
As a waterproofing agent.
Ingredient of—
Compositions used for the manufacture of artificial leather.
Impregnating and finishing compositions (Brit. 332257).
Leather substitutes used for the manufacture of footwear.

Linoleum and Oilcloth
Ingredient of—
Compositions used for the manufacture of oilcloth and linoleum.

Metallurgical
Ingredient of—
Core oils.

Miscellaneous
As a binding agent in various processes and in the production of miscellaneous compositions of matter.
In calking ships.
Ingredient (U. S. 1720487) of—
Infusible asphaltic masses of high elasticity, containing also aluminum chloride, zinc chloride, and iron chloride.

Paint and Varnish
Base in making—
Chinese lacquers.
Ingredient (U. S. 1841138) of—
Liquid coating compositions, containing a phenol-furfural resin and enough chinawood oil so that the fibrous sheets impregnated with the composition and subsequently dried will not adhere to each other at ordinary temperature, but will be free from dust.
Ingredient of—
Baking enamels.
Paint and varnish bases containing tetramethylthiuram disulphide (Brit. 321689).
Prime coaters, putty, spar varnishes containing rosin, transparent varnishes.
Varnishes for automobile hoods and other surfaces exposed to extremes in temperature.
Waterproof paints, varnishes, and enamels.
Roofing compositions.

Paper
As an impregnating agent for treating paper and pasteboard.
Ingredient of—
Waterproofing compositions for paper and pulp and paper products (Brit. 9023/1911).
Waterproofing agent for—
Treating papier-mache.

Pharmaceutical
In compounding and dispensing practice.

Plastics
Ingredient of various plastic compositions.
Lubricant in—
Molding plastic compositions.

Rubber
Ingredient of various rubber compounds.

Soap
As a soapstock.

Textile
As an impregnating agent.
As a waterproofing agent for treating cotton and silk.
Ingredient of—
Compositions used in the manufacture of wax cloth and oiled fabrics.

Woodworking
As an impregnating agent.
Ingredient of—
Compositions used for keeping parquet flooring, wainscoating, and paneling in good condition.
Preparations used in working ebony and other fine woods.
Preservative in—
Treating oil, worm-eaten furniture.

Chloracetophenone

Military
As a chemical warfare gas.

Miscellaneous
Official denaturant in—
Industrial alcohol.

8′-Chlor-1-alphanaphthylaminoanthraquinone

Chemical
Starting point in—
Organic syntheses.

Dye
Starting point (Brit. 443958 and 443959) in making—
Vat dyestuffs.

4-Chlor-5-amino-2-acetamidoanisole

Dye
Starting point (Brit. 447905, 447906, and 448016) in making—
Monoazo dyes for leather, particularly chrome leather.

1-Chlor-2-aminoanthraquinone

Chemical
Starting point in—
Organic syntheses.
Starting point (U. S. 1999996) in making—
Seleno ethers by reacting with 1:1′-dibenzanthronyl diselenide.

6-Chlor-1-aminoanthraquinone.

Chemical
Starting point in—
Organic syntheses.
Starting point (U. S. 1999996) in making—
Seleno ethers by reacting with 1:1′-dibenzanthronyl diselenide.

2-Chlor-5-aminobenzotrifluoride-4-sulphonic Acid

Dye
Intermediate (Brit. 446532) in making dyestuffs.

4-Chlor-5-aminobenzotrifluoride-2-sulphonic Acid

Dye
Intermediate (Brit. 446532) in making dyestuffs.

5-Chlor-2-aminobenzotrifluoride-3-sulphonic Acid

Dye
Intermediate (Brit. 446532) in making dyestuffs.

2-Chlor-3-amino-1:5-bistrifluoromethylbenzene-6-sulphonic Acid

Dye
Intermediate (Brit. 446532) in making various dyestuffs.

8-Chlor-1-benzothiazylbetaparatoluenesulphonylethyl Sulphide

Chemical

Intermediate (Brit. 444262 and 444501) in—
Organic syntheses.

Insecticide

Insecticide (Brit. 444262 and 444501) for—
Animal pests, vegetable pests.

3-Chlor-10-betadiallylaminoethylaminoacridin Dihydrochloride

Pharmaceutical

Claimed (Brit. 441007, 441132, and addition to 363392) as—
New pharmaceutical.

6-Chlor-9-betadiethylaminoethoxyethyl-2-methylthiolacridin

Pharmaceutical

Claimed (Brit. 363392 and 437953) as—
New pharmaceutical.

6-Chlor-9-betadiethylaminoethoxyethyl-2-methylthiolacridin Methylenedisalicylate

Pharmaceutical

Claimed (Brit. 363392 and 437953) as—
New pharmaceutical.

6-Chlor-9-betadiethylaminoethylthiolethyl-2-methylthiolacridin

Pharmaceutical

Claimed (Brit. 363392 and 437953) as—
New pharmaceutical.

6-Chlor-9-betadiethylaminoethylthiolethyl-2-methyl thiolacridin Methylenedisalicylate

Pharmaceutical

Claimed (Brit. 363392 and 437953) as—
New pharmaceutical.

3-Chlor-10-betadimethylaminoethoxyethylaminoacridin Dihydrochloride

Pharmaceutical

Claimed (Brit. 441007, 441132, and addition to 363392) as—
New pharmaceutical.

2-Chlor-6-brom-4-nitroanilin

Dye

Starting point (Brit. 429936 and 430079) in making—
Orange-brown dyes for acetate rayon and animal fibers by diazotizing and coupling with normal-ethylbetasulphatoethylanilin.
Orange-brown dyes for acetate rayon and animal fibers by diazotizing and coupling with normal-methylbetasulphatoethylanilin.
Orange-brown dyes for acetate rayon and animal fibers by diazotizing and coupling with normal-gammasulphato-normal-propylanilin.
Red-brown dyes for acetate rayon and animal fibers by diazotizing and coupling with 3-betasulphatoethylaminoparatolylmethyl ether.

1:3-Chlor-2-butadiene

Synonyms: Chloroprene.

Miscellaneous

Protective coating (Brit. 426708) for—
Metals, synthetic resins, plastics, fibrous materials, and other articles against attack by corrosive liquors (said to be of particular application for centrifugal devices, rayon spindles, acid-holding vessels and pipes).

Rubber

Starting point in making—
Polymerized product constituting a synthetic rubber said to be (a) of very high quality, (b) superior to natural rubber in certain respects; namely resistance to the deteriorating effect of crude oil, refined petroleum products, coaltar solvents, animal and vegetable oils, and other oily materials; also said to withstand heat better than natural rubber and to be less inflammable.

3-Chlor-4'-butylthioldiphenylamine-6-carboxylic Acid

Pharmaceutical

Claimed (Brit. 363392 and 437953) as—
New pharmaceutical.

3-Chlor-2-(chloromethyl)-1-phenylpropane-3

Petroleum

Solvent (Brit. 437573) in—
Refining mineral oils.

3-Chlor-10-deltadiethylaminoalphamethylbutylaminoacridin

Pharmaceutical

Claimed (Brit. 441007, 441132, and addition to 363392) as—
New pharmaceutical.

3-Chlor-10-deltadiethylaminoalphamethylbutylaminoacridin Bromide

Pharmaceutical

Claimed (Brit. 441007, 441132, and addition to 363392) as—
New pharmaceutical.

3-Chlor-10-deltadiethylaminoalphamethylbutylaminoacridin Trihydrochloride

Pharmaceutical

Claimed (Brit. 441007, 441132, and addition to 363392) as—
New pharmaceutical.

3-Chlor-10-deltadiethylaminoalphamethylbutylamino-3-ethylacridin Dihydrochloride

Pharmaceutical

Claimed (Brit. 441007, 441132, and addition to 363392) as—
New pharmaceutical.

6-Chlor-9-deltadiethylaminoalphamethylbutylamino-2-ethylthiolacridin

Pharmaceutical

Claimed (Brit. 363392 and 437953) as—
New pharmaceutical.

6-Chlor-9-deltadiethylaminoalphamethylbutylamino-2-ethylthiolacridin Bromide

Pharmaceutical

Claimed (Brit. 363392 and 437953) as—
New pharmaceutical.

6-Chlor-9-deltadiethylaminoalphamethylbutylamino-2-ethylthiolacridin Citrate

Pharmaceutical

Claimed (Brit. 363392 and 437953) as—
New pharmaceutical.

6-Chlor-9-deltadiethylaminoalphamethylbutylamino-2-ethylthiolacridin Hydrochloride

Pharmaceutical

Claimed (Brit. 363392 and 437953) as—
New pharmaceutical.

6-Chlor-9-deltadiethylaminoalphamethylbutylamino-2-ethylthiolacridin Methylenedisalicylate

Pharmaceutical

Claimed (Brit. 363392 and 437953) as—
New pharmaceutical.

6-Chlor-9-deltadiethylaminoalphamethylbutylamino-2-isooctylthiolacridin

Pharmaceutical

Claimed (Brit. 363392 and 437953) as—
New pharmaceutical.

6-Chlor-9-deltadiethylaminoalphamethylbutylamino-2-isooctylthiolacridin Citrate

Pharmaceutical

Claimed (Brit. 363392 and 437953) as—
New pharmaceutical.

3-Chlor-10-deltadiethylaminoalphamethylbutylamino-3-methylacridin Dihydrochloride

Pharmaceutical

Claimed (Brit. 441007, 441132, and addition to 363392) as—
New pharmaceutical.

6-Chlor-9-deltadiethylaminoalphamethylbutylamino-2-methylthiolacridin

Pharmaceutical

Claimed (Brit. 363392 and 437953) as—
New pharmaceutical.

6-Chlor-9-deltadiethylaminoalphamethylbutylamino-2-methylthiolacridin Citrate

Pharmaceutical

Claimed (Brit. 363392 and 437953) as—
New pharmaceutical.

6-Chlor-9-deltadiethylaminoalphamethylbutylamino-2-normal-butylthiolacridin

Pharmaceutical
Claimed (Brit. 363392 and 437953) as—
New pharmaceutical.

6-Chlor-9-deltadiethylaminoalphamethylbutylamino-2-normal-butylthiolacridin Citrate

Pharmaceutical
Claimed (Brit. 363392 and 437953) as—
New pharmaceutical.

6-Chlor-9-deltadiethylaminoalphamethylbutyl-2-methylthiol-6-methylacridin

Pharmaceutical
Claimed (Brit. 363392 and 437953) as—
New pharmaceutical.

6-Chlor-9-deltadiethylaminoalphamethylbutyl-2-methylthiol-6-methylacridin Citrate

Pharmaceutical
Claimed (Brit. 363392 and 437953) as—
New pharmaceutical.

6-Chlor-9-deltadiethylamino-normal-butyl-2-methylthioacridin

Pharmaceutical
Claimed (Brit. 363392 and 437953) as—
New pharmaceutical.

6-Chlor-9-deltadiethylamino-normal-butyl-2-methylthiolacridin Citrate

Pharmaceutical
Claimed (Brit. 363392 and 437953) as—
New pharmaceutical.

2-Chlor-2':4'-diaminoazobenzene

Disinfectant
Bactericide and Bacteriostatic (U. S. 2030897).

3-Chlor-2':4'-diaminoazobenzene

Disinfectant
Claimed (U. S. 2030897) to be—
Bactericide, bacteriostatic.

3'-Chlor-4':6'-diethoxyanilide

Dye
In dye syntheses.
Starting point (U. S. 1984739) in making—
Cardinal-red dyes with 3-chloropara-anisidin.

2-Chlor-4-diethylparaphenylenediamine

Dye
Starting point (Brit. 447905, 447906, and 448016) in making—
Monoazo dyes for leather, particularly chrome leather.

3-Chlor-4-diethylparaphenylenediamine

Dye
Starting point (Brit. 447905, 447906, and 448016) in making—
Monoazo dyes for leather, particularly chrome leather.

3-Chlor-2-dihydroxydiphenyl

Disinfectant
Claimed (U. S. 2014720) to be—
Antiseptic, germicide.

4-Chlor-2:5-dimethoxyanilide

Dye
Starting point (Brit. 434209 and 434433) in making—
Red water-insoluble dyestuffs by coupling (in substance or on the fiber) with phenyl-2:4-dichloroanilin 5-sulphonate.

5-Chlor-2:4-dimethoxyanilide

Dye
Starting point (Brit. 434209 and 434433) in making—
Red dyestuffs (water-insoluble) by coupling (in substance or on the fiber) with phenylorthoanisidin 4-sulphonate.
Bluish-red dyestuffs (water-insoluble) by coupling (in substance or on the fiber) with meta-4-xylidin-6-sulphobenzylmethylamide.
Reddish-bordeaux dyestuffs (water-insoluble) by coupling (in substance or on the fiber) with 6-chlorometatoluidin 5-sulphonpiperidide.

2-Chlor-4-dimethylparaphenylenediamine

Dye
Starting point (Brit. 447905, 447906, and 448016) in making—
Monoazo dyes for leather, particularly chrome leather.

2-Chlor-4-dimethylparaphenylenediamine-6-sulphonic Acid

Dye
Starting point (Brit. 447905, 447906, and 448016) in making—
Monoazo dyes for leather, particularly chrome leather.

1-Chlor-2:4-dinitrobenzene

Dye
Starting point in making—
Chrome-printing reddish-yellow azo dyes wtih paranitroanilin-2-sulphonic acid and salicylic acid (Brit. 435513).
Chrome-printing yellow azo dyes with metaphenylenediamine-4-sulphonic acid (Brit. 435513).
Chrome-printing yellowish-brown azo dyes with metanitroanilin-4-sulphonic acid, metatoluidin, and salicylic acid (Brit. 435513).
Chrome-printing brown azo dyes with 5-amino-salicylic acid and Cleve's acid (Brit. 435513).

3'-Chlor-4':6'-diphenoxyanilide.

Dye
In dye syntheses.
Starting point (U. S. 1984739) in making—
Red dyes with 4:4'-diaminodiphenyl ether.

6-Chlor-9-e-dimethylaminoamylamino-2-methylthiol-6-methylacridin

Pharmaceutical
Claimed (Brit. 363392 and 437953) as—
New pharmaceutical.

6-Chlor-9-e-dimethylaminoamylamino-2-methylthiol-6-methylacridin Citrate

Pharmaceutical
Claimed (Brit. 363392 and 437953) as—
New pharmaceutical.

3-Chlor-10-epsilondiethylaminoamylaminoacridin Citrate

Pharmaceutical
Claimed (Brit. 441007, 441132, and addition to 363392) as—
New pharmaceutical.

3-Chlor-4-ethoxy-2':4'-diaminoazobenzene Hydrochloride

Disinfectant
Claimed (U. S. 2009086) to be—
Bactericide.

3-Chlor-4'-ethylthiodiphenylamine-6-carboxylic Acid

Pharmaceutical
Claimed (Brit. 363392 and 437953) as—
New pharmaceutical.

4-Chlor-2-gammachlorodeltabetabutenylphenol

Disinfectant
Claimed (Brit. 443113 and 389514) to be—
Disinfectant free of odor.

4-Chlor-3-gammachlorodeltabetabutenylphenol

Disinfectant
Claimed (Brit. 443113 and 389514) to be—
Disinfectant free of odor.

6-Chlor-9-gammadiethylaminobetabetadimethylpropylamino-2-methylthiol-6-methylacridin

Pharmaceutical
Claimed (Brit. 363392 and 437953) as—
New pharmaceutical.

6-Chlor-9-gammadiethylaminobetabetadimethylpropylamino-2-methylthiol-6-methylacridin Citrate

Pharmaceutical
Claimed (Brit. 363392 and 437953) as—
New pharmaceutical.

3-Chlor-10-gammadiethylaminobetahydroxypropylaminoacridin Dihydrochloride

Pharmaceutical
Claimed (Brit. 441007, 441132, and addition to 363392) as—
New pharmaceutical.

3-Chlor-10-gammadiethylaminobutylaminoacridin Dihydrochloride

Pharmaceutical
Claimed (Brit. 441007, 441132, and addition to 363392) as—
New pharmaceutical.

3-Chlor-10-gammadiethylaminoethylaminoacridin Dihydrochloride

Pharmaceutical
Claimed (Brit. 441007, 441132, and addition to 363392) as—
New pharmaceutical.

3-Chlor-10-gammadiethylaminoethylthiolpropylaminoacridin Dihydrochloride

Pharmaceutical
Claimed (Brit. 441007, 441132, and addition to 363392) as—
New pharmaceutical.

3-Chlor-10-gammadimethylaminopropylaminoacridin Dihydrochloride

Pharmaceutical
Claimed (Brit. 441007, 441132, and addition to 363392) as—
New pharmaceutical.

3-Chlor-4-hydroxy-2′:4′-diaminoazobenzene Hydrochloride

Disinfectant
Claimed (U. S. 2009086) to be—
Bactericide.

3-Chlor-2-hydroxydiphenyl

Disinfectant
As a germicide.

Fungicide
As a fungicide.

3-Chlor-4-hydroxydiphenyl

Synonyms: 3-Chlor-4-phenylphenol.

Disinfectant
As a germicide.

5-Chlor-2-hydroxydiphenyl

Disinfectant
As a bactericide (U. S. 1989081).

3-Chlor-4-hydroxydiphenylmethane

Disinfectant
Claimed (U. S. 1967825) to be—
Bactericide.

4′-Chlor-4-hydroxydiphenylmethane

Disinfectant
Claimed (U. S. 1967825) to be—
Bactericide.

5-Chlor-2-hydroxydiphenylmethane

Disinfectant
Claimed (U. S. 1967825) to be—
Bactericide.

3-Chlor-2-hydroxy-5-normal-propyldiphenyl

Disinfectant
Claimed (U. S. 2014720) to be—
Antiseptic, germicide.

Chlorinated Copperas

Synonyms: Chlorinated iron sulphate.
French: Sulphate de fer chloré, Sulphate ferreux chloré.
German: Chloriertes eisensulfat, Chloriertes ferrosulfat.

Miscellaneous
Reagent in—
Purification of various waste waters from chemical and other plants.

Sanitation
Reagent in—
Purification of sewage.

Water
Reagent in—
Purification of water.

Chlorinated Rubber

French: Caoutchouc chloré.
German: Chlorkautschuk.

Ceramics
Ingredient of—
Compositions for coating various ceramic wares.

Construction
Ingredient of—
Compositions for coating concrete and stucco.

Leather
Ingredient of—
Compositions used to stimulate leather.

Metallurgical
Ingredient of—
Compositions for coating metals.

Miscellaneous
Ingredient of—
Compositions for coating various articles.
Compositions for fireproofing curtains and the like.

Paint and Varnish
Ingredient of—
Acid-resistant varnishes, antirust paints, elastic paints, elastic varnishes.

Plastics
Ingredient of—
Plastic compositions.

Rubber
Ingredient of—
Compositions for coating rubber goods.

Stone
Ingredient of—
Compositions for coating artificial and natural stone.

Woodworking
Ingredient of—
Coating compositions, fireproofing compositions.

Chlorinated Train Oil

Synonyms: Chlorinated whale oil.
French: Huile de baleine chlorée, Huile de cétaces chlorée.
German: Chlorinertes walfischtran.

Abrasives
Binding agent (Brit. 323801) in making—
Abrasive products.

Chemical
General binding agent (Brit. 323801).

Glues and Adhesives
Ingredient (Brit. 323801) of—
Binders and adhesive preparations.

Leather
Binder (Brit. 323801) in making—
Composition leather substitutes.

Miscellaneous
Binding agent in making various compositions (Brit. 323801).

Paint and Varnish
Ingredient (Brit. 323801) of—
Paint, varnishes.

Plastics
Binder (Brit. 323801) in making various compositions.

Rubber
Binder (Brit. 323801) in making various compositions.

Woodworking
Binder (Brit. 323801) in making—
Compositions containing ground wood, sawdust, and the like.

Chlorine

French: Chlore.
German: Chlor, Chlorin.
Spanish: Cloro.
Italian: Clorine.

Ceramics
Reagent in—
Treating metallic oxides, contained in the under-glaze, for the purpose of producing colored effects on various products.

Chemical
Catalyst in making—
Cellulose acetate from hydrocellulose by the action of acetic anhydride, acetyl chloride, and other acetylating agents.
General chlorinating agent for making organic and inorganic compounds of great variety.
General oxidizing agent.
General reducing agent.
Reagent and starting point in making—
Acetic anhydride from acetic acid.
Acetyl chloride.
Acetylene tetrachloride by reaction with acetylene and subsequent distillation.

Chlorine (Continued)

Reagent and starting point in making—(*Continued*)

- Alloxan.
- Alumina in pure state.
- Aluminum chloride from aluminum carbide in the presence of aluminum metal (German 25474) and from bauxite after roasting by direct chlorination.
- Aluminum-sodium chloride from alumina, coal, and sodium chloride (German 52770).
- Ammonium chlorostannate.
- Amyl acetate by chlorination of pentane.
- Antimony pentachloride by chlorination of metallic antimony with excess chlorine.
- Antimony trichloride by reaction with metallic antimony.
- Arsenic acid by the oxidizing action of chlorine on arsenious acid (U. S. 1515079).
- Arsenic trichloride by the action of dry chlorine gas on metallic arsenic.
- Barium chlorate.
- Barium perchlorate.
- Benzal chloride.
- Benzoic acid by chlorination of hot toluene and subsequent treatment.
- Benzotrichloride by chlorination of boiling toluene.
- Benzoyl chloride by chlorination of benzaldehyde.
- Benzyl chloride by passing chlorine over boiling toluene and subsequent treatment.
- Benzyl dichloride by chlorination of toluene.
- Bismuth chloride by chlorination of pulverized bismuth metal.
- Bismuth pentoxide from bismuthic acid.
- Bleaching powder by chlorination of slaked lime.
- Boric acid by action on various raw materials (German 118073).
- Boron trichloride by union of the elements; also by the action of chlorine on an incandescent mixture of boric acid anhydride and carbon.
- Bromine by action on potash liquors.
- Butyl alcohol, butyl chloride, butyl chlorohydrate.
- Butylchloral by action on cooled paraldehyde.
- Cadmium chloride by action on metallic cadmium.
- Calcium chlorate by action on calcium hydroxide.
- Calcium chloride, calcium hypochlorite.
- Carbon tetrachloride by action on methane in the presence of cuprous chloride, or by action on carbon bisulphide.
- Carbon trichloride by the action of sunshine on chlorine and ethyl chloride and ethylene chloride, and from acetylene tetrachloride and sulphur chloride by the action of chlorine in the presence of iron powder as a catalyst (German 174068).
- Carbonyl chloride by action on carbon monoxide in the presence of a catalyst.
- Chloral by chlorination of ethyl alcohol and subsequent distillation.
- Chloranil from anilin by chlorination in the presence of chlorosulphonic acid.
- Chlorates of various bases, chlorides of various bases.
- Chloracetic acid by action on acetic acid in the presence of acetic anhydride.
- Chloroacetone by chlorination of acetone.
- Chloroacetyl chloride by action on acetyl chloride in sunlight.
- 5-Chloro-2-aminotoluene from acetoorthotoluide.
- Chlorinated benzene derivatives.
- Chlorinated naphthalene derivatives.
- Chlorine monoxide from chlorine and yellow oxide of mercury.
- Chlorine-sulphur compounds.
- Chlorobenzanthrone by action on benzanthrone in acetic acid solution.
- Chlorobenzene by action on benzene in the presence of molybdenum chloride.
- Chlorocosane by passing chlorine through melted paraffin.
- Chloroform, 5-chloroisatin, chloromethyl ether.
- Chloronitrobenzenes by chlorinating benzene in the presence of iodine.
- Chloroparanitroanilin by chlorination of paranitroanilin in acid solution.
- Chlorophthalic acid by chlorination of phthalic acid.
- Chloropicrin.
- Chlorotolueneparasulphonic acid, ortho, by chlorination of toluene-parasulphonic acid.
- Chromium sesquichloride by chlorination of a mixture of chromic oxide and carbon.
- Compounds from sulphite cellulose waste liquor.
- Cupric chloride by chlorination of metallic copper.
- Copper oxychloride.
- Cyanogen chloride by action on moist sodium cyanide suspended in carbon tetrachloride.
- Dichloroacetic acid, dichlorobenzal chloride.
- Dichlorobenzaldehyde by chlorination of benzaldehyde in the presence of iodine or antimony.
- Dichlorobenzidin by chlorination of diacetylbenzidin.
- Dichloroethyl oxide by chlorination of ethyl ether.
- Dichloroethylene by chlorination of acetylene.
- 5:7-Dichloroisatin, dichloromethyl ether.
- 1:2-Dichloro-4-nitrobenzene by chlorination of chloronitrobenzene in the presence of ferric chloride.
- Dichlorophthalic acid anhydride by chlorination of a solution of phthalic acid anhydride in fuming sulphuric acid.
- Dimethyl sulphate.
- 6-Dichlorotoluene from 2-amino-6-chlorotoluene.
- Dinitrochlorobenzene by chlorination of dinitrobenzene.
- Diphenylchloroarsine.
- Ethyl chloride.
- Ethyldichloroamine.
- Ethylene chloride by chlorination of ethylene and subsequent distillation.
- Ethylene chlorobromide by chlorination of ethylene bromide.
- Ethylene chlorochloride by chlorination of ethylene chloride.
- Ethylene chlorohydrin.
- Ethylene dichloride by chlorination of ethylene and subsequent distillation.
- Ethylidene chloride.
- Ethylsulphonic chloride.
- Ferric chloride by chlorination of solution of ferrous chloride.
- Ferricyanides from ferrocyanides.
- Gadolinium chloride, glucinum chloride, gold chloride.
- Hexachlorobenzene.
- Hydrochloric acid by burning chlorine in an atmosphere of hydrogen or causing hydrogen and chlorine to unite in the presence of catalysts.
- Hypochlorites of various bases.
- Iodine monochloride by the action of dry chlorine on iodine.
- Iodine trichloride by the interaction of chlorine and iodine.
- Lanthanum chloride, lead chloride, lead peroxide, lithium chloride, magnesium chloride, manganese chloride.
- Mercuric chloride by the direction combination of chlorine and mercury heated to the point of volatilization; also by reaction of mercury and chlorine in the presence of a small quantity of hypochlorous acid (German 379493).
- Mercurous chloride by reaction between chlorine and excess mercury.
- Metadichlorobenzene by chlorination of monochlorobenzene.
- Methanol, methyl chloride, methyl chlorosulphonate.
- Methylene chloride by chlorination of methyl chloride and subsequent distillation.
- Methylene chlorofluoride, monochloro ether, naphthalene tetrachloride.
- Nickel chloride by the ignition of very finely divided nickel in a current of chlorine.
- Nitrogen pentoxide by action of chlorine on sliver nitrate.
- Omegadichlorobetamethylanthraquinone.
- Orthochlorobenzal chloride.
- Orthochloronaphthylamine.
- Orthochlorophenol by chlorination of phenol (German 155631).
- Orthochlorotoluene from paratoluene sulphochloride.
- Orthodichlorobenzene from monochlorobenzene by chlorination.
- Orthonitrobenzaldehyde.
- Orthonitrobenzyl chloride.
- Orthotoluene sulphochloride.
- Palladium chloride.
- Parachlorobenzaldehyde.
- Parachlorophenol by chlorination of phenol.
- Parachlorotoluene.
- Paradichlorobenzene from monochlorobenzene by chlorination.
- Paranitrobenzyl chloride.
- Paratoluene sulphochloride by chlorination of paratoluenesulphonic acid.
- Paris blue.
- Pelargonidin chloride.
- Pentachloroethane by chlorination of ethyl chloride or ethylene chloride.

Chlorine (Continued)

Reagent and starting point in making—(*Continued*)
- Perchlorates of various bases.
- Perchloromethyl ether.
- Phosphorus pentachloride by the action of chlorine on phosphorus or phosphorus trichloride.
- Phosphorus trichloride by passing a current of dry chlorine gas over gently heated phosphorus.
- Phosphorus trichloride from ferrophosphorus (French 669099).
- Phthalchloroimide.
- Platinum bichloride by heating platinum sponge in the presence of dry chloride.
- Potassium cholrate by the action of chlorine on potassium hydroxide solution.
- Potassium ferricyanide by passing chlorine gas into a solution of potassium ferrocyanide.
- Propylene dichloride by the action of chlorine on propylene.
- Samarium chlorohydrate.
- Silicon hexachloride by the action of chlorine on ferrosilicon.
- Silicon tetrachloride by the action of chlorine on an electrically heated mixture of silica and carbon.
- Sodium chlorate.
- Sodium ferricyanide by the action of chlorine on a solution of sodium ferrocyanide.
- Sodium permanganate by passing a current of chloride through a solution of sodium manganate.
- Stannic chloride by chlorination of metallic tin or stannous chloride.
- Stannous chloride by the action of chlorine on stannus oxide (German 33925).
- Strontium chlorate by passing chlorine gas into a warmed solution of strontium hydroxide.
- Strontium chloride by heating strontium sulphite in a current of chlorine gas (German 162913).
- Sulphur chloride by passing chlorine over molten sulphur.
- Sulphur dichloride by passing chlorine into sulphur monochloride to saturation.
- Sulphur tetrachloride by the action of chloride on sulphur.
- Sulphuryl chloride by the action of chlorine gas on sulphur dioxide.
- Tertiary butyl chloride, tetrachloro ether, tetrachloroethylene.
- Tetrachlorophthalic acid by passing a stream of chlorine gas through a mixture of phthalic anhydride and antimony pentachloride.
- Tetrachlorophthalic acid anhydride by the action of chlorine on phthalic acid anhydride (German 50177).
- Thionyl chloride by chlorination of a mixture of sulphur dioxide and phosphorus (U. S. 1753754).
- Thorium tetrachloride by heating thorium dioxide in a current of chlorine containing sulphur chloride vapors.
- Titanium tetrachloride by heating titanium dioxide and carbon to redness in a current of chlorine gas.
- Trichloroacetic acid by the action of chlorine on glacial acetic acid in the presence of sunlight, ultraviolet radiation, or catalysts.
- Trichloro ether.
- Trichloroethylene by chlorination of ethylene and subsequent distillation.
- Trichloroisopropyl alcohol.
- Trichloromethylchloro formate.
- Trichloronitromethane.
- 2:4:6-Trichloro-1:3:5-triazin from hydrocyanic acid and chlorine under the influence of sunlight.
- Tungsten hexachloride by chlorination of metallic tungsten.
- Tungsten oxychloride by the action of chlorine on metallic tungsten in the presence of oxygen.
- Vanadium chloride.
- Vanadium oxytrichloride by the action of chlorine on vanadium pentoxide.
- Vanadium tetrachloride by chlorination of ferrovanadium, or by the action of chlorine on vanadium carbide.
- Various intermediates, pharmaceuticals, and synthetic aromatics.
- Xylene chlorinated derivatives.
- Zinc chloride by the action of chlorine gas on metallic zinc.
- Zinc ferricyanide from zinc ferrocyanide.
- Zirconium chloride by the action of chlorine gas on zirconium carbide.

Dye

Reagent in making—
- Alizarin.
- Brilliant indigo by action on a suspension of indigo and crystallized sodium acetate in glacial acetic acid.
- Indanthrene golden orange R paste.
- Tetrabromoindigo from indigo suspensed in nitrobenzene.
- Various synthetic dyestuffs.

Disinfectant

As a disinfectant and germicide.

Gas

Reagent in—
- Purifying crude benzene (U. S. 1674472 and 1729543).

Ink

Reagent in making various inks.

Insecticide

As an insecticide.

Reagent in making—
- Lead arsenate.

Metallurgical

Reagent in—
- Extracting copper, lead, and zinc from mixed ores.

Reagent in purifying—
- Lead by treatment in the molten state (U. S. 1920211).
- Nickel liquors to remove oxides of the iron group of metals.

Reagent in recovering—
- Gold from its ores.
- Nickel from its ores by the wet method.
- Platinum from its ores by the wet method.
- Silver from its ores by the chloridizing roast process.
- Titanium from its ores.
- Zinc and lead from complex ores.
- Zirconium by action on zirconium carbide.

Reagent in reducing—
- Cobalt and nickel.

Reagent in separating—
- Tungsten and vanadium from their ores.

Reagent in treating—
- Scrap galvanized iron with simultaneous production of chloride of zinc free from iron when gaseous chlorine absolutely free from water is used.
- White cast-iron scrap for the recovery of tin.

Military

As a poison gas.

Ingredient of—
- Various mixtures used as poison gases.

Miscellaneous

As a bleach for various purposes.

Reagent for—
- Bleaching sponges.
- Treating asphalt, pitches, and tars to obtain harder and higher-melting products (German 406689 and Brit. 186861).

Paper

Reagent for—
- Bleaching various raw materials used in the manufacture of paper, such as flax fibers, hemp fibers, wood pulp, rag pulp.

Reagent for treating—
- Esparto grass to make straw pulp.
- Straw, wood, and other raw materials to obtain pure cellulose.

Pharmaceutical

In compounding practice.

Recommended for use (in very dilute state) in treating cold.

Resins and Waxes

Reagent in—
- Making resinous products from crude anthracene.
- Treating resins to improve their color and quality (German 426283).

Rubber

Reagent in making—
- Chlorinated rubber, rubber substitutes.

Starch

Reagent in making—
- Solubilized starch by action on starch milk mixed with nitric acid (German 103399).

Textile

Reagent in—
- Bleaching cotton and linen fabrics and yarns.

Water and Sanitation

Reagent for various sanitary purposes.

Chlorine (Continued)
Reagent in—
Purifying drinking water.
Treating waste liquors of various origin.

3-Chlor-4'-iso-octylthioldiphenylamine-6-carboxylic Acid
Pharmaceutical
Claimed (Brit. 363392 and 437953) as—
New pharmaceutical.

5-Chlor-7-methoxy-4-methylisatin Alphachloride
Dye
Starting point (Brit. 441548) in making—
Dyestuffs by condensing with 4-chlor-2-hydroxy-6-methoxy-3-methylthionaphthen.

4-Chlor-2-methyl-6-gammachlorodeltabetabutenylphenol
Disinfectant
Claimed (Brit. 443113 and 389514) to be—
Disinfectant free of odor.

7-Chlor-4-methylisatin Alphachloride
Dye
Starting point (Brit. 443275) in making—
Blue dyestuffs by condensation with 4-chloroalphanaphthol.

3-Chlor-4'-methylthioldiphenylamine-6-carboxylic Acid
Pharmaceutical
Claimed (Brit. 363392 and 437953) as—
New pharmaceutical.

9-Chlor-2-methylthiol-6-methylacridin
Pharmaceutical
Claimed (Brit. 363392 and 437953) as—
New pharmaceutical.

8-Chlor-1-naphthylbetaparatoluenesulphonylethyl Sulphide
Chemical
Intermediate (Brit. 444262 and 444501) in—
Organic syntheses.
Insecticide
Insecticide (Brit. 444262 and 444501) for—
Animal pests, vegetable pests.
Textile
As a dyestuff (when employing suitable initial materials) (Brit. 444262 and 444501).
Assistant (Brit. 444262 and 444501) in—
Textile processing.

8-Chlor-1-naphthylbetaparatolylthioethyl Sulphoxide
Chemical
Intermediate (Brit. 444262 and 444501) in—
Organic syntheses.
Insecticide
Insecticide (Brit. 444262 and 444501) for—
Animal pests, vegetable pests.
Textile
As a dyestuff (when employing suitable initial materials) (Brit. 444262 and 444501).
Assistant (Brit. 444262 and 444501) in—
Textile processing.

2-Chlor-6-nitrobenzaldimercuri Oxide
Disinfectant
Germicide (U. S. 1996006).

3-Chlor-5-nitrobenzoxazolone
Chemical
Starting point in making—
3-Chloro-2:1-benzoxazolone-5-arsinic acid (Brit. 261133).

3-Chlor-1-nitro-4:6-diethoxybenzene
Chemical
In organic syntheses.
Dye
In dye syntheses.
Starting point (U. S. 1984739) in making—
Anilins and anilides used in making ice colors.

3-Chlor-1-nitro-4:6-dimethoxybenzene
Chemical
In organic syntheses.
Dye
In dye syntheses.
Starting point (U. S. 1984739) in making—
Anilins and anilides used in making ice colors.

3-Chlor-1-nitro-4:6-diphenoxybenzene
Chemical
In organic syntheses.
Dye
In dye syntheses.
Starting point (U. S. 1984739) in making—
Anilins and anilides used in making ice colors.

2-Chlor-4-normal-amylphenol
Sanitation
As a bactericide (U. S. 1980966).

2-Chlor-4-normal-butylphenol
Sanitation
As a bactericide (U. S. 1980966).

2-Chlor-4-normal-heptyl phenol
Sanitation
As a bactericide (U. S. 1980966).

2-Chlor-4-normal-hexylphenol
Sanitation
As a bactericide (U. S. 1980966).

2-Chlor-4-normal-propylphenol
Sanitation
As a bactericide (U. S. 1980966).

Chloroacetaldehyde
Petroleum
Solvent (Brit. 437573) in—
Refining mineral oils.

Chloroacetic Acid Cyclohexylester
Detergent
Starting point (Brit. 408754) in making—
Saponaceous products by reaction with tertiary amines, which may be used alone or with other soaps, fillers, or compounds giving off oxygen.

Chloroacetic Acid Dodecylester
Detergent
Starting point (Brit. 408754) in making—
Saponaceous products by reaction with tertiary amines, which may be used alone or with other soaps, fillers, or compounds giving off oxygen.

Chloroacetic Acid Hexadecylester
Soap
Starting point (Brit. 403883) in making—
Saponaceous products by reaction with amines, such as anilin, piperidin bases, hydroxyethylanilin, dihydroxyethylanilin, paratoluidin (these products may be used alone, or with soaps, fillers, or compounds giving off oxygen).

Chloroacetic Acid Octadecylester
Detergent
Starting point (Brit. 408754) in making—
Saponaceous products by reaction with tertiary amines, which may be used alone or with other soaps, fillers, or compounds giving off oxygen.

Chloroacetic Acid Tetradecylester
Soap
Starting point (Brit. 403883) in making—
Saponaceous products by reaction with amines, such as anilin, piperidin bases, hydroxyethylanilin, dihydroxyethylanilin, paratoluidin (these products may be used alone, or with soaps, fillers, or compounds giving off oxygen).

Chloroacetodidodecylamide
Chemical
Starting point (Brit. 443265) in making—
Scouring and wetting agents for textile and other purposes by condensation with a degradation product of albumin or an albuminous substance.

4-Chlor-2-aminodiphenyl Ether
French: Éther 4-chloro-2-aminodiphénylique, Éther de 4-chloro-2-aminodiphényle.
German: 4-Chlor-2-aminodiphenylaether.

Dye
Starting point (Brit. 248946) in making azo dyestuffs with 2:3-oxynaphthol derivatives of—
Meta-m'-diaminoazoxybenzene, meta-m'-diamino-para-p'-dimethylazoxybenzene, meta-m'-diamino-para-p'-dimethoxyazobenzene, meta-m'-diamino-para-p'-dimethoxyazoxybenzene, para-p'-diaminoazobenzene, para-p'-diaminoazoxybenzene.

2-Chloro-4-amino-5-sulphobenzoic Acid
French: Acide de 2-chloro-4-amino-sulfobenzoique.
German: 2-Chlor-4-amino-5-sulfobenzoesaeure.

Dye
Starting point (Brit. 275220) in making monoazo dyestuffs with—
2:8:6-Aminonaphtholsulphonic acid, betanaphthol, pyrazolones.

2-Chloro-5-amino-4-sulphobenzoic Acid
French: Acide de 2-chloro-5-amino-4-sulfobenzoique.
German: 2-Chlor-5-amino-4-sulfobenzoesaeure.

Dye
Starting point (Brit. 275220) in making monoazo dyestuffs with—
2:8:6-Aminonaphtholsulphonic acid, betanaphthol, pyrazolones.

4-Chloroanilin-3-sulphonic Acid
French: Acide sulphonique de chloroaniline, 4:3.
German: Chloranilinsulfosaeure.

Dye
Starting point in making—
Soluble chromium compounds of azo dyestuffs (Brit. 260830).

5-Chloro-2-anisidide
German: 5-Chlor-2-anisidid.

Dye
Reagent (Brit. 274128) in making azo dyestuffs with—
1:3-Dimethyl-4-amino-6-bromobenzene.
1:3-Dimethyl-4-amino-6-chlorobenzene.
1:3-Dimethyl-4-amino-2:6-dibromobenzene.
1:3-Dimethyl-4-amino-2:6-dichlorobenzene.

1-Chloroanthraquinone

Chemical
Starting point in—
Organic syntheses.
Starting point (U. S. 1999996) in making—
Seleno ethers by reacting with 1:1′-dibenzanthronyl diselenide.

3-Chloro-1:2-benzanthraquinone
German: 3-Chlor-1:2-benzanthrachinon.

Chemical
Starting point in making—
Intermediates, pharmaceuticals.

Dye
Starting point (Brit. 340524) in making dyestuffs with the aid of—
Alpha-amino-4-benzoylanthraquinone.
Alpha-amino-5-benzoylanthraquinone.

4-Chlorobenzene Sulphanilide
Synonyms: 4-Chlorobenzenethioanilide.
French: Sulphanilide de 4-chlorobenzène, Thioanilide de 4-chlorobenzène.
German: 4-Chlorbenzolsulfanilid, 4-Chlorbenzolthioanilid.

Chemical
Starting point in making—
Intermediates, pharmaceuticals.

Miscellaneous
Reagent in—
Simultaneous dyeing and mothproofing of fur, hair, and feathers.

Textile
Reagent in—
Simultaneous dyeing and mothproofing of wool and felt.

2-Chlorobenzoic Acid Benzylester

Detergent
Starting point (Brit. 408754) in making—
Saponaceous products by reaction with tertiary amines, which may be used alone or with other soaps, fillers, or compounds giving off oxygen.

2-Chlorobenzoic Acid Betaphenylethylester

Detergent
Starting point (Brit. 408754) in making—
Saponaceous products by reaction with tertiary amines, which may be used alone or with other soaps, fillers, or compounds giving off oxygen.

2-Chlorobenzoic Acid Dodecylester

Soap
Starting point (Brit. 403883) in making—
Saponaceous products by reaction with amines, such as anilin, piperidin bases, hydroxyethylanilin, dihydroxyethylanilin, para-toluidin (these products may be used alone or with other soaps, fillers, or compounds giving off oxygen).

2-Chlorobenzoic Acid Hexadecylester

Soap
Starting point (Brit. 403883) in making—
Saponaceous products by reaction with amines, such as anilin, piperidin bases, hydroxyethylanilin, dihydroxyethylanilin, para-toluidin (these products may be used alone or with other soaps, fillers, or compounds giving off oxygen).

2-Chlorobenzoic Acid Tetradecylester

Soap
Starting point (Brit. 403883) in making—
Saponaceous products by reaction with amines, such as anilin, piperidin bases, hydroxyethylanilin, dihydroxyethylanilin, para-toluidin (these products may be used alone or with other soaps, fillers, or compounds giving off oxygen).

5-Chlorobenzothiazole-1-carboxylic Chloride

Dye
Starting point (Brit. 441915) in making—
Greenish-yellow vat dyes of good fastness to light, chlorine, and alkali, by condensing with an orthoaminothiol of the benzene, naphthalene, or anthraquinone series.
Greenish-yellow vat dyes of good fastness to light, chlorine, and alkali, by condensing with an arylamine and the orthothiol group subsequently introduced and the product cyclized.

Chlorobenzoylaminoanthraquinone
French: Chlorobenzoyleaminoanthraquinone.
German: Chlorbenzoylaminoanthrachinon.

Chemical
Starting point in making—
Intermediates, pharmaceuticals.

Dye
Starting point (Brit. 298696) in making anthraquinone vat dyestuffs with—
Aminohydroxybenzoylaminoanthraquinone.
Benzoylaminoanthraquinones.

Chlorobenzoyl Chloride
French: Chlorure de chlorobenzoyle, Chlorure chlorobenzoylique.
German: Chlorbenzoylchlorid, Chlorchlorbenzoyl.

Fats and Oils
Reagent for bleaching—
Various fats and oils of animal and vegetable origin (used with oleyl chloride, lauryl chloride, and caproyl chloride).
Various oilseed meals (used in conjunction with oleyl chloride, lauryl chloride, and caproyl chloride).

Food
Reagent for bleaching—
Various foodstuffs, such as flour, various milling products, egg yolk, food preparations of animal and vegetable origin (used in conjunction with oleyl chloride, lauryl chloride, and caproyl chloride).

Soap
As a bleaching agent (used in conjunction with oleyl chloride, caproyl chloride, and lauryl chloride).

Waxes and Resins
Reagent for bleaching—
Various resins (used in conjunction with oleyl chloride, caproyl chloride, and lauryl chloride).

2-Chloro-4-bromobenzoic Acid

Chemical
Starting point in making—
Esters and salts, intermediates, pharmaceuticals.
Starting point (Brit. 353537) in making acridin derivatives with—
4-Anisidin, 4-cresidin, 4-phenetidin, 4-toluidin, 4-xylidin.

5-Chloro-7-bromo-8-hydroxyquinolin

Pharmaceutical
Suggested for use (Brit. 351605) as—
Antiseptic.

Chlorobutanone

Petroleum
Solvent (Brit. 437573) in—
Refining mineral oils.

Chlorocyclohexane

German: Chlorzyklohexan.
Chemical
Starting point (Brit. 261764) in making cyclohexylamines with—
Alpha-aminoanthraquinone, anilin, beta-aminoanthraquinone, carbazole, diaminoanthraquinone, 1:4-diaminochloroanthraquinone, diphenylamine, monochloroanilin, monoethylanilin, monomethyl anilin, naphthylamine, toluidin, xylidin.

Chlorodibenzoyl Dimethylsulphide

Lubricant
Extreme pressure agent (Brit. 454552) in—
Extreme pressure lubricants.

Chlorodi-isobutylpyrocatechol

Disinfectant
As a germicide (U. S. 2023160).

Chlorodi-isobutylquinol

Disinfectant
As a germicide (U. S. 2023160).

Chlorodi-isobutylresorcinol

Disinfectant
As a germicide (U. S. 2023160).

2-Chloro-4-dimethylaminobenzaldehyde

Dye
Starting point (Brit. 262141) in making dyestuffs with—
Alphanaphthaldehyde.
Alphanaphthaldehydesulphonic acid.
4-Aminobenzaldehyde.
4-Aminobenzaldehydesulphonic acid.
Aryltetrahydronaphthalene-1-aldehyde.
Aryltetrahydronaphthalene-1-aldehydesulphonic acid.
4-Dimethylaminobenzaldehyde.
4-Dimethylaminobenzaldehydesulphonic acid.

6-Chloro-2:4-dinitroanilin

Chemical
Starting point in—
Organic synthesis.
Dye
Starting point in making various dyestuffs, including—
Light-fast and readily discharged violet dyestuffs for acetate rayon by diazotizing and coupling with di-(betahydroxyethyl)metatoluidin (Brit. 421975).
Light-fast and readily discharged navy-blue dyestuffs for acetate rayon by diazotizing and coupling with normal-betahydroxyethyl-N-N-butylcresidin (Brit. 421975).
Light-fast and readily discharged red-violet dyestuffs for acetate rayon by diazotizing and coupling with normal-ethyl-N-betagammadihydroxypropylanilin (Brit. 421975).
Light-fast and readily discharged blue-violet dyestuffs for acetate rayon by diazotizing and coupling with normal-betahydroxyethyl-N-butylmetatoluidin (Brit. 421975).
Starting point (Brit. 429936 and 430079) in making—
Blue dyestuffs for acetate rayon and animal fibers by diazotizing and coupling with 2:5-dimethoxybutylbetasulphatoethylanilin.
Navy-blue dyestuffs for acetate rayon and animal fibers by diazotizing and coupling with the alphabutyl, betabutyl, or betapropyl derivative of 3-betasulphatoethylaminoparatolylmethyl ether.
Red-violet dyestuffs for acetate rayon and animal fibers by diazotizing and coupling with ethylbetasulphatoethylanilin.
Violet dyestuffs for acetate rayon and animal fibers by diazotizing and coupling with methylbetasulphatoethylanilin.

Chloroform

Synonyms: Methenyl trichloride, Trichloromethane.
Latin: Chlorofirmium, Chloroformum-chloroformum purificatum, Formylum trichloratum.
French: Chloroforme officinale.
German: Chloroform, Reines chloroform.
Spanish: Chloridoformico, Cloroformo.
Italian: Cloroformio.

Agriculture
As a stimulant of plant growth.
Analysis
Extracting medium for various purposes.
Solvent for the extraction and assay of—
Alkaloids, drugs.
Solvent in analyzing and testing—
Alkaloids, animal oils, ashes, breadstuffs, butter, cakes, cheese, chocolate, cocoa, essential oils, fats, flour, hops, meals, meat, milk, mineral phosphates, resins, rosin, rosin oil, rubber, soaps, vegetable oils.
Solvent in making—
Toxicological examinations.
Automotive
Degreasing agent for—
Automobile bodies, automobile parts.
Dewaxing agent in—
Manufacturing operations.
Beverage
Ingredient of—
Cider flavor, containing also amyl alcohol, amyl acetate, amyl butyrate, and amyl valerate.
Ceramics
Solvent in—
Coating compositions, containing cellulose acetates, as well as resins, waxes, and gums, used for protecting and decorating ceramic ware.
Chemical
Extractant for—
Acid gases from gaseous mixtures (Austrian 135047).
Alkaloids, drug principles.
Ingredient of solvent mixtures containing—
Acetone, alcohol, benzene, chlorinated hydrocarbons, turpentine.
Noninflammable ingredient of—
Solvents mixtures.
Reagent (U. S. 1891415) in making—
Brominated hydrocarbons from aluminum bromide.
Solvent for—
Acetylsalicylic acid.
Acid in concentration of acetic acid (Brit. 400169).
Cellulose acetate.
Solvent in making—
C. P. chemicals, drugs, inorganic chemicals, intermediates, organic chemicals, pharmaceuticals, substituted alkyl chlorides (Brit. 402159), U. S. P. chemicals.
Dry Cleaning
Ingredient of—
Noninflammable cleaning fluid, containing also carbon tetrachloride and deodorized gasoline.
Noninflammable cleaning fluid, containing also carbon tetrachloride, deodorized naphtha and benzene.
Solvent for—
Removing oils, fats, waxes, tar, and other stains and impregnated substances.
Spotting agent for—
All textiles except cellulose acetate fabrics.
Dye
Reagent and solvent in making—
Synthetic dyestuffs of various classes.
Electrical
Solvent for—
Cellulose acetate used as a coating for battery electrodes (Brit. 395456).
Cleaning electric motors and other electrical machinery.
Solvent in—
Compositions, containing cellulose acetate and at times resins, gums, and the like, used for insulating cables, wiring, and electrical machinery and equipment.
Fats and Oils
Extractant for—
Animal oils, essential oils, fats, greases, vegetable oils.
Solvent for—
Animal oils, essential oils, fats, greases, vegetable oils.
Solvent for—
Recovering oils from fuller's earth and other substances used in bleaching.
Fertilizer
Solvent for—
Degreasing fish scrap.
Food
Extractant of soluble substances from—
Berries, fruits, seeds.
Solvent for—
Decaffeinizing coffee extracts (Brit. 397323).

Chloroform (Continued)
Decaffeinizing coffee (Brit. 314059).
Detheinizing tea extracts (Brit. 397323).
Making food flavors, purifying foodstuffs.

Glass
Solvent for—
Degreasing glass.
Solvent in—
Compositions, containing cellulose acetate and artificial or natural resins, waxes, and gums, used in the manufacture of nonscatterable glass and for the decoration and protection of glassware.

Glues and Adhesives
Ingredient of—
Special adhesive compositions containing cellulose acetate.
Solvent for—
Degreasing bones and hides preparatory to the manufacture of glue and gelatin.

Gums
Solvent for various gums.

Ink
Solvent in making—
Printing inks.

Insecticide
Ingredient of—
Insecticidal compositions.
Preparations for exterminating parasites.
Vermicidal compositions.

Leather
Solvent for—
Cleansing spotted leathers.
Removing natural oils and greases from hides and skins before tanning, so as to prevent staining thereafter and insure evenness of the leather finish and tan.
Solvent in—
Compositions, containing cellulose, acetate, as well as artificial or natural resins, gums, and waxes, used in the manufacture of artificial leather and for the protection and decoration of leather goods.

Mechanical
Solvent for—
Cleansing and degreasing machinery of various sorts.
Cleansing drive wheels of compression pumps and other mechanical equipment.
Degreasing automobile brakebands.

Metallurgical
Solvent for—
Cleansing and degreasing metallic surfaces preparatory to painting or other coating.
Degreasing die castings, metal stampings, metals to be electroplated, nuts and bolts.
Preparing metals for pickling, plating, shellacking, sherardizing, varnishing.
Solvent and diluent in—
Compositions, containing cellulose acetate, used for protecting and decorating metallic articles.

Miscellaneous
As a dental solvent.
As a general solvent.
Degreasing agent in treating—
Furs (also acts as a parasiticide).
Ingredient of—
Compositions of clay, for cleansing ivory, horn and bone.
Polishing compositions of various sorts.
Preparations used for the removal of stains from celluloid articles.
Preparations used for cleansing typewriters.
Solvent and diluent in—
Compositions, containing cellulose acetate, used for decorating and protecting various articles.

Oilcloth and Linoleum
Solvent in making—
Coating compositions.

Paint and Varnish
Diluent (Brit. 395478) in—
Lacquer composed of vinyl chloroacetate and vinyl stearate, polymerized in acetone solution and the resulting solution diluted.

Paper
Solvent for—
Removing oil from paper and paperstock.
Solvent in—
Compositions containing cellulose acetate, with gums, waxes, and natural or artificial resins, used in the manufacture of coated paper and for coating and decorating paper and pulp products.

Perfume
Solvent for—
Extracting aromatic principles from flowers, particularly those alterable by heat.
Solvent in making—
Nail-whitening preparation, containing also zinc white, paraffin, and oil of neroli.

Petroleum
Solvent for—
Degreasing light mineral oils.
Extracting wax from mineral oil distillates.

Pharmaceutical
Anesthetic.
Extractant for—
Alkaloids, drugs.
In compounding and dispensing practice.
Ingredient of—
Antiseptic toothache drops, containing also beechwood creosote, oil of clove, cinnamic aldehyde or oil of cassia, and ethyl aminobenzoate.
Inhalant for colds, containing also formaldehyde, ether, menthol, oils of eucalyptus and lavender, and isopropyl alcohol.
Inhalant for colds, containing also isopropyl alcohol, oils of sassafras, clove, and eucalyptus, thymol, camphor, menthol, and phenol.
Liniment, containing also oil of mustard, oil of rosemary, powdered camphor, ethyl aminobenzoate, oleoresin of capsicum, oils of laurel and camphor.
Mouthwash, containing also oils of peppermint and cinnamon, alcohol, phenol, benzoic acid, and glycerin.
Psoriasis preparation, containing also oil of mace, olive oil, ammonia, essence of rosemary, rose water, lecithin, and an aromatic.
Refrigerant counter-irritant, containing also menthol, iodine, and tincture of aconite.
Ingredient of, and process material in making—
External and internal pharmaceutical preparations.
Rubbing liniments, salves.
Preservative for—
Serums, vegetable drugs.
Suggested for use as—
Anthelmintic, antiseptic, antispasmodic, analgesic, antidote, counter-irritant, sedative in cough remedies, stimulant, vermicide.

Photographic
Solvent for—
Cleansing and degreasing motion picture film.
Solvent in making—
Motion picture film.

Plastics
Degreasing solvent.
Solvent and diluent in making—
Compositions containing cellulose acetate, with gums, waxes and artificial or natural resins.
Films and insulating materials from acetone-soluble cellulose acetate, dimethylanilin, and tetrachloropyrimidin (Brit. 393914).
Insulating materials from unsaponified cellulose acetates containing a small amount of radicals of other organic or inorganic acids (French 749575).

Printing
Solvent for—
Cleansing engraved plates, lithographic stones, printing machinery, type.

Resins and Waxes
Solvent for various resins and waxes.

Rubber
Ingredient of—
Rubber cements, rubber mastics, rubber compositions used in the manufacture of rubberized cloth.
Solvent for—
Rubber.
Solvent in—
Coating compositions, containing cellulose acetate, with gums and waxes, used for decorating and protecting rubber goods.

Soap
Ingredient of—
Dry-cleaning compositions, spotting fluids.

Stone
Solvent in—
Compositions, containing cellulose acetate, with arti-

Chloroform (Continued)
ficial or natural resins, gums, and waxes, used for the decoration and protection of artificial and natural stone.
Sugar
Solvent for—
Extracting waxes from filter press "mud" in refining.
Textile
——, *Finishing*
Solvent in—
Coating compositions containing cellulose acetate.
——, *Manufacturing*
Shrinking agent (Brit. 403106) in making—
Filaments, threads, ribbons, and the like from organic derivatives of cellulose.
Solvent for—
Cleaning knitting machine needles.
Cleaning silk and silk hosiery.
Degreasing textiles.
Degreasing wool.
Degumming silk.
Solvent and diluent in making—
Compositions, containing cellulose acetate, used for making coated textiles.
Threads from acetone-soluble cellulose acetate, dimethylanilin and tetrachloropyrimidin (Brit. 393914).
Scouring compositions.
Tobacco
Solvent for—
Extracting nicotine.
Woodworking
Solvent and diluent in—
Compositions, containing cellulose acetate, used for decorating and protecting woodwork.

2-Chlorohydrazin-5-nitropyridin
Chemical
Starting point (Brit. 259982) in making derivatives with—
Acetone, acetoacetic ester, benzaldehyde, propionic aldehyde.

2-Chloro-5-hydrazopyridin
Chemical
Starting point in making—
Intermediates with acetoacetic ester and the like (Brit. 259982).

4-Chloro-1-hydroxy-3:5-dimethylbenzene
Miscellaneous
Improver (Brit. 431645) of—
Absorbent properties of materials of various kinds; for example, wood flour and leather.
Textile
Imparter (Brit. 431645) of—
Antiseptic, germicidal, and deodorant properties to textiles (applied by impregnation during (1) manufacturing operations, (2) laundering operations).
Improver (Brit. 431645) of—
Absorbent properties of textile fibers.

4′-Chloro-3-hydroxydiphenylaminecarboxylic Acid
French: Acide de 4′-chloro-3-hydroxyediphényleaminecarbonique, Acide de 4′-chloro-3-hydroxyediphényleaminecarboxylique.
German: 4-Chlor-3-hydroxydiphenylamincarbonsäure.
Chemical
Starting point in making—
Esters, intermediates, salts.
Starting point (Brit. 336420) in making intermediates with the aid of—
Alpha-aminoanthraquinone, 3-aminocarbazole, 6-amino-3-hydroxyl-1-methylbenzene, anilin, beta-aminoanthraquinone, betanaphthylamine, 5-chloro-2-aminoanisole, 4-chloro-2-aminotoluene, 5-chloro-2-aminotoluene, 2-chloroparanitranilin, 6-chloroparanitranilin, 1:5-diaminonaphthalene, dianisidin, 2:5-dichloroanilin, meta-aminophenol, meta-anisidin, metachloroanilin, metanitranilin, metaphenetidin, metatoluidin, orthoaminophenol, orthoanisidin, orthochloroanilin, orthonitranilin, orthophenetidin, orthotoluidin, 4-nitro-2-aminoanisole, 5-nitro-2-aminotoluene, para-aminophenol, para-amisidin, parachloroanilin, paranitranilin, paraphenetidin, paratoluidin.

7-Chloro-2-hydroxy-4-methylquinolin
Pharmaceutical
Suggested for use (Brit. 351605) as—
Antiseptic.

7-Chloro-2-hydroxy-4-methylquinolin Methanesulphonate
Pharmaceutical
Suggested for use (Brit. 351605) as—
Antiseptic.

5-Chloro-8-hydroxyquinolin
Pharmaceutical
Suggested for use (Brit. 351605) as—
Antiseptic.

5-Chloro-8-hydroxyquinolin Betadicyclohexylaminoethyl Ether
Pharmaceutical
Suggested for use (Brit. 351605) as—
Antiseptic.

2-Chloro-5-hydroxytoluene
Miscellaneous
Improver (Brit. 431645) of—
Absorbent properties of materials of various kinds; for example, wood flour and leathers.
Textile
Imparter (Brit. 431645) of—
Antiseptic, germicidal, and deodorant properties to textiles (applied by impregnation during (1) manufacturing operations, (2) laundering operations).
Improver (Brit. 431645) of—
Absorbent properties of textile fibers.

2-Chloro-4-iodobenzoic Acid
Chemical
Starting point in making—
Esters and salts, intermediates, pharmaceuticals.
Starting point (Brit. 353537) in making acridin derivatives with—
4-Anisidin, 4-cresidin, 4-phenetidin, 4-toluidin, 4-xylidin.

5-Chloro-7-iodo-8-hydroxyquinolin
Pharmaceutical
Suggested for use (Brit. 351605) as—
Antiseptic.

Chloromercury Chloride
Agriculture
For control of—
Bottom rust of lettuce.
Covered smut and stripe disease of barley.
Kernel smut of sorghum.
Loose and covered smuts of oats.
Soil-borne parasitic fruit.
Stinking smut of wheat.
Woodworking
For control of—
Blue stain and sap stain in sapwood of freshly sawed lumber.

6-Chlorometatoluidin 5-Sulphonpiperidide
Dye
Coupling agent (Brit. 434209 and 434433) in making—
Water-insoluble reddish bordeaux dyestuffs with 5-chlor-2:4-dimethoxyanilide.

4-Chloro-7-methoxyisatin Chloride
French: Chlorure de 4-chloro-7-méthoxyeisatine, Chlorure de 4-chloro-7-méthoxyeisatinique.
German: 4-Chlor-7-methoxyisatinchlorid.
Chemical
Starting point in making various intermediates.
Dye
Starting point (Brit. 309379) in making thioindigoid dyestuffs with—
5-Chloro-3-oxythionapththene.
5:7-Dichloro-3-oxythionaphthene.
4:7-Dimethyl-5-chloro-3-oxythionaphthene.
4-Methyl-6-chloro-3-oxythionaphthene.
5-Methyl-6:7-dichloro-3-oxythionaphthene.
4-Methyl-5:7-dichloro-3-oxythionaphthene.
5:6:7-Trichloro-3-oxythionaphthene.

4-Chloro-1-methylanthraquinone
German: 4-Chlor-1-methylanthrachinon.
Chemical
Starting point in making—
4-Anilino-1-methylanthraquinone.
1-Chloro-4-methylanthranol.
4-Chloro-1-methylanthracene.
9:10-Dihydro-4-chloro-1-methylanthracene.
Dye
Starting point in making various dyestuffs.

5-Chloro-2-methylindole

Textile

Starting point (Brit. 396893) in—
Producing violet shades in dyeing acetate with rayon.

Chloromethylorthocresotinic Acid
French: Acide de chlorométhylecrésotinique.
German: Chlormethylorthocresotinsaeure.

Chemical

Starting point (Brit. 265203) in making aminodiarylmethane derivatives with—
Anilin, alphanaphthylamine, betanaphthylamine, metatoluidin, metaxylidin, orthotoluidin, orthoxylidin, paratoluidin, paraxylidin.

Dye

Starting point (French 627521) in making mordant azo dyestuffs with—
Alphanaphthylamine, anilin, benzylamine, betanaphthylamine, diethylamine, dimethylamine, meta-anisidin, metaphenylenediamine, metatoluidin, metaxylidin, monoethylanilin, monomethylanilin, orthoanisidin, orthophenylenediamine, orthotoluidin, orthoxylidin, para-anisidin, paraphenylenediamine, paratoluidin, paraxylidin, phenylamin.

6-Chloro-7-methyl-3-oxythionaphthalene
German: 6-Chlor-7-methyl-3-oxythionaphtalin.

Dye

Starting point (Brit. 267177) in making thioindigo dyestuffs from—
Acenaphthenequinone.
Alphaisatin chloride.
5-Bromo-2:1-thionaphthisatin.
5-Chloro-7-methylthionaphthenequinoneparadimethylaminoanil.
6-Chloro-4-methylthionaphthenequinoneparadimethylaminoanil.
6-Chlorothionaphthenequinone.
1-Chloro-2:3-thionaphthisatin.
5:7-Dibromoisatin.
5:7-Dibromoisatin chloride.
6-Ethoxy-4-methylthionaphthenequinoneparadimethylaminoanil.
6-Ethoxy-7-methylthionaphthenequinoneparadimethylaminoanil.
Paradimethylaminoanil of thionaphthenequinone.
Paranitrosodimethylanilin.
1:2-Thionaphthisatin.
2:3-Thionaphthisatinparadimethylaminoanil.

5-Chloro-7-methyl-3-oxythionaphthene
French: 5-Chloro-7-méthyle-3-oxyesulphonaphthène, 5-Chloro-7-méthyle-3-oxyethionaphthène.
German: 5-Chlor-7-methyl-3-oxysulfonaphten, 5-Chlor-7-methyl-3-oxythionaphten.

Dye

Starting point (Brit. 309379) in making thioindigoid dyestuffs with—
4:5-Dichloro-7-methoxyisatin chloride.
4-Methyl-5-chloro-7-methoxyisatin chloride.
4-Methyl-7-methoxyisatin chloride.

5-Chloro-7-methyl-1-thionaphthene-2:3-carboxylic Acid
French: Acide de 5-chloro-7-méthyle-1-thio-naphthène-2:3-carbonique.
German: 5-Chlor-7-methyl-1-thionaphten-2:3-carbonsaeure.

Dye

Starting point (Brit. 261384) in making thionaphthene dyestuffs with—
Benzene, cymene, anthracene, mesitylene, naphthalene, naphthylmethane, tolyldiphenylmethane, toluene.

Chloronaphthalenes
Synonyms: Chlorinated naphthalene.
French: Chloronaphthalène, Naphthalène chlorée.
German: Chlornaphthalin.
Spanish: Chloronaftalino.
Italian: Cloronaftalena.
(Note: As indicated by the title these are products of the chlorination of naphthalene. According to the degree of chlorination the physical state varies from a thinly fluid, mobile liquid to a crystalline, amorphous wax. The degree of chlorination is indicated by chemical nomenclature, thus:—1-chloronaphthalene, monochloronaphthalene, alphachlornaphthalene, dichloronaphthalene, trichloronaphthalene, tetrachloronaphthalene, hexachloronaphthalene, polychloronaphthalenes. Commercially they are marketed (1) under the chemical name, or (2) under a trade-name or brand-name, such as "Halowax" oils and waxes, or "Seekay" waxes and oils, "I. G." waxes and oils; or "Haftax" waxes and oils. For obvious reasons the uses below are not indicated for the degree of chlorination.)

Adhesives

Ingredient of—
Adhesive composed also of natural resins, rubber latex, and castor oil (French 691293).
Adhesive mixtures with glycerol ester of rosin (French 691293).
Cement for uniting glass, porcelain, pottery, metals, wood, and other substances (U. S. 1945803).

Agriculture

Ingredient (French 649853) of—
Smoke-screen compositions for treating vegetables.

Analysis

Standard in—
Testing index of refraction.

Building and Construction

Flameproofing agent for—
Fibrous materials, rubber tile.
Impregnating agent (U. S. 1941769) for—
Celotex building block.
Protective coating (against corrosive action of acid and alkaline liquids and acid fumes) for—
Asphalt-coated building materials (Brit. 209727).
Metal surfaces, other surfaces, stone surfaces, wood surfaces.

Chemical

Condensing agent in making—
Aminoaralkylarylcarboxylic acids (U. S. 1936090).
Aralkylarylcarboxylic acids (U. S. 1937963).
Starting point in making—
Alphachloronaphthalenesulphonic acid (Brit. 362016).
Alphanaphthol (U. S. 1996745).
Ammonium 1:5-chloronaphthalenesulphonate (Brit. 263873 and 280262).
Ammonium 1:6-chloronaphthalenesulphonate (Brit. 263873 and 280262).
Benzanthrone and its derivatives.
Betanaphthol (U. S. 1996745).
1:5-Chloronaphthalene sulphonate (Brit. 263873 and 280262).
1:6-Chloronaphthalene sulphonate (Brit. 263873 and 280262).
4-Chloroalphanitronaphthalene and other nitration derivatives.
1-Chloronaphthalene-2-thioglycollic acid (Brit. 284288).
1:2-Chloronaphthoyl chloride (German 432579).
8-Cyano-4-chloronaphthalenealphasulphonic acid (Brit. 276126).
Derivatives used in making intermediates such as—
2:3-Dichloroalphanaphthaquinone.
3:4-Dichloroalphanaphthol.
5:8-Dichloroalphanitronaphthalene.
Nitronaphthalene tetrachloride.
Phthalic acid.
Halogen derivatives, such as (Brit. 341926; French 683792)—
1:4-Chlorobromonaphthalene.
1:8-Chlorobromonaphthalene.
1:4-Chlorobromonaphthalene-8-sulphonic acid.
1-Chlor-4:8-dibromonaphthalene.
1:8-Dichloro-4-bromonaphthalene.
1:4-Dichlor-8-bromonaphthalene.
Halonaphthalene ketones, such as (German 495332)—
1:4-Dichlor-8-alphanaphthoylnaphthalene.
1:4-Dichlor-8-benzoylnaphthalene.
1:5-Dichlor-8-benzoylnaphthalene.
1:4-Dichlor-8-orthochlorobenzoylnaphthalene.
1:4-Dichlor-8-orthotoluylnaphthalene.
1:4-Dichlor-8-orthochlorobenzoylnaphthalene.
1:4-Dichlor-8-parachlorobenzoylnaphthalene.
Isopropylchloronaphthalene sulphonate (Brit. 252392).
Naphthol derivatives.
Naphthylamine derivatives.
Potassium 1:5-chloronaphthalenesulphonate (Brit. 263873 and 280262).
Potassium 1:6-chloronaphthalenesulphonate (Brit. 263873 and 280262).
Potassium isopropylchloronaphthalenesulphonate (Brit. 252392).
Sodium 1:5-chloronaphthalenesulphonate (Brit. 263873 and 280262).
Sodium 1:6-chloronaphthalenesulphonate (Brit. 263873 and 280262).
Sodium isopropylchloronaphthalenesulphonate (Brit. 252392).
Sulphonic acids.

Chloronaphthalenes (Continued)
Tanning agents by sulphonating and condensing with hydroxybenzyl alcohol (French 614661).

Coke By-Products
Stabilizing agent (French 698554) in—
Tar emulsions.

Dye
Solvent for—
Anilin dyes, other dyes.
Starting point in making—
1-Chloronaphthalene-2-thioglycollic acid in making thioindigoid dyes (Brit. 284288) with—
Acenaphthenequinone.
Alphaisatinanilide.
5:7-Dibromoisatin.
Isatin homologs, derivatives, and substitution products.
Ortho-diketones.
Quinoneimidin dyes.

Electrical
Bonding agent in—
Magnetic cores (Brit. 404544).
Rubber-textile insulations for wires and cables.
Flameproofing agent for—
Wire insulations.
Impregnating and coating agent for—
Condensers in radio, telegraphy, telephone, transmission, electric machinery, and installations of all kinds.
Coils in radio, telegraphy, telephone transmission, electrical machinery and installations of all kinds.
Ingredient of—
Insulating compositions.
Sealing compositions for dry batteries.
Protective agent (against corrosive action of acid and alkaline liquids and acid fumes) for—
Cables, wires.
Softening and flexibilizing agent in—
Rubber-textile insulations for wires and cables.
Starting point (Brit. 418557) in making—
Insulating materials with rubber and polymerized hydrcoarbons of the polyvinyl group.

Explosives and Pyrotechnics
Ingredient of—
Smoke-screen compositions.
Dampproofing absorbent in—
Explosives.

Fats, Oils, and Waxes
Ingredient of—
Wax emulsions, waxlike bodies (Brit. 406355).
Plasticizer for—
Waxes.
Solvent for—
Vegetable oils, waxes.
Starting point in making—
High-melting wax products (U. S. 1928438).
Substitute for—
Paraffin and other waxes.

Fuel
Claimed as fuel (French 642681) for—
Diesel engines, internal-combustion motors, semidiesel engines.

Gums
Plasticizer for—
Gums.
Solvent for—
Gums.

Ink and Related Products
Ingredient (Brit. 275747; U. S. 1608742, 1608743, 1639080, 1645141, and many others) of—
Stencil coating compositions.

Insecticide
Ingredient of—
Composition for spraying peach trees to combat Oriental fruit moth.
Compositions for spraying trees and other plants.
Emulsified compositions for destroying flies and their larvae (Brit. 261055).
Emulsified compositions for destroying parasites on sheep (Brit. 261055).
Hydrocarbon oil solutions for destroying flies and their larvae (Brit. 261055).
Hydrocarbon oil solutions for destroying parasites on sheep (Brit. 261055).
Insecticidal oil containing also rotenone, and viscous petroleum oil (U. S. 2013028).
Insecticidal spray compositions for codling moth.
Insecticidal spray compositions for red spider.
Insecticidal spray compositions for scale.
Insecticidal compositions with anilin sulphocyanide (French 654416).
Insectproofing compositions for wood and other fibrous materials.
Mothproofing compositions for textiles and other fabrics.
Verminproofing compositions for wool.

Lubricant
As a lubricant (German 302986).
Extreme pressure agent in—
Cup greases used as lubricants in wire-drawing operations (said to improve efficiency).

Mechanical
As a lubricant (German 302986).
Ingredient of—
Carbon removers of various compositions.
Carbon-removing composition (U. S. 1949588).
Top-cylinder lubricant for—
Preventing carbon and gum deposits on valves in internal-combustion motors.

Metallurgical
Ingredient (Brit. 413519) of—
Soldering fluxes for aluminum alloys.

Miscellaneous
Acidproofing agent for—
Fibrous materials of all kinds.
Alkaliproofing agent for—
Fibrous materials of all kinds.
As a heat-transfer medium.
Flameproofing agent for—
Fibrous materials of all kinds.
Ingredient of—
Emulsified compositions for killing weeds (Brit. 261055).
Hydrocarbon oil compositions for killing weeds (Brit. 261055).
Impregnating compositions.
Insectproofing agent for—
Fibrous materials of all kinds.
Moistureproofing agent for—
Fibrous materials of all kinds.
Plasticizer for—
Many products in various industries.
Preservative for—
Manuscripts, books, and bindings in libraries.
Protective coating (against corrosive action of acid and alkaline liquids and acid fumes) for—
Metal surfaces, other surfaces, stone surfaces, wood surfaces.
Solvent for various products.
Starting point in making—
Polishes of many kinds.

Paint and Varnish
Ingredient of—
Insectproof distemper containing also gum arabic, flour, zinc oxide, and iron oxide (Brit. 447753).
Putty containing also clay, rosin, and rosin oil (Brit. 420528).
Water-resistant varnishes.
Plasticizer in—
Lacquers.

Paper
Ingredient (Brit. 428873) of—
Flameproofing compositions for paper, containing also chlorinated rubber and polyvinyl chloride.

Petroleum
Solvent for—
Mineral oils.

Photographic
Plasticizer in making—
Film.
Substitute for—
Camphor in nitrocellulose film.

Plastics
Hardening agent (French 616506) for—
Phenol-formaldehyde molded products.
Ingredient of—
Imitation porcelain plastic, insulating compositions, molding compositions.
Plasticizer in—
Plastic compositions.
Starting point in making—
Plastics from phenols.

Rayon
Delustering agent for—
Cellulose acetate rayon.
Viscose rayon (Brit. 399512).

Chloronaphthalenes (Continued)
Ingredient (French 706709) of—
Compositions for protecting rayon against short-wave light rays.

Refrigeration
Solvent (U. S. 1991188) in—
Methyl chloride absorption type refrigeration plants.

Resins
Ingredient of—
Resins made from chinawood oil, cresol, and formaldehyde (French 688303).
Plasticizer for—
Moldable resin composed of phenol, orthocresol, and hexamethylenetetramine (U. S. 1975884).
Resins.
Solvent for—
Resins.
Starting point in making—
Resin substitutes (German 332725).
Resins from phenols.
Synthetic resins (Brit. 392382).
Varnish and lacquer resins.

Rubber
Bonding agent for rubber in making—
Rubberized cloth.
Imparter of—
Flameresisting properties (to a marked degree) to rubber.
Ingredient (Brit. 448093) of—
Chlorinated rubber compositions suitable for use as lacquers and coating materials.
Penetration promoter in making—
Rubberized cloth.
Solvent for—
Caoutchouc, gutta-percha, rubber.
Softening agent for rubber in making—
Rubberized cloth.

Textile
Acidproofing agent for—
Fabrics.
Acidfumes-proofing agent for—
Fabrics.
Alkaliproofing agent for—
Fabrics.
Flameproofing agent for—
Fabrics.
Ingredient of—
Flameproofing compositions for wool, cotton and silk, containing also chlorinated rubber and polyvinyl chloride (Brit. 428873).
Waterproofing compositions for textiles.
Moistureproofing agent for—
Fabrics.
Protective (French 623555) in—
Localizing delustring effects.

Wood
Acidproofing agent for—
Wood.
Acidfumes-proofing agent for—
Wood.
Alkaliproofing agent for—
Wood.
Flameproofing agent for—
Wood.
Impregnating agent (French 697496) for—
Weaver's wooden shuttles.
Ingredient (Brit. 428873) of—
Flameproofing and waterproofing compositions for wood, containing also chlorinated rubber and polyvinyl chloride.
Insectproofing agent for—
Wood.
Moistureproofing agent for—
Wood.
Protective coating (against corrosive chemical action) for—
Wood.

1:5-Chloronaphthalenesulphonic Acid
Miscellaneous
As an emulsifying agent (Brit. 263873).
For uses, see under general heading: "Emulsifying agents."

1:6-Chloronaphthalenesulphonic Acid
French: Acide de 1:6-chloronaphthalènesulphonique.
German: 1:6-Chlornaphthalinsulfosaeure.

Chemical
Reagent (Brit. 263873) in making—
Emulsions containing aromatic hydrocarbons.
Terpene emulsions.

Fats and Oils
Reagent in making—
Emulsions of various oils and fats.

Leather
Reagent in making—
Emulsions containing tanning agents.

Miscellaneous
Reagent in making—
Cleansing emulsions, washing emulsions.

Paper
Reagent for treating—
Cardboard and paper in order to increase their absorbing and wetting properties.

Petroleum
Reagent in making—
Emulsions containing mineral oils.

Textile
——, *Dyeing*
Reagent in making—
Dye liquor emulsions.

——, *Finishing*
Reagent in making—
Cleansing and washing emulsions.

——, *Manufacturing*
Reagent in making—
Wool carbonizing compositions.

Waxes and Resins
Reagent in making—
Resin emulsions, wax emulsions.

1:2-Chloronaphthoyl Chloride
French: Chlorure de 1:2-chloronaphthoyle, Chlorure 1:2-chloronaphthoylique.
German: Chlor-1:2-chlornaphtoyl, 1:2-Chlornaphtoylchlorid.

Dye
Starting point in making—
Anthraquinone vat dyestuffs with 1:4-diaminoanthraquinone (German 432579).

4-Chloro-2-nitranilin
French: 4-Chlorure-2-nitraniline.
German: 4-Chlor-2-nitranilin, 4-Chlor-2-nitroanilin.

Dye
Intermediate in making various dyestuffs.

Paint and Varnish
Coloring agent (Brit. 390649) for—
Cellulose nitrate or acetate varnishes.

4-Chloro-2-nitro-4′-aminodiphenylamine
German: 4-Chlor-2-nitro-4-aminodiphenylamin.

Chemical
Starting point in making—
Intermediates, pharmaceuticals.

Dye
Starting point in making various synthetic dyestuffs.
Starting point (Brit. 323792) in making azo dyestuffs for rayons with the aid of—
Alkylaryl anilins, allylaminophenol, allylnaphthylamine, alphanaphthylamine, aminonaphthoic acids, aminonaphthols, amylaminophenol, amylnaphthylamine, betanaphthylamine, butylaminophenol, butylnaphthylamine, cresols and derivatives, dimethylmeta-aminophenol, ethylaminophenol, ethylnaphthylamine, gammachlorobetaoxypropionylnaphthylamine, meta-aminophenol, meta-anisidin, metacresidin, metaphenetidin, metaphenylenediamine, metatoluidin, metaxylidin, methylaminophenol, methylnaphthylamine, naphthylamine ethers, omegaoxyethylalphanaphthylamine, orthoaminophenol, orthoanisidin, orthocresidin, orthophenetidin, orthophenylenediamine, orthotoluidin, orthoxylidin, para-aminophenol, para-anisidin, paracresidin, paranitrometaphenylenediamine, paraphenylenediamine, paratoluidin, paraxylidin, phenols and derivatives, resorcinol.

1-Chloro-4-nitrobenzene Sulphonate
Photographic
Reagent (Brit. 385522) for—
Coating back of ferroprussiate paper to produce oxidation in developing.

4-Chloro-3-nitrobenzene-1-sulphondodecylamide

Miscellaneous

As a wetting agent (Brit. 436862).
For uses, see under general heading: "Wetting agents."

Chloronitrobenzoyl Chloride

French: Chlorure de chloronitrobenzoyle, Chlorure chloronitrobenzoylique.
German: Chlornitrobenzoylchlorid.

Chemical

Starting point (Brit. 314909) in making derivatives with—
3-Carboxyphenylthiocarbamide.
Diphenylurea-3:3′-dicarboxylic acid.
4-Quinolylphenylurea-3:6′-dicarboxylic acid.
Symmetrical diphenylurea-3:3′-dicarboxylic acid.
Thiourea.
Thiourea-3:3′-dicarboxylic acid.
Urea.

4-Chloro-3-nitrobenzoylorthobenzoic Acid

French: Acide de 4-chloro-3-nitrobenzolortho-benzoique.
German: 4-Chlor-3-nitrobenzoylortho-benzoesaeure.

Chemical

Starting point (Brit. 265545) in making—
4-Allyl-3-nitrobenzoylorthobenzoic acid.
4-Amino-3-nitrobenzoylorthobenzoic acid.
4-Amyl-3-nitrobenzoylorthobenzoic acid.
4-Benzyl-3-nitrobenzoylorthobenzoic acid.
4-Butyl-3-nitrobenzoylorthobenzoic acid.
4-Ethyl-3-nitrobenzoylorthobenzoic acid.
4-Hexyl-3-nitrobenzoylorthobenzoic acid.
4-Isoamyl-3-nitrobenzoylorthobenzoic acid.
4-Isobutyl-3-nitrobenzoylorthobenzoic acid.
4-Isopropyl-3-nitrobenzoylorthobenzoic acid.
4-Methyl-3-nitrobenzoylorthobenzoic acid.
4-Naphthyl-3-nitrobenzoylorthobenzoic acid.
4-Phenyl-3-nitrobenzoylorthobenzoic acid.
4-Phthalyl-3-nitrobenzoylorthobenzoic acid.
4-Propenyl-3-nitrobenzoylorthobenzoic acid.
4-Propyl-3-nitrobenzoylorthobenzoic acid.
4-Tolyl-3-nitrobenzoylorthobenzoic acid.
4-Xylyl-3-nitrobenzoylorthobenzoic acid.

Starting point (U. S. 1614584) in making derivatives with—
Ammonia, alphanaphthylamine, anilin, benzidin, benzylamine, betanaphthylamine, butylamine, isoamylamine, isobutylamine, isopropylamine, diphenylamine, metatoluidin, metaxylidin, monoethylamine, monoethylanilin, monomethylamine, monomethylanilin, orthotoluidin, orthoxylidin, paraphenylenediamine, paratoluidin, paraxylidin, propylamine.

4-Chloro-orthoaminophenol

Dye

Intermediate in—
Dye synthesis.

Pharmaceutical

Suggested for use (Brit. 351605) as—
Antiseptic.

5-Chloro-orthotoluidide

Dye

Starting point (Brit. 434416) in making—
Bordeaux red, water-insoluble dyestuffs by coupling, in substance or on the fiber, with 2:4-dichlor-2′-amino-4′-methylazobenzene.

5-Chloro-orthotolylbetaparatoluenesulphonylethyl Sulphide

Chemical

Intermediate (Brit. 444262 and 444501) in—
Organic syntheses.

Insecticide

Insecticide (Brit. 444262 and 444501) for—
Animal pests, vegetable pests.

Textile

As a dyestuff (when employing suitable initial materials) (Brit. 444262 and 444501).
Assistant (Brit. 444262 and 444501) in—
Textile processing.

3-Chloro-3-oxybenzyl-1-arsinic Acid

French: Acide de 3-chloro-3-oxybenzyl-1-arsinique.
German: 3-Chloro-3-oxybenzyl-1-arsinsaeure.

Chemical

Starting point in making—
Normal acidyl derivatives of amino-3-chloro-4-oxybenzene-1-arsinic acid (German 441004).

3-Chloro-2-oxypropylphthalimide

French: 3-Chloro-2-oxypropylephthalimide.
German: 3-Chlor-2-oxypropylphthalimid.

Chemical

Starting point (Brit. 276012) in making therapeutic compounds, such as—
Alpha-amino-3-diallylamino-2-propanol.
Alpha-amino-3-diamylamino-2-propanol.
Alpha-amino-3-dibutylamino-2-propanol.
Alpha-amino-3-diethylamino-2-propanol.
Alpha-amino-3-dimethylamino-2-propanol.
Alpha-amino-3-dipropylamino-2-propanol.
Alpha-amino-3-phenylmethylamino-2-propanol.
Alpha-amino-3-piperidin-2-propanol.

4-Chlorophenol Betadicyclohexylaminoethyl Ether

Pharmaceutical

Suggested for use (Brit. 351605) as—
Antiseptic.

4-Chloro-1-phenol-4′-chloroanilide

French: 4′-Chloroanilide de 4-chloro-1-phénole.
German: 4-Chlor-1-phenol-4′-chloranilid.

Chemical

Starting point in making—
Intermediates, pharmaceuticals.

Dye

Starting point in making various synthetic dyestuffs.

Miscellaneous

Reagent in—
Mothproofing furs, feathers, and hair while they are being dyed.

Textile

Reagent in—
Mothproofing wool and felt while they are being dyed.

4-Chloro-1-phenol-3:5-disulphoanilide

Synonyms: 4-Chloro-1-phenol-3:5-dithioanilide.
French: 3:5-Disulphoanilide de 4-chloro-1-phénole, 3:5-Thioanilide de 4-chloro-1-phénole.
German: 4-Chlor-1-phenol-3:5-disulfoanilid, 4-Chlor-1-phenol-3:5-dithioanilid.

Chemical

Starting point in making—
Intermediates, pharmaceuticals.

Dye

Starting point in making various synthetic dyestuffs.

Miscellaneous

Reagent in—
Mothproofing furs, feathers, and hair while they are being dyed.

Textile

Reagent in—
Mothproofing wool and felt while they are being dyed.

2-Chlorophenoxyacetylamino-8-hydroxynaphthalenedisulphonic Acid

French: Acide de 2-chlorophenoxyeacetylamino-8-hydroxyenaphthalenedisulphonique.
German: 2-Chlorophenoxyacetylamino-8-oxynaphtalindisulfonsäure.

Chemical

Starting point in making—
Intermediates and other derivatives.

Dye

Starting point (Brit. 313710) in making dyestuffs with—
Anilin derivatives, 4-aminoacetanilide, betaacetamino-5-aminoanisol, betaaminobenzoic acid, paraxylidin.

4-Chlorophenoxyacetylamino-8-hydroxynaphthalene-3:6-disulphonic Acid

French: Acide de 4-chlorophénoxyacétylamino-8-hydroxynaphtalène-3:6-disulphonique.
German: 4-Chlorphenoxyacetylamino-8-hydroxynaphtalin-3:6-disulfonsäure.

Chemical

Starting point in making various derivatives.

Dye

Starting point (Brit. 313710) in making dyestuffs with—
Anilin, beta-acetamino-5-aminoanisol, beta-aminobenzoic acid, 4-aminoacetanilide, paraxylidin.

Chlorophenyl Mercaptostearate

Lubricant

Extreme pressure agent (Brit. 454552) in—
Extreme pressure lubricants.

4-Chlorophthalic Acid
Cellulose Products
Plasticizer (Brit. 390541) for—
Cellulose esters and ethers.
For uses, see under general heading: "Plasticizers."

Chlorophyll (Oil Soluble)
French: Chlorophylle soluble à l'huile.
German: Oelloslicheschlorophyll.
Chemical
Starting point in making—
Alcohol-soluble products and pigments.
Chlorosan (chlorophyll plus iron and lime salts).
Copper pheophytin.
Oil-soluble products and pigments.
Water-soluble products and pigments.
Zinc pheophytin.
Food
Ingredient of—
Food compositions, confectionery, and the like, added for the purpose of hiding their true colors.
Fats and Oils
Bleaching agent in treating—
Caraway seed oil, cottonseed oil, linseed oil, olive oil, rapeseed oil, peanut oil, perilla oil, poppyseed oil, sesame oil, teaseed oil, wormseed oil.
Fuel
Coloring for—
Fancy stearin candles.
Leather
Coloring for—
Leather and leather goods.
Perfumery
Coloring for—
Cosmetics, perfumes.
Petroleum
Ingredient of—
Mineral oil products, added for the purpose of hiding their true color.
Pharmaceutical
In compounding and dispensing practice.
Resins and Waxes
Coloring for—
Resins and waxes.
Soap
Coloring for—
Soap, added for the purpose of hiding the yellow color of the soap, to give a brighter look and greenish color.
Reagent in treating—
Olive oil foots, so as to bring back the color in a product bleached with age by processing with sulphuric acid.

Chlorophyll (Water-Soluble)
French: Chlorophylle soluble à l'eau.
German: Wasserloslischeschlorophyll.
Miscellaneous
Coloring for—
Preparations which consist of neutral and alkaline liquors free from metallic salts.
Pharmaceutical
In compounding and dispensing practice.

Chloroplatinic Acid
Synonyms: Platinic chloride.
French: Chlorure de platine.
German: Chlorplatinsäure, Platinchloridsäure.
Analysis
Reagent.
Ceramics
Ingredient of—
Batches and glazes for producing fine iridescent effects.
Chemical
Platinizing agent for—
Pumice and other carriers (to coat them with a platinum film for the production of catalysts used in various chemical processes).
Starting point in making—
Platinum-ammonium chloride (ammonium chloroplatinate), platinum bichloride (patinous chloride), platinum black, platinum resinate, platinum tetrachloride, potassium chloroplatinate, platinum sponge.
Ink
Ingredient of—
Indelible inks.
Metallurgical
Electrolyte in—
Platinum-plating baths.
Miscellaneous
Reagent in—
Microscopical work.
Reagent in making—
Platinum mirrors.
Photographic
Reagent in—
Toning baths.
Printing
Etching agent for—
Zinc plates.

5-Chloro-1:9-pyrazolanthrone
Chemical
Starting point (Brit. 264503) in making dye intermediates with—
Butyl sulphate, ethyl alcohol, ethyl sulphate, isobutyl alcohol, isobutyl sulphate, isopropyl alcohol, isopropyl sulphate, propyl alcohol, propyl sulphate, toluenesulphonicbutylester, toluenesulphonicethylester, toluenesulphonicmethylester, toluenesulphonicpropylester.

2-Chloroquinaldin
German: 2-Chlorchinaldin.
Chemical
Starting point (Brit. 305589) in making pharmaceutical phenoxyquinolin carboxylic acids and esters from—
Aromatic hydroxycarboxylic acids and esters.
Parahydroxybenzoic acid and its esters.
Pharmaceutical
In compounding and dispensing practice.

2-Chloroquinazolin
French: 2-Chloroquinazoléine.
German: 2-Chlorochinazolin.
Chemical
Starting point in making—
Intermediates, pharmaceuticals.
Dye
Starting point (Brit. 310076) in making dyestuffs with—
Aminoanisylpyrazolone.
Aminoanthranylpyrazolone.
Aminobenzoylpyrazolone.
Aminobenzylpyrazolone.
Aminocinnamylpyrazolone.
Aminocresylpyrazolone.
Aminogallylpyrazolone.
Aminometanylpyrazolone.
Aminonaphtholsulphonic acid.
Aminonaphthylpyrazolone.
Aminophenylpyrazolone.
Aminophthalylpyrazolone.
Aminosalicylpyrazolone.
Aminosuccinylpyrazolone.
Aminosulphanylpyrazolone.
Aminotolylpyrazolone.
Aminovalerylpyrazolone.
Aminoxylylpyrazolone.
Benzidin-3-sulphonic acid.
Metaphenylenediamine-4-sulphonic acid.
Paraphenylenediaminesulphonic acid.

4-Chloroquinazolin
French: 4-Chloroquinazoléine.
German: 4-Chlorchinazolin.
Chemical
Starting point in making—
Intermediates, pharmaceuticals.
Dye
Starting point (Brit. 310076) in making dyestuffs with—
Alpha-aminoanthraquinone.
Alpha-amino-5-benzoylaminoanthraquinone.
Aminoanthraquinonesulphonic acids.
1:5-Diaminoanthraquinone.
H acid.
Paraphenylenediamine.
See also 2-Chloroquinazolin.

2-Chloro-4-quinolincarboxylic Acid Chloride
Chemical
Starting point in making—
Pharmaceutical derivatives.
Starting point (Brit. 294118) in making therapeutic preparations with the aid of—
Diethylaminoethanol, sodium diethylaminoethanol.

5-Chlorosalicyl Anilide
French: Anilide de 5-chlorosalicyle, Anilide 5-chlorosalicilique.
German: 5-Chlorsalicylanilid.

Agricultural
Fungicide for the treatment of—
Seeds and grains.

Chemical
Starting point in making—
Intermediates and other derivatives.

Leather
Preservative in the treatment of—
Skins, to avoid the formation of mildew and the growth of fungi.

Paper
Preservative to prevent the formation of mildew and the growth of fungi.

Rubber
Preservative to prevent the formation of mildew and the growth of fungi.

Textile
Preservative in the treatment of—
Cotton yarns and fabrics to prevent the formation of mildew and the growth of fungi.

Woodworking
Preservative to prevent the formation of mildew and the growth of fungi.

Chlorostearic Acid Nitrate

Petroleum
Primer (Brit. 436027) for—
Diesel engine fuels (lowers ignition point).

4-Chloro-2-styrylquinolin
French: 4-Chloro-2-styrylequinoléine.
German: 4-Chlor-2-styrylchinolin.

Chemical
Starting point (Brit. 282143) in making pharmaceuticals with—
Allylamine, allylenediamine, alphanaphthylamine, ammonia, amylamine, amylenediamine, benzylamine, benzylenediamine, betanaphthylamine, butylamine, butylenediamine, cumylamine, cumylenediamine, ethylamine, ethylenediamine, heptylamine, heptylenediamine, hexylamine, hexylenediamine, metaphenylenediamine, metatoluylenediamine, methylamine, methylenediamine, orthophenylenediamine, orthotoluylenediamine, paraphenylenediamine, paratoluylenediamine, propylamine, propylenediamine, toluylamine.

Chlorosulphonic Acid
Synonyms: Sulphuryl oxychloride, Sulphuric chlorohydrin.
French: Acide chlorosulphonique, Acide chlorosulphurique, Chlorohydrine sulphurique, Chlorure de sulphurylhydroxyle, Chlorure sulphurylehydroxylique, Oxychlorure de sulphuryle, Oxychlorure sulphurylique.
German: Sulfuryloxychlorid, Sulphurylhydroxychlorid.
Spanish: Acido clorosulfonico.
Italian: Acido clorosolfonico.

Chemical
Catalyst in acetylating—
Cellulose to produce cellulose acetate suitable for the manufacture of lacquers and plastics.
Reagent in—
Absorbing ethylene from gases which contain it in the manufacture of ethyl alcohol by synthesis from this gas (French 516668).
Reagent in making—
Acetic acid, acetyl chloride, alphachloronaphthalenesulphonic acid, benzyl chloride, benzyl sulphonchloride, brominated thiobenzanthrones, derivatives with the aid of methyl chloride, dimethylanilinsulphonchloride, diethylamine, ethyl sulphate, metanitrobenzenesulphonic acid, methyl chlorosulphonate, methyl sulphate, naphthoxythiophene, nitrobenzene metasulphonchloride, nitrobenzenemetasulphonic acid, orthotoluene sulphonchloride, paratoluene sulphonchloride, persulphuric acid, saccharine, sulphon mono peracid, sulphuryl chloride, thionyl chloride.
Various organic chemicals, intermediates, pharmaceutical chemicals, and aromatic chemicals.
Reagent (Brit. 281290) in making—
Bromoamylbenzen mercaptan.
Bromoamylbenzene sulphonchloride.
Bromoamylbenzenethioglycollic acid.
Bromoethylbenzene mercaptan.
Bromoethylbenzene sulphonchloride.
Bromoethylbenzenethioglycollic acid.
Bromoethylbenzene mercaptan.
Bromoethylbenzene sulphonchloride.
Bromoethylbenzenethioglycollic acid.
Bromopropylbenzene mercaptan.
Bromopropylbenzene sulphonchloride.
Bromopropylbenzenethioglycollic acid.
Chloroallylbenzene mercaptan.
Chloroallylbenzene sulphonchloride.
Chloroallylbenzenethioglycollic acid.
Chlorobutylbenzene mercaptan.
Chlorobutylbenzene sulphonchloride.
Chlorobutylbenzenethioglycollic acid.
Chloroethylbenzene mercaptans.
Chloroethylbenzene sulphonchloride.
Chloroethylbenzenethioglycollic acid.
Chloromethylbenzene mercaptan.
Chloromethylbenzene sulphonchloride.
Chloromethylbenzenethioglycollic acid.
Chloropropylbenzene mercaptan.
Chloropropylbenzene sulphonchloride.
Chloropropylbenzenethioglycollic acid.
Reagent in making—
Wetting agents from naphthalene and anthracene.

Dye
Reagent in making—
Halogenated dimethylthioindigo coloring matters (Brit. 254340).
Soluble vat dyestuffs from indanthrene, flavanthrene, and thioindigo (Brit. 271533).
Sulphonic acid compounds of rosanilin, alizarin, and purpurin.
Tetrachlorothioindigo colors (Brit. 251321).
Vat colors of the dibenzanthrone series.
Vat red B.
Reagent for—
Converting vat coloring matters into soluble form (Brit. 251491).

Military
As a poison gas.
In admixture with sulphur trioxide to form smoke screens.

Petroleum
Reagent (Brit. 309042) for refining—
Mineral oils, ozocerite, paraffin.
Reagent (U. S. 1538287) for—
Deodorizing burning oil.

Textile
Catalyst in making—
Cellulose acetate rayon.

Chloro-tertiary-amylpyrocatechol

Disinfectant
As a germicide (U. S. 2023160).

Chloro-tertiary-amylquinol

Disinfectant
As a germicide (U. S. 2023160).

Chloro-tertiary-amylresorcinol

Disinfectant
Germicide (U. S. 2023160).

Chloro-tertiary-butylpyrocatechol

Disinfectant
As a germicide (U. S. 2023160).

Chloro-tertiary-butylquinol

Disinfectant
As a germicide (U. S. 2023160).

Chloro-tertiary-butylresorcinol

Disinfectant
As a germicide (U. S. 2023160).

Chlorothymol

Chemical
Starting point in making—
Dental disinfectant with camphor (German 433293).

2-Chlorothymol Betadiethylaminoethyl Ethers

Pharmaceutical
Suggested for use (Brit. 351605) as—
Antiseptic.

4-Chloro-2-toluidin
Chemical
Starting point in making—
Aromatics, intermediates, pharmaceuticals.
Starting point (Brit. 324041) in making intermediates and insecticides with the aid of—
2-Amino-4-sulphobenzoic acid.
3-Amino-5-sulphobenzoic acid.
2-Amino-5-sulphobenzoic acid.
4-Amino-3-sulphobenzoic acid.
3-Amino-5-sulpho-4-hydroxybenzoic acid.
5-Amino-3-sulpho-2-hydroxybenzoic acid.
4-Sulpho-4-methyl-3-aminobenzoic acid.
Dye
Starting point in making various synthetic dyestuffs.

5-Chloro-2:3-tolylenethiazothionium Chloride
Dye
Starting point (Brit. 399583) in making—
Sulphur dyestuffs.

6-Chlor-9-para-aminomethylphenyl-2-methylthiolacridin
Pharmaceutical
Claimed (Brit. 363392 and 437953) as—
New pharmaceutical.

6-Chlor-9-para-aminomethylphenyl-2-methylthiolacridin Bromide
Pharmaceutical
Claimed (Brit. 363392 and 437953) as—
New pharmaceutical.

6-Chlor-9-para-aminomethylphenyl-2-methylthiolacridin Hydrochloride
Pharmaceutical
Claimed (Brit. 363392 and 437953) as—
New pharmaceutical.

6-Chlor-9-parabetadiethylaminoethylthiolphenylamino-2-methylthiolacridin
Pharmaceutical
Claimed (Brit. 363392 and 437953) as—
New pharmaceutical.

6-Chlor-9-parabetadiethylaminoethylthiolphenylamino-2-methylthiolacridin Bromide
Pharmaceutical
Claimed (Brit. 363392 and 437953) as—
New pharmaceutical.

6-Chlor-9-parabetadiethylaminoethylthiolphenylamino-2-methylthiolacridin Hydrochloride
Pharmaceutical
Claimed (Brit. 363392 and 437953) as—
New pharmaceutical.

6-Chlor-9-paradiethylaminoethoxyphenyl-2-methylthiolacridin
Pharmaceutical
Claimed (Brit. 363392 and 437953) as—
New pharmaceutical.

6-Chlor-9-paradiethylaminoethoxyphenyl-2-methylthiolacridin Bromide
Pharmaceutical
Claimed (Brit. 363392 and 437953) as—
New pharmaceutical.

6-Chlor-9-paradiethylaminoethoxyphenyl-2-methylthiolacridin Hydrochloride
Pharmaceutical
Claimed (Brit. 363392 and 437953) as—
New pharmaceutical.

2-Chlor-5-paraxylylbeta-9-carbazolylethyl Sulphide
Chemical
Intermediate (Brit. 444262 and 444501) in—
Organic syntheses.
Pharmaceutical
Claimed (Brti. 444262 and 444501) to have—
Value for pharmaceutical purposes.
Rubber
Accelerator (Brit. 444262 and 444501) in—
Vulcanizing.

4-Chlor-1-phenylbenzothiazole
Insecticide
Exterminant for—
Culicine mosquito larvae.

5-Chlor-1-phenylbenzothiazole
Insecticide
Exterminant for—
Culicine mosquito larvae.

4-Chlor-2-propyl-6-gammachlorodeltabetabutenylphenol
Disinfectant
Claimed (Brit. 443113 and 389514) to be—
Disinfectant free of odor.

Cholesterin Cinnamate
French: Cinnamate de cholestérine.
German: Zimtsäurecholesterinester, Zimtsäurescholesterin.
Chemical
Starting point in making—
Pharmaceuticals and other derivatives.
Pharmaceutical
In compounding and dispensing practice.

Cholic Acid
Synonyms: Cholalic acid.
French: Acide cholalique, Acide cholique.
German: Cholalsaeure.
Chemical
Starting point in making—
Cholates of various bases.
Kotarin salt and kotarnin superoxide salts.
Starting point (Brit. 282356) in making antiparasitic agents with—
Dihydrocuprein ethyl ether.
Dihydrocuprein ethyl ether hydrochloride.
Dihydrocuprein isoamyl ether.
Dihydrocuprein isoamyl ether hydrochloride.
Dihydrocuprein normal octyl ether.
Dihydrocuprein normal octyl ether hydrochloride.
Dihydroquinone.
Pharmaceutical
In compounding and dispensing practice.

Chrome Alum
Synonyms: Chromium alum, Chromium-potassium sulphate, Sulphate of chrome and potash.
Latin: Alumen chromicum, Chromalaun.
French: Alun de chrome, Sulfate de chromium et de potassium, Sulfate chromique-potassique, Sulfate double de chromium et de potassium.
German: Schwefelsäureschrompotassche.
Spanish: Alumbre de cromo, Solfato de cromo y de potassio.
Italian: Allume di cromio, Solfato di cromio et di potassio.
Ceramics
Ingredient of—
Glazes and coating compositions for various ceramic products.
Chemical
Ingredient of catalytic preparations used in the manufacture of—
Acenaphthylene, acenaphthaquinone, bisacenaphthylidenedione, naphthaldehydic acid, naphthalic anhydride, and hemimellitic acid from acenaphthene (Brit. 295270).
Acetaldehyde from ethyl alcohol (Brit. 281307).
Acetic acid from ethyl alcohol (Brit. 281307).
Alcohols from aliphatic hydrocarbons (Brit. 281307).
Aldehydes and acids by the reduction of the corresponding esters (Brit. 306471).
Aldehydes and acids from toluene, orthochlorotoluene, orthonitrotoluene, orthobromotoluene, parachlorotoluene, paranitrotoluene, parabromotoluene, metachlorotoluene, metanitrotoluene, metabromotoluene, dichlorotoluenes, dinitrotoluenes, dibromotoluenes, chlorobromotoluenes, chloronitrotoluenes, bromonitrotoluenes (Brit. 295270).
Aldehydes and acids from xylenes, pseudocumenes, mesitylene, and paracymene (Brit. 281307).
Alphanaphthaquinone from naphthalene (Brit. 295270).
Anthraquinone from naphthalene (Brit. 295270).
Benzaldehyde and benzoic acid from toluene (Brit. 281307).
Benzoquinone from phenanthraquinone (Brit. 281307).
Benzyl alcohol from benzaldehyde by reduction (Brit. 306471).
Benzyl alcohol, benzaldehyde, or benzyl phthalide by the reduction of phthalic anhydride (Brit. 306471).
Butyl alcohol by the reduction of crotonaldehyde (Brit. 306471).

Chrome Alum (Continued)
Chloroacetic acid from ethylenechlorohydrin (Brit. 295270).
Diphenic acid from ethyl alcohol (Brit. 295270).
Ethyl alcohol by the reduction of acetaldehyde (Brit. 306471).
Fluorenone from fluorene (Brit. 295270).
Formaldehyde by the reduction of carbon dioxide or carbon monoxide (Brit. 306471).
Formaldehyde by the reduction of methanol or methane (Brit. 306471).
Hydroxyl compounds by the reduction of anthraquinone, benzoquinone, and similar compounds (Brit. 306471).
Isopropyl alcohol by the reduction of acetone (Brit. 306471).
Maleic acid and fumaric acid by the oxidation of toluene, benzene, phenols, tar phenols, or furfural, or from benzoquinone or phthalic anhydride (Brit. 306471).
Methane by the reduction of carbon dioxide or carbon monoxide (Brit. 306471).
Methanol by the reduction of carbon dioxide or carbon monoxide (Brit. 306471).
Naphthaldehydic acid, acenaphthaquinone, or bisacenaphthylidenedione from acenaphthylene (Brit. 281-307).
Phenanthraquinone from phenanthrene or diphenic acid (Brit. 295270).
Phthalic acid and maleic acid from naphthalene (Brit. 295270).
Primary alcohols by the reduction of the corresponding aldehydes (Brit. 306471).
Propionic acid and butyric acid and higher alcohols, ketones, and acids by the reduction of carbon dioxide or carbon monoxide (Brit. 306471).
Reduction products of ketones, aldehydes, acids, esters, alcohols, ethers, and other organic compounds which contain oxygen (Brit. 306471).
Salicylic acid and salicylic aldehyde from cresol (Brit. 295270).
Secondary butyl alcohol by the reduction of methylethyl ketone (Brit. 306471).
Valeryl alcohol by the reduction of valeraldehyde (Brit. 306471).
Vanillin and vanillic acid by the oxidation of eugenol or isoeugenol (Brit. 295270).
Ingredient (Brit. 306460) of catalytic mixtures which are used in the production of various aromatic and aliphatic compounds, including—
Alphanaphthylamine from alphanitronaphthalene.
Amines from aliphatic nitro compounds, such as allyl nitriles or nitromethane.
Amino compounds from the corresponding nitrophenols.
3-Aminopyridin from 3-nitropyridin.
Amylamine from pyridin.
Cyclohexamine, dicyclohexamine, and cyclohexylanilin from nitrobenzene.
Anilin, azo-oxybenzene, azobenzene, and hydrazobenzene from benzene by reduction.
Piperidin from pyridin.
Pyrrolidin from pyrrol.
Tetrahydroquinolin from quinolin.
Starting point in making—
Chromium salts, pigments for ceramic products.

Glues and Adhesives
Reagent in—
Treating glues and adhesive preparations to render them insoluble in water.

Gums
Reagent in—
Treating gums to render them insoluble in water.

Ink
Reagent in making—
Writing inks.

Leather
Reagent in—
Chrome tanning process.
Tanning in a bath containing stannic chloride and silicate of soda neutralized with hydrochloric acid (French 631109).

Miscellaneous
Reagent in waterproofing—
Various fibrous compositions of matter.

Plastics
Reagent (French 601297) in—
Treating cellulose acetate plastics in order to preserve their luster, transparency and appearance when treated with hot or boiling liquids.

Photographic
Fixative in the photographic process.
Reagent in—
Hardening gelatin on plates, films and papers.

Textile
——, *Dyeing and Printing*
As a mordant.
——, *Finishing*
Reagent in—
Waterproofing fabrics.

Chrome Cake
French: Gateaux de chrome.
German: Chromkuchen.

Chemical
Starting point in making—
Glauber's salt, or pure sodium sulphate, anhydrous and hydrous.
Sodium acetate, sodium carbonate, sodium hypochlorite, sodium silicate or waterglass, sodium thiosulphate, washing sodas.

Fuel
Ingredient (U. S. 1618465) of—
Fuel preparations (acting as a fuel economizer).

Glass
Ingredient of—
Batch in making low grades of glass.

Glue
Reagent in making—
Glues.

Insecticide
Ingredient of—
Germicidal compositions, insecticidal compositions.

Leather
Reagent in—
Tanning.

Paint and Varnish
Ingredient of—
Paint and varnish removers.

Paper
Reagent in making—
Pulp.

Refrigeration
Ingredient of—
Freezing mixtures.

Soap
Ingredient of—
Detergent compositions.

Chrome Nitroacetate
Synonyms: Chromium nitroacetate.
French: Nitroacétate de chrome, Nitroacétate chromique.
German: Chromnitroacetat, Chromnitroazetat, Nitroessigsäureschrom, Nitroessigsäureschromoxyd.

Perfume
Ingredient of—
Carnation odors, perfumes, rose odors.
Perfume in—
Cosmetics.

Soap
Perfume in—
Toilet soaps.

Textile
Mordant in dyeing and printing.

Chrome Sulphoacetate

Textile
Mordant in dyeing and printing.

Chromic Acetate
Synonyms: Chrome acetate.
French: Acétate de chrome, Acétate chromique.
German: Chromiacetat, Essigsäureschrom, Essigsäureschromoxyd.

Chemical
Starting point in making—
Chromium salts.
Ingredient of catalytic mixtures used in the manufacture of—
Acenaphthylene, acenaphthaquinone, bisacenaphthylidenedione, naphthaldehydic acid, naphthalic anhydride, and hemimellitic acid from acenaphthene (Brit. 295270).
Acetaldehyde from ethyl alcohol (Brit. 281307).
Acetic acid from ethyl alcohol (Brit. 281307).
Alcohols from aliphatic hydrocarbons (Brit. 281307).

Chromic Acetate (Continued)
Aldehydes or alcohols by the reduction of esters (Brit. 306471).
Alphacampholid by the reduction of camphoric acid (Brit. 306471).
Aldehydes and acids from toluene, orthochlorotoluene, orthobromotoluene, orthonitrotoluene, parachlorotoluene, paranitrotoluene, parabromotoluene, metanitrotoluene, metachlorotoluene, metabromotoluene, dichlorotoluenes, dibromotoluenes, dinitrotoluenes, nitrochlorotoluenes, nitrobromotoluenes, chlorobromotoluenes (Brit. 295270).
Aldehydes and acids from xylenes, pseudocumene, mesitylene, and paracymene (Brit. 281307).
Alphanaphthaquinone from naphthalene (Brit. 281307).
Anthraquinone from naphthalene (Brit. 295270).
Benzaldehyde and benzoic acid from toluene (Brit. 281307).
Benzoquinone from phenanthraquinone (Brit. 281307).
Benzyl alcohol by the reduction of benzaldehyde (Brit. 306471).
Benzyl alcohol or benzaldehyde or phthalide by the reduction of phthalic anhydride (Brit. 306471).
Butyl alcohol by the reduction of crotonaldehyde (Brit. 306471).
Chloroacetic acid from ethylenechlorohydrin (Brit. 295270).
Diphenic acid from ethyl alcohol (Brit. 281307).
Ethyl alcohol by the reduction of acetaldehyde (Brit. 306471).
Fluorenone from fluorene (Brit. 295270).
Formaldehyde by the reduction of methane or methanol (Brit. 306471).
Formaldehyde by the reduction of carbon monoxide or carbon dioxide (Brit. 306471).
Hydroxyl reduction compounds of anthraquinone, benzoquinone, and the like (Brit. 306471).
Isopropyl alcohol by the reduction of acetone (Brit. 306471).
Maleic and fumaric acids by the oxidation of toluene, benzene, phenols, tar phenols, or furfural, or from benzoquinone or phthalic anhydride (Brit. 295270).
Methane by the reduction of carbon dioxide or carbon monoxide (Brit. 306471).
Methanol by the reduction of carbon dioxide or carbon monoxide (Brit. 306471).
Naphthaldehydic acid, acenaphthaquinone or bisacenaphthylidenedione from acenaphthylene (Brit. 281307).
Phenanthraquinone from phenanthrene or diphenic acid (Brit. 295270).
Phthalic acid and maleic acid from naphthalene (Brit. 295270).
Primary alcohols by the reduction of the corresponding aldehydes (Brit. 306471).
Propionic acid and butyric acid and higher alcohols, ketones, and acids by the reduction of carbon monoxide or carbon dioxide (Brit. 306471).
Reduction products of ketones, aldehydes, acids, esters, alcohols, ethers and other organic compounds containing oxygen (Brit. 306471).
Salicylic acid and salicylic aldehyde from cresol (Brit. 295270).
Secondary butyl alcohol by the reduction of methylethyl ketone (Brit. 306471).
Valeryl alcohol by the reduction of valeraldehyde (Brit. 306471).
Vanillin and vanillic acid from eugenol or isoeugenol (Brit. 295270).
Ingredient (Brit. 304640) of catalytic preparations used in the production of various aromatic and aliphatic compounds, including—
Alphanaphthylamine from alphanitronaphthalene.
Amines from aliphatic nitro compounds, such as allyl nitriles or nitromethane.
Amylamine from pyridin.
Anilin, azo-oxybenzene, azobenzene, and hydrazobenzene from nitrobenzene by reduction.
Aminophenols from nitrophenols.
3-Aminopyridin from 3-nitropyridin.
Amino compound from the corresponding nitroanisole.
Amines from oximes, Schiff's base, and nitriles.
Cyclohexamine, dicyclohexamine, and cyclohexylanilin from nitrobenzene.
Piperidin from pyridin.
Pyrollidin from pyrrol.
Tetrahydroquinolin from quinolin.

Dye
Ingredient of—
Dye preparation, known as indigo substitute, containing logwood.

Leather
Reagent in tanning.

Textile
—, *Dyeing and Printing*
As a mordant.

—, *Manufacturing*
Catalyst in making—
Acetate rayon.

Chromite
Synonyms: Chrome ore.
French: Chromite, Minérale de chrome.
German: Chromerz, Chromit.

Chemical
Starting point in making—
Chromium salts, chromic acid.

Dye
Reagent in making various dyestuffs.

Glass
Ingredient of—
Glass batch (added for the purpose of obtaining a distinctive color).

Leather
Reagent in—
Tanning.

Metallurgical
Starting point in making—
Chrome steels, chromic iron ore, chromium metal, ferrochromium.

Miscellaneous
Binder for—
Furnace linings.
Lining for—
Furnaces.

Paint and Varnish
Starting point in making—
Chromium pigments, such as chrome green, chrome oxide green, chrome yellow, Pennettier's green, emerald green.

Paper
Lining for—
Digesters used in making sulphite pulp.

Refractory
Starting point in making—
Chromate binders, chromate brick.

Textile
Mordant in dyeing—
Dyeing and printing.

Chromium
Synonyms: Chromium metal.
French: Chrome.
German: Chrom.
Spanish: Cromo.
Italian: Cromo.

Chemical
Starting point in making—
Chromium salts.
Pigments for coloring porcelains, potteries, chinaware, glass, and the like.
Ingredient of catalytic preparations used in making—
Acenaphthylene, acenaphthaquinone, bisacenaphthylidenedione, naphthaldehydic acid, naphthalic aldehyde, and hemimellitic acid from acenaphthene (Brit. 295270).
Acetaldehyde from ethyl alcohol (Brit. 281307).
Acetic acid from ethyl alcohol (Brit. 281307).
Alcohols from aliphatic hydrocarbons (Brit. 281307).
Aldehydes and acids from toluene, orthochlorotoluene, orthonitrotoluene, orthobromotoluene, parachlorotoluene, paranitrotoluene, parabromotoluene, metachlorotoluene, metanitrotoluene, metabromotoluene, chlorobromotoluene, chloronitrotoluene, bromonitrotoluene, dichlorotoluenes, dibromotoluenes, dinitrotoluenes (Brit. 295270).
Aldehydes and acids from xylenes, pseudocumenes, mesitylenes, and paracymene (Brit. 295270).
Alphanaphthaquinone from naphthalene (Brit. 281307).
Anthraquinone from anthracene (Brit. 295270).
Benzaldehyde and benzoic acid from toluene (Brit. 281307).
Benzoquinone from phenanthraquinone (Brit. 281307).
Chloroacetic acid from ethylenechlorohydrin (Brit. 295270).
Diphenic acid from ethyl alcohol (Brit. 281307).
Fluorenone from fluorene (Brit. 295270).
Formaldehyde from methanol or methane (Brit. 295270).
Maleic acid and fumaric acid by the oxidation of ben-

Chromium (Continued)
zene, toluene, phenol, tar phenols, or furfural, or from benzoquinone or phthalic anhydride (Brit. 295270).
Naphthaldehydic acid, acenaphthaquinone, or bisacenaphthylidenedione from acenaphthylene (Brit. 281307).
Phenanthraquinone from phenanthrene or diphenic acid (Brit. 281307).
Phthalic acid and maleic acid from naphthalene (Brit. 295270).
Salicylic acid and salicylic aldehyde from cresol (Brit. 295270).
Vanillin or vanillic acid from eugenol or isoeugenol (Brit. 295270).

Miscellaneous
Metal or ingredient of metallic compositions for making various instruments and apparatus.

Metallurgical
Raw material in making—
Aluminum alloys, cobalt alloys, copper alloys, ferrochromium, nickel alloys, silicon alloys, stainless steel, tungsten alloys.

Paint and Varnish
Starting point in making—
Chrome green, chrome yellow, emerald green, pennettier's green.

Chromium-Ammonium Chloride
French: Chlorure de chrome et ammonium.
German: Chlorchromammonium, Chromammoniumchlorid.

Miscellaneous
Reagent (Brit. 271026) in carrotting—
Furs and felts.

Chromium Benzenesulphonate
Synonyms: Chromium benzolsulphonate.
French: Benzènesulphonate de chrome.
German: Benzolsulfonsaeureschromium, Chromiumbenzolsulfonat.

Plastics
Catalyst in making—
Cellulose acetate.

Textile
——, *Manufacture*
Catalyst in making—
Acetate rayon (Brit. 265267).

Chromium Betabenzoylpropionate

Plastics
Starting point (U. S. 2001380) in making—
Films.

Chromium Bromide
French: Bromure de chrome, Bromure chromique.
German: Chrombromid.
Spanish: Bromuro de cromo.
Italian: Bromuro di cromo.

Rubber
Thermoplasticizing agent (French 615195) for—
Rubber.

Chromium Butyrate
French: Butyrate de chrome.
German: Buttersaeureschrom, Chrombutyrat.

Plastics
Catalyst in making—
Cellulose acetate plastics (Brit. 265267).

Textile
——, *Manufacturing*
Catalyst in making—
Acetate rayon (Brit. 265267).

Chromium Chlorate
French: Chlorate de chrome, Chlorate chromique, Chrome chlorique.
German: Chlorsäureschrom, Chromchlorat.
Spanish: Clorato de cromo.
Italian: Clorato di cromo.

Textile
As a mordant.
Starting point in producing—
Nongreening blacks in printing processes.
Orange colors by double decomposition with lead salts on the fiber.
Yellows by double decomposition with lead salts on the fiber.

Chromium Chloroacetate
French: Chloracétate de chrome.
German: Chloressigsaeureschrom, Chromchloracetat.

Plastics
Catalyst in making—
Cellulose acetate plastics (Brit. 265267).

Textile
——, *Manufacturing*
Catalyst in making—
Acetate rayon (Brit. 265267).

Chromium Chromate
French: Chromate de chrome, Chromate d'oxyde de chrome.
German: Chromchromat, Chromsaeures-chromoxid.

Dye
Reagent in making—
Soluble chromium compounds of azo dyestuffs from 4-chloroanilin-3-sulphonic acid and 1:2-aminonaphthol-4-sulphonic acid.

Paint and Varnish
Reagent in making—
Rust-preventing coatings in admixture with resins, solvents, oils, varnishes and waxes (German 425900).

Textile
——, *Dyeing*
Mordant with—
Alizarin on cotton fabrics.

——, *Printing*
Mordant with—
Alizarin on cotton fabrics.

Chromium Fluoride-Sodium-Antimony Fluoride

Insecticide
Mothproofing agent (Brit. 454458) for—
Animal fibers.

Textile
Antirotting agent (Brit. 454458) for—
Animal fibers.
Mold inhibitor (Brit. 454458) for—
Animal fibers.

Chromium Formate
French: Formiate de chrome.
German: Ameisensaeurechromoxyd, Ameisensaeureschrom, Chromformiat.

Dye
Reagent in making—
Soluble chromium compounds of the azo dyestuffs made from 4-nitro-2-aminophenol-6-sulphonic acid or 1-amino-2-naphthol-4-sulphonic acid (Brit. 262418).

Leather
Reagent in tanning and finishing.

Textile
——, *Dyeing*
Mordant in dyeing yarns and fabrics.

——, *Printing*
Mordant in printing various fabrics.

Chromium-Gammabutylacetylacetone
French: Gammabutyleacétyleacétone chromique.
German: Chrom-gammabutylacetylaceton.

Chemical
Reagent (Brit. 289493) in making—
Aromatics, intermediates, pharmaceuticals.

Dye
Reagent (Brit. 289493) in making—
Synthetic dyestuffs.

Petroleum
Antidetonant (Brit. 289493) in—
Motor fuels.

Chromium-Gammaethylacetylacetone
French: Gamma éthyleacétyleacétone chromique.
German: Chrom-gammaethylacetylaceton.

Chemical
Reagent (Brit. 289493) in making—
Aromatics, intermediates, pharmaceuticals.

Dye
Reagent (Brit. 289493) in making—
Synthetic dyestuffs.

Petroleum
Antidetonant (Brit. 289493) in—
Motor fuels.

Chromium Naphthalenesulphonate
French: Naphthalènesulphonate de chrome.
German: Chromnaphtalinsulfonat, Naphtalinsulfonsaeureschrom.

Plastics
Catalyst in making—
Cellulose acetate plastics (Brit. 265267).

Textiles
——, *Manufacturing*
Catalyst in making—
Acetate rayon (Brit. 265267).

Chromium Oleate

Petroleum
Inhibitor (Brit. 431066) of—
Sludge formation in lubricating oils.

Chromium Resinate
Synonyms: Chrome resinate, Resinate of chromium.
French: Résinate de chrome.
German: Chromresinat.

Ceramics
Pigment in producing green colorations on—
Chinaware, porcelains, potteries.

Paint and Varnish
Drier in making—
Enamels, lacquers, paints, varnishes.

Chromium Salt of Coconut Oil Fatty Acids

Chemical
Catalyst (Brit. 396311) in making—
High molecular alcohols from fats, oils, waxes, fatty acids, and the like.

Chromium Stannate
French: Stannate de chrome.
German: Chromstannat, Zinnsaeureschromoxyd.

Ceramics
Pigment for—
Porcelains and chinaware.

Paint and Varnish
Pigment for—
Oil colors.
Starting point in making—
Pink pigments.

Chromium Stearate
French: Stéarate de chrome, Stéarate chromique.
German: Chromistearat, Chromstearat, Stearinsäureschrom, Stearinsäureschromoxyd.
Spanish: Estearato de cromo.

Miscellaneous
Reagent in making—
Phonograph records.

Perfume
Ingredient of—
Dental pastes, mouth washes.

Chromium Stearotoluenesulphonate
French: Stéarotoluènesulphonate de chrome, Stéarotoluènesulphonate chromique.
German: Chromstearotoluolsulfonat, Stearotoluolsulfonsäureschrom, Stearotoluolsulfonsäureschromoxyd.

Paper
Ingredient (Brit. 269917) of—
Pastes used in printing wallpaper (added to obtain level shades and effects).

Textile
Ingredient (Brit. 269917) of—
Printing pastes (added to enhance the saturating of the fabric and to equalize effects).

Chromogene Red
French: Rouge de chromogène.
German: Chromogenrot.

Dye
Starting point in making—
Chromogene blue R.

Textile
——, *Dyeing*
Dyestuff for various yarns and fabrics.
——, *Printing*
Coloring for various fabrics.

Chromous Acetate
Synonyms: Chrome acetate, Chromium acetate.
French: Acétate de chrome, Acétate chromeux.
German: Essigsäureschrom, Essigsäureschromoxydul, Chromoacetat, Chromoazetat.

Analysis
Reagent in—
Gas analysis (for absorption of oxygen).

Chemical
Starting point in making—
Chromous compounds.
Ingredient of catalytic preparations used in the manufacture of—
Acenaphthylene, acenaphthaquinone, bisacenaphthylidenedione, naphthaldehydic acid, naphthalic anhydride, and hemimellitic acid from acenaphthene (Brit. 295270).
Acetaldehyde from ethyl alcohol (Brit. 281307).
Acetic acid from ethyl alcohol (Brit. 281307).
Alcohols from aliphatic hydrocarbons (Brit. 281307).
Aldehydes or alcohols by the reduction of esters (Brit. 306471).
Alphacampholid by the reduction of camphoric acid (Brit. 306471).
Aldehydes and acids from toluene, parachlorotoluene, parabromotoluene, paranitrotoluene, orthochlorotoluene, orthonitrotoluene, orthobromotoluene, metachlorotoluene, metanitrotoluene, metabromotoluene, dichlorotoluenes, dibromotoluenes, dinitrotoluenes, chlorobromotoluenes, chloronitrotoluenes, bromonitrotoluenes (Brit. 295270).
Aldehydes and acids from xylenes, pseudocumene, mesitylene, and paracymene (Brit. 281307).
Alphanaphthaquinone from naphthalene (Brit. 281307).
Anthraquinone from naphthalene (Brit. 295270).
Benzaldehyde and benzoic acid from toluene (Brit. 281307).
Benzoquinone from phenanthraquinone (Brit. 281307).
Benzyl alcohol by the reduction of benzaldehyde (Brit. 306471).
Benzyl alcohol, or benzaldehyde or phthalide by the reduction of phthalic anhydride (Brit. 306471).
Butyl alcohol by the reduction of crotonaldehyde (Brit. 306471).
Chloroacetic acid from ethylenechlorohydrin (Brit. 295270).
Diphenic acid from ethyl alcohol (Brit. 281307).
Ethyl alcohol by the reduction of acetaldehyde (Brit. 306471).
Fluorenone from fluorene (Brit. 295270).
Formaldehyde by the reduction of methane or methanol (Brit. 306471).
Formaldehyde by the reduction of carbon dioxide or carbon monoxide (Brit. 306471).
Hydroxyl reduction compounds of anthraquinone, benzoquinone, and the like (Brit. 306471).
Isopropyl alcohol by the reduction of acetone (Brit. 306471).
Maleic acid and fumaric acid by the oxidation of toluene, benzene, phenols, tar phenols, or furfural, or from benzoquinone or phthalic anhydride (Brit. 295270).
Methane by the reduction of carbon dioxide or carbon monoxide (Brit. 306471).
Methanol by the reduction of carbon dioxide or carbon monoxide (Brit. 306471).
Naphthaldehydic acid, acenaphthaquinone, or bisacenaphthylidenedione from acenaphthylene (Brit. 281-307).
Phenanthraquinone from phenanthrene or diphenic acid (Brit. 295270).
Phthalic acid and maleic acid from naphthalene (Brit. 295270).
Primary alcohols by the reduction of the corresponding aldehydes (Brit. 306471).
Propionic acid and butyric acid and higher alcohols, ketones, and acids by the reduction of carbon dioxide or carbon monoxide (Brit. 306471).
Reduction products of ketones, aldehydes, acids, esters, alcohols, ethers, and other organic compounds containing oxygen (Brit. 306471).
Salicylic acid and salicylic aldehyde from cresol (Brit. 295270).
Secondary butyl alcohol by the reduction of methylethyl ketone (Brit. 306471).
Valeryl alcohol by the reduction of valeraldehyde (Brit. 306471).
Vanillin and vanillic acid from eugenol or isoeugenol (Brit. 306471).
Ingredient (Brit. 304640) of catalytic preparations used in making various aromatic and aliphatic compounds, particularly amino compounds, such as—
Alphanaphthylamine from alphanitronaphthalene.
Amines from aliphatic nitro compounds, such as allyl nitriles, or nitromethane.

Chromous Acetate (Continued)
Amylamine from pyridin.
Anilin, azo-oxybenzene, azobenzene, hydrazobenzene, and the like from nitrobenzene by reduction.
Aminophenols from nitrophenols.
3-Aminopyridin from 3-nitropyridin.
Amino compound from the corresponding nitroanisole.
Amines from oximes, Schiff's base, and nitriles.
Cyclohexamine, dicyclohexamine, and cyclohexylanilin from nitrobenzene.
Piperidin from pyridin.
Pyrrolidin from pyrrol.
Tetrahydroquinolin from quinolin.

Leather
Reagent in tanning.

Textile
Mordant in dyeing and printing.

Cinchonidine
German: Cinchonidin.

Chemical
Starting point in making—
Cinchonidine acetate, cinchonidine arsenate, cinchonidine arsenite, cinchonidine benzoate, cinchonidine bisulphate, cinchonidine bitartrate, cinchonidine borate, cinchonidine carbolate, cinchonidine citrate, cinchonidine dihydrobromide, cinchonidine dihydrochloride, cinchonidine ferrocyanide, cinchonidine formate, cinchonidine glycerophosphate, cinchonidine, hydrobromide, cinchonidine hydrochloride, cinchonidine hydroiodide, cinchonidine hypophosphite, cinchonidine lactate, cinchonidine phosphate, cinchonidine salicylate, cinchonidine sulphate, cinchonidine sulphocarbolate, cinchonidine tannate, cinchonidine tartrate, cinchonidine valerate.

Insecticide
Ingredient of—
Moth-proofing compositions for treating furs and feathers (Brit. 263092).

Pharmaceutical
In compounding and dispensing practice.

Textile
—, *Miscellaneous*
Ingredient of—
Moth-proofing compositions for treating woolen fabrics (Brit. 263092).

Cinchonine
German: Cinchonin.

Chemical
Starting point in making—
Cinchonine acetate, cinchonine arsenate, cinchonine arsenite, cinchonine benzoate, cinchonine bisulphate, cinchonine bitartrate, cinchonine borate, cinchonine carbolate, cinchonine citrate, cinchonine dihydrobromide, cinchonine dihydrochloride, cinchonine ferrocyanide, cinchonine formate, cinchonine glycerophosphate, cinchonine hydrobromide, cinchonine hydrochloride, cinchonine hydroiodide, cinchonine hypophosphite, cinchonine lactate, cinchonine phosphate, cinchonine salicylate, cinchonine sulphate, cinchonine sulphocarbolate, cinchonine tannate, cinchonine tartrate, cinchonine valerate.

Insecticide
Ingredient of—
Moth-proofing compositions for treating furs and feathers (Brit. 263092).

Pharmaceutical
In compounding and dispensing practice.

Textile
—, *Miscellaneous*
Ingredient of—
Moth-proofing compositions for treating woolen fabrics (Brit. 263092).

Cinchonine-Ethyl Carbonate
French: Carbonate de cinchonine-éthyle, Carbonate cinchoninique-éthylique.
German: Cinchonin aethylkarbonat.
Spanish: Carbonato de cinchonine-etile.
Italian: Carbonato di cinchonine-etil.

Chemical
Starting point (Brit. 27952-1911) in making—
Hydrocinchonine-ethyl carbonate.

Pharmaceutical
In compounding and dispensing practice.

Cinchoninic Acid
Synonyms: Quinolin-4-carboxylic acid, Quinolingammacarboxylic acid.
French: Acide cinchonique, Acide quinoléine-4-carbonique, Acide quinoléinegammacarbonique.
German: Cinchonsäure, Chinolin-4-carbonsäure, Chinolingammacarbonsäure.

Chemical
Starting point in making—
Bismuth cinchoninate (German 411051).
Esters and salts used as pharmaceuticals.
Hydrogenated derivatives (German 351464).
Silver cinchoninate (German 410365).

Pharmaceutical
In compounding and dispensing practice.

Cinchophen
Synonyms: Atophan, Betaphenylquinolin-4-carboxylic acid, Phenoquin, Phenylcinchoninic acid, Phenylquinolincarboxylic acid, Quinophan.
Latin: Acidum phenylcinchoninum.
French: Acide phénylecinchoninique, Acide de phényle-quinoline-4-carboxylique.
German: Betaphenylchinolin-4-carbonsäure, 2-Phenylchinolin-4-carbonsäure, Phenylcinchoninsäure.

Chemical
Starting point in making—
Pharmaceutical derivatives.

Pharmaceutical
In compounding and dispensing practice.

Cinnamic Aldehyde
Synonyms: Cinnamyl aldehyde.
French: Aldéhyde cinnamique, Aldéhyde de cinnamyle, Cinnamaldéhyde.
German: Cinnamaldehyd, Zimtaldehyd.

Chemical
Starting point (Brit. 263853) in making vulcanization accelerators with—
Anilin, N-butylamine, diethylamine, dimethylamine, ethylamine, ethylanilin, ethylenediamine, guanidin, methylamine, methylanilin, methylenediamine, naphthylamine (alpha and beta), naphthylenediamine, orthotolyldiguanide.

Fats and Oils
Ingredient of—
Artificial oil of cinnamon.

Food
Ingredient of—
Flavoring extracts.

Perfumery
Ingredient of—
Cosmetics, perfumes.

Pharmaceutical
In compounding and dispensing practice.

Cinnamyl Acetate
French: Acétate de cinnamyle, Acétate cinnamylique.
German: Cinnamylacetat, Cinnamylazetat, Essigsäurecinnamylester, Essigsäurescinnamyl.

Miscellaneous
As a perfume for various purposes.

Perfume
Ingredient of artificial essence of—
Hyacinth, jasmine, lilac, lily of the valley.
Perfume in—
Cosmetics.

Soap
Perfume in—
Toilet soaps.

Cinnamyl Carboxethylate
French: Carboxéthylate de cinnamyle.
German: Cinnamylcarboxaethylat.
Spanish: Carboxetilato de cinamil.
Italian: Carbossietilato di cinnamile.

Perfume
Ingredient (French 650100) of—
Perfumes.

4-Cinnamyl Chloride
French: Chlorure de 4-cinnamyle, Chlorure 4-cinnamylique.
German: 4-Cinnamylchlorid.

Chemical
Reagent (Brit. 278037) in making synthetic drugs with—
Alkoxynaphthylaminesulphonic acids.
Alphanaphthylamine-4:8-disulphonic acid.

4-Cinnamyl Chloride (Continued)
Alphanaphthylamine-3:6:8-trisulphonic acid.
Alphanaphthylamine-4:6:8-trisulphonic acid.
4-Aminoacenaphthene-3:5-disulphonic acid.
4-Aminoacenaphthene-3-sulphonic acid.
4-Aminoacenaphthene-5-sulphonic acid.
4-Aminoacenaphthenetrisulphonic acid.
2:8-Aminonaphthol-3:6-disulphonic acid.
1:8-Aminonaphthol-3:6-disulphonic acid.
1:5-Aminonaphthol-7-sulphonic acid.
Bromonaphthylaminesulphonic acid.
Chloronaphthalenesulphonic acid.
Iodonaphthalenesulphonic acid.

Cinnamyl Cinnamate
French: Cinnamate de cinnamyle, Cinnamate cinnamylique.
German: Cinnamylcinnamat, Zimtsäurecinnamylester, Zimtsäurescinnamyl.

Perfume
Fixative in making—
Flower odors.
Ingredient of—
Champaca perfumes, hyacinth perfumes.
Perfume in—
Cosmetics, toilet waters.

Soap
Perfume in—
Toilet soaps.

Cinnamylidene Acetone

Chemical
Starting point in making intermediates for perfumes (Brit. 264830).

Perfumery
Ingredient (Brit. 264830) of—
Cosmetics, perfumes.

Cinnamyl Phosphate
French: Phosphate de cinnamyle, Phosphate cinnamylique.
German: Cinnamylphosphat, Phosphorsäurecinnamylester, Phosphorsäurescinnamyl, Zimtphosphat.

Chemical
Starting point in making various derivatives.

Miscellaneous
Mothproofing agent (U. S. 1748675) for treating—
Furs, feathers, and the like.

Textile
Mothproofing agent (U. S. 1748675) for treating—
Woolen yarns and fabrics.

Citral

Chemical
Starting point (Brit. 249113) in making vulcanization accelerators with—
Anilin, diethylanilin, ethylamine, ethylanilin, ethylenediamine, guanidin, methylamine, methylanilin, normal butylamine, orthotoluidin, orthotolyldiguanide.

Food
As a flavoring.

Perfume
Ingredient of—
Artificial citronella, artificial rose, artificial violet.

Citric Acid
Synonyms: Oxytricarballylic acid.
Latin: Acidum citricum.
French: Acide citrique, Acide oxytricarbolique.
German: Citronsäure, 3-Methylsäurepentanol-(3)-disäure, Oxytricarballylsäure, Zitronensäure.
Spanish: Acido de citrico, Acido de oxicarballico.
Italian: Acido di citrico, Acido di ossicarballico.

Analysis
Reagent for—
Analyzing superphosphates.
Superphosphate fertilizers.
Differentiating between mucin and albumen.
Determining albumen.
Biliary pigments, citrate-soluble phosphoric acid, glucose, mucin.
Separating iron oxide and aluminum oxide.

Chemical
Reagent in making—
Effervescent salts, light-sensitive ammonium-ferric citrate.
Starting point in making—
Ammonium citrate by reaction with aqua ammonia.
Bismuth citrate by boiling with bismuth nitrate.
Calcium citrate.
Ferric citrate by reaction with ferric hydroxide.
Lithium citrate by reaction with lithium carbonate.
Magnesium citrate by reaction with magnesium hydroxide.
Manganous citrate by reaction with manganese hydroxide.
Potassium citrate by reaction with potassium carbonate.
Silver citrate, sodium citrate, strontium citrate, various esters of citric acid, various derivatives of alkaloids, zinc citrate.

Food
Bleaching agent for—
Vegetable foods.
Ingredient of—
Confectioneries, flavoring extracts, fruit juices, lemonades, jams, pastries, soft drinks, various food compositions.
Reagent for—
Disinfecting milk and milk products.
Improving taste of rapeseed oil used for food purposes (U. S. 1004891).
Making carbonated beverages.
Preserving various foods.
Treating teas to improve their flavor (U. S. 1750768).
Substitute for—
Vinegar in various food compositions.

Gas
Ingredient of—
Iron oxide purifier mass for use in purifying coal gas, cokeoven gas, and water gas (used for the purpose of preventing precipitation of iron hydroxide)

Glass
Ingredient of—
Compositions used for silvering mirrors.

Ink
Ingredient of—
Special inks, various printing inks, various writing inks.

Leather
Reagent in—
Deliming pelts and hides before tanning.

Linoleum and Oilcloth
Ingredient of—
Compositions used for making linoleum and oilcloth.

Metallurgical
Ingredient of—
Baths used for the deposition of bright coatings of nickel (U. S. 1837835).
Baths, containing nickel sulphate, zinc sulphate, ammonium sulphate, and amonium sulphocyanide, used for coloring zinc metal black.
Platinum plating baths containing platinum chloride.

Miscellaneous
Ingredient of—
Ink eradicators, floor cements, floor polishes, metal polishes, various polishing compositions (U. S. 1774221).

Perfume
Ingredient of—
Lemon rinses, sunburn preparations, skin creams and lotions.

Photographic
Ingredient of—
Emulsions for making silver chloride developing paper.
Emulsions for making silver chloride-gelatin (artists) paper.
Preparation containing ferric-ammonium citrate and potassium ferricyanide for making blueprint paper.
Toning baths.

Pharmaceutical
Ingredient of—
Acetoform, citrophen, citrovanilla, kephaldol (quinine preparation), urecidin.
Various antipyretic mixtures containing antipyrine.
Suggested for use as antipyretic, mild astringent.
Used for correcting taste of various pharmaceutical preparations, such as lecithin mixtures.

Printing
Reagent in—
Photomechanical printing.

Resins and Waxes
Reagent in making—
Synthetic resins from glycerin.

Citric Acid (Continued)
Sugar
Reagent for—
Preventing crystallization of sugar in refining.
Textile
——, *Dyeing*
Reagent in—
Deepening shades on dyed fabrics, dyeing fabrics and yarns.
——, *Printing*
Reserve in—
Calico printing.
Wine
Reagent for—
Correcting acid content of wine.

Citric Acid Ester of Grapeseed Alcohol
Bituminous
Solvent (Brit. 445223) for—
Asphalt and other bituminous bodies.
Dye
Solvent (Brit. 445223) for—
Dyestuffs, particularly oil-soluble coaltar dyes.
Fats, Oils, and Waxes
Solvent (Brit. 445223) for—
Fats, oils, waxes.
Resins
Solvent (Brit. 445223) for—
Oil-soluble glycerol-phthalic acid resins.
Polymerized vinyl compounds.
Synthetic resins.
Rubber
Solvent (Brit. 445223) for—
Rubber.

Citric Acid Ester of Ricinoleic Alcohol
Bituminous
Solvent (Brit. 445223) for—
Asphalt and other bituminous bodies.
Dye
Solvent (Brit. 445223) for—
Dyestuffs, particularly oil-soluble coaltar dyes.
Fats, Oils, and Waxes
Solvent (Brit. 445223) for—
Fats, oils, waxes.
Resins
Solvent (Brit. 445223) for—
Oil-soluble glycerol-phthalic acid resins.
Polymerized vinyl compounds.
Synthetic resins.
Rubber
Solvent (Brit. 445223) for—
Rubber.

Citronella Oil
Synonyms: Lana batu, Verbena oil.
Latin: Andropogon nardi.
French: Huile de citronnelle.
German: Zitronelloel.
Agricultural
To keep insects from cattle.
Chemical
Starting point in making—
Aromatics, citronellol, geraniol.
Glues and Adhesives
Ingredient of—
Adhesive preparations containing casein (U. S. 1604307).
Insecticide
As an insectifuge.
Ingredient of—
Fungicidal compositions.
Insecticidal compositions.
Ink
Ingredient of—
Printing inks (U. S. 1724603).
Miscellaneous
As a disguiser of odors.
In veterinary medicine.
Perfume
Ingredient of—
Perfume preparations.
Perfume in—
Cosmetics.
Pharmaceutical
In compounding and dispensing practice.
Soap
Perfume in—
Toilet soaps.

Citronellyl Acetate
Synonyms: Citronellol acetate, Citronellylacetic ether.
French: Acétate de citronellole, Acétate citronellylique, Acétate de citronellyle, Éther citronellylacétique.
German: Citronellylacetat, Citronellylacetoaether, Citronellylazetat, Essigsäurecitronellolester, Essigsäurecitronellylester, Essigsäurescitronellol, Essigsäurescitronellyl.
Miscellaneous
As a perfume for various purposes.

Citronellyl Carboxethylate
French: Carboxéthylate de citronellyle.
German: Citronellylcarboxaethylat.
Spanish: Carboxetilato de citronelil.
Italian: Carbossietilato di citronellile.
Perfume
Ingredient (French 650100) of—
Perfumes.

Citryl Carbamide
French: Carbamide de citryle, Carbamide citrylique.
German: Citrylcarbamid.
Chemical
Starting point in making various derivatives.
Resins and Waxes
Starting point (Brit. 292912) in making synthetic resins with—
Acetylsalicylic acid, aliphatic dibasic acids, ammonium salicylate, anthranilic acid, benzoic acid, gallic acid, hydroxynaphthoic acid, magnesium salicylate, oxalic acid, phenolic acids, phthalic acid, salicylamide, salicylic acid, strontium salicylate, succinic acid.

Coaltar Creosote
Synonyms: Creosote.
French: Brai de créosote, Créosote de houille, Huile de créosote de houille.
German: Kohlenteerkresot, Kreosot, Kreosotoel.
Agriculture
Sterilizer for soils.
Ceramics
Ingredient of—
Mass for producing a blue-colored brick in the kiln.
Lubricant for brick-making machinery.
Chemical
Starting point in making—
Lampblack.
Fuel
As a fuel.
Gas
Starting point in making—
Artificial illuminant.
Washing agent in removing—
Benzene from coal gas and coke-oven gas.
Insecticide
Ingredient of—
Insecticides and fungicides.
Mechanical
Ingredient of—
Cart or axle grease in admixture with lime.
Metallurgical
Material for working and finishing—
Iron and steel.
Miscellaneous
Diesel engine fuel.
Ingredient of—
Disinfecting compositions, sheep dips.
Preservative for treating—
Paving blocks made of wood, railroad ties, telegraph and similar poles.
Paint and Varnish
Ingredient of—
Preservative paints, shingle stains.
Woodworking
As a preservative.

Cobalt Acetate
Synonyms: Cobaltous acetate.
Latin: Cobaltum acetatum.
French: Acétate de cobalt, Acétate cobalteux.
German: Essigsäureskobalt, Kobaltacetat, Kobaltazetat.
Analysis
Reagent for the detection of potassium.
Chemical
Starting point in making various salts.

Cobalt Acetate (Continued)
Ingredient of catalytic preparations used in making—
Acenaphthalene, acenaphthaquinone, naphthaldehydic acid, naphthalic anhydride, and hemimellitic acid from acenaphthene (Brit. 295270).
Acetaldehyde from ethyl alcohol (Brit. 281307).
Acetic acid from ethyl alcohol (Brit. 281307).
Alcohols from aliphatic hydrocarbons (Brit. 281307).
Aldehydes or acids by the reduction of the corresponding ester (Brit. 306471).
Aldehydes and acids from toluene, orthochlorotoluene, orthonitrotoluene, orthobromotoluene, parachlorotoluene, parabromotoluene, paranitrotoluene, metachlorotoluene, metabromotoluene, metanitrotoluene, dichlorotoluenes, dinitrotoluenes, dibromotoluenes, chlorobromotoluenes, chloronitrotoluenes, bromonitrotoluenes (Brit. 295270).
Aldehydes and acids from xylenes, pseudocumenes, mesitylene, and paracymene (Brit. 295270).
Alphacampholid by the reduction of camphoric acid (Brit. 306471).
Alphanaphthaquinone from naphthalene (Brit. 281307).
Anthraquinone from naphthalene (Brit. 295270).
Benzaldehyde and benzoic acid from toluene (Brit. 281307).
Benzoquinone from phenanthraquinone (Brit. 281307).
Benzyl alcohol or benzaldehyde or phthalide by the reduction of phthalic anhydride (Brit. 306471).
Benzyl alcohol by the reduction of benzaldehyde (Brit. 306471).
Butyl alcohol by the reduction of crotonaldehyde (Brit. 306471).
Chloroacetic acid from ethylenechlorohydrin (Brit. 295270).
Diphenic acid from ethyl alcohol (Brit. 281307).
Ethyl alcohol by the reduction of acetaldehyde (Brit. 306471).
Fluorenone from fluorene (Brit. 295270).
Formaldehyde from methanol or methane (Brit. 295270).
Formaldehyde by the reduction of carbon dioxide or carbon monoxide (Brit. 306471).
Hydroxyl reduction compounds of anthraquinone, benzoquinone, and the like (Brit. 306471).
Isopropyl alcohol by the reduction of acetone (Brit. 306471).
Maleic acid and fumaric acid by the oxidation of toluene, benzene, phenols, tar phenols, or furfural, or from benzoquinone or phthalic anhydride (Brit. 295270).
Methane by the reduction of carbon dioxide or carbon monoxide (Brit. 306471).
Methanol by the reduction of carbon dioxide or carbon monoxide (Brit. 306471).
Naphthaldehydic acid, acenaphthaquinone, or bisacenaphthylidenedione from acenaphthylene (Brit. 281307).
Phenanthraquinone from phenanthrene or diphenic acid (Brit. 295270).
Phthalic acid and maleic acid from naphthalene (Brit. 295270).
Primary alcohols by the reduction of the corresponding aldehydes (Brit. 306471).
Propionic acid and butyric acid and higher alcohols, ketones, and acids by the reduction of carbon dioxide or carbon monoxide (Brit. 306471).
Reduction products of carbon dioxide and carbon monoxide (Brit. 306471).
Reduction products of ketones, aldehydes, acids, esters, ethers, alcohols, and other organic compounds containing oxygen (Brit. 306471).
Salicylic acid and salicylic aldehyde from cresol (Brit. 295270).
Secondary butyl alcohol by the reduction of methylethyl ketone (Brit. 306471).
Valeryl alcohol by the reduction of valeraldehyde (Brit. 306471).
Vanillin and vanillic acid from eugenol or isoeugenol (Brit. 295270).
Ingredient (Brit. 304640) of catalytic preparations used in making—
Alphanaphthylamine from alphanitronaphthalene.
Amines from aliphatic nitro bodies, such as allyl nitrites or nitromethanes.
Amines from oximes, Schiff's base, and nitrites.
Amine compounds from nitroanisole.
3-Aminopyridin from 3-nitropyridin.
Amylamine from pyridin.
Aminophenols from nitrophenols.
Anilin from nitrobenzene.
Azoxybenzene, azobenzene, and hydrazobenzene from nitrobenzene.
Cyclohexamine, dicyclohexamine, and cyclohexylanilin from nitrobenzene.
Piperidin from pyridin.
Pyrrolidin from pyrrol.
Tetrahydroquinolin from quinolin.
Ink
Ingredient of—
Sympathetic inks.
Oilcloth and Linoleum
Reagent in making—
Varnish and lacquer coatings (added to prevent the yellowing of the product).
Paint and Varnish
Bleaching agent and drier in making—
Lacquers, paints, varnishes.

Cobalt Albuminate
French: Albuminate de cobalt, Albuminate cobaltique.
German: Albuminsaeureskobalt, Kobaltalbuminat.
Rubber
Reagent (U. S. 1640817) in—
Reclaiming rubber.

Cobalt-Ammonium Chloride
French: Chlorure de cobalt et ammonium.
German: Kobaltammoniumchlorid.
Miscellaneous
Carrotting agent (Brit. 271026) in treating—
Felt, furs.

Cobalt Betabenzoylpropionate
Plastics
Starting point (U. S. 2001380) in making—
Films.

Cobalt Bismuthide
French: Bismuthide de cobalt.
German: Cobaltwismuthid.
Chemical
Catalyst in making—
Acetone from isopropyl alcohol.
Isobutyraldehyde from isobutyl alcohol.
Isobutyronitrile from isobutylamine.
Paracymene from turpentine oil.
Naphthalene from tetrahydronaphthalene.

Cobaltic Gammamethylacetylacetone
French: Gammaméthyleacétyleacétone de cobalt, Gammaméthyleacétyleacétone cobaltique.
German: Kobaltgammamethylaceton.
Chemical
Reagent (Brit. 289493) in making—
Aromatics, intermediates, pharmaceuticals.
Dye
Reagent (Brit. 289493) in making various synthetic dyestuffs.
Petroleum
Ingredient (Brit. 289493) of—
Motor fuels, to improve their combustion.

Cobalt Resinate
French: Résinate de cobalt.
German: Kobaltresinat.
Ceramics
Reagent for producing lustrous coatings on—
Chinaware, porcelains, potteries.
Glass
Reagent for producing lustrous effects on glassware.
Paint and Varnish
Drier in making—
Clear paints, enamels, lacquers, varnishes.
Textile
——, *Finishing*
Ingredient of—
Compositions used to produce waxed fabrics without changing colors dyed thereon.

Cobalt Selenide
French: Séléníde de cobalt.
German: Kobaltselenid.
Chemical
Catalyst in making—
Acetone from isopropyl alcohol.
Isobutylaldehyde from isobutyl alcohol.
Isobutyronitrile from isobutylamine.
Naphthalene from tetrahydronaphthalene.
Paracymene from turpentine oil.

Cobalt Tungate
German: Kobalttungat.
Paint and Varnish
Drier (Brit. 270387) in making—
Enamels, lacquers, paints, varnishes.
Photographic
Ingredient of—
Light-sensitive varnishes.

Cocaine
Chemical
Starting point in making—
Cocaine albuminate, cocaine camphorate, cocaine ferrocyanide, cocaine salts with various acids, guanadin compounds, phenylurethane compounds.
Pharmaceutical
In compounding and dispensing practice.

Coconut Oil
Synonyms: Coconut butter, Coconut oil, Palm oil.
French: Beurre de coco, Huile de coco.
German: Coconussfett, Cocosbutter, Cocosfett, Cocosoel, Koprafett.
Chemical
Starting point in making—
Caprinic acid, caprylic acid, lauric acid.
Fats and Oils
Ingredient of—
Edible oil mixtures.
Food
Ingredent of—
Candies, chocolate coatings, lard substitutes, margarines, pastries, vegetarian foods.
Fuel
Burning agent in—
Night lights.
Ingredient of—
Candles.
Glass
Ingredient (U. S. 1638272) of—
Detergent agents.
Perfumery
Ingredient of—
Cosmetics, hair oils, pomades.
Soap
Starting point in making—
Curd soaps, lime soaps, peroxide soaps, medicinal soaps, saltwater soaps, shaving soaps, toilet soaps.
Textile
——, *Dyeing*
Assistant in dyeing—
Cotton yarns, warp, fabrics.
——, *Finishing*
Reagent in obtaining—
Soft handle on mercerized cotton.

Coconut Oil Alcohols Alphabromolauricester
Detergent
Starting point (Brit. 408754) in making—
Saponaceous products by reaction with tertiary amines, which may be used alone or with other soaps, fillers, or compounds giving off oxygen.

Coconut Oil Alcohols Bromosuccinicester
Detergent
Starting point (Brit. 408754) in making—
Saponaceous products by reaction with tertiary amines, which may be used alone or with other soaps, fillers, or compounds giving off oxygen.

Coconut Oil Alcohols Chloraceticester
Detergent
Starting point (Brit. 408754) in making—
Saponaceous products by reaction with tertiary amines, which may be used alone or with other soaps, fillers, or compounds giving off oxygen.

Coconut Oil Alcohols Dichloroaceticester
Detergent
Starting point (Brit. 408754) in making—
Saponaceous products by reaction with tertiary amines, which may be used alone or with other soaps, fillers, or compounds giving off oxygen.

Coconut Oil Alcohols Gammachlorobutyricester
Detergent
Starting point (Brit. 408754) in making—
Saponaceous products by reaction with tertiary amines, which may be used alone or with other soaps, fillers, or compounds giving off oxygen.

Coconut Oil Alcohols Gammachlorovalericester
Detergent
Starting point (Brit. 408754) in making—
Saponaceous products by reaction with tertiary amines, which may be used alone or with other soaps, fillers, or compounds giving off oxygen.

Coconut Oil Fatty Acid
French: Acide gras d'huile de copra, Acide gras d'huile de coprah, Acide gras d'huile de copre.
German: Kopraoelfettsaeure.
Chemical
Starting point in making various salts and esters.
Food
Ingredient of—
Prepared foods, hydrogenated oil products.
Fuel
Compound of—
Candles.
Miscellaneous
Ingredient of—
Cleansing compositions with alkaline hypochlorites (Brit. 280193).
Polishing compositions.
Paint and Varnish
Starting point in making—
Driers.
Pharmaceutical
In compounding and dispensing practice.
Soap
Raw material in soapmaking.
Textile
——, *Bleaching*
Ingredient of—
Bleaching compositions containing alkaline hypochlorites (Brit. 280193).
——, *Finishing*
Ingredient of—
Finishing compositions.
Washing compositions containing alkaline hypochlorites (Brit. 280193).
Waterproofing compositions.

Codeine Hydrobromide
German: Bromwasserstoffsäureskodein, Codeinbromhydrat, Codeinhydrobromid, Kodeinbromhydrat, Kodeinhydrobromid.
Spanish: Bromhidrato de codeina.
Italian: Bromidrato di codeina.
Pharmaceutical
Suggested for use as—
Sedative in nervousness and coughs.

Codliver Oil
French: Huile de foie morue, médicinale.
German: Dorschleberoel, Dorschlebertran, Kabeljauleberoel, Lebertran, Stockfischleberoel.
Pharmaceutical
In compounding and dispensing practice.
Suggested for use as—
Source of vitamins.

Cod Oil
Synonyms: Banks oil, Brown codliver oil.
French: Huile de foie de morue, industrielle, Huile de morue.
German: Dorschoel.
Spanish: Aciete de merluza.
Italian: Olio di merluzzo.
Fats and Oils
Ingredient of—
Lubricants.
Starting point in making—
Fatty acids.
Hardened oil by treatment with hydrogen in the presence of nickel or other catalyst.
Food
Ingredient of—
Food preparations, oleomargarins.

Cod Oil (Continued)

Ink
Ingredient of—
Printers' ink.

Leather
Ingredient of—
Dressing compositions, enamelling compositions.
Reagent in—
Currying leathers of various sorts.
Tanning chamois leathers.

Mechanical
As a lubricant.

Metallurgical
Quenching oil in—
Hardening steels.

Miscellaneous
Ingredient of—
Preparations used for the treatment of cloth to make tarpaulins.
Shoe polishes.

Oilcloth and Linoleum
Ingredient of—
Coating compositions (used either with linseed oil or as a substitute for it).

Paint and Varnish
Ingredient of—
Paints, varnishes, enamels, and other preparations (used with linseed oil) or as a substitute.

Pharmaceutical
In compounding and dispensing practice.

Soap
Raw material in making—
Laundry soaps, industrial soaps.

Cod Oil Fatty Acids
French: Acide gras d'huile de foie de morue, industrielle, Acide gras d'huile de morue.
German: Dorschoelfettsäure.

Chemical
Starting point in making—
Various salts and esters.

Food
Ingredient of—
Food preparations (used in purified form).
Halogenated oil products.

Fuel
Component of—
Candles.

Miscellaneous
Ingredient of—
Cleansing compositions (used with the addition of alkaline hypochlorites, such as sodium hypochlorite) (Brit. 280193).
Polishing compositions.

Paint and Varnish
Starting point in making—
Driers.

Pharmaceutical
In compounding and dispensing practice.

Soap
Raw material in making—
Laundry soaps.

Textile
——, *Bleaching*
Ingredient (Brit. 280193) of—
Bleaching compositions containing alkaline hypochlorites.

——, *Finishing*
Ingredient of—
Finishing compositions.
Washing compositions containing alkaline hypochlorites (Brit. 280193).
Waterproofing compositions.

Cod Oil Soap
French: Savon d'huile de foie de morue, industrielle, Savon d'huile de morue.
German: Dorschoelseife.

Fats and Oils
Ingredient of—
Lubricating compositions.

Miscellaneous
Ingredient of—
Detergent preparations.

Soap
Base for—
Shampoos, shaving soaps.

Coke Pitch
French: Brai de coke.
German: Kokspech.

Chemical
Lining for—
Furnaces and other apparatus used in the chemical industries.

Electrical
Lining for—
Electric furnaces.

Fuel
As a fuel for household and industrial use.

Gas
Raw material in making—
Carburetted watergas, watergas.

Metallurgical
As a fuel.
Reagent in making—
Steel in the open-hearth furnace (used for carburizing purposes).

Condurango
Synonyms: Condor vine, Cundurango, Eagle vine.
French: Écorce de condurango.
German: Condurangorinde, Condurangobast, Condurangoborke, Condurangolohe.

Fats and Oils
Starting point in extracting—
Oil.

Food
Ingredient of—
Flavoring compositions.

Pharmaceutical
In compounding and dispensing practice.

Coniine Hydrobromide
Synonyms: Alphapropylpiperidine hydrobromide, Conine hydrobromide.
Latin: Coninum bromhydricum.
French: Bromhydrate de conine.
German: Conicinbromhydrat.
Spanish: Bromhidrato de conicina.
Italian: Bromidrato di conicina.

Pharmaceutical
In compounding and dispensing practice.
Suggested for use as—
Antineuralgic, sedative.
Suggested for use in treating—
Whooping cough.

Copal
Synonyms: Anime, Cowrie, Gum copal.
French: Gomme copal, Résine copal.
German: Kopal, Resincopal, Resinkopal.

Miscellaneous
Ingredient of—
Amber substitutes, cement compositions, compositions for closing punctures in tires (Brit. 253113), coatings for under-fabric of oil-cloth and linoleum, linoleum cements, rosin cements.

Paint and Varnish
Ingredient of—
Asphalt lacquers, spirit lacquers, varnishes.
Starting point in making—
Sulphur-lime fusion product used for manufacturing paints for fishing nets (U. S. 1617426).

Pharmaceutical
In compounding and dispensing practice.

Resins and Waxes
Hardener for—
Rosin compositions (Brit. 252656).

Textile
——, *Finishing*
Ingredient of—
Waterproofing compositions.

——, *Manufacture*
Ingredient of—
Spinning solution in the manufacture of nitro rayon.

Copellidin

Photographic
Starting point in making—
Sensitizing agents (Brit. 262816).

Copper
Synonyms: Red metal.
Latin: Cuprum.
French: Cuivre.
German: Kupfer.
Spanish: Cobre.
Italian: Rame.

In Common Commercial Forms

(Billets, Cakes, Cathodes, Ingots, Ingot Bars, Plates, Rods, Sheets, Shot, Slabs, Strips, Wedge Cakes, Wire, Wire Bars, and Others.)

Brewing
Coppersmithing material in fabricating—
Coils, cookers, coolers, false bottoms, fittings, hop tanks, kettles, mash tanks, piping, valves, water tanks, yeast equipment.

Building Construction
Material in fabricating—
Downspouts, electrical installations, flashings, gutters. hardware, pipes and fittings, plumbing and heating equipment, pumps, roofing, screens, tanks, valves, ventilators, water heaters, weather strips, window sash.

Chemical
Base material in making—
Copper salts.
Coppersmithing material in fabricating—
Agitators, autoclaves, baffles, belts, blades, blow cases, burner tips, burners, chlorinators, chutes, coils, condensers, containers, conveyors, coolers, crystallizers, dephlegmators, digesters, dryers, extractors, evaporators, fans, filling machines, filters, fittings, fusion pots, gas-scrubbers, heating coils, heating equipment, hoppers, jacketed kettles, kettles, knives, laboratory apparatus, linings, mixers, percolators, pipelines, pots, preheaters, pumps, pump rods, screens, shafts, sifters, solvent recovery apparatus, springs, stills, tanks, trays, trucks, vacuum pans, valves, vats.

Distilling
Coppersmithing material in fabricating—
Blenders, coils, condensers, cookers, coolers, fermenters, fittings, piping, recovery equipment, separators, stills, tanks, valves, yeast equipment.

Electrical
Material in fabricating—
Cables, conductors.
Parts for motors, dynamos, generating sets, lighting fixtures, switches and most other devices and services operated by electricity.
Wires.

Metallurgical
Ingredient of—
Alloys, such as brasses, bronzes, German or nickel silvers.
Electroplating solutions.

Minting
Base material in—
Coinage.

Miscellaneous
Coppersmithing material in fabricating—
Airplane equipment, automobile equipment, bearings, bowls, coils, condensers, converters, cookers, cooking utensils, coolers, dephlegmators, digesters, evaporators, expansion joints, extractors, false bottoms, farm machine parts, filters, fittings, gaskets, heat-interchangers, hotwater heaters, kettles, laundry equipment, lighting fixtures, liners, marine machinery, mixers, oil-burning equipment, pans, percolators, pipe coils and bends, piping, preheaters, pumps, radio apparatus, railroad equipment, rectifiers, recuperators, refrigerators, screens, scrubbers, separators, sifters, solvent-recovery apparatus, stills, tanks, vacuum pans, washing machines, yeast equipment.
Coppersmithing material in fabricating apparatus for—
Bakeries, canneries, confectionery plants, cosmetic plants, dairies and creameries, dye works, dyeing extract plants, extract plants of various sorts, flavoring extract plants, food products plants, hospitals, hotels, laboratories, laundries, milk condenseries, paint factories, perfumery plants, petroleum refineries, printing plants, pulp and paper mills, restaurants, salt works, soap factories, sugar mills and refineries, tanneries, tanning extract plants, textile plants, turpentine and rosin plants, varnish plants, vinegar works, wood-distillation plants.

Pharmaceutical
Coppersmithing material in fabricating—
Pill-coating equipment, tablet-coating equipment.
Various equipment (see under Chemical).

Printing
In making electrotypes and halftone plates.

In Finely Divided Forms

Ceramics
Decorative material in—
Coating ceramic products.

Chemical
Catalytic reduction agent in—
Organic syntheses.
Starting point in making—
Copper salts.

Dye
Dye syntheses.
Catalytic reduction agent in—

Fats and Oils
Catalytic reduction agent in—
Hydrogenation processes.

Glass
Decorative material in—
Coating glassware.

Paint and Varnish
Pigment in—
Decorative coatings, protective coatings, ship's-bottom paints.

Pharmaceutical
In colloidal form in compounding and dispensing practice.

Soap
Catalytic reduction agent in—
Hydrogenation treatment of soapstocks.

Copper Acetate, Basic
Synonyms: Copper acetate, Copper subacetate, Green verdigris.
Latin: Cuprum subacetum.
French: Acétate de cuivre, Acétate de cuivre, brut; Acétate cuivrique, Sousacétate de cuivre, Verdet boule, Verdet gris, Verdet de Montpellier, Vert de gris.
German: Basisch essigsäureskupfer, Basisch gruenspan, Gruenspan.

Agricultural
Alone and in mixtures for fighting insect pests in orchards and fields.

Analysis
Reagent for detecting glucose in the presence of dextrin.

Ceramics
As a pigment in coating.

Chemical
Ingredient of catalytic preparations used in making—
Acenaphthylene, acenaphthaquinone, bisacenaphthylidenedione, naphthaldehydic acid, naphthalic anhydride, and hemimellitic acid from acenaphthene (Brit. 295270).
Acetaldehyde from ethyl alcohol (Brit. 281307).
Acetic acid from ethyl alcohol (Brit. 281307).
Alcohols from aliphatic hydrocarbons (Brit. 281307).
Aldehydes and acids from the reduction of the corresponding esters (Brit. 306471).
Alphacampholid by the reduction of camphor in acid (Brit. 306471).
Aldehydes and acids from toluene. orthochlorotoluene, metachlorotoluene, parachlorotoluene, orthobromotoluene, metabromotoluene, parabromotoluene, orthonitrotoluene, paranitrotoluene, metanitrotoluene, dichlorotoluenes, dinitrotoluenes, dibromotoluenes, chlorobromotoluenes, chloronitrotoluenes, bromonitrotoluenes (Brit. 295270).
Aldehydes and acids from xylenes, pseudocumene, mesitylene, and paracymene (Brit. 281307).
Alphanaphthaquinone from naphthalene (Brit. 281307).
Anthraquinone from naphthalene (Brit. 295270).
Benzaldehyde and benzoic acid from toluene (Brit. 281307).
Benzoquinone from phenanthraquinone (Brit. 281307).
Benzyl alcohol by the reduction of benzaldehyde (Brit. 306471).
Benzyl alcohol or benzaldehyde or phthalide by the reduction of phthalic anhydride (Brit. 306471).
Butyl alcohol by the reduction of crotonaldehyde (Brit. 306471).

Copper Acetate, Basic (Continued)
Chloroacetic acid from ethylenechlorohydrin (Brit. 306471).
Diphenic acid from ethyl alcohol (Brit. 281307).
Ethyl alcohol by the reduction of acetaldehyde (Brit. 281307).
Fluorenone from fluorene (Brit. 295270).
Formaldehyde by the reduction of carbon dioxide or carbon monoxide (Brit. 306471).
Formaldehyde by the reduction of methane or methanol (Brit. 306471).
Hydroxyl reduction compounds of anthraquinone, benzoquinone and the like (Brit. 306471).
Isopropyl alcohol by the reduction of acetone (Brit. 306471).
Maleic acid and fumaric acid by the oxidation of toluene, benzene, phenols, tar phenols, or furfural, or from benzoquinone or phthalic anhydride (Brit. 295270).
Methane by the reduction of carbon dioxide or carbon monoxide (Brit. 306471).
Methanol by the reduction of carbon dioxide or carbon monoxide (Brit. 306471).
Naphthaldehydic acid, acenaphthaquinone, or bisacenaphthylidenedione from acenaphthylene (Brit. 281307).
Phenanthraquinone from phenanthrene or diphenic acid (Brit. 295270).
Phthalic acid and maleic acid from naphthalene (Brit. 295270).
Primary alcohols by the reduction of the corresponding aldehydes (Brit. 306471).
Propionic acid and butyric acid and higher alcohols, ketones and acids by the reduction of carbon dioxide or carbon monoxide (Brit. 306471).
Reduction products of ketones, aldehydes, acids, esters, alcohols, ethers and other organic compounds containing oxygen (Brit. 306471).
Salicylic acid and salicylic aldehyde from cresol (Brit. 295270).
Secondary butyl alcohol by the reduction of methylethyl ketone (Brit. 306471).
Valeryl alcohol by the reduction of valeraldehyde (Brit. 306471).
Vanillin and vanillic acid from eugenol or isoeugenol (Brit. 295270).
Ingredient (Brit. 304640) of catalytic preparations used in the production of various aromatic and aliphatic amines, including—
Alphanaphthylamine from alphanitronaphthalene.
Amines from aliphatic nitro compounds, such as alkyl nitriles, or nitromethane.
Amylamine from pyridin.
Anilin, azo-oxybenzene, azobenzene, and hydrazobenzene, by reduction of nitrobenzene.
Aminophenols from nitrophenols.
3-Aminopyridin from 3-nitropyridin.
Amino compounds from the corresponding nitroanisoles.
Amines from oximes, Schiff's base, and nitriles.
Cyclohexamine, dicyclohexamine, and cyclohexylanilin from nitrobenzene.
Piperidin from pyridin.
Pyrrolidin from pyrrol.
Tetrahydroquinolin from quinolin.
Starting point in making various copper salts.
Reagent in the isolation of—
Phytin.

Dye
Oxidizing agent in making—
Indigo and vat dyes.

Ink
Ingredient of—
Inks used on metals, glass and similar surfaces.

Insecticide
Ingredient of—
Insecticidal and fungicidal compositions.
Preparations used in the place of Bordeaux mixtures.

Linoleum and Oilcloth
Ingredient of—
Coating compositions.

Metallurgical
Ingredient of—
Baths used in the electrodeposition of copper.

Miscellaneous
In veterinary medicine.
Ingredient of—
Enamelling compositions used in the preparation of miniatures.
Gilder's wax preparations used in fire gilding.
Reagent in making—
Artificial flowers.

Paint and Varnish
As a pigment.
Reagent in making—
Schweinfurt green and other pigments.

Paper
Pigment in making—
Wallpaper.

Pharmaceutical
In compounding and dispensing practice.
Ingredient of—
Corn plasters and salves.

Textile
Mordant in—
Dyeing wool with blacks.
General dyeing and printing.

Copper Acetoarsenite
Synonyms: Cupric acetoarsenite, Emerald green, Emperor green, Imperial green, Kaiser green, King's green, Meadow green, Moss green, New green, Paris green, Parrot green, Patent green, Schweinfurt green.
French: Acétoarsénite de cuivre, Acétoarsénite cuivrique, Vert de Paris, Vert de schweinfurt.
German: Englishgruen, Kaisergruen, Kasselgruen, Kupfer acetatarsenit, Kupfer arseniacetat, Kupfer arseniazetat, Mitisgruen, Neuwiedergruen, Papagegruen, Patentgruen, Schweinfurtergruen, Wienergruen.
Spanish: Arseniacetato de cobre.
Italian: Acetoarsenito di cobre.

Insecticide
General insecticide.
Ingredient of—
Insecticidal and fungicidal compositions.

Miscellaneous
Preservative for various purposes.

Paint and Varnish
Ingredient of—
Paints for preserving ships' bottoms.
Paints for submarine work.

Woodworking
As a preservative.

Copper Acetylacetonate

Chemical
Reagent in—
Organic syntheses.

Fuel
Primer (Brit. 404682) in—
Diesel engine fuels (used in conjunction with alkyl nitrates, having two to four atoms in the molecule, whose function is that of reducing the delay period).
Reducer (Brit. 404682) of—
Spontaneous ignition temperature of diesel engine fuels.

Copper Albuminate
French: Albuminate de cuivre, Albuminate cuivrique.
German: Albuminsaeurekupferester, Albuminsaeureskupfer, Kupfersalbuminat.

Rubber
Reagent in—
Reclaiming rubber from old tires and the like (U. S. 1640817).

Copper-Ammonium Alginate

Ceramics
Ingredient of—
Compositions used for the waterproofing of various ceramic wares, porcelains, potteries, chinaware, stoneware, earthenware.

Chemical
Emulsifying agent in making—
Dispersions of various chemicals.
Ingredient of—
Various chemical solutions (added for the purpose of increasing their viscosity).

Construction
Ingredient of—
Compositions used for treating cement and concrete for the purpose of preventing deterioration when exposed to the action of alkalies and seawater.
Waterproofing compositions used for treating plaster of Paris, wallboard, cement, stucco, concrete, and masonry.

Copper-Ammonium Alginate (Continued)
Fats and Oils
Reagent for—
Stabilizing emulsions of various animal and vegetable fats and oils.
Fuel
Binder in—
Coal-dust compositions used as fuel briquettes.
Glues and Adhesives
Ingredient of—
Adhesive preparations.
Ink
Thickener in—
Printing inks.
Leather
Ingredient of—
Compositions for sizing leather.
Mechanical
Ingredient of—
Compositions for covering steel tubes.
Miscellaneous
Binder in—
Compositions containing powdered mica, asbestos, coal, carbon, graphite, minerals and the like.
Sizing compositions used on various articles.
Emulsifying agent for various products.
Ingredient of—
Compositions used for treating rope and twine.
Waterproofing compositions.
Paint and Varnish
Ingredient of—
Compositions used for proofing interior walls and ceilings.
Paper
Binder in—
Sizing compositions (used to cement the fibers more closely).
Wood-flour compositions.
Ingredient of—
Finishing compositions, waterproofing compositions.
Petroleum
Dispersing agent in—
Emulsions of petroleum and petroleum distillates.
Plastics
Ingredient of—
Various plastic compositions, containing such substances as horn, ebonite, celluloid, ivory, bone, shell, gelalith, formaldehydephenol condensation products, urea-formaldehyde condensation products, and other artificial resins.
Rubber
Ingredient of—
Products obtained with rubber latex.
Soap
Ingredient of—
Detergent preparations.
Textile
——, *Dyeing*
Dispersing agent in—
Dye baths.
——, *Finishing*
Ingredient of—
Sizing compositions, waterproofing compositions.
——, *Printing*
Ingredient of—
Printing pastes.
Waxes and Resins
Dispersing agent in—
Preparations of waxes and resins, both artificial and natural.

Copper-Ammonium Silicates (Complex)
Insecticide
As fungicides (Brit. 427128).

Copper-Ammonium Sulphate
Synonyms: Ammoniated copper sulphate, Ammoniocupric sulphate, Ammonium-cupric sulphate, Copper ammoniosulphate, Cupric-ammonium sulphate.
Latin: Cuprum ammoniatum.
French: Sulfate de cuivre ammoniacal.
German: Ammoniakalischeskupfersulfat, Cuprisulfatammoniak, Kupferammoniaksulfat, Kupferammoniumsulfat, Kupferammonsulfat, Schwefelsaeureskupferoxydammoniak.
Chemical
Starting point in making—
Copper arsenate.
Explosives
Ingredient of—
Colored lights, fireworks.
Insecticide
Ingredient of—
Insecticidal compositions, for example, azurin.
Miscellaneous
Coloring for—
Druggists' show-globe solutions.
Pharmaceutical
In compounding and dispensing practice.
Textile
——, *Dyeing and Printing*
Mordant for—
Yarns and fabrics, particularly calicoes.

Copper Borotungstate
Synonyms: Copper tungstoborate.
French: Borotungstate de cuivre.
German: Kupferborwolframat.
Metallurgical
Ingredient of (French 600774) antioxidation coating (electrically deposited) for—
Bismuth, copper, nickel, steel, tin, zinc.

Copper Bromide
Synonyms: Cupric bromide.
French: Bromure de cuivre.
German: Cupribromid, Kuperbromid.
Spanish: Bromuro cobrico.
Italian: Bromuro ramico.
Chemical
Catalyst (Brit. 398527) in making—
Esters from lower aliphatic acids and olefins.
Reagent in—
Organic synthesis to replace the iodine radical by bromine.
Petroleum
Ingredient (Brit. 406963) of—
Catalytic mixtures used in the purifying of mineral oils by hydrogenation.
Photographic
Intensifier in—
Photographic processes.
Reagent (Brit. 382320) in—
Oxidizing action in image layer of a differential treatment of images obtained in different depths of an emulsion.

Copper Carbonate
French: Carbonate de cuivre, Carbonate cuivrique.
German: Kohlensäureskupferoxyd, Kupfercarbonat, Kupricarbonat.
Spanish: Carbonato de cobre.
Italian: Carbonato di rame.
Ceramics
As a pigment.
Chemical
Starting point in making—
Copper salts.
Explosives
Ingredient of—
Pyrotechnic compositions.
Insecticide
As an insecticide
Ingredient of—
Insecticidal compositions.
Metallurgical
Pickling agent for—
Imparting black color to brass.
Paint and Varnish
Pigment in—
Paints and varnishes.
Pharmaceutical
In compounding and dispensing practice.

Copper Chlorate
Synonyms: Cupric chlorate.
French: Chlorate de cuivre, Chlorate cuivrique.
German: Chlorsäureskupfer, Kupferchlorat.
Spanish: Clorato de cobre.
Italian: Clorato di rama.
Textile
Mordant in—
Dyeing processes, printing processes.

Copper Chromite

Chemical

Catalyst (Brit. 395198) in making—
Dodecyl alcohol from borolauric anhydride.
Dodecyl alcohol from silicolauric anhydride.
Mixed higher alcohols from mixed anhydrides of boric acid and coconut oil fatty acids.
Mixed higher alcohols from mixed anhydrides of silicic acid and coconut oil fatty acids.
Octodecyl alcohol from borostearic anhydride.
Octodecyl alcohol from silicostearic anhydride.
Phenylethyl alcohol from silicophenylacetic anhydride.

Ingredient of—
Catalytic mixture, containing also chromites of zinc and cadmium, used in reduction of aliphatic acids to alcohols and esters; for example, in making alcohols from lauric, butyric, acetic, ricinoleic, oleic, stearic, and coconut oil fatty acids (Brit. 397938).
Catalytic mixture used in converting esters of aliphatic carboxylic acids into alcohols by hydrogenation; for example, (1) ethyl acetate, into ethanol; (2) ethyl normal-butyrate, normal butyl acetate, ethyl laurate, ethylphenyl acetate, normal butyl normal-butyrate, ethyl adipate into the corresponding alcohols (Brit. 385625).
Catalytic mixture used in dehydrogenation of partly or completely hydrogenated polynuclear hydrocarbons to produce aromatic hydrocarbons having the same number of carbon atoms in the molecule; for example, dehydrogenating, tetrahydronaphthalene, betaphenyldecahydronaphthalene, naphthyltetrahydronaphthalene, betacyclohexyl tetrahydronaphthalene, dicyclohexylbenzene, betabenzyltetrahydronaphthalene, cyclohexyldiphenyl, methyl cyclohexylbenzene (Brit. 406808).

Fats and Oils

Catalyst (Brit. 394073) in making—
Unsaturated hydrocarbons from fatty oils, the unsaturated products then being polymerized in the presence of condensing agents to yield lubricating oils.

Copper Cupricyanide

Chemical

Catalyst (Brit. 446411) in—
Halogenating unsaturated hydrocarbons.

Starting point (Brit. 446411) in making—
Catalysts with metal chlorides for halogenating unsaturated hydrocarbons.

Copper Cyanide

Synonyms: Cuprocyanid, Cuprous cyanide.
French: Cyanure de cuivre, Cyanure cuivrique.
German: Cyankupfer, Cyanwasserstoffsaeureskupfer, Kupfercyanid, Kupferzyanid, Zyankupfer, Zyanwasserstoffsaeureskupfer.

Chemical

Ingredient of catalytic preparations used in making—
Acenaphthylene, acenaphthaquinone, bisacenaphthylidenedione, naphthaldehydic acid, naphthalic anhydride, and hemimellitic acid from acenaphthene (Brit. 295270).
Acetaldehyde from ethyl alcohol (Brit. 295270).
Acetic acid from ethyl alcohol (Brit. 295270).
Alcohols from aliphatic hydrocarbons (Brit. 281307).
Aldehydes and acids from toluene, orthonitrotoluene, orthobromotoluene, orthochlorotoluene, parabromotoluene, parachlorotoluene, paranitrotoluene, metanitrotoluene, metachlorotoluene, metabromotoluene, dichlorotoluenes, dibromotoluenes, dinitrotoluenes, chlorobromotoluene, chloronitrotoluene, bromonitrotoluene (Brit. 295270).
Aldehydes and acids from xylenes, pseudocumene, mesitylene, and paracymene (Brit. 295270).
Alphanaphthaquinone from naphthalene (Brit. 281307).
Anthraquinone from anthracene (Brit. 295270).
Benzaldehyde and benzoic acid from toluene (Brit. 281307).
Benzoquinone from phenanthraquinone (Brit. 281307).
Chloroacetic acid from ethylenechlorohydrin (Brit. 295270).
Diphenic acid from ethyl alcohol (Brit. 281307).
Fluorenone from fluorene (Brit. 295270).
Formaldehyde by methanol or methane (Brit. 295270).
Maleic acid and fumaric acid by the oxidation of benzene, toluene, phenol, tar phenols, or furfural, or from benzoquinone or phthalic anhydride (Brit. 295270).
Naphthaldehydic acid, acenaphthaquinone, or bisacenaphthylidenedione from acenaphthylene (Brit. 281307).
Phenanthraquinone from phenanthrene or diphenic acid (Brit. 295270).
Phthalic acid and maleic acid from naphthalene (Brit. 295270).
Primary alcohols from aldehydes by reduction (Brit. 306471).
Propionic acid and butyric acid and higher alcohols, ketones, and acids from carbon dioxide or carbon monoxide by reduction (Brit. 306471).
Salicylic acid and salicylic aldehyde by reduction of cresol (Brit. 306471).
Secondary buty alcohol by the reduction of methylethyl ketone (Brit. 306471).
Valeryl alcohol by the reduction of valeraldehyde (Brit. 306471).
Vanillin and vanillic acid from eugenol or isoeugenol (Brit. 295270).

Ingredient (Brit. 306471) of catalytic preparations used in the reduction of—
Acetaldehyde to ethyl alcohol.
Acetone to isopropyl alcohol.
Anthraquinone, benzoquinone, and the like to the corresponding hydroxyl compounds.
Benzaldehyde to benzoic acid.
Camphoric acid to alphacampholide.
Carbon dioxide or carbon monoxide to formaldehyde, methane, methanol, and other products.
Crotonaldehyde to butyl alcohol.
Ketones, aldehydes, acids, esters, alcohols, ethers, and other organic compounds containing oxygen.
Phthalic anhydride to benzyl alcohol, benzaldehyde or phthalide.

As a general cyanogenating agent.

Reagent (Brit. 261422) in making—
1-Amino-2-bromo-4-anthraquinonenitrile.
Anthraquinone-1-nitrile.
Anthraquinone-1:2-dinitrile.
Anthraquinone-1:3-dinitrile.
Anthraquinone-1:4-dinitrile.
Anthraquinone-1:5-dinitrile.
1:2:3:4-Anthraquinone tetranitrite.
1:4:5:8-Anthraquinone tetranitrite.
Cyanophenylthioglycollic acid.
1:3-Dibromo-2-aminoanthraquinonenitrile.
1:4:5-Tricyano-8-chloroanthraquinone.

Dye

Reagent to introduce the cyanogen radicle in making dyestuffs.

Metallurgical

Ingredient of—
Electrolytic bath for the deposition of copper in galvanoplastic work.

Pharmaceutical

In compounding and dispensing practice.

Copper Erucate

Synonyms: Cupric erucate.
French: Érucate de cuivre, Érucate cuivrique.
German: Erucinsäureskupfer, Erucinsäureskupferoxyd, Kupfererucat.

Fats and Oils

Ingredient of—
Lubricating greases.

Leather

Ingredient of—
Waterproofing compositions.

Mechanical

Ingredient of—
Compositions used for lubricating purposes.

Miscellaneous

Ingredient of—
Compositions used for the preservation of fishing gear, fishing nets, and twine, and also to prevent the formation of mildew.
Waterproofing compositions used for various purposes.

Oilcloth and Linoleum

Drier in—
Coating compositions.

Paint and Varnish

Drier in making—
Enamels, lacquers, paints, varnishes.

Paper

Ingredient of—
Compositions used in the waterproofing of paper, pulp, and products made from them.

Petroleum

Ingredient of—
Lubricating compositions.

Copper-Glucinium Alloys

(Alloys having the hardness of steel but characterized by the property of inability to produce sparks on subjection to blows or shocks.)

Automotive
As a nonferrous alloy having safety features.

Aviation
As a nonferrous alloy having safety features.

Metal Fabricating
Base material in making—
Carbide gas-burning devices.
Chemical manufacturing plant equipment.
Electrodes for soldering machines.
Explosive plant manufacturing equipment.
Friction surfaces.
Helicoid gears for high-speed sewing machines.
Small springs, soldering irons, surgical instruments, valve guides.

Copper Hypochlorite
French: Hypochlorure de cuivre.
German: Kupferhypochlorit, Unterchlorigesaeureskupfer.

Petroleum
Purifying agent in treating hydrocarbon oils (U. S. 1627055).

Copper Laurate
Synonyms: Cupric laurate.
French: Laurate de cuivre, Laurate cuivrique.
German: Kupferlaurat, Laurinsäureskupfer, Laurinsäureskupferoxyd.

Fats and Oils
Ingredient of—
Lubricating greases.

Leather
Ingredient of—
Waterproofing compositions.

Mechanical
Ingredient of—
Compositions used for lubricating purposes.

Miscellaneous
Ingredient of—
Compositions used for the preservation of fishing gear, fishing nets and twine, and also to prevent the formation of mildew.
Waterproofing compositions used for various purposes.

Oilcloth and Linoleum
Drier in—
Coating preparations.

Paint and Varnish
Drier in making—
Enamels, lacquers, paints, varnishes.

Paper
Ingredient of—
Compositions used in the waterproofing of paper, pulp, and products made from them.

Petroleum
Ingredient of—
Lubricating compounds.

Copper Nitrate
Synonyms: Cupric nitrate.
French: Azotate de cuivre, Nitrate de cuivre, Nitrate cuivrique.
German: Kupfernitrat, Salpetersaeureskupfer.

Analysis
Reagent in various processes.

Ceramics
Ingredient of enamels for—
Chinaware, porcelains, potteries.

Chemical
Starting point in making the following salts of copper—
Abietate, ammonium nitrate, arsenite, borate, carbonate, chlorate, chloride, chromate, cyanide, oleate, oxide (black), resinate, silicofluoride, stearate, sulphide.
Catalyst in making—
Methanol (Brit. 271538).
Reagent in making—
Benzaldehyde, paranitrobenzaldehyde.

Dye
Reagent in making—
Catechu brown.

Explosives and Matches
Ingredient of—
Pyrotechnic compositions.

Ink
Ingredient of—
Ink for writing on white iron.

Insecticide
Ingredient of—
Compositions with copper sulphate and calcium nitrate for use in viniculture.

Metallurgical
Reagent in—
Burnishing iron, coloring copper black.
Ingredient of—
Nickel electroplating bath.

Paint and Varnish
Ingredient of—
Enamels, paints, varnishes.
Reagent in making—
Copper pigments.

Paper
Reagent in making various products.

Photographic
Ingredient of—
Sensitive coatings on reproductive paper.

Textile
——, Dyeing
Mordant in—
Dyeing textiles with indigoes, general dyeing practice.
——, Printing
Mordant in general practice.
Reserve in—
Printing textiles with indigoes.

Copper Oleate
Synonyms: Cupric oleate.
French: Oléate de cuivre, Oléate cuivrique.
German: Kupferoleat, Oleinsäureskupfer, Oleinsäureskupferoxyd.

Fats and Oils
Reagent in promoting—
Intimate contact between the catalyst and the oil in the hydrogenation of vegetable oils.

Insecticide
Ingredient of—
Fungicidal sprays.

Miscellaneous
Ingredient of—
Compositions used for the preservation of fish nets and lines.

Pharmaceutical
In compounding and dispensing practice.

Paint and Varnish
Ingredient of paints used on ships' bottoms.

Copper Oxide, Black
Synonyms: Copper monoxide, Cupric oxide.
Latin: Cupri oxidum nigrum, Cuprum oxydatum.
French: Oxyde de cuivre, Oxyde noir de cuivre, Safran de venus.
German: Kupferoxyd.

Analysis
Reagent in—
Analytical work.

Ceramics
Pigment in—
Enamels, faience, glazes, porcelain, stoneware.

Chemical
Catalyst in—
Hydrolysis of chlorinated diphenyls (U. S. 1925367).
Reduction of organic compounds.
Catalyst in making—
Acetic acid from ethyl alcohol by oxidation (U. S. 1911315).
Fatty alcohols (having eight to twenty carbon atoms) by hydrogenation of fatty or naphthenic acids or their derivatives, for example esters, amides, chlorides; alcohols specified are octyl, nonyl, decyl, undecyl, lauryl, tridecyl, myristyl, pentadecyl, palmityl, margaryl, linoleyl, oleyl, hypogael, ricinoleyl, stearyl, and nonadecyl (Brit. 424283).
Phenols from halogenated hydrocarbons (U. S. 1961834).
Ingredient of catalytic mixtures used in making—
Acetic acid from carbon monoxide, methanol, and steam (Brit. 405282).
Alcohols by hydrogenation of esters of aliphatic carboxylic acids; examples describe the conversion into the corresponding alcohols of (1) ethyl acetate, (2) ethyl normal-butyrate, (3) normal-butyl acetate, (4) ethyl laurate, (5) ethyl phenylacetate, (6) normal-butyl normal-butyrate, (7) ethyl adipate (Brit. 385625).

Copper Oxide, Black (Continued)
Ethylene oxide, particularly for the preparation of ethyleneglycol and its derivatives (Brit. 402438).
Higher alcohols by hydrogenation of a mixture of methyl and ethyl alcohols; normal-propyl alcohol, isobutyl alcohol, normal-butyl alcohol, methylethylcarbin carbinol, hexyl, heptyl, octyl, and nonyl alcohols are formed (Brit. 381185).
Higher ketones form lower alcohols and lower ketones (Brit. 400384).
Reagent in—
Purification of hydrogen.
Starting point in making—
Copper catalyst used in making methylamines by hydrogenation of hydrocyanic acid (Brit. 398502 and 398504).
Copper salts.
Hard, granular, porous gels having catalytic or absorbent properties (Brit. 398517).

Electrical
As a rectifier (U. S. 1905724).
Ingredient of—
Dry batteries.
Magnetic core compound, in admixture with ferric trioxide and a binder (U. S. 1946964).
Rectifier (U. S. 1901563).

Explosives and Matches
Ingredient (U. S. 1903814) of—
Pyrotechnic starter containing also calcium silicide, lead peroxide, and fused silica.

Fats and Oils
Catalyst in making—
Alcohols by hydrogenation of fatty oils or wax esters (including sperm and similar oils) or the corresponding fatty acids or other esters thereof (Brit. 433549).
Fatty alcohols (having eight to twenty carbon atoms) by hydrogenation of fatty acids or their derivatives, for example, esters, amides, chlorides; alcohols specified are octyl, nonyl, decyl, undecyl, lauryl, tridecyl, myristyl, pentadecyl, palmityl, margaryl, linoleyl, oleyl, hypogael, ricinoleyl, stearyl, and nonadecyl (Brit. 424283).
Ingredient of—
Catalytic mixtures used in making lubricating oils by converting animal or vegetable fatty substances into unsaturated products, practically free from oxygen, and polymerizing or condensing these products in presence of condensing agents; the oils may be used to improve the viscosity curves of other lubricating oils (Brit. 394073).

Glass
Ingredient of—
Compounds for producing colored effects in glassware.
Pigment for marking quartz thermometers, in admixture with sand and glycerin.

Insecticide
Insecticide for—
Potato plant.

Metallurgical
Ingredient of—
Electrolytes in electroplating.
Flux for welding bronze, containing also boric acid, borax, and sodium silicate.
Flux (with soda ash) used in reverberatory refining of copper to remove arsenic and sulphur (U. S. 1921180).
Reagent in—
Coating aluminum with copper, dissolving chromic iron ores.

Miscellaneous
Exciter (Brit. 403233) in making—
Weatherproof, luminous substances from oxides of aluminum, calcium, beryllium, magnesium, and zinc.
Ingredient of—
Metal cleaner, containing also powdered zinc, sodium acid tartrate, and mineral oil.
Starting point in making—
Imitation precious stones.

Paint and Varnish
As a pigment.
Ingredient of—
Antifouling paints for ships' bottoms.

Petroleum
Catalytic agent in desulphurizing.
Hydrocarbon products (U. S. 1937113 and 1943583).
Petroleum.
Ingredient of—
Catalytic mixtures used in purifying mineral oils and obtaining refinery products by hydrogenation (Brit. 405736 and 406963).

Copper Oxide, Red
Synonyms: Copper hemioxide, Copper protoxide, Copper suboxide, Cuprous oxide.
French: Oxyde cuivreux, Oxyde rouge de cuivre.
German: Kupferoxydul, Kupferprotoxyd.

Analysis
Reagent in—
Analytical work.

Ceramics
Ingredient of—
Red glazes for porcelain, potteries, chinaware, and the like.

Chemical
Catalyst in making—
Aliphatic alcohols and ethers by absorbing olefins at elevated temperatures and pressure in aqueous solutions of acids which are weaker than sulphuric acid and hydrolyzing the product (Brit. 397187).
4:4′-Diaminodiphenyl ether (U. S. 1890256).
Organic compounds of the anthraquinone series (U. S. 1892302).
Organic amines (Brit. 402063).
Orthodihydroxybenzenes by the hydrolysis of orthodihalogenobenzenes and their alkyl, alkoxyl, hydroxyl, and nitro derivatives (Brit. 425230).
Ingredient (Brit. 400384) of—
Catalytic mixture use in making higher ketones from lower alcohols and lower ketones.
Starting point in making—
Copper salts.

Electrical
Ingredient (U. S. 1920151) of—
Anodes in grid-bias batteries.

Insecticide and Fungicide
Fungicide for—
Fungoid growths on plants and vegetables, such as hop cones and succulent leaves of tomato and rose.

Glass
Ingredient of—
Red glassware.

Metallurgical
Electrolyte in—
Electroplating.

Paint and Varnish
As a pigment.
Ingredient of—
Antifouling paints for ships' bottoms.

Tobacco
Fungicide for inhibiting—
Downy mildew.

Copper Palmitate
Synonyms: Cupric palmitate.
French: Palmitate de cuivre, Palmitate cuivrique.
German: Kupferpalmitat, Palmitinsäureskupfer, Palmitinsäureskupferoxyd.

Fats and Oils
Ingredient of—
Lubricating greases.

Leather
Ingredient of—
Waterproofing compositions.

Mechanical
Ingredient of—
Compositions used for lubricating purposes.

Miscellaneous
Ingredient of—
Compositions used for the preservation of fishing gear, fishing nets and twine, and also to prevent the formation of mildew.
Waterproofing compositions used for various purposes.

Oilcloth and Linoleum
Drier in—
Coating compositions.

Paint and Varnish
Drier in making—
Enamels, lacquers, paints, varnishes.

Paper
Ingredient of—
Compositions used in the waterproofing of paper, pulp, and products made from them.

Petroleum
Ingredient of—
Lubricating compounds.

Copper Palmitobenzenesulphonate
French: Palmitobenzènesulfonate de cuivre, Palmitobenzènesulfonate cuivrique.
German: Kupferpalmitobenzolsulfonat, Palmitobenzolsulfonsaeureskupfer.

Textile
——, *Printing*
Ingredient of—
Printing pastes, added for the purpose of enhancing the absorption of the color by the textile fiber and the levelness of the printed design on the fabric.

Copper Parachlorobenzenesulphonate
Synonyms: Copper parachlorobenzolsulphonate.
French: Parachlorobenzènesulphonate de cuivre.
German: Kupferparachlorbenzolsulfonat, Parachlorbenzolsulfonsaeureskupfer.

Chemical
Reagent (Brit. 265985) in separating from their solutions diazo compounds of—
Metachloronitranilin.
Metachlorotoluidine.
Metadichloronitranilin.
Metanitranilin.
Metanitrotoluidin.
Nitroaminophenol allyl ether.
Nitroaminophenol butyl ether.
Nitroaminophenol ethyl ether.
Nitroaminophenol heptyl ether.
Nitroaminophenol hexyl ether.
Nitroaminophenol methyl ether.
Nitroaminophenol propyl ether.
Nitroaminophenol valeryl ether.
Orthoaminophenol anthranyl ether.
Orthoaminophenol benzoyl ether.
Orthoaminophenol benzyl ether.
Orthoaminophenol naphthyl ether.
Orthoaminophenol phenyl ether.
Orthoaminophenol tolyl ether.
Orthoaminophenol xylyl ether.
Orthochloronitranilin.
Orthochlorotoluidin.
Orthodichloronitranilin.
Orthonitranilin.
Orthonitrotoluidin.
Parachloronitranilin.
Parachlorotoluidin.
Paradichloronitranilin.
Paranitranilin.
Paranitrotoluidin.

Copper Phosphide
Synonyms: Cuprous phosphide.

Electric
Getter (U. S. 1989790) for—
Incandescent lamps (in admixture with sodium-aluminum fluoride).

Metallurgical
Source of phosphorus in making—
Phosphor-bronze.

Copper Platinate
French: Platinate de cuivre, Platinate cuivrique.
German: Kupferplatinat, Platinsaeureskupfer.

Chemical
Reagent for various chemical purposes.
Ingredient of catalytic preparations used in making—
Acenaphthylene, acenaphthaquinone, bisacenaphthylidenedione, naphthaldehydic acid, naphthalic anhydride, and hemimellitic acid (Brit. 295270).
Acetaldehyde from ethyl alcohol (Brit. 281307).
Acetic acid from ethyl alcohol (Brit. 281307).
Alcohols from aliphatic hydrocarbons (Brit. 281307).
Aldehydes or alcohols by the reduction of esters (Brit. 306471).
Alphacampholid by the reduction of camphoric acid (Brit. 306471).
Aldehydes and acids from toluene, orthochlorotoluene, orthonitrotoluene, orthobromotoluene, metachlorotoluene, metabromotoluene, metanitrotoluene, parachlorotoluene, parabromotoluene, paranitrotoluene, dichlorotoluenes, dinitrotoluenes, dibromotoluenes, chlorobromotoluenes, chloronitrotoluenes, bromonitrotoluenes (Brit. 295270).
Aldehydes and acids from xylenes, pseudocumene, mesitylene, and paracymene (Brit. 295270).
Alphanaphthaquinone from naphthalene (Brit. 281307).
Anthraquinone from naphthalene (Brit. 295270).
Benzaldehyde and benzoic acid from toluene (Brit. 281307).
Benzoquinone from phenanthraquinone (Brit. 281307).
Benzyl alcohol by the reduction of benzaldehyde (Brit. 306471).
Benzyl alcohol or benzaldehyde or phthalide by the reduction of phthalic anhydride (Brit. 306471).
Butyl alcohol by the reduction of crotonaldehyde (Brit. 306471).
Chloroacetic acid from ethylenechlorohydrin (Brit. 295270).
Diphenic acid from ethyl alcohol (Brit. 281307).
Ethyl alcohol from acetaldehyde (Brit. 306471).
Fluorenone from fluorene (Brit. 295270).
Formaldehyde from methanol or methane (Brit. 306471).
Formaldehyde by the reduction of carbon monoxide or carbon dioxide (Brit. 306471).
Isopropyl alcohol by the reduction of acetone (Brit. 306471).
Maleic acid and fumaric acid by the oxidation of benzene, toluene, phenol, tar phenols, or furfural, or from benzoquinone or phthalic anhydride (Brit. 295270).
Methane by the reduction of carbon dioxide or carbon monoxide (Brit. 306471).
Methanol by the reduction of carbon dioxide or carbon monoxide (Brit. 306471).
Naphthaldehydic acid, acenaphthaquinone, or bisacenaphthylidenedione from acenaphthylene (Brit. 281307).
Phenanthraquinone from phenanthrene or diphenic acid (Brit. 295270).
Phthalic acid and maleic acid from naphthalene (Brit. 295270).
Primary alcohols by the reduction of aldehydes (Brit. 306471).
Propionic acid and butyric acid and higher alcohols, ketones, and acids from carbon dioxide or carbon monoxide by reduction (Brit. 306471).
Reduction of anthraquinone, benzoquinone, and the like to the corresponding hydroxyl compounds (Brit. 306471).
Reduction of carbon dioxide and carbon monoxide (Brit. 306471).
Reduction of ketones, aldehydes, acids, esters, alcohols, ethers, and other organic compounds containing oxygen (Brit. 306471).
Salicylic acid and salicylic aldehyde from cresol (Brit. 295270).
Secondary butyl alcohol by the reduction of methylethyl ketone (Brit. 306471).
Valeryl alcohol by the reduction of valeraldehyde (Brit. 306471).
Vanillin and vanillic acid from eugenol or isoeugenol (Brit. 295270).

Copper Propionylacetonate

Chemical
Reagent in—
Organic syntheses.

Fuel
Primer (Brit. 404682) in—
Diesel engine fuels (used in conjunction with alkyl nitrates, having two to four atoms in the molecule, whose function is that of reducing the delay period).
Reducer (Brit. 404682) of—
Spontaneous ignition temperature of Diesel engine fuels.

Copper-Pyridin Chloride
French: Chlorure de cuivre et de pyridine, Chlorure cuivrique-pyridinique.
German: Chlorkupferpyridin, Chlorwassersaeureskupferpyridin, Kupferpyridinchlorid.

Chemical
As a general reagent.

Dye
Reagent (Brit. 306859) in making azo dyestuffs with—
Acetyl H acid.
Alphahydroxynaphthalene-4-sulphonic acid.
Alphaethoxy-8-hydroxynaphthalene-3:6-disulphonic acid.
3-Aminobenzaldehyde.
2-(4′-Aminobenzoyl)amino-5-naphthol-7-sulphonic acid.
2-(3′-Aminobenzoyl)amino-5-naphthol-7-sulphonic acid.
Anthranilic acid.
Benzidin-3:3′-dicarboxylic acid.
Beta-aminobenzaldehyde.
Beta-aminobenzene-5-sulphonic acid.
Beta-aminobenzoic acid.
Beta-amino-1-hydroxybenzene.
Beta-aminonaphthalene-3-carboxylic acid.
Betanaphthol.

Copper-Pyridin Chloride (Continued)
Betaphenylamino-4-hydroxynaphthalene-77-sulphonic acid.
4-Chloro-2-chloro-2-aminobenzoic acid.
4:4′-Diaminodiphenylurea-3:3′-dicarboxylic acid.
4:6-Dichloro-2-amino-1-hydroxybenzene.
5:5′-Dihydroxy-2:2′-dinaphthylamine-7:7′-disulphonic acid.
J acid.
5-Nitro-2-aminobenzoic acid.

Copper Resinate
Synonyms: Copper soap, Resinate of copper.
French: Résinate de cuivre.
German: Kupferresinat.

Ceramics
Pigment in producing brownish-red shade on—
Chinaware, porcelains, potteries.

Insecticide
Ingredient of—
Gasoline solutions used for fungicidal, germicidal and insecticidal purposes.

Paint and Varnish
Drier in making—
Enamels, lacquers, paints, varnishes.
Ingredient of—
Ships' bottoms paints, submarine paints.

Copper Selenite
French: Sélénite de cuivre.
German: Kupferselenit.

Metallurgical
Reagent in burnishing—
Iron.

Copper Stearate
Synonyms: Cupric stearate.
French: Stéarate de cuivre, Stéarate cuivrique.
German: Stearinsäureskupfer, Stearinsäureskupferoxyd.

Fats and Oils
Reagent in promoting—
Intimate contact between the catalyst and the oil in the hydrogenation of vegetable oils.

Insecticide
Ingredient of—
Insecticidal preparations, spraying compounds for fungicidal purposes.

Miscellaneous
Ingredient of—
Compositions used for bronzing statues.

Paint and Varnish
As a drier.
Ingredient of—
Paints and varnishes used for painting ships' bottoms.

Pharmaceutical
In compounding and dispensing practice.

Sanitation
Ingredient of—
Disinfectants, germicides.

Copper Stearotoluenesulphonate
French: Stéarotoluènesulphonate de cuivre, Stéarotoluènesulphonate cuivrique.
German: Kupferstearotoluolsulfonat, Stearotoluolsulfonsäureskupfer.

Paper
Ingredient (Brit. 269917) of—
Pastes used in printing wallpaper (added to produce level shades and effects).

Textile
Ingredient (Brit. 269917) of—
Printing pastes (added to enhance the saturating of the fabric and to equalize effects).

Copper Sulfate
Synonyms: Blue stone, Blue vitriol, Cupric sulphate, Roman vitriol.
Latin: Cupri sulphas, Cuprum sulfuricum, Cuprum vitriolatum.
French: Coupérose bleu, Sulphate de cuivre, Vitriol bleu.
German: Blauer galitzenstein, Blauvitriol, Kupfersulfat, Kupfervitriol, Schwefelsäureskupfer.
Spanish: Sulfato cuprico, Vitrolo azul.
Italian: Solfato di rame, Vitriolo di rame.

Analysis
As a reagent.

Chemical
As a dehydrating agent (in the anhydrous form).
Catalyst (Brit. 398527) in making—
Esters from lower aliphatic acids and olefines.
Reagent in—
Saccharification of carbohydrates (Brit. 400168).
Starting point in making—
Copper-ammonium sulphate, copper arsenite (Scheele's green), copper carbonate, copper cyanide, copper hydroxide, copper oleate, copper resinate, copper stearate.

Dye
Reagent (Brit. 388332) in making—
Azo dyes.

Electrical
Ingredient of—
Battery electrolytes.

Glues and Adhesives
Improver for—
Casein glues.

Insecticide
Ingredient (U. S. 1903626) of—
Fungicide and insecticide containing also clay, diatomaceous earth, and an alkali.
Starting point in making—
Bordeaux mixture from caustic soda.
Bordeaux mixture from slaked lime.
Insecticide or fungicide consisting of a voluminous, light-green, insoluble copper compound containing 50 to 55 percent of metallic copper (U. S. 1937524).
Low density paris green from sodium arsenite and acetic acid (U. S. 1928771).
Paris green.

Leather
Reagent in—
Tanning processes.

Metallurgical
Electrolyte in—
Copper-coating dust of a magnetic metal or alloy in making magnetic cores (U. S. 1919806).
Electrolytic refining of brass (U. S. 1920819).
Ingredient of—
Acid electrolytes in copperplating.
Pickling agent (Brit. 399685) for—
Removing irregularities or projections in copper wire prior to enameling for use in the electrical industry.
Precipitation promoter (U. S. 1920442) in—
Freeing zinc sulphate solutions (from leaching roasted zinc ores) from cobalt, nickel, cadmium, and germanium.

Miscellaneous
Electrolyte in making—
Master records for phonographs.
Emulsification agent (Brit. 380065 and 380052) in making—
Stable emulsions of fats, oils, paraffin, neatsfoot oil, benzene, trichloroethylene.
Ingredient (U. S. 1881128) of—
Motion picture projection screen coating (containing also glue, sodium fluoride, glycerin, casein, borax, cobalt blue, and water) said to have properties of non-stickiness, permanence, and adaptability to climatic conditions.

Paint and Varnish
As a pigment.
Reagent in—
Removing chlorine from zinc sulphate solutions used in making lithopone (U. S. 1901925).
Starting point in making—
Scheele's green.

Paper
Preservative for—
Ground pulp, pulp wood.

Perfume
Reagent in making—
Hair dyes.

Petroleum
Catalyst (Brit. 367848) in—
Purifying hydrocarbon oils with ozonized air.

Pharmaceutical
In compounding and dispensing practice.

Printing
Electrolyte in—
Electrotyping.
Reagent in—
Process engraving, photoengraving.

Copper Sulfate (Continued)

Textile
As a mordant.
Water and Sanitation
Reagent for—
Destroying algae and low forms of animal life in ponds.
Woodworking
As a preservative.

Copper Tungstomolybdate
Synonyms: Copper molybdotungstate.
French: Tungstomolybdate de cuivre.
German: Kupfermolybdenumwolframat.
Metallurgical
Ingredient of (French 600774) antioxidation coating electrically deposited) for—
Bismuth, copper, nickel, steel, tin, zinc.

Cork, Ground
French: Farine de liége, Liége broyé, Liége poudrée.
German: Korkmehl.
Chemical
Source of—
Suberin.
Starting point in making—
Cork black.
Suberic acid by reaction with nitric acid.
Construction
Ingredient of—
Artificial stone floorings.
Corkstone, used as a fireproof material insulating sound and heat.
Fireproof constructional materials.
Heat-insulating compositions and materials.
Resilient composition floorings.
Resilient floor tile containing also mineral fillers, pigments, nitrocellulose, and ester gum (U. S. 1876289).
Sound-deadening compositions and materials.
Ceramics
Suggested filler for—
Ceramic products of various kinds.
Food
Packing and conserving agent for—
Eggs, fruits, vegetables.
Linoleum and Oilcloth
Filler in—
Linoleum.
Miscellaneous
As a filler in many products where any of the following properties may be desirable:—Elasticity, ductility, high resistance to heat, cold, sound, and penetration by gases and liquids at normal and elevated pressures.
Filler for—
Life preservers, lifeboat floating media.
Sound-deadening filler for—
Telephone booths (used alone or in various mixtures).
Paint and Varnish
Cold-insulating filler in—
Paints.
Corrosion-resisting filler in—
Paints.
Heat-insulating filler in—
Paints.
Ingredient of—
Paints applied to the under side of automobile engine hoods to protect the outside lacquer films from the radiating heat of the motor.
Paints applied to the surfaces of airplane cabins to provide a certain amount of insulation against motor noise.
Moisture-resisting filler in—
Paints.
Rust-resisting filler in—
Paints.
Sound-deadening filler in—
Paints.
Paper
Ingredient of—
Mixture with paper pulp, known as corkboard and used for sound and heat insulation.
Plastics
Suggested filler in—
Plastic compositions used for insulating purposes.
Refrigeration
Insulating medium for—
Domestic refrigerators, industrial refrigerators.

Corn Oil
Synonyms: Maize oil.
French: Huile de mais.
German: Kornoel, Kukuruzoel, Maisoel.
Chemical
Starting point in making—
Fatty acids, glycerin.
Food
Baker's oil for greasing pans.
Frying oil.
Ingredient of—
Lard compound, oleomargarin, salad oils.
Raw material in making—
Cakes, biscuits, and other baked products.
Salad oil.
Substitute for lard in hydrogenated or solid form.
Electrical
Ingredient of—
Insulating compositions.
Fats and Oils
Ingredient of—
Compositions containing animal oils, used as a filler.
Lubricating compositions.
Starting point in making—
Vulcanized oil, water-soluble oils.
Substitute for codliver oil and cottonseed oil.
Leather
Ingredient of—
Dressing compositions.
Reagent in—
Finishing, tanning.
Linoleum and Oilcloth
Ingredient of—
Compositions used in making coatings.
Fuel
Illuminant.
Ingredient of—
Illuminating compositions.
Mechanical
As a lubricant.
Paint and Varnish
Grinding oil for—
Pigments, used along with linseed oil.
Ingredient of—
Paints, varnishes.
Pharmaceutical
In compounding and dispensing practice.
Rubber
Ingredient (in vulcanized condition) of—
Compositions used in making rubber bands, rubber boots, sole rubber, surgical instruments, solid rubber truck tires, bicycle and carriage tires, buffers, artificial sponges.
Reagent in making—
Imitation rubber.
Soap
Raw material in making—
Soap powders, soft soaps, textile soaps.
Textile
——, *Finishing*
Ingredient of—
Cotton softening compositions.
Rainproofing compositions.

Corn Oil Fatty Acid
Synonyms: Maize oil fatty acid.
French: Acide gras d'huile de mais.
German: Maisoelfettsaeure.
Chemical
Starting point in making various salts and esters.
Food
Ingredient of—
Prepared foods, halogenated oil products.
Fuel
Component of—
Candles.
Miscellaneous
Ingredient of—
Cleansing compositions with alkaline hypochlorites (Brit. 280193), polishing compositions.
Paint and Varnish
Starting point in making—
Driers.
Pharmaceutical
In compounding and dispensing practice.

Corn Oil Fatty Acid (Continued)

Soap
Raw material in soapmaking.
Textile
——, *Bleaching*
Ingredient of—
Bleaching compositions containing alkaline hypochlorites (Brit. 280193).
——, *Finishing*
Ingredient of—
Finishing compositions, washing compositions containing alkaline hypochlorites (Brit. 280193), waterproofing compositions.

Cornstarch

Latin: Amylum zeae.
French: Fécule de mais.
German: Maisstarke.

Agriculture
Ingredient of—
Cattle foods.
Analysis
Reagent in testing for—
Chlorine, copper, iodine, nitrous acid.
Brewing
Starting point in making—
Beer, fermented liquors.
Chemical
Ingredient of—
Colloidal preparations (added for the purpose of preventing precipitation).
Starting point in making—
Acetone by bacterial fermentation.
Acetylmethylcarbinol by fermentation (U. S. 1899094).
Alcoylated products (French 640174).
Dextrin and dextrin products, fusel oil by fermentation, lactic acid, levulinic acid, starch glycollate, starch iodide, solubilized starch.
Tanning agent by sulphonation with sulphuric acid (French 544253).
Dye
Ingredient (U. S. 1889491) of—
Household dye compositions for silk.
Distilling
Starting point in making various types of distilled liquors.
Electrical
Carrier and filler (Brit. 398638) for—
Exciting salts used in the manufacture of electrolytes used for rechargeable dry cells.
Explosives
Ingredient of—
Gelatin dynamites, permissibles for coal mining, regular nitroglycerin dynamites.
Starting point in making—
Nitro-starch explosives, nitro-starch dynamites.
Food
As a foodstuff.
Ingredient of—
Baking powders, candies, cocoa powders, cake powders, custard preparations, chocolate preparations, ice cream preparations and powders.
Sauces of various sorts (to make them thick).
Various culinary and food preparations.
Vegetarian foods.
Raw material in—
Biscuit, pastry, baking, and confectionery industries.
Fuel
Binder in making—
Fuel briquets.
Reagent (German 389401) in combination with muriatic acid for treating—
Non-floatable constituents of coal.
Glues and Adhesives
Ingredient of—
Cold-water glues, various adhesive paste preparations, wallpaper pastes, xanthate adhesive preparations.
Starting point (French 648019) in making—
Glues in bead form.
Insecticide
Ingredient (U. S. 1891750) of—
Seed-treating insecticide.
Leather
Ingredient of—
Cleansing compositions.
Compositions used in the manufacture of artificial leather (French 558630).
Compositions containing lime, calcium phenolate, and sodium hydroxide, used for softening and dehairing hides and skins (French 612409).
Vehicle for—
Holding tanning extract in the drum-tanning process.
Mechanical
Ingredient (U. S. 1720565) of—
Compositions used for the purpose of preventing incrustation of scale in boilers.
Miscellaneous
Ingredient of—
Compositions used in laundries for the dressing and sizing of fabrics after washing.
Compositions used for coating purposes, prepared by the action of calcium chloride, calcium nitrate, zinc chloride, and magnesium chloride on the starch (French 557085).
Compositions in emulsified form (French 599908).
Compositions used for stiffening fabrics.
Compositions containing coloring matter, such as azo dyestuffs.
Compositions, colored black and containing naphthalene and its derivatives (French 641442).
Compositions containing pitch, rosin soap (such as potassium resinate), oil, flour, used for road surfacing purposes.
Dental impression material (U. S. 1897034).
Starch glazes.
Starting point in making—
Starch tablets.
Paint and Varnish
Fixative (French 616204) in making—
Whitewashes and starch coating compositions with the addition of sodium carbonate and nitrobenzene.
Paper
Ingredient of—
Compositions used for sizing different qualities of paper, particularly writing paper.
Compositions used in the manufacture of surface-coated paper.
Compositions used in the manufacture of pasteboard.
Perfume
Ingredient of—
Massaging compositions (French 616204).
Perfumes, pomades, sachets, toilet powders.
Pharmaceutical
Binder in tablet mixtures, diluent, dusting powder.
In compounding and dispensing practice.
Printing
In bookbinding practice.
Rubber
Ingredient (Brit. 397279) of—
Compositions for coating surface of rubber articles to produce a smooth matt finish.
Soap
Ingredient of—
Compositions containing carbon tetrachloride, glycerin, and the like, used for the dry cleaning of hands which have become stained with crankcase oil, tar, grease, paint (French 611895).
Detergent preparations containing potassium silicate.
Soapstock in making special grade of soap.
Soft soaps (used as a filler).
Sugar
Starting point in making—
Burnt sugar or carmel, malt sugar, various syrups and mixtures, white glucose.
Textile
——, *Dyeing*
Ingredient of—
Dye bath for various yarns and fabrics.
——, *Finishing*
Ingredient of—
Compositions used for sizing cotton fabrics.
Compositions used for starching knitted merchandise, such compositions also containing glucose, sodium silicate, glycerin, olive oil, and borax (French 649899).
Fireproofing compositions, containing ammonium sulphate, sodium carbonate, boric acid, sodium biborate, used for treating rayons (French 595286).
Sizing compositions containing sodium resinate (French 523282).
Weighting compositions for treating calicoes, lace curtains, and other textiles.

Cornstarch (Continued)
——, *Manufacturing*
Ingredient of—
Spinning bath in making viscose rayon.
Size for—
Cotton yarns before weaving.
——, *Printing*
Ingredient of—
Printing pastes (added to thicken them).

Corundum
German: Diamonospat, Korund.
Abrasives
Abrasive for general purposes.
Component of—
Emery cloth, emery paper.
Ingredient of—
Abrasive compositions, abrasive stones, abrasive wheels.
Miscellaneous
Raw material in making chemical apparatus of various sorts.
Refractory
Ingredient of refractory compositions.
Raw material in making—
Refractory apparatus, refractory furnaces, refractory parts.

Cottonseed Oil
Synonyms: Cotton oil.
Latin: Oleum gossypii, Oleum gossypii seminis.
French: Huile de coton, Huile de semences de cotonier.
German: Baumoel, Baumwollsamenoel.
Spanish: Aceite de semilla de algodon.
Animal Husbandry
Ingredient of—
Cattle feeds.
Abrasives
Starting point in making—
Hydrogenated products used in making buffing and grinding compositions.
Building Construction
Ingredient of—
Coating and waterproofing compositions for concrete.
Chemical
Ingredient of—
Turkey red oils.
Process material in—
Recovering cresols.
Starting point in making—
Fatty acids, glycerin.
Cosmetic
Base for—
Cosmetic compositions.
Fats and Oils
Starting point in making—
Blown cottonseed oil, hydrogenated oil, stearin.
Food
Cooking oil.
Ingredient of—
Bread doughs (various patents, cooking oils, egg mixtures, food products of various kinds, lard compounds, olive oils, salad dressings, salad oils).
Liquid packing medium in—
Canning sardines and other fish.
Minimizer of—
Evaporation losses on fish in cold storage.
Preservative for—
Eggs, fish.
Process material in making—
Egg substitutes.
Salad oil.
Starting point in making—
Butter substitutes.
Hydrogenated products used for various purposes in the food industry.
Lard substitutes, oleomargarin, shortenings.
Gum
Filler for—
Art gums, chicle gums, gum substitutes.
Ink
Ingredient of—
Printing inks.
Leather
Ingredient of—
Dressings and finishing compositions.
Lubricant
Ingredient of—
Lubricating compositions.
Process material in making—
Cutting oils.
Mechanical
Lubricant.
Metal Fabricating
Coating and rustproofing agent for—
Iron.
Miscellaneous
Ingredient of—
Belt dressings.
Phonograph record compositions.
Waterproofing compositions for various purposes.
Solvent for—
Amber.
Starting point in making—
Hydrogenated products used for various purposes in industry.
Pharmaceutical
In compounding and dispensing practice.
Rubber
Filler for—
Gutta-percha.
Soap
Soapstock in making—
Laundry soaps, scouring powders, toilet soaps, washing powders, wool-washing soaps.
Textile
Ingredient of—
Dressing compositions.

Cottonseed Pitch
French: Poix des semences de coton.
German: Baumwollesaatpech, Baumwollesamenpech.
Chemical
Ingredient (Brit. 263520) of—
Emulsions for various chemical purposes.
Ink
Ingredient of—
Printing inks.
Insecticide
Ingredient of—
Insecticidal and germicidal emulsions.
Paint and Varnish
Ingredient of—
Paints, varnishes.

Cotton Spirits
(A name given to various acetate solutions of tin, analogous to, but distinct from tin spirits; cotton spirits are stannic compounds; tin spirits are principally stannous compounds).
Textile
Mordant in—
Dyeing processes.

Crackling Grease
Lubricant
Raw material in making—
Cup and other greases.

4-Cresidin
Synonyms: 4-Cresidine.
French: 4-Crésidine.
Chemical
Starting point in making various derivatives.
Starting point (Brit. 353537) in making acridin derivatives with—
2-Chloro-4-bromobenzoic acid.
2-Chloro-4-iodobenzoic acid.
2:4-Dichlorobenzoic acid.
Dye
Starting point (Brit. 398163) in making—
Claret shades fast to kier-boiling and chlorine.

2-Cresol-3:5-disulphobis-4′-chloroanilide
French: 2-Crésole-3:5-disulphobis-4′-chloroanilide.
German: 2-Cresol-3:5-disulfobis-4′-chloranilid.
Chemical
Starting point in making various derivatives.
Miscellaneous
Dyestuff and mothproofing agent in treating—
Feathers, hair, fur.
Textile
Dyestuff and mothproofing agent in treating—
Felt, wool.

Cresol-Mercury Chloride

Agriculture
For control of—
Bottom rust of lettuce.
Covered smut and stripe disease of barley.
Kernel smut of sorghum.
Loose and covered smut of oats.
Soil-borne parasitic fungi.
Stinking smut of wheat.

Woodworking
For control of—
Blue stain and sap stain in sapwood of freshly sawed lumber.

Cresolsulphuric Acid
French: Acide de crésole et sulfurique.
German: Kresolschwefelsaeure.

Insecticide
Solvent (Brit. 265131) in making—
Sulphur dioxide compositions.

Miscellaneous
Solvent (Brit. 265131) in making—
Sulphur dioxide antiseptics and disinfectants.

Cresotinic Acid Sulphochloride
French: Sulfochloride de crésotinique acide.
German: Kresotinsaeuressulfochlorid.

Chemical
Starting point in making—
Water-soluble tanning agents with chlorosulphonic acid (Brit. 266697).

Cresylphenyl-Aluminum

Petroleum
Addition agent (Brit. 433257) in—
Lubricating oils or greases, especially for use at high temperatures, such as cylinder oils, hydrogenated oils, or oils refined by treatment with sulphuric acid, clay, or extraction solvents.

Cresylphenyl-Bismuthine

Petroleum
Addition agent (Brit. 433257) in—
Lubricating oils or greases, especially for use at high temperatures, such as cylinder oils, hydrogenated oils, or oils refined by treatment with sulphuric acid, clay, or extraction solvents.

Cresylphenyl-Cadmium

Petroleum
Addition agent (Brit. 433257) in—
Lubricating oils or greases, especially for use at high temperatures, such as cylinder oils, hydrogenated oils, or oils refined by treatment with sulphuric acid, clay, or extraction solvents.

Cresylphenyl-Mercury

Petroleum
Addition agent (Brit. 433257) in—
Lubricating oils or greases, especially for use at high temperatures, such as cylinder oils, hydrogenated oils, or oils refined by treatment with sulphuric acid, clay, or extraction solvents.

Cresylphenyl-Stibine

Petroleum
Addition agent (Brit. 433257) in—
Lubricating oils or greases, especially for use at high temperatures, such as cylinder oils, hydrogenated oils, or oils refined by treatment with sulphuric acid, clay, or extraction solvents.

Cresylphenyl-Thallium

Petroleum
Addition agent (Brit. 433257) in—
Lubricating oils or greases, especially for use at high temperatures, such as cylinder oils, hydrogenated oils, or oils refined by treatment with sulphuric acid, clay, or extraction solvents.

Cresylphenyl-Zinc

Petroleum
Addition agent (Brit. 433257) in—
Lubricating oils or greases, especially for use at high temperatures, such as cylinder oils, hydrogenated oils, or oils refined by treatment with sulphuric acid, clay, or extraction solvents.

Cresylphenyl-Zinc Sulphide

Petroleum
Addition agent (Brit. 433257) in—
Lubricating oils or greases, especially for use at high temperatures, such as cylinder oils, hydrogenated oils, or oils refined by treatment with sulphuric acid, clay, or extraction solvents.

Cresyl Phosphate
French: Phosphate de crésyle, Phosphate crésylique.
German: Kresylphosphat, Phosphorsäurecresylester, Phosphorsäurescresyl.

Miscellaneous
Mothproofing agent (U. S. 1748675) in treating—
Feathers, furs, hair.

Textile
Mothproofing agent (U. S. 1748675) in treating—
Wool and felt.

Cresylthioglycolic Acid
French: Acide de crésylethioglycolique.
German: Cresylthioglykolsaeure.

Dye
Starting point (Brit. 284288) in making thioindigoid dyestuffs with—
Acenaphthenequinone, alphaisatinanilide, 57-dibromoisatin, isatin.
Isatin homologs, substitution products, alpha derivatives.
Orthodiketones.

Crotonic Acid
French: Acide crotonique, Acide de crotonyle.
German: Crotonsaeure.

Chemical
Starting point in making—
Oxybutyric acid (Brit. 441003).

Crotonic Aldehyde
Synonyms: Crotonaldehyde.
French: Aldéhyde crotonique.
German: Krotonaldehyd.

Chemical
Starting point in making—
Butyl alcohol by catalysis, intermediates, organic compounds, pharmaceuticals, quinaldin.
Starting point (Brit. 325669) in making synthetic perfumes with the aid of—
1:3-Cyclohexadiene, 1:1-dimethylbutadiene, 1:3-dimethylbutadiene, 1:4-dimethylbutadiene, 2:3-dimethylbutadiene, 2:4-dimethylbutadiene, 1-methylbutadiene.
Starting point (Brit. 249113) in making rubber vulcanization accelerators with—
Anilin, diethylamine, ethylamine, ethylanilin, ethylenediamine, guanidin, methylamine, methylanilin, normal butylamine, orthotoluidin, orthotolyldiguanidin.

Fats and Oils
Solvent for—
Fats, vegetable oils.

Paint and Varnish
Solvent for—
Shellac.
Solvent in making—
Varnishes.

Petroleum
Solvent for—
Oils and distillates.

Resins and Waxes
Solvent for—
Rosin, uncured resins, waxes, wood-distillation resins.
Starting point (Brit. 270433) in making artificial resins with—
Alphanaphthylamine, anilin, benzidin, benzylamine, betanaphthylamine, dianisidin, dibenzylamine, dimethylanilin, diphenylamine, metaphenylenediamine, metatoluidin, methylethylanilin, monoethylanilin, monomethylanilin, naphthylenediamine, orthophenylenediamine, orthotoluidin, paraphenylenediamine, paratoluidin, phenylamine, phenyldimethylamine, phenylmethylamine, toluylenediamine, xylidin, xylylenediamine.

Rubber
As a solvent.

Crotonyl Peroxide
French: Peroxyde de crotonyle, Peroxyde crotonylique.
German: Crotonylperoxyd.

Chemical
Reagent and starting point in making various organic compounds.

Fats and Oils
Bleaching agent (Brit. 328544) in treating—
Vegetable and animal oils (used together with hydrogen peroxide).

Food
Bleaching agent (Brit. 328544) used together with hydrogen peroxide in treating—
Egg yolk, flour, meal.

Soap
Bleaching agent (Brit. 328544) in treating—
Soapmakers' raw materials (used together with hydrogen peroxide).

Waxes and Resins
Bleaching agent (Brit. 328544) in treating—
Waxes (used together with hydrogen peroxide).

Crotylsorbitol
Synonyms: Crotylsorbite.

Miscellaneous
Plasticizer (U. S. 1936093) for—
Cellulose acetate, cellulose esters and ethers, cellulose nitrate, natural resins, synthetic resins.
For uses, see under general heading: "Plasticizers."

Crystal Violet

Chemical
Starting point (Brit. 295605) in making bacteriological preparations, bactericides, therapeutic compounds, and biological stains, with the aid of—
Cresol, guaiacol, hydroquinone, phenol, phloroglucinol, pyrocatechol, pyrogallol, resorcinol.

Miscellaneous
Dyestuff for—
Various substances.

Textile
For dyeing and printing yarns and fabrics.

Cumenedisulphonic Acid
French: Acide de cumènedisulphonique.
German: Cumendisulfonsaeure.

Chemical
Starting point in making—
Esters and salts, intermediates, pharmaceuticals.
Ingredient (Brit. 262873) of—
Aromatic hydrocarbon emulsions.
Fat solvents in emulsified form.
Terpene emulsions.

Fats and Oils
Ingredient (Brit. 263873) of—
Emulsified preparations.

Leather
Ingredient (Brit. 263873) of—
Impregnating compositions.
Tanning preparations in emulsified form.

Miscellaneous
Ingredient (Brit. 263873) of—
Washing and cleansing compositions in emulsified form.

Paper
Ingredient (Brit. 263873) of—
Emulsified preparations for treating paper and cardboard.

Petroleum
Reagent (Brit. 263873) in making—
Emulsions containing petroleum and petroleum distillates.

Resins and Waxes
Reagent (Brit. 263873) in making—
Emulsified resin preparations.

Textile
——, *Dyeing*
Ingredient (Brit. 263873) of—
Acid dye baths.
Emulsified finishing and wetting compositions.

——, *Finishing, Manufacturing*
Ingredient (Brit. 263873) of—
Wool carbonizing liquors.

Cup Grease
(The uses given under the several names of industries may be considered as unique to the particular industry. Uses that may be considered as common to many industries, such as materials handling, power generation, power transmission, will be found under those operation headings instead of repeated under many industries.)

Agriculture
Lubricant for—
Cotton gin parts, disc harrows, farm implements, grain harvesters, plows, tractor parts, wagon axles, wheel bearings.

Automobile
Lubricant for—
Chassis parts, distributor, fan bearings, rear end gears, transmission gears, speedometer shafts, steering gear parts, suspension springs, water pump, wheel bearings.

Aviation
Lubricant for—
Various parts.

Beverage
Lubricant for—
Bottling machine bearings.

Brick and Refractories
Lubricant for—
Bearings, brick cutters, cars, skip hoists, tile cutters.

Cement
Lubricant for—
Ball mills, crushers, dryers, granulators, rotary kilns.

Chemical
Lubricant for—
Air compressors, motor bearings.

Construction and Building
Lubricant for—
Air compressors, cable car pulleys and bearings, concrete mixers, cranes, elevator bearings and slides and guides, gas engine parts, hoisting machinery, motor bearings.
Pneumatic tools, such as drills, concrete breakers, wood and stone-working tools, riveting hammers.
Pumps, tractors.

Electrical
Lubricant for—
Motors.

Food
Lubricant for—
Dough dividers in baking plants.
Mill bearings.

Laundry
Lubricant for—
Overhead trolley systems, roller ironing machines, rotary driers, plant tracking systems, washing machines.

Lumbering
Lubricant for—
Chassis bearings on trucks, donkey engines, lime blocks, slides.

Materials Handling
Lubricant for—
Belt conveyor bearings and other parts, buggies, cable car pulleys and reel bearings.
Cars of various kinds used for transporting materials in factories, mills, and quarries.
Car loaders, conveyor moving parts, coal and ash handling equipment, cranes, elevating machinery, grain and ore handling equipment, freight elevators, hoisting machinery, loading machinery, overhead trolley systems, plant tracking systems, stacking machines, tractor, truck, and trolley parts.

Materials Treating
Lubricant for—
Crushers, cutters, disintegrators, dryers, grinders, kilns, millers, mixers, pulverizers, screeners, shredders, sievers, sifters.

Mechanical
Lubricant for—
Air compressors, ball bearings.
Elevator bearings, passenger and freight.
Fittings, gear trains, lathes, moving parts generally.
Pistons, valves, and other moving parts.
Pneumatic tools, such as drills, riveting hammers, chipping and caulking tools, breakers, wood and stone working tools, stoppers, augers, hammers, and other tools.
Roll machines, roller bearings, speed reducers.

Cup Grease (Continued)

Metallurgical
Lubricant for—
Blast furnace trolley bearings and line shafting, charging machines, cold rolls, converter manipulating parts, conveyors, ladle cranes, open-hearth furnace door plungers, pot trucks, wire-drawing operations.

Milling
Lubricant for—
Bagging machines, bran dusters, collar and other bearings, flour cleaning screens, flour dressers, grain elevators, line shafting, purifiers, reels, scourers, sifters, wheat rolls.

Mining
Lubricant for—
Air compressors, air hoists, cable car pulleys and bearings, conveyors.
Elevator bearings, skids, slides.
Mine cars, motors, pneumatic machinery, pumps, trolley systems.

News Publishing
Lubricant for—
Paper hoists, printing machinery, trucks.

Paper
——, *Logging*
Lubricant for—
Chassis bearings on trucks, donkey engines, crank pins, slides.

——, *Paper Box Plants*
Lubricant for—
Auto box machine, cornering machines, die presses, paper slitters, scouring machines.

——, *Pulp and Paper Mills*
Lubricant for—
Bag and box machinery, chip screens, creping and corrugating machines.
Envelope, tube, cup, and cap machinery.
Folders, foudrinier machine parts, layboys, paper coating and saturating machines, pulp screens, rewinders, roll wrapping machines, rotary cutters, splitters, trimmers, tumbling drums.

Petroleum
Lubricant for—
Drilling machinery, pumps.

Power Generation
Lubricant for—
Ash grate roll bearings, blowers, coal conveyor rolls, coal hoisting machinery, draft fan bearings, fans, pump bearings, scraper chain sheaves, steam engine crank pins, mechanical stokers.

Power Transmission
Lubricant for—
Bearings, chains, fly wheels, gear sets, hawlers, line shafts, pulleys, speed reducers, speed transmissions, sprockets, wheels.

Railroading
Lubricant for—
Air brakes, general shop purposes, moving parts, signalling systems, trackage.

Shipping
Lubricant for—
Davits, freight handling and miscellaneous deck equipment, power generation and transmission equipment, winches.

Shipyards
Lubricant for—
Cranes, general shop use, launching skidways, pneumatic tools.

Shoe Factories
Lubricant for—
Brushing machines, buffing machines, burnishing machines, butting machines, channeling machines, clicking machines, counter moulders, crimping machines, gang brushing machines, heeling machines, heel seat nailers, heel slicing machines, heel trimmers and breast scourers, inking machines, inseam trimmers, inseam welt stitchers, insole and heel seat trimmers, insole tackers, jack rollers, large splitting machines, marking machine, nigger heads, outsole stitchers, pullovers, rollers, rounding machines, sanders, skivers, slugging machines, sole cutters, sole grading machines, sole leveling machines, soling machines, stamping machines, stapling machines, tip scouring machines, toe trimmers, tree machines.

Street Railways
Lubricant for—
Air brake cylinders, ball and roller bearings.
Controller contacts, trips, fingers, drums.
Door engines, slides.
Motor bearings, motormen's valves.
Signals, interlocking (cylinders).
Trackage, trolley bases, trolley wheels.

Sugar
Lubricant for—
Crushing machinery, pumps.

Tanneries
Lubricant for—
Coloring machines, jack rollers, setting machines, shaving machines, splitters.

Textile
Lubricant for—
Carding machinery comb boxes, bearings.
Cleansing equipment bearings.
Combing machine bearings, cams.
Drawing and spinning machine bearings, gears.
Finishing machinery bearings, such as fullers, washers, raisers, nappers, croppers, pressing machines.
Gills and backwashers (fallers, screws, slides, gears).
Scouring machinery bearings, cams, and other moving parts.
Tapestry machinery.
Weaving equipment bearings, gears, chains.

Woodworking
Lubricant for—
Sawmill machines, such as carriers, planers, saws.

Cuprammonium Carbonate
Synonyms: Copper-ammonium carbonate.
French: Carbonate cupro-ammoniaque, Carbonate de cupro-ammonium.
German: Cupraammoniumcarbonat, Kohlensaeures-cuprammonium.

Chemical
Reagent (Brit. 286212) in making catalysts with sodium aluminate and copper nitrate and kieselguhr, quartz, or pumice meal, used in making—
Anilin by the reduction of nitrobenzene.
Camphor from borneol.
Crotonaldehyde from acetaldehyde.
Crotonic alcohol from crotonaldehyde.
Chlorine carriers in the chlorination of methane or thiophenes and aliphatic hydrocarbons present as impurities in benzol.
Cyclohexanone by hydrogenation of cyclohexanol.
Naphthylamine from nitronaphthalene.
Reduction compounds from nitroaromatic compounds.

Cuprein

Miscellaneous
Ingredient of—
Mothproofing compositions for treating furs and feathers (Brit. 263092).

Textile
——, *Miscellaneous*
Ingredient of mothproofing compositions for treating woolens (Brit. 263092).

Cuprene
French: Cuprène.
German: Cupren.

Chemical
As a carrier of catalysts (used in the place of kieselguhr).

Electrical
Starting material in making—
Electrodes.

Explosives
Substitute for kieselguhr in making—
Dynamites, gelatins, permissibles.

Linoleum and Oilcloth
Substitute for cork in making coatings.

Miscellaneous
Substitute for cork in making—
Various compositions of matter.

Plastics
Starting point in making—
Highly resistant plastic products.

Rubber
As a filler.

Cupric Chloride
Synonyms: Copper bichloride, copper chloride, copper dichloride.
French: Chlorure cuivrique.
German: Chlorkupfer, Kupferbichlorid, Kupferchlorid, Kuprichlorid.
Spanish: Cloruro cobrico.
Italian: Cloruro ramico.

Analysis
As a reagent.

Chemical
Catalyst in—
Deacon chlorine process.
Catalyst in making—
Acids, esters, and ethers from alcohols and carbon monoxide (Brit. 397852).
Cellulose esters (French 660623).
Esters from lower aliphatic acids and olefins (Brit. 398527).
Organic chemicals by various processes.
Phthalyl chloride or its homologs by reacting phthalic anhydride, or its homologs, or its nuclear substitution products, with benzilidene chloride or benzyl chloride (Brit. 414570).
Crystallizing accelerator (French 689040) for—
Ammonium chloride solutions.
Ingredient of—
Catalytic mixture used in making chlorobenzene from benzene, air, and hydrochloric acid gas (Brit. 362817).
Catalytic mixture used in making ketenes (such as acetic ketene, ethyl ketene and propyl ketene) from an aliphatic ketone or a secondary alcohol (Brit. 396568).
Oxidizing agent in various manufacturing operations.
Starting point in making—
Copper chromate.

Dye
Oxidizing agent in making various dyestuffs.

Explosives and Matches
Ingredient of—
Pyrotechnic compositions.

Ink
Ingredient of two-solutions in making—
Indelible inks, laundry marking inks.
Reagent in—
Synthetic inks.

Insecticide
Ingredient of—
Insecticides.

Metallurgical
Ingredient of—
Baths for coloring iron and tin.
Electrotype for plating copper on aluminum.

Miscellaneous
Etching agent for—
Galvanized iron prior to painting.

Paint and Varnish
Starting point in making—
Chrome brown pigment.
Pigments with portland cement, casein, oil and glue (French 573338).

Paper
Preservative for—
Pulp.

Petroleum
Catalyst (French 671035) in—
Transforming mineral oils into hydrocarbons of lower boiling point with simultaneous decoloration.
Deodorizing agent in processing—
Distillates.
Desulphurizing agent in processing—
Distillates.
Impregnating agent (U. S. 1965821) for—
Fuller's earth used in sweetening processes for gasoline.
Ingredient (Brit. 406963) of—
Catalytic mixtures used in manufacturing and refining operations involving hyrogenation.
Purifying agent (U. S. 1963555, 1963556, and 1914953; Brit. 398794) for—
Hydrocarbon oils.

Photographic
Fixing agent (Brit. 401340) in making—
Color pictures from silver pictures.
Reagent in—
Photographic processes.

Textile
Catalyst in making—
Diphenyl and anilin blacks in printing cotton goods.
Ingredient of—
Discharge baths containing also nickel and cobalt.

Veterinary Medicine
Suggested for use as a drug.

Water and Sanitation
As a disinfectant.

Woodworking
As a wood preservative.
Starting point (French 629145) in making—
Wood preservatives by admixture with arsenic and other products.

Cupric Normalbutylhydrogenphthalate
French: N-Butylehydrogènephthalate de cuivre, N-Butylebiphthalate cuivrique.
German: N-Butylphtalsaeureskupfer, Kupfer-n-butylsaeuresphtalat.

Resins and Waxes
Reagent (Brit. 250265) in making—
Synthetic resins.

Plastics
Reagent in making—
Plastic compositions.

Cuprous Chloride
Synonyms: Copper chloride, Copper protochloride, Copper subchloride.
French: Chlorure de cuivre, Chlorure cuivreux, Protochlorure de cuivre.
German: Kupferchlorur, Kuprochlorid.
Spanish: Protocloruro de cobre.
Italian: Cloruro rameoso.

Analysis
Absorbent for carbon monoxide in—
Gas analysis.
Absorbent for oxygen in—
Gas analysis.
Reagent in—
Analytical work.

Chemical
Absorbent for—
Butadiene and carbon monoxide (French 705214).
Butadiene and derivatives from gases (French 669337).
Carbon monoxide.
Carbon monoxide (using ammoniacal solution) (French 512542).
Carbon monoxide in making formic acid by reacting carbon monoxide with water or steam in the presence of an acid or acid substance (Brit. 396375).
Carbon monoxide in process for eliminating it from gaseous mixtures by absorption in ammoniacal solution containing also copper sulphate (French 629743).
Oxygen (oxychloride is formed and oxygen can be liberated by heating).
Water, carbon dioxide, and carbon monoxide from gaseous mixtures of hydrogen and nitrogen used in the synthesis of ammonia (French 628138).
Catalyst in making—
Chlorine from hydrochloric acid and oxygen.
Diluents or solvents for pyroxylin or resin compositions by treating unsaturated hydrocarbons with carbon monoxide and steam (U. S. 1973662).
Nonbenzenoid hydrocarbons from acetylene (Brit. 401678, 384654, and 390179).
Phenol or alphanaphthol by reaction between chlorobenzene or alphachloronaphthalene and steam (French 709184).
Synthetic organic chemicals
Reagent in—
Organic synthesis, for example, Sandmeyer reactions.

Fats and Oils
Condensing agent (Brit. 398474) in making—
Polymerized products from glycerin, chlorinated glycerin, or a mixture of glycerin and higher alcohols containing more than three hydroxy groups; such products are used (1) in compounding lubricants (explosion-proof) for use in compressors, valves, and other apparatus; (2) as softeners for shellac and other resins, rendering them soluble in water or alcohol.

Insecticide and Fungicide
Ingredient of—
Insecticidal preparations.
In viniculture.

Cuprous Chloride (Continued)

Metallurgical
Electrolyte (French 611598) in—
Copper refining.

Petroleum
Catalyst (U. S 1973662) in making—
Diluents or solvents for pyroxylin or resin compositions from vaporphase-cracked petroleum products and carbon monoxide or steam.
Decolorizing agent for—
Cracking products (French 610498 and 610499).
Shale oils (French 610498 and 610499).
Desulphurizing agent for—
Petroleum and cracking products (French 611890).

Pharmaceutical
In compounding and dispensing practice.

Soap
Condensing agent (Brit. 398474) in making—
Polymerized products from glycerin, chlorinated glycerin, or a mixture of glycerin and higher alcohols containing more than three hydroxy groups; such products are used (1) in compounding lubricants (explosion-proof) for use in compressors, valves, and other apparatus; (2) as softeners for shellac and other resins, rendering them soluble in water or alcohol.

Textile
Reagent in—
Denitration of rayon.

Cyanamide
Synonyms: Calcium cyanamide, Lime nitrogen.
French: Cyanamide calcique.
German: Cyanamidcalcium, Kalkstickstoff, Stickstoffkalk.
Spanish: Cianamide de calcio.
Italian: Cianamide di calcio.

Chemical
Starting point in making—
Aluminum carbides, ammonia (gaseous), dicyandiamine, nitrogen products, cyanides, urea.
Starting point (Brit. 279884) in making—
Allylguanidin, amylguanidin, butylguanidin, decamethyleneguanidin, ethylguanidin, heptylguanidin hexamethyleneguanidin, hexylguanidin, isoallylguanidin, isoamylguanidin, isobutylguanidin, isopropylguanidin, methylguanidin, pentamethyleneguanidin, phenylethylguanidin, propylguanidin.

Dye
Reagent in making various dyestuffs.

Explosives
Starting point in making various explosives.

Fertilizer
As a fertilizer.
Ingredient of—
Fertilizing compositions for various horticultural and agricultural purposes.

Metallurgy
In case-hardening steel.

8-Cyannaphthalene-1:5-disulphonic Acid
French: Acide de 8-cyanonaphthalène-1:5-disulfonique.
German: 8-Cyannaphtalin-1:5-disulfonsaeure.

Chemical
Starting point (Brit. 276126) in making intermediates with—
Ethoxy derivatives of 5-oxynaphthostyril.
Methoxy derivatives of 5-oxynaphthostyril.
5-Oxynaphthostyril.
5-Oxy-8-naphthamide-1-sulphonic acid.

8-Cyano-4-chloronaphthalenealphasulphonic Acid
French: Acide de 8-cyano-4-chloronaphthalènealphasulfonique.
German: 2-Cyan-4-chlornaphtalinalphasulfonsaeure.

Chemical
Starting point (Brit. 276126) in making—
4-Chloronaphthostyril.

Cyanocresolmercury Chloride

Agriculture
For control of—
Bottom rust of lettuce.
Covered smut and stripe disease of barley.

Cyanogen Bromide
French: Bromure de cyanogène.
German: Bromcyan.
Spanish: Bromuro de cianogeno.
Italian: Bromuro di cianogeno.

Insecticide
As a parasiticide (German 351894).
Ingredient of—
Fumigating composition, containing also hydrocyanic acid and bromoacetophenone, or chloropicrin, or bromoacetic ester (U. S. 1949466).
Rat exterminant, in admixture with hydrocyanic and oxalic acids (French 694139).

Metallurgical
Cyaniding reagent in—
Gold extraction from minerals.

Textile
Reagent for—
Treating cellulose in presence of alkalies and solvents, such as benzene, xylene and organic bases (French 689557).

8-Cyanonaphthalenealphasulphonic Acid
French: Acide de 8-cyanonaphthalènealphasulfonique.
German: 8-Cyannaphtalinalphasulfonsaeure.

Chemical
Starting point (Brit. 276126) in making—
1:8-Aminonaphthoic acid, naphthostyril, 1:8-oxynaphthoic acid, 1-sulphonaphthalene-8-carboxylic acid.

1-Cyano-2-sulphocyano-4-chlorobenzene
Synonyms: Alphacyano-2-sulphocyano-4-chlorobenzene, 1-Cyano-2-sulfocyano-4-chlorbenzol.

Chemical
Starting point (Brit. 305140) in making—
Orthoanthranylthioglycollic acid.
Orthobenzylthioglycollic acid.
Orthocinnamylthioglycollic acid.
Orthocresylthioglycollic acid.
Orthometanylthioglycollic acid.
Orthonaphthylthioglycollic acid.
Orthophenylthioglycollic acid.
Orthophthalylthioglycollic acid.
Orthosalicylthioglycollic acid.
Orthosulphanylthioglycollic acid.
Orthotolylthioglycollic acid.
Orthoxylylthioglycollic acid.

Dye
Starting point (Brit. 305140) in making—
Thioindigoid dyestuffs.

1-Cyano-2-sulphoncyano-4-ethoxybenzene
Synonyms: Alphacyano-2-sulphocyano-4-ethoxybenzene.
French: 1-Cyano-2-sulphocyano-4-éthoxyebenzène.
German: 1-Cyano-2-sulfocyano-4-aethoxybenzol.

Chemical
Starting point (Brit. 305140) in making—
Orthoanthranylthioglycollic acid.
Orthobenzylthioglycollic acid.
Orthocinnamylthioglycollic acid.
Orthocresylthioglycollic acid.
Orthometanylthioglycollic acid.
Orthonaphthylthioglycollic acid.
Orthophenylthioglycollic acid.
Orthophthalylthioglycollic acid.
Orthosalicylthioglycollic acid.
Orthosulphanylthioglycollic acid.
Orthotolylthioglycollic acid.
Orthoxylylthioglycollic acid.
Starting point (Brit. 305140) in making—
Thioindigoid dyestuffs.

Cyclocitral
Synonyms: Delta-1-cyclocitral, Delta-2-cyclocitral, 3:2:2:6-Trimethyldelta-5-tetrahydrobenzaldehyde, 3:2:2:6-Trimethyldelta-6-tetrahydrobenzaldehyde, 1:1:3-Trimethyl-2-methylalcyclohexene-2.

Chemical
Starting point in making—
Aromatic derivatives.

Food
Ingredient of—
Beverages, flavorings.

Perfumery
Ingredient of—
Cosmetics, perfumes.

Cyclogeranyl Acetate

Chemical
Starting point in making various derivatives.

Miscellaneous
Odor for various purposes.

Perfume
Ingredient of—
Cosmetics, perfumes.

Soap
Ingredient of—
Toilet soaps.

Cyclohexamine Selenite
French: Sélénite de cyclohexamine, Sélénite cyclohexaminique.
German: Cyklohexaminselenit, Selenigsäurecyklohexaminester, Selenigsäurescyklohexamin, Selenigsäurezyklohexaminester, Selenigsäureszyklohexamin, Zyklohexaminselenit.
Spanish: Selenito de ciclohexamine.
Italian: Selenito ciclohexaminico.

Miscellaneous
Reagent (Brit. 340318) in—
Mothproofing furs, feathers, hair.

Textile
Reagent (Brit. 340318) in—
Mothproofing wool and felt.

Cyclohexane
French: Cyclohexane.
German: Zyklohexan.

Chemical
Solvent in making—
Fine chemicals (used in the recrystallization process).

Fats and Oils
Solvent for various fats and oils.

Perfume
Solvent in—
Extracting essential oils.

Petroleum
Solvent for—
Paraffin.

Resins and Waxes
Solvent for—
Waxes.

Rubber
As a solvent.

Cyclohexanediacetic Acid Ester of Grapeseed Alcohol
(Uses same as those given for the item following).

Cyclohexanediacetic Acid Ester of Ricinoleic Alcohol

Bituminous
Solvent (Brit. 445223) for—
Asphalt and other bituminous bodies.

Dye
Solvent (Brit. 445223) for—
Dyestuffs, particularly oil-soluble coaltar dyes.

Fats, Oils, and Waxes
Solvent (Brit. 445223) for—
Fats, oils, waxes.

Resins
Solvent (Brit. 445223) for—
Oil-soluble glycerol-phthalic acid resins.
Polymerized vinyl compounds.
Synthetic resins.

Rubber
Solvent (Brit. 445223) for—
Rubber.

Cyclohexanol Acetate
Synonyms: Adronal acetate, Adronol acetate, Hexalin acetate.
French: Acétate d'adronol, Acétate adronolique, Acétate de cyclohexanol, Acétate cyclohexanolique, Acétate d'hexaline, Acétate hexalinique.
German: Adronolacetat, Adronolazetat, Essigsäureadronalester, Essigsäuresadronal, Essigsäurecyclohexanolester, Essigsäurehexalinester, Essigsäurescyclohexanol, Essigsäureshexalin, Hexalinacetat.

Cellulose Products
Solvent for—
Cellulose acetate, cellulose esters and ethers, cellulose nitrate.
For uses, see under general heading: "Solvents."

Chemical
Starting point in making various derivatives.

Cyclohexanol Butyrate
French: Butyrate de cyclohexanol, Butyrate de cyclohexyle.
German: Cyclohexylbutyrat.
Spanish: Butirato de ciclohexil.
Italian: Butirato di cicloessile.

Rubber
Regenerating agent (French 636641).

Cyclohexanol Oxalate

Cellulose Products
Plasticizer for—
Cellulose nitrate (nitrocellulose).
For uses, see under general heading: "Plasticizers."

Cyclohexanol Phthalate

Cellulose Products
Plasticizer for—
Cellulose nitrate (nitrocellulose).
For uses, see under general heading: "Plasticizers."

Cyclohexanoneoxime

Fuel
Primer (Brit. 429763) for—
Diesel engine fuel oils produced by the hydrogenation of coal.

Petroleum
Primer (Brit. 429763) for—
Diesel oils containing a high proportion of aromatic bodies.

Cyclohexanyl Cinnamate
French: Cinnamate de cyclohéxanyle, Cinnamate cyclohéxanylique.
German: Cyklohexanylcinnamat, Zimtsäurescyklohexanylester, Zimtsäurescyklohexanyl, Zimtsäureszyklohexanyl, Zimtsäurezyklohexanylester, Zyklohexanylcinnamat.

Chemical
Starting point in making—
Aromatics and other derivatives.

Perfume
Ingredient of—
Synthetic perfumes.
Perfume in—
Cosmetics.

Soap
Perfume in—
Toilet soaps.

Cyclohexanyl Formate
French: Formiate de cyclohexanyl.
German: Ameisensaeurescyklohexanyl, Cyklohexanylformiat.

Paint and Varnish
Solvent (Brit. 254041) in making—
Nitrocellulose enamels, lacquers, and varnishes.
See also: "Solvents."

Cyclohexyl Adipate
French: Adipate de cyclohexyle, Adipate cyclohexylique.
German: Adipinsäurecyklohexylester, Adipinsäurescyklohexyl, Adipinsäureszyklohexyl, Adipinsäurezyklohexylester, Cyklohexyladipat, Zyklohexyladipat.

Cellulose Products
Solvent (Brit. 330909) for—
Cellulose esters and ethers, cellulose nitrate, synthetic resins.
For uses, see under general heading: "Solvents."

Chemical
Solvent for various purposes.
Starting point in making various derivatives.

Cyclohexylamine
French: Cyclohexyleamine.
German: Cyklohexylamin.

Chemical
Starting point in making—
Intermediates, pharmaceuticals.

Cyclohexylamine (Continued)

Dye
Starting point (Brit. 340495) in making dyestuffs, which are used for dyeing and printing rayons and cellulose acetate, with the aid of—
1-Chloro-2:4-dinitrobenzene.
1-Chloro-2:4-dinitrobenzene-4-sulphonic acid.
1-Chloro-2:6-dinitrobenzene-6-sulphonic acid.
1-Chloro-2:4-dinitronaphthalene.
1-Chloro-4-nitrobenzene-2-carboxylic acid.
1-Chloro-2-nitrobenzene-4-sulphonic acid.
1-Chloro-4-nitrobenzene-2-sulphonic acid.
1:4-Dichloro-2-nitrobenzene.

Cyclohexylamineformaldehyde

Glass
Stabilizer (Brit. 437304) for—
Halogenated rubber derivatives used as cements for laminated glass.

Miscellaneous
Inhibitor (Brit. 437304) of—
Photochemical action.

Paper
Stabilizer (Brit. 437304) for—
Halogenated rubber derivatives used for impregnating or coating wrapping paper.

Rubber
Promoter (Brit. 437304) of—
Resistance to the deteriorating action of light on chlorinated rubber.
Stabilizer (Brit. 437304) for—
Coating and impregnating agents made from halogenated rubber derivatives and used for treating fabrics to be used as wrapping materials.
Transparent films or sheets made from halogenated rubber derivatives.

Cyclohexylaminoacetonitrile

Glass
Stabilizer (Brit. 437304) for—
Halogenated rubber derivatives used as cements for laminated glass.

Miscellaneous
Inhibitor (Brit. 437304) of—
Photochemical action.

Paper
Stabilizer (Brit. 437304) for—
Halogenated rubber derivatives used for impregnating or coating wrapping paper.

Rubber
Promoter (Brit. 437304) of—
Resistance to the deteriorating action of light on chlorinated rubber.
Stabilizer (Brit. 437304) for—
Coating and impregnating agents made from halogenated rubber derivatives and used for treating fabrics to be used as wrapping materials.
Transparent films or sheets made from halogenated rubber derivatives.

Cyclohexylanilin

Chemical
Starting point (Brit. 261747) in making—
Cyclohexylethylanilin, cyclohexylmethylanilin.

Cyclohexyl Bromide

Chemical
Reagent in—
Organic syntheses.

Fuel
Primer (Brit. 404682) in—
Diesel engine fuels (used in conjunction with alkyl nitrates, having two to four atoms in the molecule, whose function is that of reducing the delay period).
Reducer (Brit. 404682) of—
Spontaneous ignition temperature of diesel engine fuels.

Cyclohexyl Carbonate
French: Carbonate de cyclohexyle.
German: Cyclohexylcarbonat.
Spanish: Carbonato de ciclohexil.
Italian: Carbonate di cicloessile.

Cellulose Products
Plasticizer for—
Cellulose acetate, cellulose esters and ethers, cellulose nitrate.
For uses, see under general heading: "Plasticizers."

Cyclohexylcresol

Chemical
Starting point (Brit. 444351) in making—
Fat-splitting catalysts and emulsifying agents, useful in dyeing, laundering, bleaching, and various other purposes, by reacting with formaldehyde and non-aromatic secondary amines (the salts of the products with water-soluble acids, or water-insoluble acids, or the quaternary ammonium salts, are claimed to be valuable for the purposes named).

Cyclohexylcyclohexanol Sulphonate

Miscellaneous
As an emulsifying agent (Brit. 449607 and 425239).
For uses, see under general heading: "Emulsifying agents."

1-Cyclohexyl-2:3-dimethyl-5-pyrazolone

Pharmaceutical
Suggested (Brit. 433053) for use as—
Febrifuge, sedative.

Cyclohexylethanolamine
French: Cyclohexyle-éthanolamine.
German: Cyklohexylaethanolamin, Zyklohexylaethanolamin.

Ceramics
Plasticizer and solvent (Brit. 297484) in—
Coating compositions containing cellulose esters or ethers.

Chemical
Emulsifying agent for various chemicals.

Glass
Plasticizer and solvent (Brit. 297484) in—
Compositions, containing cellulose acetate, nitrocellulose, or other esters or ethers of cellulose, used for making non-scatterable glass and in coating glass.

Insecticide
Ingredient of—
Anticryptogamic compositions, germicidal compositions, insecticidal compositions.

Leather
Plasticizer and softener (Brit. 297484) in—
Compositions, containing cellulose acetate, nitrocellulose, or other esters or ethers of cellulose, used in making artificial leathers.

Miscellaneous
Preservative in treating—
Proteins.
Reagent in making—
Emulsions of miscellaneous materials.

Paint and Varnish
Plasticizer (Brit. 297484) in making—
Lacquers, enamels, varnishes, and paints containing cellulose acetate, nitrocellulose, or other esters or ethers of cellulose.

Paper
Plasticizer and solvent (Brit. 297484) in making—
Coating compositions containing cellulose acetate, nitrocellulose, or others esters or ethers of cellulose.

Photographic
Plasticizer (Brit. 297484) in—
Films from cellulose acetate, nitrocellulose, or other esters and ethers of cellulose.

Plastics
Plasticizer (Brit. 297484) in making—
Compositions containing cellulose acetate, nitrocellulose, or other esters or ethers of cellulose (used in the place of camphor, for example, in the manufacture of celluloid).

Resins and Waxes
Emulsifying agent.

Rubber
Plasticizer and solvent (Brit. 297484) in—
Coating compositions containing cellulose acetate, nitrocellulose, or other esters or ethers of cellulose.

Stone
Plasticizer and solvent (Brit. 297484) in—
Coating compositions containing cellulose acetate, nitrocellulose, or other esters or ethers of cellulose.

Textile
——, Dyeing
Wetting agent in—
Dye baths (used to secure better penetration of the color into the dyed yarn and fabric).

Cyclohexylethanolamine (Continued)
Softener (Brit. 297484) in—
Finishing baths (used to obtain a better finish on various textiles).
——, *Finishing*
Plasticizer and solvent (Brit. 297484) in—
Compositions containing cellulose acetate, nitrocellulose, or other esters or ethers of cellulose, used in the production of coated fabrics.
——, *Printing*
Ingredient (Brit. 302252) of—
Printing pastes (to obtain better impregnation of the color into the printed fabric).
Woodworking
Plasticizer and solvent (Brit. 297484) in—
Compositions containing cellulose acetate, nitrocellulose, or other esters or ethers of cellulose, used in the finishing of wood and wood products.

Cyclohexylglucamine
Dye
Coupling agent (Brit. 429618) in making—
Dyestuffs with diazotized arylamines (color being developed on the fiber by acid treatment).

Cyclohexylidenecyclohexanone
Chemical
Starting point (Brit. 397883) in making—
Cyclohexylcyclohexanol by hydrogenation.

Cyclohexylisoamyl Phthalate
Synonymns: Isoamylcyclohexylphthalate.
French: Phthalate de cyclohexyle et de isoamyle, Phthalate cyclohexylique et isoamylique.
German: Cyklohexylisoamylphtalat, Isoamylcyklohexylphtalat, Isoamylzyklohexylphtalat, Phtalsaeurecyklohexylisoamylester, Phtalsaeureisoamylcyklohexylester, Phtalsaeurescyklohexylisoamyl, Phtalsaeuresisoamylcyklohexyl, Phtalsaeureszyklohexylisoamyl, Phtalsaeurezyklohexylisoamylester.
Paint and Varnish
Solvent and plasticizer (Brit. 302961) in making nitrocellulose lacquers containing—
Copal, coumarone resins, cyclic ketone resins, dammar, elemi, ester resins, indene resins, manila gum, mastic, sandarac, urea-aldehyde condensation products, vinyl resins.
Plastics
Solvent and plasticizer (Brit. 302961) in making nitrocellulose products containing—
Copal, coumarone resins, cyclic ketone resins, dammar, elemi, ester resins, indene resins, manila gum, mastic, sandarac, urea-aldehyde condensation products, vinyl resins.

1-Cyclohexyl-3-methyl-5-pyrazolone
Pharmaceutical
Suggested (Brit. 433053) for use as—
Febrifuge, sedative.

Cyclohexyl Montanate
Resins and Waxes
Modifying agent (Brit. 390534) in—
Polishing waxes (replaces part of the wax constituents.

Cyclohexylnaphthalenesulphonic Acid
French: Acide cyclohexylenaphthalènesulfonique.
German: Cyklohexylnaphtalinsulfonsaeure, Zyklohexylnaphtalinsulfonsaeure.
Miscellaneous
Ingredient (Brit. 277391) of—
Stain-removing compositions.
Washing and cleansing compositions.
Textile
——, *Finishing*
Ingredient of—
Fulling compositions (Brit. 277391).

Cyclohexyl Naphthenate
Miscellaneous
As an emulsifying agent.
Ingredient (Brit. 390534) of—
Metal cleansing composition containing also silicious chalk and a volatile solvent.
See also: "Emulsifying agents."

Cyclohexylphenol
Chemical
Starting point (Brit. 444351) in making—
Fat-splitting catalysts and emulsifying agents, useful in dyeing, laundering, bleaching, and various other purposes, by reacting with formaldehyde and non-aromatic secondary amines (the salts of the products with water-soluble acids, or water-insoluble acids, or the quaternary ammonium salts, are claimed to be valuable for the purposes named).

Cyclohexylphenyl Ether
Chemical
Starting point in—
Organic synthesis.
Lubricant
Starting point (Brit. 440916) in making—
Products useful as lubricating oils or as pour-point depressors for paraffin base lubricating oils by condensation with halogenated derivatives of aliphatic hydrocarbons, such as paraffin oils, paraffin, petrolatum, ceresin, ozokerite, or others contained in the middle to higher fractions of petroleum.

Cyclohexylresorcinol
Chemical
Starting point (Brit. 444351) in making—
Fat-splitting catalysts and emulsifying agents for use in dyeing, laundering, bleaching, and various other purposes, by reacting with formaldehyde and non-aromatic secondary amines (the salts of the products with water-soluble or water-insoluble acids, and the quaternary ammonium salts are claimed to be valuable for the purposes named).

Cyclohexyl Thiocyanate
Insecticide
As an insecticide.
Ingredient (Brit. 361900) of—
Insecticidal and germicidal compositions containing soaps and organic solvents.

Cyclohexyl Xanthate
Metallurgical
Reagent (U. S. 1823316) in recovering—
Mineral from ores by broth flotation.

Cyclopentadene
French: Cyclopentadiene.
German: Cyklopentaden, Zyklopentaden.
Chemical
Starting point in making—
Derivatives by condensation with the sodium compound of a suitable malonic acid derivative (Brit. 400452).
5:5-Dideltacyclopentylallylbarbituric acid.
Methyl bicyclopentenylacetate.
Terpineol cyclopentenylacetate.
Thymol cyclopentanylacetate.

Cyclopentanone
German: Zyklopentanon.
Chemical
Starting point in making—
Piperidinomethylcyclohexanone hydrochloride (German 422916).

Cyclopentanoneoxime
Fuel
Primer (Brit. 429763) for—
Diesel engine fuel oils produced by the hydrogenation of coal.
Petroleum
Primer (Brit. 429763) for—
Diesel oils containing a high proportion of aromatic bodies.

Cymene Sulphonylchloride
Miscellaneous
Viscosity increaser (Brit. 438413 and 438415) for—
Tars.

Dammar
Synonyms: Gum dammar, Resin dammar.
Latin: Dammargummi.
French: Gomme dammar, Résine dammar.
German: Dammarharz.
Adhesives
Ingredients of special products.

Dammar (Continued)
Chemical
Reagent in making—
Dry color preparations.
Explosives
Ingredient of—
Match head compositions, pyrotechnic compositions.
Ink
Ingredient of—
Printing inks, writing inks.
Miscellaneous
Ingredient of—
Plaster preparations, shoeblackings.
Reagent in—
Mounting microscopical specimens.
Oilcloth and Linoleum
Ingredient of—
Compositions used in making coatings.
Paint and Varnish
Raw material in making—
Lacquers, light-colored transparent varnishes, varnishes in general.
Paper
Ingredient of—
Special coatings.
Pharmaceutical
In compounding and dispensing practice.
Printing
Reagent in lithography and process engraving.
Resins and Waxes
Reagent (Brit. 303386) in making—
Synthetic resins with glycerin, glycol, or glucose plus phthalic anhydride or other polybasic aromatic acids or anhydrides.
Rubber
Ingredient of—
Rubber batch.
Soap
Raw material in making—
Special grades of soaps.
Textile
Ingredient of—
Printing pastes.

Deacetylated Chitin
Adhesives
Adhesive (Brit. 458818 and 458839) for—
Asbestos and its products, canvas, cement, cloth, cork and its products, furniture, glass, lacquered surfaces, laminated paper, leather and its products, mica and its products, painted surfaces, paper and its products, plaster, plywood surfaces, porcelain, regenerated cellulose, rubber, safety glass, veneers, wood surfaces.

Deacetylated Chitin Acetate
Adhesives
Adhesive (Brit. 458818 and 458839) for—
Asbestos and its products, canvas, cement, cloth, cork and its products, furniture, glass, lacquered surfaces, laminated paper, leather and its products, mica and its products, painted surfaces, paper and its products, plaster, plywood surfaces, porcelain, regenerated cellulose, rubber, safety glass, veneers, wood surfaces.

Deacetylated Chitin Formate
Adhesives
Adhesive (Brit. 458818 and 458839) for—
Asbestos and its products, canvas, cement, cloth, cork and its products, furniture, glass, lacquered surfaces, laminated paper, leather and its products, mica and its products, painted surfaces, paper and its products, plaster, plywood surfaces, porcelain, regenerated cellulose, rubber, safety glass, veneers, wood surfaces.

Deacetylated Chitin Malate
Adhesives
Adhesive (Brit. 458818 and 458839) for—
Asbestos and its products, canvas, cement, cloth, cork and its products, furniture, glass, lacquered surfaces, laminated paper, leather and its products, mica and its products, painted surfaces, paper and its products, plaster, plywood surfaces, porcelain, regenerated cellulose, rubber, safety glass, veneers, wood surfaces.

Decahydronaphthalene
Synonyms: Decalin.
French: Décalin, Décahydronaphthalène.
German: Dekalin, Dekahydronaphtalin.
Adhesives
Solvent in—
Casein glue compositions.
Analysis
As a solvent.
Ceramics
Solvent in—
Coating compositions for potteries and porcelains.
Chemical
As a solvent.
Explosives
Solvent in—
Fireworks manufacture.
Fats and Oils
As a general solvent.
Solvent in making—
Belting greases, lubricating compositions.
Germicide
Solvent in—
Germicidal compositions.
Glass
Solvent in—
Waterproof mastics.
Ink
Ingredient of—
Lithographic inks, printing inks.
Insecticide
Vehicle in—
Liquid insecticides (used in place of turpentine).
Leather
Solvent in—
Finishing and dressing compositions, leather cements, leather polishes, patent leather finishes, shoe polishes, waterproofing compositions and finishes.
Linoleum and Oilcloth
Solvent in—
Linoleum and oilcloth cements.
Mechanical
Cleansing agent for—
Machinery.
Metallurgical
As a flotation agent (used in place of turpentine).
Solvent for—
Waterproof mastics in metal work.
Miscellaneous
Ingredient of—
Compositions for transferring pictures and prints.
Floor polishes, furniture polishes, glass cements.
Pigment preparations used as drawing crayons.
Stove polishes, waterproofing compositions.
Solvent in—
Compositions for cleansing firearms, ivory, substances attacked by chlorine.
Stain remover.
Substitute for—
Turpentine.
Paint and Varnish
Ingredient of—
Auto top dressing.
Solvent and thinner in—
Coach finishes, driers, enamels, glazing putty, lacquers, paint removers, paints of all kinds, piano rubbing varnishes, resins, roofing cements, stain removers, stains, varnishes, varnish removers, wax color-binding compositions.
Substitute for—
Turpentine.
Paper
Cleansing agent for—
Paper machine wires.
Perfume
Substitute for turpentine in—
Cosmetics, emollients.
Printing
As a general solvent and cleanser.
Solvent in—
Color process printing.
Resins and Waxes
Solvent for—
Resins, waxes.

Decahydronaphthalene (Continued)
Solvent in wax compositions for—
Grafting, modelling, sealing, various purposes.
Rubber
Solvent in—
General processing, rubber cements.
Soap
Ingredient of—
Detergent compositions, grease-removing soaps, household soaps, medicated soaps, washing compounds.
Textile
Solubilizing agent for various dyestuffs.
Solvent for—
Removing paint and oil stains from fabrics.
Woodworking
Impregnating agent.
Preservative agent.
Solvent and thinner in—
Fillers, polishes.
Waterproofing agent.

1'-Decahydronaphthyl-2-methylcyclohexanol Sulphonate
Miscellaneous
As an emulsifying agent (Brit. 449607).
For uses, see under general heading: "Emulsifying agents."

2'-Decahydronaphthyl-2-methylcyclohexanol Sulphonate
Miscellaneous
As an emulsifying agent (Brit. 449607).
For uses, see under general heading: "Emulsifying agents."

1'-Decahydronaphthylmethylmethylcyclohexanol Sulphonate
Miscellaneous
As an emulsifying agent (Brit. 449607).
For uses, see under general heading: "Emulsifying agents."

Decyl Acetate
Synonyms: Decylic acetate, Normal decylic acetate.
French: Acétate de décyle, Acétate décylique, Acétate de N-décyle, Acétate de N-décylique.
German: Decylacetat, Decylazetat, Essigsäuredecylester, Essigsaeuresdecyl, N-Decylacetat, N-decylazetat.
Food
Base in making—
Fruit flavorings.
Perfume
Ingredient of—
Fancy perfumes.
Perfume in—
Cosmetics, toilet waters.
Soap
Perfume in—
Toilet soaps.

Decyl Chloride
French: Chlorure d'alcool décylique, Chlorure de décyle, Chlorure de décyle alcool.
German: Chlordecyl, decylchlorid.
Chemical
Agent in—
Recovering volatile solvents from gases.
Emulsifiable higher fatty alcohol derivative, more readily emulsifiable in water than the usual hydrocarbons.
Solvent for—
Aromatic hydrocarbons, coaltar constituents, fatty acids.
Reagent for—
Introducing long-chain alkyl residues into the most varied types of organic substances.
Dye
Reagent in making—
Fat-soluble colors.
Fats and Oils
Solvent for—
Fatty acids, oils.
Insecticide
As an insecticide (potent in toxicity to lower organisms, but nontoxic to the human organism).
Carrier for—
Insecticides generally, nicotine, pyrethrum extracts.
Leather
Starting point in making—
Protective agents.
Miscellaneous
Ingredient of—
Shoe creams and polishes.
Solvent for—
Bitumens.
Resins and Waxes
Solvent for—
Resins, waxes.
Textile
Starting point in making—
Textile soaps.

Decylene
Miscellaneous
As an emulsifying agent (Brit. 360602).
For uses, see under general heading: "Emulsifying agents."

Decylguanidin Chloride
Textile
Assistant (Brit. 421862) in—
Aqueous baths for treating textiles.
Promoter (Brit. 421862) of—
Uniform dyeing with basic dyestuffs.
Wetting and washing agent (Brit. 421862) in—
Textile processes.

Decylguanidin Hydrochloride
Miscellaneous
As an emulsifying agent (Brit. 422461).
For uses, see under general heading: "Emulsifying agents."

Decylpyrocatechol
French: Pyrocatechole décylique.
German: Decylbrenzcatechin, Decylpyrocatechin.
Chemical
Starting point in making various derivatives.
Starting point (Brit. 330519) in making drugs with—
Betaine, hexamethylenetetramine, piperazine, sarcosine anhydride.

Decylresorcinol
French: Résorcinole décylique.
German: Decylresorcin.
Chemical
Starting point in making various derivatives.
Starting point (Brit. 330519) in making synthetic drugs with—
Betaine, hexamethylenetetramine, piperazine, sarcosine anhydride.

Decyl Rhodanate, Sodium Salt
Insecticide
Insecticide of high toxicity for use in sprays.

Dehydrothioparatoluidin
Synonyms: Aminobenzenylorthotoluidinthiocresol.
French: Aminobenzènyleorthotoluidinethiocrésole.
German: Aminobenzenylorthotoluidinthiocresol.
Chemical
Starting point in making—
Aromatics, dehydrothioparatoluidinsulphonic acid, intermediates, pharmaceuticals.
Dye
Starting point in making—
Azo dyestuffs, brilliant geranin, chlorophenin, chromin G, diamine rose, diamine rose extra R, dianil rose BD, dianil yellow, direct chloramine yellow, direct rose G, erica 2GN, flavin, geranin BB, geranin G, methylene yellow H, rhodulin yellow T, thiazo dyestuffs, thiazol yellow, thioflavin T, thiorubin.
Starting point (Brit. 306981) in making dyestuffs for dyeing cellulose acetate with the aid of—
2:3-Aminonaphthoic acid, betanaphthol, betanaphthylamine, 4-chlorophenol, 2:4-dichloronitrobenzene, dimethylanilin, 2:5-dinitrochlorobenzene, 2-ethoxy-1-naphthylamine, ethyl-1-naphthylamine, ethyl-2-naphthylamine, metatoluidin, metatoluylenediamine, 2:3-oxynaphthoic acid.

Dehydrothiotoluidinorthomonosulphonic Acid
French: Acide de déhydrothiotoluidine-orthomonosulphonique.
German: Dehydrothiotoluidinorthomonosulfonsaeure.

Dehydrothiotoluidinorthomonosulphonic Acid (Cont'd)
Chemical
Starting point in making—
Aromatics, intermediates, pharmaceuticals, salts and esters.
Dye
Starting point (Brit. 306981) in making dyestuffs for dyeing cellulose acetate with the aid of—
2:3-Aminonaphthoic acid, betanaphthol, betanaphthylamine, 4-chlorophenol, 2:4-dichloronitrobenzene, dimethylanilin, 2:5-dinitrochlorobenzene, 2-ethoxy-1-naphthylamine, ethyl-1-naphthylamine, ethyl-2-naphthylamine, metatoluidin, metatoluylenediamine, 2:3-oxynaphthoic acid.
Starting point (Brit. 310354) in making azo dyestuffs and lakes with the following acetoacetic compounds—
Alphanaphthalide, betanaphthalide, meta-anilide, meta-anisidide, metachloroanilide, metanaphthalide, metaphenetidide, metatoluidide, metaxylidide, orthoanilide, orthoanisidide, orthochloroanilide, orthonaphthalide, orthophenetidide, orthotoluidide, orthoxylidide, para-anilide, para-anisidide.
Starting point in making—
Alkali brown, alkali yellow, azidin yellow 5G, azo dyestuffs, benzamin fast yellow B, benzo brown 3R, benzoin fast red AF, brilliant geranin, chloramin yellow, chlorophenin, chromin G, clayton cloth red, clayton yellow, chlorophosphin, columbia yellow, cotton yellow R, curcuphenin, diamine fast yellow, diamine rose, diamine rose extra R, dianil pure yellow HS, dianil rose BD, direct chloramin yellow, direct fast yellow, direct rose G, erica 2GN flavin, flavin, geranin BB, geranin G, methylene yellow H, mimosa, naphthamine yellow, naphthylamine pure yellow, oriol yellow, oriole yellow, oxydiamine yellow, oxyphenin gold, rhodulin yellow T, terra cotta F, thiazin red GN, thiazin red R, thiazo dyestuffs, thiazol yellow, titan rose 3B, thiazol yellow, thioflavin S, thioflavin T, thiophosphin, thiorubin, triazo fast yellow, vigoreux yellow.

Dehydrothioxylidin
Synonyms: Aminotoluenylorthoaminothioxylenol.
Dye
Starting point in making various dyestuffs and intermediates.
Petroleum
Reagent for—
Imparting fluorescence to hydrocarbon oils or liquids.

Dekanaphthene
German: Dekanaphten.
Chemical
Solvent (Brit. 269960) in various processes and for various purposes.
Miscellaneous
Solvent for various purposes.
Textile
——, *Dyeing and Printing*
Solvent in making—
Dye liquor for textiles, paste for printing or stenciling.

Delta-alpha-aminoalphaphenylbutyramide
Chemical
Starting point (U. S. 1861458) in making—
Delta-5-phenyl-5-ethylhydantoin, suggested for use in hypnotics.

Delta-alpha-aminoalphaphenylbutyric Acid
Chemical
Starting point (U. S. 1861458) in making—
Delta-5-phenyl-5-ethylhydantoin.
Suggested for use as a hypnotic.

Delta-alphacyanoalphaphenylbutyramide
Chemical
Starting point (U. S. 1861458) in making—
Delta-5-phenyl-5-ethylhydantoin.
Suggested for use as a hypnotic.

Deltacamphoroxime
Analysis
As a reagent.
Chemical
Reagent in—
Organic synthesis.

4-Delta^2-cyclohexenylamino-1-phenyl-2:3-dimethyl-5-pyrazolone
Pharmaceutical
Suggested (Brit. 433053) for use as—
Febrifuge, sedative.

Dextrin
Synonyms: Artificial gum, British gum, Dextrine, Starch gum, Vegetable gum.
French: Dextrine, Gommeline, Léiocome, Léiogomme.
German: Starkegummi, Starkemehlgummi, Starkemehlschleim, Starkeschleim.
Ceramics
Ingredient of clay batch for—
Bricks, porcelains, potteries, tiles.
Ingredient of decorative effects for—
Porcelains, potteries.
Chemical
Starting point in making—
Emulsifying agents.
Dye
Thickener in—
Dye pastes.
Explosives and Matches
Absorbent in—
Explosives, matchhead compositions, pyrotechnic compositions.
Food
Ingredient of—
Bakery products, confectionery, various food products.
Polishing agent for—
Barley, coffee, rice.
Glass
Ingredient of—
Silvering compounds.
Glues and Adhesives
Alone as an adhesive.
Ingredient of—
Adhesive preparations, envelope adhesives, glues, glues for leather and leather substitutes, label glues, library pastes, mucilages, postage stamp adhesives.
Substitute for—
Gum arabic, gum tragacanth, other gums.
Gums
Raw material in making—
Liquid gums.
Substitute for—
Gum arabic, gum tragacanth, other gums.
Ink
Thickener in—
Lithographic inks, marking inks, printing inks, stamping inks, writing inks.
Leather
Ingredient of—
Flesh pastes, leather finishes, tanning extracts (to increase viscosity), weighting preparations.
Miscellaneous
As a binder.
As a filler in many products.
Binder, filler, size, and stiffener in making—
Felt.
Ingredient of—
Briquetting composition (U. S. 1800875), emulsions, metal polishes, shoe polishes, solder (U. S. 1844287).
Stiffening agent in—
Preparation of fibrous materials.
Substitute for—
Gum arabic, gum tragacanth, other gums.
Oilcloth and Linoleum
As a binder.
Paper
Glossing agent for—
Cardboard, paper.
Ingredient of—
Color batch in wallpaper printing.
Stiffener for—
Paper and pulp (mixed with rye meal and slaked lime).
Sizing agent for—
Boxboard, cardboard, paper, wallpaper.
Pharmaceutical
In compounding and dispensing practice.
Ingredient of—
Excipients.
Photographic
Ingredient of—
Pastes for mounting prints.

Dextrin (Continued)

Reagent in—
Reproduction processes.

Printing
Ingredient of—
Bookbinding adhesives (with alum and phenol).
Reagent in—
Process engraving and lithography.

Textile
——, *Finishing*
Ingredient of—
General textile sizes, lace-sizing compositions, tulle-sizing compositions, silk-sizing compositions, stiffening compositions for various fibers.

——, *Printing*
Ingredient—
Color pastes for calicoes.

Diacetic Acid Ester of Grapeseed Alcohol

Bituminous
Solvent (Brit. 445223) for—
Asphalt and other bituminous bodies.

Dye
Solvent (Brit. 445223) for—
Dyestuffs, particularly oil-soluble coaltar dyes.

Fats, Oils, and Waxes
Solvent (Brit. 445223) for—
Fats, oils, waxes.

Resins
Solvent (Brit. 445223) for—
Oil-soluble glycerol-phthalic acid resins, polymerized vinyl compounds, synthetic resins.

Rubber
Solvent (Brit. 445223) for—
Rubber.

Diacetic Acid Ester of Ricinoleic Alcohol

Bituminous
Solvent (Brit. 445223) for—
Asphalt and other bituminous bodies.

Dye
Solvent (Brit. 445223) for—
Dyestuffs, particularly oil-soluble coaltar dyes.

Fats, Oils, and Waxes
Solvent (Brit. 445223) for—
Fats, oils, waxes.

Resins
Solvent (Brit. 445223) for—
Oil-soluble glycerol-phthalic acid resins, polymerized vinyl compounds, synthetic resins.

Rubber
Solvent (Brit. 445223) for—
Rubber.

Diacetone

Synonyms: Diacetone alcohol.
French: Alcool de diacétone, Alcool diacétonique.
German: Diacetonalkohol.

Chemical
As a solvent.
Solvent for—
Cellulose acetate, nitrocellulose.

Dye
As a solvent.

Electrical
Solvent for—
Cellulose acetate in the production of insulating coatings on wires and parts of electric machinery.

Explosives
Solvent for—
Nitrocellulose.

Fats and Oils
Solvent for—
Fats, oils.

Glass
Solvent for—
Cellulose acetate in coating glass to prevent fogging by condensed moisture.

Jewelry
Solvent for—
Cellulose acetate in heightening the luster of artificial pearls.

Leather
Solvent for—
Cellulose acetate in rendering leather non-inflammable and impermeable.

Mechanical
Ingredient of—
Hydraulic compression fluids.

Miscellaneous
Ingredient of—
Antifreeze solutions, preparations for removing ink from printers' rollers.

Paint and Varnish
Solvent in making—
Cellulose acetate lacquers, nitrocellulose lacquers, stains that do not raise grain of wood.

Pharmaceutical
As a preservative.

Photographic
Solvent for—
Cellulose acetate in the production of noninflammable film.

Plastics
Sealing agent for—
Transparent waterproof wrappings.
Solvent for—
Cellulose acetate, nitrocellulose, resins.

Resins and Waxes
Intermediate in making—
Synthetic resins.
Solvent for various resins.

Textile
Ingredient of solvents in making—
Cellulose acetate.
Ingredient of—
Stripping agents for cellulose ester fabrics.

Woodworking
Ingredient of—
Wood preservatives.

Diacetoneanil

Rubber
Age-resisting agent (U. S. 1958928).

3:5-Diacetoxymercuri-4-nitroguaiacol

Pharmaceutical
Suggested (U. S. 1974506) for use as—
Bactericide.

3-Diacetoxymercuri-4-nitro-2-oxy-1-methylbenzene

Chemical
Starting point in making—
Alkali salts which are used as pharmaceuticals.

Pharmaceutical
Suggested for use as a strong bactericide.

Diacetyl

Synonyms: Biacetyl, Butanedione, Diketobutane, Dimethyldiketone, Dimethylglyoxal.

Dairying
Odorant for—
Butter, cream, milk.

Fats and Oils
Odorant for—
Butter substitutes, such as hydrogenated fats and oils.

Food
Odorant for—
Butter, butter substitutes, cheese, coffee.
Confectionery of various kinds, such as the so-called "rum and butter" taffies.
Fats, honey, margarine, other food products, vinegar.
Odorant in—
Essences, flavoring agents.

Glue and Adhesives
Hardening agent (Brit. 444289) for—
Gelatin (the hardening effect is greatest at a pH value of 8).

Perfumery
Odorant in—
Blended perfumes, perfume materials.

Photographic
Hardening agent (Brit. 444289) for—
Gelatin (the hardening effect is greatest at a pH value of 8).

Soft Beverages and Ice Cream
Odorant in—
Essences, flavoring agents, ice cream mixes.

Diacetylethylenediamine

Chemical
In organic syntheses.

Electrical
Stabilizer (Brit. 423938) for—
Transformer oils.

Fats and Oils
Stabilizer (Brit. 423938) for—
Vegetable oils.

Fuel
Stabilizer (Brit. 423938) for—
Coal-carbonization spirits.

Lubricant
Stabilizer (Brit. 423938) for—
Lubricants, lubricating oils.

Petroleum
Stabilizer (Brit. 423938) for—
Petroleum oils, shale oils.

Diacetyltannin
Synonyms: Acetannin, Tanacetin, Tanacetine, Tanigen, Tanigene, Tannigen, Tannigene.

Chemical
Starting point in making—
Pharmaceutical derivatives.

Pharmaceutical
In compounding and dispensing practice.

1:1′-Diallyl-4:4′-tricarbocyanin Iodide

Photographic
Sensitizer (Brit. 436941 and 437017) for—
Photographic emulsions to infrared light with maxima at 800 to 1000 mu.

1:2-Dialphanaphthylaminoethane

Chemical
Starting point in making—
Intermediates and other derivatives.

Rubber
Antioxidant (Brit. 314756) in—
Vulcanizing.

1:4-Diaminoanthraquinone
German: 1:4-Diaminoanthrachinon.

Chemical
Starting point in making—
Methylomegasulphonate derivatives (Brit. 252992).

Dye
Starting point in making anthraquinone dyestuffs with—
Betanaphthoylchloride (German 432579).
1:2-Chloronaphthoyl chloride (German 432579).
Metabenzamidobenzoic acid (French 604347).
Meta-m′-diphenyldicarboxylic acid (French 604347).
Metamethoxybenzoyl chloride (French 604347).
5-Methoxyisophthalic acid (French 604347).
2:3-Methoxynaphthoyl chloride (German 432579).
3-Methylthiolbenzoic acid (French 604347).
Starting point in making—
Algol red 5G, algol red R, various other dyestuffs.

1:5-Diaminoanthraquinone
German: 1:5-Diaminoanthrachinon.

Dye
Starting point (French 604347) in making anthraquinone dyestuffs with—
Metabenzamidobenzoic acid, meta-m′-diphenyldicarboxylic acid, metamethoxybenzoyl chloride, 5-methylisophthalic benzoic acid, 3-methylthiolbenzoic acid.
Starting point in making—
Dianthraquinone carboxylaminoanthraquinone.
Indanthrene bordeaux B, indanthrene maroon, various other dyestuffs.

Fats, Oils, and Waxes
Coloring agent (Brit. 432867) for—
Stearic acid, tallow, waxes.

1:8-Diaminoanthraquinone
German: 1:8-Diaminoanthrachinon.

Dye
Starting point (Brit. 282854) in making dyestuffs with—
Acetaldehyde, benzaldehyde, butyraldehyde, crotonaldehyde, cinnamaldehyde, formaldehyde, heptaldehyde, hexaldehyde, paraformaldehyde, propionaldehyde, succinaldehyde.
Starting point in making—
Various synthetic dyestuffs.

4:8-Diaminoanthrarufin

Dye
Starting point (French 604347) in making anthraquinone vat dyestuffs with—
Metamethoxybenzoyl chloride.
Metabenzamidobenzoic acid.
Meta-m′-diphenyldicarboxylic acid.
5-Methylisophthalic acid.
3-Methylthiolbenzoic acid.
Starting point in making—
Various synthetic dyestuffs.

1:4-Diamino-5:8-dihydroxyanthraquinone

Chemical
Starting point (Brit. 396976) in making—
Triaminohydroxyanthraquinones.
Starting point in making—
Various synthetic dyestuffs.

3:3′-Diamino-4:4′-dimethyldiphenylmethane

Chemical
Starting point in making—
Intermediates, pharmaceuticals.

Dye
Starting point making various synthetic dyestuffs.

Metallurgical
Ingredient (Brit. 313134) of—
Liquid soldering fluxes, pickling baths for metals.
Reagent (Brit. 313134) in cleansing—
Rust from metals.

Diaminodinaphthylmethane

Chemical
Starting point in making—
Intermediates, pharmaceuticals.

Dye
Starting point in making various synthetic dyestuffs.

Metallurgical
Ingredient (Brit. 313134) of—
Liquid soldering fluxes, pickling baths for metals.
Reagent (Brit. 313134) in cleansing—
Rust from metals.

1:5-Diamino-4:8-dinitroanthraquinone
German: 1:5-Diamino-4:8-dinitroanthrachinon.

Dye
Starting point (Brit. 282854) in making dyestuffs with—
Acetaldehyde, benzaldehyde, butyraldehyde, cinnamaldehyde, crotonaldehyde, formaldehyde, hexaldehyde, heptaldehyde, paraformaldehyde, propionaldehyde, succinaldehyde.
Starting point in making—
Various synthetic dyestuffs.

1:8-Diamino-4:5-dinitroanthraquinone
German: 1:8-Diamino-4:5-dinitroanthrachinon.

Dye
Starting point (Brit. 282854) in making dyestuffs with—
Acetaldehyde, benzaldehyde, butyraldehyde, cinnamaldehyde, crotonaldehyde, formaldehyde, heptaldehyde, hexaldehyde, paraformaldehyde, propionaldehyde, succinaldehyde.
Starting point in making—
Various synthetic dyestuffs.

4:4′-Diamino-3:3′-dinitrobenzophenenone
French: 4:4′-Diamino-3:3′-dinitrobenzophénone.
German: 4:4′-Diamino-3:3′-dinitrobenzophenon.

Chemical
Starting point in making—
Intermediates, pharmaceuticals.

Dye
Starting point (Brit. 323792) in making azo dyestuffs for rayons, with the aid of—
Alkylaryl amines, allylaminophenol, allylnaphthylamine, alphanaphthylamine, aminonaphthoic acids, aminonaphthols, amylaminophenol, amylnaphthylamine, betanaphthylamine, butylnaphthylamine, cresols and their derivatives, dimethylmeta-aminophenol, ethylaminophenol, ethylnaphthylamine, gamma-chlorobetaoxypropionylnaphthylamine, meta-aminophenol, meta-anisidin, metacresidin, metaphenylenediamine, metaphenetidin, metatoluidin, metaxylidin, methylaminophenol, methylnaphthylamine, naphthylamine ethers, orthoaminophenol, orthoanisidin, orthocresidin, orthophenylenediamine, orthophenetidin, orthotoluidin, orthoxylidin, para-aminophenol, para-anisidin, paracresidin, paraphenylenediamine, para-

4:4′-Diamino-3:3′-dinitrobenzophenone (Cont'd)
nitrometaphenylenediamine, paratoluidin, paraxylidin, phenols and their derivatives, resorcinol, omega-oxyethylalphanaphthylamine.
Starting point in making—
Various synthetic dyestuffs.

4:4′-Diamino-3:3′-dinitrodiphenylmethane
Chemical
Starting point in making—
Intermediates, pharmaceuticals.
Dye
Starting point (Brit. 323792) in making azo dyestuffs for rayons, with the aid of—
Alkylaryl anilins, allylaminophenol, allylnaphthylamine, alphanaphthylamine, aminonaphthoic acids, aminonaphthols, amylaminonaphthylamine, betanaphthylamine, butylnaphthylamine, cresols and their derivatives, dimethylmeta-aminophenol, ethylaminophenol, ethylnaphthylamine, gammachlorobetaoxypropionylnaphthylamine, meta-aminophenol, meta-anisidin, metacresidin, metaphenetidin, metaphenylenediamine, metatoluidin, metaxylidin, methylaminophenol, methylnaphthylamine, naphthylamine ethers, omega-oxyethylalphanaphthylamine, orthoaminophenol, orthoanisidin, orthocresidin, orthophenylenediamine, orthophenetidin, orthotoluidin, orthoxylidin, para-aminophenol, para-anisidin, paracresidin, paranitrometaphenylenediamine, paraphenylenediamine, paratoluidin, paraxylidin, phenols and their derivatives, resorcinol.
Starting point in making various synthetic dyestuffs.

2:4′-Diaminodiphenyl
Dye
Starting point (Brit. 285504) in making nitro dyestuffs with—
Alphachloro-2:6-dinitrobenzene-4-sulphonic acid.
Alphachloro-2:4-dinitrobenzene-6-sulphonic acid.
Alphachloro-2-nitrobenzene-4-sulphonic acid.
Potassium alphachloro-2:6-dinitrobenzene-4-sulphonate.
Potassium alphachloro-2:4-dinitrobenzene-6-sulphonate.
Potassium alphachloro-2-nitrobenzene-4-sulphonate.
Starting point in making various synthetic dyestuffs.

2:4-Diaminodiphenylamine
Rubber
Age-resisting agent (U. S. 1959110).

4:4′-Diaminodiphenylamine-2-sulphonic Acid
French: Acide de 4:4′-diaminodiphényleamine-2-sulfonique.
German 4:4′-Diaminodiphenylaminsulfonsaeure.
Dye
Starting point (Brit. 282111) in making dyestuffs for animal fibers, pelts, and acetate rayon with the aid of—
Alphanaphthol, alphanaphthylamine, betanaphthol, betanaphthylamine, 1:5-dioxynaphthalene, 2:7-dioxynaphthalene.
Starting point in making various synthetic dyestuffs.

2:7-Diaminodiphenylene Oxide
Rubber
Antiaging agent (Brit. 422191).

Diaminodiphenylmethane
Chemical
Starting point in making—
Intermediates, pharmaceuticals.
Dye
Starting point in making various synthetic dyestuffs.
Metallurgical
Ingredient (Brit. 313134) of—
Liquid soldering fluxes, pickling baths for metals.
Reagent (Brit. 313134) in cleansing—
Rust from metals.

3:3′-Diamino-4:4-ditolyl Ketone
Dye
Starting point (Brit. 279146) in making azo dyestuffs with—
2:3-Oxynaphthoicanilide.
2:3-Oxynaphthoicbetanaphthalide.
2:3-Oxynaphthoic-4-chloroanilide.
2:3-Oxynaphthoic-3-phenetidide.
2:3-Oxynaphthoic-4-toluidide.
Starting point in making various synthetic dyestuffs.

Diaminoditolylmethane
Chemical
Starting point in making—
Intermediates, pharmaceuticals.
Dye
Starting point in making various synthetic dyestuffs.
Metallurgical
Ingredient (Brit. 313134) of—
Liquid soldering fluxes, pickling baths for metals.
Reagent (Brit. 313134) in cleansing—
Rust from metals.

1:5-Diamino-4-hydroxyanthraquinone
German: 1:5-Diamino-4-hydroxyanthrachinon.
Dye
Starting point (French 604347) in making anthraquinone dyestuffs with—
Metabenzamidobenzoic acid.
Metamethoxybenzoyl chloride.
Meta-diphenyldicarboxylic acid.
5-Methylisophthalic acid.
3-Methylthiolbenzoic acid.
Starting point in making various synthetic dyestuffs.

Diamino-2-hydroxynaphthalene
German: Diamino-2-hydroxynaphtalin.
Miscellaneous
Reagent in dyeing—
Furs, skins, hairs, and feathers (U. S. 1643246).

Diaminoisopropanol
Chemical
Absorbent (U. S. 1985885) for—
Acidic gases, such as carbon dioxide and hydrogen sulphide, from gaseous mixtures.
Metallurgical
Absorbent (U. S. 1985885) for—
Acidic gases, such as carbon dioxide and hydrogen sulsulphide, from gaseous mixtures.
Miscellaneous
As an emulsifying agent.
For uses, see under general heading: "Emulsifying agents."

1:5-Diamino-4-methoxyanthraquinone
German: 1:5-Diamino-4-methoxyanthrachinon.
Dye
Starting point (French 604347) in making anthraquinone dyestuffs with—
Metabenzamidobenzoic acid.
Metamethoxybenzoyl chloride.
Meta-m′-diphenyldicarboxylic acid.
5-Methylisophthalicbenzoic acid.
3-Methylthiobenzoic acid.
Starting point in making various synthetic dyestuffs.

3:6-Diamino-10-methylacridinium Chloride
Veterinary Medicine
Starting point (U. S. 1999750) in making—
Therapeutical products by dissolving in water in the presence of an excess of a sulphonated dyestuff of the group consisting of trypan blue, trypan red, and acid fuchsin (claimed especially useful for injections; for example, when treating certain infectious diseases of cattle and dogs).

4:4′-Diamino-2-nitrodiphenyl
Dye
Starting point (Brit. 285504) in making nitro dyestuffs with—
Alphachloro-2:6-dinitrobenzene-4-sulphonic acid.
Alphachloro-2:4-dinitrobenzene-6-sulphonic acid.
Alphachloro-2-nitrobenzene-4-sulphonic acid.
Potassium salts of the above acids.
Starting point in making various synthetic dyestuffs.

4:4′-Diamino-2-nitrodiphenylmethane
French: 4:4′-Diamino-2-nitrodiphénylemèthane.
German: 4:4′-Diamino-2-nitrodiphenylmethan.
Chemical
Starting point in making—
Intermediates, pharmaceuticals.
Dye
Starting point (Brit. 323792) in making azo dyestuffs for dyeing and printing various rayons, with the aid of—
Alkylarylanilins, allylaminophenol, allylnaphthylamine, alphanaphthylamine, aminonaphthoic acids, aminonaphthols, amylaminophenol, amylnaphthylamine,

4:4′-Diamino-2-nitrodiphenylmethane
betanaphthylamine, butylaminophenol, butylnaphthylamine, cresols and their derivatives, dimethylmeta-aminophenol, ethylaminophenol, ethylnaphthylamine, gammachlorobetaoxypropionylnaphthylamine, heptylaminophenol, heptylnaphthylamine, hexylaminophenol, hexylnaphthylamine, meta-aminophenol, meta-anisidin, metacresidin, metaphenylenediamine, metaphenetidin, metatoluidin, metaxylidin, methylaminophenol, methylnaphthylamine, naphthylamine ethers, ortho-aminophenol, orthoanisidin, orthocresidin, orthophenylenediamine, orthophenetidin, orthotoluidin, orthoxylidin, para-aminophenol, para-anisidin, paracresidin, paraphenylenediamine, paratoluidin, paraxylidin, pentylaminophenol, pentylnaphthylamine, phenols and their derivatives, propylaminophenol, propylnaphthylamine, resorcinol, omegaoxyethylalphanaphthylamine.
Starting point in making various synthetic dyestuffs.

2:4-Diamino-4′-oxydiphenylsulphone-3′-carboxylic Acid
French: Acide de 2:4-diamino-4′-oxydiphénylesulphone-3′-carbonique.
German: 2:4-Diamino-4′-oxydiphenylsulphone-3′-carbonsaeure.
Dye
Starting point (Brit. 262143) in making dyestuffs with—
Diazotized sulphanilic acid, metanilic acid, naphthionic acid, paranitranilinorthosulphonic acid.

Diaminopropanolamine Borate
Metallurgical
Absorbent (U. S. 1964808) for—
Hydrogen sulphide and carbon dioxide in extracting these gases from air or flue gas.

3:5-Diaminopyridin
Chemical
Starting point in—
Organic synthesis.
Disinfectant
Starting point (Brit. 442190) in making—
Bactericidal azo dyestuffs by coupling with diazotized arylamines or their substitution products.

Diammonium Undecoate
Miscellaneous
As a wetting agent.
For uses, see under: "Wetting agents."

Diamylalphanaphthylaminesulphonic Acid
French: Acide de diamylealphanaphthyleaminesulfonique.
German: Diamylalphanaphtylaminsulfonsaeure.
Chemical
Dispersing agent (Brit. 277048) in making—
Sulphur dispersions, soot dispersions.
Dye
Dispersing agent (Brit. 277048) in making—
Dispersions with indigoid dyestuffs.
Ingredient of various dyestuff preparations (Brit. 252392).
Miscellaneous
Ingredient of—
Washing and cleansing compositions (Brit. 278752).
Paint and Varnish
Dispersing agent (Brit. 277048) in making—
Fine dispersions of mineral pigments, barytes, and the like.
Textile
——, *Dyeing*
Ingredient of—
Dye liquors, to increase the absorption of the dyestuffs by the textile fiber (Brit. 278752).
——, *Finishing*
Ingredient of—
Finishing compositions (Brit. 278752).

Diamylamine
Cellulose Products
Solvent for—
Cellulose esters and ethers.
For uses, see under general heading: "Solvents."

Diamyl Phthalate
Cellulose Products
Plasticizer for—
Cellulose esters or ethers, cellulose nitrate (nitrocellulose).
For uses, see under general heading: "Plasticizers."
Resins
Plasticizer for—
Resins.

Diamyl Sulphide
Metallurgical
Flotation reagent.
Miscellaneous
Stench-producing agent.

2:5-Dianilidobenzoquinone
Synonyms: Quinonanilide.
German: 2:5-Dianilidobenzochinon.
Dye
Starting point in making—
Various vat dyestuffs (U. S. 1576678).

1-Diazo-5-nitroanthraquinone-2-carboxylic Acid
French: Acide de 1-diazo-5-nitroanthraquinone-2-carbonique.
German: 1-Diazo-5-nitroanthraquinone-2-carbonsaeure.
Chemical
Starting point in making—
Alphachloro-5-nitroanthrachinon-2-carboxylic acid (Brit. 262119).
Dye
Starting point in making various azo dyestuffs (Brit. 262119).

Diazo-orthoanisol
Synonyms: Azophor pink A.
Chemical
Starting point in making—
Guaiacol (U. S. 1623949).
Textile
——, *Dyeing*
Dyestuff for—
Betanaphtholated fabrics and yarns.

1-Diazo-2-oxynaphthalene-4-sulphonic Acid
Dye
Intermediate in making various dyestuffs.
Starting point (Brit. 404198) in making—
Dyestuffs (for coloring bones and bone objects rose tints) by nitrating and then reacting with 1-(32-sulphamido) phenyl-3-methyl-5-pyrazolone and a chromium salt.
Dyestuffs (for coloring bones and bone objects black tints) by nitrating and then reacting with betanaphthol and a chromium salt.

1;2:5:6-Dibenzanthracene
Miscellaneous
Reagent for—
Promoting cancerous growths on animals in pathologic research.

1:1′-Dibenzanthronyl Diselenide
Chemical
Starting point (U. S. 1999996) in making—
1-anthraquinonyl-1′-benzathronyl selenide by reacting with 1-chloroanthraquinone, sodium acetate, and naphthalene.
Various seleno ethers by reacting with 2-bromoanthraquinone, 1-chlor-2-aminoanthraquinone, 6-chlor-1-aminoanthraquinone, and 2-brom-1-aminoanthraquinone respectively.

5:6-Dibenzothiazole-1-carboxylic Chloride
Dye
Starting point (Brit. 441915) in making—
Greenish-yellow vat dyes of good fastness to light, chlorine, and alkali, by condensing with an ortho-aminothiol of the benzene, naphthalene, or anthraquinone series.
Greenish-yellow vat dyes of good fastness to light, chlorine, and alkali, by condensing with an arylamine and the orthothiol group subsequently introduced and the product cyclized.

Dibenzoyl
Glue and Adhesives
Hardening agent (Brit. 444289) for—
Gelatin (the hardening effect is greatest at a pH value of 8).
Photographic
Hardening agent (Brit. 444289) for—
Gelatin (the hardening effect is greatest at a pH value of 8).

1:5-Dibenzoyl-2:6-dioxynaphthalene
German: 1:5-Dibenzoyl-2:6-dioxynaphtalin.

Dye
Starting point (Brit. 249147) in making halogenated dibenzopyrenequinones with—
Phosphorus oxychloride, phosphorus pentabromide, phosphorus pentachloride.

3:9-Dibenzoylperylene

Petroleum
Reagent for—
Imparting fluorescence to hydrocarbon oils or liquids.

Dibenzoyl-1:4:5:8-tetraaminoanthraquinone
German: Dibenzoyl-1:4:5:8-tetraaminoanthrachinon.

Dye
Starting point (Brit. 282854) in making dyestuffs with—
Acetaldehyde, benzaldehyde, butyraldehyde, cinnamaldehyde crotonaldehyde, heptaldehyde, hexaldehyde, paraformaldehyde, propionaldehyde, succinaldehyde.

Dibenzthiazyl Disulphide

Rubber
Accelerator in—
Vulcanization processes.
Starting point in making—
Delayed-action accelerators by reaction with accelerators of the amine type, such as diphenylguanidine, the constituents of which, on fission, can give rise to the "two-accelerator effect."

Dibenzylanilin
French: Dibenzylaniline.
German: Dibenzilanilin.
Spanish: Dibenzilanilina.
Italian: Dibenzilanilina.

Ceramics
Stabilizing agent (Brit. 342288) in—
Compositions, containing various esters or ethers of cellulose, used for the production of protective and decorative coatings on ceramic products.

Chemical
Starting point in making—
Intermediates, pharmaceuticals, other derivatives.

Dye
Starting point in making various dyestuffs.

Glass
Stabilizing agent (Brit. 342288) in—
Compositions, containing various esters or ethers of cellulose, used in the manufacture of nonscatterable glass and for the production of decorative and protective coatings on glassware.

Leather
Stabilizing agent (Brit. 342288) in—
Compositions, containing various esters or ethers of cellulose, such as benzylcellulose, butylcellulose, and the like, used in the manufacture of artificial leather and for the production of decorative and protective coatings on leather goods.

Metallurgical
Stabilizing agent (Brit. 342288) in—
Compositions, containing various esters or ethers of cellulose, used for the production of decorative and protective coatings on metal ware.

Miscellaneous
Stabilizing agent (Brit. 342288) in—
Compositions, containing various esters or ethers of cellulose, used for the production of protective and decorative coating on miscellaneous compositions of matter.

Paint and Varnish
Stabilizing agent (Brit. 342288) in making—
Paints, varnishes, enamels, lacquers, and dopes containing various esters or ethers of cellulose, such as benzylcellulose, butylcellulose, and the like.

Paper
Stabilizing agent (Brit. 342288) in—
Compositions, containing various esters or ethers of cellulose, used in making coated paper and for the production of decorative and protective finishes on paper and pulp products.

Plastics
Stabilizing agent (Brit. 342288) in making—
Plastic compositions containing various esters or ethers of cellulose, such as butylcellulose, benzylcellulose, and the like.

Rubber
Stabilizing agent (Brit. 342288) in—
Compositions, containing various esters or ethers of cellulose, such as butylcellulose, benzylcellulose and the like, used for the production of decorative and protective coatings on rubber goods.

Stone
Stabilizing agent (Brit. 342288) in—
Compositions, containing various esters or ethers of cellulose, used for the production of decorative and protective coatings on artificial and natural stone.

Textile
Stabilizing agent (Brit. 342288) in—
Compositions, containing various esters or ethers of cellulose, such as butylcellulose, benzylcellulose, and the like, used for the production of decorative and protective coatings on woodwork.

Woodworking
Stabilizing agent (Brit. 342288) in—
Compositions, containing various esters or ethers of cellulose, such as butylcellulose, benzylcellulose, and the like, used for the production of decorative and protective coatings on woodwork.

Dibenzylanilinsulphonic Acid
French: Acide dibenzylanilinesulphonique.
German: Dibenzylanilinsulfonsäure.
Spanish: Acido dibenzilanilinasolfonico.

Chemical
Starting point in making—
Intermediates, pharmaceuticals, salts and esters, synthetic aromatic chemicals.

Dye
Starting point in making—
Eriocyanin.

Dibenzyldiphenylethylenediamine
German: Dibenzyldiphenylaethylendiamin.

Dye
Condensing agent in making—
Triarylmethane series dyestuffs (Brit. 249160).

Dibenzyl Disulphide

Paper
Stabilizing agent (U. S. 1963489) for—
Paraffin wax in papercoating.

Dibenzyl Ether

Textile
Delustring agent (Brit. 419477) for—
Viscose rayon.

Dibenzylmethylene Ether

Electrical
Starting point (Brit. 399868) in making—
Plastic materials with benzyl or ethyl cellulose, fillers, and coloring matter used as a component of insulated conductors.

Dibenzylnaphthalene

Chemical
Ingredient (U. S. 1897773) of—
Colloidal suspension used in tanning.

Dibenzylphenol

Chemical
In organic syntheses.

Dibetabutoxyethyl Sebacate

Cellulose Products
Plasticizer (U. S. 1991391) for—
Cellulose esters and ethers.
For uses, see under general heading: "Plasticizers."

Dibeta-9-carbazolyldiethyl Sulphide

Chemical
Intermediate (Brit. 444262 and 444501) in—
Organic syntheses.

Pharmaceutical
Claimed (Brit. 444262 and 444501) to have—
Value for pharmaceutical purposes.

Rubber
Accelerator (Brit. 444262 and 444501) in—
Vulcanizing.

Dibetaethoxyethyl Adipate

Cellulose Products
Solvent and plasticizer (U. S. 1991391) for—
Cellulose esters and ethers.
For uses, see under general heading: "Solvents."

Di(betahydroxyethyl)metatoluidin

Chemical
Reagent in—
Organic synthesis.

Dye
Coupling agent (Brit. 421975) in making—
Light-fast and readily discharged violet dyestuffs for acetate rayon with diazotised 6-bromo-2:4-dinitroanilin or 6-chloro-2:4-dinitroanilin.

4-Di(betahydroxyethyl)paraphenylenediamine-2-sulphonic Acid

Dye
Starting point (Brit. 447905, 447906, and 448016) in making—
Monoazo dyes for leather, particularly chrome leather.

Dibeta-1-indolyldiethyl Sulphide

Chemical
Intermediate (Brit. 444262 and 444501) in—
Organic syntheses.

Pharmaceutical
Claimed (Brit. 444262 and 444501) to have—
Value for pharmaceutical purposes.

Rubber
Accelerator (Brit. 444262 and 444501) in—
Vulcanizing.

Di(beta)methylphenoxyethyl Phthalate
French: Phthalate de dibétaméthylephénoxye-éthyle, Phthalate dibétaméthylephénoxye-éthylique.
German: Dibetamethylphenoxyaethylphtalat, Phtalsaeuredibetamethylphenoxyaethylester, Phtalsaeuresdibetamethylphenoxyaethyl.

Leather
Solvent and plasticizer (Brit. 306911) in—
Cellulose acetate compositions for coating artificial leather.

Paint and Varnish
Solvent and plasticizer (Brit. 306911) in making—
Cellulose acetate paints, varnishes, lacquers, and enamels.

Photographic
Solvent and plasticizer (Brit. 306911) in making—
Cellulose acetate films.

Plastics
Solvent and plasticizer (Brit. 306911) in making—
Cellulose acetate compositions.

Textile
Solvent and plasticizer (Brit. 306911) in making—
Cellulose acetate compositions for coating fabrics.

2:7-Di(betanaphthylamino)diphenylene Oxide

Rubber
Antiaging agent (Brit. 422191).

Dibetanaphthylnitrosoamine
German: Dibetanaphtylnitrosamin.

Chemical
Starting point in making—
Intermediates and other derivatives.

Rubber
Reagent (U. S. 1734633) in—
Controlling the action of vulcanizing and accelerating agents.

Dibetaparatoluenesulphonyldiethyl Sulphide

Chemical
Intermediate (Brit. 444262 and 444501) in—
Organic syntheses.

Insecticide
Insecticide (Brit. 444262 and 444501) for—
Animal pests, vegetable pests.

Textile
As a dyestuff (when employing suitable initial materials) (Brit. 444262 and 444501).
Assistant (Brit. 444262 and 444501) in—
Textile processing.

Di(beta)phenoxyethyl Phthalate
French: Phthalate de dibétaphénoxye-éthyle, Phthalate dibétaphénoxye-éthylique.
German: Dibetaphenoxyaethylphtalat, Phtalsaeuredibetaphenoxyaethylester, Phtalsaeuresdibetaphenoxyaethyl.

Leather
Softener and plasticizer (Brit. 306911) in—
Cellulose acetate compositions for coating artificial leather.

Paint and Varnish
Softener and plasticizer (Brit. 306911) in making—
Cellulose acetate paints, varnishes, lacquers, and enamels.

Photographic
Softener and plasticizer (Brit. 306911) in making—
Cellulose acetate films.

Plastics
Softener and plasticizer (Brit. 306911) in making—
Cellulose acetate compositions.

Textile
Softener and plasticizer (Brit. 306911) in making—
Cellulose acetate compositions for coating fabrics.

Di-b′-ethoxybetaethoxyethyl Adipate

Cellulose Products
Plasticizer (U. S. 1991391) for—
Cellulose esters and ethers.
For uses, see under general heading: "Plasticizers."

2:6-Dibrom-1:5-bis-2′-methylalphanaphthylaminoanthraquinone

Chemical
Starting point in—
Organic syntheses.

Dye
Starting point (Brit. 443958 and 443959) in making—
Vat dyestuffs.

Dibrom-1:2-chrysenequinone

Dye
Intermediate (Brit. 438609) in making—
Synthetic dyes.

Dibrom-2:8-chrysenequinone

Dye
Intermediate (Brit. 438609) in making—
Synthetic dyes.

5:7-Dibrom-7-methoxy-4-methylisatin Alphachloride

Dye
Starting point (Brit. 441548) in making—
Dyestuffs by condensing with 4-chlor-2-hydroxy-6-methoxy-3-methylthionaphthen.

2:6-Dibrom-4-nitroanilin

Dye
Starting point (Brit. 429936 and 430079) in making—
Orange-brown dyes for acetate rayon and animal fibers by diazotizing and coupling with normal-ethylbetasulphatoethylanilin.
Orange-brown dyes for acetate rayon and animal fibers by diazotizing and coupling with normal-methylbetasulphatoethylanilin.
Orange-brown dyes for acetate rayon and animal fibers by diazotizing and coupling with normal-gammasulphato-normal-propylanilin.
Red-brown dyes for acetate rayon and animal fibers by diazotizing and coupling with 3-betasulphatoethylaminoparatolylmethyl ether.

Dibromoacraldehyde

Photographic
Hardening agent (Brit. 406750) for—
Gelatin emulsions (reduces "fogging" tendency as compared with ordinary aldehydes).

1:3-Dibromoanthraquinone

Dye
In dye synthesis.
Starting point (Brit. 399241, addition to Brit. 381920) in making—
Vat dyes of the anthraquinone series.

Dibromoflavanthrone
German: Dibromflavanthron.

Chemical
Starting point in making various derivatives.

Dye
Starting point (Brit. 325550) in making vat dyestuffs with—
Alpha-aminoanthraquinone, aminodibenzanthrone, cyclohexamine.

3:5-Dibromohydroxydiphenyl
Disinfectant
As a bactericide (U. S. 1989081).

5:7-Dibromo-8-hydroxyquinolin
Pharmaceutical
Suggested for use (Brit. 351605) as—
Antiseptic.

5:7-Dibromoisatin Anilide
French: Anilide de 5:7-dibromoisatine, Anilide 5:7-dibromoisatinique.
German: 5:7-Dibromisatinanilid.
Chemical
Starting point in making various intermediates and other derivatives.
Dye
In dye synthesis.
Starting point (Brit. 291825) in making indigoid dyestuffs with—
4-Allymercapto-1-naphthol.
4-Amylmercapto-1-naphthol.
4-Benzylmercapto-1-naphthol.
4-Butylmercapto-1-naphthol.
4-Ethylmercapto-1-naphthol.
4-Heptylmercapto-1-naphthol.
4-Hexylmercapto-1-naphthol.
4-Isoallylmercapto-1-naphthol.
4-Isoamylmercapto-1-naphthol.
4-Isobutylmercapto-1-naphthol.
4-Isopropylmercapto-1-naphthol.
4-Methylmercapto-1-naphthol.
4-Naphthylmercapto-1-naphthol.
Paratolylmercapto-1-naphthol.
4-Pentylmercapto-1-naphthol.
4-Phenylmercapto-1-naphthol.
4-Tolylmercapto-1-naphthol.
4-Xylylmercapto-1-naphthol.

Dibromonitromethane
Fuel
Primer (Brit. 461320) for—
Diesel fuels.

1:2-Dibromopropanol-3
Petroleum
Solvent (Brit. 437573) in—
Refining mineral oils.

3:5-Dibromosalicylaldehyde
Photographic
Purification agent (U. S. 1973472) for—
Methylpara-aminophenol (developing agent).

Dibutenylanilin, Normal
Chemical
Starting point in making—
Intermediates, pharmaceuticals.
Insecticide
As an insecticide, alone and in compositions (Brit. 313934).
Soap
Ingredient (Brit. 313934) of—
Insecticidal soaps.

Dibutylamine
Chemical
Catalyst (Brit. 252870) in making—
Normal butyl para-aminobenzoate.
Normal butyl paranitrobenzoate.

Dibutylanilinsulphonic Acid
French: Acide de dibutyleanilinesulfonique.
German: Dibutylanilinsulfonsaeure.
Dye
Ingredient of various dyestuff preparations (Brit. 252392).
Soap
Ingredient of—
Detergents (Brit. 280110).
Textile
——, *Bleaching*
Ingredient of—
Bleach liquors for wool (Brit. 280110).
——, *Dyeing*
Ingredient of—
Dye liquors (Brit. 280110).
——, *Finishing*
Ingredient (Brit. 280110) of—
Fulling baths.
Wetting preparations for felt-like fabrics.
——, *Manufacturing*
Ingredient (Brit. 280110) of—
Wool-carbonizing preparations.
Wool-degreasing preparations.

Dibutyl Ether
French: Éther de butanole, Éther butilique, Éther de butyle.
German: Butanolaether, Butylaether.
Chemical
Solvent for—
Organic acids, such as acetic, propionic, benzoic, salicylic, stearic.
Dye
As a solvent.
Fats and Oils
Solvent for—
Essential oils, fatty oils.
Food
Solvent in making and purifying—
Flavoring materials.
Gums
As a solvent.
Miscellaneous
As a general solvent.
Perfume
Extractant for—
Perfume materials.
Petroleum
Solvent for—
Dewaxed petroleum oils.
Resins and Waxes
Solvent for—
Resins, rosin, waxes.

Dibutyl Malate
Cellulose Products
Plasticizer (U. S. 1942843) for—
Cellulose acetate.
For uses, see under general heading: "Plasticizers."

Dibutyl Mesotartrate
Cellulose Products
Plasticizer (U. S. 1659906) for—
Cellulose nitrate (nitrocellulose).
For uses, see under general heading: "Plasticizers."

Dibutylnaphthalenesulphonic Acid
Agriculture
Wetting agent (Brit. 422350) for—
Green fodder preservatives, such as dilute solution of formic or hydrochloric acid, added to the fodder before placing in silos.

Di-2-butyloctyl Phthalate
Gums and Resins
Plasticizer (Brit. 442643) for—
Natural and artificial gums and resins.

Di-2-butyloctyl Succinate
Gums and Resins
Plasticizer (Brit. 442643) for—
Natural and artificial gums and resins.

Dibutyl Phthalate
Synonyms: Butyl phthalate.
French: Phthalate de butyle, Phthalate butylique, Phthalate de dibutyle, Phthalate dibutylique.
German: Butylphtalat, Dibutylphtalat, Phtalsäurebutylester, Phtalsäuredibutylester, Phtalsäuresbutyl, Phtalsäuresdibutyl.
Cellulose Products
Plasticizer and solvent for—
Cellulose acetate, cellulose esters and ethers, cellulose nitrate.
For uses, see under general heading: "Plasticizers."

Dibutyl Ricinoleicsulphonate
Miscellaneous
As a dispersing agent (Brit. 362195).
For uses, see under general heading: "Dispersing agents."

Dibutyl Selenide

Lubricant
Starting point (Brit. 440175) in making—
Addition agents for high-pressure lubricating oils or greases by mixing and reacting with organo-metallic compounds.

Dibutylstearamide

Cellulose Products
Plasticizer (U. S. 1986854) for—
Cellulose esters and ethers.
For uses, see under general heading: "Plasticizers."

Dibutyl Succinate

Cellulose Products
Plasticizer for—
Cellulose acetate, cellulose esters or ethers, cellulose nitrate (nitrocellulose).
For uses, see under general heading: "Plasticizers."

Dibutyl Tartrate

Synonyms: Butyl tartrate.
French: Tartrate de butyle, Tartrate dibutylique.
German: Weinsaeurebutylester, Weinsaeuredibutyl.

Chemical
Starting point in making—
Butyl stearate.

Miscellaneous
See also "Plasticizers."

Paint and Varnish
Plasticizer in making—
Cellulose ester and ether lacquers.

Photographic
Substitute for camphor in making films.

Plastics
Substitute for camphor as a plasticizer.

Dicalcium Phosphate

Synonyms: Bibasic calcium phosphate, Bicalcic phosphate, Dibasic calcium phosphate, Dicalcium orthophosphate, Secondary calcium phosphate.

Ceramics
Raw material in making—
Bone china.

Cosmetic
Polishing agent in—
Dentifrices.

Fertilizer
As a fertilizer (has two-fold action: [a] source of phosphoric acid, [b] soil sweetener).
Ingredient of—
Fertilizer mixtures.

Food
Ingredient of—
Baking powder (usually present as an impurity in the monobasic phosphate used as a leavening agent; the presence of the dibasic salt is advantageous in that it insures the absence of free phosphoric acid).

Miscellaneous
As a relatively soft abrasive.

4-(2':4'-Dicarboxyphenyl)-7:8-phthaloyl-2-acridone Acid Chloride

Chemical
In organic syntheses.

Dye
In dye syntheses.
Starting point (Brit. 449263) in making—
Yellow vat dyes with 1-amino-5-benzamidoanthraquinone.

Dicetylpiperidinium Bromide

Dry-Cleaning
Addition agent (Brit. 453523) to—
Solvents, such as trichloroethylene, carbon tetrachloride, and benzene.

Textile
Addition agent (Brit. 453523) to—
Solvents, such as trichloroethylene, carbon tetrachloride, and benzene.

2:5-Dichlor-3-aminobenzotrifluoride-6-sulphonic Acid

Dye
Intermediate (Brit. 446532) in making dyestuffs.

4:6-Dichlor-3-aminobenzotrifluoride-2-sulphonic Acid

Dye
Intermediate (Brit. 446532) in making dyestuffs.

2:4-Dichlor-2'-amino-4'-methoxy-5-methylazobenzene

Dye
Coupling agent (Brit. 434416) in making—
Red-brown, water-insoluble dyestuffs with anilides.

2:4-Dichlor-2'-amino-4'-methylazobenzene

Dye
Coupling agent (Brit. 434416) in making—
Bordeaux red, water-insoluble dyestuffs with 5-chloro-orthotoluidide.

6:9-Dichlor-2-butylthiolacridin

Pharmaceutical
Claimed (Brit. 363392 and 437953) as—
New pharmaceutical.

Dichlor-2:8-chrysenequinone

Dye
Intermediate (Brit. 438609) in making—
Synthetic dyes.

Dichlordifluoromethane

French: Dichlorure et difluorure de méthane.
German: Dichlordifluormethan.

Refrigeration
Refrigerant for—
Use in freezing machines, particularly in small domestic machines and in all units where it is vital that the refrigerant does not attack metals.

4:6-Dichlor-1:2-diketo-3-methyldihydrothionaphthen-1-paradimethylaminoanil

Dye
Starting point (Brit. 441548) in making—
Dyestuffs by condensing with 4-chlor-2-hydroxy-6-methoxy-3-methylthionaphthen.

6:9-Dichlor-2-ethylthiolacridin

Pharmaceutical
Claimed (Brit. 363392 and 437953) as—
New pharmaceutical.

Dichlorhydrin

Synonyms: Dichloroisopropyl alcohol, 1:3-Dichloropropanol-2.
French: Dichlorhydrine.
German: Alphapropenyldichlorhydrin, Glyceroldichlorhydrin.
(The commercial product is a mixture of the two isomers 1:3-dichlor-2-hydroxypropane and 1:2-dichlor-3-hydroxypropane, of which the former is in a dominant amount.)

Cellulose Products
Solvent for—
Cellulose acetate, ethylcellulose, nitrocellulose.

Ceramic
Solvent in—
Compositions, containing natural or synthetic resins, nitrocellulose, cellulose acetate, or other cellulose esters or ethers, used as coatings for protecting and decorating ceramic products.

Chemical
Intermediate in—
Organic syntheses.
Solvent miscible with—
Most organic solvents, vegetable oils.

Cosmetic
Solvent in—
Nail enamels and lacquers containing natural or synthetic resins, nitrocellulose, cellulose acetate, or other cellulose esters or ethers as base material.

Electrical
Solvent in—
Insulating compositions, containing natural or synthetic resins, nitrocellulose, cellulose acetate, or other cellulose esters or ethers, used for covering wire and in making electrical machinery and equipment.

Glass
Solvent in—
Compositions, containing natural or synthetic resins, nitrocellulose, cellulose acetate, or other cellulose esters or ethers, used in the manufacture of non-scatterable glass and as coatings for decorating and protecting glassware.

Dichlorhydrin (Continued)

Glue and Adhesives
Solvent in—
Adhesive compositions containing natural or synthetic resins, nitrocellulose, cellulose acetate, or other cellulose esters or ethers.

Gums
Solvent for—
Copal, copal-ester, dammar, elemi, manila, mastic, other gums.

Leather
Solvent in—
Compositions, containing natural or synthetic resins, nitrocellulose, cellulose acetate, or other cellulose esters or ethers, used in the manufacture of artificial leathers and as coatings for decorating and protecting leathers and leather goods.

Metal Fabricating
Solvent in—
Compositions, containing natural or synthetic resins, nitrocellulose, cellulose acetate, or other cellulose esters or ethers, used as coatings for protecting and decorating metallic articles.

Miscellaneous
Solvent miscible with—
Most organic solvents, vegetable oils.
Solvent in—
Coating compositions, containing natural or synthetic resins, nitrocellulose, cellulose acetate, or other cellulose esters or ethers, used for protecting and decorating various articles.

Paint and Varnish
Binder for—
Water colors.
Solvent in—
Paints, varnishes, lacquers, enamels, and dopes containing natural or synthetic resins, nitrocellulose, cellulose aacetate, or other cellulose esters or ethers.

Paper
Solvent in—
Compositions, containing natural or synthetic resins, nitrocellulose, cellulose acetate, or other cellulose esters or ethers, used in the manufacture of coated papers and as coatings for decorating and protecting products made of paper or pulp.

Photographic
Solvent in making—
Films from nitrocellulose, cellulose acetate, or other esters or ethers of cellulose.

Plastics
Solvent in making—
Plastics from or containing natural or synthetic resins, nitrocellulose, cellulose acetate, or other cellulose esters or ethers.

Resins
Solvent for—
Benzyl abietate, cumar resins, ester gums, glyptal resins, shellac, resins in general.

Rubber
Solvent in—
Compositions, containing natural or synthetic resins, nitrocellulose, cellulose acetate, or other cellulose esters or ethers, used as coatings for protecting and decorating rubber goods.

Stone
Solvent in—
Compositions, containing natural or synthetic resins, nitrocellulose, cellulose acetate, or other cellulose esters or ethers, used as coatings for decorating and protecting artificial and natural stone.

Textile
Solvent in—
Compositions, containing natural or synthetic resins, nitrocellulose, cellulose acetate, or other cellulose esters or ethers, used in the manufacture of coated fabrics.

Wood
Solvent in—
Compositions, containing natural or synthetic resins, nitrocellulose, cellulose acetate, or other cellulose esters or ethers, used as protective and decorative coatings on woodwork.

6:9-Dichlor-2-iso-octylthiolacridin

Pharmaceutical
Claimed (Brit. 363392 and 437953) as—
New pharmaceutical.

Dichlor-1-ketotetrahydronaphthalene

French: Dichlor-1-cétotétrahydronaphthalène.
German: Dichlor-1-ketotetrahydronaphthalin.
Spanish: Dichlor-1-cetotetrahidronaftoleno.
Italian: Dichlor-1-cetotetraidronaftalene.

Chemical
Intermediate (German 377587) in making—
Synthetic aromatics, synthetic chemicals, synthetic pharmaceuticals.

Dye
Intermediate (German 377587) in making—
Synthetic dyestuffs.

Insecticide
As an insecticide (German 377587).

5:7-Dichlor-7-methoxy-4-methylisatin Alphachloride

Dye
Starting point (Brit. 441548) in making—
Dyestuffs by condensing with 4-chlor-2-hydroxy-6-methoxy-3-methylthionaphthen.

5:7-Dichlor-4-methylindoxyl

Dye
Starting point (Brit. 443275) in making—
Blue dyestuffs by oxidation.

5:7-Dichlor-4-methylisatin

Dye
Starting point (Brit. 443275) in making—
Brown dyestuffs by condensation with 2:1-naphthathioindoxyl.

5:7-Dichlor-4-methylisatin Alphachloride

Dye
Starting point (Brit. 443275) in making—
Blue dyestuffs by condensation with 5-chlor-2-hydroxy-4-methoxyalphanaphthol.
Red-blue dyestuffs by condensation with 2-hydroxy-3-methylthionaphthen.
Violet dyestuffs by condensation with 5-chlor-2-hydroxy-3-methylthionaphthen.

1:3-Dichlor-2-methylpropanol-2

Petroleum
Solvent (Brit. 435096) in—
Refining mineral oils.

6:9-Dichlor-2-methylthiolacridin

Pharmaceutical
Claimed (Brit. 363392 and 437953) as—
New pharmaceutical.

2:6-Dichlor-4-nitroanilin

Dye
In dye synthesis.
Starting point (Brit. 429936 and 430079) in making—
Orange-brown dyes for acetate rayon and animal fibers by diazotizing and coupling with normal-ethylbetasulphatoethylanilin.
Orange-brown dyes for acetate rayon and animal fibers by diazotizing and coupling with normal-methylbetasulphatoethylanilin.
Orange-brown dyes for acetate rayon and animal fibers by diazotizing and coupling with normal-gammasulphato-normal-propylanilin.
Red-brown dyes for acetate rayon and animal fibers by diazotizing and coupling with 3-betasulphatoethylaminoparatolylmethyl ether.

Dichloroacetic Acid

French: Acide acétique, dichloré; Acide dichloracétique, Dichlorure d'acide acétique.
German: Dichloressigsäure.

Chemical
Reagent in making—
Aminoesters having saponaceous properties and used as addition agents to soaps and as bases for washing compounds (Brit. 403883).
Intermediates.
Organic thiosulphates used as wetting agents and as bases for washing compounds (Brit. 397445).
Pharmaceutical chemicals.
Synthetic organic chemicals.

Dichloroacetic Acid Cyclohexylester

Detergent
Starting point (Brit. 408754) in making—
Saponaceous products by reaction with tertiary amines, which may be used alone or with other soaps, fillers, or compounds giving off oxygen.

Dichloroacetic Acid Dodecylester

Detergent
Starting point (Brit. 408754) in making—
Saponaceous products by reaction with tertiary amines, which may be used alone or with other soaps, fillers, or compounds giving off oxygen.

Dichloroacetic Acid Hexadecylester

Soap
Starting point (Brit. 403883) in making—
Saponaceous products by reaction with amines, such as anilin, piperidin bases, hydroxyethylanilin, dihydroxyethylanilin, paratoluidin (these products may be used alone, or with soaps, fillers, or compounds giving off oxygen).

Dichloroacetic Acid Octadecylester

Detergent
Starting point (Brit. 408754) in making—
Saponaceous products by reaction with tertiary amines, which may be used alone or with other soaps, fillers, or compounds giving off oxygen.

Dichloroacetic Acid Tetradecylester

Soap
Starting point (Brit. 403883) in making—
Saponaceous products by reaction with amines, such as anilin, piperidin bases, hydroxyethylanilin, dihydroxyethylanilin, paratoluidin (these products may be used alone, or with soaps, fillers, or compounds giving off oxygen).

1:3-Dichloro-2-aminoanthraquinone
German: 1:3-Dichlor-2-aminoanthrachinon.

Dye
In dye synthesis.
Starting point (Brit. 278417) in making dyestuffs for wool, silk, and acetate rayon, with the aid of—
Allylamine, amylamine, anilin, benzylamine, butylamine, cresidin, diallylamine, diamylamine, dibenzylamine, dibutylamine, diethylamine, dimethylamine, diphenylamine, dipropylamine, ethylamine, ethylanilin, formylamine, isoallylamine, isoamylamine, isobutylamine, isopropylamine, metaphenylenediamine, metatoluidin, metaxylidin, methylamine, methylanilin, naphthylamine (alpha and beta), orthophenylenediamine, orthotoluidin, orthoxylidin, paraphenylenediamine, paratoluidin, paraxylidin, propylamine, tolylamine.

4:6-Dichloro-2-aminophenol

Dye
Starting point in making—
Azarin S.

2:5-Dichloroanilin
French: 2:5-Dichloroaniline.
German: 2:5-Dichloranilin.
Italian: 2:5-Dicloroanilino.

Chemical
Starting point in making—
Aromatics, intermediates, pharmaceuticals, various other derivatives.

Dye
Starting point in making—
Chloramine black N, chloramine blue 3G, chloramine blue HW, chloramine green B, nigrophor BASF.
Various dyestuffs.
Starting point (Brit. 347113) in making azo dyestuffs right on the fiber, with the aid of—
7-Hydroxyalphanaphthocarbazol-6-carboxylic beta-anilide.
7-Hydroxyalphanaphthocarbazol-6-carboxylic 5-chloro-orthoanisidide.
7-Hydroxyalphanaphthocarbazol-6-carboxylic 2:5-dimethoxyanilide.
7-Hydroxyalphanaphthocarbazol-6-carboxylic 3:4-dimethoxyanilide.
7-Hydroxyalphanaphthocarbazol-6-carboxylic meta-nitranilide.
7-Hydroxyalphanaphthocarbazol-6-carboxylic ortho-anisidide.
7-Hydroxyalphanaphthocarbazol-6-carboxylic ortho-toluidide.
7-Hydroxyalphanaphthocarbazol-6-carboxylic ortho-methylpara-anisidide.
7-Hydroxyalphanaphthocarbazol-6-carboxylic para-anisidide.

Dichloroazodicarbonamidin, Normal

Disinfectant
As a bactericide (Brit. 436093).

Dichlorobenzenesulphamide

Insecticide and Fungicide
Essential ingredient (U. S. 1997918) of—
Agent for destroying rust on cultivated plants.

1:2-Dichlorobenzene-4-sulphanilide
French: 4-Sulphanilide de 1:2-dichlorobenzène.
German: 1:2-Dichlorbenzol-4-sulphanilid.

Chemical
Starting point in making—
Pharmaceuticals and other derivatives.

Dye
Starting point in making various dyestuffs.

Miscellaneous
Reagent for—
Treating furs, hair, and feathers in order to dye them simultaneously with mothproofing.

Textile
Reagent for—
Treating wool and felt in order to dye them simultaneously with mothproofing.

2:4-Dichlorobenzoic Acid

Chemical
Starting point in making—
Intermediates, salts, esters, and other derivatives.
Starting point (Brit. 353537) in making acridin derivatives with the aid of—
4-Anisidin, 4-cresidin, 4-phenetidin, 4-toluidin, 4-xylidin.

4:4′-Dichlorobenzophenone

Dye
Starting point (Brit. 439815 and 417014) in making—
Blue dyestuffs by condensing with (1) ethylbutylmetatoluidin, (2) a primary 4-alkoxy- or 4-aryloxyarylamine and sulphonating the product.
Very greenish-blue dyestuffs by condensing with (1) dinormal-butylmetaxylidin, (2) a primary 4-alkoxy- or 4-aryloxyarylamine and sulphonating the product.

1:4-Dichlorobutanol-2

Petroleum
Solvent (Brit. 435096) in—
Refining mineral oils.

Dichloroctyl Alcohol

Petroleum
Solvent (Brit. 435096) in—
Refining mineral oils.

Dichlorodiethylenediamino-Cobaltic Chloride

Miscellaneous
Restrainer (Brit. 415672) of—
Corrosion of alkali-sensitive metals and alloys by alkaline cleansing compositions, such as mixtures of trisodium phosphate, soda ash, sodium metasilicate.

Dichlorodiethylenediamino-Cobaltic Nitrate

Miscellaneous
Restrainer (Brit. 415672) of—
Corrosion of alkali-sensitive metals and alloys by alkaline cleansing compositions, such as mixtures of trisodium phosphate, soda ash, sodium metasilicate.

4:4′-Dichloro-2:2′-dimethyloxacarbocyanin Iodide

Photographic
Sensitizing agent (Brit. 430357) for—
Emulsions.

5:5′-Dichloro-6:6′-dimethylthioindigo
French: 5:5′-Dichloro-6:6′-diméthylethioindigo.
German: 5:5′-Dichlor-6:6′-dimethylthioindig.

Dye
Starting point (Brit. 277398) in making soluble derivatives by treatment with chlorosulphonic acid, methylchlorosulphonate or sulphur trioxide in the presence of—
Acetyl chloride, benzoyl chloride, carbonyl chloride, chloroformic ester, paratoluenesulphonic chloride, phthalic anhydride, phthalimid.

Dichlorodiphenylmethane

Electrical

Cooling medium (Brit. 413596, 433070, 433071, and 433072) in—
- Electrical apparatus, such as transformers, switches, capacitors, cables, bushings, and junction boxes (may be employed in admixture with trichlorobenzene, chlorinated diphenyl, and the like).

Dielectric (Brit. 413596, 433070, 433071, and 433072) in—
- Electrical apparatus, such as transformers, switches, capacitors, cables, bushings, and junction boxes (may be employed in admixture with trichlorobenzene, chlorinated diphenyl, and the like).

Dichloroethylene

Synonyms: Dichlorethylene, Ethylene dichloride.
French: Chlorure d'éthylène, Éthylène, dichlorée.
German: Aethylenchlorid, Aethylendichlorid, Chloraethylen, Dichloraethylen.
(Note: In some cases where dichloroethylene is used as a commercial solvent it is necessary to dilute, or mix it with other solvents.)

Analysis

Extracting medium for various purposes.
Solvent for the extraction and assay of—
- Drugs.

Solvent in analyzing and testing—
- Animal oils, ashes, breadstuffs, butter, cakes, cheese, chocolate, cocoa, essential oils, fats, flour, hops, meals, meat, milk, mineral phosphates, resins, rosin, rosin oil, rubber, soaps, vegetable oils.

Automotive

Degreasing agent for—
- Automobile bodies, automobile parts.

Dewaxing agent in—
- Manufacturing operations.

Ceramics

Solvent in—
- Coating compositions, containing nitrocellulose, cellulose acetates, or other esters of cellulose, as well as resins, waxes, and gums, used for protecting and decorating ceramic ware.

Chemical

Catalyst in—
- Acetylation of cellulose in making cellulose acetate (U. S. 1823359).
- Hydrolysis of an organic acid solution of cellulose acetate (U. S. 1857190).

Extractant for—
- Alkaloids, drug principals.

Ingredient of solvent mixtures containing—
- Acetone, alcohol, benzene, chlorinated hydrocarbons, turpentine.

Reagent for—
- Introducing chlorine in the manufacture of inorganic and organic compounds.

Reagent in making—
- Intermediates, organic chemicals, pharmaceuticals.

Solvent for—
- Acid in concentration of acetic acid (Brit. 400169).
- Inorganic chemicals, organic chemicals.

Solvent (in admixture with alcohol) for—
- Cellulose acetate, cellulose nitrate.

Solvent in—
- Processes for separating isomers, particularly nitrophenols, dioxybenzenes, and the like.

Starting point in making—
- Chlorinated fats, chlorinated chemical compounds.
- Ethyleneglycol (by heating with a solution of an alkali carbonate or bicarbonate under pressure).
- Ethyleneglycol (by heating with sodium formate in methanol solution).
- Ethyleneglycol (German 574064).
- Ethylidene chloride (U. S. 1900276).

Construction

Solvent for—
- Washing tiles, tiled fronts of buildings.

Dye

Reagent and solvent in making—
- Synthetic dyestuffs of various classes.

Electrical

Solvent for—
- Cleaning electric motors and other electrical machinery.

Solvent in—
- Compositions, containing cellulose acetate, nitrocellulose, or other esters or ethers of cellulose, and at times resins, gums and the like, used for insulating cables, wiring, and electrical machinery and equipment.

Fats and Oils

Extractant for—
- Animal oils, essential oils, fats, greases, vegetable oils.

Solvent for—
- Animal oils, essential oils, fats, greases, vegetable oils.

Solvent in—
- Recovering oils from fuller's earth and other substances used in bleaching.

Dry Cleaning

Solvent.
Spotting agent.

Fertilizer

Solvent for—
- Degreasing fish scrap.

Food

Extractant of soluble substances from—
- Berries, fruits, seeds.

Solvent for—
- Making food flavors, purifying foodstuffs.

Glass

Solvent for—
- Degreasing glass.

Solvent in—
- Compositions, containing cellulose acetate, nitrocellulose, or other esters or ethers of cellulose, and artificial or natural resins, waxes and gums, used in the manufacture of non-scatterable glass and for the decoration and protection of glassware.

Glues and Adhesives

Ingredient of—
- Glues.
- Special adhesive composition containing cellulose acetate, nitrocellulose, or other esters or ethers of cellulose.

Reagent in—
- Preparing gelatins.

Solvent for—
- Degreasing bones and hides preparatory to the manufacture of glue and gelatin.

Gums

Solvent for various gums.

Ink

Solvent in making—
- Printing inks.

Insecticide

As a general insecticide.
Extractant (U. S. 1915662) for—
- Pyrethrum flowers in manufacture of a spray type insecticide.

Ingredient of—
- Fumigant said to be very effective against fur, grain, and household insects and parasites (mixed with carbon tetrachloride, 25 percent).
- Fumigating compositions, insecticidal compositions, preparations for exterminating parasites, vermicidal compositions.

Leather

Solvent for—
- Cleansing spotted leathers.
- Removing natural oils and greases from hides and skins before tanning so as to prevent staining thereafter and insure evenness of the leather finish and tan.

Solvent in—
- Compositions, containing cellulose acetate, nitrocellulose, or other esters or ethers of cellulose, as well as artificial or natural resins, gums, and waxes, used in the manufacture of artificial leather and for the protection and decoration of leather goods.

Mechanical

Penetrating and softening agent (U. S. 1909200) in—
- Carbon-removing agent.

Solvent for—
- Cleansing and degreasing machinery of various sorts.
- Cleansing drive wheels of compression pumps and other mechanical equipment.
- Degreasing automobile brake bands.

Suggested ingredient of—
- Motor fuel, now under test in France.

Metallurgical

Solvent for—
- Cleansing and degreasing metallic surfaces preparatory to painting or coating.
- Degreasing die castings, metal stampings, metals to be electroplated, nuts and bolts.
- Preparing metals for pickling, plating, shellacking, sherardizing, varnishing.

Dichloroethylene (Continued)
Solvent and diluent in—
Compositions, containing cellulose acetate, or other esters or ethers of cellulose, used for protecting and decorating metallic articles.
Miscellaneous
As a general solvent.
Degreasing agent in treating—
Furs (also acts as a parasiticide).
Ingredient of—
Compositions, containing clay, for cleansing ivory, horn, and bone.
Polishing compositions of various sorts.
Preparations used for the removal of stains from celluloid articles.
Preparations used for cleansing typewriters.
Solvent for—
Degreasing dishes, kitchenware, hardware, metal furniture, safety razor blades.
Solvent and diluent in—
Compositions, containing cellulose acetate or other esters or ethers of cellulose, used for decorating and protecting various products.
Preservative for—
Biological products.
Oilcloth and Linoleum
Solvent in making—
Coating compositions.
Paint and Varnish
Diluent (Brit. 395478) in—
Lacquer composed of vinyl chloracetate and vinyl stearate, polymerized in acetone solution and the resulting solution diluted.
Ingredient of—
Paint, lacquer, and varnish removers.
Solvent in making—
Nitrocellulose composition (U. S. 1915163).
Nitrocellulose lacquers (Brit. 390867).
Paints, varnishes, lacquers, enamels and dopes containing cellulose acetate, nitrocellulose, or other esters or ethers of cellulose, with waxes, gums and artificial or natural resins.
Paper
Solvent for—
Removing oil from paper.
Solvent in—
Compositions, containing cellulose acetate, nitrocellulose, or other esters or ethers of cellulose, with gums, waxes, and natural or artificial resins, used in the manufacture of coated paper and for coating and decorating paper and pulp products.
Perfume
Solvent for—
Extracting aromatic principles from flowers, particularly those altered by heat.
Petroleum
Ingredient of—
Compounded solvent preparations containing petroleum distillates.
Solvent for—
Degreasing light mineral oils.
Extracting wax from mineral oil distillates.
Photographic
Solvent for—
Cleaning and degreasing motion picture film.
Solvent in making—
Motion picture film.
Pharmaceutical
Extractant for—
Cocaine.
Solvent in—
Iodide solution used for disinfecting skin prior to surgical operations.
Pharmaceutical products.
Plastics
Degreasing solvent.
Ingredient (U. S. 1896145) of—
Solvent mixture (with methanol) used in making flexible films from cellulose acetate.
Solvent and diluent in making—
Compositions containing cellulose acetate, nitrocellulose, or other esters or ethers of cellulose, with gums, waxes, and artificial or natural resins.
Solvent (Brit. 390867) in making—
Plastics containing nitrocellulose.
Printing
Solvent for—
Cleaning engraved plates, lithographic stones, printing machinery, type.
Resins and Waxes
Solvent for various resins and waxes.
Rubber
Ingredient of—
Rubber cements, rubber mastics.
Rubber compositions used in the manufacture of rubberized cloth.
Solvent for—
Rubber, chlorinated rubber.
Solvent in—
Coating compositions, containing cellulose acetate, nitrocellulose, or other esters or ethers of cellulose, with gums, waxes, and artificial or natural resins, used for decorating and protecting rubber goods.
Sanitation
Solvent for—
Degreasing garbage.
Soap
Ingredient of—
Cleansing compositions, dry-cleaning compositions, spotting fluids.
Solvent in making—
Gelatinous water-soluble soaps from sulphonated oils and resins.
Paste soaps for removing grease stains.
Soaps with sodium ricinoleate.
Textile soaps from linseed oil or castor oil.
Stone
Solvent in—
Compositions containing cellulose acetate, nitrocellulose, or other esters or ethers of cellulose, with artificial or natural resins, gums and waxes, used for the decoration and protection of artificial and natural stone.
Sugar
Solvent for—
Extracting waxes from filter press mud in refining.
Textile
——, *Dyeing*
Ingredient of—
Preparations, containing turkey red oil and chlorinated hydrocarbons, used for dyeing and wetting.
——, *Finishing*
Ingredient of—
Scouring compositions containing sulphonated oil soaps.
Solvent in—
Coating compositions containing cellulose acetate, nitrocellulose, or other ethers or esters of cellulose.
——, *Manufacturing*
Solvent for—
Cleaning knitting machine needles, cleaning silk and silk hosiery, degreasing textiles, degreasing wool, degumming silk.
Solvent and diluent in—
Compositions containing cellulose acetate, nitrocellulose, or other esters or ethers of cellulose, used for making coated textiles.
Scouring compositions.
Tobacco
Solvent for—
Extracting nicotine.
Woodworking
Solvent and diluent in—
Compositions, containing cellulose acetate, nitrocellulose, or other esters or ethers of cellulose, used for decorating and protecting woodwork.

Dichloroethyl Ether
Synonyms: Beta-b′-dichlorethyl ether, Dichlordiethyl ether, 2:2′-Dichloroethyl ether, Symmetrical dichlorethyl ether.
French: Éther dichlordiéthylique, Éther dichloréthylique.
German: Dichloraethylaether, Doppeltchloraethylaether.
Analysis
Extracting medium for various purposes.
Solvent in analyzing and testing—
Animal oils, ashes, breadstuffs, butter, cakes, cheese, chocolate, cocoa, essential oils, fats, flour, hops, meals, meat, milk, mineral phosphates, resins, rosin, rosin oil, rubber, soaps, vegetable oils.
Cellulose Products
As a solvent which is soluble in organic solvents.
As a solvent which is very resistant to hydrolysis.
As a stable, high boiling-point solvent.

Dichloroethyl Ether (Continued)

Solvent, in conjunction with alcohol, for—
Cellulose acetate, cellulose esters, cellulose ethers, nitrocellulose.

Ceramic
Solvent or diluent in—
Compositions, containing natural or synthetic resins, nitrocellulose, cellulose acetate, or other cellulose esters or ethers, used as coatings for protecting and decorating ceramic products.

Chemical
Activating medium in—
Chemical reactions, sulphonation.
As a stable, high boiling-point solvent.
As a solvent which is soluble in organic solvents.
As a solvent which is insoluble in water.
As a solvent which is very resistant to hydrolysis.
Intermediate in—
Syntheses.

Cosmetic
Solvent or diluent in—
Nail enamels and lacquers containing natural or synthetic resins, nitrocellulose, cellulose acetate, or other cellulose esters or ethers as base material.

Dry Cleaning
Ingredient of—
Cleaning solutions, spotting agents.
Solvent for—
Fats, greases, gums, insoluble soap, oils, paint and varnish stains, tars, waxes.

Dye
Solvent in making various dyestuffs.

Electrical
Solvent or diluent in—
Insulating compositions, containing natural or synthetic resins, nitrocellulose, cellulose acetate, or other cellulose esters or ethers, used for covering wire and in making electrical machinery and equipment.

Explosives and Matches
Solvent, in conjunction with alcohol, for—
Nitrocellulose.

Fats, Oils, and Waxes
Solvent for—
Essential oils, fats, greases, vegetable oils, waxes.
Solvent in—
Recovering oils from fuller's earth and other substances used in bleaching.

Fertilizer
Solvent in—
Degreasing fish scrap.

Food
Solvent for—
Edible oils, fats, pectin.

Glass
Solvent or diluent in—
Compositions, containing natural or synthetic resins, nitrocellulose, cellulose acetate, or other cellulose esters or ethers, used in the manufacture of nonscatterable glass and as coatings for decorating and protecting glassware.

Glue and Adhesives
Solvent or diluent in—
Adhesive compositions containing natural or synthetic resins, nitrocellulose, cellulose acetate, or other cellulose esters or ethers.

Gums
Solvent for—
Gums.

Leather
Solvent in—
Cleansing spotted leathers.
Removing natural oils and greases from hides and skins before tanning so as to prevent staining thereafter and insure evenness of the leather finish and tan.
Solvent or diluent in—
Compositions, containing natural or synthetic resins, nitrocellulose, cellulose acetate, or other cellulose esters or ethers, used in the manufacture of artificial leathers and as coatings for decorating and protecting leathers and leather goods.

Metal Fabricating
Solvent or diluent in—
Compositions, containing natural or synthetic resins, nitrocellulose, cellulose acetate, or other cellulose esters or ethers, used as coatings for protecting and decorating metallic articles.

Miscellaneous
Solvent in—
Cleaning solutions, polishes.
Solvent or diluent in—
Coating compositions, containing natural or synthetic resins, nitrocellulose, cellulose actate, or other cellulose esters or ethers, used for protecting and decorating various articles.

Paint and Varnish
As a stable, high boiling-point solvent.
As a solvent which is soluble in organic solvents.
As a solvent which is very resistant to hydrolysis.
Ingredient of—
Paints, varnishes, lacquers, enamels, and dopes containing natural or synthetic resins, nitrocellulose, cellulose acetate, or other cellulose esters or ethers.
Paint and varnish removers.
Solvent, in conjunction with alcohols, for—
Cellulose acetate, cellulose esters, cellulose ethers, nitrocellulose.
Solvent for—
Natural resins, synthetic resins.

Paper
Solvent or diluent in—
Compositions, containing natural or synthetic resins, nitrocellulose, cellulose acetate, or other cellulose esters or ethers, used in the manufacture of coated papers and as coatings for decorating and protecting articles made of paper or pulp.

Plastics
Solvent or diluent in making—
Laminated fiber products, molded products.
Solvent for plastics from or containing natural or synthetic resins, nitrocellulose, cellulose acetate, or other cellulose esters or ethers.

Petroleum
Selective solvent in making—
High-grade lubricating oils from low-grade products.
High-grade lubricating oils from Mid-continent crudes by the "Chlorex" process.

Resins
Solvent for—
Natural resins, synthetic resins.
Solvent or diluent in making—
Artificial resins from or containing nitrocellulose, cellulose acetate, or other cellulose esters or ethers.

Rubber
Solvent or diluent in—
Compositions, containing natural or synthetic resins, nitrocellulose, cellulose acetate, or other cellulose esters or ethers, used as coatings for protecting and decorating rubber goods.

Soap
Ingredient of—
Cleansing compositions, dry-cleaning compositions, grease-removing soaps, penetrating agents, spotting agents, scouring agents, textile soaps, wetting agents.

Stone
Solvent or diluent in—
Compositions, containing natural or synthetic resins, nitrocellulose, cellulose acetate, or other cellulose esters or ethers, used as coatings for decorating and protecting artificial and natural stone.

Textile
As a conditioning agent.
As a desizing agent.
As a softening agent.
As a spotting agent.
Assist in—
Kier boiling, mercerizing, scouring operations.
Assister of—
Soap solutions where high temperatures are involved.
Compounding agent in making—
Dewaxing agents for cotton.
Penetrating agents with soaps and sulphonated oils.
Wetting agents with soaps and sulphonated oils.
Ingredient of—
Cleaning solutions, fulling preparations and soaps, scouring soaps, spotting preparations and soaps.
Penetrating agent in—
Peroxide bleaching, textile processes.
Scouring agent in removing—
Fat from raw wool, grease from raw wool, oil from raw wool, paint brands from raw wool, tar brands from raw wool.
Solvent for—
Fats, greases, oils, waxes.

Dichloroethyl Ether (Continued)
Substitute for—
Caustic alkalies.
Ethylene dichloride, especially where high temperatures are required.
Solvent or diluent in—
Compositions, containing natural or synthetic resins, nitrocellulose, cellulose acetate, or other cellulose esters or ethers, used in the manufacture of coated fabrics.
Wood
Solvent or diluent in—
Compositions, containing natural or synthetic resins, nitrocellulose, cellulose acetate, or other cellulose esters or ethers, used as protective and decorative coatings on woodwork.

Dichlorofluoromethane
German: Dichlorfluormethan.
Chemical
Starting point in making various derivatives.
Miscellaneous
As medium for extinguishing fires.
Refrigeration
Active medium in industrial refrigerating systems.

3:5-Dichlorohydroxydiphenyl
Disinfectant
As a bactericide (U. S. 1989081).

5:7-Dichloroisatin Anilide
French: Anilide de 5:7-dichloroisatine, Anilide 5:7-dichloroisatinique.
German: 5:7-Dichlorisatinanilid.
Chemical
Starting point in making—
Intermediates, pharmaceuticals.
Dye
Starting point (Brit. 291825) in making indigoid dyestuffs with—
4-Allylmercapto-1-naphthol.
4-Amylmercapto-1-naphthol.
4-Benzylmercapto-1-naphthol.
4-Butylmercapto-1-naphthol.
4-Ethylmercapto-1-naphthol.
4-Formylmercapto-1-naphthol.
4-Gallylmercapto-1-naphthol.
4-Heptylmercapto-1-naphthol.
4-Hexylmercapto-1-naphthol.
4-Isoallylmercapto-1-naphthol.
4-Isoamylmercapto-1-naphthol.
4-Isobutylmercapto-1-naphthol.
4-Isopropylmercapto-1-naphthol.
4-Lactylmercapto-1-naphthol.
4-Methylmercapto-1-naphthol.
4-Naphthylmercapto-1-naphthol.
Paracresylmercapto-1-naphthol.
Paratolylmercapto-1-naphthol.
Paraxylylmercapto-1-naphthol.
4-Pentylmercapto-1-naphthol.
4-Phenylmercapto-1-naphthol.
4-Propylmercapto-1-naphthol.
4-Tolylmercapto-1-naphthol.
4-Xylylmercapto-1-naphthol.

5:7-Dichloroisatin Chloride
French: Chlorure de 5:7-dichloroisatine, Chlorure 5:7-dichloroisatinique.
German: Chlor-5:7-dichlorisatin, 5:7-dichlorisatinchlorid.
Chemical
Starting point in making various intermediates and other derivatives.
Dye
Starting point (Brit. 291825) in making indigoid dyestuffs with—
4-Allylmercapto-1-naphthol.
4-Amylmercapto-1-naphthol.
4-Benzylmercapto-1-naphthol.
4-Butylmercapto-1-naphthol.
4-Ethylmercapto-1-naphthol.
4-Heptylmercapto-1-naphthol.
4-Hexylmercapto-1-naphthol.
4-Isoallylmercapto-1-naphthol.
4-Isoamylmercapto-1-naphthol.
4-Isobutylmercapto-1-naphthol.
4-Isopropylmercapto-1-naphthol.
4-Methylmercapto-1-naphthol.
4-Naphthylmercapto-1-naphthol.
Paratolylmercapto-1-naphthol.
4-Pentylmercapto-1-naphthol.
4-Phenylmercapto-1-naphthol.
4-Propylmercapto-1-naphthol.
4-Tolylmercapto-1-naphthol.
4-Xylylmercapto-1-naphthol.

Dichloroketoanthraquinone-2:1-dihydrothiazin
German: Dichlorketoanthrachinon-2:1-dihydrothiazin.
Dye
Starting point (German 430901) in making anthraquinone vat dyestuffs with—
1-Amino-2-paratolylaminoanthraquinone.
1:2-Diaminoanthraquinone.
4:5-Diaminometaxylene.
Orthophenylenediamine.

Dichloroketoparatolyldihydroparathiazin
Dye
Starting point (German 430901) in making anthraquinone vat dyestuffs with—
1:2-Diaminoanthraquinone.
1:2-Diaminoanthraquinone and ethylation with ethylparatoluene sulphonate.

Dichloromethane
Synonyms: Methylene bichloride, Methylene chloride, Methylene dichloride.
French: Chlorure de méthylène, Dichlorure de méthane, Dichlorure de méthylène.
German: Chlormethylen, Dichlormethan, Dichlormethylen, Methandichlorid, Methylenchlorid, Methylendichlorid.
(Note: For certain uses as a solvent dichloromethane must be mixed with other solvents; for example, for dissolving cellulose acetate, dichloromethane must be mixed with methanol.)
Analysis
Extracting medium for various purposes.
Solvent in—
Extraction and assay of drugs.
Solvent in analyzing and testing—
Animal oils, ashes, breadstuffs, butter, cakes, cheese, chocolate, cocoa, coffee, essential oils, fats, flour, hops, meals, meat, milk, mineral phosphates, resins, rosin, rosin oil, rubber, soaps.
Various industrial products containing oils, fats, resins, waxes, rubber, cellulose derivatives.
Vegetable oils.
Solvent in making toxicological assays for—
Atropine, berberine, brucine, cocaine, codeine, corydaline, cryptopine, ecgonine, emetine, eucaine, hydrastine, hyoscine (scopolamine), hyoscyamine, laudanine, laudanosine, morphine, narceine, narcotine, nicotine, protopine, quinine, strychnine, thebaine, tropinone.
Ceramics
Solvent in—
Compositions containing nitrocellulose, cellulose acetate, or other esters or ethers of cellulose, as well as resins, waxes and gums, used for coating and decorating ceramic ware.
Chemical
Extractant for—
Alkaloids, drug principles.
Ingredient of—
Solvent mixtures containing acetone, alcohol, benzene, chlorinated hydrocarbons, turpentine.
Reagent for—
Introducing chlorine in the manufacture of inorganic and organic compounds.
Reagent in making—
Acetic anhydride (Brit. 402462).
Intermediates, organic chemicals, pharmaceuticals.
Solvent for—
Cellulose acetate, cellulose nitrate, inorganic chemicals, organic chemicals.
Solvent (Brit. 400169) in—
Concentration of acetic acid.
Starting point in making—
Chlorinated fats, chlorinated chemical compounds.
Solvent for—
Washing tiles, tiled fronts of buildings.
Dye
Reagent and solvent in making—
Synthetic dyestuffs of various classes.
Electrical
Solvent for—
Cleaning motors and other electrical machinery.

Dichloromethane (Continued)
Solvent in—
Compositions containing cellulose acetate, nitrocellulose, or other esters or ethers of cellulose, and at times resins, gums, and the like, used for insulating cables, wiring, and electrical machinery and equipment.

Fats and Oils
Extractant for—
Animal oils, essential oils, fats, greases, vegetable oils.
Solvent for—
Animal oils, essential oils, fats, greases, vegetable oils.
Solvent in—
Recovering oils from fuller's earth and other substances used in bleaching.

Fertilizer
Solvent in—
Degreasing fish scrap.

Food
Extracting medium in obtaining soluble substances from—
Berries, fruits, seeds.
Solvent in—
Extracting caffeine from coffee (Brit. 404228).
Making food flavors, purification of foodstuffs.

Glass
Solvent for—
Degreasing glass.
Solvent in—
Compositions containing cellulose acetate, nitrocellulose, or other esters or ethers of cellulose, and artificial or natural resins and waxes and gums, used in the manufacture of nonscatterable glass and for the decoration and protection of glassware.

Glues and Adhesives
Ingredient of—
Glues.
Special adhesive compositions containing cellulose acetate, nitrocellulose, or other esters or ethers of cellulose.
Reagent in—
Preparing gelatins.
Solvent for—
Degreasing bone and hide gluestocks.

Gums
Solvent for various gums.

Ink
Solvent in making—
Printing inks.

Insecticide
As a general insecticide.
Ingredient of—
Fumigating compositions.
Insecticidal compositions.
Preparations for the extermination of mosquitoes.
Preparations for exterminating parasites.
Preparations for combatting grape lice.
Vermicidal compositions.

Leather
Solvent for—
Cleansing spotted leathers.
Removing natural oils and greases from hides and skins before tanning, so as to prevent staining and insure evenness of the leather finish and tan.
Solvent in—
Compositions containing cellulose acetate, nitrocellulose, or other esters or ethers of cellulose, as well as artificial or natural resins, gums and waxes, used in the manufacture of artificial leather and for the protection and decoration of leather goods.

Mechanical
Solvent for—
Cleansing and degreasing machinery of various sorts.
Cleansing automobile engines and gears.
Cleansing drive wheels for compression pumps and other mechanical equipment.
Cleansing and degreasing metallic surfaces prior to painting and coating.
Degreasing automobile brakebands.
Recovering oil and grease from cotton and wool waste and rags from factory machinery, machine shops, pipe-fitting shops, engine and pumping stations, and similar mechanical departments.
Removing oils and greases from leather belting and the like.

Metallurgical
Solvent for—
Degreasing die castings, metal stampings, metals to be electroplated, nuts and bolts.
Preparing metals for pickling, plating, shellacking, sherardizing, varnishing.
Solvent and diluent in—
Compositions containing cellulose acetate, nitrocellulose, or other esters or ethers of cellulose, used for protecting and decorating metallic articles.

Miscellaneous
As a general solvent.
Cleansing agent for—
Furs.
Dry-cleaning solvent and spotting agent.
Ingredient of—
Compositions containing clay, for cleansing ivory, horn, and bone.
Polishing compositions of various sorts.
Preparations used for the removal of stains from celluloid articles.
Preparations used for cleansing typewriters.
Preservative for—
Biological products.
Solvent for—
Degreasing dishes, kitchenware, hardware, metal furniture, safety-razor blades.
Solvent and diluent in—
Compositions containing cellulose acetate or other esters or ethers of cellulose, used for decorating and protecting various products.

Oilcloth and Linoleum
Solvent in making—
Coating compositions.

Paint and Varnish
Ingredient of—
Paint, lacquer, and varnish removers.
Solvent in making—
Nitrocellulose lacquers (Brit. 390867).
Paints, varnishes, lacquers, enamels, and dopes containing cellulose acetate, nitrocellulose, or other esters or ethers of cellulose, with waxes, gums, and artificial or natural resins.

Paper
Solvent for—
Removing oil from paperstock.
Solvent in—
Compositions containing cellulose acetate, nitrocellulose, or other esters or ethers of cellulose, with gums, waxes, and natural or artificial resins, used in the manufacture of coated paper and for coating and decorating paper and pulp products.

Perfume
Solvent for—
Extracting essential oils and other aromatic substances from flowers.

Petroleum
Ingredient of—
Compounded solvent preparations containing mineral oil distillates.
Solvent for—
Degreasing light mineral oils.
Extracting wax from mineral oil (acting simultaneously as a dewaxing solvent and refrigerant).

Photographic
Solvent for—
Degreasing and cleaning motion-picture film.
Making motion-picture film.

Pharmaceutical
Suggested as anesthetic in—
Dental work, general work.
Solvent for—
Atropine, berberine, brucine, cocaine, codeine, corydaline, cryptopine, ecgonine, emetine, eucaine, hydrastine, hyoscine (scopolamine), hyoscyamine, laudanine, laudanosine, morphine, narceine, narcotine, nicotine, protopine, quinine, strychnine, thebaine, tropinone.

Plastics
Solvent for—
Degreasing plastics.
Solvent and diluent in making—
Compositions containing cellulose acetate, nitrocellulose, or other esters or ethers of cellulose, with gums, waxes, and artificial or natural resins.
Plastics containing nitrocellulose (Brit. 390867).

Printing
Solvent for—
Cleaning engraved plates, lithographic stones, printing machinery, type.

Dichloromethane (Continued)

Refrigeration
Refrigerant in—
Air-conditioning machines.
Low-pressure ice machines.

Resins and Waxes
Solvent for various resins and waxes.

Rubber
Ingredient of—
Rubber cements, rubber mastics.
Rubber compositions used in the manufacture of rubberized cloth.
Solvent for—
Rubber.
Solvent in—
Compositions containing cellulose acetate, nitrocellulose, or other esters or ethers of cellulose, with gums, waxes, and artificial or natural resins, used for decorating and coating rubber goods.

Sanitation
Solvent for—
Degreasing garbage.

Soap
Ingredient of—
Cleansing compositions, dry-cleaning compositions, spotting fluids.
Solvent in making—
Gelatinous water-soluble soaps from sulphonated oils and resins.
Paste soaps for removing grease stains.
Soaps with sodium ricinoleate.
Textile soaps from linseed oil and castor oil.

Stone
Solvent in—
Compositions containing cellulose acetate, nitrocellulose, or other esters or ethers of cellulose, with artificial or natural resins, gums, and waxes, used for the decoration and protection of artificial and natural stone.

Sugar
Solvent for—
Extracting waxes from filter press mud in sugar refinery.

Textile
——, *Dyeing*
Ingredient of—
Preparations containing turkey red oil and chlorinated hydrocarbons used for dyeing and wetting.

——, *Finishing*
Ingredient of—
Scouring compositions containing sulphonated-oil soaps.
Solvent in—
Coating compositions containing cellulose acetate, nitrocellulose, or other ethers or esters of cellulose.
Scouring compositions.

——, *Manufacturing*
Solvent for—
Cleaning knitting machine needles.
Cleaning silk and silk hosiery.
Degreasing textiles, degreasing wool, degumming silk.
Solvent, shrinking and softening agent (Brit. 403106) for—
Organic derivatives of cellulose used as rayon filaments, threads, ribbons.

Tobacco
Solvent for—
Extracting nicotine.

Woodworking
Solvent and diluent in—
Compositions containing cellulose acetate, nitrocellulose, or other esters or ethers of cellulose, used for decorating and protecting woodwork.

1:4-Dichloronaphthalene
German: 1:4-Dichlornaphtalin.

Chemical
Starting point in making—
Alphachloronaphthalene.
5:8-Dichloroalphanitronaphthalene.

1:4-Dichloro-2-nitrobenzene
Synonyms: 1:4-Dichloro-2-nitrobenzol.
German: 1:4-Dichlor-2-nitrobenzol.

Chemical
Starting point in making—
Chloronitroanisole (French 602977).

2:5-Dichloroparaphenylenediamine

Dye
Starting point (Brit. 397034) in making—
Tetrazo compounds.

2:3-Dichloroparatoluidin

Dye
As an intermediate.
Starting point (Brit. 397016) in making—
Blue-red water-insoluble dyes.

Dichloropentanes
Note: Isomeric mixture of the dichlorides of the pentanes.

Adhesives
Solvent for—
Rubber in cements and other adhesives.

Construction
Solvent for—
Bituminous materials used in water-proofings for buildings and construction projects.

Electrical
Solvent for—
Rubber, resins, and bituminous materials used in electrical insulation.

Miscellaneous
Solvent for—
Bituminous materials used in impregnating various compositions.

Paint and Varnish
Solvent for—
Rubber, resins, and bituminous materials used in paints, lacquers, varnishes, enamels, roof cements, and other products.

Paper
Solvent for—
Bituminous materials used in impregnation of paper for roofing and insulating purposes.

Plastics
Solvent for—
Resins and bituminous materials.

Rubber
Solvent for—
Rubber.

Stone
Solvent for—
Rubber and bituminous materials in making paving materials, artificial stone, composition flavoring, and the like.

2:6-Dichlorophenolindo-orthocresol

Analysis
Indicator in—
Oxidation-reduction potential determinations (of particular interest to biologists and physiologists and in investigations of various materials, such as soils, wines, cheese, gasoline antiknock compounds).

2:6-Dichlorophenolindophenol

Analysis
Indicator in—
Oxidation-reduction potential determinations (of particular interest to biologists and physiologists and in investigations of various materials, such as soils, wines, cheese, gasoline antiknock compounds).

3:6-Dichlorophthalic Acid

Cellulose Products
Solvent (Brit. 390541) for—
Cellulose esters and ethers.
For uses, see under general heading: "Solvents."

2:4-Dichloroquinazolin
German: 2:4-Dicnlorchinazolin.

Chemical
Starting point (Brit. 309102) in making derivatives with—
Alpha-aminoanthraquinone.
Alpha-amino-4-benzoylaminoanthraquinone.
Alpha-amino-5-benzoylaminoanthraquinone.
Alphanaphthol.
Aminoarylacetopyrazolones.
Aminoarylazopyrazolones.
Aminonaphtholsulphonic acid.
5-Aminosalicylic acid.
Benzyl alcohol.
Beta-aminoanthraquinone.
Betadimethylaminoethylethylamin.

2:4-Dichloroquinazolin (Continued)
Chloronaphthylamines.
Cyclohexanol.
2:7-Dihydronaphthalene.
Diethylglycol.
G acid.
H acid.
Halogenated anilins.
J acid.
Metabenzoic acid-5-azo-orthoanisidin.
Metaphenylenediamine-4-sulphonic acid.
Methanol.
Naphthylamines.
Nitranilins.
Nitronaphthylamines.
4-Nitronaphthylamine-6-sulphonic acid.
Para-aminoacetanilide-2-sulphonic acid.
Para-aminobenzoic-5-salicylic acid.
Paranitranilin-3-sulphonic acid.
Parapara'-diaminodiphenylamineorthosulphonic acid.
Paraphenylenediaminecarboxylic acid.
Paraphenylenediaminesulphonic acid.
Parathiocresol.
Paratoluenesulphonic acid.
Phenol.
S acid.
Salicylic-5-sulphonic acid.
Toluidins.

Dichlororetene
Petroleum
Imparter (Brit. 431508) of—
High-film strength, adhesion power, and abrasion resistance to lubricants for use with extreme pressures (blended with mineral lubricating oil).

3:5-Dichlorosalicylaldehyde
Photographic
Purification agent (U. S. 1973472) for—
Methylpara-aminophenol (developing agent).

Dichloro-tertiary-amyl Alcohol
Petroleum
Solvent (Brit. 435096) in—
Refining mineral oils.

Dichloro-tertiary-butyl Alcohol
Petroleum
Solvent (Brit. 435096) in—
Refining mineral oils.

Dichlortetrafluorethane
French: Dichlorure et tétrafluorure d'éthane.
German: Dichlortetrafluoraethan.
Refrigeration
Refrigerant for—
Use in freezing machines, particularly in small domestic machines and in all units where it is vital that the refrigerant does not attack metals.

Dicresoldithiophosphoric Acid
See: Dicresyldithiophosphoric acid.

Dicresyl Carbonate
French: Carbonate, dicrésylique.
German: Dicresylcarbonat, Dikresylkarbonat.
Spanish: Carbonato dicresilico.
Italian: Carbonato dicresilico.
Explosives and Matches
Substitute (German 302361) for glycerin in making—
Explosives.
Low-freezing nitrated compounds for low-freezing dynamites.
Fats and Oils
As a lubricant (German 302361).
Ingredient (German 302361) of—
Lubricating compositions.
Ink
Substitute (German 302361) for glycerin in making—
Inks.
Leather
Substitute (German 302361) for glycerin in—
Finishing processes.
Mechanical
Substitute (German 302361) for glycerin as—
Lubricant for delicate machinery.
Miscellaneous
As a solvent (German. 302361).
As a waterproofing agent (German 302361).
Substitute (German 302361) for glycerin as—
Antimolding agent for cork stoppers.
Antishrinkage agent for wooden molds and vessels.
Ingredient of litharge cement and similar cements for pipe joints.
Ingredient of compositions for making rubber stamps.
Lubricant and plasticizer in clay modeling.
Lubricant and softening agent in shoe polishes.
Reagent in manufacture of felt.
Softening agent in millinery.
Paint and Varnish
Substitute (German 302361) for glycerin as—
Softening agent in artist's colors.
Paper
Substitute (German 302361) for glycerin in making—
Marbled papers, parchment papers, surface-coated papers, waterproofed paper.
Perfume
Substitute for glyerin (German 302361).
Photographic
Substitute for glyerin (German 302361).
Plastics
Substitute (German 302361) for glycerin in making—
Plastic compositions such as are used for making hectograph pads and printing rollers.
Rubber
Substitute for glyerin (German 302361).
Textile
Substitute (German 302361) for glycerin in—
Dyeing and printing fabrics.
Sizing and lubricating fabrics.

Dicresyldithiophosphoric Acid
Insecticide and Fungicide
Wetting agent (U. S. 2019443) in—
Insecticidal compositions.
Mining
Flotation agent (Brit. 455224) in—
Froth flotation of minerals.
Miscellaneous
As a wetting agent (U. S. 2019443).

Dicresyldithiophosphoric Acid, Ammonium Salt
Mining
Flotation agent (Brit. 455224) in—
Froth flotation of minerals.

Dicyandiamide
German: Dicyandiamin, Dizyandiamid.
Chemical
Starting point in making various derivatives.
Metallurgical
Ingredient (Brit. 311588) of—
Case-hardening preparations.

Dicyan Diselenide
French: Disélénure de dicyane.
German: Dicyandiselenid, Dizyandiselenid.
Spanish: Diselenuro de dician.
Italian: Diselenuro di diciano.
Automotive
Ingredient (U. S. 1920766) of—
Motorfuel compositions with high compression values.

Dicyclohexylamine
Dye
Starting point in making lakes with—
Alpha-amino-4-para-acetaminoanilidoanthraquinone-2-sulphonic acid.
Alphahydroxy-4-paratoluidoanthraquinonesulphonic acid.
Anthrapyrimidin-4-toluidosulphonic acid.
Azo dyestuffs.
1:4-Diamino-2-phenoxyanthraquinonesulphonic acid.
1:4-Dihydroxy-5:8-diparatoluidoanthraquinonedisulphonic acid.
1:5-Dihydroxy-5:8-diparatoluidoanthraquinonedisulphonic acid.
1:5-Diparatoluidoanthraquinonedisulphonic acid.
4:8-Diparatoluidoanthraquinonedisulphonic acid.
Dyestuffs derived from orthotoluidin and fluorescein anilide.
Methylanthrapyridin-4-arylsulphonic acids.
Paranitrophenylazosalicylic acid.
Patent blue A.
Sodium alpha-amino-4-anilidoanthraquinone-2-sulphonate.

Dicyclohexylamine (Continued)

Glass
Stabilizer (Brit. 437304) for—
Halogenated rubber derivatives used as cements for laminated glass.

Miscellaneous
Inhibitor (Brit. 437304) of—
Photochemical action.

Paper
Stabilizer (Brit. 437304) for—
Halogenated rubber derivatives used for impregnating or coating wrapping paper.

Rubber
Promoter (Brit. 437304) of—
Resistance to the deteriorating action of light on chlorinated rubber.
Stabilizer (Brit. 437304) for—
Coating and impregnating agents made from halogenated rubber derivatives and used for treating fabrics to be used as wrapping materials.
Transparent films or sheets made from halogenated rubber derivatives.

Dicyclohexylcyclohexanol Sulphonate

Miscellaneous
As an emulsifying agent (Brit. 449607).
For uses, see under general heading: "Emulsifying agents."

Dicyclohexyl Hexahydrophthalate

Cosmetic
Fixative (U. S. 2015239) in—
Perfumes.

Dicyclohexyl Malate

Cellulose Products
Plasticizer (Brit. 432404) for—
Cellulose esters and ethers.
For uses, see under general heading: "Plasticizers."

1:2-Dicyclohexyl-3-methyl-5-pyrazolone

Pharmaceutical
Suggested (Brit. 433053) for use as—
Febrifuge, sedative.

Didecahydrobetanaphthyl Tartrate

Cellulose Products
Plasticizer (Brit. 432404) for—
Cellulose acetate, cellulose esters and ethers.
For uses, see under general heading: "Plasticizers."

4-Didelta²-cyclohexenylamino-1-phenyl-2:3-dimethyl-5-pyrazolone

Pharmaceutical
Suggested (Brit. 433053) for use as—
Febrifuge, sedative.

5:5-Dideltacyclopentenylallylbarbituric Acid

Pharmaceutical
Claimed to have—
Combined properties as analgesic, sedative, and hypnotic.

Didimethylcyclohexanol Oxalate
French: Oxalate de didiméthylecyclohéxanole.
German: Didimethylcyklohexanoloxalat, Didimethylzyklohexanoloxalat, Oxalsäuredidimethylzyklohexanolester.

Cellulose Products
Plasticizer for—
Cellulose acetate, cellulose esters or ethers, nitrocellulose.
For uses, see under general heading: "Plasticizers."

Didymium Oxide
French: Oxyde de didymium.
German: Didymoxyd.

Chemical
Catalyst (Brit. 254819) in making—
Alcohols, aldehydes, amines, carboxylic acid esters, nitric acid, oxygenated organic compounds, sulphuric acid.

Miscellaneous
Ingredient in making—
Incandescent mantles for gas-burning.

Glass
Pigment for coloring and decorating.

Didymium Sulphate
French: Sulfate de didyme, Sulfate didymique.
German: Didymsulfat.

Chemical
Catalyst in oxidizing—
Sulphur trioxide to sulphuric acid.

Disinfectant
Ingredient of—
Disinfectants and germicides.

Glass
Coloring matter in—
Decorating fine glassware.

Miscellaneous
Ingredient of—
Compositions used in making gas mantles.

Pharmaceutical
In compounding and dispensing practice.

Diethanolamine Citrate

Textile
De-electrifying agent (Brit. 430221) for—
Yarns, films, fabrics, and the like, subject to charging by static electricity (applied in admixture with all usual lubricating agents as vehicle).

Diethanolamine Gallate

Textile
De-electrifying agent (Brit. 430221) for—
Yarns, films, fabrics, and the like, subject to charging by static electricity (applied in admixture with all usual lubricating agents as vehicle).

Diethanolamine Lactate

Textile
De-electrifying agent (Brit. 430221) for—
Yarns, films, fabrics, and the like, subject to charging by static electricity (applied in admixture with all usual lubricating agents as vehicle).

Diethanolamine Mucate

Textile
De-electrifying agent (Brit. 430221) for—
Yarns, films, fabrics, and the like, subject to charging by static electricity (applied in admixture with all usual lubricating agents as vehicle).

Diethanolamine Saccharate

Textile
De-electrifying agent (Brit. 430221) for—
Yarns, films, fabrics, and the like, subject to charging by static electricity (applied in admixture with all usual lubricating agents as vehicle).

Diethanolamine Salicylate

Textile
De-electrifying agent (Brit. 430221) for—
Yarns, films, fabrics, and the like, subject to charging by static electricity (applied in admixture with all usual lubricating agents as vehicle).

Diethanolamine Tannate

Textile
De-electrifying agent (Brit. 430221) for—
Yarns, films, fabrics, and the like, subject to charging by static electricity (applied in admixture with all usual lubricating agents as vehicle).

Diethanolamine Tartrate

Textile
De-electrifying agent (Brit. 430221) for—
Yarns, films, fabrics, and the like, subject to charging by static electricity (applied in admixture with all usual lubricating agents as vehicle).

Diethylacetylquinine Hydrochloride

Pharmaceutical
Claimed (Brit. 433261) as—
Water-soluble and practically tasteless form of quinine.

Diethylamine
German: Diaethylamin.

Chemical
Catalyst (Brit. 252870) in making—
Normal butyl paranitrobenzoate from sodium paranitrobenzoate and butyl bromide.
Reagent (Brit. 310534) in making rubber-vulcanization accelerators with the aid of—
Allylenethiourea, amylenethiourea, butylenethiourea, ethylenethiourea, heptylenethiourea, hexylenethiourea,

Diethylamine (Continued)
isoallylenethiourea, isoamylene thiourea, isobutylenethiourea, isopropylenethiourea, methylenethiourea, propylenethiourea.
Starting point in making—
Diethylbenzylamine, diethylphenylamine, symmetrical; diethylglycocollguaiacol hydrochloride (Guaiasanol), nirvanine, methylenetetraethyldiamine, novocaine.
Dye
Starting point in making—
New phosphin G, tannin orange.

4-Diethylaminobetahydroxyethylaminoanilin
Dye
Starting point (Brit. 447905, 447906, and 448016) in making—
Monoazo dyes for leather, particularly chrome leather.

Diethylaminoethanol Antimoniate
Disinfectant
Germicide, claimed (U. S. 1988632) to be valuable against infectious diseases.

Diethylaminoethyl Chloride
French: Chlorure de diéthyleaminoéthyle, Chlorure diéthyleaminoéthylique.
German: Chlordiaethylaminoaethyl, Diaethylaminoaethylchlorid.
Chemical
Starting point (Brit. 274058) in making—
3:4-Diethoxy-n-mono (diethylaminoethyl) anilin.
3-4-Diethoxy-n-di (diethylaminoethyl) anilin.
1-Di (diethylaminoethyl) amino-4-dimethylamino-2-methylthiophenol.
3-N (ethyldiethylaminoethyl) amino-4-methyl-1-oxybenzene.
Meta-amino-n-diethylaminoethylanilin.
Metaoxy-n-diethylaminoethylanilin.
Metaoxy-n-dimethylaminoethylanilin.
Paraoxy-n-methyldiethylaminoethylanilin.

Diethylaminoethyloleylamide Lactate
Perfume
Ingredient of—
Dentrifice, containing also gum tragacanth, pectin, glycol, water, titanium, dioxide, pepsin, glycerin, and a flavoring material.

Diethylaminoethyloleylamide Salts
Synonyms: Sapamine salts.
Perfume
Detergents and wetting agents in—
Shampoos (two such formulas are: (1) sapamine citrate, citronellic acid, saponin, glycerin, and alcohol; (2) sapamine acetate, boric acid, perfume, and water).

Diethylaminoethyloleyl Phosphate
Perfume
Emulsifying agent in—
Cold cream, containing also glyceryl monostearate, beeswax; white petrolatum, lard, mineral oil, sweet almond oil, glycerin, and water.

5-Diethylaminomethyl-1:3:2-xylenol
Rubber
Anti-ager (Brit. 459045) for—
Rubber mixes.

5:5-Diethylbarbituric Acid
Synonyms: Barbital, Barbitalum, Barbitone, Diethylmalonylurea, Malonal, Malonurea, Veronal.
French: Acide de 5:5-diéthylebarbiturique.
German: 5:5-Diaethylbarbiturinsaeure.
Chemical
Starting point (Swiss 113251) in making synthetic drugs with—
Allylamine, amylamine, butylamine, diallylamine, diamylamine, dibutylamine, diethylamine, dimethylamine, dipropylamine, ethylamine, isoallylamine, isoamylamine, isobutylamine, isopropylamine, methylamine, propylamine.
Starting point (Brit. 255434) in making—
Codiene diethylbarbiturate, pyrazolone barbituric acid, quinine diethylbarbiturate (Chineonal), sodium diethylbarbiturate.
Pharmaceutical
In compounding and dispensing practice.

Diethylbetanaphthylamine
French: Diéthylebétanaphthylamine.
German: Diaethylbetanaphtylamin.
Dye
Reagent in making color lakes with—
Alpha-amino-4-para-acetaminoanilidoanthraquinone-2-sulphonic acid.
Anthrapyrimidin-4-paratoluidosulphonic acid.
Azo dyestuffs.
1:4-Diamino-2-phenoxyanthraquinonesulphonic acid.
1:4-Dihydroxy-5:8-diparatoluidoanthraquinonedisulphonic acid.
1:5-Dihydroxy-5:8-diparatoluidoanthraquinonedisulphonic acid.
4:8-Diparatoluidoanthraquinonedisulphonic acid.
1:5-Diparatoluidoanthraquinonedisulphonic acid.
Dyes derived from orthotoluidin and fluorescein chloride.
1-Hydroxy-4-paratoluidoanthraquinonesulphonic acid.
Methylanthrapyridone-4-arylsulphonic acids.
Paranitrophenylazosalicylic acid.
Patent blue A.
Sodium 1-amino-4-anilidoanthraquinone-2-sulphonic acid.

2:2′-Diethyl-4:5:4′:5′-bisethylenedioxyselenadicarbocyanin Iodide
Photographic
Sensitizer (Brit. 425417) for—
Photographic emulsions.

2:2′-Diethyl-4:5:4′:5′-bisethylenedioxyselenatricarbocyanin Iodide
Photographic
Sensitizer (Brit. 425417) for—
Photographic emulsions.

2:2′-Diethyl-4:5:4′:5′-bismethylenedioxythiadicarbocyanin Iodide
Photographic
Sensitizer (Brit. 425417) for—
Photographic emulsions.

2:2-Diethyl-4:5:4′:5′-bismethylenedioxythiatricarbocyanin Iodide
Photographic
Sensitizer (Brit. 425417) for—
Photographic emulsions.

Diethyl Carbinol
Synonyms: 3-Pentanol, Secondary normal amyl alcohol.
French: Carbinol de diéthyle, Carbinol diéthylique.
German: Diaethylchlorid.
Chemical
General solvent for various purposes.
Starting point in making—
Phenylmethane.
Miscellaneous
General solvent for various purposes.
Paint and Varnish
Solvent in making—
Cellulose ester and ether varnishes and lacquers.
Plastics
Solvent in making—
Cellulose ester and ether compositions.

Diethyl Carbonate
English Synonym: Carbonic acid diethyl ester.
French: Carbonate de diéthyle, Carbonate diéthylique.
German: Diaethylcarbonat, Diaethylkarbonat, Kohlenstoffsäureaethylester, Kohlenstoffsäuresaethyl.
Spanish: Carbonata dietilico.
Ceramics
Solvent in—
Coating compositions containing nitrocellulose and/or resins.
Chemical
Absorbent (German 413037) for—
Acetylene.
Solvent for—
Nitrocellulose.
Fuel
Absorbent (German 413037) for—
Acetylene.
Glass
Solvent in—
Compositions, containing nitrocellulose, used for the coating of glassware.

Diethyl Carbonate (Continued)

Leather
Solvent in—
Compositions, containing nitrocellulose, used for coating leather and leather goods and in the manufacture of artificial leather.

Metallurgical
Solvent in—
Compositions, containing nitrocellulose, used for coating metallic ware.

Miscellaneous
Solvent in—
Compositions, containing nitrocellulose, used for various coating purposes.

Paint and Varnish
Solvent in making—
Dopes, paints, varnishes, enamels, and lacquers containing nitrocellulose and/or resins, particularly products for coating the surfaces of woodwork and automobile bodies.
(Note: Diethyl carbonate gives nitrocellulose solutions of high viscosity; but, when nitrocellulose is dissolved in mixtures of diethyl carbonate and alcohol, the viscosity is lowered until it is almost twice as low as that of nitrocellulose dissolved in mixtures of amyl acetate and alcohol).

Paper
Solvent in—
Coating compositions containing nitrocellulose.

Perfume
Solvent in—
Nail-coating lacquers containing nitrocellulose.

Petroleum
Ingredient (U. S. 1917910) of—
Precipitating mixture, containing also benzene and ethyl or methyl formates, used in dewaxing lubricating oils.

Pharmaceutical
Suggested for use as—
External anesthetic in dental work.

Photographic
Solvent in making—
Films from nitrocellulose.

Plastics
Solvent for—
Celluloid.
Solvent in making—
Compositions containing nitrocellulose and/or resins.

Resins and Waxes
Solvent for—
Artificial and natural resins.

Rubber
Solvent in—
Coating compositions containing nitrocellulose.

Stone
Solvent in—
Coating compositions containing nitrocellulose.

Textile
Degreasing agent (Brit. 282164) for—
Raw wool.
Solvent for—
Nitrocellulose.
Solvent in—
Coating compositions containing nitrocellulose.

Woodworking
Solvent in—
Coating compositions containing nitrocellulose.

2:2′-Diethyl-8-cyclohexylselenacarbocyanin Iodide

Dye
Dye possessing (Brit. 439359)—
Abnormally high solubility in organic solvents.

2:2′-Diethyl-8-cyclohexylthiacarbocyanin Iodide

Dye
Dye possessing (Brit. 439359)—
Abnormally high solubility in organic solvents.

2:2′-Diethyl-4:4′-diethoxy-5:5′-dimethylthiadicarbocyanin Iodide

Photographic
Sensitizer (Brit. 425417) for—
Photographic emulsions.

2:2′-Diethyl-4:4′-dimethoxy-5:5′-dimeththiothiacarbocyanin Iodide

Photographic
Sensitizer (Brit. 420971) in—
Photographic emulsions.

2:2′-Diethyl-4:4′-dimethoxy-5:5′-dimeththiothiadicarbocyanin Iodide

Photographic
Sensitizer (Brit. 425417) for—
Photographic emulsions.

2:2′-Diethyl-4:4′-dimethoxy-5:5′-dimeththiothiatricarbocyanin Iodide

Photographic
Sensitizer (Brit. 425417) for—
Photographic emulsions.

2:2′-Diethyl-5:5′-dimethselenothiacarbocyanin Bromide

Photographic
Sensitizer (Brit. 420971) in—
Photographic emulsions.

2:2′-Diethyl-5:5′-dimeththiacarbocyanin Bromide

Photographic
Sensitizer (Brit. 420971) in—
Photographic emulsions.

Diethyl-1:4-dioxane

French: 1:4-Dioxane d'éthyle, 1:4-Dioxane diéthylique.
German: Diaethyl-1:4-dioxan.

Ceramics
Solvent in—
Coating compositions containing cellulose acetate, nitrocellulose, or other esters or ethers of cellulose.

Chemical
General solvent.
Solvent in making—
Emulsions containing starches, dextrins, glues, resins, waxes, gelatin, casein, vegetable gums, and the like.
Solvent for—
Benzyl cellulose, cellulose acetate, ethyl cellulose, nitrocellulose.

Dye
Solvent for—
Oil-soluble dyestuffs.
Solvent in making—
Dyestuff preparations containing starches, dextrins, glues, casein, gelatin, vegetable gums, and the like.

Fats and Oils
Solvent for—
Certain vegetable oils.

Glass
Solvent in—
Compositions containing cellulose acetate, nitrocellulose, or other esters or ethers of cellulose, used for coating glassware and in the manufacture of nonscatterable glass.

Glues and Adhesives
Solvent in making—
Adhesive preparations containing glues, gelatins, casein, starches, dextrin, and vegetable gums.

Ink
Solvent (Brit. 326824) in making—
Printing inks.

Leather
Solvent in making—
Compositions containing cellulose acetate, nitrocellulose, or other esters or ethers of cellulose, used for coating leather goods and in the production of artificial leathers.
Dressing compositions.
Treating compositions containing starches, dextrins, gelatin, glue, casein, vegetable gums.

Miscellaneous
Ingredient of—
Dyeing and staining solutions, polishing compositions.
Solvent for—
Fats, oils, and greases.

Paint and Varnish
Solvent (Brit. 326824) in making—
Paints, varnishes, enamels, dopes, lacquers, primers, containing cellulose acetate, nitrocellulose, or other esters or ethers of cellulose, as well as oils, such as

Diethyl-1:4-dioxane (Continued)
perilla oil, and resins, such as sandarac, mastic, congo copal, kauri.
Paint and varnish removers.

Paper
Solvent in—
Coating compositions containing cellulose acetate, nitrocellulose, or other esters or ethers of cellulose.
Treating compositions containing starches, dextrins, casein, glue, vegetable gums.

Petroleum
Solvent for—
Mineral oils, paraffin.

Plastics
Solvent (Brit. 326824) in making—
Compositions containing cellulose acetate, nitrocellulose, or other esters or ethers of cellulose.

Resins and Waxes
General solvent.
Solvent for—
Beeswax, carnauba wax, montan wax.
Solvent in making—
Emulsions of waxes and resins containing glues, gelatins, vegetable gums, casein, starches, and the like.

Rubber
Solvent in—
Coating compositions containing cellulose acetate, nitrocellulose, or other esters or ethers of cellulose.
Emulsions containing starches, glues, gelatins, casein, vegetable gums, dextrins, and the like.

Soap
Solvent in making—
Detergent and cleansing preparations.

Stone
Solvent in—
Coating compositions containing cellulose acetate, nitrocellulose, or other esters or ethers of cellulose.

Textile
——, *Dyeing*
Ingredient of—
Dye bath (as an assistant in dyeing and solvent for the dyestuff).
Dye liquors containing starches, dextrins, vegetable gums, and the like.
——, *Finishing*
Solvent in—
Compositions containing cellulose acetate, nitrocellulose, or other esters or ethers of cellulose, used in the finishing of textiles.

Woodworking
Solvent in—
Coating compositions containing cellulose acetate, nitrocellulose, or other esters or ethers of cellulose.

Diethyldiphenylethylenediamine
German: Diaethyldiphenylaethylendiamin.

Dye
Condensing agent in making—
Triarylmethane series dyestuffs (Brit. 249160).

Diethyldiphenylurea

Cellulose Products
Solvent for—
Cellulose esters and ethers.
For uses, see under general heading: "Solvents."

Chemical
Starting point in making—
Intermediates, pharmaceuticals, various other derivatives.

Diethyleneglycol
French: Glycole de diéthylène.
German: Diaethylenglykol, Glykoldiaethylen.

Adhesives
Ingredient (U. S. 1786417) of—
Adhesive, containing also starch and a water-soluble alkaline boron compound.
Moistening and softening agent for—
Adhesives, casein, gelatin, glues, pastes.

Analysis
Extracting medium for various purposes.

Ceramics
Solvent for—
Nitrocellulose and resins in coating compositions used for protecting and decorating ceramic ware.

Chemical
Ingredient of solvent mixtures with—
Acetone, ethyl alcohol, ethyleneglycol, water.
Solvent for—
Nitrocellulose, organic compounds.
Starting point in making—
Chemical derivatives, such as diethyleneglycol phthalate.

Dyes
Solvent for—
Dyestuffs of many types.

Electrical
Solvent for—
Nitrocellulose and resins in coating compositions used for insulating cables, wiring, and electrical machinery and equipment.

Explosives and Matches
Starting point in making—
Dinitrate derivative valued for its low freezing point and excellent solvent power for the nitrocotton used in the manufacture of blasting gelatins.

Fats and Oils
Extractant for—
Oils.
Solvent for—
Oils.

Fire Protection
Antifreeze in—
Sprinkler systems.

Food
Ingredient of—
Plastic sealing composition for glass jars and the like. (This composition withstands the action of oils and fats and contains also edible glue, casein, talc, titanium dioxide, paraformaldehyde, ammonium hydroxide, and water.)

Fuel
Antifreeze in—
Water seals for gas tanks and the like.
Moistening agent for—
Jute or hemp joint-packing to prevent leakage in bell and spigot joints in low pressure gas distribution systems where "dry" gas is distributed.

Leather
Solvent for—
Nitrocellulose and resins in making artificial leather and protective and decorative coatings for leather goods.

Linoleum and Oilcloth
Ingredient of—
Linoleum backing cement, containing also ethyleneglycol, glycerin, phthalic anhydride, and drying oil acids, dissolved in ethyleneglycol monoethylether or similar low-boiling solvent.

Miscellaneous
Moistening and softening agent for—
Composition cork.

Paint and Varnish
Starting point (Brit. 389914) in making—
Condensation products with oleic acid or phthalic anhydride or its homologs, such products forming lacquers for flexible materials when formulated with (1) soluble polymerized vinyl compounds, such as polymerized vinyl chloride, polymerized mixture of vinyl chloride and vinyl acetate, mixture of polymerized vinyl chloride and polymerized vinyl acetate; (2) plasticizers, such as diethoxyphthalate, diethyl phthalate, diamyl phthalate, dibutyl phthalate; (3) waxes such as ozokerite, paraffin, spermaceti, ceresin, candelilla, beeswax, palm, synthetic waxes; (4) solvent consisting of a mixture of ethyl acetate, butyl acetate, toluene; (5) optional constituents, such as oleic acid, butyl oleate, other oleates, gums, resins, condensation products of fatty acids with phthalic anhydride; (6) nitrocellulose; (7) pigments.

Paper
Moistening and softening agent for—
Paper.

Perfume
Ingredient of—
Plastic sealing composition for glass jars and the like. (This composition withstands the action of oils and fats and contains also edible glue, casein, talc, titanium dioxide, paraformaldehyde, ammonium hydroxide, and water.)

Diethyleneglycol (Continued)

Pharmaceutical
Ingredient of—
Plastic sealing composition for glass jars and the like. (This composition withstands the action of oils and fats and contains also edible glue, casein, talc, titanium dioxide, paraformaldehyde, ammonium hydroxide, and water.)

Plastics
Solvent for—
Nitrocellulose and resins.

Printing
Antiwarping agent in—
Bookbinding pastes, particularly those used in connection with artificial leather covers.
Moistening and softening agent for—
Bookbinding pastes, particularly those used in connection with artificial leather covers.

Refrigeration
Antifreeze in—
Refrigerators.

Resins and Waxes
Process material in making—
Synthetic resins.
Solvent for—
Resins.
Starting point in making synthetic resins from—
Coconut oil, glycerin, and phthalic anhydride (Brit. 397554).
Adipic and phthalic acids, with glycerin, mannitol, or pentaerythritol (Brit. 396354).
Azelaic and phthalic acids, with glycerin, mannitol, or pentaerythritol (Brit. 396354).
Fumaric and phthalic acids, with glycerin, mannitol, or pentaerythritol (Brit. 396354).
Glutaric and phthalic acids, with glycerin, mannitol, or pentaerythritol (Brit. 396354).
Maleic and phthalic acids, with glycerin, mannitol, or pentaerythritol (Brit. 396354).
Malic and phthalic acids, with glycerin, mannitol, or pentaerythritol (Brit. 396354).
Pimelic and phthalic acids, with glycerin, mannitol, or pentaerythritol (Brit. 396354).
Sebacic and phthalic acids, with glycerin, mannitol, or pentaerythritol (Brit. 396354).
Suberic and phthalic acids, with glycerin, mannitol, or pentaerythritol (Brit. 396354).
Succinic and phthalic acids, with glycerin, mannitol, or pentaerythritol (Brit. 396354).

Rubber
Solvent for—
Nitrocellulose and resins in making decorative and protective coatings for rubber goods.

Soap
Coupling agent in—
Textile soaps.
Ingredient of—
Concentrated liquid soap, containing also caustic potash and oleic acid, for treating silk stockings and other silk goods.
Soaps for use with dry-cleaning solvents, especially carbon tetrachloride or trichloroethylene, consisting of a fatty acid soap (with a content of a polyglycol) with or without a chlorinated aliphatic hydrocarbon (Brit. 407088).

Textile
Agent for increasing stretching properties of—
Cotton, mohair, rayon, silk, wool.
Conditioning agent for—
Yarns.
Desizing agent in—
Peroxide bleaching of cotton (dissolves natural wax).
Flexibilizing agent for—
Cotton, mohair, rayon, silk, wool.
Ingredient (Brit. 401350) of—
Discharge paste, containing also an alkali-formaldehyde sulphoxylate and zinc sulphocyanide.
Lubricant for—
Cotton, mohair, rayon, silk, wool.
Moistening agent in making—
Nondrying dye pastes.
Plasticizing agent for—
Warp sizes (used to overcome entangling of protruding fibers in weaving hairy and fuzzy worsted warps of "singles" yarn).
Reagent for—
Setting twist of yarns.
Reagent in—
Kier boiling.
Softening agent for—
Cotton, mohair, rayon, silk, wool.
Solvent for dyes in—
Dyeing and printing processes.

Diethyleneglycol Monobutyl Ether
French: Éther de diéthylèneglycolemonobutylique.
German: Diaethylenglykolmonobutylaether.

Chemical
Solvent for—
Cellulose esters and ethers, cellulose nitrate, various oils.
Solvent in making—
Special textile oils.
For uses, see under general heading: "Solvents."

2:1′-Diethyl-4-ethseleno-6′-methylthiapsicyanin Iodide

Photographic
Sensitizer (Brit. 420971) in—
Photographic emulsions.

2:2′-Diethyl-8-furylselenacarbocyanin Iodide

Dye
Dye possessing (Brit. 439359)—
Abnormally high solubility in organic solvents.

2:2′-Diethyl-8-furylthiacarbocyanin Iodide

Dye
Dye possessing (Brit. 439359)—
Abnormally high solubility in organic solvents.

Diethylguanylthiourea
Synonyms: Diethylguanylsulphourea.
French: Sulphourée de diéthyleguanyle, Sulphourée diéthyliqueguanylique, Thiourée de diéthyleguanyle, Thiorurée diéthyliqueguanylique.
German: Diaethylguanylsulfoharnstoff, Diaethylguanylthioharnstoff.

Chemical
Starting point in making—
Intermediates, pharmaceuticals.
Starting point (Brit. 286749) in making vulcanization accelerators with—
Butyl mercaptan, ethyl mercaptan, 2-mercaptobenzimidazole, mercaptobenzothioazole, mercaptobenzoxazole, 2-mercaptoiminazole, 2-mercaptothiazolin, mercaptotolylthiazole, meta-aminothiophenol, naphthylthiazole, orthoaminothiophenol, para-aminothiphenol, thioammelin, thioamides, thioanilides, thiocresol, thioxyindole, thiphenol.

Diethylhydrazin

Chemical
Starting point in making—
Tetraethyltetrazone, triethylazonium iodide.

Diethyl Hydrophthalate
French: Hydrophthalate de diéthyle, Hydrophthalate diéthylique.
German: Diaethylhydrophtalat, Hydrophtalsäurediaethyl, Hydrophtalsäuresdiaethylester.

Cellulose Products
Plasticizer for—
Cellulose acetate, cellulose esters and ethers, cellulose nitrate.
For uses, see under general heading: "Plasticizers."

Diethylisorosindulin-1:6:13-trisulphonic Acid

Dye
Starting point (Brit. 431708 and 431709) in making—
Greenish-blue dyestuffs with 4-aminodiphenylamine-2-sulphonic acid.

Diethyl Ketone Peroxide

Fuel
Ignition improver (Brit. 444544) for—
Diesel engine fuels.
Reducer (Brit. 444544) of—
Inflammability hazards in diesel engine fuels.

Diethylmalonic Acid

Chemical
Starting point (Brit. 410385) in making—
Aliphatic esters by esterification in the presence of a neutral solvent, such as benzin, benzene hydrocarbons, chlorinated hydrocarbons, and ethers, and of an esterification catalyst, such as sulphuric, hydrochloric, phosphoric, or a sulphonic acid or an acid sulphuric acid ester.

Diethylmeta-aminophenol
Chemical
As an intermediate in organic syntheses.
Dye
As an intermediate.
Rubber
As an antioxidant (U. S. 1899120).

Diethyloctadecylamine Hydrochloride
Textile
Reagent (Brit. 390553) for—
Increasing fastness to water of cellulosic materials dyed with substantive colors.

Diethyloctylamine Oxide
Chemical
Starting point (Brit. 460710) in making—
Cleansing, disinfecting, and wetting agents by reacting with alkylene oxides.
Emulsifying agents for soaps, glue, gelatin, gums, and mucilages.
Textile stripping agents for vat dyestuffs by reacting with alkylene oxides and mixing with hydrosulphites.

Diethyloleylamine Oxide
Chemical
Starting point (Brit. 460710) in making—
Cleansing, disinfecting, and wetting agents by reacting with alkylene oxides.
Emulsifying agents for soaps, glue, gelatin, gums, and mucilages.
Textile stripping agents for vat dyestuffs by reacting with alkylene oxides and mixing with hydrosulphites.

Diethyl Oxalate
Cellulose Products
Solvent and plasticizer for—
Cellulose esters or ethers.
For uses, see under general heading: "Solvents."
Chemical
Starting point in making—
Aromatic chemicals, ethylbenzyl malonate from benzyl acetate (German 427856), intermediates, organic chemicals, pharmaceutical chemicals, trimethylamine.
Perfume
As a solvent.

2:2′-Diethyl-8-pentadecylselenacarbocyanin Iodide
Dye
Dye possessing (Brit. 439359)—
Abnormally high solubility in organic solvents.

2:2′-Diethyl-8-pentadecylthiacarbocyanin Iodide
Dye
Dye possessing (Brit. 439359)—
Abnormally high solubility in organic solvents.

Diethyl Phthalate
Synonyms: Phthalic ether.
French: Phthalate de diéthyle, Phthalate diéthylique.
German: Diaethylphtalat, Phtalsaeuresdiaethyl, Phtalsaeurediaethylester.
Abrasives
Plasticizing agent in making—
Grinding wheels, whetstones, sand paper, emery paper, and cloth (Brit. 281711).
Chemical
Reagent in—
Making various products and in various chemical processes, in which its high resistance to heat is of advantage.
Fats and Oils
Ingredient of various mixtures.
Paint and Varnish
Solvent in making—
Cellulose acetate dopes, varnishes, and lacquers.
Perfumery
Fixative in making—
Cosmetics, perfumes.
Solvent in making—
Cosmetics, perfumes.
Starting point in making—
Synthetic perfumes.
Plastics
Ingredient of—
Moulding powders containing cellulose esters and ethers (Brit. 282723).
Substitute for camphor in making—
Celluloid and other plastics.
Resins and Waxes
Solvent in making—
Synthetic ester-condensation products (Brit. 252394).
Textile
——, *Finishing*
Ingredient of—
Compositions used in oiling fabrics.

2:2′-Diethyl-8-propylselenacarbocyanin Iodide
Dye
Dye possessing (Brit. 439359)—
Abnormally high solubility in organic solvents.

2:2′-Diethyl-8-propylthiacarbocyanin Iodide
Dye
Dye possessing (Brit. 439359)—
Abnormally high solubility in organic solvents.

Diethyl Sebacinate
Synonyms: Diethyl sebacate, Ethyl sebacate, Ethyl sebacinate, Sebacic ether, Sebacinic ether.
French: Éther sébacique, Éther sébacinique, Sébacate de diéthyle, Sébacate diéthylique, Sébacate d'éthyle, Sébacate éthylique, Sébacinate de diéthyle, Sébacinate diéthylique, Sébacinate d'éthyle, Sébacinate éthylique.
German: Aethylsebacat, Aethylsebacinat, Diaethylsebacat, Diaethylsebacinat, Sebacinsäurediaethylester, Sebacinsäureaethylester, Sebacinsäuresdiaethyl, Sebacinsäuresaethyl.
Food
Flavoring in—
Confectionery, food preparations.
Ingredient of—
Fruit flavors.
Miscellaneous
See also: "Plasticizers."
Perfume
Ingredient of—
Perfumes.
Perfume in—
Cosmetics.

Diethyl Succinate
Synonyms: Diethyl ethanealphabetadicarboxylate, Succinic ester.
French: Éthanealphabétadicarbonate de diéthyle, Éthanealphabétadicarboxylate diéthylique, Éther succinique, Succinate de diéthyle, Succinate diéthylique.
German: Aethanalphabetadicarbonsäurediaethylester, Aethanalphabetadicarbonsäuresdiaethyl, Bernsteinaether, Bernsteinsäurediaethylester, Bernsteinsäuresdiaethyl, Butandisäurediaethylester, Butansäuresdiaethyl, Diaethylsuccinat.
Food
Ingredient of—
Currant flavoring, various flavoring compositions.
Perfume
Fixative in various perfumes.
Ingredient of—
Flower-oil preparations, nonalcoholic perfumes.
Miscellaneous
As an odorizer in various preparations.

Diethyl Sulphate
French: Diéthyle sulfate, Sulfate de diéthyle, Sulfate diéthylique.
German: Diaethylschwefelsäuresester, Schwefelsäuresdiaethyl.
Chemical
Ethylating agent (noninflammable and noncorrosive) in making—
Amines, esters, ethers, imides, synthetic organic chemicals.

Diethyl Telluride
French: Tellurure de diéthyle, Tellurure diéthylique.
German: Diaethyltellurid, Telluridiaethyl.
Petroleum
Ingredient of—
Gasoline motor fuels (added to reduce motor knock).

2:2'-Diethyl-4:5:4':5'-tetraethoxyselenatricarbocyanin Iodide

Photographic
Sensitizer (Brit. 425417) for—
Photographic emulsions.

2:2'-Diethyl-4:5:4':5'-tetraethoxythiatricarbocyanin Iodide

Photographic
Sensitizer (Brit. 425417) for—
Photographic emulsions.

2:2'-Diethyl-4:5:4':5'-tetramethoxythiadicarbocyanin Iodide

Photographic
Sensitizer (Brit. 425417) for—
Photographic emulsions.

2:2'-Diethyl-4:5:4':5'-tetramethoxythiatricarbocyanin Iodide

Photographic
Sensitizer (Brit. 425417) for—
Photographic emulsions.

3:3'-Diethylthiazolinotricarbocyanin Iodide

Photographic
Sensitizer (Brit. 436941 and 437017) for—
Photographic emulsions to infrared light with maxima at 710 mu.

1:1'-Diethyl-4:4'-tricarbocyanin Iodide

Photographic
Sensitizer (Brit. 436941 and 437017) for—
Photographic emulsions to infrared light with maxima at 980 mu.

Diethyl-Zinc

Lubricant
Starting point (Brit. 440175) in making—
Addition agents for high-pressure lubricating oils or greases, by reacting with oil-soluble organic compounds.

4:4'-Difluorodiphenyl

Miscellaneous
Reagent (Brit. 333583) in—
Mothproofing furs, feathers, hair.
Textile
Reagent (Brit. 333583) in—
Mothproofing wool and felt.

Digammachlorobetahydroxypropylpiperidinium Chloride

Textile
Assistant (Brit. 454320) in—
Textile processes.

Digammaethylheptenylcarbinol Hydrogensulphate, Calcium Salt

Miscellaneous
As a general detergent (Brit. 440539).
As a general emulsifying agent (Brit. 440539).
As a general wetting agent (Brit. 440539).
Textile
As a detergent (Brit. 440539).
As an emulsifying agent (Brit. 440539).
As a textile assistant (Brit. 440539).
As a wetting agent (Brit. 440539).

Digammaethylheptenylcarbinol Hydrogensulphate, Magnesium Salt

Miscellaneous
As a general detergent (Brit. 440539).
As a general emulsifying agent (Brit. 440539).
As a general wetting agent (Brit. 440539).
Textile
As a detergent (Brit. 440539).
As an emulsifying agent (Brit. 440539).
As a textile assistant (Brit. 440539).
As a wetting agent (Brit. 440539).

Digammaethylheptenylcarbinol Hydrogensulphate, Sodium Salt

Miscellaneous
As a general detergent (Brit. 440539).
As a general emulsifying agent (Brit. 440539).
As a general wetting agent (Brit. 440539).
Textile
As a detergent (Brit. 440539).
As an emulsifying agent (Brit. 440539).
As a textile assistant (Brit. 440539).
As a wetting agent (Brit. 440539).

Diglycerol Tetra-acetate

Ceramic
Plasticizer (Brit. 364807) in—
Compositions, containing cellulose esters or ethers, used as coatings for protecting and decorating ceramic products.
Chemical
Plasticizer (Brit. 364807) for—
Cellulose esters or ethers.
Cosmetic
Plasticizer (Brit. 364807) in—
Nail enamels and lacquers containing cellulose esters or ethers as a base material.
Electrical
Plasticizer (Brit. 364807) in—
Insulating compositions, containing cellulose esters or ethers, used for covering wire and in making electrical machinery and equipment.
Glass
Plasticizer (Brit. 364807) in—
Compositions, containing cellulose esters or ethers, used in the manufacture of nonscatterable glass and as coatings for protecting and decorating glassware.
Glues and Adhesives
Plasticizer (Brit. 364807) in—
Adhesive compositions containing cellulose esters or ethers.
Leather
Plasticizer (Brit. 364807) in—
Compositions, containing cellulose esters or ethers, used in the manufacture of artificial leather and as coatings for protecting and decorating leather and leather goods.
Metallurgical
Plasticizer (Brit. 364807) in—
Coating compositions, containing cellulose esters or ethers, used for protecting and decorating metallic articles.
Miscellaneous
Plasticizer (Brit. 364807) in—
Coating compositions, containing cellulose esters or ethers, used for protecting and decorating various articles.
Paint and Varnish
Plasticizer (Brit. 364807) in—
Paints, varnishes, lacquers, enamels, and dopes containing cellulose esters or ethers.
Paper
Plasticizer (Brit. 364807) in—
Compositions, containing cellulose esters or ethers, used in the manufacture of coated papers and as coatings for protecting and decorating products made of paper or pulp.
Photographic
Plasticizer (Brit. 364807) in making—
Films from cellulose esters or ethers.
Plastics
Plasticizer (Brit. 364807) in making—
Laminated fiber products.
Molded products.
Plastics from or containing cellulose esters or ethers.
Resins
Plasticizer (Brit. 364807) in making—
Artificial resins from or containing cellulose esters or ethers.
Rubber
Plasticizer (Brit. 364807) in—
Compositions, containing cellulose esters or ethers, used as coatings for protecting and decorating rubber goods.
Stone
Plasticizer (Brit. 364807) in—
Compositions, containing cellulose esters or ethers, used as coatings for decorating and protecting artificial and natural stone.
Textile
Plasticizer (Brit. 364807) in—
Compositions, containing cellulose esters or ethers, used in the manufacture of coated textile fabrics.
Woodworking
Plasticizer (Brit. 364807) in—
Compositions, containing cellulose esters or ethers, used as protective and decorative coatings on woodwork.

Diglycerylamine
French: Diglycérylamine.
German: Diglycerylamin.
Soap
Starting point in making—
Soaps, when warmed with fatty acids, soluble in organic liquids and suitable for making dry-cleaning preparations.

Diglycol Laurate
Miscellaneous
As an emulsifying agent.
For uses, see under general heading: "Emulsifying agents."

Diguaiacolisatin
Chemical
Starting point (Brit. 278672) in making drugs with—
Benzyl bromide, benzyl iodide, butyl bromide, butyl iodide, dimethyl sulphate, ethyl bromide, ethyl iodide, methyl bromide, methyl iodide, phenyl bromide, phenyl iodide, propyl bromide, propyl iodide.

Diheptoxybenzoic Acid
French: Acide de diheptoxyebenzoique.
German: Diheptoxybenzoesaeure.
Chemical
Starting point in making—
Esters and salts, intermediates, pharmaceuticals.
Dye
Starting point (Brit. 291361) in making thioindigoid dyestuffs with—
Anthracene and derivatives.
Benzene and members of the benzene series.
Naphthalene and derivatives, such as the naphthalene and naphthol sulphonic acids.

Dihexoxybenzoic Acid
French: Acide de dihexoxyebenzoique.
German: Dihexoxybenzoesaeure.
Chemical
Starting point in making—
Esters and salts, intermediates, pharmaceuticals.
Dye
Starting point (Brit. 291361) in making thioindigoid dyestuffs with—
Anthracene and derivatives.
Benzene and members of the benzene series.
Naphthalene and naphthalene derivatives, such as the naphthalene and naphthol sulphonic acids.

Dihydrobenzene
Synonyms: Dihydrobenzol.
Chemical
Ingredient (Brit. 263873) of—
Aromatic hydrocarbon emulsions, terpene emulsions.
Fats and Oils
Ingredient (Brit. 263873) of—
Emulsions.
Leather
Ingredient (Brit. 263873) of—
Emulsified tanning compositions.
Miscellaneous
Ingredient (Brit. 263873) of—
Emulsified washing and cleansing compositions.
Paper
Reagent (Brit. 263873) in treating—
Paper and cardboard to increase their absorbing and wetting capacities.
Petroleum
Ingredient (Brit. 263873) of—
Mineral oil emulsions.
Resins and Waxes
Ingredient (Brit. 263873) of—
Emulsified compositions.
Textile
——, *Dyeing*
Ingredient (Brit. 263873) of—
Emulsified dye liquors.
——, *Finishing*
Ingredient (Brit. 263873) of—
Emulsified washing and cleansing compositions.
——, *Manufacturing*
Ingredient (Brit. 263873) of—
Wool-carbonizing compositions.

Dihydro-1:2:3:9-benzisotetrazole
Pharmaceutical
Claimed (U. S. 2008536) to have—
Valuable therapeutic properties and solubility in water.

Dihydrocarveole
Synonyms: Alphamethyl-4-isopropenylcyclohexanole-2, Alphamethyl-4-isopropenylcyclohexanol-2, Dihydrocarveol.
Food
For giving food a caraway flavor.
Perfume
Ingredient of—
Elder perfume preparations.
Hyacinth preparations.
Lily-of-the-valley preparations.
Perfume in—
Cosmetics.
Soap
Perfume in—
Toilet soaps.

Dihydrocarveyl Acetate
French: Acétate de dihydrocarveyle, Acétate dihydrocarveylique.
German: Dihydrocarveylacetat, Dihydrocarveylazetat, Essigsäuredihydrocarveylester, Essigsäuresdihydrocarveyl.
Food
Ingredient of—
Flavoring compounds, fruit essences and extracts.
Flavoring agent in—
Food preparations.
Perfume
Ingredient of—
Fancy perfumes.
Perfume in—
Cosmetics, toilet waters.

Dihydrocarvone
Synonyms: Alphamethyl-4-isopropenylcyclohexanone-2.
French: Dihydrocarvone.
German: Alphamethyl-4-isopropenylcyclohexanon-2, Dihydrocarvon.
Food
As a flavoring.
Ingredient of—
Fruit essences and extracts.
Perfume
Ingredient of—
Perfume preparations.
Perfuming and flavoring ingredient in making—
Cosmetics.

Dihydrocuprein Ethyl Ether Hydrochloride
French: Chlorohydrate de dihydrocupreine éthyle éther, Hydrochlorure d'éther dihydrocupreinéthylique.
German: Dihydrocupreinaethylaetherchlorhydrat, Dihydrocupreinaethylaetherhydrochlorid.
Chemical
Starting point (Brit. 282356) in making antiparasitic agents with the following acids or their sodium and potassium salts—
Apocholic, cholic, dehydrocholic, desoxycholic, glycocholic, taurocholic.

Dihydrocupreinisoamyl Ether
French: Éther de dihydrocupreineisoamyle, Éther de dihydrocupreineisoamylique.
German: Dihydrocupreinisoamylaether.
Chemical
Starting point (Brit. 282356) in making antiparasitic agents with the following acids or their sodium and potassium salts—
Apocholic, cholic, dehydrocholic, desoxycholic, glycocholic, taurocholic.

Dihydrocupreinisoamyl Ether Hydrochloride
French: Chlorohydrate d'éther de dihydrocupreineisoamyle, Chlorohydrate d'éther dihydrocupreineisoamylique, Hydrochlorure d'éther de dihydrocupreineisoamyle, Hydrochlorure d'éther dihydrocupreineisoamylique.
German: Dihydrocupreinisoamylaetherchlorhydrat, Dihydrocupreinisoamylhydrochlorid.
Chemical
Starting point (Brit. 282356) in making antiparasitic agents with the following acids or their sodium and potassium salts—
Apocholic, cholic, dehydrocholic, desoxycholic, glycocholic, taurocholic.

Dihydrocuprein-normal-octyl Ether
French: Éther de dihydrocupreine-N-octyle, Éther dihydrocupreine-N-octylique.
German: Dihydrocuprein-N-octylaether.

Chemical
Starting point (Brit. 282356) in making antiparasitic agents with the following acids or their sodium and potassium salts—
Apocholic, cholic, dehydrocholic, desoxycholic, glycocholic, taurocholic.

Dihydrocuprein-normal-octylether Hydrochloride
French: Éther de dihydrocupreine-N-octylechlorhydrique, Éther de dihydrocupreine-N-octylehydrochlorique.
German: Chlorwasserstoffsaeuresdihydrocupreine-N-octylaether, Dihydrocupreine-N-octylaetherchlorhydrat, Dihydrocupreine-N-octylaetherhydrochlorid.

Chemical
Starting point (Brit. 282356) in making antiparasitic agents with the following acids or their sodium and potassium salts—
Apocholic, cholic, dehydrocholic, desoxycholic, glycocholic, taurocholic.

Dihydro-6:8-dimethyl-1:2:3:9-benzisotetrazole

Pharmaceutical
Claimed (U. S. 2008536) as having—
Valuable therapeutic properties and solubility in water.

Dihydrodioxymorphine-D

Pharmaceutical
Claimed (U. S. 1980972) as—
Preparation having physiological properties of morphine, but less toxic.

Dihydro-8-methyl-1:2:3:9-benzisotetrazole

Pharmaceutical
Claimed (U. S. 2008536) to have—
Valuable therapeutic properties and solubility in water.

Dihydroquinone
German: Dihydrochinon.

Chemical
Starting point (Brit. 282356) in making parasiticides with—
Apocholic acid, cholic acid, dehydrocholic acid, desoxycholic acid, glycocholic acid, potassium salts of these acids, sodium salts of these acids, taurocholic acid.

Dihydrothebaine

Chemical
Starting point in making derivatives used as drugs (German 437451).

Pharmaceutical
In compounding and dispensing practice.

2:4-Dihydroxybenzimidothiophenylether Hydochloride
Synonyms: 2:4-Dihydroxybenzimidophenylsulphide-hydrochloride.

Fungicide and Insecticide
Larvicide for—
Culicine mosquito larvae.

2:6-Dihydroxy-1:5-dibenzoylnaphthalene
German: 2:6-Dihydroxy-1:5-dibenzoylnaphtalin.

Dye
Starting point in making—
Dyestuffs of the halogenated dibenzopyrenequinone type (Brit. 249147).

5:5′-Dihydroxy-2:2′-dimethyloxacarbocyanin Perchlorate

Photographic
Sensitizing agent (Brit. 430357) for—
Emulsions.

8:8′-Dihydroxy-2:2′-dinaphthylamine-6:6′-disulphonic Acid
French: Acide de 8:8′-dihydroxy-2:2′-dinaphthylamine-6:6′-disulphonique.
German: 8:8′-Dihydroxy-2:2′-dinaphtylamin-6:6′-disulphonsaeure.

Dye
Starting point (Brit. 270446) in making azo dyestuffs for viscose rayon with—
Allphanaphthylamine, aminoazobenzene sulphonic acid, aminosalicylic acid, meta-aminobenzoic acid, metaxylidene, naphthionic acid, orthoanisidin, para-aminoacetanilide, parachloroanilin, paranitranilin.

8:8′-Dihydroxy-2:2′-dinaphthylaminetetrasulphonic Acid
French: Acide de 8:8-dihydroxy-2:2′-dinaphthylamine-tétrasulfonique.
German: 8:8′-Dihydroxy-2:2′-dinaphtylamintetrasulfonsaeure.

Dye
Starting point (Brit. 270446) in making azo dyestuffs for viscose rayon with—
Dihydrothioparatoluidinsulphonic acid, metanitranilin, metaxylidenesulphonic acid, para-aminoacetanilide, paranitranilinorthosulphonic acid, salicylicazoalphanaphthylamine.

Dihydroxydiphenylpropane

Glass
Stabilizer (Brit. 437304) for—
Halogenated rubber derivatives used as cements for laminated glass.

Miscellaneous
Inhibitor (Brit. 437304) of—
Photochemical action.

Paper
Stabilizer (Brit. 437304) for—
Halogenated rubber derivatives used for impregnating or coating wrapping paper.

Rubber
Promoter (Brit. 437304) of—
Resistance to the deteriorating action of light on chlorinated rubber.
Stabilizer (Brit. 437304) for—
Coating and impregnating agents made from halogenated rubber derivatives and used for treating fabrics to be used as wrapping materials.
Transparent films or sheets made from halogenated rubber derivatives.

2:4′-Dihydroxydiphenyl Sulphide
Synonyms: 2:4′-Dihydroxybisphenyl sulphide.

Fungicide and Insecticide
As a fungicide (Brit. 349004).
As an insecticide (Brit. 349004).

Sanitation
As a bactericide (Brit. 349004).

Dihydroxymethylcetylamine Oxide

Chemical
Starting point (Brit. 460710) in making—
Cleansing, disinfecting, and wetting agents by reacting with alkylene oxides.
Emulsifying agents for soaps, glue, gelatin, gum, and mucilages.
Textile stripping agents for vat dyestuffs by reacting with alkylene oxides and mixing with hydrosulphites.

1:2-Dihydroxynaphthalene

Chemical
Starting point in making—
Aromatics, intermediates, pharmaceuticals.

Coaltar
Inhibitor (Brit. 432121) of—
Gums or sludges in crude benzene.
Stabilizer (Brit. 432121) for—
Crude benzene.

Dye
Starting point in making various synthetic dyestuffs.

Petroleum
Inhibitor (Brit. 432121) of—
Formation of gummy, resinous products or sludge in liquid hydrocarbons, such as cracked gasoline, diesel oil, transformer oil.
Stabilizer (Brit. 432121) for—
Liquid hydrocarbons such as cracked gasoline, diesel oil, transformer oil.

1:4-Dihydroxynaphthalene

Chemical
Starting point in making—
Aromatics, intermediates, pharmaceuticals.

Coaltar
Inhibitor (Brit. 432121) of—
Gums or sludges in crude benzene.
Stabilizer (Brit. 432121) for—
Crude benzene.

Dye
Starting point in making various synthetic dyestuffs.

1:4-Dihydroxynaphthalene (Continued)
Petroleum
Inhibitor (Brit. 432121) of—
Formation of gummy, resinous products or sludge in liquid hydrocarbons, such as cracked gasoline, diesel oil, transformer oil, etc.
Stabilizer (Brit. 432121) for—
Liquid hydrocarbons such as cracked gasoline, diesel oil, transformer oil.

1:5-Dihydroxynaphthalene
German: 1:5-Dihydroxynaphtalin.
Chemical
Starting point in making—
Aromatics, intermediates, pharmaceuticals.
Dye
Starting point in making various synthetic dyestuffs.
Rubber
Ingredient (Brit. 342502) of—
Rubber batch (added in admixture with the following chemicals, for the purpose of retarding the deterioration of the rubber):
Acetaldehyde, allylthiourea, alphaphenyldibetahydroxyethylthiourea, anilin, betanaphthylamine, butyraldehyde, dialphanaphthylurea, dianisidin, dibenzylamine, dibenzylanilin, diethanolamine, diethylenetriamine, dimethylalphanaphthylamine, dinaphthylbenzidin, diphenylamine, ethylalphanaphthylamine, ethylenediamine, formaldehyde, dicyandiamine, heptaldehyde, metanitromethylanilin, methylalphanaphthylamine, methylbenzylanilin, methylphenylhydrazin, monoethanolamine, naphthyldiaminodiphenylmethane, paraminodimethylanilin, paraphenylenediamine, pentamethyldiethylenetriamine, phenylalphanaphthylamine, polyethylenepolyamine, triethanolamine, triethylamine, triethyltrimethylenetetramine.

1:7-Dihydroxynaphthalene
Chemical
Starting point in making—
Aromatics, intermediates, pharmaceuticals.
Coaltar
Inhibitor (Brit. 432121) of—
Gums or sludges in crude benzene.
Stabilizer (Brit. 432121) for—
Crude benzene.
Dye
Starting point in making various synthetic dyestuffs.
Petroleum
Inhibitor (Brit. 432121) of—
Formation of gummy or resinous products or sludge in liquid hydrocarbons, such as cracked gasoline, transformer oil, diesel oil.
Stabilizer (Brit. 432121) for—
Liquid hydrocarbons, such as cracked gasoline, diesel oil, transformer oil.

2:3-Dihydroxynaphthalene
Chemical
Starting point in making—
Aromatics, intermediates, pharmaceuticals.
Coaltar
Inhibitor (Brit. 432121) of—
Gums or sludges in crude benzene.
Stabilizer (Brit. 432121) for—
Crude benzene.
Dye
Starting point in making various synthetic dyestuffs.
Petroleum
Inhibitor (Brit. 432121) of—
Formation of gummy or resinous products or sludge in liquid hydrocarbons, such as cracked gasoline, diesel oil, transformer oil.
Stabilizer (Brit. 432121) for—
Liquid hydrocarbons, such as cracked gasoline, diesel oil, transformer oil.

2:8-Dihydroxynaphthalene-6-sulphonic Acid
Dye
Coupling agent (Brit. 421421) in making—
Red-grey colors (on wool) with orthoaminophenol-4-sulphonic acid chromium salt.
Blue-violet colors (on leather) with 6-nitro-orthoaminophenol-4-sulphonic acid copper salt.
Intermediate in—
Dye manufacture.

3:4-Dihydroxyphenylalphapropanolamine Hydrochloride
Synonyms: Hydrochloride of betaaminoalpha-3:4-dihydroxyphenyl-normal-propyl alcohol.
Pharmaceutical
Suggested for use (Brit. 439168) as—
Vaso-constrictor in conjunction with betadiethylaminoethyl para-aminobenzoate hydrochloride as anesthetic (greater stability and less toxicity than adrenalin are claimed).

Dihydroxypropyl-N′-Butylthiourea, Normal
Textile
Wetting agent (Brit. 436660) in—
Mercerizing processes (used in conjunction with phenols).

Dihydroxystearic Acid
French: Acide de dihydroxyestearique, Acide de dioxyestearique.
German: Dihydroxystearinsaeure, Dioxystearinsaeure.
Chemical
Ingredient (Brit. 303379) of—
Emulsified preparations.
Starting point in making—
Esters and salts, stearic acid compositions.
Miscellaneous
Ingredient (Brit. 303379) of—
Bleaching compositions, cleansing compositions.
Perfume
Ingredient of—
Cosmetics.
Soap
Ingredient of—
Saponaceous cleansing and bleaching compositions.
Textile
——, *Finishing*
Ingredient (Brit. 303379) of—
Finishing, bowking, and softening baths.
——, *Manufacturing*
Ingredient (Brit. 303379) of—
Oiling compositions.

2:4-Di-iodoanisol
German: 2:4-Dijodanisol.
Chemical
Starting point (Brit. 275313) in making iodo derivatives of cyanophenol ethers with—
Metallic salts of iodo-oxybenzo nitriles.
Metallic salts of iodophenol ethers.

5:7-Di-iodo-8-hydroxyquinolin
Pharmaceutical
Suggested for use (Brit. 351605) as—
Antiseptic.

2:6-Di-iodophenol-4-sulphonic Acid
Synonyms: Sozoiodolic acid.
French: Acide 2:6-di-iodophénol-4-sulphonique, Acide sozoiodolique.
German: 2:6-Dijodphenol-4-sulfonsäure, 2:6-Dijodphenol-4-sulfosäure, Sozojodolsäure.
Chemical
Starting point in making—
Esters, pharmaceuticals, various salts.
Pharmaceutical
In compounding and dispensing practice.

Di-iodoricinostearolic Acid
Synonyms: Ricinstearolic di-iodide.
French: Acide di-iodoricinostearolique, Di-iodure ricinostearolique.
German: Dijodricinstearolinsäure, Ricinstearolindijodid.
Chemical
Starting point in making—
Esters, pharmaceuticals, various salts.
Pharmaceutical
In compounding and dispensing practice.

Di-iodosalicylic Acid
Latin: Acidum di-iodosalicylicum.
French: Acide de di-iodosalicyle, Acide di-iodosalicilique.
German: Dijodsalicylsäure.
Chemical
Starting point in making—
Esters, pharmaceuticals, various salts.
Pharmaceutical
In compounding and dispensing practice.

Di-isobutylamine

Insecticide

Suspension promoter for—
Insoluble powdered insecticides.

Di-isobutyl Phthalate

French: Phthalate de di-isobutyle, Phthalate di-isobutylique.
German: Di-isobutylphtalat, Phtalsäuredi-isobutylester, Phtalsäuresdi-isobutyl.

Cellulose Products

Plasticizer for—
Cellulose acetate, cellulose esters and ethers, cellulose nitrate.
For uses, see under general heading: "Plasticizers."

Chemical

Starting point in making various derivatives.

Perfume

As a fixative.
Solvent for—
Aromatic oils.

Di-isopropylcarbazole

Resins

Treating agent (German 578039) for—
Raising softening point of natural resins.

Di-isopropyldithiophosphoric Acid

Chemical

Starting point (U. S. 1949629) in making—
Vulcanization accelerators by reaction with sulphur chloride.

Di-isopropyldithiophosphoric Acid Sodium Salt

Chemical

Starting point (U. S. 1949629) in making—
Vulcanization accelerators by reaction with sulphur chloride.

Di-isopropyl Ketone Peroxide

Fuel

Ignition improver (Brit. 444544) for—
Diesel engine fuels.
Reducer (Brit. 444544) of—
Inflammability hazards in diesel engine fuels.

Di-isopropyl Sulphite

French: Sulphite de di-isopropyle, Sulphite di-isopropylique.
German: Di-isopropylsulfit, Schwefeligsäuredi-isopropylester, Schwefeligsäuresdi-isopropyl.

Agriculture

Reagent (Brit. 340685) in destroying—
Grain weevils.

Chemical

Starting point in making various derivatives.

Insecticide

As an insecticide.
Ingredient (Brit. 340685) of—
Insecticidal preparations.

Dilauryl Dithiocarbamate

Fungicide and Insecticide

As a fungicide (Brit. 436327).
As an insecticide (Brit. 436327).

Dilauryl Dithiocarbonate

Insecticide and Fungicide

As an anticryptogamic (Brit. 436327).
As an insecticide (Brit. 436327).

Diluents

See: "Solvents."

Dimenthyl Malate

Cellulose Products

Plasticizer (Brit. 432404) for—
Cellulose acetate, cellulose esters and ethers.
For uses, see under general heading: "Plasticizers."

Dimercaptodiphenyl

Petroleum

Antioxidant (Brit. 425569) for—
Lubricating, transformer, and switch oils, particularly solvent-extracted oils and others of a paraffinic nature, in which the natural inhibitor content may have been reduced during refining.

2:5-Dimethoxyanilide

Dye

Starting point (Brit. 434209 and 434433) in making—
Yellowish-red water-insoluble dyestuffs by coupling (in substance or on the fiber) with meta-4-xylidin-6-sulphondiethylamide.

2:4-Dimethoxyanilin-5-sulphonbenzylmethylamide

Dye

Coupling agent (Brit. 434209 and 434433) in making—
Water-insoluble bordeaux dyestuffs with 5-methoxyorthotoluidide.

3:4′-Dimethoxydiphenylamine

Rubber

As an antioxidant (Brit. 435024).

6:6′-Dimethoxy-1:3:3:1′:3′:3′-hexamethylindocarbocyanin Chloride

Dye

Starting point (Brit. 448508) in making—
Color lakes which are especially fast to light, oil, and alcohols, and are claimed to be superior to the corresponding lakes from triarylmethane dyes.

2:5-Dimethoxyparaphenylenediamine

Dye

As an intermediate.
Starting point (Brit. 397034) in making—
Tetrazo compounds.

Dimethoxyphenylguanadin

Chemical

Starting point in making—
Dipara-anisylmonophenetylguanadin hydrochloride (a coin).

Dimethoxystrychnine

Chemical

Starting point in making—
Delta-5-phenyl-5-ethylhydantoin, used as a hypnotic.
Hydrochloride, nitrate, and sulphate, used as paralyzants to the sensory nerves.

Dimethylamine

German: Dimethylamin.

Agricultural

Reagent for—
Attracting boll weevils in order to exterminate them.

Chemical

Starting point (Brit. 310534) in making rubber vulcanization accelerators with the aid of—
Allylenethiourea, amylenethiourea, butylenethiourea, ethylenethiourea, heptylenethiourea, hexylenethiourea, isoallylenethiourea, isoamylenethiourea, isobutylenethiourea, isopropylenethiourea, methylenethiourea, propylenethiourea.
Starting point in making—
Aromatics, dimethylamine hydrochloride, dimethylaminocarbinol, diphenylhydrazin, intermediates, pharmaceuticals.
Starting point (Brit. 270334) in making pharmaceutical chemicals with the aid of—
Alphabromoquinolingammacarboxylic acid.
Alphachloroquinolingammacarboxylic acid.
Alphaiodoquinolingammacarboxylic acid.

Dye

Starting point in making—
New methylene blue GG.

Rubber

Accelerator in—
Vulcanization.

Dimethylaminebenzoylbenzoic Acid

French: Acide de diméthyleaminebenzoylbenzoique.
German: Dimethylamin-benzoylbenzolsaeure.

Chemical

Starting point in making—
3-Dimethylaminoanthraquinone.

1:4-Di(methylamino)anthraquinone

Oils, Fats, Waxes

Coloring agent (Brit. 432867) for—
Paraffin and other mineral waxes, stearic acid, tallow and other solid triglycerides, beeswax, carnauba wax, and others.

Dimethylaminoantipyrin
Synonyms: Amidopyrin, Aminopyrin, Pyramidon.
Chemical
Starting point in making—
Dimethylaminoantipyrin acetate.
Dimethylaminoantipyrin benzoate.
Dimethylaminoantipyrin borate.
Dimethylaminoantipyrin camphorate.
Dimethylaminoantipyrin cinnamate.
Dimethylaminoantipyrin citrate.
Dimethylaminoantipyrin glycolate.
Dimethylaminoantipyrin glycocholate.
Dimethylaminoantipyrin glycerinate.
Dimethylaminoantipyrin glycerophosphate.
Dimethylaminoantipyrin lactate.
Dimethylaminoantipyrin phosphate.
Dimethylaminoantipyrin salicylate.
Dimethylaminoantipyrin sulphate.
Pyramidon butylchloral.
Pharmaceutical
In compounding and dispensing practice.

Dimethylaminobenzaldehyde
Analysis
Reagent in testing for—
Salvarsan, tryptophan.
Chemical
Starting point in making—
Derivative with paratoluidinsulphonic acid.
Dye
Starting point in making—
Acid violet 6B, naphthalene green V.
Insecticide
Starting point in making—
Insecticidal compounds with 2:4:6-trimethoxypyridin chloride (German 438241).

Dimethylaminobenzoyl Chloride
French: Chlorure de diméthyleaminobenzoyle, Chlorure diméthyleaminobenzoylique.
German: Dimethylaminobenzoylchlorid.
Dye
Starting point in making—
Acid violet 7B.

Dimethylaminobenzyl Alcohol
French: Alcool de diméthyleaminobenzyle, Alcool dimethylaminbenzylique.
Chemical
Starting point in making—
Paradimethylaminobenzaldehyde.

1:3-Dimethyl-4-amino-6-bromobenzene
Synonyms: 1:3-Dimethyl-4-amino-6-brombenzol.
Dye
Starting point (Brit. 274128) in making azo dyestuffs with—
Alphanaphthalide, betanaphthalide, 4-chloro-2-anisidide, 4-chloro-2-toluidide.

1:3-Dimethyl-5-amino-4-bromobenzene
French: 1:3-Diméthyle-5-amino-5-bromobenzène.
German: 1:3-Dimethyl-5-amino-4-bromobenzol.
Chemical
Starting point in making—
Aromatics, pharmaceuticals, intermediates.
Dye
Starting point (Brit. 300504) in making azo dyestuffs with—
2:3-Oxynaphthoic alpha-anilide.
2:3-Oxynaphthoic alphanaphthalide.
2:3-Oxynaphthoic 4-anisidide.
2:3-Oxynaphthoic anthroxyanilide.
2:3-Oxynaphthoic benzoxyanilide.
2:3-Oxynaphthoic betanaphthalide.
2:3-Oxynaphthoic 2-chloroanilide.
2:3-Oxynaphthoic 4′-chloro-2-anisidide.
2:3-Oxynaphthoic 5′-chloro-2-anisidide.
2:3-Oxynaphthoic 4-chloro-2-anisidide.
2:3-Oxynaphthoic 4′-chloro-2′-toluidide.
2:3-Oxynaphthoic cresoxyanilide.
2:3-Oxynaphthoic diacetoaceticanilide.
2:3-Oxynaphthoic 2′:5′-dimethoxy-1′-anilide.
2:3-Oxynaphthoic 4-methoxyanilide.
2:3-Oxynaphthoic naphthoxyanilide.
2:3-Oxynaphthoic 1-naphthylamide.
2:3-Oxynaphthoic 2-naphthylamide.
2:3-Oxynaphthoic orthotoluidide.
2:3-Oxynaphthoic phenoxyanilide.
2:3-Oxynaphthoic toloxyanilide.
2:3-Oxynaphthoic 3-toluidide.
2:3-Oxynaphthoic 4-toluidide.

1:3-Dimethyl-5-amino-2-bromo-4-chlorobenzene
French: 1:3-Diméthyle-5-amino-2-bromo-4-chlorobenzène.
German: 1:3-Dimethyl-5-amino-2-brom-4-chlorbenzol.
Chemical
Starting point in making—
Aromatics, intermediates, pharmaceuticals.
Dye
Starting point (Brit. 300504) in making azo dyestuffs with the aid of—
2:3-Oxynaphthoic alpha-anilide.
2:3-Oxynaphthoic alphanaphthalide.
2:3-Oxynaphthoic alphanaphthylamide.
2:3-Oxynaphthoic 4-anisidide.
2:3-Oxynaphthoic anthroxyanilide.
2:3-Oxynaphthoic benzoxyanilide.
2:3-Oxynaphthoic betanaphthalide.
2:3-Oxynaphthoic betanaphthylamide.
2:3-Oxynaphthoic 2-chloroanilide.
2:3-Oxynaphthoic 4-chloro-2-anisidide.
2:3-Oxynaphthoic 4′-chloro-2-anisidide.
2:3-Oxynaphthoic 5′-chloro-2-anisidide.
2:3-Oxynaphthoic 4′-chloro-2′-toluidide.
2:3-Oxynaphthoic cresoxyanilide.
2:3-Oxynaphthoic diacetoaceticanilide.
2:3-Oxynaphthoic 2′:5′-dimethoxy-1′-anilide.
2:3-Oxynaphthoic 4-methoxyanilide.
2:3-Oxynaphthoic naphthoxyanilide.
2:3-Oxynaphthoic orthotoluidide.
2:3-Oxynaphthoic phenoxyanilide.
2:3-Oxynaphthoic toloxyanilide.
2:3-Oxynaphthoic 3-toluidide.
2:3-Oxynaphthoic 4-toluidide.

1:3-Dimethyl-4-amino-6-chlorobenzene
Synonyms: 1:3-Dimethyl-4-amino-6-chlorbenzol.
Dye
Starting point (Brit. 274128) in making azo dyestuffs with—
Alphanaphthalide, betanaphthalide, 5-chloro-2-anisidide, 4-chloro-2-toluidide.

1:3-Dimethyl-5-amino-2-chloro-4-bromobenzene
German: 1:3-Dimethyl-5-amino-2-chlor-4-brombenzol.
Chemical
Starting point in making—
Aromatics, intermediates, pharmaceuticals.
Dye
Starting point (Brit. 300504) in making azo dyestuffs with the aid of—
2:3-Oxynaphthoic alpha-anilide.
2:3-Oxynaphthoic alphanaphthalide.
2:3-Oxynaphthoic alphanaphthylamide.
2:3-Oxynaphthoic 4-anisidide.
2:3-Oxynaphthoic anthroxyanilide.
2:3-Oxynaphthoic benzoxyanilide.
2:3-Oxynaphthoic betanaphthalide.
2:3-Oxynaphthoic betanaphthylamide.
2:3-Oxynaphthoic 2-chloroanilide.
2:3-Oxynaphthoic 4-chloro-2-anisidide.
2:3-Oxynaphthoic 4′-chloro-2-anisidide.
2:3-Oxynaphthoic 5′-chloro-2-anisidide.
2:3-Oxynaphthoic 4′-chloro-2′-toluidide.
2:3-Oxynaphthoic cresoxyanilide.
2:3-Oxynaphthoic diacetoaceticanilide.
2:3-Oxynaphthoic 2′:5′-dimethoxy-1′-anilide.
2:3-Oxynaphthoic 4-methoxyanilide.
2:3-Oxynaphthoic naphthoxyanilide.
2:3-Oxynaphthoic orthotoluidide.
2:3-Oxynaphthoic phenoxyanilide.
2:3-Oxynaphthoic toloxyanilide.
2:3-Oxynaphthoic 3-toluidide.
2:3-Oxynaphthoic 4-toluidide.

1:3-Dimethyl-4-amino-2:6-dibromobenzene
German: 1:3-Dimethyl-4-amino-2:6-dibrombenzol.
Dye
Starting point (Brit. 274128) in making azo dyestuffs with—
Alphanaphthalide, betanaphthalide, 5-chloro-2-anisidide, 4-chloro-2-toluidide.

1:3-Dimethyl-4-amino-2:6-dichlorobenzene
German: 1:3-Dimethyl-4-amino-2:6-chlorbenzol.
Dye
Starting point (Brit. 274128) in making azo dyestuffs with—
Alphanaphthalide, betanaphthalide, 5-chloro-2-anisidide, 4-chloro-2-toluidide.

1:3-Dimethyl-5-amino-2:4-dichlorobenzene
French: 1:3-Diméthyle-5-amino-2-chlorobenzène.
German: 1:3-Dimethyl-5-amino-2-chlorbenzol.
Chemical
Starting point in making—
Aromatics, intermediates, pharmaceuticals.
Dye
Starting point (Brit. 300504) in making azo dyestuffs with—
2:3-Oxynaphthoic-4-anilide.
2:3-Oxynaphthoic-4′-chloro-2-anisidide.
2:3-Oxynaphthoic-5′-chloro-2-anisidide.
2:3-Oxynaphthoic diacetoaceticanilide.
2:3-Oxynaphthoic-2′:5′-dimethoxy-1′-anilide.
2:3-Oxynaphthoic-1-naphthylamide.
2:3-Oxynaphthoic-2-naphthylamide.
2:3-Oxynaphthoic-2′-toluidide.

1:3-Dimethyl-5-amino-4:6-dichlorobenzene
French: 1:3-Diméthyle-5-amino-4:6-dichlorobenzène.
German: 1:3-Dimethyl-5-amino-4:6-dichlorbenzol.
Chemical
Starting point in making—
Aromatics, intermediates, pharmaceuticals.
Dye
Starting point (Brit. 300504) in making azo dyestuffs with—
2:3-Oxynaphthoic alpha-anilide.
2:3-Oxynaphthoic alphanaphthalide.
2:3-Oxynaphthoic 4-anisidide.
2:3-Oxynaphthoic anthroxyanilide.
2:3-Oxynaphthoic benzoxyanilide.
2:3-Oxynaphthoic betanaphthalide.
2:3-Oxynaphthoic 2-chloroanilide.
2:3-Oxynaphthoic 4′-chloro-2-anisidide.
2:3-Oxynaphthoic 5′-chloro-2-anisidide.
2:3-Oxynaphthoic 4-chloro-2-anisidide.
2:3-Oxynaphthoic 4′-chloro-2′-toluidide.
2:3-Oxynaphthoic cresoxyanilide.
2:3-Oxynaphthoic diacetoaceticanilide.
2:3-Oxynaphthoic 2′:5′-dimethoxy-1′-anilide.
2:3-Oxynaphthoic 4-methoxyanilide.
2:3-Oxynaphthoic naphthoxyanilide.
2:3-Oxynaphthoic 1-naphthylamide.
2:3-Oxynaphthoic 2-naphthylamide.
2:3-Oxynaphthoic orthotoluidide.
2:3-Oxynaphthoic phenoxyanilide.
2:3-Oxynaphthoic toloxyanilide.
2:3-Oxynaphthoic 3-toluidide.
2:3-Oxynaphthoic 4-toluidide.
2:3-Oxynaphthoic xyloxyanilide.

5-Dimethylaminomethyl-1:3:2-xylenol
Rubber
Anti-ager (Brit. 459045) for—
Rubber mixes.

Dimethylaminonaphthophenazoxonium Chloride
Chemical
Ingredient (Brit. 364046) of—
Preparations, containing substituted amide of a fatty acid, for sterilizing seeds.
Preparations for treating infected soils.
Insecticide
Ingredient (Brit. 364046) of—
Fungicidal and insecticidal compositions containing substituted amides of fatty acids.
Miscellaneous
Dyestuff for various products.
Paper
Dyestuff for paper and pulp.
Textile
Coloring for dyeing and printing yarns and fabrics.

4-Dimethylamino-1-phenyl-2:3-dimethyl-5-pyrazolone
Chemical
Starting point (U. S. 1881317) in making—
Medicinal products with para-aminobenzoic acid.

Dimethylaminoquinaldin Ethiodide
French: Éthiodure diméthyleaminoquinaldinique.
German: Aethjoddimethylaminochinaldin, Dimethylaminochinaldinaethjodid.
Insecticide
Starting point in making—
Insecticidal compounds with cinnamaldehyde.

1:3-Dimethyl-5-amino-2:4:6-tribromobenzene
French: 1:3-Diméthyle-5-amino-2:4:6-tribromobenzène.
German: 1:3-Dimethyl-5-amino-2:4:6-tribrombenzol.
Chemical
Starting point in making—
Aromatics, intermediates, pharmaceuticals.
Dye
Starting point (Brit. 300504) in making azo dyestuffs with the aid of—
2:3-Oxynaphthoic alpha-anilide.
2:3-Oxynaphthoic alphanaphthalide.
2:3-Oxynaphthoic alphanaphthylamide.
2:3-Oxynaphthoic 4-anisidide.
2:3-Oxynaphthoic anthroxyanilide.
2:3-Oxynaphthoic benzoxyanilide.
2:3-Oxynaphthoic betanaphthalide.
2:3-Oxynaphthoic betanaphthylamide.
2:3-Oxynaphthoic 2-chloroanilide.
2:3-Oxynaphthoic 4-chloro-2-anisidide.
2:3-Oxynaphthoic 4′-chloro-2-anisidide.
2:3-Oxynaphthoic 5′-chloro-2-anisidide.
2:3-Oxynaphthoic 4′-chloro-2′-toluidide.
2:3-Oxynaphthoic cresoxyanilide.
2:3-Oxynaphthoic diacetoaceticanilide.
2:3-Oxynaphthoic 2′:5′-dimethoxy-1′-anilide.
2:3-Oxynaphthoic 4-methoxyanilide.
2:3-Oxynaphthoic naphthoxyanilide.
2:3-Oxynaphthoic orthotoluidide.
2:3-Oxynaphthoic phenoxyanilide.
2:3-Oxynaphthoic toloxyanilide.
2:3-Oxynaphthoic 3-toluidide.
2:3-Oxynaphthoic 4-toluidide.

1:3-Dimethyl-5-amino-2:4:6-trichlorobenzene
Synonyms: 1:3-Dimethyl-5-amino-2:4:6-trichlorobenzol.
French: 1:3-Diméthyle-5-amino-2:4:6-trichlorobenzène.
Chemical
Starting point in making—
Aromatics, intermediates, pharmaceuticals.
Dye
Starting point (Brit. 300504) in making azo dyestuffs with—
2:3-Oxynaphthoic alpha-anilide.
2:3-Oxynaphthoic alphanaphthalide.
2:3-Oxynaphthoic alphanaphthylamide.
2:3-Oxynaphthoic 4-anisidide.
2:3-Oxynaphthoic anthroxyanilide.
2:3-Oxynaphthoic benzoxyanilide.
2:3-Oxynaphthoic betanaphthalide.
2:3-Oxynaphthoic betanaphthylamide.
2:3-Oxynaphthoic 2-chloroanilide.
2:3-Oxynaphthoic 4-chloro-2-anisidide.
2:3-Oxynaphthoic 4′-chloro-2-anisidide.
2:3-Oxynaphthoic 5′-chloro-2-anisidide.
2:3-Oxynaphthoic 4′-chloro-2′-toluidide.
2:3-Oxynaphthoic cresoxyanilide.
2:3-Oxynaphthoic diacetoaceticanilide.
2:3-Oxynaphthoic 2′:5′-dimethoxy-1′-anilide.
2:3-Oxynaphthoic 4-methoxyanilide.
2:3-Oxynaphthoic naphthoxyanilide.
2:3-Oxynaphthoic orthotoluidide.
2:3-Oxynaphthoic phenoxyanilide.
2:3-Oxynaphthoic toloxyanilide.
2:3-Oxynaphthoic 3-toluidide.
2:3-Oxynaphthoic 4-toluidide.

Dimethylanilin
Synonyms: Dimethylaniline.
French: Aniline diméthyle, Aniline diméthylique.
German: Dimethylanilin.
Chemical
Starting point in making—
Benzotrichloride, diaspirin, diethyl carbonate with ethyl alcohol and ethyl chloroformate, dimethylmeta-aminophenol, michler's ketone, nitrosodimethylanilin, novaspirin, paradimethylaminobenzaldehyde, paradimethylaminobenzene, tetramethylaminobenzophenone, tetramethyldiaminodiphenylmethane, thyresol, vanillin.

Dimethylanilin (Continued)

Dye
Catalyst (Brit. 251491) in—
Conversion of vat colors into soluble form by means of chlorosulphonic acid.
Reagent (Brit. 401137) in making—
Lithium salts of acid disulphuric esters of leuco-vat dyes useful for printing purposes.
Starting point in making—
Auramine, benzal green 00, betadimethylsafranin, butter yellow, chrystal violet, dahlia B, ethylene blue, helianthin, indanthrene red BN, malachite green, methyl green, methyl red, methyl violet, methylene blue, new solid green 2B, patent blue, phenylauramine, safranin, tetramethylsafranin.

Explosives
Starting point in making—
Tetranitromethylanilin, trinitrophenylmethylnitramine.

Fuel
Washing agent (Brit. 371888) in—
Treating industrial gases for the recovery of sulphuric anhydride.

Petroleum
Stabilizing agent (Brit. 406658) for—
Motor fuels (lowers the rate of gum formation).

Plastics
Reagent (Brit. 393914 and 342167) in—
Treating cellulose acetate (in conjunction with chloroform and cyanuric chloride) in making thread, films, insulating material, and other products.
Cellulose acetate (in conjunction with tetrachloropyrimidin and chloroform) in making threads, films, insulating material, and other products.
Ethyl cellulose (in conjunction with cyanuric chloride and benzene) in making threads, films, insulating material, and other products.

Rubber
Accelerator in—
Vulcanization process.

Soap
Starting point (Brit. 391435) in making—
Cleansing agents for textile use and other purposes from sulphuric esters of long-chain unsaturated alcohols prepared by treatment of the alcohols with a compound of a tertiary amine and sulphur trioxide in equimolecular proportions.

Textile
Solubilizing agent (Brit. 276100) in making—
Dye liquors and printing pastes containing acridin dyestuffs.
Aminoanthraquinones, reduced and unreduced.
Anthraquinone dyestuffs, azins, azo dyestuffs, basic diarylmethane dyestuffs, basic triarylmethane dyestuffs, benzoquinone anilides, chrome mordant dyestuffs, indigoids, naphthoquinones, reduced and unreduced, naphthoquinone anilides, nitroarylamines, nitroarylphenols, nitrodiarylamines, nitrodiarylphenols, oxazins, pyridin dyestuffs, quinolin dyestuffs, quinoneimides, reduced and unreduced, sulphur dyestuffs, thiazins, xanthenes.

Dimethylanthrarufin

Textile
——, *Dyeing*
Pigment in dyeing various yarns and fabrics.
——, *Printing*
Pigment in dyeing various fabrics.

2:3′-Dimethylazobenzene-4:6-disulphonic Acid
French: Acide de 2:3′-diméthyleazobenzène-4:6-disulphonique.
German: 2:3′-Dimethylazobenzol-4:6-disulfosaeure.
Dye
Starting point (U. S. 165550-1) in making—
Tetrakosazo dyestuffs, trisazo dyestuffs.

6:8-Dimethyl-1:2:3:9-Benzisotetrazole
Pharmaceutical
Claimed (U. S. 2008536) to have—
Valuable therapeutic properties and solubility in water.

2:4-Dimethylbenzylphthalamide
German: 2:4-Dimethylbenzylphtalimid.
Chemical
Starting point (Brit. 249883) in making—
2:4-Dimethyl-1:5-di(omegaphthalimidemethyl)benzene.
2:4-Dimethyl-1:5-di(omega-aminomethyl)benzene.

Dimethylbetahydroxyethyldodecylthiomethyl-Ammonium Chloride
Disinfectant
Claimed (Brit. 436725 and 436726) to be—
Bactericide, disinfectant.
Insecticide and Fungicide
Claimed (Brit. 436725 and 436726) to be—
Fungicide.

1:1-Dimethylbutadiene
French: 1:1-Diméthylebutadiène.
German: 1:1-Dimethylbutadien.
Chemical
Starting point in making—
Intermediates, pharmaceuticals.
Starting point (Brit. 309911) in making synthetic perfumes with—
Acrolein, crotonaldehyde, tetrolic aldehyde.

2:3-Dimethylbutadiene
French: 2:3-Diméthylebutadiène.
German: 2:3-Dimethylbutadien.
Chemical
Starting point in making—
Intermediates, pharmaceuticals.
Starting point (Brit. 309911) in making synthetic perfumes with—
Acrolein, crotonaldehyde, propargylaldehyde.

Dimethylcarbinol
Chemical
Starting point in making pharmaceutical chemicals.
Pharmaceutical
In compounding and dispensing practice.

Dimethylcetylsulphonium Bromide
Textile
Mordant (Brit. 436592) in—
Dyeing natural or regenerated cellulosic textile materials with chrome dyestuffs.

1:4-Dimethyl-2-chlorobenzene
French: 1:4-Diméthyle-2-chlorobenzène.
German: 1:4-Dimethyl-2-chlorbenzol.
Chemical
Starting point (Brit. 281290) in making—
Dimethyl-2-chlorobenzene-5-mercaptan.
Dimethyl-2-chlorobenzene-5-sulphochloride.
Dimethyl-2-chlorobenzene-5-thioglycollic acid.

4:7-Dimethyl-5-chloroxythionaphthene
German: 4:7-Dimethyl-5-chloroxythionaphten.
Dye
Starting point (Brit. 274527) in making thioindigoid dyestuffs with—
6-Chloro-7-methylisatin anhydride, 5:7-dibromoisatin arylide, 5:7-dibromoisatin chloride, 5:7-dichloroisatin arylide, 5:7-dichloroisatin chloride, isatin alpha-anilide.

Dimethylcyclohexanol Phthalate
French: Phthalate de diméthylecyclohexanole.
German: Dimethylcyklohexanolphtalat, Dimethylzyklohexanolphtalat, Phthalsäuredimethylcyklohexanolester, Phthalsäuredimethylzyklohexanolester, Phthalsäuresdimethylcyklohexanol, Phthalsäuresdimethylzyklohexanol.
Cellulose Products
Plasticizer for—
Cellulose acetate, cellulose esters and ethers, cellulose nitrate.
For uses, see under general heading: "Plasticizers."

Dimethylcyclohexyldimethylcyclohexanol, Sulphonated
Miscellaneous
As an emulsifying agent (Brit. 425239).
For uses, see under general heading: "Emulsifying agents."

Dimethylcyclohexyl Tartrate
Cellulose Products
Plasticizer (Brit. 432404) for—
Cellulose esters and ethers.
For uses, see under general heading: "Plasticizers."

1:4-Dimethyldiaminoanthraquinone
German: 1:4-Dimethyldiaminoanthrachinon.
Dye
Starting point (Brit. 251139) in making dyestuffs with—
Dimethylanilin, pyridin, quinolin.

Dimethyldibenzanthrone

Dye
Starting point (Brit. 277398) in making soluble vat dyestuffs by treatment with sulphuric acid or sulphur trioxide, in the presence of—
Acetyl chloride, benzoyl chloride, carbonyl chloride, chloroformic ester, paratoluenesulphonic chloride, phthalic anhydride, phthalimide, succinic anhydride, succinimide.

2:2′-Dimethyl-4:5:4′:5′-dibenzoxacarbocyanin Bromide

Photographic
Sensitizer (Brit. 432969) for—
Silver halide emulsions (sensitizing maxima: 550 mu).

5:5-Dimethyl-1:1-dicarboxyhexane

Disinfectant
Claimed (U. S. 2032159) as having—
High bactericidal action.

Dimethyldicetyl-Ammonium Bromide

Dry-Cleaning
Addition agent (Brit. 453523) to—
Solvents, such as trichloroethylene, carbon tetrachloride, and benzene.

Textile
Addition agent (Brit. 453523) to—
Solvents, such as trichloroethylene, carbon tetrachloride, and benzene.

2:4-Dimethyldimethylene Dioxide

Cellulose Products
Solvent and softener (Brit. 391769) for—
Cellulose esters and ethers.
For uses, see under general heading: "Solvents."

Dimethyl-1:4-dioxane

French: 1:4-Dioxane de diméthyle, 1:4-Dioxane diméthylique.
German: Dimethyl-1:4-dioxan.

Ceramics
Solvent in—
Coating compositions containing cellulose acetate, nitrocellulose, or other esters or ethers of cellulose.

Chemical
Solvent for—
Cellulose esters, general use, organic and inorganic chemicals.
Solvent in making—
Emulsions containing starches, dextrins, glues, resins, waxes, gelatin, casein, vegetable gums, and the like.

Dye
Solvent for—
Oil-soluble dyestuffs.
Solvent in making—
Dyestuff preparations containing starches, dextrins, glues, casein, gelatin, vegetable gums, and the like.

Fats and Oils
Solvent for—
Certain vegetable oils.

Glass
Solvent in—
Compositions containing cellulose acetate, nitrocellulose, or other esters or ethers of cellulose, used for coating glassware and in the manufacture of non-scatterable glass.

Glues and Adhesives
Solvent in making—
Preparations containing glues, gelatin, casein, starches, dextrin, or vegetable gums.

Ink
Solvent (Brit. 326824) in making—
Printing inks.

Leather
Solvent in making—
Compositions containing cellulose acetate, nitrocellulose, or other esters or ethers of cellulose, used for coating leather goods and in the production of artificial leathers.
Dressing compositions.
Treating compositions containing dextrins, gelatin, glue, starches, vegetable gum, casein, and the like.

Miscellaneous
Solvent in—
Dyeing and staining solutions, polishing compositions.

Paint and Varnish
Solvent in making—
Paints, varnishes, lacquers, enamels, dopes and primers containing cellulose acetate, nitrocellulose, or other esters or ethers of cellulose, together with oils, such as perilla oil, and resins, such as sandarac, mastic, copal, and kauri (Brit. 326824).
Paint and varnish removers, polishing compositions.

Paper
Solvent in—
Coating compositions containing cellulose acetate, nitrocellulose, or other esters or ethers of cellulose.
Sizing compositions containing starches, dextrins, vegetable gums, casein, and the like.

Petroleum
Solvent for—
Mineral oils, paraffin.

Plastics
Solvent (Brit. 326824) in making—
Compositions containing cellulose acetate, nitrocellulose, or other esters or ethers of cellulose.

Resins and Waxes
Solvent for—
Beeswax, carnauba wax, general use, montan wax.
Solvent in making—
Emulsions containing glues, gelatin, vegetable gums, casein, starches, and the like.

Rubber
Solvent in—
Coating compositions containing cellulose acetate, nitrocellulose, or other esters or ethers of cellulose.
Emulsions containing starches, glues, gelatin, casein, vegetable gums, dextrins, and the like.

Soap
Solvent in making—
Detergent and cleansing preparations.

Stone
Solvent in—
Coating compositions containing cellulose acetate, nitrocellulose, or other esters or ethers of cellulose, used on artificial and natural stones.

Textile
——, *Dyeing*
As an assist.
Ingredient of—
Dye liquor containing starches, dextrins, vegetable gums, and the like.
Solvent for various dyestuffs.

——, *Finishing*
Solvent in—
Finishing compositions containing cellulose acetate, nitrocellulose, or other esters or ethers of cellulose.

Woodworking
Solvent in—
Coating compositions containing cellulose acetate, nitrocellulose, or other esters or ethers of cellulose.

Dimethyldodecylamine Oxide

Chemical
Starting point (Brit. 460710) in making—
Cleansing, disinfecting, and wetting agents by reacting with alkylene oxides.
Emulsifying agents for soaps, glue, gelatin, gum, and mucilages.
Textile stripping agents for vat dyestuffs by reacting with alkylene oxides and admixing with hydrosulphites.

Dimethyldodecylthiomethyl-Ammonium Chloride

Disinfectant
Claimed (Brit. 436725 and 436726) to be—
Bactericide, disinfectant.

Insecticide and Fungicide
Claimed (Brit. 436725 and 436726) to be—
Fungicide.

2:7-Di(methyleneamino)diphenylene Oxide

Rubber
Antiaging agent (Brit. 422191).

Dimethylene Dioxide

Cellulose Products
Solvent and softener (Brit. 391769) for—
Cellulose esters or ethers.
For uses, see under general heading: "Solvents."

Dimethyl Ether

Analysis
Solvent and reaction medium in various laboratory processes.

Chemical
General extracting medium.
General reaction medium.
General solvent for various purposes.
Starting point in making compounds with—
Acetylene, aluminum chloride, ammonia, antimony trichloride, bismuth chloride, boron fluoride, calcium chloride, carbon dioxide, ethylene, ferric chloride, hydriodic acid, hydrobromic acid, hydrochloric acid, nitrogen monoxide, phosphoric acid, stannic chloride, sulphur dioxide, sulphuric acid, titanium tetrachloride, zinc chloride.

Miscellaneous
Solvent for various purposes.

Pharmaceutical
Suggested for use as an anesthetic.

Refrigeration
As a refrigerating medium.

Dimethylether-anthraflavinic Acid
German: Anthraflavinsaeuresdimethylester.

Chemical
Starting point in making—
2:6-Dimethoxyanthracene (Brit. 260000).

2:6′-Dimethyl-1′-ethyl-4:5-benzoxaisocyanin

Photographic
Sensitizer (Brit. 432969) for—
Silver halide emulsions (sensitizing maxima: 515 mu).

2:6′-Dimethyl-1′-ethyl-4:5-benzoxa-psi-cyanin

Photographic
Sensitizer (Brit. 432969) for—
Silver halide emulsions (sensitizing maxima: 485 mu).

2:6′-Dimethyl-1′-ethyl-3:4-benzoxa-psi-cyanin Iodide

Photographic
Sensitizer (Brit. 423827) for—
Photographic emulsions to blue-green light.

2:6′-Dimethyl-1′-ethyl-5:6-benzoxa-psi-cyanin Iodide

Photographic
Sensitizer (Brit. 423827) for—
Photographic emulsions to blue-green light.

1:5-Dimethyl-1′-ethyl-5′:6′-benz-2:2′-pyrazinopyridocyanin Iodide

Photographic
As a dyestuff (Brit. 435542).

Dimethylethylcarbinol
Synonyms: Amylene hydrate, Tertiary amyl alcohol.
French: Alcool d'amyle tertiaire, Alcool amylique tertiaire, Carbinole de diméthyleéthyle, Carbinole diméthyleéthylique, Hydrate d'amylène, Hydrate amylènique.
German: Amylenhydrat, Dimethylaethylcarbinol, Tertiaeramylalkohol.

Chemical
Solvent for various purposes.
Starting point in making—
Intermediates used in the synthesis of dyestuffs, drugs, and perfumes.

Food
Ingredient of—
Fruit essences.

Miscellaneous
Solvent for various purposes.

Paint and Varnish
Solvent in making—
Dopes, lacquers, and varnishes containing cellulose esters and ethers.

Pharmaceutical
In compounding and dispensing practice.

Plastics
Solvent in making—
Cellulose ester and ether compounds.

2:2′-Dimethyl-8-ethyl-4:5:4′:5′-dibenzoxacarbocyanin Bromide

Photographic
Sensitizer (Brit. 432969) for—
Silver halide emulsions (sensitizing maxima: 580 mu).

2:2′-Dimethyl-8-ethyloxacarbocyanin Iodide

Photographic
Sensitizer (U. S. 1962123, 1962124, and 1962133) for—
Blue-green light.

2:6-Dimethyl-1′-ethyloxa-psi-cyanin Iodide

Photographic
Sensitizer (Brit. 423827) for—
Photographic emulsions to blue-green light.

3:5-Dimethylfurodiazole

Chemical
Starting point (Brit. 396778) in making triazoles with—
4-Amino-2-butoxypyridin, methylamine, phenylamine.

Dimethylheptenol Acetate
French: Acétate de diméthyleheptenol, Acétate diméthyleheptenolique.
German: Dimethylheptenolacetat, Dimethylheptenolazetat, Essigsäuredimethylheptenolester, Essigsäuresdimethylheptenol.
Spanish: Acetato de dimetilheptenol.
Italian: Acetato di dimetilheptenole.

Chemical
Starting point in making various derivatives.

Perfume
Ingredient of—
Artificial perfumes.
Perfume in—
Cosmetics.

Soap
Perfume in—
Toilet soaps.

Dimethylhydroquinone
Synonyms: Hydroquinone dimethyl ether, Hydroquinone methyl ether.
French: Diméthylehydroquinone, Éther diméthylehydroquinone, Éther de diméthylehydroquinone, Éther méthylehydroquinonique.
German: Dimethylhydrochinon, Hydrochinondimethylaether, Hydrochinonmethylaether.

Chemical
Starting point in making—
Aromatics and other derivatives.

Miscellaneous
Perfume for various industrial and other purposes.

Perfume
Ingredient of artificial essence of—
Clover, hawthorne, heliotrope, hyacinth, new mown hay, narcissus, ylang-ylang.
Perfume in—
Cosmetics.

Soap
Perfume in—
Toilet soaps.

Dimethylmeta-aminophenol

Chemical
As an intermediate.

Dye
As an intermediate.

Rubber
As an antioxidant (U. S. 1899120).

Dimethylmetanilic Acid
French: Acide diméthylemétanilique, Acide de diméthylemétanyle.
German: Dimethylmetanilsaeure.

Chemical
Dispersing agent (Brit. 277048) in making—
Dispersion with sulphur, soot, and the like.
Starting point in making various organic compounds.

Dye
Dispersing agent (Brit. 277048) in making—
Indigoid dyestuff compositions.
Starting point in making various synthetic dyestuffs.

Paint and Varnish
Dispersing agent (Brit. 277048) for—
Fine dispersion of mineral pigments, barytes, and the like.

Dimethyloctylamine Oxide

Chemical

Starting point (Brit. 460710) in making—

Cleansing, disinfecting, and wetting agents by reacting with alkylene oxides.

Emulsifying agents for soaps, glue, gelatin, gum, and mucilages.

Textile stripping agents for vat dyestuffs by reacting with alkylene oxides and admixing with hydrosulphites.

Dimethyloleylamine Oxide

Chemical

Starting point (Brit. 460710) in making—

Cleansing, disinfecting, and wetting agents by reacting with alkylene oxides.

Emulsifying agents for soaps, glue, gelatin, gums, and mucilages.

Textile stripping agents for vat dyestuffs by reacting with alkylene oxides and admixing with hydrosulphites.

2:6-Dimethylolparacresol

Resins

Starting point (Brit. 434850) in making—

Synthetic resins with a phenol, which may be dihydric-dinuclear; the product may be modified with a vegetable oil or resin acids.

Dimethylolthiourea

Synonyms: Dimethylolsulphourea.

French: Sulphourée de diméthylole, Thiourée de diméthylole.

German: Dimethylolsulfoharnstoff, Dimethylolthioharnstoff.

Chemical

Starting point in making various derivatives.

Resins and Waxes

Starting point (Brit. 338937) in making artificial resins with the aid of paraformaldehyde or trioxymethylene, in the presence of—

Dibromobenzyl alcohol, dichlorobenzyl alcohol, dioxane, ethyleneglycol bromophenyl ether, ethyleneglycol bromosalicylic ether, ethyleneglycol monomethyl ether, ethyleneglycol monoethyl ether, glycol monohalogenaryl ethers, glycol monobromobenzoic ethers, metachlorobenzyl alcohol, monobromobenzyl alcohol, orthochlorobenzyl alcohol, parachlorobenzyl alcohol.

Dimethylolurea

French: Urée de diméthylole.

German: Dimethylolharnstoff.

Chemical

Starting point in making various derivatives.

Resins and Waxes

Starting point (Brit. 338937) in making artificial resins with the aid of paraformaldehyde or trioxymethylene, in the presence of—

Dibromobenzyl alcohol, dichlorobenzyl alcohol, dioxane, ethyleneglycol bromophenol ether, ethyleneglycol bromosalicylic ether, ethyleneglycol monoethyl ether, ethyleneglycol monomethyl ether, glycol monobromobenzoic ethers, glycol monohalogen-aryl ethers, metachlorobenzyl alcohol, monobromobenzyl alcohol, orthochlorobenzyl alcohol, parachlorobenzyl alcohol.

Dimethyl Phthalate

French: Phthalate de diméthyle, Phthalate diméthylique.

German: Dimethylphtalat, Phtalsäuredimethylester, Phtalsäuresdimethyl.

Italian: Ftalato-dimetilica.

Cellulose Products

Solvent and Plasticizer for—

Cellulose acetate, cellulose esters and ethers, cellulose nitrate.

For uses, see under general heading: "Plasticizers."

Dimethylpyrrolidene Methyliodide

French: Méthyleiodure de diméthylepyrrolidène.

German: Dimethylpyrrolidenmethyljodid.

Chemical

Starting point in making—

Butadiene 1:3.

1:1′-Dimethyl-4:4′-tricarbocyanin Iodide

Photographic

Sensitizer (Brit. 436941 and 437017) for—

Photographic emulsions to infrared light with maxima at 980 mu.

Dimethylurea

French: Urée de diméthyle, Urée diméthilique.

German: Dimethylharnstoff.

Chemical

Starting point in making various derivatives.

Resins and Waxes

Starting point (Brit. 338937) in making artificial resins with the aid of paraformaldehyde or trioxymethylene, in the presence of—

Dibromobenzyl alcohol, dichlorobenzyl alcohol, dioxane, ethyleneglycol bromophenyl ether, ethyleneglycol bromosalicylic ether, ethyleneglycol monoethyl ether, ethyleneglycol monomethyl ether, glycol monobromobenzoic ethers, metachlorobenzyl alcohol, monobromobenzyl alcohol, orthochlorobenzyl alcohol, parachlorobenzyl alcohol.

Dinaphthalene Dioxide

French: Dioxyde dinaphthalonique.

German: Dinaphtalindioxyd.

Dye

Starting point (Swiss 114913) in making vat dyestuffs such as—

Brominated quinone derivative.

Nitrated and reduced quinone derivative.

Normal amyl derivative of aminoquinone.

Normal benzyl derivative of aminoquinone.

Normal benzoyl derivative of polyaminoquinone.

Normal butyl derivative of aminoquinone.

Normal ethyl derivative of aminoquinone.

Normal methyl derivative of aminoquinone.

Normal phenyl derivative of aminoquinone.

Normal propyl derivative of aminoquinone.

Normal tolyl derivative of aminoquinone.

Normal xylyl derivative of aminoquinone.

Dinaphthyl-Bismuth-Dicresyl-Arsenic Compound

Lubricant

Addition agent (Brit. 445813) in—

Lubricants for motors, turbines, flushing and high-temperature work generally.

Dinaphthylene Dioxide

Petroleum

Fluorescence imparter (Brit. 420371) for—

Gasoline.

Gum inhibiter (Brit. 420371) for—

Gasoline.

Dinaphthylene Oxide

Petroleum

Fluorescence imparter (Brit. 420371) for—

Gasoline.

Gum inhibiter (Brit. 420371) for—

Gasoline.

Dinaphthyl Ether

Chemical

Starting point in—

Organic synthesis.

Lubricant

Starting point (Brit. 440916) in making—

Products useful as lubricating oils or as pour-point depressors for paraffin base lubricating oils by condensation with halogenated derivatives of aliphatic hydrocarbons, such as paraffin oils, paraffin, petrolatum, ceresin, ozokerite, or others contained in the middle to high fractions of petroleum.

Di-1-naphthylethylenediamine

German: Di-1-naphtylaethylendiamine.

Dye

Starting point in making triarylmethane dyestuffs with—

Tetra-amyl-4:4′-diaminobenzophenone.

Tetrabutyl-4:4′-diaminobenzophenone.

Tetraethyl-4:4′-diaminobenzophenone.

Tetraisoamyl-4:4′-diaminobenzophenone.

Tetraisobutyl-4:4′-diaminobenzophenone.

Tetraisopropyl-4:4′-diaminobenzophenone.

Tetramethyl-4:4′-diaminobenzophenone.

Tetrapropyl-4:4′-diaminobenzophenone.

Dinaphthylthiourea

German: Dinaphtylthioharnstoff, Dinaphtylthiourea.

Dye

Starting point (Brit. 270883) in making dyestuffs for viscose with—

Anilin, meta-aminobenzoic acid, para-aminoacetanilide.

Dinaphthylurea
French: Dinaphthylurée.
German: Dinaphtylharnstoff.
Dye
Starting point (Brit. 270883) in making dyestuffs for viscose rayon with—
Anilin, meta-aminobenzoic acid, metaxylidin, napthionic acid, orthoanisidin, para-aminoacetanilide, parachloroanilin, paranitranilin, sulphanilic acid.

4:5-Dinitroalphanaphthylamine
German: 4:5-Dinitroalphanaphtylamin.
Dye
Starting point (Brit. 270428) in making azo dyestuffs with—
Alphanaphthylamine, 1:2-aminoaphthol, anilin, metatoluidin, paranitranilin, cresidin, 2:4-dinitranilin, metanitranilin, metaphenylenediamine, picramic acid.
Starting point (Brit. 252957) in making—
Diazo dyestuffs.

3:5-Dinitro-2-aminobenzylsulphonic Acid
French: Acide de 3:5-dinitro-2-aminobenzylsulphonique.
German: 3:5-Dinitro-2-amino-benzylsulphonsaeure.
Dye
Starting point (Brit. 265767) in making monoazo dyestuffs with—
Beta-amino-8-naphthol.
Beta-amino-8-naphthol-6-sulphonic acid.
Betamethylaminonaphthalene-7-sulphonic acid.
Betamethylamino-8-naphthol-6-sulphonic acid.
Betanaphthylamine-3-carboxylic acid.
Betanaphthylamine-6-sulphonic acid.
Betaphenylamino-8-naphthol-6-sulphonic acid.
Ethylbenzylanilin.
Ethylbetanaphthylamine.
Phenylbetanaphthylamine-6-sulphonic acid.

3:5-Dinitro-5-aminobenzylsulphonic Acid
French: Acide de 3:5-dinitro-5-aminobenzylsulphonique.
German: 3:5-Dinitro-5-aminobenzylsulfonsaeure.
Dye
Starting point (Brit. 265767) in making monoazo dyestuffs with—
Betamethylaminonaphthalene-6-sulphonic acid.
Betanaphthylamine.

2:4-Dinitro-3′-aminodiphenylamine
Chemical
Starting point in making—
Intermediates, pharmaceuticals.
Dye
Starting point (Brit. 323792) in making azo dyestuffs for rayons, with the aid of—
Alkyl-aryl anilins, allylaminophenol, allylnaphthylamine, alphanaphthylamine, aminonaphthoic acids, aminonaphthols, amylaminophenol, amylnaphthylamine, betanaphthylamine, butylaminophenol, butylnaphthylamine, cresols and derivatives, dimethyl-meta-aminophenol, ethylaminophenol, ethylnaphthylamine, gammachlorobetaoxypropionylnaphthylamine, meta-aminophenol, meta-anisidin, metacresidin, metaphenylenediamine, metaphenetidin, metatoluidin, metaxylidin, methylaminophenol, methylnaphthylamine, naphthylamine ethers, orthoaminophenol, orthoanisidin, orthocresidin, orthophenylenediamine, orthophenetidin, orthotoluidin, orthoxylidin, para-aminophenol, para-anisidin, paracresidin, paraphenylenediamine, paranitrometaphenylenediamine, paratoluidin, paraxylidin, phenols and derivatives, resorcinol, omegaoxyethylalphanaphthylamine.

2:4-Dinitro-4′-aminodiphenylamine
Chemical
Starting point in making—
Intermediates, pharmaceuticals.
Dye
Starting point (Brit. 323792) in making azo dyestuffs for various rayons with the aid of—
Alkyl-aryl anilins, allylaminophenol, allylnaphthylamine, alphanaphthylamine, aminonaphthoic acids, aminonaphthols, amylaminophenol, amylnaphthylamine, betanaphthylamine, butylaminophenol, butylnaphthylamine, cresols and derivatives, dimethyl-meta-aminophenol, ethylaminophenol, ethylnaphthylamine, gammachlorobetaoxypropionylnaphthylamine, meta-aminophenol, meta-anisidin, metacresidin, metaphenylenediamine, metatoluidin, metaxylidin, methylaminophenol, methylnaphthylamine, naphthylamine ethers, orthoaminophenol, orthoanisidin, orthocresidin, orthophenylenediamine, orthophenetidin, orthotoluidin, orthoxylidin, para-aminophenol, para-anisidin, paracresidin, paraphenxylenediamine, paranitrometaphenylenediamine, paratoluidin, paraxylidin, phenols and derivatives, propylaminophenol, propylnaphthylamine, resorcinol, omegaoxyethylalphanaphthylamine.

3:5-Dinitro-4-chlorobenzoic Acid
French: Acide de 3:5-dinitro-4-chlorobenzoique.
German: 3:5-Dinitro-4-chlorbenzoesaeure.
Dye
Starting point (Brit. 279133) in making dinitroarylaminodiarylamine dyestuffs with—
4-Amino-2-carboxy-4′-methoxydiphenylamine.
4-Aminodiphenylamine.
4-Amino-2-sulpho-2′-carboxyldiphenylamine.

Dinitrodibenzyldisulphonic Acid
French: Acide de dinitrodibenzyledisulfonique.
German: Dinitrodibenzyldisulfonsaeure.
Chemical
Starting point in making various intermediates.
Dye
Starting point in making—
Diphenyl citronin G, diphenyl fast yellow, mikado yellow, stilbene yellow.

3:5-Dinitro-2:4-dimethyl-6-tertiarybutylacetophenone
Mechanical
Improver (Brit. 404046) of—
Exhaust odors from internal combustion engines (added to fuels not derived from petroleum, either alone or in conjunction with (1) acetophenone, methylacetophenone, 4-methoxyacetophenone, 1-naphthylmethyl ketone, 2-naphthylmethyl ketone, or (2) any of the ketones listed under (1) and any of the following: Camphor, waste camphor oil, borneol, bornyl acetate, clove oil, ionone, coumarin, indole, skatole, paracresyl acetate, methyl anthranilate, isopropylmethylhydrocinnamic aldehyde).
Petroleum
Reagent (Brit. 404046) for—
Improving exhaust odors from internal combustion engines (added to gasoline or diesel oil, either alone or in conjunction with (1) acetophenone methylacetophenone, 4-methoxyacetophenone, 1-naphthylmethyl ketone, 2-naphthylmethyl ketone, or (2) any of the ketones listed under (1) and any of the following: Camphor, waste camphor oil, borneol, bornyl acetate, clove oil, ionone, coumarin, indole, skatole, paracresyl acetate, methyl anthranilate, isopropylmethylhydrocinnamic aldehyde).

2:5-Dinitrodiphenylamine-3′:4-disulphonic Acid
French: Acide de 2:5-dinitrodiphényleamine-3′:4-disulfonique.
German: 2:5-Dinitrodiphenylamin-3′:4-disulfonsaeure.
Dye
Starting point in making—
Agalma green B.

Dinitrodiphenylethane
Cellulose Products
Plasticizer (U. S. 1891601) for—
Cellulose acetate, cellulose esters or ethers, nitrocellulose.
For uses, see under general heading: "Plasticizers."

Dinitrohydroquinone Acetate
French: Acétate de dinitrohydroquinone.
German: Dinitrohydrochinonacetat, Dinitrohydrochinonazetat, Essigsäuredinitrohydrochinonester, Essigsäuresdinitrohydrochinon.
Spanish: Acetato de dinitrohidroquinona.
Italian: Acetato di dinitroidrochinone.
Analysis
Reagent in carrying out hydrogen ion determinations for pH 4 to 5 and 9 to 10.
Chemical
Starting point in making various derivatives.

2:4-Dinitro-4′-hydroxydiphenylamine
Dye
Starting point in making—
Immedial black, immedial dark brown A, immedial dark brown B, pyrogene blue, pyrogene direct blue.

3:5-Dinitrometatoluidin

Chemical

Starting point in making—
Aromatics, intermediates, pharmaceuticals.

Dye

Starting point (Brit. 319390) in making azo dyestuffs with the aid of—
Acetoacetic alphanaphthylide, acetoacetic anilide, acetoacetic anisidide, acetoacetic arylides, acetoacetic ester, acetoacetic naphthylide, acetoacetic phenetidide, acetoacetic toluidide, acetoacetic xylidide, aliphatic derivatives of anilin, alkylnaphthylamines, allylanilin, allylnaphthylamine, alpha-amino-2-ethoxynaphthalene, alpha-aminonaphthol, alphanaphthylamine, amylanilin, amylnaphthylamine, anilin, butylanilin, butylnaphthylamine, ethylanilin, ethylnaphthylamine, methylanilin, methylnaphthylamine, 4-nitro-1:3-phenylenediamine, omegaoxyethylalphanaphthylamine, orthoaminophenol, para-aminophenol, parachlorobetaoxypropylalphanaphthylamine, propylanilin, propylnaphthylamine, pyrazolones.

3:5-Dinitro-orthoanisidin

Chemical

Starting point in making—
Intermediates, pharmaceuticals, and other derivatives.

Dye

Starting point (Brit. 313390) in making azo dyestuffs with—
Acetoacetic alphanaphthylide, acetoacetic anilide, acetoacetic anisidide, acetoacetic arylides, acetoacetic betanaphthylide, acetoacetic ester, acetoacetic phenetidide, acetoacetic toluidide, acetoacetic xylidide, aliphatic derivatives of anilin, alkyl naphthylamines, allylanilin, allylnaphthylamine, alpha-amino-2-ethoxynaphthalene, alpha-aminonaphthol, alphanaphthylamine, amylanilin, amylnaphthylamine, anilin, butylnaphthylamine, ethylnaphthylamine, methylnaphthylamine, 4-nitro-1:3-phenylenediamine, omegaoxyethylalphanaphthylamine, orthoaminophenol, para-aminophenol, parachlorobetaoxypropylalphanaphthylamine, propylnaphthylamine, pyrazolones.

6:6′-Dinitro-orthoanisidin

Chemical

Starting point in making—
Intermediates, pharmaceuticals.

Dye

Starting point (Brit. 323792) in making azo dyestuffs for rayons with the aid of—
Alkylaryl amines, allylaminophenol, allylnaphthylamine, alphanaphthylamine, aminonaphthoic acids, aminonaphthols, amylaminophenol, amylnaphthylamine, betanaphthylamine, butylaminophenol, butylnaphthylamine, cresols and their derivatives, dimethylmeta-aminophenol, ethylaminophenol, ethylnaphthylamine, gammachlorobetaoxypropionylnaphthylamine, meta-aminophenol, meta-anisidin, metacresidin, metaphenylenediamine, metaphenetidin, metatoluidin, metaxylidin, methylaminophenol, methylnaphthylaamine, naphthylamine ethers, orthoaminophenol, orthoanisidin, orthocresidin, orthophenylenediamine, orthotoluidin, orthoxylidin, para-aminophenol, para-anisidin, paracresidin, paraphenylenediamine, paranitrometaphenylenediamine, paratoluidin, paraxylidin, phenols and their homologs, propylaminophenol, propylnaphthylamine, resorcinol, omegaoxyethylalphanaphthylamine.

3:5-Dinitro-orthocresidin

Chemical

Starting point in making—
Intermediates, pharmaceuticals, and other derivatives.

Dye

Starting point (Brit. 319390) in making azo dyestuffs with the aid of—
Acetoacetic alphanaphthylide, acetoacetic anilide, acetoacetic anisidide, acetoacetic arylides, acetoacetic ester, acetoacetic naphthylide, acetoacetic phenetidide, acetoacetic toluidide, acetoacetic xylidide, aliphatic derivatives of anilin, alkylnaphthylamines, alkylanilins, allylnaphthylamine, alpha-amino-2-ethoxynaphthalene, alpha-aminophenol, alphanaphthylamine, amylanilin, amylnaphthylamine, anilin, butylanilin, butylnaphthylamine, ethylanilin, ethylnaphthylamine, methylanilin, methylnaphthylamine, 4-nitro-1:3-phenylenediamine, heptylanilin, heptylnaphthylamine, hexylanilin, hexylnaphthylamine, isoallylanilin, isoallylnaphthylamine, isoamylanilin, isoamylnaphthylamine, isobutylanilin, isobutylnaphthylamine, isopropylanilin, isopropylnaphthylamine, omegaoxyethylalphanaphthylamine, orthoaminophenol, para-aminophenol, parachlorobetaoxypropylalphanaphthylamine, propylanilin, propylnaphthylamine, pyrazolones, meta-aminophenol.

Dinitro-orthocresol

Woodworking

Ingredient of—
Compositions used for the preservation of wood (U. S. 1616468).

3:5-Dinitro-orthotoluidin

Chemical

Starting point in making—
Aromatics, intermediates, pharmaceuticals.

Dye

Starting point (Brit. 319390) in making azo dyestuffs with the aid of—
Acetoacetic alphanaphthylide, acetoacetic anilide, acetoacetic anisidide, acetoacetic arylides, acetoacetic ester, acetoacetic phenetidide, acetoacetic toluidide, acetoacetic xylidide, aliphatic derivatives of anilin, alkyl naphthylamines, allylanilin, allylnaphthylamine, alpha-amino-2-ethoxynaphthalene, alpha-aminonaphthol, alphanaphthylamine, amylanilin, amylnaphthylamine, anilin, butylnaphthylamine, ethylnaphthylamine, methylnaphthylamine, 4-nitro-1:3-phenylenediamine, omegaoxyethylalphanaphthylamine, orthoaminophenol, para-aminophenol, parachlorobetaoxypropylalphanaphthylamine, propylnaphthylamine, pyrazolones.

2:2-Di-3-nitro-4-oxyphenylpropane

Rubber

Reagent (French 757442) for—
Restraining premature vulcanization of rubber at low temperatures (in the neighborhood of 127°).

3:5-Dinitropara-anisidin

Chemical

Starting point in making—
Aromatics, intermediates, pharmaceuticals.

Dye

Starting point (Brit. 319390) in making azo dyestuffs with the aid of—
Acetoacetic alphanaphthylide, acetoacetic anilide, acetoacetic anisidide, acetoacetic ester, acetoacetic naphthylide, acetoacetic phenetidide, acetoacetic toluidide, acetoacetic xylidide, aliphatic derivatives of anilin, alkylnaphthylamines, alkylanilins, allylanilin, allylnaphthylamine, alpha-amino-2-ethoxynaphthalene, alpha-aminonaphthol, alphanaphthylamine, amylanilin, amylnaphthylamine, anilin, butylanilin, butylnaphthylamine, ethylanilin, ethylnaphthylamine, methylanilin, methylnaphthylamine, 4-nitro-1:3-phenylenediamine, omegaoxyethylalphanaphthylamine, orthoaminophenol, para-aminophenol, parachlorobetaoxypropylalphanaphthylamine, propylanilin, propylnaphthylamine, pyrazolones.

3:5-Dinitroparatoluidin

Chemical

Starting point in making—
Intermediates, pharmaceuticals.

Dye

Starting point (Brit. 313390) in making azo dyestuffs with the aid of—
Acetoacetic alphanaphthylide, acetoacetic anilide, acetoacetic anisidide, acetoacetic arylides, acetoacetic betanaphthylide, acetoacetic ester, acetoacetic phenetidide, acetoacetic toluidide, acetoacetic xylidide, aliphatic derivatives of anilin, alkyl naphthylamines, allylanilin, allylnaphthylamine, alpha-amino-2-ethoxynaphthalene, alpha-aminonaphthylamine, amylanilin, amylnaphthylamine, anilin, butylnaphthylamine, ethylnaphthylamine, methylnaphthylamine, 4-nitro-1:3-phenylenediamine, omegaoxyethylalphanaphthylamine, orthoaminophenol, para-aminophenol, parachlorobetaoxypropylalphanaphthylamine, propylnaphthylamine, pyrazolones.

Dinitrosoresorcinol

Chemical

Intermediate in—
Organic synthesis.

Petroleum

Inhibitor (U. S. 1982277, 1982267, and 1982618) of—
Gum formation in gasoline, particularly in vapor-phase cracked gasoline.

Dinitrostilbenedisulphonic Acid
French: Acide de dinitrostilbènedisulphonique.
German: Dinitrostilbendisulfonsaeure.

Chemical
Starting point in making various organic chemicals.

Dye
Starting point (Brit. 263192) in making dyestuffs with—
Metanilic acid azometa-amidocresol-methyl ether.
Sulphanilic acid azoalphanaphthylamine.
Starting point in making—
Azidin fast yellow G, curcumin S, diamine fast yellow A, diphenyl fast yellow, direct orange G, direct yellow G, direct yellow R, polychromin B, renol yellow G, stilbene yellow G, sun yellow G, sun yellow GG.

Dinitrotertiarybutylparacymene

Mechanical
Improver (Brit. 404046) of—
Exhaust odors from internal combustion engines (added to fuels not derived from petroleum, either alone or in conjunction with (1) acetophenone, methylacetophenone, 4-methoxyacetophenone, 1-naphthylmethyl ketone, 2-naphthylmethyl ketone, or (2) any of the ketones listed under (1) and any of the following: Camphor, waste camphor oil, borneol, bornyl acetate, clove oil, ionone, coumarin, indole, skatole, paracresyl acetate, methyl anthranilate, isopropylmethylhydrocinnamic aldehyde).

Petroleum
Reagent (Brit. 404046) for—
Improving exhaust odors from internal combustion engines (added to gasoline or diesel oil, either alone or in conjunction with (1) acetophenone, methylacetophenone, 4-methoxyacetophenone, 1-naphthylmethyl ketone, 2-naphthylmethyl ketone, or (2) any of the ketones listed under (1) and any of the following: Camphor, waste camphor oil, borneol, bornyl acetate, clove oil, ionone, coumarin, indole, skatole, paracresyl acetate, methyl anthranilate, isopropylmethylhydrocinnamic aldehyde).

Di-normal-butylmetaxylidin

Dye
Starting point (Brit. 439815 and 417014) in making—
Blue dyestuffs by condensing with (1) a 4:4′-dihalogeno- or 4:4′-dialkoxybenzophenone, (2) a primary 4-alkoxy- or 4-aryloxyarylamine and sulphonating the product.
Greenish-blue dyestuffs by condensing with (1) a 4:4′-dihalogeno- or 4:4′-dialkoxybenzophenone, (2) a primary 4-alkoxy- or 4-aryloxyarylamine and sulphonating the product.

Dioctyl Sulphosuccinate

Miscellaneous
As a wetting agent (Brit. 446568).
For uses, see under general heading: "Wetting agents."

Diorthocresol

Dye
Coupling agent.
Intermediate.

Diorthoformylalkylaminodiphenyl Bisulphide

Photographic
Antifogging agent (U. S. 1962123, 1962124, and 1962133) in—
Photographic emulsions.

1:4-Dioxane
Synonyms: Diethylene dioxide.
French: Dioxyde de diéthylène, Dioxyde diéthylènique.
German: Diaethylendioxyd.

Chemical
Reagent and solvent (Brit. 307079) in making—
Emulsions with starches, dextrins, glues, gelatin, casein, vegetable gums, and the like.

Dye
Solvent (Brit. 307079) in making—
Dyestuff preparations containing starches, dextrins, glues, gelatin, vegetable gums, casein, and the like.

Glues and Adhesives
Reagent and solvent (Brit. 307079) in making—
Adhesive preparations containing starches, dextrins, glues, gelatin, casein, vegetable gums, and the like.

Leather
Solvent (Brit. 307079) in making—
Compositions in emulsion form, containing starches, vegetable gums, glues, gelatins, casein, dextrins, for treating leather.

Paper
Solvent (Brit. 307079) in making—
Compositions containing starches, dextrins, glues, gelatins, casein, vegetable gums, for treating paper.

Resins and Waxes
Solvent (Brit. 307079) in making—
Emulsions of waxes or resins containing glues, gelatins, casein, starches, vegetable gums, dextrins, and the like.

Rubber
Solvent (Brit. 307079) in making—
Rubber emulsions containing starches, dextrins, vegetable gums, casein, glues, gelatins, and the like.

Textile
Solvent (Brit. 307079) in making—
Dye liquors containing glues, gelatins, casein, vegetable gums, starches, dextrins, and the like.

2:4-Dioxybenzene-1-carboxylic Acid
French: Acide de 2:4-dioxyebenzène-1-carbonique, Acide de 2:4-dioxyebenzène-1-carbolique.
German: 2:4-Dioxybenzol-1-carbonsaeure.

Chemical
Starting point in making—
Esters and salts, intermediates, pharmaceuticals.

Dye
Starting point (Brit. 306447) in making azo dyestuffs with—
Alphahydroxy-2-amino-6-carboxybenzene-4-sulphonic acid.
Alphanaphthylamine-4-sulphonic acid.
Aminoazobenzenedisulphonic acid.
Anilin, benzidin.
Betanaphthol-6:8-disulphonic acid.
Betanaphthylamine-3:6-disulphonic acid.
4-Chloroanilin-3-sulphonic acid.
1:5-Dioxynaphthalene.
1:5-Dioxynaphthalenesulphanilic acid.
Sulphanilic acid.

1:5-Dioxy-4:8-diaminoanthraquinone

Oils, Fats, and Waxes
Coloring agent (Brit. 432867) for—
Paraffin and other mineral waxes, stearic acid, tallow and other solid triglycerides, beeswax, carnauba wax, and others.

1:5-Dioxy-4:8-di(benzoylamino)anthraquinone

Oils, Fats, and Waxes
Coloring agent (Brit. 432867) for—
Paraffin and other mineral waxes, stearic acid, tallow and other solid triglycerides, beeswax, carnauba wax, and others.

Dioxydiethylanilin
German: Dioxydiaethylanilin.

Dye
Reagent (Brit. 274823) in making dyestuffs for acetate rayon with—
Paranitranilindiazobenzene, 3:4:5-trichloroanilin.

Dioxydiethylmetatoluidin

Dye
Reagent (Brit. 274823) in making dyestuffs for acetate rayon from—
2:4-Dimethylanilin, diazotized paranitranilin.

4:4′-Dioxydiphenyldimethylmethane-3:3′-disulphonic Acid

Chemical
Starting point (Brit. 425037) in making—
Tanning agents by combination with urea-formaldehyde condensation products, used for removing the green shade of chrome leather and for tanning reptile hides.

1:5-Dioxynaphthalene
German: 1:5-Dioxynaphtalin.

Chemical
Starting point in making—
1:5-Diaminonaphthalene.

1:5-Dioxynaphthalene (Continued)

Dye
Starting point in making—
Azin green, blue benzidin dyestuffs, chrome azo dyestuffs, diamond black PV.
Starting point (Brit. 282111) in making arylaminonaphthalene derivatives for dyeing acetate rayon, pelts and animal fibers with the aid of—
4:4′-Amino-oxydiphenylamine, 4:4′-diaminodiphenylamine-2-sulphonic acid, meta-aminophenol, metaphenylenediamine, orthoaminophenol, orthophenylenediamine, para-aminophenol, paraphenylenediamine.
Sulphonic, carboxylic, and other substitution products of leucoindophenols and leucoindamines.

2:7-Dioxynaphthalene
German: 2:7-Dioxynaphtalin.

Dye
Starting point in making—
Dioxin, gambin G, muscarin.
Starting point in making arylaminonaphthalene derivatives for dyeing animal fibers, acetate rayon, and pelts with the aid of—
4:4′-Amino-oxydiphenylamine.
Carboxylic, sulphonic, and other substitution products of leucoindophenols and leucoindamines.
4:4-Diaminodiphenylamine-2-sulphonic acid, meta-aminophenol, metaphenylenediamine, orthoaminophenol, orthophenylenediamine, para-aminophenol, paraphenylenediamine.

2:6-Dioxynaphthalene-3-carboxylic Acid
French: Acide de 2:6-dioxynaphthalène-3-carbonique.
German: 2:6-Dioxynaphtalin-3-carbonsaeure.

Dye
Starting point (Brit. 270308) in making azo dyestuffs with—
2-Aminochloro-6-nitrophenol.
2-Aminophenol-4-sulphonic acid.
4-Chloro-2-aminophenol-6-sulphonic acid.
4-Nitro-2-aminophenol.
4-Nitro-2-aminophenol-6-sulphonic acid.

1:3-Dioxyphenol Diacetate

Chemical
Starting point in making—
Synthetic tanning agents (Brit. 242694).

1:3-Dioxyquinolin
French: 1:3-Dioxyequinoline.
German: 1:3-Dioxychinolin.

Chemical
Starting point in making—
Intermediates, pharmaceuticals.

Dye
Starting point (Brit. 298518) in making azo dyestuffs with—
1-Amino-2:7-dimethoxynaphthalene.
1-Amino-2:7-dioxynaphthalene glycolate.
1-Amino-2-ethoxynaphthalene-6-sulphonic acid.
2-Amino-1-methoxy-4-sulphonic acid.
1-Amino-2-methoxynaphthalene.
1-Aminonaphthalene-6-sulphonic acid.
1-Aminonaphthalene-7-sulphonic acid.
1-Amino-2-naphthoxybetapropionic acid.
1-Amino-2-oxyethoxynaphthalene sulphonic acid.
2-Amino-5-sulphobenzoic acid.
Anilin.
Anilin-3-chloro-6-sulphonic acid.
Anilin-2:4-disulphonic acid.
Anilin-2:6-disulphonic acid.
Anilin-4-nitro-2:5-disulphonic acid.
Anilin-3-sulphonic acid.

2:4-Dioxyquinolin

Dye
Starting point (Brit. 431649) in making—
Dyestuffs with anilin or halogen anilins, toluidins, xylidins, and the like, for coloring organic solvents, lacquers, fats, oils, resins, and waxes; in clear yellow, greenish-yellow, or reddish shades, fast to sublimation and other influences.
Starting point (Brit. 404198) in making—
Dyestuffs (for coloring bones and bone objects red tints) by reaction with 2-amino-1-phenol-4:6-disulphonic acid and a chromium salt.

Dipara-anisylethylene
Dye
Starting point (Brit. 435449) in making—
Dyestuffs by coupling with paranitranilin.

1:4-Diparatoluidinoanthraquinone
Oils, Fats, Waxes
Coloring agent (Brit. 432867) for—
Paraffin and other mineral waxes, stearic acid, tallow and other solid triglycerides, beeswax, carnauba wax, and others.
Petroleum
Coloring agent (Brit. 420371) for—
Gasoline, transformer oils, and turbine oils.

1:8-Diparatoluidinoanthraquinone
Oils, Fats, Waxes
Coloring agent (Brit. 432867) for—
Paraffin and other mineral waxes, stearic acid, tallow and other solid triglycerides, beeswax, carnauba wax, and others.

1:4-Diparatolylaminoanthraquinone
German: 1:4-Diparatolylaminoanthrachinon.
Dye
Starting point (Brit. 261139) in making dyestuffs with—
Dimethylanilin, pyridin, quinolin.

Diparatolylethylenediamine
Rubber
Antioxidant (U. S. 1941012).

Dipentamethylenethiuram Polysulphide
Rubber
Accelerator (Brit. 443219).

Dipentamethylenethiuram Sulphide
Disinfectant
As a bactericide (Brit. 406979, U. S. 1972961).
Insecticide and Fungicide
As a fungicide (Brit. 406979, U. S. 1972961).
As an insecticide (claimed effective against aphids) (Australian 8103/32, Brit. 406979, U. S. 1972961).

Dipentene
Synonyms: Cinen, Cinene, Cajeputene, Diamylene, Dipenten, Dipentin, Inactive limonene.
French: Cinène, Diamylène, Dipentène, Limonène inactif.
German: Cinen, Dipenten, Dipentin, Kautschin.
Chemical
Ingredient of synthetic mandarin orange oil.
Solvent in various processes.
Fats and Oils
Extracting medium in producing various fats and oils.
Food
Solvent in making—
Extracts of various sorts.
Miscellaneous
General solvent in various industries.

Diphenolisatin
Chemical
Starting point in making—
0-0-Diacetyldiphenolisatin (U. S. 1624164).

Diphenyl
French: Diphényle.
German: Diphenyl.
Chemical
Starting point in making various intermediates used in making organic chemicals, plastics, dyestuffs, and insecticides.
Abrasives
Plasticizing agent in making—
Grinding wheel compositions, sandpaper (Brit. 281711).
Mechanical
Heat transfer agent in—
Power plants and various heating and cooking operations in the process industries.
Ingredient of—
Heat transfer agent consisting of admixture with diphenyl oxide (U. S. 1,882,809).
Paint and Varnish
Plasticizing agent in making—
Cellulose ester and ether paints, varnishes, lacquers, and dopes.
Plastics
Plasticizing agent in making—
Cellulose ester and ether compounds.

Diphenylamine Blue
French: Bleu de diphényleamine.
German: Diphenylamin blau.
Dye
Starting point in making—
Anilin blue.
Textile
——, *Dyeing and Printing*
In dyeing and printing silk and other textiles.

Diphenylaminechlorarsin
Military
As a chemical warfare gas.

Diphenylamine Trichloroacetate
French: Trichloroacétate de diphényleamine.
German: Diphenylamintrichloracetat, Diphenylamintrichlorazetat, Trichloressigsaeurediphenylaminester, Trichloressigsaeuresdiphenylamin.
Chemical
Reagent in making intermediates.
Rubber
Reagent (Brit. 282778) in making conversion products with—
Alphanaphthol, betanaphthol, catechol, cresol, parachlorophenol, phenol, resorcinol.

1:2-Diphenylaminoethane
Chemical
Starting point in making—
Intermediates, pharmaceuticals.
Dye
Starting point in making various synthetic dyestuffs.
Rubber
Antioxidant (Brit. 314756) in—
Vulcanizing.

Diphenylanthracene
German: Diphenylanthracen.
Chemical
Starting point in making—
9:10:10-Triphenyl-9-hydroxydihydroanthracene.

Diphenyl-Bismuth-Di-isopropyltindicresyl-Arsenic Compound
Lubricant
Addition agent (Brit. 445813) in—
Lubricants for motors, turbines, flushing, and high-temperature work generally.

Diphenyl Bisulphide
Petroleum
Antioxidant (Brit. 425569) for—
Lubricating, transformer, and switch oils, particularly solvent-extracted oils and others of a paraffinic nature, in which the natural inhibitor content may have been reduced during refining.

Diphenylbromodiphenyl
Spanish: Difenilbromodifenil.
Italian: Difenilebromodifenile.
Chemical
Reagent (U. S. 1853818) for treating—
Sulphur to render it fireproof.
Starting point in making—
Derivatives used as pharmaceuticals, etc.
Intermediates.

Diphenyl Carbonate
French: Carbonate de diphényle, Carbonate diphénylique.
German: Kohlensaeuresdiphenyl, Kohlensäurediphenylester.
Chemical
Reagent in making—

Diquinnicarboxylic acid ester (Aristochin).
Plastics
Substitute for camphor in making celluloid.

Diphenyl, Chlorinated
French: Diphényle chlorée.
German: Chlordiphenyl.
Paint and Varnish
Base material (German 563080) in making—
Varnishes and lacquers.
Paper
Impregnating material (U. S. 1889088) (in admixture with sulphur) for—
Paper material.
Textile
Delustering agent (Brit. 409521 and 409625) for—
Rayons.

Diphenyl-Chloroarsine
Military
As a poison gas (blue cross gas).
Petroleum
Addition agent (Brit. 433257) in—
Lubricating oils or greases, especially for use at high temperatures, such as cylinder oils, hydrogenated oils, or oils refined by treatment with sulphuric acid, clay, or extraction solvents.

Diphenylchlorodiphenyl
Spanish: Difenilclorodifenil.
Italian: Difenileclorodifenile.
Chemical
Reagent (U. S. 1853818) for treating—
Sulphur to render it fireproof.
Starting point in making—
Derivatives used as pharmaceuticals, etc.
Intermediates.

Diphenylcyanoarsin
German: Diphenylcyanarsin.
Explosives
Ingredient of—
Nitro-starch explosive compositions used for filling gas shells (U. S. 1588277).

Diphenyldisazo-orthoethoxyaminophenolorthoaminobenzoic Acid Sodium Salt
Pharmaceutical
Ingredient (U. S. 2010512) of—
Antiseptic, consisting of admixture in equal parts with orthohydroxyquinolin sulphate.

Diphenylene Oxide
Chemical
Starting point in—
Organic synthesis.
Lubricant
As a high-temperature lubricant (U. S. 1867968).
Starting point (Brit. 440916) in making—
Products useful as lubricating oils or as pour-point depressors for paraffin base lubricating oils by condensation with halogenated derivatives of aliphatic hydrocarbons, such as paraffin oils, paraffin, petrolatum, ceresin, ozokerite, or others contained in the middle to high fractions of petroleum.
Mechanical
Ingredient (U. S. 1874258) of—
Stabilized heating fluid, containing also diphenyl oxide.

Diphenylethane
Cellulose Products
Plasticizer (U. S. 1891601) for—
Cellulose acetate, cellulose esters or ethers, nitrocellulose.
For uses, see under general heading: "Plasticizers."

Diphenylethyl Phosphate
French: Éthylediphényle phosphate, Phosphate de éthylediphényle.
German: Diphenylaethylphosphat, Phosphatischesdiphenylaether, Phosphoraetherdiphenyl.
Photographic
Reagent (French 606969) for—
Reducing inflammability in making film from cellulose derivatives.
Solvent (French 606969) in making—
Film from cellulose derivatives.
Plastics
Reagent (French 606969) for—
Reducing inflammability in making plastics from cellulose derivatives.
Solvent (French 606969) in making—
Plastics from cellulose derivatives.
Textile
Reagent (French 606969) for—
Reducing inflammability in making fibers from cellulose derivatives.
Solvent (French 606969) in making—
Fibers from cellulose derivatives.

Diphenylethylstibin
French: Stibine de diphényle et éthyle, Stibine diphénylique et éthylique.
German: Diphenylaethylstibin.

Chemical
Starting point in making various derivatives.

Miscellaneous
Mothproofing agent (Brit. 303092) for treating—
Furs and hair.

Textile
Mothproofing agent (Brit. 303092) for treating—
Wool and felt.

Diphenylguanidin

Chemical
Stabilizer (Brit. 397914) for—
Chlorinated hydrocarbons.
Starting point in making—
Derivatives.

Paint and Varnish
Ingredient (Brit. 370699) of—
Lacquers applied as overcoats to protect main coating on metals or other materials.

Paper
Ingredient (U. S. 1911774) of—
Diphenyl-base impregnating or varnishing agents for protecting checks and the like against chemical erasure.

Rubber
Accelerator in—
Vulcanizing processes.

Diphenylguanidin Oleate

Paper
Ingredient (U. S. 1911774) of—
Diphenyl-base impregnating or varnishing agents for protecting checks and the like against chemical erasure.

Diphenylguanidin Palmitate

Paper
Ingredient (U. S. 1911774) of—
Diphenyl-base impregnating or varnishing agents for protecting checks and the like against chemical erasure.

Diphenylguanidin Resinate

Paper
Ingredient (U. S. 1911774) of—
Diphenyl-base impregnating or varnishing agents for protecting checks and the like against chemical erasure.

Diphenylguanidin Stearate

Paper
Ingredient (U. S. 1911774) of—
Diphenyl-base impregnating or varnishing agents for protecting checks and the like against chemical erasure.

Diphenyliododiphenyl
Spanish: Difenilyododifenil.
Italian: Difenileiododifenile.

Chemical
Reagent (U. S. 1853818) for treating—
Sulphur to render it fireproof.
Starting point in making—
Derivatives used as pharmaceuticals, etc.
Intermediates.

Diphenyl-Mercury

Lubricant
Starting point (Brit. 440175) in making—
Addition agents for high-pressure lubricating oils or greases by reacting with oil-soluble organic compounds.

Diphenylmeta-m′-dicarboxylic Acid
French: Acide de diphénylemétа-m′-dicarboxylique.
German: Diphenyl-meta-m′-dicarbonsaeure.

Dye
Starting point (Brit. 264561) in making vat dyestuffs with—
1-Aminoanthraquinone.
1-Benzylamino-4-aminoanthraquinone.
1-Benzylamino-5-aminoanthraquinone.
3-Bromo-1-aminoanthraquinone.

Diphenylmethane
Synonyms: Benzylbenzene.
French: Méthane de diphényle.
German: Methandiphenyl.

Chemical
Reagent in—
Organic synthesis.
Solvent (U. S. 1467095) in—
Compositions based on ethyl cellulose.

Dye
Reagent in synthesis of—
Dyestuffs, intermediates.

Perfume
Reagent in making—
Synthetic perfumes.

Diphenylmethyl-Aluminum

Petroleum
Addition agent (Brit. 433257) in—
Lubricating oils or greases, especially for use at high temperatures, such as cylinder oils, hydrogenated oils, or oils refined by treatment with sulphuric acid, clay, or extraction solvents.

Diphenylmethyl-Bismuthine

Petroleum
Addition agent (Brit. 433257) in—
Lubricating oils or greases, especially for use at high temperatures, such as cylinder oils, hydrogenated oils, or oils refined by treatment with sulphuric acid, clay, or extraction solvents.

Diphenylmethyl-Cadmium

Petroleum
Addition agent (Brit. 433257) in—
Lubricating oils or greases, especially for use at high temperatures, such as cylinder oils, hydrogenated oils, or oils refined by treatment with sulphuric acid, clay, or extraction solvents.

Diphenylmethylene Ether

Electrical
Starting point (Brit. 399868) in making—
Plastic materials with benzyl or ethyl cellulose, fillers, and coloring matter used as a component of insulated conductors.

3:3′-Di(2-phenyl-1-methylindolyl) Ketone

Dye
Starting point (Brit. 428468) in making—
Blue dyestuffs for wool or silk by condensing with 3′-ethoxy-4-methyldiphenylamine.
Reddish-violet dyestuffs for wool, silk, and lacquers by condensing with trisulphonated 2-phenyl-1-methylindole.

Diphenylmethyl-Mercury

Petroleum
Addition agent (Brit. 433257) in—
Lubricating oils or greases, especially for use at high temperatures, such as cylinder oils, hydrogenated oils, or oils refined by treatment with sulphuric acid, clay, or extraction solvents.

Diphenylmethyl Phosphate
French: Méthylediphényle phosphate, Phosphate de méthylediphényle.
German: Diphenylmethylphosphat, Phosphatischesmethyldiphenyl, Phosphormethyldiphenyl.

Photographic
Reagent (French 606969) for—
Reducing inflammability in making film from cellulose derivatives.
Solvent (French 606969) in making—
Film from cellulose derivatives.

Plastics
Reagent (French 606969) for—
Reducing inflammability in making plastics from cellulose derivatives.
Solvent (French 606969) in making—
Plastics from cellulose derivatives.

Textile
Reagent (French 606969) for—
Reducing inflammability in making fibers from cellulose derivatives.
Solvent (French 606969) in making—
Fibers from cellulose derivatives.

Diphenylmethyl-Stibine

Petroleum
Addition agent (Brit. 433257) in—
Lubricating oils or greases, especially for use at high temperatures, such as cylinder oils, hydrogenated oils, or oils refined by treatment with sulphuric acid, clay, or extraction solvents.

Diphenylmethyl-Thallium

Petroleum
Addition agent (Brit. 433257) in—
Lubricating oils or greases, especially for use at high temperatures, such as cylinder oils, hydrogenated oils, or oils refined by treatment with sulphuric acid, clay, or extraction solvents.

Diphenylmethyl Thiophosphate

Miscellaneous
Plasticizer (U. S. 1982903) for—
Cellulose esters and ethers, synthetic resins.
For uses, see under general heading: "Plasticizers."

Diphenylmethyl-Zinc

Petroleum
Addition agent (Brit. 433257) in—
Lubricating oils or greases, especially for use at high temperatures, such as cylinder oils, hydrogenated oils, or oils refined by treatment with sulphuric acid, clay, or extraction solvents.

Diphenyl-2:6-naphthylenediamine, Normal-N′

Rubber
Age resister (Brit. 427495).

Diphenyl-2:7-naphthylenediamine, Normal-N′

Rubber
Age resister (Brit. 427495).

Diphenylnitrosoamine

Paint and Varnish
Ingredient of—
Drying oil compositions, used to prevent rapid oxidation.

Diphenylolbutane

French: Butane de diphénylole, Butane diphénylolique.
German: Diphenylolbutan.

Miscellaneous
Plasticizer (Brit. 313133) for—
Cellulose esters and ethers, natural resins, synthetic resins.
For uses, see under general heading: "Plasticizers."

Diphenylolcyclohexane

Cellulose Products
Plasticizer (Brit. 342429) for—
Cellulose acetate, cellulose esters or ethers, cellulose nitrate (nitrocellulose).
For uses, see under general heading: "Plasticizers."

Diphenylolpropane

French: Propane de diphénylole, Propane diphénylolique.
German: Diphenylolpropan.

Miscellaneous
Plasticizer (Brit. 313133) for cellulose esters and ethers, natural resins, synthetic resins.
For uses, see under general heading: "Plasticizers."

Diphenyl Oxide

French: Oxyde de diphényle.
German: Diphenyloxyd.

Chemical
Reagent in organic synthesis.

Dye
Synthesis of dyestuffs.

Mechanical
Heat transfer agent in—
Power plants and various heating and cooking operations in the process industries.
Ingredient of—
Heat transfer medium consisting of admixture with diphenyl (U. S. 1,882,809).
Heat-energy transfer medium, containing also naphthalene, pyrene or parahydroxydiphenyl (U. S. 1893051).
Stabilized heating fluid, containing also diphenylene oxide (U. S. 1874258).

Perfume
Odorant in—
Cosmetics, synthetic perfumes, toilet waters.

Soap
Odorant in—
Toilet soaps.

Diphenylpropane

French: Diphénylepropane, Propane diphénylique.
German: Diphenylpropan.

Cellulose Products
Plasticizer (Brit. 313133) for—
Cellulose esters and ethers.
For uses, see under general heading: "Plasticizers."

Chemical
Starting point in making various intermediates.

Resins
Plasticizer (Brit. 313133) for—
Resins.

Diphenyl Sulphide

Insecticide and Fungicide
Fungicide (French 702703) for—
Puccinia graminis (wheat rust).
Ingredient (French 702703) of—
Dusting agent for destroying wheat rust, containing also a wetting agent or an adhesive and an inert material, such as prepared chalk, talc, or kieselguhr.

Paper
Ingredient (U. S. 1911774) of—
Diphenyl-base impregnating or varnishing agents for protecting checks and the like against chemical erasure.

Diphenylsulphone

Insecticide
Exterminant for—
Culicine mosquito larvae.

Diphenyl Sulphoxide

Insecticide
Exterminant for—
Culicine mosquito larvae.

2:2-Diphenyltetramethylene 1:3-Disulphide

Insecticide
Exterminant for—
Culicine mosquito larvae.

Diphenylthiourea

French: Sulfourée de diphényle, Sulfourée diphénylique, Thiourée de diphényle, Thiourée diphénylique.
German: Diphenylsulfoharnstoff, Diphenylthioharnstoff.

Chemical
Starting point in making—
Pharmaceuticals and other derivatives.

Insecticide
Ingredient of—
Insecticidal preparations (used in conjunction with starch mixture) (U. S. 1734519).

Diphenylurea-3:3′-dicarboxylic Acid

Chemical
Starting point (Brit. 314909) in making derivatives with—
Alpha-amino-5-naphthol-7-sulphonic acid.
Alkoxyalphanaphthalenesulphonic acids.
Alphanaphthylamine-4:8-disulphonic acid.
Alphanaphthylamine-4:6:8-trisulphonic acid.
4-Aminoacenaphthene-3:5-disulphonic acid.
4-Aminoacenaphthene-3-sulphonic acid.
4-Aminoacenaphthene-5-sulphonic acid.
4-Aminoacenaphthenetrisulphonic acid.
Aminoarylcarboxylic acids.
Aminoheterocyclic-carboxylic acids.
1:8-Aminonaphthol-3:6-disulphonic acid.
Bromonitrobenzoyl chlorides.
Chloroalphanaphthalenesulphonic acids.
Chloronitrobenzoyl chlorides.
Iodonitrobenzoyl chloride.
Nitroanisoyl chlorides.
Nitrobenzene sulphochlorides.
Nitrobenzoyl chlorides.
2-Nitrocinnamyl chloride.
3-Nitrocinnamyl chloride.
4-Nitrocinnamyl chloride.
1-Nitronaphthalene-5-sulphochloride.
1:5-Nitronaphthoyl chloride.
2-Nitrophenylacetyl chloride.
4-Nitrophenylacetyl chloride.
Nitrotoluyl chloride.

Dipotassium Glutamate
French: Glutamate dipotassique.
German: Dikaliumglutamat, Glutamsaeuresdikalium.
Brewing
Ingredient of—
Flavoring extracts used in making beer (Brit. 279985).
Food
Reagent (Brit. 279985) in making—
Flavoring extracts for foods and drinks.
Food preparations from fish, meat, starches, casein, egg yolk, wheat, maize, and the like.
Pharmaceutical
Ingredient of—
Flavored vehicles (Brit. 279985).
Wine
Ingredient of—
Flavoring extracts used in making wines (Brit. 279985).

Dipropanolamine Citrate
Textile
De-electrifying agent (Brit. 430221) for—
Yarns, films, fabrics, and the like, subject to charging by static electricity (applied in admixture with all usual lubricating agents as vehicles).

Dipropanolamine Gallate
Textile
De-electrifying agent (Brit. 430221) for—
Yarns, films, fabrics, and the like, subject to charging by static electricity (applied in admixture with all usual lubricating agents as vehicles).

Dipropanolamine Lactate
Textile
De-electrifying agent (Brit. 430221) for—
Yarns, films, fabrics, and the like, subject to charging by static electricity (applied in admixture with all usual lubricating agents as vehicles).

Dipropanolamine Mucate
Textile
De-electrifying agent (Brit. 430221) for—
Yarns, films, fabrics, and the like, subject to charging by static electricity (applied in admixture with all usual lubricating agents as vehicles).

Dipropanolamine Saccharate
Textile
De-electrifying agent (Brit. 430221) for—
Yarns, films, fabrics, and the like, subject to charging by static electricity (applied in admixture with all usual lubricating agents as vehicles).

Dipropanolamine Salicylate
Textile
De-electrifying agent (Brit. 430221) for—
Yarns, films, fabrics, and the like, subject to charging by static electricity (applied in admixture with all usual lubricating agents as vehicles).

Dipropanolamine Tannate
Textile
De-electrifying agent (Brit. 430221) for—
Yarns, films, fabrics, and the like, subject to charging by static electricity (applied in admixture with all usual lubricating agents as vehicles).

Dipropanolamine Tartrate
Textile
De-electrifying agent (Brit. 430221) for—
Yarns, films, fabrics, and the like, subject to charging by static electricity (applied in admixture with all usual lubricating agents as vehicles).

Dipropargyl
Synonyms: 1:5-Hexadine.
Chemical
Solvent (Brit. 398561) in making—
Betaphenylethyl alcohol from chlorobenzene, magnesium, and ethylene oxide.
Miscellaneous
See also: "Solvents."
Petroleum
Extractant (U. S. 1897979) in—
Removing solvents from mineral lubricating oils.
Plastics
Extraction agent (Brit. 394244) for—
Retained softeners and solvents in sheets and films made from polymerized polyvinyl chloride.

Dipropylamine
Chemical
Catalyst (Brit. 252870) in making—
Normal butyl para-aminobenzoate.
Normal butyl paranitrobenzoate.

Disilicon Hexachloride
French: Hexachlorure de disilicium.
German: Disiliciumhexachlorid.
Construction
Hardening and preserving agent (Brit. 260031) in treating—
Concretes, stone, stuccos.

Disodium Carbimide
French: Carbimide disodique.
Miscellaneous
As a household cleansing agent (French 753038).

Disodium Glutamate
French: Glutamate disodique, Glutamate de disoude.
German: Dinatriumglutamat, Glutaminsaeuresdinatrium.
Brewing
Ingredient of—
Beers and ales, added to improve the taste (Brit. 279985).
Food
Reagent (Brit. 279985) in making—
Flavoring extracts.
Food products from fish, meat, starches, casein, wheat, maize, egg yolk.
Pharmaceutical
In compounding and dispensing practice (Brit. 279985).
Wine
Ingredient of—
Flavorings for wines (Brit. 279985).

Disodium Phosphate
Synonyms: Dibasic sodium phosphate, Disodium hydrogen phosphate, Disodium orthophosphate, Hydrosodium phosphate, Phosphate of soda, dibasic.
Latin: Natrium phosphoricum, Phosphas natricus, Phosphas sodicus, Sal mirabile perlatum, Sodii phosphas.
French: Phosphate disodique, Phosphate de soude.
German: Dinatriumphosphat, Phosphorsäuresdinatron.
Spanish: Fosfato sodico.
Italian: Fosfato bisodico.
Analysis
As a reagent.
Ceramics
Ingredient of—
Glazes for chinaware, potteries, and porcelains.
Chemical
Reagent in making—
Aluminum phosphate from aluminum sulphate.
Ammonium phosphate from ammonia.
Calcium phosphate (dibasic) from a solution of a calcium salt.
Ferric phosphate.
Sodium resinate (U. S. 1881858).
Reagent in making—
Fireproof starches.
Dye
Reagent in making—
Dyes, such as Schnitzler's green.
Explosives and Matches
Ingredient of—
Matchhead compositions.
Fertilizer
Ingredient of—
Fertilizer compositions.
Food
Ingredient of—
Baking powders.
Glass
Substitute for—
Bone ash as an opacifying agent in opaque and translucent glasses.
Leather
Reagent in—
Tanning processes.
Metallurgical
Ingredient of—
Baths in galvanoplastic work, fluxes in soldering and tinning.

Disodium Phosphate (Continued)

Miscellaneous
As a—
Boiler-scale removing agent.
Boiler-water softening agent.
Boiler water softening agent (U. S. 1247833, 1273857).
Boiler-water softening agent, followed by adding a soluble soap and agitating to cause a froth which will hold the precipitated salts in suspension (U. S. 1333393).
Ingredient of—
Boiler compounds.
Boiler compounds containing also borax and sodium or calcium carbonate (U. S. 1162024).
Fireproofing compositions.
Fireproofing mixtures with boric acid (U. S. 1501895, 1501911).
Fireproofing compositions for wood and metal, containing also sodium silicate, powdered asbestos, magnesium sulphate, saponified resin, glycerin, and water (U. S. 1397028).

Paint and Varnish
Reagent in making—
Pigments, such as cobalt phosphate.

Pharmaceutical
In compounding and dispensing practice.
Suggested for use as—
Mild purgative.
Suggested for use in treating—
Constipation in children.
Deficiency of phosphates in human system.
Infantile diarrhea.
Jaundice.

Photographic
Ingredient of—
Photographic emulsion, containing also silver nitrate, potassium chlorate, citric acid, and chrome alum.
Reagent in making—
Silver phosphate.
Retarding agent in—
Developing solutions.
Source of alkali in—
Gold toning baths for printing-out papers.

Textile
Buffer (U. S. 1912345) in—
Extraction of tannin substances from tanner's wool prior to bleaching.
Fireproofing agent for various textiles.
Impregnating agent in—
Dyeing, calico printing.
Reagent (Brit. 388044) in—
Vat dye baths.
Weighting agent for—
Silk (Brit. 403239, U. S. 1902226).

Woodworking
Ingredient of—
Fireproofing compositions.

Disodium Undecoate

Miscellaneous
As a wetting agent (U. S. 2020999).
For uses, see under general heading: "Wetting agents."

Dispersing Agents

See: "Emulsifying agents."

Ditetrahydrofurfurylamine

Glass
Stabilizer (Brit. 437304) for—
Halogenated rubber derivatives used as cements for laminated glass.

Miscellaneous
Inhibitor (Brit. 437304) of—
Photochemical action.

Paper
Stabilizer (Brit. 437304) for—
Halogenated rubber derivatives used for impregnating or coating wrapping paper.

Rubber
Promoter (Brit. 437304) of—
Resistance to the deteriorating action of light on chlorinated rubber used in the production of flexible, transparent films suitable for wrappings, paper-coatings, or the like, or in the manufacture of laminated glass.
Stabilizer (Brit. 437304) for—
Coating and impregnating agents made from halogenated rubber derivatives and used for treating fabrics to be used as wrapping materials.
Transparent films or sheets made from halogenated rubber derivatives.

Dithiocarbazide

Chemical
Starting point in making various derivatives.

Metallurgical
Promoter (U. S. 1852109) in—
Recovering minerals from ores by the froth flotation process.

Dithioglycollic Acid Disulphide Dimethylester

Oils, Fats, and Waxes
Starting point (Brit. 440175) in making—
Addition agents for high-pressure lubricating oils or greases, by mixing and reacting with organo-metallic compounds.

Dithymol Di-iodide

Synonyms: Aristol, Dithymol bi-iode, Thymol iodide.
German: Dithymol jodid.

Miscellaneous
Ingredient of—
Dental cements (U. S. 1613532).

Pharmaceutical
In compounding and dispensing practice.

Ditolyl-Aluminum

Lubricant
Addition agent (Brit. 433257) to—
Lubricating oils or greases, especially for use at high temperatures, such as cylinder oils, hydrogenated oils, or oils refined by treatment with sulphuric acid, clay, or extraction solvents.

Ditolyl-Bismuthine

Petroleum
Addition agent (Brit. 433257) in—
Lubricating oils or greases, especially for use at high temperatures, such as cylinder oils, hydrogenated oils, or oils refined by treatment with sulphuric acid, clay, or extraction solvents.

Ditolyl-Cadmium

Petroleum
Addition agent (Brit. 433257) in—
Lubricating oils or greases, especially for use at high temperatures, such as cylinder oils, hydrogenated oils, or oils refined by treatment with sulphuric acid, clay, or extraction solvents.

Ditolylethane

Cellulose Products
Plasticizer (U. S. 1891601) for—
Cellulose acetate, cellulose esters or ethers, nitrocellulose.
For uses, see under general heading: "Plasticizers."

Ditolyl-Mercury

Petroleum
Addition agent (Brit. 433257) in—
Lubricating oils or greases, especially for use at high temperatures, such as cylinder oils, hydrogenated oils, or oils refined by treatment with sulphuric acid, clay, or extraction solvents.

Ditolyl-Mercury Sulphide

Petroleum
Addition agent (Brit. 433257) in—
Lubricating oils or greases, especially for use at high temperatures, such as cylinder oils, hydrogenated oils, or oils refined by treatment with sulphuric acid, clay, or extraction solvents.

Ditolyl-Stibine

Lubricant
Addition agent (Brit. 433257) to—
Lubricating oils or greases, especially for use at high temperatures, such as cylinder oils, hydrogenated oils, or oils refined by treatment with sulphuric acid, clay, or extraction solvents.

Ditolyl-Thallium
Lubricant
Addition agent (Brit. 433257) to—
Lubricating oils or greases, especially for use at high temperatures, such as cylinder oils, hydrogenated oils, or oils refined by treatment with sulphuric acid, clay, or extraction solvents.

Ditolyl-Zinc
Petroleum
Addition agent (Brit. 433257) in—
Lubricating oils or greases, especially for use at high temperatures, such as cylinder oils, hydrogenated oils, or oils refined by treatment with sulphuric acid, clay, or extraction solvents.

Divi Divi
Synonyms: Libidibi, Libidivi, Lilidibi.
German: Gerbschotra.
Leather
Tanning agent.
Textile
——, *Dyeing*
Reagent in dyeing textiles in black shades.

Dixylylethane
French: Éthane de dixylyle, Éthane dixylylique.
German: Dixylylaethan.
Cellulose Products
Plasticizer (U. S. 1891601) for—
Cellulose acetate, cellulose esters or ethers, nitrocellulose.
For uses, see under general heading: "Plasticizers."

Dodecene
Miscellaneous
As an emulsifying agent (Brit. 343872).
For uses, see under general heading: "Emulsifying agents."

Dodecyl Alcohol
French: Alcool de dodecyl, Alcool dodecylique.
German: Dodecylalkohol.
Chemical
Starting point in making—
Activators for flotation reagents by reacting with pyridin and sulphuric acid (Brit. 410956).
Dodecyl bromide (Brit. 401707).
Dodecylbenzyl ether (Brit. 378454, 393937).
Dodecylchloracetic ester (Brit. 397445).
Dodecylpyridinium bromide (Brit. 397553, 404969).
Dodecylsulphobenzyl ether sodium salt (Brit. 378454).

Dodecylbenzyl Ether
French: Benzyle éther de dodecyl, Benzyle éther dodecylique, Éther benzilique de dodecyl.
German: Dodecylbenzilaether.
Soap
Starting point (Brit. 378454) in making—
Sulphonated derivatives used as cleansing agents.

Dodecylbetagammadihydroxypropylamine
Soap
Emulsifying agent (Brit. 421490 and 411295) in—
Shaving creams, superfatted soaps, and the like.

Dodecylbetagammadihydroxypropylsulphone
Soap
Emulsifying agent (Brit. 421490 and 411295) in—
Shaving creams, superfatted soaps, and the like.

Dodecyl Bromide
Insecticide
Reagent (Brit. 401707) in making—
Insecticides, by reaction with nicotine or with its salts, such as nicotine hydrobromide and hydrochloride.

Dodecylchloracetic Ester
Textile
Starting point (Brit. 397445) in making—
Wetting agents by reacting with sodium thiosulphate.

Dodecylchloromethyl Ether
Chemical
Starting point (Brit. 434911) in making—
Dodecyldodecoxymethylpiperidinium chloride by reacting with normal dodecylpiperidin.

Dodecylcresol
Chemical
Starting point (Brit. 444351) in making—
Fat-splitting catalysts and emulsifying agents, useful in dyeing, laundering, bleaching, and various other purposes, by reacting with formaldehyde and non-aromatic secondary amines (the salts of the products with water-soluble acids, or water-insoluble acids, or the quaternary ammonium salts, are claimed to be valuable for the purposes named).

Dodecyldimethylamine
Firefighting
Basic ingredient (Brit. 460649) in—
Air-foaming compositions for fire-extinguishing purposes.

Dodecyldimethylamine Formate
Firefighting
Basic ingredient (Brit. 460649) in—
Air-foaming compositions for fire-extinguishing purposes.

Dodecyldimethylbetaine
Firefighting
Basic ingredient (Brit. 460649) in—
Air-foaming compositions for fire-extinguishing purposes.

Dodecyldodecoxymethylpiperidinium Chloride
Textile
Increaser (Brit. 434911) of—
Fastness to water of dyeings on textile fibers.
Softener (Brit. 434911) of—
Dyed textile fibers.

Dodecylguanidin Chloride
Textile
Assistant (Brit. 421862) in—
Aqueous baths for treating textiles.
Promoter (Brit. 421862) of—
Uniform dyeing with basic dyestuffs.
Wetting and washing agent (Brit. 421862) in—
Textile processes.

Dodecylguanidin Hydrochloride
Miscellaneous
As an emulsifying agent (Brit. 422461).
For uses, see under general heading: "Emulsifying agents."

Dodecylphenol
Chemical
Starting point (Brit. 444351) in making—
Fat-splitting catalysts and emulsifying agents, useful in dyeing, laundering, bleaching, and various other purposes, by reacting with formaldehyde and non-aromatic secondary amines (the salts of the products with water-soluble acids, or water-insoluble acids, or the quaternary ammonium salts are claimed to be valuable for the purposes named).

Dodecylpiperidin Normal Oxide
Miscellaneous
As a general wetting agent (Brit. 437566).
Textile
As a dyeing assistant (Brit. 437566).
As a general wetting agent (Brit. 437566).
Wetting agent (Brit. 437566) in—
Wool washing.

Dodecylpyridinium Bromide
French: Bromure de dodécylpyridinium.
German: Bromdodecylpyridinium, Dodecylpyridiniumbromid.
Spanish: Bromuro de dodecylpyridinium.
Italian: Bromuro di dodecylpyridinium.
Metallurgical
Inhibitor (Brit. 397553) of—
Corrosion of metal by sulphuric acid in pickling baths for steel.
Miscellaneous
Agent (Brit. 404969) for—
Pretreating furs to be dyed by a chrome dye, a direct cotton dye, an acid dye, or a vat dye, or a mixture of such dyes.

Dodecylresorcinol

Chemical
Starting point (Brit. 444351) in making—
Fat-splitting catalysts and emulsifying agents for use in dyeing, laundering, bleaching, and various other purposes, by reacting with formaldehyde and non-aromatic secondary amines (the salts of the products with water-soluble acids or water-insoluble acids, and the quaternary ammonium salts, are claimed to be valuable for the purposes named).

Dodecylsulphobenzyl Ether Sodium Salt

Soap
Ingredient (Brit. 378454) of—
Cleansing composition, containing also sodium sulphate and sodium salt of cetylsulphobenzyl ether.

Dodekanaphthene
German: Dodekanaphten.

Chemical
As a general solvent (Brit. 269960).

Miscellaneous
Solvent for various substances (Brit. 269960).

Textile
——, *Dyeing and Printing*
Solvent (Brit. 269960) in preparing—
Dye liquors and printing paste.
——, *Stenciling*
Solvent (Brit. 269960) in making—
Stenciling compositions.

Dolomitic Magnesite
French: Magnésite dolomitique.
German: Dolomitsch Magnesitspat.

Construction
For building purposes.

Cement
Raw material in making—
Rapid-setting magnesium oxychloride cement.

Chemical
Raw material in making—
Magnesium and calcium chemicals.

Metallurgical
Lining of furnaces.

Paper
Raw material in making—
Pulp digestion liquor.

Refractories
Raw material in making—
Refractory brick.

Dragon's Blood
Latin: Sanguis draconis.
French: Sang-dragon.
German: Drachenblut.
Spanish: Sangre de drago.

Ceramics
Ingredient of—
Pigment preparations for chinaware, porcelains, potteries, stoneware.

Construction
Pigment for—
Plasters, stuccoes.

Jewelry
Ingredient of—
Gold lacquering preparations.

Leather
Ingredient of—
Tanning compositions.

Miscellaneous
Ingredient of—
Compositions for treating tobacco pipes.

Paint and Varnish
Red Pigment in—
Enamels, lacquers, fine paints, fine varnishes.

Paper
Pigment for—
Paper, pulp.

Perfumery
Ingredient of—
Cosmetics, dentifrices.

Pharmaceutical
In compounding and dispensing practice.

Photographic
Ingredient of—
Compositions used in making photographic papers.

Printing
In process engraving and the litho trades.

Stone
Pigment for—
Artificial stones, marbles, natural stones.

Woodworking
Ingredient of—
Polishing compositions.

Duodecylene

Chemical
Solvent for various purposes.

Miscellaneous
Solvent for various purposes.

Textile
——, *Dyeing, Printing and Stenciling*
Solvent in decorating or coloring acetate rayon (Brit. 269960).

Dutch Pink
French: Stil de grain.
German: Schuettgelb.

Paint and Varnish
Pigment in making—
Paints, lacquers, varnishes.

Elaidic Acid Chloride

Chemical
Starting point (Brit. 407956) in making pour-point improvers for machine oils, gear oils, and other lubricants by condensing with—
Anilin, anthracene oil.
Aromatics obtained by destructive hydrogenation or by dehydrogenation.
Benzene.
Cracking gases containing gaseous olefins (ethylene, propylene, and butylene).
Cyclic terpenes, ethylnaphthalene, liquid olefins, middle oil, naphthalene, naphthols, naphthylamines, nitrated aromatics, phenols, tars, toluene, xylene.

Elemi Gum
Synonyms: Manila elemi.
French: Gomme elemi.
German: Elemiharz, Oelbaumharz.

Adhesives
Ingredient of—
Elastic spirit adhesives.

Ink
Ingredient of—
Lithographic inks, printing inks.

Insecticide
Ingredient of insecticidal and germicidal preparations.

Miscellaneous
Reagent in processing felt materials.
Stiffening agent for felt hats.

Oils and Fats
Starting point in making—
Essential oil.
Ingredient of—
Essential oil compositions.

Paint and Varnish
Ingredient of—
Elastic varnishes, high-luster varnishes, lacquers.

Pharmaceutical
In compounding and dispensing practice.

Resins and Waxes
Ingredient of—
Special resinous compositions (Brit. 252656).

Eleostearic Acid
French: Acide d'éléostearique.
German: Eleostearinsäure.

Paint and Varnish
Starting point (Brit. 284389) in making—
Paint bases, varnish bases.

Resins and Waxes
Starting point (Brit. 284349) in making—
Synthetic resins.

Emetine
French: Emétine.
German: Emetin.
Chemical
Starting point (Brit. 283533) in making the following salts of emetine—
Apocholate, cholate, choleinate, dihydrocholate, disoxycholate, glycocholate, taurocholate.
Miscellaneous
In dental work.
Pharmaceutical
In compounding and dispensing practice.
Rubber
Accelerator in vulcanizing.

Emulsifying Agents
Also includes applications for products commonly referred to as "Dispersing agents" or "Suspending agents."
Building and Construction
Emulsifying agent in making—
Emulsified waterproofing compositions.
Chemical
Emulsifying agent in making—
Emulsions of various chemicals.
Textile lubricants in emulsified form.
Wetting compositions in emulsified form.
Cosmetic
Emulsifying agent in making—
Emulsified cosmetics.
Disinfectant
Emulsifying agent in making—
Emulsified germicidal and disinfecting compositions.
Dye
Emulsifying agent in making—
Emulsified color lakes.
Fats, Oils, and Waxes
Emulsifying agent in making—
Emulsified boring oils.
Emulsified drilling oils.
Emulsified fat-splitting preparations.
Emulsified fatty acids of animal or of vegetable origin.
Emulsified greasing compositions.
Emulsified greasing and lubricating compositions containing various vegetable and animal fats and oils.
Emulsified preparations of natural and synthetic waxes.
Emulsified sulphonated oils.
Emulsified wire-drawing oils.
Emulsions of animal and vegetable fats and oils.
Glue and Adhesives
Emulsifying agent in making—
Emulsified adhesive preparations.
Ink
Emulsifying agent in making—
Emulsified printing and writing inks.
Insecticide
Emulsifying agent in making—
Emulsified insecticidal and fungicidal compositions.
Horticultural sprays.
Leather
Emulsifying agent in making—
Emulsified compositions for softening hides.
Emulsified dressing compositions.
Emulsified fat-liquoring baths.
Emulsified finishing compositions.
Emulsified soaking compositions.
Emulsified tanning compositions containing formocresylic or coumarone resins.
Emulsified waterproofing compositions.
Miscellaneous
As a dispersing or emulsifying agent not precipitable by electrolytes and stable with respect to lime and magnesia.
Emulsifying agent in making—
Automobile polishes in emulsified form.
Emulsified cleansing compositions.
Emulsified compositions for cleansing painted and metallic surfaces.
Emulsified degreasing compositions.
Emulsified furniture polishes.
Emulsified greasing compositions.
Emulsified metal polishes.
Emulsions of various substances.
Waterproofing compositions in emulsified form.
Paint and Varnish
Emulsifying agent in making—
Emulsified shellac preparations.
Waterproofing compositions in emulsified form.
Paper
Emulsifying agent in making—
Emulsified compositions for sizing paper and pulp products.
Emulsified compositions for waterproofing paper and pulp compositions and paperboard.
Waxing compositions in emulsified form.
Petroleum
Emulsifying agent in making—
Emulsified cutting oils for screwpress and lathe work.
Emulsified mineral oils.
Kerosene emulsions.
Naphtha emulsions.
Petroleum pitch emulsions.
Petroleum tar emulsions.
Textile oils in emulsified form, such as rayon oils.
Soluble greases in emulsified form.
Solubilized emulsified oils and distillates.
Plastics
Emulsifying agent in making—
Emulsified plastic compositions.
Resins
Emulsifying agent in making—
Emulsified preparations of natural and synthetic resins.
Rubber
Emulsifying agent in making—
Emulsified rubber cements and compositions.
Soap
Emulsifying agent in making—
Dry-cleaning soaps.
Emulsified detergents, containing soaps, used for various purposes.
Emulsified hand-cleansing compositions containing soap.
Emulsified textile soaps.
Spotting fluids for the laundry and textile industries.
Textile
——, *Bleaching*
Emulsifying agent in making—
Emulsified bleaching baths.
——, *Dyeing*
Emulsifying agent in making—
Dye baths in emulsified form.
——, *Finishing*
Emulsifying agent in making—
Emulsified coating compositions.
Emulsified scouring compositions.
Emulsified sizing compositions.
Emulsified washing compositions.
Emulsified waterproofing compositions.
Emulsified waxing compositions.
——, *Manufacturing*
Emulsifying agent in making—
Emulsified baths for the carbonization of wool.
Emulsified baths for degumming and boiling-off silk.
Emulsified baths for soaking silks.
Emulsified bowking baths.
Emulsified compositions used for degreasing raw wool.
Emulsified fulling baths.
Emulsified keir-boiling baths for cotton.
Emulsified mercerization baths.
Emulsified spinning compositions.
Oiling emulsions for various textile purposes.
——, *Printing*
Emulsifying agent in making—
Emulsified printing pastes.

Ephedrine Erucate
Pharmaceutical
Stabilizer in making—
Colloidal solutions or organo-mercurials in mineral or vegetable oils.

Ephedrine Oleate
Pharmaceutical
Stabilizer in making—
Colloidal solutions or organo-mercurials in mineral or vegetable oils.

Epichlorhydrin
Synonyms: Chloropropylene oxide, Glycid hydrochloride.
French: Épichlorhydrine.
German: Epichlorhydrin, Glycidchlorhydrat, Salzsäuresglycid.

Epichlorhydrin (Continued)

Cellulose Products
As a solvent miscible with—
Alcohols, aliphatic halogen derivatives, esters, ethers, ketones, plasticizers.
As a solvent immiscible with—
Petroleum hydrocarbons, water.
Powerful solvent for—
Cellulose esters, cellulose ethers.

Ceramic
Solvent in—
Compositions, containing natural or synthetic resins or cellulose esters or ethers, used as coatings for protecting and decorating ceramic products.

Chemical
As a solvent miscible with—
Alcohols, aliphatic halogen derivatives, esters, ethers, ketones.
As a solvent immiscible with—
Petroleum hydrocarbons, water.
Starting point in making—
Allyl alcohol.
Betachlorolactic acid.
Chlorohydroxypropylmalonamide.
Condensation products with anilin.
Condensation products with prussic acid.
Condensation products with salicylic acid.
Glyceryl dialkylethers, glyceryl diarylethers, ketolactonic acids, trichlorohydroxypropylamine.

Cosmetic
Solvent in—
Nail enamels and lacquers containing natural or synthetic resins or cellulose esters or ethers as base material.

Electrical
Solvent in—
Insulating compositions, containing natural or synthetic resins or cellulose esters or ethers, used for covering wire and in making electrical machinery and equipment.

Glass
Solvent in—
Compositions, containing natural or synthetic resins or cellulose esters or ethers, used in the manufacture of nonscatterable glass and as coatings for decorating and protecting glassware.

Glue and Adhesives
Solvent in—
Adhesive compositions containing natural or synthetic resins or cellulose esters or ethers.

Gums
Solvent for—
Gums.

Leather
Solvent in—
Compositions, containing natural or synthetic resins, or cellulose esters or ethers, used in the manufacture, of artificial leathers and as coatings for decorating and protecting leathers and leather goods.

Metal Fabricating
Solvent in—
Compositions, containing natural or synthetic resins or cellulose esters or ethers, used as coatings for protecting and decorating metallic articles.

Miscellaneous
Solvent in—
Coating compositions, containing natural or synthetic resins or cellulose esters or ethers, used for protecting and decorating various articles.

Paint and Varnish
Solvent in—
Paints, varnishes, lacquers, enamels, and dopes containing natural or synthetic resins or cellulose esters or ethers.

Paper
Solvent in—
Compositions, containing natural or synthetic resins or cellulose esters or ethers, used in the manufacture of coated papers and as coatings for decorating and protecting products made of paper or pulp.

Plastics
Solvent in making—
Plastics from or containing natural or synthetic resins or cellulose esters or ethers.

Resins
Solvent for—
Natural resins, synthetic resins.
Solvent in making—
Artificial resins from or containing cellulose esters or ethers.

Rubber
Solvent in—
Compositions, containing natural or synthetic resins or cellulose esters or ethers, used as coatings for protecting and decorating rubber goods.

Stone
Solvent in—
Compositions, containing natural or synthetic resins or cellulose esters or ethers, used as coatings for decorating and protecting artificial and natural stone.

Textile
Solvent in—
Compositions, containing natural or synthetic resins or cellulose esters or ethers, used in the manufacture of coated fabrics.

Wood
Solvent in—
Compositions, containing natural or synthetic resins or cellulose esters or ethers, used as protective and decorative coatings on woodwork.

Epiethylin

Cellulose Products
Solvent for—
Cellulose acetate, cellulose esters and ethers, cellulose nitrate (nitrocellulose).
For uses, see under general heading: "Solvents."

Chemical
Starting point in making—
Derivatives, especially amino-ethers, by the action of various bases.

Gums
Solvent for—
Artificial and natural gums of all sorts.

Resins and Waxes
Solvent for—
Copals, coumaroneresins, shellac.

Epiphenylin

Ceramics
Solvent in—
Compositions, containing various esters or ethers of cellulose, used to improve the water-resisting properties of coatings and decorations for ceramic products.

Chemical
Starting point in making—
Derivatives, especially amino derivatives.
Intermediates.

Electrical
Solvent in—
Insulating compositions, containing various esters or ethers of cellulose, as well as gums and resins, used to improve the water-resisting properties of the covering on electrical wires and electrical equipment and machinery.

Glass
Solvent in—
Compositions, containing various esters or ethers of cellulose, such as benzylcellulose and nitrocellulose, used in the manufacture of non-scatterable glass and for the decoration and protection of glassware.

Glues and Adhesives
Solvent in—
Adhesive compositions, containing various esters or ethers of cellulose, such as benzylcellulose and nitrocellulose, as well as gums and resins (added in order to increase the water-resisting properties of the product).

Gums
Solvent for—
Shellac, various gums.

Leather
Solvent in—
Compositions, containing various esters or ethers of cellulose, such as benzylcellulose and nitrocellulose, as well as gums and resins, used in the manufacture of artificial leather and for decorating and protecting leather goods (added to increase the water-resisting properties of the film).

Epiphenylin (Continued)

Metallurgical

Solvent in—

Compositions, containing various esters or ethers of cellulose, such as benzylcellulose and nitrocellulose, as well as gums and resins, used for the decoration and protection of metallic ware (added for the purpose of increasing the water-resisting properties of the film).

Miscellaneous

Solvent in—

Compositions, containing various esters or ethers of cellulose, as well as gums and resins, used for the decorations and protection of various compositions of matter (added to improve the water-resisting qualities of the film).

Paint and Varnish

Solvent in making—

Paints, varnishes, lacquers, dopes, and enamels containing various esters or ethers of cellulose, such as benzylcellulose and nitrocellulose, as well as gums and resins (added to improve the water-resistance of the film).

Paper

Solvent in—

Compositions, containing various esters or ethers of cellulose, as well as gums and resins, used in the manufacture of coated paper and for the decoration and protection of paper and pulp products (added to increase the water-resisting properties of the film).

Plastics

Solvent in making—

Compositions containing various esters or ethers of cellulose, such as cellulose nitrate and benzylcellulose, as well as gums and resins.

Rubber

Solvent in—

Compositions, containing various esters or ethers of cellulose, as well as gums and resins, used for the decoration and protection of rubber goods (added to increase the water-resisting properties of the film).

Stone

Solvent in—

Compositions, containing various esters or ethers of cellulose, as well as gums and resins, used for the decoration and protection of artificial and natural stone (added to improve the water-resistant properties of the film).

Textile

Solvent in—

Compositions, containing various cellulose esters or ethers, such as nitrocellulose and benzylcellulose, used in making coated textile fabrics.

Woodworking

Solvent in—

Compositions, containing various esters or ethers of cellulose, such as cellulose nitrate and benzylcellulose, as well as gums and resins, used for the decoration and protection of woodwork (added for the purpose of increasing the water-resistant properties of the film).

Ergothioneine

Chemical

Starting point in making—

Trimethylhistidin.

Pharmaceutical

In compounding and dispensing practice.

Erucic Acid Chloride

Chemical

Starting point (Brit. 407956) in making pour-point improvers for machine oils, gear oils, and other lubricants by condensing with—

Anilin, anthracene oil.

Aromatics obtained by destructive hydrogenation or by dehydrogenation.

Benzene.

Cracking gases containing gaseous olefins (ethylene, propylene, and butylene).

Cyclic terpenes, ethylnaphthalene, liquid olefins, middle oil, naphthalene, naphthols, napthylamines, nitrated aromatics, phenols, tars, toluene, xylene.

Erythritol Tetranitrate

French: Tétranitrate de érythritole.

German: Erythritoltetranitrat, Tetranitroerythritol.

Explosives

Ingredient (U. S. 1744693) of—

Detonating charges, together with fulminate of mercury, in blasting caps.

Esculin

Synonyms: Aesculin, Bicolarin, Polychrome.

French: Acide ésculinique.

German: Esculinsäure.

Perfumery

Ingredient of—

Ointments for protecting the skin against sunburn.

Pharmaceutical

In compounding and dispensing practice.

Rubber

Ingredient (Brit. 325312) of—

Rubber compositions.

Ethanolamine

Synonyms: 2-Hydroxyethylamine, Monoethanolamine.

German: Aethanolamin.

Chemical

Absorbent for—

Acid gases, carbon dioxide, hydrochloric acid in gaseous form, hydrogen sulphide, sulphur dioxide.

Absorbent in—

Recovering and purifying gases.

Amine useful as—

Moderately viscid liquid.

Base useful as—

Active chemically.

Somewhat stronger than ammonia.

Emulsifying agent (commonly used in the form of one of its soaps) with—

Fatty acids, oleic acid, stearic acid.

Solvent for—

Some organic substances.

Starting point in making—

Dispersing agents, emulsifying agents, soaps having valuable properties, various derivatives.

Substitute for—

Triethanolamine (q. s.) in applications where advantage can be taken of its lower combining weight.

Gases

Absorbent for—

Acid gases, carbon dioxide, hydrochloric acid in gaseous form, hydrogen sulphide, sulphur dioxide.

Absorbent in—

Recovering and purifying gases.

Miscellaneous

Emulsifying agent (commonly used in the form of one of its soaps).

Substitute for—

Triethanolamine in applications where advantage can be taken of its lower combining weight.

Ethanolamine Borate

Metallurgical

Absorbent (U. S. 1964808) for—

Hydrogen sulphide and carbon dioxide in extracting these gases from air or flue gas.

Ethanolamine Oleate

French: Oléate d'éthanolamine, Oléate éthanolaminique.

German: Aethanolaminoleat, Oelsäureaethanolaminester, Oelsäuresaethanolamin.

Ceramics

Stabilizing agent in making—

Aqueous suspensions of clay and finely divided mineral matter.

Chemical

Reagent in making—

Stable emulsions.

Fats and Oils

Ingredient of—

Drilling oils, in emulsified form, containing mineral oils and fatty oils, such as linseed oil and other ingredients such as glycerin or alcohol.

Grinding and cutting oils, in emulsified form, containing rosin soaps, mineral oils, and fatty oils.

Reagent in making—

Emulsions of animal or vegetable oils.

Miscible compositions of animal or vegetable oils.

Insecticide

Ingredient of—

Preparations containing mineral oils, alkali soaps, calcium caseinate, glue, copper hydroxide, ferric hy-

Ethanolamine Oleate (Continued)
droxide, miscible oils, vegetable oils, sulphonated mineral oils, phenol.

Miscellaneous
Ingredient of—
Spotting fluids containing mineral oils.
Special detergent compositions containing mineral oils and a solvent such as carbon tetrachloride or ethylene dichloride used for cleansing automobile bodies, parts of machinery and the like.
Reagent in making—
Stable emulsions of various substances.

Paint and Varnish
Ingredient of—
Compositions containing mineral oils, and such solvents as carbon tetrachloride or dichloroethylene, used for cleansing walls.
Stabilizer in making—
Paints, varnishes, enamels, lacquers (added for the purpose of obtaining a more stable suspension or emulsion of the pigment).

Perfume
Emulsifying agent and stabilizer in making—
Creams, lotions, ointments.

Petroleum
Ingredient of—
Lubricating compositions, in stabilizing emulsified form, containing mineral oils.
Reagent in making—
Mineral oil preparations, such as paraffin oil, miscible in water.
Reagent in refining—
Mineral oils and mineral oil distillates (added for the purpose of eliminating colloidal materials and materials held in suspension).

Resins and Waxes
Emulsifying agent and stabilizer in making—
Emulsions.

Soap
Emulsifying agent and stabilizer in making—
Dry-cleaning agents, containing mineral oil and such solvents as carbon tetrachloride or dichloroethylene.
Shaving soap and creams.

Textile
——, *Dyeing*
Wetting agent in making—
Dye baths (added for the purpose of obtaining better penetration of the color into the fabric and yarn).

——, *Finishing*
Ingredient of—
Finishing preparations, impregnating compositions.

——, *Manufacturing*
Ingredient of—
Wool-oiling compositions containing mineral oils and such solvents as carbon tetrachloride or dichloroethylene.

——, *Printing*
Ingredient (Brit. 302252) of—
Printing paste (added for the purpose of securing better penetration of the color into the fabric).

Woodworking
Ingredient of—
Cleansing compositions.

Ethanolamine Palmitate
French: Palmitate d'éthanolamine, Palmitate éthanolaminique.
German: Aethanolaminpalmoleat, Palmitinsäureaethanolaminester; Palmitinsäuresaethanolamin.

Miscellaneous
As an emulsifying agent.
For uses, see under general heading: "Emulsifying agents."

Ethanolamine Stearate
French: Stéarate éthanolaminique.
German: Aethanolaminstearat, Stearinsäuresaethanolamin, Stearinsäureaethanolaminester.

Ceramics
Stabilizing agent in making—
Aqueous suspensions of clay and finely divided mineral matter.

Chemical
Reagent in making—
Stable emulsions.

Fats and Oils
Ingredient of—
Drilling oils, in emulsified form, containing mineral oils and fatty oils, such as linseed oil, and other ingredients, such as glycerin or alcohol.
Grinding and cutting oils, in emulsified form, containing rosin soaps, mineral oils, and fatty oils.
Reagent in making—
Emulsions of animal or vegetable oils.
Miscible compositions of animal or vegetable oils.

Insecticide
Ingredient of—
Preparations containing mineral oils, alkali soaps, calcium caseinate, glue, copper hydroxide, ferric hydroxide, miscible oils, vegetable oils, sulphonated mineral oils, phenols.

Miscellaneous
Ingredient of—
Spotting fluids containing mineral oils.
Special detergent compositions containing mineral oils and a solvent, such as carbon tetrachloride or ethylene dichloride, used for cleansing automobile bodies, machine parts, and so on.
Reagent in making—
Stable emulsions of various substances.

Paint and Varnish
Ingredient of—
Compositions containing mineral oils, and such solvents as carbon tetrachloride or dichloroethylene, used for cleansing walls.
Stabilizer in making—
Paints, varnishes, enamels, lacquers, and the like (added for the purpose of obtaining a more stable suspension or emulsion of the pigment).

Perfume
Emulsifying agent and stabilizer in making—
Creams, lotions, ointments.

Petroleum
Ingredient of—
Lubricating compositions (added for the purpose of obtaining stable emulsions).
Reagent in making—
Mineral oil preparations, such as paraffin oil, miscible with water.
Reagent in refining—
Mineral oils and mineral oil distillates (added for the purpose of eliminating colloidal matters and materials held in suspension).

Resins and Waxes
Emulsifying agent and stabilizer in making—
Emulsions.

Textile
——, *Dyeing*
Wetting agent in making—
Dye baths (added for the purpose of obtaining better penetration of the color into the fabric and yarn).

——, *Finishing*
Ingredient of—
Finishing preparations, impregnating preparations.

——, *Manufacturing*
Ingredient of—
Wool-oiling preparations, containing mineral oils, and such solvents as carbon tetrachloride and dichloroethylene.

——, *Printing*
Ingredient (Brit. 302252) of—
Printing pastes (added for the purpose of securing better penetration of the color into the fabric).

Woodworking
Ingredient of—
Cleansing preparations.

Ethenylphenylenediamine
French: Éthenylephénylènediamine.
German: Aethenylphenylendiamin.

Dye
Starting point in making lakes with—
1-Amino-4-para-acetaminoacetanilidoanthraquinone-2-sulphonic acid.
Anthrapyrimidin-2-paratoluidoanthraquinone-2-sulphonic acid.
Azo dyestuffs.
1:4-Diamino-2-phenoxyanthraquinonesulphonic acid.
1:4-Dihydroxy-5:8-diparatoluidoanthraquinonedisulphonic acid.
1:5-Dihydroxy-5:8-diparatoluidoanthraquinonedisulphonic acid.

Ethenylphenylenediamine (Continued)
1:5-Diparatoluidoanthraquinonedisulphonic acid.
4:8-Diparatoluidoanthraquinonedisulphonic acid.
Dyestuffs derived from orthotoluidin and fluorescein chloride.
1-Hydroxy-5-paratoluidoanthraquinonesulphonic acid.
Methylanthrapyridin-2-arylsulphonic acids.
Paranitrophenylazosalicylic acid.
Patent blue A.
Sodium-1-amino-4-anilidoanthraquinone-2-sulphonate.

Ether
Synonyms: Ethoxyethane, Ethyl oxide, Ethylene hydrate, Hydric ether, Purified ether, Sulfuric ether.
Latin: Aether, Aether purificatus, Aether sulphuricus.
French: Éther hydrique, Éther officinal, Éther pur; Éther sulphurique, Éther vinique.
German: Aether, Reiner aether, Schwefelaether.
Spanish: Eter, Eter sulfurico.
Italian: Etere.

Agriculture
Stimulant for—
Plant growth.

Analysis
Extracting medium for various purposes in institutional, industrial research, and control work.
Solvent in the extraction and assay of—
Alkaloids, drugs.
Solvent in analyzing and testing—
Alkaloids, animal oils, breadstuffs, butter, cakes, cheese, chocolate, cocoa, essential oils, fats, flour, hops, meals, meat, milk, resins, rosin, rosin oil, soaps, vegetable oils.
Solvent in making—
Toxicological examinations.

Automotive
Degreasing agent for—
Automobile bodies, automobile parts.
Dewaxing agent in—
Manufacturing operations.

Ceramics
Solvent in—
Coating compositions, containing nitrocellulose as well as resins, waxes, and gums, used for protecting and decorating ceramic ware.

Chemical
Denaturant in—
Industrial alcohol.
Extractant for—
Acetic acid from crude pyroligneous acid, alkaloids, chemicals, drug principles.
Extractant in—
Purification of chemicals by extraction and crystallization.
Preparing catalysts for production of synthetic formic acid (Brit. 406244 and 406345).
Ingredient of solvent mixtures containing also—
Acetone, alcohol, benzene, turpentine.
Solvent for—
Nitrocellulose.
Solvent in making—
C. P. chemicals, drugs.
Emulsifying, wetting, and dispersing agents (French 750647).
Inorganic chemicals, intermediates, organic chemicals, pharmaceuticals, U. S. P. chemicals.

Dry Cleaning
Ingredient of—
Grease spot removing creams.
Solvent in—
Removing oils, fats, waxes, gums, resins, and other stains and impregnated substances.
Spotting agent for—
Textiles and hats.

Dye
Reagent and solvent in making synthetic dyestuffs of various classes.

Electrical
Solvent in—
Cleaning electric motors and other electrical machinery.
Compositions, containing nitrocellulose and, at times, resins, gums, and the like, used for insulating cables, wiring, and electrical machinery and equipment.

Explosives
Solvent for—
Nitrocellulose.

Fats, Oils, and Waxes
Extractant for—
Animal oils, essential oils, fats, greases, vegetable oils.
Solvent for—
Animal oils, essential oils, fats, greases, vegetable oils, waxes.
Solvent in—
Recovering oils from fuller's earth and other substances used in bleaching.

Fertilizer
Solvent in—
Degreasing fish scrap.

Food
Extractant of soluble substances from—
Berries, fruits, seeds.
Ingredient of—
Nonalcoholic vanilla flavor, containing also vanillin, coumarin, glycerin, syrup, color, and water.
Solvent for—
Fats, oils.

Glass
Solvent in—
Degreasing glass.
Compositions, containing nitrocellulose and artificial or natural resins, waxes, and gums, used in the manufacture of nonscatterable glass and for the decoration and protection of glassware.

Glues and Adhesives
Ingredient of—
Special adhesive compositions, containing also nitrocellulose, or gums, resins, oils, or waxes.
Solvent in—
Degreasing bones and hides preparatory to the manufacture of glue and gelatin.

Gum
Solvent for various gums.

Insecticide
Ingredient (U. S. 1954517) of—
Insecticidal composition.

Leather
Solvent in—
Cleansing spotted leathers.
Removing natural oils and greases from hides and skins before tanning, so as to prevent staining thereafter and insure evenness of the leather finish and tan.
Compositions, containing nitrocellulose, as well as artificial or natural resins, gums, and waxes, used in the manufacture of artificial leather and for the protection and decoration of leather goods.

Mechanical
Ingredient (Brit. 411904) of—
Fuel-addition agent containing also acetone and ammonia.
Priming agent for—
Internal combustion motors.
Solvent in—
Cleansing and degreasing machinery of various sorts.
Cleansing drive wheels of compression pumps and other mechanical equipment.
Degreasing automobile brakebands.

Metallurgical
Solvent in—
Cleansing and degreasing metallic surfaces preparatory to painting or other coating.
Degreasing die-castings, metal stampings, metals to be electroplated, nuts and bolts.
Preparing metals for pickling, plating, shellacking, sherardizing, varnishing.
Solvent and diluent in—
Compositions, containing nitrocellulose, or gums, resins, or waxes, used for protecting and decorating metallic articles.

Miscellaneous
As a general solvent.
Degreasing agent in treating—
Furs (also acts as parasiticide), hats.
Ingredient of—
Biological fixing fluids, compositions of clay, for cleansing ivory, horn, and bone.
Preparations used for the removal of stains from celluloid articles.
Preparations used for cleansing typewriters.
Solvent and diluent in—
Compositions, containing nitrocellulose, or gums, resins, or waxes, used for decorating and protecting various articles.

Ether (Continued)

Oilcloth and Linoleum
Solvent in making—
Coating compositions containing nitrocellulose, gums, resins, or waxes.

Paint and Varnish
Solvent in—
Paints, varnishes, lacquers, enamels, and dopes containing nitrocellulose, oils, waxes, gums, and resins.

Paper
Solvent in—
Removing oil from paper and paperstock.
Compositions, containing nitrocellulose with gums, waxes, and natural or artificial resins, used in the manufacture of coated paper and for coating and decorating paper and pulp products.

Perfume
Solvent in—
Extracting aromatic principles from flowers, particularly those alterable by heat.
Solvent in making—
Nail enamels and lacquers containing nitrocellulose as a base material.

Petroleum
Solvent in—
Degreasing light mineral oils.
Extracting wax from mineral oil distillates.

Pharmaceutical
As an anesthetic.
Ingredient of—
Proprietary preparations.
In compounding and dispensing practice.

Photographic
Solvent in—
Cleansing and degreasing motion picture film.
Preparing squeegee plates.
Solvent in making—
Films from nitrocellulose, photographic emulsions.

Plastics
As a degreasing solvent.
Extractant (Brit. 394244) for—
Retained softeners and solvents in sheets and films made from polymerized polyvinyl chlorides (mechanical properties of the sheets are improved by this extraction).
Solvent and diluent in making—
Compositions containing nitrocellulose with gums, waxes, and artificial or natural resins.
Laminated fiber products, molded products.

Printing
Solvent in—
Photoengraver's collodion.
Solvent in cleansing—
Engraved plates, lithographic stones, printing machinery, type.

Resins
Solvent for resins of various kinds.

Rubber
Solvent in—
Coating compositions, containing nitrocellulose, with gums and waxes, used for decorating and protecting rubber goods.

Soap
Ingredient of—
Dry-cleaning compositions, spotting fluids.
Solvent for—
Fats, oils.
Solvent (Brit. 388485) in making—
Sulphonated cleansing and emulsifying agents from the unsaturated alcohols which are produced by removing water from 7:18-stearicglycol.

Stone
Solvent in—
Compositions containing nitrocellulose, with artificial or natural resins, gums, and waxes, used for the decoration and protection of artificial and natural stone.

Sugar
Solvent in—
Extracting waxes from filter press "mud" in refining.

Textile
——, Finishing
Solvent in—
Coating compositions containing nitrocellulose.

——, Manufacturing
Solvent in—
Cleaning knitting machine needles, cleaning silk and silk hosiery, degreasing textiles, degreasing wool, degumming silk, preparing nitrocellulose.
Solvent and diluent in making—
Compositions, containing nitrocellulose, used for making coated textiles.
Scouring compositions.

Tobacco
Solvent in—
Extracting nicotine.

Woodworking
Solvent in—
Compositions, containing nitrocellulose, gums, resins, and waxes, used for decorating and protecting woodwork.
Plastic compositions, containing nitrocellulose, used for many filling and repairing purposes on wood.

1-Ethinylcyclohexanol
German: 1-Aethinylcyklohexanol.
Chemical
Starting point in making—
Cyclohexylideneacetaldehyde (Brit. 267954).

Ethinyldimethylcarbinol
Chemical
Starting point in making—
Isopropyleneacetaldehyde (Brit. 267954).

Ethinylmethylethylcarbinol
German: Aethinylmethylaethylcarbinol.
Chemical
Starting point in making—
Secondary butylideneacetaldehyde.

Ethinylmethylphenylcarbinol
German: Aethinylmethylphenylcarbinol.
Chemical
Starting point in making—
Betamethylcinnamic aldehyde (Brit. 267954).

2-Ethoxy-5-acetylaminodiphenyl
Disinfectant
Intermediate (U. S. 2073683) in making—
Bactericides.

2-Ethoxybenzeneazoalphanaphthylamine
German: 2-Aethoxybenzolazoalphanaphtylamin.
Dye
Starting point (Brit. 263164) in making azo dyestuffs with sulphonated derivatives of—
Anilide, anilide 5-chlorotoluidide, anilide 5-chloro-2-anisidide, anilide betanaphthylamide, anilide 2:3-oxynaphthoic acid.

Ethoxybenzidin
German: Aethoxybenzidin.
Dye
Starting point in making—
Diamin blue B.

2-Ethoxy-5-chlor-2′:4′-diaminoazobenzene Hydrochloride
Disinfectant
Claimed (U. S. 2009086) to be—
Bactericide.

Ethoxynitrochloroacridin
German: Aethoxynitrochloracridin.
Chemical
Reagent (Brit. 283510) in making bactericidal compositions with—
Alphabetadiethylaminoethylamine-2-hydroxy-3-para-aminophenylaminopropane.
Alphadiethylamino-2-hydroxy-3-(para-aminophenylamino)propane.
Gammadiethylaminobetahydroxypropylamine.
Normal diethyl-N′-(para-aminophenyl)ethylenediamine.

4-Ethoxyphenylmalonamic Acid
French: Acide 4-éthoxyphénylemalonamique.
German: 4-Aethoxyphenylmalonaminsäure.
Spanish: Acido 4-etoxifenilomalonamico.
Italian: Acido 4-etossifenilomalonamico.
Chemical
Starting point in making—
Pharmaceutical caffeine compounds soluble in water.
Pharmaceutical
In compounding and dispensing practice.

Ethoxyquinaldin Ethiodide
French: Éthiodure éthoxyquinaldinique.
German: Aethoxychinaldinaethjodid.

Insecticide
Starting point (German 438241) in making—
Fungicide and bactericide from glyoxal for treating diseased seeds.

Ethyl Abietate
Synonyms: Ethyl resinate.
French: Abietate d'éthyle, Abietate éthylique, Résinate d'éthyle, Résinate éthylique.
German: Harzsäureaethylester, Harzsäuresaethyl.

Cellulose Products
Plasticizer (Brit. 313133) for—
Cellulose acetate, cellulose esters and ethers, cellulose nitrate.
For uses, see under general heading: "Plasticizers."

Ethyl Acetate
Synonyms: Acetic ester, Acetic ether, Vinegar naphtha.
Latin: Aether aceticus.
French: Acétate d'éthyle, Acétate éthylique, Éther acétique, Éther ethylacétique, Naphthe acétique.
German: Aethylacetat, Aethylazetat, Aethylaether, Essigaether, Essignaphta, Essigsäureaethylester, Essigsäuresaethyl, Essigsäuresaethyloxyd.
Spanish: Acetato de etil.
Italian: Acetato di etile.

Analysis
Solvent and reagent for laboratory use.

Ceramic
Solvent in—
Compositions, containing nitrocellulose or other esters or ethers of cellulose, used for coating and protecting ceramic ware.

Chemical
Reagent in—
Concentrating acetic acid (French 665412).
Dehydrating alcohol by rectification (French 558875).
Extracting various organic acids from dilute solutions, to obtain concentrated products (used along with benzene in admixture for extracting acetic acid, butyric acid, propionic acid, and other aliphatic acids) (Brit. 302174).
Making catechin.
Solvent for—
Phosgene, pyroxylin.
Various chemicals and chemical bodies.
Solvent in making—
Ketene.
Starting point in making—
Acetamide, acetoacetic ester, acetylethylamide, dimethyl ketone, intermediates, methylheptenone, organic chemicals, synthetic perfumes, synthetic pharmaceuticals.

Dye
Reagent in making various dyestuffs.
Solvent in separating various dyestuffs.

Electrical
Solvent in—
Compositions, containing nitrocellulose or other esters or ethers of cellulose, used for insulating purposes and in the manufacture of electrical machinery and equipment.

Explosives
Solvent in making—
Guncotton, smokeless powder, various explosive compositions.

Fats and Oils
Solvent for various fats and oils.

Food
Flavoring in—
Bakery products, beverages, candies.
Ingredient of—
Apple flavoring compositions.
Artificial fruit essences and flavors.
Flavoring compositions.
Fruit essences (to produce aromatic odor and flavor).
Peach flavor, strawberry flavor, yellow plum flavor.
Reagent in extracting—
Caffeine from coffee.

Glues and Adhesives
Solvent in making—
Adhesive preparations containing nitrocellulose, cellulose acetate, or other esters or ethers of cellulose.

Insecticide
Ingredient (Brit. 234456) of—
Insecticidal compositions containing carbon tetrachloride, used for the fumigation of wheat and destruction of weevils.

Leather
Solvent in—
Compositions, containing cellulose acetate, nitrocellulose, or other esters or ethers of cellulose, used in the manufacture of artificial leathers and for coating and decorating leathers and leather goods.

Metallurgical
Solvent in—
Compositions, containing cellulose acetate, nitrocellulose, or other esters or ethers of cellulose, used for decorating and protecting metallic articles.

Military
Waterproofing agent in—
Filling hand grenades.

Miscellaneous
Reagent in making—
Artificial bristles, artificial horsehair.
Solvent in—
Compositions, containing cellulose acetate, nitrocellulose, or other esters or ethers of cellulose, used for decorating and protecting various articles.

Paint and Varnish
Ingredient of—
Brushing lacquers (U. S. 1744085).
Paint and varnish removers.
Solvent in making—
Lacquers and varnishes with synthetic resins of the vinyl ester type (Brit. 312049).
Lacquers, varnishes, paints, dopes, and enamels containing cellulose acetate, nitrocellulose, or other esters or ethers of cellulose.

Paper
Solvent in—
Compositions, containing cellulose acetate, nitrocellulose, or other esters or ethers of cellulose, used in the manufacture of coated papers and for decorating and protecting products made from paper or pulp.

Perfume
Ingredient of—
Cosmetics, perfumes.

Pharmaceutical
As a solvent for various purposes.
In compounding and dispensing practice.

Photographic
Solvent in making—
Films from cellulose acetate, nitrocellulose, or other esters or ethers of cellulose.

Plastics
Solvent in making—
Colloidal cements.
Plastic products containing cellulose acetate, nitrocellulose, or other esters or ethers of cellulose.

Resins and Waxes
Solvent for various resins and waxes.

Rubber
Solvent for—
Removing resinous matters from balata gum and guttapercha.
Solvent in—
Compositions, containing cellulose acetate, nitrocellulose, or other esters or ethers of cellulose, used for decorating and protecting rubber goods.

Stone
Solvent in—
Compositions, containing cellulose acetate, nitrocellulose, or other esters or ethers of cellulose, used for decorating and protecting artificial and natural stone.

Textile
——, *Dyeing*
Reagent (Brit. 308605) in preparing—
Woolen fabrics for dyeing.

——, *Finishing*
Solvent in—
Cleansing operations.

——, *Manufacturing*
Solvent in—
Compositions, containing cellulose acetate, nitrocellulose, or other esters or ethers of cellulose, used in making coated textiles.

Ethyl Acetate (Continued)
Solvent in making
Rayon yarns.

Woodworking
Solvent in—
Compositions, containing cellulose acetate, nitrocellulose, or other esters or ethers of cellulose, used for decorating and protecting woodwork.

Ethyl Acetoacetate
Synonyms: Acetoacetic ether, Diacetic ester, Diacetic ether.
French: Acétoacétate d'éthyle, Acétoacétate éthylique, Acétylacétate d'éthyle, Acétylacétate éthylique, Acide éthyldiacétique, Éther acétoacétique.
German: Acetessigester, Acetessigsäureaethylester, Acetessigsäuresaethyl.
Spanish: Acetilacetato de etil.
Italian: Acetilacetato di etile.

Ceramics
Solvent in—
Compositions, containing nitrocellulose, used for the decoration and protection of ceramic ware.

Chemical
Solvent for—
Nitrocellulose.
Starting point in making—
Acetoacetic amide, acetoacetic anilide, acetoacetic naphthylide, acetoacetic phenylamide, acetoacetic toluide, acetoacetic xylidide, acids with strong alkalies, amidopyrine, amino-acids with ammonia, antipyrine, chloro-acids with phosphorus pentachloride, dehydroacetic acid, diaceticsuccinic acid by hydrolysis, diethylmalonic ether, dimethylglyoxime, fatty acids, hydropyridins with aldehydes and ammonia, hydroxy acids, ionone, jasmone, ketohydrobenzenes with aldehydes, ketones with dilute alkalies, methylheptenone, nitriles with hydrocyanic acid, organic chemicals, parasulphonphenyl-3-methylpyrazolone, pharmaceuticals, 1-phenyl-5-methyl-3-pyrazolone, ring compounds, salipyrin, uracels with urea, various aromatic chemicals, various derivatives by acetoacetic synthesis.

Dye
Starting point in making—
Anthracene yellow, azo colors of the pyrazolone series, coumarins with phenols, cumarins with quinones, dianil yellow 3G, dianil yellow 2R, fast light yellow, flavazin L, pyrimidins with anisidins, pyridins, pyrones, quinolins.
Various dyestuffs of the phenylpyrazolone derivatives class.
Xylene light yellow, xylene yellow 3G.

Food
Ingredient of—
Fruit essences.

Glass
Solvent in—
Compositions, containing nitrocellulose, used in the manufacture of nonscatterable glass and for the decoration and protection of glassware.

Leather
Solvent in—
Compositions, containing nitrocellulose, used in the manufacture of artificial leather and for the protection and decoration of leather goods.

Metallurgical
Solvent in—
Compositions, containing nitrocellulose, used for the decoration and protection of metallic articles.

Miscellaneous
Solvent in—
Compositions, containing nitrocellulose, used in the decoration and protection of various articles.

Paint and Varnish
Solvent in making—
Lacquers, enamels, dopes, varnishes, and paints, containing nitrocellulose.

Paper
Solvent in—
Compositions, containing nitrocellulose, used for the decoration and protection of paper and pulp products and in the manufacture of coated paper.

Perfume
Ingredient of—
Eau de cologne, perfume compositions, toilet water (to lend a freshness to the odor).

Plastics
Solvent in making—
Plastic compositions containing nitrocellulose.

Rubber
Solvent in—
Compositions, containing nitrocellulose, used for the decoration and protection of rubber goods.

Stone
Solvent in—
Compositions, containing nitrocellulose, used for the decoration and protection of natural and artificial stone.

Textile
Solvent in—
Compositions, containing nitrocellulose, used in the manufacture of coated fabrics.

Ethyl Acetylglycollate
Synonyms: Acetylglycollic ether.
French: Acétyleglycollate d'éthyle, Acétyleglycollate éthylique, Éther d'acétyleglycollique.
German: Acetylglykolsäureaethylester, Acetylglykolsäuresaethyl, Aethylacetylglykolat, Aethylazetylglycolat.

Cellulose Products
Solvent for—
Cellulose acetate, cellulose esters and ethers, cellulose nitrate.
For uses, see under general heading: "Solvents."

Chemical
Starting point in making various derivatives.

Fats and Oils
Solvent for certain fats and oils.

Ethyl Acrylate
Synonyms: Acrylic acid ethyl ester.
French: Acrylate d'éthyle, Acrylate éthylique, Éther acrylique.
German: Acrylsäureaethylester, Acrylsäuresaethyl, Aethylacrylat.

Cellulose Products
Plasticizer (Brit. 321258) for—
Cellulose acetate, cellulose esters and ethers, cellulose nitrate, rubber.
For uses, see under general heading: "Plasticizers."

Chemical
Starting point in making various derivatives.

Ethyl Adipate
Synonyms: Ethyl adipinate.
French: Adipate d'éthyle, Adipate éthylique, Adipinate d'éthyle, Adipinate éthylique.
German: Adipinsäureaethylester, Adipinsauresaethyl, Aethyladipat, Aethyladipinat.
Spanish: Adipato de etil.
Italian: Adipato di etile.

Cellulose Products
Plasticizer for—
Cellulose acetate, cellulose esters and ethers, cellulose nitrate.
For uses, see under general heading: "Plasticizers."

Chemical
Starting point in making—
Esters, salts.

Ethyl Alphacrotonate
Synonyms: Alphacrotonic ethyl ester.
French: Alphacrotonate d'éthyle, Alphacrotonate éthylique.
German: Alphacrotonaethylester, Alphacrotonsäureaethylester, Alphacrotonsäuresaethyl, Aethylalphacrotonat.

Cellulose Products
Plasticizer (Brit. 321258) for—
Cellulose acetate, cellulose esters and ethers, cellulose nitrate, rubber.
For uses, see under general heading: "Plasticizers."

Chemical
Starting point in making various derivatives.

Ethyl Aminoacetate
French: Aminoacétate éthylique.
German: Aethylaminoacetat, Aminoessigsaeuresaethylester.

Paint and Varnish
Ingredient of—
Cellulose acetate lacquers and varnishes, added for stabilizing purposes (Brit. 243722).

7-Ethylaminoalphanaphtholsulphonic Acid
French: Acide de 7-Éthyleaminoalphanaphthole, 7-Éthyleaminoalphanaphtholique.
German: 7-Aethylaminoalphanaphtolsulfonsaeure, Sulfonsaeure-7-aethylaminoalphanaphtolester.

Dye
Starting point in making—
Diphenyl blue black.

4-Ethylaminobetahydroxyethylaminoanilin

Dye
Starting point (Brit. 447905, 447906, and 448016) in making—
Monoazo dyes for leather, particularly chrome leather.

Ethyl Aminoformate
French: Aminoformiate éthylique.
German: Aethylaminoformat, Aminoameisensaeuresaethlyester.

Paint and Varnish
Ingredient of—
Cellulose acetate lacquers and varnishes, added for stabilizing purposes (Brit. 243722).

Ethylanilinmetasulphonic Acid

Miscellaneous
As an emulsifying agent (Brit. 341053).
For uses, see under general heading: "Emulsifying agents."

5-Ethyl-5-anilinobarbituric Acid Hydrochloride

Pharmaceutical
Suggested for use (Brit. 414293) as—
Hypnotic with low toxic properties.

5-Ethylbarbituric Acid Hydrochloride

Pharmaceutical
Suggested for use (Brit. 414293) as—
Hypnotic with low toxic properties.

Ethylbenzoylhydroquinone

Petroleum
Stabilizing agent (Brit. 406195) for—
Cracked gasolines and other motor fuels.

Ethylbenzoylphloroglucinol

Petroleum
Stabilizing agent (Brit. 406195) for—
Cracked gasolines and other motor fuels.

Ethylbenzoylpyrocatechol

Petroleum
Stabilizing agent (Brit. 406195) for—
Cracked gasolines and other motor fuels.

Ethylbenzoylpyrogallol

Petroleum
Stabilizing agent (Brit. 406195) for—
Cracked gasolines and other motor fuels.

Ethylbenzoylresorcinol

Petroleum
Stabilizing agent (Brit. 406195) for—
Cracked gasolines and other motor fuels.

Ethylbenzylanilin
German: Aethylbenzylanilin.

Dye
Starting point (Brit. 265767) in making monoazo dyestuffs with—
Ammonium 5-nitro-2-aminobenzylsulphonate.
3:5-Dinitro-2-aminobenzylsulphonic acid.
Starting point in making—
Acidol green, brilliant acid blue A, formyl violet.

Ethylbenzylanilinsulphonic Acid
French: Acide d'éthylebenzylanilinesulphonique.
German: Aethylbenzylanilinsulfosaeure.

Dye
Starting point in making—
Acid violet 5B, azo cardinal, benzyl green B, erioglaucin A, B, G, BB, JJ, RB extra, P. V. super X conc; formyl violet 5BN, Guinea green B, night green, patent green AGL.

Ethylbeta-amylbarbituric Acid

Pharmaceutical
Ingredient (U. S. 1928346) of—
Anesthetic composition for rectal administration in obstetrics, containing also mineral oil and ethyl ether.

Ethylbetabutoxyethyl Sebacate

Cellulose Products
Plasticizer (U. S. 1991391) for—
Cellulose esters and ethers.
For uses, see under general heading: "Plasticizers."

Ethylbetagammadihydroxypropylanilin, Normal

Chemical
Reagent in—
Organic synthesis.

Dye
Coupling agent (Brit. 421975) in making—
Light-fast and readily discharged red-violet dyestuffs for acetate rayon with diazotised 6-bromo-2:4-dinitroanilin or 6-chloro-2:4-dinitroanilin.

4-Ethylbetahydroxyethylaminoanilin

Dye
Starting point (Brit. 447905, 447906, and 448016) in making—
Monoazo dyes for leather, particularly chrome leather.

Ethylbetahydroxyethylparaphenylenediaminesulphonic Acid

Dye
Starting point (Brit. 447905, 447906, and 448016) in making—
Monoazo dyes for leather, particularly chrome leather.

Ethylbetanaphthylamine
German: Aethylbetanaphtylamin.

Chemical
Starting point in making—
Developer B for primulin dyestuffs.

Dye
Starting point (Brit. 265767) in making monoazo dyestuffs with—
3:5-Dinitro-2-aminobenzylsulphonic acid.
3:5-Dinitro-4-aminobenzylsulphonic acid.
Starting point in making—
Primulin bordeaux.

Textile
—, *Dyeing*
Developing agent in dyeing with polychromin colors.

Ethylbetaparatoluenesulphonylethyl Sulphide

Chemical
Intermediate (Brit. 444262 and 444501) in—
Organic syntheses.

Insecticide
Insecticide (Brit. 444262 and 444501) for—
Animal pests, vegetable pests.

Textile
As a dyestuff (when employing suitable initial materials) (Brit. 444262 and 444501).
Assistant (Brit. 444262 and 444501) in—
Textile processing.

Ethyl Betaphenylmethylglycidate
French: Bétaphénylemétylhyleglycidate d'éthyle, Bétaphénylemétylhyleglycidate éthylique.
German: Aethylbetaphenylmethylglycidat, Betaphenylmethylglycinsäureaethylester, Betaphenylmethylglycidinsäuresaethyl.
Spanish: Betafenilmetilglicidato de etil.
Italian: Betafenilmetilglicidato di etile.

Food
Used in various food preparations and flavors to give them a strawberry taste.

Perfume
Ingredient of—
Perfume compositions (added for the purpose of freshening the odor).

Ethyl Borate
French: Borate d'éthyle, Borate éthylique.
German: Aethylborat, Borsäureaethylester, Borsäuresaethyl.
Spanish: Borato de etil.
Italian: Borato di etile.

Petroleum
Ingredient (Brit. 334181) of—
Motor fuels (added to prevent knock).

Ethylbourbonal
French: Bourbonale el d'éthyle, Bourbonale éthylique.
German: Aethylbourbonal.

Chemical
Starting point in making—
Aromatic chemicals.

Perfume
Ingredient of—
Artificial perfumes.
Perfume in—
Cosmetics.

Soap
Perfume in—
Toilet soaps.

Ethyl Bromide
Synonyms: Bromic ether, Hydrobromic ether.
Latin: Aethyliumbromatum.
French: Bromure d'éthyle, Bromure éthylique.
German: Aethylbromid, Bromaethyl.
Spanish: Bromuro de etil.
Italian: Bromuro di etile.

Chemical
Reagent in—
Organic synthesis.
Reagent in making—
Pharmaceutical chemicals.
Starting point (Brit. 207499) in making—
Vulcanizing accelerators with ethyl iodide and hexamethylenetetramine.

Dye
Reagent in making—
Dyestuffs.

Miscellaneous
Starting point (French 636714) in making—
Chemical fire-extinguishers by admixture with carbon tetrachloride.

Pharmaceutical
Suggested for use as—
Local anesthetic.

Refrigeration
As a refrigerant.

Ethyl Bromoacetate
French: Bromoacétate d'éthyle.
German: Aethylbromoacetat, Bromessigsaeuresaethyl.

Chemical
Starting point in making—
Ethyl acetoacetate, ethyl acetosuccinate, ethyl citrate, ethyl gammabromoacetate, ethylphenylethylglycin, methyl duodecaldehyde, methyl nonylaldehyde.

Ethylbutenylanilin, Normal

Chemical
Starting point in making—
Intermediates and other derivatives.

Insecticide
As an insecticide, alone and in compositions (Brit. 313934).

Soap
Ingredient (Brit. 313934) of—
Insecticidal and germicidal soaps.

5:5-Ethylbutylbarbituric Acid
French: Acide de 5:5-éthylebutylebarbiturique.
German: 5:5-Aethylbutylbarbiturinsaeure.

Chemical
Starting point (Swiss 113251) in making synthetic drugs with—
Allylamine, amylamine, butylamine, diallylamine, diamylamine, dibutylamine, diethylamine, dimethylamine, dipropylamine, isoallylamine, isoamylamine, isobutylamine, isopropylamine.

Ethylbutyl Carbonate
Synonyms: Butylethyl carbonate.
French: Carbonate de butyle et d'éthyle, Carbonate butylique-éthylique, Carbonate d'éthyle et de butyle, Carbonate éthylique-butylique.
German: Aethylbutylkarbonat, Butylaethylkarbonat, Kohlenstoffsäureaethylbutylester, Kohlenstoffsäurebutylaethylester, Kohlenstoffsäuresaethylbutyl, Kohlenstoffsäuresbutylaethyl.

Chemical
Starting point in making various derivatives.

Miscellaneous
Solvent for—
Cellulose esters and ethers, cellulose nitrate, natural resins, synthetic resins.
For uses, see under general heading: "Solvents."

Ethylbutylmetatoluidin

Dye
Starting point (Brit. 439815 and 417014) in making—
Blue dyestuffs by condensing with (1) a 4:4′-dihalogeno- or 4:4′-dialkoxybenzophenone (2) a primary 4-alkoxy- or 4-aryloxyarylamine and sulphonating the product.
Greenish-blue dyestuffs by condensing with (1) a 4:4′-dihalogeno- or 4:4′-dialkoxybenzophenone (2) a primary 4-alkoxy- or 4-aryloxyarylamine and sulphonating the product.

Ethyl Butyrate
Synonyms: Butyric ester, Butyric ether, Oil of pineapple (artificial).
French: Butyrate d'éthyle, Butyrate éthylique, Éther butyrique, Huile d'ananas artificielle.
German: Aetherbutyl, Aethylbutyrat, Butylsäureaethylester, Butylsäuresaethyl, Synthetisches ananasoel, Synthetische fichtenapfeloel, Synthetisches fichtenzapfenoel.

Chemical
As a starting point in making—
Aromatics, intermediates, other organic chemicals.

Food
Flavoring agent in making—
Candies, desserts.
Ingredient of—
Flavorings, liqueurs.

Perfumery
Ingredient of—
Artificial odors.
Perfume in—
Cosmetics, mouth washes.

Soap
Perfume in—
Toilet soaps.

Ethylcellulose
French: Cellulose de éthyle, Cellulose éthylique.
German: Aethylzellulose.
Spanish: Cellulosa de etil.
Italian: Cellulosa di etile.

Adhesives
Ingredient of—
Heat-sealing adhesives.

Ceramic
Ingredient of—
Coating compositions, containing artificial resins and used for the decoration and protection of ceramic products.

Construction
Ingredient of—
Coating compositions, containing artificial resins and the like, used for the protection of brickwork and other construction.

Electrical
Ingredient of—
Coating compositions, containing artificial resins and the like, used for insulating electrical apparatus, wire, and other articles.

Glass
Ingredient of—
Compositions, containing artificial resins and the like, used in making nonscatterable glass and for coating glassware.

Leather
Ingredient of—
Compositions, containing artificial resins and the like, used in the manufacture of artificial leathers and for coating leather goods.

Metallurgical
Ingredient of—
Coating compositions, containing artificial resins and the like, used for decorating and protecting metalware.
Thermoplastic coatings.

Miscellaneous
As a transparent wrapping film.
As a waterproofing agent.
Ingredient of—
Moulding powders.

Ethylcellulose (Continued)

Paper
Ingredient of—
Compositions, containing artificial resins and the like, used in the manufacture of coated papers and in the coating of paper and pulp products for protective and decorative purposes.

Paint and Varnish
Ingredient (used together with artificial resins and the like) of—
Dopes, lacquers, enamels, paints, priming compositions, varnishes, wax finishes.

Plastics
Ingredient of—
Compositions, containing artificial resins and the like.
Moulding powders.

Rubber
Ingredient of—
Compositions, containing artificial resins and the like, used for coating rubber goods.

Stone
Ingredient of—
Compositions, containing artificial resins and the like, used for coating artificial and natural stones.

Textile
Ingredient of—
Compositions, containing artificial resins and the like, used in making coated textiles.
Waterproofing agent for—
Fabrics.

Wood
Ingredient of—
Compositions, containing artificial resins and the like, used as coatings for decorating and protecting wood.
Wax polishes.

Ethyl Chloride
Synonyms: Chloroethyl, Hydrochloric ether, Monochlorethane.
Latin: Aether chloratus, Aethylis chloridum, Ethyl chloridum, Ethylum chloratum.
French: Chlorure d'éthyle, Éther hydrochlorique.
German: Aethylchlorid, Chloraethyl, Chlorwasserstoffaether.

Analysis
As a reagent.

Chemical
Reagent in making—
Synthetic organic chemicals for pharmaceutical and other purposes.
Solvent for—
Phosphorus, sulphur, various products.
Starting point (U. S. 1907701) in making—
Tetraethyl lead.

Dye
Reagent in making various synthetic dyestuffs.

Fats, Oils and Waxes
Solvent for—
Fats, mixed oils, volatile oils, waxes.

Insecticide
Ingredient of—
Insecticidal preparations.

Miscellaneous
As a general solvent.

Pharmaceutical
Suggested for use as—
General anesthetic, local anesthetic.

Refrigeration
As a refrigerant.

Resins
As a solvent.

Ethyl Chloroacetate
French: Chloroacétate d'éthyle.
German: Aethylchlorazetat, Chloressigsaeuresaethyl.

Dye
Reagent (Brit. 263898) in making vat dyestuffs from—
Dibenzanthrone, dimethoxydibenzanthrone, flavanthrone, indanthrone, indigo.

Ethylchloroformic Acid
French: Acide d'éthylechloroformique.
German: Aethylchlorameisensaeure.

Chemical
Starting point in making—
Quinine ethylcarbonate.

Ethyl Chlorosulphonate
French: Chlorosulphonate éthylique.
German: Chlorsulfosaeuresaethyl.

Chemical
Reagent in making—
Sodium compound of glutaconaldehyde (German 438009).

Ethyl Cinnamate
Synonyms: Cinnamic ether, Ethyl cinnamylic ester.
Latin: Aether cinnamylicus.
French: Cinnamate de éthyle, Cinnamate éthylique.
German: Aethylcinnamat.

Cellulose Products
Solvent and plasticizer (Brit. 321258) for—
Cellulose acetate, cellulose esters and ethers, cellulose nitrate.
For uses, see under general heading: "Solvents."

Food
Ingredient of—
Cherry flavor, fruit essences.

Miscellaneous
Ingredient (Brit. 321258) of—
Compositions, containing rubber and cellulose acetate, nitrocellulose, or other esters or ethers of cellulose, used for decorating and protecting various materials.

Perfume
Ingredient of—
Eau de cologne, perfumes (used to produce a "sweet" effect).
Perfume in—
Cosmetics.

Ethyl Cinnamate Bromonitrate Derivative

Petroleum
Primer (Brit. 436027) for—
Diesel engine fuels (lowers ignition point).

4-Ethyl-5:7-dichloro-oxynaphthene
German: 4-Aethyl-5:7-dichlorhydroxynaphten.

Dye
Starting point (Brit. 274527) in making thioindigo dyestuffs with—
6-Chloro-7-methylisatin chloride.
5:7-Dibromoisatin arylide.
5:7-Dibromoisatin chloride.
5:7-Dichloroisatin arylide.
5:7-Dichloroisatin chloride.
Isatinalpha anilide.

Ethyl Dicresylphosphate

Cellulose Products
Solvent for—
Cellulose esters or ethers.
For uses, see under general heading: "Solvents."

Ethyldihydrocollidin Dicarboxylate

Photographic
Plate emulsions to the extreme ultra violet.

Ethyldihydrocuprein Ether
French: Éther d'éthyledihydrocupreine.
German: Aethyldihydrocupreinaether.

Chemical
Starting point (Brit. 282356) in making antiparisitic agents with—
Apocholic acid, cholic acid, dehydrocholic acid, desoxycholic acid, glycocholic acid, taurocholic acid.
Sodium and potassium salts of the above acids.

Ethyl Dimethyldithiocarbamate

Disinfectant
As a bactericide (Australian 8103/32, Brit. 406979, U. S. 1972961).

Insecticide and Fungicide.
As a fungicide (claimed effective against barley spores) (Australian 8103/32, Brit. 406979, U. S. 1972961).
As an insecticide (Australian 8103/32, Brit. 406979, U. S. 1972961).

Ethyl Diphenylphosphate

Cellulose Products
Solvent for—
Cellulose esters or ethers.
For uses, see under general heading: "Solvents."

Ethyldodecylguanidin Chloride

Textile

Assistant (Brit. 421862) in—
Aqueous baths for treating textiles.
Promoter (Brit. 421862) of—
Uniform dyeing with basic dyestuffs.
Wetting and washing agent (Brit. 421862) in—
Textile processes.

Ethylene

Synonyms: Bicarburetted hydrogen, Elayl, Ethene, Etherin, Heavy carburetted hydrogen, Olefiant gas.
French: Éthylène.
German: Aethylen.

Chemical

Reagent in making—
Mustard gas.
Starting point in making—
Acenaphthene, aldehyde, anthracene, chlorethyl chloroacetate, ethane, ethyl alcohol, ethylene bromide, ethylenebromhydrin, ethylene chloride, ethylene chloroiodide, ethylene iodide, ethylene nitrate, ethylene nitrosite, ethylsulphonic acid, formaldehyde, naphthalene, pyrazolon, styrol, sulphuric ether.

Horticultural

Ingredient of—
Gaseous mixtures, in combination with formaldehyde, for ripening and preserving citrous fruits (Australian 17327).

Pharmaceutical

In compounding and dispensing practice.

Refrigeration

As a refrigerating medium.

Ethylenechlorhydrin

Synonyms: 2-Chloroethyl alcohol.
German: Glycolchlorhydrin.

Agriculture

Promoter of—
Early sprouting of dormant potatoes.

Cellulose Products

Solvent for—
Cellulose acetate (tolerates the addition of water at the same time).
Ethylcellulose (used in admixture with methanol).

Chemical

As a solvent miscible with—
Alcohol, benzene, methanol, water.
Intermediate in synthesis of—
Novacaine.
Introducer of—
Hydroxyethyl group in organic syntheses.
Reagent in making—
Malonic acid.
Starting point in making—
Glycol esters (used with salts of organic acids).
Phenylethyl alcohol.

Dye

Intermediate in making—
Synthetic indigo.

Glass

Suggested ingredient of—
Solvent mixtures for cellulose acetate or ethylcellulose in making safety glass.

Leather

Suggested ingredient of—
Solvent mixtures for cellulose acetate or ethylcellulose in making artificial leather or flexible coatings for leather.

Paint and Varnish

Suggested ingredient of—
Solvent mixtures for cellulose acetate.
Solvent mixtures for ethylcellulose.

Paper

Suggested ingredient of—
Solvent mixtures for cellulose acetate or ethylcellulose in making flexible coatings for paper.

Textile

Suggested ingredient of—
Solvent mixtures for cellulose acetate.
Solvent mixture for ethylcellulose.

Ethylene Chlorobromide

Synonyms: Symmetrical chlorobromoethane.
French: Chlorure et bromure d'éthylène, Chlorure et bromure éthylènique.
German: Aethylenchlorbromid, Chlorbromaethan, Chlorbromaethylen.
Spanish: Clorobromuro de etileno.
Italian: Clorobromuro di etilene.

Cellulose Products

Solvent for—
Cellulose acetate, cellulose esters and ethers, cellulose nitrate.
For uses, see under general heading: "Solvents."

Ethylenediamine

Synonyms: Ethylenediamin.
French: Éthylènediamine, Diamine d'éthylène.
German: Aethylendiamin.
Spanish: Etilendiamine.
Italian: Etilenediamina.

Analysis

Reagent in—
Precipitating uranium salts.

Chemical

Reagent in making—
Intermediates, organic chemicals, pharmaceuticals.
Reagent for—
Protecting dissolved albumen against coagulation.
Starting point in making—
Argentamine (with silver nitrate).
Camphor compound with camphoric acid anhydride (German 408183).
Compounds with various metallic salts.
Diethylenediamine.
Ethylenediamine-silver phosphate.
Ethylenediamine perchlorate.
Euphyllin (with theophyllin).
Lycetol (dimethylpiperazin tartrate).
Lysidine (ethylene-ethenyldiamine).
Mercury compounds (German 496801).
Pharmaceuticals for treating gout.
Pharmaceuticals with silver sulphate.
Sedative with alphabromoisovalerianic acid (French 543912).
Sublamin.
Various mothproofing compounds, such as ethylenediamine selenite and ethylenediamine selenate (Brit. 340318).
Solvent for—
Albumen, casein, fibrin, sulphur.

Disinfectant

Ingredient of various disinfecting and germicidal compositions.

Explosive

Starting point in making—
Explosive compounds, such as ethylenediamine nitrate and ethylenediamine chlorate.

Leather

Reagent in—
Dehairing hides.

Miscellaneous

Reagent in—
Dehairing fur skins.

Petroleum

Reagent in making—
Emulsions of petroleum and petroleum distillates.

Pharmaceutical

Suggested as medicament in gout.
Suggested as noncorrosive solvent for dissolving false diphtheritic membranes.
Suggested for use with 10 percent solution of calcium chloride for intravenous injection to stop hemorhages.

Rubber

Accelerator in—
Vulcanization (U. S. 1503702 and 1592820).

Ethylenediamine-Mercury Sulphate

French: Sulfate éthylènediamine de mercure.
German: Aethylendiaminequecksilbersulfat.

Miscellaneous

Disinfectant for seeds.

Pharmaceutical

In compounding and dispensing practice.
Nonirritant germicide for—
Disinfecting hands and skin.
Suggested for use in treating—
Venereal diseases.

Ethylenediamine Selenate
French: Séléinate d'éthylènediamine.
German: Aethylendiaminselenat, Selensäuresaethylendiaminester, Selensäuresaethylendiamin.

Miscellaneous
Reagent (Brit. 340318) in—
Mothproofing furs, feathers, and hair.

Textile
Reagent (Brit. 340318) in—
Mothproofing wool and felt.

Ethylene Dibromide
Synonyms: 1:2-Dibromoethene, Ethylene bromide.
French: Bromoéthylène, Bromure d'éthylène, Bromure éthylènique, Dibromure d'éthylène, Dibromure éthylènique.
German: Aethylenbromid, Aethylendoppeltebromid, Bromaethylen, Dibromaethen, Doppeltebromaethylen.
Spanish: Bromuro de etileno, Dibromuro de etileno.
Italian: Bromuro di etilene, Dibromuro di etilene.

Analysis
Solvent in—
Analytical processes involving control and research work.

Cellulose Products
Ingredient of—
Solvents for cellulose esters and ethers.
Solvent for—
Nitrocellulose.

Chemical
As a carrier for—
Tetraethyl lead in the manufacture of antiknock agents to be added to motor fuel.
Solvent miscible with most other solvents.
Starting point in making—
Aromatics, diethyleneglycol, diethylenetetramine, dioxyethylene, ethylene chlorobromide, ethylene cyanide, ethylene oxide, ethylenediamine, ethyleneglycol, ethylenemercaptan, intermediates, piperazin, pharmaceuticals, symmetrical diethylenediethylamine, synthetic organic chemicals, tetraethylenetriamine, triethylenetriamine.

Dye
Reagent in making various dyestuffs.

Explosives
Solvent for—
Nitrocellulose.

Fats, Oils, and Waxes
Solvent for—
Fats, oils, waxes.

Gums
Powerful solvent for—
Gums.

Miscellaneous
Solvent for waxes in—
Polishes, waterproofing preparations.
Solvent for—
Nitrocellulose in miscellaneous coating agents.
Resins in miscellaneous coating agents.
Solvent miscible with most other solvents.

Paint and Varnish
Solvent for—
Nitrocellulose, resins.
Solvent miscible with—
Most other solvents, thinners.

Plastics
Solvent for—
Cellon, celluloid, cellulose derivatives, resins.

Petroleum
As a carrier for—
Tetraethyl lead in the manufacture of antiknock agents to be added to motor fuel.
Solvent for gums and waxes in—
Lubricating gasolines.

Resins
Solvent for—
Resins.

Rubber
Reactant in making—
Elastic bodies resembling caoutchouc.
Starting point in making—
Elastic bodies resembling caoutchouc by polymerization.

Ethylene Difluoride

Petroleum
Solvent (Brit. 436044) in—
Flushing oil composition for internal-combustion engines; flushing oil is based on light lubricating oil of either paraffinc or naphthenic origin and contains various other products; naphtha, isopropyl alcohol, or acetone may be added to reduce the viscosity; practice is to flush (1) with oil containing a high proportion of solvent to remove most of the sludge, (2) with oil containing a lower proportion of solvent.

Ethylenedinitroamine

Explosives
As an explosive with high resistance to detonation by shock (U. S. 2011578).
As an explosive with relatively low ignition temperature (U. S. 2011578).
As an initiating explosive (U. S. 2011578).
Substitute (U. S. 2011578) for—
Nitroglycerin or nitrocellulose in propellent powders.

Ethylenediphenylphosphonium Bromide
French: Bromure d'éthylènediphénylephosphonium.
German: Aethylendiphenylphosphoniumbromid, Bromaethylendiphenylphosphonium.

Miscellaneous
Mothproofing and moldproofing agent (Brit. 312163) in treating—
Hair, fur, feathers, felt, and the like.

Textile
Mothproofing and moldproofing agent (Brit. 312163) in treating—
Wool and other products.

Ethylene-Ferrous Chloride
French: Chlorure éthylènique et ferreux, Chlorure d'éthylène et fer.
German: Aethylenferrochlorid.

Chemical
Reagent in making various organic compounds.

Ethyleneglycol
Synonyms: Ethylene alcohol, Glycol, Glycol alcohol.
French: Alcool éthylènique, Éthanediol, Glycol d'éthylène, Glycol éthylènique.
German: Aethylenglycol, Glykol.

Chemical
Moistening agent in making—
Non-fermentable compositions and preparations.
Preservative in making—
Various chemical and pharmaceutical compositions.
Reagent in making—
Plasticizers and softening agents.
Solvent in making—
Pharmaceutical preparations.
Starting point in making—
Ethylenechlorohydrin, glycol diacetate, glycol diformate (Brit. 255887), glycol formate, quinaldin, spirosal (ethyleneglycol monosalicylate).
Substitute for glycerin in organic synthesis and for various chemical purposes.

Dye
Ingredient of—
Stable leuco compounds of indigo, thioindigo and anthraquinone dyestuffs (Brit. 260253).
Solvent in making dye preparations.

Explosives and Matches
Ingredient of—
Low-freezing dynamite.
Starting point in making—
Ethyleneglycol dinitrate.

Fats and Oils
Reagent in purifying—
Fats and oils by esterification (German 315222).

Food
Ingredient of—
Canned goods, confectionery, food pastes, food preparations of various sorts, ketchups, mincemeats, salad dressings.
Preservative in making—
Concentrated fruit essences, flavoring extracts, soda fountain supplies.

Gas
Lubricant in gas meters.

Ethyleneglycol (Continued)

Ink
Ingredient of—
Stamping inks, writing inks.

Leather
Ingredient of—
Compositions used for preserving the softness and flexibility of leather during working.

Mechanical
Anti-freeze agent for filling—
Exposed dashpots in Corliss engines and the like.
Exposed gages and other instruments.
Radiators of airplanes and automobiles.
Ingredient of—
Lubricating compositions used in machinery employed for producing liquefied products, such as liquid air (Brit. 277378).

Miscellaneous
General solvent for various purposes.
Ingredient of—
Compositions used in treating and preserving skins and furs, printers' rollers mass.
Preservative for treating—
Anatomical and biological specimens.
Substitute for glycerin for various purposes.

Perfumery
Ingredient of—
Cosmetics.

Pharmaceutical
In compounding and dispensing practice.

Refrigeration
Ingredient of—
Low-freezing solutions.
Lubricant in ice machines.

Resins and Waxes
Solvent for various resins and waxes.
Solvent in making—
Phenol-formaldehyde synthetic resins (Brit. 260253).

Textile
——, *Dyeing*
Assist in making—
Dye liquors for acetate rayon.
Solubilizing or dispersing agent (Brit. 276100) in making dye liquors containing—
Acridines.
Aminoanthraquinones, reduced or unreduced.
Anthraquinones, reduced or unreduced.
Azines, azo dyestuffs, basic diarylmethane dyestuffs, basic triarylmethane dyestuffs, benzoquinoneanilides, chrome mordant dyestuffs, indigoids.
Naphthoquinones, reduced and unreduced.
Naphthoquinoneanilides, nitroarylamines, nitroarylphenols, nitrodiarylamines, nitrodiarylphenols, oxazines, pyridines.
Quinoneimides, reduced and unreduced.
Quinolines, sulphur dyestuffs, thiazines, xanthenes.
——, *Finishing*
Ingredient of—
Finishing compositions for yarns and fabrics.
Softening agent for hydrogroscopic salts in textiles.
——, *Printing*
Assist in making pastes.

Tobacco
Ingredient of—
Compositions for moistening and treating tobacco.

Ethyleneglycol Chlorotolylether

Chemical
Starting point (Brit. 416943) in making—
Wetting, foaming, detergent, emulsifying, and dispersing agents by condensation with butyl alcohol and sulphonation with sulphuric acid.

Ethyleneglycol Diformate
Synonyms: Ethylene glycol biformate.
French: Biformiate d'éthylène glycole, Biformiate éthylèneglycollique, Diformiate d'éthylèneglycole, Diformiate éthylèneglycollique.
German: Aethylenglykolbiformiat, Aethylenglykoldiformiat, Biameisensaeureaethylenglykolester, Biameisensaeuresaethylenglykol, Diameisensaeureaethylenglykolester.

Cellulose Products
Plasticizer (Brit. 311795) for—
Cellulose acetate.
For uses, see under general heading: "Plasticizers."

Dye
Ingredient (Brit. 311795) of—
Dye pastes.

Ink
Ingredient (Brit. 311795) of—
Printing inks.

Ethyleneglycol Ditolylether

Chemical
Starting point (Brit. 416943) in making—
Wetting, foaming, detergent, emulsifying, and dispersing agents by condensation with butyl alcohol and sulphonation with sulphuric acid.

Ethyleneglycol Isopropylether

Petroleum
Solvent (Brit. 436044) in—
Flushing oil composition for internal-combustion engines; flushing oil is based on light lubricating oil of either paraffinic or naphthenic origin and contains various other products; naphtha, isopropyl alcohol, or acetone may be added to reduce the viscosity; practice is to flush (1) with oil containing a high proportion of solvent to remove most of the sludge, (2) with oil containing a lower proportion of solvent.

Ethyleneglycol Monoacetate
Synonyms: Glycol monoacetate.
French: Monoacétate de glycole, Monoacétate glycollique.
German: Glykolmonoacetat, Glykolmonoazetat, Monoessigsaeureglykolester, Monoessigsauresglykol.

Chemical
Starting point in making various derivatives.

Paint and Varnish
Solvent and plasticizer in making—
Products containing cellulose acetate, nitrocellulose, and other cellulose esters and ethers.
See also: "Solvents."

Plastics
Solvent and plasticizer in making—
Artificial horn products from albuminous substances.
Compounds of nitrocellulose, cellulose acetate, and other cellulose esters and ethers.

Resins and Waxes
Solvent for—
Formaldehyde condensation resins, glyptal resins, urea-formaldehyde resins.

Ethyleneglycol Monoamyl Ether
French: Éther d'éthylèneglycole monoamylique.
German: Aethylenglykolmonoamylester.

Ceramics
Solvent in—
Compositions, containing cellulose acetate, nitrocellulose, benzylcellulose, or other esters or ethers of cellulose, used for the decoration and protection of ceramic products.

Chemical
Dispersing agent in making—
Emulsions of hydrocarbons of various groups of the aliphatic and aromatic series.
Emulsions of various chemicals, terpene emulsions.
Solvent for—
Cellulose acetate, nitrocellulose.
Starting point (Brit. 302258) in making—
Cleansing agents, dispersive agents, dissolving compositions, emulsifiers, foam-producing compositions, lathering agents, textile lubricating and oiling compositions, washing agents, wetting agents.

Dye
Dispersive agent in making—
Color lakes.

Electrical
Solvent in—
Compositions, containing nitrocellulose, cellulose acetate, benzylcellulose, cellluose butyrate, or other cellulose ethers or esters and also resins, used for insulating electrical wiring and equipment.

Fats and Oils
Dispersing agent in making—
Boring oils, drilling oils, greasing compositions.
Lubricating compositions of animal or vegetable oils.
Solvent for fats (Brit. 302258).
Stabilized emulsions of animal and vegetable fats and oils.
Wire-drawing oils.

Ethyleneglycol Monoamyl Ether (Continued)

Gas
Solvent for—
Bitumen.

Germicide
Dispersing agent (Brit. 302258) in making—
Germicidal and deodorizing compositions.

Glass
Solvent in—
Compositions, containing various esters or ethers of cellulose and resins used in the manufacture of non-scatterable glass and for the decoration and protection of glassware.

Insecticide
Dispersing agent (Brit. 302258) in making—
Emulsified insecticidal and fungicidal compositions.

Leather
Dispersing agent (Brit. 302258) in making—
Emulsified tanning preparations, emulsified leather dressings, emulsified fat-liquoring baths, emulsified soaking compositions, emulsified waterproofing compositions.

Metallurgical
Solvent in—
Compositions, containing various esters or ethers of cellulose and resins, used in the manufacture of artificial leather and for the decoration and protection of leather goods.
Solvent in—
Compositions, containing various esters or ethers of cellulose and resins, used for the protection and decoration of metallic ware.

Miscellaneous
Dispersing agent in making—
Cleansing compositions of various types.
Metal polishes and other polishing compositions.
Scouring compositions.
Waterproofing compositions in emulsified form.
Solvent in—
Compositions containing various esters or ethers of cellulose and resins, used for the decoration and protection of fibrous compositions.

Paint and Varnish
Solvent in making—
Quick-drying paints, varnishes, enamels, dopes, and lacquers containing various esters or ethers of cellulose, such as cellulose acetate, cellulose butyrate, nitrocellulose, benzylcellulose, and resins.

Paper
Dispersing agent in making—
Sizing compositions in emulsified form.
Waterproofing compositions for paper and pulp compositions and paperboard.
Waxing compositions for treating paper and paperboard.
Solvent in—
Compositions, containing cellulose acetate, cellulose butyrate, nitrocellulose, or other esters or ethers of cellulose, and resins, used in the manufacture of coated paper and for the decoration and protection of pulp and paper compositions.

Petroleum
Ingredient of—
Emulsified cutting oils for lathe and screwpress work.
Kerosene emulsions, naphtha emulsions, soluble greases, soluble lubricating oils, soluble oils for lubricating textile machinery, rayon oils, various textile oils.

Plastics
Solvent in making—
Compositions containing various esters or ethers of cellulose, such as cellulose acetate, cellulose butyrate, nitrocellulose, benzylcellulose, and natural or artificial resins.

Resins and Waxes
Dispersing agent in making—
Emulsions of natural and artificial resins.
Emulsions of natural and artificial waxes.

Rubber
Solvent in—
Compositions, containing various esters or ethers of cellulose and resins, used for the decoration and protection of rubber products.

Soap
Dispersing agent (Brit. 302258) in making—
Hand-cleansing compositions.
Various emulsified cleansing and lathering compositions.

Stone
Solvent in—
Compositions, containing various cellulose esters or ethers and resins, used for the decoration and protection of natural and artificial stone.

Textile
—, *Bleaching*
Dispersing agent (Brit. 302258) in—
Emulsified bleaching baths.

—, *Dyeing*
Dispersing agent (Brit. 302258) in—
Dye baths.

—, *Finishing*
Ingredient (Brit. 302258) of—
Emulsified coating compositions containing various esters or ethers of cellulose, such as nitrocellulose, cellulose acetate, benzylcellulose, or cellulose butyrate.
Emulsified sizing compositions.
Emulsified washing compositions.
Emulsified compositions used for impregnation purposes.

—, *Manufacturing*
Ingredient (Brit. 302258) of—
Emulsified carbonizing baths for wool.
Emulsified degreasing compositions for treating raw wool.
Emulsified mercerizing baths.
Emulsified oiling compositions.
Emulsified preparations for bast scouring silk.
Emulsified preparations for fulling operations.
Emulsified spinning preparations.

—, *Printing*
Ingredient (Brit. 302258) of—
Emulsified printing pastes.

Woodworking
Solvent in—
Compositions, containing various esters or ethers of cellulose, such as cellulose acetate and nitrocellulose, and resins, used for the decoration and protection of woodwork.

Ethyleneglycol Monobutyl Ether

Cellulose Products
Solvent for—
Cellulose acetate, cellulose esters and ethers, cellulose nitrate (nitrocellulose).
For uses, see under general heading: "Solvents."

Gas
Solvent for—
Bitumen.

Ethyleneglycol Monoethyl Ether

Synonyms: Glycol monoethyl ether.
French: Éther de éthylène glycolemonoéthylique.
German: Aethylenglykolmonoaethylaether.

Paint and Varnish
Plasticizer and solvent in making—
Products containing nitrocellulose, cellulose acetate, and other cellulose esters and ethers.
See also: "Plasticizers."

Plastics
Plasticizer and solvent in making—
Compounds of nitrocellulose, cellulose acetate, and other cellulose esters and ethers.

Ethyleneglycol Monoformate

French: Monoformiate d'éthylèneglycole, Monoformiate éthylèneglycollique.
German: Aethylenglykolmonoformiat, Monoameisensaeureaethylenglykolester, Monoameisensaeuresaethylenglykol.

Cellulose Products
Plasticizer (Brit. 311795) for—
Cellulose acetate.
For uses, see under general heading: "Plasticizers."

Dye
Ingredient (Brit. 311795) of—
Dye pastes.

Ink
Ingredient (Brit. 311795) of—
Printing inks.

Ethyleneglycol Monomethyl Ether
Synonyms: Glycol monomethyl ether.
French: Éther de éthylèneglycolemonométhyle.
German: Aethylenglykolmonomethylaether.

Miscellaneous
See also: "Plasticizers."

Paint and Varnish
Plasticizer and solvent in making—
Products containing nitrocellulose, cellulose acetate, and other cellulose esters and ethers.

Plastics
Plasticizer and solvent in making—
Compounds of nitrocellulose, cellulose acetate, and other cellulose esters and ethers.

Ethyleneglycolmonomethylether Acetate
French: Monométhyle-étheracétate d'éthylèneglycole.
German: Aethylenglykolmonomethylaetheracetat.

Paint and Varnish
Plasticizer in making—
Cellulose acetate varnishes and lacquers (Brit. 278735).

Plastics
Plasticizer in making—
Cellulose acetate compositions (Brit. 278735).

Ethyleneglycolmonomethylether Formate
French: Formiate d'éthylèneglycolemonométhyleéther.
German: Aethylenglykolmonomethylaetherformat, Ameisensaeuresaethylenglykolmonomethylaether.

Paint and Varnish
Plasticizer in making—
Cellulose acetate varnishes and lacquers (Brit. 278735).

Plastics
Plasticizer in making—
Cellulose acetate compositions (Brit. 278735).

Ethyleneglycol Monotolylether

Chemical
Starting point (Brit. 416943) in making—
Wetting, foaming, detergent, emulsifying, and dispersing agents by condensation with butyl alcohol and sulphonation with sulphuric acid.

Ethylene Oxide
French: Oxyde éthylènique, Oxide d'éthylène.
German: Aethylenoxid.

Chemical
Reagent (Brit. 265233) in making—
Butyl alcohol, glycol monoacetate, glycol mononitrate, glycol monosulphate, glycol diacetate, glycol dinitrate, glycol disulphate.

Ethylene Oxide, Polymerized

Miscellaneous
As an emulsifying agent (Brit. 353926).
For uses, see under general heading: "Emulsifying agents."

1-Ethyleneoxy-4-aminoanthraquinone
Synonyms: Alphaethylenehydroxy-4-aminoanthraquinone.
French: Alphaéthylènehydroxye-4-anthraquinone, 1-Éthylèneoxye-4-anthraquinone.
German: 1-Aethylenhydroxy-4-aminoanthrachinon, 1-Aethylenoxyl-4-aminoanthrachinon, Alpha-aethylenhydroxy-4-aminoanthrachinon, Alpha-aethylenoxy-4-aminoanthrachinon.

Chemical
Starting point in making—
Intermediates, pharmaceuticals.

Dye
Starting point (Brit. 285096) in making dyestuffs in the presence of dimethylanilin, nitrobenzene, orthodichlorobenzene, naphthalene, and the like, with the aid of—
Acetylparaphenylenediamine, 5-amino-2-methylbenzimidazole, benzidin and its derivatives and homologs, dimethylparaphenylenediamine, metanaphthylenediamine, metaphenylenediamine, metatoluylenediamine, metaxylidenediamine, orthonaphthylenediamine, orthophenylenediamine, orthotoluylenediamine, orthoxylidenediamine, paranaphthylenediamine, paraphenylenediamine, paratoluylenediamine, paraxylidenediamine.

Ethylenethiodiglycol
German: Aethylenthiodiglykol.

Dye
Reagent (Brit. 276023) in making—
Stable leuco compounds of indigo, thioindigo, and anthraquinone colors, and other vat dyestuffs.

Ethylenethiourea
Synonyms: Ethylenesulphourea.
French: Éthylènesulphourée, Éthylènethiourée, Sulphourée d'éthylène, Sulphourée éthylènique, Thiourée d'éthylène, Thiourée éthylique.
German: Aethylensulfoharnstoff, Aethylenthioharnstoff.

Chemical
Starting point (Brit. 310534) in making vulcanization accelerators with—
Alphanaphthylamine, anilin, benzylamine, betanaphthylamine, cyclohexylanilin, meta-anisidin, metacresidin, metanaphthylenediamine, metaphenylamine, metaphenylenediamine, metatoluidin, metatoluylenediamine, metaxylenediamine, metaxylidin, monoethylanilin, monomethylanilin, orthoanisidin, orthocresidin, orthonaphthylenediamine, orthophenylamine, orthophenylenediamine, orthotoluidin, orthotolylenediamine, orthoxylenediamine, orthoxylidin, para-anisidin, paracresidin, paranaphthylenediamine, paraphenylamine, paraphenylenediamine, paratoluidin, paratolylenediamine, paraxylidin, paraxylenediamine.

Ethylenetriphenylphosphonium Bromide
French: Bromure d'éthylènetriphénylephosphonium.
German: Aethylentriphenylphosphoniumbromid, Bromaethylentriphenylphosphonium.

Miscellaneous
Mothproofing and moldproofing agent (Brit. 312163) in treating—
Hair, fur, feathers, felt, and the like.

Textile
Mothproofing and moldproofing agent (Brit. 312163) in treating—
Wool and other products.

Ethyl Ethylisopropylmalonate
French: Malonate d'éthyle éthyleisopropyle.
German: Aethylaethylisopropylmalonat, Malonsaeuresaethylaethylisopropylester.

Chemical
Starting point in making—
5:5-Ethylisopropylbarbituric acid (U. S. 1576014).

Ethyl Formate
French: Formiate d'éthyle, Formiate éthylée, Formiate éthylique.
German: Formiataethyl, Formylsäuresaethylester.

Food
Larvicide in—
Grain milling, packaged dried fruits.
Larvicide for treating—
Cereals.

Insecticide and Fungicide
As a larvicide.

Tobacco
Larvacide for treating—
Tobacco.

Ethylfurol
French: Furole d'éthyle, Furole éthylique.
German: Aethylfurol.

Cellulose Products
Solvent for—
Cellulose acetate, cellulose esters and ethers, cellulose nitrate.
For uses, see under general heading: "Solvents."

Chemical
General solvent.
Starting point in making—
Intermediates, pharmaceuticals.

Gums, Resins and Waxes
Solvent for—
Guaiac resin, ester resins, coumarone resin, other natural and synthetic resins.

Ethylglycol Acetate
French: Acétate d'éthyleglycole, Acétate éthylique-glycollique.
German: Aethylglykolacetat, Aethylglykolazetat, Essigsäureaethylglykol, Essigsäuresaethylglykol.

Cellulose Products
Solvent for—
Cellulose acetate, cellulose esters and ethers, cellulose nitrate.
For uses, see under general heading: "Solvents."

Ethylhydrocuprein
Pharmaceutical
Bactericide (U. S. 1997440) for—
Pneumococcus infections and other diseases.

Ethylidene Acetobenzoate
French: Acétobenzoate d'éthylidène, Acétobenzoate éthylidènique.
German: Acetobenzoesäureaethylidenester, Acetobenzoesäuresaethyliden, Aethylidenacetobenzoat.
Miscellaneous
Solvent for—
Cellulose acetate, cellulose esters and ethers, cellulose nitrate, natural and artificial resins.
For uses, see under general heading: "Solvents."

Ethylidene Diacetate
French: Diacétate d'éthylidène, Diacétate d'éthylidènique.
German: Aethylidendiacetat, Aethylidendiazetat, Essigsäurediaethylidenester, Essigsäuresdiaethyliden.
Spanish: Diacetato de etilideno.
Italian: Diacetato di etilidene.
Cellulose Products
Solvent for—
Cellulose acetate, cellulose esters and ethers, cellulose nitrate.
For uses, see under general heading: "Solvents."
Chemical
Starting point in making—
Acetaldehyde, acetic acid anhydride, acetic acid (German 284996).
Resins and Waxes
Solvent for—
Artificial resins, natural resins.

Ethylideneglycerol
French: Glycérole d'éthylidène, Glycérole éthylidènique.
German: Aethylidenglycerol.
Miscellaneous
Solvent for—
Cellulose esters and ethers.
Various gums and resins.
Various organic substances.
For uses, see under general heading: "Solvents."

Ethylidene Iodide
French: Iodure d'éthylidène, Iodure éthylidènique.
German: Aethylidenjodid, Jodaethyliden.
Spanish: Yoduro de etiliden.
Italian: Ioduro di etilidene.
Chemical
Starting point in making—
Pharmaceuticals and other derivatives.
Starting point (Brit. 353477) in making contrast media used in X-ray photography with the aid of—
Ammonium sulphite, magnesium sulphite, monomethylamine sulphite, piperazin sulphite, piperidin sulphite, sodium sulphite.

Ethylindene 3-Carboxylate
Cellulose Products
Plasticizer (U. S. 1975697) for—
Cellulose acetate, cellulose esters and ethers, cellulose nitrate.
For uses, see under general heading: "Plasticizers."

Ethylisothiourea Sulphate
French: Sulfate d'éthylisothiourée.
German: Aethylisothioharnstoffsulphat, Schwefelsaeureaethylisothioharnstoffester.
Chemical
Reagent (Brit. 272686) in making—
Aminoamyleneguanidin sulphate, aminobutyleneguanidin sulphate, aminoethyleneguanidin sulphate, aminoheptyleneguanidin sulphate, aminohexyleneguanidin sulphate, aminopentyleneguanidin sulphate, aminopropyleneguanidin sulphate.

Ethylisovanillin
German: Aethylisovanillin.
Food
As a flavoring agent.
Ingredient of—
Flavoring compositions.
Perfume
Ingredient of—
Perfumes.
Perfume in—
Cosmetics.

Ethyl Lactate
French: Lactate d'éthyle, Lactate éthylique.
German: Aethyllactat, Milchsaeuresaethylester.
Chemical
Reagent in the purification of—
Lactic acid.
Dye
Solvent in making—
Indulin dyestuffs, nigrosin dyestuffs.
Paint and Varnish
Solvent in making—
Cellulose acetate lacquers, cellulose acetate varnishes, nitrocellulose lacquers, nitrocellulose varnishes, pyroxylin varnishes.
Plastics
Solvent in making—
Celluloid, cellulose acetate plastics, nitrocellulose plastics.
Textile
——, *Dyeing*
Mordant in dyeing various textile fibers and fabrics.
——, *Manufacturing*
Solvent in making—
Cellulose acetate rayon.
——, *Printing*
Mordant in printing various textile fabrics.

Ethyl Malonate
Chemical
Starting point in making—
Diallylbarbituric acid (dial), diethylmalonic ether, ethyldiethyol malonate, intermediates, phenylbarbituric acid (luminal), dipropylbarbituric acid (propanal), synthetic aromatic chemicals, diethylbarbituric acid (barbital, veronal).

Ethyl Mandelate
French: Mandélate d'éthyle, Mandélate éthylique.
German: Aethylmandelat, Mandelsaeureaethylester, Mandelsaeuresaethyl.
Paint and Varnish
Plasticizer (Brit. 270650) in making—
Lacquers, varnishes.
Plastics
Plasticizer in making—
Nitrocellulose plastics.

Ethylmercaptan
Synonyms: Ethyl sulphhydrate.
French: Éthyle mercaptan, Mercaptan éthylique, Mercaptan d'éthyle.
German: Aethylmercaptan.
Chemical
Starting point in making—
Diethylsulphondimethylmethane (sulphonal).
Diethylsulphonethylmethylmethane (trional).
Tetranal.
Various other pharmaceutical products.
Starting point (Brit. 286749) in making vulcanization accelerators with the aid of—
Dibenzylamine, diethylguanylthioureas, diphenyl biguanide, ditolyl biguanide, ethanolamine, guanylureas, isothioureas, isoureas, monophenyl biguanide, monophenylguanylthiourea, monotolyl biguanide, pentaphenyl biguanide, pentatolyl biguanide, piperidin, piperazin, tetramethylammonium hydroxide, tetraphenyl biguanide, tetratolyl biguanide, thioureas, trimethylsulphonium hydroxide.
Gas
As a leak detector for natural gas.
Insecticide and Fungicide
Attractant for—
House flies (*Musca domestica* L.).
Screw-worm flies (*Cochliomyia macellaria* Fab.).
Fumigant and insecticide for—
Flour weevils (*Tribolium confusum* Fab.).
Granary weevils (*Sitophilus granarius* L.).
House flies (*Musca domestica* L.).
Rice weevils (*Sitophilus oryza* L.).
Larvicide for—
Larvae of the Indian-meal moth (*Plodia interpunctella* Hbn.).
Mild repellant to—
Green bottle flies (*Lucilia* spp.).

Ethylmercaptan (Continued)

Mining
As a warning agent in mines.
Miscellaneous
As an aid to the detection of noxious vapors and wasteful leaks.

Ethylmercapto-1-naphthol

Chemical
Starting point in making—
Intermediates, pharmaceuticals.
Dye
Starting point (Brit. 291825) in making synthetic indigoid dyestuffs with—
5:7-Dibromoisatin anilide.
5:7-Dibromoisatin chloride.
5:7-Dichloroisatin anilide.
5:7-Dichloroisatin chloride.
Isatin anilide.
Isatin chloride.
Reactive alpha derivatives of isatin.

Ethyl-Mercury Chloride

Disinfectant
Starting point (Brit. 450256) in making—
Disinfectants with water-glass and other reactive silicon compounds.
Insecticide and Fungicide
Starting point (Brit. 450256) in making—
Seed immunizers with water-glass and other reactive silicon compounds.
Leather
Preventer of—
Slime and molds in tanning liquors.

Ethyl-Mercury Hydroxide

French: Hydroxyde d'éthyle et de mercure, Hydroxyde éthylique et mercurique.
German: Aethylmerkurhydroxid.
Agricultural
Disinfectant for—
Seed grains.
Rubber
Dispersive agent in making—
Rubber compositions.
Sanitation
Dispersive agent in making—
Disinfectants, germicides.
Soap
Dispersive agent (German 371293) in making—
Special soap preparations.
Various scouring preparations.
Textile
——, *Dyeing and Printing*
Dispersive agent (German 371293) in making—
Dye liquors and printing pastes.
——, *Manufacturing*
Dispersive agent (German 371293) in making—
Wool-degreasing preparations.
Woodworking
Dispersive agent (German 371293) in making—
Wood preservatives.

Ethyl-Mercury Oleate

Lubricant
Starting point (Brit. 440175) in making—
Addition agents for high-pressure lubricating oils or greases, by reacting with oil-soluble organic compounds.
Insecticide and Fungicide
Controller of—
Melanose on citrus trees, scab on citrus trees.
Reducer of—
Stem-end fungous infections.

Ethylmercury Phosphate

Agriculture
For control of—
Bottom rust of lettuce.
Covered smut and stripe disease of barley.
Kernel smut of sorghum.
Loose and covered smuts of oats.
Soil-borne parasitic fungi.
Stinking smut of wheat.
Woodworking
For control of—
Blue stain and sap stain in sapwood of freshly sawed lumber.

Ethylmercury Sulphate

French: Sulphate d'éthyle et de mercure, Sulphate éthyliquemercurique.
German: Aethylmercurisulfat, Aethylquecksilbersulfat, Schwefelaethylquecksilber, Schwefelsäuresaethylquecksilber.
Agricultural
For disinfecting and preserving seed grains (Brit. 330258).

Ethylmethyl Ketone Peroxide

Fuel
Reducer (Brit. 444544) of—
Inflammability hazards in diesel engine fuels.

Ethyl Monobromoacetate

Chemical
Starting point in making—
Tear gases.

Ethyl Monochloroacetate

Chemical
As a solvent.
Miscellaneous
As a solvent.

Ethyl Monoiodoacetate

French: Monoiodoacétate d'éthyle, Monoiodoacétate éthylique.
German: Aethylmonojodacetat, Monojodoessigsaeuresaethylester.
Chemical
Starting point in making—
Monoiodoacetic acid.

Ethylnaphthalenesulphonic Acid

French: Acide d'éthylenaphthalènesulphonique.
German: Aethylnaphtalinsulfonsäure.
Chemical
Starting point in making—
Salts, esters, and other derivatives.
Miscellaneous
As a dispersing agent (Brit. 322005).
For uses, see under general heading: "Emulsifying agents."

Ethyl Nitrate

Chemical
Reagent in—
Organic syntheses.
Fuel
Primer (Brit. 404682) in—
Diesel engine fuels (used in conjunction with other primers, consisting of organic bromides or organic copper compounds, whose function is that of reducing the spontaneous ignition temperature).
Reducer (Brit. 404682) of—
Delay period in diesel engine fuels.

Ethyl Nonylate

Food
As a flavoring.
Ingredient of—
Flavoring preparations.
Perfume
Ingredient of—
Cosmetics, such as lipsticks.

Ethylnormalbutylbarbituric Acid

Pharmaceutical
Ingredient (U. S. 1928346) of—
Anesthetic composition for rectal administration in obstetrics, containing also mineral oil and ethyl ether.

Ethyloctadecenyl Sulphide

Chemical
Starting point (Brit. 422937) in making—
Textile assistants by oxidation and subsequent sulphonation.

Ethylolamine

German: Aethylolamin.
Chemical
Starting point in making—
Pharmaceuticals and other derivatives.
Electrical
Dispersive agent (Brit. 340294) in making—
Special lubricating compositions for use in electric switches.

Ethylolamine (Continued)

Mechanical
Dispersive agent (Brit. 340294) in—
Non-freezing lubricating compositions, containing animal and vegetable oils and fats, as well as ethyleneglycol or its esters, borax, benzyl alcohol.
Special lubricating compositions of the above type, for use on locomotive axles, railway switches, hydraulic presses and hydraulic brakes.
Miscellaneous
Ingredient (Brit. 340294) of—
Compositions, containing vegetable, animal, and mineral oils and greases, used as rust preventives.
Petroleum
Ingredient (Brit. 340294) of—
Special lubricating compositions containing mineral oils and greases.

Ethyl Orthosilicate

Glue and Adhesives
Ingredient (Brit. 428548) of—
Cellulose acetate or nitrocellulose base adhesives for safety glass.
Metallurgical
Binder (Brit. 441639) in—
Electric welding fluxes containing also (a) magnetic iron ore, ferromanganese, titanium dioxide, kaolin; or (b) pyrolusite, silica, magnetic iron oxide.

Ethyloxyphenylparabutylamine

Chemical
Antioxidant and stabilizer (Brit. 430335) for—
Unstable organic substances.
Petroleum
Antioxidant and stabilizer (Brit. 430335) for—
Petroleum products.
Rubber
As an antioxidant (Brit. 430335).

Ethyl Paraoxybenzoate

French: Paraoxyebenzoate d'èthyle, Paraoxyebenzoate éthylique.
German: Aethylparaoxybenzoat, Paraoxybenzoesäureaethyl, Paraoxybenzoesäuresaethyl.
Chemical
Starting point in making various derivatives.
Pharmaceutical
In compounding and dispensing practice.
Sanitation
As a general disinfectant.

5-Ethyl-5-paraphenetidinobarbituric Acid Chloride

Pharmaceutical
Suggested for use (Brit. 414293) as—
Hypnotic with low toxic properties.

Ethylparaphenol Sulphopara-aminobenzoate

Pharmaceutical
Ingredient of—
Analgesic mouthwashes.

Ethylparatoluenesulphonamide

French: Éthyleparatoluènesulphonamide.
German: Aethylparatoluolsulfonamid.
Chemical
Starting point in making various derivatives.
Miscellaneous
Plasticizer (Brit. 313133) for—
Cellulose acetate, cellulose esters and ethers, cellulose nitrate, resins.
For uses, see under general heading: "Plasticizers."

Ethyl Pelargonate

Synonyms: Ethyl nonylate.
French: Éther pélargonique, Nonylate d'éthyle, Pélargonate d'éthyle, Pélargonate éthylique.
German: Nonylsaeureaethylester, Pelargonsaeureaethylester.
Food
Flavor for beverages and confections.

Ethylpentaerythritol

Cellulose Products
Solvent, softener, and plasticizer (Brit. 358393) for—
Cellulose acetate, cellulose esters or ethers, nitrocellulose.
For uses, see under general heading: "Plasticizers."

1:2-Ethylphenylaminoethane

Chemical
Starting point in making—
Intermediates, pharmaceuticals.
Dye
Starting point in making various synthetic dyestuffs.
Rubber
Antioxidant (Brit. 314756) in—
Vulcanizing.

Ethyl-propenyl Ether

French: Éther éthylepropénylique, Éther d'éthyle et propényle.
German: Aethylpropenylaether.
Chemical
Starting point in making—
Intermediates and other derivatives.
Reagent in regulating (Brit. 340474)—
Polymerization of diolefins.

1-Ethylpropylalphanaphthol

Disinfectant
Claimed (U. S. 2073996 and 2073997) to be—
Germicide combining high efficiency toward *staphylococcus aureus* and low toxicity.

Ethylpropyl Sulphide

Synonyms: Thioethylpropyl ether.
Fungicide and Insecticide
As a fungicide (German 363656).
As an insecticide (German 363656).

Ethyl Salicylate

French: Éther salicylique, Salicylate d'éthyle.
German: Aethylsalicylat, Salicylsaeuresaethylester.
Chemical
Starting point in making—
Synthetic perfumery materials.
Perfumery
Ingredient of—
Cosmetics and perfumes.
Pharmaceutical
In compounding and dispensing practice.

Ethyl Silicate

Synonyms: Silicic acid ethylester.
French: Silicate d'éthyle, Silicate éthylée.
German: Aethylsilikat, Siliciumwasserstoffsaeuresaethylester.
Building and Construction
Binder and cavity filler in—
Preservative paints and compositions for protecting and impregnating porous building and construction surfaces.
Ingredient of—
Special preservative mortars and plasters.
Preservative (German 568545) for—
Bricks, cement, plaster, porous construction materials, stone, stucco.
Chemical
Agglomerating agent for—
Activated carbon.
Starting point in making—
Colloidal silica, silica gel.
Electrical
Binder and cementing and agglomerating agent in—
Molded electrical insulation.
Cavity-filling agent for—
Molded electrical insulation.
Surface-hardening agent for—
Molded electrical insulation.
Leather
Cavity-filling agent for—
Leather, leather products.
Surface-hardening agent for—
Leather, leather products.
Metallurgical
Binder and cementing and agglomerating agent in—
Crucibles used in the fusion of difficultly fusible alloys.
Cementing agent in—
Ferrochroming sand moulds.
Surface-hardening agent for—
Graphite moulds used in tapping special metals.
Sand moulds.
Miscellaneous
Binder and cementing and agglomerating agent for—
Asbestos products, cork products, loose materials, porous materials, sawdust.

Ethyl Silicate (Continued)
Cavity-filling agent for—
Asbestos products, cork products, fibrous materials, porous materials.
Surface-hardening agent for—
Asbestos products, cork products, soft, porous, or crumbly surfaces, straw products.

Paint and Varnish
Cavity filler in—
Preservative paints for porous materials and surfaces.
Impregnating agent in—
Preservative paints and coating agents for porous materials and surfaces.
Increaser of—
Color brightness and reflecting properties of asbestos cement paints for theatrical decorations.
Corrosion resistance of paints to the action of acid atmospheres, gasoline, and oil.
Heat resistance of paints.
Heat resistance and corrosion resistance of paints for metals exposed to high temperatures.
Heat resistance and corrosion resistance of paints for automobile exhaust systems.
Resistance to alkali washing agents by asbestos cement paints for theatrical decorations.
Surface hardness of asbestos cement paints for theatrical decorations.
Wearing properties of asbestos cement paints for theatrical decorations.

Paper
Cavity-filling agent for—
Paper products.
Surface-hardening agent for—
Paper products.

Refractories
Binder and cementing and agglomerating agent in—
Acid-resisting brick.
Acid-resisting mortars and cements.
Refractory brick.
Refractory mortars and cements.
Surface-hardening agent for—
Acid-resisting brick.
Acid-resisting mortars and cements.
Refractory brick.
Refractory mortars and cements.
Silica brick.

Stone
Binding, cementing and agglomerating agent for—
Stone products.
Cavity-filling agent for—
Stone products.
Impregnating agent and preservative for—
Stone products.
Surface-hardening agent for—
Stone products.

Sugar
Binder and cementing and agglomerating agent for—
By-products made from bagasse.

Textile
Cavity-filling agent for—
Textile products.
Producer of—
Cloudy effects in artificial silk.
Surface-hardening agent for—
Textile products.

Woodworking
Cavity-filling agent for—
Wood products.
Surface-hardening agent for—
Wood products.

Ethylsorbitol
Synonyms: Ethylsorbite.

Miscellaneous
Plasticizer (U. S. 1936093) for—
Cellulose acetate, cellulose esters and ethers, cellulose nitrate, natural resins, synthetic resins.
For uses, see under general heading: "Plasticizers."

Ethylsulphuric Acid Chloride
French: Chlorure d'acide éthylesulphurique.
German: Aethylschwefelsaeureschlorid.

Dye
Starting point (Brit. 271533) in making vat dyestuffs with—
Anthraquinone-1:2, flavanthrone, indanthrone, naphthacridin, thioindigo.

Ethyl Thiosalicylate
Synonyms: Ethyl sulphosalicylate.
French: Sulfosalicylate d'éthyle, Sulfosalicylate éthylique, Thiosalicylate éthylique.
German: Aethylsulfosalicylat, Aethylthiosalicylat, Sulfosalicylsaeureaethylester, Sulfosalicylsaeuresaethyl, Thiosalicylsaeureaethylester, Thiosalicylsaeuresaethyl.

Chemical
Starting point (Brit. 262427) in making synthetic drugs with—
Gold, silver, arsenic, antimony, and bismuth oxides and salts.

Ethyltoluene Sulphonate
French: Toluènesulphonate éthylique.
German: Toluolsulfosaeuresaethylester.

Chemical
Starting point in making—
Cyclohexylethylanilin (Brit. 261747).

Ethyltolyl Sulphide
Synonyms: Thioethylcresyl ether.

Fungicide and Insecticide
As a fungicide (German 363656).
As an insecticide (German 363656).

Ethyl Triacetylgallate

Chemical
Starting point in making—
Pharmaceuticals with resorcinolbenzoyl carbonate.

Pharmaceutical
In compounding and dispensing practice.

Ethyl Trifluoracetate

Chemical
Starting point (Brit. 416653) in making—
Trifluorodimethyl acetone (a new refrigerant) by condensation with ethyl acetate in the presence of sodium in an ethereal solution to form sodium trifluoracetoacetate, which is then decomposed by excess diluted sulphuric acid.

Ethyltriphenylphosphonium Bromide
French: Bromure d'éthyletriphénylephosphonium.
German: Bromaethyltriphenylphosphonium, Aethyltriphenylphosphoniumbromid.

Miscellaneous
Reagent (Brit. 312163) for mothproofing—
Furs, feathers and the like.

Textile
Reagent (Brit. 312163) for—
Mothproofing.

Ethyltritolyltriphenylphosphonium Iodide
French: Iodure d'éthyletritolyletriphénylephosphonium.
German: Aethyltritolyltriphenylphosphoniumjodid, Jodaethyltritolyltriphenylphosphonium.

Miscellaneous
Mothproofing and moldproofing agent (Brit. 312163) in treating—
Hair, fur, feathers, felt, and the like.

Textile
Mothproofing and moldproofing agent (Brit. 312163) in treating—
Wool and other products.

Ethyl Undecylenate
French: Undecylenate d'éthyle, Undecylenate éthylique.
German: Undecylensäureaethylester, Undecylensäuresaethyl.
Spanish: Undecilenato de etil.
Italian: Undecilenato di etile.

Chemical
Starting point (French 615959) in making—
Aluminum, zinc, manganese, and bismuth undecylenates.

Leather
Reagent (French 615959) for—
Weighting and polishing leather.

Ethylvanillin
German: Aethylvanillin.
Food
As a flavoring agent.
Ingredient of—
Flavoring compositions.
Perfume
Ingredient of—
Perfumes.
Perfume in—
Cosmetics.

Ethylxylylphosphonium Iodide
Chemical
Starting point in making various derivatives.
Miscellaneous
Mothproofing and moldproofing agent (Brit. 312163) in treating—
Hair, furs, feathers, and the like.
Textile
Mothproofing and moldproofing agent (Brit. 312163) in treating—
Wool, felt, and other products.

Eucalyptus Oil
French: Essence d'eucalyptus, Huile d'eucalypte, Huile d'eucalyptus.
German: Eukalyptusoel.
Spanish: Aceite esencial de eucalipto.
Italian: Olio di eucalitto.
Agriculture
Application for keeping insects from livestock.
Chemical
Denaturant for—
Ethyl alcohol.
Solvent for—
Aluminum stearate.
Starting point in making—
Aromatic chemicals, cineol, citronellol, fixatives for various industrial purposes, piperitone, synthetic menthol, synthetic thymol.
Fats and Oils
Solvent for—
Greases.
Gas
Solvent for—
Tar.
Germicide
Ingredient of—
Disinfecting and germicidal compositions.
Emulsified rosin soap disinfectants.
Various deodorant compositions.
Sanitary sweeping powders.
Insecticide
Ingredient of—
Fruit sprays, to combat scale, fungus, and insect pests.
Insect repellents used by sportsmen.
Various insecticidal preparations.
Metallurgical
Flotation oil in separating—
Gangue from minerals and ores.
Ingredient of—
Flotation mixtures for concentrating sulphide minerals and ores.
Miscellaneous
Ingredient of—
Dental preparations.
Deodorizing, asepticizing preparations (with rosin soap) for use in theatres.
Preparations for the prevention of scale in boilers.
Preparations for use in motor car radiators.
Mothproofing compositions for feathers, furs, skins.
Rubbing oils, shoe polishes, various floor polishes.
Paint and Varnish
Ingredient of—
Paint and varnish removers.
Solvent in making—
Lacquers containing cellulose acetate, nitrocellulose, or other cellulose esters or ethers.
Perfume
Ingredient of—
Special perfume preparations.
Perfume in making—
Cosmetics, dentifrices.
Pharmaceutical
In compounding and dispensing practice.
Resins and Waxes
Solvent for—
Artificial and natural resins, ceresin, carnauba wax, various waxes.
Rubber
Ingredient of—
Rubber cements.
Sanitation
As a disinfectant.
Soap
Ingredient of—
Compositions for cleaning upholstery of motor cars, furniture covers, tapestries, clothing.
Creams for cleansing hands (Polish 9083).
Perfume in—
Toilet soaps.
Textile
Ingredient of—
Moth eradicators for wools and felts.

Eucupinotoxin
Chemical
Disinfectant and preservative (Brit. 399602) in treating—
Adrenalin, digestive ferments, injection solutions, local anaesthetics, morphine, novocaine, pancreatin, pepsin, vegetable extracts and residues.
Food
As a preservative (Brit. 339602).
Glues and Adhesives
Preservative (Brit. 339602) in treating various products.
Perfume
Preservative and disinfectant (Brit. 339602) in making—
Ointments, pomades.
Pharmaceutical
In compounding and dispensing practice.
Sanitation
As a general disinfectant.
Preservative, sterilizing agent, and disinfectant (Brit. 339602) in treating—
Rinsing liquids, surgical gut, surgical dressings and bandages.
Starch
Preservative (Brit. 339602) in treating—
Dextrin solutions, starch solutions.
Textile
Preservative (Brit. 339602) in treating—
Sewing silk, yarn-sizing preparations.

Eugenol Acetamide
German: Eugenolacetamid.
Chemical
Starting point in making—
Pharmaceuticals.
Pharmaceutical
In compounding and dispensing practice.

Eugenol Cinnamate
French: Cinnamate d'eugénole, Cinnamate eugénolique.
German: Eugenolcinnamat, Zimtsäureeugenolester, Zimtsäureseugenol.
Chemical
Starting point in making—
Pharmaceuticals and other derivatives.
Pharmaceutical
In compounding and dispensing practice.

Feldspar
French: Feldspath.
German: Feldspat.
Building
Ingredient of—
Compositions used for surfacing concrete.
Cement
Ingredient of various cements.
Ceramics
Flux in making—
Porcelains.
Ingredient of—
Enamels, glazes.
Raw material in making—
Chinaware, porcelains, potteries.

Feldspar (Continued)

Chemical
Raw material in making—
Aluminum silicate, potassium salts, silicon nitride, sodium salts.

Electrical
Ingredient of—
Insulating compositions.
Raw material in making various electrical goods.

Fertilizer
As a plant food (in the powdered form).
Raw material in making—
Potash fertilizers.

Glass
Ingredient in making—
Cryolite glass, opalescent glasses, polishing compounds.

Jewelry
Certain varieties are the moonstone, amazon stone, and other semiprecious stones.

Miscellaneous
Ingredient in making—
Grinding wheels, poultry grit, sandpaper, scouring powders, sharpening stones, tarred roofing papers.
Raw material in making—
Artificial teeth.

Soap
Ingredient in making—
Scouring soaps.

Stone
Ingredient in making—
Artificial stone.

Fenchone
Synonyms: 1:3:3-Trimethylbicyclo-(1:2:2)-heptanone-(2).
French: Fenchone.
German: Fenchon.

Perfume
Ingredient of—
Fancy perfume preparations.

Fenchyl Alcohol
French: Alcool de Fenchyle, Alcool Fenchylique.
German: Fenchylalkohol.

Chemical
Starting point in making—
Fenchone and various other organic chemicals.

Miscellaneous
To give the odor of old pine wood to various products.

Fenugreek
Synonyms: Greek hay seed, Trigonella.
Latin: Foenugraecum, Semen foenugraeci.
French: Fénugrec.
German: Bockshornsamen, Hornkleesamen, Siebenzeitsamen.

Agricultural
Condiment for cattle.

Food
Ingredient in making—
Cheeses, condiments.

Pharmaceutical
In compounding and dispensing practice.

Ferric Acetate
Synonyms: Acetate of iron, Iron acetate, Vinegar martial.
Latin: Ferri aceticum.
French: Acétate de fer, Acétate ferrique, Extrait de mars, Vinaigre chalybé.
German: Eisenacetat, Eisenazetat, Essigsäureseisen, Essigsäureseisenoxyd, Ferriacetat, Ferriazetat.

Chemical
Ingredient of catalytic mixtures used in the manufacture of—
Acenaphthylene, acenaphthaquinone, bisacenaphthylidenedione, naphthaldehydic acid, naphthalic anhydride, and hemimellitic acid from acenaphthene (Brit. 295270).
Acetaldehyde from ethyl alcohol (Brit. 281307).
Acetic acid from ethyl alcohol (Brit. 281307).
Alcohols from aliphatic hydrocarbons (Brit. 281307).
Aldehydes and acids by the reduction of the corresponding esters (Brit. 306471).
Aldehydes and acids from toluene, orthochlorotoluene, orthobromotoluene, orthonitrotoluene, parachlorotoluene, parabromotoluene, paranitrotoluene, metachlorotoluene, metabromotoluene, metanitrotoluene, dichlorotoluenes, dibromotoluenes, dinitrotoluenes, chlorobromotoluenes, chloronitrotoluenes, bromonitrotoluenes (Brit. 295270).
Aldehydes and acids from xylenes, pseudocumenes, mesitylene, and paracymene (Brit. 281307).
Alphanaphthaquinone from naphthalene (Brit. 281307).
Anthraquinone from naphthalene (Brit. 295270).
Benzaldehyde and benzoic acid from toluene (Brit. 281307).
Benzoquinone from phenanthraquinone (Brit. 281307).
Benzyl alcohol from benzaldehyde by reduction (Brit. 306471).
Benzyl alcohol or benzaldehyde or phthalide by the reduction of phthalic anhydride (Brit. 306471).
Butyl alcohol by the reduction of crotonaldehyde (Brit. 306471).
Chloroacetic acid from ethylenechlorohydrin (Brit. 295270).
Diphenic acid from ethyl alcohol (Brit. 281307).
Ethyl alcohol by the reduction of acetaldehyde (Brit. 306471).
Fluorenone from fluorene (Brit. 295270).
Formaldehyde by the reduction of methane or methanol (Brit. 306471).
Formaldehyde by the reduction of carbon dioxide or carbon monoxide (Brit. 306471).
Hydroxyl compounds by the reduction of anthraquinone, benzoquinone, and similar compounds (Brit. 306471).
Isopropyl alcohol by the reduction of acetone (Brit. 306471).
Maleic acid and fumaric acid by the oxidation of toluene, benzene, phenols, tar phenols, or furfural, or from benzoquinone or phthalic anhydride (Brit. 295270).
Methane by the reduction of carbon dioxide or carbon monoxide (Brit. 306471).
Methanol by the reduction of carbon dioxide or carbon monoxide (Brit. 306471).
Naphthaldehydic acid, acenaphthaquinone, or bisacenaphthylidenedione from acenaphthylene (Brit. 281307).
Phenanthraquinone from phenanthrene or diphenic acid (Brit. 295270).
Phthalic acid and maleic acid from naphthalene (Brit. 295270).
Primary alcohols by the reduction of the corresponding aldehydes (Brit. 306471).
Propionic acid and butyric acid and higher alcohols, ketones, and acids by the reduction of carbon dioxide or carbon monoxide (Brit. 306471).
Reduction products of ketones, aldehydes, acids, esters, alcohols, ethers, and other organic compounds which contain oxygen (Brit. 306471).
Salicylic acid and salicylic aldehydes from cresol (Brit. 295270).
Secondary butyl alcohol by the reduction of methylethyl ketone (Brit. 306471).
Valeryl alcohol by the reduction of valeraldehyde (Brit. 306471).
Vanillin and vanillic acid by oxidation from eugenol or isoeugenol (Brit. 295270).
Ingredient (Brit. 306460) of catalytic preparations which are used in the production of various aromatic and aliphatic compounds, including—
Alphanaphthylamine from alphanitronaphthalene.
Amines, from aliphatic nitro compounds, such as allyl nitriles or nitromethane.
Amino compounds from the corresponding nitroanisoles.
Amylamine from pyridin.
Anilin, azo-oxybenzene, azobenzene, and hydrazobenzene from benzene by reduction.
Aminophenols from nitrophenols.
3-Aminopyridin from 3-nitropyridin.
Cyclohexamine, dicyclohexamine, and cyclohexylanilin from nitrobenzene.
Piperidin from pyridin.
Pyrrolidin from pyrrol.
Tetrahydroquinolin from quinolin.
Starting point in making—
Ferric chromate, ferric tannate, ferric tantalate, ferric tungstate, ferric valeriate, ferric vanadate.
Various iron salts of complex character.

Leather
Mordant in—
Dyeing to black shades.

Ferric Acetate (Continued)

Miscellaneous
Mordant in—
Dyeing hats, furs, and other articles to dark shades.

Pharmaceutical
In compounding and dispensing practice.

Printing
Reagent in making—
Printing ink rollers.

Textile
Mordant in dyeing—
Awnings, black and other dark shades, khaki shades on fabrics.
Mordant in printing—
Calicoes, violet shades with alizarin.

Woodworking
As a preservative.
Mordant in dyeing—
Black shades on wood.

Ferric-Butyryl Acetone
French: Butyryleacétone ferrique.
German: Eisenbutyrylaceton, Ferributyrylaceton.

Chemical
Starting point (Brit. 289493) in making—
Aromatics, intermediates, pharmaceuticals.

Dye
Starting point (Brit. 289493) in making various synthetic dyestuffs.

Petroleum
Antidetonant (Brit. 289493) in—
Motor fuels.

Ferric Carbonate
Synonyms: Iron carbonate.
French: Carbonate de fer, Carbonate ferrique.
German: Eisenkarbonat, Ferrikarbonat, Karbonsaeureseisen, Karbonsaeureseisenoxyd.

Chemical
Starting point in making various iron salts.

Pharmaceutical
In compounding and dispensing practice.

Ferric Chloride
Synonyms: Chloride of iron, Ferric trichloride, Iron chloride, Iron perchloride, Iron sesquichloride, Iron trichloride, Sesquichloride of iron.
Latin: Chloridum ferricum, Chloruretum ferricum, Ferri chloridum, Ferri perchloridum, Ferrum sesquichloratum, Ferrum muriaticum oxydatum, Flores martis.
French: Chlorure ferrique, Perchlorure de fer.
German: Chloreisen, Eisenchlorid.
Spanish: Cloruro ferrico.
Italian: Cloruro ferrico.

Abrasives
Protein-insolubilizing agent (Brit. 417177 and 417234) in—
Compositions for sandpaper, consisting of a protein, a soluble silicate, and/or gelatinous silica, a flexibility improver, and a modifying agent.

Analysis
As a reagent.

Ceramics
Coloring agent (Brit. 410651) for—
Bricks, tiles, earthenware, pottery (coloring is effected by sublimations, the actual sublimation of the metallic chloride may be effected outside the kiln, the vapors being blown in through the firing holes, or the material may be preheated outside the kiln and introduced into the kiln before sublimation has taken place).

Chemical
Accelerator (Brit. 405371) for—
Hydrogen formation from water in hydrogenation of carbonaceous materials, such as benzenes, petroleum residues, coaltar.
Catalyst in—
Friedel-Crafts synthesis using phenol, chlorobutanol, or tertiary amyl chloride (Brit. 409131).
Organic synthesis.
Catalyst in making—
Alkyl-substituted aromatic hydroxy compounds (U. S. 1892990).
Chlorinated derivatives of benzene from benzene and chlorine (Brit. 388818).
Chlorinated derivatives of benzene from partially chlorinated benzene and chlorine (Brit. 388818).
Orthoamylbenzoylbenzoic acid (U. S. 1889347).
Coagulating agent (U. S. 1911273) for—
Precipitated silver iodide in recovering iodine from seawater.
Electrolyte (U. S. 1915039) in—
Flocculation of negatively and positively charged emulsions.
Starting point in making—
Catalysts (molybdenum-iron oxides) used in preparation of formaldehyde by oxydation of methanol (U. S. 1913404).
Chlorine and pure hydrochloric acid with iron sulphate (German 568239).
Pharmaceutical chemicals and preparations.
Various iron salts.

Dye
Oxidizing agent in making—
Dyes.
Starting point (Brit. 408492) in making—
Brownish-black or blackish-brown metalliferous dyestuffs with 5-nitro-2-aminophenol, resorcinol, and lactic acid.

Fats and Oils
Catalyst (Brit. 397136) in making—
Oils from natural or synthetic rubber by hydrogenation (such oils are said to be suitable for use as impregnating agents for paper, cloth, leather, and other substances, and as vehicles in paint and varnish).

Fuel
Purifying agent (Brit. 397460) (in combination with sulphuric acid) for treating—
Crude benzene, low-temperature tars.

Glass
Coloring agent for—
Glass.

Insecticide
Ingredient (Brit. 396365) of—
Insecticide containing also arsenic acid and caustic soda.

Metallurgical
Ingredient of—
Gold-plating solution containing also potassium ferrocyanide, sodium carbonate, gold fulminate, and sodium hydroxide.
Pickling solutions, containing also hydrochloric acid, used on aluminum to dissolve oxide film prior to nickel-plating.

Miscellaneous
Catalyst (Brit. 398474) in making—
Explosion-proof lubricants for oxygen cylinders, welding burners, bearings.
Ingredient of—
Composition for generating heat on addition of water (U. S. 1901313).
Composition, used for generating heat on addition of water, which contains also powdered iron, manganese hydroxide, graphitic carbon, ferrous sulphate, manganese chloride, and manganese sulphate.
Heat storage and transfer medium, containing also salt, and aluminum chloride, for use in connection with fusion and calcination operations, also domestic heating systems and hot water storage systems (German 519062).

Paint and Varnish
Resinifying agent (Brit. 402759) in making—
Resinous varnish ingredients from water-soluble sulphite compounds of oxidized drying oils.

Petroleum
Condensing agent (Brit. 397169) in making—
Condensation or polymerization products of high molecular paraffin hydrocarbons for use in accelerating the separation of waxes from hydrocarbon oils.
Purifying agent (Brit. 397460) (in combination with sulphuric acid) for treating—
Hydrocarbons and freeing them of sulphur-containing compounds, colloidal asphaltic bodies, and unstable unsaturated substances (application is to such products as petroleum, shale oil, vaporphase cracked spirit, motor spirit, lubricating oil).
Purifying agent (Brit. 367848, 387447, and 413719) in conjunction with sulphuric acid and following ozonizing) for treating—
Liquid hydrocarbons in motor fuel production.
Reagent (Brit. 398794) in—
Purifying light hydrocarbon oils, especially those obtained by cracking, by treatment with phosphorus pentoxide.

Ferric Chloride (Continued)
Regenerating agent (Brit. 397460) (in combination with sulphuric acid) for—
Used lubricating oils.
Pharmaceutical
In compounding and dispensing practice.
Ingredient of—
Amethyst-colored solution (tincture plus sodium salicylate) for filling window display bottles (does not fade on exposure to sunlight).
Starting point in making—
Tincture, albuminate, and other preparations.
Suggested for use as—
Styptic for bleeding surfaces.
Photographic
Reagent in photographic processes.
Printing
Etching agent in—
Photoengraving processes.
Resins
Catalyst (U. S. 1846247) in making—
Resins from rubber.
Rubber
Catalyst (Brit. 390097) in—
Chlorination of rubber latex.
Textiles
Mordant in—
Dyeing processes, printing processes.
Water and Sanitation
Coagulant in—
Sewage purification (U. S. 191520), water purification.
Conditioning agent for—
Sewage sludge (U. S. 1928163), sewage sludge (before filtration).
Ingredient (U. S. 1747177) of—
Ferric-alumina (peptized hydrous alumina) coagulant for clarifying aqueous liquids.

Ferric Dimethyldithiocarbamate
Disinfectant
As a bactericide (Australian 8103/32, British 406979, U. S. 1972961).
Insecticide and Fungicide
As a fungicide (claimed effective against *Aspergillus niger* and *Fomes Annonsus*) (Australian 8103/32, Brit. 406979, U. S. 1972961).
As an insecticide (Australian 8103/32, Brit. 406979, U. S. 1972961).

Ferric Gammabutylacetylacetone
Synonyms: Iron gammabutylacetylacetone.
French: Gammabutyleacétyleacétone ferrique.
German: Eisengammabutylacetylaceton, Ferrigammabutylacetylaceton.
Chemical
Starting point (Brit. 289493) in making—
Aromatics, intermediates, pharmaceuticals.
Dye
Starting point (Brit. 289493) in making various synthetic dyestuffs.
Petroleum
Antidetonant (Brit. 289493) in motor fuels.

Ferric Gammaethylacetylacetone
Synonyms: Iron gammabutylacetylacetone.
French: Gammaéthyleacétyleacétone ferrique.
German: Eisengammaaethylacetylaceton, Ferrigammaaethylacetylaceton.
Chemical
Starting point (Brit. 289493) in making—
Aromatics, intermediates, pharmaceuticals.
Dye
Starting point (Brit. 289493) in making various synthetic dyestuffs.
Petroleum
Antidetonant (Brit. 289493) in motor fuels.

Ferric Normalbutylhydrogenphthalate
Synonyms: Ferric normalbutylacidphthalate, Iron normalbutylacidphthalate.
French: N-Butylehydrogènephthalate de fer, N-Butylehydrogènephthalate ferrique.
German: N-Butylsaeuresphtalsaeureseisen, Eisennormalbutylsaeuresphtalat, Ferrinormalbutylsaeuresphtalat.

Paint and Varnish
Ingredient (Brit. 250265) of—
Lacquers, enamels, varnishes.
Plastics
Ingredient of—
Plastic compositions.
Photographic
Starting point (Brit. 270387) in making—
Light-sensitive varnishes.

Ferric Oleate
Synonyms: Iron oleate.
French: Oléate de fer, Oléate ferrique.
German: Eisenoleat, Ferrioleat, Oleinsäureseisen, Oleinsäureseisenoxyd.
Fats and Oils
Ingredient of—
Solidified oils.
Reagent in promoting—
Contact between the catalyst and the oil in the hydrogenation of oils.
Leather
Ingredient of—
Dressing compositions, waterproofing compositions.
Mechanical
Ingredient of—
Cutting compounds, metal-working preparations and lubricants.
Paint and Varnish
Ingredient of—
Paints, varnishes.
Reagent in coloring—
Varnishes.
Starting point in making—
Driers.
Pharmaceutical
In compounding and dispensing practice.
Petroleum
Ingredient of—
Cylinder oils (used in place of fats), cup greases, steam turbine oils.
Textile
Ingredient of—
Softening preparations.
Waterproofing compositions for treating canvas and other heavy fabrics.

Ferric Palmitate
Synonyms: Iron palmitate.
French: Palmitate de fer, Palmitate ferrique.
German: Eisenpalmitat, Ferripalmitat, Palmitinsäureseisen, Palmitinsäureseisenoxyd.
Fats and Oils
Ingredient of—
Solidified oils.
Reagent in promoting—
Contact between the catalyst and the oil in the hydrogenation of oils.
Leather
Ingredient of—
Dressing compositions, waterproofing compositions.
Mechanical
Ingredient of—
Cutting compounds, metal-working preparations and lubricants.
Paint and Varnish
Ingredient of—
Paints, varnishes.
Reagent in coloring—
Varnishes.
Starting point in making—
Driers.
Pharmaceutical
In compounding and dispensing practice.
Petroleum
Ingredient of—
Cylinder oils (used in place of fats), cup greases, steam turbine oils.
Textile
Ingredient of—
Softening preparations.
Waterproofing compositions for treating canvas and other heavy fabrics.

Ferric Stearate
Synonyms: Iron stearate.
French: Stéarate de fer, Stéarate ferrique.
German: Eisenstearat, Ferristearat, Stearinsäureseisen, Stearinsäureseisenoxyd

Fats and Oils
Ingredient of—
Solidified oils.
Reagent in promoting—
Contact between the catalyst and the oil in the hydrogenation of oils.

Leather
Ingredient of—
Dressings, waterproofing compositions.

Mechanical
Ingredient of—
Cutting compounds, metal-working preparations.

Paint and Varnish
Ingredient of—
Paints, varnishes.
Reagent in coloring—
Varnishes.
Starting point in making—
Driers.

Pharmaceutical
In compounding and dispensing practice.

Petroleum
Ingredient of—
Cylinder oils (used in place of fats), cup greases, steam turbine oils.

Textile
Ingredient of—
Softening preparations.
Waterproofing compositions for treating canvas and other heavy fabrics.

Ferric Sulphate
Synonyms: Iron persulphate, Iron sesquisulphate.
Latin: Ferri sulphas.
French: Sulphate ferrique.
German: Schwefelsäuereseisen.

Analysis
As a reagent.

Chemical
Catalyst (in conjunction with copper sulphate) in—
Oxidation of N_2H_4 by hydrogen peroxide.
Promoter (U. S. 1914835 and 1914458) for—
Platinum-magnesium sulphate catalyst used in the oxidation of sulphur dioxide to sulphur trioxide.
Reagent in making—
Ethylidene diacetate (Brit. 252640).
Tetraglucosan from dextrose.
Starting point in making—
Black oxide of iron.
Ferric acetate and other ferric salts.
Iron alum and iron-ammonium alum.

Fats and Oils
Pickle (U. S. 1909676) in—
Fish fat recovery process.
Protein coagulating agent (U. S. 1909676) in—
Fish fat recovery process.

Disinfectant
Ingredient (Brit. 388149) of—
Cleansing and disinfectant agent, containing also sodium bisulphate, used on lavatory pans, sinks, drains, and the like.
Purifying agent (U. S. 1644250) for—
Fats, oils.
Reagent (Brit. 380065 and 380052) in making—
Stable emulsions of fats, oils, paraffin, neatsfoot oil, benzene, trichloroethylene.

Fertilizer
Promoter of—
Black alkali soil (Fresno, Cal.) productivity reclamation.

Fire-Prevention
Ingredient of—
Chemical fire extinguishers.

Fuel
Reagent in—
Determination of hydrogen sulphide in illuminating gas.
Hydrogen sulphide removal from fuels.
Purifying hydrocarbon products—crude benzene, low-temperature tars, and the like (Brit. 397460).

Ink
Ingredient of—
Tannin writing inks.

Meat-Packing
Ingredient of—
Fused mixture with anhydrous sodium sulphate used for coagulating blood and preventing nauseous odors in abattoirs.

Metallurgical
Etching reagent in—
Working with aluminum.
Flotation reagent for—
Galena, separating galena from sphalerite.
Purifying agent (U. S. 1316909) for—
Salt solutions, such as zinc sulphate solutions, used in the electrolytic production of zinc (must be basic ferric sulphate soluble in dilute sulphuric acid).
Reagent in—
Production of copper from ores by the wet process.

Miscellaneous
Reagent in making—
Gas-mask fillers (hopcalites).

Paint and Varnish
Starting point in making—
Berlin blue and similar pigments.

Petroleum
Reagent in—
Purifying hydrocarbon oils—petroleum, shale oil, used lubricating oils—by freeing them from sulphur-containing compounds, colloidal asphaltic bodies, and unstable unsaturated substances (used in conjunction with dilute sulphuric acid) (Brit. 397460).
Purifying hydrocarbon oils by treating them in conjunction with sulphur trioxide, and steam (U. S. 1897582).
Purifying hydrocarbon oils by treating them in conjunction with sulphuric acid and fuller's earth (Brit. 413412).
Purifying hydrocarbon oils by treating them in conjunction with sulphuric acid following treatment with ozonized air in presence of a catalyst (Brit. 413719, 387447, and 367848).
Purifying paraffin oil (preferred substitute for sulphuric acid).

Pharmaceutical
In compounding and dispensing practice.

Photographic
Reagent (Brit. 382320) in—
Process based on differential treatment of images obtained in different depths of an emulsion.

Resins
Polymerizing agent (in conjunction with fuller's earth) (U. S. 1894934) for—
Coaltar naphtha fractions, containing coumarone and indene constituents, used in making synthetic resins.

Sanitation
As a disinfectant.
Promoter (in conjunction with copper sulphate) of—
Germicidal activity of hydrogen peroxide on *Bacillus coli* and *Staphylococcus aureus*.
Purifying agent in—
Treatment of drinking water.

Soap
Bleaching agent for—
Glycerin.

Textile
Mordant in—
Dyeing dark colors on cotton and wool.
Reagent in—
Calico printing, cotton dyeing.

Ferro-Columbium
Note: Alloys containing 50-60 percent columbium.

Metallurgical
Ductility promoter in—
Chrome-nickel steels.
Reducer of—
Intergranular corrosion in chrome-nickel steels, especially when they are exposed simultaneously to heat and chemical attack.

Ferrous Acetate
Synonyms: Acetate of iron, Black liquor, Ferroacetate.
French: Acétate ferreux.
German: Essigsäureseisenoxydul, Ferroazetat.

Analysis
Reagent in carrying out reductions.

Ferrous Acetate (Continued)

Chemical
General reducing agent.
Reagent in making—
Aminobenzaldehyde, anilin, primary aromatic amines.

Dye
Reagent in carrying out—
Reductions in manufacturing processes.

Leather
Mordant in dyeing in black shades.

Miscellaneous
Mordan in dyeing—
Miscellaneous products, such as hats and furs, in black shades.

Textile
Mordant in dyeing—
Khaki colors on textiles.
Violet, black, blue, and brown effects.
Mordant in printing—
Iron buffs on textiles.

Woodworking
Mordant in dyeing and staining.

Ferrous Chloride
Synonyms: Ferrous protochloride, Iron dichloride, Iron protochloride.
French: Chlorure ferreux.
German: Eisenchlorur.

Analysis
As a reagent.

Chemical
Starting point in making—
Ferric chloride.

Dye
Reducing agent in making—
Dyestuffs.

Metallurgical
In metallurgy.
Ingredient of—
Electrolytes for iron-plating.

Pharmaceutical
In compounding and dispensing practice.

Textile
Mordant in—
Dyeing processes, printing processes.

Ferrous Iodide
Synonyms: Ferrous protoiodide, Iron iodide.
French: Iodure de fer, Iodure ferreux.
German: Eisenjoduer, Jodeisen.

Chemical
Catalyst in iodating organic compounds.
Starting point in making—
Barium iodide, calcium iodide, lithium iodide, magnesium iodide, strontium iodide.

Pharmaceutical
In compounding and dispensing practice.

Fireclay

Ceramics
Ingredient of—
Architectural terra cotta, art pottery, chemical stoneware, high-grade tile, stoneware.
White-bodied ware, including china, porcelain, general ware, chemical porcelain, porcelain electrical supplies, sanitary ware.

Construction
Ingredient of—
Plaster and plaster products, refractory cements and mortars.

Miscellaneous
Ingredient of—
Artificial abrasives, asbestos products.

Paper
As a filler (very small amounts only, compared with kaolin).

Paint
Ingredient of—
Calcimines.

Refractories
Raw material in making—
Cements and mortars, crucibles, firebrick, blocks and shapes, furnace lining, glass factory pots and tanks, pins, stilts and spurs for potters' use, retorts, saggers for potters, wads.

Fish Berries
Synonyms: Cockles, Indian berries, Oriental berries.
Latin: Cocculi indici.
French: Coque du levant.
German: Fischkoerner, Fischkörner, Kockelbeeren, Kockelkörner, Kokkelskörner, Lauesekoerner, Lauesekörner, Tollkoerner, Tollkörner.
Italian: Cocculi di levante.

Chemical
Starting point in the production of—
Picrotoxin.

Insecticide
Ingredient of—
Insecticide compositions, vermin killers.

Pharmaceutical
In compounding and dispensing practice.

Fish Glue. See Isinglass

Fish Meal
French: Farine de poisson.
German: Fischmehl.

Agricultural
Ingredient of—
Cattle feeds, poultry feeds.

Fertilizer
Alone or in mixtures as an ammoniate.

Paint and Varnish
Ingredient of—
Lacquers (U. S. 1245975 to 1245984).

Plastics
Ingredient of—
Plastics for making knobs, door handles, molded articles, buttons, and the like (German 352534).

Rubber
Ingredient of—
Artificial rubber compounds with furfural (Brit. 230629).

Flavanthrone

Dye
Starting point (Brit. 271533) in making soluble vat dyestuffs with—
Butylsulphuric acid chloride.
Chlorosulphonic acid chloride.
Methylsulphonic acid chloride.
Starting point (Brit. 271537) in making—
Leucoflavanthrones.

Flaxseed
Synonyms: Linseed.
Latin: Linum, Lini semina, Semen lini.
French: Graines de lin, Semences de lin.
German: Flachssamen, Leinsamen, Leinsaat.
Spanish: Lino, Linaza, Semilla de lino.
Italian: Lino, Semi di lino.

Fats and Oils
Starting point in extracting—
Linseed oil.

Miscellaneous
Ingredient (U. S. 1641006) of—
Leak-stopping compositions for automobile radiators.

Pharmaceutical
In compounding and dispensing practice.

Textile
——, *Printing*
Thickener in making—
Printing pastes.

Fluorescein
Synonyms: Diresorcinolphthalein, Resorcinolphthalein, Tetraoxyphthalophenonanhydride.
French: Fluorane de la résorcine, Phthaleine de la résorcine.
German: Diresorcinolphtalein, Resorcinolphtalein, Resorcinphtalein.

Dye
Starting point in making—
Coerulein B, coerulein BR, eosines, erythrosin G, erythrosin BB.

Miscellaneous
Dyestuff in coloring—
Pine oil preparations (Brit. 271555).

Perfume
Color for—
Bath salts, cosmetics.

Fluorescein (Continued)

Textile
——, *Dyeing and Printing*
Dyestuff for—
Wool, silk, and other textiles.

Fluorinated Paraffin

Electrical
Claimed (Brit. 443340) to be—
Chemically stable material having chemical and physical properties rendering it especially adapted for use as a dielectric material; in particular its high dielectric constant is of special importance in capacitor construction; by using liquid fluorinated wax in place of ordinary mineral oil in capacitors, the bulk of a capacitor of a given capacity rating may be decreased as much as 50 to 75 percent.
(1) It is claimed that fluorinated paraffin is chemically stable and nonvolatile at ordinary temperature; containing about 25 percent of fluorine by weight, it has a pour point of about minus 14° C., a viscosity at 100° C. of 50 seconds, and a specific gravity of 0.99 at 15.5° C. (referred to water at 15.5° C.). Fluorinated paraffin containing 45 percent of fluorine by weight has a pour point of minus 3° C., a viscosity at 100° C. of 94 seconds, and a specific gravity of 1.14 at 15.5° C. (referred to water at 15.5°).
(2) It is claimed that a fluorinated paraffin in which the proportion of combined fluorine is at least approximately equal to the proportion of combined hydrogen is non-inflammable; its dielectric constant varies from about 5 to 7 (the dielectric constant of mineral oil is about 2.2).
(3) It is claimed that by impregnating paper for dielectric cable insulation with fluorinated paraffin it is both rendered non-inflammable and improved in other respects; the impregnated cable is resistant to moisture, is more resistant to electrical breakdown, and these properties are not subject to deterioration due to ageing.
(4) It is claimed that fluorinated paraffine has a high viscosity at the operating temperature of electric transformers, or similar apparatus, in which insulating liquids are used also as circulating cooling fluids. To produce a liquid of lower viscosity it is associated with a more highly mobile liquid, such, for example, as trichlorobenzene or tetrachloroethylene. A mixture of 50 parts by weight of fluorinated paraffin and 50 parts of either of the above liquids has been found suitable for use as an insulating and cooling liquid.

Fluorine
French: Fluor.

Chemical
Reagent in making—
Inorganic chemicals, intermediates, organic chemicals.
Starting point in making—
Fluorides.

4-Fluorocumarin

Insecticide
Repellant (U. S. 1995247) in—
Salves used to protect against the incursions of insects transmitting parasites.

1-Fluoronaphthalene-4-sulphonic Acid
Synonyms: Alpha-fluoronaphthalene-4-sulphonic acid.
French: Acide d'alphafluoronaphthalène-4-sulphonique, Acide de 1-fluoronaphthalène-4-sulphonique.
German: Alphafluornaphtalin-4-sulfonsäure, 1-Fluornaphtalin-4-sulfonsäure.

Chemical
Starting point in making various derivatives.

Miscellaneous
Mothproofing reagent (Brit. 333583) for treating—
Furs and feathers.

Textile
Mothproofing reagent (Brit. 333583) for treating—
Wool and felt.

Fluoropseudocumene
German: Fluorpseudocumen.

Miscellaneous
Mothproofing agent (Brit. 333583) in treating—
Feathers, furs, and other articles.

Textile
Mothproofing agent (Brit. 333583) in treating—
Wool and felt.

Formaldehyde
Synonyms: Formalin, Formalith, Formic aldehyde, Formol, Methanal, Methyl aldehyde, Oxymethylene.
French: Aldéhyde formique, Aldéhyde méthylique, Formaline.
German: Ameisenaldehyd, Ameisensäurealdehyd.
Spanish: Aldehido formico, Aldehido metil.
Italian: Aldeide formica, Aldeide metile, Formaldeide.

Agriculture
Fumigant for various purposes on the farm and in the dairy.
Fungicide for various purposes in the orchard.
Reagent in—
Combating root knot disease.
Disinfecting and cleansing chicken coops.
Dairy containers and other utensils and equipment.
Kennels, pig pens, spraying tables, eradicating cutworms.
Preventing mildew on wheat and spelt, rot in oats.
Sterilizing grains, particularly wheat.
Treating old soil in greenhouses and cold frames.
Soil for growing vegetables.

Analysis
Reagent in—
Analyzing blood, milk, peppermint oil, sesame oil.
Determining, detecting, and analyzing abrastol, albumen, alkaloids, alphanaphthol, benzoyl, benzoyl peroxide, bile pigments, cholesterin, cinchona alkaloids, copper, diacetic acid, glucose, guaiacol, hydrogen peroxide, indol, morphine, methylamine, nicotine, oxydimorphine, phenol, resorcinol, salicylic acid.
Making colloidal gold solutions, diabetes tests.
Titrating emetine, nitric acid.
Treating and preserving anatomical specimens, botanical preparations and specimens, collyria.
Reducing agent in—
Determining gold, mercury, metals, protein, silver.

Brewing
Reagent in—
Aiding fermentation of beer (French 551494).
Manufacturing beer.
Stimulating action of yeast in the fermentation process.

Chemical
Reagent in—
Converting toluene into a mixture of orthoxylene and paraxylene (French 639252).
Disinfecting reaction media, thus allowing the manifestation of the phenomena of autolysis and heterolysis in nitrogenous substances, particularly yeast (French 580481 and 580482).
Making acetaldehyde.
Acetoneformaldehyde.
Alcohol by fermentation of amylaceous substances (French 580481 and 580482).
Allyl methylthioisocyanate.
Alphamethylaminoanthraquinone.
Aluminum-formaldehyde sulphite.
Amaltein.
Amidol.
Amines (Brit. 208779).
Aminoacetic acid from acetic acid
Amphotropin.
Amyloform.
Anilin and acetaldehyde (French 603889).
Anilodiphenylamine.
Barium carbonate by reduction of barium sulphide (French 622486).
Cellulose esters.
Colloidal solutions of various metals, chemicals, and other substances, by reduction.
Dimethyldiaminodiorthotolylmethane.
Empyroform.
Emulsifying agents by reaction with aromatic hydrocarbons and subsequent sulphonation (French 624843).
Euguform.
External disinfectants with the addition of calcium lactate, the products being white solids containing 12 percent of formaldehyde and easily decomposed by hot water and used with the addition of lactose (Brit. 191551 and German 372284).
Formals from alcohols.
Formicin.
Glycin.
Halogenated chlorosulphonic acid esters by reaction with chlorosulphonic acid, its derivatives, or esters (Brit. 299064).
Hexamethylenetetramine.
Hydrosulphite derivatives.

Formaldehyde (Continued)

Intermediate compounds.
Light hydrocarbons by the dissociation of heavy hydrocarbons, the reagent being used in combination with aluminum chloride (Brit. 315991).
Lysoform.
Menthylchloromethyl ether.
Methylal.
Methylenedictotoine.
Methylenephenylglycol ether (Jasmal).
Methylenetetramethyldiamine.
Methylphthalimide.
Naphthaleneformaldehyde.
Neraltein.
Organic chemicals.
Rongalite.
Rosin-formaldehyde.
Sodium formaldehyde-sulphoxylate.
Sodium paraethoxyphenylaminomethylsulphonate.
Solid products, easily soluble in water, by the addition of calcium lactate at 90° C. to a 35 percent solution of formaldehyde and cooling (German 345145).
Solidified formaldehyde by the addition of calcium lactate (French 547976).
Tanning agents by reaction with aromatic compounds (French 512549), naphthalene (Brit. 251294), phenols (French 515714 and 515715), sulphosalicylic acid and sulphocresylic acid (French 515267), various substances (German 420647); by admixture with hydrochloric acid and treatment with sulphuretted hydrogen, or by conversion into thioldehyde (French 546074).
Tannoformformaldehyde.
Tetramethyldiaminediphenylmethane.
Therapeutic agents from wood-distillation oils.
Trimethylamine.
Trimethylenesulphide.
Ureaformaldehyde.
Various synthetic pharmaceuticals and aromatic chemicals.
Veroformformaldehyde.
Woodtar-formaldehyde.
Stabilizing hydrosulphites.
Treating sulphonated mineral oil products to render them odorless and tasteless.
Starting point in making—
Formic acid, oxymethylene, paraformaldehyde, urea.

Disinfectant
As a germicide.
Ingredient of—
Compositions, containing hydrogen peroxide, phenol, and pine oil, used for disinfecting and antiseptic purposes (French 640647).
Compositions used for deodorizing.
Disinfecting compositions used for treating rooms, stables, cellars, utensils, books, clothing, furs, linen, sponges, walls, ships, laundry utensils, refrigerators, cupboards, sinks, potato bins.
Disinfecting compositions containing inorganic derivatives, aliphatic derivatives, or tannins.
Preparations, containing starch, rice flour, and potassium permanganate, used for germicidal and disinfecting purposes (French 627192).

Distilling
Preservative in—
Treating barrels, other containers, and apparatus used in the manufacture of distilled liquors.

Dye
Reagent in making—
Acid violet 6B, acridin dyestuffs, acridin orange, acridin orange NO, acridin yellow, alizarin celestol, anhydroformaldehydeanilin, anthracene dyestuffs, auramin dyestuffs.
Brown coloring matters, fast to washing, alkalies, chlorine, and light, by condensation with anilin and subsequent oxidation (French 595705).
Chrome bordeaux, chrome violet, coriophosphin dyestuffs.
Dyestuffs for use in making printing ink, lithographic inks, writing inks (German 431369).
Formyl violet 5BN, gallocyanin, indigo, methylanilin, methyl blue, naphthalene green, quinolin dyestuffs, tetramethyldiaminodiphenylmethane, triphenylmethane dyestuffs, turquoise blue, wool green B5.

Electrical
Ingredient of—
Electrolytes for storage batteries.
Insulating composition with glycerin, pitch, and other ingredients.
Reagent (French 622963) in making—
Insulation materials from cashew nut oil.

Explosives
Reagent in—
Dissolving nitrocotton and pyrocotton.
Gelatinizing nitrocotton.
Making diphenyleneanilodihydrotriazol (nitron).
Explosive compound by treating a solution of formaldehyde with hydrochloric acid and then nitrating with nitric acid.

Fats and Oils
Ingredient (Brit. 321690) of—
Mixtures containing fatty oils, fatty acids, fats, resins, and naphthenic acids.

Fertilizer
Ingredient of—
Special fertilizing compounds.

Food
Reagent for—
Disinfecting cereals, nuts, seeds (French 491097).
Macerating seed fruits.
Preserving (illegal in many countries) frozen meat on ships.
Fruit, milk, other foods.
Treating foods to cure them.
Reagent in the cold refrigeration of meats.

Fuel
As a fuel in certain countries.

Glass
Reagent in—
Reducing silver salts to produce silver coating on the back of mirrors.

Glues and Adhesives
Ingredient of—
Various glues, gelatins, and adhesive preparations of animal and vegetable origin (added to preserve them).
Reagent in—
Making adhesive products by reaction with urea and condensation products of resorcinol and formaldehyde (Brit. 316194).
Glues that harden spontaneously (French 501465).
Treating glues and gelatins in powdered form to make adhesives for inert substances, the formaldehyde being used in vapor form (French 621179).

Ink
Ingredient of—
Printing inks (U. S. 1621543).
Printing inks containing colors fast to water (French 608903).
Writing inks.

Insecticide
Fumigant and constituent of fumigating compositions.
Fungicide and ingredient of fungicidal compositions for treating plants and vegetables.
Ingredient of—
Preparations for destroying flies and other insects.
Preparations for controlling blackleg disease of potatoes and stinking smut of winter wheat.
Preparations for spraying trees.
Preparations for treating pear tree cancel.
Preparations, containing alkaline earth peroxides, used for various insecticidal purposes.
Larvacide and ingredient of larvacidal preparations for use on trees, plants, and other articles.

Leather
For hardening leather and leather goods.
For preserving and stiffening grain of hides.
Ingredient of—
Compositions used for hardening leather and leather goods.
Compositions used for giving body to surface of leather during tanning.
Compositions for preserving leather.
Compositions for preserving hides by vaporizing the formaldehyde on the hair side of the hides and then covering with sodium sulphate (French 556386).
Compositions for preliminary treatment of hides before tanning (Brit. 253549).
Compositions, containing condensation products of formaldehyde and phenolsulphonic acid or naphthalenesulphonic acids, used for the pretanning of leather.
Compositions used for stiffening the grain of leather.
Compositions for tanning.

Formaldehyde (Continued)
Compositions, containing irontrichloride, used for tanning (French 514586).
Compositions used for waterproofing tanned hides.

Linoleum and Oilcloth
Ingredient (Brit. 321690) of—
Compositions, containing fatty oils, fatty acids, fats, resins, or naphthenic acids for use as a base in making linoleum.

Metallurgical
Reducing agent in—
Recovering gold and silver.

Construction
Reagent (French 600749) in—
Aiding the painting of stone in a process replacing decalcomanias for mural decoration.

Miscellaneous
Antiseptic for general purposes.
Deodorizing agent for general purposes.
Disinfecting agent for general purposes.
Embalming agent.
Ingredient of—
Antiseptic compositions.
Compositions used for coloring gypsum by treatment with solutions of metallic salts.
Compositions used for coating rugs, mats, ornaments, to prevent them from slipping (Brit. 278785).
Compositions for cleansing floors and linoleum (Brit. 255101).
Preparations for fixing hair on fur skins.
Preparations for polishing floors and linoleum (Brit. 255101).
Preparations for hardening anatomical and microscopical specimens.
Preparations for preserving botanical, zoological, and bacteriological specimens.
Preparations for preserving animal and vegetable substances.
Waterproofing compositions for treating straw hats.
Insolubilizing agent for general purposes.
Preservative for general purposes.
Reagent in—
Making phonograph record blanks.
Wax recording blanks from compositions containing fatty oils, fats, fatty acids, resins, or naphthenic acids (Brit. 321690).
Treating pelts and fur skins to preserve them, by vaporizing the formaldehyde against the hair side of the pelts and then covering with sodium sulphate (French 556386).
Rosin to make a molding powder in construction work of various kinds.
Spangles and other articles made of gelatin, for the purpose of making them insoluble.

Paint and Varnish
Ingredient of—
Disinfectant whitewashes, lacquers, limewashes, paints, varnishes.
Reagent in making—
Dry colors.
Varnish bases starting from oil of cashew nut (French 622963).
Varnish bases from compositions containing fatty oils, fats, fatty acids, resins, or naphthenic acids (Brit. 321690).

Paper
Ingredient of—
Compositions used in the manufacture of greaseproof and waterproof paper and paperboard (U. S. 1723581).
Waterproofing compositions containing glycerin pitch (Brit. 276100).
Reagent for—
Treating cellulose products to improve them, the formaldehyde being used in alkaline solution (French 584904).
Waterproofing paper sized with albumens or albumenoids.
Paper, paperboard, and paper and pulp products containing gelatin or glue.
Wallpaper.

Perfume
Ingredient of—
Antiperspiration products, deodorizing preparations.

Petroleum
Ingredient of—
Cutting oils.

Pharmaceutical
For various sterilizing operations.
Ingredient of—
Preparations for allaying the itching of insect bites.
Preparations for treating the feet.
Various other pharmaceutical preparations.
Suggested for use as—
Antiseptic, antihydrotic, bactericide, disinfectant, inhalant.

Photographic
Developer for—
Films and plates (used in conjunction with hydroquinone).
Hardening agent for—
Negatives and prints.
Reagent for—
Rendering double transfer paper insoluble.
Toning gelatin chloride papers.
Reagent in—
Chrome printing.

Plastics
Ingredient (French 603452) of—
Plastic compositions containing albumenoids.
Reagent in—
Making bone-like galalith plastics.
Plastic compositions from cashew nut oil (French 622963).
Plastic compositions from aromatic hydrocarbons and subsequent sulphonation (French 624843).
Treating casein plastics for the purpose of insolubilizing and stabilizing them.

Resins and Waxes
Reagent in—
Dyeing artificial resins made from aromatic amines (French 573837).
Making artificial resins of the cyclohexanone type.
Artificial resins from anilin, toluidins, and naphthylamines (French 628650).
Artificial resins of the phenol, cresol, and urea types.
Artificial resins from vinyl esters, such as vinyl acetate (French 643419).

Rubber
Accelerator in the vulcanization process.
Reagent for—
Coagulating rubber latex.
Deodorizing rubber deposited by the electrophoresis process (Brit. 312443).
Preserving rubber latex.

Sanitation
Disinfectant for—
Houses and other premises.
Sewage and other wastes.
Ships and other carriers.

Soap
Ingredient of—
Disinfecting soaps.
Reagent (French 624843) in making—
Detergents with aromatic hydrocarbons and subsequent sulphonation.

Starch
Reagent in treating—
Dextrins.
Various starches and starch products for the purpose of hardening and preserving them.

Sugar
Reagent for preserving—
Beet sugar juices.
Cane sugar juices.
Syrups of various sorts.

Textile
——, *Bleaching*
As a bleaching agent.
As a mordant.
Ingredient of—
Baths for bleaching raw wool, wool waste and all textile materials of animal and vegetable origin (French 571298).
Baths for bleaching silk.

——, *Dyeing and Printing*
Ingredient of—
Dye bath containing methylene blue.
Dye bath (to aid in the penetration of the dyestuff) (French 633505).
Dye bath (to increase the fastness of the dyestuff).
Dye baths containing substantive dyestuffs (to increase the fastness of the color to washing and light).

Formaldehyde (Continued)
Solubilizing or dispersive agent (Brit. 276100) in dyeing and printing yarns and fibers with—
Acridin dyestuffs.
Aminoanthraquinone dyestuffs, reduced or unreduced.
Anthraquinone dyestuffs, reduced or unreduced.
Azins, azo dyestuffs, basic diarylmethane dyestuffs, basic triarylmethane dyestuffs, benzoquinoneanilides, indigoids.
Naphthoquinones, reduced or unreduced.
Naphthoquinoneanilides, nitroarylamines, nitroarylphenols, nitrodiarylmethanes, nitrodiarylphenols, oxazins, pyridin dyestuffs, quinolins.
Quinoneimides, reduced or unreduced.
Sulphur dyestuffs, thiazonins, xanthenes.

——, *Finishing*
Ingredient of—
Compositions for softening raw wool, wool waste, and all textile materials of animal or vegetable origin (French 571298).
Compositions for improving cellulosic products, the formaldehyde being used in alkaline solution (French 584904).
Compositions for producing effect threads in fabrics (German 423602).
Compositions for obtaining surface finishes on textiles.
Compositions for sizing fabrics and yarns.
Compositions for glossing fabrics.
Compositions for stiffening fabrics.
Compositions for waterproofing fabrics impregnated with glue, gelatin, or albumens.
Compositions for waterproofing sailcloth.
Compositions for weighting silk.
Reagent for—
Treating fabrics sized with albumens or albumenoids in order to insolubilize the size.

——, *Manufacturing*
Ingredient of—
Baths for degreasing and removing the suint from raw wool, wool waste (French 571298).
Baths containing alkalies for improving the rayon filament (French 571460).

——, *Miscellaneous*
Reagent for—
Protecting wool against the action of hot water.
Stripping colors from dyed and printed textile yarns and fabrics.

Wine
Reagent for—
Disinfecting casks, preserving vinous liquors.

Woodworking
Reagent (French 604897) for—
Preserving wood and wood products.

Formaldoxime

Fuel
Primer (Brit. 429763) for—
Diesel engine fuel oils produced by the hydrogenation of coal.

Petroleum
Primer (Brit. 429763) for—
Diesel oils containing a high proportion of aromatic bodies.

Formamide
Synonyms: Methanamide, Methanamine.
French: Formamide.
German: Formamid.

Chemical
Ionizing solvent in chemical reactions.
Reagent in making—
Various intermediate chemicals.
Solvent for inorganic salts.
Starting point in making—
Chloral, chloralformamide, formamide sulphate, methylamine.
Various compounds with formaldehyde, paraldehyde, and trioxymethylene.

Textile
Reagent in—
Retting flax and similar vegetable fibers.

Formamide Acid Sulphate
Synonyms: Formamide bisulphate.
French: Bisulphate de formamide.
German: Formamidbisulfat, Doppelte schwefelsacuresformamid.

Chemical
Starting point (U. S. 1584907) in making—
Amylformamide acid sulphate, butylformamide acid sulphate, cinnamylformamide acid sulphate, ethylformamide acid sulphate, formylformamide acid sulphate, methylformamide acid sulphate, phthalylformamide acid sulphate, propylformamide acid sulphate.

Formic Acid
Synonyms: Aminic acid, Formylic acid, Hydrogencarboxylic acid, Methane acid.
Latin: Acidum formicarum, Acidum formicum.
French: Acide formique.
German: Ameisensäure, Formylsäure.
Spanish: Acido formico.

Adhesives
Process material in making—
Adhesive cement (U. S. 1231519).
Casein glue.

Agricultural
Reagent (U. S. 1271591) for—
Treating banana plants.

Analysis
Reagent in—
Analytical methods used in control and research work.

Beverage
Preservative (U. S. 1401700) in—
Beverage.
Starting point in making—
Secondary alcohol esters for use in cordials.
Secondary alcohol esters for use as flavoring agents.

Brewing
Antiseptic for—
Yeast mash.

Cellulose Products
Addition agent (U. S. 1467493) to—
Cellulose acetate coagulating bath.
Precipitant for—
Viscose.
Process material in making—
Cellulose acetate (U. S. 1457131).
Cellulose acetonitrate, cellulose esters.
Solvent for cellulose esters (U. S. 1283183).
Solvent for cellulose ethers (U. S. 1217027 and 1217028).
Solvent for—
Cellulose acetate, ethyl cellulose.
Starting point in making—
Cellulose formate, cellulose formylphosphate.
Solvent for cellulose nitrate (U. S. 1260977 and 1283183).
Solvent for cellulose acetate (U. S. 1260977).
Solvent for cellulose formate (U. S. 1260977).
Solvent for cellulose sulphoacetate (U. S. 1260977).

Chemical
Absorbent (U. S. 1212199) for—
Sulphur dioxide, sulphur trioxide.
Catalyst in making—
Lead arsenate and other chemicals.
Extractant for—
Aluminum from clays used for clarifying oils and filling paper (Brit. 404991).
Ergot.
2-Phenylquinolyl-4-piperidoethanone hydrochloride (U. S. 1434306).
Liberating agent (U. S. 1418356) for—
Potassium salts from leucite.
Peptizing agent (Brit. 398517) in making—
Adsorbent gels, catalysts.
Process material in making—
Allyl alcohol.
2-Amino-1-(2′-phenyl-4′-quinolyl)ethanol (U. S. 1434306).
Bis- (N-ethyl-N-hydroxyethylaminophenyl) methane-omegasulphonic acid (U. S. 1483084).
Borneol (U. S. 1415340).
Cholic acid compound (U. S. 1218209).
Dehydrogenation catalysts (U. S. 1271013).
Dihydrodiethyl sulphide formic esters (U. S. 1422869).
Dinitrophenyl formate (U. S. 1198040).
Ethyl acetate (U. S. 1425624 and 1425625).
Formates, such as amyl formate, copper formate, ethyl formate, methyl formate, lead formate, nickel formate, zinc formate.
Formylisoborneol (U. S. 1420399).
Furfural (U. S. 1322054).
Hydrogenation catalysts (U. S. 1271013, 1482740, and 1511520).
Isobutyric acid (Brit. 417496).
Lead chromate.

Formic Acid (Continued)
Limonene.
Nickel catalysts (U. S. 1482740).
Nickel formylcarbonate.
Organic esters.
Para-amino-N-methylformanilide (U. S. 1273901).
Paracymene-5-sulphonic acid (U. S. 1332680).
2-Piperidyl-1-(2'-phenyl-4'-quinolyl)ethanol (U. S. 1434306).
Radium compound.
Sodium hyposulphite.
Soluble starch (U. S. 1207177).
4,4'-Tetramethyldiaminotriphenylcarbinol (U. S. 1483233).
Thymol (U. S. 1332680).
Reagent (U. S. 1503229) for removing—
Arsenic compounds from copper sulphate.
Aluminum compounds from copper sulphate.
Iron compounds from copper sulphate.
Reagent for treating—
Camphene (U. S. 1420399).
Pine oil (U. S. 1433666).
Reagent in various manufacturing processes.
Reducing agent in various manufacturing processes.
Revivifying agent (U. S. 1431982) for—
Nickel catalysts.
Solvent (U. S. 1350820) for—
Ethylstarch.
Solvent in various manufacturing processes.
Starting point in making—
Amyl formate, benzyl formate, butyl formate, citronellyl formate, ethyl formate, menthyl formate, methyl formate, propyl formate, rhodinyl formate, terpenyl formate.

Cosmetic
Starting point in making—
Aromatic formates (see under "Chemical").

Dye
Ingredient of—
Printing pastes.
Reagent in making—
Dyestuffs.
Dyestuffs for cellulose acetate (U. S. 1483797).
Reducing agent for—
Cymidinsulphonic acid diazo compounds (U. S. 1332680 and 1432298).
Dyestuffs.

Electrical
Process material (U. S. 1474482) in making—
Electrical insulation.

Explosives and Matches
Stabilizing agent (U. S. 1504986) in making—
Nitrodextrin, nitrostarch.

Food
Antiseptic for—
Yeast.
Preservative for—
Foodstuffs, fruit juices, honey, sugar, syrups.
Starting point in making—
Secondary alcohol esters for use as flavors.

Glass
Bonding agent (U. S. 1478862) for—
Celluloid and glass.
Process material in—
Silvering glass mirrors.

Glue and Gelatin
Hydrolizing agent (U. S. 1206189) for—
Glue.
Preservative for—
Glue, gelatin.
Process material (U. S. 1217027) in making—
Gelatin substitute.
Solvent (U. S. 1210987) for—
Gelatin.

Heat and Power
Inhibitor (U. S. 1405783) of—
Boiler-scale formation.

Insecticide
Ingredient (U. S. 1381586) of—
Insecticidal mixtures with hydrocyanic acid.
Starting point (U. S. 1494085 and 1515182) in making—
Moth-repellants.

Laundering
Sour in treating—
Washroom liquors.

Leather
As a tanning agent (U. S. 1426322 and 1413488).
Dearmoring agent (U. S. 1395773 and 1412968) for—
Shark skin dermal armor.
Deliming agent for—
Hides, pelts, skins.
Disinfectant for—
Hides, pelts, skins.
Process material in dyeing—
Hides, pelts, rabbit skins (Brit. 404960), skins.
Process material (U. S. 1245977) in making—
Artificial leather.
Preservative for—
Hides, pelts, skins.
Soaking agent for—
Hides, pelts, skins.
Softening agent for—
Hides, pelts, skins.

Metallurgical
Etching agent for—
Brass, copper, steel, zinc.
Ingredient (Brit. 410323) of—
Rust removing compositions, rust preventing compositions.
Precipitating agent (U. S. 1472115) for—
Copper.
Reagent (U. S. 1452662) in—
Lead ore sulphidizing, zinc ore sulphidizing.

Miscellaneous
As a preservative.
As a solvent.
Process material in—
Dyeing feathers, hair.
Sizing and dyeing straw hats (U. S. 1206189).
Process material in making—
Artificial hair (U. S. 1217028).
Containers for food products, such as biscuits, candy, chocolate, fruit (U. S. 1488634).
Hat sizings (U. S. 1206189 and 1224125).
Linoleum substitutes (U. S. 1245978 and 1245984).
Tile (U. S. 1245984).

Paint and Varnish
Process material in making—
Paint and varnish removers, varnish (U. S. 1280861).

Paper
Process material (U. S. 1500500) in making—
Paper.

Petroleum
Process material (Brit. 417496) in making—
Isobutyric acid from petroleum cracking gases.

Pharmaceutical
Extractant for—
Ergot.
In compounding and dispensing practice.
Process material in making—
Opium extracts, pharmaceutical chemicals.
Suggested for use as—
Local astringent and counterirritant.
Treating agent (U. S. 1460832) for—
Adrenal glands.

Photographic
Process material (U. S. 1214940) in—
Dyeing films, dyeing plates.

Plastics
Process material in making—
Casein, celluloid substitutes, cellulose formate, ivory substitutes, phenol-aldehyde substitutes, plastics (U. S. 1474482).

Resins
Starting point in making—
Synthetic resins with the aid of anilin, formaldehyde, and woodflour (Brit. 401965).
Synthetic resins with the aid of anilin formaldehyde, and paraformaldehyde (Brit. 404469).

Rubber
Coagulant for—
Rubber latex.
Process material in making—
Rubber substitutes (U. S. 1471059).
Synthetic rubber (U. S. 1185654, 1161904, 1289444, and 1436819).

Textile
Degumming agent for—
Vegetable fibers, such as cotton, hemp, esparto, flax, straw.
Felting agent for—
Silk.
Mordant in—
Dyeing operations.

Formic Acid (Continued)
Process material in—
Dyeing cellulose acetate fabrics (U. S. 1378443, 1517581, and 1517709).
Dyeing cotton.
Dyeing cotton fabrics (U. S. 1517709).
Dyeing and printing fabrics.
Dyeing silk.
Dyeing woolen goods with acid dyes.
Waterproofing rayon fabrics (U. S. 1377110).
Retting agent for—
Vegetable fibers, such as cotton, hemp, esparto, flax, straw.
Substitute for—
Acetic or sulphuric acid in dyeing and printing fabrics.

Formic Acid Ester of Grapeseed Alcohol
Bituminous
Solvent (Brit. 445223) for—
Asphalt and other bituminous bodies.
Dye
Solvent (Brit. 445223) for—
Dyestuffs, particularly oil-soluble coaltar dyes.
Fats, Oils, and Waxes
Solvent (Brit. 445223) for—
Fats, oils, waxes.
Resins
Solvent (Brit. 445223) for—
Oil-soluble glycerol-phthalic acid resins.
Polymerized vinyl compounds.
Synthetic resins.
Rubber
Solvent (Brit. 445223) for—
Rubber.

Formic Acid Ester of Ricinoleic Alcohol
Bituminous
Solvent (Brit. 445223) for—
Asphalt and other bituminous bodies.
Dye
Solvent (Brit. 445223) for—
Dyestuffs, particularly oil-soluble coaltar dyes.
Fats, Oils, and Waxes
Solvent (Brit. 445223) for—
Fats, oils, waxes.
Resins
Solvent (Brit. 445223) for—
Oil-soluble glycerol-phthalic acid resins.
Polymerized vinyl compounds.
Synthetic resins.
Rubber
Solvent (Brit. 445223) for—
Rubber.

Formyl-2-aminoanthraquinone
French: Formyle-béta-aminoanthraquinone.
German: Formyl-2-aminoanthrachinon.

Dye
Starting point (Brit. 282854) in making dyestuffs with—
Acetaldehyde, benzaldehyde, butyraldehyde, cinnamaldehyde, crotonaldehyde, formaldehyde, heptaldehyde, hexaldehyde, paraformaldehyde, propionaldehyde, succinaldehyde.

2-(Formylamino)diphenylene Oxide
Rubber
Antiaging agent (Brit. 422191).

Formyl Carbamide
French: Carbamide de formyle, Carbamide formylique.
German: Formylcarbamid.

Chemical
Reagent in making—
Pharmaceuticals and other derivatives.

Resin and Waxes
Starting point (Brit. 292912) in making synthetic resins with—
Acetylsalicylic acid, aliphatic dibasic acids, ammonium salicylate, anthranilic acid, benzoic acid, gallic acid, hydronaphthoic acid, magnesium salicylate, oxalic acid, phenolic dibasic acids, phthalic acid, salicylamide, salicylic acid, strontium salicylate, succinic acid.

Formylphenylhydrazin
Chemical
In organic syntheses.
Electrical
Stabilizer (Brit. 423938) for—
Transformer oils.
Fats and Oils
Stabilizer (Brit. 423938) for—
Vegetable oils.
Fuel
Stabilizer (Brit. 423938) for—
Coal-carbonization spirits.
Lubricant
Stabilizer (Brit. 423938) for—
Lubricants, lubricating oils.
Petroleum
Stabilizer (Brit. 423938) for—
Petroleum oils, shale oils.

Fuchsin
Synonyms: Anilin red, Aniline red, Azaleine, Erythrolbenzin, Fuchsiacin, Harmaline, Magenta, Magenta red, Rosein, Rubin, Solferino.
French: Rouge d'aniline.
German: Anilinrot, Fuchsiacin.
Dye
Starting point in making—
Acid magenta, alizarin yellow FS, fuchsin lakes, fuchsin scarlet, lime pink.
Fats and Oils
As a coloring agent.
Ink
Color in making—
Printing inks, stamp-pad inks, writing inks.
Leather
As a coloring.
Miscellaneous
Coloring for—
Feathers, hemp, jute, straw.
Paint and Varnish
Color for—
Lacquers, varnishes.
Paper
As a coloring.
Pharmaceutical
In compounding and dispensing practice.
Textile
——, *Dyeing and Printing*
Dyestuff for—
Silks, cottons, half-silks and other mixed fabrics.
Wool yarns and fabrics.
Waxes and Resins
Color for—
Waxes and resins.

Fuchsin Hydrochloride
French: Chlorohydrate de fuchsine, Hydrochlorure de fuchsine.
German: Chlorwasserstoffsaeurefuschinester, Fuchsinchlorhydrat, Fuchsinhydrochlorid.
Dye
Starting point (Brit. 298101) in making triarylmethane dyestuffs with—
2:3:6-Naphthol dicarboxylates.
Sodium 2:3-hydroxynaphthoate.

Fuller's Earth, Activated
French: Terre à foulon activée.
German: Aktivierte fullererde, Aktivierte walkerde, Aktivierte walkererde.
Chemical
Reagent in—
Clarifying and decolorizing aqueous solutions of various chemical and pharmaceutical products.
Explosives
Filler for—
Dynamites and permissibles.
Fats and Oils
Reagent in—
Decolorizing and purifying various animal and vegetable fats and oils.
Food
Reagent in—
Clarifying and decolorizing lard and other edible fats and oils.

Fuller's Earth, Activated (Continued)

Petroleum
Reagent in—
Clarifying, decolorizing, and purifying petroleum distillates, oils, and waxes.

Waxes and Resins
Reagent in—
Clarifying and decolorizing various waxes and resins.

Fumaryl Chloride
French: Chlorure de fumaryle, Chlorure fumarylique.
German: Chlorfumaryl, Fumarylchlorid.

Chemical
Starting point in making various derivatives.

Fats and Oils
Bleaching agent (Brit. 328544) in admixture with hydrogen peroxide.

Food
Bleaching agent (Brit. 328544) in admixture with hydrogen peroxide in treating—
Egg yolk, flour, meal.

Soap
Bleaching agent (Brit. 328544) in admixture with hydrogen peroxide.

Waxes and Resins
Bleaching agent (Brit. 328544) in admixture with hydrogen peroxide.

Furfural
Synonyms: Artificial oil of ants, Fulfuraldehyde, Furol, Pyromucic aldehyde.
French: Aldéhyde pyromucique, Furanaldéhyde.
German: Furanaldehyd.

Abrasive
Ingredient (Brit. 260354) of—
Grinding compositions.

Agricultural
Reagent and ingredient of—
Compositions used in dressing the wounds of trees.
Compositions used in treating seeds to prevent growth of fungi.

Analysis
Reagent for—
Sesame oil identification.

Chemical
As a general solvent.
Reagent in making—
1:2-Amyleneglycol, 1:5-amyleneglycol, anesthetics, antioxidants, antiseptics, maleic acid, normal amyl alcohol, pyromucic acid, succinic acid.
Reagent in making products used in printing cotton and silk and in dyeing acetate rayon, with the aid of—
Alpha-amino-4(4'-aminophenylamino)-anthraquinone.
Alpha-amino-4-hydroxyanthraquinone.
Alpha-methylamino-4-aminoanthraquinone.
Alphaphenylamino-4-aminoanthraquinone.
1:5-Diamino-4:8-diphenyldiaminoanthraquinone.
1:4-Diaminoanthraquinone.
1:5-Diaminoanthraquinone.
1:8-Diaminoanthraquinone.
1:5-Diamino-4:8-dihydroxyanthraquinone.
1:5-Diamino-4-phenylaminoanthraquinone.
5-Chloro-1:4-diaminoanthraquinone.
Reagent (Brit. 275862) in purifying—
Rosin.
Solvent (Brit. 295335) in making—
Impregnating solutions, used for various chemical purposes, containing phenol-aldehyde condensation products.
Starting point in making—
Amyl furoate, allyl furoate, butyl furoate, dithiofuroic acid, ethyl furoate, furacrolein, furacrylic acid, furan, furfuryl acetate, furfuryl acetone, furfuryl alcohol, furfuryl butyrate, furfuryl propionate, furfuramide, furil, furoin, furoic acid, furyol chloride, furyl alcohol, hydrofuramide, methyl furan, methyl furoate, propyl furoate, sodium furacrylate, tetrahydrofurfuryl alcohol.

Dye
Reagent and starting point in making various synthetic dyestuffs.

Electrical
Ingredient (U. S. 1697870) of—
Insulating compositions.

Explosives
Solvent for—
Nitrocellulose in the manufacture of military and commercial explosives.

Glues and Adhesives
Preservative in making—
Glues and other adhesives.

Gums
Solvent for—
Gums and gum compositions.

Insecticides
Ingredient of various insecticidal preparations.

Leather
Antiseptic in—
Tanning skins.
Ingredient of—
Extracts obtained from drum tannage, added for the purpose of preventing the grain of leather from drawing up.
Vegetable tanning solutions and liquors, added to reduce the astringency of the tannins.
Reagent used to lighten the color of leather.

Miscellaneous
Ingredient of—
Impregnating compositions containing phenol-aldehyde condensation products (Brit. 295335).
Polishing compositions (Canadian 260384).
Preservative compositions used for biological specimens.
Shoe polishing and dyeing compositions.

Paint and Varnish
Reagent (U. S. 1596413) in making—
Paint and varnish removers.
Solvent in making—
Lacquers and varnishes containing phenol-aldehyde condensation products.
Varnishes, along with turpentine.

Pharmaceutical
In compounding and dispensing practice.

Plastics
Solvent in making—
Compositions containing nitrocellulose, cellulose acetate, and other cellulose esters and ethers.
Compositions containing phenolaldehyde condensation products (Brit. 295335).

Resins and Waxes
Solvent for—
Resins and in compositions containing them.
Starting point in making—
Artificial resins with anilin, acetone, phenol.
Photosensitive resins.

Rubber
Solvent in making—
Rubber cements.

Sanitation
As a general antiseptic and germicide.

Furfural Acetone
French: Acétonne de furfural, Acétone furfuralique.
German: Furfuralaceton.

Chemical
Starting point in making various derivatives.
Solvent and plasticizer (Brit. 313133) for—
Cellulose acetate, cellulose esters and ethers, cellulose nitrate.
For uses, see under general heading: "Solvents."

Furfuraldehydecyanohydrin

Chemical
Starting point in making—
Ethyl ester of furfurylglycollic acid (Brit. 264143).

Furfuramide

Rubber
Accelerator in—
Vulcanization.

Furfuramide Chloride

Agriculture
Disinfectant for—
Seeds, soil, and plants.

Woodworking
For treating lumber to control sap stain and blue stain.

Fusel Oil

Synonyms: Amyl alcohol, Amylic alcohol, Fermentation amyl alcohol, Fousel oil, Grain oil, Hydrate of amyl, Hydrated oxide of amyl, Potato oil, Potato spirit oil.

Latin: Alcohol amylicum.

French: Alcool amylique, Huile de fousel, Huile fouselique, Huile de grain, Huile de pommes de terre.

German: Fuselöl.

Note: A by-product of alcoholic fermentation; the commercial product (refined fusel oil) is an oily compound consisting, essentially, of isoamyl alcohol (isobutyl carbinol or 3-methylbutanol) with a small percentage of active amyl and lower alcohols.

Analysis
General solvent in—
Analytical processes involving control and research.
Solvent for—
Alkaloids.

Aviation
Constituent (U. S. 1420006 and 1420007) of—
Airplane fuel.

Beverage
Solvent in making—
Fruit flavoring syrups and extracts.

Cellulose Products
Ingredient of solvent mixtures for—
Cellulose acetate, cellulose esters and ethers, nitrocellulose.
Solvent for—
Nitrocellulose.

Ceramic
Solvent in—
Coating compositions, containing nitrocellulose and resins, used for the decoration and protection of ceramic ware.

Chemical
General solvent.
Process material in—
Organic syntheses.
Solvent miscible with—
Ethyl alcohol, ether, essential oils.
Solvent for—
Alkaloids, camphor, fats, iodine, phosphorus, resins, sulphur.
Starting point in making—
Amyl acetate, amyl butyrate, amyl formate, amyl oleate, amyl oxalate, amyl phthalate, amyl propionate, amyl tartrate, amyl valerate, pharmaceutical chemicals, synthetic flavorings.

Cosmetic
Solvent for—
Aromatic agents, cellulosic bases.

Electrical
Solvent in making—
Compositions, containing nitrocellulose, as well as resins, used for insulating and coating electrical equipment and wiring.

Explosives
Gelatinizing agent.
Solvent for—
Nitrocellulose.

Fats, Oils, and Waxes
Solvent for—
Essential oils, fats, waxes.

Food
Solvent in making—
Fruit flavoring syrups and extracts.

Glass
Solvent in—
Compositions, containing nitrocellulose and resins, used in the manufacture of nonscatterable glass and as coatings for the decoration and protection of glassware.

Leather
Solvent in—
Compositions, containing nitrocellulose and resins, used in the manufacture of artificial leather and as coatings for the protection and decoration of leather goods.

Mechanical
Constituent (various patents) of—
Fuels for internal-combustion engines.

Metal Fabricating
Solvent in—
Coating compositions, containing nitrocellulose and resins, used for protection and decoration of metal articles.

Miscellaneous
Solvent in—
Coating compositions, containing nitrocellulose and resins, used for the decoration and protection of various fibrous compositions.

Paint and Varnish
Gloss imparter in—
Dopes, enamels, lacquers, paints, varnishes.
Promoter of—
Good flowing properties in dopes, lacquers, enamels, paints, varnishes.
Solvent having good blending properties.
Solvent in—
Paint and varnish removers.
Solvent in making—
Paints, varnishes, dopes, enamels, and lacquers containing nitrocellulose and resins.

Paper
Solvent in—
Compositions, containing nitrocellulose and resins, used in the manufacture of coated paper and as a coating for the decoration and protection of paper and pulp products.

Pharmaceutical
In compounding and dispensing practice.
Solvent for—
Alkaloids, camphor, iodine.
Starting point in making—
Amyl compounds for pharmaceutical and medical use, such as amyl nitrite and amylbarbital.

Photographic
Solvent in making—
Films from nitrocellulose.

Plastics
Solvent in making—
Compositions containing nitrocellulose and resins.

Resins
Solvent for resins of many types.

Rubber
Solvent in—
Coating compositions, containing nitrocellulose and resins, used for the decoration and protection of rubber goods.

Stone
Solvent in—
Coating compositions, containing nitrocellulose and resins, used for the decoration and protection of artificial and natural stone.

Textile
Solvent in—
Compositions, containing nitrocellulose and resins, used in the production of coated textile fabrics.

Wood
Solvent in—
Coating compositions, containing nitrocellulose and resins, used for the protection and decoration of woodwork.
Plastic compositions used for decorating, filling, and repairing woodwork.

Gallamide

Synonyms: Gallamid, Gallic acid amide.

French: Amide d'acide gallamique, Amide d'acide gallique.

German: Gallussaeureamid.

Dye
Starting point in making—
Amide gallamin blue, coelestin blue B, corein RR, corein AR, cyanazurin, gallamin blue, modern cyanin.

Gallic Acid

Latin: Acidum gallicum.

French: Acide gallique.

German: Gallussäure.

Spanish: Acido galico.

Italian: Acido gallico.

Analysis
Reagent for—
Analyzing alkaloids.
Detecting small quantities of iron (ferric) salts, for example, in mineral waters.
Small proportions of free mineral acids.
Determining dioxyacetone.

Chemical
Starting point in making—
Anthragallol.
Bismuth oxyiodogallate (airol).

Gallic Acid (Continued)
Bismuth subgallate (dermatol).
Compounds with acetaldehyde and benzaldehyde, as well as other aldehydes of the aromatic and aliphatic series.
Compounds with acetic acid and acetic anhydride.
Dimethylanthragallol, ellagic acid, flavellagic acid, galloformin, hexamethylenetetramine gallate, intermediate chemicals, methyl gallate (gallicine), methylenedigallic acid, organic chemicals, purpurogallincarboxylic acid, pyrogallol, rufigallic acid.
Salicylic-acid pharmaceutical (salitannol).
5:6:7-Trihydroxy-2-methylanthraquinone.
Various aromatic chemicals.
Various pharmaceutical chemicals.

Dye
Starting point in making—
Alizarin, alizarin brown, anthracene brown, anthraquinone dyestuffs, benzoin yellow, blue 1900 TC, chrome heliotrope, chromocyanin, coerulein S, delphin blue B, dihydroxyanthragallol, gallazin A, gallein, gallocyanin dyestuffs, gallocyanin MS, galloflavin W, gallogreen DH, hexaoxyanthraquinone, indalizarin R, indalizarin green, leuco gallothionone DH, modern blue, modern violet N, oxazin dyestuffs, phenocyanin TC, phenocyanin TV, thiazin dyestuffs, thionin dyestuffs, ultracyanin B, ultra-violet LGP, xanthone dyestuffs.

Ink
Ingredient of—
Writing inks.

Leather
Reagent in making—
Tannin compounds.
Tanning agent.

Metallurgical
Ingredient of—
Baths used for the production of brown colorations on various metals.

Miscellaneous
Ingredient (U. S. 1752933) of—
Wax baths used for coating various products (added for the purpose of prolonging the life of the coating).

Paint and Varnish
Ingredient of—
Paints and varnishes used for the production of a mellowing coat in obtaining decorative finishes on wood.

Paper
Reagent in the manufacture of certain papers.

Pharmaceutical
Suggested for use as—
Hemostatic and astringent and in treating various diseases, such as hematemesis, hematuria, diarrhoea, albuminuria.

Photographic
As a developer in certain processes.

Printing
In process engraving and litho work.
Ingredient of—
Discharge printing pastes used in lithography.

Textile
Ingredient of—
Baths, containing ammonia and olein in mordant preparations used in dry dyeing process with carbon tetrachloride.
Various dye baths.

Gamma-2-benzyl Piperidinopropylbenzoate

Pharmaceutical
Claimed (U. S. 1997828) as—
Local anesthetic.

Gammachlorobetahydroxypropylpiperidin, Normal

Textile
Assistant (Brit. 454320) in—
Textile finishing processes.

Gammachlorobutyric Acid Cyclohexylester

Detergent
Starting point (Brit. 408754) in making—
Saponaceous products by reaction with tertiary amines, which may be used alone or with other soaps, fillers, or compounds giving off oxygen.

Gammachlorobutyric Acid Dodecylester

Detergent
Starting point (Brit. 408754) in making—
Saponaceous products by reaction with tertiary amines, which may be used alone or with other soaps, fillers, or compounds giving off oxygen.

Gammachlorobutyric Acid Octadecylester

Detergent
Starting point (Brit. 408754) in making—
Saponaceous products by reaction with tertiary amines, which may be used alone or with other soaps, fillers, or compounds giving off oxygen.

Gammachlorovaleric Acid Cyclohexylester

Detergent
Starting point (Brit. 408754) in making—
Saponaceous products by reaction with tertiary amines, which may be used alone or with other soaps, fillers, or compounds giving off oxygen.

Gammachlorovaleric Acid Dodecylester

Detergent
Starting point (Brit. 408754) in making—
Saponaceous products by reaction with tertiary amines, which may be used alone or with other soaps, fillers, or compounds giving off oxygen.

Gammachlorovaleric Acid Octadecylester

Detergent
Starting point (Brit. 408754) in making—
Saponaceous products by reaction with tertiary amines, which may be used alone or with other soaps, fillers, or compounds giving off oxygen.

Gammadiethylaminopropyldiphenylacetamide

Pharmaceutical
Claimed (Brit. 438659) to possess—
Physiological properties resembling those of atropine.

4-Gamma-dinormal-butylaminopropoxy-3-carbogamma-dinormal-butylaminopropoxydiphenyl

Pharmaceutical
Claimed (U. S. 1976921, 1976922 and 1976924) as—
Anesthetic.

4-Gammahydroxypropylaminoanthraquinone

Textile
Dyestuff (Brit. 447090 and 447037) for imparting—
Deep-blue shades to acetate rayon, either by dyeing or printing.

Gamma-4-normal-butylcyclohexylbutyric Acid

Miscellaneous
As a wetting agent (Brit. 449865).

Gammaphenylenediamine

Dye
Starting point in making—
Azidin black F extra.

Gamma-2-phenylethyl Piperidinopropylbenzoate

Pharmaceutical
Claimed (U. S. 1997828) as—
Local anesthetic.

Garnet Lac

Leather
Ingredient of—
Dressing compositions.

Miscellaneous
Ingredient of—
Compositions used for making phonograph records.
Shoe polishes.

Gas Oil
French: Huile de gaz.
German: Gasoel.

Gas
Raw material in making—
Carburetted water gas by admixture with blue gas.
Oil gas.

Insecticide
Ingredient of—
Sulphuric acid mixtures.

Paint and Varnish
Starting point in making—
Varnish ingredient by treatment with sulphuric acid.

Gas Oil (Continued)

Textile

——, *Bleaching*

Starting point in making—
 Wetting agent by treatment with sulphuric acid.

——, *Dyeing*

Starting point in making—
 Wetting agent with sulphuric acid.

Geranyl Acetate

French: Acétate de géranyle, Acétate géranylique, Éther géranylacétique.
German: Essigsäuresgeranylester, Essigsäuresgeranyl, Geranylacetat, Geranylazetat.
Spanish: Acetato de geranil.
Italian: Acetato di geranile.

Perfume

Ingredient of—
 Geranium essence, lavender preparations, tuberose preparations, ylang-ylang preparations.
Perfume in making—
 Cosmetics.

Soap

Perfume in making—
 Toilet soaps.

Geranyl Carboxethylate

French: Carboxéthylate de géranyle.
German: Geranylcarboxaethylat.
Spanish: Carboxetilato de geranil.
Italian: Carbossietilato di geranile.

Perfume

Ingredient (French 650100) of—
 Perfumes.

Gilsonite

Synonyms: Gilsonit, Uintahite, Uintahit, Uintaite, Uintait.

Building

As a waterproofing, wearproofing, and weatherproofing agent.
Ingredient of—
 Waterproofing compositions, wearproofing compositions, and weatherproofing compositions, used for treating various building materials, such as concretes, stuccos, and masonry (Brit. 335247).

Electrical

Ingredient of—
 Insulating compositions for various electrical purposes.

Ink

Ingredient (U. S. 1725649) of—
 Quick-drying intaglio printing inks.

Miscellaneous

Binder in—
 Paving roads with cement.
Ingredient of—
 Compositions used in the manufacture of insulating tape.
 Paving compositions.
 Pressed and molded compositions used as insulation.
 Waterproofing compositions.
 Weatherproofing compositions.
 Wearproofing compositions.

Paint and Varnish

Ingredient of—
 Coach varnishes, japans, paints, roofing compositions, roof cements, tree paints (U. S. 1730724), varnishes.

Paper

Ingredient (Brit. 335247) of—
 Waterproofing compositions used in the treatment of paper, pulp, and products made from them.

Rubber

Ingredient of—
 Bath in compounding (used to aid the rubber to resist oxidation and changes in temperature).

Textile

Ingredient (Brit. 335247) of—
 Waterproofing, wearproofing, and weatherproofing compositions, used in the treatment of various textiles, such as bast, cotton, wool, and cotton and wool mixtures.

Woodworking

Ingredient (Brit. 335247) of—
 Waterproofing, wearproofing, and weatherproofing compositions.

Glass Wool

Synonyms: Glass silk.
French: Laine de verre, Soie de verre, Verre de laine, Verre soyeux.
German: Glaswolle, Wollartigesglas.
(A fibrous silk-line or wool-like material composed of fine filaments of glass intermingled like ordinary wool; available (1) in the form of large or small mattresses suitable for covering extensive areas, (2) in strips for covering small diameter pipes, (3) in shapeless form.)

Analysis

As a filtering medium.

Automotive

Sound insulator in—
 Automobile mufflers, motorcycle mufflers.

Construction

Fireproofing construction material in buildings.
Sound-insulator in buildings.

Electrical

Ingredient of—
 Storage battery separator compositions (Brit. 412625 and 412884).
Separator in—
 Storage batteries.

Metallurgical

Dust-collecting medium in various processes.

Mechanical

Dust-collecting medium in—
 Drying installations in various industries.
 Grinding operations on products such as stone, cement, gypsum, coal, leather, carbon, soap, cocoa, lime, milling products.
 Pneumatic conveying systems.
 Sand-blasting operations in various industries.

Miscellaneous

Collection of fly ash and dust from—
 Flue gases, stack gases.
Dust-collecting medium in—
 Coal cleaning, breaking, and grinding, and general processing installations.
 Factory and other power plants.
 Gasworks.
 Producer-gas plants.
 Heat-insulating medium for most stringent requirements of modern steam and heat engineering.
Ingredient of—
 Heat-insulating medium containing also asbestos and plaster or strong glue.

Refrigeration

As an insulating medium.

Sanitation

Collector in—
 Air-filtration installations for removal of dust, dirt, lint, pollen, bacteria, and other harmful impurities.

Gliadin

Chemical

Starting point (Brit. 311382) in making spinal anesthetics with—
 Diethylaminopropyl cinnamate.
 Diethylaminopropylcinnamate hydrochloride.
 Para-aminobenzoyldiethylaminoethanol.
 Para-aminobenzoyldiethylaminoethanol hydrochloride.
 Para-aminobenzoylidimethylaminomethylisobutanol.
 Para-aminobenzoylidimethylaminomethylisobutanol hydrochloride.

Pharmaceutical

In compounding and dispensing practice.

Glucinum

Synonyms: Beryilium, Beryllium, Glucinium.

Chemical

Starting point in making various salts.
Reagent (Brit. 281307) in making zeolite catalysts used in making—
 Acenaphthylene from acenaphthene.
 Acetaldehyde from ethyl alcohol.
 Acetic acid from ethyl alcohol.
 Alcohols from aliphatic hydrocarbons.
 Aldehydes from toluene, xylene, mesitylene, pseudocumene, and cymene.
 Aldehydes and acids by the oxidation of orthochlorotoluene, parachlorotoluene, orthobromotoluene, parabromotoluene, dichlorotoluene, chlorobromotoluenes, nitrotoluenes, chloronitrotoluenes, bromonitrotoluenes.
 Alpha-anthraquinone from naphthalene.
 Anthraquinone from anthracene.

Glucinum (Continued)

Benzaldehyde and benzoic acid from toluene.
Benzoquinone from phenanthraquinone.
Chloroacetic acid from ethylenechlorohydrin.
Diphenic acid from ethyl alcohol.
Fluorenone from fluorene.
Formaldehyde from methane or methanol.
Hemimellitic acid from acenaphthene.
Maleic acid and fumaric acid from benzene, toluene, phenol, or tar acids, or from benzoquinone or phthalic anhydride.
Naphthalic anhydride.
Naphthaldehydic acid, acenaphthaquinone, or bisacenaphthylideneione from acenaphthene or acenaphthylene.
Phenanthraquinone from phenanthrene.
Phthalic anhydride from naphthalene.
Salicyl aldehyde or salicylic acid from cresol.
Vanillin or vanillic acid from eugenol or isoeugenol.

Metallurgical
Ingredient of—
Copper alloys.

Miscellaneous
In place of aluminum for structural purposes, for example, in airplanes.

Glucinum Oxide

Synonyms: Beryllium oxide.
French: Oxyde de beryllium, Oxyde de glucinum.
German: Beryllerde.

Chemical
Catalyst (Brit. 254819) in making—
Alcohols, aldehydes, amines, carboxylic acids, carboxylic acid esters, oxygenated organic compounds.
Catalyst in making—
Acetic esters, allyl esters, amyl esters, butyl esters, ethyl esters, methyl esters, propyl esters.
Catalyst in the dehydration of various organic compounds.
Starting point in making beryllium salts of acids and halogens.

Jewelry
Ingredient of—
Precious stones with molten quartz base.
Synthetic alexandrite, synthetic emerald.

Glucinum Propionate

Synonyms: Beryllium propionate.
French: Propionate de beryllium, Propionate de glucinum.
German: Berylliumpropionat, Glucinumpropionat, Propionsäuresberyllium, Propionsäuresglucinum.

Petroleum
Ingredient (Brit. 334181) of—
Motor fuels.

Gluconic Acid

Chemical
Starting point in making—
Bismuth-sodium gluconate (U. S. 1906666).
Calcium gluconate, salts of various bases, various esters.

Pharmaceutical
Suggested for use in treating diabetic coma.

Glue

French: Colle, Colle d'os, Colle de peau.
German: Gluten, Leim.
Spanish: Ajicola, Cola.
Italian: Colla.

Abrasives
Adhesive and binder in—
Abrasive compositions, emery paper, garnet paper, sandpaper.

Adhesives
As an adhesive.
Ingredient of—
Adhesive compositions.

Construction
Binding agent in—
Insulating materials, containing also cork or wood waste, either in powder or shavings.
Wallboard size for—
Plaster walls.
Stabilizing agent for—
Bituminous emulsions.
Water-resistance promoter for—
Cements.

Electrical
Starting point in making—
Insulating materials, by dissolving in organic liquids such as phenols. (The substances obtained are similar physically to artificial resins, but present a much greater insulating resistance than other materials usually employed.)

Explosives and Matches
Binder in—
Matchhead compositions.

Fats and Oils
Stabilizing agent (Brit. 380052) in making—
Fat emulsions, oil emulsions.

Food
Suggested source of nitrogen in making—
Yeast.

Ink
Ingredient of various inks.

Insecticide
Adhesive in—
Oil-water emulsions used as plant insecticides, either alone or as carriers of insecticidal agents in suspension.
Ingredient (U. S. 1898673) of—
Spreader, containing also casein and hydrated lime, for insecticidal sprays.
Stabilizer in—
Oil-water emulsions used as plant insecticides, either alone or as carriers of insecticidal agents in suspension.

Leather
Ingredient of—
Cements, finishes, sizes.

Linoleum and Oilcloth
Binder in making—
Linoleum, oilcloth.

Metallurgical
Flotation reagent (U. S. 1906029) in—
Copper and lead separation.
Ingredient (U. S. 1914532) of—
Foundry cores, containing also sand, hydrated rubber mixture, sodium and ammonium soaps, and an extract of quince seed.
Restrainer in—
Scaling compositions (U. S. 1904445).
Sulphuric and hydrochloric acid pickling baths (reduces by 50 percent the attack of the iron).

Miscellaneous
Adhesive, size, and stiffener in—
Hat making.
Brilliance improver in—
Polishes.
Cost-reducing agent in—
Polishes.
Dispersing assistant for—
Waxes in paste polishes.
Ingredient (U. S. 1881128) of—
Motion picture projection screen coating, containing also sodium fluoride, copper sulphate, casein, glycerin, borax, cobalt blue, and water, said to have properties of non-stickiness, permanence, and adaptability to climatic conditions.
Size for various purposes.
Size in—
Cordage and rope making.
Stabilizing and dispersing agent for—
Basic emulsions in polishes.

Paint and Varnish
Base for—
Paints, lacquers and varnishes. (These products can be used either alone or as undercoats for paints and varnishes. The glue contained in these paints or varnishes is, either at the time of its application or in preparation, made insoluble with a bicromate or formaldehyde. Not only does it involve a big saving in the preparation of these products, but it also renders the paints insoluble in any solvent and much less permeable.)
Ingredient of—
Calcimines.
Coating compositions, containing also glycerin, alcohol, and water, used as an intermediate coating to prevent wood stains diffusing into the finish coat (U. S. 1908180).
Dry color compositions used in the preparation of leather body colors, water colors, and distempers (Brit. 404041).

Glue (Continued)
Mural paints, varnishes, wallpaper adhesives.
Stabilizing agent for—
Bituminous emulsions.

Paper
Adhesive in making—
Paper products, papier mache, pulp products.
Dispersing agent (U. S. 1903787) in making—
Waxed paper products.
Ingredient of—
Coating compositions containing also cellulose (U. S. 1910406).
Color batches for wallpapers.
Impregnating compositions containing also glycerin, rubber latex, and triethanolamine (U. S. 1913017).
Impregnating medium, containing also soap and alum, used in making waterproof paper bags.
Paper sizes and coatings.
Size for—
Papers.
Starting point in making—
Partly insolubilized glue base useful for sizing paper pulp.

Petroleum
Caulking and sizing agent for—
Wooden barrels.

Photographic
Coating agent for—
Non-sensitized side of printing papers (to prevent curling in rapid drying).

Plastics
Base material (U. S. 1862969) in making—
Phonograph records.
Binder in—
Plastic compositions.
Plastic compositions containing chalk, dextrin, rosin, and turpentine.
Plastic composition containing wood meal mixed with silicates.

Printing
Adhesive in—
Bookbinding.
In process engraving and lithographic processes.
Ingredient of—
Printing roller compositions.

Resins and Waxes
Stabilizing agent (Brit. 380052) in making—
Wax emulsions.

Rubber
Anti-coagulant in—
Emulsions.
Coagulation restrainer (Brit. 397997) in making—
Rubberized fabrics.
Improver of—
Durability, homogeneity, tenacity.
Ingredient of—
Rubber compounds.
Rubberized and fibrous plastic material unattackable by oil (composed of rubber fillers, fibers, glycerin, a vulcanization accelerator, and sulphur) (U. S. 1907231).
Tire-filling compositions.
Promoter of—
Thicker coatings (when making rubber objects by the steeping method).
Stabilizer of—
Emulsions.
Starting point in making—
Impregnating agent (from ammonium resinate and other products) for rubberized horsehair used for padding motor-coach seats.

Soap
Process material for—
Improving detergent power, improving lather.
Increasing hardness of the base in making household soaps, thus reducing the time of cooling in the moulds and facilitating the stamping of the soap.
Increasing solidity of the soap base and thus facilitating plodding, which is rendered possible with a much higher moisture content than is usual with flaked soap.

Textile
Ingredient of—
Bucking (scouring) baths (Brit. 398958).
Finishing compositions.
Scouring compositions (Brit. 388877).
Sizing compositions.

Woodworking
Adhesive in—
Carpentry, cabinet making, furniture making, piano making, plywood making.

Glutamic Acid
French: Acide glutamique.
German: Glutaminsaeure.

Brewing
Ingredient (Brit. 279985) of—
Beer flavors.

Food
Reagent (Brit. 279985) in making—
Flavoring extracts.
Food products from fish, meat, starch, casein, egg yolk, grains.

Pharmaceutical
Reagent (Brit. 279985) in making—
Flavored preparations.

Wine
Ingredient (Brit. 279985) of—
Flavors for wines.

Glutamic Acid Hydrochloride
French: Chlorhydrate d'acide glutamique, Hydrochlorure d'acide glutamique.
German: Glutaminsaeureschlorhydrat.

Brewing
Ingredient of—
Beers and ales, added to improve the taste (Brit. 279985).

Food
Reagent (Brit. 279985) in making—
Flavoring extracts.
Food products from fish, meat, starch, casein, egg yolk, grains.

Pharmaceutical
Reagent (Brit. 279985) in making—
Flavoring preparations.

Wine
Ingredient (Brit. 279985) of—
Flavors.

Glycerin
Synonyms: Glycerine, Glycerol, Glyceryl hydroxide, Glycyl alcohol, Propane-1:2:3-triol, Propenyl alcohol.
Latin: Glycerinum.
French: Glycérine.
German: Glycerin, Glyceryloxyhydrat, Oelsuss, Scheelesches suss.
Spanish: Glicerina.
Italian: Glicerina.

C.P. GRADE

Analysis
Ingredient of—
Grinding paste (admixture with emery powder) used for grinding and refitting glass parts.
Special lubricating mixtures with bentonite, offering the following advantages:—(1) Adjustable viscosity, (2) unaffected by nonaqueous solvents, (3) long-time stabilization even in presence of water, (4) viscosity unaffected by temperature of 100° C.
Phenol-burn antidote (admixture with bromine).
Lubricant for—
Stopcocks, interchangeable ground-glass parts.
Lubricant in—
Boring holes in rubber stoppers.
Inserting glass tubing through holes in rubber stoppers.
Reagent in analytical methods and processes involving control and research.
Softening and condition agent for—
Rubber articles (following washing and soaking in ammoniated water).

Brewing
Clarifying agent.

Cosmetic
Antiseptic, bactericide, carrier, emollient, humectant, hygroscopic agent.
Ingredient of—
Almond creams, buttermilk lotions, creams, cuticle removers, greaseless lip-rouge, jellies, liquid face powders, nail bleaches, nail polishes, skin creams, sunburn lotions, vanishing creams.
Penetration promoter, promoter of miscibility with water, softening agent, soothing agent, solvent, sterilizer, vehicle.

Glycerin (Continued)

Food

Humectant in—
Bread, cakes, confectionery, chocolate, food products, packed grain products.
Hygroscopic agent in—
Infant foods, invalid foods, various food products.
Ingredient of—
Fish preservatives.
Meat-curing mixtures with pyroligneous acid.
Shelled egg preservative mixture with succinic or phosphoric acid.
Inhibitor of—
Odor development in food products.
Lubricant for—
Beaters, choppers, and other power-driven kitchen equipment (leaves no after-taste in the food).
Preservative for—
Bread, cakes, confectionery, food products, packed grain products.
Promoter of—
Assimilation of foodstuffs.
Ductility and swelling (texture and volume) without adverse fermentation in bread doughs.
Retarder of drying in—
Bread, cakes, confectionery, packed grain products.
Retarder of mould formation in—
Bread, cakes, confectionery, packed grain products.
Sterilizer for—
Infant foods, invalid food, meat products, mustard preparations, shelled eggs, various food products.
Sweetening agent in—
Cakes, confectionery, infant foods, invalid foods, various food products.
Vehicle for—
Flavoring agents used in food products and confectionery.

Oral Hygiene

Antiseptic, sterilizer, and vehicle in—
Dentifrices, gargles, nasal douches, mouthwashes.

Pharmaceutical

Constituent of—
Biological serums.
Boric acid mixtures used as preservatives.
Gelatin bases for pastes, pastiles, and suppositories.
Gargles, glycerin suppositories.
Glycerites of alum, boric acid, lead subacetate, pepsin, phenol, starch, tannic acid, and other drugs.
Hexylresorcinol solutions.
Phenol solutions suggested for treatment of suppurative otitis media.
Picric acid dressings for wounds, sores, and burns.
Zinc oxide-calamine lotions for sunburn.
Dehydrating agent for—
Micro-organisms.
Emollient.
Excipient for—
Pills, tablets.
Humectant, hygroscopical agent.
Promoter of—
Miscibility of various drugs with water.
Reagent in making—
Glycerophosphates.
Solvent for—
Antiseptics used in surgery and dentistry (affords a means of preparing highly concentrated solutions).
Solvent for—
Iodoform and other antiseptic agents used for intra-articular and parenchymatous injection.
Phenol in making local anesthetics for the tympanum.
Soporifics of the barbiturate type.
Various drugs.
Suggested for use as—
Antidote for trichinae (intestinal phase only).
Antiseptic (said to approach ideal closely).
Bactericide (desirable for low destructive action on tissues).
Laxative, lymphagog.
Promoter of penetration of various drugs.
Softening agent for crusts and necrotic tissue in wounds.
Soothing agent in throat irritations.
Sterilizing agent for surgical instruments and gloves.
Substitute for sugar in diabetes.
Treating agent for septic conditions of uterine tract.
Sweeting agent in—
Preparations containing ferric chloride, cascara sagrada, cinchona.
Various medicinal preparations.
Vehicle in various classes of medicinal preparations.

Soft Drink

Process material in making—
Flavorings, smoothing agent, sweetening agent.

Wine

Clarifying and settling agent.
Imparter of—
Oiliness to wines.
Palatability and smoothness to cheap dry wines.
Maturing agent.
Promoter of—
Extractant action of alcohol for flavoring ingredients in cordial (liqueur) manufacture.
Sterilizing agent.
Suppressor of—
After-fermentations in wines.

OTHER GRADES

Adhesives

Ingredient of—
Label gums and adhesives.
Office and library adhesives.

Air-Conditioning

Hygroscopic agent.

Cellulose Products

Humectant for—
Transparent wrapping materials.
Hygroscopic agent for—
Transparent wrapping materials.
Ingredient of—
Adhesives for transparent wrapping materials.

Chemical

Dehydrating agent for—
Alcohol.
Liquid seal for—
Pure hydrogen (temporary storage only).
Starting point in—
Organic syntheses.
Starting point in making—
Allyl alcohol.
Chlorhydrius.
Esters, such as the nitric, sulphuric, phosphoric, glyceroboric, tartaric, succinic, malic, maleic, fumaric, citric.
Ethers, such as the monomethyl, monoethyl, dimethyl, diethyl.
Quinolin.

Dye

Process material and starting point in making various dyestuffs.

Explosives and Matches

Starting point in making—
Nitrated compounds for low-freezing dynamites (in admixture with sugar).
Nitroglycerin.

Florist

Conditioning agent for—
Plant leaves.
Imparter of—
Fresh and glossy appearance to plant leaves.
Retarder of—
Drying-out of plant leaves.

Glass

Ingredient of—
Etching agent, containing also ammonium bifluoride, calcium sulphate, and water.

Illuminating Gas

Antifreeze in—
Gas meters
Drying agent for—
Gas generated at municipal gasworks.

Ink

Adhesion promoter for—
Inks for printing on glossy paper.
Antioxidant.
Antiseptic.
Tygroscopic agent.
Ingredient of—
Autographic inks, hectographic inks, copying inks, chromolithographic inks, lithographic inks, plate-printing inks, printing inks, stamping inks, stencil inks, typewriting inks, writing inks.
Offset preventer.
Opacity promoter.
Reducer for—
Inks.
Restrainer of—
Quick-drying.

Glycerin (Continued)
Solvent for—
Anilin dyes.
Other ink ingredients.
Spreading agent.
Sterilizer in—
Preventing mould formation.
Thickener.
Toner.
Leather
Ingredient of—
Leather substitutes.
Preventer of—
Drying out of chrome leather between tanning and printing operations.
Reversion of the colloidal constituents of prepared leather.
Process material in—
Leather printing.
Softening and flexibilizing agent for—
Leather prior to dressing.
Mechanical
Antifreeze for—
Automobiles, hydraulic jacks, pumps.
Fluid medium in—
Pressure gauges.
Humectant for—
Belting.
Lubricant for—
Air-compressor pistons, ball bearings, clock mechanisms, delicate machinery.
Low-temperature work (in admixture with graphite).
Machinery for processing and pumping gasoline.
Refrigeration machinery, roller bearings, shafts in coal mines, shock absorbers.
Pressure-transmission agent in—
Testing machines.
Recoil-energy absorber in—
Stamping machinery.
Starting point in making—
Plumbers cements (with litharge) useful for many purposes.
Metallurgical
Intermediate quenching agent for—
Steel
Lubricant for—
Moulds.
Miscellaneous
Ingredient of—
Ammonia mixtures used to recondition or give new life to typewriter rollers.
Antitarnish varnishes for metalware, containing also rosin, sandarac, and alcohol.
Bottle-sealing compounds, containing also gelatin and zinc oxide.
Cements and lutes (with litharge)
Gelatin mixtures used for embedding microscopic specimens for examination.
Hat dressings and sizes.
Razor-sharpening compositions, containing also glue and gum.
Shoe polishes, waterproofing agents.
Lubricant for—
Rubber rings used in bottling and canning.
Shrinkage inhibitor for—
Wooden moulds and vessels.
Skin-conditioning agent for—
Mechanics, metal workers, and other workers whose hands become impregnated with dirt, grime, and abrasive materials.
Softener, flexibilizer, and reconditioner for—
Rubber covers for keys on typewriters which have become brittle and hardened and have lost their resiliency.
Softener, plasticizer, and lubricant in—
Clay modeling.
Solvent for various purposes.
Starting point in making—
Substitutes for india rubber stamps (with glue and molasses); claimed to be just as flexible as and superior to rubber stamps for some purposes.
Sterilizer for—
Cork stoppers (to prevent moulding).
Paint and Varnish
Softener in—
Artist's colors.
Paper
Bodying agent.
Flexibilizer.
Ingredient of—
Coatings for making marbled and other surface-coated effects.
Compositions for producing parchmented effects on papers.
Grease-proofings for paper, sizing and coatings for paper, waterproofings for paper.
Shrinkage preventer.
Softening agent.
Sterilizer.
Perfume
Extractant for—
Odorous constituents of flowers.
Sterilizer and vehicle in—
Perfume preparations.
Photographic
Anti-curling agent for—
Film.
Assister in—
Fine focusing in plant photography (used to coat ground glass which is too course).
Brittling and cracking preventer for—
Film.
Ingredient of—
Emulsions.
Preserver of—
Flexibility of film.
Process material in making—
Quinolin dye sensitizers, varnish solvents.
Varier of effects and improver for—
Deteriorated negatives, hard, sharp negatives.
Printing
Base material (with glue) in making—
Hectographic plates, printers' rollers.
Conditioning agent for—
Printers' rollers.
Crack and wrinkle preventer for—
Printers' rollers.
Offset preventer and treating agent for—
Tympan sheets.
Refrigeration
Refrigerant offering the advantages of (1) freedom from corrosive properties, (2) freedom from evaporation losses in open systems.
Resins
Starting point in making—
Ester gums by reaction with rosin.
Resins with—
Aleuritic acid and phthalic anhydride, malic acid, malic acid and sulphur, phenol, phenol and formaldehyde, phenol and sulphuric acid, phthalic anhydride, phthalic anhydride and oleic acid, phthalic anhydride and succinic acid, phthalic anhydride and malic acid.
Rubber
Center-filling agent in—
Golf balls.
Devulcanizer and modifying agent.
Improver of—
Ageing properties of rubber.
Ingredient of—
Mixes, preservative coatings for vulcanized rubber.
Preserver of elasticity of—
Rubber.
Process material in making—
Rubber substitutes.
Starting point in making—
Coatings for air bags (with sodium hydrosulphite) to prevent adhesion, sulphur migration, and ageing.
Soap
Ingredient of—
Toilet soaps, transparent soaps, shaving creams and sticks.
Textile
In bleaching processes.
In dyeing and printing processes.
In felt manufacture.
Increaser of—
Hygroscopic properties of textile fabrics, tenacity and resistance of rayon to friction in weaving.
Ingredient of—
Fireproofing compositions, gasproofing compositions, waterproofing compositions.
Sizing agent.
Solvent for—
Anilin dyes.
Tobacco
Humectant and conditioning agent.

Glycerin Monoacetate
French: Monoacétate de glycérine.
German: Glyzerinmonoacetat, Glyzerinmonoazetat. Monoessigsäureglyzerinester, Monoessigsäuresglyzerin.
Cellulose Products
Plasticizer (Brit. 311795) for—
Cellulose acetate, cellulose esters and ethers, cellulose nitrate.
For uses, see under general heading: "Plasticizers."
Textile
—, *Dyeing and Printing*
Solubilizing or dispersing agent (Brit. 276100) in printing and dyeing with—
Acridin dyestuffs, aminoanthraquinone dyestuffs, reduced and unreduced, anthraquinone dyestuffs, reduced and unreduced, azins, azo dyestuffs, basic diarylmethane dyestuffs, basic triarylmethane dyestuffs, benzoquinone anilides, chrome mordant dyestuffs, indigoids, naphthoquinanilides, naphthoquinones, reduced or unreduced, nitroarylamines, nitroarylphenols, nitrodiarylamines, nitrodiarylphenols, oxazines, pyridin dyestuffs, quinolins, quinoneimides, reduced and unreduced, sulphur dyestuffs, thioazonins, xanthenes.
—, *Finishing*
Plasticizer (Brit. 311795) in—
Coating compositions containing cellulose acetate, nitrocellulose, or other esters or ethers of cellulose.

Glycerin Monochlorohydrin
Dye
Reagent in making dyestuffs from—
Sodium-alpha-aminoanthraquinone-2-mercaptan.
Sodium-alpha-amino-4-paratoluidoanthraquinone-2-mercaptan.
Sodium-2-amino-3-bromoanthraquinone-1-mercaptan.
Sodium-1:4-diamino-3-chloroanthraquinone-2-mercaptan.
Sodium-1:5-diaminoanthraquinone-2-mercaptan.
Sodium-1:8-diaminoanthraquinone-2-mercaptan.
Sodium-2:6-diaminoanthraquinone-1:5-dimercaptan.
Sodium-2:7-diaminonathraquinone-1:8-dimercaptan.
Sodium-4:5:8-tetra-aminoanthraquinone-2-mercaptan.

Glycerin Monoformate
Synonyms: Glyceryl monoformate.
French: Monoformiate de glycérine, Monoformiate de glycéryle, Monoformiate glycérylique.
German: Glyzerinmonoformiat, Glyzerylmonoformiat, Monoameisensäureglyzerinester, Monoameisensäureglyzerylester, Monoameisensäuresglyzerin, Monoameisensäuresglyzeryl.
Cellulose Products
Plasticizer (Brit. 311795) for—
Cellulose acetate, cellulose esters and ethers, cellulose nitrate.
For uses, see under general heading: "Plasticizers."
Textile
—, *Finishing*
Plasticizer (Brit. 311795) in—
Coating compositions containing cellulose acetate, nitrocellulose, or other esters or ethers of cellulose.
—*Dyeing and Printing*
Solubilizing or dispersing agent (Brit. 276100) in printing and dyeing with—
Acridin dyestuffs, aminoanthraquinone dyestuffs, reduced and unreduced, anthraquinone dyestuffs, reduced and unreduced, azins, azo dyestuffs, basic diarylmethane dyestuffs, basic triarylmethane dyestuffs, benzoquinoneanilides, chrome mordant dyestuffs, indigoids, naphthoquinanilides, naphthoquinones, reduced or unreduced, nitroarylamines, nitroarylphenols, nitrodiarylamines, nitrodiarylphenols, oxazins, pyridin dyestuffs, quinolins, quinoneimides, reduced and unreduced, sulphur dyestuffs, thioazonins, xanthenes.

Glycerin Pitch
French: Brai de glycérine.
German: Glycerinpech.
Electrical
Ingredient of
Insulating compositions.
Paint and Varnish
Ingredient of—
Waterproofing compositions.
Paper
Impregnating agent in making—
Felts, specially treated papers.
Textile
—, *Dyeing and Printing*
Solubilizing or dispersive agent (Brit. 276100) in dyeing and printing textile yarns and fabrics with—
Acridin dyestuffs, aminoanthraquinone dyestuffs, reduced or unreduced, anthraquinone dyestuffs, reduced or unreduced, azines, azo dyestuffs, basic diarylmethane dyestuffs, basic triarylamine dyestuffs, benzoquinoneanilides, indigoids, naphthoquinones, reduced or unreduced, naphthoquinoneanilides, nitroarylamines, nitroarylphenols, nitrodiarylmethanes, nitrodiarylphenols, oxazines, pyridin dyestuffs, quinolines, quinoneimides, reduced or unreduced, sulphur dyestuffs, thiazonines, xanthenes.

Glycerol Alphanaphthylether
Chemical
Starting point (Brit. 416943) in making—
Wetting, foaming, detergent, emulsifying, and dispersing agents by condensation with butyl alcohol and sulphonation with sulphuric acid.

Glycerolbetacetylether Sulphonate
Miscellaneous
As a wetting agent (Brit. 436209).
For uses, see under general heading: "Wetting agents."

Glycerol Dichlorohydrin
Ceramics
Solvent in—
Compositions, containing aldehydeamine condensation products, used for coating and decorating ceramic products.
Chemical
Starting point in making various derivatives.
Electrical
Solvent (Brit. 343031) in making—
Compositions, containing aldehydeamine condensation products, used as insulating coatings.
Leather
Solvent (Brit. 343031) in making—
Compositions, containing aldehydeamine condensation products, used in the manufacture of artificial leather and for coating and decorating leather goods.
Miscellaneous
Solvent (Brit. 343031) in making—
Compositions, containing aldehydeamine condensation products, used for coating and decorating various fibrous compositions of matter.

Glycerol Dixylylether
Chemical
Starting point (Brit. 416943) in making—
Wetting, foaming, detergent, emulsifying, and dispersing agents by condensation with butyl alcohol and sulphonation with sulphuric acid.

Glycerol Monophenyl Ether
Cellulose Products
Plasticizer for—
Cellulose, acetate, cellulose esters or ethers, cellulose nitrate (nitrocellulose).
For uses, see under general heading: "Plasticizers."

Glycin. See Glycocoll.

Glycocholic Acid
French: Acide de glycocholique.
German: Glykocholsaeure.
Chemical
Reagent (Brit 282356) in making anti-parasitic agents with—
Dihydrocuprein ethyl ether.
Dihydrocuprein ethyl ether hydrochloride.
Dihydrocuprein isoamyl ether.
Dihydrocuprein isoamyl ether hydrochloride.
Dihydrocuprein normal octyl ether.
Dihydrocuprein normal octyl ether hydrochloride.
Dihydroquinone.
Starting point in making—
Bismuth glycocholate, hexamethylenetetramine, lithium glycocholate, potassium glycocholate, sodium glycocholate.
Pharmaceutical
In compounding and dispensing practice.
Sanitation
As a general antiseptic.

Glycocoll
Synonyms: Aminoacetic acid, Glycin.
French: Acide d'aminoacétique.
German: Aminoessigsaeure.
Chemical
Starting point in making—
Anthraquinone-2-glycin-3-carboxylic acid (Swiss 109067).
Pharmaceutical and other organic chemicals.
Dye
Starting point in making various synthetic dyestuffs.
Paint and Varnish
Stabilizer in making—
Cellulose acetate lacquers and varnishes.
Nitrocellulose lacquers and varnishes.
Pharmaceutical
In compounding and dispensing practice.
Photographic
Reagent for—
Reducing silver images, developing agent in place of caustic alkalies and pyrogallol.
Starting point in making—
Developing agents.

Glycocoll-Copperdiamine
French: Diamine de glycocolle et de cuivre.
German: Glykokollkupferdiamin.
Chemical
Reagent in making various substances.
Dye
Reagent (Brit. 306859) in making azo dyestuffs with—
Acetyl-H acid.
Alphaethoxy-8-hydroxynaphthalene-3:6-disulphonic acid.
Alphahydroxynaphthalene-4-sulphonic acid.
3-Aminobenzaldehyde.
2-(4′-Aminobenzoyl)amino-5-naphthol-7-sulphonic acid.
2-(3′-Aminobenzoyl)amino-5-naphthol-7-sulphonic acid.
Anthranilic acid.
Benzidin-3:3′-dicarboxylic acid.
Beta-aminobenzaldehyde.
Beta-aminobenzene-5-sulphonic acid.
Beta-aminobenzoic acid.
Beta-amino-1-hydroxybenzene.
Beta-aminonaphthalene-3-carboxylic acid.
Betanaphthol.
Betaphenylamino-4-hydroxynaphthalene-7-sulphonic acid.
4-Chloro-2-chloro-2-aminobenzoic acid.
4:4′-Diaminodiphenylurea-3:3′-dicarboxylic acid.
4:6-Dichloro-2-amino-1-hydroxybenzene.
5:5′-Dihydroxy-2:2′-dinaphthylamine-7:7′-disulphonic acid.
J acid.
5-Nitro-2-aminobenzoic acid.

Glycol Diacetate
French: Diacétate de glycole, Diacétate glycollique, Glycole diacétique.
German: Diessigsäureglycolester, Diessigsäureglykolester, Diessigsäuresglycol, Diessigsäuresglykol, Glycoldiacetat, Glycoldiazetat, Glykoldiacetat, Glykoldiazetat.
Cellulose Products
Solvent for—
Cellulose esters and ethers, cellulose nitrate.
For uses, see under general heading: "Solvents."
Fats and Oils
Solvent in extracting—
Essential oils.
Plastics
Gelatinizing agent and plasticizer (Brit. 230025) in making—
Artificial horn, plastic compositions.
Solvent in making—
Compositions containing nitrocellulose or other esters or ethers of cellulose.
Resins and Waxes
Solvent (Brit. 273748) in making resins of the—
Phenol-formaldehyde type, polyhydric alcohol-polybasic acid type, urea-aldehyde type.
Solvent (Brit. 252394) in making—
Ester condensation and polymerization products.

4-Glycollylaminophenylarsinic Acid
French: Acide de 4-glycollylaminophénylarsinique.
Spanish: Acido de 4-glicollilaminofenilarsinico.
Italian: Acido di 4-glicollileaminofenilearsinico.
Chemical
Starting point in making various derivatives.
Starting point (Brit. 347083) in making therapeutic products by reaction with—
Acetic acid, crotonic acid, isovaleric acid.

Glycol Mono-oleate
French: Mono-oléate de glycole, Mono-oléate glycollique.
German: Glykolmono-oleat, Oleinsäuremonoglykolester, Oleinsäuresmonoglykol.
Miscellaneous
As an emulsifying agent (Brit. 329266).
For uses, see under general heading: "Emulsifying agents."

Glycol Phthalate
Cellulose Products
Plasticizer for—
Cellulose acetate, cellulose esters or ethers, cellulose nitrate (nitrocellulose).
For uses, see under general heading: "Plasticizers."

Glyoxal
Chemical
Starting point in making—
Glycollic acid.
Plastics
Ingredient (Brit. 279863) of—
Casein compositions, artificial horn buttons, and the like.

Glyoxime
Fuel
Primer (Brit. 429763) for—
Diesel engine fuel oils produced by the hydrogenation of coal.
Petroleum
Primer (Brit. 429763) for—
Diesel oils containing a high proportion of aromatic bodies.

Goat Hair
Lubricant
Ingredient of—
High-cohesion greases.
Fertilizer
Source of nitrogen in making—
Wet base goods.
Furniture
Filling material in—
Upholstered furniture.

Gold Bromide
Synonyms: Aurous bromide.
French: Bromure aureux, Bromure d'or.
German: Goldbromuer, Goldbromid.
Chemical
Starting point (Brit. 261048) in making—
Aluminum-gold thiosulphate, ammonium-gold thiosulphate, barium-gold thiosulphate, bismuth-gold thiosulphate, cadmium-gold thiosulphate, calcium-gold thiosulphate cobalt-gold thiosulphate, copper-gold thiosulphate, iron-gold thiosulphate, lead-gold thiosulphate, magnesium-gold thiosulphate, nickel-gold thiosulphate potassium-gold thiosulphate, sodium-gold thiosulphate, strontium-gold thiosulphate, tin-gold thiosulphate, zinc-gold thiosulphate.
Pharmaceutical
In compounding and dispensing practice.

Gold Chloride
Synonyms: Chlorauric acid.
Latin: Auri chloridum.
French: Chlorure d'or.
German: Goldchlorid.
Italian: Cloruro di oro.
Analysis
As a reagent.
Ceramics
Gilding agent for—
Porcelain.
Ingredient of—
Enamels.
Chemical
Starting point in making—
Purple of Cassius.

Gold Chloride (Continued)

Glass
Coloring agent in making—
Ruby glass.
Reagent in—
Gilding glass.

Ink
Ingredient of—
Special inks.

Metallurgical
Ingredient of—
Goldplating electrolytes.
Starting point in making—
Finely divided gold.

Miscellaneous
Ingredient (Brit. 407039) of—
Antiseptic washing and cleansing agents prepared by incorporating water-soluble metal salts, which dissociate into metal ions, with aliphatic compounds having strong wetting and washing power, containing at least eight carbon atoms, having an acid sulphuric or phosphoric ester group or sulphonic acid group in an end position, and forming water-soluble salts with said metals.

Paint and Varnish
Starting point in making—
Purple of Cassius.

Pharmaceutical
In compounding and dispensing practice.

Photographic
As a toning agent.

Gold Iodide
Synonyms: Aurous iodide.
French: Iodure aureux, Iodure d'or.
German: Goldioduer, Goldjodid, Jodgold.

Chemical
Starting point (Brit. 261048) in making—
Aluminum-gold thiosulphate, ammonium-gold thiosulphate, barium-gold thiosulphate, bismuth-gold thiosulphate, cadmium-gold thiosulphate, calcium-gold thiosulphate, cobalt-gold thiosulphate, copper-gold thiosulphate, iron-gold thiosulphate, lead-gold thiosulphate, magnesium-gold thiosulphate, nickel-gold thiosulphate, potassium-gold thiosulphate, sodium-gold thiosulphate, strontium-gold thiosulphate, tin-gold thiosulphate, zinc-gold thiosulphate.

Gold Resinate
Synonyms: Resinate of gold.
French: Resinate d'or.
German: Goldresinat.

Ceramics
Pigment in admixture with aluminum resinate for producing light-purple shades on—
Chinaware, porcelains, potteries.

Gold-Thioglucose

Chemical
Starting point (Brit. 398020) in making—
Complex double compounds of organic heavy metal mercapto compounds.

Graphite
Synonyms: Black lead, Carbon, Plumbago.
French: Carbone.
German: Kohlenstoff.

Chemical
Raw material for—
Electrodes used in electrochemical processes.

Dye
Ingredient of—
Felt hat dye compositions.

Electrical
Raw material for—
Anodes for electric cells, arc-light carbons, commutator brushes, generator parts, motor parts, electrical machine parts.

Explosives and Matches
Glazing agent for—
Blasting powders, heavy ordnance powders.
Ingredient of—
Sporting powders.
Protective agent against dampness in—
Blasting powders, heavy ordnance powders.

Fats and Oils
Ingredient of—
Lubricant compositions composed of graphite, oil, and water.
Special lubricant for internal combustion motor crank-cases (U. S. 1879874).
Wire-drawing lubricant (U. S. 1724134).

Fertilizer
Ingredient of—
Fertilizer compositions.

Food
Glazing agent for—
Coffee beans, tea leaves.

Glass
Lubricant on—
Windowglass rolling tables.

Ink
Ingredient of—
Printing inks.

Mechanical
Anti-scale agent for—
Boilers.
Ingredient of—
Compositions for coating pipe joints.
Lubricant for various purposes.

Metallurgical
Electrode in—
Electro-metallurgical operations.
Facing agent for—
Foundry molds.
Ingredient (U. S. 1901409) of—
Composition for coating foundry molds.
Raw material for—
Crucibles used in steel melting and refining.
Retorts.

Miscellaneous
Core for—
Lead pencils.
Ingredient of—
Compositions for repairing stoves, ranges and boilers.
Heat-producing composition (U. S. 1901313).
Metal polishes.
Metallic packing consisting of graphite, lead, and wool grease (U. S. 1847796).
Shoe polishes, stove polishes.

Paint and Varnish
As a pigment.
Ingredient of—
Acid-resisting paints, rust-preventing paints, weather-resisting paints.

Paper
Ingredient of—
Compositions for treating carbon paper.

Printing
Coating agent for—
Molds for electrotypes.

Refractories
Ingredient of—
Refractory cement.
Raw material in making—
Small retorts, various shapes.

Rubber
Ingredient of—
Hard rubber compositions.
Rubber valve discs and washers for steam and hot water connections.

Guaiac
Synonyms: Guaiacum, Guaiacum resin, Gum guaiac.
Latin: Resina guajaci.
French: Gomme de guaiac, Résine de gayac.
German: Guajak, Guajakharz, Gummiguajak, Gummiguajakum.
Spanish: Resina de guayaco.
Italian: Resina di guajaco.

Electrical
Ingredient in making—
Electron-emitting cathodes (U. S. 1625776).

Paint and Varnish
Ingredient in making—
Paints, varnishes.

Pharmaceutical
In compounding and dispensing practice.

Guaiacol Acetate
Chemical
Starting point in making various derivatives.
Pharmaceutical
In compounding and dispensing practice.

Guaiac Saponin
French: Saponine du gaiac.
Fats and Oils
Emulsifying agent.
Food
Ingredient of—
Sparkling drinks.
Pharmaceutical
In compounding and dispensing practice.

Guanidin
Synonyms: Iminourea.
French: Guanidine, Urée iminique.
German: Iminoharnstoff.
Analysis
Reagent in analyzing—
Complex acids, molybdicarsenic acid, molybdicphosphoric acid.
Chemical
Starting point in making—
Amidophenylguanidin, barbital, dicyandiamidin, dicyandiamide, diphenylguanidin, intermediates, pharmaceuticals, rubber vulcanization accelerators with carbon bisulphide.
Dye
Starting point (French 612382) in making—
Azo dye compounds.
Fertilizer
Ingredient of—
Fertilizing compositions.
Miscellaneous
Ingredient of—
Fire-extinguishing compositions (German 485400).
Solutions used to prevent freezing (German 485012).
Resins and Waxes
Starting point in making—
Artificial resins with furfural (U. S. 1496792).
Artificial resins with formaldehyde and urea (used in the form of guanidin carbonate) (U. S. 1658597).
Textile
Ingredient of—
Viscose solutions (added to improve them) for spinning rayon.

Guanidin Polysulphide
French: Polysulphure de guanidine.
German: Guanidinpolysulfid.
Rubber
Accelerating agent in vulcanization (U. S. 1606321).

Guanin
Synonyms: 2-Amino-6-oxypurin, 2-Aminohypoxanthin.
Chemical
In organic syntheses.
Photographic
Defogging agent (Brit. 442731) for—
Gelatin having a strong tendency to cause fog.

Guano
German: Vogelduenger.
Chemical
Starting point in making—
Uric acid.
Fertilizer
As a plant food, alone or in compositions.

Guanylnitrosoaminoguanyltetracene
Explosives
Ingredient (U. S. 1889116) of—
Priming mixtures.

Gum Anime
Synonyms: Anime.
French: Gomme animé, Résine animé, Résine de courbaril.
German: Animeharz, Flüssharz, Gummianime, Gummiharz.
Linoleum and Oilcloth
Ingredient of—
Coating compositions.
Miscellaneous
Fumigant, alone and in mixtures.
Paint and Varnish
Ingredient of—
Cements, coach finishes, lacquers, varnishes.
Pharmaceutical
In compounding and dispensing practice.

Gum Arabic
Synonyms: Gum acacia, Gum senegal.
Latin: Gummi arabicum.
French: Gomme arabique, Gomme d'acacia, Gomme de sénégal.
German: Akaziengummi, Arabischergummi, Kordofangummi, Mimosengummi, Senegalgummi.
Ceramics
Ingredient of clay batch for—
Bricks, porcelains, potteries, tiles.
Chemical
Starting point in making—
Emulsifying agents (Brit. 252476).
Dye
Ingredient of—
Lakes with basic dyestuffs (Brit 270750).
Explosives and Matches
Ingredient of—
Match head compositions, pyrotechnic compositions.
Food
Ingredient of—
Bakery products, candies.
Glues and Adhesives
Ingredient of—
Mucilages, pastes.
Ink
Ingredient of—
Lithographic inks, printing inks, writing inks (as body drier).
Miscellaneous
Ingredient of—
Emulsions, metal polishes, shoes polishes, tire repairing compositions (Brit. 252113).
Stiffening agent in preparing fibrous materials.
Oilcloth and Linoleum
As a binder.
Paint and Varnish
Ingredient of—
Bronze compositions, paints, varnishes, water colors.
Paper
Sizing agent for—
Paper, cardboard, and other products.
Pharmaceutical
In compounding and dispensing practice.
Photographic
Ingredient of—
Pastes for mounting prints.
Reagent in—
Reproduction processes.
Printing
Reagent in—
Process engraving and lithographic arts.
Textile
——, *Finishing*
Ingredient of—
General textile sizes, lace-sizing compositions, tulle-sizing compositions, silk-sizing compositions, textile fiber stiffening compositions.
——, *Printing*
Ingredient of—
Color pastes for calicoes.

Gumbo Clay
Miscellaneous
Starting point in making—
Railroad ballasts.

Gum Sandarac
Synonyms: Gum juniper.
French: Gomme sandaraque.
German: Gummi sandarak.
Food
Ingredient of—
Candy, chewing gums, custard powders, ice cream, pie fillers.

Gum Sandarac (Continued)
Glue and Adhesives
Ingredient of—
Adhesive compositions for envelopes.
Ink
Ingredient of—
Inks
Powder for—
Rubbing on paper after erasures to prevent spreading of ink.
Linoleum and Oilcloth
Filler in—
Linoleum, oilcloth.
Miscellaneous
Ingredient of—
Dental cements, erasers, shoe polishes.
Paint and Varnish
Ingredient of—
Lacquers, varnishes.
Perfume
Ingredient of—
Incense compositions.
Pharmaceutical
In compounding and dispensing practice.
Ingredient of—
Ointments, plasters.
Photographic
Ingredient of—
Paper coatings.
Rubber
As a filler.

Guttapercha
Synonyms: Gutta pertscha, Gutta gettania, Gutta taban.
Latin: Gummi plasticum.
Chemical
Ingredient of—
Solutions that are used in place of collodion.
Electrical
Insulator in making—
Electric wiring, submarine cables.
Ingredient of—
Compositions used in fastening incandescent lamps in their sockets.
Glues and Adhesives
Ingredient of—
Special adhesive compositions for fixing metal or wood to leather, metal to metal, metal to glass.
Leather
Ingredient of—
Waterproofing compositions.
Mechanical
Making transmission belts.
Metallurgical
Raw material in making—
Moulds in galvano plastic work for making deposits on metals.
Miscellaneous
Raw material in making—
Acid-resistant containers and tubes, golf balls, cutlery handles, pump and hydraulic press valves, surgical instruments.
In dentistry.
Textile
Ingredient of—
Waterproofing compositions.

Hakuunboku Seed Oil
French: Huile de semences d'hakuum boku.
German: Hakuunbokusamenoel.
Fuel
As an illuminant.
Paint and Varnish
Ingredient of—
Paints, varnishes.
Starting point in making—
Boiled oil.
Soap
As a soapstock.

Hematite
Synonyms: Red hematite.
French: Hematite, rouge.
German: Blutstein, Blisterz, Eisenglanz, Eisenglimmer, Haematit, Roteisenstein.
Ceramics
Pigment in—
Enamels for porcelains and potteries.
Chemical
Reagent in making—
Sodium hydroxide, hydrogen.
Gas
Purifying agent in treating manufactured central station gas.
Glass
Ingredient of—
Polishing agents.
Metallurgical
As a source of iron.
Miscellaneous
Polishing agent for general purposes.
Paint and Varnish
As a pigment.
Perfumery
Ingredient of—
Cosmetics, theatrical makeups.
Rubber
As a coloring filler.

Hempseed Oil
Synonyms: Hemp oil.
Latin: Oleum cannabis.
French: Huile de canvre, Huile de chénévis.
German: Hanfoel.
Spanish: Aceite de canamo.
Italian: Olio di canapa.
Fats and Oils
Starting point in making—
Boiled oil, hardened oils, oil mixtures.
Substitute for other vegetable oils.
Food
As a food and salad oil in certain countries, especially in eastern Europe.
Fuel
As a burning oil.
Ingredient of—
Burning oil compositions containing rapeseed oil.
Gas
Starting point in making—
Oil gas (in certain countries only where other materials are costly).
Glues and Adhesives
Ingredient (Brit. 332257) of—
Adhesive compositions.
Leather
Ingredient (Brit. 332257) of—
Compositions used in the manufacture of artificial leather.
Compositions used as substitutes for leather in making footwear.
Compositions used for finishing leather goods.
Compositions used for impregnating leather to render it better resistant to wear and water.
Linoleum and Oilcloth
Ingredient of—
Compositions used in the manufacture of various types of floor coverings.
Miscellaneous
Binder in making—
Compositions of fibrous matter.
Ingredient (Brit. 332257) of—
Roofing compositions, wall coverings.
Paint and Varnish
Binder in making—
Artist's colors.
Substitute for linseed oil.
Vehicle in making—
Special paints, varnishes, and primers.
White paints (used in place of linseed oil to reduce the yellowing caused by the latter).
Paper
Ingredient (Brit. 332257) of—
Finishing and impregating compositions for treating paper, pasteboard, and pulp compositions.

Hempseed Oil (Continued)

Pharmaceutical
Starting point in making—
Galenicals.
Suggested for the treatment of gallstones.

Plastics
Ingredient (Brit. 332257) of—
Plastic compositions used for making pressed articles.

Rubber
Ingredient of—
Compositions used as rubber substitutes.

Soap
Ingredient of—
Mixed soapstocks.
Starting point in making—
Green soft soap.

Textile
Ingredient (Brit. 332257) of—
Compositions used for impregnating and finishing various fabrics.
Compositions for making waxed cloth.

Woodworking
Ingredient (Brit. 332257) of—
Compositions used for finishing and impregnating wood.

Henna
Synonyms: Egyptian privet, Flower of paradise.
French: Henne.
German: Mehnde.

Leather
Dyestuff in coloring leathers.

Perfumery
Ingredient of—
Hair coloring preparations.

Pharmaceutical
In compounding and dispensing practice.

Heptachloropropane
German: Heptachlorpropan.

Leather
Ingredient of—
Compositions used in making leather cloth (Brit. 279139).

Miscellaneous
Ingredient of—
Impregnating compositions used for various purposes (Brit. 279139).

Paint and Varnish
Ingredient (Brit. 279139) of—
Insulating varnishes and lacquers for electrical wiring and the like.
Paints and varnishes of various sorts.

Plastics
Ingredient of—
Compositions used in making molded articles, sheets and blocks (Brit. 279139).

Textile
——, *Finishing*
Ingredient of—
Compositions used in treating chemical fibers (Brit. 279139).

Heptadecylamine

Rubber
Activating agent (Brit. 412635) for—
Vulcanization accelerators, particularly such as the arylenethiazole mercaptans and disulphides and thiuramsulphides.

Heptadecylbisbetagammadihydroxypropylamine

Soap
Emulsifying agent (Brit. 421490 and 411295) in—
Shaving creams, superfatted soaps, and the like.

Heptaldehyde
Synonyms: Amylacetaldehyde, Heptanal, Heptoic aldehyde, Heptylaldehyde, Oenantaldehyde, Oenanthal, Oenanthic aldehyde, Oenanthol.
French: Aldéhyde d'héptyle, Aldéhyde héptylique.
German: Amylacetaldehyd, Heptaldehyd, Oenantaldehyd.
Italian: Aldeide etillica.

Chemical
Starting point in making—
Amylcinnamic aldehyde, heptinecarboxylic acid, heptoic acids, heptyl alcohol, heptin, heptyl heptoate, methylheptincarbonate, methylnonyl aldehyde, hydrazobenzene, nonylaldehyde, secondary caprylic alcohol, various esters used as aromatics.
Starting point in making derivatives with—
Acetone, anilin, cyanacetic acid, malonic acid, oxalic acid, oxalacetic acid.
Starting point in making—
Accelerators of vulcanization of rubber, with the aid of ammonia (French 553971).
Starting point (French 546516) in making derivatives with—
Anilin, benzylamine, diethylamine, naphthylamine, paratoluidin.
Starting point (French 613140) in making vulcanization accelerators with the aid of—
Ethylamine, metabutylamine, orthotolylbiguanide.

Heptaldoxime

Fuel
Primer (Brit. 429763) for—
Diesel engine fuel oils produced by the hydrogenation of coal.

Petroleum
Primer (Brit. 429763) for—
Diesel oils containing a high proportion of aromatic bodies.

Heptane
Synonyms: Dipropylmethane, Heptyl hydride, Methyl hexane, Normal heptane.
French: Hexane méthylique, Hydrure de héptyle, Hydrure héptylique.
German: Dipropylmethan, Heptylhydrid, Methylhexan.

Chemical
Solvent for various chemicals and in various chemical processes.

Fats and Oils
Solvent for fats and oils.

Miscellaneous
As an anesthetic.
Solvent for various substances.

Resins and Waxes
Solvent for resins and waxes.

Heptyl Alcohol
French: Alcool de héptyle, Alcool héptylique.
German: Heptylalkohol.

Chemical
Starting point in making—
Heptyl acetate, heptyl esters, intermediates, pharmaceuticals, synthetic aromatic chemicals.

Fats and Oils
Emulsifying agent (Brit. 277357) in making—
Emulsions, lubricants.

Fuel
Reagent in making—
Emulsified fuels (Brit. 277357).

Leather
Reagent in making—
Emulsified dressing compositions (Brit. 277357).

Petroleum
Reagent in making—
Motor fuel compositions.
Stable emulsions of petroleum and petroleum distillates (Brit. 277357).

Sanitation
Ingredient of—
Disinfecting compositions (German 273408).

Soap
Reagent in making—
Cleansing and detergent compositions in emulsified form (Brit. 277357).

Textile
——, *Finishing*
Reagent in making—
Cleansing and washing compositions (Brit. 277357).

Heptyl Bisulphide
Synonyms: Heptyl disulphide.
French: Bisulphure de héptyle, Bisulphure héptylique, Disulphure de héptyle, Disulphure héptylique.
German: Bischwefelheptyl, Dischwefelheptyl, Dischwefelwasserstoffsaeuresheptyl, Heptylbisulfid, Heptyldisulfid.
Chemical
Reagent in making—
Intermedaites, pharmaceuticals, salts and esters.
Reagent (Brit. 298511) in treating—
Albumens and albumenoids.
Glues and Adhesives
Reagent (Brit. 298511) in treating—
Vegetable proteins, such as soya bean flour, linseed protein, and peanut protein, to make adhesive preparations.
Miscellaneous
Reagent (Brit. 298511) in treating—
Vegetable proteins, such as soya bean flour, linseed protein, and peanut protein, to make sizing and finishing compositions.

Heptylcresol
Chemical
Starting point (Brit. 444351) in making—
Fat-splitting catalysts and emulsifying agents, useful in dyeing, laundering, bleaching, and various other purposes, by reacting with formaldehyde and nonaromatic secondary amines (the salts of the products with water-soluble acids, or water-insoluble acids, or the quaternary ammonium salts are claimed to be valuable for the purposes named).

2-Heptylcyclohexanone-1, Normal
Cosmetic
Odorant (Brit. 430930 and 449211) in—
Perfume mixtures.

Heptylenethiourea
Synonyms: Heptylenesulphourea.
French: Sulphourée de héptylène, Sulphourée héptylènique, Thiourée de héptylène.
German: Heptylensulfoharnstoff, Heptylenthioharnstoff.
Chemical
Starting point in making—
Intermediates, pharmaceuticals.
Starting point (Brit. 310534) in making rubber vulcanization accelerators with the aid of—
Alphanaphthylamine, anilin, betanaphthylamine, cyclohexylanilin, diphenylamine, ethylanilin, meta-anisidin, metacresidin, metanaphthylenediamine, metaphenyldiamine, metatoluidin, metatoluylenediamine, metaxylenediamine, metaxylidin, monomethylanilin, orthoanisidin, orthocresidin, orthonaphthylenediamine, orthophenylenediamine, orthotoluidin, orthotoluylenediamine, orthoxylenediamine, orthoxylidin, para-anisidin, paracresidin, paranaphthylenediamine, paraphenylenediamine, paratoluidin, paratoluylenediamene, paraxylenediamine, paraxylidin.

Heptylic Acid, Normal
Synonyms: Heptoic acid, Oenanthic acid, Oenanthylic acid.
French: Acide de héptyle, Acide héptylique, Acide oenanthique, Acide oenanthylique.
German: Heptylsäure, Oenanthansäure.
Spanish: Acido enantilico, Acido n-eptilico, Acido n-heptilico.
Italian: Acido enantilico, Acido n-eptilico, Acido n-heptilico.
Chemical
Starting point in making—
Esters and salts, used in perfumery, such as ethyl heptylate, methyl heptylate, isoamyl heptylate, octyl heptylate.
Intermediates, pharmaceuticals, synthetic aromatics.

Heptylphenol
Chemical
Starting point (Brit. 444351) in making—
Fat-splitting catalysts and emulsifying agents, useful in dyeing, laundering, bleaching, and various other purposes, by reacting with formaldehyde and nonaromatic secondary amines (the salts of the products with water-soluble acids, or water-insoluble acids, or the quaternary ammonium salts are claimed to be valuable for the purposes named).

Heptyl Phthalate, Secondary
French: Phthalate de héptyle, Phthalate héptylique.
German: Heptylphtalat, Phtalsäuresheptylester.
Cellulose Products
Plasticizer for—
Nitrocellulose.
For uses, see under general heading: "Plasticizers."

Heptylresorcinol
Chemical
Starting point (Brit. 444351) in making—
Fat-splitting catalysts and emulsifying agents for use in dyeing, laundering, bleaching, and various other purposes, by reacting with formaldehyde and nonaromatic secondary amines (the salts of the products with water-soluble acids or water-insoluble acids, and the quaternary ammonium salts, are claimed to be valuable for the purposes named).

Heptylylhydroquinone
Petroleum
Stabilizing agent (Brit. 406195) for—
Cracked gasolines and other motor fuels.

Heptylylphloroglucinol
Petroleum
Stabilizing agent (Brit. 406195) for—
Cracked gasolines and other motor fuels.

Heptylylpyrocatechol
Petroleum
Stabilizing agent (Brit. 406195) for—
Cracked gasolines and other motor fuels.

Heptylylpyrogallol
Petroleum
Stabilizing agent (Brit. 406195) for—
Cracked gasolines and other motor fuels.

Heptylylresorcinol
Petroleum
Stabilizing agent (Brit. 406195) for—
Cracked gasolines and other motor fuels.

Hernandia Seed Oil
French: Huile de semences d'hernandia.
German: Hernandiaoel, Hernandiasamenoel.
Spanish: Aceite de hernandia.
Italian: Olio di hernandia.
Fats and Oils
Starting point in making—
Boiled oil.
Fuel
As a burning oil and illuminant.
Leather
Ingredient of—
Compositions used in making artificial leather.
Miscellaneous
Ingredient of—
Various compositions of matter (used as a binder).
Oilcloth and Linoleum
Ingredient of—
Compositions used in the manufacture of linoleum and oilcloth.
Rubber
Ingredient of—
Rubber substitute compositions.
Soap
As a soapstock.

Hexachloroanthraquinone-1:2:5:6-diacridone
Dye
Starting point (U. S. 1972094) in making—
Reddish-grey vat dyes with 1-aminoanthraquinone.

Hexachloroanthraquinone-1:2:7:8-diacridone
Dye
Starting point (U. S. 1972094) in making—
Reddish-grey vat dyes with 1-aminoanthraquinone.

Hexachloroethane
Synonyms: Carbon hexachloride, Carbon trichloride, Perchloroethane, Tetrachloroethylene dichloride.
French: Dichlorure de tetrachloroéthylène, Dichlorure tetrachloroéthylènique, Hexachlorure de carbone, Hexachlorure carbonique.
German: Dichlortetrachloraethylen, Hexachloraethan, Kohlenstoffhexachlorid, Perchloraethan, Tetrachloraethylendichlorid.

Chemical
Starting point in making various intermediates and other derivatives.

Glass
Plasticizer in—
Compositions containing cellulose esters or ethers, used in the manufacture of non-scatterable glass and for coating glassware.

Insecticide
As an insecticide.
Ingredient of—
Insecticides, bactericides, germicides.

Leather
Plasticizer in—
Compositions containing cellulose esters or ethers, used in the manufacture of artificial leather and for coating leather and leather goods.

Match
Reagent in making—
Safety match head compositions.

Miscellaneous
Ingredient of—
Fireproofing compositions.
Retarding agent in—
Fermentation processes.

Paint and Varnish
Ingredient of—
Anti-cryptogamic submarine paints.
Plasticizer in making—
Dopes, paints, varnishes, enamels and lacquers from cellulose esters or ethers.

Pharmaceutical
In compounding and dispensing practice.

Plastics
Plasticizer in making—
Celluloid and other compositions (used in the place of camphor).

Rubber
Accelerator in vulcanization.
Plasticizer in—
Coating compositions containing cellulose acetate, nitrocellulose, or other esters or ethers of cellulose.

Stone
Plasticizer in—
Coating compositions containing various cellulose esters or ethers.

Textile
Ingredient of—
Fireproofing compositions.
Plasticizer in—
Coating compositions containing various esters or ethers of cellulose.

Woodworking
Ingredient of—
Fireproofing compositions.
Plasticizer in—
Coating compositions containing various cellulose esters or ethers.

Hexachloropropane.
French: Héxachlorure de propane.
German: Hexachlorpropan.

Electrical
Ingredient of—
Insulating varnishes for electric wiring (Brit. 279139).

Leather
Ingredient of—
Compositions used in making leather cloth (Brit. 279139).

Miscellaneous
Ingredient of—
Impregnating compositions used for various purposes (Brit. 279139).

Paint and Varnish
Ingredient (Brit. 279139) of—
Paints, varnishes.

Plastics
Ingredient (Brit. 279139) of—
Compositions for making molded articles, sheets, blocks and the like.

Textile
——, *Manufacturing*
Ingredient of—
Compositions used in making chemical fibers.

Hexachlororetene

Petroleum
Imparter (Brit. 431508) of—
High-film strength, adhesion power, and abrasion resistance to lubricants for use with extreme pressures (blended with mineral lubricating oil).

Hexadecylamine

Insecticide
Suspension promoter for—
Insoluble powdered insecticides.

Hexadecylcresol
Synonyms: Cetylcresol.

Chemical
Starting point (Brit. 444351) in making—
Fat-splitting catalysts and emulsifying agents, useful in dyeing, laundering, bleaching, and various other purposes, by reacting with formaldehyde and non-aromatic secondary amines (the salts of the products with water-soluble acids, or water-insoluble acids, or the quaternary ammonium salts, are claimed to be valuable for the purposes named).

Hexadecylguanidin Chloride

Miscellaneous
As an emulsifying agent (Brit. 422461).
For uses, see under general heading: "Emulsifying agents."

Textile
Assistant (Brit. 421862) in—
Aqueous baths for treating textiles.
Promoter (Brit. 421862) of—
Uniform dyeing with basic dyestuffs.
Wetting and washing agent (Brit. 421862) in—
Textile processes.

Hexadecylphenol

Chemical
Starting point (Brit. 444351) in making—
Fat-splitting catalysts and emulsifying agents, useful in dyeing, laundering, bleaching, and various other purposes, by reacting with formaldehyde and non-aromatic secondary amines (the salts of the products with water-soluble acids, or water-insoluble acids, or the quaternary ammonium salts, are claimed to be valuable for the purposes named).

Hexadecylresorcinol
Synonyms: Cetylresorcinol.

Chemical
Starting point (Brit. 444351) in making—
Fat-splitting catalysts and emulsifying agents, useful in dyeing, laundering, bleaching, and various other purposes, by reacting with formaldehyde and non-aromatic secondary amines (the salts of the products with water-soluble acids, or water-insoluble acids, or the quaternary ammonium salts, are claimed to be valuable for the purposes named).

Hexaethyl-Plumbane
Synonyms: Hexaethyl lead.

Lubricant
Starting point (Brit. 440175) in making—
Addition agents for high-pressure lubricating oils or greases by reacting with oil-soluble organic compounds.

1:3:3:1':3':3'-Hexamethylindocarbocyanin Chloride

Dye
Starting point (Brit. 448508) in making—
Color lakes which are especially fast to light, oil, and alcohols, and are claimed to be superior to the corresponding lakes from triarylmethane dyes.

1:3:3:1′:3′:3′-Hexamethylindocyanin Chloride

Dye
Starting point (Brit. 448508) in making—
Yellow lakes constituting clear shades fast to oil, spirit, and light.

Hexamethylmonoethylpararosanilin

Ink
Starting material (U. S. 1899452) in making—
Special ink for protection and authentification of checks and the like, which has the characteristic that the color is a function of the hydrogen ion concentration.

Hexane

Synonyms: Caproyl hydride, Hexyl hydride, Normal hexane.
French: Hydrure de caproyle, Hydrure caproylique, Hydrure de hexyle, Hydrure hexylique.
German: Caproylhydrid, Hexylhydrid.

Analysis
Reagent in determination of—
Refractive index of minerals.

Chemical
Solvent for various chemicals and in various chemical processes.

Fats and Oils
Solvent for fats and oils.

Miscellaneous
As a filler for thermometer tubes.
Solvent for various substances.

Resins and Waxes
Solvent for resins and waxes.

Hexaphenyl-Lead

Lubricant
Addition agent (Brit. 445813) in—
Lubricants for motors, turbines, flushing, and high-temperature work generally.

Hexaphenyl-Mercury

Lubricant
Addition agent (Brit. 445813) in—
Lubricants for motors, turbines, flushing, and high-temperature work generally.

Hexaphenyl-Tin

Lubricant
Addition agent (Brit. 445813) in—
Lubricants for motors, turbines, flushing, and high-temperature work generally.

Hexapyridin-Copper Sulphate

French: Sulphate de héxapyridine et de cuivre, Sulphate héxapyridinique et cuivrique.
German: Hexapyridinkupfersulfat, Kupferhexapyridinsulfat, Schwefelsaeureshexapyridinkupfer, Schwefelsaeuereskupferhexapyridin.

Chemical
Reagent in making various substanees.

Dye
Reagent (Brit. 306859) in making azo dyestuffs with—
Acetyl-H acid.
Alphaethoxy-8-hydroxynaphthalene-3:6-disulphonic acid.
Alphahydroxynaphthalene-4-sulphonic acid.
3-Aminobenzaldehyde.
2-(3′-Aminobenzoyl)amino-5-naphthol-7-sulphonic acid.
Anthranilic acid.
Benzidin-3:3′-dicarboxylic acid.
Beta-aminobenzaldehyde.
Beta-aminobenzene-5-sulphonic acid.
Beta-aminobenzoic acid.
Beta-amino-1-hydroxybenzene.
Beta-aminonaphthalene-3-carboxylic acid.
Betanaphthol.
Betaphenylamino-4-hydroxynaphthalene-7-sulphonic acid.
4-Chloro-2-chloro-2-aminobenzoic acid.
4:4′-Diaminodiphenylurea-3:3′-dicarboxylic acid.
4:6-Dichloro-2-amino-1-hydroxybenzene.
5:5′-Dihydroxy-2:2′-dinaphthylamine-7:7′-disulphonic acid.
J acid.
5-Nitro-2-aminobenzoic acid.

Hexenylpiperidin

Insecticide
As an insecticide.
Ingredient (Brit. 313934) of—
Insecticidal, germicidal, and vermicidal preparations.

Soap
Ingredient (Brit. 313934) of—
Insecticidal and germicidal soaps.

Hexone

Synonyms: Methylisobutyl ketone, 2-Methyl-4-pentanone.

Analysis
As an extractant.
Solvent for—
Camphor, cellulose derivatives, fats, gums, oils, resins, waxes.

Cellulose Products
Solvent for—
Cellulose acetate (with ethylene dichloride).
Cellulose ethers (certain types).
Nitrocellulose.

Ceramic
Solvent in—
Compositions, containing natural or synthetic resins, nitrocellulose, cellulose acetate, or other cellulose esters or ethers, used as coatings for protecting and decorating ceramic products.

Chemical
As an extractant.
Ketone in—
Organic syntheses.
Solvent for—
Camphor.
Cellulose acetate (with ethylene dichloride).
Cellulose ethers (certain types).
Fats, gums, oils, nitrocellulose, waxes.
Solvent miscible with most other organic solvents.

Cosmetic
Solvent in—
Fats, oils.
Solvent for—
Nail enamels and lacquers containing natural or synthetic resins, nitrocellulose, cellulose acetate, or other cellulose esters or ethers as base material.

Dry-Cleaning
Spotting agent for—
Fats, greasy stains, gums, oils, resins, waxes.

Electrical
Solvent in—
Insulating compositions, containing natural or synthetic resins, nitrocellulose, cellulose acetate, or other cellulose esters or ethers, used for covering wire and in making electrical machinery and equipment.

Fats, Oils, and Waxes
Solvent in—
Blown oils, essential oils, fats, sulphonated oils, synthetic oils, vegetable oils, waxes.

Food
Solvent for—
Fats, oils.

Glass
Solvent in—
Compositions, containing natural or synthetic resins, nitrocellulose, cellulose acetate, or other cellulose esters or ethers, used in the manufacture of nonscatterable glass and as coatings for decorating and protecting glassware.

Glue and Adhesives
Solvent in—
Adhesive compositions containing gums, natural or synthetic resins, nitrocellulose, cellulose acetate, or other cellulose esters or ethers.

Gums
As a solvent.

Leather
Solvent in—
Compositions, containing natural or synthetic resins, nitrocellulose, cellulose acetate, or other cellulose esters or ethers, used in the manufacture of artificial leathers and as coatings for decorating and protecting leathers and leather goods.

Hexone (Continued)

Metal Fabrication
Solvent in—
Compositions, containing natural or synthetic resins, nitrocellulose, cellulose acetate, or other cellulose esters or ethers, used as coatings for protecting and decorating metallic articles.

Miscellaneous
Solvent in—
Coating compositions, containing natural or synthetic resins, nitrocellulose, cellulose acetate, or other cellulose esters or ethers, used for protecting and decorating various articles.
Solvent miscible with most other organic solvents.

Paint and Varnish
Ingredient of—
Paint removers.
Solvent for—
Cellulose acetate (with ethylene dichloride).
Cellulose ethers (certain types).
Gums, nitrocellulose, oils, resins.
Solvent in—
Paints, varnishes, lacquers, enamels, and dopes containing natural or synthetic resins, nitrocellulose, cellulose acetate, or other cellulose esters or ethers.

Paper
Solvent in—
Compositions, containing natural or synthetic resins, nitrocellulose, cellulose acetate, or other cellulose esters or ethers, used in the manufacture of coated papers and as coatings for decorating and protecting products made of paper and pulp.

Petroleum
As a solvent.

Pharmaceutical
Solvent for—
Camphor, essential oils, fats, gums, mineral oils, vegetable oils, waxes.

Photographic
Solvent in making—
Films from nitrocellulose, cellulose acetate, or other esters or ethers of cellulose.

Plastics
Solvent in making—
Laminated fiber products, molded products.
Plastics from or containing natural or synthetic resins, nitrocellulose, cellulose acetate, or other cellulose esters or ethers.

Resins
Solvent for—
Natural and synthetic resins.
Solvent in making—
Artificial resins from or containing nitrocellulose, cellulose acetate, or other cellulose esters or ethers.

Rubber
Solvent in—
Compositions, containing natural or synthetic resins, nitrocellulose, cellulose acetate, or other cellulose esters or ethers, used as coatings for protecting and decorating rubber goods.

Soap
Solvent for—
Fats, oils.

Stone
Solvent in—
Compositions, containing natural or synthetic resins, nitrocellulose, cellulose acetate, or other cellulose esters or ethers, used as coatings for decorating and protecting artificial and natural stone.

Textile
Degreasing, defatting, and dewaxing agent for—
Textile fibers.
Solvent in—
Compositions, containing natural or synthetic resins, nitrocellulose, cellulose acetate, or other cellulose esters or ethers, used in the manufacture of coated fabrics.

Wood
Solvent in—
Compositions, containing natural or synthetic resins, nitrocellulose, cellulose acetate, or other cellulose esters or ethers, used as protective and decorative coatings on woodwork.

Hexyl Acetate, Secondary
French: Acétate de héxyle, sécondaire; Acétate héxylique, sécondaire.
German: Essigsaeuressekundärhexylester, Essigsaeuressekundärhexyl, Sekundär hexylacetat, Sekundar hexylazetat.

Cellulose Products
Solvent for—
Cellulose esters and ethers, cellulose nitrate.
For uses, see under general heading: "Solvents."

Resins
Solvent for—
Resins.

Hexyl Alcohol
Synonyms: Hexylic alcohol.
French: Alcool de hexyle, Alcool hexylique.
German: Hexylalkohol.

Chemical
Starting point in making—
Capronic acid, synthetic perfumes.

Fats and Oils.
Reagent (Brit. 277357) in making—
Emulsified lubricants, emulsions of various sorts.

Fuel
Reagent in making—
Emulsified fuels (Brit. 277357).

Leather
Reagent in making—
Dressings (Brit. 277357).

Petroleum
Reagent (Brit. 277357) in making—
Emulsified motor fuels.
Emulsions of petroleum and petroleum distillate.

Soap
Reagent in making—
Emulsions containing soap (Brit. 277357).

Textile
——, *Finishing.*
Reagent in making—
Washing and cleansing compositions (Brit. 277357).

Hexylcresol

Chemical
Starting point (Brit. 444351) in making—
Fat-splitting catalysts and emulsifying agents, useful in dyeing, laundering, bleaching, and various other purposes, by reacting with formaldehyde and non-aromatic secondary amines (the salts of the products with water-soluble acids, or water-insoluble acids, or the quaternary ammonium salts are claimed to be valuable for the purposes named).

Hexylenethiourea
Synonyms: Hexylenesulphourea.
French: Sulphourée d'hexylène, Sulphourée hexylènique, Thiourée d'hexylène, Thiourée hexylènique.
German: Hexylensulfoharnstoff, Hexylenthioharnstoff.

Chemical
Starting point in making—
Pharmaceuticals and other derivatives.
Starting point (Brit. 310534) in making rubber vulcanization accelerators with—
Alphanaphthylamine, anilin, benzylamine, betanaphthylamine, cyclohexylanilin, meta-anisidin, metacresidin, metanaphthylenediamine, metaphenylamine, metaphenylenediamine, metatoluidin, metatoluylenediamine, metaxylenediamine, metaxylidin, monoethylanilin, monomethylanilin, orthoanisidin, orthocresidin, orthonaphthylenediamine, orthophenylamine, orthophenylenediamine, orthotoluidin, orthotoluylenediamine, orthoxylenediamine, orthoxylidin, paraanisidin, paracresidin, paranaphthylenediamine, paraphenylamine, paraphenylenediamine, paratoluidin, paratoluylenediamine, paraxylidin, paraxylenediamine.
Starting point (Brit. 314909) in making derivatives with—
Alkoxyalphanaphthalenesulphonic acid.
Alpha-amino-5-naphthol-7-sulphonic acid.
Alphanaphthylamine-4:8-disulphonic acid.
Alphanaphthylamine-4:6:8-trisulphonic acid.
4-Aminoacenaphthene-3:5-disulphonic acid.
4-Aminoacenaphthene-3-sulphonic acid.
4-Aminoacenaphthenetrisulphonic acid.
Aminoarylcarboxylic acids.

Hexylenethiourea (Continued)
Aminoheterocyclic chlorides.
1:8-Aminonaphthol-3:6-disulphonic acid.
Bromonitrobenzoyl chlorides.
Chloroalphanaphthalenesulphonic acids.
Chloronitrobenzoyl chlorides.
2-Cinnamyl chloride.
Iodonitrobenzoyl chlorides.
Nitroanisoyl chlorides.
3-Nitrocinnamyl chloride.
4-Nitrocinnamyl chloride.
1-Nitronaphthalene 5-sulphochloride.
1:5-Nitronaphthoyl chloride.
2-Nitrophenylacetyl chloride.
4-Nitrophenylacetyl chloride.
Nitrotoluyl chlorides.

Hexylhydroquinone
Textile
Inhibitor (Brit. 446404) of—
Acidity and mould development in textile lubricants during storage of the lubricant or fabric.

Hexylnaphthyl-Aluminum
Lubricant
Addition agent (Brit. 433257) to—
Lubricating oils or greases, especially for use at high temperatures, such as cylinder oils, hydrogenated oils, or oils refined by treatment with sulphuric acid, clay, or extraction solvents.

Hexylnaphthyl-Antimony
Lubricant
Addition agent (Brit. 433257) to—
Lubricating oils or greases, especially for use at high temperatures, such as cylinder oils, hydrogenated oils, or oils refined by treatment with sulphuric acid, clay, or extraction solvents.

Hexylnaphthyl-Bismuth
Lubricant
Addition agent (Brit. 433257) to—
Lubricating oils or greases, especially for use at high temperatures, such as cylinder oils, hydrogenated oils, or oils refined by treatment with sulphuric acid, clay, or extraction solvents.

Hexylnaphthyl-Cadmium
Lubricant
Addition agent (Brit. 433257) to—
Lubricating oils or greases, especially for use at high temperatures, such as cylinder oils, hydrogenated oils, or oils refined by treatment with sulphuric acid, clay, or extraction solvents.

Hexylnaphthyl-Mercury
Lubricant
Addition agent (Brit. 433257) to—
Lubricating oils or greases, especially for use at high temperatures, such as cylinder oils, hydrogenated oils, or oils refined by treatment with sulphuric acid, clay, or extraction solvents.

Hexylnaphthyl-Thallium
Lubricant
Addition agent (Brit. 433257) to—
Lubricating oils or greases, especially for use at high temperatures, such as cylinder oils, hydrogenated oils, or oils refined by treatment with sulphuric acid, clay, or extraction solvents.

Hexylnaphthyl-Zinc
Lubricant
Addition agent (Brit. 433257) to—
Lubricating oils or greases, especially for use at high temperatures, such as cylinder oils, hydrogenated oils, or oils refined by treatment with sulphuric acid, clay, or extraction solvents.

Hexyl Paraoxybenzoate
Synonyms: Hexyl parahydroxybenzoate.
French: Parahydroxyebenzoate d'hexyle, Parahydroxyebenzoate hexylique, Paraoxyebenzoate d'hexyle, Paraoxyebenzoate hexylique.
German: Hexylparahydroxybenzoat, Parahydroxybenzoesäurehexylester, Parahydroxybenzoesäureshexyl, Paraoxybenzoesäurehexylester, Paraoxybenzoesäureshexyl.
Chemical
Starting point in making various derivatives.
Food
As a preservative.
Miscellaneous
As a general preservative and disinfectant.
Pharmaceutical
In compounding and dispensing practice.
Sanitation
As a disinfectant.

Hexylphenol
Chemical
Starting point (Brit. 444351) in making—
Fat-splitting catalysts and emulsifying agents, useful in dyeing, laundering, bleaching, and various other purposes, by reacting with formaldehyde and nonaromatic secondary amines (the salts of the products with water-soluble acids, or water-insoluble acids, or the quaternary ammonium salts are claimed to be valuable for the purposes named).

Hexyl Phthalate, Secondary
French: Phthalate d'hexile, Phthalate hexilique.
German: Hexylphtalat, Phtalsäureshexylester.
Cellulose Products
Plasticizer for—
Nitrocellulose.
For uses, see under general heading: "Plasticizers."

Hexylpyrocatechol
Textile
Inhibitor (Brit. 446404) of—
Acidity and mould development in textile lubricants during storage of the lubricant or fabric.

Hexylresorcinol
German: Hexylresorcin.
Miscellaneous
Ingredient (U. S. 1649671) of—
Antiseptic solutions in olive oil.
Pharmaceutical
In compounding and dispensing practice.

Hippuric Acid
Latin: Acidum hippuricum.
French: Acide hippurique.
German: Hippursäure, Pferdeharnsäure.
Chemical
Ingredient (Brit. 310934) of—
Insulin preparations.
Starting point in making—
Salts and other derivatives.
Pharmaceutical
In compounding and dispensing practice.

Homatropine Hydrobromide
Latin: Homatropinae hydrobromidum.
French: Bromhydrate d'homatropine.
German: Bromwasserstoffsäureshomatropin, Homatropinhydrobromid.
Spanish: Bromhidrato de homatropina.
Italian: Bromidrato di omatropina.
Pharmaceutical
In compounding and dispensing practice.
Suggested for use as—
Cycloplegic, mydriatic.

Homovanillin
Synonyms: 4-Oxy-3-methoxyphenylacetaldehyde.
French: 4-Oxy-3-méthoxyphényleacetaldéhydate.
Chemical
Starting point in making—
Aromatics.
Food
Ingredient of—
Artificial vanilla essence.
Synthetic flavoring agent in making—
Beverages, food preparations.
Perfume
Synthetic vanilla odor in making—
Cosmetics, perfumes.

Horse Fat

Fuel
Raw material in—
Candle making.

Lubricant
Raw material in making—
Cup and other greases.

Soap
As a soapstock.

Horse Hair

Fertilizer
Source of nitrogen in making—
Tankage, west base goods.

Furniture
Filling material in—
Upholstered furniture.

Lubricant
Ingredient of—
Elastic greases (used to produce a more resilient grease, preventing a heavy, sodden condition in the journal box).

Textile
Ingredient of—
Haircloth.

Horse Oil

Soap
As a soapstock.

Humic Acid

Petroleum
Viscosity decreaser (U. S. 1999766) of—
Fluid clay mud encountered in oil well drilling (used in conjunction with a small amount of caustic alkali).

Hydrazin Hydrate
German: Hydrazinhydrat.

Chemical
Reagent in making—
Colloidal copper solutions.
Colloidal gold solutions.
Colloidal platinum solutions.
Colloidal rhodium solutions.
Starting point in making—
Ethylideneazin.

Hydrazin Sulphate
French: Sulfate d'hydrazine.
German: Hydrazinsulfat, Schwefelsaeureshydrazin.

Analysis
Reagent in—
Analysis of minerals, slags, fluxes.
Determination of arsenic in metallurgical laboratories.

Chemical
Starting point in making—
Hydrazin anhydride, hydrazin hydrate, hydrazin ureas.

Metallurgical
Reagent in separating—
Polonium from tellurium.

Textile
——, *Manufacturing*
Catalyst in making—
Acetate rayon (Brit. 27228-1912).

Hydrazoic Acid
French: Acide hydrazoique.
German: Hydroazosaeure.

Chemical
Catalyst (Brit. 252460) in making—
Acetonitrile from acetaldehyde.
Amines from aromatic hydrocarbons.
Amines from organic carbonyl compounds.
Anilin from benzene.
Benzanilide from benzophenone.
Benzonitrile from acetaldehyde.
Formanilide from benzaldehyde.
Epsilonleucinlactamcyclopentamethylenetetrazole from cyclohexanone.
Methylacetamide from acetone.
Methyl formate from acetaldehyde.
Starting point in making—
Benzonitrile (Brit. 250897).
Phenyltetrazole (Brit. 250897 and 251266).

Hydrobromic Acid
Synonyms: Hydrogen bromide.
Latin: Acidum hydrobromicum.
French: Acide bromhydrique, Acide hydrobromique.
German: Bromwasserstoff, Bromwasserstoffsäure.
Spanish: Acido bromhidrico, Acido hidrobromico.
Italian: Acido bromidrico, Acido idrobromico.

Analysis
Reagent for—
Detecting palm oil in oleomargarin.
Detecting and determining sulphur in free state or in combination in the form of sulphides.
Solvent for—
Mercury, lead, copper and their sulphides.

Chemical
Reagent in making.
Aconitine.
Allyl bromide.
Alphabromobetagammadiacetylglycerol.
Aluminum bromide.
Ammonium bromide by reaction with ammonia.
Antimony bromide.
Arecoline hydrobromide.
Apoatropine hydrobromide.
Apomorphine hydrobromide.
Barium bromide by action on barium sulphide.
Benzyl bromide.
Bismuth bromide.
Bromobehenic acid.
Calcium bromide by action on calcium oxide, calcium carbonate, or calcium hydroxide.
Cinchonidine hydrobromide.
Cinchonine hydrobromide.
Cobaltous bromide by reaction on cobalt metal.
Cocaine hydrobromide.
Copper bromide.
Ethylene dibromide.
Glycolbromohydrin.
Homatropine hydrobromide.
Hydrastine hydrobromide.
Hyoscine hydrobromide.
Inorganic compounds.
Intermediate chemicals.
Lithium bromide by reaction with lithium hydroxide.
Magnesium bromide.
Methyl bromide.
Nickel bromide by reaction with nickel oxide.
Organic compounds.
Pharmaceutical chemicals.
Pilocarpine hydrobromide.
Quinine hydrobromide.
Quinidine hydrobromide.
Strontium bromide by reaction with strontium carbonate.
Strychnine hydrobromide.
Theobromine hydrobromide.
Tropoacaine hydrobromide.
Tropine hydrobromide.

Dye
Reagent in making various synthetic dyestuffs.

Pharmaceutical
Suggested for use as nerve sedative in various diseases.

Photographic
Reagent in the bromide process.

Miscellaneous
Reagent for—
Fixing microscopic preparations.

Hydrochloric Acid
Synonyms: Chlorhydric acid, Hydrogen chloride, Marine acid, Muriatic acid, Spirit of salt, Spirit of sea salt.
Latin: Acidum chlorhydricum, Acidum hydrochloratum, Acidum hydrochloricum, Acidum muriaticum.
French: Acide chlorhydrique, Acide hydrochlorique, Acide muriatique.
German: Chlorwasserstoffsäure.
Spanish: Acido clorhidrico, Acid muriatico.
Italian: Acido cloridrico, Acido muriatico.

Analysis
Reagent in—
Analytical processes involving control and research in science and industry.

Ceramic
Process material in making—
Porcelain, potteries.
Purifying agent for—
Iron in clay.

Hydrochloric Acid (Continued)

Chemical
Acidifying agent in—
Chemical processing.
Catalyst in making—
Esters, such as methyl anthranilate, ethyl cinnamate, ethyl citrate, ethyl lactate.
Hydrolyzing agent for—
Carbohydrates.
Neutralizing agent for—
Alkalies.
Process material in making—
Acetic acid, adipic acid, alginic acid, arsenic acid, chlorhydrates of organic bases, chlorhydrins, chlorosulphonic acid, chromic acid, citric acid, fatty acids from lime soaps, lead oxychloride (from galena-Pattison process), resorcinol, salicylic acid, salvarsan, selenium oxychloride, silica gel, various other chemicals.
Purifying agent in—
Manufacturing processes.
Reducing agent (with tin) in—
Organic syntheses (used with iron, stannous chloride, tin, or zinc).
Solvent for various chemicals and raw materials.
Starting point in making—
Aqua regia, inorganic chemicals.
Metal chlorides, such as aluminum chloride, barium chloride, bismuth chloride, cesium chloride, chromium chloride, magnesium chloride, manganese chloride.
Organic chemicals.
Organic chlorides, such as methyl chloride, ethyl chloride.

Dye
Process material in making—
Dyestuffs.

Fats and Oils
Bleaching agent for—
Fats, oils.

Fertilizer
Reactant in making—
Phosphate from bones.

Food
Hydrolizing agent for—
Carbohydrates.
Neutralizing agent for—
Alkalies.

Fuel
In coke purification.

Glue and Gelatin
Neutralizing agent for limed leaching solutions in making—
Glue, gelatin.
Solvent for mineral matter in making—
Glue, gelatin.

Glass
Purifying agent for—
Sand.

Ink
Reactant in making various inks.

Leather
Process material in—
Chrome tanning operations, dehairing skins and hides.

Metallurgical
Etching agent for—
Metals.
Ingredient of—
Soldering solutions.
Fluxing bath in galvanizing processes.
Pickling agent for iron in—
Cleaning processes, galvanizing processes, tinning processes.
Solvent for zinc in—
Zinc chemical manufacture from residual flux skimmings of galvanizing processes.
Zinc reclamation from galvanized scrap.
Starting point in making—
Aqua regia (gold solvent).

Mechanical
In pipefitting.

Paint and Varnish
Reactant in making—
Pigments.

Pharmaceutical
In compounding and dispensing practice.

Photographic
In general processes.

Perfume
In making synthetic perfumes.

Printing
In lithographic processess.
In process engraving.

Rubber
Process material in—
Rubber reclamation processes.

Soap
Reactant in—
Purifying soapstock.

Sugar
Diffusory auxiliary in—
Beet sugar manufacture.
Purifying agent for—
Animal charcoal.

Textile
Acidifying agent.
Neutralizing agent for—
Alkalies.
Process material in—
Dyeing processes, mercerizing processes, printing processes.
Sour for—
Fabrics.

Hydrocuprein

Chemical
Starting point (Brit. 27952-1911) in making—
Benzoylhydrocuprein, dibenzoylhydrocuprein, hydrocuprein-ethyl carbonate.

Hydrocyanic Acid

Synonyms: Cyanhydric acid, Formonitril, Formonitrile, Hydrogen cyanide, Prussic acid.
Latin: Acidum borussicum, Acidum cyanhydricum, Acidum hydrocyanicum, Acidum zooticum.
French: Acide de cyanhydrique, Acide de hydrocyanique, Acide de prussique, Cyanure de hydrogène.
German: Blausäure, Cyanwasserstoff, Cyanwasserstoffsäure, Preussichesäure.
Spanish: Acido cianhidrico.

Agriculture
Disinfectant for the soil, general parasiticide.

Analysis
Reagent in carrying out various processes.

Chemical
Reagent in making—
Aromatics, butylenecyanhydrin, ethylenecyanhydrin, intermediates, methylenecyanhydrin.
Oxy-acids from aldehydes and ketones.
Pharmaceuticals, propylenecyanhydrin.

Dye
Reagent in making various synthetic dyestuffs.

Food
Disinfectant for flour and other foodstuffs.
Fumigant in treating—
Grain elevators, storage chambers.

Insecticide
General plant fumigant and insecticide.
Ingredient of—
Fumigating compositions, containing zinc chloride (U. S. 1620365).
Fumigating compositions which contain liquefied sulphur dioxide (added for the purpose of stabilizing the preparation) (German 435714).
Insecticide agent in treating—
Citrus and other fruit trees.

Metallurgical
Reagent in the cyanide process of metal smelting.

Miscellaneous
Fumigant in—
Ridding clothes and storage warehouses of moths.
Disinfecting railroad cars.
Ridding ships of rats and other vermin.

Military
As a poison gas.

Rubber
Reagent (Brit. 300719) in treating—
Rubber latex for the purpose of accelerating coagulation, improving the quality of the rubber obtained.

Hydrocyanic Acid (Continued)

Sanitation
General disinfecting and fumigating agent.

Textile
Fumigant in treating—
Raw cotton.

Hydrofluoric Acid
Synonyms: Etching acid, Fluorhydric acid, Hydrogen fluoride.
Latin: Acidum fluorhydricum, Acidum hydrofluoricum.
French: Acide fluorhydrique, Acide fluorique, Acide hydrofluorique.
German: Fluorwasserstoffsäure, Flusssäure, Flusspathsäure.
Spanish: Acido fluorhidrico, Acido hidrofluorico.
Italian: Acido fluoridrico, Acido idrofluorico.

Analysis
Reagent for—
Analyzing alloys containing various metals.
Various minerals.
Disintegrating silicates to determine the silica content.
Separating niobium and tantalum.
Lead and copper from copper and antimony in electrolysis.
Volatilizing vanadic acid.

Brewing
Reagent for—
Controlling fermentation process in making beer so that secondary injurious bye-reactions are prevented and only alcoholic fermentation takes place.

Ceramics
Reagent for—
Frosting enamel for making porcelains and potteries.
Increasing porosity of porcelains and potteries.

Chemical
Reagent for—
Destroying bacteria which cause undesirable secondary reactions in alcoholic fermentation; for example, so that it can be carried out without the interference of lactic acid, acetic acid, and butyric acid secondary fermentations.
Reagent for making—
Aluminum fluoride by action on aluminum hydroxide.
Ammonium bifluoride by action on ammonium hydroxide.
Ammonium fluoride by action on ammonium hydroxide.
Antimony-ammonium fluoride.
Antimony pentafluoride.
Antimony trifluoride.
Arsenic pentafluoride.
Arsenic trifluoride.
Barium fluoride by reaction with barium sulphide.
Bismuth fluoride.
Cadmium fluoride.
Calcium fluoride by reaction with a soluble calcium salt.
Chlorates by electrolysis (added for the purpose of increasing the potential of the electrolyzing solution).
Chromic fluoride.
Cobalt fluoride.
Cupric fluoride by reaction with copper carbonate.
Ferrous fluoride.
Hydrogen peroxide, starting from sodium peroxide.
Lithium fluoride by reaction with lithium hydroxide.
Magnesium fluoride by reaction with solution of a magnesium salt.
Manganous fluoride by reaction with manganous hydroxide.
Nickel fluoride.
Persulphates by electrolysis (added for the purpose of increasing potential of the electrolyzing solution).
Potassium fluoride by reaction with potassium carbonate.
Silicon tetrafluoride by reaction with silica or silicates.
Sodium bifluoride.
Sodium fluoride by reaction with sodium carbonate.
Strontium fluoride.
Vanadium fluoride.
Yeast.
Zinc fluoride by reaction with zinc hydroxide.

Construction
Reagent for—
Removing efflorescence from brick and stone.

Distilled Liquors
Reagent for—
Controlling alcoholic fermentation so as to prevent injurious secondary reactions.
Making spirits from cereals by the Effront process.

Fuel
Reagent for—
Treating anthracite to make it suitable for use in the manufacture of coal gas.

Gas
Reagent for—
Refining crude lignite oil and benzene obtained from the gas carbonization process.

Glass
Ingredient of—
Mixtures, containing ammonium fluoride, used for making ground glass and for etching glass.
Reagent for—
Frosting glass, making etched glass, ground glass.
Removing iron skin in the treatment of waste products recovered from the manufacture of glass (French 601440).

Metallurgical
Ingredient (French 493295) of—
Electrolytic zinc-plating baths.
Reagent for—
Disintegrating rocks in the metallurgical process.
Making metallic boron.
Removing sand from metal castings.

Miscellaneous
Reagent for—
Cleaning copper and brass.
Engraving marks and scales on glass thermometers and other glass chemical and physical apparatus.
Microscopic work.
Preserving anatomical specimens.
Purifying graphite.
Iron and copper vessels.

Paper
Reagent in—
Making filter paper suitable for use in gravimetric chemical analysis.

Petroleum
Reagent in—
Gasoline refining.

Pharmaceutical
Suggested for use as antiseptic, antitubercular, and antifermentative.

Rubber
Reagent (French 532769) for—
Removing sand from rubber and gutta-percha.

Sugar
Reagent in—
Making sugar from beets (added to destroy *clostridium butyricum*).

Textile
Reagent for—
Working over silk which has been too heavily weighted.

Hydrofluosilicic Acid
Synonyms: Hydrofluorsilicic acid, Hydrosilicofluoric acid, Sand acid, Silicofluoric acid.
French: Acide fluosilicique, Acide hydrofluosilicique.
German: Kieselfluorwasserstoffsaeure, Kieselflussaeure.

Analysis
Reagent in laboratory work.

Brewing
Disinfectant cleaner for brass and copper apparatus (not for iron apparatus).

Chemical
Reagent in making—
Barium peroxide from barium phosphate (German 426034).
Chloric acid, platinum catalyst contact masses.
Starting point in making—
Aluminum hydrofluosilicate, ammonium hydrofluosilicate, barium hydrofluosilicate, bismuth hydrofluosilicate, cadmium hydrofluosilicate, calcium hydrofluosilicate, chromium hydrofluosilicate, copper hydrofluosilicate, iron hydrofluosilicate, lead hydrofluosilicate, magnesium hydrofluosilicate, manganese hydrofluosilicate, nickel hydrofluosilicate, potassium hydrofluosilicate, sodium hydrofluosilicate, strontium hydrofluosilicate, tin hydrofluosilicate (French 606541), titanium hydrofluosilicate, zinc hydrofluosilicate.

Hydrofluosilicic Acid (Continued)

Ceramics
Reagent for increasing hardness of—
Ceramic ware, chinaware, porcelains, potteries.

Construction
Ingredient of—
Concrete flooring compositions.
Preservative for—
Masonry.
Reagent for hardening—
Cement, plaster of paris.

Leather
Reagent for the preliminary treatment of hides and skins (Brit. 256628).

Miscellaneous
Ingredient of—
Disinfecting compositions for general purposes.

Paint and Varnish
Ingredient of technical paints and varnishes.
Preservative for pigments in oil.

Sanitation
As a general disinfectant, alone or in mixtures.

Starch
Reagent in making—
Dextrin.

Textile
——, *Printing*
Assist with—
Anilin black.

Woodworking
Ingredient of—
Impregnating compositions, preserving compositions.

Hydrogen
French: Hydrogène.
German: Wasserstoff.

Analysis
Reagent in various analytical processes, particularly reductions.

Chemical
Hydrogenating agent in making—
Amines from nitro compounds.
Decalin from naphthalene.
Formates from bicarbonates.
Hexalin from phenol.
Tetralin from naphthalene.
Various organic compounds and intermediates.
Reagent for soldering lead work in erecting sulphuric acid chambers, concentrators, and other apparatus used in making acids and chemicals.
Reagent in making—
Ammonia synthetically from the nitrogen of the air.
Barium sulphide.
Hydrochloric acid from chlorine.
Methylparatolylanthracene.
Prussic acid.
Synthetic pharmaceuticals.
Reducing agent in making—
Alcohol from acetaldehyde.
Methane from carbon dioxide.

Electrical
Filling material in making—
Tungsten-filament incandescent lamps.
Heating agent in making—
Mercury-vapor arc lamps from quartz glass.

Fats and Oils
Reagent in making—
Hydrogenated or hardened fats from oils.

Fuel
For the production of liquid fuels from coal by a special carbonization process.

Glass
Heating agent in making—
Quartz glass.

Jewelry
Reagent in making—
Synthetic rubies, synthetic sapphires.

Metallurgical
As a reducing atmosphere in pouring special castings.
Heating agent in autogeneous welding of—
Aluminum and alloys, cast iron, copper and alloys, steel, wrought iron.
Heating agent in fusing—
Tungsten powder to obtain the metal in the form of rods.
Various refractory metals.
Heating agent in making—
Laboratory utensils.
Various articles and vessels from platinum and other metals possessing high melting points.
Heating agent in the oxyhydrogen flame.
Reagent in—
Reducing tungstic acid to obtain metallic tungsten.
Removing sulphur from coke in the manufacture of high-grade steel products.
Treating steel in order to remove the oxygen dissolved in it.

Miscellaneous
In balloons and airships, in the oxyhydrogen limelight.

Petroleum
For the production of gasoline and other derivatives from crude petroleum by special cracking processes.

Photographic
Reagent in certain photographic processes.

Soap
Reagent in making soaps by certain processes.

Hydrogen Peroxide Solution
Synonyms: Peroxide of hydrogen, Solution of hydrogen dioxide.
Latin: Liquor hydrogenii dioxidi.
French: Péroxide d'hydrogène.
German: Wasserstoffhyperoxyd, Wasserstoffhyperoxydlosung.
Spanish: Aqua oxigenada.
Italian: Acqua ossigenata.

Agriculture
Disinfectant for—
Soils.
Pickling agent (Brit. 393808) for—
Seeds.

Brewing
Bactericide for—
Unfavorable ferments and moulds in the wort.
Preservative agent for—
Beer.
Sterilizing agent for—
Casks, filter pulp.

Chemical
General oxidizing agent in many processes.
Oxidizing agent in making—
Lead sulphate.

Dye
General oxidizing agent in making—
Intermediates, synthetic dyestuffs.

Explosives
Oxidizing agent (Brit. 397600) in making—
New explosive.

Fats and Oils
Oxidizing agent in—
Refining and bleaching fats and oils.

Food
Bleaching agent for—
Flour, gelatin.
General bleaching agent.
General preservative.
Hydrolytic agent for—
Starch (in breaking it down to obtain dextrin, dextrose, and other products of starch hydrolysis).
Preservative for—
Butter, milk.
Reagent for—
Removing last traces of sulphur dioxide used in bleaching various foods.

Glues and Adhesives
Bleaching agent for—
Glue.
Solvent for—
Indian gum.

Insecticide
Ingredient (Brit. 399938) of—
Insecticidal composition, consisting of quassin, quinine, gasoline, cade oil, and kerosene, used as a protective against moths.
Insecticidal composition consisting of quassin, quinine, hydrobromic acid, alcohol, and glycerol, used as a protective against moths.

Leather
Disinfectant for—
Hides subjected to long storage.

Hydrogen Peroxide Solution (Continued)
Ingredient (U. S. 1844018) of—
Tanning agent made from sulphite cellulose waste liquor.
Preservative for—
Tannins, tanning extracts.
Reagent in—
Tanning.
Metallurgical
Inhibiting agent (Brit. 375599) in the pickling of—
Chromium rustless steels, high-carbon steels, highly polished steels.
Reagent in—
Tinting metals.
Miscellaneous
Bleaching agent for—
Bones, feathers, hair, parchment, straw, teeth (in dentistry).
General antiseptic.
General bactericide in many fermentation industries.
General bleaching agent.
General hydrolytic agent.
General oxidizing agent.
Reagent in—
Restoration of old paintings.
Perfume
Bleach for the hair (used with a small amount of ammonia water or other alkali).
Ingredient of—
Dentifrices.
Hair wash (in combination with a small amount of nitric acid).
Various cosmetic creams, lotions, and other preparations.
Ingredient of—
Various cosmetic preparations.
Pharmaceutical
In compounding and dispensing practice.
Photographic
Bleaching agent.
Coating agent (Brit. 385522) for—
Ferroprussiate paper.
Eliminant for hypo.
Oxidizing agent.
Soap
Ingredient of—
Medicinal soaps, toilet soaps.
Textile
Antichlor for—
Removing last traces of sulphur dioxide from bleached wool and silk.
Reagent in—
Bleaching cotton and wool.
Bleaching, dyeing and printing processes.
Bleaching laces.
Water and Sanitation
Disinfectant and bactericide for—
Drinking water.
Ingredient of—
Sanitary compositions.
Reagent in—
Sanitary composition (with turpentine).
Wine
Bactericide for—
Unfavorable ferments and moulds in the must.
Preservative agent for—
Finished wines.
Sterilizing agent for—
Casks, filter pulp.
Woodworking
Bleaching agent.

Hydrophthalic Acid Ester of Grapeseed Alcohol
Bituminous
Solvent (Brit. 445223) for—
Asphalt and other bituminous bodies.
Dye
Solvent (Brit. 445223) for—
Dyestuffs, particularly oil-soluble coaltar dyes.
Fats, Oils, and Waxes
Solvent (Brit. 445223) for—
Fats, oils, waxes.
Resins
Solvent (Brit. 445223) for—
Oil-soluble glycerol-phthalic acid resins.
Polymerized vinyl compounds.
Synthetic resins.
Rubber
Solvent (Brit. 445223) for—
Rubber.

Hydrophthalic Acid Ester of Ricinoleic Alcohol
Bituminous
Solvent (Brit. 445223) for—
Asphalt and other bituminous bodies.
Dye
Solvent (Brit. 445223) for—
Dyestuffs, particularly oil-soluble coaltar dyes.
Fats, Oils, and Waxes
Solvent (Brit. 445223) for—
Fats, oils, waxes.
Resins
Solvent (Brit. 445223) for—
Oil-soluble glycerol-phthalic acid resins.
Polymerized vinyl compounds.
Synthetic resins.
Rubber
Solvent (Brit. 445223) for—
Rubber.

Hydroquinone
Synonyms: Hydrochinone, Hydroquinol, Methylhydrocupreine, Parahydroxybenzene, Parahydroxybenzol, Quinole, Quinone.
French: Chinone, Dihydroxyebenzène, Hydrochinone, Hydroquinole, Méthylehydrocupreine, Parahydroxyebenzène, Quinole, Quinone.
German: Chinol, Hydrochinol, Hydrochinon, Hydroquinol, Parahydroxybenzol.
Ceramics
Ingredient of—
Compositions used in the printing of ceramic ware.
Stabilizer (U. S. 1720992) in—
Coating compositions, containing nitrocellulose, used in decorating ceramic ware.
Chemical
Reagent in making—
5:8-Dihydroxy-2-methylanthraquinone.
Intermediates.
Starting point in making—
Adurol, brominated derivatives, chlorinated derivatives, iodinated derivatives, quinazarin, synthetic pharmaceuticals.
Dye
Starting point in making various dyestuffs.
Glass
Reagent in—
Printing on glassware.
Stabilizer (U. S. 1720992) in—
Coating compositions containing nitrocellulose.
Miscellaneous
Reagent in—
Printing various materials in colors and designs.
Stabilizer (U. S. 1720992) in—
Compositions, containing nitrocellulose, used for coating and decorating various materials.
Paint and Varnish
Stabilizer (U. S. 1720992) in making—
Nitrocellulose paints, varnishes, lacquers, enamels, and dopes.
Paper
Stabilizer (U. S. 1720992) in—
Coating compositions containing nitrocellulose.
Pharmaceutical
In compounding and dispensing practice.
Photographic
Developer for—
Films and plates, motion picture films.
Ingredient of—
Developing compositions.
Petroleum
Ingredient (Brit. 313155 and Brit. 312697) of—
Motor fuels.
Reagent (Brit. 312774) in treating—
Petroleum distillates, such as gasoline and kerosene, to prevent or remove discoloration of the product.
Plastics
Reagent in—
Printing on celluloid.
Rubber
Stabilizer (U. S. 1720992) in—
Coating compositions containing nitrocellulose.

Hydroquinone (Continued)

Stone
Stabilizer (U. S. 1720992) in—
Coating compositions containing nitrocellulose.

Textile
Stabilizer (U. S. 1720992) in—
Coating compositions containing nitrocellulose.

Waxes and Resins
Reagent (U. S. 1725933) in making—
Immersion wax baths, containing paraffin wax and the like (added for the purpose of prolonging the life of the bath).

Woodworking
Stabilizer (U. S. 1720992) in—
Coating compositions containing nitrocellulose.

2-Hydroxyalphanaphthacarbazole.
German: 2-Hydroxyalphanaphtacarbazol.

Dye
Starting point (French 617211) in making dyestuffs with—
Diazotized metanitranilin, 4-nitro-orthoanisidin.

4-Hydroxyalphanaphthacarbazole
German: 4-Hydroxyalphanaphtacarbozol.

Dye
Starting point in making dyestuffs with—
4-Nitro-orthoanisidin (French 617211).

7-Hydroxyalphanaphthacarbazol-6-carboxylic Orthoanisidide
French: Orthoanisidide de 7-hydroxyealphanaphthacarbazole-6-carboxylique.
German: 7-Hydroxyalphanaphtocarbazol-6-carbonylorthoanisidid.

Textile
Reagent (Brit. 347113) in producing azo dyestuffs on textile fibers with the aid of—
5-Chloro-2-toluidin.
2:5-Dichloroanilin.
5-Nitro-2-anisidin.
5-Nitro-orthotoluidin.

7-Hydroxyalphanaphthacarbazol-6-carboxylic Orthotoluidide
French: Orthotoluidide de 7-hydroxyalphanaphthacarbazole-6-carbonylique, Orthotoluidide de 7-hydroxyalphanaphthacarbazole-6-carboxylique.
German: 7-Hydroxyalphanaphtacarbazol-6-carbonylorthotoluidid.

Chemical
Starting point in making various derivatives.

Textile
Ingredient (Brit. 347113) of baths used for producing azo dyestuffs directly on the fiber with the aid of—
5-chloro-2-toluidin.
2:5-Dichloroanilin.
5-Nitro-2-anisidin.
5-Nitro-orthotoluidin.

6-Hydroxyalphanaphthaquinolin

Dye
Starting point (Brit. 394416, 410106) in making—
Brown-violet dyes; specially suitable for after-chroming on wool, by coupling with 4-nitro-2-aminophenol-6-sulphonic acid.
Blue dyes, specially suitable for after-chroming on wool, by coupling with 6-nitro-2-aminophenol-4-sulphonic acid.
Red-violet dyes, specially suitable for after-chroming on wool, by coupling with 5-amino-3-sulphosalicylic acid.

3-Hydroxy-1:2-benzofluorenone

Dye
In dye syntheses.
Starting point (German 589527) in making—
Water-insoluble dyes for coloring pigment pastes, by coupling with diazotized 1-amino-4-benzoylamino-5-methoxy-2-chlorobenzene.

2-Hydroxy-5-bromodiphenyl

Disinfectant
Bactericide claimed (U. S. 1989081) to have high efficiency.

Hydroxycitronellal
Synonyms: Citronellal hydrate, Citronellal hydroxide, Dihydroxycitronellal.
French: Hydrate de citronellale, Hydrate citronellalique, Hydroxyde de citronellale, Hydroxide citronellalique.
German: Citronellalhydrat, Citronellalhydroxyd.

Perfume
Ingredient of synthetic odors resembling—
Hayacinth, lily of the valley, lilac, linden flower, mimosa, neroli.
Perfume in—
Cosmetics.

Soap
Perfume in—
Toilet soaps.

2-Hydroxy-5-cyclohexylanilin

Dye
In dye syntheses
Starting point (Brit. 448872) in making—
Cobalt dyes.

2-Hydroxy-3:5-diaminopyridin

Chemical
Starting point in—
Organic synthesis.

Disinfectant
Starting point (Brit. 442190) in making—
Bactericidal azo dyestuffs by coupling with diazotized arylamines or their substitution products.

6-Hydroxy-2:4-dimethylquinolin
German: 6-Oxy-2:4-dimethylchinolin.

Chemical
Starting point (German 243206) in making—
5:6:7:8-Tetrahydro-6-hydroxy-2:4-dimethylquinolin.
5:6:7:8-Tetrahydro-6-hydroxy-2:4-dimethylquinolin hydrochloride.
5:6:7:8-Tetrahydro-6-hydroxy-2:4-dimethylquinolin picrate.
5:6:7:8-Tetrahydro-6-hydroxy-2:4-dimethylquinolin methiodide.
5:6:7:8-Tetrahydro-6-hydroxy-2:4-dimethylquinolin orthobenzoyl derivative.

2-Hydroxydiphenyl
Synonyms: Orthophenylphenol, 2-Phenylphenol.

Disinfectant
As a germicide.

Fungicide
As a fungicide.

Rubber
Reagent in compounding.

4-Hydroxydiphenyl
Synonyms: Paraphenylphenol, 4-Phenylphenol.

Resins
Intermediate in making—
Synthetic resins

Rubber
Reagent in compounding.

3-Hydroxydiphenylaminecarboxylic Acid
French: Acide de 3-hydroxyediphényleamine-carbonique, Acide de 3-hydroxyediphényleamine-carboxylique.
German: 3-Hydroxydiphenylamin carbonsäure.

Chemical
Starting point in making—
Intermediates and other derivatives.
Starting point (Brit. 336428) in making intermediates with the aid of—
Anilin, betanaphthylamine, 1:5-diaminonaphthalene, dianisidin, orthoanisidin, orthotoluidin, paranitranilin, paratoluidin.

Hydroxyethoxydiethyl Monoacetate

Cellulose Products
Plasticizer for—
Cellulose esters or ethers, cellulose nitrate (nitrocellulose).
For uses, see under general heading: "Plasticizers."

Hydroxyethoxydiethyl Phthalate

Cellulose Products
Plasticizer for—
Cellulose acetate, cellulose esters or ethers, cellulose nitrate (nitrocellulose).
For uses, see under general heading: "Plasticizers."

Hydroxyethylanilin Citrate

Textile
De-electrifying agent (Brit. 430221) for—
Yarns, films, fabrics, and the like, subject to charging by static electricity (applied in admixture with all usual lubricating agents as vehicles).

Hydroxyethylanilin Gallate

Textile
De-electrifying agent (Brit. 430221) for—
Yarns, films, fabrics, and the like, subject to charging by static electricity (applied in admixture with all usual lubricating agents as vehicles).

Hydroxyethylanilin Lactate

Textile
De-electrifying agent (Brit. 430221) for—
Yarns, films, fabrics, and the like, subject to charging by static electricity (applied in admixture with all usual lubricating agents as vehicles).

Hydroxyethylanilin Mucate

Textile
De-electrifying agent (Brit. 430221) for—
Yarns, films, fabrics, and the like, subject to charging by static electricity (applied in admixture with all usual lubricating agents as vehicles).

Hydroxyethylanilin Saccharate

Textile
De-electrifying agent (Brit. 430221) for—
Yarns, films, fabrics, and the like, subject to charging by static electricity (applied in admixture with all usual lubricating agents as vehicles).

Hydroxyethylanilin Salicylate

Textile
De-electrifying agent (Brit. 430221) for—
Yarns, films, fabrics, and the like, subject to charging by static electricity (applied in admixture with all usual lubricating agents as vehicles).

Hydroxyethylanilin Tannate

Textile
De-electrifying agent (Brit. 430221) for—
Yarns, films, fabrics, and the like, subject to charging by static electricity (applied in admixture with all usual lubricating agents as vehicles).

Hydroxyethylanilin Tartrate

Textile
De-electrifying agent (Brit. 430221) for—
Yarns, films, fabrics, and the like, subject to charging by static electricity (applied in admixture with all usual lubricating agents as vehicles).

Hydroxyethyldihydrocuprein

Disinfectant
Claimed (U. S. 1997719) to have—
Bactericidal value, pneumococcocidal value.

Hydroxyethyltriphenylphosphonium Chloride

French: Chlorure d'hydroxyéthyletriphénylephosphonium.
German: Chlorhydroxyaethyltriphenylphosphonium, Hydroxyaethyltriphenylphosphoniumchlorid.

Miscellaneous
Mothproofing and moldproofing agent (Brit. 312163) in treating—
Hair, fur, feathers, felt, and the like.

Textile
Mothproofing and moldproofing agent (Brit. 312163) in treating—
Wool and other products.

2-Hydroxy-4′-hydroxydiphenylamine

Chemical
Starting point in making—
Intermediates and other derivatives.

Disinfectant
As a germicidal agent.
Ingredient (Brit. 358508) of—
Germicidal compositions containing petrolatum, soap, glycerin, talc, wool-fat, paraffin or other waxes, and other components.

Insecticide
As an insecticide and fungicide.
Ingredient (Brit. 358508) of—
Compositions, containing talc, soap, glycerin, wool-fat, petrolatum, paraffin or other waxes, and other components, used for treating domestic animals to remove pests and for general insecticidal and fungicidal purposes.
Compositions for treating plants and seeds to disinfect them.

Miscellaneous
Ingredient (Brit. 358508) of—
Preparations, containing talc, soap, glycerin, wool-fat, petrolatum, paraffin or other waxes, and other components, used for treating catgut and other articles to preserve them.
Preservative preparations.
Preservative preparations for treating furs and skins.
Special polishing preparations.

Perfume
Ingredient (Brit. 358508) of—
Cosmetic ointments, dentifrices.

Resins and Waxes
Ingredient (Brit. 358508) of—
Antiseptic wax preparations.

Textile
Ingredient (Brit. 358508) of—
Compositions, containing petrolatum, talc, soap, wool-fat, glycerin, paraffin or other waxes, and other components, used for the treatment of fabrics to preserve them.

Woodworking
Ingredient (Brit. 358508) of—
Compositions, containing wool-fat, talc, glycerin, soap, petrolatum, paraffin or other waxes, and other components, used for preserving wood.

2-Hydroxy-2-isobutoxydibenzanthrone

Dye
Starting point (Brit. 434132) in making—
Green vat dyes which are probably cyclic ethers, and may be sulphonated, nitrated, nitrated and reduced, halogenated, hydroxylated, oxidized, or condensed with acid halides in presence of aluminum chloride.

2-Hydroxy-2-isopropoxydibenzanthrone

Dye
Starting point (Brit. 434132) in making—
Green vat dyes which are probably cyclic ether, and may be sulphonated, nitrated, nitrated and reduced, halogenated, hydroxylated, oxidized, or condensed with acid halides in presence of aluminum chloride.

Hydroxylamine Benzoate

Fats and Oils
Antiseptic for—
Fatty acids.
Oxidation inhibitor for—
Fatty acids.
Preservative for—
Fatty acids.

Soap
Antiseptic for—
Soap.
Oxidation inhibitor for—
Soap.
Preservative for—
Soap.

Hydroxylamine Hydrochloride

Chemical
Reagent in—
Organic syntheses.
Reducing agent in—
Organic syntheses.

Pharmaceutical
In compounding and dispensing practice.
Suggested for use in treating—
Skin diseases.

Photographic
In developing processes.

Hydroxylamine Sulphate

Chemical
Reagent in—
Organic syntheses.
Reducing agent in—
Organic syntheses.

Pharmaceutical
In compounding and dispensing practice.
Suggested for use in treating—
Skin diseases.

Photographic
In developing processes.

4-Hydroxy-3-methoxyphenyltrichloromethylcarbinol
Chemical
Starting point (Brit. 399723) in making—
Vanillin.

1-Hydroxy-6-methyl-4-chloro-2-benzoic Acid
Synonyms: Alphahydroxy-6-methyl-4-chloro-2-benzoic acid.
French: Acide d'alphahydroxye-6-méthyle-4-chloro-2-benzoique, Acide de 1-hydroxye-6-méthyle-4-chloro-2-benzoique.
German: Alphahydroxy-6-methyl-4-chlor-2-benzoesäure, 1-Hydroxy-6-methyl-4-chlor-2-benzoesäure.
Chemical
Starting point in making—
Acids, esters and other derivatives.
Miscellaneous
Mothproofing agent (U. S. 1734682) in treating—
Feathers, furs, skins, hair.
Textile
Mothproofing agent (U. S. 1734682) in treating—
Felt and woolen materials.

Hydroxymethylphenol
Chemical
Starting point (Brit. 399723) in making—
Ortho and para hydroxybenzaldehyde.

2:3-Hydroxynaphthalenemetachlorotolylamide
German: 2:3-Hydroxynaphtalinmetachlorotolylamid.
Dye
Starting point (German 430579) in making azo dyestuffs with—
4-Amino-4′-hydroxydiphenyl, 4-amino-4′-hydroxyditolyl, 4-chloro-6-amino-3:3′-dimethyldiphenyl, 4-chloro-6-amino-3:3′-dimethylditolyl, 4-chloro-2-aminodiphenyl, 4-chloro-3-aminoditolyl, 4:4′-dichloro-6-amino-3:3′-dimethyldiphenyl, 4:4′-dichloro-3-aminodiphenyl, 4:4′-dichloro-6-amino-3:3′-dimethylditolyl, 4:4′-dichloro-3-aminoditolyl.

2:3-Hydroxynaphthalenemetachoroxylylamide
German: 2:3-Hydroxynaphtalinmetachloroxylylamid.
Dye
Starting point (Brit. 430579) in making azo dyestuffs with—
4-Amino-4′-hydroxydiphenyl, 4-amino-4′-hydroxyditolyl, 4-chloro-6-amino-3:3′-dimethyldiphenyl, 4-chloro-6-amino-3:3′-dimethylditolyl, 4-chloro-3-aminodiphenyl, 4-chloro-3-aminoditolyl, 4:4′-dichloro-6-amino-3:3′-dimethyldiphenyl, 4:4′-dichloro-6-amino-3:3′-dimethylditolyl, 4:4′-dichloro-3-aminodiphenyl, 4:4′-dichloro-3-aminoditolyl.

2-Hydroxy-1:4-naphthaquinone
German: 2-Hydroxy-1:4-naphtachinon.
Dye
Starting point (German 433192) in making vat dyestuffs with—
Alphabetanaphthaphenazin-4:5-sultam, 6-chloro-5-hydroxy-alphabetanaphthaphenazin, 5-hydroxyalphabetanaphthaenazin.

2:3-Hydroxynaphthoic Acid
Synonyms: 2:3-Oxynaphthoic acid.
French: Acide 2:3-hydroxynaphthoique.
German: Betaoxynaphtoesaeure.
Chemical
Starting point in making—
2:3-Aminonaphthoic acid (Brit. 250598), arylamide derivatives, betaoxynaphthoic acid anilide, betaoxynaphthoic acid anisidide, betaoxynaphthoic acid naphthalidide, betaoxynaphthoic acid toluidide, betaoxynaphthoic acid xylidide, naphthol-3-carboxylic-6:8-disulphonic acid.
Dye
Starting point in making—
Brilliant lake red paste R, brilliant scarlet red R paste, lake bordeaux B paste, lithol rubin B, pigment dyestuffs, nigrosin dyestuffs.
Textile
——, *Dyeing*
Assist in forming—
Ground color for dyeing with anisidin blue on various textile fibers and fabrics.
——, *Printing*
Assist in forming—
Ground color in printing various textile fabrics with anisidin blue.

2:3-Hydroxynaphthoic-3′-chlor-2′-ethoxyanilide
Textile
Starting point (Brit. 453953) in making—
Dyed or printed red colors of fine purity of shade and fastness to kier-boiling with—
2-Amino-4-acetyl-1:1′-diphenyl ether or its 4′-chloro derivative, 2-amino-4-acetyl-3′-methyl-1:1′-diphenyl ether or its 4′ chloro derivative, 2-amino-4-acetyl-4′-methyl-1:1′-diphenyl ether or its 4′-chloro derivative.

2:3-Hydroxynaphthoic-4′-chlor-2′-methoxyanilide
Textile
Starting point (Brit. 453953) in making—
Dyed or printed red colors of fine purity of shade and fastness to kier-boiling with—
2-Amino-4-acetyl-1:1′-diphenyl ether or its 4′-chloro derivative, 2-amino-4-acetyl-3′-methyl-1:1′-diphenyl ether or its 4′-chloro derivative, 2-amino-4-acetyl-4′-methyl-1:1′-diphenyl ether or its 4′-chloro derivative.

2:3-Hydroxynaphthoic-5′-chlor-2′-methylanilide
Textile
Starting point (Brit. 453953) in making—
Dyed or printed red colors of fine purity of shade and fastness to kier-boiling with—
2-Amino-4-acetyl-1:1′-diphenyl ether or its 4′-chloro derivative, 2-amino-4-acetyl-3′-methyl-1:1′-diphenyl ether or its 4′-chloro derivative, 2-amino-4-acetyl-4′-methyl-1:1′-diphenyl ether or its 4′-chloro derivative.

2-Hydroxy-3-naphthoic-5-chlor-3-methylthiolortho-toluidide
Dye
Intermediate (U. S. 2025116) in—
Dye manufacture.

2:3-Hydroxynaphthoic-3′:4′-dichloroanilide
Textile
Starting point (Brit. 453953) in making—
Dyed or printed red colors of fine purity of shade and fastness to kier-boiling with—
2-Amino-4-acetyl-1:1′-diphenyl ether or its 4′-chloro derivative, 2-amino-4-acetyl-3′-methyl-1:1′-diphenyl ether or its 4′-chloro derivative, 2-amino-4-acetyl-4′-methyl-1:1′-diphenyl ether or its 4′-chloro derivative.

2:3-Hydroxynaphthoic-2′:5′-dimethoxyanilide
Textile
Starting point (Brit. 453953) in making—
Dyed or printed red colors of fine purity of shade and fastness to kier-boiling with—
2-Amino-4-acetyl-1:1′-diphenyl ether or its 4′-chloro derivative, 2-amino-4-acetyl-3′-methyl-1:1′-diphenyl ether or its 4′-chloro derivative, 2-amino-4-acetyl-4′-methyl-1:1′-diphenyl ether or its 4′-chloro derivative.

2:3-Hydroxynaphthoicdodecylamide
Dye
Starting point (Brit. 434243) in making—
Orange-red dyestuffs for lacquers, waxes, and the like, by coupling with anilin.
Reddish-blue dyestuffs for lacquers, waxes, and the like, by coupling with dianisidin.
Blue dyestuffs for lacquers, waxes, and the like, by coupling with 4-benzamido-2:5-diethoxyanilin.

2:3-Hydroxynaphthoicmetachloroanilide
French: Chloroanilide de 2:3-hydroxynaphthoique.
Dye
Starting point (German 430579) in making azo dyestuffs with—
4-Amino-4′-hydroxydiphenyl, 4-amino-4′-hydroxyditolyl, 4-chloro-3-aminodiphenyl, 4-chloro-3-aminoditolyl, 4-chloro-6-amino-3:3′-dimethyldiphenyl, 4-chloro-6-amino-3:3′-dimethylditolyl, 4:4′-dichloro-2-aminodiphenyl, 4:4′-dichloro-2-aminoditolyl, 4:4-dichloro-6-amino-3:3′-dimethyldiphenyl, 4:4′-dichloro-6-amino-3:3′-dimethylditolyl.

2:3-Hydroxynaphthoicmetachlorobenzamide
German: 2:3-Hydroxynaphtoemetachlorbenzamid.
Dye
Starting point (German 430579) in making azo dyestuffs with—
4-Amino-4′-hydroxydiphenyl, 4-amino-4′-hydroxyditolyl, 4-chloro-3-aminoditolyl, 4-chloro-6-amino-3:3′-dimethyldiphenyl, 4-chloro-6-amino-3:3′-dimethylditolyl, 4-chloro-3-aminodiphenyl, 4:4′-dichloro-3-aminodiphenyl, 4:4′-dichloro-6-amino-3:3′-dimethyldiphenyl, 4:4′-dichloro-6-amino-3:3′-dimethylditolyl, 4:4′-dichloro-3-aminoditolyl.

2:3-Hydroxynaphthoic-2'-methoxy-5'-methylanilide
Textile
Starting point (Brit. 453953) in making—
Dyed or printed red colors of fine purity of shade and fastness to kier-boiling with—
2-Amino-4-acetyl-1:1'-diphenyl ether or its 4'-chloro derivative, 2-amino-4-acetyl-3'-methyl-1:1'-diphenyl ether or its 4'-chloro derivative, 2-amino-4-acetyl-4'-methyl-1:1'-diphenyl ether or its 4'-chloro derivative.

2:3-Hydroxynaphthoic-4'-methoxy-2'-methylanilide
Textile
Starting point (Brit. 453953) in making—
Dyed or printed red colors of fine purity of shade and fastness to kier-boiling with—
2-Amino-4-acetyl-1:1'-diphenyl ether or its 4'-chloro derivative, 2-amino-4-acetyl-3'-methyl-1:1'-diphenyl ether or its 4'-chloro derivative, 2-amino-4-acetyl-4'-methyl-1:1'-diphenyl ether or its 4'-chloro derivative.

2:3-Hydroxynaphthoic-2'-methylanilide
Textile
Starting point (Brit. 453953) in making—
Dyed or printed red colors of fine purity of shade and fastness to kier-boiling with—
2-Amino-4-acetyl-1:1'-diphenyl ether or its 4'-chloro derivative, 2-amino-4-acetyl-3'-methyl-1:1'-diphenyl ether or its 4'-chloro derivative, 2-amino-4-acetyl-4'-methyl-1:1'-diphenyl ether or its 4'-chloro derivative.

2:3-Hydroxynaphthoic-3'-methylanilide
Textile
Starting point (Brit. 453953) in making—
Dyed or printed red colors of fine purity of shade and fastness to kier-boiling with—
2-Amino-4-acetyl-1:1'-diphenyl ether or its 4'-chloro derivative, 2-amino-4-acetyl-3'-methyl-1:1'-diphenyl ether or its 4'-chloro derivative, 2-amino-4-acetyl-4'-methyl-1:1'-diphenyl ether or its 4'-chloro derivative.

2:3-Hydroxynaphthoic-4'-methylanilide
Textile
Starting point (Brit. 453953) in making—
Dyed or printed red colors of fine purity of shade and fastness to kier-boiling with—
2-Amino-4-acetyl-1:1'-diphenyl ether or its 4'-chloro derivative, 2-amino-4-acetyl-3'-methyl-1:1'-diphenyl ether or its 4'-chloro derivative, 2-amino-4-acetyl-4'-methyl-1:1'-diphenyl ether or its 4'-chloro derivative.

2-Hydroxy-3-naphthoicparamethylthiolanilide
Dye
Intermediate (U. S. 2025116) in—
Dye manufacture.

2(2'-Hydroxy-3'-naphthoyl)-amino-3-naphthol Methylether
French: Éther méthylique de 2(2'-hydroxye-3'-naphthoyle)-amino-3-naphthole.
German: 2(2'-Hydroxy-3'-naphtoyl)-amino-3-naphtolmethylaether.
Textile
Reagent (Brit. 309879) in dyeing textiles with the aid of—
4-Amino-3-chlorobenzanilin, 2-amino-5-chlorophenylbetanaphthyl ether, 4-amino-1:3-dimethylbenzene, 4-amino-2:5'-dichloro-5:2'-dimethoxyazobenzene, 3-amino-4-nitro-6-methoxy-1-methylbenzene, 2-amino-5:2':5'-trichlorodiphenyl ether, 5-bromo-2-toluidin, 4-chloroanilin, 5-chloro-2-nitranilin, 2-chloro-5-nitranilin, 1:5-diaminoanthraquinone, 2:5-dibromoanilin, 2:5-dichloroanilin, 2:3-dichloroanilin, 4'-nitro-4-aminoazobenzene, 3-nitro-4-toluidin, orthoaminoazotoluene, 2:3:4-trichloroanilin.

2:3-Hydroxynaphthoylaminosulpho Alphanaphthalide
Paper
Impregnating agent and absorbent for ultraviolet light (Brit. 436891) in—
Treating paper and like products to be used as food containers.

2:3-Hydroxynaphthoylmetanitranilin
Textile
——, *Printing*
Ingredient (German 433276) of—
Mixtures used in printing fast shades on cottons.

2-Hydroxy-3-naphthyl-4-ethoxy-2-methylthiolanilide
Dye
Intermediate (U. S. 2025116) in—
Dye manufacture.

2-Hydroxy-3-naphthylorthomethylthiolanilide
Dye
Intermediate (U. S. 2025116) in—
Dye manufacture.

2-Hydroxy-5-normal-butyldiphenyl
Disinfectant
Intermediate (U. S. 2073683) in making—
Bactericides.

2-Hydroxy-5-N-Valeryldiphenyl
Disinfectant
Intermediate (U. S. 2073683) in making—
Bactericides.

15-Hydroxypentadecane-1-carboxylic Acid Lactone
Perfume
Claimed (Brit. 440416) to be—
Useful new perfumery product.

1-Hydroxyphenyl-4-hydrazinsulphonic Acid
Insecticide and Fungicide
Seed grain disinfectant (U. S. 2054062).
Starting point (U. S. 2054062) in making—
Seed grain disinfectants.

4-Hydroxyphenyl-3-Hydroxyl-1-naphthyl Sulphide
Disinfectant
As a bactericide (Brit. 349004).
Insecticide and Fungicide
As a fungicide (Brit. 349004).
As an insecticide (Brit. 349004).

2-Hydroxy-5-secondaryamylbenzoic Acid
Disinfectant
Claimed to be a very powerful disinfectant (U. S. 1998750).
Insecticide and Fungicide
Claimed to be a very powerful fungicide (U. S. 1998750).
Pharmaceutical
Claimed to be a very powerful therapeutic antiseptic (U. S. 1998750).

2-Hydroxy-5-secondaryhexylbenzoic Acid
Disinfectant
Claimed to be a very powerful disinfectant (U. S. 1998750).
Insecticide and Fungicide
Claimed to be a very powerful fungicide (U. S. 1998750).
Pharmaceutical
Claimed to be a very powerful therapeutic antiseptic (U. S. 1998750).

Hydroxystearic Diglyceride
French: Diglycéride hydroxystéarique.
German: Hydroxystearodiglycerid.
Miscellaneous
As a dispersing agent (Brit. 329266).
For uses, see under general heading: "Emulsifying agents."

14-Hydroxytetradecane-1-carboxylic Acid Lactone
Perfume
Claimed (Brit. 440416) to be—
Useful new perfumery product.

Hydroxythionaphthene-6-carboxylic Acid
Chemical
Starting point in making—
Intermediates and other derivatives.
Dye
Starting point (Brit. 354716) in making dyestuffs with the aid of—
Acenaphthenequinone.
Benzyl-4-chloro-6:7-benzohydroxythionaphthene.
1-Chloro-2:3-naphthisatin.
5:7-Dichloroisatin.
Isatin-7-carboxylic acid.
Monobromo-2:1-naphthisatin chloride.

Hyoscine Hydrobromide
Synonyms: Hydrobromate of hyoscine.
Latin: Hyoscinae hydrobromidum, Hyoscinum hydrobromicum, Scopolaminum hydrobromicum.
French: Bromhydrate de scopolamine, Bromhydrate d'hyoscine.
German: Bromwasserstoffsäureshyoscin, Hyoscinhydrobromid, Skopolaminhydrobromid.
Spanish: Bromhidrato de hioscina.
Italian. Bromidrato d'ioscina.

Pharmaceutical
Suggested for use as—
Cerebral sedative.

Hyoscyamine

Chemical
Starting point in making—
Hyoscyamine salts with acids and halogens.
Starting point (Brit. 273279) in making therapeutic compounds with—
Camphorates, malonates, meconates, phthalates, phosphates, saccharates, sulphates, sulphites, tartrates, terephthalates.

Pharmaceutical
In compounding and dispensing practice.

Hyoscyamine Hydrobromide
Synonyms: Hyoscyamine bromide, Hyoscyamine hydrobromate.
Latin: Hyoscyaminae hydrobromas, Hyoscyaminae hydrobromidum, Hyoscyaminum hydrobromicum.
French: Bromhydrate d'hyoscyamine.
German: Bromwasserstoffsäureshyoscyamin, Hyoscyaminhydrobromid.

Pharmaceutical
Suggested for use as—
Substitute for atropine wherever the action on peripheral nerves is desirable; for instance, to check excessive secretion, to dilate the pupil, or to allay intestinal spasm.

Hyoscyamine Sulphate
Latin: Hyoscyaminae sulphas, Hyoscyaminum sulfuricum.
French: Sulfate d'hyoscyamine.
German: Hyoscyaminsulfat, Schwefelsäureshyoscyamin.

Pharmaceutical
Suggested for use as—
Substitute for atropine wherever the action on peripheral nerves is desirable; for instance, to check excessive secretion, to dilate the pupil, or to allay intestinal spasm.

Iceland Moss
Synonyms: Cetraria.
German: Iceland lichen.

Food
Foodstuff for Lapps and Icelanders.

Miscellaneous
Emulsifying agent.

Paint and Varnish
Underlying medium (Brit. 406048) in—
Coating articles by dipping them in a bath containing a film of coating liquid floated upon an underlying medium, suitable coating liquids being cellulose ester lacquers, oil varnishes, and synthetic resin lacquers.

Pharmaceutical
Suggested for use in treating—
Chronic catarrhs (especially of the pulmonary mucous membrane).
Phthisis.

Ichthyol
Synonyms: Ammonium ichthyolsulphonate, Ammonium sulphoichthyolate.

Chemical
Starting point in making—
Ichthyol albuminate, ichthyol-formaldehyde derivative (ichthoform).

Pharmaceutical
In compounding and dispensing practice.

Indanthrene
Synonyms: Indanthren, Indanthrone.

Dye
Starting point (Brit. 271533) in making soluble vat dyestuffs with—
Butylsulphonic acid chloride, chlorosulphonic acid, methylsulphuric acid chloride, sulphur trioxide.
Starting point (Brit. 271537) in making—
Leuco flavanthrones.
Starting point in making—
Indanthrene blue R, indanthrene blue RS, indanthrene blue 2GS, indanthrene blue GCD, indanthrene blue GC, indanthrene blue WB.

Indanthrenesulphonic Acid
SSynonyms: Indanthronesulphonic acid.
French: Acide d'indanthrènesulfonique, Acide d'indanthronesulfonique.
German: Indanthrensulfonsaeure, Indanthronsulfonsaeure.

Dye
Starting point in making—
Indanthrene blue.

Indigo Disulphonate

Analysis
Indicator in—
Oxidation-reduction potential determinations (of particular interest to biologists and physiologists and in investigations of various materials, such as soils, wines, cheese, gasoline antiknock compounds).

Indigo Tetrasulphonate

Analysis
Indicator in—
Oxidation-reduction potential determinations (of particular interest to biologists and physiologists and in investigations of various materials, such as soils, wines, cheese, gasoline antiknock compounds).

Indigo Trisulphonate

Analysis
Indicator in—
Oxidation-reduction potential determinations (of particular interest to biologists and physiologists and in investigations of various materials, such as soils, wines, cheese, gasoline antiknock compounds).

Indirubin

Dye
Ingredient (Brit. 250251) in making dye mixtures with—
Ammonium borate, carbonate, and phosphate.
Potassium carbonate, borate, and phosphate.
Sodium borate, carbonate, and phosphate.

Textile
——, *Dyeing and Printing*
Dyestuff for yarns and fabrics.

Inulin
Synonyms: Alant starch, Alantin, Helenin.
German: Alantstaerke.

Chemical
Starting point in making—
Levulose.

Food
Ingredient of—
Diabetic bread and food preparations.

Pharmaceutical
In compounding and dispensing practice.

Inulin Nitrate

Explosives and Matches
Ingredient (U. S. 1922123) of—
Detonator compositions.
Sensitizer (U. S. 1922123) in—
Dynamites.

Paints and Varnish
Ingredient (U. S. 1922123) of—
Lacquers.

Invert Sugar
French: Sucre interverti, Sucre inverti.
German: Invertzucker.

Chemical
Softening agent in making various products and compositions.

Invert Sugar (Continued)
Food
Ingredient of—
Beverages, candies, honey substitutes, infant foods, invalid foods.
Softening agent in making—
Baked products in order to prevent thickening.
Miscellaneous
Softening agent in making various products.
Substitute for—
Glycerin for the purpose of holding moisture.
Tobacco
Ingredient of—
"Casing" for cigaret tobaccos to hold short tobacco with long tobacco and to retain moisture.
Heavy dips for making highly sweetened and flavored cigaret tobaccos.
Reagent in treating—
Cigaret tobacco, plug tobacco, smoking tobacco.
Wine
Added to improve the flavor of wines.

3-Iod-10-deltadiethylaminoalphamethylbutylaminoacridin Dihydrochloride
Pharmaceutical
Claimed (Brit. 441007, 441132, and addition to 363392) as—
New pharmaceutical.

3-Iod-10-deltadimethylaminobutylaminoacridin Dihydrochloride
Pharmaceutical
Claimed (Brit. 441007, 441132, and addition to 363392) as—
New pharmaceutical.

Iodic Acid
Latin: Acidum iodicum.
French: Acide iodique.
German: Iodsäure, Jodsäure.
Spanish: Acido yodico.
Italian: Acido iodico.
Analysis
As a general oxidizing agent in analytical work.
Reagent in organic analysis.
Reagent in the volumetric determination of—
Acetoacetic acid, adrenalin, bile pigments, emetine, guaiacol, mercury, morphine, naphthols, strychnine, sulphocyanic acid.
Chemical
Oxidizing agent in making—
Aromatics, intermediates, pharmaceuticals, salts and other compounds.
Dye
Oxidizing agent in making—
Synthetic dyestuffs.
Pharmaceutical
In compounding and dispensing practice.

Iodine
Latin: Iodinium, Iodum, Jodum.
French: Iode, Iode sublimé.
German: Jod.
Spanish: Yodo.
Italian: Jodo.
Analysis
As a reagent.
Chemical
Catalyst in—
Alkylation of primary aromatic amines, especially anilin and alphanaphthylamine, by the direct action of alcohols.
Bromination of benzene.
Chlorination of acetic acid.
Condensation of aromatic amines with naphthols or naphthylamines.
Condensation of aromatic alcohols with ketones.
Condensation of glycols to polyglycols.
Reactions involving elimination of hydrogen chloride.
Catalyst in making—
Chlorinated derivatives of benzene (Brit. 388818).
Synthetic organic chemicals.
Thiodiarylamines from sulphur and diarylamine.
Unsaturated compounds by heating hydroxy compounds —unsaturated hydrocarbons from alcohols, unsaturated ketones from ketonic alcohols, and unsaturated aldehydes from aldols.
Starting point in making—
Hydriodic acid, iodates, iodic acid, iodides, iodine cyanide, iodine monobromide, iodine monochloride.
Medicinal chemicals used as (1) antiseptics and dressings, (2) internal remedies, (3) antisyphilitics.
Periodic acid.
Disinfectant
Ingredient of—
Disinfecting candle compositions (Brit. 397238).
Disinfecting compositions (U. S. 1925135).
Disinfecting solution containing also iodides of sodium, potassium, and calcium (U. S. 1903614).
Dye
Reagent in making—
Synthetic dyestuffs.
Dry Cleaning
Spotting agent for—
Lead compounds (stain with tincture of iodine; let dry; and dissolve with concentrated potassium iodide solution).
Electrical
Ingredient of—
Fluorescent screens containing also calcium tungstate and an inorganic binder (U. S. 1909365).
Selenium photoelectric cell (U. S. 1730505).
Insecticide
Suggested for use as—
Ingredient of fungicides.
Leather
In leather manufacture.
Miscellaneous
As a germicide.
Paint and Varnish
Tinting agent for—
Lacquers and shellac (used to produce fast shades varying from light-yellow to a ruby-red, according to concentration).
Paper
Ingredient of—
Impregnating agent for safety paper, containing also alcohol, water, cobalt nitrate, and sodium thiosulphate.
In paper testing.
Pharmaceutical
In compounding and dispensing practice.
Ingredient of—
Ointments, salves, tinctures.
Styptic preparation containing also alcoholic solution of ferrous iodide and a powdered material having a pectin content of at least 10 percent.
Photographic
Starting point in making—
Iodides in sensitive coatings.
Printing
In process engraving and lithography.
Sanitation
Testing agent for—
Sulphur dioxide content of air in and around industrialized centers of population (used in combination with starch).
Soap
Ingredient of—
Special soaps.

Iodine Monochloride
French: Monochlorure d'iode.
German: Jodmonochlorid.
Analysis
Reagent for the determination of iodine number in fats and oils.
Chemical
Reagent (Brit. 244443) in making—
5:7-Di-iodoisatin, iodolecithin, 4:5:6:7-tetraiodoisatin.
Pharmaceutical
In compounding and dispensing practice.

Iodine Trichloride
French: Trichlorure d'iode.
German: Jodtrichlorid.
Chemical
Catalyst in making—
Alphachloronaphthalene from naphthalene by action of chlorine.
Reagent in making—
Polyiodinated isatin (U. S. 1592386).
Dye
Reagent in making—
Acridin yellow, halogenated derivatives of anthracene (Brit. 260998).

Iodine Trichloride (Continued)

Pharmaceutical
In compounding and dispensing practice.

Sanitation
As an antiseptic and disinfectant.
Ingredient of—
Antiseptic compositions, disinfecting compositions.

Iodoform
Synonyms: Methane tri-iodide, Tri-iodomethane.
French: Iodoforme.
German: Jodoform, Tri-jodmethan.

Chemical
Starting point in making various pharmaceuticals.

Pharmaceutical
In compounding and dispensing practice.

Printing
Sensitizing agent (Brit. 270386) in preparing compositions for—
Color record intaglio.
Halftone printing plates.
Line engraving on copper, zinc, and other materials.
Monochrome intaglio.
Relief printing plates.
Screenless grained litho plates.

8-Iodo-1:2-naphthisatin
German: 8-Jod-1:2-naphtisatin.

Dye
Starting point in making indigoid dyestuffs with—
Alpha-anthrol, alphanaphthol, acenaphthene, alphaoxyanthranol, carbazole, indoxyl, oxindole, oxythionaphthene.

4-Iodostyryl-2-quinolin
French: 4-Iodo-2-styrylequinoléine.
German: 4-Jod-2-styryl-2-chinolin.

Chemical
Starting point (Brit. 282143) in making pharmaceuticals with—
Allylamine, allylenediamine, alphanaphthylamine, ammonia, amylamine, amylenediamine, benzylamine, benzylenediamine, betanaphthylamine, butylamine, butylenediamine, cumylamine, cumylenediamine, ethylamine, ethylenediamine, heptylamine, heptylenediamine, hexylamine, hexylenediamine, metaphenylenediamine, metatoluylenediamine, methylamine, mthylenediamine, orthophenylenediamine, orthotoluylenediamine, paraphenylenediamine, paratoluylenediamine, propylamine, propylenediamine, toluylamine.

Iridium

Metallurgical
Ingredient of—
Alloys with precious and common metals.

Miscellaneous
Metal for making scientific instruments, thermocouples, fountain pen points, surgical instruments.

Irish Moss
Synonyms: Alga perlada, Carageen, Caragahen, Carragheen, Gigartina mamillosa, Hen's dulse, Killeen, Pearl moss, Pigwrack, Rocksalt moss.
Latin: Chondrus, Chondrus crispus, Fucus crispus, Lichen Hibernicus.
French: Mousse marine perlée, Mousse perlée.
German: Irlandische moos, Knorpeltang, Perlmoos.
Spanish: Caragaen, Musgo branco, Musgo d'irlande, Musgo marino perlado.
Italian: Fuco carageo, Fuco carrageo, Fuco crispo, Fuco crispa, Musco d'irlanda, Musco marien perlado.

Chemical
Clarifying agent for precipitating—
Finely suspended matter in liquids.

Food
As a nutrient in place of barley, sago, tapioca, and the like.

Glue and Adhesives
Substitute for—
Acacia (under name of imitation gum arabic).

Leather
As a dressing.

Miscellaneous
As an emulsifier, as a demulcent, as a size.

Pharmaceutical
In compounding and dispensing practice.
Suggested for use in treating—
Chronic pectoral affections, diarrhea, disorders of kidneys and the bladder, dysentery, scrofulous complaints.

Printing
Mottling agent for—
Book papers.

Soap
Ingredient in making—
Special soaps.

Textile
Size.
Thickening agent for—
Dyestuffs used in printing processes.

Iron Acetate Liquor
Synonyms: Iron liquor, Iron pyrolignite.
French: Bouillon noir, Pyrate de fer, Pyrolignite de fer.
German: Eisenpyrat, Eisenpyrolignit.

Leather
Mordant in dyeing—
Black and other dark shades.

Miscellaneous
Mordant in dyeing—
Black shades on furs, hats, and other articles.

Textile
Mordant in dyeing—
Awnings, black and other dark colors, khaki colors on various fabrics.
Mordant in printing—
Calicoes, violet shades with alizarin.

Woodworking
As a preservative.
Mordant in dyeing—
Black shades on wood.

Iron Acetylacetonate
French: Acétyleacétonate de fer, Acétyleacétone ferrique.
German: Acetylacetonsäureseisen, Acetylacetonsäureseisenoxyd, Eisenacetylacetonat.

Automotive
Ingredient (U. S. 1765692) of—
Motor fuel compositions.

Iron Albuminate
French: Albuminate de fer, Albuminate ferrique.
German: Albuminsaeureseisen, Eisenalbuminat.

Pharmaceutical
In compounding and dispensing practice.

Rubber
Reagent in—
Reclaiming rubber (U. S. 1640817).

Iron Benzoate
Synonyms: Ferric benzoate.
Latin: Ferri benzoas, Ferrum benzoicum.
French: Benzoate de fer, Benzoate ferré, Benzoate ferrique.
German: Benzoesäureseisen, Eisenbenzoat, Ferribenzoat.
Spanish: Benzoato de hierrico.
Italian: Benzoato di ferrico.

Pharmaceutical
In compounding and dispensing practice.

Rubber
Retarding agent (U. S. 1929561) in—
Vulcanizing processes employing an ultra-accelerator.

Iron Betabenzoylpropionate

Plastics
Starting point (U. S. 2001380) in making—
Films.

Iron Borotungstate
French: Borotungstate de fer.
German: Borwolframsaeureseisen, Eisenborwolframat.

Metallurgical
Ingredient of—
Electrically deposited insulating coatings on steel and iron and other metals, affording protection against oxidation (French 600774).

Iron Carbide
Synonyms: Carbide of iron, Ferric carbide.
French: Carbure de fer.
German: Eisencarbid.

Chemical
Starting point in making—
Potassium cyanide, sodium cyanide.

Iron Chlorosulphate
Synonyms: Ferric chlorosulphate.
French: Chlorosulfate de fer, Chlorosulfate ferrique, Chlorure et sulfate de fer, Chlorure et sulfate ferrique, Sulfate et chlorure de fer, Sulfate et chlorure ferrique.
German: Chlorschwefelsäureseisen.
Spanish: Chlorosulfato de hierro.
Italian: Chlorosolfato di ferro.

Leather
Tanning agent (French 521850) for—
Skins and pelts.

Iron-Copper Sulphate
French: Sulfate de fer et de cuivre, Sulfate ferrique et cuivrique, Vitriol d'almonde, Vitriol de Salzburg.
German: Adlervitriol, Admontervitriol, Doppelvitriol, Kupfereisensulfat, Kupfereisenvitriol, Salzbuergervitriol.

Paint and Varnish
Starting point in making—
Mineral pigments.

Textile
——, Dyeing and Printing
Mordant in—
Dyeing and printing yarns and fabrics.

Iron Pentacarbonyl
French: Fer pentacarbonyle, Pentacarbonyle de fer.
German: Eisenpentacarbonyl.

Explosives and Matches
As a military explosive having both incendiary and highly toxic properties (acts on blood hemoglobin and nerve centers).

Iron Resinate
Synonyms: Ferric resinate, Iron soap, Resinate of iron.
French: Résinate de fer.
German: Eisenresinat, Harzsaeureseisen.

Ceramics
Pigment in producing red shades in—
Chinaware, porcelains, potteries.
Pigment in admixture with bismuth resinate in producing dull tones in ceramic ware.

Paint and Varnish
Drier in making—
Enamels, lacquers, paints, varnishes.

Iron Stearotoluenesulphonate
Synonyms: Ferric stearotoluenesulphonate.
French: Stéarotoluènesulphonate de fer, Stéarotoluènesulphonate ferrique.
German: Eisenstearotoluolsulfonat, Ferristearotoluolsulfonat, Stearotoluolsulfonsäureseisen.

Chemical
Starting point in making various derivatives.

Leather
Ingredient (Brit. 269917) of—
Printing pastes and dye liquors (used to obtain better saturation of the leather with the color and more evenness of the dyed or printed shade).

Miscellaneous
Ingredient (Brit. 269917) of—
Dye liquors, used in the dyeing of furs, feathers, and the like (added for the purpose of obtaining better penetration of the color into the product and more level shades).

Paper
Ingredient (Brit. 269917) of—
Dye liquors (used for the purpose of obtaining better penetration of the color into the product and more level shades).

Textile
Ingredient (Brit. 269917) of—
Dye liquors and printing pastes (added to enhance the saturation of the textile with the color and to obtain equalization of the printed color).

Iron Tungstomolybdate
French: Tungstomolybdate de fer.
German: Eisen wolframmolybdat, Wolfram molybdaensaeureisen.

Metallurgical
Ingredient of—
Electrically deposited insulating coatings on steel and iron and other metals, affording protection against oxidation (French 600774).

Isatinalpha-anil

Chemical
Starting point in making various derivatives.

Dye
Starting point in making—
Alizarin indigo 3R, alizarin indigo B, helindon blue 3GN, 2-thionaphthene-2-indolindigo.
Starting point (Brit. 291825) in making indigoid dyestuffs with—
4-Allymercaptoalphanaphthol.
4-Amylmercaptoalphanaphthol.
4-Benzylmercaptoalphanaphthol.
4-Butylmercaptoalphanaphthol.
4-Ethylmercaptoalphanaphthol.
4-Heptylmercaptoalphanaphthol.
4-Hexylmercaptoalphanaphthol.
4-Isoallylmercaptoalphanaphthol.
4-Isoamylmercaptoalphanaphthol.
4-Isobutylmercaptoalphanaphthol.
4-Isopropylmercaptoalphanaphthol.
4-Methylmercaptoalphanaphthol.
4-Naphthylmercaptoalphanaphthol.
4-Paratolylmercaptoalphanaphthol.
4-Pentylmercaptoalphanaphthol.
4-Propylmercaptoalphanaphthol.
4-Xylylmercaptoalphanaphthol.

Isatinbenzylcarboxylic Acid
French: Acide d'isatinbenzylcarbonique, Acide d'isatinbenzylcarboxilique.
German: Isatinbenzylcarbonsäure.
Spanish: Acido de isatinbenzilcarbonico.
Italian: Acido di isatinbenzilcarbonico.

Chemical
Starting point in making—
Esters, salts, and other derivatives.

Dye
Starting point (Brit. 354716) in making dyestuffs with the aid of—
Alpha-aminoanthraquinone.
5:6-Benzo-7-chlorohydroxythionaphthene.
4:5-Benzohydroxythionaphthene.
Benzyl-4-chloro-6:7-benzohydroxythionaphthene.
5-chlorohydroxythionaphthene.
4:6-Dimethylhydroxythionaphthene.
4:7-Dimethyl-5-chlorohydroxythionaphthene.
6-Ethoxyhydroxythionaphthene.
4-Methyl-6-bromohydroxythionaphthene.
4-Methyl-6-chlorohydroxythionaphthene.
6-Methoxyhydroxythionaphthene.
5:6:7-Trichlorohydroxythionaphthene.

Isatinbeta-anil

Chemical
Starting point in making various derivatives.

Dye
Starting point in making—
Vat dyestuffs.
Starting point (Brit. 291825) in making indigoid dyestuffs with—
Allylmercaptoalphanaphthol.
Amylmercaptoalphanaphthol.
Benzylmercaptoalphanaphthol.
Butylmercaptoalphanaphthol.
Ethylmercaptoalphanaphthol.
Heptylmercaptoalphanaphthol.
Hexylmercaptoalphanaphthol.
Isoallylmercaptoalphanaphthol.
Isoamylmercaptoalphanaphthol.
Isobutylmercaptoalphanaphthol.
Isopropylmercaptoalphanaphthol.
4-Methylmercaptoalphanaphthol.
4-Naphthylmercaptoalphanaphthol.
4-Paratolylmercaptoalphanaphthol.
4-Pentylmercaptoalphanaphthol.
4-Phenylmercaptoalphanaphthol.
4-Propylmercaptoalphanaphthol.
4-Xylylmercaptoalphanaphthol.

Isatin Bromide

Chemical
Starting point in making—
Intermediates, pharmaceuticals.

Dye
Starting point (Brit. 291825) in making indigoid dyestuffs with—
Allylmercaptoalphanaphthol.
Amylmercaptoalphanaphthol.

Isatin Bromide (Continued)
Benzylmercaptoalphanaphthol.
Butylmercaptoalphanaphthol.
Ethylmercaptoalphanaphthol.
Heptylmercaptoalphanaphthol.
Hexylmercaptoalphanaphthol.
Isoallylmercaptoalphanaphthol.
Isoamylmercaptoalphanaphthol.
4-Methylmercaptoalphanaphthol.
4-Naphthylmercaptoalphanaphthol.
4-Paratolylmercaptoalphanaphthol.
4-Pentylmercaptoalphanaphthol.
4-Phenylmercaptoalphanaphthol.
4-Propylmercaptoalphanaphthol.
4-Xylylmercaptoalphanaphthol.

Isatin-7-carboxylic Acid
Chemical
Starting point in making—
Esters, salts, and other derivatives.
Dye
Starting point (Brit. 354716) in making dyestuffs with the aid of—
Alpha-aminoanthraquinone.
5:6-Benzo-7-chlorohydroxythionaphthene.
4:5-Benzohydroxythionaphthene.
Benzyl-4-chloro-6:7-benzohydroxythionaphthene.
5-chlorohydroxythionaphthene.
4:7-Dimethyl-5-chlorohydroxythionaphthene.
4:6-Dimethylhydroxythionaphthene.
6-Ethoxyhydroxythionaphthene.
4-Methyl-6-bromohydroxythionaphthene.
4-Methyl-6-chlorohydroxythionaphthene.
6-Methoxyhydroxythionaphthene.
5:6:7-Trichlorohydroxythionaphthene.

Isatoic Anhydride
French: Anhydride d'isatoique.
German: Isatoinanhydrid.
Textile
——, *Dyeing*
Reagent for treating cotton or cellulose derivatives to color them (German 433147).

Isinglass
Synonyms: Fish glue.
Latin: Ichthyocolla, Colla piscium.
French: Colle de poisson.
German: Fischleim, Hausenblase, Mundleim.
Food
Clarifying agent in making—
Cereal beverages, ciders, malt vinegar, wines.
Ingredient of—
Confectionery, pastries.
Thickener in—
Jellies, milk preparations, soups.
Glue and Gelatin
Ingredient of adhesive compositions for—
China and pottery, glass, leather.
Ingredient of hide glue mixtures.
Ink—
Ingredient of—
India ink.
Miscellaneous
Clarifying agent for various purposes.
Ingredient of court plaster.
Protective colloid for various purposes.
Paint and Varnish
Ingredient of—
Artists' colors.
Printing
Reagent in photo-mechanical processes.
Textile
——, *Dyeing*
Ingredient of dye liquors for various yarns and fabrics.
——, *Finishing*
Ingredient of—
Gum compositions used to impart luster and stiffness to linens and silks.
Pyroxylin-acetic acid waterproofing compositions.
——, *Manufacturing*
Reagent in making—
English taffetas.
——, *Printing*
Ingredient of paste for—
Calicoes.

Isoallyl Alphacrotonate
Cellulose Products
Plasticizer (Brit. 321518) for—
Cellulose acetate, cellulose esters or ethers, cellulose nitrate (nitrocellulose).
For uses, see under general heading: "Plasticizers."

Isoallyl Carbamide
French: Carbamide d'isoallyle, Carbamide isoallylique.
German: Isoallylcarbamid.
Chemical
Starting point in making—
Intermediates, pharmaceuticals.
Dye
Reagent in making various synthetic dyestuffs.
Resins and Waxes
Starting point (Brit. 292912) in making synthetic resins with—
Acetylsalicylic acid, aliphatic dibasic acids, ammonium salicylate, anthranilic acid, benzoic acid, gallic acid, hydroxynaphthoic acid, magnesium salicylate, oxalic acid, phenolic acids, salicylamide, salicylic acid, strontium salicylate, succinic acid.

Isoallylenethiourea
Synonyms: Isoallylenesulphourea.
French: Sulphourée d'isoallylène, Sulphourée isoallylènique, Thiourée d'isoallylène, Thiourée isoallylènique.
German: Isoallylensulfoharnstoff, Isoallylenthioharnstoff.
Chemical
Starting point in making—
Intermediates, pharmaceuticals.
Starting point (Brit. 310534) in making rubber vulcanization accelerators with the aid of—
Allylanilin, alphanaphthylamine, amylanilin, anilin, betanaphthylamine, cyclohexylanilin, diphenylamine, ethylanilin, meta-anisidin, metacresidin, metaphenylenediamine, metatoluidin, metatoluylenediamine, metaxylenediamine, metaxylidin, monomethylanilin, orthoanisidin, orthocresidin, orthonaphthylenediamine, orthophenylenediamine, orthotoluidin, orthotoluylenediamine, orthoxylenediamine, orthoxylidin, para-anisidin, paracresidin, paranaphthylamine, paranaphthylenediamine, paraphenylenediamine, paratoluidin, paratoluylenediamine, paraxylenediamine, paraxylidin, phenylamine, tolylamine, xylylamine.

Isoallyl Mercaptan
Synonyms: Isoallyl sulphydrate.
French: Mercaptane d'isoallyle, Mercaptane isoallylique, Sulphydrate d'isoallyle, Sulphydrate isoallylique.
German: Isoallylmerkaptan, Isoallylsulfhydrat.
Chemical
Starting point in making—
Intermediates, pharmaceuticals.
Starting point (Brit. 286749) in making rubber vulcanization accelerators with—
Dibenzylamine, diethylguanylthioureas, diphenyl biguanide, ditolyl biguanide, ethanolamines, guanylureas, isothioureas, isoureas, monophenyl biguanide, monophenylguanylthioureas, monotolyl biguanide, pentaphenyl biguanide, pentatolyl biguanide, piperidin, piperazin, tetramethylammonium hydroxide, tetraphenyl biguanide, thioureas, trimethylsulphonium hydroxide.

Isoamyl alcohol, primary. See: Isobutyl carbinol.

Isoamylcarbamide
French: Carbamide d'isoamyle, Carbamide isoamylique.
Chemical
Starting point in making—
Intermediates, pharmaceuticals.
Dye
Starting point in making various synthetic dyestuffs.
Resins and Waxes
Starting point (Brit. 292912) in making synthetic resins with—
Acetylsalicylic acid, aliphatic dibasic acids, ammonium salicylate, anthranilic acid, benzoic acid, gallic acid, hydroxynaphthoic acid, magnesium salicylate, oxalic acid, phenolic acids, phthalic acid, salicylamide, salicylic acid, strontium salicylate, succinic acid.

Isoamyl Cinnamate
Synonyms: Isoamyl betaphenylacrylate.
French: Betaphényleacrylate d'isoamyle, Betaphényle-acrylate isoamylique, Cinnamate d'isoamyle, Cinnamate isoamylique.
German: Amylcinnamat, Betaphenylacrylsäureisoamylester, Betaphenylacrylsäuresisoamyl, Isoamylcinnamat, Zimtsäureisoamylester, Zimtsäuresisoamyl.

Food
Ingredient of—
Cocoa essences.
Flavoring agent.

Perfume
Fixative in perfumes.
Ingredient of—
Cosmetics.

Isoamylenethiourea
French: Sulphourée d'isoamylène, Sulphourée isoamylènique, Thiourée d'isoamylène, Thiourée isoamylènique.
German: Isoamylensulfoharnstoff, Isoamylenthioharnstoff.

Chemical
Starting point in making—
Intermediates, pharmaceuticals.
Starting point (Brit. 310534) in making rubber accelerators with the aid of—
Allylanilin, alphanaphthylamine, amylanilin, anilin, betanaphthylamine, cyclohexylanilin, diphenylamine, ethylanilin, meta-anisidin, metacresidin, metanaphthylenediamine, metaphenylenediamine, metatoluidin, metatoluylenediamine, metaxylenediamine, metaxylidin, monomethylanilin, orthoanisidin, orthocresidin, orthonaphthylenediamine, orthophenylenediamine, orthotoluidin, orthotoluylenediamine, orthoxylenediamine, orthoxylidin, para-anisidin, paracresidin, paraphenylenediamine, paratoluidin, paratoluylenediamine, paraxylenediamine, paraxylidin, phenylamine.

Isoamylmercaptan
French: Mercaptane d'isoamyle, Mercaptane isoamylique.
German: Isoamylmerkaptan.

Chemical
Starting point in making various derivatives.
Reagent (Brit. 286749) in making vulcanization accelerators with—
Dibenzylamine, diethylguanylthioureas, diphenyl guanide, ditolyl biguanide, ethanolamines, guanylureas, isothioureas, isoureas, monophenyl biguanide, monotolyl biguanide, pentaphenyl biguanide, pentatolyl biguanide, piperidin, piperazin, tetramethylammonium hydroxide, tetraphenyl biguanide, tetratolyl biguanide, thioureas, trimethylsulphonium hydroxide.

Insecticide and Fungicide
Fumigant and insecticide for—
Rice weevils.

Isoamylmercapto-1-naphthol
French: Alphanaphtol d'isoamylmercaptique.
German: Isoamylmerkaptoalphanaphtol.
Spanish: Alphanaftol de isoamilmercapto.
Italian: Alphanaftole di isoamilmercapto.

Chemical
Starting point in making—
Intermediates and other derivatives.

Dye
Starting point (Brit. 291825) in making indigoid dyestuffs with the aid of—
Isatin anilide, isatin chloride, reactive alpha derivatives of isatin.

Isoamyl Nitrite
French: Nitrite d'isomylique.
German: Isoamylnitrat.

Chemical
Starting point in making—
Di-isoamylamine, isoamylamine, methyl nitrite.

Isoamyl Resinate
Synonyms: Isoamyl abietate.
French: Abiétate d'isoamyle, Abiétate isoamylique, Résinate d'isoamyle, Résinate isoamylique.
German: Abietinsaeureisoamylester, Abietinsaeuresisoamyl, Isoamylabietat.

Paint and Varnish
Plasticizer in making—
Paints and varnishes, lacquers and dopes, containing nitrocellulose, cellulose acetate or other cellulose esters or ethers.

Plastics
Plasticizer in making—
Compositions containing cellulose acetate, nitrocellulose or other cellulose esters or ethers.

Isoborneol Acetate
Synonyms: Isobornyl acetate.
French: Acétate d'isobornéol, Acétate d'isobornyle, Acétate isobornylique.
German: Essigsaeureisoborneolester, Essisaeureisobornylester, Essigsaeuresisobornyl, Isoborneolacetat, Isobornylacetat.

Paint and Varnish
Plasticizer in making—
Cellulose ester and ether varnishes, lacquers and dopes (Brit. 283619).

Perfumery
Ingredient of various preparations.

Plastics
Plasticizer in making—
Cellulose ester and ether compounds (Brit. 283619).

Isobornyl Phthalate
Synonyms: Isoborneol phthalate.
French: Phthalate d'isobornéole, Phthalate d'isobornyle, Phthalate isobornylique.
German: Isoborneolphtalat, Isobornylphtalat, Phtalsaeureisoborneolester, Phtalsaeureisobornylester, Phtalsaeuresisoborneol, Phtalsaeuresisobornyl.

Cellulose Products
Solvent and plasticizer (Brit. 283619) for—
Cellulose esters and ethers.
For uses, see under general heading: "Solvents."

Isobourbonal

Chemical
Starting point in making—
Aromatic chemicals.

Perfume
Ingredient of—
Artificial perfumes.
Perfume in—
Cosmetics.

Soap
Perfume in—
Toilet soaps.

Isobutoxydiphenylamine

Rubber
Aging retardant (Brit. 424461).
Promoter (Brit. 424461) of—
Resistance to crack formation on flexing.

Isobutyl Acetate

Cellulose Products
Solvent for—
Cellulose esters and ethers, cellulose nitrate (nitrocellulose).
For uses, see under general heading: "Solvents."

Chemical
Solvent for—
Celluloid, nitrocellulose.
Starting point in making—
Synthetic aromatics, synthetic pharmaceuticals, various derivatives.

Food
Ingredient of—
Various artificial fruit essences.

Perfumery
Ingredient of—
Rose perfumes.
Perfume in—
Cosmetics.

Isobutyl Carbamide
French: Carbamide d'isobutyle, Carbamide isobutylique.
German: Isobutylcarbamid.

Chemical
Reagent in making—
Pharmaceuticals and other compounds.

Isobutyl Carbamide (Continued)
Resins and Waxes
Starting point (Brit. 292912) in making synthetic resins with—
Acetylsalicylic acid, aliphatic dibasic acids, ammonium salicylate, anthranilic acid, benzoic acid, gallic acid, hydroxynaphthoic acid, magnesium salicylate, oxalic acid, phenolic acids, phthalic acid, salicylamide, salicylic acid, strontium salicylate, succinic acid.

Isobutyl Carbinol
Synonyms: Primary isoamyl alcohol.
French: Carbinole de isobutyle, Carbinole isobuylique.
German: Isocarbinol.
Chemical
General solvent in various processes.
Starting point in making—
Amyl acetate and other amyl compounds.
Miscellaneous
General solvent for various purposes.
See also: "Solvents."
Paint and Varnish
Solvent in making—
Cellulose ester and ether dopes, varnishes, and lacquers.
Plastics
Solvent in making—
Cellulose ester and ether compounds.

Isobutyl Chloroacetate
French: Chloroacétate de isobutyle.
German: Chloressigsaeureisobutylester, Chloressigsaeuresisobutyl, Isobutylchloracetat.
Dye
Reagent in making—
Stable, water-soluble vat dyestuff derivatives (Brit. 263898).

Isobutylene Dibromide
Chemical
Reagent in—
Organic syntheses.
Fuel
Primer (Brit. 404682) in—
Diesel engine fuels (used in conjunction with alkyl nitrates, having two to four atoms in the molecule, whose function is that of reducing the delay period).
Reducer (Brit. 404682) of—
Spontaneous ignition temperature of diesel engine fuels.

Isobutylethyl Ketone
French: Kétone d'isobutyle-éthyle, Kétone isobutylique-éthylique.
German: Isobutylaethylketon.
Cellulose Products
Solvent (Brit. 330725) for—
Cellulose esters and ethers.
For uses, see under general heading: "Solvents."
Chemical
Solvent for various chemicals.
Starting point in making—
Intermediates and other derivatives.

Isobutyl Formate
French: Formiate d'isobutyle, Formiate isobutylique.
German: Ameisensäureisobutylester, Ameisensäuresisobutyl, Methansäureisobutylester, Methansäuresisobutyl, Isobutylformiat.
Food
As a flavoring.
Ingredient of—
Fruit essences.
Perfume
Ingredient of
Perfumes.
Perfume in—
Cosmetics.
Soap
Perfume in—
Toilet soaps.

Isobutyl Glycollate
French: Glycollate d'isobutyle, Glycollate isobutylique.
German: Glykolsäureisobutylester, Glykolsäuresisobutyl, Isobutylglykolat.
Cellulose Products
Plasticizer (Brit. 311669) for—
Cellulose acetate, cellulose esters and ethers, cellulose nitrate.
For uses, see under general heading: "Plasticizers."
Chemical
Starting point in making various derivatives.

Isobutyl Mandelate
French: Mandélate isobutylique, Mandélate de isobutyle.
German: Isobutylmandelat, Mandelsaeureisobutylester, Mandelsaeuresisobutyl.
Miscellaneous
See also: "Plasticizers."
Paint and Varnish
Plasticizer in making—
Cellulose ester and ether varnishes and lacquers (Brit. 270650).
Plastics
Plasticizer in making—
Nitrocellulose plastics.

Isobutyl Mercaptan
Synonyms: 2-Methylpropanthiol-1.
Chemical
Starting point in making—
Intermediates, pharmaceuticals.
Reagent (Brit. 286749) in making rubber vulcanization accelerators with—
Dibenzylamine, diethylguanylthioureas, diphenylbiguanide, ditolylbiguanide, ethanolamines, guanylureas, isothioureas, isoureas, monophenylbiguanide, monophenylguanylthiourea, monotolylbiguanide, pentaphenylbiguanide, pentatolylbiguanide, piperidin, piperazin, tetramethylammonium hydroxide, tetraphenylbiguanide, tetratolylbiguanide, thioureas, trimethylsulphonium hydroxide.
Insecticide and Fungicide
Fumigant and insecticide for—
Rice weevils (*Sitophilus oryza* L.).

Isobutylnaphthalenesulphonic Acid
French: Acide d'isobutylenaphthalènesulfonique.
German: Isobutylnaphtalinsulfonsaeure.
Dye
Dispersive agent (Brit. 252392) in making—
Dyestuff preparations.
Soap
Dispersive agent (Brit. 252392) in making—
Detergent compositions.
Textile
——, *Bleaching*
Dispersive agent (Brit. 280110) in making—
Bleach liquors for wool.
——, *Finishing*
Dispersive agent (Brit. 280110) in making—
Fulling baths and finishing liquors for wetting felt-like fabrics.
——, *Printing and Dyeing*
Dispersive agent (Brit. 280110) in making—
Printing pastes and dye liquors.
——, *Manufacturing*
Dispersive agent (Brit. 280110) in making—
Wool-carbonizing solutions.

Isobutylphenyl Acetate
Synonyms: Eglanteria, Eglantin, Isobutylalphatoluylate.
French: Acétate d'isobutylephényle, Acétate isobutylique et phénylique, Alphatoluylate d'isobutyle, alphatoluylate isobutylique, Phénylacétate d'isobutyle, Phénylacétate isobutylique.
German: Alphatoluylsaeuresisobutyl, Isobutylphenylazetat, Phenylessigsaeureisobutylester, Phenylessigsaeuresisobutylester.
Perfume
Ingredient of the following artificial essences—
Garden pink, rose, tuberose, wild rose.
Perfume in—
Cosmetics.
Soap
Perfume for—
Toilet soaps.

Isobutyl Salicylate
Synonyms: Isobutyl-ortho-oxybenzoate, Isobutyl 2-phenolmethylate.
French: Ortho-oxybenzoate d'isobutyle, Ortho-oxybenzoate isobutylique, 2-Phénoleméthylate d'isobutyle, 2-Phénoleméthylate isobutylique, Salicylate d'isobutyle, Salicylate isobutylique.

Isobutyl Salicylate (Continued)
German: Ortho-oxybenzoesaeureisobutylester, Ortho-oxybenzoesaeuresisobutyl, 2-Phenolmethylsaeure-1-isobutylester, 2-Phenolmethylsaeures-1-isobutyl, Salicylsaeureisobutylester, Salicylsaeuresisobutyl.

Perfume
Ingredient of the following artificial odors—
Cassia, cloves, fern, orchid.
Perfume for—
Cosmetics.

Soap
Perfume for—
Toilet soaps.

Isobutyramide

Analysis
Laboratory reagent.

Chemical
Reagent in—
Chemical synthesis.

Isodithiocyanic Acid
Synonyms: Isodisulphonic acid.
French: Acide isodisulphocyanique, Acide isodithiocyanique.
German: Isodisulfocyansäure, Isodithiocyansäure.

Chemical
Starting point in making—
Salts, esters, and other derivatives.

Metallurgical
Flotation agent (Brit. 314822) in treating—
Oxidized ores in order to effect their separation.

Isodurene

Analysis
Reagent.

Chemical
Reagent in—
Chemical syntheses.

Iso-octane
Synonyms: Trimethylisobutylmethane.

Petroleum
Compounding agent for—
Aviation gasoline.

Isopentane

Chemical
General solvent for various purposes.
Starting point in making—
Chlorinated and hydrogenated derivatives.

Miscellaneous
General solvent for various processes.

Paint and Varnish
Solvent in making—
Cellulose ester and ether lacquers and varnishes.

Plastics
Solvent in making—
Cellulose ester and ether compositions.

Isopentoxydiphenylamine

Rubber
Aging retardant (Brit. 424461).
Promoter (Brit. 424461) of—
Resistance to crack formation on flexing.

Isopersulphocyanic Acid

Chemical
Starting point in making various derivatives.

Miscellaneous
Flotation agent (Brit. 314822) in the separation of—
Oxidized oils.

Isoprene
Synonyms: Isopren.
French: Isoprène.
German: Isopren.

Chemical
Starting point in making—
Intermediates, pharmaceuticals.
Starting point (Brit. 309311) in making synthetic aromatics with—
Acrolein, acrylic acid, alphamethylbetaethylacrolein, crotonaldehyde, crotonic acid, 2:2-dimethylacrolein, 2-ethylacrolein, tetrolicaldehyde.

Rubber
Starting point in making—
Synthetic rubber.

Isopropyl Acetate
French: Acétate d'isopropyle, Acétate isopropylique.
German: Essigsäureisopropylester, Essigsäuresisopropyl, Isopropylacetal, Isopropylazetat.
Spanish: Acetato de isopropil.
Italian: Acetato di isopropile.

Analysis
Solvent in the chemical laboratory (used in place of ethyl acetate).

Ceramics
Solvent in—
Compositions, containing nitrocellulose or other esters or ethers of cellulose, used for the coating and protection of ceramic ware.

Chemical
Reagent (Brit. 302174) in extracting—
Various organic acids from dilute solutions, to obtain concentrated products (used along with benzene in admixture for extracting acetic acid, butyric acid, propionic acid, and other aliphatic acids).
Solvent (used in place of ethyl acetate) for—
Phosgene, pyroxylin, various chemicals and chemical products.
Solvent (used in place of ethyl acetate) in making—
Ketene.

Dye
Solvent (used in the place of ethyl acetate) for—
Separating dyestuffs.

Electrical
Solvent in—
Compositions, containing nitrocellulose or other esters or ethers of cellulose, used for insulating purposes, and in the manufacture of electrical machinery and equipment.

Explosives
Solvent (used in place of ethyl acetate) in making—
Guncotton, smokeless powder, various explosive compositions.

Fats and Oils
Solvent for various animal and vegetable fats and oils.

Food
Solvent (used in place of ethyl acetate) for—
Extracting caffeine from coffee.

Glues and Adhesives
Solvent in making—
Adhesive preparations containing nitrocellulose or other esters or ethers of cellulose.

Leather
Solvent in—
Compositions, containing nitrocellulose or other esters or ethers of cellulose, used in the manufacture of artificial leathers and for the decoration and protection of leathers and leather goods.

Metallurgical
Solvent in—
Compositions, containing nitrocellulose or other esters or ethers of cellulose, used for the decoration or protection of metallic ware.

Miscellaneous
Reagent in making—
Artificial bristles, artificial horsehair.
Solvent in—
Compositions, containing nitrocellulose or other esters or ethers of cellulose, used for the decoration and protection of various compositions of matter.

Paint and Varnish
Ingredient of—
Brushing lacquers, paint and varnish removers.
Solvent (used in place of ethyl acetate) in making—
Lacquers and varnishes containing synthetic resin bases of the vinyl ester type.
Solvent in making—
Lacquers, varnishes, paints, dopes, and enamels containing nitrocellulose or other esters or ethers of cellulose.

Paper
Solvent in—
Compositions, containing nitrocellulose, cellulose acetate, or other esters or ethers of cellulose, used in the manufacture of coated papers and for the decoration and protection of products manufactured from paper or pulp.

Perfume
Solvent (used in place of ethyl acetate) in making—
Perfumes, cosmetics.

Isopropyl Acetate (Continued)

Photographic
Solvent in making—
Films from cellulose acetate, nitrocellulose, or other esters or ethers of cellulose.

Plastics
Solvent for—
Celluloid.
Solvent (used in place of ethyl acetate) in making—
Colloidal cements.
Solvent in making—
Plastic products containing cellulose acetate, nitrocellulose, or other esters or ethers of cellulose.

Resins and Waxes
Solvent for various resins and waxes (used in place of ethyl acetate).

Rubber
Solvent in—
Compositions, containing cellulose acetate, nitrocellulose, or other esters or ethers of cellulose, used for the decoration and protection of rubber merchandise.
Solvent (used in place of ethyl acetate) for—
Removing resinous matters from balata gum and gutta-percha.

Stone
Solvent in—
Compositions, containing cellulose acetate, nitrocellulose, or other esters or ethers of cellulose, used for the decoration and protection of artificial and natural stone.

Textile
——, *Finishing*
Solvent (used in place of ethyl acetate) for—
Cleansing textile fabrics.

——, *Manufacturing*
Solvent in—
Compositions, containing cellulose acetate, nitrocellulose, or other esters or ethers of cellulose, used for the decoration of textile fabrics.
Solvent (used in place of ethyl acetate) in making—
Rayon.

Woodworking
Solvent in—
Compositions, containing cellulose acetate, nitrocellulose, or other esters or ethers of cellulose, used for the decoration and protection of woodwork.

Isopropyl Acrylate
Synonyms: Acrylic acid isopropyl ester.
French: Acrylate d'isopropyle, Acrylate isopropylique.
German: Acrylsäureisopropylester, Acrylsäuresisopropyl, Isopropylacrylat.

Miscellaneous
Solvent (Brit. 321258, German 367294) for—
Cellulose acetate, cellulose esters and ethers, cellulose nitrate, natural and synthetic resins, rubber.
For uses, see under general heading: "Solvents."

Isopropyl Alcohol
Synonyms: Isopropanol.

Analysis
Extracting medium for various purposes in institutional, industrial research, and control work.
Solvent in the extraction and assay of—
Alkaloids, essential oils, gums, shellac.

Abrasives
Solvent for—
Shellac when used as a binder.

Automotive
Drying agent for—
Metal objects preparatory to electroplating.
Solvent for—
Shellac in gasket cement and similar products.

Chemical
Coupling agent in making—
Soluble oils.
Denaturant ("marker") in—
Alcohol formulas 39, 39A, 40, and 40M.
Extractant for—
Alkaloids, chemicals.
Reactant for—
Introducing isopropyl group in organic synthesis.
Solvent for—
Alkaloids, essential oils, gums, inorganic compounds, organic compounds.
Solvent in—
Organic synthesis.
Stabilizer for—
Oil emulsions.

Dry Cleaning
Solvent and spotting agent for—
Various organic substances.

Dye
Solvent in—
Dye synthesis.

Electrical
Solvent for shellac in making—
Electrical condensers.
Cements for lamp bases and caps.
Damp-proofing and insulating coatings for electrical appliances, coils and windings, motors, generators.
Insulators, seals.

Fats, Oils, and Waxes
Stabilizer in—
Emulsified oil preparations.

Germicide
Claimed as—
Effective germicide.

Glue and Adhesives
Solvent for—
Gums, shellac.

Gums
Solvent for—
Gums.

Inks
Solvent in making—
Inks of various kinds.

Leather
Solvent in—
Dressings, finishes, polishes, waterproofings.

Metallurgical
Drying agent for—
Metal objects preparatory to electroplating.

Miscellaneous
Rubbing alcohol in—
Massaging.
Solvent for shellac and gums in—
Binding various products, coating various products, filling various products, glazing various products.
Stiffening various products; for example, felt and straw hats.
Thickening various products.
Waterproofing various products, such as cordage, fishing tackle, rope.

Paint and Varnish
Solvent for—
Gums, shellac.

Paper
Solvent for—
Gums and shellac in making sizings, glazings, and coatings for paper products, such as art paper, boxboard, cartons, paper boxes, paper, playing cards, visiting cards.

Perfume
Ingredient of—
Denatured alcohol formulas for use in perfumes and cosmetics.
Extractant for—
Essential oils.
Solvent for—
Essential oils.

Pharmaceutical
Suggested solvent in making—
Antiseptic solutions, pharmaceutical preparations, liniments, lotions.
Solvent for—
Essential oils.

Photographic
Solvent for shellac and gums in making—
Photographic papers.

Plastics
Solvent for shellac in making—
Plastics.

Printing
Solvent for—
Gums, shellac.

Soap
Ingredient of—
Liquid soaps.

Isopropyl Alcohol (Continued)
Solvent for—
Essential oils.
Textile
Ingredient of—
Textile oil preparations.

5:5-Isopropylbromoallylbarbituric Acid
French: Acide de 5:5-isopropylbromoallylbarbiturique.
German: 5:5-Isopropylbromallylbarbiturinsaeure.
Chemical
Starting point (Swiss 113251) in making synthetic drugs with—
Allylamine, amylamine, butylamine, diallylamine, diamylamine, dibutylamine, diethylamine, dimethylamine, dipropylamine, ethylamine, isoallylamine, isoamylamine, isobutylamine, isopropylamine, methylamine, propylamine.

Isopropyl Carbamide
French: Carbamide d'isopropyle, Carbamide isopropylique.
German: Isopropylcarbamid.
Chemical
Starting point in making—
Intermediates, pharmaceuticals.
Dye
Starting point in making various synthetic dyestuffs.
Resins and Waxes
Starting point (Brit. 292912) in making synthetic resins with—
Acetylsalicylic acid, aliphatic dibasic acids, ammonium salicylate, anthranilic acid, benzoic acid, gallic acid, hydroxynaphthoic acid, magnesium salicylate, oxalic acid, phenolic acids, phthalic acid, salicylamide, salicylic acid, strontium salicylate, succinic acid.

Isopropyl Chloride
Analysis
Extractant for—
Fats.
Solvent for—
Fats.
Food
Extractant for—
Fats.
Solvent for—
Fats.
Glue and Gelatin
Defatting agent in—
Treating bones.
Leather
Solvent for—
Fats in compounded dressings.
Miscellaneous
Extractant for—
Fats.
Solvent for—
Fats.
Soap
Solvent for—
Fats.

Isopropyl Chloroacetate
French: Chloroacétate de isopropyle.
German: Chloressigsaeureisopropylester, Chloressigsaeuresisopropyl, Isopropylchloracetat.
Dye
Reagent in making—
Stable, water-soluble vat dyestuffs derivatives (Brit. 263898).

Isopropylenethiourea
Synonyms: Isopropylenesulphourea.
French: Sulphourée d'isopropylène, Sulphourée isopropylènique, Thiourée d'isopropylène, Thiourée isopropylènique.
German: Isopropylensulfoharnstoff, Isopropylenthioharnstoff.
Chemical
Starting point in making—
Intermediates, pharmaceuticals.
Starting point (Brit. 310534) in making rubber accelerators with the aid of—
Allylanilin and other allylamines, alphanaphthylamine, amylanilin and other amylamines, anilin and derivatives, betanaphthylamine, cyclohexylanilin and other cycloamines, diphenylamine, ethylanilin and other ethylamines, meta-anisidin, metacresidin, metaphenylenediamine, metatoluidin, metatoluylenediamine, metaxylenediamine, metaxylidin, monomethylanilin and derivatives, orthoanisidin, orthocresidin, orthonaphthylenediamine, orthophenylenediamine, orthotoluidin, orthotoluylenediamine, orthoxylenediamine, orthoxylidin, para-anisidin, paracresidin, paranaphthylamine, paranaphthylenediamine, paraphenylenediamine, paratoluidin, paratoluylenediamine, paraxylenediamine, paraxylidin, phenylamine, tolylamine, xylylamine.

Isopropyl Ether
French: Éther isopropylique.
German: Isopropylaether.
Analysis
Extractant for—
Acetic acid, nicotine, waxes.
Solvent for—
Fats, oils, natural resins, synthetic resins, waxes.
Solvent, in admixture with alcohols, for—
Dyes, ethylcellulose, nitrocellulose.
Cellulose Products
Solvent, in admixture with alcohols, for—
Cellulose nitrate, ethylcellulose.
Ceramics
Solvent, in admixture with alcohols, in—
Compositions, containing natural or synthetic resins, nitrocellulose, or ethylcellulose, used as coatings for protecting and decorating ceramic products.
Chemical
Extractant for—
Acetic acid, nicotine.
Solvent barely miscible with water.
Solvent miscible with most other organic solvents.
Substitute for—
Ethyl ether where advantage can be taken of higher boiling point and lower vapor pressure which is of distinct advantage for extraction purposes.
Solvent, in admixture with alcohols, for—
Cellulose nitrate, ethylcellulose.
Cosmetic
Extractant for—
Acetic acid.
Solvent for—
Fats, fixed oils, volatile oils, waxes.
Solvent, in admixture with alcohols, in—
Nail enamels and lacquers containing natural or synthetic resins, nitrocellulose, or ethylcellulose as base material.
Dye
Solvent, in admixture with alcohols, for various dyestuffs.
Electrical
Solvent, in admixture with alcohols, for—
Insulating compositions, containing natural or synthetic resins, nitrocellulose, or ethylcellulose, used for covering wire and in making electrical machinery and equipment.
Explosives
Solvent barely miscible with water.
Solvent miscible with most other organic solvents.
Substitute for—
Ethyl ether when advantage can be taken of higher boiling point and lower vapor pressure which is of distinct advantage for extraction purposes.
Solvent, in admixture with alcohols, for—
Nitrocellulose.
Dry-Cleaning
Ingredient of—
Spotting agents.
Spotting agent for—
Acetic acid, fats, greasy stains, nicotine stains, oils, resins, waxes.
Fats, Oils, and Waxes
Solvent for—
Fats, oils, waxes.
Food
Solvent for—
Fats, fixed oils, volatile oils, waxes.
Glass
Solvent, in admixture with alcohols, in—
Compositions, containing natural or synthetic resins, nitrocellulose, or ethylcellulose, used in the manufacture of nonscatterable glass and as coatings for decorating and protecting glassware.

Isopropyl Ether (Continued)

Glue and Adhesives
Ingredient of—
Rubber cements.
Solvent, in admixture with alcohols, in—
Adhesive compositions containing natural or synthetic resins, nitrocellulose, or ethylcellulose.

Leather
Solvent, in admixture with alcohols, in—
Compositions, containing natural or synthetic resins, nitrocellulose, or ethylcellulose, used in the manufacture of artificial leathers and as coatings for decorating and protecting leathers and leather goods.

Metal Fabricating
Solvent, in admixture with alcohols, in—
Compositions, containing natural or synthetic resins, nitrocellulose, or ethylcellulose, used as coatings for protecting and decorating metallic articles.

Miscellaneous
Solvent barely miscible with water.
Solvent miscible with most other organic solvents.
Solvent, in admixture with alcohols, in—
Coating compositions, containing natural or synthetic resins, nitrocellulose, or ethylcellulose, used for protecting and decorating various articles.
Substitute for—
Ethyl ether where advantage can be taken of higher boiling point and lower vapor pressure which is of distinct advantage for extraction purposes.

Paint and Varnish
Ingredient of—
Paint and varnish removers.
Solvent for—
Oils, natural resins, synthetic resins, waxes.
Solvent, in admixture with alcohols, in—
Paints, varnishes, lacquers, enamels, and dopes containing natural or synthetic resins, nitrocellulose, or ethylcellulose.

Paper
Solvent, in admixture with alcohols, in—
Compositions, containing natural or synthetic resins, nitrocellulose, ethylcellulose, waxes, used in the manufacture of coated papers and as coatings for decorating and protecting products made of paper or pulp.

Petroleum
Blending agent and improver (Brit. 445503) for—
Gasoline motor fuels.
Dewaxing agent for—
Paraffin base oils (used in admixture with isopropanol).
Solvent for—
Mineral oils.

Pharmaceutical
Solvent for—
Essential oils, mineral oils, vegetable oils.

Photographic
Extractant for—
Acetic acid.
Solvent, in admixture with alcohols, in making—
Films from nitrocellulose.

Plastics
Extractant for—
Acetic acid.
Solvent, in admixture with alcohols, in making—
Plastics from or containing natural or synthetic resins, nitrocellulose, or ethylcellulose.

Resins
Solvent for—
Natural resins, synthetic resins.
Solvent, in admixture with alcohols, in making—
Artificial resins from or containing nitrocellulose or ethylcellulose.

Rubber
Ingredient of—
Rubber cements.
Solvent, in admixture with alcohols, in—
Compositions, containing natural or synthetic resins, nitrocellulose, or ethylcellulose, used as coatings for protecting and decorating rubber goods.

Soap
Solvent for—
Fats, oils.

Stone
Solvent, in admixture with alcohols, in—
Compositions, containing natural or synthetic resins, nitrocellulose, or ethylcellulose, used as coatings for decorating and protecting artificial and natural stone.

Textile
Degreasing and defatting agent for—
Textile fibers.
Solvent, in admixture with alcohols, for—
Dyestuffs, ethylcellulose, fats, nitrocellulose, oils, waxes.

Tobacco
Extractant for—
Nicotine.

Wood
Solvent, in admixture with alcohols, in—
Compositions, containing natural or synthetic resins, nitrocellulose, or ethylcellulose, used as protective and decorative coatings on woodwork.

Isopropylideneglycerol
French: Isopropylidène de glycérole, Isopropylidène glycérolique.
German: Isopropylidenglycerol.

Miscellaneous
Solvent for—
Cellulose esters and ethers, gums, resins, various organic substances.
For uses, see under general heading: "Solvents."

Isopropyl Mandelate
French: Mandélate isopropylique, Mandélate de isopropyle.
German: Mandelsaeureisopropylester, Mandelsaeuresisopropyl.

Paint and Varnish
Plasticizer in making—
Cellulose ester and ether varnishes and lacquers (Brit. 270650).

Plastics
Plasticizer in making—
Nitrocellulose plastics.

Isopropylmercaptan
Synonyms: Propanthiol-2, Secondary propylmercaptan.

Insecticide and Fungicide
Fumigant and insecticide for—
Rice weevils (*Sitophilus oryza* L.).

Isopropylnaphthalenesulphonic Acid
French: Acide d'isopropylenaphthalène.
German: Isopropylnaphtalinsulfonsaeure.

Chemical
Ingredient of—
Emulsifying compositions (Brit. 260243).

Soap
Ingredient of—
Cleansing compositions (Brit. 260243).

Textile
——, *Finishing*
Ingredient of—
Soap solutions used in fulling woolen materials (Brit. 253105).

Isopropyl Resinate
Synonyms: Isopropyl abietate.
French: Abiétate d'isopropyle, Abiétate isopropylique, Résinate d'isopropyle, Résinate isopropylique.
German: Abietinsaeureisopropylester, Abietinsaeuresisopropyl, Isopropylabietat, Isopropylresinat.

Paint and Varnish
Plasticizer in making—
Paints and varnishes, lacquers and dopes, containing cellulose nitrate, cellulose acetate or other cellulose esters or ethers.

Plastics
Plasticizer in making—
Compositions containing cellulose acetate, nitrocellulose or other cellulose esters or ethers.

Isopropyl Ricinolsulphonate

Miscellaneous
As an emulsifying agent (German 561495).
For uses, see under general heading: "Emulsifying agents."

Isopropylsuccinic Acid
French: Acide d'isopropylsuccinique.
German: Isopropylbernsteinsaeure.

Chemical
Starting point in making—
Terebic acid.

Isopulegol
Synonyms: Isopulegole, 1-Methyl-4-isopropenylcyclohexanol-3, Paramenthenol-3, Paramentheneole-3.
French: 1-Méthyle-4-isopropenylecyclohexanole-3.
Perfumery
Ingredient of—
Cosmetics, toilet waters.
Soap
Perfume in—
Toilet soaps.

Japan Wax
Abrasives
Ingredient of—
Emery paste, containing also double pressed saponified stearic acid, oleostearin, petrolatum, paraffin, emery, and flint.
Ingredient of various adhesive compositions.
Agriculture
Ingredient of—
Compositions for curing brown bast in rubber trees.
Grafting dressing, in admixture with rosin.
Grafting dressing, containing also lanolin, rosin, rosin oil, pine oil, ceresin, and beeswax.
Brewing
Impregnating agent for—
Treating interior of barrels.
Chemical
Ingredient of—
Coating compositions for acid tanks and chemical apparatus.
Construction
Ingredient of—
Waterproofing compositions for brickwork, concrete, masonry, piles, shingles, and other porous structural materials.
Electrical
As a general insulating agent.
Binding, coating, and insulating agent in—
Electrical condensers.
Boiling-out agent for—
Treating cables and other materials to remove moisture and improve their electrical properties.
Coating and insulating agent for—
Dry-cell batteries.
Household light wires, radio wires, telephone wires, wires in all kinds of domestic electrical appliances.
Industrial electrical cables and industrial electrical machinery.
Radio coils and other electrical coils.
Utility cables and machinery.
Ingredient of—
Insulating compositions containing rubber.
Insulating compositions for wires of all kinds.
Insulating compositions for industrial electrical cables and industrial electrical machinery.
Insulating compositions for electric utility cables and machinery.
Insulating and sealing compositions for dry-cells.
Molded insulations.
Waterproofing agent for—
Electrical instruments, electrical machinery.
Explosives
Coating agent for—
Stems of paper or vesta matches.
Stems of wooden matches (used to provide a smooth, shiny surface).
Ingredient of—
Matchhead compositions.
Waterproofing agent for—
Explosives, matches.
Fuel
Component of—
Candles, night-lights.
Ink
Ingredient of—
Lithographic inks.
Non-offset compound, containing also No. 1 lithographic varnish, soft cup grease, and paraffin.
Offset compound, containing also amber petrolatum, mutton tallow, paraffin oil, kerosene, and high-flash naphtha.
Printing inks, stamping inks.
Leather
Ingredient of—
Dressings.
Dressing, containing also tallow, petrolatum, diglycol stearate, rosin, and water.
Finishing preparations.
Military paste polish, containing also turpentine and other waxes, such as carnauba, candelilla, paraffin.
Polishing compositions.
Preservative containing also 20° cold test neatsfoot oil, anhydrous lanolin, water, and soap chips.
Waterproofing agent.
Mechanical
As a coating against rust.
Ingredient of—
Drawing oil, containing also tallow, thin mineral oil, and 40° caustic soda.
Belt dressing, containing also asphalt, white lead, neatsfoot oil, tallow, and citronella oil.
Lubricating compositions.
Metallurgical
Ingredient of—
Compositions used for covering metals to provide protection against moisture, acids, alkalies, and other corrosive substances.
Corrosion-resisting compositions used as coating for metals, containing also petrolatum, oxidized petroleum bitumen, asbestos, and powdered shale.
Miscellaneous
Coating for—
Barrels.
Ingredient of—
Automobile polish, containing also carnauba wax, rosin, turpentine substitute, and potash solution.
Automobile paste polish, containing also turpentine, beeswax, paraffin, and carnauba wax.
Compositions for making dental impressions.
Compositions for making anatomical specimens.
Compositions for lining barrels and kegs.
Compositions for painting old timber to prevent attack of death watch beetle.
Compositions for waterproofing automobile tops and tarpaulins.
Floor polishes, furniture polishes.
Furniture polish, containing also yellow ceresin, beeswax, raw linseed oil, turpentine, paraffin oil, potassium carbonate, animal-fat soap chips, and water.
Furniture polish, containing also bleached carnauba wax, paraffin, turpentine, white curd soap, pale rosin, water, and an aromatic oil.
Linoleum polishes, polishes of various sorts.
Preparations for making imitation alabaster statues.
Shoe polishes, ski polishes, wood polishes.
Raw material in making—
Grease crayons, imitation fruit and flowers, oil crayons.
Spotting pencils, in admixture with stearic acid and oil dyes, for dry cleaners and textile manufacturers, used for restoring original shades to textiles which have been bleached by stain-removing chemicals.
Toys.
Wax figures for exhibition purposes and for window display.
Waterproofing agent for—
Cloth liners for automobile tires.
Pasteboard signs exposed to the weather.
Soda straws.
Oils, Fats, Waxes
Base of various lubricating compositions.
Ingredient of—
Axle greases.
Beeswax substitute, containing beeswax and glyceryl stearate.
Compounded waxes, electrotypers' wax.
Lubricating grease compound with castor oil, mineral oil, and aluminum stearate (U. S. 1881591).
Sealing wax, shoemakers' wax, special lubricants.
Substitute for—
Beeswax.
Paint and Varnish
Ingredient of—
Special floor waxes.
Varnish, containing also rosin, ceresin, barytes or other pigments, and alcohol, used for bottles and also for cork capping.
Wood fillers.
Paper
Coating for—
Waxed paper.
Ingredient of—
Coating compositions.
Compositions used in the manufacture of carbon paper.

Japan Wax (Continued)
Preparations used in making waxed paper.
Sizing for high-gloss paper.
Waterproofing agent for—
Boxboard, cardboard, cartons, paper, paper drinking cups.
Pharmaceutical
Base for—
Cerates.
In compounding and dispensing practice.
Printing
Process material in—
Lithography, photoengraving, process engraving.
Soap
Ingredient of—
Special soaps.
Stone
Ingredient of—
Waterproofing composition for treating natural and artificial stone.
Textile
Glazing agent in—
Hot calendering.
Ingredient of—
Compositions used for finishing.
Compositions used for softening.
Compositions used for sizing.
Compositions used in the manufacture of waxed cloth.
Viscose solution for producing dull-lustered rayon (U. S. 1902529).
Waterproofing coating, along with castor oil, rubber, and petrolatum.
Polishing agent for—
Weaving machine rollers.
Stiffening ("starching") agent for—
Linen.
Waterproofing agent in—
Treating yarns and fabrics.
Wax for—
Hosiery stitching threads.
Tobacco
Waterproofing agent for—
Packagings for various products.
Winemaking
Coating and impregnating agent for—
Cheap wine casks.
Ingredient of—
Compositions used for coating interior of tankcars used for transporting wine in bulk.
Woodworking
Coating and impregnating agent for—
Artificially dried wood (to prevent reabsorption of moisture).
Log ends (to prevent splitting and infection by borers).
Ingredient of—
Compositions used in the finishing of furniture and of lumber used for parquet flooring.

Juniper Oil
Synonyms: Juniper berry oil, Oil of juniper.
Latin: Oleum juniperi.
French: Essence de genièvre, Huile de genèvrier, Huile de genièvre.
German: Wachholderbeeroel, Wachholdereoel.
Chemical
Starting point in making—
Pharmaceutical products and compositions.
Gin and liqueurs.
Food
Ingredient of—
Confectionery, prepared foods.
Perfumery
Ingredient of—
Cosmetics, perfumes.
Pharmaceutical
In compounding and dispensing practice.
Soap
Perfume for—
Toilet soaps.

Kaolin
Synonyms: Argilla, China clay, Porcelain clay, White bole.
Latin: Bolus alba, Terra porcellanea.
French: Terre à porcelaine.
German: Porcellan erde, Porcellanthoncaolino.
Italian: Terra porcellana.
Abrasives
Binder in—
Emery wheels.
Adhesives
Ingredient of—
Linoleum cement, containing also iron oxide and dextrin.
Automotive
Ingredient of—
Brake-lining composition, containing also asbestos fiber, magnesia, rubber, sulphur, graphite, litharge, and iron oxide.
Friction material (for brake-lining), containing also black clay, zirconium oxide, feldspar, agalmatolite, and magnesite.
Ceramic
Ingredient of—
Floor tile, glazes, porcelains, potteries, slips, stoneware, wall tile, white earthenware.
Chemical
Carrier (Brit. 397901) for catalysts in making—
Aromatic hydrocarbon from aromatic hydroxy compounds by hydrogenation.
Catalyst in making—
Ethylene from ethyl alcohol, both for the ultimate production of ethylene and also as a step in making ethylene bromide from seawater.
Raw material in making—
Aluminum sulphate, alums, ceramic colors, ultramarine.
Construction
Filler in—
Underground metal pipes protective coating, containing also an artificial resin, rubber, and tar oil.
Wall plasters.
Raw material in making—
White portland cement, colored cements of fine tints.
Disinfectant
Ingredient of—
Disinfectant powders.
Dye
Inert base for—
Color lakes.
Electrical
Raw material in making—
Insulators, sparkplugs.
Explosives
Filler in—
Explosives, fuses.
Fats and Oils
Absorbent (Brit. 393108) in—
Purification of oils and fats by treatment with (1) alcohol-acetone, (2) phosphoric, hydrochloric, or sulphuric acid.
Clarification agent in—
Refining animal and vegetable oils.
Filler in—
Lubricants.
Fertilizer
Filler in—
Fertilizer mixtures.
Food
Ingredient of—
Patent foods.
Fuel
Carrier (Brit. 400628) for catalysts in hydrogenating—
Creosote oil, gas oil, low-temperature tar oil.
Glass
Ingredient of—
Glass batches.
Insecticide
Ingredient (U. S. 1890774) of—
Insecticidal dusting powder for agricultural purposes.
Ink
Filler in—
Lithographic inks, printing inks, writing inks.
Inert base for—
Color lakes.
Leather
Filler in—
Imitation leather.
Reagent in—
Finishing processes, tanning.
Linoleum and Oilcloth
As a filler.

Kaolin (Continued)

Metallurgical
Absorbent (French 755709) in making—
Compositions from monocalcium phosphate, acetic acid, and a metallic salt less basic than iron, for depositing a corrosion-resisting coating on metals.
Compositions from monocalcium phosphate, ethyl alcohol, and a metallic salt less basic than iron, for depositing a corrosion-resisting coating on metals.
Compositions from monocalcium phosphate, methanol, and a metallic salt less basic than iron, for depositing a corrosion-resisting coating on metals.
Ingredient of—
Annealing powders.
Self-hardening sand for foundry work (U. S. 1879272).
Welding-rod coating, containing also sodium silicate, glass, ferromanganese, and soda ash (U. S. 1903620).

Miscellaneous
As an absorbent.
As a clarification agent.
As an inert base, diluent, filler, loading agent.
Cleaner for—
White canvas goods and shoes.
Filler in—
Asbestos goods, picture frames, rope string.
Ingredient of—
Cleansing and scouring preparations.
Crayons.
Dance floor dusting compositions.
Detergent for carpets (Brit. 319084).
Lead pencils, lubricating compositions, metal polishes, shoe dressings, stove polishes.
Raw material for—
Bas relief ornaments, molded picture frames.

Paint and Varnish
Filler in—
Enamels, paints.
Inert base for—
Color lakes.
Pigment in—
White paints.

Paper
Filler in—
Paper, paperboard, wallpaper.
Ingredient of—
Coating compositions, sizes.
Inert base for—
Color lakes for wallpaper printing.

Perfume
Body material in—
Cosmetics.
Deodorant pencil, containing also aluminum chloride, mineral oil, and glyco wax.
Face clay, containing also tincture of benzoin, perfume, and water.
Face powders of various kinds, such as (1) those of medium weight, containing also talc, precipitated chalk, zinc oxide, zinc stearate, and perfume oil; (2) those of riceflour base, containing also rice starch, talc, zinc oxide, zinc stearate, and perfume oil; (3) those of light weight, containing also talc, light precipitated chalk, zinc oxide, zinc stearate, and perfume oil; (4) those of heavy weight for night wear, containing also (a) talc, zinc oxide, titanium oxide, zinc stearate, and perfume oil, or (b) titanium oxide, talc, magnesium carbonate, magnesium stearate, heliotropin and perfume oil.
Lipstick, containing also talc, ponceau 3R amaranth, ocher, zinc oxide, paraffin, beeswax, carnauba wax, sulphonated oil, and petrolatum.
Toilet powders, tooth powders, toothpastes.

Petroleum
Clarification agent in—
Refining mineral oils.
Ingredient of—
Asphalt coatings to protect pipelines from corrosion.

Pharmaceutical
As an absorbent.
As a cleaner for surgeons' hands.
As a dry dressing.
As a pill base and tablet filler.
Suggested for use in the treatment of alkaloid poisoning and as a poulticing agent.

Plastics
Absorbent and stiffener in—
Celluloid goods.

Refractories
Raw material in making—
Refractory blocks, bricks, and the like.
Refractory cements.

Rubber
Filler in—
Rubber goods.
Rubber road-surfacing compositions.

Soap
Filler in various soaps.
Ingredient of—
Cleansing powders, scouring soaps.

Textile
Filler in various fabrics.
Thickener in—
Calico printing.
Stiffener (mixed with size) for—
Cloths and other textile fabrics.

Water and Sanitation
Reagent in—
Sewage purifying, water purifying.

Winemaking
Clarifying agent for—
Wines.

Kauri Gum

Synonyms: Kauri copal, Kauri resin.
Latin: Kaurigummi.
French: Gomme de kauri, Résine de kauri.
German: Kauriharz.
Spanish: Goma de kauri.

Chemical
Ingredient of—
Linoleum cements, rosin cements.

Explosives
Ingredient of—
Match-head compositions, pyrotechnic preparations.

Glues and Adhesives
Ingredient (Brit. 332257) of—
Adhesive preparations.

Ink
Ingredient of—
Printing ink.

Leather
Ingredient of—
Compositions used in the manufacture of artificial leather.
Finishes for treating leather.
Leather substitutes used for footwear.
Preparations used for impregnating leather.

Linoleum and Oilcloth
Ingredient of—
Compositions used to coat the textile under-fabric.

Miscellaneous
Ingredient of—
Cements.
Roofing preparations (Brit. 332257).
Wall coverings (Brit. 332257).

Paint and Varnish
Ingredient of—
Asphalt lacquers, dry colors, light-colored transparent varnishes, paints, priming compositions, varnishes.

Paper
Ingredient (Brit. 332257) of—
Impregnating and finishing compositions for the treatment of paper, paperboard, cardboard, pasteboard, and various products made from paper and pulp.

Pharmaceutical
In compounding and dispensing practice.

Plastics
Ingredient (Brit. 332257) of—
Compositions used in making pressed articles.

Resins and Waxes
Ingredient of—
Rosin preparations (added to increase the hardness).

Rubber
Ingredient of—
Rubber compositions.

Textile
Ingredient (Brit. 332257) of—
Compositions used in the manufacture of floor coverings.
Compositions used for finishing fabrics.

Kauri Gum (Continued)
Compositions used for making waxed cloth.
Impregnating preparations.

Woodworking
Ingredient (Brit. 332257) of—
Impregnating and finishing compositions.

Kelp
French: Alghe marine.
German: Meertang.

Chemical
Starting point in making—
Algin and alginic acid, alginates.
Crude gelose (French 586692—addition 31868).
Cellulose compounds (French 552241).
Fermentable products.
Iodine by extraction with the aid of calcium bisulphite.
Liquid hydrocarbons (French 578564 and 643534).
Potash salts (French 578564 and 643534).
Sodium alginate (French 579381).

Glues and Adhesives
Ingredient (French 651552) of—
Adhesive preparations, in admixture with carob seeds or lichen seeds by treatment with steam at 80 deg. to 130 deg. C.

Gums
Starting point (French 633121) in extracting—
Gum.

Miscellaneous
Starting point in making—
Hydrosols used with rubber for making dental plates.
Preparations for molding sculptures (French 623547).

Paint and Varnish
Ingredient (Brit. 625087) of—
Compositions for coating cement, mortars, such products containing colloidal substances of rubber base and drying oils and aluminum sulphate, and used for protecting the coated material against penetration of very mobile or volatile liquids, such as gasoline, crude petroleum, fuel oil, other mineral oils, vegetable oils, alcohol, and turpentine.

Ketene

Chemical
Reagent (U. S. 1604472) in making—
Acetylsalicylic acid, cellulose acetate, cellulose formate.

Kola Nuts
Synonyms: Bissy, Cola, Cola nuts, Gooroo, Guru nuts, Kola, Kola seeds.
Latin: Semen coloe.
French: Noix de cola, Noix de gourou, Cafe du soudan.
German: Colanuesse, Gurunuesse, Kolanuesse.
Spanish: Nueces de cola.
Italian: Noces di cola.

Chemical
Starting point in extracting—
Caffeine, theobromine.

Food
As a food.
In making soft drinks.

Pharmaceutical
Suggested for use as stimulant, tonic, nervine, diuretic, masticatory, aphrodisiac, and astringent.

Kordophan Gum
Synonyms: Gum kordophan.
French: Gomme de cordofan.
German: Kordofangummi.

Ceramics
Reagent in decorating potteries and porcelains.

Food
Ingredient of—
Confectionery, pastries.

Glues and Adhesives
Ingredient of various adhesive compositions.

Ink
Ingredient of—
Lithographic inks, printing inks.

Paint and Varnish
Ingredient of—
Bronze color compositions, water color compositions.

Paper
Size in making various grades of paper.

Pharmaceutical
In compounding and dispensing practice.

Textile
——, *Finishing*
Ingredient of sizes for—
Cotton, silk.

——, *Printing*
Ingredient of paste for fabric printing.

Krypton

Analysis
Inert gas for laboratory work.

Electrical
Ingredient of—
Gaseous mixtures used in the so-called "Neon Signs."

Kuromoji Seed Oil
French: Huile des semences de kuromoji.
German: Kuromojisamenoel.

Fuel
As an illuminant.

Soap
As a soapstock.

Laccaic Acid
German: Laccainsaeure.

Textile
——, *Dyeing*
Color in dyeing—
Cotton with mordants.
Wool direct and with tin and aluminum mordants.
Yarns and fabrics with the aid of mordants.

Lactic Acid
Synonyms: Alpha-oxypropionic acid, 2-Propanolic acid.
French: Acide alphaoxypropionique, Acide lactique, Acide 2-propanolique.
German: Alphaoxypropionsäure, Milchsäure, 2-Propanolsäure.
Spanish: Acido de alfaoxipropionico, Acido lactico, Acido de 2-propanolico.
Italian: Acido d'alfaossipropionico, Acido lattico, Acido di 2-propanolico.

Agriculture
Ingredient of—
Poultry foods, stockfeed compositions.

Analysis
Ingredient of—
Copper salt solution used for various purposes in the laboratory.
Reagent for—
Asepticizing wort.
Decalcifying operations.
Detecting and analyzing glucose.
Phenol, pyrogallol, salicylic acid, savin oil.
Reagent in—
Electrolytic determination of cobalt and nickel.
Laboratories in alcohol-distillation plants and yeast factories.
Microscopy.

Brewing.
Reagent for—
Acidulating wort.
Reagent in—
Making low-alcohol content beers.
Malt beverages.
Treating mash, to check bacterial growth.
Water used in making beer.

Ceramics
Solvent and plasticizer in—
Compositions, containing cellulose acetate, used for the protection and decoration of ceramic products (German 146106 and 151918).

Chemical
Dispersing agent (Brit. 343899) in making—
Emulsions and dispersions of various chemicals.
Emulsions of hydrocarbons of various groups of the aliphatic and aromatic series.
Terpene emulsions.
Wetting compositions in emulsified form.
Ingredient of—
Mash and wort (added for the purpose of increasing the yield of alcohol) (German 249331).
Medium for—
Growing yeast cells.

Lactic Acid (Continued)

Reagent for—
- Making betaeucaine lactate.
  - Glycerin lactate (dianol), lactates of various alkaloids, lactol, lactyltropine, methylenesulphonic acid, paralactylphenetidin (lactophenine), quinaldine, quinine lactate, santalyl lactate.
  - Silver lactate (actol) from silver carbonate.
  - Various pharmaceuticals and other organic chemicals.
- Removing *clostridium butyricum* in the manufacture of yeast.
- Treating yeast in alcohol manufacture, to sour it and to prevent the development and growth of undesirable fermentations.

Solvent for—
- Cellulose acetate (German 146106 and 151918).
- Cellulose formate.

Starting point in making—
- Acetaldehyde by heating with dilute sulphuric acid.
- Ammonium lactate.
- Antimony lactate by action on salt of antimony.
- Antimony lactate from antimony oxide and sodium lactate.
- Barium lactate, beryllium lactate.
- Bismuth lactate by action on bismuth hydroxide.
- Calcium lactate, cadmium lactate, cerium lactate, cobalt lactate, ethyl lactate.
- Ferric lactate by action on ferric hydroxide.
- Ferrous lactate by action on iron filings or ferrous carbonate.
- Iron lactate (pharmaceutical) by action on ferric sulphate.
- Magnesium lactate by action on magnesium oxide or magnesium carbonate.
- Manganese lactate.
- Mercuric lactate by action on red oxide of mercury.
- Potassium lactate by action on a solution of potassium carbonate.
- Sodium lactate by action on a solution of sodium carbonate.
- Strontium lactate by action on strontium carbonate.
- Titanium lactate by action on titanic acid.
- Zinc lactate by action on zinc hydroxide.

Disinfectant

Dispersing agent (Brit. 343899) in making—
- Emulsified germicides and deodorizing preparations.

Distilled Liquor

Reagent in making various distilled liquors.

Dye

Dispersing agent (Brit. 343899) in making—
- Color lakes in emulsified form.
- Dye preparations in emulsified form.

Reagent in making—
- Lakes.

Starting point in making—
- Alcohol-soluble indulin, night blue, nigrosin.

Fats and Oils

Dispersing agent (Brit. 343899) in making—
- Boring oils in emulsified form.
- Emulsified drilling oils.
- Greasing compositions in emulsified form.
- Lubricating compositions in emulsified form, containing various vegetable and animal fats and oils.
- Solvents for fats and oils in emulsified form.
- Stabilized emulsions of various animal and vegetable fats and oils.
- Wetting compositions in emulsified form, containing vegetable oils and fats.
- Wire-drawing oils in emulsified form.

Food

Acidulating agent in—
- Essences, extracts, fruitades.
- Lemonades (added to improve the taste and make the product more resistant to spoiling).
- Syrups, various soft drinks.

Ingredient of—
- Food preparations, infants' foods, meat extracts.

Preservative for—
- Fruits.

Reagent in making—
- Bread (U. S. 1170474), cheese, confectionery, pickles, sauerkraut.

Starting point in making—
- Pectin.

Gas

Reagent (German 181063) for—
- Washing manufactured gas.

Glass

Solvent and plasticizer in—
- Compositions, containing cellulose acetate or cellulose formate, used for the decoration and protection of glassware.

Glues and Adhesives

Dispersing agent (Brit. 343899) in making—
- Emulsified adhesive preparations.

Ink

Dispersing agent (Brit. 343899) in making—
- Printing inks, writing inks.

Insecticide

Dispersing agent (Brit. 343899) in making—
- Insecticidal preparations in emulsified form.
- Vermin exterminators in emulsified form.

Leather

Dispersing agent (Brit. 343899) in making—
- Emulsified dressing compositions.
- Emulsified fat-liquoring baths.
- Emulsified soaking compositions.
- Emulsified waterproofing compositions.

Ingredient of—
- Baths used for bating hides.
- Baths used for dehairing hides.
- Baths used for deliming hides in the dehaired state (the lactic acid used being free from sulphuric acid and iron).
- Baths used for plumping hides.
- Baths used in the vegetable tanning of leather.
- Deliming hides (used in place of bran drench).
- Solutions containing sulphite cellulose waste liquor used in the preliminary treatment of hides and skins before tanning (Brit. 255566).

Mordant in—
- Dyeing hides and pelts.

Reagent for—
- Souring bark liquors and extract liquors.

Reagent in—
- Chrome tanning.
- Preliminary treatment of hides (Brit. 253549).

Solvent and plasticizer (German 146106 and 151918) in—
- Compositions, containing cellulose acetate or cellulose formate, used in the manufacture of artificial leather and also for the decoration and protection of leather goods.

Mechanical

Flux in—
- Soft soldering.

Metallurgical

Ingredient of—
- Baths used for the preliminary treatment of iron plate before tinning.
- Baths used for the preliminary treatment of tin, lead, or britannia metal articles before coating them with metals.

Solvent and plasticizer (German 146106 and 151918) in—
- Compositions, containing cellulose acetate or cellulose formate used for the decoration and protection of metallic goods.

Miscellaneous

Dispersing agent (Brit. 343899) in making—
- Cleansing compositions in emulsified form.
- Emulsified furniture polish.
- Emulsified detersive and greasing compositions.
- Metal polishes in emulsified form.
- Scouring compositions in emulsified form.
- Various emulsified compositions for use in wetting, washing and dispersing.
- Waterproofing compositions in emulsified form.

In laundries for counteracting the alkalinity of materials which are subjected to after-treatment with alkalies.

In veterinary medicine.

Mordant in—
- Dyeing hats.

Preservative for various purposes.

Reagent for—
- Treating dextrose, levulose, or other monosaccharides in making flour improvers.

Solvent and plasticizer (German 146106 and 151918) in making—
- Compositions, containing cellulose acetate or cellulose formate, used for the decoration and protection of various compositions of matter.

Paint and Varnish

Dispersing agent (Brit. 343899) in making—
- Emulsified paints and varnishes, pigment emulsions.

Lactic Acid (Continued)
Reagent in making—
Pigments.

Paper
Dispersing agent (Brit. 343899) in making—
Sizing compositions in emulsified form for use in tub and machine processes.
Waterproofing compositions for paper and pulp compositions and paperboard.
Waxing compositions in emulsified form.
Solvent and plasticizer (German 146106 and 151918) in—
Compositions, containing cellulose acetate or cellulose formate, used in the manufacture of coated paper and for the decoration and protection of paper and pulp compositions.

Perfume
Dispersing agent in making—
Creams, lotions, lanolin preparations, latherless shampoos, sunburn preparations.
Various ointments in emulsified form.
Ingredient of—
Corn removers, freckle remover.
Preparations for removing tartar from teeth.
Skin bleachers.
Various cosmetics and toilet preparations.

Petroleum
Dispersing agent (Brit. 343899) in making—
Emulsified cutting oils.
Emulsions of medicinal oils.
Kerosene emulsions, naphtha emulsions.
Solubilized greases in emulsified form.
Reagent (German 181063) in—
Refining petroleum.

Pharmaceutical
Reagent in—
Preparing vaccines.
Substitute for glycerin in various pharmaceutical preparations.
Suggested for use as astringent, caustic, digestive, sedative and antidiabetic; also for treating diarrhoea in infants, internal diseases of adults.

Plastics
Solvent and plasticizer in making—
Plastic compositions containing cellulose acetate or cellulose formate.

Resins and Waxes
Dispersing agent in making—
Emulsions of natural and artificial waxes.
Emulsions of natural and artificial resins.
Starting point in making—
Artificial resins (Brit. 316322).
Artificial resins for use along with cellulose acetate in the manufacture of lacquers and dopes (Brit. 311657).

Rubber
Dispersing agent in making—
Emulsified rubber compositions, such as cements and coatings.
Solvent and plasticizer in—
Compositions, containing cellulose acetate or cellulose formate, used for the decoration and protection of rubber goods.

Soap
Dispersing agent (Brit. 343899) in making—
Emulsions of soaps and alkaline earth soaps.

Stone
Solvent and plasticizer in—
Compositions, containing cellulose acetate or cellulose formate, used for the decoration and protection of artificial and natural stone.

Textile
——, *Dyeing*
Assistant (French 595705) in—
Baths used for dyeing cottons with developed colors in fast brown shades.
Dispersing agent (Brit. 343899) in making—
Dye baths in emulsified wool.
Ingredient of—
Baths for dyeing wool.
Baths for dyeing fabrics containing silk.
Baths for dyeing various textile fabrics with vegetable colors, for example, madder, logwood, yellow-wood, redwood, orchil, cochineal, and the like.
Baths containing mineral colors, such as prussian blue, used for dyeing various fabrics and yarns.
Baths containing synthetic colors, such as anilin black, indocyanins, metachrome yellows, for dyeing various fabrics and yarns.
Baths for dyeing woolen yarns and fabrics with acid colors.
Baths containing alizarin dyestuffs for dyeing various yarns and fabrics.
Mordanting liquors, containing chromium salts, alums, antimony salts, and tin salts, used on wool.
Mordanting liquors used for various purposes in dyeing textiles.
Mordanting liquors containing potassium bichromate (added to assist the chromate in the mordanting process).
Reagent in—
Dyeing fabrics and yarns by the oxidation of anilin black.
Reducing agent in—
Chrome mordanting of wool (used in place of tartaric acid).
Solvent for—
Water-insoluble dyestuffs in making dye liquors.

——, *Dyeing and Printing*
Solubilizing agent (Brit. 276100) in making dye liquors and printing pastes containing the following dyestuffs:—
Acridin dyestuffs.
Aminoanthraquinones, reduced and unreduced.
Anthraquinone dyestuffs, azines, azo dyestuffs, basic diarylmethane dyestuffs, basic triarylmethane dyestuffs, benzoquinone anilides, chrome mordant dyestuffs, indigoids, naphthoquinoneanilides.
Naphthaquinones, reduced and unreduced.
Nitroarylamines, nitrodiarylamines, nitroarylphenols, nitrodiarylphenols, oxazines, pyridin dyestuffs, quinolin dyestuffs.
Quinoneimides, reduced and unreduced.
Sulphur dyestuffs, xanthene dyestuffs.

——, *Finishing*
Dispersing agent (Brit. 343899) in making—
Emulsified coating compositions.
Emulsified scouring compositions.
Emulsified sizing compositions.
Emulsified washing compositions containing soaps.
Ingredient of—
Baths used for sofetening silk and cotton fabrics and giving them the appearance of velvet.
Baths used in finishing fabrics for collars and cuffs.

——, *Manufacturing*
Dispersing agent (Brit. 343899) in making—
Dispersions used for fulling operations.
Dispersions used for the carbonization of wool.
Emulsified mercerizing baths.
Emulsions for kier-boiling cotton.
Emulsions for degumming silks.
Emulsions for soaking silks.
Oiling emulsions for treating fabrics.
Ingredient of—
Baths used for producing scroop on rayon filament.
Viscose rayon precipitating baths (German 274550).
Reagent for—
Accelerating copper solution in making ammoniacol-copper solvent for use in the manufacture of cuprammonium rayon.
Treating cuprammonium rayon to preserve it and increase its strength.
Various rayon filaments to increase their strength (German 197965).
Viscose rayon filaments (added as a lactate) to remove all traces of sulphuric acid from the filaments (the sulphuric acid reacts with the lactate to form lactic acid which does not harm the filament).

——, *Printing*
Ingredient of—
Pastes containing sulphuric acid and tartaric acid for printing wool with acid dyestuffs.
Pastes for printing basic colors, especially indulins, with the aid of tannin and tartar emetic mordants.
Pastes used for printing cotton fabrics with logwood.
Pastes used for printing blacks on double satin-finished fabrics.
Pastes used for printing thick fabrics (the lactic acid being useful in allowing the color to penetrate the interior of the fabric).
Printing pastes used alone without a mordant.
Printing pastes containing chromotrope dyestuffs, basic greens, diamines, and the like.
Printing pastes containing anilin black for use on cotton fabrics.
Mordant in various printing processes.

Lactic Acid (Continued)
Reagent for—
Discharging turkey red in fabric printing.
Making thickener used instead of starch preparations.
Reducing agent in—
Chrome mordanting in wool printing (used in place of tartaric acid).
Solvent for—
Water-insoluble printing colors.
Substitute for—
Glycerin and tartaric acid in printing processes.

Lactic Acid Ester of Grapeseed Alcohol

Bituminous
Solvent (Brit. 445223) for—
Asphalt and other bituminous bodies.
Dye
Solvent (Brit. 445223) for—
Dyestuffs, particularly oil-soluble coaltar dyes.
Fats, Oils, and Waxes
Solvent (Brit. 445223) for—
Fats, oils, waxes.
Resins
Solvent (Brit. 445223) for—
Oil-soluble glycerol-phthalic acid resins.
Polymerized vinyl compounds.
Synthetic resins.
Rubber
Solvent (Brit. 445223) for—
Rubber.

Lactic Acid Ester of Ricinoleic Alcohol

Bituminous
Solvent (Brit. 445223) for—
Asphalt and other bituminous bodies.
Dye
Solvent (Brit. 445223) for—
Dyestuffs, particularly oil-soluble coaltar dyes.
Fats, Oils, and Waxes
Solvent (Brit. 445223) for—
Fats, oils, waxes.
Resins
Solvent (Brit. 445223) for—
Oil-soluble glycerol-phthalic acid resins.
Polymerized vinyl compounds.
Synthetic resins.
Rubber
Solvent (Brit. 445223) for—
Rubber.

Laevo-Adrenaline-Coumarin 3-Carboxylate

Pharmaceutical
Claimed (Brit. 440968) to be—
Less excitant in action than the bases from which it is derived.

Laevo-Linalyl Butyrate
Synonyms: Linalylbutyric ether.
French: Butyrate de 1-linalyle, Éther linalylbutyrique,
German: 1-Linalylbutyrat.
Spanish: Butirato de 1-linalil.
Italian: Butirato di 1-linalile.

Perfume
Imparter of lavender odor to—
Perfumes, lotions, toilet waters, cosmetics.
Soap
Imparter of lavender odor to—
Soaps.

Lampblack
French: Noire de fumée, Noire de lampe.
German: Lampenruss, Lampenschwarz.
Spanish: Hallin de lampara.
Italian: Nero di lampada.

Cement
Black pigment in making—
Dark-colored cement mixtures.
Ceramics
Black pigment in making—
Tile and other ceramic products.
Construction
Black pigment in making—
Mortars, stuccos, concretes.
Electrical
Ingredient of—
Insulating compositions used in the manufacture of electrical machinery and equipment, as well as cables and wiring.
Explosives
Ingredient of—
Liquid air explosive compositions.
Preparations used for making matches.
Fertilizer
Ingredient of—
Fertilizing compositions.
Ink
Black pigment in making—
Chinese inks, India inks, lithographic inks, marking inks, printing inks, stenciling inks.
Jewelry
Black pigment in coloring—
Artificial stones.
Leather
Black pigment in making—
Artificial leather, black leather, patent leather.
Linoleum and Oilcloth
Black pigment in making—
Oilcloth and linoleum.
Mechanical
Ingredient of—
Furnace lutes.
Metallurgical
Black pigment in making—
Compositions for coating mechanical apparatus.
Reagent in treating—
Steel by the cementation process.
Miscellaneous
Ingredient of—
Auto-top dressings.
Compositions for making black buttons.
Compositions for making phonograph records.
Compositions for making typewriter ribbons.
Crayons, shoe polishes, stove polishes.
Paint and Varnish
Black pigment for various purposes.
For shading oil colors.
Pigment in making—
Automobile lacquers, black paints, black varnishes, black enamels, black lacquers, casein paints, glue paints, Japan varnishes, oil paints, paints for scenery.
Paper
Pigment in making—
Black coated paper, bookbinders' board, carbon copying paper, gray coated paper, paperboard products.
Petroleum
Ingredient of—
Lubricating compositions containing mineral oil distillates or mixture of the same with other oils (added in place of graphite to increase the viscosity).
Plastics
Black pigment in making—
Colored cellulose and other plastic compositions.
Printing
In process engraving and the litho trades.
Soap
Pigment in—
Soaps.
Stone
Black pigment in making—
Artificial stone.
Textile
Black pigment in making—
Carriage cloth, tarpaulins, waxed colored cloth.
Woodworking
Black pigment for impregnating—
Furniture, ornamental work, musical instruments, picture frames, tops of desks and the like.

Lauric Acid Ester of Grapeseed Alcohol

Bituminous
Solvent (Brit. 445223) for—
Asphalt and other bituminous bodies.
Dye
Solvent (Brit. 445223) for—
Dyestuffs, particularly oil-soluble coaltar dyes.

Lauric Acid Ester of Grapeseed Alcohol (Continued)
Fats, Oils, and Waxes
Solvent (Brit. 445223) for—
Fats, oils, waxes.
Resins
Solvent (Brit. 445223) for—
Oil-soluble glycerol-phthalic acid resins.
Polymerized vinyl compounds.
Synthetic resins.
Rubber
Solvent (Brit. 445223) for—
Rubber.

Lauric Acid Ester of Ricinoleic Alcohol
Bituminous
Solvent (Brit. 445223) for—
Asphalt and other bituminous bodies.
Dye
Solvent (Brit. 445223) for—
Dyestuffs, particularly oil-soluble coaltar dyes.
Fats, Oils, and Waxes
Solvent (Brit. 445223) for—
Fats, oils, waxes.
Resins
Solvent (Brit. 445223) for—
Oil-soluble glycerol-phthalic acid resins.
Polymerized vinyl compounds.
Synthetic resins.
Rubber
Solvent (Brit. 445223) for—
Rubber.

Lauroyl Peroxide
Synonyms: Dodecanoyl peroxide.
Chemical
Catalyst in—
Polymerization processes.
Cosmetic
Bleaching agent for—
Fats, oils, waxes.
Fats and Oils
Bleaching agent for—
Fats, greases, oils.
Food
Bleaching agent for—
Fats, greases, oils.
Miscellaneous
Bleaching agent in—
Processing various products.
Resins and Waxes
Catalyst in—
Polymerization processes.
Bleaching agent for—
Waxes.

Lauryl Acetate
French: Acétate de dodécyle, Acétate dodécylique, Acétate de lauryle, Acétate laurylique.
German: Duodecylacetat, Duodecylazetat, Essigsäureduodecylester, Essigsäurelaurinester, Essigsäuresduodecyl, Essigsäureslaurin, Laurinacetat, Laurinazetat.
Spanish: Acetato de dodecil, Acetato de lauril.
Italian: Acetato di dodecile, Acetato di laurile.
Perfume
Ingredient of—
Perfumes.
Perfume in—
Cosmetics.
Soap
Perfume in—
Toilet soaps.

Lauryl Adipate
Paint and Varnish
Gelatinizing or softening agent (Brit. 387534) in making—
Varnishes and similar compositions having a base of cellulose esters or ethers, in particular nitrocellulose and cellulose acetate.

Laurylamine
Chemical
Starting point (Brit. 436327) in making—
Laurylurea by reacting with phosgene in toluene to give urea chloride and reacting with aqueous ammonia.
Laurylthiourea by reacting similarly with thiophosgene.
Dilauryldithiocarbamates by reacting with carbon bisulphide and an alkali.
Trilaurylamine by reacting with lauryl bromide.

Laurylaminoethanesulphonic Acid, Normal
Paper
Remover (Brit. 438403) of—
Printing ink, oily impurities, and other matter in process for reclaiming used or waste paper.

Lauryl Chloride
French: Chlorure de lauryle, Chlorure laurylique.
German: Chlorlauryl, Laurylchlorid.
Chemical
Starting point in making various derivatives.
Fats and Oils
Bleaching agent (Brit. 328544) in treating—
Various vegetable and animal oils (used together with hydrogen peroxide).
Food
Bleaching agent (Brit. 328544) used together with hydrogen peroxide in treating—
Egg yolk, flour, meal.
Soap
Bleaching agent (Brit. 328544) in treating—
Raw materials for soapmaking.
Waxes and Resins
Bleaching agent (Brit. 328544) in treating—
Waxes (used together with hydrogen peroxide).

Laurylcresol
Chemical
Starting point (Brit. 444351) in making—
Fat-splitting catalysts and emulsifying agents for use in dyeing, laundering, bleaching, and various other processes, by reacting with formaldehyde and nonaromatic secondary amines (the salts of the products with water-soluble acids, or water-insoluble acids, or the quaternary ammonium salts, are claimed to be valuable for the purposes named).

Lauryl Cyanate
Insecticide and Fungicide
As an insecticide (Brit. 436327).
As a fungicide (Brit. 436327).

Lauryl Cyanide
Insecticide and Fungicide
As an insecticide (Brit. 436327).
As an anticryptogamic (Brit. 436327).

Lauryl Diethyldithiocarbamate
Insecticide and Fungicide
Anticryptogamic (Brit. 436327).
Insecticide (Brit. 436327).

Lauryl Hexahydrophenylenediacetate
Paint and Varnish
Gelatinizing or softening agent (Brit. 387534) in making—
Varnishes and similar compositions having a base of cellulose esters or ethers, in particular nitrocellulose and cellulose acetate.

Lauryl Hydrophthalate
Paint and Varnish
Gelatinizing or softening agent (Brit. 387534) in making—
Varnishes and similar compositions having a base of cellulose esters or ethers, in particular nitrocellulose and cellulose acetate.

Lauryl Isoselenocyanate
Disinfectant
Parasiticide (U. S. 1993040).

Lauryl Isotellurocyanate
Disinfectant
Parasiticide (U. S. 1993040).

Lauryl Isothiocyanate
Disinfectant
Parasiticide (U. S. 1993040).

Lauryl Phthalate
Paint and Varnish
Gelatinizing or softening agent (Brit. 387534) in making—
Varnishing and similar compositions having a base of cellulose esters or ethers, in particular nitrocellulose and cellulose acetate.

Laurylpyridinium Sulphate

Fire-Prevention
Starting point (Brit. 434856) in making—
Fire-extinguishing air foam by admixture with water, especially suitable for alcohol fires.

Fuel
Activator (Brit. 410956) in—
Flotation of coal.

Metallurgical
Activator (Brit. 410956) for—
Flotation reagents in ore separation.

Laurylresorcinol

Chemical
Starting point (Brit. 444351) in making—
Fat-splitting catalysts and emulsifying agents for use in dyeing, laundering, bleaching, and various other purposes, by reacting with formaldehyde and non-aromatic secondary amines (the salts of the products with water-soluble acids or water-insoluble acids, and the quaternary ammonium salts, are claimed to be valuable for the purposes named).

Lauryl Rhodanate Sodium Salt

Insecticide
Insecticide of high toxicity for use in sprays.

Lauryl Sebacate

Paint and Varnish
Gelatinizing or softening agent (Brit. 387534) in making—
Varnishes and similar compositions having a base of cellulose esters or ethers, in particular nitrocellulose and cellulose acetate.

Lauryl Selenocyanate

Disinfectant
Parasiticide (U. S. 1993040).

Lauryl Succinate

Paint and Varnish
Gelatinizing or softening agent (Brit. 387534) in making—
Varnishes and similar compositions having a base of cellulose esters or ethers, in particular nitrocellulose and cellulose acetate.

Lauryl-1-sulphuric Acid (Normal) Ester

Chemical
As an emulsifying agent.
Reagent in—
Organic synthesis.
Starting point (Brit. 440575) in making—
Emulsifying agents with salts of lead, aluminum, iron, tin, or barium (such emulsifying agents are said to form water-in-oil emulsions and are, preferably, produced in situ by (1) dissolving the sulphuric acid ester in the oil and (2) agitating with an aqueous solution of the metal salt, for example, lead acetate; they are said to be useful for treating medicinal paraffin oil, neatsfoot oil, olive oil, castor oil, cottonseed oil, linseed oil, and petroleum lubricating oils; a heavy paraffin oil, so treated on the basis of 50 parts by weight of oil to 48.75 parts of water, is said to yield a heavy grease that has good lubricating properties and may readily be extended with oil; a water-linseed oil type emulsion is offered as suitable for use as a paint base).

Lauryl Tellurocyanate

Disinfectant
Parasiticide (U. S. 1993040).

Lauryl Thiocyanate

Disinfectant
Parasiticide (U. S. 1993040).

Oils, Fats, and Waxes
Starting point (Brit. 440175) in making—
Addition agents for high-pressure lubricating oils or greases, by mixing and reacting with organo-metallic compounds.

Laurylthiourea

Fungicide and Insecticide
As a fungicide (Brit. 436327).
As an insecticide (Brit. 436327).

Laurylurea

Fungicide and Insecticide
As a fungicide (Brit. 436327).
As an insecticide (Brit. 436327).

Lauryl Xanthate

Fungicide and Insecticide
As a fungicide (Brit. 436327).
As an insecticide (Brit. 436327).

Lead Acetate

French: Acétate de plomb, Acétate plombique.
German: Bleiacetat, Bleiazetat, Essigsäuresblei, Essigsäuresbleioxyd.
Spanish: Acetato de plombo.
Italian: Acetato di piombo.

Analysis
Clarifying agent in carrying out optical determinations.
Reagent in preparing—
Lead test paper for sulphuretted hydrogen.
Reagent in determining or testing—
Albumin and protein matters, chromium trioxide and chromium salts, cottonseed oil, dextrin and other degraded starch products, glucose, gallic acid and gallates, malic acid and malates, molybdic acid and molybdates, liquid petrolatum (test for sulphur compounds), oxalic acid and oxalates, picric acid and picrates, picrolexin, tannic acid, saccharose, wool and silk fibers, sensitive reagent for sulphuretted hydrogen.

Cement
Ingredient of—
Ferrite cements.

Chemical
Ingredient of catalytic mixtures used in the manufacture of—
Acenaphthylene, acenaphthaquinone, bisacenaphthylidenedione, naphthaldehydic acid, naphthalic anhydride, and hemimellitic acid from acenaphthene (Brit. 295270).
Acetaldehyde from ethyl alcohol (Brit. 281307).
Acetic acid from ethyl alcohol (Brit. 281307).
Alcohols from aliphatic hydrocarbons (Brit. 281307).
Aldehydes and acids by the reduction of the corresponding esters (Brit. 306471).
Aldehydes and acids from toluene, orthochlorotoluene, orthobromotoluene, orthonitrotoluene, parachlorotoluene, parabromotoluene, paranitrotoluene, metachlorotoluene, metabromotoluene, metanitrotoluene, dinitrotoluenes, dibromotoluenes, dichlorotoluenes, chloronitrotoluenes, chlorobromotoluenes, nitrobromotoluenes (Brit. 295270).
Aldehydes and acids from xylenes, pseudocumenes, mesitylene, and paracymene (Brit. 281307).
Alphanaphthaquinone from naphthalene (Brit. 281307).
Anthraquinone from naphthalene (Brit. 295270).
Benzaldehyde and benzoic acid from toluene (Brit. 281307).
Benzoquinone from phenanthraquinone (Brit. 281307).
Benzyl alcohol from benzaldehyde by reduction (Brit. 306471).
Benzyl alcohol or benzaldehyde or benzyl phthalide by the reduction of phthalic anhydride (Brit. 306471).
Butyl alcohol by the reduction of crotonaldehyde (Brit. 306471).
Chloroacetic acid from ethylenechlorohydrin (Brit. 295270).
Diphenic acid from ethyl alcohol (Brit. 281307).
Ethyl alcohol by the reduction of acetaldehyde (Brit. 306471).
Fluorenone from fluorene (Brit. 295270).
Formaldehyde by the reduction of methane or methanol (Brit. 306471).
Formaldehyde by the reduction of carbon dioxide or carbon monoxide (Brit. 306471).
Hydroxyl compounds by the reduction of anthraquinone, benzoquinone, and similar compounds (Brit. 306471).
Isopropyl alcohol by the reduction of acetone (Brit. 306471).
Maleic acid and fumaric acid by the oxidation of toluene, benzene, phenols, tar phenols, or furfural, or from benzoquinone or phthalic anhydride (Brit. 295270).
Methane by the reduction of carbon dioxide or carbon monoxide (Brit. 306471).
Methanol by the reduction of carbon dioxide or carbon monoxide (Brit. 306471).
Naphthaldehydic acid, acenaphthaquinone, or bisacenaphthylidenedione from acenaphthylene (Brit. 281307).
Phenanthraquinone from phenanthrene or diphenic acid (Brit. 295270).

Lead Acetate (Continued)
Phthalic acid and maleic acid from naphthalene (Brit. 295270).
Primary alcohols by the reduction of the corresponding aldehydes (Brit. 306471).
Propionic acid and butyric acid and higher alcohols, ketones, and acids by the reduction of carbon dioxide or carbon monoxide (Brit. 306471).
Reduction products of ketones, aldehydes, acids, esters, alcohols, ethers, and other organic compounds which contain oxygen (Brit. 306471).
Salicylic acid and salicylaldehyde from cresol (Brit. 295270).
Secondary butyl alcohol by the reduction of methylethylketone (Brit. 306471).
Valeryl alcohol by the reduction of valeraldehyde (Brit. 306471).
Vanillin and vanillic acid by the oxidation of eugenol or isoeugenol (Brit. 295270).
Ingredient (Brit. 306460) of catalytic preparations used in the production of various aromatic and aliphatic compounds, including—
Alphanaphthylamine from alphanitronaphthalene.
Amines from aliphatic nitro compounds, such as allyl nitriles or nitromethane.
Amino compounds from the corresponding nitroanisoles.
Amylamine from pyridin.
Anilin, azo-oxybenzene, azobenzene, and hydrazobenzene from benzene by reduction.
Aminophenols from nitrophenols.
3-Aminopyridin from 3-nitropyridin.
Cyclohexamine, dicyclohexamine, and cyclohexylanilin from nitrobenzene.
Piperidin from pyridin.
Pyrrolidin from pyrrol.
Tetrahydroquinolin from quinolin.
Reagent in making—
Acetone, alum mordants, aluminum sulphacetate, aluminum-potassium sulphocyanate, catechin, diastase, ethyl isothiocyanate, lead acetate paper, malic acid and malates, oleic acid and oleates, peristalin.
Starting point in making—
Aluminum acetate, basic lead acetate, copper acetate, ferrous acetate, lead salts of acids and halogens, lead soaps, tin acetate, various other metallic acetates.

Dye
Ingredient of—
Lakes.

Ink
Ingredient of—
Printing inks.

Insecticide
Ingredient of—
Compositions containing arsenicals.
Starting point in making—
Lead arsenate insecticides.

Leather
Reagent in—
Dyeing, tanning, tawing.

Metallurgical
Ingredient of—
Compositions used to produce a light steel-blue coloration on copper, a grayish-blue coloration on iron, an iridescent coloration on nickel, and a violet-bluish coloration on silver.

Miscellaneous
Clarifying and decolorizing agent for various purposes.
Ingredient of—
Compositions used to produce various colors on stone and similar substances.

Paint and Varnish
As a drier.
Ingredient of—
Paints and varnishes used on keels and bottoms of ships.
Starting point in making—
Chrome orange, chrome yellow, chrome red, lead driers, white lead.

Paper
Reagent in making—
Pulp and paper.

Perfumery
Ingredient of—
Cosmetics, hair dyes.

Pharmaceutical
In compounding and dispensing practice.

Sugar
Clarifying agent in—
Refining molasses and sugar.

Textile
——, *Dyeing*
Mordant on—
Cotton fabrics and other yarns and fabrics.
Resist in—
Indigo dyeing.
Mordant on—
Calicoes and other fabrics.

——, *Finishing*
Ingredient of—
Waterproofing compositions.

——, *Manufacturing*
In weighting silks and rayons.

——, *Printing*
Ingredient of—
Pastes used for the production of colored designs on indigo-dyes fabrics.
Paste containing anilin black.

Wine
As a clarifying agent.

Lead Albuminate
French: Albuminate de plomb, Albuminate plombique.
German: Albuminsaeuresblei, Bleialbuminat.

Rubber
Reagent in—
Reclaiming rubber from old tires and other manufactures (U. S. 1640817).

Lead Alizarate
French: Alizarate de plomb, Alizarate plombique.
German: Bleializarat.

Textile
——, *Dyeing*
Pigment in dyeing various yarns and fabrics.

——, *Printing*
Pipment in printing various fabrics.

Lead-Ammonium Chloride
French: Chlorure de plomb et ammonium.
German: Bleiammoniumchlorid.

Miscellaneous
Carotting agent in treating—
Furs and felt (Brit. 271026).

Lead Anacardate
French: Anacardate de plomb, Anacardate plombique.
German: Anacardsäuresblei, Anacardsäuresbleioxyd, Bleianacardat.

Insecticide
As a vermifuge.

Lead Antimoniate
Synonyms: Antimony yellow, Naples yellow.
French: Plomb antimonique.
German: Bleiantimon, Spiessglanzblei.

Ceramics
Ingredient of—
Compositions used for decorating porcelain and other ceramic wares.

Glass
Staining agent.

Paint and Varnish
As a pigment.

Lead Carbonate
Synonyms: White lead.
French: Carbonate de plomb, Carbonate plombique.
German: Bleicarbonat, Kohlenstoffsaeuresblei, Kohlensaeuresblei.

Ceramics
Ingredient of—
Glazes for potteries and porcelains.

Chemical
Catalyst (Brit. 291419) in purifying—
Anthracene, coaltar ammonia.
Stabilizer (Brit. 291419) in catalytic mixtures used in making—
Acenaphthylene, acenaphthaquinone, bisacenaphthylidenedione, naphthaldehydic acid, naphthalic anhydride, and hemimellitic acid from acenaphthene.
Acetaldehyde from ethyl alcohol (Brit. 281307).
Acetic acid from ethyl alcohol.
Alcohols from aliphatic hydrocarbons (Brit. 281307).

Lead Carbonate (Continued)

Aldehydes and corresponding acids from toluene, orthonitrotoluene, orthobromotoluene, orthochlorotoluene, chlorobromotoluene, chloronitrotoluene, bromonitrotoluene, dichlortoluene, dinitrotoluenes, dibromotoluenes, metachlortoluene, metachlorotoluene, metabromotoluene, paranitrotoluene, parabromotoluene, parachlorotoluene.
Aldehydes and corresponding acids from xylenes, pseudocumenes, mesitylene, paracymene, and other derivatives.
Alphanaphthaquinone from anthracene (Brit. 281307).
Anthraquinone from anthracene.
Benzaldehyde and benzoic acid from toluene (Brit. 281307).
Benzoquinone from phenanthraquinone (Brit. 281307).
Chloroacetic acid from ethylenechlorohydrin.
Diphenic acid from phenanthrene.
Fluorenone from fluorene.
Formaldehyde from methanol or methane.
Maleic acid from naphthalene.
Maleic acid and furmaric acid from benzol, toluol, phenols, tar phenols, or furfural, or from benzoquinone or phthalic anhydride.
Naphthaldehydic acid, acenaphthaquinone, or bisacenaphthylidenedione from acenaphthylene (Brit. 281307).
Naphthaquinone from naphthalene.
Phenanthraquinone from phenanthrene or diphenic acid.
Phthalic anhydride from naphthalene.
Phthalic acid and maleic acid from naphthalene (Brit. 295270).
Salicylic aldehyde and salicylic acid from cresols.
Vanillin and vanillic acid from eugenol or isoeugenol.
Reagent in making—
Trichloromethylsulphonic acid.
Starting point in making—
Litharge.

Miscellaneous
Ingredient (U. S. 1606662) of—
Transfer compositions.

Paint and Varnish
Component of—
Putties.
Pigment in—
Paints, varnishes.

Pharmaceutical
In compounding and dispensing practice.

Lead Chloride

Synonyms: Horn lead.
French: Chlorure de plomb.
German: Bleichlorid.

Chemical
Reagent in purifying—
Alcohols (U. S. 1601404).
Starting point in making—
Lead aceate, lead arsenate, lead borate, lead chromate, lead hydroxide, lead iodide, lead linoleate, lead molybdate, lead betanaphthalenesulphonate, lead oleate, lead resinate, lead silicate, lead stearate, lead sulphate, lead tetrachloride, lead tetraethyl, lead thiosulphate, lead tungstate, zinc chloride, anhydrous (U. S. 1590229).

Paint and Varnish
Ingredient of—
Cassel yellow, Paris yellow, Turner's yellow.
Starting point in making—
Chrome orange, chrome red, chrome yellow.

Lead Diamyldithiocarbamate

Rubber
Accelerator (Brit. 439215) for—
Vulcanization.

Lead Dibenzyldithiocarbamate

Rubber
Accelerator (Brit. 439215) in—
Vulcanization.

Lead Dibutyldithiocarbamate

Rubber
Accelerator (Brit. 439215) for—
Vulcanization.

Lead Diethylthiocyanate

Synonyms: Lead diethylsulphocyanate, Lead diethylsulphocyanide.
French: Diéthylesulphocyanate de plomb, Diéthylethiocyanate de plomb.
German: Bleidiaethylsulfocyanid, Bleidiaethylthiocyanat, Diaethylsulfocyansaeuresblei.

Chemical
Starting point in making—
Ethyl phosphate.

Lead 3: 5-Dinitrobenzoate

Explosives
Ingredient (U. S. 1887919) of—
Priming mixtures.

Lead Dipentamethylenethiuramdisulphide

Rubber
Secondary activator in—
Vulcanizing processes (for use with mercaptabenzthiazole).

Lead Dipentamethylenethiurammonosulphide

Rubber
Secondary activator in—
Vulcanizing processes (for use with mercaptabenzthiazole).

Lead Dipentamethylenethiuramtetrasulphide

Rubber
Secondary activator in—
Vulcanizing processes (for use with mercaptabenzthiazole).

Lead Erucate

French: Erucate de plomb, Erucate plombique.
German: Bleierucat, Erucinsäuresblei, Erucinsäuresbleioxyd.

Building
Reagent in waterproofing—
Concrete, stucco.

Leather
Reagent in waterproofing—
Leather and leather goods.

Mechanical
As a lubricant.

Metallurgical
Ingredient of—
Metal-coating compositions.

Miscellaneous
Ingredient of—
Compositions used in the manufacture of insulating tape.
Compositions used in the dressing and finishing of leather.
Insulating compositions.
Reagent in treating—
Fishing gear, nets, and tackle to prevent marine growths and to protect them against mildew.

Oils and Fats
Ingredient of—
Lubricating compositions.
Substitute for fats in making—
Cup greases, cylinder oils, steam turbine oils.

Paint and Varnish
Drier in making—
Enamels, lacquers, paints, varnishes.
Starting point in making—
Paint and varnish driers.

Paper
Ingredient of—
Compositions used in the waterproofing of paper, pulp, and various products made from them.

Pharmaceutical
In compounding and dispensing practice.

Lead Laurate

French: Laurate de plomb, Laurate plombique.
German: Bleilaurat, Laurinsäuresblei, Laurinsäuresbleioxyd.

Building
Ingredient of—
Compositions used in the waterproofing and dampproofing of concrete, stucco, plaster, and other porous surfaces in walls, cellars, and other parts of buildings.

Lead Laurate (Continued)

Fats and Oils
Ingredient of—
Preparations used for various purposes.
Special lubricating compositions.
Substitute for fats in making—
Cup greases, cylinder oils, steam turbine oils.

Leather
Ingredient of—
Compositions used for the waterproofing and softening of leather and leather goods.

Miscellaneous
Ingredient of—
Compositions used in the manufacture of adhesive tape.
Compositions used for insulating purposes.
Compositions used in various waterproofing processes.
Compositions containing black and colored leads, used in the manufacture of pencils, crayons, and the like.
Compositions for preventing mildew.
Compositions for treating fishing gear and tackle to prevent sea growths thereon.
Compositions containing starch and boric acid, used for medical purposes.
Compositions used in cutting metals.

Paint and Varnish
Drier in making—
Lacquers, paints, varnishes, special roofing preparations.

Paper
Ingredient of—
Compositions used in the waterproofing of paper, pulp, cardboard, and paper board and products made therefrom.

Petroleum
Ingredient of—
Lubricating compositions containing petroleum oils or distillates, mineral oil greases, solid lubricants.

Pharmaceutical
In compounding and dispensing practice.

Linoleum and Oilcloth
Drier in making—
Coating compositions.

Textile
Ingredient of—
Softening compositions, waterproofing compositions.

Lead 2-Mononitroresorcinate

Explosives and Matches
Ingredient (Brit. 428872) of—
Flash composition for use in electric igniters for blasting fuses and the like (used in admixture with finely divided zirconium in a solution of nitrocellulose).

Lead Oleate
French: Oléate de plomb, Oléate plombique.
German: Bleioleat, Oleinsäuresblei, Oleinsäuresbleioxyd.

Building
Reagent in waterproofing—
Concrete, stucco.

Fats and Oils
Ingredient of—
Lubricating compositions, metal-coating compositions.
Substitute for fats in making—
Cup grease, cylinder oils, steam turbine oils.

Leather
Ingredient of—
Dressing compositions.
Reagent in wraterproofing—
Leather and leather goods.

Mechanical
As a lubricant.

Miscellaneous
Ingredient of—
Compositions used in the manufacture of insulating tape.
Insulating compositions.
Reagent in treating—
Fishing gear, nets, and tackle to prevent marine growths and to protect them against mildew.

Paint and Varnish
Drier in making—
Enamels, lacquers, paints, varnishes.
Starting point in making—
Paint and varnish driers.

Paper
Ingredient of—
Compositions used in the waterproofing of paper, pulp, and various products made from these.

Perfume
Ingredient of—
Cosmetics.

Pharmaceutical
In compounding and dispensing practice.

Rubber
Accelerator in vulcanization.

Lead Palmitate
French: Palmitate de plomb, Palmitate plombique.
German: Bleipalmitat, Palmitinsäuresblei, Palmitinsäuresbleioxyd.

Building
Ingredient of—
Compositions used in the waterproofing and dampproofing of concrete, stucco, plaster and other porous surfaces.

Fats and Oils
Ingredient of—
Fat and oil preparations used for various purposes.

Leather
Ingredient of—
Waterproofing and softening compositions.

Linoleum and Oilcloth
Drier in making—
Coating compositions.

Miscellaneous
Ingredient of—
Adhesive tape coatings.
Black and colored compositions used in the manufacture of pencils and crayons.
Compositions for treating fishing nets and lines to prevent marine growths on them.
Insulating compositions, mildew preventives, waterproofing compositions.

Mechanical
Ingredient of—
Metal-cutting compositions, special lubricating compositions.

Paint and Varnish
Drier in making—
Lacquers, paints, varnishes.
Special roofing preparations, such as asbestos-creosote tar cements.

Paper
Ingredient of—
Compositions used in the waterproofing of paper, pulp, paperboard, cardboard, and their products.

Perfume
Ingredient of—
Pomades.

Petroleum
Ingredient of—
Lubricating compositions, mineral oil greases, solid lubricants.
Substitute for fats in making—
Cup greases, cylinder oils, steam turbine oils.

Pharmaceutical
In compounding and dispensing practice.

Rubber
Ingredient of—
Dusting powders.
Substitute for gum rubber.

Textile
Ingredient of—
Softening compositions, waterproofing compositions.

Lead Pentamethylenedithiocarbamate

Rubber
Secondary activator in—
Vulcanizing processes (for use with mercaptabenzthiazole).

Lead Resinate
Synonyms: Lead soap, Resinate of lead.
Freneh: Résinate de plomb.
German: Bleiresinat, Harzsaeuresblei.

Paint and Varnish
Drier in making—
Enamels, lacquers, paints, varnishes.

Lead Resinate (Continued)

Textile

——, *Finishing*

Ingredient of—

Waterproofing compositions for textile yarns and fabrics.

Lead Silicate

French: Silicate de plomb.

German: Bleisilikat, Kieselsaeuresblei.

Ceramics

Ingredient of—

Enamels, glazes.

Glass

Raw material in glassmaking.

Paint and Varnish

White pigment in combination with lead sulphate.

Textile

——, *Finishing*

Ingredient of—

Fireproofing compositions, waterproofing compositions.

Lead Stearate

French: Stéarate de plomb, Stéarate plombique.

German: Bleisterat, Stearinsäuresblei, Stearinsäuresbleioxyd.

Building

Ingredient of—

Compositions used in the waterproofing and dampproofing of concrete, stucco, plaster, and other porous surfaces.

Fats and Oils

Ingredient of—

Fat and oil preparations used for various purposes.

Leather

Ingredient of—

Compositions used for the waterproofing and softening.

Linoleum and Oilcloth

Drier in—

Coating compositions.

Miscellaneous

Ingredient of—

Adhesive tape coatings.

Black and colored compositions used in the manufacture of pencils and crayons.

Compositions for treating fishing nets and lines to prevent sea growths thereon.

Insulating compositions, mildew preventives, waterproofing compositions.

Mechanical

Ingredient of—

Metal-cutting compositions, special lubricating compositions.

Paint and Varnish

Drier in making—

Lacquers, paints, varnishes.

Special roofing preparations, such as asbestos-creosote tar cements.

Paper

Ingredient of—

Compositions used in the waterproofing of paper, pulp, cardboard, paperboard, and their products.

Perfume

Ingredient of—

Pomades.

Petroleum

Ingredient of—

Lubricating compositions, mineral oil greases, solid lubricants.

Substitute for fats in making—

Cup greases, cylinder oils, steam turbine oils.

Pharmaceutical

In compounding and dispensing practice.

Rubber

Ingredient of—

Dusting powders.

Substitute for gum rubber.

Textile

Ingredient of—

Softening compositions, waterproofing compositions.

Lead Sulphate

Synonyms: Sublimed white lead.

French: Sulfate de plomb neutre.

German: Bleisulfat, Metallweiss, Muehlhausenerweiss, Normales bleisulfat, Schwefelsaeuresblei.

Ceramics

Ingredient of—

Glazes used on chinaware and porcelains.

Chemical

Catalyst (Brit. 291419) in purifying—

Anthracene, calicoes.

Stabilizer (Brit. 291419) in catalytic mixtures used in making—

Acenaphthylene, acenaphthaquinone, bisacenaphthylidenedione, naphthaldehydic acid, naphthalic anhydride, and hemimellitic acid from acenaphthene.

Acetaldehyde from ethyl alcohol (Brit. 281307).

Acetic acid from ethyl alcohol.

Alcohols from aliphatic hydrocarbons (Brit. 281307).

Aldehydes and corresponding acids from toluol, orthonitrotoluene, paranitrotoluene, metanitrotoluene, orthochlorotoluene, parachlorotoluene, metachlorotoluene, parabromotoluene, orthobromotoluene, metabromotoluene, dinitrotoluenes, dichlorotoluenes, dibromotoluenes, bromonitrotoluenes, chloronitrotoluenes, chlorobromotoluenes.

Aldehydes and corresponding acids from xylenes, pseudocumenes, mesitylene, paracymene, and other intermediates.

Alphanaphthaquinone from anthracene (Brit. 281307).

Anthraquinone from anthracene.

Benzaldehyde and benzoic acid from toluene (Brit. 281307).

Benzoquinone from phenanthraquinone (Brit. 281307).

Chloroacetic acid from ethylenechlorohydrin.

Diphenic acid from phenanthrene.

Fluorenone from fluorene.

Formaldehyde from methanol or methane.

Maleic acid from naphthalene.

Maleic acid and fumaric acid from benzol, toluol, phenols, tar phenols, or furfural, or from benzoquinone or phthalic anhydride.

Naphthaldehydic acid, acenaphthaquinone, or bisacenaphthylidenedione from acenaphthylene (Brit. 281307).

Naphthaquinone from naphthalene.

Phenanthraquinone from phenanthrene or diphenic acid.

Phthalic anhydride from naphthalene.

Phthalic acid and maleic acid from naphthalene (Brit. 295270).

Salicylic aldehyde and salicylic acid from cresols.

Vanillin and vanillic acid from eugenol or isoeugenol.

Starting point in making—

Lead salts, metallic lead.

Dye

Substratum in—

Lakes.

Electrical

Starting point in making—

Active material for positive electrodes of storage batteries.

Ink

Substratum in—

Lithographic inks.

Paint and Varnish

Pigment in—

Paints, rapid drying oil varnishes.

Rubber

Ingredient of—

Batches.

Textile

——, *Dyeing*

As an assist.

——, *Printing*

Assist on.

Lead Titanate

French: Plomb de titane, Plomb titanique, Titanate de plomb.

German: Titansäuresblei.

Paint and Varnish

Pigment in—

Paints, varnishes, lacquers; it is of pale-yellow color, and is said to have a very high hiding power, to be inert toward all paint mediums and resistant to chalking, to absorb ultraviolet light to the extent of practically 100 percent, placing it in the same category as carbon black and giving it a protective effect on other tints; it is claimed that a film of lead

Lead Titanate (Continued)
titanate in oil exposed to outside weather conditions showed remarkable superiority over the usual lead-zinc oxide extender paint, and after three years at 45° south was still sound; it is further claimed that: "many uses suggest themselves for a pigment with the characteristics of lead titanate; it will undoubtedly find a place in the formulation of exterior house paint tints for the purpose of increasing durability and giving better control of type of failure and tint retention; its properties and behavior in linseed oil also suggest possible advantages as a pigment constituent for exterior primers on wood; it is also apparent that it will be a useful material for finishing coats on steel structures such as bridges, gasholders, and other industrial units where long life and protection from corrosion are essential factors; lead titanate has also rust-inhibitive properties to a marked extent when applied as the first coat to iron and steel; comparisons with red lead, for example, have been made over a four-year period, and for this period lead titanate appears to be equal to red lead as a rust-inhibitor; this period of test is not sufficiently long nor are the tests sufficiently numerous for making an arbitrary statement with respect to this property, but it is at least indicative of some rust-inhibitive value; in enamels and other gloss finishes lead titanate contributes not only to long life, but to gloss and color retention."

Lead-Titanium Tungstate-Resinate
Miscellaneous
As an emulsifying agent (Brit. 395406).
For uses, see under general heading: "Emulsifying agents."

Lead Triethyliodide
French: Triéthyleiodure de plomb.
German: Bleitriaethyljodid, Triaethyljodblei.
Printing
Ingredient (as stabilizer) (Brit. 270386) of preparations for—
Color record intaglio, halftone printing plates, line engraving on zinc and copper, monochrome intaglio, relief printing plates, screenless grained litho plates.

Lead Tungate
French: Tungate de plomb.
German: Bleitungat.
Paint and Varnish
Drier (Brit. 270387) in making—
Enamels, lacquers, paints, stains, varnishes.
Photographic
Ingredient (Brit. 270387) in making—
Light-sensitive varnishes.

Lemon Oil
Latin: Oleum citri, Oleum limonis.
French: Essence de citron, Huile de citron, Huile volatile de citron.
German: Citronenöl, Zitronenoel.
Spanish: Essencia de limon.
Italian: Olio di limone.
Chemical
Ingredient of—
Artificial raspberry essence.
Starting point in making—
Citral, limonene, phellandrene.
Food
Flavoring in—
Bakery products, beverages, candies.
Perfumery
Ingredient of—
Cosmetics, mouthwashes, perfumes, skin-bleaching powder (U. S. 1620269).
Soap
Perfume for—
Toilet soap.

Leptospermum Citratum Oil
Latin: Oleum leptospermum citratum.
Chemical
Source of—
Citral, citronellol, geraniol.

Leuco-5-aminoindole-2:1′-anthraceneindigo Dihydrogendisulphate
Dye
Starting point (U. S. 2000133) in making—
Water-soluble, azo dyes, which are said to form insoluble dyes of deeper shade and good washing fastness when oxidized on the fiber.

Leuco-5-amino-4′-methoxy-4:7-dimethylindole-2:2′ naphthaleneindigo Dihydrogendisulphate
Dye
Starting point (U. S. 2000133) in making—
Water-soluble azo dyes which form insoluble dyes of deeper shade and good washing fastness when oxidized on the fiber.

Leuco-5-amino-4′-methoxyindole-2:2′-naphthaleneindigo Dihydrogendisulphate
Dye
Starting point (U. S. 2000133) in making—
Water-soluble azo dyes which form insoluble dyes of deeper shade and good washing fastness when oxidized on the fiber.

Leuco-4′-chlor-5-aminoindole-2′-naphthaleneindigo Dihydrogendisulphate
Dye
Starting point (U. S. 2000133) in making—
Water-soluble, azo dyes, which are said to form insoluble dyes of deeper shade and good washing fastness when oxidized on the fiber.

Leuco-4′-chlor-5-aminoindole-2:1′-thionaphthenindigo Dihydrogendisulphate
Dye
Starting point (U. S. 2000133) in making—
Water-soluble, azo dyes, which are said to form insoluble dyes of deeper shade and good washing fastness when oxidized on the fiber.

Leuco-5-chlor-4′-amino-7-methoxy-4-methylindole-2:2′-thionaphthenindigo Dihydrogendisulphate
Dye
Starting point (U. S. 2000133) in making—
Water-soluble azo dyes which form insoluble dyes of deeper shade and good washing fastness when oxidized on the fiber.

Leuco-9-chloro-5-amino-1:2-naphthindole-2:1′-(3:4-benzo) thionaphthenindigo Dihydrogendisulphate
Dye
Starting point (U. S. 2000133) in making—
Water-soluble azo dyes which form insoluble dyes of deeper shade and good washing fastness when oxidized on the fiber.

Leuco-6:4′-dichlor-5-amino-7-methylindole-2:2′-naphthaleneindigo Dihydrogendisulphate
Dye
Starting point (U. S. 2000133) in making—
Water-soluble azo dyes which form insoluble dyes of deeper shade and good washing fastness when oxidized on the fiber.

Leuco Dimethylphenylene Green
French: Vert de leuco diméthylephénylène.
German: Leukodimethylphenylengruen.
Dye
Starting point (Brit. 282111) in making dyestuffs for animal fibers, pelts, and acetate rayon with the aid of—
Alphanaphthol, alphanaphthylamine, betanaphthol, betanaphthylamine, 1:5-dioxynaphthalene, 2:7-dioxynaphthalene.

Leuco Quinonephenolimide
German: Leukochinonphenolimid.
Dye
Starting point (Brit. 282111) in making dyestuffs for animal fibers, pelts, and acetate rayon with the aid of—
Alphanaphthol, alphanaphthylamine, betanaphthol, betanaphthylamine, 1:5-dioxynaphthalene, 2:7-dioxynaphthalene.

Leucotetrabromoindigo
German: Leukotetrabromindig.
Dye
Starting point (Brit. 267952) in making ester derivatives (dyestuffs) with—
Dimethylanilin, pyridin.

Leuco-1:4:5:8-tetrahydroxyanthraquinone

Chemical
Starting point (Brit. 396976) in making—
Triaminohydroxyanthraquinones.

Leucothioindigo
German: Leukothioindig.

Dye
Starting point (Brit. 267952) in making ester derivatives (dyestuffs) with—
Dimethylanilin, pyridin.

Leuco Toluylene Blue
French: Leuco bleu de tolylène.
German: Leucotoluylenblau.

Dye
Starting point (Brit. 282111) in making dyestuffs for animal fibers, pelts, acetate rayon with—
Alphanaphthol, alphanaphthylamine, betanaphthol, betanaphthylamine, 1:5-dioxynaphthalene, 2:7-dioxynaphthalene.

Levant Wormseed Oil
Synonyms: Oil of santonica.
French: Essence de semencontra, Essence de semencine.
German: Wurmsamenoel, Zitwersamenoel.

Chemical
Starting point in making—
Santonin.

Pharmaceutical
In compounding and dispensing practice.

Levulic Acid
Synonyms: Beta-acetylpropionic acid, Laevulinic acid, Levulinic acid, Pentanone-4-oic-1 acid.
French: Acide béta-acétylepropionique, Acide lévulinique, Acide lévulique, Acide pentanone-4-oique-1.
German: Betylpropionsäure, Laevulinsäure, Levulinsäure, 4-Pentanon-1-säure.
Spanish: Acido beta-acetilpropionico.
Italian: Acido beta-acetilepropionico.

Chemical
Starting point in making—
Antithermin, intermediates, organic chemicals, pharmaceutical chemicals, synthetic aromatic chemicals.

Textile
——, *Dyeing*
Mordant in—
Dyeing yarns and fabrics (used in place of acetic acid).

——, *Printing*
Mordant in—
Printing fabrics (used in place of acetic acid).
Solvent for—
Indulins and nigrosin in printing cottons.

Lignic Acid

Petroleum
Viscosity decreaser (U. S. 1999766) of—
Fluid clay mud encountered in oil well drilling (used in conjunction with a small amount of caustic alkali).

Lignite Pitch
French: Brai d'houille, brune; Brai de lignite.
German: Braunkohlepech.

Chemical
Starting point in making—
Lignite pitch coke.

Electrical
Ingredient of—
Insulating compositions used for various electrical purposes.

Fuel
As a fuel.
Binder in—
Fuel briquettes
Ingredient of—
Artificial fuels.

Gas
Raw material in making gas for illumination and industrial use.

Miscellaneous
Ingredient of—
Fillers used for paving purposes.
Paving compositions.
Various asphaltic compositions.
Various compositions used for coating pipes and the like.
Waterproofing, weatherproofing, and wearproofing compositions used in treating various materials (Brit. 335247).

Paint and Varnish
Ingredient of—
Roof cements, roofing papers, roofing preparations.
Special varnishes and paints.
Waterproofing compositions for treating concrete, building stone, and the like.

Paper
Ingredient of—
Compositions used for treating paper and pulp products in order to render them waterproof, weatherproof, and wearproof (Brit. 335247).
Compositions used in making heavy papers.

Textile
Ingredient (Brit. 335247) of—
Compositions used in treating cotton, woolen, and other textiles to render the products waterproof, weatherproof, and wearproof.

Woodworking
Ingredient of—
Preserving compositions.
Waterproofing, wearproofing, and weatherproofing compositions (Brit. 335247).

Linalyl Acetate
Synonyms: Artificial oil of bergamot, Bergamiol, Linalyl methanecarboxylate.
French: Acétate de linalyle, Acétate linalylique, Essence de bergamote, artificielle; Methanecarboxylate de linalyle, Methanecarboxylate linalylique.
German: Essigsäurelinanylester, Essigsäurelinalyl, Kuenstliche bergamotoel, Linalylmethancarboxylat, Methancarbonsäurelinalylester.

Food
Flavor in—
Candies and other food products.

Perfume
Ingredient of the following artificial essences:—
Bergamot, clover, curomoji, gardenia, jasmine, lavender, lemon, lilac, linden, lily of the valley, neroli, orange, petitgrain, ylang-ylang.
Perfume in—
Cosmetics, toilet waters.

Soap
Perfume in—
Toilet soaps.

Linalyl Carboxethylate
French: Carboxéthylate de linalyle.
German: Linalylcarboxaethylat.
Spanish: Carboxetilato de linalil.
Italian: Carbossietilato di linalile.

Perfume
Ingredient (French 650100) of—
Perfumes.

Linalyl Propionate
Synonyms: Linalyl methylacetate.
French: Méthyleacétate de linalyle, Méthyleacétate linalylique, Propionate de linalyle, Propionate linalylique.
German: Aethancarbonsäurelinalylester, Aethancarbonsäureslinalyl, Linalylmethylacetat, Linalylmethylazetat, Linalylpropionat, Methylessigsäurelinalylester, Methylessigsäureslinalyl, Propansäurelinalylester, Propansäureslinalyl, Propionsäurelinalylester, Propionsäureslinalyl.

Oils and Fats
Odor-enhancer for—
Bergamot oil, lavender oil.

Perfume
Ingredient of—
Eau de cologne, lily perfumes, lily of the valley perfumes.
Perfume in—
Cosmetics.

Soap
Perfume in—
Toilet soaps.

Linoleic Acid Chloride

Chemical

Starting point (Brit. 407956) in making pour-point improvers for machine oils, gear oils, and other lubricants by condensing with—
- Anilin, anthracene oil.
- Aromatics obtained by destructive hydrogenation or by dehydrogenation.
- Benzene.
- Cracking gases containing gaseous olefins (ethylene, propylene, and butylene).
- Cyclic terpenes, ethylnaphthalene, liquid olefins, middle oil, naphthalene, naphthols, naphthylamines, nitrated aromatics, phenols, tars, toluene, xylene.

Liquid Sulphur Dioxide

French: Acide sulfureux, liquefié, Acide sulfureux, liquide, Oxyde sulfureux, liquefié, Oxyde sulfureux, liquide.

German: Verfluessigte schwefligesäure, Verfluessigtes schwefeldioxyd.

Agriculture

General fumigant and disinfectant on the farm and in the dairy.

Brewing

General fumigant for—
- Beer barrels, apparatus, and containers.

Preservative for—
- Hops, beer, and porter, particularly for preventing the growth of fungi in hops.

Chemical

Extracting agent in various processes.
Oxidizing agent in various processes.
Purifying agent in various processes.
Reagent for—
- Continuous treatment of hydrocarbons (French 553546).
- Deodorizing organic solvents.

Reagent in—
- Extracting bituminous matters from lignite.
- Liquefying nitrous oxide in admixture with carbon dioxide.
- Purifying crude tannin extracts.
  - Mashes used in the manufacture of alcohol.
  - Organic chemicals.
  - Waste organic matter.
- Recovering volatile solvents and other volatile products.

Reagent in making—
- Alum from alum shale.
- Aluminum sulphite from aluminum oxide or aluminum hydroxide.
- Ammonium sulphite from ammonium salts.
- Bismuth sulphite from bismuth chloride.
- Calcium bisulphite by action on calcium hydroxide.
- Calcium hyposulphite from calcium hydroxide and sulphur.
- Calcium sulphite by action on calcium carbonate.
- Chromium bisulphate from chromium hydroxide.
- Chromium alum from chromium sulphate and potassium sulphate.
- Compounds made with phenols and the like and used as photographic developers (German 164664).
- Calcium hydrosulphite.
- Cuprous chloride from copper sulphate and sodium chloride.
- Cuprous iodide from copper sulphate and potassium iodide.
- Cuprous sulphocyanide from a solution of a cupric salt, such as cupric sulphate, and potassium sulphocyanide or ammonium sulphocyanide.
- Cuprous bromide from copper sulphate and potassium bromide or sodium bromide.
- Dicalcium phosphate from tricalcium phosphate obtained from the treatment of bones.
- Dithionic acid as manganese salt by action on suspensions of manganese dioxide in water.
- Double salts with acetates of various metals, such as sodium acetate, potassium acetate, lead acetate, nickel acetate, copper acetate, magnesium acetate, strontium acetate, calcium acetate, zinc acetate (Brit. 212902).
- Glauber's salt from sodium chloride (German 17409).
- Hydrosulphites of various metals of the alkali, alkaline earth, earth, rare, and heavy metals series.
- Iodine by action on the natural mother liquors obtained from the ashes of seaweed or from Chile saltpeter.
- Intermediate chemicals.
- Hydroquinone from quinone.
- Lead sulphite by reaction with a solution of a lead salt, such as lead nitrate.
- Lead thiosulphate by reaction with a solution of a lead salt, such as lead nitrate.
- Liquid air.
- Lithium sulphite by reaction with a solution of a lithium salt, such as lithium hydroxide.
- Mangesium hydrosulphite.
- Magnesium sulphite by action on a solution of magnesium nitrate.
- Manganese sulphite by reaction with a solution of a manganese salt, such as manganese chloride.
- Mercurous chloride from mercuric chloride.
- Metanitranilin from metadinitrobenzene.
- Metabisulphites of various alkali metals, alkaline earth metals, and earth metals.
- Nickel sulphite by reaction on a solution of a nickel salt, such as nickel nitrate.
- 4-Nitro-2-aminophenol from 2:4-dinitrophenol (German 289454).
- Organic chemicals.
- Paraphenylenediaminesulphonic acid from quinonediimide.
- Pharmaceutical chemicals.
- Phosphoric acid.
- Potassium hydrosulphite by reaction on a solution of a potassium salt.
- Potassium metabisulphite by reaction with a potassium salt.
- Potassium sulphite.
- Potassium sulphate and ammonium chloride from potassium chloride and ammonia (French 627299).
- Saltcake by the Hargreave's process.
- Sodium hydrosulphite by action on solutions of sodium salts.
- Sodium metabisulphite.
- Sodium nitrite by reduction of sodium nitrate.
- Sodium sulphate, sodium sulphite.
- Sodium thiosulphate from sodium sulphite mother liquor.
- Sulphuryl chloride by reaction with gaseous chlorine.
- Tartaric acid.
- Thionyl chloride with the aid of phosphorus pentoxide.
- Thiosulphates of various elements, such as heavy metals, alkali metals, alkaline earth metals, and earth metals.
- Trithionic acid from potassium thiosulphate or potassium bisulphite.
- Para-aminophenolalphadisulphonic acid and para-aminophenolsulphonic acid from paranitrophenol.
- Various pharmaceutical chemicals, such as alkylhydroxy-alkyl and dihydroxy-alkyl-arsinic acids (French 585970).
- Zinc sulphite.

Reducing agent in various processes.
Solvent for—
- Acids, such as chloroacetic, dichloroacetic, alphabromobutyric, benzoic, salicylic, metaoxybenzoic, betanaphtholic.
- Ammonium iodide.
- Ammonium sulphocyanide.
- Bases, such as formamide, acetnaphthalide, diethylamine, anilin, diphenylamine, benzylamine, paratoluidin, alphanaphthylamine, betanaphthylamine, phenylbetanaphthylamine, benzidin, chrysanilin, carbazol, quinolin, pyridin, acetanalide.
- Esters, such as ethyl acetate, diethyl succinate, diethyl isopropylacetoacetate, diamyl bromoaleinate, diethyl cinnamate, dimethyl malate, diethyl mandelate.
- Fatty alcohols, including those from methyl to capryl, benzyl alcohol, menthol, borneol, orthocresol, betanaphthol, hydroquinone, phenol, trinitroresorcinol.
- Hydrocarbons, such as benzene, toluene, diphenyl, fluorene, phenanthrene, naphthalene, nitrobenzene, limonene, pinene, anthraquinone.
- Nitrocellulose (French 553546).
- Picric acid, potassium bromide, potassium iodide, sodium bromide, sodium iodide.
- Various chemicals and chemical products.

Starting point in making—
- Stripping compounds.

Dye

Reagent in making—
- Sulphur dyestuffs.

Fats and Oils

Reagent in bleaching—
- Fatty acids derived from animal and vegetable fats and oils.
- Vegetable and animal fats and oils.

Liquid Sulphur Dioxide (Continued)
Reagent in making—
Corn oil.
Reagent in treating—
Animal and vegetable fats and oils, for the purpose of removing bad odors.
Animal substances, oilseeds, and the like, for the purpose of removing their fat and oil content.
Solvent for—
Fats and oils.

Food
Bleaching agent in treating—
Edible gelatin.
Flour, such as wheat flour and rye flour.
Fruits, such as cherries, plums, grapes.
Malt, mushrooms, nuts, oats and other grains.
Various other natural and artificial food products.
Preservative and disinfectant in treating—
Asparagus in glass bottles, dry meats, grapes, mutton, plums, potatoes, sausage casings, various other natural and prepared foods, vegetables.
Reagent in making—
Cider.

Gas
Reagent in—
Refining oils and other products obtained from coal-tar, brown tar, and the like, by distillation (Brit. 275884).

Glues and Adhesives
Reagent in—
Bleaching bone glue, gelatin, isinglass.
Extracting gelatin from bones (German 50360).
Preserving bone stock, gluestock.

Gums
Bleaching agent for—
Gum arabic.

Leather
As a bleaching agent.
Reagent in—
Deliming hides.
Reducing chrome tan liquors.
Soaking and pickling hides in the chrome tanning process.

Mechanical
As a lubricant in ice machines.

Metallurgical
Reagent in various smelting and other processes.
Reagent in extracting—
Copper from certain ores.
Copper and lead from roasted ores.
Copper and other metals from sulphide ores.
Gold from its ores, selenium from its ores, silver from its ores, tellurium from ores, titanium from ores.
Vanadium from its ores (French 580094).
Various metals from their ores, zinc from its ores.
Reagent in—
Reducing ores and minerals (U. S. 1528206).

Miscellaneous
As a fire extinguisher.
As a preservative.
As a rat-killer.
Bleaching agent in treating—
Animal and vegetable matter of various sort.
Basketware, catgut, cork, feathers, hog bristles, plumes, sponges, straw hats (French 618007).
Woven work of rattan and similar material.
Disinfectant for—
Barrels and casks (French 609849 and 613615).
Cotton, wool, gauze, and the like, for the manufacture of bandages (Brit. 14813-1893).
General purposes.
Miscellaneous products (French 597622).
Rooms and ships.
Reagent for—
Extracting and decomposing bitumens (German 437210).
Recovering volatile products.
Removing wine and fruit stains from fabrics.
Solvent for—
Nitrocellulose, to form films (French 553546).
Sterilizing agent for various purposes (French 597622).

Paper
Antichlor in bleaching process.
Bleaching agent in treating—
Rag stock, wood pulp.
Reagent in—
Direct digestion of wood by the sulphite process.
Making sulphite liquor.

Petroleum
Reagent in—
Purifying mineral oils (French 550758).
Refining kerosene.
Light lubricating oils (Brit. 275433).
Petroleum, transformer oils.
Treating cracked oils to remove unsaturated hydrocarbons.

Pharmaceutical
Suggested for the treatment of skin diseases.

Photographic
Reagent (French 553546) in—
Dissolving nitrocellulose in the manufacture of films.

Refrigeration
As a refrigerant.
Ingredient (Canadian 272902) of—
Refrigerating mixtures with ether.

Resins and Waxes
Reagent (German 219570) in making—
Artificial resins by the condensation of phenol.

Sanitation
Fumigant (French 623395) for—
Rooms and clothing.

Starch
As a bleaching agent.
Reagent in making—
Cornstarch.

Sugar
Bleaching agent in treating—
Sugars and sugar juices.
Reagent for—
Making sugar by the saccharification of starch.
Saturating sugar juices.
Treating beet juice after saturation.

Textile
——, *Bleaching*
Antichlor in—
Bleaching with chlorine.
Bleaching agent for—
Silk and wool.

——, *Finishing*
Reagent (French 553546) for—
Dissolving nitrocellulose in the process of making coated fabrics.

——, *Manufacturing*
Reagent in—
Purifying crude viscose for the manufacture of viscose rayon.

Wine
Disinfectant for—
Barrels, apparatus, and containers.

Liquorice Juice
Latin: Succus liquiritiae.
French: Jus de réglisse, Suc de réglisse.
German: Baerenzuckersaft, Lakritzensaft, Suessholzsaft.

Food
Ingredient of—
Beverages, confectionery.

Pharmaceutical
In compounding and dispensing practice.

Tobacco
Ingredient of—
Chewing tobacco.

Litharge
Synonyms: Lead monoxide, Lead oxide.
Latin: Lithargyrum, Plumbi oxidum, Plumbum oxydatum.
French: Mono-oxide de plomb, Proto-oxide de plomb.
German: Bleiglätte, Bleioxyd.
Spanish: Litargirio.
Italian: Litargirio.
Note: An oxide of lead corresponding to the formula PbO. Massicot is the unfused and litharge the fused compound; the tendency at the present time is to drop the use of the term, "Massicot," and to use "Litharge" for all varieties of lead monoxide.

Ceramics
Base material in making lead glazes for—
Insides of saggers, insulating porcelain, ornamental tile, stoneware.
Yellow ware, such as bowls, tubs, crocks, household utensils.

Litharge (Continued)
Starting point in making—
Acid-resisting cements, stoneware cements.

Chemical
Starting point in making—
Chrome yellow, lakes, lead chemicals.

Electrical
Base material in making—
Storage battery plates.

Glass
Substitute for red lead in making—
Automobile lamp lenses, camera lenses, cut glassware, flint glass.
Glass of brilliance, clearness, and quality.
Lead glass, microscope lenses, optical lenses, searchlight lenses, tableware of good quality, telescope lenses.

Insecticide
Starting point in making—
Lead arsenate.

Linoleum and Oilcloth
Drier.
Starting point in making—
Driers.

Mechanical
Starting point in making—
Pipe-joint cements.

Metallurgical
Flux in assaying—
Gold ores, silver ores.
Ingredient of—
Enamel frits for enameled iron sanitary ware, stove parts, signs, and various other enameled iron products (but not enameled cooking utensils).

Paint and Varnish
Drier.
Pigment in—
Paints.
Starting point in making—
Chrome yellow, driers, lakes.

Petroleum
Starting point in making—
Sodium plumbite used as a sulphur-removing agent.

Pharmaceutical
In compounding and dispensing practice.

Rubber
Accelerator in—
Curing processes, chiefly for footwear, mechanical and molded goods.
Toughener in—
Curing processes, chiefly for footwear, mechanical and molded goods.

Lithium Acetate
French: Acétate de lithine, Acétate de lithium.
German: Essigsäureslithium, Essigsäureslithiumoxyd, Lithiumacetat, Lithiumazetat.

Chemical
Starting point in making various salts of lithium.

Pharmaceutical
In compounding and dispensing practice.

Textile
Delustering agent (Brit. 260312) in making—
Dull rayons.

Leather
Ingredient (Brit. 299395) of—
Tanning compositions, containing chromic acid (added for the purpose of obtaining more uniform tanning).

Lithium Bromide
Synonyms: Bromide of lithia.
Latin: Lithium bromatum.
French: Bromure de lithium, Lithium bromique.
German: Bromlithium, Lithiumbromid.
Spanish: Bromure de litio.
Italian: Bromuro di litio.

Metallurgical
Ingredient (French 671501) of—
Soldering compositions for magnesium alloys, containing also lithium chloride and potassium fluoride.

Pharmaceutical
Suggested for use as—
Sedative.
Suggested for use in treating—
Nervous conditions.

Photographic
Ingredient of—
Film emulsions.

Lithium Carbonate
French: Carbonate de lithium.
German: Kohlensaeureslithium.

Chemical
Starting point in making the following salts of lithium:—
Acetylsalicylate, benzoate, bromide, chloride, citrate, fluoride, iodide.
Reagent in making—
Apyron.

Food
Ingredient in making—
Mineral waters.
Reagent in treating—
Citrous fruits to prevent decay and decomposition.

Jewelry
Ingredient in making—
Synthetic aquamarines (Brit. 270316).
Synthetic emeralds (U. S. 1579033).

Paint and Varnish
Ingredient in making—
Luminescent paints and varnishes.

Pharmaceutical
In compounding and dispensing practice.

Lithium Chromate
French: Chromate de lithium.
German: Chromsaeureslithium, Lithiumchromat.

Textile
——, *Miscellaneous*
Ingredient of solutions for—
Delustering fabrics and threads made of rayon (Brit. 260312).

Lithium Citrate
Synonyms: Citrate of lithia.
Latin: Lithae citras, Lithium citricum.
French: Citrate de lithane, Citrate de lithium.
German: Lithiumcitrat.
Spanish: Citrato de litio.
Italian: Citrato di litio.

Beverage
Ingredient of—
Effervescent beverages.

Pharmaceutical
In compounding and dispensing practice.
Suggested for use as—
Diaphoretic, diuretic, substitute of other citrate salts.

Lithium Iodide
French: Iodure de lithium.
German: Lithiumjodid.

Food
Reagent in making—
Mineral waters.

Pharmaceutical
In compounding and dispensing practice.

Photographic
Reagent in making photographic papers.

Lithium Oxalate
French: Oxalate de lithium.
German: Lithiumoxalat, Oxalsaeureslithium.

Textile
——, *Miscellaneous*
Ingredient of solutions for—
Delustering rayon fabrics or threads (Brit. 260312).

Lithium Selenide
German: Selenlithium.

Insecticide
Ingredient of—
Compositions used against chestnut blight fungus.

Lithium Sulphate
French: Sulphate de lithium.
German: Lithiumsulfat, Schwefelsaeureslithium.

Textile
——, *Miscellaneous*
Ingredient of solutions for—
Delustering fabrics and threads made of rayon (Brit. 260312).

Logwood
Synonyms: Hematine.
Latin: Haematoxylon, Lignum campechianum, Lignum coeruleum.
French: Bois de campêche, Boise de sang, Hematine.
German: Blauholz, Blutholz, Campecheholz, Haematein, Hematein, Kampeschenholz.

Ink
Ingredient of—
Printing inks, stencil inks, writing inks.

Leather
Mordant in dyeing—
Leather.
Tanning agent.

Paper
Reagent in making—
Paper (needle).

Pharmaceutical
In compounding and dispensing practice.

Textile
——, *Dyeing and Printing*
Mordant for—
Yarns and fabrics.

——, *Finishing*
Reagent in—
Weighing silk.

Stone
Ingredient of—
Artificial stones.

Wine
Coloring agent for—
Cheap grades of wine (U. S. 1643272).

Madder
Synonyms: Ruria, Turkey red.
Latin: Radix rubiae tinctorium.
French: Garance.
German: Faerberroete, Krapp.

Paint and Varnish
Ingredient of—
Artists' pigments, fine paints, stains.

Paper
Pigment in printing—
Wallpaper.

Pharmaceutical
In compounding and dispensing practice.

Textile
——, *Dyeing*
Pigment for—
Wool.
Ingredient of—
Fermentation indigo vat liquors.

——, *Printing*
Pigment for—
Calicoes.

Mafura Tallow
French: Buerre de mafouraire, Buerre de mafura, Suif de mafouraire, Suif de mafura.

Food
Ingredient of—
Edible fat compositions.

Fuel
Constituent of—
Candles.

Soap
Raw material.

Magenta
See Fuchsin.

Magnesite

Cement
Lining for portland cement kilns.
Raw material in making—
Sorel cement.

Ceramics
As a raw material.

Chemical
Lining material in—
Chemical furnaces.
Starting point in making—
Carbon dioxide, magnesium salts.

Metallurgical
Ingredient of—
Compositions used to line furnaces and other equipment.
Starting point in making—
Ferromagnesium, magnesium metal.
Substitute for—
Dolomite in making iron and steel.

Miscellaneous
Ingredient of—
Disinfecting powders.
Floor-treating compositions (Brit. 277444).
Fireproofing compositions.

Paint and Varnish
Filler in—
Paints, pigments.

Paper
Lining for pulp digesters, whitening agent.

Plastics
Stabilizer in making—
Celluloid.

Refractories
Ingredient of—
Highly refractory firebrick, refractory crucibles.

Stone
Raw material in making—
Artificial building stone, brick, plaster tiles.

Textile
——, *Finishing*
Reagent—
To obtain whitened effect on wool fabrics.

Woodworking
Ingredient of—
Fireproofing compositions.

Magnesium Albuminate
French: Albuminate de magnésie, Albuminate magnésique.
German: Albuminsaeuresmagnesium, Magnesiumalbuminat.

Rubber
Reagent in—
Reclaiming rubber (U. S. 1640817).

Magnesium-Aluminum-Iron Cyanide

Chemical
Catalyst (Brit. 446411) in—
Halogenating unsaturated hydrocarbons.

Magnesium-Ammonium Sulphate
Synonyms: Ammonium-magnesium sulphate.
French: Sulfate magnésique et ammoniaque, Sulfate de magnésium et d'ammonium.
German: Ammoniummagnesiumsulfat, Magnesiumammoniumsulfat, Schwefelsaeuresmagnesiumammonium.

Analysis
Reagent in various laboratory operations.

Paper
Ingredient of—
Compositions used for fireproofing.

Textile
——, *Finishing*
Ingredient of—
Compositions used for fireproofing fabrics and yarns.

Woodworking
Ingredient of—
Compositions used for fireproofing.

Magnesium-Anilin

Dye
Starting point (German 436533) in making anthracene dyestuffs from—
3:9-Dichlorobenzanthrone.
11:3-Dichlorobenzanthrone.

Magnesium Bromide
Synonyms: Bromide of magnesia.
Latin: Magnesii bromidum.
French: Bromure magnésique, Bromure de magnésium, Magnésium bromure.
German: Brommagnesium, Magnesiumbromid.
Chemical
Reagent in—
Organic synthesis.
Pharmaceutical
In compounding and dispensing practice.

Magnesium-Cadmium Cyanide
Chemical
Catalyst (Brit. 446411) in—
Halogenating unsaturated hydrocarbons.
Starting point (Brit. 446411) in making—
Catalysts with metal chlorides for halogenating hydrocarbons.

Magnesium-Calcium-Copper-Boron Alloy
Metallurgical
Degasifying and oxidizing agent for—
Metals (principally nonferrous metals).

Magnesium Carbonate
Synonyms: Light magnesium carbonate, Magnesia alba, Magnesia alba levis.
Latin: Magnesii carbonas.
French: Carbonate de magnésie, Carbonate magnésique, Carbonate de magnésium.
German: Kohlensäuresmagnesia, Kohlensäuresmagnesium, Kohlenstoffsäuresmagnesia, Kohlenstoffsäuresmagnesium, Magnesiumcarbonat.
Spanish: Carbonato de magnesio.
Italian: Carbonato di magnesio.
Analysis
Clarifying agent in—
Filtering liquids in chemical analyses.
Source of—
Carbon dioxide for analytical purposes.
Cement
Reagent in making—
Oxychloride cement, Sorel cement.
Ceramics
Ingredient of—
Ceramic compositions.
Chemical
Filtering medium in—
Treating solutions of various chemicals and chemical liquids for the purpose of clarification.
Ingredient of catalytic mixtures used in making—
Acenaphthylene, acenaphthaquinone, bisacenaphthylidenedione, naphthaldehyde acid, naphthalic anhydride and hemimellitic acid from acenaphthene (Brit. 295270).
Acetaldehyde from ethyl alcohol (Brit. 281307).
Acetic acid from ethyl alcohol (Brit. 281307).
Alcohols from aliphatic hydrocarbons (Brit. 281307).
Aldehydes or acids by the reduction of the corresponding esters (Brit. 306471).
Aldehydes and acids from toluene, orthochlorotoluene, orthonitrotoluene, orthobromotoluene, parachlorotoluene, paranitrotoluene, parabromotoluene, metachlorotoluene, metanitrotoluene, metabromotoluene, dichlorotoluenes, dibromotoluenes, dinitrotoluenes, chloronitrotoluenes, chlorobromotoluenes, bromonitrotoluenes (Brit. 295270).
Aldehydes and acids from xylenes, pseudocumenes, mesitylenes, and paracymene (Brit. 281307).
Alphacampholide from camphoric acid by reduction (Brit. 306471).
Alphanaphthaquinone from naphthalene (Brit. 281307).
Benzaldehyde and benzoic acid from toluene (Brit. 281307).
Anthraquinone from naphthalene (Brit. 281307).
Benzoquinone from phenanthaquinone (Brit. 281307).
Benzyl alcohol by the reduction of benzaldehyde (Brit. 306471).
Benzyl alcohol or benzaldehyde or benzyl phthalide by the reduction of phthalic anhydride (Brit. 306471).
Butyl alcohol by the reduction of crontonaldehyde (Brit. 306471).
Diphenic acid from ethyl alcohol (Brit. 306471).
Ethyl alcohol by the reduction of acetaldehyde (Brit. 306471).
Fluorenone from fluorene (Brit. 295270).
Formaldehyde by the reduction of methane or methanol (Brit. 306471).
Formaldehyde by the reduction of carbon dioxide or carbon monoxide (Brit. 306471).
Hydroxyl compounds by the reduction of anthraquinone, benzoquinone, or the like (Brit. 306471).
Isopropyl alcohol by the reduction of acetone (Brit. 306371).
Maleic acid and fumaric acid by the oxidation of toluene, benzene, phenols, tar phenols, or furfural, or from benzoquinone or phthalic anhydride (Brit. 295270).
Methane by the reduction of carbon dioxide or carbon monoxide (Brit. 306471).
Methanol by the reduction of carbon dioxide or carbon monoxide (Brit. 306471).
Naphthaldehyde acid, acenaphthaquinone, bisacenaphthylidenedione from acenaphthylene (Brit. 281307).
Phenanthraquinone from phenanthrene or diphenic acid (Brit. 295270).
Primary alcohols by the reduction of the corresponding aldehydes (Brit. 306471).
Propionic acid and butyric acid and higher alcohols, ketones, and acids by the reduction of carbon dioxide or carbon monoxide (Brit. 306471).
Reduction products of ketones, aldehydes, esters, ethers, alcohols, and other organic compounds which contain oxygen (Brit. 306471).
Salicylic acid and salicylic aldehyde from cresol (Brit. 306471).
Secondary butyl alcohol by the reduction of methylethyl ketone (Brit. 306471).
Valeryl alcohol by the reduction of valeraldehyde (Brit. 306471).
Vanillin and vanillic acid from eugenol or isoeugenol (Brit. 295270).
Ingredient (Brit. 304640) of catalytic preparations used in the manufacture of various aromatic and aliphatic amines, including—
Alphanaphthylamine from alphanitronaphthalene.
Amines from aliphatic nitro compounds, such as alkyl nitriles, or nitromethane.
Amylamine from pyridin.
Anilin, azo-oxybenzene, azobenzene, and hydrazobenzene from nitrobenzene by reduction.
Aminophenols from nitrophenols.
3-Aminopyridin from 3-nitropyridin.
Amino compounds from the corresponding nitroanisoles.
Amines from oximes, Schiff's base, and nitriles.
Cyclohexamine, dicyclohexamine, and cyclohexylanilin from nitrobenzene.
Piperidin from pyridin.
Pyrrolidin from pyrrol.
Tetrahydroquinolin from quinolin.
Reagent in making—
Benzoic acid from benzonitrile.
Hydrogen peroxide in concentrated solutions from barium dioxide and phosphoric acid (German 428707).
Iron oxide (ferric) (Brit. 313999).
Source of—
Carbon dioxide.
Starting point in making—
Magnesium citrate solution and powders, magnesium fluoride, magnesium hydroxide, magnesium oxide, magnesium silicofluoride, magnesium sulphate, magnesium-ammonium phosphate.

Dye
Reagent in making—
Vat dyestuffs from 2:5-diarylidoparabenzoquinone.
Various other dyestuffs.

Fats and Oils
Filtering agent for—
Clarifying animal and vegetable fats and oils.
Reagent in—
Making powdered oil preparations.
Splitting fats (used in the place of zinc oxide).

Fertilizer
As a fertilizer.
Ingredient of—
Fertilizing compositions used particularly in the cultivation of the sugar beet.

Magnesium Carbonate (Continued)

Food
Ingredient of—
Free-running table salt.

Glass
Ingredient of—
Batch for making high-grade glass, such as Pyrex glass.

Ink
Ingredient of—
Magnesia inks, printing inks, writing inks.

Insecticide
Ingredient of—
Fungicidal preparations (Brit. 251330).
Insecticidal powders (Brit. 278816).

Linoleum and Oilcloth
Pigment and filler in making—
Coating compositions.

Mechanical
As a heat insulator.
Ingredient of—
Compositions used for heat-insulating boilers, pipes, and the like.
Compositions used for removing boiler scale.

Metallurgical
Reagent in—
Making open-hearth steel, smelting copper ores.

Miscellaneous
Ingredient of—
Fireproofing compositions.
Flooring compositions.
Heat-insulating compositions.
Insulating compositions (U. S. 1597093).
Polishing compositions for metals and the like.
Preparations for making typewriter ribbons.
Reagent in—
Filtering various liquids.
Making products from quillaia bark and lupin or broom seeds for mothproofing textile fabrics and yarns, as well as furs, feathers and felt.

Paint and Varnish
Pigment and filler in making—
Lacquers, paints.
Special fire-retarding paint (in admixture with magnesium chloride).
Varnishes.
Reagent in making—
Dry colors.

Paper
Filler in making—
Cigaret paper and other paper of high quality.
Special grades of paper (U. S. 1595416).
Reagent in making—
Pulp.

Perfume
Ingredient of—
Combined face powder and skin bleach (U. S. 1620269).
Dentifrices, shaving creams, toilet powders, various cosmetics.

Pharmaceutical
In compounding and dispensing practice.

Plastics
Filler and pigment in making—
Various plastic compositions.

Refractory
Ingredient of—
Refractory products.

Rubber
Filler in—
Compounds for making rubber goods.

Textile
Filler in—
Textile fabrics.
In dry cleaning processes.

Magnesium Chlorate
French: Chlorate de magnésium.
German: Chlorsäuresmagnesium, Magnesiumchlorat.
Spanish: Clorato de magnesio.
Italian: Clorato di magnesio.

Miscellaneous
In solution form as a nontoxic herbicide (French 659957).

Magnesium Chloride
Synonyms: Chloride of magnesia.
Latin: Magnesii chloras, Magnesii chloridum.
French: Chlorure magnésique, Chlorure de magnésium, Magnésium chlorée.
German: Chlormagnesium, Magnesiumchlorid.
Spanish: Cloruro de magnesio.
Italian: Cloruro di magnesio.

Analysis
As a reagent.

Automotive
Starting point in making—
Resilient flooring for buses.

Chemical
Accelerator (Brit. 405371) for—
Hydrogen formation from water in hydrogenation of carbonaceous materials, such as benzenes, petroleum residues, coaltar.
Catalyst in—
Hydration of olefins (Brit. 396107, 394375, and 394674).
Oxidation processes.
Dehydrating agent (Brit. 400169) in—
Concentrating acetic acid.
Starting point in making—
Magnesium salts.

Construction
Ingredient of—
Hydraulic cement (U. S. 1904639).
Wall plaster compositions.
Starting point in making—
Magnesium oxychloride cements, known variously as sorel cement, magnesia cement, and used for various purposes in building and construction (as stucco, as artificial building stone, in the manufacture of artificial marble for interior decoration of buildings, of sanitary, resilient stone flooring, of light-weight construction units, of decorative and flooring tile).

Disinfectant
Ingredient of—
Disinfecting compositions.

Electrical
Ingredient (U. S. 1908792) of—
Thermionic tube heater element.

Fireproofing
Ingredient of—
Fireproofing compositions.

Fuel
Plasticizer (U. S. 1899811) in making—
Liquid fuel from coal.

Metallurgical
Flux (U. S. 1913929) in—
Refining crude zinc.
Starting point in making—
Magnesium metal and its lightweight alloys.

Mining
Chilling agent for—
Drilling tools in drilling for saline deposits (used to prevent the dissolution of the salts).
Ingredient of—
Fireproofing and preservative compositions for impregnating mine timbers.
Starting point in making—
Air-humidifying solution for laying dust in gold mines.

Miscellaneous
Ingredient of—
Antifreeze compositions containing also magnesium acetate and magnesium chromate (U. S. 1823216).
Bath salts, artificial sea salts. and the like.
Dust-laying compositions for use on roads and railways.
Floor-sweeping compositions (mixed with various other cheap materials, such as sand, sawdust, talc, kieselguhr, mineral oil).
Explosion-proof lubricants for oxygen cylinders, welding burners, valves, compressors, bearings (Brit. 398474).
Spraying composition for restoring color to artificial grass, containing also malachite green, auramine, alcohol, and turkey red oil (U. S. 1897900).

Paper
Ingredient (U. S. 1894566, 1894567, and 1894959) of—
Waterproofing compositions.

Magnesium Chloride (Continued)
Reagent in making—
Glassine or imitation parchment (U. S. 1914798, and 1914799).
Moisture-resistant, non-fibrous sheet (Brit. 391153).
Railroading
Impregnating and fireproofing agent for—
Ties and timbers.
Ingredient of—
Fireproofing and impregnating compositions.
Starting point in making—
Composition (with alkali chromate) for melting snow and ice from switches.
Resilient floorings for cars.
Refrigeration
Ingredient of—
Brines.
Textile
Dressing and filling agent for—
Cotton fabrics, woolen fabrics.
Ingredient of—
Sizes.
In wool carbonizing.
Thread lubricant in—
Weaving processes.
Woodworking
Fireproofing agent, impregnating agent.
Ingredient (U. S. 1852900) of—
Preserving composition.

Magnesium-Cupro Cyanide
Chemical
Catalyst (Brit. 446411) in—
Halogenating unsaturated hydrocarbons.
Starting point (Brit. 446411) in making—
Catalysts with metal chlorides for halogenating hydrocarbons.

Magnesium-Ethyl Chloride
French: Chlorure de magnésium et d'éthyle, Chlorure magnésioéthylique.
German: Chlormagnesiumaethyl, Magnesiumaethylchlorid.
Chemical
Reagent in making—
Lead tetraethyl (Brit. 279106).

Magnesium Fluoride
French: Fluorure de magnésium.
German: Fluormagnesium, Fluorwasserstoffsaeuresmagnesium.
Ceramics.
Ingredient of fluxes for—
Chinaware, porcelains, potteries.
Metallurgical
Ingredient to protect molten baths of easily oxidized metals, such as magnesium (Brit. 257221).
Starting point in making—
Metallic magnesium.

Magnesium Glycyrrhizate
Pharmaceutical
Ingredient (U. S. 1976668) of—
Laxative (in admixture with epsom salt).
Laxative consisting of mixed glycyrrhizates and epsom salt.

Magnesium Iodate
French: Iodate magnésique.
German: Jodsaeuresmagnesium, Magnesiumjodat.
Food
Preservative (Brit. 274164) in treating—
Butter, cream, eggs, fish, fruit preserves, margarin, milk, meat.

Magnesium Linoleate
French: Linoléate de magnésie, Linoléate magnésien, Linoléate magnésique.
German: Leinoelsaeuresmagnesium.
Paint and Varnish
Ingredient of—
Coating compositions for wood, stone, brick and other porous substances (Brit. 275610).
Enamels, paints, varnishes.

Magnesium-Nickel Cyanide
Chemical
Catalyst (Brit. 446411) in—
Halogenating unsaturated hydrocarbons.
Starting point (Brit. 446411) in making—
Catalysts with metal chlorides for halogenating unsatuarted hydrocarbons.

Magnesium Nitrate
French: Azotate de magnésie, Azotate magnésique, Azotate de magnésium, Nitrate de magnésie, Nitrate magnésique, Nitrate de magnésium.
German: Magnesiumnitrat, Salpetersäuresmagnesium, Salpetersäuresmagnesiumoxyd.
Spanish: Nitrato de magnesio.
Italian: Nitrato di magnesio.
Analysis
Reagent in various operations.
Chemical
Ingredient of catalytic mixtures used in the manufacture of—
Acenaphthylene, acenaphthaquinone, bisacenaphthylidenedione, naphthaldehydic acid, naphthalic anhydride, and hemimellitic acid from acenaphthene (Brit. 295270).
Acetaldehyde from ethyl alcohol (Brit. 281307).
Acetic acid from ethyl alcohol (Brit. 281307).
Alcohols from aliphatic hydrocarbons (Brit. 281307).
Aldehydes or alcohols by the reduction of the corresponding esters (Brit. 306471).
Alphacampholide from camphoric acid by reduction (Brit. 306471).
Aldehydes and acids from toluene, orthochlorotoluene, orthonitrotoluene, orthobromotoluene, parachlorotoluene, paranitrotoluene, parabromotoluene, metachlorotoluene, metanitrotoluene, bromotoluene, dichlorotoluenes, dinitrotoluenes, dibromotoluenes, chlorobromotoluenes, chloronitrotoluenes, bromonitrotoluenes (Brit. 295270).
Aldehydes and acids from xylenes, pseudocumenes, mesitylene, and paracymene (Brit. 281307).
Alphanaphthaquinone from naphthalene (Brit. 281307).
Anthraquinone from naphthalene (Brit. 295270).
Benzaldehyde and benzoic acid from toluene (Brit. 281307).
Benzoquinone from phenanthraquinone (Brit. 281307).
Benzyl alcohol by the reduction of benzaldehyde (Brit. 306471).
Benzyl alcohol or benzaldehyde for benzyl phthalide by the reduction of phthalic anhydride (Brit. 306471).
Butyl alcohol by the reduction of crotonaldehyde (Brit. 306471).
Chloroacetic acid from ethylenechlorohydrin (Brit. 295270).
Diphenic acid from ethyl alcohol (Brit. 281370).
Ethyl alcohol by the reduction of acetaldehyde (Brit. 306471).
Fluorenone from fluorene (Brit. 295270).
Formaldehyde by the reduction of methane or methanol (Brit. 306471).
Formaldehyde by the reduction of carbon dioxide or carbon monoxide (Brit. 306471).
Hydroxyl compounds by the reduction of anthraquinone, benzoquinone, and similar compounds (Brit. 306471).
Isopropyl alcohol by the reduction of acetone (Brit. 306471).
Maleic acid and fumaric acid by the oxidation of toluene, benzene, tar phenols, phenols, or furfural, or from benzoquinone or phthalic anhydride (Brit. 295270).
Methane by the reduction of carbon dioxide or carbon monoxide (Brit. 306471).
Methanol by the reduction of carbon dioxide or carbon monoxide (Brit. 306471).
Naphthaldehydic acid, acenaphthaquinone, or bisacenaphthylidenedione from acenaphthylene (Brit. 281307).
Phenanthraquinone from phenanthrene or diphenic acid (Brit. 295270).
Phthalic acid and maleic acid from naphthalene (Brit. 295270).
Primary alcohols by the reduction of the corresponding aldehydes (Brit. 306471).
Propionic acid and butyric acid and higher alcohols, ketones, and acids by the reduction of carbon dioxide or carbon monoxide (Brit. 306471).

Magnesium Nitrate (Continued)
Reduction products of ketones, aldehydes, acids, esters, alcohols, ethers, and other compounds which contain oxygen (Brit. 306471).
Salicylic acid and salicylic aldehyde from cresol (Brit. 295270).
Secondary butyl alcohol by the reduction of methylethyl ketone (Brit. 306471).
Valeryl alcohol by the reduction of valeraldehyde (Brit. 306471).
Vanillin and vanillic acid from eugenol or isoeugenol (Brit. 295270).
Ingredient (Brit. 306460) of catalytic preparations used in the production of various aromatic and aliphatic compounds, including—
Alphanaphthylamine from alphanitronaphthalene.
Amines from aliphatic nitro compounds, such as alkyl nitriles, or nitromethane.
Amino compounds from the corresponding nitroanisoles.
Amylamine from pyridin.
Anilin, azo-oxybenzene, azobenzene, and hydrazobenzene from nitrobenzene by reduction.
Aminophenols from nitrophenols.
3-Aminopyridin from 3-nitropyridin.
Cyclohexamine, dicyclohexamine, and cyclohexylanilin from nitrobenzene.
Piperidin from pyridin.
Pyrrolidin from pyrrol.
Tetrahydroquinolin from quinolin.
Reagent in making—
Acetal.
Starting point in making—
Magnesium-ammonium phosphate, magnesium benzoate, magnesium biphosphate, magnesium citrate, magnesium fluoride, magnesium glycerinophosphate, magnesium hydroxide, magnesium oleate, magnesium silicate, magnesium tungstate.

Explosives
Ingredient of—
Pyrotechnic compositions.

Gas
Ingredient of—
Compositions used in the manufacture of gas mantles (added for the purpose of increasing the resistance of the mantle to shock in transportation and handling).
Reagent in—
Treating heads of gas mantles in order to harden them.

Plastics
Ingredient of—
Plastic compositions of the Sorel cement (magnesium oxychloride cement) type.

Magnesium Nitride

Miscellaneous
Indicator (U. S. 1925905) in—
Gas mask absorbent compositions, to react with carbon monoxide and warn the wearer by the odor of ammonia that the absorbent is nearing exhaustion.

Magnesium Oleate
French: Oléate de magnésie, Oléate magnésique.
German: Oelsaeuresmagnesium.

Miscellaneous
Ingredient of—
Compositions used in dry cleaning fabrics, added to prevent benzin burns.

Paint and Varnish
Drier and flattener in—
Enamels, lacquers, paints, varnishes.
Ingredient of—
Coating compositions for treating wood, stone, brick, plaster, and the like (Brit. 275610).

Magnesium Oxide
Synonyms: Burnt magnesia, Calcined magnesia, Heavy calcined magnesia, Heavy magnesia, Light calcined magnesia, Magnesia.
Latin: Magnesia usta, Magnesia usta levis, Magnesia usta ponderosa, Magnesium oxydatum.
French: Magnésie, Magnésie calcinée, Magnésie décarbonatée, Oxyde de magnésium.
German: Bitterde, Gebrannte magnesia, Magnesiumoxyd.
Spanish: Oxido de magnesio.
Italian: Oxido di magnesio.

Analysis
Neutralizing reagent for various analytical purposes.
Reagent in determining—
Sulphur in iron and steel.
Sulphur in organic substances and the like.
Substitute for platinum (used in the form of rods) for various analytical purposes.

Cement
Ingredient of—
Batch in the manufacture of Sorel cement and oxychloride cement.

Ceramics
Raw material in making—
Various ceramic products, such as firebrick, muffles, crucibles, and the like.

Chemical
Compound of—
Refractory linings for chemical furnaces and other apparatus.
Ingredient of catalytic mixtures used in the manufacture of—
Acenaphthylene, acenaphthaquinone, bisacenaphthylidenedione, naphthaldehydic acid, naphthalic anhydride, and hemimellitic acid from acenaphthene (Brit. 295270).
Acetaldehyde from ethyl alcohol (Brit. 281307).
Acetic acid from ethyl alcohol (Brit. 281307).
Alcohols from aliphatic hydrocarbons (Brit. 281307).
Aldehydes and acids by the reduction of the corresponding aldehydes (Brit. 306471).
Aldehydes and acids from toluene, orthochlorotoluene, orthonitrotoluene, orthobromotoluene, parachlorotoluene, paranitrotoluene, parabromotoluene, metachlorotoluene, metabromotoluene, metanitrotoluene, dichlorotoluenes, dibromotoluenes, dinitrotoluenes, chlorobromotoluenes, chloronitrotoluenes, bromonitrotoluenes (Brit. 295270).
Aldehydes and acids from xylenes, pseudocumenes, mesitylene, and paracymene (Brit. 281307).
Alphanaphthaquinone from naphthalene (Brit. 281307).
Anthraquinone from naphthalene (Brit. 295270).
Benzaldehyde and benzoic acid from toluene (Brit. 281307).
Benzoquinone from phenanthraquinone (Brit. 281307).
Benzyl alcohol from benzaldehyde by reduction (Brit. 306471).
Benzyl alcohol or benzaldehyde or benzyl phthalide by the reduction of phthalic anhydride (Brit. 306471).
Butyl alcohol by the reduction of crotonaldehyde (Brit. 306471).
Chloroacetic acid from ethylenechlorohydrin (Brit. 295270).
Diphenic acid from ethyl alcohol (Brit. 281307).
Ethyl alcohol by the reduction of acetaldehyde (Brit. 306471).
Fluorenone from fluorene (Brit. 295270).
Formaldehyde by the reduction of methane or methanol (Brit. 306471).
Formaldehyde by the reduction of carbon dioxide or carbon monoxide (Brit. 306471).
Hydroxyl compounds by the reduction of anthraquinone, benzoquinone, and similar compounds (Brit. 306471).
Isopropyl alcohol by the reduction of acetone (Brit. 306471).
Maleic acid and fumaric acid by the oxidation of toluene, benzene, phenols, tar phenols, or furfural, or from benzoquinone or phthalic anhydride (Brit. 295270).
Methane by the reduction of carbon dioxide or carbon monoxide (Brit. 306471).
Methanol by the reduction of carbon dioxide or carbon monoxide (Brit. 306471).
Naphthaldehydic acid, acenaphthaquinone, or bisacenaphthylidenedione from acenaphthylene (Brit. 281307).
Phenanthraquinone from phenanthrene or diphenic acid (Brit. 295270).
Phthalic acid and maleic acid from naphthalene (Brit. 295270).
Primary alcohols by the reduction of the corresponding aldehydes (Brit. 306471).

Magnesium Oxide (Continued)
Propionic acid and butyric acid and higher alcohols, ketones, and acids by the reduction of carbon dioxide or carbon monoxide (Brit. 306471).
Reduction products of ketones, aldehydes, acids, esters, alcohols, ethers, and other organic compounds which contain oxygen (Brit. 306471).
Salicylic acid and salicylic aldehyde from cresol (Brit. 295270).
Secondary butyl alcohol by the reduction of methylethyl ketone (Brit. 306471).
Valeryl alcohol by the reduction of valeraldehyde (Brit. 306471).
Vanillin and vanillic acid by the oxidation of eugenol or isoeugenol (Brit. 295270).
Ingredient (Brit. 306460) of catalytic preparations used in the production of various aromatic and aliphatic compounds, including—
Alphanaphthylamine from alphanitronaphthalene.
Amines from aliphatic nitro compounds, such as allyl nitriles or nitromethane.
Amino compounds from the corresponding nitroanisoles.
Amylamine from pyridin.
Anilin, azo-oxybenzene, azobenzene, and hydrazobenzene from benzene by reduction.
Aminophenols from nitrophenols.
3-Aminopyridin from 3-nitropyridin.
Cyclohexamine, dicyclohexamine, and cyclohexylanilin from nitrobenzene.
Piperidin from pyridin.
Pyrrolidin from pyrrol.
Tetrahydroquinolin from quinolin.
Neutralizing agent for various chemical purposes.
Reagent in making—
Alkali and alkaline earth cyanides and ammonia (Brit. 250182).
Anthraquinone-2-glycin-3-carboxylic acid (Swiss 109063).
Starting point in making—
Magnesium borate, magnesium bromide, magnesium chloride, magnesium formate, magnesium hypophosphite, magnesium lactate, magnesium nitrate, magnesium perborate, magnesium peroxide, magnesium phosphate, magnesium salicylate, magnesium silicofluoride, magnesium sulphate, magnesium sulphite.
Substitute for platinum (used in the form of rods) for various chemical purposes.

Dye
Reagent in making—
Benzylauramine.
Vat dyestuffs from 2:5-diarylidoparabenzoquinone.

Electrical
Ingredient for—
Compositions used for making linings for electric furnaces.

Fats and Oils
Reagent in making—
Powdered oils.
Reagent in splitting—
Fats in autoclaves (used in place of zinc oxide).

Fertilizer
Ingredient of—
Fertilizing compositions used for the cultivation of sugar beet.

Food
Ingredient of—
Mineral waters, artificially prepared.

Glass
Ingredient of—
Glass batch.

Glues and Adhesives
Ingredient of—
Casein glues.
Adhesive preparations for special purposes.
Cements of various kinds.

Leather
Filler in making—
Artificial leathers.

Linoleum and Oilcloth
Filler in making—
Linoleum, oilcloth, and various other floor coverings.

Mechanical
Ingredient of—
Compositions used for heat-insulation purposes, especially for covering steam pipes and boilers.
Compositions used as steam packings and the like.

Metallurgical
Ingredient of—
Mixtures used for the formation of investments suitable for casting metals and alloys of high melting point (U. S. 1719276).
Compositions used for lining metallurgical furnaces.
Reagent in making—
Steel by the open-hearth process.
Reagent in smelting—
Various copper ores.

Miscellaneous
Ingredient of—
Compositions used for fireproofing various fibrous materials.
Compositions used in the manufacture of typewriter ribbons.
Compositions used for the permanent filling of the root canals in teeth.
Dental cements (U. S. 1613532).
Various cleansing compositions.

Paint and Varnish
Filler in making—
Lacquers, paints, varnishes.
Starting point in making—
Dry colors.

Paper
Ingredient of—
Fireproofing compositions used in treating paper and pulp products.
Starting point in making—
Digestion liquor for the manufacture of chemical pulp.

Perfume
Ingredient of—
Baby powders, cosmetics, dentifrices.

Pharmaceutical
In compounding and dispensing practice.

Plastics
Filler in making—
Compositions which contain pulp and sawdust.

Refractory
Raw material in making—
Firebrick and pipe which are resistant to the action of alkalies.
Refractory cements.
Refractory products, such as crucibles.

Rubber
Accelerator in—
Rubber goods.

Soap
Ingredient of—
Dry-cleaning preparations, spotting fluids containing benzene.

Stone
Ingredient of—
Artificial stone.

Textile
As a grease and oil remover.
Ingredient of—
Fireproofing compositions, sizing compositions.

Magnesium Perchlorate, Anhydrous

Chemical
Dehydrating agent for—
Air in manufacture of oxygen.

Magnesium Propylbromide
French: Propylebromure de magnésie.
German: Brompropylmagnesium.

Chemical
Reagent in making—
Methyl normal-propylcarbinol.

Magnesium Resinate
Synonyms: Resinate of magnesia.
French: Résinate de magnésium.
German: Harzsaeuresmagnesium, Magnesiumresinat.

Dye
Substratum in making—
Lakes in admixture with basic anilin dyestuffs.

Miscellaneous
Ingredient of—
Sealing wax compositions.

Magnesium Resinate (Continued)

Paint and Varnish
Clarifying agent in making—
Oil varnishes.
Neutralizing agent in making—
Oil varnishes.
Hardening agent in making—
Oil varnishes.

Magnesium Ricinoleate

Pharmaceutical
Claimed (U. S. 2019933) to be—
Intestinal detoxification agent suitable for oral administration.

Magnesium Salicylate

French: Salicylate de magnésie, Salicylate magnésique, Salicylate de magnésium.
German: Magnesiumsalicylat, Salicylsäuresmagnesium, Salicylsäuresmagnesiumoxyd.

Resins and Waxes
Reagent (Brit. 292912) in making synthetic resins with the aid of—
Acetylcarbamide, allylcarbamide, amylcarbamide, benzoylcarbamide, butylcarbamide, cinnamylcarbamide, citrylcarbamide, cyanamide, ethylcarbamide, formylcarbamide, gallylcarbamide, heptylcarbamide, hexylcarbamide, isoallylcarbamide, isomylcarbamide, isobutylcarbamide, isopropylcarbamide, lactylcarbamide, methylcarbamide, pentylcarbamide, phenylcarbamide, propionylcarbamide, propylcarbamide, resorcinoylcarbamide, toluoylcarbamide.

Pharmaceutical
In compounding and dispensing practice.

Magnesium Selenide

French: Sélénure de magnésie, Sélénure magnésique, Sélénure de magnésium.
German: Magnesiumselenid, Selenmagnesium.
Spanish: Selenuro de magnesio.
Italian: Selenuro di magnesio.

Chemical
Catalyst (Brit. 263877) in making—
Acetone from isopropyl alcohol.
Dehydrogenation products of cyclohexane.
Isobutyraldehyde from isobutyl alcohol.
Isobutyronitrile from isobutylamine.
Naphthalene from tetrahydronaphthalene.
Paracymene from turpentine.
Catalyst (Brit. 262120) in making—
Isovaleraldehyde from isoamyl alcohol.

Magnesium Silicofluoride

French: Fluosilicate de magnésium, Silicofluorure de magnésium.
German: Fluorsiliciummagnesium, Magnesiumfluorsilikat, Magnesiumsiliciumfluorid, Siliciumfluorstoffsaeuresmagnesium, Siliciumfluorwasserstoffsaeuresmagnesium.

Construction
Agent (Brit. 271203) for—
Hardening cement, concrete, stucco, and other materials.
Waterproofing cement, concrete, stucco, brickwork, and other materials.

Ceramics
Ingredient of—
Glazes.

Woodworking
Ingredient of—
Preserving compositions.

Magnesium Stearate

Latin: Stearopodis.
French: Stéarate de magnésie, Stéarate magnésique.
German: Magnesiumsterat, Stearinsaeuresmagnesium.

Paint and Varnish
As a drier.
Ingredient (Brit. 275610) of—
Coating compositions for brick, plaster, stone and wood.

Magnesium Sulphate

Synonyms: Epsom salt.
Latin: Magnessi sulphas, Magnesium sulfuricum, Sal amarum, Sal anglicum, Sal epsomense, Sal sedlicense, Sulfas magnesicus.
French: Sel amer, Sel de sedlitz, Sel d'epsom, Sulphate de magnésie.
German: Bittersalz, Magnesiumsulfat, Schwefelsäuresmagnesia.
Spanish: Sulfato magnesico.
Italian: Solfato di magnesio.

Agriculture
Ingredient of—
Stockfeeds.

Animal Remedy
Ingredient of—
Animal conditioner, containing also sulphur, rosin, fenugreek seed, flaxseed meal, African ginger, gentian root, copperas, sodium bicarbonate, antimony, salt, and potassium nitrate.
Worm-expeller, containing also calcium sulphate, calcium silicate, Venetian red, sand, and nicotine.

Beverage
Ingredient of—
Artificial mineral waters.

Ceramics
Ingredient of—
Glazes.
Mill addition, containing also clay and tin oxide.

Chemical
Dehydrating agent (U. S. 1912585) in—
Concentrating dilute aqueous solutions of acetic acid.
Dispersing agent (Brit. 415972) in making—
Sodium fluosilicate solutions.
Ingredient (U. S. 1914835) of—
Catalytic mixture for oxidation of sulphur dioxide to sulphur trioxide.
Reagent (Brit. 376080) in—
Decolorizing barytes.
Reagent (U. S. 1929476) in making—
Alkali or ammonium phosphates.
Starting point in making—
Magnesium hydrate (French 755409, Brit. 403860).
Magnesium salts, such as magnesium-ammonium phosphate, magnesium bromate, magnesium carbonate, magnesium hydroxide, magnesium peroxide, magnesium tungstate.

Construction
Ingredient of—
Hardening preparation, containing also sodium silicate and fused calcium chloride, used for impregnating statues and decorations of gypsum and alabaster.
High-early-strength hydraulic cement composition (U. S. 1904640).
Oxychloride cement, containing also calcium chloride, calcined magnesite, and casein.
Plastic magnesia cements (added to increase their water resistance and reduce expansion).
Wall plaster, containing also magnesium chloride, hydrated lime, and plaster of paris.

Explosives and Matches
Ingredient of—
Explosive compositions, matchhead compositions.

Fertilizer
Ingredient of—
Fertilizer compositions.

Fireproofing
Ingredient of—
Fireproofing compositions for balloon fabrics.
Fireproofing compositions for fibrous wallboard, containing also ammonium phosphate and boric acid.

Food
Ingredient of—
Yeast preparation for candy mixtures, containing also glycerin, citric acid, dried yeast, and tapioca starch.

Insecticide
Dispersing agent (Brit. 415972) in making—
Highly concentrated sodium fluosilicate solutions used for pest-destroying purposes.

Leather
As a tanning agent.
Ingredient (U. S. 1800776) of—
Finishing composition.

Metallurgical
Activator for—
Gold-bearing pyritic-quartz ores.
Marmatite in flotation of zinc ore (used in conjunction with copper sulphate and a suitable froth-

Magnesium Sulphate (Continued)
ing agent, such as eucalyptus oil, thereby replacing some of the more costly reagents now employed).
Ingredient of—
Electrolyte, containing also magnesium hydroxide and potassium bromate, for producing green patina on copper.
Electrolyte for nickel plating, containing also nickel sulphate, nickel-ammonium sulphate, and boric acid.
Reagent (Brit. 409636) in recovering—
Lithium from silicious lithium-bearing minerals.
Miscellaneous
Ingredient of—
Compound for melting snow and ice melting, containing also sal ammoniac and silica sand.
Fat-reducing baths, fireproofing compositions.
Silver cleaning and polishing composition, containing also sodium chloride, quinine hydrochloride, and indigo (U. S. 1795676).
Snow for use in—
Motion pictures, window displays.
Paper
Ingredient of—
Emulsified waterproofing compositions (U. S. 1894566, 1882212, and 1894959).
Sizes.
Protective (U. S. 1916606) for—
Impregnated safety paper.
Perfume
Ingredient of—
Cosmetic lotions.
Facial and body reducing lotion, containing also camphor, isopropyl alcohol, tincture of iodine, water, and perfume.
Petroleum
Coating agent (U. S. 1921116) for—
Contact material in neutralization of acid-treated oils.
Pharmaceutical
In compounding and dispensing practice.
Ingredient of—
Weight-reducing bath salts.
Suggested for use as—
Cathartic.
Local application in treating bruises, sprains, erysipelas, cellulitis, epididymitis and other localized inflammatory conditions.
Rubber
Coagulating agent (Brit. 397997) for—
Highly diluted dispersions used in coating fabrics.
Soap
Ingredient of—
Soap powder, containing also sodium silicate, soda ash, soap, and sodium perborate.
Textile
Conditioning agent in—
Compositions for finishing cotton, calico, linen, fancy woven goods, ticking, heavy woolen cloth.
Delustering agent in—
Rayon manufacture.
Ingredient of—
Baths in dyeing calicos.
Baths in dyeing with aniline black.
Buffer solutions in dyeing wool with ice colors (Brit. 401938).
Fireproofing compositions.
Preservative composition for knitted textile fibers, containing also alum and sodium chloride (U. S. 1781730).
Sizing compositions.
Spinning baths for viscose rayon.
Mordant in—
Dyeing wool with certain basic colors.
Promoter of—
Resistance to water action by direct cotton colors.
Weighting agent for—
Flanelettes, cottons, calicoes, linen.

Magnesium Tantalate
French: Tantalate de magnésia, Tantalate magnésique, Tantalate de magnésium.
German: Magnesiumtantalat, Tantalsaeuresmagnesium.
Chemical
Ingredient of catalytic preparations used in making—
Acenaphthylene, acenaphthaquinone, bisacenaphthylidenedione, naphthaldehydic acid, naphthalic anhydride, and hemimellitic acid from acenaphthene (Brit. 295270).
Acetaldehyde from ethyl alcohol (Brit. 281307).
Acetic acid from ethyl alcohol (Brit. 281307).
Alcohols from aliphatic hydrocarbons (Brit. 281307).
Aldehydes and acids from toluene, orthochlorotoluene, orthonitrotoluene, orthobromotoluene, metachlorotoluene, metabromotoluene, metanitrotoluene, parachlorotoluene, parabromotoluene, paranitrotoluene, dichlorotoluenes, dinitrotoluenes, dibromotoluenes, nitrochlorotoluenes, nitrobromotoluenes, chlorobromotoluenes (Brit. 295270).
Aldehydes and acids from xylenes, pseudocumene, mesitylene, and paracymene (Brit. 295270).
Alphanaphthaquinone from naphthalene (Brit. 281307).
Anthraquinone from anthracene (Brit. 295270).
Benzaldehydes and benzoic acid from toluene (Brit. 281307).
Benzoquinone from phenanthraquinone (Brit. 281307).
Chloroacetic acid from ethylenechlorohydrin (Brit. 295270).
Diphenic acid from ethyl alcohol (Brit. 281307).
Fluorenone from fluorene (Brit. 295270).
Formaldehyde from methanol or methane (Brit. 295270).
Maleic acid and fumaric acid by the oxidation of benzene, toluene, phenol, tar phenols, or furfural, or from benzoquinone or phthalic anhydride (Brit. 295270).
Naphthaldehydic acid, acenaphthaquinone, or bisacenaphthylidenedione from acenaphthylene (Brit. 281307).
Phthalic acid and maleic acid from naphthalene (Brit. 295270).
Salicylic acid and salicylic aldehyde from cresol (Brit. 295270).
Vanillin and vanillic acid from eugenol or isoeugenol (Brit. 295270).
Ingredient (Brit. 304640) of catalytic preparations used in making—
Alphanaphthylamine from alphanitronaphthalene.
Amines from aliphatic nitrobodies, such as allyl nitrites or nitromethane.
Amines from oximes, Schiff's bases, and nitrites.
Amylamine from pyridin.
Anilin from nitrobenzene.
3-Aminopyridin from 3-nitropyridin.
Aminophenols from nitrophenols.
Aminoanisole from nitroanisole.
Azobenzene from nitrobenzene.
Azoxybenzene from nitrobenzene.
Cyclohexamine, dicyclohexamine, and cyclohexylanilin from nitrobenzene.
Hydrazobenzene from nitrobenzene.
Piperidin from pyridin.
Pyrrolidin from pyrrol.
Tetrahydroquinolin from quinolin.

Magnesium Titanate
French: Titanate de magnésie, Titanate magnésique.
German: Magnesiumtitanat, Titansaeuresmagnesium.
Chemical
Reagent for general chemical purposes.
Ingredient of catalytic preparations used in making—
Acenaphthylene, acenaphthaquinone, bisacenaphthylidenedione, naphthaldehydic acid, naphthalic anhydride, and hemimellitic acid from acenaphthene (Brit. 295270).
Acetaldehyde from ethyl alcohol (Brit. 281307).
Acetic acid from ethyl alcohol (Brit. 281307).
Alcohols from aliphatic hydrocarbons (Brit. 281307).
Aldehydes or alcohols by reduction of esters (Brit. 306471).
Alphacampholid by the reduction of camphoric acid (Brit. 306471).
Aldehydes and acids from toluene, orthochlorotoluene, orthonitrotoluene, orthobromotoluene, metachlorotoluene, metabromotoluene, metanitrotoluene, parachlorotoluene, paranitrotoluene, parabromotoluene, dichlorotoluene, dibromotoluenes, dinitrotoluenes, dichlorotoluenes, chloronitrotoluenes, chlorobromotoluenes, bromonitrotoluenes (Brit. 295270).
Aldehydes and acids from xylenes, pseudocumenes, mesitylene and paracymene (Brit. 295270).
Alphanaphthaquinone from naphthalene (Brit. 281307).
Anthraquinone from naphthalene (Brit. 295270).

Magnesium Titanate (Continued)
Benzaydehyde and benzoic acid from toluene (Brit. 281307).
Benzoquinone from phenanthraquinone (Brit. 281307).
Benzyl alcohol by the reduction of benzaldehyde (Brit. 306471).
Benzyl alcohol or benzaldehyde or phthalide by the reduction of phthalic anhydride (Brit. 306471).
Butyl alcohol by the reduction of crotonaldehyde (Brit. 306471).
Chloroacetic acid from ethylenechlorohydrin (Brit. 295270).
Diphenic acid from ethyl alcohol (Brit. 281307).
Ethyl alcohol from acetaldehyde (Brit. 306471).
Fluorenone from fluorene (Brit. 295270).
Formaldehyde from methanol or methane (Brit. 295270).
Formaldehyde by the reduction of carbon monoxide or carbon dioxide (Brit. 306471).
Isopropvl alcohol by the reduction of acetone (Brit. 306471).
Maleic acid and fumaric acid by the oxidation of benzene, toluene, phenol, tar phenols, or furfural, or from benzoquinone or phthalic anhydride (Brit. 295270).
Methane by the reduction of carbon dioxide or carbon monoxide (Brit. 306471).
Methanol by the reduction of carbon dioxide or carbon monoxide (Brit. 306471).
Naphthaldehydic acid, acenaphthaquinone, or bisacenaphthylidenedione from acenaphthylene (Brit. 281307).
Phenanthraquinone from phenanthrene or diphenic acid (Brit. 295270).
Phthalic acid and maleic acid from naphthalene (Brit. 295270).
Primary alcohols by the reduction of aldehydes (Brit. 306471).
Propionic acid and butyric acid and higher alcohols, ketones, and acids from carbon dioxide or carbon monoxide by reduction (Brit. 306471).
Salicylic acid and salicylic aldehyde by the reduction of methylethyl ketone (Brit. 306471).
Valeryl alcohol by the reduction of valeraldehyde (Brit. 306471).
Vanillin and vanillic acid from eugenol or isoeugenol (Brit. 295270).
Ingredient of catalytic preparations used in—
Reduction of anthraquinone, benzoquinone, and the like to the corresponding hydroxyl compounds (Brit. 306471).
Reduction of carbon dioxide or of carbon monoxide (Brit. 306471).
Reduction of ketones, aldehydes, acids, esters, alcohols, ethers, and other organic compounds containing oxygen (Brit. 306471).

Magnesium Vanadate
French: Vanadate de magnésie, Vanadate magnésique.
German: Magnesiumvanadat, Vanadinsaeuresmagnesium.

Chemical
Reagent for general purposes.
Ingredient of catalytic preparations used in making—
Acenaphthylene, acenaphthaquinone, bisacenaphthylidenedione, naphthaldehydic acid, naphthalic anhydride, and hemimellitic acid from acenaphthene (Brit. 295270).
Acetaldehyde from ethyl alcohol (Brit. 295270).
Acetic acid from ethyl alcohol (Brit. 295270).
Alcohols from aliphatic hydrocarbons (Brit. 281307).
Aldehydes and acids from toluene, orthochlorotoluene, orthonitrotoluene, orthobromotoluene, metanitrotoluene, metachlorotoluene, metabromotoluene, parachlorotoluene, parabromotoluene, paranitrotoluene, dichlorotoluenes, dibromotoluenes, dinitrotoluenes, chlorobromotoluenes, chloronitrotoluenes, bromonitrotoluenes (Brit. 295270).
Aldehydes and acids from xylenes, pseudocumene, mesitylene, and paracymene (Brit. 295270).
Alphanaphthaquinone from naphthalene (Brit. 281307).
Anthraquinone from anthracene (Brit. 295270).
Benzaldehyde and benzoic acid from toluene (Brit. 281307).
Benzoquinone from phenanthraquinone (Brit. 281307).
Chloroacetic acid from ethylenechlorohydrin (Brit. 295270).
Diphenic acid from ethyl alcohol (Brit. 281307).
Fluorenone from fluorene (Brit. 295270).
Formaldehyde from methanol or methane (Brit. 295270).
Maleic acid and fumaric acid by the oxidation of benzene, toluene, phenol, tar phenols, or furfural, or from benzoquinone or phthalic anhydride (Brit. 295270).
Naphthaldehydic acid, acenaphthaquinone, or bisacenaphthylidenedione from acenaphthylene (Brit. 281307).
Phenanthraquinone from phenanthrene or diphenic acid (Brit. 295270).
Phthalic acid and maleic acid from naphthalene (Brit. 295270).
Primary alcohols from aldehydes by reduction (Brit. 306471).
Propionic acid and butyric acid and higher alcohols, ketones, and acids from carbon dioxide or carbon monoxide (Brit. 306471).
Salicylic acid and salicylic aldehyde from cresol (Brit. 306471).
Secondary butyl alcohol by the reduction of methylethyl ketone (Brit. 306471).
Valeryl alcohol by the reduction of valeraldehyde (Brit. 306471).
Vanillin and vanillic acid from eugenol or isoeugenol (Brit. 295270).
Ingredient of catalytic preparations used in—
Reduction of camphoric acid to form alphacampholid (Brit. 306471).
Reduction of benzaldehyde to form benzoic acid (Brit. 306471).
Reduction of acetaldehyde to form ethyl alcohol (Brit. 306471).
Reduction of phthalic anhydride to form benzyl alcohol, benzaldehyde or phthalide (Brit. 306471).
Reduction of crotonaldehyde to form butyl alcohol.
Reduction of carbon monoxide or carbon dioxide to form formaldehyde (Brit. 306471).
Reduction of acetone to form isopropyl alcohol (Brit. 306471).
Reduction of carbon dioxide or carbon monoxide to form methanol (Brit. 306471).
Reduction of carbon dioxide or carbon monoxide to form methane (Brit. 306471).
Reduction of anthraquinone, benzoquinone, and the like to the corresponding hydroxyl compounds (Brit. 306471).
Reduction of carbon dioxide and carbon monoxide (Brit. 306471).
Reduction of ketones, aldehydes, acids, esters, alcohols, ethers, and other organic compounds containing oxygen (Brit. 306471).

Magnesium-Zinc Cyanide

Chemical
Catalyst (Brit. 446411) in—
Halogenating unsaturated hydrocarbons.
Starting point (Brit. 446411) in making—
Catalysts with metal chlorides for halogenating hydrocarbons.

Malabar Tallow
Synonyms: Piney tallow, White dammar of South India.
French: Suif de malabar, Suif de piney.
German: Malabartalg, Pineytalg, Pflanzentalg, Valeriatalg.

Fuel
Raw material in making—
Candles.

Soap
As a raw material.

Maleic Acid
French: Acide maléique.
German: Maleinsäure.

Chemical
Starting point in making—
Acrylic acid, aspartic acid, hydracyclic acid, lactic acid, malic acid, malonic acid, propionic acid, succinic acid, tartaric acid.

Maleic Acid (Continued)
Fats and Oils
Rancidity retardant for—
Fats, oils.
Food
Rancidity retardant for—
Butter, caramels, milk powder, oleomargarin, pastry.

Malic Acid
French: Acide malique.
German: Apfelsäure.
Beverage
As a flavoring agent.
As an acidulant.
Stabilizing agent (U. S. 1427903) in making—
Grape juice.
Chemical
Process material (U. S. 1491465) in making—
Succinic acid.
Starting point in making—
Coumarines, esters, salts.
Starting (U. S. 1421604) point in making—
Ethyl malate, glycerol malate, glycol malate, propyl malate.
Cosmetic
Ingredient of—
Dentifrice (U. S. 1516206).
Mouthwash (U. S. 1275275).
Electrical
Ingredient (U. S. 1412514) of—
Electrolyte for electrolytic condensers.
Electrolyte for electrolytic lightning arresters.
Food
Acidulant for—
Candy, jellies.
Flavoring agent for—
Candy, jellies.
Extractant (U. S. 1385525) for—
Pectin.
Metallurgical
Ingredient (U. S. 1965682, 1965683, and 1965684) of—
Electrolyte used in oxidizing aluminum by electrolytic methods.
Miscellaneous
Process material in making—
Celluloid substitute (U. S. 1245976, 1245984, and 1280862).
Floor covering (U. S. 1245984).
Pharmaceutical
In compounding and dispensing practice.
Plastics
Process material in making—
Ivory substitute (U. S. 1245976).
Molded article (U. S. 1489744).
Phonograph record (U. S. 1424137).
Plastics (U. S. 1489744).
Resins
Process material in making—
Phenol-formaldehyde condensate substitute (U. S. 1245976).
Synthetic resins useful as shellac substitutes (U. S. 1413144 and 1413145).
Starting point (U. S. 1443935 and 1443936) in making—
Synthetic resin, suitable for molded electric insulation, by condensing with glycerin or polyglycerol.
Synthetic resin by condensing with glycol.
Rubber
Process material (U. S. 1245984) in making—
Rubber substitute.

Malonic Acid
Synonyms: Methanedicarboxylic acid, Propane diacid.
French: Acide malonique, Acide propanedioique.
German: Malonsäure, Methandicarbonsäure, Propandisäure.
Chemical
Starting point in making—
Diallylbarbituric acid (Dial).
Diethylbarbituric acid (Veronal).
Ethyl derivatives.
Phenylbarbituric acid (Luminal).
Propylbarbituric acid (Propanol).
Various intermediate, pharmaceutical, and aromatic chemicals.
Tribromoacetic acid.

Malonyl Chloride
French: Chlorure de malonyle, Chlorure malonylique.
German: Chlormalonyl, Malonylchlorid.
Chemical
Reagent in making—
1:4:5:8-Naphthalenetetracarboxylic acid (German 439511).

Maltose
Synonyms: Malt sugar.
French: Sucre de malt.
German: Malzzucker.
Agriculture
Ingredient (U. S. 1511856) of—
Sterilized bee food.
Beverages and Soft Drinks
Process material in making—
Beverages (U. S. 1461808).
Carbonated beverages.
Brewing
Process material in making—
Beer worts.
Coloring for beer.
Dealcoholized beers and ales (U. S. 1487842).
Malt extracts (U. S. 1515108).
Malt syrups.
Chemical
Process material in making—
Acetaldehyde (U. S. 1511754).
Acetone.
Alcohol (U. S. 1511754 and 1472344).
Glycerin (U. S. 1511754).
Glycerin substitute.
Lactic acid.
Polycarboxylic acids (U. S. 1425605).
Propanetriol (U. S. 1368023).
Reagent in making—
Stable glycerinophosphoric acid preparations.
Stabilizing agent for—
Calcium polysulphides.
Starting point in making—
Citric acid, pharmaceutical preparations, vaccines, yeast preparations (U. S. 1650738).
Dye
Ingredient of—
Indigoid vat dye pastes.
Reducing agent (U. S. 1375972) in making—
Anthranol from anthraquinone.
Fertilizer
Ingredient (U. S. 1254908) of—
Fertilizer composition.
Food
Dehydrating agent (U. S. 1361238 and 1361239) for—
Citrous fruit juices, grape juice, loganberry juice, orange juice, pineapple juice, raspberry juice, strawberry juice.
Ingredient of—
Bread doughs (U. S. 1438441 and 1505236), candy (U. S. 1450865), confectionery, infant foods, invalid foods, jams, malted milk products (U. S. 1446120), milk-iron preparation (U. S. 1393049), milk substitutes.
Process material in making—
Coffee substitutes.
Tea extracts (U. S. 1520122).
Yeast (U. S. 1434462 and 1306569).
Fungicide and Insecticide.
Process material in making—
Fungicides, insecticides.
Leather
Ingredient of—
Depilatory composition containing also glucose, lactic acid, and sodium sulphide.
Treating agent (U. S. 1419497) for—
Skins.
Miscellaneous
Ingredient of—
Snuff, stamp pad composition.
Pharmaceutical
In compounding and dispensing practice.
Ingredient of—
Lozenges, pills, and tablets (U. S. 1450865).
Printing
Ingredient (U. S. 1268135) of—
Printers' roller composition.

Maltose (Continued)

Rubber
Sugar (Brit. 393600) in making—
Polymerized products, useful as rubber substitutes, from uncracked hydrocarbon distillates, cellulosic materials, and sugars.

Textile
——, *Dyeing*
Ingredient (U. S. 1419497) of—
Dyeing solutions.

——, *Printing*
Ingredient of—
Pastes containing indigoid vat dyes.

——, *Treating*
Ingredient (U. S. 1419497) of—
Wool-washing solution.

Manganese
French: Mangane, Manganèse.
German: Mangan.

Chemical
Catalyst in—
Hydrogenation of various chemicals.
Manufacture of sulphuric acid from sulphur trioxide by the contact process.
Various organic syntheses.
Starting point in making—
Manganese salts.

Fats and Oils
Catalyst in—
Hydrogenation processes.

Metallurgical
Agent for—
Deoxidizing and desulphurizing copper, bronze, nickel, and other castings, particularly to avoid the inclusion of air bubbles during casting.
Raw material in making—
Duralumin, ferromanganese, manganese alloys.

Miscellaneous
Catalyst in hydrogenating—
Coal, tars, pitches, and so on, to produce oils used for lubricating purposes and as motor fuels.

Petroleum
Catalyst in—
Hydrogenating petroleum, distillates, and pitches.

Manganese Acetate
French: Acétate de manganèse, Acétate manganeux.
German: Essigsäuresmangan, Essigsäuresmanganoxydul, Manganacetat, Manganazetat.

Analysis
Reagent in testing for glucose and albumoses.

Chemical
Catalyst in carrying out various reactions, particularly oxidation reactions.
Ingredient of catalytic mixtures used in the manufacture of—
Acenaphthylene, acenaphthaquinone, bisacenaphthylidenedione, naphthaldehydic acid, naphthalic anhydride, and hemimellitic acid from acenaphthene (Brit. 295270).
Acetaldehyde from ethyl alcohol (Brit. 281307).
Acetic acid from ethyl alcohol (Brit. 281307).
Alcohols from aliphatic hydrocarbons (Brit. 281307).
Aldehydes or alcohols by the reduction of the corresponding esters (Brit. 306471).
Alphacampholide by the reduction of camphoric acid (Brit. 306471).
Aldehydes and acids from toluene, orthochlorotoluene, orthonitrotoluene, orthobromotoluene, parachlorotoluene, parabromotoluene, paranitrotoluene, metachlorotoluene, meatabromotoluene, metanitrotoluene, dichlorotoluenes, dinitrotoluenes, dibromotoluenes, chlorobromotoluenes, chloronitrotoluenes, bromonitrotoluenes (Brit. 295270).
Aldehydes and acids from xylenes, pseudocumenes, mesitylene, and paracymene (Brit. 281307).
Alphanaphthaquinone from naphthalene (Brit. 281307).
Anthraquinone from naphthalene (Brit. 295270).
Benzaldehyde and benzoic acid from toluene (Brit. 281307).
Benzoquinone from phenanthraquinone (Brit. 281307).
Benzyl alcohol by the reduction of benzaldehyde (Brit. 306471).
Benzyl alcohol or benzaldehyde or phthalide by the reduction of phthalic anhydride (Brit. 306471).
Butyl alcohol by the reduction of crotonaldehyde (Brit. 306471).
Chloroacetic acid from ethylenechlorohydrin (Brit. 295270).
Diphenic acid from ethyl alcohol (Brit. 281307).
Ethyl alcohol by the reduction of acetaldehyde (Brit. 306471).
Fluorenone from fluorene (Brit. 295270).
Formaldehyde by the reduction of methane or methanol (Brit. 306471).
Formaldehyde by the reduction of carbon dioxide or carbon monoxide (Brit. 306471).
Hydroxyl compounds by the reduction of anthraquinone, benzoquinone, and similar compounds (Brit. 306471).
Isopropyl alcohol by the reduction of acetone (Brit. 306471).
Maleic acid and fumaric acid by the oxidation of toluene, benzene, phenols, tar phenols, or furfural, or from benzoquinone or phthalic anhydride (Brit. 295270).
Methane by the reduction of carbon dioxide or carbon monoxide (Brit. 306471).
Methanol by the reduction of carbon dioxide or carbon monoxide (Brit. 306471).
Naphthaldehydic acid, acenaphthaquinone, or bisacenaphthylidenedione from acenaphthylene (Brit. 281307).
Phenanthraquinone from phenanthrene or diphenic acid (Brit. 295270).
Phthalic acid and maleic acid from naphthalene (Brit. 295270).
Primary alcohols by the reduction of the corresponding aldehydes (Brit. 306471).
Propionic acid and butyric acid and higher alcohols, ketones, and acids by the reduction of carbon dioxide and carbon monoxide (Brit. 306471).
Reduction products of ketones, aldehydes, acids, esters, alcohols, ethers and other organic compounds containing oxygen (Brit. 306471).
Salicylic acid and salicylic aldehyde from cresol (Brit. 295270).
Secondary butyl alcohol by the reduction of methylethyl ketone (Brit. 306471).
Valeryl alcohol by the reduction of valeraldehyde (Brit. 306471).
Vanillin and vanillic acid from eugenol or isoeugenol (Brit. 295270).
Ingredient (Brit. 306460) of catalytic preparations used in the production of various aromatic and aliphatic compounds, including—
Alphanaphthylamine from alphanitronaphthalene.
Amines from aliphatic nitro compounds, such as allyl nitriles or nitromethane.
Amino compounds from the corresponding nitroanisoles.
Amylamine from pyridin.
Anilin, azo-oxybenzene, azobenzene, and hydrazobenzene from nitrobenzene by reduction.
Aminophenols from nitrophenols.
3-Aminopyridin from 3-nitropyridin.
Cyclohexamine, dicyclohexamine and cyclohexylanilin from nitrobenzene.
Piperidin from pyridin.
Pyrrolidin from pyrrol.
Tetrahydroquinolin from quinolin.
Starting point in making—
Bister.
Driers for paints, varnishes, oil compositions.
Manganese salts.

Fertilizer
Ingredient of—
Fertilizer compositions.

Leather
Reagent in—
Tanning and finishing various kinds of leather.

Paint and Varnish
Drier in making—
Enamels, paints, varnishes.
Reagent in treating—
Linseed oil for the manufacture of boiled oil.

Manganese Acetate (Continued)
Textile
——, *Dyeing*
As a mordant.
Reagent in dyeing bister shades on yarns and fabrics.
——, *Finishing*
Ingredient of finishing compositions.
——, *Printing*
As a mordant.

Manganese Albuminate
French: Albuminate de manganèse, Albuminate manganèsique.
German: Albuminsaeuresmangan, Manganalbuminat.
Rubber
Reagent (U. S. 1640817) in—
Reclaiming rubber.

Manganese-Ammonium Sulphate
Synonyms: Ammonium-manganese sulphate.
French: Sulfate de manganèse et d'ammonium, Sulfate manganoso-ammoniaque.
German: Ammoniummangansulfat, Manganammoniumsulfat, Manganammonsulfat.
Chemical
Starting point in making—
Manganese sulphate.
Textile
——, *Finishing*
Ingredient of—
Compositions used in fireproofing fabrics.
Woodworking
Ingredient of—
Compositions used in fireproofing.

Manganese Betabenzoylpropionate
Plastics
Starting point (U. S. 2001380) in making—
Films.

Manganese Borate
French: Borate de manganèse, Borate manganique.
German: Borsäuresmangan, Borsäuresmanganoxyd.
Chemical
Ingredient of catalytic mixtures used in the manufacture of—
Acenaphthylene, acenaphthaquinone, bisacenaphthylidenedione, naphthaldehydic acid, naphthalic anhydride, and hemimellitic acid from acenaphthene (Brit. 295270).
Acetaldehyde from ethyl alcohol (Brit. 281307).
Acetic acid from ethyl alcohol (Brit. 281307).
Alcohols from the corresponding aliphatic hydrocarbons (Brit. 281307).
Aldehydes and alcohols by the reduction of the corresponding esters (Brit. 306471).
Aldehydes and acids from toluene, orthochlorotoluene, orthonitrotoluene, orthobromotoluene, parachlorotoluene, parabromotoluene, paranitrotoluene, metachlorotoluene, metabromotoluene, metanitrotoluene, dichlorotoluenes, dibromotoluenes, dinitrotoluenes, chloronitrotoluenes, chlorobromotoluenes, bromonitrotoluenes (Brit. 295270).
Aldehydes and acids from xylenes, pseudocumenes, mesitylene, and paracymene (Brit. 281307).
Alphanaphthaquinone from naphthalene (Brit. 281307).
Anthraquinone from napthalene (Brit. 295270).
Benzaldehyde and benzoic acid from toluene (Brit. 281307).
Benzoquinone from phenanthraquinone (Brit. 281307).
Benzyl alcohol by the reduction of benzaldehyde (Brit. 306471).
Benzyl alcohol or benzaldehyde or phthalide by the reduction of phthalic anhydride (Brit. 306471).
Butyl alcohol by the reduction of crotonaldehyde (Brit. 306471).
Chloroacetic acid from ethylene chlorohydrin (Brit. 295270).
Diphenic acid from ethyl alcohol (Brit. 281307).
Ethyl alcohol by the reduction of acetaldehyde (Brit. 306471).
Fluorenone from fluorene (Brit. 295270).
Formaldehyde by the reduction of methane or methanol (Brit. 306471).
Formaldehyde by the reduction of carbon dioxide or carbon monoxide (Brit. 306471).
Hydroxyl compounds by the reduction of anthraquinone, benzoquinone, and similar compounds (Brit. 306471).
Isopropyl alcohol by the reduction of acetone (Brit. 306471).
Maleic acid and fumaric acid by the oxidation of toluene, benzene, phenols, tar phenols, or furfural, or from benzoquinone or phthalic anhydride (Brit. 295270).
Methane by the reduction of carbon dioxide or carbon monoxide (Brit. 306471).
Methanol by the reduction of carbon dioxide or carbon monoxide (Brit. 306471).
Naphthaldehydic acid, acenaphthaquinone, bisacenaphthylidenedione from acenaphthylene (Brit. 281307).
Phenanthraquinone from phenanthrene or diphenic acid (Brit. 295270).
Phthalic acid and maleic acid from naphthalene (Brit. 295270).
Primary alcohols by the reduction of the corresponding aldehydes (Brit. 306471).
Propionic acid and butyric acid and higher alcohols, ketones, and acids by the reduction of carbon dioxide or carbon monoxide (Brit. 306471).
Reduction products of ketones, aldehydes, acids, esters, alcohols, ethers, and other organic compounds which contain oxygen (Brit. 306471).
Salicylic acid and salicylic aldehyde from cresol (Brit. 295270).
Secondary butyl alcohol by the reduction of methylethyl ketone (Brit. 295270).
Valeryl alcohol by the reduction of valeraldehyde (Brit. 306471).
Vanillin and vanillic acid from eugenol or isoeugenol (Brit. 295270).
Ingredient (Brit. 306460) of catalytic preparations used in the production of various aromatic and aliphatic compounds, including—
Alphanaphthylamine from alphanitronaphthalene.
Amines from aliphatic nitro compounds, such as allyl nitriles or nitromethane.
Amino compounds from the corresponding nitroanisoles.
Amylamine from pyridin.
Anilin, azo-oxybenzene, azobenzene, and hydrazobenzene from nitrobenzene by reduction.
Aminophenols from nitrophenols.
3-Aminopyridin from 3-nitropyridin.
Cyclohexamine, dicyclohexamine, and cyclohexylanilin from nitrobenzene.
Piperidin from pyridin.
Pyrrolidin from pyrrol.
Tetrahydroquinone from quinolin.
Construction
Ingredient (Brit. 250439) of—
Coating compositions for treating cement and concrete structures to render them impermeable to mineral oils.
Leather.
Ingredient of—
Compositions, containing linseed oil and rosin, used for impregnating leather.
Paint and Varnish
Drier for oils.
Drier in—
Paints, varnishes.
Stone
Ingredient (Brit. 250439) of—
Coating compositions for treating stone to render it impermeable to mineral oils.
Wood
Ingredient (Brit. 250439) of—
Coating compositions for treating wood to render it impermeable to mineral oils.

Manganese-Boron Alloy
Metallurgical
Degasifying and oxidizing agent for—
Metals (principally nonferrous metals).

Manganese Chloride
Synonyms: Manganese protochloride, Manganous chloride.
Latin: Manganum chloratum.
French: Chlorure de manganèse, Chlorure manganeux, Protochlorure de manganèse.
German: Chlormangan, Manganchlorür, Salzsäuresmanganozydul.
Spanish: Cloruro de manganesa.
Italian: Cloruro di manganese.

Manganese Chloride (Continued)

Adhesives
Ingredient (U. S. 1273571) of—
Adhesive compositions.

Building and Construction
Starting point (French 655911) in making—
Cements for coatings, stuccos, mosaics, by reacting with zinc oxide.

Cellulose Products
Catalyst (French 660623) in making—
Cellulose esters.

Chemical
Catalyst in—
Cellulose saccharification (U. S. 1428217).
Chlorination of organic compounds.
Catalyst (U. S. 1428217) in making—
Alcohol, 2-furfuraldehyde, methanol.
Ingredient of—
Catalytic mixtures used in making ammonia.
Process material in making—
Acetal (U. S. 1312186).
Lead chloride (U. S. 1441063).
Manganese dioxide (U. S. 1289707).
Manganese oleate, manganese oxalate.
Manganese oxide (U. S. 1327536 and 1520305).
Manganese phosphate (U. S. 1206075).
Promoter (French 689040) of—
Ammonium chloride crystallization from its solutions.
Starting point in making—
Catalysts (U. S. 1520305).

Dye
Catalyst in—
Chlorination processes.
Process material in making—
Chrome brown.

Electrical
Process material (U. S. 1221991) in making—
Depolarizers for dry batteries.
Substitute for—
Sal ammoniac in charging electric batteries, the Lenelanche cells.

Fertilizer
Stimulant in—
Fertilizer compositions.

Firefighting and Fireproofing
Ingredient of—
Fire extinguisher (U. S. 1421436).
Fireproofing composition for wood (U. S. 1126132).

Fuel
Accelerator (Brit. 405371) of—
Hydrogen formation in destructive hydrogenation of coal or petroleum.

Metallurgical
Addition agent (U. S. 1960700) to electrolyte in making—
Magnesium alloys.
Ingredient (U. S. 1269443) of—
Iron-pickling solution.
Process material in producing—
Copper (U. S. 1441063), gold (U. S. 1236236), iron (U. S. 1236236), lead (U. S. 1441063 and 1485909), zinc (U. S. 1236236).
Rustproofing agent (U. S. 1206075) for—
Iron and steel.
Source (U. S. 1377374) of—
Manganese in manganese-magnesium alloys.
Treating agent for—
Argentite (U. S. 1441063), chalcocite (U. S. 1441063), copper ores, galena (U. S. 1441063), lead ores, silver ores.

Miscellaneous
Enrichener (U. S. 1142153) for—
Radium substances.
Ingredient of—
Compositions added to water to preserve cut flowers.
Composition for producing heat upon the addition of water (U. S. 1901313).
Motion picture screen composition (U. S. 1166569).
Stamp pad composition (U. S. 1268135).

Paint and Varnish
Ingredient (U. S. 1291186) of—
Drier.
Process material in making—
Chrome browns.

Paper
Reagent (U. S. 144469) in making—
Photographic paper.

Petroleum
Accelerator (Brit. 405371) of—
Hydrogen formation in destructive hydrogenation processes.

Pharmaceutical
In compounding and dispensing practice.

Textile
——, *Dyeing*
Mordant in—
Dyeing processes.
Processing agent in—
Dyeing cotton to brown or bronze.
Promoter (Brit. 404005) of—
Membrane formation when using aluminum salts as resists in azo-dyeing.
——, *Printing*
Mordant in—
Printing process.

Water and Sanitation
Process material (U. S. 1455363) in making—
Artificial zeolites.

Manganese Dioxide

Synonyms: Battery manganese, Black manganese, Black oxide of manganese, Deutoxide of manganese, Glassmaker's soap, Manganese binoxide, Manganese black, Manganese peroxide, Peroxide of manganese, Pyrolusite.
Latin: Mangani dioxidum, Manganum hyperoxydatum, Oxydum manganicum.
French: Oxyde(bi) de manganèse, Oxyde noir de manganèse.
German: Braunstein, Mangandioxid, Manganoxyd.
Spanish: Bioxido de manganesa, Pyrolusita.
Italian: Biossido di manganese.

Adhesives
Ingredient (U. S. 1336055 and 1299663) of—
Sizing compositions.

Analysis
Oxidizing agent in—
Analytical processes involving control and research in science and industry.

Automotive
Coloring agent in making—
Enamels for coach work (U. S. 1221561 and 1221562).

Building Construction
Ingredient of—
Cements (U. S. 1269116).
Coating for insulation paper (U. S. 1374885).
Reinforced cement (U. S. 1230475).
Shingle-coating compositions (U. S. 1373217).
Pigment in—
Cement, concrete.
Waterproofing agent (U. S. 1519286) for—
Cement, concrete.

Ceramic
Ingredient of—
Black enamels, brown glazes, dark-violet enamels, metallic-like enamels.
Intensifying agent for—
Cobalt colors (used to lower costs).

Chemical
Accelerator in making—
Oxygen from perchlorate.
Catalyst (U. S. 1379221) in decomposing—
Hydrogen peroxide.
Catalyst in making—
Anthraquinone (U. S. 1466683), calcium chloride (U. S. 1153502), chlorine (U. S. 1255020 and 1166524).
Deodorizing agent (U. S. 1491916) for—
Isopropyl alcohol.
Hydrolyzing agent (U. S. 1890590) in making—
Glutamic acid and derivatives, such as sodium glutamate, potassium glutamate (used in conjunction with hydrochloric acid).
Ingredient of—
Catalytic mixtures, gas absorbent (U. S. 1422211).
Oxidizing agent in various chemical processes.
Oxidizing agent in making—
Inorganic chemicals, synthetic organic chemicals.
Process material in making—
Alkali benzoates (U. S. 1463255).
Alkali iodides (U. S. 1249863).
Anthraquinone (U. S. 1324715).
Benzaldehyde, benzoic acid.
Butylene chloride (U. S. 1308763).
Butylene dichloride (U. S. 1308763).

Manganese Dioxide (Continued)
Butylene oxide (U. S. 1253617).
Butylenechlorhydrin (U. S. 1308763 and 1253617).
Butyleneglycol (U. S. 1253617).
Camphor (U. S. 1518732).
1-Chloranthraquinone-2-carboxylic acid (U. S. 1504164).
Chlorine (U. S. 1456590 and 1483256).
Chlorobenzene (U. S. 1468220).
Decolorizing carbons (U. S. 1286187).
Diaminoaryloxyanthraquinone sulphonic acids (Brit. 405632 and 363027).
Dibenzanthrone derivatives (Brit. 405706).
Ethylene dichloride (U. S. 1308763).
Ethylene oxide (U. S. 1308797 and 1308796).
Ethylenechlorhydrin (U. S. 1308707, 1308796, and 1308763).
Hydrocyanic acid (U. S. 1242264).
Ethyleneglycol (U. S. 1253617).
Hydrogen (U. S. 1506323).
Iodine (U. S. 1249863).
Methane (U. S. 1242264).
Nitrogen dioxides (U. S. 1242264).
Phthalic acid (U. S. 1365956).
Propylene chloride (U. S. 1308763).
Propylene dichloride (U. S. 1308763).
Propylene oxide (U. S. 1253617).
Propylenechlorhydrin (U. S. 1308763).
Process material in purification of—
Acetylene.
Acetic anhydride (U. S. 1467074).
Barium sulphide (U. S. 1256593).
Cadmium sulphate (U. S. 1264802).
Iron oxide (U. S. 1318432).
Solvent (U. S. 1360271) for—
Bismuth (used with sulphuric acid).
Starting point in making—
Catalysts, manganates, manganese salts, permanganates.

Distilling
Catalyst (U. S. 1396009) in making—
Organic acids from distillery waste.

Dye
Oxidizing agent in making—
Dyestuffs, intermediates.

Electrical
Depolarizer for—
Batteries.
Ingredient of—
Dry cells for batteries, electric insulations.

Explosives and Matches
Ingredient of—
Explosives, friction surface compositions for matchboxes, matchhead compositions, pyrotechnic compositions, signal flares.

Fertilizer
Fertilizer
Ingredient of—
Fertilizer compositions.

Gas
Process material (U. S. 1506323) in making—
Water-gas.

Glass
Coloring agent for—
Glass batches (violet, black, toning other colors).
Neutralizing agent for—
Yellow effects produced by iron impurities in the glass batch.
Oxidizing agent (U. S. 1449793) in making—
Borosilicate glass.

Ink
Drier (U. S. 1421125 and 1342638) in—
Printing ink.

Metal Fabrication
Coloring agent in making—
Enameled ironware.
Ingredient of—
Arc-welding compositions (U. S. 1467825, 1460476, 1374711, and 1451392).

Metallurgical
Oxidizing agent in—
Case-hardening processes (U. S. 1480230).
General processes.
Process material in making—
Alloy steels, electrodes for electroplating purposes.
Reagent for removing—
Cobalt from zinc solutions (U. S. 1336386).
Molybdenum from molybdenite (U. S. 1401924 and 1401932).
Rustproofing agent for—
Iron (used with phosphoric acid in "Parkerizing").
Source of manganese in making—
Manganese steels, alloys, and other products.

Miscellaneous
Ingredient of—
Antileak composition (U. S. 1343150).
Arc-welding compositions.
Artificial spinel compositions (Brit. 403233).
Chemical heat-producing compositions (U. S. 1506323 and 1488656).
Light filter (U. S. 1331937).
Slag-forming and gas-forming coatings for welding electrodes (U. S. 1902948).
Oxidizing agent.

Paint and Varnish
Drier in—
Dopes, enamels, lacquers, paints, varnishes.
Ingredient of—
Drying oils, driers.
Pigment in—
Enamels, lacquers, paints, varnishes.
Starting point in making—
Barium manganate, manganese, linoleate, manganese-lead resinate, manganese oleate, manganese oxalate, manganese resinate, water colors (with gums in paste form).

Pharmaceutical
In compounding and dispensing practice.

Rubber
Tackiness increaser in—
Rubber batches.

Textile
Mordant in—
Dyeing and printing processes.
Process material in producing—
Brown shades, khaki effects.

Water and Sanitation
Reagent for removing—
Iron from water.

Manganese Dioxysulphate
French: Sulfate de bioxyde de manganèse.
German: Mangandioxysulfat.

Chemical
Oxidizing agent in making—
Orthonitrobenzoic aldehyde from orthotoluene.

Dye
Oxidizing agent in making various synthetic dyestuffs.

Manganese Polysulphide
French: Polysulfure de manganèse.
German: Manganpolysulfid.
Spanish: Polisulfurato de manganesa.
Italian: Polisulfurato di manganese.

Fertilizer
Ingredient of—
Fertilizing compositions used as top dressing.

Manganese Resinate
French: Résinate de manganèse.
German: Harzsaeuresmangen, Manganresinat.

Paint and Varnish
Drier in making—
Enamels, hard resin varnishes, lacquers, paints, rosin varnishes, titanium white paints, varnishes.

Manganese-Silicon-Boron Alloy

Metallurgical
Degasifying and oxidizing agent for—
Metals (principally nonferrous metals).

Manganese Tungate
French: Tungate de manganèse.
German: Mangantungat.

Paint and Varnish
Drier (Brit. 270387) in making—
Enamels, lacquers, paints, stains, varnishes.

Photographic
Ingredient (Brit. 270387) in making—
Light-sensitive varnishes.

Manila Gum
French: Gomme manila.
German: Manilgummi.

Adhesives
Ingredient of various products.

Manila Gum (Continued)

Electrical
Ingredient (Brit. 303386) of—
Coating compositions containing synthetic resins made from glycerin, glycol, or glucose with phthalic anhydride or other polybasic acids or anhydrides.

Explosives
Ingredient of—
Match-head preparations, pyrotechnic preparations.

Ink
Ingredient of—
Printing and lithographic inks.

Miscellaneous
Ingredient (Brit. 303386) of—
Compositions containing synthetic resins made from glycerin, glycol, or glucose with phthalic anhydride or other polybasic acids or anhydrides.
Ingredient of—
Shoe blackings.

Oilcloth and Linoleum
Ingredient of—
Coating compositions.

Paint and Varnish
Ingredient (Brit. 303386) of—
Paints, lacquers, and varnishes containing synthetic resins made from glycerin, glycol, or glucose with phthalic anhydride or other polybasic acids or anhydrides.
Ingredient of—
Enamels, spirit lacquers and varnishes.

Plastics
Ingredient (Brit. 303386) of—
Compositions containing synthetic resins made from glycerin, glycol, or glucose with phthalic anhydride or polybasic acids or anhydrides.

Printing
Reagent in—
Process engraving and the lithographic arts.

Rubber
Ingredient of various compositions.

Textile
Ingredient of—
Printing pastes.

Manioc
Synonyms: Manihot utilissima, Maniok.

Chemical
Starting point in making—
Alcohol by fermentation.

Food
Starting point in making—
Special dietetic flour.

Starch
Starting point in making—
Special starch.

Manjak

Mechanical
General lubricant.

Oils and Fats
Ingredient of—
Lubricating compositions, lubricants for gear cases, lubricants for sprocket wheels and roller bearings, thread greases for piping and casings.

Miscellaneous
Ingredient of—
Preparations for protecting underground and surface piping against corrosion and hydrolysis.
Preparations for repairing leaks in tanks.
Roofing compositions.

Paint and Varnish
Ingredient of—
Auto fender paints, high heat resistant paints.
Paints and varnishes used on boilers, pipelines, chimneys, bridges, machinery, tanks.
Paints and varnishes containing mineral oils.
Quick-drying paints and varnishes.

Petroleum
Ingredient of—
Pipe cements for use in rotary drilling for oil, to obtain tight connection between casings sections.

Mannitol
Synonyms: Mannite.

Analysis
Reagent in—
Boron determinations.

Chemical
Reagent in—
Organic syntheses.
Stabilizing agent (Brit. 413043) for—
Catalysts used in processes involving hydration of olefins.
Starting point in making—
Acetals, anhydrides, esters, ethers, fructose, mannide, mannitan, mannose, nitromannite, other derivatives, saccharic acid, secondary hexyliodide.

Electrical
Ingredient of—
Pastes used in the "dry" type of electrolytic condensers.

Explosives and Matches
Starting point in making—
Nitromannite, special detonants.

Paint and Varnish
Starting point (Brit. 383764 and 385139) in making—
Softening agents for cellulose lacquers, with cyclohexanone or 2-methylcyclohexanone.

Paper
Process material in making—
Fancy papers.

Petroleum
Stabilizing agent (Brit. 413043) for—
Catalysts used in processes involving hydration of olefins.

Pharmaceutical
In compounding and dispensing practice.

Resins
Process material in making—
Synthetic resins.
Starting point in making synthetic resins with—
Diabasic aliphatic acids, such as malic, fumaric, glutaric, pimelic, suberic, azalaic, maleic and sebacic (Brit. 407914, 396354).
Dihydric alcohols, such as ethyleneglycol, diethyleneglycol, propyleneglycol, tetramethyleneglycol (Brit. 407914, 407965, and 396354).
Fatty acids of drying or semidrying oils, such as linseed, chinawood, cottonseed, perilla, and soybean (Brit. 407914 and 407965).
Phenols, such as hydroxydiphenyls, amylphenols, butylphenol, benzylphenol, salicylic acid, and resorcinol (Brit. 407965).
Polybasic acids, such as phthalic, adipic, sebacic, succinic, and maleic (Brit. 407914, 407965, and 396354).

Margine
Synonyms: Marchies, Sanse.

Agriculture
As a fertilizer.

Insecticide
As an insecticide.

Maripa Fat
French: Graisse de maripa, Huile de maripa.
German: Maripafett.
Spanish: Aciete de maripa.
Italian: Sego di maripa.

Food
As a food fat.
Ingredient of—
Food preparations.

Pharmaceutical
In compounding and dispensing practice.

Marjoram
Synonyms: Knotted marjoram, Sweet marjoram.
Latin: Origanum majorana.
French: Marjolaine, Marjolaine sauvage.
German: Majoran, Meiran.

Food
As a condiment and flavor.

Oils and Fats
Starting point in making a volatile oil.

Pharmaceutical
In compounding and dispensing practice.

Mastic
Synonyms: Gum mastic, Pistachia galls.
Latin: Resina lentisci.
French: Gomme mastic, Mastich, Mastix, Mastic, Résine mastic.
German: Mastiche, Resina mastiche.
Spanish: Almaciga, Mastic.

Food
Ingredient of—
Condiments.

Glues and Adhesives
Ingredient of—
Ceramic cements, cements containing fish glue, cements for jewelry and precious stones, dental cements with wax, glass cements, special adhesives.

Miscellaneous
Ingredient of—
Chewing gum with gum sandarac, incense, plasters, sealing waxes, sizing compositions for twine and cordage.

Paint and Varnish
Ingredient of—
Spirit varnishes and lacquers, alone or in admixture with other resins.

Paper
Sizing agent for—
Paper and paper products.

Perfumery
Ingredient of—
Cosmetics.

Pharmaceutical
In compounding and dispensing practice.

Photographic
Ingredient of—
Retouching varnishes.

Plastics
Ingredient of—
Special compositions.

Printing
Reagent in—
Lithographic printing.

Resins and Waxes
Ingredient of—
Rosin compositions, added for the purpose of hardening them (Brit. 252656).

Textile
——, *Finishing*
Ingredient of—
Sizing compositions.

Woodworking
Ingredient of—
Compositions used in covering wood with metal leaf.

Meldola Blue
French: Bleu de meldola.
German: Meldolablau.

Insecticide
Ingredient (Brit. 303932) of—
Insecticides, fungicides, and vermin-destroying compositions containing arsenic acid, arsenous acid, or salts of these acids.

Miscellaneous
Dyestuff for coloring various substances.

Sanitation
Ingredient (Brit. 303932) of—
Disinfecting and bactericidal compositions containing arsenous acid, arsenic acid, or the salts of these acids.

Textile
Dyestuff in dyeing and printing yarns and fabrics.

Menhaden Oil
Synonyms: Mossbunk oil, Pogy oil.
French: Huile d'alose, Huile de menhaden.
German: Amerikanisches fischoel, Maifischoel, Menhadenoel.
Spanish: Aciete de menhaden.

Fats and Oils
Starting point in making—
Fats by hydrogenation, sod oil, various treated oils.
Substitute for—
Tallow oil.

Food
As a food (in the hydrogenated or hardened form).

Ink
Substitute for linseed oil in making—
Lithographic inks, printing inks.

Leather
Reagent in—
Currying.
Substitute for linseed oil in making—
Patent leather.
Tanning agent in making—
Chamois leather.

Linoleum and Oilcloth
Starting material (Brit. 9023-1911) in making—
Vehicles for making oilcloth and linoleum (used in conjunction with chlorinated hydrocarbons, and silica, zinc silicate, calcium silicate, obsidian, lead sulphite, white and ferruginous clays, rosin, zinc oxide, lead sulphate, and other materials).
Substitute for—
Linseed oil.

Metallurgical
Reagent in—
Tempering steel.

Miscellaneous
Ingredient (Brit. 9023-1911) of—
Fatty cements, paving compositions, and other products (used in conjunction with chlorinated hydrocarbons and silica, calcium silicate, obsidian, lead sulphite, white and ferruginous clays, rosin, zinc oxide, lead sulphate, and other materials).
Ingredient of—
Cork flooring compositions, rope-treating compositions.
Various compositions in which quick drying and binding are requirements.
Waterproofing compositions.

Paint and Varnish
Ingredient of—
Japans and paints used for smokestacks, boiler fronts, and other ironwork which is subjected to high temperatures.
Ingredient (Brit. 9023-1911) (used in conjunction with chlorinated hydrocarbons and silica, zinc silicate, calcium silicate, obsidian, lead sulphite, white and ferruginous clays, rosin, zinc oxide, lead sulphate and other materials) of—
Paints, putty, roofing compositions, varnishes.
Substitute for linseed oil in making—
Paints, varnishes.

Rubber
Ingredient of—
Rubber substitutes.

Soap
As a soapstock.

Woodworking
Ingredient (Brit. 9023-1911) of—
Artificial lumber (used in conjunction with chlorinated hydrocarbons and silica, lead sulphite, calcium silicate, zinc silicate, white or ferruginous clays, rosin, zinc oxide, lead sulphate, and other materials).

Menthol
Synonyms: Hexahydrothymol, Methylhydroxyisopropylcyclohexaneparamentheneol, Methylpropylphenyl hexahydride, Mint camphor, Peppermint camphor, Pipmenthol, 3-Terpanol.
Latin: Mentholum.
French: Camphre de menthe.
German: Menthakampher, Mentholum, Pfefferminzkampher.
Spanish: Mentol.
Italian: Mentolo.

Chemical
Starting point in making—
Betamethyladipic acid, cymene, derivatives.
Esters, such as acetic, cinnamic ester, salicylic.
Hexahydrocymene, menthene, menthone, methyl menthyloxybenzoate (U. S. 1133832), stereo-isomerides, thymol.

Cosmetic
Ingredient of—
Creams, lotions, pomades, powders.

Disinfectant
Ingredient of—
Disinfecting preparations (U. S. 1420634).

Firefighting
Ingredient (U. S. 1270396) of—
Fire extinguishing mixture with carbon tetrachloride.

Menthol (Continued)

Food
Flavoring agent.
Ingredient of—
Chewing gum (U. S. 1171392).

Miscellaneous
Ingredient of—
Deodorizing agent (U. S. 1346337), deodorizing sticks (French 742307).

Oral Hygiene
Ingredient of—
Dentifrices (many patents), nasal douche (U. S. 1471987), styptic (U. S. 1420634).
Solvent (U. S. 1471987) for—
Mucin.

Paint and Varnish
Ingredient (U. S. 1189804) of—
Paint remover.
Odorizer for—
Paints.

Pharmaceutical
Claimed to have value in treating—
Boils, carbuncles, coryza, gastrodynia, headache, laryngitis, nausea, neuralgia, pharyngitis, skin diseases accompanied by itching.
In compounding and dispensing practice.
Ingredient of—
Anodyne (U. S. 1420634), antiseptic (U. S. 1471987), cathartic (U. S. 1212888), cough drops, dental anesthetic (U. S. 1420634), salves, sanitary douches (U. S. 1471987), sprays.
Medicating agent for—
Air (U. S. 1409364).
Sterilizing agent (U. S. 1495180) for—
Surgical ligatures.
Suggested for use as—
Antiseptic, anodyne, bactericide, carminative, counter irritant, ingredient of healing compounds, local anesthetic, stimulant.

Menthol Salicylate
Synonyms: Menthyl salicylate, Salimenthol.
French: Salicylate de camphre de menthe.
German: Menthakamphersalicylat, Mentholsalicylat, Pfefferminzkamphersalicylat, Pipmentholsalicylat.

Perfume
Analgesic in—
Cosmetics used as protection against sunburn.
Absorbent of ultra-violet rays in—
Cosmetics used as protection against sunburn.

Pharmaceutical
Suggested analgesic in—
Muscular rheumatism, acute neuralgia.
Suggested for use in treating hayfever.

Menthyl Acetate
Synonyms: Menthyl methane carboxylate.
French: Acétate de menthyle, Acétate menthylique, Éther menthylacétique, Méthanecarboxylate de menthyle, Méthanecarboxylate menthylique.
German: Aethansaeurementhylester, Aethansaeuresmenthyl, Essigsaeurementhylester, Essigsaeuresmenthyl, Menthylacetat, Menthylazetat, Menthylmethancarbonat, Methancarbonsaeurementhylester, Methancarbonsaeuresmenthyl.

Chemical
Starting point in making—
Aromatics.

Perfume
Ingredient of—
Cosmetics, odorous sprays, perfumes.

Soap
Ingredient of—
Toilet soaps.

Mercaptobenzothiazole

Rubber
Accelerator in—
Vulcanizing processes.
Starting point in making—
Delayed-action vulcanization accelerators by condensation with (1) itself, (2) chloroketones, (3) dinitrochlorobenzene, (4) dinitrochloronaphthalene, (5) aromatic acyl derivatives, (6) cyanuric chloride.
Delayed-action vulcanization accelerators by treatment with a deactivating ketone followed by heating to a temperature of reactivation (Brit. 420852).

Metallurgical
Ingredient (U. S. 1736934) of—
Caustic alkali solution added as inhibitor in metal-pickling baths (acid).
Starting point (U. S. 1932553) in making—
Inhibitors for metal-pickling baths (sulphuric acid) by heating with a fully saturated aliphatic amine at 90° to 135° until reaction ceases.

Mercaptobenzothiazole Dinitrophenylester

Lubricant
Starting point (Brit. 440175) in making—
Addition agents for high-pressure lubricating oils or greases by mixing and reacting with organo-metallic compounds.

1-Mercapto-benzoxazole
Synonyms: 2-Thiobenzoxazole.

Insecticide and Fungicide
Larvicide for—
Culicine mosquito larvae.

Mercurated Camphoric-alpha-allylamides

Pharmaceutical
Starting point (Brit. 447877) in making—
Diuretics by partially neutralizing with sodium alcoholate (these products are claimed to give clear, stable aqueous solutions which are not strongly alkaline, and to be suitable for rectal administration).

Mercuric Acetate
Synonyms: Mercury acetate.
French: Acétate de mercure, Acétate mercurique, Deutoacétate de mercure.
German: Essigsäuresmerkur, Essigsäuresmerkurioxyd, Essigsäuresquecksilber, Essigsäuresquecksilberoxyd, Mercuriacetat, Mercuriazetat, Merkuriacetat, Merkuriazetat, Quecksilberacetat, Quecksilberazetat.

Analysis
As a reagent for testing—
Turpentine oil, wine coloring matters.

Chemical
Catalyst in making—
Acetaldehyde from acetylene (French 479656).
Ethylidene diacetate.
Ingredient of catalytic preparations used in making—
Acenaphthylene, acenaphthaquinone, bisacenaphthylidenedione, naphthaldehydic acid, naphthalic anhydride, and hemimellitic acid from acenaphthene (Brit. 295270).
Acetaldehyde from ethyl alcohol (Brit. 281307).
Acetic acid from ethyl alcohol (Brit. 281307).
Alcohols from aliphatic hydrocarbons (Brit. 281307).
Aldehydes or acids by the reduction of the corresponding esters (Brit. 306471).
Alphacampholide by the reduction of camphoric acid (Brit. 306471).
Aldehydes and acids from toluene, orthochlorotoluene, orthobromotoluene, orthonitrotoluene, metachlorotoluene, metabromotoluene, metanitrotoluene, parachlorotoluene, parabromotoluene, paranitrotoluene, dichlorotoluenes, dibromotoluenes, dinitrotoluenes, chlorobromotoluenes, chloronitrotoluenes, bromonitrotoluenes (Brit. 295270).
Aldehydes and acids from xylenes, pseudocumenes, mesitylene, and paracymene (Brit. 281307).
Alphanaphthaquinone from naphthalene (Brit. 281307).
Anthraquinone from naphthalene (Brit. 295270).
Benzaldehyde and benzoic acid from toluene (Brit. 281307).
Benzoquinone from phenanthraquinone (Brit. 281307).
Benzyl alcohol by the reduction of benzaldehyde (Brit. 306471).
Benzyl alcohol or benzaldehyde or phthalide by the reduction of phthalic anhydride (Brit. 306471).
Butyl alcohol by the reduction of croton aldehyde (Brit. 306471).
Chloroacetic acid from ethylenechlorohydrin (Brit. 295270).
Diphenic acid from ethyl alcohol (Brit. 281307).
Ethyl alcohol by the reduction of acetaldehyde (Brit. 306471).
Fluorenone from fluorene (Brit. 295270).
Formaldehyde by the reduction of methane or methanol (Brit. 306471).
Formaldehyde by the reduction of carbon dioxide or carbon monoxide (Brit. 306471).
Hydroxyl reduction compounds of anthraquinone, benzoquinone and the like (Brit. 306471).

Mercuric Acetate (Continued)

Isopropyl alcohol by the reduction of acetone (Brit. 306471).
Maleic acid and fumaric acid by the oxidation of toluene, benzene, phenols, tar phenols, or furfural, or from benzoquinone or phthalic anhydride (Brit. 295270).
Methane by the reduction of carbon dioxide or carbon monoxide (Brit. 306471).
Methanol by the reduction of carbon dioxide or carbon monoxide (Brit. 306471).
Naphthaldehydic acid, acenaphthaquinone, or bisacenaphthylidenedione from acenaphthylene (Brit. 281307).
Phenanthraquinone from phenanthrene or diphenic acid (Brit. 295270).
Primary alcohols by the reduction of the corresponding aldehydes (Brit. 306471).
Propionic acid and butyric acid and higher alcohols, ketones, and acids by the reduction of carbon dioxide or carbon monoxide (Brit. 306471).
Reduction products of ketones, aldehydes, acids, esters, alcohols, ethers, and other organic compounds containing oxygen (Brit. 306471).
Salicylic acid and salicylic aldehyde from cresol (Brit. 295270).
Secondary butyl alcohol by the reduction of methylethyl ketone (Brit. 306471).
Valeryl alcohol by the reduction of valeraldehyde (Brit. 306471).
Vanillin and vanillic acid from eugenol or isoeugenol (Brit. 295270).

Ingredient (Brit. 304640) of catalytic preparations used in the production of various aromatic and aliphatic amines, including—
Alphanaphthylamine from alphanitronaphthalene.
Amines from aliphatic nitro compounds, such as alkyl nitriles or nitromethane.
Amylamine from pyridin.
Anilin, azo-oxybenzene, azobenzene, and hydrazobenzene from nitrobenzene by reduction.
Aminophenols from nitrophenols.
3-Aminopyridin from 3-nitropyridin.
Amino compounds from the corresponding nitroanisoles.
Amines from oximes, Schiff's base, and nitriles.
Cyclohexamine, dicyclohexamine, and cyclohexylanilin from nitrobenzene.
Piperidin from pyridin.
Pyrrolidin from pyrrol.
Tetrahydroquinolin from quinolin.

Reagent in making—
Mercuriated hydroaromatic hydrocarbons (Austrian 100723).
Pharmaceuticals.

Starting point in making—
Mercuric benzoate, bromide, iodide, and other salts and derivatives.

Perfumery
Ingredient of—
Cosmetics.

Pharmaceutical
In compounding and dispensing practice.

Mercuric Chloride

Synonyms: Corrosive mercuric chloride, Corrosive sublimate, Mercury bichloride, Sublimate.
Latin: Hydrargyri chloridum corrosivum.
French: Bichlorure de mercure, Bichlorure mercurique, Chlorure de mercure, Chlorure mercurique, Chlorure de mercure corrosif.
German: Chlormerkur, Chlorquecksilber, Chlorwasserstoffsaeuresmerkur, Chlorwasserstoffsaeuresquecksilber, Merkurichlorid, Quecksilberchlorid, Sublimat.

Agricultural
Reagent for treating—
Lawns.

Analysis
Reagent in various processes.

Chemical
Catalyst in making—
Intermediates, pharmaceuticals.

Catalyst in various chemical processes, such as bromination, sulphonation, nitration, diazotization, reduction.

Ingredient of catalytic preparations used in making—
Acenaphthylene, acenaphthaquinone, bisacenaphthylidenedione, naphthaldehydic acid, naphthalic anhydride, and hemimellitic acid from acenaphthene (Brit. 295270).
Acetaldehyde from ethyl alcohol (Brit. 281307).
Acetic acid from ethyl alcohol (Brit. 281307).
Alcohols from aliphatic hydrocarbons (Brit. 281307).
Aldehydes and acids from toluene, orthochlorotoluene, orthobromotoluene, orthonitrotoluene, parachlorotoluene, parabromotoluene, paranitrotoluene, metachlorotoluene, metanitrotoluene, metabromotoluene, dinitrotoluenes, dibromotoluenes, dichlorotoluene, chlorobromotoluenes, chloronitrotoluenes, bromonitrotoluenes (Brit. 295270).
Aldehydes and acids from xylenes, pseudocumene, mesitylene, and paracymene (Brit. 295270).
Alphanaphthaquinone from naphthalene (Brit. 281307).
Anthraquinone from anthracene (Brit. 295270).
Benzaldehyde and benzoic acid from toluene (Brit. 281307).
Benzoquinone from phenanthraquinone (Brit. 281307).
Chloroacetic acid from ethylenechlorohydrin (Brit. 295270).
Diphenic acid from ethyl alcohol (Brit. 281307).
Fluorenone from fluorene (Brit. 295270).
Formaldehyde from methanol or methane (Brit. 295270).
Maleic acid and fumaric acid by the oxidation of benzol, toluol, phenol, tar phenols, or furfural, or from benzoquinone or phthalic anhydride (Brit. 295270).
Naphthaldehydic acid, acenaphthaquinone, or bisacenaphthylidenedione from acenaphthylene (Brit. 281307).
Phenanthraquinone from phenanthrene or diphenic acid (Brit. 295270).
Phthalic acid and maleic acid from naphthalene (Brit. 295270).
Salicylic acid and salicylic aldehyde from cresol (Brit. 295270).
Vanillin and vanillic acid from eugenol or isoeugenol (Brit. 295270).

Reagent in making—
Albuminous pharmaceuticals, aminophenylmercuric arsenate (aspirochyl), ethylene chlorobromide, ethylene isothiocyanate, rubber vulcanization accelerators from amino compounds and acetylene vinyl chloride.

Starting point in making—
Arsenic trichloride, calomel, mercuric-ammonium chloride, mercuric salts.

Electrical
Depolarizing reagent in making—
Batteries and cells.

Insecticide
Ingredient of—
Bedbug killers, germicidal preparations, preparations for the removal of fly in sheep, worm-killing compositions.

Leather
Ingredient of—
Compositions used in dressing skins.

Tanning agent in making—
Special leathers.

Metallurgical
Ingredient of—
Compositions used for coating metals of various sorts.
Compositions used for coloring metals.
Preparations used in the electroplating of aluminum.
Steel-bronzing compositions.

Reagent in making—
Zinc and tin alloys of fine metallographic characteristics.

Starting point in making—
Aluminum amalgam.

Miscellaneous
Ingredient of—
Coating compositions which contain metals (German 424658).
Embalming fluids.
Preparations used for general antiseptic purposes.

Mordant in treating—
Rabbit and beaver hair in the manufacture of hats.

Preservative for—
Anatomical specimens.

Reagent in making—
Yeast preparations and preparations of other microorganisms.

Reagent in dressing—
Furs.

Mercuric Chloride (Continued)
Paint and Varnish
Ingredient of—
Antiseptic and germicidal paints and varnishes, fruit tree paints.
Reagent (Brit. 292168) in making—
Lacquer and varnish bases from amino compounds and acetylene.
Pharmaceutical
In compounding and dispensing practice.
Perfumery
Ingredient of various cosmetics.
Photographic
As an intensifier.
Printing
Reagent in—
Process engraving and in lithographic work.
Textile
——, *Dyeing and Printing*
Mordant on various textiles.
Reagent (Brit. 292168) in making—
Reserve compositions used in dyeing and printing.
——, *Finishing*
Reagent (Brit. 292168) in making—
Finishing, wetting out and fiber protecting compositions.
Woodworking
Ingredient of—
Impregnating compositions, preservative applications.

Mercuric Iodide
Synonyms: Deutoiodide of mercury, Mercury biniodide, Red iodide of mercury.
Latin: Deutoioduretum hydrargyri, Hydrargyrum iodatum, Hydrargyrum bijodatum, Ioduretum hydrargyricum, Mercurius iodatus.
French: Bi-iodure de mercure, Deutoiodure de mercure, Iodure mercurique, Iodure rouge de mercure.
German: Quecksilberjodid, Mercurijodid, Rothesjodquecksilber.
Spanish: Yoduro, mercurico.
Italian: Bijoduro di mercurio.
Analysis
Ingredient of—
Nessler's reagent for detecting and estimating ammonia in water.
Chemical
Starting point in making—
Compounds with iodine fatty acids (German 215664).
Paint pigment with the aid of cuprous iodide.
Mechanical
Reagent for—
Revealing overheating of machine parts and bearings by change in color.
Miscellaneous
Reagent for—
Distinguishing between precious stones.
Paint and Varnish
Pigment in—
Artists' colors.
Paints for indicating excess heat.
Pharmaceutical
In compounding and dispensing practice.
Photographic
As an intensifier.

Mercuric Oxide, Red
Synonyms: Red oxide of mercury, Red precipitate.
Latin: Hyrarayrum oxidatum rubrum.
French: Oxyde de mercure, Oxyde mercurique, Oxyde de mercure rouge.
German: Quecksilberoxyd, Rotes quecksilberoxyd.
Analysis
Oxidizing agent in nitrogen determination.
Ceramics
Pigment in coloring and decorating—
Chinaware, porcelains, potteries.
Chemical
Catalyst in making—
Acetone from acetylene.
Desulphurizing agent in making various organic compounds.
Oxidizing agent in making—
Amino and acetylene compounds by sulphonation, reduction, nitration (Brit. 292168).
Cyanamide from sulphourea.
Di-iodosalicylic acid.
Hypochlorous anhydride.
Starting point in making—
Mercuric salts, mercury parasiticides, mercury pharmaceuticals.
Electrical
Depolarizer in admixture with graphite in—
Batteries containing manganese dioxide and sulphuric acid.
Insecticide
Ingredient of—
Parasiticides.
Miscellaneous
Ingredient of—
Metal polishes.
Paint and Varnish
Pigment in making—
Anti-fouling paints and varnishes.
Marine paints and varnishes.
Reagent (Brit. 292168) in making—
Lacquer and varnish bases.
Perfumery
Ingredient of—
Grease paints, pomades.
Pharmaceutical
In compounding and dispensing practice.
Rubber
Reagent (Brit. 292168) in making—
Vulcanization assistants.
Textile
——, *Dyeing and Printing*
Reagent (Brit. 292168) in making—
Reserves.
——, *Finishing*
Reagent (Brit. 292168) in making—
Fiber-protecting agents, wetting agents.

Mercuric Oxide, Yellow
Synonyms: Yellow mercury oxide, Yellow oxide of mercury, Yellow precipitate.
Latin: Hydrarayri oxidum flavum, Hydragyrum oxydatum.
French: Oxyde de mercure, Oxyde de mercure jaune, Oxyde mercurique.
German: Gelbes quecksilberoxyd, Quecksilberoxyd.
Analysis
Reagent in various processes.
Ceramics
Pigment for various wares.
Chemical
Catalyst in making—
Acetone from acetylene.
Oxdizing agent in making—
Acetaloxime, allyl ether, cacodylic acid.
Derivatives from amino compounds and acetylene by sulphonation, reduction, nitration, etc. (Brit. 292168).
Diphenyleneanilidodihydrotriazole.
Reagent (Brit. 292168) in making—
Rubber-vulcanization assistants.
Starting point in making—
Mercury salts, parasiticides, pharmaceuticals.
Electrical
Depolarizer in—
Batteries containing manganese dioxide and sulphuric acid.
Miscellaneous
Ingredient of—
Polishing compositions.
Paint and Varnish
Pigment in—
Special paints.
Reagent (Brit. 292168) in making—
Lacquer and varnish bases.
Perfumery
Ingredient of various cosmetics.
Pharmaceutical
In compounding and dispensing practice.
Textile
——, *Dyeing and Printing*
Reagent (Brit. 292168) in making—
Reserves.
——, *Finishing*
Reagent (Brit. 292168) in making—
Fiber protecting agents, wetting agents.

Mercuric-Potassium Iodide
Synonyms: Mayer's reagent, Mercury and potassium iodide, Potassium iodohydrargyrate.
Analysis
As a reagent in various processes.
Miscellaneous
Ingredient of—
Dental cements (U. S. 1613532).
Pharmaceutical
In compounding and dispensing practice.

Mercurous Chloride
Synonyms: Calomel, Mercury chloride, mild; Mild chloride of mercury, Mercury monochloride, Mercury protochloride, Mercury subchloride, Muriate of mercury, Submuriate of mercury.
Latin: Calomelas, Chloruretum hydrargyrosum, Hydrargyri chloridum, Hydrargyrum chloratum, Mercurius dulcis.
French: Calomel, Chlorure mercureux, Protochlorure de mercure, Protochlorure ou sous-muriate de mercure.
German: Kalomel, Mercurochlorid, Quecksilberchlorür.
Spanish: Cloruro mercurioso sublimado.
Italian: Protocloruro di mercurio.
Electrical
Electrolyte in batteries.
Explosives
Ingredient of—
Pyrotechnic compositions.
Miscellaneous
Ingredient of—
Dental cements (U. S. 1613532).
Paint and Varnish
Ingredient of—
Ships-bottoms paints.
Paper
Ingredient of—
Compositions for making safety paper.
Perfumery
Ingredient of—
Cosmetic creams and lotions.
Pharmaceutical
In compounding and dispensing practice.

Mercury
Synonyms: Quicksilver.
Latin: Argentum vivum, Hydrargyrum, Hydrargyrum vivum, Mercurius vivus.
French: Mercure purifié, Vif argent.
German: Quecksilber.
Spanish: Mercurio.
Italian: Mercurio.
Aviation
Process material in making—
Manifolds for airplane engines (U. S. 1282266 and 1282269).
Propellors for airplanes (U. S. 1335846, 1343191, 1282265, 1282268, and 1282270).
Chemical
Base material in making—
Double salts, such as ammonium chloride, potassium cyanide, potassium iodide, potassium thymolsulphonate, silicylarsenate.
Mercuric compounds, such as acetate, benzoate, bromide, chloride, cyanide, formate, iodide, lactate, mercaptide, nitrate, oleate, linoleate, oxides, oxycyanide, salicylate, succinimide, sulphates, sulphides, sulphocyanate.
Mercurous compounds, such as bromide, chloride, formate, iodide, nitrate, oxide, sulphate, tannate.
Base material in making—
Mercury fulminate.
Pharmaceuticals, such as thymegol, enesol, mercurochrome, mercurol, mercurosal, and the like.
Catalyst in making—
Acetaldehyde (U. S. 1151928, 1151929, 1436550, 1184177, 1247270, 1319365, 1431301, 1489915, and 1501502).
Acetic anhydride (U. S. 1425500).
Acetic acid (U. S. 1128780, 1159376, and 1431301).
Acetylene oxidation products (U. S. 1355299).
Ammonia (U. S. 1495655, and 1396557).
Anilin (U. S. 1239822).
Anthraquinonesulphonic acids (U. S. 1437571).
Benzene nitration or oxidation products (U. S. 1380185, and 1380186).
Para-aminophenol (U. S. 1239822).
Various other organic chemicals.
Vinyl esters (U. S. 1425130).
Cathode in—
Electrochemical processes.
Process material in making—
Activated carbons (U. S. 1520437).
Aurintricarboxylic acid derivatives (U. S. 1412440).
Ammonium formate (U. S. 1185028).
Calcium formate (U. S. 1185028).
Carbon blacks (U. S. 1498924).
Cinnabar (U. S. 1137467).
Dinitrophenol-2:4 (U. S. 1320076).
Formic acid (U. S. 1185028).
Hydrochloric acid (U. S. 1498924).
Hydrogen peroxide (U. S. 1904101).
4-Hydroxymeta-arsanilic acid (U. S. 1232373).
Lampblack (U. S. 1498924).
Lead sulphate (U. S. 1485794).
Lithopone (U. S. 1455963).
Nitrobenzene (U. S. 1320076 and 1320077).
Nitrophenol (U. S. 1320076 and 1320077).
Phthalic anhydride (U. S. 1261022, and 1443094).
Starting point in making—
Albumen compounds, alkali amalgams, alkali-earth amalgams, colloidal suspensions.
Dibromofluorescein derivatives (U. S. 1455495).
Glucosides (U. S. 1354105).
Hydroquinolphthalein derivatives (U. S. 1455495).
Methylfluorescein derivatives (U. S. 1455495).
Picric acid (U. S. 1320076 and 1320077).
Vermilions and scarlets.

Disinfectant
Process material in making—
Bactericides (U. S. 1145634).

Electrical
As such in—
Arc lamps of the mercury type.
Electric devices of various kinds.
Rectifiers of the electric current.
Ultraviolet ray equipment.
Bath (U. S. 1328530) in—
Sealing glass bulbs.
Color varier in—
Neon sign gaseous mixtures.
Ingredient (U. S. 1352331, 1138220, and 1138221) of—
Electrolytes.
Plating agent (U. S. 1366489) for—
Electrodes.
Preventer (U. S. 1393739) of—
Corroding of zinc electrodes (in combination with bismuth).
Process material in making—
Batteries (wet and dry) (U. S. 1174798, 1134093, 1211388, 1486613, 1138220, 1139213, 1342953, 1299693, 1497160, 1370119, and 1342953).

Explosives and Matches
Base material in making—
Fulminate of mercury.
Process material in making—
Picric acid (U. S. 1320076 and 1320077).

Glass
Silvering agent in amalgamation with tin for—
Mirrors and reflectors.

Lubricant
Mercurating agent (Brit. 433257) in making—
Addition agents for high-temperature lubricants.

Mechanical
Heat-transfer medium in—
Mercury boilers.
Ingredient of—
Boiler compounds (U. S. 1181562, and 1210965).
Starting point in making—
Boiler compounds with castor oil.

Metallurgical
As a rustproofing agent (U. S. 1518622).
Current carrier (Brit. 403404) in—
Electrical heat-treatment of metals.
Electrode in—
Electroplating processes.
Hardening agent (U. S. 1360346, 1360347, and 1360348) for—
Lead.
Ingredient of—
Alloys.
Bearing metals (U. S. 1360272, 1360346, and 1360347).
Dental alloys, special solders, special soldering fluxes

Mercury (Continued)
Starting point in making amalgams with—
Bismuth, copper, gold, lead, potassium, silver, sodium, tin, zinc.
Mining
Amalgamating agent in extracting—
Gold from its ores.
Precious metals from lead, gold, and silver ores.
Silver from its ores.
Miscellaneous
Amalgamating agent in—
Dentistry.
Indicator in—
Barometers, hydrometers, thermometers.
Ingredient of—
Dental fillings.
Process material in—
Chinese gilding process.
Sharpening agent (U. S. 1314450) for—
Files.
Paint and Varnish
Starting point in making—
Mercury linoleates, resinates, palmitates, and other soaps.
Vermilion pigments.
Paper
Drying agent (U. S. 1147808 and 1147809).
Petroleum
Process material (U. S. 1373653) in making—
Gasoline.
Pharmaceutical
Ingredient of—
Ointments, pills, powders.
Mercurating agent in making—
Antisyphilitics.
Diuretics (Brit. 447877).
Photographic
Drying agent (U. S. 1232077) for—
Motion picture film.
Intensifier (U. S. 1433806) for—
Photographic images.
Printing
Process material (U. S. 1377517) in making—
Printing plates.
Resins
Catalyst (U. S. 1377517) in making—
Phenol-aldehyde condensates.
Soap
Ingredient of—
Medicinal soaps and soap ointments.
Textile
Drying agent (U. S. 1147808 and 1147809) for—
Cotton fabrics.

Mercury-Amino Chloride
Agriculture
For control of—
Bottom rust of lettuce.
Covered smut and stripe disease of barley.
Kernel smut of sorghum.
Loose and covered smut of oats.
Soil-borne parasitic fungi.
Stinking smut of wheat.
Disinfectant for—
Seeds and soil.
Woodworking
For control of—
Blue stain and sap stain in sapwood of freshly-cut lumber.

Mercury-Anilin Hydrochloride
French: Hydrochlorure de mercure et aniline.
German: Quecksilberanilinchlorhydrat.
Agriculture
Disinfectant in treating—
Seeds (Brit. 274974).

Mercury Benzenesulphonate
French: Benzènesulphonate de mercure.
German: Benzolsulfonsaeuresmerkur, Quecksilberbenzolsulfonat.
Chemical
Catalyst in making—
Ethylidene diacetate (Brit. 252632).

Mercury Benzotrifluoride
Pharmaceutical
Claimed (U. S. 2050075) to be—
Antiseptic and valuable for other therapeutic purposes in which mercury compounds are employed.

Mercury Bisulphate
Synonyms: Mercuric sulphate, Mercury persulphate, Mercury sulphate, Normal mercury sulphate.
French: Sulphate mercurique.
German: Mercurisulfat, Merkurisulfat, Quecksilbervitriol.
Chemical
Catalyst in making—
Acetaldehyde from acetylene.
Ethylene diacetate (Brit. 252632).
Starting point in making—
Ethylenediamine-mercury sulphate (Sublamin), mercuric benzoate, mercuric bromide, mercuric chloride, mercuric cyanide, mercuric iodide, mercuric sulphide (black), mercuric sulphocyanate, mercurous chloride, mercurous oxide (black), mercury sulphate (basic).
Electrical
Ingredient of active agent in electric batteries.
Metallurgical
Reagent in extraction of—
Gold and silver from roasted pyrites.
Pharmaceutical
In compounding and dispensing practice.

Mercury Dinaphthylnaphthenate
Lubricant
Addition agent (Brit. 433257) in—
Lubricating oils or greases, especially for use at high temperatures, such as cylinder oils, hydrogenated oils, or oils refined by treatment with sulphuric acid, clay, or extraction solvents.

Mercury Erucate
Synonyms: Mercuric erucate.
French: Érucate de mercure, Érucate mercurique.
German: Erucinsäuresmerkur, Erucinsäuresmerkuroxyd, Merkurerucat.
Insecticide
Ingredient of—
Insecticidal compositions.
Spraying compounds for fungicidal purposes.
Paint and Varnish
Ingredient of—
Antifouling paints and varnishes.
Perfume
Ingredient of—
Cosmetics.
Pharmaceutical
In compounding and dispensing practice.
Sanitation
Ingredient of—
Disinfectants, germicides.

Mercury Fulminate
French: Fulminate de mercure.
German: Knallquecksilber, Merkurfulminat, Quecksilberfulminat.
Explosives
Active agent in—
Detonators, fuses.

Mercury Laurate
Synonyms: Mercuric laurate.
French: Laurate de mercure, Laurate mercurique.
German: Laurinsäuresmerkur, Laurinsäuresmerkuroxyd, Merkurlaurat.
Insecticide
Ingredient of—
Insecticidal compositions.
Spraying compounds for fungicidal preparations.
Paint and Varnish
Ingredient of—
Antifouling paints and varnishes.
Perfume
Ingredient of—
Cosmetics.
Pharmaceutical
In compounding and dispensing practice.
Sanitation
Ingredient of—
Disinfectants, germicides.

Mercury Naphthalenesulphonate
French: Naphthalènesulphonate de mercure.
German: Naphtalinsulfonsaeuresmerkur, Quecksilbernaphtalinsulfonat.
Chemical
Catalyst in making—
Ethylidene diacetate (Brit. 252632).

Mercury Oleate
Latin: Hydrargyrum oleatum.
French: Oléate de mercure, Oléate mercurique, Oléate de vif argent.
German: Merkurioleat, Oleinsäuresmerkur, Oleinsäuresmerkuroxyd, Quecksilberoleat.
Insecticide
Ingredient of—
Fungicidal sprays, insecticidal preparations.
Paint and Varnish
Ingredient of—
Antifouling paints and varnishes.
Perfume
Ingredient of—
Cosmetics of various sorts.
Pharmaceutical
In compounding and dispensing practice.
Sanitation
Ingredient of—
Disinfectants.

Mercury Palmitate
French: Palmitate de mercure, Palmitate mercurique, Palmitate de vif argent.
German: Merkuripalmitat, Palmitinsäuresmerkur, Palmitinsäuresmerkuroxyd, Quecksilberpalmitat.
Insecticide
Ingredient of—
Fungicidal sprays, insecticidal compositions.
Paint and Varnish
Ingredient of—
Antifouling paints and varnishes.
Perfume
Ingredient of—
Cosmetics of various sorts.
Pharmaceutical
In compounding and dispensing practice.
Sanitation
Ingredient of—
Disinfectants.

Mercury Phenylacetate
Petroleum
Addition agent (Brit. 433257) in—
Lubricating oils or greases, especially for use at high temperatures, such as cylinder oils, hydrogenated oils, or oils refined by treatment with sulphuric acid, clay, or extraction solvents.

Mercury Salicylate
Synonyms: Mercuric salicylate.
French: Salicylate de mercure.
German: Mercurisalicylat, Quecksilbersalicylat, Salicylsaeuresquecksilber.
Chemical
Starting point in making—
Enesol, mercury salicylarsinate.
Pharmaceutical
In compounding and dispensing practice.

Mercury Stearate
French: Stéarate de mercure, Stéarate mercurique, Stéarate de vif argent.
German: Merkuristearat, Quecksilberstearat, Stearinsäuresmerkur, Stearinsäuresmerkuroxyd.
Insecticide
Ingredient of—
Fungicidal sprays, insecticidal preparations.
Paint and Varnish
Ingredient of—
Antifouling paints and varnishes.
Perfume
Ingredient of—
Cosmetics of various sorts.
Pharmaceutical
In compounding and dispensing practice.
Sanitation
Ingredient of—
Disinfectants.

Mesitylenedisulphonic Acid
French: Acide de mésitylènedisulphonique.
German: Mesitylendisulfonsaeure.
Chemical
Emulsifying agent (Brit. 263873) for—
Aromatic hydrocarbons, solvents for fats, terpenes.
Starting point in making—
Acids and salts, intermediates, pharmaceuticals.
Dye
Starting point in making various synthetic dyestuffs.
Fats and Oils
As an emulsifying agent (Brit. 263873).
Leather
Emulsifying agent (Brit. 263873) in making—
Impregnating compositions, tanning compositions.
Miscellaneous
Emulsifying agent (Brit. 263873) in making—
Cleansing and washing compositions.
Paper
Emulsifying agent (Brit. 263873) in making—
Impregnating compositions.
Petroleum
As an emulsifying agent (Brit. 263873).
Resins and Waxes
As an emulsifying agent (Brit. 263873).
Textile
——, *Dyeing*
Emulsifying agent (Brit. 263873) in—
Acid dye liquors.
——, *Finishing*
Emulsifying agent (Brit. 263873) in making—
Wetting-out preparations.
——, *Manufacture*
Emulsifying agent (Brit. 263873) in making—
Wool-carbonizing liquors.

Mesityl Oxide
French: Éther mésitylique, Oxyde mésitylique, Mésityle.
German: Mesityloxyd.
Chemical
Starting point in making—
Methylisobutyl ketone.
Synthetic organic chemicals.
Miscellaneous
As a solvent.
Paint and Varnish
Solvent in making—
Lacquers and varnishes with sulphuretted condensation products of phenols and fatty aldehydes (Brit. 273756).
Lacquers, varnishes and enamels.
Resins and Waxes
Reagent in making—
Artificial resins with formaldehyde (German 433853).
Solvent for—
Vinyl resins.

Mesodibutylacridane
Fats and Oils
Antioxidant (Brit. 405797) for—
Fats, oils.
Petroleum
Antioxidant (Brit. 405797) for—
Petroleum derivatives.
Soap
Antioxidant (Brit. 405797) for—
Fats, oils, soaps.

Mesodiethylacridane
Fats and Oils
Antioxidant (Brit. 405797) for—
Fats, oils.
Petroleum
As an antioxidant (Brit. 405797).
Soap
Antioxidant (Brit. 405797) for—
Fats, oils, soaps.

Mesodimethylacridane
Fats and Oils
Antioxidant (Brit. 405797) for—
Fats, oils.
Petroleum
Antioxidant (Brit. 405797) for—
Petroleum derivatives.
Soap
Antioxidant (Brit. 405797) for—
Fats, oils, soaps.

Mesodimethylethylacridane
Fats and Oils
Antioxidant (Brit. 405797) for—
Fats, oils.
Petroleum
Antioxidant (Brit. 405797) for—
Petroleum derivatives.
Soap
Antioxidant (Brit. 405797) for—
Fats, oils, soaps.

Mesodimethylnaphthacridane
Fats and Oils
Antioxidant (Brit. 405797) for—
Fats, oils.
Petroleum
Antioxidant (Brit. 405797) for—
Oils.
Soap
Antioxidant (Brit. 405797) for—
Soaps, soapstocks.

Mesodiphenylacridane
Fats and Oils
Antioxidant (Brit. 405797) for—
Fats, oils.
Petroleum
Antioxidant (Brit. 405797) for—
Petroleum derivatives.
Soap
Antioxidant (Brit. 405797) for—
Fats, oils, soaps.

Mesodiphenyldinaphthacridane
Fats and Oils
Antioxidant (Brit. 405797) for—
Fats, oils.
Petroleum
Antioxidant (Brit. 405797) for—
Oils.
Soap
Antioxidant (Brit. 405797) for—
Soaps, soapstocks.

Mesodipropylacridane
Fats and Oils
Antioxidant (Brit. 405797) for—
Fats, oils.
Petroleum
Antioxidant (Brit. 405797) for—
Petroleum derivatives.
Soap
Antioxidant (Brit. 405797) for—
Fats, oils, soaps.

Mesoditolylacridane
Fats and Oils
Antioxidant (Brit. 405797) for—
Fats, oils.
Petroleum
Antioxidant (Brit. 405797) for—
Petroleum derivatives.
Soap
Antioxidant (Brit. 405797) for—
Fats, oils, soaps.

Mesoethylphenylacridane
Fats and Oils
Antioxidant (Brit. 405797) for—
Fats, oils.
Petroleum
Antioxidant (Brit. 405797) for—
Petroleum derivatives.
Soap
Antioxidant (Brit. 405797) for—
Fats, oils, soaps.

Mesomethylbutylacridane
Fats and Oils
Antioxidant (Brit. 405797) for—
Fats, oils.
Petroleum
Antioxidant (Brit. 405797) for—
Petroleum derivatives.
Soap
Antioxidant (Brit. 405797) for—
Fats, oils, soaps.

Mesomethylphenylacridane
Fats and Oils
Antioxidant (Brit. 405797) for—
Fats, oils.
Petroleum
Antioxidant (Brit. 405797) for—
Petroleum derivatives.
Soap
Antioxidant (Brit. 405797) for—
Fats, oils, soaps.

Mesothorium
German: Mesothor.
Chemical
As a catalyst.
Miscellaneous
As a substitute for radium.
Paint and Varnish
Ingredient of—
Luminous paints and varnishes, in admixture with zinc sulphide.
Pharmaceutical
In compounding and dispensing practice.

Meta-acetamidedimethylanilin
German: Meta-acetamiddoppeltemethylanilin.
Dye
Starting point in making—
Flaveosin.

Meta-aminobenzaldehyde
Dye
Starting point (Brit. 263164) in making azo dyestuffs with sulphonated derivatives of the following derivatives of 2:3-napthoic acid—
Anilide, betanaphthylamide, 5-chloro-2-anisidide, 5-chloro-5-toluidide, orthotoluidide.

Meta-aminobenzoic Anilide
French: Anilide de méta-aminobenzoique.
German: Meta-aminobenzoesaeuresanilid.
Dye
Starting point (Brit. 263164) in making azo dyestuffs as sulphonated derivatives of 2:3-oxynaphthoic acid as—
Anilide, betanaphthylamide, 5-chloro-5-toluidide, 5-chloro-2-anisidide, orthotoluidide.

Meta-aminoparacresol
Synonyms: 2-Aminoparacresol.
Chemical
Starting point in making—
Aromatics, intermediates, organic chemicals, pharmaceuticals.
Dye
Starting point in making—
Erichrome dyes.
Starting point (Brit. 347099) in making azo dyestuffs for cotton, with the aid of—
Acetoacetanilide, barbituric acid.
Benzoyl betanaphthylamine-5:7-disulphonic acid.
H acid, 3-methyl-5-pyrazolone.
2-Naphthylamine-4:8-disulphonic acid.
1-Naphthylamine-8:4-disulphonic acid.
Paratoluene-5-sulphonic acid.
Phenyl-2-naphthylamine-8:6-disulphonic acid.

Meta-aminophenyltriethylammonium Chloride
French: Chlorure de méta-aminophényletriéthyleammoniaque.
German: Meta-aminophenyltriaethylammoniumchlorid.
Dye
Starting point in making—
Janus red B.

Meta-aminophenyltrimethylammonium Chloride
French: Chlorure de méta-aminophényletriméthyle-ammonium.
German: Meta-aminophenyltrimethylammonium-chlorid.
Spanish: Cloruro de meta-aminofeniltrimetilamonio.
Italian: Cloruro di meta-aminofeniltrimetilammonio.

Dye
Intermediate in making—
Azophosphin, janus yellow G and R, janus red B, janus blue B.

Meta-aminosalicylic Acid Hydrochloride
French: Hydrochlorure de acide méta-amino salicylique.
German: Meta-aminosalicylsäurechlorhydrat, Meta-aminosalicylsäurehydrochlorid.

Chemical
Starting point in making—
Intermediates, pharmaceuticals, various other derivatives.

Dye
Starting point in making—
Azo-dyestuffs, sulphur dyestuffs.

Paper
Reagent in making—
Light-sensitive paper.

Pharmaceutical
In compounding and dispensing practice.

Metabenzamidobenzoic Acid
French: Acide de métabenzamidobenzoique.
German: Metabenzamidobenzoesaeure.

Dye
Reagent (French 604347) in making vat dyestuffs with—
Alpha-amino-4-methoxyanthraquinone.
1:4-Diaminoanthraquinone.
1:5-Diaminoanthraquinone.
4:8-Diaminoanthrarufin.
1:5-Diamino-4-hydroxyanthraquinone.
1:5-Diamino-4-methoxyanthraquinone.

Metabenzoylaminobenzoic Acid
French: Acide de métabenzoylaminobenzoique.
German: Metabenzoylaminobenzoesaeure.

Dye
Starting point in making—
Vat dyestuffs with 1:5-diaminoanthraquinone (Brit. 264561).

Metabromophenolindophenol

Analysis
Indicator in—
Oxidation-reduction potential determinations (of particular interest to biologists and physiologists and investigations of various materials, such as soils, wines, cheese, gasoline antiknock compounds).

Metachloro-orthobetachlorodeltabetaisopentenylphenol

Disinfectant
Claimed (Brit. 443113 and 389514) to be—
Disinfectant free of odor.

Metacresyl Acetate
French: Acétate metacrésylique, Acétate de metacrésyle.
German: Essigsäuremetakresylester, Essigsäuresmetakresyl, Metakresylacetat, Metakresylazetat.

Chemical
Starting point in making various derivatives.

Pharmaceutical
In compounding and dispensing practice.

Metacresylmethyl Ether

Chemical
Raw material in making—
Artificial musk.

Metadibenzylaminophenol

Petroleum
Gum inhibitor (U. S. 1980200 and 1980201) in—
Motor fuels.

Metaethyloxyphenoltolylamine

Dye
Starting point in making—
Acid violet 6BN.

Meta-m'-diaminoazoxybenzene
Synonyms: Meta-m'-diaminoazoxybenzol.
German: Meta-m'-diaminoazoxybenzol.

Dye
Starting point (Brit. 248946) in making azo dyestuffs with—
Alpha-aminoanthraquinone, 4-chloro-2-aminodiphenylether, 4-chloro-2-anisidin, 4-chloro-2-nitranilin, dianisidin, 2:4-dichloroanilin, 2:5-dichloroanilin, metachloroanilin, metanitranilin, 4-nitro-2-anisidin, 5-nitro-2-anisidin, 4-nitro-2-toluidin, 5-nitro-2-toluidin, 3-nitro-4-toluidin, orthoaminoazotoluene, orthoaminodiphenylether, orthophenetoleazoalphanaphthylamine, xylidin.

Meta-m'-diaminopara-p'-dimethyloxyazobenzene
Synonyms: Meta-m'-diaminopara-p'-dimethyloxyazobenzol.
German: Meta-m'-diaminepara-p'-dimethyloxyazobenzol.

Dye
Starting point (Brit. 248946) in making azo dyestuffs with—
Alpha-aminoanthraquinone, 4-chloro-2-aminodiphenyl ether, 4-chloro-2-anisidin, 4-chloro-2-nitranilin, dianisidin, 2:4-dichloroanilin, 2:5-dichloroanilin, metachloroanilin, metanitranilin, 4-nitro-2-anisidin, 5-nitro-2-anisidin, 3-nitro-4-toluidin, 4-nitro-2-toluidin, 5-nitro-2-toluidin, orthoaminodiphenyl ether, orthoaminoazotoluene, orthophenetoleazoalphanaphthylamine, xylidin.

Meta-m'-diphenyldicarboxylic Acid
French: Acide de méta-m'-diphényledicarboxyle.
German: Meta-m'-diphenyldicarbonsaeure.

Dye
Starting point (French 604347) in making anthraquinone vat dyestuffs with—
Alpha-amino-4-methoxyanthraquinone.
1:4-Diaminoanthraquinone.
1:5-Diaminoanthraquinone.
4:8-Diaminoanthrarufin.
1:5-Diamino-4-hydroxyanthraquinone.
1:5-Diamino-4-methoxyanthraquinone.

Metamethoxybenzaldehyde

Dye
As an intermediate.
Starting point (Brit. 398163) in making—
Claret shades fast to kier-boiling and chlorine.

Metamethoxybenzoyl Chloride
French: Chlorure de métaméthoxybenzoyle, Chlorure métaméthoxybenzoylique.
German: Chlormetamethoxybenzol.

Dye
Reagent (French 604347) in making vat dyestuffs with—
Alpha-amino-4-methoxyanthraquinone.
1:4-Diaminoanthraquinone.
1:5-Diaminoanthraquinone.
4:8-Diaminoanthrarufin.
1:5-Diamino-4-hydroxyanthraquinone.
1:5-Diamino-4-methoxyanthraquinone.

Metamethylbenzaldehyde

Chemical
Reagent in—
Organic synthesis.

Cosmetic
Ingredient of—
Perfumes.

Metamethylphenylethylamine
German: Meta-methylphenylaethylamin.

Chemical
Starting point in making—
6-Methyltetrahydroisoquinolin (German 423027).

Metanitroparatoluyl Chloride
French: Chlorure de métanitroparatoluyle.
German: Chlormetanitroparatoluyl, Metanitroparatoluylchlorid.

Chemical
Starting point (Brit. 278037) in making synthetic drugs with—
Alkoxynaphthylaminesulphonic acid.
Alphanaphthylamine-4:8-disulphonic acid.
Alphanaphthylamine-3:6:8-trisulphonic acid.
Alphanaphthylamine-4:6:8-trisulphonic acid.

Metanitroparatoluyl Chloride (Continued)
4-Aminoacenaphthene-3:5-disulphonic acid.
4-Aminoacenaphthene-3-sulphonic acid.
4-Aminoacenaphthene-5-sulphonic acid.
4-Aminoacenaphthenetrisulphonic acids.
1:5-Aminonaphthol-3:6-disulphonic acid.
1:8-Aminonaphthol-3:6-disulphonic acid.
1:5-Aminonaphthol-7-sulphonic acid.
Bromonaphthylaminesulphonic acid.
Chloronaphthylaminesulphonic acid.
Iodonaphthylaminesulphonic acid.

Metaphenylenediaminedisulphonic Acid
French: Acide de métaphénylènediaminedisulfonique.
German: Metaphenylendiamindisulfonsaeure.

Dye
Starting point in making—
Cotton orange G, cotton orange R, pyramin orange 3G.

Metaxylene
Synonyms: 1:3-Dimethylbenzene, 1:3-Dimethylbenzol, Metadimethylbenzene, Metaxylol.
German: 1:3-Dimethylbenzol.

Chemical
Solvent in making—
Artificial musk, bornyl acetate, diethylallylacetonitrile from diethylacetonitrile (Brit. 253950).
Starting point in making—
2:4-Dimethylbenzylamine (Brit. 249883), 2:4-dimethylbenzylphthalimide (Brit. 249883), metaxylidin, tetramethylanthracene.

Dye
Starting point in making—
Color lakes, various dyestuffs of red, scarlet, blue, and green shades.

Explosives
Starting point in making—
Trinitroxylene.

Gas
Solvent in removing—
Naphthalene from illuminating gas, preventing stopping-up of pipes with naphthalene.

Miscellaneous
Reagent in microscopical work.
Solvent for general purposes.

Paint and Varnish
Solvent in making—
Enamels, lacquers, varnishes.

Rubber
Solvent in making—
Rubber cements.

Textile
——, *Finishing*
Solvent in making—
Sizing compositions for rayons.

Meta-4-xylidin-6-sulphonbenzylmethylamide

Dye
Coupling agent (Brit. 434209 and 434433) in making—
Water-insoluble bluish-red dyestuffs with 5-chlor-2:4-dimethoxyanilide.

Meta-4-xylidin-6-sulphondiethylamide

Dye
Coupling agent (Brit. 434209 and 434433) in making—
Yellowish-red, water-insoluble dyestuffs with 2:5-dimethoxyanilide.

Methanol
Synonyms: Acetone alcohol, Carbinol, Colonial spirits, Columbian spirits, Columnian spirits, Green wood spirits, Manhattan spirits, Methyl alcohol, Methyl hydrate, Methyl hydroxide, Methylic alcohol, Pyroligneous spirit, Pyroxylic spirit, Standard wood spirits, Wood alcohol, Wood naphtha, Wood spirit.
Latin: Spiritus pyroxylicus rectificatus.
French: Alcool de bois, Alcool méthylique, Esprit de bois, Esprit pyroligneux, Méthanol, Méthanole, Méthyle alcool.
German: Holzalkohol, Holzgeist, Methanol, Methylalkohol.

Adhesives
Solvent in making—
Cements of various kinds.

Analysis
Reagent and solvent in—
Analytical processes involving control and research in science and industry.

Aviation
Nitrocellulose solvent in making—
Airplane dopes.

Cellulose Products
Solvent for—
Nitrocellulose in making various products from pyroxylin and the like.

Chemical
Denaturant for—
Ethyl alcohol.
Extractant in—
Manufacturing processes.
Methylating agent in making—
Esters of various kinds, such as methyl acetate, methyl benzoate, methyl chloride, methyl cinnamate, methyl formate, methyl salicylate.
Halogenation products, such as methyl bromide, methyl iodide, and methyl chloride.
Intermediates used in the manufacture of drugs, chemicals, and pharmaceutical products.
Organic chemicals, such as methylacetanilide, methylal, paramethylaminophenol, methyl anthranilate, methylanthraquinone, methyl sulphide, methylthionin chloride.
Solvent for—
Fats, nitrocellulose, oils, raw materials used in chemical manufacture, resins, various chemicals.
Solvent miscible with—
Ethyl alcohol, many other organic compounds, water.
Solvent having a fairly high tolerance for—
Benzene, ethyl ether, isopropyl ether.
Stabilizing agent (Brit. 427423) for—
Aqueous formaldehyde solutions.
Starting point in making—
Acetic acid (U. S. 1953905, 1961736, 1961737, and 1961738).
Derivatives, dimethyl ether (U. S. 1949344), formaldehyde.

Cosmetic
Extractant for—
Perfume components.
Solvent for—
Cosmetic ingredients, fats, hair tonic ingredients, nitrocellulose, oils.
Solvent in making—
Nail enamels and lacquers, synthetic perfumes.

Disinfectant
Solvent medium for—
Disinfectants, germicides.

Dye
Methylating agent in making—
Dimethylanilin, dyestuffs, intermediates, methylanthraquinone.
Solvent in—
Organic syntheses.

Explosives and Matches
Process material in making—
Poisonous gases.
Solvent for nitrocellulose in making—
Explosives.

Fats, Oils, and Waxes
Extractant for—
Oils.
Solvent for—
Fats, oils.

Fuel and Light
Fuel for—
Chafing dishes, cigarlighters, miners' lamps, small stoves, soldering torches.
Ingredient of—
Admixtures with ethyl alcohol for various heating and lighting purposes.
Fuel compositions used for heating and lighting purposes.
Starting agent for—
Gasoline lamps (either gasoline only or lamps for burning either gasoline or kerosene).
Starting point in making—
Solid alkohol fuels.

Glass
Degreasing and cleansing agent.
Diluent and solvent for—
Materials used in making nonscatterable glass.

Ink
Solvent for—
Ink ingredients.

Insecticide
Solvent (U. S. 1945235) in making—
Colorless pyrethrum spray products for household use.

Methanol (Continued)

Laundry and Dry Cleaning
Dry-cleaning agent.
Ingredient of—
Dry-cleaning solutions, spotting agents.
Solvent for—
Fats, oils.
Spotting agent.

Leather
Diluent and solvent for nitrocellulose in making—
Artificial leather, decorative effects on leather.

Lubricant
Solvent for—
Fats and oils.

Mechanical
Antifreeze for—
Internal combustion engine radiators.
Ingredient of—
Antifreeze preparations, fuels for internal combustion engines.

Metallurgical
Degreasing agent.

Miscellaneous
Cleansing agent for—
Various purposes where adequate ventilation is available or where means are provided to prevent overexposure to its vapors.
Diluent in—
Furniture polishes, metal polishes, lacquers used for various decorative effects, special polishes for various purposes.
Ingredient of—
Methylated spirits.
Solvent miscible with—
Ethyl alcohol, many other organic compounds, water.
Solvent having a fairly high tolerance for—
Benzene, ethyl ether, isopropyl ether.
Taxidermy agent.

Paint and Varnish
Diluent or solvent in—
Dopes, enamels, lacquers, paints, paint-removers, stains, varnishes.
Solvent for—
Colors, nitrocellulose, oils, resins.

Paper
Solvent for—
Nitrocellulose used in coatings and decorative effects.

Petroleum
Solvent in—
Refining processes.

Photographic
Drying agent.
Solvent for—
Nitrocellulose.

Plastics
Solvent for—
Nitrocellulose, resins.
Solvent in making—
Pyroxylin plastics.

Refrigeration
Starting point in making—
Methyl chloride.

Resins
Solvent.
Stabilizing agent (Brit. 427423) for—
Aqueous formaldehyde solutions.
Starting point in making—
Formaldehyde.

Rubber
Process material in making—
Vulcanization accelerators.

Soap
Solvent for—
Fats, oils.
Solvent in making—
Disinfecting soaps, special soaps, textile soaps, transparent soaps.

Textile
Cleaning agent.
Solvent in making—
Textile soaps.
Solvent for—
Nitrocellulose.

3-Methoxyacetamido-4:6-dimethoxyanilin

Dye
Starting point (Brit. 435711) in making—
Violet dyestuffs by coupling with parachloranilide.
Violet dyestuffs by coupling with 4-methoxy-2-methylanilide.

4-Methoxyacetamido-2:5-dimethoxyanilin

Dye
Starting point (Brit. 435711) in making—
Reddish-blue dyestuffs by coupling with 2:3 hydroxynaphthoic-2:5-dimethoxyanilide.

4-Methoxy-5-acetamino-2-amino-1-methylbenzene
Synonyms: 4-Methoxy-5-acetamino-2-amino-1-methylbenzol.
French: 4-Méthoxye-5-acétamino-2-amino-1-méthylebenzène.
German: 4-Methoxy-5-acetamino-2-amino-1-methylbenzol.

Chemical
Starting point in making—
Intermediates, pharmaceuticals.

Dye
Starting point (Brit. 307303) in making monoazo dyestuffs with—
Betachloropropionyl H acid, carbethoxy acetyl H acid, chloroacetyl H acid, normal acetyl H acid, phenylacetyl H acid.

4-Methoxyacetophenone

Mechanical
Improver (Brit. 404046) of—
Exhaust odors from internal combustion engines (added to fuels not derived from petroleum, either alone or in conjunction with (1) artificial musk compounds, or (2) artificial musk compounds and any of the following: Camphor, waste camphor oil, borneol, bornyl acetate, clove oil, ionone, coumarin, indole, skatole, paracresyl acetate, methyl anthranilate, isopropylmethylhydrocinnamic aldehyde).

Perfume
Aromatic in—
Cosmetics, perfumes.

Petroleum
Reagent (Brit. 404046) for—
Improving exhaust odors from internal combustion engines (added to gasoline or diesel oil, either alone or in conjunction with (1) artificial musk compounds, or (2) artificial musk compounds and any of the following: Camphor, waste camphor oil, borneol, bornyl acetate, clove oil, ionone, coumarin, indole, skatole, paracresyl acetate, methyl anthranilate, isopropylmethylhydrocinnamic aldehyde).

Soap
Aromatic in—
Soaps.

5-Methoxy-2:6′-dimethyl-1′-ethyloxa-psi-cyanin Iodide

Photographic
Sensitizer (Brit. 423827) for—
Photographic emulsions to blue-green light.

3-Methoxy-4:6-di-tertiarybutyltoluene
Synonyms: 3-Methoxy-4:6-di-tertiary-butyltoluol.

Aromatics
Starting point (U. S. 1926080) in making—
2:5-Dinitro derivatives used as an artificial musk odor.

Methoxyethyl-Mercury Acetate

Disinfectant
Starting point (Brit. 450256) in making—
Disinfectants with water-glass and other reactive silicon compounds.

Insecticide and Fungicide
Starting point (Brit. 450256) in making—
Seed immunizers with water-glass and other reactive silicon compounds.

5-Methoxyisophthalic Acid
French: Acide de 5-méthoxyisophthalique.
German: 5-Methoxyisophtalinsaeure.

Dye
Starting point (French 604347) in making anthraquinone dyestuffs with—
Alpha-amino-4-methoxyanthraquinone, 1:4-diaminoanthraquinone, 1:5-diaminoanthraquinone, 4:8-diaminoanthrarufin, 1:5-diamino-4-hydroxyanthraquinone, 1:5-diamino-4-methoxyanthraquinone.

3-Methoxy-4'-methyldiphenylamine

Rubber
As an antioxidant (Brit. 435024).

Methoxynaphthoyl Chloride
French: Chlorure de méthoxynaphthoyle, Chlorure méthoxynaphthoylique.
German: Chlormethoxynaphtoyl, Methoxynaphtoylchlorid.

Dye
Starting point in making—
Anthraquinone vat dyestuffs with 1:4-diaminoanthraquinone (German 432579).

5-Methoxyorthotoluidide

Dye
Starting point (Brit. 434209 and 434433) in making—
Yellowish-red, water-insoluble dyestuffs by coupling (in substance or on the fiber) with parachlorophenylorthotoluidin-4-sulphonate.
Bordeaux-red, water-insoluble dyestuffs by coupling (in substance or on the fiber) with phenyl-3-chloropara-anisidin-6-sulphonate.
Bordeaux, water-insoluble dyestuffs by coupling (in substance or on the fiber) with 2:4-dimethoxyanilin-5-sulphonbenzylmethylamide.

Methoxyphenylethylmethyl Ketone
German: Methoxyphenylaethylmethylketon.

Perfumery
Ingredient of—
Hair restorers, pomades.

Methyl Abietate
Synonyms: Methyl resinate.
French: Abiétate de méthyle, Abiétate méthylique, Résinate de méthyle, Résinate méthylique.
German: Abietinsaeuremethylester, Abietinsaeuresmethyl, Harzsaeuremethylester, Harzsaeuresmethyl, Methylabietat, Methylresinat.

Paint and Varnish
Plasticizer (Brit. 308524) in making—
Nitrocellulose varnishes, lacquers, and dopes.

Plastics
Plasticizer (Brit. 308524) in making—
Nitrocellulose plastics containing wood flour or cork.

Methyl Acetate
Latin: Aether lignosus, Methylum acetatum, Spiritus pyroaceticus.
French: Acétate de méthyle, Acétate méthylique.
German: Essigsäuremethylester, Essigsäuresmethyl, Methylacetat, Methylazetat.
Spanish: Acetato de metil.
Italian: Acetato di metile.

Ceramics
Solvent in—
Compositions containing cellulose acetate, nitrocellulose, or other esters or ethers of cellulose, used for the decoration and protection of ceramic products.

Chemical
Reagent in making—
Aromatics, intermediates, organic chemicals, pharmaceuticals.
Solvent for—
Nitrocellulose and cellulose acetate.
Various purposes (used in place of acetone).

Dye
Reagent in making various synthetic dyestuffs.
Solvent in making various synthetic dyestuffs.

Electrical
Solvent in—
Compositions containing cellulose acetate, nitrocellulose, or other esters or ethers of cellulose, used for insulating purposes in the manufacture of wire, cable, and electrical machinery and equipment.

Fats and Oils
Solvent for extracting—
Fats and oils from vegetable and other sources.

Food
Ingredient of—
Artificial fruit flavors.

Glass
Solvent in—
Compositions containing cellulose acetate, nitrocellulose, or other esters or ethers of cellulose, used in the manufacture of nonscatterable laminated glass and for the decoration and protection of glassware.

Glues and Adhesives
Solvent in making—
Adhesive compositions containing cellulose acetate, nitrocellulose, or other esters or ethers of cellulose.

Leather
Solvent in—
Compositions containing cellulose acetate, nitrocellulose, or other esters or ethers of cellulose, as well as resins and the like, used in the manufacture of artificial leather and for the decoration and protection of leather goods.

Metallurgical
Solvent in—
Compositions containing cellulose acetate, nitrocellulose, or other esters or ethers of cellulose, used for the decoration and protection of metallic wares.

Miscellaneous
Solvent in—
Compositions containing cellulose acetate, nitrocellulose, or other esters or ethers of cellulose, used for the decoration and protection of various articles.
Solvent in various processes.

Paint and Varnish
Solvent in making—
Paints, varnishes, dopes, enamels, lacquers, containing cellulose acetate, nitrocellulose, or other esters or ethers of cellulose, with natural and artificial resins (used in place of acetone).

Paper
Solvent in—
Compositions containing cellulose acetate, nitrocellulose, or other esters or ethers of cellulose, used in the manufacture of coated paper and for the protection and decoration of pulp and paper products.

Perfume
Ingredient of—
Cosmetics, perfumes.

Photographic
Solvent in making—
Films from cellulose acetate, nitrocellulose, or other esters or ethers of cellulose.

Plastics
Solvent in—
Compositions containing cellulose acetate, nitrocellulose, or other esters or ethers of cellulose (used in place of acetone).

Rubber
Solvent in—
Compositions containing cellulose acetate, nitrocellulose, or other esters or ethers of cellulose, used for the decoration and protection of rubber goods.

Stone
Solvent in—
Compositions containing cellulose acetate, nitrocellulose, or other esters or ethers of cellulose, used for the decoration and protection of natural and artificial stone.

Textile
Solvent in—
Compositions containing cellulose acetate, nitrocellulose, or other esters or ethers of cellulose, used for the decoration of textile fabrics.

Woodworking
Solvent in—
Compositions containing cellulose acetate, nitrocellulose, or other esters or ethers of cellulose, used for decoration and protection of woodwork.

Methyl Acetylglycollate
French: Acétyleglycollate de méthyle, Acétyleglycollate méthylique.
German: Acetylglykolsäuresmethylester, Acetylglykolsäuresmethyl, Methylacetylglykolat.

Miscellaneous
Solvent for—
Cellulose acetate, cellulose esters and ethers, cellulose nitrate, fats and oils, natural resins, synthetic resins.
For uses, see under general heading: "Solvents."

Methyl Acrylate
Synonyms: Acrylic acid methyl ester.
French: Acrylate de méthyle, Acrylate méthylique.
German: Acrylsäuresmethylester, Acrylsäuresmethyl, Akrylsäuresmethylester, Akrylsäuresmethyl, Methylacrylat, Methylakrylat.

Methyl Acrylate (Continued)
Spanish: Acrylato de metil.
Italian: Acrylato di metile.

Adhesives
Starting point in making—
Polymerization products which are waterproof, elastic, alcohol-resistant, gasoline-resistant, turpentine oil-resistant; such products are used in making colorless, stable, tenaceous, adhesives suitable for joining wood to wood, glass to glass, fibrous materials to leather, paper to paper, metallic foil to metallic foil, paper to metal.

Cellulose Products
Solvent (Brit. 321258) for—
Cellulose acetate, cellulose esters and ethers, cellulose nitrate.
For such uses, see under general heading: "Solvents."

Chemical
Starting point in making various derivatives.

Leather
Starting point (Brit. 387736) in making—
Polymerized products used for priming leather prior to coating it with cellulose lacquer or synthetic resin varnish.

Textile
——, *Dyeing and Printing*
Ingredient (Brit. 321258) of—
Compositions, containing cellulose acetate, nitrocellulose, or other esters or ethers of cellulose, as well as rubber, used as softeners in dye liquors and printing pastes.
——, *Finishing*
Ingredient (Brit. 321258) of—
Compositions, containing cellulose acetate, nitrocellulose, or other esters or ethers of cellulose, as well as rubber, used in the manufacture of coated fabrics.

Water and Sanitation
Breaker (U. S. 1964444) of—
Emulsoids in sewage.

Methyladipic Acid
French: Acide de méthyleadipinique, Acide de méthyleadipique.
German: Methyladipinsäure.
Spanish: Acido metiladipico.
Italian: Acido metiladipico.

Chemical
Starting point in making—
Esters and salts.
Esters with hydroaromatic alcohols.
Glyceryl methyladipate.
Methylcyclohexanol betamethyladipate.
Perfume fixatives.
Pharmaceutical chemicals.
Sodium oleomethyladipate.

Dye
Ingredient of various dyestuffs.

Food
Ingredient of—
Mineral yeast (used in the place of tartaric acid, cream of tartar, and bisphosphates for the purpose of making a more stable product and one that is nonhygroscopic).
Reagent for removing—
Bitter matters and principles from pressed yeast.

Leather
Resist in—
Dyeing leather goods.

Metallurgical
Reagent in—
Coloring metals and producing bronze effects.

Pharmaceutical
In compounding and dispensing practice.

Photographic
Reagent in making—
Photographic papers.

Plastics
In galvanoplastic work.

Textile
Mordant and resist in—
Dyeing and printing cottons, rayons, silks.

Methylal
Synonyms: Formal, Methylene dimethylate, Methylene dimethylester.

Chemical
Solvent for—
Organic acids.
Substitute for—
Ether as solvent where advantage is taken of its properties (1) solubility in water, (2) solvent for organic acids.
Formaldehyde in carrying out condensation reactions.

Essential Oil
Extractive for—
Aromatic principles.

Pharmaceutical
Suggested for use as—
Hypnotic.

Methyl Alphacrotonate
Synonyms: Alphacrotonic methyl ester.
French: Alphacrotonate de méthyle, Alphacrotonate méthylique.
German: Alphacrotonsäuremethylester, Alphacrotonsaeuresmethyl, Methylalphacrotonat.

Cellulose Products
Plasticizer (Brit. 321258) for—
Cellulose acetate, cellulose esters and ethers, cellulose nitrate, rubber.
For uses, see under general heading: "Plasticizers."

Chemical
As a general reagent.
Starting point in making various derivatives.

1-Methylaminoanthraquinone

Miscellaneous
Dyestuff (U. S. 1989133) for—
Cellulose acetate products (imparts shades of red).

Textile
Dyestuff (U. S. 1989133) for—
Cellulose acetate products (imparts shades of red).

1-Methylamino-4-betahydroxyethylaminoanthraquinone

Textile
Dyestuff (Brit. 447090 and 447037) for imparting—
Deep-blue shades to acetate rayon, either by dyeing or printing.

1-Methylamino-4-bromoanthraquinonyl-2-methanesulphonic Acid

Dye
Starting point (Brit. 440208) in making—
Acid wool dyes by condensation with organic bases having at least one hydrogen atom attached to the nitrogen atom.

1-Methylamino-4-chloroanthraquinonyl-2-methanesulphonic Acid

Dye
Starting point (Brit. 440208) in making—
Acid wool dyes by condensation with organic bases having at least one hydrogen atom attached to the nitrogen atom.

2-Methylaminonaphthalene-7-sulphonic Acid
French: Acide de 2-méthyleaminonaphthalène-7-sulphonique.
German: 2-Methylaminonaphtalin-7-sulfonsaeure.

Dye
Starting point (Brit. 265767) in making monoazo dyestuffs with—
3:5-Dinitro-2-aminobenzylsulphonic acid.

1-Methylamino-4-normalbutylaminoanthraquinone

Textile
Dyestuff (Brit. 447090 and 447037) for imparting—
Deep-blue shades to acetate rayon, either by dyeing or printing.

1-Methylamino-4-paratoluidinoanthraquinone

Oils, Fats, Waxes
Coloring agent (Brit. 432867) for—
Paraffin and other mineral waxes, stearic acid, tallow and other solid triglycerides, beeswax, carnauba wax, and others.

Petroleum
Dye (U. S. 1969249) for—
Gasoline.

Methyl-Ammonium Chloride
Automotive
Ingredient (Brit. 334181) of—
Motor fuels.
Chemical
Starting point in making various derivatives.

Methylamyl Alcohol
Synonyms: Methylisobutylcarbinol.
Chemical
As a medium high-boiling alcohol in chemical processes.
Starting point in making—
Esters, such as methylamyl acetate.
Dye
Solvent for—
Dyestuffs.
Fats, Oils, and Waxes
Solvent for—
Oils, waxes.
Gums
Solvent for—
Gums.
Miscellaneous
As a solvent, miscible with most common organic solvents, in industrial processes.
Paint and Varnish
Imparter of—
Good "flow-out" in lacquer formulation.
High gloss in lacquer formulation.
Resistance to blushing in lacquer formulation.
Medium high-boiling solvent in—
Lacquer formulation.
Solvent for—
Dyes, gums, oils, resins, waxes.
Resins
Solvent for—
Resins.

Methyl Anthranilate
Food
As a flavoring.
Ingredient of—
Artificial grape flavors.
Cosmetic
Aromatic in—
Cosmetic preparations.
Compounding agent for—
Synthetic perfumes.

8-Methyl-1:2:3:9-benzisotetrazole
Pharmaceutical
Claimed (U. S. 2008536) to have—
Valuable therapeutic properties and solubility in water.

1-Methylbenzothiazole-5-carboxylic Chloride
Dye
Starting point (Brit. 441915) in making—
Greenish-yellow vat dyes of good fastness to light, chlorine, and alkali, by condensing with an orthoaminothiol of the benzene, naphthalene, or anthraquinone series.
Greenish-yellow vat dyes of good fastness to light, chlorine, and alkali, by condensing with an arylamine and the orthothiol group subsequently introduced and the product cyclized.

Methylbetaphenoxyethyllauramide
Chemical
Starting point (Brit. 443902) in making—
Sulphonated sodium salts, stable to calcium chloride and acids, which are used as scouring agents for raw wool.

Methylbetaphenoxyethylstearamide
Chemical
Starting point (Brit. 443902) in making—
Sulphonated sodium salts, stable to calcium chloride and acids, which are used as scouring agents for raw wool.

Methyl Biscyclopentenylacetate
Food
Agent for—
Producing pineapple aroma and flavor.

Methyl Borate
French: Borate de méthyle, Borate méthylique.
German: Borsäuremethylester, Borsäuresmethyl.
Spanish: Borato de metil.
Italian: Borato di metile.
Petroleum
Ingredient (Brit. 334181) of—
Motor fuels (added to prevent knock).

Methyl Bromide
French: Bromure de méthyle.
German: Methylbromid.
Spanish: Bromuro de metail.
Italian: Bromuro di metile.
Chemical
As a noninflammable solvent.
Ingredient of—
Noninflammable solvent mixture with methyl chloride (French 530052).
Noninflammable solvent mixture with ethyl chloride alone and with methyl chloride (French 531293).
Solvent mixtures.
Methylating agent in—
Organic synthesis.
Starting point in making—
Atropine methylbromide, codeine methylbromide, morphine methylbromide.
Dry Cleaning
Ingredient (French 531293) of—
Noninflammable solvent mixture with ethyl chloride alone and with methyl chloride.
Miscellaneous
As an extinguishing fluid for—
Airplane fires, automobile fires, chemical flames, domestic fires, factory fires, gasoline flames.
Ingredient of—
Fire-extinguishing fluid comprising a mixture with pentachlorethane in conjunction with compressed nitrogen as a propellant (Brit. 369003).
Fire-extinguishing fluid comprising a mixture with ethyl chloride alone and with methyl chloride (French 531293).
Fire-extinguishing fluid comprising a mixture with carbon tetrachloride (French 636714).
Fire-extinguishing fluid comprising a mixture with ethyl bromide (French 636714).
Refrigeration
Ingredient of—
Noninflammable refrigerating fluid comprising a mixture with methyl chloride (French 530052).
Noninflammable refrigerating fluid comprising a mixture with ethyl chloride alone and with methyl chloride (French 531293).

4-Methyl-5-bromo-3-oxythionaphthene
German: 4-Methyl-5-brom-3-oxythionaphten.
Spanish: 4-Metil-5-brom-3-oxisolfonaftene.
Italian: 4-Metile-5-bromo-3-oxisulfonaftena.
Chemical
Starting point in making—
Intermediates, pharmaceuticals.
Dye
Starting point (Brit. 271906) in making thioindigoid dyestuffs with the aid of—
Acenapththaquinones, dichloroisatin anilide, dichloroisatin chloride, diketones and derivatives, isatin and derivatives.

1-Methylbutylalphanaphthol
Disinfectant
Claimed (U. S. 2073996 and 2073997) to be—
Germicide combining high efficiency toward *staphylococcus aureus* and low toxicity.

Methylbutyleneglycol Acetate
French: Acétate de méthylebutylène glycol.
German: Essigsäuresmethylbutylenglykolester, Methylbutylenglykolacetat, Methylbutylenglykolazetat.
Cellulose Products
High-boiling-point solvent for—
Benzylcellulose, cellulose derivatives, nitrocellulose.
For uses, see under general heading: "Solvents."

Methylcarbamide
French: Carbamide de méthyle, Carbamide méthylique.
German: Methylcarbamid.

Methylcarbamide (Continued)

Chemical
Starting point in making—
Intermediates, pharmaceuticals.

Resins and Waxes
Reagent (Brit. 292912) in making synthetic resins with the aid of—
Acetylsalicylic acid, aliphatic dibasic acids, ammonium salicylate, anthranilic acid, benzoic acid, gallic acid, hydronaphthoic acid, magnesium salicylate, oxalic acid, phenolic dibasic acids, phthalic acid, salicylamide, salicylic acid, strontium salicylate, succinic acid.

Methyl Cellulose
French: Cellulose méthylique.

Ceramics
Ingredient of—
Coating compositions used for the decoration and protection of ceramic ware.

Glass
Ingredient of—
Compositions used for coating glass and also in the manufacture of nonscatterable glass.

Leather
Ingredient of—
Compositions used for the decoration of leather goods and also in the manufacture of artificial leather.

Metallurgical
Ingredient of—
Coating compositions for decorating metalware.

Paint and Varnish
Ingredient of—
Dopes, enamels, lacquers, paints, varnishes.

Paper
Ingredient of—
Coating compositions for treating paper and pulp products as well as in making coated paper.

Plastics
Ingredient of—
Compositions, threads, films, and sheets.

Rubber
Ingredient of—
Coating compositions.

Stone
Ingredient of—
Coating compositions for natural and artificial stone.

Textile
Ingredient of—
Coating compositions.

Woodworking
Ingredient of—
Coating compositions for decorating and protecting wood.

Methylcetylglucamine Hydrochloride

Miscellaneous
Detergent (Brit. 428142 and 428148) in—
Cleansing operations, particularly in hard water or acids.

Methyl Chloride
Synonyms: Chloromethane.
Latin: Methylium chloratum.
French: Chlorure de méthyle.
German: Chlormethyl, Methylchlorid.
Spanish: Cloruro de metil.
Italian: Cloruro di metile.

Animal Products
Ingredient (Brit. 152550) of—
Solvent mixture, with ethyl chloride, for selective extraction of oils and greases, odorous materials, and other derivatives.

Chemical
Chlorinating agent in—
Organic syntheses.
Ingredient (French 530052) of—
Noninflammable solvent mixture with methyl bromide.
Methylating agent in—
Organic syntheses.
Starting point in making—
Acetyl chloride by reacting with carbon monoxide (Brit. 308666).
Ethyl chloride (French 564641).
Ethylene (French 564641).
Ethylidene chloride (French 564641).
Hydrocarbons boiling between 120° and 200° C. by reacting with ethylene in presence of aluminum chloride (French 695125).
Methylene chloride.
Various organic chemicals containing chlorine or methyl groups.

Dry Cleaning
Ingredient (German 584515) of—
Noninflammable spotting and staining agents consisting of various mixtures with gasoline and various noninflammable solvents.

Dye
Chlorinating agent in—
Dye syntheses.
Methylating agent in—
Dye syntheses.

Fats and Oils
Ingredient (Brit. 152550) of—
Solvent mixture, with ethyl chloride, for selective extraction of oils, greases, and odorous materials.

Miscellaneous
Ingredient (French 530052) of—
Noninflamable solvent mixture with methyl bromide.

Perfume
Extractant for—
Essential oils, odorous principles, perfume materials.
Ingredient (Brit. 152550) of—
Solvent mixture, with ethyl chloride, for selective extraction of odorous materials.
Solvent for—
Essential oils, odorous principles, perfume materials.

Petroleum
Solvent (Brit. 423303) for—
Coloring matters and asphaltic compounds in processes of dewaxing hydrocarbon oils, such as residium stocks, overhead distillates, and crude petroleum or shale oils.

Pharmaceutical
Claimed as—
Local anesthetic.

Refrigeration
Ingredient of—
Refrigerant mixture with methyl bromide (French 530052).
Refrigerant mixture with chloropicrin (U. S. 1879893).
Refrigerant in—
Air-conditioning systems, baking industry units, candy industry units, dairy products units, dispensing units, flower storage units, frozen food industry units, fur storage units, household units, ice cream units, ice cream plant systems, motor truck cooling units, multiple unit systems, refrigerator car units, room coolers, water coolers.

Resins
Ingredient (Brit. 152550) of—
Solvent mixture, with ethyl chloride, for selective extraction of resins.

Soap
Ingredient (Brit. 152550) of—
Solvent mixture, with ethyl chloride, for selective extraction of oils, fats, greases.

4-Methyl-5-chloro-7-methoxyisatin Chloride
French: Chlorure de 4-méthyle-5-chloro-7-méthoxyeisatine, Chlorure de 4-méthyle-5-chloro-7-méthoxyeisatinique.
German: Chlor-4-methyl-5-chlor-7-methoxyisatin, 4-Methyl-5-chloro-7-methoxyisatinchlorid.

Chemical
Starting point in making various intermediates.

Dye
Starting point (Brit. 309379) in making thioindigoid dyestuffs with—
5-Bromo-3-oxythionaphthene.
5-Chloro-7-methyl-3-oxythionaphthene.
5-Chloro-3-oxythionaphthene.
5:7-Dibromo-3-oxythionaphthene.
5:7-Dichloro-3-oxythionaphthene.
4:7-Dimethyl-5-chloro-3-oxythionaphthene.
5-Methyl-7-chloro-3-oxythionaphthene.
4-Methyl-5:7-dichloro-3-oxythionaphthene.
5:6:7-Trichloro-3-oxythionaphthene.

4-Methyl-5-chloro-3-oxythionaphthene
Synonyms: 4-Methyl-5-chloro-3-oxysulphonaphthene.
French: 4-Méthyle-5-chloro-3-oxythionaphthène.
German: 4-Methyl-5-chloro-3-oxysulfonaphten.

Chemical
Starting point in making—
Intermediates.

Dye
Starting point (Brit. 271906) in making thioindigoid dyestuffs with the aid of—
Acenaphthenequinones, dichloroisatin anilide, dichloroisatin chloride, diketones and derivatives, isatins.

6-Methyl-4-chloro-2:1-phenylenethiazonium Chloride
French: Chlorure de 6-méthyle-4-chloro-2:1-phénylènethiazonique.
German: 6-Methyl-4-chlor-2:1-phenylenthiazoniumchlorid.

Chemical
Starting point (Brit. 265545) in making 4-arylaminoarylene-2:1-thioazonium compounds with—
Alphanaphthylamine, aminophenol ethers, anilin, anthanilic acid, benzidin, benzylamine, betanaphthylamine, bromoanilin, bromotoluidin, diethylanilin, dimethylanilin, metatoluidin, metaxylidin, methylparatoluidin, monochloroanilin, monochlorotoluidin, monoethylanilin, monomethylanilin, mononitroanilin, mononitrotoluidin, orthotoluidin, orthoxylidin, paratoluidin, paraxylidin.

3-Methyl-4-chlorophenyl-1-thioglycollic Acid
French: Acide de 3-méthyle-4-chlorophènyle-1-thioglycollique.
German: 3-Methyl-4-chlorphenyl-1-thioglykolsaeure.

Dye
Starting point (Brit. 271906) in making—
Thioindigo dyestuffs.

6-Methyl-4-chloroquinazolin
French: 6-Méthyle-4-chloroquinazoléine.
German: 6-Methyl-4-chlorchinazolin.

Chemical
Starting point in making—
Intermediates, pharmaceuticals.

Dye
Starting point (Brit. 310076) in making dyestuffs with—
Aminonaphtholsulphonic acid, H acid.
1-Para-aminophenyl-5-pyrazolone-3-carboxylic acid.

8-Methyl-4-chloroquinazolin

Chemical
Starting point in making—
Intermediates, pharmaceuticals.

Dye
Starting point (Brit. 310076) in making dyestuffs with—
Aminonaphtholsulphonic acids, H acid.
1-Para-aminophenyl-5-pyrazolone-3-carboxylic acid.

Methyl Chlorosulphonate
French: Chlorosulphonate de méthyle.
German: Chlorsulfonsaeuresmethyl, Methylesterchlorsulfonat.

Chemical
Reagent in making—
Sodium compound of glutaconaldehyde (German 438009).

Methyl Cinnamate
Synonyms: Methyl betaphenylacrylate.
French: Bétaphényleacrylate de méthyle, Bétaphényleacrylate méthylique, Cinnamate de méthyle, Cinnamate méthylique.
German: Betaphenylacrylsäuremethylester, Betaphenylacrylsäuresmethyl, Methylbetaphenylacrylat, Methylcinnamat, Zimtsäuremethylester, Zimtsäuresmethyl.

Cellulose Products
Solvent and plasticizer (Brit. 321258) for—
Cellulose acetate, cellulose esters and ethers, cellulose nitrate.
For uses, see under general heading: "Solvents."

Food
Flavoring ingredient to give strawberry flavor to—
Confectionery, food preparations, liqueurs.
Ingredient of—
Peach essences, strawberry essences, various other essences.

Perfume
Fixative in making—
Perfumed salts, perfumes, toiletries.
Ingredient of the following synthetic odors—
Cherry blossom, eau de cologne, lilac, lavender, oriental bouquet, rose.
Perfume in making—
Cosmetics, dentifrices.

Soap
Perfume and fixative in making—
Toilet soaps.

1-Methyl-2-cyano-3-sulphocyano-5-chlorobenzene
Synonyms: Alphamethyl-2-cyano-3-sulphocyano-5-chlorobenzene.
German: 1-Methyl-2-cyano-3-sulfocyano-5-chlorbenzol.

Chemical
Starting point (Brit. 305140) in making—
Orthoanthranylthioglycollic acid, orthobenzylthioglycollic acid, orthocinnamylthioglycollic acid, orthocresylthioglycollic acid, orthometanylthioglycollic acid, orthonaphthylthioglycollic acid, orthophenylthioglycollic acid, orthophthalylthioglycollic acid, orthosalicylthioglycollic acid, orthosulphanylthioglycollic acid, orthotolylthioglycollic acid, orthoxylylthioglycollic acid.

Dye
Starting point (Brit. 305140) in making—
Thioindigoid dyestuffs.

Methylcyclohexanol
French: Cyclohexanole de méthyle, Cyclohexanole méthylique, Méthylecyclohexanole.
German: Methylhexalin, Methylzyklohexalin.

Analysis
Reagent or solvent in various analytical operations.

Ceramics
Solvent in—
Coating compositions containing cellulose acetate, nitrocellulose, or other esters or ethers of cellulose.

Chemical
As a general solvent.
Reagent in making various organic chemicals.
Starting point in making—
Cyclohexanol.

Fats and Oils
Solvent for various fats, oils, greases.

Glass
Solvent in—
Compositions containing cellulose acetate, nitrocellulose, or other esters or ethers of cellulose, used in coating glassware and in making non-scatterable glass.

Insecticide
Ingredient of—
Parasiticides of various sorts.

Leather
Ingredient of—
Compositions used for glazing.
Solvent in—
Compositions containing cellulose acetate, nitrocellulose, or other esters or ethers of cellulose, used in making artificial leather and in coating natural leather and leather goods.

Metallurgical
Ingredient of—
Compositions used in the treatment of metals.
Solvent in—
Coating compositions containing cellulose acetate, nitrocellulose, or other esters or ethers of cellulose.

Miscellaneous
Ingredient of—
Boring oils, cutting oils and pastes, drilling oils and pastes, lubricating greases, oils, and compounds, machine oil and paste compositions, dry-cleaning compositions, gun oils, wax and encaustic compositions.
Solvents for various substances, particularly in coatings.

Paint and Varnish
Ingredient of—
Paints, lacquers, enamels, and dopes containing cellulose acetate or nitrocellulose and various artificial and natural resins and gums.

Paper
Ingredient of—
Compositions employed in removing ink from printed paper.

Methylcyclohexanol (Continued)
Solvent in—
Coating compositions containing cellulose acetate, nitrocellulose, or other esters or ethers of cellulose.

Photographic
Solvent in making—
Films from compositions containing cellulose acetate or nitrocellulose.

Plastics
Solvent in making—
Cellulose acetate and nitrocellulose compositions.
Solvent for celluloid.
Substitute for camphor in making—
Celluloid and other plastics.

Resins and Waxes
Solvent for various resins and waxes.

Rubber
Ingredient of—
Rubber compounded with celluloid.
Solvent in—
Regenerated and reworking rubber.
Solvent in making—
Coating compositions containing cellulose acetate, nitrocellulose, or other esters or ethers of cellulose.
Starting point in making—
Synthetic rubber.

Sanitation
Ingredient of—
Disinfectants.

Soap
Ingredient of—
Detergent compositions.
Soap solutions used for dissolving greases, oils, hydrocarbons and colors.
Solid soaps containing benzin, benzene, gasoline, tetralin, carbon tetrachloride, or trichloroethylene.
Textile soaps and special soaps containing various ingredients.

Stone
Solvent in—
Coating compositions containing cellulose acetate, nitrocellulose, or other esters or ethers of cellulose.

Textile
——, *Bleaching*
Reagent in bleaching textiles.
——, *Finishing*
Reagent in finishing textiles.
Solvent in—
Coating compositions containing cellulose acetate, nitrocellulose, or other esters or ethers of cellulose.
——, *Manufacturing*
Ingredient (U. S. 1693788) of—
Compositions used in improving the retting of flax.

Woodworking
Ingredient of—
Preservative agents, vitrifying agents.
Solvent in—
Coating compositions containing cellulose acetate, nitrocellulose, or other esters or ethers of cellulose.

Methylcyclohexanol Acetate

Cellulose Products
Solvent for—
Cellulose acetate, cellulose esters and ethers, cellulose nitrate (nitrocellulose).
For uses, see under general heading: "Solvents."

Methylcyclohexanol Betamethyladipate
French: Bétaméthyleadipate de méthylecyclohexanole, Bétaméthyleadipinate de méthylecyclohexanole, Bétaméthyleadipinate méthylecyclohexanolique.
German: Betamethyladipinsäuremethylcyklohexanolester, Betamethyladipinsäuremethylzyklohexanolester, Betamethyladipinsäuresmethylcyklohexanol, Betamethyladipinsäuresmethylzyklohexanol, Methylcyklohexanolbetamethyladipat, Methylzyklohexanolbetamethyladipat.

Cellulose Products
Plasticizer (German 406013) for—
Cellulose acetate, cellulose nitrate.
For uses, see under general heading: "Plasticizers."

Plastics
Plasticizer and solvent (German 406013) in making—
Compositions containing cellulose acetate, nitrocellulose.
Compositions of the celluloid type (used to render the product more flexible at temperatures below zero centigrade and to make them better able to withstand mechanical fatigue and shock).

Rubber
Solvent for—
Crepe rubber.

Textile
——, *Dyeing*
Solvent in making—
Dye baths.
——, *Finishing*
Plasticizer and solvent (German 406013) in—
Compositions, containing cellulose acetate, nitrocellulose, used for making coated textiles.
——, *Printing*
Solvent in making—
Color pastes (added to increase the resistance of the printed fabric to washing and friction).
——, *Manufacturing*
Reagent for increasing—
Luster of rayons.

Methylcyclohexanol Oxalate
French: Oxalate de méthylecyclohexanole.
German: Methylcyklohexanoloxalat, Methylzyklohexanoloxalat, Oxalsäuremethylzyklohexanolester.

Cellulose Products
Plasticizer for—
Cellulose acetate, cellulose esters and ethers, cellulose nitrate.
For uses, see under general heading: "Plasticizers."

Methylcyclohexanol Stearate
French: Stéarate de méthylecyclohexanole.
German: Methylcyklohexanolstearat, Methylzyklohexanolstearat, Stearinsäuremethylcyklohexanolester, Stearinsäuremethylzyklohexanolester, Stearinsäuresmethylcyklohexanol, Stearinsäuresmethylzyklohexanol, Talgsäuremethylcyklohexanolester, Talgsäuremethylzyklohexanolester, Talgsäuresmethylcyklohexanol, Talgsäuresmethylzyklohexanol.

Ceramics
Plasticizer in—
Compositions, containing cellulose acetate, nitrocellulose, or other esters or ethers of cellulose, used for protecting and decorating ceramic products.

Electrical
Plasticizer in—
Insulating compositions, containing cellulose acetate, nitrocellulose, or other esters or ethers of cellulose, used for covering wire and in making electrical machinery and equipment.

Fats and Oils
Solvent for—
Essential oils, fats, mineral oils, vegetable oils of all classes.

Glass
Plasticizer in—
Compositions, containing cellulose acetate, nitrocellulose, or other esters or ethers of cellulose, used in the manufacture of non-scatterable glass and for coating and decorating glassware.

Glues and Adhesives
Plasticizer in—
Adhesive compositions containing cellulose acetate, nitrocellulose, or other esters or ethers of cellulose.

Ink
Ingredient of—
Cellulose ether inks, cellulose nitrate inks, intaglio inks, letterpress inks of the quick process type, lithographic inks of the quick process type.
Offset inks which are required to dry in the shortest possible time and yet be capable of being used after a thin film has stood on the machine overnight or a week-end.
Printing inks, spirit inks.
Solvent for—
Dyes.

Leather
Plasticizer in—
Compositions, containing cellulose acetate, nitrocellulose, or other esters or ethers of cellulose, used in the manufacture of artificial leathers and for coating and decorating leathers and leather goods.

Metallurgical
Plasticizer in—
Compositions, containing cellulose acetate, nitrocellulose, or other esters or ethers of cellulose, used for coating and decorating metallic articles.

Methylcyclohexanol Stearate (Continued)

Miscellaneous
Ingredient of—
French polishes, furniture polishes, polishes for resinous finishes.
Plasticizer in—
Compositions, containing cellulose acetate, nitrocellulose, or other esters or ethers of cellulose, used for protecting and decorating various products.

Paint and Varnish
Plasticizer in making—
Paints, varnishes, lacquers, enamels, and dopes containing cellulose acetate, nitrocellulose, or other esters or ethers of cellulose.
Resin-nitrocellulose compositions used as finishes, where the chief requirement is a hard, flexible film of good gloss, good adhesion, and high resistance to marking by hot articles or attack by mild alkalies.

Paper
Plasticizer in—
Compositions, containing cellulose acetate, nitrocellulose, or other esters or ethers of cellulose, used in the manufacture of coated papers and for coating and decorating products made of paper.

Photographic
Plasticizer in making—
Films from cellulose acetate, nitrocellulose, or other esters or ethers of cellulose.

Plastics
Plasticizer in making—
Laminated fiber products, molded products, plastics from nitrocellulose, cellulose acetate, or other esters or ethers of cellulose.

Resins and Waxes
Plasticizer for—
Resins, natural and synthetic.
Resin-nitrocellulose compositions and solutions.
Solvent for—
Waxes.

Rubber
Plasticizer in—
Compositions, containing cellulose acetate, nitrocellulose, or other esters or ethers of cellulose, used for decorating and protecting rubber products.

Stone
Plasticizer in—
Compositions, containing cellulose acetate, nitrocellulose, or other esters or ethers of cellulose, used for decorating and protecting artificial and natural stone.

Textile
Plasticizer in—
Compositions, containing cellulose acetate, nitrocellulose, or other esters or ethers of cellulose, used in the manufacture of coated fabrics.

Woodworking
Plasticizer in—
Compositions, containing cellulose acetate, nitrocellulose, or other esters or ethers of cellulose, used for coating and decorating woodwork.

Methylcyclohexanone
German: Methylzyklohexanon.

Paint and Varnish
Solvent (Brit. 263175) in making—
Lacquers, varnishes.
See also: "Solvents."

Methylcyclohexylmethylcyclohexanol, Sulphonated

Miscellaneous
As an emulsifying agent (Brit. 425239).
For uses, see under general heading: "Emulsifying agents."

Methylcyclohexylnaphthalenesulphonic Acid
French: Acide de méthylecyclohexylenaphthalènesulfonique.
German: Methylhexycyclonaphtalinsulfonsaeure.

Miscellaneous
Ingredient (Brit. 277391) of—
Spot-removing compositions, washing compositions.

Textile
——, *Finishing*
Ingredient of—
Fulling compositions (Brit. 277391).

Methylcyclohexyl 4-Sulphophthalate

Miscellaneous
As an emulsifying agent (Brit. 418334).
For uses, see under general heading: "Emulsifying agents."

Methyl Cyclopentenylacetate

Food
Agent for—
Producing pineapple aroma and flavor.

Methyldibutenylamine
French: Méthyledibuténylеamine.
German: Methyldibutenylamin.

Chemical
Starting point in making various derivatives.

Insecticide
As an insecticide.
Ingredient (Brit. 313934) of—
Insecticidal and germicidal preparations.

Soap
Ingredient (Brit. 313934) of—
Insecticidal and germicidal soaps.

1-Methyl-2:4-dichlorobenzene
Synonyms: Alphamethyl-2:4-dichlorobenzene.
German: 1-Methyl-2:4-dichlorbenzol.

Chemical
Starting point (Brit. 281290) in making—
Alphamethyl-2:4-dichlorobenzene-5-mercaptan.
Alphamethyl-2:4-dichlorobenzene-5-sulphochloride.
Alphamethyl-2:4-dichlorobenzene-5-thioglycollic acid.

1-Methyl-2:6-dichlorobenzene
Synonyms: Alphamethyl-2:6-dichlorobenzene.
French: 1-Méthyle-2:6-chlorobenzène.
German: 1-Methyl-2:6-dichlorbenzol.

Chemical
Starting point (Brit. 281290) in making—
1-Methyl-2:6-dichlorobenzene-3-mercaptan.
1-Methyl-2:6-dichlorobenzene-3-sulphochloride.
1-Methyl-2:6-dichlorothioglycollic acid.

8-Methyl-2:2′-diethyl-5:5′-dimethselenothiacarbocyanin Iodide

Photographic
Sensitizer (Brit. 420971) in—
Photographic emulsions.

Methyldiethyldodecylthioethyl-Ammonium Iodide

Disinfectant
Claimed (Brit. 436725 and 436726) to be—
Bactericide, disinfectant.

Insecticide and Fungicide
Claimed (Brit. 436725 and 436726) to be—
Fungicide.

8-Methyl-2:2′-diethyloxacarbocyanin Iodide

Photographic
Sensitizer (U. S. 1962123, 1962124, and 1962133) for—
Blue-green light.

11-Methyldiethylrosindulin-1:6-disulphonic Acid

Dye
Starting point (Brit. 431708 and 431709) in making—
Blue dyestuffs with 2:5-tolylenediamine.
Blue dyestuffs with 4-amino-2′-dimethoxydiphenylamine-2-sulphonic acid.
Blue dyestuffs with 4-aminomethylanilin-2-sulphonic acid.
Blue dyestuffs with 4-aminocyclohexylanilin-2-sulphonic acid.
Greenish-blue dyestuffs with 2:6-dichloroparaphenylenediamine.
Greenish-blue dyestuffs with 2:5-diaminometaxylene-4-sulphonic acid.

5′-Methyl-2:1′-diethylthia-2′-pyrazinocarbocyanin Iodide

Photographic
As a dyestuff (Brit. 435542).

4-Methyl-5:7-di-iodo-3-oxythionaphthene
French: 4-Méthyle-5:7-di-iodo-3-oxysulphonaphthène.
4-Méthyle-5:7-di-iodo-3-oxythionaphthène.
German: 4-Methyl-5:7-di-iodo-3-oxysulfonaphten.
4-Methyl-5:7-di-iodo-3-oxythionaphten.

4-Methyl-5:7-di-iodo-3-oxythionaphthene (Continued)

Chemical
Starting point in making various intermediates.

Dye
Starting point (Brit. 271906) in making thioindigoid dyestuffs with—
Acenaphthenequinones, dichloroisatin chlorides, dichloroisatin anilides, diketones and derivatives, isatins.

Methyl Disulphide

Synonyms: Methyl bisulphide.
French: Bisulphure de méthyle, Bisulphure méthylique, Disulphure de méthyle, Disulphure méthylique.
German: Bischwefelmethyl, Bischwefelwasserstoffsaeuremethylester, Dischwefelmethyl, Dischwefelwasserstoffsaeuremethylester, Dischwefelwasserstoffsauremethyl.

Chemical
General chemical reagent.
Reagent (Brit. 298511) in treating—
Albumenoids and albumens.

Glues and Adhesives
Reagent (Brit. 298511) in treating—
Vegetable proteins, such as soya bean flour, flaxseed protein, and peanut protein, to make adhesives.

Miscellaneous
Reagent (Brit. 298511) in making—
Sizes and finishes from vegetable proteins, such as soya bean flour, flaxseed protein, and peanut protein.

Methyldodecylguanidin Chloride

Textile
Assistant (Brit. 421862) in—
Aqueous baths for treating textiles.
Promoter (Brit. 421862) of—
Uniform dyeing with basic dyestuffs.
Wetting and washing agent (Brit. 421862) in—
Textile processes.

Methyleneaminoacetonitrile

Glass
Stabilizer (Brit. 437304) for—
Halogenated rubber derivatives used as cements for laminated glass.

Miscellaneous
Inhibitor (Brit. 437304) of—
Photochemical action.

Paper
Stabilizer (Brit. 437304) for—
Halogenated rubber derivatives used for impregnating or coating wrapping paper.

Rubber
Promoter (Brit. 437304) of—
Resistance to the deteriorating action of light on chlorinated rubber.
Stabilizer (Brit. 437304) for—
Coating and impregnating agents made from halogenated rubber derivatives and used for treating fabrics to be used as wrapping materials.
Transparent films or sheets made from halogenated rubber derivatives.

2-Methyleneaminodiphenylene Oxide

Rubber
Antiaging agent (Brit. 422191).

Methylenebisbetanaphthol, Nitrated

Rubber
Antiscorching agent (Brit. 418376 and 418445) in—
Rubber compounding.

Methylene Blue

French: Bleu de méthylène.
German: Methylenblau.

Analysis
Indicator in the laboratory for acidmetric and alkalimetric purposes.

Chemical
Ingredient (Brit. 295605) of bacteriological preparations, therapeutic preparations and biological stains containing—
Cresol, guaiacol, hydroquinone, phenol, phloroglucinol, pyrocatechol, pyrogallol, resorcinol.

Dye
Ingredient of—
Dye pastes, lakes.

Miscellaneous
Coloring matter in making—
Bacteriological and histological slides.
Pathological reagent.

Pharmaceutical
In compounding and dispensing practice (special medicinal quality).

Textile

——, *Dyeing*
Coloring matter in dyeing—
Cotton yarns, woolen and cotton fabrics.

——, *Printing*
Coloring matter in printing—
Calicoes with tannin and tartar emetic.

Methylene-Ethylene Ether

French: Éther méthylène-éthylènique.
German: Methylenaethylenaether.

Ceramics
Low-boiling solvent (Brit. 407709) in—
Compositions, containing cellulose acetate having a high acetyl content (particularly an acetyl content of 56 up to 62.5 percent) or other esters or ethers of cellulose, used as coatings for protecting and decorating ceramic products.

Chemical
Low-boiling solvent (Brit. 407709) for—
Cellulose acetate having a high acetyl content (particularly an acetyl content of 56 up to 62.5 percent) and other esters or ethers of cellulose.

Electrical
Low-boiling solvent (Brit. 407709) in—
Insulating compositions, containing cellulose acetate having a high acetyl content (particularly an acetyl content of 56 up to 62.5 percent) or other esters or ethers of cellulose, used for covering wire and in making electrical machinery and equipment.

Glass
Low-boiling solvent (Brit. 407709) in—
Compositions, containing cellulose acetate having a high acetyl content (particularly an acetyl content of 56 up to 62.5 percent) or other esters or ethers of cellulose, used in the manufacture of non-scatterable glass and as coatings for protecting and decorating glassware.

Glues and Adhesives
Low-boiling solvent (Brit. 407709) in—
Adhesive compositions containing cellulose acetate having a high acetyl content (particularly an acetyl content of 56 up to 62.5 percent) or other esters or ethers of cellulose.

Leather
Low-boiling solvent (Brit. 407709) in—
Compositions, containing cellulose acetate having a high acetyl content (particularly an acetyl content of 56 up to 62.5 percent) or other esters or ethers of cellulose, used in the manufacture of artificial leathers and as coatings for protecting and decorating leathers and leather goods.

Metallurgical
Low-boiling solvent (Brit. 407709) in—
Compositions, containing cellulose acetate having a high acetyl content (particularly an acetyl content of 56 up to 62.5 percent) or other esters or ethers of cellulose, used as coatings for protecting and decorating metallic articles.

Miscellaneous
Low-boiling solvent (Brit. 407709) in—
Compositions, containing cellulose acetate having a high acetyl content (particularly an acetyl content of 56 up to 62.5 percent) or other esters or ethers of cellulose, used as coatings for protecting and decorating various products.

Paint and Varnish
Low-boiling solvent (Brit. 407709) in—
Paints, varnishes, lacquers, enamels, and dopes containing cellulose acetate having a high acetyl content (particularly an acetyl content of 56 up 62.5 percent) or other esters or ethers of cellulose.

Paper
Low-boiling solvent (Brit. 407709) in—
Compositions, containing cellulose acetate having a high acetyl content (particularly an acetyl content of 56 up to 62.5 percent) or other esters or ethers of cellulose, used in the manufacture of coated papers and as coatings for protecting and decorating products made of paper or pulp.

Methylene-Ethylene Ether (Continued)
Photographic
Low-boiling solvent (Brit. 407709) in making—
Films from cellulose acetate having a high acetyl content (particularly an acetyl content of 56 up to 62.5 percent) or other esters or ethers of cellulose.
Plastics
Low-boiling solvent (Brit. 407709) in making—
Laminated fiber products, molded products.
Plastics from cellulose acetate having a high acetyl content (particularly an acetyl content of 56 up to 62.5 percent) or other esters or ethers of cellulose.
Rubber
Low-boiling solvent (Brit. 407709) in—
Compositions, containing cellulose acetate having a high acetyl content (particularly an acetyl content of 56 up to 62.5 percent) or other esters or ethers of cellulose, used as coatings for decorating and protecting rubber products.
Stone
Low-boiling solvent (Brit. 407709) in—
Compositions, containing cellulose acetate having a high acetyl content (particularly an acetyl content of 56 up to 62.5 percent) or other esters or ethers of cellulose, used as coatings for decorating and protecting artificial and natural stone.
Textile
Low-boiling solvent (Brit. 407709) in—
Compositions, containing cellulose acetate having a high acetyl content (particularly an acetyl content of 56 up to 62.5 percent) or other esters or ethers of cellulose, used in the manufacture of coated fabrics.
Woodworking
Low-boiling solvent (Brit. 407709) in—
Compositions, containing cellulose acetate having a high acetyl content (particularly an acetyl content of 56 up to 62.5 percent) or other esters or ethers of cellulose, used as protective and decorative coatings on woodwork.

Methylene Iodide
French: Iodure de méthylène, Iodure méthylènique.
German: Jodmethylen, Methylenjodid.
Spanish: Yoduro de metilen.
Italian: Ioduro di metilene.
Chemical
Starting point in making—
Pharmaceutical chemicals and other derivatives.
Starting point (Brit. 353477) in making contrast media for X-Ray photography with the aid of—
Ammonium sulphite, magnesium sulphite, monomethylamine sulphite, piperazin sulphite, piperidin sulphite, sodium sulphite.

Methylenethiourea
Synonyms: Methylenesulphourea.
French: Méthylènesulphourée, Méthylènethiourée, Sulphourée de méthylène, Sulphourée méthylènique, Thiourée de méthylène, Thiourée méthylique.
German: Methylensulphoharnstoff, Methylenthioharnstoff.
Chemical
Starting point (Brit. 310534) in making vulcanization accelerators with—
Alphanaphthylamine, anilin, benzylamine, betanaphthylamine, cyclohexylanilin, meta-anisidin, metacresidin, metanaphthylenediamine, metaphenylamine, metaphenylenediamine, metatoluidin, metatolylenediamine, metaxylenediamine, metaxylidin, monoethylanilin, monomethylanilin, orthoanisidin, orthocresidin, orthonaphthylenediamine, orthophenylamine, orthotoluidin, orthotolylenediamine, orthoxylenediamine, orthoxylidin, para-anisidin, paracresidin, paranaphthylenediamine, paraphenylamine, paraphenylenediamine, paratoluidin, paratolylenediamine, paraxylenediamine, paraxylidin.
Starting point in making—
Synthetic pharmaceuticals.

1-Methyl-5-ethylbarbituric Acid Hydrochloric
Pharmaceutical
Suggested for use (Brit. 414293) as—
Hypnotic with low toxic properties.

Methylethylbenzhydroxylamate
German: Methylaethylbenzhydroxylamat.
Chemical
Starting point in making—
Alphamethylhydroxylamine.

Methylethylcyclohexylbetahydroxygammadodecoxypropyl-Ammonium Iodide
Disinfectant
Claimed (Brit. 436725 and 436726) to be—
Bactericide, disinfectant.
Insecticide and Fungicide
Claimed (Brit. 436725 and 436726) to be—
Fungicide.

Methylethyleneglycol Monopalmitate
French: Monopalmitate de méthyléthylèneglycole.
German: Methylaethylenglykolmonopalmitat, Methylaethylenglycolmonopalmitinsäureester, Monopalmitinsäuremethylaethylenglykolester, Monopalmitinsäuresmethylaethylenglykol.
Spanish: Monopalmitato de metiletileneglycol.
Italian: Monopalmitato di metiletileneglycol.
Ceramics
Solvent in—
Compositions, containing cellulose acetate, nitrocellulose, or other esters or ethers of cellulose, used for the purpose of decorating and protecting ceramic products (produces dull films).
Chemical
Dispersing agent in making—
Emulsions of hydrocarbons of various groups of the aliphatic and aromatic series.
Emulsions of various chemicals.
Terpene emulsions.
Starting point (German 582106) in making—
Cleansing compositions.
Dispersing compositions.
Impregnating compositions.
Waterproofing compositions.
Wetting compositions.
Dye
Dispersing agent in making—
Color lakes.
Electrical
Solvent in—
Compositions, containing nitrocellulose, cellulose acetate or other esters or ethers of cellulose, used for insulating purposes in the manufacture of electrical machinery and equipment.
Fats and Oils
Dispersing agent (German 582106) in making—
Boring oil emulsions.
Drilling oil emulsions.
Greasing compositions in emulsified form.
Lubricating compositions in emulsified form, containing various vegetable and animal fats and oils.
Various fat and oil emulsions.
Wire-drawing oils in emulsified form.
Food
Dispersing agent (German 582105) in making—
Margarin dispersions.
Milk dispersions.
Various dispersed food products.
Germicide
Dispersing agent (German 582106) in making—
Germicidal and deodorizing compositions in emulsified form.
Glass
Solvent in—
Compositions, containing cellulose acetate, nitrocellulose, or other esters or ethers of cellulose, used for the production of dull coatings on glass products for decorative and protective purposes, and in the manufacture of nonscatterable glass.
Glues and Adhesives
Dispersing agent (German 582106) in making—
Glue and gelatin dispersions.
Insecticide
Dispersing agent (German 582106) in making—
Emulsified insecticidal and fungicidal preparations.
Leather
Dispersing agent (German 582106) in making—
Emulsified tanning compositions.
Emulsified waterproofing compositions.
Emulsified finishing compositions.
Emulsified dressing compositions.
Emulsified fat-liquoring baths.
Emulsified soaking compositions.

Methylethyleneglycol Monopalmitate (Continued)
Solvent in—
Compositions, containing cellulose acetate, nitrocellulose, or other esters or ethers of cellulose, used for the production of dull coatings on leather goods for their protection and decoration, and in the manufacture of artificial leather.

Metallurgical
Solvent in—
Compositions, containing cellulose acetate, nitrocellulose, and other esters or ethers of cellulose, used for the protection of dull coatings for the protection and decoration of metallic ware.

Miscellaneous
Dispersing agent (German 582106) in making—
Automobile polishes in emulsified form.
Cleansing compositions in emulsified form.
Furniture polishes in emulsified form.
Metal polishes in emulsified form.
Shoe polishes in emulsified form.
Scouring compositions and detergent preparations in emulsified form.
Various emulsified preparations for use in wetting, washing, and dispersing operations.
Waterproofing compositions in emulsified form.
Solvent in—
Compositions, containing cellulose acetate, nitrocellulose, or other esters or ethers of cellulose, used for the production of dull coatings for the protection and decoration of various articles.

Paint and Varnish
Solvent in making—
Paints, varnishes, dopes, lacquers, and enamels containing esters or ethers of cellulose.

Paper
Dispersing agent (German 582106) in making—
Emulsified preparations used for the treatment of paper and pulp products.
Sizing compositions in emulsified form.
Waterproofing compositions in emulsified form, for treating paper and paperboard and other pulp products.
Waxing compositions in emulsified form.
Solvent in—
Compositions, containing various esters or ethers of cellulose, used for the production of dull coatings on paper and pulp products for their protection and decoration, and in the manufacture of coated paper.

Perfume
Dispersing agent (German 582106) in making—
Creams in emulsified form.
Lotions.
Lanolin preparations in emulsified form.
Latherless shaving cream emulsions.
Shampoos in emulsified form.
Sunburn preparations.
Various emulsified cosmetics and perfumes.

Petroleum
Dispersing agent (German 582105) in making—
Emulsified cutting oils for lathe and screwpress work.
Kerosene emulsions.
Emulsions containing petroleum or heavy petroleum distillates.
Emulsified medicinal mineral oil.
Naphtha emulsions.
Soluble lubricating oils in emulsified form.
Soluble greases in emulsified form.
Stabilized emulsions containing paraffin oil or other petroleum oils and distillates.
Various textile oils in emulsified form.

Pharmaceutical
Dispersing agent (German 582106) in making—
Emulsified pharmaceutical preparations.

Plastics
Solvent in making—
Compositions, of various esters or ethers of cellulose.

Resins and Waxes
Dispersing agent (German 582106) in making—
Emulsions of natural and artificial resins.
Emulsions of natural and artificial waxes.

Rubber
Dispersing agent (German 582106) in making—
Emulsified rubber compositions, such as rubber cements and rubber coatings.
Solvent in—
Compositions, containing various esters or ethers of cellulose, used for the production of dull coatings on rubber articles for decorative and protective purposes.

Soap
Dispersing agent (German 582106) in making—
Emulsions of ordinary coaps and alkaline earth soaps.
Hand-cleansing compositions in emulsified form.
Various emulsified cleansing and lathering compositions.
Various emulsified scouring compositions.
Various superfatted soaps.

Stone
Solvent in—
Compositions, containing various esters or ethers of cellulose, used for the production of dull coatings on artificial and natural stone for protective and decorative purposes.

Textile
——, *Bleaching*
Dispersing agent (German 582106) in making—
Emulsified bleaching baths.

——, *Dyeing*
Dispersing agent (German 582106) in making—
Dye baths in emulsified form.

——, *Finishing*
Dispersing agent (German 582106) in making—
Emulsified coating compositions.
Emulsified coating compositions containing various esters or ethers of cellulose.
Emulsified sizing compositions.
Emulsified dressing compositions.
Emulsified finishing compositions.
Emulsified impregnating compositions.
Emulsified scouring compositions.
Emulsified washing compositions.
Emulsified waterproofing compositions.
Emulsified waxing compositions.

——, *Manufacture*
Dispersing agent (German 582106) in making—
Emulsified bowking baths.
Emulsified fulling baths.
Emulsified baths for the carbonization of wool.
Emulsified baths for washing wool.
Emulsified baths for degreasing and treating raw wool.
Emulsified spinning baths.
Emulsified mercerization baths.
Emulsified oiling compositions.
Emulsified baths for use in the kier-boiling of cotton.
Emulsified baths for soaking silk.
Emulsified baths for degumming and boiling-off raw silk.

——, *Printing*
Dispersing agent (German 582106) in making—
Emulsified printing compositions.

Woodworking
Solvent in—
Compositions, containing various esters or ethers of cellulose, used for the production of dull coatings on woodwork for decorative and protective purposes.

Methylethyl Ketone Peroxide

Fuel
Ignition improver (Brit. 444544) for—
Diesel engine fuels.

Methylethylketoxime

Fuel
Primer (Brit. 429763) for—
Diesel engine fuel oils produced by the hydrogenation of coal.

Petroleum
Primer (Brit. 429763) for—
Diesel oils containing a high proportion of aromatic bodies.

Methylformamide

Fats and Oils
Deterioration retardant (Brit. 423938) for—
Vegetable oils.

Fuel
Deterioration retardant (Brit. 423938) for—
Coal-carbonization spirits.

Methylformamide (Continued)

Petroleum

Deterioration retardant (Brit. 423938) for—
Cracked petroleum oils, lubricating oils, shale oils, transformer oils.

Methyl Formate

French: Éther méthyleformique, Formiate de méthyle, Formiate méthylique.
German: Ameisensaeuresmethyl, Ameisensaeuremethylester.

Chemical

Reagent in making various organic and intermediate chemicals.
Starting point in making—
Carbonyl chloride.
Formamide.
Hydrocyanic acid (Swiss 115702).
Methanol by catalytic reduction (French 581175).

Miscellaneous

See also: "Solvents."

Paint and Varnish

Solvent in making—
Cellulose acetate airplane dopes.
Cellulose acetate varnishes.

Plastics

Solvent in making—
Cellulose acetate plastics.

Textile

——, *Manufacture*

Solvent in making—
Cellulose acetate rayon.

Methylfurol

French: Furole méthylique.
German: Methylfurol.

Miscellaneous

Solvent for—
Cellulose esters and ethers, cellulose nitrate, coumarone resin, ester resins.
For uses, see under general heading: "Solvents."

Methylgammaphenoxypropyllauramide

Chemical

Starting point (Brit. 443902) in making—
Sulphonated sodium salts, stable to calcium chloride and acids, which are used as scouring agents for raw wool.

Methylgammaphenoxypropylstearamide

Chemical

Starting point (Brit. 443902) in making—
Sulphonated sodium salts, stable to calcium chloride and acids, which are used as scouring agents for raw wool.

Methylglucamine

Dye

Coupling agent (Brit. 429618) in making—
Dyestuffs with diazotized arylamines (color being developed on the fiber by acid treatment).

Methylglucamine Stearate

Miscellaneous

Detergent (U. S. 1994467).

Insecticide

Emulsifying agent (U. S. 1994467) for—
Insecticides.

Methylglycol Acetate

French: Acétate de glycole et de méthyle, Acétate glycollique-méthylique, Acétate de méthyle et de glycole, Acétate méthylique-glycollique.
German: Essigsäuremethylglykolester, Essigsäuresmethylglykol, Methylglykolacetat, Methylglykolazetat.

Cellulose Products

Solvent for—
Cellulose acetate, cellulose esters and ethers, cellulose nitrate.
For uses, see under general heading: "Solvents."

Chemical

Solvent for—
Various chemicals.

Methyl Glycolate

French: Glycolate de méthyle, Glycolate méthylique.
German: Glykolsaeuremethylester, Glykolsaeuresmethyl, Methylglykolat.

Cellulose Products

Plasticizer (Brit. 311664) for—
Cellulose esters and ethers.
For uses, see under general heading: "Plasticizers."

Chemical

Starting point in making—
Intermediates, pharmaceuticals.

Methylheptin Carbonate

French: Carbonate de méthyleheptine, Carbonate méthylique et heptinique, Vert de violette artificiel.
German: Kohlensäuremethylheptinester, Kohlensäuresmethylheptin, Methylheptinkarbonat.

Perfume

Ingredient of—
Cassia essence, mimosa essence, violet essence.
Perfume in—
Cosmetics.

Soap

Perfume in—
Toilet soaps

3-Methyl-5-heptylcyclopentanone-1

Cosmetic

Odorant (Brit. 430930 and 449211) in—
Perfume mixtures.

2-Methyl-3-hydroxyquinolin

Dye

Starting point (Brit. 429176) in making—
Yellow dyes for wool, by fusing with phthalic anhydride and sulphonating the product.

Methylideneglycerol

French: Glycérole de méthylidène, Glycérole méthylidènique.
German: Methylidenglycerol.

Miscellaneous

Solvent for—
Cellulose esters and ethers.
Various gums and resins.
Various organic substances.
For uses, see under general heading: "Solvents."

Methylisoeugenol

Synonyms: Propenylveratrol.
French: Isoeugénol méthylique, Propénylevératrole.
German: Isoeugenolmethylaether, Propenylveratrol.

Perfume

Ingredient of—
Artificial essence of pinks.
Ylang-ylang odors.
Perfume in—
Cosmetics.

Methylisopropylcyclohexanone

French: Cyclohexanone de méthyle et de'isopropyle, Méthyleisopropylecyclohexanone.
German: Methylisopropylcyklohexanon, Methylisopropylzyklohexanon.

Cellulose Products

Plasticizer for—
Cellulose acetate, cellulose esters and ethers, cellulose nitrate.
For uses, see under general heading: "Plasticizers."

3-Methyl-6-isopropylenephenol

Chemical

Starting point (Brit. 273685) in making—
Menthol, thymol.

4-Methyl-6-isopropylenephenol

French: 4-Méthyle-6-isopropylènephénole.
German: 4-Methyl-6-isopropylenphenol.

Chemical

Starting point (Brit. 273685) in making—
4-Methyl-6-isopropylcyclohexanol.
4-Methyl-6-isopropylphenol.

Methylisothiourea Sulphate
French: Sulphate de méthyleisothiourée.
German: Methylisothioharnstoffsulfat, Schwefelsaeuresmethylisothioharnstoff.

Chemical
Reagent (Brit. 272686) in making—
Aminobutyleneguanidin.
Aminoethyleneguanidin.
Aminohexyleneguanidin.
Aminomethyleneguanidin.
Aminopentyleneguanidin.
Aminopropyleneguanidin.

Methylisovanillin
French: Isovanilline de méthyle, Isovanilline méthylique.

Chemical
Starting point in making—
Aromatics.

Perfume
Ingredient of—
Artificial perfume preparations.
Perfume in—
Cosmetics.

Soap
Perfume in—
Toilet soaps.

Methyllaurylcetylglucamine Hydrochloride

Miscellaneous
Detergent (Brit. 428142 and 428148) in—
Cleansing operations, particularly in hard water or acids.

Methyllaurylglucamine Hydrochloride

Miscellaneous
Detergent (Brit. 428142 and 428148) in—
Cleansing operations, particularly in hard water or acids.

Methyl Mandelate

Cellulose Products
Plasticizer (Brit. 270650) for—
Cellulose esters or ethers.
For uses, see under general heading: "Plasticizers."

4-Methylmercaptoalphanaphthol

Chemical
Starting point in making—
Intermediates, pharmaceuticals.

Dye
Starting point (Brit. 291825) in making indigoid dyestuffs with—
Isatin anilide.
Isatin chloride.
Reactive alpha derivatives of isatin.

3-Methyl-4′-methylthioldiphenylamine-6-carboxylic Acid

Pharmaceutical
Claimed (Brit. 363392 and 437953) as—
New pharmaceutical.

Methylnaphthalene Sulphonate
French: Naphthalènesulphonate de méthyle, Naphthalènesulphonate méthylique.
German: Methylnaphtalinsulfonat, Naphtalinsulfonsäuremethylester, Naphtalinsulfonsäuresmethyl.

Chemical
Starting point in making various derivatives.

Miscellaneous
As a dispersing agent (Brit. 322005).
For uses, see under general heading: "Emulsifying agents."

3-Methyl-5-nitrobenzoxazolone

Chemical
Starting point in making—
3-Methyl-2:1-benzoazolone-5-arsinic acid (Brit. 261133).

Methyl Nonylate

Food
As a flavoring.
Ingredient of—
Flavoring preparations.

Perfume
Ingredient of—
Cosmetics, particularly lipstick.

Methylolamine
German: Methylolamin.

Chemical
Starting point in making—
Pharmaceuticals and other derivatives.

Electrical
Dispersive agent (Brit. 340294) in making—
Special lubricating compositions for use in electric switches.

Fats and Oils
Dispersive agent (Brit. 340294) in making—
Non-freezing lubricating compositions, containing animal and vegetable oils and fats, as well as ethyleneglycol, borax, benzyl alcohol, or esters of ethyleneglycol in place of the latter.
Special lubricating compositions of the above type, for use on locomotive axles, railway switches, hydraulic presses, and hydraulic brakes.
Ingredient (Brit. 340294) of—
Compositions, containing vegetable, animal and mineral oils and greases, used as rust preventives.

Petroleum
Ingredient (Brit. 340294) of—
Special lubricating compositions containing mineral oils and greases.

Methyl Oleate

Miscellaneous
As an emulsifying agent (Brit. 343899).
For uses, see under general heading: "Emulsifying agents."

Methyl Oleicsulphonate

Miscellaneous
As an emulsifying agent (Brit. 343098).
For uses, see under general heading: "Emulsifying agents."

Methylolformamide

Textile
——, *Printing*
Solvent for basic and other dyestuffs (German 433153).

Methyl Oxide
Synonyms: Dimethyl ether.
French: Oxyde méthylique, Oxyde de méthyle.
German: Methyloxyd.

Chemical
Reagent in making—
Chloromethyl ether, dichloromethyl ether, perchloromethyl ether.

1-Methyloxy-4-amino-anthraquinone
Synonyms: Alphamethyloxy-4-aminoanthraquinone, Alphamethoxy-4-aminoanthraquinone.
French: 1-Méthyleoxy-4-amino-anthraquinone.
German: 1-Methyloxy-4-aminoanthrachinon.

Dye
Starting point (Brit. 285096) in making dyestuffs in the presence of dimethylanilines, nitrobenzene, orthodichlorobenzene, or naphthalene, with the aid of—
Acetylparaphenylenediamine.
5-Amino-2-methylbenzimideazole.
Benzidin and derivatives and homologs.
Dimethylparaphenylenediamine.
Metaphenylenediamine.
Metatoluylenediamine.
Naphthylenediamine.
Orthophenylenediamine.
Orthotoluylenediamine.
Paraphenylenediamine.
Paratoluylenediamine.

Methyl-4-oxy-2-quinolin, Normal

Dye
Starting point (Brit. 431649) in making—
Dyestuffs with anilin or 4-nitroanilin, the cyclohexyl ester of 3-aminobenzoic acid, halogen anilins, toluidins, xylidins, and the like, for coloring organic solvents, lacquers, fats, oils, resins and waxes in clear yellow, greenish yellow or reddish shades, fast to sublimation, and other influences.

Methylpara-aminobenzoate

Analysis
Reagent.

Chemical
Reagent in—
Organic synthesis.

Methylparahydroxydiphenyl Sulphide

Pharmaceutical
As a germicide (U. S. 2011582).

Methyl Paraoxybenzoate

Synonyms: Methylparahydroxybenzoate.
French: Parahydroxyebenzoate de méthyle, Parahydroxyebenzoate méthylique, Paraoxyebenzoate de méthyle, Paraoxyebenzoate méthylique.
German: Methyl parahydroxybenzoat, Methylparaoxybenzoat, Parahydroxybenzoesäuremethylester, Parahydroxybenzoesäuresmethyl, Paraoxybenzoesäuremethylester, Paraoxybenzoesäuresmethyl.

Food
Preservative for various preparations.

Pharmaceutical
In compounding and dispensing practice.

Sanitation
Antiseptic and disinfectant for various purposes.

Soap
Ingredient of—
Antiseptic and disinfectant soaps.

Methylpentaerythritol

Cellulose Products
Solvent, softener and plasticizer (Brit. 358393) for—
Cellulose acetate, cellulose esters or ethers, nitrocellulose.
For uses, see under general heading: "Plasticizers."

Methylphenoxyethyl Laurate

As a softening agent (U. S. 1874310).
For uses, see under general heading: "Softening agents."

Methylphenoxyethyl Phthalate

French: Phthalate de méthylephénoxye-éthyle, Phthalate méthylephénoxye-éthylique.
German: Methylphenoxyaethylphtalat, Phtalsaeuremethylphenoxyaethylester, Phtalsaeuresmethylphenoxyaethyl.

Leather
Softener (Brit. 306911) in—
Cellulose acetate compositions for coating artificial leather.

Miscellaneous
See also: "Softening agents."

Paint and Varnish
Plasticizer and softener (Brit. 306911) in making—
Cellulose acetate paints, varnishes, lacquers and enamels.

Plastics
Softener and plasticizer (Brit. 306911) in making—
Cellulose acetate compositions.

Photographic
Softener (Brit. 306911) in making—
Cellulose acetate films.

Textile
Softener and plasticizer (Brit. 306911) in making—
Cellulose acetate compositions for coating fabrics.

6-Methyl-4-phenylamino-1:2-aminothiophenol

Dye
Starting point (Brit. 265641) in making thioindigoid dyestuffs with—
Dichloroquinone, trichloroquinone, quinone.

Methylphenylcarbinol Acetate

Synonyms: Phenylmethylcarbinol acetate, Styrolyl acetate.
French: Acétate de méthylephénylecarbinol, Acétate méthylephénylecarbinolique, Acétate de phénylemétylecarbinol, Acétate phénylemétylecarbinolique, Acétate de styrolyle, Acétate styrolylique.
German: Essigsäuremethylphenylcarbinolester, Essigsäuresmethylphenylcarbinol, Methylphenylcarbinolacetat, Methylphenylcarbinolazetat, Phenylmethylcarbinolacetat, Phenylmethylcarbinolazetat, Styrolylacetat, Styrolylazetat.
Spanish: Acetato de fenilmetilcarbinol, Acetato de metilfenilcarbinol, Acetato de stirolil.
Italian: Acetato di fenilmetilcarbinole, Acetato di metilfenilcarbinole, Acetato di stirolile.

Chemical
Starting point in making various derivatives.

Perfume
Ingredient of—
Artificial lily of the valley compositions.
Perfume in—
Cosmetics.

Soap
Perfume in—
Toilet soaps.

Methylphosphoric Dichloride

French: Dichlorure de méthylephosphorique.
German: Methylphosphordichlorid.

Dye
Reagent in making—
Vat dyestuffs (Brit. 248802).
Starting point in making—
Leuco compounds of vat dyestuffs.

Methylpropyl Carbinol

French: Carbinol de méthyle et d'ethyle, Carbinol méthylique et propylique.
German: Methylaethylcarbinol.

Chemical
General solvent for various purposes.

Explosives
Stabilizer in making—
Nitrocellulose explosives.

Miscellaneous
General solvent for various purposes.

Paint and Varnish
Solvent in making—
Cellulose acetate lacquers and varnishes.

Paper
Solvent in extracting—
Resin from wood for making pulp.

Plastics
Solvent in making—
Cellulose acetate compositions.

Textile
——, *Manufacturing*
Solvent in purifying—
Raw cotton.

Methyl-2-pyridonebenzylimide, Normal

Resins
Starting point (Brit. 425435) in making—
Yellow to pale-brown resins (soluble in water and inert organic solvents) with benzyl chloride or stearyl chloride.

Methyl-2-pyridonebetanaphthylimide, Normal

Resins
Starting point (Brit. 425435) in making—
Yellow to pale-brown resins (soluble in water and inert organic solvents) with benzyl chloride, stearyl chloride, palmityl chloride, or oleyl chloride.

Methyl-2-pyridonemethylimide, Normal

Resins
Starting point (Brit. 425435) in making—
Yellow to pale-brown resins (soluble in water and inert organic solvents) with acetyl chloride, benzyl chloride or stearyl chloride.

Methyl-2-pyridonepara-anisylimide, Normal

Resins
Starting point (Brit. 425435) in making—
Yellow to pale-brown resins (soluble in water and inert organic solvents) with benzyl chloride, stearyl chloride, palmityl chloride, or oleyl chloride.

Methyl-2-pyridoneparaphenetylimide, Normal

Resins
Starting point (Brit. 425435) in making—
Yellow to pale-brown resins (soluble in water and inert organic solvents) with benzyl chloride, stearyl chloride, palmityl chloride, or oleyl chloride.

Methyl-2-pyridonephenylimide, Normal

Resins
Starting point (Brit. 425435) in making—
Yellow to pale-brown resins (soluble in water and inert organic solvents) with benzyl chloride, stearyl chloride, palmityl chloride, or oleyl chloride.

Methylpyrogallol Ethyl-ether

Photographic
Starting point (U. S. 2017295) in making—
Developers having no tendency to become oxidized.

Methylpyrogallol Methyl-ether

Photographic
Starting point (U. S. 2017295) in making—
Developers having no tendency to become oxidized.

6-Methylpyronone

Dye
Starting point (Brit. 419447) in making—
Blue-red dyes for wool by coupling with diazotized 1-amino-2-naphthol-4-sulphonate and after-treating with bichromate.
Orange-yellow dyes for acetate rayon and lacquers by coupling with diazotized 2-amino-5-nitroanisole.
Orange dyes for wool by coupling with diazotized 1-amino-2-methoxynaphthalene-6-sulphonate.
Red dyes for wool by coupling with diazotized 1-amino-2-methoxynaphthalene-6-sulphonate and after-treating with chromium formate.
Red-brown dyes for wool by coupling with diazotized 1-amino-2-naphthol-4-sulphonate.
Yellow dyes for lacquers, waxes, and oils by coupling with diazotized anilin.
Yellow dyes for wool by coupling with diazotized 3-amino-6-chlorotoluene-4-sulphonate.

Methyl-2-quinolonephenylimide, Normal

Resins
Starting point (Brit. 425435) in making—
Yellow to pale-brown resins (soluble in water and inert organic solvents) with benzyl chloride or stearyl chloride.

Methyl Salicylate

Synonyms: Artificial oil of wintergreen.
French: Éther méthylesalicylique, Salicylate méthylique, Salicylate de méthyle.
German: Gaultheriaoel, Kuenstlisches wintergruenoel, Salicylsaeuremethylester.

Fats and Oils
Ingredient of—
Synthetic cassia flower oil, synthetic ylang-ylang oil.

Food
Ingredient of—
Artificial peach essence, artificial fruit essence, artificial strawberry essence, beverages, confectionery, food compositions.

Miscellaneous
Ingredient of—
Disinfecting compositions.

Perfumery
Ingredient of—
Cosmetics, dentifrices, perfumes.

Pharmaceutical
In compounding and dispensing practice.

Methyl Silicate

French: Silicate de méthyle, Silicate méthylique.
German: Kieselsäuremethylester, Kieselsäuresmethyl, Methylsilikat.

Ceramics
Ingredient of—
Compositions used for coating ceramic ware and for filling the pores in such ware to provide a smooth surface for further treatment.

Chemical
Starting point in making—
Silicic acid.

Construction
Binding agent in—
Compositions used for coating concrete, cement, and masonry work for the purpose of obtaining a smooth surface for further treatment.

Metallurgical
Ingredient of—
Compositions used for coating metallic surfaces.

Miscellaneous
Ingredient of—
Compositions used in coating various materials, to produce smooth surfaces and to fill porous bodies.

Paint and Varnish
Binding agent in making—
Paints, varnishes, and various filling and coating compositions containing pigments such as titanium white and the like, as well as asbestos and other products.

Stone
Ingredient of—
Compositions used for producing smooth surfaces on stone, gypsum, and artificial stones and for preserving both natural and artificial stone.

Woodworking
Ingredient of—
Compositions used for producing smooth coverings on wood products.

Methylsulphuric Acid Chloride

French: Chlorure d'acide méthylesulphurique.
German: Methylschwefelsaeureschlorid.

Dye
Starting point (Brit. 271533) in making soluble vat dyestuffs with—
Anthraquinone-1:2, flavanthrone, indanthrone, naphthacridone, thioindigo.

Methyl-tertiary-amyl Ether

Petroleum
Blending agent and improver (Brit. 445503) for—
Gasoline motor fuels (the blended fuel can also contain a small amount of tetraethyl lead or tetramethyl lead).

Methyl-tertiary-butyl Ether

Petroleum
Blending agent and improver (Brit. 445503) for—
Gasoline motor fuels (the blended fuel can also contain a small amount of tetraethyl lead or tetramethyl lead).

3-Methylthiobenzoic Acid

French: Acide de 3-méthylethiobenzoique.
German: 3-Methylthiobenzoesaeure.

Dye
Starting point (French 604347) in making anthraquinone vat dyestuffs with—
1-Amino-4-methoxyanthraquinone.
1:4-Diaminoanthraquinone.
1:5-Diaminoanthraquinone.
1:5-Diamino-4-hydroxyanthraquinone.
1:5-Diamino-4-methoxyanthraquinone.
4:8-Diaminoanthrarufin.

Methyl Thiosalicylate

French: Thiosalicylate de méthyle, Thiosalicylate méthylique.
German: Methylthiosalicylat, Thiosalicylsaeuremethylester, Thiosalicylsaeuresmethyl.

Chemical
Starting point (Brit. 282427) in making synthetic drugs with oxides and other salts of—
Antimony, arsenic, bismuth, gold, silver.

Methyltoluenesulphonamide

French: Sulphonamide de méthyletoluène.
German: Methyltoluolsulfonamid.

Cellulose Products
Plasticizer (Brit. 311657) for—
Cellulose esters and ethers.
For uses, see under general heading: "Plasticizers."

Chemical
Starting point in making—
Intermediates, pharmaceuticals.

Dye
Starting point in making various synthetic dyestuffs.

Methyltriphenylphosphonium Iodide
French: Iodure de méthyletriphénylephosphonium.
German: Jodmethyltriphenylphosphonium, Methyltriphenylphosphoniumjodid.

Miscellaneous
Mothproofing and moldproofing agent (Brit. 312163) in treating—
Hair, fur, feathers, felt, and the like.

Textile
Mothproofing and moldproofing agent (Brit. 312163) in treating—
Wool and other products.

Methyl Undecylenate
French: Undecylenate de méthyle, Undecylenate méthylique.
German: Methylundecylenat, Undecylensäuremethylester, Undecylensäuresmethyl.
Spanish: Undecilenato de metil.
Italian: Undecilenato di metile.

Chemical
Starting point (French 615959) in making—
Aluminum, zinc, manganese, and bismuth undecylenates.

Leather
Reagent (French 615959) for—
Weighting and polishing leather.

Methylvanillin
Synonyms: 3:4-Dimethoxybenzaldehyde, Protocatechuic aldehyde dimethyl ether, Veratrum aldehyde.
French: Aldéhyde de vératrum, Éther méthylique de vanillin, Éther diméthylique de protocatéchuiquealdéhyde.
German: Protocatechualdehyddimethylaether. Vanillinmethylaether, Veratrumaldehyd.

Perfume
Base and fixative in making—
Perfume preparations.
Ingredient of—
Cosmetics of various sorts.

Soap
Ingredient of—
Toilet soaps.

Milk Sugar
Synonyms: Lactin, Lactose, Sugar of milk.
Latin: Sacharum lactis.
French: Sucre de lait.
German: Milchzucker.
Spanish: Azucar de leche, Lactosa.
Italian: Lattosio, Zucchero di latte.

Agriculture.
Lactic fermentation generator in—
Ensilage.

Explosives and Matches
Ingredient of—
Red smoke compound, containing also potassium chlorate and paranitranilin red.
Green smoke compound, containing also synthetic indigo, auramine yellow O, and potassium chlorate.
Stabilizer in—
Explosives.

Fats and Oils
Preservative for—
Oilcakes containing readily fermentable oils and fats (produces slight lactic fermentation which protects the oil from oxidation by preventing the development of oxidizing lipases, such as olease, and makes the cake more appetising and digestible).
Reagent for—
Increasing fermentable sugar in olives intended for oil extraction by microbiologic methods.

Food
Coating agent for—
Olives.
Preserved citrous and other fruits.
Sugared almonds.
Excipient in—
Concentrating fruit juices in vacuum.
Flavoring agent for—
Chocolate products.
Firming agent for—
Soft fruits (also increases their resistance to preservatives and diminishes discoloration).
Generator of—
Lactic acid in food products.
Ingredient of—
Baking mixes.
Biscuit mixes.
Dry coloring material for edible fats and oils (admixture with a dye) (U. S. 1921738).
Infants' foods.
Invalids' foods.
Prepared dietary milk.
Proprietary food preparations, consisting of (1) various mixtures with pure cacao; (2) various mixtures with pure cacao and a high content of readily digestible iron.
Soft cheese made from fresh cheese, water, a source of butter fat, an emulsifying agent and salt (U. S. 1879162).
Soups.
Preservative for—
Flavor, color, and consistency of pork and other meat products (more advantageous than nitrites from a health standpoint).
Slightly acidulated fruits.
Reducing agent and preservative for—
Oxidizable essences in concentrated citrous fruit juices.
Substitute for—
More readily fermentable sugars in jam manufacture.

Glass
Reducing agent in making—
Mirrors.

Perfume
Ingredient of—
Dentifrice, containing also cream of tartar, colloidal clay, flavor, and color.
Stabilizer in—
Effervescent bath salts (tablets), containing also sodium biborate, sodium sulphate, sodium bicarbonate, tartaric acid, talc, oil of pinus silvertris, oil of pinus pumilio, and a coloring matter, such as fluorescein.

Pharmaceutical
In compounding and dispensing practice.

Rubber
Preservative for—
Latex (used on account of its reducing action and because it opposes resinification).

Soap
Stabilizer for—
Emulsification.
Natural organic colors, such as chlorophyl emulsion.
Transparency.

Milori Blue
French: Bleu de milori.
German: Milorblau.

Ink
Coloring in—
Lithographic inks, printing inks.

Paint and Varnish
Coloring in—
Lacquers, paints, varnishes.
Starting point in making—
Pigments.

Paper
As a coloring.

Soap
Coloring for—
Toilet soaps.

Textile
——, *Dyeing*
Coloring for—
Yarns and fabrics.

Mineral Black
Synonyms: Oil black, Slate black.
French: Noir d'huile, Noir de schiste.
German: Mineralschwarz, Oelschwarz, Schieferschwarz.

Ink
Pigment in—
Drawing inks, printing inks.

Leather
Pigment in—
Coating compositions.

Mineral Black (Continued)

Linoleum and Oilcloth
Pigment in—
Coating compositions.
Miscellaneous
Ingredient of—
Pigments used for obtaining black effects on or in various products and compositions.
Paint and Varnish
As a black pigment.
Ingredient of—
Grayish pigments (made by the addition of gypsum flour).
Lime colors, water colors.
Paper
Pigment in printing—
Wallpaper.
Stone
Pigment in—
Compositions for treating stone.
Woodworking
Pigment in—
Wood compositions.

Mixed Pentanes
German: Pentanvermischung.
Chemical
General solvent for various purposes.
Starting point in making—
Chlorinated and hydrogenated derivatives.
Miscellaneous
General solvent for various purposes.
Reagent in—
Low-temperature thermometers.
Lubricant in—
Claude's liquid air machine.
Paint and Varnish
Solvent in making—
Cellulose derivative paints and varnishes and lacquers.
Pharmaceutical
In compounding and dispensing practice.
Plastics
Solvent in making—
Cellulose ester and ether compounds.
Refrigeration
Active medium in—
Refrigerating machines.

Molasses-Alcohol Residue
(A complex mixture of vegetable gums, unfermentable sugars, inorganic salts, and water.)
Fuel
Suggested binder in—
Briquetted fuels.
Metallurgical
Low-cost core binder in foundry practice.

Molybdenum Betabenzoylpropionate
Plastics
Starting point (U. S. 2001380) in making—
Films.

Molybdenum Oxide
Synonyms: Molybdenum sesquioxide.
French: Oxyde de molybdène, Sesquioxyde de molybdène.
German: Molybdaenoxyd, Molybdaensesquioxyd.
Chemical
Catalyst in making—
Acetaldehyde from ethyl alcohol (U. S. 1636952).
Anthraquinone (U. S. 1636856).
Maleic acid from benzene (U. S. 1636857).

Molybdenum Reds
French: Rouge molybdène.
German: Molybdanrot.
Ink
New pigment for—
Printing inks (said to have great covering power).
Paint and Varnish
New pigments for—
Emulsified paints, glue paints, linseed oil base paints, oil varnishes.

Molybdenum Trioxide
Synonyms: Molybdenic acid, Molybdenic anhydride, Molybdic acid.
French: Acide molybdènique, Anhydride molybdènique, Trioxyde de molybdène, Trioxyde molybdènique.
German: Molybdaenanhydrid, Molybdaensäure, Molybdaentrioxyd.
Spanish: Acido molibdenico, Anhidrido molibdenico, Trioxido molibdenico.
Italian: Anidrido molibdenico, Triossido molibdenico.
Analysis
Constituent of—
Froehde's reagent for analyzing alkaloids.
Reagent for—
Analyzing albumen.
Aromatic oxy-compounds.
Ethyl alcohol.
Hydrogen.
Hydrogen peroxide.
Phenol.
Phosphoric acid.
Determining arsenic.
Bismuth.
Lead.
Water in ethyl alcohol and ether.
Ceramics
Ingredient of—
Blue glazes for various ceramic products, such as chinaware, porcelains, and potteries.
Chemical
Catalyst in—
Converting hexahydrotoluene into toluene with addition of alumina (French 629838).
Dehydrogenation (Brit. 323713) of allylene to give allanol.
Amylene to give amanol.
Butylene to give isobutanol.
Ethylene to give ethanol.
Heptylene to give heptanol.
Hexylene to give hexanol.
Propylene to give isopropanol.
Hydration of acetylene to acetic acid (French 518574).
Hydrogenation (Brit. 312043) of aldehydes and ketones into alcohols.
Aldehyde-ketones into glycols.
Making alcohols from methylene, ethylene, amylene, butylene, propylene, and the like (Brit. 335551).
Reagent in making—
Calcium molybdate, with addition of water and lime (French 621640).
Pigments for water-color painting and wash designs, with addition of tin.
Starting point in making—
Molybdenum salts.
Dye
Reagent in making various synthetic dyestuffs.
Electrical
Reagent in making—
Filaments for incandescent lamps.
Fertilizer
Reagent (French 632310) in making—
Neutral calcium phosphate used as fertilizer.
Gas
Ingredient (French 667877) of—
Mixtures, containing oxides of metals of the first, second, third, and fourth groups, used as catalysts for the removal of organic sulphur compounds and thiophene from gas.
Miscellaneous
Reagent for—
Coloring various metals.
Starting point in making—
Metallic molybdenum.
Metallurgical
Catalyst for—
Washing flue gases with water to remove sulfur dioxide.
Reagent in—
Electrocoloration of metals.
Starting point in making—
Molybdenum.
Paint and Varnish
Pigment in—
Oil and water paints.

Molybdenum Trioxide (Continued)
Reagent (French 569385) for making—
Pigments in which basic and acid functions are assumed by compounds of molybdenum of various valencies.
Petroleum
Catalyst (French 632850) in—
Converting crude mineral oils into light products.
Starting point (Brit. 311251) in making—
Catalysts used for the destructive hydrogenation and cracking of oils.
Pharmaceutical
In compounding and dispensing practice.
Textile
Reagent in—
Dyeing silk yarns and fabrics.

Monoacetin
Chemical
Solvent for—
Tannins.
Starting point in making various derivatives.
Dye
Solvent for—
Basic dyestuffs.
Explosives
Ingredient of—
Low-freezing dynamites.
Smokeless powder (as gelatinizing agent).
Starting point in making—
Dinitroacetylglycerin.
Leather
Assistant in—
Tanning.
Miscellaneous
As a solvent.
Textile
——, *Manufacturing*
Ingredient (Brit. 313885) of—
Solutions of esters or ethers of cellulose (added for the purpose of facilitating spinning into yarn).
——, *Printing*
Ingredient of—
Printing pastes, particularly those containing indulins.

Monoacetylphenylhydrazin
Chemical
Starting point in making—
Pharmaceuticals and other derivatives.
Pharmaceutical
In compounding and dispensing practice.

Monoamylamine
Chemical
Solvent for—
Organic compounds.
Starting point in making—
Salts and soaps with most acids.
Leather
Ingredient of—
Leather finishes.
Miscellaneous
Solvent for—
Products of organic composition.
Solvent in—
Polishes of various kinds.
Paper
Solvent in—
Sizing compositions.
Textile
Starting point in making—
Textile lubricants.

Monobenzylpara-aminophenyl
Petroleum
Antioxidant for—
Treating cracked gasoline to inhibit the formation of gum.
Prevent material loss of initial antiknock value.
Stabilize against discoloration.

Monobrom-1:2-chrysenequinone
Dye
Intermediate (Brit. 438609) in making—
Synthetic dyes.

Monobromobenzene
Synonyms: Bromobenzene, Bromobenzol, Monobrombenzene, Monobrombenzol.
French: Bromure de benzyle, Bromure benzylique.
German: Brombenzol.
Spanish: Bromuro de benzil.
Italian. Bromuro di benzile.
Chemical
Reagent in—
Organic synthesis.
Pharmaceutical
In compounding and dispensing practice.

Monobromoisovalerylglycolurea
Synonyms: Monobromisovalerylglycolcarbamide.
French: Monobromisovaleryleglycolurée.
German: Monobromisovalerylglycolylharnstoff.
Spanish: Monobromisovalerilglicolilcarbamida, Monobromisovalerilglicolilurea.
Italian: Monobromisovalerilglicolilcarbamida, Monobromisovalerilglicolilurea.
Pharmaceutical
Suggested for use as—
Hypnotic and sedative.

Monobromonaphthalene
German: Bromnaphtalin.
Insecticide
Ingredient of—
Weed-killers and insecticides (Brit. 260055).
Miscellaneous
Reagent in determining index of refraction of crystals.

Monobromonitromethane
Fuel
Primer (Brit. 461320) for—
Diesel fuels.

Monobutyldiphenyl Phosphate, Chlorinated
Lubricant
Stabilizer (Brit. 448424) for—
Viscous oils, such as Pennsylvania or Midcontinent petroleums, used for extreme pressure work.

Monobutylnaphthalenesulphonic Acid
French: Acide monobutylenaphthalènesulfonique.
German: Monobutylnaphtalinsulfonsäure.
Spanish: Acido monobutilnaftalinasulfonico.
Italian: Acido monobutilnaftalinasolfonico.
Chemical
Dispersing agent (Brit. 266746) in making—
Emulsions of hydrocarbons of various groups of the aliphatic and aromatic series.
Emulsions of various chemicals.
Terpene emulsions.
Textile lubricants in emulsified form.
Wetting compositions in emulsified form.
Starting point in making—
Esters and salts.
Disinfectant
Dispersing agent (Brit. 266746) in making—
Emulsified germicidal and disinfecting compositions.
Dye
Dispersing agent (Brit. 266746) in making—
Emulsified color lakes.
Fats and Oils
Dispersing agent (Brit. 266746) in making—
Emulsified boring oils.
Emulsified drilling oils.
Emulsified fat-splitting preparations.
Emulsified fatty acids of animal or vegetable origin.
Emulsified greasing compositions.
Emulsified greasing and lubricating compositions containing various vegetable and animal fats and oils.
Emulsified sulphonated oils.
Emulsified wire-drawing oils.
Emulsions of animal and vegetable fats and oils.
Glues and Adhesives
Dispersing agent (Brit. 266746) in making—
Emulsified adhesive preparations.

Monobutylnaphthalenesulphonic Acid (Continued)

Ink
Dispersing agent (Brit. 266746) in making—
Emulsified printing and writing inks.

Insecticide
Dispersing agent (Brit. 266746) in making—
Emulsified insecticidal and fungicidal compositions.
Horticultural sprays.

Leather
Dispersing agent (Brit. 266746) in making—
Emulsified compositions for softening hides (Brit. 266746).
Emulsified dressing compositions (Brit. 266746).
Emulsified fat-liquoring baths (Brit. 266746).
Emulsified finishing compositions (Brit. 266746).
Emulsified soaking compositions (Brit. 266746).
Emulsified tanning compositions containing formocresylic and coumarone resins (Brit. 302938).
Emulsified waterproofing composition (Brit. 266746).

Miscellaneous
Dispersing agent (Brit. 266746) in making—
Automobile polishes in emulsified form.
Emulsified cleansing compositions.
Emulsified compositions for cleansing painted and metallic surfaces.
Emulsified degreasing compositions.
Emulsified furniture polishes.
Emulsified greasing compositions.
Emulsified metal polishes.
Emulsions of various substances.
Waterproofing compositions in emulsified form.

Paint and Varnish
Dispersing agent (Brit. 266746) in making—
Emulsified shellac preparations.
Waterproofing compositions in emulsified form.

Paper
Dispersing agent (Brit. 266746) in making—
Emulsified compositions for sizing paper and pulp products.
Emulsified compositions for waterproofing paper and pulp compositions and paperboard.
Waxing compositions in emulsified form.

Perfume
Dispersing agent (Brit. 266746) in making—
Emulsified cosmetics.

Petroleum
Dispersing agent (Brit. 266746) in making—
Emulsified cutting oils for screw-press and lathe work.
Emulsified mineral oils.
Kerosene emulsions.
Naphtha emulsions.
Soluble greases in emulsified form.
Solubilized emulsified oils and distillates.
Various petroleum pitch emulsions.
Various petroleum tar emulsions.
Various textile oils in emulsified form, such as rayon oils.

Plastics
Dispersing agent (Brit. 266746) in making—
Emulsified plastic compositions.

Resins and Waxes
Dispersing agent (Brit. 266746) in making—
Emulsified preparations of natural and artificial waxes.
Emulsified preparations of natural and artificial resins.

Rubber
Dispersing agent (Brit. 266746) in making—
Emulsified rubber cements and compositions.

Soap
Dispersing agent (Brit. 266746) in making—
Emulsified detergents, containing soaps, used for various purposes.
Emulsified hand-cleansing compositions containing soap.
Emulsified textile soaps.

Textile
—, *Bleaching*
Dispersing agent (Brit. 266746) in making—
Emulsified bleaching baths.

—, *Dyeing*
Dispersing agent (Brit. 266746) in making—
Dye baths in emulsified form.

—, *Finishing*
Dispersing agent (Brit. 266746) in making—
Emulsified coating compositions.
Emulsified scouring compositions.
Emulsified sizing compositions.
Emulsified washing compositions.
Emulsified waterproofing compositions.
Emulsified waxing compositions.

—, *Manufacturing*
Dispersing agent (Brit. 266746) in making—
Emulsified bowking baths.
Emulsified fulling baths.
Emulsified baths for the carbonization of wool.
Emulsified compositions used for degreasing raw wool.
Emulsified spinning compositions.
Emulsified mercerization baths.
Emulsified keir-boiling baths for cotton.
Emulsified baths for soaking silks.
Emulsified baths for degumming and boiling-off silk.
Oiling emulsions for various textile purposes.

—, *Printing*
Dispersing agent (Brit. 266746) in making—
Emulsified printing pastes.

Monochloracetic Acid
French: Acide acétique, monochloré; Acide monochloracétique, Monochlorure d'acide acétique.
German: Chloressigsäure.

Chemical
Dehydrating agent (Brit. 388485) in making—
Cleansing and emulsifying agents from 7:18-stearicglycol.
Peptizing agent (Brit. 398517) in making—
Hard, granular, porous gels having catalytic or adsorbent properties.
Reagent in making—
Intermediates.
3-Oxyselenonaphthene and derivatives (French 754756).
Pharmaceutical chemicals.
Photographic chemicals.
Synthetic organic chemicals.

Dye
Reagent in making—
Dyes.

Resins
Reagent (Brit. 395894) in making—
Synthetic resins.

Monochlorhydrin

Chemical
Intermediate in—
Organic syntheses.
Intermediate in making—
Novocaine.
Solvent immiscible with oils and other hydrocarbons.
Solvent miscible with water and various organic solvents.

Cellulose Products
Solvent for—
Cellulose acetate (used with water).

Explosives
Intermediate in making—
Explosives.

Resins
Partial solvent for—
Benzyl abietate, ester gum, mastic, shellac.
Solvent for—
Glyceryl phthalate resins.

Monochlor-1-ketotetrahydronaphthalene
French: Monochlor-1-cétotétrahydronaphthalène.
German: Monochlor-1-ketotetrahydronaphtalin.
Spanish: Chlor-1-cetotetrahidronaftaleno.
Italian: Chlor-1-cetotetraidronaftalene.

Chemical
Intermediate (German 377587) in making—
Synthetic aromatics, synthetic chemicals, synthetic pharmaceuticals.

Dye
Intermediate (German 377587) in making—
Synthetic dyestuffs.

Insecticide.
As an insecticide (German 377587).

Monochlorobenzene
Synonyms: Benzene chloride, Chlorobenzene, Chlorobenzol, Phenyl chloride.
French: Chlorure de benzène, Chlorure de phényle, Chlorure phénylique.
German: Chlorbenzene, Chlorbenzol, Monochlorbenzol.

Chemical
Solvent (Brit. 260623) in making—
Alphabromo-2-naphthylglycollic acid.
Alphachloro-2-naphthylglycollic acid.
Alphaiodo-2-naphthylglycollic acid.
Starting point in making—
Aromatic compounds.
Chloroanthraquinone.
Dinitrochlorobenzene.
Intermediates.
Organic compounds.
Orthochloro-2-nitrobenzene-4-sulphonic acid.
Orthodichlorobenzene.
Orthonitrochlorobenzene.
Parachlorobenzenesulphonic acid.
Paradichlorobenzene.
Paranitrochlorobenzene.
Pharmaceutical chemicals.
Phenol.
Picric acid.

Dye
Starting point in making—
Sulphur blacks, sulphur browns, various other dyestuffs.

Miscellaneous
Reagent in—
Measuring temperature by optical methods.

Paint and Varnish
Ingredient (U. S. 1596413) of—
Paint and varnish removers.
Solvent in making—
Oil lacquers, varnishes.

Monochlororetene

Petroleum
Imparter (Brit. 431508) of—
High-film strength, adhesion power, and abrasion resistance to lubricants for use with extreme pressures (consists of blends with mineral lubricating oil).

Monoethanolamine Citrate

Textile
De-electrifying agent (Brit. 430221) for—
Yarns, films, fabrics, and the like, subject to charging by static electricity (applied in admixture with all usual lubricating agents as vehicle).

Monoethanolamine Gallate

Textile
De-electrifying agent (Brit. 430221) for—
Yarns, films, fabrics, and the like, subject to charging by static electricity (applied in admixture with all usual lubricating agents as vehicle).

Monoethanolamine Lactate

Textile
De-electrifying agent (Brit. 430221) for—
Yarns, films, fabrics, and the like, subject to charging by static electricity (applied in admixture with all usual lubricating agents as vehicle).

Monoethanolamine Mucate

Textile
De-electrifying agent (Brit. 430221) for—
Yarns, films, fabrics, and the like, subject to charging by static electricity (applied in admixture with all usual lubricating agents as vehicle).

Monoethanolamine Saccharate

Textile
De-electrifying agent (Brit. 430221) for—
Yarns, films, fabrics, and the like, subject to charging by static electricity (applied in admixture with all usual lubricating agents as vehicle).

Monoethanolamine Salicylate

Textile
De-electrifying agent (Brit. 430221) for—
Yarns, films, fabrics, and the like, subject to charging by static electricity (applied in admixture with all usual lubricating agents as vehicle).

Monoethanolamine Tannate

Textile
De-electrifying agent (Brit. 430221) for—
Yarns, films, fabrics, and the like, subject to charging by static electricity (applied in admixture with all usual lubricating agents as vehicle).

Monoethanolamine Tartrate

Textile
De-electrifying agent (Brit. 430221) for—
Yarns, films, fabrics, and the like, subject to charging by static electricity (applied in admixture with all usual lubricating agents as vehicle).

Monofluoroacetic Acid
French: Acide de monofluoroacétique.
German: Monofluoressigsäure.

Miscellaneous
Mothproofing agent (Brit. 333583) in treating—
Feathers, furs and other articles.

Textile
Mothproofing agent (Brit. 333583) in treating—
Wool and felt.

Monomethyldioxyethylamine
French: Monométhyledioxye-éthyleamine.

Chemical.
Reagent (Brit. 295024) in making dispersing agents with—
Castor oil, cottonseed oil, linseed oil, oleic acid, olive oil, palmitic acid, ricinoleic acid, stearic acid, sulphoricinoleic acid.
Starting point in making—
Intermediates, salts and esters.

Monomethylorthotoluidin.

Chemical
Starting point in making—
Indol.

Dye
Starting point in making—
Auramin G.
Brilliant fern blue.
Brilliant ice blue.
Brilliant rhoduline red B.
Brilliant rhoduline red BD.
Glacier blue.

Monomethylthyolurea

Plastics
Starting point in making—
Condensation products with thiourea and other compounds (Brit. 262148).

Monomethylxylenesulphonamide
French: Monométhylexylènesulphonamide, Sulphonamide de monométhylexylène.
German: Monomethylxylolsulfonamid.

Cellulose Products
Plasticizer (Brit. 313133) for—
Cellulose acetate, cellulose esters and ethers, cellulose nitrate.
For uses, see under general heading: "Plasticizers."

Mono-oxydiphenylene

Cellulose Products
Plasticizer and softener (German 591365) for—
Cellulose acetate.
For uses, see under general heading: "Plasticizers."

Monophenylglycerin

Miscellaneous
As an emulsifying agent (Brit. 350379).
For uses, see under general heading: "Emulsifying agents."

Monophenylthiourea
Synonyms: Monophenylsulphourea.
French: Sulphourée de monophényle, Thiourée de monophényle.
German: Monophenylsulfoharnstoff, Monophenylthioharnstoff.

Chemical
Starting point in making various derivatives.

Metallurgical
Ingredient (U. S. 1779961) of—
Baths for cleaning metals, combined in the form of a protective condensation product with aldehyde ammonia.

Monotolylglycerin

Miscellaneous
As an emulsifying agent (Brit. 350379).
For uses, see under general heading: "Emulsifying agents."

Monoxylylglycerin

Miscellaneous
As an emulsifying agent (Brit. 350379).
For uses, see under general heading: "Emulsifying agents."

Montanic Acid Chloride

Chemical
Starting point (Brit. 407956) in making pour-point improvers for machine oils, gear oils, and other lubricants by condensing with—
Anilin, anthracene oil.
Aromatics obtained by destructive hydrogenation or by dehydrogenation.
Benzene
Cracking gases containing gaseous olefins (ethylene, propylene, and butylene).
Cyclic terpenes, ethylnaphthalene, liquid olefins, middle oil naphthalene, naphthols, naphthylamines, nitrated aromatics, phenols, tars, toluene, xylene.

Montanic Acid Ester of 2-Hydroxyethanesulphonic Acid

Insecticide and Fungicide
Addition agent (German 550961) to—
Bordeaux mixtures for controlling *Peronospora* and *Fusicladium.*

Montan Wax
Synonyms: Lignite wax, Mineral wax.
French: Cire de lignite, Cire de montane.
German: Lignitwachs, Montanwachs.

Chemical
Ingredient of—
Boiler compounds.
Lining for—
Acid tanks, used alone or in compositions.

Electrical
Ingredient of—
Insulating preparations.
Insulating in—
Cables, motors, generators, and other electrical apparatus.

Fats and Oils
Hardener for—
Fats and greases.
Ingredient of—
Axle greases, gun oils, lubricating greases.

Food
Ingredient of—
Compositions used for decorating confections.
Reagent in making—
Artificial honeycombs.

Glues and Adhesives
Ingredient of—
Special adhesive pastes and compositions.

Ink
Ingredient of—
Printing inks, writing inks.

Leather
Ingredient of—
Finishing and dressing compositions.
Tanning compositions.

Metallurgical
Ingredient of—
Compositions used for coating metals to protect them against moisture, acids, alkalies, and so on.
Reagent in—
Electroplating.

Miscellaneous
Ingredient of compositions used in making—
Alabaster imitations, candles, dolls, statuettes, toys, wax figures.
Ingredient of—
Preservatives for sculptures, linings for kegs and barrels, leather polishes, grease crayons, metal polishes, shoe creams, shoe dressings.

Paint and Varnish
Ingredient of—
Dry colors, encaustic points, floor waxes, paints, tar roofing compositions, varnishes, waterproof paints and varnishes, wood fillers.

Paper
Ingredient of—
Compositions for making waxed paper.
Sizing compositions for producing high gloss paper.

Perfumery
Ingredient of—
Cosmetics, pomades.

Pharmaceutical
In compounding and dispensing practice.

Printing
In lithography.
In photoengraving.
In process engraving.

Rubber
Filler in making—
Rubber products.
Ingredient of—
Vulcanizing mixtures.

Soap
Ingredient of—
Special soaps.

Textile
Ingredient of—
Compositions used for making waxed cloth.

Waxes and Resins
Hardener in—
Wax compositions.
Ingredient of—
Sealing wax, shoemaker's wax.
Substitute for—
Beeswax, carnauba wax.

Morphine Acetate
Latin: Acetas morphicus, Acetas morphinae.
French: Acétate de morphine.
German: Essigsäuresmorphin, Morphinacetat, Morphinazetat.
Spanish: Acetato de morfina.
Italian: Acetato di morfina.

Chemical
Starting point in making—
Pharmaceutical derivatives.

Pharmaceutical
In compounding and dispensing practice.

Morpholine
German: Morpholin.

Chemical
Chemical whose dilute water solutions boil or evaporate with little change in composition.
Chemical which, during evaporation or distillation, maintains a constant alkalinity both in the solution and in the distillate.
Solvent miscible with—
Many miscellaneous chemicals, most organic solvents.
Suggested starting and processing material in making—
Inhibitors, pharmaceutical chemicals, rubber chemicals, textile lubricants, various synthetic organic chemicals.

Cosmetic
Ingredient of—
Hair-waving preparations.

Morpholine (Continued)

Dye
Solvent for—
Dyes.
Suggested processing material in making—
Dyes.

Fats, Oils, and Waxes
Dispersing agent in making—
Emulsified products of fats, oils, or waxes.

Leather
Dispersing agent in making—
Dressing compositions, finishing compositions, softening compositions, waterproofing compositions.

Miscellaneous
Chemical whose dilute water solutions boil or evaporate with little change in compositions.
Chemical which, during evaporation or distribution, induces a constant alkalinity both in the solution and in the distillate.
Dispersing agent in—
Polishing compositions for furniture, automobiles, metals, wood, and other surfaces.
Various processes involving aqueous solutions.
Waterproofing compositions.
Emulsifying agent in—
Polishing compositions resistant to water spotting.
Solvent miscible with—
Most organic solvents, many miscellaneous materials.

Paint and Varnish
Dispersing agent in making—
Paints, varnishes.
Solvent for—
Casein, dyes, resins, shellac, waxes.

Paper
Dispersing agent in making—
Coatings, sizings.

Plastics
Solvent for—
Casein, resins, shellac.

Power Generation
Reducer of—
Corrosion in closed boiler systems.

Resins
Solvent for—
Resins, shellac.

Textile
Dispersing agent in—
Coating compositions, scouring compositions, sizing compositions, waterproofing compositions, waxing compositions.
Solvent for—
Dyes.

Morpholineoleicamide Phosphate
French: Morpholine-oléique-amide phosphaté, Morpholine-oléique-amide phosphatique, Phosphate de morpholine-oléique-amide.
German: Morpholinoleinamidphosphat, Phosphorsäuresmorpholinoleinamid.
Spanish: Fosfato de morfoline-oleico-amide.
Italian: Fosfato di morfoline-oleico-amido.

Miscellaneous
As an emulsifying agent (Brit. 364104).
For uses, see under general heading: "Emulsifying agents."

Morpholineoleicamide Sulphate

Miscellaneous
As an emulsifying agent (Brit. 364104).
For uses, see under general heading: "Emulsifying agents."

Morpholinomethyl-1:3:2-xylenol

Rubber
Anti-ager (Brit. 459045) for—
Rubber mixes.

Mountain Green
Synonyms: Mineral green.
French: Verte de minérale, Verte de montagne.
German: Berggruen, Mineralgruen.

Paint and Varnish
Pigment in—
Lacquers, paints, varnishes, stains.

Mucic Acid
Synonyms: Saccharolactic acid.
French: Acide mucique, Acide saccharolacétique.
German: Saccharomilchsäure, Schleimsäure.
Italian: Acido mucico, Acido sarcolatico.

Chemical
Reagent in—
Making artificial yeast by admixture with sodium bicarbonate.
Granular effervescent salt.
Treating yeast to accelerate its growth.
Starting point in making—
Adipic acid, esters and salts, intermediates, pharmaceuticals, pyromucic acid, pyrrol.

Disinfectant
Reagent in making—
Alkaloid disinfectants by synthesis.

Food
Acidulant in making—
Ice cream.
Ingredient of—
Baking powders (used in the place of tartaric acid or potassium acid tartrate).
Reagent in making—
Soft drinks.

Miscellaneous
Ingredient of—
Electroplating baths.

Plastics
Ingredient of—
Plastic compositions.

Textile
Ingredient of—
Chrome baths for dyeing wool with alizarin.
Reagent in making—
Mordant solutions for dyeing fabrics and yarns.

Musk
Synonyms: Assam and Nepaul musk, Blue pile musk, Cabardine musk, Chinese, Thibet or Tonquin musk, Deer musk, Grain musk, Yaman musk.
Latin: Moschus, Moschus orientalis, Moschus chinensis, Moschus tibetanus.
French: Bésain, Musc, Musc sanko, Musc de tonquine.
German: Bisam, Moschus, Tonchinmoschus.
Spanish: Almizcle.
Italian: Muschio.

Food
Ingredient of—
Flavoring preparations.
Flavoring agent in—
Confectionery, food preparations.

Ink
Ingredient of—
Chinese ink.

Insecticide
Ingredient of—
Insecticidal preparations.

Miscellaneous
Mothproofing agent in treating—
Furs, feathers, hair, and other articles.
Preservative agent for furs.

Perfume
Fixative in making the following odors:—
Lilac, lily of the valley, rose, violet.
Ingredient of—
Cosmetics, sachet powders.

Pharmaceutical
In compounding and dispensing practice.

Soap
Perfume in—
Toilet soaps.

Textile
Mothproofing agent in treating—
Wool and felt.

Musk Ambrette
Synonyms: Dinitrobutylmetacresolmethylether, Dinitropseudobutylmetacresolmethylether.
French: Éther de dinitrobutylemétacrésoleméthyle, Éther dinitropseudobutylemétacrésolméthyle, Moschus ambrette, Musc ambrette.

Musk Ambrette (Continued)
German: Dinitrobutylmetakresolmethylaether, Dinitropseudobutylmetakresolmethylaether, Kresolmoschus.

Food
Ingredient of—
Candies, flavorings.

Insecticide
Ingredient of—
Fumigating compositions, insecticidal compositions.

Miscellaneous
Ingredient of—
Compositions used for mothproofing furs, feathers, and the like.

Perfume
Fixative for perfumes.
Ingredient of—
Cosmetics, dentifrices, perfumes.

Pharmaceutical
In compounding and dispensing practice.

Sanitation
Ingredient of—
Germicidal preparations.

Soap
Perfume in—
Toilet soaps.

Textile
Ingredient of—
Compositions used for mothproofing wool, felt, and other products.

Musk, Ketone
Synonyms: Dinitroacetotertiarybutylxylol, 3:5-Dinitro-2:4-dimethyl-6-tertiarybutylacetophenone, 2:6-Dinitro-1:3-dimethyl-5-tertiarybutylacetylbenzol, Dinitropseudobutylxylylmethylketone.
French: Musc de kétone, Musc kétonique, Musc de mallmann.
German: Moschus keton.

Food
Flavoring in—
Cakes, candies.
Ingredient of—
Flavoring compositions.

Insecticide
Ingredient of—
Insecticidal preparations.

Miscellaneous
Mothproofing agent for—
Feathers, furs.

Pharmaceutical
In compounding and dispensing practice.

Soap
Perfume for—
Toilet soaps.

Textile
Mothproofing agent for—
Felt, wool.

Musk Xylol
Synonyms: Moschus xylol, 2:4:6-Trinitro-5-tertiarybutylmetaxylol.
French: Musc de baur, Musc xylène.
German: Moschus xylol.

Food
Flavoring in—
Confectionery, pastries.
Ingredient of—
Flavoring extracts.

Insecticides
Ingredient of—
Insecticides, germicides, vermicides, and the like.

Miscellaneous
Mothproofing agent in—
Treating furs and feathers.

Perfumery
Fixative in—
Perfume making.
Ingredient of—
Cosmetics, dentifrices, perfumes.

Pharmaceutical
In compounding and dispensing practice.

Soap
Perfume in—
Shampoos, soaps.

Textile
Mothproofing agent for—
Treating wool.

Myristyl-1-sulphuric Acid (Normal) Ester

Chemical
As an emulsifying agent.
Reagent in—
Organic synthesis.
Starting point (Brit. 440575) in making—
Emulsifying agents with salts of lead, aluminum, iron, tin, or barium (such emulsifying agents are said to form water-in-oil emulsions and are, preferably, produced in situ by (1) dissolving the sulphuric acid ester in the oil and (2) agitating with an aqueous solution of the metal salt, for example, lead acetate; they are said to be useful for treating medicinal paraffin oil, neatsfoot oil, olive oil, castor oil, cottonseed oil, linseed oil, and petroleum lubricating oils; a heavy paraffin oil, so treated on the basis of 50 parts by weight of oil to 48.75 parts of water, is said to yield a heavy grease that has good lubricating properties and may readily be extended with oil; a water-linseed oil type emulsion is offered as suitable for use as a paint base).

N'-Acetyl-N'-cyclohexylparaphenylenediamine
French: N'-acétyle-N'-cyclohexyleparaphénylènediamine.
German: N'-acetyl-N'-zyklohexylparaphenylendiamin.

Chemical
Starting point in making various derivatives.

Dye
Starting point (Brit. 340640) in making dyestuffs with the aid of—
Betanaphthol-8-sulphonic acid, benzoyl-K acid, 1-(2-chloro-5-sulphophenyl)-3-methyl-5-pyrazolone, H acid, orthoanisoylgamma acid, R acid, Schaeffer's acid, tetrahydronaphthalenebetasulphonyl-H acid.

Naphtha. See Solvent Naphtha.

Naphthalene
Synonyms: Moth camphor, Naphthalin, Tar camphor, White tar.
Latin: Naphthalenum.
French: Naphtalène, Naphthalène, Naphtaline.
German: Naphtalin.
Spanish: Naftalina.
Italian: Naftalina.

Chemical
Diluent in making—
Dihydrothiotoluidin.
Ingredient (U. S. 1893051) of—
Heat-energy transfer medium composed of diphenyl oxide, pyrene or parahydroxydiphenyl.
Starting point (Brit. 251294) in making—
Tanning agents by reaction with formaldehyde.
Starting point, either directly or indirectly, in making—
Alpha-amino-2-naphthol.
Alphanaphthalenesulphonic acid.
Alphanaphthol.
Alphanaphthylamine.
Alphanaphthylamine hydrochloride.
Alphanaphthylamine-8-sulphonic acid (peri acid).
Alphanitronaphthalene.
1-Amino-2-naphthol-4-sulphonic acid (1:2:4-hydroxy acid).
1-Amino-8-naphthol-2:4-disulphonic acid (Chicago acid, SS acid, 2S acid).
1-Amino-8-naphthol-3:6-disulphonic acid (H acid).
1-Amino-8-naphthol-4-sulphonic acid (S acid).
2-Amino-5-naphthol-7-sulphonic acid (J acid).
2-Amino-8-naphthol-6-sulphuric acid (gamma acid).
Anthraquinone (Brit. 295270 and 281307).
Betanaphthalenesulphonic acid.
Betanaphthol.
Betanaphthyl salicylate betol, naphtholsalol, naphthalol, salinaphthol, salicylicnaphthyl ester.
Betanaphthylamine.
Betanaphthyl ether (bromelia, nerolin 2).
Betanaphthylmethyl ether (nerolin, yara-yara betanaphtholmethyl ether, methyl betanaphtholate).
Decalin (dekalin: decahydronaphthalene).
Diaminonaphthalene (naphthalenediamine).
1-Diazo-2-naphthol-4-sulphonic acid.

Naphthalene (Continued)
5:8-Dichloroalphanitronaphthalene.
Dichlorophthalic acid.
1:5-Dihydroxyanthraquinone (anthrarufin).
1:8-Dihydroxyanthraquinone (chrysazine).
1:3-Dihydroxynaphthalene (naphthoresorcinol).
1:5-Dihydroxynaphthalene.
1:6-Dihydroxynaphthalene.
1:7-Dihydroxynaphthalene.
1:8-Dihydroxynaphthalene.
2:3-Dihydroxynaphthalene.
2:6-Dihydroxynaphthalene.
2:7-Dihydroxynaphthalene.
Dimethylbetanaphthylamine.
1:5-Dinitronaphthalene.
1:8-Dinitronaphthalene.
Diphenylnaphthylenediamine.
Intermediates.
Maleic acid (Brit. 295270).
Naphthalene-1:5-disulphonic acid (Armstrong's acid).
Naphthalene-2:7-disulphonic acid.
Naphthaleneformaldehyde.
1-Naphthalidoanthraquinone-2-carboxylic acid.
Naphthaquinone (Brit. 295270 and 281307).
1:8-Naphthasultam-2:4-disulphonic acid (Sultan acid).
Naphthionic acid (1-aminonaphthalene-4-sulphonic acid, 4-amino-1-naphthalenesulphonic acid).
2-Naphthol-3:6-disulphonic acid (R acid, betanaphthol-disulphonic acid).
2-Naphthol-6:8-disulphonic acid (potassium salt).
1-Naphthol-4-sulphonic acid (Nevile and Winther's acid, alphanaphtholsulphonic acid).
1-Naphthol-5-sulphonic acid (Cleve's acid, alphanaphtholsulphonic acid).
2-Naphthol-1-sulphonic acid (Tobias acid).
2-Naphthol-6-sulphonic acid (Schaeffer's acid, betanaphtholsulphonic acid).
2-Naphthol-7-sulphonic acid (Cassella's acid, monosulphonic acid F; F acid, mono acid F, betanaphtholsulphonic acid).
1-Naphthylamine-3:8-disulphonic acid (epsilon acid).
1-Naphthylamine-4:8-disulphonic acid (Schoelkopf's acid).
1-Naphthylamine-3:6:8-trisulphonic acid (Koch's acid).
1-Naphthylamine-5-sulphonic acid (Laurent's acid, L acid).
2-Naphthylamine-5:7-disulphonic acid (amino-J acid).
2-Naphthylamine-6:8-disulphonic acid (amino-G acid).
2-Naphthylaminesulphonic acid.
2-Naphthylamine-5-sulphonic acid.
2-Naphthylamine-8-sulphonic acid.
2-Naphthylamine-6-sulphonic acid (Broenner's acid).
2-Naphthylamine-7-sulphonic acid (Cassella's acid F, Bayer's acid, F acid, delta acid).
1:5-Naphthylenediamine-3:7-disulphonic acid (4:8-diamino-2:6-naphthalenedisulphonic acid).
1:8-Naphthylenediamine-3:6-disulphonic acid (4:5-diamino-2:7-naphthalenedisulphonic acid).
1:4-Naphthylenediamine-2-sulphonic acid (1:4-diamino-2-naphthalenesulphonic acid).
1:3-Naphthylenediamine-6-sulphonic acid (5:7-diamino-2-naphthalenesulphonic acid).
8-Nitro-1-diazo-2 naphthol-4-sulphonic acid.
1-Nitronaphthalene-5-sulphonic acid (Laurent's alpha acid).
Phthalic acid (Brit. 295270).
Phthalic anhydride.
Tetralin (tetrahydronaphthalene).

Disinfectant
Ingredient of—
Disinfecting agent (with nitrobenzene).

Dye
Starting point in making—
Eosin dyes, synthetic indigo.

Explosives and Matches
Ingredient of—
Ammonium nitrate explosives, liquid air explosive compositions, permissible explosives, smokeless powders of certain types.

Fuel
Binder for—
Anthracite briquets.
In candle making.

Gas
Enrichener in—
Lamps of the albocarbon type.

Insecticide
Fumigant for—
Gladiolus bulbs to free them from thrips.
Fungicide.
Ingredient of—
Flea powders, insecticidal compositions, sulphur mixtures for insecticidal purposes.
Moth-repellent.

Leather
Ingredient of—
Synthetic tannins.
Preservative for—
Hides, skins.

Lumbering
Ingredient of—
Preservative compositions.

Mechanical
Fuel in—
Internal combustion engines.
Ingredient of—
Carbon remover (U. S. 1878245), lubricating compositions, motor fuels.

Miscellaneous
Solvent for—
Asphalt.

Paint and Varnish
Ingredient of—
Fatty lacquers, rosin varnishes.

Petroleum
Condensing agent (Brit. 397169) for—
Chlorinated paraffin wax and other chlorinated waxes (the product of the condensation being useful in the dewaxing of petroleum oils).
Reagent in—
Removing efflorescence in petroleum oils and distillation products.
Substitute for—
Paraffin.

Pharmaceutical
In dispensing and compounding practice.
Suggested for use as expectorant, tenicide, vermifuge.
Suggested for use in treatment of chronic bronchitis, intestinal catarrah, intestinal inflammation, seatworms, skin diseases, typhoid.

Plastics
Ingredient (U. S. 1846356) of—
Thermoplastic molding compositions.
Plasticizer in—
Celluloid manufacture.

Resins and Waxes
Condensing agent in making—
Artificial resins from formaldehyde.
Solvent for—
Resins.
Starting point (Brit. 397096) in making—
Artificial resins from polyvalent alcohols and decomposition products of an aromatic hydrocarbon.

Rubber
Preservative packing for—
Rubber goods.
Solvent in—
Rubber manufacturing processes.

Sanitation
Disinfectant and germicide.

Naphthalenedimethylsulphonamide
Cellulose Products
Plasticizer (Brit. 417871) for—
Cellulose acetate.
For uses, see under general heading: "Plasticizers."

Naphthalene Ethylsulphonamide
Cellulose Products
Plasticizer (Brit. 417871) for—
Cellulose acetate.
For uses, see under general heading: "Plasticizers."

Naphthalenemercuric Acetate
Synonyms: Mercury-naphthalene acetate.
French: Acétate de mercure et de naphthalène, Acétate mercurique-naphthalènique.
German: Essigsäuremerkurnaphtalinester, Merkurnaphtalinacetat, Merkurnaphtalinazetat.

Chemical
Starting point in making various derivatives.

Insecticide
Ingredient (Brit. 321396) of—
Compositions for immunizing wheat and other grains.

Woodworking
Ingredient (Brit. 321396) of—
Preserving and disinfecting compositions.

Naphthalene-methyl-sulphonamide

Cellulose Products
Plasticizer (Brit. 417871) for—
Cellulose acetate.
For uses, see under general heading: "Plasticizers."

Naphthalene-1:4:5:8-tetracarboxylic Acid
French: Acide naphthalène-1:4:5:8-tétracarboxylique, Acide de naphthalène-1:4:5:8-tétracarboxyle.
German: Naphtalin-1:4:5:8-tetracarbonsaeure.

Dye
Starting point (Brit. 265232) in making naphthalene dyestuffs with—
Orthonitranilin, 3-nitro-4-amino-1-phenetole.

Miscellaneous
Reagent for treating meats to obtain a water-soluble albuminous product (German 427275).

Naphthazarin
German: Naphtazarin.

Dye
Starting point (Brit. 304804) in making dyestuffs with—
Allylamine, allylenediamine, amylamine, amylenediamine, butylamine, butylenediamine, caprylamine, citrylamine, ethylamine, ethylenediamine, formylamine, gallylamine, heptylamine, heptylenediamine, hexylamine, hexylenediamine, isoallylamine, isoamylamine, isobutylamine, isopropylamine, lactylamine, methylamine, methylenediamine, propylamine, propylenediamine.

Naphthenic Acid Ester of Grapeseed Alcohol
(Uses same as those given for item immediately following.)

Naphthenic Acid Ester Ricinoleic Alcohol

Bituminous
Solvent (Brit. 445223) for—
Asphalt and other bituminous bodies.

Dye
Solvent (Brit. 445223) for—
Dyestuffs, particularly oil-soluble coaltar dyes.

Fats, Oils, and Waxes
Solvent (Brit. 445223) for—
Fats, oils, waxes.

Resins
Solvent (Brit. 445223) for—
Oil-soluble glycerol-phthalic acid resins, polymerized vinyl compounds, synthetic resins.

Rubber
Solvent (Brit. 445223) for—
Rubber.

3:4-Naphthoheptathiocarbocyanin

Photographic
Sensitizer in—
Infra-red photography now important in long distance photography, foggy weather photography, aerial photography, night-time photography without use of visible lighting.

Naphtholdisulphonic Acid (2:3:6)

Chemical
Starting point in making—
Aluminum-naphthol disulphonate-2:3:6 (aluminol).
2:3-Dioxynaphthalene.
2:3-Dioxynaphthalene-6-sulphonic acid.
Naphthylaminedisulphonic acid-2:3:6.
2-Naphthol-3-disulphonic acid.

Dye
Starting point in making—
Acid alizarin red B, amaranthe, azo grenadin L, brilliant crocein 9B, cloth red B, congo blue 2B, fast red, naphthol black, naphthol black B, naphthol black 6B, orange III, poncea dyestuffs.

1-Naphthol-2-sulphonate-indo-3:5-dichlorophenol

Analysis
Indicator in—
Oxidation-reduction potential determinations (of particular interest to biologists and physiologists and in investigations of various materials, such as soils, wines, cheese, gasoline antiknock compounds).

1-Naphthol-2-sulphonateindophenol

Analysis
Indicator in—
Oxidation-reduction potential determinations (of particular interest to biologists and physiologists and in investigations of various materials, such as soils, wines, cheese, gasoline antiknock compounds).

1:2-Naphthoquinone Chlorimide
French: Chloroimdure de 1:2-naphthoquinone.
German: 1:2-Naphtochinonchlorimid.

Agricultural
As a seed disinfectant (Brit. 340500).

Chemical
Starting point in making various derivatives.

Naphthotetronic Acid
Synonyms: 6:7-Benzochroman-2:4-dione.

Dye
Starting point (Brit. 419447) in making—
Bordeaux shades for wool by coupling with diazotized 2-nitro-6-aminophenol-4-sulphonate and aftertreating with bichromate.
Orange-yellow dyes for wool by coupling with diazotized paranitranilin.
Violet dyes for wool by coupling with diazotized 2-nitro-6-amino-phenol-4-sulphonate.

Naphthoxybenzylbutylamine

Chemical
Antioxidant and stabilizer (Brit. 430335) for—
Unstable organic substances.

Fats, Oils, and Waxes
Antioxidant and stabilizer (Brit. 430335) for—
Fats, oils, waxes.

Petroleum
Antioxidant and stabilizer (Brit. 430335) for—
Petroleum products.
Inhibitor (Brit. 430335) of—
Gumming in petroleum products.

Rubber
As an antioxidant (Brit. 430335).

Naphthoylenebenzimidazoleperdicarboxylic Anhydride

Chemical
Starting point in making—
Intermediates and other derivatives.

Dye
Starting point (Brit. 313887) in making dyestuffs with—
4-Bromo-1:2-diaminobenzene.
4-Chloro-1:2-diaminobenzene.
3:4-Diaminoacenaphthene.
1:2-Diaminonaphthalene.
3:4-Diaminophenetole.
3:4-Diaminotoluene.
1:2-Dimethyl-4:5-diaminobenzene.
4-Nitro-1:2-diaminobenzene.
Orthophenylenediamine.

Naphthoylenebenzimidazoperdicarboxylic Acid

Chemical
Starting point in making various derivatives.

Dye
Starting point (Brit. 313887) in making dyestuffs with—
4-Bromo-1:2-diaminobenzene.
4-Chloro-1:2-diaminobenzene.
3:4-Diaminoacenaphthene.
1:2-Diaminonaphthalene.
3:4-Diaminophenetole.
3:4-Diaminotoluene.
4-Nitro-1:2-diaminobenzene.
Orthophenylenediamine.

Naphthylalphapropyl-Aluminum

Petroleum
Addition agent (Brit. 433257) in—
Lubricating oils or greases, especially for use at high temperatures, such as cylinder oils, hydrogenated oils, or oils refined by treatment with sulphuric acid, clay, or extraction solvents.

Naphthylalphapropyl-Bismuthine

Petroleum
Addition agent (Brit. 433257) in—
Lubricating oils or greases, especially for use at high temperatures, such as cylinder oils, hydrogenated oils, or oils refined by treatment with sulphuric acid, clay, or extraction solvents.

Naphthylalphapropyl-Cadmium

Petroleum
Addition agent (Brit. 433257) in—
Lubricating oils or greases, especially for use at high temperatures, such as cylinder oils, hydrogenated oils, or oils refined by treatment with sulphuric acid, clay, or extraction solvents.

Naphthylalphapropyl-Mercury

Petroleum
Addition agent (Brit. 433257) in—
Lubricating oils or greases, especially for use at high temperatures, such as cylinder oils, hydrogenated oils, or oils refined by treatment with sulphuric acid, clay, or extraction solvents.

Naphthylalphapropyl-Stibine

Petroleum
Addition agent (Brit. 433257) in—
Lubricating oils or greases, especially for use at high temperatures, such as cylinder oils, hydrogenated oils, or oils refined by treatment with sulphuric acid, clay, or extraction solvents.

Naphthylalphapropyl-Thallium

Petroleum
Addition agent (Brit. 433257) in—
Lubricating oils or greases, especially for use at high temperatures, such as cylinder oils, hydrogenated oils, or oils refined by treatment with sulphuric acid, clay, or extraction solvents.

Naphthylalphapropyl-Zinc

Petroleum
Addition agent (Brit. 433257) in—
Lubricating oils or greases, especially for use at high temperatures, such as cylinder oils, hydrogenated oils, or oils refined by treatment with sulphuric acid, clay, or extraction solvents.

Naphthyl Bisulphide

Petroleum
Antioxidant (Brit. 425569) for—
Lubricating, transformer, and switch oils, particularly solvent-extracted oils and others of a paraffinic nature, in which the natural inhibitor content may have been reduced during refining.

2-Naphthylmercaptan
Synonyms: Betathionaphthol.

Insecticide and Fungicide
Larvicide for—
Culicine mosquito larvae.

Naphthyl-Mercury Iodide

Petroleum
Addition agent (Brit. 433257) in—
Lubricating oils or greases, especially for use at high temperatures, such as cylinder oils, hydrogenated oils, or oils refined by treatment with sulphuric acid, clay, or extraction solvents.

Naphthyl-Mercury Sulphide

Petroleum
Addition agent (Brit. 433257) in—
Lubricating oils or greases, especially for use at high temperatures, such as cylinder oils, hydrogenated oils, or oils refined by treatment with sulphuric acid, clay, or extraction solvents.

1-Naphthylmethyl Ketone

Mechanical
Improver (Brit. 404046) of—
Exhaust odors from internal combustion engines (added to fuels not derived from petroleum, either alone or in conjunction with (1) artificial musk compounds, or (2) artificial musk compounds and any of the following: Camphor, waste camphor oil, borneol, bornyl acetate, clove oil, ionone, coumarin, indole, skatole, paracresyl acetate, methyl anthranilate, isopropylmethylhydrocinnamic aldehyde).

Petroleum
Reagent (Brit. 404046) for—
Improving exhaust odors from internal combustion engines (added to gasoline or diesel oil, either alone or in conjunction with (1) artificial musk compounds, or (2) artificial musk compounds and any of the following: Camphor, waste camphor oil, borneol, bornyl acetate, clove oil, ionone, coumarin, indole, skatole, paracresyl acetate, methyl anthranilate, isopropylmethylhydrocinnamic aldehyde).

2-Naphthylmethyl Ketone

Mechanical
Improver (Brit. 404046) of—
Exhaust odors from internal combustion engines (added to fuels not derived from petroleum, either alone or in conjunction with (1) artificial musk compounds, or (2) artificial musk compounds and any of the following: Camphor, waste camphor oil, borneol, bornyl acetate, clove oil, ionone, coumarin, indole, skatole, paracresyl acetate, methyl anthranilate, isopropylmethylhydrocinnamic aldehyde).

Petroleum
Reagent (Brit. 404046) for—
Improving exhaust odors from internal combustion engines (added to gasoline or diesel oil, either alone or in conjunction with (1) artificial musk compounds, or (2) artificial musk compounds and any of the following: Camphor, waste camphor oil, borneol, bornyl acetate, clove oil, ionone, coumarin, indole, skatole, paracresyl acetate, methyl anthranilate, isopropylmethylhydrocinnamic aldehyde).

Naphthyl Phosphate
French: Phosphate de naphthyle, Phosphate naphthylique.
German: Naphtylphosphat, Phosphorsäurenaphtylester, Phosphorsäuresnaphtyl.

Miscellaneous
Mothproofing agent (U. S. 1748675) in treating—
Feathers, furs, skins, and other animal products that are subject to attack by the clothes moth larvae.

Textile
Mothproofing agent (U. S. 1748675) in treating—
Woolen materials and felt.

2-Naphthylthioglycollic Acid
French: Acide de bétanaphthylethioglycollique.
German: Betanaphtylthioglykolsaeure.

Chemical
Starting point in making—
1-Bromobetanaphthylthioglycollic bromide.
1-Bromobetanaphthylthioglycollic chloride.
1-Bromobetanaphthylthioglycollic iodide.
1-Chlorobetanaphthylthioglycollic bromide.
1-Chlorobetanaphthylthioglycollic chloride.
1-Chlorobetanaphthylthioglycollic iodide.
1-Iodobetanaphthylthioglycollic bromide.
1-Iodobetanaphthylthioglycollic chloride.
1-Iodobetanaphthylthioglycollic iodide.
Naphthoxythiophene.

Naphthylthiosalicylic Acid
French: Acide de naphthylesulphosalicyle, Acide naphthylesulphosalicylique, Acide de naphthylethiosalicyle, Acide naphthylethiosalicylique.
German: Naphtylsulfosalicylsäure, Naphtylthiosalicylsäure.

Chemical
Starting point in making—
Esters, intermediates, pharmaceuticals, salts.
Starting point (Brit. 282427) in making synthetic pharmaceutical derivatives of—
Antimony, arsenic, bismuth, gold, silver.

Naphthyltriethyl Iodide
French: Iodure de naphthyletriéthyle, Iodure naphthylique et triéthylique.
German: Jodnaphthyltriaethyl, Naphthyltriaethyljodid.

Chemical
Starting point in making various derivatives.

Miscellaneous
Reagent (Brit. 312613) for treating—
Hair, feathers, furs, and other animal products to render them resistant to moths.

Textile
Reagent (Brit. 312613) for treating—
Wool and felt to mothproof them.

Neon

Analysis
Inert gas for laboratory work.
Electrical
Gaseous filler in—
Neon signs.
Ingredient of gaseous fillers for—
Antifog devices, electrical current detectors, high-voltage indicators for high tension electric lines, lightning arresters, signs, television tubes, tubes for indicating ignition sparking in automobiles, voltage indicating devices in substations, warning signals, wave meter tubes.

Nephelin

Ceramics
As a raw material.
Chemical
Raw material in making various compounds.
Fertilizer
Ingredient of—
Fertilizer preparations.
Glass
As a substitute for alkali.

Neroli Oil

Synonyms: Oil of neroli.
Latin: Oleum neroli, Oleum naphae.
French: Huile de néroli.
German: Nerolioel, Pomerantzenbluethenoel.
Food
Flavoring agent in—
Confectionery and candies.
Ingredient of—
Flavoring preparations.
Miscellaneous
Ingredient of—
Disinfectants and deodorants (Brit. 272543).
Perfume
Ingredient of—
Cosmetics, perfumes.
Petroleum
Perfume in improving odor of petroleum products.
Pharmaceutical
In compounding and dispensing practice.
Soap
Perfume in making—
Toilet soaps.

Neryl Acetate

Perfume
Ingredient of—
Neroli perfumes, orange blossom perfumes.
Perfume in—
Creams, powders, and other toilet preparations.
Soap
Perfume in—
Toilet soaps.

N-Ethylcarbazole

French: Éthyle de carbazol, Carbazol éthylique, Carbazol éthylée.
German: N-Aethylcarbazol.
Spanish: N-Etilcarbazol.
Italian: N-Etilcarbazole.
Ceramics
Ingredient (Brit. 342288) of—
Compositions containing cellulose ethers, such as butylcellulose and benzylcellulose, used for the production of decorative and protective coatings on ceramic ware (added for the purpose of stabilizing the film against ageing).
Chemical
Starting point in making intermediates and other derivatives.
Glass
Ingredient (Brit. 342288) of—
Compositions containing various cellulose ethers, such as butylcellulose and benzylcellulose, used in the manufacture of nonscatterable glass and also for the production of decorative and protective coatings on glassware (added for the purpose of stabilizing the film against ageing).
Leather
Ingredient (Brit. 342288) of—
Compositions containing various cellulose ethers, such as benzylcellulose and butylcellulose, used in the manufacture of artificial leather and for the production of decorative and protective coatings on leather goods (added for the purpose of stabilizing the film against ageing).
Metallurgical
Ingredient (Brit. 342288) of—
Compositions containing various cellulose ethers, such as benzylcellulose and butylcellulose, used for the production of decorative and protective coatings on metallic goods (added for the purpose of stabilizing the film against ageing).
Miscellaneous
Ingredient (Brit. 342288) of—
Compositions containing various cellulose ethers, such as benzylcellulose and butylcellulose, used for the production of decorative and protective coatings (added for the purpose of stabilizing the film against ageing).
Paint and Varnish
Ingredient (Brit. 342288) of—
Paints, varnishes, lacquers, dopes, and enamels made from various cellulose ethers, such as benzylcellulose and butylcellulose (added for the purpose of stabilizing the film against ageing).
Paper
Ingredient (Brit. 342288) of—
Compositions containing various ethers of cellulose, such as butylcellulose and benzylcellulose, used in the manufacture of coated paper and for the production of decorative and protective coatings on pulp and paper products (added for the purpose of stabilizing the film against ageing).
Rubber
Ingredient (Brit. 342288) of—
Compositions containing various ethers of cellulose, such as butylcellulose and benzylcellulose, used for the production of decorative and protective coatings on rubber products (added for the purpose of stabilizing the film against ageing).
Stone
Ingredient (Brit. 342288) of—
Compositions containing various ethers of cellulose, such as butylcellulose and benzylcellulose, used for the production of decorative and protective coatings on artificial and natural stone (added for the purpose of stabilizing the film against ageing).
Textile
Ingredient (Brit. 342288) of—
Compositions containing various ethers of cellulose, such as butylcellulose and benzylcellulose, used for coating textile fabrics (added for the purpose of stabilizing the film against ageing).
Woodworking
Ingredient (Brit. 342288) of—
Compositions containing various ethers of cellulose, such as butylcellulose and benzylcellulose, used for the production of decorative and protective coatings on woodwork (added for the purpose of stabilizing the film against ageing).

New Blue DA

French: Nouveau bleu DA.
German: Neues blau DA.
Chemical
Ingredient (Brit. 295605) of bacteriological and therapeutic compositions and biological stains, containing—
Cresol, guaiacol, hydroquinone, phenol, phloroglucinol, pyrocatechol, pyrogallol, resorcinol.
Miscellaneous
In various coloring and staining processes.
Textile
Coloring agent for dyeing and printing.

Nickel

Latin: Niccolum.
French: Nickel.
German: Nickel.
Spanish: Niquel.
Italian: Nickelio.

In Common Commercial Forms

(Anodes, Blocks, Ingots, Plates, Rods, Sheets, Shot, Strips, Tubes, Wires, and Others).
Brewing
Metal for making—
Tanks, vats.

Nickel (Continued)

Ceramic
Process material in—
Chinaware plating (U. S. 1444113).

Chemical
Basic material in making various nickel salts.
Metal for making—
Agitators, autoclaves, baffles, belts, blowcases, chlorinators, chutes, coils, condensers, containers, conveyors, coolers, crystallizers, digesters, dryers, evaporators, fans, filling machines, filters, fittings, fusion pots, gas scrubbers, heating coils, hoppers, jacketed kettles, kettles, laboratory apparatus, linings, mixers, pipelines, pots, pumps, pump rods, screens, shafts, sieves, stills, tanks, trays, trucks, vacuum pans, valves, vats.

Cosmetic
Metal for making—
Agitators, blenders, jacketed kettles, filling machinery, pipelines, stills, tanks, valves.

Dairy
Metal for making—
Ageing equipment, coils, coolers, dump tanks, filling machinery, filters, fittings, forewarming tanks, heaters, holding tanks, hoppers, milker pails, pasteurizers, pipelines, pumps, ripeners, tanks, trucks, truck tanks, vacuum pans, valves, vats, weigh cans.

Dye
Metal for making—
Agitators, fittings, heating coils, pipelines, pressure kettles, retorts, thermometer wells, valves.

Electrical
Metal for making—
Contacts, Edison cell parts, electrode holders, electrodes, lead pipes, lightning rod tips, loud-speaker diaphragms, magnets for telegraphic instruments, parts for various equipment and purposes, rectifier parts, sparkplug electrodes, storage battery elements, vacuum tube parts.

Fats and Oils
Metal for making—
Catalyst holders, coolers, fittings, heating coils, kettles, pipelines, truck kettles, valves.

Food
Metal for making—
Bins, blanchers, bottling machinery, canning machinery, chutes, coils, containers, conveyors, cooking coils, cooking kettles, dispensing equipment, dryers, evaporator tubes, evaporators, extractors, filling machines, filters, fittings, freezing apparatus, grinders, heating coils, hoppers, jacketed kettles, kettles, mixers, pans, pasteurizers, percolators, pipelines, pumps, screens, storage tanks, strainers, table tops, tanks, trays, trucks, utensils, vacuum pans, valves, vats.

Glass
Metal for making—
Automatic feeders, blowpipes, burner pipes, chain guides, charging ladles, conveyors, cracking irons, etching tanks, lehrs, molds, punty rods, shear blades, skimmers, stowing tools, utensils.

Glue and Adhesives
Metal for making—
Agitators, kettles, mixers, rolls.

Ink
Metal for making—
Mixers, pipelines, pump liners, tanks.

Metallurgical
Component of—
Aluminum alloys, arc-welding compositions, argentan, bearing metals, Chinese silver, coinage alloys, ferronickel, German silver, high-speed nickel steels, invar, molybdenum nickel, monel metal, new silver, nickel alloys, nickel brasses, nickel castiron, nickel silvers, nickel-chromium alloys, nickel-copper alloys, nickel-gold alloys, nickel-iron alloys, nickel-steel alloys, tiers-argent, tungsten carbide alloys.

Metal for making—
Burner parts, burning points, carbon combustion boats, carbonizing boxes, enameling racks, firebrick bolts, furnace linings, oven linings, pyrometer protection tubes.
Process material in—
Electroplating.

Minting
Base material in—
Coinage.

Miscellaneous
Metal for making—
Cylinders for blood-transfusion apparatus.
Water-softening equipment for various uses.

Paint and Varnish
Metal in making—
Fittings, kettles, valves.

Petroleum
Metal for making—
Agitators, catalyst holders, tubing, wire, well strainers.

Pharmaceutical
Metal for making—
Pill-coating equipment, tablet-coating equipment, various equipment (see under Chemical).

Photographic
Metal for making—
Developing and finishing equipment, film-coating apparatus, paper-coating apparatus, various equipment for making film bases (see under Plastics).

Plastics
Metal for making—
Agitators, cooling coils, conveyor belts, condenser tubes, containers, drying nets, evaporator tubes, filter cloth, fittings, heaters, heating coils, linings, pipelines, packaging machinery, pumps, rolls, tanks, tubing, valves, wire.

Power and Heat
Metal for making—
Coils, evaporator steam chests, fittings, heater tubes.

Printing
Process material in making—
Electrotypes.

Rayon
Metal for making—
Spinerette adapters, various equipment (see under Plastics).

Rubber
Metal for making—
Pipelines, truck tanks.

Soap
Metal for making—
Cooling frames, moulds, various equipment (see under Fats and Oils).

Water
Metal for making—
Pipelines, water-softening equipment.

In Finely Divided Forms
(Including Nickelized Catalyst Carriers)

Analysis
Reagent for—
Determination of nitrogen by hydrogenation.

Chemical
Catalyst in—
Absorbing ethylene in sulphuric acid (Brit. 336603).
Decomposition of alcohol.
Destructive hydrogenation of carbonaceous materials (Brit. 335215).
Dehalogenizing aromatic chloro derivatives.
Hydrogenation of aldehydes and ketones.
Hydrogenation of benzene compounds.
Oxidation of methane.
Catalyst in making—
Alcohols from aldehydes.
Alcohols from ketones.
Alcohols of various kinds from olefins (Brit. 335551).
Cyclohexane from phenol.
Cyclohexanol from phenol.
Dihydrofurfuryl alcohols from furfuryl alcohol.
Furfuryl alcohol from furfural.
Hexahydrodiphenylene oxide.
Menthones.
Methyltetrahydrofuran from furfuryl alcohol.
Saturated compounds from olefin derivatives.
Tetrahydrofuran from furfuryl alcohol.
Tetrahydrofurfuryl alcohol from furfuryl alcohol.
Catalyst in reducing—
Acetylene to ethane.
Acetonylacetone to the anhydride of the corresponding glycol.
Aliphatic nitriles to amines and ammonia.
Aliphatic nitro compounds to primary amines.
Aliphatic nitro compounds to paraffins and ammonia.
Alphaheptin to heptane.
Alphanitronaphthalene to ammonia and tetralin.
Aromatic hydrocarbons to their hexahydro derivatives.

Nickel (Continued)
Aromatic nitriles to ammonia and an aromatic hydrocarbon.
Benzene nuclei.
Benzil, benzoin, and benzoylacetone to the corresponding hydrocarbons.
Carbon monoxide to methane and water.
Carbon dioxide to methane and water.
Diacetyl to mixtures of hydroxyketone and glycol.
Ketones of the benzophenone type and phenylbenzyl ketone to aromatic hydrocarbons.
Naphthalene to dekahydronaphthalene (dekalin).
Naphthalene to tetrahydronaphthalene (tetralin).
Olefins to the corresponding paraffins.
Phenylacetylene to mixture of ethylcyclohexane, methylcyclohexane, and methane.
Pinene to a dihydro derivative.
Quinones to quinols and carbylamines.
Terpenes, such as limonene, sylvestrene, terpinene, menthene, to paramethylisopropylcyclohexane.
Unsaturated ketones to the corresponding saturated ketones.

Coal Processing
Catalyst in—
Hydrogenation of coal.

Fats and Oils
Catalyst in the hydrogenation of—
Animal oils, fats, fish oils, oxidized oils, polymerized oils, unsaturated oils, vegetable oils.

Gas
Catalyst in—
Freeing combustible gases from carbon monoxide (Brit. 335228).
Purifying natural gas (Brit. 335394).

Petroleum
Catalyst in—
Reducing olefins to the corresponding paraffins.
Removing sulphur from high-sulphur naphthas and mineral oils.
Catalyst in making—
Alcohols of various kinds from olefins (Brit. 335551).
Saturated compounds from olefin derivatives.

Resins
Catalyst in—
Hydrogenation processes.

Soap
Catalyst in hydrogenation of—
Animal oils, fats, fish oils, vegetable oils.

Nickel Acetosulphate
French: Acétosulphate de nickel, Acétosulphate nickelique.
German: Acetoschwefelsäuresnickel, Acetoschwefelsäuresnickeloxyd, Nickelacetosulfat.

Textile
Mordant in fixing—
Alizarin orange in dyeing and printing cottons.

Nickel-Ammonium Chloride
French: Chlorure ammoniaque et nickelique, Chlorure d'ammonium et de nickel.
German: Nickelammoniakchlorid.

Miscellaneous
Reagent (Brit. 271026) in—
Carroting furs and felts.

Nickel Bismuthide

Chemical
Catalyst in making—
Acetone from isopropyl alcohol, isobutyraldehyde from isobutyl alcohol, isobutyronitrile from isobutylamine, naphthalene from tetrahydronaphthalene, paracymene from turpentine oil.

Nickel Borotungstate

Metallurgical
Ingredient of—
Insulating coatings for steel and other metals to afford protection against oxidation (French 600774).

Nickel Carbonyl, Polymerized

Fuel
Antiknock agent (U. S. 2002805) in—
Motor fuels.

Nickel-Dimethylglyoxime

Paint and Varnish
As a light-fast pigment.

Nickel Erucate
French: Érucate de nickel, Érucate nickelique.
German: Erucinsäuresnickel, Erucinsäuresnickeloxyd, Nickelerucat.

Fats and Oils
Ingredient of—
Solidified oils, special lubricants.
Reagent in promoting—
Intimate contact between the catalyst and the oil in the hydrogenation of vegetable oils.

Leather
Ingredient of—
Dressing compositions, waterproofing compositions.

Mechanical
Ingredient of—
Cutting compounds, solidified lubricants.

Metallurgical
Reagent in—
Metal working.

Miscellaneous
Ingredient of—
Compositions used for the dry cleansing of chamois and the like.

Paint and Varnish
Ingredient of—
Special varnishes.

Petroleum
Ingredient of—
Cylinder oils, cup greases, steam turbine oils.

Soap
Ingredient of—
Lathering and detergent preparations containing benzene or similar solvents.

Textile
Ingredient of—
Dry cleansing preparations, softening compositions.

Nickel Laurate
French: Laurate de nickel, Laurate nickelique.
German: Laurinsäuresnickel, Laurinsäuresnickeloxyd, Nickellaurat.

Fats and Oils
Ingredient of—
Solidified oils, special lubricants.
Reagent in promoting—
Intimate contact between the catalyst and the oil in the hydrogenation of vegetable oils.

Leather
Ingredient of—
Dressing compositions, waterproofing compositions.

Mechanical
Ingredient of—
Cutting compounds, solidified lubricants.

Miscellaneous
Ingredient of—
Compositions used for the dry cleaning of chamois and the like.

Paint and Varnish
Ingredient of—
Special varnishes.

Petroleum
Ingredient of—
Cylinder oils, cup greases, steam turbine oils.

Soap
Ingredient of—
Lathering and detergent preparations containing benzene or similar solvents.

Textile
Ingredient of—
Dry-cleansing preparations, softening compositions.

Nickel Oleate
French: Oléate de nickel, Oléate nickelique.
German: Nickeloleat, Oleinsäuresnickel, Oleinsäuresnickeloxyd.

Fats and Oils
Catalyst in promoting—
Intimate contact between the catalyst and the oil in the hydrogenation of vegetable oils.
Ingredient of—
Cup greases, cutting compounds, cylinder oils, lubricating compositions, solidified lubricants, solidified oils, steam turbine oils.

Nickel Oleate (Continued)
Leather
Ingredient of—
Dressing compositions, waterproofing compositions.
Mechanical
As a special lubricant.
Metallurgical
Reagent in—
Metal working.
Miscellaneous
Ingredient of—
Compositions used for the dry cleaning of chamois and other articles.
Paint and Varnish
Ingredient of—
Special varnishes.
Petroleum
Ingredient of—
Lubricants of various sorts.
Soap
Ingredient of—
Lathering and detergent preparations containing benzene or other volatile solvents.
Textile
Ingredient of—
Dry-cleaning preparations, softening compounds.

Nickelous Acetylmesityloxide
French: Acétylemésityle-oxyde de nickel, Acétylemésityle-oxyde nickeleux.
German: Nickelacetylmesityloxydul, Nickeloacetylmesityloxyd.
Chemical
Starting point and reagent in making—
Aromatics, intermediates, pharmaceuticals.
Dye
Starting point and reagent (Brit. 289493) in making—
Synthetic dyestuffs.
Petroleum
Antiknock reagent (Brit. 289493) in—
Motor fuels.

Nickelous Gammamethylacetylacetone
Chemical
Reagent (Brit. 289493) in making—
Aromatics, intermediates, pharmaceuticals.
Dye
Reagent (Brit. 289493) in making various synthetic dyestuffs.
Petroleum
Ingredient (Brit. 289493) of—
Motor fuels, to improve their combustion.

Nickel Palmitate
French: Palmitate de nickel, Palmitate nickelique.
German: Nickelpalmitat, Palmitinsäuresnickel, Palmitinsäuresnickeloxyd.
Fats and Oils
Catalyst in promoting—
Intimate contact between the catalyst and the oil in the hydrogenation of vegetable oils.
Ingredient of—
Cup greases, cutting compounds, cylinder oils, lubricating compositions, solidified lubricants, solidified oils, steam turbine oils.
Leather
Ingredient of—
Dressing compositions, waterproofing compositions.
Mechanical
As a special lubricant.
Metallurgical
Reagent in—
Metal-working.
Miscellaneous
Ingredient of—
Compositions used for the dry cleaning of chamois and other articles.
Paint and Varnish
Ingredient of—
Special varnishes.
Petroleum
Ingredient of—
Lubricants of various sorts.
Soap
Ingredient of—
Lathering and detergent preparations containing benzene or other volatile solvents.
Textile
Ingredient of—
Dry-cleaning preparations, softening compositions.

Nickel Resinate
Synonyms: Nickel soap, Resinate of nickel.
French: Résinate de nickel.
German: Nickelresinat.
Ceramics
Pigment in producing light-brown colors in—
Chinaware, porcelains, potteries.
Waxes and Resins
Ingredient of—
Resin clarifying compositions, resin hardening compositions, resin neutralizing compositions.
Paint and Varnish
Drier in making—
Enamels, lacquers, paints, varnishes.

Nickel-Rhodium
(Alloys containing nickel and 25 to 80 percent of rhodium; but sometimes also some platinum, iridium, parradium, molybdenum, tungsten, copper, iron, or cobalt).
Electrical
Metal (Brit. 451823) for making—
Electrodes.
Metal Fabrication
Metal (Brit. 451823) for making—
Chemical apparatus, reflectors.
Miscellaneous
Metal (Brit. 451823) for making—
Pen points.

Nickel Selenide
French: Sélénide de nickel.
Chemical
Catalyst in making—
Acetone from isopropyl alcohol, isobutyraldehyde from isobutyl alcohol, isobutyronitrile from isobutylamine, naphthalene from tetrahydronaphthalene, paracymene from turpentine oil.

Nickel Stearate
French: Stéarate de nickel, Stéarate nickelique.
German: Nickelstearat, Stearinsäuresnickel, Stearinsäuresnickeloxyd.
Fats and Oils
Catalyst in promoting—
Intimate contact between the catalyst and the oil in the hydrogenation of vegetable oils.
Ingredient of—
Cup greases, cutting compounds, cylinder oils, lubricating compositions, solidified lubricants, solidified oils, steam turbine oils.
Leather
Ingredient of—
Dressing compositions, waterproofing compositions.
Mechanical
As a special lubricant.
Metallurgical
Reagent in—
Metal-working.
Miscellaneous
Ingredient of—
Compositions used for the dry cleaning of chamois and other articles.
Paint and Varnish
Ingredient of—
Special varnishes.
Petroleum
Ingredient of—
Lubricants of various sorts.
Soap
Ingredient of—
Lathering and detergent preparations containing benzene or other volatile solvents.
Textile
Ingredient of—
Dry-cleaning preparations, softening compositions.

Nickel Tungstomolybdate

Metallurgical
Ingredient of—
Insulating coatings for steel and other metals to afford protection against oxidation (French 600774).

Nicotine Pyrogallate

Petroleum
Antioxidant (U. S. 1970339) for—
Vapor-phase-cracked hydrocarbon distillates (inhibits usual deterioration, loss of antiknock properties, gum development on storage).

Nicotinic Acid
Synonyms: Pyridinmonocarboxylic acid.
French: Acide nicotianique, Acide nicotinique, Acide pyridinemonocarbonique.
German: Nicotinsäure, Pyridinmonocarbonsäure.

Chemical
Starting point in making—
Arecolin, esters and salts, pharmaceuticals.

Niger Oil
Synonyms: Nigerseed oil.
French: Huile de niger.
German: Nigeroel.
Italian: Olio di niger.

Oilcloth and Linoleum
Substitute for linseed oil in linoleum coatings.

Paint and Varnish
Vehicle in—
Varnishes (claimed to produce better water-resistance than is obtained with linseed oil).

Nigrosin, Spirit-Soluble
Synonyms: Nigrosin base, Spirit nigrosin.
French: Base de nigrosine, Nigrosine à l'alcool, Nigrosine 2B, 3B, G, SS, and T, Noir CNN, CBR.
German: Azodiphenylblau, Nigrosin BB blaeulich, Nigrosin B roetlich, Nigrosin fettfarbe.

Fats and Oils
As a coloring.

Leather
Coloring for—
Leather (applied with a brush).

Paints and Varnishes
Coloring in—
Spirit lacquers, spirit varnishes.

Miscellaneous
Coloring in—
Shoe polishes.

Resins and Waxes
As a coloring.

Nigrosin, Water-Soluble
Synonyms: Nigrosin B, G, K, W, CBR, CNBJ, WS, SS, 7600.
French: Nigrosine soluble à l'eau, Noir CBRS.
German: Anilingrau, Anilingrau B und R, Nigrosinwasserloeslich.

Dye
Starting point in making—
Color lakes.

Ink
Coloring matter in—
Printing inks, stamping inks, stencil inks, typewriter inks, writing inks.

Leather
As a coloring.

Paint and Varnish
Coloring matter in—
Lacquers, paints, stains, varnishes.

Paper
As a coloring.

Textile
——, *Dyeing and Printing*
Coloring for—
Coconut fibers, cotton yarns and fabrics, jute fibers, silk yarns and fabrics, wool yarns and fabrics.

Woodworking
As a stain.

Nile Blue
French: Bleu de nile.
German: Nileblau.

Insecticide
Ingredient (Brit. 303932) of—
Insecticides, fungicides, and vermin-destroying compositions containing arsenous acid, arsenic acid, or the salts of these acids.

Miscellaneous
Dyestuff for coloring various substances.

Sanitation
Ingredient (Brit. 303932) of—
Bactericidal and disinfecting compositions containing arsenous acid, arsenic acid, or the salts of these acids.

Textile
Dyestuff for dyeing and printing yarns and fabrics.

Niobium
Synonyms: Columbium.

Chemical
Starting point in making—
Niobium chemicals.

Electrical
Material in making—
Electrodes for radio uses, grid wire in vacuum tubes.

Metallurgical
Ingredient of—
Alloys with nickel, iron, and tantalum used for making resistance coils and the like (Canadian 209342).
Alloys with nickel and tantalum to give a white malleable metal.
Alloy steels, in conjunction with tantalum, tungsten, vanadium, chromium, molybdenum, or uranium.
Alloys with zirconium and tantalum, giving a product which is not attacked by hydrochloric acid, sulphuric acid, nitric acid, aqua regia, alkalies, chlorine, or nascent oxygen either in the hot or cold (U. S. 1334089).
Niobium steel (Brit. 152371).

Niobium Oxide
Synonyms: Columbium oxide.
French: Oxyde de columbium, Oxyde de niobium.
German: Nioboxyd.

Chemical
Catalyst (Brit. 254819) in making—
Alcohols, aldehydes, amines, carboxylic acid esters, oxygenated organic compounds.

Niter Cake
Synonyms: Crude bisulphate of soda.
French: Bisulfate de soude, Bisulfate de sodium brut, Gateaux de nitre.
German: Natriumsaeuresulfat, Schwefelsäuressäuresnatrium.

Agricultural
Reagent in various operations.

Ceramics
Ingredient of—
Glazes.
Reagent in making—
Slag brick.

Chemical
Neutralizing agent for—
Molasses in alcoholic distillation.
Reagent in—
Boric acid extraction from borosodium calcite.
Rare earth extraction processes.
Regenerating residual liquors recovered in the manufacture of anthraquinone, containing chromium salts, for decomposition of calcium chromate.
Saccharification of carbohydrates (Brit. 400168).
Reagent in making—
Carbon dioxide for use in baths, epsom salt, hydrochloric acid, potassium bichromate, sodium fluoride from calcium fluoride, sulphuretted hydrogen.
Organic acids from their salts; for example, formic and acetic acids from formates and acetates.
Starting point in making—
Nonhygroscopic compositions, containing also sodium carbonate and aluminum sulphate (U. S. 1905833).
Sodium sulphide (by roasting with salt and coal).
Potassium sulphate, salt cake, sodium-aluminum sulphate, sodium-ammonium sulphate, sodium sulphate, sodium sulphite.
Substitute for sulphuric acid in many chemical processes.

Distilling
Substitute for sulphuric acid in—
Neutralizing molasses prior to its distillation.

Niter Cake (Continued)

Dye

Substitute for sulphuric acid in making various dyestuffs.

Fats and Oils

Substitute for sulphuric acid in—
Recovering fatty acids and grease from wool wash liquors and other residual waters of similar character.

Fertilizer

Reagent in making—
Superphosphates.
Substitute for sulphuric acid in—
Absorbing ammonia gas to make sulphate of ammonia.
Decomposing phosphate rocks.

Food

Reagent (used in place of sulphuric acid) in making—
Aerated mineral waters and soft drinks.

Gas

Substitute for sulphuric acid in—
Absorbing ammonia from coal gas and coke-oven gas.

Glass

Ingredient of—
Batch.

Glue

Substitute for sulphuric acid in—
Treating meat scrap, hide scrap, leather cuttings, and leather dust for making glue and gelatin.

Leather

Substitute for sulphuric acid in—
Bleaching and plumping leather and cause it to swell.

Metallurgical

Substitute for hydrochloric acid in—
Cleansing and scouring metals, especially sheet irons which is to be coated with zinc or tin.
Ingredient of—
Compositions used in pickling and corroding metals.
Reagent (German 426669) in—
Decomposing materials containing selenium so as to recover the metal.
Reagent in the metallurgy of—
Copper, nickel, sulphide minerals.

Miscellaneous

General substitute for sulphuric acid in miscellaneous processes.
In thermophores.

Paint and Varnish

Reagent in making—
Permanent white.

Paper

Substitute for alum in—
Sizing operations.

Perfumery

Reagent in various processes.

Rubber

Substitute for sulphuric acid in—
Reworking old rubber.

Sanitation

Disinfectant for—
Antityphoid treatment of water.

Soap

Substitute for sulphuric acid in various operations.

Textile

——, *Bleaching*

Substitute for sulphuric acid in making—
Sour liquors.

——, *Dyeing*

Substitute for sulphuric acid in making—
Dye liquors.
Substitute for tartar in making—
Dye liquors.

——, *Manufacturing*

Ingredient of—
Wool-washing liquors.
Substitute for sulphuric acid in making—
Baths for precipitating spun viscose filament.
Wool-carbonizing solutions.

3-Nitranilin

Synonyms: 3-Nitroanilin.

Chemical

Starting point in making—
Intermediates, pharmaceuticals.

Dye

Starting point in making—
Dyestuffs.
Starting point (Brit. 306153) in making dyestuffs with—
Alphanaphthol-3:6-disulphonic acid, alphanaphthylamine-3:6-disulphonic acid.

4-Nitranilin-2-carboxylic Acid

French: Acide de 4-nitraniline-2-carbonique, Acide de 4-nitraniline-2-carboxylique.
German: 4-Nitranilin-2-carbonsaeure.

Chemical

Starting point in making—
Esters and salts, intermediates, pharmaceuticals.

Dye

Starting point (Brit. 306153) in making dyestuffs with—
Dehydrothiotoluidinsulphonic acid.

Nitric Acid

Synonyms: Spirits of nitre.
Latin: Acidum nitri, Acidum nitricum, Aqua fortis, Azoticum, Spiritus nitri acidus.
French: Acide azotique officinal, Acide nitrique.
German: Salpetersäure.
Spanish: Acido nitrico.
Italian: Acido nitrico concentrato.
Swedish: Shedwater.
Dutch: Zaltpeterzuur, Sterkwater.

Analysis

Nitrating agent.
Oxidizing agent.
Solvent.

Ceramics

Reagent in—
Manufacturing processes.

Chemical

As a nitrating agent.
As an oxidizing agent.
As a solvent.
Activating agent (Brit. 291725) in making—
Activated carbon.
Catalyst (French 606541) in making—
Tin fluosilicate from hydrofluorsilicic acid and tin.
Hydrolyzing agent (U. S. 1890590) in making—
Glutamic acid, sodium glutamate.
Nitrating agent in making—
Collodion cotton.
2:7-Dinitroanthraquinone (U. S. 1622168).
Dinitronaphthalene-1:4:5:8-tetracarboxylic acid (Brit. 400069).
Nitrosyl chloride (U. S. 1920333).
1-Para-aminophenyl-2-methylaminopropanol (U. S. 1892532).
Soluble cotton, varnish cotton.
Oxidizing agent in making—
Arsenic acid from arsenic trioxide (Brit. 255522).
Oxalic acid from sawdust (German 588159).
Sodium nitrate (Brit. 401121).
Sulphuric acid (U. S. 1912832).
Sulphuric acid by the chamber process.
Reagent in making—
Ammonium trinitrate (Brit. 403289).
Fatty acids from paraffin hydrochloride (Brit. 368869).
Solvent (Brit. 402977) in extracting—
Alumina from leucite, potash from leucite.
Solvent in making—
Magnesia (deficient in lime and calcium nitrate) from dolomite (Brit. 403054).
Starting point, nitrating agent, oxidizing agent or solvent in making—
Acetyl-1-naphthyldiamine-6-sulphonic acid, adipic acid from animal fats, alloxantin, alloxin, alpha-amino-2-naphthol.
Alpha-1:5-dinitronaphthalene from naphthalene.
Alphadinitrophenol from phenol.
Alpha-4:6:8-naphthalenetrisulphonic acid.
Alphanaphthylamine from naphthalene.
Alphanaphthylamine hydrochloride from naphthalene through alphanaphthylamine.
Alphanaphthylamine-3:6-disulphonic acid.
Alphanaphthylamine-6 (and 7)-sulphonic acid.
Alphanaphthylamine-8-sulphonic acid.
Alphanitrobeta-anthraquinone from anthraquinone.
Alphanitro-2-methylanthraquinone.
Alphanitronaphthalene from naphthalene.
Alphanitronaphthalene-5-sulphonic acid from naphthalene.
Alphanitronaphthalene-6-sulphonic acid from naphthalene.
Alphanitronaphthalene-7-sulphonic acid from naphthalene.

Nitric Acid (Continued)

Alphanitronaphthalene-8-sulphonic acid from naphthalene.

Aluminum acetonitrate.

Aluminum nitrate from aluminum.

1-Amino-8-naphthol-2:4-disulphonic acid from naphthalene through Peri and Sultam acids.

1-Amino-8-naphthol-3:6-disulphonic acid from naphthalene through Peri and 1-naphthylamine-4:8-disulphonic acids.

1-Amino-2-naphthol-4-sulphonic acid from betanaphthol through 1-nitrosobetanaphthol.

1-Amino-8-naphthol-4-sulphonic acid from naphthalene through Peri and 1-naphthylamine-4:8-disulphonic acids.

2-Amino-1-phenol-4-sulphonic acid from phenol.

Ammonium chlorostannate.

Ammonium nitrate from aqua ammonia.

Ammonium phosphomolybdate from ammonia molybdate, phosphoric acid.

Ammonium phosphotungstate from ammonium phosphate and ammonium tungstate.

Analgen, anisic acid from anethole, anisic aldehyde from anethole, anthranilic acid, anthraquinone, anthraquinonedicarboxylic acid, antimony nitrate from antimony, antimony pentachloride from antimony nitrate, antimony trichloride from antimony nitrate, antimony trioxide from antimony, arsenic acid from arsenic.

Barium nitrate from barium carbonate, oxide, or hydroxide.

Benzene, benzidin from benzene or diphenyl, betaaminoanthraquinone from anthraquinone, betamethylanthraquinone.

Betanaphthylamine-4:6-disulphonic acid.

Betanaphthylamine-4-sulphonic acid.

Betanitroanthraquinone from anthraquinone.

Bismuth basic gallate from bismuth and acetic and gallic acids.

Bismuth fluoride from bismuth and hydrofluoric acid.

Bismuth nitrate from bismuth.

Bismuth-ammonium citrate from bismuth, aqua ammonia, and citric acid.

Bismuthic acid, boron carbide, bromobenzene.

Cadmium nitrate from cadmium oxide, calcium nitrate from calcium carbonate, camphor from borneol, caproic acid, cerium nitrate from cerium oxide, cesium nitrate from cesium oxide.

1-Chloro-2:6-dinitrobenzene-4-sulphonic acid (potassium salt) from benzene.

Chloronitrobenzene from benzene.

2-Chloro-5-toluidin-4-sulphonic acid from orthotoluene-parasulphonic acid.

Chromium nitrate from chromium oxide, chrysene, cobalt nitrate from cobalt oxide, copper nitrate from copper oxide, cupric sulphide.

Diaminoazotoluene from toluidine.

5:7-Dibromoisatin chloride.

5:8-Dichloroalphanitronaphthalene from naphthalene.

2:5-Dichloroanilin from paradichlorobenzene.

Didymium nitrate from monazite sand, diethylmeta-aminophenol, dinitroanilin from anilin, dinitroanthraflavic acid from anthraflavic acid, dinitrobenzyldisulphonic acid from toluene, dinitrochlorobenzene from benzene.

2:4-Dinitro-4-hydroxydiphenylamine from chlorobenzene and para-aminophenol.

1:5-Dinitro-2-methylanthraquinone from anthraquinone.

Dinitrophenol (sodium salt) from phenol.

3:5-Dinitrosalicylic acid from salicylic acid.

Dinitrostilbene-disodium sulphonate from toluylene.

Dinitrotoluene from toluene.

Esters with alcohol, ethylbenzylanilindisulphonic acid, ethylene nitrate, ethyl nitrate from ethyl alcohol and urea nitrate, ethyl nitrite from ethyl alcohol, ethylorthoaminoparacresol, ethylorthotoluidinparasulphonic acid.

Ferric nitrate from iron or its oxide.

Glucinum nitrate from glucinum oxide, glycollic acid.

Hydroxylamine.

Lead antimoniate from lead and potassium antimoniate, lead carbonate from lead, lead nitrate from lead, lithium nitrate from lithium or its hydroxide.

Magnesium nitrate from magnesia, malic acid, manganese nitrate from manganic hydroxide, mercury nitrate from quicksilver, metanitrobenzaldehyde from benzaldehyde, metadinitrobenzene from benzene, metanitrophenol from anilin, metatolylenediamine from toluene, metatolylenediaminesulphonic acid from toluene, methylsulphonic acid, molybdenum nitrate from molybdenum.

1:8-Naphthasultam-2:4-disulphonic acid from naphthalene through Peri acid.

1-Naphthylamine-3:8-disulphonic acid from naphthalene-1:5 and 1:6-disulphonic acid.

1-Naphthylamine-4:8-disulphonic acid from naphthalene through Peri acid.

2-Naphthylamine-4:8-disulphonic acid.

1-Naphthylamine-3:6:8-trisulphonic acid from naphthalene through naphthylamine-1:3:6-trisulphonic acid.

1:5-Naphthylenediamine-3:7-disulphonic acid from 2:6-naphthalenedisulphonic acid.

1:8-Naphthylenediamine-3:6-disulphonic acid from 2:7-naphthalenedisulphonic acid.

1:4-Naphthylenediamine-2-sulphonic acid from naphthalene through naphthylaminesulphonic acid and combination with diazobenzene.

Nickel nitrate from nickel oxide, nitroanthraquinone-2-carboxylic acid from anthraquinone, nitrobenzene from benzene, nitrobenzenesulphonic acid from benzene, nitrobenzoic acid from benzoic acid, 2-nitrobromoveratol, nitrochlorobenzene from benzene.

Nitrochlorobenzenesulphonic acid (ammonium salt) from chlorobenzene.

Nitrocresol methylether from cresol.

8-Nitro-1-diazo-2-naphthol-4-sulphonic acid from 1-amino-2-hydroxynaphthalene-4-sulphonic acid.

Nitrodichlorobenzene from paradichlorobenzene, nitrohydrochloric acid by admixture with hydrochloric acid, nitrometadiaminoanisole from metadiaminoanisole, nitrometadiaminophenetole from metadiaminophenetole, nitrometatoluylenediamine from diacetyltoluenediamine, nitroparacresol from paracresol, nitroparatoluidin from paratoluidin, 4-nitro-1:3-phenylenediamine, nitrosalicylic acid from salicylic acid, nitrosodimethylanilin from dimethylanilin, nitrosonaphthol from betanaphthol, nitrotartaric acid from tartaric acid, nitrotoluene from toluene, nitrotoluene-orthosulphonic acid from toluene, nitroxylene from xylene.

Orthochloroparanitranilin from anilin, orthonitranilin from anilin, orthonitroanisole from phenol, orthonitrobenzaldehyde from benzaldehyde, orthonitrobenzidin, orthonitrobenzoyl chloride, orthonitrophenol from phenol, orthonitrophenolnitromethane, orthophenylene, oxalic acid from carbohydrates.

Palladous oxide, parachloro-orthonitrophenol (sodium salt) from paradichlorobenzene, paranitroacetanilide from anilin, paranitroanilin from anilin, paranitroanilinorthosulphonic acid from chlorobenzene, parachlorobenzenesulphonic acid from chlorobenzene.

Paranitro-orthoaminophenol from phenol, paranitro-orthoanisidin from orthoanisidin, paranitro-orthotoluidin from orthotoluidin, paranitrophenetole from phenol, paranitrophenol from phenol, paratoluidinmetasulphonic acid from toluene.

Picramic acid from phenolsulphonic acid, picric acid from phenolsulphonic acid, phenylenediamine from benzene or anilin, phenyl-1-naphthylamine-8-sulphonic acid, phosphoric acid from phosphorus, potassium nitrate from potassium chloride, saccharic acid, sodium nitroprussiate from sodium ferrocyanide, stannic nitrate from tin.

Strontium nitrate from strontium chloride, tetra-aminoditolylmethane, tetranitroanthraflavic acid, tetranitromethane, thorium nitrate from monazite sand, tolyl-1-naphthylamine-8-sulphonic acid.

Tribromoacetic acid from bromal, trichloroacetic acid from chloral hydrate, uranium nitrate from uranium oxide, vanadium nitrate, yttrium nitrate from monazite sand, zinc nitrate from zinc or its oxides, zirconium nitrate from zirconium oxide.

Dye

Reagent in making—

Alizarin brown, alizarin cardinal, alizarin orange, alizarin saphirol, amido dyestuffs, anilin dyestuffs, aurantin, azo dyestuffs, azoflavin RS, azoflavin 3R, azoflavin S, diazo dyestuffs.

Explosives and Matches

Nitrating agent in making—

Ammonium nitrate from aqua ammonia, detonators, explosives, gun cotton, mercury fulminate from alcohol and mercury, nitroglycerin, nitrostarch, picric acid, primers, smokeless powder, soluble cotton, tetranitranilin, tetryl, trinitrotoluene, various pyrotechnic chemicals.

Nitric Acid (Continued)
Oxidizing agent (Brit. 397600) in making—
New explosive.
Fertilizer
Nitrating agent in making—
Ammonium nitrate from aqua ammonia.
Ammonium sulphate—nitrate.
Highly concentrated plantfoods.
Reagent for—
Treating basic phosphatic slag of high citrate-soluble phosphates to produce ammonium nitrate, calcium carbomate, calcium nitrate, and calcium sulphate (Brit. 287439).
Treating raw phosphates to produce ammonium nitrate or a fertilizer mixture containing ammonium nitrate, monoammonium phosphate, and calcium sulphate (Brit. 396729).
Source of nitrogen in—
Compounding fertilizer mixtures.
Glass
Reagent (French 601440) for—
Treating waste products from glass manufacture to remove the iron skin (used alone or admixed with other mineral acids).
Insecticide
Reagent (U. S. 1908544) in—
Coloring lead arsenate green by treatment with prussian blue and sodium bichromate.
Making insecticidal compositions.
Leather
Nitrating agent in making—
Nitrated castor and linseed oils used in the preparation of varnishes for enameling leather.
Soluble pyroxylins for leather dopes.
Reagent in—
Felting skins.
Carroting animal fibers, hair, hairy skins.
Making artificial leather.
Tannins from wood charcoal, humus, coal, peat, and lignite.
Metallurgical
Ingredient of—
Bath used for softening scale on stainless steel (U. S. 1919624), pickling liquors.
Reagent for—
Etching metals.
Improving space factor in high-silicon transformer steels (U. S. 1902815).
Reagent in—
Bright-annealing nickel-chromium steels and alloys of the stellite type (Brit. 399049).
Engraving steel and other metals, gilding brass, gold refining, palladium refining, platinum refining, precious metal refining.
Solvent for—
Aluminum, antimony, beryllium, bismuth, cadmium, calcium, cerium, cesium, chromium, cobalt, copper, gold, iridium, iron, lanthanum, lead, lithium, magnesium, mercury, molybdenum, nickel, osmium, palladium, potassium, rubidium, silver, sodium, steel, tellurium, thallium, thorium, tin, titanium, tungsten, uranium, vanadium, yttrium, zinc.
Solvent in making—
Aluminum combinations adapted for production of aluminum (U. S. 1914768).
Tungsten filaments (U. S. 1904105).
Washing agent (U. S. 1905866) in making—
Yttrium and metals of the yttrium group.
Miscellaneous
Reagent in—
Fur dyeing, hat making.
Paint and Varnish
Digesting agent (Brit. 404007) in making—
Lead pigments from crushed lead ores, concentrates, or scrap.
Ingredient (U. S. 1865799) of—
Enamel remover.
Nitrating agent in making—
Berlin blue, colors of various kinds, lead pigments, Naples yellow.
Soluble pyroxylins used in lacquers, bronzing liquids, enamels, dopes, cements.
Paper
In general paper making processes.
Nitrating agent (U. S. 1913116 and 1914302) in making—
Nitrocellulose from wood pulp.
Reagent (Brit. 391153) in making—
Moisture-resistant, parchmentized, nonfibrous cellulose sheets or filaments.
Pharmaceutical
In compounding and dispensing practice.
Suggested for use as antiseptic, astringent, escharotic.
Suggested for use in treatment of cancrum oris, hepatitis, indigestion, poisoned wounds, rabies, venereal ulcers, warts.
Photographic
Nitrating agent in making—
Nitrocellulose films.
Plastics
Nitrating agent in making—
Pyroxylins, nitramide, celluloid.
Printing
Etching agent in—
Lithography, photoengraving.
Reagent (U. S. 1903778) for—
Treating etched printing plates or flats to prevent adhesion of varnish (to be applied later) to portions of plate requiring additional etching.
Rubber
Reagent in making—
Rubber substitutes.
Solvent for—
Compounded rubber, vulcanized rubber.
Textile
Assist in—
Silk dyeing.
Ingredient (Brit. 252064) of—
Solutions used in treating silk to reduce the mineral content.
Nitrating agent in—
Rayon manufacture.
Woodworking
As a stain.

3-Nitroacenaphthene
Analysis
Reagent.
Chemical
Reagent in—
Organic synthesis.

2-Nitro-4′-acetylaminodiphenylamine
French: 2-Nitro-4′-acétyleaminodiphényleamine.
German: 2-Nitro-4′-acetylaminodiphenylamin.
Chemical
Starting point in making—
Intermediates, pharmaceuticals.
Dye
Starting point in making various synthetic dyestuffs.
Textile
Solubilizing agent (Brit. 305560) in—
Dye liquors, printing pastes and stenciling compositions used on acetate rayon and on mixed fabrics containing acetate rayon.

2-Nitroalphanaphthylamine
Chemical
Reagent in—
Starting point in making—
Betanitronaphthalene.

4-Nitro-1-aminobenzene-2-sulphonic Acid
French: Acide de 4-nitroalpha-aminobenzène-2-sulphonique, Acide de 4-nitro-1-aminobenzène-2-sulphonique.
German: Nitroalpha-aminobenzol-2-sulfonsaeure, 4-Nitro-1-aminobenzol-2-sulfonsaeure.
Chemical
Starting point in making—
Esters and salts, intermediates, pharmaceuticals.
Dye
Starting point (Brit. 311708) in making monoazo dyestuffs with—
2-Allylaminoanthraquinone, 2-amylaminoanthraquinone, 2-butylaminoanthraquinone, 2-ethylaminoanthraquinone, 2-heptylaminoanthraquinone, 2-hexylaminoanthraquinone, 2-isoallylaminoanthraquinone, 2-isoamylaminoanthraquinone, 2-isobutylaminoanthraquinone, 2-isopropylaminoanthraquinone, 2-methylaminoanthraquinone, 2-propylaminoanthraquinone, sulphonic acid derivatives of the above.

2-Nitro-2′-aminobenzophenone
Chemical
Starting point in making—
Intermediates, pharmaceuticals.

2-Nitro-2'-aminobenzophenone (Continued)

Dye

Starting point (Brit. 323792) in making azo dyestuffs for rayons, with the aid of—

Alkylaryl anilins, allylaminophenol, allylnaphthylamine, alphanaphthylamine, aminonaphthoic acids, aminonaphthols, amylaminophenol, amylnaphthylamine, betanaphthylamine, butylnaphthylamine, cresols and their derivatives, dimethylmeta-aminophenol, ethylaminophenol, ethylnaphthylamine, gammachlorobetaoxypropionylnaphthylamine, meta-aminophenol, meta-anisidin, metacresidin, metaphenylenediamine, metaphenetidin, metatoluidin, metaxylidin, methylaminophenol, methylnaphthylamine, naphthylamine ethers, orthoaminophenol, orthoanisidin, orthocresidin, orthophenylenediamine, orthophenetidin, orthotoluidin, orthoxylidin, para-aminophenol, para-anisidin, paracresidin, paraphenylenediamine, paranitrometaphenylenediamine, paratoluidin, paraxylidin, phenols and their derivatives, resorcinol, omegaoxyethylalphanaphthylamine.

3-Nitro-4-aminobenzo Trifluoride

Dye

Starting point (Brit. 440207) in making—

Water-insoluble orange-red dyes fast to light and oils, by coupling with betanaphthol.

Water-insoluble orange dyes fast to light and oils, by coupling with 1-phenyl-3-methyl-5-pyrazolone.

2-Nitro-5-aminobenzotrifluoride-4-sulphonic Acid

Dye

Intermediate (Brit. 446532) in making various dyestuffs.

3-Nitro-4-aminobenzotrifluoride-5-sulphonic Acid

Dye

Intermediate (Brit. 446532) in making various dyestuffs.

5-Nitro-2-aminobenzylsulphonic Acid

Dye

Starting point (Brit. 265767) in making monoazo dyestuffs with—

Beta-amino-8-naphthol-6-sulphonic acid.
Betamethylamino-8-naphthol-6-sulphonic acid.
Methyldiphenylamine, oxyethylbetanaphthylamine.

5-Nitro-2-amino-4-cresolmethyl Ether

French: Éther 5-nitro-2-amino-4-crésolméthylique.
German: 5-Nitro-2-amino-4-kresolmethylaether.

Dye

Starting point (Brit. 248946) in making azo dyestuffs with—

Alpha-aminoanthraquinone, 4-chloro-2-aminodiphenyl ether, 4-chloro-2-anisidin, 4-chloro-2-nitranilin, dianisidin, 2:4-dichloroanilin, 2:5-dichloroanilin, metachloranilin, metanitranilin, 4-nitro-2-anisidin, 5-nitro-2-anisidin, 3-nitro-4-toluidin, 4-nitro-2-toluidin, 5-nitro-2-toluidin, orthoaminodiphenyl ether, orthoaminoazotoluene, orthophenetoleazoalphanaphthylamine, xylidin.

4-Nitro-4'-aminodiphenylamine

French: 4-Nitro-4'-aminodiphényleamine.
German: 4-Nitro-4'-aminodiphenylamin.

Chemical

Starting point in making—

Intermediates, pharmaceuticals.

Dye

Starting point (Brit. 323729) in making azo dyestuffs for the dyeing and printing of various rayons, with the aid of—

Alkylarylamines, alkylarylanilins, allylaminophenol, allylnaphthylamine, alphanaphthylamine, aminonaphthoic acid, aminonaphthols, amylaminophenol, amylnaphthylamine, betanaphthylamine, butylaminophenol, butylnaphthylamine, cresols and their derivatives, dimethylmeta-aminophenol, ethylaminophenol, ethylnaphthylamine, gammachlorobetaoxypropionylnaphthylamine, heptylaminophenol, heptylnaphthylamine, hexylaminophenol, hexylnaphthylamine, meta-aminophenol, meta-anisidin, metacresidin, metanaphthylenediamine, metaphenylenediamine, metaphenetidin, metaphenylamine, metatoluidin, metatoluylenediamine, metaxylenediamine, metaxylidin, methylaminophenol, methylnaphthylamine, naphthylamine ethers, omegaoxyethylalphanaphthylamine, orthoaminophenol, orthoanisidin, orthocresidin, orthophenylenediamine, orthophenetidin, orthonaphthylenediamine, orthophenylamine, orthotoluidin, orthotoluylenediamine, orthoxylenediamine, orthoxylidin, para-aminophenol, para-anisidin, paracresidin, paranaphthylenediamine, paraphenetidin, paraphenylamine, paraphenylenediamine, paratoluidin, paratoluylenediamine, paraxylenediamine, paraxylidin, pentylaminophenol, pentylnaphthylamine, phenols and their derivatives, propylaminophenol, propylnaphthylamine, resorcinol.

2-Nitro-4'-aminodiphenylamine-4-sulphonic Acid

French: Acide de 2-nitro-4'-aminodiphényleamine-4-sulfonique.
German: 2-Nitro-4'-aminodiphenylamin-4-sulfonsaeure.

Dye

Starting point in making azo dyestuffs with—

Alphachloro-2-nitro-4-sulphonic acid (Brit. 274999).

4'-Nitro-4-aminodiphenylamine-2'-sulphonic Acid

Dye

Starting point (Brit. 437657) in making—

Olive-brown dyestuffs for chrome or vegetable-tanned leather by coupling with metaphenylenediamine and sulphanilic acid and coppering.

3-Nitro-4-aminodiphenyl Ether

French: Éther de 3-nitro-4-aminodiphényle.
German: 3-Nitro-4-aminodiphenylaether.

Chemical

Starting point in making—

Intermediates, pharmaceuticals, various other derivatives.

Dye

Starting point (Brit. 323792) in making azo dyestuffs for use in dyeing and printing viscose rayon, nitro rayon and cuprammonium rayon, with the aid of—

Alkylarylamines, allylaminophenol, allylnaphthylamine, alphanaphthylamine, aminonaphthoic acids, amylaminophenol, amylnaphthylamine, betanaphthylamine, butylaminophenol, butylnaphthylamine, cresols and their derivatives, dimethylmeta-aminophenol, ethylaminophenol, ethylnaphthylamine, gammachlorobetaoxypropionylnaphthylamine, heptylaminophenol, heptylnaphthylamine, hexylaminophenol, hexylnaphthylamine, meta-aminophenol, meta-anisidin, metacresidin, metaphenetidin, metaphenylenediamine, metatoluidin, metaxylidin, methylaminophenol, methylnaphthylamine, naphthylamine ethers, omegaoxyethylalphanaphthylamine, orthoaminophenol, orthoanisidin, orthocresidin, orthophenetidin, orthophenylenediamine, orthotoluidin, orthoxylidin, para-aminophenol, para-anisidin, paracresidin, paranitrometaphenylenediamine, paraphenetidin, paraphenylenediamine, paraxylidin, phenols and their derivatives, propylaminophenol, propylnaphthylamine, resorcinol.

4-Nitro-4'-aminodiphenyl Ether

French: Éther de 4-nitro-4'-aminodiphényle, Éther 4-nitro-4'-aminodiphénylique.
German: 4-Nitro-4'-aminodiphenylaether.

Chemical

Starting point in making—

Intermediates, pharmaceuticals, and other derivatives.

Dye

Starting point (Brit. 323792) in making azo dyestuffs for use in dyeing and printing viscose rayon, nitro rayon, and cuprammonium rayon, with the aid of—

Alkylarylanilines, allylaminophenol, allylnaphthylamine, alphanaphthylamine, aminonaphthoic acids, amylaminophenol, amylnaphthylamine, betanaphthylamine, butylaminophenol, butylnaphthylamine, cresols and their derivatives, dimethylmeta-aminophenol, ethylaminophenol, ethylnaphthylamine, gammachlorobetaoxypropionylnaphthylamine, heptylaminophenol, heptylnaphthylamine, hexylaminophenol, hexylnaphthylamine, meta-aminophenol, meta-anisidin, metacresidin, metaphenylenediamine, metaphenetidin, metatoluidin, metaxylidin, methylaminophenol, methylnaphthylamine, naphthylamine ethers, orthoaminophenol, orthoanisidin, orthocresidin, orthophenylenediamine, orthophenetidin, orthotoluidin, orthoxylidin, para-aminophenol, para-anisidin, paracresidin, paranitrometaphenylenediamine, paraphenylenediamine, paraphenetidin, paratoluidin, paraxylidin, phenols and their derivatives, propylaminophenol, omegaoxyethylalphanaphthylamine, resorcinol.

5-Nitro-2-aminodiphenyl Ether
French: Éther de 5-nitro-2-aminodiphényle.
German: 5-Nitro-2-aminodiphenylaether.

Chemical
Starting point in making—
Intermediates, pharmaceuticals.
Dye
Starting point (Brit. 323792) in making azo dyestuffs for use in dyeing and printing viscose rayon; nitro rayon and cuprammonium rayon, with the aid of—
Alkylarylamines, allylalphanaphthylamine, allylaminophenol, aminonaphthoic acids, amylaminophenol, amylnaphthylamine, betanaphthylamine, butylaminophenol, butylnaphthylamine, cresols and their derivatives, dimethylmeta-aminophenol, ethylaminophenol, ethylnaphthylamine, gammachlorobetaoxypropionylnaphthylamine, heptylaminophenol, heptylnaphthylamine, hexylaminophenol, hexylnaphthylamine, meta-aminophenol, meta-anisidin, metacresidin, metaphenetidin, metaphenylenediamine, metatoluidin, metaxylidin, methylaminophenol, methylnaphthylamine, naphthylamine ethers, omegaoxyethylalphanaphthylamine, orthoaminophenol, orthoanisidin, orthocresidin, orthophenetidin, orthophenylenediamine, orthotoluidin, orthoxylidin, para-aminophenol, para-anisidin, paracresidin, paranitrometaphenylenediamine, paraphenetidin, paraphenylenediamine, paratoluidin, paraxylidin, phenols and their derivatives, propylaminophenol, propylnaphthylamine, resorcinol.

4-Nitro-4′-aminodiphenyl Sulphide
French: Sulphure de 4-nitro-4′-aminodiphényle, Sulphure 4-nitro-4′-aminodiphénylique.
German: 4-Nitro-4′-aminodiphenylsulfid, Schwefel-4-nitro-4′-aminodiphenyl.

Chemical
Starting point in making—
Intermediates, pharmaceuticals.

Dye
Starting point (Brit. 321483) in making disazo dyestuffs with—
Alphanaphthol-5-sulphonic acid.
Alphanaphthol-4-sulphonic acid.
Beta-acetylamino-8-naphthol-6-sulphonic acid.
Beta-amino-8-naphthol-3:6-disulphonic acid.
Beta-amino-8-naphthol-6-sulphonic acid.
Betabenzoylamino-8-naphthol-6-sulphonic acid.
Betanaphthol-6-sulphonic acid.
Betaphenylamino-8-naphthol-6-sulphonic acid.
Metaphenylenediamine, orthocresotinic acid.
8-Oxy-2:2′-dinaphthylamine-3:6-disulphonic acid.
8-Oxy-2-naphthylglycin-6-sulphonic acid.
Resorcinol, salicylic acid.

5-Nitro-2-aminohydroquinonedimethyl Ether
French: Éther de 5-nitro-2-aminohydroquinonediméthyle.
German: 5-Nitro-2-aminohydrochinondimethylaether.

Dye
Starting point (Brit. 248946) in making azo dyestuffs with—
Alpha-aminoanthraquinone, 4-chloro-2-anisidin, 4-chloro-2-aminodiphenyl ether, 4-chloro-2-nitranilin, 2:4-dichloroanilin, 2:5-dichloroanilin, dianisidin, metachloroanilin, metanitranilin, 4-nitro-2-toluidin, 5-nitro-2-toluidin, 3-nitro-4-toluidin, 4-nitro-2-anisidin, 5-nitro-2-anisidin, orthoaminodiphenyl ether, orthoaminoazotoluene, orthophenetoleazoalphanaphthylamine, xylidin.

2-Nitro-2′-amino-4′-methoxy-5′-methyl-4-trifluoromethylazobenzene

Dye
Coupling agent (Brit. 434416) in making—
Dark-brown water-insoluble dyestuffs with orthoanisidide.

4-Nitro-2-amino-1-methylbenzene
Synonyms: 4-Nitro-2-amino-1-methylbenzol.
German: 4-Nitro-2-amino-1-methylbenzol.

Dye
Starting point (Brit. 263164) in making azo dyestuffs with sulphonated derivatives of 2:3-oxynaphthoic acid—
Anilide, betanaphthylamide, 5-chloro-2-anisidide, 5-chloro-2-toluidide, orthotoluidide.

3-Nitro-4-amino-4-methylbenzophenone
French: Benzophénone de 3-nitro-4-amino-4′-méthyle, Benzophénone 3-nitro-4-amino-4′-méthylique.
German: 3-Nitro-4-amino-4′-methylbenzophenon.

Chemical
Starting point in making—
Intermediates, pharmaceuticals, and other derivatives.

Dye
Starting point (Brit. 323792) in making azo dyestuffs for dyeing various rayons with the aid of—
Alkylarylanilines, allylaminophenol, allylnaphthylamine, alphanaphthylamine, aminonaphthoic acids, amylaminophenol, amylnaphthylamine, betanaphthylamine, butylaminophenol, butylnaphthylamine, cresols and their derivatives, dimethylmeta-aminophenol, ethylaminophenol, ethylnaphthylamine, gammachlorobetaoxypropionylnaphthylamine, meta-aminophenol, meta-anisidin, metacresidin, metaphenylenediamine, metaphenetidin, metatoluidin, metaxylidin, methylaminophenol, methylnaphthylamine, naphthylamine ethers, orthoaminophenol, orthoanisidin, orthocresidin, orthophenylenediamine, orthophenetidin, orthotoluidin, orthoxylidin, para-aminophenol, para-anisidin, paracresidin, paraphenylenediamine, paranitrometaphenylenediamine, paraphenetidin, paratoluidin, paraxylidin, phenols and their derivatives, propylaminophenol, propylnaphthylamine, omegaoxyethylalphanaphthylamine, resorcinol.

5-Nitro-2-amino-4′-methylbenzophenone
French: Benzophénone de 5-nitro-2-amino-4′-méthyle, Benzophénone 5-nitro-2-amino-4′-méthylique.
German: 5-Nitro-2-amino-4′-methylbenzophenon.

Chemical
Starting point in making—
Intermediates, pharmaceuticals.
Dye
Starting point (Brit. 323792) in making azo dyestuffs for rayons, with the aid of—
Alkylarylanilines, allylaminophenol, allylnaphthylamine, alphanaphthylamine, aminonaphthoic acids, amylaminophenol, amylnaphthylamine, betanaphthylamine, butylaminophenol, butylnaphthylamine, cresols and derivatives, dimethylmeta-aminophenol, ethylaminophenol, ethylnaphthylamine, gammachlorobetaoxypropionylnaphthylamine, meta-aminophenol, meta-anisidin, metacresidin, metaphenylenediamine, metatoluidin, metaxylidin, methylaminophenol, methylnaphthylamine, naphthylamine ethers, omegaoxyethylalphanaphthylamine, orthoaminophenol, orthoanisidin, orthocresidin, orthophenylenediamine, orthophenetidin, orthotoluidin, orthoxylidin, para-aminophenol, para-anisidin, paracresidin, paraphenylenediamine, paranitrometaphenylenediamine, paraphenetidin, paratoluidin, paraxylidin, phenols and their derivatives, propylaminophenol, propylnaphthylamine, resorcinol.

4-Nitro-2-aminophenol
Chemical
Starting point in making—
5-Nitrobenzoxazolone (Brit. 261133).
Dye
Starting point in making—
After-chromed dyestuffs.

4-Nitro-2-aminophenol-6-sulphonic Acid
Dye
Intermediate in dye making.
Paint and Varnish
Starting point (German 572475 and 529840) in making—
Red pigments used in varnishes for metals, paper, and other materials.

4′-Nitro-4-aminostilbene-2:2′-disulphonic Acid
Dye
Starting point (Brit. 427241) in making—
Brownish-red dyes, reddish-brown dyes, yellow dyes for cotton.

5-Nitrobenzoazolon
Chemical
Starting point in making—
Benzoxazolon-5-arsinic acid (German 439606).

1:5-Nitrobenzothiazyl Cyclohexylethyldithiocarbamate
Rubber
Accelerator (Brit. 442978) for—
Vulcanization.

1:5-Nitrobenzothiazyl Dicyclohexyldithiocarbamate

Rubber
Accelerator (Brit. 442978) for—
Vulcanization.

5-Nitrobenzoxazolone

Chemical
Starting point in making—
2:1-Benzoxazolone-5-arsinic acid (Brit. 261133).

Nitrobenzoyl Chloride

French: Chlorure de nitrobenzoyle, Chlorure nitrobenzoylique.
German: Chlornitrobenzoyl, Nitrobenzoylchlorid.

Chemical
Reagent (Brit. 315200) in making acidylamino compounds, with the aid of—
Aminoacenaphthenesulphonic acids.
Aminobenzenesulphonic acids.
Aminonaphthalenesulphonic acids.
Chloroacenaphthenesulphonic acids and derivatives.
Chlorobenzenesulphonic acids and derivatives.
Chloronaphthalenesulphonic acids and derivatives.
Hydroxyacenaphthenesulphonic acids and derivatives.
Hydroxybenzenesulphonic acids and derivatives.
Hydroxynaphthalenesulphonic acids and derivatives.
Methylacenaphthenesulphonic acids and derivatives.
Methylbenzenesulphonic acids and derivatives.
Methylnaphthalenesulphonic acids and derivatives.
Starting point in making—
Intermediates and other derivatives.

4-Nitrobenzyl Chloride

French: Chlorure de 4-nitrobenzyle, Chlorure 4-nitrobenzylique.
German: Chlor-4-nitrobenzoyl.

Chemical
Reagent in making—
Intermediates—

Dye
Reagent (Brit. 323710) in making dyestuffs with—
Alphanaphthylamine-6-sulphonic acid.
Anilin-2-sulphonic acid.

2-Nitro-4-bromodiphenylamine

Chemical
Starting point in making—
Intermediates, pharmaceuticals.

Dye
Starting point in making various synthetic dyestuffs.

Textile
Solubilizing agent (Brit. 305560) in—
Dye baths, printing pastes, and stenciling compositions used on acetate rayon and fabrics containing cellulose acetate.

2-Nitro-4-bromo-4′-methoxydiphenylamine

French: 2-Nitro-4-bromo-4′-méthoxyediphénylеamine.
German: 2-Nitro-4-brom-4′-methoxydiphenylamin.

Chemical
Starting point in making various derivatives.

Dye
Starting point in making various synthetic dyestuffs.

Textile
Solubilizing agent (Brit. 305560) in—
Dye baths, printing pastes, and stenciling compositions for use on mixed textiles containing cellulose acetate rayon.

Nitrocellulose

Synonyms: Cellulose nitrate, Collodion cotton, Colloxylin, Gun cotton, Nitrated cellulose, Nitrated cotton, Nitrocotton, Pyroxylin, Pyroxylon, Soluble cotton, Soluble gun cotton.
Latin: Gossypium ignarium, Pyroxylinum.
French: Coton azotique, Fulmicoton soluble, Nitrate de cellulose.
German: Kollodiumwolle, Nitriete baumwolle, Schlessbaumwolle, Zellstoffnitrat, Zellulosenitrat.
Spanish: Piroxilana.

Ceramics
Ingredient of—
Coating compositions used for protecting and decorating ceramic products.

Chemical
Starting point in making—
Collodion, soluble pyroxylins.
Reagent for treating—
Filter cloths for use in filter presses in which acid liquors are being filtered (the function of the nitrocellulose is to render the cloth resistant to acid).

Cosmetic
Ingredient of—
Nail enamels and lacquers.

Explosives
Raw material in the manufacture of—
Cordite, gelatin dynamites, guncottons, smokeless powders, sporting powders.

Glass
Ingredient of—
Compositions used in the manufacture of nonscatterable glass and as coatings for protecting and decorating glass products.

Glue and Adhesives
Ingredient of—
Adhesive preparations containing also gums, resins, and other substances.

Leather
Ingredient of—
Compositions used in the manufacture of artificial leather and as coatings for protecting and decorating leather goods.

Metal Fabricating
Ingredient of—
Compositions used as coatings for the decoration and protection of metal articles.

Miscellaneous
Ingredient of—
Compositions used as coatings for the decoration and protection of various fibrous and other products.
Compositions for coating skins.
Solidified alcohols used as fuel.

Paint and Varnish
Raw material in making—
Bronzing lacquers, cements, dopes, enamels, lacquers.

Pharmaceutical
In compounding and dispensing practice.
Ingredient of—
Collodions.

Paper
Ingredient of—
Compositions used as coatings for the decoration and protection of products made from paper and pulp and in the manufacture of coated paper.

Photographic
Raw material in making—
Sheet and roll films.

Plastics
Raw material in making—
Celluloid and other plastic compositions.

Rayon
Base of various forms of rayon.

Rubber
Ingredient of—
Compositions used as coatings for the decoration and protection of rubber and rubber merchandise.

Stone
Ingredient of—
Compositions used as coatings for the decoration and protection of artificial and natural stone.

Textile
As a textile material in the form of nitro or Chardonnet rayon.
Ingredient of—
Coating compositions for protecting and decorating textile fabrics.

Wood
Ingredient of—
Compositions used as coatings for the decoration and protection of wood products.
Plastic compositions used for filling and decorating woodwork.

2-Nitro-4-chloro-4′-acetylaminodiphenylamine

Chemical
Starting point in making—
Intermediates.

Textile
Solubilizing agent (Brit. 305560) in—
Dye liquors, printing pastes, and stenciling compositions used on acetate rayon and fabrics containing cellulose acetate.

2-Nitro-9-chloroacridin

Chemical
Starting point in making—
Intermediates, pharmaceuticals.

Dye
Starting point (Brit. 305487) in making azo dyestuffs with—
2-Aminonaphthalene-4:8-disulphonic acid.

4-Nitro-2-chloro-1-aminobenzene

Chemical
Starting point in making—
Intermediates, pharmaceuticals, other derivatives.

Dye
Starting point (French 743041) in making azo dyestuffs, suitable for dyeing cellulose esters and ethers, with the aid of—
Benzylamine, cresylamine, orthophenylamine, orthotolylamine, orthoxylylamine, paraphenylamine, paratolylamine, paraxylylamine.

2-Nitro-4-chlorodiphenylamine

Chemical
Starting point in making—
Intermediates, pharmaceuticals.

Dye
Starting point in making various synthetic dyestuffs.

Textile
Solubilizing agent (Brit. 305560) in—
Dye liquors, printing pastes, and stenciling compositions used on acetate rayon and other fabrics containing cellulose acetate.

2-Nitro-4-chloro-4′-ethoxydiphenylamine
French: 2-Nitro-4-chloro-4′-éthoxyediphénylеamine.
German: 2-Nitro-4-chlor-4′-aethoxydiphenylamin.

Chemical
Starting point in making—
Intermediates.

Textile
Solubilizing agent (Brit. 305560) in—
Dye baths, printing pastes, and stenciling compositions used on acetate rayon and fabrics containing cellulose acetate.

2-Nitro-4-chloro-3′-methyldiphenylamine

Chemical
Starting point in making various intermediates.

Textile
Solubilizing agent (Brit. 305560) in—
Dye baths, printing pastes and stenciling compositions used on acetate rayon and fabrics containing cellulose acetate.

7-Nitro-4-chlorophenanthridin

Chemical
Starting point in making—
Intermediates, pharmaceuticals.

Dye
Starting point in making—
Azo dyestuffs with 1:4-phenylenediaminesulphonic acid (Brit. 305487).
Various synthetic dyestuffs.

6-Nitro-4-chloroquinazolin
German: 6-Nitro-4-chlorchinazolin.

Chemical
Starting point in making—
Intermediates, pharmaceuticals.

Dye
Starting point (Brit. 310076) in making dyestuffs with—
Aliphatic amines, aliphatic diamines, aminoarylpyrazolones, aminocarboxylic acids of the benzene and the naphthalene series, 1-amino-7-naphthol, aminonaphtholsulphonic acids, 5-aminosalicylic acid, aminosulphonic acids of the benzene and the naphthalene series, ammonia, anilin, beta-amino-5-naphthol, betanaphthylamine, dehydrothiotoluidinsulphonic acid, dithioglycol, J acid, metanitranilin, metaphenylenediamine, metatoluylenediamine, monomethylaniline, monoformylmetaphenylenediamine, naphthols, naphthylenediamines, nitrophenols, 4-nitro-1-naphthol-5-sulphonic acid, orthoaminophenol, orthoanisidin, paraphenylenediamine, paratoluidin, phenols, phenolsulphonic acids, salicylic-3-sulphonic acid.

2-Nitrocinnamyl Chloride
French: Chlorure de 2-nitrocinnamyle.
German: Chlor-2-nitrocinnamyl.

Chemical
Reagent (Brit. 278037) in making synthetic drugs with the aid of—
Alkoxynaphthylaminesulphonic acids.
4-Aminoacenaphthene-3:5-disulphonic acid.
4-Aminoacenaphthene-3-sulphonic acid.
4-Aminoacenaphthene-5-sulphonic acid.
4-Aminoacenaphthenetrisulphonic acids.
2:8-Aminonaphthol-3:6-disulphonic acid.
1:5-Aminonaphthol-7-sulphonic acid.
Bromonaphthylaminesulphonic acids.
Chloronaphthylaminesulphonic acids.
Iodonaphthylaminesulphonic acids.

3-Nitrocinnamyl Chloride
French: Chlorure de 3-nitrocinnamyle, Chlorure 3-nitrocinnamylique.
German: Chlor-3-nitrocinnamyl, 3-Nitrocinnamylchlorid.

Chemical
Reagent (Brit. 278037) in making synthetic drugs with—
Alkoxynaphthylaminesulphonic acid.
Alphanaphthylamine-4:8-disulphonic acid.
Alphanaphthylamine-3:6:8-trisulphonic acid.
Alphanaphthylamine-4:6:8-trisulphonic acid.
4-Aminoacenaphthene-3:5-disulphonic acid.
4-Aminoacenaphthene-3-sulphonic acid.
4-Aminoacenaphthene-5-sulphonic acid.
4-Aminoacenaphthenetrisulphonic acid.
1:5-Aminonaphthol-3:6-disulphonic acid.
1:8-Aminonaphthol-3:6-disulphonic acid.
1:5-Aminonaphthol-7-sulphonic acid.
Bromonaphthylaminesulphonic acid.
Chloronaphthylaminesulphonic acid.
Iodonaphthylaminesulphonic acid.

1-Nitro-2:4-diaminobenzene

Dye
Starting point (Brit. 270352) in making azo dyestuffs for cellulose acetate rayon with—
Anilin, betachloroanilin, orthotoluidin, para-aminomethylacetanilide, paranitranilin.

6-Nitro-2:4-dichloroquinazolin
German: 6-Nitro-2:4-dichlorchinazolin.

Chemical
Starting point in making—
Intermediates, pharmaceuticals.

Dye
Starting point (Brit. 305487) in making azo dyestuffs with—
4′-Amino-4-hydroxyazobenzene-3-carboxylic acid.
5-Amino-2-hydroxybenzoic acid.
4-Aminotoluene-3-sulphonic acid.
Dimethylamine, J acid.
1:4-Phenylenediaminesulphonic acid.
5-Sulpho-3-aminobenzoic acid.

7-Nitro-2:3-dichloroquinazolin
French: 7-Nitro-2:3-dichloroquinazoléine.
German: 7-Nitro-2:3-dichlorchinazolin.

Chemical
Starting point in making—
Intermediates, pharmaceuticals.

Dye
Starting point in making azo dyestuffs with—
Anilin, H acid, J acid, 3-nitrobenzoyl chloride.

2-Nitrodiphenylamine
French: 2-Nitrodiphénylеamine.
German: 2-Nitrodiphenylamin.

Chemical
Starting point in making—
Aromatics, intermediates, pharmaceuticals.

Dye
Starting point in making various synthetic dyestuffs.

Textile
——, *Dyeing and Printing*
Solubilizing agent (Brit. 305560) in—
Dye baths, printing pastes, and stenciling compositions for use on mixtures containing cellulose acetate rayon.

Nitroethyl-Mercury Chloride

Agriculture

For control of—

Bottom rust of lettuce, covered smut and stripe disease of barley, kernel smut of sorghum, loose and covered smut of oats, soil-borne parasitic fungi, stinking smut of wheat.

Woodworking

For control of—

Blue stain and sap stain in sapwood of freshly sawed timber.

Nitrogen

French: Nitrogène.
German: Stickstoff.

Chemical

As an atmosphere for carrying out various chemical reactions which cannot be properly accomplished in the presence of oxygen or oxidizing agents.

Starting point in making—

Barium nitride from barium chloride, copper nitride from copper chloride, synthetic ammonia, synthetic nitric acid, various metallic nitrides, various metallic cyanides, nitrogen oxides.

Electrical

Filling agent in making—

High-candlepower electric light bulbs.

Fertilizer

Raw material in making—

Cyanamid, synthetic nitrate of soda.

Food

Reagent in preserving—

Food products by preventing access of atmospheric oxygen.

Miscellaneous

Material for filling—

High-temperature thermometers and other scientific instruments.

For filling automobile tires, the nitrogen prolonging the life of the tire in that it does not have the oxidizing action of the oxygen in ordinary air.

Petroleum

As an atmosphere in storing and transferring highly inflammable petroleum distillates, such as gasoline and naphthas.

Nitroglycerin

Synonyms: Blasting oil, Glonoin oil, Glycerin trinitrate, Glyceryl trinitrate, Nitroleum.
French: Nitroglycérine.
German: Sprengoel, Trinitrin, Trinitroglycerin.

Explosives

Ingredient of—

Dynamites, gelatins, military explosives, permissibles.

Petroleum

Explosive in shooting oil wells.

Pharmaceutical

In compounding and dispensing practice.

3-Nitro-2-hydroxy-5-cyclohexylanilin

Dye

In dye syntheses.

Starting point (Brit. 448872) in making—

Cobalt dyes.

Dyes, usable on wool alone or with metachrome mordants, by coupling with 1:4-hydroxynaphtholsulphonic acid or 1-acetamido-8-naphthol-4-sulphonic acid.

Nitromannite

Synonyms: Nitromannitol.

Explosives and Matches

Substitute for—

Mercury fulminate.

6-Nitro-2-mercaptobenzothiazole

Rubber

Accelerator in vulcanization (Brit. 265920).

Nitrometadiaminoanisole

Miscellaneous

Reagent in—

Dyeing hair, fur, feathers, and other articles.

Nitrometadiaminophenetole

Miscellaneous

Reagent in—

Dyeing hair, fur, feathers, and other articles.

6-Nitrometaphenylenediamine

Dye

Starting point in making—

Pyramin orange R.

Nitrometatoluylenediamine

Miscellaneous

Reagent in—

Dyeing hair, fur, feathers, and other articles.

4-Nitro-2-methoxy-4-dimethylaminoazobenzene

Textile

Dye for—

Cellulose acetate in bath containing also turpentine, turkey red oil, and olive oil soap.

7-Nitro-4-methyl-2-chloroquinolin

French: 7-Nitro-4-méthyle-2-chloroquinoléine.
German: 7-Nitro-4-methyl-2-chlorchinolin.

Chemical

Starting point in making—

Intermediates, pharmaceuticals.

Dye

Starting point in making—

Azo dyestuffs with H acid (Brit. 305487).
Various synthetic dyestuffs.

2-Nitro-4-methyldiphenylamine

Chemical

Starting point in making—

Intermediates, pharmaceuticals.

Dye

Starting point in making various synthetic dyestuffs.

Textile

Solubilizing agent (Brit. 305560) in—

Dyeing and stenciling compositions used on acetate rayon and acetate rayon mixtures.

2-Nitro-4-methyl-4′-ethoxydiphenylamine

Chemical

Starting point in making—

Intermediates, pharmaceuticals.

Dye

Starting point in making various synthetic dyestuffs.

Textile

Solubilizing agent (Brit. 305560) in—

Dye liquors, printing pastes, and stenciling compositions used on acetate rayon and acetate rayon mixtures.

1:5-Nitronaphthoyl Chloride

French: Chlorure de 1:5-nitronaphthoyle, Chlorure 1:5-nitronaphthoylique.
German: Chlor-1:5-nitronaphtoyl, 1:5-Nitronaphtoylchlorid.

Chemical

Reagent (Brit. 278037) in making synthetic drugs with—

Alkoxynaphthylaminesulphonic acids.
Alphanaphthylamine-4:8-disulphonic acid.
Alphanaphthylamine-3:6:8-trisulphonic acid.
Alphanaphthylamine-4:6:8-trisulphonic acid.
4-Aminoacenaphthene-3:5-disulphonic acid.
4-Aminoacenaphthene-3-sulphonic acid.
4-Aminoacenaphthene-5-sulphonic acid.
4-Aminoacenaphthenetrisulphonic acid.
2:8-Aminonaphthol-3:6-disulphonic acid.
1:8-Aminonaphthol-3:6-disulphonic acid.
1:7-Aminonaphthol-7-sulphonic acid.
Bromonaphthylaminesulphonic acid, chloronaphthylaminesulphonic acid, idonaphthylaminesulphonic acid.

Starting point (Brit. 314909) in making derivatives with—

3-Carboxyphenylthiocarbimide, diphenylure-3:3′-dicarboxylic acid, 4-quinolylphenylurea-3:6′-dicarboxylic acid, symmetrical diphenylurea-3:3′-dicarboxylic acid, thiourea, thiourea-3:3′-dicarboxylic acid, urea.

Nitronaphthyl Chloride

Chemical

Reagent (Brit. 315200) in making acylamino compounds with the aid of—

Aminoacenaphthenesulphonic acids, aminobenzenesulphonic acids, aminonaphthalenesulphonic acids.

Chloroacenaphthenesulphonic acids and their derivatives, chlorobenzenesulphonic acids and their derivatives, chloronaphthalenesulphonic acids and their derivatives.

Hydroxyacenaphthenesulphonic acids and their derivatives, hydroxybenzenesulphonic acids and their deriv-

Nitronaphthyl Chloride (Continued)
atives, hydroxynaphthalenesulphonic acids and their derivatives.
Methylacenaphthenesulphonic acids and their derivatives, methylbenzenesulphonic acids and their derivatives, methylnaphthalenesulphonic acids and their derivatives.
Starting point in making—
Intermediates and other derivatives.

4-Nitro-orthoaminophenol-6-sulphonic Acid
Dye
Starting point (Brit. 431201) in making chrome brown dyestuffs with—
Phenol and acetone, phenol and cyclohexanone, orthocresol and acetone, phenol and ethylmethyl ketone.

5-Nitro-orthoanisidin
Chemical
As an intermediate.
Dye
Starting point (Brit. 397016) in making—
Bordeaux water-insoluble dyes.

5-Nitro-orthotoluenesulphonic Acid
Chemical
Starting point in making—
Diaminostilbenedisulphonic acid, intermediates, paratoluidinorthosulphonic acid, pharmaceuticals, various other derivatives.
Dye
Starting point in making—
Chicago orange G, chloramine orange G, chlorophenin, curcurphenin, diamine fast yellow A, diphenyl catechin, diphenyl citronin G, diphenyl chrysosin, diphenyl chrysosin RR, diphenyl fast brown G, diphenyl orange RR, direct brown R, direct yellow R, direct yelow RT, mikado orange, mikado yellow, naphthylamine orange, polychromin B, renol yellow R, stilbene colors, such as stilbene yellow, sun yellow.

3-Nitroparatoluidin
Chemical
As an intermediate.
Paint and Varnish
Coloring agent (Brit. 390649) for—
Cellulose acetate and nitrocellulose varnishes.

2-Nitrophenoxyacetylamino-8-hydroxynaphthalene-3:6-disulphonic Acid
Chemical
Starting point in making various derivatives.
Dye
Starting point (Brit. 313710) in making dyestuffs with—
4-Aminoacetanilide, anilin, beta-acetamino-5-aminoanisol, beta-aminobenzoic acid, paraxylidin.

2-Nitrophenylacetyl Chloride
French: Chlorure de 2-nitrophényleacétyle, Chlorure 2-nitrophényleacétylique.
German: Chlor-2-nitrophenylacetyl, 2-Nitrophenylacetylchlorid.
Chemical
Starting point (Brit. 314909) in making derivatives with—
3-Carboxyphenylthiocarbamide, diphenylurea-3:3′-dicarboxylic acid, 4-quinolylphenylurea-3:6′-dicarboxylic acid, symmetrical diphenylurea-3:3′-dicarboxylic acid, thiourea, thiourea-3:3′-dicarboxylic acid, urea.

2-(4′-Nitrophenyl)-4:6-dichloropyrimidin
French: 2(4′-Nitrophényle)-4:6-dichloropyrimidine.
German: 2-(4′-Nitrophenyl)-4:6-chlorpyrimidin.
Chemical
Starting point in making—
Intermediates, pharmaceuticals.
Dye
Starting point (Brit. 305487) in making azo dyestuffs with—
Anilin, H acid, J acid.

Nitrosalicylic Acid
Chemical
Starting point in making—
Aminosalicylic acid, aminosalicylic acid hydrochloride, para-aminophenol, para-aminophenolsulphonic acid.
Dye
Starting point in making—
Azo dyestuffs, diamond black, hydron blue, hydron colors, sulphur dyestuffs.
Pharmaceutical
In compounding and dispensing practice.

Nitrosodimethylanilin
Dye
As an intermediate.
Starting point in making—
Methylene blue, bordeaux violet shades for acetate rayon (Brit. 396893).
Rubber
As a vulcanizing accelerator.

Nitrosomethylcarbamide
Chemical
Starting point in making—
Methylhydrazin sulphate.

1:2-Nitrosonaphthol
Chemical
Intermediate in—
Organic synthesis.
Petroleum
Inhibitor (U. S. 1982277, 1982267, and 1982618) of—
Gum formation in gasoline, particularly in vapourphase cracked gasoline.

1:4-Nitrosonaphthol
Chemical
Intermediate in—
Organic synthesis.
Petroleum
Inhibitor (U. S. 1982277, 1982267, and 1982618) of—
Gum formation in gasoline, particularly in vapourphase cracked gasoline.

1:5-Nitrosonaphthol
Chemical
Intermediate in—
Organic synthesis.
Petroleum
Inhibitor (U. S. 1982277, 1982267, and 1982618) of—
Gum formation in gasoline, particularly in vapourphase cracked gasoline.

1:8-Nitrosonaphthol
Chemical
Intermediate in—
Organic synthesis.
Petroleum
Inhibitor (U. S. 1982277, 1982267, and 1982618) of—
Gum formation in gasoline, particularly in vapourphase cracked gasoline.

2:1-Nitrosonaphthol
Chemical
Intermediate in—
Organic synthesis.
Petroleum
Inhibitor (U. S. 1982277, 1982267, and 1982618) of—
Gum formation in gasoline, particularly in vapourphase cracked gasoline.

1:4-Nitrosonaphthylamine
Chemical
Intermediate in—
Organic synthesis.
Petroleum
Inhibitor (U. S. 1982277, 1982267, and 1982618) of—
Gum formation in gasoline, particularly in vapourphase cracked gasoline.

5:2-Nitrosonaphthylamine
Chemical
Intermediate in—
Organic synthesis.
Petroleum
Inhibitor (U. S. 1982277, 1982267, and 1982618) of—
Gum formation in gasoline, particularly in vapourphase cracked gasoline.

Nitroso(normal)ethylurethane
Petroleum
Priming agent (Brit. 405658) for—
Fuel oil for diesel and other compression-ignition engines.

Nitroso(normal)methylurethane

Petroleum
Priming agent (Brit. 405658) for—
Fuel oil for diesel and other compression-ignition engines.

Nitrosoparatolylaminomethylbenzothiazyl, Normal, Sulphide

Rubber
Antiscorch agent (Brit. 447458) in—
Vulcanizable rubber mixtures, which may contain an ultra-accelerator.

Nitrosophenylaminomethylbenzothiazyl, Normal, Sulphide

Rubber
Antiscorch agent (Brit. 447458) in—
Vulcanizable rubber mixtures, which may contain an ultra-accelerator.

Nitrosotriacetonamine

Petroleum
Priming agent (Brit. 405658) for—
Fuel oil for diesel and other compression-ignition engines.

Nitrotoluyl Chloride
French: Chlorure de nitrotoluyle, Chlorure nitrotoluylique.
German: Chlornitrotoluyl, Nitrotoluylchlorid.

Chemical
Reagent (Brit. 315200) in making acylamino compounds with the aid of—
Aminoacenaphthenesulphonic acids, aminobenzenesulphonic acids, aminonaphthalenesulphonic acids.
Chloroacenaphthenesulphonic acids and derivatives, chlorobenzenesulphonic acids and derivatives, chloronaphthalenesulphonic acids and derivatives.
Hydroxyacenaphthenesulphonic acids and derivatives, hydroxybenzenesulphonic acids and derivatives, hydroxynaphthalenesulphonic acids and derivatives.
Methylacenaphthenesulphonic acids and derivatives, methylbenzenesulphonic acids and derivatives, methylnaphthalenesulphonic acids and derivatives.
Starting point in making—
Intermediates and other derivatives.

Nitroxylethylenechlorhydrin

Fuel
Primer (Brit. 461320) for—
Diesel fuels.

N-Monodibutenylanilin

Chemical
Starting point in making—
Intermediates and other derivatives.

Insecticide
Ingredient (Brit. 313934) of—
Insecticidal compositions.

Soap
Ingredient (Brit. 313934) of—
Insecticidal and germicidal soaps.

N-N'-Dodecylmethylethylenediamine

Firefighting
Basic ingredient (Brit. 460649) in—
Air-foaming compositions for fire-extinguishing purposes.

N-N'-Tetrahydroxyethylethylenediamine
French: N:N'-Tétrahydroxye-éthyle-éthylènediamine.
German: N:N'-Tetrahydroxyaethylaethylendiamin.

Chemical
Starting point in making various intermediates.
Starting point (Brit. 306116) in making—
Dispersing agents, emulsifying agents, solvents for organic substances.

Leather
Ingredient (Brit. 306116) of—
Impregnating compositions.

Miscellaneous
Ingredient (Brit. 306116) of—
Cleansing agents, emulsified preparations.

Paint and Varnish
Ingredient (Brit. 306116) of—
Lacquers.

Soap
Ingredient (Brit. 306116) of—
Cleansing preparations.

Textile
——, *Bleaching*
Ingredient (Brit. 306116) of—
Bleach liquors.

——, *Dyeing*
Assist (Brit. 306116) in—
Dye baths.

——, *Finishing*
Ingredient (Brit. 306116) of—
Fulling baths, wetting preparations.

——, *Manufacturing*
Ingredient (Brit. 306116) of—
Carbonizing liquors.

Noninecarboxylic Acid
French: Acide de noninecarboxylique.
German: Nonincarbonsäure.

Chemical
Starting point in making—
Ethyl ester, methyl ester.

Nonyl Acetate
French: Acétate de nonyle, Acétate nonylique.
German: Essigsäurenonylester, Essigsäuresnonyl, Nonylacetat, Nonylazetat.
Spanish: Acetato de nonil.
Italian: Acetato di nonile.

Perfume
Ingredient of—
Perfume preparations, such as orange and orange flower odors.
Perfume in—
Cosmetics.

Soap
Perfume in—
Toilet soaps.

Nonylic Acid

Chemical
Starting point in making—
Ethyl nonylate, methyl nonylate, various other esters, salts, intermediates, and pharmaceuticals.

2-Normal-amyl-4-chlorophenol

Pharmaceutical
Bactericide (U. S. 2101595) for—
Bacillus typhosus, staphylococcus aureus, other bacteria.

2-Normal-heptylcyclopentanone-1

Cosmetic
Odorant (Brit. 430930 and 449211) in—
Perfume mixtures.

2-Normal-hexylcyclohexanone-1

Cosmetic
Odorant (Brit. 430930 and 449211) in—
Perfume mixtures.

2-Normal-hexylcyclopentanone-1

Cosmetic
Odorant (Brit. 430930 and 449211) in—
Perfume mixtures.

Normal-n'-dichloroazodicarbonamidin

Disinfectant
Bactericide, the definite characteristics of which are said to make it especially useful in the presence of oxidizable organic matter (U. S. 2016257).

Water and Sanitation
Bactericide, the definite characteristics of which are said to make it especially useful in the presence of oxidizable organic matter (U. S. 2016257).

Novocaine
Synonyms: Ethocane, Erocaine, Para-aminobenzoldiethylaminoethanol, Procaine, Syncaine.

Chemical
Starting point (Brit. 260346) in making—
Ethocaine hydrobromide, ethocaine hydrochloride, ethocaine pentoborate, ethocaine salicylate, ethocaine sulphate.

Miscellaneous
Suggested for use as—
Anaesthetic in dentistry.

Pharmaceutical
In compounding and dispensing practice.

N'-Phenyl-4-metatolylenediamine
French: N'-Phényle-4-métatolylènediamine.
German: N'-Phenyl-4-metatolylendiamin.
Dye
Starting point in making—
Rhodulin red B, rhodulin red G, rhodulin violet.

N-Propyl Disulphide
Synonyms: N-Propyl bisulphide.
French: Bisulfure de N-propyle, Bisulfure N-propylique.
German: Bischwefel-N-propyl, Dischwefel-N-propyl, N-Propylbisulfid, N-Propyldisulfid.
Chemical
Reagent in making—
Intermediates, pharmaceuticals, salts and esters.
Reagent (Brit. 298511) in treating—
Albumens and albumenoids.
Glues and Adhesives
Reagent (Brit. 298511) in treating—
Vegetable proteins, such as soya bean flour, linseed protein, and peanut protein, to make glues and adhesives.
Miscellaneous
Reagent (Brit. 298511) in treating—
Vegetable proteins to make sizes and finishes.

Nucleinic Acid
Synonyms: Nucleic acid.
French: Acide nucléinique, Acide nucléique.
German: Nucleinsäure.
Chemical
Starting point in making—
Iron nucleinate (triferrin), magnesium nucleinate, quinine nucleinate, silver nucleinate, various other salts and esters used for pharmaceutical purposes.
Pharmaceutical
In compounding and dispensing practice.

Nux Vomica
Synonyms: Bachelor's buttons, Dog's buttons, Poison nut, Quaker buttons, Vomit nut.
Latin: Semen strychnos nux vomica.
French: Noix vomique.
German: Brechnuesse, Kraehenaugen, Strychnossamen.
Chemical
Raw material for obtaining—
Brucine, strychnine.
Insecticide
Ingredient of—
Compositions used for eradicating ants, cock-roaches, rats and other vermin.
Pharmaceutical
In compounding and dispensing practice.

Octadecenylaminesulphonic Acid
Miscellaneous
As an emulsifying agent (Brit. 353232).
For uses, see under general heading: "Emulsifying agents."

Octadecyl Alcohol
French: Alcool de octadécyl, Alcool octadécylique.
German: Oktadecylalkohol, Oktodecylalkohol.
Chemical
Starting point in making—
Octadecylalphapicolinium bromide (Brit. 398175).
Octadecylbenzyl ether (Brit. 378454, 393937).
Octadecyl bromide (Brit. 401707).
Octadecyl chloride.
Octadecylpyridinium bromide (Brit. 397553, 398175, 404969).
Octadecyltrimethylammonium bromide (Brit. 397553).
Octadecyltrimethylammonium methosulphate (Brit. 396992).

Octadecylalphapicolinium Bromide
French: Bromure de octadécylalphapicolinium.
German: Bromoktadecylalphapicolinium, Bromoktadecylalphapicolinium, Oktadecylalphapicoliniumbromid, Oktodecylalphapicoliniumbromid.
Textile
Reagent (Brit. 398175) for—
Increasing the fastness of dyes on cotton textiles.

Octadecylbenzyl Ether
French: Benzyle éther de octadécyl, Benzyle éther octadécylique, Éther benzilique de octadécyl.
German: Oktadecylbenzilaether, Oktodecylbenzilaether.
Soap
Starting point (Brit. 378454) in making—
Sulphonated derivatives used as cleansing agents.

Octadecyl Bromide
Insecticide
Reagent (Brit. 401707) in making—
Insecticides, by reaction with nicotine.

Octadecyl Chloride
French: Chlorure d'alcool octadécylique, Chlorure de octadécyl, Chlorure de octadécyl alcool.
German: Oktadecylchlorid, Oktodecylchlorid.
Chemical
Agent in—
Recovering volatile solvents from gases.
Emulsifiable higher fatty alcohol derivative, more readily emulsifiable in water than the usual hydrocarbons.
Reagent for—
Introducing long-chain alkyl residues into the most varied types of organic substances.
Solvent for—
Aromatic hydrocarbons, coaltar constituents, fatty acids.
Dye
Reagent in making—
Fat-soluble colors.
Fats and Oils
Solvent for—
Fatty acids, oils.
Insecticide
As an insecticide (potent in toxicity to lower organisms, but nontoxic to the human organism).
Carrier for—
Insecticides generally, nicotine, pyrethrum extracts.
Leather
Starting point in making—
Protective agents.
Miscellaneous
Ingredient of—
Shoe creams and polishes.
Solvent for—
Bitumens.
Resins and Waxes
Solvent for—
Resins, waxes.
Textile
Starting point in making—
Textile soaps.

Octadecylchloromethyl Ether
Chemical
Starting point (Brit. 434911) in making—
Triethyloctodecoxymethyl-ammonium chloride by reacting with triethylamine.

Octadecylcresol
Chemical
Starting point (Brit. 444351) in making—
Fat-splitting catalysts and emulsifying agents, useful in dyeing, laundering, bleaching, and various other purposes, by reacting with formaldehyde and nonaromatic secondary amines (the salts of the products with water-soluble acids, or water-insoluble acids, or the quaternary ammonium salts are claimed to be valuable for the purposes named).

Octadecylphenol
Chemical
Starting point (Brit. 444351) in making—
Fat-splitting catalysts and emulsifying agents, useful in dyeing, laundering, bleaching, and various other purposes, by reacting with formaldehyde and nonaromatic secondary amines (the salts of the products with water-soluble acids, or water-insoluble acids, or the quaternary ammonium salts are claimed to be valuable for the purposes named).

Octadecylpyridinium Bromide
French: Bromure de octadécylpyridinium.
German: Bromoktadecylpyridinium, Bromoktodecylpyridinium, Oktadecylpyridiniumbromid.
Spanish: Bromuro de octadecylpyridinium.
Italian: Bromuro di octadecylpyridinium.

Octadecylpyridinium Bromide (Continued)

Metallurgical
Inhibitor (Brit. 397553) of—
Corrosion of metal by sulphuric acid in pickling baths for steel.

Miscellaneous
Pretreating agent (Brit. 404969) for—
Furs to be dyed by acid, chrome, direct vat, or mixtures of such dyes.

Textile
Reagent (Brit. 398175) for—
Increasing the fastness of dyes on cotton textiles.

Octadecylresorcinol

Chemical
Starting point (Brit. 444351) in making—
Fat-splitting catalysts and emulsifying agents, useful in dyeing, laundering, bleaching, and various other purposes, by reacting with formaldehyde and nonaromatic secondary amines (the salts of the products with water-soluble acids, or water-insoluble acids, or the quaternary ammonium salts are claimed to be valuable for the purposes named).

Octadecyltrimethylammonium Bromide
French: Bromure de octadécyltriméthyleammonium.
German: Bromoktadecyldreifachmethylammoniak, Bromoktadecyldreifachmethylammonium, Oktadecyldreifachmethylammoniumbromid.
Spanish: Bromuro de octadecyltrimetailammonio.
Italian: Bromuro di octadecyltrimetileammonio.

Metallurgical
Inhibitor (Brit. 397553) of—
Corrosion of metal by sulphuric acid in pickling baths for steel.

Octadecyltrimethylammonium Methosulphate

Paper
Reagent (Brit. 396992) for—
Increasing the fastness to water of dyestuffs on tissue paper, particularly paper dyed with direct safranin B or kiton blue A.

Octahydrobetanaphthoquinolin
French: Octahydrobétanaphthoquinoléine.
German: Octahydrobetanaphtochinolin.

Chemical
Starting point in making—
Intermediates, pharmaceuticals.

Dye
Starting point (Brit. 285382) in making indophenols and leucoindophenols with—
Dichloroquinonechlorimide, 2:6-dichloro-4-aminophenol, para-aminophenol, quinone halogenimides.

Octyl Acetate
Synonyms: Capryl acetate.
French: Acétate de capryle, Acétate caprylique, Acétate d'octyle, Acétate octylique.
German: Caprylacetat, Caprylazetat, Essigsäurecaprylester, Essigsäureoctylester, Essigsäurescapryl, Essigsäuresoctyl, Octylacetat, Octylazetat.
Spanish: Acetato de capril, Acetato de octil.
Italian: Acetato di caprile, Acetato di octile.

Perfume
Ingredient of—
Artificially prepared perfume preparations.
Perfume in—
Cosmetics.

Soap
Perfume in—
Toilet soaps.

Octyl Alcohol, Secondary
Synonyms: Methylhexylcarbinol, Normal secondary caprylic alcohol, Normal secondary octylic alcohol, Octonol-2, Octoic alcohol.
French: Alcool caprylique normal sécondaire, Alcool octylique sécondaire.
German: Sekundar normal caprylalkohol, Sucundair normal oktylalkohol.
Spanish: Metilhexilcarbinol.
Italian: Metilehexilecarbinole.

Chemical
Starting point in making—
Intermediates and other organic chemicals, pharmaceutical chemicals, synthetic aromatic chemicals.

Insecticide
Ingredient (German 237408) of—
Preparations used for the destruction of fungi and insects.

Soap
Ingredient (German 237408) of—
Disinfectant soaps and disinfectant liquors.

Octylcresol

Chemical
Starting point (Brit. 444351) in making—
Fat-splitting catalysts and emulsifying agents, useful in dyeing, laundering, bleaching, and various other purposes, by reacting with formaldehyde and nonaromatic secondary amines (the salts of the products with water-soluble acids, or water-insoluble acids, or the quaternary ammonium salts are claimed to be valuable for the purposes named).

Octyl Cyclopentenylacetate, Secondary

Food
Agent for—
Producing pineapple aroma and flavor.

Octylguanidin Chloride

Miscellaneous
As an emulsifying agent (Brit. 422461).
See under general heading: "Emulsifying agents."

Textile
Assistant (Brit. 421862) in—
Aqueous baths for treating textiles.
Promoter (Brit. 421862) of—
Uniform dyeing with basic dyestuffs.
Wetting and washing agent (Brit. 421862) in—
Textile processes.

Octyl Isoselenocyanate

Disinfectant
Paraciticide (U. S. 1993040).

Octyl Isotellurocyanate

Disinfectant
Paraciticide (U. S. 1993040).

Octyl Isothiocyanate

Disinfectant
Paraciticide (U. S. 1993040).

Octylphenol

Chemical
Starting point (Brit. 444351) in making—
Fat-splitting catalysts and emulsifying agents, useful in dyeing, laundering, bleaching, and various other purposes, by reacting with formaldehyde and nonaromatic secondary amines (the salts of the products with water-soluble acids, or water-insoluble acids, or the quaternary ammonium salts are claimed to be valuable for the purposes named)

Octyl Phthalate, Secondary
French: Phthalate d'octyle, Phthalate octylique.
German: Oktylphtalat, Phtalsäuresoktylester.

Cellulose Products
Plasticizer for—
Nitrocellulose.
For uses, see under general heading: "Plasticizers."

Octylresorcinol

Chemical
Starting point (Brit. 444351) in making—
Fat-splitting catalysts and emulsifying agents, useful in dyeing, laundering, bleaching, and various other purposes, by reacting with formaldehyde and non-aromatic secondary amines (the salts of the products with water-soluble acids, or water-insoluble acids, or the quaternary ammonium salts are claimed to be valuable for the purposes named)

Octyl Rhodanate Sodium Salt

Insecticide
Insecticide of high toxicity for use in sprays.

Octyl Selenocyanate

Disinfectant
Paraciticide (U. S. 1993040).

Octyl Tellurocyanate

Disinfectant
Paraciticide (U. S. 1993040).

Octyl Thiocyanacetate

Insecticide and Fungicide
Toxic agent (German 562672) in—
Kerosene-base flysprays.

Octyl Thiocyanate

Disinfectant
Paraciticide (U. S. 1993040).

Oenanthic Acid

Synonyms: Enanthic acid, Heptoic acid (normal), Heptylic acid, Oenanthylic acid.
French: Acide d'héptyle, Acide héptylique, Acide d'oenanthyle, Acide oenanthylique.
German: Oenanthsaeure, Oenanthylicsaeure.

Chemical
Catalyst in making—
Rubber vulcanization accelerator with heptaldehyde and orthotolyldiguanide or ethylamine (Brit. 249113).
Starting point in making—
Acetyl oenanthate, butyl oenanthate, barium oenanthate, calcium oenanthate, ethyl oenanthate, formyl oenanthate, heptyl oenanthate, isoamyl oenanthate, lactyl oenanthate, magnesium oenanthate, methyl oenanthate, methylhexylketone, octyl oenanthate, secondary, potassium oenanthate, phenyl oenanthate, propyl oenanthate, sodium oenanthate, strontium oenanthate, succinyl oenanthate, salicylyl oenanthate, strontium oenanthate, tolyl oenanthate, uranyl oenanthate, valerianyl oenanthate, xylyl oenanthate.

Oil Refinery Spent Clays

(Containing asphaltic, resinous, and polymerized bodies with no free oil; preferably containing 20 to 40 percent of petroleum products).

Construction
Addition agent (U. S. 1755638) to—
Clinker in making plastic waterproof cement of excellent quality as to strength.

Oil Shale

French: Schiste bitumineux.
German: Oelshiefer.

Chemical
Raw material in making—
Ammoniacal liquor, ammonium sulphate and other salts.

Gas
Starting point in making—
Burning and illuminating gas.

Fertilizer
Raw material for extraction of—
Potash.

Oil
Raw material in making—
Burning oils, lubricants, motor fuels, shale oil.

Paint and Varnish
Starting point in making—
Mineral pigments.

Oiticica Fatty Acids

Synonyms: Fatty acids of oiticica oil.

Miscellaneous
Ingredient of—
Polishes of various kinds, preparations containing waxes.

Paint and Varnish
Ingredient of—
Special coatings.
Plasticizer in—
Antifouling coatings for bottom of ships (claimed to be very effective).

Oleic Acid

Synonyms: Oleinic acid, Red oil.
French: Acide d'oléique.
German: Oeleinsaeure, Rotoel.

Chemical
Reagent in making—
Caprinic acid, caprylic acid, liparin, palmitic acid, vasogene.
Solvent in making—
Anthracene.
Starting point in making—
Oleates of alkaloids, alkalies, and metals.
Solubilizing agent for dyeing acetate rayon.

Construction
Ingredient of—
Emulsified asphaltic preparations used in the curing of concrete.
Asphaltic road-surfacing emulsions.

Dye
Ingredient of—
Color lakes.

Fats and Oils
Ingredient of—
Cutting oils, lubricating greases and oils, neatsfoot oil emulsions, olive oil emulsions, pine oil emulsions.
Reagent in refining.
Starting point in making—
Hardened oils, sulphonated oils, textile oils.
Thickener in making—
Viscous lubricants.

Fuel
Ingredient of—
Candles.

Ink
Ingredient of—
Carbon-paper inks, multi-tone printing inks, stamp-pad inks.

Insecticide
Ingredient of—
Insecticidal emulsions, tree-spraying emulsions.

Leather
Ingredient of—
Dressing compositions, emulsified tanning compositions containing neatsfoot oil.

Miscellaneuos
Ingredient of—
Cleansing compositions containing ethylene dichloride.
Cleansing compositions for use on woodwork.
Deodorizing preparations.
Dirt and grease removers (U. S. 1624055).
Emulsified polishes containing carnauba wax or other waxes and oils.
Metal cleansing compositions, metal polishes.
Mineral oil metal polishes in emulsified form.
Scrubbing compositions for rugs.
Spotting fluids, in emulsified form, containing ethylene dichloride or other solvents.
Shoe and leather polishes.

Paint and Varnish
Ingredient of—
Auto-top dressings, marine paints.
Reagent in making—
Driers.

Paper
Reagent in making—
Easy-bleaching pulp.

Perfume
Ingredient of—
Carriers for perfumes, cold creams, cosmetic creams, dentifrices, grease paints, hair tonics, lotions, mouthwashes, ointments, shampoos, skin lotions, soapless shaving creams.

Petroleum
Ingredient of—
Kerosene emulsions, mineral oil emulsions, paraffin emulsions, petrolatum emulsions.
Reagent in—
Separating crude petroleum from water.

Pharmaceutical
In compounding and dispensing practice.

Rubber
Accelerator in vulcanizing.
Reagent in making—
Rubber heels.

Sanitation
Ingredient of—
Disinfecting emulsions.

Soap
Ingredient of—
Cleansing and scouring preparations.
Starting point in making—
Antimony soaps for mothproofing, rayon soaps, silk soaps, textile soaps.

Textile
——, *Dyeing*
Ingredient of—
Dyeing assistants in emulsified form.

Oleic Acid (Continued)
——, *Finishing*
Ingredient of—
Finishing preparations.
Mixtures to produce scroop effect on cotton fabrics.
Various textile finishes.
Waterproofing agent in treating—
Various fabrics.

——, *Manufacturing*
Ingredient of—
Lubricating compositions containing ethanolamine.
Silk-degumming baths.
Silk-lubricating oils for spinning, weaving, and knitting.
Wool-lubricating oils for spinning, carbonizing, weaving, and knitting.
Oiling agent in treating—
Wool for spinning and weaving.

——, *Miscellaneous*
Ingredient of—
Scouring preparations, wetting-out agents.

——, *Printing*
Ingredient of—
Printing paste containing alizarin red and alizarin rose (Brit. 255148).

Waxes and Resins
Ingredient of—
Wax emulsions.

Oleic Acid Chloride

Chemical
Starting point (Brit. 407956) in making pour-point improvers for machine oils, gear oils, and other lubricants by condensing with—
Anilin, anthracene oil.
Aromatics obtained by destructive hydrogenation or by dehydrogenation.
Benzene.
Cracking gases containing gaseous olefins (ethylene, propylene, and butylene).
Cyclic terpenes, ethylnaphthalene, liquid olefins, middle oil, naphthalene, naphthols, naphthylamines, nitrated aromatics, phenols, tars, toluene, xylene.

Oleic Acid Ester of Oxyethylpyridinium Chloride

Textile
Reagent (Brit. 390553) for—
Increasing fastness to water of cellulosic materials dyed with substantive colors.

Oleic Amide

Miscellaneous
As an emulsifying agent (Brit. 343899).
For uses, see under general heading: "Emulsifying agents."

Oleic Anilide
French: Anilide oléique.
German: Oleinanilid.

Chemical
Starting point in making various derivatives.
Reagent in making—
Emulsions of various chemicals.

Fats and Oils
Reagent (Brit. 328675) in making emulsions of—
Fats, fatty acids, vegetable and animal oils.

Miscellaneous
Reagent in making—
Emulsions of various substances.

Petroleum
Reagent in making emulsions of—
Crude petroleum, petroleum distillates.

Resins and Waxes
Reagent (Brit. 329675) in making emulsions of—
Natural resins, synthetic resins, waxes.

Soap
Ingredient of—
Emulsified detergents.

Textile
——, *Bleaching*
Ingredient (Brit. 329675) of—
Bleaching baths.

——, *Finishing*
Ingredient (Brit. 329675) of—
Finishing and washing, as well as fulling, baths.

Oleic Cyclohexylamide
French: Cyclohexyleamide oléique.
German: Oleincyklohexylamid, Oleinzyklohexylamid.

Miscellaneous
As a dispersing agent (Brit. 328675).
For uses, see under general heading: "Emulsifying agents."

Oleic Diethylamide
French: Diéthyleamide oléique.
German: Oleindiaethylamid.

Chemical
Starting point (Brit. 341053) in making—
Derivatives used for emulsification and other purposes.

Fats and Oils
Reagent (Brit. 341053) in making—
Fat and oil dispersive agents.

Miscellaneous
Reagent (Brit. 341053) in making—
Dispersing and emulsifying agents, detergent preparations, colloidal sols of various kinds.

Textile
Reagent (Brit. 341053) in making—
Lubricating compositions, such as those used in weaving, knitting, winding, reeling, warping, and coning of yarns and fabrics.

Oleicdimethylamidesulphonic Acid

Miscellaneous
As an emulsifying agent (Brit. 341503).
For uses, see under general heading: "Emulsifying agents."

Oleicethylanilidesulphonic Acid

Miscellaneous
As an emulsifying agent (Brit. 341053).
For uses, see under general heading: "Emulsifying agents."

Oleicmethyl Ester Sulphuric Ester

Miscellaneous
As an emulsifying agent (Brit. 343524).
For uses, see under general heading: "Emulsifying agents."

Oleicoxyethylmorpholin

Miscellaneous
As an emulsifying agent (Brit. 364104).
For uses, see under general heading: "Emulsifying agents."

Oleic-sulphonic Methyl Ester
French: Ester de méthyle-oléique-sulphonique.
German: Oleinsulphonsäuremethylester.

Chemical
Starting point (Brit. 341053) in making dispersing agents with the aid of—
Allylanilin, allylbenzylamine, allylnaphthylamine, allylphenylamine, allyltolylamine, allylxylylamine, amylanilin, amylbenzylamine, amylnaphthylamine, amylphenylamine, amyltolylamine, amylxylylamine, benzylanilin, benzylnaphthylamine, benzylphenylamine, benzyltolylamine, benzylxylylamine, butylanilin, butylbenzylamine, butylnaphthylamine, butylphenylamine, butyltolylamine, butylxylylamine, diallylamine, diamylamine, dibenzylamine, dibutylamine, diethylamine, diheptylamine, dihexylamine, di-isoallylamine, di-isoamylamine, di-isobutylamine, di-isopropylamine, diphenylamine, dipropylamine, ethylanilin, ethylbenzylamine, ethylnaphthylamine, ethylphenylamine, ethyltolylamine, ethylxylylamine, methylanilin, methylbenzylamine, methylnaphthylamine, methylphenylamine, methyltolylamine, methylxylylamine, morpholin, piperidin, propylanilin, propylbenzylamine, propylnaphthylamine, propylphenylamine, propyltolylamine, propylxylylamine, secondary amines containing cyclohexyl, hexylcetyl, and other groups.

Oleic Toluide

Chemical
Starting point in making various derivatives.

Petroleum
Ingredient (U. S. 1853571) of—
Lubricating compositions containing mineral oils (added for the purpose of increasing the consistency of the lubricant and raising its melting point).

Olein
Synonyms: Elain, Oleine oil, Triolein.
French: Élaine.
German: Oelsaeureglycerid.
Chemical
Starting point in making—
Benzoic acid, formaldehyde-potash soap solution (lysoform), oleic acid.
Dye
Ingredient of—
Anilin dye compositions.
Fats and Oils
Ingredient (Brit. 266746) of—
Boring oils, candles, emulsifying compositions, lubricants, textile oil preparations, turkey red oil.
Ink
Ingredient of—
Printing inks.
Leather
Ingredient of—
Dressing, softening and finishing compositions (Brit. 266746).
Miscellaneous
Ingredient (Brit. 266746) of—
Cleansing compositions, metal polishes, stain-removing compositions, washing compositions, wetting compositions.
Perfumery
Ingredient of—
Cosmetics, pomades, perfume.
Petroleum
Ingredient (Brit. 266746) of—
Emulsions of petroleum and petroleum distillates.
Soap
Starting point in making various soaps.
Pharmaceutical
In compounding and dispensing practice.
Textile
——, *Dyeing and Printing*
Ingredient of—
Color compositions.
Tobacco
Reagent in the treatment of tobacco.
Waxes and Resins
Ingredient (Brit. 266746) of—
Emulsified wax and resin compositions.

Oleone
Miscellaneous
As an emulsifying agent (Brit. 343098).
For uses, see under general heading: "Emulsifying agents."

Oleyl Chloride
French: Chlorure d'oléyle, Chlorure oléylique.
German: Chloroleyl.
Fats and Oils
Ingredient of—
Bleaching preparations (used with hydrogen peroxide) (Brit. 328544).
Bleaching preparations (used with benzoyl chloride, chlorobenzoyl chloride, or bromobenzoyl chloride).
Food
Ingredient of—
Bleaching compositions containing hydrogen peroxide, used on flour, egg yolk, and meal (Brit. 328544).
Bleaching compositions containing benzoyl chloride, chlorobenzoyl chloride, or bromobenzoyl chloride, used on flour, milling products, animal and vegetable foodstuffs, oilseed meals.
Soap
Ingredient of—
Bleaching compositions containing hydrogen peroxide (Brit. 328544).
Bleaching compositions containing benzoyl chloride, bromobenzoyl chloride, or chlorobenzoyl chloride.
Waxes and Resins
Ingredient of—
Bleaching compositions containing hydrogen peroxide (Brit. 328544).
Bleaching compositions containing benzoyl chloride, bromobenzoyl chloride, or chlorobenzoyl chloride.

Oleyldiethylethylenediamine
German: Oleyldiaethylaethylendiamin.
Fats and Oils
Ingredient (Brit. 328675) of—
Fat emulsions, oil emulsions.
Miscellaneous
Ingredient (Brit. 328675) of various emulsions.
Petroleum
Ingredient (Brit. 328675) of—
Emulsions containing various petroleum distillates.
Textile
Ingredient (Brit. 328675) of—
Bleaching compositions, finishing compositions, fulling compositions, washing compositions.
Waxes and Resins
Ingredient (Brit. 328675) of—
Resin emulsions, wax emulsions.

Oleyldiethylethylenediamine Citrate
Miscellaneous
As an emulsifying agent (Brit. 361860).
For uses, see under general heading: "Emulsifying agents."

Oleyldiethylethylenediamine Hydrochloride
Miscellaneous
As an emulsifying agent (Brit. 361860).
For uses, see under general heading: "Emulsifying agents."

Oleylhydroquinone
Petroleum
Stabilizing agent (Brit. 406195) for—
Cracked gasolines and other motor fuels.

Oleylhydroxyethanesulphonic Acid
French: Acide oléylehydroxyéthanesulphonique.
Disinfectant
Starting point (French 753149) in making—
Salts useful as deodorants or disinfectants.
Salts useful as deodorants or deodorants in combination with phenol pentachlorophenol, dichloroxylenol, ichthyol, menthol, sulphur.
Sanitation
As a disinfectant or deodorant (French 753149).

Oleylhydroxymethanesulphonic Acid
French: Acide oléylehydroxyméthanesulphonique.
Disinfectant
Starting point (French 753149) in making—
Salts useful as deodorants or disinfectants.
Salts useful as deodorants or disinfectants, in combination with phenol, pentachlorophenol, dichloroxylenol, ichthyol, menthol, sulphur.
Sanitation
As a disinfectant or deodorant (French 753149).

Oleylphloroglucinol
Petroleum
Stabilizing agent (Brit. 406195) for—
Cracked gasolines and other motor fuels.

Oleylpyrocatechol
Petroleum
Stabilizing agent (Brit. 406195) for—
Cracked gasolines and other motor fuels.

Oleylpyrogallol
Petroleum
Stabilizing agent (Brit. 406195) for—
Cracked gasolines and other motor fuels.

Oleylresorcinol
Petroleum
Stabilizing agent (Brit. 406195) for—
Cracked gasolines and other motor fuels.

Oleyl-1-sulphuric Acid (Normal) Ester
Chemical
As an emulsifying agent.
Reagent in—
Organic syntheses.
Starting point (Brit. 440575) in making—
Emulsifying agents with salts of lead, aluminum, iron, tin, or barium (such emulsifying agents are said to form water-in-oil emulsions and are, preferably, produced in situ by (1) dissolving the sulphuric acid ester in the oil, and (2) agitating with an aqueous solution of the metal salts, for example, lead acetate; they are said to be useful for treating medicinal

Oleyl-1-sulphuric Acid (Normal) Ester (Continued)
paraffin oil, neatsfoot oil, olive oil, castor oil, cottonseed oil, linseed oil, and petroleum lubricating oils; a heavy paraffin oil, so-treated on the basis of 50 parts by weight of oil to 48.75 parts of water, is said to yield a heavy grease that has good lubricating properties and may readily be extended with oil; a water-linseed oil type emulsion is offered as suitable for use as a paint base).

Orangeflower Oil, Bitter
Synonyms: Neroli oil.
Latin: Oleum aurantii florum, Oleum naphae.
French: Huile de fleurs d'orange amère, Huile de néroli.
German: Bittere pomeranzenbluethenoel, Bittere pomeranzenblumenoel, Nerolioel.

Food
Flavoring agent in—
Beverages, candies, foods.
Ingredient of—
Flavoring preparations.

Miscellaneous
Flavoring and perfuming agent for various purposes.

Insecticide
Ingredient of—
Insecticidal preparations (Brit. 272543).

Perfume
Ingredient of—
Perfumes, toilet waters.
Perfume in—
Cosmetics, dentifrices.

Petroleum
Reagent in treating—
Petroleum products (used for improving their odor).

Pharmaceutical
In compounding and dispensing practice.

Soap
Perfume in—
Special detergent preparations, toilet soaps.

Orangeflower Oil, Sweet
Synonyms: Portugal neroli oil, Portugal orangeflower oil, Sweet orangeflower oil.
Latin: Oleum aurantii florum.
French: Huile de fleurs d'orange douce, Huile de néroli de Portugal.
German: Nerolioel (Portugal), Suesse pomeranzenbluethenoel, Suesse pomeranzblumenoel.

Food
Flavoring agent in—
Beverages, candies, foods.
Ingredient of—
Flavoring preparations.

Insecticide
Ingredient of—
Insecticidal compositions (Brit. 272543).

Miscellaneous
Flavoring and perfuming agent for various purposes.

Perfume
Ingredient of—
Perfumes, toilet waters.
Perfume in—
Cosmetics, dentifrices.

Petroleum
Reagent in treating—
Petroleum products (used for improving their odor).

Pharmaceutical
In compounding and dispensing practice.

Soap
Perfume in—
Toilet soaps.

Orange Flower Water
French: Eau des fleurs d'orange.
German: Orangenbluettenwasser.

Food
Flavoring agent for—
Beverages, confectionery, desserts.

Perfumery
As an odorous ingredient.

Pharmaceutical
In compounding and dispensing practice.

Orange Mineral
Synonyms: Orange red, Sandix.
Note: This is an oxide of lead corresponding to the same formula as red lead; but, it differs from red lead in color and in some of its properties and is made by a different process. It is valued for its beautiful bright, uniform orange-red color.

Chemical
Base in making—
Eosin lake.

Ink
Base in making—
Eosin lake.

Paper
Base in making—
Eosin lake.

Paint and Varnish
Base in making—
Eosin lake.
Pigment.

Textile
Base in making—
Eosin lake.

Orange Oil, Bitter
Synonyms: Oil of bitter orange peel.
Latin: Oleum aurantii amari.
French: Huile d'écorce d'orange amère, Huile d'orange amère.
German: Bittere pomeranzeoel, Bittere pomeranzschaleoel.

Food
Flavoring agent in—
Beverages, candies, foods.
Ingredient of—
Flavoring preparations.

Miscellaneous
Flavoring agent and perfuming agent for various purposes.

Perfume
Ingredient of—
Perfumes, toilet waters.
Perfuming agent in—
Cosmetics, dentifrices.

Pharmaceutical
In compounding and dispensing practice.

Soap
Perfuming agent in making—
Special detergent preparations, toilet soaps.

Orange Oil, Sweet
Synonyms: Orange peel oil, Portugal oil.
Latin: Oleum aurantii.
French: Huile d'écorce d'orange douce, Huile d'orange douce, Huile de Portugal.
German: Apfelsineoel, Apfelsineschaleoel, Pomeranzoel, Suesse pomeranzoel, Suesse pomeranzschaleoel.

Fats and Oils
Ingredient of—
Artificial banana oil.

Food
Flavoring agent in—
Beverages, candies, foods.
Ingredient of—
Flavoring preparations.

Miscellaneous
Perfume and flavoring agent for various purposes.

Perfume
Ingredient of—
Perfumes, toilet waters.
Perfuming agent in—
Cosmetics, dentifrices.

Pharmaceutical
In compounding and dispensing practice.

Soap
Perfume in making—
Special detergent preparations, toilet soaps.

Oregon Fir Balsam

Ceramics
As an adhesive in the application of decorations.
As a varnish.

Miscellaneous
Cement for—
Glassware, lenses and other optical equipment, porcelains, special purposes.

Oregon Fir Balsam (Continued)
Mounting medium for—
Histological specimens.
Substitute in various uses for—
Canada balsam, Venice turpentine.
Paint and Varnish
Ingredient of—
Fine varnishes.
Pharmaceutical
In compounding and dispensing practice.

Orris Root
Latin: Radix ireos, Rhizoma iridis.
French: Iris de Florence.
German: Iriswurzel, Florentinische violenwurzel, Veilchenwurzel.
Spanish: Lirio florentino.
Italian: Ireos.
Fats and Oils
Starting point in extraction of—
Orris oil.
Perfumery
Ingredient of—
Cosmetics, dentifrices, perfumes, sachets.
Pharmaceutical
In compounding and dispensing practice.
Tobacco
Ingredient of—
Snuff.

Orthoacetotoluide
Chemical
Starting point in making—
Aminobenzoic acid, intermediates, pharmaceuticals.
Pharmaceutical
Suggested for use as analgesic, antipyretic, sedative, antiseptic.

Orthoaldehydophenoxyacetic Acid
French: Acide d'orthoaldéhydophénoxyacétique.
German: Orthoaldehydophenoxyessigsaeure.
Chemical
Starting point in making—
Coumarone.

Orthoaminoanthraquinonethiohydrin
German: Orthoaminoanthrachinonthiohydrin.
Dye
Starting point in making—
Dyestuffs for cellulose acetate rayon (Brit. 263179).

Orthoaminoazotoluene
Synonyms: Orthoaminoazotoluol.
Chemical
Starting point in making—
Intermediates, iodine derivative (azodalen), diacetyl derivative (pelidol), monoacetyl derivative (azodemin), surhodin, synthetic pharmaceuticals.
Dye
Starting point in making—
Acidol cloth red, cloth red 3GA, cloth red 3B extra, cloth red B, cloth red G, cloth red G extra, crocein 3B, fast yellow R, safranin, safranin T extra, spirit yellow R, sudan IV, wool red B.
Pharmaceutical
In compounding and dispensing practice.
Textile
Yellow dye for—
Fabrics and yarns.

Orthoaminobenzaldehyde
Chemical
Starting point in making—
Organic arsinic compounds.

Orthoamino-4-chlorophenylmercaptan Hydrochloride
Synonyms: 2-Amino-4-chlorothiophenol hydrochloride.
Insecticide and Fungicide
Larvicide for—
Culicine mosquito larvae.

Orthoaminodiphenyl Ether
Dye
Starting point (Brit. 248946) in making azo dyestuffs with di-2:3-oxynaphthoyl derivatives of—
Meta-m'-diaminoazoxybenzene.
Meta-m'-diaminopara-p'-dimethoxyazobenzene.
Meta-m'-diaminopara-p'-dimethoxyazoxybenzene.
Meta-m'-diaminopara-p'-dimethylazoxybenzene.
Para-p'-diaminoazobenzene.
Para-p'-diaminoazoxybenzene.

Orthoaminophenylmercaptan Hydrochloride
Synonyms: Orthoaminothiophenol hydrochloride.
Insecticide and Fungicide
Larvicide for—
Culicine mosquito larvae.

Orthoaminosalicylic Acid
French: Acide d'orthoaminosalicylique.
German: Orthoaminosalicylsäure.
Chemical
Starting point in making—
Esters and salts, intermediates, pharmaceuticals.
Dye
Starting point in making—
Azo colors.
Starting point (Brit. 325485) in making dyestuffs with the aid of—
4-Acetylaminophenol, 2:4-dimethylphenol, hydroquinonemonomethyl ether, parachlorometacresol, paracresol, parahydroxydiphenylmethane.

Orthoanisidin-4-sulphonamide
Dye
Intermediate in—
Dye syntheses.
Starting point (Brit. 425839) in making—
Water-insoluble azo dyes for use as red pigments for rubber, by coupling with orthoanisidide.

Orthobenzoylbenzoic Acid
Chemical
Starting point in—
Organic synthesis.
Rubber
Retardant (Brit. 426649) of—
Vulcanization of rubber mixes containing sulphur and an accelerator, in the initial stages.

Orthobenzylmethylaminophenol
Petroleum
Gum inhibitor (U. S. 1980200 and 1980201) in—
Motor fuels.

Orthobeta-p'-toluenesulphonylethylaminothiophenyl-betaparatoluenesulphonylethyl Ether
Chemical
Intermediate (Brit. 444262 and 444501) in—
Organic syntheses.
Insecticide
Insecticide (Brit. 444262 and 444501) for—
Animal pests, vegetable pests.
Textile
As a dyestuff (when employing suitable initial materials) (Brit. 444262 and 444501).
Assistant (Brit. 444262 and 444501) in—
Textile processing.

Orthobromoanilin
Chemical
Starting point in making—
2:3-Oxynaphtholic acid derivatives.
Diamino compounds.
Parabromodiazonium chloride.

Orthobromomethylcyclohexane
German: Orthobrommethylcyclohexan.
Chemical
Starting point in making—
Methylcyclohexylanilin (Brit. 261764).

Orthochlorobenzoylbenzene
German: Orthochlorbenzoylbenzol.
Chemical
Starting point in making—
Fluorone (Brit. 263163).

1-Orthochlorobenzoyl-2:6-dimethylnaphthalene
German: Alphaorthochlorbenzoyl-2:6-dimethylnaphtalin.
Chemical
Starting point in making—
4-Benzyl-2-dimethylbenzanthrone (Brit. 263163).

1-Orthochlorobenzoyl-2-methylnaphthalene
German: Alphaorthochlorbenzoyl-2-methylnaphtalin.
Chemical
Starting point in making—
4-Methylbenzanthrone (Brit. 263163).

Orthochlorobenzoylnaphthalene
German: Orthochlorbenzoylnaphtalin.
Chemical
Starting point in making—
Benzanthrone (Brit. 263163).

Orthochlorobenzylidenemalonic Acid
French: Acide d'orthochlorobenzylidènemalonyle, Acide orthochlorobenzylidènemalonique.
German: Orthochlorbenzylidinmalonsaeure.
Chemical
Starting point in making—
Orthochlorocinnamic acid.

Orthochloroparadiethylaminobenzaldehyde
Dye
Starting point (Brit. 431652) in making—
Orange dyestuffs with 1-metasulphophenyl-3-methyl-5-pyrazolone.

Orthochloroparanitranilin
German: 2-Chlor-4-nitroanilin.
Dye
Intermediate in—
Dye manufacture.

Orthochlorophenol
French: Orthochlorophénol.
German: Orthochlorphenol.
Fungicide
Starting point (French 688209) in making—
Seed disinfectants by condensation with phenylated mercuric hydroxide.
Tree preservatives by condensation with phenylated mercuric hydroxide.
Wood preservatives by condensation with phenylated mercuric hydroxide.

Orthochlorophenolindophenol
Analysis
Indicator in—
Oxidation-reduction potential determinations (of particular interest to biologists and physiologists and investigations of various materials, such as soils, wines, cheese, gasoline antiknock compounds).

1-Orthochlorophenyl-3-methyl-5-pyrazolone
Textile
Starting point (Brit. 396893) in—
Producing reddish-yellow shades in dyeing acetate rayon.

Orthocoumaric Acid
French: Acide coumarique, ortho; Acide de coumaryle, ortho.
German: Cumarinsaeure.
Chemical
Starting point in making—
Coumarin (German 440341).

Orthocresotinic Acid
Dye
Intermediate in—
Dye manufacture.

Orthocresyl Benzoate
French: Benzoate d'orthocrésyle, Benzoate orthocrésylique.
German: Benzosaeureorthocresylester, Benzosaeuresorthocresyl.
Electrical
Dispersive agent (Brit. 273290) in making—
Insulating enamels and lacquers for electric wires.
Miscellaneous
Dispersive agent in making—
Cements for laminated mica.
See also "Emulsifying agents."
Paint and Varnish
Dispersive agent in making—
Varnish bases.
Plastics
Dispersive agent in making—
Moldable compositions.
Resins and Waxes
Dispersive agent in making—
Synthetic resins.
Solvent (Brit. 273748) in making artificial resins of—
Phenol-aldehyde type, polyhydric alcohol-polybasic acid type, urea-aldehyde type.

Orthocyanocinnamic Acid
French: Acide cyanocinnamique [ortho].
German: Ortho-cyanzimtsaeure.
Chemical
Starting point in making—
Betaphenylbetahydroxypropinorthocarboxylic anhydride.
Orthoaminocinnamic acid (German 440052).

Orthodianisidin
Chemical
Starting point in making—
Orthodianisidindisulphonic acid.
Dye
Starting point in making—
Azo violet, azidin blue BA, azidin pure blue FA, azidin wool blue B, azophor blue D, azophor black S, benzoazurin G, benzoazurin 3G, benzocyanin B, benzocyanin 3B, benzopurpurin, benzopurpurin 10B, benzo fast blue, benzo pure blue 9B, benzo sky blue 4B, brilliant azurin 5G, Chicago blue B, Chicago blue 4B, Chicago blue 6B, Chicago blue G, Chicago blue RW, chlorazol blue 3G, Columbia black B, congo blue 2B, congo fast blue B, cotton red 10B, diamine blue RW, diamine brilliant blue G, diamine pure blue, dianisidin blue, dianil blue G, diazo colors, diazamine B, diazurin B, direct black B, direct blue B, direct violet BB, heliotrope B, heliotrope 2B, indazurin B, indazurin BB, indazurin GM, indazurin 5GM, oxamine black RR, oxamine blue B, sky blue, trisulphon blue B, trisulphon brown GG.
Starting point (Brit. 285504) in making nitro dyestuffs with—
Alphachloro-2:6-dinitrobenzene-4-sulphonic acid.
Alphachloro-2:4-dinitrobenzene-6-sulphonic acid.
Alphachloro-2-nitrobenzene-4-sulphonic acid.
Potassium alphachloro-2:6-dinitrobenzene-4-sulphonate.
Potassium alphachloro-2:4-dinitrobenzene-6-sulphonate.
Potassium alphachloro-2-nitrobenzene-4-sulphonate.
Textile
——, *Dyeing and Printing*
Reagent in producing—
Azo colors on fabrics.

Orthodiazocinnamic Acid
French: Acide d'orthodiazocinnamyle, Acide orthodiazocinnamique.
German: Orthodiazozimtsaeure.
Chemical
Starting point in making—
Orthochlorocinnamic acid.

Orthodibenzylaminophenol
Petroleum
Gum inhibitor (U. S. 1980200 and 1980201) in—
Motor fuels.

Orthodichlorobenzene
German: Orthodichlorbenzol.
Chemical
Reagent in—
Freeing hydrochloric acid from arsenic.
Starting point in making various organic chemicals.
Dye
Diluent (German 439467) in making—
Condensation products of benzanthrones.
Miscellaneous
Cleansing and polishing agent for—
Brass.
Solvent for various purposes.

Orthodichlorobenzene Sulphonamide
Insecticide and Fungicide
Suggested for use as—
Fungicide, pesticide.
Miscellaneous
Suggested for use as—
Bleaching agent.
Textile
Suggested for use as—
Delustering agent, treatment assistant.

Orthodichlorobenzene Sulphonchloride
Miscellaneous
Suggested for use as—
Bleaching agent.

Orthodichlorobenzene Sulphondichloramide
Disinfectant
Suggested for use as—
Oil-soluble disinfectant.

Orthodichlorobenzene Sulphonsodiochloramide
Disinfectant
Suggested for use as—
Disinfectant.
Miscellaneous
Suggested for use as—
Bleaching agent.

Orthodinitrobenzene
Synonyms: Orthodinitrobenzol.
German: Orthodinitriertbenzol.
Chemical
Reagent in—
Organic synthesis.
Starting point in making—
Intermediates, synthetic perfumes, synthetic pharmaceuticals.
Dye
Starting point in making various synthetic dyestuffs.
Perfume
Starting point in making—

Orthoethylidene-Cyclohexanone
German: Ortho-aethylidincyclohexanon.
Chemical
Starting point in making intermediates for perfumes (Brit. 264830).
Perfumery
Ingredient (Brit. 264830) of—
Cosmetics, perfumes.

Orthohydroxydiphenylmethane
Glue and Gelatin
Preservative (Brit. 396737) for—
Glue and gelatin to prevent attack by micro-organisms.

Orthohydroxyquinolin Sulphate
Chemical
Reagent in—
Organic synthesis.
Pharmaceutical
Ingredient (U. S. 2010512) of—
Antiseptic, consisting of admixture in equal parts with the sodium salt of diphenyldisazo-orthoethoxyaminophenolorthoaminobenzoic acid.

Orthohydroxytriphenylmethane
Chemical
Starting point in making—
Intermediates, pharmaceuticals.
Dye
Starting point in making various synthetic dyestuffs.

Orthomethylaminophenol
Chemical
As an intermediate.
Stabilizing agent (Brit. 397914) for—
Chlorinated hydrocarbons.

Orthomethylcyclohexanol Adipate
Synonyms: Methylhexalin adipate, Methylhexalin adipinate, Orthomethylcyclohexanol adipin ester, Orthomethylcyclohexyl adipate, Orthomethylcyclohexyl adipinate.
French: Adipate de méthylehexaline, Adipate méthylehexalinique, Adipate de orthométhylecyclohexyle, Adipate orthométhylecyclohexylique.
German: Adipinsäureorthomethylcyklohexylester, Adipinsäureorthomethylzyklohexylester, Adipinsäuresorthomethylcyclohexyl, Adipinsäuresorthomethylzyklohexyl, Orthomethylcyklohexyladipat, Orthomethylzyklohexyladipat.
Spanish: Adipato de orthometilciclohexil.
Italian: Adipato di orthometilcicloessile.

Cellulose Products
Plasticizer (German 406013) for—
Cellulose acetate, cellulose esters and ethers, cellulose nitrate.
For uses, see under general heading: "Plasticizers."

Orthomethylethylbenzene
French: Benzène orthométhyle et éthyle, Benzène orthométhylique et éthylique.
German: Orthomethylaethylbenzol.
Chemical
Starting point in making—
Aromatics, intermediates, pharmaceuticals.
Dye
Starting point in making various synthetic dyestuffs.
Textile
——, *Dyeing and Printing*
Solvent (Brit. 269960) in making—
Dye liquors and printing pastes for use on acetate rayon and materials containing it.
——, *Finishing*
Solvent (Brit. 269960) in making—
Stenciling preparations for use on acetate rayon and materials containing it.

Orthomonobenzylaminobenzyl-w-sulphonic Acid
Dye
Intermediate (Brit. 447067) in making—
Dyes containing one or more aryl residues carrying one or more alkylsulphonic groups directly combined to the nucleus.

Orthonitrobenzidin
French: Orthonitrobenzidine.
Dye
Starting point in making—
Anthracene red.

Orthonitrodiphenyl Ether
Plastics
Ingredient (Brit. 398091) of—
Dinitrotoluene-dinitrobenzene solvent mixture used for dissolving a polymerized vinyl halide to produce a resilient, rubber-like gel.

Orthonitrometatoluyl Chloride
French: Chloro-orthonitrométatoluyle, Chlorure d'orthonitrométatoluyle, Chlorure orthonitrométatoluylique.
German: Chlornitrometatoluyl, Orthonitrometatoluylchlorid.
Chemical
Reagent (Brit. 278037) in making synthetic drugs with—
Alkoxynaphthylamine-sulphonic acids, alphanaphthylamine-4:8-disulphonic acid, alphanaphthylamine-3:6:8-trisulphonic acid, alphanaphthylamine-4:6:8-trisulphonic acid, 4-aminoacenaphthene-3:5-disulphonic acid, 4-aminoacenaphthene-3-sulphonic acid, 4-aminoacenaphthene-5-sulphonic acid, 4-aminoacenaphthenetrisulphonic acids, 2:8-aminonaphthol-3:6-disulphonic acid, 1:8-aminonaphthol-3:6-disulphonic acid, 1:5-aminonaphthol-7-sulphonic acid, bromonaphthylaminesulphonic acids, chloronaphthylaminesulphonic acids, iodonaphthylaminesulphonic acids.

Orthonitroparatoluyl Chloride
French: Chlorure d'orthonitroparatoluyle, Chlorure d'orthonitroparatoluylique.
German: Chlororthonitroparatoluyl, Orthonitroparatoluylchlorid.
Chemical
Reagent (Brit. 278037) in making synthetic drugs with—
Alkoxynaphthylamine sulphonic acids, 4-aminoacenaphthene-3:5-disulphonic acid, 4-aminoacenaphthene-3-sulphonic acid, 4-aminoacenaphthene-5-sulphonic acid, 4-aminoacenaphthenetrisulphonic acid, 1:8-aminonaphthol-3:6-disulphonic acid, 2:8-aminonaphthol-3:6-disulphonic acid, 1:5-aminonaphthol-7-sulphonic acid, bromonaphthylaminesulphonic acids, chloronaphthylaminesulphonic acids, iodonaphthylaminesulphonic acids, 1-naphthylamine-4:8-disulphonic acid, 1-naphthylamine-3:6:8-trisulphonic acid, 1-naphthylamine-4:6:8-trisulphonic acid.

Orthonitrophenyl-1-Benzothiazylselenosulphide
Rubber
Nonscorching accelerator (Brit. 441653) in—
Vulcanizing.

Ortho-oxyquinolin
Synonyms: Carbostyril, Oxy-8-quinolin, Quinophenol.
German: Ortho-oxychinolin, Oxychinolin.

Chemical
Starting point in making—
Ethoxyanabenzoylaminoquinolin (analgen).
Ortho-oxyquinolin sulphate (quinosal).
Ortho-oxyquinolinmetasulphonic acid (quinaseptol).
Oxyquinaseptol (diaphtherin).

Ortho-oxyquinolin Sulphate
Synonyms: Sunoxol.
French: Sulfate d'orthooxyquinoléine.
German: Orthohydroxychinolinsulfat, Schwefelsaeures-orthooxychinolin.

Food
Preservative in—
Food products, candies.

Pharmaceutical
In compounding and dispensing practice.
Preservative for serums.

Orthophenetoleazoalphanaphthylamine

Dye
Starting point (Brit. 248946) in making azo dyestuffs with di-2:3-oxynaphthoyl derivatives of—
Meta-m'-diaminoazoxybenzene, meta-m'-diamino-para-p'-dimethylazoxybenzene, meta-m'-diamino-para-p'-dimethoxyazobenzene, meta-m'-diamino-para-p'-dimethoxyazoxybenzene, para-p'-diaminoazobenzene, para-p'-diaminoazoxybenzene.

Orthosulphomethyl-normal-phenyltaurin

Dye
Intermediate (Brit. 447067) in making—
Dyes containing one or more aryl residues carrying one or more alkylsulphonic groups directly combined to the nucleus.

Orthosulphonbenzoic Acid

Dye
Intermediate in making—
Dyestuffs, sulphone phthalein indicators.

Pharmaceutical
Starting point (U. S. 1863268) in making—
Hydroxymercuri-derivatives of resorcinol iodinated sulphonphthaleins, suggested for use as antiseptics and germicides.

Orthosulphoparadiethylaminobenzaldehyde

Chemical
Starting point in—
Organic synthesis.

Dye
Starting point (Brit. 481652) in making—
Acid wool dyestuffs (yellow) by condensing with malonic dinitrile in the presence of piperidin.
Acid wool dyestuffs (red) by condensing with hydroxythionaphthene.
Acid wool dyestuffs (orange) by condensing with 5-methylbeta-cumaronone.

Orthotoluidinmethyleneorthocresotinic Acid
French: Acide d'orthotoluidineméthylèneorthocrésotinique.
German: Orthotoluidinmethylenorthokresotinsaeure.

Dye
Starting point (Brit. 256203) in making azo dyestuffs with—
Acetyl H acid, alphanaphthol-3:6-disulphonic acid, alphanaphthol-3:8-disulphonic acid.

Orthotoluidin-4-sulphonanilide

Dye
Intermediate in—
Dye syntheses.
Starting point (Brit. 425839) in making—
Water-insoluble azo dyes for use as red pigments for rubber, by coupling with orthoanisidide.

Orthotoluquinone Chloroimide
French: Chloroimide d'orthotoluquinone.
German: Orthotoluchinonchlorimid.

Agriculture
Disinfectant (Brit. 340500) in treating—
Various seed grains.

Chemical
Starting point in making various derivatives.

Orthotolyldiguanide

Chemical
Starting point in making—
Rubber vulcanization accelerator with heptaldehyde and oenanthic acid (Brit. 249113).

Orthotolyl-Mustard Oil

Chemical
Starting point (U. S. 1730536) in making—
Paradimethylaminophenylorthotolylguanidin from para-aminodimethylanilin.

Orthotolylphenyl Ketone

Chemical
Starting point in—
Organic synthesis.

Rubber
Retardant (Brit. 426649) of—
Vulcanization of rubber mixes containing sulphur and an accelerator, in the initial stages.

Orthotritolylphosphin Oxide

Chemical
Starting point (Brit. 326137) in making pharmaceuticals and mothproofing and insect-exterminating compounds with the aid of—
Alphahydroxyphenyl-3:4-dicarboxylic acid dibutyl ester,
Alphamethyl-3-hydroxy-6-isopropylbenzene.
Alphamethyl-3-hydroxy-4-isopropyl-6-chlorobenzene.
Alphanaphthol, 4-benzylphenol, betanaphthol, 6-chloro-2-cresol, 3-chloro-4-cresol, 2:6-dichlorophenol, 2:4-dichlorophenol, 2-isobutyl-4-chlorophenol, metacresol, metahydroxydiethylanilin, 4-normal-butylphenol, orthochlorophenol, orthocresol, parachlorophenol, paracresol, parahydroxybenzoic acid ethyl ester, parahydroxybenzaldehyde, paranitrophenol, phenol, pyrocatechin monoethyl ether, resorcinol, symmetrical xylenol, 2:4:6-trichlorophenol, ar-tetrahydrobetanaphthol, thymol.

Osmic Oxide
Synonyms: Osmic acid, Osmic anhydride, Osmic tetroxide, Perosmic anhydride, Perosmic oxide.
Latin: Acidicum osmicum.
French: Acide osmique, Anhydride osmique, Peroxyde osmique, Peroxyde d'osmium.
German: Osmiumsäure, Osmiumtetroxyd, Ueberosmiumsäureanhydrid.
Spanish: Acido osmico, Anhidrato de osmico, Peroxido osmico, Superoxido osmico.
Italian: Acido osmico, Anidrato osmico.

Analysis
Reagent in—
Analyzing adrenalin.
Making microchemical analyses and tests for fatty and nerve substances.
Testing for idican in urine.

Chemical
Ingredient of catalytic mixtures used in the manufacture of—
Acenaphthylene, acenaphthaquinone, bisacenaphthylidenedione, naphthaldehydic acid, naphthalic anhydride, and hemimellitic acid from acenaphthene (Brit. 295270).
Acetaldehyde from ethyl alcohol (Brit. 281307).
Acetic acid from ethyl alcohol (Brit. 281307).
Alcohols from aliphatic hydrocarbons (Brit. 281307).
Aldehydes and acids by the reductions of the corresponding esters (Brit. 306471).
Aldehydes and acids from toluene, orthochlorotoluene, orthobromotoluene, orthonitrotoluene, parachlorotoluene, parabromotoluene, paranitrotoluene, metachlorotoluene, metanitrotoluene, metabromotoluene, dichlorotoluene, dibromotoluenes, dinitrotoluenes, chlorobromotoluenes, chloronitrotoluenes, bromonitrotoluenes (Brit. 295270).
Aldehydes and acids from xylenes, pseudocumenes, mesitylene, and paracymene (Brit. 281307).
Alphanaphthaquinone from naphthalene (Brit. 295270).
Anthraquinone from naphthalene (Brit. 295270).
Benzaldehyde and benzoic acid from toluene (Brit. 281307).
Benzaquinone from phenanthraquinone (Brit. 281307).
Benzyl alcohol from benzaldehyde by reduction (Brit. 306471).
Benzyl alcohol, benzaldehyde, or benzyl phthalide by the reduction of phthalic anhydride (Brit. 306471).
Butyl alcohol by the reduction of crotonaldehyde (Brit. 306471).

Osmic Oxide (Continued)
Chloroacetic acid from ethylenechlorohydrin (Brit. 295270).
Diphenic acid from ethyl alcohol (Brit. 295270).
Ethyl alcohol by the reduction of acetaldehyde (Brit. 306471).
Fluorenone from fluorene (Brit. 295270).
Formaldehyde by the reduction of carbon dioxide or carbon monoxide (Brit. 306471).
Formaldehyde by the reduction of methane or methanol (Brit. 306471).
Hydroxyl compounds by the reduction of anthraquinone, benzoquinone, and similar compounds (Brit. 306471).
Isopropyl alcohol by the reduction of acetone (Brit. 306471).
Maleic acid and fumaric acid by the oxidation of toluene, benzene, phenols, tar phenols, or furfural, or from benzoquinone or phthalic anhydride (Brit. 306471).
Methane by the reduction of carbon dioxide or carbon monoxide (Brit. 306471).
Methanol by the reduction of carbon dioxide or carbon monoxide (Brit. 306471).
Naphthaldehydic acid, acenaphthaquinone, or bisacenaphthylidenedione from acenaphthylene (Brit. 281307).
Phenanthraquinone from phenanthrene or diphenic acid (Brit. 295270).
Phthalic acid and maleic acid from naphthalene (Brit. 295270).
Primary alcohols by the reduction of the corresponding aldehydes (Brit. 306471).
Propionic acid and butyric acid and higher alcohols, ketones, and acids by the reduction of carbon dioxide or carbon monoxide (Brit. 306471).
Reduction products of ketones, aldehydes, acids, esters, alcohols, ethers, and other organic compounds which contain oxygen (Brit. 306471).
Salicylic acid and salicylic aldehyde from cresol (Brit. 295270).
Secondary butyl alcohol by the reduction of methylethyl ketone (Brit. 306471).
Valeryl alcohol by the reduction of valeraldehyde (Brit. 306471).
Vanillin and vanillic acid by the oxidation of eugenol or isoeugenol (Brit. 295270).
Ingredient (Brit. 306460) of catalytic preparations which are used in the production of various aromatic and aliphatic compounds, including—
Alphanaphthylamine from alphanitronaphthalene.
Amines from aliphatic nitro compounds, such as allyl nitriles or nitromethane.
Amino compounds from the corresponding nitrophenols.
3-Aminopyridin from 3-nitropyridin.
Amylamine from pyridin.
Cyclohexamine, dicyclohexamine, and cyclohexylanilin from nitrobenzene.
Anilin, azo-oxybenzene, azobenzene, and hydrazobenzene from benzene by reduction.
Piperidin from pyridin, pyrrolidin from pyrrol, tetrahydroquinolin from quinolin.
Starting point in making—
Osmium salts.

Gas
Reagent in making—
Gas mantles.

Miscellaneous
Fixative in technical histology and microscopy.

Pharmaceutical
Suggested as a caustic in medicine and also for the treatment of neuralgia and epilepsy.

Photographic
Reagent in various processes.

Oxalic Acid
Latin: Acidum oxalicum.
French: Acide carboneux, Acide oxalique.
German: Kleesäure.
Italian: Acido ossalico.

Analysis
Used as—
Dehydrating agent in condensations, microchemical reagent, reagent in various processes, reducing agent, solvent.

Brewing
Cleansing agent for—
Brewery equipment.

Chemical
Acidifying agent for—
Hydrogen peroxide to prevent alkaline reaction in purified product.
Catalyst in making—
Alphanaphthylamine-4-sulphonic acid.
Borneol and isoborneol fatty acid esters (Brit. 250255).
Manganous nitrate.
Precipitating agent for—
Rare earth oxides.
Purification agent in making—
Aluminum sulphate, pure, bleaching powders (German 567725), cream of tartar, glycerin, synthetic acetic acid, tartaric acid.
Reagent in making—
Alpha-aminoanthraquinone-3-sulphonic acid, alphanaphthofluoran, allyl alcohol, amyl formate, benzoic anhydride, caprylene, camphor (German 134553), coumarin, dextrin, pure; ethyl formate, formic acid, glycollic acid, ionin, malonic acid, metachlorobenzaldehyde, naphthionic acid, orthochlorobenzaldehyde, pararosolic acid, phosphorus acid, phosphorus oxybromide, phenylethyl anhydride, pyrazolon, quinazarin, uranium oxide, black; vanadyl sulphate.
Reducing agent in making—
Gold pigments used as decorative agents on fine chinaware.
Solvent for—
Milori blue.
Solvent in making—
Chrome green.
Starting point in making—
Aluminum oxalate, ammonium binoxalate, ammonium oxalate, antimonyl oxalate, bornyl oxalate, cuprous oxalate, cyclohexanol oxalate, dibutyl oxalate, diisoamyl oxalate, di-isobutyl oxalate, diphenyl orthooxalate (German 226231), ferrous oxalate, ferrous, potassium oxalate.
Glyoxylic acid (German 163842, 194038, 210673, 239312, 243746).
Isobornyl oxalate, lead oxalate, magnesium oxalate, manganous oxalate, metacresol oxalate (German 229143), nickel oxalate, potassium binoxalate, potassium oxalate, potassium tetroxalate, sodium binoxalate, sodium oxalate, stannous oxalate, strontium oxalate, thorium oxalate, titanium-ammonium oxalate, zinc oxalate, zinconium oxalate.
Various other simple and compound oxalates.

Ceramics
Suspension agent for—
Glaze mixtures.

Dye
Condensing agent in making various intermediates.
Reagent in making—
Anilin blue, aurin, acridin orange NO, carbazol blue (German 134983), dyestuff mixtures (for household use), diaminoacridin, malachite green, safranin.

Explosives and Matches
Bleaching agent for—
Cotton linters.
Ingredient of—
Match-head compositions, pyrotechnic compositions.

Fats and Oils
Purifying agent in making—
Stearin.

Food
Reagent (German 108880) in making—
Degraded protein food.
Starting point in making—
Artificial apple flavor.

Fuel
Cleaning agent for—
Anthracite.
Ingredient of—
Fuel compositions (added to promote combustion).

Gas
Precipitating agent for—
Rare earth oxides in the manufacture of incandescent gas mantles.

Germicide
Ingredient of—
Cresol disinfectant (German 229143).
Phenol disinfectant (German 224812, 226231).
Seed disinfectant, various germicides.

Glues and Adhesives.
Reagent (German 121422, 122018) in making—
Pectin glues from desaccharified beet cuttings.

Oxalic Acid (Continued)

Ink
Reagent in making—
Blue ink, copying ink, hematoxylin writing inks, printing ink, writing ink.

Insecticide
Ingredient of various insecticidal compositions.

Laundering
Reagent for—
Removing old laundry markings from clothes.
Removing rust and ink stains from clothes.

Leather
Disinfectant for—
Hides, skins.
Ingredient (Brit. 255566) of—
Sulphite cellulose waste liquor solutions used in the pre-treatment of hides or pelts before tanning.
Reagent in—
Tanning.
Solvent (French 498761) in—
Recovery of chromium from chromed leather.

Metallurgical
Cleansing agent for—
Brass and other metals.
Ingredient of—
Compositions used to cleanse the surface of iron before it is lead-coated (German 591116).
Electrolytes used in electroplating aluminum.
Flux used in autogeneous soldering of aluminum bronze (French 574392).
Gilding compositions for nickel, iron, steel, and silver (German 134428).

Miscellaneous
Bleaching agent for—
Straw hats, braids, and the like.
Straw and reeds.
Ingredient of—
Fire-extinguishing composition along with sodium bicarbonate and saponin.
Floor-cleansing and polishing compositions (Brit. 255101).
Ink eradicators.
Linoleum-cleansing and polishing compositions (Brit. 255101).
Metal-polishing compositions.
Tatoo-removing solutions.

Paint and Varnish
Ingredient (French 603360) of—
Compositions used for renovating surfaces after the paint has been removed.

Paper
As a reagent.

Pharmaceutical
In compounding and dispensing practice.
Suggested for use (poisonous) as emmenagog, expectorant, sedative.

Photographic
Ingredient of—
Gallic acid-iron solution used for developing paper.
Solutions used for argentotype papers.
Standard iron solutions for making platinum photographic paper.
Solubilizing agent for—
Blue coloring matter in making blueprint paper.

Plastics
Reagent in making—
Celluloid.

Printing
Coating agent for—
Chemically treated paper in photo-mechanical printing.
Polishing agent for—
Lithographic stones.
Reagent in—
Process engraving.

Resins and Waxes
Reagent (Brit. 316323) in making—
Artificial resins.

Rubber
Ingredient of—
Rubber batches.

Sanitation
Digestant for—
Sludge.
Reagent in—
Formaldehyde disinfection (German 189960).

Sugar
Reagent in making—
Glucose from starch by the Dutch process.

Textile
——, *Bleaching*
As a bleaching agent.

——, *Dyeing*
Accelerating agent for—
Chromium salts in wool chroming.
As a mordant.
Fixing agent for—
Chromium fluoride mordant in wool dyeing.
Ingredient of—
Bath in alum-developing process, bath in cochineal scarlet dyeing.
Reducing agent in—
Chrome mordant bath.

——, *Miscellaneous*
Bleaching agent for—
Cotton linters.
Impregnating agent for—
Cotton fabric in making buckram cloth.
Rust and ink stain eradicator.

——, *Printing*
Accelerator in—
Mordant printing.
Developing agent for—
Anilin black.
Ingredient of—
Bath for discharging indigo, nitroso blue paste, nitroso blue slop-padding bath, steam blue printing paste.
Reagent in—
Calico printing.
Substitute for—
Tartaric acid in making chromic acid discharge.
Thickener for—
Pastes containing Persian berries extract, tin, and aluminum.

Woodworking
Ingredient of—
Bleaching solutions, cleansing solutions, fireproofing compositions (German 162212).

Oxalyl Chloride
French: Chlorure d'oxalyle, Chlorure oxalylique.

Chemical
Reagent (Brit. 265224) in making—
Normal paratoluenesulpho derivatives of 4-methylisatin.
Normal paratoluenesulpho derivatives of 6-methylisatin.
Normal paratoluenesulpho-5-methylisatin.
Paratoluenesulpho derivatives of 5-methyl-4-chloroisatin.
Paratoluenesulpho derivatives of 5-methyl-6-chloroisatin.
Paratoluenesulpho-1:8-naphthisatin.

Oxanilide

Chemical
Starting point in making—
Orthonitroanisidin.

2^4-Oxo-2^3-methobutylcyclohexane
German: 2^4-Oxo-2^3-methobutylzyklohexan.

Perfume
Ingredient (Brit. 347052) of perfume compositions containing—
Ambrette musk, artificial jasmine oil, benzyl acetate, benzyl alcohol, bergamot oil, cinnamic alcohol, cumarin, heliotropin, hydroxycitronellal, ionone, methylionone, phenylethyl alcohol, orange oil, sandalwood oil, ylang-ylang oil.

1-Oxy-4-aminoanthraquinone
Synonyms: Alphaoxy-4-aminoanthraquinone.
German: 1-Oxy-4-aminoanthrachinon.

Dye
Starting point (Brit. 261139) in making dyestuffs with—
Dimethylanilin, pyridin, quinolin.

3-Oxy-4-aminobenzene-1-arsinic Acid
French: Acide de 3-oxye-4-aminobenzène-1-arsénieux.
German: 3-Oxy-4-aminobenzol-1-arsinigsaeure.

Chemical
Starting point in making—
Benzoxazolonarsinic acid (German 439605).

1-Oxy-2-bromo-4-benzylaminoanthraquinone
French: 1-Oxye-2-bromo-4-benzyleaminoanthraquinone.
German: 1-Oxy-2-brom-4-benzylaminoanthrachinon.

Chemical
Starting point in making—
Intermediates, pharmaceuticals.

Dye
Starting point (Brit. 268542) in making wool dyestuffs with—
Ammonium sulphite, potassium sulphite, sodium sulphite.

Oxybutyric Acid
Synonyms: Aldol.

Chemical
Starting point in making—
Cyanin, isocyanin, 2-methylquinolin (quinaldin), rubber vulcanization accelerator by reaction with anilin (Brit. 259933).

Pharmaceutical
In compounding and dispensing practice.

4′-Oxy-3′-carboxy-2-aminodiphenylsulphonemethylpyrazolone
French: 4′-Oxye-3′-carboxye-2-aminodiphényle-sulphone-méthylepyrazolone.
German: 4′-Oxy-3′-carboxy-2-aminodiphenylsulfonmethylpyrazolon.

Chemical
Starting point in making various intermediates.

Dye
Starting point (Brit. 306843) in making azo dyestuffs with—
Alphanaphthylamine, alphanaphthylaminesulphonic acid, anilin, anilinsulphonic acid, betanaphthylamine, betanaphthylaminesulphonic acid, 4-chloro-4-toluidin, cresylsulphonamide, cresylsulphonanilide, metacresidin, metacresidinsulphonic acid, metanaphthylamine, metanaphthylaminesulphonic acid, metaphenylenediamine, metaphenylenediaminesulphonic acid, metatoluidin, metatoluidinsulphonic acid, metaxylidin, metaxylidinsulphonic acid, naphthylsulphonamide, naphthylsulphonanilide, orthocresidin, orthocresidinsulphonic acid, orthonaphthylamine, orthonaphthylaminesulphonic acid, orthophenylenediamin, orthophenylenediaminsulphonic acid, orthotoluidin, orthotoluidinsulphonic acid, orthoxylidin, orthoxylidinsulphonic acid, paracresidin, paracresidinsulphonic acid, paranaphthylamine, paranaphthylaminesulphonic acid, paraphenylenediamine, paraphenylenediaminesulphonic acid, paratoluidin, paratoluidinsulphonic acid, paraxylidin, paraxylidinsulphonic acid, phenylsulphonamide, phenylsulphonanilide, tolylsulphonamide, tolylsulphonanilide, xylylsulphonamide, xylylsulphonanilide.

4′-Oxy-3′-carboxy-2-aminodiphenylsulphonpyrazolonecarboxylic Acid
French: Acide de 4′-oxye-3′-carboxye-2-aminodiphénylesulphonpyrazolonecarbonique, Acide de 4′-oxye-3′-carboxye-2-aminodiphénylesulphonpyrazolonecarboxylique.
German: 4′-Oxy-3′-carboxy-2-aminodiphenylsulphonpyrazoloncarbonsaeure.

Dye
Starting point (Brit. 306843) in making azo dyestuffs with—
Alphanaphthylamine, alphanaphthylaminesulphonic acid, anilin, anilin-3-sulphonic acid, betanaphthylamine, betanaphthylaminesulphonic acid, 2-chloro-4-toluidin, metacresidin, metacresidinsulphonic acid, metaphenylamine, metaphenylaminesulphonic acid, metatoluidin, metatoluidinsulphonic acid, metaxylidin, metaxylidinsulphonic acid, naphthylsulphonamide, naphthylsulphonanilide, orthocresidin, orthocresidinsulphonic acid, orthophenylamine, orthophenylaminesulphonic acid, orthotoluidin, orthotoluidinsulphonic acid, orthoxylidin, orthoxylidinsulphonic acid, paracresidin, paracresidinsulphonic acid, paraphenylamine, paraphenylaminesulphonic acid, paratoluidin, paratoluidinsulphonic acid, paraxylidin, paraxylidinsulphonic acid, phenylsulphonamide, phenylsulphonanilide, tolylsulphonamide, tolylsulphonanilide, xylylsulphonamide, xylylsulphonanilide.

4′-Oxy-3′-carboxy-2-aminodiphenylsulphon-4-sulphonic Acid
French: Acide de 4′-oxye-3′-carboxye-2-aminodiphénylesulphon-4-sulphonique.
German: 4′-Oxy-3′-carboxy-2-aminodiphenylsulfon-4-sulfonsaeure.

Chemical
Starting point in making various intermediates.

Dye
Starting point (Brit. 306843) in making azo dyestuffs with—
Alphanaphthylamine, alphanaphthylaminesulphonic acid, anilin, anilin-3-sulphonic acid, betanaphthylamine, betanaphthylaminesulphonic acid, 2-chloro-4-toluidin, metacresidin, metacresidinsulphonic acid, metaphenylamine, metaphenylaminesulphonic acid, metatoluidin, metatoluidinsulphonic acid, metaxylidin, metaxylidinsulphonic acid, naphthylsulphonamide, naphthylsulphonanilide, orthocresidin, orthocresidinsulphonic acid, orthophenylamine, orthophenylaminesulphonic acid, orthotoluidin, orthotoluidinsulphonic acid, orthoxylidin, orthoxylidinsulphonic acid, paracresidin, paracresidinsulphonic acid, paraphenylamine, paraphenylaminesulphonic acid, paratoluidin, paratoluidinsulphonic acid, paraxylidin, paraxylidinsulphonic acid, tolylsulphonamide, tolylsulphonanilide, xylylsulphonamide, xylylsulphonanilide.

4-Oxy-3-carboxylbenzenesulphinic Acid
French: Acide de 4-oxye-3-carboxylbenzènesulphinique.
German: 4-Oxy-3-carbonylbenzolsulphinsaeure.

Dye
Starting point in making diarylsulphone dyestuffs with—
2:5-Diaminoanisol, 2:5-diaminophenetole, paraphenylenediamine.

Oxycholesterol

Pharmaceutical
Ingredient (U. S. 2013524) of—
Nasal medicament comprising an emulsion of water and oxycholesterol, with water in the discontinuous phase and oxycholesterol in the continuous phase and having a water-soluble medicament, such as hexylresorcinol, dissolved in the water and an oil-soluble medicament, such as trichlorobutanol or menthol, dissolved in the oxycholesterol.

4-Oxy-4:6-dichloro-2-benzoic Acid
French: Acide de 4-oxye-4:6-dichloro-2-benzoique.
German: 4-Oxy-4:6-dichlor-2-benzoesaeure.

Leather
Mothproofing agent (Brit. 274425) in treating—
Skins.

Miscellaneous
Mothproofing agent in treating—
Felt, feathers, furs, hair.

Textile
——, *Finishing*
Mothproofing agent in treating—
Wool.

Oxyethylpyridinium Chlorostearate

Paper
Reagent (Brit. 396992) for—
Increasing the fastness to water of dyestuffs on paper half-stuff, particularly half-stuff dyed a vivid green with benzyl green B.

Oxygen
French: Oxygène.
German: Sauerstoff.
Italian: Ossigeno.

Analysis
Reagent in the chemical laboratory.

Chemical
Reagent in making—
Sulphuric acid by the contact process (used in the place of air for admixture with sulphur dioxide to be catalytically converted into sulphur trioxide).
Starting point in making—
Ozone, ozonized air.
Starting point in recovering—
Argon.

Distillation
Reagent in treating—
Distilled liquors of various sorts in order to hasten their maturing.

Explosives
Ingredient of—
Explosive called "oxyliquite."

Fats and Oils
Reagent in—
Thickening oils.

Oxygen (Continued)

Food

Ingredient of—

Vinegar mash (added in small amounts for the purpose of stimulating the mycoderma bacteria so as to increase the rate of acetification).

Fuel

Reagent in—

Activating combustion of low grade fuels.

Effecting more efficient utilization of various fuels.

Gas

Reagent in making—

Illuminating gas of high calorific power.

Reagent in—

Purifying coal gas and coke oven gas (used as an assistant in the removal of the sulphur compounds contained in these gases).

Linoleum and Oilcloth

Reagent in—

Treating various oils in order to thicken them.

Metallurgical

In welding and cutting—

In combination with acetylene to give the oxyacetylene flame.

In combination with hydrogen to give the oxyhydrogen flame.

In combination with illuminating gas to give a flame of high temperature.

Reagent in—

Operating converters, roasting blends and various ores.

Miscellaneous

As a general bleaching agent—

For dental work, for revivifying the air in crowded halls, for various oxidizing purposes, in anesthesia, in diving bells and caissons, in resuscitation, in submarine vessels.

Paint and Varnish

Reagent in—

Treating varnishes and oils, used in their manufacture, in order to thicken them.

Paper

Reagent in—

Bleaching paper pulps, resulting in the production of white fibers and effecting economy in the consumption of chlorine bleach.

Pharmaceutical

In compounding and dispensing practice.

Wine

Reagent in—

Treating wines in order to hasten their maturing.

1-Oxynaphthalene-4:8-disulphonic Acid

Synonyms: Beta-oxynaphthalene-4:8-disulphonic acid.

French: Acide bétaoxynaphthalène-4:8-disulphonique.

German: Beta-oxynaphtalin-4:8-bisulphonsaeure.

Dye

Starting point (Brit. 249884) in making—

A 20 dyestuffs with beta-amino-1-oxynaphthalene-4:8-disulphonic acid or 4-nitro-2-amino-1-phenol.

2:3-Oxynaphthoic Alphanaphthylamide

French: Alphanaphthyleamide 2:3-oxynaphthoique.

German: 2:3-Oxynaphtoealphanaphtylamid.

Chemical

Starting point in making—

Intermediates.

Dye

Starting point in making various dyestuffs.

Textile

——, *Dyeing*

Coupling agent (Brit. 319247) in dyeing yarns and fabrics with the aid of—

Allyl 5-bromo-2-amino-1-benzoate.
Amyl 5-bromo-2-amino-1-benzoate.
Butyl 5-bromo-2-amino-1-benzoate.
Ethyl 5-bromo-2-amino-1-benzoate.
Heptyl 5-bromo-2-amino-1-benzoate.
Hexyl 5-bromo-2-amino-1-benzoate.
Isoallyl 5-bromo-2-amino-1-benzoate.
Isoamyl 5-bromo-2-amino-1-benzoate.
Isobutyl 5-bromo-2-amino-1-benzoate.
Isopropyl 5-bromo-2-amino-1-benzoate.
Methyl 5-bromo-2-amino-1-benzoate.
Pentyl 5-bromo-2-amino-1-benzoate.
Propyl 5-bromo-2-amino-1-benzoate.
Various alkyl esters of 5-chloro-2-aminol-1-benzoic acid.

2:3-Oxynaphthoic Anilide

French: Anilide de 2:3-oxynaphthoique.

German: 2:3-Oxynaphtoeanilid.

Textile

——, *Dyeing and Printing*

Reagent (Brit. 310779) in dyeing and printing and stenciling materials containing cellulose esters and ethers, with the aid of—

4-Acetylaminobenzeneazoalphanaphthylamine.
Alpha-amino-4-acetylaminonaphthalene.
Alpha-amino-4-acetylanthraquinone.
Alpha-aminoanthraquinone.
Alpha-amino-4-bromoanthraquinone.
Alpha-amino-2-methoxynaphthalene.
Alpha-amino-2-methylanthraquinone.
Alpha-amino-3-para-toluenesulphonaminobenzene.
Alphanaphthylamine, alpha-4-oxyanthraquinone, aminoazobenzene, aminoazonaphthalene, aminoazotoluene.
4-Aminobenzeneazoalphanaphthylamine.
4-Aminodiphenyl, 4-aminodiphenyl ether, 4-aminodiphenylmethane.
1-(4′-Amino)phenyl-3-methyl-5-pyrazolon.
Anilin, benzeneazoalphanaphthylamine, benzidin, betanaphthylamine.
8-Chloroalphanaphthylamine.
4-Chloronaphthalene-azo-alpha-amino-2-methoxynaphthalene.
4-Chloro-2-nitro-4′-aminoazobenzene.
1:5-Diaminoanthraquinone.
1:8-Diaminoanthraquinone.
4:4′-Diaminoazobenzene.
4:4′-Diaminodiphenylmethane.
Dianisidin, dihydrothioparatoluidin.
4-Dimethylamino-4′-aminoazobenzene.
2:4-Dinitrobenzeneazoalphanaphthylamine.
2:4-Dinitro-3′:6′-dimethyl-4′-aminoazobenzene.
3:5-Dinitro-2-oxybenzeneazoalphanaphthylamine.
Metanitroparatoluidin, metatoluidin.
4-Methoxy-4′-aminoazobenzene.
2-Methoxybenzeneazo-alphanaphthylamine.
1:5-Naphthylenediamine.
4-Nitro-4′-aminoazobenzene.
2-Nitro-2′:4-dimethyl-4′-aminoazobenzene.
4-Nitro-2′-methoxybenzene-azoalphanaphthylamine.
4-Nitro-1-naphthylamine, orthoanisidin, orthochloroanilin, para-aminoacetanilide, paranitranilin, paranitro-orthoanisidin.
4-Phenoxy-2′-methyl-4′-aminoazobenzene.

2:3-Oxynaphthoic Betanaphthylamide

Chemical

Starting point in making—

Intermediates.

Dye

Starting point in making various dyestuffs.

Textile

Coupling agent (Brit. 319247) in dyeing yarns and fabrics with the aid of—

Allyl 5-bromo-2-amino-1-benzoate.
Amyl 5-bromo-2-amino-1-benzoate.
Butyl 5-bromo-2-amino-1-benzoate.
Ethyl 5-bromo-2-amino-1-benzoate.
Heptyl 5-bromo-2-amino-1-benzoate.
Hexyl 5-bromo-2-amino-1-benzoate.
Isoallyl 5-bromo-2-amino-1-benzoate.
Isoamyl 5-bromo-2-amino-1-benzoate.
Isobutyl 5-bromo-2-amino-1-benzoate.
Isopropyl 5-bromo-2-amino-1-benzoate.
Methyl 5-bromo-2-amino-1-benzoate.
Pentyl 5-bromo-2-amino-1-benzoate.
Propyl 5-bromo-2-amino-1-benzoate.
Various alkyl esters or 5-chloro-2-amino-1-benzoic acid.

2:3-Oxynaphthoic 5′-Chloro-2′-anisidide

French: 5′-Chloro-2′-anisidide, 2:3-Oxynaphthoique.

German: 2:3-Oxynaphtoe-5′-chlor-2′-anisidid.

Chemical

Starting point in making—

Intermediates, pharmaceuticals.

Dye

Starting point (Brit. 301410) in making azo dyestuffs and lakes with the aid of—

Alpha-amino-2-cyano-4-chlorobenzene.
Alpha-amino-2-cyano-5-chlorobenzene.
Alpha-amino-3-cyano-6-chlorobenzene.
Alpha-amino-2-cyano-5-methylbenzene.
Alpha-amino-3-cyano-2-methylbenzene.
Alpha-amino-2:4-dimethyl-5-cyanobenzene.

2:3-Oxynaphthoic Metanitranilide
French: Métanitranilide de 2:3-oxynaphthoique.
German: 2:3-Oxynaphtoemetanitranilid.

Chemical
Starting point in making various intermediates.

Dye
Starting point in making various synthetic dyestuffs.

Textile
——, *Dyeing and Printing*
Reagent (Brit. 310779) in dyeing and printing and stenciling materials containing cellulose esters and ethers, with the aid of—
4-Acetylaminobenzeneazoalphanaphthylamine.
Alpha-amino-4-acetylaminonaphthalene.
Alpha-amino-4-acetylanthraquinone.
Alpha-aminoanthraquinone.
Alpha-amino-4-bromoanthraquinone.
Alpha-amino-2-methoxynaphthalene.
Alpha-amino-2-methylanthraquinone.
Alpha-amino-3-paratoluenesulphonaminobenzene.
Alphanaphthylamin, alpha-4-oxyanthraquinone, aminoazobenzene, aminoazonaphthalene, aminoazotoluene.
4-Aminobenzeneazoalphanaphthylamine.
4-Aminodiphenyl, 4-aminodiphenyl ether, 4-aminodiphenylamine.
1-(4′-Amino)phenyl-3-methyl-5-pyrazolon.
Anilin, benzeneazoalphanaphthylamine, benzidin.
Betanaphthylamine, 8-chloroalphanaphthylamine.
4-Chloronaphthaleneazoalpha-amino-2-methoxynaphthalene.
4-Chloro-2-nitro-4′-aminoazobenzene.
1:5-Diaminoanthraquinone, 1:8-diaminoanthraquinone, 4:4′-Diaminoazobenzene, 4:4′-Diaminodiphenylmethane, dianisidin, dihydrothioparatoluidin.
4-Dimethylamino-4′-aminoazobenzene.
2:4-Dinitrobenzeneazoalphanaphthylamine.
2:4-Dinitro-3′:6-dimethyl-4′-aminoazobenzene.
3:5-Dinitro-2-oxybenzeneazoalphanaphthylamine.
Metanitroparatoluidin, metatoluidin.
4-Methoxy-4′-aminoazobenzene.
2-Methoxybenzeneazoalphanaphthylamine.
1:5-Naphthylenediamine.
4-Nitroalphanaphthylamine.
4-Nitro-4′-aminoazobenzene.
2-Nitro-2:4-dimethyl-4′-aminoazobenzene.
4-Nitro-2-methoxybenzeneazoalphanaphthylamine.
Orthoanisidin, orthochloroanilin, para-aminoacetanilide, paranitranilin, paranitro-orthoanisidin.
4-Phenoxy-2′-methyl-4′-aminoazobenzene.

2:3-Oxynaphthoic-4′-methoxyanilide
French: 4′-Méthoxyeanilide de 2:3-oxynaphthoique.
German: 2:3-Oxynaphtoe-4-methoxyanilid.

Chemical
Starting point in making—
Intermediates, pharmaceuticals.

Dye
Starting point (Brit. 301410) in making azo dyestuffs and lakes with—
Alpha-amino-2-cyano-4-chlorobenzene.
Alpha-amino-2-cyano-5-chlorobenzene.
Alpha-amino-3-cyano-6-chlorobenzene.
Alpha-amino-2-cyano-5-methylbenzene.
Alpha-amino-3-cyano-2-methylbenzene.
Alpha-amino-2:4-dimethyl-5-cyanobenzene.

2:3-Oxynaphthoic Para-anisidide
French: Para-anisidide de 2:3-oxynaphthoique.
German: 2:3-Oxynaphtoepara-anisidid.

Chemical
Starting point in making—
Intermediates.

Dye
Starting point in making various synthetic dyestuffs.

Textile
Reagent (Brit. 310779) in dyeing, printing, and stenciling materials containing cellulose esters or ethers, with the aid of—
4-Acetylaminobenzeneazoalphanaphthylamine.
Alpha-amino-4-acetylaminonaphthalene.
Alpha-amino-4-acetylanthraquinone.
Alpha-aminoanthraquinone.
Alpha-amino-4-bromoanthraquinone.
Alpha-amino-2-methoxyanthraquinone.
Alpha-amino-2-methylanthraquinone.
Alpha-amino-3-paratoluenesulphonaminobenzene.
Alphanaphthylamine, alpha-4-oxyanthraquinone, aminoazobenzene, aminoazonaphthalene, aminoazotoluene.
4-Aminobenzeneazoalphanaphthylamine.
4-Aminodiphenyl, 4-aminodiphenyl ether, 4-aminodiphenylamine.
1-(4′-Amino)phenyl-3-methyl-5-pyrazolon.
Anilin, benzeneazoalphanaphthylamine, benzidin.
Betanaphthylamine, 8-chloroalphanaphthylamine.
4-Chloronaphthaleneazoalpha-amino-2-methoxynaphthalene.
4-Chloro-2-nitro-4′-aminoazobenzene.
1:5-Diaminoanthraquinone, 1:8-diaminoanthraquinone, 4:4′-Diaminoazobenzene, 4:4′-diaminodiphenylmethane, dianisidin.
4-Dimethylamino-4′-aminoazobenzene.
2:4-Dinitrobenzenealphanaphthylamine.
2:4-Dinitro-2-oxybenzeneazoalphanaphthylamine.
2:4-Dinitro-3′:6-dimethyl-4′-aminoazobenzene.
Metanitroparatoluidin, metatoluidin.
4-Methoxy-4′-aminoazobenzene.
2-Methoxybenzeneazoalphanaphthylamine.
1:5-Naphthylenediamine, 4-nitroalphanaphthylamine.
4-Nitro-4′-aminoazobenzene.
2-Nitro-2:4-dimethyl-4′-aminoazobenzene.
4-Nitro-2-methoxybenzeneazoalphanaphthylamine.
Orthoanisidin, orthochloroanilin, para-aminoacetanilide, paranitranilin, paranitro-orthoanisidin.
4-Phenoxy-2′-methyl-4′-aminoazobenzene.

2:3-Oxynaphthoic 2′-Toluidide
French: 2′-Toluidide de 2:3-oxynaphthoique.
German: 2:3-Oxynaphtoe-2′-toluidid.

Chemical
Starting point in making various intermediates.

Dye
Starting point (Brit. 301410) in making azo dyestuffs and lakes with—
Alpha-amino-2-cyano-4-chlorobenzene.
Alpha-amino-2-cyano-5-chlorobenzene.
Alpha-amino-3-cyano-6-chlorobenzene.
Alpha-amino-2-cyano-5-methylbenzene.
Alpha-amino-3-cyano-2-methylbenzene.
Alpha-amino-2:4-dimethyl-5-cyanobenzene.

2:3-Oxynaphthoic-3-toluidide
French: 2:3-Oxynaphthoique-3-toluidide.
German: 2:3-Oxynaphtoe-3-toluidid.

Dye
Reagent (Brit. 279146) in making dyestuffs with—
3-Amino-4-chlorodiphenylsulphone.
3′-Amino-4′-methylbenzoyl-4-chloro-2-toluidin.
2-Aminotoluene-4-sulpho(normal)-methylanilide.

2:3-Oxynaphthoic-4-toluidide
French: 4-Toluidide de 2:3-oxynaphthoique.
German: 2:3-Oxynaphtoe-4-toluidid.

Dye
Starting point (Brit. 279146) in making azo dyestuffs with—
2-Aminotolyl-4-phenylsulphone.
3′-Amino-4′-methylbenzoyl-4-chloro-2-toluidine.
3:3′-Diamino-4:4′-ditolylketone.

2:3-Oxynaphthoylamide
French: 2:3-Oxynaphthoyleamide.
German: 2:3-Oxynaphtoylamid.

Chemical
Starting point in making—
Intermediates, pharmaceuticals.

Dye
Starting point (Brit. 304441) in making azo dyestuffs with—
Dimethyl sulphate, monochloroacetic acid.

2:3-Oxynaphthoylbetanaphthylamine

Textile
——, *Dyeing*
Reagent for—
Developed dyestuffs on cellulose acetate rayon (Brit. 262830).

2:3-Oxynaphthoyl Chloride
French: Chlorure de 2:3-oxynaphthoyle.
German: Chlor-2:3-oxynaphtoyl.

Chemical
Starting point in making—
Intermediates, pharmaceuticals.
Starting point (Brit. 305763) in making—
2:3-Oxynaphthoyl 4′-anthranylketone.
2:3-Oxynaphthoyl 4′-anisylketone.
2:3-Oxynaphthoyl 4′-benzylketone.
2:3-Oxynaphthoyl 4′-cinnamylketone.
2:3-Oxynaphthoyl 4′-cresylketone.

2:3-Oxynaphthoyl Chloride (Continued)
2:3-Oxynaphthoyl 4'-gallylketone.
2:3-Oxynaphthoyl 4'-metanylketone.
2:3-Oxynaphthoyl 4'-naphthylketone.
2:3-Oxynaphthoyl 4'-phenylketone.
2:3-Oxynaphthoyl 4'-phthalylketone.
2:3-Oxynaphthoyl 4'-salicylketone.
2:3-Oxynaphthoyl 4'-sulphanylketone.
2:3-Oxynaphthoyl 4'-tolylketone.
2:3-Oxynaphthoyl 4'-valerylketone.
2:3-Oxynaphthoyl 4'-xylylketone.

Dye
Starting point in making various synthetic dyestuffs.

2:3-Oxynaphthylaminohydroquinonedimethyl Ether
German: 2:3-Oxynaphtylaminhydrochinondimethylaether.

Textile
——, *Dyeing*
Reagent for—
Developed dyestuffs on cellulose acetate rayon (Brit. 262830).

1-Oxy-4-para-toluidino-anthraquinone

Oils, Fats, Waxes
Coloring agent (Brit. 432867) for—
Paraffin and other mineral waxes, stearic acid, tallow and other solid triglycerides, beeswax, carnauba wax, and others.

4-Oxy-1-tertiarybutylbenzene
French: 4-Oxye-1-tértiairebutylebenzène.
German: 4-Oxy-1-ternaerbutylbenzol.

Cellulose Products
Plasticizer (U. S. 1740854) for—
Cellulose acetate.
For uses, see under general heading: "Plasticizers."

Ozokerite
Synonyms: Fossil wax, Mineral wax, Native paraffin.
French: Cire minérale, Cire fossile, Ozocérite.
German: Mineralwachs, Ozokerit.

Chemical
For lining acid tanks and coating apparatus to avoid the corrosive action of acids.
Raw material in making—
Bottles used for holding and shipping hydrofluoric acid.
Starting point in making—
Ceresin wax, both in the white state and in the partially purified yellow condition.
Starting point (Brit. 287514) in making—
Aldehydes, alcohols, ketones, carboxylic acids, various oxidation products.

Construction
Waterproofing agent, used alone or in combination with other suitable substances (Brit. 287514), for treating—
Brickwork, concrete, masonry.
Various structural materials characterized by porosity.

Electrical
Ingredient of—
Insulating compositions used for various purposes.
Insulating compositions containing rubber.
Insulating agent in making—
Apparatus, cables, wires.

Food
Ingredient of—
Compositions which are used for decorating fancy food products.
Raw material in making—
Artificial honeycombs.

Ink
Ingredient of—
Lithographic inks, printing inks, writing inks.

Leather
Ingredient of—
Finishing preparations, polishing compositions.

Metallurgical
Ingredient of—
Compositions used for covering metals to provide protection against moisture, acids, alkalies, and other corrosive substances.
In various electroplating processes.

Miscellaneous
Ingredient of—
Compositions used for lining barrels and kegs.
Composition used in the manufacture of incandescent gas mantles.
Floor polishes, polishes of various sorts.
Preparations for making imitation alabaster statues.
Shoe polishes.
Raw material in making—
Candles, grease crayons, toys, wax figures.
Waterproofing agent (Brit. 287514), used either alone or in compositions, for the treatment of—
Asbestos, strawboard.
Various porous materials that have to be made resistant to water.

Oils and Fats
Base of various lubricating compositions.
Ingredient of—
Axle greases, gun oils, special lubricants.

Paint and Varnish
Ingredient of—
Preparations containing dry colors, special floor waxes.
Raw material in making—
Paints, varnishes, wood-fillers.

Paper
Ingredient of—
Compositions used in the manufacture of carbon paper.
Emulsified sizing compositions (Brit. 287514).
Preparations used in making waxed paper.
Sizings for high-gloss paper.

Perfume
Raw material in making—
Pomades and other waxy products.

Pharmaceutical
In compounding and dispensing practice.

Printing
Ingredient of—
Compositions used for the preparation of acidproof coatings for plates in the electrotyping process.
Compositions used for making matrices in galvanoplastic work.
Process material in—
Lithography, photoengraving, process engraving.

Rubber
Filler in making—
Rubber compositions.

Stone
Waterproofing agent (Brit. 287514), used either alone or in admixture with other substances, for treating—
Artificial stone, natural stone.

Textile
Ingredient of—
Compositions used in the manufacture of waxed cloth.
Compositions used for sizing linen and cotton fabrics.
Emulsified dressings (Brit. 287514).
Waterproofing compositions (Brit. 287514).
Waterproofing agent in treating—
Yarns and fabrics.

Waxes and Resins
Ingredient of—
Electrotypers' wax, sealing wax, shoemaker's wax.
Substitute for—
Beeswax, carnauba wax.

Woodworking
Ingredient of—
Compositions used in the treatment of furniture and of lumber used for parquet floorings.

Palladous Oxide
French: Oxyde palladeux.
German: Palladium oxydul.

Chemical
Catalyst in hydrogenation processes.
Oxidizing agent in making various chemicals.
Reagent in making—
Leptynol.

Palmitic Acid Chloride

Chemical
Starting point (Brit. 407956) in making pour-point improvers for machine oils, gear oils, and other lubricants by condensing with—
Anilin, anthracene oil.
Aromatics obtained by destructive hydrogenation or by dehydrogenation.
Benzene
Cracking gases containing gaseous olefins (ethylene, propylene, and butylene).
Cyclic terpenes, ethylnaphthalene, liquid olefins, middle oil, naphthalene, naphthols, naphthylamines, nitrated aromatics, phenols, tars, toluene, xylene.

Palmitic Toluide

Chemical
Starting point in making various derivatives.
Petroleum
Ingredient (U. S. 1853571) of—
Lubricating compositions containing mineral oils (added for the purpose of increasing the consistency of the lubricant and raising its melting point).

Palmitone

German: Palmiton.
Chemical
Reagent (Brit. 343098) in making—
Emulsions containing sulphuric esters of high molecular weight compounds, sulphonic acids of polynuclear compounds, hydroxyalkylamines, quaternary ammonium bases or salts.
Emulsions of various chemicals.
Fats and Oils
Reagent (Brit. 343098) in making—
Dispersions and emulsions of various animal and vegetable fats and oils.
Ink
Reagent (Brit. 343098) in making—
Ink dispersions.
Leather
Reagent (Brit. 343098) in making—
Dressing compositions, polishing compositions, stuffing compositions, tanning compositions, treating compositions.
Miscellaneous
Reagent (Brit. 343098) in making—
Cleansing, dispersing, emulsifying, and wetting compositions used for various purposes.
Resins and Waxes
Reagent (Brit. 343098) in making—
Emulsions and dispersions of waxes and resins.
Soap
Ingredient (Brit. 343098) of—
Washing and scouring compositions containing soaps.
Textile
——, *Dyeing*
Ingredient (Brit. 343098) of—
Dye baths.
——, *Finishing*
Ingredient (Brit. 343098) of—
Finishing baths.
——, *Manufacturing*
Ingredient (Brit. 343098) of—
Rayon-spinning baths and the like.
Wool-carbonizing solutions.

Palmityl Chloride

French: Chlorure de palmityle, Chlorure palmitylique.
Chemical
Reagent in making—
Starch esters (U. S. 1651366).

Palmitylhydroquinone

Petroleum
Stabilizing agent (Brit. 406195) for—
Cracked gasolines and other motor fuels.

Palmitylhydroxyethanesulphonic Acid

French: Acide palmitylehydroxyéthanesulphonique.
German: Palmitylhydroxyaethansulfonsäure.
Spanish: Acido de palmitilhidroxietansulfonico.
Italian: Acido di palmitilidrossietansolfonico.
Chemical
Starting point in making—
Acids, esters, and other derivatives.
Leather
As a tanning agent and lubricating agent (French 743517).
Miscellaneous
Tanning agent and lubricating agent (French 743517) in treating furs.

Palmitylphloroglucinol

Petroleum
Stabilizing agent (Brit. 406195) for—
Cracked gasolines and other motor fuels.

Palmitylpyrocatechol

Petroleum
Stabilizing agent (Brit. 406195) for—
Cracked gasolines and other motor fuels.

Palmitylpyrogallol

Petroleum
Stabilizing agent (Brit. 406195) for—
Cracked gasolines and other motor fuels.

Palmitylresorcinol

Petroleum
Stabilizing agent (Brit. 406195) for—
Cracked gasolines and other motor fuels.

Palm Kernel Fat

French: Graisse de palmiste.
German: Palmkernfett.
Fats and Oils
Ingredient of—
Fat compositions, to render them plastic and pliable (Brit. 269384).
Food
Ingredient of—
Baked products, butter substitutes, confectionery and candy, chocolate coatings, vegetarian foods.
Perfumery
Ingredient of—
Cosmetics, pomades.
Pharmaceutical
In compounding and dispensing practice.
Soap
Starting point in making—
Fine white soaps.

Palm Oil

Synonyms: Palm butter, Palm grease.
French: Beurre de palme, Huile de palme.
German: Palmfett, Palmoel.
Chemical
Starting point in making—
Aluminum palmitate, palmitic acid, palm oil stearin, palmitates of metals, palmitic acid esters.
Fats and Oils
Ingredient of—
Cutting tools' lubricating compositions.
Lubricating compositions for the tube and metal industries.
Railway axle grease compositions.
Starting point in making—
Degras.
Food
Ingredient of—
Baked products, butter substitutes, food compositions, oleomargarin.
Fuel
Ingredient of—
Candles, illuminating compositions.
Leather
Ingredient of—
Softening and finishing compositions.
Woodworking
Ingredient of—
Compositions used in treating and preserving wood.
Mechanical
As a lubricant.
Metallurgical
Ingredient of—
Compositions used in coating iron plates used in the tinplate industry so as to protect the plates until dipped in molten tin.
Miscellaneous
As an emollient for various purposes.
Ingredient of—
Reagents used in bleaching operations.
Petroleum
Ingredient of—
Motor fuels.
Rubber
Ingredient of—
Rubber compositions.

Palm Oil (Continued)

Soap
Raw material in making—
Crude soap powders, soap lubricating compositions, toilet soaps.

Textile
—, *Finishing*
Ingredient of—
Compositions used in softening and finishing cottons.
Waterproofing compositions (U. S. 1625672).

—, *Manufacturing*
Ingredient of—
Lubricating compositions for use on apparatus em-employed in the spinning of rayons.

Papain

Synonyms: Papaina, Papaine, Papayotin, Vegetable pepoin.

Food
Ferment in making cheese.

Pharmaceutical
Suggested for aiding digestion in chronic dyspepsia, for treatment of gastric fermentation and gastritis, and in dissolving false membranes in diphtheria, croup, and cancer.

Textile
Ingredient (U. S. 1855431) of—
Composition, containing salts of hydrosulphurous acid, used for the degumming of silk.

Para-acetaminophenolallyl Ether

French: Éther de para-acétaminophénolallyle, Éther para-acétaminophénolallylique.
German: Para-acetaminophenolallylaether.

Chemical
Starting point in making—
Dialacetin.

Para-acetanisidin

Chemical
Ingredient of—
Disinfectant and deodorant preparations (Brit. 297074).

Para-acetylaminoethoxybenzene

Synonyms: Pertonal.
French: Para-acétyleaminoéthoxybenzène.
German: Para-acetylaminoaethoxybenzol.

Chemical
Starting point in making—
Pharmaceutical derivatives.

Pharmaceutical
Suggested for use as antipyretic.

Para-aminoacetanilide

Chemical
Starting point in making—
Aromatics, intermediates, pharmaceuticals.

Dye
Starting point in making—
Aminonaphthol red 6B, azo acid red B, azotol C, chromotrope 6B, coomassie wool black R, coomassie wool black S, cotton yellow G, lanafuchsin, thiocatechin, thiophor yellow bronze G, victoria violet, violet black.

Glass
Ingredient (Brit. 340101) of—
Compositions, containing cellulose acetate, nitrocellulose, or other esters or ethers of cellulose, used in the manufacture of non-shatterable, laminated glass, and for the decoration and protection of glassware.

Glues and Adhesives
Ingredient (Brit. 340101) of—
Compositions, containing cellulose acetate, nitrocellulose, or other esters or ethers of cellulose, used for adhesive purposes.

Leather
Ingredient (Brit. 340101) of—
Compositions, containing cellulose acetate, nitrocellulose, or other esters or ethers of cellulose, used in the manufacture of artificial leather and for decorating and protecting leather goods.

Metallurgical
Ingredient (Brit. 340101) of—
Compositions, containing cellulose acetate, nitrocellulose, or other esters or ethers of cellulose, used for the decoration and protection of metallic wares.

Miscellaneous
Ingredient (Brit. 340101) of—
Compositions, containing cellulose acetate, nitrocellulose, or other esters or ethers of cellulose, used for the decoration and protection of various fibrous compositions.

Paint and Varnish
Ingredient (Brit. 340101) of—
Paints, varnishes, lacquers, dopes, and enamels containing cellulose acetate, nitrocellulose or other esters or ethers of cellulose.

Paper
Ingredient (Brit. 340101) of—
Compositions, containing cellulose acetate, nitrocellulose, or other esters or ethers of cellulose, used in the manufacture of coated paper and for the decoration and protection of paper and pulp products.

Plastics
Ingredient (Brit. 340101) of—
Compositions, containing cellulose acetate, nitrocellulose, or other esters or ethers of cellulose.

Rubber
Ingredient (Brit. 340101) of—
Compositions, containing cellulose acetate, nitrocellulose, or other esters or ethers of cellulose, used for coating and decorating rubber ware.

Stone
Ingredient (Brit. 340101) of—
Compositions, containing cellulose acetate, nitrocellulose, or other esters or ethers of cellulose, used for coating and decorating natural and artificial stone.

Textile
Ingredient (Brit. 340101) of—
Compositions, containing cellulose acetate, nitrocellulose, or other esters or ethers of cellulose, used in the production of coated fabrics and in finishing fabrics.
Compositions, containing cellulose acetate, used in the production of rayon filaments.

Woodworking
Ingredient (Brit. 340101) of—
Compositions, containing cellulose acetate, nitrocellulose, or other esters or ethers of cellulose, used for the decoration and protection of woodwork.

Para-aminoacetophenone

Chemical
Starting point in making—
Para-acetophenonearsinic acid (Brit. 261133).

Para-aminoazobenzene

Synonyms: Aminoazobenzene, Aminoazobenzol, Para-aminoazobenzol.
French: Aminoazobenzène.
German: Aminoazobenzol.

Chemical
Starting point in making—
Aminoazobenzene hydrochloride.
Aminoazobenzenebetanaphthol.
Aminoazobenzeneparasulphonic acid.
Intermediates.
Pharmaceuticals and other organic chemicals.
Synthetic aromatics.

Dye
Starting point in making—
Azo acid violet, acetin blue, benzo fast scarlet, acid yellow, brilliant crocein, chrysoidin, cloth red G, crocein AX, crosein B, diazo dyestuffs, erythrin P, fast yellow, indamine blue, indamine (spirit-soluble), indulin (water-soluble), indulin dyestuffs, inulin, paraphenylene blue R, ponceau 5R, solid yellow dyestuffs, spirit yellow, sudan yellow.

Food
Reagent in—
Coloring food compositions.

Paint and Varnish
Reagent in—
Coloring spirit varnishes.

Para-aminobenzaldehyde

Chemical
Starting point in—
Organic synthesis.

Dye
Starting point in making—
Acid wool dyestuffs by condensing with compounds containing a methyl or methylene group which is reactive with an aldehyde group (Brit. 481652).
Dyestuffs.
Substitution products useful in making acid wool dyestuffs (Brit. 481652).

Para-aminobenzeneazoalphanaphthylamine

Chemical
Starting point in making—
Intermediates, pharmaceuticals.

Dye
Starting point in making various synthetic dyestuffs.

Textile
——, *Printing*
Reagent (Brit. 310773) in producing photographic pattern effects on fabrics and films with the aid of—
Acetoacetic ester, alpha-aminonaphthol, alphanaphthol, betanaphthol, betaoxynaphthoic acid, beta-aminonaphthol, dimethylanilin, metaphenylenediamine, paraxylidin.

Para-aminobenzoic Acid
French: Acide para-aminobenzoique.
German: Paraamidobenzoesäure.

Chemical
Reagent in making—
Intermediates.
Starting point in making—
Esters used in perfumery.
Ethyl ester (anaethesin).
Isobutyl ester (cycloform).
Propyl ester (propesin).

Dye
Reagent in making various synthetic dyestuffs.

Paper
Reagent in making—
Transfer papers.

Para-aminobenzoic Acid Ethylester Hydrochloride

Chemical
Starting point in making various pharmaceutical chemicals.

Para-aminobenzoic Acid Propylester
Synonyms: Propyl-para-aminobenzoate.
German: Para-aminobenzoesaeurespropylaether.

Chemical
Reagent in making various synthetic products used in medicine.

Para-aminobenzylidenephenylethylhydrazone

Dye
Starting point in making—
Chromogene blue R.

Para-aminobenzylparatoluidin

Chemical
Starting point in making—
Dihydrothioparatoluidin.

Para-aminodimethylanilin

Analysis
Reagent in various processes.

Dye
Starting point in making—
Azo acid blues, azogallein, brilliant alizarin blue, clematin, diphenyl blue R, ethyl acid blue RR, ethylene blue, fast blue, fast marine blue, fuchsia, indophenol, leucogallotrionin, methylene blue, methylene grays, methylene green, methylene violet, modern cyanin, naphthol blue, neutral red extra, neutral violet extra, new fast blue B, thionin GO, thionin O, thionin BR, thiophor indigo G, toluylene blue, toluylene red, urania blue.

Photographic
Developing agent for films and plates.

Chemical
Starting point in making—
Rubber vulcanization accelerator by reaction with heptyl aldehyde (Brit. 259933).

Para-aminomethylanilin

Dye
Starting point in making—
Azo acid blue B.

Para-aminophenol

Analysis
Reagent in testing for—
Formaldehyde.

Chemical
Starting point in making—
Intermediates, organic chemicals, pharmaceutical chemicals, salophene, sugar substitutes, synthetic aromatic chemicals.

Dye
Starting point in making—
Azo chromin, diphenyl chrysoin, immedial blacks, immedial dark brown A, immedial dark brown B, immedial indone, immedial pure blue, Italian green, pyrogene black B, pyrogene black G, pyrogene direct blue, pyrogene blue, pyrogene olive N, pyrogene yellow synthetic dyestuffs with the aid of chromotropic acid, thion blue B, ursol P, various sulphur dyestuffs, vidal black.
Starting point (Brit. 319390) in making azo dyestuffs with the aid of—
3:5-Dinitro-orthoanisidin.
3:5-Dinitro-orthotoluidin.
3:5-Dinitro-metatoluidin.
3:5-Dinitroparatoluidin.
Starting point (Brit. 323792) in making azo dyestuffs for use in dyeing viscose rayon, cuprammonium rayon, and nitro rayon, with the aid of—
2-Amino-5-nitrobenzanilide.
4-Chloro-2-nitro-4′-aminodiphenylamine.
4:4′-Diamino-3:3-dinitrobenzophenone.
4:4′-Diamino-3:3′-dinitrodiphenylamine.
4:4′-Diamino-2:2′-dinitrodiphenylurea.
2:4-Dinitro-3′-aminodiphenylamine.
2:4-Dinitro-4′-aminodiphenylamine.
2:2′-Dinitrobenzidin.
3:3′-Dinitrobenzidin.
6:6′-Dinitro-orthoanisidin.
5:5′-Dinitro-orthotoluidin.
5-Nitro-2-aminobenzophenone.
3-Nitro-4-aminobenzophenone.
2-Nitro-2′-aminobenzophenone.
2-Nitrobenzidin.
4-Nitro-4′-aminodiphenyl ether.
3-Nitro-4-aminodiphenyl ether.
5-Nitro-2-aminodiphenyl ether.
2-Nitro-4′-aminodiphenylamine.
4-Nitro-4′-aminodiphenylamine.
2-Nitro-4-aminodiphenylamine.
2-Nitro-4-amino-4′-methyldiphenylamine.
4-Nitrobenzoylparaphenylenediamine.
4′-Nitrobenzyl-2-amino-5-nitroaniline.
5-Nitro-orthotoluidin.

Leather
Reagent in dyeing.

Miscellaneous
Reagent in dyeing—
Furs, hair, hair by oxidation with hydrogen peroxide or sodium bichromate.

Photographic
As a developer.

Textile
Reagent in dyeing—
Yarns and fabrics.

Para-aminophenol Hydrochloride

Chemical
Starting point in making—
Intermediates and other organic chemicals, pharmaceutical chemicals, photographic chemicals, synthetic aromatics.

Dye
Starting point in making—
Anilin black dyestuffs, azo chromin, diphenyl chrysoin,

Para-aminophenol Hydrochloride (Continued)
immedial blacks, immedial dark brown A, immedial dark brown B, immedial indone, immedial pure blue, Italian green, monoazo dyestuffs, pyrogene black B, pyrogene black G, pyrogene blue, pyrogene direct blue, pyrogene olive N, pyrogene yellow, stilbene dyestuffs, synthetic dyestuffs with the aid of chromotropic acid, thion blue B, ursol P, various sulphur dyestuffs, vidal black.
Starting point (Brit. 319390) in making azo dyestuffs with the aid of—
3:5-Dinitrometatoluidin.
3:5-Dinitro-orthoanisidin.
3:5-Dinitro-orthotoluidin.
3:5-Dinitroparatoluidin.
Starting point (Brit. 323792) in making azo dyestuffs for use in dyeing viscose rayon, cuprammonium rayon, and nitro rayon, with the aid of—
2-Amino-5-nitrobenzanilide.
4-Chloro-2-nitro-4′-aminodiphenylamine.
4:4′-Diamino-3:3′-dinitrobenzophenone.
4:4′-Diamino-3:3′-dinitrodiphenylamine.
4:4′-Diamino-2:2′-dinitrodiphenylurea.
2:4-Dinitro-3′-aminodiphenylamine.
2:4-Dinitro-4′-aminodiphenylamine.
2:2′-Dinitrobenzidin.
3:3′-Dinitrobenzidin.
6:6′-Dinitro-orthoanisidin.
5:5′-Dinitro-orthotoluidin.
5-Nitro-2-aminobenzophenone.
3-Nitro-4-aminobenzophenone.
2-Nitro-2′-aminobenzophenone.
2-Nitrobenzidin.
4-Nitro-4′-aminodiphenyl ether.
3-Nitro-4-aminodiphenyl ether.
5-Nitro-2-aminodiphenyl ether.
2-Nitro-4′-aminodiphenylamine.
4-Nitro-4′-aminodiphenylamine.
2-Nitro-4-aminodiphenylamine.
2-Nitro-4-amino-4′-methyldiphenylamine.
4-Nitrobenzoylparaphenylenediamine.
4′-Nitrobenzyl-2-amino-5-nitroaniline.
5-Nitro-orthotoluidin.

Leather
Reagent in dying.

Miscellaneous
Reagent in dyeing—
Deep reddish shades in furs, plumes, and hair (French 549000).
Furs, hair, hair by oxidation with hydrogen peroxide or sodium bichromate.

Photographic
As a developer.

Resins and Waxes
Reagent in dyeing—
Artificial resins of the phenol-formaldehyde type (French 610108).

Textile
Reagent in making—
Yarns and fabrics.

Para-aminophentole Hydrochloride

Chemical
Starting point in making—
Acetphenetidin.

Para-aminophenylarsinic Acid
French: Acide de para-aminophénylearsinique.
German: Para-aminophenylarsinsaeure.

Chemical
Starting point in making—
Metanitroparahydroxylphenylarsinic acid (U. S. 1607299).

Para-aminophenylbetaethoxyethyl Ether

Dye
Starting point (Brit. 443104) in making—
Wool dyes having good light-fastness, by reacting with a triphenylmethane dye derived from benzaldehyde in which the paraposition is substituted by a replaceable group (nitro-, halogen, sulphonate).
Wool dyes having good light-fastness, by reaction with light green SF in the presence of hydrochloric acid.
Wool dyes having good light-fastness, by reacting with the dye from parachlorobenzaldehyde and sulphobenzylethylanilin.
Wool dyes having good light-fastness, by reacting with the dye from parachlorobenzaldehyde and metatoluidinsulphonic acid.

Para-aminophenylbetahydroxyethyl Ether

Dye
Starting point (Brit. 443104) in making—
Wool dyes having good light-fastness, by reacting with a triphenylmethane dye derived from benzaldehyde in which the paraposition is substituted by a replaceable group (nitro-, halogen, sulphonate).
Wool dyes having good light-fastness, by reacting with light green SF in the presence of hydrochloric acid.
Wool dyes having good light-fastness, by reacting with the dye from parachlorobenzaldehyde and sulphobenzylethylanilin.
Wool dyes having good light-fastness, by reacting with the dye from parachlorobenzaldehyde and metatoluidinsulphonic acid.

Para-aminophenylbetamethoxyethyl Ether

Dye
Starting point (Brit. 443104) in making—
Wool dyes having good light-fastness, by reacting with a triphenylmethane dye derived from benzaldehyde in which the paraposition is substituted by a replaceable group (nitro-, halogen, sulphonate).
Wool dyes having good light-fastness, by reacting with light green SF in the presence of hydrochloric acid.
Wool dyes having good light-fastness, by reacting with the dye from parachlorobenzaldehyde and sulphobenzylethylanilin.
Wool dyes having good light-fastness, by reacting with the dye from parachlorobenzaldehyde and metatoluidinsulphonic acid.

Para-aminophenylmercaptan Hydrochloride
Synonyms: Para-aminothiophenol hydrochloride.

Insecticide and Fungicide
Larvicide for—
Culicine mosquito larvae.

Para-aminosalicylic Acid
French: Acide de para-aminosalicylique.
German: Para-aminosalicylsäure.

Chemical
Starting point in making—
Esters and salts, pharmaceuticals, intermediates.

Dye
Starting point in making—
Azo dyestuffs.
Starting point (Brit. 325485) in making dyestuffs with the aid of—
4-Acetylaminophenol.
2:4-Dimethylphenol.
Hydroquinonemonomethyl ether.
Parachlorometacresol.
Paracresol.
Parahydroxydiphenylmethane.

Para-amylphenol, Tertiary
German: Amylphenol.

Insecticide
Suggested for use as—
Fumigant, insecticide.

Resins and Waxes
Starting point in making—
Oil-soluble varnish resins.

Sanitation
Suggested for use as—
Germicide.

Para-anisylmercuric Acetate

Chemical
Starting point in making various derivatives.

Insecticide
Ingredient (Brit. 321396) of—
Immunizing compositions used in the treatment of wheat and other grains.

Miscellaneous
Ingredient of—
Preservative and disinfectant compositions used for general purposes.

Woodworking
Ingredient (Brit. 321258) of—
Compositions used for preservation and disinfection.

Parabenzylideneaminophenol

Petroleum

Gum inhibitor (U. S. 1980200 and 1980201) in—
Motor fuels.

Parabenzylmethylaminophenol

Petroleum

Gum inhibitor (U. S. 1980200 and 1980201) in—
Motor fuels.

Parabenzylphenol

Chemical

Starting point (Brit. 444351) in making—
Fat-splitting catalysts and emulsifying agents, useful in dyeing, laundering, bleaching, and various other purposes, by reacting with formaldehyde and nonaromatic secondary amines (the salts of the products with water-soluble acids, or water-insoluble acids, or the quaternary ammonium salts are claimed to be valuable for the purposes named).

Parabetahydroxyethoxyphenylarsinic Acid

Germicide

Claimed (Brit. 444882) as—
Bactericide.

Parabromometahydroxybenzoic Acid

French: Acide de parabromométahydroxybenzoique.
German: Parabrommetahydroxybenzoesaeure.

Chemical

Starting point in making—
Protocatechuic acid.

Parabromophenylmercaptan

Synonyms: Parabromothiophenol.

Insecticide and Fungicide

Larvicide for—
Culicine mosquito larvae.

Paracarboxyphenyl-4-paratolyl-7:8-phthaloyl-2-quinolin Acid Chloride, Normal

Chemical

In organic syntheses.

Dye

In dye syntheses.
Starting point (Brit. 449263) in making—
Yellow vat dyes with 1-amino-5-benzamidoanthraquinone.

4-Paracarboxyphenyl-7:8-phthaloyl-2-quinolin Acid Chloride

Chemical

In organic syntheses.

Dye

In dye syntheses.
Starting point (Brit. 449263) in making—
Yellow vat dyes with 1:2-aminoanthraquinone.

Parachloroanilide

Dye

As an intermediate.
Starting point (Brit. 396859) in making—
Fast red-blue colors on wool.

Parachloroanilin

Chemical

Reagent in making—
Diazotization products.
Diparachloroanilidoanthraquinone (Brit. 248874).
Monoparachloroanilidoanthraquinone (Brit. 248874).
Starting point in making—
Paraphenylenediamine.

Dye

Starting point in making—
Diazotization products, quinolin yellow.

Parachlorobenzyltriphenyl Chloride

French: Chlorure de parachlorobenzyletriphényle, Chlorure de parachlorobenzyletriphénylique.
German: Chlorparachlorbenzyltriphenyl, Parachlorbenzyltriphenylchlorid.

Miscellaneous

Reagent in mothproofing—
Furs, hair, feathers, and other articles.

Textile

Reagent in mothproofing—
Wool and felt.

Parachlorometacresol

Insecticide and Fungicide

Fungicide for—
Mildew growth.
Inhibitor of—
Mildew growth.

Paint and Varnish

Sterilizer for treating—
Mildewed paint surfaces prior to repainting.

Parachlorometahydroxybenzoic Acid

French: Acide de parachlorométahydroxybenzoique.
German: Parachlormetahydroxybenzoesaeure.

Chemical

Starting point in making—
Protocatechuic acid.

Parachloronitrobenzene

Synonyms: Parachloronitrobenzol.

Chemical

Starting point in making—
1-Chloro-4-nitrobenzol-2-sulphonic acid.
1:4-Dichloro-4-nitrobenzene.
Paranitranilin.
Paranitrophenetole (French 602977).
Paranitrophenol.

Dye

Starting point in making various synthetic dyestuffs.

Parachlorophenol

French: Parachlorophénol.
German: Parachlorphenol.
Spanish: Paraclorofenole.
Italian: Paraclorofenol.

Disinfectant

As a disinfectant.

Parachlorophenylmercuric Acetate

French: Acétate de parachlorophénylemercure, Acétate parachlorophénylique-mercurique.
German: Essigsäureparachlorphenylmerkurester, Essigsäuresparachlorphenylmerkur, Parachlorphenylmerkuracetat, Parachlorphenylmerkurazetat.

Chemical

Starting point in making various derivatives.

Insecticide

Ingredient (Brit. 321396) of—
Immunizing compositions used in the treatment of wheat and other grains.

Miscellaneous

General preservative and disinfectant.

Woodworking

Preservative and disinfectant (Brit. 321396).

Parachlorophenylorthotoluidin 4-Sulphonate

Dye

Coupling agent (Brit. 434209 and 434433) in making—
Yellowish-red water-insoluble dyestuffs with 5-methoxyorthotoluidide.

Paracinnamylideneaminophenol

Petroleum

Gum inhibitor (U. S. 1980200 and 1980201) in—
Motor fuels.

Paracresol Cinnamate

French: Cinnamate de paracrésol, Cinnamate paracrésolique.
German: Parakresolcinnamat, Zimtsäureparacresolester, Zimtsäuresparakresol.

Chemical

Starting point in making—
Pharmaceuticals and other derivatives.

Pharmaceutical

In compounding and dispensing practice.
As an odorant antiseptic (phenol coefficient of over 500) (U. S. 2010318).

Paracresyl Acetate
French: Acétate paracresylique, Acétate de paracresyle.
German: Essigsäureparakresylester, Essigsäuresparakresyl, Parakresylacetat, Parakresylazetat.
Chemical
Starting point in making various derivatives.
Pharmaceutical
In compounding and dispensing practice.

Paracresylphenyl Acetate
Chemical
Starting point in making—
Aromatics and other derivatives.
Perfume
Ingredient of the following odors:—
Fern, jasmine, narcissus, tuberose.
Perfume in—
Cosmetics, toilet waters.
Soap
Perfume in—
Toilet soaps.

Paracymene
Synonyms: Cymol, Isopropylenetoluene, Paracymol.
Chemical
Ingredient of—
Synthetic oil of cinnamon.
Solvent in various processes.
Starting point in making—
Carvacrol, cymenesulphonic acid, 2:4-dinitrotoluene, methylisopropylanthraquinone, para-aminocarvacrol, paranitrotoluenesulphonic acid, thymol.
Dye
Starting point in making—
Azo colors.
Fats and Oils
Solvent in various extraction processes.
Metallurgical
Ingredient of—
Polishes for metals.
Paint and Varnish
Ingredient of—
Paint and varnish removers, with alcohols or acetone.
Rubber
Starting point in making—
Synthetic rubber.

Paracymenedisulphonic Acid
French: Acide de paracymènedisulphonique.
German: Paracymendisulfonsaeure.
Chemical
Reagent (Brit. 263873) in making—
Aromatic hydrocarbon emulsions, emulsified fat solvents, terpene emulsions.
Starting point in making—
Esters and salts, intermediates, pharmaceuticals.
Dye
Starting point in making various synthetic dyestuffs.
Fats and Oils
Reagent (Brit. 263873) in making—
Emulsions.
Leather
Reagent (Brit. 263873) in making—
Emulsified impregnating solutions, emulsified tanning solutions.
Miscellaneous
Reagent (Brit. 263873) in making—
Emulsified washing and cleansing compositions.
Paper
Reagent (Brit. 263873) in making—
Emulsified impregnating compositions.
Petroleum
Reagent (Brit. 263873) in making—
Emulsions of petroleum and distillates.
Resins and Waxes
Reagent (Brit. 263873) in making—
Emulsified compositions.
Textile
—, *Dyeing*
Ingredient (Brit. 263873) of—
Acid dyestuffs liquors.
—, *Finishing*
Ingredient (Brit. 263873) of—
Wetting agents.
—, *Manufacturing*
Ingredient (Brit. 263873) of—
Wool-carbonizing liquors.

Paradiaminoanthrarufin-2:6-disulphonic Acid
French: Acide de paradiaminoanthrarufin-2:6-sulphonique.
German: Paradiaminoanthrarufin-2:6-sulfonsaeure.
Dye
Starting point (Brit. 274211) in making anthraquinone dyestuffs with—
Alpha-aminonaphthol, alphanaphthol, alphanitronaphthol, alphachloronaphthol, anisole, beta-aminonaphthol, betachloronaphthol, betanitronaphthol, betanaphthol, diphenyl ether, dichlorophenol, guaiacol, hydroquinone, monochlorophenol, phenol, phenetole, pyrocatechol, resorcinol.

Paradibenzylaminophenol
Petroleum
Gum inhibitor (U. S. 1980200 and 1980201) in—
Motor fuels.

Paradibromobenzene
Synonyms: 1:4-Dibromobenzene, Paradibromobenzol.
French: Paradibromobenzène.
German: Paradibrombenzol.
Chemical
Reagent in—
Organic synthesis.

Paradiethylamino-orthosulphobenzaldehyde
Dye
Starting point (Brit. 431652) in making—
Orange dyestuffs with 5-methylcoumaronone.
Red dyestuffs with thioindoxyl.

Paradiethyldodecylamine Normal Oxide
Miscellaneous
As a wetting agent (Brit. 437566).
For uses, see under general heading: "Wetting agents."

Paradiethylhexadecylamine Normal Oxide
Miscellaneous
As a wetting agent (Brit. 437566).
For uses, see under general heading: "Wetting agents."

Paradimethylaminoanthrarufin-2:6-disulphonic Acid
French: Acide de paradiméthyleaminoanthrarufin-2:6-disulphonique.
German: Disulfonsaeure-2:6-paradimethylaminoanthrarufinester, Disulfosaeures-2:6-dimethylaminoanthrarufin, Paradimethylaminoanthrarufin-2:6-disulfonsaeure.
Dye
Starting point (Brit. 274211) in making anthraquinone dyestuffs with—
Alpha-aminonaphthol, alphanaphthol, alphachloronaphthol, alphanitronaphthol, anisole, beta-aminonaphthol, betachloronaphthol, betanaphthol, betanitronaphthol, dichlorophenol, diphenyl ether, guaiacol, hydroquinone, phenetole, phenol, pyrocatechol, resorcinol, veratrol.

Paradimethylaminobenzaldehyde
Chemical
As in intermediate.
Textile
Starting point (Brit. 396893) in making—
Red-yellow shades on acetate rayon.

Paradimethylaminododecylbenzene Normal Oxide
Miscellaneous
As a general wetting agent (Brit. 437566).
See also "Wetting agents."
Textile
As a dyeing assistant (Brit. 437566).
As a general wetting agent (Brit. 437566).
Wetting agent (Brit. 437566) in—
Wool washing.

Paradimethylaminolaurophenone Normal Oxide

Miscellaneous
As a general wetting agent (Brit. 437566).
See also "Wetting agents."

Textile
As a dyeing assistant (Brit. 437566).
As a general wetting agent (Brit. 437566).
Wetting agent (Brit. 437566) in—
Wool washing.

Paradimethylaminophenylnaphthyl Ether

Chemical
Antioxidant and stabilizer (Brit. 430335) for—
Unstable organic substances.

Fats, Oils, and Waxes
Antioxidant and stabilizer (Brit. 430335) for—
Fats, oils, waxes.

Petroleum
Antioxidant and stabilizer (Brit. 430335) for—
Petroleum products.
Inhibitor (Brit. 430335) of—
Gumming in petroleum products.

Rubber
As an oxidant (Brit. 430335).

Paradimethylaminophenylnaphthyl Telluride

Chemical
Antioxidant and stabilizer (Brit. 430335) for—
Unstable organic substances.

Fats, Oils, and Waxes
Antioxidant and stabilizer (Brit. 430335) for—
Fats, oils, waxes.

Petroleum
Antioxidant and stabilizer (Brit. 430335) for—
Petroleum products.
Inhibitor (Brit. 430335) of—
Gumming in petroleum products.

Rubber
As an oxidant (Brit. 430335).

Paradimethylaminostearophenone Normal Oxide

Miscellaneous
As a general wetting agent (Brit. 437566).
See also: "Wetting agents."

Textile
As a dyeing assistant (Brit. 437566).
As a general wetting agent (Brit. 437566).
Wetting agent (Brit. 437566) in—
Wool washing.

1-Paradimethylaminostyryl-4:5-benzoxazole Methiodide

Photographic
Sensitizer (Brit. 432969) for—
Silver halide emulsions (sensitizing maxima: indefinite).

Paradimethylaminostyrylmethyl Ketone

Textile
Starting point (Brit. 396893) in making—
Orange shades on acetate rayon.

Paradimethyldodecylamine Normal Oxide

Miscellaneous
As a general wetting agent (Brit. 437566).
See also "Wetting agents."

Textile
As a dyeing assistant (Brit. 437566).
As a general wetting agent (Brit. 437566).
Wetting agent (Brit. 437566) in—
Wool washing.

Paradimethylhexadecylamine Normal Oxide

Miscellaneous
See also "Wetting agents."

Textile
As a dyeing assistant (Brit. 437566).
As a general wetting agent (Brit. 437566).
Wetting agent (Brit. 437566) in—
Wool washing.

Paradinitrobenzene
Synonyms: Paradinitrobenzol.
German: Paradinitriertbenzol.

Chemical
Reagent in—
Organic synthesis.
Starting point in making—
Intermediates, synthetic pharmaceuticals.

Dye
Starting point in making various synthetic dyestuffs.

Perfume
Starting point in making—
Synthetic perfumes.

Paradioxydiphenyldimethylmethane

Chemical
Starting point in making—
Cyclohexanol.
4-Isopropylcyclohexanol.
Paraisopropylphenol (Brit. 254753).

Dye
Starting point in making various dyestuffs.

Paradiphenylyl Benzoate

Plastics
Addition agent (U. S. 1933822) for—
Cellulose acetate solutions used to make plastic sheets having a nacreous appearance.

Paraditolyl Ketone

Chemical
Starting point in—
Organic synthesis.

Rubber
Retardant (Brit. 426649) of—
Vulcanization of rubber mixes containing sulphur and an accelerator, in the initial stages.

Paraethoxyquinaldinmethyl Sulphate

Photographic
Starting point in making—
Sensitizing agents (Brit. 262816).

Paraffin
Synonyms: Hard paraffin, Solid paraffin.
Latin: Paraffinum durum, Paraffinum solidum.
French: Paraffine.
German: Festes paraffin, Gereinigtes erdwachs.
Spanish: Parafina.

Adhesives
Ingredient of—
Adhesive compositions.

Agriculture
Ingredient (U. S. 1738864) of—
Composition for treating Florida fruit products, containing also starch, a volatile hydrocarbon liquid, and paraffin oil.

Analysis
As a heating medium in baths.
As a reagent.

Brewing
Impregnating agent for—
Barrel interiors.

Chemical
Absorbent (Brit. 321239) for—
Wool-fat acids.
Coating agent for—
Acid tanks, chemical apparatus.
Preventive for—
Oxidation of protoxides.
Raw material for making—
Acid bottle stoppers, bottles for hydrofluoric acid.
Solvent.

Construction
General waterproofing agent (Brit. 287514), used alone or in combination with other substances, for treating—
Brickwork, concrete, masonry, piles, porous structural materials, shingles, walls.

Distilling
Coating agent for—
Vats.

Electrical
As a general insulating agent.
Binding, coating, and insulating agent for—
Electrical condensers.

Paraffin (Continued)
Boiling out agent for—
Cables and other materials to remove moisture and improve their electrical properties.
Coating and insulating agent for—
Dry-cell batteries.
Household light wires, radio wires, telephone wires, wires of all kinds of domestic electrical appliances.
Industrial electrical cables and industrial electrical machinery.
Radio coils and other electrical coils.
Utility cables and machinery.
Ingredient of—
Insulating compositions containing rubber.
Insulating compositions for wires of all kinds.
Insulating compositions for industrial electrical cables and industrial electrical machinery.
Insulating compositions for electric utility cables and machinery.
Insulating and sealing compositions for dry cells.
Molded insulations.
Waterproofing agent for—
Electrical instruments, electrical machinery.

Explosives and Matches
Coating agent for—
Cartridges, stems of paper or vesta matches, stems of wooden matches (to provide a smooth, shiny surface).
Ingredient of—
Coal-mine explosives, matchhead compositions.
Waterproofing agent for—
Explosives, matches.

Food
Coating for—
Molds for making display products, such as artificial jellies, chocolate, foods of various kinds.
Ingredient of—
Candies, chewing gums.
Compositions for decorating fancy food products.
Preservative and coating agent for—
Eggs.
Raw material in making—
Artificial honey combs.
Sealing agent for bottled and jarred foods, such as—
Catsup, fruits, jams, jellies, meats, preserves.

Forestry
Ingredient of—
Compositions for curing brown bast in rubber trees.
Grafting dressing (mixed with rosin).

Fuel
Component of—
Candles, night lights.
Fuel for—
Flares used in Asia and the East (in night-time construction work).
Miners' lamps.
Railway carriage lamps on Asiatic and Eastern railroads.
Ship lamps on ships traversing Eastern seas.
Outer coating for—
Chinese funeral candles.
Chinese joss candles.
Stiffener for—
Chinese domestic candles.

Ink
Ingredient of—
Lithographic inks.
Marking ink for stencilling designs on wooden boxes (French 738921).
Printing inks, stamping inks, writing inks.

Laundering
Detergent in—
Boiling operations.
Lubricant for—
Flatirons and ironing machines.
Polishing and stiffening agent for—
Collars.

Leather
Ingredient of—
Dressings (U. S. 1847629), finishing preparations, polishing compositions.
Waterproofing agent.

Mechanical
As a coating against rust.
Ingredient of—
Belt dressing containing also asphalt, white lead, neatsfoot oil, tallow, and citronella oil (U. S. 1751342).
Lubricating compositions.

Metallurgical
Coating agent for—
Foundry molds.
Ingredient of—
Compositions used for covering metals to provide protection against moisture, acids, alkalies, and other corrosive substances.
Corrosion-resisting composition used as coating for metals and containing also petrolatum, oxidized petroleum bitumen, asbestos, and powdered shale (Brit. 397267).
In various electroplating processes.
Protective agent in—
Acid etching.
Wire wax or core vent in—
Casting.

Miscellaneous
Coating for—
Barrels in which fish are packed on the Pacific coast.
Butter tub interiors, cheese box interiors, cigaret packers, multiple boxboard food containers (U. S. 1895527).
Embalming agent in China (the coffin is completely filled with liquid wax and the corpse is then immersed in it).
Filling for—
Artificial pearls.
Impregnating agent for—
Keeping sponges elastic.
Ingredient of—
Cleaning and polishing fluid (U. S. 1730654).
Compositions for coloring artificial citrous fruits (U. S. 1846143).
Compositions for making dental impressions (U. S. 1897034).
Compositions for making anatomical specimens.
Compositions for lining barrels and kegs.
Compositions for painting old timber to prevent attack of deathwatch beetle.
Compositions for waterproofing automobile tops and tarpaulin.
Compositions used in the manufacture of incandescent gas mantles.
Floor polishes, furniture polishes, linoleum polishes, polishes of various sorts, preparations for making imitation alabaster statues, shoe polishes, wood polishes, ski polishes.
Preservative for—
Antiques and also for wooden articles found in tombs (it acts as an impregating agent and renders the articles strong enough to be handled).
Cables on Egyptian railways.
Flowers (by the dipping process).
Raw material in making—
Grease crayons, oil crayons, imitation fruit and flowers, toys, wax figures for exhibition purposes and for window display.
Substitute for—
Batching oil in rope making—
Waterproofing agent for—
Cloth liners for automobile tires, pasteboard signs exposed to the weather, soda straws.
Waterproofing agent (Brit. 287514), used either alone or in compositions, for the treatment of—
Asbestos, cork, porous materials of various kinds, strawboards.

Oils and Fats
Base of various lubricating compositions.
Ingredient of—
Axle greases, gun oils.
Lubricant compound with beeswax, rosin, castor oil, and graphite (U. S. 1735368).
Lubricating grease compound with castor oil, mineral oil and aluminum stearate (U. S. 1881591).
Special lubricants.

Paint and Varnish
Ingredient of—
Paints, preparations containing dry colors, special floor waxes, varnishes, wood fillers.

Paper
Coating for—
Tracing paper, wax paper.
Ingredient of—
Emulsified sizing compositions (Brit. 287514).
Compositions used in the manufacture of carbon paper.

Paraffin (Continued)
Preparations used in making wax paper.
Sizings for high-gloss paper.
Shortening agent (U. S. 1894731) for—
Phenol-tung oil-formaldehyde resinous coating for stencil paper.
Waterproofing agent for—
Boxboard, cardboard, cartons, paper, paper drinking cups.

Perfume
Extraction agent for—
Perfumes and odors from flowers.
Ingredient of—
Mascara compositions.
Raw material in making—
Creams, pencils, pomades.

Pharmaceutical
Base for—
Balms, ointments.
Coating for—
Pills, tablets.
In compounding and dispensing practice.
Suggested for use as—
Dressings in treatment of wounds, ulcers, burns.
Filler in plastic surgery.
Ingredient of bone-waxes.
Substitute for plaster of paris for splints.
Used in Europe in the so-called "paraffin wax" bath treatments.

Photographic
Coating for—
Photographic papers.
Finishing agent for—
Glossy prints.

Plastics
Coating agent for—
Plaster casts.
Ingredient of—
Phonograph records.

Printing
Ingredient of—
Compositions used for the preparation of acidproof coatings for plates in the electrotyping process.
Compositions used for making matrices in galvanoplastic work.
Process material in—
Lithography, photoengraving, process engraving.

Resins and Waxes
Ingredient of—
Batikwax (used in natural dying processes in Java and the East).
Compounded waxes, electrotypers' wax, sealing wax, shoemakers' wax.
Substitute for—
Animal and vegetable waxes.

Rubber
Coating agent for—
Molds (to prevent sticking of the article molded).
Filler in making—
Rubber compositions.
Ingredient of—
Rubber compositions (added to give the rubber a polished or finished appearance).

Shipbuilding
Ingredient of—
Mixture with tallow for greasing ships' slipways to facilitate launching operations.

Soap
Ingredient of—
Laundry soaps.

Stone
Waterproofing agent (Brit. 287514), used either alone or in admixture with other substances, for treating—
Artificial stone, natural stone.

Sugar
Antifrothing agent in—
Sugar evaporators.
Sealing agent for—
Sugar cane pieces (to prevent dessication).

Textile
Assistant (Brit. 397881) in—
Stretching cellulose acetate filaments.
Glazing agent in—
Hot calendering.
Ingredient of—
Compositions used for finishing.
Compositions used for softening.
Compositions used for sizing.
Compositions used in the manufacture of waxed cloth.
Emulsified dressing (Brit. 287514).
Waterproofing coating, containing also blown asphalt and rubber, for cellulose fibers (U. S. 1880036).
Waterproofing coating, along with castor oil, rubber, and petrolatum.
Waterproofing compositions (Brit. 287514).
Polishing agent for—
Weaving machine rollers.
Stiffening ("starching") agent for—
Linen.
Waterproofing agent in treating—
Yarns and fabrics.
Wax for—
Hosiery stitching threads.

Tobacco
Waterproofing agent for packagings for various products.

Wine-making
Coating and impregnating agent for—
Cheap wine casks.
Ingredient of—
Compositions used for coating interiors of tankcars used for transporting wine in bulk.

Woodworking
Coating and impregnating agent for—
Artificially dried wood (to prevent reabsorption of moisture).
Log ends (to prevent splitting and infection by borers).
Ingredient of—
Compositions used in the finishing of furniture and of lumber used for parquet flooring.

Parahexadecyldiphenylamine

Rubber
Preservative (U. S. 2009480, 2009526, and 2009530) for—
Rubber.

Parahydroxyphenylacetimidophenyl Hydrochloride-Sulphide
Synonyms: Parahydroxyphenylacetimidothiophenylether hydrochloride.

Insecticide
Larvicide for—
Culicine mosquito larvae.

Parahydroxyphenylalphanaphthylamine

Dye
Starting point (Brit. 429642) in making—
Deep-black dyes by sulphurizing in absence of water by baking with sulphur and sodium sulphide.

Parahydroxyphenylbetanaphthylamine

Dye
Starting point (Brit. 429642) in making—
Deep-black dyes by sulphurizing in absence of water by baking with sulphur and sodium sulphide.

Paraisopropoxydiphenylamine

Rubber
Aging retardant (Brit. 424461).
Promoter (Brit. 424461) of—
Resistance to crack formation on flexing.

Paraisopropyldiphenylamine

Rubber
Preservative (U. S. 2009480, 2009526, and 2009530) for—
Rubber.

Paraldehyde
Synonyms: Para-aldehyde.

Chemical
Substitute for—
Acetaldehyde (on account of its high boiling point and ease in handling).
Solvent for—
Certain natural gums, fats, rosin, waxes.
Starting point in—
Organic synthesis.

Dye
Starting point in making—
Quinaldin dyes.

Paraldehyde (Continued)
Fats, Oils, and Waxes
Solvent for—
Fats, waxes.
Glue and Adhesives
Solvent for—
Some natural gums.
Leather
Degreasing agent for—
Hides and skins.
Plumping agent for—
Skins.
Miscellaneous
Solvent for—
Varnishes, fats, gums, resins, rosin, and waxes in processes in the manufacture of inks, paper, engravings and lithographings, textiles, waterproofings, cosmetics, insulations, matches, polishes, and dressings, coatings, linoleum and oilcloth, crayons, sealing compounds, lubricants, photographic products, plastics, printing, soap and other products.
Perfume
Raw material in making—
Synthetic perfumes.
Paint and Varnish
Solvent for—
Gums, resins, rosin, varnishes, waxes.
Pharmaceutical
Suggested for use as—
Antispasmodic, hypnotic, sedative.
Resins
Raw material in making—
Synthetic resins.
Solvent for—
Resins.
Rubber
Raw material for making—
Accelerators of vulcanization.

Paramethoxycinnamic Aldehyde
French: Aldéhyde paraméthoxycinamique.
German: Paramethoxyzimtaldehyd.
Spanish: Aldehido parametoxicinamico.
Italian: Aldeide parametossicinnamica.
Perfume
Ingredient of—
Perfume preparations with hawthorn odor.
Perfume in various toiletries.
Soap
Perfume in—
Toilet soaps.

Paramethoxydiphenylamine
Rubber
As an antioxidant (Brit. 435024).

2-Paramethoxyphenylbenziminazole
Cosmetic
Protective (Brit. 435811) in—
Sun-tan lotions (solution or dispersion in a compatible solvent, for example, glycerin or wool-fat, but not water, alcohol, benzene, carbon tetrachloride, chloroform, or acetone), said to prevent formation of painful erythemas whilst enabling the skin to grow brown in sunlight, by virtue of high absorption of ultraviolet rays.

Paramethylacetophenone
French: Paraméthyleacétophenone.
German: Paramethylacetophenon.
Chemical
Starting point in making—
Aromatics, intermediates, pharmaceuticals.
Perfume
Ingredient of—
Artificial essence of mimosa.
Perfume preparations.
Perfume for various preparations.
Substitute for coumarin for various purposes.
Soap
Perfume for—
Toilet soaps.

Paramethylaminophenol
Chemical
As an intermediate.
Stabilizing agent (Brit. 397914) for—
Chlorinated hydrocarbons.

Paramethylcyclohexanone
Chemical
Starting point (Brit. 313421) in making condensation products with—
Alphanaphthylamine, anilin, anisidin, benzylamine, benzylanilin, betanaphthylamine, cresidin, meta-anisidin, metacresidin, metanaphthylamine, metaphenylenediamine, metatoluylenediamine, metaxylenediamine, orthoanisidin, orthocresidin, orthonaphthylamine, orthophenylenediamine, orthotoluylenediamine, orthoxylenediamine, para-anisidin, paracresidin, paranaphthylamine, paraphenylenediamine, paratoluylenediamine, paraxylenediamine.

Paranitrazobenzene
Synonyms: Paranitrazobenzol.
German: Paranitrazobenzol.
Dye
Starting point (Austrian 105341) in making ice colors with—
Alphabromobetanaphthol.
Alphabromo-2-hydroxy-3-naphthoic acid.
Alphachlorobetanaphthol.
Alphamethylbetanaphthol.
Alphanitrobetanaphthol.
Alphasulphomethylbetanaphthol.
1:3:6-Tribromobetanaphthol.

Paranitrobenzhydrazide
Analysis
Reagent.
Chemical
Reagent in—
Organic synthesis.

Paranitrobenzyl Chloride
French: Chlorure de paranitrobenzyle, Chlorure paranitrobenzylique.
Chemical
Starting point in making—
Paradinitrobenzaldehyde, paranitrobenzyl alcohol, paranitrobenzylanilin.
Dye
Starting point in making—
New phosphin G, parafuchsin, tannin orange R.

Paranitrobenzyltriphenyl Chloride
French: Chlorure de paranitrobenzyletriphényle, Chlorure paranitrobenzyletriphénylique.
German: Chlorparanitrobenzyltriphenyl, Paranitrobenzyltriphenylchlorid.
Spanish: Cloruro de paranitrobenziltriphenil.
Italian: Cloruro di paranitrobenziletriphenile.
Miscellaneous
Reagent (Brit. 312163) for—
Making hair, feathers, and the like mothproof and moldproof.
Textile
Reagent (Brit. 312163) for—
Making wool and felt mothproof and moldproof.

Paranitrodiazobenzene
Chemical
Reagent in making—
Paranil A.
Dye
Reagent in making—
Para brown G.
Textile
—, *Dyeing*
Developing agent for—
Yellowish red shades on cellulose acetate rayon (Brit. 262830).

Paranitro-orthoaminotoluene
Synonyms: Paranitro-orthotoluidin, Paranitro-orthoaminotoluol.
Dye
Starting point in making—
Cotton red S, pigment chlorine GG, pigment orange R, St. Denis red.

Paranitro-orthoaminotoluene (Continued)
Starting point (French 601687) in making water-insoluble dyes with—
Dianilide of 2-naphthol-3:6-dicarboxylic acid.
Dialphanaphthylamide of 2-naphthol-3:6-dicarboxylic acid.
Dimetachloroanilide of 2-naphthol-3:6-dicarboxylic acid.
Diorthoanisidide of 2-naphthol-3:6-dicarboxylic acid.
Diorthotoluidide of 2-naphthol-3:6-dicarboxylic acid.

Paranitrophenol
Chemical
Starting point in—
Organic synthesis.
Dye
Starting point in making—
Dyestuffs.
Leather
Mold preventive for—
Leather, pickled skins, tan liquors.
Petroleum
Improver (U. S. 1969737 and 1788569) of—
Insulating stability of electrical insulating oils.
Photographic
Starting point in making—
Developing agents.
Rubber
Mold preventive for—
Smoked sheet.
Veterinary Medicine
Suggested for treatment of—
Ringworm in horses and calves (claimed to be effective in that (1) two applications are sufficient; (2) it is not necessary to scarify the affected spots before application).

Paranitrophenylmercaptan
Synonyms: Paranitrothiophenol.
Insecticide and Fungicide
Larvicide for—
Culicine mosquito larvae.

Paranitrosodimethylanilin
Chemical
Intermediate in—
Organic synthesis.
Dye
Intermediate in—
Dye synthesis.
Petroleum
Inhibitor (U. S. 1982277, 1982267, and 1982618) of—
Gum formation in gasoline, particularly in vapour-phase cracked gasoline.

Paranitrosodimethylanilin Hydrochloride
French: Hydrochlorure de paranitrosodimethylaniline.
German: Paranitrosodimethylanilinchlorhydrat.
Chemical
Starting point in making—
Dimethylaminonaphthaphenazin, dimethylanilin, para-aminodimethylanilin, paradimethylaminobenzaldehyde.
Dye
Starting point in making—
Azin green GB, capri blue GON, cotton blue R, cotton blue RR, fast black paste, fast neutral violet B, gallocyanin, indazin L, indazin M, indazin P, indophenol, methylene gray O, methylene gray ND, methylene gray NFD, methylene gray NFSt, methylene blue, metaphenylene blue B, metaphenylene blue BB, metaphenylene blue BBR, metaphenylene RJ, muscarin, naphthazin blue, neutral blue, nitroso blue MR, parma R, safranin, tannin brown B.
Rubber
Accelerator in vulcanization of rubber.
Textile
——, *Printing*
Reagent in developing colors on various fibers.

Paranitrosophenol
Chemical
Intermediate in—
Organic synthesis.
Petroleum
Inhibitor (U. S. 1982277, 1982267, and 1982618) of—
Gum formation in gasoline, particularly in vapour-phase cracked gasoline.

Paranitrosophenylmorpholin Hydrochloride
Chemical
Starting point in making various derivatives.
Dye
Starting point (U. S. 1908099) in making dyestuffs of the naphthophenazin series with the aid of—
Beta-allylnaphthylamine, beta-amylnaphthylamine, beta-butylnaphthylamine, beta-ethylnaphthylamine, beta-methylnaphthylamine, beta-phenylnaphthylamine, beta-propylnaphthylamine.
Various alkylphenyl and alkoxylphenyl derivatives of betanaphthylamine.

Para-normal-propoxydiphenylamine
Rubber
Aging retardant (Brit. 424461).
Promoter (Brit. 424461) of—
Resistance to crack formation of flexing.

Para-o'-chlorbenzylideneaminophenol
Petroleum
Gum inhibitor (U. S. 1980200 and 1980201) in—
Motor fuels.

Paraparadiaminodiphenylthiourea Sulphate
Dye
Diazo component and coupling agent.

Para-p'-diaminodibenzyl
Rubber
Preservative (U. S. 2009480, 2009526, and 2009530) for—
Rubber.

Para-p'-dianilinodibenzyl
Rubber
Preservative (U. S. 2009480, 2009526, and 2009530) for—
Rubber.

Para-p'-di-isopropylmesodimethylacridane
Fats and Oils
Antioxidant (Brit. 405797) for—
Fats, oils.
Petroleum
Antioxidant (Brit. 405797) for—
Oils.
Soap
Antioxidant (Brit. 405797) for—
Soaps, soapstocks.

Para-p'-dimethoxymesodimethylacridane
Fats and Oils
Antioxidant (Brit. 405797) for—
Fats, oils.
Petroleum
Antioxidant (Brit. 405797) for—
Oils.
Soap
Antioxidant (Brit. 405797) for—
Soaps, soapstocks.

Para-p'-dimethylmesodimethylacridane
Fats and Oils
Antioxidant (Brit. 405797) for—
Fats, oils.
Petroleum
Antioxidant (Brit. 405797) for—
Petroleum derivatives.
Soap
Antioxidant (Brit. 405797) for—
Fats, oils, soaps.

Para-p'-di-tertiary-butyldiphenylamine
Rubber
Preservative (U. S. 2009480, 2009526, and 2009530) for—
Rubber.

Paraphenetidin
Chemical
Reagent in—
Organic synthesis.
Starting point in making—
Synthetic pharmaceuticals, such as dulcin, phenacetin, phenosal.
Dye
Starting point in making—
Synthetic dyestuffs.

Paraphenylenediamine Picrate
French: Picrate de paraphénylènediamine.
German: Paraphenylendiaminpikrat, Pikrinsäureparaphenylendiaminester, Pikrinsäuresparaphenylendiamin.
Spanish: Picrato de parafenilenediamine.
Italian: Picrato di parafenilenediamine.

Explosives
Ingredient (U. S. 1852054) of—
Percussion cap charge containing diazodinitrophenol.

Paraphenylenediamine Salt of Meta-4-xylenol

Fuel
Agent for (Brit. 398219) resisting discoloration and gum formation in—
Benzene, cracked hydrocarbons, diesel oil, gasoline.
Stabilizing agent (Brit. 398219) for—
Benzene, cracked hydrocarbons, diesel oil, gasoline.

Paraquinone Chloroimide
French: Chloroimide de paraquinone.
German: Parachinonchlorimid.

Chemical
Starting point in making—
Pharmaceuticals and other derivatives.

Miscellaneous
Ingredient of—
Silver polishing and cleansing compositions (U. S. 1795676).

Pararosolic Acid
Synonyms: Aurin, Corallin, Coralline yellow.
French: Acide pararosolique. Aurine. Coralline jaune.
German: Aurin, Corallin, Pararosolsäure.

Analysis
General indicator in titrimetric analysis.
Indicator in—
Carbon dioxide detection in potable waters.
Caustic alkalies analysis.
Gastric contents analyses.
Mineral acid titrations, including sulphur dioxide, but not including phosphoric acid.

Chemical
Starting point in making—
Esters, salts, and other derivatives.

Dye
Starting point in making—
Anilin dyestuffs.

Food
Coloring for—
Confectionery, food preparations.

Ink
Coloring matter in making.
Printing inks.

Paper
Coloring matter (in lake form) in making—
Wallpaper.

Paint and Varnish
Coloring matter in making—
Lacquers, oil varnishes, spirit varnishes, stains.

Textile
Coloring matter in dyeing orange shades on—
Wool and silk yarns and fabrics.

Pararubber Seed Oil
Synonyms: Pararubber tree seed oil.
French: Huile des semences d'arbre à caoutchouc, Huile des semences de caoutchouc.
German: Kautschukbaumsamenoel, Kautschuksamenoel.

Glues and Adhesives
Ingredient (Brit. 332257) of—
Adhesive preparations.

Leather
Ingredient (Brit. 332257) of—
Artificial leather, finishing compositions, impregating compositions, leather substitutes used in shoe industry.

Miscellaneous
Ingredient (Brit. 332257) of—
Compositions used for impregnating and finishing various fibrous and similar products.
Roofing materials, wall coverings.

Oilcloth and Linoleum
Ingredient of—
Coating compositions.

Paint and Varnish
Ingredient (Brit. 332257) of—
Paints, priming coaters, varnishes.
Starting point in making—
Boiled oil.

Paper
Ingredient (Brit. 332257) of—
Compositions used in the impregnation and finishing of paper and pasteboard products.

Plastics
Ingredient (Brit. 332257) of—
Molding and other compositions.

Soap
As a soapstock.
Ingredient of—
Scouring compositions.

Textile
Ingredient (Brit. 332257) of—
Compositions used in the manufacture of waxed cloth.
Compositions used in impregnating and finishing.
Floor coverings.

Woodworking
Ingredient (Brit. 332257) of—
Compositions used in impregnating and finishing.

Parastearamidophenyltrimethyl-Ammonium Methylsulphate

Dry-Cleaning
Addition agent (Brit. 453523) to—
Solvents, such as trichloroethylene, carbon tetrachloride and benzene.

Leather
Reagent in—
Dyeing processes.

Textile
Addition agent (Brit. 453523) to—
Solvents, such as trichloroethylene, carbon tetrachloride, and benzene.

Parasulphobenzeneazoisatoic Anhydride
French: Anhydride de parasulphobenzèneazoisatoique.
German: Parasulfobenzolazoisatoinanhydrid.

Textile
——, *Dyeing*
Reagent for coloring—
Cotton and cellulose derivatives (German 433147).

1-Parasulphophenyl-3-methyl-5-pyrazolone

Dye
Starting point (Brit. 428535) in making—
Yellow dyestuffs, capable of being chromed, by coupling with a diazotized 3-halogenoanthranilic acid.

Paratoluenemonosulphonic Acid
Synonyms: Toluenesulphonic acid.
French: Acide de paratoluènemonosulphonique, Acide de toluènesulphonique.
German: Paratoluolmonosulfonsäure, Toluolsulphonsäure.

Chemical
Starting point in making—
Intermediates, paracresol, paratoluenesulphonamide, pharmaceuticals.
Synthetic tannins with tars that have been treated with sulphur (Brit. 302938).

Plastics
Reagent (French 599561) in making—
Thermoplastic products, resembling balata, from rubber.

Textile
Ingredient (Brit. 303379) of—
Finishing compositions, oiling compositions, softening compositions.

Paratoluenesulphonamide
German: Toluolsulfonamid.

Dye
Diluent (Brit. 399268 and 399274) of—
Lacquer dyes, such as 1-amino-4-anilinoanthraquinone, auramine, barium salt of Tobias acid, 6:6-dichloro-4:4-dimethylthioindigo, indanthrone-3:3-dicarboxylic acid, paranitropara-amino-azobenzene, safranin.

Paratoluenesulphonamidobetanaphthol-4-sulphonic Acid

Dye
In dye syntheses.
Starting point (Brit. 445999) in making—
Chromable orthohydroxy azo dyes by coupling with orthohydroxydiazonium compounds, such as those derived from 6-nitro-2-amino-paracresol or 4-chlor-2-aminophenol-6-sulphonic acid.

Paratolylalphanaphthylamine

Dye
Starting point in making—
Night blue.

2-Paratolylbenziminazole

Cosmetic
Protective (Brit. 435811) in—
Sun-tan lotions (solution or dispersion in a compatible solvent, for example, glycerin or wool-fat, but not water, alcohol, benzene, carbon tetrachloride, chloroform, or acetone), said to prevent formation of painful erythemas whilst enabling the skin to grow brown in sunlight, by virtue of high absorption of ultraviolet rays.

Paratolylbeta-9-carbazolylethyl Sulphide

Chemical
Intermediate (Brit. 444262 and 444501) in—
Organic syntheses.

Pharmaceutical
Claimed (Brit. 444262 and 444501) to have—
Value for pharmaceutical purposes.

Rubber
Accelerator (Brit. 444262 and 444501) in—
Vulcanizing.

Paratolylbetaparatoluenesulphonylethyl Sulphide

Chemical
Intermediate (Brit. 444262 and 444501) in—
Organic syntheses.

Insecticide
Insecticide (Brit. 444262 and 444501) for—
Animal pests, vegetable pests.

Textile
As a dyestuff (when employing suitable initial materials) (Brit. 444262 and 444501).
Assistant (Brit. 444262 and 444501) in—
Textile processing.

Paratolylmercaptan
Synonyms: Parathiocresol.

Insecticide and Fungicide
Larvicide for—
Culicine mosquito larvae.

Paratolylmercapto-1-naphthol

Chemical
Starting point in making—
Intermediates, pharmaceuticals.

Dye
Starting point (Brit. 291825) in making synthetic indigoid dyestuffs with—
5:7-Dibromoisatin anilide, 5:7-dibromoisatin chloride, 5:7-dichloroisatin anilide, 5:7-dichloroisatin chloride, isatin anilide, isatin chloride, reactive alpha derivatives of isatin.

Paratolyloxyphenylisopropylnitrosoamine

Chemical
Antioxidant and stabilizer (Brit. 430335) for—
Unstable organic substances.

Petroleum
Antioxidant and stabilizer (Brit. 430335) for—
Petroleum products.

Rubber
As an antioxidant (Brit. 430335).

Paratolylphenylenediamine

Rubber
Starting point (Brit. 425751) in making—
Antioxidants for rubber by condensation with paratoluene sulphonyl chloride.

2-Paratolylthiolquinolin Ethiodide

Dye
Process material (Brit. 454687) in making—
Cyanin dyes.

2-Paratolylthioquinolin Methiodide

Dye
Process material (Brit. 454687) in making—
Cyanin dyes.

Peachkernel Oil
Synonyms: Peach oil.
French: Huile de pêche, Huile persique.
Spanish: Aceite de aberchigo.
Italian: Olio di mandorle di pesco, Olio di pesco.

Fats and Oils
Ingredient of—
Cutting oils.
Substitute for—
Almond oil.

Food
As an edible oil.
For packing sardines and other fish.
Ingredient of—
Oleomargarin, pastries and confectionery, salad oils and dressings.

Leather
As a dressing and softening oil.

Mechanical
Ingredient of—
Miscellaneous lubricating compositions for special machinery.

Perfume
Ingredient of—
Cosmetic creams and pomades, sunburn preparations, various ointments.

Pharmaceutical
Base in making—
Ointments and liniments.
Suggested for use as emollient and gentle laxative.

Soap
Stock in making soap.

Textile
——, *Dyeing*
Ingredient of—
Baths for dyeing various textile fabrics and yarns with various colors.
——, *Manufacturing*
For oiling wool.
——, *Printing*
Ingredient of—
Color pastes for printing calico.

Peanut Oil Fatty Acid
Synonyms: Arachis oil fatty acid, Earthnut oil fatty acid.
French: Acide gras d'huile d'arachide.
German: Erdeicheloelfettsaeure, Erdnussoelfettsaeure, Grundnussoelfettsaeure.

Chemical
Starting point in making various salts and esters.

Food
Ingredient of—
Prepared foods, hydrogenated food products.

Fuel
Compounds of—
Candles.

Miscellaneous
Ingredient of—
Cleansing compositions with alkaline hypochlorites (Brit. 280193).
Polishing preparations.

Paint and Varnish
Starting point in making—
Driers.

Pharmaceutical
In compounding and dispensing practice.

Soap
Raw material in soapmaking.

Textile
——, *Bleaching*
Ingredient of—
Bleaching compositions containing alkaline hypochlorites (Brit. 280193).

Peanut Oil Fatty Acid (Continued)

——, *Finishing*
Ingredient of—
Finishing compositions.
Washing compositions in conjunction with alkaline hypochlorites (Brit. 280193).
Waterproofing compositions.

Peat Moss

Synonyms: Bog moss, Sphagnum.

Agriculture
As a disintegrating and humidifying addition to soils.
As a packing and protective material.

Miscellaneous
As a bedding for animals.
As a packing material.

Pharmaceutical
As a surgical dressing (particularly during wartime).

Pectin

French: Pectine.
German: Pektinstoff.

Adhesive
As an adhesive (beet pectin) (German 384772, 406539).

Beverage
Emulsifying agent for—
Essential oils.

Chemical
Dehydrating agent.
Promoter of—
Large crystal growth in saturated solutions of mineral salts.

Fats and Oils
Emulsifying agent for—
Essential oils.

Food
Gelatinizing agent in—
Food products.
Ingredient of—
Candy jellies for cast or slab work, consisting of various mixtures of sodium acetate, citric acid, glucose, corn sugar or cane sugar, color, and flavor.
Jams.
Mayonnaises; for example, mixtures of whole eggs, egg yolk, mustard powder, sugar, salt, vegetable oil, flavor, tincture of capsicum, lactic acid, vinegar, and water.
Pectin-acid mixture in sugar solution, which, on admixture with sugar, fruit flavor, and water yields a jelly (U. S. 1879697).
Sherbets (water ices).

Insecticide
Emulsifying agent for—
Insecticides for delicate foliage.
Insecticides for foliage of citrous fruit trees.
Mineral oils (German 479192).
Pine oils.

Miscellaneous
Dehydrating agent in making—
Powdered products.

Perfume
Emulsifying agent for—
Essential oils, pine needle-oil bath preparations.
Ingredient of—
Cosmetic preparation (use suggested on the claims that it has beneficial action on the skin; has affinity for the cellular structure of the skin and hair; is readily absorbed by both the skin and hair, not accompanied by any chemical reaction which accompanies the use of alkaline substances; neutralizes any excess alkali that may exist).
Latherless shaving creams.
Promoter of—
Water absorption (up to 50 percent) by petrolatum-base preparations.
Thickener (German 551888, 554084) in—
Dentifrices, shaving creams.

Petroleum
Emulsifying agent (German 479192) for—
Light mineral oils.

Pharmaceutical
Disintegrating ingredient in—
Pills, tablets.
Emulsifying agent for—
Essential oils.
Excipient in making—
Greaseless ointments.
Gelatinizing agent (French 686154) in making—
Colloidal iodine jelly.
Ingredient of—
Thickening compositions, containing also gum tragacanth, glycerin, and water, for corn removers using as an active agent (1) salicylic acid and glacial acetic acid; (2) glacial acetic, lactic, and salicylic acids; (3) formic acid and phenol.

Textile
Ingredient (French 600309) of—
Viscose spinning solutions (added to improve quality of product).

Penetrating Agents

See: "Wetting agents."

Pentachloroanthraquinone-1:2:5:6-diacridone

Dye
Starting point (U. S. 1972094) in making—
Reddish-grey vat dyes with 1-amino-anthraquinone.

Pentachloroanthraquinone-1:2:7:8-diacridone

Dye
Starting point (U. S. 1972094) in making—
Reddish-grey vat dyes with 1-amino-anthraquinone.

Pentachlorodiphenylmethane

Electrical
Cooling medium (Brit. 413596, 433070, 433071, and 433072) in—
Electrical apparatus, such as transformers, switches, capacitors, cables, bushings, and junction boxes (may be employed in admixture with trichlorobenzene, chlorinated diphenyl, and the like).
Dielectric (Brit. 413596, 433070, 433071, and 433072) in—
Electrical apparatus, such as transformers, switches, capacitors, cables, bushings, and junction boxes (may be employed in admixture with trichlorobenzene, chlorinated diphenyl, and the like).

Pentachloroethane

French: Pentachloréthane.
German: Pentachloraethan.
Spanish: Pentachloretano.
Italian: Pentachloretane.

Analysis
Solvent for—
Cellulose derivatives, fats, gums, oils, resins.

Brewing
Antiseptic for—
Yeast.

Cellulose Products
Solvent for—
Cellulose acetate, cellulose ethers.

Ceramics
Solvent in—
Compositions, containing natural or synthetic resins, cellulose acetate, or other cellulose esters or ethers, used as coatings for protecting and decorating ceramic products.

Chemical
Solvent for—
Cellulose acetate, cellulose ethers, fats, oils.
Solvent miscible with—
Alcohol, ether.
Substitute for—
Tetrachloroethane (acetylene tetrachloride) (said to be less toxic).

Cosmetic
Solvent for—
Essential oils, fixed vegetable oils.
Solvent in—
Nail enamels and lacquers containing natural or synthetic resins, cellulose acetate, or other cellulose esters or ethers as base material.

Distilling
Antiseptic for—
Yeast.

Dry-Cleaning
Spotting agent for—
Fats, greasy stains, gums, oils, resins.

Electrical
Solvent in—
Insulating compositions, containing natural or synthetic resins, cellulose acetate, or other cellulose

Pentachloroethane (Continued)
esters or ethers, used for covering wire and in making electrical machinery and equipment.

Fats, Oils, and Waxes
Solvent for—
Fats, vegetable oils.

Food
Antiseptic for—
Yeast.
Solvent for—
Fats, oils.

Glass
Solvent in—
Compositions, containing natural or synthetic resins, cellulose acetate, or other cellulose esters or ethers, used in the manufacture of nonscatterable glass and as coatings for decorating and protecting glassware.

Glue and Adhesives
Solvent in—
Adhesive compositions containing natural or synthetic resins, cellulose acetate, or other cellulose esters or ethers.

Gums
Solvent for—
Gums.

Ink
Ingredient of—
Printing ink removers.
Solvent in—
Inks.

Leather
Solvent in—
Compositions, containing natural or synthetic resins, cellulose acetate, or other cellulose esters or ethers, used in the manufacture of artificial leathers and as coatings for decorating and protecting leathers and leather goods.

Metal Fabrication
Solvent in—
Compositions, containing natural or synthetic resins, cellulose acetate, or other cellulose esters or ethers, used as coatings for protecting and decorating metallic articles.

Miscellaneous
Solvent in—
Coating compositions, containing natural or synthetic resins, cellulose acetate, or other cellulose esters or ethers, used for protecting and decorating various articles.
Substitute for—
Tetrachloroethane (said to be less toxic).

Paint and Varnish
Ingredient of—
Paint removers.
Solvent for—
Cellulose derivatives, gums, metallic naphthenates, oils, resins.
Solvent in—
Paints, varnishes, lacquers, enamels, and dopes containing natural or synthetic resins, cellulose acetate, or other cellulose esters or ethers.

Paper
Solvent in—
Compositions, containing natural or synthetic resins, cellulose acetate, or other cellulose esters or ethers, used in the manufacture of coated papers and as coatings for decorating and protecting articles made of paper or pulp.

Petroleum
Solvent for—
Mineral oils, mineral oils used in rust-removing agents.

Pharmaceutical
Solvent for—
Essential oils, gums, mineral oils, vegetable oils.

Photographic
Solvent in making—
Films from cellulose acetate or other esters or ethers of cellulose.

Plastics
Solvent in making—
Laminated fiber products, molded products.
Plastics from or containing natural or synthetic resins, cellulose acetate, or other cellulose esters or ethers.

Resins
Solvent for—
Dammar, elemi, mastic, sandarac.
Solvent in making—
Artificial resins from or containing cellulose acetate or other cellulose esters or ethers.

Rubber
Solvent in—
Compositions, containing natural or synthetic resins, cellulose acetate, or other cellulose esters or ethers, used as coatings for protecting and decorating rubber goods.

Soap
Ingredient of—
Cleaning compositions, special soaps.
Solvent for—
Fats, oils.

Stone
Solvent in—
Compositions, containing natural or synthetic resins, cellulose acetate, or other cellulose esters or ethers, used as coatings for decorating and protecting artificial and natural stone.

Textile
Degreasing and defatting agent for—
Textile fibers.
Retting agent for—
Textile fibers.
Solvent in—
Compositions, containing natural or synthetic resins, cellulose acetate, or other cellulose esters or ethers, used in the manufacture of coated fabrics.

Wood
Solvent in—
Compositions, containing natural or synthetic resins, cellulose acetate, or other cellulose esters or ethers, used as protective and decorative coatings on woodwork.

Pentadekanaphthene

Chemical
General solvent for chemicals and in various chemical processes (Brit. 269960).

Miscellaneous
General solvent (Brit. 269960).

Textile
——, *Dyeing and Printing*
Solvent in making—
Liquors and pastes for dyeing and printing fabrics and yarns (Brit. 269960).
——, *Finishing*
Solvent in making—
Dye preparation in stenciling fabrics (Brit. 269960).

Pentadichloropropane
German: Pentadichlorpropan.

Leather
Ingredient of—
Compositions used in making artificial leather (Brit. 279139).

Miscellaneous
Ingredient (Brit. 279139) of—
Impregnating compositions (Brit. 279139).
Insulating varnishes and lacquers for electrical wiring.

Paint and Varnish
Ingredient (Brit. 279139) of—
Paints, varnishes.

Plastics
Ingredient of—
Moldable compositions (Brit. 279139).

Textile
——, *Manufacturing*
Reagent in making—
Chemical fibers (Brit. 279139).

Pentadigalloylglucose

Miscellaneous
Reagent (U. S. 1922464) for—
Removal of emulsoids from water solutions (used in combination with trisodium phosphate).

Pentaerythrite Tetranitrate
French: Tétranitrate de pentaaerythrite, Tétranitropentaerythrit.
German: Pentaaerythrittetranitrat.

Pentaerythrite Tetranitrate (Continued)

Explosives
As a booster in—
Making explosive compositions and explosive shells.

Pentaerythritol

Chemical
Starting point in making—
Pentaerythritol tetra-acetate (U. S. 1583658).

Pentamethylenediphenylphosphonium Bromide
French: Bromure de pentaméthylènediphénylephosphonium.
German: Brompentamethylendiphenylphosphonium, Pentamethylendiphenylphosphoniumbromid.

Miscellaneous
Mothproofing and moldproofing agent (Brit. 312163) in treating—
Hair, fur, feathers, felt, and the like.

Textile
Mothproofing and moldproofing agent (Brit. 312163) in treating—
Wool and other products.

Pentamethylenetriphenylphosphonium Bromide
French: Bromure de pentaméthylènetriphénylephosphonium.
German: Brompentamethylentriphenylphosphonium, Pentamethylentriphenylphosphoniumbromid.

Miscellaneous
Mothproofing and moldproofing agent (Brit. 312163) in treating—
Hair, fur, feathers, felt, and the like.

Textile
Mothproofing and moldproofing agent (Brit. 312163) in treating—
Wool and other products.

Pentamethylmonoethylpararosanilin

Ink
Starting point (U. S. 1899452) in making—
Special ink for protection and authentification of checks and the like; such an ink has the characteristic that the color is a function of hydrogen ion concentration.

Pentane
Synonyms: Normal pentane, Amyl hydride, Isopentane.
French: Hydrure d'amyle, Hydrure amylique.
German: Amylhydrid, Isopentan, Pentan.

Chemical
As a solvent for various purposes.
Starting point in making—
Amyl acetate, amyl alcohol, amyl chloride and various halogenated derivatives.

Miscellaneous
As a solvent for various purposes.
Filler for—
Low-temperature thermometers.
Lubricant in operating—
Claude liquid air machine.
Reagent in making—
Standard photometric lamp.

Paint and Varnish
Solvent in making—
Cellulose ester and ether varnishes, lacquers, and dopes.

Pharmaceutical
In compounding and dispensing practice.

Plastics
Solvent in making—
Cellulose ester and ether compounds.

Refrigeration
Active medium in refrigerating systems.

Rubber
Starting point in making—
Synthetic rubber.

Pentatricontanol

Miscellaneous
As an emulsifying agent (Brit. 343872).
For uses, see under general heading: "Emulsifying agents."

Pentyl Alcohol
Synonyms: Pentylic alcohol.
French: Alcool de pentyle, Alcool pentylique.
German: Pentylalkohol.

Chemical
Starting point in making—
Esters of various acids, and intermediates.

Fats and Oils
Ingredient of—
Emulsified lubricants and other compositions (Brit. 277357).

Fuel
Ingredient of—
Emulsified mixtures (Brit. 277357).

Leather
Ingredient of—
Emulsified dressing and finishing compositions (Brit. 277357).

Petroleum
Ingredient of—
Motor fuel compositions in emulsified form.
Stable emulsions of petroleum and petroleum distillates (Brit. 277357).

Soap
Ingredient of—
Cleansing compositions.
Emulsified detergents (Brit. 277357).

Textile
——, *Finishing*
Ingredient of—
Washing and cleansing compositions (Brit. 277357).

Pentyl Alphacrotonate
French: Alphacrotonate de pentyle, Alphacrotonate pentylique.
German: Alphacrotonsäurepentylester, Alphacrotonsäurespentyl, Pentylalphacrotonat.

Miscellaneous
Solvent and plasticizer (Brit. 321258) for—
Cellulose acetate, cellulose esters and ethers, cellulose nitrate, rubber.
For uses, see under general heading: "Solvents."

Pentylolamine

Chemical
Starting point in making—
Intermediates, pharmaceuticals, and other derivatives.

Electrical
Dispersing agent (Brit. 340294) in—
Special lubricating compositions for use in electric switches.

Fats and Oils
Dispersing agent (Brit. 340294) in making—
Nonfreezing lubricating compositions, containing animal and vegetable oils and fats, as well as ethyleneglycol, borax, benzyl alcohol, or esters of ethyleneglycol in the place of the latter.
Special lubricating compositions of the above type for use on locomotive axles, railway switches, hydraulic presses, and hydraulic brakes.
Ingredient (Brit. 340294) of—
Compositions, containing vegetable and animal fats and oils and greases, used as rust preventives.

Petroleum
Dispersing agent in making—
Lubricating compositions containing various mineral oils and distillates.
Lubricating compositions in dispersed form for various machine shop operations, such as boring, drilling, cutting, planing.
Special lubricating compositions containing mineral oils and greases (Brit. 340294).

Peppermint
Synonyms: Brandy mint, Lamb mint.
Latin: Folia menthae peperitae.
French: Menthe, Menthe poivrée.
German: Pfeffermint, Pfefferminzblätter, Pfeffermünze.
Spanish: Menta piperita.
Italian: Menta piperita.

Chemical
Starting point in extracting—
Menthol.

Peppermint (Continued)

Fats and Oils
Starting point in extracting—
Peppermint oil.

Food
Flavoring for—
Beverages, candies, jellies, pastries, sauces.

Pharmaceutical
In compounding and dispensing practice.

Perchloric Acid

French: Acide perchlorique.
German: Perchlorsäure, Ueberchlorsäure.
Spanish: Acido perchlorico.
Italian: Acido perchlorico.

Analysis
Reagent in—
Assaying various alkaloids, such as morphine, codeine, cocaine.
Carrying out Kjeldahl digestions for the determination of the nitrogen content of various products.
Determining potash in various products by the formation of an insoluble potassium perchlorate.
Effecting electro-analyses (used for the purpose of destroying the organic matter contained in the product that is to be analyzed).

Chemical
Oxidizing agent in making—
Inorganic chemicals, intermediates, organic chemicals, pharmaceuticals, synthetic aromatics.
Starting point in making various salts.

Explosives
In the manufacture of matches.
Reagent in making—
Explosive compounds, such as the perchlorated esters of monochlorohydrin.

Metallurgical
Ingredient of—
Lead-plating baths (used for the purpose of facilitating the deposition of lead from baths containing lead perchlorate).

Pharmaceutical
In compounding and dispensing practice.

Perchloroethylene

Synonyms: Carbon bichloride, Carbon dichloride.
French: Bichlorure de carbone, Dichlorure de carbone, Tétrachloroéthane, Tétrachloroéthylène.
German: Bichlorkohlenstoff, Dichlorkohlenstoff, Kohlenstoffbichlorid, Kohlenstoffdichlorid, Perchloraethylen, Tetrachloraethan, Tetrachoraethylen.

Ceramics
Solvent in—
Compositions, containing cellulose acetate, nitrocellulose, or other esters or ethers of cellulose, used for decorating and protecting ceramic ware.

Chemical
Solvent for various purposes.
Starting point in making—
Intermediates, organic chemicals, pharmaceuticals.

Electrical
Solvent in making—
Insulating compositions containing cellulose acetate, nitrocellulose, or other esters or ethers of cellulose.

Glass
Solvent in—
Compositions, containing cellulose acetate, nitrocellulose, or other esters or ethers of cellulose, used in the manufacture of nonscatterable glass and decorating and protecting glassware.

Glues and Adhesives
Solvent in making—
Adhesive preparations containing cellulose acetate, nitrocellulose, or other esters or ethers of cellulose.

Leather
Solvent in—
Compositions, containing cellulose acetate, nitrocellulose, or other esters or ethers of cellulose, used in the manufacture of artificial leather and for decorating and protecting leather goods.

Metallurgical
Solvent in—
Compositions, containing cellulose acetate, nitrocellulose, or other esters or ethers of cellulose, used for decorating and protecting metalware.

Miscellaneous
Solvent in—
Compositions, containing cellulose acetate, nitrocellulose, or other esters or ethers of cellulose, used for decorating and protecting various fibrous products.

Paint and Varnish
Solvent in making—
Paints, varnishes, lacquers, dopes, and enamels containing cellulose acetate, nitrocellulose, or other esters or ethers of cellulose.

Paper
Solvent in—
Compositions, containing cellulose acetate, nitrocellulose, or other esters or ethers of cellulose, used in the manufacture of coated paper and for decorating and protecting paper and pulp products.

Petroleum
Solvent in treating—
Crude petroleum to obtain petrolatum oils, petrolatum, and paraffin.

Pharmaceutical
In compounding and dispensing practice.

Photographic
Solvent in making—
Films from cellulose acetate, nitrocellulose, or other esters or ethers of cellulose.

Plastics
Solvent in making—
Compositions containing cellulose acetate, nitrocellulose, or other esters or ethers of cellulose.

Rubber
Solvent in—
Compositions, containing cellulose acetate, nitrocellulose, or other esters or ethers of cellulose, used for decorating and protecting rubber products.

Soap
Solvent in making—
Detersive preparations, dry-cleansing soaps, special solvent soaps.

Stone
Solvent in—
Compositions, containing cellulose acetate, nitrocellulose, or other esters or ethers of cellulose, used for decorating and protecting artificial and natural stone.

Textile
Solvent in—
Compositions, containing cellulose acetate, nitrocellulose, or other esters or ethers of cellulose, used for coating textile fabrics.

Woodworking
Solvent in—
Compositions, containing cellulose acetate, nitrocellulose, or other esters or ethers of cellulose, used for decorating and protecting woodwork.

Perchloromethylmercaptan

Insecticide and Fungicide
Fumigant and insecticide for—
Granary weevils (*Sitophilus granarius* L.).
Ladybird beetles (*Hippodamia convergens Guerin*) (used in conjunction with hydrocyanic acid gas).

Perilla Oil

French: Huile de perilla.
German: Perillaoel.

Chemical
Starting point in making—
Cobalt driers.

Fats and Oils
Ingredient of—
Edible oil compounds.

Food
As an edible oil and in oil compounds.

Glues and Adhesives
Ingredient (Brit. 332257) of—
Adhesive preparations.

Ink
Ingredient of—
Lithographic inks, printing inks.

Perilla Oil (Continued)

Leather
Ingredient of—
Coating compositions used in the manufacture of artificial leathers.
Coating compositions containing linoxyn (Brit. 332257).
Leather substitutes for making footwear.
Reagent for—
Impregnating and finishing leather.

Linoleum and Oilcloth
Ingredient of—
Coating compositions.

Miscellaneous
Ingredient of—
Emulsified cements (Brit. 273290).
Emulsified electrical insulating compositions (Brit. 273290).
Roofing materials (Brit. 332257).
Wall coverings (Brit. 332257).

Paint and Varnish
Ingredient of—
Paints and varnishes.
Paints, varnishes, and primers, containing linoxyn (Brit. 332257).
Starting point (Brit. 273290) in making—
Varnish bases.

Paper
Ingredient of—
Impregnating compositions for treating paper lanterns, paper umbrellas, various other products, etc.
Impregnating and finishing compositions containing linoxyn, used in the treatment of pulp and paper products (Brit. 332257).

Plastics
Ingredient (Brit. 332257) of—
Compositions containing linoxyn, used in the manufacture of pressed articles.

Resins and Waxes
Reagent for extracting—
Residual Japan wax or white wax contained in press cakes.
Starting point in making—
Artificial resins.

Textile
——, *Dyeing*
Ingredient of—
Bath in dyeing textiles red.
——, *Finishing*
Ingredient of—
Compositions used in the manufacture of waxed cloth (Brit. 332257).
Compositions for making floor coverings (Brit. 332257).
Impregnating and finishing compositions used for treating textile fibers and fabrics (Brit. 332257).
Waterproofing compositions.

Woodworking
Ingredient (Brit. 332257) of—
Impregnating and finishing compositions.

Perylene-3:9-dicarboxylic Chloride
French: Chlorure de pérylène-3:9-dicarbonique, Chlorure de pérylène-3:9-dicarboxylique.
German: Perylen-3:9-dicarbonchlorid.

Chemical
Starting point in making various derivatives.

Dye
Starting point (Brit. 347099) in making vat dyestuffs for cotton, with the aid of—
Alpha-aminoanthraquinone.
Beta-aminoanthraquinone.
1-Chloro-2-aminoanthraquinone.
1-Chloro-4-aminoanthraquinone.
1:5-Diaminoanthraquinone.

Perylenetetracarboxyldi-imide
Ingredient (U. S. 1914509) of—
Gum inhibitor for motor fuels, containing also phthalide and/or its derivatives and an amylamine.

Perylenetetracarboxylicdiphenyldi-imide

Dye
Starting point (Brit. 428770) in making—
Scarlet dye of improved vatting properties by trichlorination.

Petrolatum
Synonyms: Alboline, Cosmoline, Glycolin, Hard petroleum ointment, Paraffin jelly, Petroleum jelly, Petroline, Petroleum ointment, Pimeleine, Saxoline, Soft paraffin, Soft petroleum ointment, Soft petrolatum, Vaseline.
Latin: Adeps petrolei, Gelatum petroli, Paraffinum molle, Paraffinum spissum, Paraffinum unguinosum, Petrolatum molle, Petrolatum spissum, Petrolinum, Unguentum parafinium, Unguentum petroleis, Vaselinum.
French: Graisse minérale, Pétroléine.
German: Paraffinsalbe, Vaselin, Weiches paraffin.
Spanish: Petrolato, Vaselina.
Italian: Petrolato, Vaselina.

Abrasives
Binder in—
Abrasive compound (U. S. 1353979).
Abrasive stone (U. S. 1195246).
Abrasive wheel (U. S. 1146884).
Emery pastes, grinding compounds.

Adhesives
Ingredient of—
Adhesive composition (U. S. 1137043).
Adhesive paste for joining decorative metal leaf to metal bases (U. S. 1906168).

Building and Construction
Treating agent (U. S. 1246827) for—
Concrete floors.

Cellulose Products
Solvent (U. S. 1217027 and 1217028) for—
Cellulose ethers, ethylcellulose.

Ceramics
Process material (U. S. 1295466) in making—
Baking dishes.

Chemical
Process material in making—
Catalysts for hydrogenation processes (U. S. 1519088).
Nickel catalysts (U. S. 1329322).

Cosmetic
Absorbent in extracting—
Perfumes from flowers.
Base in—
Cosmetic creams, cuticle salve (U. S. 1513233), ointments, pomades, salves, solid brilliantines.
Ingredient of—
Hair dressing (U. S. 1368758).
Hair tonic (U. S. 1368758).

Dairy Products
Coating agent (U. S. 1260899) for—
Bacillus bulgaricus (lactic acid bacteria).
Ingredient of—
Milk powders (U. S. 1202130).

Dental
Treating agent (U. S. 1516140) for—
Pyorrhea.

Disinfectant
Ingredient of—
Disinfecting tablets (U. S. 1340661).
Germicide (U. S. 1275162).

Dye
Alkali-proofing agent (U. S. 1349265) for—
Dyes.

Electrical
Acidproofing agent (U. S. 1369783) for—
Battery containers.
Ingredient of—
Electrical insulation, said to be suitable for transformers, capacitators, cables (in admixture with scale wax and mineral oil).
Protective coatings for lead coverings for cable (French 716148).
Process material in making—
Electrical insulation (U. S. 1159257).
Storage battery vent plugs (U. S. 1506216).

Explosives and Matches
Binder (U. S. 1192678, 1202712, 1276537, 1309014, 1334303, 1335788, 1335789, 1349983, 1411674, and 1509393) in—
Explosive compositions.
Coating agent (U. S. 1360397 and 1360398) for—
Sodium nitrate in explosives.
Ingredient of—
Pyrotechnic red fire.
Stabilizing agent (U. S. 1391796) for—
Trinitrotoluene.

Petrolatum (Continued)

Food
Preservative (U. S. 1174008, 1177105, and 1245294) for—
Eggs.
Vehicle (U. S. 1307090) in—
Cooking peanuts.

Fuel
Stabilizer (Brit. 417352) of—
Liquid state in fuels consisting of a suspension of coal in fuel oil.

Gum
Ingredient (U. S. 1134073) of—
Chicle substitute.

Ink
Drying-retardant in—
Inks.
Ingredient of—
Antismutting composition for printers' inks (U. S. 1273361).
Ink eradicators for tracing cloth, containing also turpentine, pumice dust, and paraffin.
Transfer inks.
Typewriter inks.
Reducer of—
Tackiness in inks.
Softener in—
Inks.

Insecticide and Fungicide
Ingredient of—
Insecticidal compositions (U. S. 1248977).

Leather
Ingredient of—
Emulsions with egg (U. S. 1302487).
Dressings and greases.
Finishing compositions (U. S. 1453723).
Oiling agent (U. S. 1847629).
Leather substitute (U. S. 1310624).
Tanning composition (U. S. 1402283).

Lubricant.
As a lubricant.
Ingredient of—
Gum lubricant (in admixture with bone oil).
Upper cylinder lubricant.
Raw material in making—
Lubricating greases.
Reviving agent (U. S. 1352502) for—
Lubricating oils.

Mechanical
As a lubricant.
Protective coating for—
Metallic parts of machinery.

Metallurgical
Ingredient of—
Metal-coating composition (U. S. 1457169 and 1472239).
Metal polishes.
Protective coatings for iron and steel (U. S. 1410391).
Rust-preventive compositions.
Preventer (U. S. 1395413) of—
Adhering of comminuted metals.

Mining and Ore Treatment
Ingredient (U. S. 1448927 and 1448928) of—
Ore concentrating agent.

Miscellaneous
Ingredient of—
Animal bait (U. S. 1366509).
Antidimming compositions (U. S. 1201440).
Coating compositions (U. S. 1415282).
Coating composition containing also paraffin and alcohol (U. S. 1292964).
Cleaning cloth composition (U. S. 1143614).
Decorative composition (U. S. 1388518).
Etching reserve composition (U. S. 1407301).
Furniture polish (U. S. 1350537).
Gas-check pad for breech blocks (U. S. 1229662).
Gasket compositions for bottles (U. S. 1322823).
Gasket compositions for cans and fruit jars (U. S. 1322823).
Gear composition (U. S. 1506239).
Gelatin (technical) substitute (U. S. 1217027).
Impregnating admixture with rosin (U. S. 1386711).
Razor strop compositions (U. S. 1353979 and 1360343).
Shoe filler (U. S. 1136459).
Shoe polishes.
Shoe waterproofings (U. S. 1167328).
Shoe waterproofings (in admixture with paraffin and wool-grease).
Sizing composition (U. S. 1299663).
Soldering fluxes (U. S. 1401154, 1444946 and 1472281).
Stencil compositions (U. S. 1168223).
Stove polish (U. S. 1403758).
Tire puncture closing composition (U. S. 1137461).
Waterproofing compositions (U. S. 1307373, 1327239, 1376553, and 1915301).
Welding composition (U. S. 1472781).
Retainer for—
Leaves and petals in lacquer and dye dipping processes in making artificial flowers.

Paint and Varnish
Ingredient of—
Antifouling composition (in admixture with heavy lubricating oil, rosin, paraffin, and salt).

Paper
Ingredient of—
Carbon paper coating (in admixture with gutta-percha, lampblack, and a wax).
Process material in making—
Translucent paper (U. S. 1345184).
Writing paper (U. S. 1234045).

Pharmaceutical
In compounding and dispensing practice.
Ingredient of—
Antiseptic (U. S. 1275162), emulsions, ointments, expectorants, salves.
Suggested for use as—
Emollient.

Photographic
Ingredient of—
Photographic film (U. S. 1345184).
Protective compound for inhibiting action (effects) of fluorescent substances (U. S. 1511874).
Protective compound for filtering ultraviolet light (U. S. 1511874).
Protective compound for inhibiting photochemical action (U. S. 1511874).
Restorative compounds for blemished motion picture films (U. S. 1139679, 1139682, 1139683, and 1192424).

Plastics
Raw material in making—
Plastic compositions (U. S. 1322823).
Solvent (U. S. 1217027) in making—
Celluloid substitutes.

Printing
In making—
Electrotypes (U. S. 1210872).
Halftones (U. S. 1507049).

Rayon
Ingredient of—
Addition agent (emulsion with casein, oleic acid, and triethanolamine) for spinning solution in making low-luster rayon (U. S. 1984303).
Addition agent (emulsion with casein and turpentine) for spinning solution in making low-luster rayon either by viscose or cuprammonium process (U. S. 1967206).
Process material (U. S. 1958238) in making—
Cellulose acetate rayon possessing enhanced toughness, pliability, and ready delustring properties.

Resins
Starting point (U. S. 1271392 and 1271393) in making—
Resins with phenol-aldehyde condensates.

Rubber
Ingredient of—
Coating agent (in admixture with mica) (U. S. 1455544).
High-grade hose tubing.
Imitation rubber compositions (U. S. 1242886 and 1363229).
Light-sensitive rubber (U. S. 1309703).
Pencil erasers.
Rubber sheeting compositions for hospitals.
Soft rubber sponge composition.
White tubing compositions.

Soap
Raw material in making—
Soap (U. S. 1342783 and 1408650).
Special soaps.

Textile
Ingredient of—
Antisizing dressing for threads (in admixture with zinc dust).
Stripping agent for dyed fabrics.

Petrolatum (Continued)

Thallium-base composition for rendering tent canvas proof against insects, mildew, and water.

Thread greases (in admixture with lanolin and camphor).

Waterproofings for textiles.

Wood

Impregnating agent (U. S. 1429288).

Petroleum

Synonyms: Crude oil, Hydrocarbon oil, Mineral oil, Naphtha, Rock oil.

French: Napthe, Petrole brut.

German: Bergoel, Erdnaphta, Erdoel, Mineraloel, Naphta, Rohoel, Rohpetroleum.

Ceramics

Ingredient of—

Electrical conducting coatings on porcelains, chinaware, terracotta, stoneware, and other ceramic products for galvanoplastic plating.

Mold lubricants used in the manufacture of electrochemical ceramic products.

Chemical

Reagent in making—

Graphite and graphitic products.

Solvent in making—

Ammonium oleate, barium cyanide.

Starting point in making—

Hexane, isopropyl alcohol, liquefied gases for metal cutting and illuminating, pentane, secondary amyl alcohol, secondary butyl alcohol, secondary hexyl alcohol.

Explosives and Matches

Reagent in making—

Chlorate explosives and dynamites, liquid air explosive compositions.

Fuel

Fuel for—

Burning, Diesel engines.

Ingredient of—

Candles.

Gas

Reagent in treating—

Illuminating gas in order to remove sulphur compounds.

Solvent in liquefying—

Oil gas.

Starting point in making—

Fuel gas.

Glass

Ingredient of—

Compositions used in obtaining conducting coatings on glass in galvanoplastic plating.

Glues and Adhesives

Ingredient of—

Marine glues, rosin cements.

Insecticide

Ingredient of—

Emulsions, sprays.

Mechanical

As a lubricant.

Metallurgical

Flotation agent in treating—

Minerals to separate the gangue.

Miscellaneous

Ingredient of—

Metal polishes.

For laying the dust on roads.

Paint and Varnish

Starting point in making—

Gas black, lampblack, V. M. & P naphtha.

Perfumery

Diluent for—

Bay oil, cajeput oil, lemongrass oil, rue oil, palmarosa oil, ylang-ylang oil.

Petroleum

Starting point in making—

Acid coke, aviation gasoline, bakers' machinery oil, benzin, benzol wash oil, binder oils, black oils, blending naphtha, boiler fuel oil, briquetting asphalts and pitches, candle wax, candymakers' oil, candymakers' wax, carbon brush coke, carbon electrode coke, cardboard wax, chewing gum wax, coach and ship illuminants, coke fuel, cold patch oils, compressor oils, cup grease, cutting oils, cylinder oils, de-emulsifying agents, detergent wax, Diesel fuel oil, dust-laying oils, dyers' and cleansers' benzine, egg packers' oils, etching wax, flotation oils, fruit packers' oils, gas absorption oils, gas machine gasoline, gas oils, gasoline, gasoline recovery oil, gear grease, grease compounding oils, ice machine oils, illuminating oils, ink oils, iron wax, journal oils, kerosene, lamp oil, lighthouse oils, light spindle oils, match wax, medicinal oil, metallurgical fuels, motor fuels, motor oils, naphtha, natural gasoline, oil gas oil, paper wax, paving felt, saturating asphalts and pitches, plastic composition asphalts and pitches, petrolatum, petroleum ether, petroleum jelly, pitch, railroad oils, residual oils and pitches, rubber-making asphalts and pitches, saponification agents, sealing wax, signal oils, slab oil, spindle oils, steam cylinder oils, still wax, stove oil, switch grease, switch oils, tractor oils, transformer oils, transmission oils, turbine oils, twine oils, valve oils, V. M. & P. naphtha, water-soluble oils, wool oils.

Pharmaceutical

In compounding and dispensing practice.

Rubber

Reagent in—

Reclaiming rubber.

Soap

Ingredient of—

Special soaps.

Woodworking

As a preservative.

Ingredient of—

Impregnating compositions.

Petroleum Ether

Synonyms: Benzin, Benzine, Canadol, Light ligroin, Ligroin, Naphtha, Petroleum spirit.

Latin: Aether petrolei, Benzinum petrolei, Benzinum purificatum.

French: Esprit de pétrole, Éther de pétrole, Naphte.

German: Canadoel, Naphta, Petrolaether, Petroleumaether, Petroleumbenzin.

Italian: Benzina del petrolio.

Analysis

Solvent for analytical purposes.

Reagent for detecting water in organic compounds.

Reagent in forensic analytical work.

Chemical

General solvent for chemicals and the like.

Solvent for alkaloids.

Solvent in—

Extractions.

Removing phenols and their homologs from various liquids, such as waste waters and the like.

Fats and Oils

Solvent for various vegetable and animal fats and oils.

Solvent in extracting—

Vegetable and animal oils and fats from seeds and other natural products.

Fuel

As a general illuminant in lamps and the like.

As a fuel for special uses.

Glues and Adhesives

For degreasing bones.

Ingredient of—

Adhesive compositions.

Ink

Solvent in making—

Lithographic inks, printing inks.

Insecticides

As an insecticide

Ingredient of—

Insecticidal preparations.

Leather

Ingredient of—

Compositions used in the manufacture of patent leathers.

Solvent for removing—

Fats and greases from hides.

Mechanical

As a motor fuel.

Ingredient of—

Automotive fuels.

Solvent in removing—

Grease and oil from various parts of machines.

Petroleum Ether (Continued)

Mining
Illuminant in miners' lamps.
Miscellaneous
As a general solvent.
For various domestic uses.
In the manufacture of brake linings.
Cleansing agent for clothing and gloves.
Carrier of various substances used for impregnating purposes.
Ingredient of—
Cleansing compositions, metal polishes, spreader compounds.
Transferring compositions (U. S. 1606662).
Transparentizing compositions (U. S. 1744767).
Wiping compositions, waterproofing compositions used for various purposes.
Saturating solution in impregating—
Asbestos board, in dry cleansing, in dental technic, in veterinary medicine.
Oilcloth and Linoleum
Solvent in making—
Coating compositions.
Paint and Varnish
Ingredient of—
Turpentine substitutes.
Thinner and solvent in making—
Insulating varnishes, lacquers, paints, varnishes, removers, specialties.
Paper
Solvent in making—
Compositions for the treatment of wallboard.
Pharmaceutical
For removing fatty constituents prior to extraction of vegetable drugs.
In compounding and dispensing practice.
Perfume
Solvent for perfume materials.
Solvent in extracting—
Perfume bases from natural products.
Photographic
Solvent for wax used for treating ferrotype glossing plates.
Solvent in various processes.
Printing
As a general solvent and cleanser in the print shop.
Solvent in cleaning—
Plates and rollers.
Resins and Waxes
As a solvent.
Rubber
Solvent in rubber technology.
Solvent in making—
Rubber cements, rubber shoes.
Solvent in treating—
Waste rubber materials for the purpose of recovering the "gum" from the fabric.
Soap
Ingredient of—
Detergent preparations, dry-cleansing soaps, spotting fluids.
Textile
In the dyeing process.
For cleansing textile fabrics.

Phellandrene

Chemical
Starting point in synthesis of—
Cymene.
Perfume
Ingredient of—
Perfumes.
Aromatic in—
Lotions, shampoos.
Miscellaneous
Aromatic in—
Dental products.
Soap
Aromatic in—
Soaps.

Phenanthrene

Synonyms: Ortho diphenylene ethylene, Phenanthrin.
Chemical
Starting point in various organic syntheses.
Dye
Starting point in various dye syntheses.
Explosives and Matches
Ingredient of—
Nitroglycerin explosives.
Stabilizer in—
Nitrocellulose explosives.

4-Phenetidin

Chemical
Starting point in making various derivatives.
Dye
Starting point (Brit. 353537) in making acridin dyestuffs with the aid of—
2-Chloro-4-bromobenzoic acid.
2-Chloro-4-iodobenzoic acid.
2:4-Dichlorobenzoic acid.

Phenol

Synonyms: Benzenol, Benzophenol, Carbolic acid, Hydrated oxide of phenyl, Hydroxybenzene, Phenic acid, Phenic alcohol, Phenyl hydrate, Phenylic acid, Phenylic alcohol.
Latin: Acidum carbolicum, Acidum phenicum, Acidum phenylicum, Phenolum, Phenylicum.
French: Acide carbolique, Acide phénylique, Hydrate de phényle.
German: Karbolsäure, Phenylalkohol, Phenylsäure.
Spanish: Acido fenico, Fenol.
Italian: Acido carbolico, Fenolo.
Abrasives
Process material in making—
Binders for abrasives (U. S. 1468960).
Waterproof abrasive belts, discs, papers, wheels (U. S. 1484759).
Adhesives
Preservative in—
Mucilages and pastes.
Process material in making—
Adhesives.
Analytical
Reagent in—
Processes involving control and research in industry.
Animal Husbandry
General disinfectant.
Ingredient (U. S. 1498639) of—
Cattle dips, sheep dips.
Automotive
Process material (U. S. 1418607 and 1429267) in making—
Brake linings.
Aviation
Fireproofing agent (U. S. 1309581) for—
Airplane dopes.
Ingredient (U. S. 1389084) of—
Airplane fabric mending composition.
Brewing
Cleanser and disinfectant for—
Equipment, plant generally.
Building Construction
Ingredient (U. S. 1340855) of—
Roofing compositions.
Road-paving compositions.
Cellulose Products
Ingredient of—
Cellulose acetate solvent mixtures.
Nitrocellulose solvent mixtures.
Waterproofings (with formaldehyde) for rayon.
Plasticizer for—
Cellulose acetate, nitrocellulose.
Process material (U. S. 1509035) in making—
Colloidal suspensions of cellulose.
Solvent for—
Cellulose butyrate, cellulose esters and ethers, ethylcellulose.
Chemical
Dispersing agent (U. S. 1395729) for—
Proteins.
Process material in making—
Acetic anhydride (U. S. 1326040).
Acetyl chloride (U. S. 1326040).
Benzene (U. S. 1430585).
Benzoyl chloride (U. S. 1326040).
Betabeta-bis-(4-hydroxyphenyl)ethylamine (U. S. 1432291).
2:2-Bis(hydroxyphenyl)propane (U. S. 1225750).

Phenol (Continued)
Butyric anhydride (U. S. 1326040).
Cresyl phosphates (U. S. 1425392).
Cresyldiphenyl phosphate (U. S. 1462306).
Diphenyltolyl phosphate (U. S. 1425392).
2-Hydroxy-3-naphthoic acid (U. S. 1470039).
Organic anhydrides (U. S. 1326040).
Parahydroxybenzyl alcohol (U. S. 1317276).
Phenolcholeic acid (U. S. 1252212).
Pyrocatechol (U. S. 1488278).
Quinazarin (U. S. 1465689).
Purifying agent for—
Anthracene (U. S. 1326515).
Anthraquinone (U. S. 1461745).
Reagent in making—
Organic and inorganic chemicals.
Solvent for—
Anthracene, carbazole, other chemicals.
Solvent-recovering agent (U. S. 1315700, 1315701, and 1439128) for—
Acetone, amyl acetate, amyl alcohol, benzene, carbon bisulphide, carbon tetrachloride, chloroform, ether, ethyl acetate, ethyl alcohol, ethylene dichloride, ethylene perchloride, ethylene trichloride, methanol, pentachloroethane, petroleum ether.
Starting point in making—
Bromophenols, chlorophenols, cyclohexanol.
Esters, such as phenyl acetate.
Ethers, such as anisole (methyl ether), phenetole (ethylether), diphenyl oxide.
Intermediate compounds used in the manufacture of many synthetic chemicals.
Iodophenols, nitrophenols, para-aminophenol, phenates (phenoxides), phenolphthalein, phenolsulphonic acids, salicylic acid, thio compounds, triphenyl phosphates.
Starting point in making synthetic tannins with—
Cresol, oleum, soda, and formaldehyde (Brit. 425527, 416191, and 375160).
Formaldehyde, urea compounds, alkylene oxides, alkylene halohydrins, and sulphonating agents (Brit. 447417).
Sulphuric acid and formaldehyde.
Sulphuric acid and sulphites.

Cosmetic
Starting point (Brit. 427147) in making—
Disinfectants for cosmetic preparations and skin creams with the aid of the acid chlorides of capric, lauric, myristic, and palmitic acids.

Dentistry
Antiseptic, bactericide, disinfectant, germicide.

Disinfectant
Antiseptic, bactericide, disinfectant, germicide.
Ingredient of—
Germicidal and disinfectant preparations.
Standard for comparison of disinfectant power.

Distilling
Cleanser and disinfectant for—
Equipment, plant generally.

Dye
Starting point in making—
Intermediate chemicals, synthetic dyestuffs.

Electrical
Agent (Brit. 406586) for—
Removing sludge coatings on surfaces in oil-filled electric transformers.

Explosives and Matches
Starting point in making—
Nitrophenols, such as picric acid (trinitrophenol).
Pyrotechnics (U. S. 1500844).

Fats, Oils, and Waxes
Solvent (U. S. 1277904) for—
Ceresin.

Firefighting
Ingredient of—
Carbon tetrachloride fire-extinguisher (U. S. 1243149).
Fireproofing paint (U. S. 1269980).

Food
Cleanser and disinfectant for—
Equipment, plant generally.

Gas
Impregating agent (U. S. 1398613) for—
Gas-purifying sponge.

Glue and Gelatin
Disinfectant for—
Gluestocks.
Hydrolyzing agent (U. S. 1323951) for—
Gelatin, glue.
Preservative.

Gum
Starting point (U. S. 1448556) in making—
Synthetic gums with acetone.

Ink
Ingredient of—
Check-writing inks (U. S. 1514222).
Document-printing inks (U. S. 1514222).
Lithographic inks (U. S. 1406837).
Marking inks (U. S. 1420289).
Mimeograph inks.
Printing inks (U. S. 1420289).
Safety inks (U. S. 1439658).

Insecticide and Fungicide
Exterminant for—
Ants, moths, termites.
Fungicide.
Insecticide.

Leather
Disinfectant for—
Hides and skins.

Mechanical
Ingredient of—
Diesel engine fuel (U. S. 1340855).
Reagents for removing carbon from internal-combustion engines (U. S. 1368965).

Metallurgical
Detinning agent (U. S. 1379237).
Ingredient of—
Electrolytes for tinplating (U. S. 1426678).
Pickling solutions for iron and steel (U. S. 1493205).

Mining
Flotation agent (U. S. 1438436, 1457708, 1317945).

Miscellaneous
Cleanser, deodorant, disinfectant, germicide for—
Factory purposes, household purposes.
Desizing agent (U. S. 1421613) for—
Chinagrass, ramie.
Impregnating agent for—
Bags (U. S. 1367177).
Barrel linings (U. S. 1323528).
Medicating agent (U. S. 1409364) for—
Atmospheres.
Preservative (U. S. 1421613 and 1460736) for—
Enzyme extracts.
Reagent in—
Solvent-recovery processes.
Starting point (Brit. 427147) in making—
Disinfectants for floor and other wax polishes with the aid of the acid chlorides of capric, lauric, myristic, and palmitic acids.

Optical
Process material (U. S. 1386046) in making—
Cements for lenses.

Paint and Varnish
Disinfectant and germicide in—
Special paints and varnishes.
Ingredient of—
Antifouling paints, dopes, lacquers, paints, paint and varnish removers, varnishes, wood-impregnating agents.

Perfume
Starting point in making—
Aromatic esters and ethers.

Petroleum
Refining agent in processing of—
Lubricating oils.

Pharmaceutical
In compounding and dispensing practice.
Precipitant for—
Albumens.
Preservative (U. S. 1476233) for—
Antitoxins, bacterial extracts, vaccines.
Process material in making—
Colloidal blood (U. S. 1395729).
Blood antitoxins (U. S. 1270270).
Reagent in making—
Hog-cholera antitoxins.
Standard for comparison of disinfectant power.
Starting point in making—
Chemical drugs, such as phenacetin, phenobarbiturates, phenolphthalein, salol.

Phenol (Continued)
Suggested for use as—
Local anesthetic in burns and other painful ulcerations.
Photographic
Starting point in making—
Developing agents.
Plastics
Solvent for—
Casein.
Starting point in making—
Plastics of various compositions including casein.
Resins
Process material in making—
Acetaldehyde condensates, phenol-formaldehyde resins, phenol-furfural resins, urea-formaldehyde resins.
Rubber
Ingredient (U. S. 2004156) of—
Preservative for rubber latex, containing also soap, ammonia, and alkali hydroxide.
Starting point in making—
Butadiene.
Rubber latex preservatives (by alkalization) (U. S. 1447930).
Sanitation
Bactericide, disinfectant, germicide.
Ingredient of—
Disinfectant and germicidal preparations.
Process material (U. S. 1491277) in—
Aeration of sewage.
Soap
Germicide in—
Disinfectant soaps.
Starting point (Brit. 427147) in making—
Disinfectants for soaps and shaving creams with the aid of the acid chlorides of capric, lauric, myristic, and palmitic acids.
Textile
Developing agent in—
Dyeing yellow colors on fabrics and yarns.
Process material in—
Degumming silk (U. S. 1421613).
Dyeing processes.
Mercerizing processes (U. S. 1343139).
Printing processes.
Waterproofing silk (U. S. 1377110).
Process material (U. S. 1389274) for—
Treating wool.
Starting point (Brit. 447417) in making—
Dispersing agents with formaldehyde, urea compounds, alkylene oxides, alkylene halohydrins, and sulphonating agents.
Wine
Cleanser and disinfectant for—
Equipment, plant generally.

Phenol-Mercurio Chloride
French: Chlorure de phénole et mercure.
German: Phenolquecksilberchlorid.
Sanitation
Disinfectant for general use.
Pharmaceutical
In compounding and dispensing practice.

Phenosafranin
Synonyms: Safronin B extra.
Dye
Starting point in making—
Alkylated derivatives, indoine blue.
Photographic
Desensitizer for—
Films and plates (ordinary, orthochromatic, panchromatic) developed by the light of candles.
Textile
——, *Dyeing*
Dyestuff for—
Cotton, silk, wool.
——, *Printing*
Ingredient of color paste for—
Cotton fabrics, silk fabrics, wool fabrics.

Phenothiazin
Insecticide
Insecticide for—
Codling moth (claimed to be as effective as lead arsenate).

Phenothioxin
Insecticide
Toxicant (U. S. 2049725) for—
Codling moths (claimed to be more effective than lead arsenate).
Houseflies (applied in kerosene sprays).
Mosquitoes (claimed to be as effective as rotenone).

Phenoxyethyl Acetate
Cellulose Products
Plasticizer (U. S. 1804503) for—
Cellulose esters or ethers.
Cellulose nitrate (nitrocellulose).
For uses, see under general heading: "Plasticizers."

Phenoxyethyl Phthalate
French: Phthalate de phénoxye-éthyle, Phthalate phénoxyéthylique.
German: Phenoxyaethylphtalat, Phtalsaeurephenoxyaethylester, Phtalsaeuresphenoxyaethyl.
Leather
Softener (Brit. 306911) in—
Cellulose acetate compositions for making artificial leather.
Paint and Varnish
Plasticizer and softener (Brit. 306911) in making—
Cellulose acetate paints, varnishes, lacquers, and enamels.
Plastics
Plasticizer and softener (Brit. 306911) in making—
Cellulose acetate compositions.
Photographic
Softener (Brit. 306911) in making—
Cellulose acetate films.
Textile
Softener (Brit. 306911) in making—
Compositions used for coating fabrics.

Phenylacetic Acid
Synonyms: Alphatoluic acid.
French: Acide phényleacétique.
German: Phenylessigsaeure.
Chemical
Starting point in making—
Condensation products with aldehydes.
Perfume bases, pharmaceutical chemicals, phenylacetaldehyde, phenylacetamide, phenylacetyl anhydride, phenylacetylazonide, phenylacetyl chloride, phenylacetylhydrazide, phenylacetylmethane, phenylacetyluronitrile, phenyldiethylamide, phenyldiphenylamide, phenylethyl alcohol, phenylethylacetonitrile.
Perfumery
Starting point in making—
Synthetic perfumes.
Pharmaceutical
In compounding and dispensing practice.

Phenylacetimidoparatolyl Hydrochloride-Sulphide
Synonyms: Phenylacetimido-thio-para-tolylether hydrochloride.
Insecticide
Larvicide for—
Culicine mosquito larvae.

Phenylacetimidophenyl Hydrochloride-Sulphide
Synonyms: Phenylacetimidothiophenyl hydrochloride.
Insecticide
Larvicide for—
Culicine mosquito larvae.

Phenylacetylhydroquinone
Petroleum
Stabilizing agent (Brit. 406195) for—
Cracked gasolines and other motor fuels.

Phenylacetyl Peroxide
French: Peroxyde de phényleacétyle, Peroxyde phénylique et acétylique.
German: Phenylacetylperoxyd.
Fats and Oils
Bleaching agent (Brit. 328544) in treating—
Various oils (used along with hydrogen peroxide).
Food
Bleaching agent (used with hydrogen peroxide) (Brit. 328544) in treating—
Egg yolk, flour, meal.

Phenylacetyl Peroxide (Continued)
Resins and Waxes
Bleaching agent (Brit. 328544) in treating—
Various waxes (used with hydrogen peroxide).
Soap
Bleaching agent (Brit. 328544) in treating—
Soap (used along with hydrogen peroxide).

Phenylacetylphloroglucinol
Petroleum
Stabilizing agent (Brit. 406195) for—
Cracked gasolines and other motor fuels.

Phenylacetylpyrocatechol
Petroleum
Stabilizing agent (Brit. 406195) for—
Cracked gasolines and other motor fuels.

Phenylacetylpyrogallol
Petroleum
Stabilizing agent (Brit. 406195) for—
Cracked gasolines and other motor fuels.

Phenylacetylresorcinol
Petroleum
Stabilizing agent (Brit. 406195) for—
Cracked gasolines and other motor fuels.

Phenylalphanaphthylamine
Dye
Starting point in making—
Azin dyestuffs, benzophenone dyestuffs, black jet R, Fuller's blue, neutral blue, sulphonazurin D, Victoria blue B, Victoria blue R.

Phenylalphanaphthylamine-8-sulphonic Acid
Synonyms: 8-Anilino-1-naphthalene-sulphonic acid, Phenyl-peri-acid.
French: Acide de 8-anilinoalphanaphthalènesulphonique, Acide de phényle-1-naphthylamine-8-sulfonique, Acide de phényle-péri.
German: 8-Anilinoalphanaphtalinsulfonsaeure, Phenylalphanaphtylamin-8-sulfonsaeure, Phenylperisaeure.
Dye
Starting point in making—
Omega chrome black PV, sulphon acid blue R, sulphur black 3B, sulphoncyanin, sulphocyanin black B, tolyl blue SR.

Phenylalphanaphthyl Ketone
Chemical
Starting point in making—
Aromatics, intermediates, other derivatives.
Perfume
Fixative in—
Cosmetics, perfumes.
Soap
Fixative for odor in—
Toilet soaps.

4-Phenylamino-6-methyl-1:2-phenylenethiazonium Chloride
French: Chlorure de 4-phényleamino-6-méthyle-1:2-phénylènethiazonique.
Dye
Starting point (U. S. 1588384) in making vat dyestuffs with—
Anilin, betahydroxyalphanaphthaquinone, chloranil, chlorobenzoquinone, 2:3-dichloroalphanaphthaquinone, quinone.

2-Phenylamino-8-naphthol-6-carboxylic Alphanaphthalide
French: Alphanaphthalide de 2-phényleamino-8-naphthole-6-carboxyle, Alphanaphthalide 2-phényleamino-8-naphthole-6-carboxylique.
German: 2-Phenylamino-8-naphtol-6-carbonyl alphanaphtalid.
Chemical
Starting point in making—
Intermediates, pharmaceuticals.
Dye
Starting point (Brit. 302773) in making azo dyestuffs with—
Ammonium-paranitranilin orthosulphonate, anilin, 3-chloroanilin, 4-chloro-2-nitranilin, 4-chloro-2-toluidin, 4:4′-diaminodiphenylamine, dianisidin, 2:5-dichloroanilin, meta-aminobenzoic acid, 5-nitro-2-toluidin, orthoaminoazotoluene, orthophenetoleazoalphanaphthylamine.

2-Phenylamino-8-naphthol-6-carboxylic Betanaphthalide
Synonyms: Betaphenylamino-8-naphthol-6-carboxylic betanaphthalide.
French: Bétanaphthalide de bétaphényleamino-8-naphthole-6-carboxylique, Bétanaphthalide de 2-phényleamino-8-naphthole-6-carbonique, Bétanaphthalide de 2-phényleamino-8-naphthole-6-carboxylique.
German: Betaphenylamino-8-naphtol-6-carbonyl-betanaphtalid, 2-Phenylamino-8-naphtol-6-carbonyl-betanaphtalid.
Chemical
Starting point in making—
Intermediates, pharmaceuticals.
Dye
Starting point (Brit. 302773) in making azo dyestuffs with—
Ammonium-paranitranilin orthosulphonate, anilin, 3-chloroanilin, 4-chloro-2-nitranilin, 4-chloro-2-toluidin, 4:4′-diaminodiphenylamine, dianisidin, 2:5-dichloroanilin, meta-aminobenzoic acid, 5-nitro-2-toluidin, orthoaminoazotoluene, orthophenetoleazoalphanaphthylamine.

2-Phenylamino-5-naphthol-7-sulphonic Acid
Synonyms: Betaphenylamino-5-naphthol-7-sulphonic acid.
French: Acide de phényleamino-5-naphthol-7-sulfonique.
German: 2-Phenylamino-5-naphtol-7-sulfonsaeure.
Dye
Starting point (Brit. 280320) in making dyestuffs for viscose rayon with the aid of diazotized—
Anilin, dihydroparatoluidinsulphonic acid, orthoanisidin, paraminoacetanilide.

2-Phenylamino-8-naphthol-6-sulphonic Acid
Synonyms: Betaphenylamino-8-naphthol-6-sulphonic acid.
French: Acide de 2-phényleamino-8-naphthol-6-sulfonique.
German: Betaphenylamino-8-naphtol-6-sulfonsaeure, 2-Phenylamino-8-naphtol-6-sulfonsaeure.
Dye
Starting point (Brit. 281767) in making dyestuffs for viscose with the aid of—
Alphanaphthylamine-6-sulphonic acid.
Aminoazobenzene.
Aminosalicylic acid.
Betanaphthylamine-4:8-disulphonic acid.
Diazotized alphanaphthylamine.
Diazotized para-aminobenzenesulphonic acid.
Meta-aminoparacresolmethyl ether.
Metaxylidinsulphonic acid.
Para-aminoazobenzene-p′-carboxylic acid.
Starting point in making—
Crumpsall direct fast brown O, diamine brown B, diphenyl fast yellow.

Phenylbetanaphthylamine
Chemical
Intermediate in—
Organic synthesis.
Dye
Intermediate in—
Dye synthesis.
Rubber
As an antioxidant.

Phenyl-3-chloropara-anisidin 6-Sulphonate
Dye
Coupling agent (Brit. 434209 and 434433) in making—
Bordeaux red water-insoluble dyestuffs with 5-methoxy-orthotoluidide.

Phenylcinchoninic Acid
Chemical
Starting point in making—
Isatophan.
Methylphenylcinchinone acid.
6-Methyl-2-phenylcinchonin-4-carboxylic acid (paratophan).
Strontium phenylcinchoninate.
Pharmaceutical
In compounding and dispensing practice.

Phenylcresol

Chemical
Starting point (Brit. 444351) in making—
Fat-splitting catalysts and emulsifying agents for use in dyeing, laundering, bleaching, and various other processes, by reacting with formaldehyde and non-aromatic secondary amines (the salts of the products with water-soluble acids, or water-insoluble acids, or the quaternary ammonium salts, are claimed to be valuable for the purposes named).

Phenyl-2:4-dichloroanilin 5-Sulphonate

Dye
Coupling agent (Brit. 434209 and 434433) in making—
Red, water-insoluble dyestuffs with 4-chloro-2:5-dimethoxyanilide.

Phenyldiethyl Phosphate
French: Diéthylephényle phosphate, Phosphate de diéthylephényle.
German: Phenyldiaethylphosphat, Phosphatisches-phenyldiaethyl.

Photographic
Reagent (French 606969) for—
Reducing inflammability in making film from cellulose derivatives.
Solvent (French 606969) in making—
Film from cellulose derivatives.

Plastics
Reagent (French 606969) for—
Reducing inflammability in making plastics from cellulose derivatives.
Solvent (French 606969) in making—
Plastics from cellulose derivatives.

Textile
Reagent (French 606969) for—
Reducing inflammability in making fibers from cellulose derivatives.
Solvent (French 606969) in making—
Fibers form cellulose derivatives.

Phenyldimethyl Phosphate
French: Diméthylephényle phosphate, Phosphate de diméthylephényle.
German: Phenyldimethylphosphat, Phosphatisches-phenyldimethyl.

Photographic
Reagent (French 606969) for—
Reducing inflammability in making film from cellulose derivatives.
Solvent (French 606969) in making—
Film from cellulose derivatives.

Plastics
Reagent (French 606969) for—
Reducing inflammability in making plastics from cellulose derivatives.
Solvent (French 606969) in making—
Plastics from cellulose derivatives.

Textile
Reagent (French 606969) for—
Reducing inflammability in making fibers from cellulose derivatives.
Solvent (French 606969) in making—
Fibers form cellulose derivatives.

1-Phenyl-2:3-dimethyl-5-thiopyrazolone

Photographic
Fog inhibitor (U. S. 1954334) in—
Photographic emulsions.

Phenyl Disulphide
French: Disulphure de phényle, Disulphure phénylique.
German: Dischwefelphenyl, Phenyldisulfid, Schwefelwasserstoffsäurephenylester, Schwefelwasserstoffsäuresphenyl.

Chemical
Reagent (Brit. 298511) in treating—
Albumenoids, albumens.
Starting point in making various derivatives.

Glues and Adhesives
Reagent (Brit. 298511) in making adhesive preparations with—
Linseed protein, peanut protein, soybean flour, vegetable proteins of various sorts.

Miscellaneous
Reagent (Brit. 298511) in making sizes and finishing preparations by treating—
Linseed protein, peanut protein, soybean flour, vegetable proteins of various sorts.

1:3-Phenylenediamine-5-sulphonic Acid
French: Acide de 1:3-Phénylènediamine-5-sulphonique.
German: 1:3-Phenylendiaminsulfonsaeure.

Chemical
Starting point in making—
Intermediates, pharmaceuticals, salts and esters.

Dye
Starting point (Brit. 310343) in making azo dyestuffs with—
4-Nitro-2-aminophenol-6-sulphonic acid.
Picramic acid.

Phenylethyl Acetate
Synonyms: Benzylcarbinyl acetate, Betaoxyalphaphenylethane acetate, Ethylphenyl acetate.
French: Acétate de benzylecarbinyle, Acétate de bétaoxyalphaphényléthane, Acétate de phényléthyle, Acétate phénylique-éthylique.
German: Aethansäurephenylaethylester, Aethansäuresphenylaethyl, Benzylcarbinylacetat, Benzylcarbinylazetat, Essigsäurephenylaethylester, Essigsäuresphenylaethyl, Methancarbonsäurephenylaethylester, Methancarbonsäuresphenylaethyl, Phenylessigsäureaethylester, Phenylessigsäuresaethyl.

Beverages
Flavoring agent in making—
Soft drinks.

Food
Flavoring agent in making—
Food preparations.
Ingredient of—
Fruit essences.

Perfume
Ingredient of—
Hedge rose odors, rose odors.
Perfume for producing honey-like odor in—
Cosmetics.

Resins and Waxes
Perfume for producing honey-like odor in—
Waxes.

Soap
Perfume in—
Honey soaps.

Tobacco
Perfume for producing honey-like odor in—
Tobacco and tobacco products.

Phenylethyl Alcohol
Synonyms: Ethylphenyl alcohol.
French: Alcool d'éthyle et de phényle, Alcool éthylique et phénylique, Alcool de phényle et d'éthyle, Alcool phénylique et éthylique.
German: Aethylphenylalkohol, Phenylaethylalkohol.

Cellulose Products
Plasticizer for—
Cellulose acetate, cellulose esters and ethers, cellulose nitrate.
For uses, see under general heading: "Plasticizers."

Chemical
Starting point in making—
Aromatics, esters, intermediates, pharmaceuticals.

Perfume
Ingredient of various artificial odors, including—
Hyacinth, jasmine, lilac, lily of the valley, narcissus, neroli, rose.
Perfume in reproducing the rose odor in—
Cosmetics, perfumes, toilet waters.

Soap
Perfume to reproduce the rose odor in—
Shampoos, soap creams, soap powders, toilet soap.

Phenylethylbarbituric Acid
French: Acide de phényle-éthylbarbiturique.
German: Phenylaetherbarbiturinsaeure.

Chemical
Starting point (Brit. 301727) in making pharmaceutical chemicals with—
1-Phenyl-2:3-dimethyl-4-diallylamino-5-pyrazolone.
1-Phenyl-2:3-dimethyl-4-diamylamino-5-pyrazolone.

Phenylethylbarbituric Acid (Continued)
1-Phenyl-2:3-dimethyl-4-dibutylamino-5-pyrazolone.
1-Phenyl-2:3-dimethyl-4-diethylamino-5-pyrazolone.
1-Phenyl-2:3-dimethyl-4-diheptylamino-5-pyrazolone.
1-Phenyl-2:3-dimethyl-4-dihexylamino-5-pyrazolone.
1-Phenyl-2:3-dimethyl-4-di-isoallylamino-5-pyrazolone.
1-Phenyl-2:3-dimethyl-4-di-isoamylamino-5-pyrazolone.
1-Phenyl-2:3-dimethyl-4-di-isobutylamino-5-pyrazolone.
1-Phenyl-2:3-dimethyl-4-di-isopropylamino-5-pyrazolone.
1-Phenyl-2:3-dimethyl-4-dimethylamino-5-pyrazolone.
1-Phenyl-2:3-dimethyl-4-dipentylamino-5-pyrazolone.
1-Phenyl-2:3-dimethyl-4-dipropylamino-5-pyrazolone.
Starting point (Swiss 113251) in making pharmaceuticals with—
Allylamine, amylamine, butylamine, diallylamine, diamylamine, dibutylamine, diethylamine, dimethylamine, dipropylamine, ethylamine, isoallylamine, isoamylamine, isobutylamine, isopropylamine, methylamine, propylamine.
Starting point (Brit. 255434) in making synthetic drugs with—
Pyrazolone.
Pharmaceutical
In compounding and dispensing practice.

Phenylethyl Butyrate
Miscellaneous
Odor for various purposes.
Perfume
Ingredient of—
Cosmetics, jasmine perfumes, rose perfumes.
Soap
Perfume in—
Toilet soaps.

Phenylethyl Carboxethylate
French: Carboxéthylate de phényléthyle.
German: Phenylaethylcarboxaethylat.
Spanish: Carboxetilato de feniletil.
Italian: Carbossietilato di feniletile.
Perfume
Ingredient (French 650100) of—
Perfumes.

2-Phenylethyl-4-chlorophenol
Pharmaceutical
Bactericide (U. S. 2101595) for—
Bacillus typhosus, staphylococcus aureus, other bacteria.

Phenylethyldimethyl Carbinol
Synonyms: Dimethylbetaphenylethyl carbinol, Gammaoxyisoamylbenzene.
French: Carbinole de diméthylebétaphényleéthyle.
German: Dimethylbetaphenylaethylcarbinol.
Perfume
Fixative for fine perfumes.
Ingredient of—
Cosmetics, jasmine perfumes.
Soap
Ingredient of—
Toilet soaps.

Phenylethyldimethylcarbinyl Acetate
Perfume
Ingredient of artificial extracts of—
Hedge rose, hyacinth, jasmine, lily, morning-glory, orchid, rose.
Perfume in—
Cosmetics.
Soap
Perfume in—
Toilet soaps.

Phenylethylethyl Ketone
Perfumery
Ingredient (Brit. 264862) of—
Hair restorers, pomades.

Phenylethyl Formate
Perfume
Ingredient of artificial extracts of—
Chrysanthemum, hedge rose, lilac, lily, orchid, rose.
Perfume in—
Cosmetics.
Soap
Perfume in—
Toilet soaps.

Phenylethyl Isovaleriate
Food
Ingredient of artificial essences of—
Apricot, banana, cherry.
Flavoring in—
Beverages, cakes, candies.
Perfume
Ingredient of—
Rose perfumes.
Perfume in—
Cosmetics.
Soap
Perfume in—
Toilet soaps.

Phenylethyl-Mercury Dithiocarbamate
Oils, Fats, and Waxes
Addition agent (Brit. 440175) for—
Lubricating oils or greases used under high-pressure working conditions.

5:5-Phenylethyl-normal-allylbarbituric Acid
Pharmaceutical
Starting point (Brit. 398132) in making—
5:5-Phenylethyl-normal-propylbarbituric acid, suggested for use in treatment of epilepsy.

Phenylethylphenyl Acetate
French: Acétate de phényléthylphényle, Acétate phényléthylphénylique.
German: Phenylaethylphenylacetat, Phenylaethylphenylazetat.
Spanish: Acetato de feniletilfenilico.
Italian: Acetato di feniletilefenilico.
Perfume
Fixative in making—
Honeysuckle perfumes, hyacinth perfumes, jonquil perfumes, linden perfumes, narcissus perfumes, rose perfumes.
Ingredient of—
Perfume compositions (added for the purpose of giving "leafy" effects in floral odors).
Perfuming agents in cosmetics.
Substitute for—
Methylheptine carbonate in various perfumes.
Soap
Perfume in—
Toilet soaps.

Phenylethylphenyl Ketone
Perfumery
Ingredient (Brit. 264862) of—
Hair restorers, pomades.

Phenylglycinorthocarboxylic Acid
French: Acide de phényleglycinorthocarbonique, Acide phényleglycineorthocarboxyle, Acide de phényleglycineorthocarboxylique.
German: Phenylglycinorthocarbonsaeure.
Dye
Starting point in making—
Brilliant indigo BASF-B, brilliant indigo BASF-2B, brilliant indigo BASF-G, bromo indigo, bromo indigo FB, ciba blue 2B, ciba blue G, ciba yellow G, dianthrene blue 2B, helindon blue BB, indigo, indigo KG, indigo MLB, indigo MLB-RR, indigo RB, indigo white, indigo yellow 3G, indigotin, indigotin P.

Phenyl Glycol Acetate
Perfume
Ingredient of—
Jasmine perfumes, lily of the valley perfumes.
Perfume in—
Cosmetics.
Soap
Perfume in—
Toilet soaps.

Phenylglycollic Acid
Synonyms: Amygdalic acid, Amygdalinic acid, Paramandelic acid.
French: Acide amygdalique, Acide paramandelique, Acide de phényleglycole, Acide phényleglycolique.
German: Mandelsäure, Phenylglycolsäure.

Phenylglycollic Acid (Continued)

Chemical
Starting point in making—
Antipyrine phenylglycolate (tussol).
Insulin preparations (Brit. 310934).
Other derivatives (esters, pharmaceuticals, salts).

Pharmaceutical
In compounding and dispensing practice.

Phenylhydrazin

Analysis
Reagent in sugar laboratories.

Chemical
Starting point in making—
Acetylphenylhydrazin, alphanaphthocarbazol, antipyrin, antipyrin salicylate (salipyrin), azobenzene, migranin, 3:2-naphthocarbazol, paraphenylenehydrazinsulphonic acid, phenylhydrazinpyrazolon, phenylhydrazon, phenylpyridazoanthron.
Synthetic sensitizers for brom-gelatin photographic papers.

Dye
Starting point in making—
Dianil yellow R, erichrome red B, guinea fast yellow G, pigment chrome yellow L, pigment fast yellow G, pigment fast yellow R.

Electrical
Dispersive agent in—
Insulating enamels and cements for electrical wiring (Brit. 273290).

Explosives
Starting point in making—
Explosive stabilizer (nitron).

Miscellaneous
Dispersive agent in—
Cements for laminated mica (Brit. 273290).

Paint and Varnish
Dispersive agent in—
Varnish bases (Brit. 273290).

Plastics
Dispersive agent in—
Moldable compositions (Brit. 273290).

Resins and Waxes
Dispersive agent in—
Artificial resins (Brit. 273290).

Rubber
Accelerator in vulcanizing.
Reagent for—
Decreasing rate of vulcanization under certain conditions.

Phenylhydrazin Hydrochloride

Analysis
Reagent in various processes.

Chemical
Starting point in making various organic compounds.

Dye
Starting point in making various synthetic dyestuffs.

Phenylhydrazinparasulphonic Acid
Synonyms: Parahydrazinobenzenesulphonic acid.
French: Acide de phénylehydrazineparasulfonique, Acide de parahydrazinobenzènesulfonique.
German: Parahydrazinobenzolsulfonsaeure, Phenylhydrazinparasulfonsaeure.

Chemical
Starting point in making—
Phenyl-3-methylpyrazolonesulphonic acid.

Dye
Starting point in making—
Dianil yellow 2R, fast light yellow, fast wool yellow, flavazin L, flavazin S, tartrazin.

Textile
——, *Printing*
Ingredient of—
Printing paste, used as resist, in printing fabrics with naphthol azo colors.

2-Phenylindole

Cosmetic
Protective (Brit. 435811) in—
Sun-tan lotions (solution or dispersion in a compatible solvent, for example, glycerin or wool-fat, but not water, alcohol, benzene, carbon tetrachloride, chloroform, or acetone), said to prevent formation of painful erythemas whilst enabling the skin to grow brown in sunlight, by virtue of high absorption of ultraviolet rays.

Phenylionone

Chemical
Starting point in making—
Aromatics and other derivatives.

Perfume
Ingredient of various perfumes.
Odoriferous ingredient of—
Cosmetics.

Soap
Odoriferous ingredient of—
Toilet soaps, cleansing and detergent preparations.

Phenyl Isothiocyanate

Lubricant
Starting point (Brit. 440175) in making—
Addition agents for high-pressure lubricating oils or greases by mixing and reacting with organometallic compounds.

Phenyllauryl-Zinc Telluride

Lubricant
Addition agent (Brit. 440175) for—
Lubricating oils or greases used in high-pressure working conditions.

Phenyl-Magnesium Chloride
Catalyst (Brit. 398561) in making—
Betaphenylethyl alcohol from chlorobenzene, magnesium, and ethylene oxide.

Phenyl Mandelate
French: Mandélate de phényle, Mandélate phénylique.
German: Mandelsaeurephenylester, Mandelsaeuresphenyl, Phenylmandelat.

Paint and Varnish
Plasticizer (Brit. 270650) in making—
Lacquers, varnishes.

Plastics
Plasticizer in making—
Nitrocellulose plastics.

Phenylmercaptan
Synonyms: Thiophenol.

Insecticide and Fungicide
Fumigant and insecticide for—
Ladybird beetles (*Hippodamia convergens Guerin*) (alone or in conjunction with hydrocyanic acid gas).

Phenylmercaptoalphanaphthol
French: Phénylemercaptoalphanaphtole.
German: Phenylmerkaptoalphanaphtol.

Chemical
Starting point in making—
Intermediates, pharmaceuticals.

Dye
Starting point (Brit. 291825) in making indigoid dyestuffs with—
5:7-Dibromoisatin anilide, 5:7-dibromoisatin bromide, 5:7-dibromoisatin chloride, 5:7-dichloroisatin anilide, 5:7-dichloroisatin bromide, 5:7-dichloroisatin chloride.
Isatin anilide, isatin alpha-anil, isatin beta-anil, isatin bromide, isatin chloride, reactive derivatives of isatin.

Phenylmercuric Acetate
Synonyms: Mercury-phenyl acetate.
French: Acétate mercurique-phénylique, Acétate de mercure et de phényle.
German: Essigsäuremerkurphenylester, Merkurphenylacetat, Merkurphenylazetat.

Chemical
Starting point in making various derivatives.

Insecticide
Ingredient (Brit. 321396) of—
Compositions for immunizing wheat and other grains.

Woodworking
Ingredient (Brit. 321396) of—
Preserving and disinfecting compositions.

Phenyl-Mercuric Adipate

Soap
Germicide (Brit. 427324) in—
Nontoxic germicidal soaps (stable solutions in standard soaps).

Phenyl-Mercuric Citrate
Soap
Germicide (Brit. 427324) in—
Nontoxic germicidal soaps (stable solutions in standard soaps).

Phenyl-Mercuric Glycollate
Soap
Germicide (Brit. 427324) in—
Nontoxic germicidal soaps (stable solutions in standard soaps).

Phenylmercuric Hydroxide
French: Hydroxye de phényle et de mercure, Hydroxye phénylique-mercurique.
German: Phenylmerkurhydroxyd.
Chemical
Starting point in making various derivatives.
Starting point (Brit. 329987) in making organic mercury derivatives used for the immunization of grain and made with the aid of—
Alphahydroxynaphthoic acid, alphanaphthol, betahydroxynaphthoic acid, betanaphthol, gallic acid, isothymol, mercaptobenzothioazole, metahydroxybenzaldehyde, orthochlorophenol, paracresols, parahydroxybenzoic acid, parathiocresol, salicylic acid, thioglycollic acid, thiophenol, thiosalicylic acid.

Phenyl-Mercuric Salicylate
Soap
Germicide (Brit. 427324) in—
Nontoxic germicidal soaps (stable solutions in standard soaps).

Phenyl-Mercuric Succinate
Soap
Germicide (Brit. 427324) in—
Nontoxic germicidal soaps (stable solutions in standard soaps).

Phenyl-Mercuric Tartrate
Soap
Germicide (Brit. 427324) in—
Nontoxic germicidal soaps (stable solutions in standard soaps).

Phenylmercury Chloride
Agriculture
For control of—
Bottom rust of lettuce, covered smut and stripe disease of barley, kernel smut of sorghum, loose and covered smuts of oats, soil-borne parasitic fungi, stinking smut of wheat.
Insecticide
Ingredient of—
Seed, plant, and soil disinfectants.
Woodworking
For control of—
Blue stain and sap stain in sapwood and freshly sawed lumber.

Phenyl-Mercury Fuorate
Disinfectant
Claimed (U. S. 2022997) to be—
Germicide.

Phenyl-Mercury Gallate
Disinfectant
Germicide (U. S. 2074040).

Phenyl-Mercury Nitrate
Germicide
Germicidal agent.
Insecticide
Fungicidal agent.
Pharmaceutical
Antiseptic agent.

Phenyl-Mercury 2-Phenylquinolin-4-carboxylate
Disinfectant
Claimed (U. S. 2022997) to be—
Germicide.

Phenyl-Mercury Protocatechuate
Disinfectant
Germicide (U. S. 2074040).

Phenyl-Mercury Quinolinate
Disinfectant
Claimed (U. S. 2022997) to be—
Germicide.

Phenyl-Mercury Salicylate
Disinfectant
Germicide (U. S. 2074040).

5-Phenyl-3-methylfurodiazole
Chemical
Starting point (Brit. 396778) in making—
Triazoles by condensation with either methylamine or phenylamine.

2-Phenyl-1-methylindole
Chemical
Starting point (Brit. 438278) in making—
1:2-Dimethylindole-3-aldehyde.
6-Nitro-2-phenyl-1-methylindole-3-aldehyde.
2-Phenyl-1-methylindole-3-aldehyde.
1:3:4-Trimethyl-2-methyleneindoline-2-w-aldehyde.

1-Phenyl-2-methyl-5-pyrazolone
Chemical
Starting point in making—
Intermediates, pharmaceuticals.
Dye
Starting point in making azo colors for wool with—
4-Aminosalicylic acid, 2-chloro-4-toluidin, meta-aminobenzoic acid, metanilic acid, sulphanilic acid.

1-Phenyl-3-methyl-5-pyrazolone
Chemical
Starting point in making—
2-Chloro-1-phenyl-5-methylpyrazole, pharmaceuticals and various synthetic organic chemicals.
Dye
Starting point in making—
Diphenylmethane dyestuffs, dianil yellow 3G, diazo gold yellow, diazo light green BL, diazo bordeaux B, diazo bordeaux G, diazo bordeaux R, diazo bordeaux V, erichrome red B, tetrakosazo dyestuffs (U. S. 1655550-1), trisazo dyestuffs (U. S. 1655550-1).
Plastics
Starting point (German 584479) in making—
Artificial films from a cellulose ester solution.
Textile
——, *Dyeing*
Starting point (Brit. 396893) in—
Dyeing acetate silk violet bordeaux.
——, *Manufacturing*
Starting point (German 584479) in making—
Artificial fibers from a cellulose ester solution.

1-Phenyl-3-methyl-5-pyrazolonesulphonic Acid
Dye
Intermediate in making various dyestuffs.
Photographic
Fog inhibitor (U. S. 1954334) in—
Photographic emulsions.

Phenylorthoanisidin 4-Sulphonate
Dye
Coupling agent (Brit. 434209 and 434433) in making—
Water-insoluble red dyestuffs with 5-chlor-2:4-dimethoxyanilide.

Phenylparadiphenylaminoparaphenylenediamine
Chemical
Starting point in making various derivatives.
Rubber
Reagent for preserving rubber.

Phenyl Paratoluenesulphonate
French: Paratoluènesulphonate de phényle, Paratoluènesulphonate phénylique.
German: Phenylparatoluolsulfonat, Paratoluolsulfonsaeurephenylester, Paratoluolsulfonsaeurespheny1.
Cellulose Products
Solvent and plasticizer (Brit. 312688) for—
Cellulose acetate, cellulose esters and ethers, cellulose nitrate.
For uses, see under general heading: "Solvents."
Chemical
Starting point in making various derivatives.

Phenylpropyl Cinnamate
Synonyms: Hydrocinnamyl cinnamate.
French: Cinnamate de hydrocinnamyle, Cinnamate hydrocinnamylique, Cinnamate de phénylepropyle, Cinnamate phénylique et propylique.
German: Hydrocinnamylcinnamat, Phenylpropylcinnamat, Zimtsäurehydrocinnamylester, Zimtsäureshydrocinnamyl, Zimtsäurephenylpropylester, Zimtsäuresphenylpropyl.

Perfume
Ingredient of—
Fancy perfumes, oriental perfumes.

Phenylresorcinol

Chemical
Starting point (Brit. 444351) in making—
Fat-splitting catalysts and emulsifying agents for use in dyeing, laundering, bleaching, and various other purposes, by reacting with formaldehyde and non-aromatic secondary amines (the salts of the products with water-soluble acids or water-insoluble acids, and the quaternary ammonium salts, are claimed to be valuable for the purposes named).

Phenyl Salicylate
Synonyms: Salol.
Latin: Phenylis salicylas, Phenylum salicylum.
French: Salicylate de phénole, Salicylate de phényle.
German: Phenylsalicylat, Salicylsäurephenylester.
Spanish: Salicilato de fenol.

Cosmetic
Ingredient of—
Dentifrices.

Pharmaceutical
Enteric coating for—
Pills, tablets.
In compounding and dispensing practice.

Resins
Ingredient (German 364044) of—
Catalyzed condensation product with formaldehyde, useful in lacquers and similar products.

Phenyl Stearate

Cellulose Products
Plasticizer (U. S. 1901129) for—
Nitrocellulose.
For uses, see under general heading: "Plasticizers."

Phenylthioglycollic Acid
Synonyms: Phenylsulphoglycollic acid.
French: Acide de phénylesulfoglycollique, Acide de phénylethioglycollique.
German: Phenylsulfoglykolsaeure, Phenylthioglykolsaeure.

Dye
Reagent (Brit. 284288) in making thioindigoid dyestuffs with—
Acenaphthenequinone, alphaisatinanilide, 5:7-dibromoisatin, isatin, isatin homologs and substitution products, orthodiketones.

4-Phenyl-5-thioketo-2-mercapto-1:3:4-thiodiazole

Metallurgical
Promoter (U. S. 1852108) in—
Recovering mineral from ores by the froth flotation process.

2-Phenylthiolquinolin Ethiodide

Dye
Process material (Brit. 454687) in making—
Cyanin dyes.

2-Phenylthiolquinolin Methiodide

Dye
Process material (Brit. 454687) in making—
Cyanin dyes.

Phloroglucinol
German: Phloroglucin.
Spanish: Floroglucina.
Italian: Floroglucine.

Analysis
Reagent in—
Color reactions with phosphotungstic acid.
Kreis test.
Testing paper and pulp to determine the presence of pentosans and mechanical wood, as well as straw pulp.

Chemical
Reagent in making—
Condensation products with aldehydes and vanillin.
Pharmaceutical chemicals.
Starting point in making—
Ammonia derivatives, diacetyl derivatives.
Organic compounds with formic acid and potassium hydroxide.
Thiocyanic acid ester derivatives.

Miscellaneous
As a general preservative.

Phosphoric Acid
Synonyms: Orthophosphoric acid.
Latin: Acidum phosphoricum, Acidum phosphoricum concentratum.
French: Acide phosphorique.
German: Phosphorsäure.
Spanish: Acido fosforico.
Italian: Acido fosforico.

Abrasives
Etching agent (U. S. 1482793) for—
Abrasives, corundum, emery, garnet, quartz, sand, silicon carbide.

Adhesives
Ingredient of—
Cloth cement (U. S. 1482357).
Paper cement (U. S. 1482357).

Agricultural
As a weed-killer (French 770858).
Starting point (Brit. 430417) in making—
Green fodder preservatives for use in silos, by admixture with powdered coal, lignite, or peat.

Analysis
Reagent in—
Analytical methods and processes involving pure science, process control, and research.

Animal Husbandry
Ingredient of—
Cattle feeds (U. S. 1515968).

Beverage
Ingredient of—
Carbonated beverages, cola beverages, fruit-flavored beverages, phosphate beverages, soft drinks.
Sterilizing agent for—
Beverages.
Substitute for—
Citric acid, tartaric acid.

Brewing
Sterilizing agent (U. S. 1140717) for—
Beer.

Building Materials
Ingredient of—
Acid-resisting cements (Brit. 416966).
Hydraulic cement mortar (U. S. 1908636 and 1908637).
Process material in making—
Acidproof cement (U. S. 1237078).
Cement (U. S. 1507379).
Waterproof cement (U. S. 1237078).
Retardant of—
Setting of plaster of paris (used in conjunction with citric acid or its salts).
Starting point in making—
Phenol esters useful in making or treating roofing materials (U. S. 1167195).

Cellulose Products
Catalyst in making—
Cellulose acetate (U. S. 1296847, 1445382, and 1466329; Brit. 400249, 405825, and 415052).
Cellulose esters (U. S. 1296847 and 1355415; Brit. 398626 and 415052).
Cellulose formate (U. S. 1296847).
Cellulose propionate (U. S. 1296847).
Hydrolyzing agent for—
Cellulose acetate (Brit. 403554).
Wood (U. S. 1323540).
Ingredient of—
Cellulose hydrate (U. S. 1355415, 1218954, and 1242030).
Rayon (U. S. 1242030 and 1355415).
Solvent for cellulose (U. S. 1296847).
Solvent for cellulose esters (U. S. 1283183).
Parchmentizing agent (U. S. 1430163) for—
Cotton.
Reagent in making—
Artificial wool from proteids and cellulose (U. S. 1400381).
Cellulose acetate (Brit. 407759).

Phosphoric Acid (Continued)
Cellulose formylphosphate (U. S. 1153596).
Cellulose nitrate plasticizer (U. S. 1370853).
Cellulose nitrate solvent (U. S. 1283183 and 1365052).
Cotton solvent (U. S. 1218954 and 1242030).
Cellulose hydrate (U. S. 1242030).
Solvent for—
Cellulose (used in admixture with acetic acid) (U. S. 1296847).
Ceramics
As a flux.
As a vitrifying agent.
Increaser of—
Color, translucency.
Chemical
Absorption agent in making—
Aliphatic alcohols and ethers from ethylene (Brit. 397187).
Aliphatic alcohols and ethers from propylene (Brit. 397187).
Aliphatic alcohols and ethers from other olefins (Brit. 397187).
Hydration products of ethylene (Brit. 389133).
Hydration products of propylene (Brit. 389133).
Hydration products of other olefins (Brit. 389133).
Catalyst (U. S. 1429650) in decomposing—
Ethylidene diacetate.
Catalyst in making—
Acetyl chloride from acetic acid and carbonyl chloride (Brit. 402328 and 402335).
Alcohols from olefins and water vapor (Brit. 413043).
Aliphatic alcohols and organic esters thereof from ethane, or propane and organic acids, such as lower aliphatic acids (Brit. 402060).
Aliphatic amines (Brit. 399201).
Aliphatic anhydrides from the corresponding acids (Brit. 407367).
Alkyl halides (U. S. 1937269).
Esters (U. S. 1400849 and 1421605).
Esters from lower aliphatic acids and olefins (Brit. 398527).
Ethyl alcohol from ethylene and steam (Brit. 368051, 368935, 370136, and 408006).
Ethylene (U. S. 1421640, 1372736, 1402329, and 1402336).
Halides, such as benzyl, chloride and acetyl chloride, from organic halides and carboxylic acids (U. S. 1921767).
Phenol ethers (U. S. 1469709).
Wetting agents from ethers of polyalcohols (French 753752).
Catalyst deterioration inhibitor (U. S. 1967189) in making—
Acetic acid from carbon monoxide and methanol.
Catalyst revivifier (U. S. 1967189) in making—
Acetic acid from carbon monoxide and methanol.
Drying agent (U. S. 1338831) for—
Sulphur dioxide, sulphur trioxide.
Drying agent in making—
Intermediate chemicals.
Ingredient (Brit. 392289 and 392685) of—
Catalytic mixtures with compounds of uranium, iron, or cobalt used in making ethyl alcohol from ethylene and steam.
Peptizing agent (Brit. 409361) in—
Stabilizing hydrogen peroxide solutions.
Polymerizing agent (Brit. 447973, 450592, and 450668) in making—
Liquids of low molecular weight from olefins.
Process material in making—
Acetaldehyde (U. S. 1213486, 1213487, 1319365, 1384842, 1431301, and 1471058).
Acetates (U. S. 1365050 and 1365052).
Acetic acid (U. S. 1159376 and 1174250).
Acetic acid series (U. S. 1283183).
Acetone (U. S. 1497817).
Alcohol (U. S. 1218954, 1245818, 1283183, 1438123, and 1517968).
Alcohols (secondary) (U. S. 1497817).
Alkali-earth peroxides (U. S. 1169703).
Alkyl acetates (U. S. 1365050 and 1365052).
Aluminum peroxide (U. S. 1169703).
Aluminum sulphate (U. S. 1126408).
5-(Aminoethyl)-imidazole hydrochloride (U. S. 1178720).
Ammonia (U. S. 1221505).
Ammonium metaphosphate (U. S. 1514912 and 1194077).
Ammonium phosphate (U. S. 1369763, 1142068, 1151074, 1151633, 1167788, 1191615, 1194077, 1208877, 1264513, 264514, 1276870, 1369763, and 1514912).
Amyl alcohol (U. S. 1438123).
Barium carbonate (U. S. 1235664).
Barium peroxide (U. S. 1169703).
Barium nitrate (U. S. 1273824).
Barium phosphate (U. S. 1213330 and 1273824).
Barium sulphate (U. S. 1235664).
Boron derivatives (U. S. 1336974).
Butyl alcohol (U. S. 1438123).
Butyric acid (U. S. 1283183).
Butyrone (U. S. 1283183).
Calcium acetate (U. S. 1283183).
Calcium acid phosphate (U. S. 1413048).
Calcium butyrate (U. S. 1283183).
Calcium peroxide (U. S. 1169703).
Calcium silicate (U. S. 1126408).
Caproic acid (U. S. 1283183).
Carbon monoxide (U. S. 1514912).
Cesium sulphate (U. S. 1126408).
Chromopyrophosphoric acid (U. S. 1323878).
Decolorizing carbons (U. S. 1308826 and 1438113).
Disodium phosphate (U. S. 1351672).
Esters (U. S. 1365050, 1365052, and 1400852).
Ethyl acetate (U. S. 1400852).
Ethyl alcohol (U. S. 1323540 and 1438123).
Ethylene (U. S. 1295339, 1438123, and 1372736).
Ethyl butyrate (U. S. 1400852).
Ethyl propionate (U. S. 1400852).
Ethylmethyl ketone (U. S. 1497817).
Fatty acids (U. S. 1517968).
Glycerol synthesis (U. S. 1466665).
Hydrogen (U. S. 1169703).
Hydrogen peroxide (U. S. 1139774, 1210651, 1235664, 1271611, 1273824, 1262589, and 1364558).
Isoborneol (U. S. 1478690).
Isoborneol acetate (U. S. 1420399 and 1478690).
Isopropyl alcohol (U. S. 1438123 and 1497817).
Ketones (U. S. 1283183 and 1497817).
Lactose (U. S. 1500770).
Lithium sulphate (U. S. 1126408).
Magnesium dioxide (U. S. 1169703).
Manganese dioxide (U. S. 1330738).
Methyl acetate (U. S. 1400852).
Methyl butyrate (U. S. 1400852).
Methyl propionate (U. S. 1400852).
Organic acids (U. S. 1497817 and 1517968).
Oxalic acid (U. S. 1446012).
3-Pentanone (U. S. 1497817).
Perborates (U. S. 1169703).
Phthalic acid (U. S. 1516756).
Potassium chloride (U. S. 1317524, 1456831, and 1450850).
Potassium phosphate (U. S. 1456831 and 1456850).
Potassium compounds (U. S. 1514912).
Potassium nitrate (U. S. 1317524).
Potassium sulphate (U. S. 1126408 and 1317524).
Potassium-aluminum sulphate (U. S. 1126408).
Propionic acid (U. S. 1517968).
Sodium acid phosphate (U. S. 1150899 and 1150900).
Sodium chloride (U. S. 1317524).
Sodium nitrate (U. S. 1317524).
Sodium perborate (U. S. 1169703).
Sodium salt (U. S. 1162617).
Sodium sulphate (U. S. 1126408, 1262589, and 1317524).
Sodium sulphide (U. S. 1213330).
Solvent (U. S. 1283183).
Terpineol (U. S. 1408462).
Terpinolene (U. S. 1408462).
Titanium compounds (U. S. 1504672).
Titanium hydroxide (U. S. 1504672).
Titanium oxide (U. S. 1504672).
Urea (U. S. 1275276).
Urea phosphate (U. S. 1275276).
Purifying agent for—
Catalysts used in decomposing organic substances, such as alcohols, by dehydration processes (U. S. 1913938).
Crude borneol (Brit. 394979).
Hydroaromatic alcohols (Brit. 394979).
Methylcyclohexanone (Brit. 394979).
Mixtures of monohydric terpene alcohols and hydroaromatic alcohols (Brit. 394979).
Monohydric terpene alcohols (Brit. 394979).
Naphthalene (U. S. 1201601).
Phenols (U. S. 1201601).
Reagent (U. S. 1172062) in making—
Nickel catalyst.
Reagent (U. S. 1451786) in—
Removing fluorine.
Removing agent for—
Nitrogen from oxygen (U. S. 1166294).
Potassium from feldspar (U. S. 1317524).
Potassium from fluedust (U. S. 1317524).
Potassium from glauconite (U. S. 1317524).
Potassium from greensand (U. S. 1317524).

Phosphoric Acid (Continued)
Potassium from leucite (U. S. 1317524).
Potassium from mica (U. S. 1317524).
Potassium from silicate (U. S. 1317524).
Sodium from feldspar (U. S. 1317524).
Sodium from fluedust (U. S. 1317524).
Sodium from glauconite (U. S. 1317524).
Sodium from greensand (U. S. 1317524).
Sodium from leucite (U. S. 1317524).
Sodium from mica (U. S. 1317524).
Sodium from silicate (U. S. 1317524).
Separating agent (Brit. 394979) for—
Borneol from admixtures with camphene.
Borneol from admixtures with fenchyl alcohol.
Camphene from the products of the hydration of camphene.
Camphor from admixtures with camphene.
Crude borneol from other substances.
Cyclohexanol from admixtures with cyclohexanone.
Hydroaromatic alcohols from other substances.
Isoborneol from admixtures with borneol.
Methylcyclohexanone from other substances.
Mixtures of monohydric terpene alcohols and hydroaromatic alcohols from other substances.
Monohydric terpene alcohols from other substances.
Terpineol from commercial pine oil.
Solvent (U. S. 1399604) for—
Silver phosphate.
Stabilizing agent (U. S. 1275765) for—
Hydrogen peroxide, perborates.
Starting point in making—
Acid esters used as polymerizing agents for olefins in the production of low boiling point liquids (Brit. 447973, 450592, and 450668).
Albumen derivatives (U. S. 1381295).
Calcium phosphates.
Catalysts for the manufacture of benzaldehyde (U. S. 1487020).
Catalysts for the manufacture of acetaldehyde (U. S. 1487020).
Catalysts for dehydrogenation processes (U. S. 1215335).
Catalysts for the manufacture of ethylene (U. S. 1372736).
Catalysts for the manufacture of formaldehyde (U. S. 1487020).
Catalysts for hydrogenation processes (U. S. 1172062 and 1215335).
Catalysts for the manufacture of nitrogen oxides (U. S. 1207706, 1207707, and 1207708).
Catalysts for the manufacture of ethyl alcohol from ethylene (Brit. 396724, and 392289).
Catalysts (with oxides of calcium, barium, strontium, or magnesium) for the production of alcohols from olefins and steam (Brit. 415417, and 407722).
Catalysts (with strontium carbonate) used in the production of ethyl alcohol from ethylene by hydration (Brit. 407944).
Catalysts (with strontium carbonate) used in the production of ethylene from ethyl alcohol by dehydration (Brit. 407944).
Catalysts (with zinc oxide) used in the production of synthetic formic acid from carbon monoxide and steam (Brit. 406244, and 406345).
Colloidal phosphates (U. S. 1458542).
Dispersing agents with sugars and lauryl alcohol (Brit. 404684).
Dispersing agents with ricinoleic acid and sugars (Brit. 404684).
Emulsifying agents with higher alcohols and boric acid (Brit. 409598).
Emulsifying agents with oleyl and cetyl alcohol mixtures and boric acid (Brit. 409598).
Emulsifying agents with oleyl alcohol, boric acid, and hydrogen peroxide (Brit. 409598).
Emulsifying agents with lauryl alcohol, boric acid, and hydrogen peroxide (Brit. 409598).
Emulsifying agents with octadecenol, boric acid, and hydrogen peroxide (Brit. 409598).
Esters with aryl chlorides (U. S. 1425392).
Ethyl esters (U. S. 1421640).
Iron phosphate (U. S. 1428087).
Nascent hydrochloric acid from metal chlorides, used in the production of chlorhydrins from dihydric alcohols (Brit. 404938).
Nascent hydrochloric acid from metal chlorides, used in the production of chlorhydrins from polyhydric alcohols (Brit. 404938).
Polysilicophosphoric acids (U. S. 1408960).
Pyrophosphoboric acid (U. S. 1323878).
Rubidium sulphate (U. S. 1126408).
Sodium phosphates.
Various phosphates.
Wetting agents with sugars and lauryl alcohol (Brit. 404684).
Wetting agents with ricinoleic acid and sugars (Brit. 404684).
Wetting agents with higher alcohol and boric acid (Brit. 409598).
Wetting agents with oleyl and cetyl alcohol mixtures and boric acid (Brit. 409598).
Wetting agents with oleyl alcohol, boric acid, and hydrogen peroxide (Brit. 409598).
Wetting agents with lauryl alcohol, boric acid, and hydrogen peroxide (Brit. 409598).
Wetting agents with octadecenol, boric acid, and hydrogen peroxide (Brit. 409598).
Zinc dihydrogen phosphate (U. S. 1926266).

Clay Products
Cleaning agent (U. S. 1438588) for—
Canadian d'Amherst clay, Canadian china clay, Fraddon clay, Wotter clay.

Coal Processing
Catalyst (Brit. 414445) in making—
Oils by hydrogenation of coal.
Deterioration inhibitor (Brit. 401131) of—
Emulsifying agent in emulsions of tar, pitch, oils.

Disinfectant
As a bactericide (French 763508).

Distilling
Treating agent (U. S. 1423042) for—
Distillery slop.

Dye
Drying agent in—
Dye syntheses.
Process material in making—
Anilin black (U. S. 1350600).
Titanium compounds suitable as mordants (Brit. 419522).
Purifying agent (U. S. 1201601) for—
Anthracene.
Solubilizing agent (Brit. 396177) for—
Basic dyestuffs, methylene blue, salts of basic dyestuffs, victoria blue R.
Starting point in making—
Catalysts for the manufacture of anthraquinone (U. S. 1487020).

Electrical
Ingredient (U. S. 1908039) of—
Electrolyte for electrolytic rectifier.
Process material in making—
Arclight electrode (U. S. 1134148).
Electric heater (U. S. 1507379).
Electric insulation (various patents).
Electrical resistance element (U. S. 1349053).
Electrolytic condenser (U. S. 1141402).
Electrolytic rectifier (U. S. 1141402).
Incandescent light filaments.
Storage battery electrolyte (U. S. 1433136).

Explosives and Matches
Process material in making—
Match splint (U. S. 1191544 and 1191545).
Potassium nitrate (U. S. 1317524).
Sodium nitrate (U. S. 1317524).

Fats and Oils
Extractant (Brit. 410813) for—
Piperitone from essential oils.
Purifying agent for—
Vegetable oils (U. S. 1170868; Brit. 393108 and 377336).
Starting point in making—
Extractant for fat from fish meal by reacting with a fat solvent and Irish moss (Brit. 405906).

Fertilizer
Ingredient of—
Fertilizer compositions (many patents).
Starting point in making—
Ammonium metaphosphate (U. S. 1194077).
Ammonium phosphate (U. S. 1369763).
Ammonium pyrophosphate (U. S. 1194077).
Calcium acid phosphate (U. S. 1252318, 1383911, and 1383912).
Dicyandiamide (U. S. 1275276).
Double phosphate.
Fertilizer compositions (many patents).
Urea phosphate (U. S. 1440056).
Treating agent for—
Phosphate rock (U. S. 1313379).

Phosphoric Acid (Continued)

Fireproofing
Starting point in making—
 Fireproofed fiber board (U. S. 1928805).
 Fireproofing solutions (U. S. 1382618).
 Phosphate salts, such as ammonium phosphates, sodium phosphates, used as fireproofing agents.

Food
Acid flavoring agent in—
 Jams, jellies.
Extractant for—
 Pectin from fruit.
Ingredient of—
 Bread dough (U. S. 1500545).
 Cake flour (U. S. 1266202).
 Yeast stimulant (U. S. 1447054).
Peeling agent (U. S. 1453781) for—
 Fruit.
Process material in making—
 Artificial milk (U. S. 1200782).
 Bran extract (U. S. 1189023).
 Dextrose (U. S. 1218954 and 1242030).
 Milk serum powder (U. S. 1246858).
Purifying agent for—
 Glucose (U. S. 1314203).
 Lactose (U. S. 1314203).
 Maltose (U. S. 1314203).
 Sorghum (U. S. 1314203 and 1314204).
Saccharifying agent (U. S. 1431525) for—
 Cereal germs.
Sterilizing agent for—
 Cream (U. S. 1140717).
 Fruit pulp (U. S. 1140717).
 Grape juice (U. S. 1140717).
 Milk (U. S. 1140717).
Stimulant (U. S. 1449127) in—
 Yeast culture.
Substitute for—
 Citric acid, tartaric acid.
Treating agent (U. S. 1189023 and 1222830) for—
 Flour, wheat.

Fuel
Process material in making—
 Fuel briquets (U. S. 1507673, 1507674, 1507675, and 1507676).

Glass
Ingredient of—
 Opaque glass batches, optical glass batches, ornamental glass batches, translucent glass batches.
Process material in making—
 Crown glass, double objective lenses, glasses transparent to ultraviolet rays.

Glue and Gelatin
Acidifying agent (U. S. 1289053) in making—
 Gelatin, glue.

Insecticide and Fungicide
Inhibitor (U. S. 1318174) of—
 Mould growth.
Parasiticide (French 763508) for treating—
 Mushrooms.
Process material in making—
 Fungicide (U. S. 1515803).
 Insecticide (U. S. 1515803).

Laundry
Starting point in making—
 Souring compositions (U. S. 1514067).

Leather
Process material in making—
 Artificial leather (U. S. 1245818, 1245977, 1275324, and 1427645).
 Tanning compounds (U. S. 1323878 and 1375975).
 Titanium compounds useful in tanning (Brit. 419522).

Linoleum and Oilcloth
Process material in making—
 Linoleum substitute (U. S. 1245978, 1245984, and 1427645).

Metallurgical
Cleaning agent for—
 Iron and steel (U. S. 1211138, 1221441, 1221442, 1503443, and 1872091; Brit. 403373).
Ingredient of—
 Bath for zinc-coating iron (U. S. 1221046).
 Brass cement (U. S. 1359137).
 Electrolyte for nickel-plating iron (U. S. 1397514 and 1211218).
 Electrolyte for lead-plating iron (U. S. 1397514).
 Electrolyte for tin-plating iron (U. S. 1397514).
 Etching solution (U. S. 1362159).
 Glacial cement for joining of iron (U. S. 1261750).
 Iron-cleaning solutions (U. S. 1268237, 1387645, 1428084, 1387645, and 1398507).
 Iron-pickling bath (U. S. 1321182).
 Iron and steel cleaning composition containing also ethyl alcohol and a water-soluble oil solvent (U. S. 1897813).
 Iron and steel cleaning composition, for use prior to painting, containing also sulphurized pyridin bases, furfural, and organic solvents (Brit. 396053).
 Metal-cleaning compositions, for use prior to painting, containing also monobutyl ether or monoethyl ether of ethyleneglycol with or without ethylmethyl ketone, saponin, oleic acid, water, and sugar base (Brit. 404819).
 Metallic oxide briquettes (U. S. 1507673 and 1507674).
 Parkerizing agent (various patents).
 Polish for iron (U. S. 1280939).
 Rust-preventing solution containing also glucose (U. S. 1329573).
 Rust-preventing compositions (U. S. 1428085, 1291352, 1341100, and 1381112; Brit. 420461).
 Rust-preventing and rust-removing compositions containing also tannic or gallic acid, cellulose or other varnish, tin chloride, and inert pigments (Brit. 410323).
 Rust-preventing composition containing also linseed oil acids, triethanolamine, mineral spirits, and varnish (Brit. 407008).
 Rust-proofing solutions and agents (various patents).
 Rust-proofing composition containing also iron sulphide, sodium carbonate, and water (Brit. 419487).
 Rust-removing composition containing also ethyleneglycol butylester, oleic acid, saponin, and water (U. S. 1935911).
 Rust-resisting coating composition containing also a varnish base, a hydrocarbon solvent, and a saturated aliphatic monohydric alcohol (U. S. 1995954).
 Rust-resisting coating composition containing also ethyl alcohol, water, and a propyl derivative (U. S. 1949921).
Pickling agent for—
 Aluminum and its alloys prior to production of firmly adherent plated coatings of zinc or copper (Brit. 404251 and 385067).
 Iron and steel (U. S. 1279101, 1279331, and 1872091).
Reagent in—
 Detinning process (U. S. 1202149).
Rustproofing agent for—
 Constructional steel, galvanized iron (U. S. 1273358), iron (many patents), pipes, plates, steel (many patents), tubing, wrought iron.
Rust-removing agent for—
 Iron and steel.
Scale-removing agent for—
 Iron and steel.
Starting point in making—
 Ferrophosphor (U. S. 1265076).
 Foundation coatings or coverings of metal before painting.

Mining
Treating agent (U. S. 1151117) for—
 Hematite ore, lead mineral ores, limonite ore, zinc mineral ores.

Miscellaneous
As a grease-removing agent (U. S. 1240395).
Ingredient of—
 Dental cements (many patents).
 Dentifrice (U. S. 1386252).
 Polishes (U. S. 1474133).
 Metal polishes (U. S. 1280939).
Process material in making—
 Crucibles (U. S. 1512801).
 Molds (U. S. 1239152).
 Ornaments (U. S. 1482357 and 1482358).
 Proteid (U. S. 1275324).
 Proteid products (U. S. 1245818, 1245976, and 1245981).
 Size (U. S. 1289053).
 Thermal insulator (U. S. 1435416).
Recovering agent for—
 Potash from flue dust (U. S. 1317524).
 Potash from mica dust (U. S. 1317524).
 Soda from flue dust (U. S. 1317524).
 Soda from mica dust (U. S. 1317524).
Treating agent for—
 Asbestos (U. S. 1427911).
 Decolorizing carbons (U. S. 1447461).

Phosphoric Acid (Continued)

Paint and Varnish

Process material in making—
Turpentine oil substitute (U. S. 1131939).
Varnish (U. S. 1245818, 1275324, 1280861, 1427645, and 1482357).
Reagent (German 609982) for—
Increasing opacity, whiteness, and light-resistance of lithopone.
Starting point in making—
Finish remover (U. S. 1167462).
Lacquers (U. S. 1245818, 1245981, 1245982, 1275324, and 1427645).
Paint (U. S. 1213330 and 1367597).
Pigments (U. S. 1213330 and 1220973).
Titanium pigment (U. S. 1410056 and 1412027).

Paper

Ingredient of—
Ligno-cellulose solvent (U. S. 1218954).
Treating composition for unsized paper in making high-grade vulcanized fiber (U. S. 1894907).
Process material in making—
Ethyl alcohol from cellulose sulphite liquor (U. S. 1320043).
Sulphur dioxide from cellulose sulphite liquor (U. S. 1253854).
Treating agent for—
Cellulose sulphite liquor (U. S. 1155256 and 1467321).

Perfume

Process material in making—
Toilet preparations (U. S. 1482358).

Petroleum

Absorbent for—
Olefins.
Absorption agent in making—
Aliphatic alcohols and ethers from ethylene (Brit. 397187).
Aliphatic alcohols and ethers from propylene (Brit. 397187).
Aliphatic alcohols and ethers from higher olefins (Brit. 397187).
Hydration products of ethylene (Brit. 389133).
Hydration products of propylene (Brit. 389133).
Hydration products of higher olefins (Brit. 389133).
Catalyst in making—
Alcohols from olefins and steam (Brit. 413043).
Aliphatic alcohols and organic esters thereof from ethane or propane and organic acids, such as lower aliphatic acids (Brit. 402060).
Ingredient (Brit. 392289 and 392685) of—
Catalytic mixtures with compounds of uranium, iron, or cobalt, used in making ethyl alcohol from ethylene and steam.
Polymerizing agent (Brit. 450592 and 450668) in making—
Liquids of low molecular weight from olefins.
Process material in—
Cracking hydrocarbons (U. S. 1362127).
Process material in making—
Diolefins (U. S. 1179408).
Solid gasolene (U. S. 1262809).
Purifying agent for—
Hydrocarbons (U. S. 1201601; Brit. 398794).
Mineral oils (U. S. 1170868; Brit. 398794).
Starting point in making—
Catalysts, with oxide of calcium, barium, strontium, or magnesium, for the production of alcohols from olefins and steam (Brit. 415417 and 407722).

Pharmaceutical

In compounding and dispensing practice.
Starting point in making—
Pharmaceutical phosphates, glycerophosphates.

Photographic

Ingredient of—
Printing paper coatings in the "anilin process" for the reproduction of line subjects.
Starting point in making—
Phosphate salts for various purposes.

Plastics

Precipitating agent for—
Casein (U. S. 1341040 and 1360356).
Process material in making—
Billiard balls, cigaret holders, door handles and knobs, films, formaldehyde-urea condensates, handles, horn substitutes, imitation ivory, ivory substitutes, phenol-aldehyde condensates, phonograph records, pipe bowls, pipe stems.
Plastics (U. S. 1242030, 1245976, 1245981, 1245894, 1360356, 1482357, 1482358, and 1507379).
Starting point in making—
Esters useful as plastic softening agents (U. S. 1425392 and 1425393).

Printing

Process material in making—
Lithographic plate (U. S. 1162168), process engravings.

Refractory

Bonding agent for—
Alumina products (U. S. 1949038).
Zircon refractory claimed to be highly resistant to heat and suitable as a cylinder lining for internal-combustion engines and in making moulds for metal die-casting (U. S. 1872876).
Process material in making—
Fire brick (U. S. 1512801 and 1491224).
Silica brick (U. S. 1420284).

Resins

Catalyst in making—
Artificial oleoresins (U. S. 1469709).
Resins from hexahydroxycyclohexane and polybasic acids or their anhydrides (Brit. 408597).
Resins from the monomethyl ether of hexahydroxy-cyclohexane and polybasic acids or their anhydrides (Brit. 408597).
Resins from the dimethyl ether of hexahydroxycyclo-hexane and polybasic acids or their anhydrides (Brit. 408597).
Resins from quebrachitol and phthalic anhydride, linseed oil fatty acids, cyclohexanol, and tetrahydro-naphthalene (Brit. 408597).
Resins from quebrachitol, rosin, and phthalic anhydride (Brit. 408597).
Resins from inositol and phthalic anhydride (Brit. 408597).
Resins from pinite and phthalic anhydride (Brit. 408597).
Resins from dambonite and phthalic anhydride (Brit. 409597).
Process material in making—
Formaldehyde-urea condensates (U. S. 1482357 and 1482358).
Resins oil (U. S. 1131939 and 1133994).

Rubber

Coagulant for—
Latex.
Process material in making—
Artificial rubber (U. S. 1245818, 1245976, 1245979, 1245984, 1275324, and 1427645).

Sugar

Clarifying agent for—
Beet sugar, cane sugar.
Defacating agent for—
Beet juice, cane juice.
Inverting agent (U. S. 1402615) for—
Sugar.
Purifying agent for—
Beet sugar, cane sugar, decolorizing carbons (U. S. 1269080).
Molasses (U. S. 1314203, 1314204, and 1449134).
Sucrose (U. S. 1314203, 1493967, and 1269080).

Soap

Starting point in making—
Cleansing and emulsifying agents from 7:18-stearic-glycol (Brit. 308824, 317039, and 388485).
Cleansing and wetting agents from sperm oil fatty alcohols (Brit. 391610).
Cleansing, wetting, and dispersing agents with sugars and lauryl alcohol (Brit. 404684).
Cleansing, wetting, and dispersing agents with ricinoleic acid and sugars (Brit. 404684).
Emulsifying, cleansing, bleaching, and wetting agents with higher alcohols and boric acid (Brit. 409598).
Emulsifying, cleansing, bleaching and wetting agents with oleyl and cetyl alcohols mixtures and boric acid (Brit. 409598).
Emulsifying, cleansing, bleaching, and wetting agents with oleyl alcohol, boric acid, and hydrogen peroxide (Brit. 409598).
Emulsifying, cleansing, bleaching, and wetting agents with lauryl alcohol, boric acid, and hydrogen peroxide (Brit. 409598).
Emulsifying, cleansing, bleaching, and wetting agents with octadecenol, boric acid, and hydrogen peroxide (Brit. 409598).
Washing and foaming agents suitable for soaps by reacting with mono- or di-saccharides and alcohols.

Phosphoric Acid (Continued)

Textile
Improver (Brit. 434599) of—
Peroxide bleaching bath (said to give better and clearer whites).
Process material in treating—
Cotton fabric (U. S. 1439513, 1439514, 1439515, 1439516, 1439518, 1439520, and 1439521).
Cotton fabric to produce organdie effects (U. S. 1519376).
Cotton fabric to produce transparent effects (U. S. 1519376).
Cotton fabric to produce wool-like effects (U. S. 1518931 and 1519376).
Process material in—
Vat dyeing, calico printing.
Reagent for—
Brightening the colors of silk.

Water and Sanitation
Sterilizing agent for—
Aerated water (U. S. 1140717), water (U. S. 1170868).

Wood By-Products
Process material in making—
Acetic acid from woodtar (U. S. 1271071).
Methanol from woodtar (U. S. 1271071).
Turpentine oil.
Wood solvent (U. S. 1218954 and 1242030).
Purifying agent (U. S. 1183749 and 1201601) for—
Tar, tar distillates, tar oil.

Phosphoric Acid Ester of Grapeseed Alcohol

Bituminous
Solvent (Brit. 445223) for—
Asphalt and other bituminous bodies.

Dye
Solvent (Brit. 445223) for—
Dyestuffs, particularly oil-soluble coaltar dyes.

Fats, Oils, and Waxes
Solvent (Brit. 445223) for—
Fats, oils, waxes.

Resins
Solvent (Brit. 445223) for—
Oil-soluble glycerol-phthalic acid resins.
Polymerized vinyl compounds.
Synthetic resins.

Rubber
Solvent (Brit. 445223) for—
Rubber.

Phosphoric Acid Ester of Ricinoleic Alcohol

Bituminous
Solvent (Brit. 445223) for—
Asphalt and other bituminous bodies.

Dye
Solvent (Brit. 445223) for—
Dyestuffs, particularly oil-soluble coaltar dyes.

Fats, Oils, and Waxes
Solvent (Brit. 445223) for—
Fats, oils, waxes.

Resins
Solvent (Brit. 445223) for—
Oil-soluble glycerol-phthalic acid resins.
Polymerized vinyl compounds.
Synthetic resins.

Rubber
Solvent (Brit. 445223) for—
Rubber

Phosphorite

Cement
Ingredient of—
Hydraulic cements (U. S. 1628872).

Fertilizer
Ingredient of—
Fertilizing compositions (Brit. 270957).

Glass
Ingredient of batch for special glass.

Insecticide
Ingredient of—
Compositions for use against plant and animal pests (German 438006).

Phosphorus
French: Phosphore.
German: Phosphor.
Spanish: Fosforo.
Italian: Fosforo.

Note. Phosphorus appears in commerce as either yellow or red phosphorus; the yellow variety takes fire readily in the air and is poisonous; the red variety is not nearly so inflammable and is nonpoisonous; it is made from yellow phosphorus and has supplanted the latter in its principal industrial use as a component of match-head compositions.

Analysis
In gas analysis.

Chemical
Catalyst in making—
Alcohols.
In organic synthesis.
Process material in making—
Hydrogen (U. S. 1506323).
Starting point in making—
Phosphine (phosphuretted hydrogen), phosphoric acid, phosphoric anhydride, phosphorus pentachloride, phosphorus trichloride.

Electrical
Process material (U. S. 1205002) in making—
Tungsten lamp filaments.

Explosives and Matches
Ingredient of—
Match-head compositions, pyrotechnic compositions.
Process material in making—
Incendiary shells, smoke bombs, tracer bullets.
Starting point in making—
Phosphine (phosphuretted hydrogen).

Fertilizer
Starting point in making—
Ammonium metaphosphate (U. S. 1284200).
Ammonium phosphate (U. S. 1510179).

Lighting
Process material in making—
Incandescent lights.

Mining
In mine lamps.

Metallurgical
Alloying agent in making—
Alloy steels, bearing metals, electric welding alloys, ferrophosphorus, phosphor bronze and other bronzes, phosphor copper, phosphor tin.

Miscellaneous
Ingredient of—
Chemical heating compositions (U. S. 1506322 and 1506323).
Light-sensitive compositions (U. S. 1430484).

Pesticide
Poisoning agent in—
Pesticidal compositions for insects, rodents, and other vermin.

Pharmaceutical
In compounding and dispensing practice.

Textile
Decorating agent (U. S. 1496743) for—
Chiffon, cloth, felt, silk.

Phosphorus Sesquisulphide
Synonyms: Tetraphosphorus trisulphide.
French: Sesquisulfure de phosphore.
German: Phosphorsesquisulfid.

Chemical
Reagent in making various organic chemicals.

Dye
Reagent in making—
Dyestuffs derived from perylene (U. S. 1615646).

Explosives
Ingredient of—
Compositions for making heads of matches.

Phosphorus Trichloride
Synonyms: Phosphorus chloride.
French: Chlorure de phosphore, Trichlorure de phosphore.
German: Phosphorchlorid, Phosphortrichlorid, Trichlorphosphor.

Chemical
Condensing agent (Brit. 311208) in making—
Aldehydes from N-ethylcarbazol, pyridin, and quinolin.

Phosphorus Trichloride (Continued)
Alphachloroanthraquinone-9-aldehyde.
Anthracene-9-aldehyde.
9-Chloroanthracene-10-aldehyde.
3-Chloro-6-ethoxythionaphthene-2-aldehyde.
3:10-Dichloroanthracene-9-aldehyde.
1:3-Dimethylbenzene-4-aldehyde.
2:5-Dimethyl-4-oxybenzaldehyde.
1:2-Dimethoxy-10-chloroanthracene-9-aldehyde.
2:7-Dioxynaphthalene-1-aldehyde.
4:8-Dioxynaphthalene-1-aldehyde.
2-Ethoxy-1-naphthaldehyde.
6-Ethoxy-3-oxythionaphthene-2-aldehyde.
1-Methoxy-5:6:7:8-tetrahydronaphthalene-4-aldehyde.
2-Methoxy-5:6:7:8-tetrahydronaphthalene-1-aldehyde.
4-Methyl-3:6-dichlorothionaphthene-2-aldehyde.
4-Methyl-6-chloro-3-oxythionaphthalene-2-aldehyde.
Naphthostyrilaldehyde.
2-Oxy-1-naphthaldehyde.
4-Oxy-1-naphthaldehyde.
2-Oxynaphthalene-1-aldehyde-3-carboxylic acid.
Para-anisaldehyde.
1:5:10-Trichloroanthracene-9-aldehyde.
Vanillin.
Reagent in making—
Acetyl chloride.
Alphabromo-2-naphthylthioglycollic chloride (Brit. 260623).
Alphachloro-2-naphthylthioglycollic chloride (Brit. 260623).
Alphaiodo-2-naphthylthioglycollic chloride (Brit. 260623).
Benzoyl chloride, benzylidene chloride, chlorides of various acids, citronellol, ethylene dichloride.
1-Phenyl-3-pyrazolonecarboxylic acid.
Pentachloroethane, saccharin, toluene sulphochloride, trimethyl phosphate, general condensing agent, general chlorinating agent, solvent for phosphorus.
Starting point in making—
Phosphorus oxychloride, phosphorus pentachloride.

Dye
Condensing agent in making—
Crystal violet.
2-Hydroxynaphthalene-3-carboxylic acid metanitroaniline (Swiss 111922).
Sulphur dyestuffs by condensation from carbazole-2-carboxylic acid.
Victoria blue.

Metallurgical
Reagent in the production of—
Iridescent effects in the form of metallic deposits.

Paint and Varnish
Reagent in making—
Linseed oil substitutes.

Phosphorus Trichlorobromide

Chemical
Reagent in making—
Ethylidene bromide.

Phosphotungstic Acid

Analysis
Reagent in alkaloidal assays.

Dye
Ingredient of—
Color lakes made with basic dyestuffs (Brit. 270750).

Phosphotungstomolybdic Acid

Dye
Ingredient (Brit. 270750) of—
Color lakes with basic dyestuffs.

Phthalamide

Fats and Oils
Deterioration retardant (Brit. 423938) for—
Vegetable oils.

Fuel
Deterioration retardant (Brit. 423938) for—
Coal-carbonization spirits.

Petroleum
Deterioration retardant (Brit. 423938) for—
Cracked petroleum oils, lubricating oils, shale oils, transformer oils.

Phthalic Acid Ester of Grapeseed Alcohol

Bituminous
Solvent (Brit. 445223) for—
Asphalt and other bituminous bodies.

Dye
Solvent (Brit. 445223) for—
Dyestuffs, particularly oil-soluble coaltar dyes.

Fats, Oils, and Waxes
Solvent (Brit. 445223) for—
Fats, oils, waxes.

Resins
Solvent (Brit. 445223) for—
Oil-soluble glycerol-phthalic acid resins, polymerized vinyl compounds, synthetic resins.

Rubber
Solvent (Brit. 445223) for—
Rubber.

Phthalic Acid Ester of Ricinoleic Alcohol

Bituminous
Solvent (Brit. 445223) for—
Asphalt and other bituminous bodies.

Dye
Solvent (Brit. 445223) for—
Dyestuffs, particularly oil-soluble coaltar dyes.

Fats, Oils, and Waxes
Solvent (Brit. 445223) for—
Fats, oils, waxes.

Resins
Solvent (Brit. 445223) for—
Oil-soluble glycerol-phthalic acid resins, polymerized vinyl compounds, synthetic resins.

Rubber
Solvent (Brit. 445223) for—
Rubber.

Phthalide

Explosives
Gelatinizing agent (Brit. 252978) in making—
Nitrocellulose explosives, nitroglycerin explosives.

Plastics
Gelatinizing agent (Brit. 252978) in making—
Nitrocellulose plastics, celluloid.

1:2-Phthaloyl-5:6-benzocarbazole

Textile
As a vat dye (Brit. 443958 and 443959).

1:2-Phthaloylcarbazole

Textile
As a vat dye (Brit. 443958 and 443959).

1:2-Phthaloyl-6-methylcarbazole

Textile
As a vat dye (Brit. 443958 and 443959).

2:3-Phthaloyl-6-methylcarbazole

Textile
As a vat dye (Brit. 443958 and 443959).

1:2-Phthaloyl-6-phenylcarbazole

Textile
As a vat dye (Brit. 443958 and 443959).

7:8-Phthaloyl-2-quinolone-3-carboxylic Acid

Chemical
In organic syntheses.

Dye
In dye syntheses.
Starting point (Brit. 449263) in making—
Orange vat dyes with 1-amino-4-benzamidoanthraquinone.
Reddish-brown vat dyes with 4:8-diaminoanthrarufin.
Yellow vat dyes with 1-aminoanthraquinone.

7:8-Phthaloyl-2-quinolone-5-carboxylic Chloride

Chemical
In organic syntheses.

Dye
In dye syntheses.
Starting point (Brit. 449263) in making—
Yellow vat dyes with 1-aminoanthraquinone.

8:9-Phthaloyl-4:5-trimethinacridin

Textile
As a vat dye (Brit. 443958 and 443959).

Phthalyl Peroxide
French: Peroxyde de phthalyle, Peroxyde phthalique.
German: Phtalylperoxyd.

Chemical
Starting point in making—
Intermediates, pharmaceuticals.

Dye
Starting point (Brit. 314825) in making xanthene dyestuffs with the aid of—
Alphachloronaphthalene, betachloronaphthalene, 4-chlorometaxylene, metachloroanilin, metachloroanisole, metachlorobenzylamine, metachlorocresidin, metachlorophenylamine, metachlorotoluene, metachlorotoluidin, metachloroxylene, metachloroxylidin, orthochloroanilin, orthochloroanisole, orthochlorobenzylamine, orthochlorocresidin, orthochlorophenylamine, orthochlorotoluene, orthochlorotoluidin, orthochloroxylene, orthochloroxylidin, parachloroanilin, parachloroanisole, parachlorobenzylamine, parachlorocresidin, parachlorophenylamine, parachlorotoluene, parachlorotoluidin, parachloroxylene, parachloroxylidin.
Various acetyl-aralkyl, thioether derivatives of aromatic halogen compounds.

Fats and Oils
Bleaching agent (Brit. 328544) in treating—
Various fats and oils of animal and vegetable origin (used in conjunction with hydrogen peroxide).

Food
Bleaching agent (Brit. 328544) in treating—
Various food preparations, such as flours, other milled products, egg yolk, meals, and various animal and vegetable foodstuffs (used in conjunction with hydrogen peroxide).

Soap
As a bleaching agent (Brit. 328544) (used in conjunction with hydrogen peroxide).

Waxes and Resins
Bleaching agent (Brit. 328544) in treating—
Various waxes (used in conjunction with hydrogen peroxide).
3:7-Tetraisopropyldiaminoxanthone.

Picramic Acid
French: Acide de picramique.
German: Picraminsaeure.

Abrasive
Catalyst (Brit. 295335) in making—
Binders from phenolic resins for use in making grinding discs and the like.

Cement
Catalyst (Brit. 295335) in making—
Fillers and cements from phenolic-formaldehyde resins.

Chemical
Catalyst (Brit. 295335) in making—
Impregnating solutions of phenolic-formaldehyde resins, used for various chemical purposes.
Starting point in making—
Esters and salts, intermediates, pharmaceuticals.

Dye
Starting point in making various synthetic dyestuffs.

Miscellaneous
Catalyst (Brit. 295335) in making—
Solutions of phenolic-formaldehyde resins used for various impregnating purposes.

Paint and Varnish
Catalyst (Brit. 295335) in making—
Lacquers and varnishes, as well as special dopes, containing phenolic-aldehyde resins.

Plastics
Catalyst (Brit. 295335) in making—
Molding mixtures and press mixtures which contain phenolic-formaldehyde resins.

Pig's-Foot Grease

Lubricant
Raw material in making—
Gear and other greases.

Pilchard Oil

Agricultural
Ingredient of—
Dips for sheep, cattle, and other domestic animals.
Source of vitamin D for—
Animal foods, poultry foods.

Construction
Ingredient of—
Asbestos cements, bitumistic compounds, protective coatings, roofing products, waterproofing coatings, weatherproofing coatings.

Fats and Oils
Ingredient of—
Fish-oil emulsions, lubricating compositions, pipe-threading dope, wire-rope greases.
Starting point in making—
Hardened oil, stearin, tallow mixtures.

Food
Ingredient of—
Lard substitutes, oleomargarin.

Ink
Ingredient of—
Lithographic inks, marking inks, printing inks.

Insecticide
Ingredient of—
Insecticidal compounds and preparations.
Insecticidal soaps, sprays.

Leather
Ingredient of—
Dressing compositions, finishing compositions.
Reagent in—
Making chamois leather, oil tanning.
Substitute for linseed oil in making—
Patent leather.

Linoleum and Oilcloth
Substitute for—
Linseed oil.

Mechanical
As a lubricant.

Metallurgical
Quenching agent in—
Steel tempering.

Miscellaneous
Ingredient of—
Caulking compounds.
Cordage waterproofing compounds and preservatives.
Fish-net preservatives and waterproofing compounds.
Oil-clothing dopes, pipe-thread cements.
Various compositions in which quick drying and binding are advantageous, such as composition flooring and powdered cork products.

Paint and Varnish
As a vehicle which will throw tough, flexible film without the necessary addition of driers or hardening substances.
As a waterproof film-forming medium.
Checking oil in—
Enamel liquids, metal paints, spar varnishes, white undercoats.
Heat-resisting vehicle in—
Enamels.
Heat-resisting paints for use on smokestacks, boiler fronts, furnaces, drying cabinets, and other structures which are subjected to high temperatures.
Japans.
Ingredient of—
Exterior coating (to improve wearing and weather-resistance properties of linseed oil).
Putty.
Vehicle in—
Aluminum paints, baking japans, barn paints, canvas paints, enamels, exterior paints, flat wall paints, interior coatings, mill whites, oil tank paints, pigmented lacquers, roof paints and roofing products, shingle stains and other stains, structural iron paints, tank paints, varnishes, waterproof paints, white house paints, white pastes.

Rubber
Ingredient of—
Rubber substitutes.

Soap
Soapstock for—
Soft soaps.

Textile
Dressing for oiled fabrics.
Oiling and softening agent for—
Fibers, prior to spinning and weaving.

Woodworking
Ingredient of—
Impregnating and waterproofing compounds.

Pilocarpine

Chemical
Starting point in making the following derivatives:
Acetate, arsenate, arsenite, benzoate, bisulphate, bitartrate, borate, carbolate, citrate, dihydrobromide, dihydrochloride, ferrocyanide, formate, glycerophosphate, hydrobromide, hydrochloride, hydroiodide, hypophosphite, lactate, phosphate, salicylate, sulphate, sulphocarbolate, tannate, tartrate, valerate.

Pharmaceutical
In compounding and dispensing practice.

Pimelic Acid

Cellulose Products
Solvent (Brit. 341447) for—
Cellulose acetate, cellulose esters and ethers, cellulose nitrate (nitrocellulose).
For uses, see under general heading: "Solvents."

Pimento

Synonyms: Allspice, Clove pepper, Jamaica pepper.
Latin: Semen anomi.
French: Piment, Piment de la Jamaïque, Piment couronné, Piment des anglais, Poivre de la Jamaïque, Poivre giroflée, Touteépice.
German: Allerlei wuerze, Allerleigewuerz, Englischer gewürz, Indianische pfeffer, Jamaikapfeffer, Nelkenpfeffer, Neugewürz.
Spanish: Pimienta de la Jamaica.
Italian: Pimenti.

Food
As a general flavoring.
Ingredient of—
Condiments, pickles, sauces.

Oils and Fats
Source of the essential oil of pimento.

Perfumery
As an aromatic in preparations of various sorts.

Pharmaceutical
In compounding and dispensing practice.

Pine Oil

Latin: Oleum pini.
French: Huile de pin.
German: Fichtenoel, Kienoel, Russiches terpentinoel.

Agricultural
General disinfectant around the farm and dairy.
Ingredient of—
Cattle washes.
Preparation used for washing barns and dairies.

Chemical
Emulsifying agent for various purposes.
General solvent.
Starting point in making—
Aromatic chemicals, borneol, fencyl alcohol, alphaterpineol.
Reagent (Brit. 274611, 311885, 399537) in making—
Wetting agent for textiles.
Starting point (U. S. 1893802) in making—
Paracymene.

Fats and Oils
Solvent for grease.

Gums
Solvent for various gums, including most of the fossil gums.

Insecticide
As an insecticide.
Ingredients of—
Fly-repellants, insecticidal preparations of the kind used for combatting the white fly, purple scale, and aphis.
Mosquito-repellants.

Metallurgical
Flotation oil in treating—
Iron sulphides to separate them from molybdenite and graphite.
Ingredient of—
Preparations used for cleaning metals before they are electroplated.
Preparations used for cleaning zinc and copper in preparation for etching with acid.
Ingredient (U. S. 1902317) of—
Mixture with dixanthogen used in froth flotation of ores.
In the flotation process of separating mineral from gangue.

Miscellaneous
Dry-cleaning agent.
General cleansing agent for laundry and household purposes.
Ingredients of—
Cleansing compositions used in the treatment of rugs, upholstery, wood, cement, porcelain, tile, papered and painted walls.
Metal polishes.
Rustproofing compositions for use in the treatment of various metals (U. S. 1592102).
Waterproofing compositions used in the treatment of different fibrous substances, such as paper and pulp (Brit. 251961).

Paper
Ingredient of—
Compositions used in the treatment of rag stock, to remove the dirt and grease and prepare it for the digestion process.
Compositions used in the treatment of old newsprint and the like for the removal of the ink by the emulsifying action of the oil.
Compositions (also used alone) as foam reducers added to the batch in the paper-coating machine, to prevent foaming and faulty deposition of the coating on the paper web.

Paint and Varnish
Deodorant in the manufacture of—
Paints and other products of the paint and varnish industry (added for the purpose of hiding the odor of oils and solvents).
Ingredient of—
Enamels, encaustic preparations, lacquers, shingle stains, ships' bottoms paints, slow-drying paints, nitrocellulose lacquers (U. S. 1746895).
Nondrying oil for coating unfilled wooden floors and the like, containing also mineral oil, fatty acids, and an alkaline solution (U. S. 1860372).
Rubbing compound for painted, varnished, lacquered, or enameled surfaces, containing also an aqueous soap suspension and a finely divided abrasive (U. S. 1927872).
Reagent in the grinding of—
Enamel paints.

Petroleum
Ingredient of—
Gasoline-resisting cements.

Perfume
Ingredient of—
Bath odorants, cosmetics.

Pharmaceutical
In compounding and dispensing practice.

Printing
As a cleansing agent in the preparation of the metal surface in making line cuts and halftones.

Rubber
As a solvent.
Ingredient (U. S. 1875552) of—
Solution for cleaning vulcanizing molds, containing also cresol.
Reagent in—
Reclaiming rubber, rubber technology.
Solvent in making—
Rubber cements.

Sanitation
As a deodorant.
As a germicide.

Soap
Ingredient of—
Cleansing preparations, detergents of various sorts, disinfectant soaps, scouring soaps.

Textile
——, *Dyeing*
Assistant and penetrant in—
Dye liquors used in the dyeing of cotton, rayon, wool (added for the purpose of preventing spotty dyeing).
Reagent in aiding the solution of dyestuffs in preparing certain dye liquors.

——, *Finishing*
Ingredient of—
Mixture with phenols used to increase the moistening power of mercerising solution (Brit. 385977, 405492).
Proofing solution for textiles, containing also triethanolamine, a diamine, oleic acid, chlorinated naphthalene wax, paraffin, aluminum acetate, and sodium silicate (Brit. 401282).

Pine Oil (Continued)

Scouring agent, containing also sulphuric linoleic acid and sodium silicate (Brit. 401282).
Scouring baths, washing solutions.

——, *Manufacturing*
Ingredient of—
Baths used for the boiling out of cotton yarns and loose cotton.
Baths used for the degumming of raw silk.
Baths used for the degreasing of wool.
Baths used for the softening of rayons.
Textile oil preparations for use in the spinning, winding, reeling, warping, knitting, and weaving.

Waxes and Resins
Solvent for various resins and waxes.

Woodworking
Ingredient of—
Compositions used for cleansing wood.

Pine Oil Foots

Miscellaneous
Ingredient (U. S. 1840989) of—
Impregnating compound, containing also rubber and rosin.

Piperidine-1-carbothionalate
French: Carbothionate de 1-pipéridine.
German: Piperidin-1-carbothionalat.

Chemical
Starting point (Brit. 340083) in making rubber vulcanization accelerator with the aid of—
4-Chloro-1:3-dinitrobenzene.

Piperidine Ethylxanthate

Rubber
As a vulcanizing accelerator (U. S. 1875943).

Piperidine Hydrochloride

Chemical
Starting point in making—
Piperidinomethylcyclohexanone hydrochloride (German 422916).

Pharmaceutical
In compounding and dispensing practice.

Piperidine Pentamethylenedithiocarbamate

Rubber
Secondary activator in—
Vulcanizing processes (for use with mercaptabenzthiazole).

5-Piperidinomethyl-1:3:2-xylenol

Rubber
Anti-ager (Brit. 459045) for—
Rubber mixes.

Plasticizers
French: Plastifiers.

Ceramics
Plasticizer in—
Compositions used for protecting and decorating ceramic products.

Chemical
Plasticizer for—
Cellulose derivatives.

Cosmetic
Plasticizer in—
Nail enamels and lacquers.

Electrical
Plasticizer in—
Insulating compositions used for covering wire and in making electrical machinery and equipment.

Glass
Plasticizer in—
Compositions used in the manufacture of nonscatterable glass and for protecting and decorating glassware.

Glues and Adhesives
Plasticizer in—
Adhesive compositions.

Leather
Plasticizer in—
Compositions used in the manufacture of artificial leathers and for protecting and decorating leathers and leather goods.

Metallurgical
Plasticizer in—
Compositions used for protecting and decorating metallic articles.

Miscellaneous
Plasticizer in—
Compositions used for protecting and decorating various products.

Paint and Varnish
Plasticizer in—
Paints, varnishes, lacquers, enamels, and dopes.

Paper
Plasticizer in—
Compositions used in the manufacture of coated papers and for protecting and decorating products made of paper or pulp.

Photographic
Plasticizer in making—
Films.

Plastics
Plasticizer in making—
Laminated fiber products, molded products, plastics.

Resins
Plasticizer for—
Resin and/or cellulose derivative compositions and solutions.

Rubber
Plasticizer in—
Compositions used for decorating and protecting rubber products.

Stone
Plasticizer in—
Compositions used for decorating and protecting artificial and natural stone.

Textile
Plasticizer in—
Compositions used in the manufacture of coated fabrics.

Woodworking
Plasticizer in—
Compositions used as protective and decorative coatings on woodwork.
Plastic compositions used for many filling and repairing purposes on wood.

Platinum Resinate
Synonyms: Resinate of platinum.
French: Résinate de platine.
German: Platinresinat.

Ceramics
Pigment in producing iridescent effects on—
Chinaware, porcelains, potteries.

Polychlororetene

Petroleum
Imparter (Brit. 431508) of—
High-film strength, adhesion power, and abrasion resistance to lubricants for use with extreme pressures (consists of blends with mineral lubricating oil).

Polyethylstyrene

Chemical
Starting point in making various derivatives.

Miscellaneous
Ingredient (Brit. 367126) of—
Compositions used for impregnating and stiffening felt.

Polyglycol

Chemical
Solvent (Brit. 272908) in making—
Various chemical products.

Dye
Solvent (Brit. 272908) in making—
Soluble metallic compounds of azo dyestuffs.

Miscellaneous
Solvent in various processes.

Polymerized Coumarone
French: Coumarone polymerizée.
German: Polymerizierte cumaron.

Electrical
Ingredient of—
Insulating compositions.

Polymerized Coumarone (Continued)

Miscellaneous
Ingredient (Brit. 335247) of—
Waterproofing, weatherproofing and wearproofing compositions.

Paper
Ingredient (Brit. 335247) of—
Compositions used for treating paper and pulp products to render them waterproof, wearproof, and weatherproof.

Textile
Ingredient (Brit. 335247) of—
Compositions used in treating various textiles to render them waterproof, wearproof and weatherproof.

Polynitrodiphenyl Sulphide

Lubricant
Starting point (Brit. 440175) in making—
Addition agents for high-pressure lubricating oils or greases by mixing and reacting with organo-metallic compounds.

Polyricinoleic Acid
French: Acide de polyricinoléique.
German: Polyricinoelsaeure, Polyricinusoelsaeure, Polyrizinoelsaeure, Polyrizinusoelsaeure.

Chemical
Ingredient of—
Emulsions (Brit. 303379).
Starting point in making—
Salts and esters.

Miscellaneous
Ingredient (Brit. 303379) of—
Washing compositions.

Soap
Ingredient (Brit. 303379) of—
Saponaceous cleansing compositions.

Textile
——, *Bleaching*
Ingredient (Brit. 303379) of—
Bleaching preparations.
——, *Finishing*
Ingredient (Brit. 303379) of—
Bowking, oiling, softening, and finishing compositions.

Polystyrene
French: Polystyrène.
German: Polystyren.
Spanish: Polisteren.
Italian: Polistirene.

Chemical
Starting point in making—
Intermediates and other derivatives.

Miscellaneous
Ingredient (Brit. 367126) of—
Compositions for impregnating and stiffening felt.

Polyvinyl Acetate
French: Acétate de polyvinyle, Acétate polyvinylique.
German: Essigsäurepolyvinylester, Essigsäurespolyvinyl, Polyvinylacetat, Polyvinylazetat.

Chemical
Starting point in making—
Intermediates and other derivatives.

Electrical
Starting point (Brit. 322517) in making—
Compositions used in making telephone receivers, and other electrical apparatus, parts of motors, and so on.

Mechanical
Starting point (Brit. 322517) in making—
Compositions used in the manufacture of brake bands, cog wheels, and other mechanical equipment.

Miscellaneous
Ingredient of—
Compositions used in waterproofing, stiffening and making materials, such as felts, felt hats, straw plait, and the like, capable of being shaped and pressed at an elevated temperature.
Starting point (Brit. 322517) in making—
Polymerized compositions used in making buttons, umbrella handles, and other devices and equipment.

Paint and Varnish
Starting point (Brit. 322517) in making—
Polymerized compositions used as bases in the manufacture of paints, varnishes, dopes, enamels, lacquers, and the like.

Paper
Starting point (Brit. 322517) in making—
Compositions used in the impregnation of paper and pulp and products made therefrom.

Photographic
Starting point (Brit. 322517) in making—
Compositions used in making films and plates.

Plastics
Starting point (Brit. 322517) in making—
Polymerized compositions.

Textile
Ingredient of—
Compositions used in waterproofing and stiffening nitro rayon, viscose rayon, cuprammonium rayon, acetate rayon products, and linings, capable of being shaped and pressed at an elevated temperature.

Woodworking
Starting point (Brit. 322517) in making—
Compositions used in impregnation of wood and wood products.

Polyvinyl Alcohol Benzyl Ether
French: Éther benzylique de polyvinyle alcool.
German: Polyvinylalkohol-benzylaether.

Chemical
Starting point in making—
Intermediates and other derivatives.

Electrical
Starting point (Brit. 322517) in making—
Polymerized compositions used in the manufacture of telephone receivers and other electrical apparatus, parts of motors, and so on.

Mechanical
Starting point (Brit. 322517) in making—
Compositions used in the manufacture of cog wheels, brake bands, and other mechanical apparatus and equipment.

Miscellaneous
Starting point (Brit. 322517) in making—
Compositions used in the manufacture of buttons, umbrella handles, and other articles and equipment.

Paint and Varnish
Starting point (Brit. 322517) in making—
Polymerized compositions used as bases in the manufacture of paints, varnishes, enamels, lacquers, and dopes.

Paper
Starting point (Brit. 322517) in making—
Compositions used in the impregnation and coating of paper and pulp and products made from them.

Photographic
Starting point (Brit. 322517) in making—
Compositions used in making films and plates.

Plastics
Starting point (Brit. 322517) in making—
Polymerized compositions.

Woodworking
Starting point (Brit. 322517) in making—
Compositions used for the impregnation of wood and wood products.

Polyvinyl Butyrate
French: Butyrate de polyvinyle, Butyrate polyvinylique.
German: Buttersäurepolyvinylester, Buttersäurespolyvinyl, Polyvinylbutyrat.

Chemical
Starting point in making various derivatives.

Miscellaneous
Reagent in—
Waterproofing, stiffening, and treating various materials that are capable of being shaped and pressed at an elevated temperature.
Reagent in treating—
Felt, felt hats, straw plait.

Textile
Reagent in treating—
Lining fabrics, rayon fabrics.

Polyvinyl Chloroacetate
French: Chloroacétate de polyvinyle, Chloroacétate polyvinylique.
German: Chloressigsäurepolyvinylester, Chloressigsäurespolyvinyl, Polyvinylchloracetat, Polyvinylchlorazetat.

Polyvinyl Chloroacetate (Continued)

Chemical
Starting point in making various derivatives.

Miscellaneous
Ingredient of—
Compositions employed in waterproofing, stiffening, and making such materials as felts, felt hats, straw plait, and the like, capable of being shaped and pressed at an elevated temperature.

Textile
Ingredient of—
Compositions employed in waterproofing and stiffening nitro rayon, acetate rayon, viscose rayon, cuprammonium rayon, and linings, capable of being shaped and pressed at an elevated temperature.

Poppyseed
Synonyms: Mawseed.
French: Graines de pavot, Semences de pavot.
German: Mohnsamen.

Fats and Oils
Starting point in making—
Poppyseed oil (maw oil).

Food
Ingredient of—
Culinary dishes, sweetmeats, bakery products.

Poppyseed Oil
Synonyms: Maw oil.
French: Huile de graines de pavots, Huile de pavots, Huile de semences de pavots.
German: Mohnoel.

Food
As a condiment (almond paste).
As a salad oil.
Ingredient of—
Olive oil mixtures.
Substitute for—
Olive oil.

Fuel
As a burning oil.

Paint and Varnish
Starting point (German 576939) in making—
Homogeneous drying extracts readily soluble in drying oils and volatile organic solvents, by mixture with heavy or alkaline earth metal salts of naphthenic acids.
Vehicle in—
Artists' oil colors, varnishes.

Pharmaceutical
In compounding and dispensing practice.

Soap
Soapstock in making—
Olive oil soaps, potash soaps.

Poppyseed Oilcake
Synonyms: Mawseed oilcake.
French: Tourteau de pavots.
German: Mohenoelkuchen.

Animal Husbandry
As a cattlefeed (contains 8 percent oil).

Porpoise Body Oil
Synonyms: Dolphin oil, Porpoise blubber oil.

Fuel
As an illuminant.

Leather
Ingredient of—
Dressing compositions.

Mechanical
As a lubricant.
Ingredient of—
Lubricating compositions.

Soap
As a soap stock.

Porpoise Jaw Oil

Mechanical
Lubricant for—
Delicate machinery, such as clocks, chronometers, and the like.

Porpoise Junk Oil
Synonyms: Porpoise face blubber oil.

Leather
Ingredient of—
Dressing compositions.

Mechanical
As a lubricant.
Ingredient of—
Lubricating compositions.

Potash Alum
Synonyms: Alum, Alum flour, Alum meal, Aluminite, Aluminum and potassium sulphate, Common alum, Cube alum, Double sulphate of aluminum and potassium, Octoheydral alum salt, Potassic-aluminic sulfate, Potassium alum, Potassium-aluminium sulphate, Roman alum, Sulphate of aluminum and potassium.
Latin: Alumen, Alumen potassicum, Aluminii et potassi sulphas, Potassa alum, Sulphas aluminico-potassicus.
French: Alun, Alun de potasse, Alun potassique, Alun de potassium, Sulphate d'alumine et de potasse, Sulphate aluminique et potassique, Sulphate d'aluminium et de potassium.
German: Alaun, Aluminiumkaliumsulfat, Kalialaun, Kaliumaluminiumsulfat, Schwefelsaeuresaluminiumkalium, Schwefelsaeureskaliumaluminium.
Spanish: Alumbre.
Italian: Allume.

Cement
Hardener for—
Plaster casts.

Ceramics
Ingredient of various wares.

Chemical
Catalyst in—
Synthesis of ammonia.
Ingredient of catalytic preparations used in making—
Acenaphthylene, acenaphthaquinone, bisacenaphthylidenedione, naphthaldehydic acid, naphthalic anhydride and hemimellitic acid from acenaphthene (Brit. 295270).
Acetaldehyde from ethyl alcohol (Brit. 281307).
Acetic acid from ethyl alcohol (Brit. 295270).
Alcohols from aliphatic hydrocarbons (Brit. 281307).
Aldehydes and acids from toluene, orthochlorotoluene, orthobromotoluene, orthonitrotoluene, metanitrotoluene, metabromotoluene, metachlorotoluene, parabromotoluene, paranitrotoluene, parachlorotoluene, dinitrotoluenes, dibromotoluenes, dichlorotoluenes, chlorobromotoluene, chloronitrotoluene, bromonitrotoluene (Brit. 295270).
Aldehydes and acids from xylenes, pseudocumene, mesitylene, and paracymene (Brit. 295270).
Alphanaphthaquinone from naphthalene (Brit. 281307).
Anthraquinone from anthracene (Brit. 295270).
Benzaldehyde and benzoic acid from toluene (Brit. 281307).
Benzoquinone from phenanthraquinone (Brit. 281307).
Chloroacetic acid from ethylenechlorohydrin (Brit. 295270).
Diphenic acid from ethyl alcohol (Brit. 281307).
Fluorenone from fluorene (Brit. 295270).
Formaldehyde from methanol or methane (Brit. 295270).
Naphthaldehydic acid, acenaphthaquinone, or bisacenaphthylidenedione from acenaphthylene (Brit. 281307).
Phenanthraquinone from phenanthrene or diphenic acid (Brit. 295270).
Phthalic acid and maleic acid from naphthalene (Brit. 295270).
Salicylic acid and salicylic aldehyde from cresol (Brit. 295270).
Vanillin or vanillic acid from eugenol or isoeugenol (Brit. 295270).
Maleic acid and fumaric acid by the oxidation of benzol, toluol, phenol, tar phenols, or furfural, or from benzoquinone or phthalic anhydride (Brit. 295270).
Starting point in making—
Aluminum driers, aluminum salts.

Coaltar
Catalyst (Brit. 295270) in purifying—
Ammonia, anthracene, coaltar, crude naphthalene.

Construction
Hardening agent in—
Plastering.

Dye
Ingredient of—
Color lakes.

Explosives
Ingredient of—
Matchhead compositions, picric acid explosives.

Potash Alum (Continued)

Food
Ingredient of—
Baking powders, candy, margarines.

Ink
Ingredient of—
Writing inks.

Leather
Ingredient of—
Tanning compositions for making white leather.
Tanning compositions along with common salt.

Miscellaneous
As a general astringent.
Ingredient of—
Compositions used in lining safes, mineral yeasts.
In taxidermy.

Paint and Varnish
Ingredient of—
Lake pigment compositions, enamels, paints, varnishes.
Reagent in making—
Mars yellow.

Paper
Ingredient of—
Sizes.

Pharmaceutical
In compounding and dispensing practice.

Photographic
Ingredient of—
Fixing baths, hardening baths.

Printing
In process engraving and lithographic work.

Sanitation
Reagent in treating—
Sewage, water.

Sugar
Reagent in refining.

Textile
——, *Dyeing*
Mordant in dyeing—
With colors sensitive to iron (alizarin red and the like).
——, *Finishing*
Ingredient of—
Fireproofing compositions, waterproofing compositions.

Woodworking
Ingredient of—
Fireproofing compositions.

Potassium Acetate
Synonyms: Acetate of potash, Diuretic salt.
Latin: Acetas potassicas, Kalium aceticum, Potassii acetas, Sal kalicus, Terra foliata tartari.
French: Acétate de potasse, Acétate potassique, Acétate de potassium, Sel de sylvius, Terre foliée de tartre, Terre foliée végétale.
German: Essigsäureskalium, Essigsäurespotasche, Kaliumacetat, Kaliumazetat.
Spanish: Acetato de potasa, Acetato potasico.
Italian: Acetato di potassio.

Analysis
Desiccating agent in various operations.
Reagent in—
Analyzing alcohol and tartaric acid, buffer solutions.

Chemical
Desiccating reagent in various processes.
Reagent in making—
Acetic anhydride, acetone, acetyl chloride, aluminum-potassium acetate (alkalsol), benzyl acetate, bismuth acetate.
Cacodylic derivatives, such as cacodylic acid and cacodylates.
Ethylene monoacetate, ethylidene diacetate, isobutyl acetate, methyl acetate, orthonitrobenzyl acetate, paranitrobenzyl acetate, titanium acetate.
Starting point in making—
Methane.

Dye
Dehydrating agent in making—
Synthetic dyestuffs.

Glass
Ingredient of—
Batch used in making crystal glass.

Metallurgical
Ingredient of—
Compositions used in connection with bronze powder for coloring metals.

Paint and Varnish
Starting point in making—
Cobalt yellow (aureolin).

Pharmaceutical
Suggested for use as diuretic, antiarthritic, alterative, eperient, diaphoretic, antipyretic, and cathartic.

Potassium Acid Adipinate

Leather
Buffer (Brit. 444184) in—
Obtaining level dyeings with acid or substantive dyes (the dyed effects are claimed to have great resistance to soap and alkalies).

Potassium Acid Diglycollate

Leather
Buffer (Brit. 444184) in—
Obtaining level dyeings with acid or substantive dyes (the dyed effects are said to have great resistance to soap and alkalies).

Potassium Acid Phthalate

Leather
Buffer (Brit. 444184) in obtaining—
Level dyeings with acid or substantive dyes (the dyed effects are claimed to have great resistance to soap and alkalies).

Potassium Acid Saccharate

Leather
Buffer (Brit. 444184) in obtaining—
Level dyeings with acid or substantive dyes (the dyed effects are claimed to have great resistance to soap and alkalies).

Potassium Allylate
French: Allylate de potasse, Allylate potassique, Allylate de potassium.
German: Kaliumallylat.

Chemical
Reagent (Brit. 304118) in making ketonic acid esters with the aid of allyl, amyl, butyl, heptyl, hexyl, and propyl esters of the following acids:—
Acetic, anthranilic, benzoic, butyric, camphoric, caproic, caprylic, chloroacetic, cinnamic, citric, cresylic, gallic, lactic, maleic, malic, malonic, metanilic, mucic, naphthionic, oxalic, palmitic, phenylacetic, phthalic, picramic, propionic, pyrogallic, salicylic, succinic, sulphanilic, tartaric, trichloroacetic, valeric.
Starting point in making—
Aromatics, intermediates, pharmaceuticals, various salts.

Dye
Reagent in making various synthetic dyestuffs.

Potassium Alphachloro-2-nitrobenzene-4-sulphonate
French: Alphachloro-2-nitrobenzène-4-sulphonate de potasse, Alphachloro-2-nitrobenzène-4-sulphonate potassique, Alphachloro-2-nitrobenzène-4-sulphonate de potassium.
German: Alphachlor-2-nitrobenzol-4-sulfonsaeureskalium, Alphachlor-2-nitrobenzol-4-sulfonsaeurespotasche, Kaliumalphachlor-2-nitrobenzol-4-sulfonat.

Chemical
Starting point in making various intermediates.

Dye
Starting point (Brit. 285504) in making nitro dyestuffs with—
Benzidin, 2:4′-diaminodiphenyl, 4:4′-diamino-2-nitrodiphenyl, orthoanisidin.

Potassium-Aluminum-Iron Cyanide

Chemical
Catalyst (Brit. 446411) in—
Halogenating unsaturated hydrocarbons.

Potassium Amylate
French: Amylate de potasse, Amylate potassique, Amylate de potassium.
German: Amylsaeureskalium, Kaliumamylat.

Chemical
Reagent (Brit. 304118) in making ketonic acid esters with esters of the following acids:—
Acetic, anthranilic, benzoic, butyric, camphoric, capric, caproic, caprylic, chloracetic, cinnamic, citric, cresylic, gallic, lactic, maleic, malic, malonic, metanilic, mucic, naphthionic, oxalic, palmitic, phenylacetic, phthalic,

Potassium Amylate (Continued)
picramic, picric, propionic, pyrogallic, salicylic, succinic, sulphanilic, tartaric, trichloroacetic, valeric.
Starting point in making—
Aromatics, intermediates, pharmaceuticals, various salts.
Dye
Reagent in making various synthetic dyestuffs.

Potassium Amylnaphthalenesulphonate
French: Amylenaphthalènesulphonate de potasse, Amylenaphthalènesulphonate potassique, Amylenaphthalènesulphonate de potassium.
German: Amylnaphtalinsulfonsaeureskalium, Amylnaphtalinsulfonsaeurespotasche, Kaliumamylnaphtalinsulfonat.
Chemical
Starting point in making various intermediates.
Starting point (Brit. 298823) in making—
Synthetic drugs.
Dye
Dispersive agent (Brit. 264860) in making—
Color lakes.
Miscellaneous
As an emulsifying agent (Brit. 298823).
For uses, see under general heading: "Emulsifying agents."

Potassium-Anthraquinone Betasulphonate
French: Bétasulphonate de potasse et anthraquinone.
German: Betasulfonsaeuresanthrachinonkalium, Kaliumanthrachinonbetasulfonat.
Chemical
Starting point in making—
Betachloroanthraquinone.

Potassium Auribromide
French: Auribromure de potasse.
German: Auribromkalium, Kaliumauribromid.
Chemical
Reagent (Brit. 265777) in making organic auromercapto acids and salts with—
4-Amino-2-mercaptobenzene-1-carboxylic acid.
Sodium-gammamercaptoglycerin alphasulphonate.
Sodium-paramercaptobenzene sulphonate.

Potassium Betatetrahydronaphthalenesulphonate
French: Bétatétrahydronaphthalènesulphonate de potasse, Bétatétrahydronaphthalènesulphonate potassique, Bétatétrahydronaphthalènesulphonate de potassium.
German: Betatetrahydronaphtalinsulfonsäureskalium, Betatetrahydronaphtalinsulfonsäurespotasche, Kaliumbetatetrahydronaphtalinsulfonat.
Miscellaneous
As an emulsifying agent (Brit. 371293).
For uses, see under general heading: "Emulsifying agents."

Potassium Bichromate
Synonyms: Bichromate of potash, Bichromate of potassa, Potassium dichromate, Red chromate of potash, Red chromate of potassa.
Latin: Kali bichromicum, Kali chromicum rubrum, Kalium dichromicum, Potassii bichromas, Potassii dichromas.
French: Bichromate de potasse, Chromate(Bi) de potasse.
German: Doppeltchromsäureskali, Kaliumdichromat, Zweifachchromsäureskali.
Spanish: Bicromato potasico.
Italian: Bicromato di potassio.
Analysis
As a reagent in various processes.
Ceramics
Starting point in making—
Chromium stannate pigment for various uses in ceramic manufacturing operations.
Chemical
Oxidizing agent in various chemical processes.
Reagent in—
Purification of pyroligneous acid.
Starting point in making—
Chromates, such as lead, zinc, barium.
Chromic oxide, chrome alum (chromium potassium sulphate).
Construction
Controlling agent (Brit. 405508) for—
Setting of mortars, cements, concretes, betons, and the like.
Dye
Oxidizing agent in making—
Alizarin from anthracene, dyestuffs, intermediates.
Electrical
Ingredient of—
Battery electrolytes.
Reagent for—
Various electrotechnical purposes.
Explosives and Matches
Ingredient of—
Composition for producing yellow smoke (U. S. 1920254).
Dynamites, matchhead compositions, pyrotechnic compositions.
Fats and Oils
Bleaching agent (in conjunction with sulphuric acid) for—
Animal oils, fats, fatty substances, fish oils, vegetable oils.
Glass
Ingredient of—
Certain glass batches.
Glues and Adhesives
Ingredient of—
Adhesive composition containing soybean flour, caustic soda, and water (U. S. 1897469).
Chrome adhesives, chrome glues.
Glue having a moderate alkali content and a thinning or cutting agent prepared from a non-protein polysaccharide carbohydrate having the characteristics of starch (e.g., tapioca flour), calcium oxide, caustic soda, and water (U. S. 1904619).
Waterproof adhesives.
Ink
Ingredient of various inks.
Jewelry
Coloring agent in making—
Artificial rubies by the Verneuil process.
Leather
Oxidizing agent in—
Chrome tanning.
Metallurgical
Ingredient of—
Brass-pickling solutions, electrolytes in electroplating.
Miscellaneous
Bleaching agent in many processes.
Hardening and preservative agent for—
Anatomical specimens.
Hardening agent for—
Moulds in galvanoplastic work.
Ingredient of—
Poisonous compositions used on flypaper.
Oxidizing agent in many processes.
Paint and Varnish
Starting point in making—
Arnaudon's green.
Chromates used as yellow pigments and known under various names, such as lead chromate, barium chromate, zinc chromate, Leipsig yellow, Cologne yellow, Paris yellow, primrose chrome, pale chrome, middle chrome, deep chrome, citron yellow, lemon chrome, lemon yellow, baryta yellow, permanent yellow, yellow ultramarine, Steinbuller yellow, jaune d'outremer.
Chromaventurine, chromic oxide green, emerald green, Guignet's green, lead chromate orange.
Lead chromate red, known under various names, such as chrome red, Chinese red, American vermilion, Austrian cinnabar, Derby red, Persian red, Victoria red.
Leaf green, Plessy's green, Schnitzer's green, Turkish green.
Paper
As a bleaching agent.
Perfume
Reagent in making—
Synthetic perfumes.
Petroleum
Reagent in—
Testing oil-field water for iron (according to Macfadyen method).
Pharmaceutical
In compounding and dispensing practice.
Suggested for external use in treatment of—
Syphilis, rodent ulcer.
Suggested for use as—
Irritant caustic.

Potassium Bichromate (Continued)

Photographic
Anticlouding agent in—
Gelatino-bromid emulsions.
Ingredient of—
Reducers.
Reagent in—
Bichromate gum process, carbon process.

Printing
Reagent in—
Lithographic work, process engraving.

Resins and Waxes
Bleaching agent (in conjunction with sulphuric acid) for—
Waxes.
Reagent (Brit. 397096) in making—
Synthetic resins from aromatic hydrocarbons and polyvalent alcohols.

Stone
Reagent in—
Producing colored effects in alabaster to give it the appearance of onyx and agate.

Textile
Discharge for—
Turkey red.
Reagent in—
Dyeing processes, especially dyeing woolen goods with alizarin dyestuffs or with logwood black.
Printing processes.
Preparation of ammonium-copper chromate solution used in dyeing cotton and woolen fabrics olive colors with logwood and in combination with fustic as well as with buckthorn.

Woodworking
Ingredient of—
Compositions used in staining wood and wood products.

Potassium Biformate
Synonyms: Potassium acid formate, Potassium diformate, Potassium hydrogen formate.
French: Biformate de potasse, Biformiate de potasse, Biformiate de potassium.
German: Ameisensaeuressaeurekalium, Kaliumbiformiat.

Chemical
Reagent (German 439289) in making—
Ethyl formate, geranyl formate, glycol formate, mixed anhydrides of formic and acetic acids plus nitric acid, phenyl formate.

Potassium Bromate
French: Bromate de potasse, Bromate de potassium.
German: Kaliumbromat.
Spanish: Bromato de potasa.
Italian: Bromato di potassio.

Analysis
Reagent in—
Analytical work.

Chemical
Purifying agent in making—
Bromine.

Food
Leavening agent in—
Baking.

Metallurgical
Ingredient of—
Electrolyte, containing also magnesium sulphate and hydroxide, for producing green patina on copper.

Potassium Bromide
Synonyms: Bromide of potash.
Latin: Bromuretum potassicum, Kalium bromatum, Potassii bromidum.
French: Bromure de potasse, Bromure de potassium, Potasse bromique.
German: Bromkalium, Kaliumbromid.
Spanish: Bromuro potasico.
Italian: Bromuro di potassio.

Electrical
Ingredient (French 682814) of—
Battery electrolytes.

Pharmaceutical
In compounding and dispensing practice.
Suggested for use as—
Nerve sedative.

Photographic
Ingredient of—
Photographic developing fixer, containing also metol, hydroquinone, sodium sulphite, sodium carbonate, caustic soda, sodium hyposulphite, and ammonium picrate.
Reagent in making—
Bromide papers, bromide plates.
Photographic emulsions.

Printing
Bromide in—
Process engraving, lithography.

Resins
Catalyst (French 707433) in making—
Phenol-formaldehyde resins.

Soap
Ingredient of—
Special soaps.

Potassium Butylnaphthalenesulphonate
French: Butylenaphthalènesulphonate de potasse, Butylenaphthalènesulphonate potassique, Butylenaphthalènesulphonate de potassium.
German: Butylnaphtalinsulfonsäureskalium, Butylnaphtalinsulfonsäurespotasche, Kaliumbutylnaphtalinsulfonat.

Miscellaneous
As an emulsifying agent (Brit. 330896).
For uses, see under general heading: "Emulsifying agents."

Potassium Butyrate
French: Butyrate de potasse, Butyrate potassique, Butyrate de potassium.
German: Kaliumbutyrat.
Spanish: Butirato potasico.
Italian: Butirato di potassio.

Dye
Reagent (Brit. 388043) in making—
Indanthrene dyes.

Rubber
Accelerator (French 629661) in—
Vulcanizing processes.

Potassium-Cadmium Cyanide

Chemical
Catalyst (Brit. 446411) in—
Halogenating unsaturated hydrocarbons.
Starting point (Brit. 446411) in making—
Catalysts with metal chlorides for halogenating unsaturated hydrocarbons.

Potassium Carbonate
Synonyms: Carbonate of potash, Salt of tartar.
Latin: Carbonas kalicus, Carbonas potassicus, Kali carbonicum, Kalicum carbonicum, Potassii carbonas, Sal tartari.
French: Carbonate de potasse, Sel de tartre.
German: Kaliumcarbonat, Kohlensäureskali.
Spanish: Carbonato potasico.
Italian: Carbonato di potassio.

Beverage
Ingredient of—
Artificial Vichy waters, effervescent beverages.

Brewing
Alkali for process work.

Ceramics
In the process.

Chemical
Alkali for various processes.
Dehydrating agent.
Neutralizing agent for acids.
Starting point, either directly or indirectly, in making—
Potassium acetate, potassium alginate, potassium allylate.
Potassium alphachloro-2:4-dinitrobenzene-4-sulphonate.
Potassium alphachloro-2-nitrobenzene-4-sulphonate.
Potassium alphanaphtholate, potassium amylate, potassium amylnaphthalenesulphonate, potassium anthraquinonate, potassium anthraquinonebetasulphonate, potassium apocholate, potassium auribromide, potassium benzylthioglycollate, potassium betanaphthalate, potassium betatetrahydronaphthalenesulphonate.
Potassium bicarbonate, potassium bichromate, potassium biformate, potassium binoxalate, potassium bisulphate, potassium bisulphite, potassium bromate, potassium bromide, potassium butylnaphthalenesulphonate, potassium carbazolate, potassium catecholate, potassium cholate.

Potassium Carbonate (Continued)
Potassium 1-chloro-2:6-dinitrobenzene-4-sulphonate.
Potassium 1:5-chloronaphthalenesulphonate.
Potassium 1:6-chloronaphthalenesulphonate.
Potassium chloroplatinate, potassium chloroplatinite, potassium chlorostannate, potassium chromate, potassium chromitesilicate, potassium citrate, potassium cresolate, potassium cresylthioglycollate, potassium cyanide, potassium cyclohexylxanthate.
Potassium desoxycholate, potassium dichromoparaphenolsulphonate, potassium difluorodisulphonate, potassium ethylxanthate, potassium ferricyanide, potassium ferritartrate, potassium ferrocyanide, potassium fluoride, potassium fluorostannate, potassium fluorotantalate, potassium fluorozirconate, potassium fluortitanate, potassium glycerophosphate.
Potassium glycocholate, potassium guaiacolsulphonate.
Potassium heptylnaphthalenesulphonate, potassium hexylate, potassium hexylnaphthalenesulphonate, potassium hydrogenphthalate, potassium hydroxide, potassium hypochlorite, potassium hypophosphite, potassium iodide, potassium isatin-5-sulphonate.
Potassium isoallylnaphthalenesulphonate.
Potassium isoamylnaphthalenesulphonate.
Potassium isobutylnaphthalenesulphonate.
Potassium isopropylchloronaphthalenesulphonate.
Potassium isopropylnaphthalenesulphonate.
Potassium lactate, potassium naphthalenetrisulphonate, potassium naphthionate, potassium naphthylthioglycollate, potassium nitrate, potassium oxalate, potassium oxide, potassium pentylnaphthalenesulphonate, potassium perchlorate, potassium permanganate, potassium peroxide, potassium persulphate, potassium phenate, potassium phenylthioglycollate, potassium phosphates, potassium phosphotungstate, potassium phosphotungstomolybdate, potassium polysulphide.
Potassium propylnaphthalenesulphonate, potassium resorcinate, potassium selenate, potassium silicofluoride, potassium silicomolybdate, potassium silicotungstate, potassium sozoiodolate, potassium sulphate, potassium sulphide, potassium sulphite, potassium sulphocyanate, potassium sulphoricinoleate.
Potassium taurocholate, potassium telluride, potassium thioglycollate, potassium vanadate, potassium vanadite, potassium xylenolate, potassium uranate, potassium-silver cyanide, potassium-sodium tartrate.

Fertilizer
Ingredient of—
Compounded fertilizers.
Source of potash.

Food
Ingredient of—
Confectionery.

Glass
Ingredient of batch in making—
Bohemian glass, hard-to-fuse glass.

Leather
Alkali in—
Finishing processes, tanning processes.

Metallurgical
Ingredient of—
Electroplating baths.

Miscellaneous
As a detergent.
Reagent in—
Treating permanent wave papers.

Paper
Reagent (U. S. 1900967) in—
Treating paper coated with potassium ferricyanide to prevent discoloration on exposure to light and air, without impairing the ink-setting properties of the paper.

Paint and Varnish
Ingredient of—
Shellac-drying oil combination.
Titanium pigments (U. S. 1892693).

Perfume
Ingredient of—
Non-greasy creams, shampoo preparations.

Pharmaceutical
Antiacid in compounding and dispensing.
Ingredient of—
Alkaline sulphur ointment, N.F.
Carminative mixture, N.F.
Effervescent salts.
Suggested for use in treatment of—
Cutaneous affections, dropsy, dyspepsia, uric acid gravel.

Photographic
Reagent (Brit. 384770) in—
Developing bichromate prints.

Printing
Alkali in—
Lithographic work, process engraving.

Soap
Saponification agent in—
Liquid soaps, shaving creams, shaving soaps, soft soaps, toilet soaps.

Rubber
Reagent (Brit. 397136) in—
Lightening the color of drying oils produced by the catalytic hydrogenation of natural or synthetic rubber in the presence of solvents, such as petroleum derivatives.

Textile
Alkali in—
Dyeing fabrics.
Antichlor.
Cleansing agent in—
Washing woolens and silks.
Saponifying agent in—
Washing woolens and silks.

Potassium Chlorate
Synonyms: Chlorate of potash, Potash oxymuriate, Potassium oxymuriate.
Latin: Kalium chloratum.
French: Chlorate de potasse, Chlorate potassique, Chlorate de potassium, Oxymuriate de potasse, Oxymuriate potassique, Oxymuriate de potassium.
German: Chlorsäureskalium, Chlorsäurespotasche, Kaliumchlorat, Kaliumoxymuriat.

Analysis
Oxidizing agent in—
Forensic and ultimate analysis.
Reagent in analyzing—
Alkaloids, aspidospermine, atropine, cocaine, phenols, tryosine.
Reagent in determining—
Histidine bases, indican, purine bases.
Sulphur by means of the Parr calorimeter.

Automotive
Ingredient of—
Compositions for removing and preventing carbon deposits in internal combustion engines.

Chemical
As a general oxidizing agent.
Reagent in making—
Barium peroxide, boron carbide, dry colors, naphthalene tetrachloride, phosphorus oxychloride, trichloroacetic acid.
Various intermediates and other organic and inorganic chemicals.
Reagent (U. S. 1733776) in making—
Di-iodofluorescein and other dihalogenated fluoresceins.
Source of oxygen for laboratory and other purposes.

Dye
Oxidizing agent in making—
Alizarin, anilin black, bengal rose B.
Various other synthetic dyestuffs.

Explosives
Ingredient of—
Explosive compositions of various sorts, including dynamites and military explosives.
Fulminating compositions, fuses, match-head compositions, percussion cap compositions, pyrotechnical compositions, safety match compositions.

Ink
Reagent in making—
Printing inks.

Insecticide
Ingredient (Brit. 335203) of weed-killing compositions in admixture with—
Acids, such as hydrochloric, sulphuric, nitric, boric, oxalic, tartaric.
Acid salts, such as sodium acid sulphate, sodium bitartrate, calcium hydrogen-phosphate.
Acid-reacting salts.
Chlorides, such as ammonium chloride, aluminum chloride, iron chloride, copper chloride, zinc chloride, and mercuric chloride.
Sodium bichromate, sodium fluosilicate.

Miscellaneous
Oxidizing agent for various purposes.

Potassium Chlorate (Continued)

Paint and Varnish
Oxidizing agent in making Mars yellow.

Paper
Reagent in the manufacture of paper.

Perfume
Ingredient of—
Cosmetics, dentifrices.

Pharmaceutical
In compounding and dispensing practice.

Textile
——, *Dyeing*
Reagent in—
Dyeing cotton and wool in black shades and in other processes.
——, *Printing*
Ingredient of—
Printing pastes.

Sanitation
As a disinfectant in sanitary work.

Potassium Chloride
Synonyms: Chloride of potash, Muriate of potash.
Latin: Kali chloratum, Kalium chloratum, Potasii chloridum, Sal digestivum, Seyvii.
French: Chlorure de potasse, Sel digestif.
German: Chlorkalium, Kaliumchlorid.
Spanish: Cloruro potasico.
Italian: Chloruro di potassio.

Analysis
Reagent.

Beverage
Ingredient of—
Mineral waters.

Chemical
Starting point, either directly or indirectly, in making—
Potassium acetate, potassium alginate, potassium allylate.
Potassium alphachloro-2:4-dinitrobenzene-4-sulphonate.
Potassium alphachloro-2-nitrobenzene-4-sulphonate.
Potassium alphanaphtholate, potassium amylate, potassium amylnaphthalene sulphonate, potassium anthraquinonate, potassium anthraquinonebetasulphonate, potassium apocholate, potassium auribromide, potassium, benzylthioglycolate, potassium betanaphthalate.
Potassium betatetrahydronaphthalenesulphonate.
Potassium bicarbonate, potassium bichromate, potassium biformate, potassium binoxalate, potassium bisulphate, potassium bisulphite, potassium bromate, potassium bromide.
Potassium butylnaphthalenesulphonate, potassium carbazolate, potassium catecholate, potassium celenate, potassium cholate, potassium chlorate.
Potassium 1-chlor-2:6-dinitrobenzene-4-sulphonate.
Potassium 1:5-chloronaphthalenesulphonate.
Potassium 1:6-chloronaphthalenesulphonate.
Potassium chloroplatinate, potassium chloroplatinite, potassium chlorostannate, potassium chromate, potassium chromitesilicate, potassium citrate, potassium cresolate, potassium cresylthioglycolate, potassium cyanide, potassium cyclohexylxanthate.
Potassium desoxycholate, potassium dibromoparaphenolsulphonate, potassium difluorodisulphonate, potassium ethylxanthate, potassium ferricyanide, potassium ferritartrate, potassium ferrocyanide, potassium fluoride, potassium fluorostannate, potassium fluorotantalate, potassium fluorozirconate, potassium fluotitanate, potassium glycerophosphate.
Potassium glycocholate, potassium guaiacolsulphonate, potassium heptylnaphthalenesulphonate, potassium hexylate, potassium hexylnaphthalenesulphonate, potassium hydrogenphthalate, potassium hydroxide, potassium hypochlorite, potassium hypophosphite.
Potassium iodide, potassium isatin-5-sulphonate.
Potassium isoallylnaphthalenesulphonate.
Potassium isoamylnaphthalenesulphonate.
Potassium isobutylnaphthalenesulphonate.
Potassium isopropylchloronaphthalenesulphonate.
Potassium isopropylnaphthalenesulphonate.
Potassium lactate, potassium naphthalenetrisulphonate, potassium naphthionate, potassium naphthylthioglycolate, potassium nitrate, potassium oxalate, potassium oxide, potassium pentylnaphthalenesulphonate, potassium perchlorate, potassium permanganate.
Potassium peroxide, potassium persulphate, potassium phenate, potassium phenylthioglycolate, potassium phosphates, potassium phosphotungstate, potassium phosphotungstomolybdate, potassium polysulphide.
Potassium propylnaphthalenesulphonate, potassium resorcinate, potassium silicofluoride, potassium silicomolybdate, potassium silicotungstate, potassium sozoiodolate, potassium sulphate, potassium sulphide, potassium sulphite, potassium sulphocyanate.
Potassium sulphoricinoleate, potassium taurocholate, potassium telluride, potassium thioglycolate, potassium vanadate, potassium vanadite.
Potassium xylenolate, potassium uranate, potassium-silver cyanide, potassium-sodium tartrate.

Electrical
Ingredient (Brit. 395456) of—
Battery electrode coating (added for the purpose of increasing the porosity of the dried coating).

Explosives and Matches
Ingredient (Brit. 315232) of—
Fuel-igniting compositions.

Fertilizer
Ingredient of—
Fertilizer mixtures.
Source of potash.

Metallurgical
Ingredient of—
Coating compositions for welding rods used in the autogenous welding of aluminum and the like (U. S. 1844969).
Saltbath for heat-treating metals (4 parts potassium chloride and 1 part anhydrous sodium borate) (U. S. 1724551).
Reagent (Brit. 403469) in making—
Finely divided metals by precipitation of their salts by means of metals in powdered form.

Miscellaneous
Ingredient (Brit. 278785) of—
Compositions for coating ornaments to prevent them from slipping.

Pharmaceutical
In compounding and dispensing practice.
Ingredient of—
Artificial Kissingen salt.

Photographic
Reagent in—
Various processes.

Soap
Ingredient of—
Liquid soaps, potash soaps, shaving creams, soft soaps, toilet soaps.

Textile
Ingredient (Brit. 278785) of—
Composition for coating rugs and mats to prevent them from slipping.

Potassium 1:5-Chloronaphthalenesulphonate
French: 1:5-Chloronaphthalènesulfonate de potasse.
German: 1:5-Chlornaphtalinsulfonsaeureskalium, Kalium-1:5-chlornaphtalinsulfonat.

Chemical
Starting point (Brit. 263873) in making—
Emulsifying agents for terpenes and aromatic hydrocarbons.

Fats and Oils
Starting point (Brit. 263873) in making—
Emulsifying agents.

Leather
Starting point (Brit. 263873) in making—
Emulsified tanning compositions.

Miscellaneous
Starting point (Brit. 263873) in making—
Emulsifying agents for washing and cleansing compositions.

Paper
Starting point (Brit. 263873) in making—
Reagents for increasing the absorbing and wetting qualities of paper and cardboard.

Petroleum
Starting point (Brit. 263873) in making—
Emulsifying agents for mineral oils.

Resins and Waxes
Starting point (Brit. 263873) in making—
Emulsifying agents.

Textile
——, *Dyeing*
Starting point (Brit. 263873) in making—
Dye liquor emulsifiers.

Potassium 1:5-Chloronaphthalenesulphonate (Cont'd)

——, *Finishing*
Starting point (Brit. 263873) in making—
Emulsified washing and cleansing compositions.

——, *Manufacturing*
Starting point (Brit. 263873) in making—
Emulsifying agents for wool carbonizing liquors.

Potassium 1:6-Chloronaphthalenesulphonate

French: 1:5-Chloronaphthalènesulfonate de potasse, 1:6-Chlornaphthalènesulfonate potassique.
German: Kalium-1:6-chlornaphtalinsulfonat.

Chemical
Reagent (Brit. 263873) in making—
Aromatic hydrocarbon emulsions, terpene emulsions.

Fats and Oils
Reagent (Brit. 263873) in making—
Emulsions.

Leather
Reagent (Brit. 263873) in making—
Emulsified tanning compositions.

Miscellaneous
Reagent (Brit. 263873) in making—
Emulsified washing and cleansing compositions.

Paper
Reagent (Brit. 263873) in making—
Compositions that are used for increasing the absorbing and wetting capacities of cardboard, paper, and paper products.

Petroleum
Reagent (Brit. 263873) in making—
Emulsions of petroleum or its products.

Resins and Waxes
Reagent (Brit. 263873) in making—
Emulsions.

Textile

——, *Dyeing*
Reagent (Brit. 263873) in making—
Dye liquors for yarns and fabrics.

——, *Finishing*
Reagent (Brit. 263873) in making—
Washing and cleansing compositions.

——, *Manufacturing*
Reagent (Brit. 263873) in making—
Wool-carbonizing baths.

Potassium Chloroplatinite

French: Chloroplatinite de potasse, Chloroplatinite potassique, Chloroplatinite de potassium.
German: Kaliumplatinchlorur, Platinkaliumchlorur.
Spanish: Cloruro de platina y de potasa.
Italian: Chloroplatinito di potassio.

Photographic
As a toning agent in printing processes.

Potassium Chlorostannate

French: Chlorostannate de potasse.
German: Chlorzinnsaeureskalium, Kaliumchlorstannat.

Chemical
Catalyst (Brit. 250897) in making—
Amines, nitriles, substituted amines.

Potassium Cholate

French: Cholate de potasse, Cholate potassique.
German: Cholinsaeureskalium, Cholinsaeurespotasche, Kaliumcholat.

Chemical
Reagent (Brit. 282356) in making parasiticides with—
Dihydrocupreine-ethyl ether.
Dihydrocupreinethyl ether hydrochloride.
Dihydrocupreineisoamyl ether.
Dihydrocupreineisoamyl ether hydrochloride.
Dihydrocupreine normal octyl ether.
Dihydrocupreine normal octyl ether hydrochloride.
Dihydroquinone.

Pharmaceutical
In compounding and dispensing practice.

Potassium Chromitesilicate

French: Chromitesilicate de potasse, Chromitesilicate potassique, Chromitesilicate de potassium.
German: Kaliumchromitsilikat.

Chemical
Catalytic reagent in making—
Acetic acid from aldehyde.
Aldehyde from alcohol.
Benzoic acid from benzaldehyde.
Sodium chloride from sodium hypochlorite.
Sodium bisulphate from sodium bisulphite.
Reagent in converting—
Manganese protoxide into permanganic acid.
Reagent in oxidizing—
Iron and manganese compounds with the aid of atmospheric oxygen.

Dye
Reagent in converting—
Leuco-malachite hydrochloride into malachite.

Metallurgical
Reagent in recovering—
Gold from seawater, metal from liquids.
Radium from wells containing radio-active water.

Miscellaneous
Reagent in sterilizing—
Liquids by means of ozone, chlorine, hydrogen peroxide, or potassium permanganate.

Sugar
Reagent in recovering—
Potash and other bases from sugar juices and molasses.

Water
Purifying reagent.
Reagent in removing—
Iron and manganese compounds from mineral waters containing carbon dioxide by oxidizing the iron and manganese by means of atmospheric oxygen.
Oxygen from water by the addition of sodium sulphite, which is converted into sodium sulphate.
Softening reagent.

Potassium Cresolate

French: Crésolate de potasse, Crésolate potassique.
German: Cresolsaeureskalium, Kaliumcresolat.

Dye
Reagent (Brit. 388043) in making—
Indanthrene dyes.

Leather
Ingredient (Brit. 263473) of—
Dyeing compositions.

Miscellaneous
Ingredient (Brit. 263473) of—
Dye liquors used on hair, feathers, and the like.

Textile

——, *Dyeing and Printing*
Ingredient (Brit. 263473) of—
Liquors and pastes containing vat dyestuffs and used in dyeing or printing fabrics and yarns containing acetate rayon, viscose, cuprammonium rayon, and nitro rayon, as well as wool-rayon and silk-rayon mixtures.

Potassium-Cupro Cyanide

Chemical
Catalyst (Brit. 446411) in—
Halogenating unsaturated hydrocarbons.
Starting point (Brit. 446411) in making—
Catalysts with metal chlorides for halogenating unsaturated hydrocarbons.

Potassium Cyclohexylxanthate

Metallurgical
Flotation agent (U. S. 1823316) in separating—
Minerals from ores (added to aid in the froth flotation process).

Potassium Dibromoparaphenolsulphonate

French: Dibromoparaphénolesulphonate de potasse.
German: Dibromparaphenolsulfonsaeureskalium.

Chemical
Starting point in making various pharmaceutical chemicals.

Potassium Dicresyldithiophosphate

Mining
Flotation agent (Brit. 455224) in—
Froth flotation of minerals.

Potassium Eleostearicsulphonate

Miscellaneous
As an emulsifying agent (Brit. 361732).
For uses, see under general heading: "Emulsifying agents."

Potassium Ethylxanthate
French: Xanthate de potasse-éthyle.
German: Kaliumaethylxanthogenat, Xanthogensaeureskaliumaethyl.

Analysis
Reagent in determining carbon disulphide.

Chemical
Reducing agent in various processes.
Starting point in making—
Rubber vulcanization accelerator with sulphur monochloride (Brit. 265169).
Thiophenols from diazonium compounds.

Insecticide
Ingredient of—
Insecticidal compositions.

Potassium Ferritartrate
Synonyms: Ferric-potassium tartrate.
Latin: Kalium ferrotartaricum.
French: Ferritartrate de potasse, Tartrate de fer et de potasse, Tartrate ferricopotassique, Tartre chalybé, Tartre martial.
German: Eisenweinstein, Kalium ferritartrat.

Pharmaceutical
In compounding and dispensing practice.

Photographic
Reagent in making—
Blueprint papers in combination with potassium ferricyanide.

Textile
——, *Dyeing*
Mordant in dyeing yarns and fabrics.
——, *Printing*
Mordant in printing fabrics.

Potassium Fluorostannate
Synonyms: Potassium stannifluoride.
French: Fluorostannate de potasse, Fluorostannate potassique, Fluorostannate de potassium.
German: Fluorzinnsaeureskalium, Fluorzinnsaeurespottasche, Stannifluorwasserstoffsaeureskalium, Stannifluorwasserstoffsaeurespottasche.

Adhesives
Ingredient of—
Acidproof cement containing water glass (Brit. 283471).

Leather
Ingredient of—
Liquors, containing dyewoods, used in dyeing leather with the aid of alum or tannin mordant.

Potassium Fluorotantalate
French: Fluorotantalate de potasse, Fluorotantalate potassique, Fluorotantalate de potassium.
German: Fluortantalsäureskalium, Fluortantalsäurespotasche, Kaliumfluortantalat.

Glues and Adhesives
Ingredient (Brit. 283471) of—
Acidproof cements made with the addition of sodium silicate.

Potassium Fluotitanate
French: Fluotitanate de potasse.
German: Kaliumfluotitanat, Titanfluorwasserstoffsaeureskalium.

Leather
Ingredient of—
Dyeing liquor in admixture with tinctorial woods.

Potassium Guaiacolsulphonate

Pharmaceutical
In compounding and dispensing practice.

Potassium Heptylnaphthalenesulphonate

Miscellaneous
As an emulsifying agent (Brit. 298823).
For uses, see under general heading: "Emulsifying agents."

Potassium Hexylate
French: Hexylate de potasse, Hexylate potassique.
German: Kaliumhexylat.

Chemical
Reagent (Brit. 304118) in making ketonic acid esters with the aid of allyl, amyl, butyl, heptyl, hexyl, and propyl esters of the following acids—
Acetic, anthranilic, benzoic, butyric, camphoric, caproic, caprylic, chloroacetic, cinnamic, citric, cresylic, gallic, lactic, maleic, malic, malonic, metanilic, mucic, naphthionic, oxalic, palmitic, phenylacetic, phthalic, picramic, propionic, pyrogallic, salicylic, succinic, sulphanilic, tartaric, trichloroacetic, valeric.
Starting point in making—
Aromatics, intermediates, pharmaceuticals, salts and esters.

Dye
Reagent in making various synthetic dyestuffs.

Potassium Hexylnaphthalenesulphonate
French: Hexylenaphthalènesulphonate de potasse, Hexylenaphthalènesulphonate potassique, Hexylenaphthalènesulphonate de potassium.
German: Hexylnaphtalinsulfonsäureskalium, Hexylnaphtalinsulfonsäurespotasche, Kaliumhexylnaphtalinsulfonat.

Chemical
Emulsifying agent (Brit. 298823) in making—
Pharmaceuticals.
Starting point in making—
Intermediates, pharmaceuticals, and other derivatives.

Miscellaneous
As an emulsifying agent (Brit. 298823).
For uses, see under general heading: "Emulsifying agents."

Potassium Hydroxide
Synonyms: Caustic potash, Caustic potassa, Hydrate of potassa.
Latin: Kali causticum fusum, Kali hydricum fusum, Kali purum, Lapis causticus chirurgorum, Potassae hydras, Potassii hydras, Potassii hydroxidum.
French: Pierre à cautère, Potasse caustique, Potasse fondue.
German: Aetzkali, Kaliumhydrat, Kaliumhydroxyd, Kaustischeskali.
Spanish: Hidrato potasico.
Italian: Potassa caustica.

Analysis
Neutralizing agent for acids.
Reagent.
Source of potash.

Chemical
Alkali for various processes.
Neutralizing agent for acids.
Reagent in making—
4-Aminopyridin from pyridin, thionyl bromide, and calcium oxide (Brit. 382327).
Salts of adenylpyrophosphoric acids (Brit. 396647).
Salts of aliphatic acids (Brit. 405846).
Reagent (Brit. 398807) in—
Recovering alcohols from sperm oil and spermaceti.
Saponifying agent (U. S. 1912440) in making—
Sterol from yeast.
Starting point, either directly or indirectly, in making—
Potassium acetate, potassium alginate, potassium allylate.
Potassium alphachloro-2:4-dinitrobenzene-4-sulphonate.
Potassium alphachloro-2-nitrobenzene-4-sulphonate.
Potassium alphanaphtholate, potassium amylate, potassium amylnaphthalenesulphonate, potassium anthraquinonate, potassium anthraquinonebetasulphonate.
Potassium apocholate, potassium auribromide, potassium benzylthioglycollate, potassium betanaphthalate.
Potassium betatetrahydronaphthalenesulphonate.
Potassium bicarbonate, potassium bichromate, potassium biformate, potassium binoxalate, potassium bisulphate, potassium bisulphite, potassium bromate, potassium bromide.
Potassium butylnaphthalenesulphonate, potassium carbazolate, potassium catecholate, potassium cholate.
Potassium 1-chloro-2:6-dinitrobenzene-4-sulphonate.
Potassium 1:5-chloronaphthalenesulphonate.
Potassium 1:6-chloronaphthalenesulphonate.
Potassium chloroplatinate, potassium chloroplatinite, potassium chlorostannate, potassium chromate, potassium chromitesilicate, potassium citrate, potassium cresolate, potassium cresylthioglycollate, potassium cyanide, potassium cyclohexylxanthate.
Potassium desoxycholate, potassium dibromoparaphenolsulphonate, potassium difluorodisulphonate, potassium ethylxanthate, potassium ferricyanide, potassium ferritartrate, potassium ferrocyanide, potassium fluoride, potassium fluorostannate, potassium fluorotantalate.
Potassium fluorozirconate, potassium fluotitanate, potassium glycerophosphate, potassium glycocholate,

Potassium Hydroxide (Continued)
potassium guaiacolsulphonate, potassium heptylnaphthalenesulphonate, potassium hexylate.
Potassium hexylnaphthalenesulphonate, potassium hydrogenphthalate, potassium hypochlorite, potassium hypophosphite, potassium iodide, potassium isatin-5-sulphonate.
Potassium isoallylnaphthalenesulphonate.
Potassium isoamylnaphthalenesulphonate.
Potassium isobutylnaphthalenesulphonate.
Potassium isopropylchloronaphthalenesulphonate.
Potassium isopropylnaphthalenesulphonate.
Potassium lactate, potassium naphthalenetrisulphonate, potassium naphthionate, potassium naphthylthioglycollate, potassium nitrate, potassium oxalate, potassium oxide, potassium pentylnaphthalenesulphonate, potassium perchlorate, potassium permanganate.
Potassium peroxide, potassium persulphate, potassium phenate, potassium phenylthioglycollate, potassium phosphates, potassium phosphotungstate, potassium phosphotungstomolybdate, potassium polysulphide.
Potassium propylnaphthalenesulphonate, potassium resorcinate, potassium selenate, potassium silicofluoride, potassium silicomolybdate, potassium silicotungstate, potassium sozoiodolate, potassium sulphate.
Potassium sulphide, potassium sulphite, potassium sulphocyanate, potassium sulphoricinoleate, potassium taurocholate, potassium telluride, potassium thioglycollate, potassium vanadate.
Potassium vanadite, potassium xylenolate, potassium uranate, potassium-silver cyanide, potassium-sodium tartrate.

Construction
Ingredient (Brit. 387825) of—
Emulsifying composition for bitumen employed in road construction (used to produce an emulsion in wet weather).

Dye
Component of—
Dyes for various uses.
Reagent in making—
Sulphonic acids of terphenyls and their conversion products for use as dye intermediates (Brit. 404381).
Vat dyes of the anthraquinone series (Brit. 399724).

Explosives and Matches
Ingredient of—
Matchhead compositions.

Fertilizer
Source of potash.

Insecticide
Ingredient (Brit. 381290) of—
Water-soluble insecticide prepared from nicotine, lauric acid, caprylic acid, and sodium carbonate.

Metallurgical
Reagent (U. S. 1908473) in—
Separating tantalum from columbium.

Perfume
Ingredient (Brit. 394949) of—
Cuticle-removing preparation.

Petroleum
Reagent (U. S. 1905383) in making—
Chromic oxide gel used as catalyst in hydrocarbon dehydrogenation and hydrogenation processes.

Pharmaceutical
In compounding and dispensing practice.
Ingredient of several official preparations.
Suggested as—
Caustic, escharotic.

Printing
Alkali in—
Lithography, process engraving.

Rubber
Coagulation restrainer (Brit. 397997) in making—
Rubber-coated fabrics.

Sanitation
As a bactericide.

Soap
Saponifying agent (Brit. 398807) for—
Sperm oil.
Saponification agent in making—
Liquid soaps, shaving creams of brushing type, shaving creams of brushless type, shaving soaps, soft soaps, toilet soaps.

Textile
Mercerizing agent for—
Cotton.

Woodworking
Ingredient (U. S. 1909241) of—
Non-bleaching, fireproofing compositions for wood.

Potassium 2-Hydroxydiphenyldisulphonate

Cosmetic
Protectant (U. S. 2015005) in—
Oils, creams, and lotions against harmful effects of light of short wave length (sunburn).

Potassium 4-Hydroxydiphenyldisulphonate

Cosmetic
Protectant (U. S. 2015005) in—
Oils, creams, and lotions against harmful effects of light of short wave length (sunburn).

Potassium 2-Hydroxydiphenylmonosulphonate

Cosmetic
Protectant (U. S. 2015005) in—
Oils, creams, and lotions against harmful effects of light of short wave length (sunburn).

Potassium 4-Hydroxydiphenylmonosulphonate

Cosmetic
Protectant (U. S. 2015005) in—
Oils, creams, and lotions against harmful effects of light of short wave length (sunburn).

Potassium Hypochlorite
French: Hypochlorite de potassium.
German: Kaliumhypochlorit, Unterchloridsaeureskalium.

Chemical
Reagent in making—
Orthoaminocinnamic acid from orthocyanocinnamic acid (German 440052).

Miscellaneous
Bleaching agent in the treatment of—
Bone, feathers, fur, horn, straw.

Textile
——, *Bleaching*
Reagent in bleaching fabrics and yarns.

Potassium Iodide
Synonyms: Iodide of potash.
Latin: Ioduretum potassicum, Ioduretum kalicum, Kali hydriodicum, Kalium jodatum, Potassii hydriodas, Potassii iodidum.
French: Iodure de potasse, Iodure de potassium.
German: Jodkalium, Kaliumjodid.
Spanish: Yoduro potasico.
Italian: Joduro di potassio.

Analysis
As a reagent.

Dry Cleaning
Reagent for removing—
Argyrol stains (used in solution, followed with solution of sodium thiosulphate).
Copper stains.
Iodine stains (10 percent solution, followed with 10 percent sodium thiosulphate solution, followed with water).
Lead compounds (stain with tincture of iodine; remove with solution of potassium iodide).
Photographic developer stains (in combination treatment with (1) iodine and (2) sodium thiosulphate).

Leather
Ingredient (Brit. 399725) of—
Ammoniacal solution used for processing cellulosic material in the production of translucent, moldable, and highly absorptive products useful as leather-like paper materials.

Miscellaneous
Ingredient of—
Cattlefeeds.

Paper
Ingredient (Brit. 399725) of—
Ammoniacal solution used for processing cellulosic material in the production of translucent and highly absorptive products which, on grinding, give material suitable for the production of sanitary pads or filter paper.

Pharmaceutical
In compounding and dispensing practice.

Photographic
As a reagent.

Potassium Iodide (Continued)
Plastics
Ingredient (Brit. 399725) of—
Ammoniacal solution used for processing cellulose material in the production of translucent, moldable, and highly absorptive products used as cellulose plastics.
Textile
Ingredient (Brit. 400180) of—
Spinning solution in manufacture of rayon threads, yarns, and the like.

Potassium Isoallylnaphthalenesulphonate
Synonyms: Isoallylnaphthalenesulphonate of potash.
French: Isoallylenaphthalènesulphonate de potasse, Isoallylenaphthalènesulphonate potassique, Isoallylenaphthalènesulphonate de potassium.
German: Isoallylnaphtalinsulfonsaeureskalium, Isoallylnaphtalinsulfonsaeurespotasche, Kaliumisoallylnaphtalinsulfonat.
Chemical
Emulsifying agent (Brit. 298823) in making—
Pharmaceuticals.
Dye
Emulsifying agent (Brit. 264860) in making—
Color lakes.
Fats and Oils
Dispersive agent (Brit. 298823) in making—.
Lubricating and greasing compositions.
Solvents for fats.
Ink
Dispersive agent (Brit. 264860) in making—
Printing inks.
Insecticide
Dispersive agent (Brit. 298823) in making—
Insecticidal and germicidal compositions.
Miscellaneous
Emulsifying agent (Brit. 298823) in making—
Washing compositions.
Paint and Varnish
Dispersive agent (Brit. 264860) in making—
Paints, varnishes.
Perfumery
Emulsifying agent (Brit. 264860) in making—
Cosmetics, perfumes.
Plastics
Dispersive agent (Brit. 298823) in making—
Compounds of cellulose esters and ethers.
Resins and Waxes
Dispersive agent (Brit. 264860) in making—
Artificial resin compositions, natural resin compositions.
Rubber
Dispersive agent (Brit. 264860) in making various compositions.
Textile
Dispersive agent (Brit. 298823) in making—
Dye liquors containing dyestuffs, indigoes, and anthraquinone vat dyestuffs.
Dye liquors for rayon, wool, cotton, and silk.

Potassium Isoamylnaphthalenesulphonate
French: Isoamylenaphthalènesulphonate de potasse, Isoamylenaphthalènesulphonate potassique, Isoamylenaphthalènesulphonate de potassium.
German: Kaliumisoamylnaphtalinsulfonat, Isoamylnaphtalinsulfonsaeureskalium, Isoamylnaphtalinsulfonsaeurespotasche.
Chemical
Starting point in making—
Intermediates, pharmaceuticals.
Dye
Dispersive agent (Brit. 264860) in making—
Color lakes.
Fats and Oils
Dispersive agent (Brit. 298823) in making—
Lubricating and greasing compositions.
Solvents for fats.
Ink
Dispersive agent (Brit. 264860) in making—
Printing inks.
Insecticide
Emulsifying agent (Brit. 298823) in making—
Insecticidal and germicidal compositions.
Miscellaneous
Emulsifying agent (Brit. 298823) in making—
Washing compositions.
Paint and Varnish
Dispersive agent (Brit. 264860) in making—
Lacquers, paints, varnishes.
Perfume
Emulsifying agent (Brit. 298823) in making—
Cosmetics, perfumes.
Plastics
Emulsifying agent (Brit. 298823) in making—
Compounds of cellulose nitrate, cellulose acetate, and other esters and ethers of cellulose.
Resins and Waxes
Emulsifying agent (Brit. 264860) in making—
Artificial resin preparations, natural resin preparations.
Rubber
Dispersive agent (Brit. 264860) in making—
Rubber compositions.
Textile
Dispersive agent (Brit. 264860) in making—
Liquors containing sulphur dyestuffs, indigoes, and anthraquinone vat dyestuffs.
Liquors for dyeing rayon, wool, cotton, and silk.

Potassium Isobutenylxanthate
Insecticide
As an insecticide (Brit. 425192).
Ingredient (Brit. 425192) of—
Insecticidal compositions.
Metallurgical
Flotation agent (Brit. 425192) in—
Mining.
Rubber
Accelerator (Brit. 425192) in—
Vulcanizing processes.

Potassium Isobutylnaphthalenesulphonate
French: Isobutylenaphthalènesulphonate de potasse, Isobutylenaphthalènesulphonate potassique, Isobutylenaphthalènesulphonate de potassium.
German: Isobutylnaphtalinsulfonsaeureskalium, Isobutylnaphtalinsulfonsaeurespotasche, Kaliumisobutylnaphtalinsulfonat.
Chemical
Starting point in making various intermediates.
Starting point (Brit. 298823) in making—
Synthetic drugs.
Miscellaneous
As an emulsifying agent (Brit. 298823).
For uses, see under general heading: "Emulsifying agents."

Potassium Isopropylnaphthalenesulphonate
French: Isopropylenaphthalènesulphonate de potasse, Isopropylenaphthalènesulphonate potassique, Isopropylenaphthalènesulphonate de potassium.
German: Kaliumisopropylnaphtalinsulfonat, Isopropylnaphtalinsulfonsaeureskalium, Isopropylnaphtalinsulfonsaeurespotasche.
Chemical
Starting point in making—
Intermediates, pharmaceuticals.
Emulsifying agent (Brit. 298823) in making—
Pharmaceuticals.
Dye
Emulsifying agent (Brit. 264860) in making—
Color lakes.
Fats and Oils
Emulsifying agent (Brit. 298823) in making—
Lubricating and greasing compositions.
Solvents for fats.
Ink
Dispersive (Brit. 264860) in making
Printing inks.
Insecticide
Dispersive agent (Brit. 298823) in making—
Insecticidal and germicidal compositions.
Miscellaneous
Emulsifying agent (Brit. 298823) in making—
Washing compositions.
Paint and Varnish
Emulsifying agent (Brit. 264860) in making—
Lacquers, paints, varnishes.
Perfume
Emulsifying agent (Brit. 298823) in making—
Cosmetics, perfumes.

Potassium Isopropylnaphthalenesulphonate (Cont'd)

Plastics
Dispersive agent (Brit. 298823) in making—
Compounds of cellulose nitrate, cellulose acetate, and other cellulose esters and ethers.

Resins and Waxes
Emulsifying agent (Brit. 264860) in making—
Artificial resin preparations, natural resin preparations.

Rubber
Emulsifying agent (Brit. 264860) in making—
Rubber compositions.

Textile
Dispersive agent (Brit. 264860) in making—
Liquors containing sulphur dyestuffs, indigoes, and anthraquinone vat dyestuffs.
Liquors for rayon, wool, cotton, and silk.

Potassium Naphthalenetrisulphonate
French: Naphthalènetrisulfonate de postasse, Naphthalènetrisulfonate potassique.
German: Kaliumnaphtalintrisulfonat, Naphtalintrisulfonsaeureskalium.

Chemical
Reagent (Brit. 280945) in making derivatives with diazotized—
Anilin, azoxyanilin, benzidin, 2:5-chlorotoluidines, 4-chloro-2-toluidin, 5-chloro-2-toluidin, dianisidin, 2:5-dichloroanilin, meta-anisidin, metachloroanilin, metanitranilin, metanitroparatoluidin, metatoluidin, 4-nitro-2-anisidin, 5-nitro-2-anisidin, orthoanisidin, orthochloroanilin, orthonitranilin, orthotoluidin, para-anisidin, parachloroanilin, paranitranilin, paranitro-orthotoluidin, paratoluidin.

Potassium Naphthionate
French: Naphthionate de potasse.
German: Kaliumnaphthionat.

Miscellaneous
Dust-laying substance for treating roads (French 599497).

Potassium Naphthylthioglycolate

Chemical
Starting point in making various derivatives.

Dye
Reagent (Brit. 284288) in making thioindigoid dyestuffs with the aid of—
Acenaphthenequinone, alphaisatinanilide, 5:7-dibromoisatin.
Isatin, its homologs, substitution products, and alpha derivatives.
Orthodiketones.

Potassium-Nickel Cyanide

Chemical
Catalyst (Brit. 446411) in—
Halogenating unsaturated hydrocarbons.
Starting point (Brit. 446411) in making—
Catalysts with metal chlorides for halogenating unsaturated hydrocarbons.

Potassium 3-Nitrophthalimide

Analysis
Reagent for—
Reacting with organic halides to form crystalline compounds with definite melting points, by means of which organic halogen derivatives may be identified.

Potassium Pentamethylenedithiocarbamate

Disinfectant
As a bactericide (Australian 8103/32, Brit. 406979, U. S. 1972961).

Insecticide and Fungicide
As a fungicide (Australian 8103/32, Brit. 406979, U. S. 1972961).
As an insecticide (claimed effective against aphids) (Australian 8103/32, Brit. 406979, U. S. 1972961).

Potassium Pentylnaphthalenesulphonate
French: Pentylenaphthalènesulphonate de potasse, Pentylenaphthalènesulphonate potassique, Pentylenaphthalènesulphonate de potassium.
German: Kaliumpentylnaphtalinsulfonat, Pentylnaphtalinsulfonsäureskalium, Pentylnaphtalinsulfonsäurespotasche.

Miscellaneous
As an emulsifying agent (Brit. 264860).
For uses, see under general heading: "Emulsifying agents."

Potassium Permanganate
Synonyms: Permanganate of potash.
Latin: Hypermanganas kalicus, Hypermanganas potassicus, Kali hypermanganicum, Kalium permanganicum, Potassae permanganas, Potassii permanganas.
French: Permanganate de potasse.
German: Chamaeleon, Kaliumpermanganat, Übermangansäureskali.
Spanish: Permanganato potasico.
Italian: Permanganato di potassio.

Agriculture
Soil disinfectant.

Analysis
Oxidizing agent, reagent.

Beverage
Purifying agent for—
Carbon dioxide used in the manufacture of effervescent drinks.

Chemical
General oxidizing agent in many processes.
Oxidizing agent in—
Making saccharin, making synthetic pharmaceuticals, purifying synthetic methanol (U. S. 1744180), purifying vanillin (Brit. 319747).
Purifying agent (Brit. 398136) in making—
Aromatic alcohols, such as phenylethyl alcohol.

Construction
Setting-control agent (Brit. 405508) in making—
Betons, cements, concretes, mortars.

Disinfectant
Ingredient (Brit. 319776) of—
Disinfecting, deodorizing, and antiseptic compound.

Dye
Oxidizing agent in making—
Dyestuffs, intermediates.

Fats and Oils
Bleaching agent, decolorizing agent, deodorant, disinfectant, oxidizing agent.

Glues and Adhesives
Ingredient (U. S. 1833527) of—
Adhesive composition.

Leather
Bleaching agent, decolorizing agent, deodorant, disinfectant, oxidizing agent.

Metallurgical
Addition agent (U. S. 1878837) for—
Limestone and water used as a scouring composition for surfaces of metals to be electroplated.
Reagent in—
Producing gray colors on copper.

Miscellaneous
Absorbent in—
Military gas masks.
Bleaching agent, decolorizing agent, deodorant, disinfectant, germicide.

Paint and Varnish
Addition agent to—
Lithopone (to increase the whiteness of the pigment).

Paper
Oxidizing agent (Brit. 398730) in making—
Cottonlike fabric from sulphite cellulose.

Perfume
Oxidizing agent in making—
Synthetic perfumes.

Petroleum
Reagent in—
Testing oil-field water for lime (according to Macfadyen method).

Pharmaceutical
Caustic, deodorant, disinfectant, germicide, oxidizing agent.
Suggested in treatment of—
Alkaloid poisonings, bacterial affections, insect stings, snake bites.

Photographic
Oxidizing agent in various processes.

Resins and Waxes
Bleaching agent, decolorizing agent, oxidizing agent.

Soap
Activating agent (Brit. 395570) for—
Silver in disinfecting soaps.

Potassium Permanganate (Continued)

Textile
Bleaching agent for—
Cotton, flax, linen, silk.
Textile fabrics (in conjunction with sodium nitrite).
Bleaching agent (U. S. 1915952) in—
Finishing viscose rayon.
Reagent in—
Obtaining brown shades in dyeing and printing.
Reagent (U. S. 1903828) in making—
Artificial wool from jute.

Water and Sanitation
Bactericide, decolorizing agent, deodorant, disinfectant, germicide, oxidizing agent.
Oxidizing agent (U. S. 1915240) in—
Sewage treatment process.
Purifying agent.

Woodworking
As a preservative.
Reagent in—
Coloring wood brown shades.

Potassium Persulphate
French: Persulfate de potasse, Persulfate potassique.
German: Kaliumpersulfat, Perschwefelsaeureskalium.

Chemical
Oxidizing agent in making—
Aldehydes from alcohols, alizarin from hydroxyanthraquinone, dihydroxybenzoic acid from salicylic acid, ferric salts from ferrous salts, nitrohydroquinone from orthonitrophenol, pharmaceutical chemicals, purpurin from alizarin.
Starting point in making—
Oxysulphuric acid.

Dye
Oxidizing agent in making various dyestuffs.

Fats and Oils
As a bleaching agent.

Food
Bleaching agent in treating—
Flour.

Glue and Gelatin
As a bleaching agent.

Miscellaneous
As a bleaching agent.
As an oxidizing agent.
As a disinfectant for the hands and other sanitary purposes.

Perfumery
Ingredient of—
Hair dyes.

Pharmaceutical
In compounding and dispensing practice.

Photographic
As a hypo eliminator and reducer.

Soap
Bleaching agent in making—
Soft soaps.

Starch
Bleaching agent in treating—
Dextrins, starches.

Textile
Bleaching agent in treating—
Cottons and other textiles.

Potassium Phenate
Synonyms: Potassium phenolate, Potassium phenoxide.
French: Phénate de potasse, Phénate potassique, Phénolate de potasse, Phénolate potassique, Phénolate de potassium, Phénoxyde de potasse, Phénoxyde potassique, Phénoxyde de potassium.
German: Kaliumphenat, Kaliumphenolat, Phenolkalium, Phenolsaeureskalium.

Dye
Reagent (Brit. 388043) in making—
Indanthrene dyes.

Leather
Ingredient (Brit. 263473) of—
Liquors for dyeing.

Miscellaneous
Ingredient (Brit. 263473) of—
Liquors for dyeing hair and feathers.

Pharmaceutical
In compounding and dispensing practice.

Textile
——, *Dyeing and Printing*
Ingredient (Brit. 263473) of—
Liquors and pastes containing vat dyestuffs which are used in dyeing and printing acetate and other rayons in fabric or yarn form, and also mixtures of rayon with wool and silk.

Potassium Phosphate, Dibasic
Synonyms: Dipotassium orthophosphate, Potassium acid phosphate, Potassium hydrogen phosphate, Potassium monophosphate.
German: Dikaliumphosphat.

Construction
Ingredient of—
Mortars used for ornamental purposes.

Food
Ingredient of—
Baking powders.

Pharmaceutical
In compounding and dispensing practice.

Potassium Phosphotungstate
French: Phosphotungstate de potasse, Phosphotungstate potassique, Tungstophosphate de potasse, Tungstophosphate potassique.
German: Kaliumphosphowolfromat, Kaliumwolframphosphat, Phosphowolframsaeureskalium, Wolframphosphorsaeureskalium.

Chemical
Reagent (Brit. 275943) in making lakes with—
Para-aminobenzaldehyde.
4:4′-Tetramethyldiaminobenzhydrol.
4:4′-Tetramethyldiaminobenzophenone.
4:4′-Tetramethyldiaminodiphenylmethane.

Paint and Varnish
Ingredient (Brit. 275969) of—
Cellulose ester or ether, oil or spirit lacquers colored with basic dyestuffs.

Potassium Polysulphide
Synonyms: Liver of sulphur, Polysulphide of potash.
Latin: Hepar sulfuris, Kali sulfuratum, Potassi sulphuretum, Potassium sulphuret, Trisulfuretum potassicum.
French: Foie de soufre, Polysulfure de potasse, Polysulfure potassique, Polysulfure de potassium.
German: Kaliumpolysulfid, Kaliumsulfuret, Schwefelleber.
Spanish: Higado de azufre, Polisulfurato de potasa, Polisulfurato potasico.
Italian: Polisulfurato di potassio.

Chemical
Reducing agent in various processes.

Dye
Reducing agent in making various dyestuffs.

Fats and Oils
Reagent (Brit. 271553) in making—
Vulcanized oils.

Insecticide
As an insecticide and fungicide.
Ingredient of—
Insecticidal and fungicidal compositions.

Leather
Reagent in—
Dehairing hides.

Paper
Ingredient (Brit. 271553) of—
Compositions, containing rubber latex, used for treating paper and pulp.

Pharmaceutical
Ingredient of—
Parasitic pomades, sulphur baths, sulphurized lotions.

Rubber
Reagent (Brit. 271553) in treating—
Rubber latex.

Textile
Reagent in—
Denitrating nitro rayons, removing sulphur from viscose rayon filament.

Potassium Resinate

Miscellaneous
As a wetting agent (Brit. 411908).
For uses, see under general heading: "Wetting agents."

Potassium Resorcinate
French: Résorcinate de potasse, Résorcinate de potassium.
German: Kaliumresorcinat, Resorcinsaeureskalium.

Leather
Ingredient of vat liquors for dyeing various leathers (Brit. 263473).

Miscellaneous
Ingredient of vat liquors for dyeing and stenciling furs and hair (Brit. 263473).

Textile
——, *Dyeing and Printing*
Ingredient of (Brit. 263473) vat liquors for—
Cellulose acetate fabrics and yarns, cuprammonium rayon yarns and fabrics, nitro rayon yarns and fabrics, silk rayon yarns and fabrics, viscose rayon yarns and fabrics, wool rayon yarns and fabrics.

Potassium Ricinoleicsulphonate

Miscellaneous
As an emulsifying agent (Brit. 361732).
For uses, see under general heading: "Emulsifying agents."

Potassium Salt of Cholesteryl Sulphoacetate

Metallurgical
Frothing agent in—
Flotation concentration of minerals said to closely approach the ideal properties of a reagent for these purposes; namely:—(1) the formation of an abundant froth, but one not too persistent, at low concentrations; (2) as effective in acid mediums as in alkaline mediums; (3) insensitive to salts, even in high concentrations; (4) absolutely inert as a collector in regard to both sulphurized and nonsulphurized minerals; (5) its froth-forming properties should not be affected by the collecting agents, including the soap; (6) it should emulsify rapidly and have a dispersive action on all collecting reagents that are usually employed; by the use of this reagent the employment of new collectors, such as the insoluble paraffin oils and butyl sulpholeate, is practicable).

Potassium Selenate

Miscellaneous
Reagent in
Mothproofing feathers, furs, hair, and other articles.

Textile
Reagent in—
Mothproofing woolens, felts, carpets, rugs, and other textiles.

Potassium Silicofluoride
Synonyms: Potassium fluosilicate.
French: Fluosilicate de potassium, Silicofluorure de potassium.
German: Fluorsiliciumstoffsaeureskalium, Kaliumfluorsilicat, Kaliumsilicofluorid, Silicofluorstoffsaeureskalium, Silicofluorwasserstoffsaeureskalium.

Chemical
Starting point in making—
Sodium fluoride (U. S. 158189).

Metallurgical
Reagent in the treatment of—
Difficultly decomposable minerals, especially rare earth minerals, titanium minerals, zirconium minerals, and monazite sand (German 440274).

Pharmaceutical
In compounding and dispensing practice.

Potassium Silicomolybdate
Synonyms: Potassium molybdosilicate.
French: Molybdosilicate de potasse, Molybdosilicate potassique, Silicomolybdate de potasse, Silicomolybdate potassique.
German: Molybdokieselsaeureskalium, Siliciummolybdaensaeureskalium.

Dye
Reagent (Brit. 275943) in making color lakes with—
Para-aminobenzaldehyde.
4:4′-Tetramethyldiaminobenzhydrol.
4:4′-Tetramethyldiaminobenzophenone.
4:4′-Tetramethyldiaminodiphenylmethane.

Metallurgical
Starting point in making—
Metallic silicon.

Paint and Varnish
Ingredient of—
Cellulose ester or ether oil or spirit lacquers containing basic dyestuffs (Brit. 275969).

Potassium-Silver Cyanide
French: Cyanure de potasse-argent.
German: Kaliumsilbercyanid, Hydrocyansaeureskaliumsilber.

Sanitation
Disinfectant and ingredient of disinfecting compositions (U. S. 1606359).

Potassium Sulphate
Synonyms: Sulphate of potash, Salt of lemery, Vitriolated tartar.
Latin: Arcanum duplicatum, Kalium sulfuricum, Potasii sulphas, Sulfas potassicus, Sal kalicus, Tartarum vitrolatum, Sal duobus, Sal polychrestum glaseri.
French: Potasse vitrolée, Sulphate de potasse.
German: Kaliumsulfat, Schwefelsäureskali.
Spanish: Sulfato potasico.
Italian: Solfato di potassio.

Analysis
Reagent.

Chemical
Starting point in making—
Potash alum, potassium salts of acids and halogens.

Fertilizer
Ingredient of—
Fertilizer mixtures.
Source of potash.

Glass
Ingredient of—
Baths used in making frosted glass, flint optical glass.

Metallurgical
Ingredient (U. S. 1844969) of—
Coating compositions for welding rods used in the autogenous welding of aluminum and the like.

Pharmaceutical
In compounding and dispensing practice.
Ingredient of—
Artificial Carlsbad salts.
Suggested for use as—
Agent for drying up the milk, aperient.

Potassium-Sulphoricinoleate
Synonyms: Potassium sulphoricinate, Potassium thioricinate, Potassium thioricinolate.
French: Sulforicinoléate de potasse, Sulforicinolate potassique, Thioricinolate de potasse.
German: Kaliumsulforicinat, Kaliumsulforicinoleat, Kaliumsulforizinat, Kaliumthioricinat, Sulforicinoelsaeureskalium, Sulforicinusoelsaeureskalium, Sulforizinusoelsaeureskalium, Thioricinusoelsaeureskalium, Thiorizinoelsaeureskalium.

Textile
——, *Dyeing and Printing*
General assistant in dyeing and printing yarns and fabrics.
Ingredient of—
Liquor or paste for dyeing, printing, or stenciling acetate rayon threads and films and fabrics containing acetate rayon with the aid of—
4-Chloro-2-nitrophenylbenzylamine.
3:3′-Dinitrobenzidin.
3:3′-Dinitro-4:4′-diaminodiphenylmethane.
2:2′-Dinitro-4:4′-di(dimethylamine)-6:6′-ditorylmethane.
3:3′-Dinitro-4:4′-di(dimethylamine) diphenyl ketone.
3:3′-Dinitro-orthotoluidin.
2:4′-Dinitrophenylbenzylamine.
3-Nitro-4-aminodiphenyl ether.
3-Nitrobenzidin.
2-Nitrophenylbenzylamine.
4-Nitrophenylbenzylamine.
Various nitrodiphenyls, nitrobenzidines, nitrotolidines, nitrophenylbenzylamines, nitrophenylethers, nitrodiphenylmethanes, nitrobenzophenones.

Potassium Thioglycollate

Dye
Starting point (Brit. 284288) in making thioindigoid dyestuffs with the aid of—
Acenaphthaquinone, alphaisatinanilide, 5:7-dibromoisatin, isatin, isatin homologs, substitution products, orthodiketones.

Potassium Xanthate
French: Xanthate de potasse, Xanthate potassique.
German: Kaliumxanthogenat, Xanthogensäureskalium.
Metallurgical
Flotation agent in—
Ore concentration processes.
Textile
Reagent in—
Indigo printing.

Potassium-Zinc Cyanide
Chemical
Catalyst (Brit. 446411) in—
Halogenating unsaturated hydrocarbons.
Starting point (Brit. 446411) in making—
Catalysts with metal chlorides for halogenating unsaturated hydrocarbons.

Potato Starch
French: Fécule de pommes de terre.
German: Kartoffelstarke.
Agriculture
Ingredient of—
Cattle foods.
Analysis
Reagent in testing for—
Chlorine, copper, iodine, nitrous acid.
Brewing
Starting point in making—
Beer, fermented liquors.
Chemical
Ingredient of—
Colloidal preparations (added for the purpose of preventing precipitation).
Starting point in making—
Acetone by bacterial fermentation, acetylmethylcarbinol by fermentation (U. S. 1899094), alcoylated products (French 640174), dextrin and dextrin products, fusel oil by fermentation, lactic acid, levulinic acid, starch glycollate, starch iodide, solubilized starch.
Tanning agent by sulphonation with sulphuric acid (French 544253).
Construction
Emulsifying agent (Brit. 387657) in making—
Bituminous emulsions of coal, coaltar, water gas tar, tar oils or their distillates, and like substances, and water, used in the production of coating compositions, road-making compositions, and compositions mixable with fibrous materials to form pressed goods.
Ingredient of—
Compositions containing pitch, rosin soap (such as potassium resinate), oil, flour, used for road-surfacing purposes.
Dye
Ingredient (U. S. 1889491) of—
Household dye compositions for silk.
Distilling
Starting point in making various types of distilled liquors.
Electrical
Carrier and filler (Brit. 398638) for—
Exciting salts used in the manufacture of electrolytes used for rechargeable dry cells.
Ingredient (U. S. 1911400) of—
Coating compositions, containing ammonium and zinc chlorides, for paper used for lining electrical dry cells.
Explosives
Crystallizing and binding promoter (U. S. 1913344) in—
Moulded black powder explosive.
Ingredient of—
Gelatin dynamites, igniting composition for matches and other purposes (U. S. 1831760), permissibles for coal mining, regular nitroglycerin dynamites.
Starting point (U. S. 1908857) in making—
Nitrostarch dynamites, nitrostarch explosives.
Food
Absorbent carrier (U. S. 1913776) for—
Mixtures of organic peroxides used in flour bleaching.
As a foodstuff.
Filler (U. S. 1913044) in—
Bread-dough improving and bleaching agent.
Ingredient of—
Baking powders, candies, cocoa powders, cake powders, custard preparations, chocolate preparations, ice cream preparations and powders, sauces of various sorts (to make them thick), various culinary preparations, vegetarian foods.
Raw material in—
Biscuit, pastry, baking, and confectionery industries.
Fuel
Addition agent for—
Slurry from coal washing (to increase settling rate).
Binder in making—
Fuel briquets.
Reagent (German 389401) for—
Treating non-floatable constituents of coal (in combination with hydrochloric acid).
Glues and Adhesives
Ingredient of—
Cold-water glues, various adhesive pastes, wallpaper pastes, xanthate adhesives.
Starting point (French 648019) in making—
Glues in bead form.
Insecticide
Carrier for various vermin-killing substances.
Ingredient (U. S. 1891750) of—
Seed-treating insecticide.
Leather
Ingredient of—
Cleansing compositions.
Compositions used in the manufacture of artificial leather (French 558630).
Compositions, containing lime, calcium phenolate, and sodium hydroxide, used for softening and dehairing hides and skins (French 612409).
Vehicle for—
Holding tanning extract in the drum-tanning process.
Mechanical
Ingredient (U. S. 1720565) of—
Compositions used for the purpose of preventing incrustation of scale in boilers.
Miscellaneous
Emulsifying agent (Brit. 387657) in making—
Bituminous emulsions of coal, coaltar, water gas tar, tar oils or their distillates, and like substances, and water, used in the production of coating compositions and compositions mixable with fibrous materials to form pressed goods.
Ingredient of—
Compositions used in laundries for the dressing and sizing of fabrics after washing.
Compositions used for coating purposes prepared by the action of calcium chloride, calcium nitrate, zinc chloride, and magnesium chloride on the starch (French 557085).
Compositions in emulsified form (French 599908).
Compositions used for stiffening fabrics.
Compositions containing coloring matter, such as azo dyestuffs.
Compositions colored black and containing naphthalene and its derivatives (French 641442).
Dental impression material (U. S. 1897034).
Starch glazes.
Starting point in making—
Starch tablets.
Paint and Varnish
Fixative (French 616204) in making—
Whitewashes and starch coating compositions with the addition of sodium carbonate and nitrobenzene.
Starting material (U. S. 1833526) in making—
Nitrostarch lacquer compositions.
Reagent (Brit. 385139) in making—
Condensation products used as softening agents in cellulose lacquers.
Paper
Dispersing agent (U. S. 1903787) in making—
Waxed paper.
Ingredient of—
Compositions used for sizing different types of paper, particularly writing paper.
Compositions used in the manufacture of surface-coated paper.
Compositions used in the manufacture of pasteboard.
Neutralizing agent (U. S. 1903236) for—
Paper pulp while on the Fourdrinier wire.
Perfume
Ingredient of—
Massaging compositions (French 616204).
Perfumes, pomades, sachets, toilet powders.

Potato Starch (Continued)

Pharmaceutical
Binder in tablet mixtures.
Diluent, dusting powder.
In compounding and dispensing practice.

Plastics
Starting material (U. S. 1908485) in making—
Glycerin-carbohydrate plastic.

Printing
In bookbinding practice.

Rubber
Dispersing agent (Brit. 397270 and 397997) for—
Rubber, in coating articles with a smooth matt finish.
Ingredient (Brit. 397279) of—
Compositions for coating surface of rubber articles to produce a smooth matt finish.

Soap
Ingredient of—
Compositions containing carbon tetrachloride, glycerin, and the like, used for the dry cleaning of hands which have become stained with crankcase oil, tar, grease, paints (French 611895).
Detergent preparations containing potassium silicate.
Soapstock for special grades of soap.
Soft soaps (used as a filler).

Sugar
Starting point in making—
Burnt sugar or caramel, malt sugar, various syrups and mixtures, white glucose.

Textile
——, *Dyeing*
Ingredient of—
Dye bath for various yarns and fabrics.
——, *Finishing*
Ingredient of—
Compositions used for sizing cotton fabrics.
Compositions containing glucose, sodium silicate, glycerin, olive oil, and borax, used for starching knitted articles (French 649899).
Fireproofing compositions containing ammonium sulphate, sodium carbonate, boric acid, sodium biborate, used for treating rayons (French 595286).
Sizing compositions containing sodium resinate (French 523282).
Weighting compositions for treating calicoes, lace curtains, and other textiles.
——, *Manufacturing*
Ingredient of—
Spinning bath in making viscose rayon.
Size for—
Cotton yarns before weaving.
——, *Printing*
Ingredient of—
Printing pastes (added to thicken them).

6-Propaminobetanaphthol-4-sulphonic Acid

Dye
In dye syntheses.
Starting point (Brit. 445999) in making—
Chromable orthohydroxy azo dyes by coupling with orthohydroxydiazonium compounds, such as those derived from 6-nitro-2-aminoparacresol or 4-chlor-2-aminophenol-6-sulphonic acid.

Propandione-1:3-dioxime

Fuel
Primer (Brit. 429763) for—
Diesel engine fuel oils produced by the hydrogenation of coal.

Petroleum
Primer (Brit. 429763) for—
Diesel oils containing a high proportion of aromatic bodies.

Propane

Note: Propane, according to the purpose, may be used either alone or in admixture with butane or air.

Agriculture
Fuel for—
Farming machinery, gas refrigerators, heating and cooking equipment.
Orchard heating equipment used to prevent damage by frosting of citrous fruits and other crops.
Poultry equipment, such as incubators, brooders, disinfecting burners.
Stationary engines running pumps, lighting units, power units.

Analysis
As an extractant.
As a solvent.
Fuel for—
Burners, hot-plates, water stills, flashpoint testers, sterilizers, ovens, and other heating and heated equipment in laboratories.

Animal Products
Fuel for—
Cooking equipment in packing plants.

Automotive
Internal combustion fuel for—
Automobile engines in block testing and running-in operations.

Aviation
Ingredient of—
Zeppelin engine fuels, in admixture with hydrogen (U. S. 1936155).
Zeppelin engine fuels, in admixture with hydrogen or natural gas.

Bituminous Products
Precipitating agent (Brit. 409278) for—
Asphalts in hydrogenation residues obtained from coal, tars, and other materials.

Brewing
Fuel for—
De-pitching burners, keg-branding irons.

Ceramics
Fuel for—
China kilns, testing furnaces.

Chemical
As a low-boiling extracting solvent.
Starting point in making—
Aliphatic alcohols or organic esters thereof by subjecting to thermal decomposition in the presence of the vapor of an organic acid, preferably a lower aliphatic acid, such as acetic or propionic acid, and in the presence or absence of steam (Brit. 402060).
Organic chemicals.

Construction
Internal-combustion fuel for—
Ditch-diggers, excavating machinery, hauling equipment, hoisting equipment, power shovels, road-graders, trucks.

Fats and Oils
Extracting agent for—
Vegetable oils (claim is made that high yields of good quality, pale oils are obtained and the meal is easily freed from solvent).

Food
Fuel in—
Bakery plant equipment, canning plant equipment, coffee roasters, confectionery apparatus.
Cooking equipment in homes, hotels, restaurants, roadstands.
Dairy equipment, peanut roasters, ripening heaters for bananas and other fruits.

Fuel
Fuel for—
Battery and radiator torches, bench furnaces.
Burners of various types, such as ring burners, bar burners, jet burners, ribbon burners, cluster burners, furnace burners, furnace kindlers.
Industrial or domestic heating where artificial or natural gas is not available or where the supply is limited or of high cost, or not used for various reasons; can also be used as standby fuel or temporary substitute because the same burners or burning equipment is adaptable to all these fuels.
Industrial or domestic heating where dust and dirt incidental to the use of coal is not desirable.
Industrial or domestic heating where adequate coal-storage space is not available or must be avoided for various reasons.
Internal-combustion engines.
Internal-combustion power equipment operating mostly on full throttle.
Water-heaters of various kinds.

Glass
Fuel for—
Burners, glass furnaces, glassworking machinery.

Hospitals
Fuel for—
Bandage incinerators, coffee urns, constant burning devices, diet kitchen equipment, hot-plates, main kitchen equipment, steam-tables, sterilizers, stoves.

Propane (Continued)

Laundering
Fuel for—
Dryers, irons, mangles, pressing machines, small steam boilers.

Mechanical
Fuel for—
Stationary engines connected direct to generators as sources of regular power, boosters, or standby units.
Stationary engines running compressors, lighting units, pumps.

Metallurgical
Fuel for—
Blow torches, brazing torches, crucible furnaces, cutting torches, enameling ovens, japanning ovens, lead-melting pots, preheating torches, welding torches.
Gaseous fuel in—
Annealing processes, carburizing processes, heat-treating processes.
Substitute for—
Acetylene in steel industry.

Miscellaneous
Fuel for—
Barber shop equipment, beauty-shop equipment.
Cleaning, pressing, and tailoring equipment.
Dental equipment, doctors' office equipment, household incinerators.
Illuminating equipment, such as household lights, portable lanterns, gas floodlights.

Motor Transportation
Combination internal-combustion fuel and refrigerant for—
Refrigerated trucks.
Internal combustion fuel for—
Buses, industrial plant jitneys, trackless vehicles, trucks.

Paint and Varnish
Fuel for—
Varnish kettles.
Solvent in—
Lacquer formulation.
Starting point in making—
Black pigment by incomplete combustion.

Petroleum
Fuel for—
Internal-combustion engines running pumps on pipelines.
Stationary engines connected direct to generators as sources of regular power, boosters, or standby units.
Stationary engines running compressors or lighting units.
Precipitating agent for—
Asphalts in crude petroleum, or distillation, cracking, or destructive hydrogenation residues obtained from tars or mineral oils (Brit. 409278, U. S. 1948163 and 1948164).
Solvent for—
High molecular weight constituents in making high-quality lubricating oils (Brit. 422471).
Paraffinic fractions in refining mineral oils and making lubricating oils (Brit. 421123).
Paraffin in refining mineral oils (Brit. 390222, 408947, 408948, and 423303; U. S. 1977054, 1977055, 1948346, and 1943236).
Standby gas for—
Fuel purposes (in admixture with air).

Printing
Fuel for—
Glue pots, linotype burners, intertype burners, monotype burners, typemetal melting pots.

Railroading
Fuel for—
Brazing torches, buffet stoves, building heating, cooking and dining-car equipment, cutting equipment, engine-driven lighting and power generators, gas-fired switch heaters, gas refrigerators, gas service in restaurants and lunchrooms, heating passenger sections in cold weather, prime-movers, soldering torches.
Stationary engines operating electric generators, air compressors, water pumps, shop shafting.
Stationary power units on switching locomotives, construction locomotives, rail cars, trains, and locomotives (propane is especially suitable and is used as refrigerant in air-conditioning trains).
Steaming-up locomotives in terminals and stations where use of oil burners for this purpose is objectionable or impracticable and where high-pressure steam is not available around the yard or powerhouse.
Thawing torches, water heaters.

Refractories
Fuel for—
Furnaces.

Refrigeration
As a refrigerant.
Fuel in—
Gas refrigerators.

Rubber
Fuel for—
Burners for cleaning tire molds, vulcanizing equipment.

Textile
Fuel in—
Calendering processes, drying processes, singeing processes.

Utilities

——, *Gasmaking*
Enrichener for—
Manufactured gas in recarburation of domestic and industrial gases.
Heating agent for—
Underfired cokeovens (to reduce coke production).
Increaser of—
Gas production in coalgas, watergas, or oilgas plants.
Substitute for—
Gas oil for the carburetion of watergas.
Standby gas (in admixture with air) for—
Peak loads, utilities supplying natural gas.

——, *Power*
Fuel for—
Stationary engines connected direct to generators.
Stationary engines running compressors, lighting units.
Stationary engines as sources of regular power, boosters, or standby units.

Propanone Oxime
German: Propanonoxim.

Chemical
Starting point (Brit. 282083) in making—
2-Aminopropane, secondary amines.

Propenylguaethol

Chemical
Antioxidant for—
Sulphonated oils.

Fats and Oils
Antioxidant for—
Animal oils, fats, fatty substances, fish oils, vegetable oils.

Insecticides
Suggested as oxidation-retarding agent for—
Insecticidal oils and compositions containing oils.

Leather
Antioxidant for—
Dressing oils, sulphonated oils.

Mechanical
Antioxidant for—
Lubricants of all types.

Metallurgical
Suggested as antioxidant for—
Quenching oils.

Miscellaneous
Antioxidant for—
Sulphonated oils used in fur-dyeing.

Perfume
Antioxidant for—
Oils and fats used in cosmetic creams, pomades, lotions, and the like.

Petroleum
Antioxidant for—
Lubricating compositions, lubricating greases, lubricating oils of various kinds.

Soap
Antioxidant for—
Fats, fatty substances, fish oils, vegetable oils.

Textile
Antioxidant for—
Sulphonated oils.

1-Propionamido-4-aminoanthraquinone

Miscellaneous
Dyestuff (U. S. 1989133) for—
Cellulose acetate products (imparts shades of red).

Textile
Dyestuff (U. S. 1989133) for—
Cellulose acetate products (imparts shades of red).

Propionyl Carbamide
French: Carbamide de propionyle, Carbamide propionylique.
German: Propionylcarbamid.

Chemical
Reagent in making—
Pharmaceuticals and other derivatives.

Resins and Waxes
Starting point (Brit. 292912) in making synthetic resins with—
Acetylsalicylic acid, aliphatic dibasic acids, ammonium salicylate, anthranilic acid, benzoic acid, gallic acid, hydronaphthoic acid, magnesium salicylate, oxalic acid, phenolic dibasic acids, phthalic acid, salicylamide, salicylic acid, strontium salicylate, succinic acid.

Propionylhydroquinone

Petroleum
Stabilizing agent (Brit. 406195) for—
Cracked gasolines and other motor fuels.

Propionylphloroglucinol

Petroleum
Stabilizing agent (Brit. 406195) for—
Cracked gasolines and other motor fuels.

Propionylpyrocatechol

Petroleum
Stabilizing agent (Brit. 406195) for—
Cracked gasolines and other motor fuels.

Propionylpyrogallol

Petroleum
Stabilizing agent (Brit. 406195) for—
Cracked gasolines and other motor fuels.

Propionylresorcinol

Petroleum
Stabilizing agent (Brit. 406195) for—
Cracked gasolines and other motor fuels.

Propyl Acetate, Normal

Food
Ingredient of—
Fruit essences.

Perfume
Ingredient of—
Cosmetics and pomades, perfume preparations.

Soap
Perfume in—
Toilet soaps.

Propyl Aldehyde
Synonyms: Proprionic aldehyde, Propylic aldehyde.
French: Aldéhyde proprionique.
German: Propionaldehyd.

Chemical
Starting point (Brit. 263853) in making aldehyde-amine condensation products (vulcanization accelerators) with—
Anilin, ethylamine, ethylanilin, ethylenediamine, normalbutylamine, orthotolyldiguanide.
Starting point in making—
Methylcinnamic aldehyde.

Miscellaneous
Antiseptic and preservative.

Pharmaceutical
In compounding and dispensing practice.

Propyl Alphacrotonate
Synonyms: Alphacrotonic propyl ester.
French: Alphacrotonate de propyle, Alphacrotonate propylique, Éther d'alphacrotoniquepropylique.
German: Alphacrotonpropylaether, Alphacrotonsäurepropylester, Alphacrotonsäurespropyl, Propylalphacrotonat.

Chemical
Starting point in making various derivatives.

Miscellaneous
Plasticizer and solvent (Brit. 321258) for—
Cellulose acetate, cellulose esters and ethers, cellulose nitrate, rubber.
For uses, see under general heading: "Plasticizers."

Propylbenzene
Synonyms: Propylbenzol.
German: Propylbenzol.

Textile
——, *Dyeing and Printing*
Solvent in making—
Color compositions used in the dyeing, printing and stenciling of materials composed of, or containing, cellulose acetate (Brit. 269960).

Propyl Chloroacetate
French: Chloroacétate de propyle.
German: Chloressigsaeurepropylester, Chloressigsaeurespropyl, Propylchloracetat.

Dye
Reagent in making—
Stable, water-soluble vat dyestuffs derivatives (Brit. 263898).

Propyl Chlorosulphonate

Chemical
Starting point in making—
Sodium compound of glutaconaldehyde (German 438009).

Propyl Dimethylaminoisovaleryloxyisobutyrate Hydrobromide
French: Bromhydrate de diméthylamino-isovaléryloxyisobutyrate de propyle.
German: Bromwasserstoffsäuresdimethylaminoisovaleryloxyisobuttersäurespropyl.
Spanish: Bromhidrate de dimetilamino-isovalerioloxi-isobutirate de propil.
Italian: Bromidrato di dimetilamino-isovalerilisobutirato di propile.

Pharmaceutical
Suggested for use as—
Sedative.

Propylenechlorohydrin

Cellulose Products
Ingredient of—
Solvent mixtures for cellulose acetate.

Chemical
Starting point in making—
Propylene oxide.

Dye
Starting point (Brit. 263178) in making dyestuffs for acetate rayon, which are sodium salts of—
Alpha-aminoanthraquinone-2-mercaptan.
Alpha-amino-4-paratoluidoanthraquinone-2-mercaptan.
2:6-Diaminoanthraquinone-1:5-dimercaptan.
2:7-Diaminoanthraquinone-1:8-dimercaptan.
1:5-Diaminoanthraquinone-2-mercaptan.
1:8-Diaminoanthraquinone-2-mercaptan.
1:4-Diamino-3-chloroanthraquinone-2-mercaptan.
1:4:5:8-Tetraminoanthraquinone-2-mercaptan.

Propylene Dibromide
French: Dibromure de propylène.
German: Propylendibromid.

Chemical
Starting point in making—
Propylenediamine, propyleneglycol, propaldehyde.

Propylene Dichloride
Synonyms: 1:2-Dichloropropane, Dichloropropylene.
French: Dichloropropylène.
German: Dichlorpropylen.
Spanish: Dicloropropilano.
Italian: Dicloropropilene.

Chemical
As a solvent.
Intermediate in—
Organic synthesis.
Reagent in making—
Acids, alcohols, amines, nitriles.
Starting point in making—
Propyleneglycol.

Dry Cleaning
As a spot-removing agent.
As a solvent.
Ingredient of—
Cleaning compounds, dry-cleaning soaps, scouring compounds, spot-removing agents.

Fats, Oils, Waxes
Solvent for—
Fats, oils, waxes.

Glue and Adhesives
Solvent for—
Gums.

Propylene Dichloride (Continued)
Mechanical
Ingredient of—
Carbon remover, containing also isopropyl ether, ethylene dichloride, and chloronaphthalene, for treating internal combustion engines.
Miscellaneous
As a general solvent.
Ingredient of—
Fumigants.
Paint and Varnish
Solvent in—
Paint and varnish removers.
Resins
Solvent for—
Resins.
Rubber
As a solvent.
Soap
Solvent in—
Cleaning compounds, scouring compounds, special soaps.
Textile
Solvent for—
Brand marks on woolens and other textiles.
Fats, greases, gums, oils, paint, resins, waxes.

Propyleneglycol Monoformate
French: Monoformiate de propylèneglycole, Monoformiate propylèneglycollique.
German: Monoameisensäurepropylenglykolester, Monoameisensäurespropylenglykol, Propylenglykolformiat.
Cellulose Products
Solvent and plasticizer (Brit. 311795) for—
Cellulose acetate.
For uses, see under general heading: "Solvents."
Chemical
Starting point in making various derivatives.
Dye
Ingredient (Brit. 311795) of—
Dye pastes.
Ink
Ingredient (Brit. 311795) of—
Printing inks.

1-Propylenehydroxy-4-aminoanthraquinone
Synonyms: Alphapropyleneoxy-4-aminoanthraquinone.
French: Alphapropylènehydroxye-4-aminoanthraquinone, Alphapropylèneoxye-4-aminoanthraquinone.
German: Alphapropylenhydroxy-4-aminoanthrachinon, Alphapropylenoxy-4-aminoanthrachinon.
Chemical
Starting point in making—
Intermediates, pharmaceuticals.
Dye
Starting point (Brit. 285096) in making dyestuffs in the presence of dimethylanilin, nitrobenzene, orthodichlorobenzene, naphthalene, and the like, with the aid of—
Acetylparaphenylenediamine, 5-amino-2-methylbenzimidazole, benzidin and its derivatives and homologs, dimethylparaphenylenediamine, metanaphthylenediamine, metaphenylenediamine, metatoluylenediamine, metaxylidenediamine, orthonaphthylenediamine, orthophenylenediamine, orthotoluylenediamine, orthoxylidenediamine, paranaphthylenediamine, paraphenylenediamine, paratoluylenediamine, paraxylidenediamine.

Propylenethiourea
French: Thiourée de propylène, Thiourée propylènique.
German: Propylenthioharnstoff.
Chemical
Starting point in making—
Intermediates, pharmaceuticals.
Starting point (Brit. 314909) in making derivatives with—
Alkoxyalphanaphthalenesulphonic acid.
Alpha-amino-5-naphthol-7-sulphonic acid.
Alphanaphthylamine-4:8-disulphonic acid.
Alphanaphthylamine-4:6:8-trisulphonic acid.

Propylether Ethyleneglycol
Paint and Varnish
Ingredient of—
Preparations for removing lacquers and lacquer-enamels (U. S. 1618482).

Propylfurol
French: Furole de propyle, Furole propylique.
Cellulose Products
Solvent for—
Cellulose acetate, cellulose esters and ethers, cellulose nitrate.
For uses, see under general heading: "Solvents."
Chemical
General solvent.
Starting point in making—
Intermediates, pharmaceuticals.
Gums, Resins, Waxes
Solvent for various varnish gums and artificial and natural resins.

Propylidene Iodide
Chemical
Starting point in making intermediates.
Starting point (Brit. 353477) in making contrast mediums for x-ray photography with the aid of—
Ammonium sulphite, magnesium sulphite, monomethylamine sulphite, piperidin sulphite, piperazin sulphite, sodium sulphite.

Propyl Mandelate
French: Mandélate de propyle, Mandélate propylique.
German: Mandelsaeurepropylester, Mandelsaeurespropyl, Propylmandelat.
Paint and Varnish
Plasticizing agent (Brit. 270650) in making—
Cellulose ester and ether lacquers and varnishes.
Plastics
Plasticizing agent (Brit. 270650) in making—
Nitrocellulose plastics.

Propylmercaptan, Normal
Synonyms: Propanthiol-1, Primary propylmercaptan.
Insecticide and Fungicide
Fumigant and insecticide for—
Rice weevils (*Sitophilus oryza L.*).

Propylmercaptoalphanaphthol
Chemical
Starting point in making—
Intermediates, pharmaceuticals.
Dye
Starting point (Brit. 291825) in making indigoid dyestuffs with the aid of—
5:7-Dibromoisatin anilide, 5:7-dibromoisatin bromide, 5:7-dibromoisatin chloride, 5:7-dichloroisatin anilide, 5:7-dichloroisatin bromide, 5:7-dichloroisatin chloride, isatin anilide, isatin alpha-anil, isatin beta-anil, isatin bromide, isatin chloride, reactive derivatives of isatin.

Propyl Naphthenate, Normal
Miscellaneous
As an emulsifying agent (Brit. 359116).
For uses, see under general heading: "Emulsifying agents."

Propyl Nitrate
Chemical
Reagent in—
Organic syntheses.
Fuel
Primer (Brit. 404682) in—
Diesel engine fuels (used in conjunction with other primers consisting of organic bromides or organic compounds whose function is that of reducing the spontaneous ignition temperature).
Reducer (Brit. 404682) of—
Delay period in diesel engine fuels.

Propylolamine
German: Propylolamin.
Chemical
Starting point in making—
Pharmaceuticals and other derivatives.
Fats and Oils
Dispersive agent (Brit. 340294) in making—
Nonfreezing lubricating compositions, containing animal or vegetable oils and fats, as well as ethyleneglycol or its esters, borax, benzyl alcohol.
Special lubricating compositions of the above type for use on locomotive axles, railway switches, hydraulic presses, and hydraulic brakes.

Propylolamine (Continued)

Electrical

Dispersive agent (Brit. 340294) in making—
Special lubricating compositions for use in electric switches.

Miscellaneous

Ingredient (Brit. 340294) of—
Compositions, containing vegetable, animal or mineral oils and greases, used as rust preventives.

Petroleum

Ingredient (Brit. 340294) of—
Special lubricating compositions containing mineral oils and greases.

Propyl Parahydroxybenzoate

Synonyms: Propyl paraoxybenzoate.
French: Parahydroxyebenzoate de propyle, Parahydroxyebenzoate propylique, Paraoxyebenzoate de propyle, Paraoxyebenzoate propylique.
German: Parahydroxybenzoesäurepropylester, Parahydroxybenzoesäurespropyl, Paraoxybenzoesäurepropylester, Paraoxybenzoesäurespropyl.

Food

Preservative for various preparations.

Pharmaceutical

In compounding and dispensing practice.

Sanitation

Antiseptic and disinfectant for various purposes.

Soap

Ingredient of—
Antiseptic and disinfectant soaps.

Propylphenyl Acetate

Synonyms: Hydroxycinnamyl acetate.
French: Acétate de hydrocinnamyle, Acétate hydrocinnamylique, Acétate de propyle et de phényle, Acétate propylique-phénylique.
German: Essigsäurehydroxycinnamylester, Essigsäurephenylpropylester, Essigsäureshydroxycinnamyl, Essigsäuresphenylpropyl, Hydroxycinnamylacetat, Hydroxycinnamylazetat, Phenylessigsäurepropylester, Phenylessigsäurespropyl, Phenylpropylacetat, Phenylpropylazetat, Propylphenylacetat.
Spanish: Acetato de propil y phenil.
Italian: Acetato di propile ed phenile.

Perfume

Ingredient of the following odors—
Hyacinth, lily of the valley, mignonette, rose.
Perfume in—
Various cosmetics.

Soap

Perfume in—
Toilet soaps.

Propylresorcinol

Textile

Inhibitor (Brit. 446404) of—
Acidity and mould development in textile lubricants during storage of the lubricant or fabric.

Propylsulphuric Acid Chloride

French: Chlorure d'acide propylesulphurique.
German: Propylschwefelsaeureschlorid.

Dye

Starting point (Brit. 271533) in making vat dyestuffs with—
Anthraquinone-1:2, flavanthrone, indanthrone, naphthacridin, thioindigo.

Propyltetrahydronaphthalenecarboxylic Acid

French: Acide de propyletétrahydronaphthalènecarbonique, Acide de propyletétrahydronaphthalènecarboxylique.
German: Propyltetrahydronaphtalincarbonsaeure.

Chemical

Ingredient of—
Emulsifying and dissolving mediums used in various chemical processes (German 432942).

Miscellaneous

Ingredient of—
Emulsifying and dissolving mediums used in various processes (German 432942).

Propyl Thiosalicylate

Synonyms: Propyl sulphosalicylate.
French: Sulfosalicylate de propyle, Sulfosalicylate propylique, Thiosalicylate propylique.
German: Propylsulfosalicylat, Propylthiosalicylat, Sulfosalicylsaeurespropyl, Thiosalicylsaeurepropylester, Thiosalicylsaeurespropyl.

Chemical

Starting point (Brit. 282427) in making synthetic drugs with—
Oxides and other salts of gold, silver, arsenic, bismuth, and antimony.

Protocatechuic Aldehyde-3-ethyl Ether

French: Éther de protocatéchuique-aldéhyde-3-éthyle.
German: Protokatechualdehyd-3-aethylaether.

Chemical

Starting point in making—
Aromatics.

Food

Used in the place of vanillin for flavoring foods and in making flavoring compositions.

Perfume

Ingredient of—
Perfumes.
Perfume for—
Cosmetics.

Prussian Blue

Synonyms: Berlin blue, Ferric ferrocyanide, Insoluble iron cyanide, Iron blue, Iron ferrocyanide.
French: Bleu de Berlin, Bleu de Prusse, Ferrocyanure de fer insoluble, Ferrocyanure ferrique.
German: Berlinerblau, Eisenferrocyanid, Eisenferrozyanid, Unloslicheseisenferrocyanid, Unloslichesferriferrocyanid, Unloslichespreussischblau.

Chemical

Starting point in making—
Pigments.

Explosives

Blue pigment for—
Match heads.

Ink

Blue pigment in making—
Blue inks (in oxalic acid solution).

Fertilizer

Ingredient of—
Fertilizing compositions.

Miscellaneous

Pigment in making—
Laundry blue (used in oxalic acid solution).

Paint and Varnish

As a dry color.
Ingredient of—
Oil colors, water colors.

Paper

Pigment in making—
Colored paper.

Pharmaceutical

In compounding and dispensing practice.

Soap

Pigment for—
Coloring and mottling soaps.

Textile

——, *Dyeing*

Pigment for—
Yarns and fabrics.

——, *Printing*

Printing pigment for—
Cottons.

Pseudocumenedisulphonic Acid

French: Acide de pseudocumènedisulphonique.
German: Pseudocumendisulfonsaeure.

Chemical

Reagent (Brit. 263873) in making—
Emulsions with aromatic hydrocarbons.
Fat-solvents in emulsified form.
Terpene emulsions.
Starting point in making—
Esters and salts.

Fats and Oils

Reagent (Brit. 263873) in making—
Emulsions.

Leather

Reagent (Brit. 263873) in making—
Emulsified impregnating compositions.
Emulsified tanning preparations.

Pseudocumenedisulphonic Acid (Continued)

Miscellaneous
Reagent (Brit. 263873) in making—
Emulsified washing and cleansing compositions.

Paper
Reagent (Brit. 263873) in making—
Emulsified preparations for treating paper and cardboard.

Petroleum
Reagent (Brit. 263873) in making—
Emulsions of petroleum or its distillates.

Resins and Waxes
Reagent (Brit. 263873) in making—
Emulsions containing natural and synthetic resins.

Textile
——, *Dyeing*
Ingredient (Brit. 263873) of—
Acid dye baths.

——, *Finishing*
Ingredient (Brit. 263873) of—
Finishing and wetting compositions.

——, *Manufacturing*
Ingredient (Brit. 263873) of—
Wool-carbonizing liquors.

Pumice
Synonyms: Obsidian, Pumice stone.
Latin: Lapis pumicis, Pumex.
French: Pierre-ponce, Ponce.
German: Bimstein.
Spanish: Piumis, Pumis.
Italian: Tomice.

Abrasives
Abrasive in—
Wheels, discs, stones, buffers, and the like.
Ingredient of—
Abrasive powders, knife-polishing compounds, knife sharpeners, razor hones.

Analysis
Dehydrating agent for—
Viscous organic liquids.

Automotive
Smoothing agent in—
Automobile coachwork painting and decorating.

Building Construction
Filler for—
Walls and partitions.
Ingredient of—
Artificial granite, artificial stone, bricks, building blocks, heat insulations, resilient floorings, roofing compositions, sound insulations, special cements and concretes, stone floorings, stuccos, tiles, wood-finishing compositions.
Polishing agent for—
Marble, stone, wood.
Preventer of—
Dust formation by concrete and cement.

Cellulose Products
Catalyst carrier in making—
Solvents for celluloid, solvents for cellulose esters, nitrocellulose.

Chemical
Carrier for—
Catalysts.
Catalyst carrier in making—
Acetaldehyde, acetic acid, acetic anhydride, acetone, acetone oils, alcohols, aldehydes, ammonia, ammonia oxidation products, ammonium nitrate, ammonium sulphate, aromatic hydrocarbons, anilin, anthraquinone, benzene, butylmethyl ketone, chlorhydrins, chlorinated hydrocarbons, ethane, ethylene, ethylmethyl ketone, formaldehyde.
Hydrogenation products of acetylene, olefins, linoleic acid, linolein.
Ketones, maleic acid, maleic anhydride, methane, methyl ketone, methylpropyl ketone, naphthalene, nitric acid, nitrogen, nitrogen oxides, nitrous acid, olefins, oleic acid, olein, phthalaldehyde, phthaldehydic acid, phthalic acid, phthalic anhydride, phthalimide, propylene, succinic acid, sulphur trioxide.
Decolorizing agent for—
Sulphur dichloride, sulphur monochloride.
Purifying agent for—
Sulphur dioxide.

Coal Processing
Remover of—
Hydrogen sulphide from coke-oven gas.
Hydrogen sulphide from illuminating gas.
Hydrogen sulphide from producer gas.

Electrical
Process material in making—
Wet batteries.

Explosives
Absorbent for—
Explosive materials.

Fats and Oils
Catalyst carrier in hydrogenation processes for—
Fats of fish, animal, or vegetable origin.
Fatty acids.
Oils of fish, animal, or vegetable origin.
Olein.
Process material in removing—
Arsenic compounds from marine animal oils.
Catalyst poisons from fats and oils.
Chlorine compounds from marine animal oils.
Cyanide compounds from marine animal oils.
Sulphur compounds from marine animal oils.

Firefighting
Ingredient of—
Fireproofing preparations and insulations.

Food
Catalyst carrier in making—
Synthetic flavoring extracts.

Glass
Ingredient of—
Cheap glassware batches.
Polishing agent for—
Glassware.

Insecticide and Fungicide
Ingredient of repellents for—
Croton bugs, flies, insects, roaches.

Leather
Process material in making—
Patent leather.

Linoleum and Oilcloth
Smoothing agent for—
Linoleum, oilcloth.

Mechanical
Ingredient of—
Nonconducting packings.

Military
Absorbent in—
Gas masks.

Miscellaneous
Absorbent for—
Gases.
General inert filler.
Ingredient of—
Metal polishes, polishing pastes, repellents for rodents, smoothing pastes.
Preservative for—
Graphite crucibles.

Oral Hygiene
Ingredient of—
Dentifrices.

Paint and Varnish
Ingredient of—
Enamels, paints, varnishes.
Catalyst carrier in making—
Turpentine substitutes.

Paper
Process material in making—
Abrasive paper, glass paper.

Petroleum
Catalyst carrier in hydrogenation of—
Olefins, petroleum.
Remover of—
Hydrogen sulphide from natural gas, hydrogen sulphide from oil gas.

Plastics
Filler in—
Plastic masses.

Printing
Process material in making—
Lithographic stones.
Smoothing agent in—
Engraving processes, plating work.

Pumice (Continued)
Soap
Catalyst carrier in hydrogenation processes for—
Fats of fish, animal, or vegetable origin.
Fatty acids.
Oils of fish, animal, or vegetable origin.
Olein.
Ingredient of—
Hand soaps, scouring compositions, soap powders.
Process material in removing—
Arsenic compounds from marine animal oils.
Catalyst poisons from fats and oils.
Chlorine compounds from marine animal oils.
Cyanide compounds from marine animal oils.
Sulphur compounds from marine animal oils.

Wood
Ingredient of—
Wood finishing and polishing compositions.
Polishing abrasive in—
Wood finishing.

Purple of Cassius
Synonyms: Gold tin precipitate, Gold tin purple.
Ceramics
Coloring agent for—
Porcelain and chinaware.
Ingredient of—
Enamels.
Glass
As a coloring agent.
Ink
As a pigment.
Paint and Varnish
As a pigment.

Pyrazolanthrone
Chemical
Starting point in making—
Intermediates, pharmaceuticals.
Various other derivatives.
Starting point (Brit. 282375) in making alkylpyrazoleanthrones with—
Dibutyl sulphate, diethyl sulphate, diheptyl sulphate, dihexyl sulphate, dimethyl sulphate, dipropyl sulphate, ethyl bromide.
Dye
Starting point (Brit. 345728) in making dyestuffs with the aid of—
Alpha-aminoanthraquinone.
Alpha-amino-4-benzoylaminoanthraquinone.
Alpha-amino-5-benzoylaminoanthraquinone.
Alpha-aminoanthraquinone-2-aldehyde.
Benzyl-1-aminobenzanthrone.
Beta-aminoanthraquinone.
Carbazole.
1:4-Diaminoanthraquinone.
1:2-Diaminoanthraquinone.
Starting point (Brit. 263494) in making pyrazoleanthrone red vat dyestuffs by heating with the following compounds of paratoluene sulphonate—
Acetyl, allyl, anthranyl, benzoyl, benzyl, butyl, ethyl, heptyl, hexyl, lactyl, methyl, naphthyl, nonyl, octyl, phenyl, phthalyl, propionyl, propyl, salicyl, succinyl, sulphanyl, toluyl, valeryl, xylyl.

Pyrazolanthrone-2-carboxylic Bromide
French: Bromure de pyrazolanthrone-2-carbonique, Bromure de pyrazolanthrone-2-carboxylique.
German: Pyrazolanthron-2-carbonsäurebromid.
Spanish: Bromuro de pirazolantrone-2-carbonico, Bromuro de pirazolantrone-2-carboxilico.
Italian: Bromuro di pirazolantrone-2-carbossilico.
Chemical
Starting point in making various derivatives.
Dye
Starting point (Brit. 340334) in making vat dyestuffs with the aid of—
Alphamonoaminoanthraquinone.
Aminoanthrones.
Aminoanthrimides and their carboxylic derivatives.
Aminodibenzanthrones.
Aminodibenzpyrenequinones.
Aminopyranthrones.
Diaminoanthraquinone.

Pyrazolanthrone-2-carboxylic Chloride
French: Chlorure pyrazolanthrone-2-carbonique, Chlorure pyrazolanthrone-2-carboxylique.
German: Pyrazolanthron-2-carbonsäurechlorid.
Chemical
Starting point in making various derivatives.
Dye
Starting point (Brit. 340334) in making vat dyestuffs with the aid of—
Alphamonoaminoanthraquinone.
Aminoanthranthones.
Aminoanthrimides.
Aminodibenzanthrones.
Aminodibenzopyrenequinones.
Aminopyranthrones.
Diaminoanthraquinones.
Carbazolic derivatives of the above compounds.

5-Pyrazolone-3-carboxylic Allylester
French: Allyle-5-pyrazolone-3-carboxylate, Éther de 5-pyrazolone-3-carbonyleallylique, Éther de 5-pyrazolone-3-carboxyleallylique, 5-Pyrazolone-3-carboxylate allylique, 5-pyrazoline-3-carboxylate d'allyle.
German: Allyl-5-pyrazolin-3-carboxylat, 5-Pyrazolon-3-carbonsaeureallylester, 5-Pyrazolon-3-carbonsaeuresallyl.
Chemical
Starting point in making—
Intermediates, pharmaceuticals.
Dye
Starting point (Brit. 294583) in making dyestuffs with—
Alphanaphthylamine, diazotized.
Aminoazobenzenesulphonic acid.
Aminoazotoluenesulphonic acid.
1:2-Aminonaphthol-4-sulphonic acid.
Anilin, anilinsulphonic acid, anthranilic acid.
Benzidin, diazotized.
Betanaphthylamine, diazotized.
4-Chloro-2-aminonaphthol-5-sulphonic acid.
4-Chloro-2-aminophenol.
4-Chloro-2-aminophenol-5-carboxylic acid.
4-Chloro-2-aminophenol-6-sulphonic acid.
4:4′-Diaminodiphenylureadisulphonic acid.
Diaminodiphenylureas, tetrazotized.
Dianisidin, tetrazotized.
Dihydrotoluidin-2-sulphonic acid.
Metachloroanilin, diazotized.
Metadichloroanilin, diazotized.
Metanitranilin, diazotized.
Metanitroparatoluidin, diazotized.
Metaxylidin, diazotized.
2-Naphthylamine-6-sulphonic acid.
4-Nitro-2-aminophenol-5-sulphonic acid.
4-Nitro-2-aminophenol-6-sulphonic acid.
6-Nitro-2-aminophenol-4-sulphonic acid.
Orthoanilin, diazotized.
Orthochloroanilin, diazotized.
Orthonitranilin, diazotized.
Orthonitranilinparasulphamide, diazotized.
Orthonitroparatoluidin, diazotized.
Orthotoluidin, diazotized.
Orthoxylidin, diazotized.
Parachloroanilin, diazotized.
Parachloro-orthonitranilin, diazotized.
Paradichloroanilin, diazotized.
Paranitranilin, diazotized.
Paratoluidin, diazotized.
Paraxylidin, diazotized.
Picramic acid.
3-Sulpho-2-aminophenol-6-carboxylic acid.

5-Pyrazolone-3-carboxylic Amylester
Synonyms: Amyl 5-pyrazolone-3-carboxylate.
French: Éther de 5-pyrazolone-3-carbonyleamylique, Éther de 5-pyrazolone-3-carboxyleamylique, 5-Pyrazolone-3-carboxylate d'amyle, 5-Pyrazolone-3-carboxylateamylique.
German: Amyl-5-pyrazolon-3-carboxylat, 5-Pyrazolon-3-carbonsaeureamylester, 5-Pyrazolon-3-carbonsaeures-amyl.
Chemical
Starting point in making—
Intermediates, pharmaceuticals.
Dye
Starting point (Brit. 294583) in making dyestuffs with—
Alphanaphthylamine, diazotized.
Aminoazobenzenesulphonic acid.
Aminoazotoluenesulphonic acid.
1:2-Aminonaphthol-4-sulphonic acid.
Anilin, anilinsulphonic acid, anthranilic acid.
Benzidin, diazotized.
Betanaphthylamine, diazotized.
4-Chloro-2-aminonaphthol-5-sulphonic acid.

5-Pyrazolone-3-carboxylic Amylester (Continued)
4-Chloro-2-aminophenol.
4-Chloro-2-aminophenol-5-carboxylic acid.
4-Chloro-2-aminophenol-6-sulphonic acid.
4:4′-Diaminodiphenylureadisulphonic acid.
Diaminodiphenylureas, tetrazotized.
Dianisidin, tetrazotized.
Dihydrotoluidin-2-sulphonic acid.
Metachloroanilin, diazotized.
Metadichloroanilin, diazotized.
Metanitranilin, diazotized.
Metanitroparatoluidin, diazotized.
Metaxylidin, diazotized.
2-Naphthylamine-6-sulphonic acid.
4-Nitro-2-aminophenol-5-sulphonic acid.
4-Nitro-2-aminophenol-6-sulphonic acid.
6-Nitro-2-aminophenol-4-sulphonic acid.
Orthochloroanilin, diazotized.
Orthonitranilin, diazotized.
Orthonitranilinparasulphamide, diazotized.
Orthonitroparatoluidin, diazotized.
Orthotoluidin, diazotized.
Orthoxylidin, diazotized.
Parachloroanilin, diazotized.
Parachloro-orthonitranilin, diazotized.
Paradichloroanilin, diazotized.
Paranitranilin, diazotized.
Paratoluidin, diazotized.
Paraxylidin, diazotized.
Picramic acid.
3-Sulpho-2-aminophenol-6-carboxylic acid.

5-Pyrazolone-3-carboxylic Ethylester
Synonyms: Ethyl 5-pyrazolone-3-carboxylate.
French: Éther de 5-pyrazolone-3-carbonique-éthylique, Éther de 5-pyrazolone-3-carboxylique-éthylique, 5-Pyrazolone-3-carbonate d'éthyle, 5-Pyrazolone-3-carbonate éthylique, 5-Pyrazolone-3-carboxylate d'éthyle, 5-Pyrazolone-3-carboxylate éthylique.
German: Aethyl-5-pyrazolon-3-carboxylat, 5-Pyrazolon-3-carbonsaeureaethylester, 5-Pyrazolon-3-carbonsaeuresaethyl.

Chemical
Starting point in making various intermediates.

Dye
Starting point (Brit. 294583) in making dyestuffs with—
Alphanaphthylamine, diazotized.
Aminoazobenzenesulphonic acid.
Anilin, anilinsulphonic acid, anthranilic acid.
Benzidin, diazotized.
Betanaphthylamine, diazotized.
4-Chloro-2-aminophenol.
4-Chloro-2-aminophenol-5-carboxylic acid.
4-Chloro-2-aminophenol-5-sulphonic acid.
4-Chloro-2-aminophenol-6-sulphonic acid.
4:4′-Diaminodiphenylureadisulphonic acid.
Diaminodiphenylureas, tetrazotized.
Dianisidin, tetrazotized.
Dihydrothiotoluidin-2-sulphonic acid.
Metachloroanilin, diazotized.
Metadichloroanilin, diazotized.
Metanitranilin, diazotized.
Metanitroparatoluidin, diazotized.
Metaxylidin, diazotized.
2-Naphthylamine-6-sulphonic acid.
4-Nitro-2-aminophenol-5-sulphonic acid.
4-Nitro-2-aminophenol-6-sulphonic acid.
6-Nitro-2-aminophenol-4-sulphonic acid.
Orthochloroanilin, diazotized.
Orthonitroanilin, diazotized.
Orthonitranilinparasulphamide, diazotized.
Orthonitroparatoluidin, diazotized.
Orthotoluidin, diazotized.
Orthoxylidin, diazotized.
Parachloroanilin, diazotized.
Parachloro-orthonitranilin, diazotized.
Paradichloroanilin, diazotized.
Paranitranilin, diazotized.
Paratoluidin, diazotized.
Paraxylidin, diazotized.
Picramic acid.
3-Sulpho-2-aminophenol-6-carboxylic acid.

5-Pyrazolone-3-Carboxylic Heptylester
Synonyms: Heptyl-5-pyrazolone-3-carboxylate.
French: Éther de 5-pyrazolone-3-carboxyleheptylique, 5-Pyrazolone-3-carboxylate de heptyle, 5-Pyrazolone-3-carboxylateheptylique.
German: Heptyl-5-pyrazolon-3-carboxylat, 5-Pyrazolon-3-carbonsaeureheptylester, 5-Pyrazolon-3-carbonsaeuresheptyl.

Chemical
Starting point in making various intermediates.

Dye
Starting point (Brit. 294583) in making dyestuffs with—
Alphanaphthylamine, diazotized.
Aminoazobenzenesulphonic acid.
Aminoazotoluenesulphonic acid.
1:2-Aminonaphthol-4-sulphonic acid.
Anilin, anilinsulphonic acid, anthranilic acid.
Benzidin, diazotized.
Betanaphthylamine, diazotized.
4-Chloro-2-aminonaphthol-5-sulphonic acid.
4-Chloro-2-aminophenol.
4-Chloro-2-aminophenol-5-carboxylic acid.
4:4′-Diaminodiphenylureadisulphonic acid.
Diaminodiphenylureas, tetrazotized.
Dianisidin, tetrazotized.
Dihydrotoluidin-2-sulphonic acid.
Metachloronitranilin, diazotized.
Metadichloroanilin, diazotized.
Metanitranilin, diazotized.
Metanitroparatoluidin, diazotized.
Metaxylidin, diazotized.
2-Naphthylamine-6-sulphonic acid.
4-Nitro-2-aminophenol-5-sulphonic acid.
4-Nitro-2-aminophenol-6-sulphonic acid.
6-Nitro-2-aminophenol-4-sulphonic acid.
Orthochloroanilin, diazotized.
Orthonitranilin, diazotized.
Orthonitranilinparasulphamide, diazotized.
Orthonitroparatoluidin, diazotized.
Orthotoluidin, diazotized.
Orthoxylidin, diazotized.
Parachloroanilin, diazotized.
Parachloro-orthonitranilin, diazotized.
Paradichloroanilin, diazotized.
Paranitranilin, diazotized.
Paratoluidin, diazotized.
Paraxylidin, diazotized.
Picramic acid.
3-Sulpho-2-aminophenol-6-carboxylic acid.

5-Pyrazolone-3-carboxylic Hexylester
Synonyms: Hexyl-5-pyrazolone-3-carboxylate.
French: Éther de 5-pyrazolone-3-carboxylhexylique, 5-Pyrazolone-3-carboxylate d'hexyle, 5-Pyrazolone-3-carboxylatehexylique.
German: Hexyl-5-pyrazolon-3-carboxylat, 5-Pyrazolon-3-carbonsaeurehexylester, 5-Pyrazolon-3-carbonsaeureshexyl.

Chemical
Starting point in making various intermediates.

Dye
Starting point (Brit. 294583) in making dyestuffs with—
Alphanaphthylamine, diazotized.
Aminoazobenzenesulphonic acid.
Aminoazotoluenesulphonic acid.
1:2-Aminonaphthol-4-sulphonic acid.
Anilin, anilinsulphonic acid, anthranilic acid.
Benzidin, diazotized.
Betanaphthylamine, diazotized.
4-Chloro-2-aminonaphthol-5-sulphonic acid.
4-Chloro-2-aminophenol.
4-Chloro-2-aminophenol-5-carboxylic acid.
4-Chloro-2-aminophenol-6-sulphonic acid.
4:4′-Diaminodiphenylureadisulphonic acid.
Diaminodiphenylureas, tetrazotized.
Dianisidin, tetrazotized.
Dihydrotoluidin-2-sulphonic acid.
Metachloroanilin, diazotized.
Metadichloroanilin, diazotized.
Metanitranilin, diazotized.
Metanitroparatoluidin, diazotized.
Metaxylidin, diazotized.
2-Naphthylamine-6-sulphonic acid.
4-Nitro-2-aminophenol-5-sulphonic acid.
4-Nitro-2-aminophenol-6-sulphonic acid.
6-Nitro-2-aminophenol-4-sulphonic acid.
Orthochloroanilin, diazotized.
Orthonitranilin, diazotized.
Orthonitranilinparasulphamide, diazotized.
Orthonitroparatoluidin, diazotized.
Orthotoluidin, diazotized.
Orthoxylidin, diazotized.
Parachloroanilin, diazotized.
Parachloro-orthonitranilin, diazotized.

5-Pyrazolone-3-carboxylic Hexylester (Continued)
Paradichloroanilin, diazotized.
Paranitranilin, diazotized.
Paratoluidin, diazotized.
Paraxylidin, diazotized.
Picramic acid.
3-Sulpho-2-aminophenol-6-carboxylic acid.

5-Pyrazolone-3-carboxylic Isoallyl Ester

Chemical
Starting point in making various intermediates.

Dye
Starting point (Brit. 294583) in making dyestuffs with—
Alphanaphthylamine, diazotized.
Aminoazobenzenesulphonic acid.
Anilin, anilinsulphonic acid, anthranilic acid.
Benzidin, diazotized.
Betanaphthylamine, diazotized.
4-Chloro-2-aminophenol.
4-Chloro-2-aminophenol-5-carboxylic acid.
4-Chloro-2-aminophenol-5-sulphonic acid.
4-Chloro-2-aminophenol-6-sulphonic acid.
4:4′-Diaminodiphenylureadisulphonic acid.
Diaminodiphenylureas, tetrazotized.
Dianisidin, tetrazotized.
Dihydrotoluidin-2-sulphonic acid.
Metachloroanilin, diazotized.
Metadichloroanilin, diazotized.
Metanitranilin, diazotized.
Metanitroparatoluidin, diazotized.
Metaxylidin, diazotized.
2-Naphthylamine-6-sulphonic acid.
4-Nitro-2-aminophenol-5-sulphonic acid.
4-Nitro-2-aminophenol-6-sulphonic acid.
6-Nitro-2-aminophenol-4-sulphonic acid.
Orthochloroanilin, diazotized.
Orthonitroanilin, diazotized.
Orthonitranilinparasulphamide, diazotized.
Orthonitroparatoluidin, diazotized.
Orthotoluidin, diazotized.
Orthoxylidin, diazotized.
Parachloroanilin, diazotized.
Parachloro-orthonitranilin, diazotized.
Paradichloroanilin, diazotized.
Paranitranilin, diazotized.
Paratoluidin, diazotized.
Paraxylidin, diazotized.
Picramic acid.
3-Sulpho-2-aminophenol-6-carboxylic acid.

5-Pyrazolone-3-carboxylicmethylester
Synonyms: Methyl 5-pyrazolon-3-carboxylate.
French: Éther de 5-pyrazolone-3-carboniqueméthylique, Éther de 5-pyrazolone-3-carboxyleméthylique, 5-Pyrazolone-3-carbonate de méthyle, 5-Pyrazolone-3-carbonate méthylique, 5-Pyrazolone-3-carboxylate de méthyle, 5-Pyrazolone-3-carboxylate méthylique.
German: 5-Pyrazolon-3-carbonsaeuremethylester, 5-Pyrazolon-3-carbonsaeuresmethyl.

Chemical
Starting point in making various intermediates.

Dye
Starting point (Brit. 294583) in making dyestuffs with—
Alphanaphthylamine, diazotized.
Aminoazobenzenesulphonic acid.
Aminoazotoluenesulphonic acid.
Anilin, anilinsulphonic acid, anthranilic acid.
Benzidin, diazotized.
Betanaphthylamine, diazotized.
4-Chloro-2-aminonaphthol-5-sulphonic acid.
4-Chloro-2-aminophenol.
4-Chloro-2-aminophenol-5-carboxylic acid.
4-Chloro-2-aminophenol-6-sulphonic acid.
4:4′-Diaminodiphenylureadisulphonic acid.
Diaminodiphenylureas, tetrazotized.
Dianisidin, tetrazotized.
Dihydrothiotoluidin-2-sulphonic acid.
Metachloroanilin, diazotized.
Metadichloroanilin, diazotized.
Metanitranilin, diazotized.
Metanitroparatoluidin, diazotized.
Metaxylidin, diazotized.
2-Naphthylamine-6-sulphonic acid.
4-Nitro-2-aminophenol-5-sulphonic acid.
4-Nitro-2-aminophenol-6-sulphonic acid.
6-Nitro-2-aminophenol-4-sulphonic acid.
Orthochloroanilin, diazotized.
Orthonitranilin, diazotized.
Orthonitranilinparasulphamide, diazotized.
Orthonitroparatoluidin, diazotized.
Orthotoluidin, diazotized.
Orthoxylidin, diazotized.
Parachloroanilin, diazotized.
Parachloro-orthonitranilin, diazotized.
Paradichloroanilin, diazotized.
Paranitranilin, diazotized.
Paratoluidin, diazotized.
Paraxylidin, diazotized.
Picramic acid.
3-Sulpho-2-aminophenol-6-carboxylic acid.

Pyrethrum Flowers
Synonyms: Chrysanthrene insecticide, Insect flowers, Persian insect flowers, Persian pellitory.
French: Fleurs de pyrèthre insecticide, Pyrèthre insecticide.
German: Insekpulverbluethen.

Insecticide
Ingredient of—
Alcoholic insecticidal tinctures.
Insecticidal emulsions.
Insecticidal compositions with copper salts and sulphur for killing vermin in houses and on animals.
Insecticidal preparations for use against bedbugs, ants, flies.
Viticultural and horticultural insecticidal preparations.

Perfumery
Ingredient in making—
Hygienic lotions.

Soap
Starting point in making—
Special soap.

Resins and Waxes
Starting point in extracting—
Pyrethrum oleoresin.

Pyrethrum Oleoresin
French: Oléorésine pyrèthre.
German: Bertramoelharz, Speichelwurzeloelharz, Zahnwurzeloelharz.

Insecticide
Ingredient of—
Insecticidal compositions for domestic and animal industry use.

Pyridin
Synonyms: Pyridine, Pyridine base.

Analysis
As a reagent in various processes.

Chemical
Catalyst in making—
Acetyl compounds of phenolic groups, quinolinic anhydride from quinolinic acid, salicylosalicylic acid (displosal).
Denaturant for—
Industrial alcohol.
Reagent in making—
Carbonyl derivative of orthoaminophenol, guaiacol methylglycollate (monotal), lead chloride derivatives, picryl chloride.
Solvent in making—
Anhydrous metallic salts, beryllium chloride, diazonium derivatives, fluorene, quinine ethylcarbonate.
Solvent in purifying—
Anthracene.
Solvent in separating—
Anthracene from phenanthrene and carbazol.
1:5-Dinitronaphthalene from 1:8-dinitronaphthalene.
Starting point in making—
Addition products with carbonyl chloride-organic acid derivatives, piperidine.

Dye
Catalyst in making—
Leuco compounds of vat dyestuffs.
Solubilized products from vat dyestuffs by means of chlorosulphonic acid (Brit. 251491).
Reagent in making—
Purified indigo.

Glass
Solvent for—
Silver nitrate reagent in producing pictures, marks, and the like on glass (U. S. 1592429).

Illumination
Reagent in—
Denitration of mantles for incandescent gas lamps.

Pyridin (Continued)

Insecticide
Ingredient of various insecticidal and fungicidal compositions.

Leather
Depilating agent for preparing hides for tanning.

Metallurgical
Ingredient of—
Electrolytic bath for the deposition of platinum-nickel alloys.

Paint and Varnish
Ingredient of—
Paint and varnish removers.
Solvent in making—
Enamels, lacquers, paints, varnishes.

Plastics
Solvent in making—
Cellulose acetate by the interaction of cellulose and acetyl chloride.

Rubber
Accelerator of vulcanization.
Solvent in making—
Rubber cements, rubber solutions.

Sanitation
Ingredient of antiseptic and germicidal compositions.

Soap
Ingredient of special soaps, solvent in general processes.

Textile
——, *Dyeing*
Assist in—
Coloring of various materials.
Vat liquor to increase the dispersion of the dyestuff and to produce greater depth of color and greater fastness.
——, *Finishing*
Solvent in—
Producing pattern effects in woven goods by removing the rayon threads from rayon-cotton union fabrics (Brit. 237909).

Pyridin Oleate, Chlorinated

Lubricant
Stabilizing agent (Brit. 451412 and 453047) for—
Lubricating oils subjected to high pressures.
Top cylinder lubricating compositions.

Pyridin Stearate, Chlorinated

Lubricant
Stabilizing agent (Brit. 451412 and 453047) for—
Lubricating oils subjected to high pressures.
Top cylinder lubricating compositions.

3-Pyridylhydrazin

Chemical
Starting point (Brit. 259982) in making—
Benzaldehyde derivatives.

Pyrites Cinder
(Residues from the burning of pyrites).

Metallurgical
Source of—
Metals, such as copper, iron, zinc, silver.

Miscellaneous
As a weed-killer, spread about railway platforms and tracks to inhibit weed growth and so minimize danger of fire.

Pyrocatechinsulphonic Acid
French: Acide de pyrocatéchinsulphonique.
German: Brenzcatechinsulfonsaeure.

Chemical
Starting point in making—
Intermediates, pharmaceuticals, salts and esters.
Starting point (Brit. 295734) in making synthetic pharmaceuticals with oxides, hydroxides, or carbonates of—
Aluminum, antimony, arsenic, bismuth, cadmium, chromium, copper, iron, lead, manganese, tin, vanadium, zinc.

Pyrogallic Acid
Synonyms: Trihydroxybenzene, Pyrogallol.
Latin: Acidum pyrogallicum, Pyrogallolum.
French: Acide pyrogallique.
German: Brenzgallussäure, Pyrogallussäure.
Spanish: Acido pirogalico, Trioxibenzene.
Italian: Acido pirogallico, Triossibenzana.

Analysis
Absorbent for—
Oxygen in the analysis of flue gas, illuminating gas, coal gas, water gas, coke-oven gas, and other gases.
Active reducing agent in treating—
Salts of silver, gold, and mercury, even in the cold.
Reagent in—
Analyzing and detecting carbon monoxide (in blood), chloral hydrate, copper, diastase, lignin, lignified cell membrane, nitric acid, nitrous acid, oxygen, propeptone, sesame oil, sulphonal.
Determining nitric acid and nitrous acid.

Chemical
Protective colloid in making—
Colloidal solutions of metals.
Reagent (German 202561) in making—
Colloidal arsenic.
Reducing agent in—
Processes involving the reduction of silver and mercury salts.
Starting point in making—
Haemogallol (plus haemoglobin), pyrogallol monoacetate (eugallol) (German 104663), pyrogallol salicylate, pyrogallol triacetate, pyrogallolsulphonic acid, pyrophan, saligallol.
Sodium pyrogallolsulphonate.
Various intermediate chemicals, pharmaceutical chemicals and other salts and esters.

Dye
Starting point in making—
Alizarin yellow A, alizarin yellow C, anthracene yellow, anthraquinone dyestuffs.
Azo dyestuffs for use on yarns and fabrics mordanted with chromium salts.
Azochromin, azogallein, chrome brown RR, coerulein S, gallein (plus phthalic anhydride), monoazo dyestuffs, xanthone dyestuffs.

Electrical
Developing agent in—
Galvano-technology.

Leather
Mordant in—
Dyeing various types of leather.

Linoleum and Oilcloth
Ingredient (Brit. 321690) of—
Compositions containing hardened fatty oils, resins, naphthenic acids, or fats, used for coating purposes.

Metallurgical
Reagent in making—
Colloidal solutions of metals.

Miscellaneous
Coloring matter for—
Dyeing hair brown.
Ingredient of—
Compositions containing hardened fatty oils, resins, naphthenic acid, or fats, used as substitute for wax records (Brit. 321690).
Wax baths used for impregnating various products and compositions (added to prolong the life of the bath) (U. S. 1752933).
Reagent in—
Dyeing furs and skins to produce yellow shades.
Hair in black shades (used in conjunction with silver nitrate in alkaline solutions).

Paint and Varnish
Ingredient (Brit. 321690) of—
Compositions containing hardened fatty oil, resins, naphthenic acids, or fats, used in the manufacture of varnishes.

Perfume
Ingredient of—
Bath salts containing sassafras oil and dilute alcohol.
Hair-dyeing compositions.
Hair color restorers (used in connection with an alkaline solution of silver nitrate).

Petroleum
Reagent (Brit. 312774) in—
Treating petroleum distillates, such as kerosene and gasoline, for the purpose of preventing and removing discoloration.

Pharmaceutical
Suggested for use as antiseptic, for use in various skin diseases and as an ingredient of salves.

Photographic
Developer for—
Negatives, positives, and certain prints.

Pyrogallic Acid (Continued)

Printing
In process engraving and the litho trades.

Textile
Reagent in—
Producing indigo shades with the aid of ferrous sulphate.

Woodworking
Mordant in—
Dyeing wood.

Pyrogallol Acetate

Chemical
Starting point in making—
Pharmaceuticals and other derivatives.

Pharmaceutical
In compounding and dispensing practice.

Pyrogallol Ethylether

Photographic
Starting point (U. S. 2017295) in making—
Developers having no tendency to become oxidized.

Pyrogallol Methylether

Photographic
Starting point (U. S. 2017295) in making—
Developers having no tendency to become oxidized.

Pyroligneous Acid

Synonyms: Pyroligneous vinegar, Wood vinegar.
French: Acide pyrolignique, Vinaigre de bois.
German: Brenzessigsaeure, Holzessig.

Chemical
Starting point in making—
Acetic acid, calcium acetate, derivatives of acetic acid and methanol, methanol, potassium acetate, pyrolignite.

Dye
Reagent in making various synthetic dyestuffs.

Food
Starting point in making—
Vinegar.

Metallurgical
Starting point in making—
Iron liquor.

Miscellaneous
In veterinary practice.

Pharmaceutical
In compounding and dispensing practice.

Pyrrole Oleate, Chlorinated

Lubricant
Stabilizing agent (Brit. 451412 and 453047) for—
Lubricating oils subjected to high pressures.
Top cylinder lubricating compositions.

Pyrrole Stearate, Chlorinated

Lubricant
Stabilizing agent (Brit. 451412 and 453047) for—
Lubricating oils subjected to high pressures.
Top cylinder lubricating compositions.

Quartz

Synonyms: Quarz, Silex, Silica, Silicic oxide, Silicon dioxide.

Abrasive
Component of—
Finishing powders, flint paper, grinding pastes, grindstones, millstones, oilstones, polishing powders, sandpaper, scythestones, whetstones.

Cement
Raw material in making—
Magnesia or oxychloride cements.

Ceramics
Ingredient of—
Ceramic ware in general, added for the purpose of reducing the shrinkage in firing.
Enamels, glazes.
Raw material in making—
Art potteries, chemical porcelain, electrical porcelain, pottery bodies, sanitary ware, silica brick and similar brick, tableware in general, tiles, whiteware in general.

Chemical
Absorbent in making—
Compositions containing phenol and other coaltar products.
Absorbent for various chemical purposes.
Catalyst in making—
Alcohol from ethylene.
Clarifying agent in treating—
Various chemical products and waste water from chemical plants.
Deodorizing agent in treating—
Chemical effluents, various chemical solutions, and products and waste waters from chemical plants.
Filtering medium for treating—
Solutions of chemicals, chemical products, waste waters, effluents, and miscellaneous chemical substances.
Flux in making—
Phosphorus in free state, as an ingredient of the mixture of raw materials.
Packing for—
Chemical apparatus, particularly where corrosive acid liquors are being handled, such as towers, condensers, and absorbers.
Reagent in making—
Ultramarine.
Starting point in making—
Carborundum, as an ingredient of the raw material mixture fed to the electric furnace.
Colloidal silicon, silicate of soda (water glass) and other silicates, silicon fluoride.
As a base on which catalysts are deposited for making various organic compounds, including—
Acenaphthylene, acenaphthaquinone, bisacenaphthylidenedione, naphthaldehydic acid, naphthalic anhydride, and hemimellitic acid from acenaphthene (Brit. 295270).
Acetaldehyde from ethyl alcohol (Brit. 281307).
Acetic acid from ethyl alcohol (Brit. 281307).
Alcohols from aliphatic hydrocarbons (Brit. 281307).
Aldehydes or alcohols by the reduction of esters (Brit. 306471).
Alphacampholide by the reduction of camphoric acid (Brit. 306471).
Aldehydes and acids from toluene, orthochlorotoluene, orthobromotoluene, orthonitrotoluene, parachlorotoluene, parabromotoluene, paranitrotoluene, metachlorotoluene, metanitrotoluene, metabromotoluene, dichlorotoluenes, dinitrotoluenes, dibromotoluenes, nitrochlorotoluenes, nitrobromotoluenes, chlorobromotoluenes (Brit. 295270).
Aldehydes and acids from xylenes, pseudocumene, mesitylene, and paracymene (Brit. 281307).
Alphanaphthaquinone from naphthalene (Brit. 281307).
Anthraquinone from naphthalene (Brit. 295270).
Benzaldehyde and benzoic acid from toluene (Brit. 281307).
Benzoquinone from phenanthraquinone (Brit. 281307).
Benzyl alcohol by the reduction of benzaldehyde (Brit. 306471).
Benzyl alcohol or benzaldehyde or benzyl phthalide by the reduction of phthalic anhydride (Brit. 306471).
Butyl alcohol by the reduction of crotonaldehyde (Brit. 306471).
Chloroacetic acid from ethylenechlorohydrin (Brit. 295270).
Diphenic acid from ethyl alcohol (Brit. 281307).
Ethyl alcohol by the reduction of acetaldehyde (Brit. 306471).
Fluorenone from fluorene (Brit. 295270).
Formaldehyde by the reduction of methane or methanol (Brit. 306471).
Formaldehyde by the reduction of carbon dioxide or carbon monoxide (Brit. 306471).
Hydroxyl reduction compounds of anthraquinone, benzoquinone, and the like (Brit. 306471).
Isopropyl alcohol by the reduction of acetone (Brit. 306471).
Maleic and fumaric acids by the oxidation of toluene, benzene, phenols, tar phenols, or furfural, or from benzoquinone or phthalic anhydride (Brit. 295270).
Methane by the reduction of carbon dioxide or carbon monoxide (Brit. 306471).
Methanol by the reduction of carbon dioxide or carbon monoxide (Brit. 306471).
Naphthaldehydic acid, acenaphthaquinone, or bisacenaphthylidenedione from acenaphthylene (Brit. 281307).
Phenanthraquinone from phenanthrene or diphenic acid (Brit. 295270).
Phthalic acid and maleic acid from naphthalene (Brit. 295270).
Primary alcohols by the reduction of the corresponding aldehydes (Brit. 306471).

Quartz (Continued)
Propionic acid and butyric acid and higher alcohols, ketones, and acids by the reduction of carbon monoxide or carbon dioxide (Brit. 306471).
Reduction products of ketones, aldehydes, acids, esters, alcohols, ethers and other organic compounds containing oxygen (Brit. 306471).
Salicylic acid and salicylic aledhyde from cresol (Brit. 295270).
Secondary butyl alcohol by the reduction of methylethyl ketone (Brit. 306471).
Valeryl alcohol by the reduction of valeraldehyde (Brit. 306471).
Vanillin and vanillic acid from eugenol or isoeugenol (Brit. 295270).
As a base (Brit. 304640) on which catalysts are deposited for the production of various aromatic and aliphatic compounds, including—
Alphanaphthylamine from alphanitronaphthalene.
Amines from aliphatic nitro compounds, such as allyl nitriles or nitromethane.
Amylamine from pyridin.
Anilin, azo-oxybenzene, azobenzene, and hydrazobenzene from nitrobenzene by reduction.
Aminophenols from nitrophenols.
3-Aminopyridin from 3-nitropyridin.
Amino compound from the corresponding nitroanisole.
Amines from oximes, Schiff's base and nitriles.
Cyclohexamine, dicyclohexamine, and cyclohexylanilin from nitrobenzene.
Piperidin from pyridin, pyrrolidin from pyrrol, tetrahydroquinolin from quinolin.

Construction
General structural material.
Raw material in making—
Gypsum plaster board, stucco plaster board, stucco pebble dash finish on plaster cast surfaces.
Ingredient of—
Compositions used for lining electric furnaces and for making various electric equipment.

Explosives
Filler in—
Dynamites and permissible explosives, matchhead compositions.

Fats and Oils
Medium for—
Decolorizing, clarifying, and filtering oils, fats, and greases.

Fertilizer
Ingredient of—
Lime mixtures, nitrogenous mixtures.

Food
Ingredient of—
Ammonium persulphate compositions, used to increase the bleaching and preserving characteristics of the latter in the treatment of flour.

Glass
Abrasive for—
Grinding glass surfaces.
Polishing agent in making—
Glassware, plate glass.
Raw material in making—
Flint glass, frosted glass, glass in general, quartz glass.

Insecticide
Absorbent in making—
Various insecticidal, germicidal, and bactericidal preparations.

Leather
Reagent in certain manufacture processes.

Mechanical
Ingredient of—
Compositions used for making various electrical equipment.
Lining compositions used in electric furnaces and acid converters.
Lining compositions for paper mill equipment.
Material for making—
Ball mill linings and balls.
Linings for various grinding machines.
Linings for chemical equipment, such as digesters, evaporators, stills.

Metallurgical
Abrasive for—
Finishing metals, sand-blasting and cleansing castings.
Flux in smelting—
Basic oxides, copper ores.
Pyrites, added also for the purpose of removing the iron oxide formed.
Ingredient of—
Dusting compositions for treating molds prior to casting.
Preparations used for making molds for casting steel.
Preparations used in enamelling iron and steel ware.
Source of silicon in making—
Copper-silicon, ferro-silicon.

Miscellaneous
Ingredient of—
Buffing compositions, compositions containing asbestos.
Heat-insulating preparations for various purposes.
Marble scouring and polishing preparations.
Metal polishes.
Non-imflammable compositions, in combination with asbestos for making various products used for structural and other purposes.
Packing compositions, polishes of various sorts.
Preparations for the prevention of sticking of roofing papers.
Preparations for making roofing papers.
Preparations for finishing bone and pearl buttons.
Preparations for making sand belts, sandblasting preparations, scouring compositions, soundproof compositions for lining telephone booths.

Paint and Varnish
Filler in making—
Enamels, paints, pigments, wood fillers.
Filtering medium in treating—
Chinawood oil, linseed oil, soya bean oil.

Paper
Filler in making—
Special papers and pulp compositions, such as blotting papers.

Perfume
Ingredient of—
Dry rouges (Brit. 255713), tooth powders.

Petroleum
Filtering medium, deodorizing agent and clarifying agent in treating—
Refined products.

Plastics
Filler in making various compositions.

Refractory
Ingredient of—
Compositions used in the construction and lining of the hearths of reverberatory furnaces.
Compositions used in making coke-oven and open-hearth firebrick.
Compositions used in making foundry facings and partings.

Rubber
Filler in making—
Tires and other products.

Stone
Agent in grinding and polishing—
Marble and other stone.
Ingredient of—
Compositions used in making artificial stone.

Sugar
Filtering medium in the refining and purification of—
Beet sugar, cane sugar, molasses, syrups.

Soap
Filler and abrasive in making—
Cleansing powders and pastes, detergents of various kinds, floor cleansers, grit soaps, hand soaps, kitchen cleansers, mechanics' soaps, scouring compositions of all sorts, soap powders, wall cleansers.

Waxes and Resins
Filtering medium in refining various substances.

Water and Sanitation
Deodorizing, decolorizing, purifying, cleansing and clarifying agent in treating—
Potable waters, sewage, waste waters.

Wine
Filtering medium.

Quercitron Bark
Synonyms: Dyer's oak bark, Stone oak bark, Yellow oak bark.
Latin: Quercus tinctoria, Quercus velutina.
French: Écorce de chêne.
German: Eichenrinde.

Chemical
Starting point in making—
Quercetine, quercitrin, tannic acid, tanning extracts.

Quercitron Bark (Continued)
Pharmaceutical
In compounding and dispensing practice.
Textile
——, *Dyeing*
Dyestuff for fabrics and yarns.

Quercitron Extract
French: Extrait de quercitron.
Dye
Starting point in making—
Green lakes with basic dyestuffs.
Leather
Tanning agent in general practice.
Miscellaneous
Dyestuff for straw.
Paper
Stain in making—
Colored papers, wallpaper.
Textile
——, *Dyeing*
Dyestuff for—
Cotton yarns and fabrics, khaki uniform cloth, olive-drab uniform cloth, sails and tents, silk yarns and fabrics, wool yarns and fabrics.
——, *Printing*
Ingredient of—
Paste for producing dark-brown effects on fabrics.

Quinaldin
Synonyms: Alphamethylquinoline.
French: Alphaméthylequinoléine.
German: Chinaldin.
Spanish: Alfametilechinolina.
Ceramics
Ingredient (Brit. 371901) of—
Coating compositions containing nitrocellulose or other esters or ethers of cellulose (added for the purpose of increasing the resistance).
Chemical
Starting point in making various derivatives.
Dye
Starting point in making—
Quinaldin yellow, quinolin yellow.
Glass
Ingredient (Brit. 371901) of—
Coating compositions containing nitrocellulose or other esters or ethers of cellulose (added for the purpose of increasing the light resistance of the coating).
Metallurgical
Ingredient (Brit. 371901) of—
Coating compositions containing nitrocellulose or other esters or ethers of cellulose (added for the purpose of increasing the resistance to light).
Miscellaneous
Ingredient (Brit. 371901) of—
Coating compositions containing nitrocellulose or other esters or ethers of cellulose (added for the purpose of increasing the resistance to light).
Paint and Varnish
Ingredient (Brit. 371901) of—
Compositions containing nitrocellulose or other esters or ethers of cellulose (added for the purpose of increasing the resistance to light).

Quinazarin
Synonyms: Dihydroxyanthraquinone.
German: Chinizarin, Quinizarin.
Chemical
Starting point in making—
Diparachloroanilidoanthraquinone (Brit. 248874), hydroxychrysazin, hydroxyquinazarin, leucoquinazarin, monoparachloroanilidoanthraquinone (Brit. 248874), quinazarin acetate, quinazarinsulphonic acid.
Dye
Starting point in making—
Alizarin cyanin green, alizarin virisol, quinazarin blue, quinazarin green.

Quinhydrone
German: Chinhydron.
Analysis
Reagent in making—
Hydrogen electrode in determining concentration of hydrogen ions in liquids.

Quinidine
Synonyms: Betaquinine.
French: Quinidine.
German: Chindin, Krystallisierteschinidin.
Chemical
Starting point in making—
Quinidine salts with acids and halogens.
Insecticide
Ingredient of—
Mothproofing compositions (Brit. 263092).
Pharmaceutical
In compounding and dispensing practice.
Textile
——, *Finishing*
For mothproofing various fabrics.

Quinidine Oleate
French: Oléate de quinidine.
German: Chinidinoleat, Oleinsaeureschinidin.
Insecticide
Ingredient of—
Mothproofing compositions for treating furs and feathers (Brit. 263092).
Pharmaceutical
In compounding and dispensing practice.
Textile
——, *Miscellaneous*
Ingredient of—
Mothproofing compositions for woolen fabrics (Brit. 263092).

Quinine
Synonyms: Methylcupreine.
French: Chininum, Quinine hydratée.
German: Chinin.
Spanish: Quinina.
Italian: Chinina.
Chemical
Starting point in making—
Quinidine, quinine albuminate, quinine camphorate, quinine ferrocyanide, quinine salts of various acids.
Insecticide
Ingredient of—
Mothproofing compositions for treating furs and feathers (Brit. 263092).
Pharmaceutical
In compounding and dispensing practice.
Textile
——, *Miscellaneous*
Ingredient of—
Compositions for repelling moths (Brit. 263092).

Quinine Acetate
Latin: Acetas quinicus.
French: Acétate de quinine.
German: Chininacetat, Chininazetat, Essigsäurechininester, Essigsäureschinin.
Spanish: Acetato de quinina.
Italian: Acetato di chinina.
Chemical
Starting point in making—
Pharmaceutical derivatives.
Miscellaneous
Ingredient of—
Mothproofing compositions for hair and feathers and furs.
Pharmaceutical
In compounding and dispensing practice.
Textile
Ingredient of—
Mothproofing compositions for wool and felt.

Quinine Hydrochloride
Synonyms: Quinine chloride, Quinine muriate.
French: Chlorure de quinine, Hydrochlorure de quinine.
German: Chininchlorid, Chininhydrochlorid, Chininmuriat, Chlorchinin, Chlorwasserstoffsäurechininester, Chlorwasserstoffsäureschinin, Salzsäureschinin.
Spanish: Clorhidrato de quinina.
Italian: Cloridrato di chinina.
Analysis
Reagent in testing—
Carbon monoxide in blood, cellulose, phenol, hydrochloric acid, sodium carbonate in sodium bicarbonate.

Quinine Hydrochloride (Continued)

Miscellaneous
In veterinary medicine.
Ingredient (U. S. 1795676) of—
Silver polishing and cleaning compositions.

Pharmaceutical
In compounding and dispensing practice.

Quinine 3-Hydroxybetanaphthyloxyacetate

Pharmaceutical
Claimed (Brit. 439937) as—
Practically tasteless form of quinine.

Quinine Oleate

French: Oléate de quinine, Oléate quinique.
German: Chininoleat, Oleinsaeureschinin, Oleinsaeurechininester.

Insecticide
Ingredient of—
Mothproofing compositions for treating furs and feathers (Brit. 263092).

Pharmaceutical
In compounding and dispensing practice.

Textile
——, *Miscellaneous*
Ingredient of—
Mothproofing compositions for treating woolen fabrics (Brit. 263092).

Quinine Stearate

French: Stéarate de quinine.
German: Chininstearat, Stearinsaeurechininester, Stearinsaeureschinin.

Pharmaceutical
In compounding and dispensing practice.

Textile
Ingredient of—
Mothproofing compositions for woolen fabrics (Brit. 263092).

Quinine Sulphate

French: Sulphate de quinine.
German: Chininsulfat, Schwefelsaeureschinin.

Chemical
Starting point in making—
Quinine ethylcarbonate.

Perfumery
Ingredient of—
Hair lotions and pomades.

Pharmaceutical
In compounding and dispensing practice.

Quinoidine

French: Chinoidine, Quinoidine.
German: Chinoidin.

Metallurgical
Ingredient (Brit. 342235) of—
Pickling baths (for controlling the action of the acid in the bath on the metal).

Paint and Varnish
Ingredient (Brit. 342235) of—
Anticorrosive paints.

Pharmaceutical
In compounding and dispensing practice.

Quinoidine Borate

Metallurgical
Ingredient of—
Pickling baths (used as an inhibitor).

Paint and Varnish
Ingredient of—
Anticorrosion paints.

Pharmaceutical
Suggested for use as antipyretic, antiperiodic, astringent, and tonic.

Rubber
Accelerator in—
Vulcanization.

Quinoidine Citrate

French: Citrate de chinoidine, Citrate de quinoidine, Citrate quinoidinique.
German: Chinoidincitrat, Citronsäurechinoidinester, Citronsäureschinoidin.

Metallurgical
Ingredient (Brit. 342235) of—
Pickling baths (added for the purpose of controlling the action of the acid in the bath on the metal).

Paint and Varnish
Ingredient (Brit. 342235) of—
Anticorrosion paints.

Pharmaceutical
In compounding and dispensing practice.

Quinoidine Hydrochloride

Metallurgical
Ingredient of—
Pickling baths (used as an inhibitor).

Paint and Varnish
Ingredient of—
Anticorrosion paints.

Pharmaceutical
Suggested for use as antipyretic, astringent, antiperiodic, and tonic.

Quinoidine Sulphate

French: Sulphate de quinoidine, Sulphate quinoidinique.
German: Chinoidinsulfat, Schwefelsäurechinoidinester, Schwefelsäureschinoidin.

Metallurgical
Ingredient (Brit. 342235) of—
Pickling baths (added for the purpose of controlling the action of the acid in the bath on the metal).

Paint and Varnish
Ingredient (Brit. 342235) of—
Anticorrosion paints.

Pharmaceutical
In compounding and dispensing practice.

Quinolin

Synonyms: Leucoline.
French: Quinoléine.
German: Chinolin, Leukolin.

Chemical
Starting point in making—
Alphaoxyquinolin (carbosbyril).
Cyanine (sensitizer for photographic work).
Orthoquinolinsulphonic acid, paraquinolinsulphonic acid, quinosol (ortho-oxyquinolin sulphate).

Dye
Starting point in making lakes with—
Anthrapyrimidin-2-paratoluidosulphonic acid.
Azo dyestuffs.
1-Amino-4-para-acetaminoanilidoanthraquinone-2-sulphonic acid.
1:4-Diamino-2-phenoxyanthraquinonesulphonic acid.
1:4-Dihydroxy-5:8-diparatoluidoanthraquinonedisulphonic acid.
1:5-Dihydroxy-5:8-diparatoluidoanthraquinonedisulphonic acid.
1:5-Diparatoluidoanthraquinonedisulphonic acid.
4:8-Diparatoluidoanthraquinonedisulphonic acid.
Dyestuffs derived from orthotoluidin and fluorescein chloride.
1-Hydroxy-5-paratoluidoanthraquinonesulphonic acid.
Methylanthrapyridin-2-arylsulphonic acids.
Paranitrophenylazosalicylic acid.
Patent blue A.
Sodium-1-amino-4-anilidoanthraquinone-2-sulphonate.
Starting point in making—
Cyanin dyes, isocyanin dyes, isoquinolin dyes, quinolin dyes.

Insecticide
Ingredient of—
Insecticidal compositions, particularly those for use in viniculture.

Miscellaneous
Disinfectant in treating—
Anatomical specimens.

Pharmaceutical
In compounding and dispensing practice.

Quinolin Oleate, Chlorinated

Lubricant
Stabilizing agent (Brit. 451412 and 453047) for—
Lubricating oils subjected to high pressures.
Top cylinder lubricating compositions.

Quinolin Stearate, Chlorinated

Lubricant
Stabilizing agent (Brit. 451412 and 453047) for—
Lubricating oils subjected to high pressures.
Top cylinder lubricating compositions.

Quinolin Sulphocyanate
Synonyms: Quinolin rhodanate, Quinolin sulphocyanide.
French: Rhodanate de quinoléine, Sulfocyanate de quinoléine, Sulfocyanure de quinoléine, Thiocyanate de quinoléine.
German: Chinolinrhodanid, Chinolinsulfocyanat, Chinolinsulfocyanid, Chinolinthiocyanat, Rhodanwasserstoffsaeureschinolinester, Rhodanwasserstoffsaeureschinolin.

Chemical
Starting point in making—
Quinolin bisulphocyanate (crurin).

Pharmaceutical
In compounding and dispensing practice.

Sanitation
Antiseptic for various purposes.

4-Quinolylphenylurea-3:6-dicarboxylic Acid
French: Acide de 4-quinolinphényleurée-3:6-dicarbonique, Acide de 4-quinolinphényleurée-3:6-dicarboxylique.
German: 4-Chinolinphenylharnstoff-3:6-dicarbonsäure.

Chemical
Starting point in making—
Esters, salts, and other derivatives.
Starting point (Brit. 314909) in making pharmaceutical derivatives with the aid of—
Alkoxyalphanaphthalenesulphonic acid.
Alpha-amino-5-naphthol-7-sulphonic acid.
Alphanaphthylamine-4:6:8-trisulphonic acid.
4-Aminoacenapthene-3:5-disulphonic acid.
4-Aminoacenaphthene-3-sulphonic acid.
4-Aminoacenaphthene-5-sulphonic acid.
4-Aminoacenaphthene-trisulphonic acids.
Aminocarboxylic acids.
Aminoheterocyclic carboxylic acids.
1:8-Aminonaphthol-3:6-disulphonic acid.
Bromonitrobenzoyl chlorides.
Chloroalphanaphthalenesulphonic acids.
Chloronitrobenzoyl chlorides.
Iodonitrobenzoyl chlorides.
Nitroanisoyl chlorides.
Nitrobenzene sulphochlorides.
Nitrobenzoyl chlorides.
2-Nitrocinnamyl chloride.
3-Nitrocinnamyl chloride.
4-Nitrocinnamyl chloride.
1-Nitronaphthalene-5-sulphochloride.
1:5-Nitronaphthoyl chloride.
2-Nitrophenylacetyl chloride.
4-Nitrophenylacetyl chloride.
Nitrotoluyl chlorides.

Quinone
Synonyms: Chinone, Parabenzoquinone.

Chemical
As an oxidizing agent.
Starting point in making—
Bromides, bromoquinol, chlorides, chloranil, chloroquinol.
Colored, crystalline compounds, by reaction with phenols, phenolic ethers, amines, and complex hydrocarbons.
Fumaric acid (Brit. 295270).
Intermediates for pharmaceutical manufacture by condensations with 1:3-butadienes (French 677296).
Maleic acid (Brit. 295270).
Maleic acid, by reaction with benzene and gas containing oxygen (U. S. 1318632).
Nuclearly substituted derivatives with aromatic diazo compounds (German 508395).
Products with metallic sheen by condensation with aromatic nitroso compounds (German 563968).
Products by reaction with compounds containing the amino group.
Quinol (hydroquinone).
Sodium quinosulphonate by reduction with sodium sulphite.
Substituents with chlorine and bromine.

Dye
Starting point in making—
Dyes, intermediates.
Intermediates with substituted or unsubstituted aromatic amines (U. S. 1735432).
Intermediates by condensation with 1:3-butadienes (French 677296).

Germicide
Suggested for use as—
Germicide.

Glue and Gelatin
Insolubilizing agent for—
Gelatin (in boiling water).

Leather
As a tanning agent.

Paint and Varnish
Starting point (Brit. 277371 and 313094) in making—
Derivatives useful as pigments in nitrocellulose varnishes and lacquers.

Photographic
Reagent in—
Intensifying and toning silver images.
Photographic processes.

Textile
Increaser (U. S. 998370) of—
Strength and durability of animal textile fibers.

Radium Bromide
French: Bromure de radium.
German: Bromradium, Radiumbromid.
Spanish: Bromuro de radio.
Italian: Bromuro di radio.

Analysis
In chemical research experiments.

Chemical
Starting point in making other radium compounds.

Glass
In making special glass.

Food
In making preservative receptacles from mixtures of carnotite ore and white Portland cement, so as to prevent bacterial action through radio-activity.

Fertilizer
In making fertilizers and for other agricultural purposes.

Miscellaneous
For eliminating fire hazards in rubber works by the prevention of sparks of static electricity.
For carrying out refined scientific measurements.
For testing the minute leakage of air through rubberized fabric.
In the manufacture of drinking vessels designed to produce radio-active water.
In physical research experiments.

Paint and Varnish
As a luminous pigment, in admixture with calcium sulphide, used for painting watch and clock dials, electric switch buttons, keyholes, and so on.
Ingredient of—
Luminous paints.

Raisinseed Oil

Cosmetic
New vegetable oil for—
Cosmetic creams and lotions (said to offer advantages of lack of odor, tastelessness, and complete absence of yellow pigment).

Food
As a salad oil.

Raisinseed Presscake

Agriculture
As a cattlefeed.

Rapeseed Oil
Synonyms: Blown rapeseed oil, Colleseed oil, Collza oil, Rape oil, Rubsen oil, Rubsen seed oil.
Latin: Oleum brassicae, Oleum napi, Oleum rapae, Oleum raparum.
French: Huile de colza, Huile de navette, Huile de navette cuite, Huile de rabette, Huile de rabette cuite.
German: Colzaoel, Geblasene rapssamenoel, Geblasene repsoel, Geblasene ruebensamenoel, Kohloel, Kohlrapsoel, Rapsoel, Rappsamenoel, Repsoel, Rueboel, Ruebsenoel.
Spanish: Aciete de rabina.
Italian: Olio di colza, Olio di napi.

Chemical
Starting point in making—
Behenolic acid, erucic acid.

Rapeseed Oil (Continued)

Electrical
Ingredient (Brit. 273290) of—
Insulating enamels and compositions for wires and electrical machinery and devices.

Fats and Oils
Ingredient of—
Compounded cylinder oils.
Compounded compressor oils.
Starting point in making—
Boiled and blown rapeseed oils.
Sulphonated oils of the turkey red oil type.

Food
As a cooking and dressing oil.
Ingredient of—
Oleomargarins, various food compositions.

Glues and Adhesives
Ingredient (Brit. 273290) of—
Special cements and adhesive preparations for such use as cementing laminated mica.

Mechanical
Lubricant for—
Cylinders, steam engines.

Metallurgical
Reagent in—
Hardening steel, quenching steel plates.

Miscellaneous
Binder in making various compositions of matter.
Illuminant, especially in railway lamps and miners' safety lamps and in lamps used in churches.
Ingredient of—
Oil baths.

Paint and Varnish
Ingredient of—
Artists' colors, lacquers, paints, varnishes.
Starting point in making—
Paint and varnish bases from tetramethylthiuram disulphide (Brit. 321689).
Varnish bases (Brit. 273290).

Plastics
Ingredient (Brit. 273290) of—
Moldable plastic compositions.

Petroleum
Ingredient of—
Lubricating oils containing mineral oil and mineral distillates (added to increase the viscosity of the product).

Rubber
Ingredient of—
Compositions used as substitutes for rubber.

Soap
Starting point in making—
Soft soaps.

Sugar
Ingredient of—
Boiling mass in kettles (added to prevent foaming).

Textile
Oiling woolen yarns and fabrics.

Red Bole

Synonyms: Armenian bole, Red bolus.
Latin: Bolus armeniae.
French: Bol rouge.
German: Roter bolus, Rotkreide, Striegaver armenishe erde.

Glass
Ingredient of—
Polishing compositions.

Metallurgical
Ingredient of—
Polishing compositions.

Paint and Varnish
Pigment in making—
Enamels, paints, varnishes.

Perfumery
Coloring agent for—
Cosmetics, dentifrices.

Miscellaneous
Ingredient of—
Red crayons.

Rubber
Color for—
Mixtures, used in place of antimony sulphide.

Stone
Ingredient of—
Polishing compositions.

Red Hematite

Synonyms: Hematite rouge, Natural red oxide of iron, Red iron ore, Specular iron ore.
French: Hématite rouge, Rouge d'hématite.
German: Bluterz, Blutstein, Eisenglanz, Eisenglimmer, Haematit, Roteisenstein, Roter glaskopf.
Spanish: Hematita.

Fertilizer
Ingredient of—
Fertilizer compositions (used along with calcium cyanamid).

Gas
Reagent for—
Purifying coal gas, water gas, and coke-oven gas by the dry process.

Glass
Ingredient of—
Batch in making green-colored glass.
Compositions used for polishing glass.

Mechanical
Ingredient of—
Polishing compositions.

Metallurgical
Raw material in making—
Pig iron.

Paint and Varnish
Pigment in—
Freight-car and barn paints.
Structural iron and steel paints.
Starting point in making—
English reds, ochers.

Perfume
As a rouge.
Starting point in making—
Rouges.

Rubber
Pigment in—
Certain grades of rubber goods.

Red Lead

Synonyms: Lead oxide red, Plumbo-plumbix oxide.
Latin: Plumbi oxidum rubrum.
French: Deutoxide de plomb, Minium, Oxide rouge de plomb, Plomb rouge.
German: Bleirot, Mennige, Rotes bleioxyd.
Spanish: Minio.
Italian: Minio.

Note:—A higher oxide of lead than litharge, corresponding to the formula Pb_3O_4; its formula has also been written as Pb_2PbO_4. It is formed by the oxidation of litharge, and it is never a true red lead, but always contains some under-oxidized material—litharge.

Ceramic
Base material in making lead glazes for—
Insides of saggers, insulating porcelain, ornamental tile, stoneware.
Yellow ware, such as bowls, tubs, crocks, household utensils.
Substitute for litharge in making—
Acid-resisting cements, stoneware cements.

Chemical
Starting point in making—
Lead chemicals.

Electrical
Starting point in making—
Pastes for storage battery plates.

Glass
Base material in making—
Lead glass, flint glass.
Refractive agent in—
Automobile lamp lenses, camera lenses, cut glassware.
Glass of brilliancy, clearness, and quality.
Microscope lenses, optical lenses, searchlight lenses, tableware of good quality, telescope lenses.

Mechanical
Substitute for litharge in making—
Pipe joint cements.

Metal Fabricating
Substitute for litharge in making—
Enamel frits for enameled iron sanitary ware, stove parts, signs, and various other enamelled iron products (but not enamelled cooking utensils).

Red Lead (Continued)

Miscellaneous
Pigment in making—
Red pencils.

Paint and Varnish
Base material in making—
Red lead paints for the protection of iron, steel, and other metals.
Pigment.
Starting point in making—
Driers.

Red Oxide of Iron

Synonyms: Colcothar, English red, Ferric oxide red, Ferric trioxide, Hematite, Indian red, Iron oxide, Iron peroxide, Iron trioxide, Iron sesquioxide, Ironic oxide, Polishing crocus, Pompey red, Purple oxide, Red iron trioxide, Red oxide, Red stone, Red iron ore, Rouge, Venetian red.
Latin: Capit mortuum, Crocus martis, Crocus martis adstringens.
French: Oxyde de fer, Oxyde ferrique, Peroxyde de fer, Rouge, Rouge anglais, Rouge d'Angleterre, Rouge de Venise.
German: Eisenoxyd, Eisensesquioxyd, Eisenrot, Englischesrot, Ferrioxyd, Ferritrioxyd, Indianischesrot, Rotesstein, Venizianerrot.
Spanish: Oxido ferrico.

Analysis
Reagent in various operations.

Cement
Raw material in making—
Iron cements.

Ceramics
Abrasive for—
Polishing porcelain.
Ingredient of—
Ceramics, potteries.

Chemical
Catalyst in making—
Hydrochloric acid from chlorine and steam (German 427539).
Nitric acid by the oxidation of ammonia with oxygen or air.
Sulphuric acid or sulphur trioxide by the oxidation of sulphur dioxide with oxygen or air.
Ingredient of—
Compositions used in making chemical ware.
Reagent in making—
Hydrogen (French 606421).
Prepared calamine.
Starting point in making—
Ferrite compounds, such as calcium and copper.
Iron salts, magnetic oxide of iron.

Dye
Ingredient of—
Color compositions fitted by heating (French 604759).
Leather yellow.

Electrical
Ingredient of—
Compositions used in making electrodes.

Fertilizer
Ingredient of—
Compositions containing calcium cyanamid.

Food
Reagent in making—
Mineral waters.

Gas
Reagent for—
Purifying coal gas, water gas, and coke-oven gas by the dry process.

Glass
Abrasive for—
Polishing glass.
Ingredient of—
Batch to make green-colored glass.

Linoleum and Oilcloth
Pigment in—
Coating compositions.

Mechanical
Ingredient of—
Abrasive compositions, used for polishing precious and other metals.

Metallurgical
For metallurgical purposes, including making metallic iron.

Miscellaneous
Ingredient of—
Compositions used for polishing jewelry and precious stones.
Compositions used in making printers' rollers.
Liquid coating compositions (U. S. 1598688).

Paint and Varnish
As a dry color.
Pigment in—
Freight-car and barn paints.
Structural iron and steel paints.
Starting point in making—
English red, red ocher.
Various other red pigments.

Paper
Pigment in making—
Wallpaper.

Perfume
Ingredient of—
Grease paints, make-up preparation.

Pharmaceutical
In compounding and dispensing practice.

Plastics
Ingredient of—
Plastic fibrous compositions (Brit. 252112).

Refractory
Ingredient of—
Compositions used in making refractory products.

Rubber
Pigment in—
Compounding rubber.

Textile
As a dye.
Mordant for—
Dyeing with anilin black.

Resinate

See: Metal resinate, e.g., Silver resinate.

Resorcinol

Synonyms: Dihydroxybenzene, Dihydroxybenzol, Metadihydroxybenzene, Metadihydroxybenzol, Metadioxybenzene, Metadioxybenzol, Résorcin.
French: Métadioxybenzène, Métadihydroxybenzène, Résorcine.
German: Resorcin, Doppeltehydroxybenzol.
Spanish: Dihidroxibenzol, Dihidroxibenzene, Metadihydroxibenzol, Metadihydroxibenzene, Resorcina.
Italian: Di-idrossibenzene, Di-idrossibenzol, Metadi-idrossibenzol, Metadi-idrossibenzene, Resorcina.

Analysis
Reagent for—
Aconitine, aldehydes, aldoses, allyl alcohol, artificial honey, asparagine, beet sugar, cane sugar, caramel, chloral hydrate, chloric acid, chloroform, cocaine, cottonseed oil, formaldehyde, hydrochloric acid in gastric juice, invert sugar, iodoform, levulose, lignified cell tissue, lignin, mineral acids, naphthalene, narcine, nitric acid, nitrous acid, organic acids, phenols, quinic acid, saccharin, salvarsan, sesame oil, tartaric acid, wool, zinc.
Reagent in—
Detecting albumen.
Detecting cotton in woolen goods.
Determining cineol in essential oils.
Testing edible oils and fats.

Chemical
Reagent in making—
Antipyrin (resopyrin).
Betaiodoresorcinsulphonic acid (anusol).
Ethylmeta-aminophenol.
Meta-aminophenol (German 44792).
Metaoxydiphenylamine.
Orthobenzoic acid ether ester.
Polyformin, resorcin diacetate.
Resorcinolhexamethylenetetramine (hetralin).
Tannins with acetaldehyde (German 282313).
Various intermediates, pharmaceuticals, aromatics, and other organic chemicals.
Starting point in making—
Benzotrichloride.
Betaresorcylic acid (2:4-dioxybenzoic acid).
Butyrylresorcinol (Brit. 250893), caproylresorcinol (Brit. 250893), caprylylresorcinol (Brit. 250893), carbonyl compounds, compounds with iodoform.
Compounds with aldehydes, caffeine, acetylene.
Condensation products with isatin.
2:6-Dihydroxybenzoic acid.

Resorcinol (Continued)
Di-iodoresorcinolsulphonic acid potassium salt (picrol).
Dimethylaminophenol.
2:4-Dioxybenzaldehyde (resaldol), diphenic anhydride compounds, dodecylylresorcinol (Brit. 250893), euresol (resorcinol monoacetate), heptylylresorcinol (Brit. 250893), hexamethylenetetramine compounds.
Intermediates, isobutyrylresorcinol (Brit. 250893), isocaproylresorcinol (Brit. 250893), isovalerylresorcinol (Brit. 250893), organic chemicals.
Phenoresorcinol (made with the addition of phenol).
Photographic developers, pharmaceuticals, potassium hydroxide compounds, primary alcohol compounds, propionyl resorcinol (Brit. 250893).
Symmetrical diphenylmetaphenylenediamine.
Synthetic aromatics.
Tanning materials for all sorts of leathers by condensation with various aromatic and aliphatic aldehydes.
Tannoxyphenol, thioresorcinol, valerylresorcinol (Brit. 250893), xylidylmeta-aminophenol.

Dye
Reagent in—
Preventing precipitation of solutions of coloring matters by tannins.
Solvent for—
Basic dyestuffs.
Starting point (Brit. 278789) in making dyestuffs with—
Alpha-amino-3-acetylamino-4-phenol.
5-Chloro-1-amino-3-acetylamino-4-phenol.
Starting point (Brit. 343014) in making dyestuffs for use in varnishes, lacquers, and the like, with the aid of—
4-Chloro-2-aminophenol, 4-nitro-2-aminophenol.
Starting point in making—
Acid alizarin brown B, acid alizarin garnet R, acid eosin, acid rosanin A, acme yellow, azo corinth, azo dyes, azo phosphin, azo phosphin G, azo phosphin GO, bengal pink, bengal rose B, benzoin C, benzoin G, carmine naphtha J, chloramine orange G, chlorin, chrysolin, chrysoin, coerulein B, congo brown R, congo 4R, congo red 4R, coomassie union blacks, cotton red R, cotton red 4R, cyanosin, spirit-soluble.
Cyanosin B, dark green C, dinitrosoresorcin, diazo dyestuffs, diazogen black DE, eosin, eosin G, eosin S, eosin SP, eosin BN, eosin, spirit-soluble.
Erythrosin B, erythrosin G, fast acid violet B, fast acid violet A2R, fast blue R, fast brown, fast green, fluorescein dyestuffs, fluorescein phthalate A, Hessian brown BBN, indazin, iris blue, isodiphenyl black R, janus brown B, R and J, janus yellow G and R, lacmoid, methyl eosin, mikado orange.
Monoazo dyestuffs, new phosphin G, nitroso blue MR, nitroso dyestuffs, oxazin dyestuffs, phenocyanin R, phenocyanin TC, phenocyanin TV, phenocyanin VS, phloroglucin, phloxin P, phthalein dyestuffs, pyramidol brown BG, pyramidol brown T, pyramine brown T, pyronin colors, resazurin, resofurin.
Resorcin brown, resorcin blue, rhodamine B, rhodamine 12GF, rose bengal B, solid green O, succineins, sudan G, sulphur colors, stilbene dyestuffs, tetrakisazo dyestuffs, trisazo dyestuffs, tropaeolin, ultra-alizarin S, ultracyanin TV, B, and R.
Uranin.

Explosives
Reagent (German 282313) in making—
Detonating compounds by condensation with acetaldehyde in the presence of sulphuric acid.
Starting point (German 76511) in making—
Trinitroresorcinol.

Fats and Oils
Stabilizing agent in making—
Emulsions of various animal and vegetable fats and oils.

Leather
Reagent in—
Tanning.

Miscellaneous
Reagent in making—
Unbreakable phonograph plates and other articles by admixture with paperboard and other chemicals (French 593897).

Paint and Varnish
Preservative in—
Tempera colors containing yellow of egg.

Perfume
Ingredient of—
Antiseptic tooth powders, antiperspiration preparations, hair lotions, skin creams.

Petroleum
Reagent for—
Preventing discoloration of petroleum distillates, such as kerosene and gasoline (Brit. 312774).

Pharmaceutical
Suggested for use as antiseptic, antispasmodic, antipyretic, antiemetic, antizymotic; in treating insufflation in rhinolaryngology, vomiting, seasickness, asthma, dyspepsia, emphysema, frostbite, gastric ulcer, cholera, hay fever, diarrhoea, whooping cough, intestinal cystitis, diphtheria; as antiferment and bactericide.

Photographic
Sensitizer for—
Silver bromide-gelatin paper.

Plastics
Substitute for camphor in making celluloid.

Soap
Ingredient of—
Medicinal soaps.

Textile
——, *Dyeing*
As a developing agent.
Ingredient of—
Baths for dyeing browns with the aid of azidin orange D2R.
Baths for dyeing blacks with the aid of azidin black.
Baths for producing polychromin orange shades.
Baths containing basic colors.
Reagent in—
Preventing precipitation of dyes by tannins.
Producing nitroso solvent for basic colors.
Blue on fibers from mixtures which contain tannins.
Solvent for basic colors.

——, *Dyeing and Printing*
Solubilizing agent (Brit. 276100) in making dye liquors and printing pastes which contain the following dyestuffs—
Acridin dyestuffs.
Aminoanthraquinones, reduced and unreduced.
Anthraquinone dyestuffs, azins, azo dyestuffs, basic diarylmethane dyestuffs, basic triarylmethane dyestuffs, benzoquinoneanilides, chrome mordane dyestuffs, indigoids, naphthaquinoneanilides.
Naphthaquinones, reduced and unreduced.
Nitroarylamines, nitrodiarylamines, nitroarylphenols, nitrodiarylphenols, oxazins, pyridin dyestuffs, quinolin dyestuffs.
Quinoneimides, reduced and unreduced.
Sulphur dyestuffs, xanthene dyestuffs.

——, *Printing*
As a developer in printing pastes.
Ingredient of—
Nitroso blue printing pastes.
Nitroso blue slop-padding bath.
Pastes used in color discharge printing.
Printing pastes containing basic colors.
Printing paste used for discharge printing, containing rongalite and used in producing white discharges of basic colors in printing on cellulose acetate rayon.
Solvent in making—
Printing pastes (added for the purpose of avoiding precipitation of the color, particularly basic dyestuffs, by the tannin).
Printing pastes containing basic colors.

Resorcinol Diacetate

Cellulose Products
Plasticizer for—
Cellulose nitrate (nitrocellulose).
For uses, see under general heading: "Plasticizers."

Resorcinol Monoacetate

Synonyms: Eurosol, Resorcin acetate.
French: Acétate de résorcine, Acétate résorcinique, Monoacétate de résorcine, Monoacétate résorcinique.
German: Essigsäureresorcinester, Essigsäuresresorcin, Monoessigsäureresorcinester, Monoessigsäuresresorcin, Resorcinmonoacetat, Resorcinmonoazetat.
Spanish: Acetato de resorcina, Monoacetato de resorcina, Resorcina monoacetilata.
Italian: Acetato di resorcina, Monoacetato di resorcina.

Pharmaceutical
Suggested for use in the treatment of acne and other dermatological afflictions.

Resorcinol Monoacetate (Continued)

Plastics
Reagent (German 298806) in treating—
Cellulose acetate plastic compositions for the purpose of making them more pliable and resistant to the action of low temperatures.

Retene

Chemical
Starting point in making—
Intermediates, pharmaceuticals, and other derivatives.

Dye
Starting point (U. S. 1375238) in making—
Azo dyestuffs.

Rhizophora Bark Extract
Synonyms: Extract of Italian Somaliland mangrove bark.

Leather
New tanning agent.

Rhodinyl Acetate
Synonyms: Rhodinol acetate.
French: Acétate de rhodinol, Acétate de rhodinyle, Acétate rhodinylique.
German: Aethansäurerhodinylester, Aethansäuresrhodinyl, Methancarbonsäurerhodinylester, Methancarbonsäuresrhodinyl, Rhodinylacetat, Rhodinylazetat.
Spanish: Acetato de rodinil.
Italian: Acetato di rodinilo.

Chemical
Starting point in making various derivatives.

Perfume
Ingredient of—
Geranium perfumes, rose perfumes.
Various perfume compositions (added to freshen the composition and impart a fruity odor).
Perfume in—
Cosmetics.

Soap
Perfume in—
Toilet soaps.

Rice Starch
French: Amidon de riz, Fécule de riz.
German: Reisstaerke.

Agriculture
Ingredient of—
Cattle foods.

Analysis
Reagent in testing for—
Chlorine, copper, iodine, nitrous acid.

Brewing
Starting point in making—
Beer, fermented liquors.

Chemical
Ingredient of—
Colloidal preparations (added for the purpose of preventing precipitation).
Starting point in making—
Acetone by bacterial fermentation, alcoylated products (French 640174), dextrin and dextrin products, fusel oil by fermentation, lactic acid, levulinic acid, starch glycollate, starch iodide, solubilized starch.
Tanning agents by sulphonation with sulphuric acid (French 544253).

Distilling
Starting point in making various types of distilled liquors.

Explosives
Ingredient of—
Gelatin dynamites, permissibles for coal mining, regular nitroglycerin dynamites.
Starting point in making—
Nitro-starch explosives, nitro-starch dynamites.

Food
Ingredient of—
Baking powders, candies, cocoa powders, cake powders, custard preparations, chocolate preparations, ice cream preparations and powders.
Sauces of various sorts (to make them thick).
Various culinary and food preparations.
Vegetarian foods.
Raw material in—
Biscuit, pastry, baking, and confectionery industries.

Fuel
Binder in making—
Fuel briquettes.

Glues and Adhesives
Ingredient of—
Cold-water glues.
Various adhesive paste preparations.
Xanthate adhesive preparations.
Starting point (French 648019) in making—
Glues in bead form.

Leather
Ingredient of—
Cleansing compositions.
Compositions used in the manufacture of artificial leather (French 558630).
Compositions, containing lime, calcium phenolate, and sodium hydroxide, used for softening and dehairing hides and skins (French 612409).
Vehicle for—
Holding tanning extract in the drum tanning process.

Mechanical
Ingredient of—
Compositions used for the purpose of preventing incrustation of scale in boilers (U. S. 1720565).

Miscellaneous
Ingredient of—
Compositions used in laundries for the dressing and sizing of fabrics after washing.
Compositions used for coating purposes, prepared by the action of calcium chloride, calcium nitrate, zinc chloride, and magnesium chloride on the starch (French 557085).
Compositions in emulsified form (French 599908).
Compositions used for stiffening fabrics.
Compositions containing coloring matter, such as for example azo dyestuffs.
Compositions, colored black and containing naphthalene and its derivatives (French 641442).
Compositions containing pitch, rosin soap (such as potassium resinate), oil, flour, (used for road surfacing purposes.
Pastes for hanging wallpaper.
Starch glazes.
Starting point in making—
Starch tablets.

Paint and Varnish
Fixative in making—
Whitewashes and starch coating compositions with the addition of sodium carbonate and nitrobenzene (French 616204).

Paper
Ingredient of—
Compositions used for sizing different qualities of paper, particularly writing paper.
Compositions used in the manufacture of surface-coated paper.
Compositions used in the manufacture of pasteboard.

Perfume
Ingredient of—
Massaging compositions (French 616204).
Perfumes, pomades, sachets, toilet powders.

Pharmaceutical
In compounding and dispensing practice.

Printing
In bookbinding practice.

Soap
Ingredient of—
Compositions, containing carbon tetrachloride, glycerin, and the like, used for the dry cleaning of hands which have become stained with crankcase oil, tar, grease, paint (French 611895).
Detergent preparations containing potassium silicate.
Soapstock in making special grades of soap.
Soft soaps (used as a filler).

Sugar
Starting point in making—
Burnt sugar or caramel, malt sugar, various syrups and mixtures, white glucose.

Textile
——, *Dyeing*
Ingredient of—
Dye bath for various yarns and fabrics.
——, *Finishing*
Ingredient of—
Compositions used for sizing cotton fabrics.
Compositions used for starching knitted merchandise, such compositions also containing glucose, sodium

Rice Starch (Continued)
silicate, glycerin, olive oil, and borax (French 649899).
Fireproofing compositions, containing ammonium sulphate, sodium carbonate, boric acid, sodium biborate, used for treating rayons (French 595286).
Sizing compositions containing sodium resinate (French 523282).
Weighting compositions for treating calicoes, lace curtains, and other textiles.

——, *Manufacturing*
Ingredient of—
Spinning bath in making viscose rayon.
Size for—
Cotton yarns before weaving.

——, *Printing*
Ingredient of—
Printing pastes (added to thicken them).

Ricinoleic Acid
French: Acide ricinoléique.
German: Ricinoelsaeure, Ricinussaeure, Rizinoelsaeure, Rizinussaeure.

Chemical
Ingredient (Brit. 303379) of—
Emulsified preparations.
Starting point in making—
Salts, esters and other derivatives.

Fats and Oils
Ingredient of—
Turkey red oil.

Miscellaneous
Ingredient (Brit. 303379) of—
Cleansing compositions.

Soap
Ingredient (Brit. 303379) of—
Saponaceous cleansing and washing compositions.

Textile
——, *Bleaching*
Ingredient (Brit. 303379) of—
Bleaching compositions.

——, *Finishing*
Ingredient (Brit. 303379) of—
Sizing compositions, waterproofing compositions.

——, *Manufacturing*
Ingredient (Brit. 303379) of—
Bowking, softening, and oiling compositions.

Ricinoleic Alcohol Succinic Acid Ester

Bituminous
Solvent (Brit. 445223) for—
Asphalt and other bituminous bodies.

Dye
Solvent (Brit. 445223) for—
Dyestuffs, particularly oil-soluble coaltar dyes.

Fats, Oils, and Waxes
Solvent (Brit. 445223) for—
Fats, oils, waxes.

Resins
Solvent (Brit. 445223) for—
Oil-soluble glycerol-phthalic acid resins, polymerized vinyl compounds, synthetic resins.

Rubber
Solvent (Brit. 445223) for—
Rubber.

Ricinoleic Alcohol Tartaric Acid Ester

Bituminous
Solvent (Brit. 445223) for—
Asphalt and other bituminous bodies.

Dye
Solvent (Brit. 445223) for—
Dyestuffs, particularly oil-soluble coaltar dyes.

Fats, Oils, and Waxes
Solvent (Brit. 445223) for—
Fats, oils, waxes.

Resins
Solvent (Brit. 445223) for—
Oil-soluble glycerol-phthalic acid resins, polymerized vinyl compounds, synthetic resins.

Rubber
Solvent (Brit. 445223) for—
Rubber.

Ricinoleyl-1-sulphuric Acid (Normal) Ester

Chemical
As an emulsifying agent.
Reagent in—
Organic syntheses.
Starting point (Brit. 440575) in making—
Emulsifying agents with salts of lead, aluminum, iron, tin, or barium (such emulsifying agents are said to form water-in-oil emulsions and are, preferably, produced in situ by (1) dissolving the sulphuric acid ester in the oil and (2) agitating with an aqueous solution of the metal salt, for example, lead acetate; they are said to be useful for treating medicinal paraffin oil, neatsfoot oil, olive oil, castor oil, cottonseed oil, linseed oil and petroleum lubricating oils; a heavy paraffin oil, so treated on the basis of 50 parts by weight of oil to 48.75 parts of water, is said to yield a heavy grease that has good lubricating properties and may readily be extended with oil; a water-linseed oil type emulsion is offered as suitable for use as a paint base).

Rock Wool
(Fibrous, wool-like material composed of fine silicate filaments made by processing an argillaceous limestone.)

Construction
Acoustical improver in—
Public buildings, talking picture studios, theaters.
Anticracking ingredient in—
Wall plasters.
Antishrinkage ingredient in—
Wall plasters.
Binder in—
Wall plasters.
Fireproofing medium in buildings.
Heat-insulating medium in buildings.
Sound-insulator in buildings.

Glass
Heat-insulating medium for lehrs.

Metallurgical
Heat-insulating medium.

Miscellaneous
Heat-insulating medium for—
Airducts, boilers, furnaces, ovens, piping installations.
Water-heater shells, either gas or electric.

Refrigeration
Heat-insulating medium in—
Electric refrigerators.
Ice and cold-storage installations of all kinds.

Rosanilin

Chemical
Ingredient (Brit. 295605) of bacteriological preparations, therapeutic compositions, and biological stains containing—
Cresol, guaiacol, hydroquinone, phenol, phloroglucinol, pyrocatechol, pyrogallol.

Textile
Dyestuff for—
Fabrics, yarns.

Rosin
Synonyms: Colophony, Resin colophony.
Latin: Resina, Colophonium.
French: Arcanson, Colophone, Résine blanche, Résine jaune.
German: Fichtenharz, Geigenharz, Kolophonium.
Spanish: Cologonia ez griega.
Italian: Colofonia, Pece greca.

Agriculture
Protective agent in—
Pruning and grafting.

Adhesives
As a cement.
Ingredient of—
Casein glues, cements for laminated mica (Brit. 273290).

Brewing
Ingredient of—
Brewers' pitches.

Chemical
Reagent in making—
Aluminum resinate, bismuth resinate, benzene derivatives, calcium resinate, cobalt resinate, copper resinate, lead resinate, manganese resinate, zinc resinate.
Starting point in making—
Abietic acid.

Rosin (Continued)
Construction
Ingredient of—
Concrete waterproofing compositions, roofing cements, roofing materials, sizing for caulking oakum.
Electrical
Binder, cementing and insulating material in dry batteries.
Ingredient of—
Insulating compositions, soldering pastes.
Soldering flux.
Explosives
Ingredient of—
Fireworks, match compositions, shrapnel shell explosives.
Fats and Oils
Ingredient of—
Axle greases.
Compositions of emulsifiable cutting oils used on high-speed tools.
Lubricating compositions of various kinds.
Starting point in making—
Rosin oil by distillation.
Fuel
Binder in—
Briquettes.
Ingredient of—
Fire kindlers.
Glass
Ingredient of—
Glass cements.
Gas
Processing material in—
Direct manufacture of illuminating gas.
Ink
Ingredient of—
Plateless engraving inks, printing inks.
Insecticide
Ingredient of—
Coating for sticky fly-paper, insect powders, protective bandings for trees.
Leather
Ingredient of—
Dressings of various kinds, fillings for shoe soles, leather cements, lubricating compounds, stiffening compounds.
Linoleum and Oilcloth
Ingredient of—
Coating batch, linoleum cements.
Mechanical
Applied to belting to reduce slipping.
Ingredient of—
Belt greases.
Dusting agent for—
Foundry molds.
Metallurgical
Flux for—
General soldering and tin plating.
Ingredient of—
Core oils.
Soldering compositions (admixed with lard, suet, grease, waxes, oils).
Reagent in—
Pattern making, steel hardening.
Miscellaneous
Binder in—
Asphalt compositions.
Cement for—
Setting bristles in brushes.
Hardening agent for—
Tallow candles, wax tapers.
Ingredient of—
Alum-oil cements.
Cements for setting knife blades in handles.
Sizings, stamping powders, sweeping powders, weatherproofing compounds.
Protective coating for—
Mounted fish and other products of taxidermy.
Reagent for—
Maintaining proper contact of bow and strings in the playing of violins and similar musical instruments.
Making stencils.
Paint and Varnish
Ingredient of—
Alcohol varnishes, bases for varnishes (Brit. 273290), benzin lacquers, dark varnishes, driers, enamels for brick walls, transparent oil varnishes, tung oil varnishes, weatherproofing and waterproofing compositions, wood stains.
Paper
Component of—
Dressings for boxboard, papier mache.
Sizes for paper, paperboard, pulp compositions, and products made from them.
Waterproofing compositions.
Reagent in—
Utilization of sulphite cellulose waste liquors by the Tripp process.
Petroleum
Contact agent (U. S. 1904173) in—
Removing corrosive sulphur from hydrocarbon oils.
Pharmaceutical
Ingredient of—
Cerates, plasters, salves.
Plastics
Component of—
Plastic wood.
Reagent in making—
Artificial amber, moldable compositions (Brit. 273290), phonograph records, synthetic resins.
Resins and Waxes
Ingredient of—
Compound waxes, grafting wax, imitation burgundy pitch, sealing wax, sealing wax compositions (Brit. 252186), shellac substitutes.
Reagent (U. S. 1894580) in making—
Resinous products, in combination with phenylamine and furfuraldehyde.
Starting point in making—
Ester gums, neutral rosin, soluble resins.
Rubber
Ingredient of—
Rubber batches, rubber substitutes.
Shipbuilding
Reagent for—
Impregnating or sizing oakum in caulking.
Soap
General soapstock.
Ingredient of—
Bituminous waterproofing soaps, soap powders.
Textile
Ingredient of—
Powders for transferring designs, special sizes, waterproofing compositions.
Woodworking
Ingredient of—
Impregnating compositions.
Weatherproofing hot dip.

Rosin Oil
Synonyms: Resin oil, Rosinoil.
French: Huile de colophone, Huile de résine.
German: Harzoel, Kolophonoel.
Brewing
Reagent in brewing practice.
Electrical
Ingredient of—
Transformer oils (acting as insulating oil).
Fats and Oils
Ingredient of—
Axle greases, castor oil compositions, lubricating greases, olive oil compositions.
Ink
Ingredient of—
Lithographic inks, printing inks.
Insecticide
Ingredient of—
Mixtures for coating tree trunks to prevent depredations of caterpillars.
Leather
Ingredient of—
Dressing compositions, shoe polishes.
Linoleum and Oilcloth
Reagent in manufacturing processes.
Mechanical
Lubricant for—
Canvas belting.
Metallurgical
Flotation oil for—
Concentrating minerals and ores.

Rosin Oil (Continued)

Miscellaneous
Ingredient of—
Brewers' pitch, cements, sweeping compounds.
Waterproofing agent for—
Cordage.

Paint and Varnish
Ingredient of—
Funnel paints for ships, shingle stains, varnishes.
Starting point in making—
Lampblack.

Rubber
Ingredient of—
Cements and compositions.
Reagent in—
Reclaiming rubber.

Soap
Ingredient of various sorts of soap.

Textile
——, *Finishing*
Ingredient of—
Waterproofing compositions.

Rosin Pitch
French: Brai de colophone, Brai de résine, Poix de colophone, Poix de résine.
German: Harspech, Kolophoniumpech.

Agricultural
Ingredient of—
Grafting waxes.

Construction
Ingredient of—
Waterproof compositions for treating masonry, concrete work, brickwork, and the like.

Electrical
Ingredient of—
Insulating compositions used in dry batteries.
Wire-insulating preparations.

Fuel
Binder in making—
Briquets from coal dust.

Metallurgical
Ingredient of—
Plastic compositions used in making molds for castings.
Steel-hardening compositions.

Miscellaneous
Caulking agent in building ships and boats.
Cement for fixing brush bristles in the handle.
Ingredient of—
Shoemakers' wax, street-paving compositions.
Preservative for—
Nets, lines, and cordage.

Paint and Varnish
Ingredient of—
Bituminous paints, bituminous varnishes, roofing cements, roofing felts, roofing lutes.

Paper
Ingredient of—
Waterproofing compositions for paper and cardboard.

Rosolic Acid
Synonyms: Diphenolcresolcarbinol anhydride.
French: Acide rosolique, Anhydride de diphénolecrésolcarbinole.
German: Diphenolcresolanhydrid, Rosolsäure.

Analysis
General indicator in titrimetric analysis.
Indicator in—
Carbon dioxide detection in potable waters.
Caustic alkalies, gastric analysis.
Mineral acids, including sulphur dioxide but not including phosphoric acid.

Chemical
Starting point in making—
Esters, salts and other derivatives.

Dye
Starting point in making—
Anilin dyestuffs.

Food
Coloring for—
Candies, food preparations.

Ink
Coloring matter in—
Printing inks.

Paper
Coloring matter, in lake form, in—
Wallpaper.

Paint and Varnish
Coloring matter in—
Alcoholic varnishes, lacquers, oil varnishes.

Textile
Coloring matter in—
Dyeing orange shades on wool and silk.

Rubidium
French: Rubidium.
German: Rubidium.

Chemical
Starting point in making—
Rubidium salts of acids and halogens.
Reagent (Brit. 281307) in making zeolite catalysts used in making—
Acenaphthylene from acenaphthene.
Acetaldehyde from ethyl alcohol.
Acetic acid from ethyl alcohol.
Alcohols from aliphatic hydrocarbons.
Aldehydes from toluene, xylene, mesitylene, pseudocumene, and cymene.
Aldehydes and acids by the oxidation of orthochlorotoluene, parachlorotoluene, orthobromotoluene, parabromotoluene, dichlorotoluene, chlorobromotoluenes, nitrotoluenes, chloronitrotoluenes, bromonitrotoluenes.
Alpha-anthraquinone from naphthalene.
Anthraquinone from anthracene.
Benzaldehyde and benzoic acid from toluene.
Benzoquinone from phenanthraquinone.
Chloroacetic acid from ethylenechlorohydrin.
Diphenic acid from ethyl alcohol.
Fluorenone from fluorene.
Formaldehyde from methyl alcohol or methane.
Hemimellitic acid from acenaphthene.
Maleic and fumaric acids from benzene, toluene, phenol, or tar acids, or from benzoquinone or phthalic anhydride.
Naphthalic anhydride.
Naphthaldehydic acid, acenaphthaquinone, or bisacenaphthylidenedione from acenaphthene or acenaphthylene.
Naphthalic anhydride.
Phenanthraquinone from phenanthrene.
Phthalic anhydride from naphthalene.
Salicyl aldehyde or salicylic acid from cresol.
Vanillin or vanillic acid from eugenol or isoeugenol.

Rubidium Chromate
French: Chromate de rubidium.
German: Chromsaeuresrubidium.

Chemical
Catalyst (French 598447) in making the following alcohols—
Amyl, butyl, heptyl, hexyl, propyl.

Rubidium Manganate
French: Manganate de rubidium.
German: Mangansaeuresrubidium.

Chemical
Catalyst (French 598447) in making the following alcohols—
Amyl, butyl, heptyl, hexyl, propyl.

Rubidium Molybdate
French: Molybdate de rubidium.
German: Molybdansaeuresrubidium, Rubidiummolybdat.

Chemical
Catalyst (French 598447) in making the following alcohols—
Amyl, butyl, heptyl, hexyl, higher (aliphatic), propyl.

Rubidium Oxide
French: Oxyde de rubidium.
German: Rubidiumoxyd.

Chemical
Starting point in making—
Rubidium salts.
Reagent (Brit. 281307) in making zeolite catalysts used in making—
Acenaphthylene from acenaphthene.
Acetaldehyde from ethyl alcohol.
Acetic acid from ethyl alcohol.
Alcohols from aliphatic hydrocarbons.
Aldehydes from toluene, xylene, mesitylene, pseudocumene, and cymene.

Rubidium Oxide (Continued)
Aldehydes and acids by the oxidation of orthochlorotoluene, parachlorotoluene, orthobromotoluene, parabromotoluene, dichlorotoluenes, chlorobromotoluenes, nitrotoluenes, chloronitrotoluenes, bromonitrotoluenes.
Alphanaphthaquinone from naphthalene.
Anthraquinone from anthracene.
Benzaldehyde and benzoic acid from toluene.
Benzoquinone from phenanthraquinone.
Chloroacetic acid from ethylenechlorohydrin.
Diphenic acid from ethyl alcohol.
Fluorenone from fluorene.
Formaldehyde from methanol or methane.
Hemimellitic acid from acenaphthene.
Maleic and fumaric acids from benzene, toluene, phenol, or tar acids, or from benzoquinone or phthalic anhydride.
Naphthalic anhydride.
Naphthaldehydic acid, acenaphthaquinone, or bisacenaphthylidenedione from acenaphthene or acenaphthylene.
Phenanthraquinone from phenanthrene.
Phthalic anhydride from naphthalene.
Salicyl aldehyde or salicylic acid from cresol.
Vanillin or vanillic acid from eugenol or isoeugenol.

Rubidium Tungstate
French: Tungstate de rubidium.
German: Rubidiumwolframat, Wolframsaeuresrubidium.

Chemical
Catalyst (French 598447) in making the following alcohols—
Amyl, butyl, heptyl, hexyl, higher (aliphatic), propyl.

Rubidium Uranate
French: Uranate de rubidium.
German: Uransaeuresrubidium.

Chemical
Catalyst (French 598447) in making the following alcohols—
Amyl, butyl, heptyl, hexyl.

Rubidium Vanadate
French: Vanadate de rubidium.
German: Vanadinsaeuresrubidium.

Chemical
Catalyst (French 598447) in making the following alcohols—
Amyl, butyl, heptyl, hexyl, propyl.

Saffron
Synonyms: Crocus, French saffron, Saffran, Safran, Spanish saffron, Valencia saffran.

Food
Coloring agent in making special food compositions.
Flavoring agent in making special food compositions.

Miscellaneous
Pigment in coloring—
Artificial flowers.
In greenish-yellow shades.
Marble.

Oils and Fats
Color for special oils, particularly for cosmetic use.
Starting point in making an essential oil.

Pharmaceutical
In compounding and dispensing practice.

Textile
——, *Dyeing*
Dyestuff for—
Textiles mordanted with alumina in yellow shades.
Textiles mordanted with tin salts in orange shades.

Safranin

Dye
Starting point in making—
Diazin green, methyl indone B.

Ink
Pigment in making—
Typewriter inks, writing inks.

Photographic
Reagent in—
Color photography.

Textile
——, *Dyeing and Printing*
Dyestuff for yarns and fabrics.

Safranin T
French: Safranine T.

Chemical
Ingredient (Brit. 295605) of bactericidal, therapeutic preparations and biological stains, containing—
Cresol, guaiacol, hydroquinone, phenol, phloroglucinol, pyrocatechol, pyrogallol, resorcinol.

Miscellaneous
Dyestuffs for coloring various articles.

Textile
Color in dyeing and printing.

Safrol
Synonyms: Propyldioxybenzene methyleneester, Shikimol, Synthetic oil of sassafras.

Chemical
Starting point in making—
Heliotropin (piperonal), protocatechuic aldehyde.

Miscellaneous
Odor or disguise in metal polishes and the like.

Perfumery
Ingredient of—
Creams, hair oils, perfumes, pomades.

Pharmaceutical
In compounding and dispensing practice.

Soap
Odor and disguise in hard, soft and liquid soaps.

Salep
Synonyms: Satyrion.
French: Patte de loup, Scrotum de chien.
German: Salepknollen.

Food
As a foodstuff in the Orient.

Pharmaceutical
In compounding and dispensing practice.

Textile
——, *Finishing*
Ingredient of sizing compositions for the treatment of silks.
——, *Printing*
Thickener for printing paste.

Salicin

Pharmaceutical
In compounding and dispensing practice.

Rubber
Preservative (U. S. 1823119) in treating—
Rubber latex.

Salicylaldehyde
Synonyms: Ortho-oxybenzaldehyde, Salicylic aldehyde, Salicylous acid.

Analysis
Reagent in—
Analytical work.

Chemical
Reagent in—
Organic synthesis.
Starting point in making—
Coumarin by reaction with sodium acetate and acetic anhydride.

Cosmetic
Ingredient of—
Cosmetics, pomades.

Dye
Intermediate in—
Dye manufacture.

Salicylamide
French: Amide de salicyle, Amide salicylique.
German: Salicylamid.

Agricultural
Reagent in treating—
Seeds and grain to protect them against mildew and the action of fungi.

Chemical
Starting point in making—
Aromatics, intermediates, pharmaceuticals.

Dye
Starting point in making various synthetic dyestuffs.

Salicylamide (Continued)

Fats and Oils
Deterioration retardant (Brit. 423938) for—
Vegetable oils.

Fungicide
Starting point (Brit. 408258) in making—
Seed disinfectants by (1) mercuration and (2) dilution with talc and glycerin.

Fuel
Deterioration retardant (Brit. 423939) for—
Coal-carbonization spirits.

Leather
Reagent in treating—
Leather to protect it against the action of fungi and mildew.

Paper
Reagent in treating—
Paper, pulp and products made therefrom, to prevent the action of fungi and mildew.

Petroleum
Deterioration retardant (Brit. 423938) for—
Cracked petroleum oils, lubricating oils, shale oils, transformer oils.

Rubber
Reagent in treating—
Rubber and rubber products against mildew and the action of fungi.

Textile
Reagent in treating—
Cotton yarns and fabrics against mildew and the action of fungi.

Salicylhydroquinone

Petroleum
Stabilizing agent (Brit. 406195) for—
Cracked gasolines and other motor fuels.

Salicylic Acid

Synonyms: Orthohydroxybenzoic acid, Ortho-oxybenzoic acid.
Latin: Acidum salicylum, Acidum spiricum.
French: Acide orthohydroxyebenzoique, Acide salicylique.
German: Orthohydroxybenzoesäure, Salicoylsäure, Salicylsäure, Spiroylsäure, Spirsäure.
Spanish: Acido salicilico.
Italian: Acido salicilico.

Analysis
Reagent for—
Nitrogen determination by the Kjedahl or Gunning method, to include nitrate nitrogen.
Reagent in detecting and analyzing—
Acetone, citric acid, formaldehyde, fusel oil, lactic acid, methanol, nitrates, nitrous acid, titanium.

Brewing
Reagent in making—
Ales, beers.

Chemical
Activating agent (Brit. 291725) in making—
Activated charcoal.
Catalyst in making—
Accelerators of rubber vulcanization by reaction between amines and aldehydes.
Starting point or reagent in making—
Acetylpara-aminophenyl salicylate (salophen), acetylsalicylic acid, alkaloidal salicylates, allyl salicylate, aminosalicylic acid, amyl salicylate, ammonium salicylate, apyron, aromatics, aspirophen, barium salicylate, betanaphthyl salicylate, bismuth subsalicylate, butyl salicylate, cadmium salicylate, calcium salicylate, derivatives of phenyl-2:3-dimethyl-5-pyrazolonylimipyrin, diplosal, ethyl borosalicylate, glycol salicylate (spirosal), guaiacol salicylate (guaiacol salol), hexamethylenetetramine salicylate (saliformin), isopropylsuccinic acid (pimelic acid), lithium salicylate, magnesium salicylate, menthyl salicylate (samol), menthyl salicylic-methyl ester, mercuric salicylate, metallic salicylates, methyl salicylate (artificial oil of wintergreen), organic salicylates, pharmaceuticals, phenyl salicylate (salol), potassium salicylate, propyl salicylate, sodium pyrophosphate-salicylate, sodium salicylate, sodium-theobromine salicylate (diuretin), strontium salicylate.

Dye
Starting point in making—
Acidol chrome yellow R, alizarin yellow FS, alizarin yellow GG, alizarin yellow R, alkali yellow R, anthracene acid brown G, anthracene brown G, anthracene red, anthracene yellow C, anthracene yellow BN, aurichrome phosphin R, azidin brown M, azidin fast red F, azidin yellow G, azo alizarin black I, azo alizarin bordeaux W, azo alizarin yellow 5G, azo alizarin yellow 6G, azo green, benzamine brown 3GO, benzidin fast red F, benzo fast yellow 5G, benzo gray S extra, benzo olive, benzo orange, benzoin brown G, brilliant orange G, chlorazol deep brown G, chlorazol orange 2R, chrysamine, chrysamine G, chrysamine R, chrome fast yellow GG, chrome violet, chrome yellow D, columbia black green D, columbia green, congo brown G, congo brown R, cotton yellow G, cotton yellow R, crumpsall direct fast brown B, crumpsall direct fast brown O, crumpsall direct fast red R, crumpsall yellow, diamine bronze G, diamine brown B, diamine brown M, diamine fast red, diamond green, diamond yellow, diamond yellow N, diamond black, diamond flavin G, diamond yellow G, diazo colors from 3-amino-5-sulpho-2-oxybenzoic acid (Brit. 251637), diphenyl brown BN, diphenyl brown 3GN, diphenyl brown RN, diphenyl green 3G, dutch yellow, eboli green (eriochrome phosphin R, fast mordant yellow, flazol, Hessian yellow, milling orange mordant yellow O, naphthamine brown 4G, oriol yellow, oxamine green G, oxamine maroon, oxamine red, salicin red G, trisulphon brown B, trisulphone brown G, trisulphon brown 3G, xylene yellow 3G.
Starting point (Brit. 298518) in making azo colors with the aid of—
Alpha-amino-2:7-dimethoxynaphthalene.
Alpha-amino-2:7-dioxynaphthalene glycollate.
Alpha-amino-2-ethoxynaphthalene-6-sulphonic acid.
Alpha-amino-2-methoxynaphthalene.
Alpha-aminonaphthalene.
Alpha-aminonaphthalene-6-sulphonic acid.
Alpha-aminonaphthalene-7-sulphonic acid.
Alpha-amino-2-naphthoxybetapropionic acid.
Alpha-amino-2-oxyethoxynaphthalene sulphonate.
Anilin, anilin-3-chloro-6-sulphonic acid, anilin-2:4-disulphonic acid, anilin-2:5-disulphonic acid, anilin-4-nitro-2:5-disulphonic acid, anilin-3-sulphonic acid.
Beta-amino-1-methoxybenzene-4-sulphonic acid, beta-amino-5-sulphobenzoic acid.

Electrical
Ingredient of—
Storage battery electrolytes.

Food
Preservative in—
Cider, food preparations of various sorts, sausages, vinegar.

Gas
Ingredient of—
Oxide of iron purifier mass in the purification of coal gas and coke-oven gas (added for the purpose of preventing the precipitation of iron hydroxide).

Glues and Adhesives
Ingredient of—
Glue preparations (used to make them more adhesive).
Mucilage preparations.
Preservative in making—
Gelatin preparations, glue preparations.
Various adhesive preparations containing such substances as degraded starches, dextrins, and casein.

Leather
Preservative in the treatment of—
Hides to prevent their decomposition during the process of converting them into leather.

Miscellaneous
General preservative.
Preservative in treating—
Fur skins, various albuminous materials.
Preservative, in admixture with sodium thiosulphate, in treating—
Animal products of various sorts.
Reagent in making—
Catgut.

Paper
Reagent in making—
Parchmentized paper.

Salicylic Acid (Continued)

Perfume
Ingredient of—
Dentifrices, mouth washes.

Pharmaceutical
In compounding and dispensing practice.

Soap
Ingredient of—
Special medicated soaps.

Resins and Waxes
Starting point (Brit. 292912) in making synthetic resins and waxes with the aid of—
Acetylcarbamide, allylcarbamide, amylcarbamide, benzoylcarbamide, butylcarbamide, cinnamylcarbamide, citrylcarbamide, cyanamide, ethylcarbamide, formylcarbamide, gallylcarbamide, heptylcarbamide, hexylcarbamide, isoallylcarbamide, isoamylcarbamide, isobutylcarbamide, isopropylcarbamide, lactylcarbamide, methylcarbamide, pentylcarbamide, phenylcarbamide, propionylcarbamide, propylcarbamide, resorcinoylcarbamide, toluoylcarbamide.

Textile
Solubilizing agent (Brit. 276100) in making dye liquors and printing pastes, containing the following classes of dyestuffs—
Acridins, aminoanthraquinones, reduced or unreduced.
Anthraquinones, azins, azo, basic diarylmethanes, basic triarylmethanes, benzoquinoneanilides, chrome mordant, indigoid, naphthaquinoneanilides, naphthoquinones, reduced or unreduced.
Nitroarylamines, nitrodiarylamines, nitrodiarylphenols, oxazins, pyridins, quinolins, quinonimides, reduced and unreduced.
Sulphur, thiazins, xanthenes.

Salicylic Acid Ester of Grapeseed Alcohol
(Uses same as those given for the item immediately following.)

Salicylic Acid Ester of Ricinoleic Alcohol

Bituminous
Solvent (Brit. 445223) for—
Asphalt and other bituminous bodies.

Dye
Solvent (Brit. 445223) for—
Dyestuffs, particularly oil-soluble coaltar dyes.

Fats, Oils, and Waxes
Solvent (Brit. 445223) for—
Fats, oils, waxes.

Resins
Solvent (Brit. 445223) for—
Oil-soluble glycerol-phthalic acid resins.
Polymerized vinyl compounds, synthetic resins.

Rubber
Solvent (Brit. 445223) for—
Rubber.

Salicyl Orthoanisidide
French: Orthoanisidide de salicyle, Orthoanisidide salicylique.
German: Salicylorthoanisidid.

Agricultural
Reagent in treating—
Seeds and grains to protect them against mildew and the action of fungi.

Chemical
Starting point in making—
Intermediates, pharmaceuticals.

Dye
Starting point in making various synthetic dyestuffs.

Leather
Reagent in treating—
Leather and leather goods to protect them against mildew and the action of fungi.

Paper
Reagent in treating—
Paper, pulp, and products made therefrom against mildew and the action of fungi.

Rubber
Reagent in treating—
Rubber and rubber products against the action of mildew.

Textile
Reagent in treating—
Cotton yarns and fabrics against mildew and the action of fungi.

Woodworking
Reagent in treating—
Wood and wood products against mildew and the action of fungi.

Salicyl Orthotoluide

Agriculture
For protecting seeds and grains against decomposition and spoiling.

Chemical
Starting point in making—
Intermediates and other derivatives, pharmaceutical chemicals, sodium salicylorthotoluide.

Fungicide
Starting point (Brit. 408258) in making—
Seed disinfectants by (1) mercuration and (2) dilution with talc and glycerin.

Leather
For protecting leather against the formation of mildew.

Paper
For protecting paper against the formation of mildew and fungi.

Rubber
For protecting rubber against the formation of mildew and fungi.

Textile
For protecting cotton yarns and fabrics against the formation of fungi and mildew.

Woodworking
For protecting wood against the formation of mildew and fungi.

Salicyl Paratoluide

Fungicide
Starting point (Brit. 408258) in making—
Seed disinfectants by (1) mercuration and (2) dilution with talc and glycerin.

Salicylphloroglucinol

Petroleum
Stabilizing agent (Brit. 406195) for—
Cracked gasolines and other motor fuels.

Salicylylpyrocatechol

Petroleum
Stabilizing agent (Brit. 406195) for—
Cracked gasolines and other motor fuels.

Salicylylpyrogallol

Petroleum
Stabilizing agent (Brit. 406195) for—
Cracked gasolines and other motor fuels.

Salicylylresorcinol

Petroleum
Stabilizing agent (Brit. 406195) for—
Cracked gasolines and other motor fuels.

Salt Cake
Synonyms: Crude sodium sulphate.
French: Gateaux de sel, Sulfate sodique brut, Sulfate de sodium brut, Sulfate de soude brut.
German: Rohes natriumsulfat, Rohes schwefelsaeuresnatrium, Salzkuchen.

Ceramics
Ingredient of—
Glazes.

Chemical
Reagent in making—
Ammonium-magnesium sulphate, barium sulphate, barium-sodium sulphate, aluminum hydroxide, oxalic acid.
Starting point in making—
Glauber's salt, or pure sodium sulphate, anhydrous and hydrous.
Sodium acetate, sodium carbonate, sodium hypochlorite, sodium silicate, or waterglass, sodium thiosulphate, washing sodas.

Dye
Diluting agent in making—
Commercial dyestuff preparations.

Salt Cake (Continued)
Reagent in making—
Ultramarine blue.

Fats and Oils
Reagent in making—
Turkey red oil.

Fuel
Ingredient (U. S. 1618465) of—
Fuel preparations (acting as a fuel economizer).

Glass
Ingredient of batch in making—
Bottle glass, window glass, plate glass.

Glue
Reagent in making various glues and gelatines.

Ink
Reagent in making—
Printers' ink.

Insecticides
Ingredient of various compositions.

Leather
Reagent in—
Tanning.

Paint and Varnish
Ingredient of—
Paint and varnish removers.
Reagent in making—
Dry colors, lake pigments, mineral pigments.

Paper
Reagent in making—
Soda pulp (used in place of sodium carbonate).
Sulphate pulp.

Refrigeration
Ingredient of—
Freezing mixtures.

Sanitation
Reagent in precipitating—
Barium from salt brines in the Mills process for sewage treatment.

Soap
Ingredient of—
Detergent preparations.

Textile
——, *Bleaching*
Reagent in bleaching processes.

——, *Dyeing*
Ingredient of—
Dye liquors.

——, *Finishing*
Ingredient of various finishing preparations.

Sand
(Sea sand and other sands, but not including monazite sand).

Abrasives
Ingredient of—
Abrasives.
Starting point in making—
Silicon carbide with coke, salt, and sawdust.

Adhesives
Ingredient of—
Aquarium cements, containing also (1) plaster of paris, litharge, resin, boiled linseed oil; (2) red lead, litharge, rosin, spar varnish.

Analysis
Heat transfer medium in—
Sand-baths.

Animal Remedies
Ingredient of—
Lice and mite tablets for poultry, containing also calcium sulphide, gypsum, sugar, and starch.
Worm-expeller, containing also epsom salt, gypsum, calcium silicate, venetian red, and nicotine.

Automotive
Abrasive in—
Blast-removing old paint from automobile bodies.
Blast-leveling surfaces between paint coats.

Ceramics
As placing for—
Bisque ware.

Chemical
As a heat-transfer medium.
Starting point in making—
Sodium silicate, with calcined soda and powdered coal.

Construction
Blasting medium in—
Cleaning stone, brick, or concrete buildings and monuments.
Filler in—
Sanitary resilient stone floorings.
Ingredient of—
Bituminous cements for sealing pipes and conduits.
Bituminous mixes for roads, floors, tennis courts.
Building blocks, concretes, mortars.
Pipe joint compounds, containing also flour, portland cement, talc, and lampblack.
Sand cement (equal mix with portland cement).

Explosives and Matches
Ingredient (Brit. 404298) of—
Central core of repeatedly ignitable matches.

Fertilizer
Filler in—
Fertilizer mixtures.

Glass
Source of silica.

Metallurgical
As a moulding material.
Starting point in making—
Silicon—(a) with coke, (b) with powdered magnesium.

Miscellaneous
Blasting agent in—
Smoothing and leveling surfaces in various industrial processes.
Filler in many commercial products.
Filler and absorbent in—
Sweeping compounds (mostly admixtures with sawdust, salt, mineral oil, and coloring matter).
Ingredient of—
Vitrified stove wick, containing also pumice, charcoal, coke, grit, rosin, and sodium silicate.

Plastics
As a filler.

Sanitation
Filtering medium in—
Water purification (filter beds).

Soap
Filler and abrasive in—
Sand soaps, scouring powders, soap powders.

Stone
Abrasive in—
Finishing operations.
Filler in—
Artificial stone.

Santalol

Chemical
Starting point in making the following derivatives—
Acetate, allophanate, arsenate, arsenite, benzoate, bisulphate, bitartrate, borate, carbonate, camphorate, citrate, dihydrobromide, dihydrochloride, ferrocyanide, formaldehydesantalol, formate, glycerophosphate, hydrobromide, hydrochloride, hydroiodide, hypophosphite, lactate, methylether (threysol), phosphate, phosphite, salicylate, sulphate, sulphocarbolate, tannate, tartrate, valerate.

Pharmaceutical
In compounding and dispensing practice.

Santalyl Acetate
French: Acétate de santalyle, Acétate santalylique.
German: Aethansäuresantalylester, Aethansäuressantalyl, Essigsäuresantalylester, Essigsäuressantalyl, Methancarbonsäuresantalylester, Methancarbonsäuressantalyl, Santalylacetat, Santalylazetat.
Spanish: Acetato de santalil.
Italian: Acetato di santalilo.

Chemical
Starting point in making various derivatives.

Perfume
Ingredient of—
Flower bouquets, sandalwood perfumes.
Perfume in—
Cosmetics.

Soap
Perfume in—
Toilet soaps.

Santalyl Chloride
French: Chlorure de santalyle, Chlorure santalylique.
German: Santalylchlorid.

Chemical
Starting point in making—
Santalol methylether (threysol).

Saponin
French: Saponine.
Spanish: Saponina.
Italian: Saponina.

Chemical
Emulsifying agent (Brit. 361860) in making—
Emulsions of hydrocarbons of various groups of the aliphatic and aromatic series.
Emulsions of various chemicals, terpene emulsions.
Wetting compositions in emulsified state.

Construction
Emulsifying agent (Brit. 361860) in making—
Acoustic plaster containing calcimined gypsum and other substances.
Dispersed coating compositions for use on concrete, brick, stucco, and other construction materials (Brit. 361860).
Dispersed impregnating compositions for treating builders' felt, tar paper, and similar construction materials (Brit. 361860).

Disinfectant
Dispersing agent in making—
Emulsified germicides and deodorizing preparations.

Fats and Oils
Dispersing agent (Brit. 361860) in making—
Boring oils in emulsified form, drilling oil emulsions, greasing compositions in emulsified form.
Lubricating compositions in emulsified form, containing various vegetable and animal fats and oils.
Stabilized emulsions of various animal and vegetable fats and oils.
Wetting compositions containing various animal and vegetable fats and oils in emulsified form.
Wire-drawing oils in emulsified form.

Food
Ingredient of—
Carbonated beverages (nonpoisonous sort used for the purpose of producing foam).

Gas
Emulsifying agent (Brit. 361860) in making—
Tar emulsions.

Insecticide
Dispersing agent (Brit. 361860) in making—
Insecticidal preparations in emulsified form, for combating vegetable and animal pests.
Vermin exterminators in emulsified form.

Leather
Dispersing agent (Brit. 361860) in making—
Emulsified fat-liquoring baths, emulsified leather-dressing compositions, emulsified leather-finishing compositions, emulsified leather-softening compositions, emulsified leather-waterproofing compositions.

Miscellaneous
Detergent for various purposes.
Dispersing agent (Brit. 361860) in making—
Cleansing compositions in emulsified form, scouring compositions and detersives in emulsified form, various emulsified preparations, various emulsified wetting compositions, waterproofing emulsions.
Foam-producing agent for various purposes.
Ingredient of—
Fire-extinguishing solutions.

Paper
Dispersing agent in making—
Emulsified preparations for use in the treatment of paper and pulp products.
Sizing compositions in emulsified form.
Waterproofing compositions for paper and pulp products and paperboard.

Perfume
Dispersing agent (Brit. 361860) in making—
Creams in emulsified form, various emulsified cosmetics and toilet articles.

Ingredient of—
Toothpastes and other dentifrices (only the nonpoisonous sort should be so used).

Petroleum
Dispersing agent (Brit. 361860) in making—
Emulsified preparations, containing mineral oils, used in boring operations and other machine processes.
Emulsified lubricating compositions containing mineral oils and greases.
Emulsions containing petroleum and petroleum distillates.
Stabilized emulsions containing paraffin oil or other mineral oils and distillates.

Pharmaceutical
Dispersing agent (the nonpiosonous sort) in making various pharmaceutical products.

Plastics
Dispersing agent (Brit. 361860) in making various plastic compositions.

Rubber
Preservative for—
Rubber latex.

Soap
Dispersing agent (Brit. 361860) in making—
Emulsions of soaps, hand-cleansing compositions in emulsified form, various emulsified cleansing compositions, various emulsified scouring compositions.
Substitute for soap.

Textile
——, *Finishing*
Dispersing agent (Brit. 361860) in making—
Emulsified coating compositions, emulsified dressing compositions, emulsified finishing compositions, emulsified impregnating compositions, emulsified scouring compositions, emulsified sizing compositions.

——, *Printing*
Thickener (Brit. 314761) in making—
Printing pastes.

Wine
Ingredient (the nonpoisonous sort) of—
Wines (added for the purpose of producing foam).

Woodworking
Ingredient (Brit. 361860) of—
Emulsified coating compositions.
Emulsified impregnating compositions.

Sardine Oil
French: Huile de sardine.
German: Sardinoel.
Italian: Olio di sardella, Olio di sardina.

Agriculture
Ingredient of—
Dips for sheep, cattle, and other domestic animals.

Construction
Ingredient of—
Asbestos cements, bitumistic compounds, protective coatings, roofing products, waterproofing coatings, weatherproofing coatings.

Explosives and Matches
Ingredient of—
Matchhead compositions.

Fats and Oils
Ingredient of—
Fish oil emulsions, lubricating compositions, wire rope greases.
Starting point in making—
Hardened oil, stearin, tallow mixtures.

Food
Ingredient of—
Lard substitutes, oleomargarin.

Fuel
Ingredient of—
Compositions used in making candles.

Ink
Ingredient of—
Marking inks, printing inks.

Insecticide
Ingredient of—
Insecticidal compounds and preparations, insecticidal soaps, sprays.

Leather
Ingredient of—
Dressing compositions, finishing compositions.
Reagent in—
Making chamois leather, oil tanning.

Linoleum and Oilcloth
Substitute for—
Linseed oil.

Sardine Oil (Continued)

Mechanical
Lubricating agent (used alone or in mixtures) for—
Clocks, light machinery, screw-cutting machines, spindles.

Metallurgical
Quenching agent in—
Steel tempering.

Miscellaneous
Ingredient of—
Cordage waterproofing compounds and preservatives.
Fish net preservatives and waterproofing compounds.
Oil clothing dopes, pipe thread cements.

Paper
Ingredient of—
Impregnating compositions for treating paper, pasteboard, and papier-maché.

Paint and Varnish
As a—
Vehicle which will throw tough, flexible films without the necessary addition of driers or hardening substances.
Waterproof film forming medium.
Checking oil in—
Enamel liquids, metal paints, spar varnishes, white undercoats.
Heat-resisting agent in—
Enamels, japans, paints.
Ingredient of—
High-grade exterior coatings (to improve wearing and weather-resistance properties of linseed oil).
Putty.
Vehicle in—
Aluminum paints, baking japans, barn paints, canvas paints, enamels, exterior paints, flat wall paints, interior coatings, mill whites, oil tank paints, pigmented lacquers, roof paints, shingle stains and other stains, structural iron paints, varnishes, waterproof paints, white house paints, white pastes.

Soap
As a soapstock.

Textile
Dressing for oiled fabrics.
Oiling and softening agent for—
Fibers (prior to spinning and weaving).

Woodworking
Ingredient of—
Impregnating and waterproofing compounds.

Sarrapia

Perfume
Ingredient of—
Cosmetics, perfumes.

Tobacco
Flavoring for—
Smoking and chewing tobaccos.

Sawdust
French: Sciure de bois.
German: Holzmehl.

Abrasives
Source of reducing gases (Brit. 415392) in making—
Boron carbide.

Chemical
Starting point in making—
Activated chars, ethyl alcohol, methanol, oxalic acid.

Construction
Filler in—
Composition floorings.
Ingredient of—
Wallboard compositions.

Explosives and Matches
Absorbent in—
Dynamites, permissible explosives, pyrotechnics, safety explosives.

Food
Packing for—
Eggs, fruits, vegetables.

Fuel
Ingredient of—
Briquetted fuels.
Absorbent and combustion agent in—
Fire-kindlers (consisting mostly of mixtures with such products as paraffin, rosin, pitch, mineral oil, distillery waste, charcoal, coal dust).

Leather
Suggested absorbent for—
Drying oily leathers.

Linoleum and Oilcloth
Filler in—
Linoleum, oilcloth.

Miscellaneous
As an absorbent in many products.
As a filler in many products.
Ingredient of—
Floor-sweeping compounds (consisting of mixtures with such products as sand, salt, paraffin oil, water, heavy mineral oil, iron oxide, naphthalene flakes, odorants).
Stuffing in—
Upholstery in cheap furniture.
Substitute for—
Ground cork in sound-absorbing compositions.

Paint and Varnish
Filler in—
Wood-fillers of the plastic wood type.
Sound-absorbent (Brit. 408930) in—
Rubberized wall paints.

Paper
Filler in—
Paperboard and similar rough paper products.

Plastics
Absorbent and filler in—
Plastics and molded products.

Refrigeration
Insulating covering material for—
Ice.
Insulating filler in—
Iceboxes, storage-rooms.

Resins
Filler in—
Synthetic resins.

Woodworking
Ingredient of—
Pressed moldings.

S-Benzylisothiourea Hydrochloride
French: Hydrochlorure de S-benzylisothiourée.
German: Chlorwasserstoffsaeure-S-benzylisothioharnstoffester, Chlorwasserstoffsaeures-S-benzylisothioharnstoff, S-Benzylisothioharnstoffchlorhydrat, S-Benzylisothioharnstoffhydrochlorid.

Chemical
Starting point (Brit. 262155) in making therapeutic compounds with—
Anilin, benzylamine, diphenylamine, meta-anisidin, metaphenylenediamine, metatoluidin, metaxylidin, monoethylanilin, monomethylanilin, naphthylamine, orthoanisidin, orthophenylenediamine, orthotoluidin, orthoxylidin, para-anisidin, paraphenylenediamine, paratoluidin, paraxylidin, phenylamine.

S-Butylisothiourea Hydrochloride
French: Chlorhydrate de S-butyleisothiourée, Hydrochlorure de S-butyleisothiourée.
German: S-Butylisoharnstoffchlorhydrat, Chlorwasserstoffsaeure-S-butylisoharnstoffester.

Chemical
Starting point (Brit. 262155) in making therapeutic compounds with—
Anilin, benzylamine, diphenylamine, meta-anisidin, metaphenylenediamine, metatoluidin, metaxylidin, monoethylanilin, monomethylanilin, naphthylamine, orthoanisidin, orthophenylenediamine, orthotoluidin, orthoxylidin, para-anisidin, paraphenylenediamine, paratoluidin, paraxylidin.

Scammony Resin
Latin: Scammoniae resina.
French: Résine de scammonée.
German: Scammoniaharz, Skammonium, Windenharz.
Italian: Resina di scammonea.

Pharmaceutical
In compounding and dispensing practice.

Scatol
Synonyms: Betamethylindol, 3-Methylindol.
German: Skatol.

Perfume
As a fixative in synthetic perfumes.

Scopolamine
Synonyms: Hyoscine.
Chemical
Starting point (Brit. 273279) in making therapeutic compounds with—
Camphorates, malonates, meconates, phthalates, phosphates, saccharates, sulphates, sulphites, tartrates, terephthalates.
Pharmaceutical
In compounding and dispensing practice.

Scopolamine Camphorate
German: Kamphersaeurescopolaminester, Kamphersaeuresscopolamin, Kamphorsaeuresscopolamin.
Pharmaceutical
In compounding and dispensing practice.

Scopolamine Hydrochloride
French: Hydrochlorure de scopolamine.
German: Scopolaminchlorhydrat.
Chemical
Starting point in making—
Scopomorphine.
Pharmaceutical
In compounding and dispensing practice.

S-Diphenylurea-3:3′-dicarboxylic Acid
Chemical
Starting point in making—
Esters, salts, and other derivatives.
Starting point (Brit. 314909) in making derivatives with—
Alkoxyalphanaphthalenesulphonic acids.
Alpha-amino-5-naphthol-7-sulphonic acid.
Alphanaphthylamine-4:8-disulphonic acid.
Alphanaphthylamine-4:6:8-trisulphonic acid.
4-Aminoacenaphthene-3:5-disulphonic acid.
4-Aminoacenaphthene-3-sulphonic acid.
4-Aminoacenaphthene-5-sulphonic acid.
4-Aminoacenaphthenetrisulphonic acid.
Aminoarylcarboxylic acids.
Aminoheterocyclic carboxylic acids.
1:8-Aminonaphthol-3:6-disulphonic acid.
Bromonitrobenzoyl chloride, chloroalphanaphthalenesulphonic acids, chloronitrobenzoyl chlorides, iodonitrobenzoyl chlorides, nitroanisoyl chlorides, nitrobenzene sulphochlorides, nitrobenzoyl chlorides, 2-nitrocinnamyl chloride, 3-nitrocinnamyl chloride, 4-nitrocinnamyl chloride, 1-nitronaphthalene-5-sulphochloride, 1:5-nitronaphthoyl chloride, 2-nitrophenylacetyl chloride, 4-nitrophenylacetyl chloride, nitrotoluyl chlorides.

Seal Oil
French: Huile de phoque, Huile de veau marin.
German: Robbentran, Seehundstran.
Spanish: Aceite de foca.
Italian: Olio di foca.
Fats and Oils
Starting point in making—
Degras, sod oil.
Fuel
As a fuel oil.
As an illuminant.
As a special illuminant in lamps of lighthouses, signal lamps, and the like.
Ingredient of—
Compositions used in making candles.
Ink
Ingredient of—
Marking inks (used as a vehicle), printing inks.
Jewelry
Lubricant in making—
Watches.
Leather
Ingredient of—
Dressing compositions, finishing compositions.
Reagent in—
Oil tanning.
Reagent in making—
Chamois leather.
Mechanical
As a special lubricant.
Ingredient of—
Special lubricating compositions.
Metallurgical
In special polishing work.
Ingredient of—
Oil baths used for tempering special steels.
Paint and Varnish
Ingredient of—
Oil stains, paints, varnishes.
Substitute for—
Linseed oil.
Pharmaceutical
Used in compounding and dispensing practice in the place of codliver oil.
Soap
Soapstock in making—
Hard and soft soaps.
Textile
For oiling woolen yarns and fabrics.

Sebacic Acid
Synonyms: Decan-diacid, Pyroleic acid, Sebacinic acid, Sebacylic acid.
French: Acide décandioique, Acide pyroléique, Acide sébacinique, Acide sébacique, Acide sébacylique.
German: Decan-disäure, Pyroleinsäure, Sebacinsäure, Sebacylsäure.
Spanish: Acido sebacico, Acido sebacilico, Acido sebacinico.
Italian: Acido sebacico, Acido sebacilico, Acido sebacinico.
Analysis
Reagent in testing for—
Thorium.
Ceramics
Ingredient (Brit. 341447) of—
Compositions, containing various esters or ethers of cellulose, such as cellulose acetate and nitrocellulose, as well as resins, used for coating and decorating ceramic ware.
Chemical
Reagent in making—
Emulsifying agents from amino alcohols and organic acids (Brit. 394657).
Salts and esters of various bases.
Starting point in making various esters, such as—
Amyl sebacinate, ethyl sebacinate, methyl sebacinate.
Fuel
Ingredient of—
Compositions used in making candles.
Glass
Ingredient (Brit. 341447) of—
Compositions, containing various esters or ethers of cellulose, such as nitrocellulose and cellulose acetate, and also resins, used for the decoration and protection of glassware.
Leather
Ingredient (Brit. 341447) of—
Compositions, containing various esters or ethers of cellulose, such as cellulose acetate and nitrocellulose, as well as resins, used in the manufacture of artificial leather and for the decoration and protection of leather goods.
Metallurgical
Ingredient (Brit. 341447) of—
Compositions, containing various esters or ethers of cellulose, such as cellulose acetate and nitrocellulose, as well as resins, used for the decoration and protection of metallic products.
Miscellaneous
Ingredient (Brit. 341447) of—
Compositions, containing various esters or ethers of cellulose, such as cellulose acetate and nitrocellulose, as well as resins, used for the decoration and protection of various fibrous and porous compositions.
Paint and Varnish
Ingredient (Brit. 341447) of—
Paints, varnishes, lacquers, enamels, and dopes containing various esters or ethers of cellulose, such as cellulose acetate and nitrocellulose, and also resins.
Paper
Ingredient (Brit. 341447) of—
Compositions, containing various esters or ethers of cellulose, such as cellulose acetate and nitrocellulose, as well as resins, used in the manufacture of coated paper and also for the decoration and protection of porous paper and pulp products.
Plastics
Ingredient of—
Plastic compositions containing various esters or ethers of cellulose, such as cellulose acetate and nitrocellulose, and resins.

Sebacic Acid (Continued)
Substitute for—
Camphor in making celluloid and other compositions.

Resins and Waxes
Starting point (Brit. 396354) in making—
Synthetic resins from diethyleneglycol, phthalic acid, and glycerin.
Diethyleneglycol, phthalic acid, and mannitol.
Diethyleneglycol, phthalic acid, and pentaerythritol.
Ethyleneglycol, phthalic acid, and glycerin.
Ethyleneglycol, phthalic acid, and mannitol.
Ethyleneglycol, phthalic acid, and pentaerythritol.
Propyleneglycol, phthalic acid, and glycerin.
Propyleneglycol, phthalic acid, and mannitol.
Propyleneglycol, phthalic acid, and pentaerythritol.
Tetramethyleneglycol, phthalic acid, and glycerin.
Tetramethyleneglycol, phthalic acid, and mannitol.
Tetramethyleneglycol, phthalic acid, and pentaerythritol.

Rubber
Ingredient (Brit. 341447) of—
Compositions, containing various esters or ethers of cellulose, such as cellulose acetate and nitrocellulose, as well as resins, used for the decoration and protection of rubber goods.

Stone
Ingredient (Brit. 341447) of—
Compositions, containing various esters or ethers of cellulose, such as nitrocellulose and cellulose acetate, and also resins, used for the decoration and protection of natural and artificial stone.

Textile
Ingredient (Brit. 341447) of—
Compositions, containing various esters or ethers of cellulose, such as cellulose acetate and nitrocellulose, used for the coating of woven fabrics.

Woodworking
Ingredient (Brit. 341447) of—
Compositions, containing various esters or ethers of cellulose, such as cellulose acetate and nitrocellulose, as well as resins, used for the decoration and protection of woodwork.

Sebacic Acid Ester of Grapeseed Alcohol

Bituminous
Solvent (Brit. 445223) for—
Asphalt and other bituminous bodies.

Dye
Solvent (Brit. 445223) for—
Dyestuffs, particularly oil-soluble coaltar dyes.

Fats, Oils, and Waxes
Solvent (Brit. 445223) for—
Fats, oils, waxes.

Resins
Solvent (Brit. 445223) for—
Oil-soluble glycerol-phthalic acid resins.
Polymerized vinyl compounds.
Synthetic resins.

Rubber
Solvent (Brit. 445223) for—
Rubber.

Sebacic Acid Ester of Ricinoleic Alcohol

Bituminous
Solvent (Brit. 445223) for—
Asphalt and other bituminous bodies.

Dye
Solvent (Brit. 445223) for—
Dyestuffs, particularly oil-soluble coaltar dyes.

Fats, Oils, and Waxes
Solvent (Brit. 445223) for—
Fats, oils, waxes.

Resins
Solvent (Brit. 445223) for—
Oil-soluble glycerol-phthalic acid resins.
Polymerized vinyl compounds.
Synthetic resins.

Rubber
Solvent (Brit. 445223) for—
Rubber.

Sebacylic Acid
See: Sebacic acid.

2-Secondary-octyl-4-chlorophenol

Pharmaceutical
Bactericide (U. S. 2101595) for—
Bacillus typhosus.
Other bacteria.
Staphylococcus aureus.

Selachyl Alcohol

Chemical
Starting point (Brit. 398818) in making—
Detergent, by sulphonation with sodium thiosulphate.

Selenic Anhydride
Synonyms: Selenium trioxide.
French: Anhydride sélénique.
German: Selensaeuresanhydrid, Selentrioxyd.

Paint and Varnish
Ingredient of—
Luminous preparations, to increase the phosphorescence.

Selenium
French: Sélénium.
German: Selen.

Analysis
Reagent in various analytical methods in chemical and physiological laboratories.

Ceramics
Ingredient of—
Glazes used for the purpose of producing ruby-red effects.

Chemical
Control medium in making—
Sulphuric acid by the contact catalytic process.
Reagent in making—
Pharmaceutical chemicals.
Reagent in biological chemistry.
Starting point in making—
Selenates, selenic acid, selenium oxychloride.

Electrical
As a metallic base in making—
Electrodes for arc lights.
Electric torpedoes.
Electrical instruments and apparatus of various types.
Selenium cells.
Telautograph apparatus.
Wireless telephony apparatus.
Telephotographic apparatus.
Used as a coating in flameproofing—
Electric switchboard cables and wires.

Glass
Decolorizing agent in making—
Glass (used for the purpose of neutralizing the yellowish effects which are produced by traces of iron in the raw materials).
Pigment in making—
Orange-colored glass, pink-colored glass.
Red-colored glass, particularly suitable for making railroad signal lights, sailing signal lights, automobile tail lights.
Ruby-red glass (used in the place of manganese).

Mechanical
Used for various mechanical purposes, in the form of fine-drawn wire.

Miscellaneous
As a metal in making—
Control apparatus for chimney drafts.
Octophones.
Self-lighting buoys.
Sound photographing apparatus.
Ventilation control apparatus.
Reagent in—
Bacteriology.
Microscopy (used particularly in imbedding material in making microscopical examinations).

Pharmaceutical
In compounding and dispensing practice.

Photographic
Ingredient of—
Toning baths.

Rubber
Accelerator in vulcanizing.
Vulcanizing agent in the processing of rubber.

Seleniumdiethyldithiocarbamate
French: Diéthyledithiocarbamate de sélénium.
German: Seleniumdiaethyldithiocarbamat.

Rubber
Accelerator in vulcanization (U. S. 1622534).

Selenium Oxychloride
French: Oxychlorure sélénique, Oxychlorure de sélénium.
German: Selenoxychlorid.

Chemical
Solvent for—
Products that are difficult to dissolve.

Miscellaneous
Solvent for—
Substances that are hard to dissolve.

Resins and Waxes
Solvent for—
Synthetic phenolic resins.

Selenium Sulphide

Plastics
Ingredient (Brit. 351188) of—
Thermoplastic compositions containing asbestine, slate, iron oxide, talc, clay, marble dust, ground flint, black natural slate, diatomaceous earth, woodflour, mica, and the like, used for making gears, insulating material, acidproof coating on iron, and for other purposes.

Serpolet
Synonyms: Mother of thyme, Pellamountain, Quendel, Wild thyme.
Latin: Thymus serpylum.
German: Feldkuemmel, Feldthymian, Gründling, Wilder thymian.

Oils and Fats
Raw material for an essential oil.

Pharmaceutical
In compounding and dispensing practice.

S-Ethylisothiourea Hydrobromide
French: Hydrobromure de S-éthyleisothiourée.
German: Bromwasserstoffsaeures-S-aethylisothioharnstoffester, S-aethylisothioharnstoffhydrobromid.

Chemical
Starting point (Brit. 262155) in making therapeutic compounds with—
Anilin, benzylamine, diphenylamine, meta-anisidin, metaphenylenediamine, metatoluidin, metaxylidin, monoethylanilin, monomethylanilin, naphthylamine, orthoanisidin, orthophenylenediamine, orthotoluidin, orthoxylidin, para-anisidin, paraphenylenediamine, paratoluidin, paraxylidin, phenylamine.

Shale Tar Oil
French: Huile de brai de schiste.
German: Schieferteeroel.

Agricultural
Animal dip.
Ingredient of—
Sheep dips.

Dye
Ingredient (Brit. 269942) of—
Dye preparations.

Insecticide
Ingredient of—
Vermin destroying compositions.

Paint and Varnish
Ingredient of—
Paints, varnishes.

Textile
——, *Bleaching*
Wetting agent in bleaching—
Textile yarns and fabrics.
——, *Dyeing*
Ingredient of—
Dye liquors.

Shellac
Synonyms: Bleached shellac, Button lac, Garnet lac, Gum lac, Lac, Mecca, Stick lac.
Latin: Gummi lacca, Lacca, Resina lacca.
French: Lacque, Gomme lacque.
German: Gummilack, Lack.

Abrasives
Binder for—
Abrasive compositions, grinding wheels.

Construction
Cementing agent for—
Artificial stone, marble.
Filling agent for—
Artificial stone, marble.
Polishing agent for—
Artificial stone, marble.

Electrical
Bonding and insulating agent in—
Electrical condensers.
Cementing agent for—
Electric lamp bases and caps.
Damp-proofing and insulating coating for—
Electrical appliances.
Electrical coils and windings.
Electrical motors, generators, and other machines.
Filler and binder in—
Molded insulators.
Sealing agent for—
Dry batteries.

Explosives and Matches
Coating agent for—
Shell case interiors.
Flame carrier in—
Fireworks, military signals, pyrotechnic signals.
Ingredient of—
Matchhead compositions.

Food
Glazing agent for—
Coffee, chocolates and other candies.

Glass
Cementing agent for—
Lenses for grinding.

Glues and Adhesives
As a cement.
Ingredient of—
Casein glues, gelatin glues, ordinary glue compositions.

Ink
Thickening agent in—
Embossing inks, printing ink, writing ink.

Leather
Enamelling agent.
Ingredient of—
Dressings, finishes, polishes.
Waterproofing agent.

Linoleum and Oilcloth
Filler and binder in—
Linoleum, oilcloth.

Metallurgical
Protective coating for—
Aluminum foil, tin foil.

Miscellaneous
Binder in many industrial products.
Cementing agent for many industrial products.
Coating agent for—
Crayons, many industrial products, marking chalks, writing chalks.
Filler in many industrial products.
Glazing agent for many industrial products.
Stiffener for—
Felt hats, straw hats.
Stiffener in many industrial products.
Thickener in many industrial products.
Waterproofing and tightening agent for—
Cordage, fishing tackle, rope.

Paint and Varnish
Ingredient of—
Gold size for metal joints, lacquers, stains, varnishes.
Restrainer for—
Resin in wood and tarred surfaces.

Paper
Glazing agent for—
Art paper, boxboard, cartons, paper, paper boxes, playing cards, visiting cards.
Ingredient of—
Coatings, sizes.

Pharmaceutical
Glazing agent for—
Pills, tablets.

Photographic
Cementing agent in—
Dry-mounting prints.
Protective coating (Brit. 374735 and 397740) for—
Antihalation layer on the back of photographic film to protect it from splitting.

Shellac (Continued)

Plastics
Agent for—
Preventing suction in plaster casts.
Cementing agent for—
Smooth surfaces under heat and pressure.
Filler and binder in—
Buttons, molded insulating materials, molded novelties, phonograph records.
Ingredient of—
Artificial ivory.

Printing
Process material in—
Lithography, process engraving.

Resins and Waxes
Ingredient of—
Sealing waxes.

Rubber
Filler in—
Rubber compositions.

Textile
Stiffening agent for—
Crepe and other fabrics.
Tightening agent for—
Silk.

Woodworking
Ingredient of—
Furniture polishes, wood finishes.

Shiromoji Seed Oil
Synonyms: Lindera oil.
Latin: Oleum linderae trilobae.
French: Huile de lindera, Huile de graines de shiromoji, Huile de semences de shiromoji.
German: Linderaoel, Shiromojisamenoel.
Spanish: Aciete de lindera, Aciete de semilla de shiromoji.
Italian: Olio di lindera, Olio di shiromoji.

Fuel
As an illuminant.

Soap
As a soapstock.

Silica Black
French: Noir de silice.
German: Kieselsäureschwarz.
Spanish: Negro de silex.
Italian: Negro di silice.

Chemical
Carrier for various catalysts.

Fats and Oils
Absorbent for—
Animal and vegetable fats and oils.
Carrier for—
Nickel catalyst in the hydrogenation process.

Ink
Ingredient of—
Printing inks.

Insecticide
Diluent of—
Fungicides, insecticides, and the like.

Leather
Ingredient of—
Compositions used to coat and color leather.

Linoleum and Oilcloth
Pigmenting filler in—
Compositions used in the manufacture of oilcloth and linoleum.

Miscellaneous
Pigment in—
Various compositions of matter.

Paint and Varnish
Pigment in—
Paints and varnishes.

Woodworking
Reagent in—
Producing grain in wood.

Silicochloroform
Synonyms: Silicium chloroform.

Construction
Hardening agent (Brit. 260031) in treating—
Artificial stones, concretes, natural stones.
Preserving agent (Brit. 260031) in treating—
Artificial stones, concretes, natural stones.

Silicolauric Acid Anhydride

Chemical
Starting point (Brit. 395198) in making—
Dodecyl alcohol.

Silicon
French: Silicium, Silicon.
German: Silicium.

Chemical
Reagent in making—
Hydrogen with alkaline liquors.
Reducing agent, used in place of aluminum.

Metallurgical
Ingredient in making—
Ferrosilicon, silicon bronze, silicon copper.
Reagent (German 302305) in coating iron with ferrosilicon.

Silicon Disulphide
Synonyms: Silicon bisulphide.
French: Bisulfure de silicium, Disulfure de silicium.
German: Siliciumsulfid.

Construction
Reagent (Brit. 260031) for hardening—
Concrete and like structural material.

Stone
Hardening and preservative agent in treating—
Artificial stones, natural stones.

Silicon Methane
French: Silicon méthanique, Silicon de méthane.
German: Silicium methan.

Construction
Hardening agent (Brit. 260031) in treating—
Artificial stones, concretes, natural stones.
Preserving agent (Brit. 260031) in treating—
Artificial stones, concretes, natural stones.

Silicon Methide
German: Silicium methid.

Construction
Hardening agent (Brit. 260031) in treating—
Artificial stones, concretes, natural stones.
Preserving agent (Brit. 260031) in treating—
Artificial stones, concretes, natural stones.

Silicon Oxychloride
French: Oxychlorure de silicium, Oxychlorure de silicon.
German: Siliciumoxychlorid.

Cement
Reagent (Brit. 260031) in hardening—
Concretes.

Stone
Reagent in hardening—
Artificial stones, natural stones, stuccos.

Silicon Tetrachloride
Synonyms: Silicon chloride.
French: Tétrachlorure de silicium, Tétrachlorure de silicon.
German: Siliciumtetrachlorid, Tetrachlorsilicium.

Chemical
Reagent in making—
Acetic anhydride, acetyl chloride, ethylenechlorohydrin, organic silicon derivatives.
Reagent (Brit. 343165) in making therapeutic compounds with—
Calcium ricinoleate, dihydroxystearic acid, ethyl ricinoleate, lactic acid, methyl salicylate, ricinoleic acid, ricinoleic dibromide, vinyl salicylate.
Starting point in making—
Silicon esters, such as ethyl silicate and methyl silicate.

Construction
Reagent (Brit. 260031) in hardening and preserving—
Artificial stone, concrete, stone, stucco.

Metallurgical
Starting point in making—
Metallic silicon.

Military
Reagent in making—
Smoke screens.

Miscellaneous
Reagent in sky writing with airplanes.

Silicon Tetrafluoride
Synonyms: Silicon fluoride.
French: Tétrafluorure de silicium, Tétrafluorure de silicon.
German: Siliciumtetrafluorid, Tetrafluorsilicium.
Construction
Reagent (Brit. 260031) in hardening and preserving—
Artificial stone, concrete, stone, stucco.

Silicophenylacetic Anhydride
Chemical
Starting point (Brit. 395198) in making—
Phenylethyl alcohol.

Silicotungstic Acid
Synonyms: Tungstosilicic acid.
French: Acide silicotungstique, Acide tungstosilicique.
German: Silicowolframsäure, Wolframkieselsäure.
Spanish: Acido silicotungstico, Acido tungstosilico.
Analysis
Reagent for the detection and determination of—
Aconitine, antipyrine, atropine, brucine, nicotine, pyramidon, sparteine.
Reagent in making—
Specifically heavy solutions for the separation of minerals.
Chemical
Starting point in making various salts.
Textile
Mordant (German 286467 and 289878) in—
Dyeing fabrics and yarns with basic colors (used to fix the color giving shades very fast to light).

Silver Acetate
Latin: Argenti acetas.
French: Acétate d'argent.
German: Essigsaeuressilber, Silberacetat, Silberazetat.
Chemical
Ingredient of catalytic preparations used in making—
Acenaphthylene, acenaphthaquinone, bisacenaphthylidenedione, naphthaldehydic acid, naphthalic anhydride, and hemimellitic acid from acenaphthene (Brit. 295270).
Acetaldehyde from ethyl alcohol (Brit. 281307).
Acetic acid from ethyl alcohol (Brit. 281307).
Alcohols from aliphatic hydrocarbons (Brit. 281307).
Aldehydes or alcohols by the reduction of corresponding esters (Brit. 306471).
Alphacamphalide by the reduction of camphoric acid (Brit. 306471).
Aldehydes and acids from toluene, orthochlorotoluene, orthonitrotoluene, orthobromotoluene, metachlorotoluene, metabromotoluene, metanitrotoluene, parabromotoluene, parachlorotoluene, paranitrotoluene, dichlorotoluenes, dibromotoluenes, dinitrotoluenes, chloronitrotoluenes, chlorobromotoluenes, bromonitrotoluenes (Brit. 295270).
Aldehydes and acids from xylenes, pseudocumene, mesitylene, and paracymene (Brit. 295270).
Alphanaphthaquinone from naphthalene (Brit. 281307).
Anthraquinone from naphthalene (Brit. 295270).
Benzaldehyde and benzoic acid from toluene (Brit. 281307).
Benzoquinone from phenanthraquinone (Brit. 281307).
Benzyl alcohol by the reduction of benzaldehyde (Brit. 306471).
Benzyl alcohol or benzaldehyde or phthalide from the reduction of phthalic anhydride (Brit. 306471).
Butyl alcohol by the reduction of crotonaldehyde (Brit. 306471).
Chloroacetic acid from ethylene chlorohydrin (Brit. 295270).
Diphenic acid from ethyl alcohol (Brit. 295270).
Ethyl alcohol by the reduction of acetaldehyde (Brit. 306471).
Fluorenone from fluorene (Brit. 292570).
Formaldehyde by the reduction of methanol or methane (Brit. 306471).
Formaldehyde by the reduction of carbon dioxide or carbon monoxide (Brit. 306471).
Isopropyl alcohol by the reduction of acetone (Brit. 306471).
Maleic acid and fumaric acid by the oxidation of toluene, benzene, phenols, tar phenols, furfural, or from benzoquinone or phthalic anhydride (Brit. 295270).
Methane by the reduction of carbon dioxide or carbon monoxide (Brit. 306471).
Methanol by the reduction of carbon dioxide or carbon monoxide (Brit. 306471).
Naphthaldehydic acid, acenaphthaquinone, or bisacenaphthylidenedione from acenaphthylene (Brit. 281307).
Phenanthraquinone from phenanthrene or diphenic acid (Brit. 295270).
Phthalic acid and maleic acid from naphthalene (Brit. 295270).
Primary alcohols by the reduction of the corresponding aldehydes (Brit. 306471).
Propionic acid and butyric acid and higher alcohols, ketones, and acids by the reduction of carbon dioxide or carbon monoxide (Brit. 306471).
Hydroxy derivatives of anthraquinone, benzoquinone, and the like (Brit. 306471).
Reduction products of carbon dioxide and carbon monoxide (Brit. 306471).
Reduction products of ketones, aldehydes, acids, esters, alcohols, ethers, and other organic compounds containing oxygen (Brit. 306471).
Salicylic acid and salicylic aldehyde from cresol (Brit. 295270).
Secondary butyl alcohol by the reduction of methylethyl ketone (Brit. 306471).
Valeryl alcohol by the reduction of valeraldehyde (Brit. 306471).
Vanillin and vanillic acid from eugenol or isoeugenol (Brit. 295270).
Starting point in making—
Acetates of other metals, other silver salts.
Pharmaceutical
In compounding and dispensing practice.

Silver Benzenesulphinate
Photographic
Sensitizing agent (German 622866) for—
Photographic plates.

Silver Bromate
French: Bromate d'argent.
German: Bromsaeuressilber, Silberbromat.
Photographic
Ingredient of—
Sensitizing solutions for films and plates (Brit. 253380).

Silver Chlorate
French: Chlorate d'argent.
German: Chlorsaeuressilber, Silberchlorat.
Pharmaceutical
In compounding and dispensing practice.
Photographic
Ingredient of—
Sensitizing solutions for films and plates (Brit. 253380).
Sanitation
Ingredient of—
Disinfecting compositions.

Silver Cholalate
French: Cholalate d'argent.
German: Silbercholalat.
Chemical
Starting point in making—
Complex silver salts by treatment with cyanides (German 423231).
Pharmaceutical
In compounding and dispensing practice.

Silver Erucate
French: Érucate d'argent, Érucate argentique.
German: Erucinsäuressilber, Erucinsäuressilberoxyd, Silbererucat.
Chemical
Ingredient of—
Pharmaceutical products (used in the place of silver nitrate and silver caseins, proteins, vitellins).
Pharmaceutical
In compounding and dispensing practice.

Silver Glycocholate
French: Glycocholate d'argent.
German: Silberglycocholat.
Chemical
Starting point in making—
Complex silver salts by treatment with cyanides (German 423231).
Pharmaceutical
In compounding and dispensing practice.

Silver Laurate
French: Laurate d'argent, Laurate argentique.
German: Laurinsäuressilber, Laurinsäuressilberoxyd, Silberlaurat.

Chemical
Ingredient of—
Pharmaceutical products (used in the place of silver nitrate, silver caseins, proteins, vitellins).

Pharmaceutical
In compounding and dispensing practice.

Silver Nitrate
Synonyms: Lapis caustic, Luna caustic, Lunar caustic, Nitrate of silver.
Latin: Argenti nitras, Argentum nitricum.
French: Azotate d'argent, Azotate argentique, Nitrate d'argent, Nitrate argentique.
German: Hoellenstein, Salpetersaeuressilber, Silbernitrat, Silbersalpeter.
Spanish: Nitrato de plata.
Italian: Nitrato di argentico.

Analysis
Reagent in various processes.

Ceramics
Ingredient of—
Compositions used to decorate porcelains, fine potteries, and chinaware with fire colors.
Compositions used to coat porcelains, potteries, and chinaware so that they conduct electric current.

Chemical
Ingredient of—
Yeast preparations that contain metals (German 424658).
Reagent in making—
Albargin, argochron, beta-aminopropionic acid, linalool.
Starting point in making—
Silver salts of acids and halogens.
Catalyst in oxidizing—
Toluene to benzaldehyde.

Glass
Ingredient of—
Compositions used in producing coatings on glass, that conduct electric current.
Reagent in making—
Mirrors by silvering glass.
Special sorts of glass.
Reagent in producing—
Marks, pictures, and the like on glass (U. S. 1592329).

Ink
Ingredient of—
Indelible inks.

Metallurgical
Ingredient of—
Bath used to coat various metals with silver by galvanic action.

Miscellaneous
Coloring agent for—
Marble, mother of pearl.
Disinfectant and causticizing agent.

Perfumery
Ingredient of—
Hair dyes.

Pharmaceutical
In compounding and dispensing practice.

Photographic
Ingredient of—
Compositions used in coating silver bromide gelatin plates.
Compositions used in coating photographic paper.

Silver Oleate
French: Oléate d'argent, Oléate argentique.
German: Oleinsäuressilber, Oleinsäuressilberoxyd, Silberoleat.

Chemical
Ingredient of—
Pharmaceutical products (used in place of silver nitrate, silver caseins, proteins, vitellins).

Pharmaceutical
In compounding and dispensing practice.

Silver Palmitate
French: Palmitate d'argent, Palmitate argentique.
German: Palmitinsäuressilber, Palmitinsäuressilberoxyd, Silberpalmitat.

Chemical
Ingredient of—
Pharmaceutical products (used in place of silver nitrate, silver caseins, proteins, vitellins).

Pharmaceutical
In compounding and dispensing practice.

Silver Perchlorate
French: Perchlorate d'argent.
German: Perchlorsaeuressilber, Silberperchlorat.

Pharmaceutical
In compounding and dispensing practice.

Photographic
Ingredient of—
Sensitizing solutions for films and plates (Brit. 253380).

Silver Resinate
Synonyms: Resinate of silver.
French: Résinate d'argent.
German: Silberresinat.

Ceramics
Pigment in producing brilliant colors in—
Chinaware, porcelains, potteries.

Silver Stearate
French: Stéarate d'argent, Stéarate argentique.
German: Silberstearat, Stearinsäuressilber, Stearinsäuressilberoxyd.

Chemical
Ingredient of—
Pharmaceutical products (used in place of silver nitrate, silver caseins, proteins, vitellins).

Pharmaceutical
In compounding and dispensing practice.

Slag

Agriculture
As a fertilizer (particularly valuable for grazing lands; this applies to Thomas, or basic, slag).

Construction
Aggregate in—
Concrete.
As an insulating medium.
Ballast and filling material in—
General constructional projects, highway construction, land reclamation.

Cement
Starting material in making—
Slag cements, underwater cements.

Railroading
Ballast in—
Roadbeds.

Slag Wool

Construction
Fireproofing medium in buildings.
Heat-insulating medium in buildings.
Sound-insulator in buildings.

Miscellaneous
As a filtering medium.
As a packing.
Heat-insulating medium for—
Airducts, boilers, furnaces, ovens.

Refrigeration
Heat-insulating medium in—
Electric refrigerators.
Ice and cold-storage installations of all kinds.

Slip Clay

Abrasives
Ingredient of—
Artificial abrasives.

Ceramics
Ingredient of—
Enamelings, coatings, and glazes for graniteware, stoneware, electrical porcelain, potteries.
Sealing agent for—
Kiln door wickets.

S-Methylisothiourea Hydroiodide
French: Hydroiodure de S-méthyleisothiourée.
German: Jodwasserstoffsaeures-S-methylisothiourea, S-Methylisothioharnstoffhydrojodid.

S-Methylisothiourea Hydroiodide (Continued)

Chemical
Starting point (Brit. 262155) in making therapeutic compounds with—
Anilin, benzylamine, diphenylamine, meta-anisidin, metaxylidin, monoethylanilin, monomethylanilin, orthoanisidin, orthophenylenediamine, orthotoluidin, metaphenylenediamine, metatoluidin, orthoxylidin, para-anisidin, paraphenylenediamine, paratoluidin, paraxylidin, phenylamine.

Soap Works Grease
French: Graisse des usines à savon.
German: Seifenfabrikfett.

Chemical
Ingredient (Brit. 305742) of—
Emulsions of coaltar derivatives, oils, and bituminous substances.
Insecticide
Ingredient of—
Sheep dips.
Miscellaneous
Ingredient of—
Compositions used for road building purposes.
Soap
Ingredient of—
Detergents.

Sodium Acetate
Synonyms: Acetate of soda.
Latin: Acetas sodicus, Natrium aceticum, Terra foliata tartari crystallisata, Terra foliata tartari.
French: Acétate sodique, Acétate de soude, Terre foliée minérale.
German: Essigsäuresnatrium, Essigsäuresnatron, Natriumacetat, Natriumazetat.
Spanish: Acetato sodico, Acetato de sosa.
Italian: Acetato di sodio.
Analysis
Reagent in detection of—
Creatinine, gallic acid, glucose, tannin.
Reagent in the determination of—
Narceine, narcotine, papaverine, phosphoric acid.
Reagent in precipitating—
Iron and aluminum.
Reagent in quantitative separation of opium alkaloids.
Chemical
Catalyst (German 439695) in making—
Camphene from bornyl chloride.
Reagent in making—
Acetyl-1-naphthylamine-6-sulphonic acid, acetylalphanaphthylamine-5-sulphonic acid, acetphenetidin, acetonal, aluminum-sodium acetate, aminopara-acetanilide, amyl acetate from pentane, 4-anilide-1-methylanthraquinone, barium acetate, benzyl acetate, bismuth acetate, bismuth basic gallate, bismuth oxyiodide, cadmium acetate, calcium acetate, copper acetate, copper oxychloride, coumarin, cystopurin, ethyl acetate ethylene-ethenyldiamine, hydrazotoluene, ionone, iron acetate, lead acetate, magnesium acetate, manganese acetate, menthyl acetate, mercury acetate, methylpara-aminophenol sulphonate, orthonitrobenzaldehyde, orthonitrodiphenylamine, paradimethylaminobenzaldehyde, phenanthraquinone, phenylorthophenylenediamine, salophen, sulpho-halogen-amide carboxylates, strontium acetate, tetramethyldiarsin (cacodyl), tin acetate, triacetylchrysarobin, zinc acetate.
Other intermediates, organic chemicals, pharmaceutical chemicals, and synthetic aromatic chemicals.
Reagent in—
Carrying out dehydration reactions in the synthesis of intermediates and other synthetic chemicals, for example, in the preparation of cinnamic acid by the Perkin method.
Reagent in—
Separating various alkaloids from opium.
Starting point in making—
Acetic anhydride, acetyl chloride, acetic ester, acetone, carbon monoxide, methanol, methane, pure acetic acid.
Dye
Catalyst (Brit. 252182) in making—
Azo dyestuffs from anilin-2:5-disulphonic acid and amino-4-cresol.
Reagent in making—
Algol bordeaux B, 2:5-diaryldiparabenzoquinones, greenish blue dyestuffs, immediate black V extra, paranitranilin red.
Reagent in—
Neutralizing acid in diazo solutions.
Food
Preservative of meats.
Used in cold storage of foods.
Leather
Mordant in—
Dyeing leathers.
Miscellaneous
Ingredient of—
Bleaching liquor mixtures.
Mordant in—
Dyeing various products.
Reagent in—
Foot warmers, milk thermophores, chafing dishes, and hot-water bottles, in which use is made of the heat given off by the fused chemical.
Paint and Varnish
Reagent in making—
Schweinfurt green.
Paper
Mordant in—
Dyeing paper and pulp products.
Pharmaceutical
Suggested for use as diuretic.
Photographic
Reagent in photographical processes.
Soap
Reagent in making special soaps.
Sugar
Reagent in—
Purifying glucose.
Textile
——, *Dyeing*
Assist (French 595705) in—
Dyeing cottons with developed colors, particularly brown shades.
Mordant in—
Dyeing yarns and fabrics.
Reagent in—
Dyeing yarns and fabrics with paranitranilin red (used to develop the color).
Resist in—
Dyeing with anilin black.
——, *Finishing*
Neutralizing agent in treating—
Cotton whose color has been refreshened by treatment with acid.
——, *Printing*
Ingredient of—
Printing pastes (added to protect the fibers).
Mordant in—
Printing various fabrics.

Sodium Acetone-Bisulphite

Chemical
Starting point in making—
Acetone in pure state.
Photographic
As a reagent.
Textile
Reagent in—
Dyeing and printing.

Sodium Acid Adipinate

Leather
Buffer (Brit. 444184) in—
Obtaining level dyeings with acid or substantive dyes (the dyed effects are said to have great resistance to soap and alkalies).

Sodium Acid Diglycollate

Leather
Buffer (Brit. 444184) in—
Obtaining level dyeings with acid or substantive dyes (the dyed effects are said to have great resistance to soap and alkalies).

Sodium Acid Phosphate
Synonyms: Monobasic sodium phosphate, Monosodium hydrogen phosphate, Monosodium orthophosphate, Monosodium phosphate, Sodium biphosphate, Sodium dihydrogen phosphate.
Latin: Sodii phosphas acidus.
French: Phosphate sodique acide, Phosphate de soude acide.

Sodium Acid Phosphate (Continued)
German: Natriumbiphosphat, Phosphorsäuresnatriumwasserstoff.
Italian: Bifosfato di sodio.

Analysis
As a reagent.

Food
Ingredient of—
Baking powders.

Miscellaneous
As a—
Boiler-water softening agent in conjunction with ammonia.

Pharmaceutical
In compounding and dispensing practice.
Suggested for use in—
Increasing acidity of the urine.

Sodium Acid Phthalate

Leather
Buffer (Brit. 444184) in—
Obtaining level dyeings with acid or substantive dyes (the dyed effects are said to have great resistance to soap and alkalies).

Sodium Acid Saccharate

Leather
Buffer (Brit. 444184) in—
Obtaining level dyeings with acid or substantive dyes (the dyed effects are said to have great resistance to soap and alkalies).

Sodium Alginate
French: Alginate sodique, Alginate de sodium, Alginate de soude.
German: Alginsäuresnatrium, Alginsäuresnatron, Natriumalginat.
Spanish: Alginato de sodico.
Italian: Alginato di sodio.

Ceramics
Ingredient of—
Compositions used for the waterproofing of various ceramic products.

Chemical
Emulsifying agent in making—
Dispersions of various chemicals.
Ingredient of—
Various chemical liquids (added for the purpose of increasing their viscosity).
Reagent (French 570636) in—
Treating various chemical liquids, as well as solutions of pharmaceutical products, for the purpose of purifying and clarifying them.
Stabilizing agent in—
Emulsions of various chemical products.
Starting point in making—
Iodinated pharmaceutical products.

Construction
Ingredient of—
Compositions used for treating cement and concrete for the purpose of preventing deterioration when exposed to the action of alkalies and seawater.
Waterproofing compositions used for treating plaster of paris, wallboard, cement, stucco, concrete.

Fats and Oils
Stabilizing agent in—
Emulsions of various animal and vegetable fats and oils.

Fuel
Binder in—
Coal dust composition fuel briquettes (used in place of pitch).
Non-smoking fuel briquettes (used because it burns without developing large amounts of smoke, as do the usual binders employed for this purpose).

Gas
Binder (French 518037) in—
Preparations, containing graphite, lampblack, and antiseptics, used for repairing stoves.

Glues and Adhesives
Ingredient (French 563726) of—
Adhesive preparations.
Reagent in—
Treating solutions of gelatin, glue, and other adhesives for the purpose of purifying and clarifying them.

Ink
Ingredient (French 563726) of—
Various inks.
Thickening agent in—
Printing inks.

Leather
Ingredient of—
Compositions containing various fatty substances and used in the preparation of emulsions for tanning and tawing (French 533465).
Sizing compositions (French 563726).

Mechanical
Ingredient of—
Compositions used for covering steel tubes.
Compositions containing sodium carbonate and used as boiler compounds (added for the purpose of improving the water-softening properties of the carbonate).

Metallurgical
Binder (French 518037) in—
Compositions, containing graphite, lampblack, and antiseptics used for repairing metallurgical furnaces and ovens.

Miscellaneous
Binder in—
Compositions of powdered mica, asbestos, coal, carbon, graphite, minerals, and the like.
Sizing compositions for various purposes (used in place of starches and gum tragacanth to produce a size of greater elasticity and transparency) (French 563726).
Ingredient of—
Antigrease coatings (French 563726).
Compositions used for treating rope and twine.
Compositions used for waterproofing purposes.
Stabilizing agent in—
Emulsions of various substances.

Paint and Varnish
Ingredient of—
Compositions used for proofing interior walls and ceilings.
Various paints, lacquers, and enamels (French 563726).

Paper
Binder (French 563726) in—
Sizing compositions (used in place of starches and gum tragacanth to give a more elastic and transparent product).
Ingredient of—
Compositions for finishing paper.
Compositions used for waterproofing pulp and paper products.
Compositions containing woodflour.
Reagent in—
Treating waste liquors and the like for the purpose of purifying them and clarifying them.

Petroleum
Ingredient of—
Emulsions of petroleum and petroleum distillates (added for the purpose of securing better dispersion).

Plastics
Binder in making—
Various plastic compositions containing such substances as horn, ebonite, celluloid, ivory, bone, shell, galalith, formaldehyde-phenol condensation products, urea-formaldehyde condensation products, and other artificial resins.

Rubber
Ingredient of—
Products obtained with rubber latex.

Soap
Ingredient of—
Bleaching preparations, detergent preparations.

Sugar
Defecating agent in the refining of sugar.
Reagent in—
Clarifying and purifying liquors in sugar beet refining.

Textile
—, *Dyeing*
Ingredient of—
Various dye baths (added for the purpose of increasing the dispersion of the dyestuff).
Mordant in various processes.
—, *Finishing*
Ingredient of—
Compositions used for the waterproofing of fabrics, this treatment being followed by one in a solution of a metallic salt.

Sodium Alginate (Continued)

Compositions used for treating woolen fabrics to protect them against decompositions (French 518059).

Compositions used for sizing yarns and fabrics (used in place of starches or gum tragacanth for the purpose of obtaining a more elastic and more transparent size) (French 563726).

——, *Printing*

Mordant in printing various fabrics.

Thickener in—

Printing pastes (used in place of gum tragacanth or British gum).

Resins and Waxes

Emulsifying agent in making—

Dispersions of waxes and resins, both artificial and natural (added for the purpose of increasing the dispersion).

Stabilizing reagent in—

Emulsions of natural or artificial resins and waxes.

Water and Sanitation

Reagent in—

Treating waste waters and the like for the purpose of purifying and clarifying them.

Wine

As a clarifying agent.

Sodium Allylate

French: Allylate sodique, Allylate de sodium, Allylate de soude.

German: Natriumallylat.

Chemical

Reagent (Brit. 304118) in making ketonic acid esters with esters of the following acids—

Acetic, anthranilic, benzoic, butyric, camphoric, caproic, caprylic, chloroacetic, cinnamic, citric, cresylic, gallic, lactic, maleic, malic, malonic, metanilic, mucic, naphthionic, oxalic, palmitic, phenylacetic, phthalic, picramic, propionic, pyrogallic, salicylic, succinic, sulphanilic, tartaric, trichloroacetic, valeric.

Starting point in making—

Aromatics, intermediates, pharmaceuticals, various salts.

Dye

Reagent in making various synthetic dyestuffs.

Sodium Allylnaphthalenesulphonate

French: Allylnaphthalènesulphonate de soude.

German: Allylnaphtalinsulfonsaeuresnatrium, Natriumallylnaphtalinsulfonat.

Dye

Dispersing agent in making—

Color lakes (Brit. 264860).

Ink

Dispersing agent in making—

Printing inks (Brit. 264860).

Paint and Varnish

Dispersing agent (Brit. 264860) in making—

Paints, pigments, varnishes.

Plastics

Dispersing agent in making—

Cellulose ester and other plastics (Brit. 264860).

Resins and Waxes

Dispersing agent (Brit. 264860) in making—

Artificial resin compositions, natural resin compositions.

Rubber

Dispersing agent in making—

Rubber compositions (Brit. 264860).

Textile

——, *Finishing*

Dispersing agent in making—

Finishing and dressing compositions (Brit. 264860).

——, *Dyeing*

Dispersing agent in making—

Dye liquors containing sulphur, indigo, and anthraquinone vat dyestuffs for use on rayon, wool, cotton, and natural silk (Brit. 264860).

——, *Manufacturing*

Dispersing agent in making—

Lubricating compositions for spinning fibers (Brit. 268387).

Sodium Allylphthalate

French: Allylphthalate de soude.

German: Allylphtalsaeuresnatrium, Natriumallylphtalat.

Resins and Waxes

Starting point (Brit. 250265) in making synthetic resins with soluble salts of—

Barium, calcium, lead, magnesium, strontium, zinc.

Sodium Alpha-amino-4-bromo-2-anthraquinonesulphonate

French: Alpha-amino-4-bromo-2-anthraquinone-sulfonate de sodium, Alpha-amino-4-bromo-2-anthraquinone-sulfonate de soude.

German: Alpha-amino-4-bromo-2-anthrachinonsulfonsaeuresnatrium, Natriumalpha-amino-4-brom-2-anthrachinonsulfonat.

Dye

Starting point (Brit. 282409) in making wool dyestuffs with—

Para-allylacetanilide, para-amylacetanilide, parabenzoylacetanilide, parabenzylacetanilide, parabutylacetanilide, paraethylacetanilide, paraheptylacetanilide, parahexylacetanilide, paramethylacetanilide, parapropylacetanilide.

Starting point (Brit. 282452) in making dyestuffs with—

Pentamethyleneaminobornylamine.

Sodium Alpha-amino-2-naphthol-3-sulphonate

Synonyms: Sodium 1-amino-2-naphthol-3-sulphonate.

French: Alpha-amino-2-naphthole-3-sulphonate sodique, Alpha-amino-2-naphthole-3-sulphonate de sodium, 1-Amino-2-naphthole-3-sulphonate de soude.

German: Alpha-amino-2-naphtol-3-sulfonsäuresnatrium, Alpha-amino-2-naphtol-3-sulfonsäuresnatron, Alpha-amino-2-naphtol-3-sulfonsäuressodium, 1-Amino-2-naphtol-3-sulfonsäurenatrium, Natrium-alpha-amino-2-naphtol-3-sulfonat, Natrium-1-amino-2-naphtol-3-sulfonat.

Analysis

Reagent in—

Determining potassium.

Chemical

Starting point in making various derivatives.

Photographic

As a developer.

Sodium Alpha-amino-2-oxyethoxynaphthalenesulphonate

French: Alpha-amino-2-oxyéthoxyenaphthalènesulphonate sodique, Alpha-amino-2-oxyéthoxyenaphthalènesulphonate de sodium, Alpha-amino-2-oxyéthoxyenaphthalènesulphonate de soude.

German: Alpha-amino-2-oxyaethoxynaphtalinsulfonsaeuresnatrium, Alpha-amino-2-oxyaethoxynaphtalinsulfonsaeuresnatron, Natriumalpha-amino-2-oxyaethoxynaphtalinsulfonat.

Chemical

Starting point in making various intermediates.

Dye

Starting point (Brit. 298518) in making azo dyestuffs with—

Alpha-aminonaphthalene, Alpha-aminonaphthalene-6-sulphonic acid, alpha-aminonaphthalene-7-sulphonic acid, anilin, anilin-3-chloro-6-sulphonic acid, anilin-2:4-disulphonic acid, anilin-2:5-disulphonic acid, anilin-4-nitro-2:5-disulphonic acid, anilin-3-sulphonic acid, beta-amino-1-methoxybenzene-4-sulphonic acid, beta-amino-5-sulphobenzenzoic acid, 1:3-dioxyquinolin, methyl ketol, methyl ketol-sulphonic acid, orthocresotinic acid, 1-phenyl-3-carboxy-5-pyrazolone, 1-phenyl-3-methyl-5-pyrazolone, salicylic acid, sulphazone.

Sodium Alphanaphtholate

French: Alphanaphtholate sodique, Alphanaphtholate de sodium, Alphanaphtholate de soude.

German: Alphanaphtalsaeuresnatrium, Natriumalphanaphtolat.

Leather

Ingredient (Brit. 263473) of—

Liquors for dyeing leather.

Miscellaneous

Ingredient (Brit. 263473) of—

Liquors for dyeing hair and feathers.

Textile

——, *Dyeing and Printing*

Ingredient (Brit. 263473) of—

Dye liquors and printing pastes used on fabrics and yarns containing acetate and other rayons, wool-rayon mixtures and silk-rayon mixtures in connection with vat dyestuffs.

Sodium Alphatetrahydronaphthalenesulphonate
Miscellaneous
As an emulsifying agent (Brit. 371293).
For uses, see under general heading: "Emulsifying agents."

Sodium-Aluminum-Iron Cyanide
Chemical
Catalyst (Brit. 446411) in—
Halogenating unsaturated hydrocarbons.

Sodium Amalgam
French: Amalgame sodique, Amalgame de soude.
German: Amalgamiertenatrium, Amalgamiertenatron.
Analysis
Reagent in various processes.
Chemical
Reducing agent in making—
Hydrogen, organic chemicals.

Sodium Aminoazotoluenesulphonate
French: Aminoazotoluènesulphonate de soude.
German: Aminoazotoluolsulfosaeuresnatrium, Natriumaminoazotoluolsulfonat.
Dye
Starting point in making—
Tetrakisazo dyestuffs (Brit. 265553).

Sodium Aminoethanesulphonate
Insecticide and Fungicide
Starting point (German 550961) in making—
Addition agents, with montanic acid chloride, for copper, calcium, and lead arsenate products, particularly for controlling *Peronospora* and *Fusicladium.*

Sodium-2-aminopyridin
German: Natrium-2-aminopyridin.
Chemical
Starting point (Brit. 265167) in making—
2-Allylaminopyridin, 2-amylaminopyridin, 2-butylaminopyridin, 2-cetylaminopyridin, 2-diethylaminopyridin, 2-dimethylaminopyridin, 2-ethylaminopyridin, 2-isoamylaminopyridin, 2-isobutylaminopyridin, 2-isopropylaminopyridin, 2-methylaminopyridin, 2-propylaminopyridin.

Sodium-Ammonium-1:2-dihydroxynaphthalene-3:6-disulphonate
French: 1:2-Dihydroxyenaphthalène-3:6-disulphonate de soude-ammonium.
German: Natriumammonium-1:2-dihydroxynaphtalin-3:6-disulphonat.
Starting point in making—
Antimony trioxide and antimony hydroxide compounds (German 424952).

Sodium-Ammonium Phosphate
Synonyms: Fusible salt of urine, Microcosmic salt, Phosphorus salt, Salt of phosphorus, Sodium-ammonium hydrogen phosphate.
Analysis
Reagent in various processes—for example, standardizing uranium solutions, determination of Mg and Mn, blowpipe analysis.

Sodium-Ammonium Undecoate
Miscellaneous
As a wetting agent (U. S. 2020999).

Sodium Amylate
French: Amylate sodique, Amylate de sodium, Amylate de soude.
German: Natriumamylat.
Chemical
Reagent (Brit. 304118) in making ketonic acid esters with the aid of the butyl, amyl, allyl, heptyl, hexyl, and propyl esters of the following acids—
Acetic, anthranilic, benzoic, butyric, camphoric, caproic, caprylic, chloroacetic, cinnamic, citric, cresylic, gallic, lactic, maleic, malic, malonic, metanilic, mucic, naphthionic, oxalic, palmitic, phenylacetic, phthalic, picramic, propionic, pyrogallic, salicylic, succinic, sulphanilic, tartaric, trichloroacetic, valeric.
Starting point in making—
Aromatics, intermediates, pharmaceuticals, various salts.
Dye
Reagent in making various synthetic dyestuffs.

Sodium Amylnaphthalenesulphonate
French: Amylnaphthalènesulphonate de soude.
German: Amylnaphtalinsulfonsaeuresnatrium, Natriumamylnaphtalinsulfonat.
Dye
Dispersive agent (Brit. 264860) in making—
Color lakes.
Ink
Dispersive agent in making—
Printing inks, writing inks.
Paint and Varnish
Dispersive agent in making—
Paints, pigments, varnishes.
Plastics
Dispersive agent in making—
Cellulose ester and other plastics.
Rubber
Dispersive agent in making—
Rubber compositions.
Textile
——, *Dyeing*
Dispersive agent in making—
Dye liquor, containing sulphur dyes, anthraquinone vat dyes, and the like, for use on rayon, wool, cotton, and natural silk yarns and fabrics.
——, *Finishing*
Dispersive agent in making—
Finishing and dressing compositions.
——, *Manufacturing*
Lubricant in spinning—
Textile fibers (Brit. 268387).

Sodium Amylphthalate
French: Amylphthalate de soude.
German: Amylphtalsaeuresnatrium, Natriumamylphtalat.
Waxes and Resins
Starting point (Brit. 250265) in making synthetic resins with salts of—
Barium, calcium, lead, magnesium, strontium, zinc.

Sodium Amylsulphophthalate
Textile
Wetting agent (Brit. 399319 and 399320) in—
Bleaching by means of hypochlorite liquor.

Sodium-Anilin
French: Aniline de sodium, Aniline sodique.
German: Natriumanilin.
Dye
Reagent (German 436533) in making anthracene dyestuffs from—
3:9-Dichlorobenzanthrone, 11:3-dichlorobenzanthrone.

Sodium Anthranilate
French: Anthranilate de soude.
German: Natriumanthranilat.
Chemical
Starting point in making—
Monochloride of normal betamethoxyethylanthranilic acid betapiperidine ethylester (Brit. 260605).
Miscellaneous
Ingredient (German 485012) of—
Antifreeze solutions.

Sodium-Anthraquinone-1:4-disulphonate
French: Anthraquinone-1:4-disulphonate de soude.
German: Anthrachinon-1:4-disulphosaeuresnatron, Natriumanthrachinon-1:4-disulphonat.
Textile
——, *Dyeing*
Reagent in dyeing—
Cellulose acetate rayon (U. S. 1602695).
——, *Printing*
Reagent in printing—
Cellulose acetate rayon fabrics (U. S. 1602695).

Sodium Antimonate
French: Soude antimonié.
German: Natriumantimon, Natriumspiessglanz.
Ceramics
Ingredient of—
Enamels for metalware, opaque glazes.
Glass
Ingredient of—
Opaque glass.

Sodium Arsanilate
Synonyms: Sodium aminophenolarsonate, Sodium anilinarsonate.
Chemical
Reagent in—
Organic synthesis.
Pharmaceutical
In compounding and dispensing practice.
Suggested for use in treating—
Sleeping sickness.

Sodium-Arsphenamine
Synonyms: Sodium diarsenol, Sodium salt of 3-diamino-4-dihydroxy-1-arsenobenzene.
Pharmaceutical
In compounding and dispensing practice.
Substitute for—
Arsphenamine.

Sodium 3-Aurothiosulphanilate
Chemical
Starting point (Brit. 398020) in making—
Complex double compounds of organic heavy metal mercapto compounds.

Sodium 4-Benzamido-2:5-dimethoxyphenylhydrazinbetasulphonate
Textile
Starting point (Brit. 398846) in dyeing textile fibers—
Yellow colors with biacetoacetyltoluidin.
Blue-green colors with 2-hydroxyanthracene-3-carboxylicorthotoluidide.
Blue-violet colors with 2:3-hydroxynaphthoic-5-chloroorthoanisidide.

Sodium Benzoate
Synonyms: Benzoate of soda.
French: Benzoate sodique, Benzoate de sodium, Benzoate de soude.
German: Benzoesäuresnatrium, Benzoesäuresnatron, Natriumbenzoat.
Chemical
Ingredient of—
Caffeine solutions.
Preservative in making—
Alkaloid solutions and alkaloidal preparations.
Reagent in making—
Pharmaceutical chemicals.
Starting point in making—
Benzaldehyde, benzoic anhydride, various benzoates.
Dye
Starting point in making—
Anilin blues and other synthetic dyestuffs.
Fats and Oils
As a preservative.
Food
As a preservative.
Leather
Ingredient of—
Cleansing preparations.
Miscellaneous
As a preservative.
Perfume
Ingredient of—
Cosmetics, dentifrices.
Pharmaceutical
In compounding and dispensing practice.
Paint and Varnish
Ingredient of—
Paints for making designs on textiles.
Textile
Reagent in dyeing and printing yarns and fabrics.
Tobacco
Reagent for—
Improving the taste and for preservative purposes.

Sodium Benzosulphimide
Synonyms: Soluble saccharin.
Food
Sweetening agent.
Pharmaceutical
Substitute for—
Sugar in diabetic conditions.

Sodium Benzosulphopara-aminophenylarsonate
Pharmaceutical
Suggested for use in treating—
Venereal diseases.

Sodium Benzylanilinsulphonate
French: Benzylanilinesulphonate sodique, Benzylanilinesulphonate de soude.
German: Benzylanilinsulfonsäuresnatrium, Benzylanilinsulfonsäuresnatron, Natriumbenzylanilinsulfonat, Natronbenzylanilinsulfonat.
Miscellaneous
As an emulsifying agent (Brit. 350379).
For uses, see under general heading: "Emulsifying agents."

Sodium Benzylanthranilate
Miscellaneous
As an emulsifying agent (Brit. 350379).
For uses, see under general heading: "Emulsifying agents."

Sodium Benzylchloro-orthosulphonate
French: Benzylechloro-orthosulfonate sodique, Benzylechloro-orthosulfonate de sodium.
German: Benzylchlororthosulfonsaeuresnatrium, Natriumbenzylchlororthosulfonat.
Photographic
Reagent (Brit. 277137) in making—
Non-inflammable films from mercerized cellulose.
Textile
——, *Finishing*
Reagent (Brit. 277137) in making—
Fireproofed yarns and fabrics.

Sodium Benzylchloroparasulphonate
French: Benzylechloro-parasulfonate sodique, Benzylechloro-parasulfonate de sodium, Benzylechloro-parasulfonate de soude.
German: Benzylchlorparasulfonsaeuresnatrium, Natriumbenzylchlorparasulfonat.
Photographic
Ingredient (Brit. 277317) of—
Fireproofed film made from mercerized cellulose.
Textile
——, *Finishing*
Ingredient (Brit. 277317) of—
Impregnating compositions for treating textiles.

Sodium Benzylnaphthalene Sulphonate
French: Benzylenaphthalènesulphonate sodique, Benzylenaphthalènesulphonate de sodium, Benzylenaphthalènesulphonate de soude.
German: Benzylnaphtalinsulfonsäuresnatrium, Benzylnaphtalinsulfonsäuresnatron, Natriumbenzylnaphtalinsulfonat, Natronbenzylnaphtalinsulfonat.
Miscellaneous
As a wetting agent (Brit. 411908).
For uses, see under general heading: "Wetting agents."

Sodium Benzylpara-amidobenzylanilinsulphonate
French: Benzylepara-amidobenzylaniline sulphonate sodique, Benzylepara-amidobenzylaniline sulphonate de soude.
German: Benzylpara-amidobenzylanilinsulphonsäuresnatrium, Benzylpara-amidobenzylanilinsulphonsäuresnatron, Natriumbenzylpara-amidobenzylanilinsulfonat, Natronbenzylpara-amidobenzylanilinsulfonat.
Miscellaneous
As an emulsifying agent (Brit. 350379).
For uses, see under general heading: "Emulsifying agents."

Sodium Benzylsuccinate
Pharmaceutical
In compounding and dispensing practice.
Suggested for use as—
Antispasmodic.

Sodium Benzylthioglycollate
Synonyms: Benzylsulphoglycollate.
French: Benzylesulfoglycollate sodique, Benzylesulfoglycollate de sodium, Benzylesulfoglycollate de soude, Benzylethioglycollate sodique, Benzylethioglycollate de soude.
German: Benzylsulfoglykolsaeuresnatrium, Benzylsulfoglykolsaeuresnatron, Benzylthioglykolsaeuresnatrium, Natriumbenzylsulfoglycollat, Natriumbenzylthioglycollat.

Sodium Benzylthioglycollate (Continued)

Dye
Reagent (Brit. 284288) in making thioindigoid dyestuffs with—
Acenaphthenequinone, alphaisatinanilide, 5:7-dibromoisatin, isatin, isatin homologs and derivatives, orthodiketones.

Sodium-Beryllium Fluoride

Fuel
Ingredient (Brit. 463218) of—
Automotive fuels consisting of gasoline and ethyl alcohol (added to inhibit corrosion of magnesium metal, magnesium alloys, or other metal parts).
Automotive fuels consisting of gasoline, benzol, and methanol (added to inhibit corrosion of magnesium metal, magnesium alloys or other metal parts).

Sodium Betahydroxyethyldithiocarbamate

Disinfectant
As a bactericide (Australian 8103/32, Brit. 406979, U. S. 1972961).
Insecticide and Fungicide
As a fungicide (Australian 8103/32, Brit. 406979, U. S. 1972961).
As an insecticide (claimed effective against aphids) (Australian 8103/32, Brit. 406979, U. S. 1972961).

Sodium-Betanaphthalene Sulphochloramide
German: Natriumbetanaphtalinsulphonchloramid.

Chemical
Starting point in making—
Magnesium-betanaphthalene sulphochloramide (German 422076).

Sodium Betanaphtholate
Synonyms: Microcidin.
French: Bétanaphtholate sodique, Bétanaphtholate de sodium, Bétanaphtholate de soude.
German: Natriumbetanaphtolat.

Leather
Ingredient of—
Liquors used in dyeing.
Miscellaneous
Disinfectant for various purposes.
Ingredient of—
Liquors for dyeing hair and feathers.
Pharmaceutical
In compounding and dispensing practice.
Textile
——, *Dyeing and Printing*
Ingredient of—
Liquors and pastes, containing vat dyestuffs, used for dyeing and printing fabrics and yarns containing acetate rayon and other rayons, wool-rayon mixtures, and silk-rayon mixtures.
Sanitation
As a disinfectant.

Sodium Betasulphoethyllaurate

Textile
Assistant (Brit. 440103) in—
Textile processing.

Sodium Betasulphoethyloleate

Textile
Assistant (Brit. 440103) in—
Textile processing.

Sodium Betatetrahydronaphthalenesulphonate
French: Bétatétrahydronaphthalènesulphonate sodique, Bétatétrahydronaphthalènesulphonate de sodium, Bétatétrahydronaphthalènesulphonate de soude.
German: Betatetrahydronaphtalinsulfonsäuresnatrium, Betatetrahydronaphtalinsulfonsäuresnatron, Betatetrahydronaphtalinsulfonsäuressodium, Natriumbetatetrahydronaphtalinsulfonat.

Miscellaneous
As an emulsifying agent (Brit. 371293).
For uses, see under general heading: "Emulsifying agents."

Sodium Bicarbonate
Synonyms: Acid sodium carbonate, Baking soda, Bicarbonate of soda, Sodium hydrocarbonate, Sodium hydrogen carbonate.
Latin: Bicarbonas sodicus, Natrium bicarbonicum, Natrium carbonicum acidulum, Saleratus, Sodii bicarbonas.
French: Bicarbonate sodique, Bicarbonate de soude, Carbonate acide de sodium, Sel de vichy.
German: Doppeltkohlensäuresnatron, Natriumbikarbonat.
Spanish: Bicarbonato sodico, Bicarbonato de sosa.
Italian: Bicarbonato di sodio.

Agriculture
Disinfecting agent for—
Beans, cabbage, potatoes, seeds.
Retting agent for—
Flax, hemp, jute.
Analysis
Alkali in—
Analytical processes involving control and research.
Animal Husbandry
Ingredient of—
Cattle feeds, cattle salt licks.
Beverage
Ingredient of—
Artificial mineral waters, effervescent drinks.
Ceramic
In porcelain manufacture, in pottery manufacture.
Chemical
Process material in making—
Acetaldehyde, ammonium carbonate, baking powders (many patents), benzene derivatives, carbonates of various bases, chlorhydrins, glycols, nickel carbonate, thorium salts.
Source of—
Carbon dioxide.
Neutralizing agent for—
Acids in various reactions and processes of chemical manufacturing.
Saponifying agent for—
Acetin, fats, greases, oils.
Solubilizing agent for—
Substances insoluble in nitric acid.
Starting point in making—
Sodium salts.
Clay Products
Cleaning agent for—
Canadian d'Amherst clay, Canadian china clay, Fraddon china clay, kaolins, wotter clay.
Deflocculating agent.
Floating agent.
Peptizing agent.
Purifying agent.
Cosmetic
Ingredient of—
Cosmetic and toilet specialties, such as cuticle salves, hair-treating lotions, deodorants, bath salts.
Dye
Process material in making—
Dyes.
Explosives and Matches
Process material in making—
Explosives.
Fertilizer
Ingredient of—
Fertilizer compositions.
Fire-Fighting
Ingredient of—
Chemical fire-extinguishers (many patents), fireproofing agents, fire foams.
Food
Conditioning agent in—
Large-scale cooking of foods, such as canning and baking.
Disinfecting agent in—
Grain milling.
Ingredient of—
Self-raising flours.
Neutralizing agent for—
Acidity in milk, acidity in various food products, in cooking processes.
Preservative for—
Butter, various food products, yeast.
Reagent for—
Treating peeled fruits.
Starting point (many patents) in making—
Baking powders with such chemicals as cream of tartar, tartaric acid, phosphates, and starch.
Glass
Process material in making—
Opaque glass.

Sodium Bicarbonate (Continued)

Laundering
Carrier for—
Blueing.
Neutralizer of—
Acid odors.

Leather
Alkali in—
Tanning processes.

Metallurgical
Ingredient of electrolytes in—
Gold-plating, platinum-plating.
Reagent in—
Flotation processes for galena, sphalerite.

Miscellaneous
Ingredient of—
Cleansing compositions for various purposes.
Metal polishes.

Oral Hygiene
Ingredient of—
Dentifrices, mouthwashes.

Pharmaceutical
In compounding and dispensing practice.

Plastics
Dissolving agent for—
Casein.

Sanitation and Water
Regenerating agent for—
Peat used in water-softening.

Textile
Degumming agent for—
Textile fibers.
Scouring agent for—
Textile fibers.
Washing agent for—
Textile fibers.

Wood
Preventer of—
Molding.

Sodium Bichromate

Synonyms: Bichromate of soda, Sodium acid chromate, Sodium dichromate.
French: Bichromate sodique, Bichromate de soude.
German: Doppeltchromsäuresnatrium, Doppeltchromsäuresnatron, Natriumdichromat, Natrondichromat, Zweifachchromsäuresnatrium, Zweifachchromsäuresnatron.
Spanish: Bicromato sodico, Bicromato de sosa.
Italian: Bicromato di sodio.

Analysis
Reagent in various processes.

Chemical
Oxidizing agent in making—
Aldehydes, intermediates, ketones, synthetic pharmaceuticals, various chemicals.
Oxidizing and neutralizing agent (Brit. 402529) in making—
Benzoic acid from toluene.
Starting point in making—
Chromates.
Chromic acid by reaction with hydrochloric or sulphuric acid.
Lead chromate by reaction with lead sulphate and sodium acetate, the latter being regenerated (French 752674).
Substitute for—
Potassium bichromate.

Dye
Oxidizing agent in making various synthetic dyes.

Electrical
Ingredient of—
Battery electrolytes, battery pastes.

Explosives and Matches
Ingredient of—
Matchhead compositions, pyrotechnic compositions.

Fats and Oils
Bleaching and oxidizing agent in—
Refining fats and oils.

Glues and Adhesives
Reagent for—
Rendering glue, gums, and gelatin insoluble.
Source of chromium in making—
Chrome gelatin, chrome glue.

Insecticide
Reagent (U. S. 1908544) in making—
Green-colored lead arsenate from prussian blue, lead oxide, nitric acid, and arsenate acid.

Leather
Reagent in—
Chrome tanning.

Metallurgical
Ingredient of—
Brass pickling solutions.
Etching solution (containing also nitric and sulphuric acids) for roughening zinc surface so that cellulose base lacquers will have greater adherence to the metal.
Pickling agent for—
Ornamental or other silver articles, the object being to cover them with a silver chromate coating which prevents tarnishing (German 592710).

Miscellaneous
Bleaching agent for—
Sponges.
Hardening and preservative agent for—
Anatomical specimens, oxidizing agent in many processes.

Paint and Varnish
Reagent in making—
Chrome colors.

Paper
Bleaching and oxidizing agent.

Perfume
Oxidizing agent in making—
Synthetic perfumes.

Petroleum
Refining agent for—
Petroleum products (used in conjunction with sulphuric acid).

Photographic
Hardening agent for—
Gelatin.
Reagent for—
Rendering gelatin insoluble.
Reagent in—
Gum-bichromate printing process.

Printing
Oxidizing agent in—
Electroengraving copper plates.

Resins and Waxes
Bleaching agent in—
Wax refining.
Decomposing agent (Brit. 397096) in making—
Synthetic resins from decomposition products of aromatic hydrocarbons with polyvalent alcohols.

Textile
Ingredient of—
Waterproofing compositions.
Mordant in—
Dyeing textile fabrics, dyeing wool with alizarin dyestuffs, dyeing wool with logwood black.

Sodium Bifluoride

Synonyms: Sodium acid fluoride.
French: Bifluorure sodique, Bifluorore de soude.
German: Bifluornatrium, Bifluornatron, Difluornatrium, Difluornatron, Natriumbifluorid, Natriumdifluorid, Natronbifluorid, Natrondifluorid.
Spanish: Bifluoruro sodico.
Italian: Bifluoruro di sodio.

Food
As a preservative (not permitted in certain countries).

Glass
As an etching agent.
Opacifying agent in making—
Frosted glass, opaque glasses, translucent glasses.

Miscellaneous
Preservative for—
Anatomical specimens, zoological specimens.

Pharmaceutical
In compounding and dispensing practice.

Sodium Biformate

Synonyms: Sodium hydrogen formate.
French: Biformiate de soude, Biformiate sodique.
German: Ameisensaeuressaeurenatrium, Natriumbiformiat.

Sodium Biformate (Continued)

Chemical
Reagent (German 439289) in making—
Ethyl formate, geranyl formate, glyceryl formate, glycol formate, mixed anhydrides of formic and acetic and nitric acids, phenyl formate.

Sodium Bismuthyltartrate

French: Bismuthyltartrate de soude.
German: Natriumwismuthyltartrat, Wismuthylweinsaeuresnatrium.

Chemical
Reagent (Brit. 266820) in making the following basic bismuth compounds—
Normal phenylglycinamide-para-arsinate, para-aminophenylarsinate, 3-acetylamino-4-oxyphenylarsinate, 2-oxy-5-acetylaminophenylarsinate.

Sodium Bisulphite

Synonyms: Acid sodium sulphite, Leucogen, Sodium-meta-bisulfite.
Latin: Sodii bisulphis, Natrium bisulfurosum.
French: Bisulphite de soude.
German: Doppeltschwefligsäuresnatrium, Doppeltschwefligsäuresnatron.

Analysis
Reagent in various processes.

Brewing
Antiseptic in—
Fermentations.
Sterilizing agent for—
Barrels, casks, plant equipment, vats.

Chemical
Antiseptic in—
Fermentations.
Catalyst (Brit. 398626) in making—
Cellulose esters.
Precipitant in extracting—
Iodine from Chile saltpeter.
Purifying agent (Brit. 398136) in making—
Aromatic alcohols.
Reagent in making—
Aldehydes, chromium bisulphite, hydroxylamine salts, intermediates, ketones, sodium hydrosulphite.
Reducing agent (Brit. 395405, 342690) in making—
Stable acridine salt solutions.

Distilled Liquors
Antiseptic in—
Fermentations.
Sterilizing agent for—
Barrels, casks, cookers, fermentation tanks, plant equipment.

Dye
Reagent in synthesis of—
Indigo from phenylglycinorthocarbonic acid.
Solubilizing agent for—
Alizarin blue and alizarin blue-black (by forming an unstable compound decomposed by water).

Food
Antifermentative for—
Food products.
Antiseptic for—
General purposes, grains.
Germicide for—
Food products.
Preservative for—
Egg yolk, food products, fruit juices, meats, syrups, vegetable juices.

Fuel
Reagent in making—
Copper sulphocyanide from gas purifying masses.

Glue and Adhesives
Antiseptic for—
Glue, gelatin.
Deodorant in making—
Dextrin glues.

Leather
Depilatory for—
Hides.
Reagent (Brit. 402524) in—
Quick tanning process.
Reagent in making—
Tanning extracts.
Substitute for—
Sodium hyposulphite as reducing agent for bichromate in chrome tanning.

Metallurgical
Ingredient of—
Baths in electro-depositing copper and brass in galvanoplastics.

Miscellaneous
Antiseptic in—
Many processes.
Bleaching agent in—
Many processes.
Bleaching agent for—
Cork, straw.
Cleansing agent in—
Many processes.
Disinfectant in—
Many processes.
Preservative in—
Many processes.

Paper
Antichlor in—
Bleaching operations.
Digesting liquor (U. S. 1915953) in making—
Chemical pulp.
Reagent (Brit. 400974) in making—
Cellulose materials of high alphacellulose content.

Perfume
Antiseptic and preservative for—
Creams.

Pharmaceutical
Germicide.
In compounding and dispensing practice.
Suggested for use in—
Gastric fermentation, parasitic skin diseases.

Photographic
Ingredient of—
Acid fixing baths.

Rubber
Coagulating agent in making—
Raw rubber from rubber latex.

Sanitation
Ingredient (U. S. 1840452) of—
Cleaning and disinfecting compound substantially inert at atmospheric dryness and effective when mixed with water.

Sugar
Antifermentative for—
Sugar solutions and syrups.
Antiseptic in making—
Glucose, sugar.
Bleaching agent for—
Sugar solutions and syrups.

Textile
Antichlor in bleaching—
Animal fibers, vegetable fibers.
Discharge in printing—
Textile fibers.
Mordant in—
Dyeing (especially with vat dyes).

Winemaking
Antiseptic in—
Fermentations.
Sterilizing agent for—
Barrels, casks, plant equipment, vats.

Sodium Bromate

French: Bromate sodique, Bromate de soude.
German: Natriumbromat, Natronbromat.
Spanish: Bromato de sosa.
Italian: Bromato di sodio.

Analysis
Reagent in—
Analytical work.

Chemical
Brominating agent for—
Organic compounds in synthesis (used in admixture with sodium bromide).

Metallurgical
Bromine generator in—
Gold extraction processes (used in admixture with sodium bromide).

Sodium Bromide
Synonyms: Bromide of soda.
Latin: Natrium bromidum, Sodii bromidum.
French: Bromure sodique, Bromure de sodium, Bromure de soude.
German: Bromnatrium, Bromnatron, Natriumbromid, Natronbromid.
Spanish: Bromuro de sosa.
Italian: Bromuro di sodio.

Analysis
Reagent in—
Analytical work.

Chemical
Brominating agent for—
Organic compounds in synthesis (used in admixture with sodium bromate).
Purifying agent in making—
Bromine.

Metallurgical
Bromine generator in—
Gold extraction processes (used in admixture with sodium bromate).

Pharmaceutical
In compounding and dispensing practice.
Suggested for use as—
Nerve sedative.

Photographic
Bromide in—
Photographic processes.

Textile
Conserving agent (French 601297) for—
Luster, transparency, and general appearance of cellulose acetate subjected to the action of hot or boiling liquids.

Sodium 2-Brom-4-phenylphenate

Disinfectant
As a germicide.

Sodium Butylate
French: Butylate sodique, Butylate de sodium, Butylate de soude.
German: Natriumbutylat, Butylsäuresnatrium, Butylsäuresnatron, Butylsäuressodium.

Chemical
Reagent (Brit. 304118) in making ketonic esters with the aid of—
Allyl acetate, amyl acetate, butyl acetate, ethyl acetate, heptyl acetate, hexyl acetate, methyl acetate, pentyl acetate, propyl acetate.
Various alkyl esters of butyric acid, oxalic acid, propionic acid, formic acid, and other carboxylic acids.

Sodium Butylbenzenesulphonate
French: Butylebenzènesulfonate sodique, Butylebenzènesulfonate de sodium, Butylebenzènesulfonate de soude.
German: Butylbenzolsulfonsaeuresnatrium, Natriumbutylbenzolsulfonat.

Fats and Oils
Starting point in making—
Solvent compositions (Brit. 279877).

Miscellaneous
Ingredient (Brit. 279877) of—
Detergent compositions containing soap.
Preparations used for cleansing and bleaching parquet floors.

Textile
——, *Dyeing*
Assist in making—
Wool dye liquors (Brit. 279877).
——, *Finishing*
Ingredient of—
Washing compositions (Brit. 279877).

Sodium Butyldithiocarbamate

Disinfectant
As a bactericide (Australian 8103/32, Brit. 406979, U. S. 1972961).

Insecticide and Fungicide
As a fungicide (claimed effective against barley spores) (Australian 8103/32, Brit. 406979, U. S. 1972961).
As an insecticide (claimed effective against aphids) (Australian 8103/32, Brit. 406979, U. S. 1972961).

Sodium Butylnaphthalenesulphonate
French: Butylenaphthalènesulphonate de soude.
German: Butylnaphtalinsulfosaeuresnatron, Natriumbutylnaphtalinsulfonat.

Dye
Dispersive agent in making—
Lakes and other preparations of dyestuffs (Brit. 264860).

Ink
Dispersive agent in making various inks (Brit. 264860).

Paint and Varnish
Dispersive agent (Brit. 264860) in making—
Copal varnishes, lacquers, spirit varnishes, water paints.

Plastics
Dispersive agent (Brit. 264860) in making—
Cellulose acetate solutions, cellulose nitrate solutions, cellulose ester and ether compositions.

Rubber
Dispersive agent in making—
Rubber solutions (Brit. 264860).

Textile
——, *Dyeing*
Dispersive agent (Brit. 264860) in making dye liquors with—
Anthraquinone dyestuffs, indigo, sulphur dyestuffs, vat dyestuffs.
——, *Finishing*
Dispersive agent in making—
Sizing compositions (Brit. 264860).

Sodium Butyl-3-nitrophthalate
French: Butyle-3-nitrophthalate de soude.
German: Butyl-3-nitrophtalsaeuresnatrium, Natriumbutyl-3-nitrophtalat.

Resins and waxes
Reagent in making—
Synthetic resins (U. S. 1618209).

Sodium Butyl-4-nitrophthalate
French: Butyle-4-nitrophthalate de soude.
German: Butyl-4-nitrophtalsaeuresnatrium, Natriumbutyl-4-nitrophtalat.

Resins and waxes
Reagent in making—
Synthetic resins (U. S. 1618209).

Sodium Butylphthalate
French: Butylephthalate de soude.
German: Butylphtalsaeuresnatrium, Natriumbutylphtalat.

Resins and Waxes
Starting point (Brit. 250265) in making synthetic resins with—
Barium acetate, barium bromide, barium chloride, barium nitrate, calcium acetate, calcium bromide, calcium chloride, calcium nitrate, lead acetate, lead bromide, lead chloride, lead nitrate, magnesium acetate, magnesium bromide, magnesium chloride, magnesium nitarte, strontium acetate, strontium bromide, strontium chloride, strontium nitrate, zinc acetate, zinc bromide, zinc chloride, zinc nitrate.

Sodium Cacodylate
Synonyms: Sodium dimethylarsonate.

Pharmaceutical
In compounding and dispensing practice.
Suggested for use in treating—
Anemia, asthma chronic bronchitis, leukemia, malaria, psoriasis and other skin diseases, tuberculosis.

Sodium Cadmium Cyanide

Chemical
Catalyst (Brit. 446411) in—
Halogenating unsaturated hydrocarbons.
Starting point (Brit. 446411) in making—
Catalysts with metal chlorides for halogenating unsaturated hydrocarbons.

Sodium Camphorate
French: Camphorate de soude.
German: Kamphersaeuresnatrium, Kamphorsaeuresnatrium, Natriumcamphorat.

Chemical
Starting point in making—
Atropine camphorate (Brit. 269498).

Pharmaceutical
In compounding and dispensing practice.

Sodium Carbonate

Synonyms: Carbonate of soda, Sal soda, Soda, Soda ash, Washing soda.
Latin: Carbonas sodicus, Natrii carbonas, Natrium carbonicum, Sal sodae, Sodii carbonas
French: Carbonate de soude, Sodé.
German: Einfach kohlensäuresnatron, Kohlensäuresnatron, Salzasche.
Spanish: Carbonato sodico.
Italian: Carbonato di sodio.
(Uses of modified forms of sodium carbonate, such as "Special Alkalies," are included).

Abrasives
Process material in making—
Abrasives.

Adhesives
Ingredient of—
Adhesives.
Solvent for—
Casein.

Agriculture
Retting and digumming agent for—
Bast fibres, hemp fibres, jute fibres, ramie fibres, sisal fibres.

Analysis
Reagent in—
Analytical processes involving control and research.

Animal Husbandry
Ingredient of—
Cattle feeds.
Sterilizing agent for—
Beehives.

Brewing
Cleansing agent.
Ingredient of—
Bottle-washing compounds.

Building Construction
Antifreeze agent in—
Building blocks, mortars.
Remover of carbon dioxide from—
Cement kiln gas.

Cellulose Products
Alkali in making—
Cellulose and cellulose derivatives.

Chemical
Absorbent for—
Acids, hydrogen sulphide, nitrogen oxides, phenols.
Dehydrating agent for—
Organic compounds.
Extractant in obtaining—
Alginic acid from seaweed.
Hydrolyzing agent for—
Albuminoids.
Neutralizing agent for—
Acids in various reactions and processes of chemical manufacturing.
Process material in activating—
Charcoals.
Process material in making—
Acetates, activated carbons, alkali salts, alkyl compounds, aluminum compounds, arsenic organic derivatives, amino compounds, ammonia catalysts, ammonium compounds.
Anthraquinone derivatives, such as sodium salts of its sulphonic acids.
Barium salts, benzene derivatives and substitution products, calcium salts, chlorinated organic compounds, cyanides, glycols, halogenated organic compounds, hydrogenation catalysts.
Intermediates used in making pharmaceuticals, aromatics, and other organic chemicals.
Iron salts, lead salts, magnesium salts, manganese salts, nickel catalysts, phosphates, potassium salts, thorium salts, uranium and compounds, ureas, vanadium compounds, various inorganic and organic chemicals.
Remover of carbon dioxide from—
Air, gas, inert gases.
Starting point in making—
Sodium inorganic salts, sodium organic compounds.

Clay Products
Cleaning agent for—
American kaolin, Canadian china clay, Canadian d'Amherst clay, Fraddon china clay, Mid-Cornwall china clay, Pentruff china clay, Wotter clay.
Deflocculating agent for—
Clay.
Peptizing agent for—
Clay.

Cosmetic
Ingredient of—
Bath salts.
Saponifying agent in making—
Cosmetic creams.

Dye
Process material in making—
Dyestuffs, intermediates.

Electrical
Process material in making—
Depolarizers for dry batteries, electrodes for storage batteries.

Explosives and Matches
Process material in making—
Explosives.

Fats and Oils
Process material in making—
Hydrogenation catalysts.
Saponifying agent for—
All fatty compounds.

Fertilizer
Process material in making—
Fertilizers.

Firefighting
Ingredient of—
Chemical fire-extinguisher liquids.

Food
Neutralizing agent for—
Acids in food.
Source of carbon dioxide in—
Food processing.
Sterilizing agent for—
Cabbage, grains, legumes.

Glass
Ingredient of—
Glass batches.

Insecticide and Fungicide
Process material in making—
Arsenates.

Laundering
Detergent in—
Washroom operations.
Neutralizer of—
Acid odors.

Leather
Process material in—
Finishing processes, tanning processes.

Metal Fabrication
Degreasing agent for—
Metalware generally, metalware prior to enamelling.
Ingredient of—
Frits in enamelling processes.

Metallurgical
Electrolyte ingredient in—
Brass-plating, copper-plating, electroplating aluminum, nickel-plating aluminum.
Flotation agent for—
Copper ores, such as chalcocite and chalcopyrite, galena, porphyry ores, silver ores, sulphide ores, zinc-lead ores.
Flux for—
Brass, iron ore.
Ingredient of—
Arc-welding fluxes, brazing fluxes, case-hardening compounds.
Process material in—
Chromite ore treatment, heat-treating various metals, molybdenum metallurgy.
Process material in extracting—
Radium from ore, tungsten from ore, uranium from ore, vanadium from ore.

Miscellaneous
Cleansing agent for—
Bottles, various articles.
Degreasing agent for—
Metallic surfaces, other surfaces.
Flotation agent for—
Minerals.
General cleansing agent.
Ingredient of—
Antifreeze mixtures, bottle-washing compositions.
Process material in making—
Bleaching compounds, heat-insulating materials.

Sodium Carbonate (Continued)
Remover of carbon-dioxide from—
Air, gas, inert gases, nitrogen.

Paint and Varnish
Ingredient of—
Paint removers.
Process material in making—
Lakes, ultramarine blue.

Paper and Pulp
Process material in making—
Blueprint paper, sizings.
Source of soda in—
Soda process of pulp manufacture (caustic production).

Petroleum
Flooding agent for—
Oil sands in restoring production of depleted wells.
Neutralizing agent for—
Sulphuric acid in refining processes.
Refining agent for—
Petroleum and its products.
Starting point in making—
Caustic soda for refining uses.

Pharmaceutical
General cleansing agent.
In compounding and dispensing practice.

Photographic
Reagent in various processes.

Plastics
Solvent for—
Casein.

Power and Heat
Ingredient of—
Anticorrosion compounds, boiler compounds, boiler feed-water treatment compounds, boiler scale-removing compounds.

Soap
Ingredient of—
Scouring compounds, soap powders, special detergent preparations for many cleansing operations.
Process material in making—
Catalysts for hydrogenation of fatty bases.
Saponifying agent for—
Soapstocks.
Starting point in making—
Caustic soda for saponification use.

Soft Drinks
Cleansing agent for—
Bottles, equipment, utensils.
Ingredient of—
Bottle-washing liquids, effervescent beverages, mineral waters.

Textile
Degreasing agent.
Degumming agent.
Emulsifying and saponifying agent for—
Fats, greases, oils.
Process material in—
Bleaching operations, dyeing operations, printing operations, scouring operations, washing operations.
Retting agent.
Starting point in making—
Caustic soda for various uses.
Washing agent.
Water-softening agent.

Water and Sanitation
Decomposing agent for—
Calcium soaps.
Neutralizing agent for—
Acid effluents.
Process material in making—
Artificial zeolites.
Reviving agent for—
Zeolites.
Softening agent for—
Water in laundries, textile plants, chemical works, artificial ice plants, paper mills, food product plants, canneries, beverage plants, soap plants, railroads, municipal waterworks, steamships, hospitals, hotels, large buildings, and other places.

Wood
Preventer of—
Mold in lumber and timber.
Washing agent.

Sodium-Cellulose Glycollate

Textile
Antifoaming agent (U. S. 1979469) in—
Dye suspensions.
Antisettling agent (U. S. 1979469) in—
Dye suspensions.

Sodium Cerate
French: Cérate sodique, Cérate de sodium, Cérate de soude.
German: Cerisaeuresnatrium, Natriumcerat.

Chemical
Reagent (Brit. 281307) in making zeolite catalysts used in making—
Acenaphthylene from acenaphthene, acetaldehyde from ethyl alcohol, acetic acid from ethyl alcohol, alcohols from aliphatic hydrocarbons.
Aldehydes from toluene, xylene, mesitylene, pseudocumene, and cymene.
Aldehydes and acids by the oxidation of orthochlorotoluene, parachlorotoluene, orthobromotoluene, parabromotoluene, dichlorotoluenes, chlorobromotoluenes, nitrotoluenes, chloronitrotoluenes, chlorobromotoluenes.
Alphanaphthaquinone from naphthalene, anthraquinone from anthracene, benzaldehyde and benzoic acid from toluene, benzoquinone from phenanthraquinone, chloroacetic acid from ethylenechlorohydrin, diphenic acid from ethyl alcohol, fluorenone from fluorene, formaldehyde from methyl alcohol or methane, hemimellitic acid from acenaphthene.
Maleic and fumaric acids from benzene, toluene, phenol, or tar acids, or from benzoquinone or phthalic anhydride.
Naphthaldehydic acid, acenaphthaquinone or bisacenaphthylideneione from acenaphthene or acenaphthylene.
Naphthalic anhydride, phenanthraquinone from phenanthrene, phthalic anhydride from naphthalene, salicyl aldehyde or salicylic acid from cresol, vanillin or vanillic acid from eugenol or isoeugenol.

Sodium Cetylsulphate

Building and Construction
Emulsifying agent (Brit. 437674) in making—
Aqueous emulsions of asphalt and similar bituminous materials.

Miscellaneous
As an emulsifying agent (Brit. 360539).
For uses, see under general heading: "Emulsifying agents."

Rubber
Stabilizing agent (Brit. 436243) in—
Vulcanizing processes.

Sodium Chlorate
Synonyms: Chlorate of soda, Sodium oxymuriate.
Latin: Natrium chloratum.
French: Chlorate sodique, Chlorate de sodium, Chlorate de soude, Oxymuriate sodique, Oxymuriate de sodium, Oxymuriate de soude.
German: Chlorsäuresnatrium, Chlorsäuresnatron, Chlorsäuressodium, Natriumchlorat, Natriumoxymuriat.
Spanish: Clorato sodico, Clorato de sosa.
Italian: Clorato di sodio.

Analysis
Oxidizing agent in—
Forensic and ultimate analysis.
Reagent in analyzing—
Aconitine, aspidospermine, atropine, cocaine, codeine, morphine, phenols, strychnine, tryosine.
Reagent in determining—
Histidin bases, indican in urine, purin bases, sulphur by means of the Parr calorimeter.

Automotive
Ingredient of—
Compositions for removing and preventing deposits of carbon in internal combustion engines (used in the place of potassium chlorate).

Chemical
As a general oxidizing agent.
Ingredient (Brit. 335203) of weed-killing compositions in admixture with—
Acids, such as hydrochloric, sulphuric, nitric, boric, oxalic, tartaric.
Acid salts, such as sodium bisulphate, sodium bitartrate, calcium hydrogen-phosphate.

Sodium Chlorate (Continued)
Acid-reacting salts.
Chlorides, such as ammonium chloride, aluminum chloride, iron chloride, copper chloride, zinc chloride, and mercuric chloride.
Sodium bichromate, sodium fluosilicate.
Reagent for various chemical purposes (used in place of potassium chlorate).
Reagent in making—
Barium peroxide, boron carbide.
Di-iodofluorescein and other dihalogenated fluoresceins (U. S. 1733776).
Dry colors, naphthalene tetrachloride, phenanthraquinone, tetrachloroanthraquinone, trichloroacetic acid.
Various intermediates and other organic and inorganic compounds.
Reagent in recovering—
Bromine from natural brines.
Dye
Oxidizing agent in making—
Alizarin, anilin black, bengal rose B, various other synthetic dyestuffs.
Explosives
Ingredient of—
Dynamites and military explosives of various sorts, fulminating compositions, fuses, matchhead compositions, percussion cap compositions, pyrotechnical compositions, safety-match compositions.
Electrical
Ingredient of—
Electrolytes in storage batteries.
Ink
Reagent in making—
Printing inks.
Insecticide
Ingredient (Brit. 258324) of—
Fumigating compositions, disinfecting compositions.
Ingredient of—
Weed-killing compositions, containing salt and crude oil.
Leather
Ingredient of—
Finishing compositions, tanning compositions.
Miscellaneous
Oxidizing agent for various purposes.
Paper
Reagent in the manufacture of paper.
Perfume
Ingredient of—
Dentifrices, lotions.
Pharmaceutical
In compounding and dispensing practice.
Sanitation
As a disinfectant.
Textile
——, *Dyeing*
Mordant in—
Dyeing cotton and wool in black shades, and in other processes.
Reagent in—
Baths containing indigosols.
——, *Printing*
As a mordant.

Sodium Chloride
Synonyms: Chloride of soda, Common salt, Muriate of soda, Rocksalt, Salt, Seasalt.
Latin: Chloruretum sodicum, Natrium chloratum, Sodii chloridum.
French: Chlorure de sodium, Hydrochlorate de soude, Sel culinaire, Sel, Sel commun, Sel de cuisine.
German: Chlornatrium, Kochsalz, Natriumchlorid.
Spanish: Cloruro sodico, Sal comun.
Italian: Cloruro di sodio, Sal commune.
Agriculture
As a cattle lick.
As a weed killer.
Analysis
As a reagent for various purposes.
Cement
Reagent in—
Recovery of potash salts as by-products in cement manufacture.
Ceramics
Ingredient of—
Glazes for chinaware, earthenware, stoneware, sewer-pipe, tile.
Chemical
Ingredient of—
Bleaching composition in admixture with magnesium chloride.
Raw material in making—
Bleaching powder, caustic soda, chlorates, chlorine, Glauber's salt, hydrochloric acid, hydrogen, hypochlorates, niter cake, sal ammoniac, sal soda, saltcake, soda ash, sodium (metallic), sodium salts of acids and halogens.
Dye
As a salting-out agent.
Fats and Oils
Reagent in—
Purification of fats and oils.
Fertilizer—
Ingredient of—
Fertilizer mixtures.
Food
Condiment and nutrient in—
Cooking, making various foodstuffs.
Ingredient (U. S. 1879162) of—
New soft cheese.
Pickling agent for—
Fish, meats, vegetables.
Preservative agent for—
Fish, meats, vegetables.
Reagent (U. S. 1882013) for—
Coagulating protein in the extraction of cacaobutter.
Refrigerating agent—
Directly in combination with chipped ice for close packing of containers.
Indirectly dissolved in water and used as a brine for circulating systems in refrigeration installations.
Glass
Ingredient of—
Batches, glazes.
Leather
Reagent in—
Chrome tanning, mineral tanning, pickling operations, salting hides, vegetable tanning.
Metallurgical
Ingredient of—
Mixes for enameling iron products.
Reagent in—
Copper extraction (from burnt pyrites), gold ore treatment, silver extraction (by the wet process), silver ore treatment, zinc metallurgy.
Miscellaneous
Ingredient of—
Feeds and medicines for domestic animals, weed-exterminating compositions.
Reagent in production of—
Sodium light used in polariscopic, spectroscopic, and similar work.
Perfume
Ingredient of—
Bath salts.
Pharmaceutical
In compounding and dispensing practice.
Suggested for use as an emetic and in clysters, fomentations for sprains and bruises, hemoptysis, increasing density of water for intravenous injections.
Photographic
As a reagent.
Refrigeration
As a brine (used in water solution).
Ingredient of—
Freezing mixtures.
Soap
As a salting-out agent.
Textile
Reagent in—
Dyeing and printing textile fabrics, mercerizing cotton.

Sodium 1:5-Chloronaphthalenesulphonate
French: 1:5-Chloronaphthalènesulphonate de soude.
German: 1:5-Chlornaphtalinsulfonsaeuresnatrium, Natrium-1:5-chlornaphtalinsulfonat.
Chemical
Ingredient of—
Emulsions with aromatic hydrocarbons and terpenes (Brit. 263873).
Fats and Oils
Ingredient of—
Emulsions.

Sodium 1:5-Chloronaphthalenesulphonate (Cont'd)

Leather
Ingredient of—
Emulsions for tanning.

Miscellaneous
Ingredient of—
Emulsified washing and cleansing compositions for various purposes.

Paper
Ingredient of—
Emulsified compositions or wetting agents for increasing the absorbing powers of paper and cardboard.

Petroleum
Ingredient of—
Emulsions with petroleum and petroleum distillates.

Textile
——, *Dyeing*
Ingredient of—
Dye liquors.

——, *Manufacture*
Ingredient of—
Carbonizing compositions for treating wool.

Waxes and Resins
Ingredient of—
Emulsions with waxes and resins.

Sodium 1:6-Chloronaphthalenesulphonate
French: 1:6-Chloronaphthalènesulfonate de soude.
German: 1:6-Chlornaphtalinsulfonsaeuresnatrium, Natrium-1:6-chlornaphtalinsulfonat.

Chemical
Reagent (Brit. 263873) in making—
Aromatic hydrocarbon emulsions, terpene emulsions.

Fats and Oils
Reagent (Brit. 263873) in making—
Emulsions.

Leather
Reagent (Brit. 263873) in making—
Tanning emulsions.

Miscellaneous
Reagent (Brit. 263873) in making—
Detergent and cleansing preparations.

Paper
Reagent (Brit. 263873) in treating—
Cardboard and paper to increase their wetting and absorbing capacity.

Textile
——, *Dyeing*
Ingredient (Brit. 263873) of—
Dye liquors.

——, *Finishing*
Ingredient (Brit. 263873) of—
Washing and cleansing compositions.

——, *Manufacturing*
Ingredient (Brit. 263873) of—
Wool carbonizing liquors.

Waxes and Resins
Reagent (Brit. 263873) in making emulsions.

Sodium Chloroplatinate
French: Chloroplatinate sodique, Chloroplatinate de sodium, Chloroplatinate de soude.
German: Natriumchlorplatinat, Natronchlorplatinat.
Spanish: Chloroplatinato de sosa.
Italian: Chloroplatinato di sodio.

Analysis
As a reagent.

Sodium 5-Chlorosalicylanilide

Insecticide
Starting point (Brit. 403411) in making—
Fungicides for seeds, tubers, and corms by reaction with copper sulphate (in dried form the precipitate product is used as a dusting powder; in paste form it is made into an aqueous suspension to which protective colloids, emulsifying and spreading agents, insecticides, or other fungicides may be added, and used in the form of a spray).

Sodium Chlorostannate
Synonyms: Sodium-tin chloride.
French: Chlorostannate sodique, Chlorostannate de sodium, Chlorostannate de soude, Chlorure d'étain et soude, Chlorure d'étain et de sodium, Chlorure sodique et stannique, Chlorure de sodium et d'étain, Chlorure de soude et d'étain, Chlorure stannique et sodique.
German: Chlornatriumstannat, Chlornatronstannat, Natriumchlorstannat, Natriumzinnchlorid, Natronchlorstannat, Natronzinnchlorid, Stanninchlornatrium, Stanninchlornatron, Stanninnatriumchlorid, Stanninnatronchlorid, Zinnchloronatrium, Zinnchloronatron, Zinnnatriumchlorid, Zinnnatronchlorid.
Spanish: Chlorostanate de sosa.
Italian: Chlorostanato di sodio.

Textile
Mordant in various dyeing processes.

Sodium 2-Chlor-4-phenylphenate

Disinfectant
As a germicide.

Sodium 6-Chlor-2-phenylphenate

Disinfectant
As a germicide.

Sodium Cholate
French: Cholate sodique, Cholate de sodium, Cholate de soude.
German: Cholinsaeuresnatrium, Natriumcholat.

Chemical
Reagent (Brit. 282356) in making parasiticides with—
Dihydrocupreine-ethyl ether.
Dihydrocupreine-ethyl ether hydrochloride.
Dihydrocupreineisoamyl ether.
Dihydrocupreineisoamyl ether hydrochloride.
Dihydrocupreine normal octyl ether.
Dihydrocupreine normal octyl ether hydrochloride.
Dihydroquinone.

Pharmaceutical
In compounding and dispensing practice.

Sodium Chromate
Synonyms: Chromate of soda.
French: Chromate sodique, Chromate de soude.
German: Chromsäuresnatrium, Chromsäuresnatron, Natriumchromat, Natronchromat.
Italian: Cromato di sodio.

Analysis
Reagent in various processes.

Chemical
Oxidizing agent in making various chemicals.
Oxidizing and neutralizing agent (Brit. 402529) in making—
Benzoic acid from toluene.
Starting point in making—
Chromates.
Chromic acid by reaction with hydrochloric or sulphuric acid.
Lead chromate by reaction with lead sulphate and sodium acetate, the latter being regenerated (French 752674).
Substitute for—
Sodium bichromate.

Ink
Ingredient of—
Writing inks.

Leather
Reagent in—
Chrome tanning.

Metallurgical
Pickling agent for—
Ornamental or other silver articles, the object being to cover them with a silver chromate coating which prevents tarnishing (German 592710).

Miscellaneous
Oxidizing agent in various processes.

Paint and Varnish
Starting point in making—
Mineral pigments.

Textile
Mordant in—
Dyeing and printing fabrics.

Sodium Chromitesilicate
French: Chromite-silicate de chrome, Chromite-silicate chromique.
German: Natriumchromitsilikat.

Chemical
Catalytic reagent in making—
Acetic acid from aldehyde, aldehyde from alcohol, benzoic acid from benzaldehyde, sodium bisulphate from

Sodium Chromitesilicate (Continued)
sodium bisulphite, sodium chloride from sodium hypochlorite.
Reagent in—
Converting manganese protoxide into permanganic acid.
Oxidizing iron and manganese compounds with the aid of atmospheric oxygen.
Dye
Reagent in converting—
Leuco-malachite hydrochloride into malacite.
Metallurgical
Reagent in recovering—
Metals from liquids, gold from seawater, radium from wells.
Miscellaneous
Reagent in sterilizing—
Liquids by means of ozone, chloride, hydrogen peroxide, or potassium permanganate.
Sugar
Reagent in recovering—
Potash and other bases from sugar juices and molasses.
Water
Reagent in—
Purifying water.
Removing iron and manganese from mineral water containing carbon dioxide by oxidizing the iron and manganese by means of atmospheric oxygen.
Removing oxygen from water by the addition of sodium sulphite, which is converted into sodium sulphate.

Sodium Chromoglucosate
Mechanical
Inhibitor of—
Corrosion in condenser systems.
Refrigeration
Inhibitor of—
Corrosion by oxygen depolarization in brine systems.

Sodium Citrate
Synonyms: Citrate of soda.
Latin: Citras sodicus, Natrium citricum.
French: Citrate de soude.
German: Citronsäuresnatrium, Citronsäuresnatron, Natriumcitrat.
Beverage
Ingredient of—
Soft drinks.
Pharmaceutical
In compounding and dispensing practice.
Ingredient (U. S. 1772183) of—
Pharmaceutical product, containing also sodium malate, ammonium citrate, and manganese bromide.
Suggested for use in treating—
Bronchitis, cystitis, diabetic acidosis, fevers (diuretic and diaphoretic, furunculosis, gout, nephritis, pneumonia, rheumatism, tracheitis, urinary acidosis.
Food
Reagent for—
Preventing curdling of milk (offsets action of rennin).
Ingredient (U. S. 1913044) of—
Reagent for improving and bleaching bread dough, containing also manganese succinate, iron lactate, gum arabic, sodium carbonate, and starch.
Modifying agent for—
Cow's-milk in infant feeding.
Photographic
Reagent in making—
Coatings for printing-out paper.

Sodium Cresolate
French: Crésolate de soude, Crésolate sodique, Crésylate de soude, Crésylate sodique.
German: Kresylsaeuresnatrium, Natriumkresylat.
Leather
Ingredient (Brit. 263473) of—
Dyeing liquors.
Miscellaneous
Ingredient (Brit. 263473) of—
Dye liquors for coloring hair and feathers.
Textile
——, *Dyeing and Printing*
Ingredient (Brit. 263473) of—
Liquors and pastes containing vat dyestuffs for coloring and printing fabrics and yarns containing acetate rayon, viscose rayon, silk-rayon mixtures, wool-rayon mixtures.

Sodium-Cupro Cyanide
Chemical
Catalyst (Brit. 446411) in—
Halogenating unsaturated hydrocarbons.
Starting point (Brit. 446411) in making—
Catalysts with metal chlorides for halogenating unsaturated hydrocarbons.

Sodium Cuprothiolactate
Chemical
Starting point (Brit. 398020) in making—
Complex double compounds of organic heavy metal mercapto compounds.

Sodium Cyanide
French: Cyanure sodique, Cyanure de sodium.
German: Cyannatrium.
Chemical
Ingredient of—
Catalytic mixtures used in making methylamines from hydrocyanic acid (Brit. 398502-4).
Mixture with calcium cyanamide used to make sodium-calcium cyanide (Brit. 400949).
Mixture with calcium cyanamide used to make calcium cyanide (Brit. 400949).
Reagent for introducing nitrile into—
Aromatics, intermediates, pharmaceuticals.
Reagent in making—
Aromatic aldehydes from an aromatic hydrocarbon, or an ether or a mono- or polyhydric phenol, or an aromatic halogenated hydrocarbon having one or several lateral chains, and aluminum chloride (French 750842).
Carbon tetrachloride solution of cyanogen chloride by reaction with chlorine in the presence of carbon tetrachloride and an amount of glacial acetic acid equal to about 4 percent by weight of the sodium cyanide (U. S. 1938324).
Hydrocyanic acid by reaction of an acid with a mixture comprising sodium cyanide and a metal sulphite (U. S. 1950899).
Starting point in making—
Case-hardening compounds, cyanogen, cyanogen chloride, cyanogen iodide, ferricyanides, ferrocyanides, hydrocyanic acid with sulphuric acid, metallic cyanides, sulphocyanides.
Disinfectant
Starting point (U. S. 1894041) in making—
Fumigating gas-producing compositions by reaction with calcium chloride and calcium oxychloride.
Food
Fumigating agent for—
Citrous and other fruits.
Insecticide
As an insecticide.
Metallurgical
Ingredient of—
Bath used in producing an electroplated zinc-tin alloy on steel and iron; claim being made that said coating has same properties as cadmium plate (U. S. 1904732).
Mixtures for producing nitrogen-containing cases of steel (U. S. 1920368).
Reagent in—
Case-hardening steel, cleaning steel surfaces, galvannealing steel, localized hardening of steel, mottling processes, reheating processes.
Solvent in—
Cyanide processes of extracting gold and silver from their ores.
Electroplating baths.
Miscellaneous
Fumigating agent for—
Grain elevators, railroad cars, various purposes.
Ingredient of—
Metal polishes.
Paint and Varnish
Ingredient (U. S. 1803607) of—
Marine paint containing also coaltar and cement.
Photographic
As a fixing agent.
Sanitation
As a fumigating agent.
Textile
Fumigating agent for—
Raw cotton.

Sodium Cyclohexylnaphthalenesulphonate
French: Cyclohexylenaphthalènesulfonate sodique, Cyclohexylenaphthalènesulfonate de sodium, Cyclohexylenaphthalènesulfonate de soude.
German: Cyclohexylnaphtalinsulfonsaeuresnatrium, Natriumcyclohexylnaphtalinsulfonat.

Fats and Oils
Starting point in making—
Solvents (Brit. 279877).

Miscellaneous
Ingredient (Brit. 279877) of—
Cleansing and bleaching compositions.
Compositions for treating parquet floors.
Washing compositions.

Soap
Ingredient of—
Soap compositions (Brit. 279877).

Textile
——, *Dyeing*
Assist in dyeing—
Woolen fabrics and yarns (Brit. 279877).

——, *Finishing*
Ingredient of—
Cleansing compositions (Brit. 279877).

Sodium Cyclohexylxanthate

Metallurgical
Flotation agent (U. S. 1823316) in separating—
Minerals from ores (added to aid in the froth flotation process).

Sodium Decylsulphonate

Miscellaneous
As an emulsifying agent (Brit. 360539).
For uses, see under general heading "Emulsifying agents."

Sodium Diamylalphanaphthylaminesulphonate

Miscellaneous
As an emulsifying agent (U. S. 1853415).
For uses, see under general heading "Emulsifying agents."

Sodium Dibutyldithiocarbamate

Disinfectant
As a bactericide (Australian 8103/32, Brit. 406979, U. S. 1972961).

Insecticide and Fungicide
As a fungicide (Australian 8103/32, Brit. 406979, U. S. 1972961).
As an insecticide (claimed effective against aphids) (Australian 8103/32, Brit. 406979, U. S. 1972961).

Sodium Dibutylsulphanilate
French: Dibutylesulphanilate de soude.
German: Dibutylsulfanilsaeuresnatrium, Natriumdibutylsulfanit.

Dye
Dispersing agent (Brit. 264860) in making—
Dyes, lakes.

Ink
Dispersive agent.

Paint and Varnish
Dispersive agent in making—
Copal varnishes, lacquers, spirit varnishes, water paints.

Plastics
Dispersive agent in making—
Solutions of cellulose nitrate, cellulose acetate and other cellulose esters and ethers.

Rubber
Dispersive agent in making—
Rubber solutions.

Textile
——, *Dyeing and Printing*
Dispersive agent in making dye liquors with—
Anthraquinone dyestuffs, indigoes, sulphur dyestuffs, vat dyestuffs.

——, *Finishing*
Dispersive agent in making finishing compositions for all fabrics.

Sodium-Dibutyl Sulphosebacate

Miscellaneous
As a wetting agent (Brit. 446568).
For uses, see under general heading: "Wetting agents."

Sodium Dibutyltetrahydronaphthalenesulphonate
French: Dibutyletétrahydronaphthalènesulfonate sodique, Dibutyletétrahydronaphthalènesulfonate de sodium, Dibutyletétrahydronaphthalènesulfonate de soude.
German: Dibutyltetrahydronaphtalinsulfonsaeuresnatrium, Natriumdibutyltetrahydronaphtalinsulfonat.

Fats and Oils
Starting point (Brit. 279877) in making—
Solvents.

Miscellaneous
Ingredient (Brit. 279877) of—
Cleansing and bleaching compositions for parquetry floors.
Washing compositions.

Soap
Ingredient (Brit. 279877) of—
Washing and detergent compositions.

Textile
——, *Dyeing*
Assist (Brit. 279877) in making—
Wool-dyeing liquors.

——, *Finishing*
Ingredient (Brit. 279877) of—
Cleansing and finishing compositions.

Sodium 1:4-Dichlorophthalate

Textile
Delustring agent (Brit. 425418) for—
Cellulose acetate rayon (used with aluminum formate).

Sodium Dicresoldithiophosphate

Metallurgical
Collector in—
Ore concentrating by flotation processes.

Mining
Flotation agent (Brit. 455224) in—
Froth flotation of minerals.

Sodium Dicresylphosphate
French: Dicrésylephosphate de soude.
German: Dicresylphosphorsaeuresnatrium, Natriumdicresylphosphat.

Chemical
Reagent in making—
Finishing compounds for use on textiles (Brit. 267534).

Dye
Reagent in making—
Dye pastes.

Soap
Ingredient in making—
Detergent compositions.

Sodium Diethyldithiocarbamate

Disinfectant
As a bactericide (Australian 8103/32, Brit. 406979, U. S. 1972961).

Insecticide and Fungicide
As a fungicide (Australian 8103/32, Brit. 406979, U. S. 1972961).
As an insecticide (claimed effective against aphids) (Australian 8103/32, Brit. 406979, U. S. 1972961).

Sodium Diethyldithiophosphate

Metallurgical
Collector in—
Ore concentrating by flotation processes.

Sodium Dihydroxytartrate
French: Dihydroxytartrate sodique, Dihydroxytartrate de sodium, Dihydroxytartrate de soude.
German: Dihydroxyweinsäuresnatrium, Dihydroxyweinsäuresnatron, Natriumdihydroxytartrat.

Chemical
Starting point in making—
Intermediates, pharmaceuticals, various derivatives.

Dye
Starting point (Brit. 340009) in making azo dyestuffs with the aid of—
2-Aminotoluene-4:5-disulphonic acid hydrazin.
2-Methylphenylhydrazin-4:5-disulphonic acid.
Phenylhydrazin-5-sulphonic acid, phenylhydrazin-3:5-disulphonic acid, tolylhydrazin-5-sulphonic acid, tolylhydrazin-3:5-disulphonic acid, xylylhydrazin-5-sulphonic acid, xylylhydrazin-3:5-disulphonic acid.

Sodium-Dimeta-aminobenzamidostilbene Disulphonate

Paper

Impregnating agent and absorbent for ultraviolet light (Brit. 436891) in—
Treating paper and like products to be used as food containers.

Sodium 1:3-Dimethylcaproate

Textile

Cleansing agent (Brit. 414485) for—
Wool, silk, cotton, ramie, jute, hemp, flax, and rayon fibers, by treatment in an aqueous alkaline liquor, using trisodium phosphate, soda ash, or sodium or potassium hydroxide as the alkaline constituent.

Sodium Dimethyldithiocarbamate

Chemical

Starting point (Brit. 340574) in making rubber vulcanization accelerators with the aid of—
Benzal chloride, 2:4-dinitro-1-chlorobenzene, 1:4-dichloro-2:6-dinitrobenzene.

Disinfectant

As a bactericide (Australian 8103/32, Brit. 406979, U. S. 1972961).

Insecticide and Fungicide

As a fungicide (claimed effective against barley spores) (Australian 8103/32, Brit. 406979, U. S. 1972961).
As an insecticide (claimed effective against aphids) (Australian 8103/32, Brit. 406979, U. S. 1972961).

Sodium 1:3-Dimethylvalerate

Textile

Cleansing agent (Brit. 414485) for—
Wool, silk, cotton, ramie, jute, hemp, flax, and rayon fibers, by treatment in an aqueous alkaline liquor, using trisodium phosphate, soda ash, or sodium or potassium hydroxide as the alkaline constituent.

Sodium Dinaphthylphosphate

French: Dinaphthylphosphate de soude.
German: Dinaphtylphosphorsaeuresnatrium, Natriumdinaphtylphosphat.

Chemical

Reagent in making—
Finishing compounds for use on textiles (Brit. 267534).

Dye

Reagent in making—
Dye pastes.

Soap

Ingredient in making—
Detergent compositions.

Sodium Dinitrostilbindisulphonate

French: Dinitrostilbènedisulphonate sodique, Dinitrostilbènedisulphonate de sodium, Dinitrostilbènedisulphonate de soude.
German: Dinitrostilbendisulfonsaeuresnatrium, Dinitrostilbendisulfonsaeuresnatron, Natriumdinitrostilbendisulfonat.

Chemical

Starting point in making various intermediates.

Dye

Reagent (Brit. 311384) in making azo dyestuffs with—
Alphanaphthylamine, anilin, 4-chloro-2-aminophenol.
4-Chloro-2-aminophenol-6-sulphonic acid.
2-Chloroanilin-5-sulphonic acid, 2-chloro-2-aminobenzene-5-sulphonic acid, J acid, metanilic acid, paraphenylenediamine, salicylic acid, sulphanilic acid.

Sodium Dipentamethylenethiuramdisulphide

Rubber

Secondary activator in—
Vulcanizing processes (for use with mercaptabenzthiazole).

Sodium Dipentamethylenethiurammonosulphide

Rubber

Secondary activator in—
Vulcanizing processes (for use with mercaptabenzthiazole).

Sodium Dipentamethylthiouramtetrasulphide

Rubber

Secondary activator in—
Vulcanizing processes (for use with mercaptabenzthiazole).

Sodium Diphenylmonosulphonate

Chemical

Ingredient (U. S. 1845309) of—
Wetting and penetrating agent containing also cresol and an emulsifying agent.

Sodium Dithiosalicylate

Pharmaceutical

In compounding and dispensing practice.
Suggested for use as—
Antipyretic, antirheumatic, antiseptic dusting powder.

Sodium Dodecanolsulphonate

Miscellaneous

As an emulsifying agent (Brit. 360539).
For uses, see under general heading: "Emulsifying agents."

Sodium Eleostearicsulphonate

Miscellaneous

As an emulsifying agent (Brit. 361732).
For uses, see under general heading: "Emulsifying agents."

Sodium Ethylmethylbutylbarbiturate

Synonyms: Nembutal.

Pharmaceutical

Suggested for use as—
New anesthetic (by basal narcosis).

Sodium Ethyl-1-methylbutylthiobarbiturate

Synonyms: Pentothal sodium.

Pharmaceutical

Suggested for use as—
New anesthetic (said to produce quickly deep anesthesia for short periods, with rapid recovery).

Sodium Ethylnaphthalenesulphonate

French: Éthylenaphthalènesulphonate sodique, Éthylenaphthalènesulphonate de sodium, Éthylenaphthalènesulphonate de soude.
German: Aethylnaphtalinsulfonsäuresnatrium, Aethylnaphtalinsulfonsäuresnatron, Natriumaethylnaphtalinsulfonat.

Chemicals

Reagent (Brit. 298823) in making—
Pharmaceuticals.
Starting point in making various derivatives.

Fats and Oils

Reagent (Brit. 298823) in making—
Dissolving emulsions.

Insecticide

Ingredient (Brit. 298823) of—
Insecticides, vermin exterminators.

Miscellaneous

Ingredient (Brit. 298823) of—
Cleansing, scouring, and detersive preparations.

Mechanical

Ingredient (Brit. 298823) of—
Lubricating compositions.

Perfume

Ingredient (Brit. 298823) of—
Cosmetics, perfumes.

Sanitation

Ingredient (Brit. 298823) of—
Disinfectants, germicides.

Sodium Ethyl-3-nitrophthalate

French: Éthyle-3-nitrophthalate de soude.
German: Aethyl-3-nitrophtalsaeuresnatrium, Natriumaethyl-3-nitrophtalat.

Resins and Waxes

Reagent in making—
Synthetic resins (U. S. 1618209).

Sodium Ethyl-4-nitrophthalate

French: Éthyle-4-nitrophthalate de soude.
German: Aethyl-4-nitrophtalsaeuresnatrium, Natriumaethyl-4-nitrophtalat.

Resins and Waxes

Starting point in making—
Synthetic resins (U. S. 1618209).

Sodium Ethylphthalate
French: Éthylephthalate de soude.
German: Aethylphtalsaeuresnatrium, Natriumaethylphtalat.

Resins and Waxes
Starting point (Brit. 250265) in making synthetic resins with—
Barium acetate, barium bromide, barium nitrate, calcium acetate, calcium bromide, calcium chloride, calcium nitrate, lead acetate, lead bromide, lead chloride, lead nitrate, magnesium acetate, magnesium bromide, magnesium chloride, magnesium nitrate, strontium acetate, strontium bromide, strontium chloride, strontium nitrate, zinc acetate, zinc bromide, zinc chloride, zinc nitrate.

Sodium Ethylxanthate
French: Xanthate de soude-éthyle.
German: Natriumaethylxanthogenat, Natronaethylxanthogenat, Xanthogensäuresnatriumaethyl, Xanthogensäuresnatronaethyl.

Analysis
As a reagent.

Chemical
Reducing agent in various processes.
Starting point in making—
Rubber vulcanization accelerator with sulphur monochloride (Brit. 265169).
Thiophenols from diazonium compounds.

Insecticide
Ingredient of—
Insecticidal compositions.

Metallurgical
Flotation agent in—
Froth processes of ore concentration.

Sodium Fluoride
French: Florure sodique, Florure de soude.
German: Natriumfluorid, Fluornatrium, Fluornatron.
Spanish: Floruro sodico.
Italian: Floruro di sodio.

Beverage
Antifermentative in—
Alcoholic fermentations.
Antiseptic in—
Alcoholic fermentations.

Ceramics
Ingredient of—
Enamels.

Chemical
Antifermentative in—
Alcoholic fermentations.
Antiseptic in—
Alcoholic fermentations.
Reagent (U. S. 1914135) in making—
Carbon halides (chlorofluorides).
Starting point in making—
Calcium fluoride, caustic soda by the fluoride process, magnesium fluoride, zinc fluoride.

Disinfectant
Ingredient of—
Disinfectant for plant and seed diseases, comprising a mercurized chlorophenol and hydrated lime (U. S. 1776423).

Food
Antiseptic and disinfectant for—
Egg storage.

Glass
Opacifying agent in making—
Opaque glasses, translucent glasses.

Glues and Adhesives
Ingredient (U. S. 1895979) of—
Vegetable glue, containing also powdered ivory nut, casein, lime, soda ash, and trisodium phosphate.

Insecticide
Ingredient of—
Insecticidal composition for impregnating woolen goods, containing also sodium taurocholate, sodium glycocholate, and carbon dioxide dissolved under pressure sufficient to cause the spray to penetrate the goods (U. S. 1901960).
Insecticidal powders for killing chicken lice, rat exterminants, roach exterminants, vermicides.

Metallurgical
Coating agent (U. S. 1905753) for—
Copper.
Ingredient of—
Flux used in melting magnesium metal (Brit. 403891).
Pickling mixture with nitric acid and molasses for removing scale from chrome steel (U. S. 1919624).
Soldering composition for aluminum, consisting of a mixture with zinc chloride and ammonium bromide (French 642778).
Soldering composition for metals, particularly aluminum and its alloys, consisting of a mixture with zinc chloride and ammonium bromide (U. S. 1761116).
Reagent (U. S. 1914768) in making—
Pure aluminum combinations adapted for production of aluminum.

Miscellaneous
Ingredient (U. S. 1881128) of—
Motion picture projection screen coating, containing also glue, copper sulphate, casein, glycerin, borax, cobalt blue, and water, said to have properties of nonstickiness, permanence, and adaptability to climatic conditions.

Pharmaceutical
In compounding and dispensing practice.
Suggested for use as—
Antiseptic in external lotions.

Textile
Ingredient (Brit. 403966) of—
Impregnating mixture with borax, for raising the safe ironing temperature of cellulose acetate fabrics.

Woodworking
Impregnating preservative agent for—
Electric light poles, telegraph poles, and the like.
Piling, railroad ties, underground woodwork.
Ingredient of—
Wood-impregnating mixtures with zinc chloride, or acid zinc fluoride.
Wood preservative (Brit. 394162).

Sodium Formaldehyde-sulfoxylate
Synonyms: Formaldehyde-sulphoxylate of soda.
French: Formaldéhyde-sulfoxylate sodique, Formaldéhyde-sulfoxylate de sodium, Formaldéhyde-sulfoxylate de soude, Sulfoxylate-formaldéhyde de sodium.
German: Natriumformaldehydsulfoxylat.

Textile
——, *Printing*
Discharge in printing fabrics.
Reagent (U. S. 1912008) in making—
Printing pastes used as colored discharges on cellulose acetate and similar fibers.

Sodium Formanilide
German: Natriumformanilid.

Chemical
Reagent in making—
Phenyldihydroquinazolin (orexin).

Sodium Formate
French: Formiate sodique, Formiate de soude.
German: Formylsäuresnatrium, Formylsäuresnatron, Natriumformiat, Natronformiat.

Chemical
Reducing agent in—
Organic synthesis.
Starting point in making—
Formic acid, oxalic acid.

Pharmaceutical
In compounding and dispensing practice.

Textile
Mordant in—
Dyeing, printing.

Sodium Glucosate

Mechanical
Inhibitor of—
Magnesium scale formation in boilers and hot-water systems.
Remover of—
Carbon dioxide formation in boiler waters and hot-water systems.
Reagent for—
Hydrogen ion adjustment (pH increase).

Sodium Glycocholate
French: Glycocholate sodique, Glycocholate de sodium, Glycocholate de soude.
German: Glycocholsaeuresnatrium, Natriumglycocholat.

Sodium Glycocholate (Continued)
Chemical
Reagent (Brit. 282356) in making antiparasitic agents with—
Dihydrocuprein ethyl ether, dihydrocuprein ethyl ether hydrochloride, dihydrocuprein isoamyl ether, dihydrocuprein isoamyl ether hydrochloride,, dihydrocuprein normal octyl ether, dihydrocuprein normal octyl ether hydrochloride, dihydroquinone.

Sodium Heptadecylsulphonate
Miscellaneous
As an emulsifying agent (Brit. 360539).
For uses, see under general heading: "Emulsifying agents."

Sodium Heptylate
French: Heptylate sodique, Heptylate de sodium, Heptylate de soude.
German: Heptylsaeuresnatrium, Heptylsaeuresnatron, Natriumheptylat.
Chemical
Reagent (Brit. 304118) in making ketonic acid esters with the aid of allyl, amyl, butyl, heptyl, hexyl, propyl, an other alkyl esters of the following acids—
Acetic, anthranilic, benzoic, butyric, camphoric, capric, caproic, caprylic, chloracetic, cinnamic, citric, cresylic, gallic, lactic, maleic, malic, malonic, metanilic, mucic, naphthionic, oxalic, palmitic, phenylacetic, phthalic, picramic, picric, propionic, pyrogallic, salicylic, succinic, sulphanilic, tartaric, trichloroacetic, valeric.
Starting point in making—
Aromatics, intermediates, pharmaceuticals, salts and esters.
Dye
Starting point in making various synthetic dyestuffs.

Sodium Heptylnaphthalenesulphonate
French: Heptylenaphthalènesulfonate sodique, Heptylenaphthalènesulfonate de sodium, Heptylenaphthalènesulfonate de soude.
German: Heptylnaphtalinsulfonsaeuresnatrium, Natriumheptylnaphtalinsulfonat.
Fats and Oils
Reagent (Brit. 277277) in making—
Emulsified boring oil compositions, emulsions of various sorts.
Petroleum
Reagent (Brit. 277277) in making—
Emulsions of petroleum and petroleum distillates.
Resins and Waxes
Reagent (Brit. 277277) in making—
Emulsions.
Textile
——, *Finishing*
Reagent (Brit. 277277) in making—
Emulsified bucking and felting compositions.
——, *Manufacturing*
Reagent (Brit. 277277) in making—
Compositions for removing incrustations from textile fibers, emulsified spinning oil compositions.

Sodium Hexylate
French: Hexylate sodique, Hexylate de sodium, Hexylate de soude.
German: Natriumhexylat.
Chemical
Reagent (Brit. 304118) in making ketonic acid esters with the aid of the butyl, amyl, allyl, heptyl, hexyl, and propyl esters of the following acids—
Acetic, anthranilic, benzoic, butyric, camphoric, caproic, caprylic, chloroacetic, cinnamic, citric, cresylic, gallic, lactic, maleic, malic, malonic, metanilic, mucic, naphthionic, oxalic, palmitic, phenylacetic, phthalic, propionic, pyrogallic, salicylic, succinic, sulphanilic, tartaric, trichloroacetic, valeric.
Starting point in making—
Aromatics, intermediates, pharmaceuticals, salts and esters.
Dye
Reagent in making various synthetic dyestuffs.

Sodium Hexylnaphthalenesulphonate
French: Hexylenaphthalènesulfonate de soude, Hexylenaphthalènesulfonate sodique.
German: Hexylnaphtalinsulfonsaeuresnatrium, Natriumhexylnaphtalinsulfonat.
Fats and Oils
Emulsifying agent (Brit. 277277) in making—
Boring oil compositions.
Reagent (Brit. 277277) in making—
Emulsions.
Petroleum
Reagent (Brit. 277277) in making—
Emulsions of petroleum and petroleum distillates.
Resins and Waxes
Reagent (Brit. 277277) in making—
Emulsions.
Textile
——, *Manufacturing*
Emulsifying agent (Brit. 277277) in making—
Bucking compositions, compositions for removing incrustations from fibers, felting compositions, spinning oils.

Sodium Hippurate
Latin: Natrium hippuratum.
French: Hippurate sodique, Hippurate de sodium, Hippurate de soude.
German: Hippursäuresnatrium, Hippursäuresnatron, Natriumhippurat.
Chemical
Ingredient (Brit. 310934) of—
Insulin preparations.
Pharmaceutical
In compounding and dispensing practice.

Sodium Hydrosulphide
Synonyms: Sodium sulphydrate, Sulphydrate of soda.
Analysis
As a reagent in various processes.
Chemical
As a reagent in various processes.
Leather
Solvent (Brit. 402327) for—
Sulphur dyes in dyeing leather and skins.

Sodium Hydrosulphite
Chemical
Reducing agent in making—
Stable solutions of acridin salt (Brit. 395405, 342690).
Triaminohydroxyanthraquinones (Brit. 396976).
Starting point in making—
Sodium formaldehyde-hydrosulphite.
Explosives and Matches
Sheathing agent for—
Coal-mining explosives.
Photographic
Reducing agent (Brit. 401340) for—
Azo dyes in the production of color pictures from silver pictures.
Textile
Bleaching agent for various fabrics.
Discharge in—
Dyeing.
Ingredient of—
Vat liquors.
Reagent for—
Reducing dyes.

Sodium Hydroxide
Synonyms: Caustic soda, Hydrate of soda, Hydrated oxide of sodium, Mineral alkali, Soda lye, Sodic hydrate, Sodium hydrate.
Latin: Natrium causticum, Natrium hydricum, Natriumhydroxydatum, Sodii hydroxidum.
French: Caustique de soude, Soude caustique.
German: Aetznatron, Natriumhydroxyd, Natriumoxyhydrat, Natron, Natronhydrat.
Spanish: Hidrato sodico.
Italian: Soda caustica, Sodio caustica.
Abrasives
Process material in making—
Abrasives.
Adhesives
Converting agent in making—
Starch glues.
Ingredient of—
Adhesives, casein cement, starch adhesives.
Agriculture
Ingredient of—
Cattle dips, cattlefeeds, sheep dips.

Sodium Hydroxide (Continued)

Analysis
Alkali in—
Analytical process involving control research.

Automotive
Process material in making—
Clutch facing.

Brewing—
Ingredient of—
Bottle-washing compositions, cleansing compositions.
Settling agent for—
Yeast.

Building Material
Process material in making—
Heat insulation, plaster-board, portland cement, sound insulations, wallboard.

Cellulose Products
Fermenting agent for—
Cellulose sulphite liquor.
Process material in making—
Cellulose, cellulose esters, ethers, and other derivatives such as rayon, viscose.

Chemical
Absorbent for—
Acids, carbon dioxide, chlorine, cyanogen, hydrogen sulphide, nitrogen oxides, phenols.
Activating agent for—
Charcoal.
Catalyst in making—
Aldol, ammonia, cyanogen, esters, sodamide.
Dehydrating agent for—
Air, alcohol, butyl ether, diethyl ketone, ethyl ether, ethylmethyl ketone, ethylpropyl ether, ketones, methylethyl ketone, organic solutions, propyl alcohols, propyl ether, pyridin.
Deodorizing agent for—
Isopropyl alcohol.
Neutralizing agent for—
Acids, in various reactions and processes of chemical manufacturing.
Process material in—
Regenerating catalysts.
Process material in making—
Arsenobenzene derivatives, absorbent carbons, acetates (from carbohydrates), 2-acetylamino-1-naphthyl-thioglycolic acid, aldehydes, aldehyde emulsions, alginic acid, aldol, aluminates, alkali-earth linoleates, allyl chloride, allyl para-aminobenzoate, allylthiobromine, aluminum compounds, 4-amino-2-auromercaptobenzoic acid, 2-amino-1-hydroxybenzene-4-sulphonamide, 2-amino-1-naphthylthioglycolic acid, N-(4′-amino-1-naphthyl)-N-toluenesulphonamide, aminonaphtholsulphonic acids, amino(paratolylsulphonamide)naphthalene-sulphonic acids, aminophenolsulphonic acids, ammonia, amyl acetate, amyl formate, amyl oleate, amyldextrin, anthranilic acid, anthraisoquinolin, antimony sulphides, arsanilic acid, arsenic and its compounds and derivatives, barium compounds, beechwood creosote, beeswax acids, benzaldehyde, benzene derivatives, benzenedisulphonamides, benzenedisulphonic acids, benzoates, benzoic acid, benzophenonearsenious acid, 4-benzoyl-1-hydroxy-2-naphthoic acid, benzyl alcohol, benzyl oleate, benzyldextrin, benzylsodium phthalate, benzyl-sodium succinate, bis-(3-carboxy-4-hydroxy-1-naphthyl) ketone, bismuth compounds, bismuth oxide catalysts, bismuth-mercury compounds, borax.
Borneol, bromoethylene, bromine, bromine organic compounds and their emulsions, butyl acetate, butyldextrin, butyric acid, carbazole sodium salt, calcium compounds, carbon dioxide absorbent, catalysts of various kinds, cerium oxide, chlorinated organic compounds, cresol, cresyl phosphates, cyanides, decolorizing agents, dialkylaminoalkyl compounds, diaminodihydroxyarsenobenzene alkali salts, diaminodihydroxyarsenobenzene silver salts, diaminodihydroxydibenzenedisulphonamides, dibutyl dixanthate, dichlorpentane, diglycerol, dimetatolylparatolyl phosphate, dimethyldi-isopropylbenzidin, dimethyl ether, dinitrophenol, diphenic acid, diphenyl ether, diphenylguanidin, emulsions of aliphatic hydrocarbons, emulsions of aromatic hydrocarbons, emulsions of cyclic hydrocarbons, emulsions of nitro compounds, emulsions of pyridin compounds, emulsions of quinolin compounds, ethane, ethylamine, ethylbenzene emulsions, ethyl chloride emulsions, ethyl 4-hydroxy-1-naphthoate, ethyl oleate, ethyldextrin, ethylene, ethylene oxide, ethyleneglycol, ethylhydrocuprein and derivatives, ethylmethyl ether, 5-ethyl-5-phenylhydantoin, ethylstarch, formaldehyde, formic acid, glutamic acid, glyceryl oleate, glycol, glyoxylates, glyoxylic acid, guaiacol, hexamethylenetetramine, hydrocinnamic acid, hydrocuprein derivatives, hydrocyanic acid, hydrogen, hydrogenation catalysts, hydroquinone, 4-hydroxymeta-arsanilic acid, 4-hydroxynaphthalene-1:3-dicarboxylic acid, 3-hydroxy-2-naphthoic acid, 4-hydroxy-1-naphthylphenyl ketone, 4-hydroxy-3-nitrobenzenearsonic acid, inulin, iodine compounds, ionine emulsions, iron compounds, isoborneol, isobutyl oleate, isopropylallylbarbituric acid, isopropylstarch, ketones, lactic acid, lead compounds, litharge, lysalbinic acid, magnesium compounds, maltose, manganates, mannose, mercury compounds, meta-aminobenzaldehyde, metahydroxybenzaldehyde, metahydroxybenzoic acid, metallic hydroxides, metatitanic acid, methane, methanol, methylamines, methyldextrin.
Methyl 4-hydroxynaphthalene-1:3-dicarboxylate, methyl 4-hydroxy-1-naphthoate, methyl oleate, 2-methylphloroglucinol, methylstarch, naphthalene, naphthalene derivatives, naphthalenesulphonic acids, naphthotic acid, naphthoisoquinolin, naphthols, naphtholsulphonic acids, 1:2-Naphthothioindoxyl, naphthylenediamine derivatives, N-(1′-naphthyl)paratoluenesulphonamide, natural gas chlorination products, nickel catalysts, nickel hydroxide, nickel sulphide, nitrobenzene, nitrobenzene emulsions, nitrogen, nitrosamine, 4-nitro-2-thiocyanobenzoic acid, nitroxylene, olefinglycol, orthoaminophenol-4-sulphon-(4′-amino)-anilide, orthoaminophenol-4-sulphonanilide, orthobenzyloxybenzoic acid, oxalic acid, oxanthanol, oxygen, oxymercury nitrophenolates, palmitic acid, para-acetylaminophenylstibnic acid, parahydroxybenzoic acid, parahydroxyphenylstibnic acid and salts, phenol, phenols, phenolphthalein, phenylphosphates, phenylstibnic acid, phenylpstibnic oxychloride, phosphoric acid, phosphoric acid esters, phthalic acid, phthalic anhydride, picric acid, polyglycerols, potassium compounds, propionic acid, propyl oleate, propyldextrin, propylene, propyleneglycol, propylstarch, protalbinic acid compounds, pyridin, quinine derivatives, radium and its compounds, resorcinol, salicylic acid, silicic acid, silicon compounds, sodium compounds, starch ethers, starch xanthogenate, stearic acid, succinic acid, synthetic aromatic chemicals, tartaric acid, tartrates, terpinol emulsions, tetraglycerol, thorium and compounds, titanium and compounds, toluene and derivatives, toluene emulsions, trichloroethylene, triethyltrimethylenetriamine, triglycerol, tritolylguanidin, uranium and its compounds, valeric acid, vanadium catalysts, vanadium compounds, vanillalacetone, vanillylamine, xylene, xylene emulsions, zinc salts, zirconium salts.
Promoter of—
Catalytic reactions.
Purifying agent (either directly or in the process) for—
Acenaphthene, alkali chloride solutions, aluminum sulphate, 2-amino-1-methyl-4-isopropylbenzene-5-sulphonic acid, benzene, bormine, carbazole, calcium acetate, calcium sulphate, cresol, ethyl ether, fluorene, gases (with pumice), hydrogen, iron oxide, lead arsenate, nitrogen monoxide, para-acetaldehyde, toluene.
Reagent for separating—
Carbon dioxide from air, flue gas, gases generally, nitrogen, water gas.
Carbon monoxide from hydrogen, hydrofluoric acid from phosphoric acid, hydrogen sulphide from gases, naphtholsulphonic acids, nitrogen monoxide from ozone, phenols from benzene, phosphorus from iron vanadate, red phosphorus from yellow phosphorus, sodium chloride from magnesium chloride, sulphur from hydrogen, suspended matter from organic liquids, theobromine from caffeine.
Reagent in solidifying—
Alcohols, alcohol-ether mixtures, butane, carbon bisulphide, carbon tetrachloride, ether.
Solubilizing agent for—
Materials difficult to dissolve, materials insoluble in nitric acid, materials insoluble in sulphuric acid.
Solvent for—
Starch.
Stabilizing agent for—
Hydrogen peroxide solutions, perborates.
Starting point in making—
Permanganates, sodium compounds, sodium salts of organic chemicals.

Sodium Hydroxide (Continued)

Chemical Specialty
Ingredient of—
Antifreeze compositions, belt dressings, boiler compounds.
Composition for removing carbon from internal-combustion engines.
Compositions for repairing automobile radiators.
Hat sizings, puncture-closing composition for tires.

Clay Products
Cleaning agent for—
Canadian d'Amherst clay, Canadian china clay, Fraddon china clay, kaolins, Wotter clay.
Deflocculating agent, floating agent, peptizing agent.
Process material in making—
Opaque glazes.
Purifying agent.

Coal By-Products
Extractant for—
Phenols from tar.
Ingredient of—
Binders for briquettes.
Purifying agent for—
Benzene.
Separating agent for—
Clarain from coal, sulphur from coaltar distillates, vitrain from coal.
Solvent for—
Coal humic substance, coal, peat.

Construction
Waterproofing agent for—
Cement, concrete, gypsum.

Cosmetic
Process material in making—
Greaseless creams, shampoos, shaving creams.

Disinfectant
Ingredient of—
Antiseptics, bactericides, disinfecting powders, disinfectants, germicides.

Distilled Beverage
Ingredient of—
Bottle-washing compositions, cleansing compositions.
Recovering agent for—
Potash salts from distillery waste.
Treating agent for—
Distillery waste.

Dye
Process material in making—
Alizarin, anilin dyes, 2-anilinoanthraquinone, 1-anilino-2-naphthol, anthraquinone, anthraquinonesulphonic acids (sodium salt), azo dyes, azobenzene, 2-bromoanthraquinone, bromoindigo, diethylanilin, dye soaps, green dyes, hydrazin hydrochloride, hydrazobenzene, hydrazocymene, hydrazotoluene, indigo, intermediates, indanthrene, indanthrene leuco derivatives, leuco derivatives of vat dyes, leuco compounds of hydron blue, leuco derivatives of indigo, leuco derivatives of thioindigo, methylanilin, methylanthracene, nitrosamine printing pastes, paranitranilin, printing pastes, sulphur dyes, thioindigo derivatives, vat dyes.
Purifying agent for—
Anthracene, anthraquinone, phenanthrene.

Electrical
Ingredient of—
Dry batteries, electrolytes for wet batteries.
Process material in making—
Depolarizers, electrodes, electrolytic condensers, insulations.

Explosives
Treating agent for—
Cellulose.
Process material in making—
Nitrocellulose, nitrostarch, nitrosugar, nitroglycerin, picric acid.

Fats and Oils
Catalyst in—
Hydrogenation processes.
Extractant for—
Oil from copra, oil from cottonseed.
Process material in—
Bleaching operations.
Process material in making—
Butter substitutes.
Emulsions of animal, fish, and vegetable oils.
Fatty acids, sodium salts of sulphonated oils.
Purifying agent for—
Wool-grease.
Refining agent for—
Fats of animal, fish, and vegetable origin.
Foots.
Hydrogenated products of blubber, oils, fats.
Oils of animal, fish, and vegetable origin.
Remover of fatty acids from—
Fats of animal, fish, and vegetable origin.
Oils of animal, fish, and vegetable origin.
Saponifying agent for—
Fats of animal, fish, and vegetable origin.
Oils of animal, fish, and vegetable origin.
Starting point in making—
Catalysts for hydrogenation processes.

Food
Neutralizing agent for—
Acids in food, milk.
Peeling agent for—
Fruit.
Pickling agent for—
Olives.
Process material in making—
Canned products, cocoa products, dried products, yeast.

Forest Products
Treating agent for—
Bamboo, baobab wood, bast, coconut fibers, sisal.

Glass
Process material in making—
Frosted glass, milk glass, opaque glass.

Glue and Gelatin
Hydrolyzing agent for—
Gelatin, glue.

Ink
Process material in making—
Lithographic inks, printing inks, writing inks.

Insecticide and Fungicide
Ingredient of—
Insecticides, fungicides, weed destroyers.
Process material in making—
Calcium arsenate, calcium arsenite.

Laundering
Ingredient of—
Washing compositions.

Leather
Process material in making—
Artificial leather, tanning materials.
Process material in tanning—
Fish skins.

Linoleum and Oilcloth
Treating agent for—
Linseed oil.

Lubricant
Process material in making—
Cup greases, cutting oils, emulsified oils, soda-base greases.

Metallurgical
Cleansing agent for—
Iron.
Process material in—
Coloring iron, detinning operations, dezincing lead, platinum metallurgy, purifying lead.
Process material in making—
Aluminum soldering compositions, aluminum wire electric coils, antimony, bismuth, cadmium.
Cadmium electroplatings on iron, knives, piano wires, razors, scissors, springs, steel products, tools, and many other articles.
Colorings on galvanized iron, copper coatings on iron, electrodepositions of iron, electrodepositions of lead, electrodepositions of nickel, electroplatings on aluminum, ferro-molybdenum, foundry sand molds, gold, lead coatings on iron, metallic coatings on aluminum, metallic coatings on copper, nickel coatings on iron, oxide coatings on aluminum wire, red lead, zinc coatings on iron.
Reagent for separating—
Antimony from bismuth, arsenic from bismuth, selenium from bismuth, sulphur from bismuth, tellurium from bismuth, tin from bismuth, zinc from bismuth.

Military
Ingredient of—
Gas-mask absorbents.

Milling
Disinfectant for—
Beans, cereals, grain.

Sodium Hydroxide (Continued)

Mining

Extracting reagent for—
Cinnabar treatment, copper from its ores, molybdenum from its ores, tungsten from its ores, uranium from its ores, vanadium from its ores, zinc from its ores.
Flotation agent for—
Carbonate ores, chalcocite, chalcopyrite, gold ores, lead ores, malachite, porphyry ores, silicate ores, silver ores, sulphide ores, zinc-lead ores.
Ingredient of—
Binders for ore-briquetting.
Process material in making—
Flotation agents, mercury.
Reclaiming agent for—
Huebnerite.
Reagent for separating—
Bauxite from clay, hematite, lead from pyrite, lead from zinc sulphide ores, limonite, platinum group of metals from ores.
Recovering agent for—
Cyanide gas during ore treatment, potash from minerals.

Miscellaneous

Carrotting agent for—
Furs.
General cleansing agent.
Process material in making—
Chewing gum, dental cements, dentifrices, dyed feathers, silvered mirrors.

Paint and Varnish

Ingredient of—
Cement paints, coldwater paints, paint removers, paint-removing composition in admixture with calcium hydroxide.
Process material in making—
Pigments.

Paper

Digesting agent in—
Pulp manufacture.
Fermenting agent for—
Cellulose sulphite liquor.
Parchmentizing agent for—
Cellulose products.
Process material in making—
Cymene emulsions.
Source of soda in—
Pulp manufacture.

Petroleum

Decolorizing agent for—
Petroleum products.
Deodorizing agent for—
Petroleum products.
Neutralizing agent for—
Acidified products in petroleum processing.
Reagent for—
Removing sulphur from petroleum products.
Refining agent for—
Petroleum products.
Sweetening agent for—
Gasoline and other petroleum products.

Pharmaceutical

In compounding and dispensing practice.
Process material in making—
Hydroxides of magnesium and other bases.
Saponification agent.
Solubilizing agent in—
Preparing aqueous solutions of certain slightly soluble substances.

Photographic

Disintegrating agent for—
Film.
Ingredient of—
Developing agents.
Reagent for
Recovering silver from emulsions.

Plastics

Process material in making—
Horn substitutes, ivory substitutes, plastics.
Reagent for—
Recovering camphor from celluloid.
Solvent for—
Casein.

Rayon

Liquefying agent for—
Cellulose in the viscose process.

Refractories

Process material in making—
Aluminum oxide and products, sand-lime brick.
Reagent for—
Separating bauxite from clay.
Treating agent for—
Alunite.

Resins

Catalyst in—
Acetone-phenol condensations, aldehyde-phenol condensations, aldehyde-resorcinol condensations, cresolformaldehyde condensations, furfural resin manufacture, ketone-aldehyde condensations, urea-formaldehyde condensations.
Process material in making—
Oleoresins, paracoumarone, paraindene, polymerized glycerol.

Rubber

Coagulant for—
Rubber.
Process material in—
Devulcanizing processes, metal-coating rubber, reclaiming rubber, vulcanizing processes.

Sanitation

Treating agent for—
Country sewage, effluents, factory sewage, garbage grease (in hydrogenation processes), tannery sewage.

Soap

Catalyst in—
Hydrogenation of fats and oils.
Extractant for—
Oil from copra, oil from cottonseed.
Ingredient of—
Detergent compositions for various purposes in the home and industry.
Process material in making—
Emulsions of animal, fish, and vegetable oils.
Fatty acids, glycerin, sodium salts of sulphonated oils.
Refining agent for—
Fats of animal, fish, and vegetable origin.
Foots.
Hydrogenated products of blubber, oils, fats.
Oils of animal, fish, and vegetable origin.
Remover of fatty acids from—
Fats of animal, fish, and vegetable origin.
Oils of animal, fish, and vegetable origin.
Saponifying agent for—
Fats of animal, fish, and vegetable origin.
Oils of animal, fish, and vegetable origin.
Starting point in making—
Catalysts for hydrogenation processes.

Sugar

Purifying agent for—
Spent decolorizing carbons.
Treating agent for—
Bagasse.

Textile

Bleaching agent for—
Fibers and fabrics.
Cleansing agent for—
Fibers and fabrics.
Desizing agent for—
Fibers and fabrics.
Lustering agent for—
Fibers and fabrics.
Mercerizing agent for—
Fibers and fabrics.
Parchmentizing and antiqueing agent for—
Fibers and fabrics.
Preventer of—
Carbonizing of fibers.
Process material in—
Bleaching processes, dyeing processes, mercerizing processes, mildew-proofing canvas and other textile fabrics, printing processes, rotproofing canvas and other textile fabrics.
Process material in dyeing—
Cellulose acetate and other products.

Water

Ingredient of—
Water softeners.
Process material in making—
Artificial zeolites.
Regenerating agent for—
Peat in water-softening processes, zeolites in water-softening processes.
Softening agent for—
Water.
Treating agent for—
Boiler water, water-softening silicates.

Sodium 2-Hydroxydiphenyldisulphonate

Cosmetic
Protectant (U. S. 2015005) in—
Oils, creams, and lotions, against harmful effects of light of short wave lengths (sunburn).

Sodium 4-Hydroxydiphenyldisulphonate

Cosmetic
Protectant (U. S. 2015005) in—
Oils, creams, and lotions, against harmful effects of light of short wave length (sunburn).

Sodium 2-Hydroxydiphenylmonosulphonate

Cosmetic
Protectant (U. S. 2015005) in—
Oils, creams, and lotions, against harmful effects of light of short wave length.

Sodium 4-Hydroxydiphenylmonosulphonate

Cosmetic
Protectant (U. S. 2015005) in—
Oils, creams, and lotions, against harmful effects of light of short wave length (sunburn).

Sodium 2:3-Hydroxynaphthoate

French: 2:3-Hydroxyenaphthoate sodique, 2:3-Hydroxyenaphthoate de sodium, 2:3-Hydroxyenaphthoate de soude.
German: 2:3-Hydroxynaphtoesaeuresnatrium, 2:3-Hydroxynaphtoesaeuresnatron, Natrium-2:3-hydroxynaphtoat.

Chemical
Starting point in making—
Intermediates, pharmaceuticals, salts and esters.

Dye
Starting point (Brit. 298101) in making triarylmethane dyestuffs with—
Fuchsin hydrochloride, methyl violet.

Sodium Hypobromite

Analysis
As a reagent.

Chemical
Solubilizing agent (Brit. 423286) for—
Starch (treatment in cold water results in paste of the nature of a salve-like gel which gives the characteristic reactions of pure starch with iodine or Fehling's solution).

Sodium Hypochlorite

Synonyms: Bleaching solution, Labarraque's disinfecting fluid, Labarraque's solution, Solution of chlorinated soda.
Latin: Liquor natri chlorati, Liquor natri hypochlorosi, Liquor sodae chloratae, Liquor sodae chlorinatae.
French: Chlorure de soude liquide, Eau de labarraque, Liqueur de labarraque.
German: Bleischflussigkeit, Chlornatronlosung.
Spanish: Solucion de hipoclorito sodico, Licor de labarraque.

Beverage
Deodorizing and sterilizing agent for—
Equipment in breweries, malt houses, soft drink plants, wineries.

Chemical
Chlorinating agent.
Ingredient (Brit. 393221) of—
Bleaching compound containing also sodium meta silicate, or trisodium phosphate.
Oxidizing agent.
Reagent in making—
Anthranilic acid from phthalimide.
Chloramine from ammonia.
Hydrazin from ammonia.
Solvents for acetic acid from vegetable oils, fats, and fatty acids (Brit. 390148).
Solvents for cyclohexanol from vegetable oils, fats, and fatty acids (Brit. 390148).
Solvents for essential oils from vegetable oils, fats, and fatty acids (Brit. 390148).
Solvents for formic acid from vegetable oils, fats, and fatty acids (Brit. 390148).
Solvents for higher alcohols from vegetable oils, fats, and fatty acids (Brit. 390148).
Solvents for paraffin from vegetable oils, fats, and fatty acids (Brit. 390148).
Solvents for phenols from vegetable oils, fats, and fatty acids (Brit. 390148).
Water-solubilizing agents for acetic acid from vegetable oils, fats, and fatty acids (Brit. 390148).
Water-solubilizing agents for cyclohexanol from vegetable oils, fats, and fatty acids (Brit. 390148).
Water-solubilizing agents for essential oils from vegetable oils, fats, and fatty acids (Brit. 390148).
Water-solubilizing agents for formic acid from vegetable oils, fats, and fatty acids (Brit. 390148).
Water-solubilizing agents for higher alcohols from vegetable oils, fats, and fatty acids (Brit. 390148).
Water-solubilizing agents for paraffin from vegetable oils, fats, and fatty acids (Brit. 390148).
Water-solubilizing agents for phenol from vegetable oils, fats, and fatty acids (Brit. 390148).

Dry Cleaning
Deodorizing and spotting agents for—
White goods.

Fats and Oils
Bactericide, bleaching agent, deodorant, germicide.

Foods
Bactericide, bleaching agent, deodorant.
Deodorant for—
Waste waters from vegetable cooking operations.
Disinfectant for—
Fruit, general purposes, shell fish, vegetables.
Germicide.
Spraying agent for—
Rendering atmosphere sterile and sweet.
Sterilizing agent for—
Equipment and utensils in canning plants, food product plants, milk product plants.

Fuel
Extractant for—
Sulphur from municipal gases.

Laundering
Bleaching agent in—
Washroom waters and soap solutions.
Germicide in—
Washroom waters and soap solutions.

Miscellaneous
Bactericide, bleaching agent, deodorant, disinfectant, germicide, sterilizing agent.
Sterilizing agent in—
Dishwashing operations in hotels, restaurants, industrial canteens.

Paper
Bleaching agent for—
Paper stock of all kinds.
Bleaching agent (Brit. 398730) in making—
Cotton-like fabric from sulphite cellulose.
Reagent (U. S. 1906824) in making—
Orange-colored safety paper (by treatment after impregnation with an alcoholic solution of para-p′-dihydroxydiphenyl).

Petroleum
Chlorinating agent (Brit. 364204, U. S. 1908273) in purifying—
Petroleum, petroleum distillates.

Pharmaceutical
Base of—
Carrel-Dakin's solution, Dakin's solution.
Disinfecting agent for—
Utensils.
In compounding and dispensing practice.
Reagent in—
Bacteriological work.
Suggested for use in treating—
Ulcers, wounds.

Sanitation
Bactericide, deodorant and sterilizing washing agent for—
Hospital walls and floors, hospital lavatories, hospital utensils, industrial buildings, industrial equipment, public and domestic convenience stations, public buildings.
Deodorant and disinfectant for—
Gaseous factory effluents, liquid factory effluents.
Germicide and deodorant in—
Earth closets, sewage systems.

Textile
Bleaching agent in—
Finishing viscose rayon (U. S. 1915952).
Making textiles with soft handle and full white color (Brit. 401199).
Reagent (Brit. 390148) in making—
Degreasing agents from sulphonated oils, fats, and fatty acids.

Sodium Hypochlorite (Continued)

Water
Bactericide, deodorant, and sterilizing agent in—
Emergency water supply systems.
Isolated water storage systems.
Municipal water storage and supply systems.
Ships' water storage systems.
Swimming pools.
Water mains under construction.
Destructive agent for—
Algae in condenser water for power plants and refrigerating plants.

Sodium Hyposulphate
Synonyms: Sodium dithionate.

Analysis
As a reagent in various processes.

Sodium Hyposulphite
Synonyms: Antichlor, Hypo, Hyposulphite of soda, Sodium subsulphite, Sodium thiosulphate.
Latin: Hyposulphis sodicus, Natrium hyposulfurosum, Natrium thiosulfuricum, Natrium subsulfurosum.
French: Hyposulphite de soude, Sulfite sulfure de soude.
German: Natriumthiosulfat, Unterschwefligsäuresnatron.
Spanish: Hiposulfito sodico.

Agriculture
Intestinal antiseptic in—
Poultry feeding, stock feeding.

Analysis
Reagent in various processes.

Chemical
Reagent (U. S. 1900001) in making—
2-Aminoanthraquinone.
Starting point in making—
Complex double compounds of organic heavy metal mercapto compounds (Brit. 398020).
Sodium cyanates from carbon dioxide and/or carbon oxysulphide and ammonia in the presence of a gas (other than a hydrocarbon) capable of decomposing water (preferably carbon monoxide) (Brit. 399820).

Dry Cleaning
Reagent for—
Removing iodine stains on fabrics.

Dye
Reagent in making—
Aldehyde green, anilin dyes, synthetic dyes of various types.
Reducing agent in—
Indigo reduction.

Explosives and Matches
Reagent in making—
Lead thiosulphate for the production of phosphorus-free matches.

Fats and Oils
Bleaching agent for—
Edible oils, technical oils.
Rancidity retardant for—
Fats, oils.

Fuel
Ingredient of—
Candlewick pickling solutions.

Glue and Gelatin
Bleaching agent for—
Bone stock.

Ink
In the manufacturing process.

Leather
Reducing agent for—
Bichromates in chrome tanning by Schultz process.

Metallurgical
Ingredient of—
Electrolytic baths in plating with gold or silver.
Reagent in—
Silver extraction from its ores by the wet method.

Miscellaneous
Bleaching agent for—
General purposes in various industries.
Ivory, straw.
Source of—
Synthetic ice for skating rinks and ponds shown in motion pictures.

Thermatic medium in—
Chemical hot-water bottles.

Paint and Varnish
Luminophore (Brit. 391914) in making—
Luminous compositions for paints.
Reagent in making—
Antimony cinnabar, mercury cinnabar.

Paper
Bleaching agent (U. S. 1894501) in making—
Pulp from poplar stock.
Extractant for—
Excess chlorine in bleaching processes.

Pharmaceutical
In compounding and dispensing practice.
Suggested for use in treating—
Cyanide poisoning, ringworm, skin diseases of the toes.

Photographic
Fixing agent for—
Photographic and motion-picture film after development.
Prints of various types.

Refrigeration
Refrigerant in—
Portable or camp cooling equipment.

Sugar
Antifermentative for—
Sugar syrups.

Soap
Preservative for—
Colors in high-grade soaps.
Perfumes in high-grade soaps.
Rancidity retardant for—
Fats, oils, stored hard soaps.

Textile
Antichlor in—
Bleaching processes.
Mordant in—
Chrome mordanting wool (U. S. 1735844).
Dyeing and printing fabrics.
Fixation of anilin green on fabrics.
Reagent (U. S. 1903828) in making—
Artificial wool from jute.
Reducing agent (Brit. 399559) in—
Coloration of materials made of or containing cellulose esters or ethers (to give reserve effects).

Water and Sanitation
Disinfectant for—
Water supply systems.

Sodium Iodate

Pharmaceutical
Suggested for use as—
Local disinfectant.

Sodium Iodide
French: Hydriodate de soude, Iodure de sodium.
German: Natriumiodatum, Natriumjodid.

Analysis
Reagent in various processes.

Chemical
Reagent in making—
Iodated pharmaceuticals, iodated derivatives of chemicals, iodated intermediates, iodides of organic and inorganic bases, methyl iodide, propyl iodide.
Solvent in making—
Iodine solutions.

Dye
Reagent (Brit. 271181) in making dyestuffs from halogenated indanthrones, such as—
Dichloroindanthrone, monochloroindanthrone.

Pharmaceutical
In compounding and dispensing practice.

Photographic
Ingredient of—
Emulsions.
Reagent in making—
Silver iodide.

Sodium Isoallylphthalate
French: Isoallylephthalate sodique, Isoallylephthalate de sodium, Isoallylephthalate de soude.
German: Isoallylphtalsaeuresnatrium, Natriumisoallylphtalat.

Resins and Waxes
Starting point (Brit. 250265) in making synthetic resins with—
Acetates, bromides, chlorides, and nitrates of barium, calcium, lead, magnesium, strontium, and zinc.

Sodium Isoamylnaphthalenesulphonate
French: Isoamylenaphthalènesulphonate de soude.
German: Isoamylnaphtalinsulfonsaeuresnatrium, Natriumisoamylnaphtalinsulfonat.

Chemical
Dispersive agent (Brit. 264860) for various chemical purposes.

Dye
Dispersive agent in making—
Color lakes.

Ink
Dispersive agent in making—
Printing inks.

Paint and Varnish
Dispersive agent in making—
Paints, pigment compositions.

Plastics
Dispersive agent in making—
Plastics with cellulose esters.

Resins and Waxes
Dispersive agent in making—
Compositions containing natural and artificial resins.

Rubber
Dispersive agent in making—
Rubber compositions.

Textile
——, *Dyeing*
Dispersive agent in making dye liquors, containing sulphur dyes, indigoes, anthraquinone vat dyestuffs, for rayons, wool, cotton, and natural silk.

——, *Finishing*
Dispersive agent in making—
Finishing compositions for fabrics.

——, *Manufacture*
Ingredient of—
Lubricating compositions used in spinning fibers (Brit. 268387).

Sodium Isoamylphthalate
French: Isoamylephthalate sodique, Isoamylephthalate de sodium, Isoamylephthalate de soude.
German: Isoamylphtalsaeuresnatrium, Natriumisoamylphtalat.

Resins and Waxes
Starting point (Brit. 250265) in making synthetic resins with—
Acetates, bromides, chlorides, and nitrates of barium, calcium, lead, magnesium, strontium, and zinc.

Sodium Isobutylnaphthalenesulphonate
French: Isobutylenaphthalènesulphonate de soude.
German: Isobutylnaphtalinsulfonsaeuresnatrium, Natriumisobutylnaphtalinsulfonat.

Dye
Dispersing agent in making—
Color lakes (Brit. 264860).

Ink
Dispersing agent in making—
Printing inks.

Paint and Varnish
Dispersing agent in making—
Paints, pigments, varnishes.

Plastics
Dispersing agent in making—
Compounds of cellulose esters and ethers.

Resins and Waxes
Dispersing agent in making—
Artificial resin preparations, natural resin preparations.

Rubber
Dispersing agent in making various compositions.

Textile
——, *Dyeing*
Dispersing agent in making—
Dyeing liquors for applying sulphur dyes, indigoes, anthraquinone vat dyestuffs and other dyestuffs to cotton, rayon, silk, wool.

——, *Finishing*
Dispersing agent in making—
Dressings and other finishes.

Sodium Isobutylphthalate
French: Isobutylephthalate de soude, Isobutylephthalate sodique.
German: Isobutylphtalsaeuresnatrium, Natriumisobutylphtalat.

Resins and Waxes
Starting point (Brit. 250265) in making synthetic resins with—
Chlorides, bromides, acetates, and nitrates of barium, calcium, lead, magnesium, strontium, and zinc.

Sodium Isopropylnaphthalenesulphonate
French: Isopropylenaphthalène sulphonate de soude.
German: Isopropylnaphtalinsulfonsaeuresnatrium, Natriumisopropylnaphtalinsulfonat.

Dye
Dispersive agent (Brit. 264860) in making—
Dye preparations, lakes.

Ink
Dispersive agent in making—
Lithographic inks, printing inks, writing inks.

Paint and Varnish
Dispersive agent in making—
Copal varnishes, lacquers, spirit varnishes, water paints.

Plastics
Dispersive agent in making—
Cellulose ether and ester solutions, cellulose ether and ester plastics.

Rubber
Dispersive agent in making—
Solutions of rubber.

Textile
——, *Dyeing*
Dispersive agent in making—
Anthraquinone dye liquors, indigo dye liquors, sulphur dye liquors.
Vat dye liquors for cottons, woolens, rayons.

Sodium Isopropylphthalate
French: Isopropylephthalate de soude, Isopropylephthalate sodique.
German: Isopropylphtalsaeuresnatrium, Natriumisopropylphtalat.

Resins and Waxes
Starting point (Brit. 250265) in making synthetic resins with—
Barium acetate, bromide, chloride, nitrate, calcium, lead, magnesium, strontium and zinc.
Calcium acetate, bromide, chloride, nitrate.
Lead acetate, bromide, chloride, nitrate.
Magnesium acetate, bromide, chloride, nitrate.
Strontium acetate, bromide, chloride, nitrate.
Zinc acetate, bromide, chloride, nitrate.

Sodium Isopropylxylenesulphonate
French: Isopropylexylènesulfonate sodique, Isopropylexylènesulfonate de sodium, Isopropylexylènesulfonate de soude.
German: Isopropylxylensulfonsaeuresnatrium, Natriumisopropylxylensulfonat.

Fats and Oils
Starting point (Brit. 279877) in making—
Solvents.

Miscellaneous
Ingredient (Brit. 279877) of—
Cleansing and bleaching compositions for parquetry floors.
Washing compositions.

Soap
Ingredients (Brit. 279877) of—
Washing and detergent compositions.

Textile
——, *Dyeing*
Assist (Brit. 279877) in making—
Wool-dyeing liquors.

——, *Finishing*
Ingredient (Brit. 279877) of—
Cleansing and finishing compositions.

Sodium Lactate

Pharmaceutical
In compounding and dispensing practice.
Suggested for use in treating—
Acidosis.

Sodium Laurate

Textile
Cleansing agent (Brit. 414485) for—
Wool, silk, cotton, ramie, jute, hemp, flax, and rayon fibers, by treatment in an aqueous alkaline liquor, using trisodium phosphate, soda ash, or sodium or potassium hydroxide as the alkaline constituent.

Sodium Laurylethylsulphonate

Miscellaneous
As an emulsifying agent.
For uses, see under general heading: "Emulsifying agents."

Sodium Laurylpyrophosphate

Chemical
Stabilizing agent (Brit. 421843) for—
Peroxide solutions.

Miscellaneous
Stabilizing agent and promoter of wetting and penetrating properties (Brit. 421843) for—
Peroxide solutions used in many industries for (1) bleaching, (2) sterilizing, (3) disinfecting.

Sodium Laurylsulphonate

Miscellaneous
Ingredient of—
Spirit cleaner containing also glycerin, alcohol, and water.

Soap
Antioxidant in—
Soaps.
Rancidity retardant in—
Soaps.

Textile
Starting point (Brit. 393164) in making—
Detersive mixtures with tetrahydrofurfuryl acetate, tetrahydrofurfuryl formate, tetrahydrofurfuryl valerate, or tetrahydrofurfuryl propionate, for washing raw wool.

Sodium Linoleate
French: Linoleate sodique, Linoleate de sodium, Linoleate de soude.
German: Leinoelnatrium, Leinoelnatron, Leinölnatrium, Leinölnatron, Natriumleinoelat, Natronleinoelat, Natriumleinölat, Natronleinölat.

Miscellaneous
As a wetting agent (Brit. 411908).
For uses, see under general heading: "Wetting agents."

Sodium Lysalbinate
French: Lysalbinate de soude.
German: Lysalbinsaeuresnatrium, Natriumlysalbinat.

Construction
Ingredient (Brit. 271181) of—
Bituminous compositions for waterproofing cement, concrete, stone, stucco, wood, and other structural materials.

Miscellaneous
Ingredient of—
Bituminous compositions used in roadmaking.

Sodium Manganate
French: Manganate de soude, Manganate sodique.
German: Mangansaeuresnatrium, Natriummanganat.

Chemical
Oxidizing agent in making—
Anisic acid from paracresolmethyl ether.
Benzoic acid from toluol.
Benzophenoneparadicarboxylic acid from paratolylorthobenzoic acid.
Orthoacetaminobenzoic acid from orthoacetoluidide.
Orthochlorobenzoic acid from orthochlorotoluol.
Orthonitrobenzidinanilinsulphonic acid from orthonitrobenzylanilinsulphonic acid.
Para-acetaminobenzoic acid from para-acetoluidide.
Parachlorobenzoic acid from parachlorotoluol.
Reagent in making—
Saccharin, sulphonal, tetranal, trional.
Starting point in making—
Oxygen gas, sodium permanganate.

Dye
Reagent in making—
Alizarin.
Anthraflavone G from betamethylanthraquinone.
Anthraquinone dyestuffs.
Benzanthronquinolin from betamethylanthraquinone.
Indanthrene dark blue BO from betamethylanthraquinone.
Pyranin from acridin red.

Insecticide
As an insecticide, alone or in compositions.

Miscellaneous
Antidote in poisoning with organic substances.

Pharmaceutical
In compounding and dispensing practice.

Sanitation
Disinfectant for various purposes.

Sodium Metaborate
French: Métaborate sodique, Métaborate de sodium, Métaborate de soude.
German: Metaborsäuresnatrium, Metaborsäuresnatron, Natriummetaborat.
Spanish: Metaborato de sosa.
Italian: Metaborato di sodio.

Adhesives
Ingredient of—
Casein glues, various glues and adhesive compositions.

Food
As a preservative for honey.

Perfumery
Ingredient of—
Bath salts.

Pharmaceutical
Rated highly as a noncorrosive antiseptic.
Suggested as an ingredient of eye lotions and in the treatment of chronic otorrhea.

Soap
As a "building" ingredient.

Sodium Metaphosphate
French: Métaphosphate sodique, Métaphosphate de sodium, Métaphosphate de soude.
German: Metaphosphorsäuresnatrium, Metaphosphorsäuresnatron.

Chemical
Catalyst (Brit. 407722) in—
Hydration of olefines.
Catalytic promoter (French 752270) in making—
Aliphatic anhydrides from aliphatic acids.
Ingredient (French 752270) of—
Catalytic-promoter mixtures used in making aliphatic anhydrides from aliphatic acids.

Food
Reagent (Brit. 387918) in making—
Crustless cheese.

Glass
Cleansing agent for—
Glassware.
Ingredient of—
Cleansing compositions for glassware, containing also trisodium phosphate, monohydrate, sodium metasilicate, pentahydrate, and caustic soda (dehydrated salts also used).

Laundering
Addition agent to—
Boil for final hot wash, to eliminate lime soaps completely.

Mechanical
Water-softening agent in—
Treating boilerfeed water (softens water without forming a precipitate through formation of soluble salts).

Miscellaneous
Ingredient of—
Detergent, containing also borax, used in washing domestic animals.

Perfume
Water-softening agent in—
Toilet preparations.

Pharmaceutical
Water-softening agent in—
Veterinary preparations.

Textile
Water-softening agent in—
Dyeing and finishing of silk (more level dyeing and a better handle being obtained by reason of the complete elimination of lime soap which causes uneven penetration).
Dyeing cotton and wool union fibers.
Kier boiling cotton.
Wool scouring.

Water and Sanitation
Water-softening agent for—
Industrial waters (softens water by formation of complex, soluble phosphates).

Sodium Metavanadate

Agriculture
In inoculation of plant-life.

Ink
Ingredient of various inks.

Miscellaneous
Mordant in—
Fur dyeing.

Pharmaceutical
In compounding and dispensing practice.

Photographic
Reagent for—
Imparting red tones to films and plates.

Textile
As a mordant.

Sodium Methoxide
French: Méthoxyde de soude.
German: Natronmethoxyd.

Chemical
Reagent in making—
1:6-Dihydroxyanthraquinone, trimethyl phosphate.

Sodium Methylarsonate

Pharmaceutical
As an arsenic carrier in medication with that element.

Sodium 3-Methylcaproate

Textile
Cleansing agent (Brit. 414485) for—
Wool, silk, cotton, ramie, jute, hemp, flax, and rayon fibers, by treatment in an aqueous alkaline liquor, using trisodium phosphate, soda ash, or sodium or potassium hydroxide as the alkaline constituent.

Sodium 3-Methylcaprylate

Textile
Cleansing agent (Brit. 414485) for—
Wool, silk, cotton, ramie, jute, hemp, flax, and rayon fibers, by treatment in an aqueous alkaline liquor, using trisodium phosphate, soda ash, or sodium or potassium hydroxide as the alkaline constituent.

Sodium Methylnaphthalenesulphonate
French: Méthylenaphthalènesulfonate sodique, Méthylenaphthalènesulfonate de soude.
German: Methylnaphtalinsulfonsaeuresnatrium, Natriummethylnaphtalinsulfonat.

Dye
Dispersive agent in making—
Color lakes (Brit. 264860).

Ink
Dispersive agent in making—
Printing inks (Brit. 264860).

Paint and Varnish
Reagent (Brit. 268387) in making—
Paints, pigments.

Plastics
Dispersive agent (Brit. 264860) in making—
Cellulose ester and other plastics.

Rubber
Dispersive agent (Brit. 264860) in making—
Rubber cements.

Textile
——, *Dyeing*
Dispersive agent (Brit. 264860) in making—
Dye liquors for rayon, wool, cotton, silk, with sulphur dyestuffs, indigo, anthraquinone vat dyes.

——, *Finishing*
Dispersive agent (Brit. 264860) in making—
Finishing and dressing compositions for fabrics and yarns.

——, *Manufacturing*
Dispersive agent (Brit. 268387) in making—
Lubricating compositions for spinning.

Sodium Methylphthalate
French: Méthylephthalate de soude.
German: Methylphtalsaeuresnatrium, Natriummethylphtalat.

Resins and Waxes
Starting point (Brit. 250265) in making synthetic resins with salts of—
Barium, calcium, lead, magnesium, strontium, zinc.

Sodium Methylpyrazolone
German: Natriummethylpyrazolon.

Dye
Starting point in making—
Wool dyestuff (Brit. 261770).

Sodium 1-Methylvalerate

Textile
Cleansing agent (Brit. 414485) for—
Wool, silk, cotton, ramie, jute, hemp, flax, and rayon fibers, by treatment in an aqueous alkaline liquor, using trisodium phosphate, soda ash, or sodium or potassium hydroxide as the alkaline constituent.

Sodium 3-Methylvalerate

Textile
Cleansing agent (Brit. 414485) for—
Wool, silk, cotton, ramie, jute, hemp, flax, and rayon fibers, by treatment in an aqueous alkaline liquor, using trisodium phosphate, soda ash, or sodium or potassium hydroxide as the alkaline constituent.

Sodium Molybdate
French: Molybdate de soude, Soude molybdate.
German: Molybdaensäuresnatrium, Molybdaensäuresnatron.

Analysis
As a reagent.

Glues and Adhesives
Ingredient of—
Casein glues.

Paint and Varnish
As a pigment.

Miscellaneous
Starting point (U. S. 1730702) in making—
Rubber-like material with concentrated cactus juice, sodium tungstate, boiled linseed oil, and a solution of rubber in turpentine.

Sodium Monobenzylsulphanilate

Textile
——, *Dyeing*
Solvent for leuco-products in—
Dyeing with vat dyestuffs.

——, *Finishing*
Wetting agent for—
Fabrics, yarns.

Sodium Monocresylphosphate
French: Monocrésylephosphate de soude.
German: Monocresylphosphorsaeuresnatrium, Natriummonocresylphosphat.

Chemical
Reagent in making—
Textile finishing compounds (Brit. 267534).

Dye
Reagent in making—
Pastes.

Soap
Ingredient of washing compositions.

Sodium-Monopara-aminobenzoylparaphenylenediamine Sulphonate

Paper
Impregnating agent and absorbent for ultraviolet light (Brit. 436891) in—
Treating paper and like products to be used as food containers.

Sodium Monostearinsulphoacetate

Fats and Oils
Antispattering agent (U. S. 1917273) in—
Edible fats and hydrogenated oils.

Food
Antispattering agent (U. S. 1917273) in—
Margarins.

Sodium Mucate

Metallurgical
Flux (U. S. 1947735) in—
Iron ore smelting.

Sodium Myricylsulphonate

Miscellaneous
As an emulsifying agent (Brit. 360539).
For uses, see under general heading: "Emulsifying agents."

Sodium Myristylpyrophosphate

Chemical
Stabilizing agent (Brit. 421843) for—
Peroxide solutions.

Miscellaneous
Stabilizing agent and promoter of wetting and penetrating properties (Brit. 421843) for—
Peroxide solutions used in many industries for (1) bleaching, (2) sterilizing, (3) disinfecting.

Sodium-Naphthalene-1:5-disulphonate
French: Naphthalène-1:5-disulphonate sodique, Naphthalène-1:5-disulphonate de sodium, Naphthalène-1:5-disulphonate de soude.
German: Naphtalin-1:5-disulfonsaeuresnatrium, Naphtalin-1:5-disulfonsaeuresnatron, Natriumnaphtalin-1:5-disulfonat.

Chemical
Starting point in making—
Intermediates, pharmaceuticals.
Reagent (Brit. 280945) in making diazo salts with diazotized—
Anilin, azoxyanilin, 4-chloro-2-toluidin, 5-chloro-2-toluidin, dianisidin, 2:5-dichloroanilin, meta-anisidin, metachloroanilin, metachlorotoluidin, metanitranilin, metanitroparatoluidin, metatoluidin, 4-nitro-2-anisidin, 5-nitro-2-anisidin, orthoanisidin, orthochlorotoluidin, orthonitranilin, orthotoluidin, para-anisidin, parachloroanilin, parachlorotoluidin, paranitranilin, paranitro-orthotoluidin, paratoluidin.
Stabilizing agent (French 610261) in making—
Solid diazo compounds of the aromatic series.

Dye
Starting point in making various synthetic dyestuffs.

Sodium Naphthalene-2:7-disulphonate
French: Naphthalène-2:7-disulphonate de soude.
German: Naphtalin-2:7-disulfonsaeuresnatrium, Natriumnaphtalin-2:7-disulfonat.

Chemical
Stabilizing agent in making—
Solid aromatic diazo compounds (French 610261).

Sodium Naphthalenetrisulphonate
French: Naphthalènetrisulfonate sodique, Naphthalènetrisulfonate de sodium, Naphthalènetrisulfonate de soude.
German: Naphtalintrisulfonsaeuresnatrium, Natriumnaphtalintrisulfonat.

Chemical
Reagent (Brit. 280945) in making diazo salts with diazotized—
Anilin, azoxyanilin, benzidin, 4-chloro-2-toluidin, 5-chloro-2-toluidin, dianisidin, 2:5-dichloroanilin, meta-anisidin, metachloroanilin, metachlorotoluidin, metanitranilin, metanitroparatoluidin, metatoluidin, 4-nitro-2-anisidin, 5-nitro-2-anisidin, orthoanisidin, orthochloroanilin, orthochlorotoluidin, orthonitranilin, orthotoluidin, para-anisidin, parachloroanilin, parachlorotoluidin, paranitranilin, paranitro-orthotoluidin, paratoluidin.

Sodium Naphthionate
French: Naphthionate sodique, Naphthionate de soude.
German: Naphthionsaeuresnatrium, Naphthionsaeuresnatron, Natriumnapthionat, Natronnapthionat.

Analysis
Reagent in detecting—
Nitrous acid.

Dye
Intermediate in making—
Azo dyes.

Mechanical
Ingredient (U. S. 1895014) of—
Lubricating composition, containing also graphite, gum solution, triethanolamine.

Miscellaneous
Dust-laying substance (French 599497) for—
Highways.

Sodium Naphthylacetamidedisulphonate

Textile
Reagent (French 750647) for—
Rendering rayon soft (supple) to the touch.

Sodium Naphthylthioglycolate

Chemical
Starting point in making various derivatives.

Dye
Reagent (Brit. 284288) in making thioindigoid dyestuffs with the aid of—
Acenaphthenequinone, alphaisatinanilide, 5:7-dibromoisatin.
Isatin, homologs, substitution products, and alpha derivatives.
Orthodiketones.

Sodium-Nickel Cyanide

Chemical
Catalyst (Brit. 446411) in—
Halogenating unsaturated hydrocarbons.
Starting point (Brit. 446411) in making—
Catalysts with metal chlorides for halogenating unsaturated hydrocarbons.

Sodium Nitrate
Synonyms: Chile saltpeter, Chile saltpetre, Chilisaltpeter, Chilisaltpetre, Cubic nitre, Nitrate of soda.
Latin: Azotas sodicus, Nitras sodicus, Natrium nitricum, Nitrum cubicum, Sodii nitras.
French: Azoate de soude, Nitrate de Chili, Nitrate de soude, Nitre cubique.
German: Natriumnitrat, Chilesaltpeter.
Spanish: Nitrato sodico, Nitrato de sosa.
Italian: Nitrato di sodio.

Analysis
Reagent in various processes.

Ceramics
Flux.
Ingredient of—
Enamels.

Chemical
Oxidizing agent.
Reagent in making.
Nitrous oxide, potassium nitrate by double decomposition with potassium chloride, various nitrate by double decomposition.
Starting point in making—
Sodium arsenate, sodium nitrite.

Dye
Reagent in making various synthetic dyes.

Explosives and Matches
Ingredient of—
Dynamites, fuses, low-density explosive composition (U. S. 1901126), military explosives, permissible explosives, pyrotechnic compositions, touch papers.

Fertilizer
As a general nitrogenous fertilizer.
Ingredient of—
Fertilizer compositions.
Source of—
Inorganic nitrogen.
Top dressing for—
Cotton, sugar beets, various crops.

Food
Pickling agent for—
Meats.
Preservative for—
Butter, butter products.

Glass
Ingredient of—
Glass batches.

Leather
Ingredient of—
Dressing compositions.

Metallurgical
Flux in—
Ore separation processes.
Oxidizing agent (U. S. 1911943) in—
Roasting operations in rhenium recovery.
Reagent (Brit. 400121) in making—
Silicon-aluminum alloys.

Pharmaceutical
In compounding and dispensing practice.
Suggested for use in treating—
Dysentery.

Tobacco
Impregnating agent for—
Enhancing burning properties.

Sodium Nitrite
Synonyms: Nitrite of soda.
French: Nitrite sodique, Nitrite de soude.
German: Salpetrigsäuresnatrium, Salpetrigsäuresnatron.
Analysis
Reagent in various processes.
Chemical
Reagent in—
Organic synthesis.
Dye
As a diazotizing agent.
Azotizing agent (Brit. 397034) in making—
Tetrazo compounds from metaphenylenediamine or paraphenylenediamine.
Food
As a meat-pickling agent.
Miscellaneous
As a general bleaching agent (in combination with potassium permanganate).
Pharmaceutical
In compounding and dispensing practice.
Photographic
Reagent in various processes.
Rubber
Addition agent (Brit. 395774) to—
Latex prior to coagulation in making soft rubber.
Textile
Bleaching agent for—
Fibers, such as flax, linen, and silk.
Developing agent in—
Dyeing and printing.

Sodium Nitrophenate
French: Nitrophénate sodique, Nitrophénate de sodium, Nitrophénate de soude.
German: Natriumnitrophenat.
Agricultural
Ingredient (Brit. 321396) of—
Compositions used for immunizing wheat.
Woodworking
Ingredient (Brit. 321396) of—
Preserving compositions.

Sodium Nitroprusside
Synonyms: Nitroprusside of soda, Sodium nitroprussiate.
Analysis
As a reagent in—
Roussin's photometer.
Testing for sulphur, sulphides, acetone, and other substances.
Testing silk for presence of animal hair.

Sodium Nucleinate
Synonyms: Nucleinate of soda.
Pharmaceutical
As a bactericide.

Sodium Octodecylsulphonate
Miscellaneous
As an emulsifying agent (Brit. 360539).
For uses, see under general heading: "Emulsifying agents."

Sodium Oleate
Latin: Natrium oleatum.
French: Oléate sodique, Oléate de sodium, Oléate de soude.
German: Natriumoleat, Oleinsäuresnatrium, Oleinsäuresnatron.
Ceramics
Emulsifying agent (Brit. 328657) in—
Compositions, containing nitrocellulose, cellulose acetate, or other esters or ethers of cellulose, used for decorating and protecting ceramic ware.
Chemical
Starting point in making—
Copper oleate, lead oleate, magnesium oleate, manganese oleate, mercury oleate, various metal oleates, zinc oleate.
Electrical
Emulsifying agent (Brit. 328657) in—
Insulating compositions containing cellulose acetate, nitrocellulose, or other esters or ethers of cellulose.
Fats and Oils
Starting point in making—
Olein.
Food
Stabilizing agent (French 605313) in making—
Mineral waters (added to prevent the precipitation of colloidal elements contained in these preparations).
Glass
Emulsifying agent (Brit. 328657) in—
Compositions, containing cellulose acetate, nitrocellulose, or other esters or ethers of cellulose, used in the manufacture of non-scatterable glass and for coating and decorating glassware.
Glues and Adhesives
Emulsifying agent (Brit. 328657) in—
Adhesive preparations containing cellulose acetate, nitrocellulose, or other esters or ethers of cellulose.
Leather
Emulsifying agent (Brit. 328657) in—
Compositions, containing cellulose acetate, nitrocellulose, or other esters or ethers of cellulose, used in the manufacture of artificial leather and for coating and decorating leathers and leather goods.
Metallurgical
Emulsifying agent (Brit. 328657) in—
Compositions, containing cellulose acetate, nitrocellulose, or other esters or ethers of cellulose, used for coating and decorating metallic products.
Miscellaneous
Emulsifying agent (Brit. 328657) in—
Compositions, containing cellulose acetate, nitrocellulose, or other esters or ethers of cellulose, used for decorating and coating various articles.
Ingredient of—
Special detergent preparations.
Paint and Varnish
Emulsifying agent (Brit. 328657) in making—
Paints, varnishes, lacquers, dopes, and enamels containing cellulose acetate, nitrocellulose, or other esters or ethers of cellulose.
Ingredient of—
Aquarelle paint compositions.
Fine colors (added to prevent the flocculation of the particles of the pigment).
Starting point in making—
Driers.
Paper
Emulsifying agent (Brit. 328657) in—
Compositions, containing cellulose acetate, nitrocellulose, or other esters or ethers of cellulose, used in the manufacture of coated paper and for decorating and protecting paper and pulp products.
Perfume
Ingredient of—
Cosmetics, dentifrices, shampoos.
Pharmaceutical
In compounding and dispensing practice.
Plastics
Emulsifying agent (Brit. 328657) in making—
Plastics containing cellulose acetate, nitrocellulose, or other esters or ethers of cellulose.
Rubber
Emulsifying agent (Brit. 328657) in—
Compositions, containing cellulose acetate, nitrocellulose, or other esters or ethers of cellulose, used for decorating and protecting rubber merchandise.
Stone
Emulsifying agent (Brit. 328657) in—
Compositions, containing cellulose acetate, nitrocellulose, or other esters or ethers of cellulose, used for decorating and protecting artificial or natural stones.
Textile
Emulsifying agent (Brit. 328657) in—
Compositions, containing cellulose acetate, nitrocellulose, or other esters or ethers of cellulose, used in making coated fabrics.
Waterproofing agent in treating—
Yarns and fabrics by the chemical waterproofing process.

Sodium Oleic Acid Methyltauride
Laundering
Preventer (Brit. 451342) of—
Precipitates or deposits of inorganic metal salts in washing fabrics in hard water with washing agents stable to hard water.

Sodium Oleic Acid Methyltauride (Continued)

Textile
Preventer (Brit. 451342) of—
Precipitates or deposits of inorganic metal salts in washing fabrics in hard water with washing agents stable to hard water.

Sodium Oleic Acid Sarcoside

Laundering
Preventer (Brit. 451342) of—
Precipitates or deposits of inorganic metal salts in washing fabrics in hard water with washing agents stable to hard water.

Textile
Preventer (Brit. 451342) of—
Precipitates or deposits of inorganic metal salts in washing fabrics in hard water with washing agents stable to hard water.

Sodium Oleicethylsulphonate

Miscellaneous
As an emulsifying agent.
For uses, see under general heading: "Emulsifying agents."

Sodium Oleylpyrophosphate

Chemical
Stabilizing agent (Brit. 421843) for—
Peroxide solutions.

Miscellaneous
Stabilizing agent and promoter of wetting and penetrating properties (Brit. 421843) for—
Peroxide solutions used in many industries for (1) bleaching, (2) sterilizing, (3) disinfecting.

Sodium Oleylsulphate

Building and Construction
Emulsifying agent (Brit. 437674) in making—
Aqueous emulsions of asphalt and similar bituminous materials.

Sodium-1-omega-sulphomethyl-2-oxynaphthalene

German: Natrium-1-omega-sulphomethyl-2-oxynapthalin.

Chemical
Starting point in making—
Synthetic tannins (Brit. 250398).

Sodium Orthophenylphenate

French: Orthophénylephénate sodique, Orthophénylephénate de sodium, Orthophénylephénate de soude.
German: Natriumorthophenylphenat.

Glues and Adhesives
Reagent for treating—
Glues, particularly joiner's glue, to prevent decomposition and the growth of molds.

Miscellaneous
Preservative for various purposes.

Sodium Orthovanadate

Agriculture
In inoculation of plant life.

Ink
Ingredient of various inks.

Miscellaneous
Mordant in—
Fur dyeing.

Pharmaceutical
In compounding and dispensing practice.

Photographic
Reagent for—
Imparting red tones to films and plates.

Sodium Oxalate

French: Oxalate sodique, Oxalate de soude.
German: Natriumoxalat, Natronoxalat, Oxalsäuresnatrium, Oxalsäuresnatron.

Analysis
Reagent in various processes.

Chemical
Reagent in making—
Alkali-metal salts of adenylpyrophosphoric acids (Brit. 396647), various chemicals.

Explosives and Matches
Ingredient of—
Explosives (to prevent mouth-firing), pyrotechnic compositions.

Leather
Reagent in—
Tanning and finishing processes.

Textile
Promotive reagent in—
Bleaching textile fibers with hydrogen peroxide.
Reagent in finishing processes.
Retardant of—
Acid attack on fibers when bleaching with hydrogen peroxide.

Sodium 1-Oxybenzene-2-omega-methylsulphonate

French: 1-Oxybenzène-2-oméga-méthylesulphonate de soude.
German: 1-Oxybenzol-2-omega-methylsulphosaeuresnatron.

Chemical
Starting point in making—
Synthetic tannins (Brit. 250398).

Sodium 2:3-Oxynaphtholate

French: 2:3-Hydroxynaphtholate de soude, 2:3-Oxynaphtholate sodique.
German: Natrium-2:3-oxynaphtolat, 2:3-Oxynaphtolsaeuresnatrium.

Chemical
Starting point in making—
2:3-Oxynaphthoic acid (Brit. 278463).

Dye
Starting point in making various synthetic dyestuffs.
Ingredient (Brit. 277391) of—
Stain-removing compositions, washing and cleansing compositions.

Textile
——, *Finishing*
Ingredient of—
Fulling compositions (Brit. 277391).

Sodium Palmitomonosulphonate

Miscellaneous
As an emulsifying agent (Brit. 343899).
For uses, see under general heading: "Emulsifying agents."

Sodium Palm-Nut Oil Tauride

Laundering
Preventer (Brit. 451342) of—
Precipitates or deposits of inorganic metal salts in washing fabrics in hard water with washing agents stable to hard water.

Textile
Preventer (Brit. 451342) of—
Precipitates or deposits of inorganic metal salts in washing fabrics in hard water with washing agents stable to hard water.

Sodium Parachlorophenate

Synonyms: Sodium parachlorocarbolate, Sodium parachlorophenolate.
French: Parachlorocarbolate de soude, Parachlorophenolate de soude.
German: Natriumparachlorcarbolat, Natriumparachlorphenat, Natriumparachlorphenolat, Parachlorphenolsaeuresnatrium.

Leather
Ingredient of vat dyeing liquors (Brit. 263473).

Miscellaneous
Ingredient of vat dye liquor for—
Furs and hair.

Textile
——, *Dyeing and Printing*
Ingredient of vat dye liquor for—
Cellulose acetate rayon, chardonnet rayon, cuprammonium rayon, silk-rayon mixtures, viscose rayon, wool-rayon mixtures.

Sodium Paradinitrophenoxide

Construction
Ingredient (U. S. 1921324) of—
Wood preservative composition consisting of 1:2 mixture with urea.

Woodworking
Ingredient (U. S. 1921324) of—
Wood perservative composition consisting of 1:2 mixture with urea.

Sodium-Paraethoxyphenylaminomethane Sulphonate
Pharmaceutical
Suggested for use as—
Antipyretic, antirheumatic.

Sodium Paraoxybenzoate
Synonyms: Sodium parahydroxybenzoate.
French: Parahydroxyebenzoate sodique, Parahydroxyebenzoate de sodium, Parahydroxyebenzoate de soude, Paraoxyebenzoate sodique, Paraoxyebenzoate de sodium, Paraoxybenzoate de soude.
German: Natriumparahydroxybenzoat, Natriumparaoxybenzoat, Parahydroxybenzoesäuresnatrium, Parahydroxybenzoesäuresnatron, Paraoxybenzoesäuresnatrium, Paraoxybenzoesäuresnatron.
Food
Preservative for various preparations.
Pharmaceutical
In compounding and dispensing practice.
Sanitation
Antiseptic and disinfectant for various purposes.
Soap
Ingredient of—
Antiseptic and disinfectant soaps.

Sodium Pentadecylsulphonate
Miscellaneous
As an emulsifying agent (Brit. 360539).
For uses, see under general heading: "Emulsifying agents."

Sodium Pentamethylenedithiocarbamate
Rubber
Secondary activator in—
Vulcanizing processes (for use with mercaptabenzthiazole).

Sodium Pentylnaphthalenesulphonate
French: Pentylenaphthalènesulphonate sodique, Pentylenaphthalènesulphonate de sodium, Pentylenaphthalènesulphonate de soude.
German: Pentylnaphtalinsulfonsaeuresnatrium, Pentylnaphtalinsulfonsaeuresnatron, Natriumpentylnaphtalinsulfonat.
Chemical
Emulsifying agent (Brit. 298823) in making—
Pharmaceuticals.
Dye
Emulsifying agent (Brit. 264860) in making—
Color lakes.
Fats and Oils
Dispersive agent (Brit. 264860) in making—
Lubricating and greasing compositions, solvents for fats and oils.
Ink
Dispersive agent (Brit. 264860) in making—
Printing inks.
Insecticide
Emulsifying agent (Brit. 298823) in making—
Insecticidal and germicidal preparations.
Miscellaneous
Emulsifying agent (Brit. 298823) in making—
Washing compositions.
Paint and Varnish
Dispersive agent (Brit. 264860) in making—
Paints, varnishes.
Perfumery
Dispersive agent (Brit. 298823) in making—
Cosmetics, perfumes.
Plastics
Emulsifying agent (Brit. 298823) in making—
Compounds of cellulose esters and ethers.
Resins and Waxes
Dispersive agent (Brit. 264860) in making—
Artificial resin preparations, natural resin preparations.
Rubber
Dispersive agent (Brit. 264860) in making various compositions.
Textile
Dispersive agent (Brit. 264860) in making—
Dye liquors containing sulphur dyestuffs, indigoes, and anthraquinone vat dyestuffs.
Dye liquors for rayon, cotton, wool, and silk.

Sodium Perborate
Synonyms: Perborax.
French: Perborate sodique, Perborate de sodium, Perborate de soude.
German: Natriumhyperborat, Natriumperborat, Perborin, Ueberborsäuresnatrium, Ueberborsäuresnatron.
Spanish: Perborato de sosa.
Italian: Perborato di soda, Perborato di sodio.
Analytical
Reagent in—
Analyses involving oxidations, analyzing blood, analyzing boiled milk, determining bile pigments in urine, physiological analyses.
Chemical
General oxdizing agent in carrying out inorganic and organic reactions.
Ingredient of oxygen baths containing catalysts, such as—
Bisulphites of various metals, colloidal iron compounds, dried blood, colloidal manganese dioxide, heavy metals salts in admixture wtih gum arabic, manganoborates, permanganates, saponin, tannin.
Reagent in making—
Diacyl perborates.
Starting point in making—
Hydrogen peroxide.
Metallurgical
Ingredient of—
Electroplating baths (added for the purpose of assisting in the production of smooth, pleasing plated surfaces).
Electroplating baths of alkaline or sodium stannate tin bath character (added for the purpose of assisting in the production of white tin plate).
Nickel-plating solutions (added for the purpose of preventing hydrogen pitting and permitting higher current densities, thus speeding up production).
Sulphate of zinc baths (added for the purpose of obtaining white zinc plate of good appearance).
Fats and Oils
Bleaching agent for various fats and oils of animal and vegetable origin for both technical and edible use.
Food
Ingredient (German 431749) of—
Flours (added for the purpose of improving the baking properties).
Reagent in—
Bleaching almond paste and other food products.
Glues and Adhesives
Reagent in—
Bleaching bones for use in manufacturing, bleaching gelatin.
Miscellaneous
Bleaching agent for—
Bristles and the like, ivory, both natural and artificial.
Panama and similar hats, sponges, straw.
Ingredient of—
Deodorizing preparations, general bleaching preparations, general oxidizing preparations, preparations used for antiseptic purposes, preparations used for bactericidal purposes.
Reagent in—
Dentistry operations, destroying organic matter in toxicology, preventing the growth of putrefactive organisms, various domestic operations.
Paper
Bleaching agent for—
Paper, pasteboard, and various types of pulp.
Perfume
Ingredient of—
Bleaching creams and lotions, dentifrices, deodorants, hair bleaches, mouthwashes and pastilles, shampoos.
Pharmaceutical
In compounding and dispensing practice.
Resins and Waxes
Bleaching agent for—
Artificial and natural resins, various waxes.
Rubber
Ingredient of—
Rubber mixtures (added for the purpose of oxidizing lead sulphide contained in them so as to prevent dark discoloration of the product).
Soap
Ingredient of—
Laundering compositions, skin-bleaching soaps, stain-removing textile soaps, textile industrial soaps and detergents.

Sodium Perborate (Continued)

Starch
As a bleaching agent—
Glossing starch for ironing, making soluble starch.

Textile
——, *Bleaching*
As an antichlor.
Bleaching agent for—
Absorbent cotton, colored cotton fabrics, delicate cotton and linen fabrics and yarns, fine silks, laces, raw stocks and yarns, raw wool, tussah silk, various mixed fabrics, wool yarns and wool cloths.

——, *Finishing*
Reagent in—
Removing sizing from fabrics.

——, *Printing*
Reagent in—
Printing with vat dyestuffs.

Sodium Permanganate

Synonyms: Permanganate of soda.
French: Permanganate sodique, Permanganate de soude, Soude permanganique.
German: Hypermangansäurenatrium, Hypermangansäuresnatron, Permangansäuresnatrium, Permangansäuresnatron.

Analysis
As an oxidizing agent in various processes.

Chemical
As an oxidizing agent in various processes.
Oxidizing agent in making—
Saccharin.

Miscellaneous
As a general bactericide.
As a general disinfectant.

Pharmaceutical
In compounding and dispensing practice.
Suggested for use as—
Antidote for poisoning by morphine, curare, and phosphorus.

Sodium Peroxide

Synonyms: Sodium binoxide, Sodium dioxide.
Latin: Bioxydum natri.
French: Peroxide de soude, Peroxide sodique.
German: Natriumhyperoxyd, Natronhyperoxyd.

Analysis
Reagent in various processes.
Source of oxygen.

Brewing
Bactericide for—
Unfavorable ferments and moulds in the wort.
Preservative agent for—
Beer.
Sterilizing agent for—
Casks, filter pulp.

Chemical
Oxidizing agent in making—
Inorganic peroxides and various persalts—for example, manganese peroxide, sodium perborate, sodium percarbonate, zinc peroxide.
Organic peroxides, such as benzoyl superoxide.
Pharmaceutical chemicals, selenic acid from selenic salts, various chemicals.

Dye
General oxidizing agent in making—
Intermediates, synthetic dyestuffs.

Food
General bleaching agent.
General oxidizing agent.
General preservative.

Glues and Adhesives
Bleaching agent for—
Gelatin, glue.

Leather
Disinfecting and oxidizing agent for—
Hides subjected to long storage.

Miscellaneous
Bleaching agent for—
Teeth (in dentistry).
Bleaching agent (in aqueous solution acidified with sulphuric acid or mixed with magnesium sulphate) for—
Bones, bristles, feathers, hair, ivory, parchment, sponges, straw.

Carbon dioxide absorbent and source of oxygen in—
Air-purifying apparatus, diving apparatus and diving bells, fire-fighting respiratory apparatus, life-saving apparatus, submarine vessels, subterranean operations of various kinds.
General antiseptic.
General bactericide in many fermentation industries.
General bleaching agent.
General oxidizing agent.

Paper
Bleaching agent (Brit. 398730) in making—
Cotton-like fabric from sulphite cellulose.

Perfumery
Ingredient of—
Bleaching preparations, cosmetic creams, dentifrices.

Pharmaceutical
In compounding and dispensing practice.
Ingredient of—
Disinfectant tablets.

Photographic
Bleaching agent.
Ingredient (U. S. 1844711) of—
Mixture with alcohol used for removing dye from cellulose nitrate film scrap.
Oxidizing agent.

Soap
Ingredient of—
Detersive compositions, medicinal soaps, toilet soaps.
Source (Brit. 395572) of active oxygen in making—
Strongly disinfectant silver soaps.

Textile
Bleaching agent for—
Animal and vegetable textile fibers.

Water and Sanitation
Disinfectant and bactericide for—
Drinking water (in combination with citric acid).
Ingredient of—
Sanitary compositions.
Oxidizing agent (U. S. 1915240) for—
Coagulated sewage in purification process.
Purifying agent for—
Sickroom air.

Wine
Bactericide for—
Unfavorable ferments and moulds in the must.
Preservative agent for—
Finished wines.
Sterilizing agent for—
Casks, filter pulp.

Woodworking
Bleaching agent.

Sodium Phenate

Synonyms: Sodium phenolate, Sodium phenoxide.
French: Phénate sodique, Phénate de soude, Phénolate sodique, Phénolate de sodium, Phénolate de soude, Phénoxyde sodique, Phénoxyde de soude.
German: Natriumphenat, Natriumphenolat, Phenolnatrium, Phenolsaeuresnatrium.

Leather
Ingredient (Brit. 263473) of—
Liquors for dyeing leather.

Miscellaneous
Ingredient (Brit. 263473) of—
Liquors for dyeing hair and feathers.

Pharmaceutical
In compounding and dispensing practice.

Textile
——, *Dyeing and Printing*
Ingredient (Brit. 263473) of—
Liquors and pastes containing vat dyestuffs used in dyeing and printing acetate and other rayons in fabric or yarn form, and also mixtures of rayons with wool or silk.

Sodium 2-Phenylbenziminazole Sulphonate

Cosmetic
Protective (Brit. 435811) in—
Sun-tan lotions (solution or dispersion in a compatible solvent, for example, glycerin or wool-fat, but not water, alcohol, benzene, carbon tetrachloride, chloroform, or acetone), said to prevent formation of painful erythemas whilst enabling the skin to grow brown in sunlight, by virtue of high absorption of ultraviolet rays.

Sodium 1-Phenyl-2:3-dimethyl-5-pyrazolone-4-amino-methanesulphonate

Pharmaceutical
Suggested for use as—
Analgesic, antipyretic, antirheumatic.

Sodium Phenylmethyldithiocarbamate

Disinfectant
As a bactericide (Australian 8103/32, Brit. 406979, U. S. 1972961).
Insecticide and Fungicide
As a fungicide (Australian 8103/32, Brit. 406979, U. S. 1972961).
As an insecticide (claimed effective against aphids) (Australian 8103/32, Brit. 406979, U. S. 1972961).

Sodium 2-Phenylphenate

Synonyms: Sodium orthophenylphenate.
Disinfectant
As a germicide (of particular value for the disinfecting of premises which have been contaminated by infected cattle).
Fungicide
As a fungicide.

Sodium Phosphoglucosate

Refrigeration
Inhibitor of—
Corrosion in brine systems.
Reagent for—
Hydrogen ion adjustment of brines.
Remover of—
Carbonic dioxide formation in brines.

Sodium Phosphotungstate

Synonyms: Sodium tungstophosphate.
French: Phosphotungstate de soude, Phosphotungstate sodique, Tungstophosphate de soude, Tungstophosphate sodique.
German: Natriumphosphorwolframat, Natriumwolframphosphat, Phosphorwolframsaeuresnatrium, Wolframphosphorsaeuresnatrium.
Dye
Reagent (Brit. 275943) in making lakes with—
Para-aminobenzaldehyde.
4:4′-Tetramethyldiaminobenzhydrol.
4:4′-Tetramethyldiaminobenzophenone.
4:4′-Tetramethyldiaminodiphenylmethane.
Paint and Varnish
Ingredient of—
Oil or spirit lacquers containing cellulose esters and ethers and colored with basic dyestuffs (Brit. 275969).

Sodium Phosphotungstomolybdate

Synonyms: Sodium phosphomolybdotungstate, Sodium tungstomolybdophosphate, Sodium tungstophosphomolybdate.
French: Molybdophosphoretungstate sodique, Molybdophosphoretungstate de sodium, Phosphoremolybdotungstate de sodium, Phosphoretungstomolybdate de soude.
German: Molybdaenphosphorwolframsaeuresnatrium, Molybdaenwolframphosphorsaeuresnatron, Natriummolybdaenwolframphosphat, Natriumphosphorwolframmolybdat, Natriumphosphormolybdaenwolframat, Phosphorwolframmolybdaensaeuresnatrium.
Dye
Ingredient (Brit. 275943) of coloring lakes containing—
Para-aminobenzaldehyde.
4:4′-Tetramethyldiaminobenzhydrol.
4:4′-Tetramethyldiaminobenzophenone.
4:4′-Tetramethyldiaminodiphenylmethane.
Paint and Varnish
Ingredient (Brit. 275969) of—
Cellulose ester or ether lacquers containing basic colors.

Sodium Phthalate

French: Phthalate sodique, Phthalate de soude.
German: Phtalnatrium, Phtalnatron.
Textile
Delustring agent (Brit. 425418) for—
Linen goods (used with aluminum formate).
Viscose rayon (used with aluminum sulphate and sodium acetate).

Sodium Polyacrylate

Rubber
Creaming agent (Brit. 429559) for—
Rubber latex.

Sodium Polysulphide

Synonyms: Polysulphide of soda.
French: Polysulfure sodique, Polysulfure de sodium, Polysulfure de soude.
German: Natriumpolysulfid.
Spanish: Polisulfuro de sosa.
Italian: Polisulfuro di sodio.
Chemical
Reagent in making—
Sodium thiosulphate by oxidation.
Reducing agent in making—
Derivatives of polynitro compounds.
Dye
Reducing agent in making—
Sulphur dyestuffs fast to chlorine, sulphur colors from 2:5-dinitrophenol, thional brown G, thiophor indigo CJ, vidal black.
Fats and Oils
Reagent (Brit. 271553) in making—
Vulcanized oils.
Insecticide
As an insecticide and fungicide.
Ingredient of—
Insecticidal and fungicidal compositions.
Leather
Reagent in—
Dehairing hides.
Paper
Ingredient (Brit. 271553) of—
Compositions, containing rubber latex, used for treating paper and pulp.
Pharmaceutical
Ingredient of—
Parasitic pomades, sulphur baths, sulphurized lotions.
Rubber
Reagent (Brit. 271553) in treating—
Rubber latex.
Textile
Reagent in—
Denitrating nitro rayons.
Removing sulphur from viscose rayon filament.

Sodium-Potassium Guaiacolate

French: Guaiacolate de soude-potasse.
German: Guaiakolsaeuresnatriumkalium, Natriumkaliumguaiacolat.
Leather
Ingredient of—
Vat liquors for dyeing leathers (Brit. 263473).
Miscellaneous
Ingredient of—
Vat liquors for dyeing and stenciling furs and hair (Brit. 263473).
Textile
——, *Dyeing*
Ingredient of (Brit. 263473) vat liquors for dyeing—
Cellulose acetate rayon yarns and fabrics.
Chardonnet rayon yarns and fabrics.
Cuprammonium rayon yarns and fabrics.
Silk-rayon mixtures.
Viscose rayon yarns and fabrics.
Wool-rayon mixtures.
——, *Finishing*
Ingredient of compositions for stenciling—
Rayon fabrics, silk-rayon mixtures, wool-rayon mixtures.

Sodium-Potassium Quinolate

French: Quinolate de soude et potasse.
German: Chinolinsaeuresnatriumpotassium, Natriumpotassiumchinolat.
Leather
Ingredient of—
Vat dyestuff liquor (Brit. 263473).
Miscellaneous
Ingredient of—
Vat dyestuff liquor for dyeing and stenciling fur and hair.

Sodium-Potassium Quinolate (Continued)

Textile
——, *Dyeing*
Ingredient of vat dyestuff liquor for—
Cellulose acetate yarns and fabrics.
Chardonnet rayon yarns and fabrics.
Cuprammonium yarns and fabrics.
Silk-rayon yarns and fabrics.
Viscose rayon yarns and fabrics.
Wool-rayon yarns and fabrics.
——, *Finishing*
Ingredient of stenciling compositions for—
Rayon fabrics.
——, *Printing*
Ingredient of printing pastes for—
Rayon fabrics.

Sodium Propylnaphthalenesulphonate
French: Propylenaphthalènesulfonate de soude.
German: Natriumpropylnaphtalinsulfonat, Propylnaphtalinsulfonsaeuresnatrium.

Dye
Dispersing agent in making—
Color lakes (Brit. 264860).

Ink
Dispersing agent in making—
Printing inks.

Paint and Varnish
Dispersing agent in making—
Paints, pigments.

Plastics
Dispersing agent in making—
Cellulose ester plastics, cellulose ether plastics.

Resins and Waxes
Dispersing agent in making—
Artificial resin preparations, natural resin preparations.

Rubber
Dispersing agent in making rubber compositions.

Textile
——, *Dyeing*
Dispersing agent in making—
Dye liquors containing sulphur dyestuffs, indigoes, anthraquinone, vat dyestuffs.
Dye liquors for rayon, wool, cotton, and silk.

Sodium Propyl-3-nitrophthalate
French: Propyle-3-nitrophthalate de soude.
German: Natriumpropyl-3-nitrophtalat, Propyl-3-nitrophtalsaeuresnatrium.

Resins and Waxes
Starting point in making—
Synthetic resins (U. S. 1618209).

Sodium Propylphthalate
French: Propylephthalate de soude.
German: Natriumpropylphtalat, Propylphtalsaeuresnatrium.

Resins and Waxes
Starting point (Brit. 250265) in making synthetic resins with the aid of—
Barium acetate, bromide, chloride, or nitrate.
Calcium acetate, bromide, chloride, or nitrate.
Lead acetate, bromide, chloride, or nitrate.
Magnesium acetate, bromide, chloride, or nitrate.
Strontium acetate, bromide, chloride, or nitrate.
Zinc acetate, bromide, chloride, or nitrate.

Sodium Pyroglucosate

Mechanical
Inhibitor of—
Corrosion in boilers and hot-water systems.
Remover of—
Oxygen and carbon dioxide from boilers and hot-water systems.
Treating agent for—
Water in boilers and hot-water systems.

Sodium Pyrophosphate
Synonyms: Disodium pyrophosphate.
Latin: Natrium pyrophoricum, Sodii pyrophosphas.
French: Pyrophosphate de soude.
German: Natriumpyrophosphat.
Spanish: Pirofosfato sodico.

Analysis
As a reagent in—
Electro-analysis of metals, general analysis.

Chemical
Absorbent (Brit. 369344) in making—
Solidified hydrogen peroxide product.
Ingredient (Brit. 399998) of—
Desizing preparations containing also starch-degrading enzymes.
Reagent (U. S. 1914311) in making—
Organic compounds containing active oxygen, from hydrogen peroxide, alcohols, aldehydes, or ketones.
Stabilizer (Brit. 394989) in—
Emulsifying baths containing organic persulphonates, calcined soda or soap, and sodium silicate.
Starting material in making—
Iron pyrophosphates, other pyrophosphates.

Pharmaceutical
In compounding and dispensing practice.

Soap
Oxygen-carrier (Brit. 395572, 395570) in—
Silver base disinfectant soaps.

Textile
Reagent for—
Removing ink stains from colored cotton fabrics without affecting colors.
Stabilizer (Brit. 394989) in—
Cleansing baths containing organic persulphonates, calcined soda or soap, and sodium silicate.

Sodium Pyrosulphate
Synonyms: Pyrosulphate of soda.
French: Pyrosulfate sodique, Pyrosulfate de soude.
German: Natriumpyrosulfat, Natronpyrosulfat, Pyroschwefelsäuresnatrium, Pyroschwefelsäuresnatron.
Spanish: Pyrosolfato sodico.
Italian: Pyrosolfato di sodio.

Analysis
As a reagent in various processes.

Chemical
Sulphonating agent for—
Batyl, chimyl, and selachyl alcohols in production of emulsions useful in tanning and impregnating processes and in making insecticides and fungicides (Brit. 398818).
Oleyl alcohol-pyridin mixture used in making cleansing agent (Brit. 391435).
Sperm oil alcohols-pyridin mixtures used in making cleansing agents (Brit. 391435).
Sulphonating agent in—
Organic synthesis.

Sodium Pyrovanadate

Agriculture
In inoculation of plant life.

Ink
Ingredient of various inks.

Miscellaneous
Mordant in—
Fur dyeing.

Pharmaceutical
In compounding and dispensing practice.

Photographic
Reagent for—
Imparting red tones to films and plates.

Sodium Resinate
Synonyms: Abietate of soda, Resinate of soda, Rosin soap, Sodium abietate.
French: Abiétate sodique, Abiétate de sodium, Abiétate de soude, Résinate sodique, Résinate de sodium, Résinate de soude, Savon résinique.
German: Abietinsaeuresnatrium, Abietinsaeuresnatron, Harzseife, Natriumabietat, Natriumresinat.

Metallurgical
Flotation agent in separating—
Copper sulphide from iron sulphide.

Miscellaneous
Ingredient of—
Germicidal preparations.

Paper
Sizing agent in treating—
Pulp in the beater.

Pharmaceutical
In compounding and dispensing practice.

Soap
Ingredient of various soaps.

Sodium Resorcinate
French: Résorcinate de soude.
German: Natriumresorcinat, Resorcinsaeuresnatrium, Resorzinsaeuresnatrium.
Leather
Ingredient of—
Vat dye liquors in dyeing and stenciling (Brit. 263473).
Miscellaneous
Ingredient of—
Vat dye liquors in the dyeing of fur and hair.
Textile
——, *Dyeing*
Ingredient of vat-dyeing liquors for—
Cellulose acetate fabrics and yarns.
Cuprammonium rayon fabrics and yarns.
Nitrocellulose rayon fabrics and yarns.
Silk-rayon mixtures.
Viscose rayon fabrics and yarns.
Wool-rayon mixtures.
——, *Printing*
Ingredient of vat-dye pastes for printing and stenciling—
Cellulose acetate fabrics, cuprammonium rayon fabrics, nitrocellulose rayon fabrics, viscose rayon fabrics, silk-rayon fabrics, wool-rayon fabrics.

Sodium Ricinoleate
Synonyms: Sodium ricinate.
French: Ricinoléate sodique, Ricinoléate de sodium, Ricinoléate de soude.
German: Natriumricinat, Natriumricinoeleat, Natriumrizinoeleat, Ricinoelsaeuresnatrium, Ricinusoelsaeuresnatrium, Rizinoelsaeuresnatrium, Rizinusoelsaeuresnatrium.
Miscellaneous
Ingredient of—
Fire extinguishing compositions (Brit. 260535).
Soap
Ingredient of—
Transparent soaps.
Textile
——, *Dyeing and Printing*
Ingredient (Brit. 283253) of dye liquors, printing pastes, and stenciling compositions used on acetate rayon threads, films, and fabrics and on acetate rayon mixtures, with the following dyestuff ingredients—
4-Chloro-2-nitrophenylbenzylamine.
3:3′-Dinitrobenzidin.
3:3′-Dinitro-4:4′-diaminodiphenylmethane.
3:3′-Dinitro-4:4′-di(dimethylamino)-diphenyl ketone.
2:2′-Dinitro-4:4′-di(dimethylamino)-6:6′-ditolylmethane.
2:4-Dinitro-2-nitrophenylbenzylamine.
3:3′-Dinitro-orthotoluidin.
3-Nitro-4-aminodiphenyl ether.
3-Nitrobenzidin.
2-Nitrophenylbenzylamine.
4-Nitrophenylbenzylamine.
Various other nitrodiphenyls, nitrobenzidines, nitrotolidines, nitrophenylbenzylamines, nitrophenylethers, nitrodiphenylmethanes, nitrobenzophenones.

Sodium Ricinoleic Sulphonate
Miscellaneous
As an emulsifying agent (Brit. 361732).
For uses, see under general heading: "Emulsifying agents."

Sodium Ricinoleylpyrophosphate
Chemical
Stabilizing agent (Brit. 421843) for—
Peroxide solutions.
Miscellaneous
Stabilizing agent and promoter of wetting and penetrating properties (Brit. 421843) for—
Peroxide solutions used in many industries for (1) bleaching, (2) sterilizing, (3) disinfecting.

Sodium Salicylanilide
Insecticide
Starting point (Brit. 403411) in making—
Fungicides for seeds, tubers, and corms by reaction with mercuric chloride. (In dried form the precipitate product is used as a dusting powder; in paste form it is made into an aqueous suspension to which protective colloids, emulsifying and spreading agents, insecticides, or other fungicides may be added, and used in the form of a spray.)
Fungicides for seeds, tubers and corms by reaction with mercurous nitrate, copper sulphate, lead nitrate, or zinc chloride.

Sodium Salicylate
Synonyms: Salicylate of soda.
Latin: Natrium salicylatum, Sodii salicylas.
French: Salicylate de soude.
German: Natriumsalicylat, Natronsalicylat.
Spanish: Salicilato sodico.
Italian: Salicilato di sodio.
Glue and Adhesives
Ingredient of—
Furniture glue, containing also animal glue, powdered white lead, powdered chalk, methanol, and water.
Mechanical
Ingredient (U. S. 1940041) of—
Rust and corrosion preventive for automobile radiators, particularly soldered joints (in admixture with sodium borate and sodium nitrite).
Paint and Varnish
Liquefying agent (Brit. 406048) for—
Gelatin when used as underlying medium for coating liquids consisting of cellulose ester lacquers, oil varnishes, synthetic resin lacquers.
Pharmaceutical
In compounding and dispensing practice.
Ingredient of—
Nonfading amethyst-colored water composition for druggists' window display bottles (contains also tincture of ferric chloride).
Substitute for—
Salicylic acid.
Suggested for use in treating—
Migraine, neuralgia, pleurisy, rheumatism.
Rubber
Dispersing agent (German 425770 and 556904) in making—
Microporous rubber.

Sodium Salicylorthoanisidide
French: Orthoanisididesalicylique sodique, Orthoanisididesalicylique de sodium, Orthoanisididesalicylique de soude.
German: Natriumsalicylorthoanisidid.
Agricultural
Reagent in treating—
Seeds and grains to protect them against the action of fungi and mildew.
Chemical
Starting point in making—
Intermediates, pharmaceuticals, and other derivatives.
Dye
Starting point in making various synthetic dyestuffs.
Leather
Reagent in treating—
Leather and leather goods to protect them against mildew and the action of fungi.
Paper
Reagent in treating—
Paper, pulp, and products made therefrom for protection against mildew and the action of fungi.
Rubber
Reagent in treating—
Rubber and rubber products for protection against mildew and the action of fungi.
Textile
Reagent in treating—
Cotton yarns and fabrics for protection against mildew and the action of fungi.
Woodworking
Reagent in treating—
Wood and wood products for protection against the action of fungi and mildew.

Sodium Salicylparatoluidide
Insecticide
Starting point (Brit. 403411) in making—
Fungicides for seeds, tubers, and corms by reaction with copper sulphate. (In dried form the precipitate product is used as a dusting powder; in paste form it is made into an aqueous suspension to which protective colloids, emulsifying and spreading agents, insecticides, or other fungicides may be added, and used in the form of a spray.)

Sodium Salt of Dodecylsulphuric Acid Ester
Mining
Flotation reagent (Brit. 405163) for—
Barytes from carbonate or silica gangue, with or without the presence of barium chloride.

Sodium Salt of Ortho-4-Sulphobenzoylbenzoic Acid
Chemical
Starting point (U. S. 1899957) in making—
Anthraquinonesulphonic acids.

Sodium Salt of Ricinoleic Acid Butyl Ester Sulphuric Acid Ester
Mining
Flotation reagent (Brit. 405163) for—
Barytes from carbonate or silica gangue, with or without the presence of barium chloride.
Cassiterite, with or without the presence of sodium oleate.

Sodium, Secondary, Butylbetabromoallylbarbiturate
Synonyms: Pernocton.
Pharmaceutical
Suggested for use as—
New anesthetic (by basal narcosis).

Sodium Selenate
French: Séléniate sodique, Séléniate de sodium, Séléniate de soude.
German: Natriumselenat, Selensäuresnatrium, Selensäuresnatron.
Spanish: Selenato sodico.
Italian: Selenato di sodio.
Analysis
Reagent in various laboratory operations.
Chemical
Reagent in various processes.
Miscellaneous
Moth repellent (used in 10 percent solution, with soap, for killing larvae deposited on feathers, furs, hair, and other animal products).
Pharmaceutical
Suggested for the treatment of cancer.
Textile
Moth repellent (used in 10 percent solution, with soap, for killing larvae deposited on wool and felt).

Sodium Selenite
Analysis
Reagent in various laboratory operations.
Chemical
Reagent in various processes.
Glass
Ingredient in making—
Red glass.
Ingredient in masking—
Green colors due to iron.
Miscellaneous
Moth repellent (used in 10 percent solution, with soap, for killing larvae deposited on feathers, furs, hair, and other animal products).
Textile
Moth repellent (used in 10 percent solution, with soap, for killing larvae deposited on wool and felt).

Sodium Silicate
Synonyms: Liquid glass, Silicate of soda, Sodium metasilicate, Soluble glass, Water glass.
Latin: Sodii silicas.
French: Métasilicate de sodium, Métasilicate de soude, Silicate sodique, Silicate de sodium, Silicate de soude, Soude silicate, Verre soluble.
German: Löslichesglas, Natriumsilikat, Natronsilikat, Wasserärnlichesglas, Wasserartigesglas, Wasserglas.
Abrasives
Binder in making—
Abrasive compositions, abrasive stones, abrasive wheels.
Adhesives
Alone as such or ingredient of—
Adhesive cements for miscellaneous purposes.
Container board adhesives (both solid and corrugated board).
Fiber board adhesives, fiber trunk cements, labelling adhesives for glass, parquet flooring cements, plywood cements, sealing agents for shipping containers, sealing agents for various purposes, veneering adhesives, wallboard adhesives, wood adhesives.
Beverage
Cleansing agent for—
Bottles, tanks, and other plant equipment.
Ingredient of—
Bottle-washing compounds.
Brewing
Cleansing agent for—
Bottles, tanks, pasteurizers, and other plant equipment.
Ingredient of—
Bottle-washing compounds.
Building Construction
Adhesive cement for—
Corrugated asbestos insulations for piping and heating equipment.
Binder in—
Building materials, construction materials.
Coating and hardening agent for—
Cement floors and other surfaces exposed to ordinary abrasion, or to chemical corrosion or interfactory trucking of goods and materials.
Dustproofing agent for—
Cement, concrete, brick, and other surfaces exposed to abrasion, corrosion, mechanical and other forms of crumbling or disintegration.
Fireproofing agent for—
Floors, walls.
Heat insulation for—
Roofs and walls (applied by spraying on shredded newspapers).
Improver in—
Whitewashes for coating heated surfaces.
Promoter of—
Cement penetration of porous rock in sealing to prevent water seepage into borings for tunnels, subways, and the like, and for solidifying or strengthening both old and new foundations.
Starting point in making—
Acidproof cements for setting bricks, tile, and shapes in acid plants, chemical plants, smelters, metallurgical plants, and various other industrial establishments where buildings and equipment are exposed to chemically corrosive conditions.
Cements for making gas-tight joints in boilers, furnaces, coke-ovens and the like.
Hydraulic cements used for various purposes.
Waterproofing agent for—
Floors, walls.
Ceramic
Deflocculating agent for—
Clays.
Ingredient of—
China cements, zinc glazes.
Mending agent for—
Saggers.
Chemical
(For constructional uses in this industry see under: *"Building Construction."*)
Starting point in making—
Silica gel.
Dairying
Cleansing agent for—
Bottles, cans, tanks, pasteurizers, and other plant equipment.
Ingredient of—
Bottle-washing compounds.
Distilling
Cleansing agent for—
Bottles, tanks, and other plant equipment.
Ingredient of—
Bottle-washing compounds.
Dry-Cleaning
Clarifying agent for—
Dry-cleaning solvents.
Coating agent for—
Identification tags subjected to immersion in hydrocarbon solvents.
Electrical
High-temperature insulator.
Ingredient of—
Electrolyte for rectifiers (U. S. 1748011).
Explosives and Matches
Fireproofing agent for—
Matches.
Ingredient (U. S. 1762911) of—
Sealing compound for torpedoes, flares, and other pyrotechnic, and signal devices (advantages claimed: Quick-drying, waterproof, fireproof, slight expansion and contraction).
Fats and Oils
Penetration-resistant coating for—
Tubs and barrels used to ship oily and greasy products.

Sodium Silicate (Continued)

Purifying agent for—
Fats, vegetable oils.

Fertilizer
Increaser of—
Barley yield (by action on phosphate).
Sizing agent for—
Fertilizer bags.

Firefighting
Fireproofing agent for—
Curtains, fabrics, flooring, woodwork.

Food
Candling agent for—
Eggs.
Cleansing agent for—
Bottles, jars, cans, tanks, and other plant equipment.
Penetration-resistant coating for—
Bakery containers.
Tubs and barrels used to ship oils and greasy products.
Preservative for—
Eggs.

Glass
Binder (U. S. 1752792) in—
Coating composition for imparting opalescent effects to electric lamp bulbs or similar glass objects (said to give favorable effects in light absorption, bulb strength, and flow in coating).
Ingredient of—
Glass cements.

Ink
Ingredient of—
Printing inks.

Insecticide and Fungicide
Efficiency promoter in—
Pyrethrum-soap sprays for combatting Japanese beetle, striped cucumber beetle, squash bug, and other insects.

Laundry
Ingredient of—
Detergent compounds containing also soda ash or trisodium phosphate.
Jointing agent for—
Fabric conveyors on ironing machines.
Preventer of—
"Red water" in washing and rinsing operations.
Staining by iron in water.

Leather
Accelerator (U. S. 1765199) of—
Depilatant action.
Improver (U. S. 1765199) of—
Grain, fullness.
Preventer (U. S. 1765199) of—
Brittleness, rigidity.
Soaking agent (U. S. 1765199) for—
Hides, skins, pelts (used in conjunction with a nitrogen base, such as ammonia, ethylenediamine, or pyridin; claimed that this soaking agent, used prior to the depilating treatment, prevents injury to the hair or wool and the true skin thereby increasing hide-value).

Mechanical
Adhesive cement for—
Corrugated asbestos insulations for piping and heating equipment.
Degreasing agent for—
Metal machine parts and surfaces.
Ingredient of—
Boiler compounds.
Preventer of—
"Red water" in boiling systems.

Metal Fabrication
Degreasing agent for—
Metal products.
Plugging agent for—
Steel barrel seams.

Metallurgical
Basic lining for—
Bessemer converters.
Corrosion inhibitor for—
Aluminum.
Ingredient of—
Enamels for such products as kitchen ware, signs, kitchen sinks, sanitary ware, washstands, bathtubs, fixtures, and the like.

Mining
Deflocculating agent in—
Ore flotation.
Promoter of—
Penetration of porous rock by cement in sealing such rocks in order to prevent water seepage into mine workings.

Miscellaneous
Anchoring agent for—
Fixing light machinery to floors (used in admixture with sawdust).
Binder for various articles.
Binder in—
Furnace cements, stove cements.
Binder and hardener for—
Tennis courts.
Cementing agent for—
Splints in surgery.
Coating agent for—
Tree wounds.
Continuous-jointing agent for—
String.
Ingredient of—
Bottle-cleansing compounds.
Compounds for cleansing painted surfaces.
Detergent compounds for various purposes (added because of ability to suspend dirt, to increase the amount and stability of the lather of soaps, to cleanse by wetting oily surfaces thus loosening the oil so that it can be rinsed away, to emulsify oils with soap in an economical and efficient manner, to reduce soap consumption in hard waters).
Hand-cleansing compositions for mechanics.
Metal-cleansing compounds.
Rug-cleansing compounds.
Tapestry-cleansing compounds.
Woodwork-cleansing compounds.
Jointing agent for—
Fabric conveyor belts used in factories for various transportation purposes.
Penetration-resistant coating for—
Tubs and barrels used to ship oily and greasy products.
Preventer of—
Corrosion of iron pipe by dissolved oxygen in the water.
Protective coating agent for many products.

Paint and Varnish
Ingredient of—
Coldwater paints, fresco paints, paint and varnish removers, silicate paints.

Paper
Addition agent in—
Beater operations (added to reduce time of operation, to improve quality of finished product, and to effect savings in rosin size).
Adhesive for—
Roll capping, splicing.
Antisliming and coating agent for—
Concrete beaters, Jordan chests, save-alls.
Coating and hardening agent for—
Cement floors.
Dispersing agent for—
Clays used in making book and other papers (doubles solids content of fluid mixtures).
Dustproofing agent for—
Floors.
Greaseproofing and oilproofing agent for—
Paper.
Hardening agent for—
Paper (used in conjunction with alum).
Increaser of—
Hardness and smoothness of paper surface.
Paper strength.
Ingredient of—
Acidproof cements for digester linings in sulphite mills.
Preservative coatings for paper.
Sizes for paper.
Loading agent for—
Paper (combined with other advantages).
Neutralizing agent for—
Acid in colorings for wallpaper.
Retention aid for—
Color, fillers, rosin, starch.
Starting point in making—
Adhesives for corrugated paper, fiberboard, sized paper, containers, paperboard, mailing tubes, cartons.

Sodium Silicate (Continued)
Antisliming agent for treating the white water and stock in the well (combined with solution of bleaching powder).

Petroleum
Deflocculating agent in—
Crank case oil reclamation, refining processes.

Refractory
Bonding agent in—
Cements and mortars.

Rubber
Parting layer in—
Making and interplant transporting rubber products, such as boots, gloves, hot-water bags, tires, toys, and the like.

Soap
Ingredient of—
Detergent compounds for various purposes (added because of ability to suspend dirt, to increase the amount and stability of the lather of soaps, to cleanse by wetting oily surfaces thus loosening the oil so that it can be rinsed away, to emulsify oils with soap in an economical and efficient manner, to reduce soap consumption in hard waters).
Penetration-resistant coating for—
Tubs and barrels used to ship soapstocks.

Stone
Hardening agent.
Starting point in making—
Artificial stone.

Sugar
Clarifying and refining agent (U. S. 1687561) for—
Molasses to be used as nutrient in yeast culture.

Textile
Emulsification promoter in—
Kier boiling.
Fireproofing agent for various fabrics.
Preventer of—
Fabric staining by iron in water.
"Red water" caused by rust from iron pipe and equipment, corrosion by dissolved oxygen in the water.
Process material in—
Boiling-off operations, chlorine bleaching processes, degumming processes.
Dyeing processes, particularly with direct dyes.
Mordanting processes, peroxide bleaching processes, printing processes, sizing processes.
Soaking and dyeing weighted silk (U. S. 1723183).
Tin-phosphate-silicate silk-weighting process.
Resist in—
Dyeing operations.
Silk protectant in—
Peroxide bleaching processes for mixed cotton and silk goods.
Stain preventer in—
Kier boiling.

Water and Sanitation
(See also: *"Mechanical."*)
Water-clarifying agent.
Water-softening agent.

Wine
Cleansing agent for—
Bottles, tanks, and other plant equipment.
Ingredient of—
Bottle-washing compounds.

Wood
Adhesive in—
Veneering.
Adhesive in making—
Plywood.
Ingredient of—
Stainproofing compositions.
Stainproofing agent.

Sodium Silicofluoride
Synonyms: Sodium fluosilicate.
French: Fluosilicate de soude.
German: Kieselfluornatrium, Kieselfluorwasserstoffsaeuresnatron, Natriumsilicofluorid.

Ceramics
Ingredient of—
Enamel glazes for use on chinaware.
Porcelain enamels.
Raw material in making—
Ceramic ware (used in place of cryolite).

Glass
Ingredient of—
Opalescent glassware.
Raw material in making various kinds of glass.

Insecticide
Ingredient of—
Compositions used in place of arsenicals and sodium fluoride.
Compositions used for destroying the boll weevil.

Leather
Reagent in treating—
Hides and skins to facilitate tanning (Brit. 256628).

Paint and Varnish
Reagent in making—
Zirconium oxide pigment (U. S. 1588476).

Pharmaceutical
In compounding and dispensing practice.

Sodium Silicomolybdate
Synonyms: Sodium molybdosilicate.
French: Molybdosilicate de soude, Molybdosilicate sodique, Silicomolybdate de soude.
German: Molybdaenkieselsaeuresnatrium, Silicium-molybdaensaeuresnatrium.

Dye
Reagent (Brit. 275943) in making color lakes with—
Para-aminobenzaldehyde, 4:4′-tetramethyldiaminobenzhydrol, 4:4′-tetramethyldiaminobenzophenone, 4:4′-tetramethyldiaminodiphenylmethane.

Paint and Varnish
Ingredient (Brit. 275969) of—
Cellulose ester or ether oil or spirit lacquers containing basic dyestuffs.

Sodium Silicotungstate
Synonyms: Sodium tungstosilicate.
French: Silicotungstate de soude, Silicotungstate sodique, Tungstosilicate de soude.
German: Silicowolframsaeuresnatrium, Wolframkieselsaeuresnatrium.

Dye
Reagent (Brit. 275969) in making color lakes with—
Para-aminobenzaldehyde, 4:4′-tetramethyldiaminobenzhydrol, 4:4′-tetramethyldiaminobenzophenone, 4:4′-tetramethyldiaminodiphenylmethane.

Paint and Varnish
Ingredient (Brit. 275943) of—
Cellulose ester or ether oil or spirit lacquers containing basic dyestuffs.

Sodium Stannate
Synonyms: Preparing salt.
French: Stanniate de sodium, Stanniate de soude.
German: Natriumstannat, Natriumorthostannat, Zinnsaeuresnatrium, Zinnsoda.

Ceramics
Ingredient of—
Glazes.

Glass
Ingredient of batch in making various glass products.

Metallurgical
Reagent in—
Refining lead in order to remove its arsenic content.

Textile
——, *Dyeing*
As a mordant.

——, *Finishing*
Reagent in—
Fireproofing, waterproofing, weighting silk.

——, *Printing*
As a mordant and fixative.

Sodium Stearylglycollate

Metallurgical
Frothing agent in—
Flotation concentration of minerals (said to closely approach the ideal properties of a reagent for these purposes; namely:—(1) the formation of an abundant froth, but one not too persistent, at low concentrations; (2) as effective in acid mediums as in alkaline mediums; (3) insensitive to salts, even in high concentrations; (4) absolutely inert as a collector in regard to both sulphurized and nonsulphurized minerals; (5) its froth-forming properties should not be affected by the collecting agents, including the soap; (6) it should emulsify rapidly and have a dispersive

Sodium Stearylglycollate (Continued)
action on all collecting reagents that are usually employed. By the use of this reagent the employment of new collcetors, such as the insoluble paraffin oils and butyl sulpholeate, is practicable).

Sodium Stearylsulphate

Building and Construction
Emulsifying agent (Brit. 437674) in making—
Aqueous emulsions of asphalt and similar bituminous materials.

Sodium Sulphanilate
Synonyms: Sulphanilate of soda.

Soap
Rancidity retardant (U. S. 1869469) for—
Soap materials.

Sodium Sulphate
Synonyms: Glauber's salt, Sulphate of soda, Vitriolated soda.
Latin: Natrium sulfuricum, Sal mirabile glauberi, Sodii sulphas, Sulfas natricus, Sulfas sodicus.
French: Sel de glauber, Sulphate sodique, Sulphate de soude.
German: Glaubersalz, Natriumsulfat, Natronsulfat, Schwefelsäuresnatrium, Schwefelsäuresnatron.
Spanish: Sal admirabile de glaubero, Solfato sodico, Solfato de sosa.
Italian: Solfato di sodio.

Agriculture
Ingredient (U. S. 1879777) of—
Banding, plugging and patching material for trees, containing also whiting and fish and castor oils.

Analysis
Reagent in various processes.

Beverage
Ingredient of—
Artificial mineral waters, artificial vichy water.

Chemical
Dehydrating agent (Brit. 400944) for—
Concentrating aliphatic acids by extracting with a solvent of low-boiling point together with an oxygen-containing liquid with a boiling point of at least 105° C. in which water is insoluble, or extracting with a low-boiling solvent and precipitating water from the extract by adding the high-boiling oxygen-containing liquid.
Ingredient of—
Detergent compositions (Brit. 391435).
Detergent composition, containing also the sodium salt of cetylsulphobenzyl ether and the sodium salt of dodecylsulphobenzyl ether (Brit. 378454).
Furnace charge in process for decolorizing tinted barytes (Brit. 376080).
Reagent in making—
Various chemicals; for example, blanc fixe, sodium-amonium sulphate, sodium persulphate.
Sulphiding agent (U. S. 1902203) in making—
Iron-free titanium dioxide.
Washing agent (Brit. 392568) in making—
Emulsifying agents by sulphonating hydroxy fatty acids.

Construction
Ingredient (U. S. 1904639) of—
Hydraulic cement.

Dye
As a diluent for synthetic dyestuffs.
Ingredient (U. S. 1889491) of—
Dye composition for home dyeing of silk.
Starting point in making—
Ultramarine from kaolin and charcoal or rosin.

Explosives and Matches
Ingredient of—
Dynamites, safety explosives.

Glass
Ingredient of—
Glass batches.

Metallurgical
Flux (U. S. 1890204) in making—
Phosphates from ferrophosphorus.
Ingredient of—
Electrolyte, containing also nickel sulphate, nickel chloride, iron chloride, iron sulphate, cobalt sulphate, cobalt chloride, and boric acid, used in electrodeposition of magnetic nickel-cobalt-iron alloys (U. S. 1920964).
Electrolyte, containing also nickel sulphate, sodium chloride, and boric acid, for plating nickel on zinc or die-cast metal.

Miscellaneous
As a dehydrating agent in various processes.
Ingredient (U. S. 1887618) of—
Thermophoric composition.

Paint and Varnish
Crystallization promoter (Brit. 405340) in making—
Titanium pigments.
Luminophore (Brit. 319914) in making—
Luminous pigments for paints.

Paper
Precipitant (Brit. 403116) in making—
Bituminous-base waterproofed paper and pasteboard.
Source of soda in—
Sulphate process used in making Kraft and other papers.

Pharmaceutical
In compounding and dispensing practice.
Ingredient of—
Effervescent laxative compositions.
Suggested for use as—
Aperient, cathartic, diuretic.

Photographic
Reagent (Brit. 403988) in making—
Cellulose acetate film from bagasse, wood pulp, cornstalks, or other cheap cellulosic material.

Plastics
Dehydrating agent (French 755316) in making—
Plastics by condensation of a polymerized vinyl alcohol with an aldehyde; for example, aliphatic aldehydes, cyclic aldehydes, aromatic aldehydes.

Refrigeration
Ingredient of—
Freezing mixtures (in making ice).

Soap
Ingredient of—
Soap batches.

Textile
Assist in—
Dyeing.
Bleaching agent for—
Fabrics.
Ingredient (Brit. 390553) of—
Dye bath in process for increasing the fastness to water of cotton, viscose, and other cellulosic materials dyed with substantive colors.
Leveling agent in—
Dyeing and printing.
Mordant in—
Cotton dyeing.

Water and Sanitation
Crystallizing and porosifying agent (U. S. 1906163) in making—
Base-exchanging gels for water softening.
Ingredient of—
Air-conditioning substance comprising a mixture with plaster of paris (U. S. 1907809).
Cleansing agents for toilets.

Sodium Sulphide
Synonyms: Sulphide of soda.
French: Sulfure sodique, Sulfure de sodium, Sulfure de soude.
German: Schwefelnatrium, Schwefelnatron.
Spanish: Solfuro sodico.
Italian: Solfuro di sodio.

Analysis
Reagent in—
Analytical processes involving control and research.

Cosmetic
Active ingredient of—
Hair-removing preparations.

Chemical
Catalyst in making—
Urea.
Process material in making—
Acetic anhydride, aluminum chloride, amino compounds, anthranilic acid, antimony sulphide, arsenic pentasulphide, barium sulphate, benzene derivatives, intermediate chemicals, mercury-arsenic pharmaceuticals, naphthylenediamine derivatives, sulphides of various bases, sulphonamides, sulphonanilides, sulphonic acids.

Sodium Sulphide (Continued)
Reducing agent in—
Organic synthesis.
Starting point in making—
Sodium salts.

Dye
Process material in making—
Dyes, principally sulphur dyes.

Food
Disinfectant for—
Grain and chaff in milling.

Glass
Ingredient of—
Special glass batches.

Insecticide and Fungicide
Ingredient of—
Fungicides, insecticidal compositions, sheep dips, weed-destroying compositions.

Leather
Depilatory for—
Hides, pelts, skins.
Treating agent (with calcium hydroxide) for—
Shark skins, porpoise, and other fish skins.

Metallurgical
Process material in flotation of—
Cassiterite (tin ore), chalcopyrite (copper pyrites), copper ores, galena (lead glance), lead ores, malachite (copper carbonate ore), porphyry ores, zinc sulphide ores.
Process material in making—
Antimony, cadmium, cobalt, gold, iron, lead, nickel.
Process material in separating—
Copper from nickel, copper from iron, molybdenum from wulfenite.
Reagent in—
Coloring of metals.
Sulphiding material in various processes.

Miscellaneous
Carroting agent for—
Hair and fur.
Process material in—
Sulphur dyeing vegetable fibers.
Process material in dyeing—
Feathers, furs, hair.
Solvent for—
Hair.
Treating agent for—
Vegetable fibers.

Paper
Processing material in—
Pulp manufacture.

Petroleum
Catalyst in making—
Benzin, gasoline.

Photographic
Reagent in—
Developing processes, toning processes.

Printing
Process material in—
Lithographic processes, process engraving.

Rayon
Denitrating agent.

Rubber
Ingredient of—
Rubber batches.

Soap
Process material in making—
Special soaps.

Textile
Detergent and saponifying agent in—
Cleansing silk, cotton, and other fabrics.
Process material in—
Dyeing processes, printing processes, sulphur dyeing vegetable fibers.
Solvent for—
Sulphur dyes in dyeing processes.
Treating agent for—
Vegetable fibers.

Wood
Reagent in—
Staining of wood.

Sodium Sulphite
Synonyms: Sulphite of soda.
Latin: Sodii sulphis, Natrium sulfurosum, Sulfis natricus, Sulfis sodicus.
French: Sulfite sodique, Sulfite de soude.
German: Natriumsulfit, Schwefligsäuresnatrium, Schwefligsäuresnatron.

Analysis
Reagent in various processes.

Brewing
Antiseptic in—
Fermentations.
Sterilizing agent for—
Barrels, casks, plant equipment, vats.

Chemical
Antiseptic in—
Fermentations.
Reagent in making—
Alkali cyanates (Brit. 399820).
Hydrocyanic acid (Brit. 401351).
Sodium naphthylacetamide disulphonate (French 750647).
Solvents for acetic acid from vegetable oils, fats, and fatty acids (Brit. 390148).
Solvents for cyclohexanol from vegetable oils, fats, and fatty acids (Brit. 390148).
Solvents for essential oils from vegetable oils, fats and fatty acids (Brit. 390148).
Solvents for formic acid from vegetable oils, fats, and fatty acids (Brit. 390148).
Solvents for higher alcohols from vegetable oils, fats, and fatty acids (Brit. 390148).
Solvents for paraffin from vegetable oils, fats, and fatty acids (Brit. 390148).
Solvents for phenols from vegetable oils, fats, and fatty acids (Brit. 390148).
Water-solubilizing agents for acetic acid from vegetable oils, fats, and fatty acids (Brit. 390148).
Water-solubilizing agents for cyclohexanol from vegetable oils, fats, and fatty acids (Brit. 390148).
Water-solubilizing agents for essential oils from vegetable oils, fats, and fatty acids (Brit. 390148).
Water-solubilizing agents for formic acid from vegetable oils, fats, and fatty acids (Brit. 390148).
Water-solubilizing agents for higher alcohols from vegetable oils, fats, and fatty acids (Brit. 390148).
Water-solubilizing agents for paraffin from vegetable oils, fats, and fatty acids (Brit. 390148).
Water-solubilizing agents for phenol from vegetable oils, fats, and fatty acids (Brit. 390148).
Reducing agent in making—
Intermediates and other products, stable acridin salt solutions (Brit. 395405, 342690).

Distillery
Antiseptic in—
Fermentations.
Sterilizing agent for—
Barrels, casks, cookers, fermentation tanks, plant equipment.

Dye
Reducing agent in making various synthetic dyestuffs.

Food
Antifermentative for various food products.
Antiseptic for—
General purposes, grains.
Preservative for—
Egg yolk, food products, fruit juices, meats, syrups, vegetable juices.

Glue and Adhesives
Antiseptic for—
Glue, gelatin.

Glass
Ingredient of—
Compositions used in silvering glass.

Leather
Sulphiting agent for—
Quebracho extract (to improve its effect).

Mechanical
Reagent for—
Treating boiler-water to remove dissolved oxygen.

Metallurgical
Frothing restrainer (Brit. 396053) in—
Pickling solutions.

Miscellaneous
Antiseptic in many processes.
Bleaching agent in many processes.
Bleaching agent for—
Cork, straw.
Cleansing agent in many processes.
Disinfectant in many processes.
Preservative in many processes.

Sodium Sulphite (Continued)

Paper
Antichlor in—
Bleaching operations.

Perfume
Antiseptic and preservative for—
Cosmetic creams.

Pharmaceutical
In compounding and dispensing practice.

Photographic
Preservative for—
Developing solutions (prevents oxidation).
Reducing agent (Brit. 401340, 404856) in—
Developing multi-colored pictures.
Regenerator of—
Oxidized developing solutions.
Substitute for—
Hyposulphite in fixing photographic negatives and prints.

Printing
Reagent in—
Lithography, process engraving.

Rubber
Coagulating agent in making—
Raw rubber from rubber latex.

Sugar
Antifermentative for—
Sugar solutions and syrups, glucose, sugar.
Bleaching agent for—
Sugar solutions and syrups.

Textile
Antichlor in bleaching—
Animal fibers, vegetable fibers.
Reagent (Brit. 390148) in making—
Degreasing agents from sulphonated oils, fats, and fatty acids.

Wine Making
Antiseptic in—
Fermentations.
Sterilizing agent for—
Barrels, casks, plant equipment, vats.

Sodium Sulphocarbolate

Synonyms: Sodium phenolsulphonate, Sulphocarbolate of soda.
French: Sulfocarbolique de sodium, Sulfocarbolique de soude.
German: Carbolschwefelsäuresnatrium, Carbolschwefelsäuresnatron.

Chemical
Denaturant for—
Alcohol.

Insecticide and Fungicide
Process material in making—
"Bouillie Lyonnaise" for destroying *Oidium* on vines.

Pharmaceutical
In compounding and dispensing practice.
Ingredient of—
Chicken remedies.
Suggested for use as—
Antiseptic, astringent.

Sodium Sulphoglucosate

Mechanical
Inhibitor of—
Corrosion in boilers and hot-water systems.
Remover of—
Oxygen and carbon dioxide in boilers and hot-water systems.
Treating agent for—
Water in boilers and hot-water systems.

Sodium Sulphoricinoleate

French: Sulforicinoléate sodique, Sulforicinoléate de soude, Thioricinoléate sodique, Thioricinoléate de soude.
German: Natriumsulforicinoleat, Natriumthioricinoleat, Sulforicinoelsaeuresnatrium, Sulforicinoelsaeuresnatron, Sulforicinusoelsaeuresnatrium, Sulforizinusoelsaeuresnatrium, Thioricinoelsaeuresnatrium.

Miscellaneous
Ingredient of—
Mixtures used in fire-extinguishers (Brit. 260535).

Textile
——, *Dyeing and Printing*
Ingredient (Brit. 283253) of liquors used in dyeing of acetate threads and films and in stenciling or printing fabrics containing acetate rayon, with the aid of—
4-Chloro-2-nitrophenylbenzylamine, 3:3'-dinitrobenzidin, 3:3'-dinitro-4:4'-diaminodiphenylmethane, 3:3'-dinitro-4:4'-di(dimethylamine)diphenylketone, 2:2'-dinitro-4:4'-di(dimethylamine)-6:6'-ditolylmethane, 3:3'-dinitro-orthotoluidin, 2:4'-dinitrophenylbenzylamine, 3-nitro-4-aminodiphenyl ether, 3-nitrobenzidin, 2-nitrophenylbenzylamine, 4-nitrophenylbenzylamine.
Various nitrodiphenyls, nitrobenzidines, nitrotoluidines, nitrophenylbenzylamines, nitrophenyl ethers, nitrodiphenylmethanes, and nitrobenzophenones.

Sodium Sulphosebacate

Miscellaneous
As a wetting agent (Brit. 446568).
For uses, see under general heading: "Wetting agents."

Sodium Taurine

Miscellaneous
As an emulsifying agent (Brit. 343899).
For uses, see under general heading: "Emulsifying agents."

Sodium Taurocholate

French: Taurocholate sodique, Taurocholate de sodium, Taurocholate de soude.
German: Natriumtaurocholat, Taurocholsaeuresnatrium.

Chemical
Reagent (Brit. 282356) in making antiparasitic agents with—
Dihydrocuprein ethyl ether, dihydrocuprein ethyl ether hydrochloride, dihydrocuprein isoamyl ether, dihydrocuprein isoamyl ether hydrochloride, dihydrocuprein normal octyl ether, dihydrocuprein normal octyl ether hydrochloride, dihydroquinone.

Sodium Telluride

Chemical
Reagent (Brit. 292222) in making synthetic drugs with—
Pentamethylene alphaepsilondibromide, pentamethylene alphaepsilondichloride, pentamethylene alphaepsilondifluoride, pentamethylene alphaepsilondi-iodide.

Sodium Tetrachlorophthalate

Textile
Delustring agent (Brit. 425418) for—
Viscose rayon (used with aluminum formate).

Sodium 2:4:5:6-Tetrachlorphenate

Fungicide
Fungicidal agent for—
Molds and fungi on woodwork and wood products.

Sodium Tetrahydronaphthalenesulphonate

French: Tétrahydronaphthalènesulfonate sodique, Tétrahydronaphthalènesulfonate de sodium, Tétrahydronaphthalènesulfonate de soude.
German: Natriumtetrahydronaphtalinsulfonat. Tetrahydronaphtalinsulfonsaeuresnatrium.

Mechanical
Impregnating agent in treating—
Belts, bands, friction clutches, pulleys, brakes (Brit. 278465).

Sodium-Tetrahydronaphthalene Sulphonchloramide

German: Natriumtetrahydronaphtalinsulphonchloramid.

Chemical
Starting point in making—
Magnesium-tetrahydronaphthalene sulphonchloramide (German 422076).

Sodium Tetrasulphide

French: Tétrasulphure de soude.
German: Natriumtetrasulfid, Tetraschwefelnatrium.

Chemical
Reagent (Brit. 263191) in making—
2-Amino-4-nitrophenoxyethanol, 2-amino-4-nitrophenoxypropandiol.

Dye
Reducing agent for—
Polynitro compounds.
Reagent in making various dyestuffs of sulphur group.

Sodium Tetrathionate
French: Tétrathionate de soude.
German: Natriumtetrathionat.

Chemical
Reagent in making—
Tetramethylthiuram disulphide by oxidation.
Dimethyldiphenylthiuram disulphide by oxidation (Brit. 259930) practice.

Sodium Thiodinaphthylsulphonate
French: Thionaphthylsulphonate de soude.
German: Natriumthionaphtylsulfonat, Thiodinaphtylsulphonsaeuresnatron.

Textile
——, *Dyeing*
Assistant in dyeing fabrics and yarns with substantive dyestuffs.

Sodium-Thioglucose

Chemical
Starting point (Brit. 398020) in making—
Complex double compounds of organic heavy metal mercapto compounds.

Sodium Thioglycollate
French: Thioglycollate de soude.
German: Natriumthioglycollat, Thioglycolsaeuresnatrium.

Chemical
Starting point (Brit. 262301) in making therapeutic compounds with—
Antimonyl gallate, antimonyl pyrocatechol, antimonyl pyrogallol.

Sodium Thiolactate

Chemical
Starting point (Brit. 398020) in making—
Complex double compounds of organic heavy metal mercapto compounds.

Sodium Tolylthioglycollate

Chemical
Starting point in making various derivatives.

Dye
Reagent (Brit. 284288) in making thioindigoid dyestuffs with the aid of—
Acenaphthenequinone, alphaisatinanilide, 5:7-dibromoisatin.
Isatin, homologs, substitution products, and alpha derivatives.
Orthodiketones.

Sodium Triazoate

Chemical
Starting point in making—
Azoate of lead and of other metals.

Sodium Trichlorophenate
Synonyms: Sodium trichlorphenolate, Sodium trichlorophenolate.
French: Trichlorophénate sodique, Trichlorophénate de sodium, Trichlorophénate de soude, Trichlorophénolate sodique, Trichlorophénolate de sodium, Trichlorophénolate de soude.
German: Natriumchlorphenat, Natriumtrichlorphenolat, Trichlorphenolsäuresnatrium, Trichlorphenolsäuresnatron.
Spanish: Trichlorfenato sodico, Trichlorfenato de sosa, Trichlorfenolato sodico, Trichlorfenolato de sosa.
Italian: Trichlorfenato di sodio, Trichlorfenolato di sodio.

Fungicide
Fungicidal agent for—
Molds and fungi.

Insecticide
Ingredient of—
Compositions for treating furs and feathers to prevent mildew (U. S. 1618416).

Leather
Ingredient of—
Compositions for treating leathers to prevent mildew (U. S. 1618416).
Dressing compositions (added for the purpose of preventing mildew or fungoid growths, essential oils being used to hide the smell of the sodium trichlorophenate).

Textile
——, *Miscellaneous*
Ingredient of—
Compositions for making textile fibers proof against mildrew (U. S. 1618416).

Sodium Tungstate
Synonyms: Sodium wolframate.
French: Tungstate sodique, Tungstate de sodium, Tungstate de soude, Wolframate sodique, Wolframate de sodium, Wolframate de soude.
German: Natriumwolframat, Wolframsäuresnatrium, Wolframsäuresnatron.

Analysis
Reagent for the detection and determination of—
Acetoacetic acid, alkaloids, bile pigments, blood, calcium, carbonates, glucose, nitrates, peptone, phenols, tannin, tryosin, uric acid.

Chemical
Ingredient of catalytic mixtures used in the manufacture of—
Acenaphthylene, acenaphthaquinone, bisacenaphthylidenedione, naphthaldehydic acid, naphthalic anhydride and hemimellitic acid from acenaphthene (Brit. 295270).
Acetaldehyde from ethyl alcohol (Brit. 281307).
Acetic acid from ethyl alcohol (Brit. 281307).
Aldehydes or alcohols by the reduction of the corresponding esters (Brit. 306471).
Alphacampholide from camphoric acid by its reduction (Brit. 306471).
Aldehydes and acids from toluene, orthochlorotoluene, orthobromotoluene, orthonitrotoluene, parachlorotoluene, parabromotoluene, paranitrotoluene, metachlorotoluene, metanitrotoluene, metabromotoluene, dichlorotoluenes, dibromotoluenes, dinitrotoluenes, chlorobromotoluenes, chloronitrotoluenes, bromonitrotoluenes (Brit. 295270).
Aldehydes and acids from xylenes, pseudocumenes, mesitylene, and paracymene (Brit. 281307).
Alphanaphthaquinone from naphthalene (Brit. 281307).
Anthraquinone from naphthalene (Brit. 295270).
Benzaldehyde and benzoic acid from toluene (Brit. 281307).
Benzoquinone from phenanthraquinone (Brit. 281307).
Benzyl alcohol by the reduction of benzaldehyde (Brit. 306471).
Benzyl alcohol or benzaldehyde or phthalide by the reduction of phthalic anhydride (Brit. 306471).
Butyl alcohol by the reduction of crotonaldehyde (Brit. 306471).
Chloroacetic acid from ethylenechlorohydrin (Brit. 295270).
Diphenic acid from ethyl alcohol (Brit. 281307).
Ethyl alcohol by the reduction of acetaldehyde (Brit. 306471).
Fluorenone from fluorene (Brit. 281307).
Formaldehyde by the reduction of methane or methanol (Brit. 306471).
Formaldehyde by the reduction of carbon dioxide or carbon monoxide (Brit. 306471).
Hydroxyl compounds by the reduction of anthraquinone, benzoquinone and similar compounds (Brit. 306471).
Isopropyl alcohol by the reduction of acetone (Brit. 306471).
Maleic acid and fumaric acid by the oxidation of toluene, benzene, phenols, tar phenols, or furfural, or form benzoquinone or phthalic anhydride (Brit. 295270).
Methane by the reduction of carbon dioxide or carbon monoxide (Brit. 306471).
Methanol by the reduction of carbon dioxide or carbon monoxide (Brit. 306471).
Naphthaldehydic acid, acenaphthaquinone, or bisacenaphthylidenedione from acenaphthylene (Brit. 281307).
Phenanthraquinone from phenanthrene or diphenic acid (Brit. 295270).
Phthalic acid and maleic acid from naphthalene (Brit. 295270).
Primary alcohols by the reduction of the corresponding aldehydes (Brit. 306471).
Propionic acid and butyric acid and higher alcohols, ketones, and acids by the reduction of carbon dioxide and carbon monoxide (Brit. 306471).
Reduction products of ketones, aldehydes, acids, esters, alcohols, ethers, and other organic compounds, which contain oxygen (Brit. 306471).

Sodium Tungstate (Continued)
Salicylic acid and salicylic aldehyde from cresol (Brit. 295270).
Secondary butyl alcohol by the reduction of methylethyl ketone (Brit. 306471).
Vanillin and vanillic acid from eugenol or isoeugenol (Brit. 295270).
Ingredient (Brit. 306460) of catalytic preparations used in the production of various aromatic and aliphatic compounds, including—
Alphanaphthylamine from alphanitronaphthalene.
Amines from aliphatic nitro compounds, such as allyl nitriles or nitromethane.
Amino compounds from the corresponding nitroanisoles.
Amylamine from pyridin.
Anilin, azo-oxybenzene, azobenzene, and hydrazobenzene from nitrobenzene by reduction.
Aminophenols from nitrophenols.
3-Aminopyridin from 3-nitropyridin.
Cyclohexamine, dicyclohexamine, and cyclohexylanilin from nitrobenzene.
Piperidin from pyridin, pyrrolidin from pyrrol, tetrahydroquinolin from quinolin.
Reagent in—
Decolorizing acetic acid.
Starting point in making—
Ammonium borotungstate, ammonium phosphotungstate.
Borotungstates and phosphotungstates of the alkali metals, alkaline earth metals and earth metals.
Tungstic acid, tungsten oxides.

Explosives
Ingredient of—
Compositions used in the manufacture of matches.

Metallurgical
Starting point in making—
Metallic tungsten.

Miscellaneous
Ingredient of—
Compositions used for the sterilization of tooth brushes.
Dry stencil compositions (U. S. 1720897), fireproofing and waterproofing compositions.

Textile
——, *Dyeing*
Mordant in—
Dyeing silks.
——, *Finishing*
Ingredient of—
Fireproofing and waterproofing compositions.
——, *Printing*
Mordant in—
Calico printing.

Sodium Uranate
Synonyms: Uranium yellow, Yellow uranium oxide.
French: Uranate de soude.
German: Natriumuranat, Uransaeuresnatron.

Ceramics
Ingredient of—
Compositions used in the enameling of porcelains, chinaware and potteries.
Compositions used in the painting of porcelains, chinaware and potteries.

Glass
Ingredient of—
Special glasses (to give a greenish fluorescent color).

Soap
Catalyst in making—
Soap from mixtures of palm oil, seal oil and sulphonated fish oil (Brit. 255508).

Sodium Valeriate
Synonyms: Sodium valerianate.
Latin: Sodii valeras.
French: Valérianate de soude.
German: Valeriansäuresnatrium.

Pharmaceutical
In compounding and dispensing practice.
Suggested for use as—
Nerve stimulant.

Sodium Vanadate
Synonyms: Sodium orthovanadate.
French: Vanadate de soude.
German: Natriumvanadinat, Vanadinsaeuresnatron.

Analysis
Reagent in various processes.

Dye
Reagent in making—
Anilin black.

Ink
Ingredient of special inks.

Paint and Varnish
Ingredient (U. S. 1610747) of—
Luminous lacquers, luminous paints, luminous varnish.

Sodium Xanthate
French: Xanthate sodique, Xanthate de soude.
German: Natriumxanthogenat, Natronxanthogenat, Xanthogensäuresnatrium, Xanthogensäuresnatron.

Metallurgical
Flotation agent in—
Ore concentration processes.

Sodium-Zinc Cyanide

Chemical
Catalyst (Brit. 446411) in—
Halogenating unsaturated hydrocarbons.
Starting point (Brit. 446411) in making—
Catalysts with metal chlorides for halogenating unsaturated hydrocarbons.

Softening Agents

Ceramics
Softening agent in—
Compositions, containing cellulose derivatives, as well as resins, used as coatings for protecting and decorat- ceramic products.

Chemical
Softening agent for—
Cellulose derivatives.

Cosmetics
Softening agent in—
Nail enamels and lacquers containing cellulose derivatives, as well as resins, as a base material.

Electrical
Softening agent in—
Insulating compositions, containing cellulose derivatives, as well as resins, used for covering wire and in making electrical machinery and equipment.

Glass
Softening agent in—
Compositions, containing cellulose derivatives, as well as resins, used in the manufacture of nonscatterable glass and as coatings for protecting and decorating glassware.

Glue and Adhesives
Softening agent in—
Adhesive compositions containing cellulose derivatives, as well as resins.

Leather
Softening agent in—
Compositions, containing cellulose derivatives, as well as resins, used in the manufacture of artificial leathers and as coatings for protecting and decorating leathers and leather goods.

Metallurgical
Softening agent in—
Compositions, containing cellulose derivatives, as well as resins, used as coatings for protecting and decorating metallic articles.

Miscellaneous
Softening agent in—
Coating compositions, containing cellulose derivatives, as well as resins, used for protecting and decorating various products.

Paint and Varnish
Softening agent in—
Paints, varnishes, lacquers, enamels, and dopes containing cellulose derivatives, as well as resins.

Paper
Softening agent in—
Compositions, containing cellulose derivatives, as well as resins, used in the manufacture of coated papers and as coatings for protecting and decorating products made of paper or pulp.

Photographic
Softening agent in making—
Films from cellulose derivatives.

Plastics
Softening agent in making—
Laminated fiber products, molded products, plastics from cellulose derivatives, as well as resins.

Softening Agents (Continued)

Resins
Softening agent for—
Resin-cellulose derivative compositions and solutions.

Rubber
Softening agent in—
Compositions, containing cellulose derivatives, as well as resins, used as coatings for decorating and protecting rubber products.

Stone
Softening agent in—
Compositions, containing cellulose derivatives, as well as resins, used as coatings for decorating and protecting artificial stone and natural stone.

Textile
Softening agent in—
Compositions, containing cellulose derivatives, as well as resins, used in the manufacture of coated fabrics.

Woodworking
Softening agent in—
Compositions, containing cellulose derivatives, as well as resins, used as protective and decorative coatings on woodwork.
Plastic compositions, containing cellulose derivatives, as well as resins, used for many filling and repairing purposes on wood.

Solanin

Chemical
Starting point in making the following derivatives—
Acetate, hydrobromide, hydrochloride, hydroiodide, nitrate, sulphate.

Pharmaceutical
In compounding and dispensing practice.

Solar Oil

French: Huile solaire.
German: Solaroel.

Insecticide
Ingredient (Brit. 269942) of—
Animal and plant insecticides and vermin-destroying compositions, sheep dips.

Leather
Ingredient of—
Finishing compositions, polishes.

Miscellaneous
Ingredient of—
Fuels for internal combustion engines, metal polishes.
As a fuel oil.
As a burning oil.

Paint and Varnish
Ingredient (Brit. 269942) of—
Varnishes.

Soluble Prussian Blue

Synonyms: Ferriferrocyanide of potash, Potassium ferriferrocyanide, Soluble Berlin blue, Soluble blue, Soluble iron "cyanide."
French: Bleu de Berlin soluble, Bleu de prusse soluble, Ferriferrocyanure de potasse, Ferriferrocyanure potassique, Ferriferrocyanure de potassium.
German: Ferriferrocyanwasserstoffsäurepotasche, Ferriferrocyanwasserstoffsäureskalium, Kaliumferriferrocyanid, Kaliumferriferrozyanid, Losliches Berlinerblau, Loslisches preussischblau.

Ink
Pigment in—
Blue ink.

Miscellaneous
Ingredient of—
Stains for anatomical specimens.

Paint and Varnish
Pigment in making—
Paints and stains.

Paper
Pigment in making—
Colored paper (not fast to washing).

Textile
As a dye.

Solvent Naphtha

French: Naphte.
German: Bergoel, Nafta, Steinoel.

Chemical
Solvent in extracting—
Carbazole from crude anthracene.
Phenanthrene from crude anthracene.
Starting point in making—
Xylene.

Construction
Solvent for—
Asphalt, pitches, road tars.
Solvent in—
Dampproofing compositions.
Waterproofing compositions.

Explosives and Matches
Starting point in making—
Nitrated naphtha for incorporation with dynamite compositions.

Fats and Oils
Solvent and extractive agent for—
Essential oils, fats, vegetable oils.

Fuel
As an illuminant.
Softening agent for—
Bituminous materials used in making briquetted fuels.

Glass
Solvent and softener for—
Asphalt in glass-etching.

Ink
Binder phase (U. S. 1906961) in—
Emulsified inks.
Ingredient of—
Printing inks.

Mechanical
Solvent for—
Bituminous materials used in impregnating belting.

Metallurgical
Solvent (U. S. 1913100) in making—
Hard alloys containing nickel and carbides or borides of tantalum, tungsten, or molybdenum.

Miscellaneous
As a solvent for various substances.
Solvent in—
Automobile polishes, cleansing compositions.
Compositions for waterproofing automobile tops and tarpaulins.
Degreasing compositions, floor polishes, furniture polishes, linoleum polishes, metal polishes, scouring compositions, shoe polishes, wood polishes.
Solvent in impregnating—
Asbestos board, brake linings, other products.

Paint and Varnish
Binder phase (U. S. 1906961) in—
Emulsified paints.
Solvent and vehicle in—
Acidproof paints, acidproof varnishes, bituminous raw materials, black varnishes, dark paints, enamels, japans, roof cements, varnishes, waterproof paints, waterproof varnishes, white paints.
Solvent in making—
Drier compositions composed of metallic drier, beta-naphthol, chlorphenol or phenol (Brit. 391093).
Varnish containing resinous product formed by heating a phenol-acetaldehyde condensation product, or the components thereof, with a fatty oil, such as linseed or tung oil (Brit. 392226).

Paper
Solvent for—
Bituminous materials used in impregnating roofing papers, insulating papers, building papers.

Petroleum
Ingredient (U. S. 1905087) of—
Mixture with methanol, acetone, and benzene, used for reactivating spent decolorizing clays.

Plastics
Solvent and softener for—
Bituminous materials.

Printing
Solvent in—
Lithography, process engraving.

Resins and Waxes
Raw material (Brit. 394000) in making—
Synthetic resin suitable for use in final coating of leather in producing a patent leather finish.
Solvent for—
Coumarone resins.

Solvent Naphtha (Continued)

Rubber
Ingredient of—
Rubber cements.
Solvent for—
Rubber.

Soap
Ingredient of—
Special soaps.

Textile
As a dry-cleaning agent.

Solvents

This covers applications for those solvents used principally in paints, varnishes, lacquers and various coating and decorative uses. For those solvents having a wider commercial application the uses are given in full under the product heading. The applications for "Diluents" are similar to those given below.

Ceramics
Solvent in—
Compositions containing cellulose acetate, nitrocellulose, or other esters or ethers of cellulose, as well as natural or artificial resins, used for protecting and decorating ceramic ware.

Chemical
As a general solvent.

Electrical
Solvent in—
Compositions containing cellulose acetate, nitrocellulose, or other esters or ethers of cellulose, as well as natural or artificial resins, used for insulating wire and in making electrical apparatus and equipment.

Glass
Solvent in—
Compositions containing cellulose acetate, nitrocellulose, or other esters or ethers of cellulose, as well as natural or artificial resins, used for the manufacture of nonscatterable glass and for decorating glassware.

Glues and Adhesives
Solvent in—
Adhesive preparations containing cellulose acetate, nitrocellulose, or other esters or ethers of cellulose, as well as natural or artificial resins.

Leather
Solvent in—
Compositions containing cellulose acetate, nitrocellulose, or other esters or ethers of cellulose, as well as natural or artificial resins, used in the manufacture of artificial leathers and also for decorating and coating leather goods.

Metallurgical
Solvent in—
Compositions containing cellulose acetate, nitrocellulose, or other esters or ethers of cellulose, as well as natural or artificial resins, used for coating metals and metal products.

Miscellaneous
Solvent in—
Compositions containing cellulose acetate, nitrocellulose, or other esters or ethers of cellulose, as well as natural or artificial resins, used for insulating purposes and also for coating and decorating various products.

Paint and Varnish
Solvent in making—
Lacquers, paints, varnishes, dopes, and enamels, containing nitrocellulose, cellulose acetate, or other esters or ethers of cellulose, as well as artificial or natural resins.

Paper
Solvent in—
Compositions containing cellulose acetate, nitrocellulose, or other esters or ethers of cellulose, as well as natural or artificial resins, used in the manufacture of coated papers and also in coating and decorating pulp and paper products.

Plastics
Solvent in making—
Compositions containing cellulose acetate, nitrocellulose, or other esters or ethers of cellulose, as well as natural or artificial resins.

Rubber
Solvent in—
Compositions containing cellulose acetate, nitrocellulose, or other esters or ethers of cellulose, as well as natural or artificial resins, used for coating and decorating rubber merchandise.

Stone
Solvent in—
Compositions containing cellulose acetate, nitrocellulose, or other esters or ethers of cellulose, as well as natural or artificial resins, used for protecting and decorating natural and artificial stone.

Textile
Solvent in—
Coating compositions containing cellulose acetate, nitrocellulose, or other esters or ethers of cellulose, as well as natural or artificial resins.

Waxes and Resins
Solvent for—
Artificial resins, natural resins.

Woodworking
Solvent in—
Compositions containing cellulose acetate, nitrocellulose, or other esters or ethers of cellulose, as well as natural or artificial resins, used in coating and decorating wood products.

Sorbitol

Synonyms: d-Sorbitol, Sorbite.

*"Humectant" is a term denoting affinity for water, with stabilizing action on the water content of an article; thus, a humectant keeps within a narrow range the moisture content fluctuations caused by wide-range humidity fluctuations.

Adhesives
Flexibilizer, humectant,* and plasticizer in—
Animal adhesives, dextrin adhesives, envelope adhesives, gum-base adhesives, label adhesives, library adhesives, office adhesives, stamp adhesives, starch pastes, vegetable adhesives.

Antifreeze
Ingredient (U. S. 1900040) of—
Antifreeze admixtures with propyleneglycol.

Cellulose Products
Flexibilizer for—
Cellulose products, sheet regenerated cellulose.
Humectant for—
Cellulose products, sheet regenerated cellulose.
Inhibitor of—
Loss of flexibility by cellulosic wrapping material when subjected to high heat conditions.
Reducer of—
Shrinkage of cellulosic products caused by drying out.
Stretching of cellulosic products caused by loss of tensile strength at high moisture contents.
Softener of—
Cellulosic sanitary tissues.

Chemical
Conditioning agent, or humectant, in—
Chemical products and processes.
Starting point in making—
Acetals, anhydrous products.
Anhydrous products having many uses (U. S. 1757468).
1-Ascorbic acid (vitamin C), esters, ethers.
Ethers having many uses (Brit. 317770).
Ethyleneglycol (U. S. 2004135), glycerin (U. S. 2004135), mixed acetalesters, mixed etheresters.
Pentaethylsorbite, useful as a solvent and plasticizer (German 510423).
Plasticizers for benzylcellulose with ethyl chloride, or crotyl chloride, or benzyl chloride (U. S. 1936093).
Plasticizers for cellulose acetate with ethyl chloride, or crotyl chloride, or benzyl chloride (U. S. 1936093).
Plasticizers for ethylcellulose with ethyl chloride, or crotyl chloride, or benzyl chloride (U. S. 1936093).
Plasticizers for nitrocellulose with polyvalent alcohols and polyoses (French 652383).
Plasticizers for nitrocellulose, with ethyl chloride, or crotyl chloride, or benzyl chloride (U. S. 1936093).
Propyl alcohol (German 524101).
Propyleneglycol.
1:2-Propyleneglycol (U. S. 1963997, German 524101).
1:3-Propyleneglycol (German 524101).
Softening agents for cellulosic products with 2-methylcyclohexanone (Brit. 385139).
Solvents for nitrocellulose (perfect), with polyvalent alcohols and polyoses (French 652383).
Sorbose.

Cosmetic
Humectant in—
Creams and other cosmetics.
Skin-softening agent (Brit. 294130) in—
Creams and other cosmetics.

Sorbitol (Continued)
Starting point (Brit. 294130) in making—
Dehydration products and their derivatives, valuable as humectants and softening agents in creams and other cosmetics.
Explosives and Matches
Ingredient of—
Explosives containing also nitrated mixtures of carbohydrates, polyvalent alcohols containing 4 to 6 carbon atoms, and liquid polyvalent alcohols (U. S. 1750949).
Explosives designed to replace dynamite, possessing very great stability and capable of being stored for long periods of time (U. S. 1751437).
Nitrated explosive (U. S. 1751438).
Electrical
Humectant in—
"Dry-type" electrolytic condensers.
Glue and Gelatin
Flexibilizer for—
Gelatin, glue.
Humectant for—
Gelatin, glue.
Plasticizer for—
Gelatin, glue.
Gums
Flexibilizer for—
Gums.
Humectant for—
Gums.
Plasticizer for—
Gums.
Ink
Ingredient (U. S. 1757915) of—
Copying inks.
Copying ink containing also methyl violet B extra and dextrin.
Hectographic inks.
Hectographic ink containing also methyl violet blue, acetic acid, and dextrin.
Rotogravure inks.
Stamppad ink containing also glycerin, methyl violet blue N, and alcohol.
Stamppad ink containing also methyl violet B extra, and alcohol.
Textile inks.
Leather
Flexibilizer for—
Leather.
Humectant for—
Leather.
Ingredient of—
Leather dressings.
Increaser of—
Pliability of leather, softness of leather, tensile strength of leather, tearing resistance of leather.
Process material (U. S. 2063337) in—
Finishing leathers.
Substitute for—
Glycerin in various processes.
Lubricant
Starting point (French 703792) in making—
Viscous lubricating oils with higher monobasic fatty acids.
Miscellaneous
Adjunct for glycerin in various processes and products.
Adjunct for lower alcohols in various processes and products.
Conditioning agent in various processes and products.
Flexibilizer in various processes and products.
Flexibilizer and promoter of glue gel-strength in—
Cork-binding, gasket papers.
Humectant in—
Polishes of various sorts, shoe polishes.
Various processes and products.
Starting point (French 703792) in making—
Polishing composition ingredients by condensing with white lignite wax.
Softening agents with higher monobasic fatty acids.
Substitute for glycerin in various processes and products.
Substitute for lower alcohols in various processes and products.
Paint and Varnish
Starting point in making—
Anticorrosive paint materials possessing great elasticity and stability, with linoleic acid, rosin, boric acid, and benzoyl peroxide (German 529483).
Drying oil substitutes by condensing with higher monobasic fatty acids (French 703792).
Oil characterized by faster drying properties than those of linseed oil and solubility in all organic solvents excepting alcohol (French 703792).
Paper
Humectant for—
Food-wrapping papers (claimed to offer as advantages: (1) Lack of odor, color, and other undesirable characteristics; (2) treated paper is not "tacky") (U. S. 1731679).
Paper products, parchmentized papers, wrapping papers.
Ingredient of—
Paper-treating admixtures with glycerin.
Paper-treating admixtures with glycol.
Paper-treating admixtures with sodium lactate.
Paper-treating admixtures with glycerin, calcium, chloride, calcium acetate, and sodium chloride.
Pharmaceutical
In compounding and dispensing practice.
Suggested for use as—
Diuretic, mild laxative, sugar substitute for diabetics.
Plastics
Starting point (French 703792) in making—
Plastics with higher monobasic fatty acids.
Printing
Flexibilizer for—
Printers' rollers.
Humectant for—
Printers' rollers.
Increaser of—
Body of printers' rollers.
Resistance of rollers to action of organic solvents used in modern high-speed inks.
Resistance of rollers to undue moisture pick-up at high temperatures and humidities, without sacrifice of desirable surface softness at low humidities.
Toughness of printers' rollers.
Resins
Hardening agent for—
Resins (French 664455).
Plasticizer for—
Resins (French 664455).
Starting point in making—
Air-drying alkyd type resins, alkyd type resins, ester-gum type resins.
Esters and ethers useful as hardening agents and plasticizers for phenolic resins (French 664455).
Esters and ethers useful as plasticizers for resins (U. S. 1936093).
Esters useful as plasticizers for resins (German 510423).
Modified alkyd type resins.
Resins with abietic acid, characterized by great hardness and high melting point (German 500504).
Resins with anhydrides of citric acid, characterized by great hardness and high melting point (German 500504).
Resins with anhydrides of maleic acid, characterized by great hardness and high melting point (German 500504).
Resins with anhydrides of phthalic acid, characterized by great hardness and high melting point (German 500504).
Resins with anhydrides of polybasic carboxylic acids, characterized by great hardness and high melting point (German 500504).
Resins with anhydrides of succinic acid, characterized by great hardness and high melting point (German 500504).
Resins with citric acid, characterized by great hardness and high melting point (German 500504).
Resins with maleic acid, characterized by great hardness and high melting point (German 500504).
Resins with phthalic acid, characterized by great hardness and high melting point (German 500504).
Resins with polybasic carboxylic acids, characterized by great hardness and high melting point (German 500504).
Resins with succinic acid, characterized by great hardness and high melting point (German 500504).
Rubber
Humectant in—
Rubber products.
Soap
Softening agent and humectant (Brit. 294130) in—
Toilet soaps.

Sorbitol (Continued)

Textile
Adjunct (U. S. 1955766) to—
Sulphonated oils used in textile processes.
Flexibilizer for—
Textile fabrics, textile sizes.
Humectant for—
Textile fabrics, textile sizes.

Tobacco
Humectant for—
Tobacco, tobacco papers.

Wood
Plasticizing and softening agent in—
Veneering processes.
Reducer of shrinkage and expansion phenomena in—
Veneering processes.

Soybean Lecithin
French: Léchithine d'huile de soja, Lécithine d'huile de soya.
German: Soyabohnenoelecithin.

Chemical
Reagent in purifying—
Pepsin.
Reagent in making—
Arsenic compounds of bromo- and iodolecithin.
Biocithin.
Copper compounds of bromo- and iodolecithin.
Glycocithin.
Iron compounds of bromo- and iodolecithin.
Mercury compounds of bromo- and iodolecithin.
Regenerin.
Starting point in making—
Hydrolecithin, lecithol, ovalecithin, phospholecithin.
Various derivatives obtained by halogenation.

Food
Ingredient of—
Food compositions, margarins.

Pharmaceutical
In compounding and dispensing practice.
In veterinary practice.

Textile
——, *Dyeing*
Emulsifier and softener in—
Dye baths (added to produce more level shades on yarns and fabrics).
——, *Finishing*
Emulsifier and softener in—
Compositions used for finishing fabrics and yarns.
Cotton-finishing compositions used to remove stiffening effect produced during drying.
——, *Manufacturing*
Emulsifier in—
Compositions used for softening filaments and fibers and rendering them more pliable.
Cotton-spinning oils, wet doubling solutions, wool-spinning oils.
——, *Printing*
Emulsifier in—
Printing pastes (used to make them thicker for the printing of calicoes).

Soybean Meal
Synonyms: Chinese bean meal, Soja bean meal, Soy oil meal, Soy bean flour, Soya bean flour.
French: Farine de fèves de soya, Farine de pois de soya.
German: Sojabohnmähl.

Agriculture
As a cattle feed.
Ingredient of—
Cattle feeds.

Fertilizer
Ingredient of—
Compounded fertilizers.

Food
Ingredient of—
Breakfast foods, diabetic foods, flours, infant foods, macaroni pastes.

Glues and Adhesives
Ingredient of—
Glues.

Miscellaneous
Ingredient of—
Laundry starches.

Paper
Reagent in—
Sizing paper and paper products.
Waterproofing paper and paper products.

Textile
——, *Finishing*
Ingredient of—
Sizing compositions, waterproofing compositions.

Soybean Oil
Synonyms: Bean oil, Chinese bean oil, Sojabean oil, Soj oil.
French: Huile de soja, Huile de soya.
German: Chinesisiches bohnenfett.
Spanish: Aceite de soja hispida.

Agricultural
Ingredient of—
Cattle foods.

Cement
Ingredient of—
Waterproofed cements.

Chemical
Starting point in making—
Glycerin, soybean lecithin.

Electrical
Ingredient (Brit. 273290) of—
Insulating enamels used for painting electrical wires and parts of electrical machinery.

Explosives
Ingredient of—
Explosive compositions.

Food
As a general food.
As a salad oil.
Ingredient of—
Compounded products used in the place of vegetable oil and animal fats.
Lard substitutes, margarins.
Starting point in making—
Artificial lard and margarin by hydrogenation.

Fuel
As a fuel oil.
As an illuminant.
Starting point in making—
Candles.

Glues and Adhesives
Ingredient (Brit. 273290) of—
Adhesive compositions used for the manufacture of laminated mica and other special products.

Ink
Ingredient of—
Printing inks.

Linoleum and Oilcloth
Raw material in making—
Coating compositions.

Mechanical
As a lubricant for special purposes.

Metallurgical
Binding agent in making—
Cores, consisting of refractory materials of various sorts, for use in making castings in foundries.

Miscellaneous
Waterproofing agent for various compositions of matter.

Paint and Varnish
Ingredient of—
Enamels, oil lacquers, paints, varnishes.
Ingredient of—
Paint and varnish vehicles (used in conjunction with polymerized linseed oil, polymerized perilla oil, and chinawood oil).
Starting point in making—
Varnish bases.
Substitute for linseed oil in making—
Paints and varnishes (used in the heated state, and with the addition of driers, such as the cobaltous salts of chinawood oil fatty acids).

Petroleum
Ingredient of—
Axle greases, lubricating compositions.
Starting point in making—
Artificial petroleum.

Plastics
Ingredient of—
Celluloid compositions.
Mouldable compositions (Brit. 273290).

Soybean Oil (Continued)
Rubber
Ingredient of—
Rubber substitutes.
Soap
Raw material in making—
Hard and soft soaps.
Textile
As a waterproofing agent.

Soybeans
Synonyms: Chinese beans, Soja beans, Soy beans, Soya beans.
French: Fèves de soya, Pois de soya.
German: Sojabohne.
Agriculture
Food for—
Cattle.
Food
Ingredient of—
Casein products, cheeses, condensed milk, confections, meat substitutes, milk powders, soup stocks.
Raw material in making—
Baked beans, canned beans (green), coffee substitutes, roasted beans, soy sauce.
Fats and Oils
Source of—
Soybean oil.

Sparteine
Chemical
Starting point in making—
Spartein salts with acids and halogens.
Pharmaceutical
In compounding and dispensing practice.

Sperm Oil
Synonyms: Spermaceti oil.
French: Huile de blanc de baleine, Huile de cacholot.
German: Pottfischtran, Pottwaltran, Spermacetioel, Walratoel.
Spanish: Aceite de balena.
Italian: Olio di spermaceti.
Explosives
Ingredient of—
Compositions used in making matches.
Fats and Oils
Starting point in making—
Hydrogenated hardened fats, lubricating compounds, stearin.
Food
Ingredient of—
Lard substitutes, oleomargarin.
Fuel
As a burning oil for heat and illumination.
Starting point in—
Making candles.
Ink
Ingredient of—
Marking inks, printing inks.
Insecticide
Vehicle in making—
Plant insecticides.
Leather
Ingredient of—
Currying compositions, dressing preparations.
Oiling compositions for treating special grades of leather.
Mechanical
Lubricating agent (used alone or in mixtures) for—
Clocks, light machinery, screw-cutting machines, spindles.
Metallurgical
Ingredient of—
Steel-tempering bath.
Resins and Waxes
Starting point in making—
Spermaceti.
Soap
Starting point in making—
Soft soaps, special soaps.
Textile
For oiling and softening hemp, jute, and other fibers in preparing them for spinning and weaving.

Sperm Oil Alcohol Boric Ester
Fats, Oils, and Waxes
Starting point (Brit. 448668) in making—
Emulsifying agents for fats, oils, and waxes by condensing, in the presence of a sulphonating agent, with boric acid esters of the cholesterols of woolfat and neutralizing the products.

S-Phenylisothiourea Hydrochloride
French: Chlorhydrate de S-phénylisothiourée, Hydrochlorure de S-phénylisothiourée.
German: Chlorwasserstoffsaeure-S-phenylisoharnstoffester, Chlorwasserstoffsaeures-S-phenylisoharnstoff, S-phenylisothioharnstoffchlorhydrat.
Chemical
Starting point (Brit. 262155) in making therapeutic compounds with—
Anilin, benzylamine, diphenylamine, meta-anisidin, metaphenylenediamine, metatoluidin, metaxylidin, monoethylanilin, monomethylanilin, naphthylamine, orthoanisidin, orthophenylenediamine, orthotoluidin, orthoxylidin, para-anisidin, paraphenylenediamine, paratoluidin, paraxylidin, phenylamine.

Spodumene
Synonyms: Hard spodumene.
Chemical
Starting point in making—
Lithium salts.

S-Propylisothiourea Hydrochloride
French: Chlorhydrate de S-propylisothiourée, Hydrochlorure de S-propylisothiourée.
German: Chlorwasserstoffsaeure-S-propylthioharnstoffester, Chlorwasserstoffsaeures-S-propylthioharnstoff, S-Propylisothioharnstoffchlorhydrat, Salzsaeures-S-propylisothioharnstoff.
Chemical
Starting point (Brit. 262155) in making therapeutic agents with—
Anilin, benzylamine, diphenylamine, meta-anisidin, metaphenylenediamine, metatoluidin, metaxylidin, monoethylanilin, monomethylanilin, orthoanisidin, orthophenylenediamine, orthotoluidin, orthoxylidin, para-anisidin, paraphenylenediamine, paratoluidin, paraxylidin, phenylamine.

Squill
Synonyms: Sea onion.
Latin: Bulbus scillae, Scilla.
French: Oignon marin, Scille.
German: Mauszwiebel, Meerzwiebel.
Spanish: Bulbo de escila, Cebolla albarrana.
Italian: Scilla.
Insecticide
Ingredient of—
Rat poisons.
Pharmaceutical
In compounding and dispensing practice.

Stand Oil
Synonyms: Dutch boiled linseed oil, Dutch enamel oil.
German: Standoel.
Abrasives
Ingredient (Brit. 295335) of—
Compositions used in binding abrasive materials in the manufacture of grinding discs, abrasive stones, and similar articles.
Chemical
Ingredient (Brit. 295335) of—
Impregnating compositions containing phenol-aldehyde condensation products.
Ink
Ingredient of—
Lithographic inks.
Linoleum and Oilcloth
Ingredient of—
Coating compositions.
Miscellaneous
Ingredient of—
Compositions used in making artificial wood.
Impregnating compositions containing phenol-aldehyde condensation products (Brit. 295335).
Substitute for brewers' pitch, in admixture with rosin and paraffin (German 203795).

Stand Oil (Continued)

Paint and Varnish
Ingredient of—
Enamels and varnishes.
Lacquers containing phenol-aldehyde condensation products (Brit. 295335).

Plastics
Ingredient of—
Compositions containing phenolaldehyde condensation products (Brit. 295335).

Stannic Bromide
Synonyms: Tin bromide.
French: Bromure d'étain, Bromure stannique.
German: Stannibromid, Zinnbromid.

Analysis
Reagent for—
Separating mineral from gangue.

Stannous Acetate
Synonyms: Acetate of tin.
Latin: Stannumaceticum.
French: Acétate d'étain, Acétate stanneux.
German: Essigsäureszinn, Essigsäureszinnoxydul, Stannoacetat, Stannoazetat, Zinnacetat, Zinnazetat, Zinnbeize.

Chemical
Starting point in making—
Tin salts.
Ingredient of catalytic mixtures used in the manufacture of—
Acenaphthylene, acenaphthaquinone, bisacenaphthylidenedione, naphthaldehydic acid, naphthalic anhydride, and hemimellitic acid from acenaphthene (Brit. 295270).
Acetaldehyde from ethyl alcohol (Brit. 281307).
Acetic acid from ethyl alcohol (Brit. 281307).
Alcohols from aliphatic hydrocarbons (Brit. 281307).
Aldehydes or alcohols by the reduction of the corresponding esters (Brit. 306471).
Alphacampholid by the reduction of camphoric acid (Brit. 306471).
Aldehydes and acids from toluene, orthochlorotoluene, orthonitrotoluene, orthobromotoluene, parachlorotoluene, paranitrotoluene, parabromotoluene, metachlorotoluene, metanitrotoluene, metabromotoluene, dichlorotoluenes, dibromotoluenes, dinitrotoluenes, chloronitrotoluenes, chlorobromotoluenes, nitrobromotoluenes (Brit. 295270).
Aldehydes and acids from xylenes, pseudocumene, mesitylene, and paracymene (Brit. 281307).
Alphanaphthaquinone from naphthalene (Brit. 281307).
Anthraquinone from naphthalene (Brit. 295270).
Benzaldehyde and benzoic acid from toluene (Brit. 281307).
Benzoquinone from phenanthraquinone (Brit. 281307).
Benzyl alcohol by the reduction of benzaldehyde (Brit. 306471).
Benzyl alcohol or benzaldehyde or phthalide by the reduction of phthalic anhydride (Brit. 306471).
Butyl alcohol by the reduction of crotonaldehyde (Brit. 306471).
Chloroacetic acid from ethylenechlorohydrin (Brit. 295270).
Diphenic acid from ethyl alcohol (Brit. 281307).
Ethyl alcohol by the reduction of acetaldehyde (Brit. 306471).
Fluorenone from fluorene (Brit. 295270).
Formaldehyde by the reduction of methane or methanol (Brit. 306471).
Formaldehyde by the reduction of carbon monoxide or carbon dioxide (Brit. 306471).
Hydroxyl compounds by the reduction of anthraquinone, benzoquinone, and similar compounds (Brit. 306471).
Isopropyl alcohol by the reduction of acetone (Brit. 306471).
Maleic acid and fumaric acid by the oxidation of toluene, benzene, phenols, tar phenols, or furfural, or from benzoquinone or phthalic anhydride (Brit. 295270).
Methane by the reduction of carbon dioxide or carbon monoxide (Brit. 306471).
Methanol by the reduction of carbon dioxide or carbon monoxide (Brit. 306471).
Naphthaldehydic acid, acenaphthaquinone, or bisacenaphthylidenedione from acenaphthylene (Brit. 281307).
Phenanthraquinone from phenanthrene or diphenic acid (Brit. 295270).
Phthalic acid and maleic acid from naphthalene (Brit. 295270).
Primary alcohols by the reduction of the corresponding aldehydes (Brit. 306471).
Propionic acid and butyric acid and higher alcohols, ketones and acids by the reduction of carbon monoxide and carbon dioxide (Brit. 306471).
Reduction products of ketones, aldehydes, acids, esters, alcohols, ethers, and other organic compounds containing oxygen (Brit. 306471).
Salicylic acid and salicylic aldehyde from cresol (Brit. 295270).
Secondary butyl alcohol by the reduction of methylethyl ketone (Brit. 306471).
Valeryl alcohol by the reduction of valeraldehyde (Brit. 306471).
Vanillin and vanillic acid from eugenol or isoeugenol (Brit. 295270).
Ingredient (Brit. 306460) of catalytic preparations used in the production of various aromatic and aliphatic compounds, including—
Alphanaphthylamine from alphanitronaphthalene.
Amines from aliphatic nitro compounds, such as allyl nitriles or nitromethane.
Amylamine from pyridin.
Anilin, azo-oxybenzene, azobenzene, and hydrazobenzene from nitrobenzene by reduction.
Aminophenols from nitrophenols.
3-Aminopyridin from 3-nitropyridin.
Amino compound from the corresponding nitroanisole.
Amines from oximes, Schiff's base, and nitriles.
Cyclohexamine, dicyclohexamine, and cyclohexylanilin from nitrobenzene.
Piperidin from pyridin.
Pyrrolidin from pyrrol.
Tetrahydroquinolin from quinolin.

Textile
——, *Dyeing*
Discharge in dyeing—
Cotton yarns and fabrics with substantive colors.
Mordant in dyeing—
Yarns and fabrics.

——, *Printing*
Discharge in printing—
Calicoes.

Starches. See under name of particular starch; e.g., Potato Starch, Corn Starch, etc.

Stavesacre Seed
Synonyms: Louseseed, Stavesaire seeds.
French: Graines de capuchin, Semences de staphisaigre.
German: Laeusepfeffer, Laeusekoerner, Rattenpfeffer, Stephankoerner.

Chemical
Starting point in making—
Delphinine.

Insecticide
Ingredient of—
Pice powders, rat killers.

Pharmaceutical
In compounding and dispensing practice.

Stearamidin Hydrochloride

Textile
Dispersing agent (Brit. 446976) in making—
Waterproof and crease-resisting finishes on natural and synthetic fibers (used in conjunction with sulphonated fats, albuminous derivatives, and formaldehyde or a substance yielding it).
Delustring agent (Brit. 446976) for—
Natural and synthetic fibers.
Wetting agent (Brit. 446976) in making—
Waterproof and crease-resisting finishes on natural and synthetic fibers (used in conjunction with sulphonated fats, albuminous derivatives, and formaldehyde, or a substance yielding it).

Stearic Acid
Synonyms: Cetylacetic acid, Stearinic acid, Stearophanic acid.
French: Acide de stéarique.
German: Stearinsaeure, Talgsaeure.

Chemical
Ingredient (Brit. 303379) of—
Preparations used for various wetting and cleansing purposes.

Stearic Acid (Continued)
Reagent in making various stearates.
Starting point in making—
Creosote stearate, guaiacol stearate, myristic acid, stearates of metallic and alkaline bases.
Dye
Ingredient of—
Oil-soluble colors.
Electrical
Ingredient of—
Non-conducting compositions for insulating.
Reagent in—
Galvanoplastic work for making molds.
Explosives
Ingredient of—
Pyrotechnic compositions, wax matches.
Fats and Oils
Ingredient of—
Lubricating greases, textile and boring oils.
Starting point in making—
Emulsions of fats and oils with aromatic and aliphatic alcohols (Brit. 266746).
Food
Ingredient of—
Various products.
Fuel
Ingredient of—
Candles.
Ink
Ingredient of—
Lithographic inks, printing inks, writing inks.
Leather
Ingredient of—
Tanning and dressing compositions.
Washing and cleansing compositions with aliphatic or aromatic alcohols (Brit. 266746).
Metallurgical
Ingredient of—
Buffing compositions, metal cleansing and polishing compositions.
Miscellaneous
Ingredient of—
Bone-dyeing compositions (U. S. 1594498).
Coatings for enteric medicaments.
Cloth-marking compositions (U. S. 1622353).
Crayons, phonograph records, shoe polishes, waterproofing compositions, waxed pencil leads.
Paint and Varnish
Ingredient of—
Ships' bottoms paints, wax color binding compositions.
Paper
Ingredient of—
Finishing compositions.
Perfumery
Ingredient of—
Cosmetics, greaseless creams and lotions.
Pharmaceutical
In compounding and dispensing practice.
Resins and Waxes
Ingredient of—
Beeswax compounds.
Starting point in making—
Resin and wax emulsions with aromatic and aliphatic alcohols (Brit. 266746).
Rubber
Activator of accelerators in vulcanizing.
Dispersive agent in—
Pigmenting processes.
Vulcanization assist in making—
Rubber heels.
Soap
Ingredient (Brit. 303379) of—
Saponaceous cleansing compositions and detergent preparations.
Raw material in making—
Shaving soaps, textile soaps.
Textile
——, *Bleaching*
Reagent in making—
Cotton-bleaching mixtures.
——, *Finishing*
Ingredient (Brit. 303379) of—
Bowking, softening, oiling, and finishing preparations.
Emulsified finishing and dressing compositions.
——, *Manufacturing*
Ingredient of—
Precipitating bath or coagulating bath in the manufacture of viscose rayon (Austrian 102148).

Stearic Acid Chloride
Chemical
Starting point (Brit. 407956) in making pour-point improvers for machine oils, gear oils, and other lubricants by condensing with—
Anilin, anthracene oil.
Aromatics obtained by destructive hydrogenation or by dehydrogenation.
Benzene.
Cracking gases containing gaseous olefins (ethylene, propylene, and butylene).
Cyclic terpenes, ethylnaphthalene, liquid olefins, middle oil, naphthalene, naphthols, naphthylamines, nitrated aromatics, phenols, tars, toluene, xylene.

Stearic Amide
French: Amide stéarique.
German: Stearinamid.
Fats and Oils
Emulsifying agent (Brit. 328675) in making emulsions of—
Fats, fatty acids, oils.
Petroleum
Emulsifying agent (Brit. 328675) in making—
Emulsions of mineral oils and mineral oil distillates.
Soap
Emulsifying agent (Brit. 328675) in making—
Emulsified detergents.
Textile
——, *Bleaching*
Ingredient (Brit. 328675) of—
Bleaching compositions.
——, *Finishing*
Ingredient (Brit. 328675) of—
Emulsified preparations used in washing, fulling, and finishing textiles.
Waxes and Resins
Emulsifying agent (Brit. 328675) in making—
Emulsions of waxes and resins.

7:18-Stearicglycol
Chemical
Starting point (Brit. 388485) in making—
Emulsifying agents by treating the unsaturated alcohols, which are produced by dehydrating, with a sulphonating agent, phosphoric acid or its anhydride or oxyhalide, removal of the water being effected by heating to 50 to 200° C in the presence of a strong nonoxidizing acid; for example, an organic sulphonic acid, such as naphthalenesulphonic acids, phosphoric acids, sulphuric acid, chloroacetic acid.
Soap
Starting point (Brit. 388485) in making—
Cleansing agents by treating the unsaturated alcohols, which are produced by dehydrating, with a sulphonating agent, phosphoric acid or its anhydride or oxyhalide, removal of the water being effected by heating to 50 to 200° C in the presence of a strong nonoxidizing acid; for example, an organic sulphonic acid, such as naphthalenesulphonic acids, phosphoric acids, sulphuric acid, chloroacetic acid.

Stearic Toluide
Chemical
Starting point in making various derivatives.
Petroleum
Ingredient (U. S. 1853571) of—
Lubricating compositions, containing mineral oils (added for the purpose of increasing the viscosity of the oil and raising the melting point).

Stearimidobutyl Ether Hydrochloride
Textile
Dispersing agent (Brit. 446976) in making—
Waterproof and crease-resisting finishes on natural and synthetic fibers (used in conjunction with sulphonated fats, albuminous derivatives and formaldehyde or a substance yielding it).
Delustring agent (Brit. 446976) for—
Natural and synthetic fibers.

Stearimidobutyl Ether Hydrochloride (Continued)
Wetting agent (Brit. 446976) in making—
Waterproof and crease-resisting finishes on natural and synthetic fibers (used in conjunction with sulphonated fats, albuminous derivatives and formaldehyde or a substance yielding it).

Stearimidoethyl Ether Hydrochloride
Textile
Dispersing agent (Brit. 446976) in making—
Waterproof and crease-resisting finishes on natural and synthetic fibers (used in conjunction with sulphonated fats, albuminous derivatives and formaldehyde or a substance yielding it).
Delustring agent (Brit. 446976) for—
Natural and synthetic fibers.
Wetting agent (Brit. 446976) in making—
Waterproof and crease-resisting finishes on natural and synthetic fibers (used in conjunction with sulphonated fats, albuminous derivatives and formaldehyde or a substance yielding it).

Stearin
Synonyms: Cotton stearin, Glyceryl stearic ester, Tristearin, Wool stearin.
French: Ester glycéryle-stéarique.
German: Stearin, Tristearin.
Spanish: Estearica.
Italian: Stearina.
Chemical
Ingredient (Brit. 303379) of—
Preparations used for various wetting and cleansing purposes.
Reagent in making—
Iodated pharmaceutical preparations (Brit. 310869).
Starting point in making—
Aluminum stearate, bismuth stearate, creosote stearate, guaiacol stearate, manganese stearate, myristic acid, sodium stearate, various metallic and alkaline base stearates, zinc stearate.
Construction
Ingredient of—
Compositions used for impregnating and waterproofing artificial and natural stone structures.
Compositions containing asphalt in emulsified form used for the waterproofing and impregnation of concrete.
Dye
Ingredient of—
Oil-soluble dyestuffs.
Electrical
Ingredient of—
Nonconducting compositions used for the insulation of wiring and cables and in the manufacture of electrical equipment and machinery.
Reagent in—
Galvanoplastic work for making molds.
Explosives
Ingredient of—
Pyrotechnic compositions, wax matches.
Fats and Oils
Ingredient of—
Boring oils, lubricating greases, textile oils.
Starting point in making—
Emulsions of fats and oils with aromatic and aliphatic alcohols (Brit. 266746).
Food
Ingredient of various food products.
Fuel
Ingredient of—
Compositions containing paraffin used for making candles (added for the purpose of giving the requisite "snap").
Solid combustible compositions (French 486557).
Raw material for—
Making candles.
Glues and Adhesives
Ingredient of—
Starch paste adhesives.
Ink
Ingredient of—
Compositions containing indulin base, used as indelible marking inks for laundry marking and similar purposes (French 579568).
Ingredient of—
Lithographic inks, printing inks, writing inks.
Leather
Ingredient of—
Dressing compositions containing carnauba wax in emulsified form.
Polishing compositions, stuffing compositions, tanning compositions.
Washing and cleansing compositions containing aliphatic and aromatic alcohols (Brit. 266746).
Linoleum and Oilcloth
Ingredient of—
Compositions containing carnauba wax and the like in emulsified form, used for finishing linoleum.
Mechanical
Ingredient of—
Lubricating compositions.
Metallurgical
Flux in—
Compositions containing mercuric chloride, sodium borate, oxalic acid, sodium bicarbonate, potassium bicarbonate, anhydrous phosphoric acid, potassium carbonate, and glycerin, used for soldering aluminum bronze (French 574392, addition 29095).
Soldering pure aluminum (French 524817).
Ingredient of—
Buffing compositions, cleansing and polishing compositions.
Miscellaneous
Ingredient of—
Bone-dyeing compositions (U. S. 1594498).
Cloth-marking compositions (U. S. 1622353).
Coatings for enteric medicaments.
Compositions containing asphalt and the like in emulsified form, used for the surfacing of roads.
Crayons.
Furniture polishes containing paraffin in emulsified form.
Phonograph records, shoe polishes, waterproofing compositions, waxed pencil "leads."
Paint and Varnish
Ingredient of—
Automobile polishes containing carnauba wax and the like in emulsified form.
Automobile top dressings containing asphalt and the like in emulsified form.
Asphalt paints and varnishes, ships' bottoms paints, waterproofing compositions, wax color binding compositions.
Paper
Ingredient of—
Compositions used in treating cigaret paper so that it is not discolored by the tobacco (Brit. 322149).
Compositions containing rosin and paraffin, used for making treated wrapping paper for food products (French 599870).
Impregnating compositions containing paraffin.
Waterproofing compositions containing paraffin.
Perfumery
Ingredient of—
Cosmetic creams containing paraffin and the like in emulsified form.
Creams used for use after shaving, face creams, vanishing creams in emulsified form.
Pharmaceutical
In compounding and dispensing practice.
Plastics
Ingredient of—
Artificial ivory containing calcium sulphate.
Resins and Waxes
Ingredient of—
Beeswax compositions.
Starting point in making—
Resin and wax emulsions with aromatic and aliphatic alcohols (Brit. 266746).
Rubber
Accelerator in—
Vulcanization.
Dispersive agent in—
Pigmenting process, with carbon black for example.
Softener for—
Improving rubber products.
Vulcanization assist in making—
Rubber heels.
Soap
Raw material in making—
Lime soaps, shaving soaps, textile soaps.
Substitute for tallow in making—
Household soaps.

Stearin (Continued)

Textile

——, *Bleaching*

Reagent in making—

Cotton bleaching mixtures.

——, *Finishing*

Ingredient of—

Bowking, softening, oiling, and finishing preparations (Brit. 303379).

Emulsified finishing and dressing compositions.

Impregnating compositions and sizing compositions containing paraffin.

Waterproofing compositions containing paraffin.

——, *Manufacturing*

Ingredient of—

Precipitating or coagulating bath in the manufacture of viscose rayon (Austrian 102148).

Stearin Pitch

Synonyms: Candle pitch, Candle tar, Palm pitch.

French: Brai de chandelle, Brai de palme, Brai de stearine, Goudron de chandelle.

German: Palmpech, Stearinpech.

Building

Ingredient of—

Waterproofing, weatherproofing, and wearproofing compositions used in the treatment of building materials.

Chemical

Starting point in making—

Pitch coke.

Electrical

Ingredient of—

Compositions used for insulating electrical apparatus and parts of electrical equipment.

Compositions used in making electrodes for various electrical machines and furnaces.

Compositions used in the manufacture of cables.

Fuel

As a fuel itself.

Binder in making—

Fuel briquettes.

Ingredient of—

Artificial fuel compositions.

Leather

Ingredient of—

Compositions used in the production of imitation leathers.

Linoleum and Oilcloth

Ingredient of—

Compositions used as substitutes for linoleums and oilcloth.

Mechanical

Ingredient of—

Lubricating compositions.

Metallurgical

Binder in making—

Core compositions for casting purposes.

Miscellaneous

Ingredient of—

Asphaltic compositions and preparations.

Compositions used as binders in various products.

Compositions used as cements.

Compositions used for caulking boats.

Compositions used in coating articles and parts of machines and the like that are subjected to high temperatures.

Compositions used for waterproofing, wearproofing, and weatherproofing felt.

Compositions used for making roofs.

Compositions used for impregnating purposes.

Compositions used for paving streets.

Compositions used for covering pipe.

Fillers.

Weatherproofing, waterproofing, and wearproofing compositions used for a great variety of purposes (Brit. 335247).

Paint and Varnish

Ingredient of—

Black paints, black varnishes, japans.

Preparations for acidproofing various apparatus subjected to acid liquors and fumes.

Preparations for impregnating paper to make roofing paper.

Preparations for waterproofing, wearproofing, and weatherproofing cement, concrete, and building stone.

Roof cements, rustproofing compositions, black enamels, substitutes for linseed oil varnishes, waterproof cements.

Paper

Ingredient of—

Compositions used in the manufacture of heavy papers, such as tarred paper.

Compositions used in waterproofing paper and pulp products.

Printing

Ingredient of—

Compositions used in making printers' rollers.

Rubber

Filler in compounding—

Special rubber goods.

Ingredient of—

Compositions used in the place of rubber, especially hard rubber.

Stone

Ingredient of—

Compositions for waterproofing and weatherproofing natural and artificial stone.

Textile

Ingredient of—

Compositions used in waterproofing, wearproofing, and weatherproofing cotton, wool, and cotton and wool mixtures.

Woodworking

Ingredient (Brit. 335247) of—

Compositions used in waterproofing, wearproofing, and weatherproofing wood.

Stearolic Acid Chloride

Chemical

Starting point (Brit. 407956) in making pour-point improvers for machine oils, gear oils, and other lubricants by condensing with—

Anilin, anthracene oil.

Aromatics obtained by destructive hydrogenation or by dehydrogenation.

Benzene.

Cracking gases containing gaseous olefins (ethylene, propylene, and butylene).

Cyclic terpenes, ethylnaphthalene, liquid olefins, middle oil, naphthalene, naphthols, naphthylamines, nitrated aromatics, phenols, tars, toluene, xylene.

Stearylhydroquinone

Petroleum

Stabilizing agent (Brit. 406195) for—

Cracked gasolines and other motor fuels.

Stearyl Isoselenocyanate

Disinfectant

Parasiticide (U. S. 1993040).

Stearyl Isotellurocyanate

Disinfectant

Parasiticide (U. S. 1993040).

Stearyl Isothiocyanate

Disinfectant

Parasiticide (U. S. 1993040).

Stearylphloroglucinol

Petroleum

Stabilizing agent (Brit. 406195) for—

Cracked gasolines and other motor fuels.

Stearylpyrocatechol

Petroleum

Stabilizing agent (Brit. 406195) for—

Cracked gasolines and other motor fuels.

Stearylpyrogallol

Petroleum

Stabilizing agent (Brit. 406195) for—

Cracked gasolines and other motor fuels.

Stearylresorcinol

Petroleum

Stabilizing agent (Brit. 406195) for—

Cracked gasolines and other motor fuels.

Stearyl Selenocyanate

Disinfectant

Parasiticide (U. S. 1993040).

Stearyl-1-sulphuric Acid (Normal) Ester

Chemical

As an emulsifying agent.

Reagent in—

Organic syntheses.

Starting point (Brit. 440575) in making—

Emulsifying agents with salts of lead, aluminum, iron, tin, or barium (such emulsifying agents are said to form water-in-oil emulsions and are, preferably, produced in situ by (1) dissolving the sulphuric acid ester in the oil and (2) agitating with an aqueous solution of the metal salt, for example, lead acetate; they are said to be useful for treating medicinal paraffin oil, neatsfoot oil, olive oil, castor oil, cottonseed oil, linseed oil, and petroleum lubricating oils; a heavy paraffin oil, so-treated on the basis of 50 parts by weight of oil to 48.75 parts of water, is said to yield a heavy grease that has good lubricating properties and may readily be extended with oil; a water-linseed oil type emulsion is offered as suitable for use as a paint base).

Stearyl Tellurocyanate

Disinfectant

Parasiticide (U. S. 1993040).

Stearyl Thiocyanate

Disinfectant

Parasiticide (U. S. 1993040).

Stibnite

Synonyms: Antimonite, Antimony glance, Antimony ore, Gray antimony.

Metallurgical

Source of—

Antimony metal.

Crude antimony (liquated sulphide).

St. Ignatius Bean

Latin: Faba ignatti, Faba sancti ignatti, Ignatia amara, Semen ignatiae.

French: Fèves de saint ignace, Fèves igasurique.

German: Bittere fiebernuss, Ignatiusbohne, Ignazbohnen.

Spanish: Hoba de santo ignacio.

Italian: Fava di santo ignazio.

Chemical

Starting point in extracting—

Brucine, strychnine.

Pharmaceutical

In compounding and dispensing practice.

Stilbene

Synonyms: Toluylene.

French: Diphényléthylène.

German: Diphenylaethylen.

Chemical

Starting point in making—

Dinitrostilbenesulphonic acid.

Stilbeneorthodisulphonic acid.

Dye

Starting point in making various dyestuffs.

Stilbinphenylazonium Chloride

French: Chlorure de stilbinephényleazonium.

German: Chlorstilbinphenylazonium, Stilbinphenylazoniumchlorid.

Photographic

Reagent (Brit. 315236) in preparing—

Plates, papers, and films.

Printing

Reagent (Brit. 315236) in—

Photomechanical printing.

Stoneware Clay

Ceramics

Raw material in making—

Architectural terra-cotta, art ware, chemical stoneware, earthenware, stoneware, yellow ware.

Refractories

Raw material in making—

Saggers.

Storax

Synonyms: Liquid storax, Oriental sweet gum, Styrax.

Latin: Balsamum storacis, Balsamum styracis, Balsamum styrax liquidus, Styrax, Styrax liquidus.

French: Résine liquide ambar d'orient, Storax liquide, Styrax liquide.

German: Flüssiger storax, Storax, Styrax.

Spanish: Estoraque, Estoraque liquido.

Italian: Storace, Storace liquido.

Food

Ingredient of—

Chewing gum.

Chemical

Starting point in making—

Cinnamic acid, styrol.

Insecticide

Ingredient of—

Insecticidal incenses, parasiticidal lotions.

Miscellaneous

Ingredient of—

Ointments, fumigating and deodorizing pastils.

Reagent in microscopy.

Perfumery

Fixative in—

Cosmetics, perfumes.

Ingredient of—

Cosmetics, incense.

Pharmaceutical

In compounding and dispensing practice.

Soap

Ingredient of—

Medicinal soaps, toilet soaps.

Stramonium Seed

Synonyms: Jimson weed seed, Thorn apple seed.

Latin: Semen stramonii.

French: Graines de datura, Graines de stramoine.

German: Stechapfelsamen.

Chemical

Starting point in extracting—

Atropine, hyoscine, hyoscyamine.

Pharmaceutical

In compounding and dispensing practice.

Strontium

Analysis

Reagent in various processes.

Metallurgical

Ingredient of—

Aluminum, lithium, and copper alloys.

Various alloys for hardening purposes.

Strontium Acetate

French: Acétate de strontiane, Acétate strontique, Acétate de strontium.

German: Essigsäuresstrontian, Essigsäuresstrontianerde, Essigsäuresstrontium, Strontiumacetat, Strontiumazetat.

Spanish: Acetato de estrontio.

Italian: Acetato di stronzio.

Analysis

Reagent in—

Testing for inosite.

Chemical

Starting point in making—

Strontium salts.

Pharmaceutical

Suggested for use as an anthelmintic, tonic, and vermifuge.

Strontium Albuminate

German: Albuminsaeuresstrontium.

Rubber

Reagent (U. S. 160817) in—

Reclaiming rubber.

Strontium-Anilin

Dye

Reagent (German 436533) in making anthracene dyestuffs from—

3:9-Dichlorobenzanthrone, 11:3-dichlorobenzanthrone.

Strontium Bromide

Latin: Strontii bromidum.

French: Bromure de strontium.

German: Strontiumbromid.

Spanish: Bromuro estroncico.

Pharmaceutical

In compounding and dispensing practice.

Suggested for use as—

Nerve sedative.

Suggested for use in treating—

Diabetes, epilepsy, gastric dilation and catarrh.

Strontium Chlorate
French: Chlorate de strontium.
German: Chlorsäuresstrontium.
Spanish: Clorato de estrontiana.
Italian: Clorato di stronzio.

Explosives and Matches
Ingredient of—
Red fire and such-like pyrotechnics.

Strontium Iodate
French: Iodate de strontium.
German: Jodsaeuresstrontium, Strontiumjodat.

Food
Preservative (Brit. 274164) in treating—
Butter, cream, eggs, fish, fruit preserves, margarin, milk, meat.

Strontium Salicylate
Latin: Strontii salicylas, Strontium salicylicum.
French: Salicylate de strontium.
German: Salicylsäuresstrontium, Strontiumsalicylat.

Pharmaceutical
In compounding and dispensing practice.
Suggested for use as—
Substitute for other salicylates on account of causing less stomachic disturbances.
Suggested for use in treating—
Articular rheumatism, gout.

Strontium Silicofluoride
Synonyms: Strontium fluosilicate.
French: Fluosilicate de strontium, Silicofluorure de strontium.
German: Fluosiliciumstrontium, Siliciumfluorwasserstoffsaeuresstrontium, Strontiumfluorsilikat, Strontiumsiliciumfluorid.

Construction
Preservative (Brit. 271203) for—
Artificial stone, brickwork, natural stone, stucco, wood.

Woodworking
As a preservative.

Strontium Telluride

Chemical
Reagent (Brit. 292222) in making synthetic drugs with—
Pentamethylene alphaepsilondibromide.
Pentamethylene alphaepsilondichloride.
Pentamethylene alphaepsilondiodide.

Strontium Tungstate
German: Strontiumwolframat, Wolframsaeuresstrontium.

Chemical
Catalyst (French 598447) in making the following alcohols—
Allyl, amyl, butyl, pentyl, hexyl, heptyl, propyl.

Strontium Uranate
German: Uransaeuresstrontium.

Chemical
Catalyst (French 598447) in making the following alcohols—
Allyl, amyl, hexyl, heptyl, butyl, propyl.

Strontium Vanadate
German: Vanadinsaeuresstrontium.

Chemical
Catalyst (German 598447) in making the following alcohols—
Amyl, butyl, heptyl, hexyl, propyl.

Strophanthus Seed
French: Semences de strophantus.
German: Strophantussamen.

Chemical
Starting point in extracting—
Strophanthine.

Pharmaceutical
In compounding and dispensing practice.

Textile
——, *Dyeing*
Assist in dyeing fabrics and yarns.

Strychnine

Chemical
Starting point in making—
Various strychnine salts and esters.

Insecticide
Ingredient of—
Poisons used in exterminating vermin.

Miscellaneous
Used by trappers in poisoning fur-bearing animals.

Pharmaceutical
In compounding and dispensing practice.

Styrol
Synonyms: Cinnamene, Cinnamol, Phenylethylene, Styrene, Styrolene, Vinylbenzene.
French: Éthylène de phényle, Styrolène.
German: Phenylaethylen.

Chemical
Starting point (Brit. 263873) in making emulsifying agents for—
Aromatic hydrocarbons, terpenes.

Fats and Oils
Starting point in making—
Emulsifying agents.

Leather
Starting point in making—
Tanning emulsifying reagents.

Miscellaneous
Starting point in making—
Ingredients of washing and cleansing compositions used for various purposes.

Paper
Starting point in making reagents for increasing the absorbing and wetting capacity of—
Cardboard, paper, pulp.

Perfumery
Ingredient of—
Cosmetics, perfumes.

Petroleum
Starting point in making emulsifying agents for—
Distillates and oils.

Pharmaceutical
In compounding and dispensing practice.

Soap
Perfume for—
Toilet soaps.

Textile
——, *Dyeing*
Starting point in making—
Dye liquor assistants.

——, *Finishing*
Starting point in making—
Detergents in emulsified form.

——, *Manufacturing*
Starting point in making—
Assistants used in the carbonizing of wool.

Waxes and Resins
Starting point in making—
Emulsifying agents, synthetic resins.

Succinic Acid
Synonyms: Butane diacid, Ethylenedicarboxylic acid.
Latin: Acidum succinicum, Acor succinicus.
French: Acide d'ambre, Acide butanedioque, Acide karabique, Acide de succinyle, Acide succinique, Esprit volatile de succin, Sel d'ambre, Sel essential de succin, Sel volatile de succin.
German: Aethylendicarbonsäure, Bernsteinsäure, Butandisäure.
Italian: Acido succinico.

Analysis
Reagent in making—
Standard volumetric solutions for analytical work.
Reagent in separating—
Iron from aluminum, cobalt, manganese, nickel and zinc.
Reagent for testing for—
Albumen, calcium.
Reagent in volumetric operations.

Ceramics
Plasticizer in—
Compositions containing cellulose acetate, nitrocellulose or other esters or ethers of cellulose, used for the decoration or preservation of ceramic ware, porcelains, pottery and the like.

Chemical
Starting point in making—
Acetonediacetic acid anhydride, aromatics, caprylene, derivatives of various sorts, esters for use as per-

Succinic Acid (Continued)
fumes, fumaric acid, intermediates, oxalic acid, paraethoxysuccinimide, pharmaceuticals, pyrantin, sanatyl succinate, succinates, succinimide, succinylsalicylic acid.

Dye
Starting point in making—
Algol brilliant violet R, algol yellow 3G, rhodamin S.

Electrical
Plasticizer in—
Insulating compositions containing cellulose acetate, nitrocellulose, or other esters or ethers of cellulose.

Glass
Plasticizer in—
Compositions, containing cellulose acetate, nitrocellulose, or other esters or ethers of cellulose, used in the manufacture of non-scatterable glass and for decorating and protecting glassware.

Glues and Adhesives
Plasticizer in—
Compositions, containing cellulose acetate, nitrocellulose, or other esters or ethers of cellulose, used for special adhesive purposes, such as gluing paper to glass or metal.

Leather
Plasticizer in—
Compositions, containing cellulose acetate, nitrocellulose, or other esters or ethers of cellulose, used in the manufacture of artificial leathers and for the decoration and protection of leather goods.

Metallurgical
Plasticizer in—
Compositions, containing cellulose acetate, nitrocellulose, or other esters or ethers of cellulose, used for the decoration and protection of metallic articles.

Miscellaneous
Plasticizer in—
Compositions, containing cellulose acetate, nitrocellulose, or other esters or ethers of cellulose, used for the decoration and protection of various products.

Paint and Varnish
Plasticizer in making—
Lacquers, paints, enamels, varnishes, and dopes containing cellulose acetate, nitrocellulose, or other esters or ethers of cellulose.

Paper
Plasticizer in—
Compositions, containing cellulose acetate, nitrocellulose, or other esters or ethers of cellulose, used in the manufacture of coated paper and for the decoration and protection of paper and pulp products.

Pharmaceutical
In compounding and dispensing practice.

Photographic
Plasticizer in making—
Films from cellulose acetate, nitrocellulose, or other esters or ethers of cellulose.
Reagent in—
Photographic processes.

Plastics
Plasticizer in making—
Compositions containing cellulose acetate, nitrocellulose, or other esters or ethers of cellulose.

Perfume
Ingredient of—
Mouthwashes.

Rubber
Plasticizer in—
Compositions, containing cellulose acetate, nitrocellulose, or other esters or ethers of cellulose, used for the decoration and protection of rubber goods.

Stone
Plasticizer in—
Compositions, containing cellulose acetate, nitrocellulose, or other esters or ethers of cellulose, used for the decoration and protection of artificial and natural stone.

Textile
Plasticizer in—
Compositions, containing cellulose acetate, nitrocellulose, or other esters or ethers of cellulose, used for the decoration of textile fabrics.

Woodworking
Plasticizer in—
Compositions, containing cellulose acetate, nitrocellulose, or other esters or ethers of cellulose, used for the decoration and protection of woodwork.

Succinic Acid Ester of Grapeseed Alcohol

Bituminous
Solvent (Brit. 445223) for—
Asphalt and other bituminous bodies.

Dye
Solvent (Brit. 445223) for—
Dyestuffs, particularly oil-soluble coal-tar dyes.

Fats, Oils, and Waxes
Solvent (Brit. 445223) for—
Fats, oils, waxes.

Resins
Solvent (Brit. 445223) for—
Oil-soluble glycerol-phthalic acid resins.
Polymerized vinyl compounds, synthetic resins.

Rubber
Solvent (Brit. 445223) for—
Rubber.

Succinic Anhydride
Synonyms: Succinic acid anhydride.
French: Anhydride succinique, Anhydride de succinyle.
German: Bernsteinsaeuresanhydrid.

Chemical
Reagent (Brit. 274095) in making cyclic ketones with—
Acenaphthene, alphachloronaphthalene, alphamethylnaphthalene, anthracene, naphthalene.

Dye
Reagent in making—
Rhodamine dyes.

Succinyl Peroxide
French: Péroxyde de succinyle, Péroxyde succinylique.
German: Succinylperoxyd.

Fats and Oils
Bleaching agent (Brit. 328544) for—
Fats and oils (used with hydrogen peroxide).

Food
Bleaching agent (Brit. 328544) for—
Flour, egg yolk, and meal (used with hydrogen peroxide).

Soap
Bleaching agent (Brit. 328544) for—
Fine soaps (used with hydrogen peroxide).

Waxes and Resins
Bleaching agent (Brit. 328544) for—
Waxes (used with hydrogen peroxide).

Sulphacetic Acid
French: Acide sulphoacétique.
German: Sulfessigsaeure.

Chemical
Reagent in making—
Ethylidene diacetate (Brit. 252632).

Miscellaneous
As an emulsifying agent (Brit. 343899).
For uses, see under general heading: "Emulsifying agents."

1-(3^1-Sulphamido)-phenyl-3-methyl-5-pyrazolone

Dye
Starting point (Brit. 404198) in making—
Dyestuffs (for coloring bones and bone objects rose tints) by reaction with nitrated 1-diazo-2-oxynaphthalene-4-sulphonic acid and a chromium salt.

Sulphatoethylcresidin, Normal

Dye
Starting point (Brit. 435807) in making—
Reddish-orange dyestuffs for acetate rayon, wool, silk, or tin-weighted silk, by coupling with a diazotized orthomononitranilin.

Sulphatoethylmetatoluidin, Normal

Dye
Starting point (Brit. 435807) in making—
Orange dyestuffs for acetate rayon, wool, silk, or tin-weighted silk, by coupling with a diazotized orthomononitranilin.

Sulphobenzide
German: Sulfobenzid.

Plastics
Reagent in making—
Celluloid (used in place of camphor).

5-Sulpho-2-chlorobenzoic Acid

Chemical
Reagent (Brit. 397445) in making—
Wetting agents.
Starting point in making—
Esters and derivatives.

5-Sulpho-2-chlorobenzoic Acid Benzylester

Detergent
Starting point (Brit. 408754) in making—
Saponaceous products by reaction with tertiary amines, which may be used alone or with other soaps, fillers, or compounds giving off oxygen.

5-Sulpho-2-chlorobenzoic Acid Betaphenylethylester

Detergent
Starting point (Brit. 408754) in making—
Saponaceous products by reaction with tertiary amines, which may be used alone or with other soaps, fillers, or compounds giving off oxygen.

5-Sulpho-2-chlorobenzoic Acid Dodecylester

Soap
Starting point (Brit. 403883) in making—
Saponaceous products by reaction with amines such as anilin, piperidin bases, hydroxyethylanilin, dihydroxyethylanilin, paratoluidin (these products may be used alone or with other soaps, fillers, or compounds giving off oxygen).

5-Sulpho-2-chlorobenzoic Acid Hexadecylester

Soap
Starting point (Brit. 403883) in making—
Saponaceous products by reaction with amines such as anilin, piperidin bases, hydroxyethylanilin, dihydroxyethylanilin, paratoluidin (these products may be used alone or with other soaps, fillers, or compounds giving off oxygen).

5-Sulpho-2-chlorobenzoic Acid Tetradecylester

Soap
Starting point (Brit. 403883) in making—
Saponaceous products by reaction with amines such as anilin, piperidin bases, hydroxyethylanilin, dihydroxyethylanilin, paratoluidin (these products may be used alone or with other soaps, fillers, or compounds giving off oxygen).

Sulpho-4-chloronaphthalic Acid
Synonyms: Thio-4-chloronaphthalic acid.
French: Acide de sulpho-4-chloronaphthalique, Acide de thio-4-chloronaphthalique.
German: Sulfo-4-chlornaphtalsaeure, Thio-4-chlornaphtalsaeure.

Chemical
Starting point in making—
Esters and acids, intermediates, pharmaceuticals.

Dye
Starting point (Brit. 312175) in making wool dyestuffs with—
Allylamine, allylenediamine, alphanaphthylamine, ammonia, amylamine, anilin, benzidin, benzylamine, benzylenediamine, betanaphthylamine, butylamine, butylenediamine, cresidin, diallylamine, diamylamine, dibenzylamine, dibutylamine, diethylamine, dimethylamine, diphenylamine, dipropylamine, ethylamine, ethylenediamine, dianisidin, dimethylanilin, ethylanilin, formylamine, heptylamine, heptylenediamine, hexylamine, hexylenediamine, isoallylamine, isoamylamine, isobutylamine, isopropylamine, meta-anisidin, metachloroanilin, metanitranilin, metanitroxylidin, metaphenylenediamine, metatoluidin, metatoluylenediamine, metaxylidin, methylamine, methylanilin, methylenediamine, orthoanisidin, orthochloroanilin, orthonitranilin, orthonitroxylidin, orthophenylenediamine, orthotoluidin, orthotoluylenediamine, orthoxylidin, para-anisidin, parachloroanilin, paranitranilin paranitrotoluidin, paranitroxylidin, paraphenylenediamine, paratoluidin, paratoluylenediamine, paraxylidin, phenylamine, phenyldimethylamine, phenylmethylamine, propylamine, propylenediamine, tolylamine, xylenediamine.

Sulpho-4-chloronaphthalic Anhydride
Synonyms: Thio-4-chloronaphthalic anhydride.
French: Anhydride de sulpho-4-chloronaphthalique, Anhydride de thio-4-chloronaphthalique.
German: Sulfo-4-chlornaphtalanhydrid, Thio-4-chlornaphtalanhydrid.

Chemical
Starting point in making—
Intermediates, pharmaceuticals.

Dye
Starting point (Brit. 312175) in making wool dyestuffs with—
Allylamine, allylenediamine, alphanaphthylamine, ammonia, amylamine, anilin, benzidin, benzylamine, benzylenediamine, betanaphthylamine, butylamine, butylenediamine, cresidin, dianisidin, diallylamine, diamylamine, dibenzylamine, dibutylamine, diethylamine, dimethylamine, dimethylanilin, diphenylamine, dipropylamine, ethylamine, formylamine, heptylamine, heptylenediamine, hexylamine, hexylenediamine, isoallylamine, isoamylamine, isobutylamine, isopropylamine, meta-anisidin, metachloroanilin, metanitranilin, metanitroxylidin, metaphenylenediamine, metatoluidin, metaxylidin, methylamine, methylanilin, methylenediamine, orthoanisidin, orthochloroanilin, orthonitranilin, orthonitroxylidin, orthophenylenediamine, orthotoluidin, orthotoluylenediamine, orthoxylidin, para-anisidin, parachloroanilin, paranitranilin, paranitrotoluidin, paranitroxylidin, paraphenylenediamine, paratoluidin, paratoluylenediamine, paraxylidin, phenylamine, phenyldimethylamine, phenylmethylamine, propylamine, propylenediamine, tolylamine, xylenediamine.

Sulphonated Pine Oil
French: Huile de pin sulfoné.
German: Sulfoniertes fichtenoel.

Abrasives
Ingredient (Brit. 321240) of—
Compositions used for lubricating the surface of bonded abrasive articles, such as grinding wheels or abrasive cloth or paper, in conjunction with oils, fats, resins, and waxes.

Chemical
Ingredient of—
Emulsified preparations.

Dye
Ingredient of various preparations.

Fats and Oils
Ingredient of—
Emulsions.

Glues and Adhesives
Ingredient of various preparations.

Leather
Ingredient of—
Finishing and dressing compositions.

Paper
Ingredient of various compositions for treating paper.

Soap
Ingredient of—
Detergent and cleansing preparations.

Textile
——, *Dyeing*
General assist in dyeing yarns and fabrics.

——, *Finishing*
As a softener.
Ingredient of—
Scouring preparations.

——, *Manufacturing*
Ingredient of—
Oiling preparations for use in winding, weaving, and knitting textiles.

Waxes and Resins
Ingredient of—
Emulsions.

Sulphonethylmethane

Insecticide
Essential ingredient (U. S. 1871949) of—
Insecticidal composition, rodent repellant.

Pharmaceutical
Suggested for use as—
Soporific.

1-(3'-Sulpho-6'-phenylsulphonylphenyl)-3-methyl-5-pyrazolone

Dye
Starting point (Brit. 428535) in making—
Yellow dyestuffs, capable of being chromed, by coupling with a diazotized 3-halogenoanthranilic acid.

Sulphophthalic Acid
French: Acide de sulfophthalique.
German: Sulfophtalsaeure.

Chemical
Starting point in making various intermediate chemicals.

Dye
Starting point in making various dyestuffs.

Textile
——, *Dyeing and Printing*
Solubilizing or dispersing agent (Brit. 276100) in making liquors or pastes containing—
Acridines.
Aminoanthraquinones, reduced or unreduced.
Anthraquinones, reduced or unreduced.
Azines, azo dyestuffs, basic diarylmethane dyestuffs, basic triarylmethane dyestuffs, benzoquinoneanilides, chrome mordant dyestuffs, indigoids.
Naphthoquinones, reduced or unreduced.
Naphthoquinoneanilides, nitroarylamine dyestuffs, nitroarylphenol dyestuffs, oxazines, pyridin dyestuffs, quinolines.
Quinoneimides, reduced or unreduced.
Sulphur dyestuffs, thiazines, xanthenes.

Sulphur (All Varieties)
Synonyms: Brimstone, Colloidal sulphur, Commercial sulphur, Crude sulphur, Elemental sulphur, Flotation sulphur, Flowers of sulphur, Fused sulphur, Lac sulphur, Lump sulphur, Milk of sulphur, Miner's sulphur, Native sulphur, Plastic sulphur, Precipitated sulphur, Refined sulphur, Roll brimstone, Rolled sulphur, Soft sulphur, Sublimed sulphur, Sulphur flour, Viscid sulphur, Volcanic sulphur, Washed sulphur.
Latin: Flores sulphuris, Flores sulphuris loti, Lac sulphur, Lac sulphuris, Magisterium sulphuris, Sulphur depuratum, Sulphur lotum, Sulphur praecipitatum, Sulphur sublimatum.
French: Crème de soufre, Fleurs de soufre, Lait de soufre, Soufre, Soufre precipité, Soufre sublimé, Soufre sublimé lavé.
German: Gereinigte schwefel, Gereinigte schwefelblumen, Niedergeschlagener schwefel, Schwefel, Schwefelblumen, Schwefelbluthe, Schwefellack, Schwefelmilch.
Spanish: Azufre, Azufre lavado, Azufre sublimado.
Italian: Solfo, Solfo precipitato, Solfo sublimato, Solfo sublimato e levato.

Agriculture
Disinfectant, herbicide.
Ingredient of—
Animal feeds.
Insecticide.
Treating agent for—
Animal drinking waters, cattle feeds, crops, feed storage, fungi, grain storage, greenhouses, kennels, moulds, orchards, outhouses, poultry houses, stables.

Analysis
Reagent in—
Analytical processes involving control and research.

Brewing
Starting material in sterilizing—
Cooperage, hops.

Building Construction
Ingredient of—
Acid-resistant and waterproof concretes and cements.

Chemical
Reagent in—
Organic syntheses.
Starting point in making—
Ammonium sulphocyanide, bismuth sulphide, carbon bisulphide, copper sulphate, ferrous sulphide, hydrogen sulphide, mercuric sulphide, potassium sulphocyanide, rhodanates, sodium sulphocyanide, stannic sulphide, sulphides, sulphurated potassium, sulphur chloride, sulphur dioxide, sulphuric acid, sulphur iodine, sulphurous acid, thiocyanates, vermilion.

Dye
In dye syntheses.

Electrical
Process material in making battery—
Cathodes, containers, depolarizers, electrodes, electrolytes, pastes, separators.

Explosives and Matches
Ingredient of—
Gun powders, matchhead compositions, pyrotechnic compositions.

Fertilizer
As a fertilizer.
Ingredient of—
Fertilizer compositions.
Starting point in making—
Sulphuric acid.

Food
Starting point in producing bleaching gas for—
Food products, fruits, juices.

Insecticide and Fungicide
Exterminant.
Fumigant.
Fungicide.
Germicide.
Insecticide.
Parasiticide.
Pesticide.
Ingredient of—
Insecticidal, germicidal, and fungicidal preparations.
Starting point in making—
Barium sulphide, dusting agents, lime-sulphur, soda-sulphur.
Vermicide.

Leather
Bleaching agent.
Deodorant.
Disinfectant.
Fumigant.
Material in tanning.
Preservative.
Vermicide.

Metallurgy
In flotation processes.
Starting point in making—
Sulphuric acid for pickling.

Miscellaneous
Bleaching agent.
Bleaching agent for—
Straw hats.
Deodorant.
Disinfectant.
Fumigant.
Ingredient of—
Dental casts and plates.
Material for making—
Casts, moulds.
Preservative.
Treating agent for—
Animal drinking waters, feathers, furs, hair, hospitals, kennels, outhouses, rattan, ships, sponges, stables, straw products, warehouses, wicker products, willow ware.

Paint and Varnish
Process material in making—
Ultramarine.

Paper
Starting point in making—
"Acid" in sulphite process.

Perfumery
Ingredient of—
Preparations for the skin or hair.

Petroleum
Starting point in making—
Sulphuric acid.

Pharmaceutical
Bleaching agent.
Deodorant.
Disinfectant.
Fumigant.
Germicide.
In compounding and dispensing practice.
Vermicide.

Photographic
Bleaching agent.
Deodorant.
Preservative.
Reagent.

Plastics
Filler in—
Plastics.
Process material in making—
Plastics.

Printing
Starting point in making—
Molds for electrotyping.

Sulphur (All Varieties) (Continued)

Refrigeration
Starting point in making—
Refrigerant sulphur dioxide.

Rubber
Process material in making—
Rubber substitutes, vulcanizing agents.
Vulcanizing agent.

Soap
Ingredient of—
Sulphurized shampoos and soaps.

Sugar
Bleaching agent for—
Molasses, sugar.
In making invert sugar.

Textile
Bleaching agent for—
Felt, hemp, jute, linen, silk, wool.
Stain-removing agent.

Viniculture
Dusting agent for—
Vines.

Water and Sanitation
Rodent exterminator in—
Municipal sewage systems.

Wine
Fumigant for—
Cooperage.

Sulphur Bichloride
Synonyms: Sulphur dichloride.
French: Bichlorure de soufre, Dichlorure de soufre.
German: Bichlorschwefel, Dichlorschwefel, Schwefelbichlorid, Schwefeldichlorid.

Analysis
As a solvent and reagent in the chemical laboratory.

Chemical
Reagent in the dehydration of—
Acetic acid to make acetic anhydride.
Reagent in making—
Acetyl chloride (acetyl tetrachloride), beta-b′-dichloroethyl sulphide, carbon tetrachloride from carbon bisulphide, chlorohydrins from various products, glycol chlorohydrin, glycerol chlorohydrin, methyl sulphide, pharmaceutical chemicals, thionyl chloride, various organic chemicals.
Solvent for—
Sulphur, various substances.

Fats and Oils
Reagent in making—
Linseed oil substitutes, vulcanized oils of various sorts.

Insecticide
Ingredient of—
Insecticidal preparations.
Reagent in making—
Insecticides.

Metallurgical
Reagent in extracting—
Gold from ores.

Military
As a military poison gas.
Reagent in making—
Poison gases.

Miscellaneous
Ingredient of—
Cement preparations (in combination with olive oil and carbon bisulphide).
Reagent in making—
Hard bituminous materials of high fusion point from acid resins (German 427607).
Waterproofing preparations.

Paint and Varnish
Reagent in making—
Acid-resisting substitutes for shellac.

Plastics
Reagent (German 426991) in making—
Sulphur-containing products by the treatment of distillates from pitches of all sorts.

Rubber
Reagent in making—
Rubber cements, rubber substitutes.
Solvent for rubber.
Vulcanizing agent in—
Cold vulcanization processes.

Sugar
Reagent in—
Purifying cane juice.

Textile
Reagent in—
Finishing and dyeing yarns and fabrics.

Woodworking
Reagent in treating—
Soft woods to render them hard.

Sulphur Dioxide
Synonyms: Sulphurous acid, Sulphurous anhydride.
Latin: Acidum sulfurosum.
French: Acide sulfureux, Anhydride sulfureux, Oxyde sulfureux.
German: Schwefeldioxyd, Schwefligesäure, Schwefligesäureanhydrid.
Spanish: Acido sulfuroso, Anhidrido sulfuroso.
Italian: Anidride solforosa, Ossido solforico, Ossido sulfuroso.
Note: See also uses under: Liquid Sulphur Dioxide.

Agriculture
For fumigating plants.
For killing field mice, poultry lice, and other pests.
General fumigant and disinfectant on the farm and in the dairy.

Analysis
Reagent in—
Routine analyses in breweries.
Various processes in general laboratory work.

Brewing
Fumigant for—
Beer barrels, apparatus, and containers.
Preservative for—
Beer and porter (French 484708), hops.

Ceramics
Reagent in—
Glazing ceramic ware with gold.

Chemical
As a disinfectant and antiseptic in preparing and preserving various products that are decomposed by micro-organisms.
As a general extracting medium for various purposes.
As a general oxidizing agent in various processes.
As a general purifying agent in various processes.
As a general reducing agent in various processes.
Reagent in—
Extracting bituminous matters from lignite coal.
Reagent in making—
Acetic anhydride.
Addition products obtained with meta-aminophenol or the like (German 198497).
Alum from shale.
Aluminum sulphite from aluminum oxide or aluminum hydroxide.
Ammonium sulphite from ammonium salts.
Benzidin, betabenzenesulphinic acid.
Bismuth sulphite from bismuth chloride.
Boric acid from colemanite.
Calcium bisulphite by action on calcium hydroxide.
Calcium hydrosulphite.
Calcium hyposulphite from calcium hydroxide and sulphur.
Calcium sulphite by action on calcium carbonate.
Chromium alum from chromium sulphate and potassium sulphate.
Chromium bisulphate from chromium hydroxide.
Chromium sulphite by action on chromium oxyhydrate.
Colloidal sulphur from alkali sulphides (German 164664).
Compounds made with phenols and the like and used as photographic developers (German 198497).
Cuprous bromide from copper sulphate and potassium bromide or sodium bromide.
Cuprous chloride from copper sulphate and sodium chloride.
Cuprous iodide from copper sulphate and potassium iodide.
Cuprous sulphocyanide from solution of a cupric salt, such as cupric sulphate, and potassium sulphocyanide or ammonium sulphocyanide.
Dicalcium phosphate from tricalcium phosphate obtained from treatment of bones.
Disinfectants, as various chemical compounds.
Dithionic acid as manganese salt by action on suspensions of manganese dioxide in water.
Double salts with the acetate of various metals, such as sodium acetate, potassium acetate, lead acetate,

Sulphur Dioxide (Continued)

nickel acetate, copper acetate, magnesium acetate, strontium acetate, calcium acetate, zinc acetate (Brit. 212902).
Ethylsulphuric acid.
Germicides of various sorts, as chemical compounds.
Glauber's salt from sodium chloride (German 17409).
Glycerin by the fermentation of sugar (added to control the progress and rate of fermentation) (French 611880).
Hydrosulphites of various metals of the alkali, alkaline earth, earth, rare, and heavy metal series.
Hydrogen sulphide by admixture with water vapor and passage of the mixture over incandescent coke or like material to induce chemical reaction between the water and the sulphur dioxide.
Hydroquinone from quinone, hydroxylamine.
Iodine by action on the natural mother liquors obtained from the ashes of seaweed or from Chile saltpeter.
Intermediate chemicals.
Lactose from skimmed milk (used to remove the casein by precipitation) (German 184300).
Lead sulphite by reaction with solution of a lead salt, such as lead nitrate.
Lead thiosulphate by reaction with a solution of a lead salt, such as lead nitrate.
Lithium sulphite by reaction with a solution of a lithium salt, such as lithium hydroxide.
Luminescent zinc sulphide from zinc sulphide.
Magnesium hydrosulphite.
Magnesium sulphite by action on a solution of magnesium hydroxide.
Manganese sulphite by reaction with a solution of a manganese salt, such as manganese chloride.
Mercurous chloride from mercuric chloride.
Metabisulphites from various metals, alkali metals, alkaline earth metals, and earth metals.
Metanitranilin from metadinitrobenzene.
Metasulphobenzoic acid.
Nickel sulphite by reaction on a salt of nickel.
4-Nitro-2-aminophenol from 2:4-dinitrophenol (German 289454).
Organic chemicals, orthosulphobenzoic acid, ozone from hydrogen peroxide.
Para-aminophenolalphadisulphonic acid and para-aminophenolsulphonic acid from paranitrophenol.
Paraphenylenediaminesulphonic acid from quinone diimide.
Pharmaceutical chemicals.
Phenol by the decomposition of the phenolate obtained by melting benzenesulphonic acid with sodium hydroxide.
Phosphoric acid from bones, potassium hydrosulphide, potassium metabisulphite.
Potassium sulphate and ammonium chloride from potassium chloride and ammonia (French 627299).
Potassium sulphite.
Saltcake by the Hargreave's process, sodium hydrosulphite, sodium metabisulphite.
Sodium nitrite by reducing sodium nitrate, sodium sulphate, sodium sulphite.
Sodium thiosulphate from sodium sulphide mother liquor.
Sulphuryl chloride by reaction with gaseous chlorine.
Tartaric acid.
Thionyl chloride with aid of phosphorus pentoxide.
Thiosulphates of various heavy metals, alkali metals, alkaline earth metals, and earth metals.
Trithionic acid from potassium thiosulphate or potassium bisulphite.
Various pharmaceutical chemicals, as alkylhydroxyalkyl and dihydroxyalkylarsinic acids (French 585970).
Zinc sulphite.

Preservative in—
Cultures of micro-organisms.

Reagent in—
Continuous treatment of hydrocarbons, disinfecting mash, protecting metallic magnesium (French 629603).
Purifying aldehydes.
Benzaldehyde (German 154499).
Crude tanning extracts, particularly quebracho extract.
Recovering various volatile substances.
Reducing decomposibility of physostigmine solutions.
Dinitro compounds partially, the process being carried out with the aid of iron filings.
Nitrogen oxide to nitrous oxide.
Treating grains, potatoes, and other starchy materials to increase yield of alcohol.
Mash to increase yield of alcohol, waste organic matters.
Washing precipitated cuprous bromide and chloride.

Starting point in making—
Sulphur, sulphuric acid.
Sulphuric acid by oxidation to sulphur trioxide in the presence of zeolithic masses (French 641619).

Dye

Reagent in making—
Sulphur dyestuffs.

Explosives

Reagent in making—
Gunpowder.

Fats and Oils

Bleaching agent for—
Animal and vegetable fats and oils, fatty acids.

Reagent in—
Deodorizing and purifying animal fats and oils, particularly those with bad odors.
Drying copra.
Treating oilseeds and other oil-bearing materials.

Fertilizer

Reagent in—
Disinfecting stable manure to convert it into suitable fertilizer (Brit. 265131).
Treating phosphate rock and fertilizing compositions containing such rock, also ground phosphate and sodium nitrate.

Food

Bleaching agent in treating—
Dried fruits, edible gelatin, flour, foods of various sorts and compositions, grains, molasses and other syrups, mushrooms, nuts, oatmeal, white cherries.

Disinfectant in—
Cold-storage plants.

Preservative in—
Asparagus in bottles, cider, foods of various sorts, meat in the dry state, mutton, potatoes, sausage casings, vegetables.

Reagent in—
Restoring yellow color to new grain and old barley and oats, treating corn.

Glues and Adhesives

Bleaching agent for—
Gelatin, bone glue.

Preservative for—
Bone stock, gluestock, library glues.

Reagent in—
Extracting gelatin from macerated bones.

Gums

Bleaching agent for—
Gum arabic.

Insecticide

Insecticide and parasiticide, particularly for killing lice and fleas.

Leather

As a bleaching agent.

Ingredient of—
Acid baths for treating hides in tanning.

Reagent in—
Dehairing hides, purifying oak and chestnut extracts before use in tanning, purring operation, recovering degras, reducing chrome tan liquors, softening dry hides and skins.

Mechanical

Reagent in—
Improving operation of steam engines by utilizing the heat of the exhaust steam.

Metallurgical

As a reagent for a variety of purposes in smelting processes and other metallurgical operations.

Reagent in—
Dissolving auriferous and argentiferous pig iron.
Extracting copper from certain ores.
Copper and lead from oxygenated ores, the sulphur dioxide being used in conjunction with alkaline hyposulphites or alkaline earth hyposulphites (French 648742).
Copper and lead from roasted ores.
Selenium from its ores.
Tellurium from its ores.
Titanium from its ores.
Vanadium from its ores (French 580094).
Various other metals from their ores.
Zinc from its ores.
Hydrometallurgical treatment of manganese and zinc.
Leaching copper ores.
Recovering cyanogen from spent leach solutions.

Sulphur Dioxide (Continued)
Regenerating ferric sulphate solution in detinning operations (Brit. 287592).
Removing iridium from platinum to obtain pure platinum metal.
Treating ores by the cyanide process (Brit. 278742).
Oxides or carbonate of copper in the form of ores, to bring them into solution (German 151658).
Ores containing manganese, to bring them into solution.
Sulphide iron ores, to bring them into solution (French 594470).

Mining
For extinguishing mine fires.

Miscellaneous
As a general bleaching agent, disinfectant, and preservative.
Bleaching agent in treating—
Animal and vegetable matter of various sorts, basketware, catgut, cork, feathers, hog bristles, plumes, rags (French 652696), sponges, straw hats, various products (French 597622), wickerware.
For extinguishing fires in chimneys and other confined places.
Poison for rats and other rodents.
Preservative for various products (French 597622).
Reagent in—
Recovering volatile substances, removing fruit and wine stains from fabrics.
Sterilizing agent for various purposes (French 597622).

Paint and Varnish
Aid in drying paints and varnishes.
Reagent in making—
Basic sulphate white lead.

Paper
Antichlor in bleaching process.
Bleaching agent for—
Ragstock and wood pulp.
Ingredient of—
Sulphite liquor.

Petroleum
Purifying and bleaching agent for—
Crude paraffin.

Pharmaceutical
Suggested for use as antiseptic, parasiticide, disinfectant, intestinal antiseptic; also for external applications in treating skin diseases, syphilitic swellings, diphtheria, swelling of the feet; in antiseptic solutions.

Resins and Waxes
Reagent in—
Making artificial resins by condensation of phenol (German 219570).
Treating low-grade resins to remove color and improve the quality (French 632838).

Rubber
Reagent in—
Direct vulcanization of rubber (used in conjunction with sulphuretted hydrogen).

Sanitation
As disinfectant and also in lactic acid solution (French 623395).
Disinfectant for—
Barrels and casks (French 609849 and 613615), clothing, general purposes, rooms and ships, various purposes (French 597622).

Starch
Reagent in making—
Starch from corn.

Stone
Reagent in making—
Artificial gypsum stone from dolomite (German 426760).

Sugar
Bleaching agent for—
Sugar juices.
Reagent in—
Saccharification of starch.
Treating sugar juices in various stages of refining process, to purify and decolorize them.
Sulphiting reagent.

Textile
—, *Bleaching*
Bleach for—
Silk and wool.

—, *Dyeing*
Reagent in—
Dyeing cellulose acetate yarns and fabrics.
Reducing agent in—
Dyeing processes, dyeing with chrome mordant.

—, *Manufacturing*
Reagent in making—
Viscose rayons.

—, *Miscellaneous*
Reagent in—
Decolorizing rags and the like (U. S. 1741496).

—, *Printing*
Reagent in—
Printing cellulose acetate rayon fabrics.

Wine
Fumigant for—
Wine barrels.
Preservative for—
Grape must, preventing wine from turning brown, stopping fermentation, sweet wines.
Reagent in making—
White wines.

Sulphuretted Hydrogen
Synonyms: Hydrogen sulphide, Hydrosulphuric acid.
French: Acide de hydrosulphurique, Hydrogène sulphurette, Sulfure de hydrogène.
German: Hydroschwefelsäure, Schwefelwasserstoff.
Spanish: Acido de hidrosulfurico, Sulfuro de hidrogene.
Italian: Acido d'idrosolfurico, Solfuro d'idrogene, Solfuro d'idrogeno.

Analysis
Reagent for the separation of metals by precipitation as sulphides.

Ceramics
Reagent (Austrian 102553) in treating—
Clays, bauxites, and other ceramic raw material for the removal of the iron content.

Chemical
Reagent in making—
Alloxantin, aminocarvacrol, anthragallol, antimony pentasulphide, antimony trisulphide, barium hydrosulphide, barium sulphide, boron sulphide, calcium hydrosulphide.
Concentrated colloidal solutions of arsenic trisulphide from arsenic trioxide (German 424141).
Hydrocyanic acid by the reaction of cuprous sulphocyanide and carbon dioxide (Brit. 2383—1930).
Copper sulphide, diastase, ethyl mercaptans, ethyl sulphide, ethyl sulphydrate, formic acid, ionone, magnesium sulphide, malic acid, mercury sulphide, metathioformaldehyde, methylpara-aminophenol (metol), ortho-orthodibromobenzidin.
Potassium ferrocyanide by treating the raw materials with sulphuric acid before or after decomposition with lime (German 188902).
Sodium hydrosulphide, thiocarbamide, thiophene from acetylene.
Thiourea from cyanamide (used in place of carbon bisulphide).
Trichloromethylenesulphonic acid.
Various sulphides of metals, inorganic compounds, organic sulphides, intermediates.
Reagent in—
Purifying hydrochloric acid.
Reducing nitro organic compounds in the presence of ammonia.
Removing arsenic from sulphuric acid.
Starting point in making—
Colloidal sulphur in the presence of glue or other suitable colloid protector (German 245621).
Sulphur by oxidation and finally to sulphur dioxide, as a stage in the Le Blanc soda process.
Sulphur by reaction with sulphur dioxide in the gas-purification process by means of liquid reagents.

Dye
Reagent in making—
Carmine naphtha J, methyl violet B, methylene blue from para-aminodimethylanilin, ethylene blue (German 886 and 24125), spirit yellow R.

Metallurgical
Reagent in making—
Foam bubbles in the flotation of sulphide ores from gangue.
Mesothorium.

Sulphuretted Hydrogen (Continued)
Reagent for—
Precipitating gold and silver from waste materials.
Removing arsenic from zinc electrolyte.
Removing copper from solutions of copper sulphate or other copper salts (used in place of sodium sulphide).
Saturating solutions used for pickling metals.
Separating nonsulphide ores by introducing the gas into the slime.
Smelting gold ores by the wet chloration process.

Paint and Varnish
Reagent (German 235390) in making—
Lithopone by atomizing the gas through a solution of zinc sulphate or atomizing zinc sulphate into an atmosphere of the gas.

Pharmaceutical
In compounding and dispensing practice.

Rubber
Reagent in—
Treating latex, which is then aftertreated with sulphur dioxide, to make a crepe rubber of superior quality.

Sulphuric Acid
Synonyms: Battery acid, Chamber acid, Contact acid, Dipping acid, Fortifying acid, Fuming sulphuric acid, Hydrogen sulphate, Oil of vitriol, Oleum, Oleum acid, Tower acid, Vitriolic acid.
Latin: Acidum sulfuricum, Acidum sulphuricum.
French: Acide sulfurique, Huile de vitriol.
German: Schwefelsäure, Vitriolol.
Spanish: Acido solforico, Acido sulforico.
Italian: Acido solforico, Acido sulforico.

Analysis
Reagent in—
Analytical processes involving control and research in science and industry.

Beverage
Acidifying agent.
Neutralizing agent for—
Alkalies.
Process material in making—
Carbonated beverages, mineral waters.

Brewing
Hydrolyzing agent for—
Starch.
Neutralizing agent for—
Alkaline reactions of fermenting liquors.

Cellulose Products
Catalyst in making—
Cellulose acetate.
Ingredient of—
Mixed acid (nitrating acid) used in making nitrocellulose solutions.

Ceramic
In making glazes.

Chemical
Acidifying agent in—
Chemical processing.
Catalyst in making—
Esters, such as amyl acetate, amyl salicylate, butyl acetate, benzyl acetate, bornyl acetate, ethyl acetate, ethyl benzoate, ethyl formate, ethyl succinate, methyl benzoate, methyl salicylate.
Glycol, hydrocarbon polymerization products, phenol ethers.
Concentrating agent for—
Hydrogen peroxide.
Dehydrating agent in making—
Esters with inorganic acids, olefins from alcohols.
Electrolyte in—
Electrolytic reduction processes in organic syntheses.
Hydrolyzing agent for—
Carbohydrates.
Ingredient of—
Mixed acid (sulphuric plus nitric) used (1) as an oxidizing agent, (2) as a nitrating agent.
Oxidizing agent in making—
Inorganic chemicals.
Organic chemicals, such as ethyl disulphide from ethyl mercaptan, pyridin from piperidin, phthalic acid from naphthalene.
Polymerizing agent for—
Olefins.
Purifying agent in manufacturing processes.
Reactant in making—
Acetic acid.
Additive products from olefins, such as polymerized olefins, alkyl hydrogen sulphates and alcohols.
Adipic acid, albumen, alginic acid, alums, benzoic acid, beryllium compounds, bleached barytes, boric acid, butyric acid.
Carbon dioxide from carbonates, such as limestone, dolomite.
Chloroacetic acid, chromic acid, citric acid, creosote, ethers (simple or composite), Glauber's salt, glycol derivatives, guanidin and its salts, gulonic lactone, hydrofluoric acid, hydrocyanic acid from potassium cyanide, hydrochloric acid, hydrogen (with iron filings), hydrogen peroxide, hydrogen sulphide from iron or other sulphides, inorganic chemicals, isatin-a-arylides, mucic acid, nitric acid, organic chemicals, phosphoric acid, phosphorus, potassium bichromate, propionic acid.
Saltcake (mostly sodium sulphate, with sodium bisulphate, calcium sulphate, iron sulphate, iron oxide, magnesium sulphate, silica and sodium chloride as minor constituents).
Sulphates, such as those of lead, potassium, ammonium, barium, sodium, calcium, iron, magnesium, manganese, aluminum, nickel, copper, zinc, mercury, cerium, cesium.
Tartaric acid.
Reagent and solvent medium in organic syntheses in oxidizing with—
Manganese dioxide, potassium bichromate, potassium permanganate, potassium persulphate, sodium bichromate.
Reagent and solvent medium in oxidation processes in organic synthesis in making—
Aldehydes from alcohols, aldehydes from aromatic hydrocarbons, aldehydes from complex alcohols, benzoic acid from benzene.
Chemicals containing a smaller number of carbon atoms from many compounds, such as from hydroxy acids, ketones, ketonic acids.
2:5-Dihydroxybenzoic acid from salicylic acid.
2:5-Dihydroxybenzaldehyde from salicylaldehyde.
Ketones from complex alcohols.
Methyl groups in benzene homologs into aldehydo groups.
Nitroso derivatives from aromatic primary amines.
Oxidation of various terpene derivatives.
Quinones from anilin.
Stable compounds of various kinds.
Sulphonic acids from sulphides or hydrosulphides in both the aliphatic and aromatic series.
Reducing agent in—
Organic syntheses (used with aluminum, sodium amalgum, or zinc).
Starting point in making—
Acid esters (alkyl hydrogen sulphates), such as ethyl hydrogen sulphate.
Aromatic sulphonic acids, such as benzenesulphonic acids, diazobenzenesulphonic acid, toluenesulphonic acids.
Normal esters, such as ethyl sulphate, methyl sulphate.
Sulphonating agent in making—
Organic chemicals.

Coal By-Products
Catalyst in making—
Hydrocarbon polymerization products.
Neutralizing agent in making—
Ammonium sulphate.
Polymerizing agent for—
Olefins.
Purifying agent for—
Coal gas.
Washing and dehydrating agent for—
Tar.

Distilled Liquor
Hydrolyzing agent for—
Starch.
Neutralizing agent for—
Alkaline reactions of fermenting liquors.

Dye
Process material in making—
Dyestuffs.

Electrical
Electrolyte in—
Storage batteries.

Explosives and Matches
Ingredient of mixed acid (nitrating acid) in making—
Explosives, nitrocotton, nitroglycerin, picric acid, soluble cotton, TNT.

Sulphuric Acid (Continued)
Fats, Oils, and Waxes
Process material in—
Fatty acid manufacture, stearin purification, tallow preparation prior to melting.
Refining agent for—
Waxes.
Fertilizer
Fertilizer.
Ingredient of—
Fertilizer mixtures.
Starting point in making—
Superphosphate, ammonium sulphate, and other fertilizers.
Food
Acidifying agent.
Dehydrating agent.
Fuel
In candle making.
Glue and Gelatin
Neutralizing agent in—
Removing lime used to dehair and soften hide scraps.
Insecticide and Fungicide
Fungicide.
Leather
Process material in—
Tanning operations.
Metallurgical
Cleansing agent for—
Brass, bronze, copper, iron, silver, steel.
Desilvering agent for—
Copper.
Electrolyte in—
Electrolytic processes of metallurgy, electroplating.
Pickling agent for—
Iron, steel.
Process material in metallurgy of—
Cobalt, copper, gold, iron, magnesium, nickel, platinum, silver.
Miscellaneous
Acidifying agent, cleansing agent, dehydrating agent.
Neutralizing agent for—
Alkalies.
Solubilizing agent, solvent, weed-killer.
Paint and Varnish
Ingredient of—
Mixed acid used in making nitrocellulose solutions used in lacquer, paint, and dope formulation.
Reactant in making—
Mineral pigments.
Paper
Parchmentizing agent.
Perfume
Reactant in making—
Synthetic perfumes.
Petroleum
Catalyst in making—
Hydrocarbon polymerization products.
Polymerizing agent for—
Olefins.
Reactant in making—
Additive products from olefins, such as polymerized olefins, alkyl hydrogen sulphates, and alcohols.
Refining agent for—
Petroleum, cracked products, distillates, greases.
Pharmaceutical
In compounding and dispensing practice.
Photographic
Catalyst in making—
Cellulose acetate.
Ingredient of—
Mixed acid used in making nitrocellulose film.
Reactant in various photographic processes.
Plastics
Catalyst in making—
Cellulose acetate.
Ingredient of—
Mixed acid (nitrating acid) used in making nitrocellulose used as the base for celluloid and pyroxylin plastics.
Printing
In lithographic processes, in process engraving.
Rayon
Extractant for—
Precipitated copper in skeins of cuprammonium (glanzstoff) yarn.
Rubber
Process material in—
Rubber reclamation.
Soap
Recovery agent for—
Fatty acids.
Textile
Mordant in—
Calico printing, dyeing processes.
Process material in—
Carbonizing processes, dyeing processes, mercerizing processes.
Resist for—
Woolen goods (used with acetic anhydride).
Sulphonating agent for—
Castor oil in making Turkey red oil.
White sour in—
Bleaching cotton goods.
Wine
Antiseptic.

Sulphuric Acid Ester of Grapeseed Alcohol
Bituminous
Solvent (Brit. 445223) for—
Asphalt and other bituminous bodies.
Dye
Solvent (Brit. 445223) for—
Dyestuffs, particularly oil-soluble coaltar dyes.
Fats, Oils, and Waxes
Solvent (Brit. 445223) for—
Fats, oils, waxes.
Resins
Solvent (Brit. 445223) for—
Oil-soluble glycerol-phthalic acid resins, polymerized vinyl compounds, synthetic resins.
Rubber
Solvent (Brit. 445223) for—
Rubber.

Sulphuric Acid Ester of Ricinoleic Alcohol
Bituminous
Solvent (Brit. 445223) for—
Asphalt and other bituminous bodies.
Dye
Solvent (Brit. 445223) for—
Dyestuffs, particularly oil-soluble coaltar dyes.
Fats, Oils, and Waxes
Solvent (Brit. 445223) for—
Fats, oils, waxes.
Resins
Solvent (Brit. 445223) for—
Oil-soluble glycerol-phthalic acid resins, polymerized vinyl compounds, synthetic resins.
Rubber
Solvent (Brit. 445223) for—
Rubber.

Sulphur Sesquioxide
French: Sesquioxyde de soufre.
German: Schwefelsesquioxyd.
Chemical
Reducing agent in various operations.
Dye
Reducing agent in making—
Dyestuffs which are nitrated derivatives of naphthalene and anthraquinone and their sulphonic acids.

Sulphuryl Chloride
French: Chlorure sulphurique, Chlorure de sulphuryle, Chlorure sulphurylique.
German: Sulfurylchlorid.
Chemical
Chlorinating agent in making—
Cellulose acetate.
General chlorinating and dehydrating agent in the manufacture of—
Aromatics, intermediates, organic chemicals, pharmaceuticals.
Reagent in making—
Acetic anhydride, acetyl chloride, alphachloromethylanthraquinone.
Alphachloro-2-naphtholglycollic bromide (Brit. 260623).
Alphachloro-2-naphtholglycollic chloride (Brit. 260623).
Alphachloro-2-naphtholglycollic iodide (Brit. 260623).
Benzoic anhydride, benzoyl chloride, benzyl chloride, chlorinated thiobenzenes, dichloroacetic acid, ethylsul-

Sulphuryl Chloride (Continued)
phuric acid chloride, monochloroacetic acid, methyl chlorosulphonate, parachlorophenol, trichloroacetic acid.
Starting point in making—
Thionyl chloride.

Dye
Chlorinating agent in making various synthetic dyestuffs.

Rubber
Chlorinating agent in making—
Heat-plastic materials from rubber.

Textile
Ingredient of—
Acetylating bath in the manufacture of acetone rayon.

Sunflower Seed Oil
Synonyms: Sunflower oil.
French: Huile de fleur du soleil, Huile d'hélianthe annuel, Huile de tournésol.
German: Sonnenblumenoel.
Italian: Olio di girasole.

Food
As an edible oil.
Ingredient of—
Food preparations.

Fuel
As an illuminant.

Glues and Adhesives
Ingredient (Brit. 332257) of—
Adhesive preparations.

Insecticide
Ingredient of—
Fungicidal preparations.

Leather
Ingredient (Brit. 332257) of—
Compositions for making artificial leathers.
Finishing compositions for various types of leather.
Impregnating compositions.
Substitutes for leather used in making footwear.

Miscellaneous
Ingredient (Brit. 332257) of—
Roofing compositions, wall-coverings.

Paint and Varnish
Ingredient of—
Paints, varnishes.

Paper
Ingredient (Brit. 332257) of—
Finishing and impregnating compositions for paper, pulp, and pasteboard products.

Plastics
Ingredient (Brit. 332257) of—
Compositions used in the manufacture of pressed articles.

Soap
Raw material in making—
Special grades of soaps.

Textile
Ingredient of—
Compositions used in the manufacture of waxed cloth (Brit. 332257)—
Finishing compositions for textile fabrics (Brit. 332257).
Floor coverings (Brit. 332257).
Impregnating compositions (Brit. 332257).
Oils used in spinning and similar operations.
Wool-oiling compositions.

Woodworking
Ingredient of—
Finishing compositions, impregnating compositions.

Suspending Agents
See: "Emulsifying agents."

Sweet Almond Oil
Synonyms: Expressed almond oil.
French: Huile d'amandes douces.
German: Mandeloel.

Fats and Oils
Reagent in making—
Emulsions with volatile oils.

Food
Ingredient of various preparations.

Mechanical
Ingredient of—
Lubricants for delicate machinery.

Perfumery
Ingredient of creams and lotions.

Pharmaceutical
In compounding and dispensing practice.

Printing
Ingredient of—
Compositions for asphalt photolithography.

Soap
Starting point in making—
Fine toilet soaps.

Sylvestrene
French: Sylvestrène.
German: Sylvestren.

Chemical
Solvent (Brit. 269960) in various processes.

Miscellaneous
Solvent for various purposes.

Textile
——, *Dyeing and Printing*
Solvent in making mixtures used in dyeing, printing, and stenciling textiles.

Takadiastase

Brewing
Ferment in making—
Beer and similar products.

Food
Ferment in making—
Bread.
Ingredient of—
Predigested or partially digested food and starch preparations, soy sauce.

Leather
Ingredient of—
Drench or bate bath in tanning.

Pharmaceutical
In compounding and dispensing practice.

Textile
——, *Manufacture*
Used to render soluble the starch used in sizing thread during the spinning process and afterwards found in the gray cloth, so that it can be removed prior to dyeing or printing the cloth.

Talc
Synonyms: French chalk, Hydrous magnesium silicate, Steatite Soapstone.
Latin: Talcum, Talcum venetum.
French: Craie de briancone, Creta gallica, Talc purifié, Talc de venise.
German: Gereinigter talk, Rennsaclerite, Speckstein, Talk, Talkstein.

(In ground form)

Ceramics
Filler for—
Porcelains and potteries, to give them body and density.
Flux for—
Batch in the production of high tension porcelain for spark plugs, electrical insulators and the like.
Ingredient of—
Cores for electrical heating appliances, gas burner tips, sanitary ware batches and glazes.
Special mixes with clay for use as a substitute for electrical porcelain.
Various special glazes.

Chemical
Carrier for various chemical catalysts.
Packing material for—
Metallic elements which oxidize rapidly in air and which decompose water with explosive violence. Such metals—calcium, cesium, lithium, potassium, sodium, for example—must be immersed in naphtha or other suitable liquid which does not contain water or free oxygen. The tins containing these immersed metals are surrounded with some dense packing material which excludes air and moisture.
Substitute for—
Other magnesium-bearing minerals as a source of magnesium salts. When so used, the sulphate is the salt directly produced and used as a starting point in the production of other magnesium chemicals.

Construction
As a surfacing material for cement work.
Filler for—
Asbestos shingles, blocks, slabs and other forms in which this product is marketed.

Talc (Continued)
Ingredient of—
Artificial building stones and blocks of various kinds.
Composition floorings, fireproofing compositions, marble floorings, roofing cements.
Special cement mixtures (used to give coherence, density, and smooth, dustless surface).
Various compositions used for covering steam pipes and boilers to prevent loss of heat units through radiation.
Wall plasters.
Substitute for—
Oil in terrazo and mosaic flooring.

Dye
As an absorbent for dyes and colors.

Electrical
Filler for—
Wire-insulating compositions of various kinds.

Explosives
Absorbent for—
Nitroglycerin in various explosive compositions.

Fats and Oils
As a filtering medium for fats and oils of various sorts.

Fertilizer
As an inert filler in many fertilizer compositions.

Food
Bleaching agent for—
Treating barley of inferior color.
Cleansing agent in—
Treatment of such foodstuffs as barley, beans, coffee, corn, peas, peanuts, rice.
Dusting agent—
In admixture with starch, for coating candy molds and molding tables in order to prevent sticking.
Packing and conserving agent for—
Eggs, fruits, vegetables.

Glass
Dusting agent for—
Bottle molds (used to prevent sticking of the glass).
Ingredient of—
Glass batches for the production of milky, opaque glass.
Polishing agent for—
Plate glass.

Insecticide
As an inert filler in various insecticidal preparations.

Leather
Absorbent for—
Drying oily leathers.
Ingredient of—
Cleansing and redyeing preparations.
Finishing and dressing compounds for the treatment of many kinds of leather.
Substitute for—
Wheat flour in the manufacture of glazed kid.

Linoleum and Oil Cloth
As a dusting agent, as a filler.

Mechanical
As a lubricant.

Metallurgical
Substitute for—
Graphite and in admixture with graphite as a dusting and facing agent to coat foundry molds to prevent sticking of castings or ingots.

Miscellaneous
As an absorbent in many industrial processes.
As a dusting agent in many industrial processes.
As a filler in many products.
Cleansing and glossing agent for—
All kinds of brushes and brooms.
Dusting, lubricating agent for—
Cork molds, rubber stamp molds.
Use in gloves, shoes, and boots in order to make them easier to put on.
Filler in—
Automobile polishes.
Filler and finishing agent in—
Manufacture of cordage, rope, string, and the like.
Filler and polishing agent in—
Pastes used as preservative coatings and polishes for stoves and furnaces.
Ingredient of—
Colored crayons made with chrome colors.
Compositions used for manufacture of crayons.
Marking chalks.
Lubricant for—
Wire nails used in automatic box-nailing machines.
Mild abrasive in—
Automobile cleansing preparations used not only for removing dirt and road scum but old wax surfacings which have bleached and lost their luster.
Wood polishes.
Substitute for—
Ground cork (in combination with woodflour and paper stuff).

Paint and Varnish
Filler and pigment in—
Cold water paints, enamels, fire-resistant paints, flexible roofing paints, waterproofing paints.

Paper
Absorbent and filler in—
Blotting paper.
Inert filler for—
Insulating paper, roofing paper, wrapping paper, writing paper.
Ingredient of—
Glazes and coatings, tissue paper made from sulphite pulp.
Reagent in—
Bleaching cellulose, removing resin from cellulose.

Perfumery
Absorbent material in—
Deodorizing pastes, creams and the like.
Ingredient of—
Body powders, creams, face powders, foot powders, lotions, pastes.

Pharmaceutical
As a dusting powder and as an ingredient of dusting powders.
Binder in—
Pills and tablets of all kinds.
Lubricant for—
Tablet dies.

Photographic
As a general polishing and cleansing agent.

Plastics
Ingredient of—
Casein compositions.
Compositions used in the manufacture of buttons and the like.
Compositions used in the manufacture of imitation amber.
Various plastic preparations (as an inert filler).

Rubber
Dusting agent and protective coating for—
Automobile inner tubes, crude rubber, rubber goods of all descriptions.
Inert filler in—
Rubber goods of all descriptions.
Protective packing material for—
Rubber goods of all kinds.

Soap
Ingredient of—
Toilet soaps (as an inert soft filler and an odor absorbent).

Sugar
Filtering medium in—
Refining and purification.

Textile
Dressing in—
Yarn and thread manufacture.
Filler in—
All kinds of textile fabrics.
Loading agent in—
Carpets and rugs.
Polishing and sizing agent for—
Pile fabrics.
Reagent in—
Bleaching cloths and yarns.
Cleansing silks and other fabrics.
Coating and sizing cotton fabrics.
Dyeing and printing textile fabrics.
Finishing cloths and yarns.
Processing cotton and linens.

Water and Sanitation
Filtration reagent in—
Purifying, decolorizing, and degreasing waste waters.

Woodworking
Filler and abrasive in—
Compositions used in the finishing of furniture, interior trim, and floors.

Talc (Continued)

(*In lump or cut form*)

Chemical
Construction material of—
Many kinds of chemical equipment and fittings where a material resistant to the action of acids, alkalies, or heat is required, for example, acid proof flooring, blocks and shapes, laboratory sinks, shelves, table tops, linings, packing, tanks, tubs, and vats.

Construction
Raw material in—
Manufacture of non-staining and corrosion-resisting flooring, laundry tubs, sinks, table tops, work benches, and the like.

Electrical
Construction material for—
Floors for power plants, insulating mediums, switchboards.

Gas
Raw material of—
Tips for burners for acetylene or illuminating gas.

Metallurgical
Construction material of—
Casting molds of various kinds.

Miscellaneous
Small pieces are used as chalks for marking cloth, metal, glass, slate, and the like.

Refractories
Raw material for—
Blocks, firebrick, shapes.

Woodworking
Mild abrasive for—
Polishing and smoothing small wooden articles, such as wooden handles and the like, which are ground by small pieces of talc in a tumbling barrel.

Tall Oil
Synonyms: Liquid rosin.
French: Huile de tall.
German: Talloel.

Chemical
Starting point in making—
Fatty acids, sulphonated oils.

Paper
Ingredient of—
Paper sizing (admixture with montan wax).
Rosin size for paper (U. S. 1929115).
Reducing agent (U. S. 1929115) for—
Melting point of rosin size.

Soap
Ingredient of—
Soapstock (in admixture with palm-kernel or coconut oil).
Starting point in making—
Soaps.

Tall Oil Amide

Miscellaneous
As an emulsifying agent (Brit. 340272).
For uses, see under general heading: "Emulsifying agents."

Tall Oil Normal-Butyl Ester
French: Éther N-butylique de huile de tall.
German: Talloel-N-butylester.

Miscellaneous
As an emulsifying agent (Brit. 340272).
For uses, see under general heading: "Emulsifying agents."

Tall Oil, Sulphonated

Leather
Ingredient of—
Finishing preparations.

Miscellaneous
As a wetting agent.

Petroleum
Reducing agent for—
Troublesome petroleum emulsions.

Textile
As an assist in dyeing.
As a wetting agent.
Ingredient of preparations for—
Finishing operations.
General dyeing purposes (along with other oils).
Improving dyeing, impregnating fabrics, mordanting, sizing operations, waterproofing fabrics.

Woodworking
Ingredient of—
Impregnating compositions for wood (admixture with tar and suitable driers).

Tamarind

Chemical
Starting point in making—
Alcohol during chemical processing for other derivatives, calcium tartrate.
Potassium tartrate, crude.
Tartaric acid by chemical processing.

Food
In baking and confectionery making.
Ingredient of—
Condiments, food compositions, relishes, soft drinks, syrups.

Pharmaceutical
As a flavoring.
Ingredient of—
Phenolphthalein laxatives, refrigerant potions, vegetable laxative confections.
Starting point in making—
Fluidextract.

Tobacco
As a flavoring.

Tannic Acid
Synonyms: Digallic acid, Gallotannic acid, Tannin.
Latin: Acidum tannicum.
French: Acide gallotanique, Acide tannique.
German: Gallusgerbsäure, Gerbsäure, Tannin.
Spanish: Acido tanico.
Italian: Acido tannico.

Analysis
Reagent for—
Detecting gelatin.
Detecting and determining albumens.
Alcohol (ethyl).
Alkalies, both sodium and potassium.
Alkaloids of various sorts, aloes, blood, caramel, carbon monoxide in blood, chelidonine, hydrochloric acid, iron, neurin, potable waters, true honey, wine coloring matters.
Reagent for determining—
Alkalinity of drinking water (used in conjunction with a tenth-normal solution of iodine).
Effective value of hide powders and solutions used in estimating tannin.
Reagent for testing—
Paper and pulp products for animal size.
Various dyeings for their fastness properties.
Various dyestuffs in order to separate them into two large groups, namely, the basic dyestuffs which are precipitated by tannin, and the acid dyestuffs which are not precipitated.

Brewing
Reagent for—
Preserving beer (French 484708).
Purifying beer and ale by clarification.

Ceramics
Ingredient of—
Clays used in the production of ceramic products (added to increase their plasticity).
Enamel glazes (added for the purpose of preventing the ingredients of the glaze from settling in the form of a hard deposit while standing).

Chemical
Reagent in—
Clarifying solutions of various organic and inorganic chemicals.
Denaturing alcohol.
Isolating various glucosidal drugs, such as adonidin, convallamarin, digitalin, euonymin, helleborein, periplocin, pseudobaptisin, k-strophanthin.
Various enzymes.
Reagent in making—
Altannol (mixture of basic aluminum acetate and tannin).
Antidysentery compounds in the form of acylated tannin.
Bismuth oxyiodotannate (German 101776 and 295988).
Blood-albumen tannate (German 317676 and 305693).
Brominated condensation products with urea and formaldehyde.

Tannic Acid (Continued)

Bromocoll (brominated combination of tannin and glue) (German 116645 and 120834).
Bromotan (brominated tannin-methylene-urea) (German 180864).
Calcium compounds with acylated tannin (used in treating dysentery).
Captol (tannin-chloral compound, for preventing hair from falling out) (German 98273).
Casein compounds.
Cinnamic acid compounds.
Compounds with yeast.
Compounds with digitalis glucosides.
Condensation products with formaldehyde.
Condensation products with phenols and formaldehyde.
Cutol (aluminum borotannate).
Enterosan (basic cobalt tannate) (German 307853 and 306979).
Eutannin.
Formaldehyde-tannin compounds, used in treating dysentery.
Glutannin (vegetable gluten tannate).
Glutannol (vegetable fibrin tannate).
Honthin (albumen tannate hardened with keratin) (German 126806).
Hydrosols of various noble metals, such as gold, silver, and platinum.
Inorganic colloids.
Iodine compounds used as pharmaceuticals.
Iodotannin glue (German 116659).
Hexamethylenetetramine compounds with acylated tannin (used in treating dysentery) (German 308047).
Mercury paranucelinate compounds.
Mercury-silver suboxytannate.
Metallic albumen-tannates.
Noventerol (aluminum salts plus albumen and tannin).
Optannin (basic calcium tannate).
Pancreas preparations (pankreon, pankrotannin, tannotrypsin) by precipitation of juices obtained from the pancreas.
Pharmaceutical condensation products with formaldehyde and formaldehyde and aromatic monohydroxy compounds.
Pharmaceutical products with blood albumen.
Phenyldihydroquinazolin tannate (orexin tannate).
Tanargentan (silver, albumen, and tannin) (German 198304 and 218728).
Tannal (aluminum tannate, soluble and insoluble).
Tannalbin (hardened tannin albumen).
Tannigen (diacetyltannin) (German 78879).
Tannin-formaldehyde-albumen compounds (German 104237, 122098 and 99617).
Tannin-silver-albumen compounds.
Tannin-silver nitrate compound (German 198304).
Tannipyrin (antipyrine tannate).
Tannismuth (bismuth tannate) (German 172933 and 202244).
Tannisol (German 88841 and 93593).
Tannobromin (formaldehyde derivative of dibromotannin) (German 125305).
Tannocol (glue tannate) (German 108130).
Tannoform (methylene-ditannin) (German 88082).
Tannoguaiaform (tannin, formaldehyde, and guaiacol).
Tannokresoform.
Tannon.
Tannopin (urotropin tannate) (German 95186).
Tannothymol (formed by action of formaldehyde on tannin and thymol) (German 188318).
Tannoxyl (oxychlorocasein) (German 204290).
Tannoxyphenol R for producing nitroso blue on fibers.
Tannyl (oxychlorocasein tannate) (German 204290).
Tanosal (creosote plustannin).
Uzarin tannate (uzaratan).
Whey albumen tannate (German 312602).

Reagent in preserving—
Hydrogen peroxide (German 196370).

Starting point in making—
Alkali and alkaline earth salts used as fixing agents.
Aluminum tannate, antimony tannate, barium tannate, bismuth tannate, cadmium tannate, calcium tannate, calcium tannate (basic), cinchonidine tannate, copper tannate, cobalt tannate, ethyl tannate, euchinin tannate (quinine-ethylcarboxylic tannate), ferric tannate.
Gallic acid by action of moulds on tannin solutions or by boiling the latter with strong acid or caustic soda.
Hexa-acetyltannin, lead tannate, lithium tannate, magnesium tannate, mercury tannate, methyl tannate, nickel tannate, phloroglucinol-tannin, potassium tannate, pyrogallic acid, quinidine tannate, quinine tannate, safranin tannate, silver tannate, sodium tannate, strontium tannate.
Thymolmethane derivative with acylated tannin used in treating dysentery (German 308047).
Zinc tannate.

Dye

Ingredient of—
Color lakes.

Precipitating agent in making—
Color lakes of basic dyestuffs.

Reducing agent in making—
Dianil direct yellow S, chloramine orange G, mikado brown, mikado golden yellow 2G, mikado golden yellow 8G, mikado orange 5R, mikado orange R, mikado yellow, naphthylamine orange 2R.

Glass

Reagent in—
Silvering mirrors.

Glues and Adhesives

Reagent in—
Insolubilizing casein glues, gelatin adhesive preparations.

Ink

Ingredient of—
Copying inks, permanent inks, printing inks, writing inks.

Leather

Reagent in—
Tanning skins and hides.

Metallurgical

Ingredient of—
Bath used for nickeling metals.
Baths used for coloring various metals.
Copper salt solutions for coating copper on brass.
Copper-plating baths.

Reagent for—
Hardening molds made of glue and gelatin used in galvano-technology.

Miscellaneous

Ingredient of—
Shoe polishes.

Reagent in—
Carrotting furs and skins (U. S. 1625458).
Making imitation horn or tortoise shell from glue, gelatin, and albumen.
Treating clay roads.

Paint and Varnish

Reagent (Brit. 312061) in—
Treating pigments (deposited thereon for the purpose of preventing agglomeration of the particles, particularly in the manufacture of nitrocellulose lacquers, paints, varnishes, enamels, and dopes).

Paper

Reagent in—
Mordanting paper and pulp, as well as various fibrous products containing either paper or pulp, to prepare them for dyeing.
Sizing paper and pulp and compositions containing them.
Treating paper and pulp products for the purpose of increasing their strength (used in combination with sodium silicate).

Pharmaceutical

Ingredient of—
Astringent solutions containing glycerin.
Galenical preparations.
Medicated oxygenated baths (added for the purpose of increasing the degree of saturation of water with oxygen) (German 235619).
Mouthwashes, tannin baths.

Suggested for use as collyrium hemostatic, astringent, and styptic, and for treating skin diseases, hemorrhoids, diarrhea, dysentery, cholera, and other ailments.

Photographic

Ingredient of—
Fixing baths, containing sodium acetate, used for treating positives.

Reagent for—
Developing black positives.
Making positives by the iron salt process.

Perfume

Ingredient of—
Antiperspiration preparations, hair tonic.

Petroleum

Reagent for—
Deodorizing crude oil.

Tannic Acid (Continued)

Rubber
Ingredient of—
Rubber substitutes.
Reagent for—
Coagulating rubber latex.

Textile
——, *Bleaching*
Stabilizing agent in—
Bleaching baths (German 196370).

——, *Dyeing*
Ingredient of—
Baths used for treating threads of various textiles to produce color effects (German 423602).
Baths for fixing nitroso blue on fibers and fabrics.
Etching baths containing vat dyestuffs.
Nitroso blue slop-padding baths.
Mordant in dyeing—
Various textiles with vat colors, basic dyestuffs, natural dyewoods.
Various fibers and yarns (used in combination with salts of iron, chromium, tin, and antimony).
Wool, half wool, and mixtures of wool, cotton, and silk.
Mordant in precipitating—
Metallic lakes on yarns and fabrics.
Various color lakes with antimony salts.
Reagent in—
Producing tannin resists on naphtholated fabrics.
Redyeing dresses and other articles of clothing.

——, *Finishing*
Ingredient of—
Impregnating baths containing aluminum acetate and aluminum formate.
Oxygen baths containing perborates (added to act as catalyst) (German 235619).
Silk-weighting baths.

——, *Manufacturing*
Ingredient of—
Baths used for increasing the strength of paper yarns.

——, *Miscellaneous*
Ingredient of—
Baths used for treating animal fibers, especially wool (added for the purpose of protecting them from the action of alkaline liquors) (French 562327).

——, *Printing*
Ingredient of—
Solutions used for developing and producing designs in color on fabrics (German 427505).
Mordant in—
Printing pastes containing basic colors, ice colors, anilin black, naphthol-azo dyestuffs.
Printing pastes for discharging whites.
Printing pastes for obtaining color discharges with direct dyestuffs.
Printing pastes containing also tartar emetic and molybdenum salts.
Printing pastes for producing tannin resists on naphtholated fabrics.
Printing pastes, containing various dyestuffs and other fixing agents, for printing cotton yarns, woolens, silks (with basic dyestuffs), half-wool mixtures.

Water
Reagent in—
Treating potable waters.

Wine
Reagent for—
Clarifying wines, improving wines, making artificial wines, purifying wines (used in conjunction with gelatin).

Tantalum
French: Tantale.
German: Tantal.

Analysis
Cathode in—
Electrolytic analysis of metallic salts, such as those of gold, silver, platinum, zinc, nickel, antimony, copper.

Aviation
Metal for—
Airplane gasoline lines.

Chemical
Corrosion-resisting lining for—
Chemical equipment subjected to temperatures below 350° C.
Containers, evaporating pans, piping, reaction equipment, storage tanks.
Corrosion-resisting material for use with—
Acetic acid (glacial), acetone, anilin, aqua regia, barium hydroxide, bromine, chlorine, cleaning solutions (sulphuric acid plus potassium bichromate), ferric chloride, hydrochloric acid, hydrogen peroxide, iodine, lactic acid, nitric acid, organic gases, oxalic acid, phenol, phosphoric acid, potassium chloride, silver nitrate, sodium chloride, sodium sulphate, sodium tungstate, stannic chloride, sulphur chloride, sulphur dioxide, sulphuric acid.
Corrosion-resisting metal for making—
Agitators, chlorinating equipment, condensation equipment, diaphragms, gaskets, heater tubes, mixer blades, nozzles, piping, pumps, screens, stills, valves.
Erosion-resisting material for carriers for—
Rapidly flowing gases, rapidly flowing liquids.

Electrical
Electrode in—
Rectifying A.C. to D.C. current in battery chargers.
Rectifying A.C. to D.C. current in D.C. power units.
Element material for—
Electronic tubes.
Filament material in—
Electric lamps subjected to vibration.
Getter for—
Gases in electronic tubes.

Fats and Oils
Corrosion-resisting material for—
Acid pipes installations in sulphonating processes.

Medical Equipment
Metal for making—
Dental instruments, hypodermic needles, laboratory apparatus, scale weights, spatulas, surgical instruments.

Metallurgical
Corrosion-resisting material for use with—
Chromium plating solutions.
Electrode in—
Rectifying A.C. to D.C. current in electroplating processes.

Miscellaneous
Absorbent for—
Hydrogen, nitrogen, oxygen.
Corrosion-resisting material in making—
Chlorination equipment parts, such as needle valves, nozzles, diaphragms.

Power and Heat
Corrosion-resisting material for making—
Condensation equipment, heater tubes, heat exchangers.
Erosion-resisting material for carriers of—
Rapidly flowing gases, rapidly flowing liquids.

Rayon
Material for making—
Spinarets resistant to corrosion and erosion.

Textile
Corrosion-resisting material for making—
Tin tetrachloride equipment.

Water and Sanitation
Corrosion-resisting material in making—
Chlorination equipment parts, such as needle valves, nozzles, diaphragms.

Tantalum Carbide
French: Carbure de tantale, Carbure tantalique.
German: Tantalcarbid.

Metallurgical
Ingredient (U. S. 1913100) of—
Hard alloys.

Tantalum Dioxide
French: Dioxyde de tantale, Dioxyde tantalique.
German: Tantaldioxyd.

Chemical
Reagent (Brit. 281307) in making zeolite catalysts used in making—
Acetaldehyde from ethyl alcohol.
Acetic acid from ethyl alcohol.
Alcohols from aliphatic hydrocarbons.
Aldehydes and acids by the oxidation of orthochlorotoluene, parachlorotoluene, orthobromobenzene, parabromotoluene, dichlorotoluene, chlorobromotoluenes, nitrotoluenes, chloronitrotoluenes, bromonitrotoluenes.
Alphanaphthaquinone from naphthalene.
Anthraquinone from anthracene.
Benzaldehyde and benzoic acid from toluene.
Benzoquinone from phenanthraquinone.
Chloracetic acid from ethylenechlorohydrin.

Tantalum Dioxide (Continued)
Diphenic acid from ethyl alcohol.
Fluorenone from fluorene.
Formaldehyde from methanol or methane.
Hemimellitic acid from acenaphthene.
Maleic acid and fumaric acid from benzene, toluene, phenol, or tar acids, or from benzoquinone or phthalic anhydride.
Naphthaldehydic acid, acenaphthaquinone, or bisacenaphthylidenedione from acenaphthene or acenaphthylene.
Naphthalic anhydride.
Phenanthraquinone from phenanthrene.
Phthalic anhydride from naphthalene.
Salicylic aldehyde or salicylic acid from cresol.
Vanillin or vanillic acid from eugenol or isoeugenol.
Metallurgical
Starting point in making—
Metallic tantalum.

Tartaric Acid
Synonyms: Dextroracemic acid, 2:3-Dihydroxybutanedioic acid, Dihydroxysuccinic acid.
Latin: Acidum tartaricum, Sal essentiale tartari.
French: Acide dextéroracémique, Acide tartarique, Acide tartarique droit, Acide du tartre, Acide tartrique.
German: Tartersäure, Tartrylsäure, Weinsäure, Weinsteinsäure.
Spanish: Acido tartrico.
Italian: Acido tartarico.
Adhesives
Ingredient of—
Adhesive compositions.
Agriculture
Fungicide and mould-inhibitor in—
Cattle feed, molasses feeds, pigeon feed, poultry feed.
Analysis
Reagent in—
Analytical processes involving control and research work.
Beverage
Acidulating agent.
Ingredient of—
Effervescent beverages.
Reagent in making—
Fruit esters.
Stabilizing agent (U. S. 1427902 and 1427903) for—
Grape juice.
Substitute for—
Citric acid.
Brewing
Process material in—
Clarifying beer, dealcoholizing beer, preserving beer.
Building Construction
Increaser (Brit. 405508) of—
Plasticity and strength of mortars, cements, concrete.
Ingredient of—
Building tile, cement (U. S. 1456667), cement waterproofing composition (U. S. 1418374), heat-insulating composition (U. S. 1456667), plaster finishes.
Cellulose Products
Process material in making—
Cellulose (U. S. 1509273), cellulose acetate (U. S. 1905536), nitrocellulose (U. S. 1509273).
Ceramics
Process material in making—
Porcelains, potteries.
Chemical
As an organic acid.
Ingredient of—
Iodine-producing tablet (U. S. 1429276).
Process material in making—
Adrenalin (U. S. 1423101), benzyl alcohol derivatives (U. S. 1423101), cobalt catalysts, copper catalysts, flavanthrene (U. S. 1478061), iron catalysts, monohydroxyphenyl-2-aminopropanol-1 (Brit. 396951), nickel catalysts, opium extracts, pectin extracts, propyl tartrate (U. S. 1421604), succinic acid (U. S. 1491465), sulphur dioxide (U. S. 1356029).
Stabilizing agent for—
Carbon dioxide solutions.
Starting point in making—
Cream of tartar, dinitrotartaric acid (U. S. 1506728), ergotamine tartrate (U. S. 1435187), ethyl tartrate (U. S. 1421604).
Intermediate chemicals, such as ethyltartaric acid, methyltartaric acid, the diethyl ester of tartaric acid, dioxytartaric acid.
d-Orthodioxyphenylethanolethylamine bitartrate (U. S. 1423101), tartar emetic, tartrates of various kinds.
Cosmetic
Ingredient (French 663392) of—
Hair dyes.
Disinfectant
Ingredient of—
Germicides.
Dye
Process material in making various synthetic dyestuffs.
Electrical
Ingredient of—
Dry batteries, electrolytes for condensers, electrolytes for cells, electrolytes for lightning-arresters.
Process material in making—
Electric insulation (many patents), silverings for electric lamps (U. S. 1486804).
Fats and Oils
Aromatizing agent (French 752693) for—
Fats and oils.
Starting point (French 752693) in making—
Aromatizing agents for fats and oils.
Fire-Fighting
Ingredient of—
Chemical fire extinguishers of carbon dioxide type.
Floor Coverings
Process material in making—
Linoleum substitute (U. S. 1245978 and 1245984).
Food
Acidulating agent.
Bleaching agent for—
Flour.
Ingredient of—
Baking powders (many patents), bakery products, candies, confections of various sorts, cream centers for candy, fondants, fruit esters, gelatin desserts, jellies, vinegar (U. S. 1459513), whipped creams.
Mould-preventer for treating—
Corn (shelled), cornflour, flours, oatmeal, wheat flour.
Peptizing agent (U. S. 1410920) for—
Pectin.
Preservative for—
Figs (U. S. 1510679), fruits (U. S. 1510679), gluten (U. S. 1330058), prunes (U. S. 1510679), raisins (U. S. 1510679).
Process material in making—
Synthetic apple oil, tonka bean extracts (U. S. 1515714), vanilla bean extracts (U. S. 1515714).
Saccarification agent (U. S. 1431525) for—
Cereal germs.
Treating agent (U. S. 1415469) for—
Yeast.
Fuel
Gelating modifier (Brit. 403401) in—
Solidified fuels based on nitrocellulose and alcohols.
Glass
Process material in—
Silvering mirrors.
Process material in making—
Safety glass (U. S. 1355625).
Ink
Preventer of—
Mould formation in inks.
Process material in making—
Inks (U. S. 1472067).
Insecticide
Disintegrating agent (U. S. 1923004) in—
Insecticide tablets composed of dried nicotine sulphate, dextrin, and sodium bicarbonate.
Laundering
Ingredient of—
Souring compositions (U. S. 1514067).
Leather
Process material in—
Dyeing kid leathers, tanning processes.
Reagent for—
Removing chromium compounds from leather.
Metallurgical
Electrolyte ingredient in—
Etching brass, copper, nickel, steel, zinc.
Obtaining pure metallic cobalt (Brit. 403281).
Plating with tin.
Ingredient of—
Resistants used in etching.

Tartaric Acid (Continued)
Process material in—
Coloring metals.
Reagent in—
Flotation of ores.
Miscellaneous
Ingredient of—
Aluminum-cleaning compound (U. S. 1890214).
Aluminum-polishing compound (U. S. 1491456).
Chemical heating agent (U. S. 1502744).
Hat sizings.
Metal-polishing compounds (Brit. 376711).
Process material in—
Galvanoplastic work.
Process material in making—
Mineral yeast.
Oral Hygiene
Cleaning agent (U. S. 1488315) for—
Artificial teeth.
Process material in making—
Dentifrice (U. S. 1470794), mouth-cleansing tablet (U. S. 1262888), mouthwash (U. S. 1275275), tooth-cleaning tablet, toothpaste (U. S. 1467024).
Paint and Varnish
Ingredient of—
Removers of paint, varnish, and lacquer.
Process material in making—
Schnitzler's green, varnish (U. S. 1443935).
Paper
Process material in making—
Blueprint paper (U. S. 1500433), paper (U. S. 1509273), photographic paper (U. S. 1444469).
Pharmaceutical
In compounding and dispensing practice.
Ingredient of—
Effervescent preparations.
Starting point in making—
Tartrate preparations.
Photographic
Process material in making—
Iron salts sensitive to light.
Reagent in—
Color process (U. S. 1315464), printing and developing processes.
Plastics
Process material in making—
Celluloid substitutes, ivory substitutes, molded products (various patents), phonograph record (U. S. 1424137), shellac substitutes (U. S. 1413145).
Resins
Process material in making—
Synthetic resins.
Rubber
Coagulating agent for—
Latex.
Process material in making—
Rubber substitute (U. S. 1245976, 1245979, and 1245984).
Synthetic rubber (U. S. 1248888).
Stone
Ingredient (U. S. 1456667) of—
Artificial stone.
Sugar
Purifying agent for—
Beet molasses, sugar.
Reagent for—
Removing potash from sugar and molasses.
Textile
Brightening agent for—
Silk colors after dyeing.
Fixing agent for—
Dyes of various types, nitrosamine dyes in printing (U. S. 1426299).
Ingredient of—
Dye baths, dye compositions, dye mixtures, sizings.
Process material in—
Calico printing.
Dyeing cotton, silk, and wool.
Tendering cotton fibers, turkey red dyeing.
Reagent for—
Liberating chlorine from bleaching powder in bleaching textiles.
Resist for—
Aluminum and other mordants.
Tobacco
Bleaching agent in conjunction with hydrogen peroxide. (U. S. 1437095).
Wine
Acidifying agent.

Tartaric Acid Ester of Grapeseed Alcohol
Bituminous
Solvent (Brit. 445223) for—
Asphalt and other bituminous bodies.
Dye
Solvent (Brit. 445223) for—
Dyestuffs, particularly oil-soluble coaltar dyes.
Fats, Oils, and Waxes
Solvent (Brit. 445223) for—
Fats, oils, waxes.
Resins
Solvent (Brit. 445223) for—
Oil-soluble glycerol-phthalic acid resins, polymerized vinyl compounds, synthetic resins.
Rubber
Solvent (Brit. 445223) for—
Rubber.

Taurocholic Acid
Synonyms: Choleic acid, Choleinic acid, Sulphocholic acid.
French: Acide cholénique, Acide cholique, Acide sulfocholique, Acide taurocholique.
German: Choleinsaeure, Sulfocholsaeure, Taurocholsaeure.
Chemical
Starting point (Brit. 282356) in making antiparasitic agents with—
Dihydrocuprein ethyl ether, dihydrocuprein ethyl ether hydrochloride, dihydrocuprein isoamyl ether, dihydrocuprein isoamyl ether hydrochloride, dihydrocuprein normal octyl ether, dihydrocuprein normal octyl ether hydrochloride, dihydroquinone.
Pharmaceutical
In compounding and dispensing practice.

Teaseed Oil
Synonyms: Tea oil.
French: Huile de camellia, Huile de thé.
German: Theesamenoel.
Italian: Olio di the.
Cosmetic
Ingredient of—
Hair oil preparations.
Disinfectant
As a disinfectant or germicide.
Ingredient of—
Disinfecting compositions.
Fuel
As an illuminant.
Insecticide
As an insecticide.
Ingredient of—
Insecticidal spray compositions.
Miscellaneous
As a deodorant.
Oils and Fats
Ingredient of—
Lubricating compositions.
Pharmaceutical
In compounding and dispensing practice, particularly in veterinary work.
Soap
As a soapstock.

Tellurium
French: Tellure.
German: Tellur.
Ceramics
Coloring agent in—
Chinaware and porcelains (to produce blue and brownish effects).
Enamels used on potteries, porcelains, and chinaware.
Chemical
Starting point in making—
Diethyl telluride, medicinal compounds, salts and esters.
Reagent in making—
Bactericidal compounds of the cyclic organic type.
Cyclic diketones, iodine-quinine derivatives.
2:6-Dimethyltellurocyclopentadione.
2-Methyl-4-butyltellurocyclopentadione.
2-Methyl-4-ethyltellurocyclopentadione.
2-Methyl-4-propyltellurocyclopentadione.
2-Methyltellurocyclopentadione.
Dye
Reagent in making various coloring matters.

Tellurium (Continued)

Electrical
In making crystal detectors or dry rectifiers for radio work.

Glass
Coloring agent in making—
Blue and brown glass of the usual or the soda-lime-silicate type.

Metallurgical
Added to metals to increase their hardness and durability.
Ingredient of—
Alloys made with copper, lead, iron, and aluminum.
Special alloys which possess marked electrical resistance.
Tellurium steels.
Reagent in producing—
Black finish on silverware (used in a hydrochloric acid solution).

Paint and Varnish
Reagent in making—
Various ultramarines.

Photographic
Reagent in treating—
Photographic prints (used in a solution of sodium sulphide) to produce brownish tints.

Pharmaceutical
In compounding and dispensing practice.

Soap
Ingredient of—
Medicinal soaps.

Tellurium Di-iodide
French: Di-iodure de tellure, Tellure di-iodurê.
German: Dijodtellur.

Petroleum
Catalyst and ingredient of catalytic mixtures (Brit. 406006) in—
Destructive hydrogenation processes.
Purification of hydrocarbons from sulphur, oxygen, and other impurities.
Processes for conversion of organic compounds containing oxygen and sulphur into the corresponding hydrocarbons.
Processes for conversion of unsaturated hydrocarbons into aromatic or hydroaromatic hydrocarbons.

Tellurium Oxide
French: Oxyde de tellure, Oxyde tellurique.
German: Telluroxyd.

Chemical
Starting point in making—
Tellurium salts.

Metallurgical
Reagent in treating—
Silverware for the purpose of giving it a black finish (used in hydrochloric acid solution).

Tellurium Sulphide
French: Sulphure de tellure, Sulphure tellurique.
German: Schwefeltellur, Tellursulfid.

Ceramics
Ingredient of—
Glazes used to produce pink effects.

Tellurium Tri-iodide
French: Tri-iodure de tellure, Tellure triodurê.
German: Tellurdreifachjodur.

Petroleum
Catalyst and ingredient of catalytic mixtures (Brit. 406006) in—
Destructive hydrogenation processes.
Purification of hydrocarbons from sulphur, oxygen, and other impurities.
Processes for conversion of organic compounds containing oxygen and sulphur into the corresponding hydrocarbons.
Processes for conversion of unsaturated hydrocarbons into aromatic or hydroaromatic hydrocarbons.

Templin Oil
Synonyms: Silver fir oil.
German: Edeltannenzapfenoel, Templioel.

Miscellaneous
Ingredient of—
Applications for insect bites.

Pharmaceutical
In compounding and dispensing practice.

Terpeneless Citronella Oil
French: Essence de citronelle nonterpénique.
German: Citronelloel terpenlose, Terpenfrei zitronelloel, Terpenlose zitronelloel.

Food
Flavoring agent in—
Condiments, confectionery, food preparations.
Ingredient of—
Flavoring compositions.

Leather
Ingredient of—
Compositions used in tanning.

Perfume
Ingredient of—
Perfumes.
Perfume in—
Cosmetics.

Soap
Perfume in—
Toilet soaps.

Pharmaceutical
In compounding and dispensing practice.

Terpenyl Acetate

Cellulose Products
Solvent and plasticizer (French 552722) for—
Cellulose esters or ethers, cellulose nitrate (nitrocellulose).
For uses, see under general heading: "Solvents."

Perfume
Ingredient of—
Cosmetics.
Substitute for lavender oil.

Pharmaceutical
In compounding and dispensing practice.

Soap
Substitute for lavender in—
Perfuming toilet soaps.

Terpinemaleic Anhydride

Glass
Adhesive agent (U. S. 1882298) for—
Binding glass and resilient transparent material in making safety glass.

Terpineol
Synonyms: Lilacine, Terpilenol.
French: Terpinéole, Terpinyle.
German: Terpinil.
Spanish: Terpinile.
Italian: Terpinile.

Chemical
Denaturant for—
Alcohol.
Starting point in making—
Terpineol acetate (lavender and bergamot odor) terpineol acid phthalate, terpineol benzoate, terpineol butyrate (eucalyptus odor), terpineol caprylate (neroli-eucalyptus odor), terpineol cinnamate, terpineol citronellate, terpineol cyclopentenylacetate, terpineol formate (jasmin and bergamot odor), terpineol isobutyrate, terpineol isovalerate (sweet orange oil odor), terpineol phthalate, terpineol propionate, terpineol salicylate, terpineol xanthate (U. S. 1886587).
Textile wetting, cleansing, and emulsifying agents by sulphonating with either sulphuric or chlorosulphonic acids (Brit. 398086).
Textile wetting, cleansing, and emulsifying agents (Brit. 274611, 311885, 399537).

Disinfectant
Ingredient of—
Disinfectant compositions.

Glues and Adhesives
Solvent in—
Cabinetmaker's glue containing glue No. 2 and 3, glycerin, water, and betanaphthol.
Case-making machine glue containing glue No. 2, glycerin, water, and betanaphthol.
Flexible bindery adhesives comprising mixtures of various grades of glue with glycerin, water, and betanaphthol.
Tablet-binding glue containing glue No. 1, zinc oxide, glycerin, water, and betanaphthol.

Gum
Solvent for—
Hard gums (by heating), semi-hard gums.

Terpineol (Continued)

Ink
Solvent (U. S. 1752462) in—
Printing ink comprising a metallic pigment (coated with pyroxylin) and a resinous substance.

Insecticide
Ingredient of—
Insecticides.

Paint and Varnish
Antidulling agent in—
Varnishes.
Ingredient of—
Solvent mixture for varnishes (Brit. 397828).
Drying oil substitute (made by reaction with maleic anhydride and castor oil) used in varnish (Brit. 405805).
Plasticizer in—
Varnishes.
Solvent in—
Varnishes.

Perfume
Base in making—
Clover odors (in combination with amyl salicylate and phenylethyl alcohol).
Jasmin odors, lilac odors, lily of the valley odors, sweet william odors.
Cheaper substitute for—
Linalyl of rosewood oil.
Deodorant for lanolin in—
Liquid cleansing cream and hand lotion comprising emulsion of stearic acid, lanolin, white mineral oil, triethanolamine, carbitol, water and quince seed mucilage.
Sunburn creams, hand lotions, and shaving creams comprising emulsions of lanolin, stearic acid, triethanolamine, and water.
Diluent for—
Bourbon geranium oil, petitgrain oil, spike lavender oil.
Odorant for—
Depilatories.
Ingredient of—
Bois de Nice violet perfume base containing also alphaionone, methylionone, natural cassie, benzyl acetate, methylheptin carbonate, coumarin, heliotropin, linalyl acetate, geranyl acetate, cyclamen aldehyde.
Hair-setting lotion containing also rose water, isopropyl alcohol, and an emulsifying agent.
Honeysuckle perfume base containing also hydroxycitronellol, alphaionone, phenylethyl alcohol, cinnamyl alcohol, vanillin, jasmin absolute, mimosa absolute, neroli absolute, musk ketone, methylnaphthyl ketone, linalool, benzyl acetate, rhodinol, cinnamyl acetate, heliotropin, and phenylacetic aldehyde.
Jacinthe perfume base containing also phenylacetic aldehyde, phenylacetic aldehyde-dimethylacetal, hydrotropic aldehyde, bromstyrol, methyloctrin carbonate, clary sage oil, Manila ylang-ylang oil, methylionone, phenylethyl alcohol, cinnamyl alcohol, synthetic rose, phenylethyl propionate, phenylpropyl acetate, vanillin, and musk ketone.
Lilac perfume base containing also hydroxycitronellol, cinnamyl alcohol, rhodinal, heliotropin, rose absolute, phenylethyl alcohol, anisic aldehyde, phenylacetic aldehyde, musk xylene, and sandalwood oil.
Lily perfume base containing also hydroxycitronellol, methylionone, ylang-ylang oil, rose absolute, jasmin absolute, heliotropin, cyclamen aldehyde, phenylethyl alcohol, vanillin, methylphenyl acetate, nerol, rhodinol, and linalool.
Narcisse perfume base containing also Bourbon ylang-ylang oil, benzyl acetate, hydroxycintronellol, cinnamyl alcohol, rose synthetic, coumarin, jasmin synthetic, paracresylphenyl acetate, paracresyl acetate, and methylparacresol.
Sandalwood perfume base containing also sandalwood oil, cedarwood oil, geraniol, hydroxycitronellol, artificial musk, and styrax resin.
Sweet pea perfume base containing also phenylethylphenyl acetate, dimethylacetophenone, ethylvanillin, benzyl acetate, musk ketone, Manila ylang-ylang oil, benzyl salicylate, synthetic rose, cinnamyl alcohol, hydroxycitronellol, linalool, hydrotropic aldehyde, and neroli petale.
Softening agent for—
Mimosa odors (in combination with methylpara-acetophenone).

Soap
Aromatic in—
Toilet soaps.

Terpineol Cyclopentylacetate

Food
Agent for producing—
Pineapple aroma and flavor.

Terpineol Formate

Synonyms: Terpinyl formate.
French: Formiate de terpinéol, Formiate terpinéolique, Formiate de terpinol, Formiate terpinylique.
German: Ameisensaeuresterpeneol, Ameisensaeuresterpenyl, Ameisensaeureterpeneolester, Ameisensaeureterpenylester, Terpeneolformiat, Terpenylformiat.

Paint and Varnish
Plasticizer in making—
Cellulose ester and ether lacquers, varnishes, and dopes (Brit. 283619).

Perfumery
Ingredient of various preparations.

Plastics
Plasticizer in making—
Cellulose ester and ether compounds (Brit. 283619).

Terpineol Phthalate

Synonyms: Terpinyl phthalate.
French: Phthalate de terpinéole, Phthalate de terpinyle, Phthalate terpinylique.
German: Phtalsaeuresterpineol, Phtalsaeuresterpinyl, Phtalsaeureterpineolester, Phtalsaeureterpinylester, Terpineolphtalat, Terpinylphtalat.

Cellulose Products
Solvent and plasticizer (Brit. 283619) for—
Cellulose esters and ethers.
For uses, see under general heading: "Solvents."

Terpineol Sulphonate

French: Sulphonate de terpinéol.
German: Sulfonsäuresterpineol.

Textile
Emulsifying agent (Brit. 398086) in making—
Dressings for fabrics, lubricants for fabrics, sizes for fabrics.

Terpinolene

Disinfectants
As a disinfectant and germicide (it is claimed that in a soap emulsion, terpinolene completely inhibits the development of tuberculosis bacilla).

Paint and Varnish
Substitute for—
Turpentine.

Perfume
Ingredient of—
Synthetic perfumes.

4-Tertiary-amylmetacresol

Pharmaceutical
As an antiseptic (U. S. 1982180).

4-Tertiary-butylmetacresol

Pharmaceutical
As an antiseptic (U. S. 1982180).

Tetra-allyl-4:4′-diaminobenzophenone

Dye
Starting point (Brit. 249160) in making triarylmethane dyestuffs with—
3-Chlorophenylethylenediamine, dibenzyldiphenylethylenediamine, diethyldiphenylethylenediamine, dimethyldiphenylethylenediamine, diorthotolylethylenediamine, 3-tolylethylenediamine, xylylethylenediamine.

5:7-Tetra-allyldiaminoxanthone

Chemical
Starting point in making—
Intermediates, pharmaceuticals.

Dye
Starting point (Brit. 314825) in making xanthene dyestuffs with—
Alphachloronaphthalene, betachloronaphthalene, 4-chlorometaxylene, metachloroanilin, metachloroanisol, metachlorobenzylamine, metachlorocresidin, metachlorophenylamine, metachlorotoluene, metachlorotoluidin, metachloroxylene, metachloroxylidin, orthochloroanilin, orthochloroanisol, orthochlorobenzylamine, orthochlorocresidin, orthochlorophenylamine, orthochlorotoluene, orthochlorotoluidin, orthochloroxylene, ortho-

5:7-Tetra-allyldiaminoxanthone (Continued)
chloroxylidin, parachloroanilin, parachloroanisol, parachlorobenzylamine, parachlorocresidin, parachlorophenylamine, parachlorotoluene, parachlorotoluidin, parachloroxylene, parachloroxylidin.
Various acyl, aralkyl thioether derivatives of aromatic halogen compounds.

Tetra-amyl-4:4′-diaminobenzophenone
Dye
Starting point (Brit. 249160) in making triarylanthrone dyestuffs with—
3-Chlorophenylethylenediamine, dibenzyldiphenylethylenediamine, diethyldiphenylethylenediamine, dimethyldiphenylethylenediamine, diorthotolylethylenediamine, 3-tolylethylenediamine, xylylethylenediamine.

3:7-Tetra-amyldiaminoxanthone
French: 3:7-Tétra-amylediaminoxanthone.
German: 3:7-Tetra-amyldiaminoxanthon.
Chemical
Starting point in making—
Pharmaceuticals and other derivatives.
Dye
Starting point (Brit. 314825) in making xanthene dyestuffs with the aid of—
Alphachloronaphthalene, betachloronaphthalene, 4-chlorometaxylene, metachloroanilin, metachloroanisol, metachlorobenzylamine, metachlorocresidin, metachlorophenylamine, metachlorotoluene, metachlorotoluidin, metachloroxylene, metachloroxylidin, orthochloroanilin, orthochloroanisol, orthochlorobenzylamine, orthochlorocresidin, orthochlorophenylamine, orthochlorotoluene, orthochlorotoluidin, orthochloroxylene, orthochloroxylidin, parachloroanilin, parachloroanisol, parachlorobenzylamine, parachlorocresidin, parachlorophenylamine, parachlorotoluene, parachlorotoluidin, parachloroxylene, parachloroxylidin.
Various acyl, aralkyl, thioether derivatives of aromatic halogen compounds.

2:4:6:8-Tetrabromo-1:5-diaminoanthraquinone
German: 2:4:6:8-Tetrabrom-1:5-diaminoanthrachinon.
Dye
Starting point in making—
Anthraquinone blue SR.

Tetrabromoindigo
Dye
Starting point (Brit. 250251) in making dye mixtures with—
Alkali borates, alkali carbonates, alkali phosphates.
Textile
——, *Dyeing and Printing*
As a color.

Tetrabromophthalic Acid
Cellulose Products
Plasticizer (Brit. 390541) for—
Cellulose esters and ethers.
For uses, see under general heading: "Plasticizers."

Tetrabutyldiaminobenzophenone
Chemical
Starting point (Brit. 272321) in making intermediate chemicals with—
Alkoxybenzenes, dialkylanilines and homologs, diphenyls, halogenated benzenes, halogenated toluenes, halogenated xylenes, naphthalenes.
Dye
Starting point (Brit. 249160) in making triarylanthrone dyes with—
3-Chlorophenylethylenediamine, dibenzyldiphenylethylenediamine, diethyldiphenylethylenediamine, dimethyldiphenylethylenediamine, diorthotolylethylenediamine, 3-tolylethylenediamine, xylylethylenediamine.

3:7-Tetrabutyldiaminoxanthone
Chemical
Starting point in making various intermediates and other derivatives.
Dye
Starting point (Brit. 314825) in making xanthene dyestuffs with—
Alphachloronaphthalene, betachloronaphthalene, 4-chlorometaxylene, metachloroanilin, metachloroanisol, metachlorobenzylamine, metachlorocresidin, metachlorophenylamine, metachlorotoluene, metachlorotoluidin, metachloroxylene, metachloroxylidin, orthochloroanilin, orthochloroanisol, orthochlorobenzylamine, orthochlorocresidin, orthochlorophenylamine, orthochlorotoluene, orthochlorotoluidin, orthochloroxylene, orthochloroxylidin, parachloroanilin, parachloroanisol, parachlorobenzylamine, parachlorocresidin, parachlorophenylamine, parachlorotoluene, parachlorotoluidin, parachloroxylene, parachloroxylidin.
Various acyl, aralkyl, thioether derivatives of aromatic halogen compounds.

Tetrachlor-1:2-chrysenequinone
Dye
Intermediate (Brit. 438609) in making—
Synthetic dyes.

3:4:5:6-Tetrachloro-2-benzoylbenzoic Acid
Chemical
Starting point in making—
Esters, intermediates, pharmaceuticals, salts.
Starting point (Brit. 273347) in making—
Dichloroanthraquinonedisulphonic acid, dichlorodisulpho-2-benzoxylbenzoic acid.

3:5:3′:5′-Tetrachloro-2:2′-dihydroxytriphenylmethane-2″-sulphonic Acid Dialphabutylether
Textile
Mothproofing agent (Brit. 422923) for—
Animal fibers (capable of application from an acid dyebath).

3:5:3′:5′-Tetrachloro-2:2′-dihydroxytriphenylmethane-2″-sulphonic Acid Dibetabutylether
Textile
Mothproofing agent (Brit. 422923) for—
Animal fibers (capable of application from an acid dyebath).

3:5:3′:5′-Tetrachloro-2:2′-dihydroxytriphenylmethane-2″-sulphonic Acid Di(methylenephenyl)ether
Textile
Mothproofing agent (Brit. 422923) for—
Animal fibers (capable of application from an acid dyebath).

3:5:3′:5′-Tetrachloro-2:2′-dihydroxytriphenylmethane-2″-sulphonic Acid Dimethylether
Textile
Mothproofing agent (Brit. 422923) for—
Animal fibers (capable of application from an acid dyebath).

Tetrachlorodiphenylmethane
Electrical
Cooling medium (Brit. 413596, 433070, 433071, and 433072) in—
Electrical apparatus, such as transformers, switches, capacitors, cables, bushings, and junction boxes (may be employed in admixture with trichlorobenzene, chlorinated diphenyl, and the like).
Dielectric (Brit. 413596, 433070, 433071, and 433072) in—
Electrical apparatus, such as transformers, switches, capacitors, cables, bushings, and junction boxes (may be employed in admixture with trichlorobenzene, chlorinated diphenyl, and the like).

4:5:6:7-Tetrachloro-3-oxy-1-thionaphthene
Dye
Starting point (Brit. 262457) in making thioindigoid dyestuffs with—
2:3-diketodihydrothionaphthene, 5:7-dichloroisatin alphachloride, isatin.

2:4:5:6-Tetrachlorophenol
German: Tetrachlorphenol.
Spanish: Tetraclorofenol.
Italian: Tetraclorofenole.
Forestry
As a wood preservative.
Fungicide
As a fungicide.
Woodworking
As a wood preservative.

Tetrachlorophthalic Acid
French: Acide de tétrachlorophthalique.
German: Tetrachlorphtalsaeure.
Chemical
Starting point in making—
Formyl tetrachlorophthalate (Brit. 251147).
Dye
Starting point in making—
Cyanosin B, phloxin, rose bengal B.

Tetrachloropyrimidin

Plastics
Reagent (Brit. 393914) in making—
Films and insulating materials from acetone-soluble cellulose acetate, dimethylanilin, and chloroform.
Textile
Reagent (Brit. 393914) in making—
Threads from acetone-soluble cellulose acetate, dimethylanilin, and chloroform.

Tetracresyl-Bismuth

Lubricant
Addition agent (Brit. 445813) in—
Lubricants for motors, turbines, flushing, and high-temperature work generally.

Tetracresyl-Mercury

Lubricant
Addition agent (Brit. 445813) in—
Lubricants for motors, turbines, flushing, and high-temperature work generally.

Tetradecene

Miscellaneous
As an emulsifying agent (Brit. 343872).
For uses, see under general heading: "Emulsifying agents."

Tetradecyldioxypropyl Ether

Miscellaneous
As an emulsifying agent (Brit. 360539).
For uses, see under general heading: "Emulsifying agents."

Tetradecylguanidin Chloride

Miscellaneous
As an emulsifying agent (Brit. 422461).
For uses, see under general heading: "Emulsifying agents."
Textile
Assistant (Brit. 421862) in—
Aqueous baths for treating textiles.
Promoter (Brit. 421862) of—
Uniform dyeing with basic dyestuffs.
Wetting and washing agent (Brit. 421862) in—
Textile processes.

Tetradekanaphthene
German: Tetradekanaphten.

Chemical
Solvent in general use (Brit. 269960).
Miscellaneous
Solvent in various processes (Brit. 269960).
Textile
——, *Dyeing and Printing*
Solvent in—
Dyeing and printing fabrics and yarns (Brit. 269960).
——, *Finishing*
Solvent in—
After-treating and stenciling (Brit. 269960).

Tetraethyl-Antimony Fluoride

Oils, Fats, and Waxes
Addition agent (Brit. 440175) for—
Lubricating oils or greases used under high-pressure working conditions.

Tetraethyldiaminobenzophenone
French: Tétraéthylediaminobenzophénone.
German: Tetraaethyldiaminobenzophenon.

Chemical
Starting point (Brit. 272321) in making intermediates with—
Alkoxybenzenes, dialkylanilins and homologs, diphenyls, halogenated benzenes, halogenated toluenes, halogenated xylenes, naphthalenes.
Dye
Starting point (Brit. 249160) in making triarylanthrone dyestuffs with—
3-Chlorophenylethylenediamine, dibenzyldiphenylethylenediamine, dimethyldiphenylethylenediamine, diorthotolylethylenediamine, 3-tolylethylenediamine, xylylethylenediamine.

Tetraethyldiaminobenzophenone Chloride
French: Chlorure de tétraéthylediaminobenzophénone.
German: Tetra-aethyldiaminobenzophenonchlorid.

Dye
Starting point in making—
Night blue.

Tetraethyl Ferrocyanide

Chemical
Catalyst in treating—
Olefin hydrocarbons, especially ethylene.

Tetraethylphosphonium Iodide
French: Iodure de tétraéthylephosphonium.
German: Jodtetraaethylphosphonium, Tetraaethylphosphoniumjodid.

Miscellaneous
Reagent (Brit. 312163) in treating—
Furs, hair, feathers, and the like to render them mothproof and moldproof.
Textile
Reagent (Brit. 312163) in treating—
Wool and felt to render them mothproof and moldproof.

Tetraethylsilicon
French: Silicium de tétraéthyle.
German: Tetraaethylsilicium.

Automotive
Antiknock agent (Brit. 334181) in—
Motor fuels.

Tetraethylthiuram Monosulphide

Disinfectant
As a bactericide (Australian 8103/32, Brit. 406979, U. S. 1972961).
Insecticide and Fungicide
As a fungicide (claimed effective against barley spores) (Australian 8103/32, Brit. 406979, U. S. 1972961).
As an insecticide (Australian 8103/32, Brit. 406979, U. S. 1972961).

Tetraethyl-Tin

Lubricant
Starting point (Brit. 440175) in making—
Addition agents for high-pressure lubricating oils or greases by reacting with oil-soluble organic compounds.

Tetrahexahydrophenylthiuram Disulphide

Oils, Fats, and Waxes
Starting point (Brit. 440175) in making—
Addition agents for high-pressure lubricating oils or greases, by mixing and reacting with organo-metallic compounds.

Tetrahexyl-4:4′-diaminobenzophenone

Chemical
Starting point (Brit. 272321) in making intermediates with—
Alkoxybenzenes, diphenyls, halogenated benzenes, halogenated toluenes, halogenated xylenes, dialkylanilines and homologs, naphthalenes.
Dye
Starting point (Brit. 249160) in making triarylanthrone dyestuffs with—
3-Chlorophenylethylenediamine, dibenzyldiphenylethylenediamine, diethyldiphenylethylenediamine, dimethyldiphenylethylenediamine, diorthotolylethylenediamine, 3-tolylethylenediamine, xylylethylenediamine.

3:7-Tetrahexyldiaminoxanthone
French: 3:7-Tétrahexylediaminoxanthrone.
German: 3:7-Tetrahexyldiaminoxanthron.

Chemical
Starting point in making—
Pharmaceuticals and other derivatives.
Dye
Starting point (Brit. 313825) in making xanthene dyestuffs with the aid of—
Alphachloronaphthalene, betachloronaphthalene, 4-chlorometaxylene, metachloroanilin, metachloroanisol, metachlorobenzylamine, metachlorocresidin, metachlorophenylamine, metachlorotoluene, metachlorotoluidin, metachloroxylene, metachloroxylidin, orthochloroanilin, orthochloroanisol, orthochlorobenzylamine, orthochlorocresidin, orthochlorophenylamine, orthochlorotoluene, orthochlorotoluidin, orthochloroxylene, orthochloroxylidin, parachloroanilin, parachloroanisol,

3:7-Tetrahexyldiaminoxanthone (Continued)
parachlorobenzylamine, parachlorocresidin, parachlorophenylamine, parachlorotoluene, parachlorotoluidin, parachloroxylene, parachloroxylidin.
Various acetylaralkyl, thioether derivatives of aromatic halogen compounds.

Tetrahydrobenzene
French: Tétrahydrobenzène.
German: Tetrahydrobenzol.

Chemical
Reagent (Brit. 263873) in making—
Aromatic hydrocarbon emulsions, terpene emulsions.

Fats and Oils
Reagent (Brit. 263873) in making—
Emulsions of various fats and oils.

Leather
Reagent (Brit. 263873) in making—
Emulsified tanning compositions.

Miscellaneous
Reagent (Brit. 263873) in making—
Washing and cleansing compositions.

Petroleum
Reagent (Brit. 263873) in making—
Emulsions of petroleum and petroleum distillates.

Paper
Reagent (Brit. 263873) in making—
Cardboard and paper of higher absorbing and wetting qualities.

Textile
——, *Dyeing*
Reagent (Brit. 263873) in making—
Dye liquors of greater degree of dispersion.
——, *Finishing*
Reagent (Brit. 263873) in making—
Washing and cleansing compositions.
——, *Manufacturing*
Reagent (Brit. 263873) in making—
Carbonizing liquors.

Waxes and Resins
Reagent (Brit. 263873) in making—
Emulsions of various substances.

Tetrahydro-1:2:3:9-benzisotetrazole

Pharmaceutical
Claimed (U. S. 2008356) to have—
Valuable therapeutic properties and solubility in water.

Tetrahydrofurfuryl Acetate
French: Acétate de tétrahydrofurfuryle, Acétate tétrahydrofurfurylique.
German: Essigsäurestetrahydrofurfuryl, Essigsäuretetrahydrofurfurylester, Tetrahydrofurfurylacetat, Tetrahydrofurfurylazetat.

Cellulose Products
Solvent (U. S. 1756228) for—
Cellulose nitrate.
For uses, see under general heading: "Solvents."

Tetrahydrofurfuryl Alcohol

Cellulose Products
Solvent (Brit. 279520) for—
Cellulose esters and ethers, cellulose nitrate (nitrocellulose).
For uses, see under general heading: "Solvents."

Chemical
Starting point in making various derivatives.

Paint and Varnish
Solvent for—
Varnish gums.
Solvent in—
Compositions, containing cellulose esters or ethers, such as nitrocellulose, used in the production of varnishes, paints, enamels, dopes, and lacquers (Brit. 279520).

Resins and Waxes
Starting point (Brit. 312049) in making artificial resins with the aid of—
Vinyl acetate, vinyl chloride, other vinyl compounds.

Tetrahydrofurfuryl Butylphthalate

Cellulose Products
Plasticizer (U. S. 1989701) for—
Cellulose esters and ethers.
For uses, see under general heading: "Plasticizers."

Tetrahydrofurfuryl Formate
French: Formiate de tétrahydrofurfuryle, Formiate tétrahydrofurfurylique.
German: Formylsäurestetrahydrofurfuryl, Formylsäuretetrahydrofurfurylester, Tetrahydrofurfurylformiat.

Chemical
Starting point (Brit. 393164) in making—
Cleansing, emulsifying and dispersing agents by mixture with soaps, sulphonated oils, sulphonated higher alcohols, or aromatic sulphonic acids.

Leather
Ingredient (Brit. 393164) of—
Cleansing mixture with Marseilles soap.
Mixture with trichloroethylene for removing fat from tanned sheepskins.

Miscellaneous
Efficiency promoter (Brit. 393164) in—
Liquid cleansing and dispersing preparations for treating fibrous materials.
Plastic cleansing and dispersing preparations for treating fibrous materials.

Textile
Efficiency promoter (Brit. 393164) in—
Dispersions used for washing wool and degreasing raw wool, emulsified washing compositions, emulsions for degumming silk.
Emulsions for kier boiling cotton to aid in the removal of the natural gums, fats, waxes, and hemicellulose.
Emulsions for soaking silk, scouring preparations.
Ingredient (Brit. 393164) of—
Cleansing mixtures with Marseilles soap.
Cleansing mixtures with curd soap (particularly for textiles soiled by mineral oil).
Cleansing mixture with sodium salt of sulphonated lauryl alcohol (for raw wool).

Tetrahydrofurfuryl Propionate
French: Proprionate de tétrahydrofurfuryle, Proprionate tétrahydrofurfurylique.
German: Proprionsäurestetrahydrofurfuryl, Proprionsäurestetrahydrofurfurylester.

Chemical
Starting point (Brit. 393164) in making—
Cleansing, emulsifying and dispersing agents by mixture with soaps, sulphonated oils, sulphonated higher alcohols, or aromatic sulphonic acids.

Leather
Ingredients (Brit. 393164) of—
Cleansing mixture with Marseilles soap.
Mixture with trichloroethylene for removing fat from tanned sheepskins.

Miscellaneous
Efficiency promoter (Brit. 393164) in—
Liquid cleansing and dispersing preparations for treating fibrous materials.
Plastic cleansing and dispersing preparations for treating fibrous materials.

Textile
Efficiency promoter (Brit. 393164) in—
Dispersions used for washing wool and degreasing raw wool, emulsified washing compositions, emulsions for degumming silk.
Emulsions for kier boiling cotton to aid in the removal of the natural gums, fats, waxes, and hemicellulose.
Emulsions for soaking silk, scouring preparations.
Ingredient (Brit. 393164) of—
Cleansing mixtures with Marseilles soap.
Cleansing mixtures with curd soap (particularly for textiles soiled by mineral oil).
Cleansing mixtures with sodium salt of sulphonated lauryl alcohol (for raw wool).

Tetrahydrofurfuryl Valerate
French: Valérate de tétrahydrofurfuryle, Valérate tétrahydrofurfurylique.
German: Valeriansäurestetrahydrofurfuryl, Valeriansäuretetrahydrofurfurylester.

Chemical
Starting point (Brit. 393164) in making—
Cleansing, emulsifying and dispersing agents by mixture with soaps, sulphonated oils, sulphonated higher alcohols, or aromatic sulphonic acids.

Leather
Ingredient (Brit. 393164) of—
Cleansing mixture with Marseilles soap.
Mixture with trichloroethylene for removing fat from tanned sheepskins.

Tetrahydrofurfuryl Valerate (Continued)
Miscellaneous
Efficiency promoter (Brit. 393164) in—
Liquid cleansing and dispersing preparations for treating fibrous materials.
Plastic cleansing and dispersing preparations for treating fibrous materials.
Textile
Efficiency promoter (Brit. 393164) in—
Dispersions used for washing wool and degreasing raw wool, emulsified washing compositions, emulsions for degumming silk.
Emulsions for kier boiling cotton to aid in the removal of the natural gums, fats, waxes, and hemicellulose.
Emulsions for soaking silk, scouring preparations.
Ingredient (Brit. 393164) of—
Cleansing mixtures with Marseilles soap.
Cleansing mixtures with curd soap (particularly for textiles soiled by mineral oil).
Cleansing mixture with sodium salt of sulphonated lauryl alcohol (for raw wool).

5:6:7:8-Tetrahydro-6-hydroxy-2:4-dimethylquinolin
Chemical
Starting point (German 423026) in making the following derivatives—
Hydrochloride, methiodide, orthobenzoyl derivative, picrate.
Pharmaceutical
In compounding and dispensing practice.

Tetrahydronaphthalene Peroxide
Mechanical
Ignition quality improver (Brit. 428972) for—
Fuels for Diesel and semi-Diesel engines.

Tetrahydronaphthalenesulphonic Acid
French: Acide de tétrahydronaphthalènesulfonique.
German: Tetrahydronaphtalinsulfonsaeure.
Mechanical
Impregnating agent for treating—
Belts, bands, friction clutches, pulleys, brakes (Brit. 278465).

Tetrahydronaphthylcresol
Chemical
Starting point (Brit. 444351) in making—
Fat-splitting catalysts and emulsifying agents, useful in dyeing, laundering, bleaching, and various other purposes, by reacting with formaldehyde and non-aromatic secondary amines (the salts of the products with water-soluble acids, or water-insoluble acids, or the quaternary ammonium salts are claimed to be valuable for the purposes named).

Tetrahydronaphthylphenol
Chemical
Starting point (Brit. 444351) in making—
Fat-splitting catalysts and emulsifying agents, useful in dyeing, laundering, bleaching, and various other purposes, by reacting with formaldehyde and non-aromatic secondary amines (the salts of the products with water-soluble acids, or water-insoluble acids, or the quaternary ammonium salts are claimed to be valuable for the purposes named).

Tetrahydronaphthylresorcinol
Chemical
Starting point (Brit. 444351) in making—
Fat-splitting catalysts and emulsifying agents, for use in dyeing, laundering, bleaching, and various other purposes, by reacting with formaldehyde and non-aromatic secondary amines (the salts of the products with water-soluble acids, or water-insoluble acids, or the quaternary ammonium salts are claimed to be valuable for the purposes named).

Tetrahydroquinolin
French: Tétrahydroquinoléine.
German: Tetrahydrochinolin.
Dye
Starting point (Brit. 285382) in making—
Indophenols and leucoindophenol dyestuffs with 2:6-dichloro-2-aminophenol.
Indophenols and leucoindophenol dyestuffs with dichloroquinonechlorimide.
Indophenols and leucoindophenol dyestuffs with para-aminophenol.
Indophenols and leucoindophenol dyestuffs with quinone halogen imides.

3:7-Tetraisoamyldiaminoxanthone
French: 3:7-Tétraisoamylediaminoxanthone.
German: 3:7-Tetraisoamyldiaminoxanthon.
Chemical
Starting point in making—
Intermediates.
Dye
Starting point (Brit. 314825) in making xanthene by dyestuffs with the aid of—
Alphachloronaphthalene, betachloronaphthalene, 4-chlorometaxylene, metachloroanilin, metachloroanisol, metachlorobenzylamine, metachlorocresidin, metachlorophenylamine, metachlorotoluene, metachlorotoluidin, metachloroxylene, metachloroxylidin, orthochloroanilin, orthochloroanisol, orthochlorobenzylamine, orthochlorocresidin, orthochlorophenylamine, orthochlorotoluene, orthochlorotoluidin, orthochloroxylene, orthochloroxylidin, parachloroanilin, parachloroanisol, parachlorobenzylamine, parachlorocresidin, parachlorophenylamine, parachlorotoluene, parachlorotoluidin, parachloroxylene, parachloroxylidin.
Various acyl, aralkyl, thioether derivatives of aromatic halogen compounds.

Tetraisopropyldiaminobenzophenone
Chemical
Starting point (Brit. 272321) in making intermediates with—
Alkoxybenzenes, such as methoxybenzene, ethoxybenzene, propoxybenzene.
Dialkylanilins and homologs, such as diethylanilin, dimethylanilin, dibutylanilin.
Diphenyls.
Halogenated benzenes, such as chlorobenzenes, bromobenzenes.
Halogenated toluene, such as chlorotoluenes, bromotoluenes.
Halogenated xylenes, such as chloroxylenes, bromoxylenes.
Naphthalenes.
Dye
Starting point (Brit. 249160) in making triarylanthrone dyestuffs with—
3-Chlorophenylethylenediamine, dibenzyldiphenylethylenediamine, dimethyldiphenylethylenediamine, diorthotolylethylenediamine, 3-tolylethylenediamine, xylylethylenediamine.

3:7-Tetraisopropyldiaminoxanthone
German: 3:7-Tetraisopropyldiaminoxanthon.
Chemical
Starting point in making various derivatives.
Dye
Starting point (Brit. 314825) in making xanthene dyestuffs with the aid of—
Alphachloronaphthalene, betachloronaphthalene, 4-chlorometaxylene, metachloroanilin, metachloroanisol, metachlorobenzylamine, metachlorocresidin, metachlorophenylamine, metachlorotoluene, metachloroxylene, metachloroxylidin, orthochloroanilin, orthochloroanisol, orthochlorobenzylamine, orthochlorocresidin, orthochlorophenylamine, orthochlorotoluene, orthochlorotoluidin, orthochloroxylene, orthochloroxylidin, parachloroanilin, parachloroanisol, parachlorobenzylamine, parachlorocresidin, parachlorophenylamine, parachlorotoluene, parachlorotoluidin, parachloroxylene, parachloroxylidin.
Various acyl, aralkyl, thioether derivatives of aromatic halogen compounds.

Tetralin
French: Tétrahydronaphthalène.
German: Tetrahydronaphtalin.
Abrasive
Solvent (Brit. 277098) in making—
Compositions that are used in the manufacture of grinding discs.
Analysis
Reagent or solvent in various operations.
Automotive
Ingredient of—
Motor fuels containing alcohol and benzene.
Ceramics
Solvent in—
Compositions, containing various esters or ethers of cellulose, such as cellulose acetate and nitrocellulose, used for coating and decorating ceramic ware.

Tetralin (Continued)

Chemical

Assistant in making—
Sulphonated organic compounds.
Ingredient of—
Disinfectants, germicides.
Reagent in making—
Wetting compositions (in combination with naphthalene and anthracene).
Solvent for—
Naphthalene, sulphur, various organic chemicals, various purposes (used in place of acetone).
Solvent (Brit. 295335) in making—
Impregnating solutions with phenolformaldehyde synthetic resins.
Starting point in making—
Aromatics, foaming and emulsifying agent (Brit. 302666), intermediates, mercuriated hydroaromatic hydrocarbons (Austrian 100723), pharmaceuticals.
Resists used in the dyeing and printing of textiles.
Tanning agent (Brit. 302666), various organic chemicals.

Dye

Solvent for various dyestuffs.
Starting point in making—
Synthetic dyestuffs.

Electrical

Solvent in making—
Compositions, containing various esters or ethers of cellulose, such as cellulose acetate and nitrocellulose, used for electrical insulating purposes and in the production of electrical machinery and equipment.

Explosives

Reagent in making—
Explosive compounds.

Fats and Oils

Ingredient of—
Boring oils, cutting oils and paste, drilling oils and paste, gun oils.
Lubricating pastes, oils, and compounds.
Machine oils and paste compositions.
Solvent for various fats and oils.
Solvent, used as an extracting medium, in recovering oil and fats from original sources and waste products.

Gas

Reagent in purifying—
Coal gas.
Reagent in treating—
Spent oxide purification mass for the recovery of sulphur compounds.

Glass

Solvent in making—
Compositions, containing various cellulose esters or ethers, such as cellulose acetate and nitrocellulose, used in the manufacture of nonscatterable glass and for decorating and coating glassware.

Glues and Adhesives

Solvent in making—
Adhesive preparations, containing cellulose esters or ethers, such as cellulose acetate and nitrocellulose, as well as other substances.
Cements containing fillers (Brit. 295335).

Gums

Solvent for various gums.
Solvent, used as an extracting medium, for recovering gums from original sources and waste products.

Insecticide

Ingredient of—
Insecticides, parasiticides.

Leather

Ingredient of—
Compositions used in glazing and finishing leather and leather goods.
Solvent in making—
Compositions, containing various cellulose esters or ethers, such as cellulose acetate and nitrocellulose, used in the manufacture of artificial leather and for coating and decorating leather and leather goods.

Mechanical

As a lubricant.
Ingredient of—
Lubricating compositions.

Metallurgical

Ingredient of—
Compositions used in various treatments of metals.
Solvent in making—
Compositions, containing cellulose esters or ethers, such as nitrocellulose and cellulose acetate, used for coating and decorating metalware.

Miscellaneous

Ingredient of—
Dry-cleaning compositions, wax and encaustic compositions.
Solvent for removing—
Dry films of oil colors.
Solvent for various substances, particularly in coatings.
Solvent in making—
Compositions, containing various esters or ethers of cellulose, such as cellulose acetate and nitrocellulose, used for coating and decorating various products.
Impregnating solutions containing synthetic resins (Brit. 295335).
Shoe creams and polishes.

Paint and Varnish

Ingredient of—
Paint and varnish removers.
Varnishes, lacquers, dopes, and the like, which contain various artificial or natural gums, such as dammar, kauri, copal, and also rosin.
Solvent in making—
Lacquers and varnishes which contain phenol-aldehyde synthetic resins (Brit. 295335).
Paints, varnishes, lacquers, dopes and enamels (used in the place of turpentine and in admixture with hexalin).
Varnishes, lacquers, enamels, dopes, and paints, containing various cellulose esters or ethers, such as nitrocellulose and cellulose acetate, as well as gums, resins, and other substances.

Paper

Ingredient of—
Compositions used for removing printing ink from paper.
Solvent in making—
Compositions, containing cellulose esters or ethers of various kinds, such as cellulose acetate and nitrocellulose, used for the manufacture of coated papers and for coating and decorating paper and pulp products.

Perfume

Solvent for—
Essential oils.
Solvent for extracting—
Essential oils from original sources.

Petroleum

Solvent for—
Solid and liquid hydrocarbons.

Photographic

Solvent in making—
Films from compositions containing various cellulose esters or ethers, such as nitrocellulose and cellulose acetate.

Plastics

Solvent for—
Celluloid.
Solvent in making—
Compositions containing cellulose acetate, nitrocellulose, or other esters or ethers of cellulose.
Mixtures for molding and pressing, containing phenol-aldehyde synthetic resins as a base (Brit. 295335).
Substitute for camphor in making—
Celluloid and other plastics.

Resins and Waxes

Solvent for various resins and waxes.
Solvent for extracting—
Resins and waxes from original sources.
Starting point (Brit. 302666) in making—
Synthetic resins.

Rubber

Ingredient of—
Rubber compounded with celluloid.
Rubber mixtures (added to the latex to increase the action of protecting colloids in the manufacture of evaporated rubber (German 432894).
Solvent for rubber.
Solvent in making—
Coating compositions, containing various cellulose esters or ethers, such as cellulose acetate and nitrocellulose, used for decorating and protecting rubber merchandise.
Regenerated and reworked rubber.

Sanitation

Ingredient of—
Disinfecting compositions.

Tetralin (Continued)

Stone

Solvent in making—

Compositions, containing various cellulose esters or ethers, such as cellulose acetate and nitrocellulose, used for decorating and protecting artificial and natural stone.

Soap

Ingredient of—

Detergent preparations.

Soap solutions used for dissolving greases, oils, hydrocarbons, and colors.

Solid soaps, containing benzin, benzene, gasoline, hexalin, methylhexalin, carbon tetrachloride, trichloroethylene, and other solvents and detergent agents, such as ammonia and alcohol.

Textile soaps containing various ingredients.

Textile

——, Dyeing and Printing

Solubilizing agent (Brit. 276100) in making dye liquors and printing paste containing—

Acridin dyestuffs.

Aminoanthraquinones, reduced and unreduced.

Anthraquinone dyestuffs, azines, azo dyestuffs, basic diarylmethane dyestuffs, basic triarylmethane dyestuffs, benzoquinone-anilides, chrome mordant dyestuffs, indigoids, naphthoquinoneanilides.

Naphthoquinones, reduced and unreduced.

Nitroarylamines, nitrodiarylamines, nitroarylphenols, nitrodiarylphenols, oxazines, pyridin dyestuffs, quinolin dyestuffs.

Quinoneimides, reduced and unreduced.

Sulphur dyestuffs, xanthene dyestuffs.

——, Finishing

Reagent in finishing textiles.

Solvent in making—

Coating compositions containing cellulose acetate, nitrocellulose, or other esters or ethers of cellulose.

Woodworking

Ingredient of—

Preservative agents.

Solvent in making—

Coating compositions, containing cellulose acetate, nitrocellulose, or other esters or ethers of cellulose.

1:1:4:4-Tetramethylbutadiene

French: 1:1:4:4-Tétraméthylebutadiène.

German: 1:1:4:4-Tetramethylbutadien.

Chemical

Starting point (Brit. 309911) in making—

Intermediates, pharmaceuticals.

Starting point (Brit. 309911) in making synthetic perfumes with—

Acrolein, crotonaldehyde.

Tetramethylbutylcresol

Chemical

Starting point (Brit. 444351) in making—

Fat-splitting catalysts and emulsifying agents for use in dyeing, laundering, bleaching, and various other processes, by reacting with formaldehyde and nonaromatic secondary amines (the salts of the products with water-soluble acids, or water-insoluble acids, or the quaternary ammonium salts, are claimed to be valuable for the purposes named).

Tetramethylbutylresorcinol

Chemical

Starting point (Brit. 444351) in making—

Fat-splitting catalysts and emulsifying agents for use in dyeing, laundering, bleaching, and various other processes, by reacting with formaldehyde and nonaromatic secondary amines (the salts of the products with water-soluble acids, or water-insoluble acids, or the quaternary ammonium salts, are claimed to be valuable for the purposes named).

2:2:10:10-Tetramethyl-6-carboxyundecane

Disinfectant

Claimed (U. S. 2032159) as having—

High bactericidal action.

Tetramethyldiaminobenzhydrol

Synonyms: Michler's hydrol.

Dye

Starting point in making—

Agalina green, chrome colors, crystal violet, fast acid violet 10B, intensive blue, new fast blue, new patent blue B and 4B, Turkish blue.

Tetramethyldiaminobenzophenone

Synonyms: Michler's ketone.

Dye

Starting point in making—

Acid violet BN, acid violet 6BN, alphanaphthol blue, auramine, crystal violet, ethyl violet, rheonin A, victoria blue 4R, victoria blue R and B, wool green S.

4:4′-Tetramethyldiaminodiphenylethylene

Dye

Starting point (Brit. 435449) in making—

Dyestuffs for producing bordeaux red on wool from alkaline bath, developed to blue by acid, by coupling with betanaphthylamine-6:8-disulphonic acid.

4:4-Tetramethyldiaminodiphenylmethane

Chemical

Starting point in making—

Intermediates, pharmaceuticals.

Dye

Starting point in making various synthetic dyestuffs.

Metallurgical

Ingredient (Brit. 313134) of—

Compositions used for cleaning rust from metals.

Liquid soldering fluxes, pickling baths.

Tetramethyldiaminodiphenyl Sulphide

Chemical

Starting point in making various derivatives.

Metallurgical

Ingredient and inhibitor (U. S. 1755812) in—

Baths used for cleaning and pickling metals.

3:7-Tetramethyldiaminoxanthone

Chemical

Starting point in making various intermediates and other derivatives.

Dye

Starting point (Brit. 314825) in making xanthene dyestuffs with—

Alphachloronaphthalene, betachloronaphthalene, 4-chlorometaxylene, metachloroanilin, metachloroanisol, metachlorobenzylamine, metachlorocresidin, metachlorophenylamine, metachlorotoluene, metachlorotoluidin, metachloroxylene, metachloroxylidin, orthochloroanilin, orthochloroanisol, orthochlorobenzylamine, orthochlorocresidin, orthochlorophenylamine, orthochlorotoluene, orthochlorotoluidin, orthochloroxylene, orthochloroxylidin, parachloroanilin, parachloroanisol, parachlorobenzylamine, parachlorocresidin, parachlorophenylamine, parachlorotoluene, parachlorotoluidin, parachloroxylene, parachloroxylidin.

Various acyl, aralkyl, thioether derivatives of aromatic halogen compounds.

Tetramethylene Dicarbamide

Explosives and Matches

Ingredient (Brit. 415779) of—

Explosive compositions (added for the propagation of combustion without excessive violence of action or loss of sensitivity).

Tetramethylene Diperoxidedicarbamide

Explosives and Matches

Initiator (U. S. 1984846) for—

Detonators.

Tetramethyleneglycol

Analysis

Reagent.

Chemical

Reagent in—

Organic synthesis.

Resins and Waxes

Starting point (Brit. 396354) in making synthetic resins from—

Adipic acid, phthalic acid, and glycerin.

Adipic acid, phthalic acid, and mannitol.

Adipic acid, phthalic acid, and pentaerythritol.

Azelaic acid, phthalic acid, and glycerin.

Azelaic acid, phthalic acid, and mannitol.

Azelaic acid, phthalic acid, and pentaerythritol.

Fumaric acid, phthalic acid, and glycerin.

Fumaric acid, phthalic acid, and mannitol.

Fumaric acid, phthalic acid, and pentaerythritol.

Glutaric acid, phthalic acid, and glycerin.

Glutaric acid, phthalic acid, and mannitol.

Glutaric acid, phthalic acid, and pentaerythritol.

Tetramethyleneglycol (Continued)
Maleic acid, phthalic acid, and glycerin.
Maleic acid, phthalic acid, and mannitol.
Maleic acid, phthalic acid, and pantaerythritol.
Malic acid, phthalic acid, and glycerin.
Malic acid, phthalic acid, and mannitol.
Malic acid, phthalic acid, and pentaerythritol.
Pimelic acid, phthalic acid, and glycerin.
Pimelic acid, phthalic acid, and mannitol.
Pimelic acid, phthalic acid, and pentaerythritol.
Sebacic acid, phthalic acid, and glycerin.
Sebacic acid, phthalic acid, and mannitol.
Sebacic acid, phthalic acid, and pentaerythritol.
Suberic acid, phthalic acid, and glycerin.
Suberic acid, phthalic acid, and mannitol.
Suberic acid, phthalic acid, and pentaerythritol.
Succinic acid, phthalic acid, and glycerin.
Succinic acid, phthalic acid, and mannitol.
Succinic acid, phthalic acid, and pentaerythritol.

3:7-Tetramethylethyldiaminoxanthone
Chemical
Starting point in making various intermediates and other derivatives.
Dye
Starting point (Brit. 314825) in making xanthene dyestuffs with—
Alphachloronaphthalene, betachloronaphthalene, 4-chlorometaxylene, metachloroanilin, metachloroanisol, metachlorobenzylamine, metachlorocresidin, metachlorophenylamine, metachlorotoluene, metachlorotoluidin, metachloroxylene, metachloroxylidin, orthochloroanilin, orthochloroanisol, orthochlorobenzylamine, orthochlorocresidin, orthochlorophenylamine, orthochlorotoluene, orthochlorotoluidin, orthochloroxylene, orthochloroxylidin, parachloroanilin, parachloroanisol, parachlorobenzylamine, parachlorocresidin, parachlorophenylamine, parachlorotoluene, parachlorotoluidin, parachloroxylene, parachloroxylidin.
Various acyl, aralkyl, thioether derivatives of aromatic halogen compounds.

Tetramethyl-Lead
Lubricant
Addition agent (Brit. 445813) in—
Lubricants for motors, turbines, flushing, and high-temperature work generally.

Tetramethyl-Mercury
Lubricant
Addition agent (Brit. 445813) in—
Lubricants for motors, turbines, flushing, and high-temperature work generally.

3:4:3′:4′-Tetramethylthiazolotricarbocyanin Bromide
Photographic
Sensitizer (Brit. 436941 and 437017) for—
Photographic emulsions to infrared light with maxima at 790 mu.

Tetramethylthiuram Bisulphide
Disinfectant
As a bactericide (Australian 8103/32, Brit. 406979, U. S. 1972961).
Insecticide and Fungicide
As a fungicide (claimed effective against *Aspergillus niger* and *Fomes Annonsus*) (Australian 8103/32, Brit. 406979, U. S. 1972961).
As an insecticide (Australian 8103/32, Brit. 406979, U. S. 1972961).
Rubber
Promoter (Brit. 437304) of—
Resistance to the deteriorating action of light on chlorinated rubber used in the production of flexible, transparent films suitable for wrappings, paper-coatings, or the like, or in the manufacture of laminated glass.

Tetramethylthiuram Monosulphide
Disinfectant
As a bactericide (Australian 8103/32, Brit. 406979, U. S. 1972961).
Insecticide and Fungicide
As a fungicide (claimed effective against barley spores and pinewood fungi) (Australian 8103/32, Brit. 406979, U. S. 1972961).
As an insecticide (claimed effective against aphids) (Australian 8103/32, Brit. 406979, U. S. 1972961).

Tetramine-Copper Sulphate
French: Sulphate de cuivre et de tétramine, Sulphate cuivrique-tétraminique.
German: Schwefelsaeurestetraminkupfer, Tetraminkupfersulfat.
Dye
Reagent (Brit. 306859) in making azo dyestuffs with—
Acetyl H acid, alphahydroxynaphthalene-4-sulphonic acid.
Alphaethoxy-8-hydroxynaphthalene-3:6-disulphonic acid.
3-Aminobenzaldehyde.
2-(3′-Aminobenzoyl)amino-5-naphthol-7-sulphonic acid.
2-(4′-Aminobenzoyl)amino-5-naphthol-7-sulphonic acid.
Anthranilic acid, benzidin-3:3′-dicarboxylic acid, beta-aminobenzaldehyde, beta-aminobenzene-5-sulphonic acid, beta-aminobenzoic acid, beta-amino-1-hydroxybenzene, beta-aminonaphthalene-3-carboxylic acid, betanaphthol, betaphenylamino-4-hydroxynaphthalene-7-sulphonic acid.
4-Chloro-2-chloro-2-aminobenzoic acid.
4:4′-Diaminodiphenylurea-3:3′-dicarboxylic acid.
4:6-Dichloro-2-amino-1-hydroxybenzene.
5:5′-Dihydroxy-2:2′-dinaphthylamine-7:7′-disulphonic acid.
J acid, 5-nitro-2-aminobenzoic acid.

Tetramonomethylamine-Copper Sulphate
French: Sulphate de tétramonométhyleamine et de cuivre, Sulphate tétramonométhyleaminique et cuivrique.
German: Schwefelsaeurestetramonomethylaminkupfer, Tetramonomethylaminkupfersulfat.
Chemical
Reagent in making various intermediates.
Dye
Reagent (Brit. 306859) in making azo dyestuffs with—
Acetyl H acid.
Alphaethoxy-8-hydroxynaphthalene-3:6-disulphonic acid.
Alphahydroxynaphthalene-4-sulphonic acid.
3-Aminobenzaldehyde.
2-(4′-Aminobenzoyl)amino-5-naphthol-7-sulphonic acid.
2-(3′-Aminobenzoyl)amino-5-naphthol-7-sulphonic acid.
Anthranilic acid, benzidin-3:3′-dicarboxylic acid, beta-aminobenzene-5-sulphonic acid, beta-aminobenzaldehyde, beta-aminobenzoic acid, beta-amino-1-hydroxybenzene, beta-aminonaphthalene-3-carboxylic acid, betanaphthol, betaphenylamino-4-hydroxynaphthalene-7-sulphonic acid.
4-Chloro-2-chloro-2-aminobenzoic acid.
4:4′-Diaminodiphenylurea-3:3′-dicarboxylic acid.
4:6-Dichloro-2-amino-1-hydroxybenzene.
5:5′-Dihydroxy-2:2′-dinaphthylamine-7:7′-disulphonic acid.
J acid, 5-nitro-2-aminobenzoic acid.

Tetranitrodianthrone
Chemical
Starting point in making—
2:7-Dimethylanthraquinone (U. S. 1622168).

Tetranitrodiglycerin
Explosives and Matches
Ingredient (U. S. 1879064) of—
Low-freezing explosive compositions containing also ethyleneglycol dinitrate and nitroglycerin.

Tetranitropentaerythrite
Synonyms: Penthrite.
Explosives
As an explosive material of great destructive force and sensitivity to shock.
Ingredient of—
Detonating compositions in admixture with fulminate of mercury.
Explosives in admixture with aromatic nitro derivatives.
Mining explosives in admixture with ammonium nitrate.
Shell explosives in admixture with nitroglycerin.

Tetrapentyldiaminobenzophenone
Chemical
Starting point (Brit. 272321) in making intermediates with—
Alkoxybenzenes, dialkylanilines and homologs, diphenyls, halogenated benzenes, halogenated toluenes, halogenated xylenes, naphthalenes.
Dye
Starting point (Brit. 249160) in making triarylanthrone dyestuffs with—

Tetrapentyldiaminobenzophenone (Continued)
3-Chlorophenylethylenediamine, dibenzyldiphenylethylenediamine, diethyldiphenylethylenediamine, dimethyldiphenylethylenediamine, diorthotolylethylenediamine, 3-tolylethylenediamine, xylylethylenediamine.

3:7-Tetrapentyldiaminoxanthone
French: 3:7-Tétrapentylediaminoxanthone.
German: 3:7-Tetrapentyldiaminoxanthon.

Chemical
Starting point in making—
Intermediates.

Dye
Starting point (Brit. 314825) in making xanthene dyestuffs with the aid of—
Alphachloronaphthalene, betachloronaphthalene, 4-chlorometaxylene, metachloroanilin, metachloroanisol, metachlorobenzylamine, metachlorocresidin, metachlorophenylamine, metachlorotoluene, metachlorotoluidin, metachloroxylene, metachloroxylidin, orthochloroanilin, orthochloroanisol, orthochlorobenzylamine, orthochlorocresidin, orthochlorophenylamine, orthochlorotoluene, orthochlorotoluidin, orthochloroxylene, orthochloroxylidin, parachloroanilin, parachloroanisol, parachlorobenzylamine, parachlorocresidin, parachlorophenylamine, parachlorotoluene, parachlorotoluidin, parachloroxylene, parachloroxylidin.
Various acyl, aralkyl thioether derivatives of aromatic halogen compounds.

Tetraphenylphosphonium Bromide
French: Bromure de tétraphénylephosphonium.
German: Bromtetraphenylphosphonium, Tetraphenylphosphoniumbromid.

Miscellaneous
Reagent (Brit. 312163) in treating—
Furs, hair, feathers, and the like to render them mothproof and moldproof.

Textile
Reagent (Brit. 312163) in treating—
Wool and felt to render them mothproof and moldproof.

Tetrapropyldiaminobenzophenone

Chemical
Starting point (Brit. 272321) in making intermediates with—
Alkoxybenzenes, dialkylanilines and homologs, diphenyls, halogenated benzenes, halogenated toluenes, halogenated xylenes, naphthalenes.

Dye
Starting point (Brit. 249160) in making triarylanthrone dyestuffs with—
3-Chlorophenylethylenediamine, dibenzyldiphenylethylenediamine, dimethyldiphenylethylenediamine, diorthotolylethylenediamine, 3-tolylethylenediamine, xylylethylenediamine.

3:7-Tetrapropyldiaminoxanthone

Chemical
Starting point in making—
Intermediates, pharmaceuticals.

Dye
Starting point (Brit. 314825) in making xanthene dyestuffs with—
Alphachloronaphthalene, betachloronaphthalene, 4-chlorometaxylene, metachloroanilin, metachloroanisol, metachlorobenzylamine, metachlorocresidin, metachlorophenylamine, metachlorotoluene, metachlorotoluidin, metachloroxylene, metachloroxylidin, orthochloroanilin, orthochloroanisol, orthochlorobenzylamine, orthochlorocresidin, orthochlorophenylamine, orthochlorotoluene, orthochlorotoluidin, orthochloroxylene, orthochloroxylidin, parachloroanilin, parachloroanisol, parachlorobenzylamine, parachlorocresidin, parachlorophenylamine, parachlorotoluene, parachlorotoluidin, parachloroxylene, parachloroxylidin.
Various acyl, aralkyl thioether derivatives of aromatic halogen compounds.

Tetrapyridin-Copper Sulphate
French: Sulphate de cuivre et de pyridine, Sulphate cuivrique-pyridinique.
German: Schwefelsaeurestetrapyridinkupfer, Tetrapyridinkupfelsulfat.

Chemical
As a general reagent.

Dye
Reagent (Brit. 306859) in making azo dyestuffs with—
Acetyl H acid, alphaethoxy-8-hydroxynaphthalene-3:6-disulphonic acid, alphahydroxynaphthalene-4-sulphonic acid, 3-aminobenzaldehyde.
2-(3′-Aminobenzoyl)amino-5-naphthol-7-sulphonic acid.
2-(4′-Aminobenzoyl)amino-5-naphthol-7-sulphonic acid.
Anthranilic acid, benzidin-3:3′-dicarboxylic acid, beta-aminobenzaldehyde, beta-aminobenzene-5-sulphonic acid, beta-aminobenzoic acid, beta-amino-1-hydroxybenzene, beta-aminonaphthalene-3-carboxylic acid, betanaphthol, betaphenylamino-4-hydroxynaphthalene-7-sulphonic acid.
4-Chloro-2-chloro-2-aminobenzoic acid.
4:4′-Diaminodiphenylurea-3:3′-dicarboxylic acid.
4:6-Dichloro-2-amino-1-hydroxybenzene.
5:5′-Dihydroxy-2:2′-dinaphthylamine-7:7′-disulphonic acid.
J acid, 5-nitro-2-aminobenzoic acid.

Tetratrimethylamine-Copper Sulphate
French: Sulphate de tétratriméthyleamine et de cuivre, Sulphate tétratriméthyleaminique et cuivrique.
German: Schwefelsaeurestetratrimethylaminkupfer.

Chemical
Reagent in making various intermediates.

Dye
Reagent (Brit. 306859) in making azo dyestuffs with—
Acetyl H acid, alphaethoxy-8-hydroxynaphthalene-3:6-disulphonic acid, alphahydroxynaphthalene-4-sulphonic acid, 3-aminobenzaldehyde.
2-(3′-Aminobenzoyl)amino-5-naphthol-7-sulphonic acid.
2-(4′-Aminobenzoyl)amino-5-naphthol-7-sulphonic acid.
Anthranilic acid, benzidin-3:3′-dicarboxylic acid, beta-aminobenzaldehyde, beta-aminobenzene-5-sulphonic acid, beta-aminobenzoic acid, beta-amino-1-hydroxybenzene, beta-aminonaphthalene-3-carboxylic acid, betanaphthol, betaphenylamino-4-hydroxynaphthalene-7-sulphonic acid.
4-Chloro-2-chloro-2-aminobenzoic acid.
4:4′-Diaminodiphenylurea-3:3′-dicarboxylic acid.
4:6-Dichloro-2-amino-1-hydroxybenzene.
5:5′-Dihydroxy-2:2′-dinaphthylamine-7:7′-disulphonic acid.
J acid, 5-nitro-2-aminobenzoic acid.

Thallium

Chemical
Starting point in making various thallium salts.
Starting point (Brit. 281307) in making zeolite catalysts used in making—
Acenaphthylene from acenaphthene.
Acetaldehyde from ethyl alcohol.
Acetic acid from ethyl alcohol.
Alcohols from aliphatic hydrocarbons.
Aldehydes and acids by the oxidation of orthochlorotoluene, parachlorotoluene, orthobromotoluene, parabromotoluene, dichlorotoluene, chlorobromotoluenes, nitrotoluenes, chloronitrotoluenes, bromonitrotoluenes.
Alphaanthraquinone from naphthalene.
Anthraquinone from anthracene.
Benzaldehyde and benzoic acid from toluene.
Benzoquinone from phenanthraquinone.
Chloracetic acid from ethylenechlorohydrin.
Diphenic acid from ethyl alcohol.
Fluorenone from fluorene.
Formaldehyde from methane or methanol.
Hemimellitic acid from acenaphthene.
Maleic acid, and fumaric acid from benzene, toluene, phenol, or tar acids, or from benzoquinone or phthalic anhydride.
Naphthalic anhydride.
Naphthaldehydic acid, acenaphthaquinone, or bisacenaphthylidenedione from acenaphthene or acenaphthylene.
Phenanthraquinone from phenanthrene.
Phthalic anhydride from naphthalene.
Salicylic aldehyde and salicylic acid from cresol.
Vanillin or vanillic acid from eugenol or isoeugenol.

Metallurgical
Ingredient of various alloys.

Thallium Acetate
French: Acétate thallique, Acétate de thallium.
German: Essigsäuresthallium, Essigsäuresthalliumoxyd, Thalliacetat.

Chemical
Ingredient of catalytic mixtures used in the manufacture of—
Acenaphthylene, acenaphthaquinone, bisacenaphthylid-

Thallium Acetate (Continued)

enedione, naphthaldehydic acid, naphthalic anhydride, and hemimellitic acid from acenaphthene (Brit. 295270).
Acetaldehyde from ethyl alcohol (Brit. 281307).
Acetic acid from ethyl alcohol (Brit. 281307).
Alcohols from aliphatic hydrocarbons (Brit. 281307).
Aldehydes and alcohols by the reduction of the corresponding esters (Brit. 306471).
Aldehydes and acids from toluene, orthochlorotoluene, orthobromotoluene, orthonitrotoluene, parachlorotoluene, parabromotoluene, paranitrotoluene, metachlorotoluene, metabromotoluene, metanitrotoluene, dichlorotoluenes, dinitrotoluenes, dibromotoluenes, chlorobromotoluenes, chloronitrotoluenes, bromonitrotoluenes (Brit. 295270).
Aldehydes and acids from xylenes, pseudocumenes, mesitylene, and paracymene (Brit. 281307).
Alphanaphthaquinone from naphthalene (Brit. 281307).
Anthraquinone from naphthalene (Brit. 295270).
Benzaldehyde and benzoic acid from toluene (Brit. 281307).
Benzoquinone from phenanthraquinone (Brit. 281307).
Benzyl alcohol from benzaldehyde by reduction (Brit. 306471).
Benzyl alcohol or benzaldehyde or phthalide by the reduction of phthalic anhydride (Brit. 306471).
Butyl alcohol by the reduction of crotonaldehyde Brit. 306471).
Chloroacetic acid from ethylenechlorohydrin (Brit. 295270).
Diphenic acid from ethyl alcohol (Brit. 281307).
Ethyl alcohol by the reduction of acetaldehyde (Brit. 306471).
Fluorenone from fluorene (Brit. 295270).
Formaldehyde by the reduction of methane or methanol (Brit. 306471).
Formaldehyde by the reduction of carbon dioxide or carbon monoxide (Brit. 306471).
Hydroxyl compounds by the reduction of anthraquinone, benzoquinone and similar compounds (Brit. 306471).
Isopropyl alcohol by the reduction of acetone (Brit. 306471).
Maleic acid and fumaric acid by the oxidation of toluene, benzene, phenols, tar phenols, or furfural, or from benzoquinone or phthalic anhydride (Brit. 295270).
Methane by the reduction of carbon dioxide or carbon monoxide (Brit. 306471).
Methanol by the reduction of carbon dioxide or carbon monoxide (Brit. 306471).
Naphthaldehydic acid, acenaphthaquinone, or bisacenaphthylidenedione from acenaphthylene (Brit. 281307).
Phenanthraquinone from phenanthrene or diphenic acid (Brit. 295270).
Phthalic acid and maleic acid from naphthalene (Brit. 295270).
Primary alcohols by the reduction of the corresponding aldehydes (Brit. 306471).
Propionic acid and butyric acid and higher alcohols, ketones, and acids by the reduction of carbon dioxide or carbon monoxide (Brit. 306471).
Reduction products of ketones, aldehydes, acids, esters, alcohols, ethers, and other organic compounds which contain oxygen (Brit. 306471).
Salicylic acid and salicylic aldehyde from cresol (Brit. 295270).
Secondary butyl alcohol by the reduction of methylethyl ketone (Brit. 306471).
Valeryl alcohol by the reduction of valeraldehyde (Brit. 306471).
Vanillin and vanillic acid from eugenol or isoeugenol (Brit. 295270).

Ingredient (Brit. 306460) of catalytic preparations used in the production of various aromatic and aliphatic compounds, including—
Alphanaphthylamine from alphanitronaphthalene.
Amines from aliphatic nitro compounds, such as allyl nitriles or nitromethane.
Amino compounds from the corresponding nitroanisoles.
Amylamine from pyridin.
Anilin, azo-oxybenzene, azobenzene, and hydrazobenzene, from nitrobenzene by reduction.
Aminophenols from nitrophenols.
3-Aminopyridin from 3-nitropyridin.
Cyclohexamine, dicyclohexamine, and cyclohexylanilin from nitrobenzene.
Piperidin from pyridin, pyrrolidin from pyrrol, tetrahydroquinolin from quinolin.

Starting point in making—
Thallium salts.

Gas
Ingredient of—
Compositions used in the manufacture of gas mantles by impregnation of rayon fabric.

Perfume
Ingredient of—
Depilatory compositions.

Pharmaceutical
Suggested for the treatment of syphilis, and night sweats in tuberculosis.

Thallium Amylalcoholate
French: Amylealcoolate thallique, Amylealcoolate de thallium.
German: Thalliumalkoholat.

Petroleum
Anti-knock agent (Brit. 279560) in making—
Motor fuels.

Thallium Benzylate
French: Benzylate de thallium, Benzylate thallique.
German: Benzylsaeuresthallium.

Petroleum
Ingredient of—
Motor fuels, added to prevent knocking (Brit. 279560).

Thallium Chloride
French: Chlorure de thallium.
German: Thallium chlorid.

Chemical
Catalyst in chlorinating intermediate chemicals and other compounds.

Gas
Catalyst in treating—
Products of carbonization, such as coke, in order to modify the ignition temperature (U. S. 1576179).

Thallium Dinaphthylnaphthenate

Lubricant
Addition agent (Brit. 433257) to—
Lubricating oils or greases, especially for use at high temperatures, such as cylinder oils, hydrogenated oils, or oils refined by treatment with sulphuric acid, clay, or extraction solvents.

Thallium Ethyl
French: Thallium éthylé.
German: Thalliumaethyl.

Petroleum
Ingredient of—
Motor fuels, added for the purpose of stopping the knock (Brit. 279560).

Thallium Formate
French: Formiate de thallium.
German: Ameisensaeuresthallium, Thalliumformiat.

Analysis
Reagent for carrying out mineralogical assays.

Thallium Iodide
French: Iodure de thallium.
German: Jodthallium, Thalliumjodur.

Photographic
Ingredient of—
Emulsion with silver iodide to increase general and chromatic sensitivity.

Thallium-Mercurous Nitrate
French: Nitrate de mercurieux thallium.
German: Thalliumercuronitrat.

Analysis
Reagent for carrying out mineralogical assays.

Thallium Nitrate
French: Nitrate de thallium.
German: Thalliumnitrat.

Paint and Varnish
Starting point in making—
Deep-yellow pigments by reaction with potassium chromate (2) and potassium bichromate (1).
Lemon-yellow pigments by reaction with an ammoniacal solution of potassium chromate and mixing the precipitate with alumina.

Thallium Nitrate (Continued)
Middle-yellow pigments by reaction with potassium chromate (1), potassium bichromate (1), and ammonia.
Orange-yellow pigments by reaction with potassium chromate (1) and potassium bichromate (1).
Pale-yellow pigments by reaction with potassium chromate.
Reddish-orange pigments by reaction with bichromate of potash.
Photographic
Addition agent to—
Silver nitrate in making iodide-free emulsions to increase contrast without fog for development with hydroquinone.

Thallium Oxalate
French: Oxalate de thallium.
German: Oxalsaeuresthallium.
Photographic
As a layer sensitive to ultra-violet light (U. S. 1880503).

Thallium Oxide
French: Oxyde de thallium.
German: Thalliumoxyd.
Glass
Coloring agent for imparting—
Greenish-yellow shades to lead glass.
Reagent for—
Increasing refractive index of lead glass.
Rendering lead glass more suitable for making artificial gems.

Thallium Oxysulphide
French: Oxysulfure de thallium.
German: Thalliumoxysulfid.
Electrical
Light-sensitive material in—
Thalophide cell (a photoelectric cell claimed superior to the selenium cell).

Thallium-Phenyl Acetate
Petroleum
Addition agent (Brit. 433257) in—
Lubricating oils or greases, especially for use at high temperatures, such as cylinder oils, hydrogenated oils, or oils refined by treatment with sulphuric acid, clay, or extraction solvents.

Thallium Phenylethylate
French: Phényleéthylate thallique, Phényleéthylate de thallium.
German: Thalliumphenylaethylat.
Petroleum
Anti-knock agent in making—
Motor fuels (Brit. 279560).

Thallium Sulphate
Insecticide
Ingredient of—
Fire-ant insecticide composed of sugar, honey and water.
Rat poison composed of whole wheat, starch, glycerin, and water.
Rat poison composed of corn syrup, peanut butter, and water.
Rat poison composed of starch, sodium bicarbonate, molasses, glycerin, saccharin, rice, and water.
Rat poison with tapioca-flour paste.
Red ant insecticide composed of sugar, honey, and water.

Thallous Selenide
French: Sélénide thalleux, Sélénide thallieux, Sélénide de thallium.
German: Thalloselenid, Thalliumselenid, Selenthallium.
Insecticide
Ingredient of—
Compositions used against chestnut blight fungus and pear blight fungus.

Thallous-Silver Nitrate
French: Nitrate d'argent-thallique.
German: Thallosilbernitrat.
Analysis
Reagent for carrying out mineralogical assays.

Thebaine
Synonyms: Paramorphine.
Chemical
Starting point in making derivatives used as drugs (German 437451).
Pharmaceutical
In compounding and dispensing practice.

Theophyllin
Chemical
Starting point (U. S. 1867332) in making—
Mono- and triethanolamine salts (of theophyllin).

Thioacetanilide
Photographic
Fog-reducer in—
Metol-quinol developing processes (said to accomplish considerable fog reduction without changing the gradation).

Thioacetnaphthalide
Photographic
Fog-reducer in—
Metol-quinol developing processes (said to accomplish considerable fog reduction without changing the gradation).

Thioammelin
Chemical
Reagent (Brit. 286749) in making vulcanization accelerators with—
Dibenzylamine, diethylguanylthioureas, diphenyl biguanide, ditolyl biguanide, ethanolamines, guanylureas, isothioureas, isoureas, monophenyl biguanide, monophenylguanylthiourea, monotolyl biguanide, pentaphenyl biguanide, pentatolyl biguanide, piperazin, piperidin, tetramethylammonium hydroxide, tetraphenyl biguanide, tetratolyl biguanide, thioureas, trimethylsulphonium hydroxide.

Thiobenzamide
Metallurgical
Ingredient (U. S. 1734560) of—
Pickling bath.

Thiocresol
Petroleum
Antioxidant (Brit. 425569) for—
Lubricating, transformer, and switch oils, particularly solvent-extracted oils and others of a paraffinic nature, in which the natural inhibitor content may have been reduced during refining.

Thiocyanic Acid Aminoanisylester
Insecticide
Exterminant (German 562672) for—
Insects.

Thiocyanic Acid Aminonaphthylester
Insecticide
Exterminant (German 562672) for—
Insects.

Thiocyanic Acid Aminotolylester
Insecticide
Exterminant (German 562672) for—
Insects.

Thiocyanic Acid 2-(2′-Butoxyethoxy)ethylester
Insecticide
Powerful exterminant (German 562672) for—
Flies and other household insects.

Thiocyanic Acid Chlorobenzylester
Insecticide
Powerful exterminant (German 520330) for—
Flies and other household insects.

Thiocyanic Acid 2:4-Dinitrophenylester
Insecticide
Exterminant (German 562672) for—
Insects.

Thiocyanic Acid 2-(2-Ethoxyethoxy)ethylester
Insecticide
Powerful exterminant (German 562672) for—
Flies and other household insects.

Thiocyanic Acid Paradimethylaminophenylester
Insecticide
Exterminant (German 562672) for—
Insects.

Thiocyanic Acid Paramorpholinphenylester
Insecticide
Exterminant (German 562672) for—
Insects.

Thiocyanoacetone
German: Thiozyanaceton.
Chemical
Starting point in making various derivatives.
Insecticide
Ingredient (Brit. 361900) of—
Insecticidal preparations (used in solution in water or in an organic solvent, such as kerosene).

Thiodiglycol
German: Thiodiglykol.
Chemical
Solvent in—
Various processes (Brit. 272908).
Dye
Solvent in making—
Soluble metallic compounds of azo dyestuffs (Brit. 272908).
Miscellaneous
Solvent for various purposes (Brit. 272908).

Thioflavin T
Chemical
Ingredient (Brit. 295605) of biological stains, therapeutic and bacteriological preparations containing—
Cresol, guaiacol, hydroquinone, phenol, phloroglucinol, pyracatechol, pyragallol, resorcinol.
Textile
As a color in dyeing and printing.

Thioglycollic Acid
French: Acide de thioglycol.
German: Thioglycolsaeure.
Chemical
Starting point in making—
Complex antimony derivatives and esters used as drugs.

2-Thionaphtheneacenaphthene Indigo
Dye
Starting point in making—
Ciba orange G paste, vat dyestuffs in solid form (Brit. 250251).
Textile
——, *Dyeing*
Color for—
Cotton, silk, wool.
——, *Printing*
Color for—
Cotton fabrics.

Thionaphthene-2:3-dicarboxylic Anhydride
French: Anhydride de thionaphthène-2:3-dicarboxyle.
German: Thionaphten-2:3-dicarbonsaeureanhydrid.
Dye
Starting point (Brit. 261384) in making thionaphthene dyestuffs with—
Anthracene, benzene, cymene, mesitylene, naphthalene, toluene, triphenylmethane, tolyldiphenylmethane, xylene.

Thionyl Bromide
Chemical
Starting point (Brit. 382327) in making—
4-Aminopyridin from pyridin.

Thionyl Chloride
Chemical
Catalyst (Brit. 398064) in making—
Cinnamic boricanhydride from cinnamic and boric acids.

Thiophene Oleate, Chlorinated
Lubricant
Stabilizing agent (Brit. 451412 and 453047) for—
Lubricating oils subjected to high pressures.
Top cylinder lubricating compositions.

Thiophene Stearate, Chlorinated
Lubricant
Stabilizing agent (Brit. 451412 and 453047) for—
Lubricating oils subjected to high pressures.
Top cylinder lubricating compositions.

Thiophosgene
French: Chlorure de thiocarbonyle, Thiophosgène.
German: Thiocarbonylchlorid, Thiophosgen.
Analysis
As a reagent.
Chemical
Reagent (Brit. 264682) in making—
Diorthocarbethoxydiphenylthiourea, diorthocarbethoxyditolylthiourea, diorthocarbethoxydixylylthiourea, diorthocarbomethoxydiphenylthiourea, diorthocarbomethoxyditolylthiourea, diorthocarbomethoxydixylylthiourea, mono-orthocarbethoxydiphenylthiourea, mono-orthocarbethoxyditolylthiourea, mono-orthocarbethoxydixylylthiourea, mono-orthocarbomethoxydiphenylthiourea, mono-orthocarbomethoxyditolylthiourea, mono-orthocarbomethoxydixylylthiourea.
Dye
Reagent in making various dyestuffs.

Thiosalicylic Acid
French: Acide thiosalicylique.
German: Thiosalicylsaeure.
Chemical
Starting point in making—
Orthosulphobenzoic acid, strontium-sodium thiosalicylate (U. S. 1561535).
Dye
Starting point in making—
Thioindigo, vat red B.

Thiosinamine
Chemical
Starting point in making—
Fibrolysin.
Photographic
Ingredient of—
Mixtures used for toning sulphide-toned silver prints on developing and printing-out paper (German 422295).

3-Thiosulphanilic Acid
Chemical
Starting point (Brit. 398020) in making—
Complex double compounds of organic heavy metal-mercapto compounds.

Thiourea
Synonyms: Sulphocarbamide, Sulfourea, Thiocarbamide.
French: Sulphourée, Thiourée.
German: Sulfocarbamid, Sulfoharnstoff, Thiocarbamid, Thioharnstoff.
Chemical
Starting point in making—
Aromatic chemicals, barbital (diethylbarbituric acid) and other pharmaceuticals, intermediates.
Starting point (Brit. 314909) in making derivatives with—
Alkoxyalphanaphthalenesulphonic acid, alpha-amino-5-naphthol-7-sulphonic acid, alphanaphthylamine-4:8-disulphonic acid, alphanaphthylamine-4:6:8-trisulphonic acid, 4-aminoacenaphthene-3:5-disulphonic acid, 4-aminoacenaphthene-3-sulphonic acid, 4-aminoacenaphthene-5-sulphonic acid, 4-aminoacenaphthene-trisulphonic acids, aminoarylcarboxylic acids, aminoheterocyclic-carboxylic acids, 1:8-aminonaphthol-3:6-disulphonic acid, bromonitrobenzoyl chlorides, chloroalphanaphthalenesulphonic acids, chloronitrobenzoyl chlorides, iodonitrobenzoyl chlorides, nitroanisoyl chlorides, 2-nitrocinnamyl chloride, 3-nitrocinnamyl chloride, 4-nitrocinnamyl chloride, 1-nitronaphthalene-5-sulphochloride, 1:5-nitronaphthoyl chloride, 2-nitrophenylacetyl chloride, 4-nitrophenylacetyl chloride, nitrotoluyl chlorides.
Dye
Starting point in making various synthetic dyestuffs.
Insecticide
Ingredient of insecticidal compositions.
Miscellaneous
Mothproofing agent (U. S. 1748579) in treating—
Furs, feathers, skins.
Pharmaceutical
In compounding and dispensing practice.

Thiourea (Continued)

Photographic
Developer for plates and films.
Fixing agent in photographic work.
Ingredient of—
Compositions used for the removal of stains from negatives.
Compositions for toning sulphide-toned images on developing and printing papers (German 422295).
Resins and Waxes
Starting point in making—
Artificial resins with the aid of formaldehyde.
Textile
——, *Finishing*
Ingredient of—
Compositions used for treating rayon so as to protect the filament in after-treatment.
——, *Miscellaneous*
Ingredient of—
Compositions used for restoring the elasticity and strength of rayons which have been heavily weighted.
Compositions used in mothproofing woolen and felt fabrics (U. S. 1748579).

Thiourea-3:3'-dicarboxylic Acid
French: Acide de sulphourée-3:3'-dicarbonique, Acide de sulphourée-3:3'-dicarboxylique, Acide de thiourée-3:3'-dicarbonique, Acide de thiourée-3:3'-dicarboxylique.
German: Sulfoharnstoff-3:3'-dicarbonsäure, Thioharnstoff-3:3'-dicarbonsäure.
Chemical
Starting point in making—
Esters, salts, and other derivatives.
Starting point (Brit. 314909) in making pharmaceutical derivatives with the aid of—
Alkoxyalphanaphthalenesulphonic acid, alpha-amino-5-naphthol-7-sulphonic acid, alphanaphthylamine-4:8-disulphonic acid, alphanaphthylamine-4:6:8-trisulphonic acid, 4-aminoacenaphthene-3:5-disulphonic acid, 4-aminoacenaphthene-3-sulphonic acid, 4-aminoacenaphthene-5-sulphonic acid, 4-aminoacenaphthene-trisulphonic acids, aminoarylcarboxylic acids, aminoheterocyclic-carboxylic acids, 1:8-aminonaphthol-3:6-disulphonic acid, bromonitrobenzoyl, chlorides, chloroalphanaphthalenesulphonic acids, chloronitrobenzoyl chlorides, iodnitrobenzoyl chlorides, nitroanisoyl chlorides, 2-nitrocinnamyl chloride, 3-nitrocinnamyl chloride, 4-nitrocinnamyl chloride, 1-nitronaphthalene-5-sulphochloride, 1:5-nitronaphthoyl chloride, 2-nitrophenylacetyl chloride, 4-nitrophenylacetyl chloride, nitrotoluyl chlorides.

Thioxyindole
Synonyms: Sulfoxyindole.
German: Thioxyindol.
Chemical
Reagent (Brit. 286749) in making vulcanization accelerators with—
Dibenzylamine, diethylguanylthioureas, diphenyl biguanide, ditolyl biguanide, ethanolamines, guanylureas, isothioureas, isoureas, monophenyl biguanide, monophenylguanyl thioureas, monotolyl biguanide, pentaphenyl biguanide, pentatolyl biguanide, piperazin, piperidin, tetramethylammonium hydroxide, tetraphenylbiguanide, tetratolyl biguanide, thioureas, trimethylsulphonium hydroxide.

Thorium Dioxide
Synonyms: Thoric oxide, Thorium anhydride, Thorium oxide.
French: Oxyde de thorium.
German: Thoriumdioxyd, Thoroxyd.
Chemical
Catalyst in—
Oxidation of ammonium to nitric acid, oxidation of carbon monoxide to water gas, oxidation of sulphur dioxide to sulphur trioxide.
Reagent in making—
Butyl acetate, butyl butyrate, butyl formate, butyl propionate, ethyl acetate, ethyl butyrate, ethyl formate, ethyl propionate, methyl acetate, methyl formate, methyl butyrate, methyl propionate, propyl acetate, propyl formate, propyl butyrate, propyl propionate.
Ceramics
Ingredient of—
Fire-resisting compositions for making crucibles and the like.
Lighting
Ingredient of—
Compositions used in making incandescent gas mantles.

Thorium Nitrate
French: Azotate thorique, Azotate de thorium, Nitrate thorique.
German: Salpetersaeuresthorium.
Chemical
Starting point in making—
Thorium dioxide.
Lighting
Ingredient of—
Compositions used in making gas mantles.
Pharmaceutical
In compounding and dispensing practice.

Thymol Cinnamate
French: Cinnamate de thymole, Cinnamate thymolique.
German: Thymolcinnamat, Zimtsäuresthymol, Zimtsäurethymolester.
Chemical
Starting point in making—
Pharmaceuticals and other derivatives.
Pharmaceutical
In compounding and dispensing practice.

Thymol Cyclopentenylacetate
Food
Agent for producing—
Pineapple aroma and flavor.

Tiglic Acid
Water and Sanitation
Breaker (U. S. 1964444) of—
Emulsoids in sewage.

Tin Albuminate
French: Albuminate d'étain, Albuminate stannique.
German: Albuminsaeureszinn, Zinnalbuminat.
Chemical
Reagent for various purposes.
Starting point in making various derivatives.
Rubber
Reagent (U. S. 1640817) in—
Reclaiming old rubber.

Tin-Ammonium Chloride
Synonyms: Ammonium chlorostannate, Pink salt.
French: Chlorostannate d'ammonium, Chlorure ammoniaque et stannique, Chlorure d'ammonium et d'étain, Chlorure d'étain et d'ammonium, Sel pink.
German: Ammoniakzinnchlorid, Ammoniumzinnchlorid, Chlorammoniumzinn, Chlorzinnammonium, Pinksalz, Zinnammoniakchlorid, Zinnammoniumchlorid.
Spanish: Cloruro de estano y de ammonio.
Italian: Clorostannate di ammonio.
Miscellaneous
Reagent (Brit. 271026) in—
Carroting furs and felts.
Textile
Reagent for—
Imparting brilliance to fabrics dyed with alizarin colors.
Imparting solidity to alizarin-dyed fabrics.
Substitute for—
Tartar emetic.
Weighting agent for—
Silk.

Tin Betabenzoylpropionate
Plastics
Starting point (U. S. 2001380) in making—
Films.

Tin Dichloride
Synonyms: Stannous chloride, Tin bichloride, Tin chloride, Tin protochloride, Tin salt.
French: Chlorure d'étain, Chlorure stanneux, Étain chloreux, Étain chlorure, Protochlorure d'étain, Sel étain, Sel d'étain.
German: Bichlorzinn, Dichlorzinn, Salzsäureszinnoxyd, Salzsäureszinnoxydul, Stannochlorid, Zinnchlorid, Zinndichlorid, Zinnsalz.
Spanish: Cloruro de estano, Protocloruro de estano.
Italian: Cloruro stannoso, Protocloruro stannoso.

Tin Dichloride (Continued)

Analysis
Reagent in—
Analytical processes, both with respect to industrial control requirements and research in pure and applied chemistry.
Testing agent for—
Arsenic.

Brewing
Retarder of—
Fermentation.

Chemical
Eliminator (French 632920) of—
Antimony and lead from arsenate solutions.
Reducing agent in making—
Acridin derivatives (Brit. 319794), inorganic chemicals, intermediate chemicals, organic chemicals.
Starting point in making—
Stabilizing agent for hydrogen peroxide solutions by reacting with phosphoric acid (U. S. 2004809).
Tin salts, such as hydrated stannic chloride, stannous oxide, stannous hydroxide, stannous acetate, tin oxalate.

Coke By-Products
Catalyst in purifying—
Hydrocarbon oils (Brit. 406963 and 405736).
Hydrocarbon oils with ozonized air (Brit. 367848).
Catalyst in treating—
Carbonaceous materials with hydrogenating gases (Brit. 406963 and 405736).
Decolorizing agent (French 610498 and 610499) for treating—
Hydrocarbon oils.
Desulphurizing agent (French 611890) for treating—
Hydrocarbon oils.

Cosmetic
Starting point (U. S. 1899707) in making—
Depilatory compounds.

Distilling
Retarder of—
Fermentation.

Dye
Reducing agent in making—
Color lakes, dyestuffs, such as indigo, intermediates.

Glass
Reagent in—
Silvering processes.

Leather
Ingredient (French 552161) of—
Tanning agent for hides, pelts, and skins, such agents comprising admixtures with chlorides of sodium, potassium, calcium, and aluminum.
Tanning agent (French 526574) for—
Hides, pelts, skins.

Metallurgical
Ingredient of—
Electrolyte, containing also zinc cyanide, sodium chloride, caustic soda, sodium cyanide, and trisodium phosphate; used in producing an alloy plating of tin and zinc on iron and steel (this coating is said to have much the same properties as cadmium plating) (U. S. 1904732).
Metal plating solutions (French 494722).
Metal plating solutions, containing also sal ammoniac and trisodium phosphate, or sal ammoniac and sodium chloride, or tartaric acid and sodium carbonate, for iron articles (French 501513).
Nonelectrolytic tin-coating baths (immersion processes) for brass, iron, and also such small articles as pins, thimbles, eyelets, fasteners, chain links, safety pins, buttons, and the like.

Miscellaneous
Eradicator for—
Ink stains.
Ingredient of—
Rust-removing solution for the cold treatment of linens and other textile fabrics (French 562129 and 602474).
Regenerator (French 684060) for—
Green earth.

Paint and Varnish
Starting point in making—
Molybdenum blue, purple of cassius (also known as gold-tin purple and gold-tin precipitate).

Petroleum
Catalyst in purifying—
Oils (Brit. 406963 and 405736).
Oils, with ozonized air (Brit. 367848).
Decoloring agent (French 610498 and 610499) for treating—
Cracking products, mineral oils, shale oils.
Desulphurizing agent (French 611890) for treating—
Cracking products, mineral oils, shale oils.

Pharmaceutical
In compounding and dispensing practice.
Suggested for use as—
Mild caustic, violent irritant.

Photographic
Reducing agent for—
Silver bromide in photographic emulsions.
Reducing agent (French 503954) in making—
Colored pictures.

Rubber
Thermoplasticizing agent (French 615195).

Sugar
Bleaching agent.

Textile
Blooming agent in—
Logwood dyebath processes.
Corrector of—
Iron impurities in printing processes.
Discharge in—
Textile printing.
Mordant in—
Dyeing processes, printing processes.
Reducing agent in—
Dyeing processes.
Printing cellulose derivative fabrics in order to obtain reserve effects (Brit. 399559).
Printing processes.
Starting point in making—
Tin pulp (prussiate of tin).
Weighting agent for—
Silk.

Woodworking
Fixing and stabilizing agent for—
Coloring material in redwood (said to stabilize the color against the action of water, acid, alkali solutions, laundry soap solutions, and to prevent bleeding).

Tin Methylate

French: Méthylate d'étain.
German: Zinnmethylat.

Chemical
Catalyst in making—
Acetic acid (Brit. 259641).

Tin Palmitobenzenesulphonate

Textile
Ingredient (Brit. 269917) of—
Printing pastes employed to enhance the saturation of textiles with dyestuff and for equalizing the printed color.

Tin Pulp

Synonyms: Prussiate of tin, Tin ferrocyanide.
French: Ferrocyanure d'étain, Prussiate d'étain, Prussiate stanneux.
German: Ferrocyanzinn, Ferrocyanwasserstoffsäureszinn, Zinnferrocyanid.

Textile
Reagent in—
Steam blueing.

Tin Silicate

Soap
Stabilizer, in conjunction with stannic acid (Brit. 437128) of—
Bleaching, washing, disinfecting compositions containing percompounds, salts of pyrophosphoric or metaphosphoric acid, and alkali, addition agents, such as soap and the sodium salts of sulphonated higher fatty alcohols or of oleylmethyltaurin.

Tin Spirits

(A name given to a large variety of solutions of tin, in the preparation of which other acids besides hydrochloric, notably sulphuric, nitric, and oxalic, are used, and along with these acids inorganic salts are added; for example, sodium nitrate, sal ammoniac, or sodium chloride; varieties are known in the trade as orange, scarlet, amaranth, purple, plum, puce, anilin).

Textile
Mordant in—
Wool dyeing with natural coloring matters.

Tin Stearotoluenesulphonate
Synonyms: Stannic stearotoluenesulphonate.
French: Stéarotoluènesulphonate d'étain, Stéarotoluènesulphonate stannique.
German: Stannistearotoluolsulfonat, Stearotoluolsulfonsäureszinn, Zinnstearotoluolsulfonat.

Chemical
Starting point in making various derivatives.

Tin Sulphocyanide
Synonyms: Stannic sulphocyanide.
French: Sulfocyanure d'étain, Sulfocyanure stannique.
German: Schwefelartigcyanzinn, Schwefelcyansäureszinn, Schwefelcyanwasserstoffsaeureszinn, Schwefelcyanzinn.

Textile
Mordant in—
Dyeing processes.

Tin Tretrachloride
Synonyms: Butter of tin, Oxymuriate of tin, Stannic chloride, Tin chloride.
French: Beurre d'étain, Chlorure stannique, Deutochlorure d'étain, Étain chlorique, Étain tétrachloré, Oxymuriate d'étain, Perchlorure d'étain, Tétrachlorure d'étain.
German: Oxymuriatzinn, Salzsäureszinnoxyd, Stannichlorid, Tetrachlorzinn, Zinnbutter, Zinnchlorid, Zinntetrachlorid.
Spanish: Cloruro de estano.
Italian: Cloruro estannico.

Adhesives
Starting point (Brit. 310461) in making—
Cements (with rubber and benzene) for bonding fibrous materials to metal.

Agricultural
As a weed-killer.

Chemical
Catalyst in making—
Acetic anhydride from sodium acetate and acetic acid (French 630424).
Orthoamylbenzoylbenzoic acid (U. S. 1889347).
Decolorizing agent (French 619857) for—
Acetone oils, methylene.
Purifying agent (U. S. 1894975) for—
Removing visible and latent color compounds from rosin.
Starting point in making—
Tin chemicals.

Coke By-Products
Condensing agent (Brit. 394073) in making—
Lubricating oils by production of polymerized products from unsaturated hydrocarbons.
Treating agent (French 633643) for—
Crude tar.

Dye
Reagent in making—
Color lakes, fuchsin.

Electrical
Starting point (French 594165) in—
Depositing tin oxide coatings on filaments of thermionic tubes, x-ray tubes, current rectifying tubes.

Fats and Oils
Condensing agent (Brit. 394073) in making—
Lubricating oils, or agents to improve the viscosity curve of other lubricants, from animal or vegetable fatty substances, such as bone oils or soybean, olive, or palm oil.
Polymerizing agent (German 596192) in making—
Stand oils from linseed or poppyseed oil, stand oils from the fatty acids of linseed or poppyseed oil, stand oils from fish oils, stand oils from the fatty acids of fish oils.

Glass
Reagent for—
Producing iridescent effects on glass.

Leather
Ingredient (French 631109) of—
Tanning agent, containing also chrome alum, and sodium silicate, for hides, skins, and pelts.

Metallurgical
Ingredient of—
Electrolytic plating baths (French 490972).
Electrolytic plating baths for superimposing a tin coating on cadmium-plated iron or steel in the production of rust and acid-resisting metal products (French 607754).
Fluxes and cleansing agents for soldering block tin and analogous metals (French 556158).
Soldering agents used in tin-coating aluminum (French 515619).

Military
Ingredient of—
Range-finding compositions used in naval shells, smoke screens.

Miscellaneous
Ingredient of—
Rust-removing agent (admixed with tartaric acid in aqueous solution).
Reagent for producing—
Iridescent effects on artificial pearls (French 560091).
Iridescent effects on artificial pearls, leaves, spangles, buttons, and other novelties, either coated or uncoated with cellulosic lacquers, casein varnishes, or pearl essences (French 684958 and 684959).

Petroleum
Catalyst in—
Cracking processes.
Condensing agent (Brit. 394073) in making—
Lubricating oils by production of polymerized products from unsaturated hydrocarbons.
Reagent (U. S. 1941251) in—
Freeing low-boiling-point cracking products of unsaturated compounds.
Solidifying agent (French 683112) for—
Petroleum ether.

Pharmaceutical
In compounding and dispensing practice.

Plastics
Impregnating agent (Brit. 449651) in—
Thermoplastic compositions and moulded products having a cellulose derivative base.

Rayon
Catalyst in making—
Cellulose acetate, or other derivatives, by esterification of cellulose with the anhydride of a fatty acid (French 664674).
Cellulose acetate yarn which is highly resistant to the delustering action of hot water (Brit. 400249).

Rubber
Catalyst (French 646414) in making—
Isomers of rubber (in presence of phenol).
Hardening agent (U. S. 1948292) for—
Rubber surfaces of golf balls.
Promoter (U. S. 1948292) of—
Oil resistance of rubber, dirt resistance of rubber.

Sugar
Bleaching agent.

Textile
Brightener for—
Colors.
Inhibitor of—
Rapid decomposition in sizing compositions for cotton warps.
Mordant in—
Dyeing processes, printing processes.
Weighting agent for—
Silk.
Silk or rayon (French 609397, 609764, 634641, 648509, 654748, 656424, 666114, and 660115; Brit. 303128 and 303129; German 291009 and 295272; U. S. 1902226, 1896381, 1896858, and 1898105).

Titanium Acetate
French: Acétate de titane, Acétate titanique.
German: Essigsäurestitan, Essigsäurestitanoxyd, Titanacetat, Titanazetat, Titaniumacetat, Titaniumazetat.
Spanish: Acetato de titanio.
Italian: Acetato di titanio.

Chemical
Ingredient of catalytic mixtures used in making—
Acenaphthylene, acenaphthaquinone, bisacenaphthylidenedione, naphthaldehydic acid, naphthalic anhydride, and hemimellitic acid (Brit. 295270).
Acetaldehyde from ethyl alcohol (Brit. 281307).
Acetic acid from ethyl alcohol (Brit. 281307).
Alcohols from aliphatic hydrocarbons (Brit. 281307).
Aldehydes or acids by the reduction of the corresponding esters (Brit. 306471).
Aldehydes and acids from toluene, orthochlorotoluene, orthobromotoluene, orthonitrotoluene, parachlorotoluene, parabromotoluene, paranitrotoluene, metachlorotoluene, metanitrotoluene, metabromotoluene, dichlorotoluenes, dibromotoluenes, dinitrotoluenes, chlo-

Titanium Acetate (Continued)
robromotoluene, chloronitrotoluene, bromonitrotoluene (Brit. 295270).
Aldehydes and acids from xylenes, pseudocumenes, mesitylenes, and paracymene (Brit. 281307).
Alphacampholide from camphoric acid by reduction (Brit. 306471).
Alphanaphthaquinone from naphthalene (Brit. 281307).
Anthraquinone from naphthalene (Brit. 281307).
Benzaldehyde and benzoic acid from toluene (Brit. 281307).
Benzoquinone from phenanthraquinone (Brit. 281307).
Benzyl alcohol by the reduction of benzaldehyde (Brit. 306471).
Benzyl alcohol or benzaldehyde or benzyl phthalide by the reduction of phthalic anhydride (Brit. 306471).
Butyl alcohol by the reduction of crotonaldehyde (Brit. 306471).
Diphenic acid from ethyl alcohol (Brit. 281307).
Ethyl alcohol by the reduction of acetaldehyde (Brit. 306471).
Fluorenone from fluorene (Brit. 295270).
Formaldehyde by the reduction of methane or methanol (Brit. 306371).
Formaldehyde by the reduction of carbon dioxide or carbon monoxide (Brit. 306471).
Hydroxy compounds by the reduction of anthraquinone, benzoquinone, or the like (Brit. 306471).
Isopropyl alcohol by the reduction of acetone (Brit. 306471).
Maleic acid and fumaric acid by the oxidation of toluene, benzene, phenols, tar phenols, or furfural, or from benzoquinone or phthalic anhydride (Brit. 295270).
Methane by the reduction of carbon dioxide or carbon monoxide (Brit. 306471).
Methanol by the reduction of carbon dioxide or carbon monoxide (Brit. 306471).
Naphthaldehydic acid, acenaphthaquinone, bisacenaphthylidenedione from acenaphthylene (Brit. 281307).
Phenanthraquinone from phenanthrene or diphenic acid (Brit. 295270).
Primary alcohols by the reduction of the corresponding aldehydes (Brit. 306471).
Propionic acid and butyric acid and higher alcohols, ketones, and acids by the reduction of carbon dioxide or carbon monoxide (Brit. 306471).
Reduction products of ketones, aldehydes, esters, ethers, alcohols, and other organic compounds which contain oxygen (Brit. 306471).
Salicylic acid and salicylic aldehyde from cresol (Brit. 306471).
Secondary butyl alcohol by the reduction of methylethyl ketone (Brit. 306471).
Valeryl alcohol by the reduction of valeraldehyde (Brit. 306471).
Vanillin and vanillic acid from eugenol or isoeugenol (Brit. 295270).
Ingredient (Brit. 304640) of catalytic preparations, used in the manufacture of various aromatic and aliphatic amines, including—
Alphanaphthylamine from alphanitronaphthalene.
Amines from aliphatic nitro compounds, such as alkyl nitriles or nitromethane.
Amylamine from pyridin.
Anilin, azo-oxybenzene, azobenzene, and hydrobenzene from nitrobenzene by reduction.
Aminophenols from nitrophenols.
3-Aminopyridin from 3-nitropyridin.
Amino compounds from the corresponding nitroanisoles.
Amines from oximes, Schiff's base, and nitriles.
Cyclohexamine, dicyclohexamine, and cyclohexylanilin from nitrobenzene.
Piperidin from pyridin.
Pyrrolidin from pyrrol.
Tetrahydroquinolin from quinolin.
Starting point in making—
Titanium salts.

Leather
Mordant in—
Dyeing leather and leather goods.

Textile
——, *Dyeing and Printing*
Mordant in—
Dyeing and printing various textile yarns and fabrics.

——, *Miscellaneous*
Reagent in—
Stripping color from dyed fabrics.

Titanium Betabenzoylpropionate
Plastics
Starting point (U. S. 2001380) in making—
Films.

Titanium Butylphthalate
Miscellaneous
Preventer (U. S. 1965608) of—
Nitrocellulose coatings discoloration by ultraviolet light.

Titanium Carbide
French: Carbure titane, Carbure de titanium.
German: Titancarbid.
Electrical
Electrode material for—
Arc lamps.
Metallurgical
As a steel purifier (U. S. 1039672).

Titanium-Cobalt Linoleate-Tungstate
Chemical
Suggested dispersing agent (Brit. 395406) in making—
Emulsions of various chemicals.
Dry Cleaning
Saponifying and emulsifying agent (Brit. 395406).
Fats and Oils
Dispersing agent (Brit. 395406) in making—
Boring oils, drilling oils, greasing compositions.
Lubricating compositions containing animal or vegetable oils.
Stabilized emulsions of various animal or vegetable fats and oils.
Wire-drawing oils.
Glues and Adhesives
Suggested dispersing agent (Brit. 395406) for—
Gums and other adhesive materials.
Ink
Suggested dispersing agent (Brit. 395406) in making—
Printing inks, writing inks.
Leather
Dispersing agent (Brit. 395406) in making—
Emulsified dressings containing waxes.
Emulsified waterproofing compositions.
Miscellaneous
Dispersing agent (Brit. 395406) in making—
Automobile polishes, floor waxes and polishes, furniture polishes containing waxes, impregnating compositions containing paraffin, metal polishes, waterproofing compositions.
Linoleum and Oilcloth
Suggested dispersing agent (Brit. 395406) in—
Coating compositions used in making oilcloth.
Paint and Varnish
Dispersing and drying agent (Brit. 395406) in making—
Emulsified lacquers, emulsified paints, emulsified roofing compositions, emulsified varnishes, emulsified waterproofing compositions.
Paper
Suggested dispersing agent (Brit. 395406) in making—
Emulsified impregnating compositions containing waxes.
Emulsified sizing compositions containing waxes.
Emulsified waterproofing compositions.
Emulsified waterproofing compositions containing waxes.
Waxing compositions in emulsified form.
Perfume
Suggested dispersing agent (Brit. 395406) in making—
Creams, greases, lotions, soaps.
Petroleum
Dispersing agent (Brit. 395406) in making—
Stabilized emulsions containing petroleum or petroleum distillates such as paraffin oil and other heavy oils.
Ingredient (Brit. 395406) of—
Kerosene emulsions, lubricating compositions, paraffin emulsions, petrolatum emulsions, soluble greases.
Resins and Waxes
Dispersing agent (Brit. 395406) in making—
Emulsions of artificial waxes, emulsions of natural waxes.
Soap
Ingredient (Brit. 395406) of—
Dry-cleaning soaps, textile soaps.

Titanium-Cobalt Linoleate-Tungstate (Cont'd)

Textile

——, *Dyeing*

Ingredient (Brit. 395406) of—
 Dye baths in mordanting.

——, *Finishing*

Ingredient (Brit. 395406) of—
 Emulsified coating compositions, emulsified sizing compositions, emulsified waterproofing compositions.

——, *Manufacturing*

Ingredient (Brit. 395406) of—
 Dispersions used for washing wool.
 Emulsions for kier-boiling cotton to aid in the removal of the natural gums, fats, waxes, and hemicellulose.

Titanium Dioxide

Synonyms: Titanic acid anhydride, Titanic oxide, Titanium white.
French: Bioxide titanique, Bioxide de titanium, Blanc de titanium, Dioxide titanique, Dioxide de titanium, Oxide titanique, Oxide de titanium.
German: Titandoppelteoxyd, Titanoxyd, Titanweiss.

Cellulose Products

Filler (Brit. 416412) in—
 Cellulose ester-resin products used as coating compositions, protective films, adhesives, impregnating agents, or moulding materials.

Ceramic

Opacifying agent in—
 Enamels, glazes.
Pigment in—
 Enamels, glazes.

Chemical

Catalyst in—
 Oxidizing carbon monoxide.
Catalyst in making—
 Acetic acid esters.
Catalyst stabilizer (Brit. 381185) in making—
 Higher alcohol mixtures such as—
 (1) Normal propyl alcohol, isobutyl alcohol, normal butyl alcohol, methylethylcarbin carbinol, hexyl alcohol, heptyl alcohol, octyl alcohol, and nonyl alcohol.
 (2) Normal propyl alcohol, isobutyl alcohol, normal butyl alcohol, and higher alcohols, with methylethylcarbin carbinol.
Reducer or accelerator (French 752270) of—
 Catalytic action in making aliphatic anhydrides.
Reducing agent (U. S. 1512271) for—
 Beryllium oxide ore, boron oxide ore.
Starting point in making—
 Catalysts.
 Gels having catalytic or adsorbent properties (Brit. 398517).
 Titanium chemicals.

Cosmetic

Pigment in—
 Skin-whitening lotions.

Dental Products

Filler in—
 Artificial teeth.
Pigment in—
 Artificial teeth.

Electrical

Incandescent medium and pigment in making—
 Arc light electrodes.
 Electrodes for mercury vapor lamps (U. S. 1902936).

Glass

Opacifying agent.

Ink

Opacifying agent.
Pigment in—
 White inks.

Leather

Mordant in—
 Leather dyeing.

Linoleum and Oilcloth

Filler in—
 Linoleum.
Opacifying agent in—
 Linoleum.
Pigment in—
 Linoleum.

Metallurgical

Ingredient (U. S. 1909217) of—
 Flux-coated welding rods.

Metal Fabricating

Opacifying agent in—
 Vitrified enamels for iron and steel articles.
Pigment in—
 Vitrified enamels for iron and steel articles.

Miscellaneous

Improver (U. S. 1913480) of—
 Color, in artificial light, of synthetic spinels used as gem stones.
Ingredient of—
 Sealing composition for glassware, containing also glue, casein, talc, diethyleneglycol, paraformaldehyde, water and ammonia (U. S. 1904445).

Paint and Varnish

Pigment in—
 Paints, lacquers, enamels, varnishes, dopes of various kinds.
Starting point in making—
 Dry pigment preparations readily dispersible in aqueous mediums (Brit. 404041).
 White pigments of lithopone type with barium, calcium, and other bases.

Paper

Filler in—
 Beater cycle.
 Book papers, tissue papers, and the like.
 Wrapping paper for foodstuffs (U. S. 1946141).
Improver of—
 Printing qualities.
Increaser of—
 Brightness and opacity in tissue papers and thin papers for books, encyclopedias, bonds, ledgers, and the like.
Pigment in—
 Book papers, tissue papers, and the like.
 Coating processes.
Reducer of—
 Offset.
Substitute for—
 Bleaching operations on off-colored stock.

Petroleum

Catalyst in making—
 Gasoline (U. S. 1124333).
 Petroleum ether (U. S. 1124333).

Photographic

Sensitivity promoter (Brit. 413095) in—
 Production and reproduction of colored pictures or designs by means of visible or invisible rays on films of cellulose ethers or esters, paper, rayon, and other cellulosic mediums.

Plastics

Filler in—
 Plastic products.
Opacifying agent.
Pigment.

Rayon

Delustring agent (Brit. 409521, 409625, 426751, and 426912; U. S. 1940602).

Rubber

Accelerator (U. S. 1326319) in—
 Vulcanizing processes.
Filler in—
 Rubber batches.
Opacifier in—
 Rubber batches.
Pigment.

Soap

Whiteness increaser in—
 Shaving soaps, toilet soaps.

Textile

Mordant in—
 Dyeing processes, printing processes.
Resist (U. S. 1864582) in—
 Color-discharge printing of silk or rayon.

Titanium Lactate

French: Lactate de titane, Lactate titanique.
German: Milchsäurestitan.
Spanish: Lactico de titanio.
Italian: Lactico di titanio.

Leather

As a mordant.

Textile

As a mordant.

Titanium Nitride
French: Nitrure de titanium.
German: Titannitrid.

Chemical
Catalyst in making—
Synthetic ammonia.
Starting point in making—
Potassium cyanide, sodium cyanide.

Miscellaneous
Ingredient of—
Compositions used for making linings for crucibles and electric furnaces.

Metallurgical
Ingredient of—
Compositions used to coat molds and cores so as to give a high resistance to molten metal and prevent the molding sand from burning or uniting with the liquid steel.

Paint and Varnish
Starting point in making—
Titanium-white pigment.

Petroleum
Catalyst in hydrogenating—
Crude petroleum or petroleum distillates (Brit. 250948).

Titanium Phosphate, Basic
French: Phosphate basique de titanium.
German: Basischesphosphorsaeurestitan, Basischestitanphosphat.

Paint and Varnish
As a pigment (Brit. 261051).

Titanium Tetrabromide
French: Tétrabromure de titane.
German: Tetrabromitan, Titantetrabromid.

Petroleum
Reagent in—
Refining mineral oils (U. S. 1643272).

Titanium Tetrafluoride
French: Tétrafluorure de titane, Tétrafluorure titanique.
German: Tetrafluortitan, Titantetrafluorid.

Petroleum
Reagent in refining—
Mineral oils and petroleum distillates.

Titanium Tetraiodide
French: Tétraiodure de titane.
German: Tetrajodtitan, Titantetrajodid.

Petroleum
Reagent in refining—
Mineral oils (U. S. 1643272).

Titanium Vanadate
French: Vanadate de titane, Vanadate titanique.
German: Titanvanadat, Vanadinsäurestitan, Vanadinsäurestitanoxyd.
Spanish: Vanadato de titanio.
Italian: Vanadato di titanio.

Chemical
Ingredient of catalytic mixtures used in the preparation of—
Acenaphthylene, acenaphthaquinone, bisacenaphthylidenedione, naphthaldehydic acid, naphthalic anhydride, and hemimellitic acid from acenaphthene (Brit. 295270).
Acetaldehyde from ethyl alcohol (Brit. 281307).
Acetic acid from ethyl alcohol (Brit. 281307).
Alcohol from aliphatic hydrocarbons (Brit. 281307).
Aldehydes and acids by the reduction of the corresponding esters (Brit. 306471).
Aldehydes and acids from toluene, orthochlorotoluene, orthonitrotoluene, orthobromotoluene, paranitrotoluene, parabromotoluene, parachlorotoluene, metachlorotoluene, metabromotoluene, metanitrotoluene, dichlorotoluene, dibromotoluenes, dinitrotoluenes, bromonitrotoluenes, chloronitrotoluenes, and chlorobromotoluenes (Brit. 295270).
Aldehydes and acids from xylenes, pseudocumenes, mesitylenes, and paracymene (Brit. 281307).
Alphacampholide from camphoric acid by reduction (Brit. 306471).
Alphanaphthaquinone from naphthalene (Brit. 281307).
Anthraquinone from naphthalene (Brit. 281307).
Benzaldehyde and benzoic acid from toluene (Brit. 281307).
Benzoquinone from phenanthraquinone (Brit. 281307).
Benzyl alcohol from benzaldehyde by reduction (Brit. 306471).
Benzyl alcohol, benzaldehyde, or benzyl phthalide by the reduction of phthalic anhydride (Brit. 281307).
Butyl alcohol by the reduction of crotonaldehyde (Brit. 306471).
Diphenic acid from methyl alcohol (Brit. 281307).
Ethyl alcohol by the reduction of acetaldehyde (Brit. 306471).
Fluorenone from fluorene (Brit. 295270).
Formaldehyde by the reduction of methanol or methane (Brit. 306471).
Formaldehyde by the reduction of carbon dioxide or carbon monoxide (Brit. 306471).
Hydroxyl compounds by the reduction of anthraquinone, benzoquinone and the like (Brit. 306471).
Isopropyl alcohol by the reduction of acetone (Brit. 306471).
Maleic acid and fumaric acid by the oxidation of toluene, benzene, phenols, tar phenols, or furfural, or from benzoquinone or phthalic anhydride (Brit. 295270).
Methane by the reduction of carbon dioxide or carbon monoxide (Brit. 306471).
Methanol by the reduction of carbon dioxide or carbon monoxide (Brit. 306471).
Naphthaldehydic acid, acenaphthaquinone, bisacenaphthylidenedione from acenaphthylene (Brit. 281307).
Phenanthraquinone from phenanthrene or diphenic acid (Brit. 295270).
Primary alcohols by the reduction of the corresponding aldehydes (Brit. 306471).
Propionic acid and butyric acid and higher alcohols, ketones, and acids by the reduction of carbon dioxide or carbon monoxide (Brit. 306471).
Reduction products of ketones, aldehydes, esters, ethers, alcohols, and other organic compounds which contain oxygen (Brit. 306471).
Salicylic acid and salicylic aldehyde from cresol (Brit. 306471).
Secondary butyl alcohol by the reduction of methylethyl ketone (Brit. 306471).
Valeryl alcohol by the reduction of valeraldehyde (Brit. 306471).
Vanillin and vanillic acid from eugenol or isoeugenol (Brit. 295270).

Ingredient (Brit. 304640) of catalytic preparations used in the manufacture of various aromatic and aliphatic amines, such as—
Alphanaphthylamine from alphanitronaphthalenes.
Amines from aliphatic nitro compounds, such as alkyl nitriles or nitromethane.
Amylamine from pyridin.
Anilin, azobenzene, azo-oxybenzene, and hydrazobenzene from nitrobenzene by reduction.
Aminophenols from nitrophenols.
3-Aminopyridin from 3-nitropyridin.
Amino compounds from the corresponding nitroanisoles.
Amines from oximes, Schiff's base and nitriles.
Cyclohexamine, dicyclohexamine, and cyclohexylanilin from nitrobenzene.
Piperidin from pyridin.
Pyrrolidin from pyrrol.
Tetrahydroquinolin from quinolin.

Titanous Chloride
Synonyms: Titanium dichloride.
French: Chlorure titaneux, Dichlorure de titane.
German: Titandichlorid, Titanochlorid.

Chemical
Starting point in making—
Titanium compounds.
Ingredient of catalytic mixtures used in the manufacture of—
Acenaphthylene, acenaphthaquinone, bisacenaphthylidenedione, naphthaldehydic acid, naphthalic anhydride, and hemimellitic acid from acenaphthene (Brit. 295270).
Acetaldehyde from ethyl alcohol (Brit. 281307).
Acetic acid from ethyl alcohol (Brit. 281307).
Alcohols from aliphatic hydrocarbons (Brit. 281307).
Aldehydes or alcohols by the reduction of the corresponding esters (Brit. 306471).
Alphacampholide by the reduction of camphoric acid (Brit. 306471).
Aldehydes and acids from toluene, orthochlorotoluene, orthonitrotoluene, orthobromotoluene, parachlorotoluene, parabromotoluene, paranitrotoluene, metachloro-

Titanous Chloride (Continued)
toluene, metabromotoluene, metanitrotoluene, dichlorotoluenes, dibromotoluenes, dinitrotoluenes, chlorobromotoluenes, chloronitrotoluenes, bromonitrotoluenes (Brit. 295270).
Aldehydes and acids from xylenes, pseudocumenes, mesitylene, and paracymene (Brit. 281307).
Alphanaphthaquinone from naphthalene (Brit. 281307).
Anthraquinone from naphthalene (Brit. 295270).
Benzaldehyde and benzoic acid from toluene (Brit. 281307).
Benzoquinone from phenanthraquinone (Brit. 281307).
Benzyl alcohol by the reduction of benzaldehyde (Brit. 306471).
Benzyl alcohol or benzaldehyde or phthalide by the reduction of phthalic anhydride (Brit. 306471).
Butyl alcohol by the reduction of crotonaldehyde (Brit. 306471).
Chloroacetic acid from ethylenechlorohydrin (Brit. 295270).
Diphenic acid from ethyl alcohol (Brit. 281307).
Ethyl alcohol by the reduction of acetaldehyde (Brit. 306471).
Fluorenone from fluorene (Brit. 295270).
Formaldehyde by the reduction of methane or methanol (Brit. 306471).
Formaldehyde by the reduction of carbon monoxide or carbon dioxide (Brit. 306471).
Hydroxyl compounds by the reduction of anthraquinone, benzoquinone, and similar compounds (Brit. 306471).
Isopropyl alcohol by the reduction of acetone (Brit. 306471).
Maleic acid and fumaric acid by the oxidation of toluene, benzene, phenols, tar phenols, or furfural, or from benzoquinone or phthalic anhydride (Brit. 295270).
Methane by the reduction of carbon dioxide or carbon monoxide (Brit. 306471).
Methanol by the reduction of carbon dioxide or carbon monoxide (Brit. 306471).
Naphthaldehydic acid, acenaphthaquinone, or bisacenaphthylidenedione from acenaphthylene (Brit. 281307).
Phenanthraquinone from phenanthrene or diphenic acid (Brit. 295270).
Phthalic acid and maleic acid from naphthalene (Brit. 295270).
Primary alcohols by the reduction of the corresponding aldehydes (Brit. 306471).
Propionic acid and butyric acid and higher alcohols, ketones, and acids by the reduction of carbon dioxide and carbon monoxide (Brit. 306471).
Reduction products of ketones, aldehydes, acids, esters, alcohols, ethers and other organic compounds containing oxygen (Brit. 306471).
Salicylic acid and salicylic aldehyde from cresol (Brit. 295270).
Secondary butyl alcohol by the reduction of methylethyl ketone (Brit. 306471).
Valeryl alcohol by the reduction of valeraldehyde (Brit. 306471).
Vanillin and vanillic acid from eugenol and isoeugenol (Brit. 295270).
Ingredient (Brit. 306460) of catalytic preparations used in the production of various aromatic and aliphatic compounds, including—
Alphanaphthylamine from alphanitronaphthalene.
Amines from aliphatic nitro compounds, such as allyl nitriles or nitromethane.
Amino compound from the corresponding nitroanisole.
Amylamine from pyridin.
Anilin, azo-oxybenzene, azobenzene, and hydrazobenzene from nitrobenzene by reduction.
Aminophenols from nitrophenols.
3-Aminopyridin from 3-nitropyridin.
Cyclohexamine, dicyclohexamine, and cyclohexylanilin from nitrobenzene.
Piperidin from pyridin.
Pyrrolidin from pyrrol.
Tetrahydroquinone from quinolin.
Reducing agent for reducing acids (used for various purposes in the chemical industry).
Laundering
Reagent used in the laundry for—
Clearing articles that have run during washing.
Removing iron rust stains from clothes.
Removing mold stains from clothes.
Textile
Antichlor and sour in treating—
Textile fabrics after the chemicking operation in bleaching.
Reagent for clearing up the whites in colored goods.

Toluene
Synonyms: Methylbenzene, Methylbenzol, Phenylmethane, Toluol.
French: Benzène de méthyle, Benzène méthylique, Méthylebenzène, Phénylemétháne, Toluole.
German: Methylbenzol, Phenylmethan, Toluol.
Abrasives
Solvent (Brit. 295335) in making—
Compositions used for the production of grinding discs and other abrasive articles.
Analysis
Reagent in testing for water.
Ceramics
Solvent and diluent in—
Compositions, containing nitrocellulose or other esters or ethers of cellulose, used for decorating and coating ceramic ware.
Chemical
As a general solvent.
Solvent in extracting—
Alkaloids.
Solvent in making—
Aloemodin, eurodin.
Starting point in making—
Anthranilic acid, aromatics, artificial musk, benzaldehyde, benzoic acid, benzyl chloride, betamethylanthraquinone, intermediates, orthotoluenesulphonchloride, paraphenetidin, pharmaceuticals, saccharin, tolidins, toluidins.
Dye
Starting point in making—
Dyestuffs of various classes.
Triarylmethane colors, such as tetramethyl-4:4′-diaminobenzophenone.
Explosives
Starting point in making—
Trinitrotoluene (TNT).
Fats and Oils
Solvent for various fats and oils.
Solvent in extracting—
Animal oils, vegetable oils.
Glass
Solvent and diluent in—
Compositions, containing nitrocellulose or other esters or ethers of cellulose, used in the production of nonscatterable glass and for decorating and coating glassware.
Glues and Adhesives
Solvent (Brit. 295335) in making—
Cements and adhesive preparations containing fillers.
Solvent in degreasing—
Bones for the manufacture of bone glues.
Ink
Solvent in making—
Printing ink.
Leather
Solvent in making—
Artificial leather.
Impregnating compositions, containing phenol resins, used in the surfacing of leather (Brit. 295335).
Leather dressings.
Solvent and diluent in—
Compositions, containing nitrocellulose or other esters or ethers of cellulose, used in the manufacture of artificial leather and for coating and decorating leather and leather goods.
Miscellaneous
Ingredient of—
Dry-cleaning preparations, finishing compositions for various materials, splicing compositions, spotting fluids, spreading compositions, wiping compositions.
Solvent for various purposes.
Solvent for degreasing—
Hair.
Solvent in making—
Cements.
Impregnating solutions containing synthetic resins (Brit. 295335).
Solvent and diluent in—
Compositions, containing nitrocellulose or other esters or ethers of cellulose, used for coating and decorating various products.

Toluene (Continued)

Oilcloth and Linoleum
Solvent in making—
Impregnating compositions, containing synthetic resins (Brit. 295335).
Linoleum cements (Brit. 274300).

Paint and Varnish
Solvent in making—
Dopes, enamels, finishing compositions.
Lacquers, containing phenol-formaldehyde synthetic resins (Brit. 295335).
Paint removers, stains, stretchers, varnishes, varnish removers.
Various coating compositions containing nitrocellulose or other esters or ethers of cellulose.

Paper
Solvent in—
Coating compositions containing synthetic resins (Brit. 295335).
Coating compositions, containing nitrocellulose or other esters or ethers of cellulose, used in the production of coated paper and for decorating and coating paper and pulp products.

Pharmaceutical
In compounding and dispensing practice.

Plastics
Solvent in making—
Compositions, for molding and pressing, containing synthetic resins (Brit. 295335).
Nitrocellulose plastics.

Resins and Waxes
Solvent for—
Resins, rosin, and waxes.

Rubber
Solvent and diluent in—
Compositions, containing nitrocellulose or other esters or ethers of cellulose, used for decorating and coating rubber articles.

Sanitation
Solvent in treating—
Garbage.

Stone
Solvent and diluent in—
Compositions, containing nitrocellulose or other esters or ethers of cellulose, used for coating and decorating artificial and natural stone.

Textile
Solvent in—
Compositions, containing artificial resins, used for impregnating fabrics (Brit. 295335).
Compositions, containing nitrocellulose or other esters or ethers of cellulose, used for coating fabrics.

Woodworking
Solvent in—
Compositions, containing artificial resins, used for impregnating wood (Brit. 295335).
Solvent and diluent in—
Compositions, containing nitrocellulose or other esters or ethers of cellulose, used for decorating and protecting woodwork.

Toluenemethylenesulfonamide
French: Amide de toluèneméthylènesulfonique.
German: Toluolmethylensulfamid.

Chemical
Starting point in making—
Intermediates and other derivatives.

Miscellaneous
Softening agent in—
Coating compositions containing cellulose acetate.
For uses, see under general heading: "Softening agents."

Tolueneparahydroxylaminosulphonic Acid
French: Acide de toluèneparahydroxylaminesulphonique.
German: Toluolparahydroxylaminsulphosaeure.

Chemical
Starting point in making—
Paradimethylaminobenzaldehyde.

4-Toluidin
French: 4-Toluidine.
Spanish: 4-Toluidina.

Chemical
Starting point in making various derivatives.

Dye
Starting point (Brit. 353537) in making acridin dyestuffs with the aid of—
2-Chloro-4-bromobenzoic acid, 2-chloro-4-iodobenzoic acid, 2:3-dichlorobenzoic acid.

Toluidins (Mixed)

Chemical
Reagent in making—
Saccharin, synthetic pharmaceuticals.
Starting point in making—
Acetoacetic ether toluidins, intermediates, various organic chemicals.

Dye
Starting point in making—
Fuchsin dyes, magenta dyes, primulin dyes, safranin dyes.

Electrical
Reagent (Brit. 273290) in making—
Insulating enamels for wires.

Glues and Adhesives
Reagent (Brit. 273290) in making—
Cements for laminated wires.

Metallurgical
As a flotation oil.

Paint and Varnish
Reagent (Brit. 273290) in making—
Bases for varnishes.

Perfume
Reagent in making—
Synthetic perfumes.

Plastics
Reagent (Brit. 273290) in making—
Bases for plastics and moldable compositions.

Rubber
Accelerator in—
Vulcanizing operations.

1:4-Toluidoanthraquinone

Insecticide and Fungicide
Promoter (U. S. 2011428) of—
Light-stability in oil-soluble pyrethrum extracts and insecticidal products thereof.

Tolylacetylhydroquinone

Petroleum
Stabilizing agent (Brit. 406195) for—
Cracked gasolines and other motor fuels.

Tolylacetylphloroglucinol

Petroleum
Stabilizing agent (Brit. 406195) for—
Cracked gasolines and other motor fuels.

Tolylacetylpyrocatechol

Petroleum
Stabilizing agent (Brit. 406195) for—
Cracked gasolines and other motor fuels.

Tolylacetylpyrogallol

Petroleum
Stabilizing agent (Brit. 406195) for—
Cracked gasolines and other motor fuels.

Tolylacetylresorcinol

Petroleum
Stabilizing agent (Brit. 406195) for—
Cracked gasolines and other motor fuels.

Tolylhydroquinone

Petroleum
Stabilizing agent (Brit. 406195) for—
Cracked gasolines and other motor fuels.

Tolylmercaptan
Synonyms: Thiocresyl.

Insecticide and Fungicide
Increaser (U. S. 1942532) of—
Floatability on water of paris green for killing anopheline larvae.

Tolylmercuric Acetate
Synonyms: Mercury-tolyl acetate.
French: Acétate de mercure et de tolyle, Acétate mercurique-tolylique.
German: Essigsäuremerkurtolylester, Merkurtolylacetat, Merkurtolylazetat.

Tolylmercuric Acetate (Continued)

Chemical
Starting point in making various derivatives.
Insecticide
Ingredient (Brit. 321496) of—
Compositions for immunizing wheat and other grains.
Woodworking
Ingredient (Brit. 321396) of—
Preserving and disinfecting compositions.

Tolylphloroglucinol

Petroleum
Stabilizing agent (Brit. 406195) for—
Cracked gasolines and other motor fuels.

Tolyl Phosphate

French: Phosphate de tolyle, Phosphate tolylique.
German: Phosphorsäurestolyl, Phosphorsäuretolylester, Tolylphosphat.
Miscellaneous
Mothproofing agent in treating—
Feathers, furs, hair.
Textile
Mothproofing agent in treating—
Wool and felt.

Tolylpyrocatechol

Petroleum
Stabilizing agent (Brit. 406195) for—
Cracked gasolines and other motor fuels.

Tolylpyrogallol

Petroleum
Stabilizing agent (Brit. 406195) for—
Cracked gasolines and other motor fuels.

Tolylresorcinol

Petroleum
Stabilizing agent (Brit. 406195) for—
Cracked gasolines and other motor fuels.

Tolyl 4-Sulphophthalate

Miscellaneous
As an emulsifying agent (Brit. 418334).
For uses, see under general heading: "Emulsifying agents."

Tolylthioglycollic Acid

Synonyms: Tolylsulfoglycollic acid.
French: Acide de tolylesulfoglycollique, Acide de tolylethioglycollique.
German: Tolylsulfoglykolsaeure, Tolylthioglykolsaeure.
Dye
Reagent (Brit. 284288) in making thioindigoid dyestuffs with—
Acenaphthenequinone, alphaisatinanilide, 5:7-dibromoisatin, isatin, isatin homologs and substitution products, ortho diketones.

Tomatoseed Oil

French: Huile de tomate.
German: Tomatosamenoel.
Italian: Olio di pomodoro.
Fats and Oils
Ingredient of—
Lubricating compositions.
Food
As a cooking oil, as an edible oil for various purposes, as a salad oil.
Ingredient of—
Food compositions.
Fuel
As an illuminant.
Linoleum and Oilcloth
Drying oil in making—
Linoleum and oilcloth coatings.
Mechanical
As a special lubricant.
Paint and Varnish
Drying oil in making—
Paints and varnishes.
Soap
Soapstock in making—
Soft and laundry soaps.

Triacetin

Synonyms: Glyceryl triacetate.
French: Triacétine.
German: Triacetin, Triazetin.
Spanish: Triacetine.
Italian: Triacetino.
Cellulose Products
Solvent and plasticizer for—
Cellulose acetate, cellulose derivatives, nitrocellulose.
For uses, see under general heading: "Solvents."

Triamyl Borate

Paint and Varnish
Inhibitor of—
Chemical action between paints or lacquers and zinc (in protective coatings over zinc).
Substitute for boric acid in—
Varnish making (acts as a source of boric acid by decomposition without forming an insoluble sludge; the unused remainder, being soluble, eliminates the filtration operation necessary with the ordinary boric acid sludge).

1:3:5-Triazin-2:4:6-tricarboxyl Choride

French: Chlorure de 1:3:5-triazine-2:4:6-tricarbonique, Chlorure de 1:3:5-triazine-2:4:6-tricarboxylique.
German: 1:3:5-Triazin-2:4:6-tricarbonylchlorid.
Dye
Starting point (Swiss 111562) in making vat dyestuffs with—
1:4-Aminoanthraquinone, 1:5-aminoanthraquinone.

Tribenzylarsin

French: Arsine de tribenzyle, Arsine tribenzylique.
Chemical
Starting point (Brit. 303092) in making—
Chemicals used for various mothproofing purposes.
Textile
Reagent (Brit. 303092) for—
Mothproofing wool and felt.

Tribromanilin Hydrobromide

French: Bromhydrate de tribromaniline.
German: Bromwasserstoffsäurestribromanilin.
Spanish: Bromhidrato de tribromanilina.
Italian: Bromidrato di tribromanilina.
Pharmaceutical
Suggested for use as—
Analgesic, antineuralgic.

2:4:6-Tribromoanisole

Paint and Varnish
Ingredient (U. S. 1880419) of—
Cellulose acetate lacquer.

Tribromoethyl Alcohol

Synonyms: Avertin.
Pharmaceutical
Suggested for use as—
New anesthetic (by basal narcosis).

Tribromonitromethane

Fuel
Primer (Brit. 461320) for—
Diesel fuels.

Tribromophenol

French: Tribromure de phénole.
German: Tribromphenol.
Spanish: Tribromofenol.
Italian: Tribromofenole.
Disinfectant
As a disinfectant.
Pharmaceutical
Suggested for use as—
External antiseptic, internal antiseptic.

Tribromophenylstibin

French: Stibine de tribromophényle, Stibine tribromophénylique.
Chemical
Starting point in making various derivatives.
Miscellaneous
Mothproofing agent (Brit. 303902) for treating—
Furs and hair.
Textile
Mothproofing agent (Brit. 303902) for treating—
Wool and felt.

Tributenylamine
French: Amine de tributényle.

Chemical
Starting point in making various derivatives.

Insecticide
As an insecticide.
Ingredient (Brit. 313924) of—
Insecticidal and germicidal preparations.

Soap
Ingredient (Brit. 313924) of—
Insecticidal and germicidal soaps.

Tributylamine Oxide

Chemical
Starting point (Brit. 460710) in making—
Cleansing, disinfecting, and wetting agents by reacting with alkylene oxides.
Emulsifying agents for soaps, glue, gelatin, gums, and mucilages.
Textile stripping agents for vat dyestuffs by reacting with alkylene oxides and admixing with hydrosulphites.

Tributyl Phosphate

Cellulose Products
Plasticizer for—
Nitrocellulose.
For uses, see under general heading: "Plasticizers."

Tributyl Phosphate, Chlorinated

Lubricant
Stabilizer (Brit. 448424) for—
Viscous oils, such as Pennsylvania or Midcontinent petroleums, used for extreme pressure work.

2:3:3-Trichlor-2-methyl-1-phenylpropane

Petroleum
Solvent (Brit. 437573) in—
Refining mineral oils.

Trichloroacetic Acid
Synonyms: Trichloracetic acid.
Latin: Acidum trichloraceticum.
French: Acide trichloracétique.
German: Trichloressigsäure.

Analysis
Reagent for—
Albumin detection.

Chemical
Reagent in—
Organic synthesis.

Pharmaceutical
In compounding and dispensing practice.
Ingredient of—
Corn removers, wart removers.
Suggested for use as—
Antiseptic, caustic.
Suggested for use in treating—
Chancroids.
Inflammations of the nose, pharynx and tonsils.
Ozena, papillomata, small growths in the mouth, vascular naevi.

2:4:5-Trichloroanilin

Dye
Starting point (Brit. 397016) in making—
Orange-brown water-insoluble dyes.

Trichloroanilin Hydrochloride
French: Chlorhydrate de trichloraniline, Hydrochlorure de trichloroaniline.
German: Chlorwasserstoffsaeurestrichloranilin, Trichloranilinchlorhydrat.

Chemical
Starting point in making various intermediates.

Rubber
Reagent (Brit. 282778) in making conversion products with—
Alphanaphthol, betanaphthol, catechol, cresol, parachlorophenol, phenol, resorcinol.

Trichloroanthraquinoneacridin
French: Trichlorure de anthraquinone-acridine.
German: Trichloranthrachinonacridin.

Chemical
Starting point in making intermediates and other derivatives.

Dye
Starting point (Brit. 314899) in making dyestuffs with—
Alpha-amino-4:8-dichloronaphthalene, alphachloronaphthylamine, betachloronaphthylamine, 4-chlorometaxylene, 3:5-dibromoanilin, 2:5-dichloroanilin, 3:5-dichloroanilin, metachloroanilin, metachloroanisidin, metachlorobenzylamine, metachlorocresidin, metachlorodiphenylamine, metachloroethylanilin, metachloromethylanilin, metachlorophenylamine, metachlorotolylamine, metachlorotoluidin, metachloroxylidin, orthochloroanilin, orthochloroanisidin, orthochlorobenzylamine, orthochlorocresidin, orthochlorodiphenylamine, orthochloroethylanilin, orthochloromethylanilin, orthochlorophenylamine, orthochlorotolylamine, orthochlorotoluidin, orthochloroxylidin, parachloroanilin, parachloroanisidin, parachlorobenzylamine, parachlorocresidin, parachlorodiphenylamine, parachloroethylanilin, parachloromethylanilin, parachlorophenylamine, parachlorotolylamine, parachlorotoluidin, parachloroxylidin, 3:4:5-trichloroanilin.
Various other halogenated aromatic amines.

Trichlorobenzene
Synonyms: Trichlorobenzol.
French: Trichlorobenzène.
German: Trichlorbenzol.

Chemical
As a solvent.
Reagent in—
Organic synthesis.

Dye
Intermediate in—
Dyestuff manufacture.

Miscellaneous
As a heat-transfer medium.

Trichlorodiphenylmethane

Electrical
Cooling medium (Brit. 413596, 433070, 433071, and 433072) in—
Electrical apparatus, such as transformers, switches, capacitors, cables, bushings, and junction boxes (may be employed in admixture with trichlorobenzene, chlorinated diphenyl, and the like).
Dielectric (Brit. 413596, 433070, 433071, and 433072) in—
Electric apparatus, such as transformers, switches, capacitors, cables, bushings, and junction boxes (may be employed in admixture with trichlorobenzene, chlorinated diphenyl, and the like).

Trichloroethylacetanilide

Cellulose Products
Solvent (Brit. 344626) for—
Cellulose esters and ethers.
For uses, see under general heading: "Solvents."

Trichloroethylene
French: Trichlorure d'éthylène.
German: Trichloraethylen.

Agricultural
Reagent in—
Treating and disinfecting soil.

Analysis
Extracting medium for various purposes.
Reagent in analyzing—
Breadstuffs, butter, cakes, cheese, chocolate, cocoa, flour, meals, meat, milk, soaps.
Reagent in determining—
Fat content in foods and industrial products.
Solvent for various purposes.

Ceramic
Solvent and diluent in—
Compositions, containing esters or ethers of cellulose, such as cellulose acetate, used for decorating and protecting ceramic ware.

Chemical
Extracting medium in obtaining—
Alkaloids, drug principles.
Ingredient of solvent mixtures containing—
Acetone, alcohol, benzene, chlorinated hydrocarbons, turpentine.
Reagent (U. S. 1813636) in separating—
Acetic acid from formic acid.
Reagent in making—
Intermediates, organic chemicals, pharmaceuticals.
Solvent for—
Cellulose acetate, phosphorus, sulphur, various organic compounds.

Trichloroethylene (Continued)
Solvent in extracting—
Perfumes.
Solvent in removing—
Phenols and homologs and recovering them from liquids, such as waste waters and the like.
Starting point in making—
Amyl acetate, chloroacetic acid, chlorinated fats, glycollic acid, mercuric trichloroethylamide, pentachloroethane, perchloroethylene phenylglycin (German 437409), sulphonic acids of various sorts.

Dye
Reagent in making—
Synthetic dyestuffs of various classes, thioindigo.

Electrical
Solvent in cleaning—
Electric motors and other electrical machinery.
Solvent and diluent in—
Compositions, containing esters or ethers of cellulose, particularly cellulose acetate, used for making insulating compositions for electrical equipment, wiring, and machinery.

Explosives
Solvent in purifying—
Explosives, particularly of the nitrated aromatic type.
Solvent in recrystallizing—
Trinitrotoluene.

Fats and Oils
Extracting medium in recovering—
Fats from cocoa bean, grease from various products, oil from corn.
Oil from olives, olive husks, and press cakes.
Extracting medium in removing—
Caffeine from coffee to make a caffeine-free product.
Ingredient of—
Egg-covering compositions.
Solvent for various fats and oils and greases.
Solvent in obtaining—
Cottonseed oil, edible oils, inedible oils, linseed oil.
Oils from bones, tankage, leather, and other subtsances.
Soybean oil.

Fertilizer
Solvent in—
Degreasing fish scrap.

Foods
Extracting medium in obtaining soluble substances from—
Berries, fruits, seeds.
Solvent in—
Purification of foodstuffs.

Gas
Solvent in removing—
Sulphur from coal and coke-oven gas.
Solvent for—
Coaltar.

Glass
Solvent for—
Degreasing glass.
Solvent and diluent in—
Compositions, containing cellulose acetate or other esters or ethers of cellulose, used in the manufacture of non-scatterable glass and for decorating and protecting glassware.

Glues and Adhesives
Ingredient of—
Glues.
Reagent in preparing—
Gelatins.
Solvent in—
Degreasing bones preparatory to the manufacture of bone glue.
Solvent in making—
Adhesive compositions containing esters or ethers of cellulose, such as cellulose acetate.

Gums
Solvent for various gums.

Insecticide
As a general insecticide.
Ingredient of—
Fumigating compositions, insecticidal compositions, preparations for exterminating mosquitoes, preparations for combatting grape lice, vermicidal compositions.

Leather
Reagent for—
Degreasing leather.
Reagent in degreasing—
Goatskins, kidskins, lambskins, sheepskins.
Solvent and diluent in—
Compositions, containing cellulose acetate, or other esters or ethers of cellulose, used in the manufacture of artificial leathers and for decorating and protecting leathers and leather goods.

Mechanical
Cleansing agent in degreasing—
Machinery of various sorts, metallic surfaces prior to painting and coating, rags and waste from machine shops.
Solvent in cleansing—
Automobile engines and gears.
Drive wheels for compression pumps and other mechanical equipment.
Solvent in degreasing—
Automobile brakebands, leather belts.

Metallurgical
Solvent in degreasing—
Die castings, metal stampings, metals to be electroplated, nuts and bolts.
Solvent in preparing metals for—
Shellacking, sheradizing, plating, pickling, varnishing.
Solvent and diluent in—
Compositions, containing cellulose acetate or other esters or ethers of cellulose, used for protecting and decorating metallic articles.

Miscellaneous
Dry cleaning agent in treating—
Furs.
Ingredient of—
Polishing compositions.
Solvent for degreasing—
Dishes, kitchenware, hardware, metal furniture, safety razor blades.
Solvent for general purposes (used in the place of benzin because of the greater safety on account of higher boiling point and lower inflammability).
Solvent and diluent in—
Compositions, containing cellulose acetate or other esters or ethers of cellulose, used for decorating and protecting various porducts.

Paint and Varnish
Ingredient of—
Paint and varnish removers, waterproofing compositions.
Solvent in—
Breaking down aqueous bituminous emulsions used in the manufacture of bituminous paints and similar coating and impregnating compositions (Brit. 251323).
Solvent in making—
Paints, varnishes, lacquers, enamels, and dopes containing cellulose acetate or other esters or ethers of cellulose.
Thinner in—
Paints and varnishes.

Paper
Ingredient (Brit. 299817) of emulsified preparation for—
Cleansing wire on paper-making machines, digestion of sulphite pulp, grinding mechanical wood pulp, removing ink from paper.
Solvent and diluent in—
Compositions, containing cellulose acetate or other esters or ethers of cellulose, used in the manufacture of coated paper and for decorating and protecting paper and pulp compositions.

Petroleum
Ingredient of—
Compounded solvent preparations containing mineral oil distillates.
Solvent in degreasing—
Light mineral oils.
Solvent in extracting—
Paraffin and the like from mineral oil distillates.

Pharmaceutical
As a solvent for various purposes.
In compounding and dispensing practice.

Photographic
Solvent in degreasing and cleaning—
Motion picture film.

Plastics
Solvent in degreasing—
Bakelite, celluloid.
Solvent and diluent in making—
Plastic compositions containing cellulose acetate or other esters or ethers of cellulose.

Trichloroethylene (Continued)
Printing
Degreasing agent in engraving, printing, and litho trades.
Solvent in removing—
Inks from plates, rollers, and presses.
Refrigeration
As a refrigerating medium.
Resins and Waxes
Solvent for various resins and waxes.
Rubber
Ingredient of—
Rubber cements, rubber mastics.
Solvent for rubber.
Solvent in—
Compositions, containing cellulose acetate or other esters or ethers of cellulose, used for decorating and protecting rubber goods.
Sanitation
Reagent in extracting—
Grease from garbage.
Soap
Ingredient of—
Germicidal soaps, spotting fluids.
Ingredient (Brit. 299817) of—
Detergent emulsions with turkey red oil and other solvents, such as chlorinated hydrocarbons, used for laundry and domestic purposes.
Stone
Solvent and diluent in—
Compositions, containing cellulose acetate or other esters or ethers of cellulose, used for decorating and protecting natural and artificial stone.
Sugar
Solvent for extracting—
Waxes from filter press mud in sugar refineries.
Textile
—, *Dyeing*
Ingredient (Brit. 299817) of preparations containing turkey red oil and chlorinated hydrocarbons used for—
Dyeing cotton, wool, rayon, and mercerized cotton.
Wetting textiles before dyeing.
—, *Finishing*
Ingredient (Brit. 299817) of emulsified preparations containing turkey red oil and other solvents, such as chlorinated hydrocarbons, used for—
Scouring and finishing cotton, wool, rayon, and mercerized cotton.
Solvent for—
Cleaning acetate rayon, scouring textile yarns and fabrics.
—, *Manufacturing*
Ingredient (Brit. 299817) of emulsified preparations, containing turkey red oil and chlorinated hydrocarbons, used for—
Degumming silk.
Solvent for—
Cleaning knitting machine needles, cleaning silk and silk hosiery, degreasing textiles, degreasing wool.
Solvent and diluent in—
Compositions, containing cellulose acetate or other esters or ethers of cellulose, used for making coated textiles.
Tobacco
Solvent in—
Extracting nicotine.
Woodworking
Solvent and diluent in—
Compositions, containing cellulose acetate or other esters or ethers of cellulose, used for decorating and protecting woodwork.

5:6:7-Trichloro-8-hydroxyquinolin
Pharmaceutical
Suggested for use (Brit. 351605) as—
Antiseptic.

Trichloronitromethane
German: Trichlornitronmethan.
Chemical
Purifying agent (U. S. 1749381) in treating—
Vaccines.

Trichlorophenol
French: Trichlorophénole.
German: Trichlorphenol.
Spanish: Triclorofenol.
Italian: Triclorofenole.
Chemical
Reagent (French 545368) in making—
Borneol from turpentine.
Dyes
In the manufacturing process.
Fungicide
As a fungicide.
Insecticide
Toxicity agent.
Toxicity agent (French 732973) in—
Fumigating agent, containing also ammonium chloride, potassium oxalate, sodium oxalate, and paraffin, or ozokerite wax, with trioxymethylene as an irritant.
Leather
Inhibitor of—
Mould growth on pickled sheep pelts during transport, particularly during ocean transport.
Paint and Varnish
Addition agent (Brit. 409009) in—
Chlorinated rubber paints.

Trichlororetene
Petroleum
Imparter (Brit. 431508) of—
High-film strength, adhesion power, and abrasion resistance to lubricants for use with extreme pressures (blended with mineral lubricating oil).

Trichloro-tertiary-butyl Acetate
Cellulose Products
Plasticizer (U. S. 1946643) for—
Cellulose acetate.
For uses, see under general heading: "Plasticizers."

Trichloro-tertiary-butyl Alcohol
Petroleum
Solvent (Brit. 435096) in—
Refining mineral oils.

Trichloro-tertiary-butyl Benzoate
Cellulose Products
Plasticizer (U. S. 1946643) for—
Cellulose acetate.
For uses, see under general heading: "Plasticizers."

Trichloro-tertiary-butyl Lactate
Cellulose Products
Plasticizer (U. S. 1946643) for—
Cellulose acetate.
For uses, see under general heading: "Plasticizers."

Trichloro-tertiary-butyl Phthalate
Cellulose Products
Plasticizer (U. S. 1946643) for—
Cellulose acetate.
For uses, see under general heading: "Plasticizers."

Trichloro-tertiary-butyl Succinate
Cellulose Products
Plasticizer (U. S. 1946643) for—
Cellulose acetate.
For uses, see under general heading: "Plasticizers."

Trichloroxythionaphthene
German: Trichloroxythionaphten.
Dye
Starting point (Brit. 274527) in making thioindigoid dyestuffs with—
5:7-Dibromoisatin chloride, isatin alpha-anilide.

Tricresol-mercuric Acetate
French: Acétate de tricrésol-mercure, Acétate tricrésolique-mercurique.
German: Essigsäuretrikresolquecksilber, Trikresolquecksilberacetat, Trikresolquecksilberazetat.
Spanish: Acetato de tricresol de mercurio.
Italian: Acetato di tricresol e mercurio.
Agriculture
Reagent for—
Disinfecting soils.
Food
Reagent for—
Disinfecting grains.

Tricresol-mercuric Acetate (Continued)
Insecticide
As an insecticide.
Ingredient of—
Insecticidal and germicidal compositions containing pulverized talc (when used in the powder form), or sodium carbonate.
Sanitation
Reagent in—
Destroying mosquito larvae in stagnant water.

Tricresyl Phosphate, Brominated
Miscellaneous
As a fire-retardant (Brit. 409896).

Tricresyl Phosphate, Chlorinated
Lubricant
Stabilizer (Brit. 448424) for—
Viscous oils, such as Pennsylvania or Midcontinent petroleum, used for extreme pressure work.

Tricyclohexyl Citrate
Cellulose Products
Plasticizer (Brit. 432404) for—
Cellulose acetate, cellulose esters and ethers.
For uses, see under general heading: "Plasticizers."

Triethanolamine
Synonyms: Trihydroxyethanolamine.
French: Triéthanolamine.
German: Triaethanolamin, Triethanolamin.
(The commercial product contains approximately 15% of diethanolamine and 2.5% monoethanolamine.)
Chemical
Absorbent for—
Acid gases, carbon dioxide, hydrochloric acid in gaseous form, hydrogen sulphide, sulphur dioxide.
Absorbent in—
Recovering and purifying gases.
Amine useful as—
Viscous liquid, very hygroscopic liquid, completely soluble in alcohols, completely soluble in water, slightly soluble in hydrocarbons.
Emulsifying agent in making—
Emulsions of various chemicals, textile lubricants in emulsified form, wetting compositions in emulsified form.
Emulsifying agent (commonly used in the form of one of its soaps) with—
Fatty acids, oleic acid, stearic acid.
Solvent for—
Some organic substances.
Starting point in making—
Dispersing agents, emulsifying agents, intermediates, soaps having valuable properties.
Synthetic products by esterification of hydroxyl groups.
Synthetic products by condensation with aldehydes.
Various derivatives.
Cosmetic
Emulsifying agent in various preparations.
Disinfectant
Emulsifying agent in making—
Emulsified germicidal and disinfecting compositions.
Dye
Emulsifying agent in making—
Emulsified color lakes.
Solvent (Brit. 272908) in making—
Soluble metallic compounds of azo dyestuffs.
Dry Cleaning
Starting point in making—
Dry-cleaning soaps.
Fats, Oils, and Waxes
Emulsifying agent for—
Fats, oils, waxes.
Emulsifying agent in making—
Emulsified boring oils, emulsified drilling oils, emulsified fat-splitting preparations.
Emulsified fatty acids of animal or vegetable origin.
Emulsified greasing compositions.
Emulsified greasing and lubricating compositions containing various vegetable and animal fats and oils.
Emulsified preparations of natural or synthetic waxes.
Emulsified sulphonated oils, emulsified wire-drawing oils, emulsions of animal and vegetable fats and oils.
Gases
Absorbent for—
Acid gases, carbon dioxide, hydrochloric acid in gaseous form, hydrogen sulphide, sulphide dioxide.
Absorbent in—
Recovering and purifying gases.
Glue and Adhesives
Emulsifying agent in making various adhesive preparations.
Plasticizer for—
Glue.
Ink
Emulsifying agent in making—
Emulsified printing and writing inks.
Insecticide
Emulsifying agent in making—
Emulsified insecticidal and fungicidal compositions.
Horticultural sprays.
Leather
Emulsifying agent in making—
Emulsified compositions for softening hides, emulsified dressing compositions, emulsified fat-liquoring baths, emulsified finishing compositions, emulsified soaking compositions, emulsified tanning compositions, emulsified waterproofing compositions.
Ingredient (Brit. 306116) of—
Impregnating compositions.
Plasticizer in—
Leather coatings.
Miscellaneous
Emulsifying agent in making—
Automobile polishes in emulsified form, emulsified cleansing compositions, emulsified compositions for cleansing painted and metallic surfaces, emulsified degreasing compositions, emulsified furniture polishes, emulsified greasing compositions, emulsified metal polishes, emulsions of various substances, waterproofing compositions in emulsified form.
Promoter of—
Penetration of liquids into porous materials.
Paint and Varnish
Emulsifying agent in making—
Emulsified shellac, casein, or rubber preparations.
Waterproofing compositions in emulsified form.
Solvent for—
Casein, rubber, shellac.
Paper
Emulsifying agent in making—
Emulsified compositions for sizing paper and pulp products.
Emulsified compositions for waterproofing paper and pulp compositions and paperboard.
Waxing compositions in emulsified form.
Petroleum
Emulsifying agent in making—
Emulsified cutting oils for screwpress and lathe work, emulsified mineral oils, kerosene emulsions, naphtha emulsions, petroleum pitch emulsions, petroleum tar emulsions.
Textile oils in emulsified form, such as rayon oils.
Soluble greases in emulsified form, solubilized emulsified oils and distillates.
Plastics
Emulsifying agent in making—
Emulsified plastic compositions, emulsified casein or shellac compositions.
Resins
Emulsifying agent in making—
Emulsified preparations of natural or artificial resins.
Emulsified shellac compositions.
Rubber
Emulsifying agent in making—
Emulsified rubber cements and compositions.
Solvent for—
Rubber.
Soap
Emulsifying agent in making—
Emulsified detergents, containing soaps, used for various purposes.
Emulsified hand-cleaning compositions containing soap, emulsified textile soaps.
Textile
——, *Bleaching*
Emulsifying agent in making—
Emulsified bleaching baths.

Triethanolamine (Continued)

——, *Dyeing*
Emulsifying agent in making—
Dye baths in emulsified form.

——, *Finishing*
Emulsifying agent in making—
Emulsified coating compositions, emulsified scouring compositions, emulsified sizing compositions, emulsified washing compositions, emulsified waterproofing compositions, emulsified waxing compositions, emulsified wetting agents.
Plasticizer in—
Coatings for textiles.

——, *Manufacturing*
Emulsifying agent in making—
Emulsified baths for the carbonization of wool, emulsified baths for degumming and boiling-off silk, emulsified baths for soaking silks, emulsified bowking baths, emulsified compositions used for degreasing raw wool, emulsified fulling baths, emulsified kierboiling baths for cotton, emulsified mercerization baths, emulsified spinning compositions, oiling emulsions for various textile purposes.

——, *Printing*
Emulsifying agent in making—
Emulsified printing pastes.

Triethanolamine Citrate

Textile
De-electrifying agent (Brit. 430221) for—
Yarns, films, fabrics, and the like, subject to charging by static electricity (applied in admixture with all usual lubricating agents as vehicle).

Triethanolamine Fluosilicate

Synonyms: Triethanolamine silicofluoride.
French: Fluosilicate de triéthanolamine, Silicofluorure de triéthanolamine.
German: Fluorkieselsäuretriaethanolaminester, Fluorkieselsäurestriaethanolamin, Flusskieselsäuretriaethanolaminester, Fluskieselsäurestriaethanolamin, Silicoflussäuretriaethanolaminester, Silicoflussäurestriaethanolamin, Triaethanolaminfluorsilikat, Triaethanolaminsilicofluorid.
Spanish: Fluosilicato de trietanolamina, Silicofluoruro de trietanolamina.
Italian: Fluosilicato di trietanolamine, Silicofluoruro di trietanolamine.

Disinfectant
As a disinfectant.
Ingredient (Brit. 391141) of—
Disinfecting and deodorizing compositions.

Insecticide
As an insecticide.
Ingredient (Brit. 391141) of—
Insecticidal and fungicidal compositions.

Miscellaneous
As a mothproofing agent.
Ingredient (Brit. 391141) of—
Mothproofing compositions for treating feathers, furs, and hair.

Textile
As a mothproofing agent.
Ingredient (Brit. 391141) of—
Mothproofing compositions for treating wool and felt.

Woodworking
Ingredient (Brit. 391141) of—
Preserving compositions.

Triethanolamine Gallate

Textile
De-electrifying agent (Brit. 430221) for—
Yarns, films, fabrics, and the like, subject to charging by static electricity (applied in admixture with all usual lubricating agents as vehicle).

Triethanolamine Lactate

Textile
De-electrifying agent (Brit. 430221) for—
Yarns, films, fabrics, and the like, subject to charging by static electricity (applied in admixture with all usual lubricating agents as vehicle).

Triethanolamine Linoleate

French: Linoléate de triéthanolamine, Linoléate triéthanolaminique.
German: Leinoeltriaethanolamin, Leinoelsäurestriaethanolamin.

Chemical
Dispersing agent in making—
Emulsions of hydrocarbons of various groups of the aliphatic and aromatic series.
Emulsions of various chemicals, terpene emulsions.
Ingredient of—
Textile lubricants for carding, combing, and drawing wool in making wool yarn for raw wool.
Wetting compositions containing ethylene dichloride or pine oil.

Construction
Ingredient of—
Emulsions containing asphalt, used in the curing of concrete.
Road-surfacing compositions containing asphalt.

Disinfectant
Dispersing agent in making—
Emulsified disinfectants containing pine oils, creosote, or phenol.
Emulsified germicidal and deodorant preparations.
Ingredient of—
Pine oil disinfectants.

Dye
Dispersing agent in making—
Color lakes.

Fats and Oils
Dispersing agent in making—
Boring oils, drilling oils, greasing compositions.
Lubricating compositions containing animal or vegetable oils.
Solvents for fats.
Stabilized emulsions of various animal or vegetable fats and oils.
Wire-drawing oils.

Glues and Adhesives
Dispersing agent in making—
Casein emulsions used as adhesive.

Ink
Dispersing agent in making—
Printing inks, writing inks.

Insecticide
Dispersing agent in making—
Emulsified insecticidal and fungicidal compositions.
Orchard sprays in emulsified form (added to increase the effectiveness).

Leather
Dispersing agent in making—
Emulsified dressings containing casein, shellac, and carnauba wax.
Emulsified fat-liquoring baths.
Emulsified soaking compositions containing neatsfoot oil.
Emulsified waterproofing compositions.

Miscellaneous
Dispersing agent in making—
Automobile polishes.
Compositions for cleansing painted and metal surfaces.
Deodorizing compositions containing pine oil.
Furniture polishes containing carnauba wax and mineral oil.
Impregnating compositions containing paraffin.
Metal polishes.
Metal polishes containing orthodichlorobenzene and abrasives.
Scouring compositions for woodwork, linoleum, rugs, and the like.
Various emulsified polishes containing oleic acid, ethylene dichloride, carnauba wax.
Waterproofing compositions.
Ingredient of—
Liquid baths (added to assist in their penetration into porous materials).

Linoleum and Oilcloth
Dispersing agent in—
Coating compositions used in making oilcloth.

Paint and Varnish
Dispersing agent in making—
Asphaltic paints and varnishes.
Auto-top dressing compositions containing paraffin.
Emulsified paints and varnishes.
Roofing compositions containing asphalt.
Shellac emulsions.
Waterproofing compositions.
Waterproofing compositions containing asphalt.

Triethanolamine Linoleate (Continued)

Paper
Dispersing agent in making—
Compositions containing paraffin, used for impregnating paperboard.
Sizing compositions containing paraffin.
Sizing compositions in emulsified form containing rosin, casein, starches, glues, and paraffin.
Waterproofing compositions.
Waterproofing compositions containing paraffin.
Waxed paper coating containing paraffin.
Waxing compositions in emulsified form.
Reagent in—
Hydration of cellulose in the beating process (aids by increasing the speed of the process without injuring the strength and other qualities of the finished paper).

Perfume
Dispersing agent in making—
After-shaving creams, cosmetic creams, dentifrices, grease paints, hair tonics, latherless shaving creams, lotions, shampoos, shaving creams, skin foods, vanishing creams.

Petroleum
Dispersing agent in making—
Stabilized emulsions containing petroleum or petroleum distillates, such as paraffin oils and other heavy oils.
Ingredient of—
Emulsified cutting oils for lathe and screwpress work.
Kerosene emulsions, lubricating compositions, medicinal oils in emulsified form, naphtha emulsions, paraffin emulsions, petrolatum emulsions, soluble greases, soluble oils for lubricating textile machinery, rayon oils, various textile oils.

Pharmaceutical
Dispersing agent in making—
Emulsions of organic mercurials in petrolatum.
Lanolin emulsions.
Various emulsified pharmaceutical preparations.

Resins and Waxes
Dispersing agent in making—
Emulsions of natural and artificial waxes.
Emulsions of natural and artificial resins.
Starting point in making—
Condensation products used as artificial resins.

Rubber
Dispersing agent in making—
Rubber emulsions and compositions.
Reagent in—
Curing sponge rubber.

Soap
Dispersing agent in making—
Hand-cleansing compositions.
Shaving creams containing lanolin.
Ingredient of—
Dry-cleaning soaps.
Textile scouring soaps (to aid in removing grease, tar, and oil spots).

Textile
——, *Dyeing*
Ingredient of—
Dye baths in emulsified form (used as an assistant in dyeing various yarns and fabrics).
——, *Finishing*
Ingredient of—
Emulsified coating compositions.
Emulsified compositions for making window shade cloth.
Emulsified sizing preparations containing paraffin.
Emulsified sizing compositions containing starches and other sizes.
Emulsified washing compositions.
Emulsified waterproofing preparations containing paraffin.
——, *Manufacturing*
Ingredient of—
Dispersions used in fulling operations.
Dispersions used for carbonization of wool.
Dispersions used for washing wool and degreasing raw wool.
Emulsions for kier boiling cotton to aid in the removal of the natural gums, fats, waxes, and hemicellulose.
Emulsions for degumming silk.
Emulsions for soaking silk.
Emulsified mercerizing baths.
Oiling emulsions for treating fabrics.
Scouring preparations.
Wetting baths.
——, *Printing*
Ingredient of—
Printing pastes in emulsified form.

Triethanolamine Mucate

Textile
De-electrifying agent (Brit. 430221) for—
Yarns, films, fabrics, and the like, subject to charging by static electricity (applied in admixture with all usual lubricating agents as vehicle).

Triethanolamine Oleate
French: Oléate de triéthanolamine, Oléate triéthanolamanique.
German: Oelsäurestriaethanolamin, Oelsäurestriaethanolaminester, Triaethanolaminoleat.

Chemical
Dispersing agent in making—
Emulsions of hydrocarbons of various groups of the aliphatic and aromatic series.
Emulsions of various chemicals, terpene emulsions.
Ingredient of—
Textile lubricants for carding, combing, and drawing wool in making wool yarn for raw wool.
Wetting compositions containing ethylene dichloride or pine oil.

Construction
Ingredient of—
Emulsions, containing asphalt, used in the curing of concrete.
Road-surfacing compositions containing asphalt.

Disinfectant
Dispersing agent in making—
Emulsified disinfectants, containing pine oils, creosote, or phenol.
Emulsified germicidal and deodorant preparations.

Dye
Dispersing agent in making—
Color lakes.

Fats and Oils
Dispersing agent in making—
Boring oils, drilling oils, greasing compositions.
Lubricating compositions containing animal or vegetable oils.
Solvents for fats.
Stabilized emulsions of various animal or vegetable fats and oils.
Wire-drawing oils.

Glues and Adhesives
Dispersing agent in making—
Casein emulsions used as adhesives.

Ink
Dispersing agent in making—
Printing inks, writing inks.

Insecticide
Dispersing agent in making—
Emulsified insecticidal and fungicidal compositions.
Orchard sprays in emulsified form (added to increase the effectiveness).

Leather
Dispersing agent in making—
Emulsified dressings containing casein, shellac, and carnauba wax.
Emulsified fat-liquoring baths.
Emulsified soaking compositions containing neatsfoot oil.
Emulsified waterproofing compositions.

Miscellaneous
Dispersing agent in making—
Automobile polishes.
Compositions for cleansing painted and metal surfaces.
Deodorizing compositions containing pine oil.
Furniture polishes containing carnauba wax and mineral oil.
Impregnating compositions containing paraffin.
Metal polishes.
Metal polishes containing orthodichlorobenzene and abrasives.
Scouring compositions for woodwork, linoleum, rugs, and the like.
Various emulsified polishes containing oleic acid, ethylene dichloride, carnauba wax.
Waterproofing compositions.
Ingredient of—
Liquid baths (added to assist in their penetration into porous materials).

Triethanolamine Oleate (Continued)

Linoleum and Oilcloth
Dispersing agent in—
Coating compositions used in making oilcloth.

Paint and Varnish
Dispersing agent in making—
Asphaltic paints and varnishes.
Auto-top dressing compositions containing paraffin.
Emulsified paints and varnishes.
Roofing compositions containing asphalt.
Shellac emulsions, waterproofing compositions.
Waterproofing compositions containing asphalt.

Paper
Dispersing agent in making—
Compositions, containing paraffin, used for impregnating paperboard.
Sizing compositions containing paraffin.
Sizing compositions in emulsified form containing rosin, casein, starches, glues, and paraffin.
Waterproofing compositions.
Waterproofing compositions containing paraffin.
Waxed paper coating containing paraffin.
Waxing compositions in emulsified form.
Reagent in—
Hydration of cellulose in the beating process (aids by increasing the speed of the process without injuring the strength and other qualities of the finished paper).

Perfume
Dispersing agent in making—
After-shaving creams, cosmetic creams, dentifrices, grease paints, hair tonics, latherless shaving creams, lotions, shampoos, shaving creams, skin foods, vanishing creams.

Petroleum
Dispersing agent in making—
Stabilized emulsions containing petroleum or petroleum distillates, such as paraffin oil and other heavy oils.
Ingredient of—
Emulsified cuttings oils for lathe and screwpress work.
Kerosene emulsions, lubricating compositions, medicinal oils in emulsified form, naphtha emulsions, paraffin emulsions, petrolatum emulsions, soluble greases.
Soluble oils for lubricating textile machinery.
Rayon oils, various textile oils.

Pharmaceutical
Dispersing agent in making—
Lanolin emulsions.
Mercurochrome emulsions in petrolatum.
Various emulsified pharmaceutical preparations.

Resins and Waxes
Dispersing agent in making—
Emulsions of natural and artificial waxes.
Emulsions of natural and artificial resins.
Starting point in making—
Condensation products used as artificial resins.

Rubber
Dispersing agent in making—
Rubber emulsions and compositions.
Reagent in—
Curing sponge rubber.

Sanitation
Ingredient of—
Pine oil disinfectants.

Soap
Dispersing agent in making—
Hand-cleansing compositions, shaving creams containing lanolin.
Ingredient of—
Dry-cleaning soaps.
Textile scouring soaps (to aid in removing grease, tar, and oil spots).

Textile
——, *Dyeing*
Ingredient of—
Dye baths in emulsified form (used as an assistant in dyeing various yarns and fabrics).

——, *Finishing*
Ingredient of—
Emulsified coating compositions, emulsified compositions for making window shade cloth, emulsified sizing preparations containing paraffin, emulsified sizing compositions containing starches and other sizes, emulsified washing compositions, emulsified waterproofing preparations containing paraffin.

——, *Manufacturing*
Ingredient of—
Dispersions used in fulling operations.
Dispersions used for carbonization of wool.
Dispersions used for washing wool and degreasing raw wool.
Emulsions for kier boiling cotton to aid in the removal of the natural gums, fats, waxes, and hemicellulose.
Emulsions for degumming silk, emulsions for soaking silk, emulsified mercerizing baths, oiling emulsions for treating fabrics, scouring preparations, wetting baths.

——, *Printing*
Ingredient of—
Printing pastes in emulsified form.

Triethanolamine Saccharate

Textile
De-electrifying agent (Brit. 430221) for—
Yarns, films, fabrics, and the like, subject to charging by static electricity (applied in admixture with all usual lubricating agents as vehicle).

Triethanolamine Salicylate

Textile
De-electrifying agent (Brit. 430221) for—
Yarns, films, fabrics, and the like, subject to charging by static electricity (applied in admixture with all usual lubricating agents as vehicle).

Triethanolamine Salt of Sulphonated Cottonseed Oil

Glues and Adhesives
Ingredient (Brit. 411908) of—
Adhesive compositions comprising a solution of at least 15°Be. of an alkali metal silicate containing 1.5 to 3.5 mols. of silica per mol. of alkali oxide together with up to 2 percent of an organic wetting agent.

Triethanolamine Stearate

Synonyms: Stearate of tri(hydroxyethyl)amine, Tri-(hydroxyethyl)amine stearate.
French: Stéarate de triéthanolamine, Stéarate triéthanolaminique.
German: Stearinsäurestriaethanolamin, Stearinsäuretriaethanolaminester.

Chemical
Dispersing agent in making—
Dispersions and emulsions of various chemicals.

Construction
Dispersing agent in making—
Emulsified waterproofing and dampproofing compositions for treating brick work, concrete, masonry, piles, porous structural materials, shingles, walls.

Cosmetics and Perfumes
Dispersing agent in making—
Emulsified creams, emulsified lotions, emulsified lanolin preparations, emulsified ointments, emulsified perfumes, emulsified shaving creams, emulsified sunburn preparations.

Fats and Oils
Dispersing agent in making—
Emulsified boring oils, emulsified drilling oils, emulsified fatty acids of animal or vegetable origin, emulsified sulphonated oils, emulsions of animal or vegetable oils.
Greasing compositions in emulsified form.
Lubricating compositions in emulsified form, containing various vegetable or animal fats and oils.
Solvents for fats in emulsified form.
Stabilized emulsions of vegetable or animal fats and oils.
Wetting compositions in emulsified form, containing animal or vegetable fats and oils.
Wire-drawing oils in emulsified form.

Glues and Adhesives
Dispersing agent in making—
Emulsified adhesive preparations containing paraffin and other waxes.

Ink
Dispersing agent in making—
Ink emulsions for printing, marking, lithographic, stamping, and writing purposes.

Insecticide
Dispersing agent in making—
Emulsified insecticidal and fungicidal preparation.
Orchard sprays in emulsified form.
Vermin exterminators in emulsified form.

Triethanolamine Stearate (Continued)

Leather

Dispersing agent in making—

Emulsified dressing compositions, emulsified fat-liquoring baths, emulsified finishing compositions, emulsified polishing compositions.

Emulsified soaking compositions containing various animal or vegetable oils.

Emulsified waterproofing compositions.

Linoleum and Oilcloth

Dispersing agent in making—

Emulsified finishing compositions containing waxes.

Miscellaneous

Dispersing agent in making—

Automobile polishes in emulsified form.

Cleansing compositions in emulsified form, for use on painted and metallic surfaces.

Compositions in emulsified form for waterproofing automobile tops and tarpaulins.

Dampproofing compositions in emulsified form.

Degreasing compositions in emulsified form.

Emulsified compositions containing various substances, such as tars and pitches.

Floor polishes in emulsified form, greasing compositions in emulsified form, furniture polishes in emulsified form, linoleum polishes in emulsified form, metal polishes in emulsified form, scouring compositions in emulsified form, shoe polishes in emulsified form, special emulsified cleaning compositions.

Various emulsified compositions containing fats, oils, and miscellaneous substances, used for wetting, washing, and dispersion purposes.

Waterproofing compositions for treating various fibers and other compositions of matter.

Wood polishes in emulsified form.

Paint and Varnish

Dispersing agent in making—

Emulsified paints, varnishes, and other coating compositions.

Pigment emulsions, shellac emulsions, waterproofing compositions in emulsified form, wood-filling compositions.

Paper

Dispersing agent in making—

Coating compositions in emulsified form.

Emulsified preparations used for the treatment of paper and pulp and various products made therefrom.

Sizing compositions in emulsified form, waterproofing compositions in emulsified form, waxing compositions in emulsified form.

Petroleum

Dispersing agent in making—

Emulsified cutting compositions containing various mineral oil distillates.

Emulsified preparations containing kerosene, naphtha emulsions, petroleum distillate and residue emulsions, rayon oils in emulsified form.

Soluble oils in emulsified form, for the lubrication of textile and other machinery.

Various textile oils in emulsified form.

Resins and Waxes

Dispersing agent in making—

Emulsified compositions containing various waxes, artificial or natural.

Emulsified compositions containing various resins, artificial or natural.

Rubber

Dispersing agent in making—

Emulsified rubber compositions, emulsified rubber cement.

Soap

Dispersing agent in making—

Emulsified detergents for various purposes, hand-cleansing compositions in emulsified form, textile-scouring soaps in emulsified form.

Textile

——, *Finishing*

Dispersing agent in making—

Emulsified coating compositions, emulsified dressing compositions, emulsified finishing compositions, emulsified impregnating compositions, emulsified scouring compositions, emulsified sizing compositions, emulsified washing compositions containing soaps, emulsified waterproofing compositions.

——, *Manufacturing*

Dispersing agent in making—

Emulsified compositions for greasing operations.

Emulsified compositions for degreasing operations.

Emulsified compositions used in fulling operations.

Emulsified compositions for lubrication purposes in spininng and weaving.

Emulsified compositions for degumming silk.

Emulsified compositions for soaking silk.

Emulsified preparations for kier-boiling cotton.

Emulsified preparations for milling purposes.

Emulsified preparations for washing wool.

——, *Printing*

Dispersing agent in making—

Emulsified printing pastes.

Triethanolamine Tannate

Textile

De-electrifying agent (Brit. 430221) for—

Yarns, films, fabrics, and the like, subject to charging by static electricity (applied in admixture with all usual lubricating agents as vehicle).

Triethanolamine Tartrate

Textile

De-electrifying agent (Brit. 430221) for—

Yarns, films, fabrics, and the like, subject to charging by static electricity (applied in admixture with all usual lubricating agents as vehicle).

Triethanolamine Thiocyanate

Lubricant

Starting point (Brit. 440175) in making—

Addition agents for high-pressure lubricating oils or greases by mixing and reacting with organo-metallic compounds.

Triethyl-Aluminum

Lubricant

Starting point (Brit. 440175) in making—

Addition agents for high-pressure lubricating oils or greases by reacting with oil-soluble organic compounds.

Triethyl-Antimony

Lubricant

Starting point (Brit. 440175) in making—

Addition agents for high-pressure lubricating oils or greases by reacting with oil-soluble organic compounds.

Triethyl-Arsenic

Lubricant

Starting point (Brit. 440175) in making—

Addition agents for high-pressure lubricating oils or greases by reacting with oil-soluble organic compounds.

Triethyl Citrate

Cellulose Products

Solvent and plasticizer for—

Cellulose esters or ethers.

For uses, see under general heading: "Solvents."

2:2':8-triethyl-5:5'-dimethselenothiacarbocyanin Iodide

Photographic

Sensitizer (Brit. 420971) in—

Photographic emulsions.

Triethyleneglycol Monostearate

French: Monostéarate de triéthylèneglycole, Monostéarate triéthylènique-glycollique.

German: Monstearinsäurestriaethylenglykol, Monostearinsäuretriaethylenglykolester, Triaethylenglykolmonostearat.

Miscellaneous

As a dispersing agent (Brit. 329266).

For uses, see under general heading: "Emulsifying agents."

Triethyl-Lead Chloride

Lubricant

Starting point (Brit. 440175) in making—

Addition agents for high-pressure lubricating oils or greases by reacting with oil-soluble organic compounds.

Triethyl-Lead Hydroxide

Lubricant

Starting point (Brit. 440175) in making—

Addition agents for high-pressure lubricating oils or greases by reacting with oil-soluble organic compounds.

Triethyl-Lead-Mercaptan
Lubricant
Addition agent (Brit. 440175) for—
Lubricating oils or greases used in high-pressure working conditions.

Triethyl-Lead Thiocyanoleate
Oils, Fats, and Waxes
Addition agent (Brit. 440175) for—
Lubricating oils or greases used in high-pressure working conditions.

Triethyloctodecoxymethyl-ammonium Chloride
Textile
Increaser (Brit. 434911) of—
Fastness to water of dyeings on textile fibers.
Softener (Brit. 434911) of—
Dyed textile fibers.

2:2':8-Triethylselenacarbocyanin Iodide
Dye
Dye (Brit. 439359) possessing good solubility in organic solvents.

Triethylstilbin Dichloride
French: Dichlorure de triéthylestilbène, Dichlorure triéthylestilbène, Dichlorure triéthylestilbènique.
German: Dichlorotriaethylstilbin, Triaethylstilbinchlorid.
Chemical
Starting point (Brit. 303092) in making—
Chemicals for treating and mothproofing animal products.
Miscellaneous
Mothproofing agent (Brit. 303092) for—
Hair, felt, and furs.
Textile
Mothproofing agent (Brit. 303092) for—
Wool.

2:2':8-Triethylthiacarbocyanin Iodide
Dye
Dye (Brit. 439359) possessing good solubility in organic solvents.

Triethyl-Tin Oleate
Lubricant
Starting point (Brit. 440175) in making—
Addition agents for high-pressure lubricating oils or greases, by reacting with oil-soluble organic compounds.

Trifluorobenzene
Synonyms: Benzene trifluoride, Benzol trifluoride.
French: Trifluorure de benzène, Trifluorure de benzol.
German: Benzoltrifluorid, Trifluorbenzol.
Electrical
Starting point (Brit. 430045) in making—
Insulating liquids for electrical apparatus (by admixture with mineral or vegetable oils).

Trifluorobenzene, Chlorinated
Synonyms: Chlorinated benzene trifluoride, Chlorinated benzol trifluoride, Chlorinated trifluorbenzol, Chlorinated trifluorobenzol.
French: Trifluorure de benzène, chloré; Trifluorure de benzol, chloré.
German: Chlorhaltigbenzoltrifluorid.
Electrical
Starting point (Brit. 430045) in making—
Insulating liquids for electrical apparatus (by admixture with mineral or vegetable oils).

Trifluorodimethyl Acetone
Refrigeration
As a refrigerant (Brit. 416653).

Triglycerol Triacetate
Cellulose Products
Plasticizer (Brit. 364807) for—
Cellulose esters and ethers.
For uses, see under general heading: "Plasticizers."

Triglycerylamine
French: Triglycérylamine.
German: Triglycerylamin.
Soap
Starting point in making—
Soaps, when warmed with fatty acids, soluble in organic liquids and suitable for making dry-cleaning preparations.

2:4:6-Trihydroxybenzimidophenyl Hydrochloride-Sulphide
Synonyms: 2:4:6-Trihydroxybenzimidothiophenylether hydrochloride.
Insecticide
Larvicide for—
Culicine mosquito larvae.

Trilaurylamine
Fungicide and Insecticide
As a fungicide (Brit. 436327).
As an insecticide (Brit. 436327).

Trimethyl Antimony
Lubricant
Starting point (Brit. 440175) in making—
Addition agents for high-pressure lubricating oils or greases by reacting with oil-soluble organic compounds.

Trimethylbetahydroxygammadodecoxypropyl-Ammonium Bromide
Disinfectant
Claimed (Brit. 436725 and 436726) to be—
Bactericide, disinfectant.
Insecticide and Fungicide
Claimed (Brit. 436725 and 436726) to be—
Fungicide.

Trimethylbetahydroxygammadodecoxypropyl-Ammonium Chloride
Disinfectant
Claimed (Brit. 436725 and 436726) to be—
Bactericide, disinfectant.
Insecticide and Fungicide
Claimed (Brit. 436725 and 436726) to be—
Fungicide.

Trimethylbetamethyldecylaminoethyl-Ammonium Bromide
Disinfectant
Claimed (Brit. 436725 and 436726) to be—
Bactericide, disinfectant.
Insecticide and Fungicide
Claimed (Brit. 436725 and 436726) to be—
Fungicide.

Trimethylbetamethyldodecylaminoethyl-Ammonium Iodide
Disinfectant
Claimed (Brit. 436725 and 436726) to be—
Bactericide, disinfectant.
Insecticide and Fungicide
Claimed (Brit. 436725 and 436726) to be—
Fungicide.

1:1:3-Trimethylbutadiene
French: 1:1:3-Triméthylebutadiène.
German: 1:1:3-Trimethylbutadien.
Chemical
Starting point in making—
Intermediates, pharmaceuticals.
Starting point (Brit. 309911) in making synthetic perfumes with—
Acrolein, acrylic acid, crotonaldehyde, crotonic acid, propargylaldehyde.

1:1:4-Trimethylbutadiene
French: 1:1:4-Triméthylebutadiène.
German: 1:1:4-Trimethylbutadien.
Chemical
Starting point in making—
Intermediates, pharmaceuticals.
Starting point (Brit. 309911) in making synthetic perfumes with—
Acrolein, acrylic acid, crotonaldehyde, crotonic acid, propargylaldehyde.

Trimethylcetylammonium Bromide
Textile
Mordant (Brit. 436592) in—
Dyeing natural or regenerated cellulosic textile materials with chrome dyestuffs.

Trimethylcyclohexane

Petroleum
Solvent (Brit. 436044) in—
Flushing oil composition for internal-combustion engines; flushing oil is based on light lubricating oil of either paraffinic or naphthenic origin and contains various other products; naphtha, isopropyl alcohol, or acetone may be added to reduce the viscosity; practice is to flush (1) with oil containing a high proportion of solvent to remove most of the sludge, (2) with oil containing a lower proportion of solvent.

1:1:3-Trimethylcyclohexanone

Cellulose Products
Solvent for—
Nitrocellulose.
For uses, see under general heading: "Solvents."

1:1:2-Trimethylcyclopentene
French: 1:1:2-Triméthylecyclopentène.
German: 1:1:2-Trimethylcyklopenten, 1:1:2-Trimethylzyklopenten.

Chemical
Solvent in various chemical processes and for various chemicals (Brit. 269960).

Miscellaneous
Solvent for various purposes (Brit. 269960).

Textile
——, *Dyeing and Printing*
Solvent in making—
Liquors and pastes for dyeing, printing, and stenciling acetate rayon (Brit. 269960).

2:2′:8-Trimethyl-4:5:4′:5′-dibenzoxacarbocyanin Bromide

Photographic
Sensitizer (Brit. 432969) for—
Silver halide emulsions (sensitizing maxima: 570 mu).

Trimethyldodecylthiomethyl-Ammonium Chloride

Disinfectant
Claimed (Brit. 436725 and 436726) to be—
Bactericide, disinfectant.

Insecticide and Fungicide
Claimed (Brit. 436725 and 436726) to be—
Fungicide.

Trimethylene
Synonyms: Cyclopropane.

Chemical
Starting point in making—
Various derivatives.

Pharmaceutical
Suggested for use as—
Anesthetic characterized by rapid induction and quick recovery.

Trimethyleneglycol

Analysis
As a reagent.

Chemical
Reagent in—
Organic synthesis.

Trimethylene Phosphate, Chlorinated

Lubricant
Stabilizer (Brit. 448424) for—
Viscous oils, such as Pennsylvania or Midcontinent petroleum, used for extreme pressure work.

Trimethylglycocoll Hydrochloride

Metallurgical
Ingredient (U. S. 1882734) of—
Soldering flux.

Trimethyloctadecylammonium Bromide

Textile
Mordant (Brit. 436592) in—
Dyeing natural or regenerated cellulosic textile materials with chrome dyestuffs.

1:1:3-Trimethyl-2(2⁴-oxo-2³-ethobutyl)cyclohexane
German: 1:1:3-Trimethyl-2(2⁴-oxo-2⁸-aethobutyl)zyklohexan.

Perfume
Ingredient (Brit. 347052) of compositions, containing—
Ambrette musk, artificial jasmine oil, benzyl acetate, benzyl alcohol, bergamot oil, cinnamic alcohol, cumarin, ionone, heliotropin, hydroxycitronellal, methylionone, orange oil, phenylethyl alcohol, sandalwood oil, ylang-ylang oil.

1:3:3-Trimethyl-2-paradiethylaminostyrylindoleninium Chloride

Dye
Starting point (Brit. 448508) in making—
Violet lakes constituting clear shades fast to oil, spirit, and light.

1:3:3-Trimethyl-2-paradiethylaminostyryl-4:5-sulphobenzoindolenium Sulphate

Dye
Starting point (Brit. 448508) in making—
Violet lakes constituting clear shades fast to oil, spirit, and light.

Trimethyl Phosphate

Cellulose Products
Plasticizer for—
Nitrocellulose.
For uses, see under general heading: "Plasticizers."

Trimethyl Phosphate, Chlorinated

Lubricant
Stabilizer (Brit. 448424) for—
Viscous oils, such as Pennsylvania or Midcontinent petroleums, used for extreme pressure work.

1:3:5-Trimethyl-5-piperidinobarbituric Acid Hydrochloride

Pharmaceutical
Suggested for use (Brit. 414293) as—
Hypnotic with low toxic properties.

2:4:6-Trimethylpyridin Ethiodide
French: Éthiodure 2:4:6-triméthylepyridinique.
German: 2:4:6-Trimethylpyridinaethjodid.

Insecticide
Starting point (German 438241) in making—
Fungicide and bactericide for treating diseased seeds with dimethylaminobenzaldehyde.

Trimethylstibin Dibromide
Synonyms: Dibromotrimethylstibine.
French: Dibromure de triméthylestibinique.
German: Dibromtrimethylstibin.

Miscellaneous
Mothproofing agent (Brit. 303092) for treating—
Furs, hair, feathers.

Textile
Mothproofing agent (Brit. 303092) for treating—
Wool and felt.

Trimethylstilbin Sulphate
French: Sulphate de triméthylestilbène, Sulphate triméthylestilbinique.
German: Schwefelsaeurestrimethylstilben, Schwefelsaeuretrimethylstilbinester, Trimethylstilbinsulfat.

Chemical
Starting point (Brit. 303092) in making—
Chemicals for treating and mothproofing animal products.

Miscellaneous
Mothproofing agent (Brit. 303092) for treating—
Hair, felt, and furs.

Textile
Mothproofing agent (Brit. 303092) for treating—
Wool.

Trimethyltriphenyl-Mercury

Lubricant
Addition agent (Brit. 445813) in—
Lubricants for motors, turbines, flushings, and high-temperature work generally.

Trimethyltriphenyl-Tin

Lubricant
Addition agent (Brit. 445813) in—
Lubricants for motors, turbines, flushings, and high-temperature work generally.

Trinaphthyl Borate
French: Borate de trinaphthyle, Borate trinaphthylique.
German: Borsäurestrinaphtyl, Borsäuretrinaphtylester, Trinaphtylborat.
Spanish: Borato de trinaftil.
Italian: Borato di trinaftile.

Rubber
Ingredient (Brit. 363483) of—
Rubber batch (added for the purpose of increasing the resistance of the rubber goods to deterioration and discoloration caused by ageing).

Trinaphthyl Phosphate, Chlorinated

Lubricant
Stabilizer (Brit. 448424) for—
Viscous oils, such as Pennsylvania or Midcontinent oil, used for extreme pressure work.

2:4:6-Trinitro-1:3-dimethyl-5-tertiarybutylbenzene

Mechanical
Improver (Brit. 404046) of—
Exhaust odors from internal combustion engines (added to fuels not derived from petroleum, either alone or in conjunction with (1) acetophenone, methylacetophenone, 4-methoxyacetophenone, 1-naphthylmethyl ketone, 2-naphthylmethyl ketone, or (2) any of the ketones listed under (1) and any of the following: Camphor, waste camphor oil, borneol, bornyl acetate, clove oil, ionone, coumarin, indole, skatole, paracresyl acetate, methyl anthranilate, isopropylmethylhydrocinnamic aldehyde).

Petroleum
Reagent (Brit. 404046) for—
Improving exhaust odors from internal combustion engines (added to gasoline or diesel oil, either alone or in conjunction with (1) acetophenone, methylacetophenone, 4-methoxyacetophenone, 1-naphthylmethyl ketone, 2-naphthylmethyl ketone, or (2) any of the ketones listed under (1) and any of the following: Camphor, waste camphor oil, borneol, bornyl acetate, clove oil, ionone, coumarin, indole, skatole, paracresyl acetate, methyl anthranilate, isopropylmethylhydrocinnamic aldehyde).

2:4:6-Trinitro-3-methyl-5-tertiarybutylanisol

Mechanical
Improver (Brit. 404046) of—
Exhaust odors from internal combustion engines (added to fuels not derived from petroleum, either alone or in conjunction with (1) acetophenone, methylacetophenone, 4-methoxyacetophenone, 1-naphthylmethyl ketone, 2-naphthylmethyl ketone, or (2) any of the ketones listed under (1) and any of the following: Camphor, waste camphor oil, borneol, bornyl acetate, clove oil, ionone, coumarin, indole, skatole, paracresyl acetate, methyl anthranilate, isopropylmethylhydrocinnamic aldehyde).

Petroleum
Reagent (Brit. 404046) for—
Improving exhaust odors from internal combustion engines (added to gasoline or diesel oil, either alone or in conjunction with (1) acetophenone, methylacetophenone, 4-methoxyacetophenone, 1-naphthylmethyl ketone, 2-naphthylmethyl ketone, or (2) any of the ketones listed under (1) and any of the following: Camphor, waste camphor oil, borneol, bornyl acetate, clove oil, ionone, coumarin, indole, skatole, paracresyl acetate, methyl anthranilate, isopropylmethylhydrocinnamic aldehyde).

2:4:6-Trinitro-1-methyl-3-tertiarybutylbenzene

Mechanical
Improver (Brit. 404046) of—
Exhaust odors from internal combustion engines (added to fuels not derived from petroleum, either alone or in conjunction with (1) acetophenone, methylacetophenone, 4-methoxyacetophenone, 1-naphthylmethyl ketone, 2-naphthylmethyl ketone, or (2) any of the ketones listed under (1) and any of the following: Camphor, waste camphor oil, borneol, bornyl acetate, clove oil, ionone, coumarin, indole, skatole, paracresyl acetate, methyl anthranilate, isopropylmethylhydrocinnamic aldehyde).

Petroleum
Reagent (Brit. 404046) for—
Improving exhaust odors from internal combustion engines (added to gasoline or diesel oil, either alone or in conjunction with (1) acetophenone, methylacetophenone, 4-methoxyacetophenone, 1-naphthylmethyl ketone, 2-naphthylmethyl ketone, or (2) any of the ketones listed under (1) and any of the following: Camphor, waste camphor oil, borneol, bornyl acetate, clove oil, ionone, coumarin, indole, skatole, paracresyl acetate, methyl anthranilate, isopropylmethylhydrocinnamic aldehyde).

Trinitrophenetole

Abrasive
Ingredient (Brit. 295335) of—
Compositions used in binding the abrasive in grinding discs, stones, and other forms.

Chemical
Catalyst (Brit. 295335) in making—
Impregnating compositions containing phenol-aldehyde resins.
Starting point in making—
Aromatics, intermediates, pharmaceuticals.

Dye
Starting point in making various synthetic dyestuffs.

Glues and Adhesives
Ingredient (Brit. 295335) of—
Binders and cements containing phenol-aldehyde resins.

Miscellaneous
Ingredient (Brit. 295335) of—
Impregnating various compositions containing phenol-aldehyde resins.

Paint and Varnish
Ingredient (Brit. 295335) of—
Lacquers and varnishes containing phenol-formaldehyde resins.

Plastics
Ingredient (Brit. 295335) of—
Molding and pressing compositions containing phenol-aldehyde resins.

Trinitrophenylethylnitroamine

Explosives and Matches
Ingredient (U. S. 1975186) of—
Detonator charges, containing also trinitrophenylmethylnitroamine and, optionally, pentaerythritol tetranitrate.

Trinitrophenylmethylnitroamine

Explosives and Matches
Ingredient (U. S. 1975186) of—
Detonator charges, containing also trinitrophenylethylnitroamine and, optionally, pentaerythritol tetranitrate.

Triorthotolylstibin

Chemical
Starting point in making—
Intermediates and other derivatives.

Miscellaneous
Mothproofing agent (Brit. 303092) for treating—
Furs, hair, and feathers.

Textile
Mothproofing agent (Brit. 303092) for treating—
Wool and felt.

1:2:6-Trioxyanthraquinone

Oils, Fats, Waxes
Coloring agent (Brit. 432867) for—
Paraffin and other mineral waxes, stearic acid, tallow and other solid triglycerides, beeswax, carnauba wax, and others.

1:2:4 Trioxyphenyl Triacetate

Chemical
Starting point in making—
Synthetic tanning agents (Brit. 242694).

Triparatolylphosphin Oxide
French: Oxyde de triparatolylephosphine, Oxyde triparatolylephosphinique.
German: Triparatolylphosphinoxyd.

Miscellaneous
Reagent (Brit. 303092) for—
Mothproofing fur and hair.

Textile
Reagent (Brit. 303092) for—
Mothproofing felt and wool.

Triparatolylstibin
French: Stibine de triparatolyle, Stibine triparatolylique.
Miscellaneous
Ingredient (Brit. 303092) of—
Mothproofing compositions for furs and hair.
Textile
Ingredient (Brit. 303092) of—
Mothproofing compositions for woolens.

Triphenyl-Aluminum
Petroleum
Addition agent (Brit. 433257) in—
Lubricating oils or greases, especially for use at high temperatures, such as cylinder oils, hydrogenated oils, or oils refined by treatment with sulphuric acid, clay, or extraction solvents.

Triphenylarsin
French: Arsine de triphényle, Arsine triphénylique.
Chemical
Starting point (Brit. 303092) in making—
Chemicals used for various mothproofing purposes.
Miscellaneous
Reagent (Brit. 303092) for—
Mothproofing fur and hair.
Textile
Reagent (Brit. 303092) for—
Mothproofing wool and felt.

Triphenyl-Bismuth
Lubricant
Starting point (Brit. 440175) in making—
Addition agents for high-pressure lubricating oils or greases by reacting with oil-soluble organic compounds.

Triphenyl-Cadmium
Petroleum
Addition agent (Brit. 433257) in—
Lubricating oils or greases, especially for use at high tempratures, such as cylinder oils, hydrogenated oils, or oils refined by treatment with sulphuric acid, clay, or extraction solvents.

Triphenylchloromethane
French: Triphénylechlorométhane.
German: Triphenylchlormethan.
Chemical
Starting point in making various organic chemicals.
Rubber
Reagent (Brit. 282778) in making rubber conversion products with—
Alphanaphthol, betanaphthol, catechol, cresol, parachlorophenol, phenol, resorcinol.

Triphenylguanidin
French: Triphényleguanidine.
Spanish: Trifenilguanidina.
Italian: Trifenilguanidine.
Ceramics
Ingredient (Brit. 342288) of—
Compositions containing cellulose esters or ethers, used for coating and decorating ceramic wares (added for the purpose of stabilizing the composition against ageing).
Chemical
Starting point in making various derivatives.
Glass
Ingredient (Brit. 342288) of—
Compositions containing cellulose esters or ethers, used in the production of nonscatterable glass and for the decoration and protection of glassware (added for the purpose of stabilizing them against ageing)
Leather
Ingredient (Brit. 342288) of—
Compositions containing cellulose esters or ethers, used in the manufacture of artificial leather and for the decoration and protection of leather goods (added for the purpose of stabilizing them against ageing).
Linoleum and Oilcloth
Ingredient of—
Coating compositions.
Metallurgical
Ingredient (Brit. 342288) of—
Compositions containing cellulose esters or ethers, used for the decoration and protection of metal articles (added for the purpose of stabilizing them against ageing).
Miscellaneous
Ingredient (Brit. 342288) of—
Compositions containing various esters or ethers of cellulose, used for the decoration and protection of various compositions of matter (added for the purpose of stabilizing them against ageing).
Paint and Varnish
Ingredient (Brit. 342288) of—
Paints, varnishes, dopes, enamels, and lacquers containing various cellulose esters or ethers, such as butylcellulose and benzylcellulose (added for the purpose of stabilizing the products against ageing).
Paper
Ingredient (Brit. 342288) of—
Compositions containing cellulose esters or ethers, used in the production of coated paper and also for the decoration and protection of paper and pulp products (added for the purpose of stabilizing them against ageing).
Pharmaceutical
Suggested for use as an antiseptic.
Plastics
Ingredient (Brit. 342288) of—
Plastic compositions containing various esters or ethers of cellulose, such as benzylcellulose and butylcellulose (added for the purpose of stabilizing the products against ageing).
Rubber
Accelerator in—
Vulcanizing processes.
Ingredient of—
Compositions containing various esters or ethers of cellulose, used for the decoration and protection of rubber goods (Brit. 242288) (added for the purpose of stabilizing them against ageing).
Rubber substitutes.
Stone
Ingredient (Brit. 342288) of—
Compositions containing various esters or ethers of cellulose, used for the decoration and protection of artificial and natural stone (added for the purpose of stabilizing them against ageing).
Textile
Ingredient of—
Compositions used for impregnating fabrics.
Compositions containing various esters or ethers of cellulose, used in the production of coated textile fabrics (Brit. 342288) (added for the purpose of stabilizing them against ageing).
Woodworking
Ingredient (Brit. 342288) of—
Compositions containing various esters or ethers of cellulose, used for the protection and decoration of woodwork (added for the purpose of stabilizing them against ageing).

Triphenylguanidin Dimethyldithiocarbamate
Disinfectant
As a bactericide (Australian 8103/32, Brit 406979, U. S. 1972961).
Insecticide and Fungicide
As a fungicide (claimed effective against *Aspergillus niger* and *Fomes Annonsus*) (Australian 8103/32, Brit. 406979, U. S. 1972961).
As an insecticide (Australian 8103/32, Brit. 406979, U. S. 1972961).

Triphenyl-Mercury
Petroleum
Addition agent (Brit. 433257) in—
Lubricating oils or greases, especially for use at high temperatures, such as cylinder oils, hydrogenated oils, or oils refined by treatment with sulphuric acid, clay, or extraction solvents.

Triphenylmethanesulphonic Acid
French: Acide de triphénylméthanesulfonique.
German: Triphenylmethansulfonsaeure.
Miscellaneous
Reagent (Brit. 280262) in treating—
Kieselguhr for the purpose of increasing its absorbent powers.

Triphenylmethanesulphonic Acid (Continued)

Paper

Reagent in treating—
Paper to increase its absorbent powers in making carbon paper.

Textile

——, *Miscellaneous*

Reagent (Brit. 280262) in treating—
Cotton wadding to increase its absorbent powers.

Triphenyl Phosphate

French: Phosphate de triphényle, Phosphate phénylique.
German: Phosphorsaeurestriphenyl, Triphenylphosphat.

Mechanical

Ingredient (German 288488) of—
Lubricating compositions, in admixture with tricresyl phosphate.

Miscellaneous

Ingredient (Brit. 252162) of—
Fireproofing compositions which are used in the treatment of abrasive sheet materials, such as emery cloth, emery paper, sandpaper, sandcloth, and the like.
See also uses under general heading: "Plasticizers."

Paper

Ingredient of—
Compositions used for impregnating roofing paper.

Paint and Varnish

Ingredient of—
Airplane dopes.
Plasticizer in making—
Lacquers, varnishes, and paints which contain nitrocellulose, cellulose acetate, or other cellulose ethers and esters.
Polishing preparations containing various resins.

Photographic

Plasticizer in making—
Photographic and cinematographic films from cellulose acetate.

Plastics

Fireproofing agent in making—
Cellulose ester and ether preparations.
Plasticizer in making—
Compositions containing nitrocellulose, cellulose acetate, or other cellulose esters and ethers (Brit. 252999).
Stabilizing agent in making—
Acetylcellulose compositions, cellulose formate compositions, nitrocellulose compositions.
Substitute for camphor in making—
Celluloid.

Textile

Ingredient of—
Compositions, containing cellulose acetate and resorcinol acetate, for coating linens.

Triphenyl Phosphate, Chlorinated

Lubricant

Stabilizer (Brit 448424) for—
Viscous oils, such as Pennsylvania or Midcontinent petroleum, used for extreme pressure work.

Triphenylphosphin Oxide

French: Oxyde de triphénylephosphine, Oxyde triphénylephosphinique.
German: Oxytriphenylphosphin, Triphenylphosphinoxyd.

Chemical

Starting point (Brit. 326137) in making pharmaceuticals and mothproofing and insect-exterminating compounds with the aid of—
Alphahydroxyphenyl-3:4-dicarboxylic acid dibutyl ester.
Alphamethyl-3-hydroxy-6-isopropylbenzene.
Alphamethyl-3-hydroxy-4-isopropyl-6-chlorobenzene.
Alphanaphthol, 4-benzylphenol, betanaphthol, 6-chloro-2-cresol, 3-chloro-4-cresol, 2:6-dichlorophenol, 2:4-dichlorophenol, 2-isobutyl-4-chlorophenol, metacresol, metahydroxydiethylanilin, 4-normal-butylphenol, orthochlorophenol, orthocresol, parachlorophenol, paracresol, parahydroxybenzoic acid ethyl ester, parahydroxybenzaldehyde, paranitrophenol, phenol, pyrocatechin monoethyl ether, resorcinol, symmetrical xylenol, 2:4:6-trichlorophenol, ar-tetrahydrobetanaphthol, thymol.

Triphenylstibin

French: Stibine de triphenyle, Stibine triphenylique.

Miscellaneous

Ingredient (Brit. 303092) of—
Mothproofing compositions for furs and hair.

Petroleum

Addition agent (Brit. 433257) in—
Lubricating oils or greases, especially for use at high temperatures, such as cylinder oils, hydrogenated oils, or oils refined by treatment with sulphuric acid, clay, or extraction solvents.

Textile

Ingredient (Brit. 303092) of—
Mothproofing compositions for woolens.

Triphenyl-Thallium

Petroleum

Addition agent (Brit. 433257) in—
Lubricating oils or greases, especially for use at high temperatures, such as cylinder oils, hydrogenated oils, or oils refined by treatment with sulphuric acid, clay, or extraction solvents.

Triphenyl-Tin Chloride

Chemical

Reagent for—
Precipitation of fluorides.

Triphenyl-Tin Thiosulphate

Lubricant

Addition agent (Brit. 440175) for—
Lubricating oils or greases used in high-pressure working conditions.

Triphenyl Zinc

Petroleum

Addition agent (Brit. 433257) in—
Lubricating oils or greases, especially for use at high temperatures, such as cylinder oils, hydrogenated oils, or oils refined by treatment with sulphuric acid, clay, or extraction solvents.

Tripoli

Synonyms: Rotten stone.

Ceramics

Raw material in making—
Whiteware

Abrasive

Raw material in making—
Buffing compositions and abrasive articles.

Chemical

Filtering medium in various processes.
Raw material in making—
Sodium silicate.

Glass

Polishing agent for—
Mirrors and glassware.

Jewelry

Polishing agent for—
Precious stones.

Insecticide

Carrier in—
Insecticidal and germicidal compositions, tree-dusting compositions.

Metallurgical

Raw material in making—
Moulds for casting small objects.

Miscellaneous

Filtering medium for various purposes.
Ingredient of—
Compositions used in removing grease stains from floors, phonograph records (as a filler), wood-surfacing compositions.
Polishing agent for—
Horn, metals, shell.
Raw material in making—
Filter stones.

Paint and Varnish

Filler in—
Paints, stains.
Ingredient of—
Ready mixed fillers, transparent wood fillers.

Rubber

Filler in—
Hard rubber.

Tripoli (Continued)

Soap
Ingredient of—
Hand soaps, scouring powders, soap powders.

Stone
Polishing agent for—
Marble.

Water and Sanitation
Filtering medium

Wine
Filtering medium.

Trisodium Phosphate

Synonyms: Phosphate of soda, Tribasic sodium phosphate, Trisodium orthophosphate.
French: Phosphate sodique, Phosphate de soude, Phosphate trisodique.
German: Phosphorsäurestrinatron, Trinatriumphosphat.
Spanish: Fosfato trisodico.

Adhesives
Ingredient (U. S. 1895979) of—
Synthetic vegetable glue, containing also powdered ivory nut, casein, lime, soda ash, and sodium fluoride.

Automotive
As a cleansing agent for—
General factory uses, glass, radiators.
As a metal-degreasing agent.
Ingredient of—
Cleansing preparations.
Degreasing mixtures, with soap, soda ash, caustic soda, or sodium perborate.

Beverage
As a detergent in—
Cleansing casks, cleansing vats, washing bottles.
As a water-softener.
Ingredient of—
Detergent mixtures.
Water-softening mixtures.
Source of phosphate in carbonated beverages.

Ceramics
Reagent for—
Removing stains from ceramic ware of all sorts prior to shipment.

Chemical
As a reagent.
Catalyst (U. S. 1891514, 1894283) in making—
Diphenyl from benzene.
Ingredient of—
Bleaching composition (Brit. 393221).
Desizing preparation containing also starch-degrading enzymes (Brit. 399998).
Reagent (U. S. 1890201) in—
Purification of arylamides of 2:3-hydroxynaphthoic acid.
Starting point in making—
Phosphates.

Construction
Cleansing agent for—
Removing oil, grease, stains and dirt from metal, marble, tile, porcelain, and woodwork in new buildings.

Dye
Reagent (Brit. 396177) in making—
Methylene blue, victoria blue R.

Fats and Oils
As a cleansing agent for utensils and apparatus.
As an emulsifying agent.

Food
As a detergent for—
Canning plant equipment, dairy equipment, food product plant equipment.
As a water-softening agent in—
Canning plant, food product plants.
Emulsifier in—
Cheese-making (acting as a substitute for sodium citrate and offering the advantage of not leaving the too pronounced salty taste of the latter).
Ingredient of—
Dairy-cleansing mixture containing also sodium silicate, plain soap, and soda ash (U. S. 1879953).
Detergent mixtures, water-softening mixtures.
Neutralizing and washing agent (U. S. 1919502) for—
Churned butter.

Glass
Cleansing and degreasing agent for—
Plate glass, tableware, window glass.

Ink
Efficiency promoter in—
Ink-eradicators containing hypochlorites.

Insecticide
As an insecticide (in aqueous solution) for—
Crawling insects, flying insects, jumping insects.
Fungicide (U. S. 1774310) in—
Inhibiting blue mold decay on fresh fruit.
Inhibiting blue mold decay on fresh citrous fruit having broken skin.

Laundering
As a detergent.
Ingredient of—
Detergent mixtures with soap, soda ash, or sodium perborate.

Leather
Reagent (U. S. 1822898) in—
Soaking.
Swelling agent in—
Tanning.

Mechanical
As a general degreasing and cleansing agent.
Ingredient of—
Boiler compound (U. S. 1002603, 1078655, 1109849, 1162024, 1273857, 1333393).
Boiler compound, containing also ammonium sulphate and soda ash (U. S. 1001935).
Boiler compound, containing also sodium amalgam, tannin, kerosene emulsified with whale oil, caustic soda, dextrin, and water (U. S. 1181562).
Boiler compound, containing also soda ash, lime, silicate of soda, caustic soda and bichromate of soda (U. S. 1278435).
Boiler compound, containing also soda ash, dextrin or starch, cutch (sufficient to yield at least 2 percent of tannic acid), and water. (This is known as "Navy Standard Boiler Compound" and was developed by Lt. Com. F. Lyon, U. S. N.)

Metallurgical
As a general degreasing agent.
Ingredient of—
Bath used in producing electroplated zinc-tin alloy on steel and iron, said to have same properties as cadmium plate (U. S. 1904732).
Core-forming sand mixture (U. S. 1751482).
Degreasing mixtures, with soap, soda ash, caustic soda, or sodium perborate.
Reagent for—
Cleansing metallic articles in order to insure an absolutely clean surface prior to electroplating.

Miscellaneous
As a general water-softening agent.
Cleansing, degreasing, and deodorizing agent for—
Bathtubs, basins, and similar fixtures.
Clothing, dishes and cooking utensils, enamelware, factories, furniture, garages, glass, household use, iceboxes, linoleum, machinery, marble counters.
Hotel and restaurant table tops, other equipment and utensils.
Metal parts and fixtures, painted surfaces, rubber flooring, shelving, silverware, sinks, stone flooring, soda fountains, tableware, tiling, windows, walls, wooden flooring, workmen's hands.
Ingredient of—
Cleanser for aluminum (U. S. 1870311).
Detergent, containing also sodium silicate and calcium oxychloride (U. S. 1894207).
Detergent, containing also aluminum phosphate (Brit. 390435).
Detergent, containing also tin phosphate (Brit. 390435).
Detergent, containing also aluminum and tin phosphates (Brit. 390435).
Detergent mixtures with soap, soda ash, caustic soda, or sodium perborate.

Paint and Varnish
Ingredient (U. S. 1892980) of—
Paint glaze, containing also casein, borax, hexamethylenetetramine, sassafras oil, and nondrying oil.
Softening agent in—
Paint removing (500 grams to 4 liters of water).

Paper
As a water-softening agent.
Ingredient of—
Liquor used as preliminary boiling agent for chips in process said to reduce digesting period by two to three hours with a yield of 48 percent to 52 percent (U. S. 1910613).

Trisodium Phosphate (Continued)
Liquor used in recovery of pulp from waste paper (Brit. 400415).

Perfumery
Ingredient of—
Bath salts, shampoos.

Pharmaceutical
In compounding and dispensing practice.

Photographic
Ingredient of—
Developing baths for use in tropical climates.
Simultaneous developing and fixing bath containing sodium sulphite, sodium thiosulphate, metaquinone, and water.
Substitute for—
Alkali metal hydrates or carbonates in alkaline developers.

Plastics
Ingredient (Brit. 403988) of catalytic mixture in making—
Cellulose acetate from cheap material, such as bagasse, wood pulp, grass products, cornstalks.

Sanitation and Water
As a water-softening agent (either alone or in mixtures).

Soap
Saponifying and emulsifying agent.

Textile
Degumming agent, degreasing agent, detergent, emulsifying agent, indicator of pH in perborate bleaching.
Ingredient (Brit. 400996) of—
Hydrolyzing agent for organic esters of cellulose used as artificial filaments, yarns, and threads.
Reagent (U. S. 1903828) in making—
Artificial wool from jute.
Scouring agent for—
Acetate rayon.
Washing agent for—
Crude cotton.
Linen, prior to bleaching.
Wool (acts as substitute for soda ash and materially influences higher yields of long fibers, together with higher brilliancy, whiteness, and suppleness).
Water-softening agent (bleaching and dyeing plants require very soft water, free of iron which T.S.P. precipitates quantitatively).
Wetting agent for—
Cotton, prior to mercerizing.

Tuna Oil
Synonyms: Tuna fish oil, Tunny oil.
French: Huile de thon.
German: Thunfischoel.
Spanish: Aceite de tuna.
Italian: Olio di tonno.

Fats and Oils
Ingredient of—
Mixtures containing other animal or vegetable oils.
Starting point in making—
Hydrogenated fats.

Ink
Ingredient of—
Marking inks, printing inks.

Leather
As a currying oil.
Ingredient of—
Compositions used in dressing leather, compositions used for impregnating leather.

Miscellaneous
Ingredient of—
Roofing preparations.

Oilcloth and Linoleum
Ingredient of—
Coating compositions.

Paint and Varnish
Ingredient of—
Paints, putty, varnishes.

Paper
Ingredient of—
Impregnating compositions for treating paper, pasteboard, and papier-mache.

Plastics
Ingredient of various plastic compositions.

Soap
Stock in making soft soaps.

Textile
Impregnating agent in making—
Coated textiles of various sorts.

Woodworking
Ingredient of—
Compositions used for the impregnation of wood.

Tungsten
French: Tungstène.
German: Wolfram.

Automotive
Contact point metal for—
Ignition systems.

Chemical
Activating agent in—
Catalytical processes.
Catalyst (Brit. 400580) in producing—
Aromatics from hydrocarbons.
Splitting agent for—
Alcohol (into ethylene and water).

Electrical
Core material for—
Carbon electrodes (to increase their electrical conductivity).
Filament metal for—
Electric lamps.
Heat-conducting medium (U. S. 1902936) for—
Mercury vapor lamp.
Lead-in wire (U. S. 1902936) for—
Mercury vapor lamp.

Metallurgical
Constituent of—
Acid-resisting alloys, alloy resistant to hot concentrated sulphuric acid, alloy resistant to nitric acid, armorplate steels, bearing metal, chemical plant equipment alloys, ferro alloys, high-pressure steels, high-speed steel, magnet steel, nonferrous alloys, self-hardening steels, shell steels, spot-welding alloys, steels resistant to high temperatures.
Electrode metal in—
Arc welding, electrolysis.
Hardening ingredient of—
Alloys.
Imparter of—
Acid corrosion resistance to alloys.
Improver of—
Nickel's resistance to corrosion by sulphuric acid.
Process material (Brit. 355792) in making—
Aluminum, aluminum alloys.

Miscellaneous
Contact point metal for—
Telegraph system sending keys.
Filament metal for—
Radio tubes.
Metal for—
Bridge-work and filling in dentistry, high-temperature ovens, phonograph needles.

Resins and Waxes
Activating agent (Brit. 388864) in—
Catalytic hydrogenation of vegetable or animal waxes to improve them in color, iodine value, melting point, content of unsaponifiable matter, and solubility in turpentine or other solvent.

Tungsten Carbide
French: Carbide de tungstène.
German: Wolframcarbid.

Ceramics
Cutting metal in—
Ceramic processes.

Construction
Cutting metal for—
Asbestos, asbestos compositions, rock of all kinds.

Electrical
Cutting metal for—
Molded insulation material.

Glass
Cutting metal for—
Glass products of all kinds.

Mechanical
Cutting tip metal for—
Machine tools.
Cutting metal for—
Die metal, gage metal, knife-edge metal.

Miscellaneous
Cutting metal for—
Fibrous products.

Plastics
Cutting metal for—
Molded products.

Tungsten Trioxide
Synonyms: Tungsten oxide, Tungstic oxide.
French: Oxyde de tungstène, Oxyde tungstique, Trioxyde de tungstène.
German: Wolframtrioxyd.

Ceramics
Pigment for painting on porcelains, potteries, chinaware.

Chemical
Reagent (Brit. 281307) in making zeolite catalysts used in making—
Acenaphthylene from acenaphthene, acetaldehyde from ethyl alcohol, acetic acid from ethyl alcohol, alcohols from aliphatic hydrocarbons.
Aldeydes from toluene, xylene, mesitylene, pseudocumene, and cymene.
Aldehydes and acids by the oxidation of orthochlorotoluene, parachlorotoluene, orthobromotoluene, parabromotoluene, dichlorotoluenes, chlorobromotoluenes, nitrotoluenes, chloronitrotoluenes, bromonitrotoluenes.
Alphanaphthaquinone from naphthlene, anthraquinone from anthracene, benzaldehyde and benzoic acid from toluene, benzoquinone from phenanthraquinone, chloroacetic acid from ethylenechlorohydrin, diphenic acid from ethyl alcohol, fluorenone from fluorene, formaldehyde from methanol or methane, hemimellitic acid from acenaphthene.
Maleic and fumaric acids from benzene, toluene, phenol, or tar acids, or from benzoquinone or phthalic anhydride.
Naphthalic anhydride.
Naphthaldehydic acid, acenaphthaquinone, or bisacenaphthylidenedione from acenaphthene or acenaphthylene.
Phenanthraquinone from phenanthrene, phthalic anhydride from naphthalene, salicyl aldehyde or salicylic acid from cresol, vanillin or vanillic acid from eugenol or isoeugenol.

Metallurgical
Starting point in making—
Filaments for electric incandescent lights, metallic tungsten and tungsten wire.

Paint and Varnish
Ingredient of—
Bronze powder.

Textile
——, *Dyeing*
Mordant in dyeing—
Fabrics and yarns with the aid of anilin black and other colors.

Turkey Red Oil
Synonyms: Sulphonated castor oil.
French: Huile pour rouge turc, Huile sulphonée.
German: Sulfonierte oel, Tuerkischrotoel.

Chemical
Ingredient (Brit. 295024) of dispersing agents with—
Aminoazobenzene, azodiphenylamine, benzeneazonaphthaleneazophenol, diethylpara-aminophenol-1:4-naphthaquinonemonoimide, 2:4-dinitrobenzene-2-azodiethylanilin, oxyethylamines, paranitranilin, rosanthrene base.
Reagent (Brit. 307079) in making—
Emulsions with starches, dextrins, vegetable gums, gelatin, glue, casein.
Starting point (U. S. 1691994) in making softener for silk with guanidines.

Dye
Ingredient of—
Dye compositions and preparations containing starches, dextrin, vegetable gums, glue, gelatin, casein (Brit. 307079).
Lakes (Brit. 270750).
Preparations containing basic dyestuffs (Brit. 270750).

Fats and Oils
Starting point in making—
Emulsified preparations.

Glues and Adhesives
Ingredient (Brit. 307079) of—
Glues and other adhesive compositions, containing starches, dextrins, vegetable gums, casein, glues, gelatin.

Insecticide
Ingredient of—
Arsenical preparations.

Leather
Ingredient (Brit. 307079) of—
Dressing compositions containing starches, dextrins, vegetable gums, gelatin, glue, casein.
Reagent for finishing leathers.

Paper
Ingredient (Brit. 299817) of emulsified preparations for—
Cleansing wire on paper machines, digestion of sulphite pulp, grinding mechanical wood pulp, removing ink from paper.
Ingredient (Brit. 307079) of—
Emulsions, containing starches, glues, gelatin, casein, vegetable gums, dextrins, for treating paper.

Perfumery
Ingredient of—
Hair dressing.

Rubber
Ingredient (Brit. 307079) of—
Emulsions.

Soap
Ingredient of—
Degreasing agents, soaps containing petroleum distillates.
Starting point (Brit. 299817) in making—
Detergent emulsions with trichloroethylene and other chlorinated hydrocarbons, for laundry and domestic purposes.

Textile
——, *Dyeing*
General assist in dyeing.
Ingredient (Brit. 299817) of preparations containing trichloroethylene and other chlorinated hydrocarbons for—
Dyeing cotton, wool, rayon, and mercerized cotton.
Wetting textiles before dyeing.
Ingredient (Brit. 307079) of—
Dye liquors containing starches, vegetable gums, dextrins, glues, gelatin, casein.
Ingredient of preparations for—
General dyeing purposes, along with other oils.
Facilitating formation of azo dyestuffs on fabrics.
Facilitating the removal of oils from the wool fiber, added during processing, before dyeing.
Improving color of naphthol dyestuffs, improving color of diamine dyestuffs, improving dyeing with vat colors, mordanting with turkey red, mordanting to fix alumina on the fiber in dyeing with alizarin.
——, *Finishing*
Ingredient (Brit. 299817) of emulsified preparations containing trichloroethylene and other chlorinated hydrocarbons for—
Scouring and finishing cotton, wool, rayon, and mercerized cotton.
Ingredient (Brit. 307079) of—
Waterproofing baths containing starches, vegetable gums, dextrins, glues, gelatin, casein.
——, *Manufacturing*
Ingredient (Brit. 299817) of emulsified preparations containing trichloroethylene and other chlorinated hydrocarbons for—
Degumming silk.
Lubricant and cleansing agent for wool.

Waxes and Resins
Ingredient (Brit. 307079) of—
Emulsions.

Turpentine Oil
Synonyms: Spirits of turpentine, Turpentine, Turps.
Latin: Oleum terebenthinae.
French: Essence de térébenthine, Huile volatile de térébenthine.
German: Terpentinoel.
Spanish: Aceite volatil de trementina, Esencia de trementina.
Italian: Essenza di trementina.

Adhesives
Solvent (U. S. 1604307) in—
Casein glue compositions.

Analysis
As a reagent.
As a solvent.
Reagent (in colored form) for—
Wood and cork in biological technic.

Ceramics
Solvent in—
Coating compositions for potteries and porcelains.

Turpentine Oil (Continued)

Chemical
Reagent in preparing—
Eucalyptol.
Solvent (U. S. 1649326) in—
Acid-resisting compositions.
Solvent in making—
Benzyl chloride by catalysis.
Starting point in making—
Camphor (artificial), isoprene, pinene, terpene hydrochloride, terpene hydrate, terpineol, terpinyl acetate.

Explosives
Solvent in—
Fireworks manufacture.

Fats and Oils
As a general solvent.
Reagent in—
Neutral oil preparations.
Solvent (U. S. 1642884) in—
Belting greases, lubricating compositions.

Germicide
As a germicide, alone or in compositions.

Glass
Lubricant in—
Glass grinding.
Solvent in—
Waterproof mastics.

Ink
Ingredient of—
Lithographic inks, printing inks.

Insecticide
Alone as—
Bug exterminator, insecticide, moth-repellent.
Ingredient of—
Insecticides.

Leather
Solvent in—
Finishing and dressing compositions, leather cements, leather polishes, patent leather finishes, shoe polishes, waterproofing compositions and finishes.

Linoleum and Oilcloth
Solvent (Brit. 274300) in—
Linoleum and oilcloth cements.

Metallurgical
Flotation reagent in—
Concentration of ores.
Solvent for—
Waterproof mastics in metal working.

Miscellaneous
Ingredient of—
Compositions for transferring pictures and prints, floor polishes, furniture polishes, glass cements, pigment preparations used as drawing crayons, polishing compositions (U. S. 1758317), stove polishes, waterproofing compositions.
Solvent in—
Compositions for cleansing firearms, ivory, substances attacked by chlorine.

Paint and Varnish
Ingredient (U. S. 1614232) of—
Auto-top dressings.
Reagent for—
Accelerating oxidation of drying oils.
Solvent and thinner in—
Coach finishes, enamels, glazing putty, lacquers, paint driers, paint removers, paints of all kinds, piano-rubbing varnishes, resins, roofing cements stains, varnishes, varnish removers, waxes, wax color binding compositions.

Paper
Cleansing agent for—
Paper machine wires.

Perfume
Ingredient (Brit. 255148) of—
Cosmetics, emollients.

Petroleum
Carrier for—
Oxygen in the oxidation refining process.

Pharmaceutical
In compounding and dispensing practice.
Ingredient of—
Dentifrices, detergents, disinfectants, germicides, internal remedies, liniments, ointments.

Printing
As a general solvent and cleanser.
Solvent in—
Color process printing.

Resins and Waxes
Solvent for—
Resins, waxes.
Solvent in wax compositions for—
Grafting, modelling, sealing, various purposes.

Rubber
Ingredient (U. S. 1875552) of—
Mixture, with cresol, used for vulcanizing molds.
Reagent (French 599869) in—
Rubber reworking.
Solvent in—
General processing, rubber cements.

Soap
Ingredient of—
Detergent compositions, grease removing soaps, household soaps, medicated soaps, washing compounds.

Textile
Reagent for—
Preventing color bleeding in textile printing.
Solubilizing agent (Brit. 276160) for various dyestuffs.
Solvent for—
Removing paint and oils stains from fabrics.

Woodworking
Alone or in combination as—
Impregnating agent, preservative agent, waterproofing agent.
Solvent and thinner in—
Fillers, polishes.

Turtle Oil
French: Huile de torture.
German: Schildkroetenoel.
Italian: Olio di tartaruga.

Fats and Oils
Component of—
Linseed oil mixtures.
Ingredient of—
Special lubricating compositions.
Starting point in making—
Substitute for degras.

Fuel
Raw material in making—
Candles.

Leather
Ingredient of—
Chamoising compositions, currying compositions, dressing compositions, softening compositions, tanning compositions.

Mechanical
As a lubricant.

Perfumery
As a base in making—
Fatty creams.

Pharmaceutical
In compounding and dispensing practice.

Soap
As a soap stock in making—
Special toilet soaps.

Ulmic Acid

Petroleum
Viscosity decreaser (U. S. 1999766) of—
Fluid clay mud encountered in oil well drilling (used in conjunction with a small amount of caustic alkali).

Undecoic Acid

Chemical
Reagent in—
Organic syntheses.

Petroleum
Breaker (U. S. 2020998) of—
Petroleum emulsions.

Undecylenic Acid
French: Acide undécylénique.
German: Undecylensäure.
Spanish: Acido endecilenico.

Chemical
Starting point in making—
Butyl undecylenate, ethyl undecylenate, methyl undecylenate, nonoic acid, sebacic acid, undecalactone, various esters and salts.

Undecylenyl Acetate

Chemical

Starting point in making derivatives.

Perfume

Ingredient of—
Artificial perfume preparations.
Perfume in—
Toilet preparations.

Soap

Perfume in—
Toilet soaps.

Undekanaphthene

German: Undekanaphten.

Chemical

General solvent for various purposes (Brit. 269960).

Miscellaneous

General solvent.

Textile

——, *Dyeing and Printing*

Solvent in making—
Dye liquor and paste.

Uranic Oxide

French: Oxyde d'urane, Oxyde uranique.
German: Uranioxyd, Uranoxyd.
Spanish: Oxido d'uranio.
Italian: Oxido di uranio.

Ceramics

Ingredient of—
Glazes used in the manufacture of porcelains and chinaware.

Chemicals

Catalyst in—
Ammonia synthesis.
Catalyst (Brit. 331000) in making—
Anthraquinone, benzaldehyde from toluene, benzoic acid from toluene.
Parachlorobenzaldehyde and parachlorobenzoic acid from parachlorotoluene.
Paratoluic acid, paratoluic aldehyde from paraxylene, phthalic acid.
Dehydrating catalyst (Brit. 323713) in making—
Allylene from allanol, amylene from amanol, butylene from isobutanol, ethylene from ethanol, heptylene from heptanol, hexylene from hexanol, methylene from methanol, propylene from isopropanol.
Ingredient of catalytic mixtures used in the manufacture of—
Acenaphthylene, acenaphthaquinone, bisacenaphthylidenedione, naphthaldehydic acid, naphthalic anhydride, and hemimellitic acid from acenaphthene (Brit. 295270).
Acetaldehyde from ethyl alcohol (Brit. 281307).
Acetic acid from ethyl alcohol (Brit. 281307).
Alcohols from the corresponding aliphatic hydrocarbons (Brit. 281307).
Aldehydes and acids by the reduction of the corresponding esters (Brit. 306471).
Aldehydes and acids from toluene, orthochlorotoluene, orthonitrotoluene, orthobromotoluene, parachlorotoluene, paranitrotoluene, parabromotoluene, metachlorotoluene, metabromotoluene, metanitrotoluene, dichlorotoluenes, dinitrotoluenes, dibromotoluenes, chlorobromotoluenes, chloronitrotoluenes, bromonitrotoluenes (Brit. 295270).
Aldehydes and acids from xylenes, pseudocumenes, mesitylene, and paracymene (Brit. 281307).
Alphanaphthylamine from naphthalene (Brit. 281307).
Anthraquinone from naphthalene (Brit. 295270).
Benzaldehyde and benzoic acid from toluene (Brit. 281307).
Benzoquinone from phenanthraquinone (Brit. 281307).
Benzyl alcohol from benzaldehyde by reduction (Brit. 306471).
Benzyl alcohol or benzaldehyde or benzyl phthalide by the reduction of phthalic anhydride (Brit. 306471).
Butyl alcohol by the reduction of crotonaldehyde (Brit. 306471).
Chloroacetic acid from ethylenechlorohydrin (Brit. 295270).
Diphenic acid from ethyl alcohol (Brit. 281307).
Ethyl alcohol by the reduction of acetaldehyde (Brit. 306471).
Fluorenone from fluorene (Brit. 295270).
Formaldehyde by the reduction of methane or methanol (Brit. 306471).
Formaldehyde by the reduction of carbon dioxide or carbon monoxide (Brit. 306471).
Hydroxyl compounds by the reduction of anthraquinone, benzoquinone, and similar compounds (Brit. 306471).
Isopropyl alcohol by the reduction of acetone (Brit. 306471).
Maleic acid and fumaric acid by the oxidation of toluene, benzene, phenols, tar phenols, or furfural, or from benzoquinone or phthalic anhydride (Brit. 295270).
Methane by the reduction of carbon dioxide or carbon monoxide (Brit. 306471).
Methanol by the reduction of carbon dioxide or carbon monoxide (Brit. 306471).
Naphthaldehydic acid, acenaphthaquinone, or bisacenaphthylidenedione from acenaphthylene (Brit. 295270).
Phenanthraquinone from phenanthrene or diphenic acid (Brit. 295270).
Phthalic acid and maleic acid from naphthalene (Brit. 295270).
Primary alcohols by the reduction of the corresponding aldehydes (Brit. 306471).
Propionic acid and butyric acid and higher alcohols, ketones, and acids by the reduction of carbon dioxide or carbon monoxide (Brit. 306471).
Reduction products of ketones, aldehydes, acids, esters, alcohols, ethers, and other organic compounds, which contain oxygen (Brit. 306471).
Salicylic acid and salicylic aldehyde from cresol (Brit. 295270).
Secondary butyl alcohol by the reduction of methylethyl ketone (Brit. 306471).
Valeryl alcohol by the reduction of valeraldehyde (Brit. 306471).
Vanillin and vanillic acid by the oxidation of eugenol or isoeugenol (Brit. 295270).
Ingredient (Brit. 306460) of catalytic preparations used in the production of various aromatic and aliphatic compounding, including—
Alphanaphthylamine from alphanitronaphthalene.
Amine from aliphatic nitro compounds, such as allyl nitriles or nitromethane.
Amino compounds from the corresponding nitroanisoles, amylamine from pyridin.
Anilin, azo-oxybenzene, azobenzene, and hydrazobenzene from benzene by reduction.
Aminophenols from nitrophenols, 3-aminopyridin from 3-nitropyridin.
Cyclohexamine, dicyclohexamine, and cyclohexylanilin from nitrobenzene.
Piperidin from pyridin, pyrrolidin from pyrrol, tetrahydroquinolin from quinolin.
Starting point in making—
Salts and other compounds of uranium.

Glass

Ingredient of—
Glass batch (used to produce opalescent green effects).

Paint and Varnish

Pigment in—
Paints and varnishes.

Uranium Acetate

Synonyms: Uranyl acetate.
French: Acétate d'uranium.
German: Essigsaeuresuran, Uraniumacetat.

Analysis

Reagent in various processes.

Paint and Varnish

Starting point in making—
Compound pigment with barium sulphide (U. S. 1615816).

Pharmaceutical

In compounding and dispensing practice.

Uranium Dioxide

French: Dioxyde d'uranium.
German: Urandioxyd.

Ceramics

Ingredient of glazes for—
Chinaware, porcelains, potteries.

Chemical

Catalyst (Brit. 263201) in making—
Benzaldehyde and benzoic acid from toluene.
Maleic acid from benzene.
Phthalic anhydride and naphthoquinone from naphthalene.

Uranium Dioxide (Continued)
Glass
Pigment in making fine glassware.
Metallurgical
Starting point in making—
Ferro-uranium.
Paint and Varnish
Pigment in paints and varnishes.
Photographic
Reagent in the special processes.

Uranium Nitrate
Synonyms: Uranyl nitrate.
French: Nitrate d'urane, Nitrate uranique, Nitrate d'uranium, Nitrate d'uranyle.
German: Salpetersäuresuran, Salpetersäuresuranoxyd, Salpetersäuresuranyl, Urannitrat, Uranylnitrat.
Analysis
Detecting cinnamic acid in benzoic acid.
Determining phosphorus and sulphur.
Indicator in various titrations.
Making volumetric solutions.
Reagent for—
Albumen, alkaloids, apomorphine, cinnamic acid, glucose, hydrocyanic acid, hydrogen peroxide, mercury oxycyanide, morphine, phenols, phosphoric acid.
Separating tungstic acid from tungstates.
Ceramics
Reagent in making—
Yellow and orange glazes.
Chemical
Ingredient of catalytic preparations used in making—
Acenaphthylene, acenaphthaquinone, bisacenaphthylidenedione, naphthaldehydic acid, naphthalic anhydride, and hemimellitic acid from acenaphthene (Brit. 295270).
Acetaldehyde from ethyl alcohol (Brit. 281307).
Acetic acid from ethyl alcohol (Brit. 281307).
Alcohols from aliphatic hydrocarbons (Brit. 281307).
Aldehydes and acids from the reduction of the corresponding esters (Brit. 306471).
Alphacampholide by the reduction of camphoric acid (Brit. 306471).
Aldehydes and acids from toluene, orthochlorotoluene, orthonitrotoluene, orthobromotoluene, parachlorotoluene, parabromotoluene, paranitrotoluene, metachlorotoluene, metabromotoluene, metanitrotoluene, dichlorotoluenes, dibromotoluenes, dinitrotoluenes, chloronitrotoluenes, chlorobromotoluenes, bromonitrotoluenes (Brit. 295270).
Aldehydes and acids from xylenes, pseudocumenes, mesitylenes, and paracymene (Brit. 281307).
Alphanaphthaquinone from naphthalene (Brit. 281307).
Anthraquinone from naphthalene (Brit. 295270).
Benzaldehyde and benzoic acid from toluene (Brit. 281307).
Benzoquinone from phenanthraquinone (Brit. 281307).
Benzyl alcohol by the reduction of benzaldehyde (Brit. 306471).
Benzyl alcohol or benzaldehyde or phthalide by the reduction of phthalic anhydride (Brit. 306471).
Butyl alcohol by the reduction of crotonaldehyde (Brit. 306471).
Chloroacetic acid from ethylenechlorohydrin (Brit. 306471).
Diphenic acid from ethyl alcohol (Brit. 281307).
Ethyl alcohol by the reduction of acetaldehyde (Brit. 281307).
Flourenone from fluorene (Brit. 295270).
Formaldehyde by the reduction of carbon dioxide or carbon monoxide (Brit. 306471).
Formaldehyde by the reduction of methane or methanol (Brit. 306471).
Hydroxyl reduction compounds of anthraquinone, benzoquinone and the like (Brit. 306471).
Isopropyl alcohol by the reduction of acetone (Brit. 306471).
Maleic acid and fumaric acid by the oxidation of toluene, benzene, phenols, tar phenols, or furfural, or from benzoquinone or phthalic anhydride (Brit. 295270).
Methane by the reduction of carbon dioxide or carbon monoxide (Brit. 306471).
Methanol by the reduction of carbon dioxide or carbon monoxide (Brit. 306471).
Naphthaldehydic acid, acenaphthaquinone, or bisacenaphthylidenedione from acenaphthylene (Brit. 306471).
Phenanthraquinone from phenanthrene or diphenic acid (Brit. 295270).
Phthalic acid and maleic acid from naphthalene (Brit. 295270).
Primary alcohols by the reduction of the corresponding aldehydes (Brit. 306471).
Propionic acid and butyric acid and higher alcohols, ketones, and acids by the reduction of carbon monoxide or carbon dioxide (Brit. 306471).
Reduction products of ketones, aldehydes, acids, esters, alcohols, ethers, and other organic compounds containing oxygen (Brit. 306471).
Salicylic acid and salicylic aldehyde from cresol (Brit. 295270).
Secondary butyl alcohol by the reduction of methylethylketone (Brit. 306471).
Valeryl alcohol by the reduction of valeraldehyde (Brit. 306471).
Vanillin and vanillic acid from eugenol or isoeugenol (Brit. 295270).
Ingredient (Brit. 305640) of catalytic preparations used in the production of various aromatic and aliphatic amines, including—
Alphanaphthylamine from alphanitronaphthalene.
Amines from aliphatic nitro compounds, such as alkyl nitriles, or nitromethane.
Amylamine from pyridin.
Anilin, azo-oxybenzene, azobenzene, and hydrazobenzene by the reduction of nitrobenzene.
Aminophenols from nitrophenols, 3-aminopyridin from 3-nitropyridin.
Amino compounds from the corresponding nitroanisoles.
Amines from oximes, Schiff's base, and nitriles.
Cyclohexylamine, dicychlohexamine, and cyclohexylanilin from nitrobenzene.
Piperidin from pyridin, pyrrolidin from pyrrol, tetrahydroquinolin from quinolin.
Starting point in making—
Uranium carbonate, uranium chloride, uranium oxide, uranium sulphate.
Paint and Varnish
Ingredient of—
Phosphorescent paints and varnishes.
Reagent in making—
Luminous paints and varnishes.
Pharmaceutical
In compounding and dispensing practice.
Photographic
Reagent in obtaining—
Brown color effects on prints.
Sensitizing reagent in treating—
Photographic papers.
Textile
Mordant in dyeing and printing.
Reagent in obtaining—
Brown colors on textiles by impregnation and aftertreatment with solutions of sodium ferrocyanide, gallic acid, tannic acid, or pyrogallol.

Urea
Synonyms: Carbamide, Carbonylamid, Diamide of carbonic acid.
French: Urée.
German: Harnstoff.
Adhesives
Anticurling agent in—
Adhesives for paper, fabrics, and the like.
Fluidity promoter in—
Adhesives (permits reduction of water content without impairing fluidity of the solution).
Increaser of—
Flexibility of adhesives on drying, hygroscopicity of adhesives on drying, strength of adhesives on drying.
Liquefying agent in aqueous solutions containing—
Animal glue, casein, gelatin, starch.
Lowerer of—
Jelling (setting) temperature.
Reducer of—
Quantity of oxidizing agent used in starch adhesives.
Resolubilizing agent (U. S. 1895446) for—
Glues made insoluble by the action of formaldehyde.
Retarder of—
Inception of the quick-setting action of glues having reduced water content without interfering with the quick-setting effects produced (desirable for veneer and plywood glues).

Urea (Continued)
Stabilizer in—
Starch adhesives.
Starting point (Brit. 421942) in making—
Adhesives with formaldehyde, phenol, or saccharose (such products are suitable for glueing veneering papers, insulating board, paperboard, asbestos sheets, textile fabrics, metal foil, leather, and other products).

Analysis
Nutrient for—
Bacterial cultures.
Reagent in—
Analytical processes involving control and research.

Brewing
Nutrient for—
Yeast.

Cellulose Products
Antacid and stabilizer in—
Cellulose acetate products, nitrocellulose products.
Plasticizer in—
Cellulose acetate products, nitrocellulose products.

Chemical
Eliminator of—
Nitrous acid from reactions, nitrous acid from products.
Increaser of—
Efficiency of leaching agents, solvent power of water for various solutes, solvent power of various solvents for certain solutes.
Nutrient for—
Yeast in alcohol manufacture.
Separating agent for—
Metacresol from crude cresol mixtures.
Solubilizing agent for various purposes.
Stabilizer for—
Aluminum acetate (Brit. 444254), cellulose acetate, hydrogen peroxide, nitrocellulose.
Starting point in—
Organic syntheses.
Starting point in making—
Acyl derivatives, or ureides, by reacting with acid chlorides or anhydrides.
Starting point in making—
Alkylated ureas, such as methylurea, diethylurea.
Ammonium carbonate, ammonium-potassium cyanate.
Crystalline compounds with salts, such as urea-sodium chloride, urea-silver nitrate.
Cyclic ureides or diureides.
Dispersing, emulsifying, and wetting agents (Brit. 432356).
Intermediates, organic chemicals for pharmaceutical purposes, organic chemicals for technical purposes.
Salts with acids, such as urea nitrate, urea chloride, urea oxalate, urea phosphate, and urea sulphate.
Salts with metallic oxides, such as with mercuric oxide.
Ureido acids.
Varier of crystal form of—
Ammonium sulphate, sodium chloride.

Dental Products
Process material (U. S. 1355834) in making—
Dental fillings.

Disinfectant
Solubilizing agent for—
Various insoluble bactericides.
Stabilizer for—
Hydrogen peroxide solutions.
Starting point (German 544678) in making—
Disinfectant with metacresol, said to be very efficient and to have a talc-like powder appearance.

Distilled Liquors
Nutrient for—
Yeast.

Dye
Controller of—
Diazotization reactions.
Eliminator of—
Nitrous acid from products, nitrous acid from reactions.
Substitute for—
Dextrin or other agents in standardizing the color strength of basic dyes.

Explosives and Matches
Antacid or stabilizer for—
Nitrocellulose (declining use), nitrostarch (declining use).
Antacid or stabilizer in—
Detonating compositions, nitroglycerin explosives (declining use).
Starting point in making—
Detonation retardants.

Fats and Oils
Remover of—
Catalyst taste from hydrogenated fats and oils.

Fertilizer
As a nitrogenous fertilizer which is highly available to crops and is not readily leached from the soil by heavy rains (said to be an exceptionally good source of nitrogen for market garden and other crops, potatoes, tobacco).
Ingredient of—
Mixed fertilizers (claimed that it can be used in large quantities without adversely affecting their physical properties).
Starting point in making—
Calurea.
Urea-ammonium liquor (said to be essentially a solution of crude urea in aqua ammonia, offered for use to fertilizer factories in making many kinds of fertilizer; the liquor is said not to cause the fertilizer to stick to the walls of the mixer, nor does the added water limit the amount of liquor that may be used; other advantages claimed are (1) superior to introducing solid urea and ammonia separately, (2) gives more intimate mixing, (3) prevents segregation, (4) reduces the tendency of the fertilizer to absorb moisture from the air).
Urea nitrate, urea phosphate, urea sulphate.

Firefighting
Ingredient (German 485400) of—
Fire extinguisher in solid form.

Food
Nutrient for—
Yeast.
Starting point in making—
Flour gluten improvers with hydrogen peroxide.
Leavening agent with karaya gum.
Urea phosphate used instead of acids in baking powders.
Varier of crystal form of—
Salt.

Glue and Gelatin
Fluidity promoter in—
Aqueous gelatin solutions, aqueous glue solutions.
Liquefying agent in—
Aqueous glue solutions, aqueous gelatin solutions.
Resolubilizing agent (U. S. 1895446) for—
Glues made insoluble by the action of formaldehyde.

Insecticide and Fungicide
Solubilizing agent for—
Various insoluble fungicides.

Leather
Liming agent (with sodium sulphide) for—
Hides, pelts.
Starting point (Brit. 388475) in making—
Synthetic tanstuffs with aldehyde and phenol and sulphonates.
Tanning agent (used with formaldehyde).
Whitening agent (used with potassium thiocyanate).

Metallurgical
Case-hardening agent (Brit. 311588) for—
Steel and iron.
Increaser of—
Solvent power of electrolytes in electroplating.
Ingredient (U. S. 1976210) of—
Quenching agent for iron and steel.
Promoter (U. S. 1362739) of—
Alloying of aluminum with lead.

Miscellaneous
Increaser of—
Efficiency of leaching agents.
Solvent power of various solvents for certain products.
Solvent power of water for various products.
Ingredient of—
Soldering flux (U. S. 1882734).
Suggested (German 485012) for use as—
Antifreeze which will not clog radiator on boiling to dryness.

Oral Hygiene
Stabilizer for—
Hydrogen peroxide mouthwashes.

Urea (Continued)

Paint and Varnish
Improver of—
Abrasion resistance of shellac films, hardness of shellac films, water resistance of shellac.
Increaser of—
Solvent power of solvents.

Pharmaceutical
In compounding and dispensing practice.
Stabilizer for—
Anesthetics said to be alkamine aminobenzoates (Brit. 447679), hydrogen peroxide solutions.
Starting point in making—
Malonylureas or barbiturates.
Salts, such as urea salicylate, urea hydrobromide, urea oxalate.

Plastics
Antacid and stabilizer for—
Plastics containing or made from cellulose derivatives.
Process material in making—
Amber substitutes, celluloid substitutes.

Refrigeration
Suggested for use as—
Cooling brine (in solution).

Resins
Improver of—
Abrasion resistance of shellac films, hardness of shellac films, water resistance of shellac.
Stabilizing agent for—
Formaldehyde solutions.
Starting point in making—
Resins used in the manufacture of divers articles.
Resins with polybasic acids and polyhydric alcohols (such resins are claimed to have less tendency to polymerize) (Brit. 412172).
Resins with formaldehyde and ammonium thiocyanate (U. S. 2011573).
Resins with rubber, or chlorinated rubber, resins, and aldehyde (German 560260).
Urea-formaldehyde condensation products, urea-furfural condensation products.

Rubber
Accelerator in—
Vulcanizing processes.
Ingredient of mixes in making—
Eraser rubbers, microporous rubbers (German 557043), sponge rubbers.
Stabilizer (German 562755) for—
Latext (used in conjunction with enzymes).

Soap
Ingredient (Brit. 407039) of—
Antiseptic washing and cleansing agents.

Textile
Coagulating agent, color fastness agent, color intensifier, delustring agent, fixing agent.
Ingredient of—
Printing pastes containing vat colors (used to obtain fuller-bodied shades).
Printing pastes containing chrome colors (added (1) to obtain a better fixation of the colors, (2) to reduce steaming time, (3) to improve brightness, (4) to improve fastness to soap, (5) to permit employment of natural gums in printing on silk).
Printing pastes containing acid colors for wool.
Printing pastes containing acid or basic colors for silk.
Mordant aid, opacifier, plasticizer, retting accelerator, selective dyeing agent, softening agent.

Wine
Nutrient for—
Yeasts.

Urea Acetylsalicylate

Pharmaceutical
In compounding and dispensing practice.

Urea-3:3′-dicarboxylic Acid

Chemical
Starting point in making—
Esters, salts, and other derivatives.
Starting point (Brit. 314909) in making pharmaceutical derivatives with the aid of—
Alkoxynaphthalenesulphonic acids.
Alpha-amino-5-naphthol-7-sulphonic acid.
Alphanaphthylamine-4:8-disulphonic acid.
Alphanaphthylamine-4:6:8-trisulphonic acid.
4-Aminoacenaphthene-3:5-disulphonic acid.
4-Aminoacenaphthene-3-sulphonic acid.
4-Aminoacenaphthene-5-sulphonic acid.
4-Aminoacenaphthenetrisulphonic acids.
Aminocarboxylic acids, aminoheterocyclic carboxylic acids, 1:8-aminonaphthol-3:6-disulphonic acid.
Bromonitrobenzoyl chlorides, chloroalphanaphthalenesulphonic acids, chloronitrobenzoyl chlorides, iodonitrobenzoyl chlorides, nitroanisoyl chlorides, nitrobenzene sulphochlorides, nitrobenzoyl chlorides, 2-nitrocinnamyl chloride, 3-nitrocinnamyl chloride, 4-nitrocinnamyl chloride, 1-nitronaphthalene-5-sulphochloride, 2-nitronaphthoyl chloride, 4-nitronaphthoyl chloride, nitrotoluyl chlorides.

Urea Nitrate

French: Azotate d'urée, Nitrate d'urée.
German: Harnstoffnitrat.

Chemical
Starting point in making—
Ethyl carbamate, urethane.

Fertilizer
Plant food, alone or in compositions.

Miscellaneous
Ingredient of—
Compositions used for transferring pictures and prints.

Ursolic Acid

Cellulose Products
Plasticizer for—
Cellulose acetate, cellulose esters or ethers.
For uses, see under general heading: "Plasticizers."

Valerylhydroquinone

Petroleum
Stabilizing agent (Brit. 406195) for—
Cracked gasolines and other motor fuels.

Valeryl Peroxide

French: Peroxyde de valéryle, Peroxyde valérylique.
German: Valerylperoxyd.

Chemical
Starting point in making—
Intermediates, organic chemicals.
Pharmaceuticals, such as bactericidal compounds and internal antiseptics.

Fats and Oils
Bleaching agent (Brit. 328544) used with hydrogen peroxide) in treating—
Animal oils and fats, vegetable oils and fats.

Food
Bleaching agent (Brit. 328544) (used with hydrogen peroxide) in treating—
Egg yolk, flour, meal.

Miscellaneous
Bleaching agent for various purposes.

Perfume
Ingredient of—
Skin-bleaching creams, toothpastes, tooth powders.

Pharmaceutical
In compounding and dispensing practice.

Resins and Waxes
Bleaching agent (Brit. 328544) (used with hydrogen peroxide) for treating—
Waxes.

Soap
As a bleaching agent (Brit. 328544) (used with hydrogen peroxide).

Valerylphloroglucinol

Petroleum
Stabilizing agent (Brit. 406195) for—
Cracked gasolines and other motor fuels.

Valerylpyrocatechol

Petroleum
Stabilizing agent (Brit. 406195) for—
Cracked gasolines and other motor fuels.

Valerylpyrogallol

Petroleum
Stabilizing agent (Brit. 406195) for—
Cracked gasolines and other motor fuels.

Valerylresorcinol

Petroleum
Stabilizing agent (Brit. 406195) for—
Cracked gasolines and other motor fuels.

Vanadium Acid Oxalate
Synonyms: Vanadium binoxalate, Vanadium dioxalate.
French: Bioxalate de vanadium, Bioxalate vanadique.
German: Oxalsäuressäuresvanad, Oxalsäuressäuresvanadin, Oxalsäuressäuresvanadinoxyd, Vanadinsäuresoxalat, Vanadsäuresoxalat.

Gas
Reagent (U. S. 979887) in obtaining—
Greenish shade in gas light.

Photographic
Reagent (U. S. 979887) in treating—
Bromide paper, to give it a greenish tone.

Vanadium Butylphthalate

Miscellaneous
Preventer (U. S. 1965608) of—
Nitrocellulose coatings discoloration by ultraviolet light.

Vanadium Chromate
French: Chromate vanadique, Chromate de vanadium.
German: Chromsaeuresvanad, Chromsaeuresvanadin, Vanadchromat, Vanadinchromat.

Chemical
Reagent for various purposes.
Ingredient of catalytic preparations used in making—
Acenaphtheylene, acenaphthaquinone, bisacenaphthylidenedione, naphthaldehydic acid, naphthalic anhydride, and hemimellitic acid from acenaphthene (Brit. 295270).
Acetaldehyde from ethyl alcohol (Brit. 281307).
Acetic acid from ethyl alcohol (Brit. 281307).
Alcohols from aliphatic hydrocarbons (Brit. 281307).
Aldehydes and alcohols by the reduction of esters (Brit. 306471).
Alphacampholid by the reduction of camphoric acid (Brit. 306471).
Aldehydes and acids from toluene, metachlorotoluene, metanitrotoluene, metabromotoluene, parachlorotoluene, parabromotoluene, paranitrotoluene, orthochlorotoluene, orthonitrotoluene, orthobromotoluene, dichlorotoluenes, dinitrotoluenes, dibromotoluenes, chlorobromotoluenes, chloronitrotoluenes, bromonitrotoluenes (Brit. 285270).
Aldehydes and acids from xylenes, pseudocumene, mesitylene, and paracymene (Brit. 295270).
Alphanaphthaquinone from naphthalene (Brit. 281307).
Anthraquinone from naphthalene (Brit. 295270).
Benzaldehyde and benzoic acid from toluene (Brit. 281307).
Benzoquinone from phenanthraquinone (Brit. 281307).
Benzyl alcohol by the reduction of benzaldehyde (Brit. 306471).
Benzyl alcohol or benzaldehyde or phthalide by the reduction of phthalic anhydride (Brit. 306471).
Butyl alcohol by the reduction of crotonaldehyde (Brit. 306471).
Chloroacetic acid from ethylenechlorohydrin (Brit. 295270).
Diphenic acid from ethyl alcohol (Brit. 281307).
Ethyl alcohol by the reduction of acetaldehyde (Brit. 306471).
Fluorenone from fluorene (Brit. 295270).
Formaldehyde from methane or methanol (Brit. 295270.)
Formaldehyde by the reduction of carbon dioxide or carbon monoxide (Brit. 306471).
Isopropyl alcohol by the reduction of acetone (Brit. 306471).
Maleic acid and fumaric acid by the oxidation of toluene, benzene, phenol, tar phenols, or furfural, or from benzoquinone or phthalic anhydride (Brit. 295270).
Methane by the reduction of carbon dioxide or carbon monoxide (Brit. 306471).
Methanol by the reduction of carbon dioxide or carbon monoxide (Brit. 306471).
Naphthaldehydic acid, acenaphthaquinone, or bisacenaphthylidenedione from acenaphthylene (Brit. 281307).
Phenanthraquinone from phenanthrene or diphenic acid (Brit. 295270).
Phthalic acid and maleic acid from naphthalene (Brit. 295270).
Primary alcohols by the reduction of aldehydes (Brit. 306471).
Propionic acid and butyric acid and higher alcohols, ketones, and acid by the reduction of carbon dioxide and carbon monoxide (Brit. 306471).
Hydroxyl compounds of anthraquinone, benzoquinone, and the like by reduction (Brit. 306471).
Reduction products of ketones, aldehydes, acids, esters, alcohols, ethers, and other organic compounds containing oxygen (Brit. 306471).
Salicylic acid and salicylic aldehyde from cresol (Brit. 295270).
Secondary butyl alcohol by the reduction of methylethylketone (Brit. 306471).
Valeryl alcohol by the reduction of valeraldehyde (Brit. 306471).
Vanillin and vanillic acid from eugenol or isoeugenol (Brit. 295270).

Vanadium Molybdate
French: Molybdate vanadique.
German: Molybdaensäuresvanadin, Molybdaensäuresvanadinoxyd, Molybdaensäuresvanadium, Molybdaensäuresvanadoxyd, Vanadinmolybdat, Vanadmolybdat.

Chemical
Ingredient of catalytic preparations used in making—
Acenaphthylene, acenaphthaquinone, bisacenaphthylidenedione, naphthaldehydic acid, naphthalic anhydride, and hemimellitic acid from acenaphthene (Brit. 295270).
Acetaldehyde from ethyl alcohol (Brit. 281307).
Acetic acid from ethyl alcohol (Brit. 281307).
Alcohols from aliphatic hydrocarbons (Brit. 281307).
Aldehydes or acids by the reduction of the corresponding esters (Brit. 306471).
Alphacampholide by the reduction of camphoric acid (Brit. 306471).
Aldehydes and acids from toluene, orthochlorotoluene, orthonitrotoluene, orthobromotoluene, parachlorotoluene, paranitrotoluene, parabromotoluene, metachlorotoluene, metanitrotoluene, metabromotoluene, dichlorotoluenes, dinitrotoluenes, dibromotoluenes, chlorobromotoluenes, chloronitrotoluenes, bromonitrotoluenes (Brit. 295270).
Aldehydes and acids from xylenes, pseudocumenes, mesitylene, and paracymene (Brit. 281307).
Alphanaphthaquinone from naphthalene (Brit. 281307).
Anthraquinone from naphthalene (Brit. 295270).
Benzaldehyde and benzoic acid from toluene (Brit. 281307).
Benzoquinone from phenanthraquinone (Brit. 281307).
Benzyl alcohol by the reduction of benzaldehyde (Brit. 306471).
Benzyl alcohol or benzaldehyde or phthalide by the reduction of phthalic anhydride (Brit. 306471).
Butyl alcohol by the reduction of crotonaldehyde (Brit. 306471).
Chloroacetic acid from ethylenechlorohydrin (Brit. 295270).
Diphenic acid from ethyl alcohol (Brit. 281307).
Ethyl alcohol by the reduction of acetaldehyde (Brit. 306471).
Fluorenone from fluorene (Brit. 295270).
Formaldehyde by the reduction of methane or methanol (Brit. 306471).
Formaldehyde by the reduction of carbon dioxide or carbon monoxide (Brit. 306471).
Hydroxyl compounds by the reduction of anthraquinone, benzoquinone, and the like (Brit. 306471).
Isopropyl alcohol by the reduction of acetone (Brit. 306471).
Maleic acid and fumaric acid by the oxidation of toluene, benzene, phenols, tar phenols, or furfural, or from benzoquinone or phthalic anhydride (Brit. 295270).
Methane by the reduction of carbon dioxide or carbon monoxide (Brit. 306471).
Methanol by the reduction of carbon dioxide or carbon monoxide (Brit. 306471).
Naphthaldehydic acid, acenaphthaquinone, or bisacenaphthylidenedione from acenaphthylene (Brit. 281307).
Phenanthraquinone from phenanthrene or diphenic acid (Brit. 295270).
Primary alcohols by the reduction of the corresponding aldehydes (Brit. 306471).
Propionic acid and butyric acid and higher alcohols, ketones, and acids by the reduction of carbon dioxide or carbon monoxide (Brit. 306471).

Vanadium Molybdate (Continued)
Reduction products of ketones, aldehydes, acids, esters, alcohols, ethers, and other organic compounds containing oxygen (Brit. 306471).
Salicylic acid and salicylic aldehyde from cresol (Brit. 295270).
Secondary butyl alcohol by the reduction of methylethyl ketone (Brit. 306471).
Valeryl alcohol by the reduction of valeraldehyde (Brit. 306471).
Vanillin and vanillic acid from eugenol and isoeugenol (Brit. 295270).
Ingredient (Brit. 304640) of catalytic preparations used in the production of various aromatic and aliphatic amines, including—
Alphanaphthylamine from alphanitronaphthalene.
Amines from aliphatic nitro compounds, such as alkyl nitriles or nitromethane.
Amylamine from pyridin.
Anilin, azo-oxybenzene, azobenzene, and hydroazobenzene from nitrobenzene by reduction.
Aminophenols from nitrophenols.
3-Aminopyridin from 3-nitropyridin.
Amino compounds from the corresponding nitroanisoles.
Amines from oximes, Schiff's base, and nitriles.
Cyclohexamine, dicyclohexamine, and cyclohexylanilin from nitrobenzene.
Piperidin from pyridin.
Pyrrolidin from pyrrol.
Tetrahydroquinolin from quinolin.
Reagent for various purposes.

Vanadium Pentoxide
Synonyms: Vanadic acid, Vanadic anhydride.
French: Anhydride vanadique, Pentoxyde de vanadium.
German: Vanadinsaeure, Vanadinsaeureanhydrid.

Ceramics
Reagent in making—
Chinaware, porcelains, potteries.

Chemical
Catalyst in making—
Anthraquinone from anthracene by oxidation.
Formaldehyde from methane by oxidation.
Oxalic acid from sugar by oxidation.
Phthalic acid from naphthalene by oxidation.
Sulphuric acid by the contact process.
Reagent in making—
Organic compounds in acid solutions by electrolytic oxidation.
Starting point in making—
Ammonium vanadate, barium vanadate, cadmium vanadate, calcium vanadate, magnesium vanadate, potassium vanadate, sodium vanadate, strontium vanadate, vanadium chloride, vanadium bromide, vanadium sulphate, vanadium sulphite.

Dye
Reagent in making—
Anilin black.

Glass
Pigment in producing—
Red colorations in glassware.

Ink
Reagent in making—
Black ink.

Metallurgical
Raw material in making metallic vanadium.

Miscellaneous
Ingredient of—
Compositions used as substitutes for gold bronze.

Pharmaceutical
In compounding and dispensing practice.

Photographic
Developer for—
Films and plates.
Toner for—
Films, plates and prints.

Textile
——, *Dyeing*
Reagent in dyeing—
Fabrics and yarns with anilin black.

Vanillal Acetone

Perfumery
Ingredient in making—
Hair restorers, pomades.

Vegetable Oil Fatty Acids
French: Acides grasses d'huiles végétal.
German: Vegetabilischesoelfettsaeure.

Chemical
Reagent (Brit. 398064) in making—
Triacidyl borates.
Starting point in making—
Esters and salts of the acids.
Solvents or solubilizing agents in water for paraffin, phenols, higher alcohols, cyclohexanol, essential oils, formic or acetic acids (Brit. 390148).

Dye
Emulsifying agent in making—
Color lakes and oil colors.

Fats and Oils
Ingredient (Brit. 313453) of—
Fat and oil splitting compositions.
Lubricating and greasing compositions.
Starting point (Brit. 390148) in making—
Solvents or solubilizing agents in water for essential oils.

Food
Ingredient of—
Hydrogenated oil products such as lard substitutes, butter substitutes, etc.
Prepared foods.

Fuel
Component of—
Candles.

Ink
Ingredient of—
Inks, stencil sheet coatings.

Insecticide
Ingredient of—
Insecticidal and germicidal compositions.

Leather
Ingredient (Brit. 313453) of—
Treating and finishing compositions.

Miscellaneous
Ingredient of—
Cleansing compositions (Brit. 313453).
Cleansing compositions with alkaline hypochlorites (Brit. 280193).
Emulsifying compositions (Brit. 313453).
Polishing compositions.
Purifying compositions (Brit. 313453).
Washing compositions (Brit. 313453).
Wetting compositions (Brit. 313453).

Paint and Varnish
Starting point in making—
Driers.

Paper
Ingredient (Brit. 313453) of—
Compositions used in the treatment and coating of paper.

Perfume
Ingredient of—
Cosmetics, creams, lotions, shampoos.

Pharmaceutical
As a coating for pills.
In compounding and dispensing practice.

Plastics
Ingredient of various compositions.

Resins and Waxes
Ingredient of—
Resin and wax compositions.
Wax-splitting compositions (Brit. 313453).
Starting point in making—
Solvents or solubilizing agents in water for paraffin (Brit. 390148).
Synthetic resins by reaction with phthalic anhydride and glycerol, such resins being used in making "wrinkle finishes" (U. S. 1893611).
Synthetic resins by reaction with phthalic anhydride and glycerol, such resins being used in making linoleum cements (French 752565).
Synthetic resins by reaction with phthalic anhydride and glycerol, such resins being used in making cements for layers in non-scatterable glass (U. S. 1920619).

Vegetable Oil Fatty Acids (Continued)

Soap
Raw material in soapmaking.
Starting point in making—
Soaps used as stabilizing and dispersing agents for bituminous emulsions.
Special soaps.

Textile
—, *Dyeing and Printing*
Fixing agent (Brit. 313453) in—
Dyeing with basic dyestuffs.
Ingredient of—
Dye baths and printing pastes.
Stabilizing agent (Brit. 313453) in—
Dyeing with vat dyestuffs.

—, *Finishing*
Ingredient of—
Bleaching compositions containing alkaline hypochlorites (Brit. 280193).
Finishing compositions.
Washing compositions containing alkaline hypochlorites (Brit. 280179).
Waterproofing compositions.
Wetting compositions (Brit. 313453).

—, *Manufacturing*
Ingredient of—
Oil compositions.
Starting point (Brit. 390148) in making—
Degreasing agents for fabrics.

Vegetable Tallow
Synonyms: Chinese vegetable tallow, Tankawang fat.
Latin: Sebura stillingiae.
French: Suif végétal de Chine.
German: Chinesischertalg, Pflanzentalg, Stillingiatalg, Vegetablischertalg.

Fuel
Component of—
Candles.

Leather
Ingredient of—
Dressings and finishes.

Mechanical
As a lubricant, alone or in compositions.

Paper
Ingredient of—
Pulp and water mixture in beaters, added for the purpose of preventing foaming.

Printing
Reagent in—
Process engraving and litho operations.

Soap
As a raw material.

Venetian Red
Synonyms: English red, Indian red.
French: Rouge de Vénise.
German: Venetianerrot, Venetianischesrot.

Lapidary
Abrasive in polishing precious stones.

Metallurgical
Polishing agent for fine metals.

Miscellaneous
Ingredient of—
Polishing preparations, razor-strop pastes.
Pigment in making—
Linoleum, oil cloth.

Paint and Varnish
Pigment in—
Enamels, lacquers, paints, varnishes.

Rubber
Pigment for rubber products.

Textile
—, *Finishing*
Ingredient of—
Products used in making waxed cloths.

Venice Turpentine
Synonyms: Larch turpentine.
French: Térébenthine de Vénise.
German: Venezianerterpentin.

Miscellaneous
Ingredient (Brit. 252186) of—
Sealing wax compositions.

Linoleum and Oilcloth
Ingredient of—
Linoleum cements.

Paint and Varnish
Ingredient of—
Varnishes.

Pharmaceutical
In compounding and dispensing practice.

Vetiver Acetate

Perfume
Ingredient of—
Perfume compositions.
Perfume in
Cosmetics.

Soap
Perfume in—
Toilet soaps.

Vine Black
French: Noir de vigne.
German: Rebeschwarz.

Paint and Varnish
Pigment (Brit. 275234) in making—
Rustproof paints, varnishes.

Vinyl Acetate
French: Acétate de vinyle, Acétate vinylique.
German: Essigsäuresvinyl, Essigsäurevinylester, Vinylacetat, Vinylazetat.
Spanish: Acetato de vinil.
Italian: Acetato di vinile.

Ceramics
Adhesive (Brit. 309659) for—
Putting together articles of porcelain, pottery, and similar wares (used in polymerized form).
Plasticizer (Brit. 308657 and 308658) in—
Compositions, containing cellulose acetate or other esters or ethers of cellulose, used for the decoration and protection of ceramic products.

Chemical
Starting point in making—
Acetaldehyde by reaction with acetic acid and water in the presence of phosphoric acid (Brit. 288213).
Acetic anhydride, acetaldehyde, and ethylidene diacetate by heating with acetic acid in the presence of sulphuric acid, phosphoric acid (Brit. 288549).

Electrical
Plasticizer (Brit. 308657 and 308658) in—
Insulating compositions, containing cellulose acetate or other esters or ethers of cellulose, used in the manufacture of electrical machinery and equipment.

Fats and Oils
Starting point (Brit. 280246) in making—
Oily products by condensation with acetaldehyde or formal.

Glass
Adhesive (Brit. 308659) for—
Putting together glass articles (used in polymerized form).
Plasticizer (Brit. 308657 and 308658) in—
Compositions, containing cellulose acetate or other esters or ethers of cellulose, used in the manufacture of nonscatterable glass and for the decoration and protection of glassware.

Glues and Adhesives
Starting point in making—
Polymerized products used as adhesives (French 624493).
Polymerized products, soluble in alcohol, used as adhesives (U. S. 1784008).

Leather
Plasticizer (Brit. 308657 and 308658) in—
Compositions containing cellulose acetate or other esters or ethers of cellulose, used in the manufacture of artificial leather and for the decoration and protection of leather goods.

Metallurgical
Plasticizer (Brit. 308657 and 308658) in—
Compositions containing cellulose acetate or other esters or ethers of cellulose, used for the decoration and protection of metallic articles.

Miscellaneous
Ingredient (in polymerized form) of—
Impregnating compositions (French 624493).
Waterproofing compositions (Brit. 315228).

Vinyl Acetate (Continued)
Plasticizer (Brit. 308657 and 308658) in—
Compositions containing cellulose acetate or other esters or ethers of cellulose, used for the protection and decoration of various compositions of matter.
Paint and Varnish
Plasticizer (Brit. 308657 and 308658) in making—
Paints, varnishes, lacquers, dopes, and enamels containing cellulose acetate or other esters or ethers of cellulose.
Starting point (Brit. 312344) in making—
Polymerized products used in making lacquers with the addition of drying oils.
Polymerized products for making lacquers (French 624493).
Varnishes from polymerized products dissolved in a mixture of tetrahydrofurfuryl alcohol, ethyl alcohol, ethyl acetate, butyl acetate, toluene, and chlorobenzene (Brit. 312049).
Paper
Plasticizer (Brit. 308657 and 308658) in—
Compositions containing cellulose acetate or other esters or ethers of cellulose, used in the manufacture of coated paper and for the decoration and protection of various compositions of matter containing paper or pulp.
Plastics
Plasticizer (Brit. 308657 and 308658) in making—
Plastic compositions containing cellulose acetate or other esters or ethers of cellulose.
Starting point in making—
Polymerized products used as mastics (French 624493).
Polymerized products, mixed with casein solution or albumen, prepared under the action of ultraviolet light in the presence or absence of catalysts (Brit. 294474).
Polymerized products, used as artificial glass, by admixture with cellulose acetate (Brit. 308587).
Used as adhesive for glueing together cellulose ester plastics (Brit. 308659).
Resins and Waxes
Starting point in making—
Gummy or resinous products by condensation with aldehydes, for example, acetaldehyde, paraldehyde, or formaldehyde (Brit. 295322).
Resins for varnishes, films, molding, by polymerizing with admixture of formocarbamic resins (Brit. 309487).
Resinous products by condensation with acetaldehyde or formaldehyde (Brit. 280246).
Synthetic resins by polymerization in the presence of acetaldehyde, paraldehyde, and sodium acetate (Brit. 261406).
Rubber
Plasticizer (Brit. 308657 and 308658) in—
Compositions containing cellulose acetate or other esters or ethers of cellulose, used for the decoration and protection of rubber goods.
Stone
Plasticizer (Brit. 308657 and 308658) in—
Compositions containing cellulose acetate or ethers or esters of cellulose, used for the protection and decoration of natural and artificial stone.
Textile
Plasticizer (Brit. 308657 and 308658) of—
Compositions containing cellulose acetate or ethers or esters of cellulose, used in the manufacture of coated textile fabrics.
Woodworking
Plasticizer (Brit. 308657 and 308658) in—
Compositions containing cellulose acetate or other esters or ethers of cellulose, used for the protection and decoration of woodwork.
Used as an adhesive (in polymerized form) for glueing together wood.

Vinyl Alcohol
French: Alcool de vinyle, Alcool vinylique.
German: Vinylalkohol.
Chemical
Starting point in making—
Intermediates and other derivatives.
Electrical
Starting point (Brit. 322517) in making—
Compositions used for making telephone receivers and other electrical apparatus, parts of motors, and so on.
Mechanical
Starting point (Brit. 322517) in making—
Polymerized compositions used in making brake bands, cog wheels, and other mechanical equipment.
Miscellaneous
Starting point (Brit. 322517) in making—
Polymerized compositions used in making buttons, umbrella handles, and other devices and equipment.
Paint and Varnish
Starting point (Brit. 322517) in making—
Polymerized compositions used as bases in the manufacture of paints, varnishes, lacquers, dopes, and the like.
Paper
Starting point (Brit. 322517) in making—
Compositions used in the impregnation of paper and pulp.
Photographic
Starting point (Brit. 322517) in making—
Compositions used in making films and plates.
Plastics
Starting point (Brit. 322517) in making—
Various compositions.
Woodworking
Starting point (Brit. 322517) in making—
Compositions for impregnating wood and wood products.

Vinyl Allylate
French: Allylate de vinyle, Allylate vinylique.
German: Allylsaeuresvinyl, Allylsaeurevinylester, Vinylallylat.
Chemical
Starting point in making—
Aromatics, intermediates, pharmaceuticals.
Starting point (Brit. 288549) in making the following anhydrides—
Acetic, anthranilic, benzoic, butyric, camphoric, caproic, caprylic, chloroacetic (mono, di, and tri), cinnamic, citric, cresylic, gallic, lactic, maleic, malic, malonic, metanilic, mucic, naphthionic, oxalic, palmitic, phenylacetic, phthalic, picramic, picric, propionic, pyrogallic, salicylic, succinic, sulphanilic, tartaric, valeric.

1-Vinylnaphthalene
Resins
Starting point (U. S. 1982676) in making—
Synthetic resins (by polymerization) which are suitable for lacquers.

2-Vinylnaphthalene
Resins
Starting point (U. S. 1982676) in making—
Synthetic resins (by polymerization) which are suitable for lacquers.

Vinyl Pentylate
French: Pentylate de vinyle, Pentylate vinylique.
German: Pentylsaeuresvinyl, Pentylsaeurevinylester, Vinylpentylat.
Chemical
Starting point in making—
Aromatics, intermediates, pharmaceuticals.
Starting point (Brit. 288549) in making the following anhydrides—
Acetic, anthranilic, benzoic, butyric, camphoric, caproic, caprylic, chloroacetic (mono, di, and tri), cinnamic, citric, cresylic, gallic, lactic, maleic, malic, malonic, metanilic, mucic, naphthionic, oxalic, palmitic, phenylacetic, phthalic, picramic, picric, propionic, pyrogallic, salicylic, succinic, sulphanilic, tartaric, valeric.

Vinyl Propionate
French: Propionate de vinyle, Propionate vinylique.
German: Propionsäuresvinyl, Propionsäurevinylester, Vinylpropionat.
Chemical
Starting point (Brit. 319587) in making—
Polymerized vinyl compounds by the action of ozone.
Plastics
Reagent (Brit. 308657) in making—
Compositions containing cellulose esters and ethers as a base and various alcohols (methanol, ethyl alcohol, butyl alcohol, benzyl alcohol, diacetin alcohol), esters (ethyl acetate, butyl acetate, amyl acetate), hydrocarbons (benzene, toluene, tetrachloroethane), and plasticizers (triphenyl phosphate and tricresyl phosphate).

Vucinotoxin

Chemical
Disinfectant and preservative (Brit. 339602) in treating—
Adrenalin, digestive ferments, injection solutions, local anesthetics, morphine, novocaine, pancreatin, pepsin, vegetable extracts and residues.

Food
As a preservative (Brit. 339602).

Glues and Adhesives
Preservative (Brit. 339602) in treating—
Adhesive preparations, glues.

Perfume
Preservative and disinfectant (Brit. 339602) in making—
Ointments, pomades.

Pharmaceutical
In compounding and dispensing practice.

Sanitation
Preservative, sterilizing agent, and disinfectant (Brit. 339602) in treating—
Rinsing liquids, surgical gut, surgical dressings and bandages, washing liquids.
General disinfectant.

Starch
Preservative (Brit. 339602) in treating—
Dextrin solutions, starch solutions.

Textile
Preservative (Brit. 339602) in treating—
Sewing silk, yarn-sizing preparations.

Walnut Oil
French: Huile de noisette de noyer, Huile de noix de noyer.
German: Walnussoel.

Paint and Varnish
Vehicle in—
Artists colors, paints, varnishes.

Perfume
Ingredient of—
Skin creams.
Starting point in making—
Shaving creams.

Pharmaceutical
In compounding and dispensing practice.
Suggested for use as—
Laxative, vermifuge.

Soap
Starting point in making—
Fine soaps.

Water Gas Tar
French: Goudron de gaz à l'eau.
German: Wassergasteer.

Agriculture
Weed killer, disinfectant, and germicide.

Chemical
Starting point in extracting—
Anthracene, benzene, phenol, naphthalene, toluene, xylene.

Construction
Raw material in making—
Paving compositions, roads.

Fats and Oils
Ingredient of—
Brake dressings (U. S. 1745682), lubricating greases, lubricating oils.

Woodworking
Preservative for—
Railroad ties, timbers, poles, and lumber.

Wetting Agents
Includes also applications for "Penetrating agents."

Chemical
Wetting agent in—
Emulsions of various chemicals, textile lubricants.

Cosmetics
Wetting agent in various cosmetic preparations.

Disinfectant
Wetting agent in—
Disinfecting compositions, germicidal compositions.

Dry Cleaning
Wetting agent in—
Dry cleaning solutions.

Dye
Wetting agent.

Fats, Oils, and Waxes
Wetting agent in—
Boring oils, drilling oils, greasing compositions, lubricating compositions, sulphonated oils, wire-drawing oils, wax preparations.

Glue and Adhesives
Wetting agent in—
Adhesive preparations.

Ink
Wetting agent in—
Printing inks, writing inks.

Insecticide
Wetting agent in—
Fungicidal compositions, horticultural sprays, insecticidal compositions.

Leather
Wetting agent in—
Dressing compositions, fat-liquoring compositions, finishing compositions, soaking compositions, softening compositions, tanning compositions, waterproofing compositions.

Metallurgical
Wetting agent in—
Metal plating processes.

Miscellaneous
Modifier of—
Surface tension of liquids in various products and processes.
Wetting agent in—
Polishing compositions for furniture, automobiles, metals, wood.
Various processes involving aqueous solutions, such as soaking, washing, impregnating, penetrating, wetting.
Waterproofing compositions.

Paint and Varnish
Wetting agent in—
Paints, varnishes, and lacquers.

Paper
Wetting agent in—
Sizing compositions, waterproofing compositions, waxing compositions.

Petroleum
Wetting agent in—
Cutting oils, mineral oil compositions, pitch compositions, tar compositions.

Photographic
Wetting agent in—
Blueprint processes, photographic processes.

Rubber
Wetting agent in—
Cements.

Soap
Wetting agent in—
Detergent compositions, textile soaps.

Textile
—, *Bleaching*
Wetting agent in—
Bleaching baths.

—, *Dyeing*
Wetting agent in—
Dye baths.

—, *Finishing*
Wetting agent in—
Coating compositions, scouring compositions, sizing compositions, washing compositions, waterproofing compositions, waxing compositions.

—, *Manufacturing*
Wetting agent in—
Bowking baths, carbonizing baths, degreasing compositions, degumming and boiling-off baths, fulling baths, kier-boiling baths, mercerizing baths, oiling compositions, soaking baths, spinning compositions.

—, *Printing*
Wetting agent in—
Printing pastes.

Whale Oil
Synonyms: Blubber oil, Body oil, Train oil.
Latin: Oleum balaena.
French: Huile de baléine, Huile de cétaces.
German: Walfischtran.
Spanish: Aceite de balena.
Italian: Olio di balena.

Whale Oil (Continued)
Agriculture
Ingredient of—
Dips for sheep, cattle, and other domestic animals.
Fats and Oils
Starting point in making—
Degras, hardened oil, stearin, tallow mixtures.
Food
Ingredient of—
Lard substitutes, oleomargarins.
Fuel
As a fuel oil and illuminant.
Ingredient of—
Compositions used in making candles.
Insecticide
As a plant insecticide.
Ingredient of—
Insecticidal preparations, used in soap form.
Leather
Ingredient of—
Currying compositions.
Mechanical
As a heavy duty lubricant.
Ingredient of—
Special lubricating compositions.
Lubricant for screw-cutting machines.
Metallurgical
For quenching steel in tempering operations.
Paint and Varnish
Ingredient (Danish 8420-1905) of—
Emulsions with tar and calcium saccharate, used as protective coatings for roofs, walls, and similar surfaces.
Soap
Soapstock in making—
Hard and soft soaps, textile soaps.
Textile
Batching agent in spinning and twisting—
Hemp, jute, and like textile materials.

Whale Oil Fatty Acids
Chemical
Starting point in making various salts.
Fuel
Component of—
Candles.
Miscellaneous
Ingredient of—
Polishing compositions.
Cleansing compositions mixed with alkali hypochlorites (Brit. 280193).
Soap
Raw material in soapmaking.
Textile
——, *Bleaching*
Ingredient of—
Textile bleaching compositions (Brit. 280193).
——, *Finishing*
Ingredient of—
Finishing compositions, washing compositions (Brit. 280193), waterproofing compositions.

Wheat-Germ Extract (Oil-Free)
Agriculture
Nutrient in—
Cattle feeds.
Food
Suggested as valuable concentrated nutrient in—
Infant's modified milk foods, invalid's modified foods, other food products.

White Cake
Synonyms: High grade salt cake.
French: Gateaux blanc.
German: Weisskuchen.
Ceramics
Ingredient of—
Glazes.
Chemical
Reagent in making various sulphates and other salts.
Starting point in making—
Glauber's salt, or pure sodium sulphate, anhydrous and hydrous.
Sodium carbonate, sodium hypochlorite, sodium silicate, or waterglass, sodium thiosulphate, washing sodas.
Dye
Diluting agent in making—
Commercial dyestuff preparations.
Fats and Oils
Reagent in making—
Turkey red oil.
Fuel
Ingredient (U. S. 1618465) of—
Fuel preparations (acting as a fuel economizer).
Glass
Ingredient of—
Batches in making certain kinds of glass (low grade).
Glue
Reagent in making—
Glues.
Insecticides
Ingredient of—
Germicidal preparations, insecticidal preparations.
Leather
Reagent in—
Tanning.
Paint and Varnish
Ingredient of—
Paint and varnish removers.
Reagent in making—
Dry colors, lake pigments, mineral pigments.
Paper
Reagent in making—
Pulp.
Refrigeration
Ingredient of—
Freezing mixtures.
Soap
Ingredient of—
Detergent compositions.

White Grease
Lubricant
Raw material in making—
Cup and other greases.

Wild Marjoram
Latin: Origanum vulgare.
French: Origan marjolaine, Origan marjolaine batarde, Marjolaine sauvage.
German: Wilder majoran, Wilder meiran, Wohlgemut.
Fats and Oils
Raw material for an essential oil.
Perfume
Ingredient of—
Cosmetics, sachet perfumes.
Pharmaceutical
In compounding and dispensing practice.

Witch Hazel
Synonyms: Snapping hazel, Spotted hazel, Striped alder, Tobacco wood, Winter bloom, Wych hazel.
Latin: Hamamelis cortex.
French: Écorce de hamamelis.
German: Hamamelis, Zauberhasel.
Spanish: Hamamelis.
Miscellaneous
Starting point in making—
Witch hazel extract.
Pharmaceutical
In compounding and dispensing practice.

Witch Hazel Extract
Synonyms: Solution of hamamelis.
Latin: Aqua hamamelis.
French: Eau distillée de hamamelis.
German: Hamameliswasser.
Miscellaneous
Ingredient of—
Brush cleansing preparations.
Perfume
Ingredient of—
Toilet preparations.
Pharmaceutical
In compounding and dispensing practice.

Wood Charcoal
Synonyms: Charcoal, Vegetable carbon, Vegetable charcoal.
Latin: Carbo, Carbo ligni, Carbo e ligno, Carbo vegetabilis.
French: Charbon de bois, Charbon végétal.
German: Holzkohle, Präparirte kohle.
Spanish: Carbon de lena, Carbon vegetal.
Italian: Carbone di legno, Carbone vegetale.
Note: Covers all uses for wood charcoal in all forms, activated or otherwise. See also: Charcoal, Activated.

Agriculture
For disinfection of the soil.
For horticultural purposes.
Ingredient of—
Poultry foods, stock foods.
Soil conditioner.

Analysis
In blowpipe work.
Reagent in making routine tests in the laboratories of breweries.

Automotive
Fuel for the internal combustion engines used in automobiles, the charcoal being used directly in the motor car.

Brewing
As a deodorant for vats and other equipment.
As a water purifier.

Chemical
As a chemical intermediate.
Carrier for—
Catalysts, such as platinum in the production of sulphuric anhydride from sulphur dioxide; also for other catalytic metals and oxides and metallic compounds, such as palladium, nickel, oxides of iron, nickel, vanadium, used for the production of ammonia from the air, nitric acid from ammonia, also in organic catalyses, such as hydrogenation of oils.
Catalyst in—
Making nitric oxides and nitric acid by the oxidation of nitrous oxide, and in other catalytic purposes; for example, synthesis of methyl alcohol.
Decolorizing agent in treating—
Various chemicals and chemical products, vegetable principles.
Deodorizing agent in treating—
Various chemicals and chemical products.
Filtering medium for—
Chemical liquors of various sorts.
General absorbent in—
Recovering volatile solvents.
General gas absorbent.
Precipitating reagent for—
Iodine salts, lead salts.
Reagent in refining—
Alcohol.
Reagent in removing—
Alkaloids for infusions.
Starting point in making—
Activated carbon.
Calcium carbide.

Electrical
Ingredient of—
Electrolytic cells, insulating compositions.
Starting point in making—
Arc light electrodes.

Explosives
Ingredient of—
Black powder, chlorate explosives, dinitrophenol explosives, nitrate explosives, nitrocellulose explosives, nitroglycerin explosives, picric acid explosives, trinitrotoluene explosives, tetranitroanilin explosives.

Fats and Oils
Decolorizing and deodorizing agent in treating—
Animal oils and fats, vegetable oils and fats.

Food
Decolorizing agent for—
Food products.

Fertilizer
Ingredient of—
Artificial fertilizers.

Fuel
As a fuel.
Ingredient of—
Briquetted fuels.
Reagent in making—
Candle wicks.

Gas
Catalyst in the purification of—
Coal gas, water gas.
Starting point in making—
Fuel gas.

Glass
Reagent in the manufacture.

Glue and Gelatin
Decolorizing agent for—
Gelatin.

Leather
Absorbent in—
Recovering volatile solvents used in the manufacture of artificial leather.

Metallurgical
Combustible in making—
Charcoal iron, pig iron.
Fuel for heating—
Forges, ladles.
Flux for removing—
Arsenic and antimony from copper.
Flux in smelting—
Oxide and other ores to produce metals.
Ingredient of—
Casehardening preparations.
Precipitating reagent in—
Cyanide processes.
Protecting agent in covering—
Molten metals, to prevent their oxidation by the air.
Reagent in—
Assaying ores and minerals, carbonizing steel, making fine metal castings.
Reduction of metallic oxides, sulphates, and sulphides.
Smelting lead and silver ores.

Military
For filling gas masks, for purifying water.

Miscellaneous
Antidote for poisoning with metallic salts, phosphorus, and certain alkaloids.
Fuel for domestic and industrial uses.
General decolorizing agent, disinfectant, filtering medium, and deodorant.
Ingredient of—
Heat-insulating compositions.
Mixtures used for the manufacture of crayons.
Various special insulating compositions, such as those used for encasing small furnaces and steam pipes.
Reagent in removing—
Odors from bad smelling liquids, odors from clothing, odors from refrigerators and ice boxes, unpleasant odors from decomposing matter.
Recovering agent for—
Solvents.

Oral Hygiene
Ingredient of—
Dentifrices.

Paint and Varnish
As a pigment.
Starting point in making—
Colors.

Petroleum
Absorbent in recovering—
Gasoline from casinghead gas.

Pharmaceutical
In compounding and dispensing practice.

Plastics
Absorbent in recovering—
Volatile solvents used in the manufacture of celluloid, cellulose acetate compositions, and other products.

Refrigeration
Gas absorbent in cold storage work.

Rubber
Compounding agent in making—
Hard rubber products.

Sugar
Decolorizing agent in refining—
Sugar and molasses.

Water and Sanitation
Deodorant for—
Cesspools.
Reagent for—
Purifying highly calcareous waters.
Sweetening cisterns and other storage containers.

Wood Flour
French: Farine de bois.
German: Holzmehl.

Ceramics
Suggested filler for—
Porcelains and potteries to give them body and density.

Chemical
Process material (U. S. 1902986) in making—
Activated carbon from charcoal fines.
Starting material in making—
Oxalic acid.

Construction
Filler in—
Composition floorings.

Explosives and Matches
Absorbent in—
Dynamites, gelatin explosives.
Paper-like fuse composition made from potassium nitrate and nitrocellulose solution (U. S. 1875932).
Permissible explosives, pyrotechnics, safety explosives.

Food
Packing for—
Eggs, fruits, vegetables.

Leather
Absorbent in—
Synthetic tannery bates.
Filler in—
Artificial leathers.
Suggested absorbent for—
Drying oily leathers.

Linoleum and Oilcloth
Filler in—
Linoleum, oilcloth.

Miscellaneous
As an inert absorbent in many products.
As an inert filler in many products.
Filler in—
Duplicating stencil compositions containing a protein, the latter being used to improve distribution of the softeners (U. S. 1902914).
Duplicating stencil coating compositions containing sulphonated oil, oleyl alcohol, myricyl alcohol, gelatin, glycerin, ultramarine, an organic nitrate, and an organic phosphate (U. S. 1915904).
Upholstery in cheap furniture.
Substitute for—
Ground cork (in combination with talc and paperstuff).

Paint and Varnish
Filler in—
Wood fillers of the plastic wood type.
Wood filler comprising a rezyl (polyhydric alcohol-polybasic acid-fatty acid) resin, and pigment (U. S. 1903768).

Paper
Filler in—
Blotting paper, oatmeal paper, paper, paperboard.

Plastics
Absorbent and filler in plastics of various kinds.

Resins and Waxes
Absorbent (U. S. 1905999) in making—
Catalyzed urea resin.
Absorbent and starting material (Brit. 397690) in making—
Synthetic moldable resin.
Filler (Brit. 391364) in making—
Synthetic resins from phenolic substances, vegetable oils, fats, or fatty acids, bituminous substances, and sulphuric acid.

Woodworking
Basic material in making—
Molded imitations relief carvings.

Wood Tar
Synonyms: Hardwood tar.
Latin: Pix ligni, Pyroleum ligni.
French: Breu cru, Goudron de bois, Goudron végétal.
German: Holzteer, Teer.
Italian: Catramo di legno.

Chemical
Starting point in making—
Empyreumatic oils, medicinal creosote, pitch, pyroligneous acid.

Fats and Oils
Starting point in making—
Flotation oils, solvent oils, tar oils.

Fuel
As a fuel.
Starting point in making a coke.

Linoleum and Oilcloth
Ingredient of—
Compositions used in the manufacture of oilcloth and linoleum.

Miscellaneous
General disinfectant, general preservative.
Ingredient of—
Compositions used for paving streets and making roads.
Compositions used for caulking ships and in shipbuilding.
Compositions used for coating tarpaulins, cables, fishing nets.
Compositions for impregnating rope and twine.
Preparations for soaking tow, preparations for coating sails and masts, preparations of asphaltic character, roof cements.

Paint and Varnish
Ingredient of—
Fillers, paints, varnishes.

Plastics
Starting point in making—
Plastic condensation products with formaldehyde.

Pharmaceutical
Ingredient of—
Pharmaceutical preparations, various hygienic drinks, veterinary preparations.
Suggested for use in treating colds, fevers, diarrhea, diseases of the genito-urinary system, skin diseases, and as an antiseptic and disinfectant.

Resins and Waxes
Starting point in making—
Synthetic resins.

Soap
Ingredient of—
Tar soap, solid and liquid.

Textile
Ingredient of—
Compositions used in making coated packing cloth and brattice cloth.

Woolfat Acids
French: Acides grasses de laine.
German: Wollfettsaeure.

Chemical
Starting point (Brit. 321239) in making—
Emulsifying agents, with the aid of chlorine, for producing emulsions with cresols, higher alcohols, and hydrocarbons.

Coaltar
Starting point (Brit. 321239) in making—
Emulsifying agents, with the aid of chlorine, for making emulsions with coaltars and bitumens.

Fats and Oils
Starting point (Brit. 321239) in making—
Emulsifying agents, with the aid of chlorine.

Insecticide
Starting point (Brit. 321239) in making—
Emulsifying agents, with the aid of chlorine, for use in insecticides and vermifuges.

Miscellaneous
Starting point (Brit. 321239) in making—
Emulsions used in road construction and for general disinfecting purposes.

Resins and Waxes
Starting point (Brit. 321239) in making—
Emulsifying agents, with the aid of chlorine.

Soap
Ingredient of—
Detergent preparations.
Raw material in making—
Special soaps.

Textile
Ingredient of—
Finishing compositions.

Wool Olein
French: Oléine de laine.
German: Wolleolein.

Chemical
Emulsifying agent (Brit. 275267) for—
Chlorohydrin, hydrocarbons, hydrogenated phenols, ketones.

Wool Olein (Continued)

Dye
Emulsifying agent (Brit. 275267) for—
Dyestuffs.

Fats and Oils
Emulsifying agent (Brit. 275267).

Leather
Emulsifying agent (Brit. 275267) in—
Oiling compositions.
Ingredient of—
Belt dressings and leather stuffing compositions.

Mechanical
Emulsifying agent (Brit. 275267) in—
Boring oils.

Miscellaneous
Emulsifying agent (Brit. 275267) in—
Wetting agents.

Soap
Raw material in—
Special soaps.

Petroleum
Emulsifying agent (Brit. 275267) for—
Mineral oils.

Textile
——, *Dyeing*
Ingredient (Brit. 275267) of—
Dye liquors, to equalize the distribution of the dyestuff.

——, *Finishing*
Impregnating agent (Brit. 275267) in—
Bleaching liquors, mercerizing liquors.

——, *Manufacturing*
Ingredient of—
Spinning waxes, wool-carbonizing liquors (Brit. 275267).

Wool Waste

Fertilizer
Source of nitrogen in making—
Wet base goods.

Lubricant
Ingredient of—
Greases designed to meet severe dust and dirt conditions and to prevent excessive leakage from bearing housings which are not tight.

Wool Yarn

Lubricant
Ingredient of—
Greases designed to meet severe dust and dirt conditions and to prevent excessive leakage from bearing housings which are not tight.

Textile
As a textile fiber for various well-known purposes.

Wormseed Oil
Synonyms: American wormseed oil, Baltimore wormseed oil, Oil of chenopodium.
Latin: Oleum chenopodii anthelmintici.
French: Essence d'ansérine vérmifuge, Essence de semen-contra d'Amerique, Essence de semencine d'Amerique.
German: Amerikanisches wurmsamenoel.

Pharmaceutical
In compounding and dispensing practice.

Wormwood Leaves
Synonyms: Absinthe.
Latin: Artemisia absinthium, Herba absinthii.
French: Herbe d'absinthe.
German: Bitterer beifuss, Magenkraut, Wermut.

Chemical
Starting point in making—
Absinthin.

Food
Flavoring for—
Liqueurs.

Pharmaceutical
In compounding and dispensing practice.

Wine
Ingredient of—
Bitter wines, vermouths.

Xenon

Electrical
Ingredient of—
Gaseous mixtures used in the so-called "Neon Signs."

Xylenemethylsulphonamide
French: Amide de xylènemonométhylesulphonique.
German: Xylolmonomethylsulfonamid.

Cellulose Products
Plasticizer (Brit. 313405) for—
Cellulose acetate, cellulose esters and ethers, cellulose nitrate.
For uses, see under general heading: "Plasticizers."

Chemical
Starting point in making various derivatives.

Xylenesulphonamide
French: Sulphonamide de xylène.
German: Xylolsulfonamid.

Cellulose Products
Solvent for—
Cellulose acetate, cellulose esters and ethers, cellulose nitrate.
For uses, see under general heading: "Solvents."

Chemical
Starting point in making—
Various derivatives.

Insecticide and Fungicide
Essential ingredient (U. S. 1997918) of—
Agent for destroying rust on cultivated plants.

Resins and Waxes
Solvent for—
Natural and artificial resins.
Starting point (Brit. 340101) in making—
Synthetic resins with the aid of benzaldehyde.

1:3:5-Xylenol

Chemical
Starting point in making—
Alphabetadimethyladipic acid (Brit. 265959).

Agricultural
Ingredient of—
Cattle dips, weed killers.

Pharmaceutical
In compounding and dispensing practice.

Miscellaneous
Antiseptic for various purposes.

Soap
Ingredient in making—
Antiseptic soaps, germicidal soaps.

Xylyldiphenylphosphonium Bromide
French: Bromure de xylylediphénylephosphonium.
German: Bromxylyldiphenylphosphonium, Xylyldiphenylphosphoniumbromid.

Miscellaneous
Mothproofing and moldproofing agent (Brit. 312163) in treating—
Hair, fur, feathers, felt, and the like.

Textile
Mothproofing and moldproofing agent (Brit. 312163) in treating—
Wool and other products.

Xylyl Phosphate
French: Phosphate de xylyle, Phosphate xylylique.
German: Phosphorsäurexylol, Phosphorsäuresxylylester, Xylylphosphat.

Miscellaneous
Mothproofing agent (U. S. 1748675) in treating—
Feathers, furs, skins, felt and other animal products subject to attack by the clothes moth larvae.

Textile
Mothproofing agent (U. S. 1748675) in treating—
Woolen materials and felt.

Xylyltriphenylphosphonium Bromide
French: Bromure de xylyletriphénylephosphonium.
German: Bromxylyltriphenylphosphonium, Xylyltriphenylphosphoniumbromid.

Miscellaneous
Mothproofing and moldproofing agent (Brit. 312163) in treating—
Hair, fur, feathers, felt, and the like.

Textile
Mothproofing and moldproofing agent (Brit. 312163) in treating—
Wool and other fabrics.

1:2-Xylyl-ww-disulphonic Acid

Dye
Intermediate (Brit. 447067) in making—
Dyes containing one or more aryl residues carrying one or more alkylsulphonic groups directly combined to the nucleus.

Yohimbine

Chemical
Starting point in making—
Yohimbine-brucine sulphate (German 437923).

Pharmaceutical
In compounding and dispensing practice.

Yohimbine Hydrochloride
French: Chlorhydrate d'yohimbine.
German: Salzsaeureyohimbinester, Yohimbinchlorhydrat.

Pharmaceutical
In compounding and dispensing practice.

Photographic
As a developer.

Yohimbine Lactate
French: Lactate de yohimbine.
German: Milchsaeuresyohimbin, Milchsaeureyohimbinester.

Pharmaceutical
In compounding and dispensing practice.

Photographic
As a developing agent.

Yohimbine Nitrate
French: Nitrate d'yohimbine.
German: Salpetersaeuresyohimbin, Yohimbinnitrat.

Pharmaceutical
In compounding and dispensing practice.

Photographic
As a developer.

Yohimbine Nucleate
French: Nucléate de yohimbine.
German: Nucleinsaeureyohimbinester, Nucleinsaeuresyohimbin, Yohimbinnucleat.

Pharmaceutical
In compounding and dispensing practice.

Photographic
As a developer.

Zinc
Latin: Zincum, Speltrum.
French: Speltre, Zinc.
German: Spelter, Zink.
Spanish: Zincico.
Italian: Zinco.

In C.P. Form

Analysis
Reagent in—
Analytical processes involving control and research work.

Metallurgical
Starting point in making—
White gold.

In Dust Form

Abrasives
Ingredient (U. S. 1263709) of—
Abrasives.

Analysis
Reagent in—
Analytical processes involving control and research work.

Ceramics
Ingredient (U. S. 1903346) of—
Abrasive and polishing compound, in admixture with quartz, for removing blemishes from enameled metal surfaces.

Chemical
Catalyst in making—
Organic chemicals.
Dechromating agent (U. S. 1919721) in making—
Bromates.
Polymerizing agent (Brit. 363846) in making—
Lubricating oils from olefins (used in combination with aluminum chloride).
Process material in making—
Absorbent carbons (U. S. 1519470).
Adrenalin (U. S. 1399144).
Albumen compound (U. S. 1371381).
1-Anilino-2-naphthol (U. S. 1460774).
Anthranilic acid (U. S. 1492664).
Arseno-stibino compounds (U. S. 1422294).
N-Arylaminonaphthols (U. S. 1460774).
Dichloroethylene (U. S. 1419969).
Dimethyldi-isopropylbenzidin (U. S. 1314924, 1314925, and 1314926).
Hydrazoanisol (U. S. 1469586).
Hydrazobenzene (U. S. 1405732).
Hydrazocymene (U. S. 1314924, 1314925, and 1314926).
Hydrazotoluene (U. S. 1405732).
Hyposulphites (U. S. 1472828).
Paracymol (U. S. 1433664).
Sodium hydrosulphite (U. S. 1412755).
Zinc oxide in benzidin preparations (U. S. 1426349).
Reducing agent in making—
Alpha-(a'-methylaminoethyl)benzyl alcohol (U. S. 1356877).
N-(4'-Amino-1'-naphthyl)paratoluenesulphonamide (U. S. 1442818).
2-Amino-1-(2'-phenyl-4'-quinolyl)-ethanol U. S. 1434306).
4-Amino-1-(paratolylsulphonamido)-naphthalene-2-sulphonic acid (U. S. 1442818).
4-Amino-1-(paratolylsulphonamido)-naphthalene-6-sulphonic acid (U. S. 1442818).
4-Amino-1-(paratolylsulphonamido)-naphthalene-7-sulphonic acid (U. S. 1442818).
4-Amino-1-(paratolylsulphonamido)-naphthalene-8-sulphonic acid (U. S. 1442818).
N:N'-Bis(4-amino-1-naphthyl)-2:6-naphthalenedisulphonamide (U. S. 1442818).
N:N'-Bis(4-amino-2-sulpho-1-naphthyl)metabenzenedisulphonamide (U. S. 1442818).
N:N'-Bis(4-amino-6-sulpho-1-naphthyl)metabenzenedisulphonamide (U. S. 1442818).
N:N'-Bis(4-amino-7-sulpho-1-naphthyl)metabenzenedisulphonamide (U. S. 1442818).
N:N'-Bis(4-amino-8-sulpho-1-naphthyl)metabenzenedisulphonamide (U. S. 1442818).
2-Chloroanthracene (U. S. 1434980).
2-Chloroanthraquinone (U. S. 1434980).
4-Chlor-3-(5'-keto-3'-methyl-4'-phenyl-azo-1'-pyrazolyl)-benzenesulphonic acid (U. S. 1511074).
2:6-Dichloroanthracene (U. S. 1434980).
2:2'-Dichlor-4:4'-dimethylamino-3:5:3':5'-tetraminoarsenobenzene (U. S. 1180627).
Dimethyldiphenylurea (U. S. 1477087).
Nitro compounds and their derivatives (U. S. 1432775).
Organic chemicals.
Perylene (U. S. 1454204).
Pyrrole from succinimide.
Reduction compounds of naphthalene-2:6-disulphochloride (U. S. 1444277).
Reduction compounds of 2-phenylquinolyl-4-isonitroso ketone (U. S. 1434306).
Reduction compounds of phenylnitropropanol (U. S. 1356877).
Reduction compounds of diacetyldioxyphenylnitroethanol (U. S. 1399144).
Reducing agent (with glacial acetic acid) for—
Peroxides.
Reducing agent (with glacial acetic acid) in—
Removing 2 atoms of halogen and the conversion of saturated compounds into olefins.
Reducing agent (with dilute acetic acid) in converting—
Nitramines to hydrazins, nitrosamines to hydrazins, osones to ketoses.
Reducing agent (with dilute sulphuric acid) in converting—
Sulphonic chlorides to thiophenols.
Reducing agent (with concentrated sulphuric acid) in converting—
Nitro compounds to aminohydroxy compounds.
Reducing agent (with acids) in converting—
Nitrates of aromatic amines to diazonium salts,
Reducing agent (with alkali) in converting—
Aromatic ketones to secondary alcohols.
Reducing agent (with water or alcohol) in converting—
Azo dyes to mixtures of amines, such as chrysoidin to anilin and triaminobenzene.
Aromatic nitro compounds to the corresponding hydroxylamines.
Sulphonic chlorides to sulphinic acids.
Starting point in making—
Catalysts (U. S. 1519470 and 1221698), chromates used as anticorrosion agents (Brit. 406445), zinc chloride, zinc nitrate, zinc salts, zinc sulphate.

Zinc (Continued)

Coke By-Products
Catalyst (U. S. 1221698) in making—
Naphtha.
Dehydrogenating agent (U. S. 1991979) for—
Tar acids.

Dye
Reducing agent in making—
Synthetic dyestuffs.

Electrical
Purifying agent for—
Zinc sulphate electrolyte in zinc plating.

Explosives and Matches
Ingredient of—
Explosive primer composition, containing also potassium chlorate, antimony, and gelatin as a binder, that is stable to shock and friction.
Matchhead composition (U. S. 1360282), smoke-producing compositions, tear gar compositions.
Various explosives (U. S. 1243231, 1334303, 1360397, 1360398, and 1276537).

Fats, Oils, and Waxes
Catalyst (U. S. 1505560) in making—
Edible fats, lard, margarin.
Purifying agent (U. S. 1247516) in removing—
Arsenic compounds from marine animal oils, chlorine compounds from marine animal oils, phosphorus compounds from marine animal oils.

Metallurgical
Coating agent in—
Protecting iron by the galvanizing process, protecting iron by the shearardizing process.
Deoxidizing agent for—
Bronze, nonferrous metals (U. S. 1967810).
Reagent in making—
Bleaching agent for mineral materials (U. S. 2020132), copper (U. S. 1180765), copper powder (U. S. 1376961).
Precipitating agent for—
Cyanide solutions (U. S. 1433965), gold in cyaniding processes, silver (U. S. 1403463 and 1479542), silver from cyanide solutions (U. S. 1461807), mercury (U. S. 1479542).
Reagent in removing—
Antimony from zinc sulphate solutions (U. S. 1283077 and 1283078).
Bismuth from zinc sulphate solutions (U. S. 1283077 and 1283078).
Cadmium from zinc solutions (U. S. 1426703).
Cadmium from zinc sulphate solutions (U. S. 1920442).
Cobalt from zinc solutions (U. S. 1426703).
Cobalt from zinc sulphate solutions (U. S. 1920442).
Copper from zinc solutions (U. S. 1426703).
Copper from zinc sulphate solutions (U. S. 1427826).
Lead compounds from zinc solutions (U. S. 1380514 and 1380515).
Nickel from zinc solutions (U. S. 1336386).
Nickel from zinc sulphate solutions (U. S. 1920442).

Miscellaneous
Ingredient of—
Coated fabric (U. S. 1210375), lighter wick (U. S. 1430543), welding compound (U. S. 1338736).

Paint and Varnish
Raw material in making—
Lithopone.
Ingredient of—
Anticorrosive paints, antifouling paint (U. S. 1493930), zinc base paints (Brit. 436164).

Petroleum
Catalyst (U. S. 1221698) in making—
Benzin, gasolene.
Condensing agent (Brit. 397169) in making—
Condensation products of high molecular paraffin hydrocarbons (used to facilitate the separation of waxes from hydrocarbon oils).
Promoter (Brit. 433780) of—
Hydrogen evolution in making soaps from paraffin wax oxidation products.
Starting point (U. S. 1152765) in making—
Catalysts for hydrogenation of hydrocarbons, lamp oil, and petroleum.

Printing
Process material (U. S. 1210375) in making—
Printer's blanket.

Textile
Reducing agent in—
Dyeing processes.

In Feathered Form

Analysis
Reagent in—
Analytical processes involving control and research work.

Chemical
Reagent in making—
Photographic chemicals.

Photographic
Stripping agent for—
Photographic solutions.

In Mossy Form

Analysis
Reagent in—
Analytical processes involving control and research work.

Chemical
Catalyst in—
Organic syntheses.

Clay Products
Ingredient of—
Coloring compositions for face brick.

Miscellaneous
Ingredient of—
Chimney soot-removing compositions.

Paint and Varnish
Starting point in making—
Zinc pigments.

In Rolled Form

(Usually sold as strip, plates or sheet; strip is either plain or crimped and sheet is either plain or corrugated.)

Automotive
Material in fabricating—
Autobody lining, body molding, curtain frames, dome lamp rims, drip molding, escutcheon plates, gasoline tank caps, hub caps, magneto hoods, running board mouldings, scuff plates, tire valve nuts.

Building and Construction
Material in fabricating—
Art glass strips, clips for shingles, conductors, corner beading, expanded metal lath, fences, flashings, frames for windows, glazier's points, gutters, leaders, roofing, shingles, siding, stair treads, weather stripping, window boltguards.

Electrical
Material in fabricating—
Anodes, cable wrappings, cups for dry battery, fuses, ground plates, insulator cups, magneto hoods.

Laundering
Material in fabricating—
Corrugated washer surfaces, tags.

Mechanical
Material in fabricating—
Fittings, gaskets, hinges, washers.

Metal Work
In general sheet metal work.
Material in fabricating—
Boiler plates, hull plates, organ pipes, ornamental fittings, perforated metal screens, signs, washing machine parts.

Miscellaneous
Material in fabricating—
Addressing machine plates, bands for steampipe coverings, binding for linoleum, bottel caps, buttons, cans for various products, cases for alarm clocks, casket ends, collapsible tube clips, embossed numbers, embossed tags, etched nameplates, eyelets, ferrules for brushes, fittings, grommets, linings for boxes, nails for shoes, novelties, oils cans, pin tubes, shoe lace tips, stencils, templates, washboards.

Printing
Material in etching—
Engraver's plates, lithographer's plates.

Railroading
Material in fabricating—
Car linings.

Refrigeration
Material in fabricating—
Drains for ice boxes, linings for ice boxes.

In Slab Form

Metallurgical
Degolding agent in—
Lead refining.

Zinc (Continued)
Desilvering agent in—
Lead refining.
Socketing material for—
Wire rope.
Source of zinc in—
Galvanizing iron materials, zinc (electro) plating iron materials.
Source of zinc in making—
Anodes for electroplating, battery zincs, bearing metals, brass, bronze, crusher face backings, die casting alloys, nickel silver, nonferrous alloys, tombec imitation gold.
Starting point in making—
Slush castings, zinc rods for wet batteries.

Miscellaneous
Base metal in—
Toys.

In Wire Form

Miscellaneous
As a metal spraying agent.

Zinc Acetate
Latin: Acetas zinci, Zincum aceticum.
French: Acétate de zinc, Acétate zincique.
German: Essigsäureszink, Essigsäureszinkoxyd, Zinkacetat, Zinkazetat.

Analysis
Reagent in analyzing metals.
Reagent in analyzing and testing for—
Albumen, blood, phosphoric acid, tannin, urine, urobilin.

Ceramics
Ingredient of—
Glazes in the production of fine porcelains.

Chemical
Catalyst (Brit. 259641) in making—
Acetic acid.
Ingredient of catalytic mixtures used in making—
Acenaphthylene, acenaphthaquinone, bisacenaphthylidenedione, naphthaldehydic acid, naphthalic anhydride, and hemimellitic acid (Brit. 295270).
Acetaldehyde from ethyl alcohol (Brit. 281307).
Acetic acid from ethyl alcohol (Brit. 281307).
Alcohols from aliphatic hydrocarbons (Brit. 281307).
Aldehydes or acids by the reduction of the corresponding esters (Brit. 306471).
Aldehydes and acids from toluene, orthochlorotoluene, orthonitrotoluene, orthobromotoluene, parachlorotoluene, parabromotoluene, paranitrotoluene, metachlorotoluene, metanitrotoluene, metabromotoluene, dichlorotoluenes, dibromotoluenes, dinitrotoluenes, chlorobromotoluenes, chloronitrotoluenes, bromonitrotoluenes (Brit. 295270).
Aldehydes and acids from xylenes, pseudocumenes, mesitylene and paracymene (Brit. 281307).
Alphacampholide from camphoric acid by reduction (Brit. 306471).
Alphanaphthaquinone from naphthalene (Brit. 281307).
Anthraquinone from naphthalene (Brit. 281307).
Benzaldehyde and benzoic acid from toluene (Brit. 281307).
Benzoquinone from phenanthraquinone (Brit. 281307).
Benzyl alcohol by the reduction of benzaldehyde (Brit. 306471).
Benzyl alcohol or benzaldehyde or phthalide by the reduction of phthalic anhydride (Brit. 306471).
Butyl alcohol by the reduction of crotonaldehyde (Brit. 306471).
Diphenic acid from ethyl alcohol (Brit. 281307).
Ethyl alcohol by the reduction of acetaldehyde (Brit. 306471).
Fluorenone from fluorene (Brit. 295270).
Formaldehyde by the reduction of methane or methanol (Brit. 306471).
Formaldehyde by the reduction of carbon dioxide or carbon monoxide (Brit. 306471).
Hydroxyl compounds by the reduction of anthraquinone, benzoquinone, or the like (Brit. 306471).
Isopropyl alcohol by the reduction of acetone (Brit. 306471).
Maleic acid and fumaric acid by the oxidation of toluene benzene, phenols, tar phenols, or furfural, or from benzoquinone or phethalic anhydride (Brit. 295270).
Methane by the reduction of carbon dioxide or carbon monoxide (Brit. 306471).
Methanol by the reduction of carbon dioxide or carbon monoxide (Brit. 306471).
Naphthaldehydic acid, acenaphthaquinone, bisacenaphthylidenedione from acenaphthylene (Brit. 281307).
Phenanthraquinone from phenenthrene or diphenic acid (Brit. 295270).
Primary alcohols by the reduction of the corresponding aldehydes (Brit. 306471).
Propionic acid and butyric acid and higher alcohols, ketones, and acids by the reduction of carbon dioxide or carbon monoxide (Brit. 306471).
Reduction products of ketones, aldehydes, acids, esters, alcohols, ethers, and other organic compounds, which contain oxygen (Brit. 306471).
Salicylic acid and salicylic aldehyde from cresol (Brit. 295270).
Secondary butyl alcohol by the reduction of methylethyl ketone (Brit. 306471).
Valeryl alcohol by the reduction of valeraldehyde (Brit. 306471).
Vanillin and vanillic acid from eugenol or isoeugenol (Brit. 295270).
Ingredient (Brit. 304640) of catalytic preparations used in the production of various aromatic and aliphatic amines, including—
Alphanaphthylamine from alphanitronaphthalene.
Amines from aliphatic nitro compounds, such as alkyl nitriles, or nitromethane.
Amylamine from pyridin.
Anilin, azo-oxybenzene, azobenzene, and hydrozobenzene from nitrobenzene by reduction.
Aminophenols and nitrophenols.
3-Aminopyridin from 3-nitropyridin.
Amino compounds from the corresponding nitroanisoles.
Amines from oximes, Schiff's base, and nitriles.
Cyclohexamine, dicyclohexamine, and cyclohexylanilin from nitrobenzene.
Piperidin from pyridin, pyrrolidin from pyrrol, tetrahydroquinolin from quinolin.
Starting point in making—
Zinc bichromate, zinc carbonate, zinc chromate, zinc ethylsulphate, zinc fluoride, zinc formate, zinc glycerophosphate, zinc hypophosphite, zinc lactate, zinc malate, zinc oleate, zinc oleostearate, zinc picrate, zinc pyrophosphate, zinc stearate, zinc sulphide.

Pharmaceutical
In compounding and dispensing practice.

Sanitation
As a disinfectant.

Textile
Mordant in dyeing—
Textile materials with alizarin blue S and similar colors.
Mordant in printing—
Calicoes and other fabrics.
Resist in dyeing—
Textile fibers and fabrics with anilin black.
Substitute for tartar emetic in dyeing—
Textile fibers and fabrics with basic colors.

Woodworking
Ingredient of—
Compositions used for preserving wood.

Zinc-Aluminum-Iron Cyanide

Chemical
Catalyst (Brit. 446411) in—
Halogenating unsaturated hydrocarbons.

Zinc-Ammonium Alginate
French: Alginate de zinc et d'ammonium, Alginate zincique-ammoniaque.
German: Alginsäureszinkammoniak, Alginsäureszinkammonium, Zinkammoniumalginat.
Spanish: Alginato de zinc y de amoniaco.
Italian: Alginato di zinco e d'ammonio.

Ceramics
Ingredient of—
Compositions used for the waterproofing of various ceramic wares.

Chemical
Emulsifying agent in making—
Dispersions of various chemicals.
Ingredient of—
Various chemical products (added for the purpose of increasing their viscosity).

Construction
Ingredient of—
Compositions used for treating cement and concrete for the purpose of preventing deterioration when exposed to the action of alkalies or seawater.

Zinc-Ammonium Alginate (Continued)
Waterproofing compositions used for treating plaster of paris, wallboard, cement, stucco, concrete.
Fats and Oils
Reagent in treating—
Emulsions of various animal and vegetable fats and oils for the purpose of stabilizing them.
Fuel
Binder in—
Composition fuel briquettes containing coal dust.
Glues and Adhesives
Ingredient of—
Adhesive preparations.
Ink
Thickener in—
Printing inks.
Leather
Ingredient of—
Sizing compositions.
Mechanical
Ingredient of—
Compositions for covering steel tubes.
Miscellaneous
Binder in—
Compositions, containing powdered mica, asbestos, coal, carbon, graphite, minerals and the like.
Sizing compositions for various uses.
Emulsifying agent in making—
Emulsions of various products.
Ingredient of—
Compositions used for treating rope and twine, waterproofing compositions.
Paint and Varnish
Ingredient of—
Compositions used for proofing interior walls and ceilings.
Paper
Binder in—
Sizing compositions, woodflour products.
Ingredient of—
Compositions for finishing paper.
Compositions for waterproofing paper and paper products.
Petroleum
Ingredient of—
Emulsions of petroleum and petroleum distillates (added for purpose of securing better dispersion).
Plastics
Ingredient of—
Various plastic compositions containing such substances as horn, ebonite, celluloid, ivory, bone, shell, galalith, formaldehyde-phenol condensation products, urea-formaldehyde condensation products, and other artificial resins.
Rubber
Ingredient of—
Products containing rubber latex.
Soap
Ingredient of—
Detergent preparations.
Textile
——, *Dyeing*
Ingredient of—
Dye baths (added for the purpose of increasing the dispersions of the dyestuff).
——, *Finishing*
Ingredient of—
Compositions used for the waterproofing of fabrics.
Compositions used for sizing yarns and fabrics.
——, *Printing*
Ingredient of—
Printing pastes.
Waxes and Resins
Ingredient of—
Dispersions of waxes and resins, both artificial and natural (added for the purpose of increasing their dispersion).

Zinc-Ammonium Chloride
French: Chlorure de zinc-ammonium.
German: Zinkammoniumchlorid.
Electrical
Ingredient of—
Electrolytic solution in batteries.
Metallurgical
Ingredient of—
Galvanizing baths.
Reagent in—
Soldering metals.
Miscellaneous
Ingredient of—
Soldering fluxes and liquids.
Paint and Varnish
Ingredient of—
Luminous paints and varnishes.

Zinc Antimonide
French: Antomoinure de zinc, Antimoinure zincique.
German: Antimonzink, Zinkantimonid.
Chemical
Catalyst (Brit. 263877) in making—
Acetone from isopropyl alcohol, dehydrogenated products from cyclohexane, isobutyraldehyde from isobutyl alcohol, isobutyronitrile from isobutylamine, naphthalene from tetrahydronaphthalene, paracymene from turpentine.
Catalyst (Brit. 262120) in making—
Isoveraldehyde from isoamyl alcohol.
General chemical reagent.

Zinc Benzoate
French: Benzoate de zinc, Benzoate zincique.
German: Benzoesäureszink, Zinkbenzoat.
Spanish: Benzoato de zinc.
Italian: Benzoato di zinco.
Rubber
Retarding agent (U. S. 1929561) in—
Vulcanizing processes employing an ultra-accelerator.

Zinc Bismuthide
French: Bismuthide de zinc.
German: Zinkwismuthid.
Chemical
Catalyst in making—
Acetone from isopropyl alcohol, isobutyraldehyde from isobutylalcohol, isobutyronitrile from isobutylamine, naphthalene from tetrahydronaphthalene, paracymene from turpentine oil.

Zinc Bromide
French: Bromure de zinc.
German: Bromzink, Zinkbromid.
Spanish: Bromuro de zinc.
Italian: Bromuro di zinco.
Chemical
Catalyst (Brit. 398527) in making—
Esters from lower aliphatic acids and olefins.
Electrical
Electrolyte (French 648716) in—
Electrolytic condensers.
Process material in making—
Primary batteries.
Miscellaneous
Ingredient of—
Soldering flux (U. S. 1428088).
Pharmaceutical
In compounding and dispensing practice.
Process material in making—
Colloidal emulsions.
Photographic
Process material in making—
Antistatic films, colloidal emulsions.
Rubber
Process material (French 646414) in making—
Isomers of caoutchouc.
Thermoplasticizing agent (French 615195) for—
Rubber.

Zinc Butylxanthogenate
Synonyms: Zinc butylxanthate.
French: Butylexanthogénate de zinc, Xanthate butilique de zinc.
German: Butylxanthogensäureszink.
Spanish: Butilxantogenato de zinc.
Italian: Butilxantogenato di zinco.
Rubber
Accelerator (French 548180, 562255, and 563397) in—
Vulcanizing processes.

Zinc-Cadmium Sulphide
Electrical
Luminous agent in—
Cathode-ray tubes used in television.

Zinc Chlorate
French: Chlorate de zinc.
German: Chlorsäureszink, Zinkchlorat.
Spanish: Clorato de zinc.
Italian: Clorato di zinco.

Chemical
As an oxidizing agent.

Zinc Chloride
Synonyms: Butter of zinc.
Latin: Chloruretum zincicum, Zinci chloridum, Zincum chloratum.
French: Chlorure de zinc.
German: Chlorzink, Zinchlorid.
Spanish: Cloruro zincico.

Analysis
Reagent in—
Analytical work of various sorts.

Chemical
Catalyst in—
Friedel and Crafts' synthesis.
Hydration of olefins by reaction with water or steam.
Organic synthesis.
Saccharification of carbohydrates (Brit. 400168).
Catalyst in making—
Alcohol from ethylene and steam (Brit. 396107).
Alkyl naphthalenes from methyl chloride and naphthalene (U. S. 1879912).
Alkyl-substituted aromatic hydroxy compounds (U. S. 1892990).
2:3-Aminonaphthoic acid from 2:3-hydroxynaphthoic acid and ammonia (U. S. 1871990).
Benzoic acid, metal benzoates, alkyl benzoates from trichlorobenzene (U. S. 1866849).
Butylphenol compositions (U. S. 1887662).
Esters from lower aliphatic acids and olefins (Brit. 398527).
Ether from ethylene and steam (Brit. 396107).
Hydrochloric acid from hydrogen and chlorine.
Vulcanization accelerators from anilin and propyl aldehyde (U. S. 1915979).
Dehydrating agent in—
Concentration of acetic acid (Brit. 400169).
Organic synthesis.
Dehydrating agent in making—
Activated charcoal from charcoal fines, woodflour, hydrochloric acid, and sugar (U. S. 1902986).
Methyl and ethyl chlorides from hydrogen chloride and the corresponding alcohol.
Reagent in making—
Light zinc carbonate for the rubber industry by reaction with alkali carbonates or bicarbonates (German 564676).
Starting point in making—
Various zinc salts.

Construction
Ingredient of—
Magnesia cements.
Metallic, heat-resistant cement for dressing stone facings (in admixture with zinc oxide).

Disinfectant
As a general disinfectant.
Ingredient of—
Antiseptic preparations, deodorant preparations.

Dye
Reagent in making—
Auramine from Michler's ketone and sal ammoniac, malachite green, methylene blue, various other dyestuffs.

Electrical
Active ingredient of—
Leclanche battery.
Gelatinizing agent (U. S. 1911400) in—
Starch coating composition for dry-cell battery paper linings.

Fats and Oils
Condensing agent (Brit. 394073) in making—
Lubricating oils (which may be used to improve the viscosity curves of other lubricating oils) by converting animal (bone oil) or vegetable fatty substances (soybean, olive, or palm oil) into unsaturated products practically free from oxygen and polymerizing or condensing these products.

Fuel
Reagent in making—
Candles.

Glass
Ingredient of—
Etching compositions.

Glues and Adhesives
Ingredient of—
Adhesive preparations, cold-water glues.

Ink
Starting point (U. S. 1899452) in making—
Special ink for protection and authentification of checks and the like; such ink has the characteristic that the color is a function of pH.

Insecticide
Ingredient of—
Weed-killer, containing also either sodium or calcium chlorate (U. S. 1925628).

Metallurgical
Ingredient of—
Burnishing and polishing compositions for finishing steel.
Composition, containing also nickel chloride, ammonium chloride, ammonium sulphocyanide, and water, used for blackening zinc.
Copper and brass solder, containing also iron chloride, lard, rosin, glycerin, tin, and lead.
Flux, in admixture with ammonium chloride, used in remelting and refining crude zinc (U. S. 1913929).
Fluxes used in tinning steel plate by "coke" process.
Fluxing baths, containing also zinc-ammonium chloride or hydrochloric acid, used in hot dip galvanizing of iron pipe.
Solder for aluminum and its alloys, containing also ammonium bromide and sodium fluoride (German 554087).
Soldering composition (U. S. 1761116).
Soldering flux (U. S. 1882734).
Soldering fluxes for copper, brass, steel, terne plate, tinned steel, monel metal, and other metals; such fluxes consisting of various mixtures of which the following are typical: (1) Rosin, ammonium chloride, glycerin, water, and zinc chloride; (2) zinc chloride, glycerin, alcohol, and water; (3) zinc chloride and ammonium chloride; (4) zinc chloride and stearic acid; (5) petrolatum, ammonium chloride, zinc chloride, and water.
Soldering solution, containing also glacial acetic acid and hydrochloric acid, used on stainless steels.
Spraying agent, in solution with acetone and carbon tetrachloride, for moulding sands for magnesium and its alloy (Brit. 399124).

Miscellaneous
Ingredient of—
Dental cements, embalming fluids.
Plastic substance capable of being hardened, containing also Portland cement and triacetin or glycerin (Brit. 403230).
Plastic substance capable of being hardened, containing also gypsum and triacetin or glycerin (Brit. 403230).
Plastic substance capable of being hardened, containing also triacetin or glycerin (Brit. 403230).
Taxidermists' fluids.
Reagent (U. S. 1720487) in making—
Infusible asphaltic masses of high elasticity.

Paint and Varnish
Starting point in making—
Zinc greens.

Paper
Defibrating agent for—
Old parchment paper.
Mercerizing agent (U. S. 1913283) for—
Kraft pulp prior to impregnating with rubber, pyroxylin, resin, and the like.
Reagent in making—
Moisture-resistant cellulose, which is absorbent and does not disintegrate or fray when wetted (Brit. 391153).
Parchment papers, vulcanized fiber.

Perfume
Ingredient of—
Dentifrices.
Mouthwash, containing also tincture of myrrh, thymol, borax, oil of clove, oil of cinnamon, alcohol, and coloring matter.

Petroleum
Catalyst (Brit. 402060) in making—
Aliphatic alcohols or organic esters thereof by subjecting hydrocarbons (petroleum, petroleum fractions,

Zinc Chloride (Continued)
ethane, propane) to thermal decomposition in the presence of vapors of an organic acid, preferably a lower aliphatic acid, such as acetic or propionic acid, in the presence or absence of steam.
Catalyst (Brit. 367848) in purifying—
Hydrocarbon oils with ozonized air.
Condensing agent (Brit. 397169) in making—
Agents for facilitating the separation of waxes from hydrocarbon oils; such products consisting of condensations or polymerizations of high molecular paraffin hydrocarbons, such as hard or soft paraffin, ceresin, ozokerite, or wax-like derivatives thereof, more particularly oxygen derivatives, such as montan wax, or their halogen, oxygen, or sulphur compounds with cyclic hydrocarbons—for example, naphthalene, crude benzene, middle tar oils, anthracene oils, or hydrogenated or cracked cyclic hydrocarbons.
Purifying agent in—
Processing petroleum and petroleum products.
Refining gasoline by polymerization of unsaturated constituents to resins or gums, the purified gasoline being separated by fractional distillation (U. S. 1917648).
Starting point (U. S. 1912603) in making—
Zinc-lead oxychlorides compositions used in removing sulphur, gum, and color-forming bodies from gasoline.

Pharmaceutical
In general compounding and dispensing practice.
Suggested for use as—
Escharotic (in cancerous affections).

Photographic
Reagent (Brit. 313974) in making—
Acetate films.

Refrigeration
Ingredient of—
Noncorrosive brine, in admixture with calcium chloride.

Resins
Catalyst in making—
Oil-soluble synthetic resins by (1) causing an aromatic compound containing a readily exchangeable halogen to react with a recent natural resin and esterifying with a mono- or polyhydric alcohol; or (2) causing an aromatic compound containing a readily exchangeable halogen to react with a product obtained by esterifying a recent natural resin with a mono- or polyhydric alcohol (Brit. 392382).
Oil-soluble synthetic resins by (1) condensing polyhydric alcohols partly esterified by fatty acids, with phenols, and treating the product with formaldehyde; or (2) condensing polyhydric alcohols with a phenol, partly esterifying with a fatty acid, and reacting with formaldehyde (German 576714).

Rubber
Catalyst (Brit. 397136) in making—
Synthetic oils for paint, varnish and impregnating purposes by hydrogenation of rubber.
Ingredient of—
Batch in vulcanizing.

Textile
Carbonizing agent in—
Wool processing.
Catalyst (Brit. 400249) in making—
Cellulose acetate yarn highly resistant to the delustering action of hot water, by further esterification of acetone-soluble acetate in the presence of inert diluents, such as carbon tetrachloride and benzene.
Ingredient of—
Sizing and weighting compositions for textile fabrics, especially cotton goods.
Mercerizing agent for—
Cotton.
Mordant in—
Printing and dyeing processes.
Reagent for—
Producing crepe effects on and crimping cotton, woolen, and silk fabrics.
Separating silk from cotton, woolen, and linen fibers.
Various purposes in the cotton, silk, and woolen industries.
Reagent in making—
Acetate rayon (Brit. 313974).
Acetate rayon of improved cross-section (Brit. 400180).
Artificial textiles (Brit. 388768).
Resist in—
Dyeing textile fabrics with sulphur colors, with albumin colors, and with para red.
Swelling agent (Brit. 397878 and 397838) in—
Improving luster of silk, increasing transparency of silk, modifying dyeing properties of silk, stiffening silk.

Woodworking
Ingredient of—
Fireproofing compositions for treating wood.
Impregnating compositions, in admixture with mineral oils and distillates, for treating railroad ties.
Preservatives and impregnating compositions.
Preservative composition (U. S. 1852098).
Preservative agent for—
Wood and wooden manufactures.

Zinc Chromates
Synonyms: Buttercup yellow, Zinc yellow.
French: Chromate de zinc.
German: Chromsäureszink, Zinkchromat.
Spanish: Cromato de zinc.
Italian: Cromato di zinco.

Building Construction
Pigment in—
Colored cements, cement coating compounds.
Waterproofing agents for cements, mortars, and the like.

Cellulose Products
Process material (French 638431) in making—
Cellulose formic ester.

Chemical
Catalyst in making—
Acetaldehyde from methane and carbon monoxide (French 599588).
Acetic acid from carbon monoxide and hydrogen (French 599588).
Allyl alcohol (Brit. 275345).
Amyl alcohol (Brit. 275345).
Butyl alcohol (Brit. 275345).
Heptyl alcohol (Brit. 275345).
Hexyl alcohol (Brit. 275345).
Isoamyl alcohol (Brit. 275345).
Isobutyl alcohol (Brit. 275345).
Isopropyl alcohol (Brit. 275345).
Methanol from methane (French 599588).
Oxygenated carbon compounds (Brit. 275345).

Electrical
Coating agent for—
Zinc electrodes in dry batteries.

Miscellaneous
Ingredient of—
Antifreeze composition (U. S. 1442330).

Paint and Varnish
Component of—
Green pigments.
Pigment in—
Artists' colors, flat wall paints, interior paints, paints, varnishes.

Zinc Cyanide
French: Cyanure de zinc, Cyanure zincique.
German: Cyanwasserstoffsaeureszink, Cyanzink, Zinkcyanid, Zinkzyanid, Zyanzink.

Chemical
General chemical reagent.
Ingredient of catalytic preparations used in making—
Acenaphthylene, acenaphthaquinone, bisacenaphthylidenedione, naphthaldehydic acid, naphthalic anhydride, and hemimellitic acid from acenaphthene (Brit. 295270).
Acetaldehyde from ethyl alcohol (Brit. 295270).
Acetic acid from ethyl alcohol (Brit. 295270).
Alcohols from aliphatic hydrocarbons (Brit. 281307).
Aldehydes and acids from toluene, orthochlorotoluene, parachlorotoluene, metachlorotoluene, orthonitrotoluene, paranitrotoluene, parabromotoluene, metanitrotoluene, metabromotoluene, metachlorotoluene, dichlorotoluenes, dibromotoluenes, dinitrotoluenes, chloronitrotoluene, chlorobromotoluene, bromonitrotoluene (Brit. 295270).
Aldehydes and acids from xylenes, pseudocumene, mesitylene, and paracymene (Brit. 295270).
Alphanaphthaquinone from naphthalene (Brit. 281307).
Anthraquinone from anthracene (Brit. 295270).
Benzaldehyde and benzoic acid from toluene (Brit. 281307).
Benzoquinone from phenanthraquinone (Brit. 281307).
Chloroacetic acid from ethylenechlorohydrin (Brit. 295270).
Diphenic acid from ethyl alcohol (Brit. 281307).
Fluorenone from fluorene (Brit. 295270).

Zinc Cyanide (Continued)
Formaldehyde from methanol or methane (Brit. 295270).
Maleic acid and fumaric acid by the oxidation of benzene, toluene, phenol, tar phenols, or furfural, or from benzoquinone or phthalic anhydride (Brit. 295270).
Naphthaldehydic acid, acenaphthaquinone, or bisacenaphthylidenedione from acenaphthylene (Brit. 281307).
Phenanthraquinone from phenanthrene or diphenic acid (Brit. 295270).
Phthalic acid and maleic acid from naphthalene (Brit. 295270).
Primary alcohols by the reduction of the corresponding aldehydes (Brit. 306471).
Propionic acid and butyric acid and higher alcohols, ketones, and acids from carbon dioxide or carbon monoxide by reduction (Brit. 306471).
Salicylic acid and salicylic aldehyde by the reduction of cresol (Brit. 306471).
Secondary butyl alcohol by the reduction of methylethyl ketone (Brit. 306471).
Valeryl alcohol by the reduction of valeraldehyde (Brit. 306471).
Vanillin and vanillic acid from eugenol or isoeugenol (Brit. 295270).
Ingredient (Brit. 306471) of catalytic preparations used in the reduction of—
Acetaldehyde to ethyl alcohol, acetone to isopropyl alcohol.
Anthraquinone, benzoquinone, and the like to the corresponding hydroxyl compounds.
Benzaldehyde to benzoic acid.
Camphoric acid to alphacampholide.
Carbon dioxide or carbon monoxide to formaldehyde, methane, methanol, and other products.
Crotonaldehyde to butyl alcohol.
Ketones, aldehydes, acids, esters, alcohols, ethers, and other organic compounds containing oxygen.
Phthalic anhydride to benzyl alcohol, benzaldehyde or phthalide.

Gas
Reagent in treating—
Coal gas and the like to remove the ammonia content.

Pharmaceutical
In compounding and dispensing practice.

Zinc Diamyldithiocarbamate
Rubber
Accelerator (Brit. 439215) for—
Vulcanization.

Zinc Dibenzyldithiocarbamate
Rubber
Accelerator (Brit. 439215) for—
Vulcanization.

Zinc Dibutyldithiocarbamate
Rubber
Accelerator (Brit. 439215) for—
Vulcanization.

Zinc Dimethyldithiocarbamate
Rubber
Ultra-accelerator in—
Vulcanization processes.

Zinc Dinaphthylnaphthenate
Lubricant
Addition agent (Brit. 433257) in—
Lubricating oils or greases, especially for use at high temperatures, such as cylinder oils, hydrogenated oils, or oils refined by treatment with sulphuric acid, clay, or extraction solvents.

Zinc Dipentamethylenethiuramdisulphide
Rubber
Secondary activator in—
Vulcanizing processes (for use with mercaptabenzthiazole).

Zinc Dipentamethylenethiurammonosulphide
Rubber
Secondary activator in—
Vulcanizing processes (for use with mercaptabenzthiazole).

Zinc Dipentamethylenethiuramtetrasulphide
Rubber
Secondary activator in—
Vulcanizing processes (for use with mercaptabenzthiazole).

Zinc Ferrocyanide
Textile
Reagent (Brit. 421360) for making—
White or colored matt effects on viscose or cellulose acetate rayon.

Zinc Formaldehyde-Sulphoxylate
Synonyms: Decolorant N, Redol Z.
French: Formaldéhydesulfoxylate de zinc, Formosulfoxylate de zinc, Sulfoxylate-formaldéhyde de zinc.
German: Decrolein, Zinkformaldehydsulfoxylat.

Chemical
Starting point in making—
Sodium formaldehyde-sulphoxylate.

Textile
——, *Printing*
Discharge in printing—
Fabrics with indigo and other dyestuffs.

Zinc Hydrosulphite
German: Hydroschwefligsaeureszink, Zinkhydrosulfit.

Chemical
Starting point in making—
Calcium hydrosulphite, sodium hydrosulphite.

Textile
——, *Dyeing and Printing*
Discharge in dyeing and printing with indigoes.

Zinc Isovalerate
French: Isovalérate zincique.
German: Isovalerinsäureszink, Zinkisovalerat.

Petroleum
Ingredient (Brit. 334181) of—
Motor fuels.

Zinc Laurate
Rubber
Promoter (U. S. 1984247) of—
Dissolution and distribution of zinc oxide in rubber mixes, giving high abrasion-resistance products.

Zinc Methylate
French: Méthylate de zinc.
German: Zinkmethylat.

Chemical
Catalyst in making—
Acetic acid (Brit. 259641).

Zinc Normalbutylhydrogenphthalate
French: N-Butylehydrogènephthalate de zinc.
German: N-Butylbiphtalsaeureszink.

Paint and Varnish
Raw material (synthetic resins) (Brit. 250265) in making—
Enamels, lacquers, varnishes.

Plastics
Raw material in making—
Plastic compositions, various molded articles.

Photographic
Ingredient (Brit. 270387) in making—
Light-sensitive varnishes.

Zinc Oxide
Synonyms: Chinese white, Flowers of zinc, Zinc white.
Latin: Flores zinci, Oxydum zincicum, Zinci oxidum, Zincum oxydatum.
French: Blanc de zinc, Fleurs de zinc, Oxyde de zinc, Oxyde zinc par voie sèche, Oxyde zincique.
German: Philosophenwolle, Zinkblumen, Zinkoxyd.
Spanish: Oxido de zinc, Oxido zincico.
Italian: Ossido di zinco.

Leadfree Form

Abrasives
Process material in making—
Abrasive agents and compositions, abrasive wheels.

Adhesives
Process material in making—
Glue, paste.

Analysis
Reagent in—
Analytical processes involving control and research.

Zinc Oxide (Continued)

Ceramics
Ingredient of—
Enamels, glazes.
Pigment in—
Chinaware, floor tiles, porcelain.

Chemical
Catalyst in—
Organic syntheses.
Starting point in making—
Zinc acetate, zinc bichromate, zinc borate, zinc carbonate, zinc chloride, zinc chromate, zinc citrate, zinc hydroxide, zinc soaps, zinc stearate, zinc sulphate, zinc valerate.

Cosmetic
Ingredient of—
Creams, lotions, pastes, pomades, powders.

Dental Products
Filler and pigment in—
Dental cements, false teeth.

Explosives and Matches
Process material in making—
Matches, dynamites.

Fuel
Pigment and filler in—
Candles.

Glass
Clarifying agent in—
Glass batches.
Polishing agent.
Process material in making—
Opaque glass, optical glass.

Ink
Pigment in—
White printing and marking inks.

Leather
Filler and pigment in making—
Boots, leather findings, shoes, white leathers.

Linoleum and Oilcloth
Filler and pigment in—
Table oilcloth.
Pigment.

Miscellaneous
Powdered packing for various products.

Paint and Varnish
Pigment in—
Antifouling paints, antirust paints, casein paints, decorative paints, elastic ivory white paints, enamels, exterior paints, fire-resisting paints, hospital white paints, interior paints, japans, laboratory white paints, lacquers, marine paints, pasteboard paints, paints used in electric works, paints for metal work, primers, putties, sanitary white paints, shingle paints, ships-bottom paints, waterproof paints, whitewashes.
Starting point in making—
Composite zinc pigments, lead-zinc pigments.

Paper
Filler and pigment in—
Paper of various kinds, particularly wallpaper.

Pharmaceutical
In compounding and dispensing practice.

Plastics
Filler and pigment in—
Artificial ivory, celluloid.

Rubber
Compounding agent influencing—
Activation, antiscorch, cure, cutting, flexing, reinforcement, resilience, stiffness, tear-resistance.
Compounding agent in making—
Athletic goods, belting, inner tubes, electric insulating material, moulded goods, repair materials, surgical articles, surgical sheeting, tires, tubing.

Textile
Discharge in—
Printing processes.
Pigment in—
Printing processes.
Resist in—
Textile processes.

Leaded Form

Ceramics
Ingredient of—
Enamels and glazes.

Paint and Varnish
Pigment in paints, varnishes, and lacquers of various sorts (see under "Leadfree Form").

Zinc Palmitate
French: Palmitate de zinc, Palmitate zincique.
German: Palmitinsäureszink, Palmitinsäureszinkoxyd, Zinkpalmitat.

Building
Waterproofing agent and ingredient of waterproofing compositions for treating—
Brickwork, concrete, stonework, stucco.

Leather
Ingredient of—
Waterproofing compositions.

Mechanical
Ingredient of—
Lubricating compositions.

Miscellaneous
Ingredient of waterproofing compositions for various applications.

Paint and Varnish
Drier in making—
Flat paints, lacquers, varnishes.
Thickening agent in making—
Oil preparations, solvent compositions.

Paper
Ingredient of—
Compositions for waterproofing paper and making waterproofed products of paper and pulp.

Petroleum
Thickener in making—
Greases and other lubricants.

Textile
Ingredient of—
Waterproofing compositions.

Zinc Pentamethylenedithiocarbamate

Rubber
Secondary activator in—
Vulcanizing processes (for use with mercaptobenzothiazole).

Zinc Perborate
French: Perborate de zinc, Perborate zincique.
German: Perborsäureszink, Perborsäureszinkoxyd, Zinkperborat.
Spanish: Perborato de zinc.
Italian: Perborato di zinco.

Analysis
As an oxidizing agent.

Chemical
As an oxidizing agent.

Miscellaneous
As an antiseptic for various purposes.
Ingredient of—
Germicidal preparations for household use.

Perfume
As a deodorant.
Ingredient of—
Bleaching preparations, face powders.

Pharmaceutical
As a general antiseptic.
Suggested for use as a dusting powder on wounds.

Zinc-Phenyl Acetate

Petroleum
Addition agent (Brit. 433257) in—
Lubricating oils or greases, especially for use at high temperatures, such as cylinder oils, hydrogenated oils, or oils refined by treatment with sulphuric acid, clay, or extraction solvents.

Zinc Phosphide
French: Phosphide de zinc.
German: Phosphenzink, Zinkphosphid, Zinkphosphor.

Pharmaceutical
In compounding and dispensing practice.

Zinc Platinate
French: Platinate de zinc, Platinate zincique.
German: Platinsaeureszink.

Chemical
Reagent for various purposes.
Ingredient of catalytic preparations used in making—
Acenaphthylene, acenaphthaquinone, bisacenaphthyl-

Zinc Platinate (Continued)
idenedione, naphthaldehydic acid, naphthalic anhydride, and hemimellitic acid from acenaphthene (Brit. 295270).
Acetaldehyde from ethyl alcohol (Brit. 281307).
Acetic acid from ethyl alcohol (Brit. 281307).
Alcohols from aliphatic hydrocarbons (Brit. 281307).
Aldehydes or alcohols by reduction of esters (Brit. 306471).
Alphacampholid by reduction of camphoric acid (Brit. 306471).
Aldehydes and acids from toluene, orthochlorotoluene, orthonitrotoluene, orthobromotoluene, metachlorotoluene, metanitrotoluene, metabromotoluene, parachlorotoluene, parabromotoluene, paranitrotoluene, dichlorotoluenes, dinitrotoluenes, dibromotoluenes, chlorobromotoluenes, chloronitrotoluenes, bromonitrotoluenes (Brit. 295270).
Aldehydes and acids from xylenes, pseudocumene, mesitylene, and paracymene (Brit. 295270).
Alphanaphthaquinone from naphthalene (Brit. 281307).
Anthraquinone from naphthalene (Brit. 295270).
Benzaldehyde and benzoic acid from toluene (Brit. 281307).
Benzoquinone from phenanthraquinone (Brit. 281307).
Benzyl alcohol by reduction of benzaldehyde (Brit. 306471).
Benzyl alcohol or benzyl aldehyde or phthalide by the reduction of phthalic anhydride (Brit. 306471).
Butyl alcohol by the reduction of crotonaldehyde (Brit. 306471).
Chloroacetic acid from ethylenechlorohydrin (Brit. 295270).
Diphenic acid from ethyl alcohol (Brit. 281307).
Ethyl alcohol by the reduction of acetaldehyde (Brit. 306471).
Fluorenone from fluorene (Brit. 295270).
Formaldehyde from methane or methanol (Brit. 295270).
Formaldehyde by the reduction of carbon monoxide or carbon dioxide (Brit. 306471).
Isopropyl alcohol by the reduction of acetone (Brit. 306471).
Maleic acid and fumaric acid by the oxidation of benzene, toluene, phenol, tar phenols, or furfural, or from benzoquinone or phthalic anhydride (Brit. 295270).
Methane by the reduction of carbon dioxide or carbon monoxide (Brit. 306471).
Methanol by the reduction of carbon dioxide or carbon monoxide (Brit. 306471).
Naphthaldehydic acid, acenaphthaquinone, or bisacenaphthylidenedione from acenaphthylene (Brit. 281307).
Phenanthraquinone from phenanthrene or diphenic acid (Brit. 295270).
Phthalic acid and maleic acid from naphthalene (Brit. 295270).
Primary alcohols by the reduction of aldehydes (Brit. 306471).
Propionic acid and butyric acid and higher alcohols, ketones, and acids by the reduction of carbon dioxide or carbon monoxide (Brit. 306471).
Reduction of anthraquinone, benzoquinone and the like to corresponding hydroxyl compounds (Brit. 306471).
Reduction of ketones, aldehydes, acids, esters, alcohols, ethers, and other organic compounds containing oxygen (Brit. 306471).
Salicylic acid and salicylic aldehyde from cresol (Brit. 295270).
Secondary butyl alcohol by reduction of methylethyl ketone (Brit. 306471).
Valeryl alcohol by the reduction of valeraldehyde (Brit. 306471).
Vanillin and vanillic acid from eugenol or isoeugenol (Brit. 295270).

Zinc Selenide
French: Sélénure de zinc, Sélénure zincique.
German: Selenzink, Zinkselenid.

Chemical
Catalyst (Brit. 263877) in making—
Acetone from isopropyl alcohol.
Dehydrogenated products from cyclohexane.
Isobutyraldehyde from isobutyl alcohol.
Isobutyronitrile from isobutylamine.
Naphthalene from tetrahydronaphthalene.
Paracymene from turpentine.
Catalyst (Brit. 262120) in making—
Isovaleraldehyde from isoamyl alcohol.
General chemical reagent.

Zinc Stearate
French: Stéarate de zinc, Stéarate zincique.
German: Stearinsäureszink, Stearinsäureszinkoxyd, Zinkstearat.

Linoleum and Oilcloth
Drier in—
Coating compositions.

Miscellaneous
Ingredient of—
Black lead compositions, crayon compositions, colored lead compositions, fireproofing compositions for various purposes, lead pencil compositions, waterproofing compositions.

Paint and Varnish
Drier in making—
Oil paints, varnishes, enamels.

Perfumery
Ingredient of—
Combined face powders and skin foods (U. S. 1620269).
Dry rouges (Brit. 255713).
Face powders, talcum powders.

Pharmaceutical
In compounding and dispensing practice.

Rubber
For dusting purposes.
Substitute for gum rubber.

Soap
Ingredient of—
Shaving creams.

Textile
Ingredient of—
Fireproofing compositions, waterproofing compositions.

Zinc Sulphate
Synonyms: White vitriol, Zinc vitriol.
Latin: Sulfas zincicus, Vitriolum album, Zincum sulphuricum crudum, Zincum sulphuricum purum.
French: Couperose blanc, Sulphate de zinc, Sulphate zincique, Vitriol blanc.
German: Galitzenstein, Schwefelsäurezink, Schwefelsäurezinkoxyd, Weisser galitzenstein, Weisser vitriol, Zincsulfat, Zinkvitriol.
Spanish: Sulfato de zinc.
Italian: Sulfato di zinco.

Agriculture
Reagent for—
Treating soil to kill weeds (used in the proportion of 8 grams, in water solution, per square-foot of ground).

Analysis
Reagent in detecting and determining—
Albumoses, glucose, proteoses, sulphur dioxide, urea.
Reagent in—
Standardizing sodium sulphide solutions for zinc determinations.

Ceramics
Reagent in making—
Ceramic products.

Chemical
Ingredient of catalytic mixtures used in the manufacture of—
Acenaphthylene, acenaphthaquinone, bisacenaphthylidenedione, naphthaldehydic acid, naphthalic anhydride, and hemimellitic acid from acenaphthene (Brit. 295270).
Acetaldehyde from ethyl alcohol (Brit. 281307).
Acetic acid from ethyl alcohol (Brit. 281307).
Alcohols from aliphatic hydrocarbons (Brit. 281307).
Aldehydes and acids by the reduction of the corresponding esters (Brit. 306471).
Aldehydes and acids from toluene, orthochlorotoluene, orthonitrotoluene, orthobromotoluene, parachlorotoluene, parabromotoluene, paranitrotoluene, metachlorotoluene, metabromotoluene, metanitrotoluene, dichlorotoluenes, dibromotoluenes, dinitrotoluenes, chlorobromotoluenes, chloronitrotoluenes, bromonitrotoluenes (Brit. 295270).
Aldehydes and acids from xylenes, pseudocumenes, mesitylenes, and paracymene (Brit. 281307).
Alphanaphthylamine from naphthalene (Brit. 281307).
Anthraquinone from naphthalene (Brit. 295270).
Benzaldehyde and benzoic acid from toluene (Brit. 281307).

Zinc Sulphate (Continued)

Benzoquinone from phenanthraquinone (Brit. 281307).
Benzyl alcohol from benzaldehyde by reduction (Brit. 306471).
Benzyl alcohol or benzaldehyde or benzyl phthalide by the reduction of phthalic anhydride (Brit. 306471).
Butyl alcohol by the reduction of crotonaldehyde (Brit. 306471).
Chloroacetic acid from ethylenechlorohydrin (Brit. 295270).
Diphenic acid from ethyl alcohol (Brit. 281307).
Ethyl alcohol by the reduction of acetaldehyde (Brit. 306471).
Fluorenone from fluorene (Brit. 295270).
Formaldehyde by the reduction of methane or methanol (Brit. 306471).
Formaldehyde by the reduction of carbon dioxide or carbon monoxide (Brit. 306471).
Hydroxyl compounds by the reduction of anthraquinone, benzoquinone, and similar compounds (Brit. 306471).
Isopropyl alcohol by the reduction of acetone (Brit. 306471).
Maleic acid and fumaric acid by the oxidation of toluene, benzene, phenols, tar phenols, or furfural, or from benzoquinone or phthalic anhydride (Brit. 295270).
Methane by the reduction of carbon dioxide or carbon monoxide (Brit. 306471).
Methanol by the reduction of carbon dioxide or carbon monoxide (Brit. 306471).
Naphthaldehydic acid, acenaphthaquinone, or bisacenaphthylidenedione from acenaphthylene (Brit. 295270).
Phenanthraquinone from phenanthrene or diphenic acid (Brit. 295270).
Phthalic acid and maleic acid from naphthalene (Brit. 295270).
Primary alcohols by the reduction of the corresponding aldehydes (Brit. 306471).
Propionic acid and butyric acid and higher alcohols, ketones, and acids by the reduction of carbon dioxide or carbon monoxide (Brit. 306471).
Reduction products of ketones, aldehydes, acids, esters, alcohols, ethers, and other organic compounds which contain oxygen (Brit. 306471).
Salicylic acid and salicylic aldehyde from cresol (Brit. 295270).
Secondary butyl alcohol by the reduction of methylethyl ketone (Brit. 306471).
Valeryl alcohol by the reduction of valeraldehyde (Brit. 306471).
Vanillin and vanillic acid by the oxidation of eugenol or isoeugenol (Brit. 295270).

Ingredient (Brit. 306460) of catalytic preparations used in the production of various aromatic and aliphatic compounds, including—
Alphanaphthylamine from alphanitronaphthalene.
Amines from aliphatic nitro compounds, such as allyl nitriles or nitromethane.
Amino compounds from the corresponding nitroanisoles.
Amylamine from pyridin.
Anilin, azo-oxybenzene, azobenzene, and hydrazobenzene from benzene by reduction.
Aminophenols from nitrophenols.
3-Aminopyridin from 3-nitropyridin.
Cyclohexamine, dicyclohexamine, and cyclohexylanilin from nitrobenzene.
Piperidin from pyridin.
Pyrrolidin from pyrrol.
Tetrahydroquinolin from quinolin.

Reagent in making—
Diazotized aminophenols (German 431513).
Zinc sulphanilate (nizine).

Reagent (Brit. 370550) in making paint and varnish driers with the aid of—
Acids obtained by the destructive oxidation of paraffin hydrocarbons.
Anilin, benzylamine, benzoic acid, cinnamic acid, diethanolamine, diethylanilin, dihydroxy ether of triethanolamine, ethylamine, ethyleneamine, hexamethylenetetramine, hydrogenated benzoic acid, monoethanolamine, monohydroxy ether of triethanolamine, naphthetic acid, normal hydroxyethylmorpholine, oleic acid, palmitic acid, para-aminophenol, propanolamine, pyridin, quinolin, resinic acid.
Sulphonic acids formed by heating petroleum oils with pyrosulphuric acid or sulphur trioxide.
Triethanolamine.

Starting point in making—
Zinc bromate from barium bromate.
Zinc bromide from barium bromide.
Zinc carbonate, precipitated, from sodium carbonate.
Zinc chlorate from barium chlorate.
Zinc cyanide from potassium cyanide.
Zinc ferrocyanide from potassium ferrocyanide.
Zinc iodate from barium iodate.
Zinc iodide from barium iodide.
Zinc oleostearate from hard soap and curd soap.
Zinc oxalate from sodium oxalate.
Zinc perborate from boric acid.
Zinc peroxide from barium peroxide.
Zinc phosphate, tribasic, from trisodium phosphate.
Zinc phosphide by reaction with phosphine.
Zinc picrate by reaction with picric acid.
Zinc pyrophosphate by reaction with ammonium phosphate.
Zinc stearate by reaction with sodium stearate.
Zinc sulphide by introduction of sulphuretted hydrogen gas into the solution.
Zinc valerianate by reaction with sodium isovalerianate.

Electrical
Ingredient of—
Electrolyte in storage batteries.

Fats and Oils
Reagent in making—
Drying oils.

Glue and Adhesives
Reagent in—
Clarifying glues, preserving glues and gelatins, protecting gelatin and flour pastes.

Insecticides
Ingredient (French 596320) of—
Insecticidal compositions containing arsenic trioxide.

Leather
Astringent preservative for skins.

Mechanical
Ingredient of—
Lubricating compositions.

Metallurgical
Electrolyte in—
Electrodeposition of zinc in refining, zinc plating.

Miscellaneous
General disinfectant.
Ingredient of—
Compositions used in treating vegetable fibers (French 600476).
Compositions used for treating hair.
Preservative in treating fur skins.
Reagent in taxidermy.

Paint and Varnish
Ingredient of—
Enamels, fireproofing paints, lacquers, paints, varnishes.
Starting point in making—
Colored zinc pigments with 6 to 30 per cent of the metallic sulphate such as nickel, cobalt, iron, and manganese.
Lithopone.

Paper
Ingredient of—
Compositions, containing barium hypochlorite, used for bleaching paper.

Perfume
Ingredient of—
Lotions, mouth washes.

Pharmaceutical
Suggested for use as emetic, astringent, antiseptic, and escharotic.

Rubber
Ingredient of—
Crude rubber batch (added for the purpose of facilitating vulcanization).

Textile
——, *Dyeing*
Mordant in—
Dyeing yarns and fabrics with alizarin blue.

——, *Finishing*
Ingredient of—
Compositions used for preserving textiles from mildew.
Fireproofing compositions.

——, *Manufacturing*
Ingredient (Brit. 253953) of—
Viscose rayon precipitating bath.

Zinc Sulphate (Continued)

——, *Printing*
Mordant in—
Printing fabrics and calicoes with alizarin blue.
Resist in—
Printing pastes containing pigment colors thickened with dextrin and china clay.

Woodworking
Preservative for wood.

Zinc Sulphide
Latin: Zincum sulphuratum.
French: Sulfure de zinc, Sulfure zincique.
German: Schwefelzink, Zinksulfid.

Chemical
Catalyst (Brit. 262100) in making—
Cymene, isobutyraldehyde, isobutyronitrile, isovaleraldehyde.
Luminous agent for various chemical purposes.

Dye
Substratum in making—
Color lakes, permanent dyestuffs.

Glass
Pigments in producing—
White and opaque glass and glassware.

Glues and Adhesives
Pigment for—
Producing white, opaque products.

Leather
Pigment in—
Compositions used in the manufacture of artificial leathers and various leather substitutes.

Linoleum and Oilcloth
Pigment in—
Compositions used in the manufacture of oilcloth and linoleum.

Miscellaneous
Luminous and phosphorescent agent for various purposes.

Paint and Varnish
Pigment in—
Varnishes, luminous paints, paints, enamels, lacquers.
Starting point in making—
Lithopone.
"Mineral white" (in admixture with zinc oxide).

Plastics
Pigment in—
Compositions containing artificial resins, natural resins, cellulose derivatives, and the like.

Rubber
Pigment in—
Rubber merchandise.
Rubber compositions for dental purposes.

Textile
Reagent in dyeing—
Yarns and fabrics by the hydrosulphite process.

Zinc Sulphocarbolate
Synonyms: Zinc phenolsulphonate.
French: Sulfocarbolique de zinc.
German: Carbolschwefelsäureszink.

Chemical
Denaturant for—
Alcohol.

Insecticide and Fungicide
Process material in making—
"Bouillie Lyonnaise" for destroying *Oidium* on vines.

Pharmaceutical
In compounding and dispensing practice.
Ingredient of—
Chicken remedies.
Suggested for use as—
Antiseptic, astringent.

Zinc 2:4:5-Trichlorophenolate

Disinfectant
As an antiseptic (U. S. 1994002).

Insecticide and Fungicide
As an agricultural fungicide (U. S. 1994002).

Zinc Tungate
French: Tungate de zinc.
German: Tungsaeureszink, Zinktungat.

Electrical
Fluorescent screen material (Brit. 456561) in—
Electronic tubes.

Paint and Varnish
Drier (Brit. 270387) in making—
Enamels, lacquers, paints, varnishes.

Photographic
Ingredient of—
Light-sensitive varnishes.

Zinc Uranate
French: Uranate de zinc, Uranate zincique.
German: Uransaeureszink, Zinkuranat.

Chemical
Ingredient of catalytic preparations used in making—
Acenaphthylene, acenaphthaquinone, bisacenaphthylidenedione, naphthaldehydic acid, naphthalic anhydride, and hemimellitic acid from acenaphthene (Brit. 295270).
Acetaldehyde from ethyl alcohol (Brit. 281307).
Acetic acid from ethyl alcohol (Brit. 281307).
Alcohols from aliphatic hydrocarbons (Brit. 281307).
Aldehydes or alcohols by reduction of their esters (Brit. 306471).
Alphacampholide by the reduction of camphoric acid (Brit. 306471).
Aldehydes and acid from toluene, orthochlorotoluene, orthobromotoluene, orthonitrotoluene, parachlorotoluene, parabromotoluene, paranitrotoluene, metachlorotoluene, metanitrotoluene, metabromotoluene, dichlorotoluenes, dinitrotoluenes, dibromotoluenes, chlorobromotoluenes, chloronitrotoluenes, bromonitrotoluenes (Brit. 295270).
Aldehydes and acids from xylenes, pseudocumene, mesitylene, and paracymene (Brit. 295270).
Alphanaphthaquinone from naphthalene (Brit. 281307).
Anthraquinone from naphthalene (Brit. 295270).
Benzaldehyde and benzoic acid from toluene (Brit. 281307).
Benzoquinone from phenanthraquinone (Brit. 281307).
Benzyl alcohol by reduction of benzaldehyde (Brit. 306471).
Benzyl alcohol or benzyl aldehyde or phthalide by the reduction of phthalic anhydride (Brit. 306471).
Butyl alcohol from crotonaldehyde (Brit. 306471).
Chloroacetic acid from ethylenechlorohydrin (Brit. 295270).
Diphenic acid from ethyl alcohol (Brit. 281307).
Ethyl alcohol from acetaldehyde (Brit. 306471).
Fluorenone from fluorene (Brit. 295270).
Formaldehyde from methanol or methane (Brit. 295270).
Isopropyl alcohol by the reduction of acetone (Brit. 306471).
Formaldehyde from carbon monoxide or carbon dioxide (Brit. 306471).
Maleic acid and fumaric acid by the oxidation of benzene, toluene, phenol, tar phenols, or furfural, or from benzoquinone or phthalic anhydride (Brit. 295270).
Methane by the reduction of carbon dioxide or carbon monoxide (Brit. 306471).
Methanol from carbon dioxide or carbon monoxide (Brit. 306471).
Naphthaldehydic acid, acenaphthaquinone, or bisacenaphthylidenedione from acenaphthylene (Brit. 281307).
Phenanthraquinone from phenanthrene or diphenic acid (Brit. 295270).
Phthalic acid and maleic acid from naphthalene (Brit. 295270).
Primary alcohol from aldehydes by reduction (Brit. 306471).
Propionic acid and butyric acid and higher alcohols, ketones, and acids from carbon dioxide or carbon monoxide by reduction (Brit. 306471).
Reduction of anthraquinone, benzoquinone, and the like, to the corresponding hydroxyl compounds, such as phenanthraquinone and naphthalene (Brit. 306471).
Reduction of carbon dioxide and carbon monoxide (Brit. 306471).
Reduction of ketones, aldehydes, acids, esters, alcohols, ethers, and other organic compounds containing oxygen (Brit. 306471).
Salicylic acid and salicylic aldehyde from cresol (Brit. 295270).
Secondary butyl alcohol by reduction of methyl ketone (Brit. 306471).
Valeryl alcohol by the reduction of valeraldehyde (Brit. 306471).
Vanillin and vanillic acid from eugenol or isoeugenol (Brit. 295270).

Zinc Vanadate
French: Vanadate de zinc, Vanadate zincique.
German: Vanadinsaeureszink.

Chemical
Reagent for various purposes.
Ingredient of catalytic preparations used in making—
Acenaphthylene, acenaphthaquinone, bisacenaphthylidenedione, naphthaldehydic acid, naphthalic anhydride, and hemimellitic acid from acenaphthene (Brit. 295270).
Acetaldehyde from ethyl alcohol (Brit. 281307).
Acetic acid from ethyl alcohol (Brit. 281307).
Alcohols from aliphatic hydrocarbons (Brit. 281307).
Aldehydes or alcohols by reduction of esters (Brit. 306471).
Alphacampholid by reduction of camphoric acid (Brit. 306471).
Aldehydes and acids from toluene, orthochlorotoluene, orthonitrotoluene, orthobromotoluene, parachlorotoluene, parabromotoluene, paranitrotoluene, metachlorotoluene, metanitrotoluene, metabromotoluene, dichlorotoluenes, dinitrotoluenes, dibromotoluenes, chloronitrotoluenes, chlorobromotoluenes, bromonitrotoluenes (Brit. 295270).
Aldehydes and acids from xylenes, pseudocumene, mesitylene, and paracymene (Brit. 295270).
Alphanaphthaquinone from naphthalene (Brit. 281307).
Anthraquinone from naphthalene (Brit. 295270).
Benzaldehyde and benzoic acid from toluene (Brit. 281307).
Benzoquinone from phenanthraquinone (Brit. 281307).
Benzyl alcohol by reduction of benzaldehyde (Brit. 306471).
Benzyl alcohol or benzyl aldehyde or phthalid from phthalic anhydride (Brit. 306471).
Butyl alcohol from crotonaldehyde (Brit. 306471).
Chloroacetic acid from ethylenechlorohydrin (Brit. 295270).
Diphenic acid from ethyl alcohol (Brit. 281307).
Ethyl alcohol from acetaldehyde (Brit. 306471).
Fluorenone from fluorene (Brit. 295270).
Formaldehyde from methanol or methane (Brit. 295270).
Formaldehyde from carbon monoxide or carbon dioxide (Brit. 306471).
Isopropyl alcohol by the reduction of acetone (Brit. 306471).
Maleic acid and fumaric acid by the oxidation of benzene, toluene, phenol, tar phenols, or furfural, or from benzoquinone or phthalic anhydride (Brit. 295270).
Methane by the reduction of carbon dioxide or carbon monoxide (Brit. 306471).
Methanol from carbon dioxide or carbon monoxide (Brit. 306471).
Naphthaldehydic acid, acenaphthaquinone, or bisacenaphthylidenedione from acenaphthylene (Brit. 281307).
Phenanthraquinone from phenanthrene or diphenic acid (Brit. 295270).
Phthalic acid and maleic acid from naphthalene (Brit. 295270).
Primary alcohols from aldehydes by reduction (Brit. 306471).
Propionic acid and butyric acid and higher alcohols, ketones, and acids from carbon dioxide or carbon monoxide (Brit. 306471).
Reduction of anthraquinone, benzoquinone, and the like to corresponding hydroxyl compounds (Brit. 306471).
Reduction of carbon dioxide and carbon monoxide (Brit. 306471).
Reduction of ketones, aldehydes, acids, esters, alcohols, ethers, and other organic compounds containing oxygen (Brit. 306471).
Salicylic acid and salicylic aldehyde from cresol (Brit. 295270).
Secondary butyl alcohol by reduction of methylethyl (Brit. 295270).
Valeryl alcohol by the reduction of valeraldehyde ketone (Brit. 306471).
Vanillin and vanillic acid from eugenol or isoeugenol (Brit. 306471).

Zirconium Nitrate
French: Nitrate de zircone, Nitrate de zirconium.
German: Zirkonnitrat.

Chemical
Starting point in making—
Zirconium acetate, zirconium bromide, zirconium carbonate, zirconium chloride, zirconium formate, zirconium hydroxide, zirconium lactate, zirconium oxalate, zirconium phosphate.

Food
Preservative in various food preparations and compositions.

Photographic
Ingredient of—
Magnesium flashlight powders.

Zirconium Oxide
Synonyms: Zirconic anhydride, Zirconium anhydride, Zirconium dioxide.
French: Anhydride zirconique, Dioxyde de zirconium, Oxyde zirconique, Oxyde de zirconium.
German: Zirkonanhydrid, Zirkondioxyd, Zirkonerde, Zirkonoxyd.

Abrasive
Ingredient of—
Discs, powders, stones, wheels.

Ceramics
Ingredient of—
Enamel compositions resistant to acids.
Ground enamel coatings used on porcelains, potteries and chinaware to form ground for dark blue colorations.
Opaque enamels.
Vitreous enamels, in the place of tin oxide.

Chemical
Catalyst in making—
Aldehyde and acetic acid from alcohol.
Carbon dioxide and water from organic substances.
Nitrogen trioxide and other oxides of nitrogen from ammonia.
Sulphur trioxide and sulphuric acid by the contact process.
Water from hydrogen.
Reagent (Brit. 281307) in making zeolite catalysts used in making—
Acetaldehyde from ethyl alcohol.
Acetic acid from ethyl alcohol.
Alcohols from aliphatic hydrocarbons.
Aldehydes and acids by the oxidation of orthochlorotoluene, parachlorotoluene, orthobromotoluene, parabromotoluene, dichlorotoluenes, dibromotoluenes, dinitrotoluenes, chlorobromotoluenes, chloronitrotoluenes, bromonitrotoluenes.
Alphanaphthaquinone from naphthalene.
Anthraquinone from anthracene.
Benzaldehyde and benzoic acid from toluene.
Benzoquinone from phenanthraquinone.
Chloroacetic acid from ethylenechlorohydrin.
Diphenic acid from ethyl alcohol.
Fluorenone from fluorene.
Formaldehyde from methanol or methane.
Hemimellitic acid from acenaphthene.
Maleic acid and fumaric acid from benzene, toluene, phenol, or tar acids, or benzoquinone or phthalic anhydride.
Naphthaldehydic acid, acenaphthaquinone or bisacenaphthylidenedione from acenaphthene or acenaphthylene.
Naphthalic anhydride.
Phenanthraquinone from phenanthrene.
Phthalic anhydride from naphthalene.
Salicylic aldehyde or salicylic acid from cresol.
Vanillin or vanillic acid from eugenol or isoeugenol.

Dye
Base in making—
Color lakes.

Electrical
Incandescent body in making Nernst light.
Ingredient of—
Compositions used for general insulating purposes.
Insulator for general purposes.

Glass
Ingredient of—
Glass for making phonograph diaphragms, in admixture with titanium dioxide.
Opal glass, used in place of tin oxide.
Optical glass.
Quartz glass, added to increase the hardness of the glass and its resistance to chemical reagents.

Illuminating
Ingredient of—
Compositions used in making gas mantles.
Substitute for—
Lime in the calcium light.

Zirconium Oxide (Continued)

Metallurgical
Ingredient of—
Crucible compositions.
Lining compositions for blast furnaces.
Lining compositions for open-hearth furnaces and electric furnaces.
Refractory linings for zinc distillation furnaces.
Starting point in making—
Ferrozirconium, metallic zirconium.

Miscellaneous
Ingredient of—
Compositions containing titanium dioxide, used to produce porous surfaces employed in surface combustion work.
Compositions used in lining the walls of safes for the purpose of rendering the latter resistant to the attack of the oxyacetylene flame.
Substitute for—
Bismuth salts in X-ray photography.

Paint and Varnish
Pigment in making—
White lacquers for wood.
White paints and varnishes.
Starting point in making—
Pigments.

Refractories
Ingredient of—
Refractory cements, refractory materials.
Refractory in making—
Muffles, retorts.

Rubber
Filler in making—
Rubber goods.

Textile
——, *Dyeing*
Assist in certain processes.

Zirconium Silicate
German: Kieselsaeurezirkon, Zirkonsilikat.

Metallurgical
Starting point in making—
Metallic zirconium.

Paint and Varnish
Starting point in making—
Zirconium oxide-silica composite pigment (U. S. 1618288).

Synonyms
and
Cross References

Consult page vi for instructions
"How to use this book"

Consult page vi for instructions "How to use this book"

Synonyms and Cross References

A

Abietate of soda. See: Sodium resinate.
Abrastol. See: Calcium betanaphthol alphasulphonate.
Absinthe. See: Wormwood leaves.
Acetannin. See: Diacetyltannin.
Acetate of lime. See: Calcium acetate.
Acetate of soda. See: Sodium acetate.
Acetate of tin. See: Stannous acetate.
Acetic acid amine. See: Acetamide.
Acetic aldehyde. See: Acetaldehyde.
Acetic ester. See: Ethyl acetate.
Acetic ether. See: Ethyl acetate.
Acetoacetic ether. See: Ethyl acetoacetate.
Acetoform. See: Acetone chloroform.
Acetoin. See: Acetylmethylcarbinol.
Acetone alcohol. See: Methanol.
Acetosol. See: Acetylene tetrachloride.
Acetylbromanilide. See: Bromoacetanilide.
Acetylglycollic ether. See: Ethyl acetylglycollate.
Acetyltribromosalol. See: Acetyltribromophenyl salicylate.
Achiotte. See: Annatto.
Acid sodium carbonate. See: Sodium bicarbonate.
Acid sodium sulphite. See: Sodium bisulphite.
Acraldehyde. See: Acrolein.
Acrylic acid ethyl ester. See: Ethyl acrylate.
Acrylic acid methyl ester. See: Methyl acrylate.
Acrylic ether. See: Ethyl acrylate.
Activated Fuller's earth. See: Fuller's earth activated.
Adronal acetate. See: Cyclohexanol acetate.
Aesculin. See: Esculin.
Alabaster. See: Calcium sulphate.
Alboline. See: Petrolatum.
Aldol. See: Oxybutyric acid.
Alexandrian laurel oil. See: Calophyllum oil.
Algaroth powder. See: Antimony oxychloride.
Alkaline air. See: Ammonia.
Allspice. See: Pimento.
Allyl aldehyde. See: Acrolein.
Almond oil, bitter. See: Bitter almond oil
Alpha crotonic acid ethyl ester. See: Ethyl alphacrotonate.
Alpha-oxypropionic acid. See: Lactic acid.
Alphatoluic acid. See: Phenylacetic acid.
Alum, ammonia. See: Ammonium alum.
Alum, common. See: Potash alum.
Alum, cube. See: Potash alum.
Alum, Roman. See: Potash alum.
Amber seed. See: Abelmoschus.
Amidopyrin. See: Dimethylaminoantipyrin.
Aminic acid. See: Formic acid.
Aminobenzene. See: Anilin.
2-Aminohypoxanthin. See: Guanin.
2-Amino-6-oxypurin. See: Guanin.
6-Aminopurin. See: Adenin.
Aminopyrin. See: Dimethylaminoantipyrin.
Ammonia alum. See: Ammonium alum.
Ammonia, crystal. See: Ammonium carbonate.
Ammoniacal gas. See: Ammonia.
Ammonium chlorostannate. See: Tin-ammonium chloride.
Ammonium-magnesium sulphate. See: Magnesium-ammonium sulphate.
Ammonium-manganese sulphate. See: Manganese-ammonium sulphate.
Ammonium rhodanate. See: Ammonium sulphocyanate.
Ammonium sesquicarbonate. See: Ammonium carbonate.
Amygdalic acid. See: Phenylglycollic acid.
Amylacetaldehyde. See: Heptaldehyde.
Amylene hydrate. See: Dimethylethylcarbinol.
Amylenol. See: Amyl salicylate.
Amyl hydride. See: Pentane.
Amyl oxide. See: Amyl ether.
Amylic alcohol. See: Amyl alcohol.
Anacardia. See: Acajou balsam.
Analgesine. See: Antipyrine.
Anesin. See: Acetone chloroform.
Aneson. See: Acetone chloroform.
Anhydrite. See: Calcium sulphate.
Anhydrous ammonia. See: Ammonia.
Anhydrous ethyl alcohol. See: Alcohol.
Anilin brown. See: Bismarck brown.
Anilin oil. See: Anilin.
Anilin red. See: Fuchsin.
Animal black. See: Bone black.
Animal charcoal. See: Bone black.
Anime. See: Gum anime; also see: Copal.
Anodynine. See: Antipyrine.
Antichlor. See: Sodium hyposulphite.
Antifebrin. See: Acetanilide.
Antimonic acid. See: Antimony pentoxide.
Antimonic anhydride. See: Antimony pentoxide.
Antimonius acid. See: Antimony trioxide.
Antimonite. See: Stibnite.
Antimony, black. See: Antimony, crude.
Antimony, butter of. See: Antimony trichloride.
Antimony, caustic. See: Antimony trichloride.
Antimony chloride, basic. See: Antimony oxychloride.
Antimony, crimson. See: Antimony sulphides.
Antimony, Flowers of. See: Antimony trioxide.
Antimony glance. See: Stibnite.
Antimony, golden. See: Antimony sulphides.
Antimony matte. See: Antimony, crude.
Antimony, needle. See: Antimony, crude.
Antimony ore. See: Stibnite.
Antimony oxysulphide. See: Antimony red.
Antimony, tartrated. See: Antimony-potassium tartrate.
Antimony vermilion. See: Antimony red.
Antimony yellow. See: Lead antimoniate.
Arachis oil fatty acid. See: Peanut oil fatty acid.
Argilla. See: Kaolin.
Aristol. See: Dithymol di-iodide.
Armenian bole. See: Red bole.
Arsenic. See: Arsenic trioxide.
Arsenic glass, red. See: Arsenic disulphide.
Arsenic, red sulphide. See: Arsenic disulphide.
Arsenic, red sulphuret. See: Arsenic disulphide.
Arsenic rouge. See: Arsenic disulphide.
Arsenic, ruby. See: Arsenic disulphide.
Arsenious acid. See: Arsenic trioxide.
Arsenious oxide. See: Arsenic trioxide.
Arsenous anhydride. See: Arsenic trioxide.
Artificial barytes. See: Blanc fixe.
Artificial gum. See: Dextrin.
Artificial heavy spar. See: Blanc fixe.
Artificial oil of bergamot. See: Linalyl acetate.
Artificial oil of wintergreen. See: Methyl salicylate.
Asaprol. See: Calcium betanaphthol alphasulphonate.
Aspirin. See: Acetylsalicylic acid.
Assam and Nepaul musk. See: Musk.
Altophan. See: Cinchophen.
Aurin. See: Pararosolic acid.
Avertin. See: Tribromoethyl alcohol.

B

Bachelor's buttons. See: Nux vomica.
Badianic acid. See: Anisic acid.
Baking soda. See: Sodium bicarbonate.
Balsam. See: Acajou balsam.
Banana oil. See: Amyl acetate.
Banks oil. See: Cod oil.
Barbital. See: 5:5-Diethylbarbituric acid.
Barium sulphate. See: Barytes; also see: Blanc fixe.
Barium sulphocarbolate. See: Barium phenolsulphonate.
Battery acid. See: Sulphuric acid.
Battery manganese. See: Manganese dioxide.
Bean oil. See: Soybean oil.
Behen oil. See: Ben oil.
Behn oil. See: Ben oil.
Benzal chloride. See: Benzyl dichloride.
Benzene chloride. See: Monochlorobenzene.
Benzene trifluoride. See: Trifluorobenzene.
Benzenol. See: Phenol.
Benzidin dicarboxylic acid. See: Diaminodiphenic acid.
Benzin. See: Petroleum ether.

C

Cinnamene. See: Styrol.
Cinnamic acid ethyl ester. See: Ethyl cinnamate.
Cinnamic ether. See: Ethyl cinnamate.
Cinnamol. See: Styrol.
Cinnamon brown. See: Bismarck brown.
Citrate of soda. See: Sodium citrate.
Citronellol. See: Citronellyl.
Clove pepper. See: Pimento.
Coal naphtha. See: Benzene.
Cola nuts. See: Kola nuts.
Colcothar. See: Red oxide of iron.
Colloidal sulphur. See: Sulphur.
Collodion cotton. See: Nitrocellulose.
Colloxylin. See: Nitrocellulose.
Collza oil. See: Rapeseed oil.
Cologne spirits. See: Alcohol.
Colonial spirits. See: Methanol.
Colophony. See: Rosin.
Columbian spirits. See: Methanol.
Columbium. See: Niobium.
Columbium oxide. See: Niobium oxide.
Columnian spirits. See: Methanol.
Commercial sulphur. See: Sulphur.
Common salt. See: Sodium chloride.
Completely denatured alcohol. See: Alcohol.
Contact acid. See: Sulphuric acid.
Copperas, chlorinated. See: Chlorinated copperas.
Copper bichloride. See: Cupric chloride.
Copper protochloride. See: Cuprous chloride.
Copper subacetate. See: Copper acetate, basic.
Copper subchloride. See: Cuprous chloride.
Corallin. See: Pararosolic acid.
Cordyl. See: Acetyltribromophenyl salicylate.
Corrosive sublimate. See: Mercuric chloride.
Cosmetic bismuth. See: Bismuth oxychloride.
Cosmoline. See: Petrolatum.
Cotton oil. See: Cottonseed oil.
Cotton stearin. See: Stearin.
Cowrie. See: Copal.
Creosote. See: Coaltar creosote.
Crocus. See: Saffron.
Crude bisulphate of soda. See: Niter cake.
Crude sodium sulphate. See: Salt cake.
Crude sulphur. See: Sulphur.
Cubic nitre. See: Sodium nitrate.
Cupric acetoarsenite. See: Copper acetoarsenite.
Cupric salts. Unless specially listed, look for the equivalent copper salt.
Cuprous salts. Unless specially listed, look for the equivalent copper salt.
Cutch. See: Catechu.
Cutt. See: Catechu.
Cyanhydric acid. See: Hydrocyanic acid.
Cymol. See: Paracymene.

D

Dagget. See: Birch tar.
Decalin. See: Decahydronaphthalene.
Decan-diacid. See: Sebacic acid.
Decolorant N. See: Zinc formaldehydesulphoxylate.
Decylic acetate. See: Decyl acetate.
Deer musk. See: Musk.
de Haen's salt. See: Antimony salts.
Dehydrated alcohol. See: Alcohol.
Denatured alcohol. See: Alcohol.
Deutoiodide of mercury. See: Mercuric iodide.
Deutoxide of manganese. See: Manganese dioxide.
Dextroracemic acid. See: Tartaric acid.
Diacetic ester. See: Ethyl acetoacetate.
Diacetic ether. See: Ethyl acetoacetate.
Diamide of carbonic acid. See: Urea.
Diammonium phosphate. See: Ammonium phosphate.
Diamylene. See: Dipentene.
Diamyl ether. See: Amyl ether.
Diamyl oxide. See: Amyl ether.
1:2-Dibromoethene. See: Ethplene dibromide.
Dibromotrimethylstibine. See: Trimethylstibin dibromide.
Dichlorbenzyl. See: Benzyl dichloride.
Dichloroisopropyl alcohol. See: Dichlorhydrin.
1:2-Dichloropropane. See: Propylene dichloride.
1:3-Dichloropropanol-2. See: Dichlorhydrin.
Diethylene dioxide. See: 1:4-Dioxane.
Diethylmalonylurea. See: 5:5-Diethylbarbituric acid.
Digallic acid. See: Tannic acid.
1:2-Dihydroxyanthraquinone. See: Alizarin.
1:4-Dihydroxyanthraquinone. See: Quinazarin.
Dihydroxybenzene. See: Resorcinol.
2:3-Dihydroxybutane. See: 2:3-Butylene glycol.
2:3-Dihydroxybutanedioic acid. See: Tartaric acid.
Dihydroxysuccinic acid. See: Tartaric acid.
Di-isoamyl ether. See: Amyl ether.
Diketobutane. See: Diacetyl.
Dilo oil. See: Calophyllum oil.
3:4-Dimethoxybenzaldehyde. See: Methyl-vanillin.
1:3-Dimethylbenzene. See: Metaxylene.
Dimethyl diketone. See: Diacetyl.
Dimethylglyoxal. See: Diacetyl.
Dimethyl ether. See: Methyl oxide.
Dimethylketol. See: Acetylmethylcarbinol.
Dimethyl ketone. See: Acetone.
Diphenolcresolcarbinol anhydride. See: Rosolic acid.
Dipping acid. See: Sulphuric acid.
Dipropylmethane. See: Heptane.
Disodium pyrophosphate. See: Sodium pyrophosphate.
Diuretic salt. See: Potassium acetate.
Divinyl. See: 1:3-Butadiene.
Dodecanoyl peroxide. See: Lauroyl peroxide.
Doggert. See: Birch tar.
Dog's buttons. See: Nux vomica.
Dolphin oil. See: Porpoise body oil.
Domba oil. See: Calophyllum oil.
Drop black. See: Bone black.
Dry ice. See: Carbon dioxide (solidified).
Dutch boiled linseed oil. See: Stand oil.
Dutch enamel oil. See: Stand oil.
Dutch liquid, monochlorinated. See: Betatrichloroethane.
Dyer's oak bark. See: Quercitron bark.

E

Earthnut oil fatty acid. See: Peanut oil fatty acid.
Earth wax. See: Ceresin.
Egg albumen. See: Albumen.
Elain. See: Olein.
Elemental sulphur. See: Sulphur.
Eleuthera bark. See: Cascarilla.
Enanthic acid. See: Oenanthic acid.
English brown. See: Bismarck brown.
English red. See: Red oxide of iron; see also: Venetian red.
Epsom salts. See: Magnesium sulphate.
Ergol. See: Benzyl benzoate.
Erocaine. See: Novocaine.
Erythrene. See: 1:3-Butadiene.
Ethal. See: Cetyl alcohol.
Ethanol. See: Alcohol.
Ethine. See: Acetylene.
Ethocane. See: Novocaine.
Ethyl alcohol. See: Alcohol.
Ethylenecarboxylic acid. See: Acrylic acid.
Ethylene chlorochloride. See: Betatrichloroethane.
Ethylenedicarboxylic acid. See: Succinic acid.
Ethylene dichloride. See: Dichloroethylene.
Ethylene hydrate. See: Ether.
Ethylene naphthene. See: Acenaphthene.
Ethyl hydroxide. See: Alcohol.
Ethyl nonylate. See: Ethyl pelargonate.
Ethyl oxide. See: Ether.
Ethylic alcohol. See: Alcohol.
Ethylphenyl acetate. See: Phenylethyl acetate.
Ethyl resinate. See: Ethyl abietate.
Ethyl sebacate (or sebacinate). See: Diethylsebacinate.
Ethyl succinate. See: Diethyl succinate.
Ethyl sulphhydrate. See: Ethylmercaptan.
Euflavin. See: Acriflavin base.
Eurosol. See: Resorcinol monoacetate.
Expressed almond oil. See: Sweet almond oil.

F

Fermentation alcohol. See: Alcohol.
Fermentation amyl alcohol. See: Fusel oil.
Ferric ferrocyanide. See: Prussian blue.
Ferric oxide red. See: Red oxide of iron.
Ferriferrocyanide of potash. See: Soluble Prussian blue.
Fish glue. See: Isinglass.
Flotation sulphur. See: Sulphur.
Flowers of antimony. See: Antimony trioxide.

G

H

I

J

K

L

Lime nitrogen. See: Cyanamide.
Lime, salts of. See specific calcium salt; e.g., Calcium acetate.
Limonene, inactive. See: Dipentene.
Linalylbutyric ether. See: Laevo-linalylbutyrate.
Linalyl methanecarboxylate. See: Linalyl acetate.
Linalyl methylacetate. See: Linalyl propionate.
Lindera oil. See: Shiromoji seed oil.
Linseed. See: Flaxseed.
Liquid glass. See: Sodium silicate.
Liquid storax. See: Storax.
Litauer balsam. See: Birch tar.
Liver of sulphur. See: Potassium polysulphide.
Locust bean. See: Carob bean.
Locust bean gum. See: Carob bean gum.
Locust kernel gum. See: Carob bean gum.
Louseseed. See: Staveacre seed.
Lucidol. See: Benzoyl peroxide.
Lump sulphur. See: Sulphur.
Luna caustic. See: Silver nitrate.
Lunar caustic. See: Silver nitrate.

M

Magenta. See: Fuchsin.
Magistery of bismuth. See: Bismuth subnitrate.
Magnesia, alba. See: Magnesium carbonate.
Magnesia, alba levis. See: Magnesium carbonate.
Magnesia, burnt. See: Magnesium oxide.
Magnesia, calcined. See: Magnesium oxide.
Magnesium silicate, hydrous. See: Talc.
Maize oil. See: Corn oil.
Malarine. See: Acetophenoneparaphenetidin.
Malonal. See: 5:5-Diethylbarbituric acid.
Malonurea. See: 5:5-Diethylbarbituric acid.
Malt sugar. See: Maltose.
Manchester brown. See: Bismarck brown.
Manganese binoxide. See: Manganese dioxide.
Manganese black. See: Manganese dioxide.
Manganese peroxide. See: Manganese dioxide.
Manganese protochloride. See: Manganese chloride.
Manganous chloride. See: Manganese chloride.
Manhattan spirits. See: Methanol.
Manihot utilissima. See: Manioc.
Maniok. See: Manioc.
Manjak. See: Asphalt.
Mannite. See: Mannitol.
Marchies. See: Margine.
Marine acid. See: Hydrochloric acid.
Massicot. See: Litharge.
Maw oil. See: Poppyseed oil.
Mawseed. See: Poppyseed.
Mawseed oilcake. See: Poppyseed oilcake.
Mayer's reagent. See: Mercuric-potassium iodide.
Mecca. See: Shellac.
Menthyl methane carboxylate. See: Menthyl acetate.
Menthyl salicylate. See: Menthol salicylate.
Mercury and potassium iodide. See: Mercuric-potassium iodide.
Mercury bichloride. See: Mercuric chloride.
Mercury biniodide. See: Mercuric iodide.
Mercury protochloride. See: Mercurous chloride.
Mercury subchloride. See: Mercurous chloride.
Mercury-tolyl acetate. See: Tolylmercuric acetate.
Metadimethylbenzene. See: Metaxylene.
Methanamide. See: Formamide.
Methane acid. See: Formic acid.
Methanedicarboxylic acid. See: Malonic acid.
Methenyl trichloride. See: Chloroform.
Methozine. See: Antipyrine.
Methyl alcohol. See: Methanol.
Methylbenzene. See: Toluene.
Methylbenzol. See: Toluene.
Methylcupreine. See: Quinine.
Methylene chloride. See: Dichloromethane.
Methylene di(or bi)chloride. See: Dichloromethane.
Methylene dimethylate. See: Methylal.
Methyl hexane. See: Heptane.
Methylhexylcarbinol. See: Octyl alcohol, secondary.
Methyl hydrate. See: Methanol.
Methyl hydroxide. See: Methanol.
Methyl 1-hydroxyethyl ketone. See: Acetylmethylcarbinol.
Methylhydroxyisopropylcyclohexaneparamentheneol. See: Menthol.
Methylic alcohol. See: Methanol.
3-Methylindol. See: Scatol.
Methylisobutylcarbinol. See: Methylamyl alcohol.
Methylisobutyl ketone. See: Hexone.
2-Methyl-4-pentanone. See: Hexone.
Methylpropylphenyl hexahydride. See: Menthol.
Methyl resinate. See: Methyl abietate.
Michler's hydrol. See: Tetramethyldiaminobenzhydrol.
Michler's ketone. See: Tetramethyldiaminobenzophenone.
Microcidin. See: Sodium betanaphtholate.
Microsmic salt. See: Sodium-ammonium phosphate.
Milk albumen. See: Albumen.
Milk of sulphur. See: Sulphur.
Mineral alkali. See: Sodium hydroxide.
Mineral butter. See: Antimony trichloride.
Mineral green. See: Mountain green.
Mineral oil. See: Petroleum.
Mineral pitch. See: Asphalt.
Mineral wax. See: Ozokerite; see also: Montan wax, and Ceresin.
Miner's sulphur. See: Sulphur.
Mint camphor. See: Menthol.
Molybdenic acid. See: Molybdenum trioxide.
Molybdenic anhydride. See: Molybdenum trioxide.
Molybdenum sesquioxide. See: Molybdenum oxide.
Monobasic sodium phosphate. See: Sodium acid phosphate.
Monosodium hydrogen phosphate. See: Sodium acid phosphate.
Monosodium orthophosphate. See: Sodium acid phosphate.
Monosodium phosphate. See: Sodium acid phosphate.
Mossbunk oil. See: Menhaden oil.
Moth camphor. See: Naphthalene.
Mother of thyme. See: Serpolet.
Motor benzol. See: Benzene.
Muriate of potash. See: Potassium chloride.
Muriate of soda. See: Sodium chloride.
Muriatic acid. See: Hydrochloric acid.
Musk mallow. See: Abelmoschus.
Musk okra. See: Abelmoschus.
Musk seed. See: Abelmoschus.
Muthmann's liquid. See: Acetylene tetrabromide.

N

Naphacetene. See: Acenaphthene.
Naphtha. See: Petroleum; also Petroleum ether; also Solvent naphtha.
Naphthalene, chlorinated. See: Chloronaphthalenes.
Naples yellow. See: Lead antimoniate.
Native paraffin. See: Ozokerite.
Native sulphur. See: Sulphur.
Ndilo oil. See: Calophyllum oil.
Needle antimony. See: Antimony, crude.
Nembutal. See: Sodium ethylmethylbutylbarbiturate.
Neroli oil. See: Orangeflower oil, bitter.
Neroli oil, Portugal. See: Orangeflower oil, sweet.
Nitrated cotton. See: Nitrocellulose.
Nitrate of soda. See: Sodium nitrate.
Nitrite of soda. See: Sodium nitrite.
Nitroleum. See: Nitroglycerin.
Nitroprusside of soda. See: Sodium nitroprusside.
Njamplung oil. See: Calophyllum oil.
Nucleinate of soda. See: Sodium nucleinate.

O

Obsidian. See: Pumice.
Octoic alcohol. See: Octyl alcohol, secondary.
Octonol-2. See: Octyl alcohol, secondary.
Octyl alcohol, normal. See: Caprylic alcohol, primary.
Oenantaldehyde. See: Heptaldehyde.
Oenanthic acid. See: Heptylic acid, normal.
Oenanthic aldehyde. See: Heptaldehyde.
Oenanthol. See: Heptaldehyde.
Oil black. See: Mineral black.
Oil, crude. See: Petroleum.
Oil of . . .
Unless so printed, see specific oil, e.g., (1) Neroli oil, (2) Lavender oil, etc.
Oil of ants, artificial. See: Furfural.
Oil of chenopodium. See: Wormseed oil.
Oil of juniper. See: Juniper oil.
Oil of pineapples (artificial). See: Ethyl butyrate.
Oil of santonica. See: Levant wormseed oil.
Oil of vitriol. See: Sulphuric acid.
Olefiant gas. See: Ethylene.
Oleum. See: Sulphuric acid.

Oleum acid. See: Sulphuric acid.
Orange red. See: Orange mineral.
Orchadie. See: Amyl salicylate.
Ordeal bean. See: Calabar bean.
Oriental sweet gum. See: Storax.
Orlean or orleana. See: Annatto.
Orpiment, red. See: Arsenic disulphide.
Orthoaminobenzoic acid. See: Anthranilic acid.
Orthodiphenylene ethylene. See: Phenanthrene.
Orthohydroxybenzoic acid. See: Salicylic acid.
Ortho-oxybenzaldehyde. See: Salicylaldehyde.
Ortho-oxybenzoic acid. See: Salicylic acid.
Orthophenylphenol. See: 2-Hydroxydiphenyl.
Oxymuriate of tin. See: Tin tetrachloride.
Oxy-8-quinolin. See: Ortho-oxyquinolin.
Oxytricarballylic acid. See: Citric acid.
Ozokerite, refined. See: Ceresin.

P

Palma christi seed oil. See: Castor oil.
Palmityl alcohol. See: Cetyl alcohol.
Palm oil. See: Coconut oil.
Palm pitch. See: Stearin pitch.
Papermaker's alum. See: Aluminum sulphate.
Para-aminoacetanilide. See: Acetylparaphenylenediamine.
Para-aminobenzoldiethylaminoethanol. See: Novocaine.
Parabenzoquinone. See: Quinone.
Paraffin jelly. See: Petrolatum.
Paramandelic acid. See: Phenylglycollic acid.
Paramorphine. See: Thebaine.
Paranaphthene. See: Anthracene.
Parathiocresol. See: Paratolylmercaptan.
Paris green. See: Copper acetoarsenite.
Parodyne. See: Antipyrine.
Pearl moss. See: Irish moss.
Pearl white. See: Bismuth oxychloride.
Pear oil. See: Amyl acetate.
Pelargonic ether. See: Ethyl pelargonate.
Pellamountain. See: Serpolet.
Pental. See: Amylene.
1-Pentanol. See: Butyl carbinol, normal.
3-Pentanol. See: Diethyl carbinol.
Pentanone-4-oic-1 acid. See: Levulic acid.
Pentent. See: Amylene.
Penthrite. See: Tetranitropentaerythrite.
Pentothal sodium. See: Sodium ethyl-1-methyl-butylthiobarbiturate.
Peppermint camphor. See: Menthol.
Perborax. See: Sodium perborate.
Perchlormethane. See: Carbon tetrachloride.
Perchloroethane. See: Hexachloroethane.
Permanent white. See: Blanc fixe.
Permanganate of soda. See: Sodium permanganate.
Pernocton. See: Sodium, secondary, butylbetabromoallylbarbiturate.
Peroxide of manganese. See: Manganese dioxide.
Persian insect flowers. See: Pyrethrum flowers.
Persian pellitory. See: Pyrethrum flowers.
Pertonal. See: Para-acetylaminoethoxybenzene.
Petroleum pitch. See: Asphalt.
Petroleum spirit. See: Petroleum ether.
Petroline. See: Petrolatum.
Phenacetine. See: Acetphenetidin.
Phenazone. See: Antipyrine.
Phenic acid. See: Phenol.
Phenic alcohol. See: Phenol.
Phenoquin. See: Cinchophen.
Phenylacetamide. See: Acetanilide.
Phenylamine. See: Anilin.
Phenyl chloride. See: Monochlorobenzene.
Phenylcinchoninic acid. See: Cinchophen.
Phenylene brown. See: Bismarck brown.
Phenylethylene. See: Styrol.
Phenylformic acid. See: Benzoic acid.
Phenyl hydrate. See: Phenol.
Phenyl hydride. See: Benzene.
Phenylic acid. See: Phenol.
Phenylic alcohol. See: Phenol.
Phenylmethane. See: Toluene.
Phenylmethanic acid. See: Benzoic acid.
Phenyl-peri-acid. See: Phenylalphanaphthylamine-8-sulphonic acid.
2-Phenylphenol. See: 2-Hydroxydiphenyl.
Phosgene. See: Carbonyl chloride.
Phosphate of soda. See: Trisodium phosphate.
Phosphorus salt. See: Sodium-ammonium phosphate.
Phthalic ether. See: Diethyl phthalate.
Pigwrack. See: Irish moss.
Pine oil, sulphonated. See: Sulphonated pine oil.
Piney tallow. See: Malabar tallow.
Pink salt. See: Tin-ammonium chloride.
Pinnay oil. See: Calophyllum oil.
Pipmenthol. See: Menthol.
Pistachia galls. See: Mastic.
Plaster of Paris. See: Calcium sulphate.
Plastic sulphur. See: Sulphur.
Platinic chloride. See: Chloroplatinic acid.
Plumbago. See: Graphite.
Plumbo-plumbix oxide. See: Red lead.
Pogy oil. See: Menhaden oil.
Poison nut. See: Nux vomica.
Polychrome. See: Esculin.
Polysulphide of soda. See: Sodium polysulphide.
Poonseed oil. See: Calophyllum oil.
Porcelain clay. See: Kaolin.
Porpoise blubber oil. See: Porpoise body oil.
Porpoise face blubber oil. See: Porpoise junk oil.
Potash lye. See: Potassium hydroxide.
Potassium-aluminium sulphate. See: Potash alum.
Potassium-chromium sulphate. See: Chrome alum.
Potassium ferriferrocyanide. See: Soluble Prussian blue.
Potato oil. See: Fusel oil.
Potato spirit oil. See: Fusel oil.
Precipitated barium sulphate. See: Blanc fixe.
Precipitated sulphur. See: Sulphur.
Preparing salt. See: Sodium stannate.
Procaine. See: Novocaine.
Proof spirit. See: Alcohol.
Propane diacid. See: Malonic acid.
Propane-1:2:3-triol. See: Glycerin.
2-Propanolic acid. See: Lactic acid.
Propanthiol-1. See: Propylmercaptan, normal.
Propenal. See: Acrolein.
Propene acid. See: Acrylic acid.
Propenyl alcohol. See: Glycerin.
Propenylveratrol. See: Methylisoeugenol.
Propyldioxybenzene methylene ester. See: Safrol.
Propylformic acid. See: Butyric acid, normal.
Protocatechuic aldehyde dimethyl ether. See: Methylvanillin.
Prussiate of tin. See: Tin pulp.
Prussic acid. See: Hydrocyanic acid.
Pseudo butylene glycol. See: 2:3-Butylene glycol.
Pure alcohol. See: Alcohol.
Pure ethyl alcohol. See: Alcohol.
Pyramidon. See: Dimethylaminoantipyrin.
Pyridinmonocarboxylic acid. See: Nicotine acid.
Pyroacetic ether. See: Acetone.
Pyroacetic spirit. See: Acetone.
Pyroborate of soda. See: Borax.
Pyroleic acid. See: Sebacic acid.
Pyroligneous spirit. See: Methanol.
Pyrolusite. See: Manganese dioxide.
Pyromucic aldehyde. See: Furfural.
Pyrosulphate of soda. See: Sodium pyrosulphate.
Pyroxylic spirit. See: Methanol.
Pyroxylin. See: Nitrocellulose.

Q

Quaker buttons. See: Nux vomica.
Quendel. See: Serpolet.
Quicklime. See: Calcium oxide.
Quicksilver. See: Mercury.
Quinole. See: Hydroquinone.
Quinolin-4-carboxylic acid. See: Cinchoninic acid.
Quinolin rhodanate. See: Quinolin sulphocyanate.
Quinonanilide. See: 2:5-Dianilidobenzoquinone.
Quinone. Se: Hydroquinone.
Quinophan. See: Cinchophen.
Quinophenol. See: Ortho-oxyquinolin.

R

Racou. See: Annatto.
Rectified spirit. See: Alcohol.
Red chromate of potash. See: Potassium bichromate.
Red iron ore. See: Red hematite.
Red oil. See: Oleic acid.
Redol Z. See: Zinc formaldehyde-sulphoxylate.

Red precipitate. See: Mercuric oxide, red.
Refined sulphur. See: Sulphur.
Resinate of . . . See particular metal soap; e.g., Lead resinate.
Resin dammar. See: Dammar.
Resin ether L. See: Benzyl resinate.
Rhodazil. See: Benzyl benzoate.
Ricinus oil. See: Castor oil.
Rock oil. See: Petroleum.
Rocksalt. See: Sodium chloride.
Rocksalt moss. See: Irish moss.
Roll brimstone. See: Sulphur.
Rolled sulphur. See: Sulphur.
Roman vitriol. See: Copper sulphate.
Rosein. See: Fuchsin.
Rosin, liquid. See: Tall oil.
Rosin soap. See: Sodium resinate.
Rotten stone. See: Tripoli.
Rubin. See: Fuchsin.
Rubsen oil. See: Rapeseed oil.
Ruria. See: Madder.

S

Saccharolactic acid. See: Mucic acid.
Saffran. See: Saffron.
Saffron of antimony. See: Antimony crocus.
Safronin B extra. See: Phenosafranin.
Sal ammoniac. See: Ammonium chloride.
Salicylate of soda. See: Sodium salicylate.
Salicylic aldehyde. See: Salicylaldehyde.
Salicylic ether. See: Ethyl salicylate.
Salicylous acid. See: Salicylaldehyde.
Salimenthol. See: Menthol salicylate.
Salinaphthol. See: Betanaphthyl salicylate.
Salol. See: Phenyl salicylate.
Sal soda. See: Sodium carbonate.
Salt. See: Sodium chloride.
Salt cake, high-grade. See: White cake.
Salt of lemery. See: Potassium sulphate.
Salt of phosphorus. See: Sodium-ammonium phosphate.
Salt of tartar. See: Potassium carbonate.
Sal volatile. See: Ammonium carbonate.
Sand acid. See: Hydrofluosilicic acid.
Sandix. See: Orange mineral.
Sanse. See: Margine.
Santonica oil. See: Levant wormseed oil.
Sapamine salts. See: Diethylaminoethyloleylamide salts.
Satyrion. See: Salep.
Saxoline. See: Petrolatum.
Schweinfurt green. See: Copper acetoarsenite.
S-Dimethylethylene glycol. See: 2:3-Butylene glycol.
Sea onion. See: Squill.
Seasalt. See: Sodium chloride.
Sebacinic acid. See: Sebacic acid.
Selenium trioxide. See: Selenic anhydride.
Sesame oil, German. See: Cameline oil.
Shikimol. See: Safrol.
Silex. See: Quartz.
Silica. See: Quartz.
Silicate of soda. See: Sodium silicate.
Silicic oxide. See: Quartz.
Silicium chloroform. See: Silicochloroform.
Silicon bisulphide. See: Silicon disulphide.
Silicon chloride. See: Silicon tetrachloride.
Silicon dioxide. See: Quartz.
Silicon fluoride. See: Silicon tetrafluoride.
Silver fir oil. See: Templin oil.
Slate black. See: Mineral black.
Soapstone. See: Talc.
Soda. See: Sodium carbonate.
Soda ash. See: Sodium carbonate.
Soda lye. See: Sodium hydroxide.
Sodic hydrate. See: Sodium hydroxide.
Sodium abietate. See: Sodium resinate.
Sodium acid chromate. See: Sodium bichromate.
Sodium acid fluorid. See: Sodium bifluoride.
Sodium acid sulphate. See: Niter cake.
Sodium aminophenolarsonate. See: Sodium arsanilate.
Sodium-ammonium hydrogen phosphate. See: Sodium-ammonium phosphate.
Sodium anilinarsonate. See: Sodium arsanilate.
Sodium antimony trifluoride. See: Antimony salts.
Sodium biborate. See: Borax.
Sodium binoxide. See: Sodium peroxide.
Sodium biphosphate. See: Sodium acid phosphate.
Sodium bisulphate. See: Niter cake.
Sodium borate. See: Borax.
Sodium diarsenol. See: Sodium-arsphenamine.
Sodium dichromate. See: Sodium bichromate.
Sodium dihydrogen phosphate. See: Sodium acid phosphate.
Sodium dimethylarsonate. See: Sodium cacodylate.
Sodium dioxide. See: Sodium peroxide.
Sodium dithionate. See: Sodium hyposulphate.
Sodium fluosilicate. See: Sodium silicofluoride.
Sodium hydrate. See: Sodium hydroxide.
Sodium hydrocarbonate. See: Sodium bicarbonate.
Sodium hydrogen carbonate. See: Sodium bicarbonate.
Sodium hydrogen formate. See: Sodium biformate.
Sodium-meta-bisulfite. See: Sodium bisulphite.
Sodium meta silicate. See: Sodium silicate.
Sodium molybdosilicate. See: Sodium silicomolybdate.
Sodium nitroprussiate. See: Sodium nitroprusside.
Sodium orthophenylphenate. See: Sodium 2-phenylphenate.
Sodium orthovanadate. See: Sodium vanadate.
Sodium oxymuriate. See: Sodium chlorate.
Sodium parachlorocarbolate. See: Sodium parachlorophenate.
Sodium parachlorophenolate. See: Sodium parachlorophenate.
Sodium parahydroxybenzoate. See: Sodium paraoxybenzoate.
Sodium phenolate. See: Sodium phenate.
Sodium phenolsulphonate. See: Sodium sulphocarbolate.
Sodium phenoxide. See: Sodium phenate.
Sodium phosphate, dibasic. See: Disodium phosphate.
Sodium phosphate, tribasic. See: Trisodium phosphate.
Sodium phosphomolybdotungstate. See: Sodium phosphotungstomolybdate.
Sodium pyroborate. See: Borax.
Sodium ricinate. See: Sodium ricinoleate.
Sodium salt of 3-diamino-4-dihydroxy-1-arsenobenzene. See: Sodium-arsphenamine.
Sodium subsulphite. See: Sodium hyposulphite.
Sodium sulphydrate. See: Sodium hydrosulphide.
Sodium tetraborate. See: Borax.
Sodium thiosulphate. See: Sodium hyposulphite.
Sodium-tin chloride. See: Sodium chlorostannate.
Sodium trichlorophenolate. See: Sodium trichlorophenate.
Sodium tungstomolybdophosphate. See: Sodium phosphotungstomolybdate.
Sodium tungstophosphate. See: Sodium phosphotungstate.
Sodium tungstophosphomolybdate. See: Sodium phosphotungstomolybdate.
Sodium tungstosilicate. See: Sodium silicotungstate.
Sodium valerianate. See: Sodium valeriate.
Sodium wolframate. See: Sodium tungstate.
Sofran. See: Saffron.
Soft sulphur. See: Sulphur.
Soja bean meal. See: Soybean meal.
Sojabean oil. See: Soybean oil.
Soja beans. See: Soybeans.
Soj oil. See: Soybean oil.
Solferino. See: Fuchsin.
Soluble Berlin blue. See: Soluble Prussian blue.
Soluble cotton. See: Nitrocellulose.
Soluble glass. See: Sodium silicate.
Soluble gun cotton. See: Nitrocellulose.
Soluble iron "cyanide." See: Soluble Prussian blue.
Soluble saccharin. See: Sodium benzosulphimide.
Solution of chlorinated soda. See: Sodium hypochlorite.
Sorbite. See: Sorbitol.
Soringa oil. See: Ben oil.
Soya bean flour. See: Soybean meal.
Soya beans. See: Soybeans.
Soy bean flour. See: Soybean meal.
Soy beans. See: Soybeans.
Soy oil meal. See: Soybean meal.
Spanish saffron. See: Saffron.
Spasmodine. See: Benzyl benzoate.
Specially denatured alcohol. See: Alcohol.
Specular iron ore. See: Red hematite.
Spermaceti oil. See: Sperm oil.
Sphagnum. See: Peat moss.
Spirit of salt. See: Hydrochloric acid.
Spirit of sea salt. See: Hydrochloric acid.
Spirit of wine. See: Alcohol.
Spirits of nitre. See: Nitric acid.
Spirits of turpentine. See: Turpentine oil.
Split nut. See: Calabar bean.
Standard wood spirits. See: Methanol.
Stannic chloride. See: Tin tetrachloride.

Stannic stearotoluenesulphonate. See: Tin stearotoluenesulphonate.
Stannic sulphocyanide. See: Tin sulphocyanide.
Stannous chloride. See: Tin dichloride.
Starch gum. See: Dextrin.
Stavesaire seeds. See: Stavesacre seed.
Stearate of tri(hydroxyethyl)amine. See: Triethanolamine stearate.
Stearinic acid. See: Stearic acid.
Stearophanic acid. See: Stearic acid.
Steatite. See: Talc.
Stibnic acid. See: Antimony pentoxide.
Stibnite, concentrated. See: Antimony, crude.
Stibnite, refined. See: Antimony, crude.
Stick lac. See: Shellac.
St. John's bread. See: Carob bean.
Stone oak bark. See: Quercitron bark.
Strontium fluosilicate. See: Strontium silicofluoride.
Styrax. See: Storax.
Styrene. See: Styrol.
Styrolene. See: Styrol.
Styrolyl acetate. See: Methylphenylcarbinol acetate.
Sublimed sulphur. See: Sulphur.
Succinic acid anhydride. See: Succinic anhydride.
Succinic ester. See: Diethyl succinate.
Succinic ether. See: Diethyl succinate.
Sugar coloring. See: Caramel.
Sugar of milk. See: Milk sugar.
Sulphanilate of soda. See: Sodium sulphanilate.
Sulphate of soda. See: Sodium sulphate.
Sulphide of soda. See: Sodium sulphide.
Sulphite of soda. See: Sodium sulphite.
Sulphocarbamide. See: Thiourea.
Sulphocholic acid. See: Taurocholic acid.
Sulphonated castor oil. See: Turkey red oil.
Sulphourea. See: Thiourea.
Sulphoxyindole. See: Thioxyindole.
Sulphur dichloride. See: Sulphur bichloride.
Sulphur flour. See: Sulphur.
Sulphuric chlorohydrin. See: Chlorosulphonic acid.
Sulphurous acid. See: Sulphur dioxide.
Sulphurous anhydride. See: Sulphur dioxide.
Sulphuryl oxychloride. See: Chlorosulphonic acid.
Sulphydrate of soda. See: Sodium hydrosulphide.
Sunflower oil. See: Sunflower seed oil.
Sunoxol. See: Ortho-oxyquinolin sulphate.
Sweet bark. See: Cascarilla.
Sweet bay. See: Bayberry.
Sweet marjoram. See: Marjoram.
Sweetwood bark. See: Cascarilla.
Syncaine. See: Novocaine.
Synthetic ethyl alcohol. See: Alcohol.
Synthetic oil of sassafras. See: Safrol.

T

Tacamahac fat. See: Calophyllum oil.
Tanacetin. See: Diacetyltannin.
Tanigen. See: Diacetyltannin.
Tankawang fat. See: Vegetable tallow.
Tannigen. See: Diacetyltannin.
Tannin. See: Tannic acid.
Tar camphor. See: Naphthalene.
Tartar emetic. See: Antimony-potassium tartrate.
Tea oil. See: Teaseed oil.
3-Terpanol. See: Menthol.
Terpilenol. See: Terpineol.
Terpinyl formate. See: Terpineol formate.
Terra alba. See: Calcium sulphate.
Terra ponderosa. See: Blanc fixe.
Tetraborate of soda. See: Borax.
Tetrabromoethane. See: Acetylene tetrabromide.
Tetrachloromethane. See: Carbon tetrachloride.
Tetrahydronaphthalene. See: Tetralin.
2-Thiobenzoxazole. See: 1-Mercapto-benzoxazole.
Thiocarbamide. See: Thiourea.
Thio-4-chloronaphthalic acid. See: Sulpho-4-chloronaphthalic acid.
Thio-4-chloronaphthalic anhydride. See: Sulpho-4-chloronaphthalic anhydride.
Thiocresyl. See: Tolylmercapton.
Thioethylcresyl ether. See: Ethyltolyl sulphide.
Thoric oxide. See: Thorium dioxide.
Thorium anhydride. See: Thorium dioxide.
Thorium oxide. See: Thorium dioxide.
Thorn apple seed. See: Stramonium seed.
Thymol iodide. See: Dithymol di-iodide.
Tiff. See: Barytes.
Tin bichloride. See: Tin dichloride.
Tin bromide. See: Stannic bromide.
Tin chloride. See: Tin dichloride.
Tin chloride. See: Tin tetrachloride.
Tin ferrocyanide. See: Tin pulp.
Tin protochloride. See: Tin dichloride.
Tin salt. See: Tin dichloride.
Titanic acid anhydride. See: Titanium dioxide.
Titanic oxide. See: Titanium dioxide.
Titanium dichloride. See: Titanous chloride.
Titanium white. See: Titanium dioxide.
Toluol. See: Toluene.
Toluylene. See: Stilbene.
Tolylsulfoglycollic acid. See: Tolylthioglycollic acid.
Tower acid. See: Sulphuric acid.
Tragasol. See: Carob bean gum.
Train oil. See: Whale oil.
Trefol. See: Amyl salicylate.
Tribasic sodium phosphate. See: Trisodium phosphate.
Tribromomethane. See: Bromoform.
Trichloracetic acid. See: Trichloroacetic acid.
Trichlorobenzol. See: Trichlorobenzene.
Trichloromethane. See: Chloroform.
Triethanolamine silicofluoride. See: Triethanolamine fluosilicate.
Trigonella. See: Fenugreek.
Trihydroxybenzene. See: Pyrogallic acid.
2:4:6-Trihydroxybenzimidothiophenylether hydrochloride. See: 2:4:6-Trihydroxybenzimidophenyl hydrochloride-sulphide.
Trihydroxyethanolamine. See: Triethanolamine.
Tri(hydroxyethyl)amine stearate. See: Triethanolamine stearate.
Trimethylene. See: Amylene.
Trimethylethylene. See: Amylene.
Triolein. See: Olein.
Trisodium orthophosphate. See: Trisodium phosphate.
Tristearin. See: Stearin.
Trypaflavine. See: Acriflavin base.
Tuna fish oil. See: Tuna oil.
Tung oil. See: Chinawood oil.
Tungsten oxide. See: Tungsten trioxide.
Tungstic oxide. See: Tungsten trioxide.
Tungstosilicic acid. See: Silicotungstic acid.
Tunny oil. See: Tuna oil.
Turkey red. See: Madder.
Turpentine. See: Turpentine oil.
Turps. See: Turpentine oil.

U

Udilo oil. See: Calophyllum oil.
Ulmarene. See: Amyl salicylate.
Uranium yellow. See: Sodium uranate.
Uranyl acetate. See: Uranium acetate.
Uranyl nitrate. See: Uranium nitrate.

V

Valencia saffron. See: Saffron.
Valerene. See: Amylene.
Vanadic acid. See: Vanadium pentoxide.
Vanadic anhydride. See: Vanadium pentoxide.
Vanadium binoxalate. See: Vanadium acid oxalate.
Vanadium dioxalate. See: Vanadium acid oxalate.
Vaseline. See: Petrolatum.
Vegetable gum. See: Dextrin.
Vegetable pepoin. See: Papain.
Veratrum aldehyde. See: Methylvanillin.
Verbena oil. See: Citronella oil.
Verdigris, green. See: Copper acetate, basic.
Vermilion, antimony. See: Antimony red.
Veronal. See: 5:5-Diethylbarbituric acid.
Vinegar acid. See: Acetic acid.
Vinegar, martial. See: Ferric acetate.
Vinegar naphtha. See: Ethyl acetate.
Vinylbenzene. See: Styrol.
Vinylethylene. See: 1:3-Butadiene.
Vinyl trichloride. See: Betatrichloroethane.
Virgin drop black. See: Bone black.
Viscid sulphur. See: Sulphur.
Vitriolated soda. See: Sodium sulphate.

W

X

Y

Z

Consult page vi for instructions

“How to use this book”